A DICTIONARY OF THE
FLOWERING PLANTS
AND FERNS

A DICTIONARY OF THE FLOWERING PLANTS AND FERNS

EIGHTH EDITION

THE LATE J. C. WILLIS

REVISED BY

H. K. AIRY SHAW

Royal Botanic Gardens, Kew

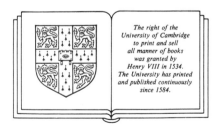

*The right of the
University of Cambridge
to print and sell
all manner of books
was granted by
Henry VIII in 1534.
The University has printed
and published continuously
since 1584.*

CAMBRIDGE UNIVERSITY PRESS

Cambridge

New York   New Rochelle

Melbourne   Sydney

Published by the Press Syndicate of the University of Cambridge
The Pitt Building, Trumpington Street, Cambridge, CB2 1RP
32 East 57th Street, New York, NY 10022, USA
10 Stamford Road, Oakleigh, Melbourne 3166, Australia

First published 1897
Second edition 1904
Third edition 1908
Reprinted 1914
Fourth edition 1919
Fifth edition 1925
Sixth edition 1931
Reprinted 1948 1951 1955 1957 1960
Seventh edition 1966
Eighth edition 1973
ELBS edition 1973
Reprinted 1980
Student edition 1985
Reprinted 1988

Printed in Great Britain at the
University Press, Cambridge

Library of Congress catalogue card number: 72–83581

ISBN 0 521 08699 X
ISBN 0 521 31395 3 student edition

FOREWORD

BY PROFESSOR J. HESLOP-HARRISON

As Sir George Taylor stated in his foreword to the last edition, the Seventh, of Dr J. C. Willis's *A Dictionary of the Flowering Plants and Ferns*, that edition marked a considerable departure from its predecessor in that certain categories of information were perforce omitted. This change was dictated by the need to do adequate justice to the subjects included within the 'reasonable compass' specified by Willis in the original preface. It is therefore gratifying that Sir George Taylor's confidence that the Seventh Edition would nevertheless find an affectionate welcome from its users has been clearly fulfilled. Between the Sixth and Seventh Editions there was a gap of thirty-five years; between the Seventh and Eighth Editions only seven years have elapsed.

This new edition embodies no new changes in policy, but as the Note to the Eighth Edition explains on p. xii, many additions and corrections have been made. The Table of 'Family Equivalents' showing the families recognised in the Eighth Edition with their equivalents according to the systems of Bentham & Hooker and Engler (ed. Melchior) will, it is hoped, be a useful innovation.

Once again the Trustees of the Bentham–Moxon Trust are most grateful to Mr H. K. Airy Shaw for the sustained care with which he has assembled the material for this new edition.

CONTENTS

PREFACE TO SEVENTH EDITION

The first edition of *A Manual and Dictionary of the Flowering Plants and Ferns*, by J. C. Willis, appeared in 1897 in two compact volumes (pp. xiv + 224, xiv + 430). The sixth and last edition, published (like the preceding four) in a single thick volume (pp. xii + 752 + lvi), appeared in 1931, with several subsequent reprintings, but there has been no revision for the past thirty years and more. In recent years the need for a complete revision of the work has become increasingly apparent.

In his original preface, Willis stated the objects of the work as follows: 'The aim with which I commenced...to prepare this book, was to supply within a reasonable compass, a summary of useful and scientific information about the plants met with in a botanical garden or museum, or in the field. The student, when placed before the bewildering variety of forms in such a collection as that at Kew, does not know where to begin or what to do to acquire information about the plants....I have endeavoured to bring together in this book as much information as is required by any but specialists, upon all plants generally met with, and upon all those points—morphology, classification, natural history, economic botany, etc.—which do not require the use of a microscope.... The principal part of the book consists of a dictionary in which the whole of the families and the important genera of flowering plants and ferns are dealt with.'

In the fourth edition, published in 1919, Willis incorporated all the separate sections into one general dictionary, and this form has been retained for all subsequent editions, including the present. It has been found desirable, however, to make certain important departures from previous editions. Whereas the original work was largely envisaged as a handy, encyclopaedic vade-mecum for students, the enormous increase in botanical knowledge over the past more than half a century has rendered it almost impossible to do justice, within the limits of a moderate-sized volume, to all the aspects of the subject which Willis sought to include. The entries in the later editions have fallen under the following main heads: (*a*) generic names, (*b*) family names, (*c*) botanical terms, (*d*) common and vernacular names, (*e*) economic products. None of these categories was covered even approximately completely. When I was asked in 1958 to prepare a revised edition of the Dictionary, my consciousness of lack of qualification for dealing adequately with headings (*c*) to (*e*), coupled with a strong sense of the advantages of a work that at least aims to cover one subject fully, decided me to confine the entries strictly to taxonomic matters—that is, headings (*a*) and (*b*), but making it as far as possible complete for these—and to exclude all others.

This, of course, changes considerably the character of the work, and some users may, I fear, find this a cause for regret. The Dictionary now bears more resemblance to Post & Kuntze's *Lexicon Generum Phanerogamarum* (Stuttgart, 1903 ['1904']), though differing from it in important respects. I trust that those who have found that work useful will not find the present one less so.

In this edition, I have aimed to include every published generic name (whether validly published or not) from 1753 onward, and every published family name from the appearance of the *Genera Plantarum* of Jussieu in 1789. In addition, a number of supra- and infra-familial taxa have been included where these have not been based on a family or generic name, for example, *Centrospermae, Tricoccae, Apetalae, Stenolobeae*, etc.; such terms have been taken almost exclusively from the systems of Bentham & Hooker and Engler & Prantl. The uninomials (or apparent uninomials) of Ehrhart and Du Petit Thouars have also been listed, since these may easily be mistaken for generic names.

Genera. So far as possible, the data for every entry have been revised. With the needs of students particularly in mind, I have here introduced a new feature. Where I have found a considerable divergence of opinion regarding the maintenance or otherwise of a given genus (*a*), I have indicated, by the use of the 'alternative' sign ($\sim$), that this genus (*a*) is by some authorities included in an older genus (*b*). Thus, '**Aclisia** E. Mey. ($\sim$ Pollia Thunb.)' indicates that some authorities treat *Aclisia* as a distinct genus, whilst others reduce it to *Pollia*. No attempt at completeness has been made in regard to this feature, but it is hoped that it may serve to put students on their guard against the too easily made assumption that taxonomy is a cut-and-dried affair, with everything in its own neat pigeon-hole. Nothing could be farther from the truth: taxonomy is very much a matter of personal opinion.

In the information provided, Willis's references to British species have been omitted. Horticultural notes are also omitted, but much of Willis's other economic information has been retained. Except in the Pteridophytes, references to literature are almost entirely excluded, owing to the impossibility of doing justice to this aspect without a prohibitive expenditure of time. Conserved generic names are indicated by means of an asterisk (*). Where a name has been changed owing to the existence of an earlier homonym in some group of Cellular Crypto-gams or fossils (not, of course, listed in this Dictionary), the authority and date of such homonym are included, since this information is not always readily accessible. Reductions to synonymy have been given as far as possible, but it is probable that many have been missed.

It should be clearly understood that the sign '=' covers a wide range of meanings, but rarely signifies 'is equivalent to' ('≡'). It usually means no more than that the type-species of the generic name preceding it is regarded as being congeneric with the type-species of the generic name following it.

A large number of inter-generic hybrids have been included. Where these represent artificially produced horticultural inter-generic crosses (especially in the *Orchidaceae*), the authority for the name is merely given as 'hort'.

A very large number of variant spellings have been listed, though no attempt has been made at complete coverage. They include many obvious slips and misprints, in addition to deliberate corrections, altera-tions or 'improvements'. In order to save space, only the author of the variant is cited: for example, '**Galactella** B. D. Jacks.' [in Ind. Kew.], rather than '**Galactella** Rafin. ex B. D. Jacks.' (= Galatella Cass.). Most of these accidental errors are indicated by the word '(sphalm.)' [= *sphalmate*] in parentheses, except in the case of authors such as Rafinesque, Steudel, Walpers, and a few others, whose habitual disregard of spelling and proof-correction is well known.

Where a generic name covers a mixture of species of other genera, for example, '**Leptolepis** Boeck. = Blysmus Panz. ex Schult. + Carex L.', such expressions almost always imply 'spp.' or '*pro parte*' understood, not that the first name embraces the entire content of the latter names.

Families. The family entries are very largely based upon the useful lists published by Bullock in *Taxon*, 7: 1–35, 158–63, 1958.[1] Every published family name has been reviewed, and where I have felt that there was or might be a case for the recognition of any such previously proposed family, the essential characters have been given. Some new families have also been proposed (see *Kew Bulletin*, 18: 249–73, 1965). The aim has been to secure, so far as possible, a greater equivalence (in morphological distinctness, not in size) between family units, and in the

---

[1] Authorities for family names are given according to these lists, except where later investigation has disclosed earlier authorities, and in a few other cases. Cf. Bullock in *Taxon*, 8: 154–81, 189–205, 1959; *Internat. Code Bot. Nomencl.* (Montreal), 187–201 (1961).

gaps separating them; but very much still remains to be done in this respect. The descriptions of all currently accepted families have been revised where necessary, and in many cases, especially in the smaller and less-known groups, greatly expanded. Brief characters of subfamilies are usually given, but only occasionally those of tribes. A few families that are widely recognised, but for which there seem to be in fact no very convincing grounds for such recognition (*Lobeliaceae*, *Hippocrateaceae*, etc.), have been reduced. Partly on these grounds, but also largely owing to the difficulty of ascertaining the correct assignment of the genera in every case, the three major subdivisions of the *Leguminosae* have not been recognised at the family level.

The treatment of the families and higher taxa of the Pteridophytes by Professor Holttum differs in some respects from that adopted for the Phanerogams. The plan followed is explained by Professor Holttum in a separate introduction (p. xiii).

The synopses of the Bentham-Hooker and Engler-Prantl systems at the end of the book have been retained, since it is believed that many workers still find this a useful feature.

It is hoped that users of the Dictionary will kindly notify the compiler of all errors and omissions that may come to their notice.

NOTE TO EIGHTH EDITION

In the present edition a large number of corrections have been made in the text. Many of these are due to colleagues and correspondents who have kindly helped by drawing attention to errors or misprints. Additional entries, the former 'Addenda' (pp. xxi–xxii of ed. 7), and much supplementary matter, have also been incorporated into the main text. These items include many further names of inter-generic hybrids in the Orchids, each of which is now accompanied by an indication of 'status' in accordance with the schedule on p. xvii. This feature owes its origin to my former colleague Mr P. F. Hunt, to whom I express my thanks for extensive and invaluable help in carrying it through. In response to requests from certain quarters, a list or 'concordance' of family equivalents as between the *Dictionary*, Melchior's edition (12th) of Engler's *Syllabus*, and Bentham & Hooker's *Genera Plantarum*, has been added on pp. liii–lxv.

As with the last edition, notification of errors or emissions will be gratefully received.

H. K. AIRY SHAW

INTRODUCTION TO PTERIDOPHYTA

BY R. E. HOLTTUM

For the *Pteridophyta* the scheme of classification here adopted is that proposed by Pichi-Sermolli in *Uppsala Univ. Årsskrift*, **6**: 70–90, 1958; this scheme includes both fossil and living genera, the former being ignored in the present treatment. The living *Pteridophyta* are divided into the classes *Lycopsida*, *Sphenopsida*, *Psilotopsida* and *Filicopsida*; subdivision of these classes is indicated under their names in the Dictionary.

The great majority of living Pteridophytes are ferns (10,000 species enumerated in Christensen's *Index Filicum* and the three supplements prepared by him up to 1934), and as there is still much uncertainty about their classification, some explanatory statement regarding the present treatment is necessary.

In his great works on the ferns of the whole world (*Species Filicum* 1844–64, *Synopsis Filicum* 1866–8), W. J. Hooker defined genera solely on the form of the sori, and the resulting arrangement was in many ways very unnatural, though attempts to define smaller, more natural genera were made during the same period by Presl, John Smith and Fée. The first later attempts at a natural classification were made by H. Christ (*Die Farnkräuter der Erde*, 1897) and L. Diels (in Engler's *Nat. Pflanzenfam.* 1899–1900), the latter being adopted as a basis for Carl Christensen's *Index Filicum* (1905). Christensen then began a series of monographic studies of several groups of genera, and from 1930 many of his ideas were taken up and extended by R. C. Ching (Peking). Simultaneously E. B. Copeland was studying ferns from the Philippines and other parts of Malaysia. These authors published new outlines of fern classification as follows: Christensen in Verdoorn's *Manual of Pteridology* (chapter 20), 1938; Ching in *Sunyatsenia*, **5**: 202–68, 1940; Copeland in *Genera Filicum*, 1947. The ideas of these authors were also considerably influenced by the morphological studies of F. O. Bower and K. von Goebel. A further outline classification, with critical notes on those of Christensen, Ching and Copeland (also on some of Bower's conclusions), was published by me in two papers (*J. Linn. Soc., Bot.* **53**: 123–58, 1947; *Biological Reviews*, **24**: 267–96, 1949). Alston then proposed a further, limited, scheme for African

FAMILY NAMES OF PTERIDOPHYTA

With authorities according to R. E. G. Pichi-Sermolli, 'A provisional catalogue of the family names of living Pteridophytes', *Webbia*, **25**: 219–97, 1970. For comments by R. E. Holttum, see 'The family names of ferns', *Taxon*, **20**: 527–31, 1971.

Actiniopteridaceae Pichi-Serm.
Adiantaceae (Presl) Ching
Angiopteridaceae Fée ex Bommer
Aspidiaceae Mett. ex Frank, nom. illeg. (*Aspidium* Sw., nom. superfl.), but likely to be conserved.
Aspleniaceae Mett. ex Frank
Athyriaceae Alston
Azollaceae Wettstein
Blechnaceae (Presl) Copel.
Cheiropleuriaceae Nakai
Christenseniaceae Ching (*Kaulfussiaceae* nom. illeg.)
Cryptogrammataceae Pichi-Serm. ('Cryptogrammaceae')
Cyatheaceae Kaulf.
Danaeaceae Agardh, nom. superfl. (*Marattia* was included), but likely to be conserved.
Davalliaceae Mett. ex Frank
Dennstaedtiaceae Pichi-Serm.
Dicksoniaceae (Hook.) Bower
Dipteridaceae (Diels) Seward & Dale
Equisetaceae L. C. Rich. ex A. P. de Candolle
Gleicheniaceae (R.Br.) Presl
Grammitidaceae (Presl) Ching
Gymnogrammaceae Herter, nom. illeg. (*Gymnogramma* nom. illeg.), = Hemionitidaceae
Hemionitidaceae Pichi-Serm. (*Gymnogrammaceae* nom. illeg.)
Hymenophyllaceae Link
Hymenophyllopsidaceae Pichi-Serm.

Isoetaceae Dumortier
Kaulfussiaceae Campbell, nom. illeg. (*Kaulfussia* Bl., non Dennst., nec Nees) = Christenseniaceae
Lindsaeaceae Pichi-Serm.
Lomariopsidaceae Alston
Lophosoriaceae Pichi-Serm.
Loxsomaceae Presl
Lycopodiaceae P. Beauv. ex Mirb.
Marattiaceae Berchtold & J. S. Presl
Marsileaceae Mirbel
Matoniaceae Presl
Negripteridaceae Pichi-Serm.
Oleandraceae Ching ex Pichi-Serm.
Ophioglossaceae (R.Br.) Agardh
Osmundaceae Berchtold & J. S. Presl
Parkeriaceae Hook.
Plagiogyriaceae Bower
Polypodiaceae Berchtold & J. S. Presl
Psilotaceae Kanitz
Pteridaceae H. L. G. Reichenbach
Salviniaceae Dumortier
Schizaeaceae Kaulf.
Selaginellaceae Willkomm, nom. superfl. (included *Lycopodium*); proposed for conservation.
Sinopteridaceae Koidz.
Thelypteridaceae Pichi-Serm.
Tmesipteridaceae Nakai
Vittariaceae (Presl) Ching

THE HYBRID GENERIC NAMES
OF ORCHIDS

The hybrid generic names of orchids fall into several categories of usage, accuracy and acceptability. In order to assist the user of such names it has been decided to categorize all orchid hybrid generic names that it has been possible to trace.

The major categories are indicated in the *Dictionary* by small roman numerals as follows:

(i) Natural hybrid generic names currently accepted and in accordance with the taxonomic and nomenclatural opinions expressed elsewhere in this book.

(ii) Natural hybrid generic names not currently accepted for reasons of taxonomy (including misidentifications) and/or of nomenclatural changes involving one or more parents.

(iii) Artificial and natural hybrid generic names currently accepted for hybrid registration purposes by the International Registration Authority for Orchid Hybrids (Royal Horticultural Society, London). These are not always necessarily 'botanically' correct.

(iv) Artificial and natural hybrid generic names not currently accepted for registration purposes because the generic names of one or more parents have been totally rejected for reasons of nomenclature (usually priority) and/or taxonomy.

(v) Artificial and natural hybrid generic names not currently accepted for registration purposes because the generic assignment of one or more parents is not in accordance with currently accepted Horticultural usage.

(vi) Natural and artificial hybrid generic names in regard to which (*a*) no plants apparently have ever existed, (*b*) plants of the alleged hybrid origin may have existed, but whose true identity was either incapable of verification or has been subsequently proved to be other than that claimed.

(vii) Synonymous natural and artificial hybrid generic names rejected for reasons of nomenclature, such as priority, homonymy or the requirements of horticultural nomenclatural conservation ('horticultural equivalents').

ACKNOWLEDGEMENTS

The list of my indebtednesses to colleagues and friends for help with this edition is very long indeed. I cannot sufficiently thank them for this help, which has so greatly enhanced whatever value the dictionary may have, and has immeasurably lightened the burden of revision. There is scarcely a colleague on the staff of the Kew Herbarium whom I have not consulted at some time or other, and it is perhaps unnecessary to list them all, but those who have made major contributions must be singled out for special mention.

The following have been *entirely* or *very largely* responsible for the entries (both family and generic) in the groups mentioned:

KEW

V. S. Summerhayes ⎫ P. F. Hunt ⎭	*Orchidaceae*; much the largest contribution
R. E. Holttum	*Pteridophyta*
N. Y. Sandwith (the late)	*Bignoniaceae*
C. Jeffrey	*Cucurbitaceae*
D. R. Hunt	*Cactaceae* *Wellstediaceae* *Gymnospermae*
A. A. Bullock	*Asclepiadaceae* *Periplocaceae*
L. L. Forman	*Fagaceae* *Pandaceae* *Podoaceae*
C. E. Hubbard	*Gramineae* (family entry)
J. P. M. Brenan	*Leguminosae–Mimosoïdeae* and *Caesalpinioïdeae* (family entry)
J. B. Gillett	*Leguminosae–Papilionoïdeae* (family entry)
S. S. Hooper (Miss)	*Cyperaceae* (family entry and major genera)
R. D. Meikle	*Salicaceae*

ACKNOWLEDGEMENTS

R. A. Blakelock (the late)   *Euonymus*

N. K. B. Robson   *Hypericum*

Mr W. D. Clayton and Dr N. L. Bor have been constantly consulted for help with the genera of *Gramineae*.

OUTSIDE KEW

L. Watson (Southampton)   { *Epacridaceae* and *Ericaceae* (family entries)

L. A. S. Johnson (Sydney)   { *Casuarinaceae* *Cycadales* *Proteaceae*

P. F. Ashton (Sarawak)   *Dipterocarpaceae*

B. L. Burtt (Edinburgh)   *Gesneriaceae* (family entry)

E. J. H. Corner (Cambridge)   *Ficus*

G. T. Prance (New York)   *Chrysobalanaceae*

J. F. Veldkamp (Leiden)   *Averrhoaceae*

Among the numerous correspondents who have kindly furnished information concerning their particular groups, either spontaneously or in response to requests, the following should be mentioned. To all of them I express my sincere thanks.

Brother Alain (*Gonianthes*)
Prof. A. L. Cabrera (miscell. *Compositae*, esp. S. American)
Dr R. C. Carolin (*Geraniaceae, Geranium, Erodium*; *Goodeniaceae, Goodenia, Scaevola*; miscell. Australian genera)
Dr K. L. Chambers (*Krigia, Microseris, Nothocalais*)
J. E. Dandy (*Androsyne*; *Magnoliaceae*; early generic names)
Dr J. W. Hardin (*Hippocastanaceae, Aesculus, Billia*)
Dr H. H. Iltis (American *Cleomaceae*)
The late Dr C. E. Kobuski (*Michoxia, Oncotheca*)
Dr P. W. Leenhouts (miscell. *Loganiaceae*)
J. McNeill (*Arenaria* and related genera)
Dr T. Morley (*Glossoma, Votomita*, etc.)
G. K. Noamesi (through H. H. Iltis) (*Meliaceae–Xylocarpeae*)
Dr S. J. van Ooststroom (*Cryptanthela, Dimerodiscus, Tridynamia*)
Prof. Dr C. G. G. J. van Steenis (much miscell. information, esp. on identity of obscure Malaysian genera, Noronha's genera, etc.)

ACKNOWLEDGEMENTS

J. H. Willis (*Reesia* and other Ewartian genera)
Dr R. E. Woodson (*Coutinia* and *Himatanthus*)
Dr J. H. Zimmermann (*Veratrum* and *Veratreae*)

I must further thank Miss P. Halliday for the immense amount of careful and accurate work that she put in, over several years, on the monotonous job of extracting names and information from standard works such as Post & Kuntze, *Lexicon Generum Phanerogamarum*; Lemée, *Dictionnaire descriptif et synonymique des genres de plantes phanérogamiques*; supplements to the *Index Kewensis*; and many others. Without her invaluable help the preparation of the dictionary would have been almost impossible. Very sincere thanks are also due to Mr W. R. Sykes (now of Christchurch, New Zealand), who gave up two or three summer vacations in order to assist in the work of revising the entries. My thanks are due, too, to the team of typists whose joint efforts succeeded in producing the final typescript in less than twelve months.

Finally, it gives me much pleasure to express to the staff of the Cambridge University Press my great appreciation of the kindness and helpfulness which they extended to me during the passage of the work through the press. Their forbearance over the countless corrections and additions made at every stage to the proofs not only earned my deepest admiration but turned for me a somewhat onerous task into a pleasant and memorable experience.

H. K. A. S.

GEOGRAPHICAL AREAS

The following may require special explanation; the remainder should be self-evident.

Medit.	The entire Mediterranean basin, incl. S. Europe from Portugal to the Balkan Peninsula, Asia Minor, Palestine, and N. Africa from Morocco to Egypt.
C. Asia	Covers approximately the Kazakh, Turkmen, Uzbek, Kirgiz and Tadzhik republics, and the Chinese province of Sinkiang.
E. Asia	Covers China, Korea and Japan.
SE. Asia	Covers Burma, Siam (Thailand) and Indochina (Vietnam, Laos, and Cambodia), and often the adjacent parts of S. China from Yunnan to Hongkong and Hainan.
W. Malaysia	Covers the Malay Peninsula, Sumatra, Java, Borneo and the Philippine Is.
E. Malaysia	Covers Celebes, the Lesser Sunda Is., the Moluccas and New Guinea.
Trop. W. Africa	Senegal to Nigeria.
W. Equatorial Africa	Cameroons and former French Equatorial Africa.
C. America	Guatemala to Panamá.

SIGNS AND ABBREVIATIONS

♂	male
♀	female
♀̣	hermaphrodite
∞	indefinite, numerous
×	(prefixed to generic name) sexual hybrid
+	(prefixed to generic name) asexual hybrid, periclinal chimaera, graft-hybrid
⋆	(prefixed to generic name) = conserved name
§	section (of a genus)
>	more than
<	less than
±	more or less
∥	parallel (to)
⊥	at right angles to, perpendicular to
~	alternatively; 'sometimes included in' or 'reduced to...'
=	equals (cf. Preface, p. xi)
K	calyx with free sepals
(K)	calyx with ± united sepals
C	corolla with free petals
(C)	corolla with ± united petals
A	androecium with free stamens
(A)	androecium with ± united stamens
$\underline{\underline{G}}$	ovary superior
$\overline{G}$	ovary inferior
G 2, 3, 4, etc.	gynoecium or ovary with 2, 3, 4, etc., free carpels
G (2), (3), (4), etc.	gynoecium or ovary with 2, 3, 4, etc., ± united carpels
P	perianth
P. & K.	Post & Kuntze, *Lexicon Generum Phanerogamarum* (Stuttgart, 1903 ['1904'])

A very large number of abbreviations of descriptive terms, geo-graphical names, etc., are employed in the Dictionary (especially in family descriptions) in order to save space. It is believed that these are self-explanatory; they are therefore not listed here.

A DICTIONARY OF THE FLOWERING
PLANTS AND FERNS

A

Aa Reichb. f. Orchidaceae. 15 trop. Am.
Aakesia Baill. = Akesia Tussac = Blighia Koenig (Sapindac.).
Aalius Kuntze = Sauropus Bl. (Euphorbiac.).
Aalius Lam. = Breynia J. R. & G. Forst. (Euphorbiac.).
Aama B. D. Jacks. = seq.
Aamia Hassk. = Adamia Wall. = Dichroa Lour. (Hydrangeac.).
Aaronsohnia Warburg & Eig. Compositae. 1 Palestine.
Ababella Comm. ex Moewes = Turraea L. (Meliac.).
Abacopteris Fée = Pronephrium Presl (Thelypteridac.).
Abacosa Alef. = Vicia L. (Legumin.).
Abalemis Rafin. = Anemone L. (Ranunculac.).
Abalon Adans. = Helonias L. (Liliac.).
Abalum Adans. = praec.
Abama Adans. = Narthecium Huds. (Liliac.).
Abamineae J. G. Agardh = Liliaceae–Narthecieae Kunth.
Abandion Adans. = Bulbocodium L. (Liliac.).
Abandium Adans. = praec.
Abaphus Rafin. = seq.
Abapus Adans. = Gethyllis L. (Amaryllidac.).
Abarema Pittier (~ Pithecellobium Mart.). Leguminosae. 50 trop. (exc. Afr.).
Abasicarpon (Andrz. ex Reichb.) Reichb. = Cheiranthus L. (Crucif.).
Abasaloa Benth. & Hook. f. = seq.
Abasoloa La Llave. Compositae. 1 Mex.
Abatia Ruiz et Pav. Flacourtiaceae. 9 trop. S. Am.
Abauria Becc. = Koompassia Maing. (Legumin.).
Abavo Risler = Adansonia L. (Bombacac.).
Abazicarpus Andrz. ex DC. = Abasicarpon (Andrz. ex Reichb.) Reichb. = Cheiranthus L. (Crucif.).
Abbevillea Berg = Campomanesia Ruiz & Pav. (Myrtac.).
Abbotia Rafin. = Triglochin L. (Juncaginac.).
Abbottia F. Muell. = Timonius DC. (Rubiac.).
Abdominea J. J. Smith. Orchidaceae. 1 Malay Penins., Java.
Abdra Greene = Draba L. (Crucif.).
Abebaia Baehni (~ Manilkara Adans.). Sapotaceae. 1 (?) E. Malaysia.
Abela Salisb. = Chamaecyparis Spach + Juniperus L. (Cupressac.).
Abelemis Britton = Abalemis Rafin. = Anemone L. (Ranunculac.).
Abelia R.Br. (~ Linnaea Gronov. ex L.). Caprifoliaceae. 30 Himal. to E. As., Mex.

ABELICEA

Abelicea Reichb. = Zelkova Spach (Ulmac.).

Abelicia Kuntze = praec.

Abeliophyllum Nakai. Oleaceae. 1 Korea.

Abelmoschus Medik. (= Laguna Cav.) (~ Hibiscus L.). Malvaceae. 15 trop. Afr. & As., Austr.

Abena Neck. = Stachytarpheta Vahl (Verbenac.).

Aberemoa Aubl. Annonaceae. 25 trop. Am.

Aberia Hochst. = Doryalis E. Mey. (Flacourtiac.).

Abesina Neck. = Verbesina L. (Compos.).

Abies Mill. Pinaceae. 50 N. temp., C. Am. The firs are evergreen trees with needle ls. borne directly on the stems. No short shoots. On the main stem the symmetry is radial, whilst on the horizontal branches the ls. twist so as to get their surfaces all much in one plane. If the top bud or leader be destroyed, however, a branch bud below it takes up the vertical growth and radial symmetry. Cones large, arranged much like *Pinus*; ♀ often brightly coloured, though wind-fertilised. The carpel scales are large and appear on the outside of the cone between the ovuliferous scales. The cone ripens in one year.

A. alba Mill. (*A. pectinata* DC.) (silver fir, mts. of S. Eur.) yields a valuable wood, Alsatian turpentine, etc. *A. balsamea* (L.) Mill. (balsam fir, E. N. Am.) yields the turpentine known as Canada balsam. Many others yield useful timbers and resins. Handsome trees; commonly cult. are *A. concolor* (Gord.) Hildebr. (NW. Am.), *A. firma* Sieb. & Zucc. (Japan), *A. procera* Rehd. (*A. nobilis* (Dougl. ex D. Don) Lindl.) (NW. Am.), *A. nordmanniana* (Steven) Spach (Caucasus), *A. pinsapo* Boiss. (Spain), *A. spectabilis* (D. Don) G. Don (*A. webbiana* Lindl.) (Himal.).

Abietaceae ('-ideae') S. F. Gray = Pinaceae Lindl.

Abietia Kent = Pseudotsuga Carr. + Keteleeria Carr. (Pinac.).

Abiga St-Lag. = Ajuga L. (Labiat.).

Abildgaardia Vahl = Fimbristylis Vahl (Cyperac.).

Abildgardia Reichb. = praec.

Abilgaardia Poir. = praec.

Abioton Rafin. = Capnophyllum Gaertn. (Umbellif.).

Ablania Aubl. = Sloanea L. (Elaeocarpac.).

Abobra Naud. Cucurbitaceae. 1 S. Am.

Abola Adans. = Cinna L. (Gramin.).

Abola Lindl. = Caucaea Schltr. (Orchidac.).

Abolaria Neck. = Globularia L. (Globulariac.).

Abolboda Humb. & Bonpl. Abolbodaceae. 20 trop. S. Am.

Abolbodaceae (Suesseng. & Beyerle) Nak. Monocots. (Farinosae). 2/20 trop. S. Am. Perenn. marsh herbs, near *Xyridac.* (*q.v.*), but differing chiefly in the usu. spirally arranged ls., the more distant bracts of the infl., the similar sepals, the united (sometimes blue or white) pets., the spinose pollen, the absence of stds., the single conspic. appendic. style, the 3-loc. ovary with axile plac., the anatr. ovules, the compressed or prismat. or obliquely trunc. ± obliquely striate seeds, and the large curved embr. Genera: *Abolboda, Orectanthe.*

Aborchis Steud. = Disa Bergius (Orchidac.).

Abortopetalum Degener. Malvaceae. 2 Hawaii.

Abramsia Gillespie = Airosperma Lauterb. & Schum. (Rubiac.).

2

Abrochis Neck. = Disa Bergius (Orchidac.).
Abrodictyum Presl. Hymenophyllaceae. 1 E. Malaysia.
Abroma Jacq. (nom. illegit.) = Ambroma L.f. (Sterculiac.).
Abroma Mart. = Theobroma L. (Sterculiac.).
Abromeitia Mez. Myrsinaceae. 1 New Guinea.
Abromeitiella Mez. Bromeliaceae. 5 trop. S. Am.
Abronia Juss. Nyctaginaceae. 35 N. Am. Anthocarp usu. winged.
Abrophaës Rafin. = Miconia Ruiz & Pav. (Melastomatac.).
Abrophyllaceae Nak. = Escalloniaceae–Cuttsiëae Engl.
Abrophyllum Hook. f. Escalloniaceae. 2 E. Austr.
Abrotanella Cass. Compositae. 20 Rodrigues, New Guinea, Austr., N.Z., Auckland Is., temp. S. Am.
Abrotanum Duham. = Artemisia L. (Compos.).
Abrus Adans. Leguminosae. 12 trop. *A. precatorius* L. has hard red seeds with black tips (crab's eyes), strung into necklaces, rosaries, etc., and used as weights (*rati*) in India (cf. *Adenanthera*). The roots are used in India as Indian liquorice.
Abryanthemum Neck. = Carpobrotus N. E. Br. (Aïzoac.).
Absinthium Mill. = Artemisia L. (Compos.).
Absintion Adans. = praec.
Absolmsia Kuntze. Asclepiadaceae. 1 SW. China, 1 Borneo.
Abulfali Adans. = Thymbra L. (Labiat.).
Abumon Adans. = Agapanthus L'Hérit. (Alliac.).
Abuta Aubl. Menispermaceae. 35 trop. S. Am. *A. rufescens* Aubl. (Guiana) yields white Pareira root.
Abutilaea F. Muell. = Abutilon Mill. (Malvac.).
Abutilodes Kuntze = Modiola Moench (Malvac.).
Abutilon Mill. Malvaceae. 100 +, trop. & subtrop. No epicalyx. Fl. mech. like *Malva sylvestris*, but some are self-sterile; the sta. do not move down, and the styles emerge through the anther-mass. Many visited by humming-birds. *A. avicennae* Gaertn. cult. in China for fibre (China jute).
Abutilothamnus Ulbr. Malvaceae. 1 trop. S. Am.
Abutua Batsch = Abuta Aubl. (Menispermac.).
Abutua Lour. = Gnetum L. (Gnetac.).
Acacallis Lindl. = Aganisia Lindl. (Orchidac.).
Acachmena H. P. Fuchs. Cruciferae. 2 C. As.
Acacia Mill. Leguminosae. 750–800 trop. & subtrop. Mostly trees (wattles); typical leaf-form bipinnate with ∞ leaflets and small scaly stips. About 300 spp., forming the § *Phyllodineae* (chiefly in Austr., where they are char., and Polynes.), have simple leaf-like phyllodes, i.e. petioles (or primary axes; cf. *Proc. Linn. Soc. N.S.W.* **45**: 24–47, 1920) flattened with their edges upwards, exposing less surface to radiation. Inspection shows the phyllode to be a l.-structure with axillary bud, but note that it is not a l. turned edgewise, though it shows no twist. Occasionally there are reversions to type (i.e. to the ancestral form) on the plant, some phyllodes occurring with leaf-blades of the ordinary bipinnate type. This is still better seen in germinating seedlings. The first ls. are typical bipinnate ls., followed by others with slightly flattened stalks and less blade, and so on, until finally only phyllodes are produced. In *A. alata* R.Br. and

3

ACANTHAMBROSIA

Acanthambrosia Rydb. Compositae. 1 Lower Calif.

Acantharia Rojas. Leguminosae. 1 Argent.

Acanthea Pharm. ex Wehmer. Quid? Acanthac.? Brazil. (N.B. Not the source of *lignum muira-puama*, which is *Ptychopetalum* Benth., Olacac.)

Acanthella Hook. f. Melastomataceae. 2 trop S. Am.

Acanthephippium Blume. Orchidaceae. 15 trop. As. to Fiji. The axial outgrowth from the base of the column, common in *O.*, is here very great and bends first downwards, then up, removing the insertion of the lateral sepals and labellum to a distance from the column.

Acanthinophyllum Burger = Clarisia Ruiz & Pav. (Morac.).

Acanthium Fabr. (uninom.) = *Onopordum* L., sp. (Compos.).

Acanthobotrya Eckl. & Zeyh. = Lebeckia Thunb. (Legumin.).

Acanthobotrys Clem. = praec.

Acanthocalycium Backeb. (∼ Echinopsis Zucc.). Cactaceae. 6 Argent.

Acanthocalyx (DC.) Van Tiegh. = Morina L. (Morinac.).

Acanthocardamum Thell. Cruciferae. 1 S. Persia.

Acanthocarpaea Dalla Torre & Harms = seq.

Acanthocarpea Klotzsch = Limeum L. (Aïzoac.).

Acanthocarpus Kuntze = praec.

Acanthocarpus Lehm. Xanthorrhoeaceae. 4 Australia.

Acanthocarya Arruda ex Endl. = Acantacaryx Arr. ex Koster = Caryocar L. (Caryocarac.).

Acanthocaulon Klotzsch = Platygyne Mercier (Euphorbiac.).

Acanthocephala Backeb. = Brasilicactus Backeb. (Cactac.).

Acanthocephalus Kar. & Kir. Compositae. 2 C. As.

Acanthocereus (Berger) Britton & Rose. Cactaceae. 8 SE. U.S. to Venez., Colombia, NE. Braz.

Acanthochiton Torr. Amaranthaceae. 1 SW. U.S.

Acanthocladium F. Muell. = Helichrysum Mill. (Compos.).

Acanthocladus Klotzsch ex Hassk. = Polygala L. (Polygalac.).

Acanthococos Barb. Rodr. Palmae. 4 S. Am.

Acanthodesmos C. D. Adams & du Quesnay. Compositae. 1 Jamaica.

Acanthodion Lem. = seq.

Acanthodium auct. = Acanthocladium F. Muell. = Helichrysum Mill. (Compos.).

Acanthodium Delile = Blepharis Juss. (Acanthac.).

Acanthodus Rafin. = Acanthus L. (Acanthac.).

Acanthoglossum Blume = Pholidota Lindl. (Orchidac.).

Acanthogonum Torr. (∼ Chorizanthe R.Br.). Polygonaceae. 2 SW. U.S.

Acantholepis Less. Compositae. 1 W. As.

Acantholimon Boiss. Plumbaginaceae. 120 E. Medit. to C. As. Deserts and stony places in mt. regions.

Acantholinum K. Koch = praec.

Acantholippia Griseb. (∼ Lippia L.). Verbenaceae. 5 subtrop. & temp. S. Am.

Acantholobivia Backeb. = Lobivia Britton & Rose (Cactac.).

Acantholobivia Y. Ito = Rebutia K. Schum. (Cactac.).

Acantholoma Gaudich. ex Baill. = Pachystroma Klotzsch (Euphorbiac.).

Acanthomintha A. Gray. Labiatae. 3 Calif.

*Acanthonema Hook. f. Gesneriaceae. 1 W. Afr.
Acanthonotus Benth. = Indigofera L. (Legumin.).
Acanthonychia (DC.) Rohrb. = Cardionema DC. (Caryophyllac.).
Acanthopale C. B. Clarke. Acanthaceae. 15 trop. Afr., SE. As.
Acanthopanax (Decne & Planch.) Miq. Araliaceae. 50 E. As., Philipp., Malay Penins.
Acanthopetalus Y. Ito = Setiechinopsis (Backeb.) De Haas = Arthrocereus A. Berger (Cactac.).
Acanthophaca Nevski = Astragalus L. (Legumin.).
Acanthophippium Blume = Acanthephippium Blume (Orchidac.).
Acanthophoenix H. Wendl. Palmae. 2 Mascarenes.
Acanthophora Merr. Araliaceae. 1 W. Malaysia, Celebes.
Acanthophyllum Hook. & Arn. = Nassauvia Juss. (Compos.).
Acanthophyllum C. A. Mey. Caryophyllaceae. 50 SW. & C. As., Siberia. Mostly desert xerophytes with prickly leaves.
Acanthophyton Sch. Bip. = seq.
Acanthophytum Less. = Cichorium L. (Compos.).
Acanthoplana C. Koch = Polylophium Boiss. (Umbellif.).
Acanthoprasium (Benth.) Spenn. = Ballota L. (Labiat.).
Acanthopsis Harv. Acanthaceae. 8 S. Afr.
Acanthopteron Britton (~ Mimosa L.). Leguminosae. 1 Mexico.
Acanthopyxis Miq. ex Lanj. = Caperonia A. St-Hil. (Euphorbiac.).
Acanthorhipsalis Britton & Rose = Rhipsalis Gaertn. (Cactac.).
Acanthorrhinum Rothm. Scrophulariaceae. 1 NW. Afr.
Acanthorrhiza H. Wendl. = Cryosophila Blume (Palm.).
Acanthosabal Proschowsky = Acoelorraphe Wendl. (Palm.).
Acanthoscyphus Small (~ Oxytheca Nutt.). Polygonaceae. 1 SW. U.S.
Acanthosicyos Welw. ex Hook. f. Cucurbitaceae. 2 S. trop. Afr. *A. horridus* Welw. ex Hook. f., the *narras*, is a remarkable plant growing on sand dunes (cf. *Welwitschia*). The thick root is very long (up to 12 m.). Above ground is a thorny shrub, without tendrils; the thorns are probably the modified tendrils.
Acanthosicyus P. & K. = praec.
Acanthosonchus (Sch. Bip.) Kirp. = Atalanthus D. Don (Compos.).
Acanthosperma Vell. = Acicarpha Juss. (Calycerac.).
Acanthospermum Schrank. Compositae. 8 W.I., S. Am., Galápagos, Madag.
Acanthosphaera Warb. (~ Ogcodeia Bur.). Moraceae. 2 Amaz. Brazil.
Acanthospora Spreng. = Tillandsia L. (Bromeliac.).
Acanthostachys Link, Klotzsch & Otto. Bromeliaceae. 3 S. Am.
Acanthostachyum Benth. & Hook. f. = praec.
Acanthostemma Blume = Hoya R.Br. (Asclepiadac.).
Acanthostyles R. M. King & H. Rob. (~ Eupatorium L.). Compositae. 2 Boliv. to Argent.
Acanthosyris Griseb. Santalaceae. 3 temp. S. Am.
Acanthothamnus T. S. Brandegee. Celastraceae. 1 Mex.
Acanthothapsus Gandog. = Verbascum L. (Scrophulariac.).
Acanthotheca DC. = Dimorphotheca Moench (Compos.).
Acanthotreculia Engl. Moraceae. 1 W. Equat. Afr.
Acanthotrichilia (Urb.) Cook & Collins = Trichilia L. (Meliac.).

Acanthura Lindau. Acanthaceae. 1 Brazil.

Acanthus L. Acanthaceae. 50 S. Eur., trop. & subtrop. As., Afr. Mostly xero. with thorny ls. (those of *A. spinosus* L. furnished, it is supposed, the pattern for the decoration of the capitals of Corinthian columns). *A. ilicifolius* L. is part of the palaeotrop. mangrove veg. Fl. a large bee-fl.; there is no upper lip to the C, and the protection of the pollen, etc., is undertaken by the K. The anthers form a box by fitting closely together at the sides, and shed their pollen sideways into it, where it is held by hairs till an insect probing for honey forces the filaments of the sta. apart and receives a shower of pollen on its head (loose-pollen mechanism, cf. many *Scrophulariaceae, Ericaceae*, etc.). In the young fl. the style is behind the anthers, later on it bends down so as to touch a visiting insect. The fr. explodes; large retinacula on the seeds.

Acanthyllis Pomel = Anthyllis L. (Legumin.).

Acareosperma Gagnep. Vitidaceae. 1 Indoch.

Acarna Boehm. = Atractylis L. (Compos.).

Acarna Hill = Cirsium Mill. + Tyrimnus Cass. (Compos.).

Acarnaceae Link = Compositae–Cynareae Spreng.

Acarpha R.Br. ex Benth. = Latrobea Meissn. (Legumin.).

Acarpha Griseb. Calyceraceae. 8 temp. S. Am.

Acarphaea Harv. & Gray ex A. Gray = Chaenactis DC. (Compos.).

Acaste Salisb. = Babiana Ker-Gawl. (Iridac.).

Acaulon N. E. Br. (1928; non C. Muell. 1847—Musci) = Aïstocaulon v. Poelln. (Aïzoac.).

Acca Berg (~ Psidium L.). Myrtaceae. 6 S. Am.

Accia A. St-Hil. = Fragariopsis A. St-Hil. (Euphorbiac.).

Accoromba Baill. = seq.

Accorombona Endl. = Galega L. (Legumin.).

Acedilanthus Benth. & Hook. f. = Acelidanthus Trautv. & Mey. = Veratrum L. (Liliac.).

Acelica Rizzini. Acanthaceae. 2 S. Am.

Acelidanthus Trautv. & Mey. = Veratrum L. (Liliac.).

Acentra Phil. = Hybanthus Jacq. (Violac.).

Acer L. Aceraceae. 200 N. temp. (esp. in hill districts) and trop. mts.; many in China and Japan; few SE. As.; 1 (*A. caesium* (Reinw. ex Bl.) Kosterm.) in W. Malaysia. Trees & shrubs, with opp. exstip. ls., decid. or evergr. Ls. often simple entire, more commonly 3- or 5-lobed, occasionally cpd. One may go through a collection of *A.* in an herbarium or elsewhere, comparing the ls. as to degree of development of the drip-tips (acum. apices to easily wetted ls., from which the water drips off rapidly after a shower, cf. *Ficus*), noting the kind of climate from which each has come. There is a correlation between length of tip and wetness of climate.

Large winter buds, covered by scale ls. In many spp. transitional forms may be seen, as the bud elongates in spring, between the scales and the green ls., showing that the scale = not the whole l., but the leaf base. In the § *Negundo* there are no scales, but the bud is protected by the base of the petiole of the l. in whose axil it arises.

The ls. commonly exhibit varnish-like smears, of sticky consistence, known as honey-dew, the excretion of aphids which live on the ls.; the insect bores into the tissues, sucks their juices, and ejects a drop of honey-dew on an average once in half an hour. In passing under a tree infested with aphids one may sometimes feel the drops falling like a fine rain (see *Pithecellobium*). The fluid is rich in sugar. When the dew falls the hygroscopic honey-dew takes it up and spreads over the l.; then later in the day evap. reduces it to a varnish on the surface. Many other trees exhibit this phenomenon, e.g. lime, beech, oak (Büsgen, 'Der Honigthau', *Jenaische Zeitschr. Naturwiss.* 25, 1891).

Fls. in racemes, sometimes contracted to corymbs or umbels, reg., polygamous, not conspic.; formula usu. K 5, C 5, A 4+4, $\underline{G}$ (2). Apetaly in some. 3 cpls. are frequent, esp. in the end fl. of a raceme. ♂ fls. protandrous; honey freely exposed on the disk, available to insects of all kinds. Fr. a samara. In germination, the long green cotyledons come above the soil almost at once.

A. saccharum Marshall (*A. saccharinum* Wangenh.) and others of E. N. Am. yield maple sugar (2-4 lb. a tree) obtained by boring holes in February and March and collecting and evaporating the juice. Many yield good timber and charcoal.

*Aceraceae Juss. Dicots. 3/200 N. temp. & trop. mts. Trees and shrubs; ls. opp., petiolate, exstip., simple, entire or more often palmately or pinnately lobed, or cpd. Infl. racemose, corymbose, or fasciculate. Fls. reg., andromonoec., androdioec., dioec., etc., 5-4-merous, usu. dichlam. Disk annular or lobed or reduced to teeth, rarely absent. A 4-10, usu. 8, hypog., perig., or on disk; ♂ fl. with rudimentary G. $\underline{G}$ (2) 2-loc., lat. compressed; styles 2, free or joined below; ov. 2 in each loc., orthotr. to anatr., with dorsal raphe. Fr. of 2 samaras, separating when ripe. Seeds usu. solitary, exalb., the cotyledons irreg. folded. Many yield good timber, sugar, etc. Largely represented in the Tertiary.

Genera: *Acer, Negundo, Dipteronia*. Related to *Sapindac.*; prob. some connection also with *Hamamelidac., Altingiac.* and *Platanac.*

× **Aceraherminium** Camus. Orchidaceae. Gen. hybr. (i) (Aceras × Herminium).

Aceranthes Reichb. = seq.

Aceranthus Morr. & Decne = Epimedium L. (Berberidac.).

Aceras R.Br. Orchidaceae. 1 Eur., N. Afr.

× **Aceras-Herminium** Gremli = × Aceraherminium Camus (Orchidac.).

Acerates Ell. (~Asclepias L.). Asclepiadaceae. 8 N. Am.

Aceratium DC. Elaeocarpaceae. 30 Malaysia, Austr.

Aceratorchis Schlechter. Orchidaceae. 2 China, Tibet.

Aceriphyllum Engl. (1890; non Fontaine 1889—gen. foss.) = Mukdenia Koidz. (Saxifragac.).

Acerophyllum P. & K. = praec.

Acerotis Rafin. = Asclepias L. (Asclepiadac.).

Acetosa Mill. (~Rumex L.). Polygonaceae. 30 N. hemisph., trop. Afr. mts.

Acetosella Kuntze = Oxalis L. (Oxalidac.).

Acetosella (Meissn.) Fourr. (~Rumex L.). Polygonaceae. 5 N. temp.

Achaemenes St-Lag. = Achimenes P.Br. (Gesneriac.).

Achaenipodium T. S. Brandegee. Compositae. 1 Mex.

Achaeta Fourn. = Calamagrostis Adans. + Koeleria Pers. (Gramin.).

Achaete Hack. = praec.

Achaetogeron A. Gray. Compositae. 20 Ariz., Mex.

Achania Sw. = Malvaviscus Guett. (Malvac.).

Achantia A. Chev. = Mansonia J. R. Drumm. (Sterculiac.).

Acharia Thunb. Achariaceae. 1 S. Afr.

*Achariaceae Harms. Dicots. 3/3 S. Afr. Habit various: shrublets, or stemless or climbing herbs, with alt., or radic., simple, cren., serr. or lobed, exstip. ls. Fls. reg., ♂ ♀, monoec., sol. or few in axill. fasc. or rac. K 3–5, open; C (3–5), in ♂ of *Guthriea* long-adnate to K, tube campan., lobes ± valv.; A 3–5 epipet., connective expanded, anth. intr.; stds. 3–5, short, fleshy, alt. with A; G̲ (3–5), 1-loc., with 2–∞ ov. on pariet. plac., styles 3–5, simple or bifid. Fr. a glob. to lin. 3–5-valved caps.; seeds arillate, with copious endosp. Genera: *Acharia, Ceratiosicyos, Guthriea*. Related to *Passiflorac.* and *Cucur-* [*bitac.*

Acharitea Benth. Dicrastylidaceae (?). 1 Madag.

Achariterium Bluff & Fingerh. = Filago L. (Compos.).

Achasma Griff. Zingiberaceae. 20 Indomal.

*Achatocarpaceae Heimerl. Dicots. (Centrospermae). 2/10 warm Am. Shrubs or small trees, often thorny, with normal growth in thickness; ls. simple, ent., alt., exstip. Fls. ♂ ♀, dioec., in small axillary branched bracteate cymes. P 4–5, imbr., persist.; A 10–20, with filif. fil. and basifixed elongate-oblong anth.; G̲ (2), 1-loc., with 1 basal campylotr. ov., and 2 conspic. simple free subul. divergent ± uncinate styles. Fr. a small 1-seeded berry; seed exarillate, with copious mealy perisp. Genera: *Achatocarpus, Phaulothamnus*. Rather closely related to *Phytolaccaceae*.

Achatocarpus Triana. Achatocarpaceae. 10 (or 1?) Mex. to Argent.

Achemora Rafin. = Archemora DC. = Oxypolis Rafin. (Umbellif.).

Achetaria Cham. & Schlechtd. Scrophulariaceae. 5 trop. Am.

Achillaea L. = seq.

Achillea L. Compositae. 200 N. temp.

Achilleopsis Turcz. = Rulingia R.Br. (Sterculiac.).

Achillios St-Lag. = Achillea L. (Compos.).

Achilus Hemsl. = Globba L. (Zingiberac.).

Achimenes P.Br. = Columnea L. (Gesneriac.).

Achimenes Pers. Gesneriaceae. 50 trop. Am.

Achimenes Vahl = Artanema D. Don (Scrophulariac.).

Achimus Poir. = Achymus Vahl ex Juss. = Streblus Lour. (Morac.).

Achiranthes P.Br. = Achyranthes L. (Amaranthac.).

Achirida Horan. = Canna L. (Cannac.).

Achiridia Bak. = praec.

Achironia Steud. = Achyronia L. (Legumin.).

Achlaena Griseb. Gramineae. 1 W.I. (Cuba, Jam.).

Achlamydosporae Benth. & Hook. f. A 'Series' of Dicot. fams. comprising the *Loranthac., Santalac.* and *Balanophorac.*

Achlyphila Maguire & Wurdack. Xyridaceae. 1 Venez.

Achlys DC. Podophyllaceae. 2 Japan & Pacif. N. Am. The perianth aborts early in development.

Achmaea Steud. = Aechmea Ruiz & Pav. (Bromeliac.).

Achmandra Wight = Aechmandra Arn. = Kedrostis Medik. (Cucurbitac.).
Achmea Poir. = Aechmea Ruiz & Pav. (Bromeliac.).
Achnatherum Beauv. (~ Stipa L.). Gramineae. 10 temp. As.
Achneria Beauv. = Eriachne R.Br. (Gramin.).
Achneria Munro = Afrachneria Sprague (Gramin.).
Achnodon Link = Phleum L. (Gramin.).
Achnodonton Beauv. = Phleum L. (Gramin.).
Achnophora F. Muell. Compositae. 1 Austr.
Achnopogon Maguire, Steyerm. & Wurd. Compositae. 3 Venez.
Achraceae Roberty = Sapotaceae Juss.
Achradelpha O. F. Cook (~ Achras L.). Sapotaceae. 1 S. Am.
Achradotypus Baill. Sapotaceae. 6 New Caled.
Achrantes Pfeiff. = seq.
Achranthes Dum. = Achyranthes L. (Amaranthac.).
Achras L. = Manilkara Adans. (Sapotac.).
Achratinis Kuntze = Arachnitis Phil. (Corsiac.).
Achratinitaceae Barkley = Corsiaceae Becc.
Achroanthes Rafin. = Malaxis Soland. ex Sw. (Orchidac.).
Achrochloa B. D. Jacks. (sphalm.) [= § Achrochloa Griseb.] = Airochloa Link
= Koeleria Pers. (Gramin.).
Achromochlaena P. & K. = seq.
Achromolaena Cass. = Cassinia R.Br. (Compos.).
Achroöstachys Benth. (sphalm.) = Athroöstachys Benth. ex Benth. & Hook. f.
(Gramin.).
Achrouteria Eyma. Sapotaceae. 1 trop. S. Am.
Achrysum A. Gray = Calocephalus R.Br. (Compos.).
Achudemia Blume. Urticaceae. 3 E. As., Java.
Achudenia Benth. & Hook. f. (sphalm.) = praec.
Achupalla Humb. = Puya Mol. (Bromeliac.).
Achymenes Batsch = Achimenes Vahl = Artanema D. Don (Scrophulariac.).
Achymus Vahl ex Juss. = Streblus Lour. (Morac.).
Achyrachaena Schau. Compositae. 1 NW. U.S. Pappus of broad, silvery
scales; fruit-heads used as 'everlastings'.
Achyracharna Walp. (sphalm.) = praec.
Achyrantes L. = seq.
Achyranthes L. Amaranthaceae. 100 trop. & subtrop., mostly Afr. & As.
(perh. only 3–5 v. variable spp.).
Achyranthus Neck. = praec.
Achyrastrum Neck. = Hyoseris L. (Compos.).
Achyrobaccharis Sch. Bip. = Baccharis L. (Compos.).
Achyrocalyx Benoist. Acanthaceae. 4 Madag.
Achyrochoma B. D. Jacks. (sphalm.) = Achyrocoma Cass. = Vernonia Schreb.
(Compos.).
Achyroclina P. & K. = seq.
Achyrocline Less. Compositae. 20 trop. Am. & Afr., Madag.
Achyrocoma Cass. = Vernonia Schreb. (Compos.).
Achyrocoma P. & K. = seq.
Achyrocome Schrank = Elytropappus Sch. Bip. (Compos.).

ACHYRODES

Achyrodes Boehm. = Lamarkia Moench (Gramin.).
Achyroma Wendl. = Achyrocoma Cass. = Vernonia Schreb. (Compos.).
Achyronia [L.] Boehm. = Aspalathus L. (Legumin.).
Achyronia Wendl. = Priestleya DC. (Legumin.).
Achyronychia Torr. & Gray. Caryophyllaceae. 2 SW. U.S., Mex.
Achyropappus Bieb. ex Fisch. = Tricholepis DC. (Compos.).
Achyropappus Kunth = Bahia Lag. + Schkuhria Roth (Compos.).
Achyrophorus Guett. = Hypochoeris L. (Compos.).
Achyropsis (Moq.) Hook. f. Amaranthaceae. 6 trop. & S. Afr.
Achyroseris Sch. Bip. = Scorzonera L. (Compos.).
Achyrospermum Blume. Labiatae. 30 trop. Afr., Madag., Himal., Malaysia
Achyrostephus G. Kunze ex Reichb. (nomen). Compositae. Quid?
Achyrothalamus O. Hoffm. Compositae. 2 trop. E. Afr.
Acia Schreb. = Acioa Aubl. + Couepia Aubl. (Chrysobalanac.).
Aciachna P. & K. = seq.
Aciachne Benth. Gramineae. 1 W. trop. S. Am.
Acialyptus B. D. Jacks. (sphalm.) = Acicalyptus A. Gray (Myrtac.).
Acianthera P. & K. = Acisanthera P.Br. (Melastomatac.).
Acianthera Scheidw. = Pleurothallis R.Br. (Orchidac.).
Acianthus R.Br. Orchidaceae. 20 Austr., N.Z., Solomon Is., New Caled.
Acicalyptus A. Gray = Cleistocalyx Bl. = Syzygium Gaertn. (Myrtac.).
Acicarpa R.Br. = Acicarpha Juss. (Calycerac.).
Acicarpa Raddi = Panicum L. (Gramin.).
Acicarpha Juss. Calyceraceae. 5 C. & S. Am.
Acicarphaea Walp. = Acarphaea Harv. & Gray ex A. Gray = Chaenactis DC (Compos.).
Acicarpus P. & K. = Acicarpa Raddi = Panicum L. (Gramin.).
Aciclinium Torr. & Gray = Bigelowia DC. (Compos.).
Acidandra Mart. ex Spreng. = Zollernia Mart. (Legumin.).
Acidanthera Hochst. Iridaceae. 40 trop. & S. Afr.
Acidanthus Clem. = Acianthus R.Br. (Orchidac.).
Acidocroton Griseb. Euphorbiaceae. 10 W.I.
Acidodendrum Kuntze (sphalm.) = Acinodendron Rafin. = Miconia L. (Melastomatac.).
Acidolepis Clem. = Acilepis D. Don = Vernonia Schreb. (Compos.).
Acidosperma Clem. = Acispermum Neck. = Coreopsis L. (Compos.).
Acidoton P.Br. = Securinega Comm. ex Juss. (Euphorbiac.).
*****Acidoton** Sw. Euphorbiaceae. 6 W.I., C.Am., N. trop. S. Am.
Aciella Van Tiegh. = Amylotheca Van Tiegh. (Loranthac.).
Acilepis D. Don = Vernonia Schreb. (Compos.).
Acinax Rafin. = Amomum L. (Zingiberac.).
Acineta Lindl. Orchidaceae. 15 trop. Am.
Acinodendron Rafin. = Miconia Ruiz & Pav. (Melastomatac.).
Acinodendrum Kuntze = praec.
Acinolis Rafin. = Miconia Ruiz & Pav. (Melastomatac.).
Acinos Mill. (~ Calamintha Lam.). Labiatae. 10 Eur., Medit., to C. As. 8 Persia.
Acinotum (DC.) Reichb. = Matthiola R.Br. (Crucif.).

Acinotus Baill. = praec.
Acioa Aubl. Chrysobalanaceae. 3 NE. S. Am., with ed. oily seed; 30 trop. Afr.
Acioja Gmel. = praec.
Aciotis D. Don. Melastomataceae. 30 trop. Am., W.I.
Acipetalum Turcz. = Cambessedesia DC. + Pyramia Cham. (Melastomatac.).
Aciphylla J. R. & G. Forst. Umbelliferae. 35 Austr., N.Z.
Aciphyllaea A. Gray = Hymenatherum Cass. (Compos.).
Aciphyllum Steud. = Chorizema Labill. (Legumin.).
Acis Salisb. = Leucojum L. (Amaryllidac.).
Acisanthera P.Br. Melastomataceae. 35 trop. Am., W.I.
Acispermum Neck. = Coreopsis L. (Compos.).
Acistoma Zipp. ex Span. = Woodfordia Sm. (Lythrac.).
Ackama A. Cunn. Cunoniaceae. 3 E. Austr., N.Z.
Acladodea Ruiz & Pav. = Talisia Aubl. (Sapindac.).
Acladodia Dalla Torre & Harms = praec.
Acleanthus Clem. = Acleisanthes A. Gray (Nyctaginac.).
Acleia DC. = Senecio L. (Compos.).
Acleisanthes A. Gray. Nyctaginaceae. 10 SW. U.S.
Acleja P. & K. = Acleia DC. = Senecio L. (Compos.).
Aclema P. & K. = Aklema Rafin. = Euphorbia L. (Euphorbiac.).
Aclinia Griff. = Dendrobium Sw. (Orchidac.).
Aclisanthes P. & K. = Acleisanthes A. Gray (Nyctaginac.).
Aclisia E. Mey. (~ Pollia Thunb.). Commelinaceae. 10 Indomal., Austr.
Acmadenia Bartl. & Wendl. f. Rutaceae. 30 S. Afr.
Acmanthera Griseb. Malpighiaceae. 3 S. Am.
Acmella Rich. ex Pers. = Spilanthes Jacq. (Compos.).
Acmena DC. = Syzygium Gaertn. (Myrtac.).
Acmenosperma Kausel = Syzygium Gaertn. (Myrtac.).
Acmispon Rafin. (~ Hosackia Dougl., ~ Lotus L.). Leguminosae. 10 N. Am.
Acmopyle Pilger. Taxaceae. 3 New Caled., Fiji.
Acmostemon Pilger = Ipomoea L. (Convolvulac.).
Acmostigma P. & K. = seq.
Acmostima Rafin. = Palicourea Aubl. (Rubiac.).
Acnadena Rafin. = Cordia L. (Ehretiac.).
Acnida L. = Amaranthus L. (Amaranthac.).
Acnide Mitch. = praec.
Acnista Durand = Acnida L. = Amaranthus L. (Amaranthac.).
Acnistus Schott. Solanaceae. 50 trop. Am.
Acocanthera P. & K. = Acokanthera G. Don (Apocynac.).
Acoeloraphe P. & K. = seq.
Acoeloraphis Durand = seq.
Acoelorraphe H. Wendl. = seq.
Acoelorrhaphe H. Wendl. corr. Becc. Palmae. *S. str.* 1, *s.l.* 7 Mex., C. Am.
Acoelorrhaphis Durand = praec.
Acoïdium Lindl. = Trichocentrum Poepp. & Endl. (Orchidac.).
Acokanthera G. Don. Apocynaceae. 15 S. Afr., trop. E. Afr., Arabia.
Acoma Adans. = Homalium Jacq. (Flacourtiac.).
Acoma Benth. = Coreocarpus Benth. (Compos.).

ACOMASTYLIS

Acomastylis Greene. Rosaceae. 13 Himal., temp. E. As., N. Am.

Acome Baker (sphalm.) = Cleome L. (Cleomac.).

Acomis F. Muell. Compositae. 3 Austr.

Acomosperma K. Schum. (nomen). Asclepiadaceae. 1 trop. S. Am. Quid?

Aconceveibum Miq. = Mallotus Lour. (Euphorbiac.).

Aconiopteris Presl = Elaphoglossum Schott (Lomariopsidac.).

Aconitella Spach (~ Delphinium L.). Ranunculaceae. 10 E. Balkan Pcnins. to C. As. & Afghan.

Aconitopsis Kemul.–Nat. = praec.

Aconitum L. Ranunculaceae. 300 N. temp. Fls. in racemes. The post. sepal forms a large hood, enclosing the two 'petals' which are repres. by nectaries on long stalks. Fl. protandrous, adapted, by its structure and its blue colour, to bees. The distribution of *A.* largely coincides with that of humble-bees (*Bombus* spp.). Humble-bees often rob the fl. of its honey by biting through the hood. Fr. of follicles, which open so far as to expose the seeds, which only escape when shaken by wind or otherwise (censer mechanism). All are poisonous; the tuberous roots contain alkaloids of the aconitin group (used in medicine). *A. ferox* Wall. (root) furnishes the *bikh* poison of Nepal.

Aconium Engl. (sphalm.) = Aeonium Webb & Berth. (Crassulac.).

Aconogonum [Meissn.] Reichb. Polygonaceae. 15 N. As., Japan, N. Am.

Acontias Schott = Xanthosoma Schott (Arac.).

Acophorum Gaudich. ex Steud. (nomen). Gramineae. Hab.? Quid?

Acoraceae ('-oïdeae') C. A. Agardh = Araceae–Acoroïdeae Eyde, Nicols. & Sherw.

Acorellus Palla ex Kneuck. = Cyperus L. (Cyperac.).

Acoridium Nees & Meyen. Orchidaceae. 60 Philippines.

Acorus L. Araceae. 2 N. temp. & subtrop. Rhiz. sympodial; ls. isobilat. Fl. ☿, protog., with P. Perh. some distant relationship with *Sparganium*, one sp. of which (*S. eurycarpum* Engelm.) is attacked by the same rust fungus (*Uromyces sparganii* Clint. & Peck) as that on *Acorus*. (Parmelee & Savile, *Mycologia*, **46**: 823–46, 1954). Should be separated from subfam. *Pothoïdeae* as subfam. *Acoroïdeae* (C. A. Ag.) Eyde, Nicols. & Sherw., *Am. Journ. Bot.* **54** (4): 478, 481, 486, 494 (1967): A introrse; multiple stylar canals; no vasc. bundles in ovary wall; no raphides.

Acosmia Benth. ex G. Don = Gypsophila L. (Caryophyllac.).

Acosmium Schott = Sweetia Spreng. (Legumin.).

Acosmus Desv. = Aspicarpa Rich. (Malpighiac.).

Acosta Adans. = Centaurea L. (Compos.).

Acosta DC. = Spiracantha Kunth (Compos.).

Acosta Lour. = Vaccinium L. (Ericac.).

Acosta Ruiz & Pav. = Moutabea Aubl. (Polygalac.).

Acostaea P. & K. = genera 4 praecc.

Acostaea Schlechter. Orchidaceae. 2 Costa Rica.

Acostia Swallen. Gramineae. 1 Ecuador. (? Digitaria × Panicum).

Acourea Scop. = seq.

Acouroa Aubl. = Dalbergia L. (Legumin.).

Acouroua Taub. = praec.

ACROCORYNE

Acourtia D. Don = Perezia Lag. (Compos.).
Acquartia Endl. = Aquartia Jacq. = Solanum L. (Solanac.).
Acrachne Wight & Arn. ex Chiov. Gramineae. 1 Abyssinia, Indoch., Indomal.,
Acradenia Kipp. Rutaceae. 1 Tasm.   [Austr.
Acraea Lindl. = Pterichis Lindl. (Orchidac.).
Acrandra Berg. Myrtaceae. 4 S. trop. Braz.
Acranthemum Van Tiegh. = Tapinanthus Bl. (Loranthac.).
*Acranthera Arn. ex Meissn. Rubiaceae. 40 Indomal., esp. Borneo.
Acranthus Clem. = Acrosanthes Eckl. & Zeyh. (Aïzoac.).
Acranthus Hook. f. (sphalm.) = Aëranthes Lindl. (Orchidac.).
Acratherum 'Hochst.' ex Rich. (pro gen. nov.) = seq.
Acratherum Link = Arundinella Raddi (Gramin.).
Acreugenia Kausel. Myrtaceae. 1 S. Brazil to N. Argent.
Acridocarpus Guill. & Perr. Malpighiaceae. 50 Afr., Madag., Arabia, New
Caled.
Acrilia Griseb. = Trichilia L. (Meliac.).
Acrilla C. DC. = praec.
Acriopsis Reinw. ex Blume. Orchidaceae. 12 Indoch., W. Malaysia, New
Guinea, Solomons.
Acrista O. F. Cook = Euterpe Gaertn. (Palm.).
Acristaceae O. F. Cook = Palmae–Areceae Benth. & Hook. f.
Acritochaeta P. & K. = seq.
Acritochaete Pilger. Gramineae. 1 trop. Afr.
Acriulus Ridl. = Scleria Bergius (Cyperac.).
Acriviola Mill. = Tropaeolum L. (Tropaeolac.).
Acroanthes Rafin. = Achroanthes Rafin. = Malaxis Soland. ex Sw. (Orchidac.).
Acroblastum Soland. ex Setchell = Balanophora J. R. & G. Forst. (Balano-
phorac.).
Acrobotrys K. Schum. & Krause. Rubiaceae. 1 Colombia.
Acrocarpidium Miq. = Peperomia Ruiz & Pav. (Peperomiac.).
Acrocarpus Nees = Cryptangium Schrad. (Cyperac.).
Acrocarpus Wight ex Arn. Leguminosae. 3 Indomal.
Acrocentron Cass. = Centaurea L. (Compos.).
Acrocentrum P. & K. = praec.
Acrocephalium Hassk. = seq.
Acrocephalus Benth. Labiatae. 100 India to China & Philipp. Is.
Acroceras Stapf. Gramineae. 15 trop. Afr., Madag., Indomal.
Acrochaene Lindl. = Monomeria Lindl. (Orchidac.).
Acrochaete Peter (~ Setaria Beauv.). Gramineae. 1 trop. E. Afr. (Tanganyika).
Acrochloa Griseb. = Airochloa Link = Koeleria Pers. (Gramin.).
Acroclinium A. Gray = Helipterum DC. (Compos.).
Acrocoelium Baill. = Leptaulus Benth. (Icacinac.).
Acrocoelium Baill. Icacinaceae. 1 Congo.
Acrocomia Mart. Palmae. 30 trop. Am., W.I.
Acrocorion Adans. = Galanthus L. (Amaryllidac.).
Acrocorium P. & K. = praec.
Acrocoryna P. & K. = seq.
Acrocoryne Turcz. = Metastelma R.Br. (Asclepiadac.).

15

ACRODICLIDIUM

Acrodiclidium Nees (~ Licaria Aubl.). Lauraceae. 30 trop. Am., W.I.
A. puchury Mez furnishes medic. *puchurim* nuts.
Acrodon N. E. Brown. Aïzoaceae. 2 S. Afr.
Acrodryon Spreng. = Cephalanthus L. (Rubiac.).
Acrodrys Clem. = praec.
Acroëlytrum Steud. = Lophatherum Brongn. (Gramin.).
Acroglochia Gérard = seq.
Acroglochin Schrad. Chenopodiaceae. 2 N. India, China. The fruit mass is prickly, many of the twigs not ending in fls.
Acroglyphe E. Mey. = Annesorhiza Cham. & Schlechtd. (Umbellif.).
Acrolasia Presl = Mentzelia L. (Loasac.).
Acrolepis Schrad. = Ficinia Schrad. (Cyperac.).
Acrolinium Engl. (sphalm.) = Acroclinium A. Gray = Helipterum DC.
Acrolobus Klotzsch = Heisteria Jacq. (Olacac.). [(Compos.).
Acrolophia Pfitz. Orchidaceae. 10 trop. & S. Afr.
Acrolophus Cass. = Centaurea L. (Compos.).
Acronema Falc. ex Edgew. Umbelliferae. 15 E. Himal., W. China.
Acronia Presl = Pleurothallis R.Br. (Orchidac.).
Acronoda Hassk. = seq.
Acronodia Blume = Elaeocarpus L. (Elaeocarpac.).
Acronozus Steud. = Acrozus Spreng. = Acronodia Bl. = Elaeocarpus L. (Elaeocarpac.).
***Acronychia** J. R. & G. Forst. Rutaceae. 50 Hainan, Malaysia, Austr., Pacif.
Acropelta Nakai = Polystichum Roth (Aspidiac.).
Acropera Lindl. = Gongora Ruiz & Pav. (Orchidac.).
Acropetalum A. Juss. = Xeropetalum Delile = Dombeya Cav. (Sterculiac.).
Acrophorus Presl. Aspidiaceae. 2 SE. As. to Fiji.
Acrophyllum Benth. Cunoniaceae. 1 New S. Wales.
Acrophyllum E. Mey. = Pappea Eckl. & Zeyh. (Sapindac.).
Acropodium Desv. = Aspalathus L. (Legumin.).
Acropogon Schlechter. Sterculiaceae. 3 New Caled.
Acropselion Spach = Acrospelion Bess. ex Roem. & Schult. = Trisetum Pers. (Gramin.).
Acropteris Link = Asplenium L. (Aspleniac.).
Acropterygium (Diels) Nakai = Dicranopteris Bernh. (Gleicheniac.).
Acroptilion Endl. = seq.
Acroptilon Cass. (~ Centaurea L.). Compositae. 2 Orient, C. Asia.
Acrorumohra Ito = Dryopteris Adans. (Aspidiac.).
Acrosanthes Eckl. & Zeyh. Aïzoaceae. 6 S. Afr.
Acrosanthes Engl. = Acrossanthes Presl = Vismia Vand. (Guttif.).
Acrosanthus Clem. = Acrosanthes Eckl. & Zeyh. (Aïzoac.).
Acroschizocarpus Gombocz = Christolea Cambess. ex Jacquem. (Crucif.).
Acrosepalum Pierre = Ancistrocarpus Oliv. (Tiliac.).
Acrosorus Copel. Grammitidaceae. 5 Malaysia to Samoa.
Acrospelion Bess. ex Roem. & Schult. = Trisetum Pers. (Gramin.).
Acrospelton Wittst. = praec.
Acrospira Welw. ex Bak. (1878; non Montagne 1857—Fungi; nec Acrospeira Berk. & Broome 1861—Fungi) = Debesia Kuntze (Liliac.).

16

Acrossanthes Presl = Vismia Vand. (Guttif.).
Acrossanthus Baill. = praec.
Acrostachys Van Tiegh. = Helixanthera Lour. (Loranthac.).
Acrostemon Klotzsch (~ Eremia D. Don). Ericaceae. 10 S. Afr.
Acrostephanus Van Tiegh. = Tapinanthus Bl. (Loranthac.).
Acrostichaceae Presl emend. Ching = Pteridaceae Gaud.
Acrostiche Dietr. = Acrotriche R.Br. (Epacridac.).
Acrostichum Linn. Pteridaceae. 3 pantrop., in mangrove. Upper pinnae reduced and covered beneath with sporangia ('acrostichoid' condition); very different ferns showing similar distribution were formerly also called *Acrostichum*, esp. *Elaphoglossum* Schott.
Acrostigma O. F. Cook & Doyle = Catostigma O. F. Cook & Doyle (Palm.).
Acrostigma P. & K. = Acmostima Rafin. = Palicourea Aubl. (Rubiac.).
Acrostoma Didr. = Macrocnemum P.Br. (Rubiac.).
Acrostylia Frappier ex Cordem. = Cynorkis Thou. (Orchidac.).
Acrostylis P. & K. = praec.
Acrosynanthus Urb. Rubiaceae. 7 Cuba.
Acrotaphros Steud. ex Hochst. = Ormocarpum Beauv. (Legumin.).
Acrothrix Clem. = Acrotriche R.Br. (Epacridac.).
Acrotiche Poir. (sphalm.) = praec.
Acrotoma P. & K. = seq.
Acrotome Benth. Labiatae. 10 trop. & S. Afr.
Acrotrema Jack. Dilleniaceae. 10 Indomal., esp. Ceylon. Herbs, sometimes with much dissected ls.
Acrotriche R.Br. (~ Styphelia J. R. & G. Forst.). Epacridaceae. 8 temp. Austr.
Acroxis Trin. ex Steud. = Muhlenbergia Schreb. (Gramin.).
Acrozus Spreng. = Acronodia Bl. = Elaeocarpus L. (Elaeocarpac.).
Acrumen Gallesio = Citrus L. (Rutac.).
Acrymia Prain. Labiatae. 1 Malay Penins.
Acryphyllum Lindl. (sphalm.) = Arcyphyllum Ell. = Tephrosia Lour. (Legumin.).
Acsmithia Hoogl. Cunoniaceae. Spp.? New Caled.
Actaea L. Ranunculaceae. 10 N. temp. Fls. in racemes. Cpl. 1. Berry.
Actaea Lour. = Tetracera L. (Dilleniac.).
Actaeogeton Reichb. = Actegeton Bl. = Azima Lam. (Salvadorac.).
Actaeogeton Steud. = Scirpus L. (Cyperac.).
Actartife Rafin. = Boltonia L'Hérit. (Compos.).
Actea Rafin. = Actaea L. (Ranunculac.).
Actegeton Blume = Azima Lam. (Salvadorac.).
Actegiton Endl. = praec.
Actephila Blume. Euphorbiaceae. 35 China, Indomal., trop. Austr.
Actephilopsis Ridley = Tylosepalum Kurz = Trigonostemon Bl. (Euphorbiac.).
Acticarnopus Rafin. = Actinocarpus Rafin. = Chaetopappa DC. (Compos.).
Actimeris Rafin. = Actinomeris Nutt. (Compos.).
Actinanthella Balle. Loranthaceae. 1 S. trop. Afr.
Actinanthus Ehrenb. (~ Oenanthe L.). Umbelliferae. 1 W. As.
Actinea Juss. (~ Hymenoxys Cass.). Compositae. 15 N. Am.

ACTINELLA

Actinella Pers. = praec.

Actinia Griff. = Dendrobium Sw. (Orchidac.).

Actinidia Lindl. Actinidiaceae. 40 E. As.

*Actinidiaceae Van Tiegh. Dicots. 3/350 trop. & E. As. to N. Austr., trop. Am. Climbing shrubs with alt. simple ls. and small hypog. fls. usu. in axillary cymose clusters, ♀ or ♂ ♀. K 5, imbr.; C 5, imbr.; A 10 or more, anthers inflexed in bud, sometimes dehisc. by pores; G̲ (5) or more, multiloc. with ∞ axile anatr. ov. Berry or caps. Endosperm. Gen.: *Actinidia, Saurauia, Clematoclethra*. Related to *Dilleniac.* and *Theac.*, esp. the former.

Actiniopteridaceae Pichi-Serm. Pteridales. Small ferns with fan-shaped ls.; sporangia in submarginal lines protected by reflexed edge of lamina. Pichi-Sermolli, *Webbia*, **17**: 1–32, 1962. 1 gen., *Actiniopteris*.

Actiniopteris Link. Actiniopteridaceae. 5 trop. Afr., As.

Actinobole Endl. = Gnaphalodes A. Gray (Compos.).

Actinocarpus R.Br. = Damasonium Juss. (Alismatac.).

Actinocarpus Rafin. = Chaetopappa DC. (Compos.).

Actinocarya Benth. Boraginaceae. 2 W. Himal., SW. China.

Actinocaryum P. & K. = praec.

Actinocheita F. A. Barkley. Anacardiaceae. 1 Mex.

Actinochloa Willd. ex Beauv. = Botelua Lag. (Gramin.).

Actinochloris Panzer = Chloris Sw. (Gramin.).

Actinocladus E. Mey. = Capnophyllum Gaertn. (Umbellif.).

Actinocyclus Klotzsch = Orthilia Rafin. (Pyrolac.).

Actinodaphne Nees. Lauraceae. 60–70 E As., Indomal.

Actinodium Schau. Myrtaceae. 1 W. Austr. Pollen strikingly different from most *Myrtac.* and resembling that of some *Santalac.* Cf. the great similarity in habit betw. spp. of *Eugenia, Syzygium*, etc., and many *Loranthac.*, another fam. of *Santalales*.

Actinokentia Dammer. Palmae. 1 New Caled.

Actinolema Fenzl. Umbelliferae. 2 E. Medit.

Actinolepis DC. = Eriophyllum Lag. (Compos.).

*Actinomeris Nutt. (~ Verbesina L.). Compositae. 13 Atl. U.S., Mex.

Actinomorphe Kuntze (sphalm.) = Actinodaphne Nees (Laurac.).

Actinomorphe Miq. = Heptapleurum Gaertn. (Araliac.).

Actinopappus Hook. f. ex A. Gray = Rutidosis DC. (Compos.).

Actinophlebia Presl = Cnemidaria Presl (Cyatheac.).

Actinophloeus Becc. (~ Ptychosperma Labill.). Palmae. 20 New Guinea, Solomons, Bismarcks.

Actinophora A. Juss. = Actinospora Turcz. = Cimicifuga L. (Ranunculac.).

Actinophora Wall. ex R.Br. = Schoutenia Korth. (Tiliac.).

Actinophyllum Ruiz & Pav. = Sciadophyllum P.Br. (Araliac.).

Actinopteris Engl. = Actiniopteris Link (Actiniopteridac.).

Actinorhytis Wendl. & Drude. Palmae. 1 Malaysia, 1 Solomon Is.

Actinoschoenus Benth. Cyperaceae. 4 Madag., Ceylon, China.

Actinoseris (Endl.) Cabr. Compositae. 5 S. Am.

Actinospermum Ell. = Baldwinia Nutt. (Compos.).

Actinospora Turcz. = Cimicifuga L. (Ranunculac.).

Actinostachys Wall. ex Hook. = Schizaea Sm. (Schizaeac.).

Actinostema Lindl. = Actinostemon Mart. ex Klotzsch (Euphorbiac.).
Actinostemma Griff. Cucurbitaceae. 1 India to Japan.
Actinostemma Lindl. = Actinostema Lindl. = seq.
Actinostemon Mart. ex Klotzsch. Euphorbiaceae. 40 trop. Am., W.I.
Actinostigma Turcz. = Seringia J. Gay (Sterculiac.).
Actinostigma Welw. = Symphonia L. f. (Guttif.).
Actinostrobaceae Lotsy = Cupressaceae Neger.
Actinostrobus Miq. Cupressaceae. 2 SW. Austr.
Actinotinus Oliv. Imaginary genus, founded through the trick of a native Chinese collector, who had carefully inserted an infl. of *Viburnum* into the terminal bud of an *Aesculus*!
Actinotus Labill. Hydrocotylaceae (~ Umbellif.). 15 Austr., Tasm., N.Z. (flannel flower).
Actipsis Rafin. = Solidago L. (Compos.).
Actispermum Rafin. = Actinospermum Ell. = Baldwinia Nutt. (Compos.).
Actogeton Clem. = seq.
Actogiton Blume = Actegeton Bl. = Azima Lam. (Salvadorac.).
Actophila P. & K. = Actephila Bl. (Euphorbiac.).
Actoplanes K. Schum. = Ilythuria Rafin. = Donax Lour. (as to type) + Schumannianthus Gagnep. (Marantac.).
Actynophloeus Becc. = Actinophloeus Becc. (Palm.).
Acuan Medik. = Desmanthus Willd. (Legumin.).
Acuania Kuntze = praec.
Acuba Link = Aucuba Thunb. (Cornac.).
Acubalus Neck. = Cucubalus L. (Caryophyllac.).
Acularia Rafin. = Scandix L. (Umbellif.).
Acuna Endl. = seq.
Acunna Ruiz & Pav. = Bejaria Mutis (Ericac.).
Acura Hill = Scabiosa L. (Dipsacac.).
Acuroa J. F. Gmel. = Acouroa Aubl. = Dalbergia L. (Legumin.).
Acustelma Baill. = Cryptolepis R.Br. (Asclepiadac.).
Acuston Rafin. = Fibigia Medik. (Crucif.).
Acylopsis P. & K. = Akylopsis Lehm. = Matricaria L. (Compos.).
Acynos Pers. = Acinos Mill. = Calamintha Moench (Labiat.).
Acyntha Medik. (~ Sanseverinia Petagna). Agavaceae. 20 trop. Afr.
Acyphilla Poir. = Aciphylla J. R. & G. Forst. (Umbellif.).
Acystopteris Nakai = Cystopteris Bernh. (Athyriac.).
Ada Lindl. Orchidaceae. 8 Colombia.
Adactylus Rolfe = Apostasia Bl. (Orchidac.).
× **Adaglossum** hort. Orchidaceae. Gen. hybr. (iii) (Ada × Odontoglossum).
Adamantogeton Schrad. ex Nees = Lagenocarpus Nees (Cyperac.).
Adamantogiton P. & K. = praec.
× **Adamara** hort. = × Yamadara hort. (Orchidac.).
Adamaram Adans. = Terminalia L. (Combretac.).
Adambea Lam. = seq.
Adamboë Adans. = Lagerstroemia L. (Lythrac.).
Adamboë Rafin. = Stictocardia Hallier f. (Convolvulac.).
Adamea Jacques-Félix = Feliciadamia Bullock (Melastomatac.).

ADAMIA

Adamia Jacques-Félix = praec.
Adamia Wall. = Dichroa Lour. (Hydrangeac.).
Adamsia Fisch. ex Steud. = Geum L. (Rosac.).
Adamsia Willd. = Puschkinia Adams (Liliac.).
Adansonia L. Bombacaceae. 10 palaeotrop. *A. digitata* L. is the baobab. Its height is not great, but the trunk may reach 9 m. in thickness. Fr. woody.
Adaphus Neck. = ? Laurus L. (Laurac.).
Adarianta Knoche (~ Pimpinella L.). Umbelliferae. 1 Balearic Is.
Adatoda Adans. = Adhatoda Mill. (Acanthac.).
Addisonia Rusby. Compositae. 2 Bolivia.
Adectum Link = Dennstaedtia Bernh. (Dennstaedtiac.).
Adelaida Buc'hoz. Quid? (Acanthac.? Verbenac.?). Réunion(?).
Adelanthus Endl. = Pyrenacantha Wight (Icacinac.).
Adelaster Lindl. = ? Pseuderanthemum Radlk. (Acanthac.).
Adelbertia Meissn. = Meriania Sw. (Melastomatac.).
Adelia P.Br. = Forestiera Poir. (Oleac.).
*Adelia L. Euphorbiaceae. 15 W.I., warm Am.
Adelia L. (*p.p.*) et auctt. mult. = Bernardia Mill. (Euphorbiac.).
Adeliodes P. & K. = seq.
Adelioïdes R.Br. ex Benth. = seq.
Adeliopsis Benth. Menispermaceae. 1 NE. Austr.
Adelmannia Reichb. = Borrichia Adans. (Compos.).
Adelmeria Ridl. = Alpinia Roxb. (Zingiberac.).
Adelobotrys DC. Melastomataceae. 20 trop. Am., W.I.
Adelocaryum Brand = Cynoglossum L. + Lindelofia Lehm. (Boraginac.).
Adeloda Rafin. = Dicliptera Juss. (Acanthac.).
Adelodypsis Becc. (~ Dypsis Nor.). Palmae. 3 Madag.
Adelonema Schott (~ Homalomena Schott). Araceae. 1 Amaz. Braz.
Adelonenga (Becc.) Hook. f. (~ Hydriastele Wendl. & Drude). Palmae. 5 New Guinea, Bismarck Is.
Adelopetalum Fitzger. = Bulbophyllum Thou. (Orchidac.).
Adelosa Blume. Verbenaceae. 1 Madag.
Adelostemma Hook. f. Asclepiadaceae. 3 Burma, China.
Adelostigma Steetz. Compositae. 2 trop. Afr.
Adeltia Mirb. (sphalm.) = Adelia L. (Euphorbiac.).
Adenacantha B. D. Jacks. (sphalm.) = Adenachaena DC. = Phymaspermum Less. (Compos.).
Adenacanthus Nees (~ Strobilanthes Blume). Acanthaceae. 4 Indomal.
Adenaceae Dulac = Droseraceae Salisb.
Adenachaena DC. = Phymaspermum Less. (Compos.).
*Adenandra Willd. Rutaceae. 25 S. Afr.
Adenandria Rafin. (nomen). Menispermaceae. Quid?
Adenanthera L. Leguminosae. 8 trop. As., Austr., Pacif. Seeds hard and bright red, or red and black (cf. *Abrus*).
Adenanthes Knight = seq.
Adenanthos Labill. Proteaceae. 20 W. & S. Austr.
Adenanthus Roem. & Schult. = praec.
Adenarake Maguire & Wurdack. Ochnaceae. 1 Venez.

Adenaria Kunth. Lythraceae. 1 Mex. to Argent.

Adenaria Pfeiff. = Adnaria Rafin. = Gaylussacia Kunth (Ericac.).

Adenarium Rafin. = Honkenya Ehrh. (Caryophyllac.).

Adeneleuterophora Barb. Rodr. = Elleanthus Presl (Orchidac.).

Adeneleuthera Kuntze = praec.

Adeneleutherophora Dalla Torre & Harms = praec.

Adenema G. Don = Enicostema Blume (Gentianac.).

Adenesma Griseb. = praec.

Adenia Forsk. Passifloraceae. 92 trop. & S. Afr., Madag., SW. Arabia, Indomal., N. Austr.

Adenia Torr. = Pilea Lindl. (Urticac.).

Adenilema Blume = Neillia D. Don (Rosac.).

Adenilemma Hassk. = praec.

Adenimesa Nieuwl. = Mesadenia Rafin. (Compos.).

Adenium Roem. & Schult. Apocynaceae. 15 trop. & subtrop. Afr., Arabia. Xerophytes with thick stems, and rather fleshy ls.

Adenleima Reichb. = Adenilema Bl. = Neillia D. Don (Rosac.).

Adenobasium Presl = Sloanea L. (Elaeocarpac.).

Adenobium Steud. = praec.

Adenocalymma Mart. ex Meissn. corr. Endl. Bignoniaceae. 40 trop. Am.

Adenocalymna Mart. ex Meissn. = praec.

Adenocalyx Bert. ex Kunth (~ Caesalpinia L.). Leguminosae. 1 Colombia.

Adenocarpum D. Don ex Hook. & Arn. = Chrysanthellum Rich. (Compos.).

Adenocarpus DC. Leguminosae. 20 Canary Is., Medit.

Adenocarpus P. & K. = Adenocarpum D. Don ex Hook. & Arn. = Chrysanthellum Rich. (Compos.).

Adenocaulon Hook. Compositae. 5 NE. temp. As.; N. temp., C. & temp. S. Am.

Adenocaulum Clem. = praec.

Adenocaulus Clem. = praec.

Adenoceras Reichb. f. & Zoll. ex Baill. = Macaranga Thou. (Euphorbiac.).

Adenochaena DC. (sphalm.) = Adenachaena DC. = Phymaspermum Less. (Compos.).

Adenochaeton Endl. = seq.

Adenocheton Fenzl = Cocculus L. (Menispermac.).

Adenochetus Baill. = praec.

Adenochilus Hook. f. Orchidaceae. 2 Austr., N.Z.

Adenochlaena Boiv. ex Baill. Euphorbiaceae. 2 Madag., Comoros, Ceylon.

Adenoclina P. & K. = seq.

Adenocline Turcz. Euphorbiaceae. 8 S. Afr.

Adenocrepis Blume = Baccaurea Lour. (Euphorbiac.).

Adenocyclus Less. = Pollalesta Kunth (Compos.).

Adenodaphne S. Moore = Litsea Lam. (Laurac.).

Adenoderris J. Sm. Aspidiaceae. 2 W.I.

Adenodiscus Turcz. = Belotia A. Rich. (Tiliac.).

Adenodolichos Harms. Leguminosae. 15 trop. Afr.

Adenodolichus P. & K. = praec.

21

ADENODUS

Adenodus Lour. = Elaeocarpus L. (Elaeocarpac.).
Adenogonum Welw. ex Hiern = Engleria O. Hoffm. (Compos.).
Adenogramma Reichb. Aïzoaceae. 7 S. Afr.
Adenogrammataceae Nak. = Aïzoaceae–Limeëae–Adenogramm[at]inae K. Müll.
Adenogyna P. & K. = Adenogyne Klotzsch = Sebastiania Spreng. (Euphorbiac.).
Adenogyna Rafin. = Saxifraga L. (Saxifragac.).
Adenogyne B. D. Jacks. = praec.
Adenogyne Klotzsch = Sebastiania Spreng. (Euphorbiac.).
Adenogynum Reichb. f. & Zoll. = Cladogynos Zipp. ex Span. (Euphorbiac.).
Adenogyras Durand = seq.
Adenogyrus Klotzsch = Scolopia Schreb. (Flacourtiac.).
Adenola Rafin. = Ludwigia L. (Onagrac.).
Adenolepis Less. = Cosmos Cav. (Compos.).
Adenolepis Sch. Bip. = Bidens L. (Compos.).
Adenolinum Reichb. = Linum L. (Linac.).
Adenolisianthus Gilg. Gentianaceae. 2 Brazil.
Adenolobus (Harv.) Torre & Hillcoat. Leguminosae. 3 S. trop. Afr.
Adenoncos Blume. Orchidaceae. 10 Malaysia.
Adenonema Bunge = Stellaria L. (Caryophyllac.).
Adenoön Dalz. Compositae. 1 Indomal.
Adenopa Rafin. = Drosera L. (Droserac.).
Adenopappus Benth. Compositae. 1 Mex.
Adenopeltis Bert. ex A. Juss. Euphorbiaceae. 2 temp. S. Am.
Adenopetalum Klotzsch & Garcke = Euphorbia L. (Euphorbiac.).
Adenopetalum Turcz. = Vitis L. (Vitidac.).
Adenophaedra Muell. Arg. Euphorbiaceae. 4 trop. S. Am.
Adenophora Fisch. Campanulaceae. 60 temp. Euras.
Adenophorus Gaud. = Amphoradenium Desv. (Grammitidac.).
Adenophyllum Pers. Compositae. 3 Mex.
Adenophyllum Thou. ex Baill. = Hecatea Thou. = Omphalea L. (Euphorbiac.).
Adenoplea Radlk. (~ Buddleja L.). Buddlejaceae. 2 Madag.
Adenoplusia Radlk. (~ Buddleja L.). Buddlejaceae. 3 E. Afr., Madag.
Adenopodia C. Presl = Entada Adans. (Legumin.).
Adenopogon Welw. = Swertia L. (Gentianac.).
Adenoporces Small (~ Tetrapteris Cav.). Malpighiaceae. 1 W.I. (S. Domingo).
Adenopus Benth. = Lagenaria Ser. (Cucurbitac.).
Adenorachis Nieuwl. (~ Aronia Medik.). Rosaceae. 4 N. Am.
Adenorhopium Reichb. = Adenoropium Pohl = Jatropha L. (Euphorbiac.).
Adenorima Rafin. = Euphorbia L. (Euphorbiac.).
Adenoropium Pohl = Jatropha· L. (Euphorbiac.).
Adenorrhopium Wittst. = praec.
Adenosachma A. Juss. = seq.
Adenosacma P. & K. = seq.
Adenosacme Wall. = Mycetia Reinw. (Rubiac.).
Adenosciadium H. Wolff. Umbelliferae. 1 SE. Arabia.
Adenoscilla Gren. & Godr. = Scilla L. (Liliac.).
Adenoselen Spach (sphalm.) = Adenosolen DC. = Marasmodes DC. (Compos.).

Adenosepalum Fourr. = Hypericum L. (Guttif.).

Adenosma R.Br. Scrophulariaceae. 15 China, Indomal., Austr.

Adenosma Nees = Synnema Benth. (Acanthac.).

Adenosolen DC. = Marasmodes DC. (Compos.).

Adenospermum Hook. & Arn. = Chrysanthellum Rich. (Compos.).

Adenostachya Bremek. (~ Strobilanthes Bl.). Acanthaceae. 2 Java.

Adenostegia Benth. (~ Cordylanthus Nutt.). Scrophulariaceae. 20 N. Am.

Adenostema Desport. = seq.

Adenostemma J. R. & G. Forst. Compositae. 30 (perhaps reducible to 5–6) trop. Am., trop. & S. Afr., 1 pantrop. Pappus glandular and sticky; fr. carried by animals.

Adenostemma Hook. f. (sphalm.) = Odontostemma Benth. = Arenaria L. (Caryophyllac.).

Adenostemon Spreng. = seq.

Adenostemum Pers. = Gomortega Ruiz & Pav. (Gomortegac.).

Adenostephanes Lindl. = seq.

Adenostephanus Klotzsch = Euplassa Salisb. (Proteac.).

Adenostoma Blume (nomen). Scrophulariaceae. Java. Quid?

Adenostoma Hook. & Arn. Rosaceae. 2 Calif. *A. fasciculatum* H. & A. is one of the shrubs forming the *chaparral* or *chamisal*.

Adenostyles Cass. Compositae. 4–5 mts. Eur., As. Min.

Adenostyles Endl. = seq.

Adenostylis Blume = Zeuxine Lindl. (Orchidac.).

Adenostylis P. & K. = Adenostyles Cass. (Compos.).

Adenostylium Reichb. = Adenostyles Cass. (Compos.).

Adenothamnus Keck. Compositae. 1 Lower Calif.

Adenotheca Welw. ex Baker = Schizobasis Baker (Liliac.).

Adenothola Lemaire = Manettia Mutis ex L. (Rubiac.).

Adenotrachelium Nees ex Meissn. = Ocotea Aubl. (Laurac.).

Adenotrias Jaub. & Spach = Hypericum L. (Guttif.).

Adenotrichia Lindl. = Senecio L. (Compos.).

Adenum G. Don = Adenium Roem. & Schult. (Apocynac.).

Adesia Eaton = Adicea Rafin. = Pilea Lindl. (Urticac.).

*****Adesmia** DC. Leguminosae. 100 S. Am. Leafstalks thorny; plants often with glandular hairs.

Adhatoda Mill. (~ Justicia L.). Acanthaceae. 20 trop. Afr. & As.

Adhunia Vell. Inc. sed. (? Myrtac.). 1 Brazil.

Adiantaceae Presl. Pteridales. Ls. variously branched, lfts. often flabellate with dichot. veins; sporangia along veins on small reflexed segments of edge of lft. (pseudo-indusia). 1 gen.: *Adiantum* L.

Adianthum Burm. = Acacia Mill. (Legumin.).

Adiantopsis Fée = Cheilanthes Sw. (Sinopteridac.).

Adiantum L. Adiantaceae. 200 cosmop., esp. trop. Am. (maidenhair).

Adicea Rafin. = seq.

Adike Rafin. = Pilea Lindl. (Urticac.).

Adina Salisb. Naucleaceae (~ Rubiac.). 20 trop. & subtrop. Afr. & As.

Adinandra Jack. Theaceae. 80 E. & SE. As., Indomal.

Adinandrella Exell = Ternstroemia L. (Theac.).

ADINANDROPSIS

Adinandropsis Pitt-Schenkel = Melchiora Kobuski (Theac.).

Adinobotrys Dunn = Whitfordiodendron Elm. (Legumin.).

× **Adioda** hort. Orchidaceae. Gen. hybr. (iii) (Ada × Cochlioda).

Adipe Rafin. = Bifrenaria Lindl. (Orchidac.).

Adipera Rafin. (~ Cassia L.). Leguminosae. 15 W.I., C. & trop. S. Am.

Adipsen Rafin. (nomen). Quid? 2 E. U.S.

Adisa Steud. = seq.

Adisca Blume = Mallotus Lour. + Sumbaviopsis J. J. Sm. (Euphorbiac.).

Adiscanthus Ducke. Rutaceae. 1 Amaz. Braz.

Adlera P. & K. = seq.

Adleria Neck. = Eperua Aubl. (Legumin.).

*****Adlumia** Rafin. ex DC. Fumariaceae. 2 Korea, E. N. Am. Leaf-climbers.

Admarium Rafin. = Adenarium Rafin. = Honkenya Ehrh. (Caryophyllac.).

Adnaria Rafin. = Styrax L. (Styracac.).

Adnula Rafin. = Pelexia Poit. (Orchidac.).

Adoceton Rafin. = Alternanthera Forsk. (Amaranthac.).

Adodendron DC. = seq.

Adodendrum Neck. = Rhodothamnus Reichb. (Ericac.).

Adoketon Rafin. = Adoceton Rafin. = Alternanthera Forsk. (Amaranthac.).

Adolia Lam. = Scutia (DC.) Comm. ex Brongn. (Rhamnac.).

Adolphia Meissn. Rhamnaceae. 2 SW. U.S., Mex.

Adonanthe Spach = Adonis L. (Ranunculac.).

Adonastrum Dalla Torre & Harms = seq.

Adoniastrum Schur = Adonis L. (Ranunculac.).

Adonidia Becc. = Veitchia H. Wendl. (Palm.).

Adonigeron Fourr. = Senecio L. (Compos.).

Adonis L. Ranunculaceae. 20 N. palaeotemp.

Adopogon Neck. = Krigia Schreb. (Compos.).

Adorion Rafin. = seq.

Adorium Rafin. = Musineon Rafin. (Umbellif.).

Adoxa L. Adoxaceae. 1 N. temp., S. to W. Himal., Colorado and Illinois, *A. moschatellina* L. Dispersal largely by snails: Müller-Schneider, *Vegetatio*, **15** (1): 27–32 (1967).

*****Adoxaceae** Trautv. Dicots. 1/1 N. temp. Small geophytic herbs. Rhiz. creeping, monopodial; flg. shoot, erect, tetrag., with 1–3 rad. ls., a pair of opp. cauline ls., and a small head of greenish fls., usu. 5 (a condensed dich. cyme). The term. fl. is usu. 4-merous, the lat. 5-merous (cf. *Ruta*, etc.). Fl. ⚥, reg., greenish, inconspic. P of 2 whorls; the outer usu. 3-merous, persistent, sometimes regarded as an invol. formed of bract and bracteoles, but quite probably a K, adnate to G; the inner (probably a C) (5–4), caduc. Sta. alt. with 'petals', divided almost to the base. Ḡ (3–5), rarely (2), at first semi-inf., with one pend. ov. in each loc. Drupe with several stones. Endosp. Chief visitors small flies, attracted by the musky smell. Only genus: *Adoxa*. No very close relationships, but indications of affinity with *Araliac.*, *Saxifragac.*, and possibly *Ranunculac.* Cf. Sprague in *J. Linn. Soc., Bot.* **47**: 471–87, 1927.

Adrastaea DC. = Hibbertia Andr. (Dilleniac.).

Adrastea Spreng. = praec.

Adriana Gaudich. Euphorbiaceae. 5 Austr.
Adriania Endl. = praec.
Adromischus Lem. Crassulaceae. 50 S. Afr.
Adrorhizon Hook. f. Orchidaceae. 1 Ceylon.
Adulpa Endl. (sphalm.) = seq.
Adupla Bosc = Mariscus Gaertn. (Cyperac.).
Aduseta Dalla Torre & Harms = seq.
Aduseton Scop. = Adyseton Adans. (Crucif.).
Adventina Rafin. = Galinsoga Ruiz & Pav. (Compos.).
Adyseton Adans. = Alyssum L. + Lobularia Desv. + Draba L. (Crucif.).
Adysetum Link = praec.
Aeceoclades Duchartre ex B. D. Jacks. (sphalm.) = Oeceoclades Lindl. = Saccolobium Bl. (Orchidac.).
Aechma C. A. Ag. = Aechmea Ruiz & Pav. (Bromeliac.).
Aechmaea Brongn. = praec.
Aechmandra Arn. = Kedrostis Medik. (Cucurbitac.).
Aechmanthera Nees. Acanthaceae. 3 Himal.
***Aechmea** Ruiz & Pav. Bromeliaceae. 150 W.I., S. Am. Epiph.
Aechmolepis Decne = Tacazzea Decne (Periplocac.).
Aechmophora Spreng. ex Steud. = Bromus L. (Gramin.).
Aectyson Rafin. = Sedum L. (Crassulac.).
Aedemone Kotschy = Herminiera Guill. & Perr. (Legumin.).
Aedesia O. Hoffm. Compositae. 3 trop. W. Afr.
Aedia P. & K. = Aïdia Lour. = Randia L. (Rubiac.).
Aedmannia Spach (sphalm.) = Oedmannia Thunb. = Rafnia Thunb. (Legumin.).
Aedula Nor. = Orophea Bl. (Annonac.).
Aeegiphila Sw. (sphalm.) = Aegiphila Jacq. (Verbenac.).
Aeëridium Salisb. = Aërides Lour. (Orchidac.).
Aegelatis Roxb. (sphalm.) = Aegialitis R.Br. (Plumbaginac.).
Aegenetia Roxb. = Aeginetia L. (Orobanchac.).
Aegeria Endl. = Ageria Adans. = Ilex L. (Aquifoliac.) + Myrsine L. (Myrsinac.).
Aegialea Klotzsch = Pieris D. Don (Ericac.).
Aegialina Schult. = Koeleria Pers. (Gramin.).
Aegialinites Presl = seq.
Aegialinitis Benth. & Hook. f. = seq.
Aegialitis R.Br. Aegialitidaceae. 2 mangrove formations Bengal & Burma, Andaman & Nicobar Is., E. Malaysia & trop. Austr.
Aegialitis Trin. = Koeleria Pers. (Gramin.).
Aegialitidaceae Lincz. (~ Plumbaginaceae Juss.). 1/2 coasts trop. As., E. Malaysia & trop. Austr. Large glabrous evergreen shrubs; trunk conical, simple; branches marked with scars of amplexicaul petioles of fallen ls. Ls. alt., ent., coriac., suborbic., with ∞ fine spreading parallel nerves; petiole long, winged, sheathing, longer than blade, amplex. at base. Fls. in few-branched rigid panic., v. shortly pedic., subtended by large concave bract and 2 small bracteoles. K (5), tubular, 5-ribbed, shortly 5-toothed; C 5, unguic., v. shortly coherent at base, linear, fleshy or subcoriac., upper portion decid., lower persistent; A 5, slightly adnate to base of pet., forming a persist. tube, anth.

AEGIALOPHILA

basifixed; pollen monomorphic. $\underline{G}$ (5), 1-loc., 1-ov., ov. pend. from basal funicle; styles 5, free; stigs small, capit. Fruit lin., elong., curved, pentag., finally splitting along angles; seed linear, filling cavity, with membr. testa; endosp. O. Only genus: *Aegialitis.*

Aegialophila Boiss. & Heldr. = Centaurea L. (Compos.).

Aegianilites C. B. Clarke (sphalm.) = Aegialinites Presl = Aegialitis R.Br. (Plumbaginac.).

Aegiatilis Griff. (sphalm.) = Aegialitis R.Br. (Plumbaginac.).

Aegiceras Gaertn. Myrsinaceae. 2 palaeotrop. *A. corniculatum* (L.) Blanco grows in mangrove swamps together with *Rhizophora*, etc., and exhibits a similar habit, vivipary, etc.

Aegicer[atac]eae Blume = Myrsinaceae R.Br.

Aegicon Adans. = seq.

Aegilops L. Gramineae. 20–25 Medit. Reg. to C. As. & Afghan.

× **Aegilosecale** Ciferri & Giacom. Gramineae. Gen. hybr. (Aegilops × Secale).

× **Aegilotrichum** G. Camus = seq.

× **Aegilotricum** Wagner ex Tschermak. Gramineae. Gen. hybr. (Aegilops × Triticum).

× **Aegilotriticum** P. Fourn. = praec.

Aeginetia Cav. = Bouvardia Salisb. (Rubiac.).

Aeginetia L. Orobanchaceae. 10 Indomal., China, Japan.

Aeginetiaceae Livera = Orobanchaceae Vent.

Aegiphila Jacq. Verbenaceae. 160 trop. Am., W.I.

Aegiphyla Steud. = praec.

*Aegle Corrêa ex Koen. Rutaceae. 3 Indomal. *A. marmelos* (L.) Corrêa is the *bael* fruit, a valuable remedy for dysentery, etc.

Aegle Dulac (sphalm.) = Aglaë Dulac = Posidonia Koen. (Posidoniac.).

Aeglopsis Swingle. Rutaceae. 5 trop. Afr.

Aegoceras P. & K. = Aegiceras Gaertn. (Myrsinac.).

Aegochloa Benth. = Navarretia Ruiz & Pav. (Polemoniac.).

Aegokeras Rafin. = Seseli L. (Umbellif.).

Aegomarathrum Steud. = Cachrys L. (Umbellif.).

Aegonychion Endl. = seq.

Aegonychon S. F. Gray = Lithospermum L. (Boraginac.).

Aegophila P. & K. = Aegiphila Jacq. (Verbenac.).

Aegopicron Giseke = Aegopricum L. = Maprounea Aubl. (Euphorbiac.).

Aegopodion St-Lag. = seq.

Aegopodium L. Umbelliferae. 7 Eur., temp. As.

Aegopogon Beauv. = Amphipogon R.Br. (Gramin.).

Aegopogon Humb. & Bonpl. ex Willd. Gramineae. 3 S. U.S. to Argent.

Aegopordon Boiss. = Jurinea Cass. (Compos.).

Aegopricon L. f. = seq.

Aegopricum L. = Maprounea Aubl. (Euphorbiac.).

Aegoseris Steud. = Crepis L. (Compos.).

Aegotoxicon Molina = Aextoxicon Ruiz & Pav. (Aextoxicac.).

Aegotoxicum Endl. = praec.

Aegylops Honck. = Aegilops L. (Gramin.).

Aeiphanes Spreng. = Aïphanes Willd. = Martinezia Ruiz & Pav. (Palm.).

Aelbroeckia De Moor = ?Aeluropus Trin. (Gramin.).
Aellenia Ulbr. Chenopodiaceae. 6 Orient.
Aeluropus Trin. Gramineae. 5 Medit. to India. Halophytes.
Aeluroschia P. & K. = Ailuroschia Stev. = Astragalus L. (Legumin.).
Aembilla Adans. = Scolopia Schreb. (Flacourtiac.).
Aenanthe Rafin. = Oenanthe L. (Umbellif.).
Aenida Scop. (sphalm.) = Acnida L. = Amaranthus L. (Amaranthac.).
Aenothera Lam. = Oenothera L. (Onagrac.).
Aeolanthus Mart. Labiatae. 50 trop. & subtrop. Afr.
Aeolianthes Spreng. = praec.
Aeolotheca P. & K. = Aiolotheca DC. (Compos.).
Aeonia Lindl. = Oeonia Lindl. (Orchidac.).
Aeonium Webb & Berth. Crassulaceae. 40 Atl. Is., Medit., Abyss., Arabia.
Aepyanthus P. & K. = Aipyanthus Stev. = Macrotomia DC. (Boraginac.).
Aera Aschers. = Aira L. (Gramin.).
Aerachne Hook. f. (sphalm.) = Acrachne Wight & Arn. = Eleusine Gaertn. (Gramin.).
Aërangis Reichb. f. Orchidaceae. 70 trop. & S. Afr., Mascarenes.
Aëranthes Lindl. Orchidaceae. 30 Mascarenes.
Aëranthus Reichb. f. = Aëranthes Lindl. *p.p.* + Macroplectrum Pfitz., etc. (Orchidac.).
Aëranthus Spreng. = Aëranthes Lindl. (Orchidac.).
Aëria O. F. Cook = Gaussia Wendl. (Palm.).
× **Aëridachnis** hort. Orchidaceae. Gen. hybr. (iii) (Aërides × Arachnis).
× **Aëridanthe** hort. (v) = × Aëridovanda hort. (Orchidac.).
Aërides Lour. Orchidaceae. 40 India, Indoch., Japan, Malaysia (exc. New Guinea).
Aëridium Pfeiff. = Aeëridium Salisb. = Aërides Lour. (Orchidac.).
Aeridium P. & K. = Airidium Steud. = Deschampsia Beauv. (Gramin.).
× **Aëridocentrum** hort. Orchidaceae. Gen. hybr. (iii) (Aërides × Ascocentrum).
× **Aëridofinetia** hort. Orchidaceae. Gen. hybr. (iii) (Aërides × Neofinetia).
× **Aëridoglossum** hort. Orchidaceae. Gen. hybr. (iii) (Aërides × Ascoglossum).
× **Aëridolabium** hort. Orchidaceae. Gen. hybr. (v, vi) (Aërides × Saccolabium).
× **Aëridopsis** hort. Orchidaceae. Gen. hybr. (iii) (Aërides × Phalaenopsis).
× **Aëridostylis** hort. Orchidaceae. Gen. hybr. (iii) (Aërides × Rhynchostylis).
× **Aëridovanda** hort. Orchidaceae. Gen. hybr. (iii) (Aërides × Vanda).
× **Aëriovanda** hort. (vii) = × Aëridovanda hort. (Orchidac.).
Aeriphracta Reichb. (sphalm.) = Aperiphracta Nees ex Meissn. = Ocotea Aubl. (Laurac.).
Aërobion Spreng. = Jumellea, Angraecum, Chameangis, Solenangis, Eulophia, Eulophidium, etc. (Orchidac.).
Aerokorion Scop. = Acrocorion Adans. = Galanthus L. (Amaryllidac.).
Aeronia Lindl. = Oeonia Lindl. (Orchidac.).
Aërope (Endl.) Reichb. = Rhizophora L. (Rhizophorac.).
Aeropsis Aschers. & Graebner = Airopsis Desv. (Gramin.).
Aerosperma P. & K. = Airosperma Lauterb. & K. Schum. (Rubiac.).

AËROVANDA

× **Aërovanda** hort. (vii) = × Aëridovanda hort. (Orchidac.).

Aerua Juss. = seq.

***Aerva** Forsk. Amaranthaceae. 10 temp. & trop. Afr. & As.

Aesandra Pierre (sphalm.) = Aïsandra Pierre corr. Airy Shaw (Sapotac.).

Aeschinanthus Endl. = Aeschynanthus Jack (Gesneriac.).

Aeschinomene Nocca = Aeschynomene L. (Legumin.).

Aeschrion Vell. (~ Picrasma Bl.). Simaroubaceae. 5 W.I., S. Brazil.

Aeschryon Pfeiff. = praec.

***Aeschynanthus** Jack. Gesneriaceae. 80 Indomal., China. Many epiphytes with fleshy leaves. Extreme protandry with movement of sta. Seeds with long hairs.

Aeschynomene L. Leguminosae. 150 trop. & subtrop. From the pith-like wood of *A. aspera* L. (*shola*, pith-plant) the solar helmets of trop. As. are made.

Aesculaceae Lindl. = Hippocastanaceae DC.

Aesculus L. Hippocastanaceae. 1 SE. Eur.; 5 India, E. As.; 7 N. Am.

Aestuaria [L.] Schaeff. = ? Diosma L. (Rutac.).

Aëtanthus (Eichl.) Engl. Loranthaceae. 10 N. Andes.

Aetenia auctt. = Atenia Hook. & Arn. = Perideridia Reichb. (Umbellif.).

Aethales P. & K. = Aïthales Webb & Berth. = Sedum L. (Crassulac.).

Aëtheilema R.Br. = Phaylopsis Willd. (Acanthac.).

Aëtheocephalus Gagnep. Compositae. 1 Indoch.

Aëtheochlaena P. & K. = seq.

Aëtheolaena Cass. = Senecio L. (Compos.).

Aëtheolirion Forman. Commelinaceae. 1 Siam. Climber. Fruit an elongate linear caps., with winged seeds.

Aëtheonema Bub. & Penzig = Aëthionema R.Br. (Crucif.).

Aëtheonema Reichb. = Gaertnera Lam. (Rubiac.).

Aëtheopappus Cass. = Centaurea L. (Compos.).

Aëtheorhiza Cass. = Crepis L. (Compos.).

Aëtheorrhiza Reichb. = praec.

Aëthephyllum N. E. Br. Aïzoaceae. 1 S. Afr.

Aetheria Endl. = Hetaeria Blume (Orchidac.).

Aëthionema R.Br. Cruciferae. 70 Medit., Orient. Fr. lomentose in some; in others, e.g. *A. heterocarpum* J. Gay, there are two kinds of fr., one many-seeded and dehiscent, the other one-seeded and indehiscent.

Aethionia auct. = Aethonia D. Don = Tolpis Adans. (Compos.).

Aethiopis (Benth.) Opiz = Salvia L. (Labiat.).

Aethiopsis Engl. = praec.

Aethonia D. Don = Tolpis Adans. (Compos.).

Aethonopogon Hackel ex Kuntze = Polytrias Hackel (Gramin.).

Aethulia A. Gray = Ethulia L. (Compos.).

Aethusa L. Umbelliferae. 1 Eur., N. Afr., W. As., *A. cynapium* L. (fool's parsley), a poisonous weed resembling parsley.

Aëtia Adans. = Combretum Loefl. (Combretac.).

Aetopsis P. & K. = Aitopsis Rafin. = Salvia L. (Labiat.).

Aëtopteron Ehrh. (uninom.) = *Polypodium aculeatum* L. = *Polystichum aculeatum* (L.) Roth (Aspidiac.).

Aetoxicon Endl. = Aextoxicon Ruiz & Pav. (Aextoxicac.).

Aëtoxylon (Airy Shaw) Airy Shaw. Thymelaeaceae. 1 Borneo.

*Aextoxicaceae Engl. & Gilg. Dicots. 1/1 Chile. Lge trees. Ls. opp. or subopp., simple, subent., lepidote, exstip. Infl. axill., racemose, lepidote. Fl. reg., ♂ ♀, envel. by bract (or outermost sep.?). K 5, much imbr.; C 5, imbr., spath.; A 5, alt. with 5 disk glands; G̲ (2), lepidote, with deflexed shortly bifid style, & 2 pend. ovules in 1 loc. only. Small dry 1-seeded drupe; seed with ruminate endosp. Only genus: *Aextoxicon.* Prob. closely related to *Monimiac.* and *Trimeniac.*

Aextoxicon Ruiz & Pav. Aextoxicaceae. 1 Chile.

Aextoxicum P. & K. = praec.

Afarca Rafin. = Sageretia Brongn. (Rhamnac.).

Affonsea A. St-Hil. Leguminosae. 8 Braz.

Affonsoa P. & K. = praec.

Afgekia Craib. Leguminosae. 2 Siam.

Aflatunia Vassilcz. = Louiseania Carr. (Rosac.).

Afrachneria Sprague. Gramineae. 10 S. Afr.

Afraegle (Swingle) Engl. Rutaceae. 4 W. Afr.

Afrafzelia Pierre = Afzelia Sm. (Legumin.).

Aframmi C. Norman. Umbelliferae. 1 Angola.

Aframomum K. Schum. Zingiberaceae. 50 trop. Afr.

Afrardisia Mez. Myrsinaceae. 16 trop. Afr.

Afraurantium A. Chev. Rutaceae. 1 W. Afr.

Afridia Duthie = Nepeta L. (Labiat.).

Afrobrunnichia Hutch. & Dalziel. Polygonaceae. 2 trop. W. Afr.

Afrocalathea K. Schum. Marantaceae. 1 W. Afr.

Afrocrania (Harms) Hutch. Cornaceae. 1 trop. Afr.

Afrodaphne Stapf (~ Beilschmiedia Nees). Lauraceae. 20 trop. Afr.

Afrofittonia Lindau. Acanthaceae. 1 trop. W. Afr.

Afroguatteria Boutique. Annonaceae. 1 trop. Afr.

Afrohamelia Wernham = Atractogyne Pierre (Rubiac.).

Afrolicania Mildbraed. Chrysobalanaceae. 1 trop. W. Afr.

Afroligusticum C. Norman. Umbelliferae. 1 trop. Afr.

Afromendoncia Gilg = Mendoncia Vell. ex Vand. (Mendonciac.).

Afropteris Alston. Pteridaceae. 2 W. Afr., Seychelles.

Afrorhaphidophora Engl. Araceae. 2 trop. W. Afr.

Afrormosia Harms = Pericopsis Thw. (Legumin.).

Afrosersalisia A. Chev. Sapotaceae. 3 trop. & subtrop. Afr.

Afrosison H. Wolff. Umbelliferae. 3 trop. Afr.

Afrostyrax Perkins & Gilg. Huaceae. 3 trop. Afr. The genus agrees with *Hua* in its very similar foliage, presence of caducous stipules, fascicled axillary flowers, valvate sepals, pubescent petals, 10 uniseriate stamens, unilocular ovary with basal placentation, and copious endosperm smelling of onion. It differs from *Hua* principally in the much larger (long and narrow) stipules, presence of a fine dense lepidote indumentum on branches, leaves, calyx, ovary and fruit, imperfect separation of some sepals at anthesis, simple non-unguiculate petals, short filiform filaments, oblong, basally pilose, apically shortly awned, unequally 4-locellate anthers (inner locelli much shorter than outer), and presence of about 6 ovules in the ovary. The suggestion of an affinity between the two genera

AFROTHISMIA

seems first to have been made by Hallier (in MS in Herb. Lugd.-Bat., *teste* van Steenis). Pierre & De Wildeman's original suggestion of a Sterculiaceous affinity for *Hua* is almost certainly correct.

Afrothismia Schlechter. Burmanniaceae. 2 W. Equat. Afr.
Afrotrewia Pax & Hoffm. Euphorbiaceae. 1 W. Equat. Afr.
Afrotrichloris Chiov. Gramineae. 2 trop. E. Afr.
Afrotrilepis (Gilly) Raynal. Cyperaceae. 2 trop. W. Afr.
Afrovivella Berger. Crassulaceae. 1 Abyss.
Afzelia J. F. Gmel. = Seymeria Pursh (Scrophulariac.).
***Afzelia** Sm. Leguminosae. 14 trop. Afr., As.
Afzeliella Gilg. Melastomataceae. 2 trop. Afr.
Agaisia Garay & Sweet (sphalm.) = Aganisia Lindl. (Orchidac.).
Agalina Hort. (sphalm.) = Agalma Miq. = Schefflera J. R. & G. Forst. (Araliac.).
***Agalinis** Rafin. = Gerardia L. emend. Benth. (Scrophulariac.).
Agallis Phil. Cruciferae. 1 Chile.
Agallochum Lam. = Aquilaria Lam. (Thymelaeac.).
Agallostachys Beer = Bromelia L. (Bromeliac.).
Agalma Miq. = Schefflera J. R. & G. Forst. (Araliac.).
Agalma Steud. = Sonchus L. (Compos.).
Agalmanthus Hombr. & Jacquinot = Metrosideros Banks (Myrtac.).
Agalmyla Blume. Gesneriaceae. 5 Borneo, Sumatra, Java.
Agaloma Rafin. = Euphorbia L. (Euphorbiac.).
Aganippa Baill. = seq.
Aganippea Moç. & Sessé ex DC. Compositae. 2 Mex.
Aganisia Lindl. Orchidaceae. 3 trop. Am., W.I.
Aganon Rafin. = Callicarpa L. (Verbenac.).
Aganonerion Pierre. Apocynaceae. 1 Indochina.
Aganope Miq. Leguminosae. 6 W. & C. Afr., India & Ceylon, 8 E. As., Malaysia.
Aganosma G. Don. Apocynaceae. 10 China, Indomal.
Agapanthaceae Lotsy = Alliaceae J. G. Agardh.
***Agapanthus** L'Hérit. Alliaceae. 5 S. Afr. Umbel cymose. Seeds winged.
Agapatea Steud. = Distichia Nees & Meyen (Juncac.).
Agapetes D. Don ex G. Don. Ericaceae. 80 E. Himal. to SE. As. & Malay
Agardhia Spreng. = Qualea Aubl. (Vochysiac.). [Penins.
Agarista DC. = Coreopsis L. (Compos.).
Agarista D. Don = Leucothoë D. Don ex G. Don + Agauria (DC.) Hook. f.
Agasillis Spreng. = Agasyllis Hoffm. (Umbellif.). [(Ericac.).
Agassizia Chavannes = Galvezia Domb. (Scrophulariac.).
Agassizia Gray & Engelm. = Gaillardia Fouger. (Compos.).
Agassizia Spach = Camissonia Link (Onagrac.).
Agassyllis Lag. = Agasyllis Hoffm. (Umbellif.).
Agasta Miers = Barringtonia J. R. & G. Forst. (Barringtoniac.).
Agastache Clayt. in Gronov. Labiatae. 30 C. As. to China, N. Am., Mex.
Fl. stalk sometimes resupinate like that of *Lobelia*.
Agastachis Poir. = seq.
Agastachys R.Br. Proteaceae. 1 Tasmania.

Agastachys Ehrh. (uninom.) = *Carex agastachys* L. f. = *Carex pendula* Huds. (Cyperac.).

Agasthosma Brongn. (sphalm.) = Agathosma Willd. (Rutac.).

Agastianis Rafin. = Sophora L. (Legumin.).

Agasulis Rafin. = Ferula L. (Umbellif.).

Agasyllis Spreng. Umbelliferae. 1 Cauc.

Agatea A. Gray. Violaceae. 12 New Guinea, New Caled., Fiji.

Agatea W. Rich ex A. Gray = Haplopetalon A. Gray = Crossostylis J. R. & G. Forst. (Rhizophorac.).

Agathaea Cass. = Aster L. + Felicia Cass. (Compos.).

Agathea Endl. = praec.

Agathelepis Reichb. (sphalm.) = seq.

Agathelpis Choisy. Scrophulariaceae. 8 S. Afr.

Agathidaceae Nak. = Araucariaceae Strasb.

Agathidanthes Hassk. = Agathisanthes Bl. = Nyssa L. (Nyssac.).

*Agathis Salisb. Araucariaceae. 20 Indoch. & W. Malaysia to N.Z. Evergr. dioec. trees; the fr. takes two years to ripen. Several give copals or animes, used for varnish, etc. *A. alba* (Lam.) Foxw. (*A. dammara* Rich., Malaysia), Manila copal. *A. australis* (D. Don) Lindl. (Austr., N.Z., kauri or cowrie pine), kauri-copal; the best pieces are dug out of the soil, often far from trees now living.

Agathisanthemum Klotzsch = Oldenlandia L. (Rubiac.).

Agathisanthes Blume = Nyssa L. (Nyssac.).

Agathodes Reichb. = Agathotes D. Don = Swertia L. (Gentianac.).

Agathomeria Baill. = seq.

Agathomeris Delaun. = Humea Sm. (Compos.).

Agathomoris Durand = praec.

Agathophora (Fenzl) Bunge (~ Halogeton C. A. Mey.). Chenopodiaceae. 1 N. Afr. to Arabia.

Agathophyllum Blume = Ocotea Aubl. (Laurac.).

Agathophyllum Juss. = Ravensara Sonnerat (Laurac.).

Agathophyton Moq. = seq.

Agathophytum Moq. = Chenopodium L. (Chenopodiac.).

Agathorhiza Rafin. = Archangelica Hoffm. (Umbellif.).

*Agathosma Willd. Rutaceae. 180 S. Afr.

Agathotes D. Don = Swertia L. (Gentianac.).

Agathyrsus D. Don = Lactuca L. (Compos.).

Agathyrus B. D. Jacks. (sphalm.) = praec.

Agati Adans. = Sesbania Scop. (Legumin.).

Agatia Reichb. = praec.

Agation Brongn. = Agatea A. Gray (Violac.).

Agatophyllum Comm. ex Thou. = Agathophyllum Juss. = Ravensara Sonnerat (Laurac.).

Agatophyton Fourr. = Agathophytum Moq. (Chenopodiac.).

Agauria (DC.) Hook. f. Ericaceae. 7 trop. Afr., Madag., Masc.

*Agavaceae Endl. Monocots. 20/670 trop. & subtrop. Robust, rhizomatous, often woody or even arborescent, sometimes scandent pl. Ls. narrow, lanceolate, crowded, often basal. Fls. as *Liliac.* or *Amaryllidac.*; P united below;

AGAVACEAE

<u>G</u> or G̅. Chief gen.: *Yucca, Cordyline, Dracaena, Sansevieria, Phormium, Nolina, Furcraea, Agave, Polianthes.* Probably a heterogeneous group.

Agavaceae ('-ineae') Dum. = Haemodoraceae R.Br.

Agave L. Agavaceae. 300 S. U.S. to trop. S. Am., incl. *A. americana* L. (century plant, maguey, American aloe). The short stem grows in thickness like *Yucca*, bearing a rosette of large fleshy ls. coated with wax; only 2 or 3 ls. form in a year. During 5 to 60 or perhaps 100 years (hence the name), depending on climate, richness of soil, etc., the plant is veg., and stores up in the ls. an enormous mass of reserves. At length it flowers, a gigantic term. infl. coming rapidly out, sometimes reaching 20 ft., and bearing many fls. When the fr. is ripe the pl. dies. Veg. repr. in two ways—by suckers from base of stem, and by formation of bulbils in place of many fls.

The rush of sap to so large and so rapidly developed an infl. is very great; the Mexicans cut off the young fl. head and collect the sap. As much as 1000 litres is said to be given by one plant. The fermented juice (*pulque*) is a national drink; from it they distil a spirit called *mescal* (cf. *Cocos*). Many yield useful fibres. The best are sisal hemp and *henequen*, given by *A. sisalana* Perrine, and *A. fourcroydes* Lem., cultivated in Yucatan, the Bahamas, India, etc. Others yield fibres variously known as *pita, istle, ixtle, lechuguilla, keratto*, etc.

Agdestidaceae Nak. Dicots. (Centrosp.). 1/1 warm Am. Twining herbs from massive globose rootstock. Near *Phytolaccac.*, but ls. cordate-ovate with pet. twisted at base; K 4, persist. and accresc.; C o; A 15–20; G (3–4), semi-inf., with 3–4 recurved stigs. and 1 basal ovule per loc. Fr. inf., 1-seeded, with 4 persist. sep. Only genus: *Agdestis.*

Agdestis Moç. & Sessé. Agdestidaceae. 1 Mex. to Braz., W.I.

Agelaea Soland. ex Planch. Connaraceae. 50 trop. Afr., Madag., SE. As., Malaysia.

Agelandra Engl. & Pax = Angelandra Endl. = Croton L. (Euphorbiac.).

Agelanthus Van Tiegh. = Tapinanthus Bl. (Loranthac.).

Agenium Nees. Gramineae. 2 S. Am.

Agenora D. Don = Hypochoeris L. (Compos.).

Ageomoron Rafin. Umbelliferae. 4 E. Medit. to Persia.

Ageratella A. Gray ex S. Wats. Compositae. 2 Mex.

Ageratina O. Hoffm. = Ageratinastrum Mattf. (Compos.).

Ageratina Spach = Eupatorium L. (Compos.).

Ageratinastrum Mattf. Compositae. 5 trop. Afr.

Ageratiopsis Sch. Bip. ex Benth. & Hook. f. = Eupatorium L. (Compos.).

Ageratium Reichb. (sphalm.) = Aceratium DC. (Elaeocarpac.).

Ageratium Steud. = seq.

Ageraton Adans. = Ageratum Mill. = Erinus L. (Scrophulariac.).

Ageratum L. Compositae. 60 trop. Am. *A. conyzoïdes* L. (goatweed), a pantrop. weed. Scaly pappus.

Ageratum Mill. = Erinus L. (Scrophulariac.).

Agerella Fourr. = Veronica L. (Scrophulariac.).

Ageria Adans. = Ilex L. (Aquifoliac.) + Myrsine L. (Myrsinac.).

Aggeianthus Wight = Porpax Lindl. (Orchidac.).

Aggeranthus Wight (sphalm.) = praec.

Aggregatae Sch. Bip. = Dipsacaceae Juss.
Agiabampoa Rose ex O. Hoffm. Compositae. 1 Mex.
Agialid Adans. = Balanites Delile (Balanitac.).
Agialida Kuntze = praec.
Agialidaceae Van Tiegh. = Balanitaceae Endl.
Agianthus Greene (~ Streptanthus Nutt.). Cruciferae. 1 Calif.
Agiella Van Tiegh. = Balanites Del. (Balanitac.).
Agihalid Juss. = Agialid Adans. = Balanites Del. (Balanitac.).
Agina Neck. = Bartonia Mühlenb. ex Willd. (Gentianac.).
Agirta Baill. = Tragia L. (Euphorbiac.).
Agistron Rafin. = Uncinia Pers. (Cyperac.).
Aglaë Dulac = Posidonia Koen. (Posidoniac.).
Aglaea P. & K. = (1) Aglaia Allam. (Cyperac.), (2) Aglaia Lour. (Meliac.), (3) Aglaia Nor. ex Thou. (Dilleniac.), *qq.v.*
Aglaea Steud. = Melasphaerula Ker-Gawl. (Iridac.).
Aglaeopsis P. & K. = Aglaiopsis Miq. = Aglaia Lour. (Meliac.).
Aglaia Allamand = ? Cyperus L. (Cyperac.).
Aglaia Dum. = Aegle Corrêa (Rutac.).
***Aglaia** Lour. Meliaceae. 250–300 China, Indomal., trop. Austr., Pacif. Some spp. have 1-foliol. ls.
Aglaia Nor. ex Thou. = Hemistemma Juss. = Hibbertia Andr. (Dilleniac.).
Aglaiopsis Miq. = Aglaia Lour. (Meliac.).
Aglaja Endl. = Aglaia Nor. ex Thou. = Hibbertia Andr. (Dilleniac.).
Aglaodendron Remy = Plazia Ruiz & Pav. (Compos.).
Aglaodendrum P. & K. = praec.
Aglaodorum Schott. Araceae. 1 Sumatra, Malay Penins., Borneo.
Aglaomorpha Schott. Polypodiaceae. 4 Formosa, Malaysia. (*Drynariopsis, Pseudodrynaria* and *Holostachyum* have been included here.)
Aglaonema Schott. Araceae. 21 Indomal. There are several infls. forming a sympodium. Fls. monoecious, naked.
Aglitheis Rafin. = Allium L. (Alliac.).
Aglossorhyncha Schlechter. Orchidaceae. 13 E.Malaysia, Palau, Bismarck and Solomon Is.
Aglotoma Rafin. = Aster L. (Compos.).
Aglycia Willd. ex Steud. = Eriochloa Kunth (Gramin.).
Agnirictus Schwantes. Aïzoaceae. 2 S. Afr.
Agnistus G. Don = Acnistus Schott (Solanac.).
Agnostus A. Cunn. = Stenocarpus R.Br. (Proteac.).
Agnus-castus Carr. = Vitex L. (Verbenac.).
Agonandra Miers ex Benth. & Hook. f. Opiliaceae. 10 Mex. to trop. S. Am.
Agoneissos Zoll. ex Niedenzu = Tristellateia Thou. (Malpighiac.).
***Agonis** (DC.) Lindl. Myrtaceae. 15 Austr.
Agonolobus Reichb. = Cheiranthus L. (Crucif.).
Agonomyrtus Schau. ex Reichb. = ? Leptospermum J. R. & G. Forst. (Myrtac.).
Agonon Rafin. = ? Ilex L. (Aquifoliac.).
Agophyllum Neck. = Zygophyllum L. (Zygophyllac.).
Agorrhinum Fourr. = Antirrhinum L. (Scrophulariac.).   [S. Am.
Agoseris Rafin. (~ Troximon Nutt.). Compositae. 9 W. N. Am., 1 temp.

AGOSTANA

Agostana Bute ex S. F. Gray = Agrostana Hill = Bupleurum L. (Umbellif.).
Agraphis Link = Endymion Dum. (Liliac.).
Agraulus Beauv. = Agrostis L. (Gramin.).
Agrestis Rafin. = Agrostis L. (Gramin.).
Agretta Eckl. = Tritonia Ker-Gawl. (Iridac.).
Agrianthus Mart. Compositae. 7 Brazil.
Agricolæa Schrank = Clerodendrum L. (Verbenac.).
Agrifolium Hill = Aquifolium Mill. = Ilex L. (Aquifoliac.).
Agrimonia L. Rosaceae. 15 N. temp. The receptacle encloses the two achenes in fr., and is covered with hooks.
Agrimoniaceae S. F. Gray = Rosaceae–Sanguisorbeae Juss.
Agrimonioïdes v. Wolf = Aremonia Neck. (Rosac.).
Agrimonoïdes Mill. = praec.
Agriodaphne Nees ex Meissn. = Ocotea Aubl. (Laurac.).
Agriodendron Endl. = Aloë L. (Liliac.).
Agriophyllum Bieb. Chenopodiaceae. 6 C. As.
Agriophyllum P. & K. = seq.
Agriphyllum Juss. = Berkheya Ehrh. (Compos.).
× **Agrocalamagrostis** Aschers. & Graebn. Gramineae. Gen. hybr. (Agrostis × Calamagrostis).
Agrocharis Hochst. = Caucalis L. (Umbellif.).
× **Agroëlymus** G. Camus ex Rousseau. Gramineae. Gen. hybr. (Agropyron × Elymus).
× **Agrohordeum** G. Camus. Gramineae. Gen. hybr. (Agropyron × Hordeum).
× **Agropogon** P. Fourn. Gramineae. Gen. hybr. (Agrostis × Polypogon).
× **Agropyrohordeum** G. Camus ex A. Camus = × Agrohordeum G. Camus (Gramin.).
Agropyron J. Gaertn. Gramineae. 100–150 temp. *A. repens* Beauv. (twitch- or couch-grass) is a troublesome weed. Its long rhizome roots at the nodes, and if broken up each node gives a new plant.
Agropyropsis A. Camus. Gramineae. 2 N. Afr., C. Verde Is.
Agropyrum Roem. & Schult. = Agropyron J. Gaertn. (Gramin.).
Agrosinapis Fourr. = Brassica L. (Crucif.).
× **Agrositanion** Bowden. Gramineae. Gen. hybr. (Agropyron × Sitanion).
Agrostana Hill = Bupleurum L. (Umbellif.).
Agrostemma L. Caryophyllaceae. 2 Euras.
Agrosticula Raddi = Sporobolus R.Br. (Gramin.).
Agrostidaceae (Kunth) Herter = Gramineae–Agrostideae Kunth.
Agrostis Adans. = Imperata Cyr. (Gramin.).
Agrostis L. Gramineae. 150–200 cosmop., chiefly N. temp. *A. stolonifera* L. (white bent or fiorin) is a valuable pasture grass.
Agrostistachys Dalz. Euphorbiaceae. 8–9 India & Ceylon to W. Malaysia.
Agrostocrinum F. Muell. Liliaceae. 1 SW. Austr.
Agrostomia Cerv. = Panicum L. (Gramin.).
Agrostophyllum Blume. Orchidaceae. 60 Seychelles to Malaysia and Poly-
× **Agrotrigia** Tsvel. Gramineae. Gen. hybr. (Agropyron) × Elytrigia. [nesia.
× **Agrotrisecale** Ciferri & Giacom. Gramineae. Gen. hybr. (Agropyron × Secale × Triticum).

× **Agrotriticum** Ciferri & Giacom. Gramineae. Gen. hybr. (Agropyron × Triticum).

Aguava Rafin. = Myrcia DC. (Myrtac.).

Aguiaria Ducke. Bombacaceae. 1 Brazil.

Agylla Phil. = Cladium P.Br. (Cyperac.).

Agylophora Neck. = Uncaria Schreb. (Naucleac.).

Agyneia aucct. = seqq.

Agyneja L. = Glochidion J. R. & G. Forst. (Euphorbiac.).

Agyneja Vent. = Synostemon F. Muell. (Euphorbiac.).

Ahernia Merr. Flacourtiaceae. 1 Hainan, Philipp. Is.

Ahouai Mill. (~ Thevetia L.). Apocynaceae. 1 C. & S. Am.

Ahovai auctt. = praec.

Ahzolia Standley & Steyerm. Cucurbitaceae. 1 C. Am.

Aïchryson Webb & Berth. Crassulaceae. 10 Macaronesia.

Aïdelus Spreng. = Veronica L. (Scrophulariac.).

Aïdia Lour. (~ Randia auctt.). Rubiaceae. ? 50 trop. Afr., trop. As. to E. Austr.

Aidomene Stopp. Asclepiadaceae. 1 Angola.

Aigiros Rafin. = Populus L. (Salicac.).

Aigosplen Rafin. = Callirhoë Nutt. (Malvac.).

Aikinia R.Br. = Epithema Blume (Gesneriac.).

Aikinia Salisb. ex A. DC. = Wahlenbergia Schrad. (Campanulac.).

Aikinia Wall. = Ratzeburgia Kunth (Gramin.).

Ailanthaceae J. G. Agardh = Simaroubaceae–Ailanthinae Engl.

Ailanthus Desf. Simaroubaceae. 10 As., Austr. Absciss layers form at base of the leaflets as well as of the petiole; the leaflets usually drop first.

Ailantopsis Gagnep. = Trichilia P.Br. (Meliac.).

Aillya De Vriese = Goodenia Sm. (Goodeniac.).

Ailuroschia Stev. = Astragalus L. (Legumin.).

Aimenia Comm. ex Planch. = Cissus L. (Vitidac.).

Aimorra Rafin. Compositae. Florida. Quid?

Ainsliaea DC. Compositae. 40 E. As. to W. Malaysia.

Ainsliea Kuntze = praec.

Ainsworthia Boiss. Umbelliferae. 2 E. Medit. to Iraq.

Aiolographis Thou. (uninom.) = Limodorum scriptum Thou. = Graphorkis scripta (Thou.) Kuntze (Orchidac.).

Aiolon Lunell. Ranunculaceae. 1 N. Am.

Aiolotheca DC. = Zaluzania Pers. (Compos.).

Aiouea Aubl. Lauraceae. 30 trop. Am., W.I.

Aïphanes Willd. Palmae. 40 trop. Am.

Aipyanthus Stev. (~ Macrotomia DC.). Boraginaceae. 1 As. Min. to Cauc. & N. Pers.

Aira L. Gramineae. 12 N. & S. temp., mts. of trop. & S. Afr., Maurit.

Airampoa Frič = Opuntia Mill. (Cactac.).

Airella Dum. = Aira L. (Gramin.).

Airidium Steud. = Deschampsia Beauv. (Gramin.).

Airochloa Link = Koeleria Pers. (Gramin.).

Airopsis Desv. Gramineae. 1 S. Eur., NW. Afr.

Airosperma Lauterb. & K. Schum. Rubiaceae. 5 New Guinea, Fiji.

AIRYANTHA

Airyantha Brummitt. Leguminosae. 1 W. & Equat. Afr., 1 NW. Borneo & SW. Philipp. Is.

Aïsandra Pierre corr. Airy Shaw (~Payena A. DC.). Sapotaceae. 1 Indoch.

Aïstocaulon v. Poelln. Aïzoaceae. 1 S. Afr.

Aïstopetalum Schlechter. Cunoniaceae. 2 New Guinea.

Aitchisonia Hemsl. ex Aitch. Rubiaceae. 1 Afghanistan.

Aïthales Webb & Berth. = Sedum L. (Crassulac.).

Aithonium Zipp. ex C. B. Clarke = Rhynchoglossum Blume (Gesneriac.).

Aititara Endl. = Atitara Juss. = Evodia J. R. & G. Forst. (Rutac.).

Aitonia Thunb. (?1780; non Aytonia J. R. & G. Forst. 1777—Hepaticae) = Nymania S. O. Lindb. (Aitoniac.).

Aitoniaceae Harv. & Sond. Dicots. 1/1 S. Afr. Much-branched rigid shrubs or small trees, with angular puberulous twigs. Ls. alt., lin.-obl., ent., obt., coriac., 1-nerved, subsess., exstip., scattered or fascic. on short shoots. Fls. sol., axill., pedic., ♀, slightly zygo. K 4, small, shortly conn., slightly imbr., caduc.; C 4, rel. large, much imbr., purplish, externally puberulous; A 8, exserted, ascending, fil. flat, shortly conn., anth. obl., intr.; disk intra-stam., cupular, fleshy, cren.; G (4), 4-lobed, pubesc., with 2 collat. axile semi-amphitr. ov. per loc., and filif. style with simple punctif. stig. Caps. inflated, membr., deeply 4-lobed, acutely 4-angled, loculic.; seeds 1–2 per loc., renif., with thick coriac. corrug. ± loose testa; embr. curved, endosp. o. Only genus: *Nymania* Lindb. (*Aitonia* Thunb.). Variously referred to the *Sapindac.*, *Meliac.*, and even *Rutac.*, but anomalous in all. From the *Sapindac.* (to which it is probably closest) it differs in the intra-staminal disk, collateral ovules and curved embr.; from the *Meliac.* in the ± zygomorphic flower, simple punctiform stigma, and

Aitopsis Rafin. = Salvia L. (Labiat.). [inflated membr. capsule.

*****Aïzoaceae** Rudolphi. Dicots. (Centrosp.). 130/1200, chiefly S. Afr., but also trop. Afr. and As., Austr., Calif., S. Am. Nearly allied to *Phytolaccac.* and *Caryophyllac.*, also to *Cactac.* Xero. herbs or undershrubs with opp. or alt. stip. or exstip. ls., often fleshy, and with cymes of ♀ reg. fls. P 4–5 or (4–5) (if 5, odd member post.); A 5 or 3 or ∞, often fasciculate; G or Ḡ (2–5–∞), rarely G 5, with 1–∞ ov. per loc. *Dédoublement* is very common in the androecium, and in these cases, e.g. *Mesembryanthemum* (*s.l.*), the outer sta. are frequently represented by petaloid stds. Ovary usu. sup. with axile plac., but in *Mesembryanthemum* (*s.l.*) inf., 4–multiloc. with often parietal plac., a very unusual feature brought about during development (see *M.*). Fr. usu. a caps.; seed with embryo curved round perisperm.

Classification and chief genera (after Pax):

    I. **Molluginoïdeae** (perianth deeply 5-lobed: 'petals' or not: ov. sup.): *Mollugo.*

    II. **Aïzoïdeae (Ficoïdeae)** (perianth tubular): (G) *Sesuvium, Trianthema, Aïzoön*; (Ḡ) *Mesembryanthemum, Delosperma.*

Aïzoanthemum Dinter ex Friedr. Aïzoaceae. 3 SW. Afr.

Aïzodraba Fourr. = Draba L. (Crucif.).

Aïzoön Andr. = Sesuvium L. (Aïzoac.).

Aïzoön Hill = Sempervivum L. (Crassulac.).

Aïzoön L. (~Sedum L.). Aïzoaceae. 15 Afr., Medit., Orient, Austr. A ∞ in bundles.

Aïzoüm L. = praec.

Ajania Poljakov (~ Tanacetum L.). Compositae. 25 C. As. & Afghan. to E. Asia.

Ajax Salisb. = Narcissus L. (Amaryllidac.).

Ajaxia Rafin. = Delphinium L. (Ranunculac.).

Ajovea Juss. = Aiouea Aubl. (Laurac.).

Ajuea P. & K. = praec.

Ajuga L. Labiatae. 40 palaeotemp. The corolla has no upper lip. Veg. repr. freq. by runners.

Ajugoïdes Makino (~ Ajuga L.). Labiatae. 1 Japan.

Ajuvea Steud. = Aiouea Aubl. (Laurac.).

Akakia Adans. = Acacia Mill. (Legumin.).

Akania Hook. f. Akaniaceae. 1 E. Austr.

*Akaniaceae Stapf. Dicots. 1/1 E. Austr. Tree with alt. imparipinnate exstip. ls. and paniculate infl.; fl. ♂, reg.; K 5, imbr.; C 5, contorted; no disk; A usu. 8, the 5 external opp. sepals; G̲ 3-loc. with 2 superposed anatr. pend. ov. in each. Loculic. caps.; fleshy endosp.; straight embryo. Only genus: *Akania*. Related to *Sapindaceae (Harpullia)*.

Akea Stokes = Akeesia Tussac = Blighia Koen. (Sapindac.).

Akebia Decne. Lardizabalaceae. 5 E. As. Infl. bisex., the lower fls. usually ♀; the ♀ much larger than the ♂. Follicles fleshy, dehiscent, ed.

Akeesia Tussac = Blighia Koen. (Sapindac.).

Akentra Benj. = Utricularia L. (Lentibulariac.).

Aker Rafin. = Acer L. (Acerac.).

Akersia Buining (~ Borzicactus Riccob.). Cactaceae. 1 Peru.

Aklema Rafin. = Euphorbia L. (Euphorbiac.).

Akylopsis Lehm. = Matricaria L. (Compos.).

Alabella Comm. ex Baill. = Turraea L. (Meliac.).

Alacospermum Neck. = Cryptotaenia DC. (Umbellif.).

Aladenia Pichon. Apocynaceae. 1 W. Afr.

Alafia Thou. Apocynaceae. 20 trop. Afr., Madag.

Alagophyla Rafin. = Gesneria L. (Gesneriac.).

Alagophylla B. D. Jacks. = praec.

Alagoptera Mart. = Allagoptera Nees (Palm.).

Alairia Kuntze = Mairia Nees (Compos.).

Alajja S. Ikonn. (~ Lamium L.). Labiatae. 3 C. As., Afghan., W. Himal.

Alalantia Corr. (sphalm.) = Atalantia Corr. (Rutac.).

Alamania La Llave & Lex. Orchidaceae. 1 Mex.

Alandina Neck. = Moringa Juss. (Moringac.).

*Alangiaceae DC. Dicots. 2/20 trop. Trees or shrubs with alt. simple sometimes lobed often asymm. exstip. ls. Fls. reg., ♂, in axill. cymes, pedic. usu. artic., bibracteol. at apex. K (4–10), with short or lanc. teeth, or truncate, persist.; C 4–10, or (5), valv., sometimes ± contorted, finally recurved or revolute; A 4–40, free or epipet., fil. often pubesc., anth. elong.-obl., intr., basifixed, more rarely dorsif. and versat. and widely sagitt.; disk usu. conspic., pulvin., more rarely o; G̅ (1–2), more rarely G̲ 1, with 1 pend. anatr. ov. per loc., and simple elong. filif. style with clavate, 2–3-lobed or punctif. stigma. Fr. a 1-seeded drupe, with crustaceous or woody endoc.; seed with fleshy

ALANGIUM

endosp., embryo with foliaceous cots. Genera: *Alangium, Metteniusa*. Relationships obscure; perhaps some connection with *Olacaceae* and *Ehretiaceae*.

*Alangium Lam. Alangiaceae. 17 trop. Afr., Madag., Comoro Is., China, SE. As., Indomal., E. Austr.

Alania Colenso (=?Thalamia Spreng.)=Dacrydium Soland. ex J. R. & G.

Alania Endl. Liliaceae. 1 SE. Austr.   Forst. (Podocarpac.).

Alarçonia DC.=Wyethia Nutt. (Compos.).

Alaternoïdes Fabr.=Phylica L. (Rnamnac.).

Alaternus Mill.=Rhamnus L. (Rhamnac.).

Alathraea Steud.=Alatraea Neck.=Phelypaea L. (Orobanchac.).

Alatoseta Compton. Compositae. 1 S. Afr.

Alatraea Neck.=Phelypaea L. (Orobanchac.).

Albersia Kunth=Amaranthus L. (Amaranthac.).

Alberta E. Mey. (1838; non *Albertia* Schimp. 1837—gen. foss.)=Ernestimeyera Kuntze (Rubiac.).

Albertia Regel ex B. & O. Fedch.=Exochorda Lindl. (Rosac.).

Albertia Regel & Schmalh.=Aulacospermum Ledeb. + Trachydium Lindl. + Kozlovia Lipsky (Umbellif.).

Albertinia DC.=Vanillosmopsis Sch. Bip. (Compos.).

Albertinia Spreng. Compositae. 1 Brazil.

Albertisia Becc. Menispermaceae. 4 Assam, Indochina, Hainan, Malay Penins., New Guinea.

Albertisiella Pierre ex Aubrév. Sapotaceae. 2 New Guinea.

Albertokuntzea Kuntze=Seguieria Loefl. (Phytolaccac.).

Albidella Pichon. Alismataceae. 1 Cuba.

Albikia J. & C. Presl=Hypolytrum Rich. (Cyperac.).

Albina Giseke=Alpinia L. (Zingiberac.).

Albinea Hombr. & Jacquinot=Pleurophyllum Hook. f. (Compos.).

Albizia Durazz. Leguminosae. 100–150 warm Old World. *A. lebbeck* (L.). Benth. (*siris*, E. Indian walnut), etc., good timber. *A. chinensis* (Osb.) Merr. (*A. stipulata* Boiv.) (*sau*), *A. moluccana* Miq., etc., as shade for tea cult. etc. (very rapid growth, about 3 m. in height, and 30 cm. in girth, a year).

Albizzia auctt.=praec.

Albolboa Hieron.=Abolboda Humb. & Bonpl. (Abolbodac.).

Albonia Buc'hoz=Ailanthus Desf. (Simaroubac.).

Albovia Schischk. Umbelliferae. 4 E. Medit. to NW. Persia.

Albowiodoxa Woron. ex Kolak.=Amphoricarpos Vis. (Compos.).

Albradia D. Dietr.=seq.

Albrandia Gaudich.=Streblus Lour. (Morac.).

Albuca L. Liliaceae. 50 Afr. Outer sta. often stds.

Albucea Reichb.=Ornithogalum L. (Liliac.).

Albuga Schreb.=Albuca L. (Liliac.).

Albugoïdes Medik.=Albuca L. (Liliac.).

Albuminosae Engl. A 'Series' within the fam. *Ochnaceae*, incl. the tribes *Euthemideae, Luxemburgieae* and *Sauvagesieae*.

Alcaea Burm. f.=Althaea Mill. (Malvac.).

Alcamaspinora 'Noronha', B. D. Jacks. (Ind. Kew., sphalm.)=*Alcanna spinosa* (L.) Gaertn.=*Lawsonia inermis* L. (Lythrac.).

Alcanna Gaertn. = Alkanna Adans. = Lawsonia L. (Lythrac.).

Alcanna Orph. = Alkanna Tausch (Boraginac.).

Alcantara Glaziou (nomen). Compositae. 2 Brazil. Quid?

Alcantarea (Morren) Harms = Vriesia Lindl. (Bromeliac.).

Alcea L. Malvaceae. 60 Medit. to C. As.

Alcea Mill. = Malva L. (Malvac.).

× **Alchamaloë** Rowley. Liliaceae. Gen. hybr. (Aloë × Chamaealoë).

Alchemilla L. Rosaceae. 250 temp., and trop. mts. Fl. inconspic., apet., with epicalyx; A 2 or 4; G 1–4 each with 1 ov. Achenes enclosed in dry receptacle. Some are apomictic; some show a kind of chalazogamy; some have an exudation of water from the ls.

Alchemill[ac]eae J. G. Agardh = Rosaceae–Sanguisorbeae Juss. ʼ

Alchimilla Mill. = Alchemilla L. (Rosac.).

Alchornea Sw. Euphorbiaceae. 70 trop.

Alchorneopsis Muell. Arg. Euphorbiaceae. 3 trop. Am., W.I.

Alcicornium Gaudich. (nom. provis.) = Platycerium Desv. (Polypodiac.).

Alcimandra Dandy. Magnoliaceae. 1 SE. As.

Alcina Cav. = Melampodium L. (Compos.).

Alcinaeanthus Merr. = Neoscortechinia Pax (Euphorbiac.).

Alciope DC. Compositae. 2 S. Afr.

Alcmene Urb. = Duguetia A. St-Hil. (Annonac.).

Alcoceratothrix Niedenzu = Byrsonima Rich. ex Juss. (Malpighiac.).

Alcoceria Fernald = Dalembertia Baill. (Euphorbiac.).

Alcytophyllum Dur. (sphalm.) = Arcytophyllum Willd. ex Schult. & Schult. f. (Rubiac.).

Aldaeëa Schlechtd. = Aldea Ruiz & Pav. = Phacelia Juss. (Hydrophyllac.).

Aldama La Llave. Compositae. 1 Mex. to N. Venez.

Aldasorea Hort. ex Haage & Schmidt = Aeonium Webb & Berth. (Crassulac.).

Aldea Ruiz & Pav. = Phacelia Juss. (Hydrophyllac.).

Aldeaea Reichb. = praec.

Aldelaster C. Koch = Adelaster Veitch = ?Pseuderanthemum Radlk. (Acanthac.).

Aldenella Greene = Cleome L. (Cleomac.).

Aldina Adans. = Brya P.Br. (Legumin.).

Aldina Engl. Leguminosae. 5 Guiana, N. Braz.

Aldina E. Mey. = Acacia Mill. (Legumin.).

Aldinia Rafin. = Croton L. (Euphorbiac.).

Aldinia Scop. = Justicia L. (Acanthac.).

Aldrovanda L. Droseraceae. 1 C. Eur., Cauc., E. & SE. As., Timor, Queensl. *A. vesiculosa* L., a rootless swimming pl. with whorls of ls. Each has a stalk portion, and a blade like *Dionaea*, working in the same way, capturing and digesting small animals. Winter buds form in cold climates.

Aldrovandaceae Nak. = Droseraceae Salisb.

Aldunatea Remy = Chaetanthera Ruiz & Pav. (Compos.).

Alectoridia A. Rich. = Arthraxon Beauv. (Gramin.).

Alectoroctonum Schlechtd. = Euphorbia L. (Euphorbiac.).

Alectorolophus Zinn = Rhinanthus L. (Scrophulariac.).

Alectorurus Makino (∼ Anthericum L.). Liliaceae. 2 Japan.

39

ALECTRA

Alectra Thunb. Scrophulariaceae. 30 trop. Am., trop. to S. Afr., trop. As.
Alectryon Gaertn. Sapindaceae. 15 Malaysia to Polynesia, trop. Austr., N.Z.
Alegria Moc. & Sessé = Luehea Willd. (Tiliac.).
Aleisanthia Ridl. Rubiaceae. 2 Malay Penins.
Aleome Neck. = Cleome L. (Cleomac.).
Alepidea La Roche. Umbelliferae. 40 trop. & S. Afr.
Alepidixia Van Tiegh. ex Lecomte = Viscum L. (Viscac.).
Alepidocalyx Piper. Leguminosae. 3 Mex.
Alepidocline Blake. Compositae. 1 C. Am.
Alepis Van Tiegh. (~Elytranthe Bl.). Loranthaceae. 1 N.Z.
× **Aleptoë** Rowley. Liliaceae. Gen. hybr. (Aloë × Leptaloë).
Alepyrum R.Br. = Centrolepis Labill. (Centrolepidac.).
Alepyrum Hieron. = Pseudalepyrum Dandy = Centrolepis Labill. (Centrolepidac.).
Aletes Coulter & Rose. Umbelliferae. 5 N. Am.
Aletris L. Liliaceae. 25 E. As., N. Am.
Aleurites J. R. & G. Forst. Euphorbiaceae. 2 trop. As., Malaysia, Pacif. *A. moluccana* (L.) Willd. yields candlenut or *lumbang* oil, a drying oil.
Aleuritia (Duby) Opiz = Primula L. (Primulac.).
Aleuritopteris Fée. Sinopteridaceae. 15 trop. & N. temp.; ± xerophytic; ls. deltoid, lr. surface covered with waxy powder.
Aleurodendron Reinw. = Melochia L. (Sterculiac.).
Alevia Baill. = Bernardia Mill. (Euphorbiac.).
Alexa Moq. Leguminosae. 7 trop. Am.
Alexandra Bunge. Chenopodiaceae. 1 C. As.
Alexandra Schomb. = Alexa Moq. (Legumin.).
Alexandrina Lindl. = praec.
Alexia Wight = Alyxia R.Br. (Apodynac.).
Alexis Salisb. = Amomum L. (Zingiberac.).
Alexitoxicon St-Lag. Asclepiadaceae. 35 temp. Euras.
Alfaroa Standley. Juglandaceae. 1 Costa Rica.
Alfonsia Kunth = Corozo Jacq. ex Giseke (Palm.).
Alfredia Cass. (~Carduus L.). Compositae. 5 C. As.
Alga Boehm. = Posidonia Koen. (Posidoniac.).
Alga Lam. = Zostera L. (Zosterac.).
Algarobia Benth. = Prosopis L. (Legumin.).
Algelanthus auctt. = Agelanthus Van Tiegh. = Tapinanthus Bl. (Loranthac.).
Algernonia Baill. Euphorbiaceae. 3 Brazil.
Alguelaguen Adans. = Sphacele Benth. (Labiat.).
Alguelagum Kuntze = praec.
Alhagi Gagnebin. Leguminosae. 5 Medit. & Sahara to C. As. & Himal. Thorny xero.; the rootstock blows about in the dry season. Honeylike sap exudes in hot weather, drying into brownish lumps (manna).
Alibertia auct. = Allibertia Marion = Agave L. (Agavac.).
Alibertia A. Rich. ex DC. Rubiaceae. 35 trop. Am., W.I.
Alibrexia Miers = Nolana L. (Nolanac.).
Alibum Less. = Liabum Adans. (Compos.).
Alicabon Rafin. = Physalis L. (Solanac.).

Alicastrum P.Br. = Brosimum Sw. (Morac.).

× **Aliceara** hort. Orchidaceae. Gen. hybr. (iii) (Brassia × Miltonia × Oncidium).

Aliciella Brand. Polemoniaceae. 1 N. Am.

Aliconia Herrera = Mikania Willd. (Compos.).

Alicosta Dulac = Bartsia L. (Scrophulariac.).

Alicteres Neck. ex Schott & Endl. = Helicteres L. (Sterculiac.).

Alifana Rafin. (Brachyotum Triana). Melastomataceae. 60 S. Am.

Alifanus Adans. = Rhexia L. (Melastomatac.).

Alifiola Rafin. = Silene L. (Caryophyllac.).

Aligera Suksdorf. Valerianaceae. 15 Pacif. N. Am.

Alina Adans. = Vimen P.Br. = Hyperbaena Miers ex Benth. (Menispermac.).

Alionia Rafin. = Allionia Loefl. (Nyctaginac.).

Alipendula Neck. = Filipendula L. (Rosac.).

Alipsa Hoffmgg. = Liparis Rich. (Orchidac.).

Alipsea auctt. = praec.

Aliseta Rafin. = Arnica L. (Compos.).

Alisma L. Alismataceae. 10 N. temp., Austr. Sta. 6 (doubling of outer whorl), coherent at base, forming nectary.

*****Alismataceae** Vent. Monocots. 13/90 cosmop. Water or marsh herbs with perenn. rhiz. Ls. rad., erect, floating or submerged, exhibiting corresponding structure (cf. *Sagittaria*). Small scales in axils. Latex. Infl. usu. much branched; primary branching racemose, secondary often cymose. Fl. ♀ or unisex., reg. K 3; C 3; A 6–∞, or 3, anth. extrorse; G̲ 6–∞, with 1 (rarely 2 or more) anatr. ov. in each. Fr. a group of achenes; seed exalb.; embryo horseshoe-shaped. Chief genera: *Alisma, Echinodorus, Sagittaria, Lophotocarpus, Damasonium*. The fam. makes a close morphol. approach to the *Ranunculac.*, but considerable anat. differences.

× **Alismodorus** Wehrh. Alismataceae. Gen. hybr. (Alisma × Echinodorus).

Alismographis Thou. (uninom.) = *Limodorum plantagineum* Thou. = *Eulophia plantaginea* (Thou.) Rolfe (Orchidac.).

Alismorchis Thou. = seq.

Alismorkis Thou. = Calanthe R.Br. (Orchidac.).

Alisson Vill. = Alyssum L. (Crucif.).

Alistilus N. E. Brown. Leguminosae. 1 S. Afr. (Bechuanaland).

Aliteria Benoist = Clarisia Ruiz & Pav. (Morac.).

Alitubus Dulac = Achillea L. (Compos.).

Alix Comm. ex DC. = Psiadia Jacq. (Compos.).

Alkanna Adans. = Lawsonia L. (Lythrac.).

*****Alkanna** Tausch. Boraginaceae. 25–30 S. Eur., Medit., to Persia. The r. of *A. tinctoria* (L.) Tausch gives the red dye alkanet or alkannin.

Alkekengi Mill. = Physalis L. (Solanac.).

Alkibias Rafin. = Aster L., Chrysocoma L., etc. (Compos.).

Allaeanthus Thw. = Broussonetia L'Hérit. (Morac.).

Allaeophania Thw. Rubiaceae. 3 Indomal.

Allaganthera Mart. = Alternanthera Forsk. (Amaranthac.).

Allagas Rafin. = Alpinia Roxb. (Zingiberac.).

Allagopappus Cass. Compositae. 1 Canaries.

Allagophyla Rafin. = Corytholoma (Benth.) Decne (Gesneriac.).

ALLAGOPHYLLA

Allagophylla auctt. = praec.

Allagoptera Nees. Palmae. 10 trop. S. Am.

Allagosperma M. Roem. = Alternasemina S. Manso = Melothria L. (Cucurbitac.).

Allagostachyum Nees = Poa L. (Gramin.).

Allamanda L. (sphalm.) = Allemanda L. (Apocynac.).

Allanblackia Oliv. Guttiferae. 8 trop. Afr. The seeds of *A. stuhlmannii* Engl. yield a tallow-like fat.

Allania Benth. = Aldina Endl. (Legumin.).

Allania Meissn. = Alania Endl. (Liliac.).

Allantodia R.Br. = Diplazium Sw. (Athyriac.).

Allantodia Wall. = Diplaziopsis C. Chr. (Athyriac.).

Allantoma Miers. Lecythidaceae. 15 Guiana, Braz.

Allantospermum Forman. Ixonanthaceae. 1 Madag., 2 Borneo.

Allardia Decne = Waldheimia Kar. & Kir. (Compos.).

Allardtia A. Dietr. = Tillandsia L. (Bromeliac.).

Allasia Lour. = Vitex L. (Verbenac.).

× **Allauminia** Rowley. Liliaceae. Gen. hybr. (Aloë × Guillauminia).

Allazia S. Manso = praec.

Alleizettea Dubard & Dop. Rubiaceae. 1 Madag.

Alleizettella Pitard. Rubiaceae. 1 Indoch.

Allelotheca Steud. = Lophatherum Brongn. (Gramin.).

Allemanda L. (corr. L.). Apocynaceae. 15 trop. S. Am., W.I. Seeds hairy.

Allemania Endl. = Allmania R.Br. ex Wight (Amaranthac.).

Allenanthus Standley. Rubiaceae. 2 C. Am.

Allendea La Llave. Compositae. 1 Mex.

Allenia Ewart = Micrantheum Desf. (Euphorbiac.).

Allenia Phillips = Radyera Bullock (Malvac.).

Allenrolfea Kuntze. Chenopodiaceae. 4 N. Am. to Patag.

Alletotheca Benth. & Hook. f. (sphalm.) = Allelotheca Steud. = Lophatherum Brongn. (Gramin.).

Allexis Pierre. Violaceae. 3 trop. W. Afr.

*****Alliaceae** J. G. Agardh. Monocots. 30/600 cosmop. (exc. Australas.). Bulbous or rhizomatous herbs, intermediate betw. *Liliaceae* and *Amaryllidaceae*, having the superior ovary of the former and the scapose umbellate infl., subtended by spathaceous ± membr. bracts, of the latter.

    1. AGAPANTHEAE (rhizome): *Agapanthus, Tulbaghia.*

    2. ALLIEAE (corm or bulb; A reg.; corona o): *Allium, Nothoscordum, Brodiaea*, etc.

    3. GILLIESIEAE (bulb; A ± zygo.; corona usu. present): *Gilliesia*, etc.

Alliaria Scop. Cruciferae. 5 Eur., temp. As.

Alliaria Kuntze = Dysoxylum Blume (Meliac.).

Allibertia Marion = Agave L. (Agavac.).

Allinum Neck. = Selinum L. (Umbellif.).

*****Allionia** L. Nyctaginaceae. 1 Am., W.I. (very variable). Anthocarp glandular (cf. *Pisonia*).

Allionia Loefl. = Mirabilis L. (Nyctaginac.).

Allioniaceae Horan. = Nyctaginaceae Juss.

Allioniella Rydb. (~ Mirabilis L.). Nyctaginaceae. 25 mostly S. Am., 1 Himal.
Allium L. Alliaceae. 450 N. Hemisph. *A. ursinum* L. (temp. Euras., in woodlands) is the wild garlic; *A. schoenoprasum* L. (N. temp.) the chive; *A. cepa* L. (Persia, etc.) the onion; *A. porrum* L. (Eur.) the leek; *A. ascalonicum* L. (Orient) the shallot; *A. sativum* L. (S. Eur.) the garlic. Mostly bulbous, sometimes rhiz. herbs with linear (or hollow centric) ls. and cymose umbels of fls. Many have collateral buds in the axils. In many the fls. are replaced by bulbils serving for veg. repr. (cf. *Lilium*). In *A. ursinum*, etc., honey is secreted by the septal glands of the ovary; fl. protandr.
Allmania R.Br. ex Wight. Amaranthaceae. 1–2 trop. As.
Allmaniopsis Suesseng. Amaranthaceae. 1 E. Afr.
Allobia Rafin. = Euphorbia L. (Euphorbiac.).
Allobium Miers = Phoradendron Nutt. (Loranthac.).
Allobrogia Tratt. = Paradisea Mazz. (Liliac.).
Alloburkillia Whitmore. Leguminosae. 1 Malay Penins.
Allocalyx Cordemoy. Scrophulariaceae. 1 Réunion.
Allocarpus Kunth = Calea L. (Compos.).
Allocarya Greene (~ Plagiobothrys Fisch. & Mey.). Boraginaceae. 3 Austr., 80 Pacif. Am.
Allocaryastrum A. Brand. Boraginaceae. 5 Calif., Chile.
Allocassine N. Robson. Celastraceae. 2 SE. trop. & S. Afr.
Alloceratium Hook. f. & Thoms. = Diptychocarpus Trautv. (Crucif.).
Allochilus Gagnep. Orchidaceae. 1 Indoch.
Allochlamys Moq. = Pleuropetalum Hook. f. (Amaranthac.).
Allochrusa Bunge (~ Acanthophyllum C. A. Mey.). Caryophyllaceae. 7 W. & S. As.
Allodape Endl. = Lebetanthus Endl. (Epacridac.).
Allodaphne Steud. = praec.
× **Alloëlla** Rowley. Liliaceae. Gen. hybr. (Aloë × Aloinella).
Alloeochaete C. E. Hubbard. Gramineae. 2 S. trop. Afr.
Alloeospermum Spreng. = Alloispermum Willd. = Calea L. (Compos.).
Allogyne Lewton = Alyogyne Alef. = Fugosia Juss. (Malvac.).
Allohemia Rafin. = Phthirusa Mart. (Loranthac.).
Alloiantheros Steud. = seq.
Alloiatheros Ell. = Gymnopogon Beauv. (Gramin.).
Alloiatheros Rafin. = Andropogon L. (Gramin.).
Alloiosepalum Gilg = Purdiaea Planch. (Cyrillac.).
Alloiozonium Kunze = Cryptostemma R.Br. = Arctotheca Wendl. (Compos.).
Alloispermum Willd. = Calea L. (Compos.).
Allolepis Soderstr. & Decker. Gramineae. 1 S. U.S.
Allomaieta Gleason. Melastomataceae. 1 Colombia.
Allomarkgrafia R. E. Woodson. Apocynaceae. 4 C. & trop. S. Am.
Allomia DC. = Alomia Kunth (Compos.).
Allomorphia Blume. Melastomataceae. 25 China, Indomal.
Alloneuron Pilger. Melastomataceae. 1 Peru.
Allopetalum Reinw. = Labisia Lindl. (Myrsinac.).
Allophyllum (Nutt.) A. & V. Grant. Polemoniaceae. 5 W. U.S.

Allophylus L. Sapindaceae. 190 trop. & subtrop. Acc. to Leenhouts, *Blumea*, **15**: 301–58, 1968, the genus contains but 1 polymorphic sp., *A. cobbe* (L.) Raeusch.

Allophyton T. S. Brandegee = Tetranema Benth. (Scrophulariac.).

***Alloplectus** Mart. Gesneriaceae. 50 trop. Am.

Allopleia Rafin. = Sibthorpia L. (Scrophulariac.).

Allopythion Schott = Thomsonia Wall. (Arac.).

Allosampela Rafin. = Ampelopsis Michx (Vitidac.).

Allosandra Rafin. = Tragia L. (Euphorbiac.).

Allosanthus Radlk. Sapindaceae. 1 Peru.

Alloschemone Schott. Araceae. 1 Braz.

Allosorus auctt. = Cryptogramma R.Br. (Cryptogrammac.).

Allosorus Bernh. = Cheilanthes Sw. (Sinopteridac.).

Allosperma Rafin. = Commelina L. (Commelinac.).

Allospondias (Pierre) Stapf = Spondias L. (Anacardiac.).

Allostis Rafin. = Baeckea L. (Myrtac.).

Alloteropsis C. Presl. Gramineae. 10 trop. Afr. & As.

Allotria Rafin. = Commelina L. (Commelinac.).

Allotropa A. Gray. Monotropaceae. 1 W. U.S.

Allouya Aubl. = Calathea G. F. W. Mey. (Marantac.).

Allowoodsonia Markgr. Apocynaceae. 1 Solomon Is.

Allozygia Naud. = Oxyspora DC. (Melastomatac.).

Alluandia Drake (sphalm.) = seq.

Alluaudia Drake. Didiereaceae. 6 Madag.

Alluaudiopsis Humbert & Choux. Didiereaceae. 2 Madag.

Allucia Klotzsch ex Petersen = Renealmia L. f. (Zingiberac.).

Allughas Steud. = Alughas Lindl. = Alpinia L. (Zingiberac.).

Almana Rafin. = Sinningia Nees (Gesneriac.).

Almeidea A. St-Hil. Rutaceae. 5 Brazil, Guiana.

Almeloveenia Dennst. = Moullava Adans. (? Legumin.).

Almideia Reichb. = Almeidea A. St-Hil. (Rutac.).

Almyra Salisb. = Pancratium L. (Amaryllidac.).

Alnaster Spach = Duschekia Opiz (Betulac.).

Alniphyllum Matsumura. Styracaceae. 8 SW. China, Indoch., Formosa.

Alnobetula Schur = Duschekia Opiz (Betulac.).

Alnus Mill. Betulaceae. 35 N. temp., S. to Assam & Indoch., and Andes. Cf. *Betula.* In the axil of each bract of the ♂ catkin are 3 fls. (cf. other genera) each with 4 sta. and 4 perianth ls. The bracteoles α, β, β', β' (in Eichler's system) are present. All these ls. are united with one another. In the ♀ catkin only two, the lat., fls. occur, and the same bracts. After fert., the ov. gives a one-seeded nut, under which is found a 5-lobed scale, the product of subsequent growth of the 5 leaves. The fl. is chalazogamic.

Aloaceae ('-ineae') J. G. Agardh = Liliaceae–Aloëae Endl.

***Alocasia** (Schott) G. Don. Araceae. 70 Indomal. Herbaceous; monoec. *A. macrorrhiza* Schott and others are cult. for ed. rhiz. (cf. *Colocasia*).

Alocasia Neck. = Dracunculus L. + Arisaema Mart. (Arac.).

Alocasia Rafin. = Arisaema Mart. (Arac.).

Alocasiophyllum Engl. = Cercestis Schott (Arac.).

Alococarpum H. Riedl & Kuber. Umbelliferae. 1 Persia.

Aloë L. Liliaceae. 275 trop. & S. Afr., 42 Madag., 12–15 Arabia. Usu. shrubby or arborescent xero., growing in thickness and branching. Ls. in dense rosettes at ends of branches, very fleshy, with thick epidermis, often waxy, and stomata in pits. They are cut across and the juice evap. to obtain the drug aloes.

Aloeatheros Endl. = Alloiatheros Ell. = Gymnopogon Beauv. (Gramin.).

Aloës Rafin. = Aloë L. (Liliac.).

Aloëxylum Lour. = Aquilaria Lam. (Thymelaeac.).

Aloïdes Fabr. (uninom.) = *Stratiotes aloïdes* L. (Hydrocharitac.).

Aloïnella (A. Berger) A. Berger ex Lemée (1939; non Cardot 1909—Musci) = Aloë L. (Liliac.).

Aloïnopsis Schwantes. Aïzoaceae. 16 S. Afr.

Aloiozonium Lindl. = Alloiozonium Kunze = Cryptostemma R.Br. = Arctotheca

Aloitis Rafin. = Gentiana L. (Gentianac.). [Wendl. (Compos.).

× **Aloloba** Rowley. Liliaceae. Gen. hybr. (Aloë × Astroloba).

Alomia Kunth. Compositae. 25 Mex. to Braz.

Alona Lindl. (∼ Nolana L.). Nolanaceae. 6 Chile.

Alonsoa Ruiz & Pav. Scrophulariaceae. 6 trop. Am.

Alopecias Stev. = Astragalus L. (Legumin.).

Alopecuropsis Opiz = seq.

Alopecuro-veronica L. = Pogostemon Desf. (Labiat.).

Alopecurus L. Gramineae. 50 temp. Eurasia, temp. S. Am. *A. pratensis* L. (foxtail) cult. for pasture. Fl. protog.

Alophia Herb. Iridaceae. 10 warm Am.

Alophium Cass. = Centaurea L. (Compos.).

Alophochloa Endl. = Lophochloa Reichb. = Koeleria Pers. (Gramin.).

Alophyllus L. = Allophylus L. (Sapindac.).

Alopicarpus Neck. = Paris L. (Trilliac.).

Aloranthus F. S. Voigt (sphalm.) = Chloranthus Sw. (Chloranthac.).

Alosemis Rafin. = Rhynchanthera DC. (Melastomatac.).

Aloseris Rafin. = praec.

Aloysia Ort. & Palau ex Pers. Verbenaceae. 37 Am.

Alpaminia O. E. Schulz. Cruciferae. 1 Peru.

Alpan Bosc ex Rafin. = Apama Lam. (Aristolochiac.).

Alphandia Baill. Euphorbiaceae. 3 New Guinea, New Hebr., New Caled.

Alphitonia Reissek ex Endl. Rhamnaceae. 20 Malaysia, Austr., Polynesia.

Alphonsea Hook. f. & Thoms. Annonaceae. 30 China, Indomal.

Alphonseopsis E. G. Baker = Polyceratocarpus Engl. & Diels (Annonac.).

Alpinia L. = Renealmia L. f. (Zingiberac.).

*Alpinia Roxb. Zingiberaceae. 250 warm As., Polynesia. K small tubular, C with short tube and 3 large teeth, big labellum; lat. stds. much reduced or absent; anther lobes divided by broad connective. *A. officinarum* Hance (China) gives *rhizoma galangae*.

Alsaton Rafin. = Siler Crantz (Umbellif.).

Alschingera Vis. = Physospermum Cusson (Umbellif.).

Alseis Schott. Rubiaceae. 15 trop. Am.

Alsenosmia Endl. = Alseuosmia A. Cunn. (Alseuosmiac.).

ALSEODAPHNE

Alseodaphne Nees (~Persea Boehm.). Lauraceae. 25 China, SE. As., Indomal.

Alseuosmia A. Cunn. Alseuosmiaceae. 8 N.Z., some hybridizing freely.

Alseuosmiaceae Airy Shaw. Dicots. 3/11 New Caled., N.Z. Shrubs with alt. or subopp. simple ent. or sin.-dent. exstip. ls. Fls. reg., ⚥ or polyg. (?dioec.), axill., sol. or fascic., rarely rac. or subterm., sometimes fragrant. K 4–5, valv. or open, decid. or persist.; C (4–7), infundib. or urc., lobes valv. or indupl.-valv. or subimbr., sometimes ± tuberc. or verruc. within, with margins ± dent. or fimbr.; A 4–7, alternipet., epipet. or free, 2 sometimes larger than remainder, anth. intr.; G̅ (2), sometimes semi-sup., crowned with flat or tumid disk, ovules 1–6–∞ per loc., axile or pariet.; style slender or stout, stig. clav. or capit., ± 2-lobed. Fr. a 2-loc. berry; seeds with endosp. Genera: *Alseuosmia, Periomphale (Pachydiscus), Memecylanthus*. Hitherto referred to *Caprifoliac.*, but the alt. ls., general habit, valvate cor.-lobes and pollen are against this. The fam. is in some respects intermediate between *Escalloniac.* and *Loganiac.* (*s.l.*). Habit sometimes recalls *Pittosporum* spp.

***Alsinaceae** Bartl. = Caryophyllaceae Juss.

Alsinanthe Reichb. = Arenaria L. (Caryophyllac.).

Alsinanthemos J. G. Gmel. = seq.

Alsinanthemum Fabr. (uninom.) = *Trientalis europaea* L. (Primulac.).

Alsinanthus Desv. = Arenaria L. (Caryophyllac.).

Alsinastrum Quer = Elatine L. (Elatinac.).

Alsine Druce = Spergularia (Pers.) J. & C. Presl (Caryophyllac.).

Alsine Gaertn. = Minuartia L. (Caryophyllac.).

Alsine L. = Arenaria L. + Stellaria L. + Delia Dum. (Caryophyllac.).

Alsinella Hill = Sagina L. (Caryophyllac.).

Alsinella Hornem. = Spergularia (Pers.) J. & C. Presl (Caryophyllac.).

Alsinella Moench = Cerastium L. (Caryophyllac.).

Alsinella Sw. = Stellaria L. + Arenaria L. (Caryophyllac.).

Alsinidendron H. Mann = Schiedea Cham. & Schlechtd. (Caryophyllac.).

Alsinodendron auctt. = praec.

Alsinoïdes Lippi ex Adans. = Jorena Adans. (?Caryophyllac.).

Alsinopsis Small = Minuartia L. (Caryophyllac.).

Alsobia Hanst. = Episcia Mart. (Gesneriac.).

Alsocydia Mart. ex J. C. Gomes Jr. = Piriadacus Pichon (Bignoniac.).

Alsodeia Thou. = Rinorea Aubl. (Violac.).

Alsodeiaceae J. G. Agardh = Violaceae–Rinoreëae Reiche & Taub.

Alsodeiïdium Engl. = seq.

Alsodeiopsis Oliv. Icacinaceae. 11 trop. Afr.

Alsolinum Fourr. = Linum L. (Linac.).

Alsomitra (Bl.) M. Roem. Cucurbitaceae. 2 Indomal.

Alsophila R.Br. = Cyathea Sm. (Cyatheac.). Has been used to include a mixture of unrelated spp. segregated from *Cyathea* owing to lack of indusium.

Alsophilaceae Presl = Cyatheaceae Reichb.

***Alstonia** R.Br. Apocynaceae. 50 Indomal., Polynesia. Ls. whorled. Bark tonic.

Alstonia Mutis = Symplocos L. (Symplocac.).

Alstonia Scop. = Pacouria Aubl. (Apocynac.).

Alstroemeria L. Alstroemeriaceae. 50 S. Am. Ls. twisted at base so that true upper surface faces down (internal anatomy also reversed). Caps. splits explosively.

*Alstroemeriaceae Dum. Monocots. 4/200 C. & S. Am. Rhizomatous, sometimes climbing herbs. Ls. alt., pet. usu. twisted through 180°. Infl. ± racemose, or capit., rarely 1-fl. P reg. or ± zygo. (at least as to colouring), tep. free; A 6, free, anth. basifixed; G (3). Caps. loculic., trunc. Seeds ∞. Genera: *Alstroemeria, Bomarea, Leontochir, Schickendantzia.*

Altamiranoa Rose (~ Villadia Rose). Crassulaceae. 18 Mex. to Peru.

Alteinia auct. = Althenia Petit (Zannichelliac.).

Altensteinia Kunth. Orchidaceae. 6 trop. Am.

Alternanthera Forsk. Amaranthaceae. 200 trop., subtrop.

Alternasemina S. Manso = Melothria L. (Cucurbitac.).

Althaea L. Malvaceae. 12 W. Eur. to NE. Siberia. *A. officinalis* L., marshmallow.

Althaeastrum Fabr. (uninom.) = *Lavatera arborea* L. (Malvac.).

Althenia Petit. Zannichelliaceae. 1 W. Medit., ? S. Afr.

Altheria Thou. = Melochia L. (Sterculiac.).

Althingia Steud. = Araucaria Juss. (Araucariac.).

Althoffia K. Schum. Tiliaceae. 6 E. Malaysia. Dioec. K, A ∞.

Altingia G. Don = Altingia Nor. (Altingiac.) + Araucaria Juss. (Araucariac.).

Altingia Nor. Altingiaceae. 7 Assam & S. China to Malay Penins., Sumatra & Java. ♂ fl. of naked sta. A form with 1-fld. ♀ infl. reported from SE. China. Large trees; good timber.

*Altingiaceae Lindl. Dicots. 2/10 As. Min., temp. & trop. SE. As., N. & C. Am. Strongly resiniferous trees, sometimes v. large. Ls. gland.-serr., ent. or palmatilobed. Fls. ♂ ♀ (the ♀ sometimes with stds.). ♂ infl. a term. rac. of globose stam.-clusters, each cluster at first envel. by a large membr. bract (arrangement of stam. prob. 'pre-floral'); P o. ♀ infl. a globose head (cf. *Platanus*); ov. united; P of ∞ minute lobes or scales, sometimes ± accresc.; G (2), with 2 stout decid. or persist. stigs.; ovules sev.–∞, *horiz.* Infruct. glob., hard, dry, of many caps. Timber often valuable. Genera: *Altingia, Liquidambar.* Somewhat intermediate between *Hamamelidac.* (*s.str.*) and *Platanac.*

Altisatis Thou. (uninom.) = *Satyrium praealtum* Thou. = *Habenaria praealta* (Thou.) Spreng. (Orchidac.).

Altora Adans. = Clutia L. (Euphorbiac.).

Alughas L. = Alpinia L. (Zingiberac.).

Alus Bub. = Coris L. (Coridac.).

Alvaradoa Liebm. Simaroubaceae. 6 W.I., C. Am. to Argent. Micropyle down.

Alvardia Fenzl = Peucedanum L. (Umbellif.).

Alvarezia Pav. ex Nees = Blechum R.Br. (Acanthac.).

Alveolina Van Tiegh. = Loranthus L. (Loranthac.).

Alvesia Welw. (1858) = Bauhinia L. (Legumin.).

*Alvesia Welw. (1869). Labiatae. 3 trop. Afr.

Alvisia Lindl. = Eria Lindl. (Orchidac.).

Alvordia T. S. Brandegee. Compositae. 3 Calif., Mex.

Alwisia Thw. ex Lindl. = Taeniophyllum Bl. (Orchidac.).

ALYCIA

Alycia Willd. ex Steud. = Eriochloa Kunth (Gramin.).

Alymeria D. Dietr. = Aylmeria Mart. = Polycarpaea Lam. (Caryophyllac.).

Alymnia [Neck.] (DC.) Spach = Polymnia L. (Compos.).

Alyogyne Alef. = Fugosia Juss. (Malvac.).

Alypaceae ('-inae') Hoffmgg. & Link = Globulariaceae Lam. & DC.

Alypum Fisch. = Globularia L. (Globulariac.).

*****Alysicarpus** Neck. ex Desv. Leguminosae. 25 warm Afr. to Austr.

Alyssoïdes Mill. Cruciferae. 4 Euras.

Alyssopsis Boiss. Cruciferae. 1 Transcauc., Persia.

Alyssopsis Reichb. = Vesicaria Lam. (Crucif.).

Alyssum L. Cruciferae. 150 Medit. reg. to Siberia.

Alytostylis Mast. = Roydsia Roxb. (Capparidac.).

*****Alyxia** Banks ex R.Br. Apocynaceae. 80 Madag., Indomal.

Alzalia F. G. Dietr. = seq.

Alzatea Ruiz et Pav. Lythraceae. 1 Peru. Plac. parietal. (Lourteig, *Ann. Missouri Bot. Gard.* **52**: 371–8, 1965.)

Alziniana F. G. Dietr. ex Pfeiff. = praec.

Amadea Adans. = Androsace L. (Primulac.).

Amagris Rafin. = Calamagrostis Adans. (Gramin.).

Amaioua Aubl. Rubiaceae. 25 N. trop. S. Am. Often dioecious.

Amalago Rafin. = Piper L. (Piperac.).

Amalia Hort. Hisp. = Tillandsia L. (Bromeliac.).

Amalia Reichb. = Laelia Lindl. (Orchidac.).

Amalias Hoffmgg. = praec.

Amalobotrya Kunth ex Meissn. = Symmeria Benth. (Polygonac.).

Amalocalyx Pierre. Apocynaceae. 3 SE. As.

Amalophyllon T. S. Brandegee. Scrophulariaceae. 1 Mex.

Amamelis Lem. = Hamamelis L. (Hamamelidac.).

Amana Honda. Liliaceae. 3 China, Japan.

Amannia Blume = Ammannia L. (Lythrac.).

Amanoa Aubl. Euphorbiaceae. 7 trop. S. Am., W.I.; 3 trop. Afr., Madag.

Amapa Steud. = Carapa Aubl. (Meliac.).

Amaraboya Linden ex Mast. Melastomataceae. 3 Colombia.

Amaracanthus Steud. (sphalm.) = seq.

Amaracarpus Blume. Rubiaceae. 60 Malaysia, Micronesia.

Amaraceae Dulac = Gentianaceae Juss.

*****Amaracus** Gled. (~ Origanum L.). Labiatae. 15 E. Medit.

Amaracus Hill = Majorana Mill. = Origanum L. (Labiat.).

Amaralia Welw. ex Benth. & Hook. f. = Sherbournia G. Don (Rubiac.).

Amarantellus Spegazz. = Amaranthus L. (Amaranthac.).

Amarantesia Hort. ex Regel = Telanthera R.Br. (Amaranthac.).

*****Amaranthaceae** Juss. Dicots. (Centrospermae). 65/850 trop. & temp. Usu. herbs or shrubs with opp. or alt. entire exstip. ls. Fls. sol. or in axillary cymes (the whole infl. racemose), ♀, rarely unisexual, usu. reg. P usu. 4–5 or (4–5), usu. membranous; A 1–5 opp. P, free or ± united to P or disk, or to one another; G̲ (2–3), free or united to P, 1-loc. with ∞–1 basal campylotr. ov. Berry, pyxidium or nut; usu. shiny testa; embryo curved; endosp.

48

AMARYLLIDACEAE

Classification and chief genera (after Schinz):
I. *Amaranthoïdeae.* Stam. 4-loc.; ovary 1–∞–ovuled.
  1. CELOSIEAE (ovary ∞- or few-ov.): *Deeringia, Celosia, Hermbstaedtia.*
  2. AMARANTHEAE (ovary 1-ov.): *Amaranthus, Sericocoma, Cyathula, Aerva, Ptilotus, Psilotrichum, Achyranthes.*
II. *Gomphrenoïdeae.* Stam. 2-loc.; ovary 1-ovuled.
  3. BRAYULINEAE (fls. sol. or fascic. in l.-axils): *Brayulinea, Tidestromia.*
  4. GOMPHRENEAE (infl. spic. or capit., sometimes also with axill. fls.): *Froelichia, Pfaffia, Alternanthera, Gomphrena, Iresine.*
Amaranthoïdes Mill. = Gomphrena L. (Amaranthac.).
Amaranthus Adans. = Celosia L. (Amaranthac.).
Amaranthus L. Amaranthaceae. 60 trop. & temp. Infl. of ∞ fl. *A. gangeticus* L., etc., are pot-herbs in India, etc.; *A. caudatus* L., *A. paniculatus* L., etc., give ed. grain, used as a cereal in trop. As.
× **Amarcrinum** Hort. Amaryllidaceae. Gen. hybr. (Amaryllis × Crinum).
Amarella Gilib. = Gentianella Moench (Gentianac.).
Amarenus C. Presl = Trifolium L. (Legumin.).
Amaria S. Mutis ex Caldas = Bauhinia L. (Legumin.).
Amaridium Ind. Kew. (sphalm.) = Camaridium Lindl. = Maxillaria Ruiz & Pav. (Orchidac.).
× **Amarine** Sealy. Amaryllidaceae. Gen. hybr. (Amaryllis × Nerine).
Amarolea Small = Cartrema Rafin. = Osmanthus Lour. (Oleac.).
Amaroria A. Gray = Soulamea Lam. (Simaroubac.).
× **Amarygia** Ciferri & Giacom. Amaryllidaceae. Gen. hybr. (Amaryllis × Brunsvigia).
*** Amaryllidaceae** Jaume St-Hil. Monocots. 85/1100, usu. trop. or subtrop. Usu. xero., often bulbous, leafing only in spring or the rains; many have rhiz. Infl. usu. on a scape, with spathe, cymose, but often umbel- or head-like by condensation; solitary fls. in some. Fl. ♀, reg. or zygo. P or (P) 3 + 3, petaloid; A 3 + 3, sometimes some staminodial, usu. introrse; G̅ (3), rarely ½-inf., 3-loc. or rarely 1-loc. with axile plac. and ∞ anatr. ov. In some (*Narcissus*, etc.) a conspic. corona looks like an extra whorl of P, between normal P and A (?combined ligules of P, or stips. of sta.), to be seen in stages in *Caliphruria, Sprekelia, Eucharis, Narcissus.* Loculic. caps., or berry; endosp., small straight embryo. Veg. repr. by bulbils common.
Classification and chief genera (after Hutchinson):
  [1. AGAPANTHEAE ⎫
  [2. ALLIEAE   ⎬ see *Alliaceae*.]
  [3. GILLIESIEAE ⎭
  4. GALANTHEAE. (G̅); corona 0; leafless scape; ovules ∞; P-tube 0 or v. short; P reg.; fls. sol. or few. *Galanthus, Leucoïum.*
  5. AMARYLLIDEAE. As 4, but P ± zygo., fls. usu. several. *Amaryllis, Brunsvigia, Nerine.*
  6. CRINEAE. As 4 and 5, but P-tube +, stam. epitep., infl. several-fld. *Crinum, Ammocharis, Cyrtanthus, Vallota,* etc.
  7. ZEPHYRANTHEAE. As 6, but fls. sol. or paired. *Zephyranthes, Sternbergia,* etc.
  8. HAEMANTHEAE. As 4–7, but ovules few. *Carpolyza, Boöphone, Haem-*
49
[*anthus,* etc.

9. IXIOLIRIEAE. As 4-8, but scape leafy below, umbel subcompound. *Ixiolirion*.
10. EUCHARITEAE. As 4-9, but corona+, conspic., of expanded often connate fil. *Pancratium, Hymenocallis, Eucharis*, etc.
11. EUSTEPHIEAE. As 10, but corona of small teeth between fil.; P-lobes not spreading. *Eustephia, Phaedranassa*, etc.
12. HIPPEASTREAE. As 11, but corona of scales, P-lobes spreading. *Hippeastrum, Sprekelia*, etc.
13. NARCISSEAE. As 10-12, but corona of sep. scales distinct from fil., or annular or tubular, distinct from fil. *Narcissus*, etc.

Amaryllis L. Amaryllidaceae. 1 S. Afr., *A. bella-donna* L.
Amaryllis Sweet = Hippeastrum Herb. (Amaryllidac.).
***Amasonia** L. f. Verbenaceae. 8 S. Am., Trinidad. Ls. alt.
Amathea Rafin. = Aphelandra R.Br. (Acanthac.).
Amatula Medik. = Lycopersicum Mill. (Solanac.).
Amauria Benth. Compositae. 3 W.N. Am. Pappus o.
Amauriella Rendle. Araceae. 1 S. Nigeria.
Amauriopsis Rydb. (~Amauria Benth.). Compositae. 1 SW. U.S.
Amauropelta Kunze = Thelypteris Schmidel (Thelypteridac.).
Amaxitis Adans. = Dactylis L. (Gramin.).
Ambaiba Adans. = Coilotapalus P.Br. = Cecropia Loefl. (Morac.).
Ambassa Steetz = Vernonia Schreb. (Compos.).
Ambavia Le Thomas. Annonaceae. 2 Madag.
Ambelania Aubl. Apocynaceae. 14 trop. S. Am. Disk o.
***Amberboa** (Pers.) Less. Compositae. 20 Medit. to C. As.
Amberboi Adans. = praec.
Amberboia Kuntze = praec.
Ambianella Willis (sphalm.) = Autranella A. Chev. = Mimusops L. (Sapotac.).
Ambinax B. D. Jacks. (sphalm.) = seq.
Ambinux Comm. ex Juss. = Vernicia Lour. (Euphorbiac.).
Amblachaenium Turcz. ex DC. = Hypochoeris L. (Compos.).
Amblatum G. Don = Anblatum Hill = Lathraea L. (Orobanchac.).
Ambleia Spach = Stachys L. (Labiat.).
Amblia Presl = Phanerophlebia Presl (Aspidiac.).
Amblirion Rafin. = Fritillaria L. (Liliac.).
Amblogyna Rafin. = Amaranthus L. (Amaranthac.).
Amblophus Merr. (sphalm.) = Amplophus Rafin. = Valeriana L. (Valerianac.).
Amblostima Rafin. = Schoenolirion Durand (Liliac.).
Amblostoma Scheidw. = Epidendrum L. (Orchidac.).
Amblyachyrum Hochst. ex Steud. = Apocopis Nees (Gramin.).
Amblyanthera Blume = Osbeckia L. (Melastomatac.).
Amblyanthera Muell. Arg. = Mandevilla Lindl. (Apocynac.).
Amblyanthopsis Mez. Myrsinaceae. 4 Indomal.
Amblyanthus A. DC. Myrsinaceae. 4 E. Himal., New Guinea.
Amblycarpum Lem. = Amblyocarpum Fisch. & Mey. (Compos.).
Amblychloa Link = Sclerochloa Beauv. (Gramin.).
Amblyglottis Blume = Calanthe R.Br. (Orchidac.).
Amblygonocarpus Harms. Leguminosae. 1 trop. Afr.

Amblygonum Reichb. (~Polygonum L.). Polygonaceae. 2 India, E. As.
Amblylepis Decne = Amblyolepis DC. = Helenium L. (Compos.).
Amblynotopsis Macbride. Boraginaceae. 8 Mex.
Amblynotus I. M. Johnston. Boraginaceae. 1 W. Siberia to Mongolia.
Amblyocalyx Benth. Apocynaceae. 1 Borneo. Ls. whorled.
Amblyocarpum Fisch. et Mey. Compositae. 1 Caspian reg.
Amblyoglossum Turcz. = Tylophora R.Br. (Asclepiadac.).
Amblyolepis DC. = Helenium L. (Compos.).
Amblyopappus Hook. & Arn. Compositae. 1 Calif., NW. Mex., Peru, Chile. Scaly pappus.
Amblyopelis Steud. (sphalm.) = Amblyolepis DC. = Helenium L. (Compos.).
Amblyopetalum (Griseb.) Malme. Asclepiadaceae. 2 S. Am. (Argent.).
Amblyopogon Fisch. & Mey. Compositae. 8 Orient.
Amblyopyrum Eig (~Aegilops L.). Gramineae. 1 E. Medit.
Amblyorhinum Turcz. = Phyllactis Pers. (Valerianac.).
Amblysperma Benth. = Trichocline Cass. (Compos.).
Amblystigma Benth. Asclepiadaceae. 7 Bolivia, Argent. Corona o.
Amblystigma P. & K. = Amblostima Rafin. = Schoenolirion Durand (Liliac.).
Amblytes Dulac = Molinia Scop. (Gramin.).
Amblytropis Kitagawa = Gueldenstaedtia Fisch. (Legumin.).
Ambongia Benoist. Acanthaceae. 1 Madag.
Ambora Juss. = Tambourissa Sonner. (Monimiac.).
Amborella Baill. Amborellaceae. 1 New Caled.
*Amborellaceae Pichon. Dicots. 1/1 New Caled. Shrubs, near *Monimiac.* and *Atherospermatac.* (*qq.v.*), but wood without vessels (cf. *Winterac.*, *Tetracentrac.*, *Trochodendrac.*), with narrow medull. rays; ls. alt., distich.; anth. dehisc. by slits; ovule orthotr. with inf. micropyle; fr. carp. compressed, stipit., closely foveolate-rugose; seed with basal embryo. Only genus: *Amborella.*
Amboroa Cabr. Compositae. 1 Bolivia.
Ambotia Rafin. = Annona L. (Annonac.).
Ambraria Cruse = Nenax Gaertn. (Rubiac.).
Ambraria Fabr. (uninom.) = *Anthospermum* L. sp. (Rubiac.).
Ambrella H. Perrier. Orchidaceae. 1 Madag.
Ambrina Spach = Chenopodium L. (Chenopodiac.).
Ambroma L. f. Sterculiaceae. 2 trop. As. to Austr.
Ambrosia B. D. Jacks. (sphalm.) = Ambrosina Bassi (Arac.).
Ambrosia L. Compositae. 35–40 cosmop. (mostly Am.). Head unisexual, ♀ one-fld. Fr. enclosed in invol.
*Ambrosiaceae Dum. & Link = Compositae–Heliantheae–Ambrosiinae O. Hoffm. [8/50, chiefly N. & C. Am., few Medit. & Afr. As *Asterac.* (*s.str.*) but fls. ♂ ♀, the latter without cor.; fil. monadelph., anth. distinct. Prob. reduced anemoph. forms. Chief gen.: *Iva, Dicoria, Ambrosia, Xanthium.*]
Ambrosina Bassi. Araceae. 1 Medit.
Ambrosinia L. = praec.
Ambuli Adans. = Limnophila R.Br. (Scrophulariac.).
Ambulia Lam. = praec.
Amburana Schwacke & Taub. Leguminosae. 3 Brazil. Good timber.
Ambuya Rafin. = Aristolochia L. (Aristolochiac.).

Amebia Repel (sphalm.) = Arnebia Forsk. (Boraginac.).
Amecarpus Benth. ex Lindl. = Indigofera L. (Legumin.).
Amechania DC. = Agarista D. Don = Leucothoë D. Don ex G. Don + Agauria Hook. f. (Ericac.).
Ameghinoa Spegazz. Compositae. 1 Patagonia.
Amelanchier Medik. Rosaceae. 25 N. temp. (chiefly N. Am.).
Amelanchus Rafin. = praec.
Amelancus Rafin. = praec.
× **Amelasorbus** Rehder. Rosaceae. Gen. hybr. (Amelanchier × Sorbus).
Ameletia DC. = Rotala L. (Lythrac.).
Amelia Alefeld = Pyrola L. (Pyrolac.).
Amelina C. B. Clarke = Aneilema R.Br. (Commelinac.).
Amellus Adans. = Aster L. (Compos.).
Amellus P.Br. = Melanthera Rohr (Compos.).
***Amellus** L. Compositae. 15 S. Afr.
Amellus Ort. ex Willd. = Tridax L. (Compos.).
Amenippis Thou. (uninom.) = *Diplecthrum amoenum* Thou. = *Satyrium amoenum* (Thou.) A. Rich. (Orchidac.).
Amentaceae Dulac = Salicaceae Mirb.
Amentaceae Juss. A ± heterogeneous assemblage of catkin-bearing families, including *Salicaceae, Juglandaceae, Betulaceae, Fagaceae, Myricaceae, Leit-*
Amentiferae H. C. Wats. = praec. [*neriaceae*, etc.
Amentiflorae Moss = praec.
Amentotaxaceae Kudo & Yamamoto = Taxaceae Lindl.
Amentotaxus Pilger. Taxaceae. 1–4 C. & S. China, Formosa, Vietnam.
Amerimnon P.Br. = Dalbergia L. f. (Legumin.).
Amerimnum P. & K. = praec.
Amerina DC. = Aegiphila Jacq. (Verbenac.).
Amerina Nor. = Aglaia Lour. (Meliac.).
Amerina Rafin. = Salix L. (Salicac.).
Amerix Rafin. = Salix L. (Salicac.).
Amerlingia Opiz = Sambucus L. (Sambucac.).
Amerorchis Hultén. Orchidaceae. 1 N. Amer.
× **Amesara** hort. (v) = × Renantanda hort. (Orchidac.).
Amesia A. Nels. & Macbr. = Epipactis Zinn (Orchidac.).
Amesiella Schltr. (nomen). Orchidaceae. Quid?
Amesiodendron Hu. Sapindaceae. 1 China.
Amesium Newm. = Asplenium L. (Aspleniac.).
Amethystanthus Nakai. Labiatae. 20 E. As.
Amethystea L. Labiatae. 1 Persia to Manchuria.
Amethystina Zinn = praec.
Ametron Rafin. = Rubus L. (Rosac.).
Amherstia Wall. Leguminosae. 1 Burma, *A. nobilis* Wall., a tree often cult. for its splendid fls. Stalk and br. as well as pets. are bright pink. Sta. united in a tube. The young ls., covered with brownish spots, hang down 'as if poured out'; later they stiffen, turn green and come to the horiz. position.
Amiantanthus Kunth = seq.
Amianthemum Steud. = seq.

*Amianthium A. Gray = Zigadenus Michx (Liliac.).
Amianthum Rafin. = praec.
Amicia Kunth. Leguminosae. 8 Mex. to Argent. In *A. zygomeris* DC. the
  large stips. protect the bud.
Amictonis Rafin. = Callicarpa L. (Verbenac.).
Amida Nutt. = Madia Molina (Compos.).
Amidena Adans. = Orontium L. (Arac.).
Amidena Rafin. = Rohdea Roth (Liliac.).
Amiris La Llave = Amyris L. (Rutac.).
Amirola Pers. = Llagunoa Ruiz & Pav. (Sapindac.).
Amischophacelus Rao & Kammathy. Commelinaceae. 2 trop.
Amischotolype Hassk. Commelinaceae. 20 trop. Afr., SE. As., Indomal.
Amischotopyle Pichon (sphalm.) = praec.
Amitostigma Schlechter. Orchidaceae. 15 India, E. As.
Ammadenia Rupr. = Honkenya Ehrh. (Caryophyllac.).
Ammandra O. F. Cook (= ?Phytelephas Ruiz & Pav.). Palmae. 1 Colombia.
Ammanella Miq. = seq.
Ammannia L. Lythraceae. 30 cosmop.
Ammanniaceae Horan. = Lythraceae J. St-Hil.
Ammanthus Boiss. & Heldr. Compositae. 5 Medit.
Ammi L. Umbelliferae. 10 Azores, Madeira, Medit., temp. W. As.
Ammiaceae Small = Apiaceae Lindl. (Umbelliferae Juss.).
Ammianthus Spruce ex Benth. & Hook. f. = Retiniphyllum Humb. & Bonpl.
Ammiopsis Boiss. Umbelliferae. 2 NW. Afr.   [(Rubiac.).
Ammios Moench = Carum L. (Umbellif.).
Ammobium R.Br. Compositae. 2 New S. Wales.
Ammobroma Torr. Lennoaceae. 1 SW. U.S. (New Mex., Calif.), 1 Mex.
× Ammocalamagrostis P. Fourn. Gramineae. Gen. hybr. (Ammophila ×
  Calamagrostis).
Ammocallis Small = Lochnera Reichb. (Apocynac.).
Ammocharis Herb. Amaryllidaceae. 5 trop. & S. Afr.
Ammochloa Boiss. Gramineae. 3 Medit.
Ammoçodon Standley = Selinocarpus A. Gray (Nyctaginac.).
Ammodaucus Coss. & Dur. Umbelliferae. 1 Algeria.
Ammodendron Fisch. ex DC. Leguminosae. 7 W. & C. As.
Ammodenia J. G. Gmel. ex Rupr. = Honkenya Ehrh. (Caryophyllac.).
Ammodia Nutt. = Chrysopsis Nutt. (Compos.).
Ammodytes Stev. = Astragalus L. (Legumin.).
Ammogeton Schrad. = Troximon Nutt. (Compos.).
Ammoïdes Adans. (~ Carum L.). Umbelliferae. 2 Medit.
Ammolirion Kar. & Kir. = Eremurus Bieb. (Liliac.).
Ammonalia Desv. = Honkenya Ehrh. (Caryophyllac.).
Ammonia Nor. = Polyalthia Bl. (Annonac.).
Ammophila Host. Gramineae. 2 Atl. N. Am., Eur., N. Afr. *A. arenaria* (L.)
  Link (marram) is common on sandy coasts and much used for sand-binding.
  After some years a light soil forms, in which other pls. take root. The ls. curl
  inwards in dry air.
Ammopiptanthus Cheng f. Leguminosae. 2 C. As.

AMMOPURSUS

Ammopursus Small (~ Laciniaria Hill). Compositae. 1 Florida.

Ammorrhiza Ehrh. (uninom.) = *Carex arenaria* L. (Cyperac.).

Ammoselinum Torr. & Gray. Umbelliferae. 3 SW. U.S.

Ammoseris Endl. = Launaea Cass. (Compos.).

Ammosperma Hook. f. Cruciferae. 1 N. Afr.

Ammothamnus Bunge. Leguminosae. 4 W. & C. As.

Ammyrsine Pursh = Leiophyllum Hedw. f. (Ericac.).

Amni Brongn. (sphalm.) = Ammi L. (Umbellif.).

Amoebophyllum N. E. Brown. Aïzoaceae. 4 S. Afr. (Namaqualand).

Amoenippis Thou. (uninom.) = *Diplecthrum amoenum* Thou. = *Satyrium amoenum* (Thou.) A. Rich. (Orchidac.).

Amogeton Neck. = Aponogeton L. f. (Aponogetonac.).

Amoleiachyris Sch. Bip. (sphalm.) = Amphiachyris Nutt. = Gutierrezia Lag. (Compos.).

Amom[ac]eae A. Rich. = Zingiberaceae Lindl.

Amomis Berg (~ Pimenta Lindl.). Myrtaceae. 10 W.I.

Amomophyllum Engl. = Spathiphyllum Schott (Arac.).

Amomum L. Zingiberaceae. 150 palaeotrop. Fls. usu. on scapes from the rhiz.

Amomyrtella Kausel. Myrtaceae. 1 N. Argent.

Amomyrtus (Burret) D. Legrand & Kausel. Myrtaceae. 2 temp. S. Am.

Amonia Nestl. = Aremonia Neck. ex Nestl. (Rosac.).

Amoora Roxb. Meliaceae. 25 Indomal.

Amooria Walp. = Amoria C. Presl = Trifolium L. (Legumin.).

Amordica Neck. = Momordica L. (Cucurbitac.).

Amorea Moq. ex Del. = Cycloloma Moq. (Chenopodiac.).

Amoreuxia Moç. & Sessé. Cochlospermaceae. 7 S. U.S. to Peru.

Amoreuxia Moq. = Cycloloma Moq. (Chenopodiac.).

Amorgine Rafin. = Achyranthes L. (Amaranthac.).

Amoria C. Presl = Trifolium L. (Legumin.).

Amorpha L. Leguminosae. 20 N. Am. Wings and keel o; standard folds round base of sta.-tube. Protog. with persistent stigma.

Amorphocalyx Klotzsch = Sclerolobium Vog. (Legumin.).

*Amorphophallus Blume ex Decne. Araceae. 100 trop. Afr. & As. Usu. corm-like rhiz., giving yearly a big l. (up to 3 m.) and infl. (in *A. titanum* Becc. c. 1 m. high), with ♂ fl. above and ♀ below. The dirty red colour and foetid smell attract carrion flies, which sometimes lay eggs on the spadix.

Amorphospermum F. Muell. (~ Lucuma Juss.). Sapotaceae. 1 trop. E. Austr.

Amorphus Rafin. = Amorpha L. (Legumin.).

Amosa Neck. = Inga Mill. (Legumin.).

Amoureuxia C. Muell. = Amoreuxia Moç. & Sessé (Cochlospermac.).

Ampacus Kuntze = Evodia J. R. & G. Forst. (Rutac.).

Ampalis Boj. Moraceae. 2 Madag.

Amparoa Schlechter. Orchidaceae. 2 Mex., Costa Rica.

Ampelamus Rafin. = Enslenia Nutt. (Asclepiadac.).

Ampelanus B. D. Jacks. = praec.

Ampelidaceae Kunth = Vitidaceae Juss.

Ampelocera Klotzsch. Ulmaceae. 9 trop. Am., W.I.

*Ampelocissus Planch. Vitidaceae. 95 trop.
Ampelodaphne Meissn. = Endlicheria Nees (Laurac.).
Ampelodesma Beauv. Gramineae. 1 Medit. When young it is used as fodder. The ls. are used like esparto (*Stipa*).
Ampelodesmos Link = praec.
Ampelodonax Lojac. = praec.
Ampeloplia Rafin. = seq.
Ampeloplis Rafin. = Sageretia Brongn. (Rhamnac.).
Ampelopsis (Rich. in) Michx. Vitidaceae. 2 temp. & subtrop. As., Am.
Ampelopsis Hort. = Parthenocissus Planch. (Vitidac.).
Ampelopteris Kunze. Thelypteridaceae. 1 Old World trop., in open swampy places. Ls. proliferous.
Ampelosicyos Thou. Cucurbitaceae. 3 Madag.
Ampelothamnus Small (~ Pieris D. Don). Ericaceae. 1 N. Am.
Ampelovitis Carr. = Vitis L. (Vitidac.).
Ampelozizyphus Ducke. Rhamnaceae. 1 Brazil.
Ampelygonum Lindl. = Polygonum L. (Polygonac.).
Amperea A. Juss. Euphorbiaceae. 6 Austr., Tasm.
Amphania Banks ex DC. = Ternstroemia L. (Theac.).
Ampherephis Kunth = Centratherum Cass. (Compos.).
Amphiachyris Nutt. Compositae. 2 Calif.
Amphianthus Torr. Scrophulariaceae. 1 SE. U.S. (Georgia).
Amphiasma Bremek. Rubiaceae. 8 trop. & SW. Afr.
Amphibecis Schrank = Centratherum Cass. (Compos.).
Amphiblemma Naud. Melastomataceae. 15 trop. Afr.
Amphiblestra Presl. Aspidiaceae. 1 Venez.
Amphibolia L. Bolus. Aïzoaceae. 3 S. Afr.
Amphibolis C. Agardh. Cymodoceaceae. 2 coasts of W. & S. Austr. & Tasm. Submerged marine aquatics.
Amphibolis Schott & Kotschy = Hyacinthus L. (Liliac.).
Amphibologyne A. Brand. Boraginaceae. 1 Mex.
Amphibromus Nees. Gramineae. 9 Austr., N.Z., S. Am.
Amphicalea (DC.) Gardn. = Geissopappus Benth. (Compos.).
Amphicalyx Blume = Diplycosia Blume (Ericac.).
Amphicarpa Ell. = seq.
*Amphicarpaea Ell. mut. DC. Leguminosae. 24 E. As., trop. & N. Am., S. Afr. Some have cleist. fls. below, which give subterranean fr. like *Arachis*.
Amphicarpon Rafin. Gramineae. 2 SE. U.S.
Amphicarpum Kunth = praec.
Amphicome Royle (~ Incarvillea Juss.). Bignoniaceae. 2 Himal.
Amphicosmia Gardn. = Cyathea Sm. (Cyatheac.).
Amphidasya Standley. Rubiaceae. 3 trop. Am.
Amphiderris Spach = Orites R.Br. (Proteac.).
Amphidesmium Schott = Metaxya Presl (Lophosoriac.) (Morton, *Am. Fern J.* 49: 151, 1959).
Amphidetes Fourn. Asclepiadaceae. 2 Brazil.
Amphidonax Nees = Arundo L. + Zenkeria Trin. (Gramin.).
Amphidoxa DC. Compositae. 7 trop. & S. Afr., Madag.

Amphiestes S. Moore = Hypoëstes Soland. ex R.Br. (Acanthac.).
Amphigena Rolfe. Orchidaceae. 2 S. Afr. (Cape Prov.).
Amphigenes Janka = Festuca L. (Gramin.).
Amphiglossa DC. Compositae. 6 S. Afr.
Amphiglottis Salisb. = Epidendrum L. (Orchidac.).
Amphilobium Loud. = Amphilophium Kunth (Bignoniac.).
Amphilochia Mart. = Qualea Aubl. (Vochysiac.).
Amphilophis Nash = Bothriochloa Kuntze (Gramin.).
Amphilophium Kunth. Bignoniaceae. 8 warm Am.
Amphimas Pierre ex Harms. Leguminosae. 4 trop. W. Afr.
Amphineurion (A. DC.) Pichon. Apocynaceae. 2 Australas.
Amphineuron Holtt. Thelypteridaceae. 12–15 E. Afr. to Polynesia.
Amphinomia DC. Leguminosae. 1 S. Afr.
Amphiodon Huber. Leguminosae. 1 Amaz. Braz.
Amphiolanthus Griseb. (~ Micranthemum Michx). Scrophulariaceae.
Amphion Salisb. = Semele Kunth (Ruscac.).   [3 Cuba.
Amphione Rafin. = Ipomoea L. (Convolvulac.).
Amphipappus Torr. & Gray (~ Amphiachyris Nutt.). Compositae. 1 SW.
U.S.
Amphiphyllum Gleason. Rapateaceae. 2 Venez.
Amphipleis Rafin. = Nicotiana L. (Solanac.).
Amphipogon R.Br. Gramineae. 7 Austr.
Amphipterum Presl. Hymenophyllaceae. 4 Malaysia.
Amphipterygium Schiede ex Standl. Julianiaceae. 4 Mex., Peru.
Amphirephis Nees & Mart. = Ampherephis Kunth = Centratherum Cass.
(Compos.).
Amphirhapis DC. = Inula L., Microglossa DC., Solidago L., etc. (Compos.).
*Amphirrhox Spreng. Violaceae. 6 trop. Am.
Amphiscopia Nees (~ Justicia L.). Acanthaceae. 20 S. Am.
Amphisiphon W. F. Barker. Liliaceae. 1 S. Afr. (Cape Prov.).
Amphisoria Trevis. = Psomiocarpa Presl (Aspidiac.).
Amphistelma Griseb. = Metastelma R.Br. (Asclepiadac.).
Amphitecna Miers. Bignoniaceae. 2 Mex., C. Am.
Amphithalea Eckl. et Zeyh. Leguminosae. 15 S. Afr.
Amphitoma Gleason. Melastomataceae. 1 Colombia.
Amphizoma Miers = Tontelaea Aubl. (Celastrac.).
Amphochaeta Anderss. = Pennisetum Pers. (Gramin.).
Amphodus Lindl. = Kennedya Vent. (Legumin.).
Amphoradenium Desvaux. Grammitidaceae. 6 Hawaii.
Amphoranthus S. Moore = Phaeoptilum Radlk. (Nyctaginac.).
Amphorchis Thou. = Cynorkis Thou. (Orchidac.).
Amphorella T. S. Brandegee = Matelea Aubl. (Asclepiadac.).
Amphoricarpos Vis. Compositae. 2 SE. Eur., Cauc.
Amphoricarpus Spruce ex Miers = Couratari Aubl. (Lecythidac.).
Amphorkis Thou. = Cynorkis Thou. (Orchidac.).
Amphorocalyx Baker. Melastomataceae. 5 Madag.
Amphorogyne Stauffer & Hürlimann. Santalaceae. 3 New Caled.
Amphymenium Kunth = Pterocarpus L. (Legumin.).

Amplophus Rafin. = Valeriana L. (Valerianac.).
Ampomele Rafin. = Rubus L. (Rosac.).
Amsinckia Lehm. Boraginaceae. 50 W. N. Am., W. temp. S. Am.
Amsonia Walt. Apocynaceae. 25 Japan, N. Am.
Amsora Bartl. = seq.
Amura Schult. = Amoora Roxb. (Meliac.).
Amydrium Schott. Araceae. 4 Siam, Malaysia.
Amyema Van Tiegh. Loranthaceae. 90 W. Malaysia to Austr. & W. Pacif.
*Amygdalaceae ('-inae') D. Don = Rosaceae–Amygdaleae Juss.
+Amygdalopersica Daniel. Rosaceae. Gen. hybr. (asex.) (Amygdalus +
Amygdalophora Néck. = Prunus L. (Rosac.).   [Persica).
Amygdalopsis Carr. = Louiseania Carr. (Rosac.).
Amygdalopsis M. Roem. = Amygdalus L. (Rosac.).
Amygdalus Kuntze = Heritiera Dryand. (Sterculiac.).
Amygdalus L. (~ Prunus L.). Rosaceae. 40 Medit. reg. to C. China.
Amylocarpus Barb. Rodr. (1902; non Currey 1857—Fungi) = Yuyba (Barb.
Rodr.) L. H. Bailey (Palm.).
Amylotheca Van Tiegh. Loranthaceae. 4 New Guinea, NE. Austr., Melanesia.
Amyrea Leandri. Euphorbiaceae. 2 Madag.
Amyrid[ac]eae R.Br. = Rutaceae–Amyridinae Engl.
Amyris P.Br. Rutaceae. 30 warm Am., W.I.
Amyrsia Rafin. Myrtaceae. 13 N. Andes.
Amyxa Van Tiegh. ex Domke. Thymelaeaceae. 1 Borneo.
Anabaena A. Juss. (1824; non *Anabaina* Bory 1821—Cyanophyceae) = seq.
Anabaenella Pax & K. Hoffm. = Romanoa Trevisan (Euphorbiac.).
Anabasis L. Chenopodiaceae. 30 Medit., C. As.
Anabata Willd. = ?Forsteronia G. F. W. Mey. (Apocynac.).
Anacampseros P.Br. = Talinum Adans. (Portulacac.).
*Anacampseros [L.] Sims. Portulacaceae. 70 S. Afr., 1 Austr. Xero. with
fleshy ls., and buds protected by bundles of hair, representing stips.
Anacampseros Mill. = Sedum L. (Crassulac.).
Anacampta Miers (~ Tabernaemontana L.). Apocynaceae. 15 trop. Am.
× **Anacamptiplatanthera** P. Fourn. Orchidaceae. Gen. hybr. (i) (Anacamptis
× Platanthera).
Anacamptis Rich. Orchidaceae. 1 Eur., N. Afr.
× **Anacampt-orchis** G. Camus. Orchidaceae. Gen. hybr. (i) (Anacamptis ×
Orchis).
Anacardia St-Lag. = Anacardium L. (Anacardiac.).
*Anacardiaceae Lindl. Dicots. 60/600, chiefly trop., but also Medit., E. As.,
Am. Trees and shrubs (rarely climbers) with alt. (opp. in *Bouea*, sometimes
ternate in *Ozoroa*) exstip. ls., and panicles of ∞ ♂ or polygamo-dioecious fls.
Resin-passages occur, but the ls. are not gland-dotted (hence they cannot be
confounded with *Rutaceae*). Recept. convex, flat, or concave; gynophores, etc.
occur. Fl. typically 5-merous, reg., hypog. to epig.; A 10–5 or other number;
G (3–1), rarely 5, each with 1 anatr. ov., often only one fertile. Usually drupe
with resinous mesocarp; embryo curved; no endosperm. The frs. of *Mangifera*,
Anacardium, *Spondias*, etc. are important. *Rhus* furnishes various useful
products. In *Drimicarpus* the ovary is inferior!

ANACARDIUM

Classification and chief genera (after Engler):
A. 5 free cpls., or 1. Ls. simple, entire:
  1. ANACARDIËAE (MANGIFEREAE): *Mangifera*, *Anacardium*.
B. Cpls. united (rarely only 1). Ls. rarely simple:
  2. SPONDIADEAE (ovule 1 in each cpl.): *Spondias*.
  3. RHOËAE (ovule in only 1 cpl., ovary free): *Rhus*.
  4. SEMECARPEAE (do., ovary sunk in axis): *Semecarpus*.
  [C. Cpl. 1. ♀ fl. naked. Ls. simple, toothed:
  5. DOBINEËAE: see *Podoaceae*.]
Anacardium Lam. = Semecarpus L. (Anacardiac.).
Anacardium L. Anacardiaceae. 15 trop. Am.; *A. occidentale* L. (cashew-nut) largely cult. Fls. polygamous. Each has 1 cpl. yielding a kidney-shaped nut with hard acrid coat. The nut (promotion nut, coffin-nail) is ed. Under it the axis swells up into a pear-like body, fleshy and ed. The stem yields a gum like arabic.
Anacharis Rich.' Hydrocharitaceae. 8 Am.
Anacheilium Hoffmgg. = Epidendrum L. (Orchidac.).
Anachortus V. Jirásek & Chrtek = Hierochloë S. G. Gmel. (Gramin.).
Anachyris Nees = Paspalum L. (Gramin.).
Anachyrium Steud. = praec.
Anacis Schrank = Coreopsis L. (Compos.).
Anaclanthe N. E. Brown. Iridaceae. 2 S. Afr.
Anaclasmus Griff. = Nenga Wendl. & Drude (Palm.).
Anacolosa Blume. Olacaceae. 1 trop. Afr., 20 Indomal., Pacif.
Anacolus Griseb. = Hockinia Gardn. (Gentianac.).
Anactinia Remy = Nardophyllum Hook. (Compos.).
Anactis Cass. = Atractylis L. (Compos.).
Anactis Rafin. = Solidago L. (Compos.).
Anactorion Rafin. = Synnotia Sweet (Iridac.).
Anacyclia Hoffmgg. Bromeliaceae. Quid?
Anacyclodon Jungh. = Leucopogon R.Br. (Epacridac.).
Anacyclus L. Compositae. 25 Medit. Some offic. (*radix pyrethri*).
Anacylanthus Steud. = Ancylanthos Desf. (Rubiac.).
Anadelphia Hackel. Gramineae. 13 W. to SE. trop. Afr.
Anademia C. A. Ag. (sphalm.) = Anadenia R.Br. = Grevillea R.Br. (Proteac.).
Anadenanthera Speg. (~ Piptadenia Benth.). Leguminosae. 2 Brazil.
Anadendron Schott (1858, text) = Anadendrum Schott (1857, text, 1858, tab.) (Arac.).
Anadendron Wight (sphalm.) = Anodendron A. DC. (Apocynac.).
Anadendrum Schott. Araceae. 9 Indomal.
Anadenia R.Br. = Grevillea R.Br. (Proteac.).
Anaectocalyx Triana = Anoectocalyx Triana (Melastomatac.).
Anaectochilus Lindl. = Anoectochilus Blume (Orchidac.).
Anafrenium Arn. = Anaphrenium E. Mey. = Heeria Meissn. (Anacardiac.).
Anagallidaceae ('-eidae') Baudo = Primulaceae Vent.
Anagallidastrum Adans. = Centunculus L. = Anagallis L. (Primulac.).
Anagallidium Griseb. (~ Swertia L.). Gentianaceae. 1 C. As.
Anagallis L. Primulaceae. 28 W. Eur., Afr., Madag., 1 pantrop., 2 S. Am.

The fl. of *A. arvensis* L. (poor man's weather-glass) closes in dull or cold weather.

Anagalloïdes Krock. = Lindernia All. (Scrophulariac.).

Anaganthos Hook. f. = Australina Gaudich. (Urticac.).

Anaglypha DC. Compositae. 3 S. Afr.

Anagosperma Wettst. = Euphrasia L. (Scrophulariac.).

Anagyris L. Leguminosae. 2 Medit.

Anagyris Lour. = Sophora L. (Legumin.).

Anagzanthe Baudo = Lysimachia L. (Primulac.).

Anaitis DC. = Sanvitalia L. (Compos.).

Analectis Juss. = Symphorema Roxb. (Symphorematac.).

Analiton Rafin. = Rumex L. (Polygonac.).

Analyrium E. Mey. ex Presl = Peucedanum Koch (Umbellif.).

Anamelis Garden = Fothergilla Murr. (Hamamelidac.).

Anamenia Vent. = Knowltonia Salisb. (Ranunculac.).

Anamirta Colebr. Menispermaceae. 1 Indomal. The fruits of *A. cocculus* Wight & Arn. are used as a fish-poison. In the angles between the big veins of the ls. are little cavities covered by hairs and inhabited by mites (*acarodomatia*).

Anamomis Griseb. (~ Eugenia L.). Myrtaceae. 10 Florida, W.I.

Anamorpha Karst. & Triana = Melochia L. (Sterculiac.).

Anamtia Koidz. = Myrsine L. (Myrsinac.).

Ananas Gaertn. = Bromelia L. (Bromeliac.).

Ananas Mill. Bromeliaceae. 5 trop. Am., incl. *A. comosus* (L.) Merr., the pineapple, largely cult. in Hawaii, Singapore, etc. Stem short and leafy, terrestrial, bearing a term. infl., which after fert. forms a common mass, fr., bracts, and axis, while the main axis grows beyond and forms a tuft of ls.—the crown of the pineapple.

Ananassa Lindl. = praec.

Anandria Less. = Leibnitzia Cass. (Compos.).

Ananthacorus Underw. & Maxon. Vittariaceae. 1 trop. Am.

Anantherix Nutt. = Asclepiodora A. Gray = Asclepias L. (Asclepiadac.).

Ananthopus Rafin. = Commelina L. (Commelinac.).

Anapalina N. E. Br. Iridaceae. 7 S. Afr.

Anapeltis J. Sm. = Microgramma Presl (Polypodiac.).

Anaphalioïdes (Benth.) Kirp. (~ Gnaphalium L.). Compositae. 1 N.Z.

Anaphalis DC. Compositae. 35 Eur., As., Am.

Anaphora Gagnep. = Malaxis Soland. ex Sw. (Orchidac.).

Anaphragma Stev. = Astragalus L. (Legumin.).

Anaphrenium E. Mey. ex Endl. = Heeria Meissn. (Anacardiac.).

Anaphyllum Schott. Araceae. 2 S. India.

Anapodophyllon Mill. = Podophyllum L. (Podophyllac.).

Anapodophyllum auctt. = praec.

Anarmodium Schott = Dracunculus Schott (Arac.).

Anarmosa Miers ex Hook. = Tetilla DC. (Francoac.).

*Anarrhinum Desf. Scrophulariaceae. 12 Medit.

Anarthria R.Br. Anarthriaceae. 5 SW. Austr.

Anarthriaceae Cutler & Airy Shaw. Monocots. (Farinosae). 1/5 Austr. Rush-

ANARTHROPHYLLUM

like herbs of swampy ground. Stems simple or branched, sometimes much compressed. Ls. basal, elong., stem-like, sheathing, equitant. Fls. reg., ♂ ♀, dioec. or rarely monoec., 1-bracteate, in lax cymes, each branch subtended by an elong. l.-like bract. P 3 + 3, glumaceous, ± equal; A 3, fil. free or conn., anth. with 2 sep. loc., dorsif. (in ♀ fl. stds. sometimes pres.); G̲ (3), 3-lobed, 3-loc., with 3 elong. free styles with decurr. stigs. (no pistillode i̲n̲ ♂ fl.). Caps. 1-seeded, sometimes indehisc. (nut-like); seed with mealy endosp. Only genus: *Anarthria*. Near *Restionac.*, but anatomy distinctive. (*Kew Bull.* **19**: 489, 1965.)

Anarthrophyllum Benth. Leguminosae. 15 Andes.

Anarthropteris Copel. Polypodiaceae(?). 1 N.Z. (Wilson, *Contr. Gray Herb.* **187**: 53–9, 1960.)

Anarthrosyne E. Mey. = Pseudarthria Wight & Arn. (Legumin.).

Anartia Miers (~ Tabernaemontana L.). Apocynaceae. 10 trop. Am.

Anaschovadi Adans. = Elephantopus L. (Compos.).

Anaspis K. H. Rechinger. Labiatae. 1 C. As.

Anasser Juss. = Geniostoma J. R. & G. Forst. (Loganiac.).

Anastatica L. Cruciferae. 1 Morocco to S. Persia, *A. hierochuntica* L. (rose of Jericho). While the seeds are ripening in the dry season the ls. fall off and the branches fold inwards, reducing the pl. to a ball of wickerwork, which rolls about with the pods closed until it reaches a wet spot, or the rains begin.

Anastrabe E. Mey. ex Benth. Scrophulariaceae. 1 S. Afr.

Anastraphia auctt. = Gochnatia Kunth (Compos.).

Anastraphia D. Don. Compositae. 1 S. Am. Quid?

Anastrephea Decne = praec.

Anastrophea Wedd. Podostemaceae. 1 Abyss.

Anastrophus Schlechtd. = Paspalum L. (Gramin.).

Anasyllis E. Mey. = Loxostylis Spreng. (Anacardiac.).

Anathallis Barb. Rodr. = Pleurothallis R.Br. (Orchidac.).

Anatherix Steud. = Anantherix Nutt. = Asclepias L. (Asclepiadac.).

Anatherum Beauv. = Andropogon L. (*s.l.*), Vetiveria Thou., etc. (Gramin.).

Anatherum Nabélĕk = Nabelekia Roshev. = Leucopoa Griseb. (Gramin.).

Anatis Sessé & Moç. ex Brongn. = Roulinia Brongn. (Liliac.).

Anatropa Ehrenb. = Tetradiclis Stev. (Zygophyllac.).

Anatropanthus Schlechter. Asclepiadaceae. 1 Borneo.

Anaua Miq. = Hemicyclia Wight & Arn. = Drypetes Vahl (Euphorbiac.).

Anaueria Kosterm. Lauraceae. 1 Brazil.

Anauxanopetalum Teijsm. & Binnend. = Swintonia Griff. (Anacardiac.).

Anavinga Adans. = Casearia Jacq. (Flacourtiac.).

Anaxagorea A. St-Hil. Annonaceae. 30 Ceylon, SE. As., W. Malaysia; [trop. Am.

Anaxeton Gaertn. Compositae. 7 SW. S. Afr.

Anaxeton Schrank = Helipterum DC. (Compos.).

Anaxetum Schott = Pessopteris Underw. & Maxon (Polypodiac.).

Anblatum Hill = Lathraea L. (Orobanchac.).

Ancalanthus I. B. Balf. = Angkalanthus I. B. Balf. (Acanthac.).

Ancana F. Muell. = Fissistigma Griff. (Annonac.).

Ancathia DC. = Cirsium Mill. (Compos.).

Anchietea A. St-Hil. Violaceae. 8 trop. S. Am.

Anchistea Presl = Woodwardia Sm. (Blechnac.).

ANCYLOBOTHRYS

Anchomanes Schott. Araceae. 10 trop. Afr.

Anchonium DC. Cruciferae. 6 W. & C. As.

Anchusa Hill = Alkanna Tausch (Boraginac.).

Anchusa L. Boraginaceae. 50 Eur., N. Afr., W. As. *A. officinalis* L. was formerly offic., and is widely scattered.

Anchusopsis Bisch. = Lindelofia Lehm. (Boraginac.).

Ancistrachne S. T. Blake. Gramineae. 2 Philipp. Is., E. Austr.

Ancistragrostis S. T. Blake. Gramineae. 1 New Guinea.

Ancistranthus Lindau. Acanthaceae. 1 Cuba.

Ancistrella Van Tiegh. = Ancistrocladus Wall. (Ancistrocladac.).

Ancistrocactus Britton & Rose. Cactaceae. 4 S. U.S., Mex.

Ancistrocarphus A. Gray = Stylocline Nutt. (Compos.).

Ancistrocarpus Kunth = Microtea Sw. (? Chenopodiac.).

*Ancistrocarpus Oliv. Tiliaceae. 5 trop. Afr.

Ancistrocarya Maxim. Boraginaceae. 1 Japan.

Ancistrochilus Rolfe. Orchidaceae. 3 trop. Afr.

Ancistrochloa Honda (~ Calamagrostis Adans.). Gramineae. 1 Japan.

*Ancistrocladaceae Planch. Dicots. 1/20 trop. Afr. to W. Malaysia. Sympodial lianes (rarely shrubs), each member of the sympodium ending in a watch-spring hook (cf. *Artabotrys*, Annonac.). Ls. alt., simple, obov. or oblanc., ent., with scattered minute glandular pits, stips. 0 or minute, caduc. Fls. reg., ⚥, v. caduc., with strongly artic. pedic., in lax (rarely condensed) dichot. cymes or apparent racemes. K 5, imbr., adnate to ovary, accresc., persist., often with conspic. dorsal glandular pits; C 5, ± fleshy, shortly conn. or cohering, contorted or imbr.; A 10, uniseriate, slightly unequal, rarely 5, fil. short, fleshy, connate below, anth. basifixed, intr. or latr.; G (3), semi-inf., 1-loc., with 1 basal erect semi-anatr. ov., and 3 free or connate artic. styles, thickened upwards, with ± hippocrepiform or punctif. stig. Fr. dry, woody, indehisc., surr. by the spreading or erect accresc. sepals, floating in water; seed large, with strongly ruminate endosp. Only genus: *Ancistrocladus*. An interesting, very isolated group, possibly related to *Dioncophyllaceae*.

*Ancistrocladus Wall. Ancistrocladaceae. 20 trop. Afr., Ceylon, E. Himal. to W. Malaysia.

Ancistrodesmus Naud. = Microlepis Miq. (Melastomatac.).

Ancistrolobus Spach = Cratoxylon Blume (Guttif.).

Ancistrophora A. Gray = Verbesina L. (Compos.).

Ancistrophyllum (Mann & Wendl.) Mann & Wendl. ex Kerchove (~ Laccosperma (Mann & Wendl.) Drude). Palmae. 6 trop. Afr.

Ancistrorhynchus Finet. Orchidaceae. 13 trop. Afr.

Ancistrostigma Fenzl = Trianthema L. (Aïzoac.).

Ancistrothyrsus Harms. Passifloraceae. 1 W. trop. S. Am.

Ancistrum J. R. & G. Forst. = Acaena L. (Rosac.).

Anclyanthus A. Juss. = Ancylanthos Desf. (Rubiac.).

Ancouratea Van Tiegh. = Ouratea Aubl. (Ochnac.).

Ancrumia Harv. ex Baker. Alliaceae. 1 Chile.

Ancylacanthus Lindau. Acanthaceae. 1 New Guinea.

Ancylanthos Desf. Rubiaceae. 9 trop. Afr.

Ancylobothrys Pierre. Apocynaceae. 10 trop. Afr., Madag., Comoro Is.

ANCYLOCALYX

Ancylocalyx Tul. = Pterocarpus L. (Legumin.).
Ancylocladus Wall. = Willughbeia Roxb. (Apocynac.).
Ancylogyne Nees = Sanchezia Ruiz & Pav. (Acanthac.).
Ancylostemon Craib. Gesneriaceae. 5 W. China.
Ancyrossemon Poepp. & Endl. (sphalm.) = seq.
Ancyrostemma Poepp. & Endl. = Sclerothrix Presl (Loasac.).
Ancystrochlora Ohwi (sphalm.) = Ancistrochloa Honda (Gramin.).
Anda Juss. = Joannesia Vell. (Euphorbiac.).
Andaca Rafin. = Lotus L. (Legumin.).
Andenea Kreuzinger = Lobivia Britton & Rose (Cactac.).
Andersonia R.Br. Epacridaceae. 22 SW. Austr.
Andersonia Buch.-Ham. = Anogeissus Wall. (Combretac.).
Andersonia Koenig = Stylidium Sw. (Stylidiac.).
Andersonia Roxb. = Aphanamixis Bl. (Meliac.).
Andersonia Willd. = Gaertnera Lam. (Rubiac.).
Anderssoniopiper Trelease. Piperaceae. 1 C. Am.
Andesia Hauman. Juncaceae. 1 S. Andes.
Andicus Vell. = Anda Juss. = Joannesia Vell. (Euphorbiac.).
****Andira** Juss. Leguminosae. 35 trop. Am., Afr. *A. inermis* Kunth (angelin) is a rain-tree (cf. *Pithecellobium*); its wood (partridge-wood) is useful.
Andouinia Reichb. (sphalm.) = Audouinia Brongn. (Bruniac.).
Andrachne L. (excl. Leptopus Decne, *q.v.*). Euphorbiaceae. 1 Peru, 1 Cuba, 15–20 C. Verde Is. & Medit. Reg. to Somalia, Socotra & W. Himal.
Andracna Marnac & Reyn. = Andrachne L. (Euphorbiac.).
Andradea Allem. Nyctaginaceae. 1 E. Braz.
Andradia T. R. Sim = Dialium L. (Legumin.).
Andrastis Rafin. ex Benth. = Cladrastis Rafin. (Legumin.).
Andrea Mez. Bromeliaceae. 1 C. Braz.
Andreoskia DC. = Dontostemon Andrz. ex Ledeb. (Crucif.).
Andreoskia Spach = Andrzeiowskya Reichb. (Crucif.).
Andresia Sleumer. Monotropaceae. 1 Malay Penins.
Andreusia Dun. = Symphysia C. B. Presl (Ericac.).
Andreusia Vent. = Myoporum Banks (Myoporac.).
Andrewsia Spreng. = Bartonia Muhlenb. ex Willd. (Gentianac.).
Andriala Decne = Andryala L. (Compos.).
Andriapetalum Pohl = Panopsis Salisb. (Proteac.).
Andrieuxia DC. = Heliopsis Pers. (Compos.).
Androcalymma Dwyer. Leguminosae. 1 Amaz. Braz.
Androcentrum Lem. Acanthaceae. 1 Mex.
Androcephalium Warb. = Lunasia Blanco (Rutac.).
Androcera Nutt. (~ Solanum L.). Solanaceae. 4 N. Am.
Androchilus Liebm. Orchidaceae. 1 Mex.
Androcoma Nees = Scirpus L. (Cyperac.).
Androcorys Schlechter. Orchidaceae. 4 India, E. As.
Androcymbium Willd. Liliaceae. 35 Medit. to S. Afr. Fls. in heads.
Androglossa Benth. = Sabia Colebr. (Sabiac.).
Andrographis Wall. Acanthaceae. 20 trop. As.
Androgyne Griff. = Panisea Lindl. (Orchidac.).

Androlepis Brongn. ex Houll. Bromeliaceae. 2 C. Am.
Andromachia Humb. & Bonpl. = Liabum Adans. (Compos.).
Andromeda L. Ericaceae. 1–2 N. temp. & cold.
Andromycia A. Rich. Araceae. 1 Cuba.
Androphilax Steud. = Androphylax Wendl. = Cocculus L. (Menispermac.).
Androphoranthus Karst. = Caperonia A. St-Hil. (Euphorbiac.).
Androphthoë Scheff. (sphalm.) = Dendrophthoë Mart. (Loranthac.).
Androphylax Wendl. = Cocculus L. (Menispermac.).
Androphysa Moq. = Halocharis Moq. (Chenopodiac.).
Andropogon L. Gramineae. 113 trop. & subtrop.
Andropogonaceae (J. Presl) Herter = Gramineae–Andropogoneae J. Presl.
Andropterum Stapf. Gramineae. 1 trop. Afr.
Andropus Brand. Hydrophyllaceae. 1 SW. U.S. (New Mexico).
Androsace L. Primulaceae. 100 N. temp. Tufted xerophytes. Often hetero-
styled like *Primula*.
Androsaces Aschers. = praec.
Androsaemum Duham. = Hypericum L. (Guttif.).
Androscepia Brongn. = Anthistiria L. (Gramin.).
Androsemum Neck. = Androsaemum Duham. = Hypericum L. (Guttif.).
Androsiphon Schlechter. Liliaceae. 1 S. Afr. (Cape Prov.).
Androsiphonia Stapf = Paropsia Nor. ex Thou. (Flacourtiac.).
Androstachydaceae Airy Shaw. Dicots. 1/5 SE. trop. Afr., Madag. Trees
of poplar-like habit, with decussate branching. Ls. opp., ent. and slightly pelt.,
or digitately 3–7-foliolate, stip., stips. large, intrapet., connate into an oblong
flattened sheath enclosing the term. bud. Fls. ♂♀, dioec. ♂ fls. amenti-
form, ± cernuous, in shortly pedunc. axill. triads: K (bracteoles?) 2–3 (in lat.
fls.) or 5 (in term. fl.), lin.-lanc., membr., ± spirally arranged; disk o; A ∞ on
an elong. axis, fil. v. short, anth. lin., biloc., loc. apiculate. ♀ fls. sol., axill.,
pedunc.: K 5–6, imbr., ± membr., acum.; disk o; stds. o; G̲ (3)(–5), with
± elong. style and 3 simple spreading stigs., and 2 apical pend. ov. per loc.
Caps. tricoccous, septi- and loculicid.; seeds with shining testa and fleshy
endosp. Only genus: *Androstachys*. Affinities obscure; perh. some connection
with *Euphorbiac.*, but anatomy is against this. The ♂ fl. and infl. are very curious.
Androstachys Prain. Androstachydaceae. 5 SE. trop. Afr., Madag.
Androstemma Lindl. = Conostylis R.Br. (Haemodorac.).
Androstephium Torr. = Bessera Schult. f. (Alliac.).
Androstoma Hook. f. = Cyathodes Labill. (Epacridac.).
Androstylanthus Ducke. Moraceae. 1 Amaz. Braz.
Androstylium Miq. = Clusia L. (Guttif.).
Androsyce Wedd. ex Hook f. = Elatostema J. R. & G. Forst. (Urticac.).
Androsynaceae Salisb. = Tecophilaeaceae Leyb.
Androsyne Salisb. = Walleria J. Kirk (Tecophilaeac.).
Androtium Stapf. Anacardiaceae. 2 Borneo.
Androtrichum Brongn. Cyperaceae. 3 E. temp. S. Am.
Androtropis R.Br. ex Wall. = Acranthera Arn. (Rubiac.).
Androya H. Perrier. Buddlejaceae. 1 Madagascar.
Andruris Schlechter. Triuridaceae. 16 Japan, Indomal., trop. Austr., Pacif.
Andryala L. Compositae. 25 Medit.

Andrzeiowskya Reichb. Cruciferae. 1 Balkans to Cauc.
Anechites Griseb. Apocynaceae. 1 W.I., trop. S. Am.
Anecio Neck. = Senecio L. (Compos.).
Anecochilus Blume = seq.
Anectochilus Blume = Anoectochilus Blume (Orchidac.).
Anectron H. Winkler = Peglera Bolus = Nectaropetalum Engl. (Erythroxylac.)
Aneilema R.Br. Commelinaceae. 100 warm, esp. Old World.
Aneimia Sw. corr. Kaulf. = Anemia Sw. (Schizaeac.).
Aneimiaebotrys Fée = Anemia Sw. (Schizaeac.).
Anelasma Miers = Abuta Aubl. (Menispermac.).
Anelsonia Macbride & Payson. Cruciferae. 1 Pacif. U.S.
Anelytrum Hackel = Avena L. (Gramin.).
Anemagrostis Trin. = Apera Adans. (Gramin.).
Anemanthus Fourr. = Anemone L. (Ranunculac.).
Anemarrhena Bunge. Liliaceae. 1 N. China.
Anemia Nutt. = Anemopsis Hook. & Arn. (Saururac.).
***Anemia** Sw. Schizaeaceae. 90 trop. & subtrop., chiefly Am. Basal pair of pinnae fertile, branched, without lamina.
Anemiaceae Presl = Schizaeaceae Mart.
Anemidictyon J. Sm. = Anemia Sw. (Schizaeac.).
Anemiopsis Endl. = Anemopsis Hook. & Arn. (Saururac.).
Anemirhiza J. Sm. = Anemia Sw. (Schizaeac.).
Anemitis Rafin. = Phlomis L. (Labiat.).
Anemoclema (Franch.) W. T. Wang (~ Anemone L.). Ranunculaceae. 1 SW. China.
Anemoisandra Pohl (nomen). 1 Brazil. Quid?
Anemonanthea (DC.) S. F. Gray = Anemone L. (Ranunculac.).
Anemonanthera Willis (sphalm.) = praec.
Anemone L. Ranunculaceae. 150 cosmop. Herbs with rhiz. and radical ls. Fls. sol. or in cymes. P petaloid; the invol. of green ls. in the hepatica (*A. hepatica* L.) is so close-to the fl. as to resemble a K. The fl. of *A. nemorosa* L. contains no honey, is white, and visited for pollen; that of *A. hepatica* is blue and bee-visited, while in *A. pulsatilla* L. there is honey secreted by stds., and the long-tubed purple fl. is visited mainly by bees. The achenes of many spp. have hairs aiding wind-dispersal.
Anemonella Spach. Ranunculaceae. 1 E. N. Am.
Anemonoïdes Mill. = Anemone L. (Ranunculac.).
***Anemopaegma** Mart. ex Meissn. Bignoniaceae. 30 trop. Am.
Anemopaegmia Mart. ex Meissn. = praec.
Anemonopsis Pritz. (sphalm.) = Anemopsis Hook. (Saururac.).
Anemonopsis Sieb. & Zucc. Ranunculaceae. 1 Japan.
Anemonospermos Boehm. = Arctotis L. (Compos.).
Anemonospermum Comm. ex Steud. = Arctotheca Wendl. (Compos.).
Anemopsis Hook. & Arn. Saururaceae. 1 SW. U.S., Mex.
Anepsa Rafin. = Stenanthium (A. Gray) Kunth (Liliac.).
Anepsias Schott. Araceae. 1 Venez.
Anerincleistus Korth. Melastomataceae. 12 Indoch., W. Malaysia.
Anerma Schrad. ex Nees = Scleria Bergius (Cyperac.).

Aneslea Reichb. = Anneslea Roxb. ex Andr. = Euryale Salisb. (Euryalac.).

Anesorhiza Endl. = Annesorhiza Cham. & Schlechtd. (Umbellif.).

Anetanthus Hiern. Gesneriaceae. 5 Mex., Braz.

Anethum L. Umbelliferae. 1 N. Afr., 3 W. As. *A. graveolens* L. is the dill; fr. a condiment.

Anetia Endl. = Byrsanthus Guillem. (Flacourtiac.).

Anetium (Kunze) Splitg. Vittariaceae. 1 trop. Am.

Aneulophus Benth. Linaceae. 2 W. trop. Afr. Ls. opp.

Aneuriscus Presl = Symphonia L. (Guttif.).

Aneurolepidium Nevski (∼ Elymus L.). Gramineae. 20 temp. Euras.

Angadenia Miers. Apocynaceae. 2 Florida, W.I.

Angeia Tidestrom = Myrica L. (Myricac.).

Angeja Vand. 1 Brazil. Quid? (Stem woody, ± tetrag.; ls. opp. trinerved; fl. v. small, in dense term. pedunc. spike; K (5), yellow, C 5, v. small, yellow; A 9 (?!), exserted; G (?), with 1 exserted style and capit. stig.; berry.)

Angelandra Endl. (1843) = Engelmannia Torr. & Gray (Compos.).

Angelandra Endl. (1850) = Croton L. (Euphorbiac.).

Angelesia Korth. = Licania Aubl. (Chrysobalanac.).

Angelica L. Umbelliferae. 80 N. hemisph. & N.Z.

Angelina Pohl ex Tul. = Siparuna Aubl. (Siparunac.).

Angelium (Reichb.) Opiz = Tommasinia Bertol. (Umbellif.).

Angelocarpa Rupr. = Angelica L. (Umbellif.).

Angelonia Humb. & Bonpl. Scrophulariaceae. 30 trop. Am., W.I.

Angelophyllum Rupr. = Angelica L. (Umbellif.).

Angelopogon Poepp. ex Poepp. & Endl. = Myzodendron Soland. ex DC.

Angervilla Neck. = Stemodia L. (Scrophulariac.).   [(Myzodendrac.).

Angianthus P. & K. = Aggeianthus Wight = Porpax Lindl. (Orchidac.).

*Angianthus Wendl. Compositae. 25 temp. Austr. Heads cpd.

Anginon Rafin. (Rhyticarpus Sonder). Umbelliferae. 3 S. Afr.

Angiopetalum Reinw. = Labisia Lindl. (Myrsinac.).

Angiopteridaceae C. Chr. Marattiales (*q.v.* for chars.). Genera: *Angiopteris, Archangiopteris, Macroglossum.*

Angiopteris Adans. = Onoclea L. (Aspidiac.).

*Angiopteris Hoffm. Angiopteridaceae. 100 (?) Madag., trop. As., Polynes. Large ferns; stem massive, not woody; sporangia not united; annulus complex.

Angiospermae A.Br. & Doell. One of the two great divisions of *Spermatophyta*, distinguished from *Gymnospermae* by the fact that the ovules (= integumented megasporangia) are enclosed in an ovary, formed by one or more modified ls. which are folded (usu. longit., more rarely transv.), and commonly connate, in such a way as to protect the ovules; furthermore by the fact that the endosperm is formed after, instead of before, fert.

In the great majority of *A.* the ovary is associated with one or more stamens (= stipitate microsporangia), both being in turn surrounded by one or more whorls of sterile protective or attractive perianth-members (sepals and/or petals); this aggregation (or ± evident derivatives of it) forms what is thought of as a typical 'flower'. Recent research, however, indicates that the concept of 'flower' is by no means of universal application, and that in a number of *Angiospermae* the ♂ 'flower' may in fact still be in a 'pre-Angiosperm' stage.

ANGKALANTHUS

(Cf. Melville, 'A new theory of the Angiosperm flower', *Kew Bulletin*, **16**(1) and **17**(1), 1962–3.)

Angkalanthus Balf. f. Acanthaceae. 1 Socotra.
Angolaea Wedd. Podostemaceae. 1 Angola.
Angolam Adans. = Alangium Lam. (Alangiac.).
Angolamia Scop. = Angolam Adans. = Alangium Lam. (Alangiac.).
Angophora Cav. Myrtaceae. 10 E. Austr.
Angorchis Thou. = Angraecum Bory (Orchidac.).
Angoseseli Chiov. Umbelliferae. 1 Angola.
Angostura Roem. & Schult. Rutaceae. 30 trop. S. Am. Sympet. & zygo.
Angostyles Benth. Euphorbiaceae. 2 trop. S. Am.
Angostylidium (Muell. Arg.) Pax & K. Hoffm. Euphorbiaceae. 1 trop. W. Afr.
Angraecopsis Kraenzl. Orchidaceae. 15 trop. Afr., Mascarene Is.
Angraecum Bory. Orchidaceae. 220 trop. & S. Afr., Madag., Masc., Philipp.
*Anguillaria R.Br. Liliaceae. 3 Austr., Tasm.
Anguillaria Gaertn. = Ardisia Sw. (Myrsinac.).
Anguillicarpus Burkill = Spirorrhynchus Kar. & Kir. (Crucif.).
Anguina Mill. = Trichosanthes L. (Cucurbitac.).
Anguinum Fourr. = Allium L. (Alliac.).
Anguloa Ruiz & Pav. Orchidaceae. 10 trop. S. Am.
× **Angulocaste** hort. Orchidaceae. Gen. hybr. (iii) (Anguloa × Lycaste).
Anguria Jacq. = Psiguria Arn. (Cucurbitac.).
Anguria Mill. = Citrullus Schrad. ex Eckl. & Zeyh. (Cucurbitac.).
Anguriopsis J. R. Johnston = Doyerea Grosourdy (Cucurbitac.).
Angusta Ellis = Gardenia Ellis (Rubiac.).
Angustinea A. Gray (sphalm.) = Augustinea A. St-Hil. & Naud. = Miconia Ruiz & Pav. (Melastomatac.).
Angylocalyx Taub. Leguminosae. 10 trop. Afr.
Anhalonium Lem. = Ariocarpus Scheidw. (Cactac.).
Ania Lindl. Orchidaceae. 11 India, China, Indoch., W. Malaysia.
Aniba Aubl. Lauraceae. 40 C. & trop. S. Am.
Anictoclea Nimmo = Tetrameles R.Br. (Tetramelac.).
Anidrum Neck. = Bifora Hoffm. (Umbellif.).
Anigosia Salisb. = seq.
Anigozanthos Labill. Haemodoraceae. 10 SW. Austr. Fl. transversely zygo.
Aniketon Rafin. = Smilax L. (Smilacac.).
Anil Mill. = Indigofera L. (Legumin.).
Anilema Kunth = Aneilema R.Br. (Commelinac.).
Aningeria Aubrév. & Pellegr. = seq.
Aningueria Aubrév. & Pellegr. corr. A. Chev. Sapotaceae. 3 trop. W. Afr.
Aniotum Soland. ex Parkinson = Inocarpus J. R. & G. Forst. (Legumin.).
Anisacantha R.Br. = Sclerolaena R.Br. (Chenopodiac.).
Anisacanthus Nees (= Idanthisa Rafin.). Acanthaceae. 15 Am.
Anisachne Keng. Gramineae. 1 SW. China.
Anisactis Dulac = Carum L. (Umbellif.).
Anisadenia Wall. ex Meissn. Linaceae. 2 Himalaya to C. China. The genus connects *Linac.* with *Plumbaginac.*
Anisandra Bartl. = Microcorys R.Br. (Labiat.).

Anisandra Planch. ex Oliv. = Ptychopetalum Benth. (Olacac.).
Anisantha C. Koch (~ Bromus L.). Gramineae. 10 N. & S. temp.
Anisanthera Griff. = Adenosma R.Br. (Scrophulariac.).
Anisanthera Rafin. (1) = Crotalaria L. (Legumin.).
Anisanthera Rafin. (2) = Caccinia Savi (Boraginac.).
Anisantherina Pennell. Scrophulariaceae. 1 trop. S. Am.
Anisanthus Sweet = Antholyza L. (Iridac.).
Anisanthus Willd. ex Roem. & Schult. = Symphoricarpos Juss. (Caprifoliac.).
Aniseia Choisy. Convolvulaceae. 5 trop., esp. Am.
Aniselytron Merr. Gramineae. 1 Philipp. Is.
Anisepta Rafin. = Croton L. (Euphorbiac.).
Aniserica N. E. Br. = Eremia D. Don (Ericac.).
Anisifolium Kuntze = Limonia L. (Rutac.).
Anisocalyx L. Bolus. Aïzoaceae. 1 SW. Afr.
Anisocalyx Hance = Brami Adans. = Bacopa Aubl. (Scrophulariac.).
Anisocampium Presl. Athyriaceae. 2 trop. As. & Malaysia.
Anisocarpus Nutt. = Madia Molina (Compos.).
Anisocentra Turcz. = Tropaeolum L. (Tropaeolac.).
Anisocentrum Turcz. = Acisanthera P.Br. (Melastomatac.).
Anisocereus Backeb. = Escontria Rose (Cactac.).
Anisochaeta DC. Compositae. 1 S. Afr.
Anisochilus Wall. Labiatae. 20 trop. Afr., As.
Anisocoma Torr. & Gray. Compositae. 1 SW. U.S.
Anisocycla Baill. Menispermaceae. 8 trop. & S. Afr., Madag.
Anisodens Dulac = Scabiosa L. (Dipsacac.).
Anisoderis Cass. = Crepis L. (Compos.).
Anisodontea Presl. Malvaceae. 19 S. Afr.
Anisodus Link & Otto (~ Scopolia Jacq.). 6 temp. E. As.
Anisogonium Presl = Diplazium Sw. (Athyriac.).
Anisolepis Steetz = Helipterum DC. (Compos.).
Anisolobus A. DC. = Odontadenia Benth. (Apocynac.).
Anisolotus Bernh. = Hosackia Dougl. (Legumin.).
Anisomallon Baill. Icacinaceae. 1 New Caled.
Anisomeles R.Br. = Epimeredi Adans. (Labiat.).
Anisomeria D. Don. Phytolaccaceae. 2 Chile.
Anisomeris Presl. Rubiaceae. 50 C. & trop. S. Am.
Anisometros Hassk. = Pimpinella L. (Umbellif.).
Anisonema A. Juss. = Phyllanthus L. (Euphorbiac.).
Anisopappus Hook. & Arn. Compositae. 20 trop. & S. Afr., S. China.
Anisopetala Walp. = Pelargonium L'Hérit. (Geraniac.).
Anisopetalon Hook. = Bulbophyllum Thou. (Orchidac.).
Anisopetalum auctt. = praec.
Anisophyllea R.Br. Anisophylleaceae. 30 trop. Afr. & As., 1 trop. S. Am.
Anisophylleaceae (Schimp.) Ridl. (~ Rhizophoraceae). 4/36 trop. As *Rhizoph.*,
  but ls. alt., exstip., often distich., sometimes anisoph., often 3–5-plinerved;
  styles free; fr. a dry or rarely fleshy drupe or winged caps. Genera: *Aniso-
  phyllea, Combretocarpus, Polygonanthus, Poga.*
Anisophyllum Boiv. ex Baill. = Croton L. (Euphorbiac.).

ANISOPHYLLUM

Anisophyllum G. Don = Anisophyllea R.Br. (Anisophylleac.).

Anisophyllum Haw. = Euphorbia L. (Euphorbiac.).

Anisophyllum Jacq. = ? Schinus L. or Astronium Jacq. (Anacardiac.).

Anisoplectus Oerst. = Alloplectus Mart. (Gesneriac.).

Anisopleura Fenzl = Heptaptera Margot & Reuter (Umbellif.).

Anisopoda Baker. Umbelliferae. 1 Madag.

Anisopogon R.Br. Gramineae. 1 Austr.

Anisoptera Korth. Dipterocarpaceae. 13 Assam, SE. As., Malaysia. Stylopodium. G semi-inf. Useful timber.

Anisopus N. E. Br. Asclepiadaceae. 4 trop. W. Afr.

Anisopyrum Gren. & Duval = Agropyron J. Gaertn. (Gramin.).

Anisora Rafin. = Helicteres L. (Sterculiac.).

Anisoramphus DC. = Crepis L. (Compos.).

Anisosciadium DC. Umbelliferae. 2 SW. As.

Anisosorus Trevis. (nomen) = Lonchitis L. (Dennstaedtiac.).

Anisosperma S. Manso. Cucurbitaceae. 1 Brazil. Seeds medicinal.

Anisostachya Nees = Justicia L. (Acanthac.).

Anisostemon Turcz. = Connarus L. (Connarac.).

Anisostichus Bureau = Bignonia L. (Bignoniac.).

Anisosticte Bartl. = Capparis L. (Capparidac.).

Anisostictus Benth. & Hook. f. = Anisostichus Bur. = Bignonia L. (Bignoniac.).

Anisostigma Schinz = Tetragonia L. (Tetragoniac.).

Anisotes Lindl. = Lythrum L. (Lythrac.).

***Anisotes** Nees. Acanthaceae. 12 trop. Afr., Madag., Arabia.

Anisothrix O. Hoffm. ex Kuntze. Compositae. 1 S. Afr.

Anisotoma Fenzl. Asclepiadaceae. 2 S. Afr.

Anisotomaria Presl = praec.

Anisotome Hook. f. Umbelliferae. 13 N.Z., Subantarctic Is.

Anistelma Rafin. = Hedyotis L. (Rubiac.).

Anistylis Rafin. = Liparis Rich. (Orchidac.).

Anisum Hill (~ Pimpinella L.). Umbelliferae. 2 Medit.

Anithista Rafin. = Carex L. (Cyperac.).

Ankylobus Stev. = Astragalus L. (Legumin.).

Ankylocheilos Summerh. = Taeniophyllum Bl. (Orchidac.).

Ankyropetalum Fenzl. Caryophyllaceae. 3 E. Medit. to Persia.

Anna Pellegr. Gesneriaceae. 1 Indoch.

Annamocarya A. Chev. (~ Carya Nutt.). Juglandaceae. 1 S. China, Indoch.

Annesijoa Pax & Hoffm. Euphorbiaceae. 1 New Guinea.

Anneslea Hook. = Anneslia Salisb. = Calliandra Benth. (Legumin.).

Anneslea Roxb. ex Andr. = Euryale Salisb. (Euryalac.).

***Anneslea** Wall. Theaceae. 4 China, Formosa, Indomal. G semi-inf.

Annesleia Hook. = Anneslia Salisb. = Calliandra Benth. (Legumin.).

Annesleya P. & K. = Anneslea Wall. (Theac.).

Anneslia Salisb. = Calliandra Benth. (Legumin.).

Annesorrhiza Cham. et Schlechtd. Umbelliferae. 15 S. Afr. *A. capensis* Ch. & Schl. has ed. roots.

Annona L. Annonaceae. 120 warm esp. Am. Fr. aggregate, often very large, made up of the individual berries derived from the separate cpls., sunk in, and united with, the fleshy recept. That of some cult. spp. is ed., e.g. of *A. cherimolia* Mill. (*cherimoyer*; trop. Am.), *A. squamosa* L. (sweet sop, custard or sugar apple; ?native in W.I.), *A. muricata* L. (sour sop; trop. Am.) and *A. reticulata* L. (custard-apple or bullock's heart; ?native in W.I.).

*Annonaceae Juss. Dicots. 120/2100, chiefly trop. (esp. Old World). Trees and shrubs (exc. one) with usu. two-ranked undivided exstip. ls. Stem sometimes sympodial, at least in infl. Oil passages present. Fls. reg., ⚥ (rarely unisex.), solitary or in infl. of various types. Usu. formula P 3 + 3 + 3 (one or two outer whorls sepaloid); A ∞ (rarely few), spiral, hypog.; G ∞ (*Monodora* is syncp.). Ovules usu. ∞, ventral or basal, anatr. Fr. commonly an aggregate of berries; when many-seeded, frequently constricted between the seeds. In *Annona*, etc., the berries coalesce with the receptacle. Ruminate endosperm (the chief character that separates *A.* from *Magnoliac.*). Many yield ed. fr., e.g. *Annona*, *Artabotrys*.

Classification and chief genera (after R. E. Fries):

I. *Annonoïdeae* (carp. spirally arranged (rarely few in 1 whorl), free (rarely united into a multiloc. syncarp)):

   1. UVARIËAE (pet. imbr., v. rarely (*Guatteria* group) ± distinctly valv.; ls. distich.): *Uvaria-, Duguetia-, Asimina-, Hexalobus-* and *Guatteria-*groups.

   2. ANNONEAE (UNONEAE) (pet. (at least the outer) valvate, v. rarely (*Trigynaea* group) imbr.; ls. distich.): *Desmos-, Polyalthia-, Unonopsis-, Xylopia-, Artabotrys-, Orophea-, Annona-, Trigynaea-* and *Monanthotaxis*-groups.

   3. TETRAMERANTHEAE. Tep. of each whorl 4(3), imbr. Stig. 3-lobed, adpr. to ov. Ls. spirally arr. Bracts 4, vertic. Only genus: *Tetrameranthus*.

II. *Monodoroïdeae* (carpels cyclically arr., united into a 1-loc. ov. with pariet. plac.):

   4. MONODOREAE. Only genera: *Monodora, Isolona*.

Annulaceae Dulac = Rosaceae Juss. (*s.str.*).

Annularia Hochst. = Voacanga Thou. (Apocynac.).

Annulodiscus Tardieu = Salacia L. (Celastrac.).

Anocheile Hoffmgg. ex Reichb. = Epidendrum L. (Orchidac.).

Anochilus (Schltr.) Rolfe = Pterygodium Sw. (Orchidac.).

Anoda Cav. Malvaceae. 10 trop. Am.

Anodendron A. DC. Apocynaceae. 20 Ceylon, Japan, Formosa, Hainan, Malaysia, Solomon Is.

Anodia Hassk. = Anoda Cav. (Malvac.).

Anodiscus Benth. Gesneriaceae. 1 Peru.

Anodontea Sweet = Alyssum L. (Crucif.).

Anodopetalum A. Cunn. Cunoniaceae. 1 Tasmania. [unequal.

Anoectocalyx Triana. Melastomataceae. 2 Venez. 6-merous. K-teeth

*Anoectochilus Blume. Orchidaceae. 25 trop. Asia, Austr., Polynesia.

× **Anoectogoodyera** hort. Orchidaceae. Gen. hybr. (vi) (Anoectochilus × Goodyera).

× **Anoectomaria** hort. Orchidaceae. Gen. hybr. (iii) (Anoectochilus × Ludisia [Haemaria]).

ANOGEISSUS

Anogeissus Wall. ex Guillem. & Perr. Combretaceae. 11 trop. Afr., Arabia, India, SE. As.

Anogra Spach = Oenothera L. (Onagrac.).

Anogramma Link. Hemionitidaceae. 7 N. & S. temp.

Anogyna Nees = Lagenocarpus Nees (Cyperac.).

Anoiganthus Baker. Amaryllidaceae. 2 trop. & S. Afr.

Anoma Lour. = Moringa Juss. (Moringac.).

Anomacanthus Good = Gilletiella De Wild. & Dur. (Mendonciac.).

Anomalanthus Klotzsch = Scyphogyne Brongn. (Ericac.).

Anomalesia N. E. Brown (~ Petamenes Salisb. ex N. E. Br.). Iridaceae.

Anomalocalyx Ducke. Euphorbiaceae. 1 trop. S. Am. [2 S. Afr.

Anomalopteris G. Don = Acridocarpus Guill. & Perr. (Malpighiac.).

Anomalosicyos Gentry = Sicyos L. (Cucurbitac.).

Anomalostemon Klotzsch = Cleome L. (Cleomac.).

Anomalostylus R. C. Foster. Iridaceae. 3 trop. S. Am.

Anomalotis Steud. = Trisetaria Forsk. (Gramin.).

Anomantha Rafin. = Verbesina L. (Compos.).

Anomanthodia Hook. f. = Randia L. (Rubiac.).

Anomantia DC. (sphalm.) = Anomantha Rafin. = Verbesina L. (Compos.).

Anomatheca Ker-Gawl. = Lapeyrousia Pourr. (Iridac.).

Anomaza Laws. ex Salisb. = Lapeyrousia Pourr. (Iridac.).

Anomeris Rafin. = Actinomeris Nutt. (Compos.).

Anomianthus Zoll. Annonaceae. 1 Siam, Indoch., Java.

Anomocarpus Miers = Calycera Cav. (Calycerac.).

Anomochloa Brongn. Anomochloaceae. 1 Brazil.

Anomochloaceae Nak. (~ Gramineae). Monocots. Trop. forest 'grasses' of Marantoid habit. L.-lam. ov.-cord., contr. into long petiole-like base and long sheath, many-nerved, with cross-nerves. Spikel. 1-fl., ♀, artic. at base, 1–3 together on short ped. in ax. of small br. and encl. in large sheath-like spathes on opp. sides of rhach. of spike-like infl. Glumes o; lemma(?) broad, many-nerved with cross-nerves, exarist.; palea (?) rigid, coriac., many-nerved, with narr. apical append.; lodic. (?) repr. by hairy disk around stam.; stam. 4; style 1. (Interpr. of parts difficult.) Only genus: *Anomochloa*.

Anomopanax Harms. Araliaceae. 9 Philipp. Is., E. Malaysia.

Anomorhegmia Meissn. = Stauranthera Benth. (Gesneriac.).

Anomosanthes Blume = Hemigyrosa Blume (Sapindac.).

Anomospermum Dalz. = Actephila Blume (Euphorbiac.).

Anomospermum Miers. Menispermaceae. 13 trop. S. Am.

Anomostachys (Baill.) Hurusawa = Excoecaria L. (Euphorbiac.).

Anomostephium DC. = Aspilia Thou. (Compos.).

Anomotassa K. Schum. Asclepiadaceae. 1 Ecuador.

Anona L. = Annona L. (Annonac.).

Anonidium Engl. & Diels. Annonaceae. 5 trop. Afr.

Anoniodes Schlechter = Sloanea L. (Elaeocarpac.).

Anonis Mill. = Ononis L. (Legumin.).

Anonocarpus Ducke = Batocarpus Karst. (Morac.).

Anonychium Schweinf. = Prosopis L. (Legumin.).

Anonymos Kuntze = Galax Rafin. (Diapensiac.).

ANTENNARIA

Anonymos Walt. A pseudo-generic term used by Walter, *Fl. Carol.* (1788), for about 42 species which he believed to be new but which he could not refer to a known genus.

Anoosperma Kunze (sphalm.) = Oncosperma Bl. (Palm.).

Anoplanthus Endl. = Anoplon Reichb. (Orobanchac.).

Anoplia Nees ex Steud. = Leptochloa Beauv. (Gramin.).

Anoplocaryum Ledeb. Boraginaceae. 1 E. Sib., Mongolia (4 Himal., China?).

Anoplon Reichb. = Phelypaea L. + Aphyllon Mitch. (Orobanchac.).

Anoplophytum Beer = Tillandsia L. (Bromeliac.).

Anopteris (Prantl) Diels. Pteridaceae. 1 trop. Am.

Anopterus Labill. Escalloniaceae. 2 E. Austr., Tasm.

Anopyxis (Pierre) Engl. Rhizophoraceae. 3 trop. Afr.

Anosmia Bernh. = Smyrnium L. (Umbellif.).

Anosporum Nees = Cyperus L. (Cyperac.).

Anota Schlechter = Rhynchostylis Bl. (Orchidac.).

Anotea Kunth. Malvaceae. 2 Mexico.

Anothea O. F. Cook. Palmae. 1 Mex.

Anotis auctt. = Neanotis W. H. Lewis (Rubiac.).

Anotis DC. = Arcytophyllum Willd. ex Schult. & Schult. f. + Hedyotis L. + Oldenlandia L. (Rubiac.).

Anotis DC. = Panetos Rafin. (Rubiac.), *q.v.*

Anotites Greene = Silene L. (Caryophyllac.).

Anoumabia A. Chev. = Harpullia Roxb. (Sapindac.).

Anplectrella Furtado = Enchosanthera King & Stapf (Melastomatac.).

Anplectrum auctt. = Dissochaeta Bl., Diplectria Reichb., Neodissochaeta Bakh. f., *p.p.* (Melastomatac.).

Anplectrum A. Gray. Melastomataceae. 2 Indoch., W. Malaysia.

Anquetilia Decne = Skimmia Thunb. (Rutac.).

Anredera Juss. Basellaceae. 5–10 S. U.S. & W.I. to Argent., Galápagos.

Anreder[ac]eae J. G. Agardh = Basellaceae–Anredereae (Endl.) Moq.

Ansellia Lindl. Orchidaceae. 2 trop. Afr., Natal.

Anselonia O. E. Schulz (sphalm.) = Anelsonia Macbr. & Payson (Crucif.).

Anserina Dum. = Chenopodium L. (Chenopodiac.).

× **Ansidium** hort. Orchidaceae. Gen. hybr. (iii) (Ansellia × Cymbidium).

Ansonia Bert. ex Hemsl. = Lactoris Phil. (Lactoridac.).

Ansonia Rafin. = Amsonia Walt. (Apocynac.).

Anstrutheria Gardn. = Cassipourea Aubl. (Rhizophorac.).

Antacanthus Rich. ex DC. = Scolosanthus Vahl (Rubiac.).

Antagonia Griseb. = Cayaponia S. Manso (Cucurbitac.).

Antaurea Neck. = Centaurea L. (Compos.).

Antegibbaeum Schwantes ex C. Weber. Aizoaceae. 1 S. Afr.

Antelaea Gaertn. = Melia L. (Meliac.).

Antennaria Gaertn. Compositae. 100 extra-trop., exc. Afr. The European *A. dioica* (L.) Gaertn. (mountain everlasting, cat's-foot) is a small creeping dioec. perenn., hairy and semi-xero., occurring chiefly on hills and at the sea-shore, but not common in intermediate places. In *A. alpina* (L.) R.Br. only ♀ plants usu. occur, and show true parthenogenesis, the ovum developing into an embryo without fert. (not to be confused with the veg. budding of *Caelebogyne*).

ANTENORON

Antenoron Rafin. (~ Persicaria Mill., ~ Polygonum L.). Polygonaceae. 4 Japan, Ryukyu Is., Philipp. Is., N. Am.

Antephora Steud. = Anthephora Schreb. (Gramin.).

Anteriscium Meyen (sphalm.) = Asteriscium Cham. & Schlechtd. (Umbellif.).

Anthacantha Lem. = Euphorbia L. (Euphorbiac.).

Anthacanthus Nees = Oplonia Rafin. (Acanthac.).

Anthactinia Bory = Passiflora L. (Passiflorac.).

Anthadenia Lem. = Sesamum L. (Pedaliac.).

Anthaea Nor. ex Thou. = Didymeles Thou. (Didymelac.).

Anthaenantia Beauv. Gramineae. 2 warm Am.

Anthaenantiopsis Mez. Gramineae. 1 Brazil.

Anthaerium Schott (sphalm.) = Anthurium Schott (Arac.).

Anthagathis Harms = Jollydora Pierre ex Gilg (Connarac.).

Anthallogea Rafin. = Polygala L. (Polygalac.).

Anthanema Rafin. = Cuscuta L. (Cuscutac.).

Anthanotis Rafin. = Asclepias L. (Asclepiadac.).

× **Anthechrysanthemum** Domin. Compositae. Gen. hybr. (Anthemis × Chrysanthemum).

Antheëischima Korth. = Gordonia Ellis (Theac.).

× **Antheglottis** hort. (vii) = × Trichovanda hort. (Orchidac.).

Antheidosorus A. Gray = Myriocephalus Benth. (Compos.).

Antheidosurus C. Muell. (sphalm.) = praec.

Antheilema Rafin. = Ruellia L. (Acanthac.).

Anthelia Schott = Epipremnum Schott (Arac.).

Antheliacanthus Ridl. Acanthaceae. 1 Siam.

Anthelis Rafin. = Helianthemum Mill., Fumana Spach, etc. (Cistac.).

Anthelmenthia P.Br. = Spigelia L. (Spigeliac.).

Anthelmenthica Pfeiff. = praec.

Anthelminthica B. D. Jacks. = praec.

Anthema Medik. = Lavatera L. (Malvac.).

× **Anthe-Matricaria** Geisenheyner = × Anthechrysanthemum Domin (Compos.).

Anthemid[ac]eae Link = Compositae–Anthemideae Cass.

× **Anthemi-Matricaria** P. Fourn. Compositae. Gen. hybr. (Anthemis × Matricaria).

Anthemiopsis Boj. = Wedelia Jacq. (Compos.).

Anthemis L. Compositae. 200 Eur., Medit. to Persia. The fr. of *A. arvensis* L. has papillae on its upper surface which become sticky when wet (cf. *Linum*). *A. nobilis* L. (chamomile) fl. offic.

Anthephora Schreb. Gramineae. 20 trop. & S. Afr.

× **Antheranthe** hort. (vii) = × Renantanda hort. (Orchidac.).

Antheric[ac]eae J. G. Agardh = Liliaceae–Asphodeloïdeae–Asphodeleae–Anthericinae Engl.

Anthericlis Rafin. = Tipularia Nutt. (Orchidac.).

Anthericopsis Engl. Commelinaceae. 1 trop. E. Afr.

Anthericum L. Liliaceae. 300 mostly trop. & S. Afr., Madag., also Eur., Am., 1 E. As.

Anthericus Aschers. & Graebn. = praec.

× **Antherisia** Wehrh. Liliaceae. Gen. hybr. (Anthericum × Paradisea).
Antherocephala DC. = Andersonia R.Br. (Epacridac.).
Antheroceras Bert. = Leucocoryne Lindl. (Liliac.).
Antherolophus Gagnep. (~ Aspidistra Ker-Gawl.). Liliaceae. 1 Indoch.
Antheropeas Rydb. Compositae. 5 N. Am.
Antheroporum Gagnep. Leguminosae. 2 Indoch.
Antherosperma Poir. ex Steud. (sphalm.) = Atherosperma Labill. (Atherospermatac.).
Antherostele Bremek. Rubiaceae. 4 Philipp. Is.
Antherostylis C. A. Gardner. Goodeniaceae. 1 W. Austr.
Antherothamnus N. E. Br. Scrophulariaceae. 1 S. Afr.
Antherotoma Hook. f. Melastomataceae. 2 trop. Afr., Madag.
Antherotriche Turcz. = Anisoptera Korth. (Dipterocarpac.).
Antherura Lour. = Psychotria L. (Rubiac.).
Antherylium Rohr & Vahl = Ginoria Jacq. (Lythrac.).
Antheryta Rafin. = Tibouchina Aubl. (Melastomatac.).
Anthesteria Spreng. = Anthistiria L. f. = Themeda Forsk. (Gramin.).
Anthillis Neck. = Anthyllis L. (Legumin.).
Anthipsimus Rafin. = Muhlenbergia Schreb. (Gramin.).
Anthirrinum Moench = Antirrhinum L. (Scrophulariac.).
Anthistiria L. f. = Themeda Forsk. (Gramin.).
Anthobembix Perkins. Monimiaceae. 8 New Guinea.
Anthobol[ac]eae Dum. = Santalaceae–Anthoboleae Spach.
Anthobolus R.Br. Santalaceae. 5 Austr. Ovary superior (cf. fam.); fruit borne on coloured fleshy receptacle.
Anthobryum Phil. Frankeniaceae. 4 Bolivia, Chile, Argent.
Anthocarpa Pierre. Meliaceae. 2 New Caled.
Anthocephalus A. Rich. (= Platanocephalus Crantz). Naucleaceae (~ Rubiac.). 3 Indomal.
Anthoceras Baker (sphalm.) = Antheroceras Bert. = Leucocoryne Lindl.
Anthocerastes A. Gray = Troxanthus Turcz. (Compos.). [(Liliac.).
Anthocercis Labill. Solanaceae. 20 Austr.
Anthochlamys Fenzl. Chenopodiaceae. 1 SW. & C. As.
Anthochloa Nees & Meyen ex Nees (corr.). Gramineae. 1 Calif., 2 Andes.
Anthochortus Nees corr. Endl. Restionaceae. 1 S. Afr.
Anthochytrum Reichb. = Crepis L. (Compos.).
Anthocleista Afzel. Potaliaceae. 11 trop. Afr., 3 Madag.
Anthoclitandra (Pierre) Pichon. Apocynaceae. 2 trop. W. Afr.
Anthocoma K. Koch = Rhododendron L. (Ericac.).
Anthocoma Zoll. & Mor. = Cymaria Benth. (Labiat.).
Anthocometes Nees = Monothecium Hochst. (Acanthac.).
Anthodendron Reichb. = Rhododendron L. (Ericac.).
Anthodiscus Endl. = Salacia L. (Celastrac.).
Anthodiscus G. F. W. Mey. Caryocaraceae. 10 trop. S. Am.
Anthodon Ruiz et Pav. Celastraceae. 2 C. & trop. S. Am.
Anthodus Mast. ex Roem. & Schult. = praec.
Anthogonium Wall. ex Lindl. Orchidaceae. 1 E. Himal. to SW. China & Siam.
Anthogyas Rafin. = Gyas Salisb. = Bletia R.Br. (Orchidac.).

ANTHOLOMA

Antholoma Labill. = Sloanea L. (Elaeocarpac.).

Antholyza L. Iridaceae. 25 Afr.

Anthomeles M. Roem. = Crataegus L. (Rosac.).

Anthonotha Beauv. = Macrolobium Schreb. (Legumin.).

Anthophyllum Steud. = Scirpus L. (Cyperac.).

Anthopogon Neck. = Gentiana L. (Gentianac.).

Anthopogon Nutt. = Gymnopogon Beauv. (Gramin.).

Anthopteropsis A. C. Sm. Ericaceae. 1 C. Am.

Anthopterus Hook. Ericaceae. 7 Andes.

Anthora DC. = Aconitum L. (Ranunculac.).

Anthosachne Steud. = Agropyron J. Gaertn. (Gramin.).

Anthosciadium Fenzl = Selinum L. (Umbellif.).

Anthoshorea Heim = Shorea Roxb. (Dipterocarpac.).

Anthosiphon Schlechter. Orchidaceae. 1 Colombia.

Anthospermum L. Rubiaceae. 35 Afr., Madag.

Anthostema Juss. Euphorbiaceae. 3 trop. Afr., Madag. Fls. in a cyathium like *Euphorbia*, but the ♂, reduced as in *E*. to 1 sta., has P where in *E*. there is a joint. The ♀ also has a P.

Anthostyrax Pierre = Styrax L. (Styracac.).

Anthotium R.Br. Goodeniaceae. 2 SW. Austr.

Anthotroche Endl. Solanaceae. 6 Austr.

Anthoxanthum L. Gramineae. 20 N. temp., & trop. mts. Afr., As. The stems of *A. odoratum* L. contain large quantities of coumarin, to which the smell char. of newly mown hay is due; it may be recognised by chewing a stalk. Fl. with 2 sta. only, protog. Awns of fr. hygroscopic.

× **Anthrichaerophyllum** P. Fourn. Umbelliferae. Gen. hybr. (Anthriscus × Chaerophyllum).

*__Anthriscus__ Pers. emend. Hoffm. Umbelliferae. 20 Eur., temp. As. *A. cerefolium* Hoffm. is the cult. chervil of France, etc.

Anthrocephalus Schlechtd. (sphalm.) = Anthocephalus A. Rich. (Naucleac.).

Anthropodium Sims (sphalm.) = Arthropodium R.Br. (Liliac.).

Anthrostylis D. Dietr. (sphalm.) = Arthrostylis R.Br. (Cyperac.).

Anthurium Schott. Araceae. 550 trop. Am., W.I. Most are sympodial herbs, with an accessory bud beside the 'continuation' bud of the sympodium. Axillary shoot often 'adnate' to the main one (cf. *Solanaceae*, etc.). Aerial roots frequent at the base of the ls. Some epiphytes. Fls. ♂, with P, protog., arranged in a dense mass upon a spadix, at whose base is a flat usu. brightly coloured spathe. Fr. a berry; when ripe it is forced out of the spadix and hangs by two threads from the P. In *A. longifolium* G. Don the root apex has been observed to develop into a shoot.

Anthyllis Adans. = Polycarpon Loefl. ex L. + Polycarpaea Lam. (Caryophyllac.).

Anthyllis L. Leguminosae. 50 Eur., N. Afr., W. As. Fl. mech. resembles *Lotus*; stigma only receptive when rubbed.

Antia O. F. Cook. Palmae. 1 Cuba.

*__Antiaris__ Lesch Moraceae. 4 trop. Afr., Madag., Indomal., incl. *A. toxicaria* Lesch. (upas-tree). The latex is poisonous. Extraordinary stories of its effects were spread abroad in the eighteenth century. The surroundings were said to be a desert, the poisonous influence emanating from the tree being fatal to life.

Antiaropsis K. Schum. Moraceae. 1–2 New Guinea.
Anticharis Endl. Scrophulariaceae. 10 SW. Afr. to Arabia & India.
Anticheirostylis Fitzg. Orchidaceae. 1 E. Austr. (N.S.W.).
Antichloa Steud. = Actinochloa Willd. = Botelua Lag. (Gramin.).
Antichorus L. = Corchorus L. (Tiliac.).
Anticlea Kunth = Zigadenus Michx (Liliac.).
Anticoryne Turcz. = Baeckea L. (Myrtac.).
Antidaphne Poepp. & Endl. Eremolepidaceae (~ Viscaceae). 2 W. trop. S. Am.
Antidesma L. Stilaginaceae. 170 Old World trop. & subtrop., esp. As.
Antidesm[atac]eae Sweet ex Endl. = Stilaginaceae C. A. Agardh.
Antidris Thou. (uninom.) = *Dryopeia oppositifolia* Thou. = *Disperis oppositifolia* [? Thou.] Sm. (Orchidac.).
Antigona Vell. = Casearia Jacq. (Flacourtiac.).
Antigonon Endl. Polygonaceae. 8 trop. Am. *A. leptopus* Hook. & Arn. is a (stem) tendril climber.
Antigramma Presl. Aspleniaceae. 2 S. Am.
Antillia R. M. King & H. Rob. (~ Eupatorium). Compositae. 1 Cuba.
Antimima N. E. Br. Aïzoaceae. 1 S. Afr.
Antimion Rafin. = Lycopersicon Mill. (Solanac.).
Antinisa (Tul.) Hutch. Flacourtiaceae. 3 Madag.
Antinoria Parl. = Aira L. (Gramin.).
Antiosorus Roemer = Lonchitis L. (Dennstaedtiac.).
Antiotrema Hand.-Mazz. Boraginaceae. 1 SW. China.
Antiphiona Merxm. Compositae. 2 trop. & SW. Afr.
Antiphyla Rafin. = Melochia L. (Sterculiac.).
Antiphylla Haw. = Saxifraga L. (Saxifragac.).
Antiphytum DC. Boraginaceae. 10 Mex. to trop. S. Am.
Antirhea Commers. ex Juss. Rubiaceae. 40 W.I., Madag., trop. E. As. to
Antirrhin[ac]eae DC. & Duby = Scrophulariaceae Juss.   [Austr.
Antirrhinum L. Scrophulariaceae. 42 Pacif. N. Am., W. Medit. The mouth of the fl. is closed and the honey thus preserved for bees, which alone are strong enough to force an entrance.
Antirrhoa Gmel. ex C. DC. = Turraea L. (Meliac.).
Antirrhoea Endl. = Antirhea Commers. ex Juss. (Rubiac.).
Antisola Rafin. = Miconia Ruiz & Pav. (Melastomatac.).
Antistrophe A. DC. Myrsinaceae. 4 Indomal.
Antitaxis Miers = Pycnarrhena Miers (Menispermac.).
Antithrixia DC. Compositae. 3 Abyss. to S. Afr.
Antitoxicum Pobedim = Alexitoxicon St-Lag. (Asclepiadac.).
Antitragus Gaertn. = Crypsis Ait. (Gramin.).
Antitypaceae Dulac = Oxalidaceae R.Br.
Antizoma Miers. Menispermaceae. 5 S. warm Afr.
Antochloa Nees & Meyen ex Nees (sphalm.) = Anthocloa Nees & Meyen (corr.) (Gramin.).
Antochortus Nees = Anthochortus Nees corr. Endl. (Restionac.).
Antodon Neck. = Leontodon L. (Compos.).
Antogoeringia Kuntze = Stenosiphon Spach (Onagrac.).
Antommarchia Colla ex Meissn. = Correa Sm. (Rutac.).

4   75   A S D

Antongilia Jum. Palmae. 1 Madag.

Antonia R.Br. = Rhynchoglossum Blume (Gesneriac.).

Antonia Pohl. Antoniaceae. 1 Brazil, Guiana.

Antoniaceae (Endl.) J. G. Agardh. Dicots. 4/8, tropics. More or less as *Loganiaceae* (*s.str.*) but cor.-lobes valv. Genera: *Antonia, Bonyunia, Norrisia, Usteria.*

Antoniana Bub. = Hesperis L. (Crucif.).

Antoniana Tuss. = Faramea Aubl. (Rubiac.).

Antonina Vved. Labiatae. 1 Cauc. to C. As.

Antopetitia A. Rich. Leguminosae. 1 mts. trop. Afr.

Antophylax Poir. (sphalm.) = Androphylax Wendl. = Cocculus DC. (Menispermac.).

Antoschmidtia Steud. Gramineae. 3 trop. & S. Afr.

Antriba Rafin. = Scurrula L. (Loranthac.).

Antriscus Rafin. = Anthriscus Hoffm. (Umbellif.).

Antrizon Rafin. = ? Antirrhinum L. (Scrophulariac.).

Antrocaryon Pierre. Anacardiaceae. 8 trop. W. Afr.

Antrolepis Welw. = Ascolepis Nees (Cyperac.).

Antrolepidaceae ('-ideas') Welw. = Cyperaceae–Scirpoïdeae.

Antrophora I. M. Johnst. Boraginaceae. 1 C. Am.

Antrophyaceae Link = Vittariaceae Presl.

Antrophyum Kaulf. Vittariaceae. 40 trop. & subtrop. Old World.

Antrospermum Sch. Bip. = Venidium Less. (Compos.).

Antschar Horsf. = Antiaris Lesch. (Morac.).

Antunesia O. Hoffm. Compositae. 1 Angola.

Antura Forsk. = Carissa L. (Apocynac.).

Anubias Schott. Araceae. 13 W. Afr.

Anulocaulis Standley. Nyctaginaceae. 5 SW. U.S., Mex.

Anumophila Link (sphalm.) = Ammophila Host (Gramin.).

Anura Tschern. (~ Puccinellia Parl.). Gramineae. 1 C. As.

Anurosperma (Hook. f.) Hallier (~ Nepenthes L.). Nepenthaceae. 1 Seychelles.

Anurus Presl = Lathyrus L. (Legumin.)

Anvillea DC. Compositae. 4 NW. Afr. to Persia.

Anvilleina Maire. Compositae. 1 Morocco.

Anychia Michx = Paronychia Adans. (Caryophyllac.).

Anychiastrum Small = praec.

Aonikena Spegazz. = Chiropetalum A. Juss. (Euphorbiac.).

Aopla Lindl. = Habenaria L. (Orchidac.).

Aostea Buscalioni & Muschler. Compositae. 2 S. trop. Afr.

Aotus Sm. Leguminosae. 15 Austr., Tasm.

Apabuta (Griseb.) Griseb. = Hyperbaena Miers ex Benth. (Menispermac.).

Apactis Thunb. = Xylosma G. Forst. (Flacourtiac.).

Apalanthe Planch. = Elodea Michx (Hydrocharitac.).

Apalantus Adans. = Hapalanthus Jacq. = Callisia L. (Commelinac.).

Apalatoa Aubl. = Crudia Schreb. (Legumin.).

Apalochlamys Cass. = Cassinia R.Br. (Compos.).

Apalophlebia Presl = Pyrrosia Mirbel (Polypodiac.).

Apaloptera Nutt. = Abronia Juss. (Nyctaginac.).

Apaloxylon Drake del Castillo. Leguminosae. 2 Madag.

Apalus DC. = Blennosperma Less. (Compos.).

Apama Lam. Aristolochiaceae. 12 Indomal., S. China.

Apargia Scop. = Leontodon L. (Compos.).

Apargidium Torr. & Gray = Microseris D. Don (Compos.).

Aparin[ac]eae Hoffmgg. & Link = Rubiaceae Juss.

Aparinanthus Fourr. = Galium L. (Rubiac.).

Aparine Guett. = Galium L. (Rubiac.).

Aparinella Fourr. = Galium L. (Rubiac.).

Aparisthmium Endl. Euphorbiaceae. 1 trop. S. Am.

Apartea Pellegr. = Mapania Aubl. (Cyperac.).

Apassalus Kobuski. Acanthaceae. 3 SE. U.S., W.I.

Apatales Blume ex Ridl. = Liparis Rich. (Orchidac.).

Apatanthus Viv. = Hieracium L. (Compos.).

Apatelia DC. = Saurauia Willd. (Actinidiac.).

Apatemone Schott = Schismatoglottis Zoll. & Mor. (Arac.).

Apatesia N. E. Brown. Aïzoaceae. 5 S. Afr.

Apation Dur. & Jacks. = Apatales Blume ex Ridl. = Liparis Rich. (Orchidac.).

Apatitia Desv. = Bellucia Neck. (Meliac.).

Apatophyllum D. McGillivray. Celastraceae. 2 Queensl., N.S.W.

Apaturia Lindl. = Pachystoma Blume (Orchidac.).

Apegia Neck. = Ceropegia L. (Asclepiadac.).

Apeiba Aubl. Tiliaceae. 10 trop. S. Am. Some have good wood.

Apeiba A. Rich. = Entelea R.Br. (Tiliac.).

Apemon Rafin. = Datura L. (Solanac.).

Apella Scop. = Appella Adans. = Laurus L. (Laurac.).

Apentostera Rafin. = Penstemon Schmidel (Scrophulariac.).

Apenula Neck. = Specularia Fabr. (Campanulac.).

Apera Adans. Gramineae. 4 Eur., W. As.

Aperiphracta Nees = Ocotea Aubl. (Laurac.).

Aperula Blume = Lindera Thunb. (Laurac.).

Apetahia Baill. Campanulaceae. 3 Society Is., Marquesas, Rapa.

Apetalae Juss. (Lindl., Benth. & Hook. f., etc.). A heterogeneous assemblage of Dicot. fams. in which the perianth is apparently single or wanting. Together with the *Polypetalae* they constitute the subclass *Archichlamydeae* of Engler.

Apetalon Wight = Didymoplexis Griff. (Orchidac.).

Apetlorhamnus Nieuwl. = Rhamnus L. (Rhamnac.).

Aphaca Mill. = Lathyrus L. (Legumin.).

Aphaenandra Miq. Rubiaceae. 2 SE. As., Sumatra, ?Java.

Aphaerema Miers. Flacourtiaceae. 1 S. Braz.

Aphanactis Wedd. Compositae. 4 C. & trop. S. Am.

Aphanamixis Blume. Meliaceae. 25 Indomal.

Aphanandrium Lindau. Acanthaceae. 1 W. trop. S. Am.

Aphananthe Link = Microtea Sw. (?Chenopodiac.).

Aphananthe Planch. Ulmaceae. 1 Madag.; 3 India & Ceylon to Japan, Indoch., Philipp., Celebes, Java, E. Austr.; 1 Mex.

4-2

APHANANTHEMUM

Aphananthemum Steud. = Helianthemum Mill. (Cistac.).
Aphanelytrum Hackel. Gramineae. 1 W. trop. S. Am.
Aphanes L. Rosaceae. 20 Am., Medit., Abyss., C. As., Austr.
Aphania Blume. Sapindaceae. 24 Indomal.; 1 W. Afr.
Aphanisma Nutt. Chenopodiaceae. 1 California.
Aphanocalyx Oliv. Leguminosae. 3 trop. W. Afr.
Aphanocarpus Steyermark. Rubiaceae. 1 Venez.
Aphanochaeta A. Gray = Pentachaeta Nutt. (Compos.).
Aphanochilus Benth. = Elsholtzia Willd. (Labiat.).
Aphanococcus Radlk. Sapindaceae. 1 Celebes.
Aphanodon Naud. = Henriettella Naud. (Melastomatac.).
Aphanomyrtus Miq. = Syzygium Gaertn. (Myrtac.).
Aphanomyxis DC. = Aphanamixis Blume (Meliac.).
Aphanopappus Endl. = Lipochaeta DC. (Compos.).
Aphanopetalum Endl. Cunoniaceae. 1 SW. & SE. Austr.
Aphanopleura Boiss. Umbelliferae. 3 Transcauc., C. As., Afghan.
Aphanostelma Malme = Melinia Decne (Asclepiadac.).
Aphanostelma Schlechter. Asclepiadaceae. 1 China. Fls. in lateral racemiform cymes.
Aphanostemma A. St-Hil. = Ranunculus L. (Ranunculac.).
Aphanostemma Willis (sphalm.) = Aphanostelma Schltr. (Asclepiadac.).
Aphanostephus DC. Compositae. 11 U.S., Mex.
Aphanostylis Pierre. Apocynaceae. 3 trop. Afr.
Aphantochaeta A. Gray = Pentachaeta Nutt. (Compos.).
Apharica Schlechtd. = Aphania Blume (Sapindac.).
Aphelandra R.Br. Acanthaceae. 200 warm Am.
Aphelandrellra Mildbraed. Acanthaceae. 1 Peru.
Aphelandros St-Lag. = Aphelandra R.Br. (Acanthac.).
Aphelexis D. Don (~ Helichrysum L.). Compositae. 10 Madag.
Aphelia R.Br. Centrolepidaceae. 1 S. Austr., Tasm.
Aphillanthes Neck. = Aphyllanthes L. (Liliac.).
Aphloia (DC.) Benn. Flacourtiaceae (?Neumanniaceae). 6 (or 1 polymorphic?) trop. E. Afr., Madag., Masc.
Aphoma Rafin. (Iphigenia Kunth). Liliaceae. 12 trop. & S. Afr., Madag., India, Austr., N.Z.
Aphomonix Rafin. = Saxifraga L. (Saxifragac.).
Aphonina Neck. = Pariana Aubl. (Gramin.).
Aphora Neck. = Virgilia Lam. (Legumin.).
Aphora Nutt. = Ditaxis Vahl ex Juss. (Euphorbiac.).
Aphragmia Nees (~ Ruellia L.). Acanthaceae. 3 trop. Am.
Aphragmus Andrz. Cruciferae. 6 Himal. to NE. Sib.
Aphylax Salisb. = Aneilema R.Br. (Commelinac.).
Aphyllangis Thou. (uninom.) = *Angraecum aphyllum* Thou. = *Solenangis aphylla* (Thou.) Summerh. (Orchidac.).
Aphyllanthaceae J. G. Agardh. Monocots. 1/1 W. Medit. Perenn. tufted herb, with a short sympodial rhiz. emitting fibr. roots. Stems rush-like, stiff, slender, usually lfless. Ls. reduced to long brownish radical sheaths (rarely bearing a short lamina). Fls. 1–2(–3) terminal, bòrne slightly obliquely, sessile

among imbr. paleac. bracts, infundib., blue or white, marcescent. P (3 + 3), petaloid, unguic.; A 6, adn. to base of P, fil. filif., anth. pelt.; G (3), loc. 1-ov., ov. amphitr.; style filif., shortly trifid. Caps. membr., seeds ovoid, with black crustaceous finely shagreened testa; embryo straight. Only genus: *Aphyllanthes*. Possibly some connection with *Xanthorrhoeaceae*.

Aphyllanthes L. Aphyllanthaceae. 1 Portugal to Italy, N. Afr., *A. monspelien-*
Aphyllarum S. Moore. Araceae. 1 Brazil. [*sis* L.
Aphylleia Champ. = Sciaphila Blume (Triuridac.).
Aphyllocalpa Cav. = Osmunda L. (Osmundac.).
Aphyllocaulon Lag. = Gerbera L. ex Cass. (Compos.).
Aphyllocladus Wedd. = Hyalis D. Don (Compos.).
Aphyllodium Gagnep. (~ Hedysarum L.). Leguminosae. 1 trop. As., Austr.
Aphyllon Mitch. = Orobanche L. (§ Gymnocaulis Nutt.) (Orobanchac.).
Aphyllorchis Blume. Orchidaceae. 20 SE. As., Formosa, Indomal.
Aphyteia L. = Hydnora Thunb. (Hydnorac.).
***Apiaceae** Lindl.: see **Umbelliferae** Juss. (nom. altern.).
Apiastrum Nutt. Umbelliferae. 2 N. Am.
Apicra *sensu* Haworth = Astroloba Uitew. (Liliac.).
Apicra Willd. = Haworthia Duval (Liliac.).
Apilia Rafin. = Fraxinus L. (Oleac.).
Apinagia Tul. Podostemaceae. 50 trop. S. Am.
Apinella [Neck.] Kuntze = Trinia Hoffm. (Umbellif.).
Apinus Neck. = Pinus L. (Pinac.).
Apiocarpus Montr. = ? Harpullia Roxb. (Sapindac.).
Apiopetalum Baill. Araliaceae. 4 New. Caled.
Apios Boehm. = Glycine L. (Legumin.).
***Apios** Medik. Leguminosae. 10 E. As., N. Am. *A. americana* Medik. (*A. tuberosa* Moench) is a climber with tuberous base. The keel of the fl. forms a tube which bends up and rests against a depression in the standard. When liberated by insects the tension of the keel makes it spring downwards, coiling up more closely, and causing the essential organs to emerge at the apex.
Apiospermum Klotzsch = Pistia L. (Arac.).
Apirophorum Neck. = Pyrus L. (Rosac.).
Apista Blume = Podochilus Blume (Orchidac.).
Apium L. Umbelliferae. 1, *A. graveolens* L., the celery, Eur. to India, N. &
Apivea Steud. = Aiouea Aubl. (Laurac.). [S. Afr.
Aplactis Rafin. = Solidago L. (Compos.).
Aplanodes Marais. Cruciferae. 2 S. Afr.
Aplarina Rafin. = Euphorbia L. (Euphorbiac.).
Aplectra Rafin. = Aplectrum (Nutt.) Torr. (Orchidac.).
Aplectrocapnos Boiss. & Reut. = Sarcocapnos DC. (Fumariac.).
Aplectrum Blume = Anplectrum A. Gray (Melastomatac.).
Aplectrum (Nutt.) Torr. Orchidaceae. 1 temp. N. Am.
Apleura Phil. Umbelliferae. 1 Chile.
Aplexia Rafin. = Leersia Sw. (Gramin.).
Aplilia Rafin. = Fraxinus L. (Oleac.).
Aplina Rafin. = Staehelina L. (Compos.).
Aploca Neck. = Periploca L. (Periplocac.).

APONOGETON

Aponogeton Hill = Zannichellia L. (Zannichelliac.).

*Aponogeton L. f. Aponogetonaceae. 30 palaeotrop., & S. Afr.

*Aponogetonaceae J. G. Ag. Monocots. 1/30 palaeotrop., & S. Afr. Water pls. with sympodial tuberous rhiz. and basal ls., usu. floating. Submerged ls. occur in some, e.g. *A.* (*Ouvirandra*) *fenestralis* (Poir.) Hook. f. The whole tissue between the veins breaks up as the l. grows, leaving a network of veins with holes between. The interior does not contain the usual intercellular spaces. The ♂ reg. fls. project above the water in spikes, sometimes divided longitudinally into 2 or 3; spathe early thrown off. P usu. 2, sometimes 3 or even 1, as in the much cult. *A. distachyon* L. f. (Cape pondweed), where it is attached by a broad base, and looks like a br. In this sp. A ∞, G 3–6, but usu. A 3 + 3, G 3, with 2 or ∞ ov. in each, basal, anatr., erect. Fr. of 3–6 free leathery dehisc. follicles. Embryo straight. Exalb. Only genus: *Aponogeton*. Distinguished from *Potamogetonaceae* by coloured P and straight embryo.

Apopetalum Pax = Brunellia Ruiz & Pav. (Brunelliac.).

Apophragma Griseb. = Curtia Cham. & Schlechtd. (Gentianac.).

Apophyllum F. Muell. Capparidaceae. 1 NE. Austr.

Apoplanesia C. Presl. Leguminosae. 1 Mex., 1 Venez.

Apopleumon Rafin. = Ipomoea L. (Convolvulac.).

Aporanthus Bromf. = Trigonella L. (Legumin.).

Aporetia Walp. = seq.

Aporetica J. R. & G. Forst. = Allophylus L. (Sapindac.).

Aporocactus Lem. Cactaceae. 6 Mex.

Aporocereus Frič & Kreuzinger = Aporocactus Lem. (Cactac.).

× **Aporodisocactus** Hort. Cactaceae. Gen. hybr. (Aporocactus × Disocactus).

× **Aporoheliocereus** Hort. Cactaceae. Gen. hybr. (Aporocactus × Heliocereus).

× **Aporophyllum** Hort. Cactaceae. Gen. hybr. (Aporocactus × Epiphyllum).

Aporosa Bl. (sphalm.) = Aporusa Bl. (Euphorbiac.).

Aporosella Chodat (~ Cicca L.). Euphorbiaceae. 1 Mex. to N. trop. S. Am., 1 Paraguay.

Aporostylis Rupp & Hatch. Orchidaceae. 1 N.Z.

× **Aporotrichocereus** Hort. Cactaceae. Gen. hybr. (Aporocactus × Tricho-

Aporrhiza Radlk. Sapindaceae. 6 trop. Afr. [cereus].

Aporuellia C. B. Clarke = Pararuellia Bremek. + Dipteracanthus Nees emend. Bremek. (Acanthac.).

Aporum Blume = Dendrobium Sw. (Orchidac.).

Aporusa Blume corr. Bl. Euphorbiaceae. 75 Indomal. to Solomon Is.

Aposeris Neck. = Hyoseris L. (Compos.).

Apostasia Blume. Orchidaceae (~ Apostasiac.). 10 trop. Asia, Malaysia, Austr.

*Apostasiaceae Bl. (~ Orchidaceae). Monocots. 3/20 SE. As., Indomal., trop. Austr. Terr. rhiz. herbs; ls. plicately nerved. Infl. rac., erect. P 3 + 3, A 3–2 (united at base and with style), Ḡ (3), ovules ∞, axile, anatr. Caps. Perhaps related also to *Hypoxidaceae*. Genera: *Apostasia, Adactylus, Neuwiedia*.

Apotaenium K.-Pol. = Conopodium K. Koch (subgen. Neoconopodium K.-Pol.) (Umbellif.).

Apoterium Blume = Calophyllum L. (Guttif.).

Apoxyanthera Hochst. = Raphionacme Harv. (Asclepiadac.).
Apozia Willd. ex Benth. = Micromeria Benth. (Labiat.).
Appella Adans. = Premna L. (Verbenac.).
Appendicula Blume. Orchidaceae. 100 trop. Asia, Polynesia.
Appendiculana Kuntze = seq.
Appendicularia DC. Melastomataceae. 1 Guiana.
Appunettia Good. Rubiaceae. 1 Angola.
Appunia Hook. f. Rubiaceae. 10 C. & trop. S. Am.
Apradus Adans. = Arctopus L. (Umbellif.).
Aprella Steud. = Asprella Schreb. = Leersia Sw. (Gramin.).
Aprevalia Baill. = Delonix Rafin. (Legumin.).
Apsanthea Jord. = Scilla L. (Liliac.).
Apseudes Rafin. = Peucedanum L. (Umbellif.).
Aptandra Miers. Olacaceae. 3 trop. S. Am.; 1 trop. W. Afr.
Aptandraceae Van Tiegh. = Olacaceae–Aptandreae Engl.
Aptandropsis Ducke. Olacaceae. 2 Brazil.
Aptenia N. E. Brown. Aïzoaceae. 1 S. Afr.
Apteranthera C. H. Wright. Amaranthaceae. 1 Mascarene Is. (Aldabra).
Apteranthes Mik. = Boucerosia Wight & Arn. (Asclepiadac.).
Apteria Nutt. Burmanniaceae. 3 warm Am., W.I.
Apterigia Galushko. Cruciferae. 1 Cauc.
Apterocaryon (Spach) Opiz = Betula L. (Betulac.).
Apteron Kurz = Ventilago Gaertn. (Rhamnac.).
Apteropteris Copel. Hymenophyllaceae. 1 Tasm., N.Z.
Apterygia Baehni (~ Bumelia Sw.). Sapotaceae. 1(?) trop. S. Am.
Apteuxis Griff. = Pternandra Jack (Melastomatac.).
Aptilon Rafin. = Serinia Rafin. (Compos.).
Aptosimum Burchell. Scrophulariaceae. 42 trop. & S. Afr.
Aptotheca Miers = Forsteronia G. F. W. Mey. (Apocynac.).
Apuleia Gaertn. = Berkheya Ehrh. (Compos.).
*****Apuleia** Mart. Leguminosae. 2 Brazil. *A. praecox* Mart. excellent timber.
Apurimacia Harms. Leguminosae. 4 N. Andes.
× **Apworthia** Poelln. Liliaceae. Gen. hybr. (Apicra × Haworthia).
Aquartia Jacq. = Solanum L. (Solanac.).
*****Aquifoliaceae** Bartl. Dicots. 2/400 trop. & temp. Trees or shrubs with alt.
or rarely opp. simple usu. coriac. stip. ls. (stips. usu. small or caduc.). Fls. reg.,
♀ or ♂ ♀, in axill. or extra-axill. cymes, or fascic., rarely rac. or sol. Usu.
K 4, imbr. (sometimes obsol.); C (4), rarely 4, imbr., usu. rounded; A 4,
alternipet., usu. adnate to C at base, anth. intr. (in ♀ fl. stds. 4, sometimes
petaloid); disk 0; G̲ (4), with sessile capit. or discoid stig., and 1 (v. rarely
2 collat.) axile pend. anatr. (rarely orthotr. or ± campylotr.) ov. per loc.;
oligomery or pleiomery, esp. of G (up to 22-merous), freq. occurs; in ♂ fl.
pistillode present. Fr. a 2–∞-pyrened drupe; seeds with copious endosp.
Genera: *Ilex*, *Nemopanthus*. Very close to *Celastrac.*
Aquifolium Mill. = Ilex L. (Aquifoliac.).
*****Aquilaria** Lam. Thymelaeaceae. 15 S. China, SE. As., Indomal. The wood
of *A. agallocha* Roxb. (*calambac*, aloe-wood, eagle-wood), in about 8% of the
trees, is saturated with resin (*agar*), used in India as a drug and perfume.

AQUILARIACEAE

Aquilariaceae ('-rineae') R.Br. ex DC. = Thymelaeaceae–Aquilarioïdeae Domke.

Aquilariella Van Tiegh. = Aquilaria Lam. (Thymelaeac.).

Aquilegia L. Ranunculaceae. 100 N. temp. Pets. with long spurs secreting honey (cf. *Delphinium*). Fl. protandrous, visited by humble-bees. Sta. often 50 or more, in whorls of 5.

Aquilicia L. = Leea Royen ex L. (Leeac.).

Aquilina Bub. = Aquilegia L. (Ranunculac.).

Arabidella (F. Muell.) O. E. Schulz. Cruciferae. 6 Austr.

Arabidium Spach = Arabis L. (Crucif.).

Arabidopsis (DC.) Heynh. Cruciferae. 13 N. Am., temp. Euras. to E. Afr.

Arabis Adans. = Iberis L. (Crucif.).

Arabis L. Cruciferae. 120 temp. Euras., Medit., trop. Afr. mts., N. Am.

Arabisa Reichb. = praec.

*Araceae Juss. Monocots. 115/2000 trop. & temp. (92 % trop.). Many types of veg. habit—herbs large and small, with aerial stems, tubers or rhiz., climbing shrubs, climbing epiph., marsh pls., one floating water pl. (*Pistia*), etc. In a few *Pothoïdeae* the stem is monopodial, but in most *A*. it is sympodial. Each joint of the sympodium begins as a rule with one or more scale ls. before bearing fol. ls. Accessory (collateral) buds often found in the leaf axils. Sometimes, as in *Anthurium, Philodendron*, etc., the axillary shoot is adnate to the main axis for some distance (cf. *Solanaceae, Zostera*, etc.). The buds usually appear in the l. axils, but often get pushed to one side, and sometimes (e.g. *Pothos*) break through the leaf-bases as in some *Commelinac., Equisetum*, etc.

Ls. are of many types. Pinnately and palmately divided ls. are frequent, but development is not like that of such leaves in Dicots. Holes develop in the ls. of *Monstera* spp. See *Monstera, Rhaphidophora, Philodendron, Helicodiceros, Dracontium, Zamioculcas*, etc.

Roots adv. and mostly formed above ground in the larger forms. Two types of aerial r.—climbing and absorbent. The former, like ivy, insensitive to gravity, show great negative heliotropism; they cling closely to the support and force their way into the crevices. The latter, insensitive to light, respond markedly to gravity; they grow down to the soil and enter it, branching out and taking up nourishment.

The larger trop. *A.* show interesting stages in the development of epiphytism. The climbing forms grow to considerable size and form longer and longer aerial r. as they grow upwards. The original r. at the base thus become of less and less importance and they often die away, together with the lower end of the stem, so that the plant thus becomes an epiph. Of course, as it still obtains its water etc. from the soil, it is not an epiph. in the sense that, e.g., many Orchids or *Bromeliaceae* are such, and it is evident that, if this method of becoming epiph. were the only one found in the family, these plants could with no more justice be classed as true epiph. than the ivy which may often be seen in the 'bowls' of pollard willows in Europe, and which has come there by climbing up the trunk and dying away below. It is found, however, that some spp. of *Philodendron, Pothos*, etc. are able to commence life as epiphytes. The fleshy fr. is eaten by birds and the seed dropped on a lofty branch. The seedling forms clasping r. and dangling aerial r. which grow steadily down to the soil, even if it be 30 m. or more away. It is hardly possible to suppose that these true epiph. spp. have been evolved in any other way than from former climbing spp. Lastly, some spp. of *Anthurium*, etc. are true epiph. without connection with the soil (e.g. *A. huegelii* Schott = *A. hookeri* Kunth); they have clasping r. and also absorbent r. which ramify amongst the humus collected by the pl. itself. The aerial r. of some *A.* possess a velamen like Orchids. The ls. of *Philodendron cannifolium* Schott have swollen petioles full of large intercellular spaces lined with mucilage. When rain falls these fill with water and act as reservoirs.

Fls. without brs., usu. massed together on a cylindrical spadix enclosed in a large spathe; the spadix usu. terminates a joint of the sympodium (the 'continuation' bud is generally in the axil of the l. next but one before the spathe), so that there is only one formed each year. Fl. ♀ or ♂ ♀, usu. monoec. (dioec. in *Arisaema*), with or without P. Sta. typically 6 but usually fewer (down to 1), often united into a synandrium (e.g. *Colocasia*, *Spathicarpa*); in *Ariopsis* the synandria again united to one another. Stds. often present, and these also may be fused into a synandrodium as in *Colocasia*. G with much variety of structure; frequently reduced to 1 cpl. Berry. Outer integument of seed often fleshy. Endosperm or none.

Fls. usu. protog. (even when monoec.). In many gen. (incl. most Eur.) the smell is disagreeable and attracts carrion flies as pollen carriers (see *Arum*, *Dracunculus*, *Helicodiceros*, etc.).

Many *A.* contain latex, which is usually poisonous but is dispelled by heat. The rhizomes of many spp. contain much starch and are used as food (*Caladium*, *Colocasia*, *Arum*, etc.).

The grouping of the *A.* is very difficult and account has to be taken of histological as well as external characters.

Classification and chief genera (after Engler):

I. **Pothoïdeae** (land pls.; no latex or raphides; ls. 2-ranked or spiral; lat. veins of 2nd and 3rd order netted; fls. usu. ♀; ov. anatr. or amphitr.): *Pothos*, *Anthurium*.

II. **Monsteroïdeae** (land pls.; no latex; raphides; lat. veins of 3rd, 4th, and sometimes 2nd orders netted; fl. ♀, usu. naked; ov. anatr. or amphitr.): *Rhaphidophora*, *Monstera*, *Spathiphyllum*, *Epipremnum*.

III. **Calloïdeae** (land or marsh pls.; latex; fl. usu. ♀; ov. anatr. or orthotr.; ls. never sagittate, usu. net-veined): *Symplocarpus*, *Calla*.

IV. **Lasioïdeae** (land or marsh pls.; latex; fl. ♀ or ♂ ♀; ov. anatr. or amphitr.; seed usu. exalbum.; ls. sagittate, often much lobed, net-veined): *Dracontium*, *Amorphophallus*.

V. **Philodendroïdeae** (land or marsh pls.; latex; fl. naked, unisex.; ov. anatr. or orthotr.; seed usu. album.; ls. usu. ‖-veined): *Philodendron*, *Zantedeschia*.

VI. **Colocasioïdeae** (land or marsh pls.; latex; fl. naked, unisex.; sta. in synandria; ov. orthotr. or anatr.; seed album. or not; ls. net-veined): *Remusatia*, *Colocasia*, *Alocasia*, *Xanthosoma*.

VII. **Aroïdeae** (land or marsh pls.; latex; ls. various, net-veined; stems mostly tuberous; fl. unisex., usu. naked; sta. free or in synandria;

ARACHIDNA

ov. anatr. or orthotr.; seed album.): *Spathicarpa, Arum, Dracunculus, Helicodiceros, Arisaema.*

VIII. **Pistioïdeae** (swimming pls.; no latex; fl. unisex., naked; ♂ fls. in a whorl, ♀ sol.): *Pistia* (only genus).

For further details of this most interesting order, see Engler in *Nat. Pfl.* and *Pfl.-R.*, from which much of the above is abridged.

Arachidna Boehm. = seq.

Arachis L. Leguminosae. 15 Brazil, Paraguay. *A. hypogaea* L. (ground-, earth-, or pea-nut), largely cult. in warm regions for its seeds, which are ed. and when pressed yield one of the many oils used in place of olive oil. The fl. after fert. bends down (cf. *Cymbalaria*) and the elongation of its stalk forces the young pod under ground, where it ripens.

Arachna Nor. = Hedychium Koen. (Zingiberac.).

Arachnabenis Thou. (uninom.) = *Habenaria arachnoïdes* Thou. (Orchidac.).

Arachnanthe Blume = Arachnis Blume (Orchidac.).

Arachne (Endl.) Pojark. = Leptopus Decne (Euphorbiac.).

Arachne Neck. = Breynia J. R. & G. Forst. (Euphorbiac.).

Arachnimorpha Desv. = Rondeletia L. (Rubiac.).

Arachniodes Bl. (*Rumohra sensu* Ching & Copel., excl. *R. aspidioïdes* Raddi). Aspidiaceae. 30 trop. & subtrop. Asia. (Tindale, *Contr. N.S.W. Nat. Herb.* **3**: 88–92, 1961; & *op. cit. Flora Ser.* **208–11**: 55–8, 1961.)

Arachnis Blume. Orchidaceae. 7 Hainan, Indoch., W. Malaysia.

Arachnites F. W. Schmidt = Ophrys L. + Chamorchis Rich. + Aceras R.Br.

Arachnit[id]aceae C. Muñoz = Corsiaceae Becc.   [(Orchidac.).

*****Arachnitis** Phil. Corsiaceae. 1 Chile.

Arachnocalyx Compton (~ Eremia D. Don). Ericaceae. 1 S. Afr.

Arachnodendris Thou. (uninom.) = *Dendrobium arachnites* Thou. = *Aëranthes arachnites* (Thou.) Lindl. (Orchidac.).

Arachnodes Gagnep. = Phyllanthodendron Hemsl. (Euphorbiac.).

× **Arachnoglottis** hort. Orchidaceae. Gen. hybr. (iii) (Arachnis × Trichoglottis).

Arachnopogon Berg ex Haberle = Hypochoeris L. (Compos.).

× **Arachnopsis** hort. Orchidaceae. Gen. hybr. (iii) (Arachnis × Phalaenopsis).

Arachnospermum Berg ex Haberle = Hypochoeris L. (Compos.).

Arachnospermum F. W. Schmidt = Podospermum DC. (Compos.).

× **Arachnostylis** hort. Orchidaceae. Gen. hybr. (iii)(Arachnis × Rhynchostylis).

Arachnothrix Walp. = seq.

Arachnothryx Planch. = Rondeletia L. (Rubiac.).

Arachus Medik. = Vicia L. (Legumin.).

Aracium [Neck.] Monnier = Crepis L. (Compos.).

Araeococcus Brongn. Bromeliaceae. 3 C. & trop. S. Am.

Aragallus Neck. = Astragalus L. (Legumin.).

Aragoa Kunth. Scrophulariaceae. 5 Andes.

Aragoaceae D. Don = Scrophulariaceae–Digitaleae Benth.

Aragus Steud. = Aragallus Neck. = Astragalus L. (Legumin.).

*****Araiostegia** Copel. Davalliaceae. 12 NE. India to S. China, Malaysia. (Included in *Leucostegia, q.v.*, by Beddome, and in C. Chr. *Ind. Fil.*)

Aralia L. Araliaceae. 35 Indomal., E. As., N. Am.

ARAUCARIACEAE

*Araliaceae Juss. Dicots. 55/700, chiefly trop., esp. Indomal., trop. Am. Usu. trees and shrubs, sometimes palm-like; some twiners, some, e.g. ivy, root-climbers. Resin-passages. Ls. alt., rarely opp. or whorled, often large and cpd., with small stips.; seedling ls. often simpler than those of mature pls. Fl. small, in umbels or heads often massed into cpd. infls., ♀, reg., usu. epig., usu. 5 (3–∞)-merous. K usu. 5, very small; C (4–)5(–10), rarely (5), often valv.; A 5 (3–∞); G̲ (5) (1–∞), rarely ½-inf. or G̲, 5-loc. with 1 anatr. pend. ov. in each, micropyle facing outwards; styles free or united in great variety. Usu. drupe with as many stones as cpls. Embryo small in rich endosperm. *Tetrapanax*, *Panax*, and others are economically important.

Classification and chief genera (after Engler):
1. SCHEFFLEREAE (C valv.): *Schefflera, Fatsia, Tetrapanax, Hedera.*
2. ARALIËAE (C ± imbr., sessile with broad base): *Aralia, Panax.*
3. MACKINLAYEAE (C valv., shortly clawed): *Mackinlaya.*

Aralidium Miq. Araliaceae. 1 Indoch., W. Malaysia. Dioec. Fr. 1-seeded. Endosp. ruminate.

*Araliopsis Engl. Rutaceae. 3 trop. W. Afr. Drupe 4-stoned.

Araliopsis Kurz = Euaraliopsis Hutch. (Araliac.).

Araliorhamnus H. Perrier. Rhamnaceae. 2 Madag.

Aralodendron Oerst. ex Marchal = Oreopanax Decne & Planch. (Araliac.).

× **Aranda** hort. Orchidaceae. Gen. hybr. (iii) (Arachnis × Vanda).

× **Arandanthe** hort. (v) = praec.

Aranella Barnhart ex Small = Utricularia L. (Lentibulariac.).

× **Aranthera** hort. Orchidaceae. Gen. hybr. (iii) (Arachnis × Renanthera).

Arapabaca Adans. = Spigelia L. (Spigeliac.).

Arariba Mart. = Sickingia Willd. = Simira Aubl. (Rubiac.).

Ararocarpus Scheff. = Meiogyne Miq. (Annonac.).

Araschcoolia Sch. Bip. = Geigeria Griesselich (Compos.).

Araucaria Juss. Araucariaceae. 18 New Guinea, E. Austr., N.Z., Norfolk I., New Caled. (many endemics), and S. Brazil to Chile. In Sect. *Colymbea* (ls. broad, fr. cpls. not winged) *A. araucana* (Mol.) K. Koch (*A. imbricata* Pav., monkey puzzle, Chile), with ed. seed; *A. angustifolia* (Bertol.) Kuntze (*A. brasiliana* A. Rich., Brazilian pine, abundant in S. Braz.); *A. bidwillii* Hook. (bunya-bunya pine, Queensland), and others. In Sect. *Intermedia* (young ls. acic., mature ls. 5–10 cm. long, scales winged, cots. epigeal) *A. klinkii* Lauterb. (N. Guin.), almost 90 m. high. In Sect. *Eutacta* (needle ls., scales winged) *A. heterophylla* (Salisb.) Franco (*A. excelsa* (D. Don) R.Br., Norfolk I. pine), *A. cunninghamii* D. Don (hoop pine, E. Austr.), etc. All have useful timber.

Araucariaceae Henk. & Höchst. Gymsp. (Conif.). 2/38 S. hemisph. (exc. Afr.) to Indoch. & Philipp. Trees with broad, or acicular and compressed, sometimes pungent ls. ♂ infl. large, catkin-like, axill. or term. on short shoots; A ∞, spirally arranged, fil. expanded to form a tough anther-scale; sporangia ∞, free, lin., borne on lower surface of scale. ♀ cones term. on short shoots, fr. cones large, ±, glob., disintegrating; carp. ∞, spirally imbr., usu. broad, sometimes narrow-conical, sometimes winged, terminally thickened, in *Araucaria* with abrupt sharp apex; ligule present in *Araucaria*, ± adnate to carpel, absent in *Agathis*; ovule 1, free or (in *Araucaria*) immersed in the ligule; nucellus free; cots. 2(–4). Genera: *Araucaria, Agathis.*

87

ARAUCASIA

Araucasia Benth. & Hook. f. (sphalm.) = Arausiaca Blume = Orania Zipp. (Palm.).

Araujia Brot. Asclepiadaceae. 2–3 S. Am.

Arausiaca Blume = Orania Zipp. (Palm.).

Arbulocarpus Tennant (= ? Staëlia Cham. & Schlechtd.). Rubiaceae. 1 trop. Afr.

Arbut[ac]eae J. G. Agardh = Ericaceae Juss.

Arbutus L. Ericaceae. 20 N. & C. Am., W. Eur., Medit., W. As. Small trees or shrubs. Fr. a dry berry. *A. menziesii* Pursh (madroña laurel, N. Am.), useful wood. *A. unedo* L. (strawberry tree), Medit. & SW. Ireland.

Arcangelina Kuntze = Kralikia Coss. & Dur. (Gramin.).

Arcangelisia Becc. Menispermaceae. 3 Malaysia.

Arcaula Rafin. = Lithocarpus Bl. (Fagac.).

Arceuthobiaceae Van Tiegh. = Loranthaceae–Viscoïdeae–Arceuthobieae Engl.

*Arceuthobium M. Bieb. Viscaceae (~ Loranthaceae). 15 N. Am., W.I., Medit., Himal., China, Malay Penins., Java.

Arceuthobium Griseb. = Dendrophthora Eichl. (Viscac.).

Arceutholobium Steud. (sphalm.) = Arceuthobium M. Bieb. (Viscac.).

Arceuthos Antoine & Kotschy. Cupressaceae. 1 Greece, As. Minor, Syria.

Archaeocarex Börner = Schoenoxiphium Nees (Cyperac.).

Archaetogeron Greenm. (sphalm.) = Achaetogeron A. Gray (Compos.).

Archangelica v. Wolf (~ Angelica L.). Umbelliferae. 12 N. temp. The petiole of *A. officinalis* (Moench) Hoffm. is offic., and is also used in confectionery.

Archangiopteris Christ & Giesenh. Angiopteridaceae. 10 SW. China, Formosa. Ls. once pinnate; sori linear, of 80–160 sporangia. (Ching, *Acta Phytotax. Sin.* **7**: 212–24, pl. 49–52, 1958.)

Archboldia E. Beer & H. J. Lam. Verbenaceae. 1 New Guinea.

Archboldiodendron Kobuski. Theaceae. 2 New Guinea.

Archemera Rafin. = seq.

Archemora DC. = Tiedemannia DC. (Umbellif.).

Archeria Hook. f. Epacridaceae. 6 Tasmania, N.Z.

Archibaccharis Heering. Compositae. 25 Mex., C. Am.

Archichlamydeae Engl. A subclass of Dicots. comprising all basically polypetalous or apetalous fams.

Archiclematis (Tamura) Tamura. Ranunculaceae. 1 Himal.

Archidendron F. Muell. (~ Pithecellobium Mart.). Leguminosae. 33 New Guinea, trop. Austr. G 2–15 free carp.

Archimedea Leandro = Lophophytum Schott & Endl. (Balanophorac.).

Archimedia Rafin. = Iberis L. (Crucif.).

Archiphyllum Van Tiegh. = Myzodendron Soland. ex DC. corr. R.Br. (Myzodendrac.).

Archiphysalis Kuang (~ Physalis, Chamaesaracha). Solanaceae. 3 E. As.

Archirhodomyrtus (Niedenzu) Burret = Rhodomyrtus Reichb. (Myrtac.).

Architaea Mart. = Archytaea Mart. (Bonnetiac.).

Archontophoenix H. Wendl. & Drude. Palmae. 3 E. Austr.

Archytaea Mart. Bonnetiaceae. 2 Brazil, Guiana.

Arcion Bub. = Arctium L. (Compos.).

Arcoa Urb. Leguminosae. 1 Haiti.

Arctagrostis Griseb. Gramineae. 6 arctic Am. & Euras.

Arcteranthis Greene (~ Oxygraphis Bunge). Ranunculaceae. 1 N. Am.
Arcterica Coville. Ericaceae. 1 NE. Sib., Sakhalin, Kamch., Japan.
Arctio(n) Lam. = Berardia Vill. (Compos.).
Arctiodracon A. Gray = Lysichitum Schott (Arac.).
Arctium Lam. = Berardia Vill. (Compos.).
Arctium L. Compositae. 5 palaeotemp. The invol. brs. become hooked and woody after the fls. wither and by clinging to fur etc. aid in dispersing the fr.
Arctocalyx Fenzl = Solenophora Benth. (Gesneriac.).
Arctocarpus Blanco = Artocarpus J. R. & G. Forst. (Morac.).
Arctocrania Nakai = Chamaepericlymenum Hill (Cornac.).
Arctogeron DC. (~ Aster L.). Compositae. 1 Siberia.
Arctomecon Torr. & Frém. Papaveraceae. 3 SW. U.S.
Arctophila Rupr. = Poa L. + Colpodium Trin. (Gramin.).
Arctopus L. Umbelliferae. 3 S. Afr.
Arctostaphyl[ac]eae J. G. Agardh = Ericaceae Juss.
*****Arctostaphylos** Adans. Ericaceae. 70 W. N. & C. Am.; 1 N. temp. & circumpolar. The fls. of *A. uva-ursi* (L.) Spreng. appear as soon as the snow melts. *A. pungens* Kunth (*manzanita,* Calif.), orn. wood.
Arctotheca Wendl. Compositae. 4 S. Afr., 2 Austr.
Arctotidaceae Bessey = Compositae–Arctotideae Cass.
Arctotis L. Compositae. 65 trop. & S. Afr., Austr. Latex!
Arctottonia Trelease. Piperaceae. 3 Mex., C. Am.
Arctoüs Niedenzu. Ericaceae. 3–4 N. circumpolar, N. temp. mts.
Arculus Van Tiegh. = Amylotheca Van Tiegh. (Loranthac.).
Arcynospermum Turcz. Malvaceae (inc. sed.). 1 Mex.
Arcyosperma O. E. Schulz. Cruciferae. 1 Himal.
Arcyphyllum Ell. = Rhynchosia Lour. (Legumin.).
Arcypteris Underw. Aspidiaceae. 4 Indoch., Malaysia. Very near *Pleocnemia, q.v.*
Arcythophyllum Schlechtd. = seq.
Arcytophyllum Willd. ex Schult. & Schult. f. Rubiaceae. 30 mts. trop. Am.
Ardernia Salisb. = Ornithogalum L. (Liliac.).
Ardinghelia Comm. ex A. Juss. = Phyllanthus L. (Euphorbiac.).
Ardinghella Thou. = Ochrocarpos Thou. = Mammea L. (Guttif.).
Ardisia Gaertn. = Cyathodes Labill. (Epacridac.).
*****Ardisia** Sw. Myrsinaceae. 400 warm countries.
Ardisiaceae Bartl. = Myrsinaceae R.Br.
Ardisiandra Hook. f. Primulaceae. 3 mts. trop. Afr.
Arduina Adans. = Kundmannia Scop. (Umbellif.).
Arduina Mill. ex L. = Carissa L. (Apocynac.).
Areca L. Palmae. 54 Indomal., Solomon & Bismarck Is., N. Austr. *A. catechu* L. largely cult. in trop. As. for its seeds (areca or betel nuts). The infl. is below the oldest living l., monoec., with the ♀ fls. at the bases of the twigs, the ♂ above. The seed, about as big as a damson, is cut into slices and rolled up in a leaf of betel pepper (*Piper betle*) with a little lime. When chewed, it turns the saliva bright red; it acts as a stimulus upon the digestive organs, and is supposed by the natives (who use it habitually) to be a preventive of dysentery.
*****Arecaceae** Schultz–Schultzenst.: see **Palmae** Juss. (nom. altern.).
Arecastrum Becc. (~ Syagrus Mart.). Palmae. 1 Brazil.

ARECHAVALETAIA

Arechavaletaia Spegazz. Flacourtiaceae. 1 Uruguay.

Aregelia Kuntze. Bromeliaceae. 40 trop. Am.

Aregelia Mez = Neoregelia L. B. Smith (Bromeliac.).

Arelina Neck. = Berkheya Ehrh. (Compos.).

***Aremonia** Neck. ex Nestler. Rosaceae. 1 SE. Eur.

Arenaria Adans. = Sagina L. (Caryophyllac.).

Arenaria L. Caryophyllaceae. 250 N. temp.

Arenbergia Mart. & Gal. = Eustoma Salisb. (Gentianac.).

***Arenga** Labill. Palmae. 11 Indomal. (exc. N. Guin.), Caroline & Christmas Is. Like *Caryota*, but spadix unisexual; sta. ∞, cpls. 3. *A. saccharifera* Labill. (*gomuti* palm) cult. for sugar (*jaggery*), obtained by wounding the young infl. and evaporating the sap. A kind of sago is obtained from the pith by washing and granulating. The tree flowers when mature, infls. appearing in descending order till it dies. An excellent fibre is obtained from the leaf-sheaths.

Arenifera Herre. Aïzoaceae. 1 S. Afr.

Arequipa Britton & Rose = Borzicactus Riccob. (Cactac.).

Arequipiopsis Kreuz. & Buining = Borzicactus Riccob. (Cactac.).

Arethusa L. Orchidaceae. 1 temp. N. Am.

Arethusantha Finet = Cymbidium (Orchidac.).

Aretia Link = Primula L. (Primulac.).

Aretia L. = Androsace L. (Primulac.).

Aretiastrum Spach. Valerianaceae. 7 Andes, Magell.

Arfeuillea Pierre ex Radlk. Sapindaceae. 1 Siam, Indoch.

Argania Roem. & Schult. Sapotaceae. 1 Morocco, *A. spinosa* (L.) Skeels. The pressed seeds yield argan oil, used like olive oil; the timber is hard and durable; the fr. is eaten by cattle.

Argelasia Fourr. = Genista L. (Legumin.).

Argelia Decne = Solenostemma Hayne (Asclepiadac.).

Argemone L. Papaveraceae. 10 W. & E. U.S., Mex., W.I.

Argenope Salisb. = Narcissus L. (Amaryllidac.).

Argentacer Small = Acer L. (Acerac.).

Argentina Hill = Potentilla L. (Rosac.).

Argeta N. E. Brown. Aïzoaceae. 1 S. Afr.

Argithamnia Sw. = Argythamnia P.Br. (Euphorbiac.).

Argocoffea (Pierre ex De Wild.) Lebrun = Coffea L. (Rubiac.).

Argocoffeopsis Lebrun = Coffea L. (Rubiac.).

Argolasia Juss. = Lanaria Ait. (Haemodorac.).

Argomuellera Pax. Euphorbiaceae. 11 trop. Afr.

Argophilum Blanco = Aglaia Lour. (Meliac.).

Argophyllum Blanco = praec.

Argophyllum J. R. & G. Forst. Escalloniaceae. 11 trop. Austr., New Caled.

Argopogon Mimeur = Ischaemum L. (Gramin.).

Argorips Rafin. = Salix L. (Salicac.).

Argostemma Wall. Rubiaceae. 100 trop. Afr., As.

Argostemmella Ridley. Rubiaceae. 2 Borneo.

Argothamnia Spreng. = Argythamnia P.Br. (Euphorbiac.).

Argusia Boehm. (Messerschmidia L. ex Hebenstr.). Boraginaceae. 1 coasts SE. U.S., C. Am. & W.I., 1 temp. Euras.; 1 islands of Indian & Pacif. oceans.

*Aristolochiaceae Juss. Dicots. 7/400, trop. & warm temp., except Austr. Herbs or shrubs, the latter usu. twining lianes. Ls. alt., stalked, often cordate, usu. simple, exstip. Fl. ♀, epig., reg. or zygo. P usu. (3), petaloid; A 6–36, free, or united with the style into a gynostemium (cf. Asclepiads, Orchids, etc.). G̅ (4–6); ov. ∞ in each loc., anatr., horiz. or pend. Caps. Embryo small in rich endosp. Genera: *Saruma, Asarum, Thottea, Apama, Holostylis, Aristolochia, Euglypha.* The *A.* are difficult to place in the system, but are probably connected with *Annonaceae* through *Thottea* and *Apama.* Some authors have even put them near *Dioscoreaceae,* though not monocot.

Aristomenia Vell. = Stifftia Mikan (Compos.).

Aristopetalum Willis (sphalm.) = Aïstopetalum Schltr. (Cunoniac.).

Aristotela Adans. = Othonna L. (Compos.).

Aristotela J. F. Gmel. = Aristotelia L'Hérit. (Elaeocarpac.).

Aristotelea Lour. = Spiranthes Rich. (Orchidac.).

Aristotelea Spreng. = Aristotelia L'Hérit. (Elaeocarpac.).

Aristotelia Comm. ex Lam. = Terminalia L. (Combretac.).

*Aristotelia** L'Hérit. Elaeocarpaceae. 5 E. Austr., Tasm., N.Z., Peru to Chile.

Aristoteliaceae Dum. = Elaeocarpaceae DC.

Arivela Rafin. = Polanisia Rafin. = Cleome L. (Cleomac.).

Arivona Steud. = Arjona Comm. ex Cav. (Santalac.).

× **Arizara** hort. Orchidaceae. Gen. hybr. (iii) (Cattleya × Domingoa × Epiden-

Arjona Comm. ex Cav. Santalaceae. 10 temp. S. Am.   [drum).

Arjonaceae Van Tiegh. corr. Bullock = Santalaceae–Thesiëae Reichb.

Arjonaea Kuntze = Arjona Comm. ex Cav. (Santalac.).

Arkezostis Rafin. = Cayaponia S. Manso (Cucurbitac.).

Arkopoda Rafin. = Reseda L. (Resedac.).

Armania Bert. ex DC. = Encelia Adans. (Compos.).

Armarintea Bub. = Cachrys L. (Umbellif.).

Armatocereus Backeb. Cactaceae. 11 Colombia, Ecuador, Peru.

Armeniaca Mill. (∼Prunus L.). Rosaceae. 10 temp. As.

Armeniastrum Lem. = Espadaea A. Rich. (Goetzeac.).

× **Armenoprunus** Janchen. Rosaceae. Gen. hybr. (Armeniaca × Prunus).

*Armeria** (DC.) Willd. Plumbaginaceae. 80 N. temp. & Andes. *A. maritima* (Mill.) Willd. (thrift, sea pink) is common on the coasts and in high mountain regions of Eur.; a fairly frequent phenomenon, due perhaps to similarity of conditions. Primary root perennial; each year's shoot dies down all but a short piece, on which the following year's shoot arises as an axillary branch. Infl. a capitulum of cincinni, surrounded by a whorl of bracts, the outer forming a reflexed sheath round the top of the peduncle. After fert. the K becomes a membranous funnel-like organ aiding seed distribution by wind.

Armeria Kuntze = Phlox L. (Polemoniac.).

Armeriaceae Horan. = Plumbaginaceae Juss.

Armeriastrum (Jaub. & Spach) Lindl. = Acantholimon Boiss. (Plumbaginac.).

× **Armodachnis** hort. Orchidaceae. Gen. hybr. (iii) (Arachnis × Armodorum).

Armodorum Breda. Orchidaceae. 3 India, China, Siam, Sumatra, Java.

Armola (Kirschl.) Montandon = Atriplex L. (Chenopodiac.).

Armoracia Fabr. (uninom.) = *Armoracia rusticana* Gaertn., Mey. & Scherb. (Crucif.).

ARMORACIA

Armoracia Gaertn., Mey. & Scherb. Cruciferae. 3 SE. Eur. to Sib. The thick root of *A. rusticana* Gaertn., Mey. & Scherb. (horse-radish) is a condiment.

Armouria Lewton. Malvaceae. 1 Haiti.

Arnaldoa Cabr. Compositae. 5 Peru.

Arnanthus Baehni. Sapotaceae. 1 (?) New Caled.

Arnebia Forsk. Boraginaceae. 25 Medit., trop. Afr., Himal. Some have black spots on the C, which fade as it grows older (see fam., and cf. *Diervilla*, **Arnebiola** Chiov. Boraginaceae. 1 Somalia.   [*Fumaria*, etc.).

Arnedina Reichb. (sphalm.) = Arundina Blume (Orchidac.).

Arnica Boehm. = Doronicum L. (Compos.).

Arnica L. Compositae. 32 N. temp. & arctic. Tincture of arnica is prepared

Arnicastrum Greenman. Compositae. 1 Mex.   [from all parts of the pl.

Arnicula Kuntze = Arnica L. (Compos.).

Arnocrinum Endl. & Lehm. Liliaceae. 3 SW. Austr.

Arnoglossium S. F. Gray = Plantago L. (Plantaginac.).

Arnoglossum Rafin. = Cacalia L. (Compos.).

Arnoldia Blume = Weinmannia L. (Cunoniac.).

Arnoldia Cass. = Dimorphotheca Moench (Compos.).

Arnoldoschultzea Mildbraed (nomen). Sapotaceae. 1 W. Equat. Afr. Quid?

Arnopogon Willd. = Urospermum Scop. (Compos.).

Arnoseris Gaertn. Compositae. 1, *A. pusilla* Gaertn., Eur. The bases of the invol. brs. enclose the ripe fr. (cf. *Rhagadiolus*).

Arnottia A. Rich. Orchidaceae. 4 Mascarene Is.

Arodendron Werth = Typhonodorum Schott (Arac.).

Arodes Kuntze = Aroïdes Fabr. = Zantedeschia Spreng. (Arac.).

Arodia Rafin. = Rubus L. (Rosac.).

Aroïdes Fabr. (uninom.) = Calla palustris L. (Arac.).

Aromadendron Andr. ex Steud. = Aromadendrum W. Anders. ex R.Br. = Eucalyptus L'Hérit. (Myrtac.).

Aromadendron Blume. Magnoliaceae. 3 Malay Penins., Borneo, Java.

Aromadendrum W. Anders. ex R.Br. = Eucalyptus L'Hérit. (Myrtac.).

Aromadendrum Blume = Aromadendron Bl. (Magnoliac.).

Aromia Nutt. = Amblyopappus Hook. (Compos.).

Aron Adans. = Colocasia Schott + Dracunculus Adans. (Arac.).

Arongana Choisy = Haronga Thou. (Guttif.).

Aronia Medik. = Amelanchier Lindl. (Rosac.).

Aronia Mitch. = Orontium L. (Arac.).

Aronicum Neck. ex Reichb. = Doronicum L. (Compos.).

Arophyton Jumelle. Araceae. 3 Madag.

Aropsis Rojas. Araceae. 1 temp. S. Am.

Arosma Rafin. = Philodendron Schott (Arac.).

Aroton Neck. = Croton L. (Euphorbiac.).

Arouna Aubl. = Dialium L. (Legumin.).

Arpitium Neck. = Pachypleurum Ledeb. (Umbellif.).

Arpophyllum La Llave & Lex. Orchidaceae. 2 C. & trop. S. Am., W.I.

Arrabidaea DC. Bignoniaceae. 70 trop. Am.

Arrabidaea Steud. = Cormonema Reissek (Rhamnac.).

Arracacha DC. = seq.

Arracacia Bancroft. Umbelliferae. 53 Mex. to Peru. *A. xanthorhiza* Bauer and others cult. for ed. tuberous r.

Arraschkoolia Hochst. = Araschcoolia Sch. Bip. = Geigeria Griess. (Compos.).

Arrhenachne Cass. = Baccharis L. (Compos.).

Arrhenatherum Beauv. Gramineae. 6 Eur., Medit.

Arrhenechthites Mattf. Compositae. 5 New Guinea.

Arrhostoxylum Nees. Acanthaceae. 10 trop. Am.

Arrhynchium Lindl. = Arachnis Bl. (Orchidac.).

Arrojadoa Britton & Rose. Cactaceae. 2 E. Brazil.

Arrojadoa Mattf. = seq.

Arrojadocharis Mattf. Compositae. 1 E. Brazil.

Arrowsmithia DC. Compositae. 1 S. Afr.

Arrostia Rafin. = Gypsophila L. (Caryophyllac.).

Arrozia Schrad. ex Kunth = Luziola Juss. (Gramin.).

Arrudea A. St-Hil. & Camb. = Clusia L. (Guttif.).

Arsace Fourr. = Erica L. (Ericac.).

Arsenia Nor. = Uvaria L. (Annonac.).

Arsenococcus Small = Lyonia Nutt. (Ericac.).

Arsis Lour. = Microcos L. (Tiliac.).

Artabotrys R.Br. Annonaceae. 100 + trop. Afr., Indomal. They usually climb by aid of recurved hooks, which are modified infl. axes, and thicken and lignify when they clasp. Some have ed. fr.

Artanacetum (Rzazade) Rzazade (~ Artemisia L.). Compositae. 1 Caucasus.

*****Artanema** D. Don. Scrophulariaceae. 4 trop. Afr., Indomal.

Artanthe Miq. = Piper L. (Piperac.).

Artaphaxis Mill. (sphalm.) = Atraphaxis L. (Polygonac.).

Artedia L. Umbelliferae. 1 W. As.

Artemisia L. Compositae. 400 N. temp., S. Afr., S. Am.; common on arid soil of the W. U.S., Russian steppes, etc. *A. tridentata* Nutt. and others form the ± halophytic 'sage-brush' of the SW. U.S. Fl.-heads small, inconspic., and wind-fert. (cf. *Poterium, Rheum* and *Rumex, Plantago, Thalictrum*, etc.). In *A. vulgaris* L. the marginal florets ♀, the rest ♂. Head pend.; the anther-tube projects beyond the C so that the dry powdery pollen is exposed to the wind. On the tips of the anthers are long bristles which together form a temporary pollen-holder. Afterwards the style emerges and the large hairy stigmas spread out. An interesting case of re-acquisition of a character not found in most higher flowering pls. The flavouring matter of absinthe is derived from wormwood (*A. absinthium* L.).

Artemisiastrum Rydb. (~ Artemisia L.). Compositae. 1 Calif.

Artemisiopsis S. Moore. Compositae. 1 S. trop. Afr.

Artenema G. Don = Artanema D. Don (Scrophulariac.).

Arthostema Neck. = Gnetum L. (Gnetac.).

Arthraerua (Kuntze) Schinz. Amaranthaceae. 1 SW. Afr.

Arthratherum Beauv. = Aristida L. (Gramin.).

Arthraxella Nakai = Psittacanthus Mart. (Loranthac.).

Arthraxon Beauv. Gramineae. 20 trop. Afr., Madag., Maurit., Indomal. to Jap.

Arthraxon Van Tiegh. = Arthraxella Nak. = Psittacanthus Mart. (Loranthac.).

Arthrobotrya J. Smith. Lomariopsidaceae. 3 E. Malaysia, Solomon Is.,

ARTHROCARPUM

Queensland. High climbing; ls. bipinnate. (*Teratophyllum* § *Polyseriatae* Holttum, *Gard. Bull. S.S.* **9**: 356, 1939.)

Arthrocarpum Balf. f. Leguminosae. 2 Somal., Socotra.

Arthrocereus A. Berger. Cactaceae. 4 Brazil.

Arthrochilium G. Beck = Epipactis Zinn (Orchidac.).

Arthrochilus F. Muell. = Spiculaea Lindl. (Orchidac.).

Arthrochlaena Boiv. ex Benth. = Sclerodactylon Stapf (Gramin.).

Arthrochloa Lorch = Dactyloctenium Willd. (Gramin.).

Arthrochloa Schultes = Holcus L. (Gramin.).

Arthrochortus Lowe = Lolium L. (Gramin.).

Arthroclianthus Baill. Leguminosae. 20 New Caled.

Arthrocnemum Moq. Chenopodiaceae. 20 coasts Medit. to Austr., warm

Arthrolepis Boiss. = Achillea L. (Compos.). [N. Am.

Arthrolobium Reichb. = Artrolobium Desv. = Ornithopus L. (Legumin.).

Arthrolobus Andrz. ex DC. = Rapistrum Crantz (Crucif.).

Arthrolobus Stev. ex DC. = Sterigmostemum M. Bieb. (Crucif.).

Arthrolophis [Trin.] Chiov. = Andropogon L. (Gramin.).

Arthromeris J. Sm. Polypodiaceae. 9 N. India to S. China.

Arthromischus Thw. = Paramignya Wight (Rutac.).

Arthrophyllum Blume. Araliaceae. 15 Indomal.

Arthrophyllum Boj. ex DC. = Phyllarthron DC. (Bignoniac.).

Arthrophytum Schrenk. Chenopodiaceae. 10 W. & C. As.

Arthropodium R.Br. Liliaceae. 1 Madag., 8 Austr., N.Z., New Caled.

Arthropogon Nees. Gramineae. 3 Brazil.

Arthropteris J. Sm. Oleandraceae. 20 Old World trop., Austr., N.Z. Petiole jointed to slender rhizome.

Arthrosamanea Britton & Rose. Leguminosae. 10 trop. Am.

Arthrosolen C. A. Mey. Thymelaeaceae. 15 trop. & S. Afr.

Arthrosprion Hassk. = Acacia Mill. (Legumin.).

Arthrostachya Link = Gaudinia Beauv. (Gramin.).

Arthrostachys Desv. = Andropogon L. (Gramin.).

Arthrostema Ruiz & Pav. Melastomataceae. 15 C. & W. trop. Am., W.I.

Arthrostemma DC. = Brachyotum Triana (Melastomatac.).

Arthrostemma Naud. = Pterolepis Miq. (Melastomatac.).

Arthrostygma Steud. = Petrophila R.Br. (Proteac.).

Arthrostylidium Rupr. Gramineae. 25 trop. Am., W.I. Climbing.

Arthrostylis Boeck. = Actinoschoenus Benth. (Cyperac.).

Arthrostylis R.Br. Cyperaceae. 3 Austr.

Arthrotaxis Endl. = Athrotaxis D. Don (Taxodiac.).

Arthrothamnus Klotzsch & Garcke = Euphorbia L. (Euphorbiac.).

Arthrotrichum F. Muell. = Trichinium R.Br. (Amaranthac.).

Arthrozamia Reichb. = Encephalartos Lehm. (Zamiac.).

Artia Guillaumin. Apocynaceae. 1 Malay Penins., 6 New Caled.

Artocarp[ac]eae R.Br. = Moraceae–Artocarpoïdeae Engl.

*★**Artocarpus** J. R. & G. Forst. Moraceae. 47 SE. As., Indomal. Many show good bud-protection by stips. Fls. monoec., the ♂ in pseudo-catkins, the ♀ in pseudo-heads. A multiple fr. is formed, the achenes being surrounded by the fleshy P and the common receptacle also becoming fleshy. Several spp. are

cult. all over the trop., e.g. *A. altilis* (Parkinson) Fosberg (bread-fruit) and
A. heterophyllus Lam. (*jak*). The fr. of *A. altilis* contains much starch, etc., and is
a valuable foodstuff. The flesh has somewhat the texture of bread and is often
roasted. The best cult. forms (cf. pear, banana, etc.) produce no seeds. The
jak and others are caulifloral. Timber useful.

Artorhiza Rafin. = Parmentiera Rafin. = Solanum L. (Solanac.).

Artorima Dressler & Pollard. Orchidaceae. 1 Mex.

Artrolobium Desv. (~ Coronilla L., Ornithopus L., etc.). Leguminosae.

Aruana Burm. f. = Myristica Boehm. (Myristicac.). [5 Medit.

Aruba Aubl. = Quassia L. (Simaroubac.).

Aruba Nees & Mart. = Almeidea A. St-Hil. (Rutac.).

Arum L. Araceae. 15 Eur., Medit. *A. maculatum* L. is a perenn. tuberous pl.
with monoec. fls.; ♀ fls. at base of spadix (each of 1 cpl., naked) and ♂ above
(each of 2–4 sta.), and above these again rudimentary ♂ fls. repres. by hairs
which project and close the mouth of the spathe. The foetid smell attracts flies,
which enter the spathe, find the stigmas ripe, and are kept prisoners till the
pollen is shed; then the hairs wither and escape is possible (cf. *Aristolochia*).
Fr. a berry. The starch of the tubers was formerly used as food under the name
Portland arrowroot, but it is difficult to get rid of the poisonous juices accom-
panying it. Other spp. are similarly used in Eur.

Aruna Schreb. = Arouna Aubl. = Dialium L. (Legumin.).

Aruncus [L.] Schaeff. Rosaceae. 12 N. temp. Sta. on inner side of axis.

Arundarbor Kuntze = Bambusa Schreb. (Gramin.).

Arundastrum Kuntze = Donax Lour. + Clinogyne Salisb. ex Benth. *p.p.* +
Marantochloa Brongn. ex Gris (Marantac.).

Arundina Blume. Orchidaceae. 1 (variable) Ceylon, SE. As., Indomal. (exc.
Philipp. & New Guin.), Tahiti.

Arundinaceae (Kunth) Herter = Gramineae–Arundineae (Kunth) Reichb.

Arundinaria Michx. Gramineae. 150 warm. Bamboos.

Arundinella Raddi. Gramineae. 55 warm. Inf. palea awned.

Arundinellaceae (Stapf) Herter = Gramineae–Arundinelleae Stapf.

Arundo Beauv. = Phragmites Trin. (Gramin.).

Arundo L. Gramineae. 12 trop. & temp. The stems of *A. donax* L. are used
for sticks, fishing-rods, etc.

Arungana Pers. = Haronga Thou. (Guttif.).

Arunia Pers. = Brunia L. (Bruniac.).

Arversia Cambess. = Polycarpon L. (Caryophyllac.).

Arviela Salisb. = Zephyranthes Herb. (Amaryllidac.).

Arytera Blume. Sapindaceae. 25 China, Indomal., Austr., Pacif.

Asacara Rafin. = Gleditsia L. (Legumin.).

Asaemia Harv. ex Benth. & Hook. f. Compositae. 2 S. Afr.

Asagraea Baill. = Dalea L. (Legumin.).

Asagraea Lindl. = Sabadilla Brandt & Ratzeburg (Liliac.).

Asaphes DC. = Toddalia Juss. (Rutac.).

Asaphes Spreng. = Morina L. (Morinac.).

Asaraceae ('-oïdeae') Vent. = Aristolochiaceae–Asareae Spach.

Asarca Poepp. ex Lindl. Orchidaceae. 20 temp. S. Am.

Asarina Mill. Scrophulariaceae. 15 N. Am., 1 Eur.

ASARUM

Asarum L. Aristolochiaceae. 70 N. temp. *A. europaeum* L. (asarabacca), formerly medic. Rhiz. below ground and creeping shoots above; the latter are sympodial, each annual joint bearing several scale ls. below, then two green ls. and a terminal fl. Fl. reg.; P (3), sometimes with 3 small teeth between the segments (perhaps remnants of a former inner whorl); A 12; G (6). The dark-brown, resinously scented fl. is visited by flies, and is very protog.; when the stigmas are ripe the sta. are all bent away but later on they move up to the centre and dehisce extr. The P lobes are bent in at first towards the centre of the fl. and form a sort of prison of it, but afterwards gradually straighten.

Ascalea Hill = Carduus L. + Cirsium Mill. (Compos.).

Ascalonicum Renault = Allium L. (Alliac.).

Ascania Crantz = Patagonula L. (Boraginac.).

Ascanica B. D. Jacks. = praec.

Ascaricida Cass. = Vernonia Schreb. (Compos.).

Ascaridia Reichb. = praec.

Ascarina J. R. & G. Forst. Chloranthaceae. 8 Malaysia, Polynesia, N.Z.

Ascarinopsis Humbert & Capuron. Chloranthaceae. 1 Madag.

Aschamia Salisb. = Hippeastrum Herb. (Amaryllidac.).

Aschenbornia Schauer. Compositae. 1 Mex.

Aschenfeldtia F. Muell. = Pimelea Banks & Soland. (Thymelaeac.).

Aschersonia F. Muell. (1878; non Endl. 1842—Fungi, nec Montaigne 1848—Fungi) = Halophila Thou. (Hydrocharitac.).

Aschersoniodoxa Gilg & Muschler. Cruciferae. 2 Andes.

Asciadium Griseb. Umbelliferae. 1 Cuba.

Ascidiogyne Cuatrec. Compositae. 1 Peru.

Ascium Schreb. = Norantea Aubl. (Marcgraviac.).

Ascleia Rafin. = Hydrolea L. (Hydrophyllac.).

*****Asclepiadaceae** R.Br. Dicots. 130/2000 trop. & subtrop., rare elsewhere. Erect or twining shrubs or perenn. herbs, sometimes fleshy and with reduced non-functional or obsolescent ls.; rootstock tuberous, fleshy, woody or sometimes absent and roots then annual from fleshy stems. Ls. opp. or whorled, ent. or v. rarely lobed or irreg. dent., exstip. Infl. cymose, often umbelliform but sometimes with fls. more or less racemosely fascic. along a simple or branched rhachis, sometimes persist. and producing successive term. crops of fls. Fls. reg., pentam.; K 5, tube short or obsolete; C (5), contorted, imbr. or valv., lobes sometimes connivent at apex; A united in a ring and adnate to style apex, the short filaments ornamented with a nectariferous corona of varied form, the whole forming a *gynostegium*; anthers provided with horny wings; pollen in tetrads united in waxy masses (*pollinia*) attached by *caudicles* of varied form to sutured *corpuscles* derived from style-apex; G of two free carpels united by their style apices; style apex peltate, with five lateral stigmatic surfaces, concave, convex or beaked above; ov. multiseriate on a single adaxial placenta. Fruit of two (or by abortion one) erect or divergent follicles, which may be linear to ovoid, membr. to woody, sometimes long-stipitate, always dehiscent, smooth or variously armed, pedic. often elongated and thickened (often kinked). Seeds usu. ∞, flattened and with a term. sess. coma of silky hairs; rarely without a coma and then not flattened; very rarely solitary or few. The fam. is readily divided into *Secamonoïdeae* with two pollinia in each anther theca, and *Asclepiadoïdeae* with solitary pollinia. The former subfam. is confined to the Old World. *Asclepiadoïdeae* occur in both hemispheres and are divisible into tribes distinguished by the orientation of the pollinia, dehiscence of the anthers and aestivation of the corolla. The remarkable *Ceropegiĕae* (including the succulent stapeliads) are confined to the Old World, and most of the spp. to Africa. The pollination mech. is unique, depending upon the trapping of insects' legs or probosces between the osmotically elastic anther wings, withdrawal entailing the capture of the pollen by means of the sutured corpuscular pollen carriers.

The Asiatic epiphytic genus *Hoya* provides a number of prized hothouse plants with large clusters of highly scented waxy flowers, whilst *Ceropegia* species are grown for their bizarre flowers; their succulent allies, *Stapelia*, *Caralluma*, *Decabelone*, etc., often have large fls. attractive to blow-flies on account of their penetrating odour of putrid carrion. *Dischidia*, another epiphytic Asiatic genus, is remarkable for the pitcher-like development of some ls., giving a means of storing free water and providing a home for species of ants.

Asclepias L. Asclepiadaceae. 120 Am., esp. U.S. (milkweeds). Herbs with umbellate infls. which spring from the stem between the petioles of the opp. ls. (cf. *Cuphea*), or above or below this; i.e. the stem is a sympodium.

Asclepiodora A. Gray = Asclepias L. (Asclepiadac.).

Asclerum Van Tiegh. = Gonystylus Teijsm. & Binnend. (Thymelaeac.).

Ascocarydion G. Taylor = Plectranthus Lour. (Labiat.).

× **Ascocenda** hort. Orchidaceae. Gen. hybr. (iii) (Ascocentrum × Vanda).

Ascocentrum Schlechter. Orchidaceae. 5 trop. SE. As. to Philipp., Celebes & Java.

Ascochilopsis C. E. Carr. Orchidaceae. 1 Malay Penins., Sumatra.

Ascochilus Blume = Geodorum Jack (Orchidac.).

Ascochilus Ridl. = Pteroceras Hassk. (Orchidac.).

× **Ascofinetia** hort. Orchidaceae. Gen. hybr. (iii) (Ascocentrum × Neofinetia).

Ascoglossum Schlechter. Orchidaceae. 2 New Guinea, Solomon Is.

*****Ascolepis** Nees. Cyperaceae. 15 warm Am., Afr.

× **Asconopsis** hort. Orchidaceae. Gen. hybr. (iii) (Ascocentrum × Phalaenopsis).

Ascopholis C. E. C. Fischer. Cyperaceae. 1 S. India.

× **Ascorachnis** hort. Orchidaceae. Gen. hybr. (iii) (Arachnis × Ascocentrum).

× **Ascorella** hort. Orchidaceae. Gen. hybr. (iii) (Ascocentrum × Renantherella).

Ascotainia Ridl. = Ania Lindl. (Orchidac.).

Ascotheca Heine. Acanthaceae. 1 W. Equat. Afr.

× **Ascovandoritis** hort. Orchidaceae. Gen. hybr. (iii) (Ascocentrum × Doritis × Vanda).

Ascra Schott = Banara Aubl. (Flacourtiac.).

Ascyroïdes Lippi ex Adans. = Bistella Adans. (Vahliac.), Bergia L. (Elatinac.), etc.

Ascyrum L. Guttiferae. 5 N. Am., W.I.

Ascyrum Mill. = Hypericum L. (Guttif.).

Ascyum Vahl = Ascium Schreb. = Norantea Aubl. (Marcgraviac.).

Asemeia Rafin. = Polygala L. (Polygalac.).
Asemanthia Ridley. Rubiaceae. 5 Malay Penins., Borneo.
Asemnantha Hook. f. Rubiaceae. 1 Mex.
Asephananthes Bory ex DC. = Passiflora L. (Passiflorac.).
Ashtonia Airy Shaw. Euphorbiaceae. 2 Malaya, Borneo.
Asiasarum Maekawa (~ Asarum L.). Aristolochiaceae. 4 E. As.
Asicaria Neck. = Persicaria Mill. (Polygonac.).
Asimia Kunth = seq.
Asimina Adans. Annonaceae. 8 E. N. Am. *A. triloba* (L.) Dun. (papaw) has
Asiphonia Griff. = Apama Lam. (Aristolochiac.).   [ed. fr.
Asisadenia Hutch. (sphalm.) = Anisadenia Wall. ex Meissn. (Linac.).
Asketanthera R. E. Woodson. Apocynaceae. 4 W.I., trop. Am.
Askidiosperma Steud. Restionaceae. 1 S. Afr.
Askofake Rafin. = Utricularia L. (Lentibulariac.).
Askolame Rafin. = Milla Cav. (Liliac.).
Asophila Neck. = Gypsophila L. (Caryophyllac.).
Aspalathoïdes K. Koch = Anthyllis L. (Legumin.).
Aspalathus Kuntze = Caragana Lam. (Legumin.).
Aspalathus L. Leguminosae. 245 S. Afr. Many are xero. with a heath-like
habit.
Aspalatus A. St-Hil. = praec.
Aspalthium Medik. = Asphalthium Medik. = Psoralea L. (Legumin.).
***Asparagaceae** ('-gi') Juss. = Liliaceae–Asparagoïdeae Engl.
Asparagopsis Kunth = Asparagus L. (Liliac.).
Asparagus L. Liliaceae. 300 Old World, mostly in dry places. Rhiz. with
aerial shoots; ls. reduced to scales with linear green shoots in axils, usu. in
tufts. These are small condensed cymes. The number of shoots that develop
varies. In the infl. the same construction holds. In the subgenus *Myrsiphyllum*
there are flat phylloclades (cf. *Ruscus*). Fr. a berry. *A. officinalis* L. cult., the
young shoots being eaten.
Aspasia Lindl. Orchidaceae. 10 C. Am., trop. S. Am.
Aspasia E. Mey. = Stachys L. (Labiat.).
Aspasia Salisb. = Ornithogalum L. (Liliac.).
× **Aspasium** hort. Orchidaceae. Gen. hybr. (iii) (Aspasia × Oncidium).
Aspazoma N. E. Brown. Aïzoaceae. 1 S. Afr.
Aspegrenia Poepp. & Endl. = Octomeria R.Br. (Orchidac.).
Aspelina Cass. = Senecio L. (Compos.).
Aspera Moench = Galium L. (Rubiac.).
Asperella Humb. (non Asprella Schreb.) = Hystrix Moench (Gramin.).
Asperifoliaceae Reichb. = Boraginaceae Juss.
Asperifoliae Batsch = praec.
× **Asperugalium** P. Fourn. Rubiaceae. Gen. hybr. (Asperula auctt. × Galium).
Asperugo L. Boraginaceae. 1 Eur.
Asperula L. (*s.str.*, i.e. quoad typum, *A. odorata* L.) = Galium L. (Rubiac.).
Asperula auctt. (Cynanchica Fourr.). Rubiaceae. 200 + Eur., As., esp. Medit.;
16 E. Austr., Tasm. The Australian spp. are dioecious and perhaps generically
distinct.
Asphalathus Burm. f. = Aspalathus L. (Legumin.).

Asphalthium Medik. = Psoralea L. (Legumin.).
Asphodelaceae ('-li') Juss. = Liliaceae–Asphodeloïdeae Engl.
Asphodeline Reichb. Liliaceae. 15 Medit.
Asphodeliris Kuntze = Tofieldia Huds. (Liliac.).
Asphodeloïdes Moench = Asphodelus L. (Liliac.).
Asphodelopsis Steud. ex Baker = Chlorophytum Ker-Gawl. (Liliac.).
Asphodelus L. Liliaceae. 12 Medit. to Himal. Ls. isobilateral; fls. protog.
Aspicarpa Rich. Malpighiaceae. 12 S. U.S. to Argentina.
Aspidalis Gaertn. = Cuspidia Gaertn. (Compos.).
Aspidandra Hassk. = Ryparosa Bl. (Flacourtiac.).
Aspidanthera Benth. = Ferdinandusa Pohl (Rubiac.).
Aspideium Zollik. ex DC. = Chondrilla L. (Compos.).
Aspidiaceae S. F. Gray. Aspidiales. Terrestrial ferns, many genera (incl.
Dryopteris, Tectaria) cosmop. Stem radially organized, usu. dictyostelic, scales
lacking superficial hairs; several small vasc. strands in petiole; hairs on ls.
mostly multicellular; veins free or variously anastomosing; sori usually round,
indusiate, on veins or at ends of veins; indusium reniform or peltate (in some
cases lacking and then sori may spread along veins).
Aspidiales. Filicidae. Fam.: *Thelypteridaceae, Aspleniaceae, Athyriaceae,
Aspidiaceae, Lomariopsidaceae.*
Aspidistra Ker-Gawl. Liliaceae. 8 E. As. The large flat style forms a lid to
the cavity made by the 6 P-leaves.
Aspidistr[ac]eae Endl. = Liliaceae–Asparagoïdeae–Convallariëae–Aspidistrinae
Engl.
Aspidium Sw. (*s.str.*) = Tectaria Cav. (Aspidiac.). The orig. genus of Swartz
was a considerable mixture, and the name *Aspidium* was later used in several
different senses (in *Syn. Fil.* mainly for *Polystichum* Roth, in C. Chr. *Ind. Fil.*
for *Tectaria*).
Aspidixia Van Tiegh. = Viscum L. (Viscac.).
Aspidocarpus Neck. = Rhamnus L. (Rhamnac.).
Aspidocarya Hook. f. & Thoms. Menispermaceae. 1 E. Himal., SE. As.
Aspidogenia Burret. Myrtaceae. 1 Chile.
Aspidoglossum E. Mey. = Schizoglossum E. Mey. (Asclepiadac.).
Aspidophyllum Ulbr. Ranunculaceae. 1 Peru.
Aspidopterys A. Juss. Malpighiaceae. 20 W. Himal. to S. China, W. Malaysia,
Celebes.
Aspidopteryx Dalla Torre & Harms = praec.
*Aspidosperma Mart. & Zucc. Apocynaceae. 80 trop. & S. Am., W.I. Wood
useful; bark (*quebracha*) used for tanning.
Aspidostigma Hochst. = Toddalia Juss. (Rutac.).
Aspidotis (Nutt. ex Hook.) Copel. Sinopteridaceae. 2 N. Am., 1 Afr. (Pichi-
Sermolli, *Webbia* 7: 326, 1950.)
Aspilia Thou. Compositae. 125 Mex. to Brazil, S. trop. Afr., Madag.
Aspiliopsis Greenman. Compositae. 1 Mex.
Aspilobium Soland. = Geniostoma J. R. & G. Forst. (Loganiac.).
Aspilotum Soland. ex Steud. = praec.
Aspitium Neck. ex Steud. = Laserpitium L. (Umbellif.).
Aspla Reichb. (sphalm.) = Aopla Lindl. = Habenaria L. (Orchidac.).

ASPLENIACEAE

Aspleniaceae S. F. Gray. Aspidiales. Mainly epiphytes or rock plants, cosmop.; sori along veins, indusiate. One very large genus, *Asplenium* L., and several smaller ones.

Asplenidictyum J. Sm. = Asplenium L. (Aspleniac.).

Aspleniopsis Mett. ex Kuhn. Hemionitidaceae. 3 New Guinea, New Caled.

Asplenium L. Aspleniaceae. 650 cosmop. *A. bulbiferum* Forst. and other spp. are 'viviparous', producing young plants on their leaves by vegetative budding. *A. nidus* L. (the bird's-nest fern) is a common pioneer epiphyte of the Old World tropics. It bears a rosette of leaves forming a nest in which humus collects; the roots ramify in this and obtain food and water. Many hybrids have been recorded between European spp. (D. E. Meyer in *Ber. Deutsch. Bot. Ges.* **70–4**, 1957–62); and many polyploids in trop. spp. (Manton & Sledge, *Phil. Trans. Roy. Soc.* B, **238**: 138, 1954).

× **Asplenoceterach** D. E. Meyer. Aspleniaceae. Gen. hybr. (Asplenium × Ceterach).

× **Asplenophyllitis** Alston. Aspleniaceae. Gen. hybr. (Asplenium × Phyllitis).

Asplundia Harling. Cyclanthaceae. 82 C. & trop. S. Am.

× **Aspoglossum** hort. Orchidaceae. Gen. hybr. (iii) (Aspasia × Odontoglossum).

Asprella Host = Psilurus Trin. (Gramin.).

Asprella Schreb. = Leersia Sw. (Gramin.).

Asprella Willd. = Asperella Humb. (Gramin.).

Aspris Adans. = Aira L. (Gramin.).

Assa Houtt. = Tetracera L. (Dilleniac.).

Assaracus Haw. = Narcissus L. (Amaryllidac.).

Assidora A. Chev. = Schumanniophyton Harms (Rubiac.).

Assonia Cav. = Dombeya Cav. (Sterculiac.).

Asta Klotzsch ex O. E. Schulz. Cruciferae. 2 Mex.

Astartea DC. Myrtaceae. 7 Austr.

*****Astelia** Banks & Soland. Liliaceae. 25 Masc., New Guinea, Austr., Tasm., Pacif. to Hawaii. Dioec.

Asteliaceae Dum. = Liliaceae–Dracaenoïdeae–Dracaeneae Reichb., or Liliac.–Milliganiëae Hutch.

Astelma R.Br. = Helichrysum Mill., Helipterum DC., etc. (Compos.).

Astelma Schltr. = Papuastelma Bullock (Asclepiadac.).

Astemma Less. Compositae. 1 Ecuador.

Astemon Regel = Lepechinia Willd. + Sphacele Benth. (Labiat.).

Astenolobium Nevski. Leguminosae. 1 C. As.

Astephananthes Bory = Passiflora L. (Passiflorac.).

Astephania Oliv. Compositae. 2 trop. E. Afr.

Astephanocarpa Baker = Syncephalum DC. (Compos.).

Astephanus R.Br. Asclepiadaceae. 2 S. Afr.

Aster L. Compositae. 500 Am., Euras., Afr.; some fleshy halophytes. (The 'China aster' of gardens is a *Callistephus*.)

Asteracantha Nees = Hygrophila R.Br. (Acanthac.).

*****Asteraceae** Dum. (*s.l.*): see **Compositae** Giseke (nom. altern.).

Asteraceae Dum. (*s.str.*) (= Compositae excl. Cichoriëae and Heliantheae–

Ambrosiinae). Cor. all tubular, or only the marginal lig.; fil. free, anth. syngen. (v. rarely distinct); pollen globose, often echin.; juice rarely milky. See *Compositae, Ambrosiaceae* and *Cichoriaceae*.

×**Asterago** Everett. Compositae. Gen. hybr. (Aster × Solidago).

Asterandra Klotzsch = Phyllanthus L. (Euphorbiac.).

Asterantha Reichb. (sphalm.) = Asteracantha Nees = Hygrophila R.Br. (Acanthac.).

***Asteranthaceae** Knuth (~ Lecythidaceae). Dicots. 1/1 Brazil. Trees with alt. entire coriaceous exstip. ls. Fl. large, sol., axill. K ± disciform, ∞-dentate, persist.; C 0; stds. 20–25, conn. into large plicate dentate membr. disk, A ∞ with slender fil. and small anth. (basifixed, introrse by slits); G (6–8) semi-inf., with ± 4 ovules pend. from apex, style filif., stig. capit. Fr. a semi-sup. crustaceous 6–8-valved and -ribbed loculic. pyram. caps., surrounded by persist. K. Only genus: *Asteranthos*. Perhaps some connection with *Punicaceae*.

Asteranthe Engl. & Diels. Annonaceae. 2 trop. E. Afr.

Asteranthemum Kunth = Smilacina Desf. (Liliac.).

Asteranthera Hanst. Gesneriaceae. 1 Chile.

Asteranthopsis Kuntze = Asteranthe Engl. & Diels (Annonac.).

Asteranthos Desf. Asteranthaceae. 1 N. Brazil.

Asteranthus Endl. = Astranthus Lour. = Homalium Jacq. (Flacourtiac.).

Asteranthus Spreng. = Asteranthos Desf. (Asteranthac.).

Asterias Borckh. = Gentiana L. (Gentianac.).

Asteriastigma Bedd. = Hydnocarpus Gaertn. (Flacourtiac.).

Asteridea Lindl. = Athrixia Ker-Gawl. (Compos.).

Asteridium Engelm. ex Walp. = Chaetopappa DC. (Compos.).

Asterigeron Rydb. (~ Aster L.). Compositae. 1 W. U.S.

Asteringa E. Mey. ex DC. = Pentzia Thunb. (Compos.).

Asteriscium Cham. et Schlechtd. Hydrocotylaceae (~ Umbellif.). 8 Mex. to Patag.

Asteriscodes Kuntze = Callistephus Cass. (Compos.).

Asteriscus Mill. Compositae. 4 N. Afr.

Asteriscus Reichb. = Asteriscium Cham. & Schlechtd. (Umbellif.).

Asteriscus Sch. Bip. = Pallenis Cass. (Compos.).

Asterocarpaceae Kerner = Resedaceae–Astrocarpeae Muell. Arg.

Asterocarpus Eckl. & Zeyh. = Pterocelastrus Meissn. (Celastrac.).

Asterocarpus Reichb. = Astrocarpa Dum. = Sesamoïdes Ortega (Resedac.).

Asterocephalus Zinn = Scabiosa L. (Dipsacac.).

Asterochaete Nees = Carpha Banks & Sol. (Cyperac.).

Asterochiton Turcz. = Thomasia J. Gay (Sterculiac.).

Asterochlaena Garcke = Pavonia Cav. (Malvac.).

Asterocytisus (Koch) Schur ex Fuss = Genista L. (Legumin.).

Asterogeum S. F. Gray = Plantago L. (Plantaginac.).

Asterogyne H. Wendl. Palmae. 2 trop. S. Am.

Asterohyptis Epling. Labiatae. 3 Mex.

Asteroïdes Mill. = Buphthalmum L. (Compos.).

Asterolasia F. Muell. Rutaceae. 12 Austr.

Asterolepidion Ducke = Dendrobangia Rusby (Icacinac.).

ASTEROLINON

Asterolinon Hoffmgg. & Link (=Borissa Rafin.). Primulaceae. 2 Medit. to Crimea, Persia, Abyss.

Asteromoea Blume. Compositae. 15 E. As.

Asteromyrtus Schau. = Melaleuca L. (Myrtac.).

Asteropeia Thou. Asteropeiaceae. 7 Madag.

Asteropeiaceae Takhtadj. Dicots. 1/7 Madag. Small trees or scrambling shrubs; ls. alt., ent., simple, exstip. Infl. in axill. or term. many-fld. thyrses. K 5, imbr., persist. & accresc. in fr.; C 5, narrow, decid.; A 9–15, connate below into a ± broad ring, persist., with dorsifixed versatile decid. anth.; G (3), more rarely (2), with 2–∞ axile pend. ov. per loc.; style 1, elongate, shortly 3-lobed at apex, or styles 3, short, reflexed, ± free, or intermediate. Fr. thick-walled, dry, indehisc., 2–∞-seeded, surr. by persist. stam. and accresc. coriaceous or membr. K. Only genus: *Asteropeia*. Affinities uncertain; possibly with *Linac.*, *Tetrameristac.* or *Flacourtiac.*

Asterophorum Sprague. Tiliaceae. 1 Ecuador.

Asterophyllum Schimp. & Spenn. = Galium L., Asperula auctt., Sherardia L., Valantia L. (Rubiac.).

Asteropsis Less. (~Podocoma Cass.). Compositae. 1 S. Brazil.

Asteropterus Adans. Compositae. 7 Medit., Afr.

Asteropus Schult. = Astropus Spreng. = Waltheria L. (Sterculiac.).

Asteropyrum J. R. Drumm. & Hutch. Ranunculaceae. 3 China.

Asteroschoenus Nees = Rhynchospora Vahl (Cyperac.).

Asterosperma Less. = Felicia Cass. (Compos.).

Asterostemma Decne. Asclepiadaceae. 1 Java.

Asterostigma Fisch. & Mey. Araceae. 5 Brazil.

Asterostoma Blume = Osbeckia L. (Melastomatac.).

Asterothamnus Novopokr. Compositae. 7 C. & E. As.

Asterothrix Cass. = Leontodon L. (Compos.).

Asterotrichion Klotzsch = Plagianthus J. R. & G. Forst. (Malvac.).

Asthenatherum Nevski. Gramineae. 2 C. As.

Asthenochloa Buese. Gramineae. 1 Philippines, Java.

Asthotheca [?sphalm. pro Athrotheca] Miers ex Planch. & Triana = Clusia L. (Guttif.).

Astianthus D. Don. Bignoniaceae. 1 Mex., C. Am.

Astiella Jovet. Rubiaceae. 1 Madag.

Astilbe Buch.-Ham. Saxifragaceae. 25 E. As., N. Am.

Astilboïdes Engl. Saxifragaceae. 1 N. China.

Astiria Lindl. Sterculiaceae (Malvaceae?). 1 Masc.

Astoma DC. Umbelliferae. 1 E. Medit.

Astomatopsis Korovin. Umbelliferae. 1 C. As.

Astorganthus Endl. = Melicope J. R. & G. Forst. (Rutac.).

Astradelphus Rémy = Erigeron L. (Compos.).

Astraea Klotzsch = Croton L. (Euphorbiac.).

Astraea Schau. = Thryptomene Endl. (Myrtac.).

Astragalina Bub. = Astragalus L. (Legumin.).

Astragaloïdes Boehm. = Phaca L. (Legumin.).

Astragaloïdes Quer = Astragalus L. (Legumin.).

Astragalus L. Leguminosae. 2000 cosmop., exc. Austr. Usu. on steppes, prairies, etc., and ± xero., often thorny; the thorns commonly form by the stiffening of the petiole or midrib of the l. when the blade falls off. *A. gummifer* Labill. and others yield gum tragacanth, obtained by wounding the stem; the gum exudes and hardens.

Astranthium Nutt. Compositae. 10 S. U.S., Mex.

Astranthus Lour. = Homalium Jacq. (Flacourtiac.).

Astrantia Ehrh. (uninom.) = *Astrantia major* L. (Umbellif.).

Astrantia L. Umbelliferae. 10 C. & S. Eur., As. Min., Cauc.

Astrapaea Lindl. = Dombeya Cav. (Sterculiac.).

Astrebla F. Muell. Gramineae. 4 Austr.

Astrephia Dufresne = Valeriana L. (Valerianac.).

Astridia Dinter & Schwantes. Aïzoaceae. 10 S. Afr.

Astripomoea A. Meeuse. Convolvulaceae. 12 trop. Afr.

Astrocalyx Merr. Melastomataceae. 2 Philipp. Is.

Astrocarpa Dum. = Sesamoïdes Ortega (Resedac.).

Astrocarpus Neck. ex DC. = praec.   [and oil.

*****Astrocaryum** G. F. W. Mey. Palmae. 50 trop. Am. Several yield fibre

Astrocasia Robinson & Millspaugh. Euphorbiaceae. 3 Mex., C. Am.

Astrocephalus Rafin. = Asterocephalus Zinn = Scabiosa L. (Dipsacac.).

Astrochlaena Hallier f. (1894; non Asterochlaena Corda 1845—gen. foss.) = Astripomoea A. Meeuse (Convolvulac.).

Astrococcus Benth. Euphorbiaceae. 2 Brazil.

Astrocodon Fedorov. Campanulaceae. 2 NE. Siberia.

Astrocoma Neck. = Staavia Dahl (Bruniac.).

Astrodaucus Drude = Ageomoron Rafin. (Umbellif.).

Astrodendrum Dennst. = Sterculia L. (Sterculiac.).

Astroglossus Reichb. f. = Trichoceros Kunth (Orchidac.).

Astrogyne Benth. = Croton L. (Euphorbiac.).

Astrogyne Wall. ex Laws. = Siphonodon Griff. (Siphonodontac.).

Astrolinon Baudo = Asterolinon Hoffmgg. & Link (Primulac.).

Astroloba Uitew. Liliaceae. 12 S. Afr. Some, e.g. *A. foliolosa* (Willd.) Uitew., show extreme superposition of ls., forming almost solid masses of tissue.

Astrolobium DC. (sphalm.) = Artrolobium Desv. = Ornithopus L. & Coronilla L. (Legumin.).

Astroloma R.Br. Epacridaceae. 25 Austr.

Astromerremia Pilger = Merremia Dennst. (Convolvulac.).

Astronia Blume. Melastomataceae. 70 Formosa, Malaysia, Polynes.

Astronia Nor. = Murraya Koen. ex L. (Rutac.).

Astronidium A. Gray (= Lomanodia Rafin.). Melastomataceae. 35 New Guinea, Pacif.

Astronium Jacq. Anacardiaceae. 15 C. & trop. S. Am., W.I.

Astropanax Seem. = Schefflera J. R. & G. Forst. (Araliac.).

Astropetalum Griff. = Swintonia Griff. (Anacardiac.).

Astrophea Reichb. = Passiflora L. (Passiflorac.).

Astrophia Nutt. = Lathyrus L. (Legumin.).

Astrophyllum Torr. & Gray = Choisya Kunth (Rutac.).

Astrophyton Lawr. & Lem. = seq.

ASTROPHYTUM

Astrophytum Lemaire. Cactaceae. 6 S. U.S., Mexico.

Astropus Spreng. = Waltheria L. (Sterculiac.).

Astroschoenus Lindl. = Asteroschoenus Nees (Cyperac.).

Astrostemma Benth. Asclepiadaceae. 1 Borneo.

Astrothalamus C. B. Robinson. Urticaceae. 1 Philipp. Is.

Astrotheca Vesque = Asthotheca Miers ex Planch. & Triana = Clusia L. (Guttif.).

Astrotricha DC. Araliaceae. 10 Austr.

Astrotrichilia (Harms) Leroy. Meliaceae. 2 Madag.

× **Astroworthia** Rowley. Liliaceae. Gen. hybr. (Astroloba × Haworthia).

Astydamia DC. Umbelliferae. 2 NW. Afr., Canaries.

Astylis Wight = Drypetes Vahl (Euphorbiac.).

Astylus Dulac = Hutchinsia R.Br. (Crucif.).

Astyposanthes Herter (~ Stylosanthus Sw.). Leguminosae. 9 S. Am.

Astyria Lindl. = Astiria Lindl. (Sterculiac.).

Asyneuma Griseb. & Schenck. Campanulaceae. 50 Medit. to Cauc., 1 E. As.

Asystasia Blume. Acanthaceae. 40 palaeotrop.

Asystasiella Lindau. Acanthaceae. 3 trop. Afr., As.

Atacca Lem. = seq.

Ataccia Presl = Tacca J. R. & G. Forst. (Taccac.).

Atadinus Rafin. = Rhamnus L. (Rhamnac.).

Ataenia Endl. = Atenia Hook. & Arn. = Perideria Reichb. (Umbellif.).

Ataenidia Gagnep. = Phrynium Willd. (Marantac.).

Atalanta Nutt. = Cleome L. (Cleomac.).

Atalanta Rafin. = praec. [to C. As.

Atalanthus D. Don (~ Sonchus L.). Compositae. ? 10 Canary Is., Medit.

*****Atalantia** Corrêa. Rutaceae. 18 trop. As., China, Austr.

Atalaya Blume. Sapindaceae. 9 S. Afr., E. Malaysia, Austr.

Atalopteris Maxon & C. Chr. Aspidiaceae. 3 W.I.

Atamasco Rafin. = Atamosco Adans. = Zephyranthes Herb. (Amaryllidac.).

Atamisquea Miers. Capparidaceae. 1 Calif., temp. S. Am.

Atamosco Adans. = Zephyranthes Herb. (Amaryllidac.).

Atanara Rafin. = Annona L. (Annonac.).

Atasites Neck. = Gerbera L. ex Cass. (Compos.).

Ataxia R.Br. = Hierochloë S. G. Gmel. (Gramin.).

Ate Lindl. = Habenaria Willd. (Orchidac.).

Atecosa Rafin. = Rumex L. (Polygonac.).

Atelandra Bello = Meliosma Blume (Meliosmac.).

Atelandra Lindl. = Hemigenia R.Br. (Labiat.).

Atelanthera Hook. f. & Thoms. Cruciferae. 3 C. As., Afgh., W. Himal. & Tibet. Some anthers monothecous.

Atelea A. Rich. = seq.

Ateleia (Moç. et Sessé ex DC.) D. Dietr. Leguminosae. 17 Mex. to trop. S. Am., W.I.

Ateleste Sond. = Doryalis E. Mey. ex Arn. (Flacourtiac.).

Atelianthus Nutt. ex Benth. = Synthyris Benth. (Scrophulariac.).

Atelophragma Rydb. = Astragalus L. (Legumin.).

Atemnosiphon Leandri. Thymelaeaceae. 1 Madag.

Atenia Hook. & Arn. = Perideria Reichb. (Umbellif.).
Ateramnus P.Br. (?Gymnanthes Sw.). Euphorbiaceae. 15 S. U.S., Mex.,
Atevala Rafin. = Aloë L. (Liliac.). [C. Am., W.I.
Athalmum Neck. = Pallenis Cass. (Compos.).
Athalmus B. D. Jacks. (sphalm.) = praec.
Athamanta L. Umbelliferae. 15 Medit.
Athamantha Rafin. = Athamanta L. (Umbellif.).
Athamus Neck. = Carlina L. (Compos.).
Athanasia L. Compositae. 50 trop. & S. Afr., Madag.
Athecia Gaertn. = ? Forstera L. f. (Stylidiac.).
Athenaea Adans. = Struchium P.Br. (Compos.).
Athenaea Schreb. = Casearia Jacq. (Flacourtiac.).
*****Athenaea** Sendtn. Solanaceae. 20 trop. Am.
Athenanthia Kunth = Anthaenantia Beauv. (Gramin.).
Atheolaena Reichb. = Aëtheolaena Cass. = Senecio L. (Compos.). [Java.
Atherandra Decne. Asclepiadaceae. 1 SE. As., Malay Penins., Sumatra,
Atheranthera Mast. = Gerrardanthus Harv. ex Hook. f. (Cucurbitac.).
Athernotus Dulac = Calamagrostis Adans. (Gramin.).
Atherocephala DC. = Andersonia R.Br. (Epacridac.).
Atherolepis Hook. f. Asclepiadaceae. 3 Burma, Siam.
Atherolepsis Willis (sphalm.) = praec.
Atherophora Willd. ex Steud. = Aegopogon Humb. & Bonpl. (Gramin.).
Atheropogon Mühlenb. ex Willd. = Botelua Lag. (Gramin.).
Atherosperma Labill. Atherospermataceae. 2 Victoria, Tasmania. The
strongly scented bark is sometimes used as a tea.
Atherospermataceae R.Br. Dicots. 5/12 New Guinea, Austr., New Caled.,
N.Z., Chile. As *Monimiaceae* (*s.str.*), *q.v.*, but medullary rays mostly narrow,
anthers opening by valves (cf. *Laurac.*), ovule anatr. or rarely orthotr., micro-
pyle inf., seed with basal embryo. Genera: *Laurelia, Nemuaron, Daphnandra,
Atherosperma, Doryphora*.
Atherostemon Blume. Asclepiadaceae. 1 Burma, Malaya.
Atherstonea Pappe = Strychnos L. (Strychnac.).
× **Athertonara** hort. (vii) = × Renanopsis hort. (Orchidac.).
Atherurus Blume = Pinella Tenore (Arac.).
Athesiandra Miers ex Benth. & Hook. f. = Ptychopetalum Benth. (Olacac.).
Athlianthus Endl. = Justicia L. (Acanthac.).
Athrixia Ker-Gawl. Compositae. 20 trop. & S. Afr., Madag., Arab., Austr.
Athroandra (Hook. f.) Pax & Hoffm. = Chloropatane Engl. (Euphorbiac.).
Athrodactylis J. R. & G. Forst. = Pandanus L. f. (Pandanac.).
Athroisma DC. Compositae. 8 trop. Afr., Indomal.
Athroisma Griff. = Trigonostemon Blume (Euphorbiac.).
Athronia Neck. = Spilanthes Jacq. (Compos.).
Athroöstachys Benth. Gramineae. 1 Brazil. Climbing.
Athrotaxidaceae Nak. = Taxodiaceae Neger.
Athrotaxis D. Don. Taxodiaceae. 3 Austr., Tasmania.
Athruphyllum Lour. = Rapanea Aubl. (Myrsinac.).
Athyana Radlk. Sapindaceae. 1 Paraguay, Argent.
Athymalus Neck. = Euphorbia L. (Euphorbiac.).

ATHYRIACEAE

Athyriaceae Alston. Aspidiales. Two vasc. strands at base of petiole, uniting upwards to U-shaped strand; sori usually asymmetric and/or elongate along veins. Doubtfully separate from *Aspidiaceae*. A new subdivision by Ching in *Acta Phytotax. Sinica* 9: 41–84, 1964. Chief genera: *Athyrium* Roth, *Diplazium* Sw., *Cystopteris* Bernh.

Athyriopsis Ching = Lunathyrium Koidz. (Athyriac.).

Athyrium Roth. Athyriaceae. 180 cosmop. Limits of genus need revision. Copeland included here *Diplazium* Sw., but see Manton & Sledge, *Phil. Tr. R. Soc.* B, **238**: 165, 1954 (in *Athyrium* n=40, in *Diplazium* n=41), and Sledge, *Bull. Brit. Mus. Nat. Hist.* **2**: 275–80, 1962.

Athyrocarpus Schlechtd. = Phaeosphaerion Hassk. (Commelinac.).

Athyrus Neck. = Lathyrus L. (Legumin.).

Athysanus Greene. Cruciferae. 1 W. U.S.

Atimeta Schott = Rhodospatha Poepp. (Arac.).

Atirbesia Rafin. = Marrubium L. (Labiat.).

Atirsita Rafin. = ? Eryngium L. (Umbellif.).

Atitara Juss. = Evodia J. R. & G. Forst. (Rutac.).

Atitara Kuntze = Desmoncus Mart. (Palm.).

Atkinsia Howard. Malvaceae. 1 Cuba.

Atkinsonia F. Muell. Loranthaceae. 1 E. Austr.

Atlantia Kurz = Atalantia Corrêa (Rutac.).

Atocion Adans. = Melandrium Roehl. (Caryophyllac.).

Atolaria Neck. = Crotalaria L. (Legumin.).

Atomostigma Kuntze. Rosaceae. 1 Brazil.

Atomostylis Steud. = Cyperus L. (Cyperac.).

Atopocarpus Cuatrec. = Clonodia Griseb. (Malpighiac.).

Atopostema Boutique. Annonaceae. 2 trop. Afr.

Atossa Alef. = Vicia L. (Legumin.).

Atractocarpa Franchet = Puelia Franch. (Gramin.).

Atractocarpus Schlechter & Krause. Rubiaceae. 10 New Caled.

Atractogyne Pierre. Rubiaceae. 3 W. Afr.

Atractylis Boehm. = Carthamus L. (Compos.).

Atractylis L. Compositae. 20 W. Medit. to Japan.

Atractylodes DC. (~ Atractylis L.). Compositae. 8 E. As.

Atragene L. = Clematis L. (Ranunculac.).

Atraphax Scop. = seq.

Atraphaxis L. Polygonaceae. 25 N. Afr., SE. Eur. to Himal. & E. Sib.

Atrategia Hook. f. = Atrutegia Bedd. = Goniothalamus Blume (Annonac.).

Atrema DC. = Bifora Hoffm. (Umbellif.).

Atrichodendron Gagnep. Solanaceae. 1 Indoch.

Atrichoseris A. Gray. Compositae. 1 SW. U.S.

Atriplex L. Chenopodiaceae. 200 temp. & subtrop. Fls. unisexual or polyg., naked or with P.

Atriplicaceae ('-ices') Juss. = Chenopodiaceae Vent.

Atropa L. Solanaceae. 4 Eur., Medit. to C. As. & Himal. *A. bella-donna* L. (deadly nightshade) contains the alkaloid atropin, the basis of the drug bella-donna used in medicine.

Atropanthe Pascher (~ Scopolia Jacq.). Solanaceae. 1 China.

Atropis Rupr. = Puccinellia Parl. (Gramin.).

Atroxima Stapf. Polygalaceae. 5 trop. Afr.

Atrutegia Bedd. = Goniothalamus Bl. (Annonac.).

Attalea Kunth. Palmae. 40 S. Am.,W.I., trop. Afr. *A. funifera* Mart. (Brazil) yields Bahia *piassaba* fibre. *A. cohune* Mart. (Honduras) yields the ivory-like *cohune* nuts.

Attractilis Hall. ex Scop. = Atractylis L. (Compos.).

Atulandra Rafin. = Rhamnus L. (Rhamnac.).

Atuna Rafin. Chrysobalanaceae. 11 Malaysia, Polynesia.

Atylosia Wight & Arn. Leguminosae. 1 trop. Afr., Madag., Masc., 20 trop. As., Austr., New Caled.

Atylus Salisb. = Petrophila R.Br. (Proteac.).

Atyson Rafin. = Aectyson Rafin. = Sedum L. (Crassulac.).

Aubentonia Domb. ex Steud. = Waltheria L. (Sterculiac.).

Aubertia Bory = ? Evodia J. R. & G. Forst. vel ? Fagara L. (Rutac.).

Aubertia Chapel. ex Baill. = Croton L. (Euphorbiac.).

Aubertiella Briq. (sphalm.) = Audibertiella Briq. = Salvia L. (Labiat.).

Aubion Rafin. = Cleome L. (Cleomac.).

Aubletella Pierre = Chrysophyllum L. (Sapotac.).

Aubletia Gaertn. = Sonneratia L. f. (Sonneratiac.).

Aubletia Le Monn. ex Rozier, mut. Dandy = Obletia Le Monn. ex Rozier = Verbena L. (Verbenac.).

Aubletia Lour. = Paliurus Mill. (Rhamnac.).

Aubletia Neck. = Ruellia L. (Acanthac.).

Aubletia Rich. = Monnieria L. (Rutac.).

Aubletia Schreb. = Apeiba Aubl. (Tiliac.).

Aubregrinia H. Heine. Sapotaceae. 1 trop. W. Afr.

Aubrevillea Pellegr. Leguminosae. 2 trop. W. Afr.

Aubrieta Adans. Cruciferae. 15 mts. Italy to Persia.

Aubrietia DC. = praec.

Aubrya Baill. = Sacoglottis Mart. (Houmiriac.).

Auchera DC. = Cousinia Cass. (Compos.).

Aucklandia Falc. = Saussurea DC. (Compos.).

Aucoumea Pierre. Burseraceae. 1 trop. W. Afr. Yields resin.

Aucuba Cham. = Aruba Nees & Mart. = Raputia Aubl. (Rutac.).

Aucuba Thunb. Aucubaceae. 3–4 Himal. to Japan. Dioec. C sometimes long-caudate.

Aucubaceae J. G. Agardh (~ Cornaceae Dum.). Dicots. 1/3 Himal. to Japan. Shrubs or small trees, glabr. or sparsely pilose, with opp. shining coriac. ent. or dent. exstip. ls. Fls. ♂ ♀, dioec., in term. or axill. dichot. thyrses. ♂: K 4, minute; C 4, valv., often with slender inflexed tip; A 4, alternipet., v. short, anth. intr.; disk fleshy; pedicel non-artic. and ebracteolate. ♀: K 4, minute; C 4, as in ♂; stds. O; disk low, fleshy; G̅ (1), 1-loc., with 1 pend. anatr. ovule; style short, thick, with conspic. oblique capit. stig.; pedicel artic. and bibracteolate. Fr. an ovoid scarlet berry; seed with minute embr. at apex of copious endosp. Only genus: *Aucuba*.

Aucubaephyllum Ahlburg = Grumilea Gaertn. (Rubiac.).

Aucuparia Medik. = Sorbus L. (Rosac.).

AUDIBERTIA

Audibertia Benth. (1829) = Mentha L. (Labiat.).
Audibertia Benth. (1831) = seq.
Audibertiella Briq. = Salvia L. (Labiat.).
Audouinia Brongn. Bruniaceae. 1 S. Afr.
Auerodendron Urb. Rhamnaceae. 7 W.I.
Auganthus Link = Primula L. (Primulac.).
*Augea Thunb. Zygophyllaceae. 1 S. Afr. Exalb.
Augea Thunb. ex Retz. = Lanaria Ait. (Haemodorac.).
Augia Lour. = Rhus L. (Anacardiac.) (+ ? Calophyllum L. [Guttif.]).
Augouardia Pellegr. Leguminosae. 1 W. Equat. Afr.
Augusta Ellis (apud Smith) = Warneria Ell. ex L. (1759) = Gardenia Ellis (1761) (Rubiac.).
Augusta Leandro = Stifftia Mikan (Compos.).
*Augusta Pohl. Rubiaceae. 1 E. Brazil.
Augustea DC. = praec.
Augustia Klotzsch = Begonia L. (Begoniac.).
Augustinea Karst. = Pyrenoglyphis Karst. (Palm.).
Augustinea A. St-Hil. & Naud. = Miconia Ruiz & Pav. (Melastomatac.).
Aukuba Koehne = Aucuba Thunb. (Cornac.).
Aulacia Lour. = Micromelum Bl. (Rutac.).
Aulacidium Rich. ex DC. = Macrocentrum Hook. f. (Melastomatac.).
Aulacinthus E. Mey. = Lotononis Eckl. & Zeyh. (Legumin.).
Aulacocalyx Hook. f. Rubiaceae. 8 trop. Afr.
Aulacocarpus Berg. Myrtaceae. 2 Brazil.
Aulacodiscus Hook. f. (1873; non Ehrenb. 1844—Diatomac.) = Pleiocarpidia K. Schum. (Rubiac.).
Aulacolepis Hackel. Gramineae. 4 Japan, Malaysia.
Aulacophyllum Regel (~ Zamia L.). Zamiaceae. 6 C. to NW. trop. S. Am.
Aulacorhynchus Nees = Tetraria Beauv. (Cyperac.).
Aulacospermum Ledeb. Umbelliferae. 4–5 E. Russia to C. As. & NE. Sib.
Aulacostigma Turcz. = Rhynchotheca Ruiz & Pav. (Ledocarpac.).
Aulacothele Monville ex Lem. = Coryphantha (Engelm.) Lem. (Cactac.).
Aulandra H. J. Lam. Sapotaceae. 3 Borneo. Staminal tube.
Aulax Bergius. Proteaceae. 3 S. Afr.
Aulaxanthus Ell. = Anthaenantia Beauv. (Gramin.).
Aulaxia Nutt. = praec.
Aulaxis Haw. = Saxifraga L. (Saxifragac.).
Aulaxis Steud. = Aulaxia Nutt. = Anthaenantia Beauv. (Gramin.).
Aulaya Harv. = Harveya Hook. (Scrophulariac.).
Auleya D. Dietr. = praec.
Aulica Rafin. = Hippeastrum Herb. (Amaryllidac.).
Auliphas Rafin. = Miconia Ruiz & Pav. (Melastomatac.).
Aulisconema Hua = Disporopsis Hance (Liliac.).
Auliza Salisb. = Epidendrum L. (Orchidac.).
Aulojusticia Lindau. Acanthaceae. 1 S. Afr.
Aulomyrcia Berg = Myrcia DC. (Myrtac.).
Aulonemia Goudot = Arthrostylidium Rupr. (Gramin.).
Aulonix Rafin. = Cytisus L. (Legumin.).

Aulosema Walp. = Astragalus L. (Legumin.).

Aulosolena K.-Pol. (~ Sanicula L.). Umbelliferae. 4 W. temp. N. & S. Am.

Aulospermum Coulter & Rose (~ Cymopterus Rafin.). Umbelliferae. 13 N. Am.

Aulostephanus Schlechter = Brachystelma R.Br. (Asclepiadac.).

Aulostylis Schlechter. Orchidaceae. 1 New Guinea.

Aulotandra Gagnep. Zingiberaceae. 1 trop. W. Afr., 5 Madag.

Aurantiaceae Juss. = Rutaceae–Aurantioïdeae Engl.

Aurantium Mill. = Citrus L. (Rutac.).

Aureilobivia Frič = Echinopsis Zucc. (Cactac.).

Aurelia Cass. = Grindelia Willd. (Compos.).

Aurelia J. Gay = Narcissus L. (Amaryllidac.).

Aureliana Boehm. = Panax L. (Araliac.).

Aureliana Lafiteau ex Catesb. = Aralia L. (Araliac.).

Aureliana Sendtn. = Bassovia Aubl. (Solanac.).

Aureolaria Rafin. Scrophulariaceae. 10 E. U.S., 1 Mex.

Auricula Hill = Primula L. (Primulac.).

Auricula-ursi Seguier = praec.

Aurila Nor. = ? Pyrenaria Bl. (Theac.).

Aurinia Desv. Cruciferae. 7 C. & S. Eur. to As. Min.

Aurora Nor. = Quisqualis L. (Combretac.).

Aurota Rafin. = Curculigo Gaertn. (Hypoxidac.).

Austerium Poit. ex DC. = Rhynchosia Lour. (Legumin.).

Australina Gaudich. Urticaceae. 5 S. Afr., Austr., N.Z.

Austroamericium Hendrych. (~ Thesium L.). Santalaceae. 3 Venez., SE. Brazil.

Austrobaileya C. T. White. Austrobaileyaceae. 2 Queensland.

*Austrobaileyaceae (Croiz.) Croiz. Dicots. 1/2 Queensl. Climbing shrubs; ls. opp., ent. (somewhat celastraceous), stips. small. Fls. sol., axill., pedic. P ∞ (± 12), free, imbr.; A 12–25, petaloid, intr., resin-dotted; G̱ ± 8, with 8–14 biseriate adaxial ovules per carpel and 2-lobed styles. Fls. unpleasantly scented. Only genus: *Austrobaileya*. Affinities v. obscure; cf. *Monimiac.*, *Eupomatic.*, *Calycanthac.*

Austrobassia Ulbr. (~ Bassia All.). Chenopodiaceae. 30 Austr.

Austrobuxus Miq. Euphorbiaceae. 2 W. Malaysia; 2 New Guinea, 1 E. Austr., 9–10 New Caled., 1 Fiji.

Austrocactus Britton & Rose. Cactaceae. 4 temp. S. Am.

Austrocedrus Florin & Boutelje. Cupressaceae. 1 temp. S. Am.

Austrocephalocereus Backeb. Cactaceae. 3 Brazil.

Austrocylindropuntia Backeb. = Opuntia Mill. (Cactac.).

Austrodolichos Verdc. Leguminosae. 1 N. Austr., Queensl.

Austroeupatorium R. M. King & H. Rob. Compositae. 12 trop. S. Am., Urug.

Austrogambeya Aubrév. & Pellegr. Sapotaceae. 1 S. trop. Afr.

Austrogramme Fourn. = Syngramma J. Sm. (Gymnogrammac.) + Grammitis Sw. (Grammitidac.).

Austromatthaea L. S. Smith. Monimiaceae. 1 NE. Queensl.

Austromimusops A. Meeuse = Vitellariopsis (Baill.) Dubard (Sapotac.).

AUSTROMUELLERA

Austromuellera C. T. White. Proteaceae. 1 Queensland.

Austromyrtus (Niedenzu) Burret. Myrtaceae. 37 E. Austr., New Caled., New Hebr.

Austropeucedanum Mathias & Constance. Umbelliferae. 1 NW. Argentina.

Austroplenckia Lundell = Plenckia Reissek (Celastrac.).

Austrotaxaceae auct. ex Florin = Taxaceae Lindl.

Austrotaxus Compton. Taxaceae. 1 New Caled.

Autogenes Rafin. = Narcissus L. (Amaryllidac.).

Autrandra Pierre ex Prain = Athroandra (Hook. f.) Pax & K. Hoffm. (Euphorbiac.).

Autranea C. Winkler & Barbey (~ Centaurea L.). Compositae. 1 Syria.

Autranella A. Chev. (~ Mimusops L.). Sapotaceae. 1 trop. Afr.

Autrania Willis = Autranea C. Winkl. & Barbey (Compos.).

Autunesia Dyer (Ind. Kew.) = Antunesia O. Hoffm. (Compos.).

Auxemma Miers. Ehretiaceae. 2 Brazil.

Auxopus Schlechter. Orchidaceae. 2 trop. Afr.

Aveledoa Pittier = Metteniusa Karst. (Alangiac.).

Avellanita Phil. Euphorbiaceae. 1 Chile.

Avellinia Parl. = Colobanthium Reichb. (Gramin.).

Avena Hall. ex Scop. = Agrostis L. (Gramin.).

Avena L. Gramineae. 70 temp. & mts. of trop. *A. sativa* L., the cult. oat, is perhaps derived from *A. fatua* L. It is cult. in Eur. to 69½° N. and forms the staple of the food of a large population. It occurs in two chief forms, the common oat with open spreading panicles, and the Tartarian oat with contracted one-sided panicles. The 2–6-flowered spikelets form a loose panicle. The lemmas bear a dorsal geniculate awn, the basal part of which is usually twisted and hygroscopic. In *A. sterilis* L. the awns cross, and when wetted try to uncurl and thus press on one another till a sort of explosion occurs, jerking away the fruits.

Avenaceae (Kunth) Herter = Gramineae–Aveneae (Kunth) Nees.

Avenaria Fabr. (uninom.) = Bromus L. sp. (Gramin.).

Avenastrum (Koch) Opiz = Helictotrichon Bess. ex Roem. & Schult. (Gramin.).

Avenella Koch = Deschampsia Beauv. (Gramin.).

Avenochloa Holub. Gramineae. 30 temp. Euras., Medit.

Avenula (Dum.) Dum. = Helictotrichon Bess. ex Roem. & Schult. (Gramin.).

Averia Leonard. Acanthaceae. 3 C. Am.

Averrhoa L. Averrhoaceae. 2 trop., long cult.; origin uncertain, but probably native in coastal forests of Brazil. *A. bilimbi* L. (*blimbing*) and *A. carambola* L. (*carambola*) cult. for fr., which is borne on the older stems.

Averrhoaceae Hutch., emend. (~ Oxalidaceae). Dicots. 3/16 Madag., W. Malaysia, ? trop. S. Am. Trees, shrubs or climbers; ls. alt., 1–3–∞-foliol., exstip., with artic. petiole and subopp. ent. lfts. Fls. reg., ☿, heterodistylous or (Malesian *Dapania*) androdioec., in axill. or ramiflorous rac. or panicled cymes; pedic. artic. K 5, shortly conn., imbr.; C 5, sometimes coherent above claw, contorted to apotact.; A 5 + 5, shortly conn., obdiplost., with 2-loc. longit. intr. anth.; disk o; G̲ (5), 5-lobed, with 5 free styles, and 1–6 axill. pend. superposed epi- and anatr. ov. per loc. (ov. sometimes o in ♂ fls. of *Dapania*). Fr. fleshy, ±5-lobed, indehisc. or (*Dapania*) widely loculic., usu. several-seeded; seeds exarillate or (*Dapania, Averrhoa carambola*) arillate; aril fleshy, bilabiate, enveloping the seed, oily; testa hard, smooth or transv. rugose; embr. pend., in variable fleshy oily endosp., cots. flat, ellipt., radicle ± straight. Genera: *Averrhoa, Sarcotheca, Dapania.* Recent research indicates that *Sarcotheca* and *Dapania* should be associated with *Averrhoa* rather than with *Lepidobotrys.* The *Averrhoaceae* would perhaps be better retained as a group in the *Oxalidaceae.*

Averrhoïdium Baill. Sapindaceae. 2 Brazil, Paraguay.

Aversia G. Don = Arversia Cambess. = Polycarpon L. (Caryophyllac.).

Avesicaria Barnh. = Utricularia L. (Lentibulariac.).

Avetra H. Perrier. Dioscoreaceae. 1 E. Madag.

Avicennia L. Avicenniaceae. 14 warm. Constituents of mangrove veg. They have aerial r. projecting out of the mud like *Sonneratia*. The seeds germinate in the fr.

***Avicenniaceae** Endl. ex Schnizl. Dicots. 2/15 trop. coasts. Shrubs or small trees, often greyish or yellowish tomentose. Ls. opp., simple, ent., exstip. Infl. cymose or thyrsif., condensed or spicif., term. and axill., bracteate. Fls. small, yellowish, reg., ♂. K (5), imbr.; C (4), imbr.; A 4; G (4), with 1-ovulate imperf. loc. and short bifid style. Fr. a broad compr. ovoid or spher. bivalved 1-seeded caps. Only genus: *Avicennia.* Perhaps related to *Salvadoraceae.*

Aviceps Lindl. Orchidaceae. 1 S. Afr.

Avicularia Steud. = Polygonum L. (Polygonac.).

Aviunculus Fourr. = Coronilla L. (Legumin.).

Avoira Giseke = Astrocaryum G. F. W. Mey. (Palm.).

Avornela Rafin. = Chamaespartium Adans. (Legumin.).

Awayus Rafin. = Spiraea L. (Rosac.).

Axanthes Blume = Urophyllum Wall. (Rubiac.).

Axanthopsis Korth. = praec.

Axenfeldia Baill. = Mallotus Lour. (Euphorbiac.).

Axia Lour. = Boerhavia L. (Nyctaginac.).

Axiana Rafin. = praec.

Axillaria Rafin. = Polygonatum Mill. (Liliac.).

Axinaea Ruiz & Pav. Melastomataceae. 25 trop. Am.

Axinandra Thw. Memecylaceae. 5 Ceylon, Malay Penins., Borneo. Habit recalls *Crypteroniaceae.*

Axinanthera Karst. = Bellucia Neck. (Melastomatac.).

Axinea Juss. = Axinaea Ruiz & Pav. (Melastomatac.).

Axiniphyllum Benth. Compositae. 2 Mex.

Axinopus Kunth (sphalm.) = Axonopus Beauv. = Paspalum L. (Gramin.).

Axiris L. = Axyris L. (Chenopodiac.).

Axiron Rafin. = Cytisus L. (Legumin.).

Axolopha Alef. = Lavatera L. (Malvac.).

Axolus Rafin. = Cephalanthus L. (Rubiac.).

Axonopus Beauv. Gramineae. 35 trop. S. Am.

Axonopus Hook. f. = Alloteropsis Presl (Gramin.).

Axonotechium Fenzl = Orygia Forsk. (Aïzoac.).

Axyris L. Chenopodiaceae. 7 S. Russia to E. Sib. & Korea.

Ayapana Spach (~Eupatorium L.). Compositae. 10 C. Am., W.I., trop. S. Am.

AYDENDRON

Aydendron Nees = Aniba Aubl. (Laurac.).
Ayenia L. Sterculiaceae. 68 trop. & subtrop. Am.
Ayensua L. B. Smith. Bromeliaceae. 1 Venez.
Aylacophora Cabrera. Compositae. 1 Patagonia.
Aylanthus Rafin. = seq.
Aylantus Juss. = Ailanthus Desf. (Simaroubac.).
Aylmeria Mart. = Polycarpaea Lam. (Caryophyllac.).
Aylostera Speg. (~ Rebutia K. Schum.). Cactaceae. 8 Boliv. to Argent.
Aylthonia N. Menezes (~ Barbacenia Vand.). Velloziaceae. 15 S. Am.
Ayparia Rafin. = Elaeocarpus L. (Elaeocarpac.).
Aytonia L. = Aitonia Thunb. = Nymania S. O. Lindb. (Aitoniac.).
Azadirachta A. Juss. Meliaceae. 2 Indomal. *A. indica* A. Juss. (*nim*) has astringent medicinal bark, and yields good timber, as well as gum.
Azalea Desv. = Rhododendron L. (subgen. Anthodendron Endl.) (Ericac.).
Azalea L. = Loiseleuria Desv. + Rhododendron L. (Ericac.).
Azaleastrum Rydb. = Rhododendron L. (Ericac.).
× **Azaleodendron** Rodigas. Ericaceae. Gen. hybr. (Azalea Desv. × Rhododendron L. *s.str.*).
Azaltea Walp. (sphalm.) = Alzatea Ruiz & Pav. (Lythrac.).
Azamara Hochst. ex Reichb. = Schmidelia L. (Sapindac.).
Azanza Alef. (~ Thespesia Soland. ex Corr.). Malvaceae. 2–3 trop. Afr., Indomal.
Azanza Moç. & Sessé ex DC. = Hibiscus L. (Malvac.).
Azaola Blanco = Payena A. DC. (Sapotac.).
Azara Ruiz & Pav. Flacourtiaceae. 10 S. Bolivia & Brazil to Chile & Argent.; 1 Juan Fernandez. Shrubs with alt. ls.; one stip. is frequently almost as large as the l. to which it belongs, giving the appearance of a pair of ls., not opp. Fl. apetalous; outer sta. often without anthers.
Azarolus Borkh. = Crataegus L., Sorbus L., etc. (Rosac.).
Azedara Rafin. = seq.
Azedarac Adans. = seq.
Azedaraca Rafin. = seq.
Azedarach Mill. = Melia L. (Meliac.).
Azeredia Arruda ex Allem. = Cochlospermum Kunth (Cochlospermac.).
Azima Lam. Salvadoraceae. 4 S. Afr. to Hainan, Philipp. Is., Lesser Sunda Is. In the axils are thorns (the ls. of an undeveloped shoot; cf. *Cactaceae*).
Azolla Lam. Azollaceae. 6 trop. & subtrop.
Azollaceae C. Chr. Salviniales. General structure like *Salvinia*. Two ls. are formed at each node, from the dorsal half of a segment of the apical cell; from the ventral half are formed roots and branches, but not at every node. The ls. are all alike; each is bilobed and has a small cavity near the base, opening by a small pore, and inhabited by the blue-green alga *Anabaena*. The r. hang freely down in the water; usually the root cap is thrown off after a time and the r. comes almost exactly to resemble the submerged l. of *Salvinia*. The sporocarps are formed in pairs (4 in *A. nilotica* Decne) on the ventral lobes of the first ls. of the branches. Each contains one sorus. The microspores are joined together into several masses in each sporangium by the hardened frothy mucilage. Each of these *massulae* has its outer surface provided with curious

barbed hairs (*glochidia*), and escapes on its own account. The megasporangium contains one spore. It sinks to the bottom; decay of the indusium frees the spore and it germinates, giving rise to a ♀ prothallus which floats about on the water and may be anchored to a floating massula by the barbs. Nitrogen-fixation by the *Anabaena* in plants of *Azolla* may be important in rice culture. Only genus: *Azolla*.

Azophora Neck. = Rhizophora L. (Rhizophorac.).
Azorella Lam. Hydrocotylaceae (~ Umbellif.). 70 N. Andes to temp. S. Am., Falkland Is., Antarctic Is. Densely tufted xero. *A. caespitosa* Vahl (balsam-bog, Falklands) forms tufts like *Raoulia*.
Azorellopsis H. Wolff = Mulinum Pers. (Hydrocotylac.).
Azorina Feer (~ Campanula L.). Campanulaceae. 1 Azores.
Aztekium Bödeker. Cactaceae. 1 Mex.
Azukia Takahashi ex Ohwi. Leguminosae. 10 pantrop., temp. E. As.
Azureocereus Akers & Johnson (~ Browningia Britton & Rose). Cactaceae.
Azurinia Fourr. = Veronica L. (Scrophulariac.). [2 Peru.

B

Babactes DC. = Chirita Buch.-Ham. (Gesneriac.).
Babbagia F. Muell. Chenopodiaceae. 4 Austr.
Babcockia Boulos. Compositae. 1 Canary Is.
Babiana Ker-Gawl. Iridaceae. 60 trop. & S. Afr., Socotra.
Babingtonia Lindl. = Baeckea L. (Myrtac.).
Babiron Rafin. = Spermolepis Rafin. (Umbellif.).
Baca Rafin. = Boea Comm. ex. Lam. (Gesneriac.).
Bacasia Ruiz & Pav. = Barnadesia Mutis (Compos.).
Baccataceae Dulac = Caprifoliaceae + Sambucaceae + Adoxaceae.
Baccaurea Lour. Euphorbiaceae. 80 Indomal., Polynesia.
Baccaureopsis Pax = Thecacoris Juss. (Euphorbiac.).
Baccharidastrum Cabrera. Compositae. 2 S. trop. & subtrop. S. Am.
Baccharis L. Compositae. 400 Am., esp. campos. Many are leafless xero. with winged or cylindrical green stems.
Baccharodes Kuntze = seq.
Baccharoïdes Moench = Vernonia Schreb. (Compos.).
Bachia Schomb. = ?Pogonia Juss. (Orchidac.).
Bachmannia Pax. Capparidaceae. 2 S. Afr.
Backebergia H. Bravo = Mitrocereus Backeb. (Cactac.).
Backeria Bakh. f. = Anplectrum A. Gray (Melastomatac.).
Backhousia Hook. & Harv. Myrtaceae. 7 E. Austr. *B. citriodora* F. Muell. gives an essential oil almost entirely citral.
Baclea E. Fourn. Asclepiadaceae. 2 Brazil.
Baclea Greene = Nemacladus Nutt. (Campanulac.).
Baconia DC. = Pavetta L. (Rubiac.).
***Bacopa** Aubl. Scrophulariaceae. 100 warm.
Bactris Jacq. Palmae. *S.l.* 250, *s.str.* 180 trop. Am., W.I. Fls. in groups of 3, one ♀ between two ♂♂. *B. minor* Jacq. (*pupunha* or peach palm, Brazil), ed.fr.

BACTYRILOBIUM

Bactyrilobium Willd. = Cassia L. (Legumin.).
Bacularia F. Muell. (~Linospadix Becc.). Palmae. 10 New Guinea, NE. Austr.
Badamia Gaertn. = Terminalia L. (Combretac.).
Badianifera L. = Illicium L. (Illiciac.).
Badiera DC. (~Polygala L.). Polygalaceae. 15 trop. Am., W.I.
Badiera Hassk. = Polygala L. (Polygalac.).
Badula Juss. Myrsinaceae. 12 Masc.
Badusa A. Gray. Rubiaceae. 3 Caroline, Fiji, Society Is., New Hebr.
Baea Juss. = Boea Comm. ex Lam. (Gesneriac.).
Baeckea Burm. f. = Brunia L. (Bruniac.).
Baeckea L. Myrtaceae. 1 China & Malaysia, 1 Borneo, 65 Austr., New Caled.
Baeckia R.Br. = praec.
Baeica C. B. Cl. = Boeica C. B. Cl. (Gesneriac.).
Baeobotrys J. R. & G. Forst. = Maesa Forsk. (Myrsinac.).
Baeochortus Ehrh. (uninom.) = *Carex humilis* Leyss. (Cyperac.).
Baeolepis Decne ex Moquin. Periplocaceae. 1 S. India.
Baeometra Salisb. Liliaceae. 1 S. Afr.
Baeoterpe Salisb. = Hyacinthus L. (Liliac.).
Baeothrion Pfeiff. = seq.
Baeothryon A. Dietr. = Scirpus L., Eleocharis R.Br., etc. (Cyperac.).
Baeothryon Ehrh. (uninom.) = *Scirpus baeothryon* L. f. = *Eleocharis quinqueflora* (F. X. Hartm.) Schwarz (Cyperac.).
Baeria Fisch. & Mey. Compositae. 10 SW. U.S. (mostly Calif.).
Baeriopsis Howell. Compositae. 1 Lower Calif.
Baeumerta Gaertn., Mey. & Scherb. = Nasturtium R.Br. = Rorippa Scop.
Baeumertia P. & K. = praec.   [(Crucif.).
Bafodeya Prance. Chrysobalanaceae. 1 trop. W. Afr.
Bafutia C. D. Adams. Compositae. 1 W. Equat. Afr.
Bagalatta Roxb. ex Reichb. = Tiliacora Colebr. (Menispermac.).
Bagassa Aubl. Moraceae. 2 Guiana, N. Brazil.
Bagnisia Becc. = Thismia Griff. (Burmanniac.).
Baguenaudiera Bub. = Colutea L. (Legumin.).
Bahamia Britton & Rose = Acacia Mill. (Legumin.).
Bahel Adans. = Artanema D. Don (Scrophulariac.).
Bahelia Kuntze = praec.
Bahia Lag. Compositae. 15 SW. U.S., Mex., Chile.
Bahianthus R. M. King & H. Rob. (~Eupatorium L.). Compositae. 1 Brazil.
Bahiopsis Kellogg = Viguiera Kunth (Compos.).
Baicalia Steller ex Gmel. = Astragalus L. + Oxytropis DC. (Legumin.).
Baikaea B. D. Jacks. (sphalm.) = seq.
Baikiaea Benth. Leguminosae. 10 trop. Afr.
Baikiea auctt. = praec.
Baileya Harv. & A. Gray ex Torr. Compositae. 4 SW. U.S., Mex.
Baileyoxylon C. T. White. Flacourtiaceae. 1 Queensland.
Baillaudea Roberty = Calycobolus Willd. ex Roem. & Schult. (Convolvulac.).
Baillieria Aubl. = Clibadium L. (Compos.).

BALANOPHORACEAE

Baillonacanthus Kuntze = Solenoruellia Baill. (Acanthac.).
Baillonella Pierre ex Dubard. Sapotaceae. 1 W. Equat. Afr.
Baillonia Bocquillon ex Baill. Verbenaceae. 1 S. Am.
Baillonodendron Heim = Dryobalanops Gaertn. (Dipterocarpac.).
Baimo Rafin. = Fritillaria L. (Liliac.).
Baissea A. DC. Apocynaceae. 40 trop. Afr., As.
Baitaria Ruiz & Pav. = Calandrinia Kunth (Portulacac.).
Bajan Adans. = Amaranthus L. (Amaranthac.).
Bakera P. & K. (1) = Bakeria André = Bakerantha L. B. Smith (Bromeliac.).
Bakera P. & K. (2) = Rosa L. (subgen. Bakeria Gandog.) (Rosac.).
Bakera P. & K. (3) = Bakeria Seem. = Plerandra A. Gray (Araliac.).
Bakerantha L. B. Smith. Bromeliaceae. 1 Colombia (?).
Bakerella Van Tiegh. (~ Taxillus Van Tiegh.). Loranthaceae. 16 Madag.
Bakeria André = Bakerantha L. B. Smith (Bromeliac.).
Bakeria Seem. = Plerandra A. Gray (Araliac.).
Bakeridesia Hochreut. Malvaceae. 4 Mex., trop. S. Am.
Bakeriella Pierre ex Dubard = Vincentella Pierre (Sapotac.).
Bakeriopteris Kuntze = Doryopteris J. Sm. (Sinopteridac.).
Bakerisideroxylon Engl. = Afrosersalisia A. Chev. (Sapotac.).
Bakerolimon Lincz. Limoniaceae (~ Plumbaginaceae). 2 Peru, N. Chile.
Bakerophyton (Léonard) Hutch. Leguminosae. 1 trop. Afr.
Bakeros Rafin. = Seseli L. (Umbellif.).
Balaka Becc. Palmae. 20 Fiji, Samoa.
Balanaulax Rafin. = Pasania Oerst. = Quercus L. (Fagac.).
Balaneikon Setchell = Balanophora J. R. & G. Forst. (Balanophorac.).
Balanghas Rafin. = Sterculia L. (Sterculiac.).
Balangue [Gaertn. pro nom. vernac.] DC. ?Rhamnaceae. 1 Madag.
Balania Nor. = Gnetum L. (Gnetac.).
Balania Van Tiegh. = Balanophora J. R. & G. Forst. (Balanophorac.).
Balaniella Van Tiegh. = praec.
Balanitaceae Endl. Dicots. 1/25 trop. Afr. & As. As *Zygophyllac.*, but ls. exstip.; spines axill. (not stip.); fr. drupaceous, with rather thin oil-containing flesh and very thick, bony, 5-angled, 1-seeded endoc. Only genus: *Balanites.*
*****Balanites** Delile. Balanitaceae. 25 trop. Afr. to Burma. Oil from seeds.
Balanocarpus Bedd. = Hopea Roxb. (Dipterocarpac.).
*****Balanopaceae** Benth. corr. Bullock. Dicots. 1/12 Queensl., New Caled., Fiji. Trees or shrubs with alt. or pseudo-vertic. simple ent. or dentic. exstip. ls. Fls. ♂ ♀, naked, dioec.; ♂ in catkins, ♀ sol. ♂ with one subtending scale and (2–)5–6(–12) subsess. latero-intr. anth.; pistillode sometimes present. ♀ subtended by ∞ imbr. bracts; G (2–3), imperf. 2–3-loc., with 2 basal ascending anatr. ov. per loc., and 2 short styles each ending in 2 long spreading subul. stigmatic arms. Fr. an acorn-like drupe, seated in the persist. invol., with 1–2 1-seeded pyrenes and persist. styles; seeds with sparse endosp. Only genus: *Balanops.* An isolated, doubtless ancient group, of quite uncertain affinity. Fls. prob. in 'pre-floral' stage.
Balanophora J. R. & G. Forst. Balanophoraceae. 80 Madag. to Japan, Malaysia, Austr. & Polynesia. Some apogamous.
*****Balanophoraceae** L. C. & A. Rich. Dicots. 18/120, all but one trop. Parasites

117

BALANOPLIS

(no chlorophyll) on tree roots, to which the tuberous rhiz. is attached by suckers. From it springs the infl. (sometimes developed within the rhiz. and breaking through it), which comes above ground as a spike or head with scaly ls. and small unisexual fls. ♂ usu. P 3–4 or (3–4), A 3–4 or more or less. ♀ usu. P 0, G̲ or G̅ (1–2, rarely 3); ovule without an integument. Nut- or drupe-like fr. Endosperm. For details and figures see Harms in Engler, *Nat. Pflfam.*, ed. 2, **16b** (1935).

Classification and genera (after Harms):
A. Rhiz. containing starch.
  I. *Mystropetaloïdeae* (♂ fl. with trimerous ± zygo. P and 2 stam.; ♀ with ± campan. 3-lobed P, G̅ with 3 pend. ov.): *Mystropetalon*.
  II. *Dactylanthoïdeae* (♂ fl. with or without P, with 1–2 free or united stam.; ♀ fl. with 2–3 narrow squamif. tep. or 3-lobed P): *Hachettea*, *Dactylanthus*.
  III. *Sarcophytoïdeae* (♂ fl. with conspic. P, ♀ fl. without P. Style 0, stig. sess. Ovary with 3 (or 1?) ov. Fls. in conspic. branched pan.): *Sarcophyte*, *Chlamydophytum*.
  IV. *Helosidoïdeae* (♂ fl. with 3-lobed or ent. P, ♀ fl. without P. Styles 2(–5). Fls. immersed in a dense layer of filif. chaffy hairs, arr. in dense, clavate, ± globose or pelt., apparently simple infl.): *Scybalium*, *Helosis*, *Corynaea*, *Rhopalocnemis*, *Exorhopala*, *Ditepalanthus*.
  V. *Lophophytoïdeae* (♂ and ♀ fls. naked. Fls. in clavate panicles, not immersed in a layer of chaffy hairs): *Lophophytum*, *Ombrophytum*, *Lathrophytum*, *Juelia*.
B. Rhiz. containing a waxy resinous substance (balanophorin).
  VI. *Balanophoroïdeae* (♀ fl. without [?in *Langsdorffia* and *Thonningia* with] P, style 1 filif. Infl. subtended by ∞ imbr. scales): *Balanophora*, *Langsdorffia*, *Thonningia*.

Van Tieghem treated most of the above subfamilies as distinct fams., and it is probable that the group is a heterogeneous one, the saprophytic or parasitic habit having brought about a superficial similarity.

Balanoplis Rafin. = Castanopsis (D. Don) Spach (Fagac.).
Balanops Baill. Balanopaceae. 12 Queensl., New Caled., Fiji.
Balanopseae Benth. = Balanopaceae Benth. corr. Bullock.
Balanopsidaceae Engl. = praec.
Balanopsis Rafin. = Ocotea Aubl. (Laurac.).
Balanopteris Gaertn. = Heritiera Ait. (Sterculiac.).
Balanostreblus Kurz = Sorocea A. St-Hil. (Morac.).
Balansaea Boiss. & Reut. (~ Chaerophyllum L.). Umbelliferae. 2 Spain, Morocco.
Balansaephytum Drake del Castillo = Poikilospermum Zipp. ex Miq. (Urticac.).
Balansochloa Kuntze = Germainia Bal. & Poitrasson (Gramin.).
Balantiaceae Dulac = Asclepiadaceae R.Br.
Balantium Desv. ex Ham. = Parinari Aubl. (Chrysobalanac.).
Balantium Kaulf. (*s.str.*) = Dicksonia L'Hérit. (Dicksoniac.).
Balantium auctt. (incl. C. Chr. *Ind. Fil.* 1905) = Culcita Presl (Dicksoniac.).
Balardia Cambess. = Spergularia Presl (Caryophyllac.).
Balaustion Hook. Myrtaceae. 2 W. Austr.

BALSAMACEAE

*Balbisia Cav. Ledocarpaceae. 8 S. Am. Undershrubs. Ov. ∞ per cpl.
Balbisia DC. = Rhetinodendron Meissn. (Compos.).
Balbisia Willd. = Tridax L. (Compos.).
Balboa Liebm. = Tephrosia Pers. (Legumin.).
*Balboa Planch. & Triana. Guttiferae. 1 Colombia.
Baldellia Parl. (~ Echinodorus Rich.). Alismataceae. 1 W. & S. Eur., N. Afr.
Baldingera Dennst. = Premna L. (Verbenac.).
Baldingera Gaertn., Mey. & Scherb. = Phalaris L. (Gramin.).
Baldingeria Neck. = Cotula L. (Compos.).
Baldingeria F. W. Schmidt = Leontodon L. (Compos.).
Baldomiria Herter = Leptochloa Beauv. (Gramin.).
*Balduina Nutt. Compositae. 3 S. U.S.
Balduinia Rafin. (1) = praec.
Balduinia Rafin. (2) = seq.
Baldwinia Rafin. = Passiflora L. (Passiflorac.).
Baldwinia Torr. & Gray = Balduina Nutt. (Compos.).
Balendasia Rafin. = Passerina L. (Thymelaeac.).
Balenerdia Comm. ex Steud. = Nanodea Banks ex Gaertn. f. (Santalac.).
Balessam Bruce = Balsamodendron Kunth (Burserac.).
Balexerdia Comm. ex Endl. = Nanodea Banks ex Gaertn. f. (Santalac.).
Balfouria R.Br. = Wrightia R.Br. (Apocynac.).
Balfourina Kuntze = Didymaea Hook. f. (Rubiac.).
Balfourodendron Mello ex Oliv. Rutaceae. 1 S. Braz., Paraguay, N. Argent.
Balfuria Reichb. = Balfouria R.Br. = Wrightia R.Br. (Apocynac.).
Balingayum Blanco = Calogyne R. Br. (Goodeniac.).
Baliospermum Blume. Euphorbiaceae. 6 India, SE. As., Malay Penins., Java, Sumbawa.
Balisaea Taub. = Aeschynomene L. (Legumin.).
Ballardia Montr. = Cloëzia Brongn. (Myrtac.).
Ballarion Rafin. = Stellaria L. (Caryophyllac.).
Ballela (Rafin.) B. D. Jacks. = Merremia Dennst. (Convulvulac.).
Ballexerda Comm. ex A. DC. = Nanodea Banks ex Gaertn. f. (Santalac.).
Ballieria Juss. = Baillieria Aubl. = Clibadium L. (Compos.).
Ballimon Rafin. = Daucus L. (Umbellif.).
Ballochia Balf. f. Acanthaceae. 3 Socotra.
Ballosporum Salisb. = Gladiolus L. (Iridac.).
Ballota L. Labiatae. 35 Eur., Medit., W. As.
Ballote Mill. = praec.
Balls-Headleya F. Muell. ex Bailey (nomen). Escalloniaceae (Cunoniaceae?). 1 Queensland. Quid?
Ballya Brenan. Commelinaceae. 1 trop. E. Afr.
Balmeda Nocca = Grewia L. (Tiliac.).
Balmea Martínez. Rubiaceae. 1 Mex.
Balmisa Lag. = Arisarum Targ.-Tozz. (Arac.).
Baloghia Endl. Euphorbiaceae. 13 E. Austr., New Caled., Norfolk I.
Balonga Le Thomas. Annonaceae. 1 W. Equat. Afr.
Baloskion Rafin. = Restio Rottb. (Restionac.).
Balsamaceae Lindl. = Altingiaceae Lindl.

119

BALSAMARIA

Balsamaria Lour. = Calophyllum L. (Guttif.).
Balsamea Gled. = Commiphora Jacq. (Burserac.).
Balsameaceae Dum. = Burseraceae Kunth.
Balsamiflua Griff. = Populus L. (Salicac.).
Balsamina Mill. = Impatiens L. (Balsaminac.).
***Balsaminaceae** DC. Dicots. 4/5–600, Euras., Afr., N. Am. Herbs with watery translucent stems and alt. ls., usu. exstip. Fl. ♀, zygo. K 5 (the 2 ant. small or aborted, the post. one spurred), petaloid; C 5 (the lat. petals united in pairs); A 5, anthers adhering to one another and forming a cap over the ovary, whose growth ultimately breaks the fil. at their bases; G (5), 5-loc., with ∞ ovules, anatr., pend. with dorsal raphe. Explosive capsule. Seed exalb. Genera: *Impatiens, Hydrocera, Semeiocardium, Impatientella.* Affinities prob. with *Geraniac., Sapindac.* and *Tropaeolac.*
Balsamita Mill. = Chrysanthemum L. (Compos.).
Balsamocarpon Clos = Caesalpinia L. (Legumin.).
Balsamocitrus Stapf. Rutaceae. 1 trop. E. Afr.
Balsamodendron DC. = seq.
Balsamodendrum Kunth = Commiphora Jacq. (Burserac.).
Balsamona Vand. = Cuphea P.Br. (Lythrac.).
Balsamophloeos O. Berg = Commiphora Jacq. (Burserac.).
Balsamorhiza Hook. Compositae. 12 W. N. Am.
Balsamus Stackh. = Commiphora Jacq. (Burserac.).
Balthasaria Verdc. Theaceae. 2–3 trop. Afr.
Baltimora L. Compositae. 4 Mex. to Colombia.
Baltimorea Rafin. = praec.
Bambekea Cogn. Cucurbitaceae. 2 trop. W. Afr.
Bamboga Baill. = Mamboga Blanco = Mitragyna Korth. (Naucleac.).
Bambos Retz. = Bambusa Schreb. (Gramin.).
Bamburanta L. Linden = Hybophrynium K. Schum. (Marantac.).
Bambus J. F. Gmel. = Bambusa Schreb. (Gramin.).
Bambusa Mutis ex Caldas = Guadua Kunth (Gramin.).
***Bambusa** Schreb. Gramineae. 70 trop. & subtrop. As., Afr., Am. The typical genus of bambôos.
Bambusaceae Nak. (∼ Gramineae–Bambusoïdeae or –Bambuseae). Monocots. About 45 genera, trop. & subtrop., a few temp. Shrubs or trees with persist. culms, v. rarely perenn. herbs; l. blades flat, many-nerved, often with cross-veins, usu. with petioloid base and artic. with sheath. Spikel. uniform, ♀, 1–∞-fld., variously arr.; gl. 2 or more; lemmas sim., 5–∞-nerved, usu. exarist.; lodic. usu. 3; stam. 3, 6 or more; styles usu. 2 or 3; fr. a nut, berry or caryopsis. Chief gen.: *Dendrocalamus, Melocanna, Bambusa, Oxytenanthera, Arundinaria, Sasa.*

The bamboos are char. by stems that become woody below and often grow to great size. The trop. forms usu. grow in clumps, which continually expand, the new shoots appearing at the outer side; the subtrop. and temp. forms are usu. continuous in their growth. There is a big rhiz. below ground and erect perenn. woody stems above, which appear in the rains (or spring) and grow rapidly to the full height, when the scale ls. fall and the leafy branches spread out. Growth is very rapid in *Dendrocalamus giganteus* Munro, reaching as

much as 41 cm. a day. Some climb. The height is often great, reaching to
36 m. in some forms.

Some fl. annually, others at longer intervals, and some are like *Agave* and
Corypha, flowering only once, all together, and then dying down. They fl. only
when in full leaf, and as the infl. grows the ls. usu. fall. The seedlings grow
for several years without forming tall shoots, producing large well-stored rhiz.
They then send up shoots increasing in length from year to year.

Bamia R.Br. ex Sims = Hibiscus L. (Malvac.).

Bamiania Lincz. Limoniaceae (~ Plumbaginaceae). 1 Afghan.

Bamlera K. Schum. & Lauterb. Melastomataceae. 2 New Guinea.

Banalia Bub. = Hedysarum L. (Legumin.).

Banalia Moq. = Indobanalia A. N. Henry & Roy (Amaranthac.).

Banalia Rafin. = Croton L. (Euphorbiac.).

Banara Aubl. Flacourtiaceae. 35 trop. Am., W.I.

Banava Juss. = Adamboë Adans. = Lagerstroemia L. (Lythrac.).

Bancalus Kuntze = Nauclea L. (Naucleac.).

Bancrofftia Steud. = seq.

Bancroftia Billb. = Arracacia Bancr. (Umbellif.).

Bancroftia Macfad. = Tovaria Ruiz & Pav. (Tovariac.).

Bandeiraea Welw. ex Benth. & Hook. f. Leguminosae. 3 trop. W. Afr.

Bandereia Baill. = praec.

Bandura Adans. = Nepenthes L. (Nepenthac.).

Banffya Baumg. = Gypsophila L. (Caryophyllac.).

Banglium Buch.-Ham. ex Wall. = Boesenbergia Kuntze (Zingiberac.).

Bania Becc. = Carronia F. Muell. (Menispermac.).

Banisteria auctt. = Banisteriopsis C. B. Rob. & Small (Malpighiac.).

Banisteria L. = Heteropterys Kunth (Malpighiac.).

Banisterioïdes Dubard & Dop. Malpighiaceae. 1 Madag.

Banisteriopsis C. B. Rob. & Small. Malpighiaceae. 100 trop. Am., W.I. Fr.
like *Acer*.

Banisterodes Kuntze = Xanthophyllum Roxb. (Xanthophyllac.).

Banium Ces. ex Boiss. = Carum L. (Umbellif.).

Banjolea Bowdich. ?Acanthaceae (inc. sed.). 1 Madeira.

Bankesia Bruce = Banksia Bruce = Brayera Kunth (Rosac.).

Banksea Koen. = Costus L. (Costac.).

Banksia Bruce = Brayera Kunth (Rosac.).

Banksia Domb. ex DC. = Cuphea P.Br. (Lythrac.).

Banksia J. R. & G. Forst. = Pimelea Banks (Thymelaeac.).

***Banksia** L. f. Proteaceae. 50 Austr. (Austr. honeysuckle). Shrubs and trees
with xero. habit. Fls. in dense spikes. Hard woody follicles enclosed in woody
cones derived from bract and bracteoles. Seeds winged.

Baobab Adans. = Adansonia L. (Bombacac.).

Baobabus Kuntze = praec.

Baoulia A. Chev. = Murdannia Royle (Commelinac.).

Baphia Afzel. Leguminosae. 65 warm Afr., Madag.; 1 Borneo. *B. nitida*
Afzel., cam-wood, used for red dye; the wood when first cut is white, but turns
red in the air.

Baphiastrum Harms. Leguminosae. 10 trop. Afr.

BAPHICACANTHUS

Baphicacanthus Bremek. Acanthaceae. 1 NE. India, S. China, Indoch.
Baphiopsis Benth. Leguminosae. 2 trop. Afr.
Baphorhiza Link = Alkanna Tausch (Boraginac.).
Baprea Pierre ex Pax & K. Hoffm. = Cladogynos Zipp. ex Span. (Euphorbiac.).
Baptisia Vent. Leguminosae. 35 N. Am. In *B. perfoliata* R.Br. there are perfoliate ls., really in two vertical ranks, but becoming one-ranked by twisting of internodes alt. right and left.
Baptistania Pfitz. (sphalm.) = seq.
Baptistonia Barb. Rodr. = Oncidium Sw. (Orchidac.).
Baranda Llanos = Barringtonia J. R. & G. Forst. (Barringtoniac.).
Barathranthus (Korth.) Miq. Loranthaceae. 4 Ceylon, W. Malaysia.
Baratostachys (Korth.) Kuntze = Phoradendron Nutt. (Loranthac.).
Barattia A. Gray & Engelm. = Encelia Adans. (Compos.).
Baraultia Spreng. = Barraldeia Thou. = Carallia Roxb. (Rhizophorac.).
Barbacenia Vand. Velloziaceae. 140 S. Am., trop. Afr., Madag.
Barbaceniopsis L. B. Smith. Velloziaceae. 2 N. Andes.
Barba-jovis Seguier = Ãnthyllis L. (Legumin.).
Barbaraea Beckm. = seq.
***Barbarea** R.Br. Cruciferae. 20 N. temp. *B. verna* (Mill.) Aschers. cult. as salad-pl. (winter-cress).
Barbarea Scop. = Dentaria L. (Crucif.).
Barbellina Cass. = Staehelina L. (Compos.).
Barberetta Harv. Haemodoraceae. 1 S. Afr.
Barberina Vell. = Symplocos Jacq. (Symplocac.).
Barbeuia Thou. Barbeuiaceae. 1 Madag.
Barbeuiaceae (Baill.) Nak. Dicots. (Centrosp.). 1/1 Madag. Large woody lianes; ls. alt., ent., pet. artic. at base, exstip. Fls. ☿, long-pedic., in short axill. rac. K 5, orbic., concave; C 0; A ∞, on annular disk, fil. short, anth. lin., sagitt., intr.; G (2), with 1 basal ovule per loc. and 2 thick styles; fr. a hard, woody, 1–2-seeded caps.; seed subglob., with conspic. fleshy aril, cots. v. large, unequal. Only genus: *Barbeuia*. The whole plant blackens on drying.
Barbeya Alboff = Amphoricarpos Visiani (Compos.).
Barbeya Schweinf. Barbeyaceae. 1 Abyssinia, Eritrea, Somalia, Arabia.
***Barbeyaceae** Rendle. Dicots. 1/1 NE. Afr., Arabia. Trees with opp. simple ent. exstip. ls. and ♂ ♀ dioec. reg. apet. fls. K 3–4; C 0; A 6–9; G 1-loc. with 1 term. plumose style and one pend. anatr. ov. Dry indeh. fr. surr. by membr. accr. K; no endosp. Ls. white-toment. beneath. ♂ fl. often rusty-toment. Only genus: *Barbeya*. Many tech. chars. of *Ulmac.*, but resembling the olive-tree (*Olea*) in gen. appearance. Perh. distantly related to *Simmondsiaceae*.
Barbeyastrum Cogn. = Dichaetanthera Endl. (Melastomatac.).
Barbiera Spreng. = seq.
Barbieria DC. Leguminosae. 1 trop. Am., W.I.
Barbilus P.Br. = Trichilia P.Br. (Meliac.).
Barbosa Becc. (~ Syagrus Mart.). Palmae. 1 E. Brazil.
Barbosella Schlechter. Orchidaceae. 12 trop. & temp. S. Am.
Barbula Lour. (1790; non Hedw. 1801—Musci) = Caryopteris Bunge (Verbenac.).
Barbylus Juss. = Barbilus P.Br. = Trichilia P.Br. (Meliac.).

Barcella Drude (~ Elaeis Jacq.). Palmae. 1 Brazil.
Barcena Dugès = Colubrina Rich. ex Brongn. (Rhamnac.).
Barckhausenia Menke = seq.
Barckhausia DC. = Barkhausia Moench = Crepis L. (Compos.).
Barclaya Wall. Barclayaceae. 3–4 Indomal., in muddy rain-forest streams.
Barclayaceae (Endl.) Li. Dicots. 1/4 Indomal. Totally submersed stemless water plants. Rhiz. villous. Ls. pet., elongate-oblong or rounded, cordate at base, membr., sometimes violet-coloured beneath. Fls. on short or long scapes, rarely emersed, not or scarcely opening, 'hydrocleistogamous'. Invol. of 5 oblong long-mucronate bracts immediately beneath fl., resembling a K. K (4–5), tube adnate to ov. C (8–∞), 2–3-ser. A ± 10 + 10 + 10 + 10 + 10, epipet., the outer ster., fil. v. short. G̅ (10), with ∞ pariet. ovules, stig. forming an obscurely radiate disk with central conical tubular projection. All internal parts of fl. (exc. anth.) dull reddish-purple. Fr. a globose fleshy berry with ∞ seeds; flesh bright rose, sweet; seeds densely covered with soft setae. Only genus: *Barclaya*. Should perhaps be merged in *Nymphaeac*.
Bardana Hill = Arctium L. (Compos.).
× **Bardendrum** hort. Orchidaceae. Gen. hybr. (iii) (Barkeria × Epidendrum).
Bareria Juss. = Barreria Scop. = Poraqueiba Aubl. (Icacinac.).
Baretia Comm. ex Cav. = Turraea L. (Meliac.).
Bargemontia Gaudich. = Nolana L. (Nolanac.).
Barhamia Klotzsch = Croton L. (Euphorbiac.).
Barjonia Decne. Asclepiadaceae. 12 Brazil.
Barkania Ehrenb. = Halophila Thou. (Hydrocharitac.).
Barkeria Knowles & Westc. Orchidaceae. 10 C. Am.
Barkerwebbia Becc. (~ Heterospathe Scheff. ex Becc.). Palmae. 3 New Guinea.
Barkhausenia Schur = Borckhausenia Gaertn., Mey. & Scherb. = Corydalis Vent. (Fumariac.).
Barkhausia Moench = Crepis L. (Compos.).
Barkhusenia Hoppe = praec.
× **Barkidendrum** hort. (vii) = × Bardendrum hort. (Orchidac.).
Barklya F. Muell. Leguminosae. 1 Queensland. Thin endosp.
× **Barlaceras** E. G. Camus. Orchidaceae. Gen. hybr. (i) (Aceras × Barlia).
Barlaea Reichb. f. = Cynorkis Thou. (Orchidac.).
Barleria L. Acanthaceae. 230 trop., many xero. Bracteoles frequently repres. by thorns. The seeds have surface hairs which swell when wetted.
Barleriacanthus Oerst. = Barleria L. (Acanthac.).
Barlerianthus Oerst. = Barleria L. (Acanthac.).
Barleriola Oerst. Acanthaceae. 6 W.I.
Barleriopsis Oerst. = Barleria L. (Acanthac.).
Barleriosiphon Oerst. = Barleria L. (Acanthac.).
Barlerites Oerst. = Barleria L. (Acanthac.).
Barlia Parl. Orchidaceae. 1 Medit. region.
× **Barliaceras** Cif. & Giac. (v) = Barlaceras E. G. Camus (Orchidac.).
Barnadesia Mutis. Compositae. 20 S. Am. Shrubs.
Barnardia Lindl. = Scilla L. (Liliac.).
Barneoudia Gay. Ranunculaceae. 3 Chile, Argent.

Basonca Rafin. = Rogeria J. Gay (Pedaliac.).
Basselinia Vieill. Palmae. 10 New Caled.
Bassia All. Chenopodiaceae. 10 C. Eur. & Medit. to C. As.; 79 Austr.
Bassia Koen. ex L. = Madhuca Gmel. (Sapotac.).
Bassovia Aubl. Solanaceae. 15 C. & S. Am.
Bastardia Kunth. Malvaceae. 8 W.I. & warm Am.
Bastardiopsis Hassler. Malvaceae. 1 S. Am.
Bastera J. F. Gmel. = seq.
Basteria Houtt. = Berkheya Ehrh. (Compos.).
Basteria Mill. = Calycanthus L. (Calycanthac.).
Bastia Steud. = Bustia Adans. = Buphthalmum L. (Compos.).
Basutica Phillips. Thymelaeaceae. 1 S. Afr.
Batania Hatusima. Menispermaceae. 1 Philipp. Is.
Batanthes Rafin. = Gilia Ruiz & Pav. (Polemoniac.).
Batatas Choisy = Ipomoea L. (Convolvulac.).
Bataprine Nieuwl. = Galium L. (Rubiac.).
Batemania Endl. = seq.
Batemannia Lindl. Orchidaceae. 5 trop. S. Am.
× **Bateostylis** hort. Orchidaceae. Gen. hybr. (iii) (Batemannia × Otostylis).
Baterium Miers = Haematocarpus Miers (Menispermac.).
Batesanthus N. E. Br. Periplocaceae. 1 W. Afr.
Batesia Spruce. Leguminosae. 1 N. Brazil.
Bathiaea Drake del Castillo. Leguminosae. 1 Madag.
Bathiea Schlechter = Neobathiea Schlechter (Orchidac.).
Bathieaea Willis (sphalm.) = Bathiaea Drake del Castillo (Legumin.).
Bathiorhamnus Capuron. Rhamnaceae. 2 Madag.
Bathmium Link = Tectaria Cav. (Aspidiac.).
Bathratherum Nees corr. Hochst. = Arthraxon Beauv. (Gramin.).
Bathysa C. Presl. Rubiaceae. 7 Brazil, Peru.
Bathysograya Kuntze = Badusa A. Gray (Rubiac.).
*****Batidaceae** Mart. ex Meissn. Dicots. (Centrosp.). 1/2 Pacif., W.I., Atl. S. Am. Coast shrubs with opp. fleshy linear ls. and spikes of dioec. fls. ♂ in axils of 4-ranked brs., with cup-like K, C 4, A 4. ♀ naked, G̲ (2), 4-loc., with 1 anatr. ov. in each. No endosp. Only genus: *Batis*. Systematic position has been disputed, but the general centrospermous affinity seems undeniable.
Batidaea Greene. Rosaceae. 16 N. Am.
Batidophaca Rydb. = Astragalus L. (Legumin.).
Batindum Rafin. = Oftia Adans. (Myoporac.).
Batis P.Br. Batidaceae. 1 New Guinea, Queensl.; 1 Hawaii, SW. U.S., W.I., Atl. coast S. Am.
Batocarpus Karst. Moraceae. 4 trop. Am.
Batocydia Mart. ex DC. = Bignonia L. (Bignoniac.).
Batodendron Nutt. (∼ Vaccinium L.). Ericaceae. 3 N. Am.
Batopedina Verdcourt. Rubiaceae. 2 W. & S. trop. Afr.
Batrachium S. F. Gray (∼ Ranunculus L.). Ranunculaceae. 30 N. & S. temp.
Batratherum Nees = Arthraxon Beauv. (Gramin.).

Batschia J. F. Gmel. = Lithospermum L. (Boraginac.).
Batschia Moench = Eupatorium L. (Compos.).
Batschia Mutis ex Thunb. = Abuta Aubl. (Menispermac.).
Batschia Vahl = Humboldtia Vahl (Legumin.).
Battandiera Maire. Liliaceae. 1 N. Afr.
Battata Hill = Solanum L. (Solanac.).
Bauchea Fourn. = Sporobolus R.Br. (Gramin.).
Baucis Phil. = Brachyclados D. Don (Compos.).
Baudinia Lesch. ex DC. (1828) = Calothamnus Labill. (Myrtac.).
Baudinia Lesch. ex DC. (1839) = Scaevola L. (Goodeniac.).
Baudouinia Baill. Leguminosae. 6 Madag.
Bauera Banks ex Andr. Baueraceae. 3 temp. E. Austr., Tasm.
Baueraceae Lindl. Dicots. 1/3 Austr. Shrubs of wet or marshy ground. Ls. opp., trifol., sess., exstip., sometimes glandular-hairy. Fls. sol., axill. K (4–)6–8 (–10), valvate or slightly imbr., C as many, A ∞ on a narrow disk, with filif. fil. and small anth., G̲ (2), sometimes semi-inf., with 2 or several, pend. or ascend. ov., and 2 distinct recurved styles. Fr. a loculic. 2-valved caps., with 1 to several seeds, testa granulate or pubesc. Only genus: *Bauera*. An isolated group, with some chars. (e.g. ovary and caps.) of *Saxifragac. (s.l.)*, but showing several features of *Lythrac.*, e.g. marshy habitat, 6–8-merous fls., valv. cal., and esp. the magenta-pink cor. The morph. of the ls. requires investigation; the lateral 'leaflets' could be stipules.
Bauerella Borzì. Rutaceae. 1 E. Austr., New Caled.
Bauerella Schindler = seq.
Baueropsis Hutch. Leguminosae. 1 Austr.
Bauhinia L. Leguminosae. 300 warm. Many lianes with stems curiously shaped, flattened or corrugated and twisted owing to a peculiar mode of growth in thickness (cf. other lianes). Some spp. have tendrils (branches). In some the young ls. droop. In the axils of the stips. are usu. found small linear trichome structures; in some they form stout interstipular thorns. Great variety in floral structure.
Baukea Vatke. Leguminosae. 1 Madag.
Baumannia DC. = Damnacanthus Gaertn. (Rubiac.).
Baumannia K. Schum. = Neobaumannia Hutch. & Dalz. (Rubiac.).
Baumannia Spach = Oenothera L. (Onagrac.).
Baumea Gaudich. Cyperaceae. 30 Madag., Masc., Indomal., Austr., Pacif.
Baumgartenia Spreng. = Borya Labill. (Liliac.).
Baumgartia Moench = Cocculus DC. (Menispermac.).
Baumia Engl. & Gilg. Scrophulariaceae. 1 trop. Afr.
Baumiella Wolff. Umbelliferae. 1 S. trop. Afr.
Baursea Hoffmgg. = Philodendron Schott (Arac.).
Baursia Schott = praec.
Bauschia Seub. ex Warm. = Aneilema R.Br. (Commelinac.).
Bauxia Neck. = Marica Ker-Gawl. (Iridac.).
Bavera Poir. = Bauera Banks (Bauerac.).
Baxtera Reichb. = Loniceroïdes Bullock (Asclepiadac.).
*****Baxteria** R.Br. ex Hook. Xanthorrhoeaceae. 1 W. Austr.
Bayonia Dugand = Onohualcoa Lundell (Bignoniac.).

BAZIASA

Baziasa Steud. = Sabazia Cass. (Compos.).
Bazina Rafin. = Lindernia All. (Scrophulariac.).
Bdallophyton Eichl. Rafflesiaceae. 4 Mex.
Bdellium Baill. ex Laness. = Commiphora Jacq. (Burserac.).
Bea C. B. Clarke = Boea Comm. ex Lam. (Gesneriac.).
Beadlea Small = Cyclopogon Presl (Orchidac.).
Bealia Scribn. ex Vasey = Muhlenbergia Schreb. (Gramin.).
Beata O. F. Cook. Palmae. 1 W.I.
Beatonia Herb. = Tigridia Ker-Gawl. (Iri c.).
Beatsonia Roxb. = Frankenia L. (Frankeniac.).
Beaua Pourr. = Boea Comm. ex Lam. (Gesneriac.).
Beaucarnea Lem. = Nolina Michx (Agavac.).
Beaufortia R.Br. Myrtaceae. 15 W. Austr.
Beauharnoisia Ruiz & Pav. = Tovomita Aubl. (Guttif.).
Beauica P. & K. = Boeica C. B. Cl. (Gesneriac.).
Beaumaria Deless. = Aristotelia L'Hérit. (Elaeocarpac.).
× **Beaumontara** hort. (vii) = × Recchara hort. (Orchidac.).
Beaumontia Wall. Apocynaceae. 15 China, Indomal.
Beaumulix Willd. ex Poir. = Reaumuria L. (Tamaricac.).
Beauprea Brongn. & Gris. Proteaceae. 10 New Caled.
Beaupreopsis Virot. Proteaceae. 1 New Caled.
Beautempsia Gaudich. = Capparis L. (Capparidac.).
Beautia Comm. ex Poir. = Thilachium Lour. (Capparidac.).
Beauverdia Herter. Amaryllidaceae. 9 temp. S. Am.
Beauvisagea Pierre = Lucuma Molina (Sapotac.).
Bebbia Greene. Compositae. 2 SW. U.S.
Beccabunga Hill = Veronica L. (Scrophulariac.).
Beccarianthus Cogn. Melastomataceae. 5 Borneo, Philipp. Is., NE. New Guinea.
Beccariella Pierre = Planchonella Pierre (Sapotac.).
Beccarimnea Pierre ex P. & K. = praec.
Beccarina Van Tiegh. = Trithecanthera Van Tiegh. (Loranthac.).
Beccarinda Kuntze. Gesneriaceae. 7 Burma to Hainan & Indoch.
Beccariodendron Warb. (~ Mitrephora Hook. f. & Thoms.). Annonaceae. 1 New Guinea.
Beccariophoenix Jumelle & Perrier. Palmae. 1 Madag.
Becheria Ridl. Rubiaceae. 1 Malay Penins., St Barbe I.
Bechium DC. = Vernonia Schreb. (Compos.).
Bechonneria Hort. ex Carr. = Beschorneria Kunth (Amaryllidac.).
Bechsteineria Muell. (sphalm.) = Rechsteineria Regel (Gesneriac.).
Becium Lindl. Labiatae. 10 trop. Afr.
Beckea Pers. = Baeckea Burm. f. = Brunia L. (Bruniac.).
Beckea A. St-Hil. = Baeckea L. (Myrtac.).
Beckera Fresen. = Snowdenia C. E. Hubbard (Gramin.).
Beckeria Bernh. = Melica L. (Gramin.).
Beckeria Heynh. = Beckera Fresen. = Snowdenia C. E. Hubb. (Gramin.).
Beckeropsis Figari & De Not. Gramineae. 6 trop. Afr.
Beckia Rafin. = Baeckea L. (Myrtac.).

Beckmannia Host. Gramineae. 2 N. temp.
Beckwithia Jepson (~Ranunculus L.). Ranunculaceae. 2 Calif.
Beclardia A. Rich. Orchidaceae. 2 Masc.
Becquerelia Brongn. Cyperaceae. 10 trop. S. Am.
Beddomea Hook. f. = Aglaia Lour. (Meliac.).
Bedfordia DC. Compositae. 2 SE. Austr., Tasm.
Bedousi Augier = Casearia Jacq. (Flacourtiac.).
Bedousia Dennst. = praec.
Bedusia Rafin. = praec.
Beehsa Endl. = Beesha Kunth = Melocanna Trin. (Gramin.).
Beën Schmidel = Limonium L. (Plumbaginac.).
Beera Beauv. = Hypolytrum Rich. (Cyperac.).
Beesha Kunth = Melocanna Trin. (Gramin.).
Beesha Munro = Ochlandra Thw. (Gramin.).
Beesia Balf. f. & W. W. Smith. Ranunculaceae. 3 Burma, SW. China.
Beethovenia Engl. = Ceroxylon Humb. (Palm.).
Befaria Mutis ex L. = Bejaria Mutis ex L. corr. Zea ex Vent. (Ericac.).
Begonia L. Begoniaceae. 900 trop. & subtrop., esp. Am. Most are perenn. herbs with thick rhiz. or tubers. Several climb by aid of roots like ivy. Ls. rad. or alt., in two ranks, with large stips. One side of the l. is larger than the other, whence the name 'elephant's-ear', by which they are sometimes known. The surface of the l. is easily wetted, and drip-tips are frequent (cf. *Ficus*). In the axils groups of little tubers are frequently found; these are not axillary branches, but are borne upon the true axillary branch, which does not lengthen. They also repr. easily by adv. buds which readily form on pieces of l. cut off and placed on the soil under suitable conditions of moisture, etc. (the common mode used in horticulture). A callus forms over the wound, and in it there develops a meristem which gives rise to one or more buds.
*Begoniaceae C. A. Agardh. 5/920 trop. Perenn. herbs, with watery stems and rad. or distich., usu. asymm. ls. and large membr. stips. Infl. axillary, dich. with a bostryx tendency. The first axes usually end in ♂, the last and sometimes the last but one in ♀ fls. ♂ fl.: P 2, valvate, or 4, decussate, corolline; A ∞, free or not, the connective often elongated and the anthers variously shaped. ♀ fl.: P 2–5; G̅ usu. (2–3), with 2–3 loc. and axile plac. often projecting far into them; ovules ∞, anatr.; styles ± free. Ovary usu. winged. Fr. a ± horny, sometimes papery, leathery or fleshy, usu. 1–3(–6)-winged caps. Seeds ∞, minute, without endosp. Genera: *Hillebrandia, Begonia, Semibegoniella, Begoniella, Symbegonia.*
Begoniella Oliv. Begoniaceae. 5 Colombia.
Beguea Capuron. Sapindaceae. 1 Madag.
Behaimia Griseb. Leguminosae. 1 Cuba.
Behen Hill = Jacea Mill. = Centaurea L. (Compos.).
Behen Moench = Oberna Adans. = Silene L. (Caryophyllac.).
Behenantha Schur = praec.
Behnia Didrichsen. Philesiaceae. 1 S. Afr.
Behria Greene. Liliaceae. 1 S. Calif.
Behrinia Sieber = Berinia Brign. = Crepis L. (Compos.).
Behuria Cham. Melastomataceae. 10 S. Brazil.

Beilia Eckl. = Watsonia Mill. (Iridac.).
Beilia Kuntze = Micranthus (Pers.) Eckl. (Iridac.).
Beilschmidtia Reichb. = seq.
Beilschmiedia Nees. Lauraceae. Over 200 trop., Austr., N.Z.
Beirnaertia J. Louis ex Troupin. Menispermaceae. 1 trop. Afr.
Bejaria Mutis ex L. corr. Zea ex Vent. Ericaceae. 30 trop. & subtrop. Am.
B. racemosa Vent. and others (Andes rose) form a consp. feature in the veg., taking the place of Rhododendrons.
Bejaudia Gagnep. Palmae. 1 Indoch.
Bejuco Loefl. = Hippocratea L. (Celastrac.).
Beketowia Krassn. = Braya Sternb. & Hoppe (Crucif.).
Belairia A. Rich. Leguminosae. 5 Cuba.
*****Belamcanda** Adans. Iridaceae. 2 E. As.
Belandra Blake. Apocynaceae. 1 C. Am.
Belangera Cambess. = Lamanonia Vell. (Cunoniac.).
Belangeraceae J. G. Agardh = Cunoniaceae–Belangereae Engl.
Belanthera P. & K. = Beloanthera Hassk. = Hydrolea L. (Hydrophyllac.).
Belantheria Nees = Brillantaisia Beauv. (Gramin.).
Belencita Karst. Capparidaceae. 1 Colombia.
Belendenia Rafin. = Tritonia Ker-Gawl. (Iridac.).
Belenia Decne = Physochlaina D. Don (Solanac.).
Belenidium Arn. = Hymenatherum Cass. (Compos.).
Beleropone C. B. Clarke (sphalm.) = Beloperone Nees (Acanthac.).
Belharnosia Adans. = Sanguinaria L. (Papaverac.).
Belia Steller = Claytonia L. (Portulacac.).
Belicea Lundell. Rubiaceae. 1 C. Am.
Belilla Adans. = Mussaenda L. (Rubiac.).
Belingia Pierre = Zollingeria Kurz (Sapindac.).
Belis Salisb. = Cunninghamia R.Br. (Taxodiac.).
Belladona Mill. = Atropa L. (Solanac.).
Belladonna Boehm. = praec.
Belladonna Sweet = Amaryllis L. (Amaryllidac.).
Bellardia All. (~ Bartsia L.). Scrophulariaceae. 3 Medit.
Bellardia Colla = Microseris D. Don (Compos.).
Bellardia Schreb. = Coccocypselum P.Br. corr. Schreb. (Rubiac.).
Bellardiochloa Chiov. (~ Poa L.). Gramineae. 1 S. Eur.
Bellendena R.Br. Proteaceae. 1 Tasmania.
Bellendenia Endl. = praec.
Bellendenia Rafin. = Tritonia Ker-Gawl. (Iridac.).
Bellermannia Klotzsch = ? Gonzalagunia Ruiz & Pav. (Rubiac.).
Bellevalia Delile = Althenia Petit (Zannichelliac.).
Bellevalia Lapeyr. (~ Hyacinthus L.) Liliaceae. 50 W. Medit. to Persia.
Bellevalia Montrouz. apud Beauvis. = Agatea A. Gray (Violac.).
Bellevalia Roem. & Schult. = Richeria Vahl (Euphorbiac.).
Bellevalia Scop. = Marurang Adans. = Clerodendrum L. (Verbenac.).
Bellia Bub. = Chaerophyllum L. (Umbellif.).
Bellida Ewart. Compositae. 2 Austr.
Bellidastrum Scop. (~ Aster L.). Compositae. 1 mts. S. Eur.

Bellidiaster Dum. = praec.
Bellidiastrum Cass. = praec.
Bellidiastrum Less. = Osmites L. (Compos.).
Bellidiopsis Spach = Osmites L. (Compos.).
Bellidistrum Reichb. = Bellidastrum Scop. (Compos.).
Bellidium Bertol. = Bellis L. (Compos.).
Bellilla Rafin. = Belilla Adans. = Mussaenda L. (Rubiac.).
Bellinia Roem. & Schult. = Saracha Ruiz & Pav. (Solanac.).
Belliolum Van Tiegh. Winteraceae. 8 Solomon Is., New Caled.
Belliopsis Pomel = Bellium L. (Compos.).
Bellis L. Compositae. 15 Eur., Medit. *B. perennis* L. (daisy) multiplies and hibernates by short rhiz. Ray ♀. Pappus usu. o. Head closes at night and
Bellium L. Compositae. 6 Medit. [in wet weather.
Belloa Remy. Compositae. 11 Andes.
Bellonia L. Gesneriaceae. 2 W.I. Often axillary thorns.
Bellota C. Gay = Beilschmiedia Nees (Laurac.).
Belluccia Adans. = Ptelea L. (Rutac.).
*****Bellucia** Neck. ex Rafin. Melastomataceae. 18 trop. Am. Trees. G 8–15-loc. Fr. ed.
Bellynkxia Muell. Arg. Rubiaceae. 6 trop. Am.
*****Belmontia** E. Mey. = Sebaea Soland. ex R.Br. (Gentianac.).
Beloakon Rafin. = Phlomis L. (Labiat.).
Beloanthera Hassk. = Hydrolea L. (Hydrophyllac.).
Beloëre Shuttlew. ex A. Gray = Abutilon Gaertn. (Malvac.).
Beloglottis Schlechter. Orchidaceae. 2 Costa Rica, Bolivia.
Belonanthus Graebn. Valerianaceae. 5 Peru, Bolivia.
Belonia Adans. = Bellonia L. (Gesneriac.).
Belonites E. Mey. = Pachypodium Lindl. (Asclepiadac.).
Belonophora Hook. f. Rubiaceae. 8 trop. W. Afr.
Beloperone Nees. Acanthaceae. 60 warm Am., W.I.
Beloperonides Oerst. = praec.
Belopis Rafin. = Salvia L. (Labiat.).
Belostemma Wall. (~ Tylophora R.Br.). Asclepiadaceae. 2 India, China.
Belosynapsis Hassk. (~ Cyanotis D. Don). Commelinaceae. 6 Indomal.
Belotia A. Rich. (~ Trichospermum Bl.). Tiliaceae. 12 Mex. to Ecuador, W.I. Androphore. Fr. 2-valved, winged.
Belotropis Rafin. = Salicornia L. (Chenopodiac.).
Belou Adans. = Aegle Corrêa (Rutac.).
Belovia Moq. = Suaeda Forsk. (Chenopodiac.).
Beltokon Rafin. = Origanum L. (Labiat.).
Beltrania Miranda. Euphorbiaceae. 1 Mex.
Belutta Rafin. = Allmania R.Br. (Amaranthac.).
Belutta-kaka Adans. = Chonemorpha G. Don (Apocynac.).
Belvala Adans. = Struthiola L. (Thymelaeac.).
Belvalia Delile = Althenia Petit (Zannichelliac.).
Belvisia Desv. = Napoleona Beauv. (Napoleonac.).
Belvisia Mirbel. Polypodiaceae. 15 Africa to Polynesia. Sporangia confined to narrow apex of l., covered when young by scales.

BELVISI[AC]EAE

Belvisi[ac]eae R.Br. = Napoleonaceae P. Beauv.
Bemarivea Choux. Sapindaceae. 1 Madag.
Bembecodium Lindl. = Athanasia L. (Compos.).
Bembicia Oliv. Flacourtiaceae. 1 Madag. G. Infl. axillary, sessile, strobiloid, with densely imbr. bract-scales.
Bembicidium Rydb. Leguminosae. 1 Cuba.
Bembicina Kuntze = Bembicia Oliv. (Flacourtiac.).
Bembiciopsis H. Perrier = Camellia L. (Theac.).
Bembicium Mart. = Eupatorium L. (Compos.).
Bembicodium P. & K. = Bembycodium Kunze = Athanasia L. (Compos.).
Bembix Lour. = Ancistrocladus Wallich (Ancistrocladac.).
Bembycodium Kunze = Athanasia L. (Compos.).
Bemsetia Rafin. = Ixora L. (Rubiac.).
Benaurea Rafin. = Musschia Dum. (Campanulac.).
Bencomia Webb & Berth. Rosaceae. 4 Canaries, ? Madeira. Hollow axis only
Benedicta Bernh. = Carbenia Adans. (Compos.). [in ♀.
Benedictella Maire. Leguminosae. 1 Morocco.
Beneditaea Toledo. Hydrocharitaceae. 1 trop. Am.
Benevidesia Saldanha & Cogn. Melastomataceae. 2 SE. Brazil.
Benguellia G. Taylor. Labiatae. 1 Angola.
Benincasa Savi. Cucurbitaceae. 1 trop. As. Fr. of *B. hispida* (Thunb.) Cogn. coated with wax; it is eaten in curries.
Benitoa Keck. Compositae. 1 Calif.
Benitzia Karst. = Gymnosiphon Blume (Burmanniac.).
Benjamina Vell. = Dictyoloma A. Juss. (Rutac.).
Benjaminia Mart. ex Benj. = Quinquelobus Benj. = Bacopa Aubl., etc. (Scrophulariac.).
Benkara Adans. = Randia L. (Rubiac.).
Bennetia DC. = Bennettia S. F. Gray = Saussurea DC. (Compos.).
Bennetia Rafin. = Sporobolus R.Br. (Gramin.).
Bennettia R.Br. = Galearia Zoll. (Pandac.).
Bennettia S. F. Gray = Saussurea DC. (Compos.).
Bennettia Miq. = Bennettiodendron Merr. (Flacourtiac.).
Bennettiaceae R.Br. = Scepaceae Lindl. (*q.v.*) + Pandaceae Pierre.
Bennettiodendron Merr. Flacourtiaceae. 3 China, Indomal.
Benoicanthus Heine & A. Raynal. Acanthaceae. 2 Madag.
Benoistia H. Perrier & Leandri. Euphorbiaceae. 2 Madag.
Bensonia Abrams & Bacigal. (1929; von Buckman 1845—gen. foss.) = seq.
Bensoniella Morton. Saxifragaceae. 1 NW. U.S.
Benteca Adans. = Hymenodictyon Wall. (Rubiac.).
Benteka Adans. = praec.
Benthamantha Alef. (~ Cracca Benth.). Leguminosae. 25 trop. Am.
× **Benthamara** hort. (iv, v) = × Trevorara hort. (Orchidac.).
Benthamia Lindl. (1830) = Amsinckia Lehm. (Boraginac.).
Benthamia Lindl. (1833) = Dendrobenthamia Hutch. (Cornac.).
Benthamia A. Rich. Orchidaceae. 25 Mascarenes.
Benthamidia Spach. Cornaceae. 3 N. & C. Am.
Benthamiella Spegazz. Solanaceae. 15 Argent., Patagonia.

Benthamina Van Tiegh. Loranthaceae. 1 E. Austr.

Benthamistella Ǩuntze = Stellularia Benth. (Scrophulariac.).

Benthamodendron Philipson (sphalm.) = Dendrobenthamia Hutch. (Cornac.).

Bentheca Neck. = Benteca Adans. = Hymenodictyon Wall. (Rubiac.).

Bentheka Neck. ex A. DC. = Willughbeia Roxb. (Apocynac.).

Bentia Rolfe. Acanthaceae. 1 S. Arabia.

Bentinckia Berry ex Roxb. Palmae. 2 India, Nicobar Is. G 3-loc. Berry 1-seeded. Sheaths 2–4.

Bentinckiopsis Becc. (~ Clinostigma H. Wendl.). Palmae. 3 Bonin & Caro-
Bentnickiopsis Becc. (sphalm.) = praec. [line Is.

Benzoë Fabr. (uninom.) = *Lindera* Thunb. sp. (Laurac.).

Benzoin Hayne = Styrax L. (Styracac.).

Benzoin Schaeff. = Lindera Thunb. (Laurac.).

Benzoina Rafin. = praec.

Benzonia Schumacher. Rubiaceae (inc. sed.). 1 W. Afr.

Bequaertia R. Wilczek. Celastraceae. 1 trop. Afr.

Bequaertiodendron De Wild. Sapotaceae. 3 trop. & subtrop. Afr.

Berardia Brongn. = Nebelia [Neck.] Kuntze (Bruniac.).

Berardia Vill. Compositae. 1 Alps. Head sessile or short-stalked.

*****Berberidaceae** Juss. Dicots. 4/575 N. temp., trop. mts., S. Am. Perennial
shrubs with alt., usu. spiny, simple, pinnate, or pinnately 1–3-ternate ls. Fls. in
racemes or fasc. or cymes, ♀, reg. Typical formula P 3 + 3 + 3 + 3, A 3 + 3, G̲ 1.
Of the outer whorls, the 2 outer are P proper, the two inner 'honey-ls.', usu.
with nectaries at base (cf. *Ranunculac.*), the former often termed K, the latter C.
Anthers intr., but usu. opening by two post. valves (cf. *Laurac.*); the valve
with the pollen upon it moves upward and turns round so that pollen faces
centre of fl. Cpl. always 1, with 1 or several basal ov. Berry. Embryo straight
in rich endosperm. Genera: *Berberis, Mahonia, Epimedium, Vancouveria*.
Related to *Podophyllac.*

Berberidopsis Hook. f. Flacourtiaceae. 1 Chile.

Berberis L. Berberidaceae. 450 N. & S. Am., Eurasia, N. Afr. Shrubs. The
ls. are simple, but usu. show a joint where the blade meets the petiole, seeming
to indicate a derivation from a cpd. l. There are also 'short' and 'long' shoots
(cf. *Coniferae*). The latter have their ls. metam. into spines (usu. tripartite);
transitions may often be seen. The former stand in the axils of the spines and
bear green ls. and racemes of fls. (afterwards sometimes elongating to 'long'
shoots). The pollination mechanism is interesting. The upper surface of the
base of each sta. is sensitive to contact, and when it is touched by an insect in
search of honey (secreted by the nectaries upon the bases of the inner P ls.)
the sta. springs violently upwards, covering the side of the visitor's head with
pollen, which it may place on the stigma in the next fl. visited. The frs. are
sometimes made into preserves.

An interesting point about the common barberry (*B. vulgaris* L.) is its connection with the disease known as black rust, which occurs on wheat and other
Gramineae. The fungus (*Puccinia graminis* Pers.) passes through two alt. stages
in its life history, one on the grass, the other on the barberry, so that if there
are no barberry plants in a district, the grass is to some extent, and in certain
countries, at least temporarily safeguarded against black rust.

BERCHEMIA

*Berchemia Neck. ex DC. Rhamnaceae. 22 palaeotrop., Atl. N. Am.
Berchemiella Nakai. Rhamnaceae. 2 China, Japan.
Berchtoldia Presl = Chaetium Nees (Gramin.).
Berckheya Pers. = Berkheya Ehrh. (Compos.).
Berebera Baker (sphalm.) = Berrebera Hochst. = Millettia Wight & Arn. (Legumin.).
Berendtia A. Gray (1868; non Goepp. 1845—gen. foss.) = seq.
Berendtiella Wettst. & Harms. Scrophulariaceae. 4 Mex., C. Am.
Berenice Salisb. = Allium L. (Alliac.).
Berenice Tul. (= ? Cephalostigma A. DC.). Campanulaceae. 1 Réunion.
Bergella Schnizlein = Bergia L. (Elatinac.).
Bergena Adans. = Lecythis Loefl. (Lecythidac.).
*Bergenia Moench. Saxifragaceae. 6 C. & E. As. Hybrids frequent.
Bergenia Neck. = Lythrum L. (Lythrac.).
Bergenia Rafin. = Cuphea P.Br. (Lythrac.).
Bergera Koen. ex L. = Murraya Koen. ex L. (Rutac.).
Bergeranthus Schwantes. Aïzoaceae. 23 S. Afr.
Bergeretia Bub. = Illecebrum L. (Caryophyllac.).
Bergeretia Desv. = Clypeola L. (Crucif.).
Bergeria Koen. = Koenigia L. (Polygonac.).
Bergerocactus Britton & Rose. Cactaceae. 1 Calif., Mex.
Bergerocereus Britt. & Rose (sphalm.) = praec.
Bergeronia M. Micheli. Leguminosae. 1 Paraguay.
Berghausia Endl. = Garnotia Brongn. (Gramin.).
Berghesia Nees. Rubiaceae (inc. sed.). 1 Mex.
Berghias Juss. = Bergkias Sonn. = Gardenia Ellis (Rubiac.).
Bergia L. Elatinaceae. 25 trop. & temp.
Bergiera Neck. = praec.
Berginia Harv. Acanthaceae. 3 NW. Mex.
Bergkias Sonner. = Gardenia Ellis (Rubiac.).
Bergsmia Blume = Ryparosa Blume (Flacourtiac.).
Berhardia C. Muell. = Berardia Vill. (Compos.).
Berhautia Balle. Loranthaceae. 1 W. Afr.
Beriesa Steud. = Siebera Presl = Anredera Juss. (Basellac.).
Beringeria [Neck.] Link = Ballota L. (Labiat.).
Berinia Brign. = Crepis L. (Compos.).
*Berkheya Ehrh. Compositae. 90 Afr.
Berkheyopsis O. Hoffm. Compositae. 14 trop. & S. Afr.
Berla Bub. = Sium L. (Umbellif.).
Berlandiera DC. Compositae. 5 S. & E. U.S.
Berliera Buch.-Ham. = Myrioneuron R.Br. (Rubiac.).
*Berlinia Soland. ex Hook. f. & Benth. Leguminosae. 15 trop. Afr.
Berlinianche (Harms) de Vattimo (~ Pilostyles Guillem.). Rafflesiaceae. 2 trop. Afr.
Bermudiana Mill. = Sisyrinchium L. (Iridac.).
Bernarda Adans. = Bernardia Mill. (Euphorbiac.).
Bernardia P.Br. = Adelia L. (Euphorbiac.).
Bernardia Endl. = Berardia Vill. (Compos.).

134

Bernardia Mill. Euphorbiaceae. 50 warm Am., W.I.
Bernardina Baudo = Lysimachia L. (Primulac.).
Bernardinia Planch. Connaraceae. 6 C. & trop. S. Am.
Berneuxia Decne. Diapensiaceae. 2 E. Tibet, SW. China.
Bernhardia P. & K. = Bernardia Mill. (Euphorbiac.).
Bernhardia Willd. = Psilotum Sw. (Psilotac.).
Berniera DC. = Gerbera L. ex Cass. (Compos.).
*Bernieria Baill. = Beilschmiedia Nees (Laurac.).
Bernoullia Neck. = Geum L. (Rosac.).
*Bernoullia Oliv. Bombacaceae. 2 C. & trop. S. Am.
Bernullia Rafin. = Bernoullia Neck. = Geum L. (Rosac.).
Berrebera Hochst. = Millettia Wight & Arn. (Legumin.).
Berresfordia L. Bolus. Aïzoaceae. 1 S. Afr.
Berria Roxb. = Berrya Roxb. mut. DC. (Tiliac.).
Berroa Beauverd (~ Lucilia Cass.). Compositae. 1 subtrop. S. Am.
Berrya Klein = Litsea Lam. (Laurac.).
*Berrya Roxb. mut. DC. Tiliaceae. 6 Indomal., Polynesia. B. cordifolia
  (Willd.) Burret (B. ammonilla Roxb.) gives a valuable timber (Trincomali
  wood, Ceylon, India).
Bersama Fres. Melianthaceae. 2 (polymorph.) trop. & S. Afr.
Bertera Steud. = Gladiolus L. (Iridac.).
Berteroa DC. Cruciferae. 8 N. palaeotemp.
Berteroa Zipp. (nomen). New Guinea. Quid?
Berteroëlla O. E. Schulz. Cruciferae. 1 temp. E. As.
Berthelotia DC. = Pluchea Cass. (Compos.).
Berthiera Vent. = Bertiera Aubl. (Rubiac.).
Bertholetia Brongn. = Bertholletia Humb. & Bonpl. (Lecythidac.).
Bertholetia Reichb. (sphalm.) = Berthelotia DC. = Pluchea Cass. (Compos.).
Bertholletia Humb. & Bonpl. Lecythidaceae. 2 trop. S. Am., W.I. Fr. a large
  woody capsule, containing seeds with hard woody testa and oily endosperm
  (Brazil nuts). The fr. is indehiscent and the seeds are procured by opening it
  with an axe. It is closed by a plug formed of the hardened calyx, and in
  germination the seedlings escape here.
Bertiera Aubl. Rubiaceae. 30 trop. Am., Afr.
Bertiera Blume = Mycetia Reinw. (Rubiac.).
Bertolonia DC. = Leucheria Lag. (Compos.).
× Bertolonia Hort. Melastomataceae. Gen. hybr. (Cassebeeria × Gravesia).
Bertolonia Moç. & Sessé = Cercocarpus Kunth (Rosac.).
*Bertolonia Raddi. Melastomataceae. 10 Brazil.
Bertolonia Rafin. = Pilopus Rafin. = Phyla Lour. (Verbenac.).
Bertolonia Spinola = Myoporum Banks (Myoporac.).
Bertolonia Spreng. = Tovomitopsis Pl. & Triana (Guttif.).
Bertuchia Dennst. = Fagraea Thunb. (Potaliac.).
Bertya Planch. Euphorbiaceae. 23 Austr., Tasm.
Bertyaceae J. G. Agardh = Euphorbiaceae–Ricinocarpoïdeae Pax.
Berula Hoffm. ex Bess. = Sium L. (Umbellif.).
Berula Koch. Umbelliferae. 3 Eur. to C. As. & Persia.
Beryllis Salisb. = Ornithogalum L. (Liliac.).

Berzelia Brongn. Bruniaceae. 11 S. Afr.

Berzelia Mart. = Hermbstaedtia Reichb. (Amaranthac.).

Berzeliaceae Nak. = Bruniaceae R.Br.

Beschorneria Kunth. Agavaceae. 10 Mex.

Besenna A. Rich. = Albizia Durazz. (Legumin.).

Besha D. Dietr. = Beesha Munro = Ochlandra Thw. (Gramin.).

Besleria L. Gesneriaceae. 150 warm Am., W.I.

Bessera Schult. = Pulmonaria L. (Boraginac.).

*Bessera Schult. f. Alliaceae. 3 S. U.S., Mex.

Bessera Spreng. = Xylosma J. R. & G. Forst. (Flacourtiac.).

Bessera Vell. = Pisonia L. (Nyctaginac.).

Besseya Rydb. Scrophulariaceae. 12 N. Am.

Bessia Rafin. = Intsia Thou. (Legumin.).

Bestram Adans. = Antidesma L. (Stilaginac.).

Beta L. Chenopodiaceae. 6 Eur., Medit. From *B. vulgaris* L. (*B. maritima* L.), the sea-beet, are derived the garden beetroot, the sugar-beet and the mangoldwurzel. The plant is a biennial and stores reserves in the root, the nonnitrogenous materials taking the form of sugar.

The sugar-beet is widely cult. in W. Eur. and elsewhere for its sugar, and has largely displaced the older industry of cane sugar. The sugar content of the roots has been continually improved by selection, and now frequently represents over 20 % of the weight.

The garden beet is a favourite vegetable; the mangold is valuable for feeding cattle, etc. The ls. are sometimes eaten like spinach.

Betaceae Burnett = Chenopodiaceae + Amaranthaceae.

Betchea Schlechter = Caldcluvia D. Don (Cunoniac.).

Betckea DC. = Valerianella Mill. (Valerianac.).

Betela Rafin. = Piper L. (Piperac.).

Betencourtia A. St-Hil. = Galactia P.Br. (Legumin.).

Bethencourtia Choisy = Senecio L. (Compos.).

Betonica L. (∼ Stachys L.). Labiatae. 15 temp. Euras.

Betula L. Betulaceae. 60 N. temp., arct. *B. alba* L., the birch, reaches to the N. limit of trees, which in much of the N. temp. zone is occupied by *B. nana* L., a creeping shrubby form. The winter buds are scaly, the scales representing stips.: the outer two or three pairs of them have no ls. Witches' brooms are very commonly to be seen as dense tufts of twigs.

Trees with catkins of fls. The ♂ catkins are laid down in autumn as large buds at the end of the year's growth, the ♀ further back, on leafy branches. In the axil of each l. of the catkin there are 3 fls. The bracts of the lateral fls. (α, β) occur, but no bracteoles. In the ♂ the bracteoles α β are joined to the bract itself. Each fl. has two sta. and a perianth, often reduced from the typical 4 ls. to the 2 median ls., or even to the single anterior l. The sta. are divided into halves nearly to the base; the lat. ones are absent. In the ♀ the bracteoles α β are free from the bract at the time of fertilisation, but afterwards they unite with it to form the 3-lobed woody scale under the fruit (or rather the tissue beneath them grows up, carrying up all together). The 2-loc. ovary gives rise to a 1-seeded nut, attached to the scale. No P. Tough wood, used for shoes, charcoal, etc. The oil from bark used in tanning Russia leather, to which it gives its fragrance. The bark of *B. papyracea* Ait. (N. Am.) used for making canoes.

*Betulaceae S. F. Gray. Dicots. 2/95, N. temp., trop. mts., Andes, Argent. Trees or shrubs with alt. undivided ls. and membranous deciduous stips. Seedling stems radial in symmetry, but in old branches ls. ± distich. Fls. monoec., anemoph., in term. catkins (or ♀ in heads); stem thus sympodial. In axils of catkin-ls. are small dich. cymes, typically of 3 fls. ♂ fl. united to br., P minute, A 2–4; ♀ P o, G (2), 2-loc. at base, each with 1 pend. anatr. ov. with one integument; 2 free styles. Some chalazogamic. 1-seeded nut; endosp. o. After fert. br. and bracteoles grow into a scale-like organ which may remain attached to fr. Genera: *Alnus, Betula.*

Betulaster Spach = Betula L. (Betulac.).
Beurera Kuntze = Bourreria P.Br. (Boraginac.).
Beureria Ehret = Calycanthus L. (Calycanthac.).
Beurreria Jacq. = Bourreria P.Br. (Boraginac.).
Bevania Bridges ex Endl. = Desfontainia Ruiz & Pav. (Potaliac.).
Beveria Collinson = Beureria Ehret = Calycanthus L. (Calycanthac.).
Beverinckia Salisb. ex DC. = Azalea Desv. = Rhododendron L. (Ericac.).
Beverna Adans. = Babiana Ker-Gawl. (Iridac.).
Bewsia Goossens. Gramineae. 1 S. Afr.
Beyeria Miq. Euphorbiaceae. 12 Austr.
Beyeriopsis Muell. Arg. = praec.
Beyrichia Cham. & Schlechtd. Scrophulariaceae. 3 Braz., W.I.
Beythea Endl. = Elaeocarpus L. (Elaeocarpac.).
Bezanilla Remy = Psilocarphus Nutt. (Compos.).
Bhesa Buch.-Ham. ex Arn. Celastraceae. 5 Indomal., Pacif.
Bhidea Stapf ex Bor. Gramineae. 1 India.
Bia Klotzsch = Tragia L. (Euphorbiac.).
Biancaea Todaro = Caesalpinia L. (Legumin.).
*Biarum Schott. Araceae. 15 Medit. Infl. appears when plant is leafless.
Biaslia Vand. = Mayaca Aubl. (Mayacac.).
Biasolettia Koch = Freyera Reichb. (Umbellif.).
Biasolettia Pohl ex Bak. = Eupatorium L. (Compos.).
Biasolettia Presl = Hernandia L. (Hernandiac.).
Biassolettia Endl. = praec.
Biatherium Desv. = Gymnopogon Beauv. (Gramin.).
Biauricula Bub. = Iberis L. (Crucif.).
Bicarpellatae Benth. & Hook. f. A 'Series' of Dicots., comprising the orders *Gentianales, Polemoniales, Personales* (*q.v.*), and *Lamiales,* and the 'anomalous'
Bicchia Parl. = Leucorchis E. Meyer (Orchidac.).   [fam. *Plantaginaceae.*
Bichea Stokes = Cola Schott (Sterculiac.).
Bichenia D. Don = Trichocline Cass. (Compos.).
Bicornaceae Dulac = Saxifragaceae Juss. (*s.str.*).
Bicornella Lindl. = Cynorkis Thou. (Orchidac.).
Bicorona A. DC. = Melodinus J. R. & G. Forst. (Apocynac.).
Bicuculla Borckh. = Adlumia Rafin. (Fumariac.).
Bicuculla auctt. = seq.
Bicucullaria Juss. = Dicentra Borckh. corr. Bernh. (Fumariac.).

BICUCULLATA

Bicucullata Marchant ex Adans. = praec.

Bicuspidaria Rydb. = Mentzelia L. (Loasac.).

Bidaria Decne. = Gymnema R.Br. (Asclepiadac.).

Bidens L. Compositae. 230 cosmop. Fr. distr. by the 2–6 barbed bristles of the pappus. *B. beckii* Torr. (N. Am.) a heterophyllous water pl.

Bidwellia Herb. = seq.

Bidwillia Herb. corr. Lindl. Liliaceae. 1 E. Austr. (N.S.W.) (? = Asphodelus L.).

Biebersteinia Stephan ex Fisch. Biebersteiniaceae. 5 Greece to C. As.

Biebersteiniaceae Endl. Dicots. 1/5 SE. Eur. to C. As. Perenn. herbs, occasionally stemless, with ± tuberous rhiz. and pinnate or pinnatipartite stip. ls. Infl. rac. or panic. K 5, imbr.; C 5, often dentic. at apex, sometimes unguic. below; A 10, v. shortly connate, with 5 alternipet. glands; G̲ (5), deeply lobed, styles arising from base of lobes and connate into a capitate stig., ovules sol., apical, pend. Fr. of 5 indeh. 1-seeded carp., separating from axis. Only genus: *Biebersteinia*. Palynological research suggests the possibility that this group should be referred to the *Rosaceae*, perhaps near the *Potentilleae*. (Bortenschlager, *Grana Palynol.* 7: 420–1, 1967).

Bielschmeidia Panch. & Sebert (sphalm.) = Beilschmiedia Nees (Laurac.).

Bielzia Schur = Centaurea L. (Compos.).

Bieneria Reichb. f. = Chloraea Lindl. (Orchidac.).

Bienertia Bunge. Chenopodiaceae. 1 W. C. As.

Biermannia King & Pantl. Orchidaceae. 2 India, China.

Bifaria (Hack.) Kuntze = Mesosetum Steud. (Gramin.).

Bifaria Van Tiegh. = Korthalsella Van Tiegh. (Viscac.).

Bifariaceae Nak. = Loranthaceae–Phoradendreae Engl.

Bifolium Gaertn., Mey. & Scherb. = Maianthemum Web. ex Wigg. (Liliac.).

Bifolium Petiver ex Nieuwl. = Listera R.Br. (Orchidac.).

*Bifora Hoffm. Umbelliferae. 2 Medit. to C. As.

Biforis Spreng. = praec.

Bifrenaria Lindl. Orchidaceae. 10 trop. S. Am.

× **Bifrinlaria** hort. Orchidaceae. Gen. hybr. (vi) (Bifrenaria × Maxillaria).

Bigamea Koen. = Ancistrocladus Wall. (Ancistrocladac.).

Bigelonia Rafin. = Bigelowia Rafin. = Stellaria L. (Caryophyllac.).

Bigelovia Sm. = Forestiera Poir. (Oleac.).

Bigelovia Spreng. (1821) = Casearia Jacq. (Flacourtiac.).

Bigelovia Spreng. (1827) = Borreria G. F. W. Mey. (Rubiac.).

Bigelowia DC. ex Ging. (1824) = Noisettia Kunth (Violac.).

Bigelowia DC. (1830) = Bigelovia Spreng. (1827) = Borreria Mey. (Rubiac.).

*Bigelowia DC. (1836). Compositae. 40 N. Am. to Ecuador.

Bigelowia Rafin. = Stellaria L. (Caryophyllac.).

Biggina Rafin. = Salix L. (Salicac.).

Biglandularia Karst. = Leiphaimos Cham. & Schlechtd. (Gentianac.).

Biglandularia Seem. = Sinningia Nees (Gesneriac.).

Bignonia L. Bignoniaceae. Now restricted to 1 sp., *B. capreolata* L., N. Am.

Bignonia auctt. = Doxantha Miers (Bignoniac.).

*Bignoniaceae Juss. Dicots. 120/650 trop.; a few temp. One genus (*Catalpa*) common to Old and New Worlds. Trees and shrubs, most commonly lianes, with opp. usu. cpd. exstip. ls. Many xero. shrubs with condensed stem, but the chief interest centres in the climbers, a very important feature in the forest veg. of S. Am., incl. twiners (e.g. *Tecomaria, Pandorea*), root-climbers (*Campsis radicans*), and tendril climbers (most *B*.). In *Eccremocarpus*, etc., the internodes and petioles are sensitive, but in most *B*. the tendrils are at the ends of the ls. (in place of leaflets, as in *Vicia*). The tendrils are frequently branched; in some cases the branched tendril occupies the place of one leaflet. Three types of tendril are found—simple twiners, tendrils provided with adhesive disks (as in *Parthenocissus*), and hooked tendrils. See *Glaziovia, Doxantha*, etc. The climbing stems exhibit many features of anatomical interest, owing to the peculiar growth in thickness.

Infl. usu. dich. with cincinnal tendency; bracts and bracteoles present. Fl. ♀, zygo., hypog. K (5); C (5), usu. bell- or funnel-shaped, descendingly imbr.; A 4, epipet., didynamous, the anther-lobes usu. one above the other, the post. std. always present; G (2) on hypog. disk, 2- (or rarely 1-) loc., with ∞ erect anatr. ov. on axile (or in 1-loc. ovaries sometimes parietal) plac. Caps. septifr. or loculic.; seed usu. flattened and with large membranous wing, exalb.

Classification and chief genera (after Schumann):

1. BIGNONIEAE (ovary completely 2-loc., compressed parallel to septum, or cylindrical; caps. septifr. with winged seeds; usu. tendrillate): *Glaziovia, Doxantha, Oroxylum.*
2. TECOMEAE (ovary 2-loc., compressed at rt. angles to septum or cylindrical; caps. loculic. with winged seeds; rarely tendrillate): *Incarvillea, Jacaranda, Catalpa, Tecoma, Spathodea.*
3. ECCREMOCARPEAE (ovary 1-loc.; caps. splits from below up; seeds winged; tendrils): *Eccremocarpus* (only gen.).
4. CRESCENTIEAE (ovary 1- or 2-loc.; fr. berry or dry indehiscent; seed not winged; usu. erect pl.): *Parmentiera, Crescentia, Phyllarthron, Kigelia.*

Bihai Mill. = Heliconia L. (Heliconiac.).
Bihaia Kuntze = praec.
Bihania Meissn. = Eusideroxylon Teijsm. & Binnend. (Laurac.).
Bijlia N. E. Brown. Aïzoaceae. 1 S. Afr.
Bikera Adans. = Tetragonotheca L. (Compos.).
***Bikkia** Reinw. Rubiaceae. 20 E. Malaysia, Polynesia.
Bikkiopsis Brongn. & Gris = praec.
Bikukulla Adans. = Dicentra Borckh. corr. Bernh. (Fumariac.).
Bilabium Miq. = Didymocarpus Wall. (Gesneriac.).
Bilabrella Lindl. = Habenaria Willd. (Orchidac.).
Bilacus Kuntze = Aegle Corrêa (Rutac.).
Bilamista Rafin. = Gentiana L. (Gentianac.).
Bilderdykia Dum. = Fallopia Adans. (Polygonac.).
Bilegnum Brand. Boraginaceae. 2 N. Persia, C. As.
Bileveillea Vaniot = Blumea DC. (Compos.).
Bilitalium Buch.-Ham. (nomen). 1 India. Quid?
Billardiera Moench = Verbena L. (Verbenac.).
Billardiera Sm. Pittosporaceae. 9 Austr.
Billardiera Vahl = Coussarea Aubl. (Rubiac.).

BILLBERGIA

Billbergia Thunb. Bromeliaceae. 50 warm Am. Epiph.

Billia Peyr. Hippocastanaceae. 2 S. Mex. to trop. S. Am.

Billiotia G. Don = Billiottia DC. = Melanopsidium Colla (Rubiac.).

Billiotia Reichb. = Billottia R.Br. = Agonis (DC.) Lindl. (Myrtac.).

Billiottia DC. = Melanopsidium Colla (Rubiac.).

Billiottia Endl. = seq.

Billotia G. Don = Billottia R.Br. = Agonis (DC.) Lindl. (Myrtac.).

Billotia Sch. Bip. = Crepis L. (Compos.).

Billottia R.Br. = Agonis (DC.) Lindl. (Myrtac.).

Billottia Colla = Calothamnus Labill. (Myrtac.).

Billya Cass. = Helichrysum.Mill. (Compos.).

Biltia Small = Rhododendron L. (Ericac.).

Bima Nor. = Castanopsis (D. Don) Spach (Fagac.) + Nephelium L. (Sapindac.).

Binaria Rafin. = Bauhinia L. (Legumin.).

Bindera Rafin. = ? Aster L. (Compos.).

Binectaria Forsk. = Imbricaria Comm. ex Juss. (Sapotac.).

Bingeria A. Chev. = Turraeanthus Baill. (Meliac.).

Binghamia Backeb. = Borzicactus Riccob. (Cactac.).

Binghamia Britton & Rose (1920; non Farlow ex Agardh, 1899—Algae) = Haageocereus Backeb. + Pseudoëspostoa Backeb. (Cactac.).

Binia Nor. ex Thou. = Noronhia Stadm. (Oleac.).

Binnendijkia Kurz = Leptonychia Turcz. (Sterculiac.).

Binotia Rolfe. Orchidaceae. 1 Brazil.

Binotia W. Watson = Worsleya W. Watson = Hippeastrum L. (Amaryllidac.).

Biolettia Greene. Compositae. 1 Calif.

Biondea Usteri = Blondea Rich. = Sloanea L. (Elaeocarpac.).

Biondia Schlechter. Asclepiadaceae. 4 China.

Bionia Mart. ex Benth. = Camptosema Hook. & Arn. (Legumin.).

Biophytum DC. Oxalidaceae. 70 trop. Many have sensitive pinnate ls.; the leaflets bend down when touched (cf. *Mimosa*). Explosive aril on the seeds (cf. *Oxalis*).

Biota (D. Don) Endl. = Platycladus Spach = Thuja L. (Cupressac.).

Biotia Cass. = Madia Molina (Compos.).

Biotia DC. = Aster L. (Compos.).

Biovularia Kamienski = Utricularia L. (Lentibulariac.).

Bipinnula Comm. ex Juss. Orchidaceae. 8 temp. S. Am.

Bipontia Blake (~ Soaresia Sch. Bip.). Compositae. 1 Brazil.

Bipontinia Alef. = Psoralea L. (Legumin.).

Biporeia Thou. = Quassia L. (Simaroubac.).

Biramella Van Tiegh. = Ochna L. (Ochnac.).

Biramia Néraud = Macleania Hook. (Ericac.).

Birchea A. Rich. = Luisia Gaudich. (Orchidac.).

Biris Medik. = Iris L. (Iridac.).

Birnbaumia Kostel. = Anisacanthus Nees (Acanthac.).

Birolia Bell. = Elatine L. (Elatinac.).

Birolia Rafin. = Clusia L. (Guttif.).

Birostula Rafin. = Scandix L. (Umbellif.).

Bisaschersonia Kuntze = Diospyros L. (Ebenac.).

Bisboeckelera Kuntze. Cyperaceae. 8 S. Am.

Bischoffia Decne = seq.

Bischofia Blume. Bischofiaceae. 1, *B. javanica* Bl., India, China & Formosa to Polynes.; good timber, bark medic.; 1 C. & SE. China.

Bischofiaceae (Muell. Arg.) Airy Shaw. Dicots. 1/2 trop. As. Large decid. trees, with alt. 3(–5)-foliol. pinn. ls. and membr. decid. stips. Fls. small, ♂ ♀, dioec. (rarely monoec.), in axill. ∞-fld. thyrses. K 5, ♂ valv., cucull., ♀ imbr.; C 0; disk 0; A 5, opp. to and encl. by the sep., fil. v. short, anth. large, intr. (Stds. in ♀ fl. 0 or v. minute.) G (3), 3-loc., with 2 pend. ov. per loc., and 1 v. short style with 3 elong. lin.-subul. ent. spreading or reflexed stigs. (Pistillode in ♂ fl. broad, pelt., shortly stipit.) Fr. a small globose fleshy drupe with thin horny endoc.; seeds 3–6, with fleshy endosp. Only genus: *Bischofia*. Probably related to *Staphyleac.* (cf. esp. *Tapiscia*), differing in the apetaly, absence of disk, few ovules and long reflexed styles. Long included in *Euphorbiac.*, but connection with this fam. probably illusory.

Biscutela Rafin. = seq.

Biscutella L. Cruciferae. 10 S. & mid-Eur.

Bisedmondia Hutch. = Calycophysum Triana (Cucurbitac.).

Biserrula L. Leguminosae. 1 Medit.

Bisetaria Van Tiegh. = Campylospermum Van Tiegh. (Ochnac.).

Bisglaziovia Cogn. Melastomataceae. 1 Brazil.

Bisgoeppertia Kuntze. Gentianaceae. 4 W.I.

Bisluederitzia Kuntze = Neoluederitzia Schinz (Zygophyllac.).

Bismalva Medik. = Malva L. (Malvac.).

Bismarckia Hildebr. & H. Wendl. (~ Medemia de Württ.). Palmae. 1 Madag.

Bisnaga Orcutt = Ferocactus Britton & Rose (Cactac.).

Bisphaeria Nor. (nomen) = Poikilospermum Zipp. ex Miq. (Urticac.).

Bisquamaria Pichon. Apocynaceae. 1 Brazil.

Bisrautanenia Kuntze = Neorautanenia Schinz (Legumin.).

Bistania Nor. = Litsea Lam. (Laurac.).

Bistella Adans. Vahliaceae. 4 trop. & S. Afr., SW. As. to NW. India.

Bistorta Scop. (~ Polygonum L.). Polygonaceae. 50 temp. N. Am., Euras.

Bistropogon auctt. = Bystropogon L'Hérit. (Labiat.).

Biswarea Cogn. Cucurbitaceae. 1 Himal.

Bitteria Börner = Carex L. (Cyperac.).

Bituminaria Fabr. (uninom.) = Psoralea L. sp. (Legumin.).

Biventraria Small (~ Asclepias L.). Asclepiadaceae. 1 N. Am.

Bivinia Tul. = Calantica Jaub. ex Tul. (Flacourtiac.).

Bivolva Van Tiegh. = Balanophora J. R. & G. Forst. (Balanophorac.).

*Bivonaea DC. Cruciferae. 1 W. Medit.

Bivonaea Moç. & Sessé = Cardionema DC. (Caryophyllac.).

Bivonea Rafin. = Jatropha L. (Euphorbiac.).

Bivonia Rafin. = praec.

Bivonia Spreng. = Bernardia Mill. (Euphorbiac.).

Biwaldia Scop. = Garcinia L. (Guttif.).

Bixa L. Bixaceae. 3–4 trop. Am., W.I. *B. orellana* L. cult. for the seed; the orange colouring matter of the outer layer of the testa (*annatto, arnotto, roucou*) is used in dyeing sweetmeats, etc.

BIXACEAE

*Bixaceae Link. Dicots. 1/4 trop. Shrubs or small trees, with alt. simple ent. long-pet. stip. ls. Fls. reg., ♀, in term. thyrses. K 5, imbr., biglandular at base; C 5, large, imbr.; disk o; A ∞, fil. elong., free, anth. oblong, biloc., thecae narrowly hippocrepif., dehisc. by short slits at the apical bend; G̱ (2–4), 2–4-loc., with 2–4 pariet. plac. bearing ∞ anatr. ov., and simple elong. style with shortly bilobed stig. Fr. a 2–4-valved setose caps., splitting between the plac.; seeds ∞, with red fleshy papillae forming an arilloid mass; endosp. starchy. Only genus: *Bixa*. Perhaps related to *Cochlospermac.*

Bixagrewia Kurz = Trichospermum Blume (Tiliac.).
Bizonula Pellegr. Sapindaceae. 1 trop. Afr.
Blabeia Baehni. Sapotaceae. 1 (?) New Caled.
Blaberopus A. DC. (~ Alstonia R.Br.). Apocynaceae. 6 China, Indomal.
Blachia Baill. Euphorbiaceae. 12 India, Andamans, SE. As., Philipp. Is.
Blackallia C. A. Gardner. Rhamnaceae. 2 W. Austr.
Blackbournea Kunth = seq.
Blackburnia J. R. & G. Forst. = Zanthoxylum L. (Rutac.).
Blackia Schrank = Myriaspora DC. (Melastomatac.).
Blackiella Aellen (~ Atriplex L.). Chenopodiaceae. 3 Austr.
Blackstonia Huds. Gentianaceae. 5–6 Eur., Medit.
Blackstonia A. Juss. = Blakstonia Scop. = Moronobea Aubl. (Guttif.).
Blackwellia Comm. ex Juss. = Homalium Jacq. (Flacourtiac.).
Blackwellia Gaertn. = Blakwellia Gaertn. = Palladia Lam. = ? Escallonia Mutis ex L. f. (Escalloniac.).
Blackwellia Sieber ex Pax & Hoffm. = Claoxylon Juss. (Euphorbiac.).
Blackwelliaceae Sch. Bip. = Flacourtiaceae–Homalieae Reichb.
Bladhia Thunb. = Ardisia Sw. (Myrsinac.).
Blaeria L. Ericaceae. 70 trop. & S. Afr., Madag.
Blainvillea Cass. Compositae. 10 pantrop., incl. Madag.
Blairia Adans. = Priva Adans. (Verbenac.).
Blairia Spreng. = Blaeria L. (Ericac.).
Blakburnia J. F. Gmel. = Blackburnia J. R. & G. Forst. = Zanthoxylum L. (Rutac.).
Blakea P.Br. Melastomataceae. 70 C. & S. Am., W.I. Ed. fr.
Blakiella Cuatrec. Compositae. 1 Venez.
Blakstonia Scop. = Moronobea Aubl. (Guttif.).
Blakwellia Comm. ex Juss. = Homalium Jacq. (Flacourtiac.).
Blakwellia Gaertn. = Palladia Lam. = ? Escallonia Mutis ex L. f. (Escalloniac.).
Blakwellia Scop. = Leea L. (Leeac.).
Blanchea Boiss. = Iphiona Cass. (Compos.).
Blanchetia DC. Compositae. 1 NE. Brazil. Raises perspiration.
Blanchetiastrum Hassler. Malvaceae. 1 Brazil.
Blanckia Neck. = Conobea Aubl. (Scrophulariac.).
Blancoa Blume (1836) = Didymosperma H. Wendl. (Palm.).
Blancoa Blume (1847) = Harpullia Roxb. (Sapindac.).
Blancoa Lindl. (1839) = Conostylis R. Br. (Haemodorac.).
Blandfordia Andr. = Galax Rafin. (Diapensiac.).
*Blandfordia Sm. Liliaceae. 5 E. Austr.
Blandfortia Poir. = Blandfordia Andr. = Galax Rafin. (Diapensiac.).

Blandibractea Wernh. Rubiaceae. 1 Brazil.
Blandina Rafin. = Leucas R.Br. (Labiat.).
Blandowia Willd. ?Podostemaceae (gen. dub.). 1 S. Am.
Blanisia Pritz. = Polanisia Rafin. = Cleome L. (Cleomac.).
Blastania Kotschy & Peyr. = Ctenolepis Hook. f. (Cucurbitac.).
Blastemanthus Planch. Ochnaceae. 5 N. Brazil, Guiana. K 5 + 5.
Blastocaulon Ruhland. Eriocaulaceae. 4 Brazil.
Blastotrophe F. Didrichs. = Alafia Thou. (Apocynac.).
Blastus Lour. Melastomataceae. 6 Indomal.
Blattaria Kuntze = Pentapetes L. (Sterculiac.).
Blattaria Mill. = Verbascum L. (Scrophulariac.).
Blatti Adans. = Sonneratia L. f. (Sonneratiac.).
Blattiaceae Niedenzu = Sonneratiaceae Engl.
Blaxium Cass. = Dimorphotheca Moench (Compos.).
Bleasdalea F. Muell. = Grevillea R.Br. (Proteac.).
Blechnaceae Presl. Blechnales. Terrestrial or scandent ferns, cosmop. Sori continuous or broken, on veins parallel to midrib of leaflet, indusium opening toward midrib (acrostichoid in *Brainea*). Genera: *Blechnum, Sadleria, Woodwardia, Salpichlaena, Doodia, Lorinseria, Brainea, Stenochlaena*.
Blechnales. Filicidae. 1 fam.: Blechnaceae.
Blechnidium Moore = Blechnum L. (Blechnac.).
Blechnopsis Presl = Blechnum L. (Blechnac.).
Blechnopteris Trevis. (nom. nud.) = Blechnum L. (Blechnac.).
Blechnum L. Blechnaceae. 220 cosmop., but mainly in S. hemisph. Division into 2 genera (*Blechnum* & *Lomaria*) on uniformity or dimorphism of ls. is not sharp.
Blechum P.Br. Acanthaceae. 10 trop. Am., W.I.
Bleekeria Hassk. Apocynaceae. 10 Mascarene Is. to Hawaii & New Caled.
Bleekeria Miq. = Alchornea Sw. (Euphorbiac.).
Bleekrodea Blume = Streblus Lour. (Morac.).
Blencocoës B. D. Jacks. = Blenocoës Rafin. = Nicotiana L. (Solanac.).
Blennoderma Spach = Oenothera L. (Onagrac.).
Blennodia R.Br. Cruciferae. 15 Austr.
Blennosperma Less. Compositae. 2 California, 1 Chile.
Blennospora A. Gray = Calocephalus R.Br. (Compos.).
Blenocoës Rafin. = Nicotiana L. (Solanac.).
Blepetalon Rafin. = Scutia Comm. ex Brongn. (Rhamnac.).
Blephanthera Rafin. = Bulbine L. (Liliac.).
Blepharacanthus Nees = Blepharis Juss. (Acanthac.).
Blepharaden Dulac = Swertia L. (Gentianac.).
Blepharandra Griseb. Malpighiaceae. 3 Guiana.
Blepharanthemum Klotzsch = Plagianthus J. R. & G. Forst. (Malvac.).
Blepharanthera Schlechter = Brachystelma R.Br. (Asclepiadac.).
Blepharanthes Sm. = Modecca Lam. (Passiflorac.).
Blepharidachne Hackel. Gramineae. 2 W. N. Am.
Blepharidium Standley. Rubiaceae. 2 Mex., C. Am.
Blephariglottis Rafin. Orchidaceae. 10 N. Am.
Blepharipappus Hook. Compositae. 1 W. U.S.

Blepharis Juss. Acanthaceae. 100 palaeotrop., Medit., S. Afr., Madag. The seeds have hairs which swell up when wetted.

Blepharispermum Benth. = Blepharophyllum Klotzsch = Scyphogyne Brongn. (Ericac.).

Blepharispermum Wight ex DC. Compositae. 15 trop. Afr. & As.

Blepharistemma Wall. ex Benth. Rhizophoraceae. 1 India.

Blepharitheca Pichon. Bignoniaceae. 2 trop. Am.

Blepharizonia Greene. Compositae. 2 California.

Blepharocalyx Berg. Myrtaceae. 25 warm S. Am.

Blepharocarya F. Muell. Blepharocaryaceae. 2 N. Austr., E. trop. Austr.

Blepharocaryaceae Airy Shaw. Dicots. 1/2 N. & E. Austr. Tall buttressed trees; young parts densely pubesc. Ls. opp., paripinn., exstip., lfts. opp., ent. Fls. ♂ ♀, dioec., minute, in glomerules in term. and axill. opp.-branched panicles. ♂: K 4, shortly connate, subimbr.; C 4, slightly larger, imbr.; A 4+4, exserted, at edge of cupular pilose disk; pistillode columnar, pubesc. ♀: K 4-5, imbr., ± persist.; C 4-5, slightly smaller, imbr. (?); stds. o; disk annular, undulate, glabr.; G 1, compressed, 1-loc., 1-ov., with short filif. style and capit. stig. Fr. dry, indehisc., much compressed, with densely ciliate margin; seed filling the cavity, endosp. o. ♀ fl. and fr. borne in the interior of a many-valved, woody, bracteate, externally and internally puberulous, involucriform cupule (cf. *Castanea*), formed by the flattened, externally grooved and partly concrescent ultimate branches of the infl.; valves at first connivent and closed, bearing the fls. on their inner surface (always at or below the junction of two adjacent valves), subsequently separating and exposing the fr. Only genus: *Blepharocarya*. An interesting type, closely related to *Anacardiac.*, differing in its opposite pinnate ls. and in the remarkable ♀ infl., with which the cupule of *Fagaceae* is apparently almost exactly homologous. Cf. also *Julianiaceae*.

Blepharochlamys Presl = Mystropetalon Harv. (Balanophorac.).

Blepharochloa Endl. = Leersia Sw. (Gramin.).

Blepharodachna P. & K. = Blepharidachne Hack. (Gramin.).

Blepharodon Decne. Asclepiadaceae. 30 Mex. to Chile.

Blepharoglottis auctt. = Blephariglottis Rafin. (Orchidac.).

Blepharolepis Nees (1836) = Polpoda Presl (Aïzoac.).

Blepharolepis Nees (1843) = Scirpus L. (Cyperac.).

Blepharoneuron Nash. Gramineae. 1 SW. U.S., Mex.

Blepharopappus P. & K. = Blepharipappus Hook. (Compos.).

Blepharophyllum Klotzsch = Scyphogyne Brongn. (Ericac.).

Blepharospermum P. & K. = Blepharispermum Wight ex DC. (Compos.).

Blepharostemma Fourr. = Asperula auctt. (Rubiac.).

Blepharostemma P. & K. = Blepharistemma Wall. ex Benth. (Rhizophorac.).

Blepharozonia P. & K. = Blepharizonia Greene (Compos.).

Blepheuria Rafin. = Campanula L. (Campanulac.).

Blephilia Rafin. Labiatae. 2 N. Am.

Blephiloma Rafin. = Phlomis L. (Labiat.).

Blephistelma Rafin. = Passiflora L. (Passiflorac.).

Blephixeta Rafin. = ?Pycnanthemum Michx (Labiat.).

Blephixis Rafin. = Chaerophyllum L. (Umbellif.).

Bletia Ruiz & Pav. Orchidaceae. 45 trop. Am., W.I.

Bletiana Rafin. = praec.

*Bletilla Reichb. f. Orchidaceae. 9 E. As.

Bletti Steud. = Blatti Adans. = Sonneratia L. f. (Sonneratiac.).

Blexum Rafin. = Blechum P.Br. (Acanthac.).

Blighia Koenig. Sapindaceae. 7 trop. Afr. *B. sapida* Koenig (*akee*, vegetable marrow) cult. for ed. fr. (fleshy arillate seed stalk).

Blighiopsis Van de Veken. Sapindaceae. 1 trop. Afr.

Blinkworthia Choisy. Convolvulaceae. 3 Burma, S. China.

Blismus Montand. = Blysmus Panz. ex Roem. & Schult. (Cyperac.).

Blitanthus Reichb. = Acroglochin Schrad. (Chenopodiac.).

Blitoïdes Fabr. (uninom.) = *Amaranthus* L. sp. (Amaranthac.).

Bliton Adans. = seq.

Blitum Fabr. = Amaranthus L. (Amaranthac.).

Blitum Hill (non L.) = Chenopodium L. (Chenopodiac.).

Blitum L. = Chenopodium L. (Chenopodiac.).

Blochmannia Reichb. = Triplaris L. (Polygonac.).

Blomia Miranda. Sapindaceae. 1 Mex.

Blondea Rich. = Sloanea L. (Elaeocarpac.).

Blondia Neck. = Tiarella L. (Saxifragac.).

× **Bloomara** hort. Orchidaceae. Gen. hybr. (iii) (Broughtonia × Laeliopsis × Tetramicra).

Bloomeria Kellogg. Alliaceae. 3 S. California.

Blossfeldia Werderm. Cactaceae. 2 Argentina.

Blossfeldiana Megata (sphalm.) = Frailea Britton & Rose (Cactac.).

Blotia Leandri. Euphorbiaceae. 5 Madag.

Blotiella R. Tryon. Dennstaedtiaceae. 15 trop. Am., Afr., Masc. Is. (R. Tryon, *Contrib. Gray Herb.* **191**: 96–100, 1962).

Bluffia Nees = Panicum L. (Gramin.).

*Blumea DC. Compositae. 50 trop. & S. Afr., Madag., India & E. As. to Austr. & Pacif. *Ai* or *ngai* camphor is distilled from *B. balsamifera* DC. (SW. China).

Blumea G. Don = Blumia Spreng. = Saurauia Willd. (Actinidiac.).

Blumea P. & K. = Blumia Meyen, Nees, Spreng.

Blumea Reichb. = Neesia Blume (Malvac.).

Blumea Zipp. ex Miq. = Didymosperma Wendl. & Drude (Palm.).

Blumella Van Tiegh. = Elytranthe Bl. + Macrosolen Bl. (Loranthac.).

Blumenbachia Koeler = Sorghum Moench (Gramin.).

*Blumenbachia Schrad. Loasaceae. 3 temp. S. Am. Fr. very light, twisted, covered with grapnel hairs.

Blumeodendron (Muell. Arg.) Kurz. Euphorbiaceae. 6 Andaman Is., Malaysia.

Blumeopsis Gagnep. Compositae. 1 India to Hainan, Nicobar Is., Malay Penins.(?), Sumatra.

Blumia Meyen ex Endl. = ? Podochilus Bl. (Orchidac.).

Blumia Nees = Talauma Juss. (Magnoliac.).

Blumia Spreng. = Saurauia Willd. (Actinidiac.).

Blutaparon Rafin. = Philoxerus R.Br. (Amaranthac.).

BLYSMOCAREX

Blysmocarex Ivanova (~ Kobresia Willd.). Cyperaceae. 1 Tibet.

Blysmoschoenus Palla. Cyperaceae (inc. sed.). 1 Bolivia.

***Blysmus** Panz. ex Schult. (~ Scirpus L.). Cyperaceae. 4 temp. Euras.

Blyttia Arn. = Vincetoxicum v. Wolf (Asclepiadac.).

Blyttia Fries = Cinna L. (Gramin.).

Blyxa Noronha. Hydrocharitaceae. 10 trop. Afr. & Madag. to trop. Austr.

Blyxaceae Nak. = Hydrocharitaceae–Blyxeae Aschers. & Gürke.

Blyxopsis Kuntze = Blyxa Nor. (Hydrocharitac.).

Boadschia All. = Bohadschia Crantz = Peltaria Jacq. (Crucif.).

Boaria DC. = Maytenus Molina (Celastrac.).

Bobaea A. Rich. = Bobea Gaudich. (Rubiac.).

Bobartella Gaertn. = Mariscus Gaertn. (Cyperac.).

Bobartia L. = Cyperus L. (Cyperac.).

***Bobartia** Salisb. Iridaceae. 17 S. Afr. Ls. sword-like or centric.

Bobea Gaudich. Rubiaceae. 5 Hawaii.

Boberella E. H. L. Krause (genus aggreg.) = Physalis, Lycium, Atropa, Nicandra (Solanac.).

Bobu Adans. = Symplocos L. (Symplocac.).

Bobua DC. = praec.

Boca Vell. = Banara Aubl. (Flacourtiac.).

Bocagea A. St-Hil. Annonaceae. 2 trop. Am.

Bocageopsis R. E. Fries. Annonaceae. 3 trop. Am.

Bocco Steud. = Bocoa Aubl. = Swartzia Schreb. (Legumin.)

Bocconia L. Papaveraceae. 10 warm As., W.I. Apetalous; seeds arillate.

Bockia Scop. = Mouriri Aubl. (Memecylac.).

Bocoa Aubl. = Swartzia Schreb. (Legumin.).

Bocquillonia Baill. Euphorbiaceae. 6 New Caled.

Bodinieria Léveillé & Vaniot = Boenninghausenia Reichb. ex Meissn. (Rutac.).

Bodinieriella Léveillé = Enkianthus Lour. (Ericac.).

Bodwichia Walp. (sphalm.) = Bowdichia Kunth (Legumin.).

Boea Comm. ex Lam. Gesneriaceae. 25 trop. As. & Austr.

Boebera Willd. = Dyssodia Cav. (Compos.).

Boeberastrum Rydb. (~ Dyssodia Cav.). Compositae. 2 SW. U.S.

Boeckeleria T. Durand. Cyperaceae. 1 S. Afr.

Boeckhia Kunth = Hypodiscus Nees (Restionac.).

Boehmeria Jacq. Urticaceae. 100 trop. & N. subtrop. *B. nivea* (L.) Gaudich. is cult. in China for the fibre (China grass, *rhea*) obtained from the inner bark (cf. *Linum*), perhaps the longest, toughest, and most silky of all veg. fibres, but very difficult to prepare. In the trop. the var. *tenacissima* (Roxb.) Miq. (ramie) is cult.

Boehmeriopsis Komarov. Urticaceae. 1 Korea.

Boeica C. B. Clarke. Gesneriaceae. 6 SE. As.

Boelia Webb = Genista L. (Legumin.).

***Boenninghausenia** Reichb. ex Meissn. Rutaceae. 1 Assam to Japan.

Boenninghausia Spreng. = Chaetocalyx DC. (Legumin.).

Boerhaavia L. = seq.

Boerhavia L. Nyctaginaceae. 40 trop. & subtrop. Anthocarp often glandular, aiding in seed dispersal.

Boerlagea Cogn. Melastomataceae. 1 Borneo.

Boerlagea P. & K. = Boerlagia Pierre = seq.

Boerlagella (Pierre ex Dub.) H. J. Lam. Boerlagellaceae. 1 Sumatra.

Boerlagellaceae H. J. Lam. Dicots. 2/2 W. Malaysia. Trees with (so far as known) large obov. alt. ent. exstip. ls. Infl. axill., rac.; fls. 1–2-bracteolate. K 5, much imbr., unequal (1–2 outer smaller); C and A unknown; G̲ (5), ovules and style unknown. Fr. a 1(–10?)-seeded berry or tardily dehisc. fleshy caps.; seeds large, similar to those of some *Sapotac.*, without endosp.; cotyledons fleshy, much contorted. Genera: *Boerlagella, Dubardella.* An obscure and imperfectly known group, of very doubtful status.

Boerlagia Pierre = Boerlagella Pierre ex Boerl. (Boerlagellac.).

Boerlagiodendron Harms. Araliaceae. 30 Formosa, Malaysia, Pacif.

Boesenbergia Kuntze. Zingiberaceae. ?20 Indomal.

Bogenhardia Reichb. = Herissantia Medik. (Malvac.).

Bogoria J. J. Sm. Orchidaceae. 3 Java, New Guinea.

Bohadschia Crantz = Peltaria L. (Crucif.).

Bohadschia Presl = Turnera L. (Turnerac.).

Bohadschia F. W. Schmidt = Hyoseris L. + Leontodon L. (Compos.).

Boheravia Parodi (sphalm.) = Boerhavia L. (Nyctaginac.).

Boholia Merr. Rubiaceae. 1 Philipp. Is.

Boisduvalia Spach. Onagraceae. 8 W. N. Am., temp. S. Am.

Boissiaea Lem. (sphalm.) = Bossiaea Vent. (Legumin.).

Boissiera Domb. ex DC. = Lardizabala Ruiz & Pav. (Lardizabalac.).

Boissiera Haenseler ex Willk. = Gagea Salisb. (Liliac.).

Boissiera Hochst. & Steud. Gramineae. 1 W. As.

Boivinella A. Camus = Cyphochlaena Hack. (Gramin.).

Boivinella Pierre ex Aubrév. & Pellegr. = Bequaertiodendron De Wild. (Sapotac.).

Bojeria DC. Compositae. 6 trop. & S. Afr., Madag.

Bojeria Rafin. = Phaeomeria Lindl. (Zingiberac.).

Bolandra A. Gray. Saxifragaceae. 3 Pacif. N. Am.

Bolanosa A. Gray. Compositae. 1 Mex.

Bolanthus (Ser.) Reichb. Caryophyllaceae. 8 Greece to Palestine.

Bolax Comm. ex Juss. = Azorella Lam. (Hydrocotylac.).

Bolbidium Lindl. = Dendrobium Sw. (Orchidac.).

Bolbitis Schott. Lomariopsidaceae. 85 trop., on rocks and trees by streams.

Bolbophyllaria Reichb. f. = Bulbophyllum Thou. (Orchidac.).

Bolbophyllopsis Reichb. = Cirrhopetalum Lindl. (Orchidac.).

Bolbophyllum Spreng. = Bulbophyllum Thou. (Orchidac.).

Bolborchis Zoll. & Moritzi = Nervilia Gaud. (Orchidac.).

Bolboschoenus Palla (~ Scirpus L.). Cyperaceae. 1 cosmop.

Bolbostemma Franquet. Cucurbitaceae. 1 China.

Bolbostylis Gardn. = Eupatorium L. (Compos.).

Bolboxalis Small = Oxalis L. (Oxalidac.).

Boldea Juss. = Boldu Adans. = Peumus Molina (Monimiac.).

Boldoa Cav. Nyctaginaceae. 1 Mex., C. Am., W.I.

Boldoa Endl. = seq.

BOLDU

Boldu Adans. = Peumus Molina (Monimiac.).
Boldu Nees = Beilschmiedia Nees (Laurac.).
Bolducia Neck. = Taralea Aubl. (Legumin.).
Boldus Kuntze = Boldu Nees = Beilschmiedia Nees (Laurac.).
Boldus Schult. = Boldu Adans. = Peumus Molina (Monimiac.).
Bolelia Rafin. = Downingia Torr. (Campanulac.).
Boleum Desv. Cruciferae. 1 Spain.
Bolina Rafin. = Bertolonia Raddi (Melastomatac.).
Boliyaria Cham. et Schlechtd. = Menodora Humb. & Bonpl. (Oleac.).
Bolivariaceae Griseb. = Oleaceae–Jasminoïdeae Knobl.
Bolivicereus Cárdenas = Borzicactus Riccob. (Cactac.).
Bollaea Parl. = Pancratium L. (Amaryllidac.).
Bollea Reichb. f. Orchidaceae. 3 W. trop. Am.
× **Bolleochondrorhyncha** hort. (vii) = × Chondrobollea hort. (Orchidac.).
× **Bollwilleria** Zabel = × Sorbopyrus Schneid. (Rosac.).
Bolocephalus Hand.-Mazz. Compositae. 1 S. Tibet.
Bolophyta Nutt. = Parthenium L. (Compos.).
Bolosia Pourr. ex Willk. & Lange = Hispidella Barnad. (Compos.).
Boltonia L'Hérit. Compositae. 1 E. As., 7 N. Am.
Bolusafra Kuntze. Leguminosae. 1 S. Afr.
Bolusanthemum Schwantes = Bijlia N. E. Br. (Aïzoac.).
Bolusanthus Harms. Leguminosae. 1 S. Afr.
Bolusia Benth. Leguminosae. 4 S. trop. & S. Afr.
Bolusiella Schlechter. Orchidaceae. 6 W. & SE. Afr.
Bomarea Mirb. Alstroemeriaceae. 150 Mex., W.I., trop. Am. Often climbing. Umbels cymose.
Bomaria Kunth = praec.
*****Bombacaceae** Kunth (~ Malvaceae Juss.). Dicots. 20/180 trop., esp. Am. Trees, often very large, with thick stems, sometimes egg-shaped owing to formation of water storage tissue; ls. entire or digitate, with deciduous stips. Fl. ⚥, often large, usu. reg. K (5), often with epicalyx; C 5, conv., pets. asymmetric; A 5–∞, free or united into a tube, pollen usu. smooth; G̲ (2–5), in latter case the cpls. opp. the pets., multiloc.; style simple, lobed or capitate; ovules 2–∞ in each loc., erect, anatr. Capsule; seeds smooth, often embedded in hairs springing from wall; endosp. little or o. The *Adansonieae* are ± myrmecophilous (cf. *Acacia*), with extrafloral nectaries on l., K, or fl. stalk. Chief genera: *Adansonia, Bombax, Eriotheca, Chorisia, Durio*.
Bombacopsis Pittier. Bombacaceae. 22 C. & trop. S. Am.
Bombax L. Bombacaceae. 8 trop. Afr., As. *B. ceiba* L. (*B. malabaricum* DC.) (cotton-tree, India, Ceylon) drops its ls. in Dec. and remains leafless till Apr., but fls. in Jan. The cotton is used for cushions, etc. Dug-out canoes are made of the soft wood.
Bombix Medik. = Hibiscus L. (Malvac.).
Bombycella Lindl. = praec.
Bombycidendron Zoll. & Mor. (~ Hibiscus L.). Malvaceae. 4 W. Malaysia.
Bombycilaena (DC.) Smoljaninova. Compositae. 2 N. Afr., S. & C. Eur. to Afghan.
Bombycodendrum P. & K. = Bombycidendron Zoll. & Mor. (Malvac.).

Bombycospermum Presl = Ipomoea L. (Convolvulac.).
Bombynia Nor. = Elaeagnus L. (Elaeagnac.).
Bombyx Moench = Bombix Medik. = Hibiscus L. (Malvac.).
Bommeria Fournier. Hemionitidaceae. 4 N. Am.
Bona Medik. = Vicia L. (Legumin.).
Bonafidia Neck. = Amorpha L. (Legumin.).
Bonafousia A. DC. (~ Tabernaemontana L.). Apocynaceae. 15 trop. Am.
Bonaga Medik. = Ononis L. (Legumin.).
*****Bonamia** Thou. Convolvulaceae. 40 trop.
Bonamica Vell. = Mayepea Aubl. = Linociera Sw. (Oleac.).
Bonamiopsis Roberty = Bonamia Thou. (Convolvulac.).
Bonamya Neck. = Stachys L. (Labiat.).
Bonania A. Rich. Euphorbiaceae. 10 W.I.
*****Bonannia** Guss. Umbelliferae. 1 SE. Eur.
Bonannia Presl = Brassica L. (Crucif.).
Bonannia Rafin. = Blighia Konig (Sapindac.).
Bonanox Rafin. = Ipomoea L. (Convolvulac.).
Bonapa Larrañaga = ? sphalm. pro Bonapartea Ruiz & Pav. = Tillandsia L. (Bromeliac.).
Bonapartea Haw. = Agave L. (Agavac.).
Bonapartea Ruiz & Pav. = Tillandsia L. (Bromeliac.).
Bonarota Adans. = Paederota L. (Scrophulariac.).
Bonatea Willd. Orchidaceae. 20 trop. & S. Afr., Arabia.
Bonatia Schlechter & Krause. Rubiaceae. 1 New Caled.
Bonatoa P. & K. = Bonatea Willd. (Orchidac.).
Bonaveria Scop. = Securigera DC. = Coronilla L. (Legumin.).
Bondtia Kuntze = Bontia L. + Eremophila R.Br. + Pholidia R.Br. (Myoporac.).
Bonduc Mill. = Caesalpinia L. (Legumin.).
Bonellia Bert. ex Colla = Jacquinia Jacq. (Theophrastac.).
Bonetiella Rzedowski (~ Pseudosmodingium Engl.). Anacardiaceae. 1 Mex.
Bongardia C. A. Mey. Leonticaceae. 1 E. Medit. to Afghan. Ls. and rhiz. ed.
Bonia Balansa = Bambusa Schreb. (Gramin.).
Bonifacia S. Manso ex Steud. = Augusta Pohl (Rubiac.).
Bonifazia Standl. & Steyerm. = Disocactus Lindl. (Cactac.).
Boninia Planch. Rutaceae. 2 Bonin Is. 4-merous.
Boniniella Hayata = Asplenium L. (Aspleniac.).
Boninofatsia Nakai. Araliaceae. 2 Bonin Is.
Boniodendron Gagnep. Sapindaceae. 1 Indoch.
Bonjeanea Reichb. = seq.
Bonjeania Reichb. = Dorycnium L. (Legumin.).
Bonnaya Link & Otto = Lindernia All. (Scrophulariac.).
Bonnayodes Blatter & Hallb. Scrophulariaceae. 1 India (Bombay).
Bonneria B. D. Jacks. (sphalm.) = Bonniera Cordem. (Orchidac.).
*****Bonnetia** Mart. Bonnetiaceae. 17 trop. Am., 1 W.I.
Bonnetia Neck. = Buchnera L. (Scrophulariac.).
Bonnetia Schreb. = Mahurea Aubl. (Guttif.).
Bonnetiaceae Beauvis. Dicots. 3/22 trop. As. & Am. Trees or shrubs, mostly of sandy (incl. coastal) or swampy ground, mostly glabrous. Ls. ent., alt.,

exstip., obov. or oblanc., sometimes distinctly asymm., v. shortly and broadly
pet., with close ascend. nerves, crowded toward tips of branches. Infls. axill.
toward tips of branches, 1–∞-fld., distal fls. sometimes crowded into false
umbels, bracts conspic. Fls. rather large, red, fragrant, 2-bracteolate. K 5,
imbr., decid. or persist.; C 5, contorted; A ∞, decid. or persist., sometimes
united below into 5 epipet. bundles; anth. small, versat.; G̲ (3) or (5), styles
free or partly or wholly united, ovules ∞, axile. Caps. 3- or 5-loc., dehisc.
from base or apex, with (rarely without) persist. columella; seeds ∞, with or
without endosp. Genera: *Bonnetia*, *Ploiarium*, *Archytaea*. Closely related to
Pelliceriac., *Tetrameristac.* and *Foetidiac.*

Bonniera Cordemoy. Orchidaceae. 2 Réunion.

Bonnierella Viguier. Araliaceae. 2 Tahiti.

Bonplandia Cav. Polemoniaceae. 2 Mex. Fl. ± zygo.

Bonplandia Willd. = Angostura Roem. & Schult. (Rutac.).

×**Bonstedtia** Wehrh. Berberidaceae. Gen. hybr. (Aceranthus × Epimedium).

Bontia P.Br. = Avicennia L. (Avicenniac.).

Bontia L. Myoporaceae. 1 W.I., N. trop. S. Am. (?introd.).

Bontiaceae Horanin. = Myoporaceae R.Br.

Bonyunia Schomb. Antoniaceae. 5 trop. Am.

Bonzetia P. & K. (sphalm.) = Bouzetia Montr. (Rutac.).

Boöphone Herb. Amaryllidaceae. 5 S. & E. Afr.

Boöpidaceae Cass. = Calyceraceae R.Br.

Boöpis Juss. Calyceraceae. 13 Andes, Argent., S. Braz.

Boothia Dougl. ex Benth. = Platystemon Benth. (Papaverac.).

Bootia Adans. = Borbonia L. = Aspalathus L. (Legumin.).

Bootia Bigel. = Potentilla L. (Rosac.).

Bootia Neck. = Saponaria L. (Caryophyllac.).

Boötrophis Steud. = Botrophis Rafin. = Cimicifuga L. (Ranunculac.).

Boottia Ayres ex Baker = Pleurostylia Wight & Arn. (Celastrac.).

Boottia Wall. = Ottelia Pers. (Hydrocharitac.).

Bopusia Presl = Graderia Benth. (Scrophulariac.).

Boquila Decne. Lardizabalaceae. 1 Chile. Dioec. 6 honey-ls.

Borabora Steud. = Mariscus Gaertn. (Cyperac.).

Boraeva Boiss. = Boreava Jaub. & Spach (Crucif.).

*****Boraginaceae** Juss. Dicots. 100/2000, trop. and temp., esp. Medit. Mostly
herbs, often perenn. by fleshy r., rhiz., etc., a few shrubs; sometimes climbing.
Ls. usu. alt. (rarely opp.), exstip., usu., like rest of pl., covered with stout hairs.
Infl. usu. a coiled cincinnus, sometimes double, with marked dorsiventrality,
uncoiling as fls. open, so that newly opened fls. face always in same direction.
The morphology of this infl. is not fully clear; adnation or concrescence
occurs, and possibly dichotomy at apex. General agreement favours the view
that the 'boragoid', as it is sometimes called, is composed of dorsiventral
monopodia.

Fl. ♀, usu. reg., hypog., usu. 5-merous; K (5), imbr. or open, rarely valv.,
odd sep. post.; C (5), imbr. or conv., funnel-shaped or tubular, limb usu. flat;
A 5, epipet., alt. with pets., intr.; G̲ (2), on hypog. disk, usu. 4-loc. (rarely 2- or
10-loc.) with 'false' septum (cf. *Labiatae*), usu. with gynobasic entire or lobed
style from base of ovary; ov. 1 in each loc., erect, ascending or horiz., anatr.

Fr. of 4 achenes (nutlets) (8–10 in *Trigonotis* spp.); or a drupe; seed with straight or curved embryo in usu. slight endosp.; radicle directed upwards. Most have a short tube, partly concealing the honey; many (esp. tribes 1 and 3 of subfam. II) have scales projecting inwards from the throat of the C, concealing and protecting the honey, and narrowing the entrance, so that visiting insects must take a definite track. 'Many spp., in the course of their individual development, seem to recapitulate to us the evolution of their colours—white, rosy, blue in several spp. of *Myosotis*; yellow, bluish, violet in *M. versicolor*; and red, violet, blue in *Pulmonaria, Echium*, etc. Here, white and yellow seem to have been the primitive colours' (Müller). Many *B.* are heterostyled, e.g. *Pulmonaria*. The fls. of many spp. are pendulous (and thus bee-flowers), e.g. *Borago, Symphytum. Echium* is gynodioecious.

Classification and chief genera:

I. **Heliotropioïdeae** (style terminal; drupe): *Tournefortia, Heliotropium.*

II. **Boraginoïdeae** (style gynobasic; achenes).

    1. CYNOGLOSSEAE (fl. reg.; base of style more or less conical; tips of achenes not projecting above point of attachment): *Omphalodes, Cynoglossum, Rindera.*

    2. ERITRICHIEAE (do., but tips projecting above point of attachment): *Echinospermum, Eritrichium, Cryptanthe.*

    3. BORAGINEAE (fl. reg.; base of style flat or slightly convex; achenes with concave attachment surface): *Symphytum, Borago, Anchusa, Alkanna, Pulmonaria.*

    4. LITHOSPERMEAE (do., but surface of attachment flat): *Myosotis, Lithospermum, Arnebia, Cerinthe.*

    5. ECHIEAE (fl. zygomorphic): *Echium.*

Boraginella Kuntze = Trichodesma R.Br. (Boraginac.).

Boraginodes P. & K. = Borraginoïdes Boehm. = praec.

Borago L. Boraginaceae. 3 Medit., Eur., As. *B. officinalis* L. (borage) cult. for bee-feeding. It has a typical bee-fl. The blue pendulous fl. secretes honey below the ovary; the elastic sta. form a cone and dehisce introrsely from apex to base, the pollen ripening gradually and trickling into the tip of the cone. Insects probing for honey dislocate the sta., receiving a shower of pollen (cf. *Erica, Galanthus, Cyclamen*). In older fls. the stigma, now ripe, projects beyond the sta. so as to be touched first.

Borassaceae O. F. Cook = Palmae–Borassoïdeae Mart.

Borassodendron Becc. Palmae. 1 Malay Penins.

Borassus L. Palmae. 8 palaeotrop. *B. flabellifer* L. (Palmyra palm) cult. in Ceylon, India, etc. Dioecious. Its uses are legion; an old Tamil song enumerates 801. The wood of the trunk is very hard and durable, and resists salt water; it is also used for rafters, well-sweeps, etc. The large fan-shaped ls. are used as thatch, and made into *olas* or writing 'paper' sheets, the writing being done upon them with a stylus. From the base of the ls. Palmyra fibre is collected, and used for making brushes, etc. The split ls. are woven into mats, baskets, etc. The fr. is eaten roasted, and the infl. is tapped for toddy (cf. *Cocos, Agave*) from which sugar or jaggery is made, as well as vinegar, etc. The young seedlings are also eaten and yield a good flour when ground, and there are many other uses.

BORBASIA

Borbasia Gandog. = Dianthus L. (Caryophyllac.).

Borbonia Adans. = Ocotea Aubl. (Laurac.).

Borbonia L. = Aspalathus L.

Borbonia Mill. = Persea Mill. (Laurac.).

Borboraceae Dulac = Scheuchzeriaceae + Juncaginaceae + Alismataceae.

Borboya Rafin. = Hyacinthus L. (Liliac.).

Borckhausenia Gaertn., Mey. & Scherb. = Corydalis Vent. (Fumariac.).

Borckhausenia Roth = Teedia Rudolphi (Scrophulariac.).

Borderea Miègeville = Dioscorea L. (Dioscoreac.).

Borea Meissn. (sphalm.) = Bovea Decne = Lindenbergia Lehm. (Scrophu-
Borea Zipp. ex Mackl. (nomen). New Guinea. Quid?   [lariac.).

Boreava Jaub. & Spach. Cruciferae. 2 E. Medit.

Borellia Neck. = Cordia L. (Ehretiac.).

Boretta [Neck.] Kuntze = Daboecia D. Don (Ericac.).

Borissa Rafin. (Asterolinon Hoffmgg. & Link). Primulaceae. 2 Medit. to
Crimea, Persia, Abyss.

Borith Adans. = Anabasis L. (Chenopodiac.).

Borkhausia Nutt. = Barkhausia Moench = Crepis L. (Compos.).

Bormiera P. & K. (sphalm.) = Bonniera Cordem. (Orchidac.).

Borneacanthus Bremek. Acanthaceae. 6 Borneo.

Borneodendron Airy Shaw. Euphorbiaceae (?). 1 N. Borneo. Ls. and ♂ fls.
in whorls of 3.

Bornmuellera Hausskn. Cruciferae. 6 As. Min.

Bornoa O. F. Cook = Attalea Kunth (Palm.).

Borodinia Busch. Cruciferae. 1 E. Siberia.

Borojoa Cuatrec. Rubiaceae. 6 trop. Am.

Boronella Baill. Rutaceae. 4 New Caled.

Boronia Sm. Rutaceae. 70 Austr.

Boroniaceae J. G. Agardh = Rutaceae–Boronieae Bartl.

Borrachinea Lavy. Boraginaceae. 1 NW. Italy. Quid?

Borraginoïdes Boehm. = Trichodesma R.Br. (Boraginac.).

Borrago auctt. = Borago L. (Boraginac.).

Borrera Spreng. = seq.

***Borreria** G. F. W. Mey. Rubiaceae. 150 warm.

Borrichia Adans. Compositae. 7 warm Am., W.I.

Borsczowia Bunge. Chenopodiaceae. 1 C. As.

Borthwickia W. W. Smith. Capparidaceae. 1 Burma.

Borya Labill. Liliaceae. 3 W. Austr., Queensland.

Borya Montrouz. ex Beauvis. = Oxera Labill. (Verbenac.).

Borya Willd. = Forestiera Poir. (Oleac.).

Borzicactella Johns. = seq.

Borzicactus Riccobono. Cactaceae. 17 Andes.

Borzicereus Frič & Kreuzinger = praec.

Boscheria Carr. = Bosscheria De Vriese & Teijsm. = Ficus L. (Morac.).

Boschia Korth. = Durio Adans. (Bombacac.).

Boschniakia C. A. Mey. Orobanchaceae. 2 N. & arct. Russia & Asia to Japan
and NW. N. Am.

Bosca Vell. = Daphnopsis Mart. & Zucc. (Thymelaeac.).

152

*Boscia Lam. Capparidaceae. 37 trop. & S. Afr.
Bosciopsis B. C. Sun = Hypselandra Pax & Hoffm. (Capparidac.).
Boscoa P. & K. = Bosca Vell. = Daphnopsis Mart. & Zucc. (Thymelaeac.).
Bosea L. Amaranthaceae. 3 Canary Is., Cyprus, India.
Bosia Mill. = praec.
Bosistoa F. Muell. Rutaceae. 4 E. Austr.
Bosleria A. Nelson. Solanaceae. 1 SW. U.S.
Bosqueia Thou. ex Baill. Moraceae. 15 trop. Afr., Madag.
Bosqueiopsis De Wild. & Durand. Moraceae. 4 trop. Afr.
Bosquiea B. D. Jacks. (sphalm.) = Bosqueia Thou. ex Baill. (Morac.)
Bosscheria De Vriese & Teijsm. = Ficus L. (Morac.).
Bossekia Neck. = Rubus L. (Rosac.).
Bossekia Rafin. = Waldsteinia Willd. (Rosac.).
Bossera Leandri. Euphorbiaceae. 1 Madag.
Bossiaea Vent. Leguminosae. 35 Austr. Several xero. spp. have flattened
green stems (phylloclades) with minute scale ls. As in *Acacia*, etc., seedlings
show transitions from ls.
Bossiena B. D. Jacks. (sphalm.) = praec.
Bostrychanthera Benth. Labiatae. 1 China.
Boswellia Roxb. ex Colebr. Burseraceae. 24 trop. Afr., Madag., trop. As.
B. carteri Birdw. (Somaliland, etc.) and other spp. yield the resin frankincense
or gum olibanum, formerly offic., now used in incense. Other spp. also yield
fragrant resin. *B. serrata* Roxb., an important tree on dry hills in India.
Botelua Lag. Gramineae. 40 Canada to S. Am., mainly in SW. U.S. (mesquitᵉ
grasses, grama, side-oats). They form a large proportion of the herbage of the
prairie, and are valuable as fodder.
Botherbe Steud. ex Klatt = Calydorea Herb. (Iridac.).
Bothriaceae Dulac = Isoëtaceae Reichb.
Bothriochilus Lem. Orchidaceae. 4 C. Am.
Bothriochloa Kuntze. Gramineae. 20 warm.
Bothriocline Oliv. ex Benth. Compositae. 14 trop. Afr.
Bothriopodium Rizzini = Urbanolophium Melchior (Bignoniac.).
Bothriospermum Bunge. Boraginaceae. 5 trop. & NE. As.
Bothriospora Hook. f. Rubiaceae. 1 trop. Am.
Bothrocaryum (Koehne) Pojark. Cornaceae. 3 Himal. to NE. As., NE. Am.
Botor Adans. = Psophocarpus Neck. ex DC. (Legumin.). [Ls. alt.
Botria Lour. = Ampelocissus Planch. (Vitidac.).
Botrophis Rafin. = Cimicifuga L. (Ranunculac.).
Botrya Juss. = Botria Lour. = Ampelocissus Planch. (Vitidac.).
Botryadenia Fisch. & Mey. = Myriactis Less. (Compos.).
Botryanthe Klotzsch = Plukenetia L. (Euphorbiac.).
Botryanthus Kunth = Muscari Mill. (Liliac.).
Botryarrhena Ducke. Rubiaceae. 1 Brazil.
Botrycarpum (A. Rich.) Opiz = Ribes L. (Grossulariac.).
Botryceras Willd. = Laurophyllus Thunb. (Anacardiac.).
Botrychiaceae Nakai = Ophioglossaceae Presl.
Botrychium Sw. Ophioglossaceae. 40 cosmop. Habit like *Ophioglossum*, but
the sterile as well as the fertile part of the l. is branched.

BOTRYCOMUS

Botrycomus Fourr. = Muscari Mill. (Liliac.).
Botrydendrum P. & K. = Botryodendrum Endl. = Meryta J. R. & G. Forst. (Araliac.).
Botrydium Spach = Chenopodium L. (Chenopodiac.).
Botrylotus P. & K. = Botryolotus Jaub. & Spach = Trigonella L. (Legumin.).
Botrymorus Miq. = Pipturus Wedd. (Urticac.).
Botryocarpium (A. Rich.) Spach = Ribes L. (Grossulariac.).
Botryocytinus (E. G. Baker) Watanabe = Cytinus L. (Rafflesiac.).
Botryodendraceae J. G. Agardh = Araliaceae Juss.
Botryodendrum Endl. = Meryta J. R. & G. Forst. (Araliac.).
Botryogramme Fée = Llavea Lagasca (Cryptogrammac.).
Botryoïdes v. Wolf = Muscari Mill. (Liliac.).
Botryoloranthus (Engl. & K. Krause) Balle. Loranthaceae. 1 trop. E. Afr.
Botryolotus Jaub. & Spach = Trigonella L. (Legumin.).
Botryomeryta R. Viguier. Araliaceae. 1 New Caledonia.
Botryopanax Miq. = Polyscias J. R. & G. Forst. (Araliac.).
*****Botryophora** Hook. f. (1888; non Bompard 1867—Algae). Euphorbiaceae. 1 Lower Burma, Penins. Siam, Malay Penins., Sumatra, W. Java, Borneo.
Botryopleuron Hemsl. = Veronicastrum Fabr. (Scrophulariac.).
Botryopsis Miers = Chondrodendron Ruiz & Pav. (Menispermac.).
Botryopteris Presl = Helminthostachys Kaulf. (Ophioglossac.).
Botryoropis Presl = Barringtonia J. R. & G. Forst. (Barringtoniac.).
Botryosicyos Hochst. = Dioscorea L. (Dioscoreac.).
Botryospora B. D. Jacks. (sphalm.) = Botryophora Hook. f. (Euphorbiac.).
Botryostege Stapf. Ericaceae. 1 Japan, Kurile Is.
Botrypanax P. & K. = Botryopanax Miq. = Polyscias J. R. & G. Forst.
Botryphile Salisb. = Muscari Mill. (Liliac.). [(Araliac.).
Botryphora P. & K. = Botryophora Hook. f. (Euphorbiac.).
Botrypus Rich. = Botrychium Sw. (Ophioglossac.).
Botryropis P. & K. = Botryoropis Presl = Barringtonia J. R. & G. Forst.
Botrys Fourr. = Teucrium L. (Labiat.). [(Barringtoniac.)
Botrys Reichb. ex Nieuwl. = Chenopodium L. (Chenopodiac.).
Botschantzevia Nabiev. Cruciferae. 1 C.As. (Kazakhstan).
Bottegoa Chiov. Sapindaceae. 1 Somaliland.
Bottionea Colla. Liliaceae. 1 Chile.
Boucerosia Wight & Arn. = Caralluma R.Br. (Asclepiadac.).
Bouchardatia Baill. Rutaceae. 2 New Guinea, E. Austr.
*****Bouchea** Cham. Verbenaceae. 16 warm.
Bouchetia DC. ex Dun. Solanaceae. 3 S. U.S. to Brazil.
Bouea Meissn. Anacardiaceae. 3–4 (? or 1 polymorph.) SE. As., W. Malaysia. (exc. Philipp.), Moluccas. Ls. opp.
Bouetia A. Chev. = Hemizygia (Benth.) Briq. (Labiat.).
*****Bougainvillea** Comm. ex Juss. mut. Choisy. Nyctaginaceae. 18 S. Am. Each group of 3 fls. is surrounded by 3 lilac or red persistent bracts.
Bougainville[ac]eae J. G. Agardh = Nyctaginaceae–Mirabileae–Bougainvilleinae Heimerl.
Bougueria Decne. Plantaginaceae. 1 Andes.
Boulardia F. Schultz = Orobanche L. (Orobanchac.).

Boulaya Gandog. = Rubus L. (Rosac.).
Bouphone Lem. = Boöphone Herb. (Amaryllidac.).
Bourdaria A. Chev. Melastomataceae. 1 W. Afr.
Bourdonia Greene = Keerlia A. Gray (Compos.).
Bourgaea Coss. = Cynara L. (Compos.).
Bourgia Scop. = Salimori Adans. = Cordia L. (Ehretiac.).
Bourjotia Pomel = Heliotropium L. (Boraginac.).
Bourlageodendron K. Schum. = Boerlagiodendron Harms (Araliac.).
Bournea Oliv. Gesneriaceae. 1 China.
***Bourreria** P.Br. Ehretiaceae. 50 warm Am., W.I.
Bousigonia Pierre. Apocynaceae. 2 Indoch.
Boussingaultia Kunth = Anredera Juss. (Basellac.).
Bouteloua Lag. mut. P. Beauv. (= Botelua Lag., *q.v.*), Gramineae. 40 Canada to S. Am., esp. SW. U.S.
Boutiquea Le Thomas. Annonaceae. 1 W. Equat. Afr.
Boutonia Boj. ex Baill. = Mallotus Lour. (Euphorbiac.).
Boutonia DC. Acanthaceae. 1 Madag.
Boutonia Hort. = Goodenia Sm. (Goodeniac.).
Bouvardia Salisb. Rubiaceae. 50 trop. Am. Some heterostyled like *Primula*.
Bouzetia Montr. Rutaceae (inc. sed.). 1 New Caled.
Bovea Decne = Lindenbergia Lehm. (Scrophulariac.).
Bovonia Chiov. Labiatae. 1 trop. Afr.
Bowdichia Kunth. Leguminosae. 3 trop. S. Am. Timber.
Bowenia Hook. Zamiaceae. 2 N. & NE. Australia. Easily recognised by the bipinnate ls. The upper part of the main r. gives rise to curiously branched apogeotropic r., which contain *Anabaena* (*Cyanophyceae*, blue algae) living in symbiosis, and branch exogenously.
***Bowiea** Harv. ex Hook. f. Liliaceae. 2 S. & E. Afr. *B. volubilis* Harv., a xero. like *Dioscorea* § *Testudinaria*, with a large partly underground stock (corm), giving off each year a much-branched climbing stem. This bears small ls., but they soon drop, and assim. is carried on by the green stem.
Bowiea Haw. = Aloë L. (Liliac.).
Bowkeria Harv. Scrophulariaceae. 8 S. Afr.
Bowlesia Ruiz & Pav. Hydrocotylaceae. 14 Am. (esp. S.).
Bowmania Gardn. = Trixis P.Br. (Compos.).
Bowringia Champ. ex Benth. Leguminosae. 4 trop. Afr., Madag., Borneo, Hongkong.
Bowringia Hook. = Brainea J. Sm. (Blechnac.).
Boyania Wurdack. Melastomataceae. 1 Guiana.
Boykiana Rafin. = Boykinia Rafin. = Rotala L. (Lythrac.).
***Boykinia** Nutt. Saxifragaceae. 8 Japan, N. Am.
Boykinia Nutt. ex Rafin. = Cayaponia S. Manso (Cucurbitac.).
Boykinia Rafin. = Rotala L. (Lythrac.).
Boymia A. Juss. = Evodia J. R. & G. Forst. (Rutac.).
Bozea Rafin. = Bosea L. (Amaranthac.).
Brabejaria Burm. f. = seq.
Brabejum L. Proteaceae. 1 S. Afr., *B. stellatifolium* L. (*wilde castanjes*), whose seeds are eaten roasted.

BRABILA

Brabila P.Br. Jamaica. Quid?
Brabyla L. = Brabejum L. (Proteac.).
Bracea Britton = Neobracea Britton (Apocynac.).
Bracea King = Sarcosperma Hook. f. (Sarcospermatac.).
Bracera Engelm. = Brayera Kunth (Rosac.).
Brachanthemum DC. (~ Chrysanthemum L.). Compositae. 7 C. As.
Brachatera Desv. = Sieglingia Bernh. (Gramin.).
Bracheilema R.Br. = Vernonia Schreb. (Compos.).
Brachiaria Griseb. Gramineae. 50 warm.
Brachiolobos All. = Rorippa Scop. (Crucif.).
Brachionidium Lindl. Orchidaceae. 12 W.I., trop. S. Am.
Brachionostylum Mattf. Compositae. 1 New Guinea.
Brachiostemon Hand.-Mazz. = Ornithoboea Parish ex C. B. Cl. (Gesneriac.).
Brachistepis Thou. (uninom.) = Epidendrum brachistachion Thou. = Beclardia brachystachya (Thou.) A. Rich. (Orchidac.).
Brachistus Miers. Solanaceae. 35 C. & S. Am., Galápagos.
Brachoneuron P. & K. = Brochoneura Warb. (Myristicac.).
***Brachtia** Reichb. f. Orchidaceae. 4 Venezuela, Colombia, Ecuador.
Brachyachaenium Baker. Compositae. 1 Madag.
Brachyachne (Benth.) Stapf. Gramineae. 10 trop. Afr., Austr.
Brachyachyris Spreng. = Brachyris Nutt. = Gutierrezia Lag. (Compos.).
Brachyactis Ledeb. Compositae. 5 N. As., N. Am.
Brachyandra Naud. = Pterolepis Miq. (Melastomatac.).
***Brachyandra** Phil. Compositae. 2 Chile.
Brachyanthes Cham. ex Dunal = Petunia Juss. (Solanac.).
Brachyapium (Baill.) Maire. Umbelliferae. 4 Medit. reg.
Brachyaster Ambrosi = Aster L. (Compos.).
Brachyathera P. & K. = Brachatera Desv. = Sieglingia Bernh. (Gramin.).
Brachybotrys Maxim. ex Oliv. Boraginaceae. 1 Manchuria, Far E. Sib.
Brachycalycium Backeb. = Gymnocalycium Pfeiff. (Cactac.).
Brachycalyx Sweet ex DC. = Rhododendron L. (Ericac.).
Brachycarpaea DC. Cruciferae. 1 S. Afr.
Brachycentrum Meissn. = Centronia D. Don (Melastomatac.).
Brachycereus Britton & Rose. Cactaceae. 1 Galápagos.
Brachychaeta Torr. & Gray. Compositae. 1 S. U.S.
Brachycheila Harv. ex Eckl. & Zeyh. = Euclea L. = Diospyros L. (Ebenac.).
Brachychilum (R.Br. ex Wall.) O. G. Petersen. Zingiberaceae. 2 Java.
Brachychilus P. & K. (1) = Brachycheila Harv. ex Eckl. & Zeyh. = Diospyros L. (Ebenac.).
Brachychilus P. & K. (2) = Brachychilum (R.Br. ex Wall.) O. G. Petersen (Zingiberac.).
Brachychiton Schott & Endl. Sterculiaceae. 11 Austr. *B. rupestris* K. Schum. (bottle tree) has swollen stems, *B. acerifolius* F. Muell. (flame tree) very fine fls.
Brachychlaena P. & K. = Brachylaena R.Br. (Compos.).
Brachyclados D. Don. Compositae. 4 temp. S. Am.
Brachycladus P. & K. = praec.
Brachycodon Fedorov. Campanulaceae. 1 Medit. to C. As.
Brachycodon Progel = Pagaea Griseb. (Gentianac.).

Brachycome Cass. Compositae. 75 Austr., N.Z., N. Am., Afr.
Brachycome Gaudich. = Vittadinia Rich. (Compos.).
Brachycorys Schrad. = Lindenbergia L. (Scrophulariac.).
Brachycorythis Lindl. Orchidaceae. 32 trop. & S. Afr., trop. As.
Brachycyrtis Koidz. Liliaceae. 1 Japan.
Brachyderea Cass. = Crepis L. (Compos.).
Brachyelytrum Beauv. Gramineae. 4 warm Am., Afr.
Brachyglottis J. R. & G. Forst. Compositae. 2 N.Z.
Brachygyne Cass. = Eriocephalus L. (Compos.).
Brachygyne (Benth.) Small = Dasistoma Rafin. (Scrophulariac.).
Brachyilema P. & K. = Bracheilema R.Br. = Vernonia Schreb. (Compos.).
Brachylaena R.Br. Compositae. 23 trop. & S. Afr., Mascarene Is. Shrubs.
Brachylepis Hook. & Arn. = Melinia Decne (Asclepiadac.).
Brachylepis C. A. Mey. Chenopodiaceae. 8 S. Russia to C. As.
Brachylepis Wight & Arn. = Baeolepis Decne ex Moquin (Asclepiadac.).
Brachylobus Dulac = Melilotus L. (Legumin.).
Brachylobus Link = Brachiolobos All. = Rorippa Scop. (Crucif.).
Brachyloma Hanst. = Isoloma Decne (Gesneriac.).
Brachyloma Sond. Epacridaceae. 7 Austr.
Brachylophon Oliv. Malpighiaceae. 3 E. Afr., Siam, Malay Penins., Sumatra.
Brachymeris DC. Compositae. 6 S. trop. & S. Afr.
*Brachynema Benth. Olacaceae. 1 N. Brazil.
Brachynema Griff. = Sphenodesma Jack (Symphorematac.).
Brachynema F. Muell. = Abrophyllum Hook. f. (Escalloniac.).
Brachyoglotis Lam. = Brachyglottis J. R. & G. Forst. (Compos.).
Brachyolobos DC. = Brachiolobos All. = Rorippa Scop. (Crucif.).
Brachyotum Triana (= Alifana Rafin.). Melastomataceae. 60 S. Am.
Brachypappus Sch. Bip. = Senecio L. (Compos.).
Brachypetalum Nutt. ex Lindl. = Smilacina Desf. (Liliac.).
Brachyphragma Rydb. = Astragalus L. (Legumin.).
Brachypoda Rafin. = Eclipta L. (Compos.).
Brachypodandra Gagnep. = ? Vatica L. (Dipterocarpac.).
Brachypodium Beauv. Gramineae. 10 temp.; trop. mts. Leaf reversed (cf.
    Alstroemeria).
Brachypremna Gleason. Melastomataceae. 1 Guiana.
Brachypterum Benth. = Derris Lour. (Legumin.).
Brachypterys A. Juss. Malpighiaceae. 3 trop. S. Am., W.I.
Brachypteryx Dalla Torre & Harms = praec.
Brachypus Ledeb. = Lunaria L. (Crucif.).
Brachyramphus DC. = Lactuca L. (Compos.).
Brachyrhynchos Less. = Senecio L. (Compos.).
Brachyridium Meissn. = Lepidophyllum Cass. (Compos.).
Brachyris Nutt. = Gutierrezia Lag. (Compos.).
Brachyscome Cass. = Brachycome Cass. (Compos.).
Brachyscypha Baker = Lachenalia Jacq. (Liliac.).
Brachysema R.Br. Leguminosae. 15 Austr.
Brachysiphon A. Juss. Penaeaceae. 11 S. Afr.
Brachysorus Presl = Athyrium Roth (Athyriac.).

BRACHYSPATHA

Brachyspatha Schott = Amorphophallus Blume (Arac.).
Brachystachys Klotzsch = Croton L. (Euphorbiac.).
Brachystachyum Keng. Gramineae. 1 China.
Brachystegia Benth. Leguminosae. 30 trop. Afr.
Brachystele Schlechter. Orchidaceae. 15 warm S. Am., Trinidad.
Brachystelma R.Br. Asclepiadaceae. 30 trop. & S. Afr.
Brachystemma D. Don. Caryophyllaceae. 1 Himal.
Brachystemum Michx = Pycnanthemum Michx (Labiat.).
Brachystephanus Nees. Acanthaceae. 10 trop. Afr., Madag.
Brachystephium Less. = Brachycome Cass. (Compos.).
Brachystepis Pritz. = Beclardia A. Rich. (Orchidac.).
Brachystigma Pennell. Scrophulariaceae. 1 SW. U.S.
Brachystylis E. Mey. ex DC. = Marasmodes DC. (Compos.).
Brachystylus Dulac = Koeleria Pers. (Gramin.).
Brachythalamus Gilg. Thymelaeaceae. 3 New Guinea.
Brachytome Hook. f. Rubiaceae. 4 S. China, Indomal.
Brachytophora Dur. (sphalm.) = Brachylophon Oliv. (Malpighiac.).
Brachytropis Reichb. = Polygala L. (Polygalac.).
Bracisepalum J. J. Smith. Orchidaceae. 1 Malaysia.
Brackenridgea A. Gray. Ochnaceae. 5 Andamans, SE. Siam, Malaysia, Queensl., Fiji.
Braconotia Godr. = Agropyron J. Gaertn. (Gramin.).
Bracteanthus Ducke. Siparunaceae. 1 E. Brazil.
Bractearia DC. = Tibouchina Aubl. (Melastomatac.).
Bracteola Swallen = Chrysochloa Swallen (Gramin.).
Bracteolanthus de Wit (~ Bauhinia L.). Leguminosae. 1 Borneo.
Bracteolaria Hochst. = Baphia Afzel. (Legumin.).
Bractillaceae Dulac = Amaryllidaceae Jaume St-Hil.
Bradburia Torr. & Gray. Compositae. 2 S. U.S., Mex.
Bradburya Rafin. = Centrosema Benth. (Legumin.).
Braddleya Vell. = Amphirrhox Spreng. (Violac.).
Bradea Standley. Rubiaceae. 5 Brazil.
Bradlaeia Neck. = Siler Mill. (Umbellif.).
Bradlea Adans. = Apios Moench + Wisteria Nutt. (Legumin.).
Bradleia Banks ex Gaertn. = Glochidion J. R. & G. Forst. (Euphorbiac.).
Bradleia Rafin. = Bradlaeia Neck. = Siler Mill. (Umbellif.).
Bradleya P. & K. (1) = Bradlea Adans. = Apios Moench (Legumin.).
Bradleya P. & K. (2) = Bradleia Banks ex Gaertn. = Glochidion J. R. & G. Forst. (Euphorbiac.).
Bradleya P. & K. (3) = Bradlaeia Neck. = Siler Mill. (Umbellif.).
Bradleya P. & K. (4) = Braddleya Vell. = Amphirrhox Spreng. (Violac.).
× **Bradriguesia** hort. (vii) = × Rodrassia hort. (Orchidac.).
Bradshawia F. Muell. = Rhamphicarpa Benth. (Scrophulariac.).
Bragantia Lour. = Apama Lam. (Aristolochiac.).
Bragantia Vand. = Gomphrena L. (Amaranthac.).
Brahea Mart. Palmae. 7 S. U.S., Mexico.
Brainea J. Sm. Blechnaceae. 1 E. As., *B. insignis* (Hook.) J. Sm., a dwarf tree-fern of open places, surviving burning; fertile leaflets narrow, acrostichoid.

Brami Adans. = Bacopa Aubl. (Scrophulariac.).
Bramia Lam. = praec.
Branciona Salisb. = Albuca L. (Liliac.).
Brandegea Cogn. Cucurbitaceae. 1 SW. U.S., Mexico.
Brandesia Mart. = Telanthera R.Br. (Amaranthac.).
Brandisia Hook. f. & Thoms. Scrophulariaceae. 13 Burma, China.
Brandonia Reichb. = Pinguicula L. (Lentibulariac.).
Brandtia Kunth = Arundinella Raddi (Gramin.).
Brandzeia Baill. Leguminosae. 1 Madag., Seychelles.
Branica Endl. = Bramia Lam. = Bacopa Aubl. (Scrophulariac.).
Branicia Andrz. = Senecio L. (Compos.).
× **Brapasia** hort. Orchidaceae. Gen. hybr. (iii) (Aspasia × Brassia).
Brasea A. Voss = Senecio L. (Compos.).
Brasenia Schreb. Cabombaceae. 1 trop. Am. & Afr., Ind., temp. E. As., Austr. A 12 or more.
Brasilettia Kuntze = Peltophorum Walp. (Legumin.).
Brasilia Barroso. Compositae. 1 Brazil. [(Simaroubac.).
Brasiliastrum Lam. = Comocladia P.Br. (Anacardiac.) + Picramnia Swartz
Brasilicactus Backeb. = Notocactus (K. Schum.) Backeb. & F. M. Knuth (Cactac.).
Brasilicereus Backeb. Cactaceae. 2–3 Brazil.
Brasiliopuntia (K. Schum.) A. Berger = Opuntia Mill. (Cactac.).
Brasilium J. F. Gmel. = Brasiliastrum Lam., *q.v.*
Brasilocalamus Nakai. Gramineae. 1 Brazil.
Brasilocactus Frič = Brasilicactus Backeb. = Notocactus (K. Schum.) Backeb. & F. M. Knuth (Cactac.).
Brasilocereus auctt. = Brasilicereus Backeb. (Cactac.).
× **Brassada** hort. Orchidaceae. Gen. hybr. (iii) (Ada × Brassia).
Brassaia Endl. = Schefflera J. R. & G. Forst. (Araliac.).
Brassaiopsis Decne & Planch. Araliaceae. 35 China, SE. As., Indomal.
× **Brassatonia** hort. Orchidaceae. Gen. hybr. (iii) (Brassavola × Broughtonia).
Brassavola Adans. = Helenium L. (Compos.).
Brassavola R.Br. Orchidaceae. 15 trop. Am.
Brassavolaea P. & K. = Brassavola Adans. = Helenium L. (Compos.).
Brassenia Heynh. = Brasenia Schreb. (Cabombac.).
Brassia R.Br. Orchidaceae. 50 trop. Am.
Brassavola Hort. (*p.p.*) = Rhyncholaelia Schltr. (Orchidac.).
Brassiantha A. C. Smith. Celastraceae. 1 New Guinea.
Brassica L. Cruciferae. Eur., Medit., As. Many forms are cult., some for the fl., others for stem, root, leaf, or seed. *B. nigra* (L.) Koch is the black mustard, whose seeds yield the condiment. *B. oleracea* L. is the cabbage, with the various races derived from it, such as cauliflower and broccoli (fleshy infl.), kale or curly greens or borecole, brussels sprouts (a form in which miniature cabbages are produced in all the leaf axils on the main stem), kohlrabi or knol-kohl (trop.) (a thickened stem, or corm, showing leaf scars on its surface), etc. *B. campestris* L. is the turnip, a biennial with thickened root, and a var. of it—*B. napus* L.—is the rape, used in salads and in the preparation of rape- or colza-oil, expressed from the seeds. It is of interest to notice the great variety

of morphology in the veg. organs, correlated with the different ways in which storage of reserve materials is effected, in the root, stem, leaf, flower-stalk, etc.

Sauerkraut, or salted cabbage, made by packing cabbage shreds in barrels with salt and pepper, and slightly fermenting, is a favourite food in Germany, esp. for winter use.

The outer coat of the seed has mucilaginous cell-walls which swell when wetted (cf. *Linum*).

*Brassicaceae Burnett: see Cruciferae Juss. (nom. altern.).

Brassicaria Pomel = Brassica L. (Crucif.).

Brassicastrum Link = Brassica L. (Crucif.).

Brassicella Fourr. ex O. E. Schulz = Rhynchosinapis Hayek (Crucif.).

× Brassidium hort. Orchidaceae. Gen. hybr. (iii) (Brassia × Oncidium).

Brassiodendron Allen = Endiandra R.Br. (Laurac.).

Brassiophoenix Burret. Palmae. 1 New Guinea.

× Brassocatlaelia hort. (vii) = × Brassolaeliocattleya hort. (Orchidac.).

× Brassocattleya hort. Orchidaceae. Gen. hybr. (iii) (Brassavola × Cattleya).

× Brasso-Cattleya–Laelia hort. (vii) = × Brassolaeliocattleya hort. (Orchidac.).

× Brassodiacrium hort. Orchidaceae. Gen. hybr. (iii) (Brassavola × Caularthron [Diacrium]).

× Brassoepidendrum hort. Orchidaceae. Gen. hybr. (iii) (Epidendrum × Rhyncholaelia [Brassavola hort.]).

× Brassolaelia hort. Orchidaceae. Gen. hybr. (iii) (Brassavola × Laelia).

× Brassolaeliocattleya hort. Orchidaceae. Gen. hybr. (iii) (Cattleya × Laelia × Rhyncholaelia [Brassavola hort.]).

× Brassoleya hort. (vii) = × Brassocattleya hort. (Orchidac.).

× Brassonitis hort. (vii) = seq.

× Brassonotis hort. (vii) = Brassophronitis hort. (Orchidac.).

× Brassophronitis hort. Orchidaceae. Gen. hybr. (iii) (Brassavola × Sophronitis).

× Brassosophrolaeliocattleya hort. (vii) = × Potinara hort. (Orchidac.).

× Brassotonia hort. (vii) = × Brassatonia hort. (Orchidac.).

× Brassovolaelia hort. (vii) = × Brassolaelia hort. (Orchidac.).

Brathydium Spach = Hypericum L. (Guttif.).

Brathys Mutis ex L. f. = Hypericum L. (Guttif.).

× Bratonia hort. (vii) = × Miltassia hort. (Orchidac.).

Braunea Willd. = Tiliacora Colebr. (Menispermac.).

Brauneria Neck. ex Britton = Echinacea Moench (Compos.).

Braunlowia A. DC. = Brownlowia Roxb. (Tiliac.).

Braunsia Schwantes. Aïzoaceae. 7 S. Afr.

Bravaisia DC. Acanthaceae. 5 trop. Am., W.I.

× Bravanthes Cif. & Giac. Agavaceae. Gen. hybr. (Bravoa × Polianthes).

Bravoa Lex. Amaryllidaceae. 7 Mexico. Rhizome with tuberous roots. Fl. zygomorphic by bending.

Braxilia Rafin. = Pyrola L. (Pyrolac.).

Braxipis Rafin. = Cola Schott & Endl. (Sterculiac.).

Braxireon Rafin. = Tapeinanthus Herb. (Amaryllidac.).

Braxylis Rafin. = Ilex L. (Aquifoliac.).

Braya Sternb. & Hoppe. Cruciferae. 20 N. circumpol., Alps, C. As., Himal.

Braya Vell. = Hirtella L. (Chrysobalanac.).

Brayera Kunth = Hagenia J. F. Gmel. (Rosac.).

Brayodendron Small = Diospyros L. (Ebenac.).

Brayopsis Gilg & Muschler = Englerocharis Muschl. (Crucif.).

Brayulinea Small. Amaranthaceae. 3 Ecuador.

Brazilocactus A. Frič = Brasilicactus Backeb. = Notocactus (K. Schum.) Backeb. & F. M. Knuth (Cactac.).

Brazoria Engelm. & Gray. Labiatae. 3 S. U.S.

Brazzeia Baill. Scytopetalaceae. 3 trop. Afr.

Brebissonia Spach = Fuchsia L. (Onagrac.).

Bredemeyera Willd. Polygalaceae. 50 N. Guinea, Austr., S. Am., W. I. Seed hairy.

Bredia Blume. Melastomataceae. 30 E. As., Indoch.

Breea Less. = Cirsium Mill. (Compos.).

Brehmia Harv. = Strychnos L. (Strychnac.).

Brehmia Schrank = Pavonia Cav. (Malvac.).

Brehnia Baker = Behnia Didrichs. (Liliac.).

Bremekampia Sreemadh. Acanthaceae. 3 India.

Bremontiera DC. Leguminosae. 1 Masc. Is.

Brenania Keay. Rubiaceae. 1 trop. W. Afr.

Brenesia Schlechter. Orchidaceae. 1 Costa Rica.

Brenierea H. Humbert. Leguminosae. 1 Madag.

Breonia A. Rich. ex DC. Naucleaceae (~ Rubiac.). 16 Madag., Mauritius.

Brephocton Rafin. = Baccharis L. (Compos.).

Breteuillia Buc'hoz = Didelta L'Hérit. (Compos.).

Bretschneidera Hemsl. Bretschneideraceae. 1–2 China.

***Bretschneideraceae** Engl. & Gilg. Dicots. 1/2 China. Trees; ls. alt., imparipinn., exstip.; infl. a term. rac. of rather large slightly zygo. fls. K (5), open; C 5, unguic., imbr., inserted on K-tube; A 8, decl., fil. hairy, anth. versat., biapic.; disk o; G (3), with 2 pend. axile ovules per loc. and long simple curved style with small capit. stig. Fr. an obov. thick-walled 3-valved dehisc. caps.; seeds without endosp., red. Only genus: *Bretschneidera*. An interesting relict, showing affinities with primitive *Capparidac.*, *Legumin.* (*Caesalpinioïdeae*) and possibly *Hippocastanac.*

Brevidens Miq. ex C. B. Cl. = Cyrtandromoea Zoll. (Scrophulariac.).

Breviea Aubrév. & Pellegr. Sapotaceae. 1 trop. W. Afr.

Breviglandium Dulac = Hottonia L. (Primulac.).

Brevipodium Á. & D. Löve = Brachypodium Beauv. (Gramin.).

Brevoortia Wood = Dichelostemma Kunth (Alliac.).

Breweria R.Br. = Bonamia Thou. (Convolvulac.).

Brewerina A. Gray = Arenaria L. (Caryophyllac.).

Breweriopsis Roberty = Bonamia Thou. (Convolvulac.).

Brewiera Roberty = Breweria R. Br. = Bonamia Thou. (Convolvulac.).

Brewieropsis Roberty = Breweriopsis Roberty = Bonamia Thou. (Convolvulac.).

Brewstera M. Roem. = Ixonanthes Jack (Ixonanthac.).

BREXIA

*Brexia Nor. ex Thou. Brexiaceae. 9 coastal lowlands trop. E. Afr., Madag., Seychelles.

Brexiaceae Lindl. Dicots. 3/11 E. Afr., Madag., Masc., N.Z. Shrubs or small trees; ls. simple, alt. or opp. or vertic., ent., serr. or dent., coriaceous, stip. (*Brexia*) or exstip. Fls. rather large, sol. or in axill. few-fld. cymes. K 4–6, imbr. or valv., decid. or persist.; C 4–6 imbr., unguic., or (4–6) valv., decid. or persist.; disk ann. or 5-lobed; A 4–6, with large anthers; G̲ (4–7), with 2–∞ apotr. ov. per loc. and 1 style with capit. or punctif. stig.; fr. a caps., drupe or berry. Genera: *Brexia, Roussea, Ixerba.*

Brexiella H. Perrier. Celastraceae. 2 Madag.

Brexiopsis H. Perrier = Drypetes Vahl (Euphorbiac.).

*Breynia J. R. & G. Forst. Euphorbiaceae. 25 China, SE. As., Indomal., Austr., New Caled.

Breynia L. = Capparis L. (Capparidac.).

Breyniopsis Beille = Sauropus Bl. (Euphorbiac.).

Brezia Moq. Chenopodiaceae. 1 SE. Eur., C. As. to Afghan.

Bricchettia Pax = Cocculus DC. (Menispermac.).

*Brickellia Ell. Compositae. 100 warm Am., W.I.

Brickellia Rafin. = Gilia Ruiz & Pav. (Polemoniac.).

Bricour Adans. = Myagrum L. (Crucif.).

Bridelia Willd. corr. Spreng. Euphorbiaceae. 60 Afr., As.

Bridgesia Backeb. (1934) = Gymnocalycium Pfeiff. (Cactac.).

Bridgesia Backeb. (1935) = Neoporteria Britton & Rose (Cactac.).

*Bridgesia Bert. ex Cambess. Sapindaceae. 1 Chile.

Bridgesia Hook. = Polyachyrus Lag. (Compos.).

Bridgesia Hook. & Arn. = Ercilla A. Juss. (Phytolaccac.).

Briedelia Willd. = Bridelia Willd. corr. Spreng. (Euphorbiac.).

Brieya De Wild. = Piptostigma Oliv. (Annonac.).

Briggsia Craib. Gesneriaceae. 14 E. Himal., Burma, S. China.

Brighamia A. Gray. Campanulaceae. 1 Hawaii. Arborescent.

Brignolia Bertol. = Kundmannia Scop. (Umbellif.).

Brignolia DC. = Isertia Schreb. (Rubiac.).

Brillantaisia Beauv. Acanthaceae. 40 trop. Afr., Madag. 2 post. sta. perfect

Brimeura Salisb. = Hyacinthus L. (Liliac.). [(only case in fam.).

Brimys Scop. (sphalm.) = Drimys J. R. & G. Forst. (Winterac.).

Brindonia Thou. = Garcinia L. (Guttif.).

Brintonia Greene = Solidago L. (Compos.).

Briquetastrum Robyns & Lebrun. Labiatae. 1 trop. W. Afr.

Briquetia Hochr. Malvaceae. 1 Paraguay.

Briquetina Macbride. Icacinaceae. 3 Peru.

Brisegnoa Remy = Oxytheca Nutt. (Polygonac.).

Briseïs Salisb. = Allium L. (Alliac.).

Brisonia Mutis apud Alba. Colombia. Quid?

Brissonia Neck. = Indigofera L. + Tephrosia Pers. (Legumin.).

Britoa Berg = Campomanesia Ruiz & Pav. (Myrtac.).

Brittenia Cogn. Melastomataceae. 1 Borneo. 5-merous. Sta. equal.

Brittonamra Kuntze = Cracca Benth. (Legumin.).

Brittonastrum Briq. = Agastache Kuntze (Labiat.).

Brittonella Rusby. Malpighiaceae. 1 Bolivia.
Brittonia Houghton ex C. A. Armstr. = Hamatocactus Britt. & Rose + ? Thelocactus Britt. & Rose (Cactac.).
Brittonia Kuntze (sphalm.) = Brissonia Neck. = Indigofera L. + Tephrosia Pers. (Legumin.).
Brittonrosea Speg. = Echinofossulocactus Lawrence (Cactac.).
Briza L. Gramineae. 20 N. temp., S. Am. (Gramin.).
Brizochloa Jirásek & Chrtek. Gramineae. 1 Balkan Penins. to Crimea & **Brizophila** Salisb. = Ornithogalum L. (Liliac.). [Caucasus.
Brizopyrum Link = Desmazeria Dum. (Gramin.).
Brizopyrum J. Presl = Distichlis Rafin. (Gramin.).
Brizopyrum Stapf = Plagiochloa Adamson & Sprague (Gramin.).
Brizula Hieron. = Aphelia R.Br. (Centrolepidac.).
Brocchia Mauri ex Ten. = Simmondsia Nutt. (Simmondsiac.).
Brocchia Vis. = Cotula L. (Compos.).
Brocchinia Schult. f. Bromeliaceae. 12 trop. S. Am.
Brochoneura Warb. Myristicaceae. 10 trop. Afr., Madag.
Brochosiphon Nees = Dicliptera Juss. (Acanthac.).
Brockmania W. V. Fitzgerald. Malvaceae. 1 W. Austr.
*****Brodiaea** Sm. Alliaceae. *S.str.* 10, *s.l.* 30, W. N. Am. Cymose umbels. (P). Sta. with projecting appendages.
Brogniartia Walp. (sphalm.) = Brongniartia Kunth (Legumin.).
Bromaceae Burnett = Sterculiaceae Vent.
Brombya F. Muell. = Melicope J. R. & G. Forst. (Rutac.).
Bromelia Adans. = Pitcairnia L'Hérit. (Bromeliac.).
Bromelia L. Bromeliaceae. 40 trop. Am., W.I. Some with ed. fr.
*****Bromeliaceae** Juss. Monocots. 44/1400 trop. Am., W.I. Many terrestrial (xero., on rocks, etc.), but most epiph. by virtue of seed distr. and xero. habit, and more char. of trop. Am. than the orchids. Stem usu. reduced, with rosette of fleshy ls. channelled above, fitting closely at base, so that the whole pl. forms a kind of funnel, usu. full of water. In this are dead ls., decaying animal matter, and other débris (certain *Utricularias* live only on these pitchers). There are many adv. r. which fasten the plant to its support, but which do not aid in its nutrition, or very little. The bases of the ls. are covered with scaly hairs by which the water in the pitcher is absorbed. Water is stored in the ls., which consist largely of water tissue. They have a thick cuticle and often bear scaly hairs that reduce transpiration. Some show a totally different habit from this, e.g. *Tillandsia usneoïdes* L. (*q.v.*).

Infl. usu. out of the centre of the pitcher; bracts coloured. Fl. usu. ♀, reg., 3-merous. P 3 + 3 or (3) + (3), the outer whorl sepaloid, persistent, the inner petaloid; A 6, introrse, often epitep.; G (3), inf., semi-inf., or sup., 3-loc., with ∞ anatr. ov. on the axile plac. in each. Style 1, stigmas 3. Berry or caps.; seeds in the latter case very light, or winged. Embryo small, in mealy endosp.

Classification and chief genera (after Harms):
I. *Navioïdeae.* Ls. spin.-dent.; ov. sup.; caps.; seed naked (without wings or hairs): *Navia.*
II. *Pitcairnioïdeae.* Ls. ent. or spinose; ov. sup. or semi-sup.; caps.; seeds variously appendaged (rarely naked): *Pitcairnia, Puya, Dyckia.*

163

III. ***Tillandsioïdeae.*** Ls. ent.; ov. sup.; caps.; seeds with plumose corona of hairs: *Tillandsia, Vriesea.*

IV. ***Bromelioïdeae.*** Ls. toothed or spinose; ov. inf.; berry; seeds naked: *Bromelia, Ananas, Billbergia, Aechmea.*

Bromelica Farwell = Melica L. (Gramin.).
Bromfeldia Neck. = Jatropha L. (Euphorbiac.).
Bromheadia Lindl. Orchidaceae. 11 Malaysia.
Bromidium Nees & Meyen = Deyeuxia Clar. (Gramin.).
× **Bromofestuca** Prodan. Gramineae. Gen. hybr. (Bromus × Festuca).
Bromopsis Fourr. = Bromus L. (Gramin.).
Bromuniola Stapf & C. E. Hubbard. Gramineae. 1 Angola.
Bromus L. Gramineae. 50 temp.; trop. mts. Of little value as pasture.
Bromus Scop. = Triticum L. (Gramin.).
Brongniartia Blume = Kibara Endl. (Monimiac.).
Brongniartia Kunth. Leguminosae. 60 trop. Am.
Brongniartikentia Becc. Palmae. 1 New Caledonia.
Bronnia Kunth = Fouquieria Kunth (Fouquieriac.).
Brookea Benth. Scrophulariaceae. 3 Borneo.
Brosimopsis S. Moore. Moraceae. 6 S. Brazil.

***Brosimum** Sw. Moraceae. 50 trop. & temp. S. Am. Infl. remarkable, a spherical pseudo-head composed of one ♀ fl. and many ♂ fls. The former is sunk into the centre of the common recept. and its style projects at the top, whilst the latter occupy the whole of the outer surface. Each ♂ fl. has a rudim. P and one sta., whose versatile anther in dehiscing passes from a shape somewhat like ∓ to one like ⫩. Achene embedded in the fleshy recept.

The achene of *B. alicastrum* Sw. is the bread-nut (not to be confused with *Artocarpus*, the bread-fruit), which is cooked and eaten in the W.I., etc. [The 'bread-nut' of Barbados is, however, a seeded var. of the bread-fruit.] *B. galactodendron* D. Don is the cow-tree or milk-tree of Venezuela. The milky latex flows in considerable quantities, tastes very like ordinary milk, and is used for the same purposes. The wood of several spp. is useful (snake-wood).

Brossaea L. = Gaultheria L. (Ericac.).
Brossardia Boiss. Cruciferae. 2 Persia.
Brossea Kuntze = Gaultheria L. (Ericac.).
Brotera Cav. = Melhania Forsk. (Sterculiac.).
Brotera Spreng. (1800) = Flaveria Juss. (Compos.).
Brotera Spreng. (1801) = Hyptis Jacq. (Labiat.).
Brotera Vell. = Luehea Willd. (Tiliac.).
Brotera Willd. = Cardopatium Juss. (Compos.).
Broteroa DC. = Brotera Spreng. (1800) = Flaveria Juss. (Compos.).
Broteroa Kuntze = Brotera Willd. = Cardopatium Juss. (Compos.).
Brotobroma Karst. & Triana = Herrania Goudot (Sterculiac.).
Broughtonia R.Br. Orchidaceae. 1 W.I.
Broughtonia Wall. ex Lindl. = Otochilus Lindl. (Orchidac.).
× **Broughtopsis** hort. (vii) = × Lioponia hort. (Orchidac.).
Brousemichea Willis (sphalm.) = seq.
Brousmichea Balansa. Gramineae. 1 Indo-China.

Broussaisia Gaudich. Hydrangeaceae. 2 Hawaii.
*Broussonetia L'Hérit. ex Vent. Moraceae. 7–8 E. As., Polynes. Dioecious; ♂ fls. in pseudo-racemes with explosive sta. like *Urtica* (unusual in *M.*); ♀ fls. in pseudo-heads. Multiple fr. (cf. *Morus*, etc.). A good fibre, used for paper, etc., is obtained from the inner bark of *B. papyrifera* Vent. (paper mulberry, Japan); in Polynes. the natives make *tapa* or *kapa* cloth from it. The ls. double upwards during the heat of the day.
Broussonetia Ortega = Sophora L. (Legumin.).
Browallia L. Solanaceae. 6 trop. Am., W.I. A 4. Caps.
Brownaea Jacq. = Brownea Jacq. corr. Murr. (Legumin.).
Brownanthus Schwantes. Aïzoaceae. 5 S. Afr.
*Brownea Jacq. corr. Murr. Leguminosae. 25 trop. Am., W.I. The young shoots emerge very rapidly from the bud and hang down on flaccid stalks, the leaflets at first rolled up, and later spread out, and pink or red speckled with white. After a time they turn green and stiffen up and spread out normally. (Cf. *Amherstia.*) *B. grandiceps* Jacq. and others have fine bunches
Browneopsis Huber. Leguminosae. 3 Brazil.   [of fls.
Brownetera Rich. = Phyllocladus Rich. (Podocarpac.).
Browningia Britton & Rose. Cactaceae. 1 Peru, Chile.
Brownleea Harv. ex Lindl. Orchidaceae. 14 trop. & S. Afr., Madag.
Brownlowia Roxb. (= Glabraria L.). Tiliaceae. 25 SE. As., Malaysia (exc. (Java & Lesser Sunda Is.), Solom. Is.
*Brucea J. F. Mill. Simaroubaceae. 10 palaeotrop. Very astringent. The seeds of *B. sumatrana* Roxb., etc., are a remedy in dysentery.
Bruchmannia Nutt. = Beckmannia Host (Gramin.).
Bruckenthalia Reichb. Ericaceae. 1 SE. Eur., Asia Minor.
Bruea Gaudich. ? Moraceae (inc. sed.), vel ? Mallotus Lour. vel ? Macaranga Thou. (Euphorbiac.).
Brueckea Klotzsch & Karst. = Aegiphila Jacq. (Verbenac.).
Bruennichia Willd. = Brunnichia Banks ex Gaertn. (Polygonac.).
Brùgmansia Blume = Rhizanthes Dumort. (Rafflesiac.).
Brugmansia Pers. (~ Datura L.). Solanaceae. 14 trop. Am.
Bruguiera Lam. Rhizophoraceae. 6 trop. E. Afr., As., Austr., Polynes. One of the mangroves. Like *Rhizophora*, but without the aerial r. from higher branches. The plant produces conspicuous erect knee-roots.
Bruguiera Pfeiff. = Bruquieria Pourr. ex Ort. = Mirabilis L. (Nyctaginac.).
Bruguiera Rich. ex DC. = Conostegia D. Don (Melastomatac.).
Bruguiera Thou. = Lumnitzera Willd. (Combretac.).
Bruinsmania Miq. = Isertia Schreb. (Rubiac.).
Bruinsmia Boerl. & Koord. Styracaceae. 1 Assam, Burma; 1 Malaysia (exc.
Brunella Mill. = Prunella L. (Labiat.).   [Malay Penins.).
Brunellia Ruiz & Pav. Brunelliaceae. 45 Mex. to Peru, W.I.
*Brunelliaceae Engl. Dicots. 1/45 Am. Tall trees, usu. toment. throughout. Ls. opp. or vertiç., simple or pinn., often dent., with small caduc. stip. Fls. reg., ♂ ♀, dioec., in axill. and term. pan. K 4–5(–7), valv., shortly conn. below; C 0; disk cupular, adn. to K, 8–10-lobed; A 8–10(–14), with filif. pubesc. fil. and dorsif. versat. intr. anth. (stds. pres. in ♀ fl.); G̲ 5–4–2, free, gradually atten. into long subul. styles with punctif. stigs., and with 2 ventr. collat. anatr.

BRUNFELSIA

ov. per loc. (small pistillode pres. in ♂ fl.). Fr. of 5–2 ventr. dehisc. 1–2-seeded follicles; seeds with mealy endosp. Only genus: *Brunellia*. Related to *Cunoniaceae*.

Brunfelsia L. Solanaceae. 30 trop. Am., W.I. The fls. change colour as they grow older (cf. *Boraginac.*).

Brunfelsiopsis Urb. Solanaceae. 1 trop. Am.

Brunia L. Bruniaceae. 7 S. Afr.

***Bruniaceae** DC. Dicots. 12/75 S. Afr. Heath-like shrubs, with small ent. alt. exstip. ls. Fl. ♀, usu. reg., 5-merous, generally perig., in dense spicate or capit. infl. K 4–5, sometimes (4–5), imbr., persist.; C 4–5, rarely conn. below, imbr., often persist.; A 4–5, free, rarely conn. with C into a tube, often persist., anth. dorsif., intr., often versat., connective sometimes produced; disk rarely present; Ḡ, rarely G̲, (2), rarely (3), with 2–4(–10) pend. anatr. ov., or 1 with 1 ov. Caps. with 2 seeds, or nut with 1. Aril. Endosp. Chief genera: *Brunia*, *Berzelia*, *Thamnea*, *Raspalia*, *Staavia*. An isolated fam., perh. distantly related to *Hamamelidac.*

Bruniera Franch. = Wolffia Horkel ex Schleid. (Lemnac.).

Brunnera Stev. Boraginaceae. 3 SW. As. Long-pet. cordate basal ls.

Brunnichia Banks ex Gaertn. Polygonaceae. 1 N. Am.

Brunonia Sm. Brunoniaceae. 1 Austr., Tasm.

***Brunoniaceae** Dum. (~ Goodeniac.). Dicots. 1/1 Austr. Herbs with rad. ent. oblanc. exstip. ls. Infl. a scapose capitate cyme. Fls. ♀, reg. K (5), lobes subul.; C (5), lobes spathul., valv.; A 5, epipet. near base, anth. coherent, intr.; G̲ 1-loc., with 1 erect ov., encl. in K-tube. Fr. a 1-seeded 'achene' encl. in persist. K. Seed exalb. Pollen-cup like *Goodeniac.* Large suspensor haustorium. Only genus: *Brunonia*.

Brunoniella Bremek. Acanthaceae. 4 Austr.

× **Brunsdonna** van Tubergen ex Worsley. Amaryllidaceae. Gen. hybr. (Amaryllis × Brunsvigia).

× **Brunserine** Traub. Amaryllidaceae. Gen. hybr. (Brunsvigia × Nerine).

Brunsfelsia L. = Brunfelsia L. (Solanac.).

Brunsvia Neck. = Croton L. (Euphorbiac.).

Brunsvigia Heist. Amaryllidaceae. 13 Afr.

Brunswigia auctt. = praec.

Brunswigiaceae Horan. = Amaryllidaceae Jaume St-Hil.

Brunyera Bub. = Oenothera L. (Onagrac.).

Bruquieria Pourr. ex Ort. = Calyxhymenia Ort. = Mirabilis L. (Nyctaginac.).

Bruschia Bertol. = Nyctanthes L. (Verbenac.).

Bruxanelia Dennst. Inc. sed. 1 S. India.

Brya P.Br. Leguminosae. 7 C. Am., W.I. *B. ebenus* DC. yields the wood Jamaica or American ebony, cocus or cocos wood, the heartwood turning black with age (cf. *Diospyros*).

Brya Vell. = Hirtella L. (Chrysobalanac.).

Bryantea Rafin. (Neolitsea (Benth.) Merr.). Lauraceae. 80 E. As., Indomal.

Bryanthus S. G. Gmel. Ericaceae. 1 Kamch., Japan.

Bryantia Webb ex Gaudich. = Pandanus L. f. (Pandanac.).

Bryaspis Duvign. Leguminosae. 1 W. Afr.

Brylkinia F. Schmidt. Gramineae. 2 Japan, Sakhalin.

Bryobium Lindl. = Eria Lindl. (Orchidac.).

Bryocarpum Hook. f. & Thoms. Primulaceae. 1 E. Himal.

Bryocles Salisb. = Funkia Spreng. (Liliac.).

Bryodes Benth. Scrophulariaceae. 3 Mascarene Is.

Bryodes Phil. = ? Quinchamalium Juss. (Santalac.).

Bryomorpha Kar. & Kir. = Thylacospermum Fenzl (Caryophyllac.).

Bryomorphe Harv. Compositae. 1 S. Afr.

Bryonia L. Cucurbitaceae. 4 Eur., As., N. Afr., Canary Is. The distr. of *B. cretica* L. (syn. *B. dioica* Jacq.) marks the N. limit of the family *Cucurbitac.* in Eur. ♂ fl. larger. Honey secreted at the base of the P.

Bryoniaceae P. & K. = Cucurbitaceae Juss.

Bryoniastrum Fabr. (uninom.) = *Sicyos* L. sp. (Cucurbitac.).

Bryonopsis Arn. = Kedrostis Medik. (Cucurbitac.).

Bryophthalmum E. Mey. = Moneses Salisb. (Pyrolac.).

Bryophyllum Salisb. = Kalanchoë Adans. (Crassulac.).

Bryopsis Reiche (1896; non Lamour. 1809—Algae) = Reicheëlla Pax (Caryophyllac.).

Bubalina Rafin. (Burchellia R.Br.). Rubiaceae. 1 S. Afr. Buffalo wood, very

Bubania Girard = Limoniastrum Moench (Plumbaginac.).   [hard.

Bubbia Van Tiegh. Winteraceae. 30 New Guinea, Queensl., New Caled.; 1 Madag.(?).

Bubon L. = Athamanta L. (Umbellif.).

Bubonium Hill = Asteriscus Mill. (Compos.).

Bubroma Ehrh. (uninom.) = *Trifolium hybridum* L. (Legumin.).

Bubroma Schreb. = Guazuma L. (Sterculiac.).

Bucafer Adans. = Ruppia L. (Ruppiac.).

Bucanion Stev. = Heliotropium L. (Boraginac.).

Buccaferrea Bubani = Potamogeton L. (Potamogetonac.).

Buccaferrea Petagna = Bucafer Adans. = Ruppia L. (Ruppiac.).

Bucco Wendl. = Agathosma Willd. (Rutac.).

Bucculina Lindl. = Holothrix Rich. (Orchidac.).

Bucephalandra Schott. Araceae. 1 Borneo.

Bucephalon L. = Trophis P.Br. (Morac.).

Bucephalophora Pau (~ Rumex L.). Polygonaceae. 1 Medit. reg.

Buceragenia Greenman. Acanthaceae. 4 Mexico to Costa Rica.

Buceras P.Br. = Bucida L. (Combretac.).

Buceras Hall. = Trigonella L. (Legumin.).

Bucerosia Endl. = Boucerosia Wight & Arn. (Asclepiadac.).

Bucetum Parnell = Festuca L. (Gramin.).

Buchanania Sm. = Colebrookea Sm. (Labiat.).

Buchanania Spreng. Anacardiaceae. 25 Indomal., trop. Austr. G 4–6, one fertile.

Buchaniana Pierre (sphalm.) = praec.

Bucharea Rafin. = Convolvulus L. (Convolvulac.).

Buchenavia Eichl. Combretaceae. 20 trop. S. Am., W.I.

Buchenroedera Eckl. & Zeyh. Leguminosae. 23 trop. & S. Afr.

Buchera Reichb. = Hornungia Reichb. (Crucif.).

Bucheria Heynh. = Thryptomene Endl. (Myrtac.).

BUCHERIA

Bucheria auct. ex Wittst. (nomen). Euphorbiaceae. Quid?
Buchholzia Engl. Capparidaceae. 3 trop. W. Afr.
Buchia D. Dietr. = Bouchea Cham. (Verbenac.).
Buchia Kunth = Perama Aubl. (Rubiac.).
Buchingera Boiss. & Hohen. Cruciferae. 1 Armenia & Persia to C. As.
Buchingera F. Schultz = Cuscuta L. (Cuscutac.).
*__Buchloë__ Engelm. Gramineae. 1, *B. dactyloïdes* Engelm., the buffalo-grass of the western prairies of the U.S., a good fodder. It is a small creeping grass.
Buchlomimus J. R. & C. G. Reeder & Rzedowski. Gramineae. 1 Mex.
Buchnera L. Scrophulariaceae. 100 trop. & subtrop., mostly Old World.
Buchnerodendron Gürke. Flacourtiaceae. 5 trop. Afr.
Bucholtzia Meissn. = seq.
Bucholzia Mart. = Alternanthera Forsk. (Amaranthac.).
Bucholzia Stadtm. ex Willem. = Combretum L. (Combretac.).
Buch'osia Vell. = Heteranthera Ruiz & Pav. (Pontederiac.).
Buchozia L'Hérit. ex Juss. = Serissa Comm. ex Juss. (Rubiac.).
Buchozia Pfeiffer = Buch'osia Vell. = Heteranthera Ruiz & Pav. (Pontederiac.).
Buchtienia Schlechter. Orchidaceae. 1 Bolivia, Peru.
*__Bucida__ L. Combretaceae. 4 S. Florida, C. Am., W.I.
Buckinghamia F. Muell. Proteaceae. 1 Queensland.
Bucklandia R.Br. ex Griff. (1836; non Presl 1825—gen. foss.) = Exbucklandia R. W. Brown (Hamamelidac.).
Bucklandi[ac]eae J. G. Ag. = Hamamelidaceae–Exbucklandioïdeae.
*__Buckleya__ Torr. Santalaceae. 3 China, Japan, S. U.S.
Bucknera Michx = Buchnera L. (Scrophulariac.).
Bucquetia DC. Melastomataceae. 3 W. trop. S. Am.
Bucranion Rafin. = Utricularia L. (Lentibulariac.).
Buda Adans. = Spergularia (Pers.) J. & C. Presl (Caryophyllac.).
Buddleia auctt. = Buddleja L. (Buddlejac.).
Buddleja L. Buddlejaceae. 100 trop. & subtrop., esp. E. As.; 1 herbaceous sp. in Assam. Food habits of certain insects (moths and beetles) indicate an affinity between *B.* and *Scrophulariac.*
Buddlejaceae Bartl. (nom. rejic.). = Buddleja L. + Scrophulariac. quaed.
*__Buddlejaceae__ Wilhelm. Dicots. 6–10/150, trop. & warm temp. Trees or shrubs, agreeing in most techn. chars. with *Loganiac.* (*s.str.*), but differing in absence of intraxylary phloem and in the glandular, stellate or lepidote indum. K (4), C (4) imbr., A 4 epipet., $\underline{G}$ (2), rarely (4). Fr. a caps., drupe or berry. Chief genera: *Buddleja, Nuxia.*
Buechnera Wettst. = Buchnera L. (Scrophulariac.).
Buechnerla Roth = praec.
Bueckia A. Rich. = Buekia Nees = Neesenbeckia Levyns (Cyperac.).
Buekia Gaeke = Alpinia Roxb. (Zingiberac.).
Buekia Nees = Neesenbeckia Levyns (Cyperac.).
Buellia Rafin. = Ruellia L. (Acanthac.).
Buelowia Schumach. & Thonn. = Smeathmannia Soland. ex R.Br. (Passiflorac.).
Buena Cav. = Gonzalagunia Ruiz & Pav. = Gonzalea Pers. (Rubiac.).
Buena Pohl = Cosmibuena Ruiz & Pav. + Cascarilla Wedd. (Rubiac.).
Buergeria Miq. = Cladrastis Rafin. (Legumin.).

Buergeria Sieb. & Zucc. = Magnolia L. (Magnoliac.).
Buergersiochloa Pilger. Gramineae. 2 New Guinea.
Buesia (Morton) Copel. Hymenophyllaceae. 5 S. Am.
Buesiella C. Schweinf. = Rusbyella Rolfe (Orchidac.).
Buettneria Murr. = Byttneria Loefl. (Sterculiac.).
Buettneriaceae auctt. = Byttneriaceae R.Br.
Buettneriaceae Barnhart = Calycanthaceae Lindl.
Buffonea Koch = seq.
Buffonia Batsch = seq.
Bufonia L. Caryophyllaceae. 20 Canaries, Medit. Not unlike *Juncus bufonius* in habit. 4-merous. Spelling of name deliberately altered by Linnaeus, from the original *Buffonia* of Boissier de Sauvages, as a malicious pun upon *bufo*, a toad.
Buforrestia C. B. Clarke. Commelinaceae. 2 trop. S. Am. (Guiana), trop.
Bugenvillea Endl. = seq. [W. Afr.
Buginvillaea Comm. ex Juss. = Bougainvillea Comm. ex Juss. mut. Choisy (Nyctaginac.).
Buglossa S. F. Gray = Lycopsis L. (Boraginac.).
Buglossaceae ['-inae'] Hoffmgg. & Link = Boraginaceae Juss.
Buglossites Bub. = Lycopsis L. (Boraginac.).
Buglossites Moris = Borago L. (Boraginac.).
Buglossoïdes Moench (~ Lithospermum L.). Boraginaceae. 15 temp. Euras.
Buglossum Mill. = Anchusa L. (Boraginac.).
Bugranopsis Pomel = Ononis L. (Legumin.).
Buguinvillaea Humb. & Bonpl. = Bougainvillea Comm. ex Juss. (Nyctaginac.).
Bugula Mill. = Ajuga L. (Labiat.).
Buhsia Bunge. Capparidaceae. 1 Persia, Transcasp.
Buinalis Rafin. = Siphonychia Torr. & Gray (Caryophyllac.).
Buiningia Buxb. Cactaceae. 1 Brazil.
Bujacia E. Mey. = Glycine L. (Legumin.).
Bukiniczia Lincz. Limoniaceae. (~ Plumbaginaceae). 1 Afghan.
Bulbedulis Rafin. = Camassia Lindl. (Liliac.).
Bulbilis Rafin. = Buchloë Engelm. (Gramin.).
Bulbillaria Zucc. = Gagea Salisb. (Liliac.).
***Bulbine** v. Wolf. Liliaceae. 55 trop. & S. Afr.
Bulbinella Kunth. Liliaceae. 15 S. Afr., N.Z., Campbell & Auckland Is.
Bulbinopsis Borzi. Liliaceae. 3 Austr.
Bulbisperma Reinw. ex Blume = Peliosanthes Andr. (Liliac.).
Bulbocapnos Bernh. = Corydalis Vent. (Fumariac.).
Bulbocastanum Lag. = Conopodium Koch (Umbellif.).
Bulbocastanum Mill. = Bunium L. (Umbellif.).
Bulbocastanum Schur = Carum L. (Umbellif.).
Bulbocod[iac]eae Salisb. = Liliaceae-Melanthoïdeae-Colchiceae Reichb.
Bulbocodium Gron. = Romulea Maratti (Iridac.).
Bulbocodium L. (~ Colchicum L.). Liliaceae. 1 Eur.
Bulbophyllaria S. Moore = Bolbophyllaria Reichb. f. = seq.
***Bulbophyllum** Thou. Orchidaceae. 900 trop., and S. temp. In *B. minutissimum* F. Muell., etc. the pseudobulbs are hollow, with stomata on inner surface (cf. l. of *Empetrum*).

BULBOSPERMUM

Bulbospermum Blume = Peliosanthes Andr. (Liliac.).
Bulbostylis DC. = Brickellia Ell. (Compos.).
Bulbostylis Gardn. = Eupatorium L. (Compos.).
*Bulbostylis Kunth. Cyperaceae. 100 warm regions.
Bulbostylis Stev. = Eleocharis R.Br. (Cyperac.).
Bulbulus Swallen. Gramineae. 1 Brazil.
Bulga Kuntze = Bugula Mill. = Ajuga L. (Labiat.).
Bulleyia Schlechter. Orchidaceae. 1 E. Himal. to SW. China.
Bulliarda DC. = Tillaea Michx = Crassula L. (Crassulac.).
Bulliarda Neck. = Annona L. f. + Xylopia L. (Annonac.).
Bulnesia C. Gay. Zygophyllaceae. 8 Venez. to Argent. and Chile. Timber.
Bulowia Hook. = Buelowia Schumach. & Thonn. = Smeathmannia Soland. ex R.Br. (Passiflorac.).
Bulwera P. & K. = seq.
Bulweria F. Muell. = Deplanchea Vieill. (Bignoniac.).
Bumalda Thunb. = Staphylea L. (Staphyleac.).
*Bumelia Sw. Sapotaceae. 60 warm Am., W.I. Exalb.
Bumeliaceae Barnh. = Sapotaceae Juss.
Bunburia Harv. = Vincetoxicum v. Wolf (Asclepiadac.).
Bunburya Meissn. ex Hochst. = Tricalysia A. Rich. (Rubiac.).
Bunchosia Rich. ex Juss. Malpighiaceae. 50 trop. Am., W.I.
Bungea C. A. Mey. Scrophulariaceae. 2 As. Min. to C. As.
Bunias L. Cruciferae. 6 Medit., As.
Buniella Schischk. Umbelliferae. 1 C. As.
Bunion St-Lag. = Bunium L. (Umbellif.).
Bunioseris Jord. = Lactuca L. (Compos.).
Buniotrinia Stapf & Wettst. ex Stapf. Umbelliferae. 1 Persia.
Bunium Koch = Pimpinella L. (Umbellif.).
Bunium L. Umbelliferae. 40 Eur. to C. As.
Bunnya F. Muell. = Cyanostegia Turcz. (Dicrastylidac.).
Bunophila Willd. ex Roem. & Schult. = Machaonia Humb. & Bonpl. (Rubiac.).
Buonapartea G. Don = Bonapartea Ruiz & Pav. = Tillandsia L. (Bromeliac.).
Bupariti Duham. = Thespesia Soland. ex Corrêa (Malvac.).
Buphane Herb. (sphalm.) = Boöphone Herb. (Amaryllidac.).
Buphthalmum L. Compositae. 6 Eur., As. Minor. *B. salicifolium* L. is a char. pl. of the chalky Alps.
Buphthalmum Mill. = Anthemis L. (Compos.).
Buplerum Rafin. = Bupleurum L. (Umbellif.).
Bupleuroïdes Moench = Phyllis L. (Rubiac.).
Bupleurum Ehrh. (uninom.) = *Bupleurum falcatum* L. (Umbellif.).
Bupleurum L. Umbelliferae. 150 Eur., As., Afr., N. Am. *B. rotundifolium* L. has perfoliate ls., whence the name 'throw-wax' or 'thorow-wax' ('through-grow') by which it is known. All spp. have entire, often parallel-nerved ls., unusu. in this family. These were formerly thought to represent expanded petioles, but later research indicates that they are prob. true laminae.
Buplevrum Rafin. = praec.
Buprestis Spreng. = praec.
Bupthalmum Neck. = Buphthalmum L. (Compos.).

170

*Buraeavia Baill. = Austrobuxus Miq.

Burasaia Thou. Menispermaceae. 4 Madag.

Burbidgea Hook. f. Zingiberaceae. 5 Borneo. C-segments large; lat. stds. absent. The small labellum and petaloid sta. stand up in the centre of the fl.

Burbonia Fabr. (uninom.) = Persea Mill. sp. (Laurac.).

Burcardia Duham. = Callicarpa L. (Verbenac.).

Burcardia Rafin. = Campomanesia Ruiz & Pav. (Myrtac.).

Burcardia Schreb. = Piriqueta Aubl. (Turnerac.).

*Burchardia R.Br. Liliaceae. 3 Austr., Tasm.

Burchardia B. D. Jacks. (sphalm.) = Burcardia Duham. = Callicarpa L. (Ver-
Burchardia Neck. = Psidium L. (Myrtac.). [benac.).

Burchellia R.Br. (= Bubalina Rafin.). Rubiaceae. 1 S. Afr. Buffalo wood, very hard.

Burckella Pierre. Sapotaceae. 11 Moluccas, New Guinea to Samoa.

Burckhardia P. & K. (quater) = Burchardia auctt. & Burghartia Scop., q.v.

Burdachia Mart. ex A. Juss. Malpighiaceae. 4 N. Brazil.

Bureaua Kuntze = Buraeavia Baill. (Euphorbiac.).

Bureava Baill. = Combretum L. (Combretac.).

Bureavella Pierre = Lucuma Molina (Sapotac.).

Burgesia F. Muell. = Brachysema R.Br. (Legumin.).

Burghartia Scop. = Piriqueta Aubl. (Turnerac.).

Burglaria Wendl. = Ilex L. (Aquifoliac.).

Burgsdorfia Moench = Sideritis L. (Labiat.).

Burkea Benth. Leguminosae. 2 W. & S. Afr.

Burkhardia Benth. & Hook. f. = Burghartia Scop. = Piriqueta Aubl. (Turnerac.).

Burkillanthus Swingle. Rutaceae. 1 Malay Penins., Sumatra.

× **Burkillara** hort. Orchidaceae. Gen. hybr. (iii) (Aërides × Arachnis × Vanda).

Burkillia Ridley (1925; non West & West 1907—Algae) = Alloburkillia Whitmore (Legumin.).

Burkilliodendron Sastry. Leguminosae. 1 Malay Penins.

Burlingtonia Lindl. = Rodriguezia Ruiz & Pav. (Orchidac.).

Burmannia L. Burmanniaceae. 57 trop. & subtrop.

*Burmanniaceae Bl. Monocots. 17/125 trop. & subtrop., N. to Japan & E. U.S., S. to Tasm. & N.Z. Ann. or perenn. often saproph. herbs. Ls. (when present) alt. or radic., simple, ent., lin., more often reduced to scales. Fls. ♀, usu. reg., sometimes zygo., either sol. and term. or in dichas. or monochas. infl. P (3 + 3), either whorl often reduced or obsol., tube often winged, segs. sometimes remarkably appendic.; A usu. 3, sometimes 6, epitep., latr. or intr., connective often appendic.; G (3), either 3-loc. with axile plac. or 1-loc. with pariet. plac.; ov. ∞, anatr.; style 1, filif. or conical, with 3 stigs. Fr. usu. caps., sometimes fleshy, dehiscing irreg. or transv., rarely with valves; seeds ∞, with endosp.

Classification and chief genera:

1. BURMANNIEAE (P persist., no annulus in mouth of tube; A 3, subsess. in P throat; thecae transv. dehisc.; style = P tube): Burmannia, Gymnosiphon, etc.

2. HAPLOTHISMIEAE (P persist. or marcesc.; no annulus; A 6, inserted below

BURMEISTERA

mouth and recurved-pend.; thecae longit. dehisc.; style v. short):
Haplothismia.
3. THISMIEAE (P circumsciss.; annulus or diaphr. in mouth of tube; A 6
(rarely 3), pend. below annulus; thecae longit. dehisc.; style v. short):
Thismia, etc.
Burmeistera Karst. & Triana. Campanulaceae. 82 trop. S. Am.
Burnatastrum Briq. (~Plectranthus Lour.). Labiatae. 3 S. Afr., Madag.
Burnatia M. Mich. Alismataceae. 3 trop. E. Afr.
Burnettia Lindl. Orchidaceae. 1 Tasmania.
Burneya Cham. & Schlechtd. = Timonius DC. + Bobea Gaudich. (Rubiac.).
Burragea Donn. Smith & Rose = Gongylocarpus Cham. & Schlechtd.
(Onagrac.).
× **Burrageara** hort. Orchidaceae. Gen. hybr. (iii) (Cochlioda × Miltonia ×
Odontoglossum × Oncidium).
Burretiodendron Rehder. Tiliaceae. 6 SE. As.
Burretiokentia Pichi-Sermolli. Palmae. 1 New Caled.
Burriela Baill. = seq.
Burrielia DC. Compositae. 1 Calif.
Burriellia Engl. = praec.
Burroughsia Moldenke. Verbenaceae. 2 Calif., Mex.
Bursa Boehm. = Capsella Medik. (Crucif.).
Bursaia Steud. = Burasaia Thou. (Menispermac.).
Bursapastoris Quer = seq.
Bursa-pastoris Seguier = Capsella Medik. (Crucif.).
Bursaria Cav. Pittosporaceae. 3 Austr. Fr. few-seeded.
*****Bursera** Jacq. ex L. Burseraceae. 80 trop. Am. *B. gummifera* L. (birch tree,
gommier, turpentine tree) furnishes the balsam resin known as American *elemi,*
chibou, cachibou, or *gomart.*
*****Burseraceae** Kunth. Dicots. 16/500 trop. Shrubs and trees with alt., usu.
cpd., dotted, sometimes stip. ls. Balsams and resins occur, in lysigenous or
schizogenous passages. Fls. small, generally unisex., with disk like *Rutaceae.*
K (3–5), imbr. or valv.; C 3–5, rarely (3–5), imbr. or valv.; A 3–5 or 6–10,
obdiplost. when both whorls are present; anth. intr.; G (3–5), ov. usu. 2 in
each loc. Ovary 2–5-loc. with one style. Drupe or caps. Seed exalb. Many
spp. are useful on account of their resins, etc. [Ls. in 1 sp. apparently whorled;
sepals sometimes free.] Chief genera: *Protium, Commiphora, Boswellia,
Bursera, Canarium, Santiria.*
Burseria Jacq. = Bursera Jacq. ex L. (Burserac.).
Burseria Loefl. = Verbena L. (Verbenac.).
Burshia Rafin. = Myriophyllum L. (Haloragidac.).
Bursinopetalum Wight = Mastixia Blume (Cornac.).
Burtinia Buc'hoz = Magnolia L. (Magnoliac.).
Burtonia R.Br. Leguminosae. 13 Austr.
Burtonia Salisb. = Hibbertia Andr. (Dilleniac.).
Burttdavya Hoyle. Naucleaceae (~Rubiac.). 1 trop. E. Afr.
Burttia E. G. Baker & Exell. Connaraceae. 1 trop. E. Afr.
Busbeckea Endl. = Capparis L. (Capparidac.).
Busbeckea Mart. = Salpichroa Miers (Solanac.).

172

Busbequia Salisb. = Hyacinthus L. (Liliac.).
Buschia Ovczinnikov (~ Ranunculus L.). Ranunculaceae. 2 W. Medit. & E. Eur. to C. As.
Busea Miq. = Cyrtandromoea Zoll. (Scrophulariac.).
Buseria Th. Dur. Rubiaceae. 1 Madag.
Bushiola Nieuwl. = Kochia Roth (Chenopodiac.).
Busipho Salisb. = Aloë L. (Liliac.).
Bussea Harms. Leguminosae. 4 trop. Afr., Madag.
Busseria Cramer = Bursera Jacq. ex L. (Burserac.).
Busseuillia Lesson = Eriocaulon L. (Eriocaulac.).
Bustamenta Alaman ex DC. = Eupatorium L.‑ (Compos.).
Bustelina B. D. Jacks. (sphalm.) = seq.
Bustelma Fourn. Asclepiadaceae. 1 Brazil.
Bustia Adans. = Buphthalmum L. (Compos.).
Bustillosia Clos (~ Asteriscium Cham.). Hydrocotylaceae (~ Umbellif.). 1 Chile.
Butayea De Wild. = Sclerochiton Harv. (Acanthac.).
***Butea** Roxb. ex Willd. Leguminosae. 30 Indomal., China. *B. monosperma* (Lam.) Taub. (*B. frondosa* Koen. ex Roxb.) (*dhak* or *palas* tree, or bastard teak), one of the handsomest of flg. trees. The red juice when dried is known as Bengal kino and used as an astringent. The fls. yield a fugitive orange-red dye. The tree also yields lac (see *Ficus*), and is very important for lac cult.
Buteraea Nees. Acanthaceae. 3 Burma.
Butia Becc. Palmae. 9 trop. & subtrop. S. Am.
Butinia Boiss. = Conopodium Koch (Umbellif.).
Butneria P.Br. = Buttneria P.Br. = Casasia A. Rich. (Rubiac.).
Butneria Duham. = Calycanthus L. (Calycanthac.).
***Butomaceae** Rich. Monocots. 1/1 temperate Euras. Aquatic herbs with erect linear ls. Infl. a scapose cymose umbel with invol. of bracts. Fl. ♀, reg., 3-merous, hypog. P 6, petaloid, persistent. A 9, with introrse anthers. G̱ 6, apocp., with ∞ anatr. ov. scattered over the inner walls (cf. *Cabomba*), except on midrib and edges. Follicles; seed exalb.; embryo straight. Only genus: *Butomus*.
Butomissa Salisb. = Allium L. (Alliac.).
Butomopsis Kunth = Tenagocharis Hochst. (Limocharitac.).
Butomus L. Butomaceae. 1 temp. Euras., *B. umbellatus* L. (flowering rush). Infl. a term. fl. surrounded by 3 bostryx-cymes.
Butonica Lam. = Barringtonia J. R. & G. Forst. (Barringtoniac.).
Butonicoïdes R.Br. = Planchonia Blume (Barringtoniac.).
Buttneria P.Br. = Casasia A. Rich. (Rubiac.).
Buttneria Schreb. = Byttneria Loefl. (Sterculariac.).
Buttonia McKen ex Benth. Scrophulariaceae. 3 trop. & S. Afr.
Butumia G. Taylor. Podostemaceae. 1 Nigeria.
Butyrospermum Kotschy. Sapotaceae. 1 N. trop. Afr. The oily seeds of *B. parkii* Kotschy when pressed yield shea butter.
***Buxaceae** Dum. Dicots. 4/100 trop. & temp., scattered. Evergreen shrubs, more rarely trees or herbs, with alt. or opp. simple ent. or dent. exstip. ls. Fls. reg., ♂ ♀, monoec. or dioec., rarely ♀, in dense rac. or heads, or the ♀ sol.

BUXANTHUS

P 2 + 2 or 3 + 3, imbr.; A 4–6, free, with ± thick fil. and large anth.; <u>G</u> (3), 3-loc., styles 3, often distant, persist., with ± decurr. stig., and 1–2 pend. anatr. ov. per loc. Fr. a loculic. caps. or drupe; seeds black and shining, sometimes with caruncle; endosp. fleshy. Genera: *Buxus, Notobuxus, Sarcococca, Pachysandra.* An interesting and apparently ancient group, showing possible relationships with *Euphorbiac.* and perhaps *Celastrac.*

Buxanthus Van Tiegh. = Buxus L. (Buxac.).

Buxella Small = Gaylussacia Kunth (Ericac.).

Buxella Van Tiegh. = Buxus L. (Buxac.).

Buxus L. Buxaceae. 70 temp. Euras., trop. & S. Afr., Madag., to Malay Penins., Borneo, Philipp. & Lesser Sunda Is.; N. & C. Am., W.I. Fls. in heads, a term. ♀ fl. surrounded by a number of ♂ fls. The fr. dehisces explosively, the inner layer of the pericarp separating from the outer and shooting out the seeds by folding into a U-shape (cf. *Viola*). The wood of box (*B. sempervirens* L.) is exceedingly firm and close-grained, and is largely used in turning, wood-engraving, etc.

*Byblidaceae Domin. Dicots. 1/2 Austr. Insectivorous herbs. Ls. alt. elongate-lin., exstip., crowded, circinate in vernation, bearing both sessile and stalked capit. glands. Fls. reg., ⚥, sol., axill., long-pedic., ebracteol. K 5, imbr., connate at base, persist.; C 5, contorted, broad-cuneate, apically fimbr.; A 5, alternipet., sometimes unequal or declinate, anth. basifixed, dehisc. by apical pores or short slits; <u>G</u> (2), with elong. filif. style and capit. stig., and ∞ ov. on axile plac. Fr. a 2-loc. 2–4-valved ∞-seeded caps.; seeds with coarsely verruc. testa, and endosp. Only genus: *Byblis.* Perhaps some remote connection with *Pittosporac.* Great superficial similarity to *Droserac.* and *Roridulac.*

Byblis Salisb. Byblidaceae. 2 New Guinea, N. & W. Austr.

Bygnonia Barcena = Bignonia L. (Bignoniac.).

Byrnesia Rose = Graptopetalum Rose (Crassulac.).

Byronia Endl. = Ilex L. (Aquifoliac.).

Byrsa Nor. = Stephania Lour. (Menispermac.).

Byrsanthes Presl = Siphocampylus Pohl (Campanulac.).

*Byrsanthus Guillem. Flacourtiaceae. 1 W. Afr.

Byrsocarpus Schumach. & Thonn. (~ Rourea Aubl.). Connaraceae. 20 trop. Afr., Madag.

Byrsonima Rich. ex Kunth. Malpighiaceae. 120 C. & S. Am., W.I. Fr. a drupe, ed. The bark of some spp. is used in tanning.

Byrsophyllum Hook. f. Rubiaceae. 2 India, Ceylon.

Byrsopteris Morton = Arachniodes Bl. (Aspidiac.).

Bystropogon L'Hérit. Labiatae. 10 Madeira, Canary Is.

Bythophyton Hook. f. Scrophulariaceae. 1 Indomal. Submerged.

Bytneria Jacq. = seq.

*Byttneria Loefl. Sterculiaceae. 70 trop.

Byttneria Steud. = Butneria Duham. = Calycanthus L. (Calycanthac.).

*Byttneriaceae R.Br. = Sterculiaceae Vent.

C

Caapeba Mill. = Cissampelos L. (Menispermac.).
Caballeria Ruiz & Pav. = Myrsine L. (Myrsinac.).
Caballeroa Font Quer (~ Limoniastrum Fabr.). Plumbaginaceae. 1 NW. Afr.
Cabanisia Klotzsch ex Schlechtd. = Eichhornia Kunth (Pontederiac.).
Cabenia Rafin. (nomen). Crassulaceae. Quid?
Cabi Ducke. Malpighiaceae. 1 Brazil.
Cabomba Aubl. Cabombaceae. 6 warm Am.
*****Cabombaceae** A. Rich. Dicots. 2/7 cosmop. (exc. Europe). Water pls. with
erect sympod. rhiz., long, slender stems, peltate floating ls. and much-divided
submerged ls. (cf. *Ranunculus, Trapa*). Fls. ♀, small, axill., 3-merous. P 3 + 3;
A 3–6–∞; G 3–∞, fully apocp.; ovules 1–3 per loc., pariet., orthotr., some-
times attached to the cpl. *midrib*. Closed indehisc. 1–3-seeded follicles. No
aril; endo- and peri-sperm. Genera: *Cabomba, Brasenia*.
Cabralea A. Juss. Meliaceae. 40 trop. S. Am.
Cabrera Lag. = Paspalum L. (Gramin.).
Cabucala Pichon. Apocynaceae. 16 Madag.
Cacabus Bernh. Solanaceae. 10 W. trop. S. Am.
Cacalia DC. = Psacalium Cass. (Compos.).
Cacalia Kuntze = Adenostyles Cass. (Compos.).
Cacalia L. Compositae. 50 E. As., 1 extending to E. Russia; 1 N. Am.
Cacalia Lour. = Crassocephalum Moench (Compos.).
Cacaliopsis A. Gray. Compositae. 2 Pacif. U.S.
Cacao Mill. = Theobroma L. (Sterculiac.).
Cacara Thou. = Pachyrhizus Rich. (Legumin.).
Cacatali Adans. = Pedalium Royen ex L. (Pedaliac.).
Caccinia Savi. Boraginaceae. 6 W. & C. As.
Cachris D. Dietr. = Cachrys L. (Umbellif.).
Cachrydium Link = Hippomarathrum Link (Umbellif.).
Cachrys L. (emend. Koch). Umbelliferae. 22 Medit., W. & C. As.
Caconapea Cham. = Mella Vand. = Bacopa Aubl. (Scrophulariac.).
Cacosmanthus Miq. = Kakosmanthus Hassk. = Payena L. (Sapotac.).
Cacosmia Kunth. Compositae. 1 Peru.
Cacotanis Rafin. = Boltonia L'Hérit. (Compos.).
Cacoucia Aubl. = Combretum L. (Combretac.).
*****Cactaceae** Juss. Dicots. Centrospermae. 50–150 gen. (see p. 172); perhaps
2000 spp.; chiefly localised in the drier regions of trop. Am. but reaching British
Columbia and Patagonia, and ascending to over 3600 m. in the Andes. In
forest regions several genera appear as epiphytes. The only representatives of
the order in the Old World are species of *Rhipsalis*, perhaps introduced, in
Afr., Madag., Mauritius, Seychelles, and Ceylon, but several spp. of *Opuntia*
are naturalised and often troublesome in S. Afr., India, Austr., etc.

Xero. of the most pronounced type, exhibiting reduction of the transpiring
surface and also storage of water, often in great quantity. R. system generally
shallow with elongated slender but fleshy roots, sometimes a well-developed
tap-root, or occasionally tuberous. Stem fleshy, of various shapes, rarely

CACTACEAE

bearing green ls., and usually provided with spines, sometimes barbed, whose functions include maintenance of a layer of still air close to the plant, retarding transpiration, promotion of dew formation at their tips, protection from insolation and from animals, and assistance in vegetative propagation by animal dispersal. We may consider briefly some of the more important types of shoot found in C. The nearest approach to the ordinary plant type is *Pereskia*, with normally developed leaves. In each axil a group of spines arises from a cushion-like structure of short hairs termed the areole. The spines are now generally regarded as modified leaves of the axillary shoot, whose stem is usually undeveloped. The next stage is found in *Opuntia*, where the stem has taken over the assimilatory functions but still bears fleshy ls.; in a few species these aid the stem functions throughout life, but in most they fall off very early. Tufts of short barbed hairs (glochids) are present in the areoles, as well as spines. In most species, the stem consists of a branching series of joints, and in many these are flattened, exposing more surface to air and light. In the remainder of the family the true leaves are rudimentary. Many of the genera (formerly combined in *Cereus* and *Echinocactus*) have a globose to cylindrical stem bearing ribs on which are the areoles at regular intervals; the rib is formed by the confluence of each expanded leaf-base with the ones vertically above and below it in the phyllotactic series. In many species of *Epiphyllum*, *Rhipsalis*, etc., some or all of the shoots exhibit a flattened leaf-like form (cladode) with areoles in notches along their edges. This form appears to be derived from the cylindrical ribbed type (which is also to be seen in these genera, especially in seedlings and etiolated plants) by abortion of some of the ribs. The leading shoots of certain spp. of *Epiphyllum*, etc., and all mature shoots of several *Rhipsalis* spp., are perfectly cylindrical. In other genera, notably of the *Mammillaria* alliance, the leaf-bases are greatly expanded into tubercles or mam-(m)illae, which may or may not be confluent. The elongate tubercles of some spp. of *Ariocarpus*, etc., give the plant an *Aloë*- or *Sempervivum*-like habit. The microscopic leaf rudiment and the areole are in most cases elevated to the tip of the tubercle, and in many of the tuberculate gen. flowers and offsets are borne at the areole or at the base of a narrow groove-like adaxial extension of it towards the 'axil' of the tubercle. In the remaining tuberculate genera, notably *Mammillaria*, flowers and offsets are normally produced actually in the 'axils' of the tubercles, but there is no adaxial groove. Developmental studies indicate that the determinate spine-producing part of the areolar meristem is cut off from the vegetative portion while the latter is still confluent with the shoot apical meristem.

The bulk of the internal tissue consists of parenchyma in which water is stored. The cell sap is commonly mucilaginous, thus further obstructing evaporation. The cuticle is thick, and the ridges of the stem are normally occupied by mechanical tissue, whilst the stomata are in the furrows. Everything thus goes to check transpiration to the utmost extent; it is very difficult to dry a cactus for the herbarium, and its vitality is very great. Its growth is slow, but spp. of *Cereus* etc. reach a great size. Veg. reproduction is frequent in the mam(m)illate forms, and opuntias, and occurs to some extent in others. In garden practice, cacti are often multiplied by cuttings—a piece cut off and stuck into the soil will usually grow. Grafting is also widely practised.

Fls. usually solitary, sessile, borne upon or near the areoles or in the 'axils' of tubercles, often large, brightly coloured, ⚥, regular or zygo. P (∞), showing gradual transition from sepaloid to petaloid ls., spirally arranged and often fused at the base to form (with the bases of the filaments) a ± elongate hypanthium. A ∞, epipet. G̅ (2–∞), uniloc., with 2–∞ parietal placentae and ∞ anatr. (circinotropous in *Opuntia*, etc.) ovules; style simple. Fr. usu. a berry, the pulp sometimes derived from the funicles. Endosp. little or none. The fr. of many spp. is edible (e.g. *Opuntia*, etc.). Several are used for hedges. Cochineal is cultivated on *Opuntia* spp. One or two genera (e.g. *Pereskia*) supply easily worked timber. *Lophophora* and other genera contain a variety of alkaloids and are used by Mexican Indians and others for ritual and therapeutic purposes.

In spite of the great diversity of form in *C.*, the close similarity in floral structure, etc., and in karyotypes, the extent to which intergrafting and cross-fertilisation are possible, and the abundant intergradation between different types, indicate a close relationship among the members of the family. Overemphasis of 'plastic' and evidently minor characters has, however, produced from some students of the group a heavy crop of genera and species, many of which are very probably untenable but nonetheless necessitate critical evaluation. For this reason many more genera than may eventually be maintained are accepted in the body of this work. The primary division into tribes (often treated with little justification as subfamilies) at least is stable (after Britton & Rose):

1. PERESKIEAE (leaves broad, flat; glochids absent; flowers usu. stalked, often clustered): *Pereskia, Maihuenia.*

2. OPUNTIEAE (leaves usually ± terete, small, early deciduous; glochids present; flowers sessile, usually rotate): *Opuntia* and a few small genera.

3. CACTEAE (CEREËAE) (leaves rudimentary; glochids absent; flowers sessile, generally with a tube): several subtribes, containing the remaining genera, e.g. *Cephalocereus, Cereus, Echinocereus, Echinopsis, Epiphyllum, Mammillaria, Melocactus, Rhipsalis.*

Cf. Britton & Rose, *The Cactaceae* (1919–23); C. Backeberg, *Die Cactaceae* (1958–62); F. Buxbaum, various publications; L. Benson, *Cacti of Arizona* (1950); N. Boke, papers in *American Journal of Botany*, 1951→. (Backeberg recognises about 220 genera. A conservative monographer might reduce this to 40–50.)

Cactodendron Bigelow = Opuntia Mill. (Cactac.).

Cactus L. = Melocactus Link & Otto + Mammillaria Haw. + Cereus Mill. + Opuntia Mill. + Pereskia Mill., etc. (Cactac.).

Cactus L. emend. Britt. & Rose (Melocactus Link & Otto). Cactaceae. 40 C. & trop. S. Am., W.I. Ribbed pl. like *Cereus*. Fls. produced at top.—The type-species of the genus *Cactus* L., as designated by Britton & Rose in their standard monograph, *The Cactaceae*, 3: 220, 1922, is *C. melocactus* L. It is also the historic type of the name *Cactus* (this name being a shortened form, via *Melocactus*, of the original *Echinomelocactus*). In 1929, Hitchcock & Green (*Prop. Brit. Bot.*, Cambridge Congr.: 158) proposed the indefensible reversal of this designation in favour of *C. mammillaris* L., the type-species of the genus *Mammillaria* Haw., segregated from *Cactus* L. by Haworth as long ago as 1812.

CACUCIA

Hitchcock & Green's proposal was, however, accepted by the Cambridge Congress (1930), the consequential conservation of *Mammillaria* Haw. against *Cactus* L. being incorporated in the *International Rules of Bot. Nomencl.*, ed. 3: 103 (1935). This particularly glaring example of *conservation de convenance* is cited here at length in order to discourage (if possible) future would-be conservators from similar deplorable action.

Cacucia J. F. Gmel. = Cacoucia Aubl. = Combretum L. (Combretac.).
Cacuvallum Medik. = Mucuna Adans. (Legumin.).
Cadaba Forsk. Capparidaceae. 30 warm Afr., Madag., SW. As. to Ceylon; 1 Java to N. Austr. Disk prolonged post. into a tube; both androphore and gynophore present.
Cadacya Rafin. = Kadakia Rafin. = Monochoria Presl (Pontederiac.).
Cadalvena Fenzl = Costus L. (Costac.).
Cadamba Sonner. = Guettarda L. (Rubiac.).
Cadelari Adans. = Achyranthes L. (Amaranthac.).
Cadelaria Rafin. = praec.
Cadelium Medik. = Phaseolus L. (Legumin.).
Cadellia F. Muell. Simaroubaceae. 2 subtrop. Austr. Ls. stip.!
Cadetia Gaudich. Orchidaceae. 55 New Guinea, Solomon Is.
Cadia Forsk. Leguminosae. 8 E. Afr., Madag., Arabia. Fl. almost reg. with free sta.
Cadiscus E. Mey. Compositae. 1 S. Afr. Water pl.
Cadsura Spreng. = Kadsura Juss. (Schisandrac.).
Caela Adans. = Torenia L. (Scrophulariac.).
Caelebogyne John Sm. corr. Reichb. (~ Alchornea Sw.). Euphorbiaceae. 2 E. Austr. The ♀ plant of *C. ilicifolia* John Sm. produces good seed in cult. in the absence of the ♂. Adventitious embr. are formed by the budding of the nucellus round the embryo-sac (cf. *Hosta* Tratt.).
Caelestina Cass. = Ageratum L. (Compos.).
Caelia G. Don = Coelia Lindl. (Orchidac.).
Caelocline auct. ex Steud. = Coelocline A. DC. = Xylopia L. (Annonac.).
Caelodepas Benth. & Hook. f. = Koilodepas Hassk. (Euphorbiac.).
Caeloglossum Steud. = Coeloglossum Hartm. (Orchidac.).
Caelogyne Wall. = Coelogyne Lindl. (Orchidac.).
Caelospermum Blume (sphalm.) = Coelospermum Blume (Rubiac.).
Caenopteris Bergius = Asplenium L. (Aspleniac.).
Caenotus Rafin. = Erigeron L. (Compos.).
Caesalpinia L. Leguminosae. 100 trop. & subtrop., often hook-climbers. The pods of *C. bonducella* Fleming (nickar bean) are brought to Eur. by the Gulf Stream. Those of *C. coriaria* Willd. (*divi-divi*) are imported from Venezuela and W.I. for tanning. *C. sappan* L. (Indomal., cult.) and several Brazilian spp. yield a red dye from the wood (sappan, Brazil, or peach wood).
*Caesalpiniaceae R.Br. = Leguminosae–Caesalpinioïdeae Kunth. (Fls. usu. ± irreg.; K imbr., rarely valv.; C. imbr.-ascend. [upper pet. innermost] in bud; A few-10, rarely ∞, pollen usu. simple; seeds usu. without areoles; ls. usu. pinn., occas. bipinn., rarely simple.)
Caesalpiniodes Kuntze = Gleditsia L. (Legumin.).
Caesarea Cambess. = Viviania Cav. (Vivianiac.).

Caesia R.Br. Liliaceae. 9 S. Afr., Austr.

Caesia Vell. = Cormonema Reissek ex Endl. (Rhamnac.).

Caesulia Roxb. Compositae. 1 NE. India.

Caeta Steud. = Caela Adans. = Torenia L. (Scrophulariac.).

Caetocapnia Endl. = Coetocapnia Link & Otto = Bravoa Lex. (Amaryllidac.).

Caffea Nor. = Coffea L. (Rubiac.).

Cahota Karst. = Clusia L. (Guttif.).

Caidbeja Forsk. = Forskohlea L. (Urticac.).

Cailliea Guillem. & Perr. = Dichrostachys (DC.) Wight & Arn. (Legumin.).

Cailliella Jacques-Félix. Melastomataceae. 1 trop. W. Afr.

Caina Panch. ex Baill. = Neuburgia Bl. (Strychnac.).

Cainito Adans. = Chrysophyllum L. (Sapotac.).

Caiophora Presl. Loasaceae. 65 S. Am.

Cajalbania Urb. Leguminosae. 1 Cuba.

Cajan Adans. = seq.

Cajanum Rafin. = seq.

*Cajanus Adans. mut. DC. Leguminosae. 1–2 trop. Afr., As. C. cajan (L.) Millsp. (dhal, pigeon pea, or Congo pea) cult. in India, etc., for its ed. seeds.

Cajophora auctt. = Caiophora Presl (Loasac.).

Caju Kuntze = Pongam Adans. = Pongamia Adans. mut. Vent. (Legumin.).

Cajuputi Adans. = Melaleuca L. (Myrtac.).

Cakile Mill. Cruciferae. 15 sea- & lake-shores N. Am., Eur., Medit., Arabia, & Austr. C. maritima Scop. has fleshy leaves and long tap-root.

Calaba Mill. = Calophyllum L. (Guttif.).

Calacanthus T. Anders. ex Benth. & Hook. f. Acanthaceae. 1 Indomal.

Calachyris P. & K. = Calliachyris Torr. & Gray = Layia Hook. & Arn. ex DC. (Compos.).

Calacinum Rafin. = Muehlenbeckia Meissn. (Polygonac.).

Caladenia R.Br. Orchidaceae. 80 Malaysia, Austr., New Caled., N.Z.

Calad[iac]eae Salisb. = Araceae Juss.

Caladiopsis Engl. Araceae. 1 Colombia.

Caladium Rafin. = seq.

Caladium Vent. Araceae. 15 trop. S. Am.

Calaena Schlechtd. = Caleana R.Br. (Orchidac.).

Calais DC. = Microseris D. Don (Compos.).

Calamagrostis Adans. (excl. Deyeuxia Clar. ex Beauv.). Gramineae. 80 temp.

Calamina Beauv. = Apluda L. (Gramin.).

Calamintha Adans. = Nepeta L. (Labiat.).

Calamintha Mill. (~ Satureia L.). Labiatae. 6–7 W. Eur. to C. As.

Calamistrum Kuntze = Pilularia L. (Marsileac.).

Calamochloa Fourn. = Sohnsia Airy Shaw (Gramin.).

Calamochloë Reichb. = Arundinella Raddi (Gramin.).

× Calamophila O. Schwartz. Gramineae. Gen. hybr. (Ammophila × Calamagrostis).

Calamophyllum Schwantes. Aïzoaceae. 3 S. Afr.

Calamosagus Griff. = Korthalsia Blume (Palm.).

Calamovilfa (A. Gray) Hack. Gramineae. 4 S. Canada, C. & E. U.S.A.

CALAMPELIS

Calampelis D. Don = Eccremocarpus Ruiz et Pav. (Bignoniac.).

Calamus L. Palmae. 375 palaeotrop. Mostly leaf-climbers with thin reedy stems. In some there are hooks on the back of the midrib, but the more common type of l. is one in which the pinnae at the distal end are repres. by stout spines pointing backwards (cf. *Desmoncus*). The l. shoots almost vertically out of the bud up among the surrounding veg., and the hooks take hold. The stem often grows to immense lengths (150–180 m.); the plants are troublesome in trop. forests because the hooks catch. The stripped stems (rattan canes) are largely used for making chair bottoms, baskets, cables, etc.

Calamus Pall. = Acorus L. (Arac.).

Calanassa P. & K. = Callianassa Webb & Berth. = Isoplexis Lindl. ex Benth. (Scrophulariac.).

Calanchoë Pers. = Kalanchoë Adans. (Crassulac.).

Calanda K. Schum. Rubiaceae. 1 trop. Afr.

Calandra P. & K. = Calliandra Benth. = Inga L. (Legumin.).

*Calandrinia Kunth. Portulacaceae. 150 Austr., W. Canada to Cnile. The fls. close very quickly in absence of sunlight.

Calandriniopsis Franz (~ Calandrinia Kunth). Portulacaceae. 4 Chile.

Calanira P. & K. = Callianira Miq. = Piper L. (Piperac.).

*Calanthe R.Br. Orchidaceae. 120 warm. [W.I.

Calanthea (DC.) Miers (~ Capparis L.). Capparidaceae. 10 Mex. to Peru,

Calanthemum P. & K. = Callianthemum C. A. Mey. (Ranunculac.).

Calanthera Kunth ex Hook. = Buchloë Engelm. (Gramin.).

× **Calanthidiopreptanthe** hort. (iv) = Calanthe R. Br. (Orchidac.).

Calanthidium Pfitz. = Calanthe R.Br. (Orchidac.).

× **Calanthodes** hort. (iv) = praec.

× **Calanthophaius** hort. (vii) = × Phaiocalanthe hort. (Orchidac.).

Calanthus Oerst. ex Hanst. = Alloplectus Mart. (Gesneriac.).

Calanthus P. & K. = Callanthus Reichb. = Watsonia Mill. (Iridac.).

Calantica Jaub. ex Tul. Flacourtiaceae. 5 E. Afr., Madag.

Calappa Steck = Cocos L. (Palm.).

Calasias Rafin. = Anisotes Nees (Acanthac.).

Calathea G. F. W. Mey. Marantaceae. 150 trop. Am., W.I. Std. β (see fam.) present in most. The tubers of *C. allouia* (Aubl.) Lindl. (*topee tampo*) are eaten like potatoes in the W.I.

Calathiana Delarbre = Gentiana L. (Gentianac.).

Calathinus Rafin. = Narcissus L. (Amaryllidac.).

Calathodes Hook. f. & Thoms. Ranunculaceae. 3 Himal., China, Formosa.

Calathostelma Fourn. Asclepiadaceae. 1 Brazil.

Calatola Standley. Icacinaceae. 7 Mex. to Ecuador.

Calaunia Grudz. = Streblus Lour. (Morac.).

Calboa Cav. = Ipomoea L. (Convolvulac.).

Calcalia Krock. (sphalm. pro Cacalia) = Adenostyles Cass. (Compos.).

Calcaramphis Thou. (uninom.) = ? *Cynorkis* sp. (Orchidac.).

Calcarunia Rafin. = Monochoria Presl (Pontederiac.).

Calcearia Blume = Corybas Salisb. (Orchidac.).

Calceolangis Thou. (uninom.) = *Angraecum calceolus* Thou. (Orchidac.).

Calceolaria Fabr. (uninom.) = *Cypripedium calceolus* L. (Orchidac.).

Calceolaria L. Scrophulariaceae. 3–400 Mexico to S. Am.
Calceolaria Loefl. = Hybanthus Jacq. (Violac.).
Calceolus Mill. = Cypripedium L. (Orchidac.).
Calcitrapa Adans. = Centaurea L. (Compos.).
Calcitrapoïdes Fabr. = Centaurea L. (Compos.).
Calcoa Salisb. = Luzuriaga Ruiz & Pav. (Philesiac.).
Caldasia Lag. = Oreomyrrhis Endl. (Umbellif.).
Caldasia Mutis ex Caldas = Helosis Rich. (Balanophorac.).
Caldasia Willd. = Bonplandia Cav. (Polemoniac.).
Caldcluvia D. Don. Cunoniaceae. 1 Chile.
Caldenbachia Pohl ex Nees = Stenandrium Nees (Acanthac.).
Calderonia Standley. Rubiaceae. 2 C. Am.
Caldesia Parl. Alismataceae. 4 palaeotrop. (rare in Malaysia).
Calea L. Compositae. 100 trop. Am., esp. campos.
Calea Sw. = Neurolaena R.Br. (Compos.).
Caleacte R.Br. = Calea L. (Compos.).
Caleana R.Br. Orchidaceae. 5 temp. Austr., N.Z.
Caleatia Mart. ex Steud. = Lucuma Molina (Sapotac.).
Calebrachys Cass. = Calea L. (Compos.).
Calectasia R.Br. Xanthorrhoeaceae. 1 W. & S. Austr.
Calectasi[ac]eae Endl. = Xanthorrhoeaceae Dum.
Calendella Kuntze = seq.
Calendula L. Compositae. 20–30 Medit. to Persia. Fr. sometimes conspic. dimorph. or trimorph.
Calendulaceae Link = Compositae–Calenduleae Cass.
Caleopsis Fedde = Goldmanella Greenman (Compos.).
Calepina Adans. Cruciferae. 1 Eur., Medit.
Calesia Rafin. = seq.
Calesiam Adans. = Lannea A. Rich. (Anacardiac.).
Calesium Kuntze = praec.
Calestania K.-Pol. = Peucedanum L. (Umbellif.).
Caletia Baill. = Micrantheum Desf. (Euphorbiac.).
Caleya R.Br. = Caleana R.Br. (Orchidac.).
Caleyana P. & K. = praec.
Calhounia A. Nels. = Nocca Cav. (Compos.).
Calia Teran & Berland. = Sophora L. (Legumin.).
Calibanus Rose. Agavaceae. 1 Mex., a xero. with remarkable tuber and a few grass-like ls.
Calibrachoa Cerv. ex La Llave & Lex. = Petunia Juss. (Solanac.).
Calicanthus Rafin. = Calycanthus L. (Calycanthac.).
Calicera Cav. = Calycera Cav. (Calycerac.).
Calicoca Rafin. = Callicocca Schreb. = Cephaëlis Sw. (Rubiac.).
Calicorema Hook. f. Amaranthaceae. 1 trop. & S. Afr.
Calicotome Link. Leguminosae. 7 Medit. Stem branches thorny.
Caligula Klotzsch = Agapetes D. Don ex G. Don (Ericac.).
Calimeris Nees = Kalimeris Cass. = Aster L. (Compos.).
Calinea Aubl. = Doliocarpus Roland. (Dilleniac.).
Calinux Rafin. = Pyrularia Michx (Santalac.).

CALIPHRURIA

Caliphruria Herb. Amaryllidaceae. 2 S. Am. Sta. with stipular appendages (see fam.).
Calipogon Rafin. = Calopogon R.Br. (Orchidac.).
Calirhoë Rafin. = Callirhoë Nutt. (Malvac.).
Calisaya Hort. ex Pav. = Cinchona L. (Rubiac.).
Calispepla Vved. Leguminosae. 1 C. As.
Calispermum Lour. = Embelia Burm. f. (Myrsinac.).
Calista Ritg. = Callista Lour. = Dendrobium Sw. (Orchidac.).
Calistachya Rafin. = Veronicastrum Fabr. (Scrophulariac.).
Calistegia Rafin. = Calystegia R.Br. (Convolvulac.).
Calisto Néraud. ?Cyperaceae. 1 Mauritius. Quid?
Calius Blanco = Streblus Lour. (Morac.).
Calixnos Rafin. = Crawfurdia Wall. = Gentiana L. (Gentianac.).
Calla L. Araceae. 1 N. temp. and subarct., *C. palustris* L. Fls. ♀ without P, borne once in two years. Aquatic.
Callaceae Bartl. = Araceae Juss.
Calladium Rafin. = Caladium Vent. (Arac.).
Callaeocarpus Miq. = Castanopsis (D. Don) Spach (Fagac.).
Callaeolepium Karst. = Fimbristemma Turcz. (Asclepiadac.).
Callaeum Small. Malpighiaceae. 1 C. Am.
Callaion Rafin. = Calla L. (Arac.).
Callanthus Reichb. = Watsonia Mill. (Iridac.).
Callaria Rafin. = Calla L. (Arac.).
Callerya Endl. = Millettia Wight & Arn. (Legumin.).
Calliachyris Torr. & Gray = Layia Hook. & Arn. ex DC. (Compos.).
Calliagrostis Ehrh. (uninom.) = Bromus inermis Leyss. (Gramin.).
Callianassa Webb & Berth. = Isoplexis Lindl. ex Benth. (Scrophulariac.).
Calliandra Benth. Leguminosae. 100 Madag., warm As., Am.
Callianira Miq. = Piper L. (Piperac.).
× **Callianthemulus** Cif. & Giac. Ranunculaceae. Gen. hybr. (Callianthemum × Ranunculus).
Callianthemum C. A. Mey. Ranunculaceae. 10 mts. of Eur. & C. As.
Callias Cass. = Kallias Cass. = Heliopsis Pers. (Compos.).
Calliaspidia Bremek. = Drejerella Lindau (Acanthac.).
Callibrachoa auctt. = Calibrachoa Cerv. ex La Llave & Lex. = Petunia Juss. (Solanac.).
Callicarpa L. Verbenaceae. 140 trop. & subtrop.
Callicephalus C. A. Mey. = Centaurea L. (Compos.).
Callichilia Stapf. Apocynaceae. 10 trop. Afr.
Callichlamys Miq. Bignoniaceae. 1 trop. Am.
Callichloë Willd. ex Steud. = Elionurus Humb. & Bonpl. ex Willd. (Gramin.).
Callichloëa Spreng. ex Steud. = praec.
Callichroa Fisch. & Mey. = Layia Hook. & Arn. ex DC. (Compos.).
Callicocca Schreb. = Cephaëlis Sw. (Rubiac.).
Callicoma Andr. Cunoniaceae. 1 E. Austr. Corolla wanting.
Callicomaceae J. G. Agardh = Cunoniaceae–Pancherieae Engl.
Callicore Link = Amaryllis L. (Amaryllidac.).
Callicornia Burm. f. = Asteropterus Adans. (Compos.).

Callicysthus Endl. = Vigna Savi (Legumin.).
Callidrynos Néraud = Molinaea Commers. ex Juss. (Sapindac.).
Calliglossa Hook. & Arn. = Layia Hook. & Arn. ex DC. (Compos.).
Calligonum L. Polygonaceae. 80 S. Eur., N. Afr., W. As.
Calligonum Lour. = Tetracera L. (Dilleniac.).
Callilepis DC. Compositae. 6 S. Afr.
Callionia Greene = Potentilla L. (Rosac.).
Calliopea D. Don = Crepis L. (Compos.).
Calliopsis Reichb. = Coreopsis L. (Compos.).
Calliparion (Link) Reichb. ex Wittst. = Aconitum L. (Ranunculac.)
Callipeltis Stev. Rubiaceae. 3 Spain & Egypt to Baluchistan.
Calliphruria Lindl. = Caliphruria Herb. (Amaryllidac.).
Calliphyllon Bub. & Penz. = Epipactis Zinn (Orchidac.).
Calliphysa Fisch. & Mey. = Calligonum L. (Polygonac.).
Calliprena Salisb. = Allium L. (Alliac.).
Calliprora Lindl. = Brodiaea Sm. (Alliac.).
Callipsyche Herb. = Eucrosia Ker-Gawl. (Amaryllidac.).
Callipteris Bory. Athyriaceae. 3 Africa to Pacific.
Callirhoë Nutt. Malvaceae. 10 N. Am.
Callisace Fisch. = Angelica L. (Umbellif.).
Callisema Steud. = seq.
Callisemaea Benth. = Platypodium Vog. (Legumin.).
Callisia L. = Asteropterus Adans. (Compos.).
Callisia Loefl. Commelinaceae. 12 trop. Am.
Callista D. Don = Erica L. (Ericac.).
Callista Lour. = Dendrobium Sw. (Orchidac.).
Callistachya Rafin. = Veronicastrum Fabr. (Scrophulariac.).
Callistachya Sm. = Callistachys Vent. = Oxylobium Andr. (Legumin.).
Callistachys Heuffel = Heuffelia Opiz = Carex L. (Cyperac.).
Callistachys Vent. = Oxylobium Andr. (Legumin.).
Callistemma Boiss. = Tremastelma Rafin. (Dipsacac.).
Callistemma Cass. = Callistephus Cass. (Compos.).
Callistemon R.Br. Myrtaceae. 25 Austr., New Caled. The axis of the infl. grows on beyond the fl. and continues to produce ls. (cf. *Eucomis*). Sta. conspicuous (bird-pollination), as is often the case in the drier parts of Austr. (cf. *Acacia*).
Callistephana Fourr. = Coronilla L. (Legumin.).
*****Callistephus** Cass. Compositae. 1 China, Japan, *C. chinensis* (L.) Nees.
Callisteris Greene = Gilia Ruiz & Pav. (Polemoniac.).
Callisthene Mart. Vochysiaceae. 10 S. Am.
Callistigma Dinter & Schwantes. Aïzoaceae. 1 S. Afr.
Callistopteris Copel. Hymenophyllaceae. 5 Malaysia to Polynesia.
Callistroma Fenzl = Oliveria Vent. (Umbellif.).
Callistylon Pittier. Leguminosae. 1 Venez.
Callithamna Herb. = Stenomesson Herb. (Amaryllidac.).
Callithronum Ehrh. (uninom.) = *Serapias rubra* L. = *Cephalanthera rubra* (L.) Rich. (Orchidac.).
*****Callitrichaceae** Link. Dicots. 1/25 cosmop. (exc. S. Afr.). Submerged ann.

herbs, with opp. ent. exstip. ls. The submerged ls. are longer and narrower than the floating, and the more so the deeper they are below the surface. Land forms also occur. Fl. unisex., naked, commonly with 2 horn-like bracteoles, protog.; ♂ of 1 sta.; ♀ of (2) cpls., transv. placed, 4-loc. by 'false' septum (cf. *Labiatae*), with 2 styles; 1 pend. anatr. ov. in each loc., with ventral raphe. Schizocarp. Fleshy endosp. Only genus: *Callitriche*. Relationships much disputed; perh. some connection with *Scrophulariac.*, *Verbenac.* and *Labiat.*

Callitriche L. Callitrichaceae. 25 cosmop. (exc. S. Afr.).

Callitris Vent. Cupressaceae. 16 Austr., New Caled. (cypress pine). Ls. and cone-scales in whorls. The cone ripens in 1 or 2 years. Wood valuable.

Callitropsis Compton = Neocallitropsis Florin (Cupressac.).

Callitropsis Oerst. = Chamaecyparis Spach (Cupressac.).

Callixene Comm. ex Juss. = Luzuriaga Ruiz & Pav. (Philesiac.).

Callobuxus Panch. ex Brongn. & Gris = Tristania R.Br. (Myrtac.).

Callogramme Fée = Syngramma J. Sm. (Gymnogrammac.).

Callopisma Mart. = Deïanira Cham. & Schlechtd. (Gentianac.).

Callopsis Engl. Araceae. 1 coast W. Equat. Afr., 1 coast E. trop. Afr.

Callosmia Presl = Anneslea Wall. (Theac.).

Callostylis Blume = Eria Lindl. (Orchidac.).

Callotropis G. Don = Galega L. (Legumin.).

Calluna Salisb. Ericaceae. 1, *C. vulgaris* (L.) Hull (heather or ling), Atl. N. Amer., Azores, NW. Morocco, Eur. (not SE. exc. Eur. Turkey), Iceland; here and there in Siberia; covering large areas, together with spp. of *Erica* and *Vaccinium*. A low evergr. shrub, with linear closely crowded decussate wiry ls. and racemes of fls. K coloured like the almost polypetalous C. The honey is more easily accessible than in *Erica* (fl. of class B) and there is a larger circle of visiting insects, including, however, many bees (heather honey is among the best). The stigma projects beyond the mouth of the fl.; insects touch it first and in probing for honey jostle the anthers. The fl. is also wind-pollinated; the loose powdery pollen blows about easily and the stigma is not covered by the C. Related to *Cassiope* rather than to *Erica* (see fam.).

Callyntranthele Ndz. = Byrsonima Rich. ex Juss. (Malpighiac.).

Calobota Eckl. & Zeyh. = Lebeckia Thunb. (Legumin.).

Calobotrya Spach = Ribes L. (Grossulariac.).

Calobuxus P. & K. = Callobuxus Panch. ex Brongn. & Gris = Tristania R.Br. (Myrtac.).

Calocarpum Pierre. Sapotaceae. 6 Mex. to trop. S. Am.

Calocarpus P. & K. = Callicarpa L. (Verbenac.).

Calocedrus Kurz. Cupressaceae. 3 N. Burma, NE. Siam, SW. China, Hainan; Formosa; Pacif. N. Am. *C. decurrens* (Torr.) Florin (Calif., 'white cedar') yields valuable timber.

Calocephalus R.Br. Compositae. 15 temp. Austr.

Calochilus R.Br. Orchidaceae. 11 New Guinea, Austr., New Caled., N.Z.

Calochilus P. & K. = Callichilia Stapf (Apocynac.).

Calochlamys P. & K. = Callichlamys Miq. (Bignoniac.).

Calochlamys Presl = Congea Roxb. (Symphorematac.).

Calochloa P. & K. = Callichloë Willd. ex Steud. = Elionurus Humb. & Bonpl. ex Willd. (Gramin.).

Calochone Keay. Rubiaceae. 2 trop. W. Afr.

Calochortaceae Dum. = Liliaceae–Lilioïdeae Engl.

Calochortus Pursh. Liliaceae. 60 temp. W. N. Am., C. Am.

Calochroa P. & K. = Callichroa Fisch. & Mey. = Layia Hook. & Arn. ex DC. (Compos.).

Calococca P. & K. = Callicocca Schreb. = Cephaëlis Sw. (Rubiac.).

Calococcus Kurz ex Teijsm. & Binnend. = Margaritaria L. f. (Euphorbiac.).

Calocoma P. & K. = Callicoma Andr. (Cunoniac.).

Calocorema P. & K. = Calicorema Hook. f. (Amaranthac.).

Calocornia P. & K. = Callicornia Burm. f. = Leysera L. (Compos.).

Calocrater K. Schum. Apocynaceae. 1 W. Equat. Afr.

Calocysthus P. & K. = Callicysthus Endl. = Vigna Savi (Legumin.).

Calodecaryia J. F. Leroy. Meliaceae. 2 Madag.

*****Calodendrum** Thunb. Rutaceae. 2 trop. & S. Afr.

Calodium Lour. = Cassytha L. (Laurac.).

Calodonta Nutt. = Tolpis Adans. (Compos.).

Calodracon Planch. = Cordyline Comm. ex Juss. (Agavac.).

Calodryum Desv. = Quivisia Comm. ex Juss. = Turraea L. (Meliac.).

Caloglossa P. & K. = Calliglossa Hook. & Arn. = Layia Hook. & Arn. ex DC. (Compos.).

Caloglossum Schlechter = Cymbidiella Rolfe (Orchidac.).

Calogonum P. & K. (1) = Calligonum L. (Polygonac.).

Calogonum P. & K. (2) = Calligonum Lour. = Tetracera L. (Dilleniac.).

Calographis Thou. (uninom.) = *Limodorum pulchrum* Thou. = *Eulophidium pulchrum* (Thou.) Summerh. (Orchidac.).

Calogyne R.Br. Goodeniaceae. 7 China, Indochina, Philipp., E. Malaysia, Austr.

Calolepis P. & K. = Callilepis DC. (Compos.).

Calolisianthus Gilg. Gentianaceae. 11 trop. Am., W.I.

Calomecon Spach = Papaver L. (Papaverac.).

Calomeria Vent. = Humea Sm. (Compos.).

Calomicta P. & K. = Kolomikta Regel = Actinidia Lindl. (Actinidiac.).

Calomorphe Kunze ex Walp. = Lennea Klotzsch (Legumin.).

Calomyrtus Blume = Myrtus L. (Myrtac.).

Caloncoba Gilg. Flacourtiaceae. 15 trop. Afr.

Calonnea Buc'hoz = Gaillardia Fouger. (Compos.).

Calonyction Choisy = Bonanox Rafin. = Ipomoea L. (Convolvulac.).

Calopanax P. & K. = Kalopanax Miq. (Araliac.).

Calopappus Meyen = Nassauvia Comm. ex Juss. (Compos.).

Caloparion P. & K. = Calliparion (Link) Reichb. ex Wittst. = Aconitum L. (Ranunculac.).

Calopeltis P. & K. = Callipeltis Steven (Rubiac.).

Calopetalon J. Drumm. ex Harv. = Marianthus Hueg. (Pittosporac.).

Calophaca Fisch. Leguminosae. 10 S. Russia to China & Burma.

Calophanes D. Don = Dyschoriste Nees (Acanthac.).

Calophanoïdes Ridley. Acanthaceae. 9 Indomal., China.

CALOPHRURIA

Calophruria P. & K. = Caliphruria Herb. (Amaryllidac.).

Calophthalmum Reichb. = Blainvillea Cass. (Compos.).

Calophylica Presl = Phylica L. (Rhamnac.).

Calophyllaceae J. G. Agardh = Guttiferae–Calophylloïdeae Engl.

Calophylloïdes Smeathm. = Eugenia L. (Myrtac.).

Calophyllum L. Guttiferae. 8 Madag., Maurit.; 100 Indomal., Indoch., Pacif., trop. Austr.; 4 trop. Am., W.I. *C. tacamahaca* Willd. and other spp. yield resins known as *tacamahac*. (See *Populus*.) The young ls. are usu. prettily coloured.

Calophyllum P. & K. (1) = Calliphyllon Bub. & Penz. = Epipactis Zinn (Orchidac.).

Calophyllum P. & K. (2) = Kalliphyllon PohJ = Symphyopappus Turcz. (Compos.).

Calophysa DC. = Clidemia D. Don (Melastomatac.).

Calophysa P. & K. = Calliphysa Fisch. & Mey. = Calligonum L. (Polygonac.).

Calopisma P. & K. = Callopisma Mart. = Deïanira Cham. & Schlechtd. (Gentianac.).

Caloplectus Oerst. = Alloplectus Mart. (Gesneriac.).

*Calopogon R.Br. Orchidaceae. 4 N. Am.

Calopogonium Desv. Leguminosae. 10 C. & S. Am., W.I.

Caloprena P. & K. = Calliprena Salisb. = Allium L. (Alliac.).

Caloprora P. & K. = Calliprora Lindl. = Brodiaea Sm. (Alliac.).

Calopsis Beauv. ex Desv. Restionaceae. 26 S. Afr.

Calopsyche P. & K. = Callipsyche Herb. = Eucrosia Ker-Gawl. (Amaryllidac.).

Calopteryx A. C. Smith. Ericaceae. 2 W. trop. S. Am.

Caloptilium Lag. = Nassauvia Juss. (Compos.).

Calopyxis Tul. Combretaceae. 23 Madag.

Calorchis Barb. Rodr. = Ponthieva R.Br. (Orchidac.).

Calorhabdos Benth. Scrophulariaceae. 4–5 E. Himal. to Formosa.

Calorhoë P. & K. = Callirhoë Nutt. (Malvac.).

Calorophus Labill. Restionaceae. 1 Queensl. to Tasm., N.Z.

Calosace P. & K. = Callisace Fisch. ex Hoffm. = Angelica L. (Umbellif.).

Calosacme Wall. = Chirita Buch.-Ham. (Gesneriac.).

Calosanthes Blume = Oroxylum Vent. (Bignoniac.).

Calosanthos Reichb. = Kalosanthes Haw. = Rochea DC. + Crassula L. *p.p.* (Crassulac.).

Calosciadium Endl. = Aciphylla J. R. & G. Forst. (Umbellif.).

Caloscilla Jord. & Fourr. = Scilla L. (Liliac.).

Caloscordum Herb. = Nothoscordum Kunth (Alliac.).

Calosemaea P. & K. = Callisemaea Benth. = Platypodium Vog. (Legumin.).

Caloseris Benth. = Onoseris DC. (Compos.).

Calosmon Bercht. & Presl = Lindera Thunb. (Laurac.).

Calospatha Becc. Palmae. 2 Malay Penins.

Calospermum Pierre = Lucuma Molina (Sapotac.).

Calosphace Rafin. = Salvia L. (Labiat.).

Calostachya P. & K. = Callistachys Rafin. = Veronica L. (Scrophulariac.).

Calostachys P. & K. (1) = Callistachys Heuffel = Carex L. (Cyperac.).

Calostachys P. & K. (2) = Callistachys Vent. = Oxylobium Andr. (Legumin.).

Calosteca Desv. (?sphalm. pro Calostega) = Calotheca Desv. = Briza L. (Gramin.).

Calostelma D. Don = Liatris Schreb. (Compos.).

Calostemma R.Br. Amaryllidaceae. 4 E. Austr. There is no embryo, but bulbils are said to be formed in the embryo-sac.

Calostemma P. & K. = Callistemma Cass. = Callistephus Cass. (Compos.).

Calostemon P. & K. = Callistemon R.Br. (Myrtac.).

Calostephana P. & K. = Callistephana Fourr. = Coronilla L. (Legumin.).

Calostephane Benth. Compositae. 6 S. warm Afr.

Calostephus P. & K. = Callistephus Cass. (Compos.).

Calostigma Decne. Asclepiadaceae. 12 Brazil.

Calostigma Schott = Philodendron Schott (Arac.).

Calostima Rafin. = Urera Gaudich. (Urticac.).

Calostroma P. & K. = Callistroma Fenzl = Oliveria Vent. (Umbellif.).

Calostrophus F. Muell. = Calorophus Labill. (Restionac.).

Calostylis Kuntze = Callostylis Blume = Eria Lindl. (Orchidac.).

Calota Harv. ex Lindl. = Ceratandra Eckl. ex Bauer (Orchidac.).

Calothamnus Labill. Myrtaceae. 25 W. Austr. The axis goes on bearing ls. beyond the fls. (cf. *Callistemon*). Sta. in bundles before the petals, the common axis of the bundle very large.

Calothauma P. & K. = Callithauma Herb. = Stenomesson Herb. (Amaryllidac.).

Calotheca Desv. = Briza L. (Gramin.).

Calotheca Spreng. = Aeluropus Trin. (Gramin.).

Calotheria Wight & Arn. = Pappophorum Schreb. (Gramin.).

Calothyrsus Spach = Aesculus L. (Hippostanac.).

Calotis R.Br. Compositae. 20 Austr.

Calotriche P. & K. = Callitriche L. (Callitrichac.).

Calotropis R.Br. Asclepiadaceae. 6 trop. Afr., As. *C. gigantea* Ait. (*madar, mudar, wara*) yields a fibre from the bark, and a floss, used like kapok (*Eriodendron*), from the seeds.

Calotropis P. & K. = Callotropis G. Don = Galega L. (Legumin.).

Caloxene P. & K. = Callixene Comm. ex Juss. = Luzuriaga Ruiz & Pav. (Philesiac.).

Calpandria Blume = Camellia L. (Theac.).

Calpicarpum G. Don = Kopsia Blume (Apocynac.).

Calpidia Thou. = Ceodes J. R. & G. Forst. (Nyctaginac.).

Calpidisca Barnh. = Utricularia L. (Lentibulariac.).

Calpidochlamys Diels. Moraceae. 2 New Guinea.

Calpidosicyos Harms = Momordica L. (Cucurbitac.).

Calpigyne Blume = Koilodepas Hassk. (Euphorbiac.).

Calpocalyx Harms. Leguminosae. 11 trop. W. Afr.

Calpocarpus P. & K. = Calpicarpum G. Don = Kopsia Bl. (Apocynac.).

Calpurnia E. Mey. Leguminosae. 6 Afr. Pod narrowly winged.

Calsiama Rafin. = Calesiam Adans. = Lannea A. Rich. (Anacardiac.).

Caltha L. Ranunculaceae. 20 Arct. & N. temp., 12 N.Z. & temp. S. Am. K coloured. No honey-ls.; honey by cpls.

Caltha Mill. = Calendula L. (Compos.).

Calthoïdes B. Juss. ex DC. = Othonna L. (Compos.).

CALUCECHINUS

Calucechinus Hombr. & Jacquinot = Nothofagus Bl. (Fagac.).
Calusia Bert. ex Steud. = Myrospermum Jacq. (Legumin.).
Calusparassus Hombr. & Jacquinot = Nothofagus Bl. (Fagac.).
Calvaria Comm. ex Gaertn. f. Sapotaceae. 15 Madag., Masc.
Calvelia Moq. Chenopodiaceae. 1 C. As.
Calvoa Hook. f. Melastomataceae. 20 trop. Afr. Connective with scale.
Calybrachoa auctt. = Calibrachoa Cerv. ex La Llave & Lex. = Petunia Juss. (Solanac.).
Calycacanthus K. Schum. Acanthaceae. 1 New Guinea.
Calycadenia DC. = Hemizonia DC. (Compos.).
Calycampe Berg = Myrcia DC. (Myrtac.).
Calycandra Lepr. ex A. Rich. = Cordyla Lour. (Legumin.).
*Calycanthaceae Lindl. 2/6 E. Asia, N. America Shrubs, usually aromatic, with opp. simple exstip. ls. and term. acyclic fls. on short shoots. P ∞, perig., spiral, with gradual transition from sepaloid to petaloid ls.; A 5–30, the inner sterile; G ∞, in hollowed axis, with 1–2 anatr. ov. in each. Achenes enclosed in axis; embryo large, with convolute cots. in slight endosp. Genera: *Calycanthus, Chimonanthus*. Prob. related to *Monimiac.*
Calycanthemeae [L.] Vent. = Lythraceae Jaume St-Hil.
Calycanthemum Klotzsch = Ipomoea L. (Convolvulac.).
*Calycanthus L. Calycanthaceae. 3–4 SW. & E. U.S.
Calycera Cav. Calyceraceae. 20 temp. S. Am. Frs. all free, dimorphic.
Calyceraceae R.Br. ex Rich. Dicots. 4/40 S. Am. Dwarf herbs with alt. exstip. ls. Fls. in heads with invol. of br., ⚥ or ♂ ♀, reg. or zygo., epig., 4–6-merous. K leafy; C valv. or open; A in one whorl, filaments united, anthers free or slightly coherent at base; G̅ 1-loc., with 1 pend. anatr. ov., and capitate stigma. Frs. achene-like, sometimes ± united, crowned by persistent K; embryo straight in slight endosp. Closely related to *Compos.* Genera: *Boöpis, Calycera, Acicarpha, Moschopsis.*
Calyciferae Hutch. A primary division of Monocots. characterized by the possession of a herbaceous (not petaloid) calyx, incl. most of the *Helobiae, Triuridales, Farinosae* and *Scitamineae* of Engler, but excluding the *Glumiflorae, Juncac., Centrolepidac.* and *Restionac.*
Calyciflorae DC. A 'Series' of Dicots. comprising (in Bentham & Hooker) the orders *Rosales, Myrtales, Passiflorales, Ficoïdales* and *Umbellales.*
Calycinae Benth. & Hook. f. A 'Series' of Monocots. comprising the fams. *Flagellariac., Juncac.* and *Palmae.*
Calycium Ell. = Heterotheca Cass. (Compos.).
Calycobolus Willd. ex Roem. & Schult. Convolvulaceae. 18 trop. Am. & Afr.
Calycocarpum Nutt. ex Torr. & Gray. Menispermaceae. 1 Atl. N. Am.
Calycocorsus F. W. Schmidt = Chondrilla L. (Compos.).
Calycodaphne Boj. = Ocotea Aubl. (Laurac.).
Calycodendron A. C. Smith = Psychotria L. (Rubiac.).
Calycodon Nutt. = Muhlenbergia Schreb. (Gramin.).
Calycodon Wendl. = Hyospathe Mart. (Palm.).
Calycogonium DC. Melastomataceae. 30 W.I.
Calycolpus Berg. Myrtaceae. 13 W.I., S. Am. K leafy, reflexed in bud.
Calycomelia Kostel. = Fraxinus L. (Oleac.).

Calycomis R.Br. = Callicoma Andr. (Cunoniac.).
Calycomis D. Don. Cunoniaceae. 1 E. Austr.
Calycomorphum Presl = Trifolium L. (Legumin.).
Calycopeplus Planch. Euphorbiaceae. 3 Austr.
Calycophisum Karst. & Triana = Calycophysum Karst. & Triana (Cucurbitac.).
Calycophyllum DC. Rubiaceae. 6 W.I., S. Am. 1 large sep. Wood useful.
Calycophysum Karst. & Triana. Cucurbitaceae. 5 NW. trop. S. Am.
Calycoplectus Oerst. = Alloplectus Mart. (Gesneriac.).
Calycopteris Lam. = Getonia Roxb. (Combretac.).
Calycopteris Rich. ex DC. = Calycogonium DC. (Melastomatac.).
Calycopteris Sieb. = Buckleya Torr. (Santalac.).
Calycorectes Berg. Myrtaceae. 15 S. Am., W.I. Like *Eugenia*, but (K).
Calycoseris A. Gray. Compositae. 2 SW. U.S., Mex.
Calycosia A. Gray. Rubiaceae. 5 Polynes.
Calycosiphonia (Pierre) Lebrun = Coffea L. (Rubiac.).
Calycosorus Endl. = Calycocorsus F. W. Schmidt = Chondrilla L. (Compos.).
Calycostegia Lem. = Calystegia R.Br. (Convolvulac.).
Calycostemma Hanst. = Isoloma Decne (Gesneriac.).
Calycostylis Hort. ex Vilm. = Beloperone Nees (Acanthac.).
Calycothrix Meissn. = Calythrix Labill. (Myrtac.).
Calycotome Link (mut. Link) = Calicotome Link (Legumin.).
Calycotome E. Mey. = Dichilus DC. (Legumin.).
Calycotomon Hoffmgg. = Calycotome Link (Legumin.).
Calycotomus Rich. = Conostegia D. Don (Melastomatac.).
Calycotropis Turcz. Caryophyllaceae? 1 Mex.
Calyctenium Greene = Rubus L. (Rosac.).
Calyculogygas Krapov. Malvaceae. 1 Uruguay.
Calydermos Lag. = Calea L. (Compos.).
Calydermos Ruiz & Pav. = Nicandra Adans. (Solanac.).
Calydorea Herb. Iridaceae. 12 S. U.S. to S. Am.
Calygogonium G. Don (sphalm.) = Calycogonium DC. (Melastomatac.).
Calylophis Spach = seq.
Calylophus Spach. Onagraceae. 4 (or more) C. U.S., N. Mex.
Calymella Presl = Gleichenia Sm. (Gleicheniac.).
Calymenia Pers. = Calyxhymenia Ort. = Mirabilis L. (Nyctaginac.).
Calymeris P. & K. = Kalimeris Cass. = Aster L. (Compos.).
Calymmandra Torr. & Gray = Evax Gaertn. (Compos.).
Calymmanthera Schlechter. Orchidaceae. 3 New Guinea.
Calymmanthium Ritter. Cactaceae. 1 N. Peru.
Calymmatium O. E. Schulz. Cruciferae. 1 C. As. (Pamirs).
Calymmodon Presl. Grammitidaceae. 25 Ceylon to Tahiti.
Calymmostachya Bremek. Acanthaceae. 1 Siam.
Calynux Rafin. = Calinux Rafin. = Pyrularia Michx (Santalac.).
Calyplectus Ruiz & Pav. = Lafoënsia Vand. (Lythrac.).
*Calypso Salisb. Orchidaceae. 1 cold N. temp.
Calypso Thou. = Johnia Roxb. = Salacia L. (Celastrac.).
Calypsodium Link = Calypso Salisb. (Orchidac.).
Calypsogyne Néraud. Inc. sed. (Hydrocharitaceae?). Mauritius.

CALYPTERIOPETALON

Calypteriopetalon Hassk. = Croton L. (Euphorbiac.).
Calypteris Zipp. ex Mackl. (nomen). New Guinea. Quid?
Calypterium Bernh. = Onoclea L. (Aspidiac.).
Calypthrantes Raeusch. = Calyptranthes Sw. (Myrtac.).
Calyptocarpus Less. Compositae. 2 S. U.S., Mex., W.I.
Calyptochloa C. E. Hubbard. Gramineae. 1 Queensland.
Calyptosepalum S. Moore = Drypetes Vahl (Euphorbiac.).
Calyptostylis Arènes. Malpighiaceae. 1 Madag.
Calyptracordia Britton = Varronia P.Br. (Boraginac.).
Calyptranthe (Maxim.) Nakai = Hydrangea L. (Hydrangeac.).
*Calyptranthes Sw. Myrtaceae. 100 trop. Am., W.I. Ed. fr.
Calyptranthus Blume = Syzygium Gaertn. (Myrtac.).
Calyptranthus Juss. = Calyptranthes Sw. (Myrtac.).
Calyptranthus Thou. = Capparis L. (Capparidac.).
Calyptraria Naud. = Centronia D. Don (Melastomatac.).
Calyptrella Naud. Melastomataceae. 10 trop. Am.
Calyptridium Nutt. ex Torr. & Gray. Portulacaceae. 7 SW. U.S.
Calyptrimalva [sphalm. 'Calyptrae-'] Krapov. Malvaceae. 1 Brazil.
Calyptrion Ging. = Corynostylis Mart. (Violac.).
Calyptriopetalum Hassk. ex Muell. Arg. = Croton L. (Euphorbiac.).
Calyptrocalyx Blume. Palmae. 18 Moluccas, New Guinea.
Calyptrocarpus Reichb. = Calyptocarpus Less. (Compos.).
Calyptrocarya Nees. Cyperaceae. 5 trop. Am.
Calyptrochilum Kraenzl. Orchidaceae. 2 trop. Afr.
Calyptrocoryne Schott = Theriophonum Blume (Arac.).
Calyptrogenia Burret. Myrtaceae. 5 W.I., trop. S. Am.
Calyptrogyne H. Wendl. Palmae. 10 C. Am.
Calyptrolepis Steud. = Rhynchospora Vahl (Cyperac.).
Calyptromyrcia Berg = Myrcia DC. (Myrtac.).   [W.I.
Calyptronoma Griseb. (~ Calyptrogyne H. Wendl.). Palmae. 8 trop. Am.,
Calyptroön Miq. = Baccaurea Lour. (Euphorbiac.).
Calyptropetalum P. & K. = Calyptriopetalum Hassk. ex Muell. Arg. = Croton
   L. (Euphorbiac.).
Calyptropsidium Berg = Psidium L. (Myrtac.).
Calyptrosciadium K. H. Rech. & Kuber. Umbelliferae. 1 Afghan.
Calyptrosicyos Keraudren = Corallocarpus Welw. ex Hook. f. (Cucurbitac.).
Calyptrospatha Klotzsch ex Baill. = Acalypha L. (Euphorbiac.).
Calyptrospermum A. Dietr. = Bolivaria Cham. & Schlechtd. = Menodora
   Humb. & Bonpl. (Oleac.).
Calyptrostegia C. A. Mey. = Pimelea Banks (Thymelaeac.).
Calyptrostigma Klotzsch = Beyeria Miq. (Euphorbiac.).
Calyptrostigma Trautv. & Mey. = Macrodiervilla Nakai = Weigela Thunb.
   (Caprifoliac.).
Calyptrostylis Nees = Rhynchospora Vahl (Cyperac.).
Calyptrotheca Gilg. Portulacaceae. 2 NE. trop. Afr.
Calysaccion Wight = Ochrocarpos Thou. = Mammea L. (Guttif.).
Calysericos Eckl. & Zeyh. = Cryptadenia Meissn. (Thymelaeac.).
Calysphyrum Bunge = Weigela Thunb. (Caprifoliac.).

*Calystegia R.Br. (~ Convolvulus L.). Convolvulaceae. 25 temp. & trop. (N.B. The hawkmoth *Sphinx convolvuli* L. is *not* the pollinator of the Eur. *C. sepium* (L.) R. Br.).

Calythrix auctt. = Calytrix Labill. (Myrtac.).

Calythropsis C. A. Gardner. Myrtaceae. 1 W. Austr.

Calytriplex Ruiz & Pav. = Brami Adans. = Bacopa Aubl. (Scrophulariac.).

Calytrix Labill. Myrtaceae. 40 Austr. Seps. long and pointed.

Calyxhymenia Ortega = Mirabilis L. (Nyctaginac.).

Camacum Steud. = Comacum Adans. = Myristica Gron. (Myristicac.).

Camaion Rafin. = Helicteres L. (Sterculiac.).

Camara Adans. = Lantana L. (Verbenac.).

Camarandraceae Dulac = Rhamnaceae Juss.

Camarea A. St-Hil. Malpighiaceae. 8 E. S. Am.

Camaridium Lindl. = Maxillaria Ruiz & Pav. (Orchidac.).

Camarilla Salisb. = Allium L. (Alliac.).

Camarinnea Bub. & Penz. = Empetrum L. (Empetrac.).

Camarotea Scott Elliot. Acanthaceae. 1 Madag.

Camarotis Lindl. Orchidaceae. 12 SE. As., Indomal., Solomon Is., Austr.

Camassia Eckl. ex Pfeiff. = Gonioma E. Mey. (Apocynac.).

*Camassia Lindl. Liliaceae. 5–6 N. Am. The bulbs (*quamash*) form a food for the Indians of N. Am.

Camax Schreb. = Ropourea Aubl. = Diospyros L. (Ebenac.).

Cambania Comm. ex M. Roem. = Dysoxylum Blume (Meliac.).

Cambderia Steud. = Campderia Rich. = Vellozia Vand. (Velloziac.).

Cambea Endl. = Cumbia Buch.-Ham. – Careya Roxb. (Barringtoniac.).

Cambessedea Kunth = Buchanania Roxb. (Anacardiac.).

Cambessedea Wight & Arn. = Bouea Meissn. (Anacardiac.).

*Cambessedesia DC. Melastomataceae. 15 S. Brazil.

Cambogia L. = Garcinia L. (Guttif.).

Cambogiaceae Horan. = Guttiferae Juss.

Camchaya Gagnep. Compositae. 4 Siam, Indoch.

Camdenia Scop. = Evolvulus L. (Convolvulac.).

Camderia Dum. = Lachnanthes Ell. (Haemodorac.).

Camelia Rafin. = Camellia L. (Theac.).

Camelina Crantz. Cruciferae. 10 Eur., Medit., C. As. *C. sativa* (L.) Cr. (gold of pleasure) is used as a source of fibre in S. Eur.

Camellia L. (*Thea* L.). Theaceae. 82 Indomal., China, Japan. *C. sinensis* (L.) Kuntze is the tea plant, largely cult. in India, Ceylon, China, Japan, etc. Var. *assamica* (Mast.) Kitam. is also cult. It has larger ls. When growing wild it forms a small tree, but in cult. is kept pruned into a small bush. The young shoots (bud and 2 or more ls.) are nipped off, withered, rolled (to express a little juice), then fermented (except for green tea), dried, and sorted into grades (pekoe, souchong, congou, etc.).

Camelliaceae Dum. (~ *Theaceae D. Don, *q.v.*). Dicots. 8/200 trop. & warm temp. As. & Am. Closely rel. to *Ternstroemiac.*, differing esp. in the mostly small versat. anth.; fr. a loculic. caps. (rarely drupe); seed with mostly straight embryo. Chief genera: *Camellia, Laplacea, Gordonia, Pyrenaria, Schima*.

Camelliastrum Nakai = Camellia L. (Theac.).

CAMELOSTALIX

Camelostalix Pfitzer. Orchidaceae. 1 Java.

Cameraria Boehm. = Hemerocallis L. (Liliac.).

Cameraria Fabr. (uninom.) = *Montia* L. sp. (Portulacac.).

Cameraria L. Apocynaceae. 4 C. Am., W.I.

Cameridium Reichb. = Camaridium Lindl. (Orchidac.).

Camforosma Spreng. = Camphorosma L. (Chenopodiac.).

Camilleugenia Frappier ex Cordem. = Cynorkis Thou. (Orchidac.).

Camirium Gaertn. = Aleurites J. R. & G. Forst. (Euphorbiac.).

Camissonia Link. Onagraceae. 55 W. N. Am., Mex., 1 temp. W. S. Am.

Cammarum Fourr. = Aconitum L. (Ranunculac.).

Cammarum Hill = Eranthis Salisb. (Ranunculac.).

Camocladia L. = Comocladia P.Br. (Anacardiac.).

*Camoënsia Welw. ex Benth. & Hook. f. Leguminosae. 2 trop. W. Afr. *C. maxima* Welw. is a magnificent flowering creeper.

Camolenga P. & K. = Benincasa Savi (Cucurbitac.).

Camomilla Gilib. = Matricaria L. (Compos.).

Camonea Rafin. = Merremia Dennst. (Convolvulac.).

Campana P. & K. = Tecomanthe Baill. (Bignoniac.).

Campanea Decne (sphalm.) = Capanea Decne (Gesneriac.).

Campanemia P. & K. = Capanemia Barb. Rodr. (Orchidac.).

Campanocalyx Valeton. Rubiaceae. 1 Borneo.

Campanolea Gilg & Schellenb. = Linociera Sw. ex Schreb. (Oleac.).

Campanopsis Kuntze = Wahlenbergia Schrad. (Campanulac.).

Campanula L. Campanulaceae. 300 N. temp., esp. Medit., and trop. mts. The pollen is shed in the bud, the sta. standing closely round the style and depositing their pollen upon the hairs. As the fl. opens the sta. wither, except the triangular bases that protect the honey, and the style presents the pollen to insects. After a time the stigmas separate and the fl. is ♀; finally the stigmas curl right back on themselves and effect self-pollin. (See fam., and cf. *Phyteuma, Jasione.*) Seeds light and contained in a caps., which if erect dehisces at the apex, if pend. at the base, so that the seeds (cf. *Aconitum*) can only escape when the plant is shaken, e.g. in strong winds.

*Campanulaceae Juss. Dicots. 60–70/2000 temp. & subtrop., & trop. mts. Mostly perennial herbs (a few trees and shrubs), with alt., exstip. ls., and usu. with latex. The infl. may term. the primary axis, or one of the second order. It is generally racemose, ending with a term. fl. in *Campanuloïdeae*. In some cases, instead of single fls. in the axils of the bracts of the raceme, small dich. occur (cf. *Labiatae*). Others have the whole infl. cymose (*Canarina*, etc.).

Fl. usu. ♀, reg. or zygo., epig., rarely hypog. (*Cyananthus*), generally 5-merous, the odd sepal post. in *Campanuloïdeae*, but anterior in the other groups. In these cases, however, a twisting of the axis through 180° takes place before the fl. opens (cf. orchids), so that the odd sepal is finally post. K 5, open; C (5), valvate; A 5, epig.; anthers intr., sometimes united; G̲ (rarely G̲) (5), (3) or (2), multiloc., with axile plac. bearing ∞ anatr. ov. Style simple; stigmas as many as cpls. Caps., dehisc. in various ways in different gen., or berry. Fleshy endosp.

The nat. history of the fl. is of interest, both in itself and as exhibiting transitions to the Composite type. Honey is secreted by a disk at base of style and covered in most cases by the triangular bases of the sta., which fit closely together and only allow of the insertion of a proboscis between them. This, taken together with the size of the fls., their frequently blue colour and pendulous position, points to their being best adapted to the visit of bees, as is the case, but there are also many other visitors of various insect classes. A few exceptions occur; the bulk of the fam. has large fls., conspicuous by themselves, but *Phyteuma* and *Jasione* have small fls. massed in heads.

The general principle of the fl. mech. is the same throughout, and agrees with that of *Compositae*. The fl. is very protandr., and the style (with the stigmas closed up against one another) has the pollen shed upon it by the anthers, either in the bud or later. Usu. there is a bunch of hairs upon the style to hold the pollen. For some time the style acts as pollen-presenter to insects; after a time the stigmas separate and the ♀ stage sets in, and finally, in many cases, the stigmas curl back so far that they touch the pollen still clinging to their own style, and thus effect self-pollin. See genera, esp. *Campanula*, *Phyteuma*, *Jasione*, *Lobelia*, and cf. *Compositae*.

Classification and chief genera (after Schönland):

I. **Campanuloïdeae** (fl. actinomorphic, rarely slightly zygomorphic; anthers usu. free): *Campanula*, *Phyteuma*, *Wahlenbergia*, *Platycodon*, *Jasione*.

II. **Cyphioïdeae** (fl. zygomorphic; sta. sometimes united; anthers free): *Cyphia*, *Nemacladus*.

III. **Lobelioïdeae** (fl. zygomorphic, rarely almost actinomorphic; anthers united): *Centropogon*, *Siphocampylus*, *Lobelia*.

Campanulastrum Small. Campanulaceae. 1 N. Am.

Campanulatae Engl. An order of Dicots. comprising the fams. *Campanulac.*, *Goodeniac.*, *Brunoniac.*, *Stylidiac.*, *Calycerac.* and *Compositae*.

Campanuloïdes Hort. Kew. ex A. DC. = Lightfootia L'Hérit. (Campanulac.)

Campanulopsis (Roberty) Roberty = Convolvulus L. (Convolvulac.).

Campanulopsis Zoll. & Mor. = Wahlenbergia Schrad. (Campanulac.).

Campanumoea Blume = Codonopsis Wall. (Campanulac.).

Campbellia Wight = Christisonia Gardn. (Orobanchac.).

Campderia Benth. = Coccoloba L. (Polygonac.).

Campderia Lag. = Kundmannia Scop. (Umbellif.).

Campderia A. Rich. = Vellozia Vand. (Velloziac.).

Campe Dulac = Barbarea R.Br. (Crucif.).

Campecarpus H. Wendl. ex Becc. Palmae. 1 New Caled.

Campecia Adans. = Caesalpinia L. (Legumin.).

Campeiostachys Drobov. Gramineae. 1 C. As.

Campelepis Falc. = Periploca L. (Periplocac.).

Campelia Kunth = Campella Link = Deschampsia Beauv. (Gramin.).

Campelia Rich. Commelinaceae. 3 trop. Am., W.I. Ed. fr.

Campella Link = Deschampsia Beauv. (Gramin.).

Campereia Engl. = Champereia Griff. (Opiliac.).

Campesia Wight & Arn. ex Steud. = Galactia P.Br. (Legumin.).

Campestigma Pierre. Asclepiadaceae. 1 Indochina.

Camphora Fabr. (uninom.) = *Cinnamomum* Schaeff. sp. (Laurac.).

CAMPHORATA

Camphorata Fabr. = Selago L. (Selaginac.).
Camphorata Zinn = Camphorosma L. (Chenopodiac.).
Camphorina Nor. = Desmos Lour. (Annonac.).
Camphoromoea Nees ex Meissn. = Ocotea Aubl. (Laurac.).
Camphoromyrtus Schau. = Baeckea L. (Myrtac.).
Camphoropsis Moq. ex Pfeiff. = Nanophyton Less. (Chenopodiac.).
Camphorosma L. Chenopodiaceae. 11 E. Medit., C. As.
Camphusia De Vriese = Scaevola L. (Goodeniac.).
Camphyleia Spreng. = Campuleia Thou. = Striga Lour. (Scrophulariac.).
Campia Domb. ex Endl. = Capia Domb. ex Juss. = Lapageria Ruiz & Pav. (Philesiac.).
Campilostachys A. Juss. = Campylostachys Kunth (Stilbac.).
Campimia Ridl. Melastomataceae. 2 Malay Penins., Borneo.
Campium Presl = Bolbitis Schott (Lomariopsidac.).
*Campnosperma Thw. Anacardiaceae. 15 trop.
Campocarpus P. & K. = Campecarpus Wendl. ex Becc. (Palm.).
Campolepis P. & K. = Campelepis Falc. = Periploca L. (Periplocac.).
Campomanesia Ruiz & Pav. Myrtaceae. 80 S. Am. Ed. fr.
Campovassouria R. M. King & H. Rob. (~ Eupatorium L.). Compositae. 1 Brazil.
Campsanthus Steud. = Compsanthus Spreng. = Tricyrtis Wall. (Liliac.).
Campsiandra Benth. Leguminosae. 3 trop. Am.
Campsidium Seem. Bignoniaceae. 1 Chile, Argent.
*Campsis Lour. Bignoniaceae. 2 E. As., E. U.S.
Campsoneura Dur. & Jacks. (sphalm.) = Compsoneura Warb. (Myristicac.).
Camptandra Ridl. Zingiberaceae. ?6 W. Malaysia.
Camptederia Steud. = Campderia A. Rich. = Vellozia Vand. (Velloziac.)
Campteria Presl = Pteris L. (Pteridac.).
*Camptocarpus Decne. Asclepiadaceae. 5 Madag., Mauritius.
Camptocarpus C. Koch = Alkanna Tausch (Boraginac.).
Camptodium Fée. Aspidiaceae. 1 W.I.
Camptolepis Radlk. Sapindaceae. 1 trop. E. Afr.
Camptoloma Benth. Scrophulariaceae. 2 trop. Afr.
Camptophytum Pierre ex A. Chev. = Tarenna Gaertn. (Rubiac.).
Camptopus Hook. f. (~ Cephaëlis Sw.). Rubiaceae. 2 trop. Afr.
Camptorrhiza Hutch. Liliaceae. 1 S. Afr.
Camptosema Hook. & Arn. Leguminosae. 15 S. Am.
Camptosorus Link. Aspleniaceae. 2 NE. As., E. N. Am.
Camptostemon Mast. Bombacaceae. 3 Philipp. Is., Aru Is., N. Austr. A (∞).
Camptostylus Gilg. Flacourtiaceae. 4 trop. W. Afr.
Camptotheca Decne. Nyssaceae. 1 China, Tibet.
Camptouratea Van Tiegh. = Ouratea Aubl. (Ochnac.).
Campuleia Thou. = Striga Lour. (Scrophulariac.).
Campuloa Desv. = Ctenium Panz. (Gramin.).
Campuloclinium DC. = Eupatorium L. (Compos.).
Campulosus Desv. = Ctenium Panz. (Gramin.).
Campydorum Salisb. = Polygonatum Adans. (Liliac.).

Campylandra Baker. Liliaceae. 9 E. Himal., China, Indoch.
Campylanthera Hook. = Pronaya Hueg. (Pittosporac.).
Campylanthera Schott = Eriodendron DC. (Malvac.).
Campylanthus Roth. Scrophulariaceae. 7 Canary Is., Cape Verde Is., Socotra, Arabia, W. Pakistan.
Campyleia Spreng. = Campuleia Thou. = Striga Lour. (Scrophulariac.).
Campyleja P. & K. = praec.
Campylia Lindl. ex Sweet = Pelargonium L'Hérit. (Geraniac.).
Campyloa P. & K. = Campuloa Desv. = Ctenium Panz. (Gramin.).
Campylobotrys Lem. = Hoffmannia Sw. (Rubiac.).
Campylocaryum DC. ex A. DC. = Alkanna Tausch (Boraginac.).
Campylocentron Benth. & Hook. f. = seq.
Campylocentrum Benth. Orchidaceae. 35 Florida to trop. S. Am., W.I.
Campylocera Nutt. = Legousia Dur. (Campanulac.).
Campylocerum Van Tiegh. = Ouratea Aubl. (Ochnac.).
Campylochinium B. D. Jacks. (sphalm.) = Campyloclinium Endl. = Eupatorium L. (Compos.).
Campylochiton Welw. ex Hiern = Combretum Loefl. (Combretac.).
Campylochnella Van Tiegh. = Ochna L. (Ochnac.).
Campyloclinium Endl. = Campuloclinium DC. = Eupatorium L. (Compos.).
Campylogramma v. Ald. v. Ros. = Microsorium Link (Polypodiac.).
Campylogyne Welw. ex Hemsl. = Combretum Loefl. (Combretac.).
Campylonema Poir. = Campynema Labill. (Hypoxidac.).
Campyloneurum Presl. Polypodiaceae. 25 trop. Am.
Campylopelma Reichb. = Hypericum L. (Hypericac.).
Campylopetalum Forman. Podoaceae. 1 Siam.
Campylopora Van Tiegh. = Brackenridgea A. Gray (Ochnac.).
Campyloptera Boiss. = Aëthionema R.Br. (Crucif.).
Campylopus Spach (1836; non Brid. 1822—Musci) = Campylopelma Reichb. = Hypericum L. (Guttif.).
Campylosiphon Benth. Burmanniaceae. 1 trop. S. Am.
Campylosiphon St-Lag. = Siphocampylus Pohl (Campanulac.).
Campylospermum Van Tiegh. (~Ouratea Aubl.). Ochnaceae. 10 trop. E. Afr., Madag., trop. As.
Campylosporus Spach = Hypericum L. (Guttif.).
Campylostachys Kunth. Stilbaceae. 1 S. Afr.
Campylostachys E. Mey. = Fimbristylis Vahl (Cyperac.).
Campylostemon Erdtm. (sphalm.) = Campylopetalum Forman (Podoac.).
Campylostemon E. Mey. = Justicia L. (Acanthac.).
Campylostemon Welw. Celastraceae. 12 trop. W. Afr.
Campylosus P. & K. = Campulosus Desv. = Ctenium Panz. (Gramin.).
Campylotheca Cass. (~Bidens L.). Compositae. 20 Polynesia.
Campylotropis Bunge. Leguminosae. 65 E. & S. As.
Campylus Lour. Generic descr. = some zygomorphic sympetalous plant; specimen + specif. descr. = Tinospora Miers (Menispermac.).
Campynema Labill. Hypoxidaceae. 2 Tasm., New Caled.
Campynemanthe Baill. Hypoxidaceae. 1 New Caled.
Campynemataceae Dum. = Hypoxidaceae R.Br.

Camunium Adans. = Trichogamila P.Br. = ? Styrax L. (Styracac.).

Camunium Kuntze = Murraya Koen. ex L. (Rutac.).

Camusia Lorch = Dactyloctenium Willd. (Gramin.).

Camusiella Bosser. Gramineae. 2 Madag.

Camutia Bonato ex Steud. = Melampodium L. (Compos.).

Canabis Roth = Cannabis L. (Cannabidac.).

Canaca Guillaumin = Austrobuxus Miq. (Euphorbiac.).

Canacomyrica Guillaumin. Myricaceae. 1 New Caled.

Canacorchis Guillaum. Orchidaceae. 1 New Caled.

Canadaea Gandog. = Campanula L. (Campanulac.).

Canahia Steud. = Kanahia R.Br. (Asclepiadac.).

Canala Pohl = Spigelia L. (Spigeliac.).

Canalia F. W. Schmidt = Gnidia L. (Thymelaeac.).

Cananga Aubl. = Guatteria Ruiz & Pav. (Annonac.).

*****Cananga** (DC.) Hook. f. & Thoms. Annonaceae. 2 trop. As. to Austr. *C. odorata* Hook. f. is cult. for its fls., which yield the perfume known as **Cananga** Rafin. = praec.   [*ylang-ylang* or Macassar oil.

Canangium Baill. = praec.

Canaria L. = Canarina L. (Campanulac.).

Canariastrum Engl. = ls. of Canarium (Burserac.) + fr. of Uapaca (Uapacac.).

Canariellum Engl. = Canarium L. (Burserac.).

Canarina L. Campanulaceae. 3 Canary Is., trop. E. Afr. Like *Campanula*, but usu. 6-merous, and with ed. berry fr.

Canarion St-Lag. = seq.

Canariopsis Miq. = seq.

Canarium L. Burseraceae. 75 trop. Afr., As., N. Austr., Pacif. *C. luzonicum* (Bl.) A. Gray has an ed. seed (Java almond); other well-known fruit-trees are *C. album* (Lour.) Raeusch. and *C. pimela* Leenh., esp. in SE. As. *C. luzonicum* furnishes the resin Manila elemi (see *Bursera*); *C. strictum* Roxb. (S. India) and others furnish some of the black dammar of commerce (cf. *Agathis*).

Canavali Adans. = seq.

*****Canavalia** Adans. mut. DC. Leguminosae. 50 trop. & subtrop., esp. Am. & Afr. *C. ensiformis* (L.) DC. (sword or sabre bean, 'overlook') cult. for ed. pods. *C. maritima* Thou. is a common pantrop. shore plant.

Canbya Parry. Papaveraceae. 2 Oregon, California, Mex.

Cancellaria (DC.) Mattei = Pavonia Cav. (Malvac.).

Cancellaria Sch. Bip. ex Oliver = Adelostigma Steetz (Compos.).

Cancrinia Kar. & Kir. Compositae. 30 C. As., Afghan.

Cancriniella Tzvelev. Compositae. 1 C. As.

Candarum Schott = Amorphophallus Blume (Arac.).

Candelabria Hochst. = Bridelia Willd. (Euphorbiac.).

Candidea Tenore = Vernonia Schreb. (Compos.).

Candjera Decne (sphalm.) = Cansjera Juss. (Opiliac.).

Candollea Baumg. = Menziesia Sm. (Ericac.).

Candollea Labill. (1805) = Stylidium Sw. (Stylidiac.).

Candollea Labill. (1806) = Hibbertia Andr. (Dilleniac.).

Candollea Mirbel = Pyrrosia Mirbel (Polypodiac.).

Candollea Steud. = Decandolia Bast. = Agrostis L. (Gramin.).

Candolleaceae Schönl. = Stylidiaceae R.Br.
Candolleodendron R. S. Cowan. Leguminosae. 1 Brazil.
Candollina Van Tiegh. = Amyema Van Tiegh. (Loranthac.).
*Canella P.Br. Canellaceae. 2 S. Florida, W.I., trop. Am. C. alba Murr. yields canella bark, used as a tonic and stimulant.
Canella Dombey ex Endl. = Drimys J. R. & G. Forst. (Winterac.).
Canella P. & K. = Cannella Schott ex Meissn. = Ocotea Aubl. (Laurac.).
*Canellaceae Mart. Dicots. 5/16 S. Am., W.I., E. Afr., Madag., with marked discontinuity. Trees with alt., leathery, entire, exstip., gland-dotted ls. Fls. sol. or in racemes or cymes, reg. K 4–5, imbr.; C 4–12, free or united, or o; A (20–40), completely united into a tube with extr. anthers; G̲ (2–5), 1-loc., with 2–∞ semi-anatr. ov. on each parietal plac. Berry. Embryo straight or slightly curved in rich endosp. Genera: Canella, Cinnamodendron, Cinnamosma, Warburgia, Pleodendron.
Canephora Juss. Rubiaceae. 5 Madag. Fls. in clusters at the top of a phyllodineous stalk within a lobed calyculus.
Canhamo Perini = Hibiscus L. (Malvac.).
Canicidia Vell. = Connarus L. (Connarac.).
Canidia Salisb. = Allium L. (Alliac.).
Caniram Thou. ex Steud. = Strychnos L. (Strychnac.).
Canistrum Morren. Bromeliaceae. 7 Brazil.
Canizaresia Britton. Leguminosae. 1 Cuba.
Cankrienia De Vriese = Primula L. (Primulac.).
Canna L. Cannaceae. 55 trop. & subtrop. Am.
Canna Nor. = (probably) Calamus L., Plectocomia Mart. & Bl., Daemonorhops Bl., etc. (Palm.).
Cannabaceae auctt. = seq.
Cannabiaceae auctt. = seq.
*Cannabidaceae Endl. Dicots. 2/3 N. temp. Herbs, sometimes climbing, closely rel. to Morac., but with free stips.; stam. short and straight; ovule apical, anatr.; fr. an achene, seed with endosp. Genera: Cannabis, Humulus. Cf. Cannabis with Datiscac. and Haloragidac.
Cannabina Mill. = Datisca L. (Datiscac.).
Cannabinaceae Lindl. = Cannabidaceae Endl.
Cannabinastrum Fabr. (uninom.) = Galeopsis L. sp. (Labiat.).
Cannabis L. Cannabidaceae. 1 C. As., C. sativa L., the hemp. Infl. like Humulus ♂, dioec. Hemp is largely cult. both in temp. and trop. regions, in the former for the fibre, in the latter for the drug. A valuable fibre, used for ropes and other purposes, is obtained from the inner bark of the stem, much as flax is prepared from Linum, and for this purpose the plant is cult. in S. Eur., the eastern U.S., and other countries. In the trop., and esp. in India, the pl. is cult. for the sake of the narcotic resin which exudes from it, and which is used much like opium, both as a drug and as a stimulant. The drug occurs in three common forms, ganja, charas, and bhang. The first is the ♀ flg. tops with resin on them, packed together, the second, which comes from rather cooler climates, is the resin knocked off the twigs, bark, etc., and the third, which is largely obtained from the wild plants, is the mature ls., with their resinous deposit, packed together. Asiatics are much addicted to the use of

197

hemp as a narcotic. It is smoked, with or without tobacco, and an intoxicating liquor, *hashish*, is made from it. The resin has an intoxicating stimulating effect. In small quantities it produces pleasant excitement, passing into delirium and catalepsy if the quantity be increased. The names given to the plant among them indicate this use of it, e.g. leaf of delusion, increaser of pleasure, cementer of friendship. The sale of *ganja* and *charas* is kept in check in India by a stringent licensing system, but that of *bhang*, which is collected from wild plants, is hard to control.

*Cannaceae Juss. Monocots. 1/55 Am. Habit like *Zingiberaceae* or *Marantaceae*, but *C.* can be distinguished even when not in fl. by possessing neither the ligule of the former nor the pulvinus of the latter. Infl. term., usu. composed of 2-fld. cincinni. The two fls. are homodromous, but the bracteole is to the right in one and to the left in the other (behind one or other of the two lat. sepals). Fl. ♀, asymmetric, epig. K 3, C (3). The A is the most conspicuous part. There is a leafy sta. bearing half an anther on one edge, and a number of petaloid structures round it, usu. 3 but sometimes 1 or 4. One of these is the labellum (not = that of *Zingiberaceae*), and is rolled back on itself outwards. The other two are often termed the wings. When a fourth std. (γ, cf. *Marantaceae*) is present it stands behind the fertile sta. Other spp. have only the labellum. G̅ (3) with petaloid style, 3-loc.; ov. in 2 rows in each loc., anatr. Caps., usu. warty. Seed with perisperm and straight embryo.

As to the morphological explanation of the A, there are two views. Eichler (*Blütendiagr.* 1: 174) regards the labellum as a lat. sta. of the inner whorl, and the fertile sta. together with all the stds. as the post. sta. of the same whorl; the other six sta. of the inner, and all the sta. of the outer, whorl are wanting. The older view looks upon β, γ, as the 2 post. sta. of the outer whorl, and the labellum, α, and the fertile sta. as the 3 sta. of the inner whorl. (Cf. this fl. with those of *Musaceae*, *Zingiberaceae* and *Marantaceae*.)

The pollen is shed upon the style in the bud; insects alight on the labellum, touch first the term. stigma and then the pollen. The rhiz. of *Canna edulis* Ker-Gawl. is ed., containing much starch.

Cannacorus Mill. = Canna L. (Cannac.).

Cannella Schott ex Meissn. = Ocotea Aubl. (Laurac.).

Cannomois Beauv. Restionaceae. 7 S. Afr.

Canonanthus G. Don = Siphocampylus Pohl (Campanulac.).

Canopholis G. Don (sphalm.) = Conopholis Wallr. (Orobanchac.).

Canophora P. & K. = Canephora Juss. (Rubiac.).

Canopodaceae ('-piaceae') C. Presl = Santalaceae–Anthoboleae Bartl.

Canopus C. Presl = Exocarpus Labill. (Santalac.).

Canothus Rafin. = Ceanothus L. (Rhamnac.).

Canotia Torr. Canotiaceae. 1 SW. U.S.

Canotiaceae Britton. Dicots. 1/1 SW. U.S. Leafless shrubs or small trees, with rigid pale green terete striate branches ending in spines; ls. repr. by minute scattered alt. deltoid scales, above each of which is a conspic. black triangular papillose gland-field. Fls. reg., ♀, in short lat. 3–7-fld. minutely bract. cymes. K 5, conn. at base, small, minutely gland.-fimbr., imbr., persist.; C 5, imbr., thickish, carinate within, decid.; A 5, alternipet., with subul. persist. fil., and ovate-obl. cord. apic. intr. anth.; G (5), with 6 biser. axile amphitr. horiz. ov. per loc., and thick simple persist. style with minute 5-toothed stig. Fr. a rel. large ovoid-fusif. woody caps., septicid. and partly loculicid. from apex; pericarp slightly fleshy; seeds 1–2 per loc., ascending, flattened, with basal membr. wing; endosp. thin, fleshy, embr. straight. Only genus: *Canotia*. Affinities much disputed. Superfic. similar to *Koeberlinia* (*Cappari-dac.*), but fls. of that tetram., A diplostem., fr. a small glob. biloc. berry, seeds not winged, embr. coiled, pollen different. The caps. of *Canotia* suggests a possible rel. with *Brexiac.*, etc., or even *Linac.* Some anat. similarities with *Koeberlinia* and *Celastrac.*; the pollen is somewhat celastraceous.

Canschi Adans. = Trewia L. (Euphorbiac.).
Canscora Lam. Gentianaceae. 30 palaeotrop.
Cansenia Rafin. = Bauhinia L. (Legumin.).
Cansiera Spreng. = Potameia Thou. (Laurac.).
***Cansjera** Juss. Opiliaceae. 5 trop. As., Austr.
Cansjeraceae J. G. Agardh = Opiliaceae Van Tiegh.
Cantalea Rafin. = Lycium L. (Solanac.).
Cantharospermum Wight & Arn. = Atylosia Wight & Arn. (Legumin.).
Canthiopsis Seem. = Randia L. (Rubiac.).
Canthium Lam. Rubiaceae. 200 palaeotrop. Some have axillary thorns.
Canthopsis Miq. = Randia L. (Rubiac.).
Cantleya Ridl. Icacinaceae. 1 W. Malaysia.
Cantua Juss. ex Lam. Polemoniaceae. 11 Andes of Ecuad., Peru, Boliv.
Cantuffa Gmel. = Pterolobium R.Br. (Legumin.).
Caonabo Turpin ex Rafin. = Columnea L. (Gesneriac.).
Caopia Adans. = Vismia Vand. (Guttif.).
Caoutchoua J. F. Gmel. = Hevea Aubl. (Euphorbiac.).
Capanea Decne. Gesneriaceae. 10 C. & S. Am.
Capanemia Barb. Rodr. Orchidaceae. 15 Brazil.
Capassa Klotzsch. Leguminosae. 1 S. trop. Afr.
Capellenia Hassk. = Capellia Blume = Dillenia L. (Dilleniac.).
Capellenia Teijsm. & Binnend. = Endospermum Benth. (Euphorbiac.).
Capellia Blume = Dillenia L. (Dilleniac.).
Caperonia A. St-Hil. Euphorbiaceae. 60 trop. Am., Afr.
Capethia Britton = Anemone L. (Ranunculac.).
Capia Domb. ex Juss. = Lapageria Ruiz & Pav. (Philesiac.).
Capillipedium Stapf. Gramineae. 10 warm Old World.
Capirona Spruce. Rubiaceae. 5 S. Am. K like *Mussaenda*.
Capitania auctt. = Capitanya Schweinf. ex Gürke (Labiat.).
Capitanopsis S. Moore. Labiatae. 2 Madag.
Capitanya Schweinf. ex Gürke. Labiatae. 1 E. Afr.
Capitellaria Naud. = Sagraea DC. (Melastomatac.).
Capitularia J. V. Suringar (~ Chorizandra R.Br.). Cyperaceae. 2 New Guinea, Solomon Is.
Capnites Dum. = Corydalis Vent. (Fumariac.).
Capnitis E. Mey. = Lotononis Eckl. & Zeyh. (Legumin.).
Capnocystis Juss. = Cysticapnos Mill. (Fumariac.).
Capnodes Kuntze = Capnoïdes Mill. = Corydalis Vent. (Fumariac.).
Capnogorium Bernh. = Corydalis Vent. (Fumariac.).

CAPNOÏDES

Capnoïdes Mill. = Corydalis Vent. (Fumariac.).

Capnophyllum Gaertn. Umbelliferae. 4 Canary Is., Medit., S. Afr.

Capnorchis Mill. = Dicentra Borckh. corr. Bernh. (Fumariac.).

Capnorea Rafin. = Hesperochiron S. Wats. (Hydrophyllac.).

*Capparidaceae Juss. (*s.str.*). Dicots. 30/650 trop. & warm temp. Many xero., with reduced, often inrolled, ls. (cf. *Empetrum*). Small trees or shrubs, occasionally lianes, v. rarely herbs, hairy or scaly, rarely glandular, with alt. simple or palmate ls., often with stips. (frequently repres. by thorns or glands). Fls. ⚥, reg., usu. in racemes, bracteate but without bracteoles. The P resembles that of *Cruciferae* (K 2+2, C 4 diagonal), but great var. occurs in the A. In some, branching of the median sta. occurs, and usu. the post. sta. is more branched than the ant. Staminody of some of the branches is frequent. Cpls. typically (2), transv. as in *Cruciferae*, with parietal plac. In many spp. of subfam. I the number rises to 10 or 12 by the addition of a second whorl of cpls. and by *dédoublement*. Ovules ∞, campylotropous.

A further complication is the presence of axial effigurations, etc., in the fls. A disk may occur between P and sta. (usually thicker at the post. side), or a gynophore between sta. and ov., or both. Or the disk may grow up in the centre to form an androphore on which the sta. are borne, and above them there may be a gynophore also. From the disk there often grow out structures of various shapes and sizes: these may be scales quite free from one another, or, as in *Cadaba*, etc., may be united into a tube. Or the scales may, as in *Steriphoma*, etc., alt. with and be joined to the sepals.

Fr. a nut, berry or drupe. Seed exalb., with embryo folded in various ways as in *Cruciferae*. Few are useful: see *Capparis*, etc.

Classification and chief genera (modified from Pax & Hoffmann):

- I. **Capparidoïdeae** (berry or drupe): *Ritchiea, Capparis, Cadaba, Maerua.*
- II. **Podandrogynoïdeae** (dehisc. caps., no replum): *Podandrogyne.*
- III. **Dipterygioïdeae** (1-2-seeded nutlet or samara): *Dipterygium.*
- IV. **Buhsioïdeae** (dry inflated many-seeded fr.): *Buhsia.*

[For genera here excluded from *Capparidac.*, see *Cleomac., Emblingiac., Pentadiplandrac.*, and *Calyptrotheca. Dipterygium* is prob. Cruciferous.]

Capparidastrum Hutch. Capparidaceae. 12 Mex. to trop. S. Am., W. I.

Capparis L. Capparidaceae. 250 warm. Many climb by recurved stip. thorns. Fertile sta. ∞. The fl. buds of *C. spinosa* L. (Medit.) are the 'capers' used in flavouring (cf. cloves, *Syzygium*).

Cappidastrum Hutch. (sphalm.) = Capparidastrum Hutch. (Capparidac.).

Capraea Opiz = Salix L. (Salicac.).

Capraria L. Scrophulariaceae. 4 warm Am., W.I. Ls. alt. C almost choripet.

Caprella Rafin. = Capsella Medik. (Crucif.).

Caprificus Gasp. = Ficus L. (Morac.).

*Caprifoliaceae Juss. Dicots. 12/450 mostly N. temp., and trop. mts. Shrubs or small trees, rarely herbs, with opp. simple usu. ent. occasionally lobed sometimes stip. ls. Fls. ⚥, reg. or zygo., usu. in cymose infl. K 5-4, imbr. or open (odd sep. post.); C (5-4), imbr., sometimes bilab.; A 4-5, epipet.; G̲ (2-5-8), with 1-∞ pend. axile ov. per loc., and simple style with capit. stig. Fr. a fleshy berry or drupe, or achene, or dehisc. or indehisc. caps.; seeds with

CARAPICHEA

fleshy endosp. Chief genera: *Viburnum, Abelia, Lonicera, Diervilla, Leycesteria, Symphoricarpos.* Closely related to *Rubiac.* and through *Viburnum* to *Valerianac.*
Caprifolium Mill. = Lonicera L. (Caprifoliac.).
Capriola Adans. = Cynodon Rich. (Gramin.).
Caproxylon Tussac = Tetragastris Gaertn. (Burserac.).
***Capsella** Medik. Cruciferae. 5 temp., subtrop. *C. bursa-pastoris* Medik. (shepherd's purse), a cosmop. weed, self-pollin. In early spring and late autumn sta. often ± aborted. Ls. variable in shape and degree of division in various situations. *C. heegeri* Solms, a form with elongate fr., which apparently arose from *C. bursa-pastoris* as a mutation, is almost generically distinct.
Capsicodendron Hoehne = Cinnamodendron Endl. (Canellac.).
Capsicum L. Solanaceae. 50 C. & S. Am. (cf. *Tubocapsicum*). *C. annuum* L. cult. for fr. (chillies or red peppers); dried and ground they form cayenne
Capura Blanco = Otophora Blume (Sapindac.). [pepper.
Capura L. = Wikstroemia Endl. (Thymelaeac.).
Capurodendron Aubrév. Sapotaceae. 22 Madag.
Capuronetta Markgraf. Apocynaceae. 1 Madag.
Capuronia Lourteig. Lythraceae. 1 Madag.
Capuronianthus Leroy. Meliaceae. 1 Madag.
Capusia Lecomte = Siphonodon Griff. (Siphonodontac.).
Capusiaceae Gagnep. = Siphonodontaceae Gagnep. & Tardieu.
Caquepiria J. F. Gmel. = Gardenia Ellis (Rubiac.).
Carabichea P. & K. = Carapichea Aubl. = Cephaëlis Sw. (Rubiac.).
Caracalla Tod. = Phaseolus L. (Legumin.).
Caracasia Szyszył. Marcgraviaceae. 2 Venez. Pets. free; A 3.
Carachera Juss. = Charachera Forsk. = Lantana L. (Verbenac.).
Caradesia Rafin. = Eupatorium L. (Compos.).
Caraea Hochst. = Euryops Cass. (Compos.).
Caragana Fabr. Leguminosae. 80 C. As., Himal., China. Petiole persistent.
Caragna Medik. = praec.
Caraguata Adans. = Tillandsia L. (Bromeliac.).
Caraguata Lindl. Bromeliaceae. 4 S. Am., W.I.
Caraipa Aubl. Guttiferae. 20 trop. S. Am. Ls. alt. Connective with term. gland. Seeds 1 per loc. Useful hard timber (*tamacoari*) and medic. balsam.
Carajaea (Tul.) Wedd. Podostemaceae (gen. dub.). 1 trop. S. Am.
***Carallia** Roxb. Rhizophoraceae. 10 Madag., Indomal., N. Austr.
Caralluma R.Br. Asclepiadaceae. 110 Afr., Medit. region to Burma.
Caramanica Tineo = Taraxacum L. (Compos.).
Carambola Adans. = Averrhoa L. (Averrhoac.).
Caramuri Aubrév. & Pellegr. Sapotaceae. 1 Amaz. Brazil.
Caranda Gaertn. = Canthium Lam. (Rubiac.).
Carandas [Rumph.] Adans. = Carissa L. (Apocynac.).
Caranga Juss. apud Vahl (sphalm.) = Curanga Juss. (Scrophulariac.).
Carania Chiov. = Tryphostemma Harv. (Passiflorac.).
Carapa Aubl. Meliaceae. 7 trop. Seeds of *C. procera* DC. and *C. guianensis* Aubl. yield a good oil (*carapa, touloucouna, andiroba, coondi*). *C. moluccensis* Lam. is a mangrove.
Carapichea Aubl. = Cephaëlis Sw. (Rubiac.).

201

CARARA

Carara Medik. = Coronopus Zinn (Crucif.).
Caratas Rafin. = Karatas Mill. (Bromeliac.).
Caraxeron Rafin. = Iresine P.Br. (Amaranthac.).
Carbeni Adans. = Cnicus L. (Compos.).
Carbenia Benth. = praec.
Carcerulaceae Dulac = Tiliaceae Juss.
Carcia Raeusch. = Garcia Rohr (Euphorbiac.).
Carcinetrum P. & K. = Karkinetron Rafin. = Polygonum L. (Polygonac.).
Carda Nor. = Aleurites Forst. (Euphorbiac.) + Pangium Reinw. (Flacourtiac.).
Cardamind[ac]eae Link = Tropaeolaceae DC.
Cardamindum Adans. = Tropaeolum L. (Tropaeolac.).
Cardamine L. Cruciferae. 160 cosmop., chiefly temp. *C. impatiens* L. has an explosive fruit like that of *Eschscholtzia*. *C. chenopodiifolia* Pers. (S. Am.) possesses two kinds of fr. Those formed on the upper part of the plant are normal siliquae; at the base, in the axils of the ls. of the rosette, cleist. fls. form which burrow into the soil and produce fr. there (cf. *Arachis*, *Trifolium*, etc.). In *C. pratensis* L. there is extensive veg. repr. by adv. buds on the radical ls., and in *C.* (*Dentaria*) *bulbifera* (L.) R.Br. by means of axillary bulbils.
Cardaminopsis (C. A. Mey.) Hayek. Cruciferae. 13 N. temp. & arct.
Cardaminum Moench = Rorippa Scop. (Crucif.).
Cardamomum Kuntze = Amomum L. (Zingiberac.).
Cardamomum Salisb. = Elettaria Maton (Zingiberac.).
Cardamon Fourr. = Lepidium L. (Crucif.).
Cardanoglyphus P. & K. = Kardamoglyphos Schlechtd. (Crucif.).
Cardanthera Buch.-Ham. ex Voigt = Synnema Benth. (Acanthac.).
Cardaria Desv. Cruciferae. 1 Medit., W. As.
Cardenanthus R. C. Foster. Iridaceae. 8 Andes, trop. Am.
Cardenasia Rusby = Bauhinia L. (Legumin.).
Cardenasiodendron F. A. Barkley. Anacardiaceae. 1 Bolivia.
Carderina Cass. = Senecio L. (Compos.).
Cardia Dulac = Veronica L. (Scrophulariac.).
Cardiaca Mill. = Leonurus L. (Labiat.).
Cardiacanthus Schau. Acanthaceae. 2 Mex.
Cardiandra Sieb. & Zucc. Hydrangeaceae. 5 China, Formosa, Japan. A ∞.
Cardianthera Hance = Cardanthera Buch.-Ham. ex Voigt (Acanthac.).
Cardinalis Fabr. (uninom.) = *Lobelia* L. sp. (Campanulac.).
Cardiobatus Greene = Rubus L. (Rosac.).
Cardiocarpus Reinw. = Soulamea Lam. (Simaroubac.).
Cardiochlaena Fée = Tectaria Cav. (Aspidiac.).
Cardiochlamys Oliv. Convolvulaceae. 2 Madag.
Cardiocrinum Endl. Liliaceae. 3 Himal., E. As.
Cardiodaphnopsis Hutch. = Caryodaphnopsis Airy Shaw (Laurac.).
Cardiogyne Bur. = Maclura Nutt. (Morac.).
Cardiolepis Rafin. = Rhamnus L. (Rhamnac.).
Cardiolepis Wallr. = Cardaria Desv. (Crucif.).
Cardiolochia Rafin. = Aristolochia L. (Aristolochiac.).
Cardiolophus Griff. = Bacopa Aubl. (Scrophulariac.).
Cardiomanes Presl. Hymenophyllaceae. 1 N.Z.

202

Cardionema DC. Caryophyllaceae. 6 Pacif. N. Am. to Chile.
Cardiopetalum Schlechtd. Annonaceae. 1 trop. S. Am.
Cardiophora Benth. = Soulamea Lam. (Simaroubac.).
Cardiophyllarium Choux. Sapindaceae. 1 Madag.
Cardiophyllum Ehrh. (uninom.) = *Ophrys cordata* L. = *Listera cordata* (L.) R.Br. (Orchidac.).
***Cardiopterid[ac]eae** Blume = Cardiopterygaceae Bl. corr. Van Tiegh.
Cardiopteris Wall. ex Blume = Peripterygium Hassk. (Cardiopterygac.).
Cardiopterygaceae Blume corr. Van Tiegh. Dicots. 1/3 SE. As. to Austr. Climbing herbs with abundant milky juice; ls. alt., exstip., cordate, ent. or lobed, membr.; cymes axill., branched, ± scorpioid, ebract. Fls. v. small. K (5), imbr.; C (5), imbr.; A 5, epipet.; disk 0; G̱ (2), 1-loc., with 2 pend. apic. ov.; styles 2, dissimilar, 1 longer, thicker, cylindr. or subclav., persistent on fruit, 1 shorter, thinner, with capit. stig.; fr. dry, indeh., flat, 2-winged, obcordate, stramineous, 2-seeded, embryo minute in fleshy endosp. Only genus: *Peripterygium*. A curious genus of disputed affinity, but probably related to the *Convolvulac.*: cf. habit, foliage, latex, floral structure, 2 free styles and capit. stig. (*Dicranostyleae*), etc. Scorpioid infls. occur in *Convolvulac.*, *Boraginac.*, *Hydrophyllac.*
Cardeopleryx Wall. ex Bl. corr. Engl. = Peripterygium Hassk. (Cardiopterygac).
Cardiospermum L. Sapindaceae. 12 trop., esp. Am.
Cardiostegia Presl = Melhania Forsk. (Sterculiac.).
Cardiostigma Baker = Sphenostigma Baker (Iridac.).
Cardioteucris C. Y. Wu. Labiatae. 1 SW. China.
Cardiotheca Ehrenb. ex Steud. = Anarrhinum Desf. (Scrophulariac.).
Cardopatium Juss. Compositae. 3 Medit., 1 C. As.
Cardosanctus Bub. = Cnicus L. (Compos.).
Carduaceae Small = Compositae Giseke.
Carduncellus Adans. Compositae. 20 Medit.
× **Carduocirsium** Sennen. Compositae. Gen. hybr. (Carduus × Cirsium).
× **Carduogalactites** P. Fourn. Compositae. Gen. hybr. (Carduus × Galactites).
Carduus L. Compositae. 100 Eur., Medit., As. (thistles). The genus should probably include the genus *Cirsium*.
Cardwellia F. Muell. Proteaceae. 1 Queensland.
Carelia Cav. = Mikania Willd. (Compos.).
Carelia Fabr. (uninom.) = *Ageratum* L. sp. (Compos.).
Carelia Less. = Radlkoferotoma Kuntze (Compos.).
Carenophila Ridl. = Geostachys Ridl. (Zingiberac.).
Careum Adans. = Carum L. (Umbellif.).
Carex L. Cyperaceae. 1500–2000 cosmop., espec. temp., in marshes, etc. Grass-like pl. with 1-fld. pseudo-spikelets in mostly long, dense spikes, which are sometimes unisexual, sometimes with both ♂ and ♀ fls. The ♀ fl. has a second glume, modified into a closed sac (utricle) enclosing the ovary. Fls. protog., wind-pollin. Much veg. repr. by offshoots. Many spp. are alpine; others, e.g. *C. arenaria* L., grow on sand-dunes, with the habit of *Ammophila*. Cf. monograph by Kükenthal in Engl., *Pflanzenr.*
***Careya** Roxb. Barringtoniaceae. 4 Indomal. *C. arborea* Roxb. (patana oak)

CARGILA

almost the only tree on the grassy expanses known as *patanas* in Ceylon. Seeds ∞; embryo like *Barringtonia*.

Cargila Rafin. = Melampodium L. (Compos.).

Cargilla Adans. = Chrysogonum L. (Compos.).

Cargillia R.Br. = Diospyros L. (Ebenac.).

Caribea Alain. Nyctaginaceae. 1 Cuba.

Carica L. Caricaceae. 45 warm Am. *C. papaya* L. (papaw) universally cult. in warm countries for ed. fr. The ls. and the unripe fr. contain a milky juice in which is the proteid-ferment papaïn, collected in Ceylon, etc., for use in digestive salts. Meat wrapped in the ls. and buried becomes tender through partial digestion of the fibres. *C. cundinamarcensis* Hook. f. (mountain papaw) also cult. for ed. fr. in trop. mts.

Caricaceae Burnett (non Dum.) = Cyperaceae Juss.

*__**Caricaceae** Dum. Dicots. 4/55 trop. Am., Afr. Small trees, branched or not, with a term. crown of palmate or digitate alt. exstip. ls., and milky juice. Fls. reg., ♂ ♀, dioec. or monoec., occasionally ⚥, in loose axillary infls. K 5, small, ± open; C (5), contorted or valv.; ♂ with long C-tube, and 2 whorls intr. epipet. sta.; ♀ with short C-tube, G̲ (5), 1- or 5-loc., with short style and 5 stigmas; ov. ∞, anatr., usu. on parietal plac. Berry; endosp. oily. Genera: *Carica, Jacaratia, Cylicomorpha, Jarilla (Mocinna)*. Related to *Passiflorac.*, and (through *Jatropha*) to *Euphorbiac.*

Caricella Ehrh. (uninom.) = *Carex capillaris* L. (Cyperac.).

Caricina St-Lag. = Carex L. (Cyperac.).

Caricinella St-Lag. = Carex L. (Cyperac.).

Caricteria Scop. = Corchorus L. (Tiliac.).

Caridochloa Endl. = Coridochloa Nees ex Grah. = Alloteropsis Presl (Gramin.).

Carigola Rafin. = Monochoria Presl (Pontederiac.).

Carima Rafin. = Adhatoda Medik. (Acanthac.).

Cariniana Casar. Lecythidaceae. 13 trop. S. Am. Timber valuable.

Carinivalva Ising. Cruciferae. 1 S. Austral.

Carinta W. F. Wight = Geophila D. Don. (Rubiac.)

Carionia Naud. Melastomataceae. 1 Philipp. Is. Small trees. 6-merous.

*__**Carissa** L. Apocynaceae. 35 warm Afr. & As. Shrubs with branch thorns. *C. carandas* L. has ed. fr.

Carissophyllum Pichon. Apocynaceae. 1 Madag.

Carlea Presl = Symplocos L. (Symplocac.).

Carlemannia Benth. Carlemanniaceae. 3 E. Himal., Assam, SE. As., Sumatra.

Carlemanniaceae Airy Shaw. Dicots. 2/5 SE. As. to Sumatra. Perenn. herbs or subshrubs, with opp. simple sometimes oblique dent. exstip. ls., petioles connected by a raised line. Fls. slightly zygo. or subreg., ⚥, in dense term. or axill. cymes. K 4-5, ± unequal, open, persist.; C (4-5), lobes imbr. or indupl.-valv.; A 2, epipet., fil. short, anth. lin.-obl., sometimes latr., cohering round the style; disk conspic., shortly cyl. or conic; G̲ (2), with elong. style and bifid clav. or fusif. stig., and ∞ ov. on axile or sub-basal plac. Fr. a membr. or fleshy ± glob. 2-loc. 4-5-valved caps.; seeds ∞, with ± fleshy endosp. Genera: *Carlemannia, Silvianthus*. A small group showing possible connections with *Rubiac.* and *Caprifoliac.*; *Silvianthus* has even distinct points of resemblance with *Clerodendrum* in the *Verbenac.* Although *Carlemannia* and *Silvianthus* differ strikingly from each other in many points, their anatomy supports their association, as well as their separation from other fams. (*Kew Bull.* **19**: 507–12, 1965.) A more probable relationship (at least for *Carlemannia*) than those suggested may be with *Hydrangeaceae,* esp. *Cardiandra* S. & Z.

Carlephyton Jumelle. Araceae. 1 Madag.

Carlesia Dunn. Umbelliferae. 1 China.

Carlina L. Compositae. 20 Eur., Medit., As. *C. acaulis* L. is the weather thistle or silver thistle of the Alps, etc. The outer brs. of invol. are prickly, the inner membranous and shining, spreading like a star in dry air, but closing [in damp.

Carlinodes Kuntze = Berkheya Ehrh. (Compos.).

Carlomohria Greene = Halesia Ellis (Styracac.).

Carlostephania Bub. = Circaea L. (Onagrac.).

Carlotea Arruda = ? Hippeastrum Herb. (Amaryllidac.).

Carlowizia Moench = Carlina L. (Compos.).

Carlowrightia A. Gray. Acanthaceae. 20 SW. U.S., Mex.

Carludovica Ruiz & Pav. (emend. Harling). Cyclanthaceae. 3 C. & NW. trop. S. Am. Habit that of a small palm with short stem and fan ls. in whose axils arise the infls. Each is a cylindrical spadix, enclosed at first in a number of brs. which fall off and leave it naked. Its surface is covered with fls. The ♂ has a rudimentary P, and ∞ sta., united below. The ♀ is sunk in and united with the tissue of the spadix. It has 4 very long stds. and 4 stigmas corresponding to the 4 plac. in the 1-loc. ov. When the spadix opens the ♀ fls. are ripe and the long stds. give a tangled appearance to the whole. After a few days the stigmas cease to be receptive and the anthers open. Afterwards the ♂ fls. drop and a multiple fr. is formed, composed of berries.

The ls. of *C. palmata* Ruiz & Pav., gathered young, cut into thin strips and bleached, form the material of Panama hats.

Carmelita C. Gay = Chaetanthera Ruiz & Pav. (Compos.).

Carmenocania Wernham. Rubiaceae. 1 trop. Am.

Carmenta Nor. = Viburnum L. (Caprifoliac.).

Carmichaelia R.Br. Leguminosae. 41 N.Z., Lord Howe I. Xero. with flat green stems (phylloclades) and no green ls. (cf. *Bossiaea*).

Carminatia Moç. ex DC. Compositae. 1 Mex.

Carmona Cav. = Ehretia L. (Ehretiac.).

Carnarvonia F. Muell. Proteaceae. 1 Queensland.

Carnegiea Britton & Rose. Cactaceae. 1, *C. gigantea* (Engelm.) Britt. & Rose, SW. U.S., Mex., the largest of the cacti; it grows to 21 m. high and 60 cm. thick, with candelabra-like branching.

Carnegiea Perkins = seq.

Carnegieodoxa Perkins. Monimiaceae. 1 New Caled.

Carolifritschia P. & K. = Carolofritschia Engl. (Gesneriac.).

Caroli-Gmelina Gaertn., Mey. & Scherb. = Rorippa Scop. (Crucif.).

Carolinea L. f. = Pachira Aubl. (Bombacac.).

Carolinella Hemsl. = Primula L. (Primulac.).

Carolinia Néraud (nomen). Quid? (Flacourtiac.?). 1 Mauritius.

Carolofritschia Engl. = Acanthonema Hook. (Gesneriac.).

Caromba Steud. = Cabomba Aubl. (Cabombac.).

CAROPODIUM

Caropodium Stapf & Wettst. (~ Grammosciadium DC.). Umbelliferae. 5 SW. As.

Caropyxis Benth. & Hook. f. (sphalm.) = Calopyxis Tul. = Combretum L. (Combretac.).

Caroselinum Griseb. = Peucedanum L. (Umbellif.).

Carota Rupr. = Daucus L. (Umbellif.).

Caroxylon Thunb. = Salsola L. (Chenopodiac.).

Carpacoce Sond. Rubiaceae. 4 S. Afr.

Carpangis Thou. (uninom.) = *Angraecum carpophorum* Thou. (Orchidac.).

Carpanthe Rafin. = Carpanthus Rafin., *q.v.*

Carpanthea N. E. Brown. Aïzoaceae. 2 S. Afr.

Carpanthus Rafin. = Azolla Lam. (Azollac.), sec. Copeland; = ? Myriophyllum L. (Haloragidac.), sec. Pennell.

Carpentaria Becc. Palmae. 1 NE. Austr.

Carpenteria Torr. Philadelphaceae. 1 Calif. Like *Philadelphus*, but ov. sup.; sta. ∞, cpls. 5–7.

Carpentia Ewart = Cressa L. (Convolvulac.).

Carpentiera Steud. = Charpentiera Gaudich. (Amaranthac.).

Carpesium L. Compositae. 10 S. Eur., temp. As., esp. Japan. Pappus o.

Carpha Banks & Soland. Cyperaceae. 11 temp. S. Am., trop. & S. Afr., Mascarene Is., Japan, New Guinea, Austr., N.Z.

Carphalea Juss. Rubiaceae. 9 Madag.

Carphephorus Cass. Compositae. 7 SE. U.S.

Carphobolus Schott = Piptocarpha R.Br. (Compos.).

Carphochaete A. Gray. Compositae. 4 SW. U.S., Mex.

Carpholoma D. Don = Lachnospermum Willd. (Compos.).

Carphopappus Sch. Bip. = Iphiona Cass. (Compos.).

Carphophorus P. & K. = Carphephorus Cass. (Compos.).

Carphostephium Cass. = Tridax L. (Compos.).

Carpidopterix Karst. = Thouinia Poit. (Sapindac.).

Carpinaceae (Spach) Kuprianova. Dicots. 3/47 N. temp. Distinguished from the *Betulaceae* and *Corylaceae* by the following characters: Lvs. plicate in vernation, parallel to lateral nerves (cf. *Fagus*). ♂ fls. without bracteoles. Pollen grains 3–5-porate, without arci, pores with an operculum or 'plug'. ♀ fls. in catkins, each fl. with a large membranous invol. formed of the bract and 2 bracteoles. Ḡ (2), transverse. Fr. a small nutlet, attached to the accresc. invol.; cotyledons free from nutlet at germination. Vessels of the wood with simple perforations. Genera: *Carpinus, Ostrya, Ostryopsis*.

Carpinum Rafin. = seq.

Carpinus L. Carpinaceae. 35 N. temp., chiefly E. As. The young ls. hang downwards as the shoot expands. The ♀ catkins are term. on long shoots, the ♂ are themselves short shoots. In the axil of each scale of the latter are 4–12 sta., each split almost to the base. No bracteoles are present, so that it is doubtful how many fls. of the possible 3 are repres. In the ♀ there are the 2 lat. fls. with all 6 bracteoles. On the top of the 2-loc. ovary is a small P. Fr. a 1-seeded nut with a 3-lobed leafy wing on one side, whose centre lobe corresponds to the bract α or β, the lat. lobes to the bracteoles α′, β′; these unite and grow large after fert. The timber is little used.

Carpiphea Rafin. = Cordia L. (Ehretiac.).
Carpobrotus N. E. Brown. Aïzoaceae. 24 S. Afr., Austr., N.Z., Pacif.
Carpocalymna Zipp. = Epithema Blume (Gesneriac.).
Carpoceras Link = Thlaspi L. (Crucif.).
Carpoceras A. Rich. = Martynia L. (Martyniac.).
Carpodet[ac]eae Fenzl = Escalloniaceae–Argophylleae Engl.
Carpodetes Herb. = Stenomesson Herb. (Amaryllidac.).
Carpodetus J. R. & G. Forst. Escalloniaceae. 10 New Guinea, N.Z.
Carpodinopsis Pichon. Apocynaceae. 4 trop. W. Afr.
Carpodinus R.Br. ex G. Don (~ Landolphia Beauv.). Apocynaceae. 50 trop.
  Afr. Rubber is obtained by grating and boiling from the rhiz. of C. lanceolatus
  K. Schum., etc. (cf. Clitandra).
Carpodiptera Griseb. = Berrya Roxb. (Tiliac.).
Carpodontos Labill. = Eucryphia Cav. (Eucryphiac.).
Carpogymnia (H. P. Fuchs) Löve & Löve = Gymnocarpium Newm. (Aspi-
diac.).
Carpolobia G. Don. Polygalaceae. 10 trop. W. Afr.
Carpolobium P. & K. = praec.
Carpolyza Salisb. Amaryllidaceae. 1 S. Afr.
Carponema (DC.) Eckl. & Zeyh. = Heliophila Burm. f. ex L. (Crucif.)
Carpophillus Neck. = Pereskia Mill. (Cactac.).   [(galled fr.).
Carpophora Klotzsch = Silene L. (Caryophyllac.).
Carpophyllum Miq. = Sterculia L. (Sterculiac.).
Carpophyllum Neck. = Carpophillus Neck. = Pereskia Mill. (Cactac.).
Carpopodium Eckl. & Zeyh. = Heliophila Burm. f. ex L. (Crucif.).
Carpopogon Roxb. = Mucuna Adans. (Legumin.).
Carpothalis E. Mey. = Kraussia Harv. (Rubiac.).
Carpotroche Endl. Flacourtiaceae. 15 trop. Am.
Carpoxis Rafin. = Forestiera Poir. (Oleac.).
Carpoxylon H. Wendl. & Drude. Palmae. 1 New Hebrides.
Carptotepala Moldenke. Eriocaulaceae. 1 Venez., Guiana.
Carpunya Presl = Piper L. (Piperac.).
Carpupica Rafin. = Piper L. (Piperac.).
Carradoria A. DC. = Globularia L. (Globulariac.).
× Carrara hort. = × Vascostylis hort. (Orchidac.).
Carregnoa Boiss. = Tapeinanthus Herb. (Amaryllidac.).
Carria Gardn. = Gordonia Ellis (Theac.).
Carrichtera Adans. = Vella L. (Crucif.).
*Carrichtera DC. Cruciferae. 1 Canary Is., Medit. to Persia.
Carrichtera P. & K. = Caricteria Scop. = Corchorus L. (Tiliac.).
Carrichteria Wittst. = Carrichtera DC. (Crucif.).
Carrierea Franch. Flacourtiaceae. 3 S. & W. China, Indochina.
Carrissoa E. G. Baker. Leguminosae. 1 Angola.
Carroa Presl = Dalea Juss. (Legumin.).
Carronia F. Muell. Menispermaceae. 3 New Guinea to New S. Wales.
Carruanthus Schwantes ex N. E. Brown. Aïzoaceae. 1 S. Afr.
Carruthersia Seem. Apocynaceae. 6 Philippines, Solomon Is., Fiji.
Carruthia Kuntze = Nymania S.O. Lindberg (Aitoniac.).

CARSONIA

Carsonia Greene (~ Cleome L.). Cleomaceae. 1 SW. U.S.

Cartalinia Szov. ex Kunth = Paris L. (Trilliac.).

× **Carterara** hort. Orchidaceae. Gen. hybr. (iii) (Aërides × Renanthera × **Carteretia** A. Rich. = Sarcanthus Lindl. (Orchidac.). [Vandopsis).

Carteria Small = Basiphyllaea Schlechter (Orchidac.).

Carterothamnus R. M. King. Compositae. 1 Lower Calif.

Cartesia Cass. = Stokesia L'Hérit. (Compos.).

Carthamodes Kuntze = Carduncellus Adans. (Compos.).

Carthamoïdes v. Wolf = Carduus + Carduncellus + Carthamus + Centaurea + Cnicus (Compos.).

Carthamus L. Compositae. 13 Medit., Afr., As. *C. tinctorius* L. (safflower) cult. in Asia, etc.; its fls. are used in dyeing; powdered and mixed with talc they form rouge.

Cartiera Greene. Cruciferae. 6 W. N. Am.

Cartodium Soland. ex R.Br. = Craspedia G. Forst. (Compos.).

Cartonema R.Br. Cartonemataceae. 6 N. Guinea (Aru Is.), trop. Austr.

***Cartonemataceae** Pichon. Monocots. 1/6 Austr. Erect perenn. non-succ. glandular-hairy herbs, with habit of *Xanthorrhoea*; stem leafy, with cortical bundles. Ls. alt., lin., sess., sheathing. Fls. in simple or branched spikes or rac. K 3; C 3, marcescent; A 3 + 3, equal, glabr., anth. introrse, deh. by a simple longit. slit; G̲ (3), with 2 superp. axill. ov. per loc., style with capit. stig.; fr. a loculic. caps., with 2 superp. seeds per loc., seeds exarill., with retic. testa, copious mealy endosp., minute embr. and embryostega. Only genus: *Cartonema*. Near *Commelinac.*, but differing in non-succulent habit, rac. infl., cortical bundles, and in the absence of calcium oxalate in the tissues, in these chars. (exc. infl.) agreeing with *Mayacac.*

Cartrema Rafin. = Osmanthus Lour. (Oleac.).

Caruelia Parl. = Ornithogalum L. (Liliac.).

Caruelina Kuntze = Chomelia Jacq. (Rubiac.).

Carui Mill. = seq.

Carum L. Umbelliferae. 30 temp. & subtrop. *C. carvi* L. is cult. for its fr. (caraway seeds).

Carumbium Kurz = Sapium P.Br. (Euphorbiac.).

Carumbium Reinw. = Homalanthus Juss. (Euphorbiac.).

Caruncularia Haw. = Stapelia L. (Asclepiadac.).

Carusia Mart. ex Niedenzu = Burdachia Mart. ex A. Juss. (Malpighiac.).

Carvalhoa K. Schum. Apocynaceae. 2 trop. E. Afr.

Carvi Bernh. = Selinum L. (Umbellif.).

Carvi Bub. = Carui Mill. = Carum L. (Umbellif.).

Carvia Bremek. Acanthaceae. 1 Penins. India.

Carvifolia Vill. = Selinum L. (Umbellif.).

***Carya** Nutt. Juglandaceae. 25 E. As., E. N. Am., the hickory trees, cultivated for their wood, which is very tough and elastic, and for the edible fruit (pecans, like walnuts).

Caryella Bourn. ex Parment. Ebenaceae? 1 Madag.

Caryocar Allam. ex L. Caryocaraceae. 20 trop. Am. The wood is very durable and is used in shipbuilding. The fruit is a large 4-stoned drupe; the seeds are the *souari-* or butter-nuts of commerce.

*Caryocaraceae Szyszyl. Dicots. 2/25 trop. Am. Trees and shrubs with digitately 3–5-foliol. opp. or alt. ls. with deciduous stips. Fl. reg., ♀, in racemes. K (5–6), imbr. or open; C 5–6, imbr., sometimes calyptrate; A ∞, united into a ring or in 5 bundles; G̲ 4- or 8–20-loc., with as many styles, and 1 pend. ov. per loc. Usu. drupe with oily mesocarp, and woody endocarp which splits into 4 mericarps; sometimes a leathery schizocarp. Little or no endosp.; plumule thick and fleshy or elongate and spirally convolute. Genera: *Caryocar*, *Anthodiscus*. Allied to *Bombacac.*, *Theac.* and *Barringtoniac.*

Caryochloa Spreng. = Oryzopsis Michx (Gramin.).

Caryochloa Trin. = Luziola Juss. (Gramin.).

Caryococca Willd. ex Roem. & Schult. = Gonzalagunia Ruiz & Pav. (Rubiac.).

Caryodaphne Blume ex Nees = Cryptocarya R.Br. (Laurac.).

Caryodaphnopsis Airy Shaw. Lauraceae. 5 S. China, Siam, Indoch., Philipp. Is., Borneo.

Caryodendron Karst. Euphorbiaceae. 3 trop. S. Am.

Caryolobis Gaertn. (Doona Thw.). Dipterocarpaceae. 11 Ceylon. Timber,

Caryolobium Stev. = Astragalus L. (Legumin.). [resin.

Caryolopha Fisch. & Trautv. = Pentaglottis Tausch (Boraginac.).

Caryomene Barneby & Krukoff. Menispermaceae. 4–6 Brazil.

*Caryophyllaceae Juss. Dicots. (Centrospermae). 70/1750 cosmop. Mostly herbs, a few undershrubs, with opp. simple usu. entire ls., often stip.; the stem often swollen at the nodes, the branching dich. The infl. usu. term. the main axis and is typically a dich. cyme, but both in the veg. region and in the infl., of the two branches arising at any node, one (that in the axil of *β*) tends to outgrow the other, and after two or three branchings the weaker one often does not develop at all, so that a cincinnus arises. The whole infl. is very char., and such a one is often called a caryophyllaceous infl.

Fls. ♀ or ♂ ♀, and reg., but often not isomerous. K (5), imbr.; C 5 (sometimes o); A 5+5, or fewer; G̲ (2–5), with free central, axile, or basal plac., 1–5-loc. Ov. usu. ∞, in double rows corresponding to the cpls., rarely few or 1 (*Paronychieae*), usu. campylotropous. In most cases the fl. is obdiplost., as may be recognised by the cpls. (when 5) being opp. the petals. The ovary, sta., and corolla are sometimes borne on an androphore (e.g. *Lychnis*), an elongation of the axis between K and C. The petals sometimes have a ligule (e.g. *Lychnis*), and are often clawed or bifid. At the base of the ovary are often seen traces of the septa, which in the upper part do not develop.

Biologically, as well as morphologically, the fam. forms two distinct groups, a higher type, the *Caryophylloïdeae*, and a lower, the *Alsinoïdeae*. All secrete honey at the base of the sta., but while in the *A.* the fl. is wide open, so that short-tongued insects can reach the honey, in the *C.* a tube is formed by the gamosepalous K; in this stand the claws of the petals and the sta., partly filling it up, and rendering the honey inaccessible to any but long-tongued insects, esp. bees and Lepidoptera. The latter class, esp. in the Alps (see Müller's *Alpenblumen*), are the chief visitors, and many of the *C.* are adapted to them— by length of tube, red and white colours, night-flowering in many spp., or emission of scent only at night, etc. The fls. are commonly protandr. Many *A.* are gynodioec. (cf. *Labiatae*).

CARYOPHYLLATA

Fr. usu. a caps. containing several or ∞ seeds, sometimes an indeh. 1-seeded nutlet. It opens in nearly all cases by splitting from the apex into teeth which bend outwards, leaving an opening. The splitting may take place in as many, or in twice as many, lines as cpls. The seeds cannot escape from the capsule unless it be shaken, e.g. by wind or animals, and being small and light have a good chance of distr. Rarely (e.g. *Cucubalus*) the fruit is berry-like, but dry when ripe, indehisc. Embryo usu. curved round the perisperm (in a few cases nearly straight).

Classification and chief genera (after Pax):

I. **Alsinoïdeae** (fl. polysepalous; sta. often perig.).

    a. Fruit a capsule opening by teeth.

      1. ALSINEAE (styles free to base; ls. exstip.): *Stellaria, Cerastium, Sagina, Arenaria.*

      2. SPERGULEAE (do., but ls. stip.): *Spergula, Spergularia.*

      3. POLYCARPEAE (styles joined at base): *Drymaria, Polycarpon.*

    b. Fruit an achene or nut.

      4. PARONYCHIËAE (fls. all alike; stipules): *Corrigiola, Paronychia, Illecebrum, Herniaria.*

      5. SCLERANTHEAE (do., but exstip.): *Scleranthus.*

      6. PTERANTHEAE (fls. in 3's, the 2 lat. ± abortive): *Pteranthus.*

II. **Caryophylloïdeae (Silenoïdeae)** (fl. gamosepalous, hypog.):

      1. LYCHNIDEAE (calyx with commissural ribs): *Silene, Lychnis.*

      2. CARYOPHYLLEAE (DIANTHEAE) (no commissural ribs): *Gypsophila, Dianthus.*

Caryophyllata Mill. = Geum L. (Rosac.).

Caryophyllea Opiz = Aira L. (Gramin.).

Caryophyllus L. = Syzygium Gaertn. (Myrtac.).

Caryophyllus Mill. = Dianthus L. (Caryophyllac.).

Caryopitys Small = Pinus L. (Pinac.).

Caryopteris Bunge. Verbenaceae. 15 Himal. to Japan.

Caryospermum Blume = Perrottetia Kunth (Celastrac.).

Caryota L. Palmae. 12 Ceylon, Indomal., Solomon Is., NE. Austr. Stem columnar; ls. bipinnate. Infl. of a number of equal branches hanging down like a brush. They appear in descending order, the oldest in the crown, the younger lower down in the axils of the old leaf-sheaths. Fls. in groups of 3, one ♀ between two ♂♂. Sta. 9–∞. Cpl. 1. Berry. *C. urens* L. (toddy palm) cult.; it yields palm sugar (see *Arenga*), sago (see *Metroxylon*), kitul fibre, [wood, etc.

Caryotaceae O. F. Cook = Palmae–Caryoteae Mart.

Caryotaxus Zucc. ex Endl. = Torreya Arn. (Taxac.).

Caryotophora Leistn. Aïzoaceae. 1 S. Afr.

Casalea A. St-Hil. = Ranunculus L. (Ranunculac.).

Casanophorum Neck. = Castanea Mill. (Fagac.).

Casarettoa Walp. = Vitex L. (Verbenac.).

Casasia A. Rich. Rubiaceae. 10 Florida, Mex., W.I. G 1-loc.

Cascabela Rafin. = Thevetia Adans. (Apocynac.).

Cascadia A. M. Johnson (~ Saxifraga L.). Saxifragaceae. 1 W. U.S.

Cascarilla Adans. = Croton L. (Euphorbiac.).

Cascarilla Ruiz ex Steud. = Cinchona L. (Rubiac.).

Cascarilla Wedd. Rubiaceae. 25 S. Am. The bark of some resembles that of *Cinchona* (see also *Croton*), but the amount of alkaloid is small.

Cascaronia Griseb. Leguminosae. 1 Argentina.

Cascoëlytrum Beauv. = Chascolytrum Desv. = Briza L. (Gramin.).

Casearia Jacq. Flacourtiaceae. 160 trop. Pellucid 'dot-and-dash' glands in ls. *C. praecox* Griseb. (Cuba, trop. S. Am.), W.I. box (useful wood).

Caseola Nor. = Sonneratia L. f. (Sonneratiac.).

Cashalia Standley = Dussia Krug & Urb. (Legumin.).

Casia Duham. = Osyris L. (Santalac.)

Casimira Scop. = Melicocca L. (Sapindac.).

Casimirella Hassler. Icacinaceae. 1 Paraguay.

Casimiroa Domb. ex Baill. = Cervantesia Ruiz & Pav. (Santalac.).

Casimiroa La Llave. Rutaceae. 6 C. Am. Ed. fr. Exalb.

Casinga Griseb. = Laëtia Loefl. (Flacourtiac.).

Casiostega Rupr. ex Galeotti = Opizia Presl (Gramin.).

Casparea Kunth = Bauhinia L. (Legumin.).

Caspareopsis Britton & Rose = Bauhinia L. (Legumin.).

Casparya Klotzsch = Begonia L. (Begoniac.).

Caspia Scop. = Vismia Vand. (Guttif.).

Cassandra D. Don (~ Lyonia Nutt.). Ericaceae. 1 N. temp.

Cassebeera Kaulf. = Doryopteris J. Sm. (Sinopteridac.).

Cassebeeria Dennst. = Codigi Augier = Sonerila Roxb. (Melastomatac.).

Casselia Dum. = Mertensia Roth (Boraginac.).

*****Casselia** Nees & Mart. Verbenaceae. 12 trop. Am.

Cassia L. Leguminosae. 5–600 trop. & warm temp. (exc. Eur.). Trees, shrubs and herbs with paripinnate ls. and stips. of various types. Fl. zygo., but with petals almost equal in size. The sta. may be 10, but the 3 upper ones are usu. reduced to stds. or absent. The anthers usu. open by pores. The 5 upper sta. are generally short, the 2 lower are long and project outwards. In many two forms of fl. occur, one in which the lower sta. project to the left, the other in which they project to the right. It was once thought that this 'enantiostyly' was a kind of heterostylism, but both types of fl. occur on one plant. It would appear to be simply a case of variation in symmetry (cf. *Exacum*, *Saintpaulia*). In many spp. a division of labour takes place among the sta. (cf. *Heeria*); the insect visitors eat the pollen of the short sta. and carry away on their bodies that of the long. There is no honey. Fr. often chambered up by 'false' septa running across it—outgrowths from the placenta.

Many cult. for the ls., which when dried form the drug senna. Alexandrian senna from *C. acutifolia* Delile, Italian from *C. obovata* Collad., Arabian from *C. angustifolia* Vahl. *C. fistula* L. (purging cassia, pudding-pipe tree) has its seeds embedded in laxative pulp.

Cassiaceae Link = Leguminosae–Caesalpinioïdeae Kunth.

Cassiana Rafin. = Cassia L. (Legumin.).

Cassida Seguier = Scutellaria L. (Labiat.).

Cassidispermum Hemsl. Sapotaceae. 1 Solomon Is.

Cassidocarpus Presl ex DC. = Asteriscium Cham. & Schlechtd. (Umbellif.).

Cassidospermum P. & K. = Cassidispermum Hemsl. (Sapotac.).

Cassiera Raeusch. = Cansjera Juss. (Opiliac.).

CASSINE

Cassine Kuntze = Otherodendron Mak. (Celastrac.).

Cassine L. Celastraceae. 40 S. Afr., Madag., trop. As. to Pacif. Ls. alt. or opp. *C. crocea* Presl yields saffron-wood.

Cassine Loes. = Elaeodendron Jacq. f. (Celastrac.).

Cassinia R.Br. ex Ait. (1813) = Angianthus Wendl. (Compos.).

*Cassinia R.Br. (1817). Compositae. 28 trop. & S. Afr., Austr., N.Z.

Cassiniaceae Sch. Bip. = Compositae Giseke.

Cassiniola F. Muell. = Helipterum DC. (Compos.).

Cassinopsis Sond. Icacinaceae. 7 S. Afr., Madag.

Cassiope D. Don. Ericaceae. 12 circumpolar, Himal. Ls. much rolled back (see fam.; cf. *Empetrum*); in *C. redowskii* (Cham. & Schlechtd.) G. Don they are hollow.

Cassiphone Reichb. = Leucothoë D. Don (Ericac.).

Cassipourea Aubl. Rhizophoraceae. 80 trop. Am., W.I., trop. & S. Afr., Madag., Ceylon. G! (exceptional in fam.). Some affinity with *Elaeocarpac.*

Cassipoure[ace]ae J. G. Agardh = Rhizophoraceae–Macarisieae Baill.

Cassitha Hill = Cassytha L. (Laurac.).

Cassumbium Benth. & Hook. f. = Cussambium Buch.-Ham. = Schleichera Willd. (Sapindac.).

Cassumunar Colla = Zingiber Boehm. (Zingiberac.).

Cassupa Humb. & Bonpl. (~ Isertia Schreb.). Rubiaceae. 7 NW. S. Am.

Cassutha Des Moul. = Cuscuta L. (Cuscutac.).

Cassuvi[ac]eae R.Br. = Anacardiaceae R.Br.

Cassuvium Lam. = Anacardium L. (Anacardiac.).

Cassyta L. = Cassytha L. (Laurac.).

Cassyta J. Miller = Rhipsalis Gaertn. (Cactac.).

Cassytha S. F. Gray = Cuscuta L. (Cuscutac.).

Cassytha L. Lauraceae. 20 palaeotrop. Parasites with the habit of *Cuscuta*.

*Cassythaceae Bartl. ex Lindl. = Lauraceae–Cassythoïdeae Kosterm.

Castalia Salisb. = Nymphaea L. (Nymphaeac.).

Castalis Cass. Compositae. 3 S. Afr.

Castanea Mill. Fagaceae. 12 North temp. *C. sativa* Mill. (*C. vulgaris* Lam.) is the sweet chestnut; useful wood; bark used in tanning.

Castaneaceae Link = Hippocastanaceae DC.

Castanella Spruce ex Benth. & Hook. f. Sapindaceae. 1 Brazil.

Castanocarpus Sweet = Castanospermum A. Cunn. (Legumin.).

× **Castanocasta** Cif. & Giac. Fagaceae. Gen. hybr. (Castanea × Castanopsis).

Castanola Llanos = Agelaea Soland. ex Planch. (Connarac.).

Castanophorum Pfeiff. = Casanophorum Neck. = Castanea Mill. (Fagac.).

*Castanopsis (D. Don) Spach. Fagaceae. 120 trop. & subtrop. As. (See also *Chrysolepis* Hjelmq.)

Castanospermum A. Cunn. Leguminosae. 1 subtrop. Austr., *C. australe* A. Cunn. (Australian chestnut). Seeds ed.

Castanospora F. Muell. Sapindaceae. 1 warm E. Austr.

*Castela Turp. Simaroubaceae. 15 W.I., S. U.S. to Argent., Galápagos Is. Cf. Moran & Felger, *Trans. S. Diego Soc. Nat. Hist.* **15** (4): 33–40, 1968.

Castel[ac]eae J. G. Agardh = Simaroubaceae–Picrasmeae–Castelinae Engl.

Castelaria Small = Castela Turp. (Simaroubac.).

Castelia Cav. = Pitraea Turcz. (Verbenac.).

Castelia Liebm. = Castela Turp. (Simaroubac.).

Castellanoa Traub. Amaryllidaceae. 1 Argentina.

Castellanosia Cárdenas. Cactaceae. 1 Bolivia.

Castellia Tineo. Gramineae. 1 Canary Is., Medit., Sudan, Arabia, NW. India.

Castelnavia Tul. & Wedd. Podostemaceae. 9 Brazil.

Castiglionia Ruiz & Pav. = Jatropha L. (Euphorbiac.).

Castilla Cerv. Moraceae. 10 trop. Am., Cuba. The latex of *C. elastica* Cerv. yields caoutchouc (C. Am. or Panama rubber, *caucho, ulé*; cf. *Hevea*, etc.).

Castilleja Mutis. Scrophulariaceae. 2 arct. Euras., 200 N. & S. Am. (painted lady, paint-brush). The upper ls., or sometimes only their outer ends, are brightly coloured, adding to the conspicuousness of the fls. (cf. *Cornus, Poinsettia*, etc.).

Castillejoa P. & K. = praec.

Castilloa Endl. = Castilla Cerv. (Morac.).

Castorea Mill. = Duranta L. (Verbenac.).

Castra Vell. = Trixis P.Br. (Compos.).

Castratella Naud. Melastomataceae. 1 Colombia, Venez.

Castrea A. St-Hil. = Phoradendron Nutt. (Loranthac.).

Castronia Nor. = Helicia Lour. (Proteac.).

Casuarina Adans. Casuarinaceae. 45 trop. E. Afr. (? native), Masc., SE. As., Malaysia, Austr., Polynes. Known in Austr. as she-oak, forest-oak, etc. Wood of several spp. valued for hardness. The green shoots of some spp. are used as emergency fodder for sheep and cattle.

*****Casuarinaceae** R.Br. Dicots. 2/65 Masc., SE. As. to NE. Austr. & Polynesia. Trees or shrubs, often of weeping habit, with long or short slender green branches, cylindrical and deeply grooved. At the nodes are borne whorls of scale ls. like those of *Equisetum*. The stomata and green tissue are at the base and sides of the grooves, whilst the ridges are formed of sclerenchyma, so that the plant is markedly xero. Fls. ♂ ♀. The ♂ are borne in simple or cpd. spikes, usu. term. on green branchlets, but sometimes sess. The internodes are short and at every node is a cup (formed of the combined bracts) with several sta. hanging out over the edge. Each sta. repres. a ♂ fl. and has a 2-leaved P and 2 bracteoles. The ♀ fls. are borne in dense spherical or ovoid heads, sess. or terminating short usu. lat. branchlets. Each fl. is naked in the axil of a bract, has 2 bracteoles, and consists of 2 cpls., syncp., the post. loc. empty, the ant. containing 2 ov. The long styles hang out beyond the bracts and wind-fert. occurs. Afterwards the whole head becomes woody (bracteoles as well), enclosing the ripening seeds. The fr. is a single-seeded, flattened, terminally winged nut, of seed-like appearance, enclosed in the woody bracteoles, which separate at maturity. Endosp. o. Genera: *Casuarina, Gymnostoma*. A very isolated fam.; some similarity in pollen with *Betulac.* and *Myricac.*

Catabrosa Beauv. Gramineae. 4 temp.

Catabrosella (Tsvelev) Tsvelev. Gramineae. 9 SW. to C. As., Himal., W. China.

Catabrosia Roem. & Schult. = praec.

Catachaetum Hoffmgg. ex Reichb. = Catasetum Rich. (Orchidac.).

Catachenia Griseb. = Miconia Ruiz & Pav. (Melastomatac.).

CATACLINE

Catacline Edgew. = Tephrosia Pers. (Legumin.).

Catacoma Walp. = Catocoma Benth. = Bredemeyera Willd. (Polygalac.).

Catadysia O. E. Schulz. Cruciferae. 1 Peru.

Catagyna Beauv. = ? Scleria Bergius (Cyperac.).

Catagyna Hutch. & Dalz. = Coleochloa Gilly (Cyperac.).

Catakidozamia T. Hill = Macrozamia Miq. (Zamiac.).

Catalepis Stapf & Stent. Gramineae. 1 S. Afr.

Cataleuca Hort. ex K. Koch = Onoseris DC. (Compos.).

Catalissa Miers (nomen). Guttiferae. Trop. S. Am. Quid?

Catalium Buch.-Ham. ex Wall. = Carallia Roxb. (Rhizophorac.).

Catalpa Scop. Bignoniaceae. 11 E. As., Am., W.I. *C. bignonioïdes* Walt. yields a durable timber. The genus is closely related to *Paulownia* S. & Z. (*Scrophulariac.*), *q.v.*

Catalpium Rafin. = praec.

Catamixis Thoms. Compositae. 1 Himal.

× **Catamodes** hort. Orchidaceae. Gen. hybr. (iii) (Catasetum × Mormodes).

Catanance St-Lag. = seq.

Catananche L. Compositae. 5 Medit. *C. lutea* L. produces, besides its normal stalked capit., small sess. capit. in axils of rad. ls. at ground level, tightly appressed to crown of rootstock, as safeguard against grazing by goats.

Catanga Steud. = Cananga Aubl. = Guatteria Ruiz & Pav. (Annonac.).

× **Catanoches** hort. Orchidaceae. Gen. hybr. (iii) (Catasetum × Cycnoches).

Catanthera F. Muell. (Hederella Stapf). Melastomataceae. 5 Borneo, New Guinea.

Catapodium Link. Gramineae. 2 Eur., Medit.

Catappa Gaertn. = Terminalia L. (Combretac.).

Catapuntia Muell. Arg. = seq.

Cataputia Boehm. = Ricinus L. (Euphorbiac.).

Cataria Mill. = Nepeta L. (Labiat.).

Catarsis P. & K. = Katarsis Medik. = Gypsophila L. (Caryophyllac.).

Catas Domb. ex Lam. = Embothrium J. R. & G. Forst. (Proteac.).

Catasetum Rich. Orchidaceae. 70 trop. Am. Epiph. 3 widely different sexual forms occur on different (or sometimes on the same) stocks. They were formerly regarded as separate gen. The old genus *C.* is the ♂ form, *Myanthus* Lindl. the ♀, and *Monachanthus* Lindl. the ♀. The labellum is uppermost in the fl. The pollinia are ejected with violence when one of the horns of the **Cataterophora** Steud. = seq. [column is touched.

Catatherophora Steud. = Pennisetum Rich. (Gramin.).

Catatia Humbert. Compositae. 2 Madag.

Catenaria Benth. = Desmodium Desv. (Legumin.).

Catenularia Botschantsev. Cruciferae. 1 C. As.

Catepha Leschen. ex Reichb. = ? Hydrocotyle L. (Hydrocotylac.).

Catesbaea L. Rubiaceae. 10 Florida Keys, W.I.

Catevala Medik. = Aloë L. + Haworthia Duval (Liliac.).

Catha G. Don = Celastrus L. (Celastrac.).

Catha Forsk. ex Scop. Celastraceae. 1 Afr., Madag., Arabia, *C. edulis* Forsk. The ls. are used by Arabs like tea, under the name *khat* or *cafta*.

Cathanthes Rich. = Tetroncium Willd. (Juncaginac.).

Catharanthus G. Don. Apocynaceae. 5 trop., esp. Madag. *C. roseus* (L.)
G. Don (*Lochnera rosea* [L.] Reichb.), one of the commonest trop. weeds.
Cathartocarpus Pers. = Cassia L. (Legumin.).
Cathartolinum Reichb. = Mesynium Rafin. = Linum L. (Linac.).
Cathastrum Turcz. = Pleurostylia Wight & Arn. (Celastrac.).
Cathaya Chun & Kuang. Pinaceae. 2 S. & W. China.
Cathayambar (Harms) Nakai = Liquidambar L. (Altingiac.).
Cathayanthe Chun. Gesneriaceae. 1 Hainan.
Cathayeia Ohwi = Idesia Maxim. (Flacourtiac.).
Cathcartia Hook. f. = Meconopsis Vig. (Papaverac.).
Cathea Salisb. = Calopogon R.Br. (Orchidac.).
Cathedra Miers. Olacaceae. 11 trop. S. Am.
Cathedraceae Van Tiegh. = Olacaceae–Anacoloseae Engl.
Cathestecum J. Presl. Gramineae. 6 S. U.S., Mex.
Cathetostemma Blume = Hoya R.Br. (Asclepiadac.).
Cathetus Lour. = Phyllanthus L. (Euphorbiac.).
Cathissa Salisb. = Ornithogalum L. (Liliac.).
Cathormion Hassk. Leguminosae. 15 trop. (mainly Amer.).
Catimbium Holtt. = Zerumbet Wendl. (Zingiberac.).
Catimbium Juss. = Renealmia L. f. (non L.) (Zingiberac.).
Catinga Aubl. = Eugenia L. (Myrtac.).
Catis O. F. Cook = Euterpe Gaertn. (Palm.).
× **Catlaelia** hort. (vii) = × Laeliocattleya hort. (Orchidac.).
× **Catlaenitis** hort. (vii) = × Sophrolaeliocattleya hort. (Orchidac.).
Catoblastus H. Wendl. Palmae. 15 trop. S. Am.
Catocoma Benth. = Bredemeyera Willd. (Polygalac.).
Catocoryne Hook. f. Melastomataceae. 1 Peru.
Catodiacrum Dulac = Orobanche L. (Orobanchac.).
Catonia P.Br. = ? Miconia Ruiz & Pav. (Melastomatac.).
Catonia Moench = Crepis L. (Compos.).
Catonia Rafin. = Cordia L. (Ehretiac.).
Catonia Vahl = Erycibe Roxb. (Convolvulac.).
Catonia Vell. = ? Symplocos Jacq. (Symplocac.).
Catopheria Benth. Labiatae. 3 trop. Am.
Catophractes D. Don. Bignoniaceae. 1 trop. & S. Afr.
Catophyllum Pohl ex Baker = Mikania Willd. (Compos.).
Catopodium Link (sphalm.) = Catapodium Link (Gramin.).
Catopsis Griseb. Bromeliaceae. 25 Florida, Mex. to trop. S. Am., W.I.
Catosperma Benth. Goodeniaceae. 1 trop. Austr.
Catostemma Benth. Bombacaceae. 8 Guiana, Brazil.
Catostigma O. F. Cook & Doyle. Palmae. 3 Colombia.
Catsjopiri Rumph. = Gardenia L. (Rubiac.).
Cattimarus Kuntze = Kleinhovia L. (Sterculiac.).
Cattleya Lindl. Orchidaceae. 60 Mex. to trop. S. Am., W.I.
× **Cattleyodendrum** hort. (vii) = × Epicattleya hort. (Orchidac.).
Cattleyopsis Lem. Orchidaceae. 2 W.I.
× **Cattleyopsisgoa** hort. Orchidaceae. Gen. hybr. (iii) (Cattleyopsis ×
Domingoa).

CATTLEYOPSISTONIA

× **Cattleyopsistonia** hort. Orchidaceae. Gen. hybr. (iii) (Broughtonia × Cattleyopsis).

× **Cattleyovola** hort. (vii) = × Brassocattleya hort. (Orchidac.).

× **Cattleytonia** hort. Orchidaceae. Gen. hybr. (iii) (Broughtonia × Cattleya).

Cattutella (sphalm. pro Catutekka) Reichb. = Katoutheka Adans. = Wendlandia Bartl. (Rubiac.).

Catu-Adamboe Adans. = Lagerstroemia L. (Lythrac.).

Catunaregam Adans. ex v. Wolf = Randia L. (sens. lat.) (Rubiac.).

Caturus L. = Acalypha L. (Euphorbiac.).

Caturus Lour. = Malaisia Blanco (Morac.).

Catutsjeron Kuntze = Katoutsjeroe Adans. = Holigarna Buch.-Ham. (Anacardiac.).

Catyona Lindl. = Gatyona Cass. = Crepis L. (Compos.).

Caucaea Schlechter. Orchidaceae. 1 Colombia.

Caucaliopsis Wolff. Umbelliferae. 1 trop. E. Afr.

Caucalis L. Umbelliferae. 4 Eur. to C. As.

Caucaloïdes Fabr. (uninom.) = ?*Caucalis* L. sp. (Umbellif.).

Caucanthus Forsk. Malphigiaceae. 5 E. Afr., Arabia.

Caucanthus Rafin. = Sterculia L. (Sterculiac.).

Caudicia Ham. ex Wight = Parsonsia R.Br. (Apocynac.).

Caudoleucaena Britton & Rose = Leucaena Benth. (Legumin.).

Caudoxalis Small = Oxalis L. (Oxalidac.).

Caulangis Thou. (uninom.) = *Angraecum caulescens* Thou. (Orchidac.).

Caulanthus S. Wats. Cruciferae. 20 W. U.S. *C. procerus* Wats. (wild cabbage) ed.

Caularthron Rafin. Orchidaceae. 2 trop. Am.

Caulinia DC. = Posidonia Koen. (Posidoniac.).

Caulinia Moench = Kennedya Vent. (Legumin.).

Caulinia Willd. = Najas L. (Najadac.).

Caullinia Rafin. = Hippuris L. (Hippuridac.).

Caulobryon Klotzsch ex C. DC. = Piper L. (Piperac.).

Caulocarpus E. G. Baker. Leguminosae. 1 Angola.

Caulokaempferia Larsen. Zingiberaceae. 7 Himal. to SE. As.

Cauloma Rafin. = Verbesina L. (Compos.).

Caulophyllum Michx. Leonticaceae. 2 NE. As., N. Am. (cohosh). Wall of fr. early evanescent, exposing the large drupe-like seeds borne on accresc. funicles.

Caulopsis Fourr. = Arabis L. (Crucif.).

Caulotretus Rich. ex Spreng. = Bauhinia L. (Legumin.).

Caulotulis Rafin. = Ipomoea L. (Convolvulac.).

Causea Scop. = Hirtella L. (Chrysobalanac.).

Causonia Rafin. = Cayratia Juss. (Vitidac.).

Causonis Rafin. = praec.

Caustis R.Br. Cyperaceae. 10 Austr.

Cautlea Royle = seq.

Cautleya Royle. Zingiberaceae. 5 Himal.

Cavacoa J. Léonard. Euphorbiaceae. 3 trop. Afr.

Cavalam Adans. = Sterculia L. (Sterculiac.).

Cavaleriea Léveillé = Ribes L. (Grossulariac.).
Cavaleriella Léveillé = Aspidopterys Juss. (Malpighiac.) + Dipelta Maxim. (Caprifoliac.).
Cavallium Schott = Sterculia L. (Sterculiac.).
Cavanalia Griseb. = Canavalia DC. (Legumin.).
Cavanilla J. F. Gmel. = Dombeya Cav. (Sterculiac.).
Cavanilla Salisb. = Stewartia L. (Theac.).
Cavanilla Thunb. = Pyrenacantha Wight (Icacinac.).
Cavanilla Vell. = Caperonia A. St-Hil. (Euphorbiac.).
Cavanillea Desr. = Diospyros L. (Ebenac.).
Cavanillea Medik. = Anoda Cav. (Malvac.).
Cavanillesia Ruiz & Pav. Bombacaceae. 3 trop. Am.
Cavaraea Speg. Leguminosae. 1 Argentina.
Cavaria Steud. = Tovaria Ruiz & Pav. (Tovariac.).
Cavea W. W. Smith & Small. Compositae. 1 E. Himal.
*****Cavendishia** Lindl. Ericaceae. 100 trop. Am.
Cavinium Thou. = Vaccinium L. (Ericac.).
Cavoliana Rafin. = seq.
Cavolinia Rafin. = Caulinia Willd. = Najas L. (Najadac.).
*****Cayaponia** S. Manso. Cucurbitaceae. 45 warm Am., 1 trop. W. Afr. & Madag., 1 Java.
*****Caylusea** A. St-Hil. Resedaceae. 3 Cape Verde Is., N. & E. Afr. to India.
*****Cayratia** Juss. Vitidaceae. 45 Afr., Madag., Indomal., Austr., New Caled., Pacif.
Ceanothus L. Rhamnaceae. 55 N. Am.
Cearia Dum. = Eurycles Salisb. (Amaryllidac.).
Cebatha Forsk. = Cocculus DC. (Menispermac.).
Cebipira Juss. ex Kuntze = Bowdichia Kunth (Legumin.).
Cecchia Chiov. = Oldfieldia Hook. (?Euphorbiac.).
Cecidodaphne Nees = Cinnamomum Schaeff. (Laurac.).
*****Cecropia** Loefl. Urticaceae. 100 trop. Am. Trees of rapid growth, with very light wood, used for floats, etc. Infl. a very complex cyme. *C. peltata* L. is the trumpet tree, so called from the use made of its hollow stems by the Uaupés Indians. The hollows are often inhabited by fierce ants (*Azteca* spp.) which rush out if the tree be shaken, and attack the intruder. Schimper has made an investigation of this symbiosis (or living together for mutual benefit) of plant and animal, showing that there is here a true case of myrmecophily, as in *Acacia sphaerocephala* (*q.v.*). These ants protect the *C.* from the leaf-cutter ants. The internodes are hollow, but do not communicate directly with the air. Near the top of each, however, is a thin place in the wall. A gravid ♀ ant burrows through this and brings up her brood inside the stem. The base of the leaf-stalk is swollen and bears food-bodies (cf. *Acacia*) on the lower side, upon which the ants feed. New ones form as the old are eaten. Several other spp. show similar features. An interesting point, that goes to show the adaptive nature of these phenomena, is that in one sp. the stem is covered with wax which prevents the leaf-cutters from climbing up, and there are neither food-bodies nor the thin places in the internodes.—The genus *Cecropia* is somewhat intermediate between the *Urticac.* and *Morac.*

CEDRELA

Cedrela P.Br. Meliaceae. 6–7 Mex. to trop. S. Am. Some yield valuable timber, e.g. *C. odorata* L., the West Indian cedar, used in cigar-boxes.

Cedrelaceae R.Br. = Meliaceae–Cedreleae Harms.

Cedrelinga Ducke. Leguminosae. 1 Brazil.

Cedrella Scop. = Cedrela P.Br. (Meliac.).

Cedrelopsis Baill. Ptaeroxylaceae. 2 Madag.

Cedro Loefl. = Cedrela P.Br. (Meliac.).

Cedronella Moench. Labiatae. 1 Canaries, Madeira. Ls. ternate.

Cedronia Cuatrec. Simaroubaceae. 1 Colombia.

Cedrostis P. & K. = Kedrostis Medik. (Cucurbitac.).

Cedrota Schreb. = Aniba Aubl. (Laurac.).

Cedrus Duham. = Juniperus L. (Cupressac.).

Cedrus Loud. = Pinus L. (Pinac.).

Cedrus Mill. = Cedrela P.Br. (Meliac.).

*Cedrus Trew. Pinaceae. 4, *C. libani* A. Rich. (cedar of Lebanon), *C. atlantica* (Endl.) Arn. (Atlantic cedar; Algeria), *C. brevifolia* (Hook. f.) Henry (Cyprus), and *C. deodara* (Roxb. ex Lamb.) G. Don (deodar; Himal., gregarious and reaching to 12 m. in girth). Handsome evergreen trees (often planted for orn.) with needle ls. and long and short shoots; the latter may grow for several years and even develop into long shoots. Infls. sol., in the position of short shoots. The cone ripens in 2–3 years. Wood durable and valued for building, etc.

Ceiba Mill. Bombacaceae. 10 trop. Am. *C. pentandra* (L.) Gaertn. is the silk cotton, or kapok; its seeds are enveloped in silky hairs, which are used for stuffing cushions, etc.; long cult. in W. Afr.

Celaena Wedd. = Oligandra Less. (Compos.).

Celaenodendron Standley. Euphorbiaceae (?). 1 Mex.

Celasine Pritz. (sphalm.) = Gelasine Herb. (Iridac.).

*Celastraceae R.Br. Dicots. 55/850 trop. & temp. Trees or shrubs with simple, opp. or alt., stip., often leathery ls. Fl. small, reg., usu. ⚥, in cymose (rarely racemose) infl. K 3–5, free or united; C 3–5, imbr.; usu. a well-marked disk; A 3–5, alternipet.; G̲ (2–5), usu. with as many loculi, sometimes partly sunk in the disk; ovules generally 2 in each loc., usu. erect, anatr. or apotr. Fr. a loculic. caps., samara, drupe, berry or indehisc. caps. Seed often with brightly coloured aril. Endosp. usu. present. Chief genera: *Euonymus*, *Celastrus*, *Cassine*, *Elaeodendron*, *Maytenus*, *Hippocratea*, *Salacia*.

Celastrus Baill. = Denhamia Meissn. (Celastrac.).

Celastrus L. Celastraceae. 30 trop. & subtrop. Climbing shrubs with fruit like *Euonymus*.

Celebnia Nor. = Saraca L. (Legumin.).

Celeri Adans. = Apium L. (Umbellif.).

Celerina Benoist. ,Acanthaceae. 1 Madag.

Celestina Rafin. = Coelestina Cass. = Ageratum L. (Compos.).

Celianella Jabl. Euphorbiaceae. 1 Venez.

Celmisia Cass. (1817) = Alciope DC. (Compos.).

Celmisia Cass. (1825) = praec. + seq.

*Celmisia Cass. ex DC. (1836). Compositae. 65 Austr., Tasm., N.Z.

Celome Greene (~ Cleome L.). Cleomaceae. 1 W. U.S.

Celosia L. Amaranthaceae. 60 trop. & temp. In *C. cristata* L., the cock's-comb, there is a cult. (but now hereditary) monstrosity, in which fasciation of the infl. occurs.

Celsa Vell. = ? Casearia Jacq. (Flacourtiac.).

Celsia Boehm. = Bulbocodium L. (Liliac.).

Celsia Fabr. = Ornithogalum L. + Gagea Salisb. (Liliac.).

Celsia L. = Verbascum L. (Scrophulariac.).

× **Celsioverbascum** K. H. Rech. & Hub.-Mor. Scrophulariaceae. Gen. hybr. (Celsia × Verbascum).

Celtidaceae Link = Ulmaceae–Celtideae Gaud.

Celtidopsis Priemer = Celtis L. (Ulmac.).

Celtis L. Ulmaceae. 80 N. hemisph., S. Afr. Like *Ulmus*, but intr. anthers, drupe, and curved embryo. Fr. of *C. australis* L. (nettle-tree) ed.; wood useful for turning; tree used as fodder in India.

Cembra (Spach) Opiz = Pinus L. (Pinac.).

Cenarium L. = Canarium L. (Burserac.).

Cenarrhenes Labill. Proteaceae. 1 Tasmania.

Cenchropsis Nash = seq.

Cenchrus L. Gramineae. 25 trop. & warm temp. Spikelet surrounded by invol. of sterile spikelets, which in some spp. become hard and prickly, surrounding the fr. and acting as a means of distribution by animals (cf. *Tribulus*, etc.). *C. tribuloïdes* L. is a very troublesome pest in the wool-growing districts of N. Am.

Čeněkia Opiz = Campanula L. (Campanulac.).

Cenesmon Gagnep. = Cnesmone Bl. (Euphorbiac.).

Cenia Comm. ex Juss. = Lancisia Fabr. (Compos.).

Cenocentrum Gagnep. Malvaceae. 1 Indochina.

Cenocline C. Koch = Cotula L. (Compos.).

Cenolophium Koch. Umbelliferae. 1–2 E. temp. & arct. Eur., Sib., C. As.

Cenolophon Blume. Zingiberaceae. ? 10 Indoch., Malaysia.

Cenopleurum P. & K. = Kenopleurum Candargy = Ferula L. (Umbellif.).

Cenostigma Tul. Leguminosae. 5 Brazil, Paraguay.

Cenothus Rafin. = Ceanothus L. (Rhamnac.).

Cenotis Rafin. = seq.

Cenotus Rafin. = Caenotus Rafin. = Erigeron L. (Compos.).

Centaurea L. Compositae. 600 Eur. & N. Afr. to N. India & N. China; temp. N. & S. Am.; 1 Austr. In *C. scabiosa* L. and *C. cyanus* L. the outer fls. are neuter with enlarged C (cf. *Hydrangea*). *C. calcitrapa* L. (star-thistle) has long spiny invol. brs. The fl. of *C.* shows the usual construction but the sta. are sensitive to contact and when touched (e.g. by insects probing) contract, thus forcing out the pollen at the top of tube. In *C. montana* L. and others there is a nectary on each bract of the invol. Numbers of ants are thus attracted.

Centaurella Delarbre = Centaurium Hill (Gentianac.).

Centaurella Michx = Bartonia Muhlenb. (Gentianac.).

× **Centaureopappus** Hort. ex Möllers. Compositae. Gen. hybr. (Aëtheopappus × Centaurea).

Centauria L. = Centaurea L. (Compos.).

Centauridium Torr. & Gray = Xanthisma DC. (Compos.).

CENTAURION

Centaurion Adans. = Centaurium Hill (Gentianac.).
Centaurium Borckh. = Canscora Lam. (Gentianac.).
Centaurium Cass. = seq.
Centaurium Haller = Centaurea L. (Compos.).
Centaurium Hill. Gentianaceae. 40–50 cosmop. (excl. trop. & S. Afr.).
Centaurium Pers. = Centaurella Michx = Bartonia Muhlenb. (Gentianac.).
Centaurodendron Johow. Compositae. 1 Juan Fernandez.
Centauropsis Boj. ex DC. Compositae. 10 Madag.
× **Centaurserratula** Arènes. Compositae. Gen. hybr. (Centaurea × Serratula).
Centella L. Hydrocotylaceae (~ Umbellif.). 40 Afr., Austr., N.Z., Am.
Centema Hook. f. Amaranthaceae. 2 trop. Afr.
Centemopsis Schinz. Amaranthaceae. 6 trop. Afr.
Centhriscus Spreng. ex Steud. (sphalm.) = Anthriscus Pers. (Umbellif.).
Centinodia (Reichb.) Reichb. = Polygonum L. (Polygonac.).
Centinodium (Reichb.) Montandon = praec.
Centipeda Lour. Compositae. 6 Madag., Afgh., E. As., Indomal., Austr., N.Z., Polynesia, Chile.
Centopodium Burch. = Emex Neck. ex Campd. (Polygonac.).
Centotheca Desv. Gramineae. 4 trop. Afr., As., Polynesia.
Centrachena Schott = Chrysanthemum L. (Compos.).
Centradenia G. Don. Melastomataceae. 7 Mexico, C. Am. *C. rosea* Lindl. shows anisophylly.
Centradeniastrum Cogn. Melastomataceae. 2 W. trop. S. Am.
Centrandra Karst. = Julocroton Mart. (Euphorbiac.).
Centranthera R.Br. Scrophulariaceae. 9 China, Indomal., Austr.
Centranthera Scheidw. = Pleurothallis R.Br. (Orchidac.).
Centrantheropsis Bonati. Scrophulariaceae. 1 China.
Centranthus DC. Valerianaceae. 12 Eur., Medit. C spurred at the base; at the end of the spur honey is secreted. The tube of the C has a partition dividing it into two, one containing the style, the other, lined with downward-pointing hairs, leading to the spur. Fl. protandr.; only long-tongued insects can obtain honey.
Centrapalus Cass. = Vernonia Schreb. (Compos.).
Centratherum Cass. Compositae. 20 trop.
Centridobotryon Klotzsch ex Pfeiff. = seq.
Centridobryon Klotzsch ex Pfeiff. = Piper L. (Piperac.).
Centrilla Lindau = Justicia L. (Acanthac.).
Centrocarpha D. Don = Rudbeckia L. (Compos.).
Centrochilus J. C. Schau. = Platanthera Rich. (Orchidac.).
Centrochloa Swallen. Gramineae. 1 Brazil.
Centrochrosia P. & K. = Kentrochrosia Lauterb. & Schum. (Apocynac.).
Centroclinium D. Don = Onoseris DC. (Compos.).
Centrodiscus Muell. Arg. = Caryodendron Karst. (Euphorbiac.).
Centrogenium Schlechter. Orchidaceae. 11 trop. Am., W.I.
Centroglossa Barb. Rodr. Orchidaceae. 7 Brazil, Peru, Paraguay.
Centrogonium Willis (sphalm.) = Centrogenium Schltr. (Orchidac.).
Centrogyne Welw. ex Benth. & Hook. f. = Bosqueia Thou. ex Baill. (Morac.).
***Centrolepidaceae** Desv. Monocots. 5/40 SE. As. to Australas. Small ann.

or perenn., rush-, sedge-, or grass-like herbs, with scapose distich. spikes or heads of fls. subtended by glume-like bracts. Fls. small, ♂ or ♂ ♀, naked or surr. by 1–3 trichomatous bracts. P o; A 1–2, fil. often ± twisted, anth. 1- or 2-thecous; G̲ 1–∞, ± stipit., often concresc. to varying heights and obliquely or spirally superposed, each carp. membr., saccate, with 1 pend. orthotr. ov., styles filif., simple, free or variously united. Genera: *Centrolepis, Aphelia, Trithuria, Hydatella, Gaimardia*. For an interpretation of the floral morphology see Hamann, *Ber. Deutsch. Bot. Ges.* **75**: 165–9, 1962. The 'flowers' are probably pseudanthia.

Centrolepis Labill. Centrolepidaceae. 25 Hainan, Indoch., Malaysia (mts.), Australas.

Centrolobium Mart. ex Benth. Leguminosae. 7 trop. Am. Pod winged. *C. robustum* Mart. yields good timber (zebra wood).

Centromadia Greene = Hemizonia DC. (Compos.).

Centronia Blume = Aeginetia L. (Orobanchac.).

Centronia D. Don. Melastomataceae. 15 C. & W. trop. Am., 1 Guiana.

Centronota DC. = Aeginetia L. (Orobanchac.).

Centropappus Hook. f. = Senecio L. (Compos.).

Centropetalum Lindl. Orchidaceae. 10 W. trop. S. Am.

Centrophorum Trin. = Chrysopogon Trin. (Gramin.).

Centrophyllum Dum. = Carthamus L. (Compos.).

Centrophyta Reichb. = Kentrophyta Nutt. = Astragalus L. (Legumin.).

Centroplacus Pierre. Pandaceae. 1 trop. W. Afr.

Centropodia (R.Br.) Reichb. = Danthonia DC. (Gramin.).

Centropodium Lindl. = Centopodium Burch. = Emex Campd. (Polygonac.).

Centropogon Presl. Campanulaceae. 230 trop. Am., W.I.

Centropsis Endl. = Kentropsis Moq. = Sclerolaena R.Br. (Chenopodiac.).

*Centrosema (DC.) Benth. Leguminosae. 45 Am.

Centrosepis R. Hedw. (sphalm.) = Centrolepis Labill. (Centrolepidac.).

Centrosia A. Rich. = seq.

Centrosis Sw. = Limodorum Rich. (Orchidac.).

Centrosis Thou. = Calanthe R.Br. (Orchidac.).

Centrosolenia Benth. Gesneriaceae. 7 Mex., C. Am.

Centrospermae Eichler. One of the more natural and distinctive orders of Dicotyledons, comprising the families *Achatocarpac., Agdestidac., Aïzoac., Amaranthac., Barbeuiac., Basellac., Batidac., Cactac., Caryophyllac., Chenopodiac., Didiereac., Dysphaniac., Gyrostemonac., Halophytac., Nyctaginac., Phytolaccac., Portulacac., Theligonac.*

Usu. herbs, more rarely shrubs or trees, with spiral or cyclic homo- or hetero-chlam. fls.; A usu. = and opp. P, but also ∞–1; G (∞–1) or free, rarely G̲, usu. 1-loc. with ∞–1 campylotr. ov.; perisperm. (C, when pres., represents sterile sta.) The group includes many succulents and halophytes. The red pigments are due to 'betacyanin', not to anthocyanins, which are unknown in the order. Certain other fams. may bear a 'marginal' or 'peripheral' relationship to the *Centrospermae*: e.g. *Fouquieriac., Frankeniac., Tamaricac., Elatinac., Polygonac., Plumbaginac., Primulac., Sphenocleac.*

Centrospermum Kunth = Acanthospermum Schrank (Compos.).

Centrospermum Spreng. = Chrysanthemum L. (Compos.).

CENTROSPHAERA

Centrosphaera P. & K. = Kentrosphaera Volk. = Volkensinia Schinz (Amaranthac.).

Centrostachys Wall. Amaranthaceae. 1 N. Afr., India, Java, Norfolk Is.

Centrostegia A. Gray (~ Chorizanthe R.Br.). Polygonaceae. 3 California.

Centrostemma Baill. (sphalm.) = Ceratostema Juss. (Ericac.).

Centrostemma Decne. Asclepiadaceae. 5 SE. As.

Centrostigma Schlechter. Orchidaceae. 3 trop. Afr.

Centrostylis Baill. = Adenochlaena Baill. (Euphorbiac.).

Centunculus Adans. = Cerastium L. (Caryophyllac.).

Centunculus L. = Anagallis L. (Primulac.).

Ceodes J. R. & G. Forst. Nyctaginaceae. 25 Mascarene Is., Malaysia, Austr., Polynesia.

Cepa Kuntze = Eurycles Salisb. (Liliac.).

Cepa Mill. = Allium L. (Alliac.).

Cepaea Fabr. (uninom.) = Sedum L. sp. (Crassulac.).

Cepae[ac]eae Salisb. = Alliaceae J. G. Agardh.

Cepalaria Rafin. = Cephalaria Schrad. (Dipsacac.).

*****Cephaëlis** Sw. Rubiaceae. 180 trop. C. ipecacuanha (Stokes) Baill. is the ipecacuanha (Brazil), a herb with decumbent stem, and roots thickened somewhat like rows of beads. Root used in medicine.

Cephalacanthus Lindau. Acanthaceae. 1 Peru.

Cephalandra Schrad. ex Eckl. & Zeyh. = Coccinia Wight & Arn. (Cucurbitac.).

Cephalangraecum Schlechter = Ancistrorhynchus Finet (Orchidac.).

Cephalanoplos Neck. = Cirsium Mill. + Serratula L. (Compos.).

Cephalanthera Rich. Orchidaceae. 14 N. temp. No rostellum; the pollen germinates in situ, fertilising its own stigma. The lat. stds. (see fam.) are easily seen. Darwin regarded C. as a degraded Epipactis (cf. × Cephalopactis).

Cephalantheropsis Guillaumin. Orchidaceae. 1 Indoch.

Cephalanthus L. Naucleaceae (~ Rubiac.). 17 warm Am., As.

Cephalaralia Harms. Araliaceae. 1 Austr.

*****Cephalaria** Schrad. ex Roem. & Schult. Dipsacaceae. 65 Medit. reg. to C. As., S. Afr.

Cephaleis Vahl = Cephaëlis Sw. (Rubiac.).

× **Cephalepipactis** E. G. Camus, Bergon & A. Camus (vii) = × Cephalopactis Aschers. & Graebn. (Orchidac.).

Cephalidium A. Rich. = Anthocephalus A. Rich. (Naucleac.).

Cephalina Thonn. = Sarcocephalus Afzel. (Naucleac.).

Cephalipterum A. Gray. Compositae. 1 W. & S. Austr.

Cephalobembix Rydb. Compositae. 1 Mexico.

Cephalocarpus Nees. Cyperaceae. 7 trop. S. Am. Habit of Dracaena.

Cephaloceraton Gennari = Isoëtes L. (Isoëtac.).

Cephalocereus Pfeiff. Cactaceae. Sensu Britt. & Rose, 48 Florida to Brazil; sensu Backeb., 1 Mex.

Cephalochloa Coss. & Dur. = Ammochloa Boiss. (Gramin.).

Cephalocleistocactus F. Ritter (~ Cleistocactus Lem.). Cactaceae. 3 Bolivia.

Cephalocroton Hochst. Euphorbiaceae. 8 trop. Afr.

Cephalocrotonopsis Pax. Euphorbiaceae. 1 Socotra.

Cephalodes St-Lag. = Cephalaria Schrad. ex Roem. & Schult. (Dipsacac.).

CEPHALOTAXACEAE

Cephalohibiscus Ulbr. Malvaceae. 1 New Guinea, Solomon Is.
Cephaloma Neck. = Dracocephalum L. (Labiat.).
Cephalomammillaria Frič = Epithelantha Weber (Cactac.).
Cephalomanes Presl. Hymenophyllaceae. 10 India to Polynesia.
Cephalomappa Baill. Euphorbiaceae. 5 S. China, W. Malaysia.
Cephalomedinilla Merr. = Medinilla Gaud. (Melastomatac.).
Cephalonema K. Schum. = Clappertonia Meissn. (Tiliac.).
Cephalonoplos Fourr. [Cephalanoplos Neck.] = Cirsium Mill. (Compos.).
× **Cephalopactis** A'schers. & Graebn. Orchidaceae. Gen. hybr. (i) (Cephalanthera × Epipactis).
Cephalopappus Nees & Mart. Compositae. 1 E. Brazil.
Cephalopentandra Chiov. Cucurbitaceae. 1 NE. trop. Afr.
Cephalophilum (Meissn.) Börner = Tasoba Rafin. = Polygonum L. (Poly-
Cephalophora Cav. = Helenium L. (Compos.). [gonac.).
Cephalophorus Lem. ex Boom = Cephalocereus Pfeiff. (Cactac.).
× **Cephalophrys** hort. Orchidaceae. Gen. hybr. (vi) (Cephalanthera × Ophrys).
Cephalophyllum Haw. Aïzoaceae. 70 S. Afr.
Cephalophyton Hook. f. ex Baker = Thonningia Vahl (Balanophorac.).
× **Cephalopipactis** Willis (sphalm.) = × Cephalopactis Aschers. & Graebn. (Orchidac.).
Cephalopterum P. & K. = Cephalipterum A. Gray (Compos.).
Cephalorhizum Popov & Korovin. Plumbaginaceae. 4 C. As.
Cephalorrhynchus Boiss. Compositae. 6 SW. As.
Cephaloschefflera (Harms) Merr. Araliaceae. 9 Philipp. Is.
Cephaloschoenus Nees = Rhynchospora Vahl (Cyperac.).
Cephaloscirpus Kurz = Mapania Aubl. (Cyperac.).
Cephaloseris Poepp. ex Reichb. = Polyachyrus Lag. (Compos.).
Cephalosorus A. Gray = Angianthus Wendl. (Compos.).
Cephalosphaera Warb. Myristicaceae. 1 trop. Afr.
Cephalostachyum Munro. Gramineae. 12 Indomal., Madag.
Cephalostemon R. Schomb. Rapateaceae. 5 trop. S. Am.
Cephalostigma A. DC. Campanulaceae. 10 trop.
Cephalosurus C. Muell. (sphalm.) = Cephalosorus A. Gray = Angianthus Wendl. (Compos.).
***Cephalotaceae** Dum. Dicots. 1/1 W. Austr. An interesting pl. with pitchers like *Nepenthes* or *Sarracenia*, though not nearly related to either. The lower ls. of the rosette form pitchers, the upper are flat and green (cf. this division of labour with that in *N.* and *S.*), the woody rhiz. annually producing both. The pitcher has much the structure of *N.* and catches insects in the same way. Infl. a small adpr. hairy thyrse borne on a long scape. Fl. ⚥, apetalous, reg.; K 6, valvate; A 6 + 6, connective large, globose, fungose-glandular; disk broad, thick, green, papillose; G̲ 6, or ± united, hairy, each with 1 (rarely 2) basal erect anatr. ov. with dorsal raphe, and short recurved subulate style with simple stigma. Follicle with 1 seed; embryo small in fleshy endosp. Only genus: *Cephalotus*. Perhaps distantly rel. to *Saxifragac.*
***Cephalotaxaceae** Neger. Gymnosp. (Conif.). 1/7 E. Himal. to Japan. Trees or shrubs with densely leafy twigs. All shoots of unlimited growth. Ls.

CEPHALOTAXUS

narrow-lin., distich. Fls. dioec. ♂ infl. short-pedunc., ± glob., in the axils of ls. of preceding year. A up to 12, with short fil. and usu. 3 sporangia (pollensacs). ♀ infl. 2–4 together on bracteate short shoots, wh. are borne in the axils of the subterm. or ± distal ls. of long shoots (the infl. being accompanied by a veg. branchlet which later grows out), short-pedunc., bearing several decussate pairs of biovuliferous scales; only 1–2 seeds devel. in each infl., large, with fleshy outer ('aril') and thin woody inner layer; embr. large, with 2 cots. Only genus: *Cephalotaxus*.

Cephalotaxus Sieb. & Zucc. Cephalotaxaceae. 4–7 E. As. Plum yews.
Cephalotomandra Karst. & Triana. Nyctaginaceae. 1–2 Panamá, Colombia.
Cephalotos Adans. = Thymus L. (Labiat.).
Cephalotrophis Blume = Malaisia Blanco (Morac.).
Cephalotus Labill. Cephalotaceae. 1, *C. follicularis* Labill., in damp sandy or peaty ground in southern W. Austr.
Cephaloxis Desv. = seq.
Cephaloxys Desv. = Juncus L. (Juncac.).
Ceradia Lindl. = Othonna L. (Compos.).
Ceraia Lour. = Dendrobium Sw. (Orchidac.).
Ceramanthe (Reichb.) Dum. = Scrophularia L. (Scrophulariac.).
Ceramanthus Hassk. = Phyllanthus L. (Euphorbiac.).
Ceramanthus Malme = Philibertia Kunth, Pentacyphus Schltr., Funastrum Fourn. & Sarcostemma auctt. (non R.Br.) (Asclepiadac.).
Ceramanthus P. & K. = Keramanthus Hook. f. = Adenia Forsk. (Passiflorac.).
Ceramia D. Don = Erica L. (Ericac.).
Ceramicalyx Blume = Osbeckia L. (Melastomatac.).
Ceramiocephalum Sch. Bip. = Crepis L. (Compos.).
Ceramium Blume = Apama Lam. (Aristolochiac.).
Ceramocalyx P. & K. = Ceramicalyx Bl. = Osbeckia L. (Melastomatac.).
Ceramocarpium Nees ex Meissn. = Ocotea Aubl. (Laurac.).
Ceramocarpus Wittst. = Keramocarpus Fenzl = Coriandrum L. (Umbellif.).
Ceramophora Nees ex Meissn. = Ocotea Aubl. (Laurac.).
Ceranthe (Reichb.) Opiz = Cerinthe L. (Boraginac.).
Ceranthera Beauv. = Rinorea Aubl. (Violac.).
Ceranthera Ellis. Labiatae. 2 SE. U.S.
Ceranthera Endl. = Ceratanthera Hornem. = Colebrookia Sm. (Labiat.).
Ceranthera Rafin. = Solanum L. (Solanac.).
Cerantheraceae Dulac = Ericaceae Juss. (excl. Vaccinioïdeae–Vaccinieae).
Ceranthus Schreb. = Linociera Sw. (Oleac.).
× **Cerapadus** Buia. Rosaceae. Gen. hybr. (Cerasus × Padus).
Ceraria Pearson & E. L. Stephens. Portulacaceae. 5 trop. & S. Afr.
Ceraseidos Sieb. & Zucc. = Prunus L. (Rosac.).
Ceraselma Wittst. = Keraselma Neck. = Euphorbia L. (Euphorbiac.).
Cerasiocarpum Hook. f. = Kedrostis Medik. (Cucurbitac.).
Cerasiocarpus P. & K. = praec.
Cerasites Steud. = Cerastites S. F. Gray = Papaver L. (Papaverac.).
Cerasophora Neck. = Prunus L. (Rosac.).
Cerastites S. F. Gray = Papaver L. (Papaverac.).
Cerastium L. Caryophyllaceae. 60 almost cosmop.

Cerasus Mill. (~ Prunus L.). Rosaceae. 140 N. temp.
Ceratandra Eckl. ex Bauer. Orchidaceae. 6 S. Afr.
Ceratandropsis Rolfe = praec.
Ceratanthera Hornem. = Globba L. (Zingiberac.).
Ceratanthus F. Muell. Labiatae. 6 SE. As., New Guinea, Queensl.
Ceratella Hook. f. = Abrotanella Cass. (Compos.).
Ceratia Adans. = Ceratonia L. (Legumin.).
Ceratiola Michx. Empetraceae. 1 SE. U.S.
Ceratiosicyos Nees. Achariaceae. 1 S. Afr.
Ceratistes Hort. = Eriosyce Phil. (Cactac.).
Ceratites Soland. ex Miers. Apocynaceae. 1 SE. Brazil.
Ceratium Blume = Eria Lindl. (Orchidac.).
Ceratocalyx Coss. = Orobanche L. (Orobanchac.).
Ceratocapnos Dur. Fumariaceae. 2 NW. Afr., Syria & Palest.
Ceratocarpus Dur. = praec.
Ceratocarpus L. Chenopodiaceae. 2 E. Eur., temp. As.
Ceratocaryum Nees. Restionaceae. 2 S. Afr.
Ceratocaulos Reichb. = Datura L. (Solanac.).
Ceratocephala Moench (~ Ranunculus L.). Ranunculaceae. 2 C. Eur., Medit. to C. As. & Himal.
Ceratocephalus Cass. = Bidens L. (Compos.).
Ceratocephalus Kuntze = Spilanthes Jacq. (Compos.).
Ceratocephalus Pers. = Ceratocephala Moench (Ranunculac.).
Ceratochaete Lunell = Zizania L. (Gramin.).
Ceratochilus Blume. Orchidaceae. 1 Malaysia.
Ceratochilus Lindl. = Stanhopea Frost ex Hook. (Orchidac.).
Ceratochloa Beauv. = Bromus L. (Gramin.).
Ceratocnemum Coss. & Balansa. Cruciferae. 1 Morocco.
Ceratococca Willd. ex Roem. & Schult. = Microtea Sw. (Chenopodiac.?).
Ceratococcus Meissn. = Pterococcus Hassk. (Euphorbiac.).
Ceratodactylis J. Sm. = Llavea Lagasca (Cryptogrammac.).
Ceratodes Kuntze = Ceratocarpus L. (Chenopodiac.).
Ceratodiscus Dur. & Jacks. = Corallodiscus Batalin (Gesneriac.).
Ceratogonon Meissn. = Oxygonum Burch. (Polygonac.).
Ceratogonum C. A. Mey. = praec.
Ceratogyne Turcz. Compositae. 1 W. temp. Austr.
Ceratogynum Wight = Sauropus Blume (Euphorbiac.).
Ceratoïdes Gagnebin = Axyris L. (Chenopodiac.).
Ceratolacis (Tul.) Wedd. Podostemaceae. 1 Brazil.
Ceratolepis Cass. = Pamphalea DC. (Compos.).
Ceratolobus Blume. Palmae. 5 W. Malaysia.
Ceratominthe Briq. (~ Satureja § Xenopoma). Labiatae. 10 Andes.
Ceratonia L. Leguminosae. 1 Medit., *C. siliqua* L. (carob-tree). The pods (*algaroba*, St John's bread) are full of juicy pulp containing sugar and gum, and are used for fodder. The seeds are said to have been the original of the carats of jewellers.
Ceratoniaceae Link = Leguminosae–Caesalpinioïdeae–Cassieae Bronn.
Ceratonychia Edgew. = Cometes L. (Caryophyllac.).

CERATOPETALUM

Ceratopetalum Smith. Cunoniaceae. 5 New Guinea, E. Austr. Light timber.

Ceratophorus Hassk. = Payena L. (Sapotac.).

Ceratophorus Sond. = Suregada Roxb. ex Rottl. (Euphorbiac.).

*Ceratophyllaceae S. F. Gray. Dicots. 1/10 cosmop. Water pl., rootless, with thin stems and whorls of dichotomously divided submerged ls. The pl. decays behind as it grows in front, so that veg. repr. occurs by the setting free of the branches. The old ls. are translucent and horny, whence the name. Winter buds are not formed, the pl. merely sinking in autumn and rising in spring. Fls. ♂♀, monoec., axillary, sessile, with sepaloid P. In the ♂, P about (12), subvalv.; A 12–16 on convex recept., with elongate subsess. anth. and oval non-cutinised pollen. In the ♀, P (9–10), hypog.; G 1, the midrib anterior; ovule 1, orthotr., pend. Achene crowned by the persistent style, which in *C. demersum* L. is hooked. Endosp. Fl. water-pollin.; the anthers break off and float up through the water (each has a sort of float at top of theca); the pollen is of the same specific gravity as water (cf. *Zostera*) and drifts about till it reaches a stigma.

Only genus: *Ceratophyllum*. As often with water plants, it is difficult to decide upon a position for the *C.* in the classification. The one free cpl. and several P leaves seem to place them in *Ranales*, where they are perhaps remotely related to *Cabombaceae*.

Ceratophyllum L. Ceratophyllaceae. 10 cosmop.

Ceratophytum Pittier. Bignoniaceae. 1 Venez.

Ceratopsis Lindl. = Epipogium S. G. Gmel. (Orchidac.).

Ceratopteridaceae Maxon = Parkeriaceae Hook.

Ceratopteris Brongn. Parkeriaceae. 2 trop., subtrop. Aquatic, edible; sporangia solitary.

Ceratopyxis Hook. f. Rubiaceae. 1 Cuba.

Ceratosanthes Burm. ex Adans. Cucurbitaceae. 1 W.I. to Braz.

Ceratosanthus Schur = Delphinium L. (Ranunculac.).

Ceratoschoenus Nees = Rhynchospora Vahl (Cyperac.).

Ceratoscyphus Chun = Chirita Ham. ex D. Don (Gesneriac.).

Ceratosepalum Oerst. = Passiflora L. (Passiflorac.).

Ceratosepalum Oliv. Tiliaceae. 1 trop. E. Afr.

Ceratosicyus P. & K. = Ceratiosicyos Nees (Achariac.).

Ceratospermum Pers. = Eurotia Adans. (Chenopodiac.).

Ceratostachys Blume = Nyssa Linn. (Nyssac.).

Ceratostanthus B. D. Jacks. (sphalm.) = Ceratosanthus Schur = Delphinium L. (Ranunculac.).

Ceratostema G. Don = Pellegrinia Sleumer (Ericac.).

Ceratostema Juss. Ericaceae. 18 Andes of trop. S. Am.

Ceratostemma Spreng. (sphalm.) = praec.

Ceratostigma Bunge (~Plumbago L.). Plumbaginaceae. 8 E. trop. Afr., S. & E. Tibet, Himal., China, Burma, Siam. The total infl. is racemose, the partials dichasial.

Ceratostylis Blume. Orchidaceae. 60 Indomal., Polynesia.

Ceratotheca Endl. Pedaliaceae. 9 trop. & S. Afr.

Ceratoxalis Lunell = Oxalis L. (Oxalidac.).

Ceratozamia Brongn. Zamiaceae. 4 Mexico.
Ceraunia Nor. = Aegiceras Gaertn. (Myrsinac.).
Cerbera L. Apocynaceae. 6 trop. coasts of Indian and W. Pacif. Oceans. The floating frs. are familiar on the coast. Ls. alt.
Cerbera Lour.: descr. = Cerbera L. (Apocynac.); specim. = Scaevola L. (Goodeniac.).
Cerberiopsis Vieill. Apocynaceae. 3 New Caled.
Cercaceae Dulac = Ceratophyllaceae S. F. Gray.
Cercanthemum Van Tiegh. = Ouratea Aubl. (Ochnac.).
Cercestis Schott. Araceae. 9 W. Afr.
Cercidiopsis Britton & Rose = Cercidium Tul. (Legumin.).
*Cercidiphyllaceae Van Tiegh. Dicots. 1/1 E. As. Decid. tree; ls. usu. opp. on long shoots, alt. on short shoots, ± cord.-orbic., cren.-serr., with lanc. stips. adnate to pet. Fls. dioec.: ♂ subsess., axill., sol. or fascic., K 4, C 0, A 15–20, fil. elong.-filif., anth. obl.-lin.; ♀ pedic., K 4, C & A 0, G 4–6 substipit., with filif. style and ∞ biseriate pend. anatr. ov. Fr. a cluster of dehisc. follicles, with woody endoc.; seeds compr., subquadr., winged, endosp. copious. Only genus: *Cercidiphyllum*. Allied to *Hamamelidac.*
Cercidiphyllum Sieb. & Zucc. Cercidiphyllaceae. 1 China & Japan. Useful
Cercidium Tul. Leguminosae. 10 warm Am.   [wood.
Cercidophyllum P. & K. = Cercidiphyllum Sieb. & Zucc. (Cercidiphyllac.).
Cercinia Van Tiegh. = Ouratea Aubl. (Ochnac.).
Cercis L. Leguminosae. 7 N. temp. *C. siliquastrum* L. is the Judas-tree (Judas is said to have hanged himself on one). The fls. appear before the ls., in bunches on the older twigs, and have a very papilionaceous look, the two lower pets. enclosing the essential organs. Serial buds in the axils. Good wood.
Cercocarp[ac]eae J. G. Agardh = Rosaceae–Rosoïdeae–Cercocarpeae Torr. & Gray.
Cercocarpus Kunth. Rosaceae. 20 W. & SW. U.S., Mex.
Cercocodia P. & K. = Cercodia Murr. = Haloragis J. R. & G. Forst. (Haloragidac.).
Cercocoma Miq. = Rhynchodia Benth. (Apocynac.).
Cercocoma Wall. = Strophanthus DC. (Apocynac.).
Cercodea Soland. ex Lam. = seq.
Cercodia Murr. = Haloragis J. R. & G. Forst. (Haloragidac.).
Cercodiaceae ('-anae') Juss. = Haloragidaceae R.Br.
Cercopetalum Gilg = Pentadiplandra Baill. (Pentadiplandrac.).
Cercophora Miers (1874; non Fuckel 1870—Fungi) = Strailia Th. Dur. (Lecythidac.).
Cercostylos Less. = Gaillardia Fouger. (Compos.).
Cercouratea Van Tiegh. = Ouratea Aubl. (Ochnac.).
Cercus Rafin. = Cereus L. (Cactac.).
Cerdana Ruiz & Pav. = Cordia L. (Ehretiac.).
Cerdia Moç. & Sessé. Caryophyllaceae. 4 Mex.
Cerdosurus Ehrh. (uninom.) = *Alopecurus agrestis* L. = *A. myosuroïdes* Huds.
Cerea Thou. = Elaeocarpus L. (Elaeocarpac.).   [(Gramin.)
Cerefolium Fabr. = Anthriscus Pers. emend. Hoffm. (Umbellif.).
Cereopsis Rafin. = Coreopsis L. (Compositae).

CERESIA

Ceresia Pers. = Paspalum L. (Gramin.).

Cereus Mill. Cactaceae. 50 S. Am., W.I. (Formerly incl. almost all the ribbed columnar cacti; restricted by Britton & Rose, *Cactaceae*, 1920. Segregates incl. *Carnegiea, Hylocereus, Selenicereus*, etc.)

Cerinozoma P. & K. = Kerinozoma Steud. ex Zoll. (Gramin.).

Cerinthe L. Boraginaceae. 10 Eur., Medit.

Cerinthodes Kuntze = Mertensia Roth (Boraginac.).

Cerinthopsis Kotschy ex Paine = Lindelofia Lehm. (Boraginac.).

Cerionanthus Schott ex Roem. & Schult. = Cephalaria Schrad. ex Roem. & Schult. (Dipsacac.).

Ceriops Arn. Rhizophoraceae. 2 trop. coasts of Indian & W. Pacif. oceans.

Ceriscus Gaertn. = Randia L. (Rubiac.).

Ceriscus Nees = Tarenna Gaertn. (Rubiac.).

Cerium Lour. = Lysimachia L. (Primulac.).

Cerocarpus Colebr. ex Hassk. = Syzygium Gaertn. (Myrtac.).

Cerochilus Lindl. = Hetaeria Blume (Orchidac.).

Cerochlamys N. E. Brown. Aïzoaceae. 1 S. Afr.

Cerolepis Pierre = Camptostylus Gilg (Flacourtiac.).

Ceropegia L. Asclepiadaceae. 160 Canary Is., trop. & S. Afr., Madag., trop. & subtrop. As. (Arabia to New Guin.), N. Queensl. Erect or twining herbs or undershrubs, ± xero. Many have tuberous rootstocks, others are leafless and sometimes have fleshy *Stapelia*-like stems. The fls. form a trap like *Aristolochia clematitis*. The C-tube widens at the base and at the top the teeth spread out, but in some they hold together at the tips, making a sort of umbrella. The tube is lined with downward-pointing hairs, and small flies, attracted by the colour and smell, creep into the fl. and cannot escape till the hairs wither, when they emerge with pollinia on their proboscides.

Cerophora Rafin. = Myrica L. (Myricac.).

Cerophyllum Spach = Ribes L. (Grossulariac.).

Ceropteris Link = Pityrogramma Link (Gymnogrammac.).

Cerosora (Bak.) Domin. Hemionitidaceae. 2 Sumatra, Borneo. (*Kew Bull.* 13: 450, 1959.)

Cerothamnus Tidestrom = Myrica L. (Myricac.).

Ceroxylaceae O. F. Cook = Palmae–Iriarteëae Benth. & Hook. f.

Ceroxylon Humb. & Bonpl. Palmae. 20 N. Andes. *C. andicola* Humb. & Bonpl. and others secrete wax on the stems; it is used for gramophone discs, candles, etc.

Cerqueiria Berg = Myrcia DC. (Myrtac.).

Cerquieria Benth. & Hook. f. = praec.

Cerraria Tausch (sphalm.) = Cervaria L. = Peucedanum L. (Umbellif.).

Cerris Rafin. = Quercus L. (Fagac.).

Ceruana Forsk. Compositae. 1 Egypt, trop. Afr.

Ceruchis Gaertn. ex Schreb. = Spilanthes Jacq. (Compos.).

Cervantesia Ruiz & Pav. Santalaceae. 5 Andes.

Cervaria L. = Ortegia Loefl. ex L. (Caryophyllac.).

Cervaria v. Wolf = Peucedanum L. (Umbellif.).

Cervia Rodr. ex Lag. Boraginaceae. 1 Spain, Balkan Penins.

Cervicina Delile = Wahlenbergia Schrad. ex Roth (Campanulac.).

Cervispina Ludw. = Rhamnus L. (Rhamnac.).
Cesatia Endl. = Trachymene Rudge (Umbellif.).
Cesdelia DC. ex Rafin. (= ? Cornelia Ard.) = Ammannia L. (Lythrac.).
Cespa Hill = Eriocaulon L. (Eriocaulac.).
Cespedesia Goudot. Ochnaceae. 6 trop. S. Am.
Cestichis Pfitzer [Coestichis Thou.] = Liparis Lindl. (Orchidac.).
Cestraceae ('-ineae') Schlechtd. = Solanaceae–Cestreae Schlechtd.
Cestrinus Cass. = Centaurea L. (Compos.).
Cestrum L. Solanaceae. 150 warm Am., W.I.
Ceterac Adans. = Asplenium L. (Aspleniac.).
***Ceterach** DC. Aspleniaceae. 3 Old World.
Ceterachopsis Ching. Aspleniaceae. 2 E. Himal. to W. China.
Cetra Nor. = Syzygium Gaertn. (Myrtac.) + ?
Cevallia Lag. Loasaceae. 1 SW. U.S., Mex. Connective with long process; G 1 with one pend. ov.
Cevalliaceae Griseb. = Loasaceae–Gronovieae Gilg.
Chaboissaea Fourn. = Muhlenbergia Schreb. (Gramin.).
Chabraea Adans. = Peplis L. (Lythrac.).
Chabraea DC. = Leucheria Lag. (Compos.).
Chabrea Rafin. = Peucedanum L. (Umbellif.).
Chacaya Escalante. Rhamnaceae. 1 Argentina.
Chadara Forsk. = Grewia L. (Tiliac.).
Chadsia Boj. Leguminosae. 18 Madag. Cauliflorous. Fls. scarlet.
Chaelanthus Poir. (sphalm.) = Chaetanthus R.Br. (Restionac.).
Chaelothilus Beck. = Gentiana L. (Gentianac.).
Chaenactis DC. Compositae. 40 W. U.S., NW. Mex. Ls. opp.
Chaenanthe Lindl. Orchidaceae. 1 Brazil.
Chaenanthera Rich. ex DC. = Charianthus D. Don (Melastomatac.).
Chaenarrhinum Reichb. = Chaenorhinum (DC.) Reichb. (Scrophulariac.).
Chaenesthes Miers = Iochroma Benth. (Solanac.).
Chaenocarpus Juss. [Chenocarpus Neck.] = Spermacoce L. (Rubiac.).
Chaenocephalus Griseb. Compositae. 12 W.I., S. Am.
Chaenolobium Miq. = Ormosia G. Jacks. (Legumin.).
Chaenolobus Small = Pterocaulon Ell. (Compos.).
Chaenomeles Lindl. Rosaceae. 3 E. As. (Chinese or Japanese quince).
Chaenopleura Rich. ex DC. = Miconia Ruiz & Pav. (Melastomatac.).
Chaenorrhinum (DC.) Reichb. Scrophulariaceae. 20 Eur., Medit., W. As. Post. cpl. larger.
***Chaenostoma** Benth. = Sutera Roth (Scrophulariac.).
Chaenotheca Urb. (1902; non Th. Fr. 1856—Lichenes) = Chascotheca Urb. = Securinega Juss. (Euphorbiac.). [bellif.).
Chaerefolium Haller = Cerefolium Fabr. = Anthriscus (Pers.) Hoffm. (Um-
Chaerophyllastrum Fabr. (uninom.) = Myrrhis Mill. sp. (Umbellif.).
Chaerophyllopsis Boissieu. Umbelliferae. 1 W. China.
Chaerophyllum L. Umbelliferae. 40 N. temp. *C. bulbosum* L. cult. for ed. root.
Chaetacanthus Nees. Acanthaceae. 8 S. Afr.
Chaetachlaena D. Don = Onoseris DC. (Compos.).

CHAETACHME

Chaetachme Planch. = seq.

Chaetacme Planch. corr. Planch. Ulmaceae. 4 trop. & S. Afr., Madag.

Chaetadelpha A. Gray. Compositae. 1 SW. U.S. Zigzag twigs.

Chaetaea Jacq. = Byttneria Loefl. (Sterculiac.).

Chaetaea P. & K. = Chaitaea Soland. ex Seem. = Tacca J. R. & G. Forst. (Taccac.).

Chaetagastra Crueg. = Chaetogastra DC. = Tibouchina Aubl. (Melastomatac.).

Chaetantera Less. = Chaetanthera Ruiz & Pav. (Compos.).

Chaetanthera Nutt. = Chaetopappa DC. (Compos.).

Chaetanthera Ruiz & Pav. Compositae. 50 Peru, Chile. Cushion pl.

Chaetanthus R.Br. Restionaceae. 1 SW. Austr.

Chaetaria Beauv. = Aristida L. (Gramin.).

Chaethymenia Hook. & Arn. = Jaumea Pers. (Compos.).

Chaetium Nees. Gramineae. 3 trop. Am., Cuba. Glume with callus.

Chaetobromus Nees. Gramineae. 3 S. Afr.

Chaetocalyx DC. Leguminosae. 20 warm Am., W.I. Twining herbs.

Chaetocapnia Sweet = Coetocapnia Link & Otto = Polianthes L. (Agavac.).

Chaetocarpus Schreb. = Pouteria Aubl. (Sapotac.).

*Chaetocarpus Thw. Euphorbiaceae. 10 trop. (exc. E. Malaysia, Austr., Pacif.).

Chaetocephala B. Rodr. Orchidaceae. 2 Brazil.

Chaetochilus Vahl = Schwenckia L. (Solanac.).

Chaetochlaena P. & K. = Chaetachlaena D. Don = Onoseris DC. (Compos.).

Chaetochlamys Lindau. Acanthaceae. 7 trop. S. Am.

Chaetochloa Scribn. = Setaria Beauv. (Gramin.).

Chaetocladus 'Senilis' [J. Nelson] = Ephedra L. (Ephedrac.).

Chaetocrater Ruiz & Pav. = Casearia Jacq. (Flacourtiac.).

Chaetocyperus Nees = Eleocharis R.Br. .(Cyperac.).

Chaetodiscus Steud. = Eriocaulon L. (Eriocaulac.).

Chaetogastra DC. = Tibouchina Aubl. (Melastomatac.).

Chaetolepis Miq. Melastomataceae. 17 trop. Am., W.I. No appendages to connective.

Chaetolimon (Bunge) Lincz. Plumbaginaceae. 3 C. As.

Chaetonychia (DC.) Sweet. Caryophyllaceae. 1 Medit. reg.

Chaetopappa DC. Compositae. 9 SW. U.S., Mex.

Chaetophora Nutt. ex DC. = praec.

Chaetopoa C. E. Hubb. Gramineae. 1 trop. E. Afr.

Chaetopogon Janchen. Gramineae. 2 Portugal, Spain, Dalmatia.

Chaetoptelea Liebm. (~ Ulmus L.). Ulmaceae. 1 Mex. to Panamá.

Chaetosciadium Boiss. Umbelliferae. 1 E. Medit.

Chaetospermum (M. Roem.) Swingle (1913; non Sacc. 1892—Fungi) = Swinglea Merr. (Rutac.).

Chaetospira Blake. Compositae. 1 Colombia.

Chaetospora R.Br. = Schoenus L. (Cyperac.).

Chaetospora Kunth = Rhynchospora Vahl (Cyperac.).

Chaetostachydium Airy Shaw. Rubiaceae. 1 New Guinea.

Chaetostachys Benth. = Lavandula L. (Labiat.).

Chaetostachys Valeton = Chaetostachydium Airy Shaw (Rubiac.).
Chaetostemma Reichb. = Chaetostoma DC. (Melastomatac.).
Chaetostichium C. E. Hubbard. Gramineae. 1 mts. E. Afr.
Chaetostoma DC. Melastomataceae. 20 Braz. G 3-loc.
Chaetosus Benth. Apocynaceae. 1 New Guinea.
Chaetothylax Nees. Acanthaceae. 8 C. & S. Am.
Chaetothylopsis Oerst. = praec.
Chaetotropis Kunth. Gramineae. 6 temp. S. Am. Like *Agrostis*, but axis prolonged.
Chaeturus Link = Chaetopogon Janchen (Gramin.).
Chaeturus Reichb. = Chaiturus Ehrh. ex Willd. (Labiatae).
Chaffeyopuntia Frič & Schelle = Opuntia Mill. (Cactac.).
Chailletia DC. = Dichapetalum Thou. (Dichapetalac.).
Chailletiaceae R.Br. = Dichapetalaceae Baill.
Chaitaea Soland. ex Seem. = Tacca J. R. & G. Forst. (Taccac.).
Chaitea S. Parkinson = praec.   [C. As.
Chaiturus Ehrh. ex Willd. (~ Leonurus L.). Labiatae. 1 W. Eur. to
Chaixia Lapeyr. = Ramonda Rich. (Gesneriac.).
Chakiatella DC. = Chatiakella Cass. = Wulffia Neck. ex Cass. (Compos.).
Chalarium DC. = Edusaron Medik. = Desmodium Desv. (Legumin.).
Chalarium Poit. ex DC. = Ogiera Cass. = Eleutheranthera Poir. (Compos.).
Chalarothyrsus Lindau. Acanthaceae. 1 Mex.
Chalazocarpus Hiern = Schumanniophyton Harms (Rubiac.).
Chalazogamae Engl. A division of *Angiospermae*, originally proposed by Treub as the outcome of his work upon *Casuarina* (*Ann. Buitenz.* 10; 1891). It included only the fam. *Casuarinaceae*, later distinguished by Engler as the Order *Verticillatae* of his Subclass *Archichlamydeae*.

Both in the development of the macrospores (embryo-sacs), which elongated downwards into the chalaza, and in the process of fertilisation, which took place through the chalaza, the difference from other known Angiosperms seemed to Treub so great that he proposed to rearrange them into *Chalazogamae* and *Porogamae* (fertilised in the ordinary way), but as the phenomenon has since been observed in many other genera, especially of the lower types, it cannot be regarded as of classificatory value. (For literature see fifth or earlier edition of the *Dictionary*.)

Chalcanthus Boiss. Cruciferae. 1 mts. of Persia.
Chalcas L. = Murraya Koen. ex L. (Rutac.).
Chalcitis P. & K. = Xalkitis Rafin. = ? Aster L. (Compos.).
Chalcoëlytrum Lunell = Sorghastrum Nash (Gramin.).
Chaldia Boj. = Chadsia Boj. (Legumin.).
Chalebus Rafin. = Salix L. (Salicac.).
Chalepoa Hook. = Tribeles Phil. (Tribelac.).
Chalepophyllum Hook. f. Rubiaceae. 5 Venez., Guiana. 2 bracteoles. Seps. unequal.
Chalmersia F. Muell. ex S. Moore = Dichrotrichum Reinw. ex De Vriese
Chalybea Naud. = Pachyanthus A. Rich. (Melastomatac.).   [(Gesneriac.).
Chamabainia Wight. Urticaceae. 2 Indomal., Formosa.
Chamaeacanthus Chiov. Acanthaceae. 1 Somalia.

CHAMAEALOË

Chamaealoë Berger (∼Aloë L.). Liliaceae. 1 S. Afr.
Chamaeangis Schlechter. Orchidaceae. 16 trop. Afr., Madag., Masc.
Chamaeanthus Schlechter. Orchidaceae. 10 Malaysia.
Chamaeanthus Ule. Commelinaceae. 1 Amaz. Brazil.
Chamaebatia Benth. Rosaceae. 2 Calif., Lower Calif. Glandular, aromatic.
Chamaebatiaria Maxim. Rosaceae. 1 W. U.S.
Chamaebetula Opiz = Betula L. (Betulac.).
Chamaebuxus (DC.) Spach = Polygaloïdes Haller (Polygalac.).
Chamaecalamus Meyen = Deyeuxia Clarion ex Beauv. (Gramin.).
Chamaecassia Link = Cassia L. (Legumin.).
Chamaecerasus Duham. = Lonicera L. (Caprifoliac.).
Chamaecereus Britton & Rose. Cactaceae. 1 Argentina.
Chamaechaenactis Rydb. Compositae. 1 SW. U.S.
Chamaecissos Lunell = Glechoma L. (Labiat.).
Chamaecistus Fabr. = Helianthemum Mill. (Cistac.).
Chamaecistus S. F. Gray = Loiseleuria Desv. (Ericac.).
Chamaecistus Regel = Rhododendron L. (Ericac.).
Chamaecladon Miq. = Homalomena Schott (Arac.).
Chamaeclema Boehm. = Glechoma L. (Labiat.).
Chamaeclitandra (Stapf) Pichon. Apocynaceae. 1 trop. Afr.
Chamaecnide Nees & Mart. ex Miq. = Pilea Lindl. (Urticac.).
Chamaecrinum Diels ex Diels & Pritz. = Hensmania Fitzger. (Liliac.).
Chamaecrista Moench = Cassia L. (Legumin.).
Chamaecrypta Schlechtd. & Diels. Scrophulariaceae. 1 S. Afr.
Chamaecyparis Spach (∼Cupressus L.). Cupressaceae. 7 N. Am., Japan,
Formosa; incl. *C. lawsoniana* (A. Murr.) Parl. (NW. U.S.; Lawson's cypress);
C. nootkatensis (D. Don) Spach (Alaska to Oregon; Sitka or yellow cypress,
yellow cedar); *C. thyoïdes* (L.) Britton, Sterns & Poggenb. (E. N. Am.; white
cedar); *C. obtusa* (Sieb. & Zucc.) Endl. (Japan, Formosa); *C. pisifera* (Sieb. &
Zucc.) Endl. (Japan; Sawara cypress), etc. All yield good and useful timber.
Chamaecytisus Link (∼Cytisus L.). Leguminosae. 15–20 Atl. Is., Eur.,
Medit. reg.
Chamaecytisus Vis. = Argyrolobium Eckl. & Zeyh. (Legumin.).
Chamaedactylis T. Nees = Aeluropus Trin. (Gramin.).
Chamaedaphne Kuntze = Kalmia L. (Ericac.).
Chamaedaphne Mitch. = Mitchella L. (Rubiac.).
Chamaedaphne Moench = Cassandra D. Don (Ericac.).
*Chamaedorea Willd. Palmae. 100 warm Am. Small reedy palms, often
forming suckers. Dioec. A 6 on fleshy disk.
Chamaedoreaceae O. F. Cook = Palmae–Chamaedoreëae Benth. & Hook. f.
Chamaedryfolia Kuntze = Forskohlea L. (Urticac.).
Chamaedryfolium P. & K. = praec.
Chamaedrys Mill. = Teucrium L. (Labiat.).
Chamaefilix Hill = Asplenium L. (Aspleniac.).
Chamaefistula G. Don = Cassia L. (Legumin.).
Chamaegastrodia Mak. & Maekawa. Orchidaceae. 1 Japan.
Chamaegeron Schrenk. Compositae. 3 C. As.
Chamaegigas Dinter ex Heil. Scrophulariaceae. 1 SW. Afr.

CHAMAERHODENDRON

Chamaegyne Suesseng. (~ Eleocharis R. Br.). Cyperaceae. 1 Amaz. Brazil.
Chamaeiasma Gmel. = Cymbaria L. (Scrophulariac.).
Chamaeiris Medik. = Iris L. (Iridac.).
Chamaejasme Kuntze = Stellera J. G. Gmel. (Thymelaeac.).
Chamaelauciaceae Lindl. = Myrtaceae–Chamaelaucieae DC.
Chamaelaucium Desf. Myrtaceae. 12 W. Austr. Heath-like. A 10 + 10 stds.
Chamaele Miq. Umbelliferae. 1 Japan.
Chamaelea Duham. = Cneorum L. (Cneorac.).
Chamaeledon Link = Loiseleuria Desv. (Ericac.).
Chamaeleon Cass. = Atractylis L. (Compos.).
Chamaelinum Guett. = Radiola Roth (Linac.).
Chamaelinum Host = Camelina Crantz (Crucif.).
Chamaelirium Willd. Liliaceae. 1 Atl. N. Am. Dioec.
Chamaelobivia Y. Ito. Cactaceae. 3, hab.? Quid?
Chamaelum Bak. = Chamelum Phil. (Iridac.).
Chamaemeles Lindl. Rosaceae. 1 Madeira. G 1.
Chamaemelum Mill. (~ Anthemis L.). Compositae. 2–3 W. & C. Eur., Medit. *C. nobile* (L.) All. (chamomile) fls. offic.
Chamaemelum Vis. = Matricaria L. (Compos.).
Chamaemespilus Medik. = Sorbus L. (Rosac.).
Chamaemorus Ehrh. (uninom.) = *Rubus chamaemorus* L. (Rosac.).
Chamaemorus Hill = Rubus L. (Rosac.).
Chamaemyrrhis Endl. ex Heynh. = Oreomyrrhis Endl. (Umbellif.).
Chamaenerion Seguier emend. S. F. Gray (~ Epilobium L.). Onagraceae. 10 N. temp. & arct. In *C. angustifolium* (L.) Scop. the fls. are large and autogamy almost impossible. Honey is secreted by the upper surface of the ovary. The sta. are ripe when the fl. opens, and project horiz., while the style, with its stigmas closed, is bent downwards. Afterwards the sta. bend down and the style up, and the stigmas open. This is the plant in which C. K. Sprengel (1793) made the first discovery of dichogamy.
Chamaenerium Spach = praec.
Chamaeorchis Koch = Chamorchis Rich. (Orchidac.).
Chamaepentas Bremek. Rubiaceae. 1 trop. E. Afr.
Chamaepericlymenum Hill. Cornaceae. 2–3 N. circumpol., N. temp. mts. *Ch. suecicum* (L.) Graebn. is a dwarf herb giving off annual stems from the creeping stems, with purple fls. in umbels, invol. by 4 large white brs.
Chamaepeuce DC. = Ptilostemon Cass. (Compos.).
Chamaepeuce Zucc. = Chamaecyparis Spach (Cupressac.).
Chamaephoenix A. H. Curtiss = Pseudophoenix Wendl. (Palm.).
Chamaephyton Fourr. = Potentilla L. (Rosac.).
Chamaepitys Hill = Ajuga L. (Labiat.).
Chamaeplium Wallr. = Sisymbrium L. (Crucif.).
Chamaeranthemum Nees = Chameranthemum Nees (Acanthac.).
Chamaeraphis R.Br. Gramineae. 1 N. Austr.
Chamaeraphis Kuntze = Setaria Beauv. (Gramin.).
Chaemaerepes Spreng. = Chamorchis Rich. (Orchidac.).
Chamaerhodendron Bub. (sphalm.) = Chamaerhododendron Mill. = Rhododendron L. (Ericac.).

233

CHAMAERHODIOLA

Chamaerhodiola Nakai = Sedum L. (Crassulac.).
Chamaerhododendron Mill. = Rhododendron L. (Ericac.).
Chamaerhododendros Duham. = praec.
Chamaerhodos Bunge. Rosaceae. 11 Sib. & C. As. to N. China & temp.
N. Am.
Chamaeriphe Steck = Chamaerops L. (Palm.).
Chamaeriphes Kuntze = Hyphaene Gaertn. (Palm.).
Chamaeriphes Ponted. ex Gaertn. = seq.
Chamaerops L. Palmae. 2 W. Medit. *C. humilis* L. the only Eur. palm.
Dioec. Endosp. ruminate.
Chamaesaracha (A. Gray) Benth. & Hook. f. Solanaceae. 7 SW. U.S., N.
Mex. Prostrate herbs.
Chamaeschoenus Ehrh. (uninom.) = Scirpus (Isolepis) setaceus L. (Cyperac.).
Chamaesciadium C. A. Mey. (~ Trachydium Lindl.). Umbelliferae. 1 As.
Min., Cauc., Persia.
Chamaescilla F. Muell. Liliaceae. 2 Austr., Tasm. P twisted after flowering.
Chamaesenna Pittier = Cassia L. (Legumin.).
Chamaesium H. Wolff. Umbelliferae. 7 Himal., Tibet, W. China.
Chamaespartium Adans. (~ Cytisus L.). Leguminosae. 15 Atl. Is., Eur.,
Medit.
Chamaesparton Fourr. = praec.
Chamaesphacos Schrenk. Labiatae. 1 C. As., Pers., Afghan.
Chamaesphaerion A. Gray = Chthonocephalus Steetz (Compos.).
Chamaespilus Fourr. = Chamaemespilus Medik. = Sorbus L. (Rosac.).
Chamaestephanum Willd. = Schkuhria Roth (Compos.).
Chamaesyce S. F. Gray (~ Euphorbia L.). Euphorbiaceae. 250 cosmop.
Chamaetaxus Bub. = Empetrum L. (Empetrac.).
Chamaethrinax H. Wendl. ex R. Pfister = Trithrinax Mart. (Palm.).
Chamaexeros Benth. Xanthorrhoeaceae. 2 SW. Austr.
Chamaexiphium Hochst. = Ficinia Schrad. (Cyperac.).
Chamaexyphium Pfeiff. = praec.
Chamaezelum Link = Antennaria R.Br. (Compos.).
Chamagrostis Borkh. = Mibora Adans. (Gramin.).
Chamalirium Rafin. = Chamaelirium Willd. (Liliac.).
Chamalium Cass. = Chamaeleon Cass. = Atractylis L. (Compos.).
Chamalium Juss. = Cardopatium Juss. (Compos.).
Chamarea Eckl. & Zeyh. Umbelliferae. 1 S. Afr.
Chamartemisia Rydb. = Tanacetum L. (Compos.).
Chambeyronia Vieill. Palmae. 2 New Caled.
Chamedrys Rafin. (1836) = Spiraea L. (Rosac.).
Chamedrys Rafin. (1837) = Chamaedrys Moench = Teucrium L. (Labiat.).
Chamelaea P. & K. = Chamaelea Duham. = Cneorum L. (Cneorac.).
Chamelum Phil. Iridaceae. 3 Chile, Argent. Caps. enclosed in spathe.
Chamepeuce Rafin. = Chamaepeuce DC. = Cirsium Mill. (Compos.).
Chameranthemum Nees. Acanthaceae. 8 trop. Am.
Chamerasia Rafin. = Lonicera L. (Caprifoliac.).
Chamerion (Rafin.) Rafin. ex Holub = Chamaenerion Seguier emend. S. F.
Gray (Onagrac.).

Chamerops Rafin. = Chamaerops L. (Palm.).
Chamira Thunb. Cruciferae. 1 S. Afr. Lower ls. opp.
Chamisme Rafin. = Houstonia L. (Rubiac.).
***Chamissoa** Kunth. Amaranthaceae. 7 warm Am. Aril.
Chamissomneia Kuntze = Schlechtendalia Less. (Compos.).
Chamissonia Raimann = Camissonia Link (Onagrac.).
Chamissoniophila A. Brand. Boraginaceae. 2 S. Brazil.
Chamitea Kerner = Salix L. (Salicac.).
Chamitis Banks ex Gaertn. = Azorella Lam. (Hydrocotylac.).
Chamoletta Adans. = Xiphion Mill. = Iris L. (Iridac.).
Chamomilla Godr. = Anthemis L. (Compos.).
Chamomilla S. F. Gray = Matricaria L. (Compos.).
Chamorchis Rich. Orchidaceae. 1 Europe.
Champaca Adans. = Michelia L. (Magnoliac.).
Champereia Griff. Opiliaceae. 6 Indomal.
Championella Bremek. Acanthaceae. 4 Indoch., China, Japan.
Championia C. B. Clarke = Leptoboea Benth. (Gesneriac.).
Championia Gardn. Gesneriaceae. 1 Ceylon.
Chamula Nor. = Lobelia L. (Campanulac.).
Chamysyke Rafin. = Chamaesyce S. F. Gray (Euphorbiac.).
Chanekia Lundell = Licaria Aubl. (Laurac.).
Changium H. Wolff (∼ Conopodium Koch). Umbelliferae. 1 E. China.
Changnienia Chien. Orchidaceae. 1 China.
Chapelieria A. Rich. Rubiaceae. 2 Madag.
Chapelliera Nees = Cladium P.Br. (Cyperac.).
Chapmannia Torr. & Gray. Leguminosae. 1 Florida.
Chapmanolirion Dinter = Pancratium L. (Amaryllidac.).
Chaptalia Royle = Gerbera L. ex Cass. (Compos.).
***Chaptalia** Vent. Compositae. 25 warm Am., W.I.
Chaquepiria Endl. = Caquepiria J. F. Gmel. = Gardenia Ellis (Rubiac.).
Charachera Forsk. = Lantana L. (Verbenac.).
Characias S. F. Gray = Euphorbia L. (Euphorbiac.).
Charadra Scop. = Chadara Forsk. = Grewia L. (Tiliac.).
Charadrophila Marlòth. Scrophulariaceae. 1 S. Afr.
Chardinia Desf. Compositae. 1 W. As.
Charesia E. Busch = Silene L. (Caryophyllac.).
Charia C. DC. Meliaceae. 2 trop. Afr.
Charianthus D. Don. Melastomataceae. 8 W.I.
Charidia Baill. = Savia Willd. (Euphorbiac.).
Charidion Bong. Ochnaceae. 2 Brazil.
Charieis Cass. Compositae. 2 S. Afr.
Chariessa Miq. (∼ Citronella D. Don). Icacinaceae. 7 Philipp. Is., Borneo & Java to N. Queensl., Samoa, Tonga.
Chariomma Miers = Echites P.Br. (Apocynac.).
Charistemma Janka = Scilla L. (Liliac.).
× **Charlesworthara** hort. Orchidaceae. Gen. hybr. (iii) (Cochlioda × Miltonia × Oncidium).
Charlwoodia Sweet = Cordyline Comm. ex Juss. (Agavac.).

CHARPENTIERA

Charpentiera Gaud. Amaranthaceae. 1 Hawaii. Small tree.
Charpentiera Vieill. = Ixora L. (Rubiac.).
Chartacalyx Maingay ex Mast. = Schoutenia Korth. (Tiliac.).
Chartocalyx Regel = Otostegia Benth. (Labiat.).
Chartolepis Cass. = Centaurea L. (Compos.).
Chartoloma Bunge. Cruciferae. 1 C. As.
Chasalia DC. = seq.
Chasallia Comm. ex Poir. (1817) = Chassalia Comm. ex Poir. (1812) (Rubiac.).
Chascanum E. Mey. Verbenaceae. 25 Afr., Madag., Arabia to W. India.
Chascolytrum Desv. = Briza L. (Gramin.).
Chascotheca Urb. = Securinega Juss. (Euphorbiac.).
Chasea Nieuwl. = Panicum L. (Gramin.).
Chasechloa A. Camus. Gramineae. 3 Madag.
Chaseëlla Summerh. Orchidaceae. 1 S. Rhodesia.
Chasmanthe N. E. Br. (~ Petamenes Salisb. ex N. E. Br.). Iridaceae. 7 trop. & S. Afr.
Chasmanthera Hochst. Menispermaceae. 2 trop. Afr.
Chasmanthium Link (emend. Yates). Gramineae. 5 temp. N. Am.
Chasmatocallis R. C. Foster. Iridaceae. 1 S. Afr.
Chasmatophyllum Dinter & Schwantes. Aïzoaceae. 6 S. Afr.
Chasme Salisb. = Leucadendron R.Br. (Proteac.).
Chasmia Schott ex Spreng. = Arrabidaea DC. + Tynnanthus Miers (Bignoniac.).
Chasmone E. Mey. = Argyrolobium Eckl. & Zeyh. (Legumin.).
Chasmonia Presl = Moluccella L. (Labiat.).
Chasmopodium Stapf. Gramineae. 2 trop. Afr.
Chassalia Comm. ex Poir. Rubiaceae. 42 palaeotrop., esp. Madag.
Chasseloupia Vieill. = Symplocos L. (Symplocac.).
Chastenaea DC. = Axinaea Ruiz & Pav. (Melastomatac.).
Chastoloma Lindl. (sphalm.) = Chartoloma Bunge (Crucif.).
Chataea Soland. = Chaitaea Soland. ex Seem. = Tacca J. R. & G. Forst. (Taccac.).
Chatelania Neck. = Tolpis Adans. (Compos.).
Chatiakella Cass. = Wulffia Neck. ex Cass. (Compos.).
Chatinia Van Tiegh. = Psittacanthus Mart. (Loranthac.).
Chaubardia Reichb. f. Orchidaceae. 3 trop. S. Am.
Chaubardiella Garay. Orchidaceae. 3 trop. Am.
Chauliodon Summerhayes. Orchidaceae. 1 trop. W. Afr.
Chaulmoogra Roxb. = Gynocardia R.Br. (Flacourtiac.).
Chaunanthus O. E. Schulz = Iodanthus Torr. & Gray ex Steud. (Crucif.).
Chaunochiton Benth. Olacaceae. 5 C. & trop. S. Am.
Chaunochit[on]aceae Van Tiegh. = Olacaceae–Heisterieae Engl.
Chaunostoma Donnell Smith. Labiatae. 1 C. Am.
Chauvinia Steud. = Spartina Schreb. (Gramin.).
Chavannesia A. DC. (~ Urceola Roxb.). Apocynaceae. 8 Indomal.
Chavica Miq. = Piper L. (Piperac.).
Chavinia Gandog. = Rosa L. (Rosac.).
Chaydaia Pitard. Rhamnaceae. 2 S. China, N. Indoch.

Chayota Jacq. = Sechium P.Br. (Cucurbitac.).
Cheesemania O. E. Schulz. Cruciferae. 6 Tasm., N.Z.
Cheilanthaceae Nayar = Sinopteridaceae Koidz.
*Cheilanthes Sw. Sinopteridaceae. 180 trop. & temp. Mostly xero. Copeland includes here *Notholaena* R.Br., by others regarded as distinct.
Cheilanthopsis Hieron. Aspidiaceae. 1 NE. Himal., W. China.
Cheilanthos St-Lag. = Cheilanthes Sw. (Polypodiac.).
Cheiloclinium Miers. Celastraceae. 23 C. to trop. S. Am.
Cheilococca Salisb. ex Sm. = Platylobium Sm. (Legumin.).
Cheilodiscus Triana = Pectis L. (Compos.).
Cheilogramma (Bl.) Maxon = Paltonium Presl (Polypodiac.).
Cheilolepton Fée = Lomagramma J. Sm. (Lomariopsidac.).
Cheilophyllum Pennell. Scrophulariaceae. 8 W.I.
Cheiloplecton Fée. Sinopteridaceae. 1 Mexico.
Cheilopsis Moq. = Acanthus L. (Acanthac.).
Cheilosa Blume. Euphorbiaceae. 2 W. Malaysia.
Cheilosandra Griff. ex Lindl. = Rhynchotechum Blume (Gesneriac.).
Cheilosoria Trev. = Cheilanthes Sw. (Sinopteridac.).
Cheilotheca Hook. f. Monotropaceae. 2 Assam, Malay Penins.
Cheilyctis (Rafin.) Spach = Monarda L. (Labiat.).
Cheiradenia Lindl. Orchidaceae. 2 Guiana.
Cheiranthera A. Cunn. ex Brongn. Pittosporaceae. 2 SW., 2 SE. temp. Austr.
Cheiranthera Endl. = Cheirostemon Humb. & Bonpl. (Sterculiac.).
× Cheiranthesimum D. Bois. Cruciferae. Gen. hybr. (Cheiranthus × Erysimum).
Ch[e]iranthodendr[ac]eae A. Gray = Sterculiaceae–Fremontodendreae Airy Shaw.
Cheiranthodendron Benth. & Hook. f. = Chiranthodendron Cerv. ex Cav. = Cheirostemon Humb. & Bonpl. (Sterculiac.? Bombacac.?).
Cheiranthus L. Cruciferae. 10 Medit. & N. temp.
Cheiri Ludw. = Cheiranthus L. (Crucif.).
Cheiridopsis N. E. Brown. Aïzoaceae. 100 S. Afr.
Cheirinia Link = Erysimum L. + Sisymbrium L. (Crucif.).
Cheirisanthera Hort. ex Lindl. & Paxt. = Heppiella Regel (Gesneriac.).
Cheirodendron Nutt. Araliaceae. 8 Hawaii, Marquesas Is.
Cheiroglossa Presl = Ophioglossum L. (Ophioglossac.).
Cheirolaena Benth. Sterculiaceae (Malvaceae?). 1 Mauritius.
Cheirolepis Boiss. = Centaurea L. (Compos.).
Cheiroloma F. Muell. = Calotis R.Br. (Compos.).
Cheirolophus Cass. = Centaurea L. (Compos.).
Cheiropetalum E. Fries = Silene L. (Caryophyllac.).
Cheiropleuria Presl. Cheiropleuriaceae. 1 E. As.
Cheiropleuriaceae Nakai. Polypodiales. Terrestrial in ridge forest. Stem hairy. Ls. dimorphous; fertile ones narrow, acrostichoid. Sole genus: *Cheiropleuria* Presl.
Cheiropsis Bercht. & Presl = Clematis L. (Ranunculac.).
Cheiropteris Christ (1898; non Chiropteris Kurr 1858—gen. foss. inc. sed.) = Neocheiropteris Christ (Polypodiac.).

CHEIROPTEROCEPHALUS

Cheiropterocephalus Barb. Rodr. = Malaxis Soland. ex Sw. (Orchidac.).

Cheirorchis C. E. Carr. Orchidaceae. 5 Siam, Malay Penins.

Cheirostemon Humb. & Bonpl. = Chiranthodendron Larreategui (?Sterculiac.).

Cheirostylis Blume. Orchidaceae. 22 trop. Afr., As., Pacif.

× **Cheirysimum** Janchen = × Cheiranthesimum D. Bois (Crucif.).

Chelidoniaceae Nak. = Papaveraceae–Chelidonieae Reichb.

Chelidonium L. Papaveraceae. 1 temp. & subarct. Eurasia, *C. majus* L.

Chelidospermum Zipp. ex Blume = Pittosporum Banks ex Soland. (Pittosporac.).

Chelidurus 'Willdener' ex Cothenius. Inc. sed. (cf. *Euadenia* Oliv., Capparidac.). Hab.?

Cheliusia Sch. Bip. = Vernonia Schreb. (Compos.).

Chelona P. & K. = Chelone L. (Scrophulariac.).

Chelon[ac]eae D. Don = Scrophulariaceae–Cheloneae Benth.

Chelonanthera Blume. Orchidaceae. 2 Malaysia.

Chelonanthus Gilg. Gentianaceae. 20 trop. S. Am.

Chelonanthus Rafin. = seq.

Chelone L. (∼ Penstemon Schmid.). Scrophulariaceae. 4 E. U.S.

Chelonecarya Pierre = Rhaphiostyles Planch. ex Benth. (Icacinac.).

Chelonespermum Hemsl. Sapotaceae. 4 Solomon Is., Fiji.

Chelonistele Pfitzer. Orchidaceae. 20 Burma, W. Malaysia.

Chelonopsis Miq. Labiatae. 13 Kashmir, E. Tibet, China, Japan.

Chelrostylis Pritz. (sphalm.) = Cheirostylis Blume (Orchidac.).

Chelyocarpus Dammer. Palmae. 3 trop. S. Am.

Chemnicia Scop. = Strychnos L. (Strychnac.).

Chemnitzia P. & K. = praec.

Chemnizia Fabr. (uninom.) = *Lagoecia* L. sp. (Umbellif.).

Chemnizia Steud. = Chemnicia Scop. = Strychnos L. (Strychnac.).

Chenocarpus Neck. = Chaenocarpus Juss. = Spermacoce L. (Rubiac.).

Chenolea Thunb. Chenopodiaceae. 4 Medit., S. Afr.

Chenopodina Moq. = Suaeda Forsk. (Chenopodiac.).

*Chenopodiaceae Vent. Dicots. (Centrospermae). 102/1400, with an interesting geographical distr., determined by the fact that they are nearly all halophytic. The 10 chief districts char. by their presence are (according to Bunge), (1) Austr., (2) the Pampas, (3) the Prairies, (4) and (5) the Medit. coasts, (6) the Karroo (S. Afr.), (7) the Red Sea shores, (8) the SW. Caspian coast, (9) Centr. As. (Caspian to Himalayas—deserts), (10) the salt steppes of E. As. The presence of large quantities of salt in the soil necessitates the reduction of the transpiration, so that the pls. which grow in such situations exhibit xero. characters. They are mostly herbs (a few shrubs or small trees), with roots which penetrate deeply into the soil, and with exstip. ls. of various types, usu. not large, often fleshy, and often covered with hairs, which frequently give a curious and very char. mealy feeling to the pl. In some halophytes of this fam. the ls. are altogether suppressed, and the pl. has curious jointed succulent stems like a miniature cactus (e.g. *Salicornia*). Each 'limb' embraces the next succeeding one by a sort of cup at its apex. Even more than in their external form, the *C.* show xero. structure in their internal anatomy.

Infl. often primarily racemose, but the partial infls. are always cymose, at first often dich., but with a tendency to the cincinnus form, by preference of the β-bracteole. The fls. are reg., small and inconspic., ♀ or unisex. P simple, rarely absent, persistent after flowering, 5, 3, 2 (rarely 1 or 4) ± united, imbr., sepaloid; A as many as or fewer than P segments, opp. to them, hypog. or on a disk; anthers bent inwards in bud; G (semi-inf. in *Beta*), 1-loc. with 2 (rarely more) stigmas; ov. 1, basal, campylotropous. Fr. usu. a small round nut or achene; embryo usu. surrounding the endosp., either simply bent or spirally twisted. Few are useful; see *Beta, Spinacia, Cnenopodium*, etc.

Classification and chief genera (after Ulbrich):

A. 'Cyclolobeae.' Embryo ring-shaped, horseshoe-like, condupl. or semicirc., wholly or partly enclosing endosp.

I. **Polycnemoïdeae.** Roots and stem with normal structure. Fls. ♀, solitary, bracteolate; stam. 2–5, united below. Seed pend. Ls. lin. or subul. *Polycnemum, Nitrophila.*

II. **Betoïdeae.** Roots, and usu. stem, with anomalous 'Chenopodiaceous' structure. Fr. opening at maturity or at germination by an operculum. Stigma short, usu. broad-lobed, papillose within. *Hablitzia, Beta.*

III. **Chenopodioïdeae.** As II, but fr. remaining closed, more rarely breaking up, not opening by an operculum. Fls. in glomerules, more rarely arranged in spicate infls. Fr. surrounded by P or bracteoles at maturity. *Chenopodium, Spinacia, Obione, Atriplex, Eurotia, Axyris, Camphorosma, Bassia, Austrobassia, Sclerolaena, Kochia*, etc.

IV. **Corispermoïdeae.** As III, but fr. naked at maturity. Fls. ♀, spicate, ebracteolate. *Corispermum, Agriophyllum.*

V. **Salicornioïdeae.** As III, but fls. arranged in clavate, catkin-like infls., or in hollows of apparently l.-less branches. Fls. ♀, protandr., ebracteolate; P herbaceous or membr., united. Branches articulated. *Kalidium, Halostachys, Halocnemum, Arthrocnemum, Salicornia.*

B. 'Spirolobeae.' Embryo spirally twisted; endosp. wanting or divided into two masses by embryo.

VI. **Sarcobatoïdeae.** Fls. without bracteoles, monoec. ♂ fls. consisting of sta. arranged beneath peltate scales; ♀ fls. with small P adnate to the ovary, sol., axill. *Sarcobatus.*

VII. **Suaedoïdeae.** As VI, but fls. with bracteoles, which are small and scale-like. P herbaceous or membr. Stam. 5. Stig. papillose all over. Embryo plane-spiral. Ls. glabr., without sheath of collecting hairs around nerve-bundles. *Suaeda, Brezia*, etc.

VIII. **Salsoloïdeae.** As VII, but bracteoles as large as or larger than P, v. rarely smaller. P usu. membr. Stam. 4–5, rarely fewer. Stig. papillose only on inner face. Embryo helical, rarely plane-spiral. Ls. usu. covered with filif. hairs, with a sheath of collecting hairs round the nerve-bundles. *Traganum, Salsola, Arthrophytum, Haloxylon, Noaea, Anabasis, Petrosimonia, Halimocnemis, Cornulaca, Halogeton*, etc.

Chenopodium L. Chenopodiaceae. 100–150 temp. Fr. in many dimorphic; some have horiz. seeds, some vertical (esp. on the term. twigs of the cymes). Essential oil from *C. anthelminticum (ambrosioïdes)* L. (worm-seed or Mexican tea) is used as a vermifuge in the U.S. *C. quinoa* Willd. is a food plant in AS. m.; its seeds are boiled like rice. It and other spp. are used as spinach.

Cheobula Vell. = Cleobula Vell. (Legumin.?).

Cheramela Rumph. = Cicca L. (Euphorbiac.).

Cherimolia Rafin. = Annona L. (Annonac.).

Cherina Cass. = Chaetanthera Ruiz & Pav. (Compos.).

Cherleria L. (∼ Minuartia L.). Caryophyllaceae. 1 Alps, Carpathians, Scotland. Tufted alpine.

Cherophilum Nocca = seq.

Cherophylum Rafin. = Chaerophyllum L. (Umbellif.).

Chersodoma Phil. Compositae. 5 temp. S. Am.

Chersydrium Schott = Dracontium L. (Arac.).

Chesmone Bub. = Chasmone E. Mey. = Argyrolobium Eckl. & Zeyh. (Legumin.).

Chesnea Scop. = Cephaëlis Sw. (Rubiac.).

Chesneya Bertol. = Gaytania Münter = Pimpinella L. (Umbellif.).

Chesneya Lindl. Leguminosae. 15–20 Armenia & Iraq to Mongolia.

Chesniella Borisova (∼ Chesneya Lindl.). Leguminosae. 6 C. As., Kashmir.

Chetanthera Rafin. = Chaetanthera Nutt. = Chaetopappa DC. (Compos.).

Chetaria Steud. = seq.

Chetastrum Neck. = Scabiosa L. (Dipsacac.).

Chetocrater Rafin. = Casearia Jacq. (Flacourtiac.).

Chetopappus Rafin. = Chaetopappa DC. (Compos.).

Chetropis Rafin. = Arenaria L. (Caryophyllac.).

Chetyson Rafin. = Sedum L. (Crassulac.).

Chevalierella A. Camus. Gramineae. 2 Congo.

Chevalieria Gaudich. Bromeliaceae. 5 E. Brazil, Argent.

Chevalierodendron Leroy = Streblus Lour. (Morac.).

Chevreulia Cass. Compositae. 6 S. Am., Falkland Is.

Cheynia J. Drumm. ex Harv. = Balaustion Hook. (Myrtac.).

Chianthemum Kuntze = Galanthus L. (Amaryllidac.).

Chiapasia Britton & Rose = Disocactus Lindl. (Cactac.).

Chiarinia Chiov. Sapindaceae. 1 Somalia.

Chiastophyllum (Ledeb.) Stapf. Crassulaceae. 1 Caucasus.

Chiazospermum Bernh. = Hypecoum L. (Hypecoac.).

Chibaca Bertol. f. = Warburgia Engl. (Canellac.).

Chichaea Presl = Brachychiton Schott & Endl. (Sterculiac.).

Chicharronia A. Rich. = Terminalia L. (Combretac.).

Chichipia Marn.-Lapost. = Polaskia Backeb. (Cactac.).

Chickrassia Wight & Arn. = Chukrasia Juss. (Meliac.).

Chicoca Augier = Chiococca L. (Rubiac.).

Chicoinaea Comm. ex DC. = Psathura Comm. ex Juss. (Rubiac.).

Chidlowia Hoyle. Leguminosae. 1 trop. W. Afr.

Chienia W. T. Wang (∼ Delphinium L.). Ranunculaceae. 1 C. China.

Chieniodendron Tsiang & P. T. Li (∼ Desmos Lour.). Annonaceae. 1 S. China (Hainan).

Chienodoxa Y. S. Sun = Schnabelia Hand.-Mazz. (Verbenac.).

Chienopteris Ching = Woodwardia Sm. (Blechnac.).

Chikusichloa Koidz. Gramineae. 2 SE. China, Ryukyu Is., Japan.
Childsia Childs = Hidalgoa La Llave (Compos.).
Chilechium Pfeiff. = Chilochium Rafin. = Echiochilon Desf. (Boraginac.).
Chilenia Backeb. (1935) = Neoporteria Britton & Rose (Cactac.).
Chilenia Backeb. (1939) = Nichelia Bullock (Cactac.).
Chileniopsis Backeb. = Neoporteria Britt. & Rose (Cactac.).
Chileocactus Frič = Horridocactus Backeb. (Cactac.).
Chileorebutia F. Ritter = Pyrrhocactus (A. Berger) Backeb. & F. M. Knuth (Cactac.).
Chileranthemum Oerst. Acanthaceae. 2 Mexico.
Chiliadenus Cass. = Jasonia Cass. (Compos.).
Chiliandra Griff. = Rhynchotechum Blume (Gesneriac.).
Chilianthus Burchell = Buddleja L. (Buddlejac.).
Chiliocephalum Benth. Compositae. 1 Abyssinia.
Chiliophyllum DC. = Zaluzania Pers. (Compos.).
*Chiliophyllum Phil. Compositae. 1 S. Andes.
Chiliorebutia Frič = ? Chileorebutia F. Ritter = Pyrrhocactus (A. Berger) Backeb. & F. M. Knuth (Cactac.).
Chiliotrichiopsis Cabrera. Compositae. 3 Argentina.
Chiliotrichum Cass. Compositae. 2 temp. S. Am.
Chilita Orcutt = Mammillaria Haw. (Cactac.).
Chillania Roivainen. Cyperaceae. 1 Chile.
Chilmoria Buch.-Ham. = Gynocardia R.Br. (Flacourtiac.).
Chilocalyx Hook. f. (sphalm.) = Chylocalyx Hassk. ex Miq. = Echinocaulos (Meissn. ex Endl.) Hassk. (Polygonac.).
Chilocalyx Klotzsch = Cleome L. (Cleomac.).
Chilocalyx Turcz. = Atalantia Corrêa (Rutac.).
Chilocardamum O. E. Schulz. Cruciferae. 1 Patagonia.
Chilocarpus Blume. Apocynaceae. 15 Indomal., N. Queensl.
Chilochium Rafin. = Echiochilon Desf. (Boraginac.).
Chilochloa Beauv. = Phleum L. (Gramin.).
Chilococca P. & K. = Cheilococca Salisb. = Platylobium Sm. (Legumin.).
Chilodia R.Br. = Prostanthera Labill. (Labiat.).
Chilodiscus P. & K. = Cheilodiscus Triana = Pectis L. (Compos.).
Chiloglossa Oerst. = Dianthera L. (Acanthac.).
Chiloglottis R.Br. Orchidaceae. 8 Austr., N.Z.
Chilopogon Schlechter. Orchidaceae. 4 E. Malaysia.
Chiloporus Naud. = Miconia Ruiz & Pav. (Melastomatac.).
Chilopsis D. Don. Bignoniaceae. 1 S. U.S., Mexico. Ls. alt.
Chilopsis P. & K. = Cheilopsis Moq. = Acanthus L. (Acanthac.).
Chilosa P. & K. = Cheilosa Bl. (Euphorbiac.).
Chilosandra P. & K. = Cheilosandra Griff. ex Lindl. = Rhynchotechum Bl. (Gesneriac.).
Chiloschista Lindl. Orchidaceae. 3 Indomal.
Chilostigma Hochst. = Aptosimum Burch. (Scrophulariac.).
Chilotheca P. & K. = Cheilotheca Hook. f. (Monotropac.).
Chilyanthum P. & K. = Xeilyathum Rafin. = Oncidium Sw. (Orchidac.).
Chilyctis P. & K. = Cheilyctis (Rafin.) Spach = Monarda L. (Labiat.).

CHIMANTAEA

Chimantaea Maguire, Steyerm. & Wurd. Compositae. 8 Venez., Guiana.
Chimanthus Rafin. = Lauro-Cerasus Duham. (Rosac.).
Chimaphila Pursh. Pyrolaceae. 8 Euras., N. & C. Am., W.I.
Chimarrhis Jacq. Rubiaceae. 15 W.I., trop. S. Am.
Chimaza R.Br. ex DC. = Chimaphila Pursh (Pyrolac.).
Chimocarpus Baill. = Chymocarpus D. Don = Tropaeolum L. (Tropaeolac.).
***Chimonanthus** Lindl. Calycanthaceae. 4 China. *C. praecox* (L.) Link has fragrant fls. which come out very early in the year before the ls. and show marked protogyny with movement of sta. Pollination is by beetles.
Chimonobambusa Makino. Gramineae. 14 India to Japan.
Chimophila Radius = Chimaphila Pursh (Pyrolac.).
Chincharronia A. Rich. = Chicharronia A. Rich. = Terminalia L. (Combretac.).
Chinchona Howard = Cinchona L. (Rubiac.).
Chingia Holtt. Thelypteridaceae. 12 Malaysia to Tahiti.
Chingiacanthus Hand.-Mazz. Acanthaceae. 2 S. China.
Chingithamnaceae Hand.-Mazz. = Celastraceae–Euonymeae Loes.
Chingithamnus Hand.-Mazz. = Microtropis Wall. (Celastrac.).
Chiococca P.Br. Rubiaceae. 20 S. Florida, W.I., trop. Am.
Chiogenes Salisb. ex Torr. = Gaultheria L. (Ericac.).
Chionachne R.Br. Gramineae. 5 Indomal., Indoch., E. Austr. *C. cyathopoda* F. Muell., valuable fodder-grass.
Chionanthula Börner = Carex L. (Cyperac.).
Chionanthus Gaertn. = Linociera Sw. (Oleac.).
Chionanthus L. Oleaceae. 2 E. As., E. N. Am.
Chione DC. Rubiaceae. 10 C. Am., W.I.
Chione Salisb. = Narcissus L. (Amaryllidac.).
Chionice Bunge ex Ledeb. = Potentilla L. (Rosac.).
Chionocarpium Brand = Adesmia DC. (Legumin.).
Chionocharis I. M. Johnson (~ Eritrichium Schrad.). Boraginaceae. 1 Himal.
Chionochlaena P. & K. = Chionolaena DC. (Compos.).
Chionochloa Zotov. Gramineae. 1 SE. Austr., 18 N.Z.
Chionodoxa Boiss. = Scilla L. (Liliac.).
Chionoglochin Gandog. = Carex L. (Cyperac.).
***Chionographis** Maxim. Liliaceae. 7 E. As. Fl. zygo.
Chionolaena DC. Compositae. 9 Mexico, S. Am. Shrubs with the ls. rolled back.
Chionopappus Benth. Compositae. 2 Peru.
Chionophila Benth. Scrophulariaceae. 2 Rocky Mts.
Chionophila Miers ex Lindl. = Boöpis Juss. (Calycerac.).
Chionoptera DC. = Pachylaena D. Don ex Hook. & Arn. (Compos.).
× **Chionoscilla** J. Allen ex Nicholson. Liliaceae. Gen. hybr. (Chionodoxa × Scilla).
Chionothrix Hook. f. Amaranthaceae. 3 Somalia.
Chionotria Jack = Glycosmis Corrêa (Rutac.).
Chiophila Rafin. = Gentiana L. (Gentianac.).
Chiovendaea Speg. Leguminosae. 2 Argentina.
Chiradenia P. & K. = Cheiradenia Lindl. = Zygopetalum Hook. (Orchidac.).
Chiranthera P. & K. = Cheiranthera A. Cunn. ex Brongn. (Pittosporac.).

Chiranthodendron Larreategui. ?Sterculiaceae. 1 Mex. Fls. large, apet.; sta. 5, united unilaterally below.

Chirata G. Don = Chirita Buch.-Ham. (Gesneriac.).

Chiratia Montr. = Sonneratia L. f. (Sonneratiac.).

Chiridium Van Tiegh. = Helixanthera Lour. (Loranthac.).

Chirita Buch.-Ham. Gesneriaceae. 80 Indomal., SE. As., S. China; some with epiphyllous infl.

Chirocalyx Meissn. = Erythrina L. (Legumin.).

Chirocarpus A. Braun ex Pfeiff. = Caylusea A. St-Hil. (Resedac.).

Chirochlaena P. & K. = Cheirolaena Benth. (Sterculiac.).

Chirodendrum P. & K. = Cheirodendron Nutt. (Araliac.).

Chirolepis P. & K. = Cheirolepis Boiss. = Centaurea L. (Compos.).

Chiroloma P. & K. = Cheiroloma F. Muell. = Calotis R.Br. (Compos.).

Chirolophus Cass. = Cheirolophus Cass. = Centaurea L. (Compos.).

Chironea Rafin. = Chironia L. (Gentianac.).

Chironia Gaertn., Mey. & Scherb. = Centaurium Hill (Gentianac.).

Chironia L. Gentianaceae. 30 Afr., Madag.

Chironia F. W. Schmidt = Gentiana L. (Gentianac.).

Chironiaceae Horan. = Gentianaceae Juss.

Chiropetalum A. Juss. (~ Argithamnia Sw.). Euphorbiaceae. 25 Mex. to S. Am.

Chirostemum Cerv. = Cheirostemon Humb. & Bonpl. (Sterculiac.?).

Chirostylis P. & K. = Cheirostylis Bl. (Orchidac.).

Chirripoa Suesseng. Bromeliaceae. 1 C. Am.

Chirvnia Rafin. (sphalm.) = Chironia L. (Gentianac.).

Chisocheton Blume. Meliaceae. 100 Indomal., SE. As., S. China. L. apex often showing continued or indeterminate growth. Some New Guinea spp. have epiphyllous infls.

Chithonanthus Lehm. = Acacia Mill. (Legumin.).

Chitonanthera Schlechter. Orchidaceae. 11 New Guinea.

Chitonia D. Don = Miconia Ruiz & Pav. (Melastomatac.).

Chitonia Moç. & Sessé = Morkillia Rose & Painter (Zygophyllac.).

Chitonia Salisb. = Zigadenus Michx (Liliac.).

Chitonochilus Schlechter. Orchidaceae. 1 New Guinea.

Chizocheton A. Juss. = Chisocheton Blume (Meliac.).

Chlaenaceae Thou. = Sarcolaenaceae Caruel.

Chlaenandra Miq. Menispermaceae. 1 New Guinea.

Chlaenanthus P. & K. = Chlainanthus Briq. = Lagochilus Bunge (Labiat.).

Chlaenobolus Cass. = Pterocaulon Ell. (Compos.).

Chlaenosciadium C. Norman. Hydrocotylaceae (~ Umbellif.). 1 W. Austr.

Chlainanthus Briq. = Lagochilus Bunge (Labiat.).

Chlamydacanthus Lindau = Theileamea Baill. (Acanthac.).

Chlamydanthus C. A. Mey. = Thymelaea Endl. (Thymelaeac.).

Chlamydia Banks ex Gaertn. = Phormium J. R. & G. Forst. (Agavac.).

Chlamydites J. R. Drumm. Compositae. 1 Tibet.

Chlamydobalanus (Endl.) Koidz. = Castanopsis (D. Don) Spach (Fagac. .

Chlamydoboea Stapf. Gesneriaceae. 2 Burma, China.

Chlamydocardia Lindau. Acanthaceae. 4 trop. W. Afr.

Chlamydocarya Baill. Icacinaceae. 7 trop. W. Afr.

CHLAMYDOCOLA

Chlamydocola (K. Schum.) Bodard. Sterculiaceae. 2 trop. W. Afr.

Chlamydojatropha Pax & K. Hoffm. Euphorbiaceae. 1 W. Equat. Afr.

Chlamydophora Ehrenb. = Cotula L. (Compos.).

Chlamydophytum Mildbr. Balanophoraceae. 1 W. Equat. Afr.

Chlamydosperma A. Rich. = Stegnosperma Benth. (Stegnospermatac.).

Chlamydostachya Mildbr. Acanthaceae. 1 trop. E. Afr.

Chlamydostylus Baker = Nemastylis Nutt. (Iridac.).

Chlamyphorus Klatt = Gomphrena L. (Amaranthac.).

Chlamysperma Less. = Villanova Lag. (Compos.).

Chlamysporum Salisb. = Thysanotus R.Br. (Liliac.).

Chlanis Klotzsch = Xylotheca Hochst. (Flacourtiac.).

Chlaotrachelus Hook. f. (sphalm.) = Claotrachelus Zoll. = Vernonia Schreb. (Compos.).

Chleterus Rafin. = Boea Comm. ex Lam. (Gesneriac.).

Chlevax Cesati ex Boiss. = Ferula L. (Umbellif.).

Chlidanthus Herb. Amaryllidaceae. 1 S. Am. Sta. with lat. appendages (see fam.).

Chloachne Stapf. Gramineae. 2 trop. Afr.

Chloammia Rafin. = Festuca L. (Gramin.).

Chloamnia Schlechtd. = praec.

Chloanthaceae Hutch. = Dicrastylidaceae J. Drumm. ex Harv.

Chloanthes R.Br. Dicrastylidaceae. 10 Austr., N.Z.

Chloërum Willd. ex Link = Abolboda Humb. & Bonpl. (Abolbodac.).

Chloïdia Lindl. = Corymborkis Thou. + Tropidia Lindl. (Orchidac.).

Chlonanthes Rafin. = seq.

Chlonanthus Rafin. = Chelonanthus Rafin. = Chelone L. (Scrophulariac.).

Chloöpsis Blume = Ophiopogon Ker-Gawl. (Liliac.).

Chloöthamnus Buese. Gramineae. 3 Malaysia.

Chlora Adans. = Blackstonia Huds. (Gentianac.).

Chloradenia Baill. = Cladogynos Zipp. ex Span. (Euphorbiac.).

Chloraea Lindl. Orchidaceae. 100 S. Am.

*****Chloranthaceae** R.Br. ex Sims. Dicots. 5/65 trop. & subtrop. Herbs, shrubs, or trees, with opp. stip. ls. Fls. small, perhaps in a 'pre-floral' stage, in spikes or cymes, ♀ or unisex., sometimes with sepaloid P; A 1–3, united to one another and to ovary; G or̄ G 1; ov. few, pend., orthotr. Endosp. oily; perisperm; embryo minute. Genera: *Chloranthus, Sarcandra, Hedyosmum, Ascarina, Ascarinopsis*.

Chloranthus Sw. Chloranthaceae. 15 E. As., Indomal. P 1, anterior; the centre sta. has a complete anther, the lat. each a half (cf. *Fumaria*).

Chloraster Haw. = Narcissus L. (Amaryllidac.).

Chloridaceae (Kunth) Herter = Gramineae–Chlorideae Kunth.

Chloridion Stapf (1900; non Chloridium Link 1809—Fungi) = Stereochlaena Hackel (Gramin.).

Chloridiopsis J. Gay ex Scribn. = Trichloris Fourn. ex Benth. (Gramin.).

Chloridopsis Hort. ex Hack. = praec.

Chloris Sw. Gramineae. 40 trop. & warm temp. Several are useful pasture-grasses.

Chlorita Rafin. = Chlora Adans. = Blackstonia Huds. (Gentianac.).

Chloriza Salisb. = Lachenalia Jacq. (Liliac.).
Chlorocalymma W. D. Clayton. Gramineae. 1 trop. E. Afr.
Chlorocarpa Alston. Flacourtiaceae. 1 Ceylon.
Chlorocaulon Klotzsch = Chiropetalum A. Juss. (Euphorbiac.).
Chlorocharis Rikli = Eleocharis R.Br. (Cyperac.).
Chlorochlamys Miq. = Marsdenia R.Br. (Asclepiadac.).
Chlorocodon Fourr. = Erica L. (Ericac.).
Chlorocodon Hook. f. = Mondia Skeels (Asclepiadac.).
Chlorocrambe Rydb. Cruciferae. 1 W. N. Am.
Chlorocrepis Griseb. = Hieracium L. (Compos.).
Chlorocyathus Oliv. Asclepiadaceae. 1 trop. E. Afr. (Mozamb.).
Chlorocyperus Rikli = Cyperus L. (Cyperac.).
Chlorodes P. & K. = Chloroïdes Fisch. = Chloris Sw. (Gramin.).
***Chlorogalum** (Lindl.) Kunth. Liliaceae. 7 Calif. *C. pomeridianum* Kunth has a large bulb whose inner parts are used as a substitute for soap. The outer layers yield a quantity of fibre.
Chloroïdes Fisch. = Chloris Sw. (Gramin.).
Chloroleucon (Benth.) Britton & Rose = Pithecellobium Mart. (Legumin.).
Chloroluma Baill. Sapotaceae. 4 S. trop. S. Am.
Chloromeles (Decne) Decne = Malus Mill. (Rosac.).
Chloromyron Pers. = Rheedia L. (Guttif.).
Chloromyrtus Pierre (~ Eugenia L.). Myrtaceae. 1 trop. Afr.
Chloropatane Engl. (~ Erythrococca Benth.). Euphorbiaceae. 20 W. & C. trop. Afr.
Chlorophora Gaudich. Moraceae. 12 trop. Am., Afr., Madag. The wood of *C. tinctoria* (L.) Gaudich. forms the yellow dye fustic.
Chlorophyllum Liais (sphalm.) = Chrysophyllum L. (Sapotac.).
Chlorophytum Ker-Gawl. Liliaceae. 215 S. Am., Afr., Madag., India, Austr., Tasm. In *C. comosum* Baker infl. often replaced by veg. repr.; long shoots develop in the axils of the brs., weigh the stem down to the soil and take root.
Chlorophytum Pohl ex DC. = Spermacoce Gaertn. (Rubiac.).
Chloropsis Hackel ex Kuntze = Trichloris Fourn. ex Benth. (Gramin.).
Chloropyron Behr = Cordylanthus Nutt. (Scrophulariac.).
Chlorosa Blume. Orchidaceae. 2 Borneo, Java.
Chlorospatha Engl. Araceae. 1 Colombia. ♀ fls. in whorls.
Chlorostelma Welw. ex Rendle = Asclepias L. (Asclepiadac.).
Chlorostemma Fourr. = Galium L. (Rubiac.).
Chlorostis Rafin. = Chloris Sw. (Gramin.).
***Chloroxylon** DC. Flindersiaceae. 1 SW. India, Ceylon. *C. swietenia* DC. (satinwood). Timber very lasting, largely used in veneering. The tree also yields a gum.
Chloroxylon Rafin. = Diospyros L. (Ebenac.).
Chloroxylon Scop. = seq.
Chloroxylum P.Br. = Ziziphus Mill. (Rhamnac.).
Chloroxylum P. & K. = Chloroxylon DC. (Flindersiac.).
Chloryllis E. Mey. = Dolichos L. (Legumin.).
Chloryta Rafin. = Chlora Adans. = Blackstonia Huds. (Gentianac.).

CHNOANTHUS

Chnoanthus Phil. = Gomphrena L. (Amaranthac.).
Chnoöphora Kaulf. = Cyathea Sm. (Cyatheac.).
Choananthus Rendle. Amaryllidaceae. 2 mts. trop. E. Afr.
Chocho Adans. = Sechium P.Br. (Cucurbitac.).
Chodanthus Hassler = Mansoa DC. (Bignoniac.).
Chodaphyton Minod. Scrophulariaceae. 1 Paraguay.
Chodondendron Bosc (sphalm.) = Chondrodendron Ruiz & Pav. (Meni-
Choenomeles Lindl. = Chaenomeles Lindl. (Rosac.).   [spermac.).
Choeradodia Herb. = Strumaria Jacq. (Amaryllidac.).
Choeradoplectron Schau. = Peristylus Blume (Orchidac.).
Choerophillum Neck. = seq.
Choerophyllum Brongn. = Chaerophyllum L. (Umbellif.).
Choeroseris Link = Picris L. (Compos.).
Choerospondias B. L. Burtt & A. W. Hill. Anacardiaceae. 1 NE. India to
   SE. China, N. Siam.
Choetophora Franch. & Sav. (sphalm.) = Chaetospora R.Br. = Schoenus L.
   (Cyperac.).
Choisya Kunth. Rutaceae. 6 S. U.S., Mex.
Cholisma Greene = Xolisma Rafin. = Lyonia Nutt. (Ericac.).
Chomelia Jacq. = Anisomeris C. Presl (Rubiac.).
Chomelia L. (Tarenna Gaertn.) Rubiaceae. 370 trop. Afr., Madag., Seych.,
   trop. As., Austr.
Chomelia Vell. = Ilex L. (Aquifoliac.).
Chomutowia B. Fedtsch. = Acantholimon Boiss. (Plumbaginac.).
Chona D. Don = Erica L. (Ericac.).
Chonaïs Salisb. = Hippeastrum Herb. (Amaryllidac.).
Chondilophyllum Panch. ex Guillaumin = Meryta J. R. & G. Forst. (Araliac.).
Chondodendron Benth. & Hook. f. (sphalm.) = Odontocarya Miers (Meni-
   spermac.).
Chondodendron Ruiz & Pav. = Chondrodendron Ruiz & Pav. corr. Miers
   (Menispermac.).
Chondrachne R.Br. = Lepironia Rich. (Cyperac.).
Chondrachyrum Nees = Melica L. (Gramin.).
Chondradenia Maxim. ex Mak. = Orchiṣ L. (Orchidac.).
Chondrilla L. Compositae. 25 temp. Eurasia.
× **Chondrobollea** hort. Orchidaceae. Gen. hybr. (iii) (Bollea × Chondroryncha).
Chondrocarpus Nutt. = Hydrocotyle L. (Hydrocotylac.).
Chondrocarpus Stev. = Astragalus L. (Legumin.).
Chondrochilus Phil. = Chaetanthera Ruiz & Pav. (Compos.).
Chondrochlaena P. & K. = Chondrolaena Nees = Prionanthium Desv.
   (Gramin.).
Chondrodendron Ruiz & Pav. corr. Miers. Menispermaceae. 8 Brazil, Peru.
   A 6 or (3). G 6. The root of some spp. furnishes *Radix Pereirae bravae.*
Chondrolaena Nees = Prionachne Nees = Prionanthium Desv. (Gramin.).
Chondrolomia Nees = Scleria Bergius (Cyperac.).
Chondropetalon Rafin. = Chondropetalum Rottb. (Restionac.).
× **Chondropetalum** hort. (vii) = × Zygorhyncha hort. (Orchidac.).
Chondropetalum Rottb. Restionaceae. 18 S. Afr.

Chondrophora Rafin. = Bigelowia DC. (Compos.).
Chondrophylla A. Nelson = Gentiana L. (Gentianac.).
Chondropsis Rafin. = Exacum L. (Gentianac.).
Chondrorhyncha Lindl. Orchidaceae. 11 C. & trop. S. Am.
Chondrosea Haw. = Saxifraga L. (Saxifragac.).
Chondrospermum Wall. = Myxopyrum Blume (Oleac.).
Chondrostylis Boerlage. Euphorbiaceae. 2 Indoch., W. Malaysia.
Chondrosum Desv. = Botelua Lag. (Gramin.).
Chondylophyllum Panch. ex R. Viguier = Meryta J. R. & G. Forst. (Araliac.).
*Chonemorpha G. Don. Apocynaceae. 20 SE. As., Indomal.
Chonocentrum Pierre ex Pax & K. Hoffm. Euphorbiaceae. 1 Amaz. Brazil.
Chonopetalum Radlk. Sapindaceae. 1 trop. W. Afr.
Chordorrhiza Ehrh. (uninom.) = Carex chordorrhiza L. f. (Cyperac.).
Chordospartium Cheesem. Leguminosae. 1 N.Z.
Choretis Herb. = Hymenocallis Salisb. (Amaryllidac.).
Choretrum R.Br. Santalaceae. 8 Austr.
Choriantha Riedl. Boraginaceae. 1 Iraq.
Choribaena Steud. = Chorilaena Endl. (Rutac.).
Choricarpha Boeck. = Lepironia Rich. (Cyperac.).
Choricarpia Domin. Myrtaceae. 1 E. Austr.
Choriceras Baill. (~ Dissiliaria F. Muell.). Euphorbiaceae. 1 S. New Guinea, NE. Austr.
Chorilaena Endl. Rutaceae. 2 W. Austr.
Chorilepidella Van Tiegh. = Lepidaria Van Tiegh. (Loranthac.).
Chorilepis Van Tiegh. = Lepidaria Van Tiegh. (Loranthac.).
Chorioluma Baill. = Planchonella Pierre (Sapotac.).
Choriophyllum Benth. = Longetia Baill. = Austrobuxus Miq. (Euphorbiac.).
Choriosphaera Melch. (nomen) = ?Pseudocalymma A. Sampaio & Kuhlm. (Bignoniac.).
Choriozandra Steud. = Chorizandra R.Br. (Cyperac.).
Choripetalum A. DC. = Embelia Burm. f. (Myrsinac.).
Choriptera Botsch. (~ Salsola L.). Chenopodiaceae. 1 Abd al Kuri & Semha (W. of Socotra).
Chorisandra Benth. & Hook. f. = Chorizandra R.Br. (Cyperac.).
Chorisandra Wight = Phyllanthus L. (Euphorbiac.).
Chorisandrachne Airy Shaw. Euphorbiaceae. 1 Penins. Siam.
Chorisanthera Oerst. = Pentarhaphia Lindl. (Gesneriac.).
Chorisema Fisch. = Chorizema Labill. (Legumin.).
Chorisepalum Gleason & Wodehouse. Gentianaceae. 4 Venezuela.
Chorisia Kunth. Bombacaceae. 5 trop. S. Am. *C. speciosa* A. St-Hil. (*paina de seda*) gives a useful silky cotton from the pods.
Chorisis DC. = Lactuca L. (Compos.).
Chorisiva (A. Gray) Rydb. = Iva L. (Compos.).
Chorisma D. Don = Chorisis DC. = Lactuca L. (Compos.).
Chorisma Lindl. = Pelargonium L'Hérit. (Geraniac.).
Chorispermum R.Br. = seq.
*Chorispora R.Br. ex DC. Cruciferae. 10 E. Medit., C. As.

CHORISTANTHUS

Choristanthus K. Schum. = Eleutheranthus F. Muell. (Rubiac.).
Choristea Thunb. = Didelta L'Hérit. (Compos.).
Choristega Van Tiegh. = Lepeostegeres Bl. (Loranthac.).
Choristegeres Van Tiegh. = praec.
Choristemon Williamson. Epacridaceae. 1 Victoria.
Choristes Benth. = Deppea Cham. & Schlechtd. (Rubiac.).
Choristigma Baill. = Tetrastylidium Engl. (Olacac.).
Choristigma Kurtz ex Heger = Stuckertia Kuntze (Asclepiadac.).
Choristosoria Mett. = Pellaea Link (Sinopteridac.).
Choristylis Harv. Iteaceae. 1–2 E. trop. & S. Afr. Affinity with *Itea* confirmed by char. pollen.
Choritaenia Benth. Umbelliferae. 1 S. Afr.
Chorizandra Benth. & Hook. f. = Chorisandra Wight = Phyllanthus L. (Euphorbiac.).
Chorizandra R.Br. Cyperaceae. 4 Austr., Tasm.
Chorizandra Griff. ex C. B. Clarke = Boeica C. B. Clarke (Gesneriac.).
Chorizanthe R.Br. Polygonaceae. 50 dry W. Am. Some have an ochrea, usu. absent in this group. Fls. usu. single inside the invol. (cf. *Eriogonum*).
Chorizema Labill. Leguminosae. 15 W. Austr., 1 E. Austr.
Chorizopteris Moore = Lomagramma J. Sm. (Lomariopsidac.).
Chorizospermum P. & K. = Corizospermum Zipp. ex Bl. = Casearia Jacq. (Flacourtiac.).
Chorizotheca Muell. Arg. = Pseudanthus Sieber (Euphorbiac.).
Chorobanche B. D. Jacks. = seq.
Chorobane Presl = Orobanche L. (Orobanchac.).
Chorosema Brongn. = seq.
Chorozema Sm. = Chorizema Labill. (Legumin.).
Chortolirion Berger. Liliaceae. 4 Afr.
Chosenia Nak. Salicaceae. 1 temp. & subarct. NE. As. Shrub or small tree closely resembling *Salix*. Buds covered with a single, non-calyptrate scale. Ls. simple, alt. Fls. anemoph. Scale (or bract) ent., strongly devel. and imbr. in ♂ fl. Nectary o. Sta. 5, adnate to scale. Styles 2, stigs. bifid.
Chotchia Benth. (sphalm.) = Chotekia Opiz & Corda = Pogostemon Desf. (Labiat.).
Choteckia Steud. = seq.
Chotekia Opiz & Corda = Pogostemon Desf. (Labiat.).
Chotellia Hook. f. (sphalm.) = praec.
Choulettia Pomel = Gaillonia A. Rich. (Rubiac.).
Chouxia Capuron. Sapindaceae. 1 Madag.
Chresta Vell. = Eremanthus Less. (Compos.).
Chrestienia Montrouz. = Pseuderanthemum Radlk. (Acanthac.).
Chrisanthemum Neck. = Chrysanthemum L. (Compos.).
Chrisosplenium Neck. = Chrysosplenium L. (Saxifragac.).
Christannia Presl = Pineda Ruiz & Pav. (Flacourtiac.).
Christannia Walp. (sphalm.) = Christiana DC. (Tiliac.).
Christella Lév. emend. Holtt. Thelypteridaceae. Pantropic, 40 spp. in Old World; *P. parantica* (Linn.) Lév. and a few allied spp. widely distributed in man-made clearings. Club-shaped glands present on sporangium-stalks.

Christensenia Maxon. Christenseniaceae. 1 Assam, W. Malaysia. Ls. palmate; veins anast.; synangia circular.

Christenseniaceae Ching. Marattiales (*q.v.* for chars.). 1 genus: *Christensenia*.

Christia Moench. Leguminosae. 12 Indoch., Indomal., Austr.

Christiana DC. Tiliaceae. 2 trop. S. Am., trop. Afr., Madag.

Christiania Reichb. = praec.

× **Christieara** hort. Orchidaceae. Gen. hybr. (iii) (Aërides × Ascocentrum × Vanda).

Christiopteris Copeland. Polypodiaceae. 2 SE. As., New Caled. Fertile ls. acrostichoid.

Christisonia Gardn. Orobanchaceae. 17 SW. China, SE. As., Indomal. Roots parasitic on those of bamboos or *Acanthaceae*, united to a dense meshwork. The flg. shoots spring up, die, and decay, in a fortnight.

Christmannia Dennst. = Courondi Adans. = Salacia L. (Celastrac.).

Christolea Cambess. ex Jacquem. Cruciferae. 13 C. As., W. Himal., Afgh., Tib., Kamch., Alaska.

Christolia P. & K. = Chrystolia Montr. ex Beauvis. = Soja Moench (Legumin.).

Christophoriana Kuntze = Knowltonia Salisb. (Ranunculac.).

Christophoriana Mill. = Actaea L. (Ranunculac.).

Christopteris Willis (sphalm.) = Christiopteris Copel. (Polypodiac.).

Christya Ward & Harv. = Strophanthus DC. (Apocynac.).

Chritmum Brot. (sphalm.) = Crithmum L. (Umbellif.).

Chroësthes Benoist. Acanthaceae. 1 S. China, N. Indoch.

Chroïlema Bernh. Compositae. 1 Chile.

Chromanthus Phil. = Talinum Adans. (Portulacac.).

Chromatolepis Dulac = Carlina L. (Compos.).

Chromatopogon F. W. Schmidt = Scorzonera L. (Compos.).

Chromochiton Cass. = Cassinia R.Br. (Compos.).

Chromolaena DC. (~Eupatorium L.). Compositae. 130 S. U.S., W.I., C. & trop. S. Am.

Chromolepis Benth. Compositae. 1 Mexico.

Chromolucuma Ducke. Sapotaceae. 2 Guiana, NE. Brazil.

Chromophora P. & K. = Chrozophora Neck. ex Juss. (Euphorbiac.).

Chronanthos K. Koch = Genista L. (Legumin.).

Chrone Dulac = Eupatorium L. (Compos.).

Chroniochilus J. J. Smith. Orchidaceae. 3 Siam, Malay Penins., Java, Fiji.

Chronobasis DC. ex Benth. & Hook. f. = Ursinia Gaertn. (Compos.).

Chronopappus DC. Compositae. 1 Brazil.

Chrosothamnus P. & K. = Chrysothamnus Nutt. = Aster L. (Compos.).

Chrosperma Rafin. = Amianthium A. Gray (Liliac.).

*****Chrozophora** Neck. ex Juss. corr. Benth. & Hook. f. Euphorbiaceae. 12 Medit., trop. Afr. to India. *C. tinctoria* (L.) Juss. and *C. verbascifolia* (Willd.) Juss. are characteristic plants of the Medit. region. The former, once medicinal, is still sometimes used as the source of the dye turn-sole, *tournesol*, or *bezetta rubra*.

Chrozorrhiza Ehrh. (uninom.) = *Galium tinctorium* (L.) Scop. (Rubiac.).

CHRYSA

Chrysa Rafin. = Chryza Rafin. = Coptis Salisb. (Ranunculac.).

× **Chrysaboltonia** Arends. Compositae. Gen. hybr. (Boltonia × Chrysanthemum).

Chrysactinia A. Gray. Compositae. 4 SW. U.S., Mex.

Chrysactinium Wedd. = Liabum Adans. (Compos.).

Chrysaea Nieuwl. & Lunell = Impatiens L. (Balsaminac.).

Chrysalidocarpus H. Wendl. Palmae. 20 Pemba, Madag., Comoro Is. *C. lutescens* H. Wendl. branches at the r. and forms tufts of stems.

Chrysallidosperma H. E. Moore. Palmae. 1 Peru.

Chrysamphora Greene = Darlingtonia Torr. (Sarraceniac.).

Chrysangia Link = Musschia Dum. (Campanulac.).

Chrysanthellina Cass. = seq.

Chrysanthellum Rich. Compositae. 5 trop.

× **Chrysanthemoachillea** Prodan. Compositae. Gen. hybr. (Achillea × Chrysanthemum).

Chrysanthemodes P. & K. = seq.

Chrysanthemoïdes Medik. Compositae. 2 trop. & S. Afr.

Chrysanthemopsis K. H. Rechinger = Smelowskia C. A. Mey. (Crucif.).

Chrysanthemum L. Compositae. *Sensu lato*, 200 Eur., As., Afr., Am.; *sensu stricto*, 5 Euras., Medit. *C. parthenium* (L.) Bernh. (feverfew, Eur.), a popular remedy against fever; *C. cinerariifolium* Vis. yields Dalmatian, and *C. roseum* Adam Persian, insect powder (the dried and powdered fls.). In the strict sense, the genus is limited to *C. segetum* L., *C. coronarium* L., and their immediate allies.

Chrysaspis Desv. = Trifolium L. (Legumin.).

Chrysastrum Willd. ex Wedd. = Liabum Adans. (Compos.).

Chryseïs Cass. = Centaurea L. (Compos.).

Chryseïs Lindl. = Eschscholtzia Cham. (Papaverac.).

Chrysion Spach = Viola L. (Violac.).

Chrysiphiala Ker-Gawl. = Stenomesson Herb. (Amaryllidac.).

Chrysis DC. = Helianthus L. (Compos.).

Chrysithrix L. corr. Spreng. Cyperaceae. 4 S. Afr., 1 W. Austr.

Chrysitrix L. = praec.

Chrysobactron Hook. f. = Bulbinella Kunth (Liliac.).

*Chrysobalanaceae R.Br. Dicots. 10/400 trop. & subtrop. Trees or shrubs, with alt. simple ent. stip. ls. Fls. ⚥, more rarely ♂ ♀, usu. ± zygo., in simple or cpd. rac. infl. K (5), tube turbin. or campan., ± unequal or calcarate at base, segs. free or ± conn., imbr.; C 5–0, often unequal, shortly unguic., inserted in mouth of K-tube; A 2–∞, inserted with C, often larger and fert. opp. the larger K-segs., shorter and ± ster. on opp. side, fil. filif., exserted, anth. intr.; G (1) with 2 erect basal collat. ov., v. rarely (2) with 1-ov. loc., sessile or more often stipit. at base of K-tube; style simple, filif., lat. or almost gynobasic, with simple stig. Fruit a sess. or stipit. drupe, with bony almost 2-valved endoc., more rarely a crustaceous berry; embryo with thick fleshy cots.; endosp. o. Chief genera: *Parinari, Hirtella, Couepia, Licania, Magnistipula*.

Chrysobalanus L. Chrysobalanaceae. 4 trop. Am., W.I., trop. Afr. Style basal. *C. icaco* L. (coco plum), W.I., fr. ed.

Chrysobaphus Wall. = Anoectochilus Blume (Orchidac.).

Chrysobotrya Spach = Ribes L. (Grossulariac.).
Chrysocactus Y. Ito = Notocactus (K. Schum.) A. Berger (Cactac.).
Chrysocalyx Guill. & Perr. = Crotalaria L. (Legumin.).
Chrysocephalum Walp. = Helichrysum L. (Compos.).
Chrysochamela (Fenzl) Boiss. Cruciferae. 4 E. Medit. to Armenia & Iraq.
Chrysochlamys Poepp. Guttiferae. 20 trop. Am.
Chrysochloa Swallen. Gramineae. 5 trop. Afr.
Chrysochosma (J. Sm.) Kümm. = Aleuritopteris Fée (Sinopteridac.).
Chrysocoma L. Compositae. 50 S. Am., trop. & S. Afr.
Chrysocome St-Lag. = praec.
Chrysocoptis Nutt. = Coptis Salisb. (Ranunculac.).
Chrysocoryne Endl. = Angianthus Wendl. (Compos.).
Chrysocyathus Falc. = Calathodes Hook. f. & Thoms. (Ranunculac.).
Chrysocychnis Linden & Reichb. f. Orchidaceae. 5 Colombia & Ecuador.
Chrysodendron Meissn. = Protea L. (Proteac.).
Chrysodendron Teran & Berland. = Mahonia Nutt. (Berberidac.).
Chrysodiscus Steetz = Athrixia Ker-Gawl. (Compos.).
Chrysodium Fée = Acrostichum L. (Pteridac.).
Chrysoglossella Hatusima = Hancockia Rolfe (Orchidac.).
Chrysoglossum Blume. Orchidaceae. 12 Indomal., Polynes.
Chrysogonum A. Juss. = Leontice L. (Leonticac.).
Chrysogonum L. Compositae. 1 E. U.S.
Chrysolarix H. E. Moore = Pseudolarix Gordon (Pinac.).
Chrysolepis Hjelmq. (~ Castanopsis (D. Don) Spach). Fagaceae. 2 W. U.S.
♀ fl. opp. or subopp.
Chrysoliga Willd. ex DC. = Nesaea Comm. ex Juss. (Lythrac.).
Chrysolinum Fourr. = Linum L. (Linac.).
Chrysolyga Willd. ex Steud. = Nesaea Comm. ex Juss. (Lythrac.).
Chrysoma Nutt. = Solidago L. (Compos.).
Chrysomallum Thou. = Vitex L. (Verbenac.).
Chrysomelea Tausch = Coreopsis L. (Compos.).
Chrysomelon J. R. & G. Forst. ex A. Gray = Spondias L. (Anacardiac.).
Chrysonias Benth. ex Steud. = Chrysoscias E. Mey. = Rhynchosia Lour.
Chrysopappus Takhtadj. Compositae. 1 Kurdistan.   [(Legumin.).
Chrysopelta Tausch = Achillea L. (Compos.).
Chrysophaë K.-Pol. Umbelliferae. 2 E. Medit.
Chrysophania Kunth ex Less. = Zaluzania Pers. (Compos.).
Chrysophiala P. & K. = Chrysiphiala Ker-Gawl. = Stenomesson Herb.
(Amaryllidac.).
Chrysophora Cham. ex Triana = Oxymeris DC. (Melastomatac.).
Chrysophthalmum Phil. = Grindelia Willd. (Compos.).
Chrysophthalmum Sch. Bip. Compositae. 3 W. As.
Chrysophyllum L. Sapotaceae. 150 trop., esp. Am. Serial buds form in
each leaf-axil in some spp. and the undeveloped ones subsequently give rise to
fls. borne on the old wood. *C. cainito* L. (star-apple, W.I.), cult for ed. fr.
Chrysopia Nor. ex Thou. = Symphonia L. (Guttif.).
*****Chrysopogon** Trin. Gramineae. 25 trop. & subtrop., esp. Old World.
*****Chrysopsis** (Nutt.) Ell. Compositae. 20 N. Am.

CHRYSOPTERIS

Chrysopteris Link = Phlebodium (R.Br.) J. Sm. (Polypodiac.).
Chrysorhoë Lindl. = Verticordia DC. (Myrtac.).
Chrysoscias E. Mey. = Rhynchosia Lour. (Legumin.).
Chrysosperma Dur. & Jacks. = Chrosperma Rafin. = Zigadenus Michx (Liliac.).
Chrysospermum Reichb. = Anthospermum L. (Rubiac.).
Chrysosphaerium Willd. ex DC. = Calea L. (Compos.).
Chrysosplenium L. Saxifragaceae. 55 N. temp. & arct., N. Afr., temp. S. Am. Rhiz. bears both veg. and flg. shoots. Infl. cymose. The small greenish fls. are perig. and apet., homogamous. (Cf. *Adoxa*.)
Chrysostachys Poepp. ex Baill. = Sclerolobium Vog. (Legumin.).
Chrysostachys Pohl = Combretum L. (Combretac.).
Chrysostemma Less. = Coreopsis L. (Compos.).
Chrysostemma E. Mey. ex Spach = Gorteria L. (Compos.).
Chrysostemon Klotzsch = Pseudanthus Sieber (Euphorbiac.).
Chrysostoma Lilja = Mentzelia L. (Loasac.).
Chrysothamnus Nutt. Compositae. 12 W. N. Am.
Chrysothemis Decne. Gesneriaceae. 6 W.I., C. Am., northern S. Am.
Chrysothrix Roem. & Schult. = Chrysithrix L. (Cyperac.).
Chrysoxylon Casaretto = Plathymenia Benth. (Legumin.).
Chrysoxylon Wedd. = Pogonopus Klotzsch (Rubiac.).
Chrystolia Montrouz. = Glycine L. (Legumin.).
Chrysurus Pers. = Lamarckia Moench (Gramin.).
Chrytotheca G. Don = Cryptotheca Blume = Ammannia L. (Lythrac.).
Chryza Rafin. = Coptis Salisb. (Ranunculac.).
Chthamalia Decne = Lachnostoma Kunth (Asclepiadac.).
Chthonia Cass. = Pectis L. (Compos.).
Chthonocephalus Steetz. Compositae. 4 temp. Austr.
Chucoa Cabrera. Compositae. 1 Peru.
Chukrasia A. Juss. Meliaceae. 1–2 S. China to Indomal. Timber valuable (Indian red wood, Chittagong wood, white cedar).
Chulusium Rafin. = Polygonum L. (Polygonac.).
Chumsriella Bor. Gramineae. 1 NW. Siam.
Chuncoa Pav. ex Juss. = Terminalia L. (Combretac.).
Chunechites Tsiang. Apocynaceae. 1 SE. China.
Chunia Chang. Hamamelidaceae. 1 Hainan. Near *Exbucklandia* and *Mytilaria*.
Chuniodendron Hu. Meliaceae. 2 SW. China.
Chuniophoenix Burret. Palmae. 2 S. China, N. Indoch.
Chupalon Adans. = ? Cavendishia Lindl. (Ericac.).
Chuquiraga Juss. Compositae. 40 S. Am. In each axil are thorns, probably repres. ls. of an undeveloped branch; above is a normal branch.
Churumaya Rafin. = Piper L. (Piperac.). [plateaux in S. Am.
Chusquea Kunth. Gramineae. 70 Am. Like *Bambusa* (q.v.). Char. of high
Chusua Nevski. Orchidaceae. 2 N. India to E. As.
Chydenanthus Miers. Barringtoniaceae. 1 Burma, Andam., Borneo, Sumatra to W. New Guinea.
Chylaceae Dulac = Fumariaceae DC.
Chylismia Nutt. = Camissonia Link (Onagrac.).

Chylocalyx Hassk. ex Miq. = Echinocaulos (Meissn. ex Endl.) Hassk. (Polygonac.).

Chylodia Rich. ex Cass. = Wulffia Neck. ex Cass. (Compos.).

Chylogala Fourr. = Euphorbia L. (Euphorbiac.).

Chymaceae Dulac = Papaveraceae Juss.

Chymocarpus G. Don = Tropaeolum L. (Tropaeolac.).

Chymococca Meissn. = Passerina L. (Thymelaeac.).

Chymocormus Harv. = Fockea Endl. (Asclepiadac.).

Chymsydia Alboff (~ Agasyllis Spreng.). Umbelliferae. 1 Transcauc.

Chysis Lindl. Orchidaceae. 6 trop. Am.

Chytra Gaertn. f. = Gerardia Benth. (Scrophulariac.).

Chytraculia P.Br. = Calyptranthes Sw. (Myrtac.).

Chytralia Adans. = praec.

Chytranthus Hook. f. Sapindaceae. 30 trop. Afr.

Chytroglossa Reichb. f. Orchidaceae. 3 Brazil.

Chytroma Miers. Lecythidaceae. 46 trop. S. Am.

Chytropsia Bremek. Rubiaceae. 3 Venez., Guiana.

Cianitis Reinw. = Cyanitis Reinw. = Dichroa Lour. (Hydrangeac.).

Cibotarium O. E. Schulz. Cruciferae. 5 Mexico.

Cibotium Kaulf. Dicksoniaceae. 10 trop. As., Am., Hawaii. Hairs of stem used as styptic in Asia. Chinese spec. in 17th cent. mistakenly thought to be plant which gave rise to story of Tartarian lamb. (A. F. Tryon, *Missouri Bot. Gard. Bull.* **43**: 25–8, 1955.)

Cicca L. (~ Phyllanthus L.). Euphorbiaceae. 1 trop., long cult. for ed. fr.; origin probably coastal forests of NE. Brazil.

Cicendia Adans. Gentianaceae. 1 Eur., N. Afr., As. Minor; 1 Calif., W. S. Am.

Cicendia Griseb. = Exaculum Caruel (Gentianac.).

Cicendiola Bub. = Cicendia Adans. (Gentianac.).

Cicendiopsis Kuntze = Exaculum Caruel (Gentianac.).

Cicer L. Leguminosae. 20 N. Afr., Abyss., E. Medit. to C. As. Accessory buds in axils in some. *C. arietinum* L. (chick-pea, gram), cult. for food in

Cicerbita Wallr. Compositae. 18 N. temp., esp. mts. [S. Eur., Ind.

Cicercula Medik. = Lathyrus L. (Legumin.).

Cicerella DC. = praec.

Ciceronia Urb. Compositae. 1 Cuba.

Cicheria Rafin. = Cichorium L. (Compos.).

Cichlanthus Van Tiegh. = Scurrula L. (Loranthac.).

***Cichor[i]aceae** Juss. = Compositae–Cichorieae DC.

Cichorium L. Compositae. 9 Eur., Medit., Abyss. *C. intybus* L. (chicory); the r., roasted and ground, is mixed with coffee. *C. endivia* L. (endive), a pot-herb; its ls. are blanched.

Ciclospermum La Gasca (~ Apium L.). Umbelliferae. 1 warm Am.

Ciconium Sweet = Pelargonium L'Hérit. (Geraniac.).

Cicuta L. Umbelliferae. 10 N. temp. *C. virosa* L. (cow-bane or water-

Cicuta Mill. = Conium L. (Umbellif.) [hemlock) highly poisonous.

Cicutaria Fabr. (uninom.) = *Conium* L. sp. (Umbellif.).

Cicutaria Lam. = Cicuta L. (Umbellif.).

Cicutaria Mill. = Molopospermum Koch (Umbellif.).

CICUTASTRUM

Cicutastrum Fabr. (uninom.) = *Thapsia* L. sp. (Umbellif.).

Cieca Adans. = Julocroton Mart. (Euphorbiac.).

Cieca Medik. = Passiflora L. (Passiflorac.).

Cienfuegosia Cav. Malvaceae. 20 trop. & subtrop. Am., Afr., Austr.

Cienkowskia Regel & Rach = ? Ehretia L. (Ehretiac.).

Cienkowskya Solms = Kaempferia L. (Zingiberac.).

Ciliaria Haw. = Saxifraga L. (Saxifragac.).

Ciliovallaceae Dulac = Campanulaceae–Lobelioideae Engl.

Cimbaria Hill = Cymbaria L. (Scrophulariac.).

Cimicifuga Wernischek (~ Actaea L.). Ranunculaceae. 15 N. temp. *C. foetida* L. (bugbane, Eur.) is used as preventive against vermin; r. of *C. racemosa* Nutt. (black snake-root, N. Am.) as emetic.

Ciminalis Adans. = Gentiana L. (Gentianac.).

Ciminalis Rafin. = Leiphaimos Cham. & Schlechtd. (Gentianac.).

Cinara L. = Cynara L. (Compos.).

Cinarocephalae Juss. = Cynaraceae Lindl. = Compositae–Cynareae Spreng.

Cinchona L. Rubiaceae. 40 Andes. Trees. Fl. heterostyled in some. The source of Peruvian or Jesuit's bark, from which are extracted the valuable drugs (alkaloids) quinine, cinchonidine, etc. The tree used to be cut down to obtain the bark and there was danger of extinction until cult. was started on a large scale. The Dutch in 1854, followed by the British in 1859, brought it to the East, where Java and esp. India and Ceylon took up its cult., and upon so large a scale as to reduce the price of quinine from 12s. to 1s. an ounce. Decrease in price, the lack of any improvement in the barks, and attacks of disease, made the cult. die out in Ceylon; and for a time Java, where improvement was taken in hand, almost monopolised it. India grows a great deal for supply to natives through the post offices; antimalarial doses could at one time be bought for the equivalent of ½d. each. Several spp. are used, e.g. *C. calisaya* Wedd. var. *ledgeriana* Moens (yellow bark, the richest in alkaloid), *C. cordifolia* Mutis (Cartagena bark), *C. officinalis* L. (*condaminea* H. & B.) (Loxa, crown or brown bark), *C. succirubra* Pav. (red bark). Cinchona has lost much of its importance through the widespread adoption of synthetic anti-malarials free from some disadvantages of natural quinine.

Cinchonaceae Lindl. = Rubiaceae, excl. tribe Rubieae.

Cincinalis Desv. = Cheilanthes Sw. (Sinopteridac.).

Cincinalis Gled. = Pteridium Scop. (Dennstaedtiac.).

Cincinnobotrys Gilg. Melastomataceae. 3 trop. Afr.

Cinclia Hoffmgg. = Ceropegia L. (Asclepiadac.).

Cinclidocarpus Zoll. & Mor. = Caesalpinia L. (Legumin.).

Cineraria L. [type *C. maritima* (L.) L.] = Senecio L. (Compos.).

Cineraria L. emend. Less. [type *C. geifolia* (L.) L.] (~ Senecio L.). Compositae. 50 Afr., Madag.

Cinga Nor. = Cyrtandra J. R. & G. Forst. (Gesneriac.).

Cinhona L. = Cinchona L. (Rubiac.).

Cinna L. Gramineae. 4 N. & S. Am., temp. Euras. A 1.

Cinnabarinea Frič = Lobivia Britt. & Rose (Cactac.).

Cinnagrostis Griseb. = Cinna L. (Gramin.).

Cinnamodendron Endl. Canellaceae. 7 trop. S. Am., W.I.

Cinnamomum Schaeffer. Lauraceae. 250 E. As., Indomal. Young leaves often red. *C. zeylanicum* Blume (Ceylon) is the cinnamon. The pl. is coppiced in cult., and the bark of the twigs peeled off and rolled up is the spice. *C. cassia* (L.) Blume (China, Japan) yields cassia bark, often used to adulterate cinnamon. Its fl. buds are used as a spice (cf. *Syzygium*). *C. camphora* (L.) T. Nees & Eberm. (China, Japan, Formosa) is the camphor. The old trees are felled, and the wood cut into chips and distilled with steam, but in cult. the camphor is distilled from young shoots.

Cinnamosma Baill. Canellaceae. 3 Madag.

Cinnastrum Fourn. = Cinna L. (Gramin.).

Cinogasum Neck. = Croton L. (Euphorbiac.).

Cinsania Lavy. ? Ericaceae. 1 NW. Italy. Quid?

Cionandra Griseb. = Cayaponia S. Manso (Cucurbitac.).

Cionidium Moore. Aspidiaceae. 1 New Caled.

Cionisaccus Breda = Goodyera R.Br. (Orchidac.).

Cionosicyos Griseb. corr. Hook. f. Cucurbitaceae. 3 C. Am., W.I.

Cionosicys Griseb. (sphalm.) = praec.

Cionosicyus P. & K. = praec.

Cionura Griseb. (~ Marsdenia R.Br.). Asclepiadaceae. 1 Balk. Penins. & Crete to W. Iran (& W. Afghan.?).

Cipadessa Blume. Meliaceae. 3 Madag., Indomal.

Ciponima Aubl. = Symplocos L. (Symplocac.).

Cipura Aubl. Iridaceae. 2 trop. Am.

Cipura Klotzsch ex Klatt = Herbertia Sweet = Alophia Herb. (Iridac.).

Cipuropsis Ule. Bromeliaceae. 1 Peru.

Circaea L. Onagraceae. 12 N. temp. & arctic. Fl. dimerous with one whorl of sta. Fr. covered with hooked bristles.

Circaeaceae Lindl. = Onagraceae Juss.

Circaeaster Maxim. Circaeasteraceae. 1 NW. Himal. to NW. China.

*Circaeasteraceae [Kuntze ex] Hutch. Dicots. 1/1 temp. As. Small ann. glabr. herb; cots. lin.-obl., persist.; ls. cuneate-spathulate, spinul.-dent. at apex, with open dichot. venation, rosulately crowded at summit of short slender stem (elong. hypocotyl); fls. fascic. in condensed infls., shortly pedic. K 2–3, valv., persist.; C o; A 1–3, alt. with sep., anth. introrse, thecae divergent from apex; G 1–3, free, linear, with oblique sess. stig., ov. 1, apical, pend.; fr. an achene, covered with fine uncinate setae; endosp. copious. Only genus: *Circaeaster*. A very remarkable little plant, doubtless representing an extreme reduction from the general ranunculaceous type, but too isolated to be included in *Ranunculac.* or any other fam. The dichot. l.-ven. without cross-veins is exactly paralleled by that of *Kingdonia*. (Cf. Junell, *Svenska Bot. Tidsskr.* 25: 238–270, 1931; Foster, *J. Arn. Arb.* 44: 299–321, 1963.)

Circaeocarpus C. Y. Wu. Saururaceae. 1 SW. China, Indoch. Densely pelluc.-punct.; infl. a lax elong. rac.; anth. intr.; G (3–4), with 1 ov. per loc.; fr. glob., indehisc., densely glochidiate.

Circandra N. E. Br. Aïzoaceae. 1 S. Afr.

Circea Rafin. = Circaea L. (Onagrac.).

Circinus Medik. = Hymenocarpos Savi (Legumin.).

Circis Chapm. = Cercis L. (Legumin.).

CIRCUMACEAE

Circumaceae Dulac = Carophyllaceae–Alsinoïdeae Vierh.
Cirinosum Neck. = Cereus Haw. (Cactac.).
Ciripedium Zumaglini (sphalm.) = Cypripedium L. (Orchidac.).
Cirrhaea Lindl. Orchidaceae. 7 Brazil. [Tahiti.
*Cirrhopetalum Lindl. Orchidaceae. 70 trop. Afr., Masc., Indomal. to
× **Cirrhophyllum** hort. Orchidaceae. Gen. hybr. (vi) (Bulbophyllum ×
Cirrhopetalum).
Cirsellium Gaertn. = Atractylis L. (Compos.).
× **Cirsio-Carduus** P. Fourn. Compositae. Gen. hybr. (Carduus × Cirsium).
Cirsium Mill. Compositae. 150 N. temp. The genus is probably artificially
separated from *Carduus* L.
Cissaceae Horan. = Vitidaceae Juss.
Cissampelopsis Miq. = Senecio L. (Compos.).
Cissampelos L. Menispermaceae. 30 trop.
Cissarobryon Kunze = Viviania Cav. (Vivianiac.).
Cissodendron F. Muell. = Kissodendron Seem. (Araliac.).
Cissodendrum P. & K. = praec.
Cissus L. Vitidaceae. 350 trop., rarely subtrop. Fl. ♀. C 4.
*Cistaceae Juss. Dicots. 8/200 in dry sunny places, espec. on chalk or sand,
a few S. Am., the rest N. temp. (espec. Medit.). Shrubs and herbs with opp.
rarely alt. ls., often inrolled (cf. *Ericaceae*), stip. or not. Ethereal oil and
glandular hairs usu. present. Fl. sol. or in cymes, ♀, reg. K 5, the two outer
usu. smaller than the inner (sometimes regarded as bracteoles, but these are
found lower down); C 5, 3 or o (in cleistog. fls.), conv. to right or left according
as 3 inner seps. overlap to left or right, usu. crumpled in bud; A usu. ∞ on
a hypogynous disk, developed in descending order; G̲ (5–10 or 3), 1-loc. with
parietal often projecting plac. each bearing ∞ or 2 ascending orthotr. or anatr.
ov.; styles free or not. Caps. usu. loculic.; endosp. and curved embryo.
Genera: *Cistus* (C 5, ov. ∞, caps. 10–5-valved), *Helianthemum* (do. but
3-valved), *Halimium*, *Crocanthemum* (ls. alt.), *Tuberaria*, *Fumana* (outer A ster.,
torulose, ov. pend. anatr.), *Hudsonia* (C contorted), *Lechea* (C 3). (See
monograph by Grosser in Engl., *Pflanzenr.*, 1903.)
Cistanche Hoffmgg. & Link. Orobanchaceae. 16 NW. Afr., Abyss., Medit.
to W. India and NW. China.
Cistanthe Spach = Calandrinia Kunth (Portulacac.).
Cistanthera K. Schum. = Nesogordonia Baill. (Sterculiac.).
Cistella Blume = Geodorum G. Jacks. (Orchidac.).
Cistellaria Schott. ?Rutac. ?Simaroubac. 1 Brazil.
Cisticapnos Adans. = Cysticapnos Mill. (Fumariac.).
Cistocarpium Spach = Vesicaria Adans. (Crucif.).
Cistocarpum Pfeiff. = seq.
Cistocarpus Kunth = Balbisia Cav. (Ledocarpac.).
Cistomorpha Caley ex DC. = Hibbertia Andr. (Dilleniac.).
Cistrum Hill = Centaurea L. (Compos.).
Cistula Nor. = Maesa Forsk. (Myrsinac.).
Cistus L. Cistaceae. 20 Canary Is., Medit. to Transcauc. *C. villosus* L. var.
creticus (L.) Boiss. and *C. ladanifer* L. yield the resin ladanum (not laudanum),
formerly offic.

Cistus Medik. = Helianthemum L. (Cistac.).
Citharaexylon Adans. = Citharexylum L. (Verbenac.).
Citharella Nor. = Eranthemum L. (Acanthac.).
Cithareloma Bunge. Cruciferae. 3 Persia, C. As.
Citharexylum Mill. Verbenaceae. 115 S. U.S. to Argentine. Fr. often tightly enclosed in K. Drupe with 2 stones. Timber (fiddlewood, from *bois fidèle*).
Citinus All. = Cytinus L. (Rafflesiac.).
Citrabenis Thou. (uninom.) = *Habenaria* (?*Cynorkis* sp.) *citrina* Thou. (Orchidac.).
Citraceae Drude = Aurantiaceae Juss. = Rutaceae–Aurantioïdeae Engl.
Citrangis Thou. (uninom.) = *Angraecum citratum* Thou. = *Aërangis citrata* (Thou.) Schltr. (Orchidac.).
Citreum Mill. = Citrus L. (Rutac.).
Citriobathus A. Juss. = seq.
Citriobatus A. Cunn. Pittosporaceae. 5 Malaysia (Philipp., Cel., Java), N. & E. Austr.
Citriopsis Pierre ex A. Chev. (nomen). Rutaceae. 1 trop. W. Afr. Ouid?
Citriosma Tul. = Citrosma Ruiz & Pav. = Siparuna Aubl. (Siparunac.).
Citronella D. Don. Icacinaceae. *S.str.*, 7 C. & trop. S. Am.; *s.l.*, 30 Malaysia, trop. Austr., Pacif., trop. Am. *C. gongonha* (Mart.) Howard is used like maté
Citrophorum Neck. = Citrus L. (Rutac.).   [(*Ilex*).
Citropsis Swingle & Kellermann. Rutaceae. 10 trop. Afr.
Citrosena Bosc ex Steud. (sphalm.) = seq.
Citrosma Ruiz & Pav. = Siparuna Aubl. (Siparunac.).
*Citrullus Schrad. ex Eckl. & Zeyh. Cucurbitaceae. 3 Afr., Medit., trop. As. *C. lanatus* (Thunb.) Mansf., the watermelon; *C. colocynthis* (L.) Kuntze (colocynth), fr. a drug.
Citrus L. Rutaceae. 12 S. China, SE. As., Indomal. Trees or shrubs with usu. simple ls., which show a joint at meeting of blade and stalk, indicating their derivation from cpd. ls. like those of most of the fam. (cf. *Berberis*). Axillary thorns in some = metam. ls. of branch shoot. Fls. ☿ in corymbs. K, C 4–8; A ∞ in irreg. bundles, corresponding to an outer whorl only. G̲ (∞) (6 or more); a second whorl sometimes appears. Berry with leathery epicarp, the flesh made up of large cells which grow out from inner layer of pericarp.
  Many cult. in warm countries, espec. California and Florida, the W. Indies, Brazil, the Medit. region (Israel, Italy, N. Afr.), etc., for their fr. *C. medica* L. is the citron; *C. limon* (L.) Burm. f., the lemon; *C. aurantiifolia* (Christm.) Swingle, the lime, var. *limetta* the sweet lime; *C. aurantium* L. is the orange, with its vars. *bergamia*, the Bergamot orange (from which the perfume is obtained), var. *aurantium* (*bigaradia* or *amara*) the Seville or bitter orange, used in marmalade; *C. decumana* L. the shaddock, or pomelo; *C. paradisi* Macf., the grape-fruit; *C. sinensis* (L.) Osb., the Malta, Portugal, or sweet orange; and others. *C. nobilis* Lour. is the true mandarin orange, *C. reticulata* Blanco the tangerine.
Citta Lour. = Mucuna Adans. (Legumin.).
Cittaronium Reichb. = Viola L. (Violac.).
Cittorhinchus Willd. ex Kunth = Ouratea Aubl. (Ochnac.).
Cittorhynchus P. & K. = praec.

CLADANDRA

Cladandra O. F. Cook. Palmae. 1 Colombia.

Cladanthus Cass. Compositae. 4 S. Spain, NW. Afr.

Cladapus This.-Dyer (sphalm.) = Cladopus H. Moeller (Podostemac.).

*Claderia Hook. f. Orchidaceae. 2 Malaysia (exc. Philippines).

Claderia Rafin. = Murraya Koenig ex L. (Rutac.).

Cladium P.Br. Cyperaceae. 50–60 trop. & temp., esp. Austr.; 1 almost cosmop.

Cladobium Lindl. = Scaphyglottis Poepp. (Orchidac.).

Cladobium Schlechter = Lankesterella Ames (Orchidac.).

Cladocaulon Gardn. = Paepalanthus Mart. (Eriocaulac.).

Cladoceras Bremek. Rubiaceae. 1 trop. E. Afr.

Cladochaeta DC. (~ Helichrysum L.). Compositae. 2 Caucasus.

Cladocolea Van Tiegh. = Oryctanthus (Griseb.) Eichl. (Loranthac.).

Cladoda (Cladodea) Poir. = seq.

Cladodes Lour. = Alchornea Sw. (Euphorbiac.).

Cladogelonium Leandri. Euphorbiaceae. 1 Madag.

Cladogynos Zipp. ex Span. Euphorbiaceae. 1 SE. As., Malaysia.

Cladolepis Moq. = Ofaiston Rafin. (Chenopodiac.).

Cladomischus Klotzsch ex A. DC. = Begonia L. (Begoniac.).

Cladomyza Danser. Santalaceae. 20 Borneo, New Guinea, Solomon Is.

Cladophyllaceae Dulac = Dioscoreaceae R.Br.

Cladopogon Sch. Bip. = Senecio L. (Compos.).

Cladopus H. Moeller. Podostemaceae. 3 Japan, Siam, Java, Celebes.

Cladoraphis Franch. = Eragrostis Host (Gramin.).

Cladorhiza Rafin. = Corallorhiza Chatelain (Orchidac.).

Cladoseris Spach = Onoseris DC. (Compos.).

Cladosicyos Hook. f. = Cucumeropsis Naud. (Cucurbitac.).

Cladosperma Griff. = Pinanga Blume (Palm.).

Cladostachys D. Don (Deeringia R.Br.) Amaranthaceae. 12 Madag., Indomal.

Cladostemon A.Br. & Vatke. Capparidaceae. 1 trop. E. Afr.

Cladostigma Radlk. Convolvulaceae. 2 NE. trop. Afr. Fls. ♂ ♀. Ed. fr.

Cladostyles Humb. & Bonpl. = Evolvulus L. (Convolvulac.).

Cladothamnus Bong. Ericaceae. 1 NW. N. Am. Tall shrub.

Cladotheca Steud. = Cryptangium Schrad. (Cyperac.).

Cladothrix (Moq.) Nutt. ex Benth. & Hook. f. (1880; non Cohn 1875— Bacteria) = Tidestromia Standley (Amaranthac.).

Cladotrichium Vog. = Caesalpinia L. (Legumin.).

Cladrastis Rafin. Leguminosae. 4 E. As., 1 E. N. Am. *C. tinctoria* Rafin., yellow-wood; its wood yields a yellow dye.

Clairisia Benth. & Hook. f. = Clarisia Abat = Anredera Juss. (Basellac.).

Clairvillea DC. = Cacosmia Kunth (Compos.).

Clambus Miers = Phyllanthus L. (Euphorbiac.).

Clamydanthus Fourr. = Chlamydanthus C. A. Mey. = Thymelaea L. (Thymelaeac.).

Clandestina Hill = Lathraea L. (Orobanchac.).

Clandestinaria Spach = Rorippa Scop. (Crucif.).

Claotrachelus Zoll. = Vernonia Schreb. (Compos.).

Claoxylon A. Juss. Euphorbiaceae. 80 palaeotrop. A 10–200.

Claoxylopsis Leandri. Euphorbiaceae. 1 Madag.
Clappertonia Meissn. Tiliaceae. 3 trop. W. Afr.
Clappia A. Gray. Compositae. 2 S. U.S., Mexico.
Clara Kunth (Herreria Ruiz & Pav.). Liliaceae. 1 Brazil.
Clarckia Pursh = Clarkia Pursh (Onagrac.).
Clarionea Lag. = Perezia Lag. (Compos.).
Clarionella DC. ex Steud. = praec.
Clarisia Abat = Anredera Juss. (Basellac.).
*****Clarisia** Ruiz & Pav. Moraceae. 2 Mex. to trop. S. Am.
Clarkeifedia Kuntze = Patrinia Juss. (Valerianac.).
Clarkella Hook. f. Rubiaceae. 2 Himalaya, Siam.
Clarkia Pursh. Onagraceae. 36 W. N. Am., Chile. Mech. of fl. as in *Epilobium.*
Clarorivinia Pax & K. Hoffm. = Ptychopyxis Miq. (Euphorbiac.).
Clasophyllum Néraud (nomen). Urticaceae. Quid?
Clasta Comm. ex Vent. = Casearia Jacq. (Flacourtiac.).
Clastilix Rafin. = Miconia Ruiz & Pav. (Melastomatac.).
Clastopus Bunge ex Boiss. Cruciferae. 2 N. Persia.
Clathrospermum Planch. ex Benth. & Hook. f. = Enneastemon Exell (Annonac.).
Clathrotropis Harms. Leguminosae. 4 Brazil, Guiana.
Claucena Burm. f. = Clausena Burm. f. (Rutac.).
Claudia Opiz = Beckeria Bernh. = Melica L. (Gramin.).
Clausena Burm. f. Rutaceae. 30 palaeotrop. Some ed. fr.
Clausenopsis (Engl.) Engl. = Fagaropsis Mildbr. (Rutac.).
Clausia Kornuch-Trotzky (~ Hesperis L.). Cruciferae. 5 C. & N. As.
Clausonia Pomel = Asphodelus L. (Liliac.).
Clavapetalum Pulle = Dendrobangia Rusby (Icacinac.).
Clavaria Steud. (sphalm.) = Calvaria Comm. ex Gaertn. (Sapotac.).
Clavarioïdia Kreuzinger = Opuntia Mill. (Cactac.).
Clavena DC. = Carduus L. (Compos.).
Clavenna Neck. = ?Ammannia L. ?Peplis L. (Lythrac.).
Claviga Regel (sphalm.) = Clavija Ruiz & Pav. (Theophrastac.).
Clavigera DC. = Brickellia Ell. (Compos.).
Clavija Ruiz & Pav. Theophrastaceae. 55 trop. Am. Trees of palm-like habit, often showing cauliflory.
Clavimyrtus Blume = Syzygium Gaertn. (Myrtac.).
Clavipodium Desv. ex Grüning = Beyeria Miq. (Euphorbiac.).
Clavistylus J. J. Smith = Megistostigma Hook. f. (Euphorbiac.).
Clavophylis Thou. (uninom.) = Bulbophyllum clavatum Thou. (Orchidac.).
Clavula Dum. = Eleocharis R.Br. & Scirpus L. (Cyperac.).
Clavulium Desv. = Crotalaria L. (Legumin.).
Clayomyza Whitmore (sphalm.) = Cladomyza Danser (Santalac.).
Claytonia L. Portulacaceae. 35 E. Siberia, N. Am. (incl. arctic). No stips. Fl. in sympodial cymes. Before pollin. the fl.-stalk is erect; fl. protandr., with outward movement of the sta. after dehisc. Honey, at base of each petal, accessible to short-tongued insects. After pollin., the stalk bends down through 180°, to return once more to the erect position when fr. ripe. The caps. contains

3 seeds and splits into 3 valves, the seeds lying across the lines of splitting. The inner surfaces of the valves contract as they dry and shoot out the seeds (cf. *Buxus*, *Viola*).

Claytoniella Yurtsev. Portulacaceae. 2 arct. & alp. NE. As., NW. Am.

Cleachne Roland ex Steud. = Paspalum L. (Gramin.).

Cleanthe Salisb. Iridaceae. 1 S. Afr.

Cleanthes D. Don. Compositae. 2 S. Braz., Argent.

Cleghornia Wight. Apocynaceae. 4 Indomal., Indoch.

Cleianthus Lour. ex Gomes = Clerodendrum L. (Verbenac.).

Cleidiocarpon Airy Shaw. Euphorbiaceae. 1–2 Burma, W. China.

Cleidion Blume. Euphorbiaceae. 25 trop.

Cleiemera Rafin. = Ipomoea L. (Convolvulac.).

Cleiostoma Rafin. = Ipomoea L. (Convolvulac.).

Cleisocentron Brühl. Orchidaceae. 1 Himalaya.

Cleisocratera Korth. = Saprosma Bl. (Rubiac.).

Cleisostoma Blume = Sarcanthus Lindl. (Orchidac.).

Cleisostoma B. D. Jacks. = Cleiostoma Rafin. = Ipomoea L. (Convolvulac.).

Cleissocratera Miq. = Cleisocratera Korth. = Saprosma Bl. (Rubiac.).

Cleistachne Benth. Gramineae. 4 trop. & S. Afr., India.

Cleistanthes Kuntze = Cleistanthus Hook. f. ex Planch. (Euphorbiac.).

Cleistanthium Kuntze = Gerbera L. ex Cass. (Compos.).

Cleistanthopsis Capuron = Allantospermum Forman (Ixonanthac.).

Cleistanthus Hook. f. ex Planch. Euphorbiaceae. 140 palaeotrop.

Cleistes Rich. ex Lindl. Orchidaceae. 25 temp. N. & trop. S. Am.

Cleistocactus Lemaire. Cactaceae. 30 subtrop. S. Am., mostly Andes.

Cleistocalyx Blume = Syzygium Gaertn. (Myrtac.).

Cleistocalyx Steud. = Rhynchospora Vahl (Cyperac.).

Cleistocereus Frič & Kreuzinger = Cleistocactus Lem. (Cactac.).

Cleistochlamys Oliv. Annonaceae. 1 trop. E. Afr.

Cleistochloa C. E. Hubbard. Gramineae. 2 Queensland.

Cleistogenes Keng. Gramineae. 18 temp. Eurasia.

Cleistoloranthus Merr. = Amyema Van Tiegh. (Loranthac.).

Cleistopholis Pierre. Annonaceae. 5 trop. W. Afr.

Cleistoyucca Eastw. = Clistoyucca Trelease (Liliac.).

Cleithria Steud. = seq.

Cleitria Schrad. = Venidium Less. (Compos.).

Clelandia J. M. Black. Violaceae. 1 S. Austr.

Clelia Casaretto = Calliandra Benth. (Legumin.).

Clemanthus Klotzsch = Adenia Forsk. (Passiflorac.).

Clematicissus Planch. Vitidaceae. 1 W. Austr.

Clematis L. Ranunculaceae. 250 cosmop., chiefly temp. Mostly climbing shrubs with opp., usu. cpd., ls. Lower sides of petioles sensitive to contact. The petiole bends once round the support, thickens and lignifies. Fls. in cymes; K coloured; no pets. or honey secretion. The style often remains persistent upon the fr. and becomes hairy, thus forming a mech. for wind-distr.

Clematitaria Bur. = Pleonotoma Miers (Bignoniac.).

Clematitis Duham. = Clematis L. (Ranunculac.).

Clematoclethra Maxim. Actinidiaceae. 10 W. & C. China.

Clematopsis Bojer ex Hutch. Ranunculaceae. 18 trop. & S. Afr., Madag.

Clemensia Merr. = Chisocheton Bl. (Meliac.).

Clemensia Schlechter = seq.

Clemensiella Schlechter. Asclepiadaceae. 1 Philipp. Is.

Clementea Cav. (1802) = Angiopteris Hoffm. (Angiopteridac.).

Clementea Cav. (1804) = Canavalia DC. (Legumin.).

Clementsia Rose ex Britt. & Rose = Sedum L. (Crassulac.).

Cleobula Vell. Leguminosae (inc. sed.). 1 Brazil.

Cleobulia Mart. ex Benth. Leguminosae. 3 Brazil.

Cleochroma Miers = Iochroma Benth. (Solanac.).

Cleodora Klotzsch = Croton L. (Euphorbiac.).

Cleomaceae (Pax) Airy Shaw. Dicots. 12/275 trop. & subtrop. Mostly glandular ann. herbs, but some shrubs and a few trees or climbers. Intermediate between *Capparidac.* and *Crucif.-Stanleyeae*, differing from the former in the glandular covering, and in the fr. being a 'cruciferous' siliqua with a replum; from the latter in the digitately 3–7-foliolate (rarely simple) ls., often zygo. fls. and rarely tetradynamous sta. Chief genera: *Cleome, Cleomella, Gynandropsis, Physostemon.*

Cleomaceae Horan. = Capparidaceae Juss. (*s.l.*).

Cleome [L.] DC. Cleomaceae. 150 trop., subtrop. Disk usu. more developed on post. side; may bear scales.

Cleome L. *s.str.* (type *C. gynandra* L.) = Gynandropsis DC. (Cleomac.).

Cleomella DC. Cleomaceae. 10 SW. U.S., Mex.

Cleomena Roem. & Schult. = Muhlenbergia Schreb. (Gramin.).

Cleomodendron Pax = Farsetia Turra (Crucif.).

Cleomopsideae Villani (nom. illegit. pro subfam. Crucif.) = Stanleyaceae Nutt., *q.v.*

Cleonia L. Labiatae. 2 W. Medit.

Cleopatra Panch. ex Baillon = Neoguillauminia Croizat (Euphorbiac.).

Cleophora Gaertn. (~ Latania Comm. ex Juss.). Palmae. 4 Mascarene Is.

Clercia Vell. = Salacia L. (Celastrac.).

Cleretum N. E. Br. Aïzoaceae. 10 S. Afr.

Clerkia Neck. = Tabernaemontana L. (Apocynac.).

Clermontia Gaudich. Campanulaceae. 27 Hawaii. The latex is used as bird lime. Some have ed. fr.

Clerodendranthus Kudo = Orthosiphon Benth. (Labiat.).

Clerodendron Adans. = seq.

Clerodendrum L. Verbenaceae. 400 trop., subtrop. The sta. project so as to form the landing place for insects, and when they are ripe the style is bent down. Afterwards the sta. roll up and the style takes their place. *C. fistulosum* Becc. (Borneo) has hollow internodes inhabited by ants. *C. thomsonae* Balf. (W. Afr.) has white K and red C.

Cleterus Rafin. = Chleterus Rafin. (Gesneriac.).

Clethra Bert. ex Steud. = Viviania Cav. (Vivianiac.).

Clethra Gronov. ex L. Clethraceae. 68 As., Am., Madeira.

*Clethraceae Klotzsch. Dicots. 120 As., Am. Shrubs and trees with alt. exstip. ls.; fls. in racemes or panicles, without bracteoles, ⚥, reg. K 5; C 5, imbr., polypet.; A 5 + 5, hypog.; no disk; anthers bent outwards in bud, opening by pores; pollen in single grains; G̲ 3-loc., ovules ∞ in each loc.; style crenate, or with 3 short stigmas. Caps. loculic.; endosp. Only genus: *Clethra*. Rather closely ̱related to *Ericac.* and *Cyrillac.*

Clethropsis Spach = Alnus L. (Betulac.).

Clethrospermum Planch. (sphalm.) = Clathrospermum Planch. = Enneastemon Exell (Annonac.).

Clevelandia Greene ex Brandegee = ? Orthocarpus Nutt. (Scrophulariac.).

Cleyera Adans. = Polypremum L. (Loganiac.).

***Cleyera** Thunb. emend. DC. Theaceae. 1 Himal. to Japan, 16 Mex. to Panamá, W.I.

Cleyria Neck. = Dialium L. (Legumin.).

***Clianthus** Soland. ex Lindl. Leguminosae. 2–3 Austr., N.Z.

Clibadium Allem. ex Lindl. (nomen) = Clybates Reichb. (Urticac.).

Clibadium Allem. ex L. Compositae. 50 trop. Am., W.I.

Clidanthera R.Br. = Glycyrrhiza L. (Legumin.).

Clidemia D. Don. Melastomataceae. 145 trop. Am., W.I.

Clidemiastrum Naud. = Oxymeris DC. = Leandra Raddi (Melastomatac.).

Clidium P. & K. = Cleidion Bl. (Euphorbiac.).

Cliffordia Livera = Christisonia Gardn. (Orobanchac.).

Cliffortia L. Rosaceae. 80 Afr. (mainly S.). Axis hollow in ♀ only.

Cliffortiaceae Mart. = Rosaceae–Sanguisorbeae Juss.

Cliffortioïdes Dryand. ex Hook. = Nothofagus Bl. (Fagac.).

Cliftonia Banks ex Gaertn. f. Cyrillaceae. 1 SE. U.S. (exc. penins. Florida).

Climacandra Miq. = Ardisia Sw. (Myrsinac.).

Climacoptera Botsch. (~ Salsola L.). Chenopodiaceae. 23 C. As.

Climacorachis Hemsl. & Rose = Aeschynomene L. (Legumin.).

Climedia Rafarin (sphalm.) = Clidemia D. Don (Melastomatac.).

Climacanthus Nees. Acanthaceae. 2 S. China, Indoch., Malaysia.

Clinanthus Herb. = Stenomesson Herb. (Amaryllidac.).

Clinelymus (Griseb.) Nevski = Elymus L. (Gramin.).

Clinhymenia A. Rich. & Gal. = Orchidofunckia A. Rich. & Gal. = Cryptarrhena R.Br. (Orchidac.).

Clinogyne Salisb. ex Benth. = Donax Lour. + Schumannianthus Gagnep. + Marantochloa Brongn. & Gris (Marantac.).

Clinogyne K. Schum. = Marantochloa Brongn. & Gris (Marantac.).

Clinopodium L. (~ Calamintha Mill., Satureia L.). Labiatae. 10 N. temp.

Clinosperma Becc. Palmae. 1 New Caledonia.

Clinostemon Kuhlm. & Sampaio = Licaria Aubl. (Laurac.).

Clinostigma Wendl. Palmae. 5 Samoa, Fiji.

Clinostigmopsis Becc. (~ Exorrhiza Becc.). Palmae. 2 Fiji, New Hebrides.

Clinostylis Hochst. = Gloriosa L. (Liliac.).

Clinta Griff. (sphalm.) = Chirita Buch.-Ham. (Gesneriac.).

Clintonia Dougl. = Downingia Torr. (Campanulac.).

Clintonia Rafin. Liliaceae. 6 temp. E. As., N. Am.

Cliocarpus Miers = Solanum L. (Solanac.).

Cliococca Bab. = Linum L. (Linac.).

Cliomera P. & K. = Cleiemera Rafin. = Ipomoea L. (Convolvulac.).

Clipeola Hall. = Clypeola L. (Crucif.).

Clipteria Rafin. = Eclipta L. (Compos.).

Clistanthium P. & K. = Cleistanthium Kunze = Gerbera L. ex Cass. (Compos.).

Clistanthocereus Backeb. = Borzicactus Riccob. (Cactac.).

Clistanthus Muell. Arg. = Clistranthus Poit. ex Baill. = Pera Mutis (Perac.).

Clistanthus P. & K. = Cleistanthus Hook. f. (Euphorbiac.).

Clistax Mart. Acanthaceae. 2 Brazil.

Clistes P. & K. = Cleistes Rich. ex Lindl. (Orchidac.).

Clistoyucca Trel. Agavaceae. 1 SW. U.S.

Clistranthus Poit. ex Baill. = Pera Mutis (Perac.).

Clitandra Benth. Apocynaceae. 1 W. to C. trop. Afr.

Clitandropsis S. Moore. Apocynaceae. 7 New Guinea, Palau Is.

Clitanthes Herb. = Chlidanthus Herb. & Stenomesson Herb. (Amaryllidac.).

Clitanthum Benth. & Hook. f. = Chlidanthus Herb. (Amaryllidac.).

Clithria P. & K. = Cleitria Schrad. = Venidium Less. (Compos.).

Clitocyamos St-Lag. = Quamoclit Moench = Ipomoea L. (Convolvulac.).

Clitoria L. Leguminosae. 40 trop. & subtrop. Fls. inverted; the essential organs therefore touch the backs of visiting insects.

Clitoriopsis R. Wilczek. Leguminosae. 1 trop. Afr.

× **Cliveucharis** Rodigas. Amaryllidaceae. Gen. hybr. (Clivia × Eucharis).

Clivia Lindl. Amaryllidaceae. 3 S. Afr.

Cloanthe Nees = Chloanthes R.Br. (Dicrastylidac.).

Cloëzia Brongn. & Gris. Myrtaceae. 8 New Caled.

Cloiselia S. Moore = Dicoma Cass. (Compos.).

Clomena Beauv. = Muhlenbergia Schreb. (Gramin.).

Clomenocoma Cass. = Dysodia Cav. (Compos.).

Clomenolepis Cass. (nomen). Compositae (Astereae). Hab.? Quid?

Clomium Adans. = Carduus L. (Compos.).

Clomopanus Steud. = seq.

Clompanus Aubl. = Lonchocarpus Humb. & Bonpl. (Legumin.).

Clompanus Rafin. = Sterculia L. (Sterculiac.).

Clonium P. & K. = Klonion Rafin. = Eryngium L. (Umbellif.).

Clonodia Griseb. Malpighiaceae. 6 trop. S. Am.

Clonostachys Klotzsch = Sebastiania Spreng. (Euphorbiac.).

Clonostylis S. Moore. Euphorbiaceae. 1 Sumatra. ♂ fl. unknown.

Clopodium Rafin. = Lycopodium L. (Lycopodiac.).

Closaschima Korth. = Laplacea Kunth (Theac.).

Closia Remy. Compositae. 10 Chile.

Closirospermum Neck. = Crepis L. (Compos.).

Closterandra Boiv. ex Belang. = Papaver L. (Papaverac.).

Clowesia Lindl. = Catasetum Rich. (Orchidac.).

Clozelia A. Chev. = Antrocaryon Pierre (Anacardiac.).

Clozella A. Chev. = praec.

Cluacena Rafin. = Myrtus L. (Myrtac.).

Clueria Rafin. = Eremostachys Bunge (Labiat.).

Clugnia Comm. ex DC. = Dillenia L. (Dilleniac.).

Clusia L. Guttiferae. 145 mostly warm Am., Madag., New Caled. Mostly climbing epiph., clasping the host by anastomosing aerial r., and frequently strangling it (cf. *Ficus*). Fr. fleshy, probably carried by birds.

CLUSIACEAE

*Clusiaceae Lindl.: see **Guttiferae** Juss. (nom. altern.).

Clusianthemum Vieill. = Garcinia L. (Guttif.).

Clusiella Planch. & Triana. Guttiferae. 7 Colombia.

Clusiophyllea Baill. = Canthium Lam. (Rubiac.).

Clusiophyllum Muell. Arg. = Cunuria Baill. (Euphorbiac.).

Clutia L. Euphorbiaceae. 70 Afr., Arabia.

Cluytia Ait. = praec.

Cluytia Roxb. ex Steud. = Bridelia Willd. (Euphorbiac.).

Cluytiandra Muell. Arg. = Meineckia Baill. (Euphorbiac.).

Clybates Reichb. (nomen). Urticaceae. Hab.? Quid?

Clybatis Phil. Compositae. 1 Chile.

Clymenia Swingle. Rutaceae. 1 Bismarck Archip.

Clymenum Mill. = Lathyrus L. (Legumin.).

Clynhymenia A. Rich. & Gal. = Clinhymenia A. Rich. & Gal. = Cryptarrhena R.Br. (Orchidac.).

Clypea Blume = Stephania Lour. (Menispermac.).

Clypeola Burm. ex DC. = Pterocarpus L. (Legumin.).

Clypeola Crantz = Biscutella L. + Alyssum L. (Crucif.).

Clypeola L. Cruciferae. 8 Medit.

Clypeola Neck. = Adyseton Adans. = Alyssum L. (Crucif.).

Clystomenon Muell. Arg. (sphalm.) = Cyclostemon Bl. = Drypetes Vahl (Euphorbiac.).

Clytia Stokes = Clutia L. (Euphorbiac.).

Clytoria J. S. Presl = Clitoria L. (Legumin.).

Clytostoma Miers. Bignoniaceae. 12 trop. Am.

Clytostomanthus Pichon. Bignoniaceae. 1 Ecuador, Brazil, Paraguay.

Cnema P. & K. = Knema Lour. (Myristicac.).

Cnemidaria Presl. Cyatheaceae. 12 trop. Am. Simply pinnate ls.; veins anast.; spores distinctive.

Cnemidia Lindl. = Tropidia Lindl. (Orchidac.).

Cnemidiscus Pierre. Sapindaceae. 1 S. Indoch.

Cnemidophacos Rydb. = Astragalus L. (Legumin.).

Cnemidostachys Mart. = Sebastiania Spreng. (Euphorbiac.).

Cnenamum Tausch = Crenamum Adans. = Crepis L. (Compos.).

*Cneoraceae Link. Dicots. 2/3 Cuba, Canaries, Medit. Shrubs with alt. simple ent. leathery exstip. ls. with oil-glands; fls. sol. or in corymbs (peduncle sts. adnate to leaf-base), 3–4-merous, ☿, reg., with elong. torus (?androgynoph.) or bolster-like disk; A 3–4, latr.; G (3–4), lobed, with 2 ov. in each; style 1. Schizocarp. Genera: *Cneorum, Neochamaelea*. Near *Zygophyllaceae*, but only one whorl of sta., with no ligules, and no stipules, but oil-glands in the ls.

Cneoridium Hook. f. Rutaceae. 1 S. Calif., Lower Calif.

Cneorum L. Cneoraceae. 1 Cuba, 1 W. & C. Medit.

Cnesmocarpus Zipp. ex Blume = Pometia J. R. & G. Forst. (Sapindac.).

Cnesmone Blume corr. Bl. Euphorbiaceae. 10 Assam, S. China, SE. As., W. Malaysia. Climbing shrubs.

Cnesmosa Blume = praec.

Cnestidium Planch. Connaraceae. 3 C. Am., trop. S. Am.

Cnestis Juss. Connaraceae. 40 warm Afr., Madag., Malaysia. K valv. Caps. hairy within.
Cnicothamnus Griseb. Compositae. 2 Bolivia, Argentina.
Cnicus D. Don = Theodorea Cass. (Compos.).
*Cnicus L. emend. Gaertn. Compositae. 1 Medit., *C. benedictus* L., offic. The genus was much confused with *Carduus* and *Cirsium* in the past.
Cnidium Cusson. Umbelliferae. 20 N. palaeo-temp., S. Afr.
Cnidome E. Mey. ex Walp. = seq.
Cnidone E. Mey. ex Endl. = Kissenia R.Br. ex T. Anders. (Loasac.).
Cnidoscolus Pohl. Euphorbiaceae. 75 trop. Am.
Cnopos Rafin. = Polygonum L. (Polygonac.).
Coa Mill. = Hippocratea L. (Celastrac.).
Coalisia Rafin. = seq.
Coalisina Rafin. = Cleome L. (Cleomac.).
Coatesia F. Muell. = Geijera Schott (Rutac.).
Coaxana Coulter & Rose. Umbelliferae. 2 Mexico.
Cobaea Cav. Cobaeaceae. 18 trop. Am. *C. scandens* Cav. shows very rapid growth. It climbs by aid of tendrils (leaf-structures) which are much branched, the branches ending in sharp hooks. The tendril nutates with great rapidity and is highly sensitive to contact (as may be seen by rubbing one side and watching it for 5 min.); the hooks prevent the nutation from dragging away a branch before it has had time to clasp its support (Darwin, *Climbers*, p. 106). The closed bud stands erect on an erect stalk, but when going to open, the tip of the stalk bends over. Fl. very protandr. with movement of sta. and styles. At first greenish with unpleasant smell (fly-fl.), it becomes purple with pleasant honey-like smell (bee-fl.). Afterwards the stalk goes through several contortions (cf. *Linaria*).
Cobaea Neck. = Lonicera L. (Caprifoliac.).
Cobaeaceae D. Don. Dicots. 1/18 Am. Climbing shrubs, with alt. pinnate stip. ls., the stips. large and foliaceous, the term. lfts. modif. into tendrils. Fls. large, axill., 1–3 on long pend. common pedunc. K 5 or scarcely (5) at base; C (5), campan., lobes rounded to linear; A 5, exserted, epipet. near base of C, fil. bearded below, anth. versatile; disk large, fleshy, lobed; G (3), with long filif. shortly trifid style, and 2–∞ axile ov. per loc.; fr. a septicid. 3-valved caps., seeds biseriate, medifixed, compressed, winged, embryo large, endosp. o. Only genus: *Cobaea*. Somewhat intermediate between *Bignoniac.* and *Polemoniac.*
Cobamba Blanco = Canscora Lam. (Gentianac.).
Cobresia Pers. = Kobresia Willd. (Cyperac.).
Coburgia Herb. ex Sims = ? Brunsvigia Heist. (Amaryllidac.).
Coburgia Sweet = Stenomesson Herb. (Amaryllidac.).
Coccanthera C. Koch & Hanst. = Codonanthe Hanst. (Gesneriac.).
Coccineorchis Schlechter. Orchidaceae. 1 Peru.
Coccinia Wight & Arn. Cucurbitaceae. 30 trop. & S. Afr., 1 also trop. India & Malaysia. The fr. of *C. grandis* (L.) Voigt is eaten as a veg. in India.
Coccobryon Klotzsch = Piper L. (Piperac.).
Coccoceras Miq. = Mallotus Lour. (Euphorbiac.).
Coccocipsilum P.Br. = seq.
*Coccocypselum P.Br. corr. Schreb. Rubiaceae. 20 trop. Am. Heterostyled.

COCCODERMA

Coccoderma Miers (nomen). Menispermaceae. Hab.? Quid?

Coccoglochidion K. Schum. = Glochidion J. R. & G. Forst. (Euphorbiac.).

***Coccoloba** P.Br. mut. L. Polygonaceae. 150 trop. & subtrop. Am. *C. uvifera* L., and others, ed. fr. (seaside grape).

Coccolobaceae Barkley = Polygonaceae–Coccoloboïdeae Reichb.

Coccolobis P.Br. = Coccoloba P.Br. mut. L. (Polygonac.).

Coccomelia Reinw. = Baccaurea Lour. (Euphorbiac.).

Coccomelia Ridl. = Angelesia Korth. = Licania Aubl. (Chrysobalanac.).

Cocconerion Baill. Euphorbiaceae (?). 2 New Caled. Ls. whorled.

Coccos Gaertn. = Cocos L. (Palm.).

Coccosipsilum Sw. = Coccocypselum P.Br. corr. Schreb. (Rubiac.).

Coccosperma Klotzsch = Blaeria L. (Ericac.).

Coccothrinax Sargent. Palmae. 50 S. Florida, W.I.

Cocculidium Spach = Cocculus DC. (Menispermac.).

***Cocculus** DC. Menispermaceae. 11 trop. & subtrop. (excl. S. Am.).

Coccus Mill. = Cocos L. (Palm.).

Coccyganthe Reichb. = Lychnis L. (Caryophyllac.).

Cochemiea (K. Brandegee) Walton. Cactaceae. 5 Lower Calif.

Cochlanthera Choisy = Clusia L. (Guttif.).

Cochlanthus Balf. f. = Socotranthus Kuntze (Periplocac.).

Cochleanthes Rafin. Orchidaceae. 14 trop. Am.

Cochlearia L. Cruciferae. 25 N. temp., S. to E. Himal. and Java mts. (introd.?). *C. officinalis* L., with ± fleshy ls., chiefly at the seaside and on mts.

× **Cochleatorea** hort. = × Pescoranthes hort. (Orchidac.). [(cf. *Armeria*).

× **Cochlenia** hort. Orchidaceae. Gen. hybr. (iii) (Cochleanthes × Stenia).

Cochlia Blume = Bulbophyllum Thou. (Orchidac.).

Cochlianthus Benth. Leguminosae. 2 C. Himal. to SW. China.

Cochliasanthus Trew = Phaseolus L. (Legumin.).

Cochlidiosperma (Reichb.) Reichb. = Veronica L. (Scrophulariac.).

Cochlidiospermum Opiz = Cochlidiosperma (Reichb.) Reichb. = Veronica

Cochlidium Kaulfuss. Grammitidaceae. 7 trop. Am. [L. (Scrophulariac.).

Cochlioda Lindl. Orchidaceae. 4 trop. S. Am.

Cochliopetalum Beer = Pitcairnia L'Hérit. (Bromeliac.).

Cochliospermum Lag. = Suaeda Forsk. (Chenopodiac.).

Cochliostema Lem. Commelinaceae. 2 Colombia to Bolivia. The filaments of the fertile sta. develop both laterally and beyond the anthers into large wings. Anther-loculi spiral.

***Cochlospermaceae** Planch. Dicots. 2/20–25 trop. Trees and shrubs, with alt. usu. lobed stip. ls. Fl. large, ⚥, reg. or slightly zygo., in rac. infl. K 4–5, imbr.; C 4–5, imbr. or contorted; A ∞, anth. lin., dehisc. by term. pores or short slits; G̲ (3–5), with ∞ ov. on axile or pariet. plac., and simple filif. style with minute stig. Fr. a large· 1–3-loc. 2–5-valved caps., the inner and outer layers separating and forming alternating valves; seeds usu. renif., often pilose or woolly, with curved embr., endosp. oily. Genera: *Cochlospermum, Amoreuxia*. Related to *Malvales* and perhaps *Bixaceae*.

***Cochlospermum** Kunth. Cochlospermaceae. 15–20 trop., mostly xero.; some have stout tuberous underground stems; many drop their ls. and flower in the dry season.

Cochlospermum P. & K. = Cochliospermum Lag. = Suaeda Forsk. (Cheno-
Cochlostemon P. & K. = Cochliostema Lem. (Commelinac.). [podiac.).
Cochranea Miers = Heliotropium L. (Boraginac.).
Cockaynea Zotov = Stenostachys Turcz. (Gramin.).
Cockburnia Balf. f. = Poskea Vatke (Globulariac.).
Cocoaceae Schultz-Schultzenst. = Palmae–Cocoëae Kunth.
Cococipsilum Jaume St-Hil. = Coccocypselum P.Br. corr. Schreb. (Rubiac.).
Cocoloba Rafin. = Coccoloba P.Br. mut. L. (Polygonac.).
Cocops O. F. Cook = Calyptrogyne Wendl. (Palm.).
Cocos L. Palmae. 1, *C. nucifera* L. (coconut), orig. prob. trop. As. or Polynes.,
cult. throughout trop. It grows esp. well near to the sea, and its fibrous and
woody fr. is capable of floating long distances uninjured; hence it forms a char.
feature of marine island veg., and indeed probably became widely distr. in
early times. It is a tall palm with large pinnate ls. and a dense monoec. infl.
The stem rarely stands vertically, but makes a gradual curve; this would appear
to be due to heliotropism. Fr. large, one-seeded. The outer layer of the pericarp
is fibrous, the inner very hard (the shell of the coconuts sold in shops). At the
base are 3 marks, corresponding to the 3 loc. of the ovary, two of which have
become obliterated. Under one of these is the embryo. The thin testa is lined
with white endosp., enclosing a large cavity partly filled with a milky fluid.
This palm furnishes many of the necessaries of life to the inhabitants of the
tropics, and its products are largely exported from Ceylon, the Philippines, etc.
The large ls. are woven into *cadjans* for thatching, mats, baskets, etc.; their
stalks and midribs make fences, brooms, yokes, and many other useful articles.
The bud or 'cabbage' at the apex of the stem makes an excellent vegetable and
is made into pickles and preserves. When flowering the infl.-axis is tapped for
toddy, a drink like the Mexican *pulque* (cf. *Agave*), containing sugar. Evap. of
toddy furnishes a sugar known as *jaggery*; its fermentation gives an alcoholic
drink, from which distillation produces the strong spirit known as *arrack*, while
further fermentation gives vinegar. The fr. while young contains a pint or more
of a sweetish watery fluid, a refreshing drink; it decreases as the nut ripens.
The kernels are eaten raw, or in curries, milk is expressed from them for
flavouring, and oil is extracted by boiling or by pressure, in the latter case the
kernels being first dried into what is known as *copra*; this forms the only source
of wealth of many island communities throughout the tropics. The refuse cake
or *poonac*, left after the expression of the oil, is a valuable fattening food for
cattle. The great use of the oil is for soap-making and margarine; indeed at
the present time the coconut is the world's chief producer of vegetable fat.
In recent years a large industry has sprung up in desiccated coconut, largely
used in confectionery, the kernel being sliced and dried in special desiccators.
The outer wood of the stem (porcupine wood) is used for rafters, orn. articles,
etc. The thick outer husk, rarely seen in Europe upon the nut, contains a large
number of long stout fibres running lengthwise. The nut is placed in water
till the soft tissues between these fibres decay, and the fibre (*coir*) is then
beaten out; or sometimes the fibre is obtained by special machinery. (R. Child,
Coconuts, 1964; Menon & Pandalai, *The Coconut Palm*, 1958.)
Codanthera Rafin. = Salvia L. (Labiat.).
Codaria L. ex Benn. = Lerchea Kalm (Rubiac.).

CODARIOCALYX

Codariocalyx Hassk. = Desmodium Desv. (Legumin.).
Codarium Soland. ex Vahl = Dialium L. (Legumin.).
Codazzia Karst. & Triana = Delostoma D. Don (Bignoniac.).
Coddampulli Adans. = Garcinia L. (Guttif.).
Codda-Pana Adans. = Corypha L. (Palm.).
Coddingtonia S. Bowdich = ? Lonicera L. (Caprifoliac.).
Codebo Rafin. = Codiaeum Juss. (Euphorbiac.).
Codia J. R. & G. Forst. Cunoniaceae. 13 New Caledonia.
Codiaceae Van Tiegh. = Cunoniaceae–Pancherieae Engl.
*Codiaeum A. Juss. Euphorbiaceae. 15 Malaysia, Polynes., N. Austr.
  C. variegatum (L.) Blume cult., esp. in trop., for its coloured ls.; usu. known as
  'Crotons', and also used as hedges. In some forms the ls. are curiously
  twisted, or have two blades separated by a length of midrib.
Codiaminum Rafin. = Narcissus L. (Amaryllidac.).
Codieum Rafin. = Codiaeum Juss. (Euphorbiac.).
Codigi Augier = Sonerila Roxb. (Melastomatac.).
Codiocarpus Howard. Icacinaceae. 2 Andaman Is., Philippines.
Codiphus Rafin. = Prismatocarpus L'Hérit. (Campanulac.).
Codivalia Rafin. = Pupalia Juss. (Amaranthac.).
Codochisma Rafin. = Convolvulus L. (Convolvulac.).
Codochonia Dun. = Acnistus Schott (Solanac.).
Codomale Rafin. = Polygonatum Mill. (Liliac.).
Codon Royen ex L. Hydrophyllaceae. 2 S. Afr. 10–12-merous.
Codonacanthus Nees. Acanthaceae. 2 Assam, S. China.
Codonachne Wight & Arn. ex Steud. = Chloris Sw. (Gramin.).
Codonandra Karst. = Calliandra Benth. (Legumin.).
*Codonanthe (Mart.) Hanst. Gesneriaceae. 15 trop. Am.
Codonanthe Mart. ex Steud. = Hypocyrta Mart. (Gesneriac.).
Codonanthemum Klotzsch = Eremia D. Don (Ericac.).
Codonanthes Rafin. = Pitcairnia L'Hérit. (Bromeliac.).
Codonanthopsis Mansf. Gesneriaceae. 2 Brazil.
Codonanthus G. Don = Breweria R.Br. (Convolvulac.).
Codonanthus Hassk. = Physostelma Wight (Asclepiadac.).
Codonechites Markgraf. Apocynaceae. 2 Bolivia, W. Brazil.
Codonemma Miers = Tabernaemontana L. (Apocynac.).
Codonium Vahl = Schoepfia Schreb. (Olacac.)
Codonoboea Ridl. Gesneriaceae. 4 Malay Penins.
Codonocalyx Klotzsch ex Baill. = Croton L. (Euphorbiac.).
Codonocalyx Miers = Suteria DC. = Psychotria L. (Rubiac.).
Codonocarpus A. Cunn. ex Hook. Gyrostemonaceae. 3 Austr.
Codonocephalum Fenzl. Compositae. 2 Persia to C. As.
Codonochlamys Ulbr. Malvaceae. 2 Brazil.
Codonocrinum Willd. ex Schultes = Yucca L. (Agavac.).
Codonocroton E. Mey. ex Engl. & Diels = Combretum L. (Combretac.).
Codonophora Lindl. = Paliavana Vell. ex Vand. (Gesneriac.).
Codonoprasum Reichb. = Allium L. (Alliac.).
Codonopsis Wall. (incl. Campanumoea Bl.). Campanulaceae. 30–40 C. &
  E. As., Himal., Malaysia.

Codonoraphia Oerst. = Pentarhaphia Lindl. (Gesneriac.).
Codonorchis Lindl. Orchidaceae. 3 S. trop. & temp. S. Am.
Codonosiphon Schlechter. Orchidaceae. 3 New Guinea.
Codonostigma Klotzsch = Scyphogyne Brongn. (Ericac.).
Codonura K. Schum. Apocynaceae. 1 W. Equat. Afr.
Codoriocalyx Hassk. = Codariocalyx Hassk. = Desmodium Desv. (Legumin.).
Codornia Gandog. = Helianthemum Mill. (Cistac.).
Codosiphus Rafin. = Convolvulus L. (Convolvulac.).
Codylis Rafin. = Solanum L. (Solanac.).
Coelachne R.Br. Gramineae. 10 palaeotrop.
Coelachyropsis Bor. Gramineae. 1 S. India, Ceylon.
Coelachyrum Hochst. & Nees. Gramineae. 5 N. trop. Afr., trop. SW. As.
Coeladena P. & K. = Coiladena Rafin. = Ipomoea L. (Convolvulac.).
Coelandria Fitzg. = Dendrobium Sw. (Orchidac.).
Coelanthe Griseb. = Coilantha Borckh. = Gentiana L. (Gentianac.).
Coelanthera P. & K. = Coilanthera Rafin. = Cordia L. (Ehretiac.).
Coelanthum E. Mey. Aïzoaceae. 2 S. Afr.
Coelanthus Willd. ex Schult. f. = Lachenalia Jacq. (Liliac.).
Coelarthron Hook. f. = Microstegium Nees ex Lindl. (Gramin.).
Coelas Dulac = Sibbaldia L. (Rosac.).
Coelebogyne John Sm. = Caelebogyne John Sm. Corr. Reichb. (Euphorbiac.).
Coelestina Cass. = Caelestina Cass. = Ageratum L. (Compos.).
Coelestina Hill = Amellus L. (Compos.).
Coelestinia Endl. = Caelestina Cass. = Ageratum L. (Compos.).
Coelia Lindl. Orchidaceae. 1 C. Am., W.I.
Coelidium Vog. ex Walp. Leguminosae. 15 S. Afr.
Coelina Nor. = Elaeocarpus L. (Elaeocarpac.).
Coeliopsis Reichb. f. Orchidaceae. 1 Costa Rica, Panamá, Colombia.
Coelocarpum Balf. f. Verbenaceae. 1 Socotra, 4 Madag.
Coelocarpus P. & K. = Coilocarpus F. Muell. ex Domin (Chenopodiac).
Coelocarpus Scott Elliot = Coelocarpum Balf. f. (Verbenac.).
Coelocaryon Warb. Myristicaceae. 7 trop. Afr.
Coelochloa Hochst. = Coelachyrum Hochst. & Nees (Gramin.).
Coelocline A. DC. = Xylopia L. (Annonac.).
Coelococcus H. Wendl. (~ Metroxylon Rottb.). Palmae. 2 Polynes.
Coelodepas Hassk. = Koilodepas Hassk. (Euphorbiac.).
Coelodiscus Baill. = Mallotus Lour. (Euphorbiac.).
× **Coeloglossgymnadenia** Druce (vii) = seq.
× **Coeloglosshabenaria** Druce (vii) = seq.
× **Coeloglossogymnadenia** A. Camus (vii) = × Gymnaglossum Rolfe (Orchidac.).
× **Coeloglossorchis** Guétrot (vi, vii) = Dactyloglossum P. F. Hunt & Summerh. (Orchidac.).
Coeloglossum Hartm. Orchidaceae. 2 temp. Euras., N. Am.
Coeloglossum Lindl. = Peristylus Bl. (Orchidac.).
Coelogyne Lindl. Orchidaceae. 200 W. China, Indomal., Pacif.

10-2

COELONEMA

Coelonema Maxim. Cruciferae. 1 SW. China.

Coeloneurum Radlk. Goetzeaceae. 1 W.I. (S. Domingo).

Coelonox P. & K. = Coilonox Rafin. = Ornithogalum L. (Liliac.).

Coelophragmus O. E. Schulz. Cruciferae. 2 Mexico.

× **Coeloplatanthera** Cif. & Giac. Orchidaceae. Gen. hybr. (i) (Coeloglossum × Platanthera).

Coelopleurum Ledeb. (~Angelica L.). Umbelliferae. 1 NE. As.

Coelopyrena Val. Rubiaceae. 1 Moluccas, New Guinea.

Coelopyrum Jack = Campnosperma Thw. (Anacardiac.).

Coelorachis Brongn. Gramineae. 12 trop.

Coelosperma P. & K. = Coilosperma Rafin. = Deeringia R.Br. (Amaranthac.).

Coelospermum Blume. Rubiaceae. 15 Hainan, Indoch., Malaysia, Austr., New Caled.

Coelostegia Benth. Bombacaceae. 5 W Malaysia.

Coelostelma Fourn. Asclepiadaceae. 1 Brazil.

Coelostigma P. & K. = Coilostigma Klotzsch = Salaxis Salisb. (Ericac.).

Coelostigmaceae Dulac = Berberidaceae Juss.

Coelostylis (Juss.) Kuntze = Echinopterys A. Juss. (Malpighiac.).

Coelostylis P. & K. = Coilostylis Rafin. = Epidendrum L. (Orchidac.).

Coelostylis Torr. & Gray ex Endl. = Spigelia L. (Spigeliac.).

Coelotapalus P. & K. = Coilotapalus P.Br. = Cecropia Loefl. (Morac.).

Coemansia Marchal (1879; non Van Tiegh. & Le Monn. 1873—Fungi) = Coudenbergia Marchal = Pentapanax Seem. (Araliac.).

Coenochlamys P. & K. = Coinochlamys T. Anders. ex Benth. & Hook. f. (Loganiac.).

Coenogyna P. & K. = Coinogyne Less. = Jaumea Pers. (Compos.).

Coenolophium Reichb. = Cenolophium Koch (Umbellif.).

Coenotus Benth. & Hook. f. = Caenotus Rafin. = Erigeron L. (Compos.).

Coerulinia Fourr. = Veronica L. (Scrophulariac.).

Coespiphylis Thou. (uninom.) = *Bulbophyllum caespitosum* Thou. (Orchidac.).

Coestichis Thou. (uninom.) = *Malaxis caespitosa* Thou. = *Liparis caespitosa* (Thou.) Lindl. (Orchidac.).

Coetocapnia Link & Otto = Polianthes L. (Amaryllidac.).

Cofeanthus A. Chev. = Coffea L. (Rubiac.).

Cofer Loefl. = Symplocos Jacq. (Symplocac.).

Coffea L. Rubiaceae. 40 palaeotrop., esp. Afr. *C. arabica* L. (Arabian coffee) largely cult. in S. Brazil, Colombia, C. Am., Jamaica, Java, E. Afr., etc., often under the shade of large trees; this produces coffee of highest quality. *C. liberica* Bull ex Hiern (Liberian coffee) and *C. canephora* Pierre (robusta coffee) cult. usu. at lower elevations; product is not so good. Other spp. are also used. The fr. is a 2-seeded drupe, resembling a cherry. The pulp and the endocarp (which covers the two seeds like a layer of parchment) are mechanically removed. The seed, or coffee-bean, has a deep groove on the ventral side; by soaking it in water the endosperm is softened and the embryo may be dissected out. The stimulating property depends on the presence of the alkaloid caffeine. By far the largest cult. is that of Brazil, the annual export of which in recent years has exceeded 1 million tons (over half the world's total), valued at about £300 million. Although not one of the major food-crops, coffee is of unusual importance because on its trade depend the economies of several countries in Latin America and Africa.

Coffeaceae J. G. Agardh = Rubiaceae–Ixoreae Benth. & Hook. f.

× **Cogniauxara** hort. = × Holttumara hort. (Orchidac.).

Cogniauxella Baill. = seq.

Cogniauxia Baill. Cucurbitaceae. 1 trop. Afr.

Cogniauxiocharis (Schlechter) Hoehne. Orchidaceae. 1 Brazil.

Cogsvellia Rafin. = seq.

Cogswellia Roem. & Schult. = Lomatium Rafin. (Umbellif.).

Cogylia Molina = Lardizabala Ruiz & Pav. (Lardizabalac.).

Cohautia Endl. = Kohautia Cham. & Schlechtd. = Oldenlandia L. (Rubiac.).

Cohiba Rafin. = Wigandia Kunth (Hydrophyllac.).

Cohnia Kunth. Agavaceae. 3 Mascarene Is., New Caled.

Cohnia Reichb. f. = seq.

Cohniella Pfitz. Orchidaceae. 1 C. Am.

Coiladena Rafin. = Ipomoea L. (Convolvulac.).

Coilantha Borckh. = Gentiana L. (Gentianac.).

Coilanthera Rafin. = Cordia L. (Ehretiac.).

Coilmeroa Endl. (sphalm.) = Colmeiroa Reut. = Securinega Juss. (Euphorbiac.).

Coilocarpus F. Muell. ex Domin (~ Bassia All.). Chenopodiaceae. 1 Queens-

Coilochilus Schlechter. Orchidaceae. 1 New Caled.   [land.

Coilomphis Rafin. = Melaleuca L. (Myrtac.).

Coilonox Rafin. = Ornithogalum L. (Liliac.).

Coilosperma Rafin. = Deeringia R.Br. (Amaranthac.).

Coilostigma Klotzsch = Salaxis Salisb. (Ericac.).

Coilostylis Rafin. = Epidendrum L. (Orchidac.).

Coilotapalus P.Br. = Cecropia Loefl. (Morac.).

Coincya Rouy. Cruciferae. 1 Spain.

Coinochlamys T. Anders. ex Benth. & Hook. f. Loganiaceae. 5 trop. W. Afr.

Coinogyne Less. = Jaumea Pers. (Compos.).

Coix L. Gramineae. 5 trop. As. *C. lachryma* L. (Job's tears) with inverted pear-shaped body at base of infl., the sheath of the br. of the infl., hollowed out and containing the 1-fld. ♀ spikelet; the ♂♂ project beyond the mouth. Cult. for food in Khasia Hills and Burma; used in medicine in China.

Cojoba Britton & Rose = Pithecellobium Mart. (Legumin.).

*****Cola** Schott & Endl. Sterculiaceae. 125 Afr. *C. vera* K. Schum. and *C. acuminata* Schott & Endl. (possibly identical) are the source of the kola nuts which form a principal article of trade in W. Africa. The nuts contain much caffein, and when chewed confer considerable power of sustaining fatigue; they are consequently a staple in the diet of the native population (cf. *Erythroxylum*). The tree is as yet rarely cult., but is very common in W. Afr. The nuts are skinned after keeping for a few days, and packed between ls. to keep them damp. Exalb.

Colania Gagnep. (~ Aspidistra Ker-Gawl.). Liliaceae. 1 Indoch.

Colaria Rafin. = Cola Schott & Endl. (Sterculiac.).

Colax Lindl. Orchidaceae. 5 Brazil.

Colbertia Salisb. = Dillenia L. (Dilleniac.).

Colchicaceae DC. = Liliaceae–Colchiceae Reichb.

Colchicum L. Liliaceae. 65 Eur., Medit., to C. As. & N. India. *C. autumnale* L. (Eur.) is the autumn crocus or meadow saffron. Below the soil is a large corm. In autumn the fl. projects out of the soil. The P-tube is long, and the ovary remains below ground, protected from cold, etc. The protog. fl. is visited by bees. In spring the ls. appear and the capsule is brought above ground by the lengthening of its stalk. The seeds and corms are used in medicine, in gout, etc.

Coldenella Ellis = Fibra v. Fibraurea J. Colden = Coptis Salisb. (Ranunculac.)

Coldenia L. Ehretiaceae. 20 trop. & subtrop. Am., 1 palaeotrop.

*Colea Boj. ex Meissn. Bignoniaceae. 20 Madag., Masc. Is.

Coleachyron J. Gay ex Boiss. = Carex L. (Cyperac.).

Coleactina N. Hallé. Rubiaceae. 1 W. Equat. Afr.

Coleanthera Shchegl. Epacridaceae. 3 W. Austr.

*Coleanthus Seidl. Gramineae. 1 N. temp. Euras.

Coleataenia Griseb. = Panicum L. (Gramin.).

Colebrockia Steud. (1) (sphalm.) = Colebrookea Sm. (Labiat.).

Colebrockia Steud. (2) (sphalm.) = Colebrookia Donn ex Lestib. = Globba L. (Zingiberac.).

Colebrookea Sm. Labiatae. 1 India.

Colebrookia Donn ex Lestib. = Globba L. (Zingiberac.).

Colebrookia Spreng. = Colebrookea Sm. (Labiat.).

Colema Rafin. = Corema D. Don (Empetrac.).

Colensoa Hook. f. = Pratia Gaudich. (Campanulac.).

Coleobotrys Van Tiegh. = Helixanthera Lour. (Loranthac.).

Coleocarya S. T. Blake. Restionaceae. 1 Queensland.

Coleocephalocereus Backeb. = Austrocephalocereus Backeb. (Cactac.).

Coleochloa Gilly. Cyperaceae. 7 trop. & subtrop. Afr. (chiefly E.), Madag. Several chars., e.g. compressed stem, distich. decid. ls., 'ligule' of hairs, and ventrally open l.-sheaths, reminiscent of *Gramineae*.

Coleocoma F. Muell. Compositae. 1 trop. Austr.

Coleogyn[ac]eae J. G. Agardh = Rosaceae–Cercocarpeae Torr. & Gray.

Coleogyne Torr. Rosaceae. 1 SW. U.S.

Coleonema Bartl. & Wendl. Rutaceae. 5 S. Afr.

Coleophora Miers (~ Daphnopsis Mart. & Zucc.). Thymelaeaceae. 1 Brazil.

Coleophyllum Klotzsch = Chlidanthus Herb. (Amaryllidac.).

Coleosanthus Cass. = Brickellia Ell. (Compos.).

Coleospadix Becc. = Drymophloeus Zipp. (Palm.).

Coleostachys A. Juss. Malpighiaceae. 1 N. trop. S. Am.

Coleostephus Cass. (~ Chrysanthemum L.). Compositae. 7 Canary Is., W. Medit.

Coleostylis Sond. = Levenhookia R.Br. (Stylidiac.).

Coleotrype C. B. Clarke. Commelinaceae. 5 SE. Afr., Madag.

Coletia Vell. = Mayaca Aubl. (Mayacac.).

Coleus Lour. Labiatae. 150 palaeotrop. Many forms and hybrids with varieg. and coloured leaves. *C. elongatus* Trimen is a peculiar sp. found only on the top of one mountain in Ceylon, and has presumably arisen by mutation (*Ann. Perad. Bot. Gard.* **4**: 5–9, 1907).

Colicodendron Mart. = Capparis L. (Capparidac.).

Colignonia Endl. Nyctaginaceae. 6 Andes.
Colina Greene = Mohria Sw. (Schizaeac.).
Colinil Adans. = Cracca L. = Tephrosia Pers. (Legumin.).
Coliquea Steud. = Chusquea Kunth (Gramin.).
Colla Rafin. = Calla L. (Arac.).
Collabium Blume. Orchidaceae. 8 China, Indoch., Malaysia, Polynes.
Colladoa Cav. = Ischaemum L. (Gramin.).
Colladonia DC. = Heptaptera Margot & Reuter (Umbellif.).
Colladonia Spreng. = Palicourea Aubl. (Rubiac.).
Collaea Bert. ex Colla = Ardisia Sw. (Myrsinac.).
Collaea DC. = Galactia P.Br. (Legumin.).
Collaea Endl. = Collea Lindl. = Pelexia Lindl. (Orchidac.).
Collaea Spreng. = Chrysanthellum Rich. (Compos.).
Collandra Lemaire = Columnea L. (Gesneriac.).
Collania Herb. = Bomarea Mirb. (Alstroemeriac.).
Collania Schult. f. = Urceolina Reichb. (Amaryllidac.).
Collea Lindl. = Pelexia Lindl. (Orchidac.).
Collema Anders. ex R.Br. (1810; non Wigg. 1780—Lichenes) = Goodenia Sm. (Goodeniac.).
Collenucia Chiov. = Jatropha L. (Euphorbiac.).
*Colletia Comm. ex Juss. Rhamnaceae. 17 temp. & subtrop. S. Am. Habit peculiar; in each axil are 2 serial buds; the upper gives a triangular thorn, the lower fls. or a branch of unlimited growth.
Colletia Endl. = Coletia Vell. = Mayaca Aubl. (Mayacac.).
Colletia Scop. = Celtis L. (Ulmac.).
Colletoecema E. Petit. Rubiaceae. 1 trop. Afr.
Colletogyne Buchet. Araceae. 1 Madag.
Collignonia Endl. = Colignonia Endl. (Nyctaginac.).
Colliguaja Molina. Euphorbiaceae. 5 temp. S. Am.
Collinaria Ehrh. (uninom.) = Poa cristata L. = Koeleria cristata (L.) Pers. (Gramin.).
Collinia (Mart.) Liebm. ex Oerst. Palmae. 3 C. Am.
Collinia Rafin. = seq.
Colliniana Rafin. = seq.
Collinsia Nutt. Scrophulariaceae. 18 Pacif. N. Am., 2 E. U.S. The fl. resembles, in shape and mech., that of Leguminosae.
Collinsiana Rafin. = praec.
Collinsonia L. Labiatae. 5 E. N. Am.
Colliquaja Augier = Colliguaja Mol. (Euphorbiac.).
Collococcus P.Br. = Cordia L. (Ehretiac.).
Collomia Nutt. Polemoniaceae. 15 Pacif. N. Am., temp. S. Am. The seed coat has a covering of cells with mucilaginous walls which swell when wetted (cf. Brassica, Linum, etc.).
Collomia Sieber ex Steud. = Felicia Cass. (Compos.).
Collophora Mart. = Couma Aubl. (Apocynac.).
Collospermum Skottsb. Liliaceae. 5 N.Z., Fiji, Samoa.
Collotapalus P.Br. = Coilotapalus P.Br. = Cecropia Loefl. (Morac.).
Collyris Vahl = Dischidia R.Br. (Asclepiadac.).

273

COLMANARA

× **Colmanara** hort. Orchidaceae. Gen. hybr. (iii) (Miltonia × Odontoglossum × Oncidium).

Colmeiroa F. Muell. = Paracorokia Král = Corokia A. Cunn. (Escalloniac.).

Colmeiroa Reut. = Securinega Comm. ex Juss. (Euphorbiac.).

Colobachne Beauv. = Alopecurus L. (Gramin.).

Colobandra Bartl. = Hemigenia R.Br. (Labiat.).

Colobanthera Humbert. Compositae. 1 Madag.

Colobanthium Reichb. Gramineae. 1 Medit.

Colobanthus Bartl. Caryophyllaceae. 20 Andes, temp. S. Am., Falkl., Kerguelen & New Amst. Is., Austr., N.Z., Campb. & Auckl. Is. Petals o. Sta. in one whorl.

Colobanthus Trin. = Colobanthium Reichb. (Gramin.).

Colobatus Walp. = Colobotus E. Mey. = Buchenroedera Eckl. & Zeyh. (Legumin.).

Colobium Roth = Leontodon L. (Compos.).

Colobogyne Gagnep. Compositae. 1 Indoch.

Colobogynium Schott = Schismatoglottis Zoll. (Arac.).

Colobopetalum P. & K. = Kolobopetalum Engl. (Menispermac.).

Colobotus E. Mey. = Buchenroedera Eckl. & Zeyh. (Legumin.).

Colocasia Link = Richardia Kunth = Zantedeschia Spreng. (Arac.).

Colocasia Schott. Araceae. 8 Indomal., Polynes. Tuberous herbs or small shrubs. Monoec. Sta. in synandria. *C. esculenta* (L.) Schott (*taro, coco*, or scratch-coco), cult. in trop. for its rhiz., which when boiled loses its poisonous nature and forms valuable food.

Colococca Rafin. = Collococcus P.Br. = Cordia L. (Ehretiac.).

Colocrater Dur. & Jacks. (sphalm.) = Calocrater K. Schum. (Apocynac.).

Colocynthis Mill. = Citrullus Schrad. ex Eckl. & Zeyh. (Cucurbitac.).

Cologania Kunth = Amphicarpaea Ell. mut. DC. (Legumin.).

Colomandra Neck. = Aiouea Aubl. (Laurac.).

Colona Cav. Tiliaceae. 30 S. China, SE. As., Indomal.

Colonna J. St-Hil. = praec.

Colonnea Endl. = Calonnea Buc'hoz = Gaillardia Foug. (Compos.).

Colophonia Comm. ex Kunth = Canarium L. (Burserac.).

Colophonia P. & K. = Kolofonia Rafin. = Ipomoea L. (Convolvulac.).

Colophospermum Kirk ex J. Léonard. Leguminosae. 1 S. trop. Afr.

Coloptera Coulter & Rose = Cymopterus Rafin. (Umbellif.).

Coloradoa Boissev. & C. Davidson. Cactaceae. 1 SW. U.S.

Colosanthera Pohl (nomen). 2 Brazil. Quid?

Colostephanus Harv. = Cynanchum L. (Asclepiadac.).

Colpias E. Mey. Scrophulariaceae. 1 S. Afr.

Colpodium Trin. Gramineae. *S.l.* 20, *s.str.* 3, N. temp.

Colpogyne B. L. Burtt. Gesneriaceae. 1 Madag.

Colpoön Bergius. Santalaceae. 1 S. Afr.

Colpophyllos Trew = Ellisia L. (Hydrophyllac.).

Colpothrinax Griseb. & H. Wendl. (~ Pritchardia Seem. & H. Wendl.). Palmae. 1 Cuba.

Colquhounia Wall. Labiatae. 6 E. Himal., SW. China.

Colsmannia Lehm. = Onosma L. (Boraginac.).

Colubrina Montandon = Polygonum L. (Polygonac.).

*Colubrina Rich. ex Brongn. Rhamnaceae. 16 trop. & subtrop. Am.; 1 trop. E. Afr. & Maurit.; 7 E. & SE. As., Indomal., Queensl., Pacif.

Columbaria J. & C. Presl = Scabiosa L. (Dipsacac.).

Columbea Salisb. = Araucaria Juss. (Araucariac.).

Columbia Pers. = Colona Cav. (Tiliac.).

Columbra Comm. ex Endl. = Cocculus DC. (Menispermac.).

Columella Comm. ex DC. = Pavonia Cav. (Malvac.).

Columella Lour. = Cayratia Juss. (Vitidac.).

Columella Vahl = Columellia Ruiz & Pav. (Columelliac.).

Columella Vell. = Pisonia L. (Nyctaginac.).

Columellaceae Dulac = Euphorbiaceae Juss. + Buxaceae Loisel.

Columellea Jacq. = Nestlera Spreng. (Compos.).

*Columellia Ruiz & Pav. Columelliaceae. 4 N. Andes.

*Columelliaceae D. Don. Dicots. 1/4 S. Am. Shrubs with evergr. opp. asymm. exstip. ls. Fls. in cymes, slightly zygo. K 5, subimbr., persist.; C (5), imbr.; A 2, short and thick with irreg. broad connective and 1 twisted pollen sac; no disk; $\overline{G}$ (2), imperfectly 2-loc.; ov. ∞, anatr., on subcontig. pariet. plac.; style short and thick with broad 2-4-lobed stigma. Caps., enclosed in K. Endosp. Only genus: *Columellia*. Despite the sympetaly, slight zygomorphy and curious anthers, probably related to *Escalloniac.* and *Hydrangeac.*; perhaps also to *Loganiac.*

Columnea L. Gesneriaceae. 200 trop. Am. Several climbers and epiphytes. Anisophylly is frequent.

Coluppa Adans. = Gomphrena L. (Amaranthac.).

Coluria R.Br. Rosaceae. 6 S. Siberia, China.

Colus Raeusch. = Coleus Lour. (Labiat.).

Colutea L. Leguminosae. 26 S. Eur. & Abyss. to C. As. & Himal. *C. arborescens* L. is the bladder senna. Its ls. have similar properties to senna (*Cassia* spp.) and are used to adulterate the latter. The pods are inflated, and burst on being

Coluteastrum Fabr. (uninom.) = Lessertia DC. sp. (Legumin.).   [squeezed.

Coluteocarpus Boiss. Cruciferae. 2 mts. SW. As.

Colutia Medik. = Sutherlandia R.Br. (Legumin.).

Colvillea Boj. ex Hook. Leguminosae. 1 Madag.

Colymbada Hill = Centaurea L. (Compos.).

Colymbea Steud. = Columbea Salisb. = Araucaria Juss. (Araucariac.).

Colyris Endl. = Collyris Vahl = Dischidia R.Br. (Asclepiadac.).

Colysis Presl. Polypodiaceae. 30 Africa to New Guinea & Queensland. Sporangia in lines between main veins.

Colythrum Schott = Esenbeckia Kunth (Rutac.).

Comaceae Dulac = Tamaricaceae S. F. Gray.

Comacephalus Klotzsch = Eremia D. Don (Ericac.).

Comachlinium Scheidw. & Planch. = Dyssodia Cav. (Compos.).

Comacum Adans. = Myristica Gron. (Myristicac.).

Comandra Nutt. Santalaceae. 5 N. Am.; 1 SE. Eur. & As. Min.

Comanthera L. B. Smith. Eriocaulaceae. 1 N. trop. S. Am.

Comanthosphace S. Moore. Labiatae. 1 Nepal(?), 1 E. China, 1 Japan.

Comarella Rydb. = Potentilla L. (Rosac.).

COMAROBATIA

Comarobatia Greene = Rubus L. (Rosac.).

Comaropsis Rich. = Waldsteinia Willd. (Rosac.).

Comarostaphylis Zucc. (~ Arctostaphylos Adans.). Ericaceae. 20 SW. U.S., **Comarouna** Carr. = Coumarouna Aubl. (Legumin.).   [Mex.

Comarum L. = Potentilla L. (Rosac.).

Comaspermum Pers. = Comesperma Labill. (Polygalac.).

Comastoma (Wettst.) Tokoyuni = Gentiana L. (Gentianac.).

Comatocroton Karst. = Croton L. (Euphorbiac.).

Comatoglossum Karst. & Triana = Talisia Aubl. (Sapindac.).

Combera Sandwith. Solanaceae. 2 Andes of Chile & Argent.

Combesia A. Rich. = Crassula L. (Crassulac.).

*****Combretaceae** R.Br. Dicots. 20/600 trop. & subtrop. Trees and shrubs with alt. or opp. simple entire exstip. ls.; many climbers, some twining, some with hooks (persistent bases of petioles); usu. rich in tannin. Fls. usu. sessile in racemose infls., usu. ♀ and reg., 3–8-merous. Typically K 5, usu. valv.; C 5 or 0; A 5 + 5, rarely ∞; Ḡ 1-loc. with 2–5 pend. anatr. ov. and simple style. Disk on summit of ovary, sometimes with outgrowths. Dry or drupaceous 1-seeded fr., often winged at angles; endocarp, when present (e.g. *Terminalia*), split longit. into 2 unequal halves. No endosp.; cots. usu. convolute (this char. scarcely known elsewhere in the Angiosp.; cf. *Calycanthac.*), sometimes contortupl., rarely flat (*Quisqualis, Combretum* spp., etc.). Chief genera: *Terminalia, Combretum, Quisqualis*.

Combretocarpus Hook. f. Anisophylleaceae. 1 W. Malaysia (exc. Philipp. Is. & Java). Char. of freshwater peat swamps. Recently found in S. Malay Penins.: cf. Ng, *Malayan Forester*, **29** (1): 32, 1966.

Combretodendron A. Chev. Barringtoniaceae. 1 trop. W. Afr.; 1 Philipp. Is. Fr. a dry indeh. 1-seeded caps. with 4 broad membr. wings.

Combretopsis K. Schum. = Lophopyxis Hook. f. (Lophopyxidac.).

*****Combretum** Loefl. Combretaceae. 250 trop. (exc. Austr.). K deciduous, C present, A usu. 8–10. Climbers or erect; ls. alt. or opp. *Chiquito*, a butter-like substance, from fr. of *C. butyrosum* Tul. (trop. Afr.).

Comes Buc'hoz = ? Codonanthe Hanst. (Gesneriac.).

Comesperma Labill. = Bredemeyera Willd. (Polygalac.).

Cometes L. Caryophyllaceae. 2 desert reg. NE. Afr. & Abyss. to NW. India.

Cometia Thou. ex Baill. = Drypetes Vahl + Thecacoris Juss. (Euphorbiac.).

Comeurya Baill. = Dracontomelon Blume (Anacardiac.).

Cominia P.Br. = Allophylus L. (Sapindac.).

Cominia Rafin. = Rhus L. (Anacardiac.).

Cominsia Hemsl. Marantaceae. 5 Moluccas, New Guinea, Solomons.

Commarum Schrank = Comarum L. = Potentilla L. (Rosac.).

Commelina L. Commelinaceae. 230 trop. & subtrop. Infl. with sheathing bracts; fl. horiz., sta. and style projecting. In many the upper 3 sta. sterile with cruciform anthers, but lobes juicy, bees piercing them for honey. *C. benghalensis* L. has subterranean cleist. fls. Some have ed. rhiz.

Commelinaceae R.Br. Monocots. (Farinosae). 38/500 mostly trop. & subtrop. Herbs (occasionally twining) with jointed ± succ. stems and alt. sheathing ls. Infl. usu. a cincinnus like *Boraginac.* Fl. ♀, usu. reg., commonly blue; usu. formula K 3; C 3, rarely (3), differing in colour and texture from K; A 3 + 3,

COMPERIA

often hairy, but some often absent or stds.; G (3), 3-loc., with a few orthotr.
ov. in each. Caps. loculic. (lin.-elongate in *Aëthiolirion*) or indeh. Seed often
with aril, rarely winged. Endosp. copious. Calcium oxalate present in tissues.
No cortical bundles. Glandular hairs rare (*Floscopa*). Chief genera: *Commelina*,
Aneilema, Tradescantia, Cyanotis, Dichorisandra.
Commelinantia Tharp = Tinantia Scheidw. (Commelinac.).
Commelinidium Stapf. Gramineae. 3 W. Afr.
Commelinopsis Pichon. Commelinaceae. 1 trop. Am., W.I.
Commelyna Hoffmgg. ex Endl. = Commelina L. (Commelinac.).
Commerçona Sonner. = Barringtonia J. R. & G. Forst. (Barringtoniac.).
Commerçonia F. Muell. = Commersonia J. R. & G. Forst. (Sterculiac.).
Commersis Thou. (uninom.) = *Commersorchis commersonis* Thou. = ? (Or-
chidac.).
Commersonia auctt. = Commerçona Sonner. = Barringtonia J. R. & G. Forst.
(Barringtoniac.).
Commersonia Comm. ex Juss. = Polycardia Juss. (Celastrac.).
Commersonia J. R. & G. Forst. Sterculiaceae. 1 trop. SE. As., 8 Austr. Pets.
cap-like, stds. 3-partite.
Commersophylis Thou. (uninom.) = *Bulbophyllum commersonii* Thou. (Or-
chidac.).
Commersorchis Thou. Orchidaceae (inc. sed.). 1 Mascarenes.
Commia Ham. ex Meissn. = Aporusa Bl. (Euphorbiac.).
Commia Lour. = Excoecaria L. (Euphorbiac.).
Commianthus Benth. = Retiniphyllum Humb. & Bonpl. (Rubiac.).
Commicarpus Standley. Nyctaginaceae. 16 trop. & subtrop. (incl. 4 Am.,
1 SE. Spain).
Commidendron Lem. = seq.
Commidendrum Burchell ex DC. Compositae. 5 St Helena. Trees; ls
crowded at ends of twigs. *C. gummiferum* DC. yields a gum.
Commilobium Benth. = Pterodon Vog. (Legumin.).
*****Commiphora** Jacq. Burseraceae. 185 warm Afr., Madag., Arabia to Western
India. Axis cup-like. Resin exudes and collects in lumps. Several yield myrrh,
used in medicine, incense, etc. *C. opobalsamum* Engl. is said to yield balm of
Gilead; others yield bdellium and other resins.
Commirhoea Miers = Chrysochlamys Poepp. (Guttif.).
Commitheca Bremek. Rubiaceae. 1 trop. W. Afr.
Comocarpa Rydb. = Potentilla L. (Rosac.).
Comocladia P.Br. Anacardiaceae. 20 C. Am., W.I. 3–4-merous.
Comolia DC. Melastomataceae. 30 trop. S. Am.
Comomyrsine Hook. f. = Weigeltia A. DC. (Myrsinac.).
Comoneura Pierre ex Engl. = Strombosia Blume (Olacac.).
Comopycna Kuntze = Pycnocoma Benth. (Euphorbiac.).
Comoranthus Knobl. Oleaceae. 3 Madag., Comoro Is.
Comoroa Oliv. = Teclea Delile (Rutac.).
Comosperma Poir. = Comesperma Labill. (Polygalac.).
Comostemum Nees = Androtrichum Brongn. (Cyperac.).
Comparettia Poepp. & Endl. Orchidaceae. 7 trop. Am.
Comperia C. Koch = Orchis L. (Orchidac.).

277

COMPHOROPSIS

Comphoropsis Moq. (sphalm.) = Camphoropsis Moq. = Nanophytum Less. (Chenopodiac.).

Comphrena Aubl. (sphalm.) = Gomphrena L. (Amaranthac.).

*Compositae Giseke (nom. altern. **Asteraceae** Link). Dicots. One of the largest fams. of flg. pls., comprising about 900 genera, with over 13,000 spp.—more than 10 % of the total. They are distr. over the greater part of the earth. Although so large a fam. they are well marked in their characters and cannot be confounded with any other. They are prob. related to *Goodeniac.*, *Brunoniac.*, *Stylidiac.* and *Campanulac.*, and have a superficial likeness to *Dipsacaceae.*

Living in almost every conceivable situation (though rare in trop. rain forests), they present great variety in veg. habit, often within a single genus, e.g. *Senecio (q.v.).* Water and marsh plants and climbers are rare, and so also are epiphytes. This latter fact is interesting, for the distr. mech. of these pls. is admirably suited to an epiph. existence, and xero. is not uncommon. The enormous majority are herbaceous pls.; trees and shrubs are comparatively rare (about 1½ %). It is worthy of note that the latter often form an important feature in the Composite flora of oceanic islands (see Wallace's *Island Life*).

Ls. usu. alt., frequently rad., opp. in most *Heliantheae* and *Eupatorieae*, and some others, whorled in a few cases, e.g. *Zinnia verticillata* Andr.; stips. rarely present. R. usu. a tap-root, sometimes tuberous as in *Dahlia*, etc., often thickened like that of a carrot, e.g. *Taraxacum*, *Cichorium*, etc. For details of veg. organs refer to individual gen.; e.g. *Aster, Barnadesia, Bellis, Bidens, Cichorium, Dahlia, Espeletia, Gnaphalium, Helianthus, Helichrysum, Lactuca, Mutisia, Petasites, Senecio, Silphium, Taraxacum,* etc. All tribes exc. 11 and 12 contain oil-passages in the root, stem, etc. In 12 (*Lactuceae*), and some *Arctotideae* (4), laticiferous vessels are present, commonly containing a milky white latex (e.g. *Lactuca, Taraxacum*).

Infl. of racemose types, the fls. arranged in heads (*capitula*). These heads are again arranged in many cases into larger infls.—racemes, corymbs, etc., or even into cpd. heads (*Echinops*, etc.). In this last case, however, the smaller heads contain only one fl. each. Head surrounded by an invol. of bracts, usu. green, which performs for all the fls. of the head the functions that in most plants are performed by the calyces of the individual fls., viz. protection of the bud and of the young fr. Fls. arranged upon a common receptacle—the enlarged end of the axis—of various shapes, most frequently flat, slightly convex or even spindle-shaped. The shape and surface condition of the receptacle are chars. of importance in classification of the fam. It may be smooth or hairy, etc.; there may (*Helianthus*, etc.) or may not (*Calendula*, etc.) be, upon it, scaly brs. belonging to the individual fls. In *Cynareae* these brs. are divided so as to form numerous bristles.

In the simplest case the fls. of a single head are all alike and ⚥, but there are many deviations from this type. The fls. may be all actinomorphic (tubular) or all zygo. (ligulate); see below. Very commonly, however, as in *Bellis* or *Helianthus*, there is a distinction into a *disc* of actinomorphic fls., and a marginal *ray* of zygo. fls. Or, as in *Centaurea* spp., the other florets may be actinomorphic but different in size from the central. The number of ray florets varies in different spp., but according to definite rules.

The *distribution of sexes* among the fls. of a head varies much. The common case is gynomonoecism, the ray florets ♀, the disk ☿. The very large ray florets of *Centaurea* spp. and others are completely sterile (cf. *Hydrangea, Viburnum,* etc.). Cf. also *Tussilago, Petasites,* etc.

The *flower* is fully epig., usu. 5-merous. K absent in *Ambrosia* and its allies, *Sigesbeckia,* etc.; in some cases it appears only as a slightly 5-lobed rim upon the top of the inf. ovary (cf. *Rubiaceae* and *Umbelliferae*); usu. it takes the form of hairs or bristles—the *pappus*—and enlarges after fert. into a parachute (*Taraxacum*) or into hooked bristles (*Bidens*) to aid in distr. (see below). C (5), valvate in bud; actinom. (tubular) or zygo. Of the latter form there are two varieties, labiate (lipped) and ligulate (strap-shaped). The latter term, strictly speaking, should be applied to those corollas which are strap-shaped in form with 5 teeth at the end repres. the petals, but is usu. also given to those lipped forms where the lower lip is strap-shaped and ends in 3 or fewer teeth. Sta. 5, epipet. with short filaments, alt. with the petals. Anthers intr., cohering by their edges (syngenesious), forming a tube around the style (cf. *Lobelia*). Ḡ (2), with a simple style that forks at the end into two stigmas, an ant. and a post. The construction of the style and stigma is of importance in the classification. There is often a brush of hairs on the style below the stigmas. Only the inner (upper) surfaces of the stigmas are as a rule receptive to pollen. Ovary 1-loc. with 1 erect, basal, anatr. ov., which gives an exalb. seed with straight embryo, enclosed in the dry indeh. pericarp. This fr. is usu. termed an achene, but of course is, if one adhere strictly to definitions, a pseudo-nut (*cypsela*), as its pericarp is partly axial, and there is > one cpl. It is often crowned with a *pappus* (see below), sometimes winged or spiny, v. rarely drupaceous.

Natural History of the Flower. Being massed together in heads, the individual fls. may be, and usu. are, comparatively very small, and the advantage is gained that a single insect visitor may fert. many fls. in a short time without having to fly from one to the other, while there is no loss of conspicuousness, and a considerable saving of corolla material, etc. Throughout the fam., the same type of mech. of the individual fl. is found, the differences being slight and unimportant. It is simple, but effective. Honey is secreted by a ring-shaped nectary round the base of the style, and protected from rain and from short-lipped insects by the tube of the C. The depth of the tube varies within fairly wide limits, but is never so small as to permit the shortest-lipped insects to obtain the honey. As a fam., the *C.* all belong to Müller's fl. class B′, but there is considerable variety in the depth of tube, etc., and therefore also in the composition of the group of visiting insects to each. Thus the long-tubed purple-flowered Centaureas, etc., are mainly visited by bees and Lepidoptera, while the short-tubed yellow Leontodons or white Achilleas are visited mainly by flies. Many are beetle-pollinated.

At the time when the fl. opens, the style, with its stigmas tightly closed against one another, is comparatively short, reaching up to, or projecting a small distance into, the anther tube. The pollen is shed into this, and as the style grows it presses the pollen little by little out at the upper end of the tube where it will come into contact with visiting insects. At last the style itself emerges and the stigmas separate. The fl. is now ♀. Finally, in a great many cases, the stigmas curl so far back that they touch the pollen upon their own style, so that every fl. is certain to set seed, even though it be by self-fert. In a few cases, e.g. *Senecio vulgaris* L., insect visitors are very rare, and the fl. depends entirely on self-fert. The mech. is about the simplest and most perfect that exists for attaining the desired ends. A striking contrast is seen in the orchids; they have bizarre fls. with most elaborate mechs., and an enormous number of seeds in every caps. An interesting modification of the mech. is found in *Cynareae* (see *Centaurea*) where the sta. are irritable. See also *Artemisia* (wind-fert.).

The invol. bracts, or ray florets, or both, often close up over the central fls. in cold or wet weather, thus protecting the fls.

Natural History of the Fruit. The ripening fr.-head is generally protected from injury by the invol. bracts, which bend inwards over it, performing the function of a K. The calyces of the individual fls. are thus rendered useless in this respect and are, in most C., used for purposes of distr. of the fr. In most cases, the K, after the fert. of the fl., grows into the familiar pappus, as seen in *Taraxacum, Cirsium*, etc., usu. composed of fine hairs, often branched, but in some cases, e.g. *Achyrachaena*, leafy and membranous. The hairs are hygroscopic and spread out in dry air; this often helps to lever the fr. off the receptacle. In *Adenostemma* the pappus is sticky. In *Bidens* and others the pappus is formed of stout barbed bristles; the fr. adheres to animals. In *Arctium* the invol. brs. become hooked at the tips and cling to animals. In *Xanthium* the recept. is provided with hooks. In *Sigesbeckia* the bracts are sticky. A few genera, e.g. *Helianthus, Bellis*, etc., have no special arrangements at all, and the frs. remain upon the common receptacle till jerked off by wind or otherwise.

General Considerations. The C. are often regarded as occupying the highest position in the veg. kingdom. Their success may be put down perhaps to the concurrence of several useful peculiarities, viz.,

(1) the massing of the fls. in heads, surrounded by invol. bracts: from this there results

(*a*) greater conspicuousness, especially when ray florets are developed; (*b*) a saving of material in the corollas, etc.; (*c*) the fact that one insect visitor may fertilise many fls. in a short time without having to fly from one to another;

(2) the very simple and effective floral mechanism, which ensures

(*d*) protection of honey and pollen; (*e*) exclusion of the very short-lipped (allotropous) insects, but not too great specialisation for a very narrow circle of visitors; (*f*) prevention of self- and chance of cross-fertilisation till the last possible moment; (*g*) certainty of self-fertilisation if the cross fails;

(3) the use of the calyces of individual fls. for purposes of seed-distribution, and the very perfect character of the mechanism.

These considerations should be compared with the features of rival fams., e.g. *Cruciferae, Gramineae, Rubiaceae, Leguminosae.*

Economic uses. The C. furnish but few useful plants (other than border or greenhouse pls.); 1 sp. of *Brachylaena* (*q.v.*) is a timber tree. See *Lactuca, Cichorium, Cynara, Helianthus, Carthamus, Chrysanthemum, Tanacetum*, etc.

Classification and chief genera (after Cronquist). The classification of the C. and the determination of their genera is a matter of no small difficulty; we give here only the primary groupings and their chief genera. (There are several exceptions to the characters given below.)

A. **Asteroïdeae (Tubuliflorae)**. (fls. of disc not ligulate; latex rare):

1. HELIANTHEAE (incl. *Helenieae*) (style with crown of long hairs above the division; anthers usu. rounded at base with basally inserted filaments; corolla of disc fls. actinom.; pappus not hairy; invol. bracts not membranous at margins; recept. with or without scaly brs.): *Espeletia, Silphium, Xanthium, Ambrosia, Zinnia, Sigesbeckia, Helianthus, Dahlia, Bidens, Cosmos, Tithonia, Helenium, Tagetes.*

2. ASTEREAE (capit. heterog. or homog.; all or only central fls. tub.; anthers blunt at base, with filaments inserted at base; stigmas flattened with marginal rows of papillae, and terminal hairy unreceptive portions): *Solidago, Bellis, Aster, Erigeron, Baccharis, Callistephus, Olearia.*

3. ANTHEMIDEAE (as 1, but invol. br. with membranous tip and edges; pappus o or abortive): *Achillea, Anthemis, Chrysanthemum, Matricaria, Tanacetum, Artemisia.*

4. ARCTOTIDEAE (style, below or at point of division, thickened or with circle of hairs; capit. with lig. ray fls.; anthers acute at base or with longer or shorter point and with filaments inserted above the base): *Arctotis, Gazania.*

5. INULEAE (as 2; corolla in tub. fls. with 4–5-toothed limb; anthers tailed at base; styles various): *Blumea, Filago, Antennaria, Gnaphalium, Helichrysum, Inula.*

6. SENECIONEAE (as 1, but pappus hairy): *Tussilago, Petasites, Senecio, Doronicum.*

7. CALENDULEAE (capit. with ♀ ray fls., and usu. ♂ disc fls., with undivided style; anthers pointed at base; recept. not scaly; no pappus): *Calendula.*

8. EUPATORIËAE (capit. homog.; fls. tub., never pure yellow; anthers as in 2; stigmas long, but blunt or flattened at tip, with very short hairs; stigmatic papillae in marginal rows): *Ageratum, Eupatorium, Mikania, Adenostemma.*

9. VERNONIËAE (capit. homog.; fls. tub., never yellow; anthers arrow-shaped at base, pointed or rarely tailed, with filaments inserted high above the base; stigmas semi-cylindrical, long, pointed, hairy outside; stigmatic papillae all over inner surface): *Vernonia, Elephantopus.*

10. CYNAREAE (style as in 4; capit. homog. or with neuter, rarely ♀, not ligulate, ray fls.; anthers usu. tailed; recept. usu. bristly): *Echinops, Carlina, Arctium, Carduus, Cirsium, Cynara, Centaurea, Carthamus, Saussurea, Cousinia.*

11. MUTISIËAE (capit. homog. or heterog.; ray fls. when present usu. 2-lipped; disc fls. actinom. with deeply divided limb, or 2-lipped): *Barnadesia, Mutisia, Stifftia, Gerbera.*

B. **Lactucoïdeae (Liguliflorae)** (all fls. ligulate; latex)

12. LACTUCEAE (CICHORIËAE): *Cichorium, Rhagadiolus, Picris, Crepis, Hieracium, Leontodon, Taraxacum, Lactuca, Tragopogon, Scorzonera, Sonchus.*

COMPSANTHUS

Compsanthus Spreng. = Tricyrtis Wall. (Liliac.).
Compsoa D. Don = Tricyrtis Wall. (Liliac.).
Compsoaceae Horan. = Hypoxidaceae R.Br. + Liliaceae–Tricyrteae Krause.
Compsoneura Warb. Myristicaceae. 11 trop. Am.
Comptonanthus Nordenstam. Compositae. 3 S. & SW. Afr.
Comptonella Baker f. Rutaceae. 2 New Caled.
Comptonia L'Hérit. ex Ait. (~ Myrica L.). Myricaceae. 1 E. N. Am.
Comularia Pichon. Apocynaceae. 1 W. Equat. Afr.
Comus Salisb. = Muscari Mill. (Liliac.).
Conaceae Dulac = Coniferae Juss.
Conami Aubl. (**Conamia** Rafin.) = Phyllanthus L. (Euphorbiac.).
Conamomum Ridl. Zingiberaceae. 3 Malay Penins.
Conandrium Mez. Myrsinaceae. 9 Moluccas, New Guinea.
Conandron Sieb. & Zucc. Gesneriaceae. 3 Japan, Indoch.
Conanthera Ruiz & Pav. Tecophilaeaceae. 5 Chile.
Conanthes Rafin. = Pitcairnia L'Hérit. (Bromeliac.).
Conanthodium A. Gray = Helichrysum L. (Compos.).
Conanthus S. Wats. = Nama L. (Hydrophyllac.).
Conceveiba Aubl. Euphorbiaceae. 7 trop. S. Am.
Conceveibastrum Pax & K. Hoffm. Euphorbiaceae. 2 trop. S. Am.
Conchidium Griff. = Eria Lindl. (Orchidac.).
Conchium Sm. = Hakea Schrad. (Proteac.).
Conchocarpus Mikan = Angostura Roem. & Schult. (Rutac.).
Conchochilus Hassk. = Appendicula Blume (Orchidac.).
Conchopetalum Radlk. Sapindaceae. 1 Madag.
Conchophyllum Blume. Asclepiadaceae. 10 Malaysia.
Concilium Rafin. = Lightfootia L'Hérit. (Campanulac.).
Condaea Steud. = Condea Adans. = Hyptis Jacq. (Labiat.).
*Condalia Cav. Rhamnaceae. 18 warm Am.
Condalia Ruiz & Pav. = Coccocypselum P.Br. (Rubiac.).
Condaliopsis (Weberb.) Suesseng. Rhamnaceae. 5–6 subtrop. N. Am.
Condaminea DC. Rubiaceae. 3 Andes.
Condea Adans. = Hyptis Jacq. (Labiat.).
Condgiea Baill. ex Van Tiegh. = Klainedoxa Pierre (Ixonanthac.).
Condylicarpus Steud. = Tordylium L. (Umbellif.).
Condylocarpon Desf. Apocynaceae. 15 trop. S. Am.
Condylocarpus Hoffm. = Tordylium L. (Umbellif.).
Condylocarpus Salisb. ex Lamb. = Sequoia Endl. (Taxodiac.).
Condylocarpus K. Schum. = Condylocarpon Desf. (Apocynac.).
Condylocarya Bess. ex Endl. = Rapistrum Crantz (Crucif.).
Condylostylis Piper. Leguminosae. 2 Costa Rica, Colombia.
Confluaceae Dulac = Globulariaceae Lam. & DC.
Conforata Fourr. = Achillea L. (Compos.).
Congdonia Jepson = Sedum L. (Crassulac.).
Congdonia Muell. Arg. Rubiaceae. 1 SE. Brazil.
Congea Roxb. Symphoremataceae. 10 NE. India, SE. As., Malay Penins., Sumatra.
Conghas Wall. ex Hiern = Schleichera Willd. (Sapindac.).

Congolanthus A. Raynal. Gentianaceae. 1 trop. Afr.

Coniandra Schrad. ex Eckl. & Zeyh. = Kedrostis Medik. (Cucurbitac.).

Conicosia N. E. Brown. Aïzoaceae. 13 S. Afr.

Coniferae Juss. The most important class of Gymnosperms, and like the others better represented in former ages than now. They form 6–8 fams. with nearly 50 gen. and 400 spp. Like their past history, their present geographical distr. is of interest. Most are erect evergr. trees, and grow as dense forests, forming char. features of the veg. in many regions (esp. temp. and subtrop. and mountains). Beginning in the north we find *Juniperus communis* ssp. *nana* beyond the limit of trees. This limit is largely marked by the C. and the birch. Within it, in the N. temp. zone, are broad areas covered with *Larix, Abies, Pinus,* etc.). Going S., their importance decreases, and at about 40° N. they become practically confined to the mountains. Here we find in Japan and China a region of development char. by *Cephalotaxus, Pseudolarix, Cryptomeria, Cunninghamia, Sciadopitys, Chamaecyparis, Keteleeria, Glyptostrobus, Taiwania,* etc., mostly endemic gen. In Pacific N. Am. is another region, with *Pseudotsuga menziesii, Sequoia, Taxodium, Chamaecyparis lawsoniana, Thuja plicata* and *Calocedrus decurrens,* together with endemic *Abies, Tsuga, Pinus,* etc. The Himal. forms another great centre, with many endemic spp., e.g. *Cedrus deodara, Pinus wallichiana* and others, *Picea* spp., *Tsuga* spp., etc. The C. of the N. hemisph. are separated from those of the S. hemisph. by a broad band of trop. forests, etc., partially broken by groups of C. on the mts. of the Indomal. region and Am. In Austr. we find *Araucaria, Agathis, Podocarpus, Callitris, Microcachrys, Athrotaxis, Actinostrobus,* etc. In Tasm., N.Z. and Chile appear *Phyllocladus, Fitzroya,* etc. S. Am. has *Araucaria* spp., *Podocarpus* spp., and others. Few gen. and spp. of C. appear in both N. and S. hemispheres; each sp. is limited to a well-defined area.

Trees or shrubs, usu. monopodial, often of considerable or even (*Sequoiadendron*) gigantic size. Typically, as may be seen in a fir or larch plantation, a certain amount of growth is made each year and a number of branches are also formed much at the same level, so that in trees of moderate size the number of 'whorls' of branches is an index of the age. Later on, the lower branches usu. die off and the branching near the apex becomes less reg. The main stem is radially symmetrical, but the branches, which often grow almost horiz., have a tendency to dorsiventrality, expressed in a two-ranked arrangement of the ls., twisting of the ls. on their stalks, and so on. Many C. show a difference in their shoots; some (long shoots, or shoots of unlimited growth) grow continuously onwards, except for the interruption in winter; others (short shoots, shoots of limited growth, or spurs) grow only to a definite size, usu. very small, and bear a few to many ls. (e.g. *Cedrus*). Intermediate conditions occur in *Larix, Cedrus, Taxodium,* etc. When both kinds occur the foliage ls. are often borne on the short shoots only (see *Pinus,* etc., for details). The green ls. are usu. entire and are either needle-like, flat and linear, or closely appressed scales (*Cupressus,* etc.). Mention may be made of the curious 'double needles' of *Sciadopitys* and the flat green short shoots of *Phyllocladus* (*qq.v.*).

Anatomically, the C. resemble Dicots. in many important points. A very general feature is the presence of resin passages in all parts of the pl. In many

CONILARIA

Pollen-sacs 2; pollen grains with 2–4 air bladders; female inflorescence consisting of 1–∞ sterile basal bracts and one to a few fertile scales each with one ovule **Podocarpaceae**

Pollen-sacs 3; pollen grains without bladders; female 'cone' consisting of a few opposite pairs of short fleshy scales each with two axillary ovules (usually only one seed matures) **Cephalotaxaceae**

[Pollen-sacs 2–8; pollen grains without bladders; female 'flower' consisting of several sterile bracts and a solitary terminal ovule **Taxaceae**]

Conilaria Rafin. = Lecokia DC. (Umbellif.).

Conimitella Rydb. = Heuchera L. (Saxifragac.).

Coniodictyogramme Nakai = Coniogramme Fée (Gymnogrammac.).

Coniogeton Blume = Buchanania Spreng. (Anacardiac.).

*****Coniogramme** Fée. Hemionitidaceae. 20 Old World trop., Hawaii, Mexico. Mexico.

Conioneura Pierre ex Engl. = Strombosia Blume (Olacac.).

Conioselinum Fisch. Umbelliferae. 12 temp. Euras.

Coniothele DC. = Blennosperma Less. (Compos.).

Coniphylis Thou. (uninom.) = *Bulbophyllum conicum* Thou. (Orchidac.).

Conirostrum Dulac = Brassica L. (Crucif.).

Conisa Desf. ex Steud. = Conyza L. (Compos.).

Conium L. Umbelliferae. 4 N. temp. Euras., S. Afr. *C. maculatum* L. (hemlock), very poisonous. Biennial. Stem dotted red.

Coniza Neck. = Conyza L. (Compos.).

*****Connaraceae** R.Br. Dicots. 16/300–350 trop. Mostly twining shrubs with alt. cpd. (sometimes 1-fol.) exstip. ls. and panicles of reg. or subreg. fls. K 5 or (5), imbr. or valvate; C 5, imbr. (rarely valv.); A 10 or 5, sometimes joined below; G 5 or 1 or 4, each with 2 erect orthotr. ov. Fr. usu. one follicle with one seed, album. or not, often arillate. Chief genera: *Connarus, Rourea, Cnestis.* Somewhat intermediate between *Leguminosae* and *Averrhoaceae.* K sometimes 4; A sometimes 8 or 4; G (8–) 5 (–3) or 1; ovules sometimes anatr.

Connaropsis Planch. ex Hook. f. = Sarcotheca Blume (*Averrhoac.*).

Connarus L. Connaraceae. Fewer than 100 trop. Am., Afr., As., Austr., Pacif.

Connellia N. E. Br. Bromeliaceae. 4 Venez., Guiana.

Conobaea Bert. ex Steud. = Muehlenbeckia Meissn. (Polygonac.).

Conobea Aubl. Scrophulariaceae. 7 Am., W.I.

Conocalpis Boj. ex Decne = Gymnema R.Br. (Asclepiadac.).

Conocalyx Benoist. Acanthaceae. 1 Madag.

Conocarpus Adans. = Leucadendron R.Br. (Proteac.).

Conocarpus L. Combretaceae. 2 Florida, W.I., trop. Am., NE. trop. Afr.

Conocephalopsis Kuntze = seq.

Conocephalus Blume (1825; non Neck. ex Dum. 1822—Hepaticae) = Poikilospermum Zipp. ex Miq. (Urticac.).

Conocliniopsis R. M. King & H. Rob. (~ Eupatorium L.). Compositae. 1 Colombia, Venez., Brazil.

Conoclinium DC. = Eupatorium L. (Compos.).

Conohoria Aubl. = Rinorea Aubl. (Violac.).

Conomitra Fenzl = Glossonema Decne (Asclepiadac.).

Conomorpha A. DC. Myrsinaceae. 60 trop. S. Am., W.I.

Conophallus Schott = Amorphophallus Blume (Arac.).
Conopharyngia G. Don. Apocynaceae. 25 Afr., Mascarene Is.
Conopholis Wallr. Orobanchaceae. 2 SE. U.S. to Panamá.
Conophora (DC.) Nieuwl. = Mesadenia Rafin. (Compos.).
Conophyllum Schwantes. Aïzoaceae. 20 S. Afr.
Conophyta Schumacher ex Hook. f. = Thonningia Vahl (Balanophorac.).
Conophyton Haw. = seq.
Conophytum N. E. Brown. Aïzoaceae. 270 S. Afr.
Conopodium Koch (~ Bunium L.). Umbelliferae. 20 Eur., As., N. Afr.
The tuberous roots of *C. majus* (Gouan) Loret (earth-nut) are ed. when roasted.
Conopsidium Wallr. = Platanthera Rich. (Orchidac.).
Conoria Juss. = Conohoria Aubl. = Rinorea Aubl. (Violac.).
Conosapium Muell. Arg. = Sapium P.Br. (Euphorbiac.).
Conosilene Fourr. = Silene L. (Caryophyllac.).
Conosiphon Poepp. & Endl. = Sphinctanthus Benth. (Rubiac.).
Conospermum Sm. Proteaceae. 35 Austr.
Conostegia D. Don. Melastomataceae. 50 trop. Am., W.I.
Conostemum Kunth (sphalm.) = Comostemum Nees = Androtrichum Brongn.
Conostephiopsis Shchegl. = seq. [(Cyperac.).
Conostephium Benth. Epacridaceae. 6 W. & S. Austr.
Conostomium (Stapf) Cufod. Rubiaceae. 12 trop. E. Afr.
Conostylis R.Br. Haemodoraceae. 23 SW. Austr.
Conostylus Pohl ex A. DC. = Conomorpha A. DC. (Myrsinac.).
Conothamnus Lindl. Myrtaceae. 4 W. Austr.
Conotrichia A. Rich. = Manettia Mutis ex L. (Rubiac.).
Conradia Mart. = Gesneria L. (Gesneriac.).
Conradia Nutt. = Macranthera Torr. (Scrophulariac.).
Conradia Rafin. = Tofieldia Huds. (Liliac.).
Conradina A. Gray. Labiatae. 4 SE. U.S.
Conringia Adans. Cruciferae. 7 Medit., Eur. to C. As.
Consana Adans. = Subularia L. (Crucif.).
Consolea Lem. = Opuntia Mill. (Cactac.).
Consolida Gilib. = Symphytum L. (Boraginac.).
Consolida (DC.) Opiz = Delphinium L. (Ranunculac.).
Consoligo (DC.) Opiz = Adonis L. (Ranunculac.).
Constantia Barb. Rodr. Orchidaceae. 3 Brazil.
Consuegria Mutis ex Caldas (nomen). Colombia. Quid?
Contarena Adans. = Corymbium L. (Compos.).
Contortaceae Dulac = Convolvulaceae Juss. + Cuscutaceae Dum.
Contarenia Vand. = ? Alectra Thunb. (Scrophulariac.).
Contortae Engl. An Order of Dicots. comprising the fams. *Oleac., Desfontain-iac., Loganiac. (s.l.), Gentianac., Apocynac.* and *Asclepiadac.*
Contortuplicata Medik. = Astragalus L. (Legumin.).
Contrarenia Jaume St-Hil. = Contarenia Vand. = ? Alectra Thunb. (Scrophulariac.).
Conuleum A. Rich. = Siparuna Aubl. (Siparunac.).
Convallaria L. Liliaceae. 1 N. temp., *C. majalis* L. (lily of the valley), in

CONVALLARIACEAE

woods. The stock develops a few scales and 2 green ls. annually. Fls. homogamous, self-fert.

Convallariaceae Horan. = Liliaceae–Convallarieae Endl.

*Convolvulaceae Juss. Dicots. 55/1650 trop. & temp. Herbs, shrubs or rarely trees, many climbing, some thorny xero. Some with tuberous roots or stems, others with rhiz.; latex often present. Ls. alt., usu. stalked, rarely stip., often with accessory axillary buds. Infl. dich. with tendency to cincinnus or bostryx; br. and bracteoles present, sometimes close to K. Fl. ♂, reg., hypog., usu. 5-merous. K, rarely (K), imbr., odd sep. post., seps. sometimes unequal; (C) of various shapes, usu. induplicate-valv., sometimes conv.; A 5, alt. with C, epipet. on base of C, anthers usu. intr.; disk present, honey-secreting; G̲ (2), rarely (3–5), or only joined by style, 2-loc. with axile plac.; ov. 2, rarely 1 or 4, per loc., erect, anatr. or semianatr., micropyle facing outwards and downwards; ov. with one integument. Berry, nut, caps.; endosp. Fl. usu. large and showy. Many have extrafloral nectaries on petiole. Few of economic value except for handsome fls. (cf. *Ipomoea*). Closely related to *Solanac.*, *Boraginac.*, and other *Tubiflorae*.

Classification and chief genera:

1. DICHONDREAE (G usu. divided, forming 2 or 4 1-seeded mericarps): *Dichondra* (2 meri.), *Falkia* (4) (only genera).
2. DICRANOSTYLEAE (G simple; style long ± bifid, or 2, stigma capitate or slightly lobed): *Breweria, Dicranostyles, Evolvulus*.
3. HILDEBRANDTIËAE (do., stigma irreg. lobed): *Hildebrandtia, Cladostigma, Sabaudiella*.
4. CONVOLVULEAE (do., style 1): *Jacquemontia, Calystegia, Convolvulus,* etc.
5. PORANEAE (do., do., caps. thin-membr., 1-seeded): *Porana*.
6. IPOMOEËAE (plicae of C sharply marked by 2 lat. nerves; fr. dehisc., or indehisc. chartac.; pollen spinulose): *Ipomoea, Pharbitis*, etc.
7. ARGYREIËAE (do., do., fr. indehisc. woody or fleshy): *Argyreia*.
8. ERYCIBEAE (pollen smooth; style ent.; fr. indeh., baccate or woody; lobes of C usu. bilobulate): *Erycibe* (only genus).

Convolvulaster Fabr. (uninom.) = *Convolvulus* L. sp. (Convolvulac.).

Convolvuloïdes Moench = Ipomoea L. (Convolvulac.).

Convolvulus L. Convolvulaceae. 250 cosmop., mostly temp. *C. arvensis* L. has sweetly scented fls., more visited by insects than the large but scentless fls. of *Calystegia sepium* (L.) R.Br. Smaller fls. with short sta. appear on some stocks; these are due to the action of a smut fungus (*Thecaphora seminisconvolvuli* (Duby) Liro). Veg. repr. by adv. stem buds produced on the root. From incisions in the rhiz. of *C. scammonia* L. flows a resinous purgative juice (scammony). Some yield rosewood oil.

Conysa Adans. = Conyza L. (Compos.).

Conystylus Pritz. (sphalm.) = Gonystylus Teijsm. & Binnend. (Thymelaeac.).

*Conyza Less. Compositae. 60 temp. & subtrop. Head androgynous.

Conyza L. = Inula, Gnaphalium, Callistephus, etc. (Compos.).

Conyzanthus Tamamsch. Compositae. 3–4 S. Am.

Conyzella Fabr. (uninom.) = *Conyza* Less. sp. (Compos.).

Conyzoïdes DC. = Carpesium L. (Compos.).

Conyzoïdes Fabr. (uninom.) = *Erigeron* L. sp. (Compos.).

Conzattia Rose. Leguminosae. 3 Mex.

Coockia Batsch = Cookia Sonner. = Clausena Burm. f. (Rutac.).

Cookia J. F. Gmel. = Pimelea Banks & Soland. (Thymelaeac.).

Cookia Sonner. = Clausena Burm. f. (Rutac.).

Coombea van Royen. Rutaceae. 1 W. New Guinea.

× **Cooperanthes** Lancaster. Amaryllidaceae. Gen. hybr. (Cooperia × Zephyranthes).

Cooperia Herb. Amaryllidaceae. 8 U.S., N. Mex.; 1 Brazil.

Coopernookia Carolin. Goodeniaceae. 5 SW. & SE. Austr., Tasm.

Copaiba Mill. = Copaïfera L. (Legumin.).

Copaica Dur. & Jacks. (sphalm.) = praec.

***Copaïfera** L. Leguminosae. 25 trop. Am., 5 trop. Afr. Some S. Am. spp. yield the resin Balsam of Copaiba, and resins (copals) are also obtained from the Afr. spp. Timber good (purpleheart).

Copaiva Jacq. = praec.

Copedesma Gleason. Melastomataceae. 1 Guiana.

Copernicia Mart. Palmae. 30 trop. Am., W.I. *C. cerifera* Mart. (wax- or *carna-uba*-palm, Brazil) has its ls. coated with wax, removed by shaking; it is used in making gramophone records, candles, etc. The wood, ls., etc., are also useful.

Copianthus Hill. ? Euphorbiacea monstrosa. 1 India.

Copiapoa Britton & Rose. Cactaceae. 15 Chile.

Copioglossa Miers = Ruellia L. (Acanthac.).

Copisma E. Mey. = Rhynchosia Lour. (Legumin.).

Copodium Rafin. = Lycopodium L. (Lycopodiac.).

Coppensia Dum. = Oncidium Sw. (Orchidac.).

Coppoleria Todaro = Vicia L. (Legumin.).

Coprosma J. R. & G. Forst. Rubiaceae. 90 Malaysia, Austr., Polynes., N.Z., Chile. The stipules of some are glandular; some spp. have peculiar openings (? domatia) on the backs of the ls.

Coprosmanthus Kunth = Smilax L. (Smilacac.).

Coptidipteris Nakai & Momose. Lindsaeaceae (probably). 1 China & Japan.

Coptidium Nym. = Ranunculus L. (Ranunculac.).

Coptis Salisb. Ranunculaceae. 15 N. temp. & arctic.

Coptocheile Hoffmgg. = ? Gesneria L. (Gesneriac.).

Coptophyllum Gardn. = Anemia Sw. (Schizaeac.).

***Coptophyllum** Korth. Rubiaceae. 3 W. Malaysia.

Coptosapelta Korth. Rubiaceae. 13 SE. China, Formosa, Ryu-Kyu Is., SE. As., Malaysia.

Coptospelta Dur. & Jacks. (sphalm.) = praec.

Coptosperma Hook. f. Rubiaceae. 1 trop. Afr.

Coquebertia Brongn. = Zollernia Maximil. & Nees (Legumin.).

Coralliokyphos Fleischm. & Rech. = Moerenhoutia Bl. (Orchidac.).

Coralliorrhiza Aschers. = Corallorhiza Chatelain (Orchidac.).

Corallobotrys Hook. f. = Agapetes D. Don ex G. Don (Ericac.).

Corallocarpus Welw. ex Hook. f. Cucurbitaceae. 15 trop. Afr., India & Madag.

Corallodendron Mill. = Erythrina L. (Legumin.).

CORALLODISCUS

Corallodiscus Batalin. Gesneriaceae. 18 Himal. to NW. China & Indoch.

Corallophyllum Kunth = Lennoa La Llave & Lex. (Lennoac.).

Corallorrhiza Gagnebin. Orchidaceae. 15 N. temp. Saprophytes with much-branched fleshy rhiz., no r., and scaly ls. (Cf. *Epipogium*.)

Corallospartium Armstrong. Leguminosae. 2 N.Z.

Coralluma Schrank ex Haw. = Caralluma R.Br. (Asclepiadac.).

Coralorhiza Rafin. = Corallorhiza Chatelain (Orchidac.).

Corazon Loefl. Hab.? Quid?

Corbassona Aubrév. Sapotaceae. 2 New Caled.

Corbichonia Scop. Aïzoaceae. 1 trop. Afr. to India.

Corbularia Salisb. = Narcissus L. (Amaryllidac.).

Corchoropsis Sieb. & Zucc. Sterculiaceae. 3 E. As.

Corchorus L. Tiliaceae. 100 warm. *C. capsularis* L. and *C. olitorius* L. (India, etc.) furnish the chief supply of the fibre jute or gunny; annuals about 3 m. high, little branched. The stems are cut and retted in water, and the fibre beaten out (cf. *Linum*).

Corculum Stuntz = Antigonon Endl. (Polygonac.).

Corda St-Lag. = Cordia L. (Ehretiac.).

Cordaea Spreng. = Cyamopsis DC. (Legumin.).

Cordanthera L. O. Williams. Orchidaceae. 1 Colombia.

Cordeauxia Hemsl. Leguminosae. 1 trop. Afr.

Cordemoya Baill. Euphorbiaceae. 1 Masc. (excl. Madag.).

Cordia L. Ehretiaceae. 250 warm. Trees or shrubs; fr. ed.; that of *C. myxa* L. (*sebestens*; As. Min., Palest.; cult. Egypt & trop. Afr.) formerly medic. Some have good timber, e.g. *C. gerascanthus* L. (trop. Am., W.I., prince-wood), and *C. sebestena* L. (trop. Am., W.I., aloewood).

*Cordiaceae R. Br. ex Dum. = Ehretiaceae Mart.

Cordiada Vell. = Cordia L. (Ehretiac.).

Cordiera A. Rich. = Alibertia A. Rich. (Rubiac.).

Cordiglottis J. J. Smith. Orchidaceae. 1 Sumatra.

Cordiopsis Desv. = Cordia L. (Ehretiac.).

Cordisepalum Verdc. Convolvulaceae. 1 Siam, Indoch.

Cordobia Niedenzu. Malpighiaceae. 2 S. Am.

Cordula Rafin. = Paphiopedilum Pfitz. (Orchidac.).

Cordyla Blume = Nervilia Comm. ex Gaudich. (Orchidac.).

Cordyla Lour. Leguminosae. 4 trop. & S. Afr. Apet. Pods fleshy, ed.

Cordyla P. & K. = Cordula Rafin. = Paphiopedilum Pfitz. (Orchidac.).

Cordylanthus Blume = Homalium Jacq. (Flacourtiac.).

*Cordylanthus Nutt. ex Benth. Scrophulariaceae. 40 W. N. Am.

Cordylestylis Falc. = Goodyera R.Br. (Orchidac.).

Cordylia Pers. = Cordyla Lour. (Legumin.).

Cordyline Adans. = Sanseverinia Petagna (Agavac.).

*Cordyline Comm. ex Juss. Agavaceae. 15 trop., warm temp. Habit of *Dracaena*. The ls. of some spp. yield fibre.

Cordyline Fabr. = Dracaena L. (Liliac.).

Cordyloblaste Hensch. ex Moritzi (~ Symplocos Jacq.). Symplocaceae. 7 S. India, Ceylon, S. China, N. Siam, W. Malaysia.

Cordylocarpus Desf. Cruciferae. 1 N. Afr.

CORIDOTHYMUS

Cordylocarya Bess. ex DC. = Rapistrum Crantz (Crucif.).
Cordylogyne E. Mey. Asclepiadaceae. 1 S. Afr.
Cordylophorum Rydb. = Epilobium L. (Onagrac.).
Cordylostylis P. & K. = Cordylestylis Falc. = Goodyera R.Br. (Orchidac.).
Coreanomecon Nakai = Hylomecon Maxim. (Papaverac.).
Corema Bercht. & J. S. Presl = Sarothamnus Wimm. (Legumin.).
Corema D. Don. Empetraceae. 2 E. N. Am., Azores, Canary Is., SW. Eur.
Coreocarpus Benth. Compositae. 8 Calif., Mex.
Coreopsid[ac]eae Link = Compositae–Heliantheae DC.
Coreopsis L. Compositae. 120 Am., trop. Afr.
Coreopsoïdes Moench = praec.
Coreosma Spach = Ribes L. (Grossulariac.).
Coresantha Alef. = Iris L. (Iridac.).
Coresanthe Baker = praec.
Coreta P.Br. = Corchorus L. (Tiliac.).
Corethrodendron Fisch. & Basiner = Hedysarum L. (Legumin.).
Corethrogyne DC. Compositae. 3 Calif.
Corethrostylis Endl. = Lasiopetalum Sm. (Sterculiac.).
Corethrum Vahl = Botelua Lag. (Gramin.).
Coriandraceae Burnett = Umbelliferae–Apioïdeae–Coriandreae & Smyrniëae p.p. (Astoma).
Coriandropsis H. Wolff. Umbelliferae. 1 Kurdistan.
Coriandrum L. Umbelliferae. 2 Medit. The fr. (coriander seeds) of *C. sativum* L. are used in flavouring.
Coriaria L. Coriariaceae. 15 Medit. to Japan, N.Z., Mex. to Chile. *C. myrtifolia* L. (W. Medit.) yields tan, others a black dye.
***Coriariaceae** DC. Dicots. 1/15 Euras., N.Z., C. & S. Am. Mostly shrubs with opp. or whorled parallel-veined stip. (!) ls., sometimes becoming alt. at the ends of the shoots. Stips. minute, caduc., sometimes several to each l. The inconspic. ♀ or ♂♀ protog. fls. are in racemose infls. K 5, imbr.; C 5, valv.; A 5 + 5, with large anth.; G 5–10. The petals are keeled on the inner side, and after fert. grow fleshy and enclose the cpls., forming a pseudo-drupe. Ov. 1 in each loc., pend., anatr.; raphe dorsal. Endosp. thin. Only genus: *Coriaria*. Relationships very uncertain.
Coridaceae (Reichb.) J. G. Agardh. Dicots. 1/1 Medit., NE. Afr. Small thyme-like subshrubs, stem often reddish-tinged. Ls. alt., lin., coriac., stip .(!), crowded, persist., ent. or obscurely dent. (uppermost often spinulose-marg.), with 2 rows of usu. conspic., black, immersed, marg. or submarg. glands. Fls. in crowded term. rac., zygo. to subreg. K (5), campan.-urc., membr., 10-nerved, teeth short, delt., valv., each bearing a large black dorsal gland, and an outer ring of 10–15 spreading aculei below the teeth; C (5), tub.-campan., sub-bilab. (3 upper lobes longer), lobes bifid, bright magenta to rose or white; A 5, antepet. and epipet., filif., exserted, glandular at base, anth. small globose; G (5), uniloc. with 5 semi-anatr. ov. on free centr. plac., and filif. style. Caps. glob., dehisc. by 5 valves, encl. in persist. K. Only genus: *Coris*. An interesting type, ± intermed. between *Primulac.* and *Lythrac.*
Coridochloa Nees ex Grah. = Alloteropsis Presl (Gramin.).
Coridothymus Reichb. f. = Thymus L. (Labiat.).

291

CORILUS

Corilus Nocca = Corylus L. (Corylac.).

Corindum Adans. = Paullinia L. (Sapindac.).

Corindum Mill. = Cardiospermum L. (Sapindac.).

Coringia J. & C. Presl = Conringia Fabr. (Crucif.).

Corinocarpus Poir. = Corynocarpus J. R. & G. Forst. (Corynocarpac.).

Coriocarpus Pax & Hoffm. = Coreocarpus Benth. (Compos.).

Corion Hoffmgg. & Link = Bifora Hoffm. (Umbellif.).

Corion Mitch. = Spergularia (Pers.) J. & C. Presl (Caryophyllac.).

Coriophyllus Rydb. = Cymopterus Rafin. (Umbellif.).

Coriospermum P. & K. = Corispermum L. (Chenopodiac.).

Coris L. Coridaceae. 1 (v. variable) Medit., Somal.

Corisanthera C. B. Clarke = Corysanthera Wall. = Rhynchotechum Blume (Gesneriac.).

Corisanthes Steud. = Criosanthes Rafin. (Orchidac.).

Corispermaceae Link = Chenopodiaceae–Corispermeae Moq.

Corispermum L. Chenopodiaceae. 60 N. temp.

Coristospermum Bertol. = Ligusticum L. (Umbellif.).

Corium P. & K. (1) = Corion Hoffmgg. & Link = Bifora Hoffm. (Umbellif.).

Corium P. & K. (2) = Corion Mitch. = Spergularia J. & C. Presl (Caryophyllac.).

Corizospermum Zipp. ex Blume = Casearia Jacq. (Flacourtiac.).

Cormigonus Rafin. = Bikkia Reinw. (Rubiac.).

Cormonema Reissek. Rhamnaceae. 6 trop. Am.

Cormus Spach = Sorbus L. (Rosac.).

Cormylus Rafin. = Hedyotis L. (Rubiac.).

Corna Nor. = Avicennia L. (Avicenniac.).

Cornacchinia Endl. = Baeolepis Decne ex Moq. (Periplocac.).

Cornacchinia Savi = Clerodendrum L. (Verbenac.).

*Cornaceae Dum. Dicots. 12/100 N. & S. temp., & trop. mts. Trees and shrubs, rarely herbs, with opp. or alt. simple ls., usu. petiolate, entire, exstip. Infl. dich. or pleioch., usu. condensed to corymbs or umbels, or even (*Cornus*) heads with invol. Fls. ⚥ or ♂ ♀, reg., 4–5-merous. K 4–5, small, or 0; C 4–5–0, usu. valv.; A 4–5; Ḡ (4–1, usu. 2); disk epig., style simple with lobed stigma; loc. 1–4, each with 1 pend. anatr. ov., raphe usu. dorsal. Drupe or berry, with 1–4-loc. stone or 2 separate stones. Endosp. Related to *Caprifoliac.* and *Escalloniac.* Genera: *Cornus* (*q.v.* for segregates), *Aucuba, Griselinia, Melanophylla, Curtisia, Kaliphora, Mastixia.* (Cf. monograph by Wangerin in Engl., *Pflanzenr.*, 1910.)

Cornachina Endl. (sphalm.) = Cornacchinia Endl. = Baeolepis Decne ex Moq. (Periplocac.).

Cornalia Lavy. ? Cruciferae. 1 NW. Italy. Quid?

Cornelia Ard. = Ammannia L. (Lythrac.).

Cornelia Rydb. = Chamaepericlymenum Hill (Cornac.).

Cornera Furtado. Palmae. 3 W. Malaysia.

Cornicina Boiss. = Anthyllis L. (Legumin.).

Corniculatae Sonnenb. = Crassulac. + Saxifragac. + Cunoniac. + Bruniac.

Cornidia Ruiz & Pav. = Hydrangea L. (Hydrangeac.).

Corniola Adans. = Genista L. (Legumin.).

Corniveum Nieuwl. = Dicentra Borckh. corr. Bernh. (Fumariac.).

Cornopteris Nakai. Athyriaceae. 12 trop. As. (See Holttum, *Kew Bull.* 13: 447–8, 1959.)

Cornucopiae L. Gramineae. 2 E. Medit. Fls. in small heads; when frs. ripe those of *C. cucullatum* L. bend over and break off with a sharp point; they adhere to animals, and are said to be capable of burrowing into the soil.

Cornuella Pierre. Sapotaceae (inc. sed.). 1 Venezuela.

Cornulaca Delile. Chenopodiaceae. 7 Egypt to C. As.

Cornulus Fabr. (uninom.) = *Swida* Opiz sp. (Cornac.).

Cornus L. (*s.str.*). Cornaceae. 4 C. & S. Eur. to Cauc., E. As., Calif. Preserve made from fr. of *C. mas* L. (Cornelian cherry, Eur., As. Min.), whose fls. appear in spring before ls. (For other spp. formerly included in *Cornus s.l.*, see *Afrocrania, Chamaepericlymenum, Cynoxylon, Dendrobenthamia, Swida* [*Thelycrania*].)

Cornuta L. = seq.

Cornutia Burm. f. = Premna L. (Verbenac.).

Cornutia L. Verbenaceae. 12 trop. Am., W.I. A 2; stds. filiform.

Corocephalus D. Dietr. (sphalm.) = Conocephalus Blume = Poikilospermum Zipp. ex Miq. (Urticac.).

Corokia A. Cunn. Escalloniaceae. 6 Austr., N.Z., Polynes., Rapa I.

Corolliferae Hutch. A primary division of Monocots. characterized by the possession of a petaloid outer (and inner) perianth-whorl, and frequent development of bulbs or corms, incl. most of Engler's orders *Liliiflorae, Pandanales, Spathiflorae, Principes, Microspermae*, etc.

Corollonema Schlechter. Asclepiadaceae. 1 Bolivia.

× **Coromelandrium** Graebn. Caryophyllaceae. Gen. hybr. (Coronaria × Melandrium).

Corona Fisch. ex R. Grah. = Fritillaria L. (Liliac.).

Coronanthera Vieill. ex C. B. Clarke. Gesneriaceae. 1 Queensl., 10 New Caled. Shrubs or small trees.

Coronaria Guett. (~ Lychnis L.). Caryophyllaceae. 5 temp. Euras.

Coronariae [L.] Reichb. = Liliaceae Juss.

Coronarieae Benth. & Hook. f. A 'Series' of Monocots. comprising the fams. *Roxburghiac., Liliac., Pontederiac., Philydrac., Xyridac., Mayacac., Commelinac.* and *Rapateac.*

Corone Hoffmgg. ex Steud. = Silene L. (Caryophyllac.).

Coronilla Ehrh. (uninom.) = *Coronilla coronata* L. (Legumin.).

Coronilla L. Leguminosae. 20 Eur., Medit. Fl. like *Lotus*, but honey usu. secreted by outer surface of K, insects poking between claws of pets. Buds and ripening fr. bend down, open fls. and ripe fr. upwards.

Coronocarpus Schumach. & Thonn. = Aspila Thou. (Compos.).

Coronopus Mill. = Plantago L. (Plantaginac.).

*Coronopus Zinn. Cruciferae. 10 almost cosmop.

Coropsis Adans. = Coreopsis L. (Compos.).

Corothamnus Presl = Genista L. (Legumin.).

Coroya Pierre. Leguminosae. 1 Indoch.

Corozo Jacq. ex Giseke. Palmae. 1 trop. Am.

Corpodetes Reichb. = Stenomesson Herb. (Amaryllidac.).

Corpuscularia Schwantes = Delosperma N. E. Br. (Aïzoac.).

CORRAEA

Corraea Sm. = seq.

*Correa Andr. Rutaceae. 11 temp. Austr. Sympet. and tetram.

Correa Becerra = Dialium L. (Legumin.).

Correaceae J. G. Agardh = Rutaceae–Boronieae–Correinae Engl.

Correaea P. & K. = Correia Vell. = Ouratea Aubl. (Ochnac.).

Correas Hoffmgg. = Correa Andr. (Rutac.).

Correia Vell. = Ouratea Aubl. (Ochnac.).

× **Correvonia** hort. (vii) = × Brassocattleya hort. (Orchidac.).

Corrigiola Kuntze = Illecebrum L. (Caryophyllac.).

Corrigiola L. Caryophyllaceae. 10 cosmop.

Corrigiolaceae Dum. = Caryophyllaceae–Paronychieae Pax.

Corryocactus Britton & Rose. Cactaceae. 21 Peru, Bolivia, Chile.

Corryocereus Frič & Kreuz. = praec.

Corsia Becc. Corsiaceae. 25 New Guinea, 1 Austr., 2 Solomon Is.

*Corsiaceae Becc. Monocots. 2/29 N. Guinea, etc., & Chile. Small erect rhiz. or tuberous saprophytes; ls. red. to scales. Fl. sol., term., zygo., ♀ or ♂ ♀. P 3 + 3, post. member of outer whorl large, coloured, cord.-ov., marg. laterally involute in bud, encl. remaining 5, wh. are linear-spath. and finally pend. A 3 + 3, fil. short, subequ., anth. shortly ovoid, 2-loc., extr. G̅ (3), 1-loc., glob. or elong., with 3 pariet. plac. and ∞ ov., and short thick shortly trifid style. Caps dehisc. by 3 valves. Genera: *Corsia, Achratinis (Arachnitis)*. Near *Burmanniac.*

*Cortaderia Stapf. Gramineae. 15 N.Z., S. Am. Pampas grass.

Cortesia Cav. Ehretiaceae. 2 temp. S. Am.

Corthumia Reichb. = Pelargonium L'Hérit. (Geraniac.).

Corthusa Reichb. = Cortusa L. (Primulac.).

Cortia DC. Umbelliferae. 7 Afghan. to SW. China.

Cortiella C. Norman. Umbelliferae. 2 Tibet, Himal.

Cortusa L. Primulaceae. 10 mts. of C. Eur. to Japan & Sakhalin.

Cortusina Eckl. & Zeyh. = Pelargonium L'Hérit. (Geraniac.).

Corunastylis Fitzgerald (1888; non *Corynostylis* Mart. 1824) = Anticheirostylis Fitzgerald (Orchidac.).

Corvina B. D. Jacks. = Muricaria Desv. (Crucif.). ('Corv. prostrata' B. D. Jacks. = Calepina corvini + Cal. [Muricaria] prostrata!)

Corvinia Stadtm. ex Willem. = Litchi Sonn. (Sapindac.).

Corvisartia Mérat = Inula L. (Compos.).

Corya Rafin. = Carya Nutt. (Juglandac.).

Coryanthes Hook. Orchidaceae. 17 trop. Am. Fl. pend.; seps. bent back and fairly large, pets. small. Labellum complex, forming a bucket-like organ with dome above; the mouth faces upwards, and the edges are incurved; there is also an overflow pipe projecting towards the seps. and closely covered in by the bent end of the column, with the stigma and anther. From the base of the column project two horns which secrete a thin watery fluid that drips into the bucket, keeping it full to the level of the overflow pipe. The dome (above) is composed of succulent tissue attractive to bees; these fight for places on it to drill the tissue; every now and then one gets pushed off and falls into the bucket. It can neither fly nor climb out, and has to squeeze through the overflow pipe. In so doing it first passes the stigma, fertilising it if it bears any pollen, and then, passing the anther, is loaded with new pollinia. (Cf. *Stanhopea*.)

Corybas Salisb. Orchidaceae. 50 Indomal., Solomon Is., Austr., N.Z.

Corycarpus Zea ex Spreng. = Diarrhena Rafin. (Gramin.).

Corycium Swartz. Orchidaceae. 15 S. Afr.

Coryda B. D. Jacks. (sphalm.) = Cordyla Lour. (Legumin.).

*Corydalis Vent. Fumariaceae. 320 N. temp., 1 mts. trop. E. Afr. *C. claviculata* (L.) DC. is a (leaf) tendril-climbing annual. Most are perennial herbs with underground tubers. In *C. cava* (L.) Schweigg. & Kort., and others, the main axis forms a tuber, which dies away below, each annual shoot arising from the axil of a scale-l. of older date. In *C. solida* (L.) Sw., and others, the tuber is a swollen root-structure belonging to the current annual shoot. Fl. transv. zygo.; twisting through 90° brings it vertical; only one petal is spurred and contains the honey secreted by a staminal outgrowth. Its mech. resembles that of *Leguminosae*. The inner pets., united at the tip, enclose stigma and anthers; the upper pet. covers the fl. Bees alighting push down the inner pets. and cause the essential organs to emerge. In some, e.g. *C. ochroleuca* Koch and *C. lutea* (L.) DC., the emergence is explosive (cf. *Genista*). The fls. of *C. Cava* are self-sterile.

Corydandra Reichb. = Galeandra Lindl. (Orchidac.).

*Corylaceae Mirbel. Dicots. 1/15 N. temp. Trees or shrubs with alt., simple, stip., often distich. ls. Fls. monoec.; ♂ in pend. catkins, sol. in axil of each bract; bracteoles 2, united to bract; P 0; A 3–14; ♀ in pairs in axil of each bract, bracteoles + ; P small, irreg. lobed; Ḡ (2), median, with 1 pend. anatr. ovule in each cell and 2 free styles. Fr. a 1-seeded nut, surrounded by l.-like invol. (accresc. bract and bracteoles); cotyledons retained inside nut at germination. Only genus: *Corylus*.

Corylopasania (Hickel & A. Camus) Nakai = Pasania (Miq.) Oerst. = Lithocarpus Bl. (Fagac.).

Corylopsis Sieb. & Zucc. Hamamelidaceae. 20 Himal., E. As. Fls. ♀, precocious, in usu. pend. bracteate spikes or rac.

Corylus L. Corylaceae. 15 N. temp. *C. avellana* L., shrubby (largely owing to extensive formation of suckers), with monoecious catkinate fls. (the ♀ catkin small, sessile and ellipsoid, rather resembling a bud). Both are laid down in autumn; the ♂ catkins are visible all winter, but the ♀ are not obvious until the red stigmas come out early in the year. Anemoph.; the fact of flg. before the appearance of the ls. renders their chance of fert. greater. On the inner side of the br. in the ♂ catkin are found 2 scales and, adnate to these, 4–8 sta., each branched nearly to the base. Here only the central fl. of the possible 3 is present, with its bracteoles α, β. In the ♀ catkin, on the other hand, we have the two lat. and not the central fl. At the time of fert. the ovary is minute, but the long red stigmas are easily identified. After fert. the ovary (2-loc. at first) gives a one-seeded nut, enclosed in a green leafy cup, really the combined bract and bracteoles α, α', β', very much developed. The fl. is chalazogamic (cf. *Chalazogamae*). The nuts of this and other spp. are valuable as dessert fr., etc. (hazel-nut, cob-nut, filbert), and have been cultivated from very early times. Wood elastic, but cannot be obtained in large boards. Oil from the seeds.

CORYMBIFERAE

Corymbiferae Juss. = Compositae Giseke.

Corymbis Reichb. f. = Corymborkis Thou. (Orchidac.).

Corymbis Thou. (uninom.) = *Corymborkis corymbosa* Thou. (Orchidac.).

Corymbium L. Compositae. 17 S. Afr. Ls. narrow, ||-veined.

Corymborchis Thou. = seq.

Corymborkis Thou. Orchidaceae. 20 pantrop.

Corymbostachys Lindau. Acanthaceae. 1 Madag.

Corymbula Rafin. = Polygala L. (Polygalac.).

Corynabutilon (K. Schum.) Kearney. Malvaceae. 4 Chile.

Corynaea Hook. f. Balanophoraceae. 4 Andes.

Corynandra Schrad. = Cleome L. (Cleomac.).

Corynanthe Welw. Rubiaceae. 8 trop. Afr.

Corynanthelium Kunze = Mikania Willd. (Compos.).

Corynanthes Schlechtd. = Coryanthes Hook. (Orchidac.).

Corynella DC. Leguminosae. 3 W.I.

Corynelobos R. Roem. = Brassica L. (Crucif.).

Corynemyrtus (Kiaersk.) Mattos. Myrtaceae. 1 Brazil.

*Corynephorus Beauv. Gramineae. 6 Eur., Medit.

Corynephyllum Rose = Sedum L. (Crassulac.).

Corynitis Spreng. = Corynella DC. (Legumin.).

*Corynocarpaceae Engl. Dicots. 1/5 SW. Pacif. Trees or shrubs with alt. simple ent. exstip. ls. Fls. in term. pan. K 5, imbr.; C 5, imbr., adn. to base of K; A 5, antepet. and epipet., alt. with 5 petaloid stds.; disk of 5 large glands opp. stds.; G (1–2), 1 loc. fert. with 1 pend. ov., the other ster.; styles 1–2, free; fr. a glob. fleshy drupe, seed exalb. Only genus: *Corynocarpus*. Probably related to *Anacardiac.*

Corynocarpus J. R. & G. Forst. Corynocarpaceae. 4–5 New Guinea, Queensl., New Hebrides, New Caled., N.Z. Staminodes very variable.

Corynolobus P. & K. = Corynelobos R. Roem. = Brassica L. (Crucif.).

Corynophallus Schott = Amorphophallus Blume (Arac.).

Corynophorus Kunth = Corynephorus Beauv. (Gramin.).

Corynopuntia F. M. Knuth = Opuntia Mill. (Cactac.).

Corynosicyos F. Muell. = Cucumeropsis Naud. (Cucurbitac.).

Corynostigma Presl = Ludwigia L. (Onagrac.).

Corynostylis Mart. Violaceae. 4 trop. Am.

Corynostylus P. & K. = Corunastylis Fitzg. = Anticheirostylis Fitzg. (Or-

Corynotheca F. Muell. Liliaceae. 3 trop. & W. Austr.   [chidac.).

Corynula Hook. f. Rubiaceae. 2 Colombia.

Corypha L. Palmae. 8 Ceylon, SE. As., Indomal. The gigantic infl. terminates the life of the tree. *C. umbraculifera* L. (talipot palm, Ceylon, S. Ind.) grows to a great size, up to 24 m. The ls. are used as umbrellas, and for thatching, also as writing material (a metal stylus being used).

Coryphaceae Schultz-Schultzenst. = Palmae–Coryphoïdeae Spreng.

Coryphadenia Morley = Votomita Aubl. (Memecylac.).

Coryphantha (Engelm.) Lem. Cactaceae. 64 SW. U.S., Mex., Cuba.

Coryphomia Rojas. Palmae. 1 Argentina.

Coryphopteris Holtt. Thelypteridaceae. 30 SE. Asia to Samoa, in peaty soil in forest near crests of mountain ridges.

Coryphothamnus Steyerm. Rubiaceae. 1 SE. Venez.
Corysadenia Griff. = Illigera Blume (Hernandiac.).
Corysanthera Wall. = Rhynchotechum Blume (Gesneriac.).
Corysanthes R.Br. = Corybas Salisb. (Orchidac.).
Corythacanthus Nees = Clistax Mart. (Acanthac.).
Corythanthes Lem. = Coryanthes Hook. (Orchidac.).
Corythea S. Wats. Euphorbiaceae. 2 Mexico.
Corytholobium Mart. ex Benth. = Securidaca L. (Polygalac.).
Corytholoma (Benth.) Decne = Rechsteineria Regel (Gesneriac.).
Corythophora Knuth. Lecythidaceae. 1 Brazil.
Corytoplectus Oerst. = Alloplectus Mart. (Gesneriac.).
Coryzadenia Griff. = Illigera Blume (Hernandiac.).
Cosaria J. F. Gmel. = Kosaria Forsk. = Dorstenia L. (Morac.).
Cosbaea Lem. = Schisandra Michx (Schisandrac.).
Coscinium Colebr. Menispermaceae. 8 Indomal., Indoch.
Cosentinia Todaro = Cheilanthes Sw. (Sinopteridac.).
Cosmanthus Nolte ex A. DC. = Phacelia Juss. (Hydrophyllac.).
Cosmarium Dulac = Adonis L. (Ranunculac.).
Cosmea Willd. = Cosmos Cav. (Compos.).
Cosmelia R.Br. Epacridaceae. 1 SW. Austr.
Cosmia Domb. ex Juss. = Calandrinia Kunth (Portulacac.).
Cosmianthemum Bremek. Acanthaceae. 8 W. Borneo.
Cosmibuena Ruiz & Pav. (1794) = Hirtella L. (Chrysobalanac.).
*****Cosmibuena** Ruiz & Pav. (1802). Rubiaceae. 12 C. & trop. S. Am. Good
bud-protection by the stips. of the last-opened ls.
Cosmidium Nutt. = Thelesperma Less. (Compos.).
Cosmiusa Alef. = Parochetus Buch.-Ham. (Legumin.).
Cosmiza Rafin. = Polypompholyx Lehm. (Lentibulariac.).
Cosmocalyx Standley. Rubiaceae. 1 Mexico.
Cosmoneuron Pierre = Strombosia Blume (Olacac.).
Cosmophyllum C. Koch = Podachaenium Benth. (Compos.).
Cosmos Cav. Compositae. 25 trop. & subtrop. Am., W.I.
Cosmostigma Wight. Asclepiadaceae. 3 S. China (Hainan), Indomal.
Cossignea Willd. = seq.
Cossignia Comm. ex Juss. = seq.
Cossignya Baker = seq.
Cossinia Comm. ex Lam. Sapindaceae. 4 Masc., New Caled., Fiji.
Cossonia Durieu = Raffenaldia Godr. (Crucif.).
Costa Vell. = Galipea Aubl. (Rutac.).
Costaceae (K. Schum.) Nak. Monocots. 4/200 trop. As *Zingiberaceae*, but
aerial parts not aromatic; ls. alt. or 4-ranked, with closed sheaths; no erect
nectary, but depressed supra-septal nectary present (cf. *Marantac.*); anther-loc.
often apically appendiculate; $\bar{G}$ (3) with axile plac. or (2) with pariet. plac.;
ov. 1–2-seriate. Genera: *Costus, Dimerocostus, Monocostus, Tapeinochilos.*
Costaea P. & K. (1) = Costa Vell. = Galipea Aubl. (Rutac.).
Costaea P. & K. (2) = Costia Willk. (1858) = Agropyron J. Gaertn. (Gramin.).
Costaea P. & K. (3) = Costia Willk. (1860) = Iris L. (Iridac.).
Costaea A. Rich. Cyrillaceae. 3 Cuba, Colombia.

COSTANTINA

Costantina Bullock. Asclepiadaceae. 1 Indochina.

Costaricaea Schlechter = Hexisea Lindl. (Orchidac.).

Costaricia H. Christ. ?Dennstaedtiaceae. 1 Costa Rica. Descr. from young sterile plant; probably = Dennstaedtia Bernh. (Tryon, *Rhodora*, **63**: 75–8, 1961).

Costera J. J. Smith. Ericaceae. 8 Sumatra, Borneo, Philipp.

Costia Willk. (1858) = Agropyron J. Gaertn. (Gramin.).

Costia Willk. (1860) = Iris L. (Iridac.).

Costularia C. B. Clarke. Cyperaceae. 11 S. Afr., Masc., Austr.

Costus L. Costaceae. 150 trop. Labellum very large, lat. stds. wanting, sepals and petals comparatively small. Projecting in the centre is the fertile petaloid sta. with anther on its ant. face; the style reaches just above this. The fl. mech. thus resembles *Iris*.

Cota J. Gay = Anthemis L. (Compos.).

Cotema Britton & P. Wils. = Spirotecoma Baill. ex Dalla Torre & Harms (Bignoniac.).

Cotinus Mill. (~Rhus L.). Anacardiaceae. 1 SE. U.S., 1 S. Eur. to Himal. & China, 1 SW. China. *C. coggygria* Scop. (Medit. to China) is the wig-tree, often cult. in shrubberies. Fls. polyg. The stalk of each drupe remains smooth, but the sterile parts of the panicle lengthen and become hairy. Then when ripe the stalks become detached at their joints, and the whole infl., with the fr. on it, falls to the ground and may be blown about. The wood yields the yellow dye 'young fustic'.

Cotonea Rafin. = seq.

Cotoneaster [B. Ehrh.] Medik. Rosaceae. 50 N. temp. *C. vulgaris* Lindl. in the Alps is visited solely by a wasp (*Polistes gallica*) whose nests are often attached to the rocks where the pl. grows. Fl. protog. with self-fert. in default of insects.

Cotopaxia Mathias & Constance. Umbelliferae. 1 Ecuador.

Cottaea Endl. = seq.

Cottea Kunth. Gramineae. 1 S. U.S. to C. Mex., Ecuad. to Argent.

Cottendorfia Schult. f. Bromeliaceae. 1 E. Brazil.

Cottetia Gandog. = Rosa L. (Rosac.).

Cottonia Wight. Orchidaceae. 1 S. India, Ceylon.

Cotula L. Compositae. 75 almost cosmop., esp. S. hemisph.

Cotulina Pomel = praec.

Cotylanthera Blume. Gentianaceae. 4 Himal., SW. China, Java, Polynes.

Cotylaria Rafin. = Cotyledon L. (Crassulac.).

Cotyledon L. Crassulaceae. 40 S. Afr., 1 Eritr., Arabia.

Cotylelobiopsis Heim = Copaïfera L. (Legumin.).

Cotylelobium Pierre. Dipterocarpaceae. 5 Ceylon, W. Malaysia.

Cotylephora Meissn. = Neesia Blume (Malvac.).

Cotylina P. & K. = Cotulina Pomel = Cotula L. (Compos.).

Cotyliphyllum Link = Cotyledon L. (Crassulac.).

Cotyliscus Desv. = Coronopus Zinn (Crucif.).

Cotylodiscus Radlk. Sapindaceae. 1 Madag.

Cotylolobiopsis P. & K. = Cotylelobiopsis Heim = Copaïfera L. (Legumin.).

Cotylolobium P. & K. = Cotylelobium Pierre (Dipterocarpac.).

Cotylonia Norman = Dickinsia Franch. (Hydrocotylac.).
Cotylonychia Stapf. Sterculiaceae. 1 trop. Afr.
Cotylophyllum P. & K. = Cotyliphyllum Link = Umbilicus DC. (Crassulac.).
Cotyloplecta Alef. = Hibiscus L. (Malvac.).
Coublandia Aubl. = Muellera L. f. (Legumin.).
Coudenbergia Marchal = Pentapanax Seem. (Araliac.).
Couepia Aubl. Chrysobalanaceae. 58 C. & trop. S. Am.
Coula Baill. Olacaceae. 3 trop. W. Afr.
Coulaceae Van Tiegh. = Olacaceae–Couleae Engl.
Coulejia Dennst. = Antidesma L. (Stilaginac.).
Coulterella Van Tiegh. = Pterocephalus Adans. (Dipsacac.).
Coulterella Vasey & Rose. Compositae. 1 NW. Mex. (Lower Calif.).
Coulteria Kunth = Caesalpinia L. (Legumin.).
Coulterina Kuntze = Physaria (Nutt.) A. Gray (Crucif.).
Coulterophytum Robinson. Umbelliferae. 5 Mex.
Couma Aubl. Apocynaceae. 15 Brazil, Guiana.
Coumarouna Aubl. Leguminosae. 13 trop. Am. Ls. opp. 3 lower seps.
mere teeth, 2 upper wing-like. Fr. 1-seeded, indehisc. *C. odorata* Aubl.
(Tonquin or Tonka bean) used in perfumery, snuff, etc. Some yield valuable
timber.
Coupia G. Don (sphalm.) = Goupia Aubl. (Goupiac.).
Coupoui Aubl. = Duroia L. f. (Rubiac.).
Couralia Splitg. = Tabebuia Gomez (Bignoniac.).
Courantia Lemaire = Echeveria DC. (Crassulac.).
Couratari Aubl. Lecythidaceae. 18 trop. S. Am. The bark yields a soft fibre
used for making clothing.
Courbari Adans. = seq.
Courbaril Miller = Hymenaea L. (Legumin.).
Courbonia Brongn. = Maerua Forsk. (Capparidac.).
Courimari Aubl. = Sloanea L. (Elaeocarpac.).
Couringia Adans. = Conringia Fabr. (Crucif.).
Courondi Adans. = Salacia L. (Celastrac.).
Couroupita Aubl. Lecythidaceae. 20 trop. Am., W.I. The fls. of *C. guianensis*
Aubl. are borne on the old stems and followed by large spherical woody caps.
(whence the name cannon-ball tree). Good timber.
Courrantia Sch. Bip. = Matricaria L. (Compos.).
Coursetia DC. Leguminosae. 15 S. Calif. to Brazil.
Coursiana Homolle. Rubiaceae. 1–2 Madag.
Courtenia R.Br. = Cola Schott & Endl. (Sterculiac.).
Courtoisia Nees. Cyperaceae. 2 Afr., Madag., India.
Courtoisia Reichb. = Phlox L. (Polemoniac.).
Cousinia Cass. Compositae. 400 E. Medit. to C. As. & Mong., S. to Persia,
Afgh. & W. Himal.
Cousiniopsis Nevski. Compositae. 1 C. Asia.
Coussapoa Aubl. Urticaceae. 50 trop. S. Am. Somewhat intermediate
between *Urticac.* and *Morac.*
Coussarea Aubl. Rubiaceae. 90 trop. Am.
Coutaportia Urb. = Portlandia R.Br. (Rubiac.).

COUTAREA

Coutarea Aubl. Rubiaceae. 7 trop. Am., W.I.

Couthovia A. Gray = Neuburgia Bl. (Strychnac.).

Coutinia Vell. Apocynaceae. 7 Panamá, trop. S. Am.

Coutiria Willis (sphalm.) = praec.

Coutoubea Aubl. Gentianaceae. 5 trop. S. Am., W.I.

Coutubea Steud. = praec.

Covalia Reichb. = Covolia Neck. ex Rafin. = Spermacoce L. (Rubiac.).

Covellia Gasp. = Ficus L. (Morac.).

Covilhamia Korth. = Stixis Lour. (Capparidac.).

Covillea A. M. Vail = Larrea Cav. (Zygophyllac.).

Covola Medik. = Salvia L. (Labiat.).

Covolia Neck. ex Rafin. = Spermacoce L. (Rubiac.).

Cowania D. Don. Rosaceae. 5 SW. U.S., Mex.

Cowellocassia Britton = Cassia L. (Legumin.).

Cowiea Wernham = Petunga DC. (Rubiac.).

Coxella Cheesem. & Hemsl. Umbelliferae. 1 Chatham Is.

Coxia Endl. = Lysimachia L. (Primulac.).

***Crabbea** Harv. Acanthaceae. 12 trop. & S. Afr.

Crabowskia G. Don (sphalm.) = Grabowskia Schlechtd. (Solanac.).

***Cracca** Benth. ex Oerst. Leguminosae. 8 trop. Am.

Cracca Hill = Vicia L. (Legumin.).

Cracca L. = Tephrosia Pers. (Legumin.).

Craccina Stev. = Astragalus L. (Legumin.).

Cracosna Gagnep. Gentianaceae. 1 Indoch.

Craepalia Schrank = Lolium L. (Gramin.).

Craepaloprumnon Karst. = Xylosma J. R. & G. Forst. (Flacourtiac.).

Crafordia Rafin. = Tephrosia Pers. (Legumin.).

Craibia Harms & Dunn. Leguminosae. 10 trop. Afr.

Craibiodendron W. W. Smith. Ericaceae. 7 SE. As.

Craigia W. W. Smith & Evans. Tiliaceae. 1 SW. China.

Crambe L. Cruciferae. 25 Eur., Medit., N. Atl. Is., trop. Afr., W. & C. As. *C. maritima* L. (sea-kale) has fleshy and waxy ls. The young ls. blanched form a veg.

Crambella Maire. Cruciferae. 1 Morocco.

Crameria Murr. = Krameria L. (Krameriac.).

Cranichis Sw. Orchidaceae. 34 trop. Am., W.I.

Craniolaria L. Martyniaceae. 3 S. Am.

Craniospermum Lehm. Boraginaceae. 4 temp. As.

Craniotome Reichb. Labiatae. 1 Himalaya.

Cranocarpus Benth. Leguminosae. 2 Brazil.

Crantzia DC. = Cranzia Schreb. = Toddalia Juss. (Rutac.).

Crantzia Lag. ex DC. = Conringia Adans. + Moricandia DC. (Crucif.).

Crantzia Nutt. = Crantziola F. Muell. = Lilaeopsis E. L. Greene (Umbellif.).

Crantzia Pohl (nomen). 3 Brazil. Quid?

Crantzia Scop. = Alloplectus Mart. (Gesneriac.).

Crantzia Sw. = Buxus L. (Buxac.).

Crantzia Vell. = Centratherum Cass. (Compos.).

Crantziola F. Muell. = Lilaeopsis E. L. Greene (Umbellif.).

Cranzia J. F. Gmel. = Crantzia Sw. = Buxus L. (Buxac.).
Cranzia Schreb. = Toddalia Juss. (Rutac.).
Craspedaria Link = Microgramma Presl (Polypodiac.).
Craspedia Forst. f. Compositae. 7 temp. Austr., N.Z.
Craspedodictyum Copel. = Syngramma J. Sm. (Hemionitidac.).
Craspedolepis Steud. = Restio L. (Restionac.).
Craspedolobium Harms. Leguminosae. 1 SW. China.
Craspedoneuron v. d. Bosch = Pleuromanes Presl (Hymenophyllac.).
Craspedophyllum (Presl) Copel. Hymenophyllaceae. 29 SE. Austr., N.Z.
Craspedorhachis Benth. Gramineae. 5–6 trop. Afr.
Craspedospermum Boj. ex DC. Apocynaceae. 1 Madag.
Craspedostoma Domke. Thymelaeaceae. 5 S. Afr.
Craspedum Lour. = Elaeocarpus L. (Elaeocarpac.).
Craspidospermum Boj. ex DC. = Craspedospermum Boj. ex DC. (Apocynac.).
Crassangis Thou. (uninom.) = Angraecum crassum Thou. (Orchidac.).
Crassina Shchepin (Scepin) = Zinnia L. (Compos.).
Crassipes Swallen = Sclerochloa Beauv. (Gramin.).
Crassocephalum Moench. Compositae. 30 warm Afr., Madag.; 1 sp. an aggressive weed in trop. E. As.
Crassopetalum Northrop (sphalm.) = Crossopetalum P.Br. = Myginda L. (Celastrac.).
Crassouvia Comm. ex DC. = Crassuvia Comm. ex Lam. = Bryophyllum Salisb. (Crasulac.).
Crassula L. Crassulaceae. Over 300 cosmop., esp. S. Afr. (cf. *Trans. R. S. S. Afr.* **17**: 151, 1929), usu. succulent-l. xero. In *C. lycopodioïdes* Lam. the ls. are narrow and closely packed, giving to the pl. the habit of a *Lycopodium*. In *C. (Rochea) falcata* Wendl. the connate decussate ls. stand almost edgewise, and are very fleshy; some of the epidermal cells are swollen above the rest into large bladders which meet one another over the whole surface. At first living, when the l. is mature they are dead and full of air, their walls infiltrated with quantities of silica. A protection against evap. is thus afforded. In *C. nemorosa* Endl. there is veg. repr. by the formation of young plants in the infl. in place of fls.
***Crassulaceae** DC. Dicots. 35/1500 cosmop., chiefly S. Afr., a very natural group. Most are perenn. living in dry (esp. rocky) places and exhibit xero. chars., fleshy ls. and stem, often tufted growth, close packing of ls., waxy surface, sunk stomata, etc. Veg. repr. frequent; usu. by rhiz. or offsets; some form bulbils, etc. (e.g. *Crassula*), others form adv. buds upon the ls. (e.g. *Kalanchoë*). Fls. usu. in cymes (cincinni), ♂ or rarely unisex., actinom. with very reg. construction. Formula K*n*, C*n*, A*n* + *n*, G*n*, where *n* represents any number from 3 to 30. K persistent; C sometimes (e.g. *Cotyledon*) gamopet.; A frequently obdipl. Insertion of parts usu. perig., but recept. not deeply hollowed. Cpls. frequently slightly united at the base; at the base of each commonly a honey-secreting scale; ov. usu. ∞. Fr. usu. a group of follicles with very small seeds. Endosp. none or very little. Fls. mostly protandr. and chiefly visited by flies, etc., their honey being easily obtainable. Chief genera: *Sedum, Sempervivum, Cotyledon, Kalanchoë, Crassula*.
Crassularia Hochst. ex Schweinf. = Crassula L. (Crassulac.).

CRASSUVIA

Crassuvia Comm. ex Lam. = Kalanchoë Adans. (Crassulac.).

+ Crataegomespilus Simon-Louis ex G. Bellair. Rosaceae. Gen. hybr. asex. (Crataegus + Mespilus).

× Crataegosorbus Makino ex Koidz. Rosaceae. Gen. hybr. (Crataegus × Sorbus).

Crataegus L. Rosaceae. 200 N. temp. Some hundreds of spp. have been described from the U.S., but there is some possibility that they may have arisen through hybridization. The thorns of *C. monogyna* Jacq. are modified branches. Collateral buds appear in the axils. The wood is a substitute for that of box in engraving, etc.

× Crataemespilus G. Camus. Rosaceae. Gen. hybr. (Crataegus × Mespilus).

Crataeva L. = Crateva L. (Capparidac.).

Crateola Rafin. = ?Hemigraphis Nees v. ?Dipteracanthus Nees (Acanthac.).

Crateranthus E. G. Baker. Napoleonaceae. 3 trop. W. Afr.

Crateria Pers. = Casearia Jacq. (Flacourtiac.).

Craterianthus Valeton ex K. Heyne = Pellacalyx Korth. (Rhizophorac.).

Cratericarpium Spach = Oenothera L. (Onagrac.).

Crateriphytum Scheff. ex Koord. = Neuburgia Bl. (Strychnac.).

Craterispermum Benth. Rubiaceae. 17 trop. Afr., 1 Seychelles.

Crateritecoma Lindl. = Craterotecoma Mart. ex DC. = Lundia DC. (Bignoniac.).

Craterogyne Lanj. Moraceae. 4 trop. Afr.

Craterosiphon Engl. & Gilg. Thymelaeaceae. 7 trop. Afr.

Craterostemma K. Schum. Asclepiadaceae. 1 SE. Afr.

Craterostigma Hochst. Scrophulariaceae. 20 trop. & S. Afr., Madag.

Craterotecoma Mart. ex Meissn. = Lundia DC. (Bignoniac.).

Crateva L. Capparidaceae. 9 trop. (exc. Austr. & New Caled.).

Cratochwilia Neck. = Clutia L. (Euphorbiac.).

Cratoxylon Blume = seq.

Cratoxylum Blume. Guttiferae. 6 SE. As., W. Malaysia.

Cratylia Mart. ex Benth. Leguminosae. 8 S. Am.

Cratystylis S. Moore. Compositae. 3 Austr.

Crawfurdia Wall. = Gentiana L. (Gentianac.).

Creaghia Scortech. = Mussaendopsis Baill. (Rubiac.).

Creaghiella Stapf. Melastomataceae. 2 N. Borneo.

Creatantha Standley. Rubiaceae. 1 Peru.

Cremanium D. Don = Miconia Ruiz & Pav. (Melastomatac.).

Cremanthodium Benth. Compositae. 55 Himal., China.

Cremaspora Benth. Rubiaceae. 3 trop. Afr., Madag.

Cremastogyne (H. Winkler) Czerepanov (~Betula L.). Betulaceae. 3 SW. China.

Cremastopus P. Wils. Cucurbitaceae. 2 C. Am.

Cremastosciadium Rech. f. Umbelliferae. 1 Afghan.

Cremastosperma R. E. Fries. Annonaceae. 17 trop. S. Am.

Cremastostemon Jacq. = Olinia Thunb. (Oliniac.).

Cremastra Lindl. Orchidaceae. 7 E. As.

Cremastus Miers. Bignoniaceae. 5 trop. Am.

Crematomia Miers = Bourreria P.Br. (Boraginac.).

Cremnobates Ridl. = Schizomeria D. Don (Cunoniac.).
Cremnophila Rose = Sedum L. (Crassulac.).
Cremobotrys Beer = Billbergia Thunb. (Bromeliac.).
Cremocarpon Boiv. ex Baill. Rubiaceae. 1 Comoro Is.
Cremocarpus K. Schum. = praec.
Cremocephalium Miq. = seq.
Cremocephalum Cass. = Gynura Cass. (Compos.).
Cremochilus Turcz. = Siphocampylus Pohl (Campanulac.).
Cremolobus DC. Cruciferae. 7 Andes.
Cremophyllum Scheidw. = Dalechampia L. (Euphorbiac.).
Cremopyrum Schur = Agropyron J. Gaertn. (Gramin.).
Cremosperma Benth. Gesneriaceae. 18 N. Andes.
Cremospora P. & K. = Cremaspora Benth. (Rubiac.).
Cremostachys Tul. = Galearia Zoll. & Mor. (Pandac.).
Crena Scop. = Crenea Aubl. (Lythrac.).
Crenaea Schreb. = praec.
Crenamon Rafin. = ? Leontodon L. + Picris L. (Compos.).
Crenamum Adans. = Crepis L. + Picris L. (Compos.).
Crenea Aubl. Lythraceae. 3 trop. S. Am., Trinidad.
Crenias Spreng. = Mniopsis Mart. (Podostemac.).
Crenosciadium Boiss. & Heldr. = Opopanax Koch (Umbellif.).
Crenularia Boiss. = Aëthionema R.Br. (Crucif.).
Creochiton Blume. Melastomataceae. 6 Philipp., Borneo, Java, New Guinea.
Creocome Kunze (sphalm.) = Oreocome Edgew. = Selinum L. (Umbellif.).
Creodus Lour. = Chloranthus Sw. (Chloranthac.).
Creolobus Lilja = Mentzelia L. (Loasac.).
Crepalia Steud. = Craepalia Schrank = Lolium L. (Gramin.).
Crepidaria Haw. = Pedilanthus Neck. ex Poit. (Euphorbiac.).
× **Crepidiastrixeris** Kitamura. Compositae. Gen. hybr. (Crepidiastrum ×
Crepidiastrum Nakai = Ixeris Cass. (Compos.).   [Youngia).
Crepidispermum Fries = Hieracium L. (Compos.).
Crepidium Blume = Malaxis Soland. ex Sw. (Orchidac.).
Crepidium Presl = Crepidophyllum Reed (Hymenophyllac.).
Crepidium Tausch = Crepis L. (Compos.).
Crepidocarpus Klotzsch ex Boeck. = Scirpus L. (Cyperac.).
Crepidomanes v. d. Bosch = Crepidophyllum Reed (Hymenophyllac.).
Crepidomanes Presl. Hymenophyllaceae. 20 Madag. to Japan & Tahiti.
Crepidophyllum Reed (1968; non Herzog 1926—Musci) = Reediella Pichi-
Sermolli (Hymenophyllac.).
Crepidopsis Arv.-Touv. Compositae. 1 Mex.
Crepidopteris Benth. (sphalm.) = Crepidotropis Walp. = Dioclea Kunth
(Legumin.).
Crepidopteris Copel. (1938; non Presl ex Sternb. 1838—gen. foss. Filic.)
= Crepidophyllum Reed (Hymenophyllac.).
Crepidospermum Benth. & Hook. f. = Crepidispermum Fries = Hieracium L.
(Compos.).
Crepidospermum Hook. f. Burseraceae. 2 trop. S. Am.
Crepidotropis Walp. = Dioclea Kunth (Legumin.).

× **Crepi-Hieracium** P. Fourn. Compositae. Gen. hybr. (Crepis × Hieracium).
Crepinella Marchal. Araliaceae. 1 Guiana.
Crepinia Gandog. = Rosa L. (Rosac.).
Crepinia Reichb. = Pterotheca Cass. = Crepis L. (Compos.).
Crepinodendron (Baill.) Pierre = Micropholis Pierre (Sapotac.).
Crepis L. Compositae. 200 N. hemisph., trop. & S. Afr.
Crepula Hill = Cirsium Mill. (Compos.).
Crepula Nor. = Phrynium Willd. (Marantac.).
Crescentia L. Bignoniaceae. 5 trop. Am. Fls. on old stems, succeeded by gourd-like berries; the epicarp is woody, and after removal of the pulp forms a calabash (*C. cujete* L., calabash-tree, most used).
Crescentiaceae Dum. = Bignoniaceae–Crescentieae G. Don.
Cressa L. Convolvulaceae. 5 trop. & subtrop.
Cressaria Rafin. = praec.
× **Crindonna** Ragioneri. Amaryllidaceae. Gen. hybr. (Belladonna × Crinum).
Crinipes Hochst. Gramineae. 4 trop. & S. Afr.
Crinissa Reichb. = Pyrrhopappus DC. (Compos.).
Crinita Houtt. = Pavetta L. (Rubiac.).
Crinita Moench = seq.
Crinitaria Cass. (~ Aster L.). Compositae. 5 temp. Euras.
Crinodendron Molina. Elaeocarpaceae. 2 temp. S. Am.
× **Crinodonna** Stapf = × Crindonna Ragioneri (Amaryllidac.).
Crinonia Banks ex Tul. = Hedycarya J. R. & G. Forst. (Monimiac.).
Crinonia Blume. Orchidaceae. 3 Malay Penins., Sumatra, Java.
Crinopsis Herb. = seq.
Crinum L. Amaryllidaceae. 100–110 trop. & subtrop., esp. on sea coasts. Large bulbous pl. with showy fls. The seed of *C. asiaticum* L. has a very thin corky covering and is suited to distr. by water and early germination. The ovule has no integuments, and the testa is replaced by a formation of cork at the outside of the endosp.
Crioceras Pierre (~ Tabernaemontana L.). Apocynaceae. 2 trop. Afr.
Criosanthes Rafin. Orchidaceae. 1 China, temp. N. Am.
Criosophila P. & K. = Cryosophila Bl. (Palm.).
Criptangis Thou. (uninom.) = *Angraecum inapertum* Thou. (Orchidac.).
Criptina Rafin. = Crypta Nutt. = Elatine L. (Elatinac.).
Criptophylis Thou. (uninom.) = *Bulbophyllum occultum* Thou. (Orchidac.).
Crispaceae Dulac = Balsaminaceae DC.
Cristaria Cav. Malvaceae. 40 temp. S. Am.
Cristaria Sonner. = Combretum L. (Combretac.).
Cristatella Nutt. (~ Polanisia Rafin.). Cleomaceae. 2 C. & S. U.S.
Cristella Rafin. = praec.
Critamus Bess. = Falcaria Host (Umbellif.).
Critamus Hoffm. = Apium L. (Umbellif.).
Critesia Rafin. = Salsola L. (Chenopodiac.).
Critesion Rafin. = Hordeum L. (Gramin.).
Crithmum L. Umbelliferae. 1, *C. maritimum* L., on rocky coasts, Medit., W. Eur. It has much-divided and very fleshy ls. Used for making pickles.
Critho E. Mey. = Hordeum L. (Gramin.).

Crithodium Link = Triticum L. (Gramin.).

Crithopsis Jaub. & Spach. Gramineae. 1 N. Afr. & Orient.

Crithopyrum Hort. Prag. ex Steud. = Agropyron J. Gaertn. (Gramin.).

Critonia P.Br. (~ Eupatorium L.). Compositae. 32 W.I., Mex. to Parag.

Critonia Cass. = Vernonia Schreb. (Compos.).

Critonia Gaertn. = Kuhnia L. (Compos.).

Critoniadelphus R. M. King & H. Rob. (~ Eupatorium). Compositae. 2 Mex., C. Am.

Critoniopsis Sch. Bip. = Vernonia Schreb. (Compos.).

Crobylanthe Bremek. Rubiaceae. 1 Borneo.

Crocanthemum Spach (~ Helianthemum Mill.). Cistaceae. 30 Am. (mostly SE. U.S. & Mex.), W.I.

Crocanthus L. Bolus = Malephora N. E. Br. (Aïzoac.).

Crocanthus Klotzsch ex Klatt = Crocosmia Planch. (Iridac.).

Crocaria Nor. = Microcos L. (Tiliac.).

Crocidium Hook. Compositae. 2 NW. N. Am.

Crocion Nieuwl. = Viola L. (Violac.).

Crociris Schur = Crocus L. (Iridac.).

Crociseris Fourr. = Senecio L. (Compos.).

Crockeria Greene ex A. Gray. Compositae. 1 Calif.

Crococylum Steud. = Crocoxylon Eckl. & Zeyh. (Celastrac.).

Crocodeilanthe Reichb. f. & Warsc. = Pleurothallis R.Br. (Orchidac.).

Crocodilina Bub. = Atractylis L. (Compos.).

Crocodilium Hill = Centaurea L. (Compos.).

Crocodilodes Adans. = Berkheya Ehrh. (Compos.).

Crocodiloïdes B. D. Jacks. = praec.

Crocodylium Hill = Crocodilium Hill = Centaurea L. (Compos.).

Crocopsis Pax. Amaryllidaceae. 1 Peru.

Crocosmia Planch. Iridaceae. 6 trop. & S. Afr.

Crocoxylon Eckl. & Zeyh. Celastraceae. 2 S. trop. & S. Afr.

Crocus L. Iridaceae. 75 Eur., Medit. to C. As. & W. Pakistan. The corm is covered with a few scaly ls., in whose axils may arise one or more buds, giving rise to new corms on the top of the old. The ls. are dorsiv., and curiously grooved on the back. The fl. is often single and term.; in some spp. there is a small cyme. The fl. closes at night and in dull weather. The tube of the P is so long that the ovary remains below the soil and is protected from the weather (cf. *Colchicum*). The fl. is protandr. and visited by bees and Lepidoptera. Honey is secreted by the ovary, and the anthers face outwards so as to touch any insect alighting on the petals and seeking honey. The stigmas are branched. Birds often bite off the fls. in gardens (? for honey); they seem to prefer the yellow fls., often leaving the purple and white alone.

The dried stigmas of *C. sativus* L. form saffron, once largely used as an orange-yellow dye, but now chiefly employed in flavouring (cf. Cornish saffron cake) and colouring dishes, liqueurs, etc.

Crocyllis E. Mey. ex Hook. f. Rubiaceae. 1 S. Afr.

Crodisperma Poit. ex Cass. = Wulffia Neck. ex Cass. (Compos.).

Croftia King & Prain = Pommereschea Wittmack (Zingiberac.).

Croftia Small (~ Dianthera L.). Acanthaceae. 1 SE. U.S.

CROIXIA

Croixia Pierre = Planchonella Pierre (Sapotac.).

Croizatia Steyerm. Euphorbiaceae. 1 trop. S. Am.

Crolocos Rafin. = Salvia L. (Labiat.).

Cromidon Compton. Scrophulariaceae. 1 S. Afr.

Cronquistia R. M. King. Compositae. 1 Mex.

Cronquistianthus R. M. King & H. Rob. (~Eupatorium L.). Compositae. 13 Colombia to Peru.

Cronyxium Rafin. = Lloydia Salisb. (Liliac.).

Crookea Small = Hypericum L. (Guttif.).

Croomia Torr. ex Torr. & A. Gray. Croomiaceae. 3 Japan, E. U.S.

Croomiaceae Nak. Monocots. 2/5 E. & SE. As., SE. U.S. Per. rhiz. herbs, with habit of *Stemonaceae*; ls. alt., *Smilax*-like, with parallel nerves and ∞ parallel cross-veins. Fls. reg., ♀, axill., subsol. or in crowded pedunc. rac. P 2+2, imbr. or valv., spreading; A 2+2, fil. short, thick, free or adn. to P, with shortly obl. ± horiz. thecae, sometimes divergent below; G (2–3), sometimes semi-inf., 1-loc., with several anatr. or semi-anatr. ov. pend. from apex, and 2–3 sess. stigs. Fr. (of *Croomia*) a 2-valved berry-like 1–4-seeded caps.; that of *Stichoneuron* unknown. Genera: *Croomia, Stichoneuron*. Near *Stemonac.* in the foliage and tetram. fls., but differing profoundly in the small P with rounded segments, in the short fil. and simple anth., and in the apical placentation.

Croptilon Rafin. = Haplopappus Cass. corr. Endl. (Compos.).

Crosapila Rafin. = Chaerophyllum L. (Umbellif.).

Crosperma Rafin. = Amianthium A. Gray (Liliac.).

Crossandra Salisb. Acanthaceae. 50 trop. Afr., Madag., Arabia. The seeds of many spp. are covered with scales which spread out and become sticky when wetted (cf. *Linum*).

Crossandrella C. B. Clarke. Acanthaceae. 2 trop. Afr.

Crossangis Schlechter = Diaphananthe Schlechter (Orchidac.).

Crosslandia W. V. Fitzger. Cyperaceae. 1 W. Austr.

Crossocoma Hook. (sphalm.) = Crossosoma Nutt. (Crossosomatac.).

Crossolepis Benth. = Angianthus Wendl. (Compos.).

Crossolepis Less. = Gnephosis Cass. (Compos.).

Crossonephelis Baill. Sapindaceae. 1 Madag. (Nossi Bé).

Crossopetalum P.Br. Celastraceae. 15 W.I., trop. S. Am.

Crossopetalum Roth = Gentiana L. (Gentianac.).

Crossophora Link = Chrozophora Neck. ex Juss. (Euphorbiac.).

Crossophrys Klotzsch = Clethra L. (Clethrac.).

Crossopteryx Fenzl. Rubiaceae. 1 trop. & S. Afr.

Crossosoma Nutt. Crossosomataceae. 3–4 SW. U.S., Mex.

*Crossosomataceae Engl. Dicots. 1/4 N. Am. Shrubs, sometimes spinescent, with simple ent. alt. exstip. ± glauc. ls. Fls. sol., term. on short leafy shoots. K (5), imbr.; C 5, imbr., orbic. or spath.; A ∞, on K-tube, fil. slender, anth. obl.; G 3–6, each with short style and capit. stig., ov. ∞, ± biseriate; fr. of 3–6 ± long-stipit. ventr. dehisc. follicles with persist. recurved style; seeds ∞, glob.-renif., shining, surr. by lacin. aril; endosp. thin, fleshy. Only genus: *Crossosoma*. Closely rel. to *Rosaceae–Spiraeoïdeae* (exstip.!).

Crossostemma Planch. ex Benth. Passifloraceae. 1 trop. W. Afr.

306

Crossostephium Less. (~Artemisia L.). Compositae. 4 C. & E. As., Philipp. Is., Calif.

Crossostigma Spach = Epilobium L. (Onagrac.).

Crossostoma Spach = Scaevola L. (Goodeniac.).

Crossostyles Benth. & Hook. f. = seq.

Crossostylis J. R. & G. Forst. Rhizophoraceae. 10 Polynesia.

Crossotropis Stapf = Trichoneura Anderss. (Gramin.).

Crossyne Salisb. = Boöphone Herb. (Amaryllidac.).

Crotalaria L. Leguminosae. 550 (+ at least 100 undescribed) trop. & subtrop. *C. juncea* L. (India, Austr.), an annual about 2·5 m. high, is largely cult. for the fibre obtained from its stems by maceration in water (cf. *Linum*), known as *sunn* hemp, Bombay or Madras hemp, etc. *C. retusa* L. (trop.) is also employed.

Crotalopsis Michx ex DC. = Baptisia Vent. (Legumin.).

Croton L. Euphorbiaceae. 750 trop. & subtrop. Fls. mon- or dioecious, little reduced from the type of the fam. Almost all spp. have stellate hairs or scales. *C. tiglium* L. (trop. As.) is the source of croton oil (a powerful purgative drug, expressed from the seeds). *C. cascarilla* Benn. and *C. eluteria* Benn. (Bahamas) yield cascarilla bark, used as a tonic. *C. laccifer* L. (India, Ceylon) yields a lac, used in varnish-making, and several Brazilian spp. a dragon's-blood resin.

Croton[ac]eae J. G. Agardh = Euphorbiaceae–Crotoneae Bl.

Crotonanthus Klotzsch ex Schlechtd. = Croton L. (Euphorbiac.).

Crotonogyne Muell. Arg. Euphorbiaceae. 15 trop. Afr.

Crotonogynopsis Pax. Euphorbiaceae. 1 trop. Afr.

Crotonopsis Michx. Euphorbiaceae. 2 E. N. Am.

Crototerum Desv. ex Baill. = Trachycaryon Klotzsch = Adriana Gaudich. (Euphorbiac.).

Crotularia Medik. = Crotalaria L. (Legumin.).

Crotularius Medik. = praec.

Croum Pfeiff. (sphalm.) = Ervum L. = Lens Mill. + Vicia L. (Legumin.).

Crowea Smith. Rutaceae. 4 Austr.

Crozophora A. Juss. = Chrozophora Neck. ex Juss. (Euphorbiac.).

Crozophyla Rafin. = Codiaeum Juss. (Euphorbiac.).

Cruciaceae Dulac = Cruciferae Juss.

Crucianella L. Rubiaceae. 35–40 Eur., Medit. to Persia & C. As.

Cruciata Gilib. = Gentiana L. (Gentianac.).

Cruciata Mill. Rubiaceae. 20 Eur., Medit.

Crucicaryum O. Brand. Boraginaceae. 1 New Guinea.

Cruciella Leschen. ex DC. = Xanthosia Rudge (Umbellif.).

Crucifera E. H. L. Krause = genus embracing entire family *Cruciferae*.

*Cruciferae Juss. (nom. altern. **Brassicaceae** Burnett). Dicots. 375/3200 cosmop., chiefly N. temp., esp. Medit. A very natural fam., though approaching *Papaveraceae* and *Capparidaceae*. Herbs, a few undershrubs; some ann., many perenn. forming each year a new shoot term. in the infl. Ls. usu. alt., exstip., with unicellular simple or branched hairs. For other peculiarities of veg. organs see gen., e.g. *Brassica, Anastatica, Subularia, Vella*, etc. Infl. usu. a raceme or corymb, and nearly always without bracts or bracteoles.

Fl. usu. ♀, reg. (rarely zygo.), hypog., with typical formula K 2+2, C 4,

CRUCIFERAE

A 6, $\underline{G}$ (2). The K has two whorls, the C only one, alt. with the K as a whole. The petals usu. spread out in the form of a cross and are often clawed; the sta. in two whorls, an outer of 2 short, an inner of 4 long, sta. (tetradynamous); anthers intr. The two cpls. are placed transv., and have parietal plac., but the ov. is 2-loc. on account of the presence of an antero-post. partition, the *replum* or so-called spurious septum, an outgrowth of the placentae. Stigmas 2, on short style, above the placentae (cf. *Papaveraceae*). Ovules anatr. or campylotr.

The explanation of the morphology of this fl. has given rise to much dispute.

On the bases of the sta. are the nectaries, the honey being secreted into the often gibbous bases of the inner sepals. The sepals often stand almost straight up, and the petals are then provided with claws, the limbs spreading out horiz. beyond the sepals. The honey is thus concealed to some extent and protected from rain. The majority of the order exhibit this construction more or less. In many gen. the fls. are arranged in corymbs, thus getting the advantage of many fls. massed together on one level (cf. *Umbelliferae* and *Compositae*). Insects visiting the fls. touch the anthers with one side of their bodies and the stigma with the other, and may in this way effect cross-fert., as they go sometimes to one, sometimes to the other, side of the fl. Dichogamy is frequent, but not well marked, and in almost all self-fert. ultimately occurs.

Fr. a caps. of pod-like form; if at least three times as long as broad it is called a *siliqua*, if shorter a *silicula*. It is divided into two by the replum and is usu. thin and membranous. The valves break away from below upwards, leaving the replum with the seeds pressed against it and adhering. The fr. may be flattened in two ways, either parallel or perpendicular to the replum; this char. is of systematic importance. It may also be jointed between the seeds as in a lomentum (*Leguminosae*). Achene-like one-seeded frs. occur in a few gen. Others have subterranean frs. (*Cardamine* spp., etc.).

The chars. of the seed are also of great importance in classification. The seed is exalb.; the testa is often mucilaginous, swelling up when wetted (e.g. the familiar case of mustard seed). The ovules being campylotr., the embryo sacs and embryos are curved, usu. with the radicle in one half of the seed, the cots. in the other. The shape of the embryo and the position of the radicle with regard to the cots. are important. The chief cases are: (1) radicle *incumbent* (or embryo *notorhizal*), i.e. lying on the back of one cot., the cots. not being folded on themselves; this may be shown thus o ||, the o repres. the radicle; (2) *accumbent* (or embryo *pleurorhizal*), o =, the radicle against the edges of the cots.; (3) *orthoplocous* (cots. *conduplicate*, o > > ; (4) *spirolobous*, as in (1) but cots. once folded, o || ||; (5) *diplecolobous*, ditto twice or more folded, o || || || ||.

For plants of economic value see esp. *Brassica* (which gives a number of valuable vegetables), *Sinapis*, *Nasturtium* (*Rorippa*), *Lepidium*, etc. All *C.* are harmless, and most are rich in sulphur compounds (to which the smell of boiling cabbages is due), and are thus useful in scurvy, etc.

Classification and chief genera:

The grouping of the smaller divisions of the fam. and the defining of the genera is a most difficult task. Many classifications have been devised. Prantl based his largely upon the hairs borne on the leaves, as follows:

A. Hairs simple or none; no glandular hairs.

1. THELYPODIËAE (stigma equally developed all round; style undivided or prolonged above middle of cpls., or turned back): *Pringlea, Thelypodium, Heliophila.*

2. SINAPEAE (stigma better developed over placentae): *Subularia, Lepidium, Iberis, Cochlearia, Alliaria, Sisymbrium, Cakile, Isatis, Vella, Sinapis, Brassica, Raphanus, Crambe, Rorippa, Cardamine, Lunaria.*

B. Hairs branched (a few exceptions); sometimes also glandular hairs.

3. SCHIZOPETALEAE (stigma equal all round): *Schizopetalum, Physaria.*

4. HESPERIDEAE (stigma better developed over placentae): *Capsella, Draba, Arabis, Erysimum, Cheiranthus, Alyssum, Anastatica, Malcomia, Hesperis, Matthiola, Conringia.*

O. E. Schulz (in Engl. & Harms, *Nat. Pflanzenfam.* ed. 2, **17b**; 1936) published an entirely new classif., based on a wide variety of chars., and making comparatively little use of those employed by Prantl. He divided the family into 19 tribes, the chars. of which unfortunately cannot be summarised in a limited space; their names however indicate their type genera: *Pringleëae, Stanleyeae, Romanschulzieae, Streptantheae, Cremolobeae, Chamireae, Brassiceae, Heliophileae, Schizopetaleae, Lepidieae, Euclidieae, Stenopetaleae, Lunarieae, Alysseae, Drabeae, Arabideae, Matthioleae, Hesperideae, Sisymbrieae.*

The classif. published by Janchen in *Österr. Bot. Zeitschr.* **91**: 1–28, 1942, with a criticism of Schultz's system, should also be consulted.

Crucita L. = Cruzeta Loefl. = Iresine P.Br. (Amaranthac.).
Cruciundula Rafin. = Thlaspi L. (Crucif.).
Cruckshanksia Hook. (1831) = Balbisia Cav. (Ledocarpac.).
***Cruckshanksia** Hook. & Arn. (1833). Rubiaceae. 7 Chile.
Cruckshanksia Miers (1826) = Solenomelus Miers (Iridac.).
Cruddasia Prain. Leguminosae. 1 Burma.
Crudea K. Schum. = seq.
***Crudia** Schreb. Leguminosae. 50 trop.
Crudya Batsch = praec.
Cruikshanksia Benth. & Hook. f. = Cruckshanksia Hook. = Balbisia Cav. (Ledocarpac.).
Cruikshanksia Reichb. = Cruckshanksia Miers = Solenomelus Miers (Iridac.).
Crukshanksia auct. = Cruckshanksia Hook. = Balbisia Cav. (Ledocarpac.).
Crula Nieuwl. = Acer L. (Acerac.).
Crumenaria Mart. Rhamnaceae. 6 C. Am. to Argent.
Cruminium Desv. = Centrosema DC. (Legumin.).
Crunocallis Rydb. (~ Montia L.). Portulacaceae. 2 W. U.S. & C. Am.
Crupina (Pers.) Cass. Compositae. 4 S. Eur. to Persia.
Crupinastrum Schur = Serratula L. (Compos.).
Crusea Cham. ex DC. = Mitracarpum Zucc. (Rubiac.).
Crusea Cham. & Schlechtd. Rubiaceae. 15 Mex., C. Am.
Crusea A. Rich. = Chione DC. (Rubiac.).
Cruzea A. Rich. = praec.
Cruzeta Loefl. = Iresine P.Br. (Amaranthac.).
Cruzia Phil. Labiatae. 1 Patagonia.
Cruzita L. = Cruzeta Loefl. = Iresine P.Br. (Amaranthac.).

CRYBE

Crybe Lindl. Orchidaceae. 1 C. Am.

Cryophytum N. E. Brown = Gasoul Adans. (Aïzoac.).

Cryosophila Blume. Palmae. 9 Mex., C. Am.

Cryphaea Buch.-Ham. (1825; non Mohr 1803—Musci) = Chloranthus Sw. (Chloranthac.).

Cryphia R.Br. = Prostanthera Labill. (Labiat.).

Cryphiacanthus Nees (~ Ruellia L.). Acanthaceae. 10 S. Am.

Cryphiantha Eckl. & Zeyh. = Amphithalea Eckl. & Zeyh. (Legumin.).

Cryphiospermum Beauv. = Enydra Lour. (Compos.).

Crypsinna Fourn. ex Benth. = Muhlenbergia Schreb. (Gramin.).

Crypsinus Presl. Polypodiaceae. 40 India to Malaysia & Japan (possibly a composite genus, part to be transferred to *Selliguea* Bory).

*__Crypsis__ Ait. Gramineae. 10 Medit. to N. China; trop. Afr. Seed extruded from fr.

Crypsocalyx Endl. (sphalm.) = Chrysocalyx Guill. & Perr. = Crotalaria L. (Legumin.).

Crypsophila Benth. & Hook. f. = Cryosophila Blume (Palm.).

Crypta Nutt. = Elatine L. (Elatinac.).

Cryptadenia Meissn. Thymelaeaceae. 5 S. Afr. (Cape Prov.).

Cryptadia Lindl. ex Endl. = Gymnarrhena Desf. (Compos.).

Cryptandra Sm. Rhamnaceae. 40 temp. Austr.

Cryptandraceae Barkley = Rhamnaceae Juss.

Cryptangium Schrad. Cyperaceae. 20 trop. Am.

Cryptantha Lehm. Boraginaceae. 100 Pacif. Am.

Cryptanthe Benth. & Hook. f. = praec.

Cryptanthela Gagnep. = Argyreia Lour. (Convolvulac.).

Cryptanthemis Rupp. Orchidaceae. 1 E. Austr. (N.S.W.).

Cryptanthopsis Ule. Bromeliaceae. 2 NE. Brazil.

Cryptanthus Nutt. ex Moq. = Aphanisma Nutt. (Chenopodiac.).

*__Cryptanthus__ Otto & Dietr. Bromeliaceae. 22 Brazil.

Cryptanthus Osbeck. Inc. sed. 1 S. China.

Cryptaria Rafin. = Crypta Nutt. = Elatine L. (Elatinac.).

Cryptarrhena R.Br. Orchidaceae. 3 Mex. to trop. S. Am.

× **Cryptbergia** hort. Bromeliaceae. Gen. hybr. (Billbergia × Cryptanthus).

Cryptella Rafin. = Crypta Nutt. = Elatine L. (Elatinac.).

Crypteronia Blume. Crypteroniaceae. 4 Assam, SE. As., Malaysia.

*__Crypteroniaceae__ A. DC. Dicots. 2/5 trop. As. Trees or shrubs with opp. ent. simple coriac. exstip. ls. Infl. a true panicle of small subsess. ♀ or polyg.-dioec. fls. K (4–5), valv.; C 4–5, imbr., or 0; A 10, or 4–5 alternisep., fil. short or elong., anth. small, sometimes with thickened connect. and thecae divergent downwards; disk 0; $\underline{G}$ or $\bar{G}$ (2–5), loc. sometimes incompl., with 1 style and capit. stig., and 2–∞ axile or basal ov.; fr. a loculic. 2–5-valved caps., valves usu. remaining connected by style; seeds ∞, sometimes winged, endosp. 0. Genera: *Crypteronia, Dactylocladus.* Near *Lythrac.* and *Melastomatac.* (cf. *Axinandra* Thw.).

Crypterpis Thou. (uninom.) = *Goodyera bracteata* Thou. = *Platylepis occulta* (Thou.) Reichb. f. (Orchidac.).

Cryptina Rafin. = Crypta Nutt. = Elatine L. (Elatinac.).

Cryptobasis Nevski = Iris L. (Iridac.).
Cryptocalyx Benth. = Lippia L. (Verbenac.).
Cryptocapnos Rech. f. Fumariaceae. 1 Afghan.
Cryptocaria Rafin. = Cryptocarya R.Br. (Laurac.).
Cryptocarpa Steud. = Cryptocarpha Cass. = Acicarpha Juss. (Calycerac.).
Cryptocarpa Tayl. ex Tul. = Tristicha Thou. (Podostemac.).
Cryptocarpha Cass. = Acicarpha Juss. (Calycerac.).
Cryptocarpus Kunth. Nyctaginaceae. 1 W. coast S. Am., Galápagos.
Cryptocarpus P. & K. = Cryptocarpa Tayl. ex Tul. = Tristicha Thou. (Podostemac.).
Cryptocarya R.Br. Lauraceae. 200–250 trop. (exc. C. Afr.) & subtrop. The fr. of *C. moschata* Nees & Mart. (Brazilian nutmegs) used as spice.
Cryptocentrum Benth. Orchidaceae. 14 C. Am. & N. Andes.
Cryptoceras Schott & Kotschy = Corydalis Vent. (Fumariac.).
Cryptocereus Alexander (~ Selenicereus [A. Berger] Britt. & Rose). Cactaceae. 2 Mex., C. Am.
Cryptochaete Raimondi ex Herrera = Laccopetalum Ulbr. (Ranunculac.).
Cryptochilus Wall. Orchidaceae. 2 Himalaya.
Cryptochloa Swallen. Gramineae. 4 Mex. to Colombia.
Cryptochloris Benth. = Tetrapogon Desf. (Gramin.).
Cryptocodon Fedorov. Campanulaceae. 1 C. As.
Cryptocorynaceae J. G. Agardh = Araceae–Areae Engl.
Cryptocoryne Fisch. ex Wydl. Araceae. 50 Indomal. Marsh and water plants. Some spp. are apparently 'viviparous' in their germination, like mangroves.
Cryptodia Sch. Bip. (sphalm.) = Cryptadia Lindl. = Gymnarrhena Desf. (Compos.).
Cryptodiscus Schrenk. Umbelliferae. 4 Persia to C. As.
Cryptoglochin Heuff. = Carex L. (Cyperac.).
Cryptoglottis Blume = Podochilus Blume (Orchidac.).
Cryptogramma R.Br. Cryptogrammataceae. 4 Eur., As., Am.
Cryptogrammataceae Pichi-Sermolli. Pteridales. Ls. much divided; sori protected by reflexed membranous edges of segments. Genera: *Cryptogramma, Llavea, Onychium.*
Cryptogyne Cass. = Eriocephalus L. (Compos.).
***Cryptogyne** Hook. f. Sapotaceae. 1 Madag.
Cryptolappa Kuntze = Camarea A. St-Hil. (Malpighiac.).
Cryptolepis R.Br. Periplocaceae. 12 palaeotrop.
Cryptolepis Wall. ex Lindl. = praec.?
Cryptolobus Endl. = praec.
Cryptolobus Spreng. = Amphicarpaea Ell. + Voandzeia Thou. (Legumin.).
Cryptoloma Hanst. = Isoloma Decne (Gesneriac.).
Cryptomeria D. Don. Taxodiaceae. 1 Japan, *C. japonica* (L. f.) D. Don (Japanese cedar). Timber good.
Cryptomeriaceae Hay. = Taxodiaceae Neger.
Cryptonema Turcz. = Burmannia L. (Burmanniac.).
Cryptopetalon Cass. = Pectis L. (Compos.).
Cryptopetalum Hook. & Arn. = Lepuropetalon Ell. (Lepuropetalac.).
Cryptophaseolus Kuntze = Canavalia DC. (Legumin.).

Cryptophila W. Wolf = Monotropsis Schweinitz (Monotropac.).
Cryptophoranthus Barb. Rodr. Orchidaceae. 30 trop. Am., W.I.
Cryptophragmia Benth. & Hook. f. = seq.
Cryptophragmium Nees. Acanthaceae. 40 SE. As., Indomal.
Cryptophysa Standley & Macbride. Melastomataceae. 1 C. Am.
Cryptopleura Nutt. = Troximon Nutt. (Compos.).
Cryptopodium Schrad. ex Nees = Scleria Bergius (Cyperac.).
Cryptopus Lindl. Orchidaceae. 3 Madag. & Mascarene Is.
Cryptopyrum Heynh. = Triticum L. (Gramin.).
Cryptorhiza Urb. Myrtaceae. 1 Haiti.
Cryptorrhynchus Nevski = Astragalus L. (Legumin.).
Cryptosaccus Reichb. f. = Leochilus Knowles & Westc. (Orchidac.).
Cryptosanus Scheidw. = praec.
Cryptosema Meissn. = Jansonia Kippist (Legumin.).
Cryptosepalum Benth. Leguminosae. 15 trop. Afr. One large petal only. A 3.
Cryptosorus Fée = Ctenopteris Bl. (Grammitidac.).
Cryptospermum Steud. = Cyrtospermum Rafin. = Cryptotaenia DC. (Umbellif.).
Cryptospermum Young = Opercularia Gaertn. (Rubiac.).
Cryptospora Kar. & Kir. Cruciferae. 2 C. As.
Cryptostachys Steud. = Sporobolus R.Br. (Gramin.).
Cryptostegia R.Br. Periplocaceae. 2 Madag.
Cryptostemma R.Br. = Arctotheca Wendl. (Compos.).
Cryptostemon F. Muell. ex Miq. = Darwinia Rudge (Myrtac.).
Cryptostephane Sch. Bip. = Dicoma Cass. (Compos.).
Cryptostephanus Welw. Amaryllidaceae. 4 S. trop. Afr. Berry.
Cryptostoma D. Dietr. = seq.
Cryptostomum Schreb. = Moutabea Aubl. (Polygalac.).
Cryptostylis R.Br. Orchidaceae. 20 Formosa, Indomal., Austr., Pacif.
*Cryptotaenia DC. Umbelliferae. 4 Italy, Transcauc., W. Equat. Afr. (Cameroon Mt.), E. As. & N. Am.
Cryptotaeniopsis Dunn. Umbelliferae. 22 E. As.
Cryptotenia Rafin. = Cryptotaenia DC. (Umbellif.).
Cryptotheca Blume = Ammannia L. (Lythrac.).
Cryptotonia Tausch (sphalm.) = Cryptotaenia DC. (Umbellif.).
Crypturus Link = Lolium L. (Gramin.).
Crysophila Benth. & Hook. f. = Cryosophila Bl. (Palm.).
Crystallopollen Steetz = Vernonia Schreb. (Compos.).
Cszernaëvia Endl. = Czernaëvia Turcz. = Archangelica Hoffm. (Umbellif.).
Cteisium Michx = Lygodium Sw. (Schizaeac.).
Ctenadena Prokh. = Euphorbia L. (Euphorbiac.)
Ctenanthe Eichl. = Myrosma L. f. (Marantac.).
Ctenardisia Ducke. Myrsinaceae. 1 NE. Brazil.
Ctenitis C. Chr. Aspidiaceae. 150 pantrop. Veins free; short multicellular hairs on raised upper surface of axes.
Ctenitopsis Ching = Tectaria Cav. (Aspidiac.). (Holttum, *Rev. Fl. Malaya*, 2: 501, 1924.)
*Ctenium Panz. Gramineae. 20 Am., Afr.

Ctenocladium Airy Shaw. Moraceae. 1 W. Equat. Afr.

Ctenocladus Engl. (1921; non Borzì, 1883—Algae) = praec.

Ctenodaucus Pomel = Daucus L. (Umbellif.).

Ctenodon Baill. = Aeschynomene L. (Legumin.).

Ctenolepis Hook. f. Cucurbitaceae. 2 trop. Afr. & India.

Ctenolophon Oliv. Ctenolophonaceae. 1 trop. Afr., 2 Malaysia.

Ctenolophonaceae (H. Winkl.) Exell & Mendonça. Dicots. 1/3 trop. Afr., W. Malaysia. Trees, with opp. ent. coriac. ls. with conspic. arcuate-anast. venation and ± large interpet. stips.; stell. hairs on stips., young shoots, sep. and pet. Fls. in lat. and term. cymes. K 5, imbr., v. shortly connate, persist.; C 5, imbr., thick, lin.-obl., decid.; A 10, unequal, adnate at base to inner side of tubular disk, anth. ovoid; G̲ (2), with 2 apical pend. ov. per loc. on long funicles, and 2 connate or ± free styles with capit. stigs. Fr. a dry woody indehisc. 1-seeded caps., seed with fibrous aril. Only genus: *Ctenolophon.* Affinities obscure; *Linac.* and *Malpighiac.* have been suggested.

Ctenomeria Harv. (~ Tragia L.). Euphorbiaceae. 2 S. Afr.

Ctenopaepale Bremek. Acanthaceae. 1 W. Java.

Ctenophrynium K. Schum. Marantaceae. 1 Madag.

Ctenophyllum Rydb. = Astragalus L. (Legumin.).

Ctenopsis De Not. Gramineae. 2 trop. Afr.

Ctenopsis Naud. = Ctenolepis Hook. f. (Cucurbitac.).

Ctenopteris Blume. Grammitidaceae. 200 pantrop.

Ctenorchis K. Schum. = Angraecum Bory (Orchidac.).

Ctenosachna P. & K. = Ktenosachne Steud. = Prionanthium Desv. (Gramin.).

Ctenosperma Hook. f. = Cotula L. (Compos.).

Ctenosperma F. Muell. ex Pfeiff. = Brachycome Cass. (Compos.).

Ctenospermum P. & K. = Ktenospermum Lehm. = Pectocarya DC. ex Meissn. (Boraginac.).

Cuatrecasasia Standl. Rubiaceae. 1 Colombia.

Cuatrecasasiodendron Standl. & Steyerm. Rubiaceae. 2 Colombia.

Cuatrecasea Dugand. Palmae. 1 Colombia.

Cuba Scop. = Tachigalia Aubl. (Legumin.).

Cubacroton Alain. Euphorbiaceae. 1 Cuba.

Cubaea Schreb. = Cuba Scop. = Tachigalia Aubl. (Legumin.).

Cubanthus (Boiss.) Millsp. (~ Pedilanthus Neck. ex Poit.). Euphorbiaceae. 3 Cuba, Hispaniola.

Cubeba Rafin. (1838, a) = Piper L. (Piperac.).

Cubeba Rafin. (1838, b) = Litsea Lam. (Laurac.).

Cubelium Rafin. = Hybanthus Jacq. (Violac.).

Cubilia Blume. Sapindaceae. 1 Philipp. Is., Celebes, Moluccas.

Cubincola Urb. = Cneorum L. (Cneorac.).

Cubospermum Lour. = Ludwigia L. (Onagrac.).

Cuchumatanea Seidenschnur & Beaman. Compositae. 1 C. Am. (Guatem.).

Cucifera Delile = Hyphaene Gaertn. (Palm.).

Cucubalus L. Caryophyllaceae. 1, *C. baccifer* L., N. temp. Fr. a dry berry.

Cucularia Rafin. = Dicentra Borckh. corr. Bernh. (Fumariac.).

Cuculina Rafin. = Catasetum Rich. (Orchidac.).

Cucullangis Thou. (uninom.) = *Angraecum cucullatum* Thou. (Orchidac.).

CUCULLARIA

Cucullaria Endl. = Cucularia Rafin. = Dicentra Borckh. corr. Bernh. (Fumariac.).

Cucullaria Fabr. (uninom.) = *Lychnis* L. sp. (Caryophyllac.).

Cucullaria Kuntze = Callipeltis Stev. (Rubiac.).

Cucullaria Schreb. = Vochysia Juss. (Vochysiac.).

Cucullifera Nees = Cannomois Beauv. ex Desv. (Restionac.).

Cucumella Chiov. Cucurbitaceae. 4 trop. Afr.

Cucumeroïdes Gaertn. = Trichosanthes L. (Cucurbitac.).

Cucumeropsis Naud. Cucurbitaceae. 1 trop. W. Afr.

Cucumis L. Cucurbitaceae. 25, mostly Afr., few Asia, 1 introd. trop. Am. *C. melo* L. (melon), *C. sativus* L. (cucumber) cult. from early times. Tendrils simple.

Cucurbita L. Cucurbitaceae. 15 Am., 5 long cult. For tendrils see fam. Fls. monoecious. Germination interesting. On the lower side of the hypocotyl a peg is formed which holds one side of the testa firmly while the expansion of the plumule splits off the other side. The position of the peg is determined by gravity.

C. pepo L. includes the vegetable marrow and some squashes; *C. maxima* Duch. ex Lam. includes the giant pumpkin, cult. in N. Am.

***Cucurbitaceae** Juss. Dicots. 110/640, most abundant in the trop., wanting in the colder regions. Chiefly climbing herbs with very rapid growth and abundance of sap in their stems and other tissues. Ls. alt., simple, palmately veined, roundish, entire or lobed, or pedately compound. They climb by tendrils, about whose morphological nature there has been much discussion; they have been considered by various authors as 'roots, stems, leaves, stipules, shoots, flower-stalks, or organs *sui generis*'. They probably originated as stipules, as shown by their position lateral to the leaf-base and by their rarely being paired: from the work of Sensarma (*Proc. Natl. Inst. Sci. India,* **21**: 162–9, 1955) it appears in some cases that the tendril is indeed partly vascularised in the manner of a stipule, but the supply is augmented and in other cases replaced by a direct supply from the bud trace. The tendrils of C. are very sensitive and show very well all the phenomena of tendril-climbing.

Fls. diclinous, rarely ♀, in infls. of various types. K and C typically (5) or 5 each, mostly reg.; A basically 5, but great var. is introduced by cohesions, etc.; it is almost always zygo. In *Fevillea* we find 5 sta. with dithecous anthers, the simplest type; it is noteworthy that the usual 4-thecous anther rarely occurs in C., and no trace of the missing loc. is to be found in either lobe of the anther of *Fevillea*. In other genera the A is usually more complex. In *Thladiantha* two pairs of sta. stand apart from the fifth sta. In *Sicydium* these pairs show union of their members at the base, and in others the union is more complete, until, as in *Bryonia*, etc., the A apparently has only 3 sta., of which 2 have 4-thecous anthers due to unions. The more the sta. depart from the simple type the more curved do the loculi of the anthers become, till in *Cucurbita*, etc., the pollen-sacs are twisted in a most extraordinary manner (cf. *Columellia*). In *Cyclanthera* the sta. are all united into a column with two ring-shaped pollen chambers running round the top (cf. the *flowers* of *Cyclanthus*). Ḡ 1–3-loc., with 1–∞ anatr. ov. in each loc.; the most common type is, however, a 1-loc. ovary with parietal plac. projecting deep into the cavity. Stigmas as many as cpls., usually forked. Fr. usu. fleshy, berry-like, called when firm-walled a *pepo*, e.g. melon or cucumber; sometimes a capsule. Seeds exalb. In *Zanonia, Ecballium, Cyclanthera*, etc. (*q.v.*), the mode of seed dispersal is interesting. Several have ed. frs., e.g. *Cucurbita, Cucumis, Sechium, Lagenaria*, etc.; some poisonous. The relationships of this fam. have been much disputed. It has been placed near *Passifloraceae, Loasaceae* and *Begoniaceae*, but its affinities are still obscure. *Classification and chief genera* (after C. Jeffrey in *Kew Bull.* **15**: 337–71, 1962):

A. Tendrils proximally 2–7-fid or simple; seeds unwinged
                                               I. **Cucurbitoïdeae**
  a. Receptacle-tube usu. relatively short; if long in ♂ fl. then short in ♀ fl.:
    α. Petals fringed, or with ventral scales   1. JOLIFFIËAE
             (*Momordica, Thladiantha, Telfairia*)
    β. Petals without fringe or ventral scales:
      α'. Ovules many, horizontal; pollen reticulate, 3-colporate
               2. BENINCASEAE
        (*Citrullus, Acanthosicyos, Luffa, Bryonia, Ecballium, Lagenaria, Benincasa*)
      β'. Ovules 1–2, pendulous; pollen reticuloid, 3-colporate
               3. SCHIZOPEPONEAE
               (*Schizopepon*)
      γ'. Ovules 1–many, ascending; pollen smooth, mostly poly-colporate   4. CYCLANTHEREAE
             (*Echinocystis, Cyclanthera, Marah*)
      δ'. Ovule solitary, pendulous; pollen spinulose, polycol-porate   5. SICYOËAE
             (*Sechium, Sicyos*)
      ε'. Ovules many, horizontal, or 1–few, ascending; pollen spinose, pantoporate   6. CUCURBITEAE
             (*Cucurbita, Cayaponia, Sicana*)
  b. Receptacle-tube relatively long, alike in ♂ and ♀ fl.:
    α. Pollen usu. reticulate; fl. small; sta. usu. free and with simple thecae   7. MELOTHRIËAE
      (*Melothria, Cucumis, Gurania, Dendrosicyos, Trochomeria*)
    β. Pollen striate, smooth or verrucose; fl. large; sta. united, with triplicate thecae   8. TRICHOSANTHEAE
      (*Trichosanthes, Peponium, Herpetospermum*)

B. Tendrils distally bifid; seeds mostly winged
                              II. **Zanonioïdeae**
  One tribe   9. ZANONIËAE
             (*Fevillea, Zanonia, Xerosicyos*)

Cucurbitella Walp. corr. Hook. f. Cucurbitaceae. 1 S. Am.
Cucurbitula (M. Roem.) P. & K. = Zehneria Endl. (Cucurbitac.).
Cudicia Buch.-Ham. ex G. Don = Pottsia Hook. & Arn. + Parsonsia R.Br. (Apocynac.).
*Cudrania Tréc. Moraceae. 4 Japan to Austr., New Caled.

CUDRANUS

Cudranus Kuntze = praec.

Cuellara Pers. = seq.

Cuellaria Ruiz & Pav. = Clethra L. (Clethrac.).

Cuenotia Rizzini. Acanthaceae. 1 NE. Brazil.

Cuepia J. F. Gmel. = Couepia Aubl. (Chrysobalanac.).

Cuervea Triana ex Miers (~ Hippocratea L.). Celastraceae. 3 Mex. to trop. S. Am., W.I.

Cufodontia R. E. Woodson. Apocynaceae. 4 Mex., C. Am.

Cuiavus Trew = Psidium L. (Myrtac.)

Cuiete Mill. = Crescentia L. (Bignoniac.).

Cuitlanzina Roeper = seq.

Cuitlauzina La Llave & Lex. = Odontoglossum Kunth (Orchidac.).

Cujete auctt. = Cuiete Mill. = Crescentia L. (Bignoniac.).

Cujunia Alef. = Vicia L. (Legumin.).

*****Culcasia** P. Beauv. Araceae. 20 trop. Afr.

Culcita Presl. Dicksoniaceae. 5 trop. Am. & Azores (subg. *Culcita*); Malaysia to Samoa, Austral. (subg. *Calochlaena*). Stem mostly prostrate; ls. deltoid, petiole long; sorus as *Dicksonia*. (Maxon, *J. Wash. Acad. Sci.* 12: 454, 1922; Holttum, *Fl. Males.* II, 1: 166, 1963.)

Culcitaceae Ching = Dicksoniaceae Presl.

Culcitium Humb. & Bonpl. Compositae. 35 Andes. Like *Espeletia*.

Culhamia Forsk. = Sterculia L. (Sterculiac.).

Cullay Molina ex Steud. = Quillaja Molina (Rosac.).

Cullen Medik. = Psoralea L. (Legumin.).

Cullenia Wight. Bombacaceae. 3 S. India, Ceylon. Apet.

Cullmannia C. Distefano = Wilcoxia Britt. & Rose (Cactac.).

Cullomia Juss. = Collomia Nutt. (Polemoniac.).

Cullumia R.Br. Compositae. 16 S. Afr.

Cullumiopsis Drake del Castillo = Dicoma Cass. (Compos.).

Cultridendris Thou. (uninom.) = *Dendrobium cultriforme* Thou. = *Polystachya cultriformis* (Thou.) Spreng. (Orchidac.).

Cuma P. & K. = Couma Aubl. (Apocynac.).

Cumarinia (Knuth) F. Buxb. Cactaceae. 1 Mex.

Cumarouma Steud. = seq.

Cumaruma Steud. = seq.

Cumaruna J. F. Gmel. = Coumarouna Aubl. (Legumin.).

Cumbata Rafin. = Rubus L. (Rosac.).

Cumbea Wight & Arn. = seq.

Cumbia Buch.-Ham. = Careya Roxb. (Barringtoniac.).

Cumbula Steud. = seq.

Cumbulu Adans. = Gmelina L. (Verbenac.).

Cumetea Rafin. = Myrcia DC. (Myrtac.).

Cumingia Kunth = Cummingia D. Don = Conanthera Ruiz & Pav. (Tecophilaeac.).

*****Cumingia** Vidal. Bombacaceae. 1 Philipp. Is.

Cuminia Colla. Labiatae. 1 Juan Fernandez.

Cuminia B. D. Jacks. (sphalm.) = Cuminum L. (Umbellif.).

Cuminoïdes Fabr. (uninom.) = *Lagoecia* L., sp. (Umbellif.).

CUPRESSACEAE

Cuminum L. Umbelliferae. 2 Medit. to Sudan & C. As. The frs. of *C. cyminum* L. (cummin seeds) are sometimes used like caraway seeds.

Cummingia D. Don = Conanthera Ruiz & Pav. (Tecophilaeac.).

Cumminsia King ex Prain = Meconopsis Viguier (Papaverac.).

Cuncea Buch.-Ham. ex D. Don = Knoxia L. (Rubiac.).

Cunibalu B. D. Jacks. (sphalm.) = Cumbulu Adans. = Gmelina L. (Verbenac.).

Cunigunda Bub. = Eupatorium L. (Compos.).

Cunila Mill. = Sideritis L. (Labiat.).

Cunila Royen ex L. (= Hedyosmos Mitch.). Labiatae. 15 E. N. Am. to Urug.

Cunina Clos = Nertera Banks & Soland. (Rubiac.).

*Cunninghamia R.Br. Taxodiaceae. 3 S. China, Formosa.

Cunninghamia Schreb. = Malanea Aubl. (Rubiac.).

Cunninghamiaceae Hay. = Taxodiaceae Neger.

Cunonia L. (= Oosterdyckia Boehm.). Cunoniaceae. 1, *C. capensis* L., S. Afr.; 16 New Caled. Bud protection by stipules.

Cunonia Mill. = Anomalesia N. E. Brown (Iridac.).

*Cunoniaceae R.Br. Dicots. 26/250, chiefly between 13° and 35° S., a few (*Weinmannia* spp.) N. to Philipp. Is., S. Mex. & W.I. Shrubs and trees with opp. or whorled often pinn. or trifol. leathery ls., stip. (the stips. often united in pairs as in *Rubiaceae*). Fl. small, usu. ♀, in compound infl. Receptacle usu. flat. K 4–5; C 4–5, usu. smaller than calyx, often absent; A 8–10 or ∞ or 4–5; G usu. (2), rarely 2; ovary usu. 2-loc., generally with ∞–2 ovules in 2 rows in each loc. Fruit usu. a capsule, rarely drupe or nut. Endosperm. Chief genera: *Cunonia, Weinmannia, Geissoïs, Spiraeanthemum, Pancheria, Lamanonia*.

Cunto Adans. = Acronychia J. R. & G. Forst. (Rutac.).

Cunuria Baill. Euphorbiaceae. 8 trop. S. Am.

Cupadessa Hassk. = Cipadessa Blume (Meliac.).

Cupameni Adans. = Acalypha L. (Euphorbiac.).

Cupamenis Rafin. = praec.

Cupania L. Sapindaceae. 55 warm Am. Wood of some is useful.

Cupaniopsis Radlk. Sapindaceae. 60 Austr., Polynesia.

Cuparilla Rafin. = Acacia Mill. (Legumin.).

Cuphaea Moench = seq.

Cuphea P.Br. Lythraceae. 250 Am. Ls. decussate; usu. there is one fl. at each node, standing between the two ls.; this is really the axillary fl. of the l. below, and its peduncle is 'adnate' to the main stem. Many covered with sticky glandular hairs.

Cupheanthus Seem. Myrtaceae. 5 New Caled.

Cuphocarpus Decne & Planch. Araliaceae. 1 Madag.

Cuphoea Brongn. ex Neumann = Cuphea R.Br. (Lythrac.).

Cuphonotus O. E. Schulz. Cruciferae. 3 S. & E. Austr., Tasm.

Cupi Adans. = Rondeletia L. (Rubiac.).

Cupia DC. = praec.

Cupidonia Bub. = Catananche L. (Compos.).

Cupirana Miers = Coupoui Aubl. = Duroia L. f. (Rubiac.).

Cuprespinnata 'Senilis' [Nelson] = Taxodium Rich. (Taxodiac.).

Cupressaceae Bartl. Gymnosp. (Conif.). 19/130 cosmop. Trees or shrubs, much branched, erect or prostr. Ls. decuss. or in whorls of 3–4, the young

317

CUPRESSOCYPARIS

ls. acic., mature ls. usu. small and squamiform, decurrent, often dimorphic. Fls. monoec. or dioec., small, sol., axill., or term. on short shoots (rarely the ♂♂ in axill. groups), cone-scales (sporophylls) opp. or in whorls of 3. ♂: stam. with short fil. and broad anther-scale; sporangia 3-6, ellips., free, on lower margin of anther-scale or ± covered by lower part of shield. ♀ shoots usu. with 1-several pairs of whorls of fert. or ster. scales, ov. 1-8 at base of scale, rarely with 1-3 term. ov. (*Juniperus*). Fr. cones with usu. woody, sometimes coriac., or (*Juniperus*) ± fleshy scales; scales imbr. or valv., usu. finally separating (exc. *Juniperus* and *Arceuthos*); seeds free, rarely (*Arceuthos*) united into a drupe, winged or not; cots. usu. 2, rarely 5-6. Chief genera: *Callitris, Widdringtonia, Thuja, Libocedrus, Cupressus, Chamaecyparis, Juniperus.*

× **Cupressocyparis** Dallimore. Cupressaceae. Gen. hybr. (Chamaecyparis × Cupressus).

Cupresstellata 'Senilis' [Nelson] = Fitzroya Hook. (Cupressac.).

Cupressus L. Cupressaceae. 15-20 Medit., Sahara, As., N. Am. The gen. habit is xero., the ls. being much reduced and closely appressed to the stems. *C. sempervirens* L. is the cypress of the Medit. region; *C. macrocarpa* Hartw. (Monterey cypress, Calif.) is largely planted for timber and shade in warm countries. Several yield useful timber, e.g. *C. lusitanica* Mill. (Mexico, Guatemala), *C. torulosa* D. Don (W. Himal.), *C. sempervirens* L., etc.

Cuprestellata Carr. = Cupresstellata Nelson = Fitzroya Hook. f. (Cupressac.).

Cupuia Rafin. = Coupoui Aubl. = Duroia L. f. (Rubiac.).

Cupulaceae Dulac = Cupuliferae A. Rich., *q.v.*

Cupulanthus Hutch. Leguminosae. 1 W. Austr.

Cupularia Godr. & Gren. = Inula L. (Compos.). [Fagaceae.

Cupuliferae A. Rich. = Betulaceae + Corylaceae + Fagaceae; later restricted to **Cupulissa** Rafin. = Anemopaegma Mart. ex Meissn. (Bignoniac.).

Cupuya Rafin. = Coupoui Aubl. = Duroia L. f. (Rubiac.).

Curanga Juss. Scrophulariaceae. 1 Indomal.

Curania Roem. & Schult. = praec.

Curare Kunth ex Humb. = Strychnos L. (Strychnac.).

Curarea Barneby & Krukoff. Menispermaceae. 4 trop. S. Am.

Curatari J. F. Gmel. = Couratari Aubl. (Lecythidac.).

Curatella Loefl. Dilleniaceae. 2 trop. Am., W.I.

Curbaril P. & K. = Courbaril Mill. = Hymenaea L. (Legumin.).

Curcas Adans. = Jatropha L. (Euphorbiac.).

Curcubitella Walp. (sphalm.) = Cucurbitella Walp. corr. Hook. f. (Cucurbitac.).

Curculigo Gaertn. Hypoxidaceae. 10 trop. Ovary loculi imperfect.

Curcuma L. Zingiberaceae. 5 Indomal., China. *C. angustifolia* Roxb. tubers furnish East Indian arrowroot. *C. longa* L. yields the spice and yellow dye turmeric (dried and ground rhiz.). The tubers of *C. zedoaria* Rosc. yield zedoary, used in the East as a tonic and perfume.

Curcumaceae Dum. = Zingiberaceae Lindl.

Curima O. F. Cook = Aïphanes Willd. (Palm.).

Curimari P. & K. = Courimari Aubl. = Sloanea L. (Elaeocarpac.).

Curmeria Linden & André = Homalomena Schott (Arac*).

Curnilia Rafin. = Curinila Roem. & Schult. = Leptadenia R.Br. (Asclepiadac.).

318

Curondia Rafin. = Courondi Adans. = Salacia L. (Celastrac.).
Currania Copel. = Gymnocarpium Newm. (Aspidiac.).
Curraniodendron Merr. = Quintinia A. DC. (Escalloniac.).
Curroria Planch. ex Benth. Asclepiadaceae. 4 trop. & subtrop. Afr., Socotra, S. Arabia.
Cursonia Nutt. = Onoseris DC. (Compos.).
Curtia Cham. & Schlechtd. Gentianaceae. 10 Guiana to Uruguay.
Curtisia Ait. Curtisiaceae. 1 S. Afr., *C. dentata* (Burm. f.) C. A. Sm., yielding a hard and useful timber (assegai-wood).
Curtisia Schreb. = Zanthoxylum L. (Rutac.).
Curtisiaceae (Harms) Takht. Dicots. 1/1 S. Afr. Tree, with opp. coriac. coarsely dent. exstip. ls., shining above, ±ferrug.-toment. below. Fls. ⚥, minute, in many-fld. term. dichot. toment. thyrses. K (4), open, tube turbin.; C 4, valv.; A 4, alternipet., anth. intr.; disk broad, densely barbate; Ḡ (4), 4-loc., with 1 epitr. ov. in each loc. and short subconical glabr. style with 4-lobed stig. Fr. a small subglob. drupe, 4-loc., 4-seeded; seed with elong. embr. in copious endosp. Only genus: *Curtisia*.
Curtisina Ridl. = Dacryodes Vahl (Burserac.).
Curtogyne Haw. = Crassula L. (Crassulac.).
Curtoisia Endl. = Courtoisia Reichb. = Collomia Nutt. (Polemoniac.).
Curtonus N. E. Brown. Iridaceae. 1 S. Afr.
Curtopogon Beauv. = Aristida L. (Gramin.).
Curupira G. A. Black. Olacaceae. 1 Brazil.
Curupita P. & K. = Couroupita Aubl. (Lecythidac.).
Cururu Mill. = Paullinia L. (Sapindac.).
Curvangis Thou. (uninom.) = *Angraecum recurvum* Thou. = *Jumellea recurva* (Thou.) Schltr. (Orchidac.).
Curvembryae Benth. & Hook. f. A 'Series' of Dicot. fams., almost equivalent to the Order *Centrospermae* Engl. (*q.v.*).
Curvophylis Thou. (uninom.) = *Bulbophyllum incurvum* Thou. (Orchidac.).
Cuscatlania Standley. Nyctaginaceae. 1 C. Am.
Cuscuaria Schott = Scindapsus Schott (Arac.).
Cuscuta L. Cuscutaceae. 170 trop. & temp. (dodder). Many have extended their boundaries through being carried about with their host plants. The stem twines and is sensitive to contact like a tendril so that it clasps the support tightly; it rarely makes more than three turns about the same branch of the host. At the points in close contact suckers are developed which penetrate the tissues of the host, growing into organic union with them and drawing off all the food materials required by the parasite, which has no green tissue of its own. The seeds of *C.* germinate later than those of the host plant; a very short anchorage root is formed and the stem nutates in search of a host; as soon as it has clasped one the root dies away. Much damage is often done by these plants: some confine themselves to particular host pls., but others attack a variety of pls.
*****Cuscutaceae** Dum. Dicots. 1/170 cosmop. Leafl. and rootl. total parasites, with thread-like herbaceous stems and small fls. in stalked or sess. ebracteate clusters. K 4–5 or (4–5), imbr.; C (4–5), subglob., imbr.; A 4–5, epipet. and alternipet. in throat of C, with small anth.; 5 scale-like stds. alt. with stam. in

CUSCUTINA

base of C, often lobed or fimbr.; G (2), loc. sometimes incompl., with 2 axile ov. per loc.; fr. a dry or fleshy caps., dehiscing irreg. or transv.; seeds tuberc., embryo lin., curved, surr. the endosp. Only genus: *Cuscuta*. Closely related to *Convolvulac.*
Cuscutina Pfeiff. = Cuscuta L. (Cuscutac.).
Cusickia M. E. Jones. Umbelliferae. 1 N. Am.
Cuspa Humb. = Rinorea Aubl. (Violac.).
Cusparia D. Dietr. (sphalm.) = Casparia Kunth = Bauhinia L. (Legumin.).
Cusparia Humb. = Angostura Roem. & Schult. (Rutac.).
Cuspariaceae J. G. Agardh = Rutaceae–Cusparieae DC.
***Cuspidaria** DC. Bignoniaceae. 4 S. Am.
Cuspidaria (DC.) Besser = Acachmena H. P. Fuchs (Crucif.).
Cuspidaria Fée = Dicranoglossum J. Sm. (Polypodiac.).
Cuspidia Gaertn. Compositae. 1 S. Afr.
Cuspidocarpus Spenn. = Micromeria Benth. (Labiat.).
Cussambium Lam. = Schleichera Willd. (Sapindac.).
Cussapoa P. & K. = Coussapoa Aubl. (Morac.).
Cussarea P. & K. = Coussarea Aubl. (Rubiac.).
Cusso Bruce = Brayera Kunth (Rosac.).
Cussonia Comm. ex Endl. = Eliaea Cambess. (Guttif.).
Cussonia Thunb. Araliaceae. 25 trop. & S. Afr., Madag., Masc.
Cussutha Benth. & Hook. f. (sphalm.) = Cassutha Des Moul. = Cuscuta L. (Cuscutac.).
Custenia Steud. = seq.
Custinia Neck. = Salacia L. (Celastrac.).
Cutandia Willk. Gramineae. 6 Medit.
Cutarea Jaume St-Hil. = Coutarea Aubl. (Rubiac.).
Cutaria Brign. = praec.
Cuthbertia Small = Phyodina Rafin. (Commelinac.).
Cutlera Rafin. = Gentiana L. (Gentianac.).
Cuttera Rafin. = praec.
Cuttsia F. Muell. Escalloniaceae. 1 E. Austr.
Cutubaea P. & K. = seq. (Gentianac.).
Cutubea Jaume St-Hil. = Coutoubea Aubl. (Gentianac.).
Cuveraca Jones = Cedrela P.Br. (Meliac.).
***Cuviera** DC. Rubiaceae. 14 trop. Afr. Several are ant-inhabited with hollow swellings of the stem above the nodes.
Cuviera Koeler = Hordelymus (Jessen) Jessen ex Harz (Gramin.).
Cyamopsis DC. Leguminosae. 3–4 trop. & subtrop. Afr., Arabia, India. *C. tetragonolobus* (L.) Taub. is largely cultivated in India as fodder (*guar*).
Cyamus Sm. = Nelumbo Adans. (Nelumbonac.).
Cyanaeorchis Barb. Rodr. Orchidaceae. 2 Brazil.
Cyanandrium Stapf. Melastomataceae. 4 Borneo.
Cyananthaceae J. G. Agardh = Campanulaceae–Wahlenbergiinae Schönl.
Cyananthus Griff. = Stauranthera Benth. (Gesneriac.).
Cyananthus Miers = Burmannia L. (Burmanniac.).
Cyananthus Rafin. = Centaurea L. (Compos.).

320

Cyananthus Wall. ex Benth. Campanulaceae. 30 Himal., Tibet, SW. China. Ov. sup.!

***Cyanastraceae** Engl. = Tecophilaeaceae Leyb.

Cyanastrum Cass. = Volutarella Cass. (Compos.).

Cyanastrum Oliv. Tecophilaeaceae. 6 trop. Afr. Herbs with tuber or tuberous rhiz., and racemes or panicles of ⚲ reg. fls. P (3 + 3), A (6), G̅ (3), 3-loc. with 2 ov. in each. Fr. 1-seeded. Perisperm.

Cyanea Gaudich. Campanulaceae. 60 Hawaii.

Cyanella Royen ex L. Tecophilaeaceae. 7 S. Afr.

Cyanellaceae Salisb. = Tecophilaeaceae Leyb.

Cyanitis Reinw. = Dichroa Lour. (Hydrangeac.).

Cyanobotrys Zucc. = Muellera L. (Legumin.).

Cyanocarpus F. M. Bailey = Helicia Lour. (Proteac.).

Cyanococcus Rydb. = Vaccinium L. (Ericac.).

Cyanodaphne Blume = Dehaasia Bl. (Laurac.).

Cyanophyllum Naud. = Miconia Ruiz & Pav. (Melastomatac.).

Cyanopis Blume = Vernonia Schreb. (Compos.).

Cyanopis Steud. = Cyanopsis Cass. = Volutarella Cass. (Compos.).

Cyanopogon Welw. ex C. B. Clarke = Cyanotis D. Don (Commelinac.).

Cyanopsis Cass. = Volutarella Cass. (Compos.).

Cyanopsis Endl. = Cyanopis Blume = Vernonia Schreb. (Compos.).

Cyanorchis Thou. = seq.

Cyanorkis Thou. = Phaius Lour. (Orchidac.).

Cyanoseris Schur = Lactuca L. (Compos.).

Cyanospermum Wight & Arn. = Rhynchosia Lour. (Legumin.).

Cyanostegia Turcz. Dicrastylidaceae. 4 W. Austr., Queensl.

Cyanostremma Benth. ex Hook. & Arn. = Calopogonium Desv. (Legumin.).

Cyanothamnus Lindl. = Boronia Sm. (Rutac.).

Cyanothyrsus Harms. Leguminosae. 3 trop. Afr.

***Cyanotis** D. Don. Commelinaceae. 50 palaeotrop.

Cyanotis Miers = Burmannia L. (Burmanniac.).

Cyanotris Rafin. = Camassia Lindl. (Liliac.).

Cyanthillium Blume (~ Vernonia Schreb.). Compositae. 1 China, Malaysia.

Cyanus Mill. = Centaurea L. (Compos.).

Cyathanthera Pohl = Miconia Ruiz & Pav. (Melastomatac.).

Cyathanthus Engl. Moraceae. 1 trop. W. Afr.

Cyathea Sm. (incl. Hemitelia & Alsophila). Cyatheaceae. 600 trop. & subtrop. Tree-ferns, forming a char. feature in the scenery of various regions, esp. on mts. in wet tropics. Subdivision on characters of indusium alone unnatural; better characters provided by scales (Holttum, *Fl. Males.* Ser. II, I (2): 72–6, 1963), on which subgenera *Cyathea* and *Sphaeropteris* are based.

Cyatheaceae Reichenb. Cyatheales. Tree ferns; young parts scaly; sori superficial, variously indusiate or naked. 2 gen.: *Cyathea, Cnemidaria*. (For another arrangement, see Holttum & Sen, *Phytomorphology*, 11: 406–20, 1961.)

Cyatheales. Filicidae. 2 fam.: *Cyatheaceae, Lophosoriaceae.*

Cyathidium Lindl. ex Royle = Saussurea DC. (Compos.).

Cyathiscus Van Tiegh. = Barathranthus Miq. (Loranthac.).

CYATHOBASIS

Cyathobasis Aellen. Chenopodiaceae. 1 Turkey.

Cyathocalyx Champ. ex Hook. f. & Thoms. Annonaceae. 38 Indomal.

Cyathocephalum Nakai = Ligularia Cass. (Compos.).

Cyathochaeta Nees. Cyperaceae. 5 Austr.

Cyathocline Cass. Compositae. 3 India.

Cyathocnemis Klotzsch = Begonia L. (Begoniac.).

Cyathocoma Nees = Tetraria Beauv. (Cyperac.).

Cyathodes Labill. Epacridaceae. 15 Austr., Tasm., Polynes., Micronesia.

Cyathodiscus Hochst. = Peddiea Harv. (Thymelaeac.).

Cyathoglottis Poepp. & Endl. = Sobralia Ruiz & Pav. (Orchidac.).

Cyathogyne Muell. Arg. = Thecacoris Juss. (Euphorbiac).

Cyathomiscus Turcz. = Marianthus Hueg. (Pittosporac.).

Cyathomone S. F. Blake. Compositae. 1 Ecuador.

Cyathopappus F. Muell. = Gnephosis Cass. (Compos.).

Cyathopappus Sch. Bip. = Elytropappus Sch. Bip. (Compos.).

Cyathophora Rafin. = Euphorbia L. (Euphorbiac.).

Cyathopsis Brongn. & Gris (~ Styphelia Soland.). Epacridaceae. 1 New Caled.

Cyathopus Stapf. Gramineae. 1 E. Himal.

Cyathorhachis Nees ex Steud. = Polytoca R.Br. (Gramin.).

Cyathoselinum Benth. Umbelliferae. 1 Jugoslavia (Dalmatia).

Cyathospermum Wall. ex D. Don = Gardneria Wall. (Strychnac.).

Cyathostegia (Benth.) Schery. Leguminosae. 2 trop. Am.

Cyathostelma Fourn. Asclepiadaceae. 2 Brazil.

Cyathostemma Griff. Annonaceae. 8 Burma, Malaysia.

Cyathostemon Turcz. = Baeckea L. (Myrtac.).

Cyathostyles Schott ex Meissn. = Cyphomandra Sendtn. (Solanac.).

***Cyathula** Blume. Amaranthaceae. 25–30 Afr., Madag., Ceylon, China, Malaysia.

Cyathula Lour. = Achyranthes L. (Amaranthac.).

Cybanthus P. & K. = Cybianthus Mart. (Myrsinac.).

Cybbanthera Buch.-Ham. ex D. Don = Limnophila R.Br. (Scrophulariac.).

Cybele Falc. = Herminium R.Br. (Orchidac.).

Cybele Salisb. = Stenocarpus R.Br. (Proteac.).

Cybelion Spreng. = Ionopsis Kunth (Orchidac.).

Cybianthopsis (Mez) Lundell. Myrsinaceae. 1 W.I. (Puerto Rico).

***Cybianthus** Mart. Myrsinaceae. 40 trop. S. Am., W.I.

Cybiostigma Turcz. = Ayenia L. (Sterculiac.).

Cybistax Mart. ex Meissn. Bignoniaceae. 3 trop. Am. The ls. of *C. antisyphilitica* Mart. are used as a blue dye, by boiling them with the cloth.

Cybistetes Milne-Redh. & Schweick. Amaryllidaceae. 1 S. Afr.

Cybostigma P. & K. = Cybiostigma Turcz. = Ayenia L. (Sterculiac.).

Cycadaceae Pers. (emend. L. A. S. Johnson). Gymnosp. (Cycadales). 1/20 Madag., E. As., Indomal. Woody pl., usu. unbranched, sometimes reaching 15 m. in height, trunk clothed with old frond-bases. Pinnae with single thick midrib and no lat. veins, circinately involute in vernation. ♂ sporophylls in definite cones. ♀ sporophylls not forming a determinate cone, but spirally arranged in a terminal (usu. woolly-brown) mass, and falling separately at maturity, the central axis continuing veg. growth; 'lamina' pinnatif., pectinate, or toothed; ovules 2–several, marginally inserted in notches proximal to the lamina, directed obliquely outwards. Only genus: *Cycas*.

Cycadales Engl. (chiefly after Eichler). Gymnospermae. 3 families, 10 genera, with about 100 spp., the survivors of a group of plants which in past ages figured more largely in the flora of the earth, reaching their maximum about the end of the Triassic and beginning of the Jurassic period. They represent perhaps the most primitive type of living seed-plants. In appearance and habit they bear some resemblance to tree-ferns. The stem is usu. short and stout, but may reach some height in *Cycas*, *Lepidozamia*, and *Dioön*, and is often tuberously swollen; it shows a secondary growth in thickness. It has a long primary tap-root. In some spp. a sort of felt-work of roots is formed at the base of the stem and a number of short lat. dichot. branched branches of these stand erect and may emerge from the soil. The stem has usu. a crown of leaves, and its lower portion is often covered with the scale-like remains of leaf-bases. There are two sorts of ls., foliage- and reduced scale-ls., borne spirally upon the stem, and alt. with one another, as a rule several circles of scales before each circle of foliage ls., which they protect in the bud. The foliage ls. are very char. They possess usu. a thickened, woody, ± sheathing base, which often persists after the fall of the rest of the l. There is a stout rachis or petiole, frequently thorny at the base, the thorns being 'metamorphosed' leaflets. Upon its upper side are two grooves, from which spring the leaflets, which may or may not be opp. to one another; there is usu. no term. leaflet. The leaflets may be entire or toothed and are usu. very rigid and leathery. Three types of nervature occur:

(1) midrib, no lateral nerves: Cycadaceae (*Cycas*);
(2) midrib and lateral nerves: Stangeriaceae (*Stangeria*);
(3) numerous parallel or wavy, simple or forked nerves running longitudinally: Zamiaceae.

The fls. are dioec. and usu. take the form of cones; these are term., and so the stem becomes a sympodium, except in *Cycas* ♀, where the stem 'grows through' the whorl of sporophylls. The size of the cones varies considerably. Each consists essentially of a central axis bearing a number of fertile ls. or sporophylls. In the ♂ cone, the ls. (scales) are generally of a sort of nail shape (cf. *Equisetum*), and bear *sori* upon the lower side, each of 2–6 sporangia (pollen-sacs), arranged with the lines of dehiscence radiating from the common centre. In the ♀ cone the sporophyll is of somewhat similar shape but bears as a rule only two marginal sporangia (ovules), whose apices are directed towards the axis of the cone. *Cycas* (*q.v.*) has no proper cone, but the stem bears a whorl of sporophylls in place of ordinary ls. The ovule is large, orthotr., with one integument. The pollen is carried by the wind, or perhaps in some cases by beetles, to the micropyle, where it germinates. The ov. grows into a large seed; testa two-layered, the inner woody, the outer fleshy. Endopleura on the seed. Nucellus reduced to a thin cap on the top of the seed, the bulk of which is endosp., with straight embryo in centre. 2 cots., usu. united at the tips. (See Pilger in Engl. *Pflanzenf.* ed. 2, **13**, 1926; L. A. S. Johnson in *Proc. Linn. Soc. N. S. Wales*, **84**: 64–117, 1959.)

Cycas L. Cycadaceae. 20 Madag., E. & SE. As., Indomal., Austr., Polynes. The pith of *C. circinalis* L. (trop. As., sometimes called sago-palm) and *C. revoluta* Thunb. (Japan) yields a sago.

CYCCA

Cycca Batsch = Cicca L. (Euphorbiac.).

Cyclacanthus S. Moore. Acanthaceae. 2 Indochina.

Cyclachaena Fresen. (~ Iva L.). Compositae. 4 N. Am.

Cycladenia Benth. Apocynaceae. 1 SW. U.S.

Cyclamen L. Primulaceae. 15 Eur., Medit. to Persia. There is a stout corm (cf. *Colchicum*) due to thickening of the hypocotyl. The P-lobes are bent back and the fl. is pend., with loose-pollen mechanism (cf. *Acanthus, Erica*). After fert. the stalk usu. coils up spirally, drawing the ripening fr. down to the soil (cf. *Vallisneria*); in *C. persicum* Mill. it bends over and deposits the fr. on the ground.

Cyclaminos Heldr. = praec.

Cyclaminum Bub. = praec.

Cyclaminus Haller = praec.

Cyclandra Lauterb. = Ternstroemia Mutis ex L. f. (Theac.).

Cyclandrophora Hassk. = Atuna Rafin. (Chrysobalanac.).

*__*Cyclanthaceae__** Dum. (emend. Lindl.). Monocots. 11/180 trop. Am. Perenn. rhiz. herbs or shrubs (rarely epiph.), of palm-like habit, or root-climbing lianes. Ls. alt. or distich., with sheathing pet. and usu. bifid, sometimes flabell. limb. Fls. ♂ ♀, monoec., on axill. spadices encl. by several conspic. spathes; fls. either in groups of 1 ♀ surr. by 4 ♂♂, the groups spirally arranged, or fls. in alt. ♂ and ♀ whorls or spirals. ♂: P cupular, ± lobed (rarely 0); A usu. ∞, fil. conn. & bulbous at base. ♀ fls. free or completely connate in whorls; when ± free, P 4 or (4), often accresc. and persist.; stds. 4, flex.-filif., opp. tepals; G (4), uniloc., often embedded in spadix, with 4 pariet. or apical or 1 apic. plac., bearing ∞ anatr. ov., and 4 free or conn. styles with various stigs. which sometimes persist and enlarge. Fr. a fleshy syncarp; seeds ∞, with copious fleshy endosp. The fam. is nearly related to *Palmae, Pandanaceae* and *Araceae*.

Classification and chief genera (Harling, *Acta Horti Berg.* **18**: 1–428, tt. 1–110, 1958):

I. **Carludovicoïdeae** (♂ and ♀ fls. in spirally arranged groups: fr. spadix not screw-like; ls. bifid, flabell., or ent.): *Carludovica, Asplundia, Dicranopygium, Sphaeradenia*, etc.

II. **Cyclanthoïdeae** (♂ and ♀ fls. in sep. alt. whorls, or sometimes part spirals); fr. spadix screw-like; ls. deeply 2-partite, with forked costa): *Cyclanthus* (only gen.).

Cyclanthera Schrad. Cucurbitaceae. 15 trop. Am. Sta. combined into a column, the anther-loculi fused into 2 ring-shaped loculi running completely round the top of the column. Fr. explosive: the pericarp is extremely turgid on its inner surface, and the fr. dehisces into valves, each of which rolls back on itself with a jerk.

Cyclantheropsis Harms. Cucurbitaceae. 2 trop. Afr.

Cyclanthus Poit. ex Spreng. Cyclanthaceae. 1 C. Am., trop. S. Am., W.I. The rhiz. bears large ls., forked at the top. Infl. term. on a long stalk, as a large cylindrical spadix with big bracts at base, resembling a number of discs piled upon one another, with their edges sharpened to a thin rim. In some two parallel spirals compose the spadix, each with a sharpened edge. In the former case every other disc bears ♂ fls., in the latter one of the spirals, the other being ♀. The ♂ fls. occupy a groove at the edge of the rim; each has 6 sta. and no P.

The ♀ fls. are embedded in the disc; ovaries united into a long continuous chamber running all round the disc and containing numerous placentae. P's united all round the disc; on their inner sides they bear stds. Fr. multiple, consisting of a number of seeds embedded in a general fleshy mass formed of ovaries and spadix. Cf. *Carludovica*.

Cyclas Schreb. = Crudia Schreb. (Legumin.).

Cyclea Arn. ex Wight. Menispermaceae. 30 trop. As.

Cyclium Steud. = Cycnium E. Mey. (Scrophulariac.).

***Cyclobalanopsis** (Endl.) Oerst. = Quercus L. (Fagac.).

Cyclobalanus (Endl.) Oerst. = Lithocarpus Bl. (Fagac.).

Cyclobothra D. Don ex Sweet = Calochortus Pursh (Liliac.).

Cyclocampe Benth. & Hook. f. = Lophoschoenus Stapf (Cyperac.).

Cyclocampe Steud. = Schoenus L. (Cyperac.).

Cyclocarpa Afzel. ex Bak. (emend. Urban). Leguminosae. 1 trop. Afr., Indoch., SE. Borneo, E. Java, N. Austr., Queensl.

Cyclocarpa Miq. (sphalm.) = Cyclocampe Steud. = Schoenus L. (Cyperac.).

Cyclocarpus Jungh. = Evodia J. R. & G. Forst. (Rutac.).

Cyclocarya Iljinskaja (~ Pterocarya Kunth). Juglandaceae. 1 China.

Cyclocheilon Oliv. Dicrastylidaceae? 2 Tunisia to Tanganyika, Arabia, Iraq. 2 seps. large, orbic.

Cyclochilus P. & K. = praec.

Cyclocodon Griff. = Campanumoea Blume (Campanulac.).

Cyclocotyla Stapf. Apocynaceae. 2 trop. W. Afr.

Cyclodiscus Klotzsch = Apama Lam. (Aristolochiac.).

Cyclodiscus K. Schum. = Cylicodiscus Harms (Legumin.).

Cyclodium Presl. Aspidiaceae. 2 W.I., trop. S. Am.

Cyclogramma Tagawa. Thelypteridaceae. 7 SE. As.

Cyclogyne Benth. = Swainsona Salisb. (Legumin.).

Cyclolepis Gillies ex D. Don. Compositae. 1 temp. S. Am. Char. in N. Patag.

Cyclolepis Moq. = Petermannia Reichb. = Cycloloma Moq. (Chenopodiac.).

Cyclolepsis Endl. = Cyclolepis Gillies ex D. Don (Compos.).

Cyclolobeae C. A. Mey. Chenopodiaceae. A comprehensive 'Series' of subfams. incl. the *Polycnemoïdeae*, *Betoïdeae*, *Chenopodioïdeae*, *Corispermoïdeae* and *Salicornioïdeae*. See fam.

Cyclolobium Benth. Leguminosae. 6 Brazil, Guiana.

Cycloloma Moq. Chenopodiaceae. 1 N. Am.

Cyclomorium Walp. = Desmodium Desv. (Legumin.).

Cyclonema Hochst. = Clerodendrum L. (Verbenac.).

Cyclopappus Cass. ex Sch. Bip. (nomen). Compositae–Erigeronteae. Hab.? Quid?

Cyclopeltis J. Sm. Aspidiaceae. 6 trop. E. As. & Am.

Cyclophorus Desv. = Pyrrosia Mirbel (Polypodiac.).

Cyclophyllum Hook. f. Rubiaceae. 11 New Caled., New Hebr.

Cyclopia Vent. Leguminosae. 15 S. Afr. Exstip.

Cyclopis Guill. (sphalm.) = Cyclolepis Gillies ex D. Don (Compos.).

Cyclopogon Presl. Orchidaceae. 20 Florida, W.I., trop. Am., Argent.

Cycloptera Endl. = Cyclopogon Presl (Orchidac.).

Cycloptera Nutt. ex A. Gray = Abronia Juss. (Nyctaginac.).

CYCLOPTERIS

Cyclopteris Schrad. ex S.F. Gray = Cystopteris Bernh. (Athyriac.).

Cyclopterygium Hochst. = Schouwia DC. (Crucif.).

Cycloptychis E. Mey. Cruciferae. 2 S. Afr.

Cyclosanthes Poepp. = Cyclanthus Poit. (Cyclanthac.).

Cyclosia Klotzsch = Mormodes Lindl. (Orchidac.).

Cyclosorus Link. Thelypteridaceae. Now limited to the type species (pantropic) and *C. striata* (Schum.) Ching (Africa). Distinguished from *Thelypteris* by anast. veins.

Cyclospathe O. F. Cook = Pseudophoenix Engl. & Drude (Palm.).

Cyclospermum Caruel = Ciclospermum La Gasca (Umbellif.).

Cyclostachya J. & C. Reeder. Gramineae. 1 Mex.

Cyclostegia Benth. = Elsholtzia Willd. (Labiat.).

Cyclostemon Blume = Drypetes Vahl (Euphorbiac.).

Cyclostigma Hochst. ex Endl. = Voacanga Thou. (Apocynac.).

Cyclostigma Klotzsch = Croton L. (Euphorbiac.).

Cyclostigma Phil. = Leptoglossis Benth. (Solanac.).

Cyclotaxis Boiss. = Scandix L. (Umbellif.).

Cycloteria Stapf = Coelorhachis Brongn. + Rhytachne Desv. (Gramin.).

Cyclotheca Moq. = Gyrostemon Desf. (Gyrostemonac.).

Cyclotrichium Mandenova & Scheng. (~ Calamintha Lam.). Labiatae. 6 As. Min., SW. Persia.

Cycnia Griff. = ? Parinari Aubl. (Chrysobalanac.).

Cycnia Lindl. = Prinsepia Royle (Rosac.).

Cycniopsis Engl. Scrophulariaceae. 3 trop. Afr.

Cycnium E. Mey. Scrophulariaceae. 40 trop. & S. Afr.

Cycnoches Lindl. Orchidaceae. 12 trop. Am. Fl. like that of *Catasetum* in mechanism and polymorphism.

× **Cycnodes** hort. Orchidaceae. Gen. hybr. (iii) (Cycnoches × Mormodes).

Cycnogeton Endl. = Triglochin L. (Juncaginac.).

Cycnopodium Naud. = Graffenrieda DC. (Melastomatac.).

Cycnoseris Endl. = Hypochoeris L. (Compos.).

Cycoctonum P. & K. (sphalm.) = Cynoctonum Gmel. = Mitreola L. ex Schaeff. (Spigeliac.).

Cydenis Salisb. = Narcissus L. (Amaryllidac.).

Cydista Miers. Bignoniaceae. 5 trop. Am.

Cydonia Mill. Rosaceae. 1 E. As. Min., Cauc., N. Pers., C. As., *C. oblonga* Mill. (*Pyrus cydonia* L.), the quince (ed. fr.).

Cylactis Rafin. = Rubus L. (Rosac.).

Cylastis Rafin. = praec.

Cylbanida Nor. ex Tul. = Pittosporum Banks ex Soland. (Pittosporac.).

Cylicadenia Lem. = Odontadenia Benth. (Apocynac.).

Cylichnanthus Dulac = Dianthus L. (Caryophyllac.).

Cylichnium Dulac = Gaudinia Beauv. (Gramin.).

Cylicodaphne Nees = Litsea Lam. (Laurac.).

Cylicodiscus Harms. Leguminosae. 2 trop. Afr.

Cylicomorpha Urb. Caricaceae. 2 trop. Afr.

Cylidium Rafin. (nomen). Convolvulaceae. Quid?

Cylindrachne Reichb. (sphalm.) = Cylindrocline Cass. (Compos.).

Cylindria Lour. = Linociera Sw. (Oleac.).
Cylindrocarpa Regel. Campanulaceae. 1 C. As.
Cylindrochilus Thw. = Thrixspermum Lour. (Orchidac.).
Cylindrocline Cass. Compositae. 1 Mauritius.
Cylindrokelupha Kosterm. Leguminosae. 3 Indoch., W. Malaysia.
Cylindrolelupha Hutch. (sphalm.) = praec.
Cylindrolepis Boeck. = Mariscus Gaertn. (Cyperac.).
Cylindrolobus Blume = Eria Lindl. (Orchidac.).
Cylindrophyllum Schwantes. Aïzoaceae. 6 S. Afr.
Cylindropsis Pierre. Apocynaceae. 1 trop. W. Afr.
Cylindropuntia (Engelm.) F. M. Knuth = Opuntia Mill. (Cactac.).
Cylindropus Nees = Scleria Bergius (Cyperac.).
Cylindrorebutia Frič & Kreuz. = Rebutia K. Schum. (Cactac.).
Cylindrosolen Kuntze = seq.
Cylindrosolenium Lindau. Acanthaceae. 1 Peru.
Cylindrosorus Benth. = Angianthus Wendl. (Compos.).
Cylindrosperma Ducke. Apocynaceae. 1 Amaz. Brazil.
Cylipogon Rafin. = Dalea Juss. (Legumin.).
Cylista Ait. = Rhynchosia Lour. (Legumin.).
Cylista auctt. = Paracalyx Ali (Legumin.).
Cylixylon Llanos = Gymnanthera R.Br. (Asclepiadac.).
Cylizoma Neck. = Derris Lour. (Legumin.).
Cyllenium Schott = Biarum Schott (Arac.).
Cylopogon P. & K. = Cylipogon Rafin. = Dalea Juss. (Legumin.).
Cymapleura P. & K. = Kymapleura Nutt. = Troximon Nutt. (Compos.).
Cymaria Benth. Labiatae. 3 Burma, Philipp., Java, Timor.
Cymation Spreng. = Ornithoglossum Salisb. (Liliac.).
Cymatocarpus O. E. Schulz. Cruciferae. 3 Transcauc., C. As.
Cymatochloa Schlechtd. = Paspalum L. (Gramin.).
Cymatoptera Turcz. = Menonvillea R.Br. (Crucif.).
Cymba Dulac = Tofieldia Huds. (Liliac.).
Cymba Nor. = Agalmyla Bl. (Gesneriac.).
Cymbachne Retz. = Rottboellia L. f. (Gramin.).
Cymbaecarpa Cav. = Coreopsis L. (Compos.).
Cymbalaria Hill (~ Linaria Mill.). Scrophulariaceae. 15 Medit., W. Eur.
C. muralis Gaertn., Mey. & Scherb. on walls. Before fert. the fls. are positively
heliotropic and stand erect; after it they become negatively heliotropic and
bend downward, seeking out dark crannies in the substratum, where the seeds
Cymbalariella Nappi = Saxifraga L. (Saxifragac.).   [ripen.
Cymbalina Rafin. = Cymbalaria Hill (Scrophulariac.).
Cymbanth[ac]eae Salisb. = Melanthiaceae R.Br. = Liliaceae–Anguillarieae D.
Cymbanthes Salisb. = Androcymbium Willd. (Liliac.).   [Don.
Cymbaria L. Scrophulariaceae. 4 C. & E. As.
Cymbia Standley = Krigia Schreb. (Compos.).
Cymbicarpos Stev. = Astragalus L. (Legumin.).
Cymbidiella Rolfe. Orchidaceae. 5 Madag.
Cymbidium Sw. Orchidaceae. 40 trop. Asia, Austr.
× **Cymbiphyllum** hort. (vii) = × Grammatocymbidium hort. (Orchidac.).

CYMBISPATHA

Cymbispatha Pichon. Commelinaceae. 2 trop. Am.
Cymbocarpa Miers. Burmanniaceae. 2 trop. S. Am., W.I.
Cymbocarpum DC. Umbelliferae. 5 SW. As.
Cymbochasma (Endl.) Klok & Zoz. Scrophulariaceae. 1–2 S. Russia.
Cymbolaena Smoljan. Compositae. 1 As. Min. & Syria to C. As., Afgh. & Baluchistan.
Cymbonotus Cass. (~ Arctotis L.). Compositae. 1 temp. Austr.
Cymbopetalum Benth. Annonaceae. 11 Mexico to trop. S. Am.
Cymbophyllum F. Muell. = Veronica L. (Scrophulariac.).
Cymbopogon Spreng. Gramineae. 60 trop. & subtrop. Afr. & As. Several yield essential aromatic oils, e.g. *C. nardus* (L.) Rendle and others (Ceylon; citronella), *C. citratus* (DC.) Stapf (Ceylon, S. India; lemon-grass), *C. martini* (Roxb.) Watts (India; *palma rosa* or geranium oil), used in soaps, perfumery, **Cymbosema** Benth. Leguminosae. 1 Brazil.   [etc.
Cymbosepalum Baker = Haematoxylon L. (Legumin.).
Cymboseris Boiss. = Phaecasium Cass. (Compos.).
Cymbosetaria Schweickerdt. Gramineae. 1 S. Arabia to SW. Afr. & Transvaal.
Cymbostemon Spach = Illicium L. (Illiciac.).
Cymburus Rafin. = Elytraria Michx (Acanthac.).
Cymburus Salisb. = Stachytarpheta Vahl (Verbenac.).
Cymelonema Presl = Urophyllum Wall. (Rubiac.).
Cymicifuga Reichb. = Cimicifuga L. (Ranunculac.).
Cyminon St-Lag. = Cuminum L. (Umbellif.).
Cyminosma Gaertn. = Acronychia J. R. & G. Forst. (Rutac.).
Cyminum Boiss. = Microsciadium Boiss. (Umbellif.).
Cyminum Hill = Cuminum L. (Umbellif.).
Cyminum P. & K. (sphalm.) = Cuminoïdes Moench = Lagoecia L. (Umbellif.).
*****Cymodocea** Koen. Cymodoceaceae. 4 coasts of Senegal & Canary Is. to Medit., & Indo-Pacific.
*****Cymodoceaceae** N. Taylor (~ Zannichelliaceae). Monocots. 5/16 warm seas. Submerged marine perenn. herbs, with creeping rhiz. Ls. lin., sheathing at base. Fls. monoec. or dioec., sol. or cym., naked or rarely minutely bracteate: ♂ of 2 long-pedic. 2-celled anth., with thread-like pollen; ♀ of 1 or (2) cpls., sess. or stipit., styles & stig. filif. Fr. a 1-seeded nutlet, seed pend. Genera: *Cymodocea, Syringodium, Amphibolis, Halodule, Thalassodendron.* Closely related to *Zannichelliac. (q.v.),* but pollen thread-like as in *Zosterac.*
Cymonamia Roberty = Bonamia Thou. (Convolvulac.).
Cymonetra Roberty = Gilletiodendron Vermoesen (Legumin.).
Cymophora Robinson. Compositae. 1 Mexico.
Cymophyllus Mackenzie (~ Carex L.). Cyperaceae. 1 SE. U.S.
Cymopteribus Buckl. (sphalm.) = seq.
Cymopterus Rafin. Umbelliferae. 18 W. N. Am.
Cymospermum Rafin. = ? Cyrtospermum Rafin. ex DC. (Umbellif.).
Cymothoë Airy Shaw = Costera J. J. Sm. (Ericac.).
Cynamonum Deniker = Cinnamomum Boehm. (Laurac.).
Cynanchica Fourr. (Asperula auctt., non L. *s.str.*). Rubiaceae. 200 + Eur., As., esp. Medit.; 16 E. Austr., Tasm. The Australian spp. are dioecious and perhaps generically distinct.

CYNOPSOLE

Cynanchum L. Asclepiadaceae. *S.str.*, 5 E. Eur., temp. As.; *s.l.*, ?150 trop. & temp.

Cynapium Bub. = Aethusa L. (Umbellif.).

Cynapium Nutt. ex Torr. & Gray = Ligusticum L. (Umbellif.).

Cynara L. Compositae. 14 Medit. to Kurdistan. *C. scolymus* L. is the true artichoke (cf. *Helianthus*); young fl.-heads enclosed in the invol. bracts, a valuable pot-herb. The blanched summer growth (chard) is also ed. *C. cardunculus* L. is the cardoon, whose ls. are blanched and eaten like celery; it has spread over great areas of the pampas, where it was introduced.

Cynaraceae Burnett = Compositae–Cynareae Spreng.

Cynaropsis Kuntze = Silybum Adans. (Compos.).

Cyne Danser. Loranthaceae. 4 Philipp. Is.

Cynocardamum Webb & Berth. = Lepidium L. (Crucif.).

× **Cynochloris** Clifford & Everist. Gramineae. Gen. hybr. (Chloris × Cynodon).

Cynocrambaceae Nees = Theligonaceae Dum.

Cynocrambe Gagneb. = Theligonum L. (Theligonac.).

Cynocrambe Hill = Mercurialis L. (Euphorbiac.).

Cynoctonum J. F. Gmel. = Mitreola L. ex Schaeffer (Spigeliac.).

Cynodendron Baehni. Sapotaceae. 10 trop. Am., W.I.

*****Cynodon** Rich. Gramineae. 10 trop. & subtrop. *C. dactylon* Pers. (dog's-tooth or Bermuda grass), cosmop. It grows with creeping stems on sandy soil and is used for binding dunes; useful pasture. Spikes digitate, spikelets 1-fld.

Cynogeton Kunth = Cycnogeton Endl. = Triglochin L. (Juncaginac.).

Cynoglossopsis A. Brand. Boraginaceae. 1 Abyssinia.

Cynoglossospermum Kuntze (1891) = Echinospermum Sw. (Boraginac.).

Cynoglossospermum Kuntze (1898) = Eritrichium Schrad. (Boraginac.).

Cynoglossum L. Boraginaceae. 50–60 temp. & subtrop. *C. officinale* L. (hound's tongue) formerly offic. Fr. covered with hooks.

Cynomarathrum Nutt. = Lomatium Rafin. (Umbellif.).

Cynometra L. Leguminosae. 60 trop.

Cynomora R. Hedw. = praec.

Cynomorbium Opiz = Pfundia Opiz ex Nevski = Hericinia Fourr. (Ranunculac.).

*****Cynomoriaceae** Lindl. Dicots. 1/2 Medit. reg. Totally parasitic herbs, with thick brownish rhiz. bearing ∞ short haustorial appendages, and thick simple flowering stems bearing ∞ deltoid scales and a clavate term. infl. (composed of ∞ suppressed false capitula), also interspersed with scales, with ∞ minute polygam. epig. fls. ♂ fl.: P 1–5(–8), lin.-spath.; A 1 with intr. versat. 4-loc. anth.; pistillode small. ♀ fl.: P 1–5; Ḡ 1, with 1 pend. ov. with thick integ., and 1 thick grooved term. style. ☿ with P, A and G. Fruit nut-like. Only genus: *Cynomorium*.

Cynomorium L. Cynomoriaceae (~ Balanophorac.). 2 Medit. to Mongolia Whole plant reddish-brown.

Cynomyrtus Scrivenor = Rhodomyrtus Reichb. (Myrtac.).

Cynophalla J. Presl = Capparis L. (Capparidac.).

Cynopoa Ehrh. (uninom.) = *Elymus caninus* L. = *Agropyrum caninum* (L.) Beauv. (Gramin.).

Cynopsole Endl. = Balanophora J. R. & G. Forst. (Balanophorac.).

CYNORCHIS

Cynorchis Thou. = Cynorkis Thou. (Orchidac.).

Cynorhiza Eckl. et Zeyh. = Peucedanum L. (Umbellif.).

Cynorkis Thou. Orchidaceae. 125 trop. & S. Afr., Masc.

Cynorrhynchium Mitchell = Mimulus L. (Scrophulariac.).

Cynosbata Reichb. = Pelargonium L'Hérit. (Geraniac.).

Cynosciadium DC. Umbelliferae. 1 S. Cent. U.S.

Cynosorchis Thou. = Cynorkis Thou. (Orchidac.).

Cynosurus L. Gramineae. 3–4 Eur., W. Asia, N. & S. Afr. *C. cristatus* L., a valuable pasture and fodder grass.

Cynotis Hoffmgg. = Cryptostemma R.Br. (Compos.).

Cynotoxicum Vell. = ? Connarus L. (Connarac.).

Cynoxylon (Rafin.) Small = Benthamidia Spach (Cornac.).

Cynthia D. Don = Krigia Schreb. (Compos.).

Cynura d'Orb. (sphalm.) = Gynura Cass. (Compos.).

Cyparissia Hoffmgg. = Callitris Vent. (Cupressac.).

Cypella Herb. Iridaceae. 15 Mex. to Argent. One sp. unfolds its fls. in great numbers at definite times.

Cypella Klatt = Marica Ker-Gawl. (Iridac.).

Cypellium Desv. = Styrax L. (Styracac.).

*Cyperaceae Juss. Monocots. About 90/4000, worldwide. 'Sedges'. Many perennial grass-like herbs of wet places, often with a creeping sympodial rhizome. Culms solid, often trigonous with sheathing leaves in three ranks. Sheaths entire (except *Coleochloa*), not articulated with blade. Flowers aggregated into spikelets of diverse structure. Spikelets often numerous and borne in a paniculately branched inflorescence. Branches generally bearing a basal 2-keeled glume (prophyll) which in *Carex* becomes a sac-like 'utricle' surrounding the female one-flowered spikelet. Flowers simple, borne in the axil of a glume; P of bristles (often 6), or scales, or entirely wanting; A usu. 1–3, anthers basifixed; G (2–3), 1-loc., with 1 basal anatr. ov.; style often deeply divided, branches equal in number to carpels. Fruit a trigonous or lenticular achene (except *Scirpodendron* and allies), the testa not adhering to the pericarp. Usually wind-pollinated. The Sedges are of little economic value—see *Cyperus*.

Classification and chief genera (after Engler):

    I. *Cyperoïdeae* (fls. ♀ in many-fld. spikelets, or single ♂ ♀ with or without P): *Cyperus, Eriophorum, Scirpus, Eleocharis, Fimbristylis*.

    II. *Rhynchosporoïdeae* (fls. ♀ or ♂ ♀ with or without P in few-fld. spike-like cymes aggregated into spikes or heads): *Scirpodendron, Schoenus, Cladium, Rhynchospora, Mapania, Scleria*.

    III. *Caricoïdeae* (fls. ♂ ♀, naked, usu. in many-fld. spikes; ♀ enclosed by utricle): *Carex, Uncinia*.

Cyperella MacMill. = Luzula DC. (Juncac.).

× **Cyperocymbidium** hort. (iv) = Cymbidium Sw. (Orchidac.).

Cyperoïdes Seguier = Carex L. (Cyperac.).

Cyperorchis Blume = Cymbidium Sw. (Orchidac.).

Cyperus L. Cyperaceae. 550 trop. & warm temp. Ann. or more often perenn. herbs, usu. with rhiz., stolons or tubers. Stems erect, usu. leafy at the base, with a term. umbellate or capit. infl. Stigs. 3, nut trigonous. *C. papyrus* L.

(paper-reed) is a riverside plant with stems 1–4 m. high. From the stems was made the ancient writing paper, papyrus. The stem was split into thin strips, which were pressed together while still wet. The rhiz. is ed., and also the root-tubers of several spp.; the stems (whole or split) of many are used for basket-making, etc.

Cyphacanthus Leonard. Acanthaceae. 1 Colombia.

Cyphadenia P. & K. = Kyphadenia Sch. Bip. ex O. Hoffm. = Chrysactinia A. Gray (Compos.).

Cyphaea Lem. = Cuphea P.Br. (Lythrac.).

Cyphanthe Rafin. = Orobanche L. (Orobanchac.).

Cyphanthera Miers = Anthocercis Labill. (Solanac.).

Cyphea P. & K. = Cuphea P.Br. (Lythrac.).

Cypheanthus P. & K. = Cupheanthus Seem. (Myrtac.).

Cyphella P. & K. = Cypella Herb. (Iridac.).

Cyphia Bergius. Campanulaceae. 50 Afr. (esp. S. Afr.), C. Verde Is.

Cyphiaceae A. DC. = Campanulaceae–Cyphioïdeae Reichb.

Cyphiella (Presl) Spach (1840; non Cyphella Fries 1823—Fungi) = Cyphopsis Kuntze = Cyphia Bergius (Campanulac.).

Cyphisia Rizzini. Acanthaceae. 1 Brazil.

Cyphium J. F. Gmel. = Cyphia Bergius (Campanulac.).

Cyphocalyx Gagnep. Scrophulariaceae. 1 Indoch.

Cyphocalyx Presl = Aspalathus L. (Legumin.).

Cyphocardamum Hedge. Cruciferae. 1 Afghan.

Cyphocarpa (Fenzl emend.) Lopriore. Amaranthaceae. 5 trop. & S. Afr.

Cyphocarpaceae Miers = Campanulaceae–Cyphioïdeae Reichb.

Cyphocarpus Miers. Campanulaceae. 2 Chile.

Cyphocarpus P. & K. = Cuphocarpus Decne & Planch. = Polyscias J. R. & G. Forst. (Araliac.).

Cyphochilus Schlechter. Orchidaceae. 7 New Guinea.

Cyphochlaena Hackel. Gramineae. 2 Madag.

Cyphokentia Brongn. Palmae. 2 New Caled.

Cypholepis Chiov. Gramineae. 1 S. & E. Afr. to Arabia.

Cypholophus Wedd. Urticaceae. 30 Philipp. Is. & Java to Polynesia.

Cyphomandra Mart. ex Sendtn. Solanaceae. 30 C. & S. Am., W.I. *C. betacea* Sendtn. (tree tomato) has ed. fr.

Cyphomattia Boiss. = Rindera Pall. (Boraginac.).

Cyphomeris Standley. Nyctaginaceae. 1 SW. U.S., Mexico.

Cyphonema Herb. = Cyrtanthus Ait. (Amaryllidac.).

Cyphophoenix H. Wendl. ex Benth. & Hook. f. Palmae. 1 Caroline Is., 1 New Caled.

Cyphopsis Kuntze = Cyphia Bergius (Campanulac.).

Cyphorima Rafin. = Lithospermum L. (Boraginac.).

Cyphosperma H. Wendl. ex Benth. & Hook. f. Palmae. 1 New Caled.

Cyphostemma (Planch.) Alston = Cissus L. (Vitidac.).

Cyphostigma Benth. Zingiberaceae. 15 Ceylon, Siam, W. Malaysia.

Cyphostyla Gleason. Melastomataceae. 3 Colombia.

Cyphotheca Diels. Melastomataceae. 2 SW. China.

Cyprianthe Spach = Ranunculus L. (Ranunculac.).

CYPRINIA

Cyprinia Browicz. Periplocaceae. 1 SE. Turkey, Cyprus.
Cypripediaceae (Lindl.) Lindl. = Orchidaceae–Cypripedioïdeae Lindl.
Cypripedilon St-Lag. = seq.
Cypripedium L. Orchidaceae. 50 N. temp. 'Lady's-slipper' orchids. Terrestrial acranthous plants. Lat. sepals completely united. Labellum slipperlike with inturned edge; at its base is the column, partly enclosed in it. The large std. (see fam.) is visible outside the labellum; under it are the two anthers, and lower down the flat stigma. Pollen glutinous, not united into pollinia. Insects (mostly bees) visiting the fl. get inside the labellum and cannot get out by the way they entered, so have to pass out by the openings at the base, in doing which they brush against the stigma and then the anthers.
Cyprolepis Steud. = Kyllinga Rottb. (Cyperac.).
Cypselea Turp. Aïzoaceae. 2 S. Florida, W.I., Venez.
Cypselocarpus F. Muell. Gyrostemonaceae. 1 W. Austr.
Cypselodontia DC. Compositae. 1 S. Afr.
Cypsophila Gaertn., Mey. & Scherb. (sphalm.) = Gypsophila L. (Caryophyllac.).
Cyrbasium Endl. = Cristatella Nutt. (Capparidac.).
Cyrenea Allemand. Gramineae. Quid?
Cyrilla Garden ex L. Cyrillaceae. 1 SE. U.S. to N. trop. S. Am., W.I. Shrub of marshy ground. Fls. in racemes below ls.
Cyrilla L'Hérit. = Achimenes P.Br. (Gesneriac.).
*****Cyrillaceae** Endl. Dicots. 3/13 Am. Evergr. shrubs or trees with alt., simple, ent., exstip. ls., and racemes of ⚥ reg. fls. K 5, imbr., persistent; C 5 or (5), imbr.; A 5 + 5 or 5, with intr. anthers; G̲ (5–2), multiloc., with 1 (rarely 2–4) pend. anatr. ov. in each loc.; raphe dorsal, micropyle facing upwards and inwards. Fr. dry, indehisc., sometimes 2–4-winged or encl. in the accr. K. Embryo straight, in endosp. Genera: *Cliftonia, Purdiaea, Cyrilla*. Prob. related to *Ericales*.
Cyrillopsis Kuhlm. Ixonanthaceae. 1 NE. Brazil.
Cyrilwhitea Ising. Chenopodiaceae. 1 Queensl.
Cyrta Lour. = Styrax L. (Styracac.).
Cyrtacanthus Mart. ex Nees = Ruellia L. (Acanthac.).
Cyrtandra J. R. & G. Forst. Gesneriaceae. 350 Malaysia, Polynesia.
Cyrtandraceae Jack = Gesneriaceae Nees.
Cyrtandroïdea Forest Brown. Gesneriaceae(!). 1 Marquesas Is.
Cyrtandromoea Zoll. Scrophulariaceae. 10 Burma, W. Malaysia.
Cyrtandropsis C. B. Clarke ex DC. = Tetraphyllum Griff. ex C. B. Cl. (Gesneriac.).
Cyrtandropsis Lauterb. = Cyrtandra J. R. & G. Forst. (Gesneriac.).
Cyrtanthaceae Salisb. = Amaryllidaceae–Crininae Pax.
Cyrtanthe Dur. & Jacks. (sphalm.) = Cystanthe R.Br. = Richea R.Br. (Epacridac.).
Cyrtanthemum Oerst. = Besleria L. (Gesneriac.).
Cyrtanthera Nees (~ Jacobinia Nees ex Moric.). Acanthaceae. 10 trop. Am.
Cyrtantherella Oerst. = Jacobinia Nees ex Moric. (Acanthac.).
Cyrtanthus Ait. Amaryllidaceae. 47 trop. & S. Afr.
Cyrtanthus Schreb. = Posoqueria Aubl. (Rubiac.).

Cyrtidium Schlechter. Orchidaceae. 2 trop. Am.
Cyrtocarpa Kunth. Anacardiaceae. 2 Mexico.
Cyrtoceras Benn. = Hoya R.Br. (Asclepiadac.).
Cyrtochilos Spreng. = seq.
Cyrtochilum Kunth. Orchidaceae. 115 Andes.
Cyrtocladon Griff. = Homalomena Schott (Arac.).
Cyrtococcum Stapf. Gramineae. 12 palaeotrop.
Cyrtococcus Willis (sphalm.) = praec.
Cyrtodeira Hanst. = Episcia Mart. (Gesneriac.).
Cyrtoglottis Schlechter = Mormolyca Fenzl (Orchidac.).
Cyrtogonellum Ching = Phanerophlebia Presl (Aspidiac.).
Cyrtogonium J. Smith = Bolbitis Schott (Lomariopsidac.).
Cyrtogonone Prain. Euphorbiaceae. 1 trop. Afr.
Cyrtogyna P. & K. = seq.
Cyrtogyne Reichb. = Curtogyne Haw. = Crassula L. (Crassulac.).
Cyrtolepis Less. = Anacyclus L. (Compos.).
Cyrtolobum R.Br. = Crotalaria L. (Legumin.).
Cyrtomidictyum Ching. Aspidiaceae. 4 China & Japan. (*Acta Phytotax. Sinica* **6**: 260–6, pl. 51–4, 1957.)
Cyrtomium Presl = Phanerophlebia Presl (or as subst. genus for spp. from Asia & Africa) (Aspidiac.).
Cyrtonema Schrad. ex Eckl. & Zeyh. = Kedrostis Medik. (Cucurbitac.).
Cyrtonora Zipp. (nomen). 1 New Guinea. Quid?
Cyrtopera Lindl. = Eulophia R.Br. (Orchidac.).
Cyrtophlebium (R.Br.) J. Sm. = Campyloneurum Presl (Polypodiac.).
Cyrtophyllum Reinw. = Fagraea Thunb. (Potaliac.).
Cyrtopodium R.Br. Orchidaceae. 10 trop. Am.
Cyrtopogon Spreng. = Curtopogon Beauv. = Aristida L. (Gramin.).
Cyrtorchis Schlechter. Orchidaceae. 15 trop. & S. Afr.
Cyrtorrhyncha Nutt. ex Torr. & Gray = Ranunculus L. (Ranunculac.).
Cyrtosia Blume = Galeola Lour. (Orchidac.).
Cyrtosiphonia Miq. = Rauvolfia L. (Apocynac.).
Cyrtospadix C. Koch = Caladium Vent. (Arac.).
Cyrtosperma Griff. Araceae. 18 trop. (esp. New Guinea). The rhiz. of *C. edule* Schott is ed. when cooked (cult. in Polynes.).
Cyrtospermum Benth. = Campnosperma Thw. (Anacardiac.).
Cyrtospermum Rafin. ex DC. = Cryptotaenia DC. (Umbellif.).
Cyrtostachys Blume. Palmae. 10 Malaysia (esp. New Guinea), Solomon Is.
Cyrtostemma Kunze = Clerodendrum L. (Verbenac.).
Cyrtostemma (Mert. & Koch) Spach = Scabiosa L. (Dipsacac.).
Cyrtostylis R.Br. = Acianthus R.Br. (Orchidac.).
Cyrtotropis Wall. = Apios Moench (Legumin.).
Cyrtoxiphus Harms. Leguminosae. 1 trop. Afr.
× **Cysepedium** hort. Orchidaceae. Gen. hybr. (v, vi) (Cypripedium × Selenipedium).
Cyssopetalum Turcz. = Oenanthe L. (Umbellif.).
Cystacanthus T. Anders. Acanthaceae. 10 SE. As.
Cystanche Ledeb. = Cistanche Hoffmgg. & Link (Orobanchac.).

CYSTANTHE

Cystanthe R.Br. = Richea R.Br. (Epacridac.).
Cystea Sm. = Cystopteris Bernh. (Athyriac.).
Cystibax Heynh. (sphalm.) = Cybistax Mart. ex Meissn. (Bignoniac.).
Cysticapnos Mill. Fumariaceae. 1 S. Afr.
Cysticorydalis Fedde ex Ikonn. Fumariaceae. 2 C. As., W. Tibet, W. Himal.
Cystidianthus Hassk. = Physostelma Wight (Asclepiadac.).
Cystidospermum Prokh. = Euphorbia L. (Euphorbiac.).
Cystistemon P. & K. = Cystostemon Balf. f. (Boraginac.).
Cystium Stev. = Astragalus L. (Legumin.).
Cystoathyrium Ching. Athyriaceae. 1 SW. China.
Cystocapnus P. & K. = Cysticapnos Mill. (Fumariac.).
Cystocarpum Benth. & Hook. = ? Cistocarpium Spach = Vesicaria Adans.
Cystocarpus Lam. ex P. & K. = ? praec.   [(Crucif.).
Cystochilum Barb. Rodr. = Cranichis Sw. (Orchidac.).
Cystodium J. Sm. Dicksoniaceae. 1 Malaysia, lowlands. Stock prostrate.
Cystogyne Gasp. = Ficus L. (Morac.).
Cystopora Lunell = Astragalus L. (Legumin.).
*Cystopteris Bernh. Athyriaceae. 18 temp. & subtrop. Limits of genus not
clear.
Cystopus Blume (1858; non Lévl. 1847—Fungi) = Pristiglottis Cretz. & J. J.
Smith (Orchidac.).
Cystorchis Blume. Orchidaceae. 21 China, Malaysia.
Cystostemma Fourn. Asclepiadaceae. 2 Brazil.
Cystostemon Balf. f. Boraginaceae. 1 Socotra.
Cytharexylum Jacq. = Citharexylon L. (Verbenac.).
Cytheraea (DC.) Wight & Arn. = Spondias L. (Anacardiac.).
Cytherea Salisb. = Calypso Salisb. (Orchidac.).
Cytheris Lindl. = Nephelaphyllum Blume (Orchidac.).
Cythisus Schrank = Cytisus L. (Legumin.).
Cyticus Link (sphalm.) = praec.
Cytin[ac]eae Brongn. = Rafflesiaceae–Cytineae R.Br.
*Cytinus L. Rafflesiaceae. 6 Medit., S. Afr., Madag.
Cytisanthus Lang (~ Genista L.). Leguminosae. 10 S. Eur., Medit.
Cytiso-Genista Duham. = Sarothamnus Wimm. (Legumin.).
Cytisophyllum Lang (~ Cytisus L.). Leguminosae. 1 S. Eur.
Cytisopsis Jaub. & Spach. Leguminosae. 1 W. As.
Cytisus L. Leguminosae. 25–30 Atl. Is., Eur., Medit.
Cyttaranthus J. Léonard. Euphorbiaceae. 1 trop. Afr.
Cyttarium Peterm. = Antennaria Gaertn. + Helichrysum Mill. + Gnaphalium
L. (Compos.).
Czackia Andrz. = Paradisea Mazzucato (Liliac.).
Czekelia Schur = Muscari Mill. (Liliac.).
Czernaëvia Turcz. = Angelica L. (Umbellif.).
Czerniaëvia Turcz. ex Ledeb. = Deschampsia Beauv. (Gramin.).
Czerniajevia Turcz. = Czernaëvia Turcz. = Angelica L. (Umbellif.).
Czerniajewia P. & K. = Czerniaëvia Turcz. ex Ledeb. = Deschampsia Beauv.
(Gramin.).
Czernya Presl = Phragmites Trin. (Gramin.).

334

D

Dabanus Kuntze = Pometia J. R. & G. Forst. (Sapindac.).
Dabeocia C. Koch = seq.
*****Daboecia** D. Don. Ericaceae. 2 Atl. Eur., Azores.
Dachel Adans. = Phoenix L. + Elate L. (Palm.).
Dacnopholis W. D. Clayton. Gramineae. 1 Madag.
Dacrycarpus de Laubenfels (~ Podocarpus L'Hérit.). Podocarpaceae. Spp.? Distrib.?
Dacrydium Soland. Podocarpaceae. 20–25 Indomal., Tasm., N.Z., New Caled., Fiji, Chile. Most are dioecious. Fertile scales 1 or 2 or more. Seed arillate. *D. franklinii* Hook. f. (Huon pine; Tasm.) and *D. cupressinum* Soland. ex Lamb. (red pine; N.Z.), good timber.
Dacryodes Vahl. Burseraceae. 30 trop.
Dactilis Neck. = Dactylis L. (Gramin.).
Dactilon Vill. = Cynodon Rich. (Gramin.).
Dactimala Rafin. = Chrysophyllum L. (Sapotac.).
Dactiphyllon Rafin. = Trifolium L. (Legumin.).
Dactychlaena P. & K. = Dactylaena Schrad. ex Schult. f. (Capparidac.).
× **Dactyladenia** Airy Shaw (sphalm.) = × Dactylodenia Garay & H. Sweet (Orchidac.).
Dactyladenia Welw. = Acioa Aubl. (Chrysobalanac.).
Dactylaea Fedde ex H. Wolff. Umbelliferae. 1 Tibet.
Dactylaena Schrad. ex Schult. f. Cleomaceae. 6 W.I., Brazil.
× **Dactylanthera** P. F. Hunt & [Summerhayes (vii) = × Rhizanthera P. F. Hunt & Summerhayes (Orchidac.).
Dactylanthera Welw. (nomen) = Symphonia L. f. vel Garcinia L. (Guttif.).
Dactylanthes Haw. = Euphorbia L. (Euphorbiac.).
Dactylanthocactus Y. Ito = Notocactus (K. Schum.) Backeb. & F. M. Knuth
Dactylanthus Hook. f. Balanophoraceae. 1 N.Z. [(Cactac.).
× **Dactylella** Soó (vii) = × Dactylitella P. F. Hunt & Summerh. (Orchidac.).
Dactylepis Rafin. = Cuscuta L. (Cuscutac.).
Dactylethria Ehrh. (uninom.) = *Digitalis ambigua* L. (Scrophulariac.).
× **Dactyleucorchis** Soó (ii) = × Pseudorhiza P. F. Hunt (Orchidac.).
Dactyliandra Hook. f. Cucurbitaceae. 1 SW. Afr. to India.
Dactylicapnos Wall. (~ Dicentra Borckh.). Fumariaceae. 8 Himal., W. China.
Dactyliocapnos Spreng. = praec.
Dactyliophora Van Tiegh. Loranthaceae. 6 New Guin., New Irel., Solomon Is.
Dactyliota Blume = Hypenanthe (Bl.) Bl. (Melastomatac.).
Dactylis L. Gramineae. 5 temp. Euras. *D. glomerata* L., cock's-foot, a valuable pasture grass, widely naturalised or cult. in N. Am. and S. Afr.
Dactyliscapnos B. D. Jacks. (sphalm.) = Dactylicapnos Wall. (Fumariac.).
× **Dactylitella** P. F. Hunt & Summerhayes. Orchidaceae. Gen. hybr. (i) (Dactylorhiza × Nigritella).
Dactylium Griff. = Erythropalum Blume (Erythropalac.).
× **Dactylocamptis** P. F. Hunt & Summerhayes. Orchidaceae. Gen. hybr. (ii) (Anacamptis × Dactylorhiza).

DACTYLOCERAS

× **Dactyloceras** Garay & Sweet (i). Orchidaceae. Gen. hybr. (Aceras × Dactylorhiza).

Dactylocladus Oliv. Crypteroniaceae. 1 Borneo, in freshwater peat-swamps. Branching often serial.

Dactyloctenium Willd. Gramineae. 10 warm.

× **Dactylodenia** Garay & H. Sweet (vii) = × Dactylogymnadenia Soó (Orchidac.).

Dactylodes Kuntze = Tripsacum L. (Gramin.).

× **Dactyloglossum** P. F. Hunt & Summerhayes. Orchidaceae. Gen. hybr. (i) (Coeloglossum × Dactylorhiza).

Dactylogramma Link = Muhlenbergia Schreb. (Gramin.).

× **Dactylogymnadenia** Soó. Orchidaceae. Gen. hybr. (i) (Dactylorhiza × Gymnadenia).

Dactyloïdes Nieuwl. = Saxifraga L. (Saxifragac.).

Dactylon Roem. & Schult. = Dactilon Vill. = Cynodon Rich. (Gramin.).

Dactylopetalum Benth. Rhizophoraceae. 10 trop. Afr., Madag.

Dactylophora Dur. & Jacks. = Dactyliophora Van Tiegh. (Loranthac.).

Dactylophyllum Spach = Gilia Ruiz & Pav. (Polemoniac.).

Dactylophyllum Spenn. = Potentilla L. (Rosac.).

Dactylopsis N. E. Brown. Aïzoaceae. 1 S. Afr.

Dactylorchis (Klinge) Vermeulen = seq.

Dactylorhiza Neck. ex Nevski. Orchidaceae. 30 Atl. Is., N. Afr., temp. Euras., Alaska.

Dactylorhynchus Schlechter. Orchidaceae. 1 New Guinea.

Dactylostalix Reichb. f. Orchidaceae. 1 Japan.

Dactylostegium Nees (~ Dicliptera Juss.). Acanthaceae. 2 S. U.S. to trop. S. Am., W.I.

Dactylostelma Schlechter. Asclepiadaceae. 1 Bolivia.

Dactylostemon Klotzsch = Actinostemon Klotzsch (Euphorbiac.).

Dactylostylis Scheidw. = Zygostates Lindl. (Orchidac.).

Dactylus Aschers. = Cynodon Rich. (Gramin.).

Dactylus Burm. f. = Microstegium Nees (Gramin.).

Dactylus Forsk. = Diospyros L. (Ebenac.).

Dactymala P. & K. = Dactimala Rafin. = Chrysophyllum L. (Sapotac.).

Dactyphyllum Endl. = Dactiphyllum Rafin. = Trifolium L. (Legumin.).

Dadia Vell. Compositae (inc. sed.; cf. Alomia Kunth). 1 Brazil.

Daedalacanthus T. Anders. (= Upudalia Rafin.). Acanthaceae. 15 Penins. India, Ceylon, E. Himal. to L. Burma.

Daemia Poir. = Doemia R. Br. = Pergularia L. (Asclepiadac.).

Daemonorops Blume ex Schult. f. Palmae. 100 Indomal.

Daenikera Hürlimann & Stauffer. Santalaceae. 1 New Caled.

Dahlbergia Rafin. = Dalbergaria Tussac = Columnea L. (Gesneriac.).

Dahlgrenia Steyerm. Palmae. 1 Venezuela.

Dahlia Cav. Compositae. 27 Mexico, Guatem. Perenn. herbs with tuberous [roots.

Dahlia Thunb. = Trichocladus Pers. (Hamamelidac.).

Dahlstedtia Malme. Leguminosae. 1 Brazil.

Dahuronia Scop. = Licania Aubl. (Chrysobalanac.).

Dais Royen ex L. Thymelaeaceae. 2 S. Afr., Madag.

Daiswa Rafin. = Paris L. (Trilliac.).

Daknopholis W. D. Clayton. Gramineae. 1 E. Afr., Madag., Aldabra.

Dalbergaria Tussac = Alloplectus Mart. (Gesneriac.).

Dalbergia L. f. Leguminosae. 300 trop. & subtrop., S. Afr. Many lianas. *D. variabilis* Vog., a shrub with pend. twigs in the open, becomes a liana in forest, with short lat. shoots sensitive to contact. Fr. winged, indeh. Many yield valuable wood, e.g. *D. nigra* Allem. (rosewood, Braz.) and other Am. spp.; *D. melanoxylon* Guill. & Perr. (Afr. blackwood, trop. W. Afr.); *D. latifolia* Roxb. (blackwood or E. Indian rosewood, India), and *D. sissoo* Roxb. (*shisham, sissoo,* India).

Dalbergiella E. G. Baker. Leguminosae. 2 S. trop. Afr.

Dalea P.Br. (1). [Infl. rac., ramifl.; K campan., trunc., C o, A ∞, G (1), fr. bacc.] Jamaica. Quid?

Dalea P.Br. (2) = Critonia P.Br. = Eupatorium L. (Compos.).

Dalea Cramer = Petalostemon Michx (Legumin.).

Dalea Gaertn. = Microdon Choisy (Scrophulariac.).

Dalea L. ex Juss. Leguminosae. 250 Am. Ov. collat. Claws of 4 lower pets. united to A.

Dalea Mill. = Browallia L. (Solanac.).

Dalechampia L. Euphorbiaceae. 110 warm (esp. Am.). *D. roezliana* Muell. Arg. has a complex infl. (cf. diagram), enclosed in 2 large pink or white outer brs. (the big brackets). Above these on axis is a smaller br. (small bracket), with 3-fld. cyme of ♀ fls. (F) in axil. Above is ♂ part of infl., starting with 4 brs. (asterisks); above these, ant., are 9–14 ♂ fls., and post. a yellow cushion of rudimentary ♂ fls., secreting resin in some spp. Some spp. possess urticating hairs or glands in all parts of the pl.

cushion
* male fls. *

F F F

Dalembertia Baill. Euphorbiaceae. 4 Mex. ♂ fl.: A 1 enclosed in K 1.

Dalenia Korth. Melastomataceae. 3 Borneo. K-tube with deciduous cap.

Dalhousiea R. Grah. ex Benth. Leguminosae. 2 palaeotrop.

Dalia Endl. = Dulia Adans. = Ledum L. (Ericac.).

Dalia St-Lag. = Dalea L. ex Juss. (Legumin.).

Dalibarda L. = Rubus L. (Rosac.).

Dallachya F. Muell. = Rhamnella Miq. (Rhamnac.).

Dalrympelea Roxb. = Turpinia Vent. (Staphyleac.).

Dalucum Adans. = Melica L. (Gramin.).

Dalzellia Hassk. = Cyanotis D. Don (Commelinac.).

Dalzellia Wight. Podostemaceae. 1 S. India.

Dalzielia Turrill. Asclepiadaceae. 1 W. Afr.

Damapana Adans. = Smithia Ait. (Legumin.).

Damasoniaceae Nak. = Alismataceae Vent.

Damasonium Mill. Alismataceae. 5 Eur., Medit., W. & C. As., Austr., Calif.

Damasonium Schreb. = Ottelia Pers. (Hydrocharitac.).

Damatras Reichb. = seq.

Damatrias Reichb. = seq.

Damatris Cass. = Haplocarpha Less. (Compos.).

Damburneya Rafin. = Ocotea Aubl. (Laurac.).

Dameria Endl. = Dauceria Dennst. = Embelia Burm. f. (Myrsinac.).

DAMIRONIA

Damironia Cass. = Helipterum DC. (Compos.).
Dammara Gaertn. = Protium Burm. f. (Burserac.).
Dammara Lam. = Agathis Salisb. (Araucauriac.).
Dammaraceae Link = Araucariaceae Strasb. + Taxodiaceae Neger, *p.p.*
Dammaropsis Warb. Moraceae. 1 New Guinea.
Dammera K. Schum. & Lauterb. (~Licuala Wurmb). Palmae. 2 New Guinea.
Damnacanthus Gaertn. f. Rubiaceae. 6 E. As. Thorny shrubs.
Dampiera R.Br. Goodeniaceae. 60 Austr. K small or o. Syngenesious.
Damrongia Kerr ex Craib (= ? Trisepalum C. B. Cl.). Gesneriaceae. 2 SE. As.
Danaa All. = Physospermum Cusson (Umbellif.).
Danaa Colla = Senecio L. (Compos.).
Danaë Medik. Ruscaceae. 1 SW. As., SE. Balkan Penins.(?). Erect shrub; phylloclades.
***Danaea** Sm. Danaeaceae. 30 Am. Stem dorsiventral. Synangia very long, sometimes from midrib to margin of l., opening by term. pores.
Danaeaceae Agardh. Marattiales (*q.v.* for chars.). Only genus: *Danaea.*
Danaeopsis Presl. Filices. Gen. dub.
Danaïdia Link = Danaë Medik. (Ruscac.).
Danaïs Comm. ex Vent. Rubiaceae. 40 Madag., Masc.
Danatophorus Blume = Harpullia Roxb. (Sapindac.).
Danbya Salisb. = Bomarea Mirb. (Alstroemeriac.).
Dancera Rafin. = Clidemia D. Don (Melastomatac.).
Dandya H. E. Moore. Amaryllidaceae. 1 Mex.
Dangervilla Vell. = Angostura Roem. & Schult. (Rutac.).
Danguya Benoist. Acanthaceae. 1 Madag.
Danguyodrypetes Leandri. Euphorbiaceae. 1 Madag.
Danielia Lem. = Rochea DC. (Crassulac.).
Daniella Corrêa de Mello = Mansoa DC. (Bignoniac.).
Daniella Willis (sphalm.) = seq.
Daniellia Benn. Leguminosae. 11 trop. W. Afr.
Dankia Gagnep. Flacourtiaceae. 1 Indoch.
Dansera v. Steenis (~ Dialium L.). Leguminosae. 1 Malay Penins., Sumatra, Borneo.
Danserella Balle = Odontella Van Tiegh. (Loranthac.).
Danthia Steud. = Dantia Boehm. = Ludwigia L. (Onagrac.).
Danthonia DC. Gramineae. 10 trop. & temp., esp. S. Afr.
Danthoniastrum (Holub) Holub. Gramineae. 1 Balkan Penins.
Danthonidium C. E. Hubbard. Gramineae. 1 India.
Danthoniopsis Stapf. Gramineae. 15 Afr. & Arabia.
Danthorhiza Ten. Gramineae (Aveneae). 1 Italy. Quid?
× **Danthosieglingia** Domin. Gramineae. Gen. hybr. (Danthonia × Sieglingia).
Dantia Boehm. = Ludwigia L. (Onagrac.).
Dantia Lippi ex Choisy = Boerhavia L. (Nyctaginac.).
Danubiunculus Sailer = Limosella L. (Scrophulariac.).
Danzleria Bert. ex DC. = Diospyros L. (Ebenac.).
Dapania Korth. Averrhoaceae (~ Oxalidac.). 1 Madag. (?), 2 W. Malaysia.

Dapedostachys Börner = Carex L. (Cyperac.).
Daphmanthus F. K. Ward = ?Daphne L. (Thymelaeac.).
Daphnaceae ('-oïdeae') Vent. = Thymelaeaceae Juss.
Daphnandra Benth. Atherospermataceae. 5 New Guinea, Austr. Ovule pend.
Daphne L. Thymelaeaceae. 70 Eur., N. Afr., temp. & subtrop. As., Austr., Pacif. Honey is secreted by the base of the ovary, and the depth of the tube preserves it for long-tongued insects. Bark used for paper in India.
Daphnephyllum Hassk. = Daphniphyllum Blume (Daphniphyllac.).
Daphnicon Pohl = Tontelea Aubl. (Celastrac.).
Daphnidium Nees = Lindera Thunb. (Laurac.).
Daphnidostaphylis Klotzsch = Arctostaphylos Adans. (Ericac.).
Daphnimorpha Nakai. Thymelaeaceae. 2 Japan.
*Daphniphyllaceae Muell. Arg. Dicots. 1/10 E. As., Malaysia. Trees or shrubs, with alt. (often pseudovertic.) long-pet. ent. simple exstip. ls., ± glauc. below. Infl. rac., axill., with decid. bracts. Fl. ♂ ♀, pedic. K (3–6) or o; C o; A 6–12, anth. large, obl., on short fil. (short stds. present in ♀ fl.); G (2), loc. incompl., with 2 thick subsess. recurved stigs., and 2 pend. anatr. ov. per loc. Fr. a 1-seeded glauc. drupe with persist. stigs.; embryo minute, apical, in copious endosp. Only genus: *Daphniphyllum*. Prob. related to *Hamamelidales* and/or *Magnoliales*.
Daphniphyllopsis Kurz = Nyssa Gronov. ex L. (Nyssac.).
Daphniphyllum Blume. Daphniphyllaceae. 10 China, Japan, Formosa, Indomal. Some spp. very polymorphic.
Daphnitis Spreng. = Botryceras Willd. (Anacardiac.).
Daphnobryon Meissn. = Drapetes Banks (Thymelaeac.).
Daphnophyllopsis P. & K. = Daphniphyllopsis Kurz = Nyssa Gronov. ex L. (Nyssac.).
Daphnophyllum P. & K. = Daphniphyllum Bl. (Daphniphyllac.).
Daphnopsis Mart. & Zucc. Thymelaeaceae. 46 Mex. & W.I. to E. Argent.
Daphonanthe Schrad. ex Nees = Calyptrocarya Nees (Cyperac.).
Darbya A. Gray. Santalaceae. 1 SE. U.S.
Dardanis Rafin. = Ferula L. (Umbellif.).
Darea Juss. = Asplenium L. (Aspleniac.).
Dargeria Decne = Leptorhabdos Schrenk (Scrophulariac.).
Darion Rafin. = Kundmannia Scop. (Umbellif.).
Darlingia F. Muell. Proteaceae. 1–2 Queensland.
Darlingtonia DC. = Desmanthus Willd. (Legumin.).
Darlingtonia Torr. (1851) = Styrax L. (Styracac.).
*Darlingtonia Torr. (1854). Sarraceniaceae. 1 Calif., a pitcher pl. like *Sarracenia*, but top of tube bent over and a fish-tail-shaped flap in front.
Darluca Rafin. (1815). Quid? (Monimiac.?)
Darluca Rafin. (1820) = Bouvardia Salisb. (Rubiac.).
Darluca Rafin. (1838) = ?Evolvulus Linn. (Convolvulac.).
Darmera A. Voss = Peltiphyllum Engl. (Saxifragac.).
Darniella Maire & Weiller = Salsola L. (Chenopodiac.).
Dartus Lour. = Maesa Forsk. (Myrsinac.).
Darwinia Dennst. = Litsea Lam. (Laurac.).
Darwinia Rafin. = Sesbania Scop. (Legumin.).

DARWINIA

Darwinia Rudge. Myrtaceae. 35 Austr. Heath-like shrubs.
Darwiniana Lindl. = Darwinia Dennst. = Litsea Lam. (Laurac.).
Darwiniothamnus Harling. Compositae. 2 Galápagos.
Dasanthera Rafin. = Penstemon Mitch. (Scrophulariac.).
Dasianthera Presl = Scolopia Schreb. (Flacourtiac.).
Dasicephala Rafin. = Dasycephala Kunth = Cordia L. (Ehretiac.).
Dasillipe Dubard = Madhuca J. F. Gmel. (Sapotac.).
Dasiogyne Rafin. = Prosopis L. (Legumin.).
Dasiola Rafin. = Festuca L. (Gramin.).
Dasiorima Rafin. = Solidago L. (Compos.).
Dasiphora Rafin. (~ Potentilla L.). Rosaceae. 15 N. temp.
Dasispermum Rafin. = Dasyspermum Neck. (Umbellif.), *q.v.*
Dasistema Rafin. = Dasistoma Rafin. (Scrophulariac.).
Dasistemon Rafin. = Aureolaria Rafin. (Scrophulariac.).
Dasistepha Rafin. = Gentiana L. (Gentianac.).
Dasistoma Rafin. (~ Gerardia L., ~ Seymeria Pursh). Scrophulariaceae.
1 SE. U.S.
Dasoclema J. Sincl. Annonaceae. 1 Siam, W. Malaysia.
Dasouratea Van Tiegh. = Ouratea Aubl. (Ochnac.).
Dassovia Neck. = Asclepias L. (Asclepiadac.).
Dastylepis Rafin. = Cuscuta L. (Cuscutac.).
Dasurus Salisb. = Chamaelirium Willd. (Liliac.).
Dasus Lour. = Lasianthus Jack (Rubiac.).
Dasyanthera Reichb. = Dasianthera Presl = Scolopia Schreb. (Flacourtiac.).
Dasyanthes D. Don = Erica L. (Ericac.).
Dasyanthus Bub. = Gnaphalium L. (Compos.).
Dasyaulus Thw. = Payena A. DC. (Sapotac.).
Dasycarpus Oerst. = Sloanea L. (Elaeocarpac.).
Dasycarya Liebm. = Cyrtocarpa Kunth (Anacardiac.).
Dasycephala Benth. & Hook. f. (~ Diodia L.). Rubiaceae. 5 trop. Am.
Dasycephala Borkh. ex Pfeiff. (sphalm.) = Dasystephana Adans. = Gentiana L.
(Gentianac.).
Dasychloa P. & K. = Dasyochloa Willd. ex Steud. = Sieglingia Bernh.
(Gramin.).
Dasycoleum Turcz. = Chisocheton Bl. (Meliac.).
Dasydesmus Craib. Gesneriaceae. 1 S. China.
Dasygyna P. & K. = Dasiogyna Rafin. = Prosopis L. (Legumin.).
Dasylepis Oliv. Flacourtiaceae. 10 trop. Afr.
Dasylirion Zucc. Agavaceae. 18 SW. U.S., Mex. Aloe-like, xero.; stems
woody, often tuberous; hard ls. Fls. dioec., in gigantic infl. Cf. *Cordyline.*
Dasyloma DC. = Oenanthe L. (Umbellif.).
Dasymalla Endl. = Pityrodia R.Br. (Dicrastylidac.).
Dasymaschalon (Hook. f. & Thoms.) Dalla Torre & Harms (~ Desmos Lour.).
Annonaceae. 15 S. China, Indomal.
Dasynema Schott = Sloanea L. (Elaeocarpac.).
Dasynotus I. M. Johnson. Boraginaceae. 1 NW. U.S.
Dasyochloa Willd. ex Steud. = Sieglingia Bernh. (Gramin.).
Dasypetalum Pierre ex A. Chev. = Scottellia Oliv. (Flacourtiac.).

Dasyphonion Rafin. = Aristolochia L. (Aristolochiac.).

Dasyphora P. & K. = Dasiphora Rafin. (Rosac.).

Dasyphyllum Kunth. Compositae. 36 S. Am.

Dasypoa Pilger. Gramineae. 1 Peru.

Dasypogon R.Br. Xanthorrhoeaceae. 2 SW. Austr.

Dasypogon[ac]eae Dum. = Xanthorrhoeaceae Dum.

Dasypogonia Reichb. = Dasypogon R.Br. (Xanthorrhoeac.).

Dasypyrum (Coss. & Durieu) Candargy [Kantartzis] = Haynaldia Schur (Gramin.).

Dasyranthus Rafin. ex Steud. = Gnaphalium L. (Compos.).

Dasys Lem. = Dasus Lour. = Lasianthus Jack (Rubiac.).

Dasyspermum Neck. = Conium + Tordylium + Ammi + Scandix spp. (Umbellif.).

Dasysphaera Volkens. Amaranthaceae. 2 E. Afr.

Dasystachys Baker (~ Chlorophytum Ker). Liliaceae. 15 trop. Afr.

Dasystachys Oerst. (~ Chamaedorea Willd.). Palmae. 4 trop. Afr.

Dasystemon DC. = Crassula L. (Crassulac.).

Dasystepha P. & K. = Dasistepha Rafin. = Gentiana L. (Gentianac.).

Dasystephana Adans. (~ Gentiana L.). Gentianaceae. 30 N. temp.

Dasystoma Rafin. = Dasistoma Rafin. (Scrophulariac.).

Dasytropis Urb. Acanthaceae. 1 Cuba.

Dasyurus P. & K. = Dasurus Salisb. = Chamaelirium Willd. (Liliac.).

Datisca L. Datiscaceae. 1 Medit. to Himal. & C. As., 1 SW. U.S. & NW. Mex.

*__Datiscaceae__ Lindl. Dicots. 1/2 dry W. Euras. and dry N. Am. Tall glabr. perenn. herbs with habit of *Cannabis*. Ls. alt., pinn., exstip. Infl. of crowded fascicles on long leafy branches. Fls. ♂ or ♂ ♀ dioec., apet., anemoph. ♂: K 3–9, unequal; C o; A 8–∞, with short fil. and large obl. anthers; pistillode o. ♀: K 3–8; C o; A o; Ḡ (3–5), obl., ribbed, 1-loc., with pariet. plac. and ∞ anatr. ov., and 3–5 deeply bifid free styles. ♂: as ♀, but with 3–5 stam. betw. the styles. Fr. a membr. dehisc. caps.; seeds coarsely retic.; endosp. little or o. Only genus: *Datisca*. Prob. related to *Haloragidaceae*; supposed connection with *Tetramelaceae* doubtful.

Datura L. (excl. Brugmansia Pers.). Solanaceae. 10 trop. & warm temp. (esp. trop. Am.; incl. Austr.). *D. stramonium* L. has a 4-loc. ov. (see fam.) giving a 4-valved caps. covered with spines. The ls. and seeds are medic.

Daturicarpa Stapf. Apocynaceae. 1 S. Am.

D'Aubentona Buc'hoz. ?Apocynaceae. 1 S. Am.

Daubentonia DC. (~ Sesbania Adans. mut. Scop.). Leguminosae. 8 Am.

Daubentoniopsis Rydb. (~ Aeschynomene L.). Leguminosae. 1 Mexico.

Daubenya Lindl. Liliaceae. 1 S. Afr.

Daucaceae Dostál = Umbelliferae Juss.

Daucalis Pomel = Caucalis L. (Umbellif.).

Dauceria Dennst. = Embelia Burm. f. (Myrsinac.).

Daucophyllum Rydb. = Musineon Rafin. (Umbellif.).

Daucosma Engelm. & A. Gray ex A. Gray = Discopleura DC. (Umbellif.).

Daucus L. Umbelliferae. 60 Eur., Afr., As., Am. *D. carota* L., biennial with thickened root. The cult. form has much more fleshy roots than the wild. In

DAUMALIA

the centre of the umbel is usu. a red term. fl. After fert. the peduncles all bend inwards until the frs. are ripe and then spread out again, allowing the burred mericarps to adhere to animals.

Daumaiia Arènes. Compositae. 1 Syria.

Daun-contu Adans. = Paederia L. (Rubiac.).

Dauthonia Link (sphalm.) = Danthonia DC. (Gramin.).

Dauventonia Reichb. (sphalm.) = Daubentonia DC. (Legumin.).

Davaea Gandog. = Campanula L. (Campanulac.).

Davaëlla Gandog. = Chondrilla L. (Compos.).

Davallia Sm. Davalliaceae. 40 SW. Europe & Canary Is., Madag., trop. & subtrop. Asia, Pacific. Epiphytes, several common and widely distr.; often cult.

Davalliaceae Reichb. Davalliales. Mostly epiphytes with dorsiventral scaly rhizome and articulate ls.; sori at ends of veins, usu. with pouch-shaped indus. Chief genera: *Davallia, Humata, Araiostegia, Davallodes.*

Davalliales. Filicidae. Fam. *Davalliaceae, Oleandraceae.*

Davalliopsis v. d. Bosch. Hymenophyllaceae. 1 trop. Am.

Davallodes Copel. Davalliaceae. 10 Malaysia. Ls. not deltoid, ± hairy; indusia various, some *Dryopteris*-like.

Daveana Dur. & Jacks. (spnalm.) = seq.

Daveaua Willk. ex Mariz. Compositae. 1 Portugal.

Davidia Baill. Davidiaceae. 1 SW. China.

Davidiaceae Takhtadj. Dicots. 1/1 China. Decid. trees, with alt. simple dent. exstip. ls. Fls. small, ♂ ♀, andromonoec., in dense term. capitula subtended by 2 large white bracts. ♂ fls. ∞, scarcely distinct, consisting of 5–6 sta. only, with long fil. and small anth., in dense glob. head. ♀ fl. sol. in head, obliquely term.: P 0; A (stds.) small, few to many, inserted ½ way up ovary; G̅ (6–9), with columnar style exceeding the A, with 6–9 stig. lobes, and 1 axile pend. ov. per loc. Fr. a drupe, with granular mesocarp and bony longit. sulcate 3–5-loc. endocarp. Seeds with fleshy endosp. Only genus: *Davidia*. Prob. related to *Actinidiaceae.*

Davidsonia F. Muell. Davidsoniaceae. 1 Queensl., N.S.W.

Davidsoniaceae Bange. Dicots. 1/1 NE. Austr. Small slender trees bearing irritant hairs. Ls. alt., pinn., elong., with large reniform stips., rhachis dent.-alate, lfts. dent.-serr. Infl. a large lax pedunc. panicle, or a dense pedunc. spike, axill. or supra-axill. Bracts large, amplexic. Fls. ♀. K (4), valv., thick; C 0; A 10 + 10 nectarif. scales, inserted on disk, fil. ± tumid below, anth. oblong, versatile; G̅ (2), with 2 free filif. styles genic. above, and c. 7 anatr. and epitr. axile ov. per loc. Fr. a large 2-pyrened drupe, red-velvety when young, glauc. and pruinose when ripe. Seeds 2, large, pend., exalb. Only genus: *Davidsonia*. Affinities obscure; habit of *Burserac., Sapindac.*, etc.; anat. of *Cunoniac.* and *Brunelliac.*

Daviesia Poir. = Borya Labill. (Liliac.).

Daviesia Sm. Leguminosae. 55 Austr.

Davilia Mutis apud Alba = ? Paullinia L. (Sapindac.).

Davilla Vand. Dilleniaceae. 38 trop. Am., W.I. The two inner sepals are larger; after fert. they grow woody or leathery and enclose the fr.

Davya DC. = Meriania Sw. (Melastomatac.).

Davya Moç. & Sessé ex DC. = Saurauia Willd. (Actinidiac.).
Davyella Hack. = Neostapfia Davy (Gramin.).
Dawea Sprague ex Dawe = Warburgia Engl. (Canellac.).
Daydonia Britten = Anneslea Wall. (Theac.).
Dayena Adans. = Byttneria Loefl. (Sterculiac.).
D'Ayena Monier ex Mill. = Ayenia L. (Sterculiac.).
D'Ayenia Monier ex Mill. = praec.
Dayenia Michx ex Jaub. & Spach = Biebersteinia Stephan ex Fisch. (Biebersteiniac.).
Dazus Juss. = Dasus Lour. = Lasianthus Jack (Rubiac.).
Deamia Britton & Rose = Selenicereus (A. Berger) Britt. & Rose (Cactac.).
Deanea Coulter & Rose = Rhodosciadium Coult. & Rose (Umbellif.).
Deastella Loud. = Mimetes Salisb. (Proteac.).
Debeauxia Gandog. = Cephalotos Adans. = Thymus L. (Labiat.).
Debesia Kuntze. Liliaceae. 10 trop. Afr.
Debraea Roem. & Schult. = Erisma Rudge (Vochysiac.).
Debregeasia Gaudich. Urticaceae. 5 Abyss., Arabia, Afghan., Indomal., E. As. *D. edulis* Wedd. (*janatsi*, Japan) ed. fr., useful fibre (cf. *Boehmeria*).
Decabelone Decne. Asclepiadaceae. 3 trop. & S. Afr.
Decaceras Harvey = Anisotoma Fenzl (Asclepiadac.).
Decachaena Torr. & Gray ex A. Gray (~ Gaylussacia Kunth). Ericaceae. 4 N. Am.
Decachaeta DC. Compositae. 1 Mexico.
Decachaeta Gardn. = Ageratum L. (Compos.).
Decadenium Rafin. = Adenaria Kunth (Lythrac.).
Decadia Lour. = Symplocos Jacq. (Symplocac.).
Decadianthe Reichb. ? Rutaceae. 1 Austr.
Decadon G. Don = Decodon J. F. Gmel. = Nesaea Comm. ex Juss. (Lythrac.).
Decadonia Rafin. = Decadenium Rafin. = Adenaria Kunth (Lythrac.).
Decadontia Griff. = Sphenodesma Jack (Symphorematac.).
Decagonocarpus Engl. Rutaceae. 1 Amaz. Brazil.
Decaisnea Brongn. = Prescottia Lindl. (Orchidac.).
*Decaisnea Hook. f. & Thoms. Lardizabalaceae. 1 Himal., China, *D. insignis* Hook. f. & Thoms., with pinnate ls. and ed. fr.
Decaisnea Lindl. = Tropidia Lindl. (Orchidac.).
Decaisnella Kuntze = Gyrinops Gaertn. (Thymelaeac.).
Decaisnina Van Tiegh. Loranthaceae. 30 Philipp. Is. to N. Austr. & Tahiti.
Decalepidanthus Riedl. Boraginaceae. 1 W. Pakistan.
Decalepis Boeck. = Tetraria Beauv. (Cyperac.).
Decalepis Wight & Arn. Periplocaceae. 1 Penins. Ind.
Decaloba Rafin. = Ipomoea L. (Convolvulac.).
Decaloba M. Roem. = Passiflora L. (Passiflorac.).
Decalobanthus van Ooststr. Convolvulaceae. 1 Sumatra.
Decalophium Turcz. = Chamaelaucium Desf. (Myrtac.).
Decameria Welw. = Gardenia Ellis (Rubiac.).
Decamerium Nutt. (~ Gaylussacia Kunth). Ericaceae. 7 N. Am.
Decandolia Bast. = Agrostis L. (Gramin.).
Decanema Decne. Asclepiadaceae. 3 Madag.

DECANEMOPSIS

Decanemopsis Costantin & Gallaud. Asclepiadaceae. 1 Madag.
Decaneurum DC. = Centratherum Cass. (Compos.).
Decaneurum Sch. Bip. = Leucanthemella Tsvelev (Compos.).
Decapenta Rafin. (1834) = Diodia L. (Rubiac.).
Decapenta Rafin. (1838) = Litsea Lam. (Laurac.).
Decaphalangium Melch. Guttiferae. 1 Peru.
Decaprisma Rafin. = Campanula L. (Campanulac.).
Decaptera Turcz. Cruciferae. 1 Chile.
Decaraphe Miq. = Miconia Ruiz & Pav. (Melastomatac.).
Decarinium Rafin. = Croton L. (Euphorbiac.).
Decarya Choux = Decaryia Choux (Didiereac.).
Decaryanthus Bonati. Scrophulariaceae. 1 Madag.
Decarydendron Danguy. Monimiaceae. 1 Madag.
Decaryella A. Camus. Gramineae. 1 Madag.
Decaryia Choux. Didiereaceae. 1 S. Madag. Shows points of resemblance to *Pereskia* (*Cactac.*).
Decaryochloa A. Camus. Gramineae. 1 Madag.
Decaschistia Wight & Arn. Malvaceae. 12 India to Hainan & Malay Penins.
Decaspermum J. R. & G. Forst. Myrtaceae. 30 Indomal.
Decaspora R.Br. = Trochocarpa R.Br. (Epacridac.).
Decastelma Schlechter. Asclepiadaceae. 2 W.I.
Decastemon Klotzsch = Cleome L. (Cleomac.).
Decastia Rafin. = Nigella L. (Ranunculac.).
Decastrophia Griff. = Erythropalum Blume (Erythropalac.).
Decastylocarpus Humbert. Compositae. 1 Madag.
Decateles Rafin. = Bumelia Sw. (Sapotac.).
Decatoca F. Muell. Epacridaceae. 1 New Guinea.
Decatropis Hook. f. Rutaceae. 2–3 S. Mexico, Guatem.
Decavenia (Nakai) Koidz. = Pterostyrax Sieb. & Zucc. (Styracac.).
Decazesia F. Muell. Compositae. 1 W. Austr.
Decazyx Pittier & Blake. Rutaceae. 1 C. Am.
Decemium Rafin. Hydrophyllaceae. 1 Atl. N. Am.
Dechampsia Kunth = Deschampsia Beauv. (Gramin.).
Deckenia H. Wendl. Palmae. 1 Seychelles.
Deckera Sch. Bip. = Picris L. (Compos.).
Deckeria Karst. = Iriartea Ruiz & Pav. (Palm.).
Declieuxia Kunth. Rubiaceae. 40 trop. S. Am., W.I.
Decodon J. F. Gmel. Lythraceae. 1 E. N. Am.
Decodontia Haw. = Huernia R.Br. (Asclepiadac.).
Decorima Rafin. = Karwinskia Zucc. (Rhamnac.).
Decorsea R. Viguier. Leguminosae. 4 C. & E. Afr., Madag.
Decorsella A. Chev. Violaceae. 1 trop. W. Afr. Caps. 'gymnospermous'.
Decostea Ruiz & Pav. = Griselinia J. R. & G. Forst. (Cornac.).
Dectis Rafin. = Commidendron DC. (Compos.).
Decumaria L. Hydrangeaceae. 2 China, SE. U.S.
Decussocarpus de Laubenfels (~ Podocarpus L'Hérit.) Podocarpaceae. Spp.? Distrib.?
Dedea Baill. = Quintinia A. DC. (Escalloniac.).

Deeringia R.Br. (= Cladostachys D. Don). Amaranthaceae. 12 Madag., Indomal.
Deeringia Kuntze = Deringa Adans. = Cryptotaenia DC. (Umbellif.).
Deeringiaceae J. G. Agardh = Amaranthaceae–Celosieae Endl.
Deeringothamnus Small. Annonaceae. 2 Florida.
Defforgia Lam. = Forgesia Comm. ex Juss. (Escalloniac.).
Deflersia Gandog. = Verbascum L. (Scrophulariac.).
Deflersia Schweinf. ex Penzig = Erythrococca Benth. (Euphorbiac.).
× **Degarmoara** hort. Orchidaceae. Gen. hybr. (iii) (Brassia × Miltonia × Odontoglossum).
Degeneria Bailey & Smith. Degeneriaceae. 1 Fiji.
*****Degeneriaceae** Bailey & Smith. Dicots. 1/1 Fiji. Tree with alt. ent. simple exstip. ls. Fls. sol., supra-axill., pedunc., bract. K 3, persist.; C 12–13 (3–4-ser.), fleshy, decid.; A 20 (± 3-ser.), flattened, oblong, 4-loc., loc. immersed, dehisc. extrorsely by 2 slits; stds. similar but fewer; G 1, obliquely ellipsoid, open when young, with elongate decurrent stig., and ∞ biseriate ov. on 2 ventr. plac. (1 series sess., 1 series funic.). Fr. large, leathery, asymm., indehisc., with intrusions of endocarp betw. the ∞ seeds; seeds flattened, ± sculptured. Only genus *Degeneria*. Related to *Magnoliac.*, *Winterac.*, *Himantandrac.*
Degenia Hayek. Cruciferae. 1 Croatia.
Deguelia Aubl. = Derris Lour. (Legumin.).
Dehaasia Blume. Lauraceae. 20 Malaysia.
Deherainia Decne. Theophrastaceae. 2 Mex., incl. *D. smaragdina* Decne with large green fls. (coloured by chlorophyll).
Deïanira Cham. & Schlechtd. Gentianaceae. 5 Brazil, Bolivia.
Deïdamia Nor. ex Thou. Passifloraceae. 8 trop. W. Afr., Madag.
Deilanthe N. E. Brown. Aïzoaceae. 1 S. Afr.
Deilosma Andrz. ex DC. = Hesperis L. (Crucif.).
Deina Alef. = Triticum L. (Gramin.).
Deinacanthon Mez. Bromeliaceae. 1 Argentina.
Deinandra Greene = Hemizonia DC. (Compos.).
Deinanthe Maxim. Hydrangeaceae. 2 C. China, Japan.
Deinbollia Schumacher & Thonn. Sapindaceae. 40 warm Afr., Madag.
Deinosmos Rafin. = Pulicaria Gaertn. (Compos.).
Deinostema Yamazaki (~ Gratiola L.). Scrophulariaceae. 2 Manch., Korea, Japan.
Deiregyne Schlechter. Orchidaceae. 8 Mex., C. Am.
Dejanira auctt. = Deïanira Cham. & Schlechtd. (Gentianac.).
× **Dekensara** hort. Orchidaceae. Gen. hybr. (iii) (Cattleya × Rhyncholaelia [Brassavola hort.] × Schomburgkia).
Dekindtia Gilg = Linociera Sw. ex Schreb. (Oleac.).
Dekinia Mart. & Gal. (~ Lepechinia Willd.). Labiatae. 1 Mex.
Dela Adans. = Libanotis Hill = Seseli L. (Umbellif.).
Delabechea Lindl. = Sterculia L. (Sterculiac.).
Delaetia Backeb. Cactaceae. 1 Chile.
Delairea P. & K. = Delaria Desv. = Baphia Afzel. (Legumin.).
Delairia Lem. = Senecio L. (Compos.).
Delamerea S. Moore. Compositae. 1 trop. E. Afr.
Delaportea Thorel & Gagnep. Leguminosae. 3 Indoch.

DELARBREA

Delarbrea Vieill. Araliaceae. 6 New Guinea, New Caled.

Delaria Desv. = Baphia Afzel. (Legumin.).

Delastrea A. DC. = Labramia A. DC. (Sapotac.).

Delavaya Franch. Sapindaceae. 1 SW. China.

Delia Dum. = Segetella Desv. (Caryophyllac.).

Delila Pfeiff. = praec.

Delilia Spreng. = Elvira Cass. (Compos.).

Delima L. = Tetracera L. (Dilleniac.).

Delimopsis Miq. = praec.

Delissea Gaudich. Campanulaceae. 8 Hawaii.

Delivaria Miq. = Dilivaria Juss. = Acanthus L. (Acanthac.).

Deloderium Cass. = Leontodon L. (Compos.).

Delognaea Cogn. = Ampelosicyos Thou. (Cucurbitac.).

Delonix Rafin. Leguminosae. 3 trop. Afr., Madag., As. *D. regia* (Boj.) Rafin. is the 'flamboyant'. Endosp.

Delopyrum Small (~ Polygonella Michx). Polygonaceae. 5 E. U.S.

Delosperma N. E. Brown. Aïzoaceae. 120 S. Afr.

Delostoma D. Don. Bignoniaceae. 5 trop. Andes.

Delostylis Rafin. = Trillium L. (Trilliac.).

Delpechia Montr. = Psychotria L. (Rubiac.).

Delphidium Rafin. = Delphinium L. (Ranunculac.).

Delphinacanthus Benoist. Acanthaceae. 1 Madag.

Delphinastrum (DC.) Spach = Delphinium L. (Ranunculac.).

Delphiniastrum Willis = praec.

Delphinium L. Ranunculaceae. 250 N. temp. Fls. zygo., in racemes; the post. sepal is drawn out into a spur containing the spurs of the two post. petals, in which the honey is secreted. (Cf. with *Aconitum*, which is far more frequently robbed by bumble-bees.) The fl. is protandr. with movement of sta., fert. by bumble-bees. The open fl. projects horiz., but subsequently the stalk bends up and the follicles stand erect so that the seeds can only escape if shaken, e.g. by strong wind (censer mechanism).

Delphyodon K. Schum. Apocynaceae. 1 New Guinea.

Delpinoa H. Ross = Agave L. (Agavac.).

Delpinoëlla Spegazz. (1902; non Sacc. 1899—Fungi) = seq.

Delpinophytum Spegazz. Cruciferae. 1 Patagonia.

Delpya Pierre (1895). Sapindaceae. 1 Siam, ? Malay Penins.

Delpya Pierre ex Bonati (1912) = Pierranthus Bonati (Scrophulariac.).

Delpydora Pierre. Sapotaceae. 2 trop. W. Afr.

Deltaria v. Steenis. Thymelaeaceae. 1 New Caled.

Deltocarpus L'Hérit. ex DC. = Myagrum L. (Crucif.).

Deltonea Peckolt = Theobroma L. (Sterculiac.).

Delucia DC. = Bidens L. (Compos.).

Dematophyllum Griseb. = Balbisia Cav. (Ledocarpac.).

Dematra Rafin. = Euphorbia L. (Euphorbiac.).

Demazeria Dum. = Desmazeria Dum. corr. Dum. (Gramin.).

Demetria Lag. = Grindelia Willd. (Compos.).

Demeusea De Wild. & Durand = Haemanthus L. (Amaryllidac.).

Demeusia Willis = praec.

Demidium DC. (~Amphidoxa DC.). Compositae. 1 Madag.
Demidofia Dennst. = Carallia Roxb. (Rhizophorac.).
Demidofia J. F. Gmel. = Dichondra J. R. & G. Forst. (Convolvulac.).
Demidovia Hoffm. = Paris L. (Trilliac.).
Demidovia Pall. = Tetragonia L. (Tetragoniac.).
Demnosa Frič = Cleistocactus Lem. (Cactac.).
Democrita Vell. Inc. sed. (?Rutaceae). 1 Brazil.
Democritea DC. = Serissa Comm. ex Juss. (Rubiac.).
Demosthenesia A. C. Smith. Ericaceae. 9 Andes.
Demosthenia Rafin. = Sideritis L. (Labiat.).
Denckea Rafin. = Gentiana L. (Gentianac.).
Dendragrostis Nees = Chusquea Kunth (Gramin.).
Dendranthema (DC.) Des Moul. (~Chrysanthemum L.). Compositae.
50 C. & E. As., 1, *D. zawadskii* (Herb.) Tsvel., extending W. to SE. Eur.
D. indica (L.) Des Moul., *D. morifolia* (Ramat.) Tsvel. and *D. erubescens* (Stapf)
Tsvel. are perennial 'chrysanthemums'.
Dendrema Rafin. = Bessia Rafin. = Intsia Thou. (Legumin.).
Dendriopoterium Sventenius = Sanguisorba L. (Rosac.).
Dendrium Desv. = Leiophyllum Hedw. f. (Ericac.).
Dendrobangia Rusby. Icacinaceae. 3 trop. S. Am.
Dendrobenthamia Hutch. (~Cynoxylon Rafin.). Cornaceae. 12 Himal., E. As.
*Dendrobium Sw. Orchidaceae. 1400 trop. As. to Polynesia & Australas.
Dendrobrium Agardh (sphalm.) = praec.
Dendrobryon Klotzsch ex Pax = Algernonia Baill. + Tetraplandra Baill.
(Euphorbiac.).
Dendrocacalia (Nak.) Nak. Compositae. 1 Bonin Is.
Dendrocalamus Nees. Gramineae. 20 China, Indomal. *D. giganteus* Munro
(the giant bamboo), the largest known bamboo, grows with great rapidity (see
Lock in *Ann. Perad. Bot. Gard.* 2: 211, 1904), even as much as 46 cm. a day.
D. strictus Nees (male bamboo) has solid stems, used for lances, etc. Nut fr.
× **Dendrocattleya** hort. Orchidaceae. Gen. hybr. (vi) (Cattleya × Dendro-
bium). (Doubtful.)
Dendrocereus Britton & Rose. Cactaceae. 1 Cuba.
Dendrocharis Miq. = Ecdysanthera Hook. (Apocynac.).
Dendrochilum Blume. Orchidaceae. 100 Indomal. to Philippines & New Guinea.
Dendrochloa C. E. Parkinson. Gramineae. 1 Burma.
Dendrocnide Miq. Urticaceàe. 36 India & Ceylon to Formosa, Malaysia, NE.
Austr. & Pacif.
Dendrocolla Blume = Thrixspermum Lour. + Pteroceras v. Hasselt ex Hassk.
(Orchidac.).
Dendroconche Copeland. Polypodiaceae. 2 New Guinea.
Dendrocousinia Willis (sphalm.) = seq.
Dendrocousinsia Millspaugh (~Sebastiania Spreng.). Euphorbiaceae.
3 Jamaica.
Dendrodaphne Beurl. = Ocotea Aubl. (Laurac.).
Dendroglossa Presl. Polypodiaceae. 5 India to S. China & Malaysia.
Dendroleandria Arènes. Sterculiaceae. 1 Madag.
Dendrolirium Blume = Eria Lindl. (Orchidac.).

DENDROLOBIUM

Dendrolobium Benth. Leguminosae. 17 Indomal., Austr.

Dendromecon Benth. Papaveraceae. 2 Calif., 1 Mex.

Dendromyza Danser. Santalaceae. 7 SE. As., Indomal.

Dendropanax Decne & Planch. Araliaceae. 75 trop. & subtrop.

Dendropemon (Blume) Reichb. = Phthirusa Mart. (Loranthac.).

Dendrophila Zipp. ex Blume = Tecomanthe Baill. (Bignoniac.).

Dendrophthoaceae Van Tiegh. = Loranthaceae Juss.

Dendrophthoë Mart. Loranthaceae. 30 trop. Afr., Indomal., trop. Austr.

Dendrophthora Eichl. Viscaceae (~ Loranthaceae). 55 trop. Am., W.I.

Dendrophthoraceae Dostál (sphalm.) = Dendrophthoaceae Van Tiegh. = Loranthaceae Juss.

Dendrophylax Reichb. f. Orchidaceae. 4 W.I.

Dendrophyllanthus S. Moore = Phyllanthus L. (Euphorbiac.).

Dendropogon Rafin. = Tillandsia L. (Bromeliac.).

Dendrorchis Thou. = seq.

Dendrorkis Thou. = Polystachya Hook. (Orchidac.).

Dendroseris D. Don. Compositae. *S.str.* 4, *s.l.* 10, Juan Fernandez.

Dendrosicus Rafin. = Enallagma Baill. (Bignoniac.).

Dendrosicyos Balf. f. Cucurbitaceae. 1 Socotra.

Dendrosida Fryxell. Malvaceae. 2 Mex.

Dendrosipanea Ducke. Rubiaceae. 3 N. trop. S. Am.

Dendrosma R.Br. ex Cromb. = Atherosperma Labill. (Atherospermatac.).

Dendrosma Panch. & Sebert = Geijera Schott (Rutac.).

Dendrospartium Spach = Genista L. (Legumin.).

Dendrostellera Van Tiegh. (~ Stellera J. G. Gmel.). Thymelaeaceae. 10 SW. & C. As.

Dendrostigma Gleason. Flacourtiaceae. 1 Colombia.

Dendrostylis Karst. & Triana = Mayna Aubl. (Flacourtiac.).

Dendrotrophe Miq. Santalaceae. 4 SE. As., Indomal., trop. Austr.

Denea O. F. Cook (~ Howea Becc.). Palmae. 1 Lord Howe I.

Deneckia Sch. Bip. = seq.

Denekia Thunb. Compositae. 3 trop. & S. Afr.

*Denhamia Meissn. Celastraceae. 4 trop. Austr.

Denhamia Schott = Culcasia Beauv. (Arac.).

Denira Adans. = Iva L. (Compos.).

Denisaea Neck. = Phryma L. (Phrymatac.).

Denisia P. & K. = praec.

Denisonia F. Muell. Dicrastylidaceae. 1 N. Austr.

Denisophytum R. Viguier. Leguminosae. 1 Madag.

Denmoza Britton & Rose. Cactaceae. 2 Argent.

Denneckia Steud. = Denekia Thunb. (Compos.).

Dennettia E. G. Baker. Annonaceae. 1 Nigeria.

Dennstaedtia Bernh. Dennstaedtiaceae. 70 pantrop., temp. S. Am., Austr., N.Z. Large forest-ferns or thicket-formers. Probably a composite genus; distinction from *Microlepia* (and *Hypolepis*?) not clear.

Dennstaedtiaceae Ching. Dicksoniales. Large much-divided ferns with creeping dorsiventral underground hairy rhizome (except *Orthiopteris* Copel. and *Saccoloma* Kaulf.); sori marginal or submarginal, with indusium (no

DESCHAMPSIA

indusium in *Hypolepis*); in some cases fusion-sori present (*Pteridium, Histiopteris, Lonchitis*). In *Dennstaedtia* the reflexed sorus is protected by a cup formed by union of indusium with lobe of leaf-margin.

Dens Fabr. = Taraxacum Web. ex Wigg. (Compos.).
Dens-leonis Seguier = Leontodon L. (Compos.).
Denslovia Rydb. = Gymnadeniopsis Rydb. (Orchidac.).
Densophylis Thou. (uninom.) = Bulbophyllum densum Thou. (Orchidac.).
Dentaria L. (~Cardamine L.). Cruciferae. 20 Atl. N. Am., Euras.
Dentella J. R. & G. Forst. Rubiaceae. 10 Indomal., Austr.
Dentidia Lour. = Perilla L. (Labiat.).
Dentillaria Kuntze = Knoxia L. (Rubiac.).
Dentimetula Van Tiegh. = Tapinanthus Bl. (Loranthac.).
Dentoceras Small. Polygonaceae. 1 Florida.
Deonia Pierre ex Pax = Blachia Baill. (Euphorbiac.).
Depacarpus N. E. Brown. Aïzoaceae. 1 S. Afr.
Depanthus S. Moore. Gesneriaceae. 1 New Caled.
Deparia Hook. & Grev. (*s.str.*) = Athyrium Roth (Athyriac.); Bower (*The Ferns*, 3: 257) copies *Syn. Fil.* in including mixture of unrelated spp.
Depierrea auct. ex Schlechtd. = Campanula L. (Campanulac.).
Deplachne Boiss. (sphalm.) = Diplachne Beauv. (Gramin.).
Deplanchea Vieill. Bignoniaceae. 9 Malaysia, Austr., New Caled.
Deppea Cham. & Schlechtd. Rubiaceae. 20 Mex. to Venez.
Deppia Rafin. = Lycaste Lindl. (Orchidac.).
Deprea Rafin. = Athenaea Sendtn. (Solanac.).
Depremesnilia F. Muell. = Pityrodia R.Br. (Dicrastylidac.).
Depresmenilia Willis (sphalm.) = praec.
Derderia Jaub. & Spach = Jurinea Cass. (Compos.).
Derenbergia Schwantes (~Conophytum N. E. Br.). Aïzoaceae. 15 S. Afr.
Derenbergiella Schwantes. Aïzoaceae. 1 S. Afr.
Deringa Adans. = Cryptotaenia DC. (Umbellif.).
Deringia Steud. = praec.
Derlinia Néraud = Gratiola L. (Scrophulariac.).
Dermasea Haw. = Saxifraga L. (Saxifragac.).
Dermatobotrys Bolus. Scrophulariaceae(?). 1 S. Afr. (Natal, etc.).
Dermatocalyx Oerst. Scrophulariaceae. 2 Costa Rica, Ecuador.
Dermatophlebium Presl = Sphaerocionium Presl (Hymenophyllac.).
Dermatophyllum Scheele = Sophora L. (Legumin.).
Dermophylla S. Manso = Cayaponia S. Manso (Cucurbitac.).
Deroemera Reichb. f. = Holothrix A. Rich. (Orchidac.).
Deroemeria Willis = praec.
Derosiphia Rafin. = Osbeckia L. (Melastomatac.).
Derouetia Boiss. & Bal. = Crepis L. (Compos.).
***Derris** Lour. Leguminosae. 80 trop.
Derwentia Rafin. = Veronica L. (Scrophulariac.).
Desbordesia Pierre ex Van Tiegh. Ixonanthaceae. 1 trop. W. Afr. Samaroid fr.
Descantaria Schlechtd. = Tripogandra Rafin. (Commelinac.).
Deschampsia Beauv. Gramineae. 60 temp. & frigid, trop. mts. Of tufted growth; rough fodder grasses.

349

DESCLIAEA

Descliaea [Moç. & Sessé] ex DC. = Symphoricarpos Juss. (Caprifoliac.) +
Margaritopsis Wright ex Sauvalle (Rubiac.).

*****Descurainia** Webb & Berth. Cruciferae. 55 cold & temp. Am., Euras., S. Afr.

Desdemona S. Moore. Scrophulariaceae. 1 Brazil.

Desfontaena Vell. = Chiropetalum Juss. (Euphorbiac.).

Desfontaina Steud. = praec.

Desfontainea Kunth = Desfontainia Ruiz & Pav. (Potaliac.).

Desfontainea Reichb. = Desfontaena Vell. = Chiropetalum Juss. (Euphor-

Desfontaineaceae: see Desfontainiaceae. [biac.).

Desfontainesia Hoffmgg. = Fontanesia Labill. (Oleac.).

Desfontainia Ruiz & Pav. Potaliaceae. 5 Andes. Ls. spinous; ovary 5-loc.

*****Desfontain[iac]eae** Endl. = Potaliaceae Mart.

Desforgia Steud. = Defforgia Lam. = Forgesia Comm. ex Juss. (Escalloniac.).

Desideria Pampan. = Christolea Cambess. ex Jacquem. (Crucif.).

Desmanthodium Benth. Compositae. 9 Mexico to Venez.

*****Desmanthus** Willd. Leguminosae. 40 Am., Madag.

Desmaria Van Tiegh. = Loranthus L. (Loranthac.).

Desmazeria Dum. (corr. Dum.). Gramineae. 4 Medit., S. Afr.

Desmesia Rafin. = Typhonium Schott + Sauromatum Schott (Arac.).

Desmia D. Don = Erica L. (Ericac.).

Desmidochus Reichb. = seq.

Desmidorchis Ehrenb. = Boucerosia Wight & Arn. (Asclepiadac.).

Desmiograstis Börner = Carex L. (Cyperac.).

Desmitus Rafin. = Camellia L. (Theac.).

Desmocarpus Wall. = Cadaba Forsk. (Capparidac.).

Desmocephalum Hook. f. = Elvira Cass. (Compos.).

Desmochaeta DC. = Pupalia Juss. (Amaranthac.).

Desmocladus Nees = Loxocarya R.Br. (Restionac.).

Desmodiocassia Britton & Rose = Cassia L. (Legumin.).

*****Desmodium** Desv. Leguminosae. 450 trop. & subtrop. In *D. gyrans* (L. f.)
DC. (telegraph plant), during the day, if the temperature be not below 72° F.
the two small lat. leaflets of each l. move steadily round in elliptical orbits.
At night the leaves sleep, drooping downwards. Several are useful as fodder,
and are cult.

Desmofischera Holthuis = Monarthrocarpus Merr. (Legumin.).

Desmogymnosiphon Guinea. Burmanniaceae. 1 trop. W. Afr.

Desmogyne King & Prain = Agapetes D. Don ex G. Don (Ericac.).

Desmonchus Desf. = seq.

*****Desmoncus** Mart. Palmae. 65 trop. Am. Climbing palms with reedy stems,
and hooks like *Calamus*.

Desmonema Miers = Hyalosepalum Troupin (Menispermac.).

Desmonema Rafin. = Euphorbia L. (Euphorbiac.).

Desmophyla Rafin. = Ehretia L. (Ehretiac.).

Desmophyllum Webb & Berth. = Ruta L. (Rutac.).

Desmopsis Safford. Annonaceae. 15 Mex., C. Am., 1 Cuba.

Desmos Lour. Annonaceae. 30 Indomal., Austr., Pacif. Some climb by
recurved hooks which are infl. axes. Fr. an aggregate of stalked moniliform
berries, constricted between the seeds like a lomentum.

Desmoscelis Naud. Melastomataceae. 3 trop. S. Am.
Desmoschoenus Hook. f. = Scirpus L. (Cyperac.).
Desmostachya Stapf = Stapfiola Kuntze (Gramin.).
Desmostachys Planch. Icacinaceae. 3 trop. Afr., Madag.
Desmostemon Thw. = Fahrenheitia Reichb. f. & Zoll. (Euphorbiac.).
Desmothamnus Small (~Lyonia Nutt.). Ericaceae. 1 SE. U.S.
Desmotrichum Blume = Ephemerantha P. F. Hunt & Summerhayes (Orchidac.).
Despeleza Nieuwl. = Lespedeza Michx (Legumin.).
Desplatsia Bocquillon. Tiliaceae. 7 trop. W. Afr.
Despretzia Kunth = Zeugites Schreb. (Gramin.).
Desrousseauxia Van Tiegh. = Loranthus L. (Loranthac.).
Dessenia Adans. = Gnidia L. (Thymelaeac.).
Dessenia Rafin. (1838) = Lasiosiphon Fresen. (Thymelaeac.).
Dessenia Rafin. (1840) = Struthiola L. (Thymelaeac.).
Destrugesia Gaudich. = Capparis L. (Capparidac.).
Destruguezia Benth. & Hook. f. = praec.
Desvauxia Benth. & Hook. f. = Devauxia R.Br. = Centrolepis Labill. (Centrolepidac.).
Desvauxia P. & K. = Devauxia Beauv. ex Kunth = Glyceria R.Br. (Gramin.).
Desvauxi[ac]eae Lindl. = Centrolepidaceae Desv.
Detandra Miers = Sciadotenia Miers (Menispermac.).
Detari[ac]eae Burnett = Leguminosae–Cynometreae Benth.
Detarium Juss. Leguminosae. 4 trop. Afr. Pith of pod ed.
Dethardingia Nees & Mart. = Breweria R.Br. (Convolvulac.).
Dethawia Endl. Umbelliferae. 1 Pyrenees.
Detridium Nees = Felicia Cass. (Compos.).
Detris Adans. = Felicia Cass. (Compos.).
Detzneria Schlechter ex Diels. Scrophulariaceae. 1 New Guinea.
Deuterocohnia Mez. Bromeliaceae. 7 S. Am.
Deuteromallotus Pax & K. Hoffm. Euphorbiaceae. 1 Madag.
Deutzia Thunb. Philadelphaceae. 50 Himal., E. As., Philipp. Is.; 1–3 Mex. The fruit splits septicidally into its cpls. which open each at its apex. The seed is provided with a winged testa, very light. See Zaïkonnikova, *Deitsii–Dekorativnye Kustarniki* [Deutzias as ornamental shrubs: monograph of *Deutzia*]. Moscow/Leningrad, 1966.
Deutzianthus Gagnep. Euphorbiaceae. 1 Indoch.
Devauxia Beauv. ex Kunth = Glyceria R.Br. (Gramin.).
Devauxia R.Br. = Centrolepis Labill. (Centrolepidac.).
Devauxiaceae Dum. = Centrolepidaceae Desv.
Deverra DC. = Pituranthos Viv. (Umbellif.).
Deveya Reichb. = Deweya Torr. & Gray = Tauschia Schlechtd. (Umbellif.).
Devillea Bert. ex Schult. f. = Guzmania Ruiz & Pav. (Bromeliac.).
Devillea Bub. = Ligusticum L. (Umbellif.).
Devillea Tul. & Wedd. Podostemaceae. 1 Brazil.
Dewevrea M. Micheli. Leguminosae. 2 trop. Afr.
Dewevrella De Wild. Apocynaceae. 2 trop. Afr.
Deweya Eaton = Nemopanthus Rafin. (Aquifoliac.).

DEWEYA

Deweya Rafin. = Carex L. (Cyperac.).

Deweya Torr. & A. Gray = Tauschia Schlechtd. (Umbellif.).

Dewildemania O. Hoffm. Compositae. 3 trop. Afr.

Dewindtia De Wild. Leguminosae. 1 trop. Afr.

× **DeWolfara** hort. (v) = × Shigeuraara hort. (Orchidac.).

Deyeuxia Clarion ex Beauv. (~ Calamagrostis Beauv.). Gramineae. 200 temp.

× **Diabroughtonia** hort. Orchidaceae. Gen. hybr. (iii) (Broughtonia × Caularthron [Diacrium]).

Diacaecarpium Endl. = Diacicarpium Blume = Alangium Lam. (Alangiac.).

Diacalpe Blume. Aspidiaceae. 1 trop. As.

Diacantha Lag. = Chuquiraga Juss. (Compos.).

Diacantha Less. = Barnadesia Mutis (Compos.).

Diacarpa Sim. Sapindaceae. 1 trop. E. Afr.

× **Diacatlaelia** hort. (vii) = × Dialaeliocattleya hort. (Orchidac.).

× **Diacattleya** hort. Orchidaceae. Gen. hybr. (iii) (Cattleya × Caularthron [Diacrium]).

Diacecarpium Hassk. = Diacicarpium Blume = Alangium Lam. (Alangiac.).

Diachroa Nutt. ex Steud. = Glyceria R.Br. (Gramin.).

Diachyrium Griseb. (~ Sporobolus R.Br.). Gramineae. 1 temp. S. Am.

Diacicarpium Blume = Alangium Lam. (Alangiac.).

Diacidia Griseb. Malpighiaceae. 2 trop. S. Am.

Diacisperma P. & K. = Disakisperma Steud. (Gramin.), *q.v.*

Diacles Salisb. = Haemanthus L. (Amaryllidac.).

Diacrium Benth. = Caularthron Rafin. (Orchidac.).

× **Diacrocattleya** hort. (vii) = × Diacattleya hort. (Orchidac.).

Diacrodon Sprague. Rubiaceae. 1 Brazil.

Diadenaria Klotzsch & Garcke = Pedilanthus Neck. ex Poit. (Euphorbiac.).

Diadenium Poepp. & Endl. Orchidaceae. 2 W. trop. S. Am.

Diadesma Rafin. (1834) = Modiola Moench (Malvac.).

Diadesma Rafin. (1836) = Sida L. (Malvac.).

× **Dialaelia** hort. Orchidaceae. Gen. hybr. (iii) (Caularthron [Diacrium] × Laelia).

× **Dialaeliocattleya** hort. Orchidaceae. Gen. hybr. (iii) (Cattleya × Caularthron [Diacrium] × Laelia).

× **Dialaeliopsis** hort. Orchidaceae. Gen. hybr. (iii) (Caularthron [Diacrium] × Laeliopsis).

Dialanthera Rafin. = Cassia L. (Legumin.).

Dialesta Kunth = Pollalesta Kunth (Compos.).

Dialion Rafin. = Heliotropium L. (Boraginac.).

Dialiopsis Radlk. = Zanha Hiern (Sapindac.).

Dialissa Lindl. = Stelis Sw. (Orchidac.).

Dialium L. Leguminosae. 1 trop. S. Am., 40 trop. Afr., Madag., Malaysia. Petals 2, 1, or 0; sta. 2, or rarely 3. *D. guineënse* Willd. (trop. Afr.; velvet tamarind) pod contains an ed. pulp; wood useful, resists salt water. *D. indum* L. (Java; tamarind plum) and others also have ed. fr.

Dialla Lindl. = Dicella Griseb. (Malpighiac.).

Diallobus Rafin. = Cassia L. (Legumin.).

Diallosperma Rafin. = Aspalathus L. (Legumin.).

Diallosteira Rafin. = Collinsonia L. (Labiat.).
Dialyanthera Warb. = Otoba (DC.) Karst. (Myristicac.).
Dialycarpa Mast. = Brownlowia Roxb. (Tiliac.).
Dialyceras Capuron. Sphaerosepalaceae. 1 Madag.
Dialypetalae Endl. = Polypetalae Juss.
*****Dialypetalanthaceae** Rizz. & Occh. Dicots. 1/1 trop. S. Am. Tree, with white-tom. young branches. Ls. opp., ent., simple, with large persist. intrapet. stips. (members of opp. pairs conn.). Infl. a term. ∞-fld. bracteate thyrse; fls. ♀, conspic., scented. K 4, imbr.; C 4, biseriate, white; A (16–25), biser., epig., free from C, equal, with 2 sep. lin. loc. on flattened connective, intr.; disk annular, fimbr. G̲ (2), with ∞ axile anatr. ov. and simple elong. style with shortly bifid stig. Fr. a 2-loc. loculic. ∞-seeded caps., exserted from K at apex; seeds ∞, fusif., with endosp. Only genus: *Dialypetalanthus*. An interesting monotype, related to *Rubiac.* and perhaps to *Myrtac.*
Dialypetalanthus Kuhlm. Dialypetalanthaceae. 1 E. Brazil.
Dialypetalum Benth. Campanulaceae. 5 Madag.
Dialytheca Exell & Mendonça. Menispermaceae. 1 trop. Afr.
Diamarips Rafin. = Salix L. (Salicac.).
Diamonon Rafin. = Solanum L. (Solanac.).
Diamorpha Nutt. Crassulaceae. 2 E. U.S.
Diana Comm. ex Lam. = Dianella Lam. (Liliac.).
Diandriella Engl. Araceae. 1 New Guinea.
Diandrochloa de Winter. Gramineae. 7 Am., Afr., As., Austr.
Diandrolyra Stapf. Gramineae. 1, habitat unknown (?trop. Am.).
Diandrostachya (C. E. Hubbard) Jacques-Félix. Gramineae. 4 trop. Afr.
Dianella Lam. Liliaceae. 30 trop. As., Austr., Polynes., N.Z.
Dianell[ac]eae Salisb. = Liliaceae–Asphodeleae–Dianellineae Engl.
Diania Nor. ex Tul. = Dicoryphe Thou. (Hamamelidac.).
Dianisteris Rafin. = Verbesina L. (Compos.).
Dianthaceae Drude = Caryophyllaceae Juss.
Dianthella Clauson ex Pomel = Petrorhagia (Ser.) Link (Caryophyllac.).
Dianthera Klotzsch = Cleome L. (Cleomac.).
Dianthera L. = Justicia L. (Acanthac.).
Dianthoseris Sch. Bip. Compositae. 2 Abyssinia.
Dianthus L. Caryophyllaceae. 300 Eur., As., Afr., esp. Medit., mostly in dry sunny situations. Genus readily known by the bracts under the K. Fls. very protandrous, largely visited by butterflies.
Diapasis Poir. = Diaspasis R.Br. (Goodeniac.).
Diapedium Koenig = Dicliptera Juss. (Acanthac.).
Diapensia Hill = Sanicula L. (Umbellif.)
Diapensia L. Diapensiaceae. 3 Himal. & W. China; 1 circumpolar boreal. Tufted, like many alpine and arctic pl.; fl. protog.
*****Diapensiaceae** Lindl. Dicots. 6/20 temp. Euras., E. U.S. Chiefly alpine and arctic evergr. undershrubs, with rosettes of ls. Fls. sol. or in racemes, with two bracteoles, ♀, actinom. K (5) or 5; C (5), nearly polypet.; A 5, epipet., opp. sepals, with frequently 5 stds. opp. petals, or 5 + 5, with inner series sometimes free from C; anthers usu. transv., each lobe opening by longitudinal slit; pollen simple; disk rarely present; G̲ (3), rarely (5), with axile

DIAPERIA

plac. bearing ∞ anatr. or amphitr. ov.; style simple with 3-lobed capitate stigma. Fruit a loculic. caps. Embryo cylindrical, endosp. fleshy. Genera: *Diapensia, Pyxidanthera, Shortia, Schizocodon, Berneuxia, Galax*. Connects Ericac. with *Primulac*.

Diaperia Nutt. = Evax Gaertn. (Compos.).

Diaphananthe Schlechter. Orchidaceae. 42 trop. Afr.

Diaphane Salisb. = Iris L. (Iridac.).

Diaphanoptera K. H. Rechinger. Caryophyllaceae. 1 Persia.

Diaphora Lour. = Scleria Bergius (Cyperac.).

Diaphoranthema Beer = Tillandsia L. (Bromeliac.).

Diaphoranthus Anders. ex Hook. f. = Pringlea Anders. (Crucif.).

Diaphoranthus Meyen = Polyachyrus Lag. (Compos.).

Diaphorea Pers. = Diaphora Lour. = Scleria Bergius (Cyperac.).

Diaphractanthus Humbert. Compositae. 1 Madag.

Diaphycarpus Calest. = Bunium L. (Umbellif.).

Diaphyllum Hoffm. = Bupleurum L. (Umbellif.).

Diarina Rafin. = seq.

***Diarrhena** Beauv. Gramineae. 1 N. Am.

Diarthron Turcz. Thymelaeaceae. 2 temp. As. (As. Min. & Pers. to N. China & Korea).

× **Diaschomburgkia** hort. Orchidaceae. Gen. hybr. (iii) (Diacrium × Schomburgkia).

Diascia Link & Otto. Scrophulariaceae. 42 S. Afr.

Diasia DC. = Melasphaerula Ker-Gawl. (Iridac.).

Diaspananthus Miq. = Ainsliaea DC. (Compos.).

Diaspanthus Kitamura (sphalm.) = praec.

Diaspasis R.Br. Goodeniaceae. 1 SW. Austr.

Diasperus Kuntze = Phyllanthus L., Glochidion J. R. & G. Forst., etc. (Euphorbiac.).

Diaspis Niedenzu = Triaspis Burch. (Malpighiac.).

Diastatea Scheidw. Campanulaceae. 7 Mexico to Peru.

Diastella Salisb. (~ Leucospermum R.Br.). Proteaceae. 5 S. Afr.

Diastema Benth. Gesneriaceae. 40 C. & trop. S. Am.

Diastema L. f. ex B. D. Jacks. = Dalbergia L. (Legumin.).

Diastemanthe Steud. = Stenotaphrum Trin. (Gramin.).

Diastemation C. Muell. = Diastema Benth. (Gesneriac.).

Diastemella Oerst. = praec.

Diastemma Lindl. = praec.

Diastrophis Fisch. & Mey. = Aëthionema R.Br. (Crucif.).

Diateinacanthus Lindau. Acanthaceae. 1 C. Am.

Diatenopteryx Radlk. Sapindaceae. 1 S. Am.

Diatoma Lour. (1790; non DC. 1805, nec *Bory 1824—Algae–Chrysophyta–Bacillarioph.) = Carallia Roxb. (Rhizophorac.).

Diatonta Walp. = Diadonta Nutt. = Coreopsis L. (Compos.).

Diatosperma C. Muell. = Diotosperma A. Gray = Ceratogyne Turcz. (Com-

Diatrema Rafin. = seq. [pos.).

Diatremis Rafin. = Ipomoea L. (Convolvulac.).

Diatropa Dum. = Bupleurum L. (Umbellif.).

Diaxulon Rafin. = Cytisus L. (Legumin.).
Diaxylum P. & K. = praec.
Diazeuxis D. Don = Lycoseris Cass. (Compos.).
Diazia Phil. = Calandrinia Kunth (Portulacac.).
Diberara Baill. = Nebelia Neck. ex Sweet (Bruniac.).
Diblemma J. Sm. Polypodiaceae. 1 Philippines.
Dibothrospermum Knaf = Matricaria L. (Compos.).
Dibrachia Steud. = Dibrachya Eckl. & Zeyh. = Pelargonium L'Hérit. (Geraniac.).
Dibrachion Regel = Homalanthus A. Juss. (Euphorbiac.).
Dibrachion Tul. = Diplotropis Benth. (Legumin.).
Dibrachionostylus Bremek. Rubiaceae. 1 trop. E. Afr.
Dibrachium Walp. = Dibrachion Regel = Homalanthus A. Juss. (Euphorbiac.).
Dibrachya (Sweet) Eckl. & Zeyh. = Pelargonium L'Hérit. (Geraniac.).
Dibracteaceae Dulac = Callitrichaceae Link.
Dicaelosperma E. G. O. Muell. & Pax = Dicoelospermum C. B. Cl. corr. P. & K. (Cucurbitac.).
Dicaelospermum C. B. Cl. = praec.
Dicalix Lour. = Symplocos Jacq. (Symplocac.).
Dicalymma Lem. = Podachaenium Benth. (Compos.).
Dicalyx Poir. = Dicalix Lour. = Symplocos Jacq. (Symplocac.).
Dicardiotis Rafin. = Gentiana L. (Gentianac.).
Dicarpaea Presl = Limeum L. (Aïzoac.).
Dicarpellum (Loes.) A. C. Smith. Celastraceae. 5 New Caled.
Dicarpidium F. Muell. Bombacaceae. 1 Austr.
Dicarpophora Speg. Asclepiadaceae. 1 Bolivia.
Dicaryum Willd. = Geissanthus Hook. f. (Myrsinac.).
Dicella Griseb. Malpighiaceae. 5 trop. S. Am.
Dicellandra Hook. f. Melastomataceae. 6 trop. W. Afr.
Dicellostyles Benth. Malvaceae. 1 Ceylon.
*****Dicentra** Borckh. corr. Bernh. Fumariaceae. 20 W. Himal. to E. Sib., Sakhalin, Japan & W. China; N. Am. The rhiz. of many spp. (§ *Cucullaria*) resembles a succession of bulbs, on account of the fleshiness of the scale ls. and of the sheathing bases of the fol. ls. The materials formed in the ls. during the growing season are stored up in the fleshy base, which survives the winter, while the rest of the l. dies. Fls. in racemes, pend. Each outer petal has a large pouch at its base. The inner petals are spoon-shaped and cohere at the tip, forming a hood which covers the anthers and stigma. The pend. position and complex structure of the fl. render it suited to bees, which hang on to it and probe for honey, first one side, then the other, in the pouches of the petals. In so doing they push aside the hood and touch the stigma, on which there is usu. pollen from its own sta.
Dicentranthera T. Anders. = Asystasia Blume (Acanthac.).
Dicera J. R. & G. Forst. = Elaeocarpus Burm. f. (Elaeocarpac.).
Dicera Zipp. ex Blume = Gironniera Gaudich. (Ulmac.).
Dicerandra Benth. = Ceranthera Ell. (Labiatae).
Diceras P. & K. (1) = Diceros Bl. = Vandellia L. (Scrophulariac.).
Diceras P. & K. (2) = Dicera Zipp. ex Bl. = Gironniera Gaudich. (Ulmac.).

DICERAS

Diceras P. & K. (3) = Dicera J. R. & G. Forst. = Elaeocarpus L. (Elaeocarpac.).
Diceras P. & K. (4) = Diceros Lour. = Limnophila R.Br. (Scrophulariac.).
Diceras P. & K. (5) = Diceros Pers. = Artanema D. Don (Scrophulariac.).
Diceratella Boiss. Cruciferae. 9 Cyrenaica, NE. trop. Afr., Socotra, Persia.
Diceratium Boiss. = praec.
Diceratium Lag. = Notoceras R.Br. (Crucif.).
Diceratostele Summerhayes. Orchidaceae. 1 trop. Afr.
Dicerma DC. emend. Benth. (~ Desmodium Desv.). Leguminosae. 3 Burma, New Guinea, Austr.
Dicerocaryum Boj. Pedaliaceae. 1 S. trop. Afr.
Dicerolepis Blume = Gymnanthera R.Br. (Asclepiadac.).
Diceros Blume = Lindernia All. (Scrophulariac.).
Diceros Lour. = Limnophila R.Br. (Scrophulariac.).
Diceros Pers. = Artanema D. Don (Scrophulariac.).
Dicerospermum Bakh. f. Melastomataceae. 1 New Guinea.
Dicerostylis Blume. Orchidaceae. 5 Formosa, Philippines, Indonesia.
Dicersos Lour. (sphalm.) = Diceros Lour. = Limnophila R.Br. (Scrophulariac.).
Dichaea Lindl. Orchidaceae. 40 trop. Am., W.I.
Dichaelia Harv. (~ Brachystelma R. Br.). Asclepiadaceae. 10 trop. & S. Afr.
Dichaeopsis Pfitz. = praec.
Dichaespermum Hassk. = Aneilema R.Br. (Commelinac.).
Dichaeta Nutt. = Baeria Fisch. & Mey. (Compos.).
Dichaeta Sch. Bip. = Schaetzellia Sch. Bip. (Compos.).
Dichaetandra Naud. = Ernestia DC. (Melastomatac.).
Dichaetanthera Endl. Melastomataceae. 5 trop. W. Afr., 35 Madag.
Dichaetaria Nees. Gramineae. 1 India, Ceylon.
Dichaetophora A. Gray. Compositae. 1 S. U.S.
Dichanthium Willemet. Gramineae. 15 palaeotrop.
*Dichapetalaceae Baill. Dicots. 4/200 trop. Mostly shrubs, some trees and lianes, often ± grey-pubesc., with alt. simple ent. stip. ls., with a few flat glands near base below. Infl. of axill. dichot. cymes or fascicles, the peduncle often adnate to petiole of subtending l. Fls. ⚲ or ♂ ♀, reg., or in 2 small gen. zygo. K 5 or slightly (5), imbr. C 5, or in 3 smaller gen. (5–4), pet. mostly bifid or bilobed, often drying black. A 5, oppositisep., epipet. or free, anth. intr. Disk cupular or lobed. G̲ (2–3), with 2 apical pend. anatr. ov. per loc., and 2 free or conn. styles with small capit. stig. Drupe usu. pubesc., 1–3-lobed, 1–3-loc., loc. 1-seeded; seeds exalb., sometimes with caruncle. Genera: *Dichapetalum*, *Stephanopodium*, *Tapura*, *Gonypetalum*. Related to *Euphorbiaceae* (cf. pollen and anat., and the bilobed pet., turning black on drying, of *Trigonostemon* spp.), and *Chrysobalanac.*, perhaps also to *Polygalac.* and *Trigoniac.* Lf.-glands often not confined to the basal region, sometimes also or only on upper surface, sometimes absent.
Dichapetalum Thou. Dichapetalaceae. 150–200 trop., esp. Afr. Several have epiphyllous infl.
Dichasium (A. Braun) Fée = Dryopteris Adans. (Aspidiac.).
Dichazothece Lindau. Acanthaceae. 1 E. Brazil.
Dichelachne Endl. Gramineae. 5 Austr., N.Z.
Dichelactina Hance = Emblica Gaertn. = Phyllanthus L. (Euphorbiac.).

Dichelostemma Kunth (~ Brodiaea Sm.). Alliaceae. 6 N. Am.
Dichelostemma Wood = Strophilirion Torr. (Alliac.).
Dichelostylis Endl. = Dichostylis Beauv. = Scirpus L. (Cyperac.).
Dicheranthus Webb. Caryophyllaceae. 1 Canaries.
Dichilanthe Thw. Rubiaceae. 2 Ceylon, Borneo.
Dichiloboea Stapf. Gesneriaceae. 2 S. China, Burma, Langkawi Is.
Dichilos Spreng. = seq.
Dichilus DC. Leguminosae. 5 S. Afr.
Dichismus Rafin. = Scirpus L. (Cyperac.).
Dichocarpum W. T. Wang & Hsiao (~ Isopyrum L.). Ranunculaceae. 16 Himal., E. As.
Dichodon Bartl. ex Reichb. = Cerastium L. (Caryophyllac.).
Dichoespermum Wight = Aneilema R.Br. (Commelinac.).
Dichoglottis Fisch. & Mey. = Gypsophila L. (Caryophyllac.).
Dichondra J. R. & G. Forst. Convolvulaceae. 4–5 trop. & subtrop. Some amphicarpic.
***Dichondraceae** Dum. = Convolvulaceae–Dichondreae Choisy.
Dichondropsis T. S. Brandegee. Convolvulaceae. 1 Mexico.
Dichone Laws. ex Salisb. = Tritonia Ker-Gawl. (Iridac.).
Dichopetalum F. Muell. = Dichosciadium Domin (Umbellif.).
Dichopetalum P. & K. = Dichapetalum Thou. (Dichapetalac.).
Dichopogon Kunth. Liliaceae. 2 Austr.
Dichopsis Thw. = Palaquium Blanco (Sapotac.).
Dichopus Blume = Dendrobium Sw. (Orchidac.).
Dichorexia Presl = Cyathea Sm. (Cyatheac.).
***Dichorisandra** Mikan. Commelinaceae. 35 trop. Am. Infl. racemose (cf. fam.); its branches often pierce the leaf-sheath.
Dichoropetalum Fenzl = Johrenia DC. (Umbellif.).
Dichosciadium Domin. Hydrocotylaceae (~ Umbellif.). 1 Austr.
Dichosema Benth. = Mirbelia Sm. (Legumin.).
Dichosma DC. ex Loud. = Agathosma Willd. (Rutac.).
Dichospermum C. Muell. = Dichoespermum Wight = Aneilema R.Br. (Commelinac.).
Dichospermum W. T. Wang & Hsiao [non Dichoespermum Wight]. Ranunculaceae. 20 C. & E. As.
Dichostachys Krauss (sphalm.) = Dichrostachys Wight & Arn. (Legumin.).
Dichostemma Pierre. Euphorbiaceae. 3 trop. Afr.
Dichostylis Beauv. = Scirpus L. (Cyperac.).
Dichotoma Sch. Bip. = Sclerocarpus Jacq. (Compos.).
Dichotomanthes Kurz. Rosaceae. 1 SW. China.
Dichotophyllum Moench = Ceratophyllum L. (Ceratophyllac.).
Dichotrichum S. Moore (sphalm.) = Dichrotrichum Reinw. (Gesneriac.).
Dichroa Lour. Hydrangeaceae. 13 China, SE. As., Indomal. G isomerous.
Dichroanthus Webb & Berth. = Cheiranthus L. (Crucif.).
Dichrocephala L'Hérit. ex DC. Compositae. 10 warm Afr., Madag., India, China, Java. Pappus o.
Dichrolepidaceae ('-ideas') Welw. = Eriocaulaceae Desv.
Dichrolepis Welw. = Eriocaulon L. (Eriocaulac.).

DICHROMA

Dichroma Cav. = Ourisia Comm. ex Juss. (Scrophulariac.).

Dichroma Pers. = seq.

Dichromena Michx. Cyperaceae. 60 Am., W.I. P o. Stigmas 2.

Dichromus Schlechtd. = Paspalum L. (Gramin.).

Dichronema Baker = Dichromena Michx (Cyperac.).

Dichropappus Sch. Bip. ex Kraschen. = Stenachaenium Benth. (Compos.).

Dichrophyllum Klotzsch & Garcke = Euphorbia L. (Euphorbiac.).

Dichrospermum Bremek. Rubiaceae. 1 trop. Afr.

*__*Dichrostachys__ (A. DC.) Wight & Arn. Leguminosae. 20 trop. Afr. to Austr., esp. Madag. Stips. often thorny. Spike of two colours, upper fls. ♂, lower neuter.

Dichrostylis Nakai (sphalm.) = Dichostylis Beauv. = Scirpus L. (Cyperac.).

Dichrotrichum Reinw. Gesneriaceae. 35 Malaysia (Borneo & Philipp. to New Guinea). Like *Aeschynanthus*; hair of 2 colours.

Dichylium Britton = Euphorbia L. (Euphorbiac.).

Dichynchosia C. Muell. (sphalm.) = Dirhynchosia Blume = Spiraeopsis Miq.

Dickasonia L. O. Williams. Orchidaceae. 1 Burma. [(Cunoniac.).

Dickia Scop. = Stemodia L. (Scrophulariac.).

Dickinsia Franch. Hydrocotylaceae (~ Umbellif.). 1 SW. China.

Dicksonia L'Hérit. Dicksoniaceae. 30 Malaysia (mts.), Austr., New Caled., N.Z., trop. Am., St Helena; tree-ferns.

Dicksoniaceae Presl. Dicksoniales. Mainly arborescent; young ls. hairy, not scaly; sori at ends of veins, protected by reflexed margin and by inner indusium. Genera: *Dicksonia, Cystidium, Cibotium, Culcita, Thyrsopteris*. By some authors united to *Cyatheaceae, q.v.*

Dicksoniales. Filicidae. 3 fam.: *Dicksoniaceae, Dennstaedtiaceae, Lindsaeaceae.*

Dicladanthera F. Muell. Acanthaceae. 1 W. Austr.

Diclemia Naud. = Ossaea DC. (Melastomatac.).

Diclidanthera Mart. Polygalaceae. 6 trop. S. Am.

*__*Diclidantheraceae__ J. G. Agardh = Polygalaceae–Moutabeae Chod.

Diclidium Schrad. ex Nees = Mariscus Gaertn. (Cyperac.).

Diclidocarpus A. Gray = Trichospermum Bl. (Tiliac.).

Diclidopteris Brack. = Vaginularia Fée (Vittariac.).

Diclidostigma Kunze = Melothria L. (Cucurbitac.).

Diclinanona Diels. Annonaceae. 2 E. Peru, W. Braz.

Diclinothrys Endl. = seq.

Diclinotris Rafin. = seq.

Diclinotrys Rafin. = Chamaelirium Willd. (Liliac.).

*__*Dicliptera__ Juss. Acanthaceae. 150 trop. & subtrop.

Diclis Benth. Scrophulariaceae. 9 E. trop. & S. Afr., Madag. Creeping herbs.

Diclisodon Moore = Dryopteris Adans. (Aspidiac.).

Diclythra Rafin. = seq.

Diclytra Borckh. (sphalm.) = Dicentra Borckh. corr. Bernh. (Papaverac.).

Dicneckeria Vell. = Euplassa Salisb. (Proteac.).

Dicocca Thou. = Dicoryphe Thou. (Hamamelidac.).

Dicodon Ehrh. (uninom.) = Linnaea borealis L. (Caprifoliac.).

Dicoelia Benth. Euphorbiaceae. 2–3 Malaya, Sumatra, Banka, Borneo. Pets. deeply hollowed either side of midrib.

DICRASTYLIDACEAE

Dicoelospermum C. B. Cl. corr. P. & K. Cucurbitaceae. 1 S. India.

Dicolus Phil. = Zephyra D. Don (Tecophilaeac.).

Dicoma Cass. Compositae. 48 C., E. & S. Afr., Madag., Socotra, 1 India.

Diconangia Adans. = Itea L. (Iteac.).

Dicondra Rafin. = Dichondra J. R. & G. Forst. (Convolvulac.).

Dicophe Roem. (sphalm.) = Dicoryphe Thou. (Hamamelidac.).

Dicoria Torr. & A. Gray. Compositae. 5 SW. U.S., Mex. Fr. winged, with pappus.

Dicorynia Benth. Leguminosae. 2 Guiana & Amaz. Basin. A 2.

Dicorypha R. Hedw. = seq.

Dicoryphe Thou. Hamamelidaceae. 15 Madag., Comoro Is.

Dicotyledones Juss. One of the two chief divisions (classes) of *Angiospermae*. Embryo with 2 (rarely more) cotyledons; ls. mostly pinnately and reticulately nerved; fls. most commonly pentamerous or tetramerous.

Dicraea Tul. = Dicraeia Thou. (Podostemac.).

Dicraeanthus Engl. Podostemaceae. 2 trop. Afr.

Dicraeia Thou. Podostemaceae. 5 Madag. Thallus (shoot) drifting from attached base, exogenously branched, with marginal secondary shoots. Fr. isolobous.

Dicraeopetalum Harms. Leguminosae. 1 Somaliland.

Dicrairus Hook. f. corr. Airy Shaw. Amaranthaceae. 1 S. U.S., Mex.

Dicrama Klatt (sphalm.) = Dierama C. Koch (Iridac.).

Dicranacanthus Oerst. = Barleria L. (Acanthac.).

Dicrananthera Presl = Acisanthera P.Br. (Melastomatac.).

Dicranocarpus A. Gray. Compositae. 1 S. U.S., Mex. Some frs. with no pappus.

Dicranoglossum J. Sm. Polypodiaceae. 6 trop. Am.

Dicranolepis Planch. Thymelaeaceae. 30 trop. Afr. Fls. 1-2 in axils.

Dicranopetalum Presl = Toulicia Aubl. (Sapindac.).

Dicranopteris Bernh. Gleicheniaceae. 10 pantrop., forming dense thickets. Sect. *Acropterygium* in S. Am.

Dicranopygium Harling. Cyclanthaceae. 44 C. Am. & N. trop. S. Am.

Dicranostachys Tréc. = Myrianthus Beauv. (Urticac.).

Dicranostegia (A. Gray) Pennell. Scrophulariaceae. 1 Calif.

Dicranostigma Hook. f. & Thoms. Papaveraceae. 2-3 Himal., W. China.

Dicranostyles Benth. Convolvulaceae. 13 trop. S. Am.

Dicranotaenia Finet = Microcoelia Lindl. (Orchidac.).

Dicraspidia Standley. Tiliaceae. 1 C. Am.

Dicrastyles Benth. & Hook. f. = Dicrastylis J. Drumm. ex Harv. (Dicrastylidac.).

Dicrastylidaceae J. Drumm. ex Harv. Dicots. 14/90 trop. E. Afr. (?), Madag., Masc., Austr., Pacif. Shrubs, often tomentose. Ls. opp. or vertic., rarely alt., ent., exstip. Infl. term., spicate, capit. or comp., or fls. axill. K (4-8), rarely accresc.; C (4-8), reg. or zygo., imbr.; A 4-8 or 3-7, epipet.; disk 0; G̲ (2), with 2 axile ov. per loc., and filif. shortly bifid style; fr. dry (rarely drupac.), indehisc., 1-2-seeded, seeds with endosp. Chief genera: *Pityrodia*, *Dicrastylis*, *Lachnostachys*, *Chloanthes*, *Newcastelia*. Close to *Verbenac.*, differing chiefly in the albuminous seeds.

359

DICRASTYLIS

Dicrastylis Drumm. ex Harv. Dicrastylidaceae. 15 Austr.
Dicraurus Hook. f. (sphalm.) = Dicrairus Hook. f. corr. Airy Shaw (Amaranthac.).
Dicroactis Rafin. = ? Coreopsis L. (Compos.).
Dicrobotryum Willd. ex Roem. & Schult. = Guettarda L. (Rubiac.).
Dicrocaulon N. E. Brown. Aïzoaceae. 6 S. Afr.
Dicrocephala Royle (sphalm.) = Dichrocephala L'Hérit. (Compos.).
Dicrophyla Rafin. = Ludisia A. Rich. (Orchidac.).
Dicrosperma W. Wats. (sphalm.) = Dictyosperma H. Wendl. & Drude (Palm.).
Dicrostylis P. & K. = Dicrastylis Drumm. ex Harv. (Dicrastylidac.).
Dicrurus P. & K. = Dicrairus Hook. f. (Amaranthac.).
Dicrus Reinw. = Voacanga Thou. (Apocynac.).
Dicrypta Lindl. = Maxillaria Poepp. & Endl. (Orchidac.).
Dictamn[ac]eae Trautv. = Rutaceae–Ruteae–Dictamninae Engl.
Dictamnus L. Rutaceae. 6 C. & S. Eur. to E. Sib. & N. China. *D. albus* L. (*D. fraxinella* Pers.) (dittany, candle-plant). Volatile and inflammable ethereal oil is secreted, so that on hot calm days the air round the pl. may sometimes be ignited. Fl. zygo.; unripe sta. bent down. Fr. elastic-dehisc.
Dictamnus Mill. = Amaracus Gled. (Labiat.).
Dictilis Rafin. = Otostegia Benth. (Labiat.).
Dictyaloma Walp. = Dictyoloma A. Juss. (Rutac.).
Dictyandra Welw. ex Benth. & Hook. f. Rubiaceae. 2 trop. W. Afr. K large, conv. Anther loc. chambered.
Dictyanthes Rafin. = Aristolochia L. (Aristolochiac.).
Dictyanthus Decne. Asclepiadaceae. 3 Mex.
Dictymia J. Sm. Polypodiaceae. 4 Australia to Fiji.
Dictyocalyx Hook. f. = Cacabus Bernh. (Solanac.).
Dictyocarpus Wight = Sida L. (Malvac.).
Dictyocaryum H. Wendl. Palmae. 6 trop. S. Am.
Dictyochloa Camus (~ Ammochloa Boiss.). Gramineae. 1 NW. Afr.
Dictyocline Moore. Thelypteridaceae. 1 Assam to Japan.
Dictyodaphne Blume = Endiandra R.Br. (Laurac.).
Dictyodroma Ching. Athyriaceae. 5 E. & SE. As.
Dictyoglossum J. Sm. = Elaphoglossum Schott (Lomariopsidac.).
Dictyogramme Fée = Coniogramme Fée (Gymnogrammac.).
*****Dictyoloma** A. Juss. Rutaceae. 2 Peru, Brazil.
Dictyoneura Blume. Sapindaceae. 9 Malaysia.
Dictyopetalum Fisch. & Mey. = Oenothera L. (Onagrac.).
Dictyophleba Pierre. Apocynaceae. 4 trop. Afr.
Dictyophragmus O. E. Schulz. Cruciferae. 1 Peru.
Dictyopsis Harv. ex Hook. f. = Behnia Didrichs. (Liliac.).
Dictyopteris Presl (1836; non Lamx. 1809—Algae) = Arcypteris Underw. (Aspidiac.).
Dictyosperma P. & K. = Dyctisperma Rafin. = Rubus L. (Rosac.).
Dictyosperma Regel = Pirea Dur. and Jacks. = Rorippa Scop. (Crucif.).
Dictyosperma Wendl. & Drude. Palmae. 3 Mascarenes.
Dictyospermum Wight = Aneilema R.Br. (Commelinac.).
Dictyospora Hook. f. = Dyctiospora Reinw. ex Korth. = Hedyotis L. (Rubiac.).

Dictyostega Miers. Burmanniaceae. 2 Mex. to trop. S. Am.

Dictyoxiphiaceae Ching = Aspidiaceae S. F. Gray.

Dictyoxiphium Hook. Aspidiaceae. 1 C. Am. Simple ls., marginal fusion-sorus; see *Pleuroderris*.

Dictysperma Rafin. = Dyctisperma Rafin. = Rubus L. (Rosac.).

Dicyclophora Boiss. Umbelliferae. 1 Persia.

Dicymanthes Danser. Loranthaceae. 15 Philipp. Is. & Java to New Guinea.

Dicymbe Spruce ex Benth. & Hook. f. Leguminosae. 5 trop. S. Am.

Dicymbopsis Ducke. Leguminosae. 2 trop. S. Am.

Dicypellium Nees. Lauraceae. 1 Brazil, *D. caryophyllatum* (Mart.) Nees. Wood valuable; bark (Cassia caryophyllata) smells like cloves.

Dicyrta Regel (~Achimenes P.Br.). Gesneriaceae. 2 C. Am.

Didactyle Lindl. = Bulbophyllum Thou. (Orchidac.).

Didactylon Zoll. & Moritzi = Dimeria R.Br. (Gramin.).

Didaste E. Mey. ex Harv. & Sond. = Acrosanthes Eckl. & Zeyh. (Aïzoac.).

Didelotia Baill. Leguminosae. 7 trop. Afr.

*****Didelta** L'Hérit. Compositae. 2–3 S. Afr.

Diderota Comm. ex A. DC. = Ochrosia Juss. (Apocynac.).

Diderotia Baill. = Alchornea Sw. (Euphorbiac.).

Didesmandra Stapf. Dilleniaceae. 1 Borneo.

Didesmus Desv. (~Rapistrum Crantz). Cruciferae. 2 E. Medit.

Didiciea King & Prain. Orchidaceae. 1 E. Himal.

Didiclis P. Beauv. = Selaginella P. Beauv. (Selaginellac.).

Didierea Baill. Didiereaceae. 2 Madag.

Didiereaceae Radlk. ex Drake. Dicots. (Centrospermae). 4/11 Madag. Trees or shrubs with the habit of cacti or cactiform euphorbias, often armed with sol., paired, or fascicled spines. Ls. simple, ent., alt., exstip. Fls. in cymes or fascicles, ♂ ♀ or rarely (*Decaryia*) ☿ ♀. Invol. 2-leaved, ± produced or decurrent at base; P 2 + 2, imbr.; A 8–10 (staminodial in ♀ fls.), shortly united at base; G (2–4), with 1 fert. loc., style usu. bearing an expanded 3–4-lobed stig. (stig. in *Decaryia* scarcely enlarged), and 1 basal hemicampylotr. ov. (G pistillodial in ♂ fls.). Fr. dry, indehisc., usu. loosely encl. in persist. invol.; seed 1, with curved embr. and 0 or v. scanty endosp., and small aril. Genera: *Didierea, Alluaudia, Alluaudiopsis, Decaryia*. An unusually interesting fam., of which the affinities have been long disputed, but in recent years shown beyond doubt to be Centrospermous. This is confirmed, not only by the pollen and other morphological features, but, in a remarkable manner, by the successful grafting of the 2 known spp. of *Didierea* on stocks of the Cactaceous genera *Pereskia, Pereskiopsis* and *Trichocereus*. The red and purple pigments developed in different parts of the *Did.* are found to be, not anthocyanins, but betacyanins, as in typical Centrospermae. (Cf. Rauh & Reznik in Engl. *Bot. Jahrb.* 81 (1/2): 94–105, 1961.) The habit similarity with *Fouquieriaceae* (*q.v.*) should also be noted.

Didimeria Lindl. = Correa Sm. (Rutac.).

Didiplis Rafin. (~Peplis L.). Lythraceae. 1 N. Am.

Didiscus DC. = Trachymene Rudge (Hydrocotylac.).

*****Didissandra** C. B. Clarke (emend. Ridley). Gesneriaceae. 30 India, China.

Didothion Rafin. = Epidendrum L. (Orchidac.).

361

DIDYMAEA

Didymaea Hook. f. Rubiaceae. 2 Mexico to Panamá.

Didymandra Willd. = Lacistema Sw. (Lacistematac.).

Didymanthus Endl. Chenopodiaceae. 1 W. Austr.

Didymanthus Klotzsch ex Meissn. = Euplassa Salisb. (Proteac.).

Didymaotus N. E. Brown. Aïzoaceae. 1 S. Afr.

Didymelaceae Leandri. Dicots. 1/2 Madag. Trees, with simple ent. alt. exstip. coriac. or chartac. ls. drying yellowish-green. Fls. ♂ ♀, dioec.; infls. axill. or supra-axill., the ♂ shortly panic., the ♀ simply spicate, with thickened rhachis. ♂ fl. subtended by 0–2 scales (?bracts ?seps.); A 2, anth. sessile, cuneate. ♀ fl. subtended by 0–4 scales; G 1, cylindr., with large oblique decurr. stig. with median groove, sometimes recurved at apex, ov. 1 semi-anatr., with apical embr. in copious endosp. Fr. a large 1-seeded drupe, with lateral groove (as *Prunus*). Only genus: *Didymeles*. Relationships obscure. The extreme simplicity of the flowers enhances the difficulty. Tissues full of sclereids as *Theac.*, but probably no significant affinity with this fam. Fls. perhaps not 'reduced' but primitively simple. Possibly some connection with *Buxaceae* through *Styloceratac*.

Didymeles Thou. Didymelaceae. 2 Madag.

Didymeria Lindl. = Didimeria Lindl. = Correa Andr. (Rutac.).

Didymia Phil. (~ Mariscus Gaertn.). Cyperaceae. 1 Chile.

Didymiandrum Gilly. Cyperaceae. 3 trop. S. Am.

Didymocarp[ac]eae D. Don = Gesneriaceae–Didymocarpeae Endl.

*Didymocarpus Wall. Gesneriaceae. 120 trop. Afr., Madag., SE. As., Indomal., Austr.

Didymochaeta Steud. = Deyeuxia Clarion ex Beauv. (Gramin.).

Didymocheton Blume. Meliaceae. 30 S. China, Malaysia, Pacif.

Didymochiton Spreng. = praec.

Didymochlaena Desv. Aspidiaceae. 1 trop. & Natal.

Didymochlaenaceae Ching = Aspidiaceae S. F. Gray.

Didymochlamys Hook. f. Rubiaceae. 2 trop. S. Am. Epiphytic.

Didymocistus Kuhlm. Euphorbiaceae. 1 trop. S. Am.

Didymococcus Blume = Sapindus L. (Sapindac.).

Didymodoxa E. Mey. ex Wedd. = Australina Gaudich. (Urticac.).

Didymoecium Bremek. Rubiaceae. 1 Sumatra.

Didymoglossum Desv. Hymenophyllaceae. 20 + Am., Afr., Madag., Ceylon.

Didymoglossum Prantl = Crepidomanes Presl (Hymenophyllac.).

Didymogyne Wedd. = Droguetia Gaudich. (Urticac.).

Didymomeles Spreng. = Didymeles Thou. (Didymelac.).

Didymonema Presl = Gahnia J. R. & G. Forst. (Cyperac.).

Didymopanax Decne & Planch. Araliaceae. 40 trop. Am.

Didymopelta Regel & Schmalh. = Astragalus L. (Legumin.).

Didymophysa Boiss. Cruciferae. 2 Persia to C. As. & E. Himal.

Didymoplexiella Garay. Orchidaceae. 5 Siam, Malaysia.

Didymoplexis Griff. Orchidaceae. 23 trop. E. Afr., Madag., Malaysia, Pacif.

Didymopogon Bremek. Rubiaceae. 1 Sumatra.

Didymosalpinx Keay. Rubiaceae. 5 trop. Afr.

Didymosperma H. Wendl. & Drude ex Benth. & Hook. f. (~ Arenga Labill.). Palmae. 8 Assam to Ryukyu Is., W. Malaysia.

Didymotheca Hook. f. Gyrostemonaceae. 5 NW. to S. Austr., Tasmania.

Didymotoca E. Mey. = Australina Gaudich. (Urticac.).

Didyplosandra Wight. Acanthaceae. 3 (?7) Penins. Ind., Ceylon.

***Diectomis** Kunth. Gramineae. 1 trop. Annual.

Diectonis Willis (sphalm.) = praec.

Dieffenbachia Schott. Araceae. 30 trop. Am., W.I. Fls. ♂ ♀, naked; the ♂ is a synandrium of 4 or 5 sta. *D. seguine* Schott is the 'dumb cane' of the W. Ind., formerly used in torturing slaves; it renders speechless a person who chews a piece of stem.

Diegodendraceae Capuron. Dicots. 1/1 Madag. Shrub or small tree. Ls. alt., simple, ent., lanc., densely minutely pellucid-punct., smelling of camphor when crushed, with ± large convol. caduc. intrapet. stips., leaving conspic. annular scars. Infl. term., panic., ± few-fld., branches alt., subtended by a small bract, pedic. opp. or subopp., with bract and bracteoles at base. Fls. large, reg., ♀, fragrant. K 5(–6), imbr., unequal, persist. C 5(–6), large, imbr., slightly unequal, caduc. A ∞, fil. filif., anth. ov.-obl., basif., ± latr. G 1–4, coarsely verruc. and pelt.-gland., ovoid, free, but with common central elong. gynobasic style with punctif. stig.; ov. 2 in each carp., basal, collat., ascend., anatr., with micropyle facing outwards. Fr. unknown. Only genus: *Diegodendron*. Probably related to *Sphaerosepalaceae*, differing from them in the glandular-punctate ls., pentamerous fls. and outward-facing micropyle of the ovules. The conspicuous development of the disk in *Rhopalocarpus* constitutes another point of difference. The aspect of the twigs, leaves, stipules and stipular scars is strongly reminiscent of *Irvingia* and its immediate allies.

Diegodendron Capuron. Diegodendraceae. 1 Madag. Endosp. o!

Diellia Brackenridge. Aspleniaceae. 5 Hawaii. (Wagner, *Univ. Calif. Publ. Bot.* **26**: 1–212, 1952.)

Dielsantha E. Wimm. Campanulaceae. 1 trop. Afr.

Dielsia Gilg. Restionaceae. 1 Austr.

Dielsia Kudo = Plectranthus L'Hérit. (Labiat.).

Dielsina Kuntze = Polyceratocarpus Engl. & Diels (Annonac.).

Dielsiocharis O. E. Schulz. Cruciferae. 1 Persia.

Dielsiochloa Pilger. Gramineae. 1 Peru.

Dielsiothamnus R. E. Fries. Annonaceae. 1 trop. E. Afr.

Dielytra Cham. & Schlechtd. = Dicentra Borckh. corr. Bernh. (Fumariac.).

Diemenia Korth. = Parastemon A. DC. (Chrysobalanac.).

Diemisa Rafin. = Carex L. (Cyperac.).

Dieneckeria Vell. = Euplassa Salisb. (Proteac.).

Dienia Lindl. = Malaxis Soland. ex Sw. (Orchidac.).

Dierama C. Koch. Iridaceae. 25 trop. & S. Afr.

Dierbachia Spreng. = Dunalia Kunth (Solanac.).

Diervilla Mill. Caprifoliaceae. 3 N. Am.

Diervillea Bartl. = praec.

Diesingia Endl. = Psophocarpus Neck. ex DC. (Legumin.).

Dietegocarpus Willis (sphalm.) = Distegocarpus Sieb. & Zucc. = Carpinus L. (Carpinac.).

Dieteria Nutt. = Aster L. (Compos.).

Dieterica Ser. = Caldcluvia D. Don (Cunoniac.).

DIETERICHIA

Dieterichia Giseke = Dietrichia Giseke = Zingiber Adans. (Zingiberac.).
Dietes Salisb. Iridaceae. 3 S. Afr., 1 Lord Howe I.
Dietrichia Giseke = Zingiber Adans. (Zingiberac.).
Dietrichia Tratt. = Rochea DC. (Crassulac.).
Dieudonnaea Cogn. Cucurbitaceae. 1 Peru.
Diflugossa Bremek. Acanthaceae. 16 Indomal.
Digaster Miq. = Pygeum Gaertn. = Lauro-Cerasus Duham. (Rosac.).
Digastrium (Hackel) A. Camus. Gramineae. 2 Queensland.
Digera Forsk. Amaranthaceae. 2 palaeotrop.
Digitacalia Pippen. Compositae. 5 Mex.
Digital[idac]eae J. G. Agardh = Scrophulariaceae–Digitalideae Benth.
Digitalis L. Scrophulariaceae. 20–30 Eur., Medit., Canary Is. In *D. purpurea*
  L. the racemes are one-sided by twisting of pedicels. Fert. by bees. A calcifuge
  sp. Ls. offic. for digitalin.
Digitaria Adans. = Tripsacum L. (Gramin.).
Digitaria Haller. Gramineae. 380 warm.
Digitariella de Winter. Gramineae. 1 SW. Afr.
Digitariopsis C. E. Hubbard. Gramineae. 2 trop. Afr.
Digitorebutia Frič & Kreuz. ex Buining = Rebutia K. Schum. (Cactac.).
Diglosselis Rafin. = Aristolochia L. (Aristolochiac.).
Diglossophyllum H. Wendl. ex Drude = Serenoa Hook. f. (Palm.).
Diglossus Cass. = Tagetes L. (Compos.).
Diglottis Nees & Mart. = Angostura Roem. & Schult. (Rutac.).
Diglyphis Blume = seq.
Diglyphosa Blume. Orchidaceae. 5 Indomal.
Dignathe Lindl. = Leochilus Knowles & Westc. (Orchidac.).
Dignathia Stapf. Gramineae. 4 trop. E. Afr., W. India (Kutch).
Digomphia Benth. Bignoniaceae. 2 N. trop. S. Am. Std. very long.
Digomphotis Rafin. = Peristylus Blume (Orchidac.).
Digoniopterys Arènes. Malpighiaceae. 1 Madag.
Digonocarpus Vell. = Cupania L. (Sapindac.).
Digrammaria Presl = Diplazium Sw. (Athyriac.).
Digraphis Trin. = Phalaris L. (Gramin.).
Digyroloma Turcz. = ? mixture of Justicia, Ruellia, etc. (Acanthac.).
Diheteropogon Stapf. Gramineae. 4 trop. Afr.
Diholcos Rydb. = Astragalus L. (Legumin.).
Dikylikostigma Kraenzl. = Discyphus Schlechter (Orchidac.).
Dilanthes Salisb. = Anthericum L. (Liliac.).
Dilasia Rafin. = Murdannia Royle (Commelinac.).
Dilatris Bergius. Haemodoraceae. 3 S. Afr. $\overline{G}$. Ov. 1 per loc.
Dilax Rafin. = Smilax L. (Smilacac.).
Dilema Griff. = Dillenia L. (Dilleniac.).
Dilepis Suesseng. & Merxm. Compositae. 1 NW. S. Am.
Dileptium Rafin. = Lepidium L. (Crucif.).
Dilepyrum Michx = Muhlenbergia Schreb. (Gramin.).
Dilepyrum Rafin. = Oryzopsis Michx (Gramin.).
Dileucaden Rafin. = Panicum L. (Gramin.).
Dilivaria Juss. = Acanthus L. (Acanthac.).

Dilkea Mast. Passifloraceae. 6 Peru, N. Braz. A 6, united at base.
Dillenia Fabr. (uninom.) = *Sherardia arvensis* L. (Rubiac.).
Dillenia L. Dilleniaceae. 60 Masc., SE. As., Indomal., N. Queensl., Fiji.
*Dilleniaceae Salisb. Dicots. 10/400, trop. & subtrop., well repres. in the
Austr. scrub. Usu. woody pl. (many lianas), with alt. (rarely opp.) usu.
leathery ls., stip. or not, veins of 2nd and later orders parallel; sometimes
phylloclades. Infl. cymose, often raceme-like by reduction, or fls. sol. Fl.
usu. reg., ♀. K 5, 3, 4, or ∞, spiral, imbr., persistent; C usu. 5, imbr.; A ∞,
rarely 10 or fewer, hypog., free or united at base, anthers usu. adnate; G ∞–1,
free or slightly united, styles usu. free, ov. ∞–1, ascending, anatr., with ventral
raphe, plac. unthickened. Fr. dehisc. or not; funicular aril, united to testa.
Endosp. copious; embryo small, straight. Some give useful timber; tannin.
Chief genera: *Dillenia, Hibbertia, Davilla, Tetracera.*
× Dillonara hort. Orchidaceae. Gen. hybr. (iii) (Epidendrum × Laelia ×
Schomburgkia).
Dillonia Sacleux = Catha Forsk. ex Scop. (Celastrac.).
Dillwinia Poir. = Dillwynia Sm. (Legumin.).
Dillwynia Roth = Rothia Pers. (Legumin.).
Dillwynia Sm. Legumin. 15 Austr. Exstip.
Dilobeia Thou. Proteaceae. 1 Madag.
Dilochia Lindl. Orchidaceae. 5 Malaysia.
Dilodendron Radlk. Sapindaceae. 1 trop. S. Am. Oil from seed.
Dilomilis Rafin. Orchidaceae. 5 W.I., Brazil.
Dilophia T. Thoms. Cruciferae. 5 C. As.
Dilophotriche (C. E. Hubbard) Jacques-Félix. Gramineae. 5 trop. Afr.
Dilosma P. & K. = Deilosma Andrz. ex DC. = Cheiranthus L. (Crucif.).
Dilwynia Pers. = Dillwynia Sm. (Legumin.).
Dimacria Lindl. = Pelargonium L'Hérit. (Geraniac.).
Dimanisa Rafin. = Dianthera L. = Justicia L. (Acanthac.).
Dimeiandra Rafin. = Amaranthus L. (Amaranthac.).
Dimeianthus Rafin. = praec.
Dimeiostemon Rafin. = Andropogon L. (Gramin.).
Dimeium Rafin. = Zanthoxylum L. (Rutac.).
Dimejostemon P. & K. = Dimeiostemon Rafin. = Andropogon L. (Gramin.).
Dimenops Rafin. = Krameria Loefl. (Krameriac.).
Dimenostemma Steud. = Dimerostemma Cass. (Compos.).
Dimerandra Schlechter = Epidendrum L. (Orchidac.).
Dimeresia A. Gray. Compositae. 1 W. U.S. Head 2-fld., each in br.
Dimereza Labill. = Guioa Cav. (Sapindac.).
Dimeria R.Br. [1810; non *Dimera* Fries 1825 – Fungi]. Gramineae. 40 Masc.,
SE. As., Indomal., Austr., Polynes. Spikelet 1-fld.
Dimeria Endl. (sphalm.) = Dimesia Rafin. = Hierochloë R.Br. (Gramin.).
Dimerocarpus Gagnep. = Streblus Lour. (Morac.).
Dimerocostus Kuntze. Costaceae. 8 Panamá, trop. S. Am.
Dimerodiscus Gagnep. = Ipomoea L. (Convolvulac.).
Dimerostemma Cass. Compositae. 6 trop. S. Am.
Dimesia Rafin. = Hierochloë R.Br. (Gramin.).
Dimetia Meissn. = Hedyotis L. (Rubiac.).

DIMETOPIA

Dimetopia DC. = Trachymene Rudge (Umbellif.).
Dimetra Kerr. Verbenaceae. 1 Siam.
Dimia Rafin. = Diospyros L. (Ebenac.).
Dimia Spreng. = Daemia R.Br. (Asclepiadac.).
Dimitopia D. Dietr. = Dimetopia DC. = Trachymene Rudge (Umbellif.).
Dimocarpus Lour. Sapindaceae. 5 Indomal.
Dimopogon Rydb. = Drimopogon Rafin. (Rosac.).
Dimorpha D. Dietr. = Diamorpha Nutt. (Crassulac.).
Dimorpha Schreb. = Eperua Aubl. (Legumin.).
Dimorphandra Schott. Leguminosae. 25 trop. Am. Episepalous sta. reduced to stds.
Dimorphanthera F. Muell. Ericaceae. 60 Philippines to E. Malaysia (esp. New Guinea).
Dimorphanthes Cass. = Conyza Less. (Compos.).
Dimorphanthes Meissn. = seq.
Dimorphanthus Miq. = Aralia L. (Araliac.).
Dimorphocalyx Hook. f. (sphalm.) = Dimorphochlamys Hook. f. = Momordica L. (Cucurbitac.).
Dimorphocalyx Thw. Euphorbiaceae. 12 Indomal., Austr.
Dimorphochlamys Hook. f. = Momordica L. (Cucurbitac.).
Dimorphochloa S. T. Blake. Gramineae. 1 Queensland.
Dimorphocladium Britton = Phyllanthus L. (Euphorbiac.).
Dimorphocoma F. Muell. & Tate. Compositae. 1 C. Austr.
Dimorpholepis (G. M. Barroso) R. M. King & H. Rob. = Graziela R. M. King & H. Rob. (Compos.).
Dimorpholepis A. Gray = Helipterum DC. (Compos.).
Dimorphopetalum Bertero = Tetilla DC. (Francoac.).
Dimorphopteris Tagawa & Iwatsuki = Cyclosorus Link (Thelypteridac.). Fertile pinnae much contracted.
Dimorphorchis Rolfe = Arachnis Bl. (Orchidac.).
Dimorphostachys Fourn. = Panicum L. (Gramin.).
Dimorphostemon Kitagawa = Sisymbrium L. (Crucif.).
***Dimorphotheca** Moench. Compositae. 7 S. Afr. There are two kinds of fr. on the head (cf. *Calendula*).
Dinacanthon P. & K. = Deinacanthon Mez (Bromeliac.).
Dinacria Haw. (~ Crassula L.). Crassulaceae. 3 S. Afr.
Dinaeba Delile = Dinebra DC. = Botelua Lag. (Gramin.).
Dinanthe P. & K. = Deinanthe Maxim. (Hydrangeac.).
Dineba Beauv. = seq.
Dinebra DC. = Botelua Lag. (Gramin.).
Dinebra Jacq. Gramineae. 1 trop. Afr., As.
Dinema Lindl. Orchidaceae. 1 C. Am., W.I.
Dinemagonum A. Juss. Malpighiaceae. 3 Chile.
Dinemandra A. Juss. ex Endl. Malpighiaceae. 6 Peru, Chile.
Dinetopsis Roberty (~ Porana Burm. f.). Convolvulaceae. 1 E. Himal.
Dinetus Buch.-Ham. ex D. Don = Porana Burm. f. (Convolvulac.).
Dinizia Ducke. Leguminosae. 1 Amaz. Brazil.
Dinklagea Gilg = Manotes Soland. ex Planch. (Connarac.).

Dinklageanthus Melch. ex Mildbr. Bignoniaceae. 1 trop. W. Afr. (Liberia).

Dinklageëlla Mansf. Orchidaceae. 2 trop. W. Afr.

Dinklageodoxa Heine & Sandwith. Bignoniaceae. 1 W. trop. Afr. (Liberia).

Dinocanthium Bremek. Rubiaceae. 3 trop. & S. Afr.

Dinochloa Buese. Gramineae. 20 SE. As., Indomal.

Dinophora Benth. Melastomataceae. 2 trop. W. Afr.

Dinoseris Griseb. = Hyaloseris Griseb. (Compos.).

Dinosma P. & K. = Deinosmos Rafin. = Pulicaria Gaertn. (Compos.).

Dintera Stapf. Scrophulariaceae. 1 trop. Afr.

Dinteracanthus C. B. Clarke ex Schinz. Acanthaceae. 2 S. Afr.

Dinteranthus Schwantes. Aïzoaceae. 5 S. Afr.

Dioclea Kunth. Leguminosae. 50 trop. Am., 1 trop. Afr. & As.

Dioclea Spreng. = Arnebia Forsk. (Boraginac.).

Dioctis Rafin. = Polygonum L. (Polygonac.).

Diodeilis Rafin. = Clinopodium L. (Labiat.).

Diodella Small (~ Diodia Gronov. ex L.). Rubiaceae. 1 Florida, W.I.

Diodia Gronov. ex L. Rubiaceae. 50 trop. & subtrop. *D. maritima* Schumacher & Thonn. is common to Am. and Afr.

Diodioïdes Loefl. = Spermacoce L. (Rubiac.).

Diodoïs Pohl = Psyllocarpus Mart. & Zucc. (Rubiac.).

Diodonta Nutt. = Coreopsis L. (Compos.).

Diodontium F. Muell. = Glossogyne Cass. (Compos.).

Diodontocheilis Rafin. = Diodeilis Rafin. = Clinopodium L. (Labiat.).

Diodosperma H. Wendl. = Trithrinax Mart. (Palm.).

Diogenesia Sleum. Ericaceae. 11 Andes.

Diogoa Exell & Mendonça. Olacaceae. 1 trop. Afr.

Dioicodendron Steyerm. Rubiaceae. 2 NW. trop. S. Am.

Diolena Naud. = Triolena Naud. (Melastomatac.).

Diolotheca Rafin. = Diototheca Rafin. = Phyla Lour. (Verbenac.).

Diomedea Bertol. ex Colla = Helianthus L. (Compos.).

Diomedea Cass. = seq.

Diomedella Cass. = Borrichia Adans. (Compos.).

Diomedes Haw. = Narcissus L. (Amaryllidac.).

Diomedia Willis (sphalm.) = Diomedea Cass. = Borrichia Adans. (Compos.).

Diomma Engl. ex Harms = Spathelia L. (Rutac.).

Dion Lindl. = Dioön Lindl. corr. Miq. (Zamiac.).

Dionaea Ellis. Droseraceae. 1 SE. U.S., *D. muscipula* Ellis (Venus' fly-trap), in damp mossy places on the 'pine-barrens'. Short rhiz. bearing a rosette of ls., which lie close to the soil. Each has a lower and an upper blade; the former may be regarded as a winged petiole, the latter has a quadrangular shape and the margins project as long teeth close together. The two halves of this part of the l. are bent upwards so as to present a flat V-form in section. The edge of each half is green, the inner part of the surface is covered with reddish dots, which under the microscope are seen to be digestive glands; unless stimulated, no secretion is carried on. On each half of the l. are three long hairs—the trigger-hairs—jointed at the base so that they fold downwards when the l. closes. Two or three slight touches to one of these, or a vigorous stimulus to the surface of the l., causes an immediate closing. The teeth cross one another,

DIONAE[AC]EAE

and if an insect cause the movement, it is thus captured. The closing of the l. still continues till the two halves are tightly squeezed together. Then the digestive glands commence to secrete a ferment, which acts upon the proteids of the prey and renders them soluble, when they are absorbed by the l. (cf. *Drosera*). When the process is complete the l. opens again.

Dionae[ac]eae Dum. = Droseraceae Salisb.

*Dioncophyllaceae Airy Shaw. Dicots. 3/3 W. Afr. Soft-wooded lianes, sometimes ± rusty-scurfy. Ls. alt., simple, ent. or cren., exstip., usu. with midrib excurrent into 2 recurved hooks; in 1 gen. (*Triphyophyllum*) 2 additional types of l. are produced: (1) normal, but without apical hooks, (2) partly or wholly reduced to midrib, which is beset with ∞ conspic. stalked or sessile glands, and l. is circinate in vernation; nerves close, parallel, spreading. Infl. a lax ± supra-axill. cyme, bracts large or small, fls. moderate-sized, reg., ♀. K 5, valv. or open. C 5, contorted, either thick or delicate, white. Disk o. A 10(–30), equal or unequal, short or elongate, anth. ovoid or oblong. G (2) or (5), 1-loc., opening loculic. at a very early stage and exposing the ∞ anatr. ov. on pariet. plac.; styles either 2, free, filif., with capit. stig., or 5, free, filif. with minute stig., or 5, slightly conn. at base, with plumose stig. Fr. of 2 or 5 valves, enlarging or not as seeds mature, spreading widely and bearing the few seeds peltately attached on greatly elongate, thickened, rigid funicles. Seeds large, discoid, either thick and surr. by narrower wing, or thin and surr. by broad satiny wing; embryo large, discoid-obconic, axile, mostly surr. by the copious pileiform endosp. Genera: *Dioncophyllum*, *Triphyophyllum*, *Habropetalum*. A relict group of great interest, showing relationships with *Nepenthac.*, *Droserac.* and *Ancistrocladac.* (R. Schmid, *Bot. Jahrb.* (Engl.), **83**: 1–56, 1964, and refs.)

Dioncophyllum Baill. Dioncophyllaceae. 1 W. Equat. Afr.

Dionea Rafin. = Dionaea Ellis (Droserac.).

Dioneidon Rafin. = Diodia L. (Rubiac.).

Dionycha Naud. Melastomataceae. 2 Madag.

Dionychastrum A. & R. Fernandes. Melastomataceae. 1 trop. E. Afr.

Dionychia auctt. = Dionycha Naud. (Melastomatac.).

Dionysia Fenzl. Primulaceae. 41 mts. C. As., N. Iraq, Persia, Afghan.

Dionysis Thou. (uninom.) = *Diplecthrum dionysii* Thou. = ? (Orchidac.).

Dioön Lindl. corr. Miq. Zamiaceae. 3–5 Mexico, C. Am. The seeds are ground into meal, which contains much starch.

Diopogon Jord. & Fourr. = Jovibarba Opiz = Sempervivum L. (Crassulac.).

Diorimasperma Rafin. = Cleome L. (Cleomac.).

Dioryktandra Hassk. = Rinorea Aubl. (Violac.).

Dioryktandra Hassk. ex Bakh. = Diospyros L. (Ebenac.).

Diosanthos St-Lag. = Dianthus L. (Caryophyllac.).

Dioscorea L. Dioscoreaceae. 600 trop. & subtrop.; 3 spp. in the Pyrenees, Balkan Penins. & Caucasus. They have twining annual stems arising from tubers which in different spp. are of different morphological nature. In *D. batatas* Decne, etc., the tuber arises by a lateral hypertrophy of the hypocotyl, and is variously regarded as a rhiz. or a root; in *D. sinuata* Vel., etc., it arises by lateral hypertrophy of the internodes above the cotyledon; in *D. pentaphylla* L., etc., it arises from the internode just above the cotyledon together with the hypocotyl, whilst in *D. villosa* L., *D. quinqueloba* Thunb., etc., there is a fleshy rhiz. The tubers are known as yams; they contain much starch and are largely cult. for food in trop., esp. Am. The best are perhaps *D. alata* L. (white yam), *D. cayennensis* Link (negro yam), *D. trifida* L. f. (*cush-cush*; *yampi*). They are propagated by 'eyes' like potatoes. Small axillary tubers often form on the main stem and may also be used. 3 S. Afr. spp. comprising the Sect. *Testudinaria*, incl. *D. elephantipes* Salisb. (Hottentot bread), have an enormous tuber, the swollen first internode of the stem, projecting out of the soil, with a thick outer coating of cork. From it yearly, during the wet season, springs by adv. budding the year's shoot, a long thin climbing stem with large ls. and small fls. This dies down in the dry season, and the corky covering protects the mass of the plant from drought.

*Dioscoreaceae R.Br. Monocots. 5/750, trop. & warm temp. Climbing herbs or shrubs with tubers or rhizomes at the base (morphology varied; see gen.). Ls. alt., net-veined, often arrow-shaped; infl. racemose; fls. reg., ♂♀, dioec., inconspic. P (6), tubular at base; A 6, or 3 and 3 stds.; Ḡ (3), usu. 3-loc. with axile, rarely 1-loc. with parietal, plac.; ov. usu. 2 in each loc., anatr. one above the other. Capsule or berry; embryo in horny endosp. Genera: *Dioscorea, Tamus, Rajania, Stenomeris, Avetra*. (Cf. Burkill in *J. Linn. Soc., Bot.* **56**: 319–412, 1960.)

Dioscoreophyllum Engl. Menispermaceae. 10 trop. Afr.

Dioscoreopsis Kuntze = praec.

Dioscorida St-Lag. = Dioscorea L. (Dioscoreac.).

Diosma L. Rutaceae. 15 S. Afr. Heath-like xerophytes.

Diosm[ac]eae R.Br. = Rutaceae–Diosmeae DC.

Diospermum Hook. f. (sphalm.) = Oïospermum Less. (Compos.).

Diosphaera Buser (~ Tracheliopsis Buser). Campanulaceae. 3 E. Medit.

Diospyraceae Van Tiegh. = Ebenaceae Vent.

Diospyros L. Ebenaceae. 500 warm. Many spp. yield the valuable wood ebony. The sapwood is white and soft, the heartwood hard and black. *D. reticulata* Willd. (Mauritius) and *D. ebenum* Koen. (Ceylon) yield the finest ebony. *D. quaesita* Thw. (Ceylon) yields calamander wood. *D. embryopteris* Pers. (*gaub*; India) fr. contains a sticky pulp, used for caulking. *D. kaki* L. f. (Chinese date-plum, persimmon) and *D. lotus* L. (date-plum, temp. As.) fr. is used as a sweetmeat when dried. *D. virginiana* L. (N. Am. ebony or persimmon, U.S.) cult. for both wood and fr.

Diospyros Roxb. = praec.

Diostea Miers (~ Baillonia Bocq.). Verbenaceae. 3 S. Am.

Diotacanthus Benth. Acanthaceae. 2 Indomal.

Diothilophis Schlechtd. = Dothilophis Rafin. = Epidendrum L. (Orchidac.).

Diothonea Lindl. Orchidaceae. 7 W. trop. S. Am.

Dioticarpus Dunn = Hopea Roxb. (Dipterocarpac.).

Diotis Desf. = Otanthus Hoffmgg. & Link (Compos.).

Diotis Schreb. = Eurotia Adans. (Chenopodiac.).

Diotocarpus Hochst. = Pentanisia Harv. (Rubiac.).

Diotocranus Bremek. Rubiaceae. 1 C. Afr.

Diotolotus Tausch = Argyrolobium Eckl. & Zeyh. (Legumin.).

Diotosperma A. Gray = Ceratogyne Turcz. (Compos.).

Diotostemon Salm-Dyck = Pachyphytum Link, Klotzsch & Otto (Crassulac.).

DIOTOSTEPHUS

Diotostephus Cass. = Chrysogonum L. (Compos.).
Diototheca Rafin. = Phyla Lour. (Verbenac.).
Diouratea Van Tiegh. = Ouratea Aubl. (Ochnac.).
Dioxippe M. Roem. = Glycosmis Corrêa (Rutac.).
Dipanax Seem. Araliaceae. 2–3 Hawaii.
Dipcadi Medik. Liliaceae. 55 Medit., Afr., Madag., Penins. Ind.
Dipcadioïdes Medik. = Lachenalia Jacq. (Liliac.).
Dipelta Maxim. Caprifoliaceae. 4 China.
Dipelta Regel & Schmalh. = Astragalus L. (Legumin.).
Dipentaplandra Kuntze = Pentadiplandra Baill. (Pentadiplandrac.).
Dipentodon Dunn. Dipentodontaceae. 1 E. Himal., Burma, SW. China.
***Dipentodontaceae** Merr. Dicots. 1/1 NE. India, China. Small decid. tree;
ls. alt., ovate, dentic., with decid. stips. Fls. small, ⚥, in dense globose long-
pedunc. axillary umbels, with 4–5 small decid. bracts at base, pedicels slender,
artic. mid-way. K 5–7, ± imbr., lin., shortly united at base, pubesc.; C 5–7,
± imbr., lin., slightly narrower than K or indistinguishable; disk-glands 5–7,
opp. C; A 5–7, opp. K, erect, with small anth.; G̲ (3), 1-loc. above, with 2 axile
ov. per loc. on free basal plac., and simple style with small capit. stig. Fr.
a small oblong 1-loc. 1-seeded toment. caps. with persist. style, tardily dehisc.,
surr. by persist. K, C, & A. Only genus: *Dipentodon*. Despite obvious differences, probably related to *Homalium* (*Flacourtiac.*). The shortly tomentose
caps. may be compared with those of *Poliothyrsis*, *Carrierea*, etc., in the same
fam. There is also a remarkable agreement in general habit with *Nyssa* (*Nyssac.*).
Dipera Spreng. = Disperis Sw. (Orchidac.).
Diperis Wight = praec.
Diperium Desv. = Mnesithea Kunth (Gramin.).
Dipetalanthus A. Chev. = Hymenostegia (Benth.) Harms (Legumin.).
Dipetalia Rafin. = Oligomeris Cambess. (Resedac.).
Dipetalon Rafin. = Cuphea P.Br. (Lythrac.).
Dipetalum Dalz. = Toddalia Juss. (Rutac.).
Diphaca Lour. = Ormocarpum Beauv. (Legumin.).
Diphalangium Schau. Alliaceae. 1 Mexico.
Diphasia Pierre. Rutaceae. 6 trop. Afr., Madag.
Diphasiopsis Mendonça. Rutaceae. 1 trop. E. Afr.
Diphasium Presl = Lycopodium L. (Lycopodiac.).
Dipherocarpus Llanos = Nephelium L. (Sapindac.).
***Dipholis** A. DC. (= Spondogona Rafin.). Sapotaceae. 20 Florida, W.I.
C-lobes 3-fid.
Diphorea Rafin. = Sagittaria L. (Alismatac.).
Diphragmus Presl = Spermacoce L. (Rubiac.).
Diphryllum Rafin. = Listera R.Br. (Orchidac.).
Diphyes Blume = Bulbophyllum Thou. (Orchidac.).
Diphylax Hook. f. Orchidaceae. 1 NE. India, China.
Diphyleia Rafin. = Diphylleia Michx (Podophyllac.).
Diphyllanthus Van Tiegh. = Ouratea Aubl. (Ochnac.).
Diphyllarium Gagnep. Leguminosae. 1 Indoch.
Diphylleia Michx. Podophyllaceae. 3 W. China, Japan, Atl. N. Am.
Diphylleiaceae Schultz–Schultzenst. = Podophyllaceae + Sarraceniaceae.

Diphyllopodium Van Tiegh. = Ouratea Aubl. (Ochnac.).
Diphyllum Rafin. = Diphryllum Rafin. = Listera R.Br. (Orchidac.).
Diphysa Jacq. Leguminosae. 15 Mexico, C. Am.
Diphystema Neck. = Amasonia L. f. (Verbenac.).
Dipidax Laws. ex Salisb. Liliaceae. 4 S. Afr.
Diplachne Beauv. Gramineae. 15 trop. & subtrop.
Diplachne R.Br. = Verticordia DC. (Myrtac.).
Diplachyrium Nees = Muhlenbergia Schreb. (Gramin.).
Diplacorchis Schlechter = Brachycorythis Lindl. (Orchidac.).
Diplacrum R.Br. Cyperaceae. 6 trop.
Diplactis Rafin. = Aster L. (Compos.).
Diplacus Nutt. = Mimulus L. (Scrophulariac.).
Dipladenia A. DC. Apocynaceae. 30 trop. S. Am. Most are lianes climbing
Diplandra Bert. = Elodea Michx (Hydrocharitac.).   [by hooks.
Diplandra Hook. & Arn. Onagraceae. 1 Mexico.
Diplandra Rafin. = Ludwigia L. (Onagrac.).
Diplanoma Rafin. = Abelmoschus Medik. (Malvac.).
Diplanthemum K. Schum. = Duboscia Bocq. (Tiliac.).
Diplanthera Banks & Soland. ex R.Br. = Deplanchea Vieill. (Bignoniac.).
Diplanthera Gled. = Dianthera L. = Justicia L. (Acanthac.).
Diplanthera Rafin. = Platanthera Rich. (Orchidac.).
Diplanthera Thou. = Halodule Endl. (Cymodoceac.).
Diplarchaceae Klotzsch = Diapensiaceae–Diplarcheae Airy Shaw.
Diplarche Hook. f. & Thoms. Ericaceae. 2 E. Himal., SW. China. (Cf.
    Stevens, *Bot. J. Linn. Soc.* **64**: 27–9, 1971.)
Diplaria Rafin. ex DC. = Cassandra D. Don (Ericac.).
Diplarinus Rafin. = Scirpus L. (Cyperac.).
Diplarpea Triana. Melastomataceae. 1 Colombia.
Diplarrena Labill. = seq.
Diplarrhena Labill. corr. R. Br. Iridaceae. 2 S. Austr., Tasm.
Diplarrhinus Endl. = Diplarinus Rafin. = Scirpus L. (Cyperac.).
Diplasanthera Hook. f. = seq.
Diplasanthum Desv. = Andropogon L. (Gramin.).
Diplasia Rich. Cyperaceae. 1 Indoch., 2 trop. S. Am., W.I.
Diplaspis Hook. f. = Huanaca Cav. (Umbellif.).
Diplatia Van Tiegh. Loranthaceae. 3 trop. Austr.
Diplax Soland. ex Benn. = Microlaena R.Br. (Gramin.).
Diplaziopsis C. Chr. Athyriaceae. 4 E. As., Polynesia.
Diplazium Sw. Athyriaceae. 400 trop. & N. temp. United to *Athyrium* Roth
    by Copeland (*Gen. Fil.* 147, 1947).
Diplazoptilon Ling. Compositae. 1 SW. China.
Diplecosia G. Don = Diplycosia Blume (Ericac.).
Diplecthrum Pers. = Satyrium Sw. (Orchidac.).
Diplectraden Rafin. = Habenaria Willd. (Orchidac.).
Diplectria Reichb. Melastomataceae. 4 Siam, Malaysia.
Diplectrum Thou. = Diplecthrum Pers. = Satyrium Sw. (Orchidac.).
Diplegnon P. & K. = Diplolegnon Rusby (Gesneriac.).
Dipleina Rafin. = Actaea L. (Ranunculac.).

Diplemium Rafin. = Erigeron L. (Compos.).
Diplerisma Planch. = Melianthus L. (Melianthac.).
Diplesthes Harv. = Salacia L. (Celastrac.).
Dipliathus Rafin. = Licaria Aubl. (Laurac.).
Diplicosia Endl. = Diplycosia Blume (Ericac.).
Diplima Rafin. = Salix L. (Salicac.).
Diplisca Rafin. = Colubrina Rich. (Rhamnac.).
Diploblechnum Hayata = Blechnum L. (Blechnac.).
Diplocalymma Spreng. = Thunbergia Retz. (Thunbergiac.).
Diplocalyx Presl = Mitraria Cav. (Gesneriac.).
Diplocalyx A. Rich. = Schoepfia Schreb. (Olacac.).
Diplocardia Zipp. ex Blume = Pometia J. R. & G. Forst. (Sapindac.).
Diplocarex Hayata. Cyperaceae. 1 Formosa.
Diplocaulobium Kraenzl. Orchidaceae. 70 Malaya to Austr. & Pacif. Is.
Diplocea Rafin. (1817) = Salsola L. (Chenopodiac.).
Diplocea Rafin. (1818) = Triplasis Beauv. (Gramin.).
Diploceleba P. & K. = Diplokeleba N. E. Br. (Sapindac.).
Diplocentrum Lindl. Orchidaceae. 2 India.
Diploceras Meissn. = Pavolinia Webb (Crucif.).
Diplochaete Nees = Rhynchospora Vahl (Cyperac.).
Diplochilus Lindl. = Diplomeris D. Don (Orchidac.).
Diplochita DC. = Miconia Ruiz & Pav. (Melastomatac.).
Diplochiton Spreng. = praec.
Diplochlaena Spreng. = Diplolaena R.Br. (Rutac.).
Diplochlamys Muell. Arg. = Mallotus Lour. (Euphorbiac.).
Diplochonium Fenzl = Trianthema L. (Aïzoac.).
Diploclinium Lindl. = Begonia L. (Begoniac.).
Diploclisia Miers. Menispermaceae. 4 Indomal., SE. As., China.
Diplocnema P. & K. = Diploknema Pierre (Sapotac.).
Diplococea Reichb. = Diplocea Rafin. (1818) = Triplasis Beauv. (Gramin.).
Diplocoma D. Don = Heterotheca Cass. (Compos.).
Diploconchium Schau. = Agrostophyllum Blume (Orchidac.).
Diplocos Bureau = Streblus Lour. (Morac.).
Diplocrater Benth. = Cathedra Miers (Olacac.).
Diplocrater Hook. f. = Tricalysia A. Rich. ex DC. (Rubiac.).
Diplocyatha N. E. Br. Asclepiadaceae. 1 S. Afr.
Diplocyathium H. Schmidt = Euphorbia L. (Euphorbiac.).
Diplocyclos (Endl.) P. & K. corr. C. Jeffrey. Cucurbitaceae. 3 trop. Afr., 1 trop. Afr. & Indomal.
Diplocyclus (Endl.) P. & K. = praec.
Diplodiscus Turcz. Tiliaceae. 7 W. Malaysia.
Diplodium Sw. = Pterostylis R.Br. (Orchidac.).
Diplodon DC. = Diplusodon Pohl (Lythrac.).
Diplodonta Karst. = Heterotrichum DC. (Melastomatac.).
Diplodontaceae Dulac = Lythraceae Jaume St-Hil.
Diplofatsia Nakai (~ Fatsia Decne & Planch.). Araliaceae. 1 Formosa.
Diplofractum Walp. = Diplophractum Desf. = Colona Cav. (Tiliac.).
Diplogama Opiz = Otites Adans. = Silene L. (Caryophyllac.).

DIPLOPTERYS

Diplogastra Welw. ex Reichb. = Platylepis A. Rich. (Orchidac.).
Diplogatha K. Schum. (sphalm.) = Diplocyatha N. E. Br. (Asclepiadac.).
Diplogenaea A. Juss. = seq.
Diplogenea Lindl. = Medinilla Gaudich. (Melastomatac.).
Diploglossis Benth. & Hook. f. = seq.
Diploglossum Meissn. = Cynanchum L. (Asclepiadac.).
Diploglottis Hook. f. Sapindaceae. 1–2 Austr.
Diplogon Rafin. = Chrysopsis (Nutt.) Ell. (Compos.).
Diplokeleba N. E. Br. Sapindaceae. 2 S. trop. S. Am.
Diploknema Pierre. Sapotaceae. 7 Indomal. The seeds of *D. butyracea* (Roxb.) H. J. Lam (Indian butter-tree) yield a butter-like substance, used for soap-making, etc.
Diplolabellum Maekawa. Orchidaceae. 1 Korea.
Diplolaena R.Br. Rutaceae. 6–8 W. Austr.
Diplolaen[ac]eae J. G. Agardh = Rutaceae–Boronieae–Diplolaeninae Engl.
Diplolegnon Rusby. Gesneriaceae. 1 S. Am.
Diplolobium F. Muell. = Swainsona Salisb. (Legumin.).
Diploloma Schrenk = Craniospermum Lehm. (Boraginac.).
Diplolophium Turcz. Umbelliferae. 5 trop. Afr.
Diploma Rafin. = Gentiana L. (Gentianac.).
Diplomeris D. Don. Orchidaceae. 3 Himal.
Diplomorpha Griff. = Synostemon F. Muell. (Euphorbiac.).
Diplomorpha Meissn. ex C. A. Mey. = Wikstroemia Endl. (Thymelaeac.).
Diplonema G. Don = Euclea L. (Ebenac.).
Diplonix Rafin. = seq.
Diplonyx Rafin. = Wisteria Nutt. (Legumin.).
Diploön Cronquist. Sapotaceae. 1 trop. S. Am.
Diploöphyllum v. d. Bosch = Mecodium Presl (Hymenophyllac.).
Diplopanax Hand.-Mazz. Araliaceae. 1 China.
Diplopappus Cass. = Aster L. (Compos.).
Diplopapus Rafin. = praec.
Diplopeltis Endl. Sapindaceae. 4 Austr.
Diplopenta Alef. = Pavonia Cav. (Malvac.).
Diploperianthium Ritter = Calymmanthium Ritter (Cactac.).
Diplopetalon Spreng. = Cupania L. (Sapindac.).
Diplophractum Desf. = Colona Cav. (Tiliac.).
Diplophragma Meissn. = Hedyotis L. (Rubiac.).
Diplophyllum Lehm. = Veronica L. (Scrophulariac.).
Diplopia Rafin. = Salix L. (Salicac.).
Diplopilosa Dvořák. Cruciferae. 1 SE. As. Min.
Diplopogon R.Br. Gramineae. 1 W. Austr.
Diploprion Viv. = Medicago L. (Legumin.).
Diploprora Hook. f. Orchidaceae. 5 trop. As.
Diploptera C. A. Gardner = Strangea Meissn. (Proteac.).
Diplopterygium (Diels) Nakai. Gleicheniaceae. 20 trop. & subtrop. Asia to Polynesia, 1 trop. Am. Large ferns forming thickets on mts.; l.-branches bipinnatifid, not pseudo-dichot.
Diplopterys A. Juss. Malpighiaceae. 10 warm S. Am., W.I.

373

DIPLOPTERYX

Diplopteryx Dalla Torre & Harms = praec.
Diplopyramis Welw. = Oxygonum Burch. (Polygonac.).
Diplora Baker. Aspleniaceae. 4 Malaysia, Melanesia. Like *Phyllitis* Ludw. in having simple ls. with sori in facing pairs. Copeland, *Gen. Fil.*, 165, includes in *Asplenium* L.
Diplorhipis Drude = Mauritia L. f. (Palm.).
Diplorhynchus Welw. ex Ficalho & Hiern. Apocynaceae. 1 trop. Afr.
Diplorrhiza Ehrh. (uninom.) = *Satyrium viride* L. = *Coeloglossum viride* (L.) Hartm. (Orchidac.).
Diplosastera Tausch = Coreopsis L. (Compos.).
Diploscyphus Liebm. = Scleria Bergius (Cyperac.).
Diplosiphon Decne = Blyxa Thou. (Hydrocharitac.).
Diplosoma Schwantes. Aïzoaceae. 2 S. Afr. [Austr.
Diplospora DC. (~ Tricalysia A. Rich.). Rubiaceae. 25 China, Indomal.,
Diplosporopsis Wernham = Belonophora Hook. f. (Rubiac.).
Diplostachyum P. Beauv. = Selaginella P. Beauv. (Selaginellac.).
Diplostegium D. Don = Tibouchina Aubl. (Melastomatac.).
Diplostelma Rafin. = Chaetopappa DC. (Compos.).
Diplostemma DC. = Diphystema Neck. = Amasonia L. f. (Verbenac.).
Diplostemma Steud. & Hochst. ex DC. = Geigeria Griesselich (Compos.).
Diplostemon DC. ex Steud. = Ammannia L. (Lythrac.).
Diplostephion Rafin. = Dyssodia Cav. (Compos.).
Diplostephium Kunth. Compositae. 90 trop. Andes; 1 Madag.?
Diplostigma K. Schum. Asclepiadaceae. 1 E. Afr.
Diplostylis Karst. & Triana = Rochefortia Sw. (Boraginac.).
Diplostylis Sond. = Adenocline Turcz. (Compos.).
Diplosyphon Matsum. = Diplosiphon Decne = Blyxa Thou. (Hydrocharitac.).
Diplotaenia Boiss. = Peucedanum L. (Umbellif.).
Diplotax Rafin. = Cassia L. (Legumin.).
Diplotaxis DC. Cruciferae. 27 Eur., Medit.
Diplotaxis Wall. ex Kurz = Chisocheton Bl. (Meliac.).
Diploter Rafin. = Tetracera L. (Dilleniac.).
Diplotheca Hochst. = Astragalus L. (Legumin.).
Diplothemium Mart. = Allagoptera Nees (Palm.).
Diplothorax Gagnep. = Streblus Lour. (Morac.).
Diplothria Walp. (sphalm.) = seq.
Diplothrix DC. = Zinnia L. (Compos.).
Diplotropis Benth. Leguminosae. 12 trop. Am.
Diplousodon Meissn. = Diplusodon Pohl (Lythrac.).
Diplukion Rafin. = Iochroma Benth. (Solanac.).
Diplusion Rafin. = Salix L. (Salicac.).
Diplusodon Pohl. Lythraceae. 50 Brazil.
Diplycosia Blume. Ericaceae. 60 Malaysia.
Dipodium R.Br. Orchidaceae. 22 Malaysia, Palau Is., Solomon Is., New Hebr., New Caled.
Dipodophyllum Van Tiegh. = Loranthus L. (Loranthac.).
Dipogon Durand = Diopogon Jord. & Fourr. = Sempervivum L. (Crassulac.).
Dipogon Liebm. Leguminosae. 1 S. Afr.

DIPTEROCARPACEAE

Dipogon Willd. ex Steud. = Chrysopogon Trin. (Gramin.).

Dipogonia Beauv. = Diplopogon R.Br. (Gramin.).

Dipoma Franch. Cruciferae. 2 SE. Tibet, SW. China.

Diporidium Wendl. f. ex Bartl. & Wendl. f. = Ochna L. (Ochnac.).

Diporochna Van Tiegh. = Ochna L. (Ochnac.).

Diposis DC. Hydrocotylaceae (~ Umbellif.). 3 temp. S. Am.

*Dipsacaceae Juss. Dicots. 8/150, chiefly N. temp. Euras. & trop. & S. Afr. Most are herbs with opp. exstip. ls. (connate in *Dipsacus*), and dense heads of fls. That the heads are actually cymose is indicated by the fact that the fls. open both centripetally and centrifugally. The outer fls. usu. have the corolla more or less drawn out abaxially (cf. *Compos.*, *Crucif.*, etc.). Most have an epicalyx, a cup-shaped organ springing from the base of the ovary, and usu. regarded as composed of the two united bracteoles; bracteoles of the ordinary kind are rare. K and C 5-merous, or 4-merous by union of two members; A 4, epipetalous; Ḡ (2), 1-loc. with one pend. anatr. ov. Fls. usu. protandr. Fr. an achene (cf. *Compos.*) usu. enclosed in the epicalyx; endosperm. Chief genera: *Dipsacus, Cephalaria, Knautia, Scabiosa, Pterocephalus*.

Dipsacella Opiz = Virga Hill (Dipsacac.).

Dipsacozamia Lehm. ex Lindl. = Ceratozamia Brongn. (Zamiac.).

Dipsacus L. Dipsacaceae. 15 Euras., Medit., trop. Afr. The connate leaves form troughs round the stem in which rain-water collects. The protandr. fls. are chiefly visited by bees. *D. fullonum* L. ssp. *sativus* (L.) Thell. (fuller's teasel) has hooked bracts; the fr.-heads are used for raising the nap upon cloth.

Dipseudochorion Buchen. = Limnophyton Miq. (Alismatac.).

Diptanthera Schrank ex Steud. = Diplanthera Gled. = Justicia L. (Acanthac.).

Diptera Borckh. = Saxifraga L. (Saxifragac.).

Dipteracanthus Nees (emend. Bremek.). Acanthaceae. 10 trop. E. Afr., As.,

Dipteraceae Lindl. = Dipterocarpaceae Bl. [Austr.

Dipteranthemum F. Muell. Amaranthaceae. 1 W. Austr.

Dipteranthus Barb. Rodr. Orchidaceae. 8 trop. S. Am.

Dipterella Moggi. Cruciferae. 1 NE. trop. Afr.

Dipteridaceae Seward & Dale. Polypodiales. Ls. divided into 2 flabellate halves. Only genus: *Dipteris*.

Dipteris Reinw. Dipteridaceae. 8 As., Polynes., in open places on mts. Related to Jurassic fossils (*Dictyophyllum, Haussmannia*).

Dipterix Willd. = Dipteryx Schreb. (Legumin.).

Dipterocalyx Cham. = Lippia L. (Verbenac.).

*Dipterocarpaceae Bl. Dicots. 15/580 palaeotrop., chiefly Indomal. Trees, usu. tall, with large spreading sympodially branched emergent crowns, all at first monopodial and a few remaining so. Freq. gregarious in areas with marked dry season or on poor soils; dominating as a family, but with no single species dominance, the evergreen mixed lowland tropical rain forests, often called Mixed Dipterocarp forests, of humid tropical Indomal. Ls. entire, leathery, alt., stip., usu. prominently penninerved, freq. with domatia in axils of nerves, usu. evergreen. Infl. paniculate, rarely cymose (*Upuna*, some *Vatica*), term. or axill., with caduc. br.; fls. ♀, reg., 5-merous, fragrant. Twigs, infl., K, C, G, young parts and freq. other parts usu. hairy; hairs mainly unicellular, aciculate and usu. fascicled or stellate, peltate, or emarginate and

375

DIPTEROCARPUS

single; freq. also with caduc. multicellular long-stalked or capitate hairs. K 5, imbr. or valv., freq. with short or long tube free or adnate to ov.; C 5, convol., often connate at base; A 5–10–15 or more, centrifugal, sometimes with androgynophore, usu. connate, freq. adnate to base of petals, connective usu. with term. process, anthers 2-loc., opening lengthwise; G, rarely semi-inf. (*Anisoptera*), (3), 3-loc., with 2 anatr. or pend. ov. in each; style entire or trifid, freq. on a stylopodium; stigma usu. obscure, 3- or 6-lobed. Fr. usu. a 1-seeded nut, usu. surrounded or enclosed by K with some or all seps. enlarged into wings. Usu. no endosperm. Cotyledons often twisted, lobed or laciniate and enclosing the radicle. Pericarp splitting irregularly or into 3 valves at germination. Recent work suggests a close relationship with *Tiliaceae*.

Classification and chief genera (after Gilg):

I. *Monotoïdeae* (androecium with gynophore; anther dorsifixed; wood without resin canals, with uniseriate rays): *Marquesia, Monotes* (trop. Afr.).

II. *Dipterocarpoïdeae* (androecium without gynophore; anther basi-fixed; wood with intercellular resin canals and multiseriate rays): *Dipterocarpus, Dryobalanops, Hopea, Shorea, Vateria, Vatica* (Indomal.).

Dipterocarpus Gaertn. f. Dipterocarpaceae. 76 Ceylon & India to W. Malaysia & Bali. Large amplexicaul stips. protect the young bud (cf. *Magnolia*, etc.). Stylopodium. Fr. encl. in K-tube freq. proliferated into ribs or flanges. Several yield wood-oil or *gurjun* balsam, a resin obtained by tapping, and used as a varnish. Many yield useful timber.

Dipterocome Fisch. & Mey. Compositae. 1 W. As. Fr. 2-winged.

Dipterocypsela Blake. Compositae. 1 Colombia.

Dipterodendron Radlk. Sapindaceae. 3 C. Am., Venez.

Dipteronia Oliv. Aceraceae. 2 C. & S. China. Mericarp winged all round.

Dipteropeltis Hallier f. Convolvulaceae. 2 trop. W. Afr.

Dipterosiphon Huber = Campylosiphon Benth. (Burmanniac.).

Dipterosperma Hassk. = Stereospermum Cham. (Bignoniac.).

Dipterospermum Griff. = Gordonia Ellis (Theac.).

Dipterostele Schlechter. Orchidaceae. 1 Ecuador.

Dipterostemon Rydb. = Brodiaea Sm. (Alliac.).

Dipterotheca Sch. Bip. = Aspilia Thou. (Compos.).

Dipterygia Presl = Asteriscium Cham. & Schlechtd. (Umbellif.).

Dipterygium Decne. Cruciferae (~ Capparidac.). 1 Egypt to W. Pakistan. Fr. a samara.

*Dipteryx Schreb. = Coumarouna Aubl. + Taralea Aubl. (Legumin.).

Diptychandra Tul. Leguminosae. 3 Brazil, Bolivia.

Diptychocarpus Regel & Schmalh. = Clausia Trotzky (Crucif.).

Diptychocarpus Trautv. Cruciferae. 1 C. As.

Diptychum Dulac = Sesleria Scop. (Gramin.).

Dipyrena Hook. Verbenaceae. 1 temp. S. Am. Ls. alt. Stones 2, 2-loc.

Dirachma Schweinf. ex Balf. f. Dirachmaceae. 1 Socotra.

Dirachmaceae (Reiche) Hutch. Dicots. 1/1 Socotra. Shrub with long and short shoots; ls. alt., simple, serr., with persist. stips. Fls. sol., axill., ☿, with epicalyx of 4 bracteoles. K 8, subul., valv.; C 8, contorted; A 8, opp. pets., with obl. anth.; G̲ (8), deeply 8-lobed, narrowed into style with 8 lin. stigs., and 1 basal axile ascending ov. per loc. Fr. capsular, of 8 ventr. dehisc. follicles, woolly within; seeds compr., shining; endosp. scanty. Only genus: *Dirachma*. Distant connection with *Greyiaceae*?

Diracodes Blume=Amomum L. (Zingiberac.).

Dirca L. Thymelaeaceae. 2 N. Am. K almost 0, C 0, A 8, G 1-loc.

Dircaea Decne=Corytholoma Decne (Gesneriac.).

Dirhacodes Lem.=Diracodes Blume=Amomum L. (Zingiberac.).

Dirhamphis Krapov. Malvaceae. 1 Boliv., Parag.

Dirhynchosia Blume=Spiraeopsis Miq. (Cunoniac.).

Dirichletia Klotzsch (~Carphalea Juss.). Rubiaceae. 10 trop. Afr., Madag.

Dirtea Rafin.=Commelina L. (Commelinac.).

Dirynchosia P. & K.=Dirhynchosia Bl.=Spiraeopsis Miq. (Cunoniac.).

Disa Bergius. Orchidaceae. 130 trop. & S. Afr., Madag., Masc.

Disaccanthus Greene. Cruciferae. 6 W. N. Am.

Disachoena Zoll. & Mor.=Pimpinella L. (Umbellif.).

Disadena Miq.=Voyria Aubl. (Gentianac.).

Disakisperma Steud. Gramineae. 1 Mex. (Quid?=?Eragrostis Beauv.).

Disandra L.=Sibthorpia L. (Scrophulariac.).

Disandraceae Dulac=Linaceae S. F. Gray.

Disanthaceae (Harms) Nak.=Hamamelidaceae–Disanthoïdeae Harms.

Disantheraceae Dulac=Polygalaceae Juss.

Disanthus Maxim. Hamamelidaceae. 1 C. China, Japan. Fls. small, serotinous, in back-to-back pairs. C with 2 basal nectaries on inner surface (cf. *Eustigma*); 5 small staminodes present. Mizushima, *J. Jap. Bot.* **43**: 522–3, 1968.

Disarrenum Labill.=Hierochloë R. Br. (Gramin.).

Disarrhenum P. Beauv.=praec.

Discalyxia Markgraf. Apocynaceae. 3 New Guinea.

Discanthera Torr. & Gray=Cyclanthera Schrad. (Cucurbitac.).

Discanthus Spruce=Cyclanthus Poit. (Cyclanthac.).

Discaria Hook. Rhamnaceae. 10 Austr., N.Z., S. Andes, Brazil.

Dischanthium Kunth=Dichanthium Willem.=Andropogon L. (Gramin.).

Dischema Voigt=Hitchenia Wall. (Zingiberac.).

Dischidanthus Tsiang. Asclepiadaceae. 1 S. China, Indoch.

Dischidia R.Br. Asclepiadaceae. 80 Indomal., Polynes., Austr. Epiphytic, climbing by advent. roots, and with fleshy ls. covered by wax. The curious pitcher-plant, *D. rafflesiana* Wall., besides the ordinary ls., has pitcher-ls. Each is a pitcher with incurved margin, about 10 cm. deep. Into it grows an adv. r. developed from the stem or petiole just beside it. The pitcher may hang with its mouth upwards or may stand horizontally or upside down. It usu. contains a lot of débris, largely carried into it by nesting ants. Most contain ± rainwater, so that perhaps they act as humus collectors and water reservoirs. The inner surface is waxy, so that the water cannot be absorbed by the pitcher itself, but must be taken up by the roots.

Developmental study shows the pitcher to be a l. with its lower side invaginated. The existing spp. illustrate all stages. Many, e.g. *D. bengalensis* Colebr., have bi-convex ls.; others have the under surface concave, e.g.

DISCHIDIOPSIS

D. (*Conchophyllum*) *collyris* Wall., and the roots are developed under and sheltered by the concave ls. A further invagination would lead to *D. rafflesiana*.
Dischidiopsis Schlechter. Asclepiadaceae. 9 Philipp. Is., New Guinea.
Dischidium (Ging.) Opiz = Viola L. (Violac.).
Dischimia Reichb. = seq.
Dischisma Choisy. Scrophulariaceae. 13 S. Afr.
Dischistocalyx T. Anders. ex Benth. corr. C. B. Clarke. Acanthaceae. 20 trop. Afr.
Dischistocalyx T. Anders. ex Lindau = Pseudostenosiphonium Lindau (Acan-
Dischlis Phil. (sphalm.) = Distichlis Rafin. (Gramin.).   [thac.).
Dischoriste D. Dietr. = Dyschoriste Nees (Acanthac.).
Disciflorae Benth. & Hook. f. A 'Series' of Dicots., comprising the orders *Geraniales, Olacales, Celastrales* and *Sapindales*, and the 'anomalous' fams. *Coriariaceae* and *Moringaceae*.
Disciphania Eichl. Menispermaceae. 20 Mex. to trop. S. Am., W.I.
Discipiper Trelease & Stehlé (~ Piper L.). 2 C. Am., W.I.
Discladium Van Tiegh. = Ochna L. (Ochnac.).
Discocactus Pfeiffer. Cactaceae. 9 Brazil, Paraguay.
Discocalyx Mez. Myrsinaceae. 50 Philipp. Is., New Guinea, Polynesia.
D. dissectus Hatus. (New Guinea) is a remarkable subherbaceous sp. with ls. resembling those of *Anthemis, Cotula, Matricaria*, etc.
Discocapnos Cham. & Schlechtd. Fumariaceae. 2 S. Afr.
Discocarpus Klotzsch. Euphorbiaceae. 5 Brazil, Guiana.
Discocarpus Liebm. = Discocnide Chew (Urticac.).
Discocatus Walp. (sphalm.) = Disocactus Lindl. = Phyllocactus Link (Cactac.).
Discoclaoxylon (Muell. Arg.) Pax & K. Hoffm. Euphorbiaceae. 3 trop. W. Afr.
Discocleidion (Muell. Arg.) Pax & K. Hoffm. Euphorbiaceae. 3 China,
Discocnide Chew. Urticaceae. 1 Mex., C. Am.   [Ryukyu Is.
Discocoffea A. Chev. = Coffea L. (Rubiac.).
Discocrania (Harms) Král (~ Cornus L.). Cornaceae. 2 Mex.
Discoglypremna Prain. Euphorbiaceae. 1 trop. W. Afr.
Discogyne Schlechter = Ixonanthes Jack (Ixonanthac.).
Discolenta Rafin. = Polygonum L. (Polygonac.).
Discolobium Benth. Leguminosae. 7 Brazil, Paraguay.
Discoluma Baill. = Pouteria Aubl. (Sapotac.).
Discoma O. F. Cook. Palmae. 1 Mex.
Discomela Rafin. = Helianthus L. (Compos.).
Discophora Miers. Icacinaceae. 2 Panamá, trop. S. Am.
Discophytum Miers = Calycera Cav. (Calycerac.).
Discopleura DC. Umbelliferae. 2 N. Am.
Discoplis Rafin. = Mercurialis L. (Euphorbiac.).
Discopodium Hochst. Solanaceae. 2 trop. Afr.
Discopodium Steud. = Tricostularia Nees (Cyperac.).
Discorea Miq. (sphalm.) = Dioscorea L. (Dioscoreac.).
Discoseris (Endl.) Kuntze = Gochnatia Kunth (Compos.).
Discospermum Dalz. = Diplospora DC. (Rubiac.).
Discostegia Presl = Marattia Sw. (Marattiac.).

Discostigma Hassk. = Garcinia L. (Guttif.).
Discovium Rafin. (Lesquerella S. Wats.). Cruciferae. 40 N. Am.
Discurainia Walp. = Descurainia Webb (Crucif.).
Discurea Schur = praec.
Discyphus Schlechter. Orchidaceae. 1 Venez., Trinidad.
Disecocarpus Hassk. = Commelina L. (Commelinac.).
Diseldia Rafin. = ? Pycnothymus Small (Labiat.).
Disella Greene = Sida L. (Malvac.).
Diselma Hook. f. Cupressaceae. 1 Tasmania.
Disemma Labill. = Passiflora L. (Passiflorac.).
Disepalum Hook. f. Annonaceae. 6 W. Malaysia. 2-merous.
Diseris Wight = Disperis Sw. (Orchidac.).
Diserneston Jaub. & Spach = Dorema D. Don (Umbellif.).
Disgrega Hassk. = Tripogandra Rafin. (Commelinac.).
Disinstylis Rafin. = Chlora Adans. (Gentianac.).
Disiphon Schlechter = Vaccinium L. (Ericac.).
Disisocactus G. K[unze] = Disocactus Lindl. (Cactac.).
Diskion Rafin. = Saracha Ruiz & Pav. (Solanac.).
Dismophyla Rafin. = Drosera L. (Droserac.).
Disocactus Lindl. Cactaceae. 9 C. Am.
Disocereus Frič & Kreuz. = Disocactus Lindl. (Cactac.).
Disodea Pers. = Lygodisodea Ruiz & Pav. (Rubiac.).
Disomene A. DC. = Dysemone Soland. ex Forst. f. = Gunnera L. (Gunnerac.).
Disoön A. DC. = Myoporum Banks (Myoporac.).
Disoxylon Reichb. = Dysoxylum Blume (Meliac.).
Disoxylum Benth. & Hook. f. [= Dysoxylum Bl. sec. Hook. f. & Thoms.] = Amoora Roxb. (Meliac.).
Disoxylum A. Juss. = Dysoxylum Blume (Meliac.).
Dispara Rafin. = Cristatella Nutt. (Cleomac.).
***Disparago** Gaertn. Compositae. 7 S. Afr.
Dispeltophorus Lehm. = Menonvillea R.Br. (Crucif.).
Disperis Sw. Orchidaceae. 75 trop. & S. Afr., Masc. Is., Indomal.
Disperma C. B. Clarke = Duosperma Dayton (Acanthac.).
Disperma J. F. Gmel. = Mitchella L. (Rubiac.).
Dispermotheca Beauverd = Parentucellia Viv. (Scrophulariac.).
Disphenia Presl = Cyathea Sm. (Cyatheac.).
Disphyma N. E. Brown. Aïzoaceae. 2 S. Afr., 1 temp. Austr.
Displaspis Klatt = Diplaspis Hook. f. = Huanaca Cav. (Umbellif.).
Disporocarpa A. Rich. = Crassula L. (Crassulac.).
Disporopsis Hance. Liliaceae. 4 Siam, SE. China, Formosa.
Disporum Salisb. Liliaceae. 20 N. temp. As. & Am.
Disquamia Lem. = Aechmea Ruiz & Pav. (Bromeliac.).
Dissanthelium Trin. Gramineae. 17 California to Patagonia (esp. Peruv. Andes).
Dissecocarpus Hassk. = Disecocarpus Hassk. = Commelina L. (Commelinac.).
Dissiliaria F. Muell. Euphorbiaceae. 1–2 NE. Austr.
Dissocarpus F. Muell. (~ Sclerolaena R.Br.). Chenopodiaceae. 2 SE. Austr.
Dissochaeta Blume. Melastomataceae. 20 Indomal.

379

DISSOCHONDRUS

Dissochondrus (Hillebr.) Kuntze. Gramineae. 1 Hawaii.

Dissochroma P. & K. = Dyssochroma Miers (Solanac.).

Dissolaena Lour. = Rauvolfia L. (Apocynac.).

Dissomeria Hook. f. ex Benth. Flacourtiaceae. 1 trop. W. Afr.

Dissopetalum Miers = Cissampelos L. (Menispermac.).

Dissorhynchium Schau. = Habenaria Willd. (Orchidac.).

Dissothrix A. Gray. Compositae. 2 Mex., Paraguay.

*Dissotis Benth. Melastomataceae. 140 Afr.

Distandra Link = Disandra L. = Sibthorpia L. (Scrophulariac.).

Distasis DC. = Chaetopappa DC. (Compos.).

Distaxia Presl = Blechnum L. (Blechnac.).

Disteganthus Lem. Bromeliaceae. 1 Guiana.

Distegia Klatt = Didelta L'Hérit. (Compos.).

Distegia Rafin. = Lonicera L. (Caprifoliac.).

Distegocarpus Sieb. & Zucc. = Carpinus L. (Carpinac.).

Disteira Rafin. = Martynia L. (Martyniac.).

Distemma Lem. = Disemma Labill. = Passiflora L. (Passiflorac.).

Distemon Bouché = Canna L. (Cannac.).

Distemon Ehrenb. ex Aschers. = Anticharis Endl. (Scrophulariac.).

Distemon Wedd. = Neodistemon Babu & Henry (Urticac.).

Distemonanthus Benth. Leguminosae. 1 trop. W. Afr.

Distephana [Juss.] (DC.) M. Roem. = Passiflora L. (Passiflorac.).

Distephania Gagnep. = Indosinia J. E. Vidal (Ochnac.).

Distephanus Cass. = Vernonia Schreb. (Compos.).

Distephia Salisb. ex DC. = Distephana (DC.) M. Roem. = Passiflora L.

Disterepta Rafin. = Cassia L. (Legumin.). [(Passiflorac.).

Disterigma Niedenzu ex Drude. Ericaceae. 35 trop. Andes.

Distetraceae Dulac = Thymelaeaceae Juss.

Distiacanthus Linden = Karatas Adans. (Bromeliac.).

Disticheia Ehrh. (uninom.) = Bromus pinnatus L. = Brachypodium pinnatum (L.) Beauv. (Gramin.).

Distichella Van Tiegh. (~ Dendrophthora Eichl.). Viscaceae. 3 W.I.

Distichia Nees & Meyen. Juncaceae. 3 Andes.

Distichis Lindl. = Liparis Rich. (Orchidac.).

Distichis Thou. (uninom.) = Malaxis disticha Thou. = Liparis disticha (Thou.) Lindl. (Orchidac.).

Distichlis Rafin. Gramineae. 12 Am., 1 Austr. D. spicata (L.) Greene used for binding sandy soil (cf. Ammophila, Carex).

Distichmus Endl. = Dichismus Rafin. = Scirpus L. (Cyperac.).

Distichocalyx Benth. (sphalm.) = Dischistocalyx T. Anders. ex Benth. corr. C. B. Clarke (Acanthac.).

Distichostemon F. Muell. Sapindaceae. 2 N. Austr. Apet.

Distictella Kuntze. Bignoniaceae. 10 trop. S. Am. & Tobago.

Distictis Bur. = praec.

Distictis Mart. ex Meissn. Bignoniaceae. 4 Mexico, W.I.

Distigocarpus Sargent (sphalm.) = Distegocarpus Sieb. & Zucc. = Carpinus L. (Carpinac.).

Distimake Rafin. = Ipomoea L. (Convolvulac.).

Distimum Steud. = seq.
Distimus Rafin. = Pycreus Beauv. = Cyperus L. (Cyperac.).
Distira P. & K. = Disteira Rafin. = Martynia L. (Martyniac.).
Distixila Rafin. = Myrtus L. (Myrtac.).
Distoecha Phil. Compositae. 1 Chile.
Distomaea Spenn. = Neottia Ludw. (Orchidac.).
Distomanthera Turcz. ᷉ Inc. sed. (? Melastomatac. ? Rhizophorac.). 1 Peru or Chile.
Distomischus Dulac = Festuca L. (Gramin.).
Distomocarpus O. E. Schulz. Cruciferae. 1 Morocco.
Distrepta Miers = Tecophilaea Bert. (Tecophilaeac.).
Distreptus Cass. = Elephantopus L. (Compos.).
Distrianthes Danser. Loranthaceae. 3 New Guinea.
Distyliopsis P. K. Endress. Hamamelidaceae. 5–6 Burma, E. As., Formosa, Malaysia.
Distylis Gaudich. = Calogyne R.Br. (Goodeniac.).
Distylium Sieb. & Zucc. Hamamelidaceae. 15 Assam to Japan, SE. As., Malay Penins., Java.
Distylodon Summerh. Orchidaceae. 1 trop. E. Afr.
Disynanthes Reichb. = seq.
Disynanthus Rafin. = Antennaria Gaertn. (Compos.).
Disynapheia Steud. = seq.
Disynaphia DC. (~ Eupatorium L.). Compositae. 12 Brazil to Argent.
Disynia Rafin. = Salix L. (Salicac.).
Disynoma Rafin. = Aëthionema R.Br. (Crucif.).
Disynstemon R. Viguier. Leguminosae. 1 Madag.
Disyphonia Griff. = Dysoxylum Bl. (Meliac.).
Ditassa R.Br. Asclepiadaceae. 75 S. Am.
Ditaxia Endl. (sphalm.) = Ditoxia Rafin. = Celsia L. (Scrophulariac.).
Ditaxis Vahl ex A. Juss. (~ Argithamnia Sw.). Euphorbiaceae. 50 warm Am., W.I.
Diteilis Rafin. = Liparis Rich. (Orchidac.).
Ditelesia Rafin. = Dilasia Rafin. = Murdannia Royle (Commelinac.).
Ditepalanthus Fagerl. Balanophoraceae. 1–2 Madag.
Ditereia Rafin. = Evolvulus L. (Convolvulac.).
Ditheca Miq. = Ammannia L. (Lythrac.).
Dithecina Van Tiegh. = Helixanthera Lour. (Loranthac.).
Dithrichum DC. = Ditrichum Cass. = Verbesina L. (Compos.).
Dithrix Schlechter. Orchidaceae. 1 Afghanistan, NW. India.
Dithyraea Endl. = seq.
Dithyrea Harv. Cruciferae. 5 SW. U.S., Mex.
Dithyria Benth. = Swartzia Schreb. (Legumin.).
Dithyrocarpus Kunth = Floscopa Lour. (Commelinac.).
Dithyrostegia A. Gray = Angianthus Wendl. (Compos.).
Ditinnia A. Chev. = Remusatia Schott (Arac.).
Ditmaria Spreng. = Erisma Rudge (Vochysiac.).
Ditoca Banks & Soland. ex Gaertn. = Scleranthus L. (Caryophyllac.).
Ditomaga Rafin. = Irlbachia Mart. (Gentianac.).

DITOMOSTROPHE

Ditomostrophe Turcz. = Guichenotia J. Gay (Sterculiac.).

Ditoxia Rafin. = Celsia L. (Scrophulariac.).

Ditremexa Rafin. = Cassia L. (Legumin.).

Ditrichospermum Bremek. Acanthaceae. 1 Assam, Burma.

Ditrichum Cass. (1817; non Timm 1788—Musci) = Verbesina L. (Compos.).

Ditriclita Rafin. = Saxifraga L. (Saxifragac.).

Ditrisynia Rafin. = Ditrysinia Rafin. = Sebastiania Spreng. (Euphorbiac.).

Ditritra Rafin. = Euphorbia L. (Euphorbiac.).

Ditroche E. Mey. ex Moq. = Limeum L. (Aïzoac.).

Ditta Griseb. Euphorbiaceae. 1 W.I.

Ditrysinia Rafin. = Sebastiania Spreng. (Euphorbiac.).

Dittelasma Hook. f. = Sapindus L. (Sapindac.).

Dittoceras Hook. f. Asclepiadaceae. 3 E. Himal., Siam.

Dittostigma Phil. Solanaceae. 1 Chile.

Dituilis Rafin. = Liparis Rich. (Orchidac.).

Ditulima Rafin. = Dendrobium Sw. (Orchidac.).

Ditulium Rafin. = Diascia Link & Otto (Scrophulariac.).

Diuranthera Hemsl. Liliaceae. 2 W. China.

Diuratea P. & K. = Diouratea Van Tiegh. = Ouratea Aubl. (Ochnac.).

Diuris Sm. Orchidaceae. 1 Java, 37 Austr.

Diuroglossum Turcz. = Guazuma Adans. (Sterculiac.).

Diurospermum Edgew. = Utricularia L. (Lentibulariac.).

Dizonium Willd. ex Schlechtd. = Geigeria Griesselich (Compos.).

Dizygandra Meissn. = Ruellia L. (Acanthac.).

Dizygostemon Radlk. Scrophulariaceae. 1 Brazil.

Dizygotheca N. E. Br. Araliaceae. 17 New Caled.

Djaloniella P. Taylor. Gentianaceae. 1 trop. W. Afr.

Djeratonia Pierre = Landolphia Beauv. (Apocynac.).

Dobera Juss. Salvadoraceae. 2 trop. E. Afr., S. Arabia, NW. Penins. Ind.

Dobera 'Rafin.' ex Ind. Kew. = praec.

Doberia Pfeiff. = praec.

Dobinea Buch.-Ham. Podoaceae. 2 E. Himalaya, S. China.

Dobrowskya Presl = Monopsis Salisb. (Campanulac.).

Docanthe O. F. Cook. Palmae. 1 Mex., Guatem.

Dochafa Schott = Arisaema Mart. (Arac.).

Docynia Decne. Rosaceae. 6 Himal., Burma, W. China.

Docyniopsis (C. K. Schneid.) Koidz. = Malus L. (Rosac.).

Dodartia L. Scrophulariaceae. 1 S. Russia, W. As.

Dodecadenia Nees = Litsea Lam. (Laurac.).

Dodecadia Lour. = Pygeum Gaertn. = Lauro-Cerasus Duham. (Rosac.).

Dodecas L. f. = Crenea Aubl. (Lythrac.).

Dodecasperma Rafin. = Bomarea Mirb. (Alstroemeriac.).

Dodecaspermum Först. ex Scop. = Decaspermum J. R. & G. Forst. (Myrtac.).

Dodecastemon Hassk. = Drypetes Vahl (Euphorbiac.).

Dodecastigma Ducke. Euphorbiaceae. 3 trop. S. Am.

Dodecatheon L. Primulaceae. 1 arct. NE. As., 50 Pacif. N. Am., 1 Atl. N. Am.
Like *Cyclamen*.

Dodonaea Adans. = Comocladia P.Br. (Anacardiac.).

Dodonaea Mill. Sapindaceae. 60 trop. & subtrop., esp. Austr.
***Dodonaeaceae** Link = Sapindaceae–Dodonaeëae Kunth.
Dodonea P.Br. Inc. sed. 1 Jamaica.
Doellia Sch. Bip. = Blumea DC. (Compos.).
Doellingeria Nees (~ Aster L.). Compositae. 1 E. As.
Doellochloa Kuntze = Monochaete Doell (Gramin.).
Doemia R.Br. = Pergularia L. (Asclepiadac.).
Doerpfeldia Urb. Rhamnaceae. 1 Cuba.
Doerriena Borkh. = Cerastium L. (Caryophyllac.).
Doerrienia Dennst. = Acronychia J. R. & G. Forst. (Rutac.).
Doerrienia Reichb. = Genlisea A. St-Hil. (Lentibulariac.).
Doerriera Steud. = Doerriena Borkh. = Cerastium L. (Caryophyllac.).
Dofia Adans. = Dirca L. (Thymelaeac.).
Doga (Baill.) Baill. ex Nakai = Storckiella Seem. (Legumin.).
Dolia Lindl. = Nolana L. (Nolanac.).
Dolianthus C. H. Wright = Amaracarpus Bl. (Rubiac.).
Dolichandra Cham. Bignoniaceae. 1 S. Brazil, Paraguay, Argent.
***Dolichandrone** (Fenzl) Seemann. Bignoniaceae. 9 trop. E. Afr., Madag., trop. As., Austr.
Dolichangis Thou. (uninom.) = Angraecum sesquipedale Thou. (Orchidac.).
Dolichanthera Schlechter & Krause = Morierina Vieill. (Rubiac.).
Dolichlasium Lag. = Trixis P.Br. (Compos.).
Dolichochaete (C. E. Hubbard) Phipps. Gramineae. 6 trop. & S. Afr.
Dolichodeira Hanst. = Achimenes P.Br. (Gesneriac.).
Dolichodelphys K. Schum. & Krause. Rubiaceae. 1 Peru.
Dolichogyne DC. = Nardophyllum Hook. & Arn. (Compos.).
Dolichokentia Becc. corr. B. D. Jacks. Palmae. 1 New Caled.
Dolicholasium Spreng. = Dolichlasium Lag. = Trixis P.Br. (Compos.).
Dolicholobium A. Gray. Rubiaceae. 20 Philippines, New Guinea to Fiji.
Dolicholus Medik. = Rhynchosia Lour. (Legumin.).
Dolichometra K. Schum. Rubiaceae. 1 E. trop. Afr.
Dolichonema Nees = Moldenhawera Schrad. (Legumin.).
Dolichopsis Hassler. Leguminosae. 1 Paraguay.
Dolichopterys Kosterm. Malpighiaceae. 1 Guiana.
Dolichorrhiza (Pojark.) Galushko. Compositae. 3 Cauc.
Dolichos auctt. = Lablab Adans., Vigna Savi, etc. (Legumin.).
***Dolichos** L. Leguminosae. 70 warm.
Dolichosiphon Phil. = Jaborosa Juss. (Solanac.).
Dolichostachys Benoist. Acanthaceae. 1 Madag.
Dolichostemon Bonati. Scrophulariaceae. 1 Indoch.
Dolichostylis Cass. = Barnadesia Mutis (Compos.).
Dolichostylis Turcz. = Draba L. (Crucif.).
Dolichotheca Cass. = Campylotheca Cass. (Compos.).
Dolichothele Britton & Rose = Mammillaria Haw. (Cactac.).
Dolichoura Brade. Melastomataceae. 1 Brazil.
Dolichovigna Hayata. Leguminosae. 3 Formosa.
Dolichus E. Mey. = Dolichos L. (Legumin.).
Dolicokentia Becc. = Dolichokentia Becc. corr. B. D. Jacks. (Palm.).

DOLICOTHECA

Dolicotheca Benth. & Hook. f. (sphalm.) = Dolichotheca Cass. = Campylotheca Cass. (Compos.).

Doliocarpus Roland. Dilleniaceae. 40 C. & trop. S. Am., W.I.

Dollinea P. & K. (sphalm.) = seq.

Dollinera Endl. (~ Desmodium Desv.). Leguminosae. 3 Indomal.

Dollineria Sauter = Draba L. (Crucif.).

Dolomiaea DC. Compositae. 5 Tibet, Himal.

Dolophragma Fenzl = Arenaria L. (Caryophyllac.).

Dolosanthus Klatt = Vernonia Schreb. (Compos.).

Doma Lam. = Hyphaene Gaertn. (Palm.).

Dombeia Raeusch. = Dombeya Lam. = Araucaria Juss. (Araucariac.).

*Dombeya Cav. Sterculiaceae (Malvaceae?). 50 Afr., 300 + Madag. & Masc.

Dombeya Lam. = Araucaria Juss. (Araucariac.).

Dombeya L'Hérit. = Tourretia Juss. (Bignoniac.).

Dombeyaceae Schultz-Schultzenst. = Sterculiaceae–Dombeyeae Kunth.

Dombrowskya Endl. = Dobrowskya Presl = Monopsis Salisb. (Campanulac.).

Domeykoa Phil. Hydrocotylaceae (~ Umbellif.). 4 Chile.

× **Domindendrum** hort. (vii) = × Epigoa hort. (Orchidac.).

× **Domindesmia** hort. Orchidaceae. Gen. hybr. (iii) (Domingoa × Hexadesmia).

Dominella E. Wimm. Campanulaceae. 1 W. trop. S. Am.

Domingoa Schlechter. Orchidaceae. 2 W.I.

Dominia Fedde. Umbelliferae. 1 New S. Wales.

× **Domintonia** hort. Orchidaceae. Gen. hybr. (iii) (Broughtonia × Domingoa).

Domkeocarpa Markgr. Apocynaceae. 1 trop. W. Afr.

× **Domliopsis** hort. Orchidaceae. Gen. hybr. (iii) (Domingoa × Laeliopsis).

Domohinea Leandri. Euphorbiaceae. 1 Madag.

Donacium Fries = Donax Beauv. = Arundo L. (Gramin.).

Donacodes Blume = Amomum L. (Zingiberac.).

Donacopsis Gagnep. = Eulophia R.Br. ex Lindl. (fls.) + Arundina Bl. (ls.)

Donaldia Klotzsch = Begonia L. (Begoniac.). [(Orchidac.).

Donaldsonia Baker f. = Moringa L. (Moringac.).

Donatia Bert. ex Remy = Lastarriaea Remy (Polygonac.).

*Donatia J. R. & G. Forst. Donatiaceae. 2 Tasmania, N.Z., subantarct. S. Am.

Donatia Loefl. = Avicennia L. (Avicenniac.).

*Donatiaceae (Mildbr.) Skottsb. Dicots. 1/2 antarct. & subantarct. Small subalpine cushion-plants, with densely arranged alt. ent. lin. coriac. exstip. ls. Fls. term., sol., sess. K 5–7, sometimes unequal, open, with obconic tube; C 5–10, variable, ± fleshy; A 2–3, inside low annular disk, fil. long or v. short, anth. ovoid, extrorse; $\overline{G}$ (2–3), styles free, long or v. short, ± recurved, with capit. stig., ovules ∞, on axile subapical plac. Caps. turbinate, indehisc., seeds few, with fleshy endosp. Only genus: *Donatia*. Chars. ± intermediate betw. *Saxifragac.* and *Stylidiac.*

Donatophorus Zipp. = Harpullia Roxb. (Sapindac.).

Donax Beauv. = Arundo L. (Gramin.).

Donax Lour. Marantaceae. 3 Indomal., Pacif.

Donax K. Schum. = Clinogyne Salisb. ex Benth. (as to type) = Schumannianthus Gagnep. (Marantac.).

384

Doncklaeria Hort. ex Loud. = Donkelaaria Hort. ex Lem. = Centradenia G. Don (Melastomatac.).

Dondia Adans. = Lerchia Zinn = Suaeda Forsk. (Chenopodiac.).

Dondia Spreng. = Hacquetia Neck. ex DC. (Umbellif.).

Dondisia DC. = Plectronia L. (Rubiac.).

Dondisia Reichb. = Dondia Spreng. = Hacquetia Neck. ex DC. (Umbellif.).

Dondisia Scop. = Raphanus L. (Crucif.).

Donella Pierre (~ Chrysophyllum L.). Sapotaceae. 3 trop. Afr.

Donepea Camb. corr. Backer ex O. E. Schulz. Cruciferae. 1 NW. India.

Donia R.Br. in Ait. (1813) = Grindelia Willd. (Compos.).

Donia R.Br. (1819) = Oxyria Hill (Polygonac.).

Donia G. & D. Don = Clianthus Soland. (Legumin.).

Donia Nutt. = Aster L. (Compos.).

Doniana Rafin. = Donia R.Br. (1813) = Grindelia Willd. (Compos.).

Donidsia G. Don (sphalm.) = Dondisia DC. = Plectronia L. (Rubiac.).

Doniophyton Wedd. Compositae. 3 Chile, Argentina.

Donkelaaria Hort. ex Lem. = Centradenia G. Don (Melastomatac.).

Donkelaaria Lem. = Guettarda L. (Rubiac.).

Donnellia C. B. Clarke ex Donn. Sm. (1902; non C. F. Austin 1880—Musci) = Neodonnellia Rose (Commelinac.).

Donnellsmithia Coulter & Rose. Umbelliferae. 13 Mex., C. Am.

Donningia A. Gray (sphalm.) = Downingia Torr. (Campanulac.).

Dontospermum Sch. Bip. = Odontospermum Neck. ex Sch. Bip. (Compos.).

*Dontostemon Andrz. ex Ledeb. Cruciferae. 8 C. As.

Donzella Lem. = seq.

Donzellia Tenore = Flacourtia Comm. ex L'Hérit. (Flacourtiac.).

Doodia R.Br. Blechnaceae. 11 Ceylon to Austr., N.Z., Hawaii.

Doodia Roxb. = Uraria Desv. (Legumin.).

Doona Thw. Dipterocarpaceae. 11 Ceylon. Timber, resin.

Doornia De Vriese = Pandanus L. f. (Pandanac.).

Doosera Roxb. ex Wight & Arn. = Mollugo L. (Aïzoac.).

Dopatrium Buch.-Ham. ex Benth. Scrophulariaceae. 20 trop. Afr., As., Austr.

Doraena Thunb. = Maesa Forsk. (Myrsinac.).

Doranthera Steud. = Doratanthera Benth. ex Endl. = Anticharis Endl. (Scrophulariac.).

Doranxylum Néraud (nomen) = ? Doratoxylon Thou. ex Benth. & Hook. f. (Sapindac.).

Doratanthera Benth. ex Endl. = Anticharis Endl. (Scrophulariac.).

Doratium Soland. ex Jaume St-Hil. = Zanthoxylum L. (Rutac.).

Doratolepis Schlechtd. = Leptorhynchos Less. (Compos.).

Doratometra Klotzsch = Begonia L. (Begoniac.).

Doratophora Lem. = Doryphora Endl. (Atherospermatac.).

Doratoxylon Thou. ex Benth. & Hook. f. Sapindaceae. 1 Mascarenes.

Dorcapteris Presl = Polybotrya Humb. & Bonpl. (Aspidiac.).

Dorcoceras Bunge = Boea Comm. ex Lam. (Gesneriac.).

Dorella Bub. = Camelina Crantz (Crucif.).

Dorema D. Don. Umbelliferae. 16 C. & SW. As. *D. ammoniacum* D. Don is

DORIA
the source of the gum resin, gum ammoniacum (medic.), obtained by puncturing the stem.

Doria Fabr. (uninom.) = *Solidago* L. sp. (Compos.).

Doria Thunb. = Othonna L. = Senecio L. (Compos.).

× **Doricentrum** hort. Orchidaceae. Gen. hybr. (iii) (Ascocentrum × Doritis).

Doriclea Rafin. = Leucas R.Br. (Labiat.).

× **Doriella** hort. Orchidaceae. Gen. hybr. (iii) (Doritis × Kingidium [Kingiella]).

× **Doriellaopsis** hort. Orchidaceae. Gen. hybr. (iii) (Doritis × Kingidium [Kingiella] × Phalaenopsis).

Doriena Endl. = Doeringia Dennst. = Acronychia J. R. & G. Forst. (Rutac.).

Dorisia Gillespie = Mastixiodendron Melch. (Rubiac.).

× **Doritaenopsis** hort. Orchidaceae. Gen. hybr. (iii) (Doritis × Phalaenopsis).

Doritis Lindl. Orchidaceae. 2 Indomal.

× **Doritopsis** hort. (vii) = × Doritaenopsis hort. (Orchidac.).

Dornera Heuff. ex Schur = Carex L. (Cyperac.).

Dorobaea Cass. = Senecio L. (Compos.).

Doroceras Steud. = Dorcoceras Bunge (Gesneriac.).

Doronicum L. Compositae. 35 temp. Euras., N. Afr.

Dorothea Wernham. Rubiaceae. 3 trop. W. Afr.

Dorotheanthus Schwantes. Aïzoaceae. 6 S. Afr.

Dorrienia Engl. = Doerrienia Reichb. = Genlisea A. St-Hil. (Lentibulariac.).

Dorstenia L. Moraceae. 170 trop. Herbs or shrubs with peculiar cymose infl. The common recept. of the fls. is a flat or hollowed fleshy structure, often > 2·5 cm. wide. Fls. unisexual, sometimes all of one sex on one receptacle, sometimes intermingled with several ♂ round one ♀, sunk in the receptacle round whose edge project a number of bracts. P-segments completely united. Sta. in the ♂ usu. 2. The fr. when ripe is shot out of the receptacle; the latter becomes very turgid and presses on the fr. and at length ejects it as one might flip away a bit of soap between finger and thumb.

Dortania A. Chev. = Acidanthera Hochst. (Iridac.).

Dortiguea Bub. = Erinus L. (Scrophulariac.).

Dortmania Neck. = seq.

Dortmanna Hill = Lobelia L. (Campanulac.).

Dortmannia Steud. = praec.

Dorvalia Comm. ex DC. = seq.

Dorvalla Comm. ex Lam. = Fuchsia L. (Onagrac.).

Doryalis E. Mey. corr. Warb. Flacourtiaceae. 30 Afr., Ceylon. Some (cf.

Doryanthes Corrêa. Amaryllidaceae. 3 Austr.   [*Aberia*) have ed. fr.

Dorycheile Reichb. = Cephalanthera Rich. (Orchidac.).

Dorychilus P. & K. = praec.

Dorychnium Brongn. = Dorycnium L. (Legumin.).

Dorychnium Moench = Psoralea L. (Legumin.).

Dorycinopsis Lem. = Dorycnopsis Boiss. = Anthyllis L. (Legumin.).

Dorycnium Mill. Leguminosae. 15 Medit.

Dorycnopsis Boiss. = Anthyllis L. (Legumin.).

Doryctandra Hook. f. & Thoms. = Dioryktandra Hassk. = Rinorea Aubl.
[(Violac.).

Dorydium Salisb. = Asphodeline Reichb. (Liliac.).
Doryopteris J. Sm. Sinopteridaceae. 35 trop. & subtrop.
Doryphora Endl. Atherospermataceae. 1 New S. Wales.
Dorystaechas Boiss. & Heldr. ex Benth. = Dorystoechas Boiss. & Heldr. ex Benth. (corr. Benth.) (Labiat.).
Dorystephania Warb. Asclepiadaceae. 1 Philipp. Is.
Dorystigma Gaudich. = Pandanus L. f. (Pandanac.).
Dorystigma Miers = Jaborosa Juss. (Solanac.).
Dorystoechas Boiss. & Heldr. ex Benth. corr. Benth. Labiatae. 1 W. As.
Doryxylon Zoll. Euphorbiaceae. 1 Philipp., Java, Lesser Sunda Is.
Doschafa P. & K. = Dochafa Schott = Arisaema Mart. (Arac.).
Dossifluga Bremek. Acanthaceae. 1 Siam.
Dossinia C. Morr. Orchidaceae. 1 Borneo.
× **Dossinimaria** hort. Orchidaceae. Gen. hybr. (iii) (Dossinia × Ludisia [Haemaria]).
× **Dossisia** hort. (v) = praec.
Dothicroa Rafin. = Phlogacanthus Nees (Acanthac.).
Dothilis Rafin. = Chloraea Lindl. (Orchidac.).
Dothilophis Rafin. = Barberia Knowles & Westc. (Orchidac.).
Douarrea Montr. = Psychotria L. (Rubiac.).
Douepea Camb. (sphalm.) = Donepea Camb. corr. Backer (Crucif.).
Douepia Hook. f. & Thoms. = praec.
***Douglasia** Lindl. Primulaceae. 7 arct. N. Am.
Douglassia Heist. = Imhofia Heist. = Nerine Herb. (Amaryllidac.).
Douglassia Mill. = Clerodendrum L. (Verbenac.).
Douglassia Reichb. = Douglasia Lindl. (Primulac.).
Douglassia Schreb. = Aiouea Aubl. (Laurac.).
Douma [Poir.] = Hyphaene Gaertn. (Palm.).
Doupea D. Dietr. = Douepea Camb. = Donepea Camb. corr. Backer (Crucif.).
Dovea Kunth. Restionaceae. 10 S. Afr. Used for thatch.
Dovera Ehrenb. ex Reichb. = ? Dobera Juss. (Salvadorac.).
Dovyalis E. Mey. = Doryalis E. Mey. corr. Warb. (Flacourtiac.).
Dowea Steud. = Dovea Kunth (Restionac.).
***Downingia** Torr. Campanulaceae. 11 Pacif. N. & temp. S. Am. Fls. usu. inverted.
Doxantha Miers = Bignonia L. + Doxanthemum D. R. Hunt (Bignoniac.).
Doxanthemum D. R. Hunt. Bignoniaceae. 1 N. Am.
Doxanthes Rafin. = Phaeomeria Lindl. (Zingiberac.).
Doxema Rafin. = Quamoclit Moench = Ipomoea L. (Convolvulac.).
Doxomma Miers = Barringtonia J. R. & G. Forst. (Barringtoniac.).
Doxosma Rafin. = Encyclia Hook. (Orchidac.).
Doyerea Grosourdy. Cucurbitaceae. 2 C. Am., W.I., Venezuela.
Draakesteinia P. & K. = Drakenstenia Neck. = Ecastaphyllum P.Br. (Legumin.).
Draba L. Cruciferae. 300 N. temp. & arctic, mts. of C. & S. Am.
Drabastrum O. E. Schulz. Cruciferae. 1 SE. Austr.
Drabella Fourr. = Draba L. (Crucif.).
Drabella Nábělek = Thylacodraba (Nábělek) O. E. Schulz = Draba L. (Crucif.).

DRABOPSIS

Drabopsis C. Koch. Cruciferae. 1 SW. & C. As.

Dracaena Vand. ex L. Agavaceae. 150 warm Old World. Mostly trees, whose stems branch and grow in thickness (extra-fascicular cambium). The famous dragon-tree of Teneriffe (*D. draco* L.), blown down in 1868, was 70 ft. high and 45 ft. in girth and was supposed to be 6,000 years old. A resin exudes from the trunk of this sp. (dragon's blood); the original dragon's blood appears to be that of *D. cinnabari* Balf. f. (Socotra).

Dracaen[ac]eae Salisb. = Agavaceae–Dracaeneae Reichb.

Dracaenopsis Planch. = Cordyline Comm. ex Juss. (Agavac.).

Dracamine Nieuwl. = Cardamine L. (Crucif.).

Dracena Rafin. = Dracaena Vand. ex L. (Agavac.).

Draco Crantz = Calamus L. (Palm.).

Draco Fabr. (uninom.) = *Dracaena* Vand. ex L. sp. (Agavac.).

Dracocactus Y. Ito = Pyrrhocactus (A. Berger) Backeb. & F. M. Knuth (Cactac.).

Dracocephalium Hassk. = seq.

Dracocephalon All. = seq.

Dracocephalum L. Labiatae. 45 C. Eur., temp. As., 1 N. Am.

Dracocephalus Aschers. = praec.

Draconia Fabr. (uninom.) = *Artemisia* L. sp. (Compos.).

Dracont[iac]eae Salisb. = Araceae–Dracontiëae Schott.

Dracontioïdes Engl. Araceae. 1 S. Brazil.

Dracontium Hill = Dracunculus Mill. (Arac.).

Dracontium L. Araceae. 13 Mex. to trop. S. Am. The sympodial rhiz. gives rise yearly to one enormous l. and an infl. The l. has 3 chief divisions, and the lat. ones develop dichot. at first. Fl. ⚥ with P.

Dracontocephalium Hassk. = Dracocephalum L. (Labiat.).

Dracontocephalum St-Lag. = praec.

Dracontomelon Blume. Anacardiaceae. 8 Malaysia to Fiji.

Dracontomelum auctt. = praec.

Dracontopsis Lem. = Dracopis Cass. (Compos.).

Dracophilus Dinter & Schwantes. Aïzoaceae. 4 S. Afr.

Dracophyllum Labill. Epacridaceae. 30 Austr., New Caled., N.Z. The sheathing ls. leave ring-scars when they fall.

Dracopis Cass. (~ Rudbeckia L.). Compositae. 1 N. Am., Mex.

Dracunculus Ledeb. = Artemisia L. (Compos.).

Dracunculus Mill. Araceae. 2 Medit. Fert. like *Arum*.

Drakaea Lindl. Orchidaceae. 4 W. Austr.

Drakaina Rafin. = Dracaena Vand. ex L. (Agavac.).

Drakea Endl. = Drakaea Lindl. (Orchidac.).

Drake-Brockmania Stapf. Gramineae. 1 Somaliland.

Drakebrockmania White & Sloane (non Stapf) = White-Sloanea Chiov. (Asclepiadac.).

Drakensteinia DC. = seq.

Drakenstenia Neck. = Ecastaphyllum P.Br. (Legumin.).

Draparnalda St-Lag. (1881; non Draparnaldia Bory 1808—Algae–Rhodophyc.) = seq.

Draparnaudia Montr. = Xanthostemon F. Muell. (Myrtac.).

Draperia Torr. Hydrophyllaceae. 1 California.

Drapetes Banks. Thymelaeaceae. 10 Borneo, New Guinea, Australasia, antarct. S. Am., Falkland Is.
Drapiezia Blume = Disporum Salisb. (Liliac.).
Draytonia A. Gray = Saurauia Willd. (Actinidiac.).
Drebbelia Zoll. & Mor. (1846) = Spatholobus Hassk. (Legumin.).
Drebbelia Zoll. (1857) = Olax L. (Olacac.).
Dregea Eckl. & Zeyh. = Peucedanum L. (Umbellif.).
***Dregea** E. Mey. Asclepiadaceae. 12 trop. & S. Afr. to China.
Dregeochloa Conert. Gramineae. 2 S. & SW. Afr.
Dreissenia auctt. (sphalm.) = Driessenia Korth. (Melastomatac.).
Drejera Nees. Acanthaceae. 4 trop. Am.
Drejerella Lindau. Acanthaceae. 12 Mex., W.I.
Drepachenia Rafin. = Sagittaria L. (Alismatac.).
Drepadenium Rafin. = Croton L. (Euphorbiac.).
Drepanandrum Neck. = Topobea Aubl. (Melastomatac.).
Drepananthus Maingay ex Hook. f. = Cyathocalyx Champ. (Annonac.).
Drepania Juss. = Tolpis Adans. (Compos.).
Drepanocarpus G. F. W. Mey. (~ Machaerium Pers.). Leguminosae. 15 trop.
Drepanocaryum Pojark. Labiatae. 1 C. As. [Am., Afr.
Drepanolobus Nutt. ex Torr. & Gray = Hosackia Dougl. (Legumin.).
Drepanometra Hassk. = Begonia L. (Begoniac.).
Drepanophyllum Wibel = Falcaria Host (Umbellif.).
Drepanospermum Benth. = Campnosperma Thw. (Anacardiac.).
Drepaphyla Rafin. = Acacia Mill. (Legumin.).
Drepilia Rafin. = Thermopsis R.Br. (Legumin.).
Driessenia Korth. Melastomataceae. 8 Malaysia.
Drimia Jacq. Liliaceae. 45 trop. & S. Afr.
Drimiopsis Lindl. & Paxt. Liliaceae. 22 trop. & S. Afr.
Drimophyllum Nutt. = Umbellularia Nutt. (Laurac.).
Drimopogon Rafin. = Spiraea L. (Rosac.).
Drimya Lem. = Drimia Jacq. (Liliac.).
Drimyaceae Van Tiegh. = Winteraceae Lindl.
Drimycarpus Hook. f. Anacardiaceae. 1 E. Himal. G.
Drimyphyllum Burch. ex DC. = Petrobium R.Br. (Compos.).
Drimys J. R. & G. Forst. Winteraceae. 70 Borneo, New Guinea, New Caled., Austr., Tasm., N.Z., S. Am. There is a distinction between calyx and corolla (cf. *Illicium*). The bark of *D. winteri* J. R. & G. Forst. (Winter's bark) is medicinal.
Drimyspermum Reinw. = Phaleria Jack (Thymelaeac.).
Dripax Nor. ex Thou. = Rinorea Aubl. (Violac.).
Drobowskia Brongn. = Dobrowskya Presl = Monopsis Salisb. (Campanulac.).
Drobrowskia B. D. Jacks. (sphalm.) = praec.
Droceloncia J. Léonard. Euphorbiaceae. 1 trop. Afr.
Drogouetia Steud. = seq.
Droguetia Gaudich. Urticaceae. 12 trop. & S. Afr., Madag., Arabia, S. India, Java.
Dromophylla Lindl. = Dermophylla S. Manso = Cayaponia S. Manso (Cucurbitac.).

DROOGMANSIA

Droogmansia De Wild. Leguminosae. 30 trop. Afr.

Drosace A. Nelson = Androsace L. (Primulac.).

Drosanthe Spach = Hypericum L. (Guttif.).

Drosanthemum Schwantes. Aïzoaceae. 70 S. Afr.

Drosanthus R.Br. ex Planch. = Byblis Salisb. (Byblidac.).

Drosera L. Droseraceae. 100 trop. & temp. (esp. Austr. & N.Z.). Herbs usu. with shortly creeping or tuberous rhiz. and rosettes of ls., insectivorous. The blade of the l. is circular in some spp., elongated in others, and is set with curious tentacles; these are emergences containing vascular bundles and ending in swollen reddish heads which secrete a sticky glistening fluid. Flies and other insects mistaking it for honey are held by it. The tentacles are exceedingly sensitive to continued pressure even by the lightest bodies; the result is to cause an inward and downward movement of the head of the tentacle, finally placing the fly upon the blade of the l. At the same time the stimulus passes to the surrounding tentacles, causing them also to bend downwards to the same point. The victim is thus smothered, and now the glandular heads of the tentacles secrete a ferment which acts upon the proteids and brings them into solution, when they are taken up by the l. Afterwards the tentacles expand once more and recommence the secretion of the sticky fluid. The food thus obtained is of benefit to the pl., though it can live without it. *D.* is able to live in very poor soil. The extra materials obtained are devoted chiefly to seed-production. If the stimulus produced by the capture of an insect be very powerful, the l. itself may bend into a cup form, and this feature is very marked in some spp., the l. bending almost double over the prey.

The fls. of some spp. rarely open, but self-pollinate in bud.

*****Droseraceae** Salisb. Dicots. 4/105, *Drosera* cosmopolitan, the rest more local. Herbs, usu. of acid bogs, with perenn. rhiz. and rosettes of stip. or exstip. ls.; *Aldrovanda* a submerged water-plant. All are insectivorous; *Dionaea* and *Aldrovanda* have sensitive ls. which shut up when touched, the others catch their prey by sticky tentacles upon the ls. (see genera). Fls. usu. in cincinni, rarely in racemes or sol., ♀, reg., 5–4-merous, usu. hypog. K (5); C 5, imbr.; no disk; A usu. 5, pollen in tetrads (cf. *Ericaceae*); G (2, 3, or 5); plac. usu. parietal, rarely axile or free-central; style long; stigmas simple or branched; ov. 3–∞, anatr. Loculic. caps.; seed with endosp. and small basal embryo. Genera: *Dionaea, Aldrovanda, Drosophyllum, Drosera.*

Drosocarpium Fourr. = Hypericum L. (Guttif.).

Drosodendron M. Roem. = Baeckea L. (Myrtac.).

Drosophorus R.Br. ex Planch. = Byblis Salisb. (Byblidac.).

Drosophyllum Link. Droseraceae. 1, *D. lusitanicum* Link, Portugal, S. Spain, Morocco. The ls. have glands of two kinds—stalked, secreting a sticky fluid (cf. *Drosera*), and sessile, which only secrete when stimulated by nitrogenous matter, and then secrete a digestive ferment. Insects alight on the glands and are entangled; they struggle for a while and finally sink down and die, and are digested by the ferment. The taller glands have no power of movement, but are able to secrete a ferment as well as the sessile ones.

Drossera Gled. = Drosera L. (Droserac.).

Drouguetia Endl. = Droguetia Gaudich. (Urticac.).

Drozia Cass. = Perezia Lag. (Compos.).

Drudea Griseb. = Pycnophyllum Remy (Caryophyllac.).
Drudeophytum Coulter & Rose (~ Tauschia Schlechtd.). Umbelliferae. 7 N. Am.
Drummondia DC. = Mitellopsis Meissn. = Mitella L. (Saxifragac.).
Drummondita Harv. = Philotheca Rudge (Rutac.).
Drumondia Rafin. = Drummondia DC. = Mitella L. (Saxifragac.).
Drupaceae S. F. Gray = Rosaceae–Prunoïdeae Focke.
Druparia [Clairv.] = Prunus L. (*sensu latiss.*) (Rosac.).
Druparia S. Manso = Cayaponia S. Manso (Cucurbitac.).
Drupatris Lour. = Symplocos Jacq. (Symplocac.).
Drupifera Rafin. = Camellia L. (Theac.).
Drupina L. = Curanga Juss. (Scrophulariac.).
Drusa DC. (~ Bowlesia Ruiz & Pav.). Hydrocotylaceae (~ Umbellif.). 1 Canary Is.
Dryad[ac]eae S. F. Gray = Rosaceae–Potentilleae–Dryadinae Vent.
Dryadaea Kuntze = Dryadea Rafin. = Dryas L. (Rosac.).
Dryadanthe Endl. (~ Potentilla L.). Rosaceae. 1 C. As., Himal.
Dryadea Rafin. = Dryas L. (Rosac.).
Dryadodaphne S. Moore. Monimiaceae. 1 New Guinea.
Dryadorchis Schlechter. Orchidaceae. 2 New Guinea.
*****Dryandra** R.Br. Proteaceae. 50 Austr. Like *Banksia.*
Dryandra Thunb. = Vernicia Lour. (Euphorbiac.).
Dryas L. Rosaceae. 2 arctic-alpine. *D. octopetala* L. is androdioec. in the Alps. Style feathery after fert. (cf. *Clematis, Geum*).
Drymaria Willd. ex Roem. & Schult. Caryophyllaceae. 44 Mex. to Patag., W.I.; 1 Abyss., 1 S. Afr.; 1 temp. Himal., 1 Java, 1 Austr.
Drymeia Ehrh. (uninom.) = *Carex drymeia* L. f. = *C. sylvatica* L. (Cyperac.).
Drymiphila Juss. = Drymophila R.Br. (Liliac.).
Drymis Juss. = Drimys J. R. & G. Forst. (Winterac.).
Drymispermum Reichb. = Drimyspermum Reinw. = Phaleria Jack (Thymelaeac.).
Drymoanthus W. H. Nicholson. Orchidaceae. 1 Queensland, 1 N.Z.
Drymocallis Fourr. (~ Potentilla L.). Rosaceae. 30 N. Am., 1 W. Eur.
Drymocodon Fourr. = Campanula L. (Campanulac.).
Drymoda Lindl. Orchidaceae. 2 Burma, Siam.
*****Drymoglossum** Presl. Polypodiaceae. 6 Madag., trop. As., Malaysia. (C. Chr., *Dansk Bot. Ark.* **6**, no. 3, 1929.)
Drymonactes Steud. = seq.
Drymonaetes Ehrh. (uninom.) = *Bromus giganteus* L. = *Festuca gigantea* (L.) Vill. (Gramin.).
Drymonaetes Fourr. = Festuca L. (Gramin.).
Drymonia Mart. Gesneriaceae. 35 trop. Am., W.I.
Drymophila R.Br. Liliaceae. 2 E. Austr., Tasm.
Drymophloeus Zipp. Palmae. 12 Moluccas, New Guinea, Samoa, Fiji.
Drymopogon Fabr. (uninom.) = *Aruncus* Schaeffer sp. (Rosac.).
Drymopogon Rafin. = Spiraea L. (Rosac.).
Drymoscias K.-Pol. = Notopterygium Boissieu (Umbellif.).
Drymospartum Presl = Genista L. (Legumin.).

DRYMOSPHACE

Drymosphace (Benth.) Opiz = Salvia L. (Labiat.).

Drymotaenium Makino. Polypodiaceae. 2 Japan, Formosa.

Drymyrrhizae Vent. = Zingiberaceae Lindl.

Drymys Vell. = Drimys J. R. & G. Forst. (Winterac.).

Drynaria (Bory) J. Sm. Polypodiaceae. 20 palaeotrop. Ls. dimorphic, some projecting, assimilating and spore-bearing, the others small, close to rhiz., collecting humus. Cf. *Drynariopsis.*

Drynariopsis (Copel.) Ching. Polypodiaceae. 1 Malaysia, Solomon Is., *D. heraclea* (Kunze) Ching, a large epiphyte; the wide l.-bases protect roots and collect humus.

Dryoathyrium Ching = Lunathyrium Koidz. (Athyriac.).

Dryobalanops Gaertn. f. Dipterocarpaceae. 9 Sumatra, Malay Penins., Borneo. *D. aromatica* Gaertn. and other spp. yield Borneo or Sumatra camphor, used chiefly in China. The young ls. are red, and hang down. A ∞.

Dryomenis Fée = Tectaria Cav. (Aspidiac.).

Dryopaeia Roeper = seq.

Dryopeia Thou. = Disperis Sw. (Orchidac.).

Dryopetalon A. Gray. Cruciferae. 4 Calif., Mexico. Pets. 5–7-lobed.

Dryopoa Vickery. Gramineae. 1 SE. Austr., Tasm.

Dryopolystichum Copel. Aspidiaceae. 1 New Guinea.

Dryopria Thou. = Dryopeia Thou. = Disperis Sw. (Orchidac.).

Dryopsila Rafin. = Erythrobalanus (Oerst.) O. Schwarz = Quercus L. (Fagac.).

Dryopteridaceae (Holttum) Ching = Aspidiaceae S. F. Gray *p.p.*

*Dryopteris Adans. Aspidiaceae. 150 cosmop. Terrestrial ferns; type sp. *D. filix-mas* (L.) Schott. (In C. Chr. *Ind. Fil.* the following genera were also included: *Thelypteris, Cyclosorus, Gymnocarpium, Ctenitis, Goniopteris, Hypodematium, Meniscium, Polystichopsis, Stegnogramma.*)

Dryorkis Thou. = Disperis Sw. (Orchidac.).

Dryostachyum J. Sm. = Aglaomorpha Schott (Polypodiac.).

Dryparia P. & K. = Druparia S. Manso = Cayaponia S. Manso (Cucurbitac.).

Drypetes Vahl. Euphorbiaceae. 200 trop., also S. Afr. and subtrop. E. As.

Drypis L. Caryophyllaceae. 1 E. Medit.

Drypsis Duchartre = Dypsis Nor. (Palm.).

Dryptes Kanjilal, Das & De = Drypetes Vahl (Euphorbiac.).

Dryptopetalum Arn. = Gynotroches Blume (Rhizophorac.).

Duabanga Buch.-Ham. Sonneratiaceae. 3 Indomal.

Duania Nor. = Homalanthus A. Juss. (Euphorbiac.).

Dubaea Steud. = Dubyaea DC. (1828) = Diplusodon Pohl (Lythrac.).

Dubanus Kuntze = Pometia J. R. & G. Forst. (Sapindac.).

Dubardella H. J. Lam. Boerlagellaceae. 1 Borneo. (=? Adinandra Jack, Theac.)

Dubautia Gaudich. Compositae. 30 Hawaii. Ls. ‖-veined. Hybrids.

Duboisia R.Br. Solanaceae. 2 Austr., New Caled.

Duboisia Karst. = Pleurothallis R.Br. (Orchidac.).

Dubois-Reymondia Karst. = Pleurothallis R.Br. (Orchidac.).

Duboscia Bocq. Tiliaceae. 3 trop. W. Afr.

Dubouzetia Panch. ex Brongn. & Gris. Elaeocarpaceae. 10 New Guinea, New Caled.

Dubreuilia Decne=seq.
Dubrueilia Gaudich.=Pilea Lindl. (Urticac.).
Dubyaea DC. (1828) (pro syn.)=Diplusodon Pohl (Lythrac.).
Dubyaea DC. (1838). Compositae. 10 Himal., W. China.
Ducampopinus A. Chev. (~Pinus L.). Pinaceae. 1 Indoch.
Duchartrea Decne=Pentarhaphia Lindl. (Gesneriac.).
Duchartrella Kuntze=Holostylis Duch. (Aristolochiac.).
Duchassaingia Walp.=Erythrina L. (Legumin.).
Duchekia Kostel.=Palisota Reichb. (Commelinac.).
Duchesnea P. & K.=Duchesnia Cass.=Pulicaria Gaertn. (Compos.).
Duchesnea Smith (~Fragaria L.). Rosaceae. 6 India, E. As.
Duchesnia Cass.=Pulicaria Gaertn. (Compos.).
Duchola Adans.=Omphalandria P. Br.=Omphalea L. (Euphorbiac.).
Duckea Maguire. Rapateaceae. 4 Venez., Braz.
Duckeanthus R. E. Fries. Annonaceae. 1 trop. S. Am.
Duckeëlla C. Porto & Brade. Orchidaceae. 3 trop. S. Am.
Duckeodendraceae Kuhlmann. Dicots. 1/1 Brazil. Large trees, with alt. ent. simple exstip. ls. Infl. of term. or subterm. few-fld. cymes. K (5), persist.; C (5), with long tube and short much imbr. lobes, greenish-white; A 5, alternipet. and epipet., exserted, with oblong introrse medifixed basally bilobed anth.; disk present; G (2), ± immersed in disk, with 1 anatr.(?) ov. per loc., and with elongate style and dilated shortly 2-lobed stigma. Fr. a large shining red drupe; endocarp 2-loc., bony, externally fibrous, fertile loc. U-shaped, containing U-shaped seed with U-shaped embryo in scanty oily endosp., ster. loc. ± straight. Only genus: *Duckeodendron*. Probably closely related to *Apocynaceae*; cf. anatomy (!), sagittate anthers, disk, simple style, etc.
Duckeodendron Kuhlm. Duckeodendraceae. 1 E. trop. S. Am.
Duckera F. A. Barkley. Anacardiaceae. 1 Tahiti.
Duckesia Cuatrec. Houmiriaceae. 1 Amaz. Brazil.
Ducosia Vieill. ex Guillaumin=Dubouzetia Panch. ex Brongn. & Gris (Elaeocarpac.).
Ducoudraea Bur.=Tecomaria Spach (Bignoniac.).
Ducrosia Boiss. Umbelliferae. 5 Egypt to NW. India.
× **Dudleveria** Rowley. Crassulaceae. Gen. hybr. (Dudleya × Echeveria).
Dudleya Britton & Rose. Crassulaceae. 40 SW. U.S., NW. Mex.
Dufourea Bory ex Willd.=Tristicha Thou. (Podostemac.).
Dufourea Gren.=Arenaria L. (Caryophyllac.).
Dufourea Kunth=Breweria R.Br. (Convolvulac.).
Dufrenoya Chatin. Santalaceae. 14 Indomal.
Dufresnia DC.=Valerianella Mill. (Valerianac.).
Dugagelia Juss. ex Gaudich. (nomen)=?Piper L. (Piperac.).
Dugaldia Cass.=Helenium L. (Compos.).
Dugandia Britton & Killip. Leguminosae. 1 NW. trop. S. Am.
Dugesia A. Gray. Compositae. 1 Mexico.
Dugezia Montr.=Lysimachia L. (Primulac.).
Duggena Vahl (~Gonzalea Pers.). Rubiaceae. 22 trop. S. Am., W.I.
Duglassia Houst.=Douglassia Mill.=Clerodendrum L. (Verbenac.).
Dugortia Scop.=Parinari Juss. (Chrysobalanac.).

DUGUETIA

*Duguetia A. St-Hil. Annonaceae. 70 trop. S. Am., W.I. Fr. formed of the individual berries or achenes united to the fleshy recept. *D. quitarensis* Benth., etc., furnish Jamaica and Cuba lancewood. *D. (Geanthemum) rhizantha* (Eichl.) Huber has rhiz. below the soil, bearing scale ls. only. The fls. are borne on branches of these above ground. (Cf. *Polyalthia* spp.).

Duguldea Meissn. = Dugaldia Cass. = Helenium L. (Compos.).

Duhaldea DC. = Inula L. (Compos.).

Duhamela Rafin. = Duhamelia Pers. = Hamelia Jacq. (Rubiac.).

Duhamelia Domb. ex Lam. = Myrsine L. (Myrsinac.).

Duhamelia Pers. = Hamelia Jacq. (Rubiac.).

Duidaea Blake. Compositae. 3 Venez.

Duidania Standley. Rubiaceae. 1 Venez.

Dukea Dwyer. Rubiaceae. 5 Panama, Colombia.

Dulacia Neck. = Acioa Aubl. (Chrysobalanac.).

Dulacia Vell. = Liriosma Poepp. & Endl. (Olacac.).

Dulcamara Medik. = Solanum L. (Solanac.).

Dulia Adans. = Ledum L. (Ericac.).

Dulichium Pers. Cyperaceae. 1 N. Am.

Dulongia Kunth = Phyllonoma Willd. (Dulongiac.).

Dulongiaceae J. G. Agardh. Dicots. 1/8 trop. Am. Trees or shrubs; ls. alt., simple, ent. or serr., acum., with small fimbr. caduc. stips. Infl. epiphyllous, superior, fascic. (cymose false-umbels). Fls. small, green. K 5–4, open, toothed, persist.; C 5–4, valv.; A 5–4, with short fil. and small bilobed anth.; disk large, pulvinate, intra-stam., epigyn.; $\overline{G}$ (2), 1-loc., with ∞ 2-seriate ov. on 2 pariet. plac., and 2 v. short divaric. styles with small stigs. Fr. a small incompl. 2-loc. 3–6-seeded berry; seeds rugose, with endosp. Only genus: *Phyllonoma (Dulongia* Kunth). Related to the *Escalloniac.*, and providing an obvious parallel with the not distantly related *Helwingiac.*

Dumartroya Gaudich. = Malaisia Blanco (Urticac.).

Dumasia DC. Leguminosae. 9 S. Afr., Indomal., E. As.

Dumerilia Lag. ex DC. = Jungia L. (Compos.).

Dumerilia Less. = Perezia Lag. (Compos.).

Dumoria A. Chev. = Tieghemella Pierre (Sapotac.).

Dumreichera Hochst. & Steud. = Senra Cav. (Malvac.).

Dumula Lour. ex Gomes = Severinia Ten. (Rutac.).

Dunalia R.Br. = Torenia L. (Scrophulariac.).

*Dunalia Kunth. Solanaceae. 7 Andes (Colomb. to Argent.).

Dunalia Montr. = Amorphophallus Blume (Arac.).

Dunalia Spreng. = Lucya DC. (Rubiac.).

Dunantia DC. = Isocarpha R.Br. (Compos.).

Dunbaria Wight & Arn. Leguminosae. 25 trop. As., Austr.

Duncania Reichb. = Toddalia Juss. (Rutac.).

Dunnia Tutcher. Rubiaceae. 1 S. China.

Dunstervillea Garay. Orchidaceae. 1 Venez.

Duosperma Dayton. Acanthaceae. 12 trop. & S. Afr.

Duotriaceae Dulac = Cistaceae Juss.

Duparquetia Baill. Leguminosae. 1 trop. W. Afr.

Dupatya Vell. = Paepalanthus Mart. (Eriocaulac.).

Duperrea Pierre ex Pitard. Rubiaceae. 2 India, Indoch., China.
Duperreya Gaudich. = Porana Burm. f. (Convolvulac.).
Dupineta Rafin. = Osbeckia L. (Melastomatac.).
Dupinia Scop. = Ternstroemia L. f. (Theac.).
Dupontia R.Br.(~ Graphephorum Desv.). Gramineae. 2 arctic.
Dupratzia Rafin. = ?Phlox L. (Polemoniac.).
Dupuisia A. Rich. = Sorindeia Thou. (Anacardiac.).
Duquetia G. Don = Duguetia A. St-Hil. (Annonac.).
Durandea Delarbre = Raphanus L. (Crucif.).
***Durandea** Planch. Linaceae. 15 Borneo, New Guinea, New Caled., Fiji.
Durandeëldea Kuntze = Acidoton Sw. (Euphorbiac.).
Durandia Boeck. = Scleria Bergius (Cyperaceae).
Durandoa Pomel = Carthamus L. (Compos.).
Duranta L. Verbenaceae. 36 trop. & S. Am., W.I.
Durantaceae J. G. Agardh = Verbenaceae–Verbeneae Dum.
Durantia Scop. = Duranta L. (Verbenac.).
Duravia Greene = Polygonum L. *s.str.* (Polygonac.).
Duretia Gaudich. = Boehmeria Jacq. (Urticac.).
Duria Scop. = Durio Adans. (Bombacac.).
Duriala (R. H. Anders.) Ulbr. Chenopodiaceae. 1 SE. Austr.
Durieua Boiss. & Reut. = Daucus L. (Umbellif.).
Durieura Mérat = Lafuentea Lag. (Scrophulariac.).
Durio Adans. Bombacaceae. 27 Burma, W. Malaysia. *D. zibethinus* Murr. produces the durian fr., with delicate flavour and disagreeable smell. Seed with fleshy aril.
***Duroia** L.f. Rubiaceae. 20 S. Am. Myrmecophilous (cf. *Acacia*). *D. petiolaris* Hook. f. and *D. hirsuta* K. Schum. have stems swollen just below the infl. The swollen part is hollow and entrance is obtained by two longitudinal slits. *D. saccifera* Benth. & Hook. f. has 'ant-houses' on the l. At the base, on the under side, are two pear-shaped organs formed by outgrowth of the l. The entrance is upon the upper side, protected from rain by a little flap.
Duschekia Opiz (~ Alnus L.). Betulaceae. 10 temp. & arct. Euras., Japan, E. N. Am., W. Greenl.
Dusenia O. Hoffm. ex Dusén (1900; non Brotherus 1894—Musci) = seq.
Duseniella K. Schum. [1902; non Brotherus 1906—Musci]. Compositae. 2 Argent., Patag.
Dussia Krug & Urb. Leguminosae. 10 Mex. to Peru & Guiana, W.I.
Dutaillyea Baill. Rutaceae. 5 New Caled.
Duthiea Hackel. Gramineae. 3 Afghanistan, Kashmir, Nepal.
Dutra Bernh. ex Steud. = Datura L. (Solanac.).
Duttonia F. Muell. (1852) = Helipterum DC. (Compos.).
Duttonia F. Muell. (1856) = Pholidia R.Br. (Myoporac.).
Duvalia Bonpl. = Hypocalyptus Thunb. (Legumin.).
Duvalia Haw. Asclepiadaceae. 10 trop. & S. Afr.
Duvaliella Heim = Dipterocarpus Gaertn. f. (Dipterocarpac.).
Duvaljouvea Palla (~ Cyperus L.). Cyperaceae. 7 S. Am., Euras.
Duvaua Kunth = Schinus L. (Anacardiac.).

DUVAUCELLIA

Duvaucellia Bowdich. Quid? (Loganiac.?) I trop. W. Afr.

Duvernaya Desv. ex DC. = Cuphea Jacq. (Lythrac.).

Duvernoya E. Mey. ex Nees (~Adhatoda Mill., Justicia L.). Acanthaceae.
3 trop. Am., 35 trop. & S. Afr.

Duvigneaudia J. Léonard. Euphorbiaceae. I trop. Afr.

Duvoa Hook. & Arn. = Duvaua Kunth = Schinus L. (Anacardiac.).

Dyanthus P.Br. = Dianthus L. (Caryophyllac.).

Dybowskia Stapf. Gramineae. I trop. W. Afr.

Dychotria Rafin. = Psychotria L. (Rubiac.).

Dyckia Schult. f. Bromeliaceae. 80 warm S. Am.

Dyctioloma DC. (sphalm.) = Dictyoloma A. Juss. (Simaroubac.).

Dyctiospora Reinw. ex Korth. = Hedyotis L. (Rubiac.).

Dyctisperma Rafin. = Rubus L. (Rosac.). [(Palm.).

Dyctosperma H. Wendl. (sphalm.) = Dictyosperma H. Wendl. & Drude

Dydactylon Zoll. = Didactylon Mor. = Dimeria R.Br. (Gramin.).

Dyera Hook. f. Apocynaceae. 2-3 W. Malaysia.

Dyerella Heim = Vateria L. (Dipterocarpac.).

Dyerocycas Nakai = Cycas L. (Cycadac.).

Dyerophytum Kuntze (~Plumbago L.). Plumbaginaceae. 3 S. Afr., Socotra,
Arabia, India.

Dyllwinia Nees (sphalm.) = Dillwynia Sm. (Legumin.).

Dymczewiczia Horan. = Zingiber Boehm. (Zingiberac.).

Dymondia Compton. Compositae. I S. Afr.

Dynamidium Fourr. = Potentilla L. (Rosac.).

Dyneba Lag. = Dinebra Jacq. (Gramin.).

Dyospyros Dum. = Diospyros L. (Ebenac.).

Dyplecosia G. Don = Diplycosia Blume (Ericac.).

Dyplostylis Karst. & Triana = Diplostylis Karst. & Triana = Rochefortia Sw.

Dyplotaxis DC. = Diplotaxis DC. (Crucif.). [(Boraginac.).

Dypontia Dietr. ex Steud. = Dupontia R.Br. (Gramin.).

Dypsidium Baill. (~Neophloga Baill.). Palmae. 3 Madag.

Dypsis Nor. ex Mart. Palmae. 20 Madag.

Dypterygia C. Gay = Dipterygia Presl = Asteriscium Cham. & Schlechtd.

Dyschoriste Nees. Acanthaceae. 100 trop. & subtrop. [(Umbellif.).

Dyscritogyne R. M. King & H. Rob. (~Eupatorium L.). Compositae. 2 Mex.

Dyscritothamnus B. L. Robinson. Compositae. 2 Mexico.

Dysemone Soland. ex Forst. f. = Gunnera L. (Gunnerac.).

Dysinanthus DC. = Disynanthus Rafin. = Antennaria Gaertn. (Compos.).

Dysmicodon Nutt. = Legousia Durand (Campanulac.).

Dysoda Lour. = Serissa Comm. ex Juss. (Rubiac.).

Dysodia DC. = Dyssodia Cav. (Compos.).

Dysodia Spreng. = Dyssodia Willd. = Adenophyllum Pers. (Compos.).

Dysodidendron Gardn. = Saprosma Blume (Rubiac.).

Dysodiopsis Rydb. (~Dyssodia Cav.). Compositae. I SW. U.S.

Dysodium Rich. & Pers. = Melampodium L. (Compos.).

Dysolacoïdeae Engl. (nom. illegit.) = Olacaceae-Anacolosoïdeae Airy Shaw.

Dysolobium Prain. Leguminosae. 4 Indomal.

Dysophylla Blume (s. str.) = Pogostemon Desf. (Labiat.).

Dysophylla El-Gazzar & L. Watson ex Airy Shaw (=Eusteralis Rafin.) Labiatae. 25 temp. As., Austr.

Dysopsis Baill. Euphorbiaceae. 1 Andes, Juan Fernandez.

Dysosma R. E. Woodson (~Podophyllum L.). Podophyllaceae. 3 China.

Dysosmia Miq. = Saprosma Blume (Rubiac.).

Dysosmon Rafin. = Sesamum L. (Pedaliac.).

Dysoxylon Bartl. = seq.

Dysoxylum Blume. Meliaceae. 200 Indomal. *D. fraseranum* Benth. (E. Austr.; Austr. mahogany) and others, good timber.

Dyspemptemorion Bremek. Acanthaceae. 1 trop. S. Am.

Dysphania R.Br. Dysphaniaceae. 5–6 Austr.

*****Dysphaniaceae** (Pax) Pax. Dicots. (Centrosp.). 1/5 Austr. Small branched ± prostrate perenn. herbs, with small ent. or cren. alt. exstip. ls. Fls. small, ♂ or ♀, in dense axill. fascicles, sometimes crowded into leafless false spikes. P 3(–1), valv., membr., persist., accresc.; A 3–1, opp. to P-ls., exserted, with straight fil. and ovoid intr. anth.; disk 0; G̲ (3–2), 1-loc., with 1 hemianatr. (?) ov. and 1–2 filif. styles. Fr. a small achene, surr. by broadly winged accresc. P; embr. not strongly curved. Only genus: *Dysphania*. Somewhat intermediate between *Chenopodiac.* and *Caryophyllac.*

Dyssapindaceae Radlk. (nom. illegit.) = Sapindaceae–Dodonaeoïdeae Kunth.

Dyssochroma Miers. Solanaceae. 4 trop. Am.

Dyssodia Cav. Compositae. 50 SW. U.S., Mex.

Dyssodia Willd. = Adenophyllum Pers. (Compos.).

Dystaenia Kitagawa = Ligusticum L. (Umbellif.).

Dzieduszyckia Rehm. = Ruppia L. (Ruppiac.).

E

Eadesia F. Muell. = Anthocercis Labill. (Solanac.).

Eaplosia Rafin. = Baptisia Vent. (Legumin.).

Earina Lindl. Orchidaceae. 7 N.Z., Polynesia.

Earleocassia Britton = Cassia L. (Legumin.).

Earlia F. Muell. = Graptophyllum Nees (Acanthac.).

Eastwoodia Brandegee. Compositae. 1 Calif. & Lower Calif.

Eatonella A. Gray. Compositae. 2 SW. U.S.

Eatonia Rafin. = Panicum L. + Sphenopholis Scribn. (Gramin.).

Eatonia Riddell ex Rafin. = Ludwigia L. (Onagrac.).

Eatoniopteris Bommer = Cyathea Sm. (Cyatheac.).

Ebandoua Pellegr. Anacardiaceae. 1 W. Equat. Afr.

Ebelia Reichb. = Triodon DC. (Rubiac.).

Ebelingia Reichb. = Harrisonia R.Br. (Simaroubac.).

*****Ebenaceae** Gürke. Dicots. 3/500 trop. (esp. Indomal.). Trees and shrubs with alt., opp. or whorled, simple, leathery, usu. entire, exstip. ls. Fls. axillary, sol. or in small cymes, reg., usu. dioec., bracteolate, 3–7-merous. (K) persistent, very varied; (C) convolute; A epipet. at base of tube, free or united, isomerous or in 2 whorls but frequently ∞ by branching; stds. usu. present in ♀ fls.; G̲, rarely G̅, (2–16), with 1–2 pend. anatr. ov. in each loc.; styles 2–8, free or united below (rarely completely). Fr. usu. a berry with fewer seeds than there were ovules, sometimes dehiscent. Embryo straight or slightly curved, in abundant cartilaginous endosp. Many *Diospyros* yield valuable wood. Genera: *Euclea, Diospyros, Lissocarpa.* Related to *Annonac.*

Ebenidium Jaub. & Spach = Ebenus L. (Legumin.).

Ebenopsis Britton & Rose = Siderocarpos Small = Pithecellobium Mart. (Legumin.).

Ebenoxylon Spreng. = seq.

Ebenoxylum Lour. = Maba J. R. & G. Forst. = seq.

Ebenum Rafin. = Diospyros L. (Ebenac.).

Ebenus Kuntze = Maba J. R. & G. Forst. = praec.

Ebenus L. Leguminosae. 18 E. Medit. to Baluchistan.

Eberhardtia Lecomte. Sapotaceae. 3 Indoch.

Eberlanzia Schwantes. Aïzoaceae. 11 S. Afr.

Eberlea Riddell ex Nees = Hygrophila R.Br. + Justicia L. (Acanthac.).

Ebermaiera Nees = Staurogyne Wall. (Acanthac. vel Scrophulariac.).

Ebnerella F. Buxb. = Mammillaria Haw. (Cactac.).

Ebracteola Dinter & Schwantes. Aïzoaceae. 3 S. Afr.

Ebraxis Rafin. = Silene·L. (Caryophyllac.).

Ebulum Garcke = Sambucus L. (Sambucac.).

Ebulus Fabr. (uninom.) = *Sambucus ebulus* L. (Sambucac.).

Eburnangis Thou. (uninom.) = *Angraecum eburneum* Bory (Orchidac.).

Eburnax Rafin. = Mimosa L. (Legumin.).

Eburopetalum Becc. = Anaxagorea A. St-Hil. (Annonac.).

Eburophyton A. A. Heller = Cephalanthera Rich. (Orchidac.).

Ecastaphyllum P.Br. (∼ Dalbergia L. f.). Leguminosae. 6 trop. Am., W.I.

*Ecballium A. Rich. Cucurbitaceae. 1 Medit., *E. elaterium* (L.) A. Rich. (squirting cucumber). The ripe fr. is highly turgid; as it drops from the stalk, a hole is made in its lower end, and through this the contraction of the pericarp squirts the seeds, mixed with a watery fluid. A purgative (elaterium) is prepared from the fr.

Ecbolium Kuntze = Justicia L. (Acanthac.).

Ecbolium Kurz. Acanthaceae. 19 trop. & S. Afr., Madag., Socotra, India.

Ecclinusa Mart. Sapotaceae. 20 trop. S. Am., Trinidad.

Eccoilopus Steud. Gramineae. 4 N. India to Japan & Formosa.

Eccoptocarpha Launert. Gramineae. 1 S. trop. Afr.

Eccremanthus Thw. = Pometia J. R. & G. Forst. (Sapindac.).

Eccremis Baker = Excremis Willd. (Liliac.).

Eccremocactus Britton & Rose. Cactaceae. 3 Costa Rica, Ecuador.

Eccremocarpus Ruiz & Pav. Bignoniaceae. 5 W. S. Am. The valves of the fr. hang together at the top.

Eccremocereus Frič & Kreuz. = Eccremocactus Britt. & Rose (Cactac.).

Ecdeiocolea F. Muell. Ecdeiocoleaceae. 1 SW. Austr.

Ecdeiocoleaceae Cutler & Airy Shaw. Monocots. (Farinosae). 1/1 W. Austr. Glabrous perenn. herbs with simple erect slender cylindrical wiry grooved glaucous stems from·a creeping rhizome, bearing several basal brown sheathing scales (the uppermost elongated) and a single elongate sheathing scale towards the top of the stem. Inflorescence a terminal conical or cylindr. spike, with

∞ rigid imbr. obtuse shining dark brown glumes, but without a basal sheathing bract. Fls. ♂ ♀, monoec., flattened. P 3 + 3, unequal, glumaceous; A 3–4, fil. free, ± short, anth. dorsifixed, with distinct thecae (in ♀ fl. 3 stds.); G̲ (2) with 1 pend. ov. per loc., and 2 long free styles stigmatic to near base. Fr. unknown. Only genus: *Ecdeiocolea*. Near *Restionaceae*(?), but habit of *Xyridaceae*. (*Kew Bull*. **19**: 495–9, 1965.)

Ecdysanthera Hook. & Arn. Apocynaceae. 15 China, Indomal.

Echaltium Wight = Melodinus J. R. & G. Forst. (Apocynac.).

Echeandia Ortega. Liliaceae. 12 Mexico to Guiana.

Echenais Cass. = Cirsium Mill. (Compos.).

× **Echephytum** Gossot. Crassulaceae. Gen. hybr. (Echeveria × Pachy-**Echetrosis** Phil. Compositae. 1 temp. S. Am. [phytum).

Echeveria DC. Crassulaceae. 150 S. U.S. to Argent.

Echidiocarya A. Gray = Plagiobothrys Fisch. & Mey. (Boraginac.).

Echidnium Schott. Araceae. 2 trop. S. Am.

Echidnopsis Hook. f. Asclepiadaceae. 25 E. Afr. to Arabia.

Echinacanthus Nees. Acanthaceae. 9 Himal., Siam, Java.

Echinacea Moench. Compositae. 3 Atl. N. Am.

Echinais C. Koch = Echenais Cass. = Cirsium Mill. (Compos.).

Echinalysium Trin. = Elytrophorus Beauv. (Gramin.).

Echinanthus Cerv. & Cord. = Tragus Hall. (Gramin.).

Echinanthus Neck. = Echinops L. (Compos.).

*****Echinaria** Desf. Gramineae. 2 Medit.

Echinaria Fabr. (uninom.) = Cenchrus L. sp. (Gramin.).

× **Echinobivia** Rowley. Cactaceae. Gen. hybr. (Echinopsis × Lobivia).

Echinocactus Fabr. (uninom.) = Melocactus Link & Otto sp. (Cactac.).

Echinocactus Link & Otto. Cactaceae. 10 S. U.S., Mex.

Echinocalyx Benth. = Sindora Miq. (Legumin.).

Echinocarpus Blume = Sloanea L. (Elaeocarpac.).

Echinocassia Britton & Rose = Cassia L. (Legumin.).

Echinocaulon Meissn. ex Spach (nomen) = seq.

Echinocaulos (Meissn. ex Endl.) Hassk. (∼ Polygonum L.). 1 E. Himal., E. & SE. As., Philipp., Sumatra, Java.

Echinocephalum Gardn. Compositae. 3 Brazil.

Echinocereus Engelm. Cactaceae. 75 S. U.S., Mex.

Echinochlaena Spreng. = Echinolaena Desv. (Gramin.).

Echinochlaenia Börner = Carex L. (Cyperac.).

Echinochloa Beauv. Gramineae. 30 warm.

Echinocitrus Tanaka. Rutaceae. 1 New Guinea.

Echinocroton F. Muell. = Mallotus Lour. (Euphorbiac.).

*****Echinocystis** Torr. & Gray. Cucurbitaceae. 15 Am. Tuberous climbing herbs. The tendrils of *E. lobata* Torr. & Gray are very sensitive and nutate rapidly; they become straight and erect as they come round towards the main axis, thus avoiding contact.

Echinodendrum A. Rich. = Scolosanthus Vahl (Rubiac.).

Echinodiscus Benth. = Pterocarpus L. (Legumin.).

Echinodium Poit. ex Cass. = Acanthospermum Schrank (Compos.).

Echinodorus Rich. Alismataceae. 30 Am., Afr.

ECHINOFOSSULOCACTUS

Echinofossulocactus Lawrence. Cactaceae. 32 Mexico.
Echinoglochin A. Brand. Boraginaceae. 8 Pacif. N. Am.
Echinoglossum Reichb. = Echioglossum Blume = Sarcanthus Lindl. (Orchidac.).
Echinolaena Desv. Gramineae. 6 C. & S. Am., 1 Madag.
Echinolema Jacq. f. ex DC. = Acicarpha Juss. (Calycerac.).
Echinolitrum Steud. = Echinolytrum Desv. = Fimbristylis Vahl (Cyperac.).
Echinolobium Desv. = Hedysarum L. (Legumin.).
Echinolobivia Y. Ito = Lobivia Britt. & Rose (Cactac.).
Echinoloma Steud. = Echinolema Jacq. f. ex DC. = Acicarpha Juss. (Calycerac.).
Echinolysium Benth. = Echinalysium Trin. = Elytrophorus Beauv. (Gramin.).
Echinolytrum [recte Echinelytrum] Desv. = Fimbristylis Vahl (Cyperac.).
Echinomastus Britton & Rose. Cactaceae. 10 S. U.S., Mexico.
Echinomeria Nutt. = Helianthus L. (Compos.).
Echinonyctanthus Lem. = Echinopsis Zucc. (Cactac.).
Echinop[ac]eae Link = Compositae–Cynareae–Echinopinae O. Hoffm.
Echinopaepale Bremek. Acanthaceae. 1 W. Java.
Echinopanax Decne & Planch. Araliaceae. 3 NE. As., N. Am. *E. horridus* D. & P., an obstacle to travellers.
Echinopepon Naud. = Echinocystis Torr. & Gray. (Cucurbitac.).
Echinophora L. Umbelliferae. 10 Medit. to Persia. One cpl. is aborted. The umbel has one ☿ fl. in the centre, surrounded by ♂ fls. The spiny stalks of the latter enclose the fr.
Echinopiaceae Link corr. Bullock = Compositae–Cynareae–Echinopinae O. Hoffm.
Echinopogon Beauv. Gramineae. 7 Austr., N.Z.
Echinops L. Compositae. 100 E. Eur., Afr., As. The spherical head is really cpd., formed of ∞ small 1-fld. heads, each with its own invol. The fls. are largely visited by bees.
Echinopsidaceae Link = Echinopiaceae Link corr. Bullock = Compositae–Cynareae–Echinopinae O. Hoffm.
Echinopsilon Moq. = Bassia All. (Chenopodiac.).
Echinopsis Zucc. Cactaceae. 35 S. Am.
Echinopsus St-Lag. = Echinops L. (Compos.).
Echinopterys A. Juss. Malpighiaceae. 3 Mexico. Mericarp spiny.
Echinopteryx Dalla Torre & Harms = praec.
Echinopus Mill. = Echinops L. (Compos.).
Echinorebutia Frič = Rebutia K. Schum. (Cactac.).
Echinoschoenus Nees & Meyen = Rhynchospora Vahl (Cyperac.).
Echinosciadium Zohary = Anisosciadium DC. (Umbellif.).
× **Echinosicyos** [sphalm. '-sycios'] Kamner & Topa. Cucurbitaceae. Gen. hybr. [Echinocystis × Sicyos].
Echinosophora Nakai (~ Sophora L.). Leguminosae. 1 Korea.
Echinosparton Fourr. = seq.
Echinospartum (Spach) Rothm. (~ Genista L.). Leguminosae. 3–4 SW. Eur.
Echinospermum Sw. = Lappula v. Wolf (Boraginac.).
Echinosphaera Sieber ex Steud. = Ricinocarpos Desf. (Euphorbiac.).

Echinostachys Brongn. = Aechmea Ruiz & Pav. (Bromeliac.).
Echinostachys E. Mey. = Pycnostachys Hook. (Labiat.).
Echinothamnus Engl. = Adenia Forsk. (Passiflorac.).
Echinus L. Bolus = Braunsia Schwantes (Aïzoac.).
Echinus Lour. = Mallotus Lour. (Euphorbiac.).
Echiochilon Desf. Boraginaceae. 6 N. Afr., Arabia.
Echiochilopsis Caballero. Boraginaceae. 1 NW. Afr.
Echiochilus P. & K. = Echiochilon Desf. (Boraginac.).
Echiodes P. & K. (ter) = Echioïdes Desf., Fabr., Moench, qq. v.
Echioglossum Blume = Sarcanthus Lindl. (Orchidac.).
Echioïdes Desf. = Nonea Medik. (Boraginac.).
Echioïdes Fabr. (uninom.) = Lycopsis L. sp. (Boraginac.).
Echioïdes Moench = Myosotis L. (Boraginac.).
Echioïdes Ortega = Aipyanthus Steven (Boraginac.).
Echion St-Lag. = Echium L. (Boraginac.).
Echiopsis Reichb. = Lobostemon Lehm. (Boraginac.).
Echiostachys Levyns (~ Lobostemon Lehm.). Boraginaceae. 3 S. Afr.
Echirospermum Allem. ex Saldanha da Gama = mixture of Cassia, Caesalpinia, etc. (Legumin.).
Echisachys Neck. = Tragus Hall. (Gramin.).
Echitella Pichon. Apocynaceae. 2 Madag.
Echites P.Br. Apocynaceae. 6 Florida, Mex. to NE. Colombia, W.I.
Echithes Thunb. = praec.
Echium L. Boraginaceae. 40 Canaries, Azores, N. & S. Afr., Eur., W. As.
E. vulgare L. offic. Fl. zygo., protandr., gynodioec., bee-visited.
Echthronema Herb. = Sisyrinchium L. (Iridac.).
Echtrosis Dur. (sphalm.) = Echetrosis Phil. Compos.).
Echtrus Lour. = Argemone L. (Papaverac.).
Echyrospermum Schott = ? Echirospermum Allem. ex Sald. da Gama = mixture of Cassia, Caesalpinia, etc. (Legumin.).
Eckardia Endl. = seq.
Eckartia Reichb. = Peristeria Hook. (Orchidac.).
Eckebergia Batsch = Ekebergia Sparrm. (Meliac.).
Ecklonea Steud. (1829; non Ecklonia Hornem. 1828—Algae) = Trianoptiles Fenzl (Cyperac.).
Ecklonia Schrad. = praec.
*Eclipta L. Compositae. 3–4 warm Am., Afr., As., Austr. Pappus o.
Ecliptica Kuntze = praec.
Ecliptostelma T. S. Brandegee. Asclepiadaceae. 1 Mexico.
Eclopes Gaertn. = Relhania L'Hérit. (Compos.).
Eclotoripa Rafin. = Digera Forsk. (Amaranthac.).
Eclypta E. Mey. = Eclipta L. (Compos.).
Ecphymacalyx Pohl (nomen). 1 Brazil. Quid?
Ecpoma K. Schum. Rubiaceae. 1 trop. Afr.
Ecpomanthera Pohl (nomen). 1 Brazil. Quid?
Ectadiopsis Benth. Periplocaceae. 2 trop. & S. Afr.
Ectadium E. Mey. Periplocaceae. 2 S. Afr.
Ectasis D. Don = Erica L. (Ericac.).

ECTEINANTHUS

Ecteinanthus T. Anders. = Isoglossa Oerst. (Acanthac.).
Ectemis Rafin. = Cordia L. (Ehretiac.).
Ectinanthus P. & K. = Ecteinanthus T. Anders. = Isoglossa Oerst. (Acanthac.).
Ectinocladus Benth. Apocynaceae. 1 W. Afr.
Ectosperma Swallen (1950; non Vaucher 1803—Algae) = Swallenia Soderstr.
& Decker (Gramin.).
Ectotropis N. E. Brown. Aïzoaceae. 1 S. Afr.
Ectozoma Miers = Juanulloa Ruiz & Pav. (Solanac.).
Ectrosia R.Br. Gramineae. 12 trop. Austr.
Ectrosiopsis (Ohwi) Jansen. Gramineae. 5 Philippines, E. Malaysia.
Edanthe O. F. Cook & Doyle = Chamaedorea Willd. (Palm.).
Edanyoa Copel. Lomariopsidaceae (?). 1 Philippines.
Edbakeria R. Viguier. Leguminosae. 1 Madag.
Eddya Torr. & Gray = Coldenia L. (Ehretiac.).
Edechi Loefl. = Guettarda L. (Rubiac.).
Edemias Rafin. = Conyza L. (Compos.).
Edgaria C. B. Clarke. Cucurbitaceae. 1 E. Himal.
Edgeworthia Falc. = Reptonia A. DC. (Myrsinac.).
Edgeworthia Meissn. Thymelaeaceae. 3 Himalaya to Japan.
Edisonia Small (~ Gonolobus Michx). Asclepiadaceae. 1 SE. U.S.
Editeles Rafin. = Lythrum L. (Lythrac.).
Edithcolea N. E. Br. Asclepiadaceae. 2 Somal., Socotra.
Edithea Standley. Rubiaceae. 1 Mex.
Edmondia Cass. = Helichrysum L. (Compos.).
Edmondia Cogn. = Calycophysum Triana (Cucurbitac.).
Edmonstonia Seem. = Tetrathylacium Poepp. (Violac.).
Edosmia Nutt. ex Torr. & Gray = Carum L. (Umbellif.).
Edouardia Corrêa de Mello apud Stellfeld = Melloa Bur. (Bignoniac.).
Edraianthus A. DC. (~ Wahlenbergia Schrad. ex Roth). Campanulaceae. 10
SE. Eur. to Cauc.
Edrajanthus auctt. = praec.
Edrastenia B. D. Jacks. = seq.
Edrastima Rafin. = Hedyotis L. (Rubiac.).
Edrissa Endl. = Hedyotis L. (Rubiac.).
Edritria Rafin. = Carex L. (Cyperac.).
Eduardoregelia Popov (~ Tulipa L.). Liliaceae. 1 C. As.
Edusaron Medik. = Desmodium Desv. (Legumin.).
Edusarum Steud. = praec.
Edwardia Rafin. = Cola Schott & Endl. (Sterculiac.).
Edwardsia Endl. = Edwarsia Neck. = Bidens L. (Compos.).
Edwardsia Salisb. (~ Sophora L.). Leguminosae. 6 India, Hawaii, N.Z.,
Easter I., Juan Fernandez, Chile.
Edwarsia Dum. = praec.
Edwarsia Neck. = Bidens L. (Compos.).
Edwinia A. A. Heller = Jamesia Torr. & Gray (Philadelphac.).
Eeldea Th. Dur. (= ? Weldenia Reichb.) = Candollea Labill. (1806) = Hibbertia
Andr. (Dilleniac.).
Eenia Hiern & S. Moore. Compositae. 1 SW. trop. Afr.

Efossus Orcutt = Echinofossulocactus Lawr. (Cactac.).

Efulensia C. H. Wright = Deïdamia Nor. ex Thou. (Passiflorac.).

Egania Remy = Chaetanthera Ruiz & Pav. (Compos.).

Eganthus Van Tiegh. Olacaceae. 1 Brazil.

Egassea Pierre ex De Wild. = Oubanguia Baill. (Scytopetalac.).

Egena Rafin. = Clerodendrum L. (Verbenac.).

Egenolfia Schott. Lomariopsidaceae. 10 trop. As. Habit of *Bolbitis*; veins Egeria Néraud (nomen). Rubiaceae. Mauritius. Quid?   [free.

Egeria Planch. (~ Elodea Michx). Hydrocharitaceae. 2 subtrop. S. Am.

Eggelingia Summerhayes. Orchidaceae. 2 trop. Afr.

Eggersia Hook. f. = Neea Ruiz & Pav. (Nyctaginac.).

Egleria Eiten. Cyperaceae. 1 Amaz. Brazil.

Eglerodendron Aubrév. & Pellegr. Sapotaceae. 1 Brazil.

Egletes Cass. Compositae. 10 Mexico to trop. S. Am., W.I.

Ehrardia Benth. & Hook. f. = Ehrhardia Scop. = Aiouea Aubl. (Laurac.).

Ehrartha Beauv. = Ehrharta Thunb. (Gramin.).

Ehrartia Benth. (sphalm.) = Ehrhartia Weber = Leersia Sw. (Gramin.).

Ehrenbergia Mart. = Kallstroemia Scop. (Zygophyllac.).

Ehrenbergia Spreng. = Amaioua Aubl. (Rubiac.).

Ehretia P.Br. Ehretiaceae. 50 warm, chiefly Old World. Timber.

*Ehretiaceae Lindl. Dicots. 13/400 trop. & subtrop. Trees or shrubs, rarely herbs, with alt. (rarely subopp.) simple ent. or dent. exstip. ls. Infl. cymose (sometimes spiciform or capit.), term. or axill. or leaf-opp. K (5), sometimes membr. and accresc.; C [4–](5)[–6 +], imbr., rarely valv.; A 5, alternipet. and epipet., incl. or exserted; G (2–4), with 2 erect (rarely pend.) axile ov. per loc. (loc. may be 2-locell.), and 2 free or ± united styles, stig. bifid or capit. Fr. a hard and dry or baccate drupe, often encl. in persist. K; endosp. + or −. Chief genera: *Cordia, Ehretia, Bourreria, Halgania, Coldenia*. Scarcely distinct from *Boraginac*.

Ehretiana Collinson = Periploca L. (Periplocac.).

Ehrardia Scop. = Aiouea Aubl. (Laurac.).

Ehrhardta Hedw. = seq.

*Ehrharta Thunb. Gramineae. 27 S. Afr., Masc., N.Z. Useful pasture grasses for sandy soil.

Ehrhartia P. & K. (1) = Ehrhardia Scop. = Aiouea Aubl. (Laurac.).

Ehrhartia P. & K. (2) = Ehrharta Thunb. (Gramin.).

Ehrhartia Weber = Leersia Sw. (Gramin.).

*Eichhornia Kunth. Pontederiaceae. 7 SE. U.S. to Argent., W.I. The sympodium is very complex. Each shoot in turn is pushed to one side by the axillary shoot of its last l. but one; with this shoot it is combined, however, up to the last l. of the axillary shoot. After leaving the axillary shoot, each shoot bears another l., and then ends in the infl., which is enclosed in a spathe, and at first glance appears to spring from the stalk of the last l. In *E. azurea* Kunth the fls. are dimorphic, in *E. crassipes* Solms trimorphic heterostyled. This last sp. has, when floating freely, large bladder-like swollen petioles, but in soil these are not nearly so large. They cause the plant to float high and it is easily blown about by wind, and has become a very troublesome weed (water hyacinth) in Florida, Africa, Java, Australia, etc.

EICHLERIA

Eichleria Hartog = Labourdonnaisia Boj. (Sapotac.).

Eichleria Progel. Oxalidaceae. 2 E. Brazil.

Eichlerina Van Tiegh. = Loranthus L. (Loranthac.).

Eichlerodendron Briquet. Flacourtiaceae. 7 Mex. to Brazil.

Eichornia A. Rich. = Eichhornia Kunth (Pontederiac.).

Eichwaldia Ledeb. = Reaumuria L. (Tamaricac.).

Eicosia Blume (sphalm.) = Eucosia Bl. (Orchidac.).

Eilemanthus Hochst. = Indigofera L. (Legumin.).

Einadia Rafin. = Rhagodia R.Br. (Chenopodiac.).

Einomeia Rafin. = Aristolochia L. (Aristolochiac.).

Einomeria Reichb. = praec.

Einsteinia Ducke = Kutchubaea Fisch. ex DC. (Rubiac.).

Eionitis Bremek. Rubiaceae. 2 Somaliland.

Eisenmannia Sch. Bip. = Blainvillea Cass. (Compos.).

Eisocreochiton Quisumb. & Merrill. Melastomataceae. 1 Philipp. Is.

Eizaguirrea Remy = Leuceria Lag. (Compos.).

Eizia Standley. Rubiaceae. 1 Mex.

Ekebergia Sparrm. Meliaceae. 15 S. & trop. Afr., Madag.

Ekmania Gleason. Compositae. 1 Cuba.

Ekmanianthe Urb. Bignoniaceae. 2 Cuba, Hispaniola.

Ekmaniocharis Urb. Melastomataceae. 1 Haiti.

Ekmanochloa Hitchcock. Gramineae. 2 Cuba.

Elachanthera F. Muell. (~ Luzuriaga Ruiz & Pav.). Philesiaceae. 1 Austr.

Elachanthus F. Muell. Compositae. 3 temp. Austr.

Elachia DC. = Chaetanthera Ruiz & Pav. (Compos.).

Elachocroton F. Muell. = Sebastiania Spreng. (Euphorbiac.).

Elacholoma F. Muell. & Tate. Scrophulariaceae. 1 C. Austr.

Elachopappus F. Muell. = Myriocephalus Benth. (Compos.).

Elachothamnos DC. = Minuria DC. (Compos.).

Elachyptera A. C. Smith. Celastraceae. 3 C. Am. to S. trop. S. Am.

Elaeagia Wedd. Rubiaceae. 10 C. & trop. S. Am., Cuba.

*****Elaeagnaceae** Juss. Dicots. 3/50 N. hemisph., chiefly on steppes and coasts. Much branched shrubs, often with leathery ls., entire, opp. or alt., and covered, as are all parts, with scaly hairs. Thorns are frequently present (reduced shoots). Infl. racemose; fls. ♀ or unisexual, 2- or 4-merous. In the ♂ the recept. is often flat, but in the ♀ or ♀ fl. it is tubular as in *Thymelaeaceae*, and may be fused with the ovary. No petals. Sta. as many, or twice as many, as sepals. G 1 with one erect anatr. ov. Pseudo-drupe. Seed with little or no endosp. Genera: *Hippophaë, Elaeagnus, Shepherdia.*

Elaeagnus L. Elaeagnaceae. 45 Eur., As., N. Am. (oleaster). The fr. of some is ed.

Elaeagrus Pall. = praec.

Elaeis Jacq. Palmae. 2, one trop. Am., the other, *E. guineënsis* Jacq. (oil-palm), trop. Afr., from whose fr. the palm-oil, used for railway axles, etc., is obtained by boiling.

*****Elaeocarpaceae** DC. Dicots. 12/350 trop. & subtrop. Trees and shrubs with alt. or opp. stip. ls., and racemes, panicles or dichasia of fls. K 4 or 5, free or united, valvate; C 4 or 5, rarely united, often 0, the petals often much divided

ELATINACEAE

at the ends, valv. or imbr. (never contorted; disk usu. present; A ∞, free, on the disk, which is sometimes developed to an androphore; anthers 2-loc., usu. opening by two pores (sometimes confluent) at the apex; G̲ sessile, with 2–∞ (rarely 1) loc.; ovules ∞ or 2 per loc., anatr., pend., with ventral raphe; style simple, sometimes lobed at apex. Capsule or drupe; embryo straight, in abundant endosp. The *E.* show indications of relationship with *Combretac.*, *Rhizophorac.* and *Tiliac.* Chief genera: *Elaeocarpus, Sloanea, Aristotelia.*

Elaeocarpus L. Elaeocarpaceae. 200 E. As., Indomal., Austr., Pacif.
Elaeocharis Brongn. = Eleocharis R.Br. (Cyperac.).
Elaeochytris Fenzl = Peucedanum L. (Umbellif.).
Elaeococca Comm. ex A. Juss. = Vernicia Lour. (Euphorbiac.).
Elaeococcus Spreng. = praec.
Elaeodendron J. F. Jacq. ex Jacq. Celastraceae. 16–17 trop. & subtrop.
Elaeodendrum Murr. = praec.
Elaeogene Miq. = Vatica L. (Dipterocarpac.).
Elaeoluma Baill. Sapotaceae. 2 Brazil.
Elaeophora Ducke = Plukenetia L. (Euphorbiac.).
Elaeophorbia Stapf (~ Euphorbia L.). Euphorbiaceae. 5 trop. & S. Afr.
Elaeopleurum Korovin. Umbelliferae. 1 C. As.
Elaeoselinum Koch ex DC. Umbelliferae. 10 W. Medit.
Elaeosticta Fenzl (~ Conopodium Koch). Umbelliferae. 7 SW. & C. As.
Elais L. = Elaeis Jacq. (Palm.).
Elangis Thou. (uninom.) = *Angraecum elatum* Thou. = *Cryptopus elatus* (Thou.) Lindl. (Orchidac.).
Elaphoglossaceae [Herter & Ching ex] Pichi-Sermolli = Lomariopsidaceae Alston.
*****Elaphoglossum** Schott. Lomariopsidaceae. 400 trop. & subtrop., esp. trop. Am. Acrostichoid epiphytes; ls. simple, articulated to phyllopodia.
Elaphrium Jacq. = Bursera Jacq. ex L. (Burserac.).
Elasmatium Dulac = Goodyera R.Br. (Orchidac.).
Elasmocarpus Hochst. ex Chiov. = Indigofera L. (Legumin.).
Elate L. = Phoenix L. (Palm.).
Elateriodes Kuntze = Elateriospermum Bl. (Euphorbiac.).
Elateriopsis Ernst. Cucurbitaceae. 3 S. Am.
Elateriospermum Blume. Euphorbiaceae. 1 S. Siam, W. Malaysia.
Elaterium Jacq. = Rytidostylis Hook. & Arn. (Cucurbitac.).
Elaterium Mill. = Ecballium A. Rich. (Cucurbitac.).
Elateum Rafin. = Phoenix L. (Palm.).
*****Elatinaceae** Dum. Dicots. 2/40 trop. & temp. Undershrubs, herbs, or annual water pls.; the latter are able to live on land, altering their structure to suit the changed conditions (cf. *Littorella*). Ls. opp. or whorled, simple, ent. or dent., with interpet. stip. Fls. ♀̂, reg., solitary or in dichasia, 2–6-merous. K hypog., free or united; C imbr.; A in 2 whorls, or inner aborted; G̲ syncarpous, multiloc., with free styles; plac. axile; ov. ∞, anatr. Capsule septifragal; seeds straight to strongly curved; testa usu. strongly sculptured; endosp. thin or none. Genera: *Bergia, Elatine.* A probable 'peripheral Centrosperm' group (cf. *Frankeniac.* & *Tamariac.*), but with possible connections also to *Hippuridac., Haloragidac., Lythrac.,* etc.

405

ELATINE

Elatine Hill = Kickxia Dum. (Scrophulariac.).

Elatine L. Elatinaceae. 20 trop. & temp.

Elatinella (Seub.) Opiz = praec.

Elatinoïdes Wettst. = Kickxia Dum. (Scrophulariac.).

Elatinopsis Clav. = Elatine L. (Elatinac.).

Elatosema Franch. & Sav. (sphalm.) = seq.

Elatostema J. R. & G. Forst. (as to type!) = Procris Comm. ex Juss. (Urticac.).

*Elatostema Gaudich. Urticaceae. 200 trop. Old World. *E. acuminatum* Brongn. is apogamous. Some show water secretion from the ls.

Elatostematoïdes C. B. Robinson. Urticaceae. 15 Philipp. Is.

Elatostemma Endl. = Elatostema Gaud. + Procris Comm. ex Juss. (Urticac.).

Elatostemon P. & K. = Elatostema Gaud. (Urticac.).

Elattosis Gagnep. = Tenagocharis Hochst. (Limnocharitac.).

Elattospermum Solereder. Rubiaceae. 1 Madag.

Elattostachys Radlk. Sapindaceae. 14 Malaysia, Austr., Polynesia.

Elayuna Rafin. = Piliostigma Hochst. (Legumin.).

Elbunis Rafin. = Ballota L. (Labiat.).

Elburzia Hedge. Cruciferae. 1 NW. Persia.

Elcaja Forsk. = Trichilia R.Br. (Meliac.).

Elcana Blanco = Cerbera L. (Apocynac.).

Elcismia Robinson = Celmisia Cass. ex DC. (1836) (Compos.).

Eleagnus Hill = Elaeagnus L. (Elaeagnac.).

Electra DC. = Coreopsis L. (Compos.).

Electra (Electtra) Nor. = Sapindus L. (Sapindac.).

Electra Panz. = Schismus Beauv. (Gramin.).

Electrosperma F. Muell. = Eriocaulon L. (Eriocaulac.).

Elegia L. Restionaceae. 32 S. Afr.

Eleiastis Rafin. = Tetracera L. (Dilleniac.).

Eleiodoxa (Becc.) Burret. Palmae. 5 Indomal.

Eleiosina Rafin. (Sibiraea Maxim.). Rosaceae. 5 Balkans, C. As., W. China.

Eleiotis DC. Leguminosae. 1 India, Ceylon.

Elelis Rafin. = Salvia L. (Labiat.).

Elemanthus [Schlechtd.] = Eilemanthus Hochst. = Indigofera L. (Legumin.).

Elemi Adans. = Amyris P.Br. (Rutac.).

Elemifera Burm. f. = praec.

Elengi Adans. = Mimusops L. (Sapotac.).

Eleocaris Sanguinetti = seq.

Eleocharis R.Br. Cyperaceae. 200 cosmop. In *E. palustris* (L.) Roem. & Schult. the green tissue is centric. The tubers of *E. tuberosus* Schult. (E. As.) are used as food.

Eleogenus Nees = praec.

Eleogiton Link = Scirpus L. (Cyperac.).

Eleorchis Maekawa. Orchidaceae. 2 Japan.

Elephantella Rydb. = Pedicularis L. (Scrophulariac.).

Elephantina Bertol. = Rhynchocorys Griseb. = Rhinanthus L. (Scrophulariac.).

Elephantodon Salisb. = Dioscorea L. (Dioscoreac.).

Elephantopsis A. Dietr. = Elephantosis Less. = seq.

Elephantopus L. Compositae. 32 trop. *E. scaber* L. is an abundant and troublesome weed.
Elephantorrhiza Benth. Leguminosae. 10 trop. & S. Afr.
Elephantosis Less. = Elephantopus L. (Compos.).
Elephantusia Willd. = Phytelephas Ruiz & Pav. (Palm.).
Elephas Mill. = Rhynchocorys Griseb. (Scrophulariac.).
Elettaria Maton. Zingiberaceae. 7 Indomal. Fls. on leafless shoots from the rhiz. *E. cardamomum* Maton, cult. in the mountains of Ceylon and S. India (cardamoms). The ripe frs. are picked and dried; the seeds form a strongly flavoured spice, mainly used in India.
Elettariopsis Baker. Zingiberaceae. 10 Indomal.
Eleusine Gaertn. Gramineae. 9 trop. & subtrop.; 1, *E. tristachya* (Lam.) Kunth, temp. S. Am. *E. coracana* (L.) Gaertn. (*ragi, kurakkan*) is cult. as a cereal and/or alcoholic beverage in Ceylon, India, Africa, etc.; others are useful fodders.
Eleutherandra van Slooten. Flacourtiaceae. 1 Sumatra, Borneo.
Eleutheranthera Poit. ex Bosc. Compositae. 1 trop. Am.; 1 Madag.
Eleutheranthus K. Schum. = Eleuthranthes F. Muell. (Rubiac.).
Eleutheria Triana & Planch. = Elutheria M. Roem. = Schmardaea Karst. (Meliac.).
***Eleutherine** Herb. Iridaceae. 2 C. & trop. S. Am., W.I.; 2 Indoch. Bulb.
Eleutherocarpum Schlechtd. = Osteomeles Lindl. (Rosac.).
Eleutherococcus Maxim. Araliaceae. 15 Himal. to Japan.
Eleutheropetalum (H. Wendl.) H. Wendl. ex Oerst. (~ Chamaedorea Willd.). Palmae. 2 trop. Am.
Eleutherospermum C. Koch. Umbelliferae. 2 SW. As.
Eleutherostemon Herzog = Diogenesia Sleum. (Ericac.).
Eleutherostemon Klotzsch = Philippia Klotzsch (Ericac.).
Eleutherostigma Pax & Hoffm. Euphorbiaceae. 1 Colombia.
Eleutherostylis Burret. Tiliaceae. 1 New Guinea.
Eleuthrantheron Steud. = Eleutheranthera Poit. ex Bosc (Compos.).
Eleuthranthes F. Muell. Rubiaceae. 1 W. Austr.
Eliaea Cambess. Guttiferae. 1 Madag.
Eliastis P. & K. = Eleiastis Rafin. = Tetracera L. (Dilleniac.).
Elichrys[ac]eae Link = Compositae–Inuleae Cass.
Elichrysum Mill. = Helichrysum L. (Compos.).
Elictotrichon Bess. ex Andrz. = Helictotrichon Bess. ex Roem. & Schult. (Gramin.).
Elide Medik. = Asparagus L. (Liliac.).
Elidurandia Buckl. = Fugosia Juss. (Malvac.).
Eliea G. Don = Eliaea Cambess. (Guttif.).
Eligia Dum. = Elegia L. (Restionac.).
Eligmocarpus Capuron. Leguminosae. 1 Madag.
Elimus Nocca = Elymus L. (Gramin.).
Elingamita Baylis. Myrsinaceae. 1 N.Z.
Eliokarmos Rafin. = Ornithogalum L. (Liliac.).
Elionurus Humb. & Bonpl. ex Willd., corr. Kunth. Gramineae. 25 trop. & subtrop.

ELVIRA

Elvira Cass. Compositae. 4 Mex., Galápagos Is.

Elwendia Boiss. = Carum L. (Umbellif.).

Elwertia Rafin. = Clusia L. (Guttif.).

× **Elyhordeum** Mansf. Gramineae. Gen. hybr. (Elymus × Hordeum).

Elymandra Stapf. Gramineae. 6 trop. Afr.

× **Elymopyrum** Cugnac = × Agroëlymus Cugnac (Gramin.).

× **Elymordeum** Lepage. Gramineae. Gen. hybr. (Elymus × Hordeum).

× **Elymotrigia** Hylander. Gramineae. Gen. hybr. (Elymus × Elytrigia).

× **Elymotriticum** P. Fourn. Gramineae. Gen. hybr. (Elymus × Triticum).

Elymus L. Gramineae. 70 N. temp., S. Am. *E. arenarius* L. on dunes (cf. *Ammophila*); its ls. are coated with wax.

Elymus Mitchell = Zizania L. (Gramin.).

Elyna Schrad. = Kobresia Willd. (Cyperac.).

Elynanthus Beauv. ex Lestib. Cyperaceae. Gen. dub.

Elynanthus Nees = Tetraria P. Beauv. (Cyperac.).

Elyonorus Bartl. = seq.

Elyonurus Humb. & Bonpl. ex Willd. = Elionurus Humb. & Bonpl. ex Willd. corr. Kunth (Gramin.).

× **Elysitanion** Bowden. Gramineae. Gen. hybr. (Elymus × Sitanion).

Elythranthe Reichb. = Elytranthe Bl. (Loranthac.).

Elythranthera A. S. George (sphalm.) = Elytranthera (Endl.) A. S. George (Orchidac.).

Elythraria D. Dietr. = Elytraria Michx (Acanthac. or Scrophulariac.).

Elythroblepharum auctt. = Elytroblepharum (Steud.) Schlechtd. = Digitaria

Elythrophorus Dum. = Elytrophorus Beauv. (Gramin.). [Fabr. (Gramin.).

Elythrospermum Steud. = Elytrospermum C. A. Mey. = Scirpus L. (Cyperac.).

Elythrostamna Boj. = Ipomoea L. (Convolvulac.).

Elytranthaceae Van Tiegh. = Loranthaceae–Elytranthinae Engl.

Elytranthe Blume. Loranthaceae. 10 SE. As., W. Malaysia.

Elytranthera (Endl.) A. S. George. Orchidaceae. 2 W. Austr.

*****Elytraria** Michx. Acanthaceae or Scrophulariaceae. 7 trop., subtrop.

Elytrigia Desv. (~ Agropyron J. Gaertn.). Gramineae. 50 temp. Euras.

Elytrigium Benth. = praec.

× **Elytriticum** auct. Gramineae. Gen. hybr. (Elymus × Triticum).

Elytroblepharum (Steud.) Schlechtd. = Digitaria Fabr. (Gramin.).

Elytropappus Cass. Compositae. 8 S. Afr. *E. rhinocerotis* L. is a char. pl.

Elytrophorum Poir. = seq. [of the karroo.

Elytrophorus Beauv. Gramineae. 4 trop. Afr., trop. As., Austr.

Elytropus Muell. Arg. Apocynaceae. 1 Chile. Many bracts.

× **Elytrordeum** Hylander. Gramineae. Gen. hybr. (Elytrigia × Hordeum).

Elytrospermum C. A. Mey. = Scirpus L. (Cyperac.).

Elytrostachys McClure. Gramineae. 2 C. Am., NW. trop. S. Am.

Elytrostamna Choisy = Elythrostamna Boj. = Ipomoea L. (Convolvulac.).

Embadium J. M. Black. Boraginaceae. 1 S. Austr.

Embamma Griff. = Pterisanthes Blume (Vitidac.).

*****Embelia** Burm. f. Myrsinaceae. 130 trop. & subtrop. Afr., Madag., E. As., Indomal., Pacif.

Embeli[ac]eae J. G. Agardh = Myrsinaceae R.Br. (genera choripetala).

Embelica Boj. = seq.

Embergeria Boulos. Compositae. 1 Austr., 1 Chatham Is.

Emblica Gaertn. (~Phyllanthus L.). Euphorbiaceae. 4 Madag., E. As.,

Emblingia F. Muell. Emblingiaceae. 1 W. Austr. [Indomal.

Emblingiaceae (Pax) Airy Shaw. Dicots. 1/1 Austr. Prostrate herbaceous subshrub with habit of *Scaevola* (*Goodeniac.*), stems hispidulous. Ls. opp. or subopp., simple, ent., subrhomb., cartilag.-margined, scabr., minutely stip. Fls. sol., axill., greenish or yellowish. K (5), unequal, posticously dimidiate; C (2), anticous, laterally connate into a slipper-like structure, hooded at apex, externally sericeous. Androgynophore linear, flattened, incurved, inserted in slit of K. A 8–9, short, the 4 anticous fertile, the 4–5 posticous staminodial, sericeous. G 1, apically bialate, with 1 basal ovule, and small sessile stig. Fr. dry, indeh., pendulous within the K from the apex of the androgynophore, the thin pericarp adherent to the seed. Seed ± flattened, with rugose woody-crustaceous testa and lacin. funicle; embryo linear, condupl., endosp. scanty. Only genus: *Emblingia*. A remarkable local relict, in possession of an androgynophore and a conduplicate embryo agreeing with *Capparidac.*, but differing widely in almost every other character. The habit, K, and C might suggest sympetalous fams. such as *Goodeniac.*, *Scrophulariac.*, *Verbenac.*, etc., but the pollen indicates a possible relationship with *Polygalac.*

Embolanthera Merr. Hamamelidaceae. 2 Indoch., Philipp. Is.

Embothrium J. R. & G. Forst. Proteaceae. 8 E. Austr., Andes, Chile.

Embryogonia Blume = Combretum L. (Combretac.).

Embryopteris Gaertn. = Diospyros L. (Ebenac.).

Emelia Wight (sphalm.) = Emilia Cass. (Compos.).

Emelianthe Danser. Loranthaceae. 2 trop. E. Afr.

Emelista Rafin. = Cassia L. (Legumin.).

Emeorhiza Pohl (nomen) = Emmeorhiza Pohl ex Endl. (Rubiac.).

Emericia Roem. & Schult. = Vallaris Burm. f. (Apocynac.).

Emerus Mill. = Coronilla L. (Legumin.).

Emeticaceae Dulac = Apocynaceae Juss.

Emetila (Rafin.) Rafin. ex S. Wats. = Ilex L. (Aquifoliac.).

***Emex** Neck. ex Campderá. Polygonaceae. 2 Medit., S. Afr., Austr. The fr. is surrounded by the P, 3 of whose segs. are spiny.

Emicocarpus K. Schum. & Schlechter. Asclepiadaceae. 1 SE. Afr.

Emilia Cass. Compositae. 30 palaeotrop.

Emiliella S. Moore. Compositae. 1 Angola.

Emiliomarcelia Th. & H. Durand = Trichoscypha Hook. f. (Anacardiac.).

Eminia Taub. (~Rhynchosia Lour.). Leguminosae. 6 trop. Afr.

Eminium (Bl.) Schott. Araceae. 6 E. Medit. to C. As.

Emlenia Rafin. = Enslenia Nutt. = Cynanchum L. (Asclepiadac.).

Emmenanthe Benth. Hydrophyllaceae. 1 SW. U.S.

Emmenanthus Hook. & Arn. = Ixonanthes Jack (Ixonanthac.).

Emmenopterys Oliv. Rubiaceae. 1 China.

Emmenopteryx Dalla Torre & Harms = praec.

Emmenosperma F. Muell. Rhamnaceae. 2 Austr., 1 New Caled.

Emmenospermum C. B. Clarke ex Hook. f. = Phtheirospermum Bunge (Scrophulariac.).

EMMENOSPERMUM

Emmenospermum F. Muell. = Emmenosperma F. Muell. (Rhamnac.).
Emmeorhiza Pohl ex Endl. Rubiaceae. 2 trop. S. Am.
Emmotaceae Van Tiegh. = Icacinaceae–Icacineae Benth.
Emmotium Meissn. = seq.
Emmotum Desv. Icacinaceae. 12 trop. S. Am.
Emorya Torr. Buddlejaceae. 1 S. U.S.
Empedoclea A. St-Hil. = Tetracera L. (Dilleniac.).
Empedoclesia Sleum. Ericaceae. 1 C. Am.
Empedoclia Rafin. = Sideritis L. (Labiat.).
***Empetraceae** S. F. Gray. Dicots. 3/10 (?20), N. temp., Andes, Falkl. Is.,
Tristan, occupying similar positions to *Ericaceae*; heath-like habit. The alt.
exstip. ericoid ls. are incurved downwards, forming a cavity on the under side
partly filled up by hairs into which the stomata open. Infl. racemose; fls. usu.
♂ ♀ and dioec., rarely ⚥. In all but *Corema* the fls. are on 'short shoots' which
arise lat. from the main axis and bear only scales below the infl. K 3, imbr.,
± petaloid; C 3, imbr. (or K 3 + 3, C 0); A 3, with long fil. and small anth.;
disk 0; G (2–9), with 2–9 loc.; style short, often with large flabellate stigmatic
branches; ovules 1 per loc., anatr. or nearly campylotr., erect on axile plac.,
with ventral raphe. Drupe with 2–9 stones; seed albuminous with no caruncle.
Genera: *Corema, Empetrum, Ceratiola*. Related to *Grubbiaceae* and the *Ericales*.
Empetron Adans. = seq.
Empetrum L. Empetraceae. 2 (Good) or 15–16 (Vasiliev). N. temp. & arctic,
S. Andes, Falkl., Tristan; on moors. Fl. dioec. and anemoph., but sometimes
♀ and protandr. (Cf. Good in *J. Linn. Soc., Bot.* **47**: 489, 1927.)
Emphysopus Hook. f. = Lagenophora Cass. (Compos.).
Emplectanthus N. E. Br. Asclepiadaceae. 2 S. Afr.
Emplectocladus Torr. (~ Amygdalus L.). Rosaceae. 6 NW. N. Am.
Empleuridium Sond. & Harv. Rutaceae. 1 S. Afr.
Empleurosma Bartl. = Dodonaea L. (Sapindac.).
Empleurum Ait. Rutaceae. 2 S. Afr. Ls. officinal (*buchu*).
Empodium Salisb. (~ Curculigo Gaertn.). Hypoxidaceae. 8 S. Afr.
Empogona Hook. f. Rubiaceae. 5 E. trop. & S. Afr.
Empusa Lindl. = Liparis Rich. (Orchidac.).
Empusaria Reichb. = praec.
Emuleria Rafin. = Justicia L. (Acanthac.).
Emulina Rafin. = Jacquemontia Choisy (Convolvulac.).
Emurtia Rafin. = Eugenia L. (Myrtac.).
Enaemon P. & K. = seq.
Enaimon Rafin. = Olea L. (Oleac.).
Enalcida Cass. = Tagetes L. (Compos.).
***Enallagma** Baill. Bignoniaceae. 3 C. & N. trop. S. Am., W.I. Ls. alt. Berry.
Enalus Aschers. & Guerke = Enhalus Rich. (Hydrocharitac.).
Enantia Falc. = Sabia Colebr. (Sabiac.).
Enantia Oliv. Annonaceae. 9 W. Afr., 1 trop. E. Afr.
Enantiophylla Coulter & Rose. Umbelliferae. 1 C. Am.
Enantiosparton C. Koch = Genista L. (Legumin.).
Enantiotrichum E. Mey. ex DC. = Euryops Cass. (Compos.).
Enarganthe N. E. Brown. Aïzoaceae. 1 S. Afr.

Enargea Banks = Luzuriaga Ruiz & Pav. (Philesiac.).
Enartea Steud. = praec.
Enarthrocarpus Labill. Cruciferae. 5 E. Medit., N. Afr.
Enartocarpus Poir. = Enarthrocarpus Labill. (Crucif.).
Enaulophyton v. Steenis. Melastomataceae. 1 Natuna Is. (between Malay Penins. & Borneo), 1 NW. Borneo.
Encelia Adans. Compositae. 15 W. U.S. to Chile, Galápagos. Pappus usu. o.
Enceliopsis A. Nelson. Compositae. 4 SW. U.S.
Encentrus Presl = Gymnosporia Benth. & Hook. f. (Celastrac.).
Encephalartos Lehm. Zamiaceae. 30 trop. & S. Afr. The Kaffirs prepare a meal from the pith (cf. *Cycas*).
Encephalocarpus A. Berger = Pelecyphora Ehrenb. (Cactac.).
Encephalosphaera Lindau. Acanthaceae. 2 trop. S. Am.
Encheila O. F. Cook. Palmae. 1 Mex.
Encheiridion Summerhayes. Orchidaceae. 2 trop. Afr.
Enchelya Lem. = Encelia Adans. (Compos.).
Enchidion Muell. Arg. = seq.
Enchidium Jack = Trigonostemon Blume (Euphorbiac.).
Encholirium Mart. ex Schult. f. Bromeliaceae. 8 Brazil.
Enchosanthera King & Stapf ex Guillaumin. Melastomataceae. 1 Malay Penins., Sumatra.
Enchydra F. Muell. = Enydra Lour. (Compos.).
Enchylaena R.Br. Chenopodiaceae. 6 Austr.
Enchylus Ehrh. (uninom.) = *Sedum annuum* L. (Crassulac.).
Enchysia Presl = Laurentia Michx (Campanulac.).
Encilia Reichb. = Ercilla A. Juss. (Phytolaccac.).
Enckea Kunth = Piper L. (Piperac.).
Enckianthus Desf. = Enkianthus Lour. (Ericac.).
Enckleia Pfeiff. = Enkleia Griff. (Thymelaeac.).
Encliandra Zucc. = Fuchsia L. (Onagrac.).
Encopa Griseb. = Encopella Pennell (Scrophulariac.).
Encopea Presl = Faramea Aubl. (Rubiac.).
Encopella Pennell. Scrophulariaceae. 1 Cuba.
Encrypta P. & K. (sphalm.) = Eucrypta Nutt. (Hydrophyllac.).
Encurea Walp. = Enourea Aubl. = Paullinia L. (Sapindac.).
Encyanthus Spreng. = Enkianthus Lour. (Ericac.).
Encycla Benth. (sphalm.) = Eucycla Nutt. = Eriogonum Michx (Polygonac.).
Encyclia Hook. Orchidaceae. 130 trop. Am.
Encyclia Poepp. & Endl. = Polystachya Hook. (Orchidac.).
× **Encyclipedium** hort. Orchidaceae. Gen. hybr. (vi; nomen nugax) (Cypripedium × Encyclia).
Encyclium Neum. = Encyclia Hook. (Orchidac.).
Endacanthus Baill. = Pyrenacantha Hook. ex Wight (Icacinac.).
Endallex Rafin. = Phalaris L. (Gramin.).
Endammia Rafin. = Corema D. Don (Empetrac.).
Endecaria Rafin. (1815). Menispermaceae. Inc. sed.
Endecaria Rafin. (1838) = Cuphea P.Br. (Lythrac.).
Endeisa Rafin. = Dendrobium Sw. (Orchidac.).

ENDEMA

Endema Pritz. (sphalm.) = Eudema Humb. & Bonpl. (Crucif.).

Endera Regel = Taccarum Brongn. (Arac.).

Endertia v. Steenis & de Wit. Leguminosae. 1 Borneo.

Endespermum Blume = Dalbergia L. (Legumin.).

Endiandra R.Br. Lauraceae. 1 Assam(?), 80 Malaysia, Austr., Polynesia.

Endiplus Rafin. = Phacelia Juss. (Hydrophyllac.).

Endisa P. & K. = Endeisa Rafin. = Dendrobium Sw. (Orchidac.).

Endiusa Alef. = Vicia L. (Legumin.).

Endivia Hill = Cichorium L. (Compos.).

Endlichera Presl = Emmeorhiza Pohl ex Endl. (Rubiac.).

*__Endlicheria__ Nees. Lauraceae. 40 C. & trop. S. Am.

Endocarpa Rafin. = Aiouea Aubl. (Laurac.).

Endocellion Turcz. ex Herd. = ?Nardosmia Cass. = Petasites L. (Compos.).

Endochromaceae Dulac = Phytolaccaceae R.Br.

Endocles Salisb. = Zigadenus Michx (Liliac.).

Endocodon Rafin. = Calathea G. F. W. Mey. (Marantac.).

Endocoma Rafin. = Bottionea Colla (Liliac.).

Endodaca [Schlechtd.] = seq.

Endodeca Rafin. = Aristolochia L. (Aristolochiac.).

Endodesmia Benth. Guttiferae. 1 trop. W. Afr.

Endodia Rafin. = Leersia Sw. (Gramin.).

Endogona Rafin. = Anthericum L. (Liliac.).

Endogonia Lindl. = Trigonotis Stev. (Boraginac.).

Endoisila Rafin. = Euphorbia L. (Euphorbiac.).

Endolasia Turcz. = ?Manettia Mutis (Rubiac.).

Endolepis Torr. ex A. Gray = Atriplex L. (Chenopodiac.).

Endoleuca Cass. = Metalasia R.Br. (Compos.).

Endolimna Rafin. = Heteranthera Ruiz & Pav. (Pontederiac.).

Endolithodes Bartl. = Synisoön Baill. (Rubiac.).

Endoloma Rafin. = Amphilophium Kunth (Bignoniac.).

Endomallus Gagnep. Leguminosae. 2 Indoch.

Endomelas Rafin. = Thunbergia Retz. (Thunbergiac.).

Endonema A. Juss. Penaeaceae. 2 S. Afr.

Endopappus Sch. Bip. = Chrysanthemum L. (Compos.).

Endoplectris Rafin. = Epimedium L. (Berberidac.).

Endopleura Cuatrec. Houmiriaceae. 1 Amaz. Brazil.

Endopogon Nees = Strobilanthes Blume (Acanthac.).

Endopogon Rafin. (1818) = Pagesia Rafin. (Scrophulariac.).

Endopogon Rafin. (1836) = Diodia L. (Rubiac.).

Endoptera DC. = Crepis L. (Compos.).

Endorima Rafin. (1819) = Balduina Nutt. (Compos.).

Endorima Rafin. (1836) = Helipterum DC. (Compos.).

Endosiphon T. Anders. ex Benth. & Hook. f. Acanthaceae. 1 trop. Afr.

*__Endospermum__ Benth. Euphorbiaceae. 12–13 SE. As., Malaysia, Fiji.

Endospermum Endl. = Endespermum Blume = Dalbergia L. (Legumin.).

Endosteira Turcz. Tiliaceae. 1 W.I. (St. Vincent).

Endostemon N. E. Br. Labiatae. 16 trop. & S. Afr., Arabia, 1 India.

Endostephium Turcz. = Galipea Aubl. (Rutac.).

414

Endotheca Rafin. = Endodeca Rafin. = Aristolochia L. (Aristolochiac.).
Endotis Rafin. = Allium L. (Alliac.).
Endotricha Aubrév. & Pellegr. (1935; non seq. [1821], nec Endotrichia
Suringar 1870—Algae) = Aubregrinia H. Heine (Sapotac.).
Endotriche (Bunge) Steud. = Gentianella Moench (Gentianac.).
Endotropis Endl. = Cynanchum L. (Asclepiadac.).
Endotropis Rafin. = Rhamnus L. (Rhamnac.).
Endrachium Juss. = Humbertia Comm. ex Lam. (Humbertiac.).
Endrachne Augier = praec.
Endresiella Schlechter. Orchidaceae. 1 Costa Rica.
Endressia J. Gay. Umbelliferae. 2 Pyrenees, N. Spain.
Endusa Miers. Olacaceae. 1 Peru.
Endusia Benth. & Hook. f. = Endiusa Alef. = Vicia L. (Legumin.).
Endymion Dum. Liliaceae. 10 W. Eur., W. Medit.
Endysa P. & K. = Endusa Miers (Olacac.).
Enemion Rafin. = Isopyrum L. (Ranunculac.).
Enemosyne Lehm. (sphalm.) = Eremosyne Endl. (Eremosynac.).
Eneodon Rafin. = Leucas R.Br. (Labiat.).
Enetephyton Nieuwl. = Utricularia L. (Lentibulariac.).
Engelhardia Leschen. ex Bl. = seq.
Engelhardtia Leschen. ex Bl. corr. Bl. Juglandaceae. 5 Himal. to Formosa,
SE. As., Malaysia; 3 Mex., C. Am.
Engelia Karst. ex Nees = Mendoncia Vell. (Mendonciac.).
Engelmannia A. Gray ex Nutt. corr. Torr. & Gray. Compositae. 1 SW. U.S.,
Engelmannia Klotzsch = Croton L. (Euphorbiac.). [Mex.
Engelmannia Pfeiff. = Cuscuta L. (Cuscutac.).
Englemannia A. Gray ex Nutt. = Engelmannia A. Gray ex Nutt. corr. Torr.
& Gray (Compos.).
Englera P. & K. = Engleria O. Hoffm. (Compos.).
Englerastrum Briquet. Labiatae. 20 trop. Afr.
Englerella Pierre. Sapotaceae (inc. sed.; = ? Lucuma Mol.). 1 Guiana.
Engleria O. Hoffm. Compositae. 2 trop. & S. Afr.
Englerina Van Tiegh. = Tapinanthus Bl. (Loranthac.).
Englerocharis Muschler. Cruciferae. 4 Andes.
Englerodaphne Gilg (= ? Wikstroemia Endl.). Thymelaeaceae. 3 E. Afr.
Englerodendron Harms. Leguminosae. 2 trop. Afr.
Englerodoxa Hörold = Ceratostema Juss. (Ericac.).
Englerophoenix Kuntze = Maximiliana Mart. (Palm.).
Englerophytum Krause. Sapotaceae. 1 W. Equat. Afr.
Engysiphon G. J. Lewis. Iridaceae. 8 S. Afr.
Enhalaceae Nak. = Hydrocharitaceae–Enhaleae Dandy.
Enhalus Rich. Hydrocharitaceae. 1 Indomal., Austr., in salt water. The ♀ fls.
float horiz. at low water, and catch the ♂ fls. which (cf. *Vallisneria*) break off
and float. As the tide rises, the ♀ fls. stand vertically, and the pollen, heavier
than water, sinks down upon the stigmas. The testa bursts when the seed is
ripe, and the embryo is freed.
Enhydra DC. = Enydra Lour. (Compos.).
Enhydria Kanitz = Enydria Vell. = Myriophyllum L. (Haloragidac.).

Enhydrias Ridl. = Blyxa Nor. (Hydrocharitac.).

Enicosanthum Becc. Annonaceae. 16 Ceylon, Burma, Siam, W. Malaysia.

***Enicostema** Blume. Gentianaceae. 3–4 W.I., trop. & S. Afr., Madag., India, Java, Lesser Sunda Is. Frequently coastal.

Enipea Rafin. = Salvia L. (Labiat.).

Enkea Walp. = Enckea Kunth = Piper L. (Piperac.).

Enkianthus Lour. Ericaceae. 10 Himal. to Japan.

Enkleia Griff. Thymelaeaceae. 3 Andaman Is., SE. As., Malaysia.

Enkyanthus DC. = Enkianthus Lour. (Ericac.).

Enkylia Griff. = Gynostemma Blume (Cucurbitac.).

Enkylista Benth. & Hook. f. (sphalm.) = Eukylista Benth. = Calycophyllum DC. (Rubiac.).

Enneadynamis Bub. = Parnassia L. (Parnassiac.).

Ennealophus N. E. Br. (~ Trimezia Salisb.). Iridaceae. 1 Amaz. Brazil.

Enneapogon Desv. ex Beauv. Gramineae. 30 warm.

Ennearina Rafin. = Pleea Rich. (Liliac.).

***Enneastemon** Exell. Annonaceae. 15 trop. Afr.

Enneatypus Herzog. Polygonaceae. 1 Bolivia.

Ennepta Rafin. = Ilex L. (Aquifoliac.).

Enochoria Baker f. Araliaceae. 1 New Caled.

Enodium Gaudin = Molinia Schrank (Gramin.).

Enomegra A. Nelson = Argemone L. (Papaverac.).

Enomeia Spach = Einomeia Rafin. = Aristolochia L. (Aristolochiac.).

Enosanthes A. Cunn. ex Schauer = Homoranthus A. Cunn. (Myrtac.).

Enothrea Rafin. = Octomeria R.Br. (Orchidac.).

Enourea Aubl. = Paullinia L. (Sapindac.).

Enrila Blanco = Ventilago Gaertn. (Rhamnac.).

Ensatae [L.] Ker-Gawl. = Iridaceae Juss.

Ensete Bruce. Musaceae. 7 trop. Afr., Madag., S. China, SE. As., Indomal. Bracts and fls. persistent; pollen-grains warted; seeds larger than those of *Musa*. The stalk of the infl. of *Ensete ventricosa* (Welw.) E. E. Cheesm. is cooked and eaten.

Enskide Rafin. = Utricularia L. (Lentibulariac.).

Enslemia Th. Dur. (sphalm.) = seq.

Enslenia Nutt. Asclepiadaceae. 1–2 N. Am.

Enslenia Rafin. = Ruellia L. (Acanthac.).

Ensolenanthe Schott = Alocasia Schott (Arac.).

Enstoma A. Juss. = Eustoma Salisb. (Gentianac.).

***Entada** Adans. Leguminosae. 30 warm. Seeds of *E. gigas* (L.) Fawc. & Rendle (nicker bean), a trop. climber with pods 1 m. long, are carried to Eur. by the Gulf Stream drift.

Entadopsis Britton = praec.

Entagonum Poir. = Entoganum Banks ex Gaertn. = Melicope J. R. & G. Forst. (Rutac.).

Entandrophragma C. DC. Meliaceae. 9 trop. & S. Afr.

Entasicum P. & K. = seq.

Entasikom Rafin. = seq.

Entasikon Rafin. = Trepocarpus Nutt. (Umbellif.).

EPACRIDACEAE

Entaticus S. F. Gray = Coeloglossum Hartm. + Leucorchis E. Mey. (Or-
Entelea R.Br. Tiliaceae. 1 N.Z. Wood very light. [chidac.).
Enterolobium Mart. Leguminosae. 10 trop. Am., W.I. Pod spiral.
Enteropogon Nees. Gramineae. 6 Afr., Madag., Seychelles, India, Formosa,
Austr., Pacif.
Enterosora Baker = Glyphotaenium J. Sm. (Grammitidac.).
Enterospermum Hiern. Rubiaceae. 8 E. Afr., Madag.
Enthomanthus Moç. & Sessé ex Ramírez = Lopezia Cav. (Onagrac.).
Entoganum Banks ex Gaertn. = Melicope J. R. & G. Forst. (Rutac.).
Entolasia Stapf. Gramineae. 5 trop. Afr., E. Austr.
Entoplocamia Stapf. Gramineae. 3 trop. & S. Afr.
Entosiphon Bedd. = Cyclandrophora Hassk. (Chrysobalanac.).
Entrecasteauxia Montr. = Duboisia R.Br. (Solanac.).
Entrochium Rafin. = Eupatorium L. (Compos.).
Enula Boehm. = Inula L. (Compos.).
Enurea J. F. Gmel. = Enourea Aubl. = Paullinia L. (Sapindac.).
Enydra Lour. Compositae. 10 warm. Marsh pl.
Enydria Vell. = Myriophyllum L. (Haloragidac.).
Enymion Rafin. = Enemion Rafin. = Isopyrum L. (Ranunculac.).
Enymonospermum Spreng. ex DC. = Pleurospermum Hoffm. (Umbellif.).
Eomatucana Ritter. Cactaceae. 1 Peru.
Eomecon Hance. Papaveraceae. 1 E. China.
Eopepon Naud. = Trichosanthes L. (Cucurbitac.).
Eophylon A. Gray = Cotylanthera Blume (Gentianac.).
Eophyton Benth. & Hook. f. = praec.
Eora O. F. Cook = Rhopalostylis H. Wendl. & Drude (Palm.).
Eosanthe Urb. Rubiaceae. 1 Cuba.
Eotaiwania Yendo = Taiwania Hayata (Taxodiac.).
Eothinanthes Rafin. = Etheosanthes Rafin. = Tradescantia L. (Commelinac.).
***Epacridaceae** R.Br. Dicots. 30/400 Indoch. to N.Z., Hawaii, S. Am.; chiefly
Austr. and Tasm., representing the *Ericaceae* of other continents, on heaths
and boggy ground. Mostly like *Ericac.* in habit: usu. shrubs or small trees with
narrow, entire, rigid, exstipulate ls. Ls. sometimes sheathing, then sometimes
leaving annular scars, sometimes no scars. L. venation basically palmate with
each l. supplied by a single vasc. strand (exc. *Richeoïdeae*, where venation
parallel, and each l. receives many strands). Infl. various, the fls. in most
Styphelïeae assoc. with minute abortive buds (*Pentachondra*) or bract-like
abortive shoots (*Styphelia*). Fls. often fragrant, white–red–blue, reg., usu. ♀;
usu. K 5, imbr.; C (5), imbr. or valv.; A 5, epipet. or hypogynous, anthers usu.
with one central slit (2 slits in *Conostephium, Wittsteinia*, etc.), without appendages but sometimes sterile at the tips; anthers inverting during devel.
(see *Ericac.*). Pollen in tetrads, but 1–3 members of the tetrad aborting in
many *Styphelïeae* so that the mature grains sometimes appear solitary. Hypog.
disk or scales usu. present (absent in e.g. *Sprengelia*). (G) usu. 5-loc. (1–10-loc.
in *Styphelïeae*, 3-loc. in *Wittsteinia*, 2-loc. in *Oligarrhena*); G in *Wittsteinia*;
loc. usu. opp. C-lobes. Style simple; placentation axile; ovules 1–∞ per loc.,
anatropous. Fr. a loculic. caps. or a drupe. Embryo straight, in copious
endosperm. A few yield edible fr.

417

EPACRIS

No feature absolutely distinguishes *Epacridoïdeae* from *Ericac.*; but all genera fall into the one fam. or the other on combinations of morphological chars., except *Wittsteinia*, which might be placed in either.

Classification and chief genera:

I. **Epacridoïdeae** (stems without annular scars, bracteoles usu. persistent, ls. rarely sheathing; mostly ericoid shrubs):

1. Cosmeliëae (leaves sheathing, style in depression at apex of ovary G (5), fr. loculic. caps.): *Cosmelia* (stamens epipet.), *Andersonia*, *Sprengelia*.

2. Styphelieae (ls. not sheathing, G 1–5-loc., ovary attenuated into style, fr. indehisc.): *Styphelia* (pollen warty), *Leucopogon*, *Conostephium* (anthers with 2 slits), *Monotoca*, *Trochocarpa*, *Decatoca*.

3. Epacrideae (ls. not sheathing, G (5), style in depression at apex of ovary, fr. loculic. caps.): *Epacris*, *Lysinema* (sta. free), *Rupicola*, *Lebetanthus* (sta. free, S. Am.).

4. Oligarrheneae (K4 C4 A2 G(2), infl. a cpd. rac.): *Oligarrhena*.

5. Needhamielleae (C lobes indupl., G (2)): *Needhamiella*.

6. Wittsteiniëae (ls. broad, with distinct midrib, retic. venation and crenate margins; ovary 3-loc., inferior; pollen grains solitary): *Wittsteinia*.

II. **Richeoïdeae** (stems with annular scars, ls. sheathing, bracteoles often caducous; mostly Monocot. in habit; many tall shrubs or small trees): *Richea*, *Dracophyllum*, *Sphenotoma*.

*Epacris Cav. Epacridaceae. 40 SE. Austr., Tasm., N.Z.

Epacris J. R. & G. Forst. = Leucopogon + Dracocephalum + Cyathodes + Pentachondra (Epacridac.).

Epactium Willd. ex J. A. & J. H. Schultes = ? Ludwigia L. (Onagrac.).

Epallage DC. Compositae. 1 SE. Afr., 14 Madag.

Epallageiton K.-Pol. = Aulospermum Coult. & Rose (Umbellif.).

Epaltes Cass. Compositae. 17 trop. Ls. usu. decurrent. Pappus o.

Epatitis Rafin. = Adenostyles Cass. (Compos.).

Eperua Aubl. Leguminosae. 12 trop. S. Am. *E. falcata* Aubl. (*wallaba*, Br.

Epetetiorhiza Steud. = seq.   [Guiana) yields a good timber.

Epetorhiza Steud. = Physalis L. (sect. Epeteiorhiza G. Don) (Solanac.).

Ephaeola P. & K. = seq.

Ephaiola Rafin. = Acnistus Schott (Solanac.).

Ephebepogon Nees & Meyen = Pollinia Trin. (Gramin.).

Ephedra L. Ephedraceae. 40 warm temp. N. & S. Am., Euras.

*Ephedraceae Dum. Gymnosp. (Gnetales). 1/40 N. & S. warm temp. Shrubs, much branched, with opp. connate ls. reduced to scales, so that the stem performs the work of assim. Fls. diclinous, with no trace of cpls. in ♂, or of sta. in ♀; ♂ in spikes, ♀ in pairs or solitary, usu. bracteate. The ♂ has a P of 2 antero-post. united ls., beyond which the axis is prolonged and bears 2–8 sessile 2-loc. anthers. The ♀ has a tubular P and one erect orthotr. ovule with a long micropyle projecting at the top of the fl.; the fl. or fls. are enclosed by bracts which become red and fleshy after fert. and enclose the fr. The seed is enclosed in the P, which becomes woody, and the fleshy bracts cover this again. There are two cots. in the embryo; seed album. Only genus: *Ephedra*.

structure and is repeated very closely in *Casuarina*. In several spp. the internodes of the rhiz. are swollen into tubers, which serve for hibernation and veg. repr.

The spike is very like the ♂ fl. of a Conifer, and has as much (or as little) right to the title of 'flower'. It is a terminal axis with short internodes, bearing a dense strobilus of peltate sporangiophores. Each of these bears a number of sporangia upon the under side of the head (i.e. towards the stem), arranged like the horses of a 'merry-go-round'. The spores are of one kind only; each has, running round it, two spiral cuticularised bands of membrane, formed from the outer wall and termed *elaters*. These are hygroscopic, unfolding in damp air. In the rolling up again on drying, the elaters of one spore become entangled with those of others and cause them to adhere together, so that several prothalli may be formed near to one another when they germinate. This is advantageous, for the prothalli are dioec., though so far as we can tell the spores are all alike. The prothallus is fairly large, the ♂ being smaller than the ♀.

The mechanical tissues of most spp. contain much silica, and the stems of *E. hyemale* L. (Dutch rushes) are used for polishing.

Equitiris Thou. (uninom.) = *Cymbidium equitans* Thou. = *Oberonia disticha* (Lam.) Schltr. (Orchidac.).

Eraclissa Forsk. = Andrachne L. (Euphorbiac.). ˅

Eraclyssa Scop. = praec.

Eragrostidaceae (Benth.) Herter = Gramineae–Eragrost[id]eae Benth.

Eragrostiella Bor. Gramineae. 7 India, Burma, Ceylon.

Eragrostis v. Wolf. Gramineae. 300 cosmop., mostly subtrop.

Eranthemum L. Acanthaceae. 30 trop. As.

***Eranthis** Salisb. Ranunculaceae. 7 N. palaeotemp. *E. hyemalis* Salisb. has a thick rhizome or row of tubers, one formed each year. The sol. term. fls. appear in February, before the ls.; each has an invol. of three green ls., a 'calyx' of 6 segments, and several honey-ls. or petals.

Erasma R.Br. = Lonchostoma Wikstr. (Bruniac.).

Erasmia Miq. = Peperomia Ruiz & Pav. (Peperomiac.).

Eratica Hort. ex Dipp. = Eurybia Cass. = Olearia Moench (Compos.).

Erato DC. = Liabum Adans. (Compos.).

Eratobotrys Fenzl ex Endl. = Scilla L. (Liliac.).

Erblichia Seem. (~Piriqueta Aubl.). Turneraceae. 6 C. Am., S. Afr., Madag.

Ercilia Endl. = seq.

Ercilla A. Juss. Phytolaccaceae. 2 Chile, Peru. *E. volubilis* A. Juss. climbs by adhesive disks, endogenous just above the axils.

Erdisia Britton & Rose = Corryocactus Britt. & Rose (Cactac.).

Erebennus Alef. = Hibiscus L. (Malvac.).

Erebinthus Mitch. = Tephrosia Pers. (Legumin.).

Erechthites Less. = seq.

Erechtites Rafin. Compositae. 15 Am., Austr., N.Z.

Ereicoctis (DC.) Kuntze = Arcytophyllum Willd. ex Schult. & Schult. f.

Eremaea Lindl. [non Eremia D. Don] = seq.   [(Rubiac.).

Eremaeopsis Kuntze. Myrtaceae. 8 W. Austr.

EREMALCHE

Eremalche Greene (~ Malvastrum A. Gray). Malvaceae. 4 W. U.S.

Eremanthe Spach = Hypericum L. (Guttif.).

Eremanthus Less. Compositae. 25 Brazil.

Eremanthus Royle (sphalm.) = Erismanthus Wall. ex Muell. Arg. (Euphorbiac.).

Eremia D. Don. Ericaceae. *S. str.* 12, *s. ampl.* 72, S. Afr.

Eremiastrum A. Gray. Compositae. 2 SW. U.S.

Eremiella Compton. Ericaceae. 1 S. Afr.

Ereminula Greene = Dimeresia A. Gray (Compos.).

Eremiopsis N. E. Br. = Scyphogyne Brongn. (Ericac.).

Eremites Benth. = seq.

Eremitis Doell = Pariana Aubl. (Gramin.).

Eremobium Boiss. Cruciferae. 3 N. Afr., Arabia.

Eremocallis Salisb. ex S. F. Gray = Erica L. (Ericac.).

Eremocarpus Benth. = Piscaria Piper (Euphorbiac.).

Eremocarpus Lindl. (sphalm.) = Eremodaucus Bunge = Trachydium Lindl. (Umbellif.).

Eremocarpus Spach ex Reichb. (? sphalm.) = Eremosporus Spach = Hypericum L. (Guttif.).

Eremocarya Greene. Boraginaceae. 3 Pacif. N. Am.

Eremocharis R.Br. = Clianthus Soland. (Legumin.).

Eremocharis Phil. Hydrocotylaceae (~ Umbellif.). 9 Peru, Chile.

Eremochion Gilli. Chenopodiaceae. 1 Afghanistan.

Eremochlaena auctt. = Eremolaena Baill. (Sarcolaenac.).

Eremochlamys Peter = Tricholaena Schrad. (Gramin.).

Eremochloa Buese. Gramineae. 8 India, Ceylon, S. China, SE. As., W. Malaysia, Austr.

Eremochloë S. Wats. = Blepharidachne Hack. (Gramin.).

Eremocitrus Swingle. Rutaceae. 1 N. Austr.

Eremocrinum M. E. Jones. Liliaceae. 1 W. U.S. (Utah).

Eremodaucus Bunge. Umbelliferae. 1 Cauc. to C. As. & Afghan.

Eremodendron DC. ex Meissn. = Eremophila R.Br. (Myoporac.).

Eremodraba O. E. Schulz. Cruciferae. 2 Peru, Chile.

Eremogeton Standley & L. O. Williams. Scrophulariaceae. 1 Mex., Guatem.

Eremogone Fenzl = Arenaria L. (Caryophyllac.).

Eremohylema A. Nelson (~ Polypappus Nutt.). Compositae. 1 SW. U.S.

Eremolaena Baill. Sarcolaenaceae. 2 E. Madag.

Eremolepidaceae Van Tiegh. emend. Kuijt (~ Loranthaceae). Dicots. 3/11 trop. S. Am., W. I. Parasitic shrubs on trees, sometimes with evanescent epicortical roots. Ls. alt. Fls. dioec. or monoec., in bracteate spikes or catkins. P 3–4($\male$) or 2–3($\female$) (usu. 0 in $\male$ *Antidaphne*), valv.; A 3–4, opp. P, with biloc. anth.; $\overline{G}$ 1 (semi-inf. in *Antidaphne*). Fruit without staminodial bristles, with viscid tissue; endosp. (in *Antid.*) containing chlorophyll. Genera: *Antidaphne, Eremolepis, Eubrachion*. (Cf. Kuijt, *Brittonia*, **20**: 140, 1968.)

Eremolepis Griseb. Eremolepidaceae (~ Viscaceae). 7 trop. Am., W.I.

Eremolithia Jepson = Scopulophila M. E. Jones (Caryophyllac.).

Eremolobium Aschers. ex Boiss. (sphalm.) = Eremobium Boiss. (Crucif.).

Eremoluma Baill. = Lucuma Mol. (Sapotac.).
Eremomastax Lindau. Acanthaceae. 1 trop. Afr. (?Madag.), very variable. Seeds often with toothed scales, spreading when wetted.
Eremonanus I. M. Johnston. Compositae. 1 Calif.
Eremopanax Baill. Araliaceae. 10 New Caled.
Eremopappus Takhtadjian. Compositae. 1 Cauc., Pers. to C. As.
Eremophila R.Br. Myoporaceae. 45 Austr.
Eremophyton Béguinot. Cruciferae. 1 N. Afr. (Alger. Sah.).
Eremopoa Roshev. Gramineae. 6 SW. & C. As. to Himal.
Eremopodium Trev. = Asplenium L. (Aspleniac.).
Eremopogon Stapf. Gramineae. 4 warm Old World.
Eremopyrum Jaub. & Spach (~ Agropyron J. Gaertn.). Gramineae. 8 Medit. to NW. India.
Eremopyxis Baill. = Thryptomene Endl. (Myrtac.).
Eremosemium Greene. Chenopodiaceae. 2 W. U.S.
Eremosis Gleason (~ Vernonia Schreb.). Compositae. 20 Mex., C. Am.
Eremosparton Fisch. & Mey. Leguminosae. 3 W. & C. As.
Eremospatha (Mann & Wendl.) Mann & Wendl. ex Kerchove. Palmae. 12 trop. Afr.
Eremosperma Chiov. = Hewittia Wight & Arn. (Convolvulac.).
Eremosporus Spach = Hypericum L. (Guttif.).
Eremostachys Bunge. Labiatae. *S.l.* 60, *s.str.* 5, W. & C. As.
Eremosynaceae Takhtadj. Dicots. 1/1 Austr. Small ann. herbs, branched from base, pubesc. Ls. alt., sess., ent. to pinnatif., exstip. Fls. v. small, in dichot. cymes. K (5), deeply lobed, valv.; C 5; A 5 alternipet., with subul. fil. and small anth.; G̲ (2), shortly adn. to K at base, with 1 sub-basal axile ascending ov. per loc., and 2 free slender divergent styles, opp. to loc. Fr. a semi-inf. subdidymous loculic. caps.; seeds with copious endosp. Only genus: *Eremosyne*. Near *Saxifragac.*, but structure and arrangement of ov.
Eremosyne Endl. Eremosynaceae. 1 SW. Austr.   [aberrant.
Eremothamnus O. Hoffm. Compositae. 1 SW. Afr.
Eremotropa H. Andres. Pyrolaceae. 1 SW. China.
Eremurus M. Bieb. Liliaceae. 50 alpine W. & C. As. Fl. protog.; the petals crumple up before the essential organs are ripe. The ls. of *E. aurantiacus* Baker are eaten in Afghanistan.
Erepsia N. E. Brown. Aïzoaceae. 45 S. Afr.
Eresda Spach = Reseda L. (Resedac.).
Eresimus Rafin. = Cephalanthus L. (Rubiac.).
Eretia Stokes = Ehretia L. (Ehretiac.).
Ergocarpon C. C. Townsend. Umbelliferae. 1 SW. As. (Iraq–Persia border).
Erharta Juss. = Ehrharta Thunb. (Gramin.).
Erhetia Hill = Ehretia L. (Ehretiac.).
*Eria Lindl. Orchidaceae. 375 trop. As., Polynesia, Austr.
Eriachaenium Sch. Bip. Compositae. 1 Tierra del Fuego.
Eriachna P. & K. (1) = Achneria Munro (Gramin.).
Eriachna P. & K. (2) = Eriachne Phil. = Panicum L. (Gramin.).
Eriachne R.Br. Gramineae. 40 China, Indomal., Austr.
Eriachne Phil. = Panicum L. (Gramin.).

ERIADENIA

Eriadenia Miers. Apocynaceae. 1 trop. S. Am.

Eriander H. Winkler = Oxystigma Harms (Legumin.).

Eriandra v. Royen & v. Steenis. Polygalaceae. 1 New Guinea, Solomon Is.

Eriandrostachys Baill. Sapindaceae. 1 Madag.

Erianthecium L. Parodi. Gramineae. 1 Urug.

Erianthemum Van Tiegh. Loranthaceae. 12 trop. & S. Afr.

Erianthera Benth. = Alajja S. Ikonn. (Labiat.).

Erianthera Nees = Andrographis Wall. (Acanthac.).

Erianthus Michx. Gramineae. 28 trop. Am., SE. Eur. to E. As., Indomal.,
Polynes.; 1 Sahara, 1 Madag.

Eriastrum Wooton & Standley. Polemoniaceae. 14 SW. U.S.

Eriathera B. D. Jacks. (sphalm.) = Erianthera Nees = Andrographis Wall.

Eriaxis Reichb. f. Orchidaceae. 3 New Caled.   [(Acanthac.).

Eribroma Pierre = Sterculia L. (Sterculiac.).

Erica L. Ericaceae. Over 500 Eur., Atlant. Is., N. Afr., As. Min., Syria, trop.
& esp. S. Afr. The two common European heaths, *E. cinerea* L. and *E. tetralix*
L., cover great areas of moor. In habit like *Calluna*, but prob. not closely
related. Fl. bell-shaped and pendulous, visited and fert. mainly by bees. Honey
is secreted by the disk, and insects hanging on to the fl. and probing for it must
shake the sta. and receive a shower of the loose powdery pollen from the pores
in the tips of the anthers. In the wider-mouthed spp. the anthers have horn-like
projections at the back, which ensure contact with the insect's proboscis. The
stigma projects beyond the sta. so as to be touched first.

E. scoparia L. is the heath (*bruyère*) of S. France, etc., several feet high;
its rootstocks furnish briarwood pipes. The roots of heaths possess endotrophic
mycorhiza.

*Ericaceae Juss. 50/1350, cosmop., except in deserts, usu. confined to high
altitude regions in tropics; almost absent from Australasia. Form ecol. important
communities in many parts of the world, esp. on moors, in swamps and on
peaty soils. *Ericoïdeae* confined to Africa, Medit. and Europe, in two main
masses separated by the Sahara. *Vaccinioïdeae* mainly American and Asiatic,
nearly absent from Africa. *Rhododendroïdeae* with main centres in E. Asia,
New Guinea and N. Amer., absent from Africa.

Woody; small undershrubs to large shrubs and a few small trees (e.g.
Arbutus). Ls. exstip. *Ericoïdeae* characterized by the 'ericoid' habit: evergreen, with no winter buds and no bud-scales, ls. small and needle-like,
whorled, and deeply grooved beneath. The rest have winter buds with scales;
ls. usu. evergreen and leathery, sometimes deciduous, usu. ± elliptical with
flat or ± recurved margins, alt. (rarely opp.—*Leiophyllum*, or whorled—
Ledothamnus), entire; a few have needles like *Ericoïdeae* (e.g. *Bryanthus, Ledothamnus*). Many epiphytes, esp. in *Vaccinioïdeae*.

Infl. various, term. or lat.; leafy, or leafless but bracteate. Bracteoles usu.
2 or 3. Fls. usu. ⚥ (*Epigaea* dioecious), reg. (zygo. in *Rhododendreae*); K 4–5
(*Bejaria* 6–7), sometimes accrescent; C (4) or (5), urceolate, campanulate or
hypocrateriform, or 3–5 (*Ledeae*), or 6–7 (*Bejariëae*), the lobes usu. imbr., more
rarely valv. (*Oxycoccus*, etc.); A usu. 8–10 (A 5 in *Loiseleuria*), obdiplost., free;
anthers biloc., becoming inverted during development so as to appear introrse
when mature; dehiscence by terminal pores (longit. slits in e.g. *Loiseleuria*,

Epigaea); anthers often with appendages (awns and/or spurs), but not in *Rhododendroïdeae* and scattered genera and species elsewhere. Pollen in tetrads. Nectarif. disk usu. present. G̲ (4) or (5̲) (2–3-loc. in *Sympieza, Tripetaleia*, 7-loc. in *Bejaria*), loc. opp. C̲-lobes; G̲ in many *Vaccinioïdeae*; 1–∞ axile anatr. ovules per loc. Style simple, stigma usu. capitate; fr. caps., drupe or berry; embryo cylindrical, in copious endosp. Many spp. entomophilous: *Erica* spp. perhaps pollinated by thrips; cf. *Kalmia* (explosive mechanism). Useful products include: ed. fr. from some *Vacciniëae*; oil of wintergreen from *Gaultheria*; briar pipes from roots of *Erica*. Several *Rhododendroïdeae* are poisonous to livestock, and also produce toxic pollen (andromedo-toxin).

Classification and chief genera:

I. **Rhododendroïdeae** (winter buds with scales; C caducous; stamens usu. without appendages, often having fine viscous threads among the pollen tetrads; usu. septicidal capsules, seeds often winged).

    1. LEDEAE (polypet.): *Elliottia, Ledum, Tripetaleia* (C 3).

    2. RHODODENDREAE (zygo.): *Rhododendron, Menziesia*.

    3. BEJARIËAE (C 7): *Bejaria*.

    4. PHYLLODOCEAE (fls. reg.; no viscous threads among pollen; seeds usu. not winged): a tribe of rather isolated genera and small groups of genera of doubtful affinities. *Leiophyllum, Loiseleuria* (A 5), *Kalmia* (seeds sometimes winged), *Andromeda* (loculic. caps. and appendaged stamens), *Phyllodoce, Ledothamnus* (ericoid), *Daboecia* (fls. 4-merous, ls. densely tomentose beneath).

II. **Ericoïdeae** (no winter buds or scales, and habit 'ericoid' (see above), except in *Calluneae*; C usu. persistent after flowering; stamens usu. appendaged, no threads among pollen tetrads; fr. usu. loculic. caps. or nut, seeds not winged).

    1. ERICEAE (habit 'ericoid', the leaves whorled; bracteoles 2–3; seeds ∞ per loc.): *Erica, Macnabia, Philippia*.

    2. SALAXIDEAE (like *Ericeae*, but seeds 1 per loc.): *Simocheilus* (caps. septic.), *Salaxis, Eremia*.

    3. CALLUNEAE (ls. decussate, often sagittate; 2 prs. bracteoles): *Calluna* (caps. septic.), *Cassiope* (C decid.).

III. **Vaccinioïdeae** (winter buds with scales, ls. usu. broad and flat; infl. usu. leafless racemes or panicles; C caducous; urceolate; stamens usu. appendaged; often G̲; loculic. caps., drupes or berries, seeds wingless).

    1. VACCINIËAE (G̲): *Vaccinium, Gaylussacia* (fr. a drupe), *Ceratostema, Agapetes, Thibaudia*.

    3. GAULTHERIËAE (G̲ or somewhat embedded, loculic. caps. or berries): *Gaultheria* (K fleshy around caps.), *Pernettya* (K leafy, berries); caps. in *Oxydendrum, Pieris, Leucothoë, Cassandra, Lyonia*.

    3. ARBUTEAE (G̲, fr. a berry or drupe; inversion of anthers occurring late in development, so that they articulate on the filaments and are versatile when mature): *Arbutus* (berry, loc. ∞-seeded), *Arctostaphylos* (1-seeded).

IV. **Epigaeoïdeae** (ls. cordate; dioecious; stamens without appendages, anthers with longit. slits; stigma greatly expanded, 5-lobed; ovary densely pubescent, placentae double).

ERICALA

1. EPIGAEËAE: *Epigaea.*

The above classification is supported by combinations of anatomical characters, incl. stomatal distribution and morphology. See L. Watson in *J. Linn. Soc., Bot.* 59: 111–25, 1965.

Ericala S. F. Gray = Ericoila Borckh. = Gentiana L. (Gentianac.).

'×' **Ericalluna** Krüssm. = Erica L. (Ericac.). (*E. bealeana* Krüssm. = *Erica cinerea* L.).

Ericameria Nutt. Compositae. 20 W. U.S. Ls. ericoid, often glandular and scented.

Ericaulon Lour. = Eriocaulon L. (Eriocaulac.).

Ericentrodea S. F. Blake & Sherff. Compositae. 4 Andes.

Erichsenia Hemsl. Leguminosae. 1 W. Austr.

Ericilla Steud. = Ercilla Juss. (Phytolaccac.).

Ericinella Klotzsch (~ Blaeria L.). Ericaceae. 7 trop. & S. Afr., Madag. K zygo.

Ericodes Kuntze = seq.

Ericoïdes Boehm. (non seq.) = Erica L. (Ericac.).

Ericoïdes Fabr. (uninom.) = *Erica* L. sp. (Ericac.).

Ericoila Borckh. = Gentiana L. (Gentianac.).

Ericoma Vasey = Eriocoma Nutt. = Oryzopsis Michx (Gramin.).

Ericomyrtus Turcz. = Baeckea L. (Myrtac.).

Ericopsis C. A. Gardn. = Leschenaultia R.Br. (Goodeniac.).

Erigenia Nutt. Umbelliferae. 1 E. U.S.

Erigerodes Kuntze = Epaltes Cass. (Compos.).

Erigeron L. Compositae. 200 + cosmop., esp. N. Am.

Erigone Salisb. = Crinum L. (Amaryllidac.).

Erigonia A. Juss. = Erigenia Nutt. (Umbellif.).

Erimatalia Roem. & Schult. = Erycibe Roxb. (Convolvulac.).

Erinacea Adans. Leguminosae. 1 SW. Eur. Branch thorns.

Eringium Neck. = Eryngium L. (Umbellif.).

Erinia Noulet = Campanula L. (Campanulac.).

Erinna Phil. Alliaceae. 1 Chile. A 3 with 3 stds.

Erinocarpus Nimmo ex J. Grah. Tiliaceae. 1 SW. India. Androphore. Fr. spiny.

Erinosma Herb. = Leucojum L. (Amaryllidac.).

Erinus L. Scrophulariaceae. 1 Pyrenees, Alps.

Erioblastus Honda. Gramineae. 2 E. As.

Eriobotrya Lindl. Rosaceae. 30 Himal. to Japan, SE. As., W. Malaysia (exc. Philipp.). *E. japonica* (Thunb.) Lindl. has ed. fr. (loquat), largely cult.

Eriocachrys DC. = Magydaris Koch (Umbellif.).

Eriocactus Backeb. = Notocactus (K. Schum.) Backeb. & F. M. Knuth (Cactac.).

Eriocalia Sm. = Actinotus Labill. (Umbellif.).

Eriocalyx Endl. = Eriocylax Neck. = Aspalathus L. (Legumin.).

Eriocapitella Nak. = Anemone L. (Ranunculac.).

Eriocarpaea Bertol. = Onobrychis Gaertn. (Legumin.).

Eriocarpha Cass. = Montañoa Cerv. (Compos.).

Eriocarpha Lag. ex DC. = Lasiospermum Lag. (Compos.).

Eriocarpum Nutt. = Haplopappus Cass. corr. Endl. (Compos.).

Eriocarpus P. & K. (1) = Eriocarpaea Bertol. = Onobrychis L. (Legumin.).
Eriocarpus P. & K. (2) = Eriocarpha Lag. ex DC. = Lasiospermum Lag. (Compos.).
Eriocarpus P. & K. (3) = Eriocarpum Nutt. = Haplopappus Cass. corr. Endl. (Compos.).
Eriocaucanthus Chiov. = Caucanthus Forsk. (Malpighiac.).
*Eriocaulaceae Desv. Monocots. (Farinosae). 13/1150, mostly trop. & subtrop., esp. S. Am. Perenn. herbs with often grass-like ls. Fls. in invol. heads, inconspic., unisexual, 2-3-merous, reg. or zygo. P usu. 2 whorls, differing in texture. ♂ with usu. (C), A 4–6 or 3–2 with 2- or 1-thecous anthers; ♀ with G̲ (2–3)-loc. with 1 orthotr. pend. ov. in each. Caps. loculic. Endosp. floury. Chief genera: *Eriocaulon, Paepalanthus, Syngonanthus, Leiothrix*.
Eriocaulon L. Eriocaulaceae. 400 trop. & subtrop., with ± 30 in Japan and ± 8 in N. Am., incl. *E. septangulare* With. in the eastern U.S. and also in the Scottish Hebrides and the west coast of Ireland (the only repres. of the fam. in Eur.).
Eriocephala (Backeb.) Backeb. = Eriocactus Backeb. = Notocactus (K. Schum.) Backeb. & F. M. Knuth (Cactac.).
Eriocephalus L. Compositae. 30 S. Afr. (*capok-bosch*).
Eriocereus (A. Berger) Riccob. (~ Harrisia Britt.). Cactaceae. 7–8 Brazil to Argent.
Eriochaenium C. Muell. = Eriachaenium Sch. Bip. (Compos.).
Eriochaeta Fig. & De Not. = Pennisetum Rich. (Gramin.).
Eriochaeta Torr. ex Steud. = Rhynchospora Vahl (Cyperac.).
Eriochilos Spreng. = seq.
Eriochilum Ritg. = seq.
Eriochilus R.Br. Orchidaceae. 6 Austr.
Eriochiton F. Muell. = Bassia All. (Chenopodiac.).
Eriochlaena Spreng. = Eriolaena DC. (Sterculiac.).
Eriochlamys Sond. & F. Muell. Compositae. 2 S. Austr.
Eriochloa Kunth. Gramineae. 20 trop., subtrop. Fodders.
Eriochrysis Beauv. Gramineae. 8 trop. Am., Afr.
Eriochylus Steud. = Eriochilus R.Br. (Orchidac.).
Eriocladium Lindl. = Angianthus Wendl. (Compos.).
Erioclepis Fourr. = Eriolepis Cass. = Cirsium Mill. (Compos.).
Eriocline Cass. = Osteospermum L. (Compos.).
Eriocnema Naud. Melastomataceae. 2 Braz.
Eriococcus Hassk. = Phyllanthus L. (Euphorbiac.).
Eriocoelum Hook. f. Sapindaceae. 10 W. trop. Afr.
Eriocoma Kunth = Montañoa Cerv. (Compos.).
Eriocoma Nutt. = Oryzopsis Michx (Gramin.).
Eriocoryne Wall. = Saussurea DC. (Compos.).
Eriocycla Lindl. Umbelliferae. 7 alpine, N. Persia to W. China.
Eriocyclax B. D. Jacks. (sphalm.) = seq.
Eriocylax Neck. = Aspalathus L. (Legumin.).
Eriodaphus Spach = Eriudaphus Nees = Scolopia Schreb. (Flacourtiac.).
Eriodendron DC. = Bombax L. (Bombacac.).
Eriodes Rolfe = Tainiopsis Schlechter (Orchidac.).

ERIODESMIA

Eriodesmia D. Don = Erica L. (Ericac.).
Eriodictyon Benth. Hydrophyllaceae. 10 SW. U.S., Mex.
Eriodrys Rafin. = Quercus L. (Fagac.).
Eriodyction Benth. (sphalm.) = Eriodictyon Benth. (Hydrophyllac.)
Eriogenia Steud. = Eriogynia Hook. = Luetkea Bong. (Rosac.).
Erioglossum Blume. Sapindaceae. 1 Indomal., Austr.
Eriogon[ac]eae Meissn. = Polygonaceae–Eriogonoïdeae Benth.
Eriogonella Goodman. Polygonaceae. 2 Calif., Chile.
Eriogonum Michx. Polygonaceae. 200 N. Am., esp. W. U.S. Differs from most of the fam. in having no ocreae, and cymose umbels or heads of fls. The partial infls. (of a few or many fls. with special invol. of united brs.) are combined into heads, etc.
Eriogynia Hook. = Luetkea Bong. (Rosac.).
Eriolaena DC. Sterculiaceae (Malvaceae?). 17 China, SE. As., Indomal.
Eriolepis Cass. = Cirsium Mill. (Compos.).
Eriolithis Gaertn. Inc. sed. (Rosac.? Sapotac.?). 1 trop. Am.
Eriolobus (DC.) M. Roem. (~ Sorbus L.). Rosaceae. 6 SE. Eur., Orient,
Eriolopha Ridl. Zingiberaceae. 11 New Guinea.   [E. As.
Eriolytrum Desv. ex Kunth = Panicum L. (Gramin.).
Erione Schott & Endl. = Eriodendron DC. = Ceiba Mill. (Bombacac.).
Erioneuron Nash (~ Tridens Roem. & Schult.). Gramineae. 5 N. Am.
Erionia Nor. = Ocimum L. (Labiat.).
Eriopappus Arn. = Layia Hook. (Compos.).
Eriopappus Hort. ex Loud. = Eupatorium L. (Compos.).
Eriope Humb. & Bonpl. Labiatae. 28 trop. & subtrop. S. Am.
Eriopetalum Wight = Brachystelma R.Br. (Asclepiadac.).
Eriopha Hill = Centaurea L. (Compos.).
Eriophila Reichb. = Erophila DC. (Crucif.).
Eriophoropsis Palla = seq.
Eriophorum L. Cyperaceae. 20 N. temp. & arct., 1 S. Afr.; chiefly on wet moors. The ♀ fls. are massed together; each has a P of bristles which after fert. grow out into long hairs acting as a means of distr. for the fr. The hairs are sometimes used in stuffing pillows, etc.
Eriophyllum Lag. Compositae. 11 W. N. Am.
Eriophyton Benth. Labiatae. 1 Himalaya.
Eriopodium Hochst. = Andropogon L. (Gramin.).
Eriopsis Lindl. Orchidaceae. 6 Costa Rica, trop. S. Am.
Eriopus D. Don = Taraxacum Weber (Compos.).
Eriopus Sch. Bip. ex Baker = Trichocline Cass. (Compos.).
Eriorhaphe Miq. = Pentapetes L. (Sterculiac.).
Erioscirpus Palla = Eriophorum L. (Cyperac.).
Eriosema (DC.) G. Don. Leguminosae. 140 trop. & subtrop.
Eriosemopsis Robyns. Rubiaceae. 1 S. Afr.
Eriosermum Thunb. (sphalm.) = Eriospermum Jacq. (Liliac.).
Eriosolena Blume. Thymelaeaceae. 5 E. Himal. & SW. China to Java.
Eriosorus Fée. Hemionitidaceae. 35 trop. Am. (Andes).
Eriosperm[ac]eae Endl. = Liliaceae–Asphodeleae–Eriosperminae Engl.
Eriospermum Jacq. Liliaceae. 80 Afr.

Eriosphaera F. G. Dietr. = Lasiospermum Lag. (Compos.).
Eriosphaera Less. Compositae. 1 S. Afr.
Eriospora Hochst. = Coleochloa Gilly (Cyperac.).
Eriostax Rafin. = Aechmea Ruiz & Pav. (Bromeliac.).
Eriostemon Less. = Saussurea DC. (Compos.).
Eriostemon Panch. & Sebert = Myrtopsis Engl. (Rutac.).
Eriostemon Sm. Rutaceae. 32 Austr., New Caled.
Eriostemon Sweet = Eriostomum Hoffmgg. & Link = Stachys L. (Labiat.).
Eriostemum Colla ex Steud. = Elaeocarpus L. (Elaeocarpac.).
Eriostemum Poir. = Eriostemon Sm. (Rutac.).
Eriostemum Steud. = Eriostomum Hoffmg. & Link = Stachys L. (Labiat.).
Eriostoma Boiv. ex Baill. = Hypobathrum Blume (Rubiac.).
Eriostomum Hoffmgg. & Link = Stachys L. (Labiat.).
Eriostrobilus Bremek. Acanthaceae. 1 SE. As.
Eriosyce Phil. Cactaceae. 1 Chile.
Eriosynaphe DC. (~ Ferula L.). Umbelliferae. 1 SE. Russia to C. As.
Eriotheca Schott & Endl. Bombacaceae. 19 trop. Am.
Eriothrix Cass. Compositae. 1 Réunion.
Eriothymus J. A. Schmidt = Keithia Benth. = Hedeoma Pers. (Labiat.).
Eriotrichium Lem. = Eritrichium Schrad. (Boraginac.).
Eriotrichum St-Lag. = praec.
Erioxantha Rafin. = Eria Lindl. (Orchidac.).
Erioxylum Rose & Standley. Malvaceae. 2 W. Mexico.
Eriozamia Hort. ex Schuster = Ceratozamia Brongn. (Zamiac.).
Eriphia P.Br. = Besleria L. (Gesneriac.).
Eriphilema Herb. = Sisyrinchium L. (Iridac.).
Eriphlema Baker = praec.
Erisimum Neck. = Erysimum L. (Crucif.).
Erisma Rudge. Vochysiaceae. 20 N. Brazil, Guiana.
Erismadelphus Mildbraed. Vochysiaceae. 3 trop. W. Afr.
Erismanthus Wall. ex Muell. Arg. Euphorbiaceae. 2 Hainan, SE. As., Malay
  Penins., Sumatra, E. Borneo.
Erithalia Bunge ex Steud. = Gentiana L. (Gentianac.).
Erithalis P.Br. Rubiaceae. 10 Florida, W.I.
Erithalis Forst. f. = Timonius DC. (Rubiac.).
Eritheis S. F. Gray = Inula L. (Compos.).
Erithraea Neck. = Erythraea Borckh. = Centaurium Hill (Gentianac.).
Erithrorhiza Rafin. = Erythrorhiza Michx = Galax Rafin. (Diapensiac.).
Eritrichium auctt. = seq.
Eritrichum Schrad. Boraginaceae. 65 temp.
Eritronium Scop. = Erythronium L. (Liliac.).
Eriudaphus Nees = Scolopia Schreb. (Flacourtiac.).
Erlangea Sch. Bip. Compositae. 60 trop. Afr.
Ermania Cham. = Christolea Cambess. ex Jacquem. (Crucif.).
Erndelia Neck. = Passiflora L. (Passiflorac.).
Erndlia Giseke = Curcuma L. (Zingiberac.).
× Ernestara hort. Orchidaceae. Gen. hybr. (iii) (Phalaenopsis × Renanthera
  × Vandopsis).

431

Ernestella Germain = Rosa L. (Rosac.).

Ernestia DC. Melastomataceae. 10 trop. S. Am.

Ernestimeyera Kuntze. Rubiaceae. 4 S. Afr., Madag.

Ernodea Sw. Rubiaceae. 6 SE. U.S., W.I.

Ernstamra Kuntze = Wigandia Kunth (Hydrophyllac.).

Ernstia Badillo. Compositae. 1 trop. S. Am.

Ernstingia Scop. = Ratonia DC. (Sapindac.).

Erobathos Spach = seq.

Erobatos (DC.) Reichb. = Nigella L. (Ranunculac.).

Erocallis Rydb. = Lewisia Pursh (Portulacac.).

Erochloa Steud. = seq.

Erochloë Rafin. = Eragrostis Host (Gramin.).

Erodendron Meissn. = seq.

Erodendrum Salisb. = Protea L. (Proteac.).

Erodiaceae Horan. = Geraniaceae Juss. (*sensu latiss.* Benth. & Hook. f.).

Erodion St-Lag. = Erodium L'Hérit. (Geraniac.).

Erodiophyllum F. Muell. Compositae. 2 W. & S. Austr.

Erodium L'Hérit. Geraniaceae. 90 Eur., Medit. to C. As., temp. Austr., S. trop. S. Am. Like *Geranium.* The awn twists into a corkscrew with free end and is very hygroscopic (used for weather indicators, etc.). The mericarp has a sharp point with backward-pointing hairs. When it falls, the free end of the awn often catches against surrounding objects. If dampness supervene, the awn untwists and lengthens, and the fr. is driven into the soil. When dry the awn curls up, and the process may be repeated (cf. *Stipa*). The partial infl. (unlike that of *Geranium* and *Pelargonium, q.v.*) is a cincinnal umbel.

Eroeda Levyns = Oedera L. (Compos.).

Eronema Rafin. = Cuscuta L. (Cuscutac.).

Erophaca Boiss. = Astragalus L. (Legumin.).

***Erophila** DC. Cruciferae. 10 Eur., Medit. *E. verna* (L.) Chevall. occurs in a vast number of vars. which breed true, and were distinguished by Jordan as spp. (cf. works on Mendelism). Most are tufted, ± xero., with hairy or fleshy ls.

Erosion Lunell = Eragrostis Host (Gramin.).

Erosma Booth = Ficus L. (Morac.).

Erosmia A. Juss. = Euosmia Kunth (Rubiac.).

Eroteum Blanco = Trichospermum Bl. (Tiliac.).

Eroteum Sw. = Freziera Sw. (Theac.).

Erpetina Naud. = Medinilla Gaudich. (Melastomatac.).

Erpetion DC. ex Sweet = Viola L. (Violac.).

Erporchis Thou. = seq.

Erporkis Thou. = Platylepis A. Rich. (Orchidac.).

Errazurizia Phil. (~ Dalea L.). Leguminosae. 2 Calif., 1 Chile.

Errerana Kuntze = Pleiococca F. Muell. (Rutac.).

Ertela Adans. (Moniera Loefl.). Rutaceae. 2 trop. S. Am.

Eruca Mill. Cruciferae. 6 Medit., NE. Afr. (Eritr.). Oil is obtained from the seed of *E. sativa* Mill.

Erucago Mill. = Bunias L. (Crucif.).

Erucaria Cerv. = Botelua Lag. (Gramin.).

Erucaria Gaertn. Cruciferae. 9 E. Medit., Arabia, Persia.

Erucastrum (DC.) Presl. Cruciferae. 17 Canaries, Medit., C. & S. Eur.

Ervatamia Stapf. Apocynaceae. 80 palaeotrop.

Ervilia (Koch) Opiz = Vicia L. (Legumin.).

Ervum L. = Vicia L. + Lens Mill. (Legumin.).

Erxlebenia Opiz (~ Pyrola L.). Pyrolaceae. 1 N. temp.

Erxlebia Medik. = Commelina L. (Commelinac.).

Erybathos Fourr. = Erobatos (DC.) Reichb. = Nigella L. (Ranunculac.).

Erycib[ac]eae Endl. = Convolvulaceae–Convolvuleae–Erycibinae Hoogl.

Erycibe Roxb. Convolvulaceae. 66 S. China, Formosa, Japan, Indomal., N. Queensl. Differs from other *C.* in the bifid (Y-shaped) C-lobes, and in the sessile, conical or subglob. (rarely infundibulif.), 5–10-rayed stig. There is perh. some connection with *Sapotac.*

Erycina Lindl. Orchidaceae. 2 Mexico.

Eryngiophyllum Greenman. Compositae. 1 Mexico.

Eryngium L. Umbelliferae. 230 trop. & temp. (exc. trop. & S. Afr.). Prickly herbs with thick r. and fleshy ls. coated with wax. Fls. in cymose heads, blue, visited by bees. Fibre (*caraguata* fibre) is obtained from the ls. of *E. pandanifolium* Cham. & Schlechtd. (subtrop. S. Am.).

Eryocycla Pritz. = Eriocycla Lindl. = Pituranthos Viv. (Umbellif.).

Erysibe G. Don = Erycibe Roxb. (Convolvulac.).

Erysimastrum (DC.) Rupr. = seq.

Erysimum L. Cruciferae. 100 Medit., Eur., As. Very close to *Cheiranthus*.

Erythaea Pfeiff. (sphalm.) = Erythraea Borckh. (Gentianac.).

Erythalia Delarbre = Eyrythalia Borckh. = Gentiana L. (Gentianac.).

Erythea S. Wats. Palmae. 10 SW. U.S. to C. Am.

Erytheremia Endl. = Erythremia Nutt. = Lygodesmia D. Don (Compos.).

Erythorchis Blume = Galeola Lour. (Orchidac.).

Erythracanthus Nees = Ebermaiera Nees (Acanthac.).

Erythradenia (B. L. Rob.) R. M. King & H. Rob. Compositae. 1 Mex.

Erythraea Borckh. = Centaurium Hill (Gentianac.).

Erythranthe Spach = Mimulus L. (Scrophulariac.).

Erythranthera Zotov. Gramineae. 2 Austr., N.Z.

Erythranthus Oerst. ex Hanst. = Alloplectus Mart. (Gesneriac.).

Erythremia Nutt. = Lygodesmia D. Don (Compos.).

Erythrina L. Leguminosae. 100 trop. & subtrop. In *E. crista-galli* L. the bright red fls. are inverted; the wings are nearly aborted; the keel forms at its base a honey sac. *E. indica* Lam. largely planted as shade, and as support. *E. caffra* Thunb. (S. Afr., *kaffir-boom*) very light timber.

Erythrobalanus (Oerst.) O. Schwarz = Quercus L. (Fagac.).

Erythrocarpus Blume = Suregada Roxb. ex Rottl. (Euphorbiac.).

Erythrocarpus M. Roem. = Modecca Lam. (Passiflorac.).

Erythrocephalum Benth. Compositae. 15 trop. Afr. Pappus scaly.

Erythrocereus Houghton (nomen). Cactaceae. Quid?

Erythrochaete Sieb. & Zucc. = Senecio L. (Compos.).

Erythrochilus auctt. = Erythrochylus Reinw. ex Bl. = Claoxylon A. Juss. (Euphorbiac.).

Erythrochiton Griff. = Ternstroemia L. (Theac.).

ERYTHROCHITON

Erythrochiton Nees & Mart. Rutaceae. 8 trop. Am. The infl. is sometimes borne on the lower surface of l. by adnation from lower l.-axil (unique instance?). K coloured. C sympetalous.

Erythrochlaena P. & K. = Erythrolaena Sweet = Cirsium Mill. (Compos.).

Erythrochlamys Guerke. Labiatae. 5 E. trop. Afr.

Erythrochylus Reinw. ex Blume (1823) = Claoxylon A. Juss. (Euphorbiac.).

Erythrococca Benth. Euphorbiaceae. 30 trop. & S. Afr.

Erythrocoma Greene (~ Sieversia Willd.). Rosaceae. 4 N. Am.

Erythrocynis Thou. (uninom.) = *Orchis purpurea* Thou. (non Huds.) = *Cynorkis purpurascens* Thou. (Orchidac.).

Erythrodanum Thou. = Nertera Banks & Soland. (Rubiac.).

Erythrodes Blume. Orchidaceae. 100 China, Formosa, Indomal., Polynesia, S. U.S. to Argent.

Erythrodris Thou. (uninom.) = *Dryopeia discolor* Thou. = ? *Disperis cordata* Thou. = [Sw. (Orchidac.).

Erythrogyne Vis. = Ficus L. (Morac.).   [Sw. (Orchidac.).

Erythrolaena Sweet = Cirsium Mill. (Compos.).

Erythroleptis Thou. (uninom.) = *Malaxis purpurascens* Thou. = *Liparis salassia* (Pers.) Summerh. (Orchidac.).

Erythronium L. Liliaceae. 25 S. Eur., temp. As., Pacif. & Atl. N. Am. P leaves reflexed.

*Erythropalaceae (Hassk.) Van Tiegh. Dicots. 1/2 Indomal. Slender climbing shrubs or lianes. Ls. alt., ent., simple, long-pet., exstip., 3–5-nerved at base, with axillary tendrils, shortly bifid at apex. Fls. v. small, on slender pedicels, in loose axill. dichot. cymes. K (5), lobes short and broad; C 5, valv., caduc.; A 5, oppositipet. and shortly epipet., with 2 lateral bearded scales at base, fil. short, anth. ovate, introrse; disk shortly cupular, pentagonal; G̲ (3), with 1 apical pend. anatr. ov. per loc., but early becoming 1-loc. through disappearance of thin septa; style short, thick, conical, with shortly trifid stig. Fr. drupaceous, with crustac. endoc., completely encl. in subpyrif. accresc. thin red-brown long-stipit. K, which finally splits into 3–5 reflexed valves; seed 1, blue, with fleshy endosp. Only genus: *Erythropalum*. Close to Olacaceae, in which it should perhaps be included.

Erythropalla Hassk. = seq.

Erythropalum Blume. Erythropalaceae. 2–3 E. Himal. to Celebes & Java.

Erythrophila Arn. = Erythrophysa E. Mey. (Sapindac.).

Erythrophleum Afzel. ex G. Don. Leguminosae. 17 Afr., Seychelles, trop. & E. As., Austr. 10 fertile sta. *E. guineënse* G. Don (Sierra Leone, red-water tree) has poisonous bark, used as ordeal by native tribes.

Erythrophysa E. Mey. Sapindaceae. 3 Somalia, S. Afr. Petiole winged.

Erythrophysopsis Verdcourt. Sapindaceae. 1 S. Madag.

Erythropogon DC. = Metalasia R.Br. (Compos.).

Erythropsis Lindl. ex Schott & Endl. Sterculiaceae. 6 trop. Afr., Madag., India, SE. As., W. Malaysia.

Erythropyxis Engl. = Erytropyxis Pierre = Brazzeia Baill. (Scytopetalac.).

Erythrorhipsalis Berger. Cactaceae. 1 S. Brazil.

Erythrorhiza Michx = Galax Rafin. (Diapensiac.).

Erythroropalum Blume = Erythropalum Bl. (Erythropalac.).

Erythroselinum Chiov. Umbelliferae. 2 trop. Afr.

Erythrospermaceae Van Tiegh. = Flacourtiaceae–Oncobeae Gilg (Erythrospermeae DC.).

*Erythrospermum Lam. Flacourtiaceae. 6 Madag., Ceylon, Burma, China, Malaysia, Polynesia.

Erythrostaphyle Hance = Iodes Blume (Icacinac.).

Erythrostemon Klotzsch = Caesalpinia L. (Legumin.).

Erythrostictus Schlechtd. = Androcymbium Willd. (Liliac.).

Erythrostigma Hassk. = Connarus L. (Connarac.).

Erythrotis Hook. f. = Cyanotis D. Don (Commelinac.).

*Erythroxylaceae Kunth. Dicots. 2/250 trop. Trees and shrubs with usu. alt., entire, stip. ls., often showing 2 persist. longit. folds. Fl. ⚥, reg., usu. heterost., in axill. fasc. K 5 or (5), persistent, quincuncial or valvate; C 5, conv. or imbr., often with appendages on upper side; A 5 + 5, united at base, persist.; G̲ (3–4), usu. 1 only fertile, ov. 1–2, pend., anatr.; styles 3, free or conn., with obliquely clav. stigs. Drupe; endosp. or not. Genera: *Erythroxylum*, *Nectaropetalum*.

Erythroxylon L. = seq.

Erythroxylum P.Br. Erythroxylaceae. 250 trop. & subtrop., chiefly Am. & Madag. Branches often covered with distichous scales (rudimentary ls.). Ls. often showing longit. folds. *E. coca* Lam. (Peru, coca) ls. chewed or infused, enable the user to undergo much fatigue; cocaine is made from them.

Erytrochilus Reinw. ex Blume (1825) = Erythrochylus Reinw. ex Bl. (1823) = Claoxylon A. Juss. (Euphorbiac.).

Erytrochiton [Schlechtd.] = Erythrochiton Griff. = Ternstroemia L. (Theac.).

Erytronium Scop. = Erythronium L. (Liliac.).

Erytropyxis Pierre = Brazzeia Baill. (Scytopetalac.).

Esblichia Rose (sphalm.) = Erblichia Seem. (Turnerac.).

Escallonia Mutis ex L. f. (1781). Escalloniaceae. 60 S. Am., esp. Andes.

Escallonia Mutis (1821) = Dichondra J. R. & G. Forst. (Convolvulac.).

*Escalloniaceae Dum. Dicots. 7/150 trop. & S. temp. (mostly S. Am. & Australas.). An ill-defined group of woody pl. of Saxifragaceous affinity. Ls. alt. or opp., simple, often dentate or crenate-serrate, exstip. Fls. rac. or cymose. K 4–5 or (4–5), imbr. or valv.; C 4–5, imbr. or valv., sometimes clawed; A 4–5, sometimes with 4–5 stds.; disk present; G̲ or G̅ (1–6), sometimes 1-loc., with 1–6 free styles, and ∞ axile or pariet. ov. Caps. or berry; copious endosp. Chief genera: *Escallonia*, *Polyosma*, *Quintinia*.

Eschatogramme Trevis. = Dicranoglossum J. Sm. (Polypodiac.).

Eschenbachia Moench = Conyza Less. (Compos.).

Escheria Regel = Gloxinia L'Hérit. (Gesneriac.).

Eschholtzia Reichb. = Eschholzia Cham. (in tabula) = Escholtzia Dum. = **Eschscholtzia** Bernh. = seq.

Eschscholzia Cham. Papaveraceae. 10 Pacific N. Am., very variable. Fl. perig. with concave recept. (K) falling as a cap. In dull weather each pet. rolls up on itself longit., enclosing some sta. Valves of ripe fr. tend to curl spirally, and fr. explodes.

Eschweilera Mart. Lecythidaceae. 120 trop. Am. Seed sessile. Androec. like *Lecythis*.

ESCHWEILERIA

Eschweileria Zipp. ex Boerl. = Boerlagiodendron Harms (Araliac.).
Esclerona Rafin. (Xylia Benth.). Leguminosae. 15 trop. Afr., Madag., trop. As. Good timber.
Escobaria Britton & Rose. Cactaceae. 20 SW. U.S., Mex.
Escobedia Ruiz & Pav. Scrophulariaceae. 14 trop. Am. Root used for dyeing.
Escobesseya Hest. = Escobaria Britt. & Rose (Cactac.).
Escontria Rose. Cactaceae. 3 Mex.
Esculus Rafin. = Aesculus L. (Hippocastanac.).
Esdra Salisb. = Trillium L. (Trilliac.).
Esembeckia Barb. Rodr. = Esenbeckia Kunth (Rutac.).
Esenbeckia Blume = Neesia Blume (Malvac.).
Esenbeckia Kunth. Rutaceae. 38 trop. Am., W.I. Bark of some Braz. spp. (*angostura brasiliensis, quina*) used like angostura.
Esera Neck. = Drosera L. (Droserac.).
Esfandiari Charif & Aellen. Chenopodiaceae. 1 Persia.
Esmarchia Reichb. = Cerastium L. (Caryophyllac.).
Esmeralda Reichb. f. Orchidaceae. 3 India.
Esmeraldia Fourn. = Meresaldia Bullock (Asclepiadac.).
× **Esmeranda** hort. (v) = × Aranda hort. (Orchidac.).
Esopon Rafin. = Prenanthes L. (Compos.).
Espadaea A. Rich. Goetzeaceae. 1 Cuba.
Espadea Miers = praec.
Espejoa DC. = Jaumea Pers. (Compos.).
Espeletia Mutis. Compositae. 80 Andes. Char. pl. of the alpine region (*páramo*). Aloe-like xero. with dense hairs.
Espeletia Nutt. = Balsamorhiza Hook. (Compos.).
Espeletiopsis Sch. Bip. ex Benth. & Hook. f. = Helenium L. (Compos.).
Espera Willd. = Berrya Roxb. (Tiliac.).
Espicostorus Rafin. = Epicostorus Rafin. = Physocarpus (Cambess.) Maxim. (Rosac.).
Espinosa Lag. = Eriogonum Michx (Polygonac.).
Espostoa Britton & Rose. Cactaceae. 11 trop. S. Am.
Espostoöpsis F. Buxb. Cactaceae. 1 Brazil.
Esquirolia Léveillé = Ligustrum L. (Oleac.).
Esquiroliella Léveillé = Martinella Léveillé = Neomartinella Pilger (Crucif.).
Essenhardtia Sweet = Eysenhardtia Kunth (Legumin.).
Esterhazya Mikan. Scrophulariaceae. 4 Brazil, Bolivia.
Esterhuysenia L. Bolus. Aïzoaceae. 1 S. Afr.
Esula (Pers.) Haw. = Euphorbia L. (Euphorbiac.).
Etaballia Benth. Leguminosae. 1 Guiana, Venezuela.
Etaeria Blume = Hetaeria Blume (Orchidac.).
Etericius Desv. Rubiaceae (inc. sed.). 1 Guiana.
Eteriscius B. D. Jacks. = praec.
Eteriscus Steud. = praec.
Ethanium Salisb. = Renealmia L. f. (Zingiberac.).
Etheiranthus Kostel. = Muscari Mill. (Liliac.).
Etheosanthes Rafin. = Tradescantia L. (Commelinac.).
Ethesia Rafin. (1837) = Ornithogalum L. (Liliac.).

Ethesia Rafin. (1838) = Jacobinia Moric. (Acanthac.).

Ethionema Brongn. = Aëthionema R.Br. (Crucif.).

Ethnora O. F. Cook (= ? Maximiliana Mart.). Palmae. 1 trop. S. Am.

Ethulia L. f. Compositae. 10 trop. Am., trop. & S. Afr., Mascar. Is., Assam, Sunda Is.

Ethuliopsis F. Muell. = Epaltes Cass. (Compos.).

Ethusa Ludw. = Aethusa L. (Umbellif.).

Etlingera Giseke (Geanthus Loes., non Reinw.). Zingiberaceae. 40 Malaysia (esp. New Guinea).

Etorloba Rafin. = Jacaranda Juss. (Bignoniac.).

Etornotus Rafin. = Schweinfurthia A. Braun (Scrophulariac.).

Etoxoë Rafin. = Astrantia L. (Umbellif.).

Ettlingera Giseke (sphalm.) = Etlingera Giseke (Zingiberac.).

Etubila Rafin. = Dendrophthoë Mart. + Scurrula L. (Loranthac.).

Etusa Steud. = Aethusa L. (Umbellif.).

Euacer Opiz = Acer L. (Acerac.).

Euadenia Oliv. Capparidaceae. 2 trop. Afr.

Eualcida Hemsl. (sphalm.) = Enalcida Cass. = Tagetes L. (Compos.).

Euandra P. & K. = Evandra R.Br. (Cyperac.).

Euanthe Schlechter. Orchidaceae. 1 Philipp. Is.

Euaraliopsis Hutch. (sine descr. lat.) = Brassaiopsis Decne & Planch. (Araliac.).

Euarthrocarpus Endl. (sphalm.) = Enarthrocarpus Labill. (Crucif.).

Euarthronia Nutt. ex A. Gray = Coprosma J. R. & G. Forst. (Rubiac.).

Eubasis Salisb. = Aucuba Thunb. (Cornac.).

Eubotryoïdes (Nakai) Hara (∼ Leucothoë D. Don). Ericaceae. 2 Sakhalin, Japan.

Eubotrys Nutt. (∼ Leucothoë D. Don). Ericaceae. 2 N. Am.

Eubotrys Rafin. = Muscari Mill. (Liliac.).

Eubrachion Hook. f. Eremolepidaceae (∼ Viscaceae). 4 S. Am.

Eucalia Raeusch. = Encelia Adans. (Compos.).

Eucallias Rafin. = Billbergia Thunb. (Bromeliac.).

Eucalypton St-Lag. = Eucalyptus L'Hérit. (Myrtac.).

Eucalyptopsis C. T. White. Myrtaceae. 2 Moluccas, New Guinea.

Eucalyptus L'Hérit. Myrtaceae. 500 Austr., Tasm., 2 or 3 Indomal. (blue-gum, iron-bark, stringy-bark, blood-wood, mallee, etc.). One of the most characteristic genera of the Austr. flora, easily known by the operculum of the fl. bud. Trees and shrubby trees. Some spp. reach an enormous size; *E. regnans* F. Muell. is officially recorded as reaching 97 m. in height and 7·5 m. in girth at 2 m., on Mt. Baw Baw near Melbourne (cf. *Sequoia*). The ls. at first formed are often opp. and dorsiv., the later ones alt. and isobilat., more suited to the climate. The barks vary much but, being easily recognised, are a valuable aid in the classification. The most common is smooth bark (gum trees) which exfoliates in patches; other kinds are bark scaly all over the trunk (blood-woods, etc.); bark thick and fibrous, the fibres set longitudinally (stringy-barks), or felted; bark hard and furrowed, often black with age (iron-barks). Infl. usu. an umbel which by lengthening of the axis passes to a panicle or corymb. The floral recept. is hollow and becomes woody in the fr. The K is thrown off as a lid when the fl. opens.

EUCAPNIA

On account of their rapid growth and economic value, these trees are now largely cult. in warm climates. Many yield valuable timber, e.g. *E. rostrata* Schlechtd., *E. marginata* Sm. (jarrah), *E. diversifolia* F. Muell. (karri), etc.; *E. globulus* Labill. (blue-gum) and others yield oil of eucalyptus; others yield oils, kino, etc. (Cf. Maiden, *Crit. Rev. Eucalyptus* (1909–34); Blakely, *Key Eucalypts* (1934).)

Eucapnia Rafin. = Nicotiana L. (Solanac.).
Eucapnos Bernh. = Dicentra Borckh. corr. Bernh. (Fumariac.).
Eucarpha Spach = Knightia R.Br. (Proteac.).
Eucarya T. L. Mitch. Santalaceae. 4 S. & E. Austr.
Eucentrus Endl. = Encentrus Presl = Celastrus L. (Celastrac.).
Eucephalus Nutt. (~ Aster L.). Compositae. 20 N. Am.
Euceraea Mart. Flacourtiaceae. 1 Amaz. Braz.
Euceras P. & K. = praec.
Euchaetis Bartl. & Wendl. Rutaceae. 12 S. Afr.
Eucharidium Fisch. & Mey. = Clarkia Pursh (Onagrac.).
Eucharis Planch. & Linden. Amaryllidaceae. 10 trop. S. Am. Sta. from margin of corona.
Euchidium Endl. = Enchidium Jack = Trigonostemon Blume (Euphorbiac.).
Euchilodes (Benth.) Kuntze = seq.
Euchilopsis F. Muell. Leguminosae. 1 W. Austr.
Euchilos Spreng. = seq.
Euchilus R.Br. = Pultenaea Sm. (Legumin.).
Euchiton Cass. = Gnaphalium L. (Compos.).
Euchlaena Schrad. Gramineae. 2 Mex. *E. mexicana* Schrad. (*teosinte*), used as cereal and fodder. Like *Zea* in habit and infl., but ♀ spikelets free from one another, not forming 'cob'.
× **Euchlaezea** Janaki ex Bor. Gramineae. Gen. hybr. (Euchlaena × Zea).
Euchlora Eckl. & Zeyh. Leguminosae. 1 S. Afr. Ls. simple, exstip.
Euchloris D. Don = Helichrysum Mill. (Compos.).
Euchorium Ekman & Radlk. Sapindaceae. 1 Cuba.
Euchresta Bennett. Leguminosae. 5 E. As., 1 Java.
Euchroma Nutt. = Castilleja Mutis (Scrophulariac.).
Euchylaena Spreng. = Enchylaena R.Br. (Chenopodiac.).
Euchylia Dulac = Lonicera L. (Caprifoliac.).
Euchylus Poir. = Euchilus R.Br. = Pultenaea Sm. (Legumin.).
Eucladus Nutt. ex Hook. = Schiedea Cham. & Schlechtd. (Caryophyllac.).
Euclasta Franch. Gramineae. 1 trop. Am., Afr.
Euclastaxon P. & K. = Euklastaxon Steud. = Andropogon L. (Gramin.).
Euclaste Dur. & Jacks. = Euclasta Franch. (Gramin.).
Euclea L. Ebenaceae. 20 Afr., Comoro Is., Arabia. Ls. alt., opp., or whorled. Fr. ed. *E. pseudebenus* E. Mey. (Orange R. ebony), etc., good wood.
Eucliandra Steud. = Encliandra Zucc. = Fuchsia L. (Onagrac.).
*****Euclidium** R.Br. Cruciferae. 2 E. Eur. to C. As.
Euclinia Salisb. = Randia L. (Rubiac.).
Euclisia Greene = Euklisia (Nutt.) Greene (Crucif.).
Eucnemia Reichb. = seq.
Eucnemis Lindl. = Govenia Lindl. (Orchidac.).

*Eucnide Zucc. Loasaceae. 10 SW. U.S., Mex. G (5).
Eucodonia Hanst. = Achimenes P.Br. (Gesneriac.).
× Eucodonopsis Van Houtte. Gesneriaceae. Gen. hybr. (Achimenes ×
Smithiantha).
Eucolum Salisb. = Gloxinia L'Hérit. (Gesneriac.).
Eucomea Soland. ex Salisb. = Eucomis L'Hérit. (Liliac.).
Eucom[idac]eae Salisb. = Liliaceae–Scilloïdeae K. Krause.
*Eucomis L'Hérit. Liliaceae. 14 S. trop. & S. Afr. Spike crowned by tuft
of br. Sta. broadened at base.
Eucommia Oliv. Eucommiaceae. 1 China. Yields medicinal bark and rubber.
*Eucommiaceae Van Tiegh. Dicots. 1/1 E. As. Trees with alt. simple serr.
decid. exstip. ls. and latex. Fls. ♂ ♀, dioec., naked, reg., sol. in crowded
bract-axils, shortly pedic. A 6–10, fil. v. short, anth. lin., apic. G (2), one loc.
abortive, with 2 apic. pend. anatr. ov. Fr. a samara, somewhat elm-like,
1-seeded, endosp. copious. Only genus: *Eucommia*. Possibly related to
Ulmaceae.
Eucorymbia Stapf. Apocynaceae. 1 Borneo.
Eucosia Blume. Orchidaceae. 2 Java, New Guinea.
Eucrania P. & K. = Eukrania Rafin. (Cornac.).
Eucrinum (Nutt.) Lindl. = Fritillaria L. (Liliac.).
Eucriphia Pers. = Eucryphia Cav. (Eucryphiac.).
Eucrosia Ker-Gawl. Amaryllidaceae. 4 Peru, Ecuador. Ls. stalked.
Eucryphia Cav. Eucryphiaceae. 5–6 SE. Austr., Tasm., Chile.
*Eucryphiaceae Endl. Dicots. 1/5 S. temp. Trees or shrubs with evergr. opp.
simple or pinn. ls. and large stips. Fls. sol., axill., reg., ⚥, hemicyclic, conspic.
K 4, subcalyptrately decid.; C 4, large, white, imbr., sometimes asymm.; A ∞,
fil. filif., anth. small, versat.; G (4–14), 4–14-loc. and -sulcate, each with
∞ pend. ov. Styles and ripe carpels free, but joined by threads to axis. Seeds
winged, sparse, fleshy. Only genus: *Eucryphia*. Related to *Cunoniac*.
Eucrypta Nutt. Hydrophyllaceae. 2 SW. U.S., NW. Mex.
Eucycla Nutt. = Eriogonum Michx (Polygonac.).
Eucyperus Rikli = Cyperus L. (Cyperac.).
Eudema Humb. & Bonpl. Cruciferae. 8 Andes.
Eudesmia R.Br. = Eucalyptus L'Hérit. (Myrtac.).
Eudesmis Rafin. = Colchicum L. (Liliac.).
Eudianthe Reichb. (~ Silene L.). Caryophyllaceae. 5 Medit.
Eudipetala Rafin. = Commelina L. (Commelinac.).
Eudiplex Rafin. = Tamarix L. (Tamaricac.).
Eudisanthema P. & K. = Eydisanthema Neck. = Brassavola R.Br. (Orchidac.).
Eudistemon Rafin. = Coronopus Gaertn. (Crucif.).
Eudodeca Steud. = Endodeca Rafin. = Aristolochia L. (Aristolochiac.).
Eudolon Salisb. = Strumaria Jacq. (Amaryllidac.).
Eudorus Cass. = Senecio L. (Compos.).
Eudoxia D. Don ex G. Don = Gentiana L. (Gentianac.).
Eudoxia Klotzsch = Acalypha L. (Euphorbiac.).
Euforbia Tenore = Euphorbia L. (Euphorbiac.).
Eufournia Reeder = Sohnsia Airy Shaw (Gramin.).
Eufragia Griseb. = Parentucellia Viv. (Scrophulariac.).

439

EUGAMELIA

Eugamelia DC. ex Pfeiff. = Elvira Cass. (Compos.).
Eugeissona Griff. Palmae. 8 Malay Penins., Borneo. Dioec.
Eugenia L. Myrtaceae. 1000 trop. & subtrop. K-segs. free in bud. Ov. 4–∞ per loc. Berry. Cots. fleshy, plumule small. Many have ed. fr., e.g. *E. michelii* Lam. (trop. Am., Brazil cherry), etc.
Eugeniodes Kuntze = Symplocos Jacq. (Symplocac.).
Eugeniopsis Berg = Marlieria Cambess. (Myrtac.).
Eugissona P. & K. = Eugeissona Griff. (Palm.).
Euglypha Chodat & Hassler. Aristolochiaceae. 1 Paraguay.
Eugone Salisb. = Gloriosa L. (Liliac.).
Euhemus Rafin. = Lycopus L. (Labiat.).
Euhydrobryum Koidz. = Hydrobryum Endl. (Podostemac.).
Euilus Stev. = Astragalus L. (Legumin.).
Euklasta P. & K. = Euclasta Franch. (Gramin.).
Euklastaxon Steud. = Andropogon L. (Gramin.).
Euklisia (Nutt.) Greene (*sensu* O. E. Schulz). Cruciferae. 10 Pacif. U.S.
Euklisia (Nutt.) Rydb. = Cartiera Greene (Crucif.).
Euklisia Rydb. ex Small = Icianthus Greene (Crucif.).
Eukrania Rafin. = Chamaepericlymenum Hill (Cornac.).
Eukylista Benth. = Calycophyllum DC. (Rubiac.).
Eulalia Kunth. Gramineae. 30 trop. & subtrop. Afr., As.
Eulalia Trin. = Miscanthus Anderss. (Gramin.).
Eulaliopsis Honda. Gramineae. 2 India, China, Formosa, Philipp. Is.
Eulenburgia Pax = Momordica L. (Cucurbitac.).
Eulepis (Bong.) P. & K. = Mesanthemum Koern. (Eriocaulac.).
Euleria Urb. Anacardiaceae. 1 Cuba.
Euleucum Rafin. = Corema D. Don (Empetrac.).
Eulobus Nutt. ex Torr. & Gray = Camissonia Link (Onagrac.).
***Eulophia** R.Br. ex Lindl. Orchidaceae. 200 pantrop.
Eulophidium Pfitz. Orchidaceae. 30 pantrop.
Eulophiella Rolfe. Orchidaceae. 4 Madag.
Eulophiopsis Pfitz. = Graphorkis Thou. (Orchidac.).
Eulophus R.Br. = Eulophia R.Br. ex Lindl. (Orchidac.).
Eulophus Nutt. (1834) = Peucedanum L. (Umbellif.).
Eulophus Nutt. ex DC. (1829) = Perideridia Reichb. (Umbellif.).
Eulychnia Phil. Cactaceae. 5 Chile.
Eulychnocactus Backeb. = Corryocactus Britt. & Rose (Cactac.).
Eumachia DC. = Ixora L. (Rubiac.).
Eumecanthus Klotzsch & Garcke = Euphorbia L. (Euphorbiac.).
Eumolpe Decne ex Jacques & Hérincq = Gloxinia L'Hérit. (Gesneriac.).
Eumorpha Eckl. & Zeyh. = Pelargonium L'Hérit. (Geraniac.).
Eumorphanthus A. C. Smith = Psychotria L. (Rubiac.).
Eumorphia DC. Compositae. 5 S. Afr.
Eunanus Benth. = Mimulus L. (Scrophulariac.).
Eunomia DC. Cruciferae. 8 mts. of E. Medit.
Eunoxis Rafin. = Lactuca L. (Compos.).
Euodia J. R. & G. Forst. Rutaceae. 45 trop. Afr., As., Austr., Pacif.
Euodia Gaertn. (tab.) = Evodia Gaertn. (text) = Ravensara Sonner. (Laurac.).

Euoesta P. & K. = Evoista Rafin. = Lycium L. (Solanac.).

Euonymodaphne P. & K. = Evonymodaphne Nees = Licaria Aubl. (Laurac.).

Euonymoïdes Medik. = Celastrus L. (Celastrac.).

Euonymoïdes Soland. ex A. Cunn. = Alectryon Gaertn. (Sapindac.).

Euonymus L. Celastraceae. 176 subcosmopolitan with greatest number of spp. in the Himalaya, China and Japan. Several spp. have curious outgrowths of cork on their stems. Heterostyly in some spp. Protandry in *E. europaeus* L. On the ripe seed is a bright-coloured fleshy aril, serving in bird-dispersal. Loculicidal capsules, in some spp. with spiny or wing-like outgrowths. Wood has been used in turnery, for spindles, etc., and furnishes good charcoal. Some spp. at least are poisonous.

Euonyxis P. & K. = Evonyxis Rafin. = Melanthium L. + Zigadenus Michx (Liliac.).

Euopis Bartl. = Evopis Cass. = Berkheya Ehrh. (Compos.).

Euosanthes Endl. = Enosanthes A. Cunn. ex Schauer = Homoranthus A. Cunn. (Myrtac.).

Euosma Andr. = Logania R.Br. (Loganiac.).

Euosma Willd. ex Schultes = seq.

Euosmia Kunth = Evosmia Humb. & Bonpl. = Hoffmannia Sw. (Rubiac.).

Euosmus (Nutt.) Bartl. = Lindera Thunb. + Sassafras Boehm. (Laurac.).

Euothonaea Reichb. f. = Hexisea Lindl. (Orchidac.).

Euparaea Steud. = seq.

Euparea Banks & Soland. ex Gaertn. = Anagallis L. (Primulac.).

Eupatoriaceae Link = Compositae–Eupatorieae Cass.

Eupatoriadelphus R. M. King & H. Rob. Compositae. 4 E. N. Am.

Eupatoriastrum Greenman. Compositae. 4 Mexico, Venez.

Eupatorina R. M. King & H. Rob. Compositae. 1 Hispaniola.

Eupatoriophalacron Adans. = Eclipta L. (Compos.).

Eupatoriophalacron Mill. = Verbesina L. (Compos.).

Eupatoriopsis Hieron. Compositae. 1 Brazil.

Eupatorium Bub. = Agrimonia L. (Rosac.).

Eupatorium L. Compositae. 1200 mostly Am., a few in Eur., As., Afr. The fls. of *E. cannabinum* L. are largely visited by butterflies.

Eupetalum Lindl. = Begonia L. (Begoniac.).

Euphocarpus Anders. ex Steud. = Correa Andr. (Rutac.).

Euphoebe Blume ex Meissn. = Alseodaphne Nees (Laurac.).

Euphora Griff. = Aglaia Lour. (Meliac.).

Euphorbia L. Euphorbiaceae. 2000 cosmop., chiefly subtrop. and warm temp. They differ very much in vegetative habit. Many spp. are herbs, but shrubs are also frequent. The chief interest centres in those spp. that inhabit very dry places and have consequently a xerophytic habit. Most of these forms closely resemble *Cactaceae* (*q.v.*), and sometimes when not in fl. it is very difficult to decide from the outside appearance whether one has to do with a *Euphorbia* or a cactus. The presence of latex of course distinguishes the former. It is very interesting to see how similar conditions of life have called forth, in three different fams. not nearly allied to one another, such a similarity of habit as is seen in *Euphorbia*, the *Cactaceae*, and *Stapelia* (*Asclepiadaceae*). As in the cacti, we get almost spherical forms, ridged forms, cylindrical forms, etc. Many

EUPHORBIA

are armed with thorns. In all cases it is the stem which is fleshy. The outer tissue is green and does the assimilating work of the plant; the inner portion of the stem consists mainly of parenchymatous storage tissue.

For morphology cf. Goebel (*Pflanzenbiol. Schild.* 56, 1889–91). He divides the pls. into the following groups:

I. Ls. normal, well developed, serving a long time as assim. organs. (1) Shoot not water-storing: e.g. the British spp. (2) Storage in tubers below ground: *E. tuberosa* L. (3) Stem as reserve for water, etc., but not green: *E. bupleurifolia* Jacq. (cylindrical stem covered with corky scales = l. bases; ls. borne in wet season, falling in dry). (4) Stem fleshy, green, leafy in wet season only: *E. neriifolia* L., etc.

II. Ls. abortive, dropping off early. Assim. and storage carried on in stem. Various types occur here (cf. *Cactaceae*), some approaching a perfectly spherical form. Some common ones are: (1) *E. tirucalli* L. (Zanzibar), with thin cylindrical shoots. *E. pendula* Link is very similar and resembles *Rhipsalis* in *Cactaceae*. (2) *E. xylophylloïdes* Brongn. has flattened shoots (cf. *Phyllanthus* § *Xylophylla*, and *Epiphyllum* in *Cactaceae*). (3) *E. caput-medusae* L. has a stout stock giving off a number of thinner branches at the top. These are covered with little cushion-like papillae, closely crowded, which are really l. bases; the l. proper is undeveloped. Many spp. show this structure. (4) *E. mamillaris* L. has a thorn in the axil of each cushion (= metamorphosed infl.-axis). If the cushions, as in the cacti, become 'fused', we get a ridged stem, as is seen in (5) *E. polygona* Haw., *E. grandicornis* Goebel, and many others (cf. Cereoid genera in *Cactaceae*). Most of these spp. exhibit pairs of stout thorns which are the stips. of the abortive l. By the two horizontal thorns one can tell one of these pls. from a cactus, which has a group of thorns. (6) *E. meloniformis* Ait. is nearly spherical but ribbed, whilst in (7) *E. globosa* Sims (cf. *Echinocactus*) we have an almost perfect sphere. (Cf. *Cactaceae*, and *Stapelia*, and compare all these succulent forms with one another. See also Goebel, *loc. cit.*)

Besides the above, note *E. splendens* Boj. and *E. bojeri* Hook., pls. with thick stems and green ls., the latter dropped in the dry season.

The other chief point of interest in *E.* is the *cyathium*, or infl. condensed to simulate a single fl. The resemblance is almost perfect. The general branching of the plant is cymose (dichasial). The partial infl. forms a cyathium by the non-development of its internodes, the absence of the P of the individual fls. and the reduction of each ♂ fl. to one sta. There is a perianth-like organ of 5 ls., really bracts, and between these are 4 curious horn-like bodies, which are the combined stips. of the bracts. Then follow a number of sta. arranged with the oldest nearest to the centre and each with a peculiar joint half-way up the stalk. In the middle of the cyathium is a 3-carpelled ovary on a long stalk, usu. ripe for pollination before any sta. ripen.

That this cyathium is an infl. and not a fl., consisting of many ♂ fls., each of 1 sta., round a single ♀, is shown by the centrifugal (cymose) order of ripening of organs, and the joint on the sta.; at this point in the allied gen. *Anthostema* there is a P, which shows that the sta. is really a reduced ♂ fl.

In *E.* § *Poinsettia* the infl. is rendered conspicuous by the bright red colour of the larger upper bracts.

442

EUPHORBIACEAE

As in most tricoccous *Euphorbiaceae*, the capsule explodes when ripe; the carpels split off from the central axis and open at the same moment.
*****Euphorbiaceae** Juss. Dicots. 300/5000 cosmop., except arctic. Few spp. have a very wide range; the most wide-ranging genus is *Euphorbia*. Closely related to *Malvales*, esp. *Sterculiac.*, and *Parietales*, esp. *Flacourtiac.* (habit, foliage, androecium, etc.), and to *Geraniales* (gynoecium, etc.), although separated a good deal from the other fams. of the latter order by the amount of reduction in most of its fls.

Most are shrubs or trees, a few herbaceous. Many are xero.; a number of Australian spp. are of ericoid habit; several (esp. S. Afr.) *Euphorbias* are cactus-like; others resemble *Lauraceae*, or possess phylloclades (e.g. *Phyllanthus* spp.). A few are lianes; some of these have stinging hairs. Ls. usu. alt.; some have opp. ls., some opp. ls. above and alt. below. Stips. usu. present, but may be repres. by branched hair-like bodies (*Jatropha*), glands, or thorns. Many contain latex in special laticiferous vessels.

Infl. usu. complex; almost every type occurs. Often the first branching is racemose and all subsequent ones cymose. In some cases, e.g. *Dalechampia* and *Euphorbia* (*q.v.*), the partial infls. are so condensed as to give the appearance of single fls. The fls. are always unisexual, monoec. or dioec., reg., hypog. The P may be present as two whorls, usu. 5-merous; more often there is only one (calyx) and frequently the fl. is naked. Sta. 1–∞, free or united in various ways. *Ricinus* has branched sta. *Phyllanthus cyclanthera* Baill. has the sta. united, with a ring-like common anther. G̲ usu. (3), with axile placentae, and 3 loc. Styles usu. 2-lobed. The ovules are constant throughout the family and form its best distinctive feature (but cf. *Icacinac.*); they are 1 or 2 in each loc., collateral, pendulous, anatropous, with ventral raphe. The micropyle is usu. covered by a caruncle, which is also found on the seed. The fruit is most commonly a 'schizocarp capsule', more rarely a drupe (*Drypetes*, etc.). It splits into cpls. often elastically, and at the same time each cpl. opens ventrally, letting the seed escape. Seed albuminous. The following important systematic studies on the pollen of the family have appeared in recent years: Punt, Pollen morphology of the Euphorbiaceae with special reference to taxonomy, *Wentia*, 7: 1–116, 1962; Köhler, Die Pollenmorphologie der biovulaten Euphorbiaceae und ihre Bedeutung für die Taxonomie, *Grana Palynol.* 6: 26–120, 1965.

Most *E.* are poisonous. Several are important economic plants, e.g. *Manihot* (rubber, cassava), *Hevea* (rubber), *Croton*, *Ricinus*, etc.

Classification and chief genera (after Pax):

A. 'Platylobeae' (cotyledons much broader than radicle):

  I. **Phyllanthoïdeae** (ovules 2 per loc.; no latex):

    1. PHYLLANTHEAE (embryo large, little shorter than endosp.; ♂ calyx imbricate): *Phyllanthus, Glochidion, Drypetes, Baccaurea, Aporusa.*

    2. BRIDELIËAE (do., but ♂ calyx valvate): *Bridelia, Cleistanthus.*

  II. **Euphorbioïdeae** (**Crotonoïdeae**) (ovules 1 per loc.; latex usu. present):

    1. CROTONEAE (sta. bent inwards in bud): *Croton.*

    2. ACALYPHEAE (sta. erect in bud; fl. usu. apetalous; ♂ calyx valvate; infl. a raceme, spike or panicle, axillary or term.): *Acalypha, Mallotus, Macaranga, Ricinus, Dalechampia, Tragia.*

3. JATROPHEAE (do.; infl. a dichasial thyrse): *Hevea, Jatropha*.
4. ADRIANEAE (do.; infl. a simple term. spike or raceme): *Manihot*.
5. CLUTIËAE (♂ calyx imbr.; ♂ fls. with petals, in groups or cymes, these partial infls. axillary or in complex infls.): *Codiaeum, Clutia*.
6. GELONIËAE (do. but apetalous; infl. leaf-opp.): *Suregada (Gelonium)*.
7. HIPPOMANEAE (do.; apetalous; infl. axillary or term., spike-like, the partial infl. cymes): *Stillingia, Hura, Hippomane, Excoecaria, Sapium*.
8. EUPHORBIËAE (cyathium): *Euphorbia, Monadenium*.
B. 'Stenolobeae' (cotyledons as wide as radicle):
I. *Porantheroïdeae* (ovules 2 per loc.; no latex): *Poranthera*.
II. *Ricinocarpoïdeae* (ovules 1 per loc.; latex): *Ricinocarpos*.

The classif. of the family needs drastic overhauling, some of the tribes as circumscribed by Pax & Hoffmann (1931) containing quite unrelated elements. The division into the two primary series *Platylobeae* and *Stenolobeae* (*qq.v.*) is almost certainly unnatural (as already indicated by Pax), and the two subfamilies of the former, based on the number of ovules per loculus, represent only partially natural groups. In the present work the following genera, currently included in the *Euphorbiac.*, are excluded from this family: Phyllanthoïdeae: *Androstachys, Antidesma, Bischofia, Centroplacus, Hymenocardia, Uapaca*; Euphorbioïdeae: *Galearia, Microdesmis, Pera*. A number of others (e.g. the *Paivaeusinae, Petalostigma, Pogonophora*, etc.) are doubtfully retained.

Euphorbiastrum Klotzsch & Garcke = Euphorbia L. (Euphorbiac.).
Euphorbiodendron Millspaugh = praec.
Euphorbiopsis Léveillé = praec.
Euphorbium Hill = praec.
Euphoria Comm. ex Juss. = Litchi Sonn. (Sapindac.).
Euphorianthus Radlk. Sapindaceae. 3 Philipp. Is., E. Malaysia, New Hebrid.
Euphoriopsis Radlk. (1879; non Massalongo 1852—gen. foss.) = praec.
Euphorona Steud. = Euphrona Vell. (?Rutac.).
Euphrasia L. Scrophulariaceae. 200 N. temp., mts. of Malaysia, N.Z., temp. S. Am. Semi-parasites with loose-pollen fls. (see fam.). The 4 anthers lie close under the upper lip of the fl.; the two upper cohere and also the upper to the lower on each side; the lower lobe of each has a projecting spine. Insects probing for honey shake the spines and receive a shower of pollen from among the anthers. The stigma protrudes beyond the sta. in most fls. so as to be touched first, but every stage can be found from highly protog. fls. with very protruding stigmas to almost homog. fls. whose stigma does not protrude and with self-fert.
Euphroboscis Wight = Euproboscis Griff. = Thelasis Blume (Orchidac.).
Euphrona Vell. Inc. sed. (?Rutac.). 1 Brazil.
Euphronia Mart. Trigoniaceae. 3 N. trop. S. Am.
Euphrosine Allemand. Cyperaceae? Quid?
Euphrosine Endl. = seq.
Euphrosinia Reichb. = seq.
Euphrosyne DC. Compositae. 1 Mexico.
Euphyleia Rafin. = Cordyline Adans. (Agavac.).
Euphylieu Rafin. (sphalm.) = praec.
Euplassa Salisb. Proteaceae. 25 trop. Am.

Euploca Nutt. Boraginaceae. 4 N. Am., Mex.

Eupodia Rafin. = Chironia L. (Gentianac.).

Eupodium J. Sm. = Marattia Sw. (Marattiac.).

Eupogon Desv. = Andropogon L. (Gramin.).

Eupomatia R.Br. Eupomatiaceae. 2 E. New Guinea, coast E. Austr. (Queensl. to Vict.).

*Eupomatiaceae Endl. Dicots. 1/2 New Guinea, E. Austr. Shrubs or small trees, sometimes from tuberous roots, with alt. simple ent. exstip. ls. and sol. term. fls. P o, but fl. bud covered by a bract which forms an ent. decid. calyptra. A ∞, the few outer narrow and fertile, reflexed, extrorse, the ∞ inner petaloid and sterile, ± erect, bearing small 'food-bodies', and forming a pseudo-perianth. G (∞), encl. in hollow recept., with flat top, from the areolate upper surface of which the stigmas shortly project; ovules ∞, ventrally attached. Fr. an urceolate-turbinate berry, apex truncate and margined by annular base of decid. calyptra, with 1–2 seeds per loc.; endosp. ruminate, Only genus: *Eupomatia*. An isolated and ancient group, without close relatives, except perhaps distantly to the *Calycanthaceae*. Pollination by beetles (*Elleschodes*), which feed on the 'food-bodies' on the stds.

Euporteria Kreuz. & Buining = Neoporteria Britton & Rose (Cactac.).

Euprepia Stev. = Astragalus L. (Legumin.).

Eupritchardia Kuntze = Pritchardia Seem. & H. Wendl. (Palm.).

Euproboscis Griff. = Thelasis Blume (Orchidac.).

Euptelea Sieb. & Zucc. Eupteleaceae. 2 Assam, SW. & C. China, Japan.

*Eupteleaceae Van Tiegh. Dicots. 1/2 E. As. Decid. trees or shrubs, with alt. simple dent. exstip. ls. Infl. a cluster of 6–12 fls. borne in axils of bracts on short shoots, precocious. P o; A ∞, in 1 whorl, on flattened recept., fil. filif., anth. elong.-obl., basifixed; G 6–18, in 1 whorl, free, stipit., with 1–3 ventr. ov. in each, and sess. ventr. or term. stig. Fr. a cluster of small winged stipit. samaras, with ventr. notch or indentation; seed 1, with copious oily endosp. Only genus: *Euptelea*. An isolated fam. 'of general ranalian affinity'.

Eupteris Newm. = Pteridium Scop. (Dennstaedtiac.).

Eupteron Miq. = Polyscias J. R. & G. Forst. (Araliac.).

Euptilia Rafin. = Pterocephalus Adans. (Dipsacac.).

Eupyrena Wight & Arn. = Timonius DC. (Rubiac.).

× **Eurachnis** hort. = × Aranda hort. (Orchidac.).

Euranthemum Nees ex Steud. = Eranthemum L. (Acanthac.).

Euraphis (Trin.) Kuntze = Pappophorum Schreb. (Gramin.).

Eurebutia auctt. = Rebutia K. Schum. (Cactac.).

Euregelia Kuntze = Cylindrocarpa Regel (Campanulac.).

Eureiandra Hook. f. Cucurbitaceae. 9 trop. Afr. & Socotra.

Eurhaphis Trin. ex Steud. = Euraphis̆ (Trin.) Kuntze = Pappophorum Schreb.

Eurhotia Neck. = Cephaëlis Sw. (Rubiac.).   [(Gramin.).

Euriosma Desv. (sphalm.) = Eriosema (DC.) Desv. (Legumin.).

Euriples Rafin. = Salvia L. (Labiat.).

Euroschinus Hook. f. Anacardiaceae. 10 New Guinea, NE. Austr., New Caled.

Eurostorhiza G. Don ex Steud. = Physalis L. (Solanac.).

Eurostorhiza Steud. = Caustis R.Br. (Cyperac.).

EUROTIA

Eurotia Adans. = Axyris L. (*s.l.*) (Chenopodiac.).

Eurotia Adans. emend. C. A. Meyer. Chenopodiaceae. 8 N. Afr., C. & E. Eur., temp. As., W. N. Am., N. Mex.

Eurotium B. D. Jacks. (sphalm.) = Erotium Kuntze = Eroteum Sw. = Freziera Sw. (Theac.).

Eurya Thunb. Theaceae. 130 E. As., Indomal., Pacif.

Euryalaceae J. G. Ag. Dicots. 2/3-4 trop. E. As., trop. S. Am. Large aquatic ann. or short-lived perenn. herbs. Rhiz. short, thick, erect; stem o. Ls. large or very large, floating, peltate, strongly nerved below; pet. and nerves aculeate. Scapes 1-fld.; fls. large (but small in rel. to plant), sometimes 'hydrocleistogamous'; pedunc. acul. K 4, externally acul.; C ∞; A ∞, subpetaloid, pollen in tetrads; G̅ (6-8-∞), with ∞ parietal anatr. ov. in each loc., and pelt. or campan. radiate stig. Fr. an inf. many-seeded berry, crowned by persist. K & stig. Endosp. copious, fleshy. Genera: *Euryale, Victoria*.

Euryale Salisb. Euryalaceae. 1 China, SE. As. The seeds and roots are eaten in China.

Euryales Steud. = Eurycles Salisb. (Amaryllidac.).

Euryandra J. R. & G. Forst. = Tetracera L. (Dilleniac.).

Euryandra Hook. f. (sphalm.) = Eureiandra Hook. f. (Cucurbitac.).

Euryangium Kauffm. = Ferula L. (Umbellif.).

Euryanthe Cham. & Schlechtd. = Amoreuxia Moç. & Sessé (Cochlospermac.).

Eurybia Cass. = Olearia Moench (Compos.).

Eurybia S. F. Gray = Aster L. (Compos.).

Eurybiopsis DC. = Vittadinia A. Rich. (Compos.).

Eurybropsis Willis (sphalm.) = praec.

Eurycarpus Botschantsev. Cruciferae. 1 W. Tibet.

Eurycentrum Schlechter. Orchidaceae. 5 New Guinea, Solomon Is.

Eurychaenia Griseb. = Miconia Ruiz & Pav. (Melastomatac.).

Eurychanes Nees = Ruellia L. (Acanthac.).

Eurychiton Nimmo = Limonium Mill. (Plumbaginac.).

Eurychona Willis (sphalm.) = seq.

Eurychone Schlechter. Orchidaceae. 2 trop. Afr.

Eurycles Salisb. Amaryllidaceae. 3 Malaysia, NE. Austr.

Eurycoma Jack. Simaroubaceae. 4 Indomal. Sta. 5, stds. 5.

Eurycorymbus Hand.-Mazz. Sapindaceae. 1 S. China, Formosa.

Eurydochus Maguire & Wurdack. Compositae. 1 Venez.

Eurygania Klotzsch = Thibaudia Ruiz & Pav. (Ericac.).

Eurylepis D. Don = Erica L. (Ericac.).

Eurylobium Hochst. Stilbaceae. 1 S. Afr.

Euryloma D. Don = Erica L. (Ericac.).

Euryloma Rafin. = Calonyction Choisy (Convolvulac.).

Eurymyrtus P. & K. = Euryomyrtus Schauer = Baeckea L. (Myrtac.).

Eurynema Endl. = Hermannia L. (Sterculiac.).

Eurynoma Steud. = seq.

Eurynome DC. = Coprosma J. R. & G. Forst. (Rubiac.).

Eurynotia R. C. Foster. Iridaceae. 1 Ecuador.

Euryomyrtus Schauer = Baeckea L. (Myrtac.).

Euryops Cass. Compositae. 70 S. Afr. to Socotra, Arabia.
Euryosma Walp. = Eriosema DC. (Legumin.).
Eurypetalum Harms. Leguminosae. 3 trop. Afr.
Euryptera Nutt. = Lomatium Rafin. (Umbellif.).
Eurysolen Prain. Labiatae. 1 Indomal.
Euryspermum Salisb. = Leucadendron R.Br. (Proteac.).
Eurystegia D. Don = Erica L. (Ericac.).
Eurystemon Alexander. Pontederiaceae. 1 Mex.
Eurystigma L. Bolus. Aïzoaceae. 1 S. Afr.
Eurystyles Wawra. Orchidaceae. 2 trop. Am., W.I.
Eurystylus Bouché = Canna L. (Cannac.).
Eurystylus P. & K. = Eurystyles Wawra (Orchidac.).
Eurytaenia Torr. & Gray. Umbelliferae. 2 S. U.S. (Texas).
Eurytalia Fourr. = Eyrythalia Borckh. = Gentiana L. (Gentianac.).
Eurytenia Buckl. = Eurytaenia Torr. & Gray (Umbellif.).
Eurythalia D. Don = Eyrythalia Borckh. = Gentiana L. (Gentianac.).
Eusarcops Rafin. = Hippeastrum Herb. (Amaryllidac.).
Euscapha Van Tiegh. = seq.
Euscaphia Stapf = seq.
*Euscaphis Sieb. & Zucc. Staphyleaceae. 4 Japan, S. China, Indoch. Aril.
*Eusideroxylon Teijsm. & Binnend. Lauraceae. 2 Borneo.
Eusipho Salisb. = Cyrtanthus Ait. (Amaryllidac.).
Eusiphon Benoist. Acanthaceae. 3 Madag.
Eusmia Humb. & Bonpl. (tab., sphalm.) = Evosmia Kunth (Rubiaceae).
Eusolenanthe Benth. & Hook. f. = Ensolenanthe Schott = Alocasia G. Don
(Arac.).
Eustachia Rafin. = seq.
Eustachya Rafin. = Veronicastrum Fabr. (Scrophulariac.).
Eustachys Desv. Gramineae. 12 trop. Am., W.I., trop. & S. Afr.
Eustachys Salisb. = Ornithogalum L. (Liliac.).
Eustathes Spreng. = Eystathes Lour. = Xanthophyllum Roxb. (Xantho-
phyllac.).
Eustaurophorae Villani (nom. illegit. pro subfam. Crucif.) = Cruciferae Juss.
excl. Stanleyeae (Nutt.) O. E. Schulz.
Eustaxia Rafin. = Eustachya Rafin. = Veronicastrum Fabr. (Scrophulariac.).
Eustegia R. Br. Asclepiadaceae. 5 S. Afr.
Eustegia Rafin. = Conostegia D. Don (Melastomatac.).
Eustephia Cav. Amaryllidaceae. 6 Peru, Argentina.
Eustephiopsis R. Fries = Hieronymiella Pax (Amaryllidac.).
Eusteralis Rafin. Labiatae. 25 temp. As., Austr.
Eustigma Gardn. & Champ. Hamamelidaceae. 2 S. China, Indoch. C minute,
fleshy, obcordate, with 2 basal nectaries on outer surface (cf. *Disanthus*); A sub-
sessile; $\bar{G}$ (2), with large clavate–flabellate stigmas.
Eustoma Salisb. Gentianaceae. 3 S. U.S., Mex.
Eustrephia D. Dietr. = Eustephia Cav. (Amaryllidac.).
Eustrephus R.Br. (~ Luzuriaga Ruiz & Pav.). Philesiaceae. 1 S. & E. New
Guinea, E. Austr., New Caled. & Loyalty Is. (v. polymorphic).
Eustylis Engelm. & Gray = Nemastylis Nutt. (Iridac.).

EUSTYLIS

Eustylis Hook. f. = Anisotome Hook. f. (Umbellif.).
Eustylus Baker = Eustylis Engelm. & Gray = Nemastylis Nutt. (Iridac.).
Eusynaxis Griff. = Pyrenaria Blume (Theac.).
Eusynetra Rafin. = Columnea L. (Gesneriac.).
Eutacta Link = seq.
Eutassa Salisb. = Araucaria Juss. (Araucariac.).
Eutaxia R.Br. Leguminosae. 9 W. Austr., 1 W. to SE. Austr. Ls. opp.
Eutelia R.Br. ex DC. = Ammannia L. (Lythrac.).
Euteline Rafin. = Genista L. (Legumin.).
Eutereia Rafin. = Dracontium L. (Arac.).
***Euterpe** Gaertn. Palmae. 50 trop. Am., W.I. *E. edulis* Mart. (*assai* palm) ed. fr.; a beverage is prepared by soaking it in water.
Eutetras A. Gray. Compositae. 2 Mexico.
Euthales R.Br. = Velleia Sm. (Goodeniac.).
Euthales F. G. Dietr. = Tovomita Aubl. (Guttif.).
Euthalia Rupr. = Arenaria L. (Caryophyllac.).
Euthalis Banks & Soland. ex Hook. f. = Maytenus Molina (Celastrac.).
Euthamia Ell. = Solidago L. (Compos.).
Euthamnus Schlechter = Aeschynanthus Jack (Gesneriac.).
Euthemidaceae Van Tiegh. = Ochnaceae–Euthemideae Planch.
Euthemis Jack. Ochnaceae. 2 S. Indoch., Malay Penins., Sumatra, Borneo.
Eutheta Standley. Solanaceae. 1 C. Am.
Euthodon Griff. = Pottsia Hook. & Arn. (Apocynac.).
Euthrixia D. Don = Chaetanthera Ruiz & Pav. (Compos.).
Euthyra Salisb. = Paris L. (Trilliac.).
Euthystachys A. DC. Stilbaceae. 1 S. Afr.
Eutmon Rafin. = Talinum Adans. (Portulacac.).
Eutoca R.Br. = Phacelia Juss. (Hydrophyllac.).
Eutralia (Rafin.) B. D. Jacks. = Rumex L. (Polygonac.).
Eutraphis Walp. = Euscaphis Sieb. & Zucc. (Staphyleac.).
Eutrema R.Br. Cruciferae. 15 C. & E. As., Arctic, 1 SW. U.S.
Eutriana Trin. = Botelua Lag. (Gramin.).
Eutrochium Rafin. = Eupatorium L. (Compos.).
Eutropia Klotzsch = Croton L. (Euphorbiac.).
Eutropus Falc. = Pentatropis R.Br. (Asclepiadac.).
Euxena Calest. = Arabis L. (Crucif.).
Euxenia Cham. = Podanthus Lag. (Compos.).
Euxolus Rafin. = Amaranthus L. (Amaranthac.).
Euxylophora Huber. Rutaceae. 1 Amaz. Brazil. Good wood.
Euzomodendron Coss. Cruciferae. 1 S. Spain.
Euzomum Link = Eruca Mill. (Crucif.).
Evacidium Pomel. Compositae. 1 NW. Afr.
Evacopsis Pomel = Evax Gaertn. (Compos.).
Evactoma Rafin. = Silene L. (Caryophyllac.).
Evaiezoa Rafin. = Chondrosea Haw. = Saxifraga L. (Saxifragac.).
Evallaria Neck. = Polygonatum Adans. (Liliac.).
Evalthe Rafin. = Chironia L. (Gentianac.).
Evandra R.Br. Cyperaceae. 2 SW. Austr.

Evanesca Rafin. = Pimenta Lindl. (Myrtac.).
Evansia Salisb. = Iris L. (Iridac.).
Evax Gaertn. Compositae. 25 Medit. to C. As., N. Am.
Evea Aubl. (~ Cephaëlis Sw.). Rubiaceae. 18 C. & trop. S. Am.
Eveleyna Steud. = Evelyna Poepp. & Endl. = Elleanthus Presl (Orchidac.).
Eveltria Rafin. = Orthrosanthus Sweet (Iridac.).
Evelyna Poepp. & Endl. = Elleanthus Presl (Orchidac.).
Evelyna Rafin. = Litsea Lam. (Laurac.).
Everardia Ridl. Cyperaceae. 12 Venez., Guiana.
Everettia Merr. = Beccarianthus Cogn. (Melastomatac.).
Everettiodendron Merr. = Baccaurea Lour. (Euphorbiac.).
Everion Rafin. = Froelichia Moench (Amaranthac.).
Eversmannia Bunge. Leguminosae. 1 SE. Russia, N. Persia, C. As.
Evia Comm. ex Bl. = Spondias L. (Anacardiac.).
Evodea Kunth = Euodia J. R. & G. Forst. (Rutac.).
Evodia Gaertn. = Ravensara Sonner. (Laurac.).
Evodia Scop. = Euodia J. R. & G. Forst. (Rutac.).
Evodianthus Oerst. Cyclanthaceae. 1 C. Am. & N. trop. S. Am.
Evodiella v. d. Linden. Rutaceae. 2 New Guinea, Queensl.
Evodiopanax Nakai. Araliaceae. 2 China, Japan.
Evoista Rafin. = Lycium L. (Solanac.).
Evolvolus Sw. = seq.
Evolvulus L. Convolvulaceae. 100 trop. & subtrop., chiefly Am.
Evonimoïdes Duham. = Celastrus L. (Celastrac.).
Evonimus Neck. = Euonymus L. (Celastrac.).
Evonymodaphne Nees = Licaria Aubl. (Laurac.).
Evonymoïdes Medik. = Evonimoïdes Duham. = Celastrus L. (Celastrac.).
Evonymopsis H. Perrier. Celastraceae. 4 Madag.
Evonymus L. = Euonymus L. (Celastrac.).
Evonyxis Rafin. = Melanthium L. + Zigadenus Michx (Liliac.).
Evopis Cass. = Berkheya Ehrh. (Compos.).
Evosma Steud. = Euosma Andr. = Logania R.Br. (Loganiac.).
Evosmia Humb. & Bonpl. = Hoffmannia Sw. (Rubiac.).
Evosmus (Nutt.) Reichb. = Euosmus (Nutt.) Bartl. = Lindera Thunb. + Sassafras Boehm. (Laurac.).
Evota Rolfe = Ceratandra Lindl. (Orchidac.).
Evotrochis Rafin. = Primula L. (Primulac.).
Evrardia Adans. = Bursera Jacq. ex L. (Burserac.).
Evrardia Gagnep. = Hetaeria Bl. (Orchidac.).
Evrardiella Gagnep. (~ Aspidistra Ker-Gawl.). Liliaceae. 1 Indoch.
Ewaldia Klotzsch = Begonia L. (Begoniac.).
Ewartia Beauverd. Compositae. 5 SE. Austr., Tasmania, N.Z.
Ewersmannia Gorshkova = Eversmannia Bunge (Legumin.).
Ewyckia Blume = Pternandra Jack (quoad typum) + Kibessia DC. (Melasto-
Exacantha P. & K. = Exocantha Labill. (Umbellif.).   [matac.).
Exaculum Caruel = Cicendia Griseb. (Gentianac.).
Exacum L. Gentianaceae. 40 palaeotrop. The style is bent to one side or other of the fl.; both occur on the same plant (enantiostyly).

449

EXADENUS

Exadenus Griseb. = Halenia Borckh. (Gentianac.).

Exagrostis Steud. = Eragrostis Host (Gramin.).

Exalbuminosae Engl. A 'Series' within the fam. *Ochnaceae*, incl. the tribes *Ochneae* ('Ourateëae' Engl.) and *Elvasieae*.

Exallage Bremek. Rubiaceae. 24 Indomal.

Exallosis Rafin. = Ipomoea L. (Convolvulac.).

Exandra Standley. Rubiaceae. 1 Mexico, C. Am.

Exarrhena R.Br. = Myosotis L. (Boraginac.).

Exbucklandia R. W. Brown. Hamamelidaceae. 2 E. Himal. to S. China, Malay Penins. & Sumatra. The large stips. are folded against one another, enclosing and protecting the young axillary bud or infl. Fls. in heads in groups of 4, polyg. or monoec., sunk in the axis. The 'calyx-tube' becomes visible as a ring after flowering. Wood valued.

Excavatia Markgraf. Apocynaceae. 10 New Guinea & Marianne Is. to New Caled., Fiji & Hawaii.

Excoecaria L. Euphorbiaceae. 40 trop. Afr. & As.

Excoecariopsis Pax = Spirostachys Sond. (Euphorbiac.).

Excremis Willd. Liliaceae. 1 Peru, Colombia. (A) thick.

Exechostilus Willis (sphalm.) = seq.

Exechostylus K. Schum. Rubiaceae. 1 trop. Afr.

Exellia Boutique. Annonaceae. 1 trop. Afr.

Exellodendron Prance. Chrysobalanaceae. 4 Brazil, Guiana.

Exemix Rafin. = Lychnis L. (Caryophyllac.).

Exeria Rafin. = Eria Lindl. (Orchidac.).

Exinia Rafin. = Dodecatheon L. (Primulac.).

Exiteles Miers = seq.

Exitelia Blume = Maranthes Bl. (Chrysobalanac.).

Exoacantha Labill. Umbelliferae. 2 Syria, Persia.

Exocarpaceae Gagnep. = Santalaceae–Anthoboleae (Dum.) Endl.

*Exocarpos Labill. Santalaceae. 26 Indoch., Malaysia, Australas., Pacif. to Hawaii.

Exocarpus auctt. = praec.

Exocarya Benth. Cyperaceae. 2 E. Austr.

Exochaenium Griseb. = Sebaea Soland. ex R.Br. (Gentianac.).

Exochogyne C. B. Clarke. Cyperaceae. 4 trop. S. Am.

Exochorda Lindl. Rosaceae. 5 C. As., China. Caps.

Exocroa Rafin. = Ipomoea L. (Convolvulac.).

Exodeconus Rafin. = Physalis L. (Solanac.).

Exodiclis Rafin. = Acisanthera P.Br. (Melastomatac.).

Exogonium Choisy (~ Ipomoea L.). Convolvulaceae. 25 trop. Am. *E. purga* (L.) Benth. (jalap) cult. for medic. resin.

Exohebea R. C. Foster. Iridaceae. 13 S. Afr.

Exolepta Rafin. = Cassandra D. Don (Ericac.).

Exolobus Fourn. Asclepiadaceae. 5 trop. S. Am.

Exomicrum Van Tiegh. = Ouratea Aubl. (Ochnac.).

Exomiocarpon Lawalrée. Compositae. 1 Madag.

Exomis Fenzl. Chenopodiaceae. 2–3 S. & SW. Afr., St Helena.

Exophya Rafin. = Encyclia Hook. (Orchidac.).

Exorhopala v. Steenis. Balanophoraceae. 1 Malay Penins.
Exorrhiza Becc. Palmae. 1 Fiji.
Exosolenia Baill. ex Drake = Genipa L. (Rubiac.).
Exospermum Van Tiegh. Winteraceae. 2 New Caled.
Exostegia Boj. ex Decne = Cynanchum L. (Asclepiadac.).
Exostema (Pers.) Rich. ex Humb. & Bonpl. Rubiaceae. 50 Mex. to Braz. & Peru, W.I. Febrifugal alkaloids are contained in the bark.
Exostemma DC. = praec.
Exostemon P. & K. = praec.
Exostyles Schott ex Spreng. Leguminosae. 2 Brazil.
Exostylis G. Don = praec.
Exotanthera Turcz. = Rinorea Aubl. (Violac.).
Exothamnus D. Don ex Hook. = Aster L. (Compos.).
Exothea Macfadyen. Sapindaceae. 3 Florida, Mex., C. Am., W.I.
Exotheca Anderss. Gramineae. 1 Abyssinia, E. Afr.
Exothostemon G. Don = Prestonia R.Br. (Apocynac.).
Expangis Thou. (uninom.) = *Angraecum expansum* Thou. (Orchidac.).
Exphaloschoenus Nees (sphalm.) = Cephaloschoenus Nees = Rhynchospora Vahl (Cyperac.).
Exsertanthera Pichon = Lundia DC. (Bignoniac.).
Extelma Dum. = ? Oxystelma R.Br. (Asclepiadac.).
Extinctorium Dum. = ? Eucalyptus L'Hérit. (Myrtac.).
Exydra Endl. = Glyceria R.Br. (Gramin.).
Eydisanthema Neck. = Brassavola R.Br. (Orchidac.).
Eydouxia Gaudich. = Pandanus L. f. (Pandanac.).
Eylesia S. Moore. Scrophulariaceae. 1 trop. E. Afr.
Eyrea Champ. ex Benth. = Turpinia Vent. (Staphyleac.).
Eyrea F. Muell. = Pluchea Cass. (Compos.).
Eyrythalia Borckh. = Gentiana L. (Gentianac.).
Eyselia Neck. = Galium L. (Rubiac.).
Eyselia Reichb. = Egletes Cass. (Compos.).
*Eysenhardtia Kunth. Leguminosae. 14 S. U.S. to Guatemala.
Eystathes Lour. = Xanthophyllum Roxb. (Xanthophyllac.).
Ezeria Rafin. = Renealmia R.Br. non L. = Libertia Spreng. (Iridac.).
Ezehlsia Lour. ex Gomes = ? Cordyline Comm. ex Juss. (Agavac.).

F

Faba Mill. (~ Vicia L.). Leguminosae. 1 Medit. reg., *F. vulgaris* Moench (broad bean), widely cult.
*Fabaceae Lindl.: see **Leguminosae** Juss. (nom. altern.).
Fabago Mill. = Zygophyllum L. (Zygophyllac.).
Fabera Sch. Bip. = Hypochoeris L. (Compos.).
Faberia Hemsl. Compositae. 5 SW. China.
Fabiana Ruiz & Pav. Solanaceae. 25 S. Am. (mostly temp.).
Fabrenia Nor. = Toona M. Roem. (Meliac.).

FABRICIA

Fabria E. Mey. = Ruellia L. (Acanthac.).

Fabricia Adans. = Lavandula L. (Labiat.).

Fabricia Gaertn. = Leptospermum J. R. & G. Forst. (Myrtac.).

Fabricia Scop. = Alysicarpus Neck. ex Desv. (Legumin.).

Fabricia Thunb. = Curculigo Gaertn. (Hypoxidac.).

Fabrisinapis C. C. Townsend. Cruciferae. 1 Socotra.

Fabritia Medik. = Fabricia Adans. = Lavandula L. (Labiat.).

Facchinia Reichb. = Arenaria L. (Caryophyllac.).

Facelis Cass. Compositae. 4 S. Am.

Facheiroa Britton & Rose = Espostoa Britt. & Rose (Cactac.).

Facolos Rafin. = Carex L. (Cyperac.).

Factorovskya Eig. Leguminosae. 1 E. Medit. A remarkable geocarpic annual. The gynophore may elongate below ground from 1 mm. up to as much as 10 cm. Fr. 2-seeded, densely hairy.

Fadgenia Lindl. (sphalm.) = Fadyenia Endl. = Garrya Dougl. (Garryac.).

Fadogia Schweinf. Rubiaceae. 60 trop. Afr.

Fadogiella Robyns. Rubiaceae. 4 trop. Afr.

Fadyenia Endl. = Garrya Dougl. (Garryac.).

Fadyenia Hook. Aspidiaceae. 1, *F. hookeri* (Sweet) Maxon, W.I. The sterile ls. produce buds at the tips.

Faeniculum Hill = Foeniculum L. (Umbellif.).

Faetidia Juss. = Foetidia Comm. ex Lam. (Foetidiac.).

*Fagaceae Dum. Dicots. 8/900 cosmop. (exc. trop. S. Am. & trop. & S. Afr.). *Fagus, Castanea* N. extra-trop.; *Quercus* N. extra-trop., Malaysia and C. Am.; *Trigonobalanus* Siam, W. Malaysia; *Lithocarpus* and *Castanopsis* in trop. As.; *Chrysolepis* in Calif.; *Nothofagus* in New Guinea, Austr., New Caled., N.Z. & S. Am.

Mostly monoecious trees; leaves simple, alt. or v. rarely whorled, stipulate. Fls. unisexual, anemophilous or entomophilous, in dichasia which are often arranged in 'catkins'; the dichasia sometimes reduced to solitary fls.

P with usu. 6 divisions. ♂ fls. with 5–∞ stamens, with or without pistillode. ♀ fls. usu. in dichasia of 3 in *Castanea*, 2 in *Fagus*, 1 in *Quercus*. Staminodes often present. Ḡ usu. (3) with 3 styles [(6) with 6 styles in *Castanea*]. Placentas apical, each bearing 2 pendulous anatropous ovules with 2 integuments. Fr. an achene. Seeds without endosperm.

The fruits are surrounded by a cup-like organ termed a *cupule*; in the oak there is one fr. in each cupule, in the beech two, in the chestnut three. The cupule of the beech is divided from the beginning into 4 lobes; the cupule of the chestnut splits at maturity into 4 valves. Recent work suggests that the 4-partite cupule represents a reduction from three 3-lobed cupules condensed together. Elimination of the inner lobes produces the 5-lobed cupule of *Trigonobalanus*. Elimination of the central cupule together with adjacent lobes leaves the 2 lateral cupules each represented by 2 lobes, i.e. the 4-valved cupule of *Castanea*, and of *Fagus* in which the central flower is absent. The entire cupule of *Lithocarpus* containing 1 flower is equivalent to 1 of the three theoretical 3-lobed cupules. The cupule of *Quercus* probably represents a reduction from the *Trigonobalanus*-type cupule.

Classification and genera (cf. Forman in *Kew Bull.* **17**: 388–9, 1964):

I. *Fagoïdeae*: fls. in axillary clusters, rarely solitary; fr. 2- or 3-angled.
♂ infls. ∞-flowered, globular, long-peduncled; ♀ infls. with 2 fls.,
central fl. lacking. *Fagus*
♂ infls. 1–3-flowered, ± sessile; ♀ infls. with 1, 3, or 7 fls., central
fl. present. *Nothofagus*
II. *Castaneoïdeae*: infls. catkin-like, ♂ 'catkins' rigid, stamens usu. 12;
stigma punctiform.
Cupules never lobed or showing any indication of vertical divisions.
Each flower of a ♀ dichasium surrounded by its own cupule,
which is covered with scales or lamellae. *Lithocarpus*
Cupules divided into lobes or segments which sometimes remain
completely fused, but indications of vertical divisions always
evident in young cupules. Each complete ♀ dichasium surrounded by a cupule, which is usu. spiny:
♀ fls. always on separate infls. *Castanopsis*
♀ fls. on androgynous infls.:
Lobes of the cupule at first fused, at length splitting apart; fruits
not separated by cupule-lobes; styles 6–9. *Castanea*
Lobes of the cupule free throughout (from earliest stage); fruits
separated by inner cupule-lobes; styles 3. *Chrysolepis*
III. *Quercoïdeae*: infls. catkin-like; stamens usu. 6; stigma broad.
Fruit round; margin of cupule entire; ♂ 'catkins' flexuose.

Quercus

Fruit 3-angled; cupule lobed; ♂ 'catkins' flexuose or rigid.

Trigonobalanus

The family includes several important plants chiefly valuable for their
timber, e.g. oak (*Quercus*), beech (*Fagus*), chestnut (*Castanea*).

Fagara Duham. = Zanthoxylum L. (Rutac.).
*Fagara L. (*sensu* Engl.). Rutaceae. 250 trop.
Fagaras Kuntze = Fagara Duham. = Zanthoxylum L. (Rutac.).
Fagarastrum G. Don = Clausena Burm. f. (Rutac.).
Fagaropsis Mildbr. Rutaceae. 4 trop. Afr., Somal.
Fagaster Spach = Nothofagus Bl. (Fagac.).
Fagelia Neck. ex DC. = Bolusafra Kuntze (Legumin.).
Fagelia Schwencke = Calceolaria L. (Scrophulariac.).
Fagoïdes Banks & Soland. ex A. Cunn. = Alseuosmia A. Cunn. (Alseuosmiac.).
Fagonia L. Zygophyllaceae. 40 SW. U.S., Chile, Medit., SW. Afr., SW. As.
to NW. India.
*Fagopyrum Mill. Polygonaceae. 15 temp. Euras. Fls. like *Polygonum*, but
heterostyled, with long- and short-styled forms. *F. esculentum* Moench (buck-
wheat) largely cult., esp. in N. Am., for its fr. (seed), in which there is a floury
endosp. Also used as green fodder, and a good honey-plant.
Fagraea Thunb. Potaliaceae. 35 Indomal., SE. As., N. Austr., Pacif. Often
epiphytic. Some spp. have nectaries at the outside of the base of the fl.
Fagraeopsis Gilg & Schlechter = Mastixiodendron Melch. (Rubiac.).
Faguetia L. Marchand. Anacardiaceae. 1 Madag.
Fagus L. Fagaceae. 10 N. temp., Mex. *F. sylvatica* L. (beech, large parts of
Eur.) often forms homogeneous forests, and is accompanied by a peculiar

FAHRENHEITIA

ground flora, e.g. *Galium odoratum, Lathraea squamaria, Cephalanthera dama-sonium*, etc. ♂ fls. in pendulous cymose heads, ♀ in pairs; each cupule encloses two nuts. The wood is hard, and much used in the arts; an oil is expressed from the nuts. Beech hedges in many districts; when growing low it does not drop its ls., as it does when it takes the tree form, and thus affords good shelter in winter. A variety with red sap in the cells of the epidermis (copper beech) is often cult. The beech flowers only every few years.

Fahrenheitia Reichb. f. & Zoll. (~Ostodes Blume). Euphorbiaceae. 4 S. India, Ceylon, W. Malaysia, Bali.

Faidherbia A. Chev. = Acacia Mill. (Legumin.).

Fairchildia Britton & Rose = Swartzia Schreb. (Legumin.).

Fakeloba Rafin. = Anthyllis L. (Legumin.).

*Falcaria Bernh. Umbelliferae. 4–5 C. Eur., Medit., W. & C. As.

Falcata J. F. Gmel. (~Amphicarpaea Ell.). Leguminosae. 2 Himal., NE. As.; 8 Am.

Falcatifolium de Laubenfels (~Podocarpus L'Hérit.). Podocarpaceae. 4 W.

Falcatula Brot. = Trigonella L. (Legumin.).   [Malaysia to New Caled.

Falckia Thunb. = Falkia L. f. (Convolvulac.).

Falconera Salisb. = Albuca L. (Liliac.).

Falconera Wight = Falconeria Royle = Sapium P.Br. (Euphorbiac.).

Falconeria Hook. f. Scrophulariaceae. 1 W. Himalaya.

Falconeria Royle = Sapium P.Br. (Euphorbiac.).

Faldermannia Trautv. = Ziziphora L. (Labiat.).

Falimiria Bess. ex Reichb. = Gaudinia Beauv. (Gramin.).

Falkea Koen. ex Steud. = Begonia L. (Begoniac.).

Falkia L. f. Convolvulaceae. 35 Abyss. & Eritrea to S. Afr.

Fallopia Adans. (~Polygonum L.). Polygonaceae. 9 N. temp.

Fallopia auctt. = Anredera Juss. (Basellac.).

Fallopia Bub. & Penz. = Empetrum L. (Empetrac.).

Fallopia Lour. = Microcos L. (Tiliac.).

Fallugia Endl. Rosaceae. 1 SW. U.S., Mex.

Falona Adans. = Cynosurus L. (Gramin.).

Falopia Steud. = Fallopia Adans. (Polygonac.).

Falya Descoings = Carpolobia G. Don (Polygalac.).

Famarea Vitm. (sphalm.) = Faramea Aubl. (Rubiac.).

Famatina Ravenna. Amaryllidaceae. 3 Andes of Chile & Argent.

Fanninia Harv. Asclepiadaceae. 1 S. Afr.

Faradaya F. Muell. Verbenaceae. 17 N. Borneo, New Guin., Austr., Polynes.

Faramea Aubl. Rubiaceae. 120 trop. S. Am., W.I. Dimorphic pollen.

Fareinhetia Baill. (sphalm.) = Fahrenheitia Reichb. f. & Zoll. (Euphorbiac.).

Farfara Gilib. = Tussilago L. (Compos.).

Farfugium Lindl. = Ligularia Cass. (Compos.).

Fargesia Franch. Gramineae. 2 China.

Farinaceae Dulac = Chenopodiaceae Dulac.

Farinosae Engler. An order of Monocots. comprising the fams. *Flagellariac., Restionac., Centrolepidac., Mayacac., Xyridac., Eriocaulac., Thurniac., Rapateac., Bromeliac., Commelinac., Pontederiac., Cyanastrac.* and *Philydrac.* They are characterized by mealy or floury endosp.

454

Farmeria Willis. Podostemaceae. 2 S. India, Ceylon.
Farnesia Fabr. (uninom.) = *Persea* Mill. sp. (Laurac.).
Farnesia Gasp. = Acacia Mill. (Legumin.).
Faroa Welw. Gentianaceae. 17 trop. & S. Afr.
Farobaea Schrank ex Colla = Senecio L. (Compos.).
Farquharia Hilsenb. & Boj. ex Boj. = Crateva L. (Capparidac.).
Farquharia Stapf. Apocynaceae. 1 S. Nigeria.
Farrago W. D. Clayton. Gramineae. 1 trop. E. Afr.
Farreria I. B. Balf. & W. W. Sm. = Wikstroemia Endl. (Thymelaeac.).
Farringtonia Gleason. Melastomataceae. 1 Venez.
Farsetia Turra. Cruciferae. 15 Morocco to NW. India, C. Afr.
Fartis Adans. = Zizania L. (Gramin.).
Fascicularia Mez. Bromeliaceae. 6 Chile.
Fasciculus Dulac = Spergularia (Pers.) J. & C. Presl (Caryophyllac.).
Faskia Lour. ex Gomes = Strophanthus DC. (Apocynac.).
Faterna Nor. ex A. DC. = Landolphia Beauv. (Apocynac.).
Fatioa DC. = Lagerstroemia L. (Lythrac.).
Fatoua Gaudich. Moraceae. 1 Madag.; 1 Java N. to Japan and E. to N. Austr.
Fatraea Thou. ex Juss. = Terminalia L. (Combretac.).   [& New Caled.
Fatrea Juss. = praec.
× **Fatshedera** Guillaumin. Araliaceae. Gen. hybr. (Fatsia × Hedera).
Fatsia Decne & Planch. Araliaceae. 1 Japan, 1 Formosa.
Faucaria Schwantes. Aïzoaceae. 36 S. Afr.
Faucherea Lecomte. Sapotaceae. 4 Madag.
Faucibarba Dulac = Calamintha Mill. = Satureia L. (Labiat.).
Faujasia Cass. Compositae. 5 Madag., Mascarenes.
Faulia Rafin. = Ligustrum L. (Oleac.).
Faurea Harv. Proteaceae. 18 trop. & S. Afr., Madag.
Fauria Franch. = Nephrophyllidium Gilg (Menyanthac.).
Fauriea P. & K. = praec.
Faustia Font Quer & Rothm. = Saccocalyx Coss. & Dur. (Labiat.).
Faustula Cass. = Helichrysum L. (Compos.).
Favonium Gaertn. = Didelta L'Hérit. (Compos.).
Favratia Feer = Campanula L. (Campanulac.).
Fawcettia F. Muell. Menispermaceae. 1 E. Austr.
Faxonanthus Greenman. Scrophulariaceae. 1 Mexico.
Faxonia T. S. Brandegee. Compositae. 1 Lower California.
Faya Neck. = Crenea Aubl. (Lythrac.).
Faya Webb & Berth. = Myrica L. (Myricac.).
Fayana Rafin. = praec.
Feaea Spreng. = Selloa Kunth (Compos.).
Feaëlla Blake = praec.
Feddea Urb. Compositae. 1 Cuba.
Fedia Adans. = Patrinia Juss. (Valerianac.).
*Fedia Gaertn. (emend. Moench). Valerianaceae. 3 Medit.
Fedia Kunth = Astrephia Dufr. (Valerianac.).
Fedorovia Yakovl. (~ Ormosia G. Jacks.). Leguminosae. 40–50 E. Himal., E.
& SE. As., trop. Am.

FEDTSCHENKIELLA

Fedtschenkiella Kudrjaschev = Dracocephalum L. (Labiat.).
Fedtschenkoa Regel & Schmalh. ex Regel = Malcolmia R.Br. (Crucif.).
Feea Bory. Hymenophyllaceae. 5 trop. Am.
Féea P. & K. = Feaea Spreng. = Selloa Kunth (Compos.).
Feedia Hornem. = Fedia Gaertn. (Valerianac.).
Feeria Buser (~ Trachelium L.). Campanulaceae. 1 Morocco.
Fegimanra Pierre. Anacardiaceae. 2 trop. Afr.
Feidanthus Stev. = Astragalus L. (Legumin.).
Feijoa Berg = Acca Berg (Myrtac.).
*Felicia Cass. Compositae. 60 trop. & S. Afr.
Feliciadamia Bullock (~ Miconia Ruiz & Pav.). Melastomataceae. 1 W. Afr.
Feliciana Benth. & Hook. f. = seq.
Felicianea Cambess. = Myrrhinium Schott (Myrtac.).
Felipponia Hicken (1917; non Brotherus 1912—Musci) = seq.
Felipponiella Hicken. Araceae. 1 Uruguay.
Fendlera Engelm. & Gray. Philadelphaceae. 4 SW. U.S., Mex.
Fendlera P. & K. = Fendleria Steud. = Oryzopsis Michx (Gramin.).
Fendlerella A. A. Heller. Philadelphaceae. 3 SW. U.S.
Fendleria Steud. = Oryzopsis Michx (Gramin.).
Fenelonia Rafin. = Lloydia Salisb. (Liliac.).
Feneriva Diels. Annonaceae. 1 Madag.
Fenestraria N. E. Brown. Aïzoaceae. 2 S. Afr.
Feniculum Gilib. = Foeniculum Mill. (Umbellif.).
Fenixanthes Rafin. = Salvia L. (Labiat.).
Fenixia Merr. Compositae. 1 Philipp. Is.
Fentzlia Reichb. = Fenzlia Benth. = Gilia Ruiz & Pav. (Polemoniac.).
Fenugraecum Adans. = Foenugraecum Ludw. = Trigonella L. (Legumin.).
Fenzlia Benth. = Gilia Ruiz & Pav. (Polemoniac.).
Fenzlia Endl. Myrtaceae. 6 Austr.
Feracacia Britton & Rose = Acacia Mill. (Legumin.).
Ferberia Scop. = Althaea L. (Malvac.).
Ferdinanda Benth. & Hook. f. = Fernandoa Welw. ex Seem. (Bignoniac.).
Ferdinanda Lag. = Zaluzania Pers. (Compos.).
Ferdinandea Pohl = Ferdinandusa Pohl (Rubiac.).
Ferdinandia Seem. (1865) = Fernandoa Welw. ex Seem. (Bignoniac.).
Ferdinandoa Seem. (1870) = praec.
Ferdinandusa Pohl. Rubiaceae. 20 trop. S. Am., W.I.
Ferecuppa Dulac = Tozzia L. (Scrophulariac.).
Fereira Reichb. = seq.
Fereiria Vell. ex Vand. = Hillia Jacq. (Rubiac.).
Feretia Delile. Rubiaceae. 4 trop. Afr., Abyss.
Fergusonia Hook. f. Rubiaceae. 1 S. India, Ceylon.
Fernaldia R. E. Woodson. Apocynaceae. 4 Mex., C. Am.
Fernandezia Ruiz & Pav. (1794). Orchidaceae. Quid?
Fernandezia Ruiz & Pav. (1798) = Dichaea Lindl., Centropetalum Lindl., etc. (Orchidac.).
Fernandia Baill. = seq.
Fernandoa Welw. ex Seem. Bignoniaceae. 2 trop. Afr.

Fernelia Comm. ex Lam. Rubiaceae. 2 Mascarenes.
Fernseea Baker. Bromeliaceae. 1 SE. Braz.
× **Ferobergia** Glass. Cactaceae. Gen. hybr. (Ferocactus × Leuchtenbergia).
Ferocactus Britton & Rose. Cactaceae. 35 SW. U.S., Mexico.
Ferolia Aubl. = Parinari Aubl. (Chrysobalanac.).
Feronia Corrêa = Limonia L. (Rutac.).
Feroniella Swingle. Rutaceae. 3 SE. As., Java.
Ferrandia Gaudich. = Cocculus DC. (Menispermac.).
Ferraria Burm. ex Mill. Iridaceae. 2 trop. & S. Afr.
Ferreirea Allem. Leguminosae. 2 SE. Braz., Parag.
Ferreola Koen. ex Roxb. = Diospyros L. (Ebenac.).
Ferretia Pritz. = Feretia Delile (Rubiac.).
Ferreyrella Blake. Compositae. 1 Peru.
Ferriera Bub. = Paronychia Adans. (Caryophyllac.).
Ferriola Roxb. = Ferreola Koen. ex·Roxb. = Diospyros L. (Ebenac.).
Ferrum-equinum Medik. = Hippocrepis L. (Legumin.).
Ferula L. Umbelliferae. 133 Medit. to C. As. *F. communis* L. (giant fennel, cult.) only flowers after storing up material for some years (cf. *Fagus, Agave*). *F. narthex* Boiss. and *F. assa-foetida* L. are the sources of the drug asafoetida, obtained by notching the roots; used as a condiment in Persia, etc., under the name 'food of the gods', and as a stimulant in medicine. *F. galbaniflua* Boiss. & Buhse and *F. rubricaulis* Boiss. are the sources of the medic. gum galbanum.
Ferulago Koch. Umbelliferae. 50 Medit., SE. Eur., to Persia & C. As.
Ferulopsis Kitagawa. Umbelliferae. 1 Mongolia.
Fessonia DC. ex Pfeiff. = Picramnia Sw. (Simaroubac.).
Festania Rafin. = Rhus L. (Anacardiac.).
Festuca L. Gramineae. 80 cosmop. (fescue-grass). The ls. roll inwards in dry air (cf. *Stipa*). Many good pasture-grasses. When growing on mountains often viviparous.
Festucaceae (Kunth) Herter = Gramineae–Festuceae (Kunth) Nees = Gramineae–Poëae (Nees).
Festucaria Fabr. (uninom.) = *Festuca* L. sp. (Gramin.).
× **Festulolium** Aschers. & Graebn. Gramineae. Gen. hybr. (Festuca × Lolium).
Feuillaea Gled. = Fevillea L. (Cucurbitac.).
Feuillea Kuntze = praec.
Feuilléea Kuntze = seq.
Fevillaea Neck. = Albizia + Calliandra + Inga + Pithecellobium (Legumin.).
Fevillea L. Cucurbitaceae. 9 trop. Am. 5 sta. all alike.
Fevilleaceae Pfeiff. = Cucurbitaceae–Zanonioïdeae C. Jeffr.
Fezia Pitard ex Battand. Cruciferae. 1 Morocco.
Fialaris Rafin. = Rapanea Aubl. (Myrsinac.).
Fibichia Koel. = Cynodon Pers. (Gramin.).
Fibigia Medik. Cruciferae. 14 E. Medit. to Afghan.
Fibra J. Colden (apud Schoepf) = Coptis Salisb. (Ranunculac.).
Fibraurea J. Colden (apud Ellis) = praec.
Fibraurea Lour. Menispermaceae. 5 Assam, Indoch., Philipp. Is., Borneo.
Fibraureopsis Yamamoto = ? Haematocarpus Miers (Menispermac.).

FIBROCENTRUM

Fibrocentrum Pierre (nomen). Sapotaceae. 1 Brazil. Quid?

Ficaceae Dum. = Moraceae Link.

Ficalhoa Hiern. Theaceae(?). 1 trop. Afr. An interesting type, combining certain features of *Theac.* (esp. *Eurya*) with *Actinidiac.* and *Sladeniac.* (dichas. cyme), *Sapotac.* (latex), and other related fams. It should perhaps be treated as a distinct fam. The following are the salient characters: Large trees, with latex in the bark. Ls. simple, serr., distich., exstip. Fls. ♀, reg., small, in sol. or paired axill. dichasial cymes. K 5(-6), imbr.; C 5(-6), imbr., connected at the base; A 15(-18), 1-ser., arranged in alternipet. 3-membered phalanges (triads), adnate to C, with slender fil. and ovoid didymous anth. dehisc. by apical pore and lat. slit, pollen grains not in tetrads; G̲ (5-6), with ∞ ov. per loc. on spongy axile plac., and 5(-6) subul. recurved almost free styles. Fr. a depressed loculicid. caps.; seeds ∞, with retic. testa, sparse endosp. and straight embr.

Ficaria Guett. = Ranunculus L. (Ranunculac.).

Fichtea Sch. Bip. = Microseris D. Don (Compos.).

Ficindica St-Lag. = Opuntia Mill. (Cactac.).

*****Ficinia** Schrad. Cyperaceae. 60 trop. & S. Afr.

Ficodaceae Kuntze = seq.

Ficoïd[ac]eae Juss. = Aïzoaceae J. G. Agardh.

Ficoïdes Mill. = Mesembryanthemum L. (Aïzoac.).

Ficula Fabr. = praec.

Ficus L. Moraceae. 800 warm, chiefly Indomal., Polynes., etc. Trees and shrubs of the most various habit; many root-climbers, twiners and epiphytes. Ls. alt. (spiral or distich.) or opp., ent. or toothed, with stips. which envelop the bud (acting as a protection to it against heat, etc.) and soon after their unfolding drop off. Adv. roots are very common.

F. elastica Roxb. (indiarubber tree) grows as a stout independent tree, usu. commencing epiphytically, and often reaching a great size. At its base are developed buttress-roots, radiating out in all directions; their depth is often several feet, while their thickness is only a few inches. From the branches are given off copious adv. roots which grow downwards and enter the soil. The ls. are entire, and leathery, with a glossy surface. Rubber is obtained by tapping (cf. *Hevea*).

In *F. benghalensis* L. (*F. indica* L.) (banyan) the aerial roots grow in thickness and form great pillars supporting the branches, and by their means the tree may reach immense size. (The banyan is sacred in India; the young roots are provided with tubes of bamboo to protect them, and the ground is prepared for them.) See plate in *Nat. Pfl.* of the famous tree at Calcutta. *F. benghalensis* produces a form with cup-shaped ls., known as Krishna's fig (*F. krishnae* C. DC.).

F. benjamina L. and other spp. climb up other trees, giving off aerial clasping (negatively heliotropic) roots which surround the trunk. These roots thicken and unite into a network and finally often strangle the 'host' altogether. These spp. often become epiph. by the dying away of their lower portions, but like the Aroids they maintain communication with the ground by long aerial roots. Sometimes they commence as epiphytes and send down aerial roots to the soil.

F. religiosa L. (peepul or bo-tree) is, in the wild state, like the three foregoing, a strangling fig, developing epiphytically. Its ls. have a long acuminate apex, combined with an easily wetted surface. From the apex (drip-tip) the rain drips off rapidly after a shower and the l. is soon dry. In very wet trop. forests this property is of some importance.

F. sycomorus L. (sycomore or mulberry fig), N. Afr., and *F. carica* L. (fig), Eur., Medit., are erect independent trees, starting as seedlings on the ground.

F. pumila L. (*F. repens* Rottl.) is a small climbing sp. which takes hold of its support by aerial roots (as in ivy); it is said that these secrete a gummy substance containing caoutchouc, and then absorb the fluid constituents, leaving the caoutchouc as a cement, fastening the roots to their support. (Darwin, *Climbing Plants*, p. 185.) *F. thwaitesii* Miq. and other climbing spp. are heterophyllous, the ls. on the climbing shoots being small and different in shape.

The infl. of *Ficus* is hollowed out, and consists of a number of fls. (several thousand in the largest figs, 2–3 in the smallest) inside a glob. or pear-shaped common recept., the narrow apical orifice of which is closed by small overlapping bracts. There are ♂, ♀, and gall-flowers (which are sterile ♀). The genus divides into two groups: MONOECIOUS, with all three kinds of fl. in the same fig (Subgen. *Urostigma*, *Sycomorus* and *Pharmacosycea*); DIOECIOUS, with ♂ and gall-flowers in the figs of one plant (gall-plants), and ♀ flowers in the figs of another plant (seed-plants) (Subgen. *Ficus*, mainly Asiatic and Polynesian); in this subgenus the ♂ fls. are usu. concentrated near the orifice. (See Corner in *Gardens Bull. Singapore*, **19**: 202, 1962.) In the ♂ fls. there are 1–7 stamens and perianth-segments (rarely P o). The infl. as a whole is protog. The mode of pollination is extraordinary (cf. *Yucca*), there being apparently a special gall-insect for each *Ficus* species. The gall-wasps belong to several genera of the Hymenopterous family *Agaonidae* (*Chalcidoïdea*), the best known being *Blastophaga*. The gravid ♀ enters a fig infl. and lays eggs in the ovary; the ♂ wasps thus formed fertilise the ♀'s, and these as they emerge are dusted by the pollen of the ♂ fls. and carry the pollen to new figs. The theory is that the fig-wasp stimulates only endosperm development in the ovules of the gallflowers, and on this endosperm the wasp-grub feeds. (Cf. Condit in *Hilgardia*, **6**: 443–81, 1932; v. der Vecht in *Entom. Bericht*, **16**, 99–103, 1956; Grandi in *Bol. Ist. Entom. Univ. Bologna*, **26**: I–XIII, 1962.)

Many spp. bear the fls. on old parts of the stem (cauliflory), whilst a number of others are geocarpic, fruiting on underground stolons, up to 10 m. long, from the base of the trunk, the figs being buried up to 10 cm. deep in the earth. (Cf. Corner, *Wayside Trees of Malaya*, 658–88, 1952.) Fr. multiple, composed of a number of drupes inside the common fleshy recept.; that of *F. carica* L. is the edible fig (closely allied to *F. palmata* Forsk. of Africa and India). Other edible spp. include *F. auriculata* Lour. (*F. roxburghii* Wall. ex Brandis, non Miq.) and *F. pumila* L. var. *awkeotsang* Makino (Formosa).

Lac (shellac, etc.) is produced on several by the punctures of a small hemipterous insect (cf. *Butea*). Several, esp. *F. elastica* Roxb., yield caoutchouc. The buttress-roots are used as planks. *F. sycomorus* supplied timber for Egyptian mummy-cases. The latex of several spp. is anthelmintic, the property being connected with the proteolytic enzyme *ficin*.

Fidelia Sch. Bip. = Leontodon L. (Compos.).

FIEBERA

Fiebera Opiz = Physocaulis (DC.) Tausch = Chaerophyllum L. (Umbellif.).
Fiebrigia K. Fritsch. Gesneriaceae. 1 Bolivia.
Fiebrigiella Harms. Leguminosae. 1 Bolivia.
Fiedleria Reichb. = Petrorrhagia (Ser.) Link (Caryophyllac.).
Fieldia A. Cunn. Gesneriaceae. 1 SE. Austr.
Fieldia Gaudich. = Vandopsis Pfitzer (Orchidac.).
Fierauera Reichb. (sphalm.) = Fibraurea Lour. (Menispermac.).
Figaraea Viv. = Neurada L. (Neuradac.).
Figonia Rafin. = Michelia L. (Magnoliac.).
Figuierea Montr. = Coelospermum Bl. (Rubiac.).
Filaginella Opiz = Gnaphalium L. (Compos.).
Filaginopsis Torr. & Gray = Evax Gaertn. (Compos.).
*Filago L. Compositae. 20 Eur., N. Afr., As., SW. & E. N. Am.
Filago Loefl. = Evax Gaertn. (Compos.).
Filagopsis (Battand.) Rouy = Evax Gaertn. (Compos.).
Filangis Thou. (uninom.) = Angraecum filicornu Thou. (Orchidac.).
Filarum D. H. Nicolson. Araceae. 1 Peru.
Filetia Moq. Acanthaceae. 8 Sumatra, Malay Penins.
Filicaceae Dulac = Filicopsida (excl. Marsileidae and Salviniidae).
Filicidae (leptosporangiate ferns). Here divided into 14 Orders: *Schizaeales, Pteridales, Dicksoniales, Davalliales, Hymenophyllales, Loxsomales, Gleicheniales, Cyatheales, Aspidiales, Blechnales, Matoniales, Polypodiales, Plagiogyriales,*
Filicirna Rafin. = Drosera L. (Droserac.).   [*Hymenophyllopsidales.*
Filicium Thw. ex Benth. & Hook. f. Sapindaceae. 3 trop. Afr. & As.
Filicopsida. Here divided into 6 subclasses: *Ophioglossidae, Marattiidae, Osmundidae, Filicidae, Marsileidae, Salviniidae.*
Filicula Seguier = Cystopteris Bernh. (Athyriac.).
Filifolium Kitamura (~ Artemisia L.). Compositae. 1 NE. As.
Filinguis Rafin. = Phyllitis Ludw. (Aspleniac.).
Filipedium Raiz. & Jain = Capillipedium Stapf (Gramin.).
Filipendicula Guett. = seq.
Filipendula Mill. Rosaceae. 10 N. temp.
Filix Seguier = Dryopteris Adans. (Aspidiac.).
Fillaea Guill. & Perr. = Erythrophleum Afzel. (Legumin.).
Fillaeopsis Harms. Leguminosae. 1 trop. Afr.
Fimbriaria A. Juss. = Schwannia Endl. (Malpighiac.).
Fimbrillaria Cass. = Conyza L. (Compos.).
Fimbristemma Turcz. Asclepiadaceae. 5 C. & trop. S. Am.
Fimbristilis Ritg. = Fimbristylis Vahl (Cyperac.).
Fimbristima Rafin. = Aster L. (Compos.).
*Fimbristylis Vahl. Cyperaceae. 300 trop. & subtrop., esp. Indomal. & Austr.
Fimbrolina Rafin. = Besleria L. (Acanthac.).
Fimbrystylis D. Dietr. = Fimbristylis Vahl (Cyperac.).
Finckea Klotzsch = Grisebachia Klotzsch (Ericac.).
Findlaya Bowdich = Plumbago L. (Plumbaginac.).
Findlaya Hook. f. Ericaceae. 1 Trinidad.
Finetia Gagnep. Combretaceae. 1 Indoch.
Finetia Schlechter = Neofinetia Hu (Orchidac.).

Fingalia Schrank = Eleutherantha Poit. (Compos.).
Fingerhuthia Nees ex Lehm. Gramineae. 2 S. Afr.
Finlaysonia Wall. Periplocaceae. 1 Indomal.
Finschia Warb. Proteaceae. 7 Palau Is.), New Guinea (incl. Aru Is.), Solomon Is. to New Hebr.
Fintelmannia Kunth. Cyperaceae. 5 Brazil, Madag.
Fioria Mattei = Hibiscus L. (Malvac.).
Fiorinia Parl. = Aira L. (Gramin.).
Firensia Scop. = Cordia L. (Ehretiac.).
Firenzia DC. = praec.
Firkea Rafin. = Clusia L. (Guttif.).
Firmiana Marsigli. Sterculiaceae. 15 E. Afr., Indomal., SE. & E. As.
Fischera Spreng. = Platysace Bunge = Trachymene Rudge (Umbellif.).
Fischera Sw. = Leiophyllum (Pers.) Hedw. f. (Ericac.).
Fischeria DC. Asclepiadaceae. 15 W.I., C. & trop. S. Am.
Fishlockia Britton & Rose = Acacia Mill. (Legumin.).
Fisquetia Gaudich. = Pandanus L. f. (Pandanac.).
Fissendocarpa (Haines) Bennet. Onagraceae. 1 India.
Fissenia Endl. (sphalm.) = Kissenia R.Br. (Loasac.).
Fissicalyx Benth. Leguminosae. 1 Venezuela.
Fissilia Comm. ex Juss. = Olax L. (Olacac.).
Fissipes Small = Cypripedium L. (Orchidac.).
Fissipetalum Merr. = Erycibe Roxb. (Convolvulac.).
Fissistigma Griff. Annonaceae. 60 trop. Afr., SW. China, Indomal., NE. Austr.
Fistularia Kuntze (1891; non Stackhouse, 1816—Algae) = Rhinanthus L. (Scrophulariac.).
Fitchia Hook. f. Compositae. 7 Polynesia.   [(Scrophulariac.).
Fitchia Meissn. = Grevillea R.Br. (Proteac.).
Fittingia Mez. Myrsinaceae. 2 New Guinea.
Fittonia E. Coemans. Acanthaceae. 2 Peru.
Fitzalania F. Muell. Annonaceae. 1 trop. E. Austr.
Fitzgeraldia F. Muell. (1867) = Cananga (DC.) Hook. f. & Thoms. (Annonac.).
Fitzgeraldia F. Muell. (1882) = Lyperanthus R.Br. (Orchidac.).
Fitzgeraldia Schltr. = Peristeranthus T. E. Hunt (Orchidac.).
Fitzroya Benth. & Hook. f. = seq.
Fitz-Roya Hook. f. ex Hook. Cupressaceae. 1 Chile.
Fiva Steud. = Fiwa J. F. Gmel. = Litsea Lam. (Laurac.).
Fivaldia Walp. = Friwaldia Endl. = Microglossa DC. (Compos.).
Fiwa J. F. Gmel. = Litsea Lam. (Laurac.).
Flabellaria Cav. Malpighiaceae. 1 trop. Afr.
Flabellariopsis R. Wilczek. Malpighiaceae. 1 trop. Afr.
Flabellographis Thou. (uninom.) = Limodorum flabellatum Thou. = Cymbidiella flabellata (Thou.) Rolfe (Orchidac.).
Flacourtia Comm. ex L'Hérit. Flacourtiaceae. 15 trop. & S. Afr., Masc. Is., SE. As., Malaysia, Fiji. *F. ramontchi* L'Hérit. (Madagascar plum), and others, have ed. drupes.
*****Flacourtiaceae** DC. Dicots. 93/1000 trop. & subtrop. Trees and shrubs, mostly with alt. stip. leathery ls., often ± two-ranked. Fls. sol., axill., or in

FLACURTIA

cymose or racemose mixed infls., often ∞, reg., 4–5–∞-merous, sometimes partly spiral. Peduncle often jointed near base. Axis convex; disk, or glands, scales, etc., between C and A. K 2–15 or (2–15), imbr. or valv.; C 15–0; A usu. ∞, sometimes united in antepet. groups, anthers usu. opening by lat. slits; G̲ (2–10) or ½-inf. (inf. in *Bembicia*), 1- (rarely multi-) loc. with parietal plac. which often project into ov., and sometimes unite; ov. ∞, anatr.; styles = plac., or united. Usu. berry or caps.; seeds 1 or many, often with aril; embryo straight in endosp. Some have ed. fr.; a few supply timber, or oil.

Classification and chief genera (after Gilg):

1. ERYTHROSPERMEAE (fl. ♂; P usu ∞, spiral, A 5–8 with linear anthers; caps.): *Erythrospermum, Berberidopsis.*

2. ONCOBEAE (fl. ♀; K 3–5, C 4–12 imbr., A ∞ with linear anthers, G (3–10) each with ∞ ov.; fr. not, or tardily, dehisc.): *Oncoba, Mayna, Carpotroche.*

3. PANGIEAE (dioec.; K 2–5, C 5–8 with scales at base, A ∞–5, G (2–6) each with ∞–1 ov.; berry or berry-like caps.): *Pangium, Hydnocarpus, Gynocardia.*

4. PAROPSIEAE (K 5, axis slightly tubular, with disk or gynophore, C 5, often corona, A ∞–20 or 9–5, perig. or at base of gynophore, sometimes united, G (3–5), usu. with ∞ ov.): *Barteria, Paropsia.*

5. ABATIEAE (fl. ♀; K 4, valv., C 0, A ∞–8, perig., no stds., G (2–4) with ∞ ov.; ls. opp.): *Abatia.*

6. TRICHOSTEPHANEAE (fls. ♂ ♀, monoec. or dioec.; K 2+2, C 0, A (15–20) with adnate corona of 15–20 pubesc. stds., G 1-loc., with 4 pariet. plac. with ∞ ov., and 4 short thick styles): *Trichostephanus.*

7. SCOLOPIEAE (fl. ♀; K 4–6 almost valv., C small or 0, A ∞ perig. with short anthers, G (3–6), each with ∞–1 ov., 1- or multi-loc.): *Scolopia, Prockia, Banara.*

8. HOMALIEAE (K, C 4–15, A 4–15 or ∞ in bundles, ante-pet., perig. or epig.; ls. spiral, rarely paired): *Homalium.*

9. PHYLLOBOTRYEAE (fl. ♀ or polyg., K, C 3–5, A 5–∞, hypog., G (2–4), 1-loc. with ∞ ov.; ls. alt. with epiphyllous infl.): *Phyllobotryum.*

10. FLACOURTIEAE (K 4–6 imbr., C usu. 0, A ∞ with short anthers, G (2–6) each with ∞–1 ov.; berry or caps.): *Flacourtia, Xylosma, Azara.*

11. CASEARIEAE (K 4–5 imbr., C 0, A ∞ or few, sometimes stds., perig., G (2–6), usu. (3), each with ∞–2 ov.): *Casearia, Samyda.*

12. BEMBICIEAE (fls. ♀ in condensed bracteate axill. infl.; K (7–8), tubular limb petaloid; C 0; disk annular; A ∞ on mouth of K tube, fil. filif.; G̲ 1-loc., trigon., with 3 pariet. plac. and ∞ ov.; fr. 1-seeded): *Bembicia.*

The *F.* are intimately related to the *Euphorbiac., Tiliac.* and *Passiflorac.*, and the boundaries between them need extensive and careful reappraisal.

Flacurtia Comm. ex Juss. = Flacourtia L'Hérit. (Flacourtiac.).

Fladermannia Endl. = Faldermannia Trautv. = Ziziphora L. (Labiat.).

Flagellaria L. Flagellariaceae. 3 trop. Afr., Formosa, Indomal., Austr., Pacif.

*****Flagellariaceae** Dum. Monocots. (Farinosae). 1/3 trop. Afr., As., Pacif. High-climbing lianes, from a diffuse sympodial rhiz.; stems freq. branched by equal dichotomy, internodes solid, covered by lf.-sheaths; ls. elongate, dorsiv. or isobilateral, rolled in bud, the apex coiled circinately (adaxially) and acting as

FLINDERSIACEAE

a tendril; lf.-sheath closed. No hairs or setae. Secretory cells present; silica absent. Infl. term., branched, bracteate; fls. reg., ⚥, subsess. P 3 + 3, ± petaloid, ±obtuse; A 3 + 3; G̲ (3) with 1 axile ovule per loc., styles 3, free or almost so, stigs. ±plumose. Fr. drupaceous, ± globose; seeds 1(–2), with minute embr. in copious endosp. Only genus: *Flagellaria*. Some features of anat. and pollen show much similarity to *Gramineae*.

Flagellarisaema Nakai = Arisaema Mart. (Arac.).

Flagenium Baill. Rubiaceae. 2 trop. W. Afr., 4 Madag.

Flamaria Rafin. = Macranthera Torr. (Scrophulariac.).

Flammara Hill = Anemone L. (Ranunculac.).

Flammula Fourr. = Ranunculus L. (Ranunculac.).

Flanagania Schlechter. Asclepiadaceae. 1 S. Afr.

Flaveria Juss. Compositae. 14 Am. (esp. S. U.S. & Mex.), 1 Austr. Ls. opp. No pappus.

Flavia Fabr. (uninom.) = Anthoxanthum odoratum L. (Gramin.).

Flavicoma Rafin. = Schaueria Nees (Acanthac.).

Flavileptis Thou. (uninom.) = Malaxis flavescens Thou. = Liparis flavescens (Thou.) Lindl. (Orchidac.).

Fleischeria Steud. = Sida L. (Malvac.).

Fleischeria Steud. & Hochst. ex Endl. = Scorzonera L. (Compos.).

Fleischmannia Sch. Bip. Compositae. 7 Mex.

Fleischmanniopsis R. M. King & H. Rob. Compositae. 3 Mex., C. Am.

Flemingia Hunter apud Ridley = Fagraea Thunb. (Potaliac.).

***Flemingia** Roxb. (in Ait.). Leguminosae. 35 trop.

Flemingia Roxb. ex Rottl. = Thunbergia L. f. (Thunbergiac.).

Flemingia Roxb. ex Wall. = Canscora Lam. (Gentianac.).

Flemmingia Walp. = Flemingia Roxb. (in Ait.) = Maughania Jaume St-Hil. (Legumin.).

Flessera Adans. = Agastache Clayt. in Gronov. (Labiat.).

Fleura Steud. = Fleurya Gaudich. (Urticac.).

Fleurotia Reichb. = Siebera J. Gay (Compos.).

Fleurya Gaudich. = Laportea Gaudich. (Urticac.).

Fleurydora A. Chev. Ochnaceae. 1 trop. W. Afr.

Fleuryopsis Opiz = Fleurya Gaudich. = Laportea Gaudich. (Urticac.).

Flexanthera Rusby. Rubiaceae. 2 Colombia, Bolivia.

Flexularia Rafin. Gramineae (inc. sed.; ?aff. Muhlenbergia). 1 E. U.S.

Flexuosatis Thou. (uninom.) = Satyrium flexuosum L. = Schizodium flexuosum (L.) Lindl. (Orchidac.).

Flickingeria A. D. Hawkes = Ephemerantha P. F. Hunt & Summerhayes (Orchidac.).

Flindersia R.Br. Flindersiaceae. 16 Molucc., New Guinea, E. Austr., New Caled.

Flindersiaceae (Engl.) C. T. White ex Airy Shaw. Dicots. 2/17 S. India & Ceylon, E. Malaysia, E. Austr., New Caled. Trees or shrubs, often with hard, bright yellow wood, and alt. or opp. pinn. (sometimes trifol. or simple) exstip. ls.; lfts. ent., gland-dotted. Fls. reg., ⚥, small, in axill. or term. pan. K 5 or (5), imbr. or valv.; C 5, imbr., sometimes pubesc. within; A 5 + 5 (inner sometimes stds.), fil. subul., sometimes adnate to disk, sometimes pilose, anth. intr., dorsif., with

463

produced connective; disk large, cupular, cren., enclosing the ovary; G̲ (5) or (3), sometimes with glandular lobes, with simple style and 5-lobed pelt. stig., and 2–8 biseriate ov. per loc. on axile plac. Caps. woody, smooth, spiny or verruc., septic. or loculic., valves attached to or separating from septa, with 2–8 seeds per loc.; seeds compressed, winged, imbr., without endosp.; cots. fleshy, foliac., pellucid-punct. Genera: *Flindersia, Chloroxylon.* Somewhat intermediate between *Rutac.* and *Meliac.* (Cf. *Ptaeroxylac.*)

Flipanta Rafin. = Salvia L. (Labiat.).

Floerkea Spreng. = Adenophora Fisch. (Campanulac.)

Floerkea Willd. Limnanthaceae. 1 N. Am.

Flomosia Rafin. = Verbascum L. (Scrophulariac.).

Floresia Krainz & Ritter = Weberbauerocereus Backeb. (Cactac.).

Florestina Cass. Compositae. 7 Mexico.

Florinda Nor. ex Endl. = Polycardia Juss. (Celastrac.).

Floriscopa F. Muell. = Floscopa Lour. (Commelinac.).

Florkea Rafin. = Floerkea Willd. (Limnanthac.).

Floscaldasia Cuatrec. Compositae. 1 Colombia.

Floscopa Lour. Commelinaceae. 20 trop. & subtrop.

Floscuculi Opiz = Lychnis L. (Caryophyllac.).

Flostolonia Raf. = ? Gymnostyles A. Juss. (Compos.).

Flotovia Spreng. = Dasyphyllum Kunth (Compos.).

Flotowia Endl. = praec.

Flourensia Cambess. = Thylacospermum Fenzl (Caryophyllac.).

Flourensia DC. Compositae. 30 SW. U.S. to Argentina.

Flox Adans. = Coronaria Guett. = Lychnis L. (Caryophyllac.).

Floyeria Neck. = Exacum L. (Gentianac.).

Fluckigeria Rusby = Kohlerianthus Fritsch (Gesneriac.).

Flueckigera Kuntze. Phytolaccaceae. 3 trop. S. Am., W.I.

Flueggea Rich. = Ophiopogon Ker-Gawl. (Liliac.).

Flueggea Willd. (~ Securinega Juss.). Euphorbiaceae. 5 palaeotrop.

Flueggeopsis K. Schum. = Kirganelia A. Juss. = Phyllanthus L. (Euphorbiac.).

Flueggia Benth. & Hook. f. = Flueggea Willd. (Euphorbiac.).

Flugea Rafin. = Ophiopogon Ker-Gawl. (Liliac.).

Fluggea Willd. = Flueggea Willd. (Euphorbiac.).

Fluminea auctt. = seq.

Fluminia Fries = Scolochloa Link (Gramin.).

Flundula Rafin. = Hosackia Dougl. (Legumin.).

Flustula Rafin. = Vernonia Schreb. (Compos.).

Fluviales Vent. = Potamogetonac. + Ruppiac. + Zannichelliac. + Zosterac.

Fluvialis Seguier = Najas L. (Najadac.).

Fluviatiles S. F. Gray = Fluviales Vent.

Fobea Frič = Escobaria Britton & Rose (Cactac.).

Fockea Endl. Asclepiadaceae. 10 trop. & S. Afr.

× **Fockeanthus** Wehrh. Campanulaceae. Gen. hybr. (Campanula × Phyteuma).

Foeniculum Mill. Umbelliferae. 5 Medit., Eur. The young ls. of *F. vulgare* Mill. are a good veg. when blanched like celery, and the fr. is a condiment.

Foenodorum E. H. L. Krause = Anthoxanthum L. (Gramin.).

Foenugraecum Ludw. = Trigonella L. (Legumin.).

Foenum Fabr. = Trigonella L. (Legumin.).

Foenum-graecum Seguier = Foenugraecum Ludw. = Trigonella L. (Legumin.).

Foersteria Scop. = Breynia J. R. & G. Forst. (Euphorbiac.).

Foetataxus 'Senilis' [Nelson] = Torreya Arn. (Taxac.).

Foetidia Comm. ex Lam. Foetidiaceae. 5 Pemba, Madag., Masc.

Foetidiaceae (Niedenzu) Airy Shaw. Dicots. 1/5 Masc. region. Small or medium-sized trees; ls. simple, alt., exstip., ent., usu. slightly to strongly asymmetr., crowded toward ends of branches, vernation involute. Fl. sol. axill. or in small cymes, bibracteol. at apex of pedic. K 4, valv., delt., persist.; C o; A ∞, sometimes in 4 oppositisep. groups, fil. filif., anth. v. small, pollen smooth; stam. disk large, quadrate, intrastam. disk inconspic.; G̅ (4), with slender tetraquetrous style and 4 short slender divaric. stigs.; ovules 15–20 per loc., horiz., anatr., arr. in vertical ring around thick pelt. plac. Fr. turbin., drupac., the endoc. with plac. and dissep. hardening completely, 1–4-loc., loc. 1-seeded. Only genus: *Foetidia*. A small relict group, closely related to *Tetrameristac.* and *Bonnetiac.*, differing from these principally in its inferior ovary, valvate sepals, and absence of corolla; from the former also in its numerous stamens, and from the latter in its tetramerous fls. and drupaceous fr. An endophytic moniliaceous fungus is constantly present in the tissues of *F. mauritiana* Lam.

Fokienia A. Henry & H. H. Thomas. Cupressaceae. 1–3 SW. & SE. China,

Foleyola Maire. Cruciferae. 1 W. Sahara. [Indoch.

Folianthera Rafin. = Stryphnodendron Mart. (Legumin.).

Folis Dulac = Teesdalia R.Br. (Crucif.).

Folliculigera Pasq. = Trigonella L. (Legumin.).

Folomfis Rafin. = Miconia Ruiz & Pav. (Melastomatac.).

Folotsia Costantin & Bois = Cynanchum L. (Asclepiadac.).

Fometica Rafin. = Heritiera Dryand. (Sterculiac.).

Fonkia Phil. = Gratiola L. (Scrophulariac.).

Fonna Adans. = Phlox L. (Polemoniac.).

Fontainea Heckel. Euphorbiaceae. 2 New Guinea, NE. Austr., New Caled.

Fontainesia P. & K. = Fontanesia Labill. (Oleac.).

Fontanella Kluk = Isopyrum L. (Ranunculac.).

Fontanesia Labill. Oleaceae. 2 Sicily, W. As., China. Hedge plant.

Fontbrunea Pierre = Sideroxylon L. (Sapotac.).

Fontenella Walp. = seq.

Fontenellea A. St-Hil. & Tul. = Quillaja Molina (Rosac.).

Fontquera Maire. Compositae. 1 Morocco.

Fontqueriella Rothm. = Triguera Cav. (1786) (Solanac.).

Foquiera Hemsl. = Fouquieria Kunth (Fouquieriac.).

Forasaccus Bub. = Bromus L. (Gramin.).

Forbesia Eckl. ex Nel = Empodium Salisb. (Hypoxidac.).

Forbesina Rafin. = Verbesina L. (Compos.).

Forbesina Ridley. Orchidaceae. 1 Sumatra.

Forbicina Seguier = Bidens L. (Compos.).

Forchhammeria Liebm. Capparidaceae. 10 Calif. to C. Am., W.I.

FORCIPELLA

Forcipella Baill. Acanthaceae. 5 Madag.
Forcipella Small = Paronychia Mill. (Caryophyllac.).
Fordia Hemsl. Leguminosae. 12 S. China, Siam, Malay Penins., Borneo, Philipp. Is.
Fordiophyton Stapf. Melastomataceae. 10 S. China, Indoch.
Forestiera Poir. Oleaceae. 15 Am. (esp. S. U.S. & Mex.), W.I.
Forestier[ac]eae Endl. = Oleaceae Hoffmgg. & Link.
Forexeta Rafin. = Carex L. (Cyperac.).
Forfasadis Rafin. = Torfasadis Rafin. = Euphorbia L. (Euphorbiac.).
Forficaria Lindl. Orchidaceae. 2 S. Afr.
Forgerouxa Neck. = Rhamnus L. (Rhamnac.).
Forgeruxia Rafin. = praec.
Forgesia Comm. ex Juss. Escalloniaceae. 1 Réunion.
Forgetina Bocquill. ex Baill. = Sloanea L. (Elaeocarpac.).
Formania W. W. Smith & Small. Compositae. 1 SW. China.
Formosia Pichon. Apocynaceae. 1 Formosa.
Fornasinia Bertol. = Millettia Wight & Arn. (Legumin.).
Fornea Steud. = Forneum Adans. = Andryala L. (Compos.).
Fornelia Schott (sphalm.) = Tornelia Gutierrez ex Schlechtd. = Monstera Adans. (Arac.).
Forneum Adans. = Andryala L. (Compos.).
Fornicaria Rafin. = Salmea DC. (Compos.).
Fornicium Cass. = Centaurea L. (Compos.).
Forotubaceae Dulac = Ericaceae–Vaccinioïdeae–Vaccinieae.
Forrestia Rafin. = Ceanothus L. (Rhamnac.).
Forrestia A. Rich. = Amischotolype Hassk. (Commelinac.).
Forsakhlia Ball (sphalm.) = Forsskaolea L. (Urticac.).
Forsellesia Greene. Celastraceae. 8 W. U.S.
Forsgardia Vell. = Combretum L. (Combretac.).
Forshohlea Batsch = Forskaelea Scop. = Forskahlea Agardh = Forskåhlea Webb & Berth. = Forskalea Juss. = Forskoehlea Reichb. = Forskoelea Brongn. = Forskohlea L. (1767) = Forskolea L. (1764) = Forskolia Wight = seq.
Forsskaolea ('Forsskålea') L. (1764). Urticaceae. 6 Canary Is., SE. Spain, Afr., Arabia, W. India.
Forstera L. f. Stylidiaceae. 5 Tasm., N.Z.
Forstera P. & K. = Foersteria Scop. = Breynia J. R. & G. Forst. (Euphorbiac.).
Forsteria Neck. = Forstera L. f. (Stylidiac.).
Forsteria Steud. = Foersteria Scop. = Breynia J. R. & G. Forst. (Euphorbiac.).
Forsteronia G. F. W. Mey. Apocynaceae. 50 C. & trop. S. Am., W.I.
Forsteropsis Sond. = Stylidium Sw. (Stylidiac.).
***Forsythia** Vahl. Oleaceae. 1 SE. Eur.; 6 E. As.
Forsythia Walt. = Decumaria L. (Hydrangeac.).
Forsythiopsis Baker (~ Oplonia Rafin.). Acanthaceae. 5 Madag.
Forsythmajoria Kraenzl. ex Schlechter = Cynorkis Thou. (Orchidac.).
Fortunaea Lindl. = Platycarya Sieb. & Zucc. (Juglandac.).
Fortunatia Macbride. Liliaceae. 1 Peru, Chile.
Fortunea Poit. = Fortunaea Lindl. = Platycarya Sieb. & Zucc. (Juglandac.).
Fortunearia Rehder & Wilson. Hamamelidaceae. 1 C. & E. China.

466

Fortunella Swingle. Rutaceae. 6 E. As., Malay Penins. (*cumquats*).
Fortuynia Shuttl. ex Boiss. Cruciferae. 2 Persia, Afghan., Baluch.
Forysthia Franch. & Sav. (sphalm.) = Forsythia Vahl (Oleac.).
Foscarenia Vell. ex Vand. = Randia L. (Rubiac.).
Fosselinia Scop. = Clypeola L. (Crucif.).
Fosterelia L. B. Smith. Bromeliaceae. 13 trop. Am.
Fosteria Molseed. Iridaceae. 1 Mex.
Foterghillia Dum. (sphalm.) = Fothergilla Murr. (Hamamelidac.).
Fothergilla Aubl. = Miconia Ruiz & Pav. (Melastomatac.).
Fothergilla L. Hamamelidaceae. 4 Atl. N. Am. (Am. witch elder). Fl. apet., A ∞.
Fothergill[ac]eae Link = Hamamelidaceae–Fothergilleae DC.
Fougeria Moench = Baltimora L. (Compos.).
Fougerouxia Cass. = praec.
Fouha Pomel = Colchicum L. (Liliac.).
Fouilloya Benth. & Hook. f. = seq.
Foullioya Gaudich. = Pandanus L. f. (Pandanac.).
Fouquiera Spreng. = seq.
Fouquieria Kunth. Fouquieriaceae. 10 SW. U.S., Mex. *F. splendens* Engelm. (*ocotilla*, coach-whip) used for hedges. Wax obtained from bark of some.
*****Fouquieriaceae** DC. Dicots. 2/11 warm N. Am. Shrubs with little-branched stems. Ls. simple, ent., alt., exstip., decid., the midribs of some persist. and spinescent. Fls. ☿, reg. or slightly curved, showy, in ± rac. infl. K 5, much imbr., persist.; C (5), tubular, the lobes imbr.; A 10–15, with long usu. exserted fil. and small anth.; G̲ (3), 1-loc., with 3 pariet. plac. each bearing 4–6 ov. Caps. obl. or spher.; seeds with long hairs or wings, and oily endosp. Genera: *Fouquieria, Idria*. An interesting, isolated group, probably to be regarded as a 'peripheral Centrosperm', together with *Didiereac., Tamaricac., Frankeniac.*, etc.
Fourcroea Haw. = seq.
Fourcroya Spreng. = Furcraea Vent. (Agavac.).
Fourneaua Pierre ex Pax & Hoffm. = Grossera Pax (Euphorbiac.).
Fourniera Bommer = Cyathea Sm. subg. Sphaeropteris (Bernh.) Holtt. *p.p.* (Cyatheac.). Sori protected by overlapping scales, not by true indusia (Holttum, *Fl. Males.* ser. II, 1, 1963).
Fourniera Scribner = Soderstromia Morton (Gramin.).
Fournieria Van Tiegh. = Cespedesia Goudot (Ochnac.).
Fourraea Gandog. = Potentilla L. (Rosac.).
Foveolaria (DC.) Meissn. = Sloanea L. (Elaeocarpac.).
Foveolaria Ruiz & Pav. (~ Styrax L.). Styracaceae. 2 Peru, Cuba.
Foxia Parl. = Hyacinthus L. (Liliac.).
Fracastora Adans. = ? Teucrium L. (Labiat.).
Fractiunguis Schlechter = Reichenbachanthus Barb. Rodr. (Orchidac.).
Fradinia Pomel = Mecomischus Coss. (Compos.).
Fraga Lapeyr. = Potentilla L. (Rosac.).
Fragaria L. Rosaceae. 15 N. Am., Chile, Euras. (to S. India). Veg. repr. by runners is well shown. Fl. protog., with epicalyx. Fr. a number of achenes (the so-called seeds) upon a fleshy recept. The fl. bends down after fert., while

FRAGARIACEAE

the fr. ripens. In Am. the cult. forms tend to become dioec. or polyg. Several spp. cult. for the ed. fr.

Fragariaceae Rich. ex Nestl. = Rosaceae–Rosoïdeae Reichb.

Fragariastrum Fabr. (uninom.) = *Potentilla sterilis* (L.) Garcke (Rosac.).

Fragariopsis A. St-Hil. (∼ Plukenetia L.). Euphorbiaceae. 2 trop. S. Am.

Frageria Delile ex Steud. = Leuceria Lag. (Compos.).

Fragmosa Rafin. = Erigeron L. (Compos.).

Fragosa Ruiz & Pav. (∼ Azorella Lam.). Umbelliferae. 8 Andes.

Fragrangis Thou. (uninom.) = *Angraecum fragrans* Thou. = *Jumellea fragrans* (Thou.) Schltr. (Orchidac.).

Frailea Britton & Rose. Cactaceae. 12 Andes & subtrop. S. Am.

Franca Boehm. = Frankenia L. (Frankeniac.).

Francastora Steud. = Fracastora Adans. = ?Teucrium L. (Labiat.).

Francfleurya A. Chev. & Gagnep. = Pentaphragma Wall. ex G. Don (Pentaphragmatac.).

Franchetella Kuntze = Heteromorpha Cham. & Schlechtd. (Umbellif.).

Franchetella Pierre = Lucuma Molina (Sapotac.).

Franchetia Baill. = Cephalanthus L. (Rubiac.).

Franciella Guillaumin (1922; non Frankiella Speschnew 1900—Fungi, nec Frankiella Maire & Tison 1909—Fungi) = Neofranciella Guillaumin (Rubiac.).

Franciscea Pohl = Brunfelsia L. (Solanac.).

Francisia Endl. = Darwinia Rudge (Myrtac.).

Francoa Cav. Francoaceae. 1 Chile (polymorphic).

*****Francoaceae** A. Juss. Dicots. 2/2 temp. S. Am. Perenn. rhizomatous scapose herbs. Ls. subradical, alt., lyrate or orbic., exstip. Fls. in rac. or pan., tetram. K 4(–5), valv., persist.; C 4(–5) or 2, imbr. or contorted, sometimes clawed; A 8 or 4, alternating with 8 or 4 stds., anth. oblong or subglob.; G (4) [rarely (2)], tetrag., with ∞ axile, biseriate, horiz., anatr. ov., and 4 sess. glob. or flattened commissural stigs. Caps. obl., membr., erect, 4-loc., septic., ∞-seeded; seeds v. small, with thin testa and fleshy endosp. Genera: *Francoa, Tetilla*.

Francoeuria Cass. = Pulicaria Gaertn. (Compos.).

Frangula Mill. (∼ Rhamnus L., *q.v.*). Rhamnaceae. 50 Am., W. Medit., temp. Euras.

Frangulaceae Lam. & DC. = Rhamnaceae Juss.

Franka Steud. = Franca Boehm. = seq.

Frankenia L. Frankeniaceae. 80 temp. & subtrop., sea coasts, salt lakes, salt deserts. Halophytes with usu. inrolled hairy ls.

*****Frankeniaceae** S. F. Gray. Dicots. 4/90 trop. & temp. Salt-loving herbs (more rarely subshrubs) with jointed stems; ls. opp., inrolled, stip. (?). Fls. in dichasia, ⚥, reg. K (4–7); C 4–7, persist.; A usu. 6 in two whorls, fil. slightly united at base; G̲ usu. (3), 1-loc. with parietal plac., only the lower parts of which bear ovules; ov. ∞, anatr., ascending, on long funicles; style forked. Caps. loculic. Mealy endosp.; embryo straight. Genera: *Frankenia, Hypericopsis, Anthobryum, Niederleinia, ?Petrusia*. Closely related to *Tamaricaceae*, and like them probably to be regarded as a 'peripheral Controsperm' group.

Frankeria Rafin. = Frankenia L. (Frankeniac.).

Frankia Bert. ex Steud. = Cicca L. (Euphorbiac.).

Frankia Steud. ex Schimp. = Gymnarrhena Desf. (Compos.).

Franklandia R.Br. Proteaceae. 2 W. Austr.

Franklinia Bartr. ex Marshall. Theaceae. 1 SE. U.S. (?extinct).

Frankoa Reichb. = Francoa Cav. (Francoac.).

Frankoeria Steud. = Francoeuria Cass. = Pulicaria Gaertn. (Compos.).

Franquevillea Zoll. ex Miq. = Hypoxis L. (Hypoxidac.).

Franquevillia Salisb. ex S. F. Gray = Microcala Hoffmgg. & Link (Gentianac.)

*****Franseria** Cav. = Ambrosia L. (Compos.).

Fransiella Willis (sphalm.) = Franciella Guillaum. = Neofranciella Guillaum. (Rubiac.).

Frantzia Pittier. Cucurbitaceae. 1 C. Am.

Frappieria Cordem. = Psiadia Jacq. (Compos.).

Frasera Walt. (~ Swertia L.). Gentianaceae. 15 N. Am.

Fraunhofera Mart. Celastraceae. 1 Brazil.

Fraxima Rafin. = Ipomoea L. (Convolvulac.).

Fraxin[ac]eae S. F. Gray = Oleaceae–Fraxineae Bartl.

Fraxinella Mill. = Dictamnus L. (Rutac.).

Fraxinell[ace]ae Nees & Mart. = Rutaceae–Cusparieae DC. (*pro maxima parte*).

Fraxinoïdes Medik. = seq.

Fraxinus L. Oleaceae. 70 N. hemisph., esp. E. As., N. Am. & Medit. *F. excelsior* L., European ash, has large pinnate ls. with serial accessory buds in axils. The petioles are grooved, and water is said to enter this groove and be absorbed by the l.; the hollow is usu. inhabited by acarids, forming a *domatium*. The fls. appear before the ls. in densely crowded short racemes. Each ♀ consists merely of 2 sta. and 2 cpls., and is anemoph.; but polygamy is the rule in this sp. and every possible combination of the three types of fl. (♀, ♂, ♀) occurs in various places, sometimes all on one tree, or two on one and one on another, and so on. Fr. a samara or one-seeded nut with terminal wing aiding in wind distr. *F. ornus* L., the 'flowering ash' of S. Eur., has K and C. The firm elastic wood of the ash is valuable.

The weeping ash is a variety propagated veg. from a single tree which appeared as a sport at Wimpole in Cambridgeshire, England.

Fredericia G. Don = Fridericia Mart. (Bignoniac.).

Fredolia (Coss. & Dur. ex Bunge) Ulbr. Chenopodiaceae. 1 Sahara (S. Algeria).

Freemania Boj. ex DC. = Helichrysum L. (Compos.).

Freerea Willis (sphalm.) = seq.

Freeria Merr. = Pyrenacantha Hook. ex Wight (Icacinac.).

Freesea Eckl. = Tritonia Ker-Gawl., Ixia L., etc. (Iridac.).

Freesia Klatt. Iridaceae. 20 S. Afr.

Fregea Reichb. f. Orchidaceae. 2 Costa Rica, Panamá.

Fregirardia Dun. = Cestrum L. (Solanac.).

Freira C. Gay = seq.

Freirea Gaudich. = Parietaria L. (Urticac.).

Freireodendron Muell. Arg. = Drypetes Vahl (Euphorbiac.).

Fremontea Lindl. = seq.

Fremontia Torr. (1843) = Sarcobatus Nees (Chenopodiac.).

Fremontia Torrey (1854) = Fremontodendron Coville (Sterculiac.) (?Bombacac.).

FREMONTI[AC]EAE

Fremonti[ac]eae J. G. Agardh = Cheiranthodendr[ac]eae A. Gray = Sterculiaceae–Fremontodendreae Airy Shaw (Fremontieae Torr., nom. illegit.).

Fremontodendron Coville. Sterculiaceae (?Bombacaceae). 4–6 Calif., Mex. (slippery elm).

Fremya Brongn. & Gris = Xanthostemon F. Muell. (Myrtac.).

Frenela Mirb. = Callitris Vent. (Cupressac.).

Frerea Dalz. Asclepiadaceae. 1 S. India.

Fresenia DC. Compositae. 6 S. Afr.

Fresiera Mirb. = Freziera Sw. (Theac.).

Fresnelia Steud. = Frenela Mirb. = Callitris Vent. (Cupressac.).

Freuchenia Eckl. = Moraea Mill. ex L. (Iridac.).

Freycinetia Gaudich. Pandanaceae. 100 Ceylon to N.Z. & Polynes. Usu. climbing shrubs with infl. and fl. like *Pandanus*. The bracts are fleshy and usu. brightly coloured. In Java, Burck observed pollination effected by a bat (*Pteropus edulis*) which devoured the coloured bracts; in so doing it received pollen upon its head and carried it to the ♀ fl. Fr. a berry, not, as in *Pandanus*, a drupe.

Freyera Reichb. (~ Chaerophyllum L., Bunium L.). Umbelliferae. 12 Medit. Cot. 1.

Freyeria Scop. = Mayepea Aubl. = Linociera Sw. (Oleac.).

Freylinia Colla. Scrophulariaceae. 5 trop. & S. Afr.

Freyliniopsis Engl. Scrophulariaceae. 1 SW. Afr.

***Freziera** Sw. ex Willd. Theaceae. 38 trop. Am., W.I.

Fridericia Mart. Bignoniaceae. 1 S. Brazil.

Friederichsthalia A. DC. = Friedrichsthalia Fenzl = Trichodesma R.Br. (Boraginac.).

Friedericia Reichb. = Fridericia Mart. (Bignoniac.).

Friedlandia Cham. & Schlechtd. = Diplusodon Pohl (Lythrac.).

Friedrichsthalia Fenzl = Trichodesma R.Br. (Boraginac.).

Friesea Reichb. = seq.

Friesia DC. = Aristotelia L'Hérit. (Elaeocarpac.).

Friesia Frič = Pyrrhocactus A. Berger (Cactac.).

Friesia Spreng. = Crotonopsis Michx (Euphorbiac.).

Friesodielsia van Steenis. Annonaceae. 55 trop. W. Afr., Indomal.

Frisca Spach = Thesium L. (Santalac.).

Frithia N. E. Brown. Aïzoaceae. 1 S. Afr.

Fritillaria L. Liliaceae. 85 N. temp. Large nectaries at base of P. The bud stands erect and so does the caps., but the open fl. is pend.

Fritillar[iac]eae Salisb. = Liliaceae Juss.

Fritschia Walp. = Fritzschia Cham. (Melastomatac.).

Fritschiantha Kuntze = Seemannia Regel (Gesneriac.).

Fritzschia Cham. Melastomataceae. 4 Brazil.

Frivaldia Endl. = Microglossa DC. (Compos.).

Frivaldzkia Reichb. = praec.

Froebelia Regel = Acrotriche R.Br. (Epacridac.).

Froehlichia D. Dietr. = Froelichia Vahl = Coussarea Aubl. (Rubiac.).

Froehlichia Endl. = Froelichia Moench (Amaranthac.).

Froehlichia Pfeiff. = Froelichia Wulf. = Kobresia Willd. (Cyperac.).

Froelichia Moench. Amaranthaceae. 20 warm Am., Galápagos. Fr. enclosed in the P, which forms two wings.

Froelichia Vahl = Coussarea Aubl. (Rubiac.).

Froelichia Wulf. ex Roem. & Schult. = Kobresia Willd. (Cyperac.).

Froelichiella R. E. Fries. Amaranthaceae. 1 Brazil.

Froësia Pires. Quiinaceae. 2 Amaz. Brazil.

Froësiochloa G. A. Black. Gramineae. 1 NE. Brazil.

Froësiodendron R. E. Fries. Annonaceae. 2 trop. S. Am.

Frolovia (Ledeb. ex DC.) Lipschitz (~ Saussurea DC.). Compositae. 3 Siberia, C. As.

Frommia H. Wolff. Umbelliferae. 1 trop. E. Afr.

Fropiera Bouton ex Hook. f. = Psiloxylon Thou. ex Tul. (Psiloxylac.).

Froriepia C. Koch. Umbelliferae. 1 Cauc., Persia.

Froscula Rafin. = Dendrobium Sw. (Orchidac.).

Frostia Bertero ex Guillem. = Pilostyles Guillem. (Rafflesiac.).

Fructesca DC. = Gaertnera Lam. (Rubiac.).

Frumentum Krause = Hordeum L., Secale L., Triticum L., Agropyron J. Gaertn., etc. (Gramin.).

Frutesca DC. ex A. DC. = Fructesca DC. = Gaertnera Lam. (Rubiac.).

Fuchsia L. Onagraceae. 100 N.Z., Tahiti, C. & S. Am. Many show two buds in each axil, one above the other. Fl. suited to bees, humming-birds, etc. Berry ed.

Fuchsia Sw. = Schradera Vahl (Rubiac.).

Fuernrohria C. Koch. Umbelliferae. 1 Cauc., Armenia.

Fuerstia T. C. E. Fries. Labiatae. 6 trop. Afr.

Fuertesia Urb. Loasaceae. 1 W.I. (S. Domingo).

Fuertesiella Schlechter. Orchidaceae. 2 W.I.

Fugosia Juss. = Cienfuegosia Cav. (Malvac.).

Fuirena Rottb. Cyperaceae. 40 trop. & subtrop.

Fuisa Rafin. = Patrinia Juss. (Valerianac.).

× **Fujiwaraära** hort. Orchidaceae. Gen. hybr. (iii) (Brassavola × Cattleya × Laeliopsis).

Fulcaldea Poir. Compositae. 1 W. trop. S. Am.

Fulchironia Lesch. = Phoenix L. (Palm.).

Fullartonia DC. = Doronicum L. (Compos.).

Fumana (Dunal) Spach. Cistaceae. 15 Medit., SE. Eur., SW. As.

Fumanopsis Pomel = praec.

Fumaria L. Fumariaceae. 55 Eur., Medit. to C. As. & Himal., 1 mts. & highlands E. Afr. Many climb by sensitive petiolules (not petioles) (cf. *Clematis*). Fl. like *Corydalis*. *F. capreolata* L. shows colour-change in its fl.; before pollination white, it gradually turns pink or carmine (cf. *Ribes, Diervilla*).

*Fumariaceae DC. Dicots. 16/450 N. (esp. warm) temp.; few mts. E. Afr., S. Afr. Herbs, sometimes bulbous, sometimes climbing by means of petiolules, with watery (not milky) juice. Ls. alt., usu. variously cpd., more rarely simple, exstip. Fls. ☿, zygomorphic, usu. ± racemose. K 2, often squamif., decid. C 2+2, one or both outer pet. ± saccate or spurred at base, the 2 inner ± coherent at apex. A 6, in 2 phalanges (central member of each phalanx with dithecous, laterals with monothecous anth.), with 1–2 nectaries at base.

FUMARIOLA

G̲ (2), 1-loc., with 1–∞ anatr. ov. on 2 pariet. plac., and slender style with capit. stig. Fr. from a small glob. 1-seeded nut to an inflated bladdery or slender elongate caps.; seeds mostly black and shining, often with conspic. aril or caruncle. Chief genera: *Dicentra, Corydalis, Rupicapnos, Fumaria.* Connected with *Papaveraceae* by means of *Hypecoaceae.*

Fumariola Korshinsky. Fumariaceae. 1 C. As.

Funastrum Fourn. Asclepiadaceae. 10–15 trop. Am.

Funckia Dennst. = Lumnitzera Willd. (Combretac.).

Funckia Dum. = Funkia Spreng. = Hosta Tratt. (Liliac.).

Funckia Willd. = Astelia Banks & Soland. (Liliac.).

Funifera Leandro ex C. A. Mey. Thymelaeaceae. 3–4 Brazil.

Funisaria Rafin. = Uvaria L. (Annonac.).

Funium Willem. = Furcraea Vent. (Agavac.).

Funkia Benth. & Hook. f. = Funckia Willd. = Astelia Banks & Soland. (Liliac.).

Funkia Endl. = Funckia Dennst. = Lumnitzera Willd. (Combretac.).

Funkia Spreng. = Hosta Tratt. (Liliac.).

Funkiaceae Horan. = Liliac., Agavac., Alliac., with fibrous or tuberous (not bulbous) rootstock.

Funkiella Schltr. Orchidaceae. 1 Mexico, C. Am.

Funtumia Stapf. Apocynaceae. 3 trop. Afr. *F. elastica* Stapf is the chief source of Lagos or Iré rubber.

Furarium Rizzini. Loranthaceae. 1 Brazil.

Furcaria Boiv. ex Baill. = Croton L. (Euphorbiac.).

Furcaria Desv. = Ceratopteris Brongn. (Parkeriac.).

Furcaria Kostel. (~ Hibiscus L.). Malvaceae. 5 trop.

Furcilla Van Tiegh. = Phrygilanthus Eichl. (Loranthac.).

Furcraea Vent. Agavaceae. 20 trop. Am. Like *Agave*; infl. even larger. *F. gigantea* Vent. yields fibre (Mauritius hemp).

Furcroya Rafin. = praec.

Furera Adans. = Pycnanthemum Michx (Labiat.).

Furera Bub. = Corrigiola L. (Caryophyllac.).

Furiolobivia Y. Ito = Echinopsis Zucc. (Cactac.).

Furnrohria Lindl. = Fuernrohria C. Koch (Umbellif.).

Fusaea (Baill.) W. E. Safford. Annonaceae. 3 E. Peru, Amaz. Brazil, Guiana.

Fusanus R.Br. = Eucarya T. L. Mitch. (Santalac.).

Fusanus L. = Colpoön Bergius (Santalac.).

Fusarina Rafin. = Uncinia Pers. (Cyperac.).

Fusidendris Thou. (uninom.) = *Dendrobium fusiforme* Thou. = *Polystachya fusiformis* (Thou.) Lindl. (Orchidac.).

Fusifilum Rafin. = Urginea Steinh. (Liliac.).

Fusispermum Cuatrec. Violaceae. 2 Colombia.

Fussia Schur = Aira L. (Gramin.).

Fusticus Rafin. = Chlorophora Gaudich. (Morac.).

Fuziifilix Nak. & Momose = Microlepia Presl (Dennstaedtiac.).

G

Gabertia Gaudich. = Grammatophyllum Blume (Orchidac.).
Gabila Baill. = Pycnarrhena Miers (Menispermac.).
Gabunia Pierre ex Stapf (~ Tabernaemontana L.). Apocynaceae. 10 trop. & W. Afr.
Gaedawakka Kuntze = Chaetocarpus Thw. (Euphorbiac.).
Gaeodendrum P. & K. = Gaiadendron G. Don (Loranthac.).
Gaerdtia Klotzsch = Begonia L. (Begoniac.).
*****Gaertnera** Lam. Rubiaceae. 30 trop. Afr., Madag., Masc., Ceylon, Assam to S. China & Malay Penins., Borneo, Austr. G̲!
Gaertnera Retz. = Sphenoclea Gaertn. (Sphenocleac.).
Gaertnera Schreb. = Hiptage Gaertn. (Malpighiac.).
Gaertneria Medik. = Franseria Cav. (Compos.).
Gaertneria Neck. = Gentiana L. (Gentianac.).
Gagea Salisb. Lilaceae. 70 temp. Euras. Fl. protog. In the l.-axils of some are buds which, if fert. does not occur, develop into bulbils and drop off.
Gagernia Klotzsch (nomen). Ochnaceae. 1 Guiana. Quid?
Gagia St-Lag. = Gagea Salisb. (Liliac.).
Gagnebina Neck. ex. DC. Leguminosae. 1 Madag., Mauritius.
Gagnebinia P, & K. = Guagnebina Vell. = Manettia Mutis (Rubiac.).
Gagnepainia K. Schum. Zingiberaceae. 3 Indoch.
Gaguedi Bruce = Protea R.Br. (Proteac.).
Gahnia J. R. & G. Forst. Cyperaceae. 40 China, Indoch., Malaysia, Austr., N.Z., Polynesia to Hawaii & Marquesas.
Gaiadendraceae Van Tiegh. = Loranthaceae–Gaiadendrinae Engl.
Gaiadendron G. Don (~ Phrygilanthus Eichl.). Loranthaceae. 10 Andes.
Gaillardia Fougeroux. Compositae. 26 N. Am., 2 temp. S. Am.
Gaillionia Endl. = seq.
Gaillonia A. Rich. (1830; non Gaillona Bonnemaison 1828—Rhodophyc.) = Jaubertia Guill. (Rubiac.).
Gaimarda Juss. = seq.
Gaimardia Gaudich. Centrolepidaceae. 2 N.Z., temp. S. Am.
Gaiodendron Endl. = Gaiadendron G. Don (Loranthac.).
Gaissenia Rafin. = Trollius L. (Ranunculac.).
Gajanus Kuntze = Inocarpus J. R. & G. Forst. (Legumin.).
Gajati Adans. = Aeschynomene L. (Legumin.).
Gakenia Fabr. (uninom.) = Matthiola R.Br. sp. (Crucif.).
Galacaceae ('-inae') D. Don = Diapensiaceae–Galaceae A. Gray + Francoaceae Juss.
Galacanthus Lem. = Galanthus L. (Amaryllidac.).
Galactea Wight = Galactia P.Br. (Legumin.).
Galactella B. D. Jacks. = Galatella Cass. (Compos.).
Galactia P.Br. Leguminosae. 140 trop. & subtrop. Latex, which is rare in the fam., is found in this genus.
Galaction St-Lag. = praec.
*****Galactites** Moench. Compositae. 3 Canaries, Medit.

GALACTODENDRON

Galactodendron Reichb. = seq.

Galactodendrum Kunth ex Humb. = Brosimum Sw. (Morac.).

Galactoglychia Miq. = Galoglychia Gasp. = Ficus L. (Morac.).

Galactophora R. E. Woodson. Apocynaceae. 6 trop. S. Am.

Galactoxylon Pierre = Palaquium Blanco (Sapotac.).

Galagania Lipsky = Muretia Boiss. (Umbellif.).

Galanga Nor. = Alpinia Roxb. (Zingiberac.).

Galanthaceae Salisb. = Amaryllidaceae–Galanthinae Pax.

Galanthus L. Amaryllidaceae. 20 Eur., Medit. to Cauc. Bulb with 1-fld. scape. P in two whorls. On the inner surface of the inner P-ls. are green grooves secreting honey. The bud is erect, but the open fl. pendulous, visited by bees. The sta. dehisce by apical slits and lie close against the style. Each has a process outwards from the anther. The stigma projects and is first touched by an insect; in probing for honey it shakes the sta. and receives a shower of pollen (cf. *Erica*). Autogamy may occur in old fls. The fl. remains open a long time.

Galapagoa Hook. f. = Coldenia L. (Ehretiac.).

Galardia Lam. = Gaillardia Fougeroux (Compos.).

Galarhoeus Haw. = Euphorbia L. (Euphorbiac.).

Galarips Allem. ex L. = Allemanda L. (Apocynac.).

Galasia Koch = Gelasia Cass. = Scorzonera L. (Compos.).

Galasia Sch. Bip. = Microseris D. Don (Compos.).

Galatea Cass. = Aster L. (Compos.).

Galatea Herb. = Nerine Herb. (Amaryllidac.).

Galatea Salisb. = Eleutherine Herb. (Iridac.).

Galatella Cass. (~ Aster L.). Compositae. 40 temp. Euras.

Galathea Liebm. = Marica Ker-Gawl. = Neomarica Sprague (Iridac.).

Galathea Steud. (1) = Galatea Herb. = Nerine Herb. (Amaryllidac.).

Galathea Steud. (2) = Galatea Salisb. = Eleutherine Herb. (Iridac.).

Galathenium Nutt. = Lactuca L. (Compos.).

Galax L. = Nemophila Nutt. (Hydrophyllac.).

*****Galax** Rafin. Diapensiaceae. 1 SE. U.S.

Galaxa Parkinson (nomen) = Cerbera L. (Apocynac.)

Galaxia Thunb. Iridaceae. 6 S. Afr.

Galbanon Adans. = Bubon L. = Athamanta L. (Umbellif.).

Galbanophora Neck. = Seseli L. (Umbellif.).

Galbanum D. Don = Galbanon Adans. = Athamanta L. (Umbellif.).

Galbulimima F. M. Bailey. Himantandraceae. 2–3 E. Malaysia, NE. Austr.

Galdicia Néraud (nomen). ?Rhamnaceae. Mauritius.

Gale Duham. (~ Myrica L.). Myricaceae. 1 temp. N. Am., NW. Eur., 1 NE. Siberia. *G. belgica* Dum. (*Myrica gale* L.) (sweet gale, bog myrtle) in bogs and on wet heaths. Ls. with resinous smell.

Galeaceae Bub. = Myricaceae S. F. Gray.

Galeana La Llave. Compositae. 3 Mexico, C. Am.

Galeandra Lindl. Orchidaceae. 20 trop. & temp. S. Am., W.I.

Galearia Presl = Trifolium L. (Legumin.).

Galearia Zoll. & Mor. Pandaceae. 16 SE. As., Malaysia (to Solomon Is.).

Galeariaceae Pierre = Pandaceae Pierre.

Galearis Rafin. Orchidaceae. 9 India, Tibet, E. As., temp. N. Am., Greenland.
Galeata Wendl. Inc. sed. (?Boraginac. ?Convolvulac.). 1 trop. As.
Galeatella (E. Wimm.) O. & I. Deg. Campanulaceae. 4 Hawaii.
Galedragon S. F. Gray = Virga Hill (Dipsacac.).
Galedupa Lam. = Pongamia Vent. (Legumin.).
Galedupa Prain = Sindora Miq. (Legumin.).
Galega L. Leguminosae. 3 Medit. to Persia, 3 trop. E. Afr. *G. officinalis* L. sometimes cult. as a fodder-plant (goat's rue).
Galenia L. Aïzoaceae. 25 S. fr.
Galeobdolon Adans. Labiatae. 1 W. Eur. to N. Persia.
Galeoglossa Presl = Pyrrosia Mirbel (Polypodiac.).
Galeoglossum A. Rich. & Gal. = Prescottia Lindl. (Orchidac.).
Galeola Lour. Orchidaceae. 25 Madag., Indomal., Austr.
Galeopsis L. Labiatae. 10 temp. Euras. Upper ends of internodes swollen, acting as pulvini.
Galeopsis Hill = Stachys L. (Labiat.).
Galeorchis Rydb. = Galearis Rafin. (Orchidac.).
Galeottia Nees = Glockeria Nees (Acanthac.).
Galeottia A. Rich. = Mendoncella A. D. Hawkes (Orchidac.).
Galeottia Rupr. ex Gal. = Zeugites P.Br. (Gramin.).
Galeottiella Schlechter. Orchidaceae. 1 Mexico.
Galera Blume = Epipogium R.Br. (Orchidac.).
Galiaceae Lindl. = Rubiaceae–Rubieae (Galieae Dum.).
Galianthe Griseb. ex Lorentz = Spermacoce L. (Rubiac.).
× **Galiasperula** Ronniger. Rubiaceae. Gen. hybr. (Asperula × Galium).
Galiba P. & K. = Gabila Baill. = Pycnarrhena Miers (Menispermac.).
Galilea Parl. = Cyperus L. (Cyperac.).
Galimbia Endl. (sphalm.) = Palimbia Bess. = Peucedanum L. (Umbellif.).
Galiniera Delile. Rubiaceae. 2 Abyssinia, Madag.
Galinsoga Ruiz & Pav. Compositae. 4 Mexico to Argentina. *G. parviflora* Cav. now a common weed in Eur.
Galinsogaea Zucc. = praec.
Galinsogea Kunth = Tridax L. (Compos.).
Galinsogea Willd. = Galinsoga Ruiz & Pav. (Compos.).
Galinsogeopsis Sch. Bip. = Pericome A. Gray (Compos.).
Galinsoja Roth = Galinsoga Ruiz & Pav. (Compos.).
Galinzoga Dum. = praec.
Galion St-Lag. = Galium L. (Rubiac.).
Galiopsis Fourr. = praec.
Galiopsis St-Lag. = Galeopsis L. (Labiat.).
Galipea Aubl. Rutaceae. 13 C. & S. Am.
Galisongea Willd. = Galinsoga Ruiz & Pav. (Compos.).
Galium L. Rubiaceae. 400 cosmop. Herbs with whorls of ls. and stips. (see fam.); fls. in dichasial panicles, small, with honey freely exposed on the epig. disk, usu. protandrous with ultimate self-pollination. *G. aparine* L. is a feeble hook-climber with small reflexed hooks on the stem. The schizocarp is also provided with hooks.
Galiziola Rafin. = Ardisia Sw. (Myrsinac.).

GALLAPAGOA

Gallapagoa Pritz. = Galapagoa Hook. f. = Coldenia L. (Boraginac.).
Gallardoa Hicken. Malpighiaceae. 1 Argentina.
Gallaria Schrank ex Endl. = Medinilla Gaudich. (Melastomatac.).
Gallasia Mart. ex DC. = Miconia Ruiz & Pav. (Melastomatac.).
Gallesia Casaretto. Phytolaccaceae. 2 Peru, Brazil.
Gallesioa Kuntze = praec.
Gallesioa M. Roem. = Clausena Burm. f. (Rutac.).
Galliaria Bub. = Amaranthus L. (Amaranthac.).
Galliastrum Fabr. (uninom.). = *Mollugo* L. sp. (Aïzoac.).
Gallienia Dubard & Dop. Rubiaceae. 1 Madag.
Gallinsoga Jaume St-Hil. = Galinsoga Ruiz & Pav. (Compos.).
Gallion Pohl = Galium L. (Rubiac.).
Gallitrichum Fourr. = Salvia L. (Labiat.).
Gallium Mill. = Galium L. (Rubiac.).
Galloa Hassk. = Cocculus DC. (Menispermac.).
Galoglychia Gasp. = Ficus L. (Morac.).
Galophthalmum Nees & Mart. = Blainvillea Cass. (Compos.).
Galopina Thunb. Rubiaceae. 3 S. Afr.
Galordia Raeusch. = Gaillardia Fougeroux (Compos.).
Galorhoeus Endl. = Galarhoeus Haw. = Euphorbia L. (Euphorbiac.).
Galphimia Cav. = Thryallis L. (Malpighiac.).
Galphinia Poir. = praec.
Galpinia N. E. Br. Lythraceae. 3 S. Afr.
Galpinsia Britton = Oenothera L. (Onagrac.).
Galstronema Steud. = Gastronema Herb. = Cyrtanthus Ait. (Amaryllidac.).
Galtonia Decne. Liliaceae. 4 S. Afr.
Galumpita Blume = Gironniera Gaudich. (Ulmac.).
Galurus Spreng. = Acalypha L. (Euphorbiac.).
Galvania Vand. = Psychotria L. (Rubiac.).
Galvesia J. F. Gmel. = Galvezia Domb. ex Juss. (Scrophulariac.).
Galvesia Pers. = Galvezia Ruiz & Pav. = Pitavia Molina (Rutac.).
Galvezia Domb. ex Juss. Scrophulariaceae. 6 Calif., Mex., Ecuad., Peru.
Galvezia Ruiz & Pav. = Pitavia Molina (Rutac.).
Galypola Nieuwl. = Polygala L. (Polygalac.).
Gama La Llave = ? Matricaria L. (Compos.).
Gamanthus Bunge. Chenopodiaceae. 7 Palest. to C. As. & Afghan.
Gamaria Rafin. = Disa Bergius (Orchidac.).
Gamazygis Pritz. = Gamozygis Turcz. = Angianthus Wendl. (Compos.).
Gambelia Nutt. = Antirrhinum L. (Scrophulariac.).
Gambeya Pierre. Sapotaceae. 14 trop. Am., Afr.
Gamblea C. B. Clarke. Araliaceae. 2 E. Himal., Burma.
Gamblum Rafin. = Gansblum Adans. = Draba L. (Crucif.).
Gamelythrum Nees = Amphipogon R.Br. (Gramin.).
Gamelytrum Steud. = praec.
Gamocarpha DC. (~ Boöpis Juss.). Calyceraceae. 6 temp. S. Am.
Gamochaeta Wedd. Compositae. 30 trop. S. Am., W.I.
Gamochilum Walp. = Argyrolobium Eckl. & Zeyh. (Legumin.).
Gamochilus Lestib. = Hedychium Koen. (Zingiberac.).

476

Gamochlamys Baker = Spathantheum Schott (Arac.).

Gamogyne N. E. Br. = Piptospatha N. E. Br. (Arac.).

Gamolepis Less. Compositae. 13 S. Afr.

Gamopetalae Brongn. (*Sympetalae* A. Br., *Metachlamydeae* Engl.). A collective term for a probably only partially natural assemblage of Dicot. fams. in which sympetaly predominates.

Gamoplexis Falc. = Gastrodia R.Br. (Orchidac.).

Gamopoda Baker. Menispermaceae. 1 Madag.

Gamosepalum Hausskn. (~ Alyssum L.). Cruciferae. 3 Turkey, Armenia. (K)!

Gamosepalum Schlechter (~ Spiranthes Rich.). Orchidaceae. 1 Mex.

Gamotopea Bremek. Rubiaceae. 5 trop. S. Am.

Gamozygis Turcz. = Angianthus Wendl. (Compos.).

Gampsoceras Stev. = Ranunculus L. (Ranunculac.).

Gamwellia E. G. Baker. Leguminosae. 1 trop. E. Afr.

Gandasulium Kuntze = Hedychium Koen. (Zingiberac.).

Gandola L. (1762) = Basella L. (1753) (Basellac.).

Gandola Moq. = Ullucus Lozano (Basellac.).

Gandriloa Steud. = Chenopodium L. (Chenopodiac.).

Gangila Bernh. = Sesamum L. (Pedaliac.).

Ganitrum Rafin. = seq.

Ganitrus Gaertn. = Elaeocarpus L. (Elaeocarpac.).

Ganja Reichb. = Corchorus L. (Tiliac.).

Ganophyllum Blume. Sapindaceae. 1 trop. W. Afr., 1 Philipp. Is., Andaman Is., Sumatra, Java, New Guinea, NE. Austr.

Ganosma Decne = Aganosma D. Don (Apocynac.).

Gansblum Adans. = Erophila DC. (Crucif.).

Gantelbua Bremek. Acanthaceae. 1 Penins. India.

Ganua Pierre ex Dubard. Sapotaceae. 20 Malaysia. *G. pallida* (Burck) H. J. Lam yields a gutta-percha.

Ganymedes Salisb. = Narcissus L. (Amaryllidac.).

Garacium Gren. & Godr. = Lactuca L. (Compos.).

Garadiolus P. & K. = Garhadiolus Jaub. & Spach (Compos.).

Garaleum Sch. Bip. = Garuleum Cass. (Compos.).

Garapatica Karst. = Alibertia A. Rich. (Rubiac.).

Garaventia Looser. Alliaceae. 1 Chile.

× **Garayara** hort. (iv, v) = × Laycockara hort. (Orchidac.).

Garberia A. Gray. Compositae. 1 Florida.

Garcia Rohr. Euphorbiaceae. 2 Mex. (cult. trop. S. Am., W.I.). Yields an oil. Fls. rel. large. K (2–3), C 8–12, A ∞.

Garciana Lour. = Philydrum Banks (Philydrac.).

Garcilassa Poepp. & Endl. Compositae. 1 Peru.

Garcinia L. Guttiferae. 400 trop. (esp. As.); S. Afr. Trees or shrubs with leathery ls. Sta. free or united into bundles or into a common mass. Berry; seed arillate. The resin of *G. morella* Desr. and other spp., obtained by cutting notches in the stem, forms gamboge. The fr. of many is ed., esp. that of *G. mangostana* L. (mangosteen), the aril of the seed of which is a delicacy. Some yield useful timber.

Garcini[ac]eae Burnett = Guttiferae–Garcinieae Choisy.

GARDENA

Gardena Adans. = seq.

Gardenia J. Colden ex Garden (1756) = Triadenum Rafin. (Guttif.).

Gardenia Ellis [1757] (1821) = Calycanthus L. (Calycanthac.).

Gardenia Ellis [1760a] (1821) = Gelsemium Juss. (Loganiac.).

Gardenia Ellis [1760b] (1821) = Kleinhovia L. (Sterculiac.).

*Gardenia Ellis (1761). Rubiaceae. 250 palaeotrop. Some spp. have apparently whorls of leaves, 3 in each, really a case of condensation of two whorls of 2 into one with extreme anisophylly of one whorl; the fourth l. is reduced to a minute scale. The stipules of many secrete a resinous fluid.

Gardeniaceae Dum. = Rubiaceae–Gardenieae A. Rich.

Gardeniola Cham. = Alibertia A. Rich. (Rubiac.).

Gardeniopsis Miq. Rubiaceae. 1 Sumatra, Borneo.

Gardinia Bertol. = Brodiaea Sm. (Alliac.).

Gardneria Wall. Strychnaceae. 5 India to Japan.

Gardnerodoxa Sandwith. Bignoniaceae. 1 Brazil.

Gardoquia Ruiz & Pav. = Satureja L. (Labiat.).

Gareilassa Walp. (sphalm.) = Garcilassa Poepp. & Endl. (Compos.).

Garhadiolus Jaub. & Spach (~ Rhagadiolus Juss.). Compositae. 5 Turkey to C. As. & Afghan.

Garidelia Spreng. = seq.

Garidella L. = Nigella L. (Ranunculac.).

Garnieria Brongn. & Gris. Proteaceae. 1 New Caled.

Garnotia Brongn. Gramineae. 30 E. As., NE. Austr., Pacif.

Garnotiella Stapf. Gramineae. 1 Philipp. Is.

Garosmos Mitch. Quid? (?Rubiac. ?Gentianac. ?Celastrac.) 1 E. U.S.

Garrelia Gaudich. = Dyckia Schult. f. (Bromeliac.).

Garretia Welw. = Khaya Juss. (Meliac.).

Garrettia Fletcher. Verbenaceae. 1 Siam, Java & islets.

Garrielia P. & K. = Garrelia Gaudich. = Dyckia Schult. f. (Bromeliac.).

Garrya Dougl. ex Lindl. Garryaceae. 18 W. U.S., Mex., W.I.

*Garryaceae Lindl. Dicots. 1/18 warm N. Am. Shrubs with 4-angled twigs and opp. ent. exstip. evergr. ls., petioles united at base. Fls. unisexual (♂ long-stalked), in catkin-like pend. infls., 1–3 in axil of each br. ♂ P usu. 4, A 4; ♀ naked, or with 2–4 minute K-teeth at summit of G below styles, $\overline{G}$ (2–3), uniloc. with 2 pend. anatr. ov. with dorsal raphe, on parietal plac. Fr. berry-like, with thin pericarp and 1–2 seeds. Endosp. Only genus: *Garrya*. Prob. distantly related to *Cornaceae*.

Garuga Roxb. Burseraceae. 4 Himal. to S. China, Philipp., N. Borneo, E. Java, E. Malaysia, NE. Austr., Pacif. Ls. imparipinnate. Disk cup-like.

Garugandra Griseb. = Gleditsia L. (Legumin.).

Garuleum Cass. Compositae. 8 S. Afr. No pappus.

Garumbium Blume (sphalm.) = Carumbium Reinw. = Homalanthus A. Juss. (Euphorbiac.).

Gaslondia Vieill. = Cupheanthus Seem. (Myrtac.).

Gasoub B. D. Jacks. (sphalm.) = seq.

Gasoul Adans. (~ Mesembryanthemum L.). Aïzoaceae. 50 Atl. Is., Medit. reg. to Baluch., S. Afr.

Gasparinia Bertol. = Silaum Mill. (Umbellif.).

Gasparinia Endl. = Aeginetia L. (Orobanchac.).
Gassoloma D. Dietr. (sphalm.) = Geissoloma Lindl. (Geissolomac.).
× Gasteraloë Guillaumin = × Gastrolea Walther (Liliac.).
Gasteranthopsis Oerst. = seq.
Gasteranthus Benth. = Besleria L. (Gesneriac.).
× Gasterhaworthia Guillaumin. Liliaceae. Gen. hybr. (Gasteria × Haworthia).
Gasteria Duval. Liliaceae. 70 S. Afr. Xero. with succulent ls. closely packed, but often growing in shade of grass.
Gasterolychnis Rupr. = Gastrolychnis (Fenzl) Reichb. = Melandrium Roehling (Caryophyllac.).
Gasteronema Lodd. ex Steud. = Gastronema Herb. = Cyrtanthus Ait. (Liliac.).
Gastonia Comm. ex Lam. Araliaceae. 10 trop. E. Afr., Madag., Masc., Borneo, New Guinea. 10–15-merous.
Gastorchis Thou. = seq.
Gastorkis Thou. = Phaius Lour. (Orchidac.).
Gastranthopsis P. & K. = Gasteranthopsis Oerst. = Besleria L. (Gesneriac.).
Gastranthus Moritz ex Benth. & Hook. f. Acanthaceae. 1 Venezuela.
Gastranthus F. Muell. = Parsonsia R.Br. (Apocynac.).
Gastridium Beauv. Gramineae. 2 Canaries, W. Eur., Medit. Glumes persistent on axis.
Gastridium Blume (corr. Bl.) = Grastidium Bl. = Dendrobium Sw. (Orchidac.).
Gastrilia Rafin. = Daphnopsis Mart. & Zucc. (Thymelaeac.).
× Gastrocalanthe hort. Orchidaceae. Gen. hybr. (iii) (Calanthe × Gastrorchis).
Gastrocalyx Gardn. = Prepusa Mart. (Gentianac.).
Gastrocalyx Schischkin = Schischkiniella v. Steenis (Caryophyllac.).
Gastrocarpha D. Don = Moscharia Ruiz & Pav. (Compos.).
Gastrochilus D. Don. Orchidaceae. 20 India, E. As., W. Malaysia.
Gastrochilus Wall. = Boesenbergia Kuntze (Zingiberac.).
Gastrococos Morales (~ Acrocomia Mart.). Palmae. 1 Cuba.
Gastrocotyle Bunge. Boraginaceae. 2 E. Medit. to C. As. & NW. India.
Gastrodia R.Br. Orchidaceae. 20 E. As., Indomal. to N.Z. (? trop. W. Afr.).
Gastroglottis Blume = Malaxis Soland. ex Sw. (Orchidac.).
× Gastrolea Walther. Liliaceae. Gen. hybr. (Aloë × Gasteria).
Gastrolepis Van Tiegh. Icacinaceae. 1 New Caled.
× Gastrolirion Walther. Liliaceae. Gen. hybr. (Chortolirion × Gasteria).
Gastrolobium R.Br. Leguminosae. 44 W. Austr. Ls. usu. opp or whorled.
Gastrolychnis (Fenzl) Reichb. = Melandrium Roehling (Caryophyllac.).
Gastromeria D. Don = Melasma Bergius (Scrophulariac.).
Gastronema Herb. = Cyrtanthus Ait. (Amaryllidac.).
Gastronychia Small (~ Paronychia Adans.). Caryophyllaceae. 1 SE. U.S.
× Gastrophaius hort. Orchidaceae. Gen., hybr. (iii) (Gastrorchis × Phaius).
Gastropodium Lindl. = Epidendrum L. (Orchidac.).
Gastrorchis Schlechter = Phaius Lour. (Orchidac.).
Gastrostylum Sch. Bip. = Gastrosulum Sch. Bip. = Matricaria L. (Compos.).
Gastrostylus (Torr.) Kuntze = Cneoridium Hook. f. (Rutac.).
Gastrosulum Sch. Bip. = Matricaria L. (Compos.).

Gatesia Bertol. = Petalostemon Michx (Legumin.).
Gatesia A. Gray = Yeatesia Small (Acanthac.).
Gatnaia Gagnep. = Baccaurea Lour. (Euphorbiac.).
Gattenhoffia Neck. = Dimorphotheca Moench (Compos.).
Gattenhofia Medik. = Iris L. (Iridac.).
Gatyona Cass. = Crepis L. (Compos.).
Gaudichaudia Kunth. Malpighiaceae. 15 Mex. to Bolivia. Mericarp elevated on carpophore formed from wing of cpl.
Gaudina St-Lag. = seq.
Gaudinia Beauv. Gramineae. 3 Azores, Medit. Spikelet many-fld.
Gaudinia J. Gay = Limeum L. (Aïzoac.).
Gaudiniopsis (Boiss.) Eig (~ Gaudinia Beauv.). Gramineae. 1 As. Minor, Crimea.
× **Gaulnettya** W. J. Marchant = × Gaulthettya Camp (Ericac.).
Gaulteria Adans. = seq.
Gaultheria Kalm ex L. Ericaceae. 200 circumpacif. (W. to W. Himal. & S. India); 2 spp. in E. N. Am.; 8 in E. Brazil. Shrubby. Fr. berry-like, but really a caps. enclosed in fleshy K (not adherent). Many have ed. fr., e.g. *G. procumbens* L. (E. N. Am., wintergreen, checker-berry, partridge-berry), *G. shallon* Pursh (W. N. Am., sallal, shallon). Wintergreen oil distilled from some.
× **Gaulthettya** Camp. Ericaceae. Gen. hybr. (Gaultheria × Pernettya).
Gaultiera Rafin. = Gaultheria Kalm ex L. (Ericac.).
Gaumerocassia Britton = Cassia L. (Legumin.).
Gaunia Scop. = Gahnia J. R. & G. Forst. (Cyperac.).
× **Gauntlettara** hort. Orchidaceae. Gen. hybr. (iii) (Broughtonia × Cattleyop-
Gaura Lam. = Lechea Kalm ex L. (Cistac.).   [sis × Laeliopsis).
Gaura L. Onagraceae. 18 N. Am., Mex., Argent. Sta. with scale at base; anther chambered by horiz. septa in each loc. (cf. *Circaea*). Nut.
Gaurea Reichb. = Guarea Allem. (Meliac.).
Gaurella Small (~ Oenothera L.). Onagraceae. 2 N. Am.
Gauridium Spach = Gaura L. (Onagrac.).
Gauropsis Presl (~ Clarkia Pursh). Onagraceae. 1 Mex.
Gauropsis (Torr. & Frém.) Cockerell = Gaurella Small (Onagrac.).
Gaussia H. Wendl. Palmae. 2 Cuba, Puerto Rico. Stem swollen below.
Gautiera Rafin. = Gaultheria Kalm ex L. (Ericac.).
Gavarretia Baill. Euphorbiaceae. 1 Amaz. Brazil. G 2-loc.
Gavesia Walp. = Gatesia Bertol. = Petalostemon Michx (Legumin.).
Gavilea Poepp. Orchidaceae. 17 temp. S. Am.
Gavnia Pfeiff. = Gaunia Scop. = Gahnia J. R. & G. Forst. (Cyperac.).
Gaya Gaudin = Neogaya Meissn. = Pachypleurum Ledeb. (Umbellif.).
Gaya Kunth. Malvaceae. 3 N.Z.; 20 trop. Am., W.I. Like *Sida*. No epicalyx.
Gaya Spreng. = Seringia J. Gay (Sterculiac.).
Gayacum Brongn. = Guaiacum L. (Zygophyllac.).
Gayella Pierre. Sapotaceae. 1 Chile.
*****Gaylussacia** Kunth. Ericaceae. 9 N. Am., 40 S. Am. (huckleberry). The 5 loc. of the ovary are made into 10 by partitions growing out from the midribs of the cpls., as in *Linum*.

Gayoïdes (Endl.) Small = Herissantia Medik. (Malvac.).

Gayophytum A. Juss. Onagraceae. 9 temp. W. N. Am., temp. W. S. Am.

Gaytania Münter = Pimpinella L. (Umbellif.).

Gaza Teran & Berland. = Ehretia L. (Ehretiac.).

Gazachloa Phipps. Gramineae. 1–2 S. trop. Afr.

*Gazania Gaertn. Compositae. 40 trop. & (mostly) S. Afr. Latex.

Geanthemum (R. E. Fries) Safford = Duguetia A. St-Hil. (Annonac.).

Geanthia Rafin. = Crocus L. (Iridac.).

Geanthus (Bl., Benth.) Loes. = Etlingera Giseke (Zingiberac.).

Geanthus Phil. = Speea Loes. (Liliac.).

Geanthus Rafin. = Crocus L. (Iridac.).

Geanthus Reinw. = Hornstedtia Retz. (Zingiberac.).

Gearum N. E. Br. Araceae. 1 Brazil.

Geaya Costantin & Poisson = Kitchingia Bak. (Crassulac.).

Geblera Andrz. ex Bess. = Crepis L. (Compos.).

Geblera Fisch. & Mey. = Flueggea Willd. (Euphorbiac.).

Geboscon Rafin. = Nothoscordum Kunth (Alliac.).

Geeria Blume = Eurya Thunb. (Theac.).

Geeria Neck. = Paullinia L. (Sapindac.).

Geigera Less. = Geigeria Griesselich (Compos.).

Geigera Lindl. = Geijera Schott (Rutac.).

Geigeria Griesselich. Compositae. 50 trop. & S. Afr.

Geijera Schott. Rutaceae. 7 E. New Guinea, E. Austr., New Caled., Loyalty Is.

Geisarina Rafin. = Forestiera Poir. (Oleac.).

Geiseleria Klotzsch = Croton L. (Euphorbiac.).

Geiseleria Kunth = Anticlea Kunth = Zigadenus Michx (Liliac.).

Geisenia Endl. = Gaissenia Rafin. = Trollius L. (Ranunculac.).

Geisoloma Lindl. = Geissoloma Lindl. (Geissolomatac.).

Geisorrhiza Reichb. = Geissorhiza Ker-Gawl. (Iridac.).

Geissanthera Schlechter = Microtatorchis Schlechter (Orchidac.).

Geissanthus Hook. f. Myrsinaceae. 35 W. trop. S. Am.

Geissapsis Baker (sphalm.) = seq.

Geissaspis Wight & Arn. Leguminosae. 30 Afr., 2 India to Lower Burma.

Geissois Labill. Cunoniaceae. 20 Austr., New Hebrid., New Caled., Fiji.

Geissolepis Robinson. Compositae. 1 Mexico.

Geissoloma Lindl. ex Kunth. Geissolomataceae. 1 S. Afr., G. marginatum (L.) A. Juss.

*Geissolomataceae Endl. Dicots. 1/1 S. Afr. Small xero. shrubs. Ls. opp., ent., stip. (!), evergr. Fls. reg., ☿, sol., axillary, subtended by 6 persist. bracts. K 4, petaloid., imbr., persist.; C 0; A 4 + 4, with slender fil. and ellipsoid dorsifixed anth.; G (4), with 2 pend. ov. per loc., and 4 subulate styles free below but coherent above. Caps. 4-loc., 4-seeded; seeds shining. Endosp. Only genus: Geissoloma. Related to Penaeaceae.

Geissomeria Lindl. Acanthaceae. 15 Mex. to trop. S. Am.

Geissopappus Benth. Compositae. 4 trop. S. Am.

Geissorhiza Ker-Gawl. Iridaceae. 65 S. Afr., Madag.

Geissospermum Allem. Apocynaceae. 5 trop. Brazil. G. laeve Baill. has offic. bark, cortex Pereirae.

Geitonoplesium A. Cunn. (~ Luzuriaga Ruiz & Pav.). Philesiaceae. 1 Philipp. Is. & E. Malaysia (exc. Celebes) to Solomons, E. Austr. to Fiji (very polymorphic).

Gela Lour. = Acronychia J. R. & G. Forst. (Rutac.).

Gelasia Cass. = Scorzonera L. (Compos.).

Gelasine Herb. Iridaceae. 1 subtrop. S. Am.

Geleznowia Turcz. Rutaceae. 3 W. Austr.

Gelibia Hutch. Araliaceae. 2 New Guinea, Austr.

Gelonium Gaertn. = Ratonia DC. (Sapindac.).

Gelonium Roxb. ex Willd. = Suregada Roxb. ex Rottl. (Euphorbiac.).

Gelpkea Blume = Syzygium Gaertn. (Myrtac.).

Gelseminum Pursh = Gelsemium Juss. (Loganiac.).

Gelseminum Weinm. = Tecoma Juss. (Bignoniac.).

Gelsemium Juss. Loganiaceae. 1 S. China, Indoch., N. Borneo, Sumatra; 1 SE. U.S., N. Mex. The peduncle of the latter sp. bears numerous bracteoles.

Gembanga Blume = Corypha L. (Palm.).

Gemella Hill = Bidens L. (Compos.).

Gemella Lour. = Allophylus L. (Sapindac.).

Gemellaria Pinel ex Lem. = Nidularium Lem. (Bromeliac.).

Geminaceae Dulac = Circaeaceae Lindl.

Geminaria Rafin. = Synexemia Rafin. = Phyllanthus L. + Savia Willd. (Euphor-

Gemmaria Nor. = Tetracera L. (Dilleniac.). [biac.).

Gemmaria Salisb. = Periphanes Salisb. (Amaryllidac.).

Gemmingia Fabr. = Ixia L. (1753) = Belamcanda Adans. + Aristea Soland.

Gendarussa Nees (~ Justicia L.). Acanthaceae. 2 Indomal. [(Iridac.).

Genea (Dum.) Dum. = Anisantha C. Koch (~ Bromus L.) (Gramin.).

Genersichia Heuff. = Carex L. (Cyperac.).

Genesiphyla Rafin. = seq.

Genesiphylla L'Hérit. = Xylophylla L. = Phyllanthus L. (Euphorbiac.).

Genetyllis DC. = Darwinia Rudge (Myrtac.).

Genevieria Gandog. = Rubus L. (Rosac.).

Genianthus Hook. f. Asclepiadaceae. 10 Indomal.

Geniosporum Wall. ex Benth. Labiatae. 25 trop. Afr., Madag., Indochina, S. China.

Geniostemon Engelm. & Gray. Gentianaceae. 2 Mexico.

Geniostephanus Fenzl = Trichilia L. (Meliac.).

Geniostoma J. R. & G. Forst. Loganiaceae. 60 Madag. to N.Z.

Genipa L. Rubiaceae. 6 warm Am., W.I.

Genipella Rich. ex DC. = Alibertia Rich. (Rubiac.).

Genista Duham. = Spartium L. (Legumin.).

Genista L. Leguminosae. 75 Eur., N. Afr., W. As. *G. anglica* L. has large thorns (branches). The fl. has an explosive mechanism, typical of many of the fam. (*q.v.*). In *G. tinctoria* L. there is no honey; the style and tube of sta. are enclosed in the keel, which is united along the top seam as well as the bottom. The sta. shed their pollen almost in the apex of the keel, but not so near it as to pollinate the stigma. When the fl. opens there is a tension of the sta.-tube on the lower side, tending to bend it upwards; this is resisted by an opposite one in the keel and wings, but if an insect alight on the wings and press them down, the upper seam of the keel gives way and an explosion follows. In it the style flies out, striking the under side of the insect, thus probably becoming cross-pollinated, and is followed by a shower of pollen which gives the insect a fresh coating to take to another fl. A yellow dye is obtained from the fls. of this sp., which when mixed with woad gives a fine green (Kendal green).

Genista-spartium Duham. = Ulex L. (Legumin.).

Genistella Ortega = Chamaespartium Adans. (Legumin.).

Genistidium I. M. Johnston. Leguminosae. 1 Mex.

Genistoïdes Moench = Genista L. (Legumin.).

Genlisa Rafin. = Scilla L. (Liliac.).

Genlisea Benth. & Hook. f. = Genlisia Reichb. = Aristea Ait. (Iridac.).

Genlisea A. St-Hil. Lentibulariaceae. 15 C. Am., W.I., trop. & S. Afr., Madag.

Genlisia Reichb. = Aristea Ait. (Iridac.).

Gennaria Parlat. Orchidaceae. 1 N. Atl. Is., W. Medit.

Genoplesium R.Br. = Prasophyllum R.Br. (Orchidac.).

Genoria Pers. = Ginora L. f. (Lythrac.).

Genosiris Labill. = Patersonia R.Br. (Iridac.).

Gentiana L. Gentianaceae. 400 cosmop. exc. Afr., chiefly alpine. Most are alpine pl. of tufted growth. Fls. of interest. The genus shows an ascending series of fls., adapted to higher and higher types of insects. *G. lutea* L. is a primitive type, with freely exposed honey, yellow homogamous fls. and short-tongued visitors. *G. purpurea* L., *G. pneumonanthe* L., etc., are blue long-tubed humble-bee fls. *G. verna* L. has long-tubed butterfly fls., sometimes protandr.

The gentians form one of the most striking features of the flora of the Alps, occurring in large masses and with very conspicuous fls.; *G. acaulis* L. is perhaps the most beautiful. The root of *G. lutea* furnishes a tonic.

*****Gentianaceae** Juss. 80/900 in every part of the globe and in great variety of situation—arctic and alpine pls., halophytes, saprophytes (*Voyria*, etc.), marsh pls., etc. They are mostly herbaceous (often perennial); a few shrubs. The perennial herbs have usu. a rhizome. Ls. opp., exstip., usu. entire. The infl. is usu. a dichasial cyme like *Caryophyllaceae*; as in that fam., the lat. branches often become monochasial. Other cymose infls. also occur. Bracts and bracteoles present or not. Fls. reg., ☿, 4–5-merous, rarely more (e.g. *Blackstonia*). K usu. 4(–5), imbr.; C usu. (4–5), bell- or funnel-shaped, or sometimes salver-shaped, convol. (exc. *Bartonia, Obolaria*, etc.); A as many as petals, alt. with them, epipet., sometimes strongly zygo. (declinate); anthers various, usu. introrse; G̲ (2), placed in the antero-posterior plane, with a glandular disk at base. Placentae usu. parietal, but they commonly project far into the cavity and spread out at their ends; occasionally the ovary is 2-loc. with axile plac.; ovules usu. ∞, anatr.; style simple; stigma simple or 2-lobed. Fr. usu. a septicidal caps. with ∞ seeds, rarely a berry (*Chironia*, etc.); seeds small; embryo small, in abundant endosp. Closely related to *Loganiac.*; perhaps some connection also with *Melastomatac.* (Ls. rarely alt. – *Swertia* spp.)

The flowers of *G.* are insect-fertilised. The genus *Gentiana* has been very fully studied. Chief genera: *Exacum, Sebaea, Belmontia, Sabbatia, Centaurium, Canscora, Chironia, Gentiana, Swertia, Halenia, Schultesia, Leiphaimos.*

Gentianella Moench (~ Gentiana L.). Gentianaceae. 125 N. & S. temp. (exc. Afr.).

GENTIANOPSIS

Gentianopsis Ma. Gentianaceae. 26 N. temp. As., Am.

Gentianothamnus Humbert. Gentianaceae. 1 Madag.

Gentianusa Pohl = Gentiana L. (Gentianac.).

Gentilia A. Chev. & Beille = Neogoetzea Pax = Bridelia Willd. (Euphorbiac.).

Gentlea Lundell (~ Ardisia Sw.). Myrsinaceae. 6 Mex. to Peru.

Gentrya Breedlove & Heckard. Scrophulariaceae. 1 Mex.

Genyorchis Schlechter. Orchidaceae. 6 trop. Afr.

Geobalanus Small = Licania Aubl. (Chrysobalanac.).

Geobina Rafin. = Goodyera R.Br. (Orchidac.).

Geoblasta Barb. Rodr. = Chloraea Lindl. (Orchidac.).

Geocallis Horan. = Amomum L. (Zingiberac.).

Geocardia Standley = Carinta W. F. Wright = Geophila D. Don (Rubiac.).

Geocarpon Mackenzie. Aïzoaceae. 1 Central U.S. (Missouri).

Geocaryum Coss. = Carum L. (Umbellif.).

Geocaulon Fernald. Santalaceae. 1 Alaska, Canada, NE. U.S.

Geocharis (K. Schum.) Ridl. Zingiberaceae. 4 W. Malaysia.

Geochorda Cham. & Schlechtd. Scrophulariaceae. 1 warm S. Am.

Geococcus J. Drumm. ex Harv. Cruciferae. 2 NW. Austr.

Geodorum G. Jacks. Orchidaceae. 16 India to Polynesia, Austr.

Geoffraea L. = Geoffroea Jacq. (Legumin.).

Geoffraya Bonati. Scrophulariaceae. 2 Indoch.

Geoffroea Jacq. Leguminosae. 6 trop. Am., W.I.

Geoffroya Murr. = praec.

Geogenanthus Ule. Commelinaceae. 4 trop. S. Am.

Geoherpum Willd. ex Schult. = Mitchella L. (Rubiac.).

Geolobus Rafin. = Voandzeia Thou. (Legumin.).

Geomitra Becc. Burmanniaceae. 1 Borneo.

Geonoma Willd. Palmae. *S.l.* 240, *s.str.* 150 trop. Am., W.I. Style lateral.

Geonomaceae O. F. Cook = Palmae–Geonomeae Benth. & Hook. f.

Geopanax Hemsl. Araliaceae. 1 Seychelles.

Geophila Bergeret = Merendera Ram. (Liliac.).

***Geophila** D. Don. Rubiaceae. 30 trop.

Geoprumnon Rydb. = Astragalus L. (Legumin.).

Georchis Lindl. = Goodyera R.Br. (Orchidac.).

Georgia Spreng. = Cosmos Cav. + Dahlia Cav. (Compos.).

Georgina Willd. = Dahlia Cav. (Compos.).

***Geosiridaceae** Jonker. Monocots. 1/1 Madag. Small colourless saprophytic herbs from scaly rhiz. Stems simple or branched; ls. alt., squamiform. Infl. loosely cymose, bracteate; fls. reg., ⚥. P 3+3, petaloid, bluish, subequal, shortly connate, the outer imbr., the inner contorted; A 3, opp. to the outer P-segs., with v. short fil. and large obl. basifixed extrorse anth.; G̅ (3), with ∞ ov. on dendriform axile plac. Fr. trigono-obconic, crowned with an annulus at truncate apex, prob. marcescent; seeds ∞, minute. Only genus: *Geosiris*. Perhaps related to *Iridaceae* as *Petrosaviaceae* to *Liliaceae*.

Geosiris Baill. Geosiridaceae. 1 Madag.

Geostachys (Bak.) Ridl. Zingiberaceae. 12 Malay Penins., Sumatra.

Geracium Reichb. = Crepis L. (Compos.).

Geraea Torr. & Gray (~ Encelia Adans.). Compositae. 2 SW. U.S., NW. Mex.

*Geraniaceae Juss. (*s.str.*). Dicots. 5/750 cosmop. Mostly herbs, often hairy; *Sarcocaulon* fleshy. Ls. opp. or alt., often stip. Fl. reg. or zygo., ♂, 5-merous. K 5, or (5), imbr. with valvate tips, persist.; C 5, imbr. or convol.; A as many or 2 or 3 times as many as petals, united at base, obdipl. when > 1 whorl, anther usu. versatile; G (5), with 1–2 ovules in each on axile plac.; ovules usu. pend. with ventral raphe and micropyle facing upwards; style long with 5 stigmas. Fls. usu. protandr. Fr. a schizocarp, the cpls. splitting off from a central beak (the persistent style); each takes with it a strip of the tissue of the style, forming an awn, which is usu. hygroscopic (cf. *Geranium, Erodium*). Embryo straight or folded, in endosp. Genera: *Geranium, Erodium, Monsonia, Pelargonium, Sarcocaulon*.

Geranion St-Lag. = Geranium L. (Geraniac.).

Geraniopsis Chrtek. Geraniaceae. 2 Red Sea coast, Arabia, S. Persia.

Geraniospermum Kuntze = Pelargonium L'Hérit. (Geraniac.).

Geranium L. Geraniaceae. 400 cosmop., esp. temp. The partial infl. (unlike that of *Erodium* and *Pelargonium, q.v.*) is a pair of fls. or a single fl. Branching cymose, either dich. with cincinnus-tendency, or a cincinnus alone, which is straightened out into a sympodium. The nectaries are at the base of the sta. These stand at first round the undeveloped style; after dehiscence they move away, and finally the stigmas open. The fr. explodes, the awn twisting up so that the cpls. are carried up and outwards. In many spp. the cpls. open at the same time and the seeds are shot out. The 'geraniums' of greenhouses belong to the genus *Pelargonium*.

Gerardia Benth. Scrophulariaceae. 60 Am., W.I.

Gerardia L. = Stenandrium Nees (Acanthac.).

Gerardianella Klotzsch = Micrargeria Benth. (Scrophulariac.).

Gerardiina Engl. Scrophulariaceae. 1 trop. & S. Afr.

Gerardiopsis Engl. Scrophulariaceae. 1 trop. Afr.

Gerascanthus P.Br. = Cordia L. (Ehretiac.).

Gerbera Boehm. = Arnica L. (Compos.).

Gerbera J. F. Gmel. = Gerberia Scop. = Quararibea Aubl. (Malvac.).

*Gerbera L. ex Cass. mut. Spreng. Compositae. 70 Afr., Madag., As., Indonesia (Bali).

Gerberia L. ex Cass. = praec.

Gerberia Scop. = Quararibea Aubl. (Malvac.).

Gerberia Stell. ex Choisy = Lagotis Gaertn. (Scrophulariac.).

Gerdaria Presl = Sopubia Buch.-Ham. (Scrophulariac.).

Germainia Bal. & Poitr. Gramineae. 3 Assam, SE. As.; 1 NE. Austr.

Germanea Lam. = Plectranthus L'Hérit. (Labiat.).

Germania Hook. f. = praec.

Germaria Presl = Pygeum Gaertn. = Lauro-Cerasus Duham. (Rosac.).

Gerocephalus F. Ritter = Espostoopsis F. Buxb. (Cactac.).

Gerontogea Cham. & Schlechtd. = Oldenlandia L. (Rubiac.).

Geropogon L. = Tragopogon L. (Compos.).

Gerostemum Steud. = Gonostemum Haw. = Stapelia L. (Asclepiadac.).

Gerrardanthus Harv. ex Hook f. Cucurbitaceae. 5 trop. & S. Afr.

Gerrardiana Willis (sphalm.) = seq.

Gerrardina Oliver. Flacourtiaceae(?). 2 E. trop. & S. Afr.

GERSINIA

Gersinia Néraud = Bulbophyllum Thou. (Orchidac.).

Gertrudia K. Schum. = Ryparosa Bl. (Flacourtiac.).

Geruma Forsk. ? Aïzoaceae. 1 Arabia.

Gervasia Rafin. = Poterium L. (Rosac.).

Geryonia Schrank ex Hoppe = Saxifraga L. (Saxifragac.).

Gesnera Adans. = Gesneria L. (Gesneriac.).

Gesnera Mart. = Rechsteineria Regel (Gesneriac.).

Gesneria L. Gesneriaceae. 50 trop. Am., W.I.

*Gesneriaceae Dum. Dicots. 120/2000, mostly trop. & subtrop. Herbaceous or slightly woody, rarely large shrubs or root-climbers; often epiphytes. Ls. usu. opp., more rarely whorled or alt., simple, ent. or toothed, rarely pinnatisect, exstip., some (esp. of epiphytes) thick with well-marked hypoderm. Some tuberous, e.g. *Sinningia* (the 'gloxinia' of greenhouses), *Rechsteineria*; others reproducing by scale-covered catkin-like stolons (*Smithiantha, Kohleria, Achimenes*, etc.). In nearly all Old World genera the cotyledons become unequal after germination, and in the unifoliate species of *Streptocarpus* and in *Monophyllaea* the enlarged cotyledon is the only leaf of the adult plant.

Fls. usu. cymose, sometimes sol. or rac., nearly always ⚥ and zygo. K 5, sometimes divided to base, sometimes tubular and shortly lobed. C (5), usu. 2-lipped (± regular in *Ramonda, Conandron*, etc.), lobes imbr., the lateral lobes overlapping the upper. A usu. either 2 or 4 didynamous, rarely 5. Stds. often found. Disk annular or cupular or one-sided or represented by up to 5 discrete glands. G sup. or ± inf., 1-loc., with 2 parietal bifid plac., which sometimes meet in the middle so that it becomes biloc.; ovules ∞, anatr.; style simple; stigma simple or variously bilobed. Fr. either a capsule usu. splitting loculic., sometimes at length 4-valved, rarely septicidal or a pyxis, or a hard or soft fleshy berry. Seeds ∞, small, with retic. or spiral markings, sometimes (tribe *Trichosporeae*) with hair-like appendages; endosp. slight or rarely none; funicle sometimes forming food-body. Embryo straight.

Fls. mostly protandrous and often fairly large (an inch or more long); many probably insect-pollinated, but others (esp. epiphytic *Aeschynanthus, Columnea*, etc.) have characteristic syndrome of arcuate red or orange-red corolla and copious nectar associated with bird-pollination.

None of the G. are economic plants; many are hothouse favourites and the African Violet (*Saintpaulia*) has become extremely popular as a house-plant, especially in America.

The family is very closely related to its neighbour in *Tubiflorae*: *Scrophulariaceae*. No single character provides a clear demarcation, but the unilocular ovary of G. generally serves to distinguish it from S. Where this breaks down the general affinities of the plants concerned usually solve the problem (e.g. bilocular ovaries may be found in species clearly belonging to *Streptocarpus, Petrocosmea* and *Saintpaulia*), but a few genera such as *Rehmannia* are still of uncertain family position.

Classification and chief genera (for Old World genera cf. Burtt in *Notes Bot. Gard. Edinb.* **24**: 205–20, 1963):

    I. **Cyrtandroïdeae.** Cotyledons becoming unequal after germination; ovary always superior; infl. various, usually cymose with 2 flowers at each dichotomy, never a simple raceme; disk annular, rarely divided or split on one side. Confined to Old World except for 3 species of *Rhynchoglossum* (KLUGIËAE).

Tribes: TRICHOSPOREAE (*Aeschynanthus, Agalmyla, Lysionotus*); DIDYMOCARPEAE (*Didymocarpus, Streptocarpus, Chirita, Didissandra, Ramonda, Haberlea, Saintpaulia, Boea, Briggsia, Oreocharis, Corallodiscus*); LOXONIËAE (*Loxonia, Stauranthera*); CYRTANDREAE (*Cyrtandra, Rhynchotechum*); KLUGIËAE (*Rhynchoglossum, Monophyllaea, Epithema*).

II. **Gesnerioïdeae.** Cotyledons remaining equal after germination; ovary superior or more or less inferior; infl. often racemose; disk annular or more often represented by 1–5 separate glands, sometimes absent. Confined to the New World and SE. Australasia.

Tribes: (*a*) Ovary superior: COLUMNEËAE (*Columnea, Alloplectus, Episcia, Codonanthe, Hypocyrta, Drymonia, Nautilocalyx, Chrysothemis, Nematanthus*); BESLERIËAE (*Besleria*); ANETANTHEAE (*Anetanthus*); CORONANTHEREAE (*Coronanthera, Rhabdothamnus*); MITRARIËAE (*Mitraria, Sarmienta, Fieldia, Asteranthera*).

(*b*) Ovary ± inferior: BELLONIËAE (*Bellonia, Phinaea*); GLOXINIËAE (*Gloxinia, Achimenes, Smithiantha*); KOHLERIËAE (*Kohleria, Diastema*); GESNERIËAE (*Gesneria, Rhytidophyllum*); SINNINGIËAE (*Sinningia, Rechsteineria*); SOLENOPHOREAE (*Solenophora*).

Gesnouinia Gaudich. Urticaceae. 1 Canaries.

Gesnouisia Steud. = praec.

Gessneria Dum. = Gesneria L. (Gesneriac.).

Gestroa Becc. = Erythrospermum Lam. (Flacourtiac.).

Gethosyne Salisb. = Asphodelus L. (Liliac.).

Gethyllidaceae J. G. Agardh = Amaryllidaceae–Zephyrantheae (Pax) Hutch.

Gethyllis L. Amaryllidaceae. 20 S. Afr. Some ed. fr.

Gethyonis P. & K. = Getuonis Rafin. = Allium L. (Alliac.).

Gethyra Salisb. = Renealmia L. f. (Zingiberac.).

Gethyum Phil. Alliaceae. 1 Chile.

Getonia Banks & Soland. = Cyrtandra J. R. & G. Forst. (Gesneriac.).

Getonia Roxb. Combretaceae. 1 Indomal. Like *Bucida*, but ls. opp. Climber.

Getuonis Rafin. = Allium L. (Alliac.).

Geum L. Rosaceae. 40 N. & S. temp., arctic. *G. rivale* L. has a thick rhizome and large protog. fls.; *G. urbanum* L. has smaller nearly homogamous fls. Both, with many others, have a hook on each achene aiding distr. The style in a newly opened fl. has a Z-like kink in it. The lower half of this after fert. gets larger and more woody, while the upper drops off.

Geum Mill. = Saxifraga L. (Saxifragac.).

Geuncus Rafin. = praec.

Geunsia Blume. Verbenaceae. 18 Malaysia.

Geunsia Moç. & Sessé = Calandrinia Kunth (Portulacac.).

Geunsia Neck. = Hypoëstes Soland. (Acanthac.).

Geunsia Rafin. = Geum L. (Rosac.).

Geunzia Neck. = Samyda Jacq. (Flacourtiac.).

× **Geversia** Dostál. Rosaceae. Gen. hybr. (Geum × Sieversia).

Gevuina Molina. Proteaceae. 3 New Guinea, Queensl., Chile. Ed. nut.

GHAESEMBILLA

Ghaesembilla auctt. = seq.

Ghesaembilla Adans. = Embelia Burm. f. (Myrsinac.).

Ghiesbrechtia Lindl. = Ghiesbreghtia A. Rich. & Gal. = Calanthe R.Br. (Orchidac.).

Ghiesbreghtia A. Gray = Eremogeton Standl. & L. O. Williams (Scrophu-
Ghiesbreghtia A. Rich. & Gal. = Calanthe R.Br. (Orchidac.).   [lariac.).

Ghikaea Volkens & Schweinf. Scrophulariaceae. 1 trop. Afr.

Ghinia Bub. = Cardamine L. (Crucif.).

Ghinia Schreb. Verbenaceae. 7 C. & S. Am., W.I.

Giadotrum Pichon. Apocynaceae. 2–3 Indoch., Malay Penins., ?Borneo.

Gibasis Rafin. = Tradescantia L. (Commelinac.).

Gibbaeum Haw. Aïzoaceae. 30 S. Afr.

Gibbaria Cass. Compositae. 2 S. Afr.

Gibbesia Small (~ Siphonychia Torr. & Gray, Paronychia Adans.). Caryophyllaceae. 1 E. N. Am.

Gibbsia Rendle. Urticaceae. 2 W. New Guinea.

Gibsonia Stocks = Calligonum L. (Polygonac.).

× **Giddingsara** hort. (v) = × Onoara hort. (Orchidac.).

Gieseckia Reichb. = seq.

Giesekia Agardh = Gisekia L. (Aïzoac.).

Giesleria Regel = Isoloma Decne (Gesneriac.).

× **Giflifa** Chrtek & Holub. Compositae. Gen. hybr. (Gifola × Oglifa).

Gifola Cass. = Filago L. (Compos.).

Gifolaria Pomel = praec.

Gigachilon Seidl = Triticum L. (Gramin.).

Gigalobium P.Br. = Entada Adans. (Legumin.).

Gigantabies 'Senilis' [Nelson] = Sequoia Endl. + Sequoiadendron Buchholz (Taxodiac.).

Giganthemum Welw. = Camoënsia Welw. ex Benth. & Hook. f. (Legumin.).

Gigantochloa Kurz. Gramineae. 20 Indomal. Giant bamboos, used in Java, etc., for building.

Gigasiphon Drake del Castillo. Leguminosae. 1 trop. Afr., 1 Madag., 3 Philipp. Is., Timor, New Guinea.

Gigliolia Barb. Rodr. = Octomeria R.Br. (Orchidac.).

Gigliolia Becc. Palmae. 2 Borneo.

Gigotorcya Buc'hoz = fol. Artocarpi(?) (Morac.) + fruct. Pithecellobii Mart. vel aff. (Legumin.).

Gijefa (M. Roem.) P. & K. = Kedrostis Medik. (Cucurbitac.).

Gilberta Turcz. = Myriocephalus Benth. (Compos.).

Gilbertiella Boutique. Annonaceae. 1 trop. Afr.

Gilbertiodendron J. Léonard. Leguminosae. 25 trop. W. Afr.

Gilesia F. Muell. = Hermannia L. (Sterculiac.).

Gilgia Pax = Glossonema Decne (Asclepiadac.).

Gilgiochloa Pilger. Gramineae. 2 trop. E. Afr.

Gilgiodaphne Domke = Synandrodaphne Gilg (Thymelaeac.).

Gilia Ruiz & Pav. Polemoniaceae. 120 temp. & subtrop. Am.

Giliastrum (A. Brand) Rydb. (~ Gilia Ruiz & Pav.). Polemoniaceae. 2 SW. U.S., Mex.

Giliberta St-Lag. = seq.
Gilibertia J. F. Gmel. = Quivisia Comm. ex Juss. (Meliac.).
Gilibertia Ruiz & Pav. Araliaceae. 60 trop. & E. As., trop. Am.
Gilipus Rafin. = Cephalanthus L. (Rubiac.).
Gillbeea F. Muell. Cunoniaceae. 2 New Guinea, NE. Queensl.
Gillena Adans. = Clethra L. (Clethrac.).
Gillenia Moench. Rosaceae. 2 N. Am.
Gillenia Steud. = Gillena Adans. = Clethra L. (Clethrac.).
Gillespiea A. C. Smith. Rubiaceae. 1 Fiji.
Gilletiella De Wild. & Durand. Mendonciaceae. 1 Congo, Angola.
Gilletiodendron Vermoesen. Leguminosae. 7 trop. Afr.
Gillettia Rendle = Anthericopsis Engl. (Commelinac.).
Gillia Endl. (sphalm.) = Gilia Ruiz & Pav. (Polemoniac.).
Gilliesia Lindl. Alliaceae. 3 Chile.
Gilliesi[ac]eae Lindl. = Alliaceae–Gilliesieae (Lindl.).
Gillietiella auctt. = Gilletiella De Wild. & Dur. (Mendonciac.).
Gillonia A. Juss. (sphalm.) = Gillenia Moench (Rosac.).
Gilmania Coville. Polygonaceae. 1 SW. U.S.
× **Gilmourara** hort. (v) = × Lewisara hort. (Orchidac.).
Gilruthia Ewart. Compositae. 1 W. Austr.
Gimbernatea Ruiz & Pav. = Chuncoa Pav. ex Juss. = Terminalia L. (Combretac.).
Ginalloa Korth. Viscaceae (~ Loranthaceae). 15 Indomal.
Ginalloaceae Van Tiegh. = Loranthaceae–Ginalloinae Engl.
Ginannia Bub. = Holcus L. (Gramin.).
Ginannia F. G. Dietr. = Gilibertia Ruiz & Pav. (Araliac.).
Ginannia Scop. = Palovea Aubl. (Legumin.).
Gingidium J. R. & G. Forst. Umbelliferae. 5 N.Z.; ?2 Austr.
Gingidium Hill = Ammi L. (Umbellif.).
Gingidium F. Muell. = Aciphylla J. R. & G. Forst. (Umbellif.).
Ginginsia DC. = Pharnaceum L. (Aïzoac.).
Gingko, Gingkyo, Ginkgyo, Ginkyo auctt. = Ginkgo L. (Ginkgoac.).
Gingkyoaceae, Ginkyoaceae auctt. = Ginkgoaceae Engl.
Ginkgo L. Ginkgoaceae. 1, *G. biloba* L., the maidenhair-tree, perhaps found wild in E. China (Chekiang). It is valued for its ed. seed, after removal of the nauseous flesh; the seed also yields an oil. The tree is often, but erroneously, stated to be regarded as sacred. The timber is useful.
*****Ginkgoaceae** Engl. Gymnosp. (Ginkgoales). 1/1 E. As. A tree, reaching 30 m. Ls. deciduous in autumn, resembling the lfts. of *Adiantum* on a large scale, and very often with a deep median division, *forked* in venation (cf. ferns and cycads), scattered on long shoots, or crowded at the apex of short shoots, which sometimes elongate into long. Below the ls. on the short shoot are a few scale-ls. Fls. dioecious, in the axils of the uppermost scales or lowest green ls. on short shoots (position different from that usual in *Coniferae* with long and short shoots). ♂ a stalked central axis, bearing scattered, rather loosely disposed sta., each of which is a slender filament ending in an apical scale and two or more pollen-sacs with longitudinal opening. The pollen grain forms a rudimentary prothallus of a few cells, and the generative nuclei produce two large

GINKGOALES

spirally coiled spermatozoids (cf. cycads). The ♀ has the form of a long stalk with two term. elliptical ovules enclosed at the base by a collar-like envelope repres. a reduced carpellary l. Each ov. consists of a nucellus surrounded by one integument, which in the ripe seed forms a thick fleshy aril-like covering round a hard woody shell. In the mature ov. the greater part of the nucellus tissue is reduced to a thin papery layer enclosing a large embryo-sac with usually 2 archegonia. Fert. occurs either before or *after* the ovule has fallen from the tree. The embryo has 2 cots.

Ginkgo thus represents a very old type, with relationships to the *Cycadales* and the *Filicales*. Fossil species are found in the Carboniferous, Permian, Triassic, and Jurassic, and in the Tertiary of England.

Ginkgoales Engl. Gymnosp. An order containing the single fam. *Ginkgoaceae*.
Ginko Agardh = Ginkgo L. (Ginkgoac.).
Ginnania M. Roem. = Quivisia Comm. ex Juss. (Meliac.).
Ginora L. = seq.
Ginoria Jacq. Lythraceae. 14 Mex., W.I.
Ginoria [Moç. & Sessé ex] DC. = Heimia Link & Otto (Lythrac.).
Ginsa Steud. = seq.
Ginsen Adans. = Panax L. (Araliac.).
Ginseng Wood = Aralia L. (Araliac.).
Ginura Vidal = Gynura Cass. (Compos.).
Giorgiella De Wild. = Efulensia C. H. Wright = Deïdamia Nor. ex Thou. (Passiflorac.).
Giraldia Baroni = Atractylis L. (Compos.).
Giraldiella Damm. Liliaceae. 1 NW. China.
Girardinia Gaudich. Urticaceae. 8 trop. Afr., Madag., E. As., Indomal. Stinging hairs.
Gireoudia Klotzsch = Begonia L. (Begoniac.).
Girgensohnia Bunge. Chenopodiaceae. 5 W. & C. As.
Giroa Steud. (sphalm.) = Guioa Cav. (Sapindac.).
Gironniera Gaudich. Ulmaceae. 15 Indomal., Polynesia.
Girostachys Rafin. = Spiranthes Rich. (Orchidac.).
Girtaneria Rafin. = seq.
Girtanneria Neck. = Rhamnus L. (Rhamnac.).
Gisania Ehrenb. ex Moldenke = Chascanum E. Mey. (Verbenac.).
Gisechia L. = seq.
Giseckia Willd. = seq.
Gisekia L. Aïzoaceae. 5 trop. & S. Afr. to India, Céylon & Indoch.
Gisekiaceae Nak. = Aïzoaceae–Gisekieae (Moq.) K. Müll.
Gisopteris Bernh. = Lygodium Sw. (Schizaeac.).
Gissanthe Salisb. = Costus L. (Costac.).
Gissanthus P. & K. = Geissanthus Hook. f. (Myrsinac.).
Gissaspis P. & K. = Geissaspis Wight & Arn. (Legumin.).
Gissipium Medik. = Gossypium L. (Malvac.).
Gissois P. & K. = Geissois Labill. (Cunoniac.).
Gissolepis P. & K. = Geissolepis Robins. (Compos.).
Gissoloma P. & K. = Geissoloma Lindl. ex Kunth (Geissolomac.).
Gissomeria P. & K. = Geissomeria Lindl. (Acanthac.).

Gissonia Salisb. = Leucadendron R.Br. (Proteac.).
Gissopappus P. & K. = Geissopappus Benth. (Compos.).
Gissorhiza P. & K. = Geissorhiza Ker (Iridac.).
Gissospermum P. & K. = Geissospermum Allem. (Apocynac.).
Gitara Pax & K. Hoffm. Euphorbiaceae. 2 Panamá, Venez.
Githago Adans. = Agrostemma L. (Caryophyllac.).
Githopsis Nutt. Campanulaceae. 54 W. N. Am.
Gitonoplesium P. & K. = Geitonoplesium A. Cunn. (Philesiac.).
Giulianettia Rolfe. Orchidaceae. 7 New Guinea.
Givotia Griff. Euphorbiaceae. 1 NE. trop. Afr., 2 Madag., 1 S. India, Ceylon.
Gjellerupia Lauterbach. Opiliaceae. 1 New Guinea.
Glabraria L. (Brownlowia Roxb.). Tiliaceae. 25 SE. As., Malaysia (exc. Java
    & Lesser Sunda Is.), Solomon Is.
× Gladanthera Joan M. Wright. Iridaceae. Gen. hybr. (Acidanthera ×
    Gladiolus).
Gladiangis Thou. (uninom.) = Angraecum gladiifolium Thou. = Angraecum
    mauritianum (Poir.) Frapp. (Orchidac.).
Gladiolaceae Salisb. = Iridaceae–Ixieae Benth.
Gladiolimon Mobayen. Limoniaceae (~ Plumbaginaceae). 1 Afghan.
Gladiolus Gaertn. = Antholyza L. (Iridac.).
Gladiolus L. Iridaceae. 300 Canaries, Madeira, W. & C. Eur., Medit. to SW.
    & C. As.; trop. & S. Afr. Fls. often protandrous. Ls. isobilat.
Gladiopappus Humbert. Compositae. 1 Madag.
Glandiloba Rafin. = Eriochloa Kunth (Gramin.).
Glandonia Griseb. Malpighiaceae. 2 trop. S. Am.
Glandula Medik. = Astragalus L. (Legumin.).
Glandularia J. F. Gmel. (~ Verbena L.). Verbenaceae. 28 Am.
Glandulicactus Backeb. = Hamatocactus Britt. & Rose (Cactac.).
Glandulifera Dalla Torre & Harms = Glandulifolia Wendl. = Adenandra Willd.
    (Rutac.).
Glandulifera Frič = Coryphantha (Engelm.) Lem. (Cactac.).
Glandulifolia Wendl. = Adenandra Willd. (Rutac.).
Glans Gronov. = Balanites Del. (Balanitac.).
Glaphiria Spach = seq.
Glaphyria Jack = Vaccinium L. (Ericac.) + Decaspermum J. R. & G. Forst.
    (Myrtac.).
Glaphyropteridopsis Ching. Thelypteridaceae. 4 NE. India, China.
Glaphyropteris Presl = Thelypteris Schmid. (Thelypteridac.).
Glastaria Boiss. Cruciferae. 1 Syria, Iraq.
Glaucena Vitm. = Clausena Burm. f. (Rutac.).
Glaucidiaceae (Himmelb.) Tamura. Dicots. 1/1 Japan. Near Hydrastidaceae
    (s. str.), differing in vernation, vasculature of pedicel and receptacle, sepals 4,
    persistent, ovules with very thick integuments, carpels 2, developing into ∞-
    seeded quadrangular dehiscent follicles, seeds compressed and winged;
    chromosome number 20. Only genus: Glaucidium. (Tamura, Sci. Rep. Osaka
    Univ. 11: 120–1, 1963; Bot. Mag. Tokyo, 85: 40, 1972.)
Glaucidium Sieb. & Zucc. Glaucidiaceae. 1 Japan.
Glaucium Mill. Papaveraceae. 25 Eur., SW. & C. As.

GLAUCOCARPUM

Glaucocarpum Rollins. Cruciferae. 1 SW. U.S.

Glaucocochlearia (O. E. Schulz) Pobed. Cruciferae. 1 SW. Europe, Corsica, N. Italy, Istria.

Glaucosciadium B. L. Burtt & P. H. Davis. Umbelliferae. 1 S. As. Min., Cyprus.

Glaucothea O. F. Cook = Erythea S. Watson (Palm.).

Glaux Ehrh. (uninom.) = *Glaux maritima* L. (Primulac.).

Glaux Hill = Astragalus L. (Legumin.).

Glaux L. Primulaceae. 1 N. temp. coasts, *G. maritima* L., a halophyte with fleshy l. The seedling dies after producing in the axil of one cot. a hibernating shoot, with a root of its own. From this fresh plants arise veg., the process being repeated for several years before flowering. Runners with scale ls., in whose axils renewal-shoots form, appear before the flowering period. The fl. has no C, but a coloured K.

Glayphyria G. Don = Glaphyria Jack, *q.v.*

Glaziocharis Taub. ex Warm. Burmanniaceae. 1 SE. Brazil, 1 Japan.

Glaziophyton Franch. Gramineae. 1 Brazil.

Glaziostelma Fourn. Asclepiadaceae. 1 Brazil.

Glaziova Bur. Bignoniaceae. 1 Brazil. Tendrils with discs at tip (cf. *Urbanolophium* Melch. [Bignoniac.] and *Parthenocissus* Planch. [Vitidac.]).

Glaziova Mart. ex H. Wendl. = Microcoelum Burret & Potztal (Palm.).

Glaziovanthus G. M. Barroso. Compositae. 1 Brazil.

Glaziovia Benth. & Hook. f. = Glaziova Bur. (Bignoniac.).

Gleadovia Gamble & Prain. Orobanchaceae. 6 W. Himal., W. China.

Gleasonia Standley. Rubiaceae. 3 trop. S. Am.

Glebionis Cass. = Chrysanthemum L. (Compos.).

Glechoma L. Labiatae. 10–12 temp. Euras.

Glechon Spreng. Labiatae. 6 Brazil, Paraguay.

Glechonion St-Lag. = seq.

Glecoma L. = Glechoma L. (Labiat.).

Gleditschia Scop. = seq.

Gleditsia L. Leguminosae. 11 trop. & subtrop. Stems usu. with stout branched thorns (stem structures, arising in l. axils). The thorn comes from the uppermost of a series of supra-petiolar buds one above the other in the axil. No winter buds form, and the young apex of each twig dies off in winter, the next year's growth starting late. Some used for hedges; some useful timber.

Glehnia F. Schmidt. Umbelliferae. 1–2 NE. As., Pacif. N. Am.

Gleichenella Ching = Dicranopteris subg. Acropterygium Diels (Gleicheniac.).

*Gleichenia Sm. Gleicheniaceae. 10 S. Afr., Masc., Malaysia, Austr., N.Z. Leaf-segts. v. small.

Gleicheniaceae Gaudich. 5/160 trop., subtrop. & S. temp. Thicket-forming sunferns; rhiz. long-creeping; l.-axis of indefinite growth, apex resting during growth of each pair of branches; branches often pseudo-dichotomous due to dormancy of apices; sorus of 2–15 large sporangia, no indusium. (Holttum, *Fl. Males.* ser. II, 1, 1959). Genera: *Gleichenia, Sticherus, Diplopterygium, Dicranopteris, Stromatopteris.*

Gleicheniales. Filicidae. 1 fam. (also 1 fossil fam.).

Gleicheniastrum Presl = Gleichenia Sm. (Gleicheniac.).

Gleniea Willis (sphalm.) = seq.
Glenniea Hook. f. Sapindaceae. 1 Ceylon.
Glia Sond. = Annesorhiza Cham. & Schlechtd. (Umbellif.).
Glicirrhiza Nocca = Glycyrrhiza L. (Legumin.).
Glinaceae ('-oideae') Link = Aïzoaceae–Aïzoïnae K. Müll.
Glinus L. Aïzoaceae. 12 trop. & subtrop.
Gliricidia Kunth. Leguminosae. 10 trop. Am., W.I.
Glischrocaryon Endl. = Loudonia Lindl. (Haloragidac.).
Glischrocolla (Endl.) A. DC. Penaeaceae. 1 S. Afr.
Glischrothamnus Pilger. Aïzoaceae. 1 NE. Brazil.
Glissanthe Steud. = Gissanthe Salisb. = Costus L. (Costac.).
Globba L. (1754) (nomen) = Alpinia Roxb. (Zingiberac.).
Globba L. (1771). Zingiberaceae. 50 S. China, Indomal. There is a short K;
above this is the C tube, from the end of which spring 3 petals, a large labellum
and 2 stds., also the slightly petaloid fertile sta., projecting beyond which is the
style. The ovary is 1-loc. with parietal plac. The lower cymes are usu. replaced by
bulbils; the mass of one of these consists of a root, springing lat. from the axis.
Globbaria Rafin. = praec.
Globeria Rafin. = Liriope Lour. (Liliac.).
Globeris Rafin. = praec.
Globifera J. F. Gmel. = Micranthemum Michx (Scrophulariac.).
Globimetula Van Tiegh. Loranthaceae. 14 trop. Afr.
Globocarpus Caruel = Oenanthe L. (Umbellif.).
Globularia L. Globulariaceae. 28 C. Verde Is., Canaries, S. Eur., As. Min.
***Globulariaceae** DC. Dicots. 2/30 Eur., Medit., NE. Afr. Herbs or shrubs
with alt., exstip., simple ls. and heads or spikes of fls. with or without invol. of
bracts. Fl. ♂. K (5), persist.; C (5), median-zygomorphic; the upper lip of
2 petals is shorter than the 3-petalled lower lip; A 4, didynamous, epipetalous;
G 1-loc., with 1 pend. anatr. ov. Fr. a one-seeded nut, free in base of calyx;
embryo straight, in endosp. Genera: *Globularia* (*Lytanthus*), *Poskea* (*Cock-
burnia*). Closely related to *Scrophulariac.–Selagineae* (esp. *Lagotis*).
Globulariopsis Compton. Scrophulariaceae. 1 S. Afr.
Globulea Haw. = Crassula L. (Crassulac.).
Globulostylis Wernham. Rubiaceae. 3 trop. W. Afr.
Glocheria Pritz. (sphalm.) = Glockeria Nees (Acanthac.).
Glochidinopsis Steud. = Glochidionopsis Blume = seq.
***Glochidion** J. R. & G. Forst. Euphorbiaceae. 300 Madag. (few), trop. As. to
Queensl. & Polynes. (many), trop. Am. (few).
Glochidionopsis Blume = praec.
Glochidium Wittst. = praec.
Glochidocaryum W. T. Wang. Boraginaceae. 1 NW. China.
Glochidopleurum K.-Pol. = Bupleurum L. (Umbellif.).
Glochidotheca Fenzl = Caucalis L. (Umbellif.).
Glochisandra Wight = Glochidion J. R. & G. Forst. (Euphorbiac.).
Glockeria Nees. Acanthaceae. 10 Mexico, C. Am.
Gloeocarpus Radlk. Sapindaceae. 1 Philipp. Is.
Gloeospermum Triana & Planch. Violaceae. 12 C. & trop. S. Am. Exalb.
Gloiospermum Benth. & Hook. f. = praec.

GLOMERA

Glomera Blume. Orchidaceae. 40 Malaysia, Polynesia.
Glomeraria Cav. = Amaranthus L. (Amaranthac.).
Glomeropitcairnia Mez. Bromeliaceae. 2 Venezuela, W.I.
Gloneria André = Psychotria L. (Rubiac.).
Gloriosa L. Liliaceae. 5 trop. Afr., As. They climb by aid of the ls., whose tips twine like tendrils. Fl. pendulous, with spreading sta., and style sharply deflexed at the extreme base through more than a right angle, so as to project from the fl. ± horizontally.
Glosarithys Rizzini = Saglorithys Rizzini (Acanthac.).
Glosocomia D. Don = Codonopsis Wall. (Campanulac.).
Glossanthis P. Poljakov. Compositae. 3 C. As.
Glossanthus Klein ex Benth. = Rhynchoglossum Bl. (Gesneriac.).
Glossarion Maguire & Wurdack. Compositae. 1 Guiana.
Glossarrhen Mart. = Schweiggeria Spreng. (Violac.).
Glossaspis Spreng. = Habenaria Willd. (Orchidac.).
Glossidea Van Tiegh. = Loranthus L. (Loranthac.).
Glossocalyx Benth. Siparunaceae. 4 trop. W. Afr. Remarkable anisophylly, 1 l. of each pair reduced to midrib.
Glossocardia Cass. Compositae. 2 W. & C. India.
Glossocarya Wall. ex Griff. Verbenaceae. 9 Indomal., Austr.
Glossocentrum Crueg. = Miconia Ruiz & Pav. (Melastomatac.).
Glossochilus Nees. Acanthaceae. 2 S. Afr.
Glossocoma Endl. = Glossoma Schreb. = ?Votomita Aubl. (Memecylac.).
Glossocomia Reichb. = Glosocomia D. Don = Codonopsis Wall. (Campanulac.).
Glossodia R.Br. Orchidaceae. 5 Austr.
Glossodiscus Warb. ex Sleum. = Casearia Jacq. (Flacourtiac.).
Glossogyne Cass. Compositae. 8 S. China, SE. As., Indomal., Austr.
Glossolepis Gilg (~ Chytranthus Hook. f.). Sapindaceae. 2 trop. W. Afr.
Glossoloma Hanst. = Alloplectus Mart. (Gesneriac.).
Glossoma Schreb. = ?Votomita Aubl. (Memecylac.).
Glossonema Decne. Asclepiadaceae. 4–5 trop. Afr., As.
Glossopappus Kunze = Chrysanthemum L. (Compos.).
Glossopetalon A. Gray = Forsellesia Greene (Celastrac.).
Glossopetalum Benth. & Hook. f. = praec.
Glossopetalum Schreb. = Goupia Aubl. (Goupiac.).
Glossopholis Pierre. Menispermaceae. 4 trop. Afr.
Glossophyllum Fourr. = Ranunculus L. (Ranunculac.).
Glossopteris Rafin. = Phyllitis Ludw. (Aspleniac.).
Glossorhyncha Ridl. Orchidaceae. 70 Malaysia, Polynesia.
Glossoschima Walp. = Closaschima Korth. = Laplacea Kunth (Theac.).
Glossospermum Wall. = Melochia L. (Sterculiac.).
Glossostelma Schlechter. Asclepiadaceae. 6 trop. & S. Afr.
Glossostemon Desf. Sterculiaceae. 1 Persia.
Glossostemum Steud. = praec.
Glossostephanus E. Mey. = Oncinema Arn. (Asclepiadac.).
★Glossostigma Wight & Arn. ex Arn. Scrophulariaceae. 5 India, Austr., N.Z.
Glossostylis Cham. & Schlechtd. = Melasma Bergius (Scrophulariac.).

Glossula Lindl. = Habenaria Willd. (Orchidac.).

Glossula Reichb. = Aristolochia L. (Aristolochiac.).

Glottes Medik. = Glottis Medik. = Astragalus L. (Legumin.).

Glottidium Desv. = Sesbania Pers. (Legumin.).

Glottiphyllum Haw. Aïzoaceae. 50 S. Afr.

Glottis Medik. = Astragalus L. (Legumin.).

× **Gloxinantha** R. E. Lee. Gesneriaceae. Gen. hybr. (Gloxinia × Smithiantha).

× **Gloxinera** hort. Gesneriaceae. Gen. hybr. (Rechsteineria [Gesneria hort.] × Sinningia [Gloxinia hort.]).

Gloxinia L'Hérit. Gesneriaceae. 6 trop. Am.

Gloxinia hort. = Sinningia Nees (Gesneriac.).

Gluema Aubrév. & Pellegr. (= ? Lecomtedoxa Pierre). Sapotaceae. 1 trop. W. Afr.

Glumaceae Benth. & Hook. f. A 'Series' of Monocots., comprising the fams. *Eriocaulac., Centrolepidac., Restionac., Cyperac.,* and *Gramineae.*

Glumicalyx Hiern. Scrophulariaceae. 1 S. Afr.

Glumiflorae C. A. Agardh. An order of Monocots., comprising (in Engler's system) the *Gramineae* and *Cyperaceae.*

Glumosia Herb. = Sisyrinchium L. (Iridac.).

Gluta L. Anacardiaceae. 1 Madag., 12 Indomal. The sap of *G. renghas* L. yields a good varnish.

Glutago Comm. ex Poir. (Oryctanthus Eichl.). Loranthaceae. 20 trop. Am.

Glutinaria Fabr. (uninom.) = Salvia L., sp. (Labiat.).

Glutinaria Rafin. = Salvia L. (Labiat.).

Glyaspermum Zoll. & Mor. = Pittosporum Banks ex Soland. (Pittosporac.).

Glycanthes Rafin. = Columnea L. (Gesneriac.).

Glyce Lindl. = Alyssum L. (Crucif.).

*****Glyceria** R.Br. Gramineae. 40 cosmop., esp. N. Am. Pasture grasses in wet meadows.

Glyceria Nutt. = Hydrocotyle L. (Hydrocotylac.).

Glycicarpus Benth. & Hook. f. = Glycycarpus Dalz. = Nothopegia Blume (Anacardiac.).

Glycideras DC. = Glycyderas Cass. = Psiadia Jacq. (Compos.).

Glycine L. = Apios + Wisteria + Abrus + Rhynchosia + Amphicarpaea + Pueraria + Fagelia (Legumin.).

*****Glycine** Willd. Leguminosae. 10 trop. & warm temp. Afr. & As. *G. max* (L.) Merr. (*G. hispida* [Moench] Maxim.) yields soja beans, eaten in Japan, etc., and used as green fodder. An oil is obtained from the seeds. (Cf. Hermann, Rev. Gen. Glycine, *U.S. Dept. Agric. Techn. Bull.* No. **1268**, 1962.)

Glycinopsis (DC.) Kuntze = Periandra Mart. (Legumin.).

Glyciphylla Rafin. = Chiogenes Salisb. ex Torr. = Gaultheria L. (Ericac.).

Glycosma Nutt. ex Torr. & Gray = Myrrhis Scop. (Umbellif.).

Glycosmis Corrêa. Rutaceae. 60 Indomal. Fr. ed.

Glycoxylon Ducke. Sapotaceae. 5 Brazil.

Glycoxylum Capelier ex Tul. = Dicoryphe Thou. (Hamamelidac.).

Glycycarpus Dalz. = Nothopegia Blume (Anacardiac.).

Glycydendron Ducke. Euphorbiaceae. 1 Amaz. Brazil.

Glycyderas Cass. = Psiadia Jacq. (Compos.).

GLYCYNODENDRON

Glycynodendron Pax & Hoffm. = Glycydendron Ducke (Euphorbiac.).

Glycyphylla Spach = Glyciphylla Rafin. = Chiogenes Salisb. ex Torr. = Gaultheria L. (Ericac.).

Glycyphylla Stev. = Astragalus L. (Legumin.).

Glycyrrhiza L. Leguminosae. 18 temp. & subtrop. Am., temp. Euras., N. Afr., SE. Austr. An extract of the rhiz. of *G. glabra* L. is Spanish liquorice.

Glycyrrhizopsis Boiss. Leguminosae. 2 SE. As. Min., Syria.

Glypha Lour. ex Endl. = Scaevola L. (Goodeniac.).

Glyphaea Hook. f. Tiliaceae. 2 trop. Afr.

Glyphia Cass. = Glycyderas Cass. = Psiadia Jacq. (Compos.).

Glyphosperma S. Wats. Liliaceae. 1 N. Mex.

Glyphospermum G. Don = Gentiana L. (Gentianac.).

Glyphostylus Gagnep. Euphorbiaceae. 1 Siam, Indoch. Near *Excoecaria*, but infl. greatly condensed.

Glyphotaenium J. Sm. Grammitidaceae. 4 trop. Am.

Glyptocarpa Hu (~Pyrenaria Bl.). Theaceae. 1 SW. China.

Glyptocaryopsis A. Brand. Boraginaceae. 6 Calif.

Glyptomenes Collins ex Rafin. = Asimina Adans. (Annonac.).

Glyptopetalum Thw. Celastraceae. 27 India, SE. As., Philipp. Is.

Glyptopleura Eaton. Compositae. 2 W. U.S.

Glyptospermae Vent. = Annonaceae Juss.

Glyptostrobus Endl. Taxodiaceae. 1 SE. China.

Glyscosmis D. Dietr. = Glycosmis Corrêa (Rutac.).

Gmelina L. Verbenaceae. 2 trop. Afr., Masc.; 33 E. As., Indomal., Austr.

Gnafalium Rafin. = Gnaphalium L. (Compos.).

Gnaphalion St-Lag. = praec.

Gnaphaliothamnus Kirp. Compositae. 1 Mex.

Gnaphalium Adans. = Otanthus Hoffmgg. & Link (Compos.).

Gnaphalium L. Compositae. 200 cosmop.

Gnaphalodes A. Gray. Compositae. 3 temp. Austr.

Gnaphalodes Mill. = Micropus L. (Compos.).

Gnaphalon Lowe = Phagnalon Cass. (Compos.).

Gnaphalopsis DC. = Hymenatherum Cass. (Compos.).

Gnemon Kuntze = Gnetum L. (Gnetac.).

Gneorum G. Don (sphalm.) = Cneorum L. (Cneorac.).

Gnephosis Cass. Compositae. 15 temp. Austr.

Gnetaceae Lindl. Gymnosp. (Gnetales). 1/30 trop. Most are climbing shrubs, a few erect shrubs or small trees. Ls. decussate, exstip., simple, evergr., leathery. Fls. dioec., in spikes which are frequently grouped into more complex infls. The spike bears decussate bracts, in whose axils are condensed partial infls. of a large number of fls. (cf. *Labiatae*), about 3–8 in the ♀, but more (up to 40) in the ♂. These fls. form whorls round the stem, and are intermingled with numerous hair-structures. At the top of each nodal group of the ♂ infl. in most is a single ring of ♀ fls., usu. with only 1 integument and infertile, sometimes with 2 or even 3 integuments and fertile. The ♂ has a tubular (2-leafed) P, from the top of which the axis projects; at the tip of the axis, right and left, are two sessile 1-loc. anthers. The ♀ has a tubular P like that of *Ephedra*, surrounding a single orthotr. erect ovule with two integuments; the inner of these projects at the apex of the fl. But there is much difference of opinion as to the morphology of these three envelopes. After fert. the P becomes fleshy, the outer integument woody, forming a drupe-like fr. Only genus: *Gnetum*. The *Gnetac.* are usu. associated with *Ephedrac.* and *Welwitschiac. (q.v.)* to form the *Gnetales.*

Gnetales Engl. Gymnospermae. 3 families, 3 genera, 70–75 spp. See above.

Gnetum L. Gnetaceae. 30 Indomal., Fiji, N. trop. S. Am., W. trop. Afr. *G. gnemon* L. (Malaysia) and other spp. are cult. for the ed. fr.

Gnidia L. Thymelaeaceae. 100 trop. & S. Afr., Madag., SW. Arabia, W. coast

Gnidiopsis Van Tiegh. = praec.   [India, Ceylon.

Gnidium G. Don (sphalm.) = Cnidium Cass. = Selinum L. (Umbellif.).

Gnomonia Lunell = Festuca L. (Gramin.).

Gnoteris Rafin. = Hyptis Jacq. (Labiat.).

Goadbyella R. S. Rogers. Orchidaceae. 1 W. Austr.

Gobara Wight & Arn. ex Voigt = Dysoxylum Bl. (Meliac.).

Gochnatia Kunth. Compositae. 2 Himal. to SE. As.; 64 S. U.S., Mex., W.I., S. Am.

Gocimeda Gandog. = Medicago L. (Legumin.).

Godefroya Gagnep. = Cleistanthus Hook. f. (Euphorbiac.).

Godetia Spach = Clarkia Pursh (Onagrac.).

Godia Steud. = Golia Adans. = Soldanella L. (Primulac.).

Godiaeum Boj. (sphalm.) = Codiaeum Juss. (Euphorbiac.).

Godinella T. Lestib. = Lysimachia L. (Primulac.).

Godmania Hemsl. Bignoniaceae. 1 Mex. to trop. S. Am.

Godoya Ruiz & Pav. Ochnaceae. 5 W. trop. S. Am. Ls. in 1 sp. pinnate!

Godwinia Seem. = Dracontium L. (Arac.).

Goebelia Bunge ex Boiss. (~ Sophora L.). Leguminosae. 3 As. Min. to C. As. & Baluchistan.

Goeldinia Huber. Lecythidaceae. 2 Brazil.

Goeppertia Griseb. = Bisgoeppertia Kuntze (Gentianac.).

Goeppertia Nees (1831) = Calathea G. F. W. Mey. + Maranta L. + Monotagma K. Schum. spp. (Marantac.).

Goeppertia Nees (1836) = Endlicheria Nees (Laurac.).

Goerziella Urb. (~ Amaranthus L.). Amaranthaceae. 1 Cuba.

Goethalsia Pittier. Tiliaceae. 1 C. Am., Colombia.

Goethartia Herzog. Urticaceae. 1 Bolivia.

Goethea Nees. Malvaceae. 2 Brazil. Several buds in each axil, some of which give rise years later to fls., borne on the old wood. Epicalyx brightly coloured. The C does not spread out, but the styles first emerge and afterwards the sta. (reverse of usual behaviour in *Malvaceae*). Honey is secreted at the base of the K. The styles are twice as numerous as the cpls. (cf. *Pavonia*).

Goetzea Reichb. = Rothia Pers. (Legumin.).

***Goetzea** Wydler. Goetzeaceae. 2 Puerto Rico, Haiti, S. Domingo.

Goetzeaceae Miers (corr.) ex Airy Shaw. Dicots. 5/7 Mex., W.I. Muchbranched shrubs or small trees, with alt. simple ent. coriac. exstip. closely parallel-ascending striate-nervose ls., midrib sometimes fusif.-thickened; young parts and fls. often densely ferrug.-velut. Fls. ☿, reg. or zygo., sol. and extra-axill. or in few-fld. axill. rac. or fasc. K (4–6), rarely segs. almost free, valv., persist.; C (4–6), infundib. or campan., lobes mostly short, equal or unequal, valv.; A 4–6, epipet., fil. long-exserted, equal or unequal, anth. versat., sagitt., dorsifixed near base; disk usu. large, fleshy, lobed; G̲ (2), 1–2-loc., with 2 collat. basal (rarely apical?) ov. per loc., and long simple style with capit. or bilobed stig. Fr. a fleshy or coriac. 1–2-seeded berry; seeds with plicate testa; endosp. o; embryo large, sometimes with 4 cots. Genera: *Goetzea, Espadaea, Henoonia, Coeloneurum, Lithophytum.* The fruit of *Espadaea* is ed., with a flavour of apricots. The affinities of this small relict group need further investigation; apparently near *Solanac., Convolvulac.* and *Verbenac.*, but the indumentum and leaf-shape and venation are remarkably Sapotaceous, and *Henoonia* was originally referred to that fam. Cf. the Sapotaceous facies of spp. of the Convolvulaceous genus *Erycibe.* The venation and indumentum are also similar to some *Thymelaeac.–Gonystyloïdeae,* a group possibly connected with *Sapotac.* through the genus *Microsemma.*

Goetzia Miers = Goetzea Wydler (Goetzeac.).

Goetziaceae Miers = Goetzeaceae Miers corr. Airy Shaw.

Gohoria Neck. = Ammi L. (Umbellif.).

Golaea Chiov. Acanthaceae. 1 Somalia.

Golatta Rafin. = Grafia Reichb. (Umbellif.).

***Goldbachia** DC. Cruciferae. 6 temp. As.

Goldbachia Trin. = Calamochloë Reichb. = Arundinella Raddi (Gramin.).

Goldenia Raeusch. (sphalm.) = Coldenia L. (Boraginac.).

Goldfussia Nees. Acanthaceae. 30 Himal. to Philipp. Is. & Java.

Goldmanella Greenman. Compositae. 1 Mexico.

Goldmania Greenman = praec.

Goldmania Rose ex Micheli. Leguminosae. 2 Mexico, 1 S. trop. S. Am.

Goldschmidtia Dammer = Dendrobium Sw. (Orchidac.).

Golenkinianthe K.-Pol. = Chaerophyllum L. (Umbellif.).

Golia Adans. = Soldanella L. (Primulac.).

Golionema S. Wats. Compositae. 1 Mexico.

Golowninia Maxim. = Crawfurdia Wall. = Gentiana L. (Gentianac.).

Golubiopsis Becc. ex Martelli = Gulubiopsis Becc. (Palm.).

Gomara Adans. = Crassula L. (Crassulac.).

Gomara Ruiz & Pav. Scrophulariaceae. 1 Peru.

Gomaria Spreng. = praec.

Gomarum Rafin. = Comarum L. = Potentilla L. (Rosac.).

Gomesa R.Br. Orchidaceae. 12 Brazil.

Gomesia La Llave (nomen). Compositae. Mexico. Quid?

Gomesia Spreng. = Gomesa R.Br. (Orchidac.).

Gomeza Lindl. = praec.

Gomezia Bartl. = praec.

Gomezia Benth. & Hook. f. = Gomesia La Llave (Compos.).

Gomezia Mutis = Gomozia Mutis ex L. f. = Nertera Banks & Soland. (Rubiac.).

Gomidesia Berg = Myrcia DC. (Myrtac.).

Gomidezia Benth. & Hook. f. = praec.

Gomortega Ruiz & Pav. Gomortegaceae. 1 Chile.

***Gomortegaceae** Reiche. Dicots. 1/1 temp. S. Am. Large tree, containing aromatic oil. Ls. simple, ent., opp., exstip., evergr. Fls. ☿, reg., bibracteolate, in axill. and term. rac. or pan. P (10–)9(–6), spirocyclic; A (11–)9(–2), spiro-cycl., with v. short fil., 2 basal capit. glands, and 2-locell. anth. dehiscing by valves (cf. *Laurac.*); $\overline{G}$ (2–3), with 1 pend. ov. per loc., and 2–3-lobed style; fr. drupaceous, with bony endocarp; seeds with much oily endosp. Only genus: *Gomortega*. Prob. related to *Atherospermatac.* and *Laurac.*

Gomoscypha P. & K. (sphalm.)= Gonioscypha Bak. = Tupistra Ker-Gawl. (Liliac).

Gomosia Lam. = Gomozia Mutis ex L. f. = Nertera Banks & Soland. (Rubiac.).

Gomotriche Turcz. (sphalm.) = Goniotriche Turcz. = Trichinium R.Br. (Amaranthac.).

Gomozia Mutis ex L. f. = Nertera Banks & Soland. (Rubiac.).

Gomphandra Wall. ex Lindl. (~ Stemonurus Bl.). Icacinaceae. 33 SE. As. to Solomon Is.

Gomphia Schreb. Ochnaceae. 30–35 S. trop. Afr., Madag.; 1 S. India, Ceylon, SE. As., S. China, W. Malaysia (exc. Java), Celebes.

Gomphichis Lindl. Orchidaceae. 25 mts. of S. Am.

Gomphiluma Baill. = Pouteria Aubl. (Sapotac.).

Gomphima Rafin. = Monochoria Presl (Pontederiac.).

Gomphipus (Rafin.) B. D. Jacks. = Calonyction Choisy (Convolvulac.).

Gomphocalyx Baker. Rubiaceae. 1 Madag.

Gomphocarpa van Royen (sphalm.) = Gomphandra Wall. ex Lindl. (Icacinac.).

Gomphocarpus R.Br. Asclepiadaceae. 50 trop. & S. Afr.

Gomphogyna P. & K. = seq.

Gomphogyne Griff. Cucurbitaceae. 2 E. Himal. to C. China & Indoch.

Gompholobium Sm. Leguminosae. 1 New Guinea, 24 Austr.

Gomphopetalum Turcz. = Angelica L. (Umbellif.).

Gomphopus P. & K. = Gomphipus (Rafin.) B. D. Jacks. = Calonyction Choisy (Convolvulac.).

Gomphosia Wedd. = Ferdinandusa Pohl (Rubiac.).

Gomphostema Hassk. = seq.

Gomphostemma Wall. ex Benth. Labiatae. 40 India, E. As., W. Malaysia. One of the very few rain-forest genera of *Lab.*

Gomphostigma Turcz. Buddlejaceae. 2 S. Afr.

Gomphostylis Rafin. = Zigadenus Miçhx (Liliac.).

Gomphostylis Wall. ex Lindl. = Coelogyne Lindl. (Orchidac.).

Gomphotis Rafin. (Thryptomene Endl.). Myrtaceae. 25 Austr., esp. W.

Gomphraena L. = seq.

Gomphrena L. Amaranthaceae. 100 C. & S. Am. Herbs with cymose heads of fls.; fls. ♀ with 5 hairy P-leaves and (5) sta.

Gomutus Corrêa = Arenga Labill. (Palm.).

Gonancylis Rafin. = Apios Medik. (Legumin.).

Gonantherus Rafin. = Osmorhiza Rafin. (Umbellif.).

Gonatandra Schlechtd. = Campelia Rich. (Commelinac.).

Gonatanthus Klotzsch. Araceae. 2 W. Himal. to Siam & SW. China. Pro-duces specialised branched stems with hook-bearing bulbils as in *Remusatia* Schott, *q.v.*

Gonatherus P. & K. = Gonantherus Rafin. = Osmorhiza Rafin. (Umbellif.).

Gonatia Nutt. ex DC. = Gratiola L. (Scrophulariac.).

GONATOCARPUS

Gonatocarpus Schreb. = Gonocarpus Thunb. = Haloragis J. R. & G. Forst. (Haloragidac.).

Gonatogyne Klotzsch ex Muell. Arg. = Savia Willd. (Euphorbiac.).

Gonatopus (Hook. f.) Engl. Araceae. 3 trop. E. Afr.

Gonatostemon Regel = Chirita Buch.-Ham. (Gesneriac.).

Gonatostylis Schlechter. Orchidaceae. 1 New Caled.

Gonema Rafin. = Ossaea DC. (Melastomatac.).

Gongora Ruiz & Pav. Orchidaceae. 20 C. & trop. S. Am.

Gongrodiscus Radlk. Sapindaceae. 2 New Caled.

Gongronema (Endl.) Decne. Asclepiadaceae. 15 palaeotrop.

Gongrospermum Radlk. Sapindaceae. 1 Philipp. Is.

Gongrothamnus Steetz. Compositae. 9 trop. Afr., Masc. Is.

Gongylocarpus Cham. & Schlechtd. Onagraceae. 2 Mex., C. Am. The ovary is eventually sunk in the swollen stem-tissue.

Gongylolepis R. H. Schomb. Compositae. 12 N. trop. S. Am.

Gongylosperma King & Gamble. Periplocaceae. 2 Malay Penins.

Gonianthes Blume = Burmannia L. (Burmanniac.).

Gonianthes A. Rich. = Portlandia P.Br. (Rubiac.).

Goniaticum Stokes = Polygonum L. (Polygonac.).

Gonioanthela Malme. Asclepiadaceae. 5 S. Brazil.

Goniocarpus Koen. = Gonocarpus Thunb. = Haloragis J. R. & G. Forst. (Haloragidac.).

Goniocaulon Cass. Compositae. 1 Indomal.

Goniocheton Blume = Dysoxylum Blume (Meliac.).

Goniochiton Reichb. = praec.

Goniocladus Burret. Palmae. 1 Fiji.

Goniodiscus Kuhlm. Celastraceae. 1 Brazil.

Goniodium Kunze ex Reichb. (nomen). Quid? (Thymelaeac.? Myrtac.?)

Goniogyna DC. = Crotalaria L. (Legumin.).

Goniogyne Benth. & Hook. f. = praec.

Goniolimon Boiss. Plumbaginaceae. 20 NW. Afr. to Mongolia.

Goniolobium G. Beck = Conringia Adans. (Crucif.).

Gonioma E. Mey. Apocynaceae. 1 S. Afr.

Goniophlebium (Bl.) Presl. Polypodiaceae. 20 trop. As. to Fiji.

Goniopogon Turcz. = Calotis R.Br. (Compos.).

Goniopteris Presl. Thelypteridaceae. 70 trop. Am. Branched unicellular hairs on ls.

Goniorrhachis Taub. Leguminosae. 1 SE. Brazil.

Gonioscheton G. Don = Goniocheton Blume = Dysoxylum Blume (Meliac.).

Gonioscypha Baker. Liliaceae. 2 Himalaya, Indoch.

Goniosperma Burret = Physokentia Becc. (Palm.).

Goniostachyum (Schau.) Small = Lippia L. (Verbenac.).

Goniostoma Elmer (sphalm.) = Geniostoma J. R. & G. Forst. (Loganiac.).

Goniothalamus Hook. f. & Thoms. Annonaceae. 115 Indomal.

Goniotriche Turcz. = Trichinium R.Br. (Amaranthac.).

Gonipia Rafin. = Centaurium Hill (Gentianac.).

Gonistum Rafin. = Piper L. (Piperac.).

Gonistylus Baill. = Gonystylus Teijsm. & Binnend. (Thymelaeac.).

Goniurus Presl = Pothos L. (Arac.).
Gonocalyx Planch. & Linden ex A. C. Smith. Ericaceae. 3 Colombia, W.I.
Gonocarpus Ham. = Combretum L. (Combretac.).
Gonocarpus Thunb. = Haloragis J. R. & G. Forst. (Haloragidac.).
Gonocaryum Miq. Icacinaceae. 9 Formosa, SE. As., Indomal.
Gonoceras P. & K. = Gonokeros Rafin. = Cephalaria Schrad. (Dipsacac.).
Gonocitrus Kurz = Merope M. Roem. = Atalantia Corrêa (Rutac.).
Gonocormus v. d. Bosch. Hymenophyllaceae. Few spp., Old World trop. & subtrop. Ls. proliferous by buds on rachis.
Gonocrypta Baillon. Periplocaceae. 1 Madag.
Gonocytisus Spach = Genista L. (Legumin.).
Gonogona Link = Goodyera R.Br. + Ludisia A. Rich. (Orchidac.).
Gonohoria G. Don (sphalm.) = Conohoria Aubl. = Rinorea Aubl. (Violac.).
Gonokeros Rafin. = Cephalaria Schrad. (Dipsacac.).
Gonolobium Hedw. f. = seq.
Gonolobus Michx. Asclepiadaceae. 200 N. & S. Am.
Gonoloma Rafin. = Cissus L. (Vitidac.).
Gononcus Rafin. = Polygonum L. (Polygonac.).
Gonondra Rafin. = Sophora L. (Legumin.).
Gonophylla Eckl. & Zeyh. ex Meissn. = Lachnaea L. (Thymelaeac.).
Gonoptera Turcz. = Bulnesia C. Gay (Zygophyllac.).
Gonopyros Rafin. = Diospyros L. (Ebenac.).
Gonopyrum Fisch. & Mey. ex C. A. Mey. (~ Polygonella Michx, Polygonum L.). Polygonaceae. 3 E. U.S.
Gonospermum Less. Compositae. 4 Canaries.
Gonostegia Turcz. = Pouzolzia Gaudich. (Urticac.).
Gonostemma Haw. corr. Spreng. = seq.
Gonostemon Haw. = Stapelia L. (Asclepiadac.).
Gonosuke Rafin. = Covellia Gasp. = Ficus L. (Morac.).
Gonotheca Blume ex DC. = Thecagonum Babu (Rubiac.).
Gonotheca Rafin. = Tetragonotheca L. (Compos.).
Gonsii Adans. = Adenanthera L. (Legumin.).
Gontarella Gilib. ex Steud. (sphalm.) = Fontanella Kluk = Isopyrum L.
Gontscharovia A. Borisova. Labiatae. 1 C. As.   [(Ranunculac.).
Gonufas Rafin. = Celosia L. (Amaranthac.).
Gonus Loud. = Brucea J. S. Mill. (Simaroubac.).
Gonyanera Korth. Rubiaceae. 1 Sumatra.
Gonyanthes Nees = Gonianthes Blume = Burmannia L. (Burmanniac.)
Gonyclisia Dulac = Kernera Medik. (Crucif.).
Gonypetalum Ule. Dichapetalaceae. 5 trop. S. Am.
Gonyphas P. & K. = Gonufas Rafin. = Celosia L. (Amaranthac.).
*****Gonystylaceae** Van Tiegh. = Thymelaeaceae–Gonystyloïdeae Domke.
Gonystylus Teijsm. & Binn. Thymelaeaceae. 30 Malaysia (exc. Lesser Sunda Is.), Solomon Is., Fiji. Trees with alt. entire exstip. ls. and cymes of ☿ reg. fls. K 5, valv.; C 0; disk conspic., of 7–40 deltoid or subul. cartilag. processes; A ∞; G̲ (5–3), each with 1 pend. ov. Fr. a thick-walled, tardily dehisc., 1–3-seeded caps.; seeds large, arillate. *G. bancanus* (Miq.) Kurz produces useful timber (*ramin*); grows in fresh-water peat-swamp forest.

GONZALAGUNIA

Gonzalagunia Ruiz & Pav. Rubiaceae. 35 trop. Am., W.I.

Gonzalea Pers. = praec.

Goodallia Benth. Thymelaeaceae. 1 Guiana.

Goodallia Bowd. ex Reichb. (nomen). Quid? (Crassulac.?)

Goodenia Sm. Goodeniaceae. 110 Austr. & Tasm.; 1 extending to Lesser Sunda Is., Java, Indoch. & Siam. Ovary 1-loc. above, often ± 2-loc. below. ***Goodeniaceae** R.Br. corr. Dum. Dicots. 14/300 chiefly Austr. (esp. SW.), a few N.Z., Polynes., and trop. coasts. Herbs and shrubs with rad. or alt. rarely opp. exstip. ls. and no latex. Fls. ⚥, zygo., sol. in the leaf-axils, or in cymes, racemes, or spikes. K usu. 5, small; C (5); A 5, alt. with the petals, with introrse sometimes syngenesious anthers; G (2), inf. or sup., 1- or 2-loc.; ovules 1, 2, or ∞ in each, usu. ascending, anatr.; style simple or 2–3-fid, with 'pollen-cup' close under the stigma. Into this the pollen is shed in the bud; it then closes up, leaving only a narrow opening. The style bends down to stand in the mouth of the almost horizontal fl., so that insect visitors come in contact with the cup and dust themselves with a little of the powdery pollen. As the stigmatic lobes grow up in the cup they keep forcing fresh pollen into the narrow slit, and finally emerge by it themselves and then receive the pollen of younger fls. from insect visitors. The mechanism should be carefully compared with that of *Campanulaceae* and *Compositae*. Fr. usu. caps., sometimes a nut or drupe. Embryo straight, in fleshy endosp. The *G.* are allied to *Campanulaceae*, differing chiefly in the absence of latex and the presence of the pollen-cup. They resemble *Gentianaceae* in a few points. Chief genera: *Goodenia, Leschenaultia, Scaevola, Dampiera*. Leaves in 1 sp. apparently whorled; ovules sometimes campylotr.

Goodenoughia A. Voss = Goodenia Sm. (Goodeniac.).

Goodenovi[ace]ae R.Br. = Goodeniaceae R.Br. corr. Dum.

Goodia Salisb. Leguminosae. 2 S. Austr.

Goodiera Koch = seq. [Polynes., temp. N. Am.

Goodyera R.Br. Orchidaceae. 40 temp. Euras., trop. As., Masc., Austr.,

Gooringia Williams = Arenaria L. (Caryophyllac.).

Gorceixia Baker. Compositae. 1 SE. Brazil.

*Gordonia Ellis. Theaceae. 40 China, Formosa, Indomal.; 1 SE. U.S. Sta. opp. to petals. Seeds winged. The bark of *G. lasianthus* L. (loblolly-bay, S. U.S.) is employed for tanning.

Gorenia Meissn. (sphalm.) = Govenia Lindl. (Orchidac.).

Gorgasia O. F. Cook (~ Roystonea O. F. Cook). Palmae. 2 W.I.

Gorgoglossum F. C. Lehm. = Sievekingia Reichb. f. (Orchidac.).

Gorgonidium Schott. Araceae. 1 Bolivia (not Malaysia!).

Gorinkia J. & C. Presl = Brassica L. + Conringia Fabr. (Crucif.).

Gormania Britton ex Britton & Rose (~ Sedum L.). Crassulaceae. 10 W. N. Am.

Gorodkovia Botsch. & Karav. Cruciferae. 1 NE. Siberia.

Gorostemum Steud. = Gonostemon Haw. = Stapelia L. (Asclepiadac.).

Gorskia Bolle = Guibourtia Benn. (Legumin.).

Gortera Hill = Gorteria L. (1759) (Compos.).

Gorteria L. (1759). Compositae. 4 S. Afr.

Gorteria L. (1763, p.p. et auctt. mult.) = Berkheya Ehrh., Gazania Gaertn., etc. (Compos.).

Gosela Choisy. Scrophulariaceae. 1 S. Afr.
Gossania Walp. = Gouania Jacq. (Rhamnac.).
Gossampinus Buch.-Ham. = Bombax + Ceiba (Bombacac.).
Gossampinus Buch.-Ham. emend. Schott & Endl. = Ceiba Mill. (Bombacac.).
Gossweilera S. Moore. Compositae. 2 Angola.
Gossweilerodendron Harms. Leguminosae. 1 trop. Afr.
Gossypianthus Hook. Amaranthaceae. 6 N. Am., Mex., W.I.
Gossypioïdes Skovsted ex J. B. Hutch. Malvaceae. 2 trop. Afr., Madag.
Gossypiospermum (Griseb.) Urb. (~Casearia Jacq.). Flacourtiaceae. 2 Cuba, trop. S. Am.
Gossypium L. Malvaceae. Trop. & subtrop. According to Hutchinson, Silow & Stephens (1947), 20 spp.; according to Prokhanov (also 1947), 67 spp. (incl. *Erioxylum* Rose & Standl., *Ingenhouzia* Moç. & Sessé ex DC., *Notoxylinon* Lewton, *Sturtia* R.Br., *Thurberia* A. Gray; 16 spp. in subgen. *Eugossypium*). Epicalyx of 3 ls. G (5). Loculic. caps. The seeds are covered with long hairs forming the material known as cotton. The cult. forms arise mainly from a few spp., including *G. barbadense* L. (trop. Am.), *G. hirsutum* L. (Am.), *G. arboreum* L. (Old World), and *G. herbaceum* L. (ditto). The cotton separates easily from the seed in the first sp., which is the Sea Island cotton of the U.S.; in Egypt, India, etc., the Old World spp. are most used. From the seeds an oil (cotton-seed oil) is obtained by crushing, and the oil-cake left behind is largely used for feeding cattle, etc. The fls. are visited by bees and (in Am.) by humming-birds.
Gothofreda Vent. = Oxypetalum R.Br. (Asclepiadac.).
Gouania Jacq. Rhamnaceae. 20 trop. & subtrop. Some have watch-spring tendrils. The stalks of some spp. contain saponin.
Gouarea Hedw. f. = Guarea Allem. (Meliac.).
Goudenia Vent. = Goodenia Sm. (Goodeniac.).
Goudotia Decne = Distichia Nees & Meyen (Juncac.).
Gouffeia Robill. & Cast. ex DC. = Arenaria L. (Caryophyllac.).
Goughia Wight = Daphniphyllum Blume (Daphniphyllac.).
Gouinia Fourn. Gramineae. 13 Mex. to Argent., W.I.
Goulardia Husnot = Agropyron J. Gaertn. (Gramin.).
Gouldia A. Gray. Rubiaceae. 1 New Guinea, 20 Hawaii.
Goupia Aubl. Goupiaceae. 3 Guiana, N. Brazil.
Goupiaceae Miers. Dicots. 1/3 trop. S. Am. Trees or shrubs with alt. ent. subtriplinerved transversely venose stip. ls. Fls. ⚥, reg., pedic., in axill. pedunc. false umbels. K (5), imbr., small; C 5, indupl.-valv., subulate, yellow with red base, the apical 1/3 sharply inflexed in bud and remaining geniculate or sigmoid at anthesis; A 5, on edge of disk, with v. short fil., and small ovoid anth. with apically pilose connective; disk cupular; G (5), with 5 short free styles, and ∞ basal-axile ascending ov.; fr. a small hard 2–3-loc. drupe, with ∞ seeds; endosp. fleshy. Only genus: *Goupia*. Affinities very obscure; usu. associated with *Celastrac.* Foliage and fr. somewhat Rhamnaceous; habit of *Casearia* (*Flacourtiac.*), *Glochidion* (*Euphorbiac.*), etc.
*****Gourliea** Gillies ex Hook. Leguminosae. 1 temp. S. Am. Pod ed.
Gourmania A. Chev. = Hibiscus L. (Malvac.).
Govana All. = Gouania Jacq. (Rhamnac.).

GOVANIA

Govania Wall. = Givotia Griff. (Euphorbiac.).

Govantesia Llanos = Champereia Griff. (Opiliac.).

Govenia Lindl. Orchidaceae. 20 C. & trop. S. Am., W.I.

Govindooia Wight = Tropidia Lindl. (Orchidac.).

Gowenia Lindl. = Govenia Lindl. (Orchidac.).

Goyazia Taub. Gesneriaceae. 1 Brazil.

Grabowskia Schlechtd. Solanaceae. 12 S. Am.

Grabuskia Rafin. = praec.

Gracilangis Thou. (uninom.) = *Angraecum gracile* Thou. = *Chamaeangis gracilis* (Thou.) Schltr. (Orchidac.).

Gracilicaulaceae Dulac = Illecebraceae R.Br., *q.v.*

Gracilea Koen. ex Rottl. = Melanocenchris Nees (Gramin.).

Gracilophylis Thou. (uninom.) = *Bulbophyllum gracile* Thou. (Orchidac.).

Graderia Benth. Scrophulariaceae. 3 trop. & S. Afr., Socotra.

Graeffea Seem. = Trichospermum Bl. (Tiliac.).

Graeffenrieda D. Dietr. = Graffenrieda DC. (Melastomatac.).

Graellsia Boiss. Cruciferae. 3 Persia, C. As., NW. Himal.

Graemia Hook. = Cephalophora Cav. (Compos.).

Graevia Neck. = Grewia L. (Tiliac.).

Graffenrieda DC. Melastomataceae. 40 trop. S. Am., W.I.

Grafia Reichb. = Pleurospermum Hoffm. (Umbellif.).

Grahamia Gill. Portulacaceae. 1 temp. S. Am.

Grahamia Spreng. = Cephalophora Cav. (Compos.).

Grajalesia Miranda. Nyctaginaceae. 1 Mex.

Gramen E. H. L. Krause = Festuca L. (Gramin.).

Gramen Seguier = Secale L. (Gramin.).

Gramerium Desv. = Digitaria Hall. (Gramin.).

Graminaceae Lindl. = Gramineae Juss.

Graminastrum E. H. L. Krause = Dissanthelium Trin. (Gramin.).

Gramineae Juss. (nom. altern. **Poaceae** Nash). Monocots. The grasses form one of the largest fams. of flowering plants, comprising about 620 genera and 10,000 species, and are widely dispersed in all regions of the world where plants can survive. Although so large a family, its members are well marked by distinguishing chars. and should not readily be confused with any other fam. They resemble the *Cyperaceae* superficially, being separated by usu. terete hollow jointed stems and distichous ls., and in the structure of the spikelets. The *Gramineae* are thought to be distantly related to the *Commelinaceae*.

Grasses are to be found in almost every type of habitat, from the Arctic to the Antarctic (Graham Land), often dominating the vegetation in savannas, prairies and steppes, occurring from sea level to high elevations, usu. in the open, less often in the shade, but freq. forming a part of forest undergrowth, in wet or dry places, from brackish situations on the coast to the fresh waters of lakes and rivers, and in the deserts of Africa, America, Asia and Australia.

Most grasses are herbaceous, with fibrous roots, ranging in height from less than 2 cm. to 6 m. in the savanna grasses of the tropics; rarely are they shrubs or trees, with woody stems to over 30 m. high, as in some Asiatic bamboos. Their duration extends from annuals and biennials to long-lived perennials, some bamboos having life-cycles of 15, 30, 60 or even 120 years. In habit they form loose to dense tufts or mats, the perennials developing veg. shoots (tillers or innovations) which grow up within the enveloping sheath, or burst through it at the base to give rise to stolons and rhizomes (both in *Cynodon dactylon*). The stems terminated by inflorescences are known as culms (haulms). They are simple or branched, erect, genic. or prostrate, usu. cylindrical, rarely flattened, jointed, hollowed or filled with soft tissue between the joints (nodes) and closed at the latter. The nodal tissue can continue growth on its lower side, thus enabling prostrate stems to regain the vertical position.

The leaves alternate in two rows on opp. sides of the stem, originating singly at the nodes; they are freq. crowded at the base; very exceptionally are they spirally arranged as in the Australian *Micraira*. They consist of sheath, ligule and blade.

The sheath (or vagina) forms the basal portion of the l., encircling the shoot or culm, with its margins free and overlapping or ± united to form a tube (*Bromus*, *Glyceria*, etc.); they are frequently swollen at the base to form the sheath-node.

At the junction of sheath and blade, on the inner or upper surface, a membranous outgrowth, the ligule, develops, which in many tropical genera is reduced to a hairy fringe; rarely is the ligule entirely suppressed (a few species of *Echinochloa*, etc.).

The blade (or lamina) is the upper portion of the l. It is usually long and narrow, mostly linear, rarely broad, lanceolate to oblong, ovate, or elliptic in some tropical forest grasses; it usually passes imperceptibly into the sheath or is sometimes constricted at the base; rarely is it provided with a petiole-like base, as in some bamboos, etc. Blades may be flat, convolute or involute, or even terete and solid. Their veins are normally parallel, rarely oblique from the mid-vein to the margin (*Pharus*, *Leptaspis*), and in some bamboos and certain forest and aquatic grasses are provided with transverse veinlets.

The inflorescences which terminate the culm and its branches are complex, consisting of usu. numerous basic units—the spikelets (really miniature spikes), rarely reduced to a single spikelet. They may be sparingly to profusely branched, forming dense to loose panicles, or the axis (rhachis) may be undivided and the spikelets sessile in spikes or pedicelled in racemes; the spikes or racemes being solitary, digitate or scattered along the primary axis.

The spikelets are typically composed of a series of alternating scales (modified reduced ls.), borne in two opposite rows, one above the other on a very short or minute axis (rhachilla). The two scales at the base are termed the lower and upper glumes; they vary considerably in size, shape and texture, in some genera being relatively small, in others as long as or longer than the rest of the spikelet which they help to protect (*Andropogoneae*, etc.). Above the glumes, alternating and sessile on the rhachilla, are one to many pairs of scales, the outer scale of each pair being the lemma (flowering glume, lower palea or valve) and the inner scale the palea (pale, upper palea, valvule), the two scales bearing a flower in their axil, the whole being termed a floret (false flower).

The lemmas resemble the glumes or differ from them in shape, size and texture. The paleas (modified prophylls) are generally two-nerved or two-keeled, and partially or wholly enclosed by the lemmas.

The flowers are usually bisexual, less often unisexual (*Maydeae*, *Phareae*,

505

GRAMINEAE

Olyreae, etc.), small and inconspicuous, protected by the lemma and palea or, when these are delicate, by the glumes; they consist of stamens and pistil, and of normally 2 or 3 minute fleshy or hyaline scales (lodicules) considered to represent a reduced perianth rather than a second palea. The stamens are hypogynous, 1 to 6, rarely more (numerous in *Pariana*), usually 3, with delicate filaments and 2-celled anthers; their cells open usually by a longitudinal slit and soon separate at the base so that the anthers appear versatile. The ovary is unilocular, with one anatropous, hemitropous or campylotropous ovule often adnate to the adaxial side of the carpel, the point or line of attachment being visible in the mature state as the hilum; styles usually 2, rarely 1 (*Nardus*, etc.) or 3 (some *Bambuseae*, etc.), terminating the ovary or arising laterally below an apical appendage (*Bromus*, etc.), free or united; stigmas generally plumose, sometimes only papillate, whitish or in some genera purple (*Molinia*, *Panicum*, etc.).

The fruit is 1-seeded, mostly a caryopsis with a thin pericarp adhering to the seed, rarely a nut or a berry, or utricle-like with a free membranous or gelatinous pericarp, with starchy endosperm; embryo varying in size from very small (*Festuceae* and most temperate tribes) to one third or half the length of the fruit (many tropical genera) to as long as the fruit (*Spartina*, etc.), placed on the abaxial side of the fruit; starch grains simple or compound.

The typical structure of the spikelets may be altered or modified in various ways. Thus the glumes and paleas and less often the lemmas may be so reduced in size (sometimes entirely suppressed) that the interpretation of the parts of the spikelets is not obvious. For example, in the *Paniceae*, the lower glume is usually minute in *Digitaria* and generally absent in *Paspalum*. A floret too may be reduced, sometimes by the loss of the sexual organs and of the lodicules, by a reduction in the size of or by suppression of the palea, while even the lemma may be very small and in some genera the whole floret absent and the spikelet consisting only of the glumes. Reduction of the lower or both florets is characteristic of the *Andropogoneae*.

The majority of grasses are chasmogamous, their florets opening for the exsertion of anthers and stigmas. When the flowers are completely developed and climatic conditions are favourable, the lower part or the whole of each lodicule becomes swollen with sap and exerts considerable pressure at the base of the lemma and palea, forcing them apart and thus permitting the stamens and stigmas to be exserted, the latter laterally or terminally. The stamens dehisce and the loose powdery pollen is spread by air currents. Some species are mainly self-fertile, others are mostly self-sterile and only perfect seed after cross-pollination. In quite a number of grasses, especially in relatively dry regions, cleistogamy occurs, in which the anthers are very small and pollination takes place within the closed florets. Sometimes whole inflorescences are enclosed within the leaf-sheath, all the spikelets being cleistogamous, while in the terminal exserted panicles of the same plant the spikelets may be chasmogamous. Apomixis is now known to occur in a large number of grasses.

In most grasses the 'seeds' are sufficiently light and small to be dispersed by the wind. They usually consist of the lemma and palea tightly enclosing the grain, or of the complete spikelet; only in a few genera, such as *Eragrostis* and *Sporobolus*, are the grains naked. The 'seeds' may be flattened, hollowed

GRAMINEAE

out or winged or provided with a hairy pappus, all assisting wind dispersal. In other 'seeds' the glumes or lemmas bear teeth, barbs, bristles or awns, by means of which they adhere to animals and become more widely dispersed than would otherwise be possible.

Grasses are the most useful of all flowering plants. They provide food in the form of cereals for man and beast and green herbage and dried fodder for domestic and wild animals; also a wide variety of valuable by-products such as fibres, paper, sugar, aromatic and edible oils, adhesives, plastics, starch, alcohol, beverages and liquors, and packing, thatching and building materials.

The classification of grasses into genera and larger groups has, in the past, been largely based on the structure and arrangement of their spikelets, as these organs show a wider range of visible distinguishing features than do other parts of the plant. Emphasis on such characteristics, however, led to some very unnatural groupings and, since 1930, increasing attention has been devoted to other aspects of the plants' structure. Among them have been: (1) the relative size and basic numbers (4, 5, 6, 7, 9, 10, 11, 12, 13, 17) of the chromosomes; (2) the anatomy of the leaf, particularly the arrangement and nature of the chlorenchymatous tissues, and the form, presence or absence of parenchymatous and mestome bundle sheaths; (3) the structure of the epidermis, especially the shape of the silica deposits, the type of micro and macro hairs, and the shape of the stomata; (4) the structure of the embryo; (5) the form of the first green leaf of the seedling, whether erect and linear or relatively broad and spreading; (6) the flower, especially the number, shape and size of the lodicules; (7) form of starch granules, whether simple or compound; and (8) physiological nature of the plant. These and other studies of the grasses have yielded much new evidence of relationships, resulting in more natural groupings, but further investigations are required before all genera can be satisfactorily placed in a new system. Between 50 and 60 tribes have been recognised and grouped into a varying number (2–12) of subfamilies, depending on which aspect of plant-structure is considered of major importance for purposes of classification. At present it seems preferable to group the tribes tentatively as follows:

Group I: 1, BAMBUSEAE (*Bambusa, Arthrostylidium, Arundinaria, Chusquea, Dendrocalamus, Oxytenanthera*).

II: 2, STREPTOCHAETEAE (*Streptochaeta*).

III: 3, ANOMOCHLOËAE (*Anomochloa*).

IV: 4, PHAREAE (*Pharus, Leptaspis*).

V: 5, OLYREAE (*Olyra, Diandrolyra, Lithachne, Mniochloa, Raddia*).

VI: 6, BUERGERSIOCHLOËAE (*Buergersiochloa*).

VII: 7, PARIANEAE (*Pariana*).

VIII: 8, STREPTOGYNEAE (*Streptogyna*).

IX: 9, CENTOTHECEAE (*Centotheca, Megastachya, Orthoclada, Zeugites*); 10, UNIOLEAE (*Uniola*); 11, DIARRHENEAE (*Diarrhena*); 12, BRYLKINEËAE (*Brylkinea*).

X: 13, ORYZEAE [ZIZANIEAE] (*Oryza, Hydrochloa, Leersia, Luziola, Zizania, Zizaniopsis*); 14, EHRHARTEAE (*Ehrharta, Microlaena, Tetrarrhena*); 15, PHYLLO-RHACHIDEAE (*Phyllorhachis, Humbertochloa*).

XI: 16, AMPELODESMEAE (*Ampelodesmos*); 17, PHAENOSPERMATEAE (*Phaenosperma*); 18, STIPEAE (*Stipa, Nassella, Oryzopsis, Piptochaetium*); 19, BRACHY-ELYTREAE (*Brachyelytrum, Podophorus*).

507

GRAMINISATIS

XII: 20, ARUNDINEAE (*Arundo, Phragmites*); 21, THYSANOLAENEAE (*Thysanolaena*).

XIII: 22, DANTHONIËAE (*Danthonia, Eriachne, Molinia, Pentaschistis, Triodia*); 23, ELYTROPHOREAE (*Elytrophorus*); 24, MICRAIREAE (*Micraira*); 25, ARISTIDEAE (*Aristida*); 26, PEROTIDEAE (*Perotis*); 27, ARUNDINELLEAE (*Arundinella, Danthoniopsis, Loudetia, Trichopteryx, Tristachya*); 28, GARNOTIËAE (*Garnotia*).

XIV: 29, ISACHNEAE (*Isachne, Heteranthoecia, Limnopoa*); 30, HUBBARDIËAE (*Hubbardia*).

XV: 31, BRACHYPODIËAE (*Brachypodium*); 32, BROMEAE (*Bromus, Boissiera*); 33, TRITICEAE [HORDEËAE] (*Triticum, Aegilops, Agropyron, Elymus, Hordeum, Secale*); 34, GLYCERIËAE (*Glyceria, Pleuropogon*); 35, FESTUCEAE [POËAE] (*Festuca, Briza, Dactylis, Lolium, Poa, Puccinellia*); 36, AVENEAE (*Avena, Aira, Deschampsia, Holcus, Koeleria*); 37, PHALARIDEAE (*Phalaris*); 38, AGROSTIDEAE (*Agrostis, Alopecurus, Calamagrostis, Gastridium, Mibora, Phleum*); 39, MONERMEAE (*Monerma, Parapholis, Pholiurus*); 40, COLEANTHEAE (*Coleanthus*); 41, MILIËAE (*Milium, Zingeria*); 42, SESLERIËAE (*Sesleria, Oreochloa*); 43, MELICEAE (*Melica, Lycochloa, Streblochaete*).

XVI: 44, NARDEAE (*Nardus*).

XVII: 45, LYGEËAE (*Lygeum*).

XVIII: 46, PAPPOPHOREAE (*Pappophorum, Cottea, Enneapogon, Schmidtia*); 47, ORCUTTIËAE (*Orcuttia, Neostapfia*); 48, AELUROPODEAE (*Aeluropus*); 49, POMMEREULLEAE (*Pommereulla*); 50, ERAGROSTIDEAE (*Eragrostis, Dactyloctenium, Eleusine, Leptochloa, Tridens*); 51, CHLORIDEAE (*Chloris, Botelua, Cynodon, Ctenium, Microchloa*); 52, SPARTINEAE (*Spartina*); 53, SPOROBOLEAE (*Sporobolus, Calamovilfa, Crypsis, Lycurus, Muhlenbergia, Urochondra*); 54, LEPTUREAE (*Lepturus*); 55, ZOYSIËAE (*Zoysia, Latipes, Monelytrum, Tragus*); 56, SPHAEROCARYEAE (*Sphaerocaryum*).

XIX: 57, PANICEAE [MELINIDEAE] (*Panicum, Cenchrus, Digitaria, Paspalum, Pennisetum*); 58, ANDROPOGONEAE (*Andropogon, Cymbopogon, Sorghum, Themeda, Vetiveria*); 59, MAYDEAE [TRIPSACEAE] (*Zea, Chionachne, Coix, Polytoca, Tripsacum*).

Graminisatis Thou. (uninom.) = *Satyrium gramineum* Thou. = *Cynorkis graminea* (Thou.) Schltr. (Orchidac.).

Grammadenia Benth. Myrsinaceae. 15 trop. Am., W.I.

Grammangis Reichb. f. Orchidaceae. 1 Madag.

Grammanthes DC. = Vauanthes Haw. (Crassulac.).

Grammartheon Reichb. = seq.

Grammarthron Cass. = Doronicum L. (Compos.).

Grammatocarpus Presl = Scyphanthus D. Don (Loasac.).

× **Grammatocymbidium** hort. Orchidaceae. Gen. hybr. (iii) (Cymbidium × Grammatophyllum).

Grammatophyllum Blume. Orchidaceae. 10 Malaysia, Polynesia.

Grammatopteridium v. Ald. v. Ros. Polypodiaceae. 2 Malaysia.

Grammatopteris v. Ald. v. Ros. (1922; non Renault 1891—gen. foss.) = praec.

Grammatosorus Regel = Tectaria Cav. (Aspidiac.).

Grammatotheca C. Presl. Campanulaceae. 1 S. Afr., 1 Austr.

Grammeionium Reichb. = Viola L. (Violac.).

Grammica Lour. = Cuscuta L. (Cuscutac.).

Grammitidaceae Presl. Polypodiales. Small epiphytes, esp. in cloud-forests on trop. mts.; fronds ± hairy; veins free; sori exindusiate; spores trilete. Formerly included in *Polypodium* L. (Holttum, *J. Linn. Soc., Bot.* **53**: 128, 1947; Wilson, *Contr. Gray Herb.* **185**: 97–127, 1959). Chief genera: *Grammitis, Xiphopteris, Ctenopteris, Prosaptia* (some authors include all in *Grammitis*).

Grammitis Sw. (*s.str.*). Grammitidaceae. 150 pantrop. & Austral. Ls. simple.

Grammocarpus Schur = Trigonella L. (Legumin.).

Grammopetalum C. A. Mey. ex Meinsh. = Trinia Hoffm. (Umbellif.).

Grammosciadium DC. Umbelliferae. 6 E. Medit.

Grammosperma O. E. Schulz. Cruciferae. 1 Patagonia.

Granadilla Mill. = Passiflora L. (Passiflorac.).

Granat[ac]eae D. Don = Punicaceae Horan.

Granatum Kuntze = Carapa Aubl. (Meliac.).

Granatum St-Lag. = Punica L. (Punicac.).

Grandidiera Jaub. Flacourtiaceae. 1 trop. E. Afr.

Grandiera Lefeb. ex Baill. = Sindora Miq. (Legumin.).

Grangea Adans. Compositae. 6 trop. Afr., Madag., trop. As.

Grangeopsis Humbert. Compositae. 1 Madag.

Grangeria Comm. ex Juss. Chrysobalanaceae. 2 Madag., Mauritius.

Graniera Mandon & Wedd. ex Benth. & Hook. f. = Abatia Ruiz & Pav. (Flacourtiac.).

Grantia Boiss. Compositae. 6 Arabia, Persia.

Grantia Griff. ex Voigt = Wolffia Horkel ex Schleid. (Lemnac.).

Graphandra Imlay. Acanthaceae. 1 Siam.

Graphephorum Desv. = Trisetum Pers. (Gramin.).

Graphiosa Alef. = Lathyrus L. (Legumin.).

Graphistemma Champ. ex Benth. Asclepiadaceae. 1 S. China (Hongkong).

Graphophorum P. & K. = Graphephorum Desv. = Trisetum Pers. (Gramin.).

Graphorchis Thou. = seq.

Graphorkis Thou. Orchidaceae. 5 trop. Afr., Masc.

Graptopetalum Rose (~ Echeveria DC.). Crassulaceae. 10 SW. U.S. to Mex.

Graptophyllum Nees. Acanthaceae. 10 trop. W. Afr., New Guinea, Austr., Polynesia. Some have prettily marked ls.

× **Graptoveria** Rowley. Crassulaceae. Gen. hybr. (Echeveria × Graptopetalum).

Grastidium Blume = Gastridium Bl. (corr. Bl.) = Dendrobium Sw. (Orchidac.).

Gratiola L. Scrophulariaceae. 20 N. & S. temp. zones, trop. mts. Sta. 2. The dried plant of *G. officinalis* L. was formerly offic.

Gratwickia F. Muell. Compositae. 1 Austr.

Graumuellera Reichb. = Amphibolis C. Agardh (Cymodoceac.).

Gravenhorstia Nees = Lonchostoma Wikstr. (Bruniac.).

Gravesia Naud. Melastomataceae. 100 Madag. Crossed with *Cassebeeria* they give the fancy 'Bertolonias'.

Gravesiella A. & R. Fernandes. Melastomataceae. 1 trop. E. Afr.

Gravia Steud. = Grafia Reichb. = Pleurospermum Hoffm. (Umbellif.).

GRAVISIA

Gravisia Mez. Bromeliaceae. 7 trop. Am., W.I.

Graya Arn. ex Steud. = Andropogon L. (Gramin.).

Graya Endl. = Grayia Hook. & Arn. = Eremosemium Greene (Chenopodiac.).

Graya Nees ex Steud. = Isachne R.Br. (Gramin.).

Grayia Hook. & Arn. = Eremosemium Greene (Chenopodiac.).

Grazielia R. M. King & H. Rob. (~ Eupatorium L.). Compositae. 9 Brazil, Urug.

× **Greatwoodara** hort. (v) = × Kagawara hort. (Orchidac.).

Grecescua Gandog. = Epilobium L. (Onagrac.).

Greenea P. & K. = Greenia Nutt. = Thurbera Benth. (Gramin.).

Greenea Wight & Arn. Rubiaceae. 10 Indomal.

Greeneina Kuntze = Helicostylis Tréc. (Morac.).

Greenella A. Gray. Compositae. 3 SW. U.S., NW. Mex.

Greeneocharis Gürke & Harms (~ Cryptantha Lehm.). Boraginaceae. 2 Pacif. N. Am.

Greenia Nutt. = Thurberia Benth. (Gramin.).

Greeniopsis Merr. Rubiaceae. 6 Philipp. Is.

Greenmania Hieron. = Unxia L. f. (Compos.).

Greenmaniella Sharp. Compositae. 1 Mex.

× **Greenonium** Rowley. Crassulaceae. Gen. hybr. (Aeonium × Greenovia).

Greenovia Webb & Berth. Crassulaceae. 4 Canaries.

Greenwaya Giseke = Amomum L. (Zingiberac.).

Greenwayodendron Verdc. Annonaceae. 2 trop. E. Afr.

Greevesia F. Muell. = Pavonia Cav. (Malvac.).

Greggia Engelm. = Fallugia Endl. + Cowania D. Don (Rosac.).

Greggia Gaertn. = Eugenia L. (Myrtac.).

Greggia A. Gray = Parrasia Greene = Nerisyrenia Greene (Crucif.).

Gregia Carr. (sphalm.) = Greigia Regel (Bromeliac.).

Gregoria Duby = Dionysia Fenzl + Douglasia Lindl. (Primulac.).

Greigia Regel. Bromeliaceae. 18 C. Am. to Chile, Juan Fernandez.

Grenacheria Mez. Myrsinaceae. 10 Malaysia.

Greniera J. Gay = Arenaria L. (Caryophyllac.).

Grenvillea Sweet = Pelargonium L'Hérit. (Geraniac.).

Greslania Balansa. Gramineae. 4 New Caled.

Greuia Stokes = Grewia L. (Tiliac.).

Grevea Baill. Montiniaceae. 3 trop. E. Afr., Madag.

Grevellina Baill. = Turraea L. (Meliac.).

*****Grevillea** R.Br. Proteaceae. 190 E. Malaysia, New Hebrid., New Caled., Austr. Trees and shrubs with racemose infls., 2 fls. in each axil. The style projects from the bud as a long loop, the stigma being held by the P until the pollen is shed upon it. Then the style straightens out, and the pollen may be removed; presently the female stage supervenes. Some yield useful timber, and *G. robusta* A. Cunn. (silky oak) and other spp. are now extensively employed as shade and timber trees in Ceylon and elsewhere.

Grewia L. Tiliaceae. 150 Afr., As., Austr., esp. trop.

Grewiella Kuntze = seq.

Grewiopsis De Wild. et Durand = Desplastia Bocq. (Tiliac.).

Greyia Hook. & Harv. Greyiaceae. 3 SE. S. Afr.

*Greyiaceae (Gürke) Hutch. Dicots. 1/3 S. Afr. Shrubs or medium-sized trees with somewhat bare branches. Ls. alt., simple, slightly pelt., crenate-lobed and dentate (somewhat *Pelargonium*-like), subpalmatinerved at base, petiolate, exstip. but with expanded sheathing base congenitally united with the stem, forming a pseudo-cortex, decid.; young ls. and shoots whitish-tomentose or subglabrous and minutely glandular. Infl. rac., terminating branches. Fls. ⚥, slightly zygo., conspic., bract., protandr. K 5, small, imbr., v. shortly conn., persist. C 5, imbr., cuneate-obov., scarlet, ciliol., decid. Disk (or stds.) fleshy, extra-stam., but connected by 5 inter-stam. ridges with ovary, cupular, 10-lobed, lobes alt. with sta., capit.-gland. or lacerate-dent. at apex, producing copious nectar, ± persist. A 5 + 5, obdiplost., exserted, fil. subul., ± persist., anth. shortly ovoid. G̱ (5), oblong, deeply 5-lobed, 5-loc., ov. ∞ on slightly bifid parietal plac. in each loc.; style simple, slender, subul., with small shortly 5-lobed stig. Caps. chartac., septicid., with follicular loc.; seeds minute, with membr. testa, straight embr. and copious fleshy endosp. Only genus: *Greyia*. Affinities possibly Saxifragalean: cf. leaf-shape, venation, resinous glands, petiole, etc., of *Ribes* spp. (e.g. *R. pentlandii* Britton). Cf. also *Geraniac., Dirachmac., Melianthac.* The placentation is intrusive-parietal, the placentae eventually meeting and almost coalescing in the centre.

Grias L. Lecythidaceae. 15 Panamá to Peru, W.I. *G. cauliflora* L., anchovy pear, cult. in the W.I.

Grielum L. Neuradaceae. 6 S. Afr.

Griesebachia Endl. = Grisebachia Klotzsch (Ericac.).

Grieselinia Endl. = Griselinia Forst. f. (Cornac.).

Griffinia Ker-Gawl. Amaryllidaceae. 7 Brazil.

Griffithella (Tul.) Warming. Podostemaceae. 1 W. Ghats of India. Plants with the general veg. structure of *Dicraea*, but remarkable for the extraordinary polymorphism of their shoots, which may be cup- or wineglass-shaped, creeping or erect, and of many different forms, shapes, and sizes (cf. Willis, *Ann. Perad.* 1: 364, 1902).

Griffithia J. M. Black = Helipterum DC. (Compos.).

Griffithia Maingay ex King = Griffithianthus Merr. = Enicosanthum Becc. (Annonac.).

Griffithia Wight & Arn. = Randia L. (Rubiac.).

Griffithianthus Merr. = Enicosanthum Becc. (Annonac.).

Griffithiella Prain in I.K. (sphalm.) = Griffithella (Tul.) Warming (Podostemac.).

Griffonia Baill. = Bandeiraea Welw. (Legumin.).

Griffonia Hook. f. = Acioa Aubl. (Chrysobalanac.).

Grimaldia Schrank (~ Cassia L.). Leguminosae. 18 Mex. to trop. S. Am.

Grimmeodendron Urb. Euphorbiaceae. 2 W.I.

Grindelia Willd. Compositae. 60 Am. (exc. C. Am.).

Grindeliopsis Sch. Bip. = Xanthocephalum Willd. (Compos.).

Gripidea Miers = Cajophora Presl (Loasac.).

Grischowia Karst. = Monochaetum Naud. (Melastomatac.).

Grisebachia Klotzsch (~ Eremia D. Don). Ericaceae. 25 S. Afr.

Grisebachia H. Wendl. & Drude = Howeia Becc. (Palm.).

Grisebachiella Lorentz. Apocynaceae. 1 Argent.

GRISELEA

Griselea D. Dietr. = Grislea L. = Combretum L. (Combretac.).

Griselinia Forst. f. Griseliniaceae. 6 N.Z., Chile, SE. Brazil. Rupestral or littoral shrubs or trees, sometimes epiph. or climbing.

Griselinia Scop. = Pterocarpus L. (Legumin.).

Griseliniaceae (Wang.) Takht. Dicots. 1/6 N.Z., S. Am. Trees or shrubs, often epiphytic or climbing, glabr. Ls. alt., coriac., often asymm., ent. or angled or spinose, exstip., petiole subvaginate. Fls. ♂ ♀, dioec., small, rac. or panic. ♂: K (5), minute; C 5, imbr.; A 5, alternipet., anth. intr., versat.; disk fleshy; pistillode 0. ♀: K (5), tube ovoid; C 5, imbr., or 0; staminodes 0; G̅ 1, 1-loc., with 1 pend. anatr. ov., and 3 short free or shortly united styles. Fruit an ovoid 1-loc. 1-seeded berry; seed with minute embr. in copious endosp. Only genus: *Griselinia*. (Cf. Dawson, *Tuatara* **14**: 121-9, 1966; Philipson, *New Zeal. Journ. Bot.* **5**: 134-65, 1967).

Grisia Brongn. = Bikkia Reinw. (Rubiac.).

Grislea L. = Combretum L. (Combretac.).

Grislea Loefl. = Pehria Sprague (Lythrac.).

Grisleya P. & K. = praec.

Grisollea Baill. Icacinaceae. 2 Madag., Seychelles.

Grisseea Bakh. f. Apocynaceae. 1 Java.

Grobya Lindl. Orchidaceae. 3 Brazil.

Groelandia Fourr. = seq.

Groenlandia J. Gay (~ Potamogeton L.). Potamogetonaceae. 1 W. Eur. & N. Afr. to SW. As.

Gromovia Regel = Beloperone Nees (Acanthac.).

Gromphaena St-Lag. = Gomphrena L. (Amaranthac.).

Grona Benth. & Hook. f. = Nogra Merr. (Legumin.).

Grona Lour. = Desmodium Desv. (Legumin.).

Grone Spreng. = praec.

Gronophyllum Scheff. Palmae. 13 E. Malaysia, 1 N. Austr.

Gronovia Blanco = Illigera Blume (Hernandiac.).

Gronovia L. Loasaceae. 2 Mex. to Venez. & Ecuador. A 5, no stds. G 1.

Gronovi[ac]eae Endl. = Loasaceae–Gronovieae Reichb.

Grosourdya Reichb. f. (= ? Pteroceras Hassk.). Orchidaceae. 1 Malaysia.

Grosowidya B. D. Jacks. (sphalm.) = praec.

Grossera Pax. Euphorbiaceae. 11 trop. Afr., Madag.

Grossheimia Sosn. & Takht. (~ Centaurea L.). Compositae. 3 Armenia, Cauc.

Grossostylis Pers. (sphalm.) = Crossostylis J. R. & G. Forst. (Rhizophorac.).

Grossularia Mill. (~ Ribes L.). Grossulariaceae. 50 temp. Euras., N. Afr., N. Am.

*★**Grossulariaceae** DC. Dicots. 2/150 temp. Euras., NW. Afr., N. & C. Am., Pacif. S. Am. to Fuegia. Shrubs, sometimes spiny, with alt. simple variously lobed stip. or exstip. ls., often with resinous glands. Fls. reg., ♀ or ♂ ♀ and dioec., rac. or subsol. K (4-5), imbr. or subvalv., sometimes petaloid, persist.; C (or stds.) 5, small, squamiform, obov. or subul.; A 5, mostly short, rarely long (C and A adnate to K-tube); G̅ (2), 1-loc., with a lateral or median pariet. plac. bearing few-∞ ov., and 2 ± connate styles with simple stigs. Fr. a juicy berry; seeds ∞, with endosp.; aril or arillode. Genera: *Ribes, Grossularia*.

Grotefendia Seem. = Polyscias J. R. & G. Forst. (Araliac.).
Groutia Guill. & Perr. = Opilia Roxb. (Opiliac.).
Grubbia Bergius. Grubbiaceae. 5 (or 4 excl. *Strobilocarpus*) S. Afr.
*Grubbiaceae Endl. Dicots. 2/5 S. Afr. Small heath-like shrubs with opp.
decuss. lin. or lanc. ent. exstip. ls. Fls. ♂, sess., in small axill. 3-fld. dichasia
or in many-fld. strobiloid compound dichasia, the ovaries coherent or connate.
The most prob. interpr. of the P is: K 2 bracteiform + 2 vestigial, C 2+2
diagonally disposed, valv. A 4+4, with ± ligulif. laterally compressed fil. and
2-locellate anth., dehiscing lat. with valvular reflexion of the theca-wall; disk
annular, hairy; G̅ (2), with 1 pend. anatr. ov. per loc. and simple shortly
2-lobed style. Fr. a dry, drupaceous or achenial syncarp, with woody or bony
endocarp; seed 1 per syncarp, with oily endosp. Genera: *Grubbia, Strobilo-
carpus*. Related to *Empetraceae* and the *Ericales*. The supposed affinity with
Olacac. is found to be illusory. (Fagerlind, *Sv. Bot. Tidskr.* **41**: 315–320, 1947.)
Gruenera Opiz = Salix L. (Salicac.).
Gruhlmania Neck. = Spermacoce L. (Rubiac.).
Grumilea Gaertn. = Psychotria L. (Rubiac.).
Grundelia Engl. & O. Hoffm. = Gundelia L. (Compos.).
Grundlea Poir. ex Steud. = Grumilea Gaertn. (Rubiac.).
Grusonia F. Reichb. ex Britt. & Rose = Opuntia Mill. (Cactac.).
Gruvelia A. DC. = Pectocarya DC. (Boraginac.).
Grymania Presl = Parinari Aubl. + Couepia Aubl. + Maranthes Bl. (Chryso-
balanac.).
Grypocarpha Greenman. Compositae. 3 Mex.
Guacamaya Maguire. Rapateaceae. 1 Venez., Colombia.
Guachamaca De Grosourdy = Prestonia R.Br. (vel Malouetia A. DC.?)
(Apocynac.).
Guaco Liebm. = Aristolochia L. (Aristolochiac.).
Guadella Franch. (sphalm.) = Guaduella Franch. (Gramin.).
Guadua Kunth. Gramineae. 30 trop. Am.
Guaduella Franch. Gramineae. 8 trop. Afr.
Guagnebina Vell. = Manettia Mutis (Rubiac.).
Guaiabara Mill. = Coccoloba L. (Polygonac.).
Guaiacana Duham. = Diospyros L. (Ebenac.).
Guaiacan[ace]ae Juss. = Ebenaceae Vent.
Guaiacon Adans. = seq.
Guaiacum L. Zygophyllaceae. 6 warm Am., W.I. *G. officinale* L. yields
lignum vitae wood, from which is also obtained the medicinal resin *guaiacum*.
Guaiava Adans. = seq.
Guaicaia Maguire. Compositae. 1 Venez., Brazil.
Guajava Mill. = Psidium L. (Myrtac.).
Gualteria Duham. = Gaultheria L. (Ericac.).
Gualtheria J. F. Gmel. = praec.
Guamatela Donn. Smith. Rosaceae. 1 C. Am.
Guamia Merr. Annonaceae. 1 Marianne Is.
Guanabanus Mill. = Annona L. (Annonac.).
Guandiola Steud. = Guardiola Humb. & Bonpl. (Compos.).
Guania Tul. = Gouania Jacq. (Rhamnac.).

GUAPEA

Guapea Endl. = Guapira Aubl. = Pisonia L. (Nyctaginac.).
Guapeba Gomes (~ Pouteria Aubl.). Sapotaceae. 5 trop. Am.
Guapebeira Gomes (? sphalm.) = praec.
Guapeiba Gomes = praec.
Guapina Steud. = seq.
Guapira Aubl. = Pisonia L. (Nyctaginac.).
Guapurium Juss. = Eugenia L. (Myrtac.).
Guapurum J. F. Gmel. = praec.
Guarania Wedd. ex Baill. = Richeria Vahl (Euphorbiac.).
Guararibea Cav. (sphalm.) = Quararibea Aubl. (Malvac.).
Guardiola Cerv. ex Humb. & Bonpl. Compositae. 10 Mex., SW. U.S.
*Guarea Allem. ex L. Meliaceae. 150 trop. Am., 20 Afr. The disk forms a gynophore, and the sta. are completely united into a tube.
Guaria Dumort. = praec.
Guariruma Cass. = Mutisia L. f. (Compos.).
Guaropsis Presl = Clarkia Pursh (Onagrac.).
Guatemala A. W. Hill (in I.K., sphalm.) = Guamatela Donn. Smith (Rosac.).
*Guatteria Ruiz & Pav. Annonaceae. 250 S. Mex. to S. Braz. Berry stalked.
Guatteriella R. E. Fries. Annonaceae. 1 W. Brazil.
Guatteriopsis R. E. Fries. Annonaceae. 4 trop. S. Am.
Guayaba Nor. = Psidium L. (Myrtac.).
Guayabilla Sessé & Moç. = Samyda L. (Flacourtiac.).
Guaymasia Britton & Rose. Leguminosae. 1 Mex.
Guayunia C. Gay ex Moldenke = Rhaphithamnus Miers (Verbenac.).
Guazuma Mill. Sterculiaceae. 4 trop. Am.
Gubleria Gaudich. = Periloba Rafin. = Nolana L. ex L. f. (Nolanac.).
Gueinzia Sond. ex Schott = Stylochiton Lepr. (Arac.).
Gueldenstaedtia Fisch. Leguminosae. 10 C. As., Himal., China.
Gueldenstaedtia Neck. = Eurotia Adans. emend. C. A. Mey. (Chenopodiac.).
Guenetia Sagot ex Benoist = Catostemma Benth. (Malvac.).
Guenthera Andrz. ex Bess. = Brassica L. (Crucif.).
Guenthera Regel = Xanthocephalum Willd. (Compos.).
Guentheria Spreng. = Gaillardia Fouger. (Compos.).
Guepinia Bast. = Teesdalia R.Br. (Crucif.).
Guerinia J. Sm. = Isoloma J. Sm. (Lindsaeac.).
Guerkea K. Schum. Apocynaceae. 2 trop. Afr.
Guerreroia Merr. Compositae. 1 Philipp. Is.
Guersentia Rafin. = Chrysophyllum L. (Sapotac.).
Guesmelia Walp. (sphalm.) = Quesnelia Gaudich. (Bromeliac.).
Guettarda L. Rubiaceae. 20 New Caled.; 60 trop. Am.; 1, *G. speciosa* L., common on trop. coasts. Exalb. G 4–9-loc. Cots. o.
Guettardella Champ. ex Benth. = Antirhea Comm. ex Juss. (Rubiac.).
Guettardia P. & K. = Guettarda L. (Rubiac.).
Guetzlaffia Walp. = Gutzlaffia Hance (Acanthac.).
Guevina Juss. = Gevuina Molina (Proteac.).
Guevinia Hort. Par. ex Decne = Celastrus L. (Celastrac.).
Guevuina P. & K. = Gevuina Molina (Proteac.).
Guiabara Adans. = Coccoloba L. (Polygonac.).

Guibourtia Benn. Leguminosae. 4 trop. Am.; 11 trop. Afr.
Guichenotia J. Gay. Sterculiaceae. 5 W. Austr.
Guidonia P.Br. = Laëtia Loefl. ex L. (Flacourtiac.).
Guidonia (DC.) Griseb. = Samyda Jacq. (Flacourtiac.).
Guidonia Mill. = Samyda L. = Guarea Allem. ex L. (Meliac.).
Guienzia Benth. & Hook. f. = Gueinzia Sond. = Stylochiton Lepr. (Arac.).
Guiera Adans. ex Juss. Combretaceae. 1 N. trop. Afr.
Guiina Crueg. (sphalm.) = Quiina Aubl. (Quiinac.).
Guilandia P.Br. = seq.
Guilandina L. = Caesalpinia L. (Legumin.).
Guildingia Hook. = Mouriri Aubl. (Memecylac.).
Guilelma Link = Guilielma Mart. = Bactris Jacq. (Palm.).
Guilfoylia F. Muell. (~ Cadellia F. Muell.). Simaroubaceae. 1 NE. Austr.
Guilielma Mart. (~ Bactris Jacq.). Palmae. 7 trop. S. Am.
Guillainia Vieill. Zingiberaceae. 6 New Guinea, New Hebrid., New Caled.
Guillandinodes Kuntze = Schotia Jacq. (Legumin.).
Guillauminia A. Bertrand (~ Aloë L.). Liliaceae. 1 Madag.
Guilleminea Kunth = Brayulinea Small (Amaranthac.).
Guilleminia Neck. = Votomita Aubl. (Memecylac.).
Guilleminia Reichb. = Guilleminea Kunth = Brayulinea Small (Amaranthac.).
Guillenia Greene. Cruciferae. 4 SW. U.S.
Guillimia Reichb. = Gwillimia Rottl. = Magnolia L. (Magnoliac.).
Guillonea Coss. = Laserpitium L. (Umbellif.).
Guindilia Gill. ex Hook. = Valenzuelia Bert. (Sapindac.).
Guinnalda Sessé ex Meissn. = Gelsemium Juss. (Loganiac.).
Guioa Cav. Sapindaceae. 70 Indomal., Austr., Pacif.
Guiraoa Coss. Cruciferae. 1 Spain.
Guirea Steud. = Guiera Adans. ex Juss. (Combretac.).
***Guizotia** Cass. Compositae. 12 trop. Afr. *G. abyssinica* Cass. (*rantil* or Niger-seed) is cult. in India, etc., for its seeds, from which an oil is expressed.
Guldaenstedtia A. Juss. = Gueldenstaedtia Fisch. & Mey. (Legumin.).
Guldenstaedtia Dum. = praec.
Gulielma Spreng. = Guilielma Mart. (Palm.).
Gulubia Becc. Palmae. 10 Moluccas, New Guinea, New Hebrid., Solomon Is.,
Gulubiopsis Becc. = praec.   [Austr.
Gumifera Rafin. = Acacia Mill. (Legumin.).
Gumillaea Roem. & Schult. = seq.
Gumillea Ruiz & Pav. Cunoniaceae(?). 1 Peru. Ls. alt.
Gumira Hassk. = Premna L. (Verbenac.).
Gumsia Buch.-Ham. ex Wall. = Eriolaena DC. (Sterculiac.).
Gumteolis Buch.-Ham. ex D. Don = Centranthera R.Br. (Scrophulariac.).
Gumutus Spreng. = Arenga Labill. (Palm.).
Gundelia L. Compositae. 1 As. Min., Syria, Persia.
Gundelsheimera Cass. = praec.
Gundlachia A. Gray. Compositae. 8 Cuba, S. Domingo.
Gundlea Willis (sphalm.) = Grundlea Poir. ex Steud. = Grumilea Gaertn. (Rubiac.).

GYMNACHAENA

Gymnachaena Reichb. ex DC. = Perrotriche Cass. (Compos.).

Gymnachne L. Parodi. Gramineae. 1 Chile.

Gymnacranthera Warb. Myristicaceae. 17 Indomal.

Gymnactis Cass. = Glossogyne Cass. (Compos.).

Gymnadenia R.Br. Orchidaceae. 10 NE. Canada, Newf., Greenl., temp. Euras.

Gymnadeniopsis Rydb. Orchidaceae. 3 N. Am.

× **Gymnadeniorchis** A. D. Hawkes (vii) = × Orchigymnadenia E. G. Camus (Orchidac.).

Gymnagathis Schau. = Melaleuca L. (Myrtac.).

Gymnagathis Stapf = Stapfiophyton Li (Melastomatac.).

× **Gymnaglossum** hort. Orchidaceae. Gen. hybr. (i) (Coeloglossum × Gymnadenia).

Gymnalypha Griseb. = Acalypha L. (Euphorbiac.).

Gymnamblosis Pfeiff. = Gynamblosis Torr. = Croton L. (Euphorbiac.).

× **Gymnanacamptis** Aschers. & Graebn. Orchidaceae. Gen. hybr. (i) (Anacamptis × Gymnadenia).

Gymnandra Pall. = Lagotis Gaertn. (Scrophulariac.).

Gymnandropogon (Nees) Munro ex Duthie = Dichanthium Willem. (Gramin.)

Gymnantha Y. Ito = Gymnocalycium Pfeiff. (Cactac.).

Gymnanthelia Anderss. = Andropogon L. (Gramin.).

Gymnanthemum Cass. = Vernonia Schreb. (Compos.).

Gymnanthera R.Br. Periplocaceae. 4 Malaysia.

Gymnanthes Sw. (= Ateramnus P.Br.). Euphorbiaceae. 15 S. U.S., Mex., C. Am., W.I.

Gymnanthocereus Backeb. = ? Borzicactus Riccob. (Cactac.).

Gymnanthus Endl. = Gymnanthes Sw. (Euphorbiac.).

Gymnanthus Jungh. = Trochodendron Sieb. & Zucc. (Trochodendrac.).

× **Gymnaplatanthera** L. Lambert (vii) = × Gymplatanthera Camus (1906) (Orchidac.).

Gymnarren Leandro ex Klotzsch = Actinostemon Klotzsch (Euphorbiac.).

Gymnarrhena Desf. Compositae. 1 Medit., 5 W. As.

Gymnarrhoea (Baill.) P. & K. = Gymnarren Leandro ex Klotzsch = Actinostemon Klotzsch (Euphorbiac.).

Gymnartocarpus Boerlage. Moraceae. 3 Philipp. Is., Sumatra, Java.

Gymnaster Kitamura (~ Aster L.). Compositae. 3 E. As.

Gymnelaea (Endl.) Spach = Nestegis Rafin. (Oleac.).

Gymnema R. Br. Asclepiadaceae. 25 palaeotrop., S. Afr., Austr. The leaves of *G. sylvestre* R.Br. contain gymnemic acid, and when chewed temporarily destroy the capacity of tasting sugar.

Gymnema Endl. = Gynema Rafin. = Pluchea Cass. (Compos.).

Gymnemopsis Costantin. Asclepiadaceae. 2 Siam, Indochina.

Gymnerpis Thou. (uninom.) = *Goodyera nuda* Thou. (Orchidac.).

× **Gymnigritella** G. Camus. Orchidaceae. Gen. hybr. (i) (Gymnadenia × Nigritella).

Gymnima Rafin. ex Britten = Gymnema R.Br. (Asclepiadac.).

Gymnobalanus Nees & Mart. = Ocotea Aubl. (Laurac.).

Gymnobothrys Wall. ex Baill. = Sapium R.Br. (Euphorbiac.).

Gymnocactus Backeb. (~ Thelocactus (K. Schum.) Britt. & Rose). Cactaceae. 12 Mex.

Gymnocalycium Pfeiff. Cactaceae. 60 subtrop. S. Am.

Gymnocampus Pfeiff. = Gynocampus Lesch. = Levenhookia R.Br. (Stylidiac.).

Gymnocarpium Newm. Aspidiaceae. 5 N. temp., Formosa, Philippines, New Guinea. (Ching, *Contr. Biol. Lab. Sci. Soc. China*, **9**: 30–42, 1933.)

Gymnocarpon Pers. = seq.

Gymnocarpos Forssk. Caryophyllaceae. 1 Canaries, N. Afr., E. Medit. to Baluch., used as fodder for camels; 1 Tian-shan, Tibet, Mongolia.

Gymnocarpum DC. = praec.

Gymnocarpus Thou. ex Baill. = Uapaca Baill. (Uapacac.).

Gymnocarpus Viv. = Gymnocarpos Forssk. (Caryophyllac.).

Gymnocaulis Nutt. = Aphyllon Mitch. (Orobanchac.).

Gymnocaulus Phil. = Calycera Cav. (Calycerac.).

Gymnocereus Rauh ex Backeb. = Browningia Britt. & Rose (Cactac.).

Gymnochaeta Steud. = Schoenus L. (Cyperac.).

Gymnochaete Benth. & Hook. f. = praec.

Gymnochilus Blume. Orchidaceae. 2 Mascarenes.

Gymnocladus Lam. Leguminosae. 3 Assam, Burma, China; 1 N. Am. Serial axillary buds. *G. dioicus* (L.) K. Koch (*G. canadensis* Lam.) (Kentucky coffee tree) yields good timber.

Gymnocline Cass. = Chrysanthemum L. (Compos.).

Gymnococca Fisch. & Mey. ex Fisch., Mey. & Avé-Lall. = Pimelea Banks & Soland. (Thymelaeac.).

Gymnocoronis DC. Compositae. 1 Mex., 1 subtrop. S. Am.

Gymnodes Fourr. = Luzula DC. (Juncac.).

Gymnodiscus Less. Compositae. 3 S. Afr.

Gymnogonum Parry. Polygonaceae. 1 N. Am.

Gymnogramma Desv. (*s. str.*) (*nom. illegit.*) = Gymnopteris Bernh. (Hemionitidac.). (Formerly made to include an unnatural collection of ferns with elongate naked sori.)

Gymnogrammaceae Herter = Hemionitidaceae Pichi-Serm.

Gymnogramme auctt. = Eriosorus Fée (Gymnogrammac.).

Gymnogrammitidaceae Ching = Davalliaceae Reichb. (See T. & U. Sen, *Ann. Bot.* n.s. **35**: 229–35, 1971, for new evidence that *Gymnogrammitis* belongs to *Davalliaceae.*)

Gymnogrammitis Griff. Davalliaceae. 1 NE. India to S. China. Exindusiate.

Gymnogyne F. Didr. = Boehmeria Jacq. (Urticac.).

Gymnogyne Steetz = Cotula L. (Compos.).

Gymnolaema Benth. = Sacleuxia Baill. (Periplocac.).

Gymnolaena Rydb. (~ Dyssodia Cav.). Compositae. 4 Mexico.

Gymnoleima Decne = Lithodora Griseb. + Moltkia Lehm. (Boraginac.).

Gymnolomia Kunth. Compositae. 50 SW. U.S. to trop. S. Am.

Gymnoluma Baill. = Lucuma Mol. (Sapotac.).

Gymnomesium Schott = Arum L. (Arac.).

Gymnonychium Bartl. = Agathosma Willd. (Rutac.).

Gymnopentzia Benth. Compositae. 2 S. Afr.

GYMNOPETALUM

Gymnopetalum Arn. Cucurbitaceae. 3 S. China, Indomal.

Gymnophragma Lindau. Acanthaceae. 1 NE. New Guinea. [Argent.

Gymnophyton Clos. Hydrocotylaceae (~ Umbellif.). 6 Andes of Chile & × **Gymnoplatanthera** Rouy = × Gymplatanthera Camus (1906) (Orchidac.).

Gymnopodium Rolfe. Polygonaceae. 3 C. Am.

Gymnopogon Beauv. Gramineae. 10 Am.

Gymnopogon Forst. ex Scop. Apocynaceae. Quid?

Gymnopoma N. E. Brown. Aïzoaceae. 1 S. Afr.

Gymnopsis DC. = Gymnolomia Kunth (Compos.).

Gymnopteris Bernh. Hemionitidaceae. 5 warm Am., As.

Gymnopyrenium Dulac = Cotoneaster Medik. (Rosac.).

× **Gymnorchis** Osvačilová (ii) = × Pseudadenia P. F. Hunt (Orchidac.).

Gymnoreima Endl. (sphalm.) = Gymnoleima Decne = Lithodora Griseb. + Moltkia Lehm. (Boraginac.).

Gymnorinorea Keay = Decorsella A. Chev. (Violac.).

Gymnoschoenus Nees. Cyperaceae. 6 Austr.

Gymnosciadium Hochst. (~ Pimpinella L.). Umbelliferae. 2 trop. Afr.

Gymnosiphon Blume. Burmanniaceae. 30 trop.

Gymnosperma Less. Compositae. 1 S. U.S. to C. Am.

Gymnospermae Lindl. One of the two great divisions of *Spermatophyta* or seed-plants, distinguished from *Angiospermae* by the fact that the ovules are always naked in the sense of not being enclosed in an ovary, although they usually have some form of protection. Also, the endosperm (female prothallus) is formed before fertilisation. The existing *G.* are divided into five orders: *Cycadales, Ginkgoales, Coniferae* (or *Coniferales*), *Taxales* and *Gnetales.* These differ very much from one another, so much so that it is by no means impossible that the Gymnosperms are polyphyletic. Several other Gymnosperm orders are known only from fossils, which occur as far back in geological history as the Lower Carboniferous period. By that time, two very distinct orders, the *Pteridospermales* or seed-bearing ferns, and the *Cordaïtales*, were evidently already much diversified. By the end of the Permian or Triassic, however, both had declined and the surviving orders (excluding the *Gnetales*, of which there are few fossils) had begun. The *Cycadales* are more closely related to the *Pteridospermales*, judging from their wood structure, leaves and seed, whilst the *Coniferae*, *Taxales* and *Ginkgoales* resemble more closely the *Cordaïtales*. The suggestion, popular at one time, that some of the fossil gymnosperms (*Bennettitales*) with flower-like hermaphrodite cones might be on the direct line of ascent to the Angiosperms, is not now given credence. (See *Cycadales, Coniferae, Ginkgoaceae*, and refer to K. R. Sporne, *The Morphology of Gymnosperms.* London, 1966.)

Gymnospermium Spach. Leonticaceae. 1 C. As.

Gymnosphaera Bl. (excl. sect. 3 of Copel. *Gen. Fil.* 99, 1947) = Cyathea Sm. subg. Cyathea, sect. Gymnosphaera (Cyatheac.) (Holttum, *Fl. Males.* ser. II, 1, 1963). [*c.* 30 spp. trop. As. to Fiji & Australia; mostly small tree ferns of forest. 1 sp. in New Guinea is scandent (*Thysanobotrya* v. Ald. v. Ros.).]

Gymnosporia Benth. & Hook. f. (*p.p.*) = Maytenus Mol. (Celastrac.).

*****Gymnosporia** (Wight & Arn.) Benth. & Hook. f. Celastraceae. 100 trop. & subtrop., esp. Afr. Many have branches modified into thorns.

Gymnostachium Reichb. = Gymnostachyum Nees (Acanthac.).
Gymnostachys R. Br. Araceae. 2 E. Austr.
Gymnostachyum Nees. Acanthaceae. 30 India & Ceylon to Philipp. & Java.
Gymnostechyum Spach = praec.
Gymnostemon Aubrév. & Pellegr. Simaroubaceae. 1 trop. W. Afr.
Gymnostephium Less. Compositae. 7 S. Afr.
Gymnosteris Greene. Polemoniaceae. 3 W. U.S.
Gymnostichum Schreb. = Hystrix Moench (Gramin.).
Gymnostillingia Muell. Arg. = Stillingia Garden (Euphorbiac.).
Gymnostoma L. A. S. Johnson. Casuarinaceae. 20 W. Malaysia to NE. Austr., Fiji & New Caled.
Gymnostyles Juss. = Soliva Ruiz & Pav. (Compos.).
Gymnostyles Rafin. = Pluchea Cass. (Compos.).
Gymnostylis B. D. Jacks. = praec.
Gymnoterpe Salisb. = Tapeinanthus Herb. (Amaryllidac.).
Gymnotheca Decne. Saururaceae. 2 China.
Gymnotheca Presl = Marattia Sw. (Marattiac.).
Gymnothrix Spreng. = Pennisetum Pers. (Gramin.). *L . Rich.*
× **Gymnotraunsteinera** Cif. & Giac. Orchidaceae. Gen. hybr. (i) (Gymnadenia × Traunsteinera).
Gymnotrix Beauv. = praec.
Gymnouratella Van Tiegh. = Ouratea Aubl. (Ochnac.).
Gymnoxis Steud. (sphalm.) = Gymnopsis DC. = Gymnolomia Kunth (Compos.).
× **Gymnplatanthera** Camus (1908) (vii) = Gymplatanthera Camus (1906) (Orchidac.).
Gymostyles Willd. (sphalm.) = Gymnostyles Juss. = Soliva Ruiz & Pav. (Compos.).
× **Gymplatanthera** Camus (1906). Orchidaceae. Gen. hybr. (i) (Gymnadenia × Platanthera).
Gynactis Cass. = Glossogyne Cass. (Compos.).
Gynaecocephalium Hassk. = Gynocephalum Blume = Phytocrene Wall. (Icacinac.).
Gynaecopachys Hassk. = Gynopachis Blume = Randia L. (Rubiac.).
Gynaecotrochus Hassk. = Gynotroches Blume (Rhizophorac.).
Gynaecura Hassk. = Gynura Cass. (Compos.).
Gynaeum P. & K. = seq.
Gynaion A. DC. = Cordia L. (Ehretiac.).
Gynamblosis Torr. = Croton L. (Euphorbiac.).
Gynampsis Rafin. = Downingia Torr. (Campanulac.).
Gynandriris Parl. (~ Iris L.). Iridaceae. 20 Medit., S. Afr.
*****Gynandropsis** DC. (~ Cleome L.). Cleomaceae. 1 trop. & subtrop., *G. gynandra* (L.) Briq. The seeds are used like mustard.
Gynanthistrophe Poit. ex DC. = Swartzia Schreb. (Legumin.).
Gynaphanes Steetz = Epaltes Cass. (Compos.).
Gynapteina Spach = Schefflera J. R. & G. Forst. (Araliac.).
Gynastrum Neck. = Guapira Aubl. = Pisonia L. (Nyctaginac.).

Gynatrix Alef. = Plagianthus J. R. & G. Forst. (Malvac.).

Gynema Rafin. = Pluchea Cass. (Compos.).

Gynerium Humb. & Bonpl. Gramineae. 1 Mex. to subtrop. S. Am.

Gynestum Poit. = Geonoma Willd. (Palm.).

Gynetera Rafin. = Tetracera L. (Dilleniac.).

Gyneteria Spreng. = Gynheteria Willd. = Tessaria Ruiz & Pav. (Compos.).

Gynetra B. D. Jacks. = Gynetera Rafin. = Tetracera L. (Dilleniac.).

Gynheteria Willd. = Tessaria Ruiz & Pav. (Compos.).

Gynicidia Neck. = Mesembryanthemum L. (Aïzoac.).

Gynisanthus P. & K. = Gunisanthus DC. = Diospyros L. (Ebenac.).

Gynizodon Rafin. = Miltonia Lindl. (Orchidac.).

Gynocampus Lesch. ex DC. = Levenhookia R.Br. (Stylidiac.).

Gynocardia R.Br. Flacourtiaceae. 1 Assam & Burma, *G. odorata* Br. The seed yields a medicinal oil of less value than the chaulmoogra oil of *Hydnocarpus* (*q.v.*).

Gynocephala Benth. & Hook. f. = seq.

Gynocephalium Endl. = seq.

Gynocephalum Blume = Phytocrene Wall. (Icacinac.).

Gynochthodes Blume. Rubiaceae. 14 Andamans, SE. As., W. Malaysia, Caroline Is., Samoa.

Gynocraterium Bremek. Scrophulariaceae. 1 trop. S. Am.

Gynodon Rafin. = Allium L. (Alliac.).

Gynoglossum Zipp. ex Scheff. = Rapanea Aubl. (Myrsinac.).

Gynoglottis Smith. Orchidaceae. 1 Sumatra.

Gynoisa B. D. Jacks. = seq.

Gynoisia Rafin. = Ipomoea L. (Convolvulac.).

Gynomphis Rafin. = Tibouchina Aubl. (Melastomatac.).

Gynoön A. Juss. = Glochidion J. R. & G. Forst. (Euphorbiac.).

Gynopachis Blume (1823) = Randia L. (Rubiac.).

Gynopachys Blume (1825) = praec.

Gynophoraria Rydb. = Astragalus L. (Legumin.).

Gynophorea Gilli. Cruciferae. 1 Afghanistan.

Gynopleura Cav. = Malesherbia Ruiz & Pav. (Malesherbiac.).

Gynopogon J. R. & G. Forst. = Alyxia R.Br. (Apocynac.).

Gynostemma Blume. Cucurbitaceae. 2 E. As., Indomal.

Gynothrix P. & K. = Gynatrix Alef. = Plagianthus J. R. & G. Forst. (Malvac.).

Gynotroches Blume. Rhizophoraceae. 1 Burma, Siam, Malaysia, Caroline & Solomon Is.

Gynoxys Cass. Compositae. 100 C. Am. to Peru.

*Gynura Cass. Compositae. 100 trop. Afr. & Madag. to E. As. & Malaysia.

Gypothamnium Phil. (~ Plazia Ruiz & Pav.). Compositae. 1 Chile.

Gypsocallis Salisb. = Erica L. (Ericac.).

Gypsophila L. Caryophyllaceae. 125 temp. Euras. (esp. E. Medit.), Egypt; 1 Austr., N.Z. The fls. are shorter in the tube than most *Silenoïdeae* and are visited by a greater variety of insects.

Gypsophytum Adans. = Arenaria, Minuartia, Moehringia, Cerastium (Caryophyllac.).

Gypsophytum Ehrh. (uninom.) = *Gypsophila fastigiata* L. (Caryophyllac.).

Gyptidium R. M. King & H. Rob. (~ Eupatorium L.). Compositae. 2 Braz. to Argent.

Gyptis Cass. (~ Eupatorium L.). Compositae. 7 Brazil to Argent.

Gyrandra Griseb. = Centaurium Hill (Gentianac.).

Gyrandra Moq. = Tersonia Moq. (Gyrostemonac.).

Gyrandra Wall. = Daphniphyllum Blume (Daphniphyllac.).

Gyranthera Pittier. Bombacaceae. 2 Panamá, Venez.

Gyrenia Knowles & Westc. ex Loud. = Milla Cav. (Liliac.).

Gyrinops Gaertn. Thymelaeaceae. 1 Ceylon, 7 E. Malaysia.

Gyrinopsis Decne = praec.

Gyrocarpaceae Dum. Dicots. 2/22 trop. & subtrop. Small trees, shrubs or lianes. Ls. alt., simple, ent. or palmately lobed, pubesc., exstip., with cystoliths. Fls. ♂ or ♂ ♀, in dense thyrses. P 8(-4), sometimes shortly connate below, imbr.; A 3-5, with or without staminal glands, anth. dehiscing laterally by valves; stds. 3-4 or 0; Ḡ 1, 1-ovulate, with terete style and capit. stig. Fr. dry and indehisc., sometimes winged with persist. accresc. tep.; seed exalbum., embryo with leafy convol. cots. Genera: *Gyrocarpus, Sparattanthelium*. Rather closely rel. to *Laurac., Atherospermatac.* and *Gomortegac.*; less closely to *Monimiac.* and *Hernandiac.*

Gyrocarpus Jacq. Gyrocarpaceae. 7 trop. & subtrop. The fr. is rather similar to that of a Dipterocarp.

Gyrocephalium Reichb. = Gynocephalum Blume = Phytocrene Wall. (Icacinac.).

Gyromia Nutt. = Medeola L. (Liliac.).

Gyroptera Botsch. (~ Salsola L.). Chenopodiaceae. 2 Somalia, Abd al Kuri (W. of Socotra).

Gyrosorium Presl = Pyrrosia Mirbel (Polypodiac.).

Gyrostachys Blume = Spiranthes Rich. (Orchidac.).

Gyrostelma Fourn. Asclepiadaceae. 3 Brazil, Argent.

Gyrostemon Desf. Gyrostemonaceae. 6 Austr.

*****Gyrostemonaceae** Endl. Dicots. (Centrosp.). 5/16 Austr. Trees, shrubs or undershrubs; stems with normal growth. Ls. simple, ent., alt., ± succulent; stips. v. small or 0. Fls. ♂ ♀, mostly dioec., reg., sol., axill., or racemose. P usu. discoid or cupular, ent. or ± lobed, persist.; A 6-∞, 1-several-seriate, anth. oblong to subcuneate, tetragonal, sess. or subsess., arising from edge of flat or convex recept.; G (1-2-∞), when ∞ arranged in a whorl around a central column, each loc. with 1 axile campylotr. ov., funicle and micropyle much thickened; stigs. usu. sess., forming a corona, rarely ∞, subulate. Fr. dry, each carp. usu. dehiscing dorsally or ventrally or both and separating from centr. column; seeds arillate, with rather copious oily or fleshy endosp. Genera: *Didymotheca, Gyrostemon, Codonocarpus, Tersonia, Cypselocarpus*. Rather closely related to *Phytolaccac.*

Gyrostephium Turcz. = Chthonocephalus Steetz (Compos.).

Gyrotaenia Griseb. Urticaceae. 6 W.I.

Gyrotheca Salisb. = Lachnanthes Ell. (Haemodorac.).

Gytonanthus Rafin. = Patrinia Juss. (Valerianac.).

H

Haagea Frič = Mammillaria Haw. (Cactac.).

Haagea Klotzsch = Begonia L. (Begoniac.).

Haageocactus Backeb. = seq. (?).

Haageocereus Backeb. (~ Trichocereus (A. Berger) Riccob.). Cactaceae.

Haarera Hutch. & E. A. Bruce. Compositae. 1 trop. E. Afr. [50 Peru.

Haasia Nees = Dehaasia Blume (Laurac.).

Haaslundia Schumach. & Thonn. = Hoslundia Vahl (Labiat.).

Haastia Hook. f. Compositae. 3 N.Z. *H. pulvinaris* Hook. f. forms large dense cushions ('vegetable sheep') in the subalpine and alpine zones (cf. *Raoulia, Azorella*).

Habenaria Willd. Orchidaceae. 600 trop. & warm countries, Old & New Worlds.

× **Habenariorchis** Rolfe. Orchidaceae. Gen. hybr. (vi) (Habenaria × Orchis).

Habenella Small = Platanthera Lindl. (Orchidac.).

Habenorkis Thou. = Habenaria Willd. (Orchidac.).

Haberlea Frivaldszky. Gesneriaceae. 1 Balkan Penins.

Haberlea Pohl ex Baker = Praxelis Cass. (Compos.).

Haberlia Dennst. = Odina Roxb. = Lannea A. Rich. (Anacardiac.).

Habershamia Rafin. = Brami Adans. = Bacopa Aubl. (Scrophulariac.).

Hablanthera Hochst. = Haplanthera Hochst. = Ruttya Harv. (Acanthac.).

Hablitzia M. Bieb. Chenopodiaceae. 1 Caucasus. Climbing shoot given off yearly from perenn. underground stem (cf. *Bowiea*); climbs by sensitive petioles.

Hablitzlia Reichb. = praec.

Hablizia Spreng. = praec.

Hablizlia Pritz. = praec.

Habracanthus Nees. Acanthaceae. 8 Mexico to Peru.

Habranthus Herb. Amaryllidaceae. 20 trop. & S. Am.

Habrochloa C. E. Hubb. Gramineae. 1 trop. E. Afr.

Habrodictyon v. d. Bosch = Abrodictyum Presl (Hymenophyllac.).

Habroneuron Standley. Rubiaceae. 1 Mexico.

Habropetalum Airy Shaw. Dioncophyllaceae. 1 trop. W. Afr. (Sierra Leone).

Habrosia Fenzl. Caryophyllaceae. 1 W. As.

Habrothamnus Endl. = Cestrum L. (Solanac.).

Habrozia Lindl. = Habrosia Fenzl (Caryophyllac.).

Habrurus Hochst. ex Hack. = Elionurus Kunth (Gramin.).

Habsburgia Mart. = Skytanthus Meyen (Apocynac.).

Habsia Steud. = Guettarda L. (Rubiac.).

Habzelia A. DC. = Xylopia L. (Annonac.).

Hachenbachia D. Dietr. = Hagenbachia Nees & Mart. (Haemodorac.).

Hachettea Baill. Balanophoraceae. 1 New Caled.

Hachetteaceae Van Tiegh. = Balanophoraceae–Mystropetaloïdeae (Hook. f.) Engl. + Dactylanthoïdeae (Hook. f.) Engl. (*see* Balanophorac.).

Hackela Pohl ex Welden = Hackelia Pohl ex Griseb. = Curtia Cham. & Schlechtd. (Gentianac.).

524

Hackelia Opiz. Boraginaceae. 40 As., Am., 1 Eur.

Hackelia Pohl ex Griseb. = Curtia Cham. & Schlechtd. (Gentianac.).

Hackelia Vasey ex Beal = Leptochloa Beauv. (Gramin.).

Hackelochloa Kuntze = Rytilix Rafin. (Gramin.).

Hacquetia Neck. ex DC. Umbelliferae. 1 C. Eur.

Hacub Boehm. = Gundelia L. (Compos.).

Hadestaphylum Dennst. = Holigarna Buch.-Ham. (Anacardiac.).

Hadongia Gagnep. = Citharexylum Mill. (Verbenac.).

Hadrodemas H. E. Moore. Commelinaceae. 1 C. Am. (Guatem.).

Haeckeria F. Muell. = Humea Sm. (Compos.).

Haemacanthus S. Moore. Acanthaceae. 1 Somaliland.

Haemadictyon Lindl. = Prestonia R.Br. (Apocynac.).

Haemanthaceae Salisb. = Amaryllidaceae–Haemanthinae Pax.

Haemanthus L. Amaryllidaceae. 50 trop. & S. Afr., Socotra.

Haemaria L. = Ludisia A. Rich. (Orchidac.).

× **Haemari-anoectochilus** hort. (vii) = × Anoectomaria hort. (Orchidac.).

× **Haemari-macodes** hort. (vii) = × Macomaria hort. (Orchidac.).

Haemarthria Munro = Hemarthria R.Br. = Rottboellia L. f. (Gramin.).

Haemastegia Klatt = Erythrocephalum Benth. (Compos.).

Haematobanche Presl = Hyobanche L. (Scrophulariac.).

Haematocarpus Miers. Menispermaceae. 3 E. Himal., Assam, Philipp., Borneo, Java.

Haematodes P. & K. = Hematodes Rafin. = Salvia L. (Labiat.).

Haematolepis Presl = Cytinus L. (Balanophorac.).

Haematophyla P. & K. = Hematophyla Rafin. = Columnea L. (Gesneriac.).

Haematorchis Blume = Galeola Lour. (Orchidac.).

Haematospermum Wall. = Homonoia Lour. (Euphorbiac.).

Haematostaphis Hook. f. Anacardiaceae. 2 trop. W. Afr.

Haematostemon (Muell. Arg.) Pax & Hoffm. Euphorbiaceae. 2 trop. S. Am.

Haematostrobus Endl. = Thonningia Vahl (Balanophorac.).

Haematoxyllum Scop. = seq.

Haematoxylon L. = seq.

Haematoxylum L. Leguminosae. 3 Mex., C. Am., W.I., SW. Afr. *H. campechianum* L. (logwood). Young foliage red. Thorns in the leaf-axils. The heartwood contains haematoxylin and is used for dyeing.

Haemax E. Meyer = Microloma R. Br. (Asclepiadac.).

Haemocarpus Nor. ex Thou. = Haronga Thou. (Guttif.).

Haemocharis Salisb. ex Mart. & Zucc. = Laplacea Kunth (Theac.).

*Haemodoraceae R.Br. Monocots. 14/75 S. Afr., Austr., N. & trop. Am. Herbs with rad., lin. or ensif. ls., and panicled infl. of a number of cymes arranged racemosely (cf. *Aesculus*). Fl. reg. or transv. zygo. (cf. *Anigozanthos*), ⚥, 3-merous; P 3 + 3 or (3 + 3), 1- or 2-ser., imbr. or subvalv., tube straight or curved; A 6 or 3, inserted on inner perianth-ls., with intr. anthers; G̲ or G̅ (3), ovules few–∞ in each loc., anatr. or semi-anatr.; stigma capitate. Capsule. Chief genera: *Haemodorum, Conostylis, Anigozanthos*.

Haemodoron Reichb. = Cistanche Hoffmgg. & Link (Orobanchac.).

Haemodorum Sm. Haemodoraceae. 20 Austr.

Haemospermum Reinw. = Geniostoma J. R. & G. Forst. (Loganiac.).

HAENCKEA

Haenckea Juss. = Haenkea Ruiz & Pav. (1802) = Schoepfia Schreb. (Olacac.).
Haenelia Walp. = Amellus L. (Compos.).
Haenianthus Griseb. Oleaceae. 2–4 W.I.
Haenkaea Usteri = Haenkea F. W. Schmidt = Adenandra Willd. (Rutac.).
Haenkea Ruiz & Pav. (1794) = Maytenus Mol. (Celastrac.).
Haenkea Ruiz & Pav. (1802) = Schoepfia Schreb. (Olacac.).
Haenkea Salisb. = Portulacaria Jacq. (Portulacac.).
Haenkea F. W. Schmidt = Adenandra Willd. (Rutac.).
Haenselera Boiss. ex DC. = Rothmaleria Font Quer (Compos.).
Haenselera Lag. = Physospermum Cusson (Umbellif.).
Hafunia Chiov. = Sphaerocoma T. Anders. (Caryophyllac.).
Hagaea Vent. = Polycarpaea Lam. (Caryophyllac.).
Hagea Pers. = praec.
Hagea Poir. = Hagenia J. F. Gmel. (Rosac.).
Hagenbachia Nees & Mart. Haemodoraceae. 1 Brazil.
Hagenia J. F. Gmel. Rosaceae. 1 Abyssinia to N. Malawi. The dried ♀ fls. (*koso*) are used as a remedy for tapeworm.
Hagenia Moench = Saponaria L. (Caryophyllac.).
Hagidryas Griseb. = Prunus L. (Rosac.).
Hagioseris Boiss. = Picris L. (Compos.).
Hagnothesium (A. DC.) Kuntze = Thesidium Sonder (Santalac.).
Hahnia Medik. = Torminalis Medik. = Sorbus L. (Rosac.).
Hainania Merr. Tiliaceae. 1 S. China (Hainan).
Hainardia Greuter = Monerma Beauv. (Gramin.).
Haitia Urb. Lythraceae. 2 Haiti.
Haitiella L. H. Bailey. Palmae. 1 Haiti.
Haitimimosa Britton = Mimosa L. (Legumin.).
Hakea Schrad. Proteaceae. 100 Austr. Xero. with hard woody fr. The seedlings show transition stages (cf. *Acacia*) from entire ls. to the much divided ls.
Hakoneaste Maekawa. Orchidaceae. 1 Japan.   [usu. in the genus.
Hakonechloa Makino ex Honda (~ Phragmites Trin.). Gramineae. 1 Japan.
Halacsya Doerfl. Boraginaceae. 1 W. Balkan Penins., on serpentine.
Halacsyella Janchen. Campanulaceae. 1 Greece.
Halaea Garden = Berchemia Neck. ex DC. (Rhamnac.).
Halanthium C. Koch. Chenopodiaceae. 6 W. & C. As.
Halanthus Czerep. (sphalm.) = praec.
Halarchon Bunge. Chenopodiaceae. 1 Afghanistan.
Halconia Merr. = Trichospermum Bl. (Tiliac.).
Halea L. ex Sm. (1) = Eupatorium L. (Compos.).
Halea L. ex Sm. (2) = Schlosseria Garden (Palm.).
Halea Torr. & Gray = Tetragonotheca L. (Compos.).
Halecus Rafin. = Croton L. (Euphorbiac.).
Halenbergia Dinter. Aïzoaceae. 1 SW. Afr.
Halenea Wight = seq.
***Halenia** Borckh. Gentianaceae. 3 mts. C. & E. As., S. India; 100 Am. Cleistogamic fls. frequent.
Halerpestes Greene (~ Ranunculus L.). Ranunculaceae. 7 temp. N. Am., Euras.

Halesia P.Br. = Guettarda L. (Rubiac.).
*Halesia J. Ellis ex L. Styracaceae. 6 E. As., SE. U.S. Fr. inf., winged.
Halesia Loefl. = Trichilia P.Br. (Meliac.).
Halesiaceae Link = Styracaceae Rich. (genera c. fruct. inf. alat.).
Halfordia F. Muell. Rutaceae. 4 New Guinea, E. Austr., New Caled.
Halgania Gaudich. Ehretiaceae. 15 Austr.
Halia St-Lag. = Halesia J. Ellis ex L. (Styracac.).
Halianthus Fries = Honkenya Ehrh. (Caryophyllac.).
Halibrexia Phil. = Alibrexia Miers = Nolana L. (Nolanac.).
Halicacabus (Bunge) Nevski = Astragalus L. (Legumin.).
× Halimiocistus Janchen. Cistaceae. Gen. hybr. (Cistus × Halimium).
Halimione Aellen. Chenopodiaceae. 3 W. Eur. & Medit. to SW. & C. As.
Halimiphyllum (Engl.) A. Borisova (~ Zygophyllum L.). Zygophyllaceae.
5 C. As.
Halimium (Dunal) Spach emend. Willk. Cistaceae. 14 Medit.
Halimocnemis C. A. Mey. Chenopodiaceae. 19 C. As.
Halimocnemum Lindem. = Halocnemum M. Bieb. (Chenopodiac.).
Halimodendron Fisch. ex DC. Leguminosae. 1 W. & N. As., on salt steppes.
Outer leaflets often thorny.
Halimolobos Tausch. Cruciferae. 16 Pacif. Am. (Rocky Mts. to Andes).
Halimum Loefl. = Sesuvium L. (Aïzoac.).
Halimus P.Br. = Portulaca L. (Portulacac.).
Halimus Kuntze = Sesuvium L. (Aïzoac.).
Halimus Wallr. = Atriplex L. (Chenopodiac.).
Hallackia Harv. = Huttonaea Harv. (Orchidac.).
Hallera St-Lag. = seq.
Halleria L. Scrophulariaceae. 9 trop. & S. Afr., Madag.
Hallesia Scop. = Halesia P.Br. = Guettarda L. (Rubiac.).
Hallia Dum. ex Pfeiff. = Honkenya Ehrh. (Caryophyllac.).
Hallia Jaume St-Hil. = Alysicarpus Neck. ex Desv. (Legumin.).
Hallia Thunb. Leguminosae. 6 S. Afr.
Hallieracantha Stapf. Acanthaceae. 30 W. Malaysia (esp. Borneo).
Halliophytum I. M. Johnston = Securinega Juss. + Tetracoccus Engelm. ex
Parry p.p. (Euphorbiac.).
Hallomuellera Kuntze = Crantzia Nutt. (Umbellif.).
Halloschulzia Kuntze = Stenomeris Planch. (Dioscoreac.).
Halmia M. Roem. = Crataegus L. (Rosac.).
Halmyra Herb. = Pancratium L. (Amaryllidac.).
Halocharis M. Bieb. ex DC. = Centaurea L. (Compos.).
Halocharis Moq. Chenopodiaceae. 12 SW. & C. As.
Halochlamys Ind. Nom. Gen. (cf. Wagenitz in Bot. Jahrb. 87, Litber.: 35,
1967) (sphalm.) = Hyalochlamys A. Gray = Angianthus Wendl. (Compos.).
Halochloa Griseb. = Monanthochloë Engelm. (Gramin.).
Halocnemon Spreng. = seq.
Halocnemum M. Bieb. Chenopodiaceae. 1 C. Medit. to C. As.
Halodendron DC. = Halimodendron Fisch. (Legumin.).
Halodendron Roem. & Schult. = seq.
Halodendrum Thou. = Avicennia L. (Avicenniac.).

HALODULA

Halodula Benth. & Hook. f. = seq.

Halodule Endl. Cymodoceaceae. 6 all shallow trop. seas.

Halogeton C. A. Mey. Chenopodiaceae. 3 Spain, NW. Afr., SE. Russia to C. As. *H. sativus* (L.) C. A. Mey. (W. Medit.; *barilla*), formerly burnt for soda.

Halolachna Endl. (sphalm.) = Hololachna Ehrenb. (Tamaricac.).

Halongia Jeanplong = Thysanotus R. Br. (Liliac.).

Halopegia K. Schum. Marantaceae. 6 trop. Afr., Madag., India to Siam & Java.

Halopeplis Bunge. Chenopodiaceae. 3 Medit. to C. As., S. Afr. (Cape Penins.).

Halopetalum Steud. = Banisteria L. = Heteropterys Kunth (Malpighiac.).

Halophila Thou. Hydrocharitaceae. 4 trop. coasts Indian & Pacific oceans, Caribbean.

Halophilaceae J. G. Agardh = Hydrocharitaceae–Halophiloïdeae Dandy.

Halophytaceae Soriano. Dicots. (Centrospermae). 1/1 temp. S. Am. Annual succulent herbs, with 'normal' stem anat. Ls. alt., ent., fleshy, exstip. Fls. ♂ ♀, monoec., the ♂ in strobilate term. spikes, 2-bracteolate, the ♀ sol. in 4–5 uppermost l.-axils, in cavities of the swollen stem, naked, 1-bracteate. ♂ fl.: P 4, membr., A 4, alternitep., with long filif. fil. and oblong versat. anth.; pollen spherical, hexacolpate. ♀ fl.: G̱ 1, uniloc., with 1 basal campylotr.(?) ov., and 1 style with 3 finally exserted stigs. Fr. a syncarp composed of 2–3 fert. ovaries encl. in the swollen stem-apex, forming a reddish berry-like structure; in germination the radicle pushes off an operculum formed from a mamilla immediately below the base of the fallen l. Only genus: *Halophytum*. An interesting type, combining features of several centrospermous fams., e.g. *Chenopodiac.*, *Phytolaccac.*, *Basellac.*, *Batidac.*, etc.

Halophytum Spegazz. Halophytaceae. 1 Patag.

Halopyrum Stapf. Gramineae. 1 coasts of Ind. Ocean.

*Haloragidaceae R.Br. (excl. *Gunnerac.*, *q.v.*). 6/120 cosmop., esp. Austr. Land, marsh, or water herbs of various habits, with great development of adv. r., opp., alt., or whorled usu. exstip. ls., and inconspic. fls., sol. or in infl. Fl. ♀ or ♂ ♀, usu. bracteolate, reg., epig., usu. 4-merous. P 4+4, or 4, or 0; A 4+4, obdipl., or fewer; G̱ (2–4), multiloc., usu. with 1 pend. anatr. ov. in each; styles free. Nut or drupe; embryo straight, in endosp.

Classification and genera (Schindler in *Pflanzenr.*, 1905): Fr. not a schizocarp; *Loudonia* (panicle), *Haloragis* (raceme, ♂, A 2-cyclic, usu. 4-merous), *Meziella*, *Laurembergia* (♂ ♀), *Proserpinaca* (3-merous). Fr. a schizocarp: *Myriophyllum*. The fam. is prob. related to *Datiscaceae*. Connection with *Gunnerac.* perhaps superficial.

Haloragis J. R. & G. Forst. Haloragidaceae. 1 Madag., 75 E. As., Indomal., Austr., Tasm., N.Z., Pacific to Juan Fernandez & Chile.

Halorrhag[ac]eae Lindl. = Haloragidaceae R.Br.

Halorrhena Elmer = Holarrhena R.Br. (Apocynac.).

Haloschoenus Nees = Rhynchospora Vahl (Cyperac.).

Haloscias Fries = Ligusticum L. (Umbellif.).

Halosciastrum Koidz. = Cymopterus Rafin. (Umbellif.).

Halosicyos Mart. Crovetto. Cucurbitaceae. 1 N. Argentina.

Halostachys C. A. Mey. Chenopodiaceae. 1 SE. Russia & Armenia to C. As.
Halostemma Wall. ex Benth. & Hook. f. = Pandanophyllum Hassk. = Mapania
Aubl. (Cyperac.).
Halothamnus Jaub. & Spach = Salsola L. (Chenopodiac.).
Halothamnus F. Muell. = Selenothamnus Melv. (Malvac.).
Halotis Bunge (~ Halimocnemis C. A. Mey.). Chenopodiaceae. 2 C. As. to
Persia & Afghan.
Haloxanthium Ulbr. Chenopodiaceae. 2 Austr.
Haloxylon Bunge. Chenopodiaceae. 10 W. Medit. to Mongolia, S. to Persia,
Afghan., Burma & SW. China. Steppe pl. of curious habit; twigs jointed,
apparently leafless.
Halphophyllum Mansf. Gesneriaceae. 1 Ecuador.
Halymus Wahlenb. = Halimus L. = Atriplex L. (Chenopodiac.).
Hamadryas Comm. ex Juss. Ranunculaceae. 5 Antarctic Am. Dioecious.
Hamalium Hemsl. (sphalm.) = Hamulium Cass. = Verbesina L. (Compos.).
***Hamamelidaceae** R.Br. Dicots. 22/80, chiefly subtrop. (N. & S.), with
very discontinuous distrib. areas. Trees and shrubs with usu. alt., simple or
palmate, stip. ls. Infl. racemose, often a spike or head, sometimes with invol.
of coloured brs. Fl. ♀ or ♂ ♀, often apet., rarely naked, hypo-, peri-, or epi-
gynous, usu. without a disk. K 4–5, usu. imbr.; C 4–5, open or valv., pets.
sometimes long and rolled up like a watch-spring in bud; A 4–5(–14), rarely
fewer; G̲ to G̅ (2), usu. median, rarely obliquely placed, 2-loc. with term.
divided style; ovules 1 or more in each loc., pend., anatr., with ventral or lat.
raphe. Caps. loculic. or septic.; exocarp woody, endocarp horny. Embryo
straight; endosp. Some yield useful timbers, resins, etc. Allied to *Cunoniaceae*
and thence to *Saxifragaceae*; also to *Corylaceae, Altingiaceae*, etc.
Classification and chief genera (after Harms):
  I. *Disanthoïdeae* (fls. separate, in 2-fld. capitula; pet. long, narrow,
  subulate; up to 6 ov. per loc.): *Disanthus.*
  II. *Hamamelidoïdeae* (fls., when ♀ or ♀, clearly separate from one
  another, ♂ fls. in the ♂ infl. sometimes not clearly separable; 1–2 ov.
  per loc.): *Hamamelis, Trichocladus, Dicoryphe, Corylopsis, Parrotia,
  Fothergilla, Distylium, Sycopsis.*
  [III. *Rhodoleioïdeae*—see *Rhodoleiaceae.*]
  IV. *Exbucklandioïdeae* (fls. polygamo-monoec., in capitula; pet. in the
  ♀ fls. 2–5, narrow; stam. 10–14; ls. large, cordate-ovate or 3-lobed,
  palmately nerved, with broad stips., which are closely folded against
  each other): *Exbucklandia.*
  [V. *Liquidambaroïdeae*—see *Altingiaceae.*]
Hamamelis L. Hamamelidaceae. 6 E. As., E. N. Am. *H. virginiana* L.
(N. Am., witch-hazel) flowers in late autumn and ripens its fr. in the following
year.
Hamaria Fourr. = Astragalus L. (Legumin.).
Hamaria Kunze = ? Acaena L. (Rosac.).
Hamastris Mart. ex Pfeiff. = Myriaspora DC. (Melastomatac.).
Hamatocactus Britton & Rose. Cactaceae. 2–3 S. U.S., Mex.
Hamatolobium Fenzl (1842) = Hammatolobium Fenzl (1843) (Legumin.).
Hamatris Salisb. = Dioscorea L. (Dioscoreac.).

Hambergera Scop. = Cacoucia Aubl. (Combretac.).

Hamelia Jacq. Rubiaceae. 40 Mexico to Paraguay, W.I.

Hamelinia A. Rich. = Astelia Banks (Liliac.).

Hamellia L. = Hamelia Jacq. (Rubiac.).

Hamemelis Wernischek = Hamamelis L. (Hamamelidac.).

Hamilcoa Prain. Euphorbiaceae. 1 trop. W. Afr.

Hamiltonia Harv. = Colpoön Bergius (Santalac.).

Hamiltonia Mühlenb. ex Willd. = Pyrularia Michx (Santalac.).

Hamiltonia Roxb. Rubiaceae. 1 India.

Hamiltonia Spreng. = Comandra Nutt. (Santalac.).

Hammada Iljin. Chenopodiaceae. 12 S. Spain, N. Afr., E. Medit. to C. As. & NW. India.

Hammarbya Kuntze = Malaxis Soland. ex Sw. (Orchidac.).

Hammatocaulis Tausch = Ferula L. (Umbellif.).

Hammatolobium Fenzl. Leguminosae. 2–3 NW. Afr., Greece, Syria.

Hamolocenchrus Scop. (sphalm.) = Homalocenchrus Mieg ex Haller = Leersia Sw. (Gramin.).

Hamosa Medik. = Astragalus L. (Legumin.).

Hampea Schlechtd. Malvaceae. 16 Mex. to Colombia.

Hamulia Rafin. = Utricularia L. (Lentibulariac.).

Hamulium Cass. = Verbesina L. (Compos.).

Hanabusaya Nakai. Campanulaceae. 2 Corea.

Hanburia Seem. Cucurbitaceae. 2 Mexico, Guatemala. Fr. explosive.

Hanburyophyton Corrêa de Mello apud Stellfeld = Mansoa DC. (Bignoniac.).

Hancea Hemsl. = Hanceola Kudo (Labiat.).

Hancea Pierre = Hopea Roxb. (Dipterocarpac.).

Hancea Seem. = Mallotus Lour. (Euphorbiac.).

Hanceola Kudo. Labiatae. 3 China.

Hancockia Rolfe. Orchidaceae. 1 China.

Hancornia Gomes. Apocynaceae. 1 Brazil, *H. speciosa* Gomes, the *mangabeira* rubber.

Handelia Heimerl. Compositae. 1 C. As.

Handeliodendron Rehder. Sapindaceae. 1 China.

Handroanthus Mattos = Tabebuia Gomes ex DC. (Bignoniac.).

Handschia Pohl (nomen). 1 Brazil. Quid?

Hanghomia Gagnep. & Thénint. Apocynaceae. 1 Indoch.

Hanguana Blume. Hanguanaceae. 1–2 Ceylon, Indoch., Malaysia (exc. Moluccas & Lesser Sunda Is.).

Hanguanaceae Airy Shaw. Monocots. 1/2 Ceylon, Malaysia. Robust erect herbs, often with long creeping or floating stolons. Ls. mostly radic., long-pet., longit. nerved, with many cross-nervules. Infl. panic., with large bracts. Fls. small, reg., ♂ ♀, dioec. P 3 + 3; shortly connate, persist., greenish or yellowish or the inner red-dotted, the inner larger and vaulted; A 3 + 3, fil. subul., anth. small, basifixed, intr. (in ♀ fls. stds. 3 small + 3 inner large, without anth.); G̲ (3), 3-loc., with 1 axile ov. per loc., and sess. broadly 3-lobed stig. (in ♂ fl. pistillode small, with erect stigs.). Fr. a fleshy thick-walled 1–3-seeded drupe; seeds with endosp. and thick testa. Only genus: *Hanguana*. Formerly included in *Flagellariac.*, but pollen very different. Prob. related to *Xanthorrhoeac.*

(some features of pollen and anat. similar to *Lomandra*), and perhaps even to *Palmae*.

Haniffia Holttum. Zingiberaceae. 2 Penins. Siam, Malay Penins.

Hannafordia F. Muell. Sterculiaceae. 4 Austr.

Hannoa Planch. = Quassia L. (Simaroubac.).

Hannonia Braun-Blanquet & Maire. Amaryllidaceae. 1 NW. Afr.

Hansalia Schott = Amorphophallus Blume (Arac.).

Hansemannia K. Schum. = Archidendron F. Muell. (Legumin.).

Hansenia Turcz. = Ligusticum L. (Umbellif.).

Hanslia Schindl. Leguminosae. 1 E. Malaysia, Bismarck Archipel., New Hebrid., trop. Austr.

Hansteinia Oerst. Acanthaceae. 4 Mex. to Bolivia.

Hapalanthe P. & K. = Apalanthe Planch. = Elodea Michx (Hydrocharitac.).

Hapalanthus Jacq. = Callisia L. (Commelinac.).

Hapale Schott (1857) = seq.

Hapaline Schott (1858). Araceae. 5 Indomal., Indoch.

Hapalocarpum Miq. = Ammannia L. (Lythrac.).

Hapaloceras Hassk. = Payena DC. + Ganua Pierre ex Dub. (Sapotac.).

Hapalochlamys Reichb. = Apalochlamys Cass. = Cassinia R.Br. (Compos.).

Hapaloptera P. & K. = Apaloptera Nutt. ex A. Gray = Abronia Juss. (Nyctaginac.).

Hapalorchis Schlechter. Orchidaceae. 4 trop. S. Am., W.I.

Hapalosa Edgew. = seq.

Hapalosia Wall. ex Wight & Arn. = Polycarpon L. (Caryophyllac.).

Hapalostephium D. Don ex Sweet = Crepis L. (Compos.).

Hapalus Endl. = Apalus DC. = Blennosperma Less. (Compos.).

Haplachne Presl = Dimeria R.Br. (Gramin.).

Haplanthera Hochst. = Ruttya Harv. (Acanthac.).

Haplanthera P. & K. = Aplanthera Horan. = Globba L. (Zingiberac.).

Haplanthodes Kuntze. Acanthaceae. 1 Indomal.

Haplanthus Nees. Acanthaceae. 2 Indomal.

Haplanthus Nees ex Anders. = Haplanthodes Kuntze (Acanthac.).

Haplesthes P. & K. = Haploësthes A. Gray (Compos.).

Haplocalymma Blake. Compositae. 2 Mexico.

Haplocarpha Less. Compositae. 10 Afr.

Haplocarya Phil. = Aplocarya Lindl. = Nolana L. (Nolanac.).

Haplochilus Endl. = Zeuxine Lindl. (Orchidac.).

Haplochorema K. Schum. Zingiberaceae. 1 Sumatra, 6 Borneo.

Haploclathra Benth. Guttiferae. 4 N. Brazil. Wood red.

Haplocoelum Radlk. Sapindaceae. 9 trop. Afr.

Haplodesmium Naud. = Chaetolepis Miq. (Melastomatac.).

Haplodictyum Presl. Thelypteridaceae. 2 Philippines.

Haplodiscus (Benth.) Phil. = Haplopappus Cass. (Compos.).

Haplodypsis Baill. (~ Neophloga Baill.). Palmae. 1 Madag.

Haploësthes A. Gray. Compositae. 2 Mexico.

Haploleja P. & K. = Aploleia Rafin. = Tradescantia L. (Commelinac.).

Haplolobus H. J. Lam. Burseraceae. 17 Malaysia, Solomon Is.

***Haplolophium** Cham. corr. Endl. Bignoniaceae. 1 Brazil.

HAPLOPAPPUS

*Haplopappus Cass. corr. Endl. Compositae. 150 W. Am.

Haplopetalon A. Gray. Rhizophoraceae. 4 Polynesia.

Haplopetalum Miq. = praec.

Haplophandra Pichon. Apocynaceae. 1 Brazil.

Haplophloga Baill. (~ Neophloga Baill.). Palmae. 4 Madag., Masc.

Haplophragma Dop. Bignoniaceae. 4 SE. As. to Sumatra.

*Haplophyllum A. Juss. corr. Reichb. (~ Ruta L.). Rutaceae. 70 Medit. to E. Sib.

Haplophyllum P. & K. = Aplophyllum Cass. = Mutisia L. f. (Compos.).

Haplophyton A. DC. Apocynaceae. 3 SW. U.S., Mex.; Cuba?

Haplopteris Presl = Vittaria Sm. (Vittariac.).

Haplorhus Engl. Anacardiaceae. 1 Peru.

Haplormosia Harms. Leguminosae. 2 trop. W. Afr.

Haplosciadium Hochst. Umbelliferae. 1 NE. trop. Afr.

Haploseseli H. Wolff & Hand.-Mazz. Umbelliferae. 1 W. China.

Haplosphaera Hand.-Mazz. Umbelliferae. 1 SW. China.

Haplostachys Hillebr. Labiatae. 6 Hawaii.

Haplostelis Reichb. = Aplostellis A. Rich. = Nervilia Comm. ex Gaud. (Orchidac.).

Haplostemma Endl. = Vincetoxicum v. Wolf (Asclepiadac.).

Haplostemum Endl. = Aplostemon Rafin. = Scirpus L., etc. (Cyperac.).

Haplostephium Mart. ex DC. Compositae. 3 Brazil.

Haplostichanthus F. Muell. Annonaceae. 1 Queensland.

Haplostichia Phil. = Senecio L. (Compos.).

Haplostigma F. Muell. = Loxocarya R.Br. (Restionac.).

Haplostylis Nees = Rhynchospora Vahl (Cyperac.).

Haplostylis P. & K. = Aplostylis Rafin. = Cuscuta L. (Cuscutac.).

Haplotaxis Endl. = Aplotaxis DC. = Saussurea DC. (Compos.).

Haplothismia Airy Shaw. Burmanniaceae. 1 S. India.

Haploxylon (Koehne) Komarov = Pinus L. (Pinac.).

Happia Neck. ex DC. = Tococa Aubl. (Melastomatac.).

Haptocarpum Ule. Cleomaceae. 1 E. Brazil.

Haptophyllum Vis. & Panč. = Haplophyllum A. Juss. corr. Reichb.

Haquetia D. Dietr. = Hacquetia Neck. ex DC. (Umbellif.).

Haradjania K. H. Rechinger. Compositae. 1 Syria.

Haraëlla Kudo. Orchidaceae. 2 Formosa.

Harbouria Coulter & Rose. Umbelliferae. 1 SW. U.S.

Hardenbergia Benth. Leguminosae. 2 Austr.

Hardwickia Roxb. Leguminosae. 2 India. Apet.

Harfordia Greene & Parry. Polygonaceae. 2 California.

Hargasseria A. Rich. = Linodendron Griseb. (Thymelaeac.).

Hargasseria Schiede & Deppe ex C. A. Mey. = Daphnopsis Mart. & Zucc. (Thymelaeac.).

Harina Buch.-Ham. = Wallichia Roxb. (Palm.).

Hariota Adans. = Rhipsalis Gaertn. (Cactac.).

Hariota DC. = Hatiora Britt. & Rose (Cactac.).

Harlandia Hance = Solena Lour. (Cucurbitac.).

Harlanlewisia Epling. Labiatae. 1 Ecuador.

Harleya Blake. Compositae. 1 Mex.
Harmala Mill. = Peganum L. (Zygophyllac.).
Harmandia Pierre ex Baill. Olacaceae. 2 Indoch., Malay Penins.
Harmandiaceae Van Tiegh. = Olacaceae–Aptandreae Engl.
Harmandiella Costantin. Asclepiadaceae. 1 Indoch.
Harmogia Schau. = Baeckea L. (Myrtac.).
Harmsia K. Schum. Sterculiaceae. 3 trop. Afr.
Harmsiella Briq. = Chartocalyx Regel (Labiat.).
Harmsiodoxa O. E. Schulz. Cruciferae. 3 Austr.
Harmsiopanax Warb. Araliaceae. 3 Java, E. Malaysia.
Harnackia Urb. Compositae. 1 Cuba.
Haronga Thou. = Harungana Lam. (Guttif.).
Harnieria Solms = Justicia L. (Acanthac.).
Harpachaena Bunge = Acanthocephalus Kar. & Kir. (Compos.).
Harpachne Hochst. ex A. Rich. Gramineae. 2 trop. Afr.
Harpaecarpus Nutt. = Madia Molina (Compos.).
Harpagocarpus Hutch. & Dandy (~ Fagopyrum Mill.). Polygonaceae. 1 trop.
Harpagonella A. Gray. Boraginaceae. 1 Calif., NW. Mex.   [Afr.
Harpagonia Nor. = Psychotria L. (Meliac.).
Harpagophytum DC. ex Meissn. Pedaliaceae. 8 S. Afr., Madag. *H. procumbens* (Burch.) DC. (grapple-plant) fr. is beset with large woody grapples about 2·5 cm. long, pointed and barbed. It is thus suited to animal distribution, and is troublesome to wool-growers (cf. *Xanthium*).
Harpalium Cass. = Helianthus L. (Compos.).
Harpalyce D. Don = Prenanthes L. (Compos.).
Harpalyce Moç. & Sessé ex DC. Leguminosae. 25 trop. Am., W.I.
Harpanema Decne. Periplocaceae. 1 Madag.
Harpechloa Kunth (1833) = Harpochloa Kunth (1830) (Gramin.).
Harpelema Jacq. f. = Rothia Pers. (Legumin.).
Harpephora Endl. = Aspilia Thou. (Compos.).
Harpephyllum Bernh. ex Krauss. Anacardiaceae. 1 S. Afr.
Harperella Rose = Ptilimnium Rafin. (Umbellif.).
Harperia Fitzgerald. Restionaceae. 1 Austr.
Harperia Rose = Harperella Rose = Ptilimnium Rafin. (Umbellif.).
Harperocallis McDaniel. Liliaceae. 1 Florida.
Harpocarpus Endl. = Acanthocephalus Kar. & Kir. (Compos.).
Harpocarpus P. & K. = Harpaecarpus Nutt. = Madia Mol. (Compos.).
Harpochilus Nees. Acanthaceae. 3 Brazil.
Harpochloa Kunth. Gramineae. 1 S. Afr.
Harpolema P. & K. = Harpelema Jacq. f. = Rothia Pers. (Legumin.).
Harpolyce P. & K. (1) = Harpalyce Moç. & Sessé ex DC. (Legumin.).
Harpolyce P. & K. (2) = Harpalyce D. Don = Prenanthes L. (Compos.).
Harpophora P. & K. = Harpephora Endl. = Aspilia Thou. (Compos.).
Harpophyllum P. & K. = Harpephyllum Bernh. ex Krauss (Anacardiac.).
Harpostachys Trin. = Panicum L. (Gramin.).
Harpullia Roxb. Sapindaceae. 37 Indomal., trop. Austr., Pacif.
Harrachia Jacq. f. = Crossandra Salisb. (Acanthac.).
Harrera Macfad. = Tetrazygia Rich. (Melastomatac.).

533

HARRIMANELLA

Harrimanella Coville (~ Cassiope D. Don). Ericaceae. ? Arctic (Greenl., Icel., Spitzb., arctic Eur.; Jap. & Kamch. to NW. U.S.).

Harrisella Willis (sphalm.) = Harrisiella Fawc. & Rendle (Orchidac.).

Harrisia Britton. Cactaceae. 13 Florida, W.I.

Harrisiella Fawcett & Rendle. Orchidaceae. 1 Florida, Mexico, Salvador, W.I.

*Harrisonia R.Br. Simaroubaceae. 4 trop. Afr., SE. As., Indomal., trop. Austr.

Harrisonia Hook. = Loniceroïdes Bullock (Asclepiadac.).

Harrisonia Neck. = Xeranthemum L. (Compos.).

Harrysmithia H. Wolff. Umbelliferae. 1 SW. China.

× **Hartara** hort. Orchidaceae. Gen. hybr. (iii) (Broughtonia × Laelia × Sophronitis).

Hartia Dunn = Stewartia L. (Theac.).

Hartiana Rafin. = Anemone L. (Ranunculac.).

Hartighaea Reichb. = seq.

Hartighsea A. Juss. = Dysoxylum Blume (Meliac.).

Hartigia Miq. = Miconia Ruiz & Pav. (Melastomatac.).

Hartigsea Steud. = Hartighsea A. Juss. = Dysoxylum Blume (Meliac.).

Hartleya Sleumer. Icacinaceae. 1 New Guinea.

Hartmannia Spach (~ Oenothera L.). Onagraceae. 15 N. Am. to Bolivia,

Hartmannia DC. = Hemizonia DC. (Compos.).   [W.I.

Hartogia Hochst. = Cassinopsis Sond. (Icacinac.).

Hartogia L. = Agathosma Willd. (Rutac.).

Hartogia L. f. Celastraceae. 3 S. Afr., Madag.

Hartwegia Lindl. = Nageliella L. O. Williams (Orchidac.).

Hartwegia Nees = Chlorophytum Ker-Gawl. (Liliac.).

Hartwegiella O. E. Schulz. Cruciferae. 1 Mex.

Hartwrightia A. Gray. Compositae. 1 Florida.

Harungana Lam. Guttiferae. 1 trop. Afr., Madag., Maurit.

Harveya Hook. Scrophulariaceae. 40 trop. & S. Afr., Masc. Some are root parasites, like *Euphrasia.*

Harveya R. W. Plant ex Meissn. = Peddiea Harv. (Thymelaeac.).

Harwaya Steud. = Harveya Hook. (Scrophulariac.).

Haselhoffia Lindau. Acanthaceae. 4 trop. W. Afr.

Haseltonia Backeb. = Cephalocereus Pfeiff. (Cactac.).

Hasseanthus Rose (~ Dudleya Britt. & Rose). Crassulaceae. 5 Calif., N. Lower

Hasselquistia L. = Tordylium L. (Umbellif.).   [Calif.

Hasseltia Blume = Kickxia Blume (Apocynac.).

Hasseltia Kunth. Flacourtiaceae. 12 Mexico to Andes.

Hasseltiopsis Sleum. = Pleuranthodendron L.O. Williams (?Tiliac. ?Flacourtiac.).

Hasskarlia Baill. Euphorbiaceae. 4 trop. Afr. Infl. lf.-opp. (cf. *Suregada*).

Hasskarlia Meissn. = Turpinia Vent. (Staphyleac.).

Hasskarlia Walp. = Pandanus L. f. (Pandanac.).

Hasslerella Chodat. Scrophulariaceae. 1 Argentina.

Hassleria Briquet ex Moldenke = Amasonia L. f. (Verbenac.).

Hassleropsis Chodat = Basistemon Turcz. (Scrophulariac.).

Hasteola Rafin. (~ Cacalia L.). Compositae. 1 NE. Russia, 50 E. & SE. As., 1 N. Am.

Hastifolia Ehrh. (uninom.) = *Scutellaria hastifolia* L. (Labiat.).
Hastingia Koenig ex Endl. = Abroma Jacq. (Sterculiac.).
Hastingia Koen. ex Sm. = Holmskioldia Retz. (Verbenac.).
Hastingsia P. & K. (bis) = praecc.
Hastingsia S. Wats. = Schoenolirion Durand (Liliac.).
× **Hatcherara** hort. (vii) = × Colmanara hort. (Orchidac.).
Hatiora Britton & Rose (~ Rhipsalis Gaertn.). Cactaceae. 4 SE. Brazil.
Hatschbachia L. B. Smith = Napeanthus Gardn. (Gesneriac.).
Hatschbachiella R. M. King & H. Rob. (~ Eupatorium L.). Compositae.
2 Braz. to Argent.
Haumania J. Léonard. Marantaceae. 3 trop. Afr.
Haumaniastrum Duvign. & Plancke. Labiatae. 23 trop. Afr.
Hausemannia auct. (sphalm.) = Hansemannia K. Schum. = Archidendron F.
Muell. (Legumin.).
Haussknechtia Boiss. Umbelliferae. 1 Persia.
Haussmannia F. Muell. (1864; non Hausmannia Dunker 1846—gen. foss.) = seq.
Haussmannianthes v. Steenis = Neosepicaea Diels (Bignoniac.).
Haustrum Nor. = Rhododendron L. (Ericac.).
Hauya Moç. & Sessé ex DC. Onagraceae. 14 Mex., C. Am.
Havanella Kuntze = Flaveria Juss. (Compos.).
Havardia Small (~ Pithecellobium Mart.). Leguminosae. 10 N. Am., Mex.
Havetia Kunth. Guttiferae. 1 Colombia.
Havetiopsis Planch. & Triana. Guttiferae. 5 trop. S. Am.
Havilandia Stapf = Trigonotis Stev. (Boraginac.).
× **Hawaiiara** hort. Orchidaceae. Gen. hybr. (iii) (Renanthera × Vanda × Vandopsis).
× **Hawkesara** hort. (iii). Orchidaceae. Gen. hybr. (Cattleya × Cattleyopsis × Epidendrum).
***Haworthia** Duval. Liliaceae. 150 S. Afr. Xero. with fleshy ls., similar in habit to some *Crassulaceae*.
Haxtonia Caley ex D. Don = Olearia Moench (Compos.).
Haya Balf. f. Caryophyllaceae. 1 Socotra.
Hayacka Willis (sphalm.) = Hayecka Pohl = ? Paullinia L. (Sapindac.).
Hayataëlla Masamune. Rubiaceae. 1 Formosa.
Haydonia R. Wilczek. Leguminosae. 2 trop. Afr.
Hayecka Pohl = ? Paullinia L. (Sapindac.).
Haylockia Herb. Amaryllidaceae. 6 Andes. Like *Crocus*, with fls. projecting from the soil.
Haynaldia Kanitz = Lobelia L. (Campanulac.).
Haynaldia Schur. Gramineae. 3 Medit.
× **Haynaldoticum** Cif. & Giac. Gramineae. Gen. hybr. (Haynaldia × Triticum).
Haynea Reichb. = Modiola Moench (Malvac.).
Haynea Schumach. & Thonn. = Fleurya Gaudich. (Urticac.).
Haynea Willd. = Pacourina Aubl. (Compos.).
Hazardia Greene = Haplopappus Cass. (Compos.).
Hazomalania Capuron (~ Hernandia L.). Hernandiaceae. 1 Madag.
Hazunta Pichon. Apocynaceae. 8 Madag., Comoros, Seychelles.

HEARNIA

Hearnia F. Muell. = Aglaia Lour. (Meliac.).

Hebandra P. & K. = Hebeandra Bonpl. = Monnina Ruiz & Pav. (Polygalac.).

Hebanthe Mart. = Pfaffia Mart. (Amaranthac.).

Hebe Comm. ex Juss. (~ Veronica L.). Scrophulariaceae. 100–150 New Guinea, Australas., temp. S. Am., Falkl. Is. Shrubs or small trees, with handsome spikes of fls.; often cult. In N.Z. the genus is char. alpine; about 90 spp. are endemic. Many, e.g. *H. cupressoïdes* (Hook. f.) Cockayne & Allan, are xero. with reduced ls. appressed to stem, so that the twigs resemble those of *Cupressus* and other *Coniferae*.

Hebea Hedw. f. (~ Gladiolus L.). 15 S. Afr.

Hebeandra Bonpl. = Monnina Ruiz & Pav. (Polygalac.).

Hebeanthe Reichb. = Hebanthe Mart. (Amaranthac.).

*****Hebecladus** Miers. Solanaceae. 12 W. trop. S. Am.

Hebeclinium DC. = Eupatorium L. (Compos.).

Hebecocca Beurl. = Omphalea L. (Euphorbiac.).

Hebecoccus Radlk. Sapindaceae. 2 Philippines, 1 Java.

Hebelia C. C. Gmel. = Tofieldia Huds. (Liliac.).

Hebenstreitia Murr. = Hebenstretia L. (Scrophulariac.).

Hebenstreitiaceae Horan. = Selaginaceae Choisy = Scrophulariaceae–Selagineae Reichb.

Hebenstretia L. Scrophulariaceae. 40 trop. & S. Afr. The corolla is slit open along the anterior side, and the style and sta. project through the slit.

Hebepetalum Benth. Linaceae. 6 trop. S. Am.

*****Heberdenia** Banks. Myrsinaceae. 1 Mex.; 1 Canaries, Madeira.

Hebestigma Urb. Leguminosae. 1 W.I.

Hebocladus P. & K. = Hebecladus Miers (Solanac.).

Heboclinium P. & K. = Hebeclinium DC. = Eupatorium L. (Compos.).

Hebococca P. & K. = Hebecocca Beurl. = Omphalea L. (Euphorbiac.).

Hebococcus P. & K. = Hebecoccus Radlk. (Sapindac.).

Hebokia Rafin. = Euscaphis Sieb. & Zucc. (Staphyleac.).

Hebonga Radlk. = Ailanthus Desf. (Simaroubac.).

Hebopetalum P. & K. = Hebepetalum Benth. (Linac.).

Hebostigma P. & K. = Hebestigma Urb. (Legumin.).

Hebradendron R. Grah. = Garcinia L. (Guttif.).

Hecabe Rafin. = Phaius Lour. (Orchidac.).

Hecale Rafin. = Wahlenbergia Schrad. (Campanulac.).

Hecaste Soland. ex Schumacher = Bobartia L. (Iridac.).

Hecastocleïs A. Gray. Compositae. 1 SW. U.S.

Hecastophyllum Kunth = Ecastaphyllum P.Br. = Dalbergia L. (Legumin.).

Hecatactis (F. Muell.) Mattf. (~ Keysseria Lauterb.). Compositae. 2 New Guinea.

Hecatandra Rafin. = Acacia Mill. (Legumin.).

Hecatea Thou. = Omphalea L. (Euphorbiac.).

Hecaterium Kunze ex Reichb. = praec.

Hecaterosachna P. & K. = Hekaterosachne Steud. = Oplismenus Beauv. (Gramin.).

Hecatonia Lour. = Ranunculus L. (Ranunculac.).

Hecatostemon Blake. Flacourtiaceae. 1 Venez.

Hecatounia Poir. = Ranunculus L. (Ranunculac.).
Hecatris Salisb. = Asparagus L. (Liliac.).
Hechtia Klotzsch. Bromeliaceae. 35 S. U.S. to C. Am.
Hecistocarpus P. & K. = Hekistocarpa Hook. f. (Rubiac.).
Hecistopteris J. Sm. Vittariaceae. 1 trop. Am.
Heckeldora Pierre (~ Guarea L.). Meliaceae. 6 trop. Afr.
Heckelia K. Schum. = Rhipogonum J. R. & G. Forst. (Liliac.).
Heckeria Kunth = Pothomorphe Miq. (Piperac.).
Heckeria Rafin. = Litsea Lam. (Laurac.).
Hectorea DC. = Chrysopsis Nutt. (Compos.).
Hectorella Hook. f. Hectorellaceae. 1 S. N.Z.
Hectorellaceae Philipson & Skipworth. Dicots. (Centrospermae). 2/2 Antarct. Perenn. densely caespit. herbs, with simple ent. coriaceous alt. exstip. ls. Fls. ♀, ♂ or ♀, reg., sol., axill.; bracteoles 0 or 2–3. K 2, median; C 4–5, free or shortly connate; A 3–5(–6), alternipet., free or adnate to C-tube when this is present, with versat. anth.; G̲ (2), 1-loc., with simple style and shortly bifid stig., and 4–7 ovules borne near base of free central plac. Fr. a 1–5-seeded caps.; seeds with endosp. and curved embr. Genera: *Hectorella, Lyallia.* Pollen distinctive (different from *Caryophyllac.*).
Hecubaea DC. Compositae. 2 Mexico.
Hedaroma Lindl. = Darwinia Rudge (Myrtac.).
Hedeoma Pers. Labiatae. 30 Am.
Hedeomoïdes Briq. = Pogogyne Benth. (Labiat.).
Hedera L. Araliaceae. 15 Canary Is., W. & C. Eur., Medit. to Cauc.; W. Himal. to Korea & Japan; Queensl. *H. helix* L. (ivy) is a root-climber. Ls. dimorphic, those on the climbing shoots lobed, those on the freely projecting shoots that bear the infl. not. The former form leaf-mosaics better. Fls. not very conspicuous but, coming out late in the year, largely visited for the freely exposed honey by flies and wasps.
Hederaceae Bartl. = Vitidaceae + Cornaceae + Hedera, etc.
Hederanthum Steud. = Phyteuma L. (Campanulac.).
Hederella Stapf = Catanthera F. Muell. (Melastomatac.).
Hederopsis C. B. Clarke. Araliaceae. 1 Malay Peninsula.
Hederorchis Thou. = seq.
Hederorkis Thou. = Bulbophyllum Thou. (Orchidac.).
Hederula Fabr. (uninom.) = Glechoma hederacea L. (Labiat.).
Hedichium Ritg. = Hedychium Koen. (Zingiberac.).
Hedinia Ostenf. Cruciferae. 1 C. As., Tibet, NW. Himalaya.
Hediosma L. ex B. D. Jacks. = Nepeta L. (Labiat.).
Hediosmum Poir. = Hedyosmum Sw. (Chloranthac.).
Hedisarum Neck. = Hedysarum L.
Hedona Lour. = Lychnis L. (Caryophyllac.).
Hedraeanthus Griseb. = Edraianthus A. DC. (Campanulac.).
Hedraianthera F. Muell. Celastraceae. 1 E. Austr.
Hedraiophyllum (Less.) Spach = Gochnatia Kunth (Compos.).
Hedraiostylus Hassk. = Pterococcus Hassk. (Euphorbiac.).
Hedranthera (Stapf) Pichon (~ Callichilia Stapf). Apocynaceae. 1 trop. W. Afr.
Hedranthus Rupr. = Edraianthus A. DC. (Campanulac.).

HEDSTROMIA

Hedstromia A. C. Smith. Rubiaceae. 1 Fiji.
Hedwigia Medik. = Commelina L. (Commelinac.).
Hedusa Rafin. = Dissotis Benth. (Melastomatac.).
Hedwigia Sw. = Tetragastris Gaertn. (Burserac.).
Hedyachras Radlk. Sapindaceae. 1 Philipp. Is.
Hedycapnos Planch. = Dicentra Borckh. corr. Bernh. (Fumariac.).
Hedycaria Murr. = Hedycarya J. R. & G. Forst. (Monimiac.).
Hedycarpus Jack = Baccaurea Lour. (Euphorbiac.).
Hedycaria J. R. & G. Forst. Monimiaceae. 25 SE. Austr. to Solomon Is. & Fiji.
Hedycaryopsis Danguy. Monimiaceae. 1 Madag.
Hedychium Koen. Zingiberaceae. 50 Madag., Indomal., SW. China. Rhizome often tuberous. The fl. has a long tube, at the end of which spring the narrow free parts of the petals and the larger staminodes and labellum. The stigma projects just beyond the anther.
Hedychloa B. D. Jacks. = seq.
Hedychloë Rafin. = Kyllinga Rottb. (Cyperac.).
Hedycrea Schreb. = Licania Aubl. (Chrysobalanac.).
Hedyosmon Spreng. = Hedyosmum Sw. (Chloranthac.).
Hedyosmos Mitch. (Cunila Roy. ex L.). Labiatae. 14 E. N. Am. to Urug.
Hedyosmum Sw. Chloranthaceae. 1 S. China, SE. As., Sumatra, Borneo; 40 trop. Am., W.I.
Hedyotis L. Rubiaceae. 150 trop. As.
Hedyphylla Stev. = Astragalus L. (Legumin.).
Hedypnois Mill. Compositae. 3 Canaries, Madeira, Medit.
Hedypnois Scop. = Taraxacum Weber ex Wigg. (Compos.).
Hedysa P. & K. = Hedusa Rafin. = Dissotis Benth. (Melastomatac.).
Hedysar[ac]eae J. G. Agardh = Leguminosae–Hedysareae DC.
Hedysarum L. Leguminosae. 150 N. temp.
Hedyscepe H. Wendl. & Drude (~ Kentia Bl.). Palmae. 1 Lord Howe Island.
Hedystachys Fourr. = Pseudolysimachion (Koch) Opiz = Veronica L. (Scrophulariac.).
Hedythyrsus Bremek. Rubiaceae. 2 trop. Afr.
Heeria Meissn. Anacardiaceae. 1 S. Afr.
Heeria Schlechtd. = Schizocentron Meissn. (Melastomatac.), q.v.
Hegemone Bunge ex Ledeb. (~ Trollius L.). Ranunculaceae. 2 C. As.
Hegetschweilera Heer & Regel = Alysicarpus Neck. ex Desv. (Legumin.).
Hegnera Schindl. Leguminosae. 1 Burma, Sumatra.
Heimerlia Skottsb. (1936; non v. Höhnel 1903—Fungi) = seq.
Heimerliodendron Skottsb. Nyctaginaceae. 1 Mascarenes to Polynesia.
Heimia Link & Otto. Lythraceae. 3 S. U.S. to Argent.
Heimodendron Sillans = Entandrophragma C. DC. (Meliac.).
Heinchenia Hook. f. (sphalm.) = seq.
Heinekenia Webb ex Benth. & Hook. f. = Lotus L. (Legumin.).
Heinsenia K. Schum. Rubiaceae. 5 trop. Afr.
Heinsia DC. Rubiaceae. 10 trop. Afr.
Heintzia Karst. = Alloplectus Mart. (Gesneriac.).
Heintzia Steud. = Heinzia Scop. = Coumarouna Aubl. (Legumin.).

Heinzelia Nees = Chaetothylax Nees (Acanthac.).
Heinzelmannia Neck. = Spigelia L. (Loganiac.).
Heinzia Scop. = Coumarouna Aubl. (Legumin.).
Heistera Kuntze = Heisteria Boehm. = Muraltia Jacq. (Polygalac.).
Heistera Schreb. = Heisteria Jacq. (Olacac.).
Heisteria Boehm. = Muraltia Jacq. (Polygalac.).
Heisteria Fabr. (uninom.) = *Veltheimia* Gled. sp. (Liliac.).
Heisteria Jacq. Olacaceae. 50 warm Am., W. Afr.
Heisteriaceae Van Tiegh. = Olacaceae–Heisterieae Engl.
Hekaterosachne Steud. = Oplismenus Beauv. (Gramin.).
Hekeria Endl. = Heckeria Kunth = Pothomorphe Miq. (Piperac.).
Hekistocarpa Hook. f. Rubiaceae. 1 trop. W. Afr. (S. Nig., Cam.).
Hekorima Kunth = Streptopus Michx (Liliac.).
Heladena A. Juss. Malpighiaceae. 6 trop. & subtrop. S. Am.
Helanthium (Benth. & Hook. f.) Engelm. ex Britton corr. Pichon. Alismataceae. 1 E. Canada to S. Am.
Helcia Lindl. Orchidaceae. 1 Colombia.
Heldreichia Boiss. Cruciferae. 6 As. Min. to Afghan.
Heleastrum DC. = Aster L. (Compos.).
Heleiotis Hassk. = Phylacium Benn. (Legumin.).
Helemonium Steud. = Heliopsis Pers. (Compos.).
Helena Haw. = Narcissus L. (Amaryllidac.).
Heleneum Buckl. = Helenium L. (Compos.).
Helenia Mill. = Helenium L. (Compos.).
Heleniaceae Bessey = Compositae–Helenieae Cass.
Heleniastrum Fabr. (uninom.) = *Helenium* L. sp. (Compos.).
Heleniopsis Baker (sphalm.) = Heloniopsis A. Gray (Liliac.).
Helenium L. Compositae. 40 W. Am.
Helenium Mill. = Inula L. (Compos.).
Helenomoium Willd. = Heliopsis Pers. (Compos.).
Heleocharis Lestib. = Eleocharis R.Br. (Cyperac.).
Heleochloa Beauv. = Sporobolus R.Br. + Phleum L. (Gramin.).
Heleochloa (Fries) Drejer = Glyceria L. + Puccinellia Parl. (Gramin.).
Heleochloa Host ex Roem. = Crypsis Ait. (Gramin.).
Heleogenus P. & K. = Eleogenus Nees = Scirpus L. (Cyperac.).
Heleogiton Schult. = Scirpus L. (Cyperac.).
Heleonastes Ehrh. (uninom.) = *Carex heleonastes* L. (Cyperac.).
Heleophila Schult. = seq.
Heleophyla auct. = seq.
Heleophylax Beauv. ex Lestib. = Scirpus L. (Cyperac.).
Helepta Rafin. = Heliopsis Pers. (Compos.).
Helia Benth. & Hook. f. = Helië M. Roem. = Atalantia Corrêa (Rutac.).
Helia Mart. = Lisianthus Aubl. (Gentianac.).
Heliabravoa Backeb. Cactaceae. 1 Mexico.
Heliamphora Benth. Sarraceniaceae. 6 Venez., Guiana. Pitcher plants (cf. *Sarracenia*).
Helianthaceae Bessey = Compositae–Heliantheae Cass.
Helianthella Torr. & Gray. Compositae. 10 W. U.S., Mexico.

HELIANTHEMOÏDES

Helianthemoïdes Medik. = Talinum Adans. (Portulacac.).

Helianthemon St-Lag. = Helianthemum Mill. (Cistac.).

Helianthemum S. F. Gray = Helianthus L. (Compos.).

Helianthemum Mill. (excl. Crocanthemum Spach). Cistaceae. 100 W. & C. Eur., Medit., S. to C. Verde Is., Sahara, Somal., E. to C. As. & Persia. Infl. a cincinnus. The fl. contains no honey and is homogamous, with sensitive sta., which move outwards when touched.

Helianthium Britton (sphalm.) = Helanthium (Benth. & Hook. f.) Engelm. ex Britton corr. Pichon (Alismatac.).

Helianthocereus Backeb. = Trichocereus (A. Berger) Riccob. (Cactac.).

Helianthostylis Baill. Moraceae. 2 Brazil.

Helianthum Prain in I.K. (sphalm.) = Helanthium (Benth. & Hook. f.) Engelm. ex Britton corr. Pichon (Alismatac.).

Helianthus L. Compositae. 110 Am. In *H. annuus* L. (sunflower) the number of fls. upon the head is often enormous and they show very regular spiral arrangement, probably due (largely) to pressure in the bud. Ray florets neuter. The seeds give oil. *H. tuberosus* L. (Jerusalem artichoke) has subterranean tuberous stems, like potatoes, with well-marked 'eyes' (buds in axils of scale-ls.).

× **Heliaporus** Rowley. Cactaceae. Gen. hybr. (Aporocactus × Heliocereus).

Helicandra Hook. & Arn. = Parsonsia R.Br. (Apocynac.).

Helicanthera Roem. & Schult. = Helixanthera Lour. (Loranthac.).

Helicanthes Danser. Loranthaceae. 1 Penins. India.

Helichroa Rafin. = Echinacea Moench (Compos.).

Helichrysaceae Link corr. Bullock = Elichrys[ac]eae Link = Compositae–Inuleae Cass.

Helichrysopsis Kirp. Compositae. 1 trop. E. Afr.

*****Helichrysum** Mill. corr. Pers. Compositae. 500 S. Eur., trop. & S. Afr., Madag., Socotra, SW. As., S. India, Ceyl., Austr.; 200 in S. Afr. Many xero. with hairy surface, decurrent ls., etc. The dried fl.-heads of some spp. are 'everlastings'.

Helicia Lour. Proteaceae. 90 E. & SE. As., Indomal., E. Austr.

Helicia Pers. = Helixanthera Lour. (Loranthac.).

Helicilla Moq. (? = Suaeda Forsk.). Chenopodiaceae. 1 S. China.

Heliciopsis Sleum. Proteaceae. 7 Assam to S. China, W. Malaysia.

Helicodea Lem. = Billbergia Thunb. (Bromeliac.).

*****Helicodiceros** Schott. Araceae. 1 Balearics, Corsica, Sardinia, *H. muscivorus* (L. f.) Engl. The development of the pedate leaf is cymose; the later-formed branches grow more slowly than the earlier. The name *muscivorus* is due to the number of flies captured; attracted by the foul smell of the infl. (cf. *Arum*) they collect inside the spathe in enormous numbers; it may often be seen tightly packed; when it withers the top closes and they are caught.

*****Heliconia** L. Heliconiaceae. 80 trop. Am. Fls. in cincinni; odd sep. post.

Heliconiaceae (Endl.) Nak. Monocots. 1/80 trop. Am. Large perenn. herbs with distich. ls. Infl. term., with distich. coloured bracts bearing ∞-fld. cincinni. Fls. ⚥. P 3 + 3, the outer posticous tep. large and free, remainder smaller, mostly connate into a 5-toothed cymbiform struct. A 5 with lin. anth.; std. 1, short, petaloid, opp. the postic. tep. $\overline{G}$ (3) with 1 basal anatr. ov. per

540

loc.; style filif. with clav. or capit. 3-lobed stig. Fr. a caps., dehisc. into 3 cocci, ± blue; seeds exarillate. Only genus: *Heliconia.*

Heliconiopsis Miq. = Heliconia L. (Heliconiac.).

Helicophyllum Schott (1853; non Brid. 1827–Musci) = Eminium (Bl.) Schott (Arac.).

Helicostylis Tréc. Moraceae. 12 C. & trop. S. Am.

Helicotrichum Bess. ex Reichb. = Helictotrichon Bess. ex Roem. & Schult. (Gramin.).

Helicroa Rafin. = Helichroa Rafin. = Echinacea Moench (Compos.).

Helicta Cass. = Borrichia Adans. (Compos.).

Helicta Less. = Epallage DC. (Compos.).

Helicter[ac]eae J. G. Agardh = Sterculiaceae–Helictereae DC.

Helicteres L. Sterculiaceae. 60 trop. As. & Am. The fls. become zygomorphic if they happen to be in a horiz. position.

Helicterodes (DC.) Kuntze = Caiophora Presl (Loasac.).

Helicteropsis Hochr. Malvaceae. 2 Madag.

Helictonema Pierre = Hippocratea L. (Celastrac.).

Helictonia Ehrh. (uninom.) = *Ophrys spiralis* L. = *Spiranthes spiralis* (L.) Chevall. (Orchidac.).

Helictotrichon Bess. ex Roem. & Schult. Gramineae. 90 Eur., trop. & S. Afr., As.; 1 Java, 2 W. N. Am., 1 temp. S. Am.

Helië M. Roem. = Atalantia Corrêa (Rutac.).

Helietta Tul. Rutaceae. 8 Mex., Cuba, trop. S. Am.

Heligma Benth. & Hook. f. = seq.

Heligme Blume = Parsonsia R.Br. (Apocynac.).

***Helinus** E. Mey. ex Endl. Rhamnaceae. 6 trop. & S. Afr., Arabia, Himal.

× **Heliocactus** Janse = × Heliphyllum Rowley (Cactac.).

Heliocarpus L. Tiliaceae. 17 Mex. to Paraguay. Dioec.; 4-merous.

Heliocarya Bunge = Caccinia Savi (Boraginac.).

Heliocereus (A. Berger) Britton & Rose. Cactaceae. 3–4 Mex., C. Am.

Heliocharis Lindl. = Eleocharis R.Br. (Cyperac.).

× **Heliochia** Rowley. Cactaceae. Gen. hybr. (Heliocereus × Nopalxochia).

Heliochroa Rafin. (1833) = Helichroa Rafin. (1825) = Echinacea Moench (Compos.).

Heliogenes Benth. = Jaegeria Kunth (Compos.).

Heliomeris Nutt. (~ Viguiera Kunth). Compositae. 5 W. N. Am., 1 SE.

Helionopsis Franch. & Sav. = Heloniopsis A. Gray (Liliac.). [U.S.

Heliophila Burm. f. ex L. Cruciferae. 75 S. Afr.

Heliophthalmum Rafin. = Bidens L. (Compos.).

Heliophyla Neck. = Heliophila Burm. f. ex L. (Crucif.).

Heliophylax Lestib. ex Steud. = Heleophylax Beauv. = Scirpus L. (Cyperac.).

Heliophylla Scop. = Heliophila Burm. f. ex L. (Crucif.).

Heliophyton Benth. = seq.

Heliophytum DC. = Heliotropium L. (Boraginac.).

Heliopsis Pers. Compositae. 12 Am.

Helioreos Rafin. = Pectis L. (Compos.).

Heliosciadium Bluff & Fingerh. = Helosciadium Koch = Apium L. (Umbellif.).

× **Helioselenius** Rowley. Cactaceae. Gen. hybr. (Heliocereus × Selenicereus).

HELIOSOCEREUS

× **Heliosocereus** Glass & Foster (*nom. provis.*). Cactaceae. Gen. hybr. (Heliocereus × Pilosocereus).

Hellosperma Reichb. Caryophyllaceae. 8 mts. of SE. Eur.

Heliospora Hook. f. = Helospora Jack = Timonius DC. (Rubiac.).

Heliotrichum Bess. ex Schur = Helictotrichon Bess. ex Roem. & Schult. (Gramin.).

*****Heliotropi[a]ceae** Schrad. = Boraginaceae–Heliotropiëae J. G. Agardh.

Heliotropium L. Boraginaceae. 250 trop. & temp.

× **Heliphyllum** Rowley. Cactaceae. Gen. hybr. (Epiphyllum × Heliocereus).

Helipterum DC. Compositae. 90 S. Afr., Austr. Xero. with persistent invol. of white scaly bracts. The dried flower-heads are sold as 'everlastings' (cf. *Helichrysum*, etc.).

Helisanthera Rafin. = Helixanthera Lour. (Loranthac.).

Helittophyllum Blume = Helicia Lour. (Proteac.).

Helix Dum. ex Steud. = Salix L. (Salicac.).

Helix Mitch. = Inc. sed. 1 E. U.S.

Helixanthera Lour. Loranthaceae. 50 trop. Afr., Indomal. to Celebes.

Helixyra Salisb. Iridaceae. 14 S. Afr.

Helleboraceae Loisel. = Ranunculaceae–Helleboreae DC.

Helleboraster Fabr. (uninom.) = *Adonis* L. sp. (Ranunculac.).

Helleboraster Hill = Helleborus L. (Ranunculac.).

Helleborine Ehrh. (uninom.) = *Serapias latifolia* L. = *Epipactis helleborine* (L.) Crantz (Orchidac.).

Helleborine Kuntze = Calopogon R.Br. (Orchidac.).

Helleborine Mill. = Cypripedium L., Cephalanthera Rich., Epipactis Zinn, etc. (Orchidac.).

Helleborine Pers. = Serapias L. (Orchidac.).

Helleborodes Kuntze = seq.

Helleboroïdes Adans. = Eranthis Salisb. (Ranunculac.).

Helleborus Gueldenst. = Veratrum L. (Liliac.).

Helleborus L. Ranunculaceae. 20 Eur., Medit. to Cauc.; 1 W. China. Pl. woody below, each shoot from the stock taking several years to reach maturity and flower. Fl. protog., opening early in the year. Cpls. slightly coherent at base. In *H. niger* L. (Christmas rose) the P turns green after the fl. has been fert.

Hellenia Retz. = Costus L. (Costac.).

Hellenia Willd. = Alpinia L. (Zingiberac.).

Hellenocarum H. Wolff. Umbelliferae. 2 mid Medit.

Hellera Schrad. ex Doell = Raddia Bertol. (Gramin.).

Helleranthus Small = Verbena L. (Verbenac.).

Helleria Fourn. (~ Festuca L.). Gramineae. 2 Mex., Venez.

Helleria Nees & Mart. = Vantanea Aubl. (Houmiriac.).

Helleriella A. D. Hawkes. Orchidaceae. 1 C. Am. (Nicar.).

Hellerorchis A. D. Hawkes = Rodrigueziella Kuntze (Orchidac.).

Hellmuthia Steud. = Scirpus L. (Cyperac.).

Hellwigia Warb. = Alpinia Roxb. (Zingiberac.).

Helmentia Jaume St-Hil. = Helmintia Juss. = Picris L. (Compos.).

Helmholtzia F. Muell. Philydraceae. 3 New Guinea, E. Austr.

Helmia Kunth = Dioscorea L. (Dioscoreac.).
Helminta Willd. = seq.
Helminthia Juss. corr. DC. = Picris L. (Compos.).
Helminthion St-Lag. = praec.
Helminthocarpon A. Rich. (1847; non Fée 1824—Lichenes) = Vermifrux J. B. Gillett (Legumin.).
Helminthocarpum auctt. = praec.
Helminthospermum Thw. = Gironniera Gaudich. (Urticac.).
Helminthospermum (Torr.) Durand = Phacelia Juss. (Hydrophyllac.).
Helminthostachyaceae Ching = Ophioglossaceae Presl.
Helminthostachys Kaulf. Ophioglossaceae. 1, *H. zeylanica* Hook. f., Ceylon, Himal. to Queensland. Rhiz. dorsiv. with 2-ranked ls. on the upper side. Sporangia globose, on sporangiophores from the sides of the fertile spike. (Cf. Farmer & Freeman, *Ann. Bot.* **13**: 421, 1899; Lang on prothallus, *do.* **16**: 23, 1902.)
Helminthoteca Juss. = seq.
Helminthotheca Zinn = Helminthia Juss. corr. DC. = seq.
Helmintia Juss. = Picris L. (Compos.).
Helmiopsiella Arènes. Sterculiaceae. 1 Madag.
Helmiopsis H. Perrier. Sterculiaceae. 3 Madag.
Helmontia Cogn. Cucurbitaceae. 1 Brazil, Guiana.
Helmuthia Pax = Hellmuthia Steud. = Scirpus L. (Cyperac.).
Helobiae Reichb. An Order of Monocots., including (in Engler) the *Potamogetonac.* (*s.l.*), *Najadac., Aponogetonac., Juncaginac., Alismatac., Butomac.* and *Hydrocharitac.* (*s.l.*).
Helobieae auctt. = praec.
Helodea P. & K. = Elodes Adans. = Hypericum L. (Guttif.).
Helodea Reichb. = Elodea Michx (Hydrocharitac.).
Helodes St-Lag. = Elodes Adans. = Hypericum L. (Guttif.).
Helodium Dum. = Apium L. (Umbellif.).
Helogyne Benth. = Hofmeisteria Walp. (Compos.).
Helogyne Nutt. Compositae. 8 Andes.
Helonema Suesseng. Cyperaceae. 1 SE. Brazil.
Heloni[adac]eae J. G. Agardh = Liliaceae–Heloniadeae Reichb.
Helonias Adans. = Scilla L. (Liliac.).
Helonias L. Liliaceae. 1 E. U.S.
***Heloniopsis** A. Gray. Liliaceae. 4 Japan, Formosa.
Helophyllum Hook. f. = Phyllachne J. R. & G. Forst. (Stylidiac.).
Helophytum Eckl. & Zeyh. = Tillaea L. = Crassula L. (Crassulac.).
Helopus Trin. = Eriochloa Kunth (Gramin.).
Helorchis Schlechter = Cynorkis Thou. (Orchidac.).
Heloschiadium Marsson = seq.
Heloscia Dum. = seq.
Helosciadium Koch (~ Apium L.). Umbelliferae. 5 W. & C. Eur., Medit.; 1 S. Afr.
Heloseris Reichb. ex Steud. = Senecio L. (Compos.).
Helosidaceae Van Tiegh. = Balanophoraceae–Helosidoïdeae Harms, *q.v.*
***Helosis** Rich. Balanophoraceae. 3 trop. Am.

HELOSPORA

Helospora Jack = Timonius DC. (Rubiac.).
Helothrix Nees = Schoenus L. (Cyperac.).
Helvingia Adans. = Thamnia P.Br. = Laëtia Loefl. ex L. (Flacourtiac.).
Helwingia Willd. Helwingiaceae. 4–5 E. Himal. to Japan & Formosa.
Helwingiaceae Decaisne. Dicots. 1/4 Himal., E. As. Shrubs with alt. or subopp. simple serrulate stip. ls., stips. often branched, decid. Fls. ♂ ♀, dioec., in small epiphyllous umbels from upper side of midrib, ♀ v. shortly, ♂ sometimes long-pedic. K o; C 3–4(–5), valv.; A 3–4(–5), alternipet., inserted outside flat angled disk; G̅ (3–4), with apical disk passing into short style with 3–4 recurved stigs., and 1 pend. anatr. ov. per loc. Drupe with 3–4 pyrenes, seed with endosp. Only genus: *Helwingia*. A distinct group ± intermediate between *Cornac.* and *Araliac.*, providing also a parallel with the *Dulongiaceae.*
Helxine Bub. = Parietaria L. (Urticac.).
Helxine L. = Fagopyrum Mill. (Polygonac.).
Helxine Req. = Soleirolia Gaudich. (Urticac.).
Helygia Blume = Parsonsia R.Br. (Apocynac.).
Hemandradenia Stapf. Connaraceae. 3 trop. Afr., Madag.
Hemanthus Rafin. = Haemanthus L. (Amaryllidac.).
Hemarthria R.Br. Gramineae. 10 trop. Afr., Madag., E. As., Indomal.
Hematodes Rafin. = Salvia L. (Labiat.).
Hematophyla Rafin. = Columnea L. (Gesneriac.).
Hemecyclia Wight & Arn. (sphalm.) = Hemicyclia Wight & Arn. = Drypetes Vahl (Euphorbiac.).
Hemenaea Scop. (sphalm.) = Hymenaea L. (Legumin.).
Hemeotria Merr. (sphalm.) = Hemesotria Rafin. = Astrephia Dufr. (Valerianac.).
Hemerocallid[ac]eae R.Br. = Liliaceae–Hemerocallideae C. A. Agardh.
Hemerocallis L. Liliaceae. 20 temp. Euras., esp. Jap. Infl. a double bostryx. The fls. of *H. fulva* L. are self-sterile.
Hemesotria Rafin. = Astrephia Dufr. (Valerianac.).
Hemesteum Lév. = Polystichum Roth (Aspidiac.).
Hemestheum Newm. = Thelypteris Schmid. (Thelypteridac.).
Hemiachyris DC. = Gutierrezia Lag. (Compos.).
Hemiadelphis Nees = Hygrophila R.Br. (Acanthac.).
Hemiagraphis T. Anders. = Hemigraphis Nees (Acanthac.).
Hemiambrosia Delpino = Franseria Cav. (Compos.).
Hemiandra R.Br. Labiatae. 7 SW. Austr.
Hemiandra Rich. ex Triana = Trembleya DC. (Melastomatac.).
Hemiandrina Hook. f. = Agelaea Soland. ex Planch. (Connarac.).
Hemianemia (Prantl) Reed = Anemia Sw. (Schizaeac.).
Hemiangium A. C. Smith. Celastraceae. 1 Mex. to Paraguay.
Hemianthus Nutt. Scrophulariaceae. 10 trop. Am., W.I., 1 arct. Am.
Hemiarinum Rafin. (nomen). Quid?
Hemiarrhena Benth. Scrophulariaceae. 1 trop. Austr.
Hemiarthria P. & K. = Hemarthria R.Br. (Gramin.).
Hemiarthron Van Tiegh. (~Loranthus L.). 1 trop. S. Am.
Hemibaccharis S. F. Blake. Compositae. 15 Mex., C. Am.
Hemiboea C. B. Clarke. Gesneriaceae. 8 China, Formosa, Indoch.

Hemibromus Steud. = Glyceria R.Br. (Gramin.).
Hemicardion Fée = Cyclopeltis J. Sm. (Aspidiac.).
Hemicarex Benth. = Kobresia Willd. + Schoenoxiphium Nees (Cyperac.).
Hemicarpha Nees & Arn. (~ Scirpus L.). Cyperaceae. 6 trop. & subtrop.
Hemicarpurus Nees = Pinellia Tenore (Arac.).
Hemicarpus F. Muell. = Trachymene Rudge (Umbellif.).
Hemichaena Benth. Scrophulariaceae. 1 C. Am.
Hemicharis Salisb. ex DC. = Scaevola L. (Goodeniac.).
Hemichlaena Schrad. Cyperaceae. 3 S. Afr.
Hemichoriste Nees = Justicia L. (Acanthac.).
Hemichroa R.Br. Chenopodiaceae (~ Amaranthaceae). 3 Austr.
Hemicicca Baill. (~ Phyllanthus L.). Euphorbiaceae. 1 Japan.
Hemiclidia R.Br. = Dryandra R.Br. (Proteac.).
Hemiclis Rafin. = Lyonia Nutt. (Ericac.).
Hemicoa auct. (sphalm.) = Hemieva Rafin. = Suksdorfia A. Gray (Saxifragac.).
Hemicrambe Webb. Cruciferae. 1 Morocco.
Hemicrepidospermum Swart. Burseraceae. 1 trop. S. Am.
Hemicyatheon (Domin) Copel. Hymenophyllaceae. 2 Queensl., New Caled.
Hemicyclia Wight & Arn. = Drypetes Vahl (Euphorbiac.).
Hemidemus Dum. = Hemidesmus R.Br. (Periplocac.).
Hemidesma Rafin. = seq.
Hemidesmas Rafin. = Neptunia Lour. (Legumin.).
Hemidesmus R.Br. Periplocaceae. 1 S. India, SE. As., Malaysia.
Hemidia Rafin. = Ipomoea L. vel Convolvulus L. (Convolvulac.).
Hemidictyum Presl. Athyriaceae. 1 trop. Am.
Hemidiodia K. Schum. Rubiaceae. 1 Mexico to Brazil.
Hemidis Rafin. = Hemiclis Rafin. = Lyonia Nutt. (Ericac.).
Hemidistichophyllum Koidz. = Cladopus Moell. (Podostemac.).
Hemierium Rafin. = Lloydia Salisb. (Liliac.).
Hemieva Rafin. = Suksdorfia A. Gray (Saxifragac.).
Hemifuchsia Herrera. Onagraceae. 1 Peru.
Hemigenia R.Br. Labiatae. 40 Austr. Ls. in whorls of 3.
Hemiglochidion (Muell. Arg.) K. Schum.: as to Glochidion sect. Hemiglochidion Muell. Arg. = Glochidion J. R. & G. Forst. (Euphorbiac.); as to species included by Schumann = Phyllanthus L. sect. Nymania (K. Schum.) J. J. Sm. (Euphorbiac.).
Hemigramma Christ. Aspidiaceae. 6 S. China, Malaysia.
Hemigraphis Nees. Acanthaceae. 100 S. China, Indomal., trop. Austr.,
Hemigymnia Griff. = Cordia L. (Ehretiac.). [Pacif.
Hemigymnia Stapf = Ottochloa Dandy (Gramin.).
Hemigyrosa Blume = Guioa Cav. (Sapindac.).
Hemihabenaria Finet = Pecteilis Rafin. (Orchidac.).
Hemiheisteria Van Tiegh. = Heisteria Jacq. (Olacac.).
Hemilepis Kunze = Leontodon L. (Compos.).
Hemilepis Vilm. = Heliopsis Pers. (Compos.).
Hemilobium Welw. = Apodytes E. Mey. (Icacinac.).
Hemilophia Franch. Cruciferae. 2 SW. China.
Hemimeris L. = Diascia Link & Otto (Scrophulariac.).

HEMIMERIS

Hemimeris L. f. Scrophulariaceae. 9 S. Afr.

Hemimeris Pers. = Alonsoa Ruiz & Pav. (Scrophulariac.).

Hemimunroa (L. Parodi) L. Parodi. Gramineae. 1 S. Andes.

Heminema Rafin. = Tripogandra Rafin. (Commelinac.).

Hemionitidaceae Pichi-Serm. (Gymnogrammaceae Ching). Pteridales. Mainly xeroph. ferns; some with waxy powder on lower surface; sporangia spreading along veins, without indusia. Chief genera: *Syngramma, Eriosorus, Coniogramme, Hemionitis, Gymnopteris, Pityrogramma*.

Hemionitis L. Hemionitidaceae. 1 trop. As.; 6 trop. N. Am.

Hemiorchis Ehrenb. ex Schweinf. = Lindenbergia Lehm. (Scrophulariac.).

Hemiorchis Kurz. Zingiberaceae. 3 Indomal.

Hemiouratea Van Tiegh. = Ouratea Aubl. (Ochnac.).

Hemipappus C. Koch. Compositae. 4 NE. As. Min., NW. Persia.

Hemiperis Frapp. ex Cordem. = Cynorkis Thou. (Orchidac.).

Hemiphlebium Presl = Didymoglossum v. d. Bosch (Hymenophyllac.).

Hemiphora F. Muell. Dicrastylidaceae. 1 W. Austr.

Hemiphractum Turcz. = Vateria L. (Dipterocarpac.).

Hemiphragma Wall. Scrophulariaceae. 1 W. Himalaya to Assam.

Hemiphues Hook. f. = Actinotus Labill. (Umbellif.).

Hemiphylacus S. Wats. Liliaceae. 1 N. Mexico.

Hemipilia Lindl. Orchidaceae. 12 Himal., Siam, E. As.

Hemipogon Decne. Asclepiadaceae. 10 S. Am.

Hemiptelea Planch. = Zelkova Spach (Ulmac.).

Hemipteris Rosenst. = Pteris L. (Pteridac.).

Hemiptilium A. Gray = Stephanomeria Nutt. (Compos.).

Hemisacris Steud. = Schismus Beauv. (Gramin.).

Hemisandra Scheidw. = Aphelandra R.Br. (Acanthac.).

Hemisantiria H. J. Lam = Dacryodes Vahl (Burserac.).

Hemiscleria Lindl. Orchidaceae. 1 Ecuador, Peru.

Hemiscola Rafin. = Cleome L. (Cleomac.).

Hemiscolopia van Slooten. Flacourtiaceae. 1 Indoch., Sumatra, Banka, Java.

Hemisiphonia Urb. Scrophulariaceae. 1 W.I.

Hemisodon Rafin. = Leonotis R.Br. (Labiat.).

Hemisorghum C. E. Hubbard. Gramineae. 1 SE. As.

Hemispadon Endl. = Indigofera L. (Legumin.).

Hemisphace (Benth.) Opiz = Salvia L. (Labiat.).

Hemisphaerocarya A. Brand. Boraginaceae. 8 S. U.S., Mexico.

Hemistachyum Copel. = Aglaomorpha Schott (Polypodiac.).

Hemistegia Presl = Cnemidaria Presl (Cyatheac.).

Hemistegia Rafin. = Salvia L. (Labiat.).

Hemisteirus F. Muell. = Trichinium R.Br. (Amaranthac.).

Hemistema DC. (sphalm.) = seq.

Hemistemma DC. = Hibbertia Andr. (Dilleniac.).

Hemistemma Reichb. = Leucas R.Br. (Labiat.).

Hemistemon F. Muell. = Chloanthes R.Br. (Dicrastylidac.).

Hemistephia Steud. = Hemistepta Bunge = Saussurea DC. (Compos.).

Hemistephus Drumm. ex Harv. = Hibbertia Andr. (Dilleniac.).

Hemistepta Bunge = Saussurea DC. (Compos.).

Hemistoma Ehrenb. ex Benth. = Leucas R.Br. (Labiat.).
Hemistylis Walp. = seq.
Hemistylus Benth. Urticaceae. 4 trop. S. Am.
Hemitelia R.Br. = Cyathea Sm. (Cyatheac.).
Hemithrinax Hook. f. Palmae. 4 Cuba.
Hemitome Nees = Stenandrium Nees (Acanthac.).
Hemitomes A. Gray. Monotropaceae. 1 W. U.S.
Hemitomus L'Hérit. ex Desf. = Alonsoa Ruiz & Pav. (Scrophulariac.).
Hemitria Rafin. (Phthirusa Mart.). Loranthaceae. 60 trop. Am.
× **Hemiultragossypium** Roberty. Malvaceae. Gen. hybr. (Gossypium × Neogossypium).
Hemiuratea P. & K. = Hemiouratea Van Tiegh. = Ouratea Aubl. (Ochnac.).
Hemixanthidium Delpino = Franseria Cav. (Compos.).
Hemizonella A. Gray. Compositae. 2 Pacif. U.S.
Hemizonia DC. Compositae. 31 Calif. & Lower Calif.
Hemizygia (Benth.) Briq. Labiatae. 28 trop. & S. Afr.
Hemma Rafin. ex Pfitzer = Lemna L. (Lemnac.).
Hemolepis Hort. ex Vilm. = Heliopsis Pers. (Compos.).
Hemonacanthus Nees (sphalm.) = Stemonacanthus Nees = Ruellia L. (Acan-
Hemprichia Ehrenb. = Commiphora Jacq. (Burserac.).   [thac.).
Hemsleia Kudo = Ceratanthus F. Muell. (Labiat.).
Hemsleya Cogn. Curcurbitaceae. 1 E. Himal., China.
Hemsleyna Kuntze = Thryallis Mart. (Malpighiac.).
Hemyphyes Endl. = Hemiphues Hook. f. = Actinotus Labill. (Umbellif.).
Henanthus Less. = Pteronia L. (Compos.).
Henckelia Spreng. = Didymocarpus Wall. (Gesneriac.).
Hendecandras Eschsch. = Croton L. (Euphorbiac.).
Henfreya Lindl. = Asystasia Blume (Acanthac.).
Henicosanthum Dalla Torre & Harms = Enicosanthum Becc. (Annonac.).
Henicostemma Endl. = Enicostema Blume (Gentianac.).
Henisia Walp. = Heinsia DC. (Rubiac.).
Henkelia Reichb. = Henckelia Spreng. = Didymocarpus Wall. (Gesneriac.).
Henlea Griseb. = Henleophytum Karst. (Malpighiac.).
Henlea Karst. = Rustia Klotzsch (Rubiac.).
Henleophytum Karst. Malpighiaceae. 1 Cuba.
Henna Boehm. = Lawsonia L. (Lythrac.).
Hennecartia Poisson. Monimiaceae. 1 S. Braz., Paraguay.
Henningia Kar. & Kir. = Eremurus M. Bieb. (Liliac.).
Henningsocarpum Kuntze = Neopringlea S. Wats. (? Flacourtiac.).
Henonia Coss. & Dur. = Henophyton Coss. & Dur. (Crucif.).
Henonia Moq. Amaranthaceae. 1 Madag.
Henonix Rafin. = Scilla L. (Liliac.).
Henooina Melch. (sphalm.) = seq.
Henoonia Griseb. Goetzeaceae. 3 Cuba.
Henophyton Coss. & Dur. Cruciferae. 1 Algeria.
Henosis Hook. f. = Bulbophyllum Thou. (Orchidac.).
Henrardia C. E. Hubbard. Gramineae. 2 As. Min. & Persia to C. As. & Baluch.

Henribaillonia Kuntze = Cometia Thou. (Euphorbiac.), *q.v.*
Henricea Lemaire-Lisancourt = Swertia L. (Gentianac.).
Henricia L. Bolus = Neohenricia L. Bolus (Aïzoac.).
Henricia Cass. Compositae. 1 Madag.
Henrietia Reichb. = seq.
Henrietta Macfad. = seq.
Henriettea DC. Melastomataceae. 20 trop. S. Am.
Henriettella Naud. Melastomataceae. 40 C. & trop. S. Am., W.I.
Henrincquia Benth. & Hook. f. (sphalm.) = Herincquia Decne = Pentarhaphia Lindl. (Gesneriac.).
Henriquezia Spruce ex Benth. Henriqueziaceae. 7 Amaz. Brazil.
Henriqueziaceae (Hook. f.) Bremek. Dicots. 2/13 Brazil. Small or moderate-sized trees. Ls. simple, ent., opp. or vertic., stip. Fls. large, zygo., in term. thyrses. K 4–5, decid.; C (5), campan., imbr.; A 5, unequally epipet., with curved fil. and dorsifixed sagitt.-based anth.; disk ann.; G (2), with 2–4 collat. ov. per loc., and single style with 2 apical stigs. Fr. a semi-superior or almost superior, transversely ovoid, subreniform or discoid, 2–4-seeded, loculicid. caps.; seeds flattened but not winged, without endosp. Genera: *Henriquezia, Platycarpum*. A small family related to the *Bignoniac., Pedaliac., Thunbergiac.,* etc. Formerly erroneously included in the *Rubiac.,* on account of the presence of stips.
Henrya Nees. Acanthaceae. 20 Mex., C. Am.
Henryettana A. Brand. Boraginaceae. 1 SW. China.
Henschelia Presl = Illigera Blume (Hernandiac.).
Henslera Endl. = Haenselera Lag. = Physospermum Cusson (Umbellif.).
Henslera Reichb. = Haenslera Boiss. (Compos.).
Henslevia Rafin. = seq.
Henslovia A. Juss. = Henslowia Wall. = Crypteronia Blume (Crypteroniac.).
Hensloviaceae Lindl. = Henslowiaceae Lindl. corr. Endl. = Crypteroniaceae
Henslowia Blume = Dendrotrophe Miq. (Santalac.).   [DC.
Henslowia Lowe ex DC. = Notelaea Vent. (Oleac.).
Henslowia Wall. = Crypteronia Blume (Crypteroniac.).
Henslowiaceae Lindl. corr. Endl. = Crypteroniaceae DC.
Hensmania Fitzgerald. Liliaceae. 1 W. Austr.
Heocarphus Phil. (sphalm.) = Pleocarphus D. Don = Jungia L. f. (Compos.).
Hepatica Mill. (~ Anemone L.). Ranunculaceae. 10 temp. Euras.
Hepatitis Rafin. (nomen). Quid?
Hepetis Sw. = Pitcairnia L'Hérit. (Bromeliac.).
Hepetospermum Spach (sphalm.) = Herpetospermum Wall. ex Hook. f. (Cucurbitac.).
Hephestionia Naud. = Tibouchina Aubl. (Melastomatac.).
× **Heppiantha** H. E. Moore. Gesneriaceae. Gen. hybr. (Heppiella × Smithiantha).
Heppiella Regel (~ Achimenes P.Br.). Gesneriaceae. 17 trop. S. Am.
Heptaca Lour. = Oncoba Forsk. (Flacourtiac.).
Heptacarpus Conzatti = Bejaria Mutis corr. Zea ex Vent. (Ericac.).
Heptacodium Rehder. Caprifoliaceae. 2 C. & E. China. (Cf. Golubkova, *Novit. Syst. Pl. Vasc.* **1965**: 230–6, 1965.)

Heptacyclum Engl. = Penianthus Miers (Menispermac.).

Heptallon Rafin. = Croton L. (Euphorbiac.).

Heptanis Rafin. = praec.

Heptanthus Griseb. Compositae. 6 Cuba.

Heptantra O. F. Cook = Orbignya Mart. ex Endl. (Palm.).

Heptapleurum Gaertn. = Schefflera J. R. & G. Forst. (Araliac.).

Heptaptera Margot & Reut. Umbelliferae. 6 E. Medit., SW. As.

Heptarina Rafin. = Polygonum L. (Polygonac.).

Heptarinia Rafin. = praec.

Heptas Meissn. = Septas Lour. = Bacopa Aubl. (Scrophulariac.).

Hepteireca Rafin. = Chamaecrista Moench = Cassia L. (Legumin.).

Heptoneurum Hassk. = Heptapleurum Gaertn. = Schefflera J. R. & G. Forst. (Araliac.).

Heptoseta Koidz. = Agrostis L. (Gramin.).

Heptrilis Rafin. = Leucas R.Br. (Labiat.).

Heracantha Hoffmgg. & Link = Carthamus L. (Compos.).

Heraclea Hill = Centaurea L. (Compos.).

Heracleum L. Umbelliferae. 70 N. temp., trop. mts.

Herbaceae Hutch. A major division of the *Dicotyledones*, comprising families regarded by Hutchinson as 'fundamentally herbaceous'.

× **Herbertara** hort. Orchidaceae. Gen. hybr. (iii) (Cattleya × Laelia × Schomburgkia × Sophronitis).

Herbertia Sweet (1827; non S. F. Gray 1821—Hepaticae) = Alophia Herb. (Iridac.).

Herbichia Zawadski = Senecio L. (Compos.).

Herderia Cass. Compositae. 6 trop. Afr.

Hereroa (Schwantes) Dinter & Schwantes. Aïzoaceae. 28 S. Afr.

Heretiera G. Don = Heritiera Dryand. (Sterculiac.).

Hericinia Fourr. (~ Ranunculus L.). Ranunculaceae. 1 Eur. to C. As. & Himal.

Herincquia Decne = Pentarhaphia Lindl. (Gesneriac.).

Herissantia Medik. Malvaceae. 3 trop. Am.

Heritera Stokes = Heritiera Dryand. (Sterculiac.).

Heriteria Dum. (1) = praec.

Heriteria Dum. (2) = Heritiera J. F. Gmel. = Lachnanthes Ell. (Haemodorac.).

Heriteria Schrank = Tofieldia Huds. (Liliac.).

Heritiera Dryand. (in Ait.). Sterculiaceae. 35 W. trop. Afr., Indomal., trop. Austr., Pacif.

Heritiera J. F. Gmel. = Lachnanthes Ell. (Haemodorac.).

Heritiera Retz. = Alpinia L. (Zingiberac.).

Hermannia L. Sterculiaceae. 300 + trop. & subtrop. S. Am., Afr., Arabia, Austr.

Hermanniaceae Schultz-Schultzenst. = Sterculiaceae–Hermannieae Spreng.

Hermarthria auct. = Hemarthria R.Br. (Gramin.).

Hermas L. Hydrocotylaceae (~ Umbellif.). 6 S. Afr.

Hermbstaedtia Reichb. Amaranthaceae. 17 trop. & S. Afr.

Hermesia Humb. & Bonpl. = Alchornea Sw. (Euphorbiac.).

Hermesias Loefl. = Brownea Jacq. (Legumin.).

HERMIBICCHIA

× **Hermibicchia** G. Gamus, Bergon & A. Camus (ii) = × Pseudinium P. F. Hunt (Orchidac.).

Hermidium S. Wats. Nyctaginaceae. 1 SW. U.S.

× **Hermileucorchis** Cif. & Giac. (ii) = × Pseudinium P. F. Hunt (Orchidac.).

Herminiera Guill. & Perr. (~ Aeschynomene L.). Leguminosae. 1 trop. Afr., *H. elaphroxylon* Guill. & Perr. Wood light, used for floats, canoes, etc. Cf. with the development of aerenchyma seen in other marsh plants (*Lycopus, Jussiaea,* etc.).

Herminium Guett. (= Monorchis Seguier). Orchidaceae. 30 temp. Euras., Siam, Philipp., Java.

× **Herminorchis** P. Fourn. (ii) = × Pseudinium P. F. Hunt (Orchidac.).

Hermione Salisb. = Narcissus L. (Amaryllidac.).

Hermodactylon Parl. = Hermodactylus Mill. (Iridac.).

Hermodactylos Reichb. = Colchicum L. (Liliac.).

Hermodactylum Bartl. = seq.

Hermodactylus Mill. (~ Iris L.). Iridaceae. 1 S. France to Greece. G 1-loc.

Hermodactylus Reichb. = Colchicum L. (Liliac.).

Hermstaedtia Steud. = Hermbstaedtia Reichb. (Amaranthac.).

Hermupoa Loefl. = Steriphoma Spreng. (Capparidac.).

Hernandezia Hoffmgg. = seq.

Hernandia L. Hernandiaceae. 24 C. Am., Guiana, W.I., W. Afr., Zanzibar, Masc., Indomal., Pacif.

*****Hernandiaceae** Blume. Dicots. 3/54 trop. Trees, shrubs or lianes, with alt. simple (and palminerved) or digitate exstip. ls., with oil-cells. Fls. in cymes, ♀ or ♂ ♀ and monoec., reg. P 6–10, ± biseriate, the outer ± valv.; A 3–5 (staminodial in ♀ fls.), opp. the outer P-segs., sometimes accompanied by 1–2 glands at base, anth. dehisc. by valves; G̅ (1), 1-loc., with 1 pend. anatr. ov. and unilaterally grooved style with emarg. stig. Fr. dry, indehisc., winged or encl. in inflated envelope; seed without endosp., cots. fleshy, smooth or ruminate. Genera: *Hernandia, Hazomalania, Illigera.* A small family, most closely related to the *Monimiaceae* (*s.str.*).

Hernandiopsis Meissn. = Hernandia L. (Hernandiac).

Hernandria L. (sphalm.) = Hernandia L. (Hernandiac.).

Herniaria L. Caryophyllaceae. 35 Eur., Medit. to Afghan., S. Afr. Fl.

Herodium Reichb. = Erodium L'Hérit. (Geraniac.).   [apetalous.

Herodotia Urb. & Ekman. Compositae. 1 Haiti.

Heroion Rafin. = Asphodeline Reichb. (Liliac.).

Herotium Steud. = Xerotium Bluff & Fingerh. = Filago L. (Compos.).

Herpestes Kunth = seq.

Herpestis Gaertn. f. = Bacopa Aubl. (Scrophulariac.).

Herpetacanthus Nees. Acanthaceae. 10 Panamá to Brazil.

Herpetica [Rumph.] Rafin. = Cassia L. (Legumin.).

Herpetina P. & K. = Erpetina Naud. = Medinilla Gaudich. (Melastomatac.).

Herpetium Wittst. = Erpetion DC. ex Sweet = Viola L. (Violac.).

Herpetospermum Wall. ex Hook. f. Cucurbitaceae. 1 Himal., China.

Herpolirion Hook. f. Liliaceae. 1 SE. Austr., Tasm., N.Z.

Herpophyllum Zanard. = Barkania Ehrenb. = Halophila Thou. (Hydrocharitac.).

Herpothamnus Small = Vaccinium L. (Ericac.).
Herpysma Lindl. Orchidaceae. 2 India to Philipp. Is.
Herpyza C. Wright (~ Teramnus P.Br.). Leguminosae. 1 Cuba.
Herrania Goudot. Sterculiaceae. 20 trop. S. Am.
Herraria Ritg. = Herreria Ruiz & Pav. (Liliac.).
Herrea Schwantes. Aïzoaceae. 24 S. Afr.
Herreanthus Schwantes. Aïzoaceae. 1 S. Afr.
Herrera Adans. = Erithalis P.Br. (Compos.).
Herreraea P. & K. (1) = praec.
Herreraea P. & K. (2) = seq.
Herreria Ruiz & Pav. (= Clara Kunth). Liliaceae. 8 S. Am.
Herreri[ac]eae Endl. = Liliaceae–Herrerioïdeae Engl.
Herreriopsis H. Perrier. Liliaceae. 1 Madag.
Herrickia Wooton & Standley. Compositae. 1 SW. U.S.
Herrmannia Link & Otto = Hermannia L. (Sterculiac.).
Herschelia Bowdich = Physalis L. (Solanac.).
Herschelia Lindl. = Disa Bergius (Orchidac.).
Herschellia Bartl. = Herschelia Bowdich = Physalis L. (Solanac.).
Hersilea Klotzsch = Aster L. (Compos.).
Hersilia Rafin. = Phlomis L. (Labiat.).
Hertelia Neck. = Hernandia L. (Hernandiac.).
Hertelia P. & K. = Ertela Adans. = Monnieria L. (Rutac.).
Hertia Less. Compositae. 12 N. Afr. to Baluchistan, S. Afr.
Hertrichocereus Backeb. = Lemaireocereus Britton & Rose (Cactac.).
Herya Cordemoy. Celastraceae. 1 Réunion.
Herzogia K. Schum. = Euodia J. R. & G. Forst. (Rutac.).
Hesioda Vell. = Heisteria Jacq. (Olacac.).
Hesiodia Moench (~ Sideritis L.). Labiatae. 2 Medit. to C. As. & Afghan.
Hesperalcea Greene = Sidalcea A. Gray (Malvac.).
Hesperaloë Engelm. Agavaceae. 2 S. U.S., Mexico.
Hesperantha Ker-Gawl. Iridaceae. 50 trop. & S. Afr.
Hesperanthemum (Endl.) Kuntze = Anthacanthus Nees (Acanthac.).
Hesperanthes S. Wats. = Anthericum L. (Liliac.).
Hesperanthus Salisb. = Hesperantha Ker-Gawl. (Iridac.).
Hesperaster Cockerell = Mentzelia L. (Loasac.).
Hesperastragalus A. A. Heller = Astragalus L. (Legumin.).
Hesperelaea A. Gray. Oleaceae. 1 NW. Mex. (Guadalupe I., L. Calif.).
Hesperethusa M. Roem. Rutaceae. 1 NW. Himal. to Ceylon & SE. As.
Hesperevax A. Gray = Evax Gaertn. (Compos.).
Hesperhodos Cockerell (~ Rosa L.). Rosaceae. 4 SW. U.S.
Hesperidanthus (Robinson) Rydb. Cruciferae. 1 SW. U.S., N. Mex.
Hesperideae Vent. = Rutaceae–Aurantioïdeae Engl. + Olacaceae quaed. + Theaceae (Camellia L.).
Hesperidopsis Kuntze = Dontostemon Andrz. ex DC. (Crucif.).
Hesperis L. Cruciferae. 30 Eur., Medit. to Persia, C. As., W. China.
Hesperocallis A. Gray. Liliaceae. 1 deserts SW. U.S.
*Hesperochiron S. Wats. Hydrophyllaceae. 2–6 W. U.S.
Hesperochloa Rydb. Gramineae. 1 W. U.S.

HESPEROCLES

Hesperocles Salisb. = Nothoscordum Kunth (Alliac.).
Hesperocnide Torr. Urticaceae. 2 Hawaii, California.
Hesperodoria Greene = Haplopappus Cass. (Compos.).
Hesperogenia Coulter & Rose = Tauschia Schlechtd. (Umbellif.).
Hesperogeton K.-Pol. = Sanicula L. (Umbellif.).
Hesperogreigia Skottsb. Bromeliaceae. 1 Juan Fernandez.
Hesperolaburnum Maire. Leguminosae. 1 Morocco.
Hesperolinon Small. Linaceae. 12 Pacif. U.S.
Hesperomannia A. Gray. Compositae. 7 Hawaii.
Hesperomecon Greene = Platystigma Benth. (Papaverac.).
Hesperomeles Lindl. Rosaceae. 20 C. Am. to Peru.
Hesperonia Standley = Mirabilis L. (Nyctaginac.).
Hesperonix Rydb. = Astragalus L. (Legumin.).
Hesperopeuce Lemmon = Tsuga Carr. (Pinac.).
Hesperoschordum Willis (sphalm.) = seq.
Hesperoscordum Lindl. (~ Brodiaea Sm.). Alliaceae. 1–2 W. N. Am.
Hesperoseris Skottsb. Compositae. 1 Juan Fernandez.
Hesperothamnus T. S. Brandegee. Leguminosae. 6 SW. U.S., Mex.
Hesperoxalis Small = Oxalis L. (Oxalidac.).
Hesperoxiphion Baker = Cypella Herb. (Iridac.).
Hesperoyucca Baker. Liliaceae. 1 SW. U.S.
Hesperozygis Epling. Labiatae. 8 Mex. to Brazil.
Hessea Bergius = Carpolyza Salisb. (Amaryllidac.).
*****Hessea** Herb. = Periphanes Salisb. (Amaryllidac.).
Hetaeria Blume. Orchidaceae. 20 palaeotrop.
Hetaeria Endl. = Philydrella Caruel (Philydrac.).
Heteracantha Link = Carthamus L. (Compos.).
Heteracea Steud. = Heteracia Fisch. & Mey. (Compos.).
Heterachaena Fres. = Launaea Cass. (Compos.).
Heterachaena Zoll. & Mor. = Pimpinella L. (Umbellif.).
Heterachne Benth. Gramineae. 3 N. Austr.
Heterachthia Kunze = Tradescantia L. (Commelinac.).
Heteracia Fisch. & Mey. Compositae. 1 Armenia.
Heteractis DC. = Gymnostephium Less. (Compos.).
Heteradelphia Lindau. Acanthaceae. 1 W. Afr. (São Tomé).
Heterandra Beauv. = Heteranthera Ruiz & Pav. (Pontederiac.).
Heteranthelium Hochst. Gramineae. 1 temp. As.
Heteranthemia Schott = Chrysanthemum L. (Compos.).
*****Heteranthera** Ruiz & Pav. Pontederiaceae. 10 trop. & subtrop. Am., Afr.
Ls. of two types—linear submerged and orbicular floating. Some spp. have
cleistog. fls.
Heteranther[ac]eae J. G. Ag. = Pontederiaceae–Heteranthereae O. Schwartz.
Heteranthia Nees & Mart. Solanaceae. 1 Brazil.
Heteranthoecia Stapf. Gramineae. 1 trop. Afr.
Heteranthus Bonpl. = Perezia Lag. (Compos.).
Heteranthus Borckh. = Ventenata Koel. (Gramin.).
Heterapithmos Turcz. (sphalm.) = seq.
Heterarithmos Turcz. corr. B. Fedtch. = Meliosma Bl. (Meliosmac.).

HETEROKALIMERIS

Heteraspidia Rizzini. Acanthaceae. 1 Brazil.
Heterelytron Jungh. = Anthistiria L. (Gramin.).
Heteresia Rafin. = Steiranisia Rafin. = Saxifraga L. (Saxifragac.).
Heterisia B. D. Jacks. = praec.
Heterixia Van Tiegh. = Korthalsella Van Tiegh. (Viscac.).
Heteroarisaema Nakai (~Arisaema Mart.). Araceae. 3 E. As.
Heterocalymnantha Domin = Synostemon F. Muell. (Euphorbiac.).
Heterocalyx Gagnep. = Agrostistachys Dalz. (Euphorbiac.).
Heterocanscora C. B. Clarke = Canscora Lam. (Gentianac.).
Heterocarpaea Scheele = Galactia P.Br. (Legumin.).
Heterocarpha Stapf & C. E. Hubbard. Gramineae. 1 trop. E. Afr.
Heterocarpus Phil. = Cardamine L. (Crucif.).
Heterocarpus P. & K. = Heterocarpaea Scheele = Galactia P.Br. (Legumin.).
Heterocarpus Wight = Commelina L. (Commelinac.).
Heterocaryum A. DC. Boraginaceae. 6 S. Russia & As. Min. to C. As. & Himal.
Heterocentron Hook. & Arn. Melastomataceae. 12 Mex., C. Am. Some sta. attract insects, the others pollinate them (cf. *Commelina*).
Heterochaenia A. DC. Campanulaceae. 1 Mascarenes.
Heterochaeta Bess. ex Roem. & Schult. = Ventenata Koeler (Gramin.).
Heterochaeta DC. = Aster L. (Compos.). [Medit.
Heterochiton Graebn. & Mattfeld (~Herniaria L.). Caryophyllaceae. 10
Heterochlaena P. & K. (1) = Heterolaena Sch. Bip. ex Benth. & Hook. f. = Eupatorium (Compos.).
Heterochlaena P. & K. (2) = Heterolaena C. A. Mey. ex Fisch., Mey. & Avé-Lallem. = Pimelea Banks (Thymelaeac.).
Heterochlamys Turcz. = Julocroton Mart. (Euphorbiac.).
Heterochloa Desv. = Andropogon L. (Gramin.).
Heterochloa Endl. (sphalm.) = seq.
Heterochroa Bunge = Gypsophila L. (Caryophyllac.).
Heterocladus Turcz. = Coriaria L. (Coriariac.).
Heteroclita Rafin. = Canscora Lam. (Gentianac.).
Heterocodon Nutt. Campanulaceae. 1 SW. China, 1 W. N. Am.
Heterocoma DC. Compositae. 1 Brazil.
Heterocrambe Coss. & Dur. = Sinapis L. (Crucif.).
Heterocroton S. Moore = Croton L. (Euphorbiac.).
Heterodanaea Presl = Danaea Sm. (Danaeac.).
Heterodendron Spreng. = seq.
Heterodendrum Desf. Sapindaceae. 5 Austr.
Heteroderis Boiss. Compositae. 2–6 E. Medit. to Baluchistan.
Heterodon Meissn. = Berzelia Brongn. (Bruniac.).
Heterodonta Nutt. ex Benth. & Hook. f. = Coreopsis L. (Compos.).
Heterodraba Greene. Cruciferae. 1 Pacif. U.S., L. Calif.
Heterogaura Rothrock. Onagraceae. 1 W. U.S.
Heterogonium Presl. Aspidiaceae. 12 trop. As. to Malaysia (Holttum, *Sarawak Mus. J.* **5**: 156–66, 1949; *Reinwardtia* **3**: 269–74, 1955).
× **Heterokalimeris** Kitamura. Compositae. Gen. hybr. (Heteropappus × Kalimeris).

Heterolaena C. A. Mey. ex Fisch., Mey. & Avé-Lall. = Pimelea Banks (Thymelaeac.).
Heterolaena Sch. Bip. ex Benth. & Hook. f. = Chromolaena DC. (Compos.).
Heterolamium C. Y. Wu. Labiatae. 1 China.
Heterolathus Presl = Aspalathus Linn. (Legumin.).
Heterolepis Bertero ex Endl. = Senecio L. (Compos.).
***Heterolepis** Cass. Compositae. 3 S. Afr.
Heterolepis Ehrenb. ex Boiss. = Chloris Sw. (Gramin.).
Heterolobium Peter. Araceae. 2 trop. E. Afr.
Heterolobivia Ito = ? Lobivia Britt. & Rose (Cactac.).
Heteroloma Desv. ex Reichb. = Adesmia DC. (Legumin.).
Heterolophus Cass. = Centaurea L. (Compos.).
Heterolytron Hack. = Heterelytron Jungh. = Anthistiria L. (Gramin.).
Heteromeles M. Roem. Rosaceae. 1 Calif., L. Calif.
Heteromera Montrouz. = Leptostylis Benth. (Sapotac.).
Heteromera Pomel (~ Chrysanthemum L.). Compositae. 3 N. Afr.
Heteromerae Benth. & Hook. f. A ‘Series’ of Dicots. comprising the orders *Ericales, Primulales,* and *Ebenales.*
Heteromeris Spach = Crocanthemum Spach (Cistac.).
Heteromma Benth. Compositae. 2 S. Afr. mts.
Heteromorpha Cass. = Heterolepis Cass. (Compos.).
Heteromorpha Cham. & Schlechtd. Umbelliferae. 10 trop. Afr., Madag.
Heteromorpha Viv. ex Coss. = Hypochoeris L. (Compos.).
Heteromyrtus Blume = Myrtus L. (Myrtac.).
Heteronema Reichb. = Heteronoma DC. = Arthrostema Ruiz & Pav. (Melastomatac.).
Heteroneuron Fée = Bolbitis Schott (Lomariopsidac.).
Heteroneuron Hook. f. = Loreya DC. (Melastomatac.).
Heteronoma DC. = Arthrostema Ruiz & Pav. (Melastomatac.).
Heteropanax Seem. Araliaceae. 2 India, S. China.
Heteropappus Less. Compositae. 12 temp. E. As.
Heteropetalum Benth. Annonaceae. 2 Venez., Brazil.
Heterophlebium Fée = Pteris L. (Pteridac.).
Heteropholis Hubbard. Gramineae. 2 trop. E. Afr., Madag., Ceylon.
Heterophragma DC. Bignoniaceae. 2 India, Indoch.
Heterophyllaea Hook. f. Rubiaceae. 4 Bolivia, Argentina.
Heterophylleia Turcz. = Coriaria L. (Coriariac.).
Heterophyllum Boj. = Byttneria Loefl. (Sterculiac.).
Heteropleura Sch. Bip. = Hieracium L. (Compos.).
Heteroplexis Chang. Compositae. 1 E. China.
Heteropogon Pers. Gramineae. 12 trop.
Heteroporidium Van Tiegh. = Ochna L. (Ochnac.).
Heteropsis Kunth. Araceae. 12 trop. S. Am.
Heteroptera Steud. = Heptaptera Margot & Reut. = Prangos Lindl. (Umbellif.).
Heteropteris Fée (1843) = Paltonium Presl (Polypodiac.).
Heteropteris Fée (1869) = Doryopteris J. Sm. (Sinopteridac.).

Heteropteris Kunth = seq.

*Heteropterys Kunth emend. Griseb. Malpighiaceae. 100 trop. Am., 1 W. trop. Afr. Fr. a samara (cf. *Acer, Banisteriopsis*).

Heteropteryx Dalla Torre & Harms = praec.

Heteroptilis E. Mey. ex Meissn. Umbelliferae. 1 S. Afr.

*Heteropyxidaceae Engl. & Gilg. Dicots. 1/3 S. Afr. Small trees, with alt. ent. simple stip. (!) gland-dotted ls. Fls. small, ♀, reg., in term. thyrses. K (5), open or scarcely imbr., persist.; C 5, inserted in tube of K, obov., gland-dotted; A 5, *oppositipet.*, short; G̲ (2–3), with ∞ axile ov. per loc., and 1 short persist. style with capit. stig. Fr. a small 2–3-loc. loculic. caps., with few exalb. seeds. Only genus: *Heteropyxis*. Related on the one hand to *Lythraceae* and on the other hand to *Tristania* in the *Myrtaceae*; perhaps to be incl. in the latter fam.

Heteropyxis Griff. = Boschia Korth. = Durio Adans. (Bombacac.).

*Heteropyxis Harv. Heteropyxidaceae. 3 S. Rhodesia to Natal.

Heterorhachis Sch. Bip. ex Walp. Compositae. 2 S. Afr.

Heterosamara Kuntze = Polygala L. (Polygalac.).

Heterosciadium DC. = Petagnia Guss. (Umbellif.).

Heterosciadium Lange. Umbelliferae. 1 Spain.

Heterosicyos Welw. ex Hook. f. = Trochomeria Hook. f. (Cucurbitac.).

Heterosicyus [(S. Wats.)] P. & K. = Cremastopus P. Wils. (Cucurbitac.).

Heterosmilax Kunth. Smilacaceae. 15 E. As., Indoch., W. Malaysia.

Heterosoma Guill. = Heterotoma Zucc. (Campanulac.).

Heterospathe Scheff. Palmae. 18 Philipp., Palau & Mariane Is., New Guinea, Solomon Is.

Heterosperma Cav. Compositae. 10 SW. U.S. to Argentina.

Heterosperma Tausch = Heteromorpha Cham. & Schlechtd. (Compos.).

Heterospermum Willd. = Heterosperma Cav. (Compos.).

Heterostachys Ung.-Sternb. Chenopodiaceae. 2 Mex., W.I., temp. S. Am.

Heterostalis Schott = Typhonium Schott (Arac.).

Heterosteca Desv. = seq.

Heterostega Desv. corr. Kunth = Botelua Lag. (Gramin.).

Heterostegon Schweinitz ex Hook. f. = praec.

Heterostemma Wight & Arn. Asclepiadaceae. 30 trop. As.

Heterostemon Desf. Leguminosae. 8 trop. Am.

Heterostemon Nutt. ex Torr. & Gray = Oenothera L. (Onagrac.).

Heterostigma Gaudich. = Pandanus L. f. (Pandanac.).

Heterostylaceae Hutch. = Lilaeaceae Dum.

Heterostylus Hook. = Lilaea Humb. (Lilaeac.).

Heterotaenia Boiss. = Conopodium Koch (Umbellif.).

Heterotaxis Lindl. = Maxillaria Poepp. & Endl. (Orchidac.).

Heterothalamus Less. Compositae. 2 S. Am.

Heterotheca Cass. (∼ Chrysopsis [Nutt.] Ell.). Compositae. 30 S. U.S., Mexico.

Heterothrix Muell. Arg. = Echites L. (Apocynac.).

Heterothrix (Robins.) Rydb. = Pennellia Nieuwl. (Crucif.).

Heterotis Benth. = Dissotis Benth. (Melastomatac.).

Heterotoma Zucc. Campanulaceae. 11 Mexico, C. Am.

HETEROTOMMA

Heterotomma Dur. (sphalm.) = Heteromma Benth. (Compos.).
Heterotrichum M. Bieb. = Saussurea DC. (Compos.).
Heterotrichum DC. Melastomataceae. 15 trop. Am. Some ed. fr.
Heterotristicha Tobl. Podostemaceae. 1 Uruguay.
Heterotropa Morr. & Decne = Asarum L. (Aristolochiac.).
Heterozostera (Setch.) den Hartog. Zosteraceae. 1 coasts extratrop. Austr., Chile.
Heterozygis Bunge = Kallstroemia Scop. (Zygophyllac.).
Heteryta Rafin. = Phacelia Juss. (Hydrophyllac.).
Hethingeria Rafin. = Hettlingeria Neck. = Rhamnus L. (Rhamnac.).
Hetrepta Rafin. = Leucas R.Br. (Labiat.).
Hettlingeria Neck. = Rhamnus L. (Rhamnac.).
Heuchera L. Saxifragaceae. 50 N. Am. Sometimes apet.
× Heucherella Wehrh. Saxifragaceae. Gen. hybr. (Heuchera × Tiarella).
Heudelotia A. Rich. = Commiphora Jacq. (Burserac.).
Heudusa E. Mey. = Lathriogyne Eckl. & Zeyh. (Legumin.).
Heuffelia Opiz = Carex L. (Cyperac.).
Heuffelia Schur = Helictotrichon Bess. ex Roem. & Schult. (Gramin.).
Heurckia Muell. Arg. = Rauwolfia L. (Apocynac.).
Heurlinia Rafin. = Rapanea Aubl. (Myrsinac.).
Heurnia Spreng. = Huernia R.Br. (Asclepiadac.).
Heurniopsis K. Schum. = Huerniopsis N. E. Br. (Asclepiadac.).
Hevea Aubl. Euphorbiaceae. 12 trop. Am. *H. brasiliensis* Muell. Arg. is the source of the best natural rubber (Pará rubber), which, though Brazilian in origin, is now largely produced in the Far East. The tree was originally introduced into Ceylon and esp. Malaya in 1876, since when a very large industry has developed. Latex vessels in the cortex are tapped by making sloping cuts half-way round the trunk. The wound is renewed at intervals of one or two days by shaving off a thin slice from the lower side, when there is a larger flow of milk than at first. The milk is usu. coagulated with the aid of enough acid to neutralise its alkalinity, and pressed into sheets or other forms, with or without
Hewardia Hook. = Isophysis T. Moore (Iridac.). [smoking.
Hewardia J. Smith (~ Adiantum L.). Adiantaceae. 4 trop. Am. (Ching, *Acta Phytotax. Sin.* 6: 349, 1957.)
Hewardiaceae Nak. = Iridaceae–Isophysideae Hutch.
Hewittia Wight & Arn. Convolvulaceae. 1–2 trop. Afr., E. As., Indomal.
Hexabolus Steud. (sphalm.) = Hexalobus A. DC. (Annonac.).
Hexacadica Rafin. = Hexadica Lour. = Ilex L. (Aquifoliac.).
Hexacentris Nees = Thunbergia L. f. (Thunbergiac.).
Hexacestra P. & K. = Hexakestra Hook. f. = Andrachne L. (Euphorbiac.).
Hexachlamys Berg (~ Eugenia L.). Myrtaceae. 3 temp. S. Am.
Hexactina Willd. ex Schlechtd. = Amaioua Aubl. (Rubiac.).
Hexacyrtis Dinter. Liliaceae. 1 SW. Afr.
Hexadena Rafin. = Phyllanthus L. (Euphorbiac.).
Hexadenia Klotzsch & Garcke = Pedilanthus Neck. ex Poit. (Euphorbiac.).
Hexadesmia Brongn. Orchidaceae. 15 Mex. to trop. S. Am., W.I.
Hexadica Lour. = Ilex L. (Aquifoliac.).
Hexaglochin Nieuwl. = Triglochin L. (Juncaginac.).

Hexaglottis Vent. Iridaceae. 5 S. Afr.
Hexagonotheca Turcz. = Berrya Roxb. (Liliac.).
Hexakestra Hook. f. = Andrachne L. (Euphorbiac.).
Hexakistra Hook. f. (sphalm.) = praec.
Hexalectris Rafin. Orchidaceae. 3 S. U.S., Mex.
Hexalepis Boeck. = Gahnia J. R. & G. Forst. (Cyperac.).
Hexalepis Rafin. = Vriesea Lindl. (Bromeliac.).
Hexaletris Rafin. = Hexalectris Rafin. (Orchidac.).
Hexalobus A. DC. Annonaceae. 5 trop. & S. Afr., Madag.
Hexameria R.Br. = Podochilus Blume (Orchidac.).
Hexameria Torr. & Gray = Echinocystis Torr. & Gray (Cucurbitac.).
Hexamium Rafin. (nomen). Quid?
Hexaneurocarpon Dop. Bignoniaceae. 1 Indoch.
Hexanthus Lour. = Litsea Lam. (Laurac.).
Hexaphoma Rafin. = Spatularia Haw. = Saxifraga L. (Saxifragac.).
Hexaphora Rafin. (nomen). Menispermaceae. Quid?
Hexaplectris Rafin. = Aristolochia L. (Aristolochiac.).
Hexapora Hook. f. Lauraceae. 1 Malay Penins.
Hexaptera Hook. Cruciferae. 13 temp. S. Am. Fr. winged.
Hexapterella Urb. Burmanniaceae. 1 trop. S. Am.
Hexarrhena J. & C. Presl = Hilaria Kunth (Gramin.).
Hexasepalum Bartl. ex DC. = Diodia L. (Rubiac.).
Hexaspermum Domin (? = Phyllanthus L.). Euphorbiaceae. 1 N. Queensland.
Hexaspora C. T. White. Celastraceae. 1 N. Queensland.
Hexastemon Klotzsch = Eremia D. Don (Ericac.).
Hexastylis Rafin. (1825) = Asarum L. (Aristolochiac.).
Hexastylis Rafin. (1836) = Stylexia Rafin. = Caylusea A. St-Hil. (Resedac.).
Hexatheca C. B. Clarke. Gesneriaceae. 3 Borneo.
Hexatheca Sond. ex F. Muell. = Lepilaena Drumm. (Zannichelliac.).
Hexepta Rafin. = Coffea L. (Rubiac.).
Hexisea Lindl. Orchidaceae. 5 Mex. to Colombia.
Hexodontocarpus Dulac = Sherardia L. (Rubiac.).
Hexonix Rafin. = Heloniopsis A. Gray (Liliac.).
Hexonychia Salisb. = Allium L. (Alliac.).
Hexopea Steud. = Hexopia Batem. ex Lindl. = Hexadesmia Brongn. (Orchidac.).
Hexopetion Burret (~ Astrocaryum G. F. W. Mey.). Palmae. 1 Mex.
Hexopia Batem. ex Lindl. = Hexadesmia Brongn. (Orchidac.).
Hexorima Rafin. = Streptopus Michx (Liliac.).
Hexorina Steud. = praec.
Hexostemon Rafin. = Lythrum L. (Lythrac.).
Hexotria Rafin. = Ilex L. (Aquifoliac.).
Hexuris Miers. Triuridaceae. 2 trop. S. Am.
Heyderia C. Koch = Calocedrus Kurz (Cupressac.).
Heydia Dennst. = Scleropyrum Arn. (Santalac.).
Heydusa Walp. = Heudusa Presl = Lathriogyne Eckl. & Zeyh. (Legumin.).
Heyfeldera Sch. Bip. = Chrysopsis Nutt. (Compos.).
Heylandia DC. = Goniogyna DC. = Crotalaria L. (Legumin.).
Heylygia G. Don (sphalm.) = Helygia Blume = Parsonsia R.Br. (Apocynac.).

HEYMASSOLI

Heymassoli Aubl. = Ximenia L. (Olacac.).

Heymia Dennst. = Dentella J. R. & G. Forst. (Rubiac.).

Heynea Roxb. ex Sims = Trichilia P.Br. (Meliac.).

Heynella Backer. Asclepiadaceae. 1 Java.

Heynichia Kunth = Trichilia P.Br. (Meliac.).

Heynickia C. DC. = praec.

Heywoodia Sim. Euphorbiaceae. 1 Equat. E. Afr. (NW. Tangan. & adjacent Uganda; Kenya) & S. Afr. (E. Cape Prov. to SW. Natal), Cape ebony.

Heywoodiella Svent. & Bramwell. Compositae. 1 Canary Is.

Hibanthus D. Dietr. = Hybanthus Jacq. (Violac.).

Hibbertia Andr. Dilleniaceae. 100 Madag., New Guinea, Austr., New Caled., Fiji. Mostly ericoid or climbing shrubs. Some have phylloclades. Infl. dich., but often, by reduction, coming to look like a raceme. The sta., etc., vary much in number in different spp.

Hibbertiaceae J. G. Agardh = Dilleniaceae–Hibbertieae Reichb.

Hibiscaceae J. G. Agardh = Malvaceae–Hibisceae Reichb.

Hibiscadelphus Rock. Malvaceae. 4 Hawaii.

***Hibiscus** L. Malvaceae. 300 trop. & subtrop. The 5 antesepalous sta. are repres. by teeth at the top of the stamen-tube. Several are cult., esp. *H. rosasinensis* L. (shoe-flower, fls. showy), *H. sabdariffa* L. (*rozelle*, fr. for jelly, etc.), *H. (Abelmoschus) esculentus* L. (*okra* or *bandakai*, mucilaginous young fr. in soups, etc.).

Hibiscus Mill. = Malvaviscus Fabr. = Hibiscus L. + Ketmia Mill. + Malvaviscus Adans. (Malvac.).

Hicarya Rafin. = Carya Nutt. (Juglandac.).

Hickelia A. Camus. Gramineae. 2 Madag.

Hickenia Britton & Rose = Parodia Speg. (Cactac.).

Hickenia Lillo. Asclepiadaceae. 1 Argent.

Hickoria C. Mohr = Hicoria Rafin. = Carya Nutt. (Juglandac.).

Hicksbeachia F. Muell. Proteaceae. 2 NE. Austr.

Hicoria Rafin. = Carya Nutt. (Juglandac.).

Hicorias Benth. & Hook. f. = praec.

Hicorius Rafin. = praec.

Hicorya Rafin. = praec.

Hicriopteris Copel. = Diplopterygium (Diels) Nakai (Gleicheniac.).

Hicriopteris Presl = Dicranopteris Bernh. (Gleicheniac.). (Holttum, *Fl. Males.* ser. I, 1: 32, 33; 1959.)

Hidalgoa La Llave. Compositae. 4 Mex., C. Am.

Hidrocotile Neck. = Hydrocotyle L. (Hydrocotylac.).

Hidrosia E. Mey. = Rhynchosia Lour. (Legumin.).

Hieraceum Hoppe = Hieracium L. (Compos.).

Hierachium Hill = Hypochoeris L. (Compos.).

Hieraciastrum Fabr. (uninom.) = *Picris* L. sp. (Compos.).

Hieraciodes Kuntze = seq.

Hieracioïdes Fabr. = Crepis L. (Compos.).

Hieracioïdes Moench = seq.

Hieracium L. Compositae. Perhaps 5000 apomictic micro-species, or 1000 macro-species. Temp. regions (exc. Australasia) and trop. mts.

HIMANTANDRACEAE

Hieranthemum Spach = Heliotropium L. (Boraginac.).

Hieranthes Rafin. = Stereospermum Cham. (Bignoniac.).

Hierapicra Kuntze = Cnicus L. emend. Gaertn. (Compos.).

Hiericontis Adans. = Anastatica L. (Crucif.).

Hieris van Steenis. Bignoniaceae. 1 Malay Penins. (Penang).

Hiernia S. Moore. Scrophulariaceae. 1 Angola, SW. Afr.

Hierobotana Briq. Verbenaceae. 1 Ecuad., Peru.

Hierochloa Beauv. = seq.

*Hierochloë [J. G. Gmel.] R.Br. Gramineae. 30 temp. & cold; trop. mts. Exceptionally strong odour of coumarin in *H. odorata* (L.) Beauv.

Hierochontis Medik. = Euclidium R.Br. (Crucif.).

Hierocontis Steud. = Hiericontis Adans. = Anastatica L. (Crucif.).

Hieronia Vell. = Davilla Vand. (Dilleniac.).

Hieronima Allem. Euphorbiaceae. 36 trop. Am., W.I.

Hieronyma Baill. = praec.

Hieronymiella Pax. Amaryllidaceae. 1 Argentina.

Hieronymusia Engl. Saxifragaceae. 1 S. Andes.

Hierophyllus Rafin. = Ilex L. (Aquifoliac.).

Hiesingera Endl. = Hisingera Hellen. = Xylosma J. R. & G. Forst. (Flacourtiac.).

Higgensia Steud. = Higginsia Pers. = Hoffmannia Sw. (Rubiac.).

Higginsia Blume = Petunga DC. (Rubiac.).

Higginsia Pers. = Ohigginsia Ruiz & Pav. = Hoffmannia Sw. (Rubiac.).

Higinbothamia Uline = Dioscorea L. (Dioscoreac.).

Hilacium Steud. = Hylacium Beauv. = Psychotria L. (Rubiac.).

Hilairanthus Van Tiegh. = Avicennia L. (Avicenniac.).

Hilairella Van Tiegh. = Luxemburgia A. St-Hil. (Ochnac.).

Hilairia DC. = Onoseris DC. (Compos.).

Hilaria Kunth. Gramineae. 9 SW. U.S. to C. Am.

Hilariophyton Pichon = Sanhilaria Baill. (Bignoniac.).

Hilbertia Thouin ex Reichb. (nomen). Quid?

Hildebrandtia Vatke. Convolvulaceae. 8 Afr. 2 seps. enlarged on fr.

Hildegardia Schott & Endl. Sterculiaceae. 1 Cuba; 2 trop. E. Afr., 3 Madag.; 1 SW. China, 1 Philipp., 1 Lesser Sunda Is.

Hildewintera F. Ritter. Cactaceae. 1 Bolivia.

Hildmannia Kreuz. & Buining = Pyrrhocactus (A. Berger) Backeb. & F. M. Knuth, etc. (Cactac.).

Hillebrandia Oliv. Begoniaceae. 1 Hawaii.

Hilleria Vell. Phytolaccaceae. 5 C. Am., trop. S. Am.

Hilleriaceae Nak. = Phytolaccaceae–Rivineae Reichb.

Hillia Boehm. = Halesia Ellis ex L. (Styracac.).

Hillia Jacq. Rubiaceae. 20 Mex. to trop. S. Am. Epiph. shrubs. Fls. sol., term.

Hilospermae Vent. = Sapotaceae Juss.

Hilsenbergia Boj. = Dombeya Cav. (Sterculiac.).

Hilsenbergia Tausch ex Meissn. = Ehretia L. (Ehretiac.).

Himalrandia Yamazaki. Rubiaceae. 1 Himal.

Himantandra F. Muell. = Galbulimima F. M. Bailey (Himantandrac.).

*Himantandraceae Diels. Dicots. 1/2 E. Malaysia, Austr. Trees, densely

Hippophaës Aschers. = praec.

Hippophaestum S. F. Gray = Centaurea L. (Compos.).

Hippopodium Harv. ex Lindl. = Ceratandra Eckl. ex Bauer (Orchidac.).

Hipporchis Thou. = seq.

Hipporkis Thou. = Satyrium Sw. (Orchidac.).

Hipposelinum Britton & Rose = Levisticum Koch (Umbellif.).

Hipposeris Cass. = Onoseris DC. (Compos.).

Hippothronia Benth. = Hypothronia Schrank = Hyptis Jacq. (Labiat.).

Hippotis Ruiz & Pav. Rubiaceae. 11 trop. S. Am.

Hippoxylon Rafin. = Oroxylum Vent. (Bignoniac.).

*Hippuridaceae Link. Dicots. 1/3 cosmop. Water plants with creeping rhiz. and erect sympodial shoots, whose upper parts usu. project above the water. Ls. linear, ent., exstip., vertic., the submerged ones longer and more flaccid than the aerial. Fls. sol. and sess. in the axil of each l., ♂ (or sometimes ♀ in the upper axils, or on some stocks; cf. *Labiatae*), protogynous. P o (or K repr. by a slight rim); A 1, epig., with large bilobed anth.; Ḡ (1) with 1 pend. anatr. ov. with 1 integ., and 1 long subulate style stigmatic throughout; anemoph. Fr. a smooth ovoid achene. Only genus: *Hippuris*. Affinities doubtful, but prob. connected with *Haloragidac.*, *Elatinac.*, *Lythrac.*, *Primulac.*, etc.

Hippuris L. Hippuridaceae. 2–3 almost cosmop.

Hiptage Gaertn. Malpighiaceae. 20–30 Mauritius, W. Himal. to China, Formosa, Indoch., W. Malaysia, Celebes, Timor, Fiji.

Hiraea Jacq. Malpighiaceae. 30 trop. Am.

Hirculus Haw. = Saxifraga L. (Saxifragac.).

Hirculus Rafin. = praec.

Hirnellia Cass. = Myriocephalus Benth. (Compos.).

Hirpicium Cass. Compositae. 3 S. Afr.

Hirschfeldia Moench. Cruciferae. 2 Medit., Socotra.

Hirschia Baker. Compositae. 1 S. Arabia.

Hirschtia K. Schum. ex Schwartz = Pontederia L. (Pontederiac.).

Hirtella L. Chrysobalanaceae. 95 C. & trop. S. Am., W.I.; 3 trop. E. Afr., Madag. Fl. zygo., axis deeply hollowed on one side. The sta. and cpl. are not in the hollow, but on the other side of the surface of the axis.

Hirtellaceae Horan. = Chrysobalanac. + Rhizophorac. + Vochysiac. + Dichapetalac.

Hirtellaceae Nak. = Chrysobalanaceae R.Br.

Hirtellina Cass. = Staehelina L. (Compos.).

Hirundinaria B. Ehrh. = Vincetoxicum v. Wolf (Asclepiadac.).

Hisbanche Sparrm. ex Meissn. (sphalm.) = Hyobanche L. (Scrophulariac.)

Hisingera Hellen. = Xylosma G. Forst. (Flacourtiac.).

Hispidella Barnad. ex Lam. Compositae. 1 Iberian Penins.

Hissopus Nocca = Hyssopus L. (Labiat.).

Histiopteris (Agardh) J. Sm. Dennstaedtiaceae. 7 warm, and S. hemisph. Fusion-sorus protected by reflexed edge; no indus.; anast. veins.

Hisutsua DC. = Boltonia L'Hérit. (Compos.).

Hitchcockella A. Camus. Gramineae. 1 Madag.

Hitchenia Wall. Zingiberaceae. 3 India, Malay Penins.

Hitcheniopsis (Baker) Ridley. Zingiberaceae. 10 SE. As.

Hitchinia Horan. = Hitchenia Wall. (Zingiberac.).
Hitoa Nadeaud = ?Ixora L. (Rubiac.).
Hitzera B. D. Jacks. = seq.
Hitzeria Klotzsch = Commiphora Jacq. (Burserac.).
Hladnickia Meissn. = Hladnikia Reichb. = Carum L. (Umbellif.).
Hladnickia Steud. = seq.
Hladnikia Koch. Umbelliferae. 1 Adriatic.
Hladnikia Reichb. = Carum L. (Umbellif.).
Hoarea Sweet = Pelargonium L'Hérit. (Geraniac.).
Hochberg'a [sic!] Pohl (nomen). 1 Brazil. Quid?
Hochenwarthia auctt. = seq.
Hochenwartia Crantz = Rhododendron L. (Ericac.).
Hochreutinera Krapov. Malvaceae. 1 temp. S. Am.
Hochstettera Spach = seq.
Hochstetteria DC. = Dicoma Cass. (Compos.).
Hockea Lindl. (sphalm.) = Fockea Endl. (Asclepiadac.).
Hockinia Gardn. Gentianaceae. 1 E. Brazil.
Hocquartia Dum. = Aristolochia L. (Aristolochiac.).
Hodgkinsonia F. Muell. Rubiaceae. 2 E. Austr.
Hodgsonia Hook. f. & Thoms. Cucurbitaceae. 1 Indomal.
Hodgsonia F. Muell. = seq.
Hodgsoniola F. Muell. Liliaceae. 1 SW. Austr.
Hoeckia Engl. & Graebn. = Triplostegia Wall. ex DC. (Triplostegiac.).
Hoeffnagelia Neck. = Trigonia Aubl. (Trigoniac.).
Hoehnea Epling. Labiatae. 4 E. trop. S. Am. (S. Brazil to N. Argent.)
Hoehneëlla Ruschi. Orchidaceae. 3 Brazil.
Hoehnelia Schweinf. Compositae. 2 trop. E. Afr.
Hoehnephytum Cabr. Compositae. 2 Brazil.
Hoelselia Juss. = seq.
Hoelzelia Neck. = Swartzia Schreb. (Legumin.).
Hoepfneria Vatke = Abrus L. (Legumin.).
Hoferia Scop. = Ternstroemia L. f. (Theac.).
Hoffmannella Klotzsch ex A. DC. = Begonia L. (Begoniac.).
Hoffmannia Loefl. = Duranta L. (Verbenac.).
Hoffmannia Sw. Rubiaceae. 100 Mex. to Argent.
Hoffmannia Willd. = Psilotum Sw. (Psilotac.).
Hoffmanniella Schlechter ex Lawalrée. Compositae. 1 trop. Afr.
Hoffmannseggella H. G. Jones = Laelia Lindl. (Orchidac.).
Hoffmannseggia Willd. = seq.
Hoffmannsegia Bronn = seq.
Hoffmanseggia Cav. Leguminosae. 45 Am., trop. & S. Afr.
Hofmannia Fabr. (uninom.) = Amaracus Gled. sp. (Labiat.).
Hofmannia Spreng. = Hoffmannia Sw. (Rubiac.).
Hofmeistera Reichb. f. = seq.
Hofmeisterella Reichb. f. Orchidaceae. 1 Ecuador.
Hofmeisteria Walp. Compositae. 10 SW. U.S., Mexico.
Hohenackeria Fisch. & Mey. Umbelliferae. 2 N. Afr., Caucasus.
Hohenbergia Bak. = Aechmea Ruiz & Pav. (Bromeliac.).

HOHENBERGIA

Hohenbergia Schult. f. Bromeliaceae. 35 trop. Am., W.I.
Hohenwartha Vest = Carthamus L. (Compos.).
Hohenwarthia Pacher ex A. Braun = Saponaria L. (Caryophyllac.).
Hoheria A. Cunn. Malvaceae. 5 N.Z.
Hoiriri Adans. = Aechmea Ruiz & Pav. (Bromeliac.).
Hoita Rydb. = Psoralea L. (Legumin.).
Hoitzia Juss. = Loeselia L. (Polemoniac.).
Holacantha A. Gray = Castela Turp. (Simaroubac.).
Holacanthaceae Jadin = Simaroubaceae–Castelinae Engl.
Holalafia Stapf. Apocynaceae. 3 trop. W. Afr.
Holarges Ehrh. (uninom.) = *Draba incana* L. (Crucif.).
Holargidium Turcz. = Draba L. (Crucif.).
Holarrhena R.Br. Apocynaceae. 20 trop. Afr., Madag., India, SE. As.,
    Philipp., Malay Penins.
Holboellia Wall. (1824) (~ Stauntonia Wall.). Lardizabalaceae. 10 Himal.,
    China, Indoch.
Holboellia Wall. (1831) = Lopholepis Decne (Gramin.).
Holcoglossum Schlechter. Orchidaceae. 1 Formosa.
Holcolemma Stapf & C. E. Hubbard. Gramineae. 2 E. Afr., S. India, Ceylon.
Holcophacos Rydb. = Astragalus L. (Legumin.).
Holcosorus Moore. Polypodiaceae. 3 Borneo, New Guinea.
*Holcus L. Gramineae. 8 Canaries, N. Afr., Eur., to As. Min. & Cauc.;
    1 S. Afr.
Holcus Nash = Sorghum Moench (Gramin.).
Holderlinia Neck. = Serruria Salisb. (Proteac.).
*Holigarna Buch.-Ham. ex Roxb. Anacardiaceae. 8 Indomal. Ḡ.
Hollandaea F. Muell. Proteaceae. 2 E. Austr.
Hollboellia Meissn. = Holboellia Wall. (1831) = Lopholepis Decne (Gramin.).
Hollboellia Spreng. = Holboellia Wall. (1824) (Lardizabalac.).
Hollermayera O. E. Schulz. Cruciferae. 1 Chile.
Hollia Heynh. (1840) = Noltea Reichb. (Rhamnac.).
Hollia Heynh. (1846) = Chlorophytum Ker-Gawl. (Liliac.).
Hollisteria S. Wats. Polygonaceae. 1 SW. U.S.
Hollrungia K. Schum. Passifloraceae. 2–3 Moluccas, New Guinea, Solomon
    Is.
Holmbergia Hicken (~ Rhagodia R.Br.). Chenopodiaceae. 1 Urug., Argent.
Holmia Börner = Kobresia Willd. (Cyperac.).
Holmskioldia Retz. Verbenaceae. 11 trop. Afr., Madag., Masc., India to
    W. Malaysia.
Holocalyx Micheli. Leguminosae. 1 Brazil, Paraguay, NE. Argent.
Holocarpa Baker (~ Pentanisia Harv.). Rubiaceae. 1 Madag.
Holocarpha Greene. Compositae. 4 Calif.
Holocarya Th. Dur. (sphalm.) = Holocarpa Baker (Rubiac.).
Holocheila (Kudo) S. Chow. Labiatae. 1 SW. China.
Holocheilus Cass. = Trixis P.Br. (Compos.).
Holochiloma Hochst. = Premna L. (Verbenac.).
Holochilus Dalz. = Diospyros L. (Ebenac.).
Holochilus P. & K. = Holocheilus Cass. = Trixis P.Br. (Compos.).

Holochlamys Engl. Araceae. 5 New Guinea.
Holochloa Nutt. = Heuchera L. (Saxifragac.).
Holodictyum Maxon. Aspleniaceae. 2 Mex.
*Holodiscus (C. Koch) Maxim. Rosaceae. 8 W. N. Am. to Colombia.
Hologamium Nees = Ischaemum L. (Gramin.).
Holographis Nees. Acanthaceae. 4 Mex.
Hologymne Bartl. = Lasthenia Cass. (Compos.).
Hologyne Pfitzer. Orchidaceae. 5 Borneo, Java, New Guinea.
Hololachna Ehrenb. (~ Reaumuria L.). Tamaricaceae. 2 C. As.
Hololachne Reichb. = praec.
Hololafia K. Schum. = Holalafia Stapf (Apocynac.).
Hololeion Kitam. Compositae. 2 Japan.
Hololepis DC. = Vernonia Schreb. (Compos.).
Holopeira Miers = Cocculus DC. (Menispermac.).
Holopetala Wight (sphalm.) = Holoptelea Planch. (Ulmac.).
Holopetalon Reichb. = Heteropterys Kunth (Malpighiac.).
Holopetalum Turcz. = Oligomeris Cambess. (Resedac.).
Holophyllum Less. = Athanasia L. (Compos.).
Holophyllum Meissn. (sphalm.) = Hoplophyllum DC. (Compos.).
Holophytum P. & K. = Olofuton Rafin. = Capparis L. (Capparidac.)
Holopleura Regel & Schmalh. = Hyalolaena Bunge (Umbellif.).
Holopogon Komarov & Nevski. Orchidaceae. 1 Russia.
Holoptelea Planch. Ulmaceae. 2 trop. Afr., Indomal.
Holoptelaea Planch. = praec.
Holoptolaea B. D. Jacks. (sphalm.) = praec.
Holopyxidium Ducke. Lecythidaceae. 3 Amaz. Brazil.
Holoregmia Nees = Martynia L. (Martyniac.).
Holoschoenus Link = Scirpus L. (Cyperac.).
Holosepalum Fourr. = Hypericum L. (Guttif.).
Holosetum Steud. = Panicum L. (Gramin.).
Holostachys Greene (sphalm.) = Halostachys C. A. Mey. (Chenopodiac.).
Holostachyum (Copel.) Ching. Polypodiaceae (near *Aglaomorpha* Schott).
1 New Guinea.
Holostemma R.Br. Asclepiadaceae. 1–2 Indomal., China.
Holosteum L. Caryophyllaceae. 6 temp. Euras.
Holostigma G. Don = Lobelia L. (Campanulac.).
Holostigma Spach = Oenothera L. (Onagrac.).
Holostyla DC. corr. Endl. Rubiaceae. 2 New Caled.
Holostylis Duch. Aristolochiaceae. 1 C. & SE. Brazil.
Holostylis Reichb. = Holostyla DC. corr. Endl. (Rubiac.).
Holostylon Robyns & Lebrun. Labiatae. 5 trop. Afr.
Holothamnus P. & K. = Halothamnus F. Muell. = Plagianthus J. R. & G.
Forst. (Malvac.).
*Holothrix Rich. ex Lindl. Orchidaceae. 55 trop. & S. Afr., Arabia.
Holotome Endl. = Actinotus Labill. (Umbellif.).
Holozonia Greene = Lagophylla Nutt. (Compos.).
Holstia Pax (~ Tannodia Baill.). Euphorbiaceae. 2 trop. Afr.
Holtonia Standley. Rubiaceae. 1 Colombia.

× **Holttumara** hort. Orchidaceae. Gen. hybr. (iii) (Arachnis × Renanthera × Vanda).

Holttumiella Copel. = Taenitis Willd. (Gymnogrammac.). (*Blumea* 9: 533, 1962.)

Holtzea Schindl. Leguminosae. 1 N. Austr.

Holtzendorffia Klotzsch & Karst. ex Nees = Ruellia L. (Acanthac.).

Holubia Oliv. Pedaliaceae. 1 S. Afr.

Homaïd Adans. = Biarum Schott (Arac.).

Homaïda Kuntze = praec.

Homaïda Rafin. = Arisarum Targ.-Tozz. (Arac.).

Homalachne (Benth.) Kuntze. Gramineae. 2 S. Spain.

Homaladenia Miers = Dipladenia A. DC. (Apocynac.).

Homalanthus A. Juss. corr. Reichb. Euphorbiaceae. 35 Indomal., Polynes.

Homalanthus Wittst. = Omalanthus Less. = Tanacetum L. (Compos.).

Homaliaceae ('-lineae') R.Br. = Flacourtiaceae–Homalieae Reichb.

Homaliopsis S. Moore = Tristania R. Br. (Myrtac.).

Homalium Jacq. Flacourtiaceae. 200 trop. & subtrop. After fert. the sepals or petals, or both, enlarge and form wings (often hairy) to the fr.

Homalobus Nutt. ex Torr. & Gray = Astragalus L. (Legumin.).

Homalocalyx F. Muell. Myrtaceae. 2 NE. Austr.

Homalocarpus Hook. & Arn. Umbelliferae. 6 Chile.

Homalocarpus P. & K. = Omolocarpus Neck. = Nyctanthes L. (Verbenac.).

Homalocarpus Schur = Anemone L. (Ranunculac.).

Homalocenchrus Mieg ex Hall. = Leersia Sw. (Gramin.).

Homalocephala Britton & Rose = Echinocactus Link & Otto (Cactac.).

Homalocheilos J. K. Morton. Labiatae. 2 trop. Afr.

Homalocladium (F. Muell.) L. H. Bailey. Polygonaceae. 1, *H. platycladium* (F. Muell.) L. H. Bailey (*Muehlenbeckia platyclados* Meissn.), perhaps native New Guinea, Solomons or New Caledonia. It has flat green jointed phylloclades with transverse bands at the nodes, and rhomboid or hastate green ls. which drop early.

Homaloclados Hook. f. = Faramea Aubl. (Rubiac.).

Homaloclina P. & K. = seq.

Homalocline Reichb. = Omalocline Cass. = Crepis L. (Compos.).

Homalodiscus Bunge ex Boiss. = Ochradenus Delile (Resedac.).

Homalolepis Turcz. = Quassia L. (Simaroubac.).

Homalomena Schott. Araceae. 140 trop. As. & S. Am.

Homalonema Kunth (sphalm.!) = praec.

Homalopetalum Rolfe. Orchidaceae. 1 Jamaica.

Homalosche Ehrh. (uninom.) = Lycopodium complanatum L. (Lycopodiac.).

Homalosciadium Domin. Hydrocotylaceae (~ Umbellif.). 1 W. Austr.

Homalosorus Small = Athyrium Roth (Athyriac.).

Homalospermum Schau. = Leptospermum J. R. & G. Forst. (Myrtac.).

Homalostachys Boeck. = Carex L. (Cyperac.).

Homalostoma Shchegl. = Andersonia R.Br. (Epacridac.).

Homalostylis P. & K. = Homolostyles Wall. ex Wight = Tylophora R.Br. (Asclepiadac.).

Homalotes Endl. = Omalotes DC. = Tanacetum L. (Compos.).

Homalotheca Reichb. = Omalotheca Cass. = Gnaphalium L. (Compos.).
Homanthis Kunth = Perezia Lag. (Compos.).
Hombak Adans. = Capparis L. (Capparidac.).
Hombronia Gaudich. = Pandanus L. f. (Pandanac.).
Homeoplitis Endl. = Homoplitis Trin. = Pogonatherum R.Br. (Gramin.).
Homeria Vent. Iridaceae. 37 S. Afr. Bulbils in axils of lower ls.
Homilacanthus S. Moore = Isoglossa Oerst. (Acanthac.).
Homocentria Naud. = Oxyspora DC. (Melastomatac.).
Homochaete Benth. = Macowania Oliv. (Compos.).
Homochroma DC. Compositae. 1 S. Afr.
Homocnemia Miers = Stephania Lour. (Menispermac.).
Homoeantherum Steud. = Homoeatherum Nees = Andropogon L. (Gramin.).
Homoeanthus Spreng. = Homoianthus Bonpl. ex DC. = Perezia Lag. (Compos.).
Homoeatherum Nees = Andropogon L. (Gramin.).
Homoeotes Presl = Feea Bory (Hymenophyllac.).
× **Homoglad** C. Ingram. Iridaceae. Gen. hybr. (Gladiolus × Homoglossum).
Homoglossum Salisb. Iridaceae. 20 S. Afr.
Homognaphalium Kirp. Compositae. Spp.? N. Afr.
Homogyne Cass. Compositae. 3 mts. of Eur.
Homoiachne Pilger = Homalachne (Benth.) Kuntze (Gramin.).
Homoianthus Bonpl. ex DC. = Perezia Lag. (Compos.).
Homoioceltis Blume = Aphananthe Planch. (Ulmac.).
Homolepis Chase. Gramineae. 3 trop. S. Am.
Homollea Arènes. Rubiaceae. 3 Madag.
Homolliella Arènes. Rubiaceae. 1 Madag.
Homolostyles Wall. ex Wight = Tylophora R.Br. (Asclepiadac.).
Homonoia Lour. Euphorbiaceae. 3 S. China, SE. As., Indomal. *H. riparia* Lour. a 'rheophyte'.
Homonoma Bello = Nepsera Naud. (Melastomatac.).
Homopappus Nutt. = Haplopappus Cass. (Compos.).
Homopholis C. E. Hubbard. Gramineae. 1 Queensland.
Homophyllum Merino = Blechnum. L. (Blechnac.).
Homoplitis Trin. = Pogonatherum Beauv. (Gramin.).
Homopogon Stapf. Gramineae. 1 W. Equat. Afr.
Homopteryx Kitagawa = Angelica L. (Umbellif.).
Homoranthus A. Cunn. ex Schau. Myrtaceae. 3 E. Austr.
Homoscleria P. & K. = Omoscleria Nees = Scleria Bergius (Cyperac.).
Homostyles Wall. ex Hook. f. = Homolostyles Wall. ex Wight = Tylophora R.Br. (Asclepiadac.).
Homostylium Nees = Microglossa DC. (Compos.).
Homotropa Shuttlew. ex Small = Asarum L. (Aristolochiac.).
Homotropium Nees = Ruellia L. (Acanthac.).
Homozeugos Stapf. Gramineae. 4 trop. Afr.
Honckeneja Maxim. = Honkenya Ehrh. (Caryophyllac.).
Honckeneya Steud. (1) = Honkenya Willd. = Clappertonia Meissn. (Tiliac.).
Honckeneya Steud. (2) = Honkenya Ehrh. (Caryophyllac.).
Honckenia Pers. = Honckenya Willd. = Clappertonia Meissn. (Tiliac.).

HONCKENIA

Honckenia Rafin. = Honkenya Ehrh. (Caryophyllac.).

Honckenya Bartl. = praec.

Honckenya Willd. = Clappertonia Meissn. (Tiliac.).

Honckneya Spach = praec.

Hondbesseion Kuntze = seq.

Hondbessen Adans. = Paederia L. (Rubiac.).

Honkeneja Endl. = seq.

Honkenya Ehrh. Caryophyllaceae. 2 N. temp. & circumpol., S. Patag. *H. peploïdes* (L.) Ehrh., common on sandy coasts, with long creeping underground stems with scale ls., the green ls. fleshy with water tissue.

Honomoya Scheff. = Homonoia Lour. (Euphorbiac.).

Honorius S. F. Gray = Ornithogalum L. (Liliac.).

Honottia Reichb. = Limnophila R.Br. (Scrophulariac.).

Hoodia Sweet ex Decne. Asclepiadaceae. 10 SW. trop. & S. Afr.

Hoodiopsis Luckhoff. Asclepiadaceae. 1 SW. Afr.

Hoogenia Balls (sphalm.) = Hulthemia Dum. (Rosac.).

Hooibrenckia Hort. = Staphylea L. (Staphyleac.).

Hookera Salisb. = Brodiaea Sm. (Alliac.).

× **Hookerara** hort. Orchidaceae. Gen. hybr. (iii) (Cattleya × Caularthron × Rhyncholaelia).

Hookerella Van Tiegh. = Phrygilanthus Eichl. (Loranthac.).

Hookerina Kuntze = Hydrothrix Hook. f. (Pontederiac.).

Hookia Neck. = Centaurea L. (Compos.).

Hoopesia Buckl. = Acacia Mill. + Cercidium Tul. (Legumin.).

Hoorebekia Cornelissen = Grindelia Willd. (Compos.).

Hopea Garden ex L. = Symplocos L. (Symplocac.).

***Hopea** Roxb. Dipterocarpaceae. 90 S. China, SE. As., Indomal.

Hopea Vahl = Hoppea Willd. (Gentianac.).

Hopeoïdes Cretz. = Scaphula R. N. Parker = Anisoptera Korth. (Dipterocarpac.).

Hopkinsia Fitzgerald. Restionaceae. 1 W. Austr.

Hopkirkia DC. = Schkuhria Roth (Compos.).

Hopkirkia Spreng. = Salmea DC. (Compos.).

Hoplestigma Pierre. Hoplestigmataceae. 2 W. Equat. Afr.

***Hoplestigmataceae** Gilg. Dicots. 1/2 trop. Afr. Trees with large chartac. simple ent. alt. exstip. ls. Fls. ♀, reg., shortly pedic., in term. ebracteate brown-hirsute subscorpioid cymes. K ent. in bud, globose, splitting into irreg. lobes, persist.; C 11–14, shortly connate, imbr., irreg. 3–4-seriate; A 20–35, free, irreg. 3-seriate, with filif. fil. and elongate-obl. anth.; G (2), ovoid, 1-loc., with 2 pariet. plac. each bearing 2 pend. anatr. ov., and 2 filif. genic. styles, shortly united at base, with capit. hippocrepiform stigs. Fr. drupaceous, laterally compressed, channelled on the narrow sides, with soft-leathery exocarp and bony endocarp, containing 4 seeds, each in a small chamber, and with a large empty lateral chamber in the endocarp on each side of the fruit, covered only by the channelled exocarp on the narrow sides, which eventually becomes torn by the growth of the endoc. Seeds oblong, slightly curved, with sparse endosp. and large embryo. Only genus: *Hoplestigma*. A remarkable isolated relic, probably related to primitive *Ehretiaceae*.

Hoplismenus Hassk. = Oplismenus Beauv. (Gramin.).

Hoplonia P. & K. = Oplonia Rafin. = Anthacanthus Nees (Acanthac.).

Hoplopanax P. & K. = Oplopanax (Torr. & Gray) Miq. (Araliac.).

Hoplophyllum DC. Compositae. 2 S. Afr.

Hoplophytum Beer = Aechmea Ruiz & Pav. (Bromeliac.).

Hoplotheca Spreng. = Oplotheca Nutt. = Froelichia Moench (Amaranthac.).

Hoppea Endl. (1839) = Hopea L. = Symplocos L. (Symplocac.).

Hoppea Endl. (1840) = Hopea Roxb. (Dipterocarpac.).

Hoppea Reichb. = Ligularia Cass. (Compos.).

Hoppea Willd. Gentianaceae. 2 India.

Hoppia Nees = Bisboeckelera Kuntze (Cyperac.).

Hoppia Spreng. = Hoppea Willd. (Gentianac.).

Horaninovia Fisch. & Mey. Chenopodiaceae. 6 SW. & C. As.

Horanthes Rafin. = Crocanthemum Spach (Cistac.).

Horanthus B. D. Jacks. = praec.

Horau Adans. = Avicennia L. (Avicenniac.).

Horbleria Pav. ex Moldenke = Rhaphithamnus Miers (Verbenac.).

× **Hordale** Cif. & Giac. Gramineae. Gen. hybr. (Hordeum × Secale).

Hordeaceae Burnett = Gramineae–Hordeëae (Kunth) Lindl.

× **Hordelymus** Bakhtjeev & Darevskaja = × Elymordeum Lepage (Gramin.).

Hordelymus (Jessen) Jessen ex Harz. Gramineae. 1 Eur., W. As.

× **Hordeopyrum** Simonet. Gramineae. Gen. hybr. (Agropyron × Hordeum).

× **Horderoegneria** Tsvelev. Gramineae. Gen. hybr. (Hordeum × Roegneria).

Hordeum L. Gramineae. 20 temp. Spikelets in groups of 3 on the main axis, forming a dense spike. Each is 1-flowered when perfect, but commonly either the central or the two lat. fls. are aborted. The cult. barley is *H. vulgare* L. (*H. sativum* Pers.). The most common form is the var. *distichum* or 2-rowed barley, where the central fl. of each group is fertile, but 6-rowed barley (var. *hexastichum*) and 4-rowed barley, or bere, are also grown. The last is the most hardy and is cult. as far as 70° N. (in Norway). Cf. Schiemann, *Weizen, Roggen, Gerste*, 71–94 (1948).

Horkelia Cham. & Schlechtd. Rosaceae. 30 W. U.S.

Horkelia Reichb. ex Bartl. = Wolffia Horkel ex Schleid. (Lemnac.).

Horkeliella Rydb. (~ Horkelia Cham. & Schlechtd.). Rosaceae. 3 N. Am.

Hormathophylla Cullen & Dudley. Cruciferae. 4 Iberian Penins., 1 Siberia.

Hormiastis P. & K. = Ormiastis Rafin. = Salvia L. (Labiat.).

Hormidium Lindl. ex Heynh. Orchidaceae. 7 Mex., W.I., trop. Am.

Hormilis P. & K. = Ormilis Rafin. = Salvia L. (Labiat.).

Horminum L. Labiatae. 1 mts. of S. Eur.

Horminum Mill. = Salvia L. (Labiat.).

Hormocalyx Gleason. Melastomataceae. 1 Brazil.

Hormocarpus P. & K. = Ormycarpus Neck. = Raphanus L. (Crucif.).

Hormocarpus Spreng. = Ormocarpum Beauv. (Legumin.).

Hormogyne A. DC. = Planchonella Pierre (Sapotac.).

Hormogyne Pierre = Aningueria Aubrév. & Pellegr. (Sapotac.).

Hormolotus Oliver = Ornithopus L. (Legumin.).

Hormopetalum Lauterb. = Sericolea Schlechter (Elaeocarpac.).

Hormosciadium Endl. = Ormosciadium Boiss. (Umbellif.).

Hormosia Reichb. = Ormosia G. Jacks. (Legumin.).

Hormosolevia P. & K. = Ormosolenia Tausch = Peucedanum L. (Umbellif.).

Hormuzakia Guşul. Boraginaceae. 2 E. Medit.

Hornea Baker. Sapindaceae. 1 Mauritius.

Hornea Durand & Jacks. (sphalm.) = Hounea Baill. = Paropsia Nor. ex Thou. (Flacourtiac.).

Hornemannia Benth. = Ellisiophyllum Maxim. (Ellisiophyllac.).

Hornemannia Link & Otto = Lindernia All. (Scrophulariac.).

Hornemannia Vahl = Symphysia C. B. Presl (Ericac.).

Hornemannia Willd. = Lindernia All. + Mazus Lour. (Scrophulariac.).

Hornemannia Willd. emend. Reichb. = Mazus Lour. (Scrophulariac.).

Hornera Miq. = Neolitsea (Benth.) Merr. + (?)Litsea Lam. (Laurac.).

Hornera Neck. = Mucuna Adans. (Legumin.).

Hornschuchia Blume = Cratoxylon Blume (Guttif.).

Hornschuchia Nees. Annonaceae. 3 E. Brazil. Some similarity to *Aristolochiaceae–Bragantieae*.

Hornschuchia Spreng. = ?Mimusops L. (Sapotac.).

Hornschuchi[ac]eae J. G. Agardh = Annonaceae–Annoneae Reichb. ('Trigynaea group', R. E. Fries).

Hornstedtia Retz. Zingiberaceae. 60 Indomal.

Hornungia Bernh. = Gagea Salisb. (Liliac.).

Hornungia Reichb. Cruciferae. 1 W. & C. Eur., Medit.

Horreola Nor. = Procris Comm. ex Juss.

Horridocactus Backeb. (~Neoporteria Britt. & Rose). Cactaceae. 7 Chile.

Horsfielda Pers. = Horsfieldia Willd. (Myristicac.).

Horsfieldia Bl. ex DC. = Harmsiopanax Warb. (Araliac.).

Horsfieldia Chifflot = Chirita Ham. ex D. Don (Gesneriac.).

Horsfieldia Willd. Myristicaceae. 80 S. China, SE. As., Indomal., N. Austr.

Horsfordia A. Gray. Malvaceae. 4 SW. U.S., Mex.

Horstia Fabr. (uninom.) = *Salvia* L. sp. (Labiat.) [*nec* Cestrum L. (Solanac.) *pace* Rothmaler!].

Horta Thunb. ex Steud. = Hosta Tratt. (Liliac.).

Horta Vell. = Clavija Ruiz & Pav. (Theophrastac.).

Hortegia L. = Ortegia Loefl. (Caryophyllac.).

Hortensia Comm. ex Juss. = Hydrangea L. (Hydrangeac.).

Hortia Vand. Rutaceae. 10 trop. S. Am.

Hortonia Wight. Monimiaceae. 2 Ceylon.

Hortsmania Miq. = Condylocarpon Desf. (Apocynac.).

Hortsmannia Pfeiff. = praec.

Horwoodia Turrill. Cruciferae. 1 Arabia.

Hosackia Dougl. ex Benth. apud Lindl. Leguminosae. 50 W. N. Am.

Hosangia Neck. = Maieta Aubl. (Melastomatac.).

Hosea Dennst. Inc. sed. (? Symplocos). 1 S. India.

Hosea Ridley = seq.

Hoseanthus Merr. Verbenaceae. 1 Borneo (Sarawak).

Hosiea Hemsl. & E. H. Wilson. Icacinaceae. 2 W. & C. China, Japan.

Hoslunda Roem. & Schult. = seq.

Hoslundia Vahl. Labiatae. 2–3 trop. Afr.

Hosta Jacq. = Cornutia L. (Verbenac.).
Hosta Pfeiff. = Horta Vell. = Clavija Ruiz & Pav. (Theophrastac.).
*Hosta Tratt. Liliaceae. 10 China, Japan. Embryos are formed in the seeds by outgrowth of the nucellus tissue round the embryo sac (cf. *Caelebogyne*). Seeds winged.
Hostana Pers. = Hosta Jacq. = Cornutia L. (Verbenac.).
Hostea Willd. = Matelea Aubl. (Asclepiadac.).
Hostia Moench = Crepis L. (Compos.).
Hostia P. & K. = Hostea Willd. = Matelea Aubl. (Asclepiadac.).
Hostmannia Planch. = Elvasia DC. (Ochnac.).
Hostmannia Steud. ex Naud. = Comolia DC. (Melastomatac.).
Hoteia C. Morr. & Decne = Astilbe Buch.-Ham. (Saxifragac.).
Hotnima A. Chev. = Manihot Mill. (Euphorbiac.).
Hottea Urb. Myrtaceae. 6 Santo Domingo.
Hottonia L. Primulaceae. 2, one Atl. N. Am., the other, *H. palustris* L., Eur. & W. As. Floating water pls. with finely divided submerged ls. The fls. are borne above the water; they are dimorphic like *Primula*.
Hottonia Vahl = ? Myriophyllum L. (Haloragidac.).
Hottuynia Cram. = Houttuynia Thunb. (Saururac.).
Houlletia Brongn. Orchidaceae. 12 C. & trop. S. Am.
Houmiri Aubl. Houmiriaceae. 3–4 trop. S. Am.
Houmiria Juss. = praec.
Houmiriaceae Juss. Dicots. 8/50 trop. Am., W. Afr. Trees and shrubs, with alt. simple ent. or cren. ls., stips. small or o. Fls. ⚥, reg., in axill. or rarely term. thyrses. K 5, imbr., ± connate below; C 5, contorted or imbr., sometimes persist.; A ∞ in several series, or 30–10, 1–2-ser., fil. connate below, subul. or filif.; anth. ± versat., with thickened produced connective, sometimes with 4 discrete locelli; disk intrastam., cupular or of 10–20 free scales; G̲ (5), with simple style and capit. stig., and 1–2 axile anatr. ov. per loc. Fr. drupaceous, with ± fleshy exocarp, and hard woody endoc., often with resin-filled cavities, 5-loc. but usu. only 1–2-seeded, germinating by valves or opercula; seeds with fleshy and oily endosp. Genera: *Vantanea, Duckesia, Endopleura, Hylocarpa, Houmiri (Humiria), Sacoglottis, Schistostemon, Humiriastrum*. Rather closely related to *Ixonanthac.* and *Linac.*
Houmiry Duplessy = Houmiri Aubl. (Houmiriac.).
Hounea Baill. = Paropsia Nor. ex Thou. (Flacourtiac.).
Houstonia L. Rubiaceae. 50 S. & W. U.S., Mex. Fls. heterostyled as in *Primula*; similar differences in stigma and pollen.
Houtouynia Pers. = Houttuynia Thunb. (Saururac.).
Houttea Decne = Vanhouttea Lem. (Gesneriac.).
Houttea Heynh. = Achimenes P.Br. (Gesneriac.).
Houttinia Steud. = Hovttinia Neck. = Zantedeschia Spreng. (Arac.).
Houttouynia Batsch = seq.
Houttoynia Gmel. = Houttuynia Thunb. (Saururac.).
Houttuynia Houtt. = Acidanthera Hochst. (Iridac.).
Houttuynia P. & K. = Hovttinia Neck. = Zantedeschia Spreng. (Arac.).
*Houttuynia Thunb. (1784). Saururaceae. 1 Himalaya to Japan. Parthenogenetic.
Houtuynia Thunb. (1783) = praec.

Houtuynia Thunb. (1784) = Houttuynia Houtt. = Acidanthera Hochst. (Iridac.).

Houzeaubambus Mattei = Oxytenanthera Munro (Gramin.).

Hovea R.Br. Leguminosae. 12 Austr.

Hovenia Thunb. Rhamnaceae. 5 Himal. to Japan. Fr. axis succulent, ed.

Hoverdenia Nees. Acanthaceae. 1 Mex.

Hovttinia Neck. = Zantedeschia Spreng. (Arac.).

Howardia Klotzsch = Aristolochia L. (Aristolochiac.).

Howardia Wedd. = Pogonopus Klotzsch (Rubiac.).

Howea Benth. & Hook. f. = seq.

Howeia Becc. Palmae. 2 Lord Howe Island.

Howellia A. Gray. Campanulaceae. 1 W. N. Am. A flaccid aquatic.

Howelliella Rothm. Scrophulariaceae. 1 Calif.

Howiea B. D. Jacks. (sphalm.) = Howeia Becc. (Palm.).

Howittia F. Muell. Malvaceae. 1 Austr.

Hoya R.Br. Asclepiadaceae. 200 S. China, SE. As., Indomal., Austr., Pacif. Twiners and root-climbers with fleshy ls.

Hoyella Ridley. Asclepiadaceae. 1 Sumatra.

Hoyopsis Léveillé = Tylophora R.Br. (Asclepiadac.).

Hua Pierre & De Wild. Huaceae. 2 trop. Afr.

Huaceae A. Chev. Dicots. 1/2 trop. Afr. Trees with simple ent. alt. short-pet. minutely stip. ls. Fls. ☿, reg., small, on slender pedic., in axill. few-fld. fasc. K5, valv., persist.; C 5, with slender claw and ± peltate hirsute inward-facing lamina, ± indupl.-valv., ± persist.; A 10, 1-seriate, equal, with flattened fil. and peltately attached unequally 4-locellate anth.; stds. and disk 0; G̲ (5), 1-loc., with 1 basal erect anatr. ov., and short conical style with simple stig. Fr. a relatively large thin-walled 5-valved caps.; seed 1, large, oblong, with long-setose testa, and copious endosp. smelling of onion. Only genus: *Hua*. The genus *Afrostyrax* (for chars. see ADDENDA) should also perhaps be included. Affinities probably with *Sterculiaceae*.

Hualania Phil. = Bredemeyera Willd. (Polygalac.).

Huanaca Cav. Hydrocotylaceae (~ Umbellif.). 4 Chile, Patag.

Huanuca Rafin. = Acnistus Schott (Solanac.).

Huarpea Cabrera. Compositae. 1 S. Andes.

Hubbardia Bor. Gramineae. 1 W. India.

Huberia DC. Melastomataceae. 10 trop. S. Am.

Huberodaphne Ducke = Endlicheria Nees (Laurac.).

Huberodendron Ducke. Bombacaceae. 5 C. Am. to Brazil.

Hubertia Bory = Senecio L. (Compos.).

Hudsonia L. Cistaceae. 3 Atlantic N. Am. to NW. Terr. Canada.

Hudsonia Robins. ex Lunan = Terminalia L. (Combretac.).

Huebneria Reichb. = Hypericum L. (Guttif.).

Huebneria Schlechter = Orleanesia B. Rodr. (Orchidac.).

Huegelia R.Br. ex Endl. = Decadianthe Reichb. (?Rutac.).

Huegelia P. & K. = Hugelia Benth. = Eriastrum Woot. & Standl. (Polemoniac.).

Huegelia Reichb. = Trachymene Rudge (Umbellif.).

Huegelroea P. & K. = Hugelroea Steud. = Sphaerolobium Sm. (Legumin.).

Hugueninia Reichb. = Hugueninia Reichb. (Crucif.).

Huenefeldia Walp. = Calotis R.Br. (Compos.).

Huernia R.Br. Asclepiadaceae. 30 trop. & S. Afr., S. Arabia.

Huerniopsis N. E. Br. Asclepiadaceae. 3–4 S. & SW. Afr.

Huerta Jaume St-Hil. = seq.

Huertea Ruiz & Pav. Staphyleaceae(?). 4 Cuba, S. Domingo, Colombia, Peru.

Huertia G. Don = praec.

Huertia Mutis apud Alba = Swartzia Schreb. (Legumin.).

Huetia Boiss. = Carum L. (Umbellif.).

× **Hueylihara** hort. Orchidaceae. Gen. hybr. (iii) (Neofinetia × Renanthera × Rhynchostylis).

Hufelandia Nees = Beilschmiedia Nees (Laurac.).

Hugelia Benth. = Eriastrum Wooton & Standl. (Polemoniac.).

Hugelroea Steud. = Sphaerolobium Sm. (Legumin.).

Hugeria Small = Vaccinium L. (Ericac.).

Hugonia L. Linaceae. 40 trop. Afr., Madag., Masc., Indomal., New Caled. The lower twigs of the infl. are modified into hooks for climbing.

Hugoniaceae Arn. = Linaceae–Hugonieae Meissn.

Hugueninia Reichb. Cruciferae. 2 Pyren., Balearic Is., Alps.

Huidobria C. Gay = Loasa Juss. (Loasac.).

Huilaea Wurdack. Melastomataceae. 1 Colombia.

Hulemacanthus S. Moore. Acanthaceae. 1 New Guinea.

Hulletia Willis (sphalm.) = seq.

Hullettia King. Moraceae. 2 Lower Burma, Penins. Siam, Malay Penins., Sumatra.

Hulsea Torr. & A. Gray. Compositae. 8 W. U.S.

Hultemia Reichb. = seq.

Hultenia Reichb. = Hulthemia Dum. (Rosac.).

Hulthemia Blume ex Miq. = Abrus L. (Legumin.).

Hulthemia Dum. (~ Rosa L.). Rosaceae. 2 SW. & C. As.

× **Hulthemosa** Juzepczuk. Rosaceae. Gen. hybr. (Hulthemia × Rosa).

Humata Cav. Davalliaceae. 50 Madag., trop. As. to Pacific. Small epiphytes; doubtfully distinct from *Davallia*.

Humbertia Comm. ex Lam. Humbertiaceae. 1 Madag.

*****Humbertiaceae** Pichon (~ Convolvulaceae). Dicots. 1/1 Madag. Tall trees, with simple ent. alt. exstip ls. Fls. sol., axill., zygo., ♀, bibracteolate. K 5, much imbr., persist.; C (5), campan., v. shortly lobed, lobes contorted and unilaterally indupl., tube and lobes densely adpressed-ferrugineous-pubesc. outside; A 5, alternipet., inserted at base of C-tube, long-exserted, inflexed in bud, declinate at anthesis, anthers basifixed, non-versatile, introrse; disk 0; G (2), 2-loc., on a short thick gynophore, with simple style, deflexed in bud, with small truncate stig., and ∞ axile anatr. ov. Fr. dry, crustaceous, indehisc., 4-seeded. Only genus: *Humbertia*.

Humbertianthus Hochr. Malvaceae. 1 Madag.

Humbertiella Hochr. Malvaceae. 3 Madag.

Humbertina Buchet. Araceae. 1 Madag.

Humbertiodendron Leandri. Trigoniaceae. 1 Madag.

Humbertioturraea Leroy. Meliaceae. 4–5 Madag.

Humbertochloa A. Camus & Stapf. Gramineae. 2 trop. E. Afr., Madag.

HUMBERTODENDRON

Humbertodendron Leandri = Humbertiodendron Leandri (Trigoniac.).
Humblotia Baill. = Drypetes Vahl (Euphorbiac.).
Humblotidendron H. St John = Humblotiodendron Engl. (Rutac.).
Humblotiella Tard. Lindsaeaceae. 1 Madag.
Humblotiodendron Engl. Rutaceae. 2 Comoros, Madag.
Humboldia Reichb. = Humboldtia Vahl (Legumin.).
Humboldtia Neck. = Voyria Aubl. (Gentianac.).
***Humboldtia** Vahl. Leguminosae. 6 S. India, Ceylon. *H. laurifolia* Vahl is myrmecophilous. The non-flowering twigs are normal, but those that bear fls. have hollow obconical internodes. In each of these, at the top, opposite the l., is a slit leading to the cavity, which is inhabited by ants.
Humboldtiella Harms. Leguminosae. 2 trop. S. Am., W.I.
Humboltia Ruiz & Pav. = Pleurothallis R.Br. + Stelis Sw. (Orchidac.).
Humea Roxb. = Brownlowia Roxb. (Tiliac.).
Humea Sm. Compositae. 7 Madag., S. Austr.
***Humiria** Jaume St-Hil. = Houmiri Aubl. (Houmiriac.).
***Humiriaceae**: see Houmiriaceae.
Humirianthera Huber. Icacinaceae. 3 Colombia, Brazil.
Humiriastrum (Urb.) Cuatrec. Houmiriaceae. 12 C. Am. to SE. Brazil.
Humirium Rich. ex Mart. = Houmiri Aubl. (Houmiriac.).
Humularia Duvign. Leguminosae. 40 trop. Afr.
Humulus L. Cannabidaceae. 2 N. temp., S. to Indoch. & SW. U.S. Perennial climbing herbs. Infl. cymose, dioec., the ♂ a much-branched pseudo-panicle, the ♀ a few-flowered pseudo-catkin with 2 fls. in the axil of each scale. Fl. protog., wind fert. Achene. *H. lupulus* L. is the hop, largely cult.; the fr. catkin is used in brewing, etc.
Hunefeldia Lindl. = Huenefeldia Walp. = Calotis R.Br. (Compos.).
Hunemannia A. Juss. = Hunnemannia Sweet (Papaverac.). [Loyalty Is.]
Hunga Panch. ex Prance. Chrysobalanaceae. 5 New Guinea, New Caled.,
Hunguana Maury (sphalm.) = Hanguana Blume (Hanguanac.).
Hunnemannia Sweet. Papaveraceae. 1 Mex.
Hunsteinia Lauterb. = Rapanea Aubl. (Myrsinac.).
× **Huntara** hort. (v) = × Teohara hort. (Orchidac.).
Hunteria [Moç. & Sessé ex] DC. = Porophyllum Adans. (Compos.).
Hunteria Roxb. Apocynaceae. 6 trop. Afr., S. India, Ceylon, Andam., S. China, SE. As., Malay Penins., Anambas Is. [Huntleya].
× **Huntleanthes** hort. Orchidaceae. Gen. hybr. (iii) (Cochleanthes × **Huntleya** Bateman. Orchidaceae. 10 C. & trop. S. Am., Trinidad.
Huodendron Rehder. Styracaceae. 6 S. China, Siam, Indoch.
Huonia Montrouz. = Acronychia J. R. & G. Forst. (Rutac.).
Huperzia Bernh. = Lycopodium L. (Lycopodiac.).
Huperziaceae Rothm. = Lycopodiaceae Reichb.
Hura Koen. ex Retz. = Globba L. (Zingiberac.).
Hura L. Euphorbiaceae. 2 Mex. to trop. S. Am., W.I., incl. *H. crepitans* L., the sandbox tree. Fr. with numerous hard woody cpls. Each, as the ripe fr. dries, tries to expand from the △ shape to a U shape. Presently an explosion occurs and the seeds are shot out. The frs. used to be wired together and used as sand-boxes before the era of blotting-paper.

Husangia Juss. = Hosangia Neck. = Maieta Aubl. (Melastomatac.).
Husemannia F. Muell. = Carronia F. Muell. (Menispermac.).
Husnotia Fourn. Asclepiadaceae. 1 Brazil.
Hussonia Boiss. = Erucaria Gaertn. (Crucif.).
Huszia Klotzsch = Begonia L. (Begoniac.).
Hutchinia Wight & Arn. = Caralluma R.Br. (Asclepiadac.).
Hutchinsia R.Br. Cruciferae. 3 Eur.
Hutchinsiella O. E. Schulz. Cruciferae. 1 W. Tibet.
Hutchinsonia M. E. Jones = Hymenothrix A. Gray (Compos.).
Hutchinsonia Robyns = Rytigynia Bl. (Rubiac.).
Hutera Porta = Coincya Rouy (Crucif.).
Huthamnus Tsiang. Asclepiadaceae. 1 SW. China.
Huthia Brand. Polemoniaceae. 2 Peru.
Hutschinia D. Dietr. = Hutchinia Wight & Arn. = Caralluma R.Br. (Ascle-
Huttia Drumm. ex Harv. = Hibbertia Andr. (Dilleniac.).   [piadac.).
Huttia Preiss ex Hook. = Calectasia R.Br. (Liliac.).
Huttonaea Harv. Orchidaceae. 5 S. Afr.
Huttonella T. Kirk. Leguminosae. 4 N.Z.
Huttum Adans. = Barringtonia J. R. & G. Forst. (Barringtoniac.).
Huxhamia Garden ex Sm. (1) = Trillium L. (Trilliac.).
Huxhamia Garden ex Sm. (2) = Berchemia Neck. ex DC. (Rhamnac.).
Huxhamia Garden ex Sm. (3) = Gordonia Ellis (Theac.).
Huxleya Ewart. Verbenaceae. 1 N. Austr.
Hyacinth[ac]eae J. G. Agardh = Liliaceae–Scilloïdeae K. Krause.
Hyacinthella Schur (~ Hyacinthus L.). Liliaceae. 17 SE. Eur. to C. As.
Hyacinthoïdes Fabr. (uninom.) = Endymion Dum. sp. (Liliac.).
Hyacinthorchis Blume = Cremastra Lindl. (Orchidac.).
Hyacinthus L. Liliaceae. 30 Medit., Afr. Many cult. forms of hyacinth
derived from H. orientalis L.
Hyaenachne Benth. & Hook. f. = seq.
Hyaenanche Lamb. & Vahl. Euphorbiaceae (?). 1 S. Afr. Ls. verticillate.
Perhaps some distant connection with Buxaceae.
Hyala L'Hérit. ex DC. = Polycarpaea Lam. (Caryophyllac.).
Hyalaea Benth. & Hook. f. = Hyalea Jaub. & Spach = Centaurea L. (Compos.).
Hyalaena C. Muell. = Hyalolaena Bunge = Selinum L. (Umbellif.).
Hyalea Jaub. & Spach = Centaurea L. (Compos.).
Hyalis Champ. = Sciaphila Blume (Triuridac.).
Hyalis D. Don ex Hook. & Arn. = Plazia Ruiz & Pav. (Compos.).
Hyalis Salisb. = Ixia L. (1762) (Iridac.).
Hyalisma Champ. (~ Sciaphila Bl.). Triuridaceae. 1 India.
Hyalocalyx Rolfe. Turneraceae. 1 trop. E. Afr., Madag.
Hyalochlamys A. Gray = Angianthus Wendl. (Compos.).
Hyalocystis H. Hallier. Convolvulaceae. 1 trop. Afr.
Hyalolaena Bunge. Umbelliferae. 6 C. As.
Hyalolepis DC. = Myriocephalus Benth. (Compos.).
Hyalolepis Kunze = Belvisia Mirbel (Polypodiac.).
Hyalopoa (Tsvelev) Tsvelev. Gramineae. 4 Cauc., Turkey, NW. Himal., NE.
Siberia.

HYALOSEMA

Hyalosema Rolfe = Bulbophyllum Thou. (Orchidac.).

Hyalosepalum Troupin. Menispermaceae. 8 trop. Afr., Madag.

Hyaloseris Griseb. Compositae. 7 Andes (Boliv., Argent.).

Hyalosperma Steetz = Helipterum DC. (Compos.).

Hyalostemma Wall. ex Meissn. = Miliusa Leschen. (Annonac.).

Hyalotricha Copel. ? Aspidiaceae. 1 trop. Am. (*Am. Fern J.* 43: 12, 1953.)

Hybanthera Endl. = Tylophora R.Br. (Asclepiadac.).

*****Hybanthus** Jacq. Violaceae. 150 trop. & subtrop. The roots of *H. ipecacuanha* (L.) Baill. are used in medicine (white ipecacuanha) in the same way as the true drug (*Uragoga*).

Hybericum Schrank (sphalm.) = Hypericum L. (Guttif.).

Hybidium Fourr. = Centranthus DC. (Valerianac.).

Hybiscus Dum. = Hibiscus L. (Malvac.).

Hybochilus Schlechter. Orchidaceae. 2 Costa Rica.

Hybophrynium K. Schum. = Trachyphrynium Benth. (Marantac.).

Hybosema Harms. Leguminosae. 1 Mex., C. Am.

Hybosperma Urb. Rhamnaceae. 2 W.I.

Hybotropis E. Mey. ex Steud. = Rafnia Thunb. (Legumin.).

Hybridella Cass. = Zaluzania Pers. (Compos.).

Hydastylis Steud. = seq.

Hydastylus Dryand. ex Salisb. = Sisyrinchium L. (Iridac.).

Hydatella Diels. Centrolepidaceae. 2 W. Austr., 1 N.Z.

Hydatica Neck. ex S.F. Gray = Saxifraga L. (Saxifragac.).

Hydnocarpus Gaertn. Flacourtiaceae. 40 Indomal.

Hydnophytum Jack. Rubiaceae. 80 Andamans, Indoch., Malaysia (esp. New Guinea), Pacif. Epiphytes with ant-inhabited tubers, like *Myrmecodia* (*q.v.*).

Hydnora Thunb. Hydnoraceae. 12 trop. & S. Afr., Madag.

*****Hydnoraceae** C. A. Agardh. Dicots. 2/18 S. Am., Afr. Leafless parasites, somewhat sim. to *Rafflesiaceae*, the veg. portion consisting of thick creeping branched underground 'rhizomatoids', terete or angled or ribbed, and devoid of roots. Fls. thick, fleshy, sess. on rhizomatoids, reg., ♀. P (3–5), ± tubular, lobes valv.; A (3–5), opp. the P-lobes, usu. sess., united into a thick sinuose-annular or ovoid structure, with very ∞ parallel longit. or transv. thecae; Ḡ (3–5), 1-loc., with ∞ plac., either stalagmitiform and pendulous or laminar and parietal, bearing very ∞ orthotrop. ov. of reduced structure, and simple truncate-pulvinate stig. (G wholly or partly buried in the soil). Fr. a large thick-walled berry with fleshy pulp; seeds ∞, minute, with endosp. and perisp. Genera: *Hydnora, Prosopanche*. Pollination by beetles.

Hydnostachyon Liebm. = Spathiphyllum Schott (Arac.).

Hydora Bess. = Udora Nutt. = Elodea Michx (Hydrocharitac.).

Hydragonum Kuntze = Cassandra D. Don (Ericac.).

Hydrangea L. Hydrangeaceae. 80 Himal. to Japan, Philipp. Is. & Java; Atlant. N. Am., mts. C. Am. to Chile.

*****Hydrangeaceae** Dum. Dicots. 10/115 N. temp. & subtrop.; Andes from Mex. to S. Chile. Small trees or shrubs, sometimes climbers or herbs, with opp. or alt., simple, dentate or rarely lobed, exstip. ls. Fls. reg., ♀ or polyg.-dioec., in cymose infl., sometimes corymbose or capit., or pseudo-rac. by reduction, usu. conspic., outer ones often enlarged and sterile. K (4–10), more rarely 4–10; C 4–10, valv., imbr. or contorted; A 4–∞; G̅ (2–5), with as many free or ± connate styles, and usu. ∞ anatr. ov. on axile or pariet. plac. Fr. a loculic. caps., rarely baccate; seeds ∞, with endosp. Chief genera: *Dichroa, Hydrangea, Decumaria, Schizophragma, Kirengeshoma,* etc. Closely related to *Philadelphac.* [lariac.).

Hydranthelium Kunth = Herpestis Gaertn. f. = Bacopa Aubl. (Scrophu-**Hydranthus** Kuhl & van Hasselt ex Reichb. f. = Dipodium R.Br. (Orchidac.).

Hydrastidaceae ([Torr. & Gray] Engl. & Gilg) Lemesle (*s. str.*). Dicots. 1/2 Japan, N. Am. Rhizomatous herbs, somewhat intermediate between *Paeoniac., Ranunculac.* and *Podophyllac.,* differing from *Pae.* in the palmately lobed ls., absence of C and disk, and free sta.; from the *R.* in the fls. without nectaries, in the ± fleshy carpel-walls, and in the seeds with the outer integ. longer than the inner; and from the *Pod.* in the apetalous fls., with ∞ stam. and ∞ carpels. Differs from *Glaucidiaceae* (*q.v.*) in vernation, in the vasculature of pedicel and receptacle; sepals 3, caducous; ovules with outer and inner integuments of only ± 8 cell layers altogether; carpels many, 2-ovulate, developing into 1-seeded drupelets, seeds globose; chromosome number 26. Only genus: *Hydrastis.*

Hydrastis Ellis ex L. Hydrastidaceae. 1 Japan, 1 E. N. Am. *H. canadensis* L., a tonic, almost exterminated through ruthless collection of rhiz.

Hydrastylis Steud. = Hydastylus Dryand. ex Salisb. = Sisyrinchium L. (Iridac.).

Hydriastele H. Wendl. & Drude. Palmae. 3–4 New Guinea, Austr.

Hydrilla Rich. Hydrocharitaceae. 1 Euras. & Afr. to Austr.

Hydroanzia Koidz. = Hydrobryum Endl. (Podostemac.).

Hydrobryopsis Engl. Podostemaceae. 1 S. India.

Hydrobryum Endl. Podostemaceae. 3 E. Nepal, Assam, S. Japan (? & **Hydrocalyx** Triana = Juanulloa Ruiz & Pav. (Solanac.). [SE. As.).

Hydrocarpus D. Dietr. (sphalm.) = Hydnocarpus Gaertn. (Flacourtiac.).

Hydrocaryaceae Raimann = Trapaceae Dum.

Hydrocaryes Link = praec.

Hydrocera Blume. Balsaminaceae. 1 Indomal.

Hydrocer[ac]eae Bl. = Balsaminaceae DC.

Hydroceras Hook. f. & Thoms. = Hydrocera Bl. (Balsaminac.).

Hydroceratophyllon Seguier = Ceratophyllum L. (Ceratophyllac.).

Hydrochaeris Gaertn., Mey. & Scherb. (sphalm.) = Hydrocharis L. (Hydrocharitac.).

Hydrocharella Spruce ex Rohrb. = Limnobium Rich. (Hydrocharitac.).

Hydrocharis L. Hydrocharitaceae. 6 Eur., Medit., trop. Afr., As., temp. Austr. *H. morsus-ranae* L. is a rootless water pl. with orbicular floating ls. Fls. dioec., produced upon the surface. During summer the pl. multiplies by horizontal stolons, which form new pls. at the ends. Large buds on stolons in autumn drop off, winter, sprout next year.

*Hydrocharitaceae Juss. Monocots. 16/80 trop. & temp. Water pls., subfams. II and III marine, usu. with ribbon-like submerged ls., *Hydrocharis,* etc., with floating or subaerial ls.; squamulae in axils, frequently serial buds. Infl. axillary, monoec. or dioec. (rarely ⚥), ♀ usu. 1-fld., ♂ often > 1, enclosed at first in spathe of usu. 2 or more fused ls. Fl. usu. reg., 3-merous. P usu. in two heterochlam. whorls; A 1–5, innermost often stds.; $\overline{\text{G}}$ (2–15), 1-loc. with parietal plac. and ∞ orthotr. to anatr. erect to pend. ov.; stigmas = cpls. Fr. irreg. dehisc., seeds ∞, exalb.

Classification and chief genera (after Dandy):

I. **Hydrocharitoïdeae (Vallisnerioïdeae).** Freshwater (rarely marine) plants, pollinated at or above the surface of the water; pollen sphaeroid; perianth mostly double.

Spathes of 1 or 2 free bracts: *Hydrocharis, Limnobium, Stratiotes, Enhalus* (the last named marine).

Spathes of 2 bracts connate into a tube: *Ottelia, Elodea, Egeria, Anacharis, Hydrilla, Lagarosiphon, Vallisneria, Blyxa.*

II. **Thalassioïdeae.** Marine plants, pollinated beneath the surface of the water; pollen (where known) confervoid; perianth single; styles 6–12, bifid, dissepiments well developed; ls. alt.; spathes of bracts connate at base: *Thalassia.*

III. **Halophiloïdeae.** As II, but styles 2–5, entire; dissepiments obsolete; ls. opp.; spathes of 2 free bracts: *Halophila.*

Hydrochloa Beauv. Gramineae. 1 SE. U.S. Floating grass.

Hydrochloa Hartm. = Glyceria R.Br. + Puccinellia Parl. + Molinia Schrank (Gramin.).

Hydrocleïs Reichb. = seq.

Hydrocleys Rich. Limnocharitaceae. 4 trop. S. Am., strikingly resembling *Nymphaea* or *Limnanthemum.*

Hydroclis P. & K. = praec.

Hydrocotile Crantz = Hydrocotyle L. (Hydrocotylac.).

*****Hydrocotylaceae** (Drude) Hylander (~ Umbelliferae). Dicots. 30/375 temp. (esp. S.), trop. mts. Mostly low herbs, sometimes pulvinate, sometimes erect or shrubby, often with simple ls.; differing from the *Umbellif.* (and approaching the *Araliac.*) principally in the fruits possessing a woody endoc., without free carpophore; vittae either absent, or immersed in the primary ribs, with none in the furrows. Chief genera: *Hydrocotyle, Centella, Platysace, Xanthosia, Azorella, Bowlesia, Mulinum.*

Hydrocotyle L. Hydrocotylaceae (~ Umbellif.). 100 trop. & temp. Ls. with scarious or cil. stips. Many have peltate ls. Mericarp 5-ribbed.

Hydrodea N. E. Brown. Aïzoaceae. 4 S. Afr., St Helena.

Hydrogaster Kuhlm. Tiliaceae. 1 Brazil.

Hydrogeton Lour. = Potamogeton L. (Potamogetonac.).

Hydrogeton Pers. = Aponogeton L. f. (Aponogetonac.).

Hydrogetones Link = Potamogetonaceae Dum.

Hydroglossum Willd. = Lygodium Sw. (Schizaeac.).

Hydrola Rafin. = seq.

Hydrolaea Dum. = seq.

*****Hydrolea** L. Hydrophyllaceae. 20 trop. Am., Afr., As. Some have axillary thorns (branches). Ls. alt. Styles 2; plac. large, spongy. Fl. self-fert.

Hydrole[ace]ae R.Br. = Hydrophyllaceae R.Br. ex Edw.

Hydrolia Thou. = Hydrolea L. (Hydrophyllac.).

Hydrolirion Léveillé = Sagittaria L. (Alismatac.).

Hydrolythrum Hook. f. (~ Rotala L.). Lythraceae. 1 Indomal.

Hydromestes Benth. & Hook. f. = seq.

Hydromestus Scheidw. = Aphelandra R.Br. (Acanthac.).

Hydromistria Bartl. = seq.

Hydromystria G. F. W. Mey. = Limnobium Rich. (Hydrocharitac.).

Hydropectis Rydb. Compositae. 1 Mex.

Hydropeltid[ac]eae Dum. = Cabombaceae A. Rich.

Hydropeltis Michx = Brasenia Schreb. (Cabombac.).

Hydrophaca Steud. = seq.

Hydrophace Haller = Lemna L. (Lemnac.).

Hydrophila Ehrh. (uninom.) = Tillaea aquatica L. = Crassula aquatica (L.) Schönl. (Crassulac.).

Hydrophila House = Tillaea L. = Crassula L. (Crassulac.).

Hydrophylax L. f. Rubiaceae. 3 coasts of E. Afr., Madag., India. Fr. corky, indehisc.

*Hydrophyllaceae R.Br. ex Edwards. Dicots. 18/250, cosmop. exc. Austr. Herbs or undershrubs with simple or cpd., radical, alt. or opp., exstip. ls.; usu. hairy, sometimes glandular. Fls. scattered or in 'boragoid' cincinni, usu. with no bracteoles, ☿, reg., often blue, usu. 5-merous. K usu. (5), imbr., odd sep. post.; C (5), rotate, bell- or funnel-shaped, usu. imbr.; A usu. 5, epipet., alt. with pets., often with scale-like appendages at base; G̲ (2), on disk or not, 1–2-loc., with 1–2 styles; ovules in each cpl. ∞–2, sessile or pend., anatr., often parietal. Fr. usu. a loculic. caps.; embryo small, in rich endosp. Fls. chiefly visited by bees; honey secreted below ovary, and protected by sta. appendages, which are frequently united to C, and in *Hydrophyllum* form tubes leading to honey. Fls. usu. protandrous. Chief genera: *Hydrophyllum, Nemophila, Phacelia, Nama, Hydrolea.*

Hydrophyllax Rafin. = Hydrophylax L. f. (Rubiac.).

Hydrophyllum L. Hydrophyllaceae. 10 N. Am. Fl. protandrous, with sta. appendages united to C, forming tubes to honey.

Hydrophylum Rafin. = praec.

Hydrophytum auct. (sphalm.) = Hydnophytum Jack (Rubiac.).

Hydropiper Fourr. = Elatine L. (Elatinac.).

Hydropityon Gaertn. f. = Limnophila R.Br. (Scrophulariac.).

Hydropityum Steud. = praec.

Hydropoa (Dum.) Dum. = Glyceria L. (Gramin.).

Hydropyrum Link = Zizania L. (Gramin.).

Hydropyxis Rafin. = ? Bacopa Aubl. (Scrophulariac.) + Centunculus L. (Primulac.).

Hydroryza Prat (sphalm.) = Hygroryza Nees (Gramin.).

Hydroschoenus Zoll. & Mor. = Cyperus L. (Cyperac.).

Hydrosia A. Juss. = Hidrosia E. Mey. = Rhynchosia Lour. (Legumin.).

Hydrosme Schott = Amorphophallus Bl. (Arac.).

Hydrospondylus Hassk. = Hydrilla Rich. (Hydrocharitac.).

*Hydrostachydaceae Engl. Dicots. 1/30 Afr., Madag. Submerged freshwater herbs, from a short tuberous stem. Ls. (? ster. branches) all radic., elongate, simple to tripinnatisect (segs. flattened or vesic. or filif.), ligulate at base, rhachis beset with ∞ short processes. Infl. scapose, spicate, dense-fld. Fls. small, ♂ ♀, dioec., sol. and sess. in axil of bract. K 0, C 0; A 1, fil. short,

HYDROSTACHYS

anth. obl., 2-celled; G (2), 1-loc. with 2 long filif. styles connate below, and 2 pariet. plac. with ∞ ov. Fr. a small ∞-seeded caps., bract accresc. Only genus: *Hydrostachys*. An interesting type, prob. distantly related to *Podostemac.*, and preserving like them certain primitive features of organization. Recent work suggests the probability of a relationship with the *Tubiflorae*: see Rauh & Jäger-Zürn, *Adansonia*, ser 2, **6**: 515–23, 1967.

Hydrostachys Thou. Hydrostachydaceae. 30 trop. & S. Afr., Madag.
Hydrostis Reichb. = Hydrastis L. (Hydrastidac.).
Hydrotaenia Lindl. = Tigridia Juss. (Iridac.).
Hydrothauma C. E. Hubbard. Gramineae. 1 S. trop. Afr. (N. Rhod.).
Hydrothrix Hook. f. Pontederiaceae. 2 Brazil. Submerged annuals; fls. cleistog., with 1 fert. (outer) sta. and 2–1 (inner) stds.
Hydrotiche A. Juss. (sphalm.) = seq.
Hydrotriche Zucc. Scrophulariaceae. 1 Madag. Water pl. with dimorphic ls.
Hydrotrida Small = Macuillamia Rafin. = Bacopa Aubl. (Scrophulariac.).
Hydrotrida Willd. ex Schlechtd. & Cham. = Hydranthelium Kunth = Bacopa Aubl. (Scrophulariac.).
Hydrotrophus C. B. Clarke = Blyxa Nor. (Hydrocharitac.).
Hyeronima Allem. = Hieronima Allem. (Euphorbiac.).
Hygea Hanst. Gesneriaceae. 1 Chile.
Hygea Klotzsch (nomen). Asclepiadaceae. 1 Guiana. Quid?
Hygrobi[ace]ae Rich., Dulac = Haloragidaceae R.Br.
Hygrocharis Hochst. = Nephrophyllum A. Rich. (Convolvulac.).
Hygrocharis Nees = Rhynchospora Vahl (Cyperac.).
Hygrochilus Pfitz. = Vanda R.Br. (Orchidac.).
Hygrophila R.Br. Acanthaceae. 80 trop., in marshes.
Hygrorhiza Benth. (sphalm.) = seq.
Hygroryza Nees. Gramineae. 1 Indomal.
Hylacium Beauv. = Rauvolfia L. (Apocynac.).
Hylaeanthe Jonker-Verhoef & Jonker. Marantaceae. 4 trop. S. Am.
Hylandra Á. Löve (~ Arabidopsis Heynh.). Cruciferae. 1 S. Norway, C. Sweden, Finland, Baltic States, N. Germany.
Hylas Bigel. = Myriophyllum L. (Haloragidac.).
Hylebates Chippindall. Gramineae. 1 trop. E. Afr.
Hylebia Fourr. = Stellaria L. (Caryophyllac.).
Hylenaea Miers. Celastraceae. 2 C. Am., Venez., Guiana, W.I.
Hylethale Link = Prenanthes L. (Compos.).
Hyline Herb. Amaryllidaceae. 2 Brazil.
Hylocarpa Cuatrec. Houmiriaceae. 1 Amaz. Brazil.
Hylocereus (A. Berger) Britton & Rose. Cactaceae. 23 Mex. to Peru.
Hylocharis Miq. = Oxyspora DC. (Melastomatac.).
Hylocharis Tiling ex Regel & Tiling = Clintonia Rafin. (Liliac.).
Hylococcus R.Br. ex T. L. Mitch. (sphalm.) = Xylococcus R.Br. ex Britt. & S. Moore = Petalostigma F. Muell. (Euphorbiac.).
Hylodendron Taub. Leguminosae. 1 trop. Afr.
Hylogeton Salisb. = Allium L. (Alliac.).
Hylogyne Salisb. & Knight = Telopea R.Br. (Proteac.).
Hylomecon Maxim. (~ Chelidonium L.). Papaveraceae. 1 temp. E. As.

Hylomenes Salisb. = Endymion Dum. (Liliac.).

Hylomyza Danser = Dufrenoya Chatin (Santalac.).

Hylophila Lindl. Orchidaceae. 4 Malay Penins., New Guinea.

Hylonome Webb & Berth. = Behnia Didr. (Liliac.).

Hymanthoglossum Tod. = Himantoglossum Spreng. (Orchidac.).

Hymenachne Beauv. Gramineae. 8 trop.

Hymenaea L. Leguminosae. 25 Mex., Cuba, trop. S. Am. *H. courbaril* L. (West Indian locust) has buttress roots. The wood is valuable. From the stem exudes a resin (copal or anime) which is often found in lumps underground near the trees (cf. *Agathis, Trachylobium*); it is used in varnish, etc.

Hymenandra A. DC. ex Spach. Myrsinaceae. 1 E. Himal., Assam.

Hymenanthera R.Br. Violaceae. 7 E. Austr., N.Z., Norfolk I.

Hymenantherum Cass. = Dyssodia Cav. (Compos.).

Hymenanthes Blume = Rhododendron L. (Ericac.).

Hymenanthus D. Dietr. = praec.

Hymenasplenium Hayata = Asplenium L. (Aspleniac.).

Hymendocarpum Pierre ex Pitard = Nostolachma Dur. (Rubiac.).

Hymenella (Moç. & Sessé ex) DC. = Minuartia L. (Caryophyllac.).

Hymenesthes Miers = Bourreria P.Br. (Boraginac.).

Hymenetron Salisb. = Strumaria Jacq. (Amaryllidac.).

Hymenia Griff. = Hymenaea L. (Legumin.).

Hymenidium Lindl. = Pleurospermum Hoffm. (Umbellif.).

Hymenocallis Salisb. Amaryllidaceae. 50 warm Am. The stipular appendages of the sta. are united into a tube, on the summit of which the filaments stand, and which surpasses the perianth in conspicuousness (cf. *Eucharis*).

Hymenocalyx Zenk. = Hibiscus L. (Malvac.).

Hymenocapsa J. M. Black = Gilesia F. Muell. (Sterculiac.).

Hymenocardia Wall. ex Lindl. Hymenocardiaceae. 4 trop. & S. Afr.; 1 SE. As., Malay Penins., Sumatra.

Hymenocardiaceae Airy Shaw. Dicots. 1/5 trop. Afr., SE. As. Decid. trees or shrubs, with alt. simple ent. short-pet. stip. ls., sometimes 3-nerved at base and densely red-gland-dotted below. Fls. ♂ ♀, dioec., ± precocious, in rather short axill. catkin-like spikes or rac. K 4–6, imbr., sometimes irreg. conn. in ♂, narrow and free in ♀. C o. Disk o. A 4–6 or (4–6), oppositisep., with short spreading fil. and large extrorse anth., often bearing a dorsal gland. G̲ (2), compressed at rt. angles to plane of septum, with 2 apical pend. anatr. ov. per loc., and 2 long free simple styles. (Pistillode in ♂ fls. small.) Fr. a capsule, consisting of 2 broad flattened winged or wing-like cocci, separating from the persist. central axis, mostly with 1 seed per coccus; seeds flat, with sparse endosp. Only genus: *Hymenocardia*. Currently referred to the *Euphorbiac.*, but evidence of affinity not strong. Prob. equally related to the *Urticales*, esp. *Ulmaceae*. The pollen is almost indistinguishable from that of *Celtis*: cf. Livingstone, *Ecol. Monogr.* **37** (1): 41, 1967.

*****Hymenocarpos** Savi. Leguminosae. 1 Medit.

Hymenocentron Cass. = Centaurea L. (Compos.).

Hymenocephalus Jaub. & Spach = Centaurea L. (Compos.).

Hymenochaeta Beauv. = Scirpus L. (Cyperac.).

Hymenocharis Salisb. = Ischnosiphon Koern. (Marantac.).

HYMENOCHLAENA

Hymenochlaena Bremek. Acanthaceae. 3 Assam, Malay Penins., Philipp. Is.

Hymenochlaena P. & K. = Hymenolaena DC. (Umbellif.).

Hymenoclea Torr. & A. Gray. Compositae. 4 SW. U.S., Mex.

Hymenocnemis Hook. f. Rubiaceae. 1 Madag.

Hymenocrater Fisch. & Mey. Labiatae. 12 SW. As.

Hymenocyclus Dinter & Schwantes = Malephora N. E. Br. (Aïzoac.).

Hymenocystis C. A. Mey. = Woodsia R.Br. (Aspidiac.).

Hymenodictyon Wall. Rubiaceae. 20 trop. Afr., Madag., Himal. to Celebes.

Hymenodium Fée = Elaphoglossum Schott (Lomariopsidac.).

Hymenoglossum Presl. Hymenophyllaceae. 1 Chile & Juan Fernandez.

Hymenogonium Rich. ex Lebel = Spergularia J. & C. Presl (Caryophyllac.).

Hymenogyne 'N. E. Brown' (sphalm.) = seq.

Hymenogyne Haw. Aïzoaceae. 2 S. Afr.

Hymenolaena DC. (~ Pleurospermum Hoffm.). Umbelliferae. 10 C. As. to Himal.

Hymenolepis Cass. = Metagnanthus Endl. = Athanasia L. (Compos.).

Hymenolepis Kaulf. = Belvisia Mirbel (Polypodiac.).

Hymenolobium Benth. ex Mart. (1837) = Platymiscium Vog. (Legumin.).

Hymenolobium Benth. (1860). Leguminosae. 12 N. trop. S. Am.

Hymenolobus Nutt. Cruciferae. 5 Eur., Medit., C. As., Austr., N. Am., Chile.

Hymenolophus Boerl. Apocynaceae. 1 Sumatra.

Hymenolyma Korovin. Umbelliferae. 2 C. As.

Hymenolytrum Schrad. = Scleria Bergius (Cyperac.).

Hymenomena Less. = seq.

Hymenonema Cass. Compositae. 2 Greece.

Hymenopappus L'Hérit. Compositae. 10 U.S., Mex.

Hymenopholis Gardn. = Oligandra Less. (Compos.).

Hymenophora Viv. ex Coss. = Pituranthos Viv. (Umbellif.).

Hymenophyllaceae Gaudich. Hymenophyllales. 34/600 trop. & temp. (filmy ferns), in very humid locs., esp. cloud-zones on trop. mts. Ls. 5 mm.–60 cm. long, laminae 1 cell thick (exc. *Cardiomanes*); sporangia or short or elongate receptacles continuous with ends of veins, protected by tubular or 2-lobed indusia. In *Syn. Fil.* 2 polymorphous genera, *Trichomanes* L. and *Hymenophyllum* Sm. Copeland recognises 34 genera, but cytological evidence suggests that his arrangement needs revision.

Hymenophyllales. Filicidae. Only fam.: *Hymenophyllaceae.*

Hymenophyllopsidaceae C. Chr. Hymenophyllopsidales. Stem short, scaly; ls. *Hymenophyllum*-like, no stomata; sori ± as *Lindsaeaceae*. Only genus: *Hymenophyllopsis* Goebel.

Hymenophyllopsidales. Filicidae. 1 fam.

Hymenophyllopsis Goebel. Hymenophyllopsidaceae. 2 Guiana.

Hymenophyllum Sm. Hymenophyllaceae. 25 trop. & S. temp., N. temp. only Europe & Japan.

Hymenophysa C. A. Mey. Cruciferae. 4 Persia, C. As.

Hymenopogon Wall. Rubiaceae. 3 Himal., Assam, SW. China.

Hymenopyramis Wall. ex Griff. Verbenaceae. 6 India, SE. As.

Hymenorchis Schlechter. Orchidaceae. 6 New Guinea.

Hymenorebulobivia Frič = Lobivia Britt. & Rose (Cactac.).

Hymenorebutia Frič ex Buining = praec.
Hymenosicyos Chiov. = Oreosyce Hook. f. (Cucurbitac.).
Hymenospermum Benth. = Alectra Thunb. (Scrophulariac.).
Hymenosporum R.Br. ex F. Muell. Pittosporaceae. 1 New Guinea, E. Austr.
Hymenospron Spreng. = Dioclea Kunth (Legumin.).
Hymenostachys Bory = Feea Bory (Hymenophyllac.).
Hymenostegia (Benth.) Harms. Leguminosae. 24 trop. Afr.
Hymenostemma Kunze ex Willk. = Chrysanthemum L. (Compos.).
Hymenostephium Benth. Compositae. 11 Mex. to Colombia & Venez.
Hymenostigma Hochst. = Moraea L. (Liliac.).
Hymenotheca F. Muell. = Codonocarpus A. Cunn. (Gyrostemonac.).
Hymenotheca Salisb. = Ottelia Pers. (Hydrocharitac.).
Hymenothecium Lag. = Aegopogon Willd. (Gramin.).
Hymenothrix A. Gray. Compositae. 5 SW. U.S., Mex.
Hymenoxis Endl. = seq.
Hymenoxys Cass. Compositae. 26 W. N. Am. to Argent.
Hymnostemon P. & K. = Ymnostema Neck. = Lobelia L. (Campanulac.).
Hyobanche L. Scrophulariaceae. 7 S. Afr.
Hyocyamus G. Don = Hyoscyamus L. (Solanac.).
Hyogeton Steud. (sphalm.) = Ilyogeton Endl. = Vandellia L. (Scrophulariac.).
Hyophorbe Gaertn. Palmae. 2 Réunion, Mauritius.
Hyoscarpus Dulac = seq.
Hyoschyamus Zumaglini = seq.
Hyosciamus Neck. = seq.
Hyoscyamus L. Solanaceae. 20 Eur., N. Afr., Sahara, to SW. & C. As. *H. niger* L. formerly largely cult. as a narcotic. The fls. are in cincinni. The capsule stands erect enclosed in the calyx, and opens by a lid (censer mechanism).
Hyoseris L. Compositae. 3 Medit.
Hyosicamus Hill (sphalm.) = Hyoscyamus L. (Solanac.).
Hyospathe Mart. Palmae. 19 C. & trop. S. Am.
Hypacanthium Juzepczuk (∼ Cousinia Cass.). Compositae. 1 C. As.
Hypaelyptum Vahl = Hypolytrum Rich. (Cyperac.).
Hypaelytrum Poir. = praec.
Hypagophytum A. Berger. Crassulaceae. 1 Abyssinia.
Hypanthera S. Manso = Fevillea L. (Cucurbitac.).
Hypaphorus Hassk. = Erythrina L. (Legumin.).
Hyparete Rafin. = Hermbstaedtia Reichb. (Amaranthac.).
Hypargyrium Fourr. = Potentilla L. (Rosac.).
Hyparrhenia Anderss. Gramineae. 75 Medit., Afr., Arabia (Am.?).
Hypechusa Alef. = Vicia L. (Legumin.).
Hypecoaceae (Prantl & Kündig) Barkley. Dicots. 1/15 warm temp. Euras. Low herbs, branched from base, with watery (not milky) juice. Ls. alt. (many radical), usu. much pinnatifid, exstip. Infl. ± leafless below, loosely cymose, with pinnatifid bracts. K 2; C 2+2, the outer ± rhomboid to trilobed, the inner deeply trifid, with the central segment ± spatulate, stipitate and fimbriate; A 4, fil. winged below, anth. biapic.; G̲ (2), 1-loc., with ∞ ov. on 2 pariet. plac. Fr. a linear siliquiform caps., mostly a nodose lomentum

breaking up into 1-seeded sections, more rarely a dehiscent bivalved siliqua. Only genus: *Hypecoum*. Almost exactly intermediate between *Papaverac. (s.str.)* and *Fumariac.*

Hypecoum L. Hypecoaceae. 15 Medit. to C. As. & N. China. Fl. 2-merous throughout. The inner petals are 3-sect, and the middle lobe stands erect and encloses the sta. In *H. procumbens* L. the pollen is shed in the bud into pockets on the inner surface of the inner petals, which close up before the stigma develops. When pressed by an insect the pockets open and dust it with pollen. The stigma only ripens after it has grown above the level of the pollen.

Hypelate auct. ex Pfeiff. (sphalm.) = Hyperanthera Forsk. = Moringa Adans. (Moringac.).

Hypelate P.Br. Sapindaceae. 1 Florida, W.I. White ironwood.

Hypelichrysum Kirp. Compositae. ? Spp. S. Am.

Hypelythrum D. Dietr. = seq.

Hypelytrum Poir. = Hypolytrum Rich. (Cyperac.).

Hypenanthe (Bl.) Blume. Melastomataceae. 4 Malaysia (exc. Malay Penins., Borneo, Java).

Hypenia Mast. ex Benth. = Hyptis Jacq. (Labiat.).

Hyperacanthus E. Mey. = Gardenia Ellis (Rubiac.).

Hyperanthera Forsk. = Moringa Adans. (Moringac.).

Hyperanther[ac]eae Link = Moringaceae R.Br.

Hyperanthus Harv. & Sond. = Hyperacanthus E. Mey. = Gardenia Ellis (Rubiac.).

Hyperaspis Briquet. Labiatae. 2 trop. Afr.

Hyperbaena Miers ex Benth. Menispermaceae. 40 trop. Am., W.I.

Hypergyna P. & K. = Hyperogyne Salisb. = Paradisea Mazzuc. (Liliac.).

***Hyperica[ceae]** Juss. = Guttiferae Juss.

Hypericoïdes Adans. = Ascyrum L. (Guttif.).

Hypericoïdes Cambess. ex Vesque = Garcinia L. (Guttif.).

Hypericon J. F. Gmel. = Hypericum L. (Guttif.).

Hypericophyllum Steetz. Compositae. 7 trop. Afr.

Hypericopsis Boiss. Frankeniaceae. 1 S. Persia.

Hypericopsis Opiz = seq.

Hypericum L. Guttiferae. 400 temp., & trop. mts. Nearly all perenn. herbs or shrubs, with opp. gland-dotted ls. and cymes of fls., often forming pseudo-racemes or umbels. Androecium of 5 antepetalous stamen-fascicles which are freq. united to form 3 (2+2+1) or more rarely 4 (2+1+1+1) groups, or sometimes merge to form a continuous ring of apparently free stam. In *H. elodes* L. and 2 other spp. the staminodial bodies which alternate with the 3 stam.-fascicles represent the antesepalous whorls of the androecium. The fls. contain no honey, but offer abundant pollen, and the larger are freq. visited. They are homogamous, but the stigmas project through the stam. and there is thus a chance of a cross.

Hyperixanthes Blume ex Penzig = Epirixanthes Bl. = Salomonia Lour. (Polygalac.).

Hyperogyne Salisb. = Paradisea Mazzuc. (Liliac.).

Hypertelis E. Mey. ex Fenzl. Aïzoaceae. 7 S. Afr., St Helena.

Hyperthelia W. D. Clayton. Gramineae. 6 trop. Afr.

Hyperum Presl = Wendtia Meyen (Ledocarpac.).
Hypestes P. & K. = Hypoëstes Soland. ex R.Br. (Acanthac.).
Hyphaene Gaertn. Palmae. 30 warm Afr., Madag., Arabia (doum palms). The stem is frequently branched, a rare occurrence in palms.
Hyphear Danser = Loranthus L. = Psittacanthus Mart. (Loranthac.).
Hyphipus Rafin. = Psittacanthus Mart. (Loranthac.).
Hyphydra Schreb. = Tonina Aubl. (Eriocaulac.).
Hypnoticon Reichb. = seq.
Hypnoticum Barb. Rodr. = Withania Pauq. (Solanac.).
Hypobathrum Blume. Rubiaceae. 9 Burma to Philipp. Is., Java.
Hypobrichia M. A. Curt. ex Torr. & Gray = Didiplis Rafin. (Lythrac.).
Hypobrychia Wittst. = praec.
Hypocalymma Endl. Myrtaceae. 18 W. Austr.
Hypocalyptus Thunb. Leguminosae. 3 S. Afr.
Hypocarpus A. DC. = Liriosma Poepp. (Olacac.).
Hypochaeris L. = Hypochoeris L. (Compos.).
Hypochlaena P. & K. = Hypolaena R.Br. (Restionac.).
Hypochlamys Fée = Athyrium Roth (Athyriac.).
Hypochoeris L. Compositae. 100 cosmop., esp. S. Am.
Hypocistis Mill. = Cytinus L. (Rafflesiac.).
Hypocoton Urb. = Bonania A. Rich. (Euphorbiac.).
Hypocylix Wołoszczak ex Stapf. Chenopodiaceae. 1 Persia.
Hypocyrta Mart. Gesneriaceae. 17 C. & trop. S. Am.
Hypodaeurus Hochst. = Anthephora Schreb. (Gramin.).
Hypodaphnis Stapf. Lauraceae. 1 trop. W. Afr.
Hypodema Reichb. = Cypripedium L. (Orchidac.).
Hypodematium Kunze. Aspidiaceae. 3 Old World subtrop.
Hypodematium A. Rich. (1847) = Arbulocarpus Tennant (Rubiac.).
Hypodematium A. Rich. (1850) = Eulophia R. Br. ex Lindl. (Orchidac.).
Hypoderridaceae Ching = Aspidiaceae S. F. Gray.
Hypoderris R.Br. Aspidiaceae. 1 trop. Am., W.I.
*Hypodiscus Nees. Restionaceae. 12 S. Afr.
Hypoëlytrum Kunth = Hypolytrum Rich. (Cyperac.).
Hypoëstes Soland. ex R.Br. Acanthaceae. 150 palaeotrop., esp. Madag.
Hypoglottis Fourr. = Astragalus L. (Legumin.).
Hypogomphia Bunge. Labiatae. 4 C. As., Afghan.
Hypogon Rafin. (Micheliella Briq.). Labiatae. 2 SE. U.S.
Hypogynium Nees = Andropogon L. (Gramin.).
*Hypolaena R.Br. Restionaceae. 2 SE. Austr., Tasm. G 1-loc., 1-ovuled.
Hypolepidaceae Ching = Dennstaedtiaceae Ching.
Hypolepis Beauv. ex Lestib. = Ficinia Schrad. (Cyperac.).
Hypolepis Bernh. Dennstaedtiaceae. 45 trop. & subtrop. No indus., sori protected by small reflexed lobes of leaf-edge.
Hypolepis Pers. = Cytinus L. (Rafflesiac.).
Hypolobus Fourn. Asclepiadaceae. 1 Brazil.
Hypolytrum Rich. Cyperaceae. 80 trop. & subtrop.
Hypoma Rafin. = Noltea Reichb. (Rhamnac.).
Hyponema Rafin. = Cleomella DC. (Cleomac.).

Hypopeltis Michx = Polystichum Roth (Aspidiac.).
Hypophaë Medik. = Hippophaë L. (Elaeagnac.).
Hypophialium Nees = Ficinia Schrad. (Cyperac.).
Hypophyllanthus Regel. Sterculiaceae. 1 Colombia.
Hypopithis Rafin. = Hypopitys Hill (Monotropac.).
Hypopithydes Link = Monotropaceae Nutt.
Hypopithys Adans. = Hypopitys Hill (Monotropac.).
Hypopitys Ehrh. (uninom.) = *Monotropa hypopitys* L. (Monotropac.).
Hypopitys Hill (~ Monotropa L.). Monotropaceae. 1 Himal. to NE. As., temp. N. Am. to Colombia.
Hypopogon Turcz. = Symplocos L. (Symplocac.).
Hypoporum Nees = Scleria Bergius (Cyperac.).
Hypopteron Hassk. = Chirita Buch.-Ham. (Gesneriac.).
Hypopterygiopsis Sakurai [pro Bryoph.!] = Selaginella P. Beauv. (Selaginellac.).
Hypopterygium Schlechtd. (1843; non Brid. 1827—Musci) = Amphipterygium Schiede ex Standl. (Julianiac.).
Hypopythis Rafin. = Hypopitys Hill (Monotropac.).
Hypostate Hoffmgg. = ? Rhexia L. (Melastomatac.).
× **Hypotanthus** Saylor. Gesneriaceae. Gen. hybr. (Hypocyrta × Nematanthus).
Hypothronia Schrank = Hyptis Jacq. (Labiat.).
Hypoxanthus Rich. ex DC. = Miconia Ruiz & Pav. (Melastomatac.).
*****Hypoxidaceae** R.Br. Monocots. 7/120 ± cosmop. (exc. Eur. & N. As.). Herbs from a tuberous rhiz. or corm; ls. mostly radic., conspic. nerved or plicate, often clothed with long whitish hairs. Infl. rac. or subcapit., sometimes 1-fld. Fls. ⚥, reg. P 6, equal, patent, ± persist.; A 6 or 3, anth. extr. or intr.; G (3), often prolonged into an apical beak bearing the P; styles free or united; ov. ∞ (rarely few) in each loc., axile, biseriate. Fr. either capsular and variously dehisc., or baccate and indehisc.; seeds small, often black, sometimes carunculate, with copious endosp. Genera: *Hypoxis, Curculigo, Campynema, Campynemanthe, Pauridia, Rhodohypoxis, Spiloxene*. Related on one side to *Haemodoraceae* and on the other side perhaps to *Apostasiaceae* (Hutchinson).
Hypoxidopsis Steud. ex Baker = Iphigenia Kunth (Liliac.).
Hypoxis Adans. = Upoda Adans. = Hypoxis L. (Hypoxidac.).
Hypoxis Forsk. = ? Scilla L. (Liliac.).
Hypoxis L. Hypoxidaceae. 100 Am., Afr., E. As., Indomal., Austr.
Hyppochaeris Bivona = Hypochoeris L. (Compos.).
Hyppomarathrum Rafin. = Hippomarathrum Hall. = Seseli L. (Umbellif.).
Hypsagyne Jack ex Burkill = Salacia L. (Celastrac.).
Hypsela Presl. Campanulaceae. 5 E. Austr., N.Z., Andes.
Hypselandra Pax & K. Hoffm. Capparidaceae. 1 Burma.
Hypselodelphys (K. Schum.) Milne-Redhead. Marantaceae. 4 trop. Afr.
Hypseloderma Radlk. Sapindaceae. 1 Somalia.
Hypseocharis Remy. Hypseocharitaceae. 8 Andes.
Hypseocharitaceae Weddell. Dicots. 1/8 Andes. Acaulescent perenn. herbs from thick tap-root, with rosulate pinnatif. exstip. ls. Fls. reg., ⚥, in radic. shortly pedunc. or epedunc. 1–9-fld. cymes. K 5, much imbr., persist.; C 5, contorted; A 15, 1-seriate, with persist. subul. fil. and dorsifixed intr. anth.;

$\underline{G}$ (5), attached to central column, 5-lobed, 5-loc., with simple filif. style with capit. stig., and ∞ axile biseriate anatr. ov. Fr. a tardily and irreg. loculic. caps., seeds with cochlear embr. and scanty endosp. Only genus: *Hypseocharis*. Somewhat intermediate between *Geraniac.* (*s.l.*) and *Oxalidac.* (*s.l.*), differing from the former in the lack of stipules and in the ∞ ovules, from the latter in the completely connate styles with capit. stig., from both in the possession of 15 stamens. (Some approach to *Biebersteiniac.*?)

Hypseochloa C. E. Hubbard. Gramineae. 1 trop. W. Afr.

Hypserpa Miers. Menispermaceae. 20 Indomal.

Hypsipodes Miq. = Tinospora Miers (Menispermac.).

Hypsophila F. Muell. Celastraceae. 3 Austr.

Hyptiandra Hook. f. = Quassia L. (Simaroubac.).

Hyptianthera Wight & Arn. Rubiaceae. 2 N. India, Siam.

Hyptiodaphne Urb. = Daphnopsis Mart. & Zucc. (Thymelaeac.).

***Hyptis** Jacq. Labiatae. 400 warm Am., W.I.

Hyptissa Salisb. = Gladiolus L. (Iridac.).

Hypudaeurus Reichb. (sphalm.) = Hypodaeurus Hochst. = Anthephora Schreb. (Gramin.).

Hyrtanandra Miq. (~ Pouzolzia Gaudich.). Urticaceae. 15 E. As., Indomal.

Hyssopifolia Fabr. (uninom.) = *Lythrum hyssopifolia* L. (Lythrac.).

Hyssopus L. Labiatae. 15 S. Eur., Medit. to C. As. *H. officinalis* L., the hyssop, formerly used in medicine.

Hysteria Reinw. ex Blume = Corymborkis Thou. (Orchidac.).

Hystericina Steud. = Echinopogon Beauv. (Gramin.).

Hysterionica Willd. Compositae. 12 S. Brazil, Urug., Argentina.

Hysteronica Endl. = praec.

Hysterophorus Adans. = Parthenium L. (Compos.).

Hystrichophora Mattf. Compositae. 1 trop. E. Afr.

Hystringium Trin. ex Steud. = Lasiochloa Kunth (Gramin.).

Hystrix Moench. Gramineae. 7 temp. As., N. Am., N.Z.

Hystrix Rumph. = Barleria L. (Acanthac.).

I

Iacaranda Nees = Jacaranda Juss. (Bignoniac.).

Iaera H. F. Copeland = Costera J. J. Sm. (Ericac.).

Ialapa Crantz = Jalapa Mill. = Mirabilis L. (Nyctaginac.).

Ialappa Ludw. = praec.

Iantha Hook. = Ionopsis Kunth (Orchidac.).

Ianthe Pfeiff. = Janthe Griseb. = Celsia L. (Scrophulariac.).

Ianthe Salisb. (~ Hypoxis L.). Hypoxidaceae. 25 S. Afr.

Iaravaea Scop. = Acisanthera, Desmoscelis, Microlicia, Nepsera spp. (Melastomatac.).

Iasione Moench = Jasione L. (Campanulac.).

Iasmin[ac]eae Link = Jasmin[ac]eae Juss.

Iatropha Stokes = Jatropha L. (Euphorbiac.).

Ibadja A. Chev. = Loesenera Harms (Legumin.).

IBATIA

Ibatia Decne. Asclepiadaceae. 3 W.I., trop. S. Am.

Ibbertsonia Steud. = seq.

Ibbetsonia Sims = Cyclopia Vent. (Legumin.).

Iberidella Boiss. (~ Eunomia DC.). Cruciferae. 6 Alps to Cauc. & S. Persia; ?2 Himal.

Iberis Hill = Lepidium L. (Crucif.).

Iberis L. Cruciferae. 30 Eur., As. *I. amara* L. provides a good example of the corymb. The outer petals of the fls. are longer than the rest, thus adding to the conspicuousness (cf. *Umbelliferae*).

Ibervillea Greene. Cucurbitaceae. 1 N. Am.

Ibetralia Bremek. Rubiaceae. 1 Guiana.

Ibettsonia Steud. = Ibbetsonia Sims = Cyclopia Vent. (Legumin.).

Ibicella (Stapf) Van Eselt. Martyniaceae. 3 trop. S. Am.

Ibidium Salisb. ex Small = Spiranthes Rich. (Orchidac.).

Ibina Nor. = Thunbergia Retz. (Thunbergiac.).

Iboga J. Braun & K. Schum. = Tabernanthe Baill. (Apocynac.).

Iboza N. E. Brown. Labiatae. Over 12 trop. & S. Afr., India(?).

Icacina A. Juss. Icacinaceae. 5 trop. W. Afr.

*Icacinaceae Miers. Dicots. 58/400 trop. Trees and shrubs (often lianes), or rarely herbs, with alt. (rarely opp.) exstip. ls., usu. entire and often leathery. Fls. usu. in cymose or thyrsiform infl., less freq. rac. or spic. or panic., reg., usu. ⚥. K (5) or (4), rarely 0 (*Pyrenacantha*), not enlarged when the fruit is ripe; C 5 or 4, rarely united, valvate or imbr.; A 5 or 4, alt. with petals, with usu. intr. anthers; disk rarely developed; G̲ (3), rarely (5) or (2), rarely multiloc., usu. 1-loc. by abortion of the remaining cavities; ovules 2 per loc., pendulous from its apex, anatr., with dorsal raphe and micropyle facing upwards; funicle usu. thickened above the micropyle; style simple with 3 (or 5-2) stigmas. Fr. 1-loc., 1-seeded, usu. a drupe, sometimes a samara. Endosp. usu. present; embryo straight or curved. Chief genera: *Apodytes, Rhyticaryum, Gonocaryum, Platea, Stemonurus* (*Gomphandra*), *Iodes* (ls. opp.), *Pyrenacantha, Phytocrene*. The fam. shows connections with the *Celastrac., Aquifoliac.* and *Olacac.*, also with *Stilaginac.*

Icacinopsis Roberty. Icacinaceae. 1 trop. W. Afr.

Icaco Adans. = Chrysobalanus L. (Chrysobalanac.).

Icacorea Aubl. = Ardisia Sw. (Myrsinac.).

Icaranda Pers. = Jacaranda Juss. (Bignoniac.).

Icaria Macbride = Miconia Ruiz & Pav. (Melastomatac.).

Ichnanthus Beauv. Gramineae. 1 Indomal., Austr.; 25 trop. (esp. S.) Am.

*Ichnocarpus R.Br. Apocynaceae. 18 S. China, Indomal.

Ichthyomethia P.Br. = Piscidia L. (Legumin.).

Ichthyophora Baehni = Neoxythece Aubrév. & Pellegr. (!) (Sapotac.).

Ichthyosma Schlechtd. = Sarcophyte Sparrm. (Balanophorac.).

Ichthyothere Mart. Compositae. 18 trop. S. Am.

Ichtyosma Steud. = Ichthyosma Schlechtd. = Sarcophyte Sparrm. (Balanophorac.).

Ichtyothere DC. = Ichthyothere Mart. (Compos.).

Icianthus Greene (~ Streptanthus Nutt.). Cruciferae. 3 SW. U.S.

Icica Aubl. = Protium Burm. f. (Burserac.).

Icicariba Maza = Bursera Jacq. (Burserac.).

Icicaster Ridley = Santiria Bl. (Sect. Icicopsis A. W. Benn., non Icicopsis Engl.) (Burserac.).

Icicopsis Engl. (non Santiria Sect. Icicopsis A. W. Benn.) = Protium Burm. f.

Icma Phil. Compositae. 1 Chile. [(Burserac.).

Icmane Rafin. = Hakea Schrad. (Proteac.).

Icomum Hua. Labiatae. 10 trop. Afr.

Icosandra Phil. = Cryptocarya R.Br. (Laurac.).

Icosinia Rafin. = Sterculia L. (Sterculiac.).

Icostegia Rafin. = Clusia L. (Guttif.).

Icotorus Rafin. = Physocarpus (Cambess.) Maxim. (Rosac.).

Icthyoctonum Boiv. ex Baill. = Lonchocarpus Kunth (Legumin.).

Icthyothere Baker = Ichthyothere Mart. (Compos.).

Ictinus Cass. = Gorteria L. (Compos.).

Ictodes Bigel. = Symplocarpus Salisb. (Arac.).

Idahoa A. Nelson & Macbride. Cruciferae. 1 W. U.S.

Idalia Rafin. = Convolvulus L. (Convolvulac.).

Idaneum Kuntze & Post = Adenium Roem. & Schult. (Apocynac.).

Idanthisa Rafin. (Anisacanthus Nees). Acanthaceae. 15 Am.

Ideleria Kunth = Tetraria Beauv. + Macrochaetium Steud. (Cyperac.).

Idenburgia L. S. Gibbs = Sphenostemon Baill. (Sphenostemonac.).

Idertia Farron. Ochnaceae. 3 W. & C. trop. Afr.

Idesia Maxim. Flacourtiaceae. 1 China, Japan.

Idesia Scop. = Ropourea Aubl. = Diospyros L. (Ebenac.).

Idianthes Desv. = Phaecasium Cass. = Crepis L. (Compos.).

Idicium Neck. = Gerbera L. ex Cass. (Compos.).

Idiopsis (Moq.) Kuntze = Nitrophila S. Wats. (Chenopodiac.).

Idiopteris T. G. Walker. Pteridaceae. 1 Ceylon. (*Kew Bull.* **12**: 341, 1958.)

Idiospermaceae S. T. Blake. Dicots. 1/1 Queensland. Evergreen tree; ls. opp., ent., pet., exstip. Infl. axill., 1(-3)-fl., peduncle with several decussate bracts; receptacle cup-shaped. Tepals ∞, spirally arranged, petaloid, the outer caducous, leaving persistent bases, the inner narrower, persistent, at first white, later turning bright red, finally dull brownish purple. A ∞, inserted at top of recept., tepal-like, thick, laminar, with almost no filament; anthers dorsal, shorter than stamen, with two long narrow protuberant extrorse thecae. Stds on inside of recept., similar to A, becoming smaller, more distant and hood-like towards base. Carpels 1(-2), sessile on hairy base of recept., glabrous, compressed; stig. nearly sessile, fleshy, papillose. Ovule 1, on a short funicle, or sometimes a second present on a longer funicle. Fruit depressed-globose, rather large (5-6·5 cm. diam.), composed of the much enlarged, almost closed, recept., which is externally thin and crustaceous but internally somewhat fleshy, bearing the persist. tep. and stam. Pericarp thin, free from inner wall of recept. Seed filling the cavity, with membr. testa ± adherent to pericarp; endo sp. o; embryo filling the seed, with stout plumule and hypocot. and (3-)4 massive fleshy peltate cotyledons in a single whorl, which persist for a year or more on the seedling. Only genus: *Idiospermum*. An interesting relict, related to *Calycanthus*, *Eupomatia*, *Monimiaceae*, etc. The cotyledons contain a substance toxic to cattle.

IDIOSPERMUM

Idiospermum S. T. Blake. Idiospermaceae. 1 Queensland.

Idothea Kunth = Drimia Jacq. (Liliac.).

Idothearia Presl = praec.

Idria Kellogg. Fouquieriaceae. 1 Mex., *I. columnaris* Kellogg, the 'boojum tree'.

Idriaceae Barkley = Fouquieriaceae DC.

Iebine Rafin. = Leptorkis Thou. = Liparis Rich. (Orchidac.).

Iericontis Adans. = Hiericontis Adans. = Anastatica L. (Crucif.).

Ifdregea Steud. = Peucedanum L. (Umbellif.).

Ifloga Cass. Compositae. 10 Canaries, Medit., S. Afr., India.

Ifuon Rafin. = Asphodeline Reichb. (Liliac.).

Ignatia L. f. = Strychnos L. (Strychnac.).

Ignatiana Lour. = praec.

Iguanara Reichb. = seq.

Iguanura Blume. Palmae. 20 W. Malaysia.

Ikonnikovia Lincz. Plumbaginaceae. 1 C. As., NW. China.

Ildefonsia Gardn. Scrophulariaceae. 1 Brazil.

Ildefonsia Mart. ex Steud. = Urtica L. (Urticac.).

Ilemanthus P. & K. = Eilemanthus Hochst. = Indigofera L. (Legumin.).

Ileocarpus Miers = Stephania Lour. (Menispermac.).

Ileostylus Van Tiegh. Loranthaceae. 1 N.Z.

Ilex L. Aquifoliaceae. 400 cosmop. (exc. N. Am.). In *I. aquifolium* L. the fls. are dioecious, but in the ♀ the sterile sta. are so large that the fl. appears ♂. Truly ♂ fls. sometimes occur. *I. paraguariensis* A. St-Hil. is the *maté* or Paraguay tea, largely used in S. Am. The ls. contain caffeine; they are dried, broken up and used like tea. A few Bornean spp. have opp. ls.

Ilex Mill. = Quercus L. (Fagac.).

Iliamna Greene. Malvaceae. 7 W. U.S.

Ilicaceae Brongn. = Aquifoliaceae Bartl.

Iliciodes Kuntze = seq.

Ilicioïdes Dum.-Cours. = Nemopanthus Rafin. (Aquifoliac.).

Iliogeton Benth. = Ilyogeton Endl. = Vandellia L. (Scrophulariac.).

Iljinia Korovin. Chenopodiaceae. 1 C. As.

Illa Adans. = Tomex L. = Callicarpa L. (Verbenac.).

Illairea Lenné & C. Koch = Loasa Juss. (Loasac.).

*****Illecebraceae** R.Br. = Caryophyllaceae–Paronychioïdeae Fenzl.

Illecebrella Kuntze = seq.

Illecebrum L. Caryophyllaceae. 1 Canaries, W. Eur. (incl. SW. Brit.), Medit.

Illecebrum Spreng. = Alternanthera Forsk. (Amaranthac.).

*****Illiciaceae** Van Tiegh. Dicots. 1/42 trop. SE. As., N. Am., W.I. Shrubs or small trees, with simple ent. alt. or pseudovertic. exstip. ls. Fls. ☿, reg., sol., axill. P 7–∞, pluriser., imbr., the outer and innermost smaller; A 4–∞, 1–pluri-ser., anth. obl., intr.; G̲ 5–20, 1-ser., with simple styles, and 1 ventral sub-basal ov. in each carp. Fr. a ring of spreading follicles, ventrally dehisc.; seeds glossy, with copious endosp. Only genus: *Illicium*. Relationship with *Winterac.* and *Schisandrac.*

Illicium L. Illiciaceae. 42 India, E. As., W. Malaysia, Atl. N. Am., Mex.,

W.I. There is a gradual transition in the spiral P from sepaloid to petaloid structure (cf. *Nymphaea*). The fr. is an aggregate of follicles. Pollination partly by beetles. *I. verum* Hook. f. (star-anise; China) is used for flavouring.

Illigera Blume. Hernandiaceae. 30 trop. Afr., Madag., E. As., W. Malaysia.

Illigeraceae Bl. = Hernandiaceae Bl.

Illigerastrum (Prain & Burkill) A. W. Hill in I.K. (sphalm.) = Dioscorea L. (Dioscoreac.).

Illipe Gras = Madhuca Gmel. (Sapotac.).

Illus Haw. = Narcissus L. (Amaryllidac.).

Ilmu Adans. = Romulea Maratti (Iridac.).

Ilocania Merr. = Diplocyclos (Endl.) P. & K. corr. C. Jeffrey (Cucurbitac.).

Ilogeton A. Juss. = Ilyogeton Endl. = Lindernia All. (Scrophulariac.).

Iltisia Blake. Compositae. 1 Costa Rica.

Ilyogethos Hassk. = seq.

Ilyogeton Endl. = Lindernia All. (Scrophulariac.).

Ilyphilos Lunell = Elatine L. (Elatinac.).

Ilysanthes Rafin. = Lindernia All. (Scrophulariac.).

Ilysanthos St-Lag. = praec.

Ilythuria Rafin. = Donax Lour. (Marantac.).

Imantina Hook. f. (~ Morinda L.). Rubiaceae. 1 New Caled.

Imantophyllum Hook. (1854) = Clivia Lindl. (Sect. Imantophyllum (Hook.) Benth. & Hook. f.) (Amaryllidac.).

Imatophyllum Hook. (1828) = Clivia Lindl. (Sect. Clivia) (Amaryllidac.).

Imbricaria Comm. ex Juss. (~ Mimusops L.). Sapotaceae. 7 trop. Afr., Masc. Is.

Imbricaria Sm. = Baeckea L. (Myrtac.).

Imbutis Rafin. = Ribes L. (Grossulariac.).

Imerinaea Schlechter. Orchidaceae. 1 Madag.

Imhofia Heist. = Nerine Herb. (Amaryllidac.).

Imhofia Herb. (1821) = ? Nerine Herb. (Amaryllidac.).

Imhofia Herb. (1837) = Periphanes Salisb. (Amaryllidac.).

Imhofia Zoll. ex Taub. = Rinorea Aubl. (Violac.).

Imitaria N. E. Brown. Aïzoaceae. 1 S. Afr.

Impatiens L. Balsaminaceae. 5–600 trop. & temp. Euras. & Afr., esp. Madag. & mts. of India and Ceylon; about 6 in N. & C. Am. The name is derived from the explosive fr., a caps. with fleshy pericarp; the outer layers of cells are highly turgid and thus a great strain is put upon the whole. Dehiscence is septifragal and is started by a touch when the fr. is ripe. The valves roll up inwards with violence (starting at the base) and the seeds are scattered in all directions.

Impatientaceae Van Tiegh. = Balsaminaceae DC.

Impatientella H. Perrier. Balsaminaceae. 1 Madag.

Imperata Cyr. Gramineae. 10 trop. & subtrop. *I. cylindrica* (L.) P. Beauv. (*I. arundinacea* Cyr.) (*lalang*) is a very troublesome weed in Malaya.

Imperatia Moench = Petrorrhagia (Ser.) Link (Caryophyllac.).

Imperatoria L. = Peucedanum L. (Umbellif.).

Imperialis Adans. = Petilium Ludw. = Fritillaria L. (Liliac.).

Impia Bluff & Fingerh. = Filago L. (Compos.).

Incarvillea Juss. Bignoniaceae. 11–14 C. & E. As., Himal. Herbs. Ls. alt.

IONETTIA

× **Ionettia** hort. Orchidaceae. Gen. hybr. (iii) (Comparettia × Ionopsis).

Ionia Pers. ex Steud. = Ionidium Vent. = Hybanthus Jacq. (Violac.).

Ionidiopsis Walp. = Jonidiopsis Presl = Noisettia Kunth (Violac.).

Ionidium Vent. = Hybanthus Vent. (Violac.).

Ioniris Baker = Joniris Klatt = Iris L. (Iridac.).

× **Ionocidium** hort. Orchidaceae. Gen. hybr. (iii) (Ionopsis × Oncidium).

Ionopsidium (DC.) Reichb. Cruciferae. 1 Portugal, *I. acaule* (Desf.) Reichb., with sol. fls. in the axils of radical leaves.

Ionopsis Kunth. Orchidaceae. 10 trop. Am.

Ionorchis auctt. = Jonorchis G. Beck = Limodorum Rich. (Orchidac.).

Ionosmanthus Jord. & Fourr. = Ranunculus L. (Ranunculac.).

Ionoxalis Small = Oxalis L. (Oxalidac.).

Ionthlaspi Gérard = Clypeola L. (Crucif.).

Iosotoma Griseb. (sphalm.) = Isotoma Lindl. (Campanulac.).

Iostephane Benth. Compositae. 2 Mex.

Ioxylon Rafin. = Toxylon Rafin. = Maclura Nutt. (Morac.).

Iozosmene Lindl. = seq.

Iozoste Nees = Litsea Lam. + Actinodaphne Nees (Laurac.).

Ipecacuana Rafin. = seq.

Ipecacuanha Arruda = Uragoga L. (Rubiac.).

Ipecacuanha Gars. = Gillenia Moench = Porteranthus Britton (Rosac.).

Ipheion Rafin. (~ Milla Cav.). Liliaceae. 25 Mex. to Chile.

*** Iphigenia** Kunth (= Aphoma Rafin.). Liliaceae. 12 trop. & S. Afr., Madag., India, Austr., N.Z.

Iphigeniopsis F. Buxb. (~ Iphigenia Kunth). Liliaceae. 4 trop. & S. Afr.

Iphiona Cass. Compositae. 15 Medit., trop. & S. Afr., Masc. Is., Arabia, C. As.

Iphisia Wight & Arn. = Tylophora R.Br. (Asclepiadac.).

Iphyon P. & K. = Ifuon Rafin. = Asphodeline Reichb. (Liliac.).

Ipnum Phil. = Diplachne P. Beauv. (Gramin.).

Ipo Pers. = Antiaris Lesch. (Morac.).

Ipomaea Burm. f. = Ipomoea L. (Convolvulac.).

Ipomaeëlla A. Chev. = Aniseia Choisy (Convolvulac.).

Ipomea All. = Ipomoea L. (Convolvulac.).

Ipomeria Nutt. = Gilia Ruiz & Pav. (Polemoniac.).

Ipomoea L. (incl. *Batatas, Calonyction, Pharbitis* and *Exogonium* of Choisy, and *Quamoclit* Moench). Convolvulaceae. 500 trop. & warm temp., chiefly climbing herbs or shrubs, rarely aquatic. *I. pes-caprae* (L.) R.Br. (*I. biloba* Forsk.) is a char. creeping pl. of trop. beaches. *I. batatas* (L.) Lam. (*B. edulis* Choisy) is the sweet potato, largely cultivated in warm countries for its tubers, which are used like potatoes. *I. (Exogonium) purga* (Wender.) Hayne is the jalap; its rhizome gives off turnip-like roots about the size of apples. Worm-eaten tubers are most valuable, as the non-resinous parts are eaten.

Ipomopsis Michx (~ Gilia Ruiz & Pav.). Polemoniaceae. 23 W. N. Am., Florida, 1 temp. S. Am.

Iposues Rafin. = Rhododendron L. (Ericac.).

Ipsea Lindl. Orchidaceae. 1 India, Ceylon.

Irasekia S. F. Gray = Jirasekia F. W. Schmidt = Anagallis L. (Primulac.).

594

Iraupalos Rafin. = Traupalos Rafin. = Hydrangea L. (Hydrangeac.).

Ireneis Moq. = Iresine L. (Amaranthac.).

Irenella Suesseng. Amaranthaceae. 1 Ecuador.

Ireon Burm. f. = Roridula Burm. f. ex L. (Roridulac.).

Ireon Rafin. = Prismatocarpus L'Hérit. (Campanulac.).

Ireon Scop. = Lightfootia L'Hérit. (Campanulac.).

*****Iresine** P.Br. Amaranthaceae. 80 Austr., Galápagos, Am.

Ireum Steud. = Ireon Burm. f. = Roridula Burm. f. ex L. (Roridulac.).

Iria (Rich. ex Pers.) Kuntze = Fimbristylis Vahl (Cyperac.).

Iriartea Ruiz & Pav. Palmae. 7 trop. Am. The stem is supported on aerial roots (cf. *Pandanus*). Some of the branches of these roots are thorny (cf. *Acanthorhiza*). In *I. ventricosa* Mart. (*paxiuba* palm), the stem has a peculiar egg-like thickening about half-way (cf. *Bombacaceae, Jatropha*).

Iriarteaceae O. F. Cook = Palmae–Iriarteëae Benth. & Hook. f.

Iriartella H. Wendl. Palmae. 2 trop. S. Am.

Iriastrum Fabr. (uninom.) = Iris L. sp. (Iridac.).

*****Iridaceae** Juss. Monocots. 60/800 trop. & temp.; the chief centres of distr. S. Afr. and trop. Am. Chiefly herbs with a sympodial tuber or rhizome below ground. Ls. usu. equitant in two ranks. Infl. term., cymose (1 fl. only in *Sisyrinchieae*). Fl. ⚥, reg. or zygo. P 3 + 3, petaloid, united below into a long or short tube; A 3 (the outer whorl), with extr. anthers; $\overline{G}$ or very rarely $\underline{G}$ (*Isophysis*) (3), 3-loc., with axile plac. (rarely 1-loc. with parietal plac.); style usu. trifid and frequently ± petaloid. Ovules usu. ∞, anatr. Loculic. caps. Embryo small, in hard endosp.

Classification and chief genera (after Diels):

1. SISYRINCHIEAE (spathes term. or lat., stalked, rarely sessile; fls. solitary, or more often several developed centrifugally round a central one, mostly stalked; style-branches alternating with stam.; plant small; ls. not exactly in ½ phyllotaxy): *Crocus, Romulea, Sisyrinchium*.

2. IXIEAE (spathes lateral, sessile, 1-flowered; fl. often zygomorphic): *Ixia, Tritonia, Gladiolus, Freesia*.

3. IRIDEAE (MORAEËAE) (fls. numerous, in term. or lat. stalked (rarely sess.) spathes, usu. reg.; style-branches opp. the stam.; stem distinct; ls. equitant): *Iris, Moraea, Tigridia*.

Iridaps Comm. ex Pfeiff. = ? Tridaps Comm. ex Endl. = Artocarpus J. R. & G. Forst. (Morac.).

Iridion Roem. & Schult. = Ireon Burm. f. = Roridula Burm. f. ex L. (Roridulac.).

Iridis Rafin. = Iris L. (Iridac.).

Iridisperma Rafin. = Polygala L. (Polygalac.).

Iridodictyum Rodionenko (~ Iris L.). Iridaceae. 9 Turkey, Palest., Cauc., Iraq, C. As.

Iridopsis Welw. ex Baker = Moraea L. (Iridac.).

Iridorchis Blume = Cymbidium Sw. (Orchidac.).

Iridorchis Thou. = seq.

Iridorkis Thou. = Oberonia Lindl. (Orchidac.).

Iridosma Aubrév. & Pellegr. Simaroubaceae. 1 W. Equat. Afr.

Iridrogalvia Pers. (sphalm.) = Isidrogalvia Ruiz & Pav. = Tofieldia Huds. (Liliac.).

IRIHA

Iriha Kuntze = Iria (Rich. ex Pers.) Kuntze = Fimbristylis Vahl (Cyperac.).

Irillium Rafin. = Trillium L. (Liliac.).

Irina Nor. = Sapindacea indet.

Irina Nor. ex Bl. = Pometia J. R. & G. Forst. (Sapindac.).

Irine Hassk. = praec.

Irio L. = Sisymbrium L. (Crucif.).

Iripa Adans. = Cynometra L. (Legumin.).

Iris L. Iridaceae. 300 N. temp. Most have a sympodial rhiz. with equitant isobilat. ls., and small cymes of fls. in spathes. P petaloid, the outer segments usu. bending downwards at the outer ends; opp. to them and almost resting on them are the petaloid styles, under which are the sta. with their extr. anthers. Just above the anther, on the outer side of the style, is a little flap, whose upper surface is the stigma. Bees entering the fl. to get the honey secreted by the ovary rub off their pollen upon the stigma; going further in they get fresh pollen, and when they come out close the stigma flap, which prevents self-fert. (cf. *Viola*). Many spp. have flat seeds suited to wind distr. (Dykes, *The Genus I.*, 1913; Rodionenko, *Rod Iris* [*The genus I.*], 1961.)

The dried rhiz. of *I. florentina* L. (orris root) smells like violets, and is used in perfumery; 'essence of violets' is made from it.

Irium Steud. = Ireon Burm. f. = Roridula Burm. f. ex L. (Roridulac.).

Irlbachia Mart. Gentianaceae. 1 trop. S. Am.

Irma Bouton ex A. DC. = Begonia L. (Begoniac.).

Irmischia Schlechtd. Asclepiadaceae. 5 Mex., W.I., trop. S. Am.

Iron P.Br. = Sauvagesia L. (Ochnac.).

Iroucana Aubl. = Casearia Jacq. (Flacourtiac.).

Irsiola P.Br. = Cissus L. (Vitidac.).

Irucana P. & K. = Iroucana Aubl. = Casearia Jacq. (Flacourtiac.).

Irulia Bedd. = Melocanna Trin. (Gramin.).

Irvingbaileya Howard. Icacinaceae. 1 Queensland.

Irvingella Van Tiegh. = seq.

Irvingia Hook. f. Ixonanthaceae. 10 trop. Afr., Indoch., Malay Penins., Borneo. Butters from the seeds (*cay-cay*, *dika*, etc.).

Irvingia F. Muell. = Hedera L. (Araliac.).

*Irvingiaceae Pierre = Ixonanthaceae Klotzsch.

Iryanthera Warb. Myristicaceae. 30 trop. S. Am.

Isabelia Barb. Rodr. Orchidaceae. 1 Brazil.

Isacanthus Nees = Sclerochiton Harv. (Acanthac.).

Isachne R.Br. Gramineae. 60 trop. & subtrop.

Isaloa Humbert. Scrophulariaceae. 1 Madag.

Isalus Phipps. Gramineae. 3 Madag.

Isandra F. Muell. Solanaceae. 1 SW. Austr.

Isandra Salisb. = Thysanotus R.Br. (Liliac.).

Isandrina Rafin. = Cassia L. (Legumin.).

× **Isanitella** Leinig. Orchidaceae. Gen. hybr. (i) (Isabelia × Sophronitella).

Isanthera Nees = Rhynchotechum Bl. (Gesneriac.).

Isanthina Reichb. = Commelina L. (Commelinac.).

Isanthus DC. = Homoianthus Bonpl. ex DC. = Perezia Lag. (Compos.).

Isanthus [Rich. in] Michx. Labiatae. 1 N. Am.

Isartia Dum. = Isertia Schreb. (Rubiac.).

Isatis L. Cruciferae. 45 Eur., Medit. to SW. & E. As. *I. tinctoria* L. is the woad, largely used as a dye before the introduction of indigo. It is prepared by grinding the leaves to a paste and fermenting them.

Isaura Comm. ex Poir. = Stephanotis Thou. (Asclepiadac.).

Isauxis Reichb. = Vatica L. (Dipterocarpac.).

Ischaemon Hill = Ischaemum L. (Gramin.).

Ischaemon Schmiedel = Luzula DC. (Juncac.).

Ischaemopogon Griseb. = seq.

Ischaemum L. Gramineae. 50 trop. & subtrop.

Ischaleon Ehrh. (uninom.) = *Gentiana filiformis* L. = *Cicendia filiformis* (L.) Delarbre (Gentianac.).

Ischarum Blume = Biarum Schott (Arac.).

Ischina Walp. = Ischnia DC. ex Meissn. = Tamonea Aubl. (Verbenac.).

Ischnanthus Roem. & Schult. = Ichnanthus Beauv. (Gramin.).

Ischnanthus Van Tiegh. = Tapinanthus Bl. (Loranthac.).

Ischnea F. Muell. Compositae. 5 New Guinea. Pappus 0.

Ischnia DC. ex Meissn. = Ghinia Schreb. (Verbenac.).

Ischnocarpus O. E. Schulz. Cruciferae. 1 N.Z.

Ischnocentrum Schlechter. Orchidaceae. 1 New Guinea.

Ischnochloa Hook. f. Gramineae. 1 NW. Himalaya.

Ischnogyne Schlechter. Orchidaceae. 1 China.

Ischnolepis Jumelle & Perrier. Periplocaceae. 1 Madag.

Ischnosiphon Koern. Marantaceae. 35 trop. Am., W.I.

Ischnostemma King & Gamble. Asclepiadaceae. 1 coasts Malay Penins., Philipp., Java, New Guinea, trop. Austr.

Ischnurus Balf. f. Gramineae. 1 Socotra.

Ischurochloa Buese = Bambusa Schreb. (Gramin.).

Ischyranthera Steud. ex Naud. = Bellucia Neck. ex Naud. (Melastomatac.).

Ischyrolepis Steud. = Restio Rottb. (Restionac.).

Iseia O'Donell (~ Ipomoea L.). Convolvulaceae. 1 trop. S. Am.

Iseilema Anderss. Gramineae. 20 Indomal., Austr.

Iserta Batsch = seq.

Isertia Schreb. Rubiaceae. 25 C. & trop. S. Am.

Isexima Rafin. = seq.

Isexina Rafin. = Cleome L. (Cleomac.).

Isgarum Rafin. = ? Salsola L. (Chenopodiac.).

Isias De Not. = × Orchiserapias Camus (Orchidac.).

Isica Moench = Isika Adans. = Lonicera L. (Caprifoliac.).

Isidorea A. Rich. Rubiaceae. 8 W.I.

Isidrogalvia Ruiz & Pav. = Tofieldia Huds. (Liliac.).

Isika Adans. = Lonicera L. (Caprifoliac.).

Isilema P. & K. = Iseilema Anderss. (Gramin.).

Isinia K. H. Rechinger. Labiatae. 1 Persia.

Isiphia Rafin. = Aristolochia L. (Aristolochiac.).

Isis Tratt. = Iris L. (Iridac.).

Iskandera N. Busch. Cruciferae. 1 C. As.

Islaya Backeb. Cactaceae. 9 Peru.

ISMARIA

Ismaria Rafin. = Brickellia Ell. (Compos.).
Ismelia Cass. = Chrysanthemum L. (Compos.).
Ismene Salisb. = Hymenocallis Salisb. (Amaryllidac.).
Isnardia L. = Ludwigia L. (Onagrac.).
Isoberlinia Craib & Stapf. Leguminosae. 6 trop. Afr.
Isocarpellaceae Dulac = Crassulaceae DC.
Isocarpha R.Br. Compositae. 10 S. U.S. to trop. S. Am., W.I.
Isocaulon Van Tiegh. = Loranthus L. (Loranthac.).
Isochilos Spreng. = seq.
Isochilus R.Br. Orchidaceae. 4 C. Am., W.I., trop. S. Am. to Argent.
Isochoriste Miq. (?Asystasia Bl.). Acanthaceae. 1 Java.
Isocoma Nutt. Compositae. 30 N. Am., Mex.
Isocynis Thou. (uninom.) = Cynorkis fastigiata Thou. (Orchidac.).
Isodeca Rafin. = Leucas R.Br. (Labiat.).
Isodendrion A. Gray. Violaceae. 14 Hawaii.
Isodesmia Gardn. Leguminosae. 2 Brazil. ·
Isodichyophorus A. Chev. (sphalm.) = seq.
Isodictyophorus Briq. (~ Coleus Lour.). Labiatae. 1 W. Afr.
Isodon (Schrad. ex Benth.) Kudo = Plectranthus L'Hérit. (Labiat.).
Isoëtaceae Reichenb. Isoëtales. 2 gen.: Isoëtes, Stylites.
Isoëtales. Lycopsida. Only living fam., Isoëtaceae. (Pleuromeiaceae, Nathorstianaceae, fossil.)
Isoëtes L. Isoëtaceae. 75 temp. & trop. Aquatics with short erect stems of complex structure; roots branch dichotomously; ls. awl-shaped, 1-veined, ligulate. Sporangia are sunk in leaf-bases; outer ls. have mega-, inner micro-, sporangia. Development of prothalli and young sporophyte ± as in Sela-
Isoëtes Weigel = Juncus L. (Juncac.). [ginella.
Isoëtopsis Turcz. Compositae. 1 temp. Austr.
Isoglossa Oerst. Acanthaceae. 50 trop. & S. Afr., Madag.
Isolatocereus (Backeb.) Backeb. = Lemaireocereus Britton & Rose (Cactac.).
Isolepis R.Br. = Scirpus L. (Cyperac.).
Isoleucas Schwartz. Labiatae. 1 Arabia.
Isoloba Rafin. = Pinguicula L. (Lentibulariac.).
Isolobus A. DC. = Lobelia L. (Campanulac.).
Isoloma Decne = Kohleria Regel (Gesneriac.).
Isoloma J. Sm. (Guerinia J. Sm.). Lindsaeaceae. 8 Ceylon, Malaysia.
Isoloma J. Sm. (quoad 'typum' in Hist. Fil., 1875, indicatum) = Nephrolepis
Isolona Engl. Annonaceae. 20 trop. Afr., Madag. [Schott (Oleandrac.).
Isolophus Spach = Polygala L. (Polygalac.).
Isomacrolobium Aubrév. & Pellegr. Leguminosae. 10 trop. Afr.
Isomeraceae Dulac = Elatinaceae Cambess.
Isomeria D. Don ex DC. = Vernonia Schreb. (Compos.).
Isomeris Nutt. (~ Cleome L.). Cleomaceae. 1 Calif., Mex.
Isomerocarpa A. C. Smith = Dryadodaphne S. Moore (Monimiac.).
Isometrum Craib. Gesneriaceae. 2 China.
Isonandra Wight. Sapotaceae. 10 S. India, Ceylon, Malay Penins., Borneo.
Isonema R.Br. Apocynaceae. 3 W. Afr.
Isonema Cass. = Vernonia Schreb. (Compos.).

Isopappus Torr. & Gray (~ Haplopappus Cass.). Compositae. 3 N. Am.
Isopara Rafin. = Cleomella DC. (Cleomac.).
Isopetalum Sweet = Pelargonium L'Hérit. (Geraniac.).
Isophyllum Hoffm. = Bupleurum L. (Umbellif.).
Isophyllum Spach = Ascyrum L. = Hypericum L. (Guttif.).
Isophysidaceae Takhtadj. = Iridaceae–Isophysideae Hutch.
Isophysis T. Moore. Iridaceae. 1 Tasmania. G̲!
Isoplesion Rafin. = Echium L. (Boraginac.).
Isoplexis (Lindl.) Loud. (~ Digitalis L.). Scrophulariaceae. 3 Canaries, Madeira.
*****Isopogon** R.Br. ex Knight. Proteaceae. 30 Austr.
Isoptera Scheff. ex Burck = Shorea Roxb. ex Gaertn. (Dipterocarpac.).
Isopteris Klotzsch = Begonia L. (Begoniac.).
Isopteris Wall. = Trigoniastrum Miq. (Trigoniac.).
Isopteryx Klotzsch = Isopteris Klotzsch = Begonia L. (Begoniac.).
Isopyrum Adans. = Hepatica Mill. (Ranunculac.).
Isopyrum L. Ranunculaceae. 30 N. temp.
Isora Mill. = Helicteres L. (Sterculiac.).
Isorium Rafin. = Lobostemon Lehm. (Boraginac.).
Isoschoenus Nees = Schoenus L. (Cyperac.).
Isostigma Less. Compositae. 15 subtrop. S. Am., on campos.
Isostoma D. Dietr. = Isotoma Lindl. (Campanulac.).
Isostylis Spach = Banksia Gaertn. (Proteac.).
Isotheca P. & K. = Isodeca Rafin. = Leucas R.Br. (Labiat.).
Isotheca Turrill. Acanthaceae. 1 W.I. (Trinidad).
Isothylax Baill. = Sphaerothylax Bisch. (Podostemac.).
Isotoma Lindl. (~ Laurentia Adans.). Campanulaceae. 10 Austr.; 1 C. & S. Am., W.I.
Isotrema Rafin. (~ Aristolochia L.). Aristolochiaceae. 12 Himal. to Formosa & Japan, S. U.S. to C. Am.
Isotria Rafin. Orchidaceae. 2 E. U.S.
Isotrichia (DC.) Kuntze = Vanillosmopsis Sch. Bip. (Compos.).
Isotropis Benth. Leguminosae. 12 Austr.
Isotypus Kunth = Onoseris DC. (Compos.).
Isouratea Van Tiegh. = Ouratea Aubl. (Ochnac.).
Isquierda Willd. = Izquierdia Ruiz & Pav. = Ilex L. (Aquifoliaceae).
Isquierdia Poir. = praec.
Isypus Rafin. = Ipomoea L. (Convolvulac.).
Itaculumia Hoehne = Habenaria Willd. (Orchidac.).
Itasina Rafin. = Oenanthe L. (Umbellif.).
Itatiaia Ule. Melastomataceae. 1 SE. Brazil.
Itea L. Iteaceae. 15 Himal. to Japan, W. Malaysia, Atl. N. Am. Foliage of some spp. resembles *Maesa* (*Myrsinac.*).
*****Iteaceae** J. G. Agardh. Dicots. 2/17 E. & SE. As., E. N. Am., trop. & S. Afr. Shrubs or trees with simple toothed or spinose alt. stip. ls. Fls. small, reg., ♂ or polygam. Infl. term. or axill., densely elongate-racemiform or shortly cymose. K 5, valv. or open; C 5, valv.; A 5, with subul. fil. and obl. or ov. intr. anth.; disk annular; G (2), semi-inf., often elongate, with few to many usu. biseriate axill. ov., and 2 connate styles finally free but sometimes united

by capit. stig. Fr. a narrow or ovoid septic. caps., with few to many obl. or scobiform seeds; embryo large in sparse fleshy endosp. Genera: *Itea, Choristylis.* Near *Escalloniaceae.*

Iteadaphne Blume = Lindera Thunb. (Laurac.).

Iteiluma Baill. = Planchonella Pierre (Sapotac.).

Iteria Hort. (sphalm.) = Stevia Cav. (Compos.).

Itheta Rafin. = Carex L. (Cyperac.).

Ithycaulon Copel. = Orthiopteris Copel. (Dennstaedtiac.).

Itia Molina = Lonicera L. (Caprifoliac.).

Iticania Rafin. = Elytranthe Bl. (Loranthac.).

Itoa Hemsl. Flacourtiaceae. 2 S. China, E. Malaysia.

Itoasia Kuntze = Corynaea Hook. f. (Balanophorac.).

Ittnera C. C. Gmel. = Najas L. (Najadac.).

Ituridendron De Wild. Sapotaceae. 1 trop. Afr.

Ituterion Rafin. = Salvadora L. (Salvadorac.).

Itzaea Standl. & Steyerm. Convolvulaceae. 1 C. Am.

Iuga Reichb. (sphalm.) = Inga Mill. (Legumin.).

Iulocroton Baill. = Julocroton Mart. (Euphorbiac.).

Iulus Salisb. = Allium L. (Alliac.).

Iuncago Fabr. = Juncago Seguier = Triglochin L. (Jungacinac.).

Iungia Boehm. = Dianthera L. (Acanthac.).

Iva Fabr. = Teucrium L. (Labiat.).

Iva L. Compositae. 15 N. & C. Am., W.I.

Ivania O. E. Schulz. Cruciferae. 1 N. Chile.

Ivesia Torr. & Gray (~ Potentilla L.). Rosaceae. 22 W. U.S.

Ivira Aubl. = Sterculia L. (Sterculiac.).

Ivodea R. Capuron. Rutaceae. 6 Madag.

Ivonia Vell. Inc. sed. (Sympet.). 1 Brazil.

× **Iwanagara** Hort. Orchidaceae. Gen. hybr. (Cattleya × Caularthron × Laelia × Ryncholaelia [Brassavola Hort.]).

Ixalum Forst. f. = Spinifex L. (Gramin.).

Ixanthus Griseb. Gentianaceae. 1 Canaries.

Ixauchenus Cass. = Lagenophora Cass. (Compos.).

Ixerba A. Cunn. Brexiaceae. 1 Northern N.Z.

Ixeridium (A. Gray) Tsvelev. Compositae. 20–5 temp. & trop. As. to New Guinea.

Ixeris Cass. Compositae. 50 E. & SE. As. to New Guinea.

Ixerra Pritz. (sphalm.) = Ixerba A. Cunn. (Brexiac.).

Ixia L. (1753) = Aristea Soland. + Belamcanda Adans. (Iridac.).

***Ixia** L. (1762). Iridaceae. 1 trop. Afr., 44 S. Afr.

Ixia Mühlenb. ex Spreng. = Hydrilla Rich. (Hydrocharitac.).

Ixiaceae Horan. = Iridaceae Juss.

Ixianthes Benth. Scrophulariaceae. 1 S. Afr.

Ixianthus Reichb. = praec.

Ixiauchenus Less. = Ixauchenus Cass. = Lagenophora Cass. (Compos.).

Ixidium Eichl. = Eremolepis Griseb. (Loranthac.).

Ixina Rafin. = Ixine Loefl. = Krameria Loefl. (Krameriac.).

Ixine Hill = Cirsium Mill. (Compos.).

Ixine Loefl. = Krameria Loefl. (Krameriac.).

Ixiochlamys F. Muell. & Sond. = Podocoma Cass. (Compos.).

Ixiolaena Benth. Compositae. 6 Austr.

Ixioliriaceae ('Ixiolirionaceae') Nak. = Amaryllidaceae-Ixioliriinae Pax.

Ixiolirion Fisch. Amaryllidaceae. 3 W. & C. As.

Ixionanthes Endl. = Ixonanthes Jack (Ixonanthac.).

Ixiosporum F. Muell. = Citriobatus A. Cunn. (Pittosporac.).

Ixiosporus Benth. = praec.

Ixoca Rafin. = Silene L. (Caryophyllac.).

Ixocactus Rizzini. Loranthaceae (*s. str.*). 1 Colombia, Venez. Perhaps related to *Oryctanthus*. (Kuijt, *Brittonia*, 19: 62–7, 1967.)

Ixocaulon Rafin. = Ixoca Rafin. = Silene L. (Caryophyllac.).

Ixodia R.Br. Compositae. 1 SE. Austr.

Ixodia Soland. ex DC. = Brasenia Schreb. (Cabombac.).

Ixodonerium Pitard. Apocynaceae. 1 Indoch.

Ixonanthaceae Planch. ex Klotzsch. Dicots. 8/48 trop. Trees or shrubs, with simple ent. or serr. alt. stip. ls. (sometimes distich. & asymm.), stips. sometimes very long, convolute, leaving a conspic. annular intrapet. scar. Infl. rac., panic., or corymbiform. Fls. small, reg., ♀ or ♂ ♀. K 4–5 or (4–5), imbr.; C 4–5, imbr., sometimes persist. and indurated; A 5–10–20 (sometimes unequal), fil. slender, subul., often sigmoid in bud; anth. small; disk conspic., annular or cupular; G (4–5) or (2), with simple style ± deflexed in bud, and 1–2 pend. ov. per loc. Fr. a large drupe, a broadly winged samara, or a woody or leathery 2–5-loc. septic. caps. (sometimes also loculic. by secondary false septa); seeds either ovoid, encl. in aril, or cylindric with free vestigial aril, or small and flattened with adnate vestig. aril; endosp. fleshy or o. Genera: *Irvingia, Klainedoxa, Desbordesia, Cyrillopsis, Ixonanthes, Ochthocosmus, Phyllocosmus, Allantospermum*.

Ixonanthes Jack. Ixonanthaceae. 10 Himal., S. China, Indoch., Philipp., Borneo; 1 New Guinea.

Ixophorus Nash = Setaria Beauv. (Gramin.).

Ixophorus Schlechtd. Gramineae. 2–3 Mex.

Ixora L. Rubiaceae. 400 trop. The fl. is commonly red with a long narrow tube, and probably butterfly-visited.

Ixorhea Fenzl. Ehretiaceae(?). 1 Andes of Argentina.

Ixorrhoea Willis = praec.

Ixtlania M. E. Jones. Acanthaceae. 1 Mex.

× **Ixyoungia** Kitam. Compositae. Gen. hybr. (Ixeris × Youngia).

Izabalaea Lundell. Nyctaginaceae. 1 C. Am. (Guatem.).

Izquierdia Ruiz & Pav. = Ilex L. (Aquifoliac.).

J

Jaborosa Juss. Solanaceae. 20 Mex., Bolivia to Patag.

Jabotapita Adans. = Ochna L. + Ouratea Aubl. (Ochnac.).

Jacaima Rendle. Asclepiadaceae. 1 Jamaica.

Jacaranda Juss. Bignoniaceae. 50 C. & S. Am., W.I.

JACARATIA

Jacaratia A. DC. Caricaceae. 8 trop. Am., Afr.
Jacea Haller (non seq.) = Centaurea L. (Compos.).
Jacea Mill. = Centaurea L. (Compos.).
Jacea Opiz = Grammeionium Reichb. = Viola L. (Violac.).
Jackia Blume = Jakkia Blume = Xanthophyllum Roxb. (Xanthophyllac.).
Jackia Spreng. = Eriolaena DC. (Sterculiac.).
Jackia Wall. Rubiaceae. 1 Malay Penins., Banka, Borneo.
Jacksonago Kuntze = Wiborgia Thunb. (Legumin.).
Jacksonia R.Br. Leguminosae. 40 Austr.
Jacksonia Hort. ex Schlechtd. = Jasminum L. (Jasminac.).
Jacksonia Rafin. = Polanisia Rafin. (Cleomac.).
Jacobaea Kuntze = Vicoa Cass. (Compos.).
Jacobaea Mill. = Senecio L. (Compos.).
Jacobaeastrum Kuntze = Euryops Cass. (Compos.).
Jacobanthus Fourr. = Senecio L. (Compos.).
Jacobea Thunb. = Jacobaea Mill. = Senecio L. (Compos.).
***Jacobinia** Nees ex Moric. Acanthaceae. 50 Mex. to trop. S. Am.
Jacobsenia L. Bolus & Schwantes. Aïzoaceae. 1 S. Afr.
Jacosta DC. = Phymaspermum Less. (Compos.).
Jacquemontia Bélang. = Gamolepis Less. (Compos.).
Jacquemontia Choisy. Convolvulaceae. 120 trop. (mostly Am.).
Jacquesfelixia Phipps. Gramineae. 1 trop. & SW. Afr.
Jacqueshuberia Ducke. Leguminosae. 2 Amaz. Brazil.
Jacquinia L. corr. Jacq. Theophrastaceae. 50 warm Am., W.I.
Jacquinia Mutis ex L. = Prockia P.Br. ex L. (Tiliac.).
Jacquiniella Schlechter. Orchidaceae. 3 C. Am., trop. S. Am., W.I.
Jacquinotia Hombr. & Jacquinot = Lebetanthus Endl. (Epacridac.).
×**Jacquinparis** hort. Orchidaceae. Gen. hybr. (vi; nomen nugax) (Jacquiniella × Liparis).
Jacuanga Lestib. = Costus L. (Costac.).
Jacularia Rafin. = Cunninghamia R.Br. (Taxodiac.).
Jadelotia Buc'hoz. Quid? (Rosaceae? Prunus?)
Jadunia Lindau. Acanthaceae. 1 New Guinea.
Jaegera Giseke = Zingiber Adans. (Zingiberac.).
Jaegeria Kunth. Compositae. 8 Mex. to Urug.
Jaeggia Schinz = Adenia Forsk. (Passiflorac.).
Jaeschkea Kurz. Gentianaceae. 3 Himal.
Jagera Blume. Sapindaceae. 3 E. Malaysia, Austr.
Jahnia Pittier & Blake. Rubiaceae. 1 Venez.
Jaimenostia Guinea & Gómez Moreno = Sauromatum Schott (Arac.).
Jakkia Blume = Xanthophyllum Roxb. (Xanthophyllac.).
Jalambica Rafin. = Neurelmis Rafin. = ?Pinillosia Ossa vel ?Koehneola Urban (Compos.).
Jalambicea Cerv. = Limnobium Rich. (Hydrocharitac.).
Jalapa Mill. = Mirabilis L. (Nyctaginac.).
Jalapaceae Barkley = Nyctaginaceae Juss.
Jaliscoa S. Wats. Compositae. 3 Mex.
Jalombicea Steud. = Jalambicea Cerv. = Limnobium Rich. (Hydrocharitac.).

Jaltomata Schlechtd. = Saracha Ruiz & Pav. (Solanac.).
Jaltonia Steud. = praec.
Jambolana Adans. = Jambolifera L. = Acronychia J. R. & G. Forst. (Rutac.).
Jambolifera Houtt. = Syzygium Gaertn. (Myrtac.).
Jambolifera L. = Acronychia J. R. & G. Forst. (Rutac.).
Jambos Adans. = seq.
***Jambosa** Adans. mut. DC. = Syzygium Gaertn. (Myrtac.).
Jambus Nor. = praec. (Myrtac.).
Jamesbrittenia Kuntze. Scrophulariaceae. 1 Egypt to NW. India.
Jamesia Nees = Stephanomeria Nutt. (Compos.).
Jamesia Rafin. = Dalea Juss. (Legumin.).
***Jamesia** Torr. & Gray. Philadelphaceae. 1 SE. U.S.
Jamesianthus Blake & Sherff. Compositae. 1 S. U.S.
Jamesonia Hook. & Grev. Hemionitidaceae. 17 trop. S. Am., xerophytic. Rhiz. hairy; ls. narrow, of indefinite apical growth. (A. F. Tryon, *Contrib. Gray Herb.* **191**: 109–203, 1962.)
Janasia Rafin. = Phlogacanthus Nees (Acanthac.).
Jancaea Boiss. Gesneriaceae. 1 Balkan Penins.
Jandinea Steud. = Jardinea Steud. = Thelepogon Roth (Gramin.).
Jangaraca Rafin. = Tangaraca Adans. = Hamelia Jacq. (Rubiac.).
Jania Schult. f. = Baeometra Salisb. (Liliac.).
Janipha Kunth = Manihot Mill. (Euphorbiac.).
Jankaea Boiss. = Jancaea Boiss. (Gesneriac.).
Janraia Adans. = Rajania L. (Dioscoreac.).
Jansenella Bor. Gramineae. 1 India, Assam, Ceylon.
Jansenia Barb. Rodr. = Plectrophora Focke (Orchidac.).
Jansonia Kippist. Leguminosae. 1 W. Austr. Ls. simple, opp.
Jantha Steud. = Iantha Hook. = Ionopsis Kunth (Orchidac.).
Janthe Griseb. = Celsia L. (Scrophulariac.).
Janthe Nel = Ianthe Salisb. (Hypoxidac.).
Janusia A. Juss. Malpighiaceae. 12 Calif. to Argent.
Japanobotrychium Masamune = Botrychium Sw. (Ophioglossac.).
Japarandiba Adans. = Gustavia L. (Lecythidac.).
Japonasarum Nakai = Asarum L. (Aristolochiac.).
Japonolirion Nakai = Tofieldia Huds. (Liliac.).
Japotapita Endl. = Jabotapita Adans. = Ouratea Aubl. (Ochnac.).
Jaquinia L. = Jacquinia L. corr. Jacq. (Theophrastac.).
Jaquinotia Walp. = Jacquinotia Hombr. & Jacquinot = Lebetanthus Endl. (Epacridac.).
Jaracatia Endl. = Jacaratia A. DC. (Caricac.).
Jarandersonia Kosterm. Tiliaceae. 1 Borneo. (= ?Berrya Roxb. with galled ov.)
Jarapha Steud. = Jarava Ruiz & Pav. = Stipa L. (Gramin.).
Jaraphaea Steud. = Iaravaea Scop. = Microlicia D. Don (Melastomatac.).
Jarava Ruiz & Pav. = Stipa L. (Gramin.).
Jaravaea Neck. = Iaravaea Scop. = Microlicia D. Don (Melastomatac.).
Jardinea Steud. (~ Thelepogon Roth). Gramineae. 3 trop. Afr.
Jardinia Benth. & Hook. f. = praec.

bract-like, mostly acute, persist.; A 3 + 3, anth. basifixed, sagitt., latrorse; G̲ (3), with 1 axile orthotr. ovule per loc., styles 3, free or almost so, stigs. plumose, persist. Fr. drupaceous, ± triquetr., with bony endoc.; seeds 1–3, with minute embr. in copious endosp. Only genus: *Joinvillea.* Formerly incl. in *Flagellariac.*, but recent evidence from veg. morphol., anat. and pollen clearly indicates the need for separation. See Tomlinson & A. C. Smith, *Taxon,* **19**: 887–9, 1970.

Joira Steud. = Ivira Aubl. = Sterculia L. (Sterculiac.).

Joliffia Boj. ex Delile = Telfairia Hook. (Cucurbitac.).

Jollya Pierre = Achradotypus Baill. (Sapotac.).

Jollydora Pierre ex Gilg. Connaraceae. 6 trop. W. Afr.

Joncquetia Schreb. = Tapiria Juss. (Anacardiac.).

Jondraba Medik. = Biscutella L. (Crucif.).

Jonesia Roxb. = Saraca L. (Legumin.).

Jonesiella Rydb. = Astragalus L. (Legumin.).

Jonghea Lem. = Billbergia Thunb. (Bromeliac.).

Jonia Steud. = Ionidium Vent. (Violac.).

Jonidiopsis Presl = Noisettia Kunth (Violac.).

Joniris Klatt = Iris L. (Iridac.).

Jonorchis G. Beck = Limodorum Rich. (Orchidac.).

Jonquilla Haw. = Narcissus L. (Amaryllidac.).

Jonsonia Garden = Johnsonia T. Dale ex Mill. = Callicarpa L. (Verbenac.).

Jontanea Rafin. = Tontanea Aubl. = Coccocypselum P.Br. (Rubiac.).

Jonthlaspi All. = Ionthlaspi Gérard = Clypeola L. (Crucif.).

Joosia Karst. (~ Ladenbergia Klotzsch). Rubiaceae. 3 N. Andes.

Jordania Boiss. = Gypsophila L. (Caryophyllac.).

Jorena Adans. Inc. sed. (? Amaranthac. ? Caryophyllac.). Egypt.

Josepha Benth. & Hook. f. = Josephia Wight = Sirhookera Kuntze (Orchidac.).

Josepha Vell. = Bougainvillea Comm. ex Juss. (Nyctaginac.).

Josephia Salisb. = Dryandra R.Br. (Proteac.).

Josephia Steud. = Josepha Vell. = Bougainvillea Comm. ex Juss. (Nyctaginac.).

Josephia Wight = Sirhookera Kuntze (Orchidac.).

Josephina Pers. = seq.

Josephinia Vent. Pedaliaceae. 1 NE. trop. Afr.; 4 E. Java, E. Malaysia, Austr.

Jossinia Comm. ex DC. (~ Eugenia L.). Myrtaceae. 30 trop. Afr., Masc., Madag., Indomal., Austr., Pacif.

Jouvea Fourn. Gramineae. 2 Mex., C. Am.

Jovellana Ruiz & Pav. Scrophulariaceae. 7 N.Z., Chile.

Jovibarba Opiz = Sempervivum L. (Crassulac.).

Joxocarpus Pritz. (sphalm.) = Toxocarpus Wight & Arn. (Asclepiadac.).

Joxylon Rafin. = Toxylon Rafin. = Maclura Nutt. (Morac.).

Jozoste Kuntze = Iozoste Nees = Actinodaphne Nees + Litsea Lam. (Laurac.).

Jrillium Rafin. = Trillium L. (Liliac.).

Jryaghedi Kuntze = Horsfieldia Willd. (Myristicac.).

Juania Drude. Palmae. 1 Juan Fernandez.

Juanulloa Ruiz & Pav. Solanaceae. 12 Mex. to trop. S. Am.

Jubaea Kunth. Palmae. 1 Chile, *J. spectabilis* Kunth, the coquito palm. Palm-honey is prepared by evaporation of the sap, and the tree is useful in other ways.

Jubaeopsis Becc. Palmae. 1 S. Afr. (Cape Prov.).

Jubelina A. Juss. = Diplopterys Juss. (Malpighiac.).

Jubilaria Mez = Loheria Merr. (Myrsinac.).

Jubistylis Rusby = Banisteriopsis C. B. Rob. (Malpighiac.).

Jububa Bub. = Ziziphus Mill. (Rhamnac.).

Juchia Neck. = Lobelia L. (Campanulac.).

Juchia M. Roem. = Solena Lour. (Cucurbitac.).

Jucunda Cham. = Miconia Ruiz & Pav. (Melastomatac.).

Juelia Asplund. Balanophoraceae. 3 Bolivia, Argent.

Juergensenia Schlechtd. = Jurgensenia Turcz. = Bejaria Mutis (Ericac.).

Juergensia Spreng. = Rinorea Aubl. (Violac.).

Juga Griseb. (sphalm.) = Inga Scop. (Legumin.).

Jugastrum Miers. Lecythidaceae. 7 Venez., Brazil.

*Juglandaceae A. Rich. ex Kunth. Dicots. 7/50 N. temp. & subtrop., S. to India, Indoch., & S. Am. (Andes). Trees; alt. stip. pinn. ls., with brown hairy winter buds; the buds arise rather high up in the leaf axils, and sometimes several appear in descending order. Infl. monoecious, the ♂ appearing as catkins on the twigs of the previous year. P typically 4-leaved, but often fewer by abortion. ♂ fl. with 3–40 stam. (more in the lower fls.); ♀ fl. with epig. P enclosed in adnate cupule; G̲ (2), 1-loc., with 1 erect orthotr. ov.; style short with 2 stigmas. Wind-fert.; *Juglans* is chalazogamic. Drupe or nut; testa thin, seed exalb. Genera: *Carya, Juglans, Pterocarya, Engelhardtia, Oreomunnea, Platycarya, Alfaroa.* Prob. some connection with *Anacardiac.* & *Picrodendrac.*

Juglandicarya Reid & Chandler (*sensu* Hu) = Rhamphocarya Kuang (Juglandac.).

Juglans L. Juglandaceae. 15 Medit. to E. As., Indoch., N. & C. Am., Andes. *J. regia* L. (W. As., walnut), *J. cinerea* L. (Canada, U.S., butternut) and others useful. ♂ fl. adnate to br. and bracteoles, P 5–4–3–2, A to 20 in lowest fls., to 6 in upper. Drupe, with green fleshy exocarp, and hard endocarp (the 'shell'). The 'boats' into which the shell splits are not cpls.; the splitting is down the midribs. Within is the seed with thin brown seedcoat, exalb., with basal radicle and two large cots., deeply lobed (ruminate) owing to presence of partial septa in ovary. Chalazogamic. Wood valued in cabinet-making, etc. The seeds yield an oil. Many vars. cult for fr. Many Tertiary fossils.

Julbernardia Pellegr. Leguminosae. 11 trop. Afr.

Julia Steud. = Junia Adans. = Clethra L. (Clethrac.).

Juliana Reichb. = seq.

Juliana La Llave = Choisya Kunth (Rutac.).

Juliania Schlechtd. = Amphipterygium Schiede ex Standl. (Julianiac.).

*Julianiaceae Hemsl. Dicots. 2/5 warm Am. Resinous or laticiferous trees or shrubs with the habit of *Rhus* (*Anacardiac.*). Ls. alt. (often crowded), imparipinn. (rarely simple), exstip., pubesc., with serrate lflts. Fls. ♂ ♀, dioec.: ♂ ∞, in pend. pubesc. panicles; ♀ in threes or fours, collateral, encl. in common invol. at end of short cernuous or erect pedunc. ♂: P 3–8, thin, lin., valv.(?); A 3–8, alt. with P. ♀: P 0, G̲ 1-loc., with 1 basal ov. on cup-like funicle, and exserted 3-lobed style. Fr. a dry indehisc. syncarp, consisting of the accresc. invol. with broad flattened stalk and thickened subglob. apex containing 1–2 compressed nuts, ± adnate to wall of invol., with hairy pericarp; seed exalb. Genera: *Amphipterygium* (*Juliania*), *Orthopterygium*. Closely related to *Anacardiaceae* (cf. anat., pollen, habit), with which they should perhaps be reunited.

JULIBRISIN

Julibrisin Rafin. = Albizia Durazz. (Legumin.).

Julieta Leschen. ex DC. = Lysinema R.Br. (Epacridac.).

***Julocroton** Mart. Euphorbiaceae. 45 Mex. to trop. S. Am.

Julostyles Benth. & Hook. f. = seq.

Julostylis Thw. Malvaceae. 1 Ceylon. Seed reniform. A 10. G (2).

Julus P. & K. = Iulus Salisb. = Allium L. (Alliac.).

Jumellea Schlechter. Orchidaceae. 45 E. trop. & S. Afr., Masc.

Jumelleanthus Hochr. Malvaceae. 1 Madag.

***Juncaceae** Juss. Monocots. 9/400 temp., arct., trop. mts., in damp cold places. Usu. creeping sympodial rhiz., one joint of the sympodium appearing above ground each year as a leafy shoot. The stem does not often lengthen above ground, except to bear infl.; ls. usu. narrow, sometimes centric. Infl. usu. a crowded mass of fls. in cymes of various types, usu. monochas. Fls. usu. ⚥, reg., wind-fert. P 3 + 3, usu. sepaloid, with odd l. of inner whorl post.; A 3 + 3 (or inner wanting), anthers dehisc. lat. or intr., pollen in tetrads; G (3), with axile or parietal plac. and ∞ or few anatr. ov.; style simple with 3 brush-like stigmas. Loculic. caps.; embryo straight, in starchy endosp. Chief genera: *Prionium, Juncus, Luzula, Oxychloë.*

***Juncaginaceae** Rich. Monocots. 3/25 N. & S. temp. & frig. Ann. or perenn. scapig. herbs of freshwater or salt marshes; roots sometimes tub. Ls. lin., sheathing, radic. (rarely floating). Fls. rac. or spic., small, reg., ⚥ or ♂ ♀ (dioec. or polygam.), ebract. P 3 + 3, herbaceous; A 6–4, with subsess. extr. anth.; G 6–4 or (6–4), with subsess. ± plumose stig., and 1 basal anatr. (rarely apical orthotr.) ov. per loc. Fr. cylindr. or obov.; carp. free or conn., dehisc. or indehisc., sometimes conspic. calcarate at base, sometimes 3 barren; seed erect, without endosp. Genera: *Triglochin, Tetroncium, Maundia.*

Juncago Seguier = Triglochin L. (Juncaginac.).

Juncaria DC. = Ortegia L. (Caryophyllac.).

Juncastrum Fourr. = Juncus L. (Juncac.).

Juncella F. Muell. = Trithuria Hook. f. (Centrolepidac.).

Juncellus (Griseb.) C. B. Clarke (~ Cyperus L.). Cyperaceae. 18 warm &

Juncinella Fourr. = Juncus L. (Juncac.). [temp.

Juncodes Kuntze = seq.

Juncoïdes Seguier = Luzula DC. (Juncac.).

Juncus L. Juncaceae. 300 cosmop., chiefly in cold or wet places, rare in trop. Usu. low herbaceous pl. with sympodial rhiz. giving off one leafy shoot each year. The ls. are of various types, with large sheathing bases. Some are flat and grass-like, others needle-like, and still others centric in structure and standing erect. The infl. is a dense head or panicle, of cymose construction (usu. of rhipidia or drepania). In some spp. it appears to be lat. on a leaf-like cylindrical stem, but is really only pushed to one side by the bract of the infl. Fl. protog. and wind-fert.

Rushes are largely used for making baskets, chair bottoms, etc. *J. squarrosus* L. is common on hill pastures in Eur.; it is eaten by sheep and forms a valuable part of their fodder when grass is scarce.

Jundzillia Andrz. ex DC. = Cardaria Desv. (Crucif.).

Junellia Moldenke (= ?Monopyrena Speg.) (= ?Thryothamnus Phil.). Verbenaceae. 57 S. Am.

Junghansia J. F. Gmel. = Curtisia Ait. (Cornac.).

Junghuhnia R.Br. ex De Vriese = Salomonia Lour. (Polygalac.).

Junghuhnia Miq. = Codiaeum Juss. (Euphorbiac.).

Jungia Boehm. = Dianthera L. (Acanthac.).

Jungia Fabr. (uninom.) = *Salvia* L. sp. (Labiat.).

Jungia Gaertn. = Baeckea L. (Myrtac.).

Jungia L. f. (= Trinacte Gaertn.). Compositae. 30 Mex., C. Am., Andes.

Jungia Loefl. = Ayenia L. (Sterculiac.).

Junia Adans. = Clethra L. (Clethrac.).

Junia Rafin. Inc. sed. (? Saxifragac., *s.l.*). Hab.? (Shrub; ls. opp., serr.; fls. reg., dioec., in axill. 3-fld. cymes; K 5; C 5, trunc.-spath.; ♀ fl. with 5 stds(?) opp. pets., G (2) with 2 styles.)

Juniperus L. Cupressaceae. 60 N. hemisph. The juniper, *J. communis* L. (Euras.), and *J. oxycedrus* L., etc., have needle ls. throughout life; others, such as *J. sabina* L., the savin (Eur., As.), have small ls. closely appressed, as in *Cupressus*. Seedling forms of these are known (see *Retinospora* Carr.). The cone consists of 1–4 whorls of scales, one only being fertile, as a rule. In ripening the whole becomes a fleshy mass enclosing the hard seeds, and forming a good imitation of a true berry. The fruit is eaten by birds. That of *J. communis* is used in making gin. The wood of *J. virginiana* L. is the 'red cedar' used for pencils; others also give useful timber.

Junkia Ritgen (sphalm.) = Funkia Spreng. = Hosta Tratt. (Liliac.).

Juno Tratt. = Iris L. (Iridac.).

Junodia Pax = Anisocycla Baill. (Menispermac.).

Junopsis W. Schulze (~ Iris L.). Iridaceae. 1 C. Himal. to Yunnan & Siam.

Junquilla Fourn. = Jonquilla Haw. = Narcissus L. (Amaryllidac.).

Jupica Rafin. = Xyris L. (Xyridac.).

Juppia Merr. = Zanonia L. (Cucurbitac.).

Jupunba Britton & Rose = Pithecellobium Mart. (Legumin.).

Jurgensenia Turcz. = Bejaria Mutis (Ericac.).

Jurgensia Benth. & Hook. f. = Juergensia Spreng. = Rinorea Aubl. (Violac.).

Jurgensia Rafin. = Spermacoce L. (Rubiac.).

Jurighas Kuntze = Filicium Thw. (Burserac.).

Jurinea Cass. Compositae. 100 C. Eur., Medit. to China; many C. As.

Jurinella Jaub. & Spach = praec.

Juruasia Lindau. Acanthaceae. 2 Brazil.

Jussia Adans. = seq.

Jussiaea L. = Ludwigia L. (Onagrac.).

Jussiaeia Hill = praec.

Jussiea L. ex Sm. = Potentilla L. (Rosac.).

Jussiena Reichb. (sphalm.) = seq.

Jussieua Murr. = Jussiaea L. = Ludwigia L. (Onagrac.).

Jussieuaceae Drude = Onagraceae Juss.

Jussieuaea Rottl. ex DC. = Lumnitzera Willd. (Combretac.).

Jussieuia Houst. = Jatropha L. (Euphorbiac.).

Jussieuia Thunb. = Jussiaea L. = Ludwigia L. (Onagrac.).

Jussieva Gleditsch = praec.

Justago Kuntze. Cleomaceae. 1 N. & NE. Austr.

JUSTENIA

Justenia Hiern. Rubiaceae. 1 trop. Afr.
Justica Neck. = seq.
Justicea P. & K. = seq.
Justicia L. Acanthaceae. 300 trop. & subtrop.
Juttadinteria Schwantes. Aïzoaceae. 11 S. Afr.
Juzepczukia Khrzhanovsky (~ Rosa L.). Rosaceae. 3 China, Japan.

K

Kablikia Opiz = Primula L. (Primulac.).
Kabulia Bor & C. E. C. Fischer. Caryophyllaceae. 1 Afghanistan.
Kadakia Rafin. = Monochoria Presl (Pontederiac.).
Kadali Adans. = Osbeckia L. (Melastomatac.).
Kadalia Rafin. = Dissotis Benth. (Melastomatac.).
Kadaras Rafin. = Kadurias Rafin. = Cuscuta L. (Cuscutac.).
Kadsura Juss. Schisandraceae. 22 India, China, Japan, SE. As., W. Malaysia, Moluccas. Climbing shrubs with no stipules. Fls. unisexual, spiral throughout.
Kadua Cham. & Schlechtd. (~ Hedyotis L.). Rubiaceae. 23 Hawaii.
Kadula Rafin. = seq.
Kadurias Rafin. = Cuscuta L. (Cuscutac.).
Kaeleria Boiss. = Koeleria Pers. (Gramin.).
Kaempfera Houst. = Tamonea Aubl. (Verbenac.).
Kaempfera Spreng. = seq.
Kaempferia L. Zingiberaceae. 70 trop. Afr., India to S. China & W. Malaysia.
Kaernbachia Kuntze = Microsemma Labill. (Thymelaeac.).
Kaernbachia Schlechter = Turpinia Vent. (Staphyleac.).
× **Kagawara** hort. Orchidaceae. Gen. hybr. (iii) (Ascocentrum × Renanthera × Vanda).
Kagenackia Steud. = seq.
Kageneckia Ruiz & Pav. Rosaceae. 3 Chile.
Kahiria Forsk. = Ethulia L. (Compos.).
Kaieteuria Dwyer. Ochnaceae. 1 Brit. Guiana.
Kajewskia Guillaumin (~ Carpoxylon Wendl. & Drude). Palmae. 1 New Hebrides.
Kajewskiella Merr. & Perry. Rubiaceae. 1 Solomon Is.
Kajuputi Adans. = Melaleuca L. (Myrtac.).
Kakile Desf. = Cakile L. (Crucif.).
Kakosmanthus Hassk. = Madhuca J. F. Gmel. (Sapotac.).
Kalabotis Rafin. = Allium L. (Alliac.).
Kalaharia Baill. Verbenaceae. 1 trop. & S. Afr.
Kalanchoë Adans. Crassulaceae. 200 trop. & S. Afr. to China & Java; 1 trop. S. Am.
Kalappia Kosterm. Leguminosae. 1 Celebes.
Kalawael Adans. = Rourea Aubl. (Connarac.).
Kalbfussia Sch. Bip. (~ Leontodon L.). Compositae. 1 Medit. (v. variable).
Kalbreyera Burret. Palmae. 1 Colombia.
Kalbreyeriella Lindau. Acanthaceae. 1 Colombia.

Kaleniczenkia Turcz. = Brachysema R.Br. (Legumin.).
Kaleria Adans. = Silene L. (Caryophyllac.).
Kali Mill. = Salsola L. (Chenopodiac.).
Kalidiopsis Aellen. Chenopodiaceae. 1 As. Min.
Kalidium Moq. Chenopodiaceae. 4 S. Russia, W. As.
Kalimares Rafin. = seq.
Kalimeris Cass. (~ Aster L.). Compositae. 2 E. As.
Kaliphora Hook. f. Cornaceae?? 1 Madag. Probably to be excluded from *Cornaceae*; pollen quite aberrant (cf. *Montiniac*).); facies of *Olacaceae*. Cf. reference under *Griseliniaceae*.
Kallias Cass. = Heliopsis Pers. (Compos.).
Kallophyllon Pohl ex Baker = Symphyopappus Turcz. (Compos.).
Kallstroemia Scop. Zygophyllaceae. 7 N. & NE. Austr., 16 S. U.S. & W.I. to Argent.
Kalmia L. Ericaceae. 8 N. Am., Cuba. The anthers are held in pockets of the C, and the filaments are bent like bows when the fl. is open. An insect probing for honey releases them, and the anthers strike against him, loading him with pollen.
Kalmiella Small (~ Kalmia L.). Ericaceae. 1 SE. U.S., 3 Cuba.
Kalmiopsis Rehder. Ericaceae. 1 NW. U.S.
Kalomikta Regel = Actinidia Lindl. (Actinidiac.).
Kalonymus (G. Beck) Prokh. = Euonymus L. (Celastrac.).
Kalopanax Miq. (~ Acanthopanax Decne). Araliaceae. 1 E. As.
Kalosanthes Haw. = Rochea DC. (Crassulac.).
Kalpandria Walp. = Calpandria Blume = Camellia L. (Theac.).
Kaluhaburunghos Kuntze = Cleistanthus Hook. f. ex Planch. (Euphorbiac.).
Kalymopetalon Pohl (nomen). 2 Brazil. Quid?
Kambala Rafin. = Sonneratia L. f. (Sonneratiac.).
Kameli̇a Steud. = Camellia L. (Theac.).
× **Kamemotoara** hort. (v) = × Perreiraara hort. (Orchidac.).
Kamettia Kostel. = Ellertonia Wight (Apocynac.).
Kampmania Rafin. = Zanthoxylum L. (Rutac.).
Kampmannia Steud. (= ? Arundo L.). Gramineae. 1 N.Z.
Kampochloa W. D. Clayton. Gramineae. 1 S. trop. Afr.
Kamptzia Nees = Syncarpia Tenore (Myrtac.).
Kanahia R.Br. Asclepiadaceae. 1 trop. E. Afr., Arabia.
Kandelia (DC.) Wight & Arn. Rhizophoraceae. 1 E. As. & W. Malaysia.
Kandena Rafin. = Canthium Lam. (Rubiac.).
Kandis Adans. = Lepidium L. (Crucif.).
Kania Schlechter = Metrosideros Banks ex Gaertn. (Myrtac.).
Kaniaceae (Engl.) Nak. = Myrtaceae Juss.(?).
Kanilia Guett. = Bruguiera Lam. (Rhizophorac.).
Kanimia Gardn. Compositae. 14 trop. S. Am.
Kanopikon Rafin. = Euphorbia L. (Euphorbiac.).
Kantemon Rafin. = Lonicera L. (Caprifoliac.).
Kantou Aubrév. & Pellegr. Sapotaceae. 1 trop. W. Afr.
Kantuffa Bruce = Pterolobium R.Br. (Legumin.).
Kaokochloa de Winter. Gramineae. 1 SW. Afr.

Kaoue Pellegr. Leguminosae. 2 trop. Am.

Kara-angolam Adans. = Alangium Lam. (Alangiac.).

Karaguata Rafin. = Tillandsia L. (Bromeliac.).

Karaka Rafin. = Sterculia L. (Sterculiac.).

Karamyschewia Fisch. & Mey. (~ Oldenlandia L.). Rubiaceae. 1 Balkan Penins. to NW. Persia, Egypt, Abyss.

Karamyschovia Fisch. ex Steud. = praec.

Karangolum Kuntze = Kara-angolam Adans. = Alangium Lam. (Alangiac.).

Karatas Mill. (~ Bromelia L.). Bromeliaceae. 20 Mex. to Argent.

Kardamoglyphos Schlechtd. corr. O. E. Schulz. Cruciferae. 1 Andes.

Kardanoglyphos Schlechtd. = praec.

Karekandel Adans. ex v. Wolf = Carallia Roxb. (Rhizophorac.).

Karekandelia Kuntze = praec.

Karelinia Less. (~ Pluchea Cass.). Compositae. 1 E. Russia & NE. Persia to Mongolia.

Karimbolea Descoings. Asclepiadaceae. 1 Madag.

Karina Boutique. Gentianaceae. 1 Congo.

Karivia Arn. = Solena Lour. (Cucurbitac.).

Karkandela Rafin. = Kare-Kandel Adans. = Carallia Roxb. (Rhizophorac.).

Karkinetron Rafin. = Muehlenbeckia Meissn. (Polygonac.).

Karlea Pierre = Maesopsis Engl. (Rhamnac.).

Karomia Dop. Verbenaceae. 1 Indochina.

Karos Nieuwl. & Lunell = Carum L. (Umbellif.).

Karpaton Rafin. = Triosteum L. (Caprifoliac.).

Karroochloa Conert & Türpe (~ Danthonia DC.). Gramineae. 1 S. & SW. Afr.

Karsthia Rafin. = Oenanthe L. (Umbellif.).

Karthemia Sch. Bip. = Vartheimia DC. = Iphiona Cass. (Compos.).

Karvandarina K. H. Rechinger. Compositae. 1 Persia.

Karwinskia Zucc. Rhamnaceae. 14 SW. U.S. to Bolivia, W.I.

Kasailo Dennst. = ? Ambelania Aubl. (Apocynac.).

Kaschgaria Poljakov (~ Tanacetum L.). 2 C. As. to W. China.

Kastnera Sch. Bip. = Liabum Adans. (Compos.).

Katafa Costantin & Poiss. = Cedrelopsis Baill. (Ptaeroxylac.).

Katakidozamia Haage & Schmidt ex Regel = Catakidozamia T. Hill = Macrozamia Miq. (Zamiac.).

Katapsuxis Rafin. = Cnidium Cusson (Umbellif.).

Katarsis Medik. = Gypsophila L. (Caryophyllac.).

Katharinea A. D. Hawkes (1956; non *Catharinea* Ehrh. 1780—Musci) = Epigeneium Gagnep. (Orchidac.).

Katoutheka Adans. = Ardisia Sw. (Rubiac.).

Katouthexa Steud. = praec.

Katou-Tsjeroe Adans. = Holigarna Buch.-Ham. ex Roxb. (Anacardiac.).

Katubala Adans. = Canna L. (Cannac.).

Kaufmannia Regel. Primulaceae. 1-2 C. As.

Kaukenia Kuntze = Mimusops L. (Sapotac.).

Kaulfussia Blume = Christensenia Maxon (Kaulfussiac.).

Kaulfussia Dennst. = Xanthophyllum Roxb. (Xanthophyllac.).

Kaulfussia Nees = Charieis Cass. (Compos.).
Kaulfussiaceae Presl. Marattiales (*q.v.* for chars.).
Kaulinia Nayar = Microsorium Link (Polypodiac.).
Kavalama Rafin. = Sterculia L. (Sterculiac.).
Kayea Wall. = Mesua L. (Guttif.).
Kearnemalvastrum Bates. Malvaceae. 2 Mex. to Colombia.
Keayodendron Leandri. Euphorbiaceae. 1 trop. Afr.
Keckia Straw (1967; non Glocker 1841—Algae foss.) = seq.
Keckiella Straw. Scrophulariaceae. 7 N. Am.
Kedrostis Medik. Cucurbitaceae. 35 trop. & subtrop. Afr., Madag., trop. As. & Malaysia.
Keenania Hook. f. Rubiaceae. 5 Assam, SE. As.
Keerlia DC. = Aphanostephus DC. + Xanthocephalum Willd. (Compos.).
Keerlia A. Gray & Engelm. Compositae. 2 S. U.S., Mex.
Keetia E. P. Philips = Canthium Lam. (Rubiac.).
Kefersteinia Reichb. f. Orchidaceae. 20 C. & trop. S. Am.
Kegelia Reichb. f. = Kegeliella Mansf. (Orchidac.).
Kegelia Sch. Bip. = Eleutheranthera Poit. (Compos.).
Kegeliella Mansf. Orchidaceae. 2 C. & trop. S. Am., W.I.
Keiri Fabr. = Cheiranthus L. (Crucif.).
Keiria Bowdich. Quid? 1 W. Afr.
Keiskea Miq. Labiatae. 4 E. As.
Keithia Benth. Labiatae. 4 trop. S. Am.
Keithia Spreng. ? Leguminosae. 1 Brazil.
Keitia Regel = Eleutherine Herb. (Iridac.).
Kellaua A. DC. = Euclea Murr. (Ebenac.).
Kelleria Endl. = Drapetes Banks (Thymelaeac.).
Kelleronia Schinz. Zygophyllaceae. 9 Somaliland.
Kellettia Seem. = Prockia P.Br. ex L. (Flacourtiac.).
Kelloggia Torr. ex Benth. & Hook. f. Rubiaceae. 1 SW. U.S.
Kelseya Rydb. (~ Eriogynia Hook.). Rosaceae. 1 W. U.S.
Kemelia Rafin. = Camellia L. (Theac.).
Kemoxis Rafin. = Cissus L. (Vitidac.).
Kempfera Adans. = Tamonea Aubl. (Verbenac.).
Kemulariella Tamamsch. Compositae. 6 Caucasus.
Kendrickia Hook. f. Melastomataceae. 1 S. India, Ceylon.
Kengia Packer = Cleistogenes Keng (Gramin.).
Keniochloa Melderis = Colpodium Trin. (Gramin.).
***Kennedia** Vent. Leguminosae. 15 Austr. The fls. of some are almost black.
Kennedya auctt. = praec.
Kennedynella Steud. = Leptocyamus Benth. = Glycine auctt. (Legumin.).
Kenopleurum Candargy [Kantartzis]. Umbelliferae. 1 Aegean (Lesbos).
Kensitia Fedde. Aïzoaceae. 1 S. Afr.
Kentia Adans. = Trigonella L. (Legumin.).
Kentia Blume (1830) = Schnittspahnia Reichb. = Polyalthia Bl. (Annonac.).
Kentia Blume (1838) = Gronophyllum Scheff. (Palm.).
Kentia Steud. = Fagraea Thunb. (Potaliac.).
Kentiopsis Brongn. Palmae. 1 New Caled.

Kentranthus Neck. = Centranthus DC. (Valerianac.).
Kentrochrosia Lauterb. & K. Schum. = Kopsia Bl. (Apocynac.).
Kentrophyllum Neck. ex DC. = Carthamus L. (Compos.).
Kentrophyta Nutt. = Astragalus L. (Legumin.).
Kentropsis Moq. = Sclerolaena R.Br. (Chenopodiac.).
Kentrosiphon N. E. Brown. Iridaceae. 5 S. Afr.
Kentrosphaera Volkens (1897; non Borzì 1883—Algae) = Volkensinia Schinz (Amaranthac.).
Kentrothamnus Suesseng. & Overkott. Rhamnaceae. 2 Bolivia.
Kepa Rafin. = Cepa Mill. = Allium L. (Alliac.).
Keppleria Mart. ex Endl. = Bentinckia Berry (Palm.).
Keppleria Meissn. = Oncosperma Blume (Palm.).
Keracia Calest. Umbelliferae. 1 W. Medit.
Keramanthus Hook. f. = Adenia Forsk. (Passiflorac.).
Keramocarpus Fenzl = Coriandrum L. (Umbellif.).
Kerandrenia Steud. = Keraudrenia J. Gay (Sterculiac.).
Keranthus Lour. ex Endl. = Dendrobium Sw. (Orchidac.).
Keraselma Neck. = Euphorbia L. (Euphorbiac.).
Keraskomion Rafin. = Cicuta L. (Umbellif.).
Keratephorus Hassk. = Payena A. DC. (Sapotac.).
Keratolepis Rose ex Fröderstr. = Sedum L. (Crassulac.).
Keratophorus C. B. Clarke = Keratephorus Hassk. = Payena A. DC. (Sapotac.).
Keraudrenia J. Gay. Sterculiaceae. 1 Madag., 7 Austr.
Kerbera E. Fourn. Asclepiadaceae. 1 trop. S. Am.
Kerchovea Jorissenne = Stromanthe Sond. (Marantac.).
Keria Spreng. = Kerria DC. (Rosac.).
Keringa Rafin. = Vernonia Schreb. (Compos.).
Kerinozoma Steud. = Xerochloa R.Br. (Gramin.).
Kermadecia Brongn. & Gris. Proteaceae. 10 NE. Austr., New Caled., Fiji.
Kermula Nor. = Psychotria L. (Rubiac.).
*Kernera Medik. (~ Cochlearia L.). Cruciferae. 2–3 mts. of C. & S. Eur.
Kernera Schrank = Tozzia L. (Scrophulariac.).
Kernera Willd. = Posidonia Koen. (Posidoniac.).
Kerneria Moench = Bidens L. (Compos.).
Kerrdora Gagnep. = Enkleia Griff. (Thymelaeac.).
Kerria DC. Rosaceae. 1 E. As., *K. japonica* DC.
Kerriochloa C. E. Hubbard. Gramineae. 1 Siam.
Kerstania K. H. Rechinger = Astragalus L. (Legumin.).
Kerstingia K. Schum. = Belonophora Hook. f. (Rubiac.).
Kerstingiella Harms. Leguminosae. 1 trop. W. Afr. Geocarpic.
Keteleeria Carr. Pinaceae. 4–8 E. As., Indoch.
Kethosia Rafin. = Hewittia Wight & Arn. (Convolvulac.).
Ketmia Mill. = Hibiscus L. (Malvac.).
Kettmia Medik. = praec.
Ketumbulia Ehrenb. ex v. Poelln. = Talinum Adans. (Portulacac.).
Keulia Molina = Gomortega Ruiz & Pav. (Gomortegac.).
Keura Forsk. = Pandanus L. f. (Pandanac.).
Keurva Endl. = praec.

Keyserlingia Bunge ex Boiss. Leguminosae. 2 Persia, Afghan.

Keysseria Lauterb. Compositae. 6 New Guinea.

Khadia N. E. Brown. Aïzoaceae. 6 S. Afr. (Transvaal).

Khaya A. Juss. Meliaceae. 8 trop. Afr., Madag.

Khayea Planch. & Triana = Kayea Wall. (Guttif.).

Khytiglossa Nees (sphalm.) = Rhytiglossa Nees = Dianthera L. (Acanthac.).

Kiapasia Woronow ex Grossh. = Astragalus L. (Legumin.).

Kibara Endl. Monimiaceae. 37 Nicobar Is., Malaysia, trop. Austr.

Kibaropsis Vieill. ex Guillaum. (= ?Hedycarya J. R. & G. Forst.). Monimiaceae. 1 New Caled.

Kibatalia G. Don. Apocynaceae. 25 trop. Afr., W. Malaysia.

Kibbessia Walp. = Kibessia DC. (Melastomatac.).

Kibera Adans. = Sisymbrium L. (Crucif.).

Kibessia DC. Melastomataceae. 20 W. Malaysia.

Kickxia Blume = Kibatalia G. Don (Apocynac.).

Kickxia Dum. (~ Linaria L.). Scrophulariaceae. 25 Medit. to W. India.

Kielboul Adans. = Aristida L. (Gramin.).

Kielmeyera Mart. Guttiferae. 20 S. Brazil, char. of campos.

Kielmiera G. Don = praec.

Kierschlegeria Spach = Fuchsia L. (Onagrac.).

Kiersera Dur. & Jacks. (sphalm.) = Kieseria Nees = Bonnetia Mart. & Zucc. (Bonnetiac.).

Kiesera Reinw. ex Bl. = Tephrosia Pers. (Legumin.).

Kieseria Nees = Bonnetia Mart. & Zucc. (Bonnetiac.).

Kieseria Spreng. = Kiesera Reinw. ex Bl. = Tephrosia Pers. (Legumin.).

Kigelia DC. Bignoniaceae. 1 trop. Afr. Ls. alt. The infls. are borne on old wood, hanging down on very long stalks.

Kigelianthe Baill. Bignoniaceae. 3 Madag.

Kigelkeia Rafin. = Kigelia DC. (Bignoniac.).

Kigellaria Endl. = seq.

Kiggelaria L. Flacourtiaceae. 4 trop. & S. Afr. *K. dregeana* Turcz. yields a good timber (Natal mahogany).

Kiggelariaceae Link = Flacourtiaceae–Kiggelariinae Engl.

Kilbera Fourr. = Sisymbrium L. (Crucif.).

Kiliana Sch. Bip. ex Hochst. = seq.

Kiliania Sch. Bip. ex Benth. & Hook. f. = Pulicaria Gaertn. (Compos.).

Killinga Adans. = Athamanta L. (Umbellif.).

Killinga Lestib. = Kyllinga Rottb. (Cyperac.).

Killingia Juss. = praec.

Killyngia Ham. = praec.

Killipia Gleason. Melastomataceae. 2 Colombia.

Killipiella A. C. Smith. Ericaceae. 2 Colombia.

Killipiodendron Kobuski. Theaceae. 1 Colombia.

Kindasia Blume ex Koorders = Turpinia Vent. (Staphyleac.).

Kinepetalum Schlechter. Asclepiadaceae. 1 S. Afr.

Kinetostigma Dammer = Chamaedorea Willd. (Palm.).

Kingdonia Balf. f. & W. W. Smith. Kingdoniaceae. 1 W. & N. China.

Kingdoniaceae (Janchen) A. S. Foster ex Airy Shaw. Dicots. 1/1 China.

KINGDON-WARDIA

Perenn. herb with slender branched scaly rhiz., a sol. l. and fl. arising from each bud. L. long-pet., with membr. suborbic. lamina of 5 major cuneate segs., these variously lobed and dentate, with radiating open dichot. venation. Fl. rel. small on long scape: P 4–7, petaloid; A 11–21 (outer 8–13 stds.), short, with small anth.; G 4–9, free, with 1 ventr. pend. orthotr. ov. per carpel, and short style, persist. and deflexed in fruit. Fr. a small head of oblanc. beaked glabr. ach., beak strongly decurved at base. Only genus: *Kingdonia*. Combines the remarkable dichot. venation of *Circaeaster* with many features of *Ranunculac.* (Foster, *J. Arn. Arb.* **42**: 397–415, 1961.)

Kingdon-Wardia Marquand = Swertia L. (Gentianac.).

Kingella Van Tiegh. = Trithecanthera Van Tiegh. (Loranthac.).

Kingia R.Br. Xanthorrhoeaceae. 1 W. Austr., a char. pl.

Kingiaceae Endl. = Xanthorrhoeaceae Dum.

Kingidium P. F. Hunt. Orchidaceae. 5 India to W. Malaysia.

Kingiella Rolfe (1917; non *Kingella* Van Tiegh. 1895) = Kingidium P. F. Hunt (Orchidac.).

Kinginda Kuntze = Mitrephora Blume (Annonac.).

Kingiodendron Harms. Leguminosae. 1 W. Penins. India, 1 Philipp. Is., 1 Solomon Is., 1 Fiji.

Kingsboroughia Liebm. = Meliosma Blume (Meliosmac.).

Kingstonia S. F. Gray = Saxifraga L. (Saxifragac.).

Kingstonia Hook. f. & Thoms. Annonaceae. 1 Malay Penins., Java.

Kinia Rafin. Liliaceae. 1 'Borneo' (? = Comoro Is.). Quid?

Kinkina Adans. = Cinchona L. (Rubiac.).

Kinostemon Kudo = Teucrium L. (Labiat.).

Kinugasa Tatewaki & Sutô = Paris L. (Trilliac.).

Kiosmina Rafin. = Salvia L. (Labiat.).

Kippistia Miers = Hippocratea L. (Celastrac.).

Kippistia F. Muell. = Minuria DC. (Compos.).

× **Kirchara** hort. Orchidaceae. Gen. hybr. (iii) (Cattleya × Epidendrum × Laelia × Sophronitis).

Kirchnera Opiz = Astragalus L. (Legumin.).

Kirengeshoma Yatabe. Hydrangeaceae. 1 Japan.

Kirengeshomaceae (Engl.) Nak. = Hydrangeaceae–Kirengeshomeae (Engl.).

Kirganelia Juss. = Phyllanthus L. (Euphorbiac.).

Kirilowia Bunge. Chenopodiaceae. 3 C. As., Afghanistan.

Kirkia Oliv. Kirkiaceae. 8 trop. & S. Afr.

Kirkiaceae (Engl.) Takht. Dicots. 1/8 trop. & S. Afr. Trees with alt. imparipinn. exstip. ls. crowded towards tips of branches; lfts. mostly opp. Fls. tetram., ♂ ♀ or polyg., in axill. dichasia. K (4), small; C 4, 'induplicate-imbricate', much exceeding K; A 4, alternipet., with slender fil. and intr. anth.; disk intrastam., fleshy, quadrang.; G 4, partly immersed in disk, each carpel with short erect style (stig. punctif.) and 1 pend. ovule. Fruit dry, oblong, prismatic, 4-angled, splitting longitudinally into 4 linear-obl. dorsally compressed indehisc. 1-seeded carpels with leathery endocarp, attached above to a central carpophore; seeds with thin testa, without endosp. Only genus: *Kirkia*. Perhaps related to *Ptaeroxylaceae*; cf. tetramerous fls., curious aestivation of C, and carpophore.

Kirkianella Allan. Compositae. 1 N.Z.
Kirkophytum (Harms) Allan = Stilbocarpa A. Gray (Araliac.).
Kirschlegera Reichb. = seq.
Kirschlegeria Reichb. = Kierschlegeria Spach = Fuchsia L. (Onagrac.).
Kissenia R.Br. ex Endl., corr. T. Anders. Loasaceae. 2 Arabia, Somalia, SW. Afr.
Kissodendron Seem. Araliaceae. 3–4 New Guinea, Queensl.
Kita A. Chev. (~ Hygrophila R.Br.). Acanthaceae. 2 trop. W. Afr.
Kitaibelia Willd. Malvaceae. 1 SE. Eur. (lower Danube).
Kitchingia Baker. Crassulaceae. 7 Madag.
Kittelia Reichb. = Cyanea Gaudich. (Campanulac.).
Kittelocharis Alef. = Reinwardtia Dum. (Linac.).
Kixia Blume = Kickxia Blume = Kibatalia G. Don (Apocynac.).
Kixia Meissn. = Kickxia Dum. (Scrophulariac.).
Kjellbergia Bremek. Acanthaceae. 1 Celebes.
Kjellbergiodendron Burret. Myrtaceae. 3 Celebes.
Kladnia Schur = Hesperis L. (Crucif.).
Klaineanthus Pierre ex Prain. Euphorbiaceae. 1 trop. W. Afr.
Klaineastrum Pierre ex A. Chev. Melastomataceae. 1 trop. W. Afr.
Klainedoxa Pierre. Ixonanthaceae. 10 trop. Afr.
Klanderia F. Muell. = Prostanthera Labill. (Labiat.).
Klaprothia Kunth. Loasaceae. 1 N. trop. S. Am.
Klasea Cass. = Serratula L. (Compos.).
Klattia Baker. Iridaceae. 2 S. Afr.
Klausea Endl. = Klasea Cass. = Serratula L. (Compos.).
Kleinhofia Giseke = seq.
Kleinhovia L. Sterculiaceae. 1 trop. As.
Kleinia Crantz = Quisqualis L. (Combretac.).
Kleinia Jacq. = Porophyllum Guett. (Compos.).
Kleinia Juss. = Jaumea Pers. (Compos.).
Kleinia Mill. (~ Senecio L.). Compositae. 50 trop. & S. Afr., Arabia.
Kleinodendron L. B. Smith & Downs. Euphorbiaceae. 1 Brazil.
Kleistrocalyx Steud. = Rhynchospora Vahl (Cyperac.).
Klemachloa R. N. Parker. Gramineae. 1 Burma.
Klenzea Sch. Bip. = Athrixia Ker-Gawl. (Compos.).
Klingia Schönland. Amaryllidaceae. 1 SW. Afr. (Namaqualand).
Klonion Rafin. = Eryngium L. (Umbellif.).
Klopstockia Karst. = Ceroxylon Kunth (Palm.).
Klossia Ridley. Rubiaceae. 1 Malay Penins.
Klotzschia Cham. & Schlechtd. Hydrocotylaceae (~ Umbellif.). 3 S. Brazil.
Klotzschiphytum Baill. = Croton L. (Euphorbiac.).
Klugia Schlechtd. = Rhynchoglossum Bl. (Gesneriac.).
Klugiodendron Britton & Killip = Pithecellobium Mart. (Legumin.).
Klukia Andrz. ex DC. = Malcolmia R.Br. (Crucif.).
Kmeria (Pierre) Dandy. Magnoliaceae. 2 S. China, Indoch.
Knafia Opiz = Salix L. (Salicac.).
Knantia Hill (sphalm.) = Knautia L. (Dipsacac.).
Knappia Ferd. Bauer ex Steud. = Rhynchoglossum Blume (Gesneriac.).

KNAPPIA

Knappia Sm. = Mibora Adans. (Gramin.).

Knauthia Fabr. (uninom.) = *Scleranthus* L. sp. (Caryophyllac.).

Knautia L. Dipsacaceae. 50 Eur., Medit. *K. arvensis* (L.) Coult. has a large head of fls.; the C is drawn out upon the outer side (cf. *Compositae*), and this the more the farther from the centre of the head. Honey is secreted by the upper surface of the ovary, and protected from rain by hairs. The sta. are ripe first, while the style with immature stigmas is quite enclosed in the C; later the sta. wither and the style occupies their place. The stigmas of the various fls. on the head ripen nearly together.

Knavel Seguier = Scleranthus L. (Caryophyllac.).

Knawel Fabr. = praec.

Kneiffia Spach = Oenothera L. (Onagrac.).

Knema Lour. Myristicaceae. 37 SE. As., Indomal.

Knesebeckia Klotzsch = Begonia L. (Begoniac.).

Knifa Adans. = Hypericum L. (Guttif.).

Kniffa Vent. = praec.

***Knightia** R.Br. Proteaceae. 3 New Caled., N.Z. *K. excelsa* R.Br. (*rewa*, N.Z.) furnishes a beautiful timber.

***Kniphofia** Moench. Liliaceae. 75 E. & S. Afr., Madag. Bees sometimes force their way into fls. and are unable to return.

Kniphofia Scop. = Terminalia L. (Combretac.).

Knorrea [Moç. & Sessé ex] DC. = Tetragastris Gaertn. (Burserac.).

Knowlesia Hassk. = Tradescantia L. (Commelinac.).

Knowltonia Salisb. Ranunculaceae. 10 S. Afr.

Knoxia P.Br. = Ernodea Sw. (Rubiac.).

Knoxia L. Rubiaceae. 15 Indomal.

Koanophyllon Arruda ex R. M. King & H. Rob., descr. angl. (∼ Eupatorium L.). Compositae. 21 Mex. to Parag.

Kobiosis Rafin. = Euphorbia L. (Euphorbiac.).

Kobresia Willd. Cyperaceae. 50 N. temp.

Kobresiaceae Gilly = Cyperaceae–Caricoïdeae Pax. (Fls. ♂ ♀, mostly achlamyd.; ♀ fls. encl. in 'utricle', sol. or forming the term. fl. in androgyn. rhipidium; ♂ fls. in many-fld. false spikes, or sol. Genera: *Carex, Kobresia, Uncinia, Schoenoxiphium.*)

Kobria St-Lag. = Kobresia Willd. (Cyperac.).

Kobus Nieuwl. = Magnolia L. (Magnoliac.).

Kochia Roth. Chenopodiaceae. 90 C. Eur., temp. As., N. & S. Afr., Austr.

Kochiophyton Schlechter ex Cogn. = Aganisia Lindley (Orchidac.).

Koddampuli Adans. = Garcinia L. (Guttif.).

Kodda-Pail Adans. = Pistia L. (Arac.).

Kodda-Pana Adans. = Codda-Pana Adans. = Corypha L. (Palm.).

Koeberlinia Zucc. Capparidaceae. 1 S. U.S., Mex. A leafless xerophyte with thorny twigs. The reported occurrence of resin canals in the secondary cortex appears to require confirmation.

***Koeberliniaceae** Engl. = Capparidaceae–Koeberliniëae (Engl.) Pax & Hoffm.

Koechlea Endl. (=? Staehelina L.). Compositae. 1 As. Min. (Cilicia).

Koehleria Benth. & Hook. f. = Kohleria Regel = Isoloma Decne (Gesneriac.).

Koehnea F. Muell. = Nesaea Comm. ex Juss. (Lythrac.).

Koehneago Kuntze = Euosmia Humb. & Bonpl. = Hoffmannia Sw. (Rubiac.).
Koehneola Urb. Compositae. 1 Cuba.
Koeiea K. H. Rechinger = Prionotrichon Botsch. & Vved. (Crucif.).
Koelera St-Lag. = Koeleria Pers. (Gramin.).
Koelera Willd. = Xylosma Forst. (Flacourtiac.).
Koeleria Pers. Gramineae. 60 macrospp. (100 microspp.), N. & S. temp.
Koellea Biria = Eranthis Salisb. (Ranunculac.).
Koellensteinia Reichb. f. Orchidaceae. 6 trop. Am.
Koellia Moench = Pycnanthemum Michx (Labiat.).
Koellikeria Regel. Gesneriaceae. 3 Venez., Boliv.
× **Koellikohleria** Wiehler. Gesneriaceae. Gen. hybr. (Koellikeria × Kohleria).
Koeloeria Parl. = Koeleria Pers. (Gramin.).
Koelpinia Pall. Compositae. 5 N. Afr. to E. As.
Koelpinia Scop. = Acronychia J. R. & G. Forst. (Rutac.).
Koelreutera Murr. = Gisekia L. (Aïzoac.).
Koelreutera Schreb. = seq.
Koelreuteria Laxm. Sapindaceae. 8 China, Formosa, Fiji. The capsule is large and bladdery and may be blown about by wind (cf. *Colutea*).
Koelreuteria Medik. = Marsdenia R.Br. (Apocynac.).
Koelreuteriaceae J. G. Agardh = Sapindaceae–Koelreuterieae Radlk.
Koelzella Hiroe. Umbelliferae. 1 Afghan., W. Himal.
Koelzia K. H. Rechinger = Christolea Cambess. ex Jacquem. (Crucif.).
Koeniga Benth. & Hook. f. = Konig Adans. = Lobularia Desv. (Crucif.).
Koenigia Comm. ex Cav. = Ruizia Cav. (Sterculiac.).
Koenigia Comm. ex Juss. = Dombeya Cav. (Malvac.).
Koenigia L. Polygonaceae. 7 Arctic, Himal., temp. E. As.; 1 temp. S. Am. (Fuegia).
Koenigia P. & K. = Konig Adans. = Lobularia Desv. (Crucif.).
Koerinckia B. D. Jacks. (sphalm.) = Koernickea Regel = Achimenes P.Br. (Gesneriac.).
Koernickea Klotzsch = Paullinia L. (Sapindac.).
Koernickea Regel = Achimenes P.Br. (Gesneriac.).
Kohautia Cham. & Schlechtd. = Oldenlandia L. (Rubiac.).
Kohleria Regel. Gesneriaceae. 50 trop. Am. Several spp. form runners above ground, thickly covered with scaly ls.
Kohlerianthus Fritsch. Gesneriaceae. 1 Bolivia.
Kohlrauschia Kunth (~ Petrorhagia (Ser.) Link). Caryophyllaceae. 4 Eur., Medit., W. As.
Koilodepas Hassk. Euphorbiaceae. 10 S. India, SE. As., Hainan, W. Malaysia (exc. Philipp. Is.), New Guinea.
Kokabus Rafin. = Acnistus Schott (Solanac.).
Kokera Adans. = Chamissoa Kunth (Amaranthac.).
Kokia Lewton. Malvaceae. 5 Hawaii.
Kokkia Zipp. ex Blume = Odina Roxb. = Lannea A. Rich. (Anacardiac.).
Kokonoria Keng & Keng = Lagotis J. Gaertn. (Scrophulariac.).
Kokoona Thw. Celastraceae. 5 Ceylon, Burma, W. Malaysia.
Kokoschkinia Turcz. = Tecoma Juss. (Bignoniac.).
Kolbea Reichb. = Kolbia Beauv. = Modecca Lam. (Passiflorac.).

KOLBEA

Kolbea Schlechtd. = Baeometra Salisb. (Liliac.).
Kolbia Adans. = Blaeria L. (Ericac.).
Kolbia Beauv. = Modecca Lam. (Passiflorac.).
Kolerma Rafin. = Carex L. (Cyperac.).
Kolkwitzia Graebn. Caprifoliaceae. 1 China.
Kolleria Presl = Galenia L. (Aïzoac.).
Kolobochilus Lindau. Acanthaceae. 2 C. Am. (Costa Rica).
Kolobopetalum Engl. Menispermaceae. 9 trop. Afr.
Kolofonia Rafin. = Ipomoea L. (Convolvulac.).
Kolomikta Dippel = Kalomikta Regel = Actinidia Lindl. (Actinidiac.).
Kolooratia Lestib. = seq.
Kolowratia Presl = Alpinia Roxb. (Zingiberac.).
Kolpakowskia Regel (~ Ixiolorion Herb.). Amaryllidaceae. 2 C. As.
Kolrauschia Jord. = Kohlrauschia Kunth (Caryophyllac.).
Komana Adans. = Hypericum L. (Guttif.).
Komaroffia Kuntze. Ranunculaceae. 1 C. As.
Komarovia Korovin. Umbelliferae. 1 C. As.
Kommia Ehrenb. ex Schweinf. = Pupalia Juss. (Amaranthac.).
Konig Adans. = Lobularia Desv. (Crucif.).
Koniga R.Br. = praec.
Konigia Comm. ex Cav. = Dombeya Cav. (Malvac.).
Konxikas Rafin. = Lathyrus L. (Legumin.).
Koockia Moq. = Kochia Roth (Chenopodiac.).
Kookia Pers. = Cookia Sonner. = Clausena Burm. f. (Rutac.).
Koompassia Maingay. Leguminosae. 4 Malay Penins., Borneo, New Guinea.
Koon [Gaertn.] Miers = Schleichera Willd. (Sapindac.).
Koordersina Kuntze = Koordersiodendron Engl. (Anacardiac.).
Koordersiochloa Merr. = Streblochaete Hochst. (Gramin.).
Koordersiodendron Engl. Anacardiaceae. 1 Philipp., Celebes, New Guinea.
*Kopsia Blume. Apocynaceae. 25 SE. As., W. Malaysia, Caroline Is.
Kopsia Dum. = Orobanche L. (Orobanchac.).
Kopsiopsis (G. Beck) G. Beck. Orobanchaceae. 2 W. N. Am.
Kordelestris Arruda = Jacaranda Juss. (Bignoniac.).
Kornickia Benth. & Hook. f. = Koernickea Regel = Achimenes P.Br. (Gesneriac.).
Korolkowia Regel = Fritillaria L. (Liliac.).
Korosvel Adans. = Tetracera L. (Dilleniac.).
Korovinia Nevski & Vved. Umbelliferae. 3 C. As.
Korsaria Steud. = Kosaria Forsk. = Dorstenia L. (Morac.).
Korshinskia Lipsky. Umbelliferae. 2 C. As.
Korthalsella Van Tiegh. Viscaceae (~ Loranthaceae). 45 N. trop. Afr., Madag., Masc., Himal., Jap., Indoch., Malaysia, Austr., N.Z., Pacif., W.I. (Cuba).
Korthalsia Blume. Palmae. 35 Indomal. Some, e.g. *K. horrida* Becc., are said to be myrmecophilous (cf. *Cecropia*), the ants living in the sheaths of the leaves.
Korycarpus Zea = Diarrhena Rafin. (Gramin.).
Kosaria Forsk. = Dorstenia L. (Morac.).
Kosmosiphon Lindau. Acanthaceae. 1 trop. W. Afr.

Kosopoljanskia Korovin. Umbelliferae. 1 C. As.
Kosteletskya Brongn. = seq.
Kosteletzkya C. Presl (= Thorntonia Reichb.). Malvaceae. 30 N. Am., Mex., trop. &. S. Afr., Madag.
Kostermansia Soeg. Reksod. Bombacaceae. 1 Malay Penins.
Kostermanthus Prance. Chrysobalanaceae. 2 W. Malaysia.
Kostyczewa Korshinsky = Chesneya Lindl. (Legumin.).
Kotchubaea Fisch. corr. Regel ex Benth. & Hook. f. Rubiaceae. 10 trop.
Kotschya Endl. Leguminosae. 30 trop. Afr.   [S. Am.
Kotsjiletti Adans. = Xyris L. (Xyridac.).
Kowalewskia Turcz. = Clethra L. (Clethrac.).
Kozlovia Lipsky. Umbelliferae. 1 C. As.
Kozola Rafin. = Hexonix Rafin. (Liliac.).
× **Kraenzlinara** hort. (v) = × Trichovanda hort. (Orchidac.).
Kraenzlinella Kuntze. Orchidaceae. 5 trop. Am.
Krainzia Backeb. = Mammillaria Haw. (Cactac.).
Kralikella Coss. & Durieu = Tripogon Roth (Gramin.).
Kralikia Coss. & Dur. = praec.
Kralikia Sch. Bip. = Chiliocephalum Benth. (Compos.).
Kralikiella Batt. & Trab. = Kralikella Coss. & Dur. = Tripogon Roth (Gramin.).
Kramera P. & K. = seq.
Krameria L. ex Loefl. Krameriaceae. 25 S. U.S. to Chile.
***Krameriaceae** Dum. Dicots. 1/25 warm Am. Shrubs or perenn. herbs, mostly pubesc. or seric., with simple (? 1-foliolate) or rarely 3-foliolate, ent., alt., exstip. ls. Fls. ⚥, zygo., axill. or in term. rac., usu. with 2 opp. foliac. bracts. K 4–5, imbr., unequal; C 5, very unequal, the 3 upper long-clawed, sometimes partly connate below, the 2 lower much smaller, broad, thick, sess.; A 3–4, sometimes adnate to claws of upper petals, anth. dehisc. by a pore; G 1 with 2 collat. pend. anatr. ov., and simple style with discoid stig. Fr. glob., indehisc., 1-seeded, covered with bristles or spines (often barbed); seed without endosp. Only genus: *Krameria*. Related to *Polygalaceae* and *Leguminosae*, differing from the former *inter alia* in having a pet., not a sep., posterior, and in the probably basically compound ls. (in *K. cytisoïdes* Cav. they are 3-foliolate); from the latter in the absence of stips., and in anatomical features.
Kranikofa Rafin. = Krascheninnikovia Gueldenst. = Eurotia Adans. emend. C. A. Mey. (Chenopodiac.).
Kranikovia Rafin. = praec.
Krapfia DC. (∼ Ranunculus L.). Ranunculaceae. 8 Andes.
Kraschenikofia Rafin. = Krascheninnikovia Gueldenst. = Eurotia Adans. emend. C. A. Mey. (Chenopodiac.).
Krascheninikofia Rafin. = praec.
Krascheninnikovia Turcz. ex Endl. = Pseudostellaria Pax (Caryophyllac.).
Krascheninnikovia Gueldenst. = Eurotia Adans. emend. C. A. Mey. (Chenopodiac.).
Kraschennikofia Rafin. = praec.
Kraschnikowia Turcz. ex Ledeb. = Marrubium L. (Labiat.).
Krasnikovia Rafin. = Krascheninnikovia Gueldenst. = Eurotia Adans. emend. C. A. Mey. (Chenopodiac.).

KRASNOVIA

Krasnovia Popov ex Schischk. Umbelliferae. 1 C. As.
Krassera Schwartz. Melastomataceae. 2 Borneo.
Kratzmannia Opiz = Agropyron J. Gaertn. (Gramin.).
Krauhnia Steud. = seq.
Kraunhia Rafin. = Wisteria Nutt. (Legumin.).
Kraunshia Rafin. = praec.
Krausella H. J. Lam. Sapotaceae. 4 New Guinea.
Krauseola Pax & K. Hoffm. Caryophyllaceae. 2 trop. E. Afr.
Kraussia Harv. Rubiaceae. 4 trop. & S. Afr.
Kraussia Sch. Bip. = Amellus L. (Compos.).
Krebsia Eckl. & Zeyh. = Lotononis Eckl. & Zeyh. (Legumin.).
Krebsia Harv. = Stenostelma Schlechter (Asclepiadac.).
Kreidek Adans. = Scoparia L. (Scrophulariac.).
Kreidion Rafin. = Conioselinum Hoffm. (Umbellif.).
Kremeria Coss. & Dur. = Kremeriella Maire (Crucif.).
Kremeria Dur. Compositae. 4 NW. Afr.
Kremeriella Maire. Cruciferae. 1 NW. Afr.
Kreysigia Reichb. Liliaceae. 1 SE. Austr.
***Krigia** Schreb. Compositae. 8 N. Am.
Krockeria Neck. = Xylopia L. (Annonac.).
Krockeria Steud. = Krokeria Moench = Lotus L. (Legumin.).
Krokeria Endl. = Krockeria Neck. = Xylopia L. (Annonac.).
Krokeria Moench = Lotus L. (Legumin.).
Krokia Urb. Myrtaceae. 3 Cuba.
Krombholtzia Benth. = seq.
Krombholzia Rupr. ex Fourn. = Zeugites Schreb. (Gramin.).
Kromon Rafin. = Allium L. (Alliac.).
Krubera Hoffm. = Capnophyllum Gaertn. (Umbellif.).
Kruegeria Scop. = Macrolobium Schreb. (Legumin.).
Krugella Pierre = Pouteria Aubl. (Sapotac.).
Krugia Urb. Myrtaceae. 1 W.I.
Krugiodendron Urb. Rhamnaceae. 1 W.I.
Kruhsea Regel = Streptopus Michx (Liliac.).
Krukoviella A. C. Smith. Ochnaceae. 1 Peru, Brazil.
Krylovia Schischk. Compositae. 4 C. As. to E. Siberia.
Krynitzia Reichb. = seq.
Krynitzkia Fisch. & Mey. = Cryptantha Lehm. (Boraginac.).
Ktenosachne Steud. = Prionachne Nees (Gramin.).
Ktenospermum Lehm. = Pectocarya DC. (Boraginac.).
Kua Medik. = Curcuma L. (Zingiberac.).
Kuala Karst. & Triana = Esenbeckia Kunth (Rutac.).
Kudoa Masamune = Gentiana L. (Gentianac.).
Kudoacanthus Hosokawa. Acanthaceae. 1 Formosa.
Kudrjaschevia Pojark. Labiatae. 4 C. As.
Kuekenthalia Börner = Carex L. (Cyperac.).
Kuenckelia Heim = Kunckelia Heim = Vateria L. (Dipterocarpac.).
Kuenstlera K. Schum. = Kunstleria Prain (Legumin.).
Kuestera Regel = Beloperone Nees (Acanthac.).

Kugia Lindl. (sphalm.) = Krigia Bertero ex Colla = Bellardia Colla = Microseris D. Don (Compos.).

Kuhitangia Ovczinn. Caryophyllaceae. 1 C. As.

Kuhlhasseltia J. J. Smith. Orchidaceae. 5 Malaysia.

Kuhlia Kunth = Banara Aubl. (Flacourtiac.).

Kuhlia Reinw. = Fagraea Thunb. (Potaliac.).

Kuhlmannia J. C. Gomes. Bignoniaceae. 1 Brazil.

Kuhlmanniella L. Barroso. Convolvulaceae. 4 trop. S. Am.

Kuhnia L. = Brickellia Ell. (Compos.).

Kuhnia Walt. = seq.

Kuhniastera Kuntze = Kuhnistera Lam. = Petalostemum Michx (Legumin.).

Kuhniodes (A. Gray) Kuntze = Bebbia Greene (Compos.).

Kuhnistera Lam. = Petalostemum Michx (Legumin.).

Kuhnistra Endl. = praec.

Kukolis Rafin. = Hebecladus Miers (Solanac.).

Kulmia Augier = Kalmia L. (Ericac.).

Kumara Medik. = Aloë (Liliac.).

Kumaria Rafin. = Haworthia Duval (Liliac.).

Kumbaya Endl. ex Steud. = Gardenia Ellis (Rubiac.).

Kumlienia Greene = Ranunculus L. (Ranunculac.).

Kummeria Mart. = Discophora Miers (Icacinac.).

Kummerowia Schindler. Leguminosae. 2 Japan.

Kunckelia Heim = Vateria L. (Dipterocarpac.).

Kunda Rafin. = Amorphophallus Blume (Arac.).

*****Kundmannia** Scop. Umbelliferae. 1 S. Eur., Medit.

Kunhardtia Maguire. Rapateaceae. 1 Venez.

Kuniria Rafin. = Dicliptera Juss. (Acanthac.).

Kunkeliella W. T. Stearn. Santalaceae. 2 Canary Is.

Kunokale Rafin. = Fagopyrum Moench (Polygonac.).

Kunstlera [King ex] Gage = Chondrostylis Boerl. (Euphorbiac.).

Kunstleria Prain. Leguminosae. 9 Malay Penins., Borneo, Philipp. Is., 2 Austr.

Kunstlerodendron Ridley = Chondrostylis Boerl. (Euphorbiac.).

Kunthia Dennst. = Garuga Roxb. (Burserac.).

Kunthia Humb. & Bonpl. = Morenia Ruiz & Pav. (Palm.).

Kuntia Dum. = praec.

Kunzea Reichb. Myrtaceae. 30 Austr.

Kunzia Spreng. = Purshia DC. (Rosac.).

Kunzmannia Klotzsch & Schomb. (nomen). Quid? (Rutac. vel Ochnac.?) 1 Guiana.

Kurites Rafin. = Selago L. (Scrophulariac.).

Kuritis B. D. Jacks. = praec.

Kurkas Adans. = Curcas Adans. = Jatropha L. (Euphorbiac.).

Kurkas Rafin. = Croton L. (Euphorbiac.).

Kuromatea Kudo = Pasania (Miq.) Oerst. = Lithocarpus Bl. (Fagac.).

Kurria Hochst. & Steud. = Hymenodictyon Wall. (Rubiac.).

Kurria Steud. = Hermannia L. (Sterculiac.).

Kurrimia Wall. ex Meissn. = Itea L. (Iteac.).

Kurrimia Wall. ex Thwaites = Bhesa Buch.-Ham. ex Arn. (Celastrac.).

Kurtzamra Kuntze. Labiatae. 1 temp. S. Am.
Kurzamra Kuntze = praec.
Kurzia King ex Hook. f. = Hullettia King (Morac.).
Kurzinda Kuntze = Ventilago Gaertn. (Rhamnac.).
Kurziodendron Balakrishnan = Trigonostemon Bl. (Euphorbiac.).
Kuschakewiczia Regel & Smirn. = Solenanthus Ledeb. (Boraginac.).
Kustera Benth. & Hook. f. = Kuestera Regel = Beloperone Nees (Acanthac.).
Kutchubaea Fisch. ex DC. = Kotchubaea Fisch. ex DC. corr. Regel (Rubiac.).
Kyberia Neck. = Bellis L. (Compos.).
Kydia Roxb. Malvaceae. 3 E. Himal. to SE. As.
Kylinga Roem. & Schult. = seq.
Kylingia Stokes = seq.   [aromatic roots.
*****Kyllinga** Rottb. Cyperaceae. 60 trop. & subtrop., esp. Afr. Some have
Kyllingia P. & K. = Killinga Adans. = Athamanta L. (Umbellif.).
Kymapleura Nutt. = Troximon Nutt. (Compos.).
Kyphadenia Sch. Bip. ex O. Hoffm. = Chrysactinia A. Gray (Compos.).
Kyphocarpa (Fenzl) Schinz = Cyphocarpa (Fenzl emend.) Lopriore (Amar-
Kyrstenia Neck. = Eupatorium L. (Compos.).   [anthac.).
Kyrsteniopsis R. M. King & H. Rob. (~ Eupatorium L.). Compositae. 1 Mex.
Kyrtandra J. F. Gmel. = Cyrtandra J. R. & G. Forst. (Gesneriac.).
Kyrtanthus J. F. Gmel. = Posoqueria Aubl. (Rubiac.).

L

Labatia Scop. = Ilex L. (Aquifoliac.).
*****Labatia** Sw. = Pouteria Aubl. (Sapotac.).
Labiaceae Dulac = seq.
Labiataceae Boerl. = seq.
*****Labiatae** Juss. (nom. altern. **Lamiaceae** Lindl.). Dicots. 180/3500 cosmop.;
chief centre the Medit. region. Some small groups are localised in their distri-
bution, e.g. § II in Austr. and Tasmania, III in India, Malaysia, China, etc.,
VIII in Centr. Am., whereas the large ones, such as I and IV, are cosmop.
Most *L.* are land plants, and herbs or undershrubs, similar in habit and
structure. Stem usu. square, with decussate simple exstip. ls., often hairy and
with epidermal glands secreting volatile oils, which give char. scents to many.
A few marsh plants (*Mentha, Lycopus*, etc.), a few climbers (*Stenogyne* spp.,
Scutellaria, etc.), and a few small trees (*Hyptis* spp.). Very few truly rain-forest
genera (e.g. *Gomphostemma*). Many xero. with reduced, sometimes infolded,
ls., hairiness, thick cuticles, etc., e.g. *Rosmarinus*.

The axis of the first order is not closed by a fl. but only those of later orders;
thus the primary form of the infl. is racemose, and a simple raceme actually
occurs in *Scutellaria*, etc. Usu., however, a dichasial cyme, becoming cincinnal
in its later branchings, occurs in the axil of each l. upon the upper part of the
main axis. In *Teucrium, Nepeta* spp., etc., the construction of this cyme is easily
seen; in most *L.* however it is closely 'condensed' into the axil, so that all the
fls. are sessile; but it is easily seen that the central fl. opens first and then those
on either side of it. The two condensed cymes at each node overlap the leaf-

axils and often form what looks like a whorl of fls.; this infl. is often called a *verticillaster* or false whorl.

Fl. ♀ or gynodioec., zygo., hypog., 5-merous with suppression in some whorls. Usu. formula K (5), C (5), A 4, G̲ (2). K tubular, bell- or funnel-shaped, sometimes 2-lipped, persistent in fr.; C usu. 2-lipped with no clear indication of the individual petals; A 4, didynamous, or of nearly equal length, sometimes 2, epipet., with intr. anthers. G on a nectariferous disk (often developed on anterior side only), of (2) cpls. placed antero-post. Early in development a constriction appears in the ovary in the antero-post. line, dividing each cpl. into 2 loculi, so that the ovary becomes 4-loc. as it matures. Each of the 4 portions is nearly independent of the rest, and the style springs between them from the base of the ovary (i.e., is *gynobasic*); stigma 2-lobed. Placentae axile, each with 1 basal erect anatr. ovule with ventral raphe. Fr. usu. a group of 4 achenes or *nutlets*, each containing one seed; sometimes a drupe. Seed with no endosp. or very little; the radicle of the embryo points downwards (cf. *Boraginaceae*).

The 2-lipped C ensures that a visiting insect shall take a definite position in regard to the anthers and stigma whilst probing for the honey at the base of the fl. The lower lip acts as a flag to attract, and also as a landing-place, whilst the upper lip shelters the essential organs, which are usu. placed so as to touch the insect's back. The length of the C-tube varies very much, and with it the kind of visitors. Most N. temp. spp. are bee fls., the long-tubed red fls. of *Monarda*, etc., are butterfly fls., and a few spp. of *Salvia*, etc., are humming-bird fls. The pollination mech. is usu. simple; in *Lamium*, etc., the fl. is homogamous, the stigma merely projecting beyond the anthers so as to be touched first, but usu. the fl. is dichogamous (protandr.), often with movements of the essential organs, e.g. in *Teucrium*, etc. The lever mechanism of *Salvia* is almost unique. *Thymus, Origanum*, and their allies have nearly regular fls. visited by a more miscellaneous selection of insects. In many *L.*, esp. § VI, interesting distrs. of sex appear, esp. gynodioecism.

A few disperse their fr. by aid of the persistent bladdery K, or by hooks formed from the K teeth. The stalks are often hygroscopic and move in such a way as to favour dispersal in wet weather.

Useful on account of their volatile oils; many, e.g. *Thymus, Ocimum, Origanum, Salvia*, etc., used as condiments. Oils and perfumes are obtained by distillation from *Rosmarinus, Pogostemon, Lavandula*, etc. Food products from *Stachys* spp.

Closely allied to *Verbenaceae*; from *Boraginaceae* the position of the radicle sharply separates them, whilst the similarity to *Scrophulariaceae*, etc., is largely in minor chars.

Classification and chief genera (after Briquet, from whose account much of the above is condensed):

A. Style not gynobasic. Nutlets with lateral-ventral attachment and usu. large surface of contact (often > ½ as high as ovary).
    I. *Ajugoïdeae* (seed exalb.).
        1. AJUGEAE (corolla various; upper lip if present rarely concave; sta. 4 or 2; anther 2-loc.; nutlets ± wrinkled): *Ajuga, Teucrium*.

LABIATIFLORAE

2. ROSMARINEAE (corolla strongly 2-lipped; upper lip very concave and arched; sta. 2; anthers 1-loc.; nutlets smooth): *Rosmarinus* (only genus).

II. **Prostantheroïdeae** (seed albuminous): *Prostanthera*.

B. Style perfectly gynobasic. Nutlets with basal attachment and usu. small surface of contact, rarely with ± basal-dorsal attachment.

III. **Prasioïdeae** (nutlet drupaceous with fleshy or very thick exocarp and hard endocarp): *Stenogyne, Gomphostemma.*

IV. **Scutellarioïdeae** (nutlet dry; seed ± transversal; embryo with curved radicle lying on one cot.): *Scutellaria.*

V. **Lavanduloïdeae** (nutlet dry; seed erect; embryo with short straight superior radicle; disk-lobes opp. to ovary-lobes; nutlets with ± distinct dorsal-basal attachment; sta. 4 included; anthers 1-loc. at tip through union of thecae): *Lavandula* (only genus).

VI. **Lamioïdeae** (*Stachydoïdeae*) (ditto, but disk-lobes, when distinct, alt. with ovary-lobes; nut with small basal attachment; sta. ascending or spreading and projecting straight forwards): *Marrubium, Nepeta, Dracocephalum, Prunella, Phlomis, Galeopsis, Lamium, Ballota, Stachys, Salvia, Monarda, Ziziphora, Horminum, Calamintha, Satureia, Origanum, Thymus, Mentha, Pogostemon.*

VII. **Ocimoïdeae** (as VI, but sta. descending, lying upon under lip or enclosed by it): *Hyptis, Ocimum.*

VIII. **Catopherioïdeae** (nutlet dry; seed erect; embryo with curved radicle lying against the cotyledons): *Catopheria.*

Labiatiflorae DC. (subfam.) = Compositae–Mutisieae Cass.

Labichea Gaudich. ex DC. Leguminosae. 8 Austr.

Labillardiera Roem. & Schult. = Billardiera Sm. (Pittosporac.).

***Labisia** Lindl. Myrsinaceae. 9 Malaysia. Subherbaceous; stem creeping.

Lablab Adans. Leguminosae. 1 trop. Afr., *L. niger* Medik., largely cult. in trop. for ed. pods.

Lablavia D. Don = praec.

Labordea Benth. = seq.

Labordia Gaudich. Loganiaceae. 25 Hawaii.

Laboucheria F. Muell. = Erythrophleum Afzel. (Legumin.).

Labourdonnaisia Boj. Sapotaceae. 4 Natal, Madag., Mauritius. Hardwood.

Labradia Swediaur = Mucuna Adans. (Legumin.).

Labramia A. DC. = Mimusops L. (Sapotac.).

+**Laburnocytisus** C. K. Schneid. Leguminosae. Gen. hybr. asex. (*Cytisus* L. +*Laburnum* Fabr.). +*L. adami* (Poit.) C. K. Schneid. is a curious graft-hybrid between *Cytisus purpureus* Scop. and *Laburnum anagyroïdes* Medik. The latter was used as the stock; the shoots and flowers above the graft exhibit hybrid characters.

Laburnum Medik. Leguminosae. *S.str.*, 2 C. Eur.; *s.l.*, 4 Eur., N. Afr., W. As. *L. anagyroïdes* Medik. is the common laburnum. The fl. has a simple *Trifolium* mech. There is no free honey; bees pierce the swelling at the base of the vexillum (cf. *Orchis*). All parts are poisonous.

Lacaena Lindl. Orchidaceae. 2 Mex., C. Am.

Lacaitaea Brand. Boraginaceae. 1 E. Himal.

Lacanthis Rafin. = Euphorbia L. (Euphorbiac.).

Lacara Rafin. = Campanula L. (Campanulac.).

Lacara Spreng. = Bauhinia L. (Legumin.).

Lacaris Buch.-Ham. ex Pfeiff. = Zanthoxylum L. (Rutac.).

Lacathea Salisb. = Franklinia Bartr. ex Marsh. (Theac.).

Lacaussadia Gaudich. = Egenolfia Schott (Lomariopsidac.).

Laccodiscus Radlk. Sapindaceae. 5 trop. W. Afr.

Laccopetalum Ulbrich (~ Ranunculus L.). Ranunculaceae. 1 Peru. Coarse rigid herb with large yellowish-green fls.

Laccospadix H. Wendl. & Drude (~ Calyptrocalyx Blume). Palmae. 2 New Guinea, Queensl.

Laccosperma (G. Mann & H. Wendl.) Drude (~ Ancistrophyllum Mann & Wendl.). Palmae. 1 W. Afr.

Lacellia Bub. & Penz. = Laserpitium L. (Umbellif.).

Lacellia Viv. = Volutarella Cass. (Compos.).

Lacepedea Kunth = Turpinia Vent. (Staphyleac.).

Lacepedia Kuntze = praec.

Lacerdaea Berg = Campomanesia Ruiz & Pav. (Myrtac.).

Lacerpitium Thunb. = Laserpitium L. (Umbellif.).

Lachanodendron Reinw. ex Blume = Lansium Corrêa (Meliac.).

Lachanodes DC. = Senecio L. (Compos.).

Lachanostachys Endl. = Lachnostachys Hook. (Dicrastylidac.).

Lachemilla Rydb. (~ Alchemilla L.). Rosaceae. 80 Am.

Lachenalia J. F. Jacq. ex J. A. Murr. Liliaceae. 65 S. Afr.

Lachenal[iac]eae Salisb. = Liliaceae–Scilloïdeae K. Krause.

Lachnaea L. Thymelaeaceae. 20 S. Afr.

Lachnagrostis Trin. Gramineae. 5 N.Z., Auckl., Campb., Antipodes Is.

*****Lachnanthes** Ell. Haemodoraceae. 1 N. Am., *L. caroliniana* (Lam.) Wilbur (*L. tinctorum* (J. F. Gmel.) Sprague), the paint-root. The roots yield a red dye.

Lachnastoma Korth. (1851; non Lachnostoma Kunth 1818) = Hymendocarpum Pierre ex Pitard = Nostolachma Dur. (Rubiac.).

Lachnea auctt. = Lachnaea L. (Thymelaeac.).

Lachnia Baill. = praec.

Lachnocapsa Balf. f. Cruciferae. 1 Socotra.

Lachnocaulon Kunth. Eriocaulaceae. 10 SE. U.S., W.I.

Lachnocephalus Turcz. = Mallophora Endl. (Verbenac.).

Lachnochloa Steud. Gramineae. 1 Senegambia. Quid?

Lachnocistus Duchass. ex Linden & Planch. = Cochlospermum Kunth (Cochlospermac.).

Lachnolepis Miq. = Gyrinops Gaertn. (Thymelaeac.).

Lachnoloma Bunge. Cruciferae. 1 C. As.

Lachnopetalum Turcz. = Lepidopetalum Bl. (Sapindac.).

Lachnophyllum Bunge. Compositae. 2–3 W. & C. As.

Lachnopodium Blume = Otanthera Blume (Melastomatac.).

Lachnopylis Hochst. Buddlejaceae. 40 trop. & S. Afr., Madag., Comoro Is.

Lachnorhiza A. Rich. (~ Vernonia Schreb.). Compositae. 2 Cuba.

Lachnosiphonium Hochst. (~ Randia L.). Rubiaceae. 1 trop. Afr.

LACHNOSPERMUM

Lachnospermum Willd. Compositae. 2 S. Afr. Fr. glandular, with pappus.
Lachnostachys Hook. Dicrastylidaceae. 10 W. & S. Austr.
Lachnostoma auctt. = Lachnastoma Korth. = Hymendocarpum Pierre ex Pitard = Nostolachma Dur. (Rubiac.).
Lachnostoma Kunth. Asclepiadaceae. 1 SW. U.S., 4 trop. Am.
Lachnostylis Engl. (sphalm.) = Lachnopylis Hochst. (Buddlejac.).
Lachnostylis Turcz. Euphorbiaceae. 2 S. Afr. Endosp. thin.
Lachnothalamus F. Muell. = Chthonocephalus Steetz (Compos.).
Lachryma-Job Ort. = seq.
Lachrymaria Fabr. (uninom.) = *Coix lachryma* L. (Gramin.).
Laciala Kuntze = Schizoptera Turcz. (Compos.).
Lacimaria B. D. Jacks. (sphalm.) = seq.
Lacinaria Hill = Laciniaria Hill = Liatris Schreb. (Compos.).
Laciniaceae Dulac = Resedaceae S. F. Gray.
Laciniaria Hill = Liatris Schreb. (Compos.).
Lacis Dulac = Trinia Hoffm. (Umbellif.).
Lacis Lindl. = Tulasneantha van Royen (Podostemac.).
Lacis Schreb. = Mourera Aubl. (Podostemac.).
Lacistema Sw. Lacistemataceae. 20 Mex. to Paraguay, W.I.
*****Lacistemataceae** Mart. Dicots. 2/27 trop. Am., W.I. Shrubs with distichous alt. ls. with small deciduous stips. Fls. very small, ♂ or ♂ ♀, in small dense bracteate cylindr. fascic. axill. spikes or rac. K 4–6, narrow, or 0; C 0; axis expanded into a fleshy concave disk; A 1, with 2 separate sometimes stipit. loculi; G (2–3) with parietal plac. and 1–2 pend. anatr. ov. on each. Caps. 3-valved, 1-seeded. Endosp. Genera: *Lacistema, Lozania*. Closely allied to **Lacistemon** P. & K. = Lacistema Sw. (Lacistematac.). [*Flacourtiac.*
Lacistemopsis Kuhlm. = Lozania S. Mutis ex Caldas (Lacistematac.).
Lacmellea Karst. Apocynaceae. 20 trop. S. Am. Fr. ed.
Lacmellia B. D. Jacks. = praec.
Lacostea v. d. Bosch = Trichomanes L. (Hymenophyllac.).
Lacryma Medik. = Lachrymaria Fabr. = Coix L. (Gramin.).
Lactaria Rafin. = Ochrosia Juss. (Apocynac.).
Lactomamillaria Frič = Solisia Britton & Rose (Cactac.).
*****Lactoridaceae** Engl. Dicots. 1/1 Juan Fernandez. Shrub with small simple ent. obov. glabr. minutely gland-dotted alt. distich. ls., with large membr. stips. adnate to pet. Fls. small, reg., polyg.-monoec., axill., 1–3 together. P 3, imbr.; A 3 + 3, with short extrorse anth. and shortly produced connective; G 3, free, shortly beaked, with 6 basal-ventral ov. in each carp. Fr. follicular, 4–6-seeded, endosp. copious, oily. Only genus: *Lactoris*. Affinities much disputed; connection with *Winteraceae* and other *Magnoliales* often suggested, but the habit, stipules and anatomy point rather to some relationship with *Piperaceae* or *Polygonaceae*. Cf. also *Phyllanthus* (*Euphorbiac.*).
Lactoris Phil. Lactoridaceae. 1 Juan Fernandez, *L. fernandeziana* Phil.
Lactuca L. Compositae. 100 chiefly temp. Euras., extending to trop. & S. Afr. *L. sativa* L., the lettuce. *L. scariola* L. (prickly lettuce), a compass pl. (cf. *Silphium*) in dry exposed places, spreading as a weed in the U.S.
Lactucaceae Bessey = Compositae–Cichorieae Reichb.
Lactucopsis Sch. Bip. ex Vis. = Cicerbita Wallr. (Compos.).

LAGEDIUM

Lactucosonchus (Sch. Bip.) Svent. Compositae. 1 Canary Is.
Lacuala Blume = Licuala Wurmb (Palm.).
Lacunaria Ducke. Quiinaceae. 11 trop. S. Am.
Lacuris Buch.-Ham. = Zanthoxylum L. (Rutac.).
Ladaniopsis Gandog. = Cistus L. (Cistac.).
Ladanium Spach = Cistus L. (Cistac.).
Ladanum Gilib. = Galeopsis L. (Labiat.).
Ladanum Rafin. = Cistus L. (Cistac.).
Ladenbergia Klotzsch. Rubiaceae. 30 S. Am. Bark astringent, containing alkaloids.
Ladenbergia Klotzsch ex Moq. corr. Kuntze = Flueckigera Kuntze (Phytolaccac.).
Ladoicea Miq. (sphalm.) = Lodoicea Comm. ex J. St-Hil. (Palm.).
Ladrosia Salisb. ex Planch. = Drosophyllum Link (Droserac.).
Ladyginia Lipsky. Umbelliferae. 1 C. As.
Laea Brongn. = Leea L. (Leeac.).
Laechhardtia Archer ex Gordon (sphalm.) = Leichhardtia Shepherd = Callitris
Laelia Adans. = Bunias L. (Crucif.).   [Vent. (Cupressac.).
Laelia Lindl. Orchidaceae. 30 Mex., C. Am., E. trop. S. Am.
× Laeliobrassocattleya hort. (vii) = × Brassolaeliocattleya hort. (Orchidac.).
× Laeliocatonia hort. Orchidaceae. Gen. hybr. (iii) (Broughtonia × Cattleya × Laelia).
× Laeliocattkeria hort. Orchidaceae. Gen. hybr. (iii) (Barkeria × Cattleya × Laelia).
× Laeliocattleya hort. Orchidaceae. Gen. hybr. (iii) (Cattleya × Laelia).
× Laeliodendrum hort. (vii) = × Epilaelia hort. (Orchidac.).
× Laeliopleya hort. Orchidaceae. Gen. hybr. (iii) (Cattleya × Laeliopsis).
Laeliopsis Lindl. Orchidaceae. 2 W.I.
× Laeliovola hort. (vii) = Brassolaelia hort. (Orchidac.).
× Laelonia hort. Orchidaceae. Gen. hybr. (iii) (Broughtonia × Laelia).
Laemellea Pfeiff. (sphalm.) = Lacmellea Karst. (Apocynac.).
Laemellia Jacks. = praec.
Laennecia Cass. = Conyza Less. (Compos.).
× Laeopsis hort. (vii) = Liaopsis hort. (Orchidac.).
Laërtia Gromow ex Trautv. = Leersia Sw. (Gramin.).
Laestadia Kunth. Compositae. 6 trop. Andes, W.I. No pappus.
*Laëtia Loefl. ex L. Flacourtiaceae. 10 trop. Am., W.I. Ls. often gland-
Laetji Osb. ex Steud. = Litchi Sonner. (Sapindac.).   [dotted.
Lafoënsia Pohl (nomen). 3 Brazil. Quid?
Lafoënsia Vand. Lythraceae. 12 trop. Am. Trees. Fls. 8–16-merous.
Lafuentia Lag. Scrophulariaceae. 2 S. Spain, Morocco. Woolly undershrubs.
Lagansa Rafin. = Polanisia Rafin. (Cleomac.).
Lagarinthus E. Meyer. Asclepiadaceae. 20 trop. & S. Afr.
Lagaropyxis Miq. = Radermachera Zoll. & Mor. (Bignoniac.).
Lagarosiphon Harv. Hydrocharitaceae. 15 trop. & S. Afr., Madag., India. Ls. alt. ♂ fls. floating off. A 3, stds. usu. 3.
*Lagascea Cav. Compositae. 15 Mex. to trop. S. Am., W.I. Heads cpd., 1-fld.
Lagedium Soják. Compositae. 2 temp. Euras. (exc. W. Eur.).

629

LAGENANDRA

Lagenandra Dalz. Araceae. 6 S. India, Ceylon.

Lagenantha Chiov. (~ Salsola L.). Chenopodiaceae. 1 Somalia.

Lagenanthus Gilg. Gentianaceae. 2 Panamá, Colombia.

Lagenaria Ser. Cucurbitaceae. 6: 1 pantrop., 1 trop. Afr. & Madag., 4 trop. Afr. *L. siceraria* (Molina) Standley is the calabash cucumber or bottle gourd. The woody outer pericarp makes a flask.

Lagenia E. Fourn. = Araujia Brot. (Asclepiadac.).

Lagenias E. Mey. = Sebaea R.Br. (Gentianac.).

Lagenifera Cass. (1815). Compositae. 30 Japan, Ryukyu Is., Borneo, New Caled., N.Z., Pacif., C. Am., Andes.

Lagenocarpus Klotzsch = Nagelocarpus Bullock (Ericac.).

Lagenocarpus Nees. Cyperaceae. 70 trop. S. Am., W.I.

Lagenophora Cass. (1818) = Lagenifera Cass. (1815) (Compos.).

Lagenula Lour. = Cayratia A. Juss. (Vitidac.).

Lagerstroemia L. Lythraceae. 53 trop. As. to N. Austr. Trees (sometimes large) and shrubs. Some provide good timber. Some heterostyled like *Lythrum*.

Lagerstroemi[ac]eae J. G. Agardh = Lythraceae–Lagerstroemieae DC.

Lagetta Juss. Thymelaeaceae. 4 W.I. *L. lintearia* Lam. is the lace tree. Its bast fibres on removal from the stem (by maceration, etc.) form a network for making dresses, etc.

Laggera Gandog. = Rosa L. (Rosac.).

Laggera Sch. Bip. ex Oliv. Compositae. 20 trop. Afr., Arabia, India, S. China Formosa.

Lagoa Durand. Asclepiadaceae. 1 Brazil.

Lagochilium Nees = Aphelandra R.Br. (Acanthac.).

Lagochilopsis Knorring. Labiatae. 5 C. As.

Lagochilus Bunge. Labiatae. 35 C. As. to Persia & Afghan.

Lagocodes Rafin. = Scilla L. (Liliac.).

Lagoecia L. Umbelliferae. 1 Medit. One of the usual two loc. of the ovary is aborted.

Lagonychium Bieb. (~ Prosopis L.). Leguminosae. 1 E. Medit. to C. As.

Lagophylla Nutt. Compositae. 9 W. N. Am.

Lagopsis Bunge ex Benth. (~ Marrubium L.). Labiatae. 4 W. Sib. to Japan.

Lagopus Fourr. = Plantago L. (Plantaginac.).

Lagopus Hill = Trifolium L. (Legumin.).

Lagoseriopsis Kirp. Compositae. 1 C. As.

Lagoseris M. Bieb. (~ Crepis L.). Compositae. 5 Crimea, As. Min., NW. Persia.

Lagoseris Hoffmgg. & Link = Pterotheca Cass. = Crepis L. (Compos.).

Lagothamnus Nutt. = Tetradymia DC. (Compos.).

Lagotia C. Muell. (sphalm.) = Sagotia Duchass. & Walp. = Desmodium Desv. (Legumin.).

Lagotis Gaertn. Scrophulariaceae. 20 N. & C. As., S. to Cauc., Himal. &

Lagotis E. Mey. = Carpacoce Sond. (Rubiac.).   [W. China.

Lagowskia Trautv. = Coluteocarpus Boiss. (Crucif.).

Lagrezia Moq. (~ Celosia L.). Amaranthaceae. 17 trop. E. Afr., Madag., Chagos Archip.; 1 Mex.(?)

LAMIOPHLOMIS

Laguna Cav. (Abelmoschus Medik.) (~ Hibiscus L.). Malvaceae. 15 trop. Afr., As., Austr.
Lagunaea C. Agardh = Lagunea Lour. = Amblygonum Reichb. (Polygonac.).
Lagunaea Schreb. = Laguna Cav. (Malvac.).
Lagunaena Ritgen = Lagunea Lour. = Amblygonum Reichb. (Polygonac.).
Lagunaria G. Don. Malvaceae. 1 E. Austr., Norfolk I., Lord Howe I.
Laguncularia Gaertn. f. Combretaceae. 2 trop. Am., trop. W. Afr. (mangrove).
Lagunea Lour. = Amblygonum Reichb. (Polygonac.).
Lagunea Pers. = Laguna Cav. (Malvac.).
Lagunezia Scop. = Racoubea Aubl. = Homalium Jacq. (Flacourtiac.).
Lagunizia B. D. Jacks. (sphalm.) = praec.
Lagunoa Poir. = Llagunoa Ruiz & Pav. (Sapindac.).
Lagurostemon Cass. = Saussurea DC. (Compos.).
Lagurus L. Gramineae. 1 Medit., *L. ovatus* L.
Lagynias E. Mey. Rubiaceae. 5 trop. & S. Afr.
Lahaya Roem. & Schult. = Polycarpaea Lam. (Caryophyllac.).
Lahayea Rafin. = praec.
Lahia Hassk. = Durio Adans. (Bombacac.).
Laïs Salisb. = Hippeastrum Herb. (Amaryllidac.).
Lalage Lindl. = Bossiaea Vent. (Legumin.).
Lalda Bub. = Lapsana L. (Compos.).
Lallemandia Walp. = seq.
Lallemantia Fisch. & Mey. Labiatae. 5 As. Min. to C. As. & Himal.
Lalypoga Gandog. = Polygala L. (Polygalac.).
Lamanonia Vell. Cunoniaceae. 10 S. Brazil, Paraguay.
Lamarchea Gaudich. Myrtaceae. 1 W. Austr.
Lamarckea Steud. = Lamarckia Moench mut. Koeler (Gramin.).
Lamarckia Hort. ex Endl. = Elaeodendron Jacq. f. (Celastrac.).
***Lamarckia** Moench mut. Koeler. Gramineae. 1 Medit.
Lamarckia Vahl = seq.
Lamarkea Pers. = Markea Rich. (Solanac.).
Lamarkea Reichb. = Lamarchea Gaudich. (Myrtac.).
Lamarkia G. Don = Lamarkea Pers. = Markea Rich. (Solanac.).
Lamarkia Medik. = Sida L. (Malvac.).
Lamarkia Moench = Lamarckia Moench mut. Koeler (Gramin.).
Lambertia Sm. Proteaceae. 8 Austr.
Lambertya F. Muell. = Bertya Planch. (Euphorbiac.).
Lamechites Markgraf. Apocynaceae. 1 New Guinea.
Lamellisepalum Engl. Rhamnaceae. 1 trop. Afr.
Lamia Endl. = Lemia Vand. = Portulaca L. (Portulacac.).
Lamiacanthus Kuntze. Acanthaceae. 2 Java, Lombok.
***Lamiaceae** Lindl.: *see* **Labiatae** Juss. (nom. altern.).
Lamiastrum Fabr. ex Ehrend. & Polatsch. = Galeobdolon Adans. (Labiat.).
Lamiella Fourr. = praec.
Lamiodendron v. Steen. Bignoniaceae. 1 New Guinea.
Lamiofrutex Lauterb. = Vavaea Benth. (Meliac.).
Lamiophlomis Kudo = Phlomis L. (Labiat.).

LAMIOPSIS

Lamiopsis (Dum.) Opiz = seq.

Lamium L. Labiatae. 40–50 Eur., As., extratrop. Afr. *L. album* L. has sympodial rhizomes and large white homogamous humble-bee fls. *L. amplexicaule* L. has cleist. fls. in spring and autumn; they look like ordinary buds with a small C, and are pollinated without opening.

Lamottea Pomel (1860) = Carduncellus Adans. (Compos.).

Lamottea Pomel (1870) = Munbya Pomel (Legumin.).

Lamourouxia Kunth. Scrophulariaceae. 30 Mex. to trop. S. Am.

Lampas Danser. Loranthaceae. 1 N. Borneo.

Lampaya Phil. Verbenaceae. 3 Bolivia, Chile, Argent.

Lampetia Rafin. = Mollugo L. (Aïzoac.).

Lampetia M. Roem. = Atalantia Corrêa (Rutac.).

Lampocarpya Spreng. = seq.

Lampocarya R.Br. = Gahnia J. R. & G. Forst. (Cyperac.).

Lampra Benth. = Weldenia Schult. (Commelinac.).

Lampra Lindl. ex DC. = Trachymene Rudge (Umbellif.).

Lamprachaenium Benth. Compositae. 1 Penins. India.

Lampranthus N. E. Brown. Aïzoaceae. 100 S. Afr.

Lamprocapnos Endl. = Dicentra Borckh. corr. Bernh. (Fumariac.).

Lamprocarpus Blume = Pollia Thunb. (Commelinac.).

Lamprocarya Nees = Lampocarya R.Br. = Gahnia J. R. & G. Forst. (Cyperac.).

Lamprocaulos Mast. Restionaceae. 3 S. Afr.

Lamprochlaena F. Muell. = Myriocephalus Benth. (Compos.).

Lamprochlaenia Börner = Carex L. (Cyperac.).

Lamprococcus Beer = Aechmea Ruiz & Pav. (Bromeliac.).

Lamproconus Lem. = Pitcairnia L'Hérit. (Bromeliac.).

Lamprodithyros Hassk. = Aneilema R.Br. (Commelinac.).

Lamprolobium Benth. Leguminosae. 1 Queensland.

Lamprophragma O. E. Schulz. Cruciferae. 1 S. U.S., Mex.

Lamprophyllum Miers = Rheedia L. (Guttif.).

Lamprospermum Klotzsch = Matayba Aubl. (Sapindac.).

Lamprostachys Boj. ex Benth. = Achyrospermum Blume (Labiat.).

Lamprothamnus Hiern. Rubiaceae. 2 trop. E. Afr.

Lamprothyrsus Pilger. Gramineae. 2 trop. S. Am.

Lamprotis D. Don = Erica L. (Ericac.).

Lampsana Mill. = Lapsana L. (Compos.).

Lampujang Koen. = Zingiber Boehm. (Zingiberac.).

Lamyra Cass. (~ Cirsium Mill.). Compositae. 6 Medit. to Crimea & Cauc.

Lamyropappus Knorr. & Tamamsch. Compositae. 1 C. As.

Lamyropsis (Charadze) Dittrich. Compositae. 3 Sardinia, S. Aegean, Cauc., C. As.

Lanaria Adans. = Gypsophila L. (Caryophyllac.).

***Lanaria** Ait. Liliaceae or Tecophilaeaceae. 1 S. Afr.

Lancea Hook. f. & Thoms. Scrophulariaceae. 2 Tibet, China.

Lancisia Fabr. (~ Cotula L.). Compositae. 10 S. Afr.

Lancisia Lam. = Lidbeckia Bergius (Compos.).

Lancretia Delile = Bergia L. (Elatinac.).

Landersia Macfad. = Melothria L. (Cucurbitac.).

Landesia Kuntze (sphalm.) = Londesia Fisch. & Mey. (Chenopodiac.).
Landia Comm. ex Juss. = Mussaenda Burm. f. (Rubiac.).
Landolfia D. Dietr. = seq.
*Landolphia Beauv. Apocynaceae. 55 trop. & S. Afr., Madag., Masc. Several
are lianes with curious hook tendrils like *Strychnos*. Fr. a large berry full of an
acid pulp composed of the hair-structures on the seeds. Several, e.g. *L. kirkii*
Dyer, *L. comorensis* Benth. & Hook. f., etc., yield rubber, the coagulated latex.
It is known in trade as African and Madagascar rubber.
Landtia Less. Compositae. 6 trop. & S. Afr.
Landukia Planch. = Parthenocissus Planch. (Vitidac.).
Laneasagum Bedd. = Drypetes Vahl (Euphorbiac.).
Lanesagum Pax & Hoffm. = praec.
Lanessania Baill. Moraceae. 2 N. trop. S. Am.
Langefeldia Steud. = Langeveldia Gaudich. = Elatostema Gaudich. (Urticac.).
Langerstroemia Cram. = Lagerstroemia L. (Lythrac.).
Langeveldia Gaudich. = Elatostema Gaudich. (Urticac.).
Langevinia Jacques-Félix = Mapania Aubl. (Cyperac.).
Langia Endl. = Hermbstaedtia Reichb. (Amaranthac.).
Langlassea H. Wolff. Umbelliferae. 1 Mex.
Langleia Scop. = Casearia Jacq. (Flacourtiac.).
Langloisia Greene (~ Gilia Ruiz & Pav.). Polemoniaceae. 6 SW. U.S.
Langsdorffia Fisch. ex Regel = Chloris Sw. (Gramin.).
Langsdorffia Mart. Balanophoraceae. 1 New Guinea; 1 Mex. to trop. S. Am.,
L. hypogaea Mart.
Langsdorffia Raddi = Barbosa Becc. (Palm.).
Langsdorffia Steud. = Langsdorfia Leandro = Zanthoxylum L. (Rutac.).
Langsdorffia Willd. ex Steud. = Lycoseris Cass. (Compos.).
Langsdorffiaceae Van Tiegh. = Balanophoraceae–Langsdorffieae Schott &
Endl.
Langsdorfia Leandro = Zanthoxylum L. (Rutac.).
Langsdorfia Pfeiff. = Langsdorffia Raddi = Barbosa Becc. (Palm.).
Langsdorfia Rafin. = Nicotiana L. (Solanac.).
Langsdorfia Willd. ex Less. = Lycoseris Cass. (Compos.).
Languas Koenig = Alpinia Roxb. (Zingiberac.).
Lanigerostemma Chapel. ex Endl. = Eliaea Cambess. (Guttif.).
Lanipila Burch. = Lasiospermum Lag. (Compos.).
Lanium Lindl. Orchidaceae. 4 trop. Am.
Lankesterella Ames. Orchidaceae. 15 trop. Am.
Lankesteria Lindl. Acanthaceae. 7 trop. Afr., Madag.
*Lannea A. Rich. Anacardiaceae. 70 trop. Afr., 1 Indomal.
Lanneoma Delile = praec.
Lansbergia De Vriese = Trimezia Salisb. ex Herb. (Iridac.).
Lansium Corrêa. Meliaceae. 6–7 Indomal. *L. domesticum* Jack, ed. fr.
Lantana L. Verbenaceae. 150 trop. Am., W.I., trop. & S. Afr. Shrubs, often
used for hedges, but some pernicious weeds. Some have ed. fr.
Lantanopsis Wright. Compositae. 2 W.I.
Lanthorus Presl = Helixanthera Lour. (Loranthac.).
Lanugia N. E. Brown. Apocynaceae. 3 trop. E. Afr., Madag.

LANZANA

Lanzana Stokes = Buchanania Roxb. (Anacardiac.).

Laoberdes Rafin. = Apium L. (Umbellif.).

Laothoë Rafin. = Chlorogalum Kunth (Liliac.).

Lapageria Ruiz & Pav. Philesiaceae. 1 Chile, *L. rosea* Ruiz & Pav., a climbing shrub with ed. fr.

Lapageriaceae Kunth = Philesiaceae Dum.

Lapasathus Presl = Aspalathus L. (Legumin.).

Lapathon Rafin. = seq.

Lapathum Mill. = Rumex L. (Polygonac.).

Lapeirousia Pourr. Iridaceae. 60 trop. & S. Afr.

Lapeirousia Thunb. = Peyrousea DC. (Compos.).

Lapeyrousa Poir. = seq.

Lapeyrousia Pourr. = Lapeirousia Pourr. (Iridac.).

Lapeyrousia Spreng. = Lapeirousia Thunb. = Peyrousea DC. (Compos.).

Laphamia A. Gray (~ Perityle Benth.). Compositae. 20 SW. U.S., Mex.

Lapidaria Dinter & Schwantes (~ Dinteranthus Schwantes). Aïzoaceae. 1 S. Afr. (Namaqualand).

Lapiedra Lag. Amaryllidaceae. 1 Spain.

Lapithea Griseb. = Sabatia Adans. (Gentianac.).

*Laplacea Kunth. Theaceae. 30 Malaysia, trop. Am., W.I.

*Laportea Gaudich. Urticaceae. 23 trop. & subtrop., also temp. E. As. & E.N. Am., S. Afr., Madag.

Lappa Scop. = Arctium L. (Compos.).

Lappago Schreb. = Tragus Hall. (Gramin.).

Lappagopsis Steud. = Paspalum L. (Gramin.).

Lappula v. Wolf. Boraginaceae. 50 temp. Euras., Austr., 5 N. Am. *L. myosotis* v. Wolf cult. The fls. change from white to red and blue (see fam.). Fr. hooked.

Lappularia Pomel = Caucalis L. (Umbellif.).

Lapsana L. Compositae. 9 temp. Euras. Self-pollinated. No pappus.

× **Lapsyoungia** Hiyama. Compositae. Gen. hybr. (Lapsana × Youngia).

Lapula Gilib. = Lappula v. Wolf (Boraginac.).

Larbraea Fourr. = seq.

Larbrea A. St-Hil. = Stellaria L. (Caryophyllac.).

Lardizabala Ruiz & Pav. Lardizabalaceae. 2 Chile. Tough fibre from the stems of *L. biternata* Ruiz & Pav.

*Lardizabalaceae Decne. Dicots. 8/35 Himal. to Japan, Chile. Mostly climbing shrubs with palmate ls. Rarely erect tree with pinnate ls. (*Decaisnea*). Fls. in racemes, usu. in the axils of the scale ls. at the bases of the branches, polygamous or diclinous. Usual formula P 3 + 3, A 3 + 3, $\underline{G}$ 3 or more. 2 whorls of small honey-leaves (see *Ranunculaceae*) often occur between P and A; sta. sometimes united; anthers extrorse; ovules ∞ in longitudinal rows on the lat. walls (cf. *Nymphaeaceae*), anatr. The fl. of either sex shows rudiments of the organs of the other sex. Berry. Embryo small and straight, in copious endosp. Genera: *Decaisnea, Akebia, Lardizabala, Stauntonia, Holboellia, Sinofranchetia, Boquila, Parvatia.*

Larentia Klatt = Alophia Herb. (Iridac.).

Larephes Rafin. = Echium L. (Boraginac.).

Laretia Gill. & Hook. Hydrocotylaceae (~ Umbellif.). 2 Andes of Chile.

Lariadenia [Schlechtd.] (sphalm.) = Lasiadenia Benth. (Thymelaeac.).
Laricopsis Kent = Pseudolarix Gord. (Pinac.).
Lariospermum Rafin. = Ipomoea L. (Convolvulac.).
Larix Mill. Pinaceae. 10–12 Eur., N. As., N. Am. The general chars. are those of *Cedrus*, but the ls. are deciduous, and the cones ripen in a single year. *L. decidua* Mill. (*L. europaea* DC.) is the common larch, cult. on a large scale for its wood, bark (used in tanning) and turpentine (Venice t.). Others are also important, e.g. *L. laricina* (Du Roi) Koch (tamarack).
Larmzon Roxb. = Launzan Buch.-Ham. = Buchanania Roxb. (Anacardiac.).
Larnalles Rafin. = Commelina L. (Commelinac.).
Larnandra Rafin. = Epidendrum L. (Orchidac.).
Larnastyra Rafin. = Salvia L. (Labiat.).
Larnax Miers = Athenaea Sendtn. (Solanac.).
Larochea Pers. = Crassula L. (Crassulac.).
Larradia Pritz. (sphalm.) = Lavradia Vell. (Violac.).
*Larrea Cav. Zygophyllaceae. 2 temp. S. Am.
Larrea Ortega = Hoffmanseggia Cav. (Legumin.).
Larsenia Bremek. Acanthaceae. 1 Siam.
Larysacanthus Oerst. = Ruellia L. (Acanthac.).
Lasallea Greene (~ Aster L.). Compositae. 3 N. Am.
Lascadium Rafin. = Croton L. (Euphorbiac.).
Laseguea A. DC. Apocynaceae. 10 trop. S. Am.
Lasemia Rafin. = Salvia L. (Labiat.).
Laser Borckh. ex Gaertn., Mey. & Scherb. Umbelliferae. 3 C. & S. Eur., W. As.
Laserpicium Aschers. = seq.
Laserpitium L. Umbelliferae. 35 Canaries, Medit. to SW. As.
Lasespitium Rafin. = praec.
Lasia Lour. Araceae. 3 Indomal.
Lasiacis Hitchcock. Gramineae. 30 trop. & subtrop. Am.
Lasiadenia Benth. Thymelaeaceae. 1 trop. S. Am.
Lasiagrostis Link (~ Stipa L.). Gramineae. 4 S. Russia, temp. As.
Lasiake Rafin. = Verbascum L. (Scrophulariac.).
Lasiandra DC. = Tibouchina Aubl. (Melastomatac.).
Lasiandros St-Lag. = praec.
Lasianthaea DC. = Zexmenia La Llave (Compos.).
Lasianthea Endl. = praec.
Lasianthemum Klotzsch = Talisia Aubl. (Sapindac.).
Lasianthera Beauv. Icacinaceae. 1 trop. W. Afr.
Lasianthus Adans. = Gordonia Ellis (Theac.).
*Lasianthus Jack. Rubiaceae. 1 W.I., 10–20 trop. Afr., 150 Indomal.
Lasianthus Zucc. ex DC. = Lasianthaea DC. = Zexmenia La Llave (Compos.)?
Lasiarrhenum I. M. Johnston. Boraginaceae. 2 Mex.
Lasierpa Torr. = Chiogenes Salisb. ex Torr. = Gaultheria L. (Ericac.).
Lasimorpha Schott = Cyrtosperma Griff. (Arac.).
Lasinema Steud. = Lysinema R.Br. (Epacridac.).
Lasingrostis Link (sphalm.) = Lasiagrostis Link (Gramin.).
Lasinia Rafin. = Baptisia Vent. (Legumin.).

LASIOBEMA

Lasiobema (Korth.) Miq. (~ Bauhinia L.). Leguminosae. 10 E. Himal. to Japan, Indoch., Java & Sumba.

Lasiocarphus Pohl ex Baker = Stilpnopappus Mart. (Compos.).

Lasiocarpus Banks & Soland. ex Hook. f. = Acaena Vahl (Rosac.).

Lasiocarpus Liebm. Malpighiaceae. 4 Mex.

Lasiocarys Balf. f. (sphalm.) = Lasiocorys Benth. (Labiat.).

Lasiocaryum I. M. Johnston. Boraginaceae. 7 C. As., Himal.

Lasiocephalus Willd. ex Schlechtd. = Culcitium Humb. & Bonpl. (Compos.).

Lasiocereus Ritter = ? Haageocereus Backeb. (Cactac.).

Lasiochlamys Pax & K. Hoffm. Euphorbiaceae. 1 New Caled.

Lasiochloa Kunth. Gramineae. 4 S. Afr.

Lasiocladus Boj. ex Nees. Acanthaceae. 5 Madag.

Lasiococca Hook. f. Euphorbiaceae. 3 E. Himal., India (Orissa), Hainan, Malay Penins.

Lasiococcus Small = Gaylussacia Kunth (Ericac.).

Lasiocoma Bolus. Compositae. 1 S. Afr.

Lasiocorys Benth. Labiatae. 9 trop. & S. Afr., Socotra, Arabia.

Lasiocroton Griseb. Euphorbiaceae. 6 W.I.

Lasiodendrum P. & K. = ? praec.

Lasiodiscus Hook. f. Rhamnaceae. 9 trop. Afr., Madag.

Lasiogyne Klotzsch = Croton L. (Euphorbiac.).

Lasiolepis Benn. = Harrisonia R.Br. (Rutac.).

Lasiolepis Boeck. = Eriocaulon L. (Eriocaulac.).

Lasiolytrum Steud. = Arthraxon Beauv. (Gramin.).

Lasiomorpha P. & K. = Lasimorpha Schott = Cyrtosperma Griff. (Arac.).

Lasionema D. Don = Macrocnemum R.Br. (Rubiac.).

Lasiopera Hoffmgg. & Link = Bellardia All. + Odontites Zinn + Parentucellia Viv. (Scrophulariac.).

Lasiopetalaceae J. G. Agardh = Sterculiaceae–Lasiopetaleae Gay.

Lasiopetalum Sm. Sterculiaceae. 30 Austr.

Lasiophyton Hook. & Arn. = Micropsis DC. (Compos.).

Lasiopoa Ehrh. (uninom.) = Bromus asper L. (Gramin.).

Lasiopogon Cass. Compositae. 2–3 Medit. to India, S. Afr.

Lasioptera Andrz. ex DC. = Lepidium L. (Crucif.).

Lasiopus Cass. = Gerbera L. ex Cass. (Compos.).

Lasiopus D. Don = Eriopus D. Don = Taraxacum Wigg. (Compos.).

Lasiorrhachis Stapf. Gramineae. 1 Madag.

Lasiorrhiza Lag. = Leuceria Lag. (Compos.).

Lasiosiphon Fresen. Thymelaeaceae. 50 trop. & S. Afr. & Madag. to W. Penins. India & Ceylon.

Lasiospermum Fisch. = Scorzonera L. (Compos.).

Lasiospermum Lag. Compositae. 5 S. Afr.

Lasiospora Cass. = Scorzonera L. (Compos.).

Lasiostega Rupr. ex Benth. = Buchloë Engelm. (Gramin.).

Lasiostelma Benth. Asclepiadaceae. 2–3 S. Afr.

Lasiostemon Benth. & Hook. f. = Lasiostemum Nees & Mart. = Cusparia Humb. ex R.Br. (Rutac.).

Lasiostemon Schott ex Endl. = Esterhazya Mikan (Scrophulariac.).
Lasiostemum Nees & Mart. = Angostura Roem. & Schult. (Rutac.).
Lasiostoma Benth. = Hydnophytum Jack (Rubiac.).
Lasiostoma Schreb. = Strychnos L. (Strychnac.).
Lasiostomum Zipp. ex Blume = Geniostoma J. R. & G. Forst. (Loganiac.).
Lasiostyles Presl = Cleidion Blume (Euphorbiac.).
Lasiostylis Pax & K. Hoffm. = praec.
Lasiotrichos Lehm. = Fingerhuthia Nees (Gramin.).
Lasipana Rafin. = Mallotus Lour. (Euphorbiac.).
Lasiurus Boiss. Gramineae. 3 trop. E. Afr., India.
Lasius Hassk. = Lasia Lour. (Arac.).
Lass Adans. = Pavonia Cav. (Malvac.).
Lassa Kuntze = praec.
Lassia Baill. = Tragia L. (Euphorbiac.).
Lassonia Buc'hoz = Magnolia L. (Magnoliac.).
Lastarriaca B. D. Jacks. (sphalm.) = seq.
Lastarriaea Remy = Chorizanthe R.Br. (Polygonac.).
Lasthenia Cass. Compositae. 2 SW. U.S., 1 Chile.
Lastila Alef. = Lathyrus L. (Legumin.).
Lastrea Bory = Thelypteris Schmid. (unless restricted to spp. immediately related to type sp. *L. oreopteris* (Ehrh.) Bory) (Thelypteridac.).
Lastrella (H. Ito) Nak. = Phegopteris Fée emend. Ching (Thelypteridac.).
Lastreopsis Ching. Aspidiaceae. 25 cosmop. (Included in *Ctenitis* C. Chr. by Copel. 1947; but see Tindale, *Vict. Nat.* **73**: 180–5, 1957.)
Lasynema Poir. = Lysinema R.Br. (Epacridac.).
Latace Phil. = Leucocoryne Lindl. (Liliac.).
Latana Robin = Lantana L. (Verbenac.).
Latania Comm. ex Juss. Palmae. 3 trop. E. Afr., Mascarenes.
Laterifissum Dulac = Montia L. (Portulacac.).
Lateropora A. C. Smith. Ericaceae. 1 Panamá.
Lathirus Neck. = Lathyrus L. (Legumin.).
Lathraea L. Scrophulariaceae. 7 temp. Euras. *L. squamaria* L. is a total parasite living upon the roots of hazel, beech, etc. It has a thick rhiz. bearing 4 rows of tooth-like scaly ls. The fl. shoot comes above ground and bears a raceme of purplish fls., all bent round to the same side of the infl., protogynous. The scales upon the rhiz. are hollowed, each containing a branched cavity opening to the outside by a narrow slit at the base of the back of the l. This arises by a development similar to that which forms the chambers in the ls. of *Empetrum*, *Cassiope*, etc. In the small lat. cavities opening out of the main one there are found peculiar glandular organs, resembling those of insectivorous plants. Small insects, etc., are often found in these leaves (cf. bladders of *Utricularia*, etc.) and it has been supposed that these organs absorb their proteids like the glands of *Drosera*, etc. This, however, is doubtful. *L. clandestina* L. is parasitic upon willows, beech, poplar, etc. The capsule of *L.* splits explosively.
Lathraeocarpa Bremek. Rubiaceae. 2 Madag.
Lathraeophila Hook. f. = Latraeophila Leandro ex A. St-Hil. = Helosis Rich. (Balanophorac.).
Lathriogyna Eckl. & Zeyh. Leguminosae. 1 S. Afr.

LATHRISIA

Lathrisia Sw. = Bartholina R.Br. (Orchidac.).

Lathrophytum Eichl. Balanophoraceae. 1 SE. Brazil.

Lathyraceae Burnett = Leguminosae–Sarcolobae DC. (Viciëae + Phaseoleae + Dalbergiëae).

Lathyroïdes Fabr. (uninom.) = *Lathyrus* L. sp. (Legumin.).

Lathyropteris Christ = Pteris L. (Pteridac.).

Lathyros St-Lag. = seq.

Lathyrus L. Leguminosae. 130 N. temp., and mts. of trop. Afr. & S. Am. *L. aphaca* L. has large green stipules performing assim. functions, whilst the l. is transformed into a tendril; *L. nissolia* L. has its petioles flattened into phyllodes and has no l. blade at all (see *Acacia*). *L. macrorrhizus* Wimm. has tuberous roots which may be eaten like potatoes. *L. sativus* L. (jarosse) and *L. cicera* L. are cult. in S. Eur. as fodder and are also eaten like chick-pea (*Cicer*). *L. odoratus* L. is the sweet pea; *L. latifolius* L. is the everlasting pea. The fl. is like that of *Vicia*; on the style is a tuft of hairs that brushes the pollen out of the apex of the keel, where it is shed by the anthers.

Laticoma Rafin. = Nerine Herb. (Amaryllidac.).

Latipes Kunth. Gramineae. 1–2 Senegal to Sind.

Latosatis Thou. (uninom.) = *Satyrium latifolium* Thou. = *Benthamia latifolia* (Thou.) A. Rich. (Orchidac.).

Latouchea Franch. Gentianaceae. 1 E. China.

Latouria Blume = Dendrobium Sw. (Orchidac.).

Latouria Lindl. = Leschenaultia R.Br. (Goodeniac.).

Latraeophila Leandro ex A. St-Hil. = Helosis Rich. (Balanophorac.).

Latraeophilaceae Leandro ex A. St-Hil. = Balanophoraceae–Helosidoïdeae.

Latreillea DC. = Ichthyothere Mart. (Compos.).

Latrienda Rafin. = Ipomoea L. (Convolvulac.).

Latrobea Meissn. Leguminosae. 6 W. Austr.

Latua Phil. Solanaceae. 1 Chile.

Latyrus Gren. = Lathyrus L. (Legumin.).

Laubertia A. DC. Apocynaceae. 6 C. to trop. S. Am.

Lauchea Klotzsch = Begonia L. (Begoniac.).

Laugeria Hook. f. = Terebraria Sessé ex Kuntze (Rubiac.).

Laugeria L. = seq.

Laugieria Jacq. = Guettarda L. (Rubiac.).

Launaea Cass. Compositae. 40 Canaries, Medit. to E. As., trop. & S. Afr. *L. sarmentosa* (Willd.) Alston is a char. plant of sandy trop. beaches.

Launaya Kuntze = praec.

Launea Endl. = praec.

Launzan Buch.-Ham. = Buchanania Roxb. (Anacardiac.).

Launzea Endl. = praec.

*Lauraceae Juss. Dicots. 32/2000–2500 trop. & subtrop.; chief centres of distr. SE. As. and Brazil. Trees and shrubs with usu. leathery evergr. alt. or opp. exstip. ls. The tissues contain numerous oil-cavities. *Cassytha* is an interesting parasite. Infl. racemose, cymose, umbelliform, or mixed. Fl. actinom., usu. 3-merous, ⚥ or monoec. Formula usu. P $2n$, A $4n$, G n. P in two whorls, perig.; A perig. or epig., in 3 or 4 whorls, some of which are commonly reduced to stds.; anther 2- or 4-loc., opening by valves, usu. intr., but in many cases those of the third whorl extr. The axis is ± concave, and the ovary is free from it at the sides. G (v. rarely G̅—*Hypodaphnis*) 1 or more probably (3), forming a 1-loc. ovary, with 1 pend. anatr. ov. Fr. a berry, often ± enclosed by the cup-like recept. which also becomes fleshy in these cases. Embryo straight; seed exalb. Important economic plants are found in several genera.

Classification and chief genera (after Kostermans):

I. *Lauroïdeae.* Arborescent; ls. normal.
    1. PERSEËAE (infl. panic., fl. umbels exinvolucrate; fr. without cupule): *Persea, Phoebe, Dehaasia, Beilschmiedia, Endiandra, Potameia.*
    2. CINNAMOMEAE (as 1, but fr. base embedded in a cupule): *Ocotea, Cinnamomum, Actinodaphne, Aiouea, Aniba, Endlicheria, Licaria.*
    3. LAUREAE (LITSEËAE) (fl. umbels surr. by invol. of large decuss. persist. bracts; fr. ± embedded in a cupule): *Litsea, Neolitsea, Lindera, Laurus.*
    4. CRYPTOCARYEAE (as 2; ovary superior, but fr. completely encl. in the accresc. perianth-tube): *Cryptocarya, Ravensara, Eusideroxylon.*
    5. HYPODAPHNIDEAE (as 4, but ovary inferior): *Hypodaphnis.*

II. *Cassythoïdeae.* Parasitic twiners without proper ls.; infl. indef., spicate or rac., or capit., exinvolucrate; ov. and fr. as in 4, above: *Cassytha.*

The valvular anther dehisc., taken in conjunction with the char. lat. stds. and certain anat. features, indicates that the *Laurac.* are most closely related to the *Atherospermatac., Gomortegac.* and *Gyrocarpac.*

Lauradia Vand. = Lavradia Vell. (Violac.).

Laurea Gaudich. = Bagassa Aubl. (Urticac.).

*Laurelia Juss. Atherospermataceae. 1 N.Z.; 1 Chile. *L. novae-zelandiae* A. Cunn. supplies a useful timber. The frs. of *L. aromatica* Juss. are used as a spice under the name Peruvian nutmegs.

Laurembergia Bergius. Haloragidaceae. 4 trop. S. Am., trop. Afr., Madag., Masc., India, Java.

Laurencellia Neum. = Lawrencellia Lindl. = Helichrysum L. (Compos.).

Laurenta Medik. = seq.

Laurentia Adans. Campanulaceae. 2 W. U.S.; 15 Medit., S. Afr. (See also *Isotoma* Lindl.)

Laurentia Steud. = Lorentea Ortega = Sanvitalia Gualt. (Compos.).

Laureola Hill = Daphne L. (Thymelaeac.).

Laureola M. Roem. = Skimmia Thunb. (Rutac.).

Laureria Schlechtd. = Juanulloa Ruiz & Pav. (Solanac.).

Lauridia Eckl. & Zeyh. = Elaeodendron Jacq. (Celastrac.).

Laurocerasus M. Roem. = seq.

Lauro-Cerasus Duham. (~ Padus Mill. or Prunus L.). Rosaceae. 75 trop. & temp. As. & Am., 1–2 trop. Afr., Madag., 1 Canaries to Portugal, 1 SE. Eur. to Cauc.

Lauromerrillia C. K. Allen = Beilschmiedia Nees (Laurac.).

Laurophyllus Thunb. Anacardiaceae. 1 S. Afr.

Laurus L. Lauraceae. 2, *L. nobilis* L., the true laurel or sweet bay, Medit. (ls. aromatic, used in condiments, etc., berries in veterinary medicine), and *L. canariensis* Webb & Berth., Canaries and Madeira. Fls. unisexual by

Lausonia Juss. = Lawsonia L. (Lythrac.).   [abortion.

LECHLERIA

Lechleria Phil. = Huanaca Cav. (Umbellif.).
Lechytis Augier = Lecythis Loefl. (Lecythidac.).
Leciscium Gaertn. = ? Memecylon L. (Memecylac.).
Lecocarpus Decne. Compositae. 1 Galápagos.
Lecockia Meissn. = Lecokia DC. (Umbellif.).
Lecointea Ducke. Leguminosae. 3 Brazil, Peru.
Lecokia DC. Umbelliferae. 1 Crete to Persia.
Lecomtea Koidz. (~ Mniopsis Mart. & Zucc.). Podostemaceae. 1 Indoch.
Lecomtea Pierre ex Van Tiegh. = ? Harmandia Pierre (Olacac.).
Lecomtedoxa Dubard. Sapotaceae. 2 W. Equat. Afr.
Lecomtella A. Camus. Gramineae. 1 Madag.
Lecontea Rafin. = Lecontia Cooper ex Torr. = Peltandra Rafin. (Arac.).
Lecontea A. Rich. Rubiaceae. 3 Madag.
Lecontia W. Cooper ex Torr. = Peltandra Rafin. (Arac.).
Lecoqia P. & K. = Lecokia DC. (Umbellif.).
Lecoquia Caruel = praec.
Lecostemon (Moç. & Sessé ex) DC. = Sloanea L. (Elaeocarpac.).
Lectandra J. J. Smith = Poaephyllum Ridl. (Orchidac.).
Lecticula Barnhart = Utricularia L. (Lentibulariac.).
*****Lecythidaceae** Poiteau. Dicots. 15/325 trop. Am. Trees with ls. bunched at ends of twigs, alt., simple, exstip. Fls. sol. or in racemose infl., reg. or C and A zygo., perig. or epig., always with complete fusion of recept. and ovary. Usu. intrastaminal disk as well as one under C and A. K usu. 4–6, valv.; C, rarely (C), 4–6, imbr.; A ∞ in several whorls, sta. ± united at base, anther usu. versatile, bent inwards in bud. Sometimes A of remarkable appearance. owing to one-sided development of the union, and abortion of some anthers. G (2–6) or more, multiloc.; 1–∞ anatr. ov. in each, simple style. Berry or woody caps., indehisc. or operculate ('monkey-pots'); seeds exalbum. Chief genera: *Gustavia, Couroupita, Lecythis, Chytroma, Eschweilera, Bertholletia, Couratari.*
Lecythis Loefl. Lecythidáceae. 50 trop. Am. Like *Couroupita,* but sta. of helmet sterile. Caps. with lid woody (monkey-pot; used with sugar to catch monkeys, which cannot withdraw the inserted hand). Oily seeds ed. (*sapucaia* [nuts].
Lecythopsis Schrank = Couratari Aubl. (Lecythidac.).
Leda C. B. Clarke. Acanthaceae. 10 Malay Penins.
Ledebouria Reichb. = Ledeburia Link = Pimpinella L. (Umbellif.).
Ledebouria Roth = Scilla L. (Liliac.).
Ledebouriella H. Wolff. Umbelliferae. 2 C. As.
Ledeburia Link = Pimpinella L. (Umbellif.).
Ledelia Rafin. = Pomaderris Labill. (Rhamnac.).
Ledenbergia Klotzsch ex Moq. (sphalm.) = Ladenbergia Klotzsch ex Moq. corr. Kuntze = Flueckigera Kuntze (Phytolaccac.).
Ledermannia Mildbr. & Burret = Desplatsia Bocq. (Tiliac.).
Ledermanniella Engl. Podostemaceae. 1 W. Equat. Afr.
Ledgeria F. Muell. = Galeola Lour. (Orchidac.).
Ledocarpaceae Meyen. Dicots. 2/11 Andes. Small shrubs; ls. usu. opp., ent. or deeply lobed or partite, exstip. Fls. reg., ☿, sol. or in few-fld. corymbs. K 5, imbr., often with an epicalyx; C 5 or 0, imbr. or contorted; disk 0; A 10, extr. or intr.; G (5) or (3), styles short or 0, ovules either 2 collat. pend., or

642

∞ axile 2-seriate. Caps. loculic. or septifrag., sometimes beaked, with 1–2 or ∞ seeds per loc., endosp. thin, fleshy. Genera: *Balbisia* (*Ledocarpon*), *Wendtia*. Closely related to *Geraniac*. The genus *Rhynchotheca* Ruiz & Pav. should probably be removed from this family to the *Vivianiaceae*.

Ledocarpon Desf. = Balbisia Cav. (Ledocarpac.).

Ledonia Spach = Cistus L. (Cistac.).

Ledothamnus Meissn. Ericaceae. 5 Venez., Guiana.

Ledum L. Ericaceae. 10 N. temp. & arct. *L. palustre* L. circumpolar (used as tea in Labrador).

Ledum Reichb. (sphalm.) = Sedum L. (Crassulac.).

Leea Royen ex L. Leeaceae. 70 palaeotrop.

***Leeaceae** (DC.) Dum. Dicots. 1/70 palaeotrop. Trees, shrubs or herbs, branches occasionally prickly. Ls. pinn. to tripinn., rarely ternate or simple, alt. (v. rarely opp.), usu. dent., exstip. (but petiole usu. with 2 auricles or sheathing expansions near base). Infl. usu. corymbose, many-fld., erect, term. (v. rarely pend. or axill.), often ferrug.-toment. K, C and A usu. 5 (4 in 3 spp.). K (5–4), cupular, shortly toothed; C (5–4), valv., reflexed; A (5–4), oppositipet. and adnate to C, stam. tube short or long, conic to subglob., variously 5-lobed, lobes ent. or bifid, alt. with stam., with an internal ± obconic free membr. tube pendent midway from the stam. tube; anth. extr.; G̲ (3–8), often somewhat immersed in the recept., with simple style and 1 axile ov. per loc. Fr. a ± depressed 3–8-loc. berry; seeds with ruminate endosp. Only genus: *Leea*. Related to *Vitidaceae*; ?some remote connection with *Staphyleac*.

Leeania Rafin. = Leea L. (Leeac.).

Leeria Steud. = Leria DC. = Chaptalia Vent. (Compos.).

***Leersia** Soland. ex Sw. Gramineae. 15 trop. & warm temp. Marsh grasses like *Oryza*, used as fodder in As. Glumes rudimentary. *L. oryzoïdes* (L.) Sw. (Eur.) has cleist. fls.

Leeuwenhockia Steud. = Levenhookia R.Br. (Stylidiac.).

Leeuwenhoeckia E. Mey. ex Endl. = Dombeya Cav. (Sterculiac.).

Leeuwenhoekia Spreng. = Levenhookia R.Br. (Stylidiac.).

Leeuwenhookia Reichb. = praec.

Lefeburea Endl = seq.

Lefeburia Lindl. = seq.

Lefebvrea A. Rich. Umbelliferae. 15 trop. & SW. Afr.

Lefrovia Franch. = Cnicothamnus Griseb. (Compos.).

Leganosperma P. & K. = Lecanosperma Rusby (Rubiac.).

Legazpia Blanco = Torenia L. (Scrophulariac.).

Legendrea Webb & Berth. = Turbina Rafin. (Convolvulac.).

Legenere McVaugh. Campanulaceae. 1 Calif., 1 Chile.

Leggouzia Th. Dur. & Jacks. (sphalm.) = Legousia Durand (Campanulac.).

Legnea O. F. Cook. Palmae. 1 C. Am.

Legnephora Miers. Menispermaceae. 3 New Guinea, NE. Austr.

Legnotid[ace]ae Endl. = Rhizophoraceae–Macarisieae Baill.

Legnotis Sw. = Cassipourea Aubl. (Rhizophorac.).

Legocia Livera = Christisonia Gardn. (Orobanchac.).

Legouixia Heurck & Muell. Arg. = Epigynum Wight (Apocynac.).

Legousia Durand. Campanulaceae. 15 N. temp., S. Am.

LEGOUXIA

Legouxia Gérard = praec.

Legouzia Delarbre = praec.

Legrandia Kausel. Myrtaceae. 1 Chile.

Leguminaceae Dulac = Leguminosae Juss.

Leguminaria Bur. = Memora Miers (Bignoniac.).

***Leguminosae** Juss. (nom. altern. **Fabaceae** Reichb.). Dicots. The third largest fam. of flg. pls., with 600/12000, cosmop. *Mimosoïdeae* and *Caesalpinioïdeae* are mostly trop., *Papilionoïdeae* both trop. and temp.

Living in every soil and climate they show great variety in habit—trees, shrubs, water plants, xerophytes, climbers, etc. The roots of most exhibit peculiar tubercles—metamorphosed lat. roots containing peculiar bacterial organisms (*Rhizobium* spp.). Plants provided with these are able to take up much more atmospheric nitrogen. The plant appears actually to consume the bacteria which live in its cells, after they have stored up in themselves a considerable amount of nitrogenous material. Hence the value of the *L.* as a crop on poor soil, or as preceding wheat in the rotation of crops; for instead of impoverishing the soil they enrich it, either by the nitrogen contained in their roots and liberated as these decay, or by that of the whole pl. if ploughed in as 'green manure'.

Stem commonly erect; many climbers. Some, e.g. *Vicia*, climb by leaf-tendrils, some, e.g. *Bauhinia*, by stem-tendrils, some by hooks (modified in *Caesalpinia*, etc., emergences in *Acacia*, etc.), some by twining. Creeping stems, rooting at the nodes, also occur. Thorns, usu. modified branches (e.g. *Gleditsia*) or stipules (*Acacia*) or simply emergences from the stem (*Acacia*), are common. The stems of the erect trop. spp. sometimes branch so that the branches run parallel and erect, and bear crowns of ls. at the top. The stems of many lianes are peculiarly shaped, often flat, or corrugated in various ways, owing to peculiar growth in thickness.

Ls. usually alt., stip., and usu. cpd. Some have very small ls., e.g. *Ulex*, or scale-ls. and flat stems, e.g. *Carmichaelia*. In many spp. of *Acacia* the ls. are represented by simple green phyllodes. The stipules vary much in size, etc. (see *Acacia, Lathyrus, Vicia*). The ls. usu. perform sleep-movements at night, some moving upwards, some downwards, or in other ways, but finally usu. placing the leaflet edgwise to the sky. In *Mimosa* and *Neptunia* the ls. are sensitive to a touch and at once assume their sleep-position, recovering after a time. In *Desmodium gyrans* (L. f.) DC. the lat. leaflets execute continuous spontaneous movements as long as the temperature is high enough.

Infl. apparently always basically racemose, but with variety; simple raceme very common, also panicle and spike. Heads of sessile fls. common in *Mimosoïdeae*. Single flowers sometimes occur. Dorsiventral racemes, resembling the cymes of *Boraginaceae*, also occur (e.g. *Dalbergia*). The fls. are regular (and then frequently polygamous) or zygo. (and then usu. ☿); recept. more often cupular than convex or flat, so that the fl. is usu. slightly perig. Tubular receptacles occur sometimes, e.g. in *Bauhinia*. K developed in ascending order, usu. 5-merous, the odd (oldest) sepal anterior; the sepals ± united. C polypetalous, alt. with the K; aestivation valvate (*Mimosoïdeae*), ascending imbr. (*Caesalpinioïdeae*), or descending (*Papilionoïdeae*). In many cases it is zygo. to a high degree, having a large petal posterior (*vexillum* or standard), two lateral (*alae* or wings), and two anterior ± joined to form a keel or *carina*. A typically of 10 sta., or frequently ∞ in *Mimosoïdeae*, free or united into a tube; in the latter case the tenth sta. (the posterior one) often remains free, so as to leave a slit in the tube, only covered loosely by this sta. Many variations are found. In cases where a keel is present, the sta. are enclosed in it. G typically of one cpl. with its ventral side directly posterior; long style and terminal stigma. There are two rows of ovules (alt. with one another so as to stand in one vertical rank), anatr. or amphitr., ascending or pend.

FERTILISATION (*Papilionoïdeae*). The keel encloses the essential organs, protecting them from rain, etc., and rendering the fl. complex. Honey is secreted by the inner sides of the sta. near their base, and accumulates in the stamen-tube round the base of the ovary. The tenth sta. is free of the tube, and at the base, on either side of it, are two openings leading to the honey. The honey is thus concealed and at some depth, so that a clever insect with a tongue of moderate length is required. All this points to the *P.* being bee flowers, as is in fact the case. Insects alight upon the wings and depress them by their weight, whilst they probe for honey under the standard. The wings are always joined to the keel, usually by a protuberance in the former fitting into a suitable hollow in the latter, so that the keel is thus depressed likewise. This causes the emergence of the essential organs, the stigmas usu. coming first, so that a fair chance of cross-fert. exists. Self-pollination usu. occurs when the insect flies off, leaving the keel to return to its former position.

'Four different types of structure may be distinguished (in *Papilionoïdeae*) according to the manner in which the pollen is applied to the bee: (1) *Papil.* in which the sta. and stigma emerge from the carina and again return within it. They admit repeated visits; e.g. *Trifolium, Onobrychis*. (2) *P.* whose essential organs are confined under tension and explode. In these only one insect's visit is effective; e.g. *Medicago, Genista, Ulex*. (3) *P.* with a piston mechanism which squeezes the pollen in small quantities out of the apex of the carina, and not only permits but requires numerous insect visits; e.g. *Lotus, Ononis, Lupinus*. (4) *P.* with a brush of hairs upon the style which sweeps the pollen in small portions out of the apex of the carina. They for the most part require repeated insect visits; e.g. *Lathyrus, Vicia*.' (Müller.)

Cleistogamy is fairly common. In several cases the stigma in the unvisited fl. lies in the keel among the pollen, but it has been shown that it only becomes receptive (if young) when rubbed, so that autogamy does not necessarily occur. For the phenomenon of *enantiostyly* (right- and left-styled fls.), see *Cassia*. Some have fls. which after fertilisation bury themselves in the earth and there ripen their fruit (geocarpic); e.g. *Arachis, Lathyrus, Trifolium, Vicia, Voandzeia*, etc.

Fruit extremely variable, typically a *legume* opening by both sutures, but sometimes by one suture, or quite indehiscent. The fruit may be dry or fleshy, straight, curved or spirally coiled. In some the pod is constricted between the seeds, forming a *lomentum* which breaks up into indeh. one-seeded portions. The pods frequently open explosively, the valves twisting up spirally, e.g. in *Ulex, Cytisus* spp., etc. In *Colutea*, etc., they are inflated. Some are eaten by animals, but the seed-coats are hard enough to preserve the seeds from injury. Some seeds have a coloured fleshy aril (*Afzelia* spp., etc.). Still others have

LEGUMINOSAE

hooked pods, e.g. *Medicago*, *Mimosa*. The seed is alb. or more often exalb.; usu. large store of reserves in the cot.

Economically the *L.* are most important. The seeds of many spp. form important foodstuffs, e.g. of *Arachis*, *Cajanus*, *Cicer*, *Dolichos*, *Glycine*, *Lathyrus*, *Lens*, *Lotus*, *Lupinus*, *Phaseolus*, *Pisum*, *Vicia*, *Voandzeia*, etc. The pods of *Ceratonia*, *Tamarindus*, *Phaseolus*, *Prosopis*, etc., are also eaten. A great number are valuable as fodder, e.g. *Trifolium*, *Medicago*, *Onobrychis*, *Lotus*, *Vicia*, etc. Many trop. and subtrop. spp. yield valuable timber, e.g. *Acacia*, *Albizia*, *Dalbergia*, *Gleditsia*, *Hymenaea*, *Melanoxylon*, *Pericopsis* (*Afrormosia*), *Pterocarpus*, *Robinia*, *Sophora*, etc.; *Crotalaria* and others are sources of fibre; *Acacia*, *Genista*, *Haematoxylon*, *Indigofera*, etc., yield dyes; gums and resins are obtained from *Acacia*, *Astragalus*, *Copaifera*, *Hymenaea*, etc.; oil is expressed from the seeds of *Arachis* and *Voandzeia*; kino is obtained from *Pterocarpus*, and so on.

Nearly related to *Rosaceae*, *Chrysobalanaceae* and *Connaraceae*. Sometimes made a separate order, with the 3 divisions as fams.

Classification and chief genera:

I. **Mimosoïdeae**. Fls. reg. Sepals valvate, rarely imbricate. Petals valvate in bud. Stamens 4–10 or numerous. Pollen frequently compound. Seeds generally marked with areoles on side. Lvs. often bipinnate.

II. **Caesalpinioïdeae**. Fls. usually ± irreg. Sep. imbricate, rarely valvate. Petals imbricate-ascending in bud. Stamens 10 or fewer, rarely numerous. Pollen grains usually simple. Seeds generally without areoles. Lvs. usually pinnate, occasionally bipinnate, rarely simple.

III. **Papilionoïdeae (Faboïdeae)**. Fls. usually irreg. Sep. imbricate. Petals imbricate-descending in bud. Stamens generally 10. Pollen grains simple. Seeds without areoles. Leaves pinnate, digitate, trifoliolate or simple.

These are again subdivided as follows:

I. Mimosoïdeae

A. Calyx valvate:
   a. Sta. more than 10:
      1. INGEAE (A united into a tube): *Inga* (ls. once pinnate, G 1), *Calliandra*, *Pithecellobium*, *Albizia* (ls. twice pinnate, G 1, pod flat, straight, not elastic).
      2. ACACIËAE (A free or only slightly joined at base): *Acacia* (only genus).
   b. Sta. as many or twice as many as pets.:
      3. MIMOSEAE (anther glandless): *Mimosa* (pod compressed, flat).
      4. ADENANTHEREAE (anther in bud crowned by a gland; endosperm): *Neptunia* (head of fl. ♂ or neuter below), *Prosopis*.
      5. PIPTADENIËAE (anther usually glandular, rarely without a gland; no endosp.): *Piptadenia*, *Entada*.
B. Calyx imbricate:
      6. PARKIËAE: *Parkia* (fls. in heads), *Pentaclethra* (spikes).

II. *Caesalpinioïdeae*

A. Ls. bipinnate:
a. Fls. small, in spikes or spike-like racemes. Calyx-lobes connate
below into a short tube extending beyond the hypanthium.
DIMORPHANDREAE
b. Fls. usu. medium to large, in racemes or panicles of racemes.
Sepals 5, free to hypanthium. CAESALPINIËAE
B. Ls. simply pinnate, or sometimes simple or unifoliolate:
a. Sepals distinct in bud, free to base:
x. Anthers firm in texture, dehiscing by pores, sometimes by short
slits, sometimes by slits as long as anther, which is then basifixed. Leaves normally simply pinnate. CASSIËAE
xx. Anthers dorsifixed, dehiscing by slits:
§. Ls. normally imparipinnate. SCLEROLOBIËAE
§§. Ls. paripinnate or simple:
o. Bracteoles small or large, not enclosing the flower-buds, or
enclosing them, but then never valvate. CYNOMETREAE
oo. Bracteoles well developed, enclosing the flower-buds,
valvate, usu. persistent. AMHERSTIËAE
b. Sepals ± joined above the hypanthium, or calyx entire in bud
and becoming variously lobed or slit as flower opens:
x. Leaves usually simple, bilobed or entire, sometimes with two
separate leaflets. Calyx shortly toothed or lobed in bud.
Stamens 10 or fewer. CERCIDEAE (BAUHINIËAE)
xx. Leaves simply pinnate, rarely unifoliolate. Calyx entire in bud,
closed, not divided into sepals, becoming variously lobed or
slit as flower opens. SWARTZIËAE

III. *Papilionoïdeae*

(Conspectus of tribes, after Bentham, modified.)
*Stamens free or almost so:
1. SOPHOREAE: trees, shrubs or rarely woody herbs or lianes; ls. pinnate or
1-foliolate with a joint between petiole and lamina: *Myroxylon, Ormosia,
Sophora, Baphia*, etc.
2. PODALYRIËAE: shrubs or, less often, herbs; ls. simple, with no joint
between petiole and lamina, or digitately 3- or more-foliolate: *Anagyris,
Podalyria*, etc.
** Stamens mon- or di-adelphous:
(A) Herbs or shrubs (or trees or lianes in tribes 6 & 10); pods dehiscent
unless short and 1–2-seeded, or inflated:
(a) Pods never transversely jointed:
(α) Ls. never stipellate:
3. GENISTEAE: usually shrubs; ls. simple or digitately 3- or more-foliolate,
lflts. entire; stamens usually monadelphous, anthers often of 2 sizes:
Lotononis, Crotalaria, Lupinus, Cytisus, etc.
4. TRIFOLIËAE: usually herbs; ls. pinnately or digitately 3-foliolate, lflts.
usually toothed; corolla glabrous; stamens mon- or di-adelphous:
Ononis, Medicago, Melilotus, Trifolium, etc.

LEHMANNA

5. LOTEAE: ls. pinnately 3–∞-foliolate, lflts. entire; corolla glabrous: stamens mon- or di-adelphous, alternate filaments usually dilated at the tip: *Anthyllis, Lotus, Hosackia,* etc.

(β) Ls. sometimes stipellate (except in tribe 7):

6. GALEGEAE: ls. pinnately 5–∞- or rarely 3- or 1-foliolate, or rarely digitately 3-foliolate; lflts. usually entire; rhachis never ending in a tendril; stamens usually diadelphous: *Psoralea, Indigofera, Tephrosia, Millettia, Sesbania, Astragalus, Oxytropis,* etc.

7. FABEAE (VICIËAE): herbs; ls. paripinnate, without stipels, rhachis ending in a point or tendril; stamens 10, diadelphous: *Cicer, Vicia, Lens, Lathyrus, Pisum.*

8. ABREAE: shrubs or twiners woody at the base; ls. paripinnate, rhachis ending in a point, usually stipellate; stamens 9, united: *Abrus.*

9. PHASEOLEAE: often twining; ls. pinnately 3-foliolate, usually stipellate, rarely 1- or 5–7-foliolate; lflts. entire or lobed; stamens di- or sub-monadelphous: *Glycine, Erythrina, Mucuna, Galactia, Canavalia, Rhynchosia, Eriosema, Phaseolus, Vigna, Dolichos,* etc.

(b) Pods transversely jointed:

10. HEDYSAREAE: habit of *Loteae, Galegeae* or *Phaseoleae*; ls. stipellate or not: *Coronilla, Hedysarum, Onobrychis, Aeschynomene, Kotschya, Arachis, Zornia, Desmodium.*

(B) Trees, shrubs or lianes: pods indehiscent:

11. DALBERGIËAE: ls. pinnately 5–∞-foliolate, rarely 3–1-foliolate, stipellate or not: *Dalbergia, Pterocarpus, Derris, Leptoderris, Lonchocarpus.*

Lehmanna Casseb. & Theob. = Gentiana L. (Gentianac.).
Lehmannia Jacq. ex Steud. = Moschosma Reichb. (Labiat.).
Lehmannia Spreng. = Nicotiana L. (Solanac.).
Lehmannia Tratt. = Tylosperma Botsch. (Rosac.).
Lehmanniella Gilg. Gentianaceae. 1 Colombia.
Leiachenis Rafin. = Aster L. (Compos.).
Leiachensis Merr. (sphalm.) = praec.
Leiacherus Rafin. (sphalm.) = praec.
Leiandra Rafin. = Tradescantia L. (Commelinac.).
Leianthostemon Miq. = Voyria Aubl. (Gentianac.).
Leianthus Griseb. = Lisianthus L. (Gentianac.).
Leibergia Coulter & Rose = Lomatium Rafin. (Umbellif.).
Leibnitzia Cass. Compositae. 3 E. As., 2 C. Am.
Leiboldia Schlechtd. ex Gleason = Vernonia L. (Compos.).
Leicesteria Pritz. = Leycesteria Wall. (Caprifoliac.).
Leichhardtia F. Muell. = Phyllanthus L. (Euphorbiac.).
Leichhardtia Sheph. = Callitris Vent. (Cupressac.).
Leichtlinia H. Ross = Agave L. (Agavac.).
Leidesia Muell. Arg. (~ Seidelia Baill.). Euphorbiaceae. 1 S. Afr.
Leiena Rafin. = Restio Rottb. (Restionac.).
Leighia Cass. = Viguiera Kunth (Compos.).
Leighia Scop. = Ethulia L. (Compos.).
Leimanisa Rafin. = Gentianella Moench (Gentianac.).
Leimanthemum Ritgen = seq.

Leimanthium Willd. = Melanthium L. (Liliac.).
Leinckeria Neck. = seq.
Leinkeria Scop. = Roupala Aubl. (Proteac.).
Leiocalyx Planch. ex Hook. = Dissotis Benth. (Melastomatac.).
Leiocarpodicraea (Engl.) Engl. = Leiothylax Warm. (Podostemac.).
Leiocarpus Blume = Aporusa Blume (Euphorbiac.).
Leiocarya Hochst. = Trichodesma R.Br. (Boraginac.).
Leiochilus Benth. = Leochilus Knowles & Westc. (Orchidac.).
Leiochilus Hook. f. = Buseria Th. Dur. (Rubiac.).
Leioclusia Baill. = Carissa L. (Apocynac.).
Leiodon Shuttlew. ex Sherff = Coreopsis L. (Compos.).
Leiogyne K. Schum. = Neves-Armondia K. Schum. (Bignoniac.).
Leioligo Rafin. = Solidago L. (Compos.).
Leiolobium Benth. = Dalbergia L. (Legumin.).
Leiolobium Reichb. = Rorippa Scop. (Crucif.).
Leioluma Baill. = Lucuma Molina (Sapotac.).
Leiophaca Lindau (= ? Whitfieldia Hook.). Acanthaceae. 1 trop. Afr.
Leiophyllum Ehrh. (uninom.) = *Schoenus compressus* L. = *Blysmus compressus* (L.) Panz. ex Link (Cyperac.).
Leiophyllum (Pers.) Hedw. f. Ericaceae. 1 Atl. U.S.
Leiophyltum Rafin. = praec.
Leiopoa Ohwi. Gramineae. 2 Korea, Japan.
Leiopogon Dur. & Schinz (sphalm.) = Lepipogon Bertol. f. = ? Tricalysia A. Rich. (Rubiac.).
Leioptyx Pierre ex De Wild. = Entandrophragma C. DC. (Meliac.).
Leiopyxis Miq. = Cleistanthus Hook. f. ex Planch. (Euphorbiac.).
Leiosandra Rafin. = Verbascum L. (Scrophulariac.).
Leiospermum D. Don = Weinmannia L. (Cunoniac.).
Leiospermum Wall. = Psilotrichum Blume (Amaranthac.).
Leiospora (C. A. Mey.) Vassiljeva. Cruciferae. 2 C. As.
Leiostegia Benth. = Comolia DC. (Melastomatac.).
Leiostemon Rafin. Scrophulariaceae. 1 N. Japan, Kamtsch., etc.
Leiotelis Rafin. = Seseli L. (Umbellif.).
Leiothamnus Griseb. = Lisianthus L. (Gentianac.).
Leiothrix Ruhland. Eriocaulaceae. 65 S. Am.
Leiothylax Warm. Podostemaceae. 6 trop. Afr.
Leiotulus Ehrenb. = Malabaila Hoffm. (Umbellif.).
Leiphaimos Cham. & Schlechtd. Gentianaceae. 40 C. & trop. S. Am., W.I., trop. Afr.
Leipoldtia L. Bolus. Aïzoaceae. 20 S. Afr.
Leipotyx auct. (sphalm.) = Leioptyx Pierre (Meliac.), *q.v.*
Leitgebia Eichl. Ochnaceae. 4 N. trop. S. Am.
Leitneria Chapm. Leitneriaceae. 1 SE. U.S., *L. floridana* Chapm. (corkwood).
*****Leitneriaceae** Benth. Dicots. 1/1 SE. U.S. Shrubs with resin-canals, and with simple ent. alt. long-pet. exstip. ls. Fls. ♂ ♀, dioec., precocious, in erect axill. catkin-like spikes, each fl. subtended by a bract. ♂: P 0; A 3–12, with filif. fil. and basifixed anth. ♀: P (?) of 2–4 minute unequal glandular-fimbr. scales, ± connate at base; G 1, with 1 ascend. laterally attached ov., and long

649

simple falcate unilaterally stigmatic style. Fr. drupaceous, ± ellipsoid, slightly asymm., somewhat flattened, with conspic. retic. endocarp; seed with thin fleshy endosp. Only genus: *Leitneria*. An interesting relict, of very obscure affinities; there may be distant connections with *Myricac.*, *Didymelac.*, or *Daphniphyllac.*

Lejica DC. = Lepia Hill = Zinnia L. (Compos.).

Lejocarpus (DC.) P. & K. = Anogeissus Wall. (Combretac.).

Lejochilus P. & K. = Leiochilus Benth. = Leochilus Knowles & Westc. (Orchidac.).

Lejogyna Bur. ex P. & K. = Leiogyne K. Schum. = Neves-Armondia K. Schum. (Bignoniac.).

Lejophyllum P. & K. = Leiophyllum (Pers.) Hedw. f. (Ericac.).

Lejopogon P. & K. = Leiopogon Dur. & Schinz = Lepipogon Bertol. f. = ? Tricalysia A. Rich. (Rubiac.).

Leleba Rumph. ex Schult. (as to Malaysian plants only) (~ Bambusa Schreb.). Gramineae. 6 Moluccas to Solomon Is.

Leloutria Gaudich. = Nolana L. (Nolanac.).

Lelya Bremek. Rubiaceae. 1 trop. Afr.

Lemairea De Vriese = *mixtum compositum*: Mycetia Korth. (Rubiac., *quoad ram. fructif.*) + Scaevola L. (Goodeniac., *quoad fl.*).

Lemaireocereus Britton & Rose. Cactaceae. 25 C. Am. to Venez. & Colombia, W.I.

Lemapteris Rafin. = Pteris L. (Pteridac.).

Lembertia Greene = Eatonella A. Gray (Compos.).

Lembocarpus Leeuwenb. Gesneriaceae. 1 Guiana.

Lembotropis Griseb. = Cytisus L. (Legumin.).

Lemia Vand. = Portulaca L. (Portulacac.).

Lemma Adans. = Marsilea L. (Marsileac.).

Lemmaphyllum Presl. Polypodiaceae. 5 NE. India to Malaysia & Japan.

Lemmatium DC. = Calea L. (Compos.).

Lemmonia A. Gray. Hydrophyllaceae. 1 California.

Lemna L. Lemnaceae. 15 cosmop. The plant consists of a flat green floating (e.g. *L. minor* L.) or submerged (e.g. *L. trisulca* L.) blade, or frond, which performs leaf functions. From the under side hangs down a long adv. root, with well-marked root-cap, visible to the eye. No ls. The floating fronds are oval and slightly turned up at the ends, so that if two are placed near together in water, the surface tension will cause them to run against one another and adhere by the tips. In the post. portion on either side is a groove under the edge. In this arise branches which may either (as in *L. trisulca*, etc.) remain in union with the parent shoot, or become detached and give rise to new plants. In autumn a number of these are formed ready to start growth next spring, whilst the mother pls. sink to the bottom. Fls. also borne in these grooves; spathe very reduced, with 2 ♂ fls. (each 1 sta.) and a ♀ (1 cpl.).

*Lemnaceae S. F. Gray. Monocots. 6/30 cosmop. Free-swimming, floating or submerged, small or minute, monoec. (rarely dioec.), annual water-pl., with undifferentiated thalloid fronds of various form, bearing 0, 1, or several roots, and 1 or 2 budding pouches. Infl. of 1 ♀ + 1–2 ♂ fls., naked or surr. by a spathe. P o; A 1, anther 1- or 2-loc.; G̲ 1, glob., 1-loc., with short style and

1–4 ovules (amphitr., anatr. or orthotr.). Fr. a 1–4-seeded utricle; seeds smooth or ribbed.

Classification (after den Hartog & van der Plas, *Blumea*, **18**: 355–68, 1970):

I. *Lemnoïdeae.* Roots present; budding pouches 2; infl. 1, devel. from 1 budding pouch, of 1 ♀ and 2 ♂ fls., encl. in a membr. spathe; anther biloc., transv. dehisc.; raphides present. *Spirodela* (fronds with dorsal & ventral scale, often many roots and many nerves; brown pigment cells present); *Lemna* (fronds without scales, with 1 root [rarely 0] and 1–3 nerves; no pigment cells).

II. *Wolffioïdeae.* Roots absent; budding pouch 1, never flowering; infl. dorsal, from a cavity containing 1 ♀ and 1 ♂ fl., without a spathe; anther 1-loc., apically dehisc.; raphides 0. *Wolffia* (floating, glob.; budding pouch with circular opening; stipes ['petiole'] not evident); *Pseudowolffia* (floating, flat, without pigment cells; budding pouch symmetric; stipes median, long; infl. 1); *Wolffiopsis* (frond submerged, membr., symm., with pigment cells, ±ovate, slightly curled; budding pouch symm.; stipes median, short; infl. 2); *Wolffiella* (as last, but frond asymm., ±linear, ±falc., budding pouch asymm.; stipes lat.; infl. 1).

Lemnescia Willd. = seq.

Lemniscia Schreb. = Vantanea Aubl. (Houmiriac.).

Lemniscoa Hook. = Bulbophyllum Thou. (Orchidac.).

Lemnopsis Zipp. = Utricularia L. (Lentibulariac.).

Lemnopsis Zoll. = Halophila Thou. (Hydrocharitac.).

Lemonia Lindl. = Ravenia Vell. (Rutac.).

Lemonia Pers. = Lomenia Pourr. = Watsonia Mill. (Iridac.).

Le-Monniera Lecomte (1918; non Lemonniera De Wild. 1894—Fungi) = Neolemonniera Heine (Sapotac.).

Lemotris Rafin. = Camassia Lindl. (Liliac.).

Lemotrys Rafin. = praec.

Lemphoria O. E. Schulz. Cruciferae. 1 S. Austr.

Lemuranthe Schlechter = Cynorkis Thou. (Orchidac.).

Lemurella Schlechter. Orchidaceae. 4 Madag.

Lemuropisum H. Perrier. Leguminosae. 1 Madag.

Lemurorchis Kraenzl. Orchidaceae. 1 Madag.

Lemurosicyos Keraudren. Cucurbitaceae. 1 Madag.

Lemyrea (A. Chev.) A. Chev. (~ Coffea L.). Rubiaceae. 3 Madag.

Lencymmaea Benth. & Hook. f. = seq.

Lencymmoea C. Presl. Inc. sed. (?Myrtaceae). 1 Burma.

Lenda Koidz. = Tectaria Cav. (Aspidiac.).

Lendneria Minod = Poarium Desv. (Scrophulariac.).

Lenidia Thou. = Wormia Rottb. = Dillenia L. (Dilleniac.).

Lennea Klotzsch. Leguminosae. 6 Mex. to Urug. Style coiled.

Lennoa Lex. Lennoaceae. 3 C. Mex. Sta. of two lengths.

*****Lennoaceae** Solms-Laubach. Dicots. 3/5 SW. U.S., Mex. Parasitic fleshy herbs devoid of chlorophyll; ls. reduced to short scales; hosts shrubby *Compositae, Clematis, Euphorbia*, etc. Fls. ♀, reg. or slightly zygo., in dense bracteate thyrses, cpd. spikes, or discoid heads. K 5–10, linear, persist.; C (5–8), tubular or hypocrat., persist., lobes short, imbr., veined, with ± erose margin; A 5–10, alternipet., inserted below mouth of C, with intr. anth.; G̲ (6–15), each loc. divided into two 1-ov. locelli by a false septum, ov. anatr., axile, ± horiz.; style simple with capit. crenate stig. Fr. drupaceous-capsular, encl. by persist. K and C, finally ± circumsciss., with 12–28 1-seeded stones; seeds with endosp., embryo spherical, undifferentiated. Genera: *Lennoa, Pholisma, Ammobroma.* Probably related to *Ehretiaceae.* K sometimes (5–10).

Lenophyllum Rose. Crassulaceae. 6 S. U.S., Mex.

Lenormandia Steud. (1850; non Montagne 1843—Algae) = Mandelorna Steud. = Vetiveria Thou. ex Virey (Gramin.).

*Lens Mill. Leguminosae. 10 Medit., W. As. *L. esculenta* Moench (*Ervum lens* L.), the lentil, is a food pl. of great antiquity. The seeds furnish a flour. Close to *Vicia,* but ovules 2 only.

Lens Stickm. [L.]. = Entada Adans. (Legumin.).

Lentago Rafin. = Viburnum L. (Caprifoliac.).

Lentibularia Seguier = Utricularia L. (Lentibulariac.).

*Lentibulariaceae Rich. Dicots. 4/170 cosmop. (cf. Lloyd, *Carniv. Pls.* (1942), and for details see gen.). Insectivorous herbs, of water or moist places, often without r. Infl. usu. a raceme or spike, or fls. sol. Fls. ⚥, zygo., 5-merous. K 2–5-lobed, the odd sep. post., often 2-lipped, persistent; C (5), 2-lipped, lower lip ± spurred; A 2 (anterior pair), epipet., with 1-loc. anthers; G̲ (2), 1-loc. with free-central plac. and usu. sessile 2-lobed stigma (post. lobe ± abortive); ovules ∞–2, anatr., often ± sunk in plac. Caps. with 2–4 valves, circumscissile, or indehisc.; seeds 1–∞, exalb. Genera: *Pinguicula* (K 5, fls. sol. on long stalks, land pls.), *Genlisea* (K 5, raceme, land), *Polypompholyx* (K 4, land), *Utricularia* (K 2, land or water).

Lenticula Hill = Lemna L. (Lemnac.).

Lenticularia Montand. = praec.

Lenticularia Seguier = Spirodela Schleid. (Lemnac.).

Lentilla W. F. Wight = Lens Mill. (Legumin.).

Lentiscus Mill. = Pistacia L. (Pistaciac.).

Lentzia Schinz = seq.

Lenzia Phil. Portulacaceae. 1 Chile.

Leobardia Pomel = seq.

Leobordea Delile = Lotononis Eckl. & Zeyh. (Legumin.).

Leocereus Britton & Rose. Cactaceae. 5 E. Brazil.

Leochilus Knowles & Westc. Orchidaceae. 17 C. Am., W.I., trop. & temp. S. Am.

Leocus A. Chev. Labiatae. 1 trop. W. Afr.

Leonardendron Aubrév. Leguminosae. 1 W. Equat. Afr.

Leonardia Urb. = Thouinia Poit. (Sapindac.).

Leonardoxa Aubrév. Leguminosae. 3 trop. Afr.

Leonhardia Opiz = Nepa Webb (Legumin.).

Leonia Cerv. ex La Llave & Lex. = Salvia L. (Labiat.).

Leonia Mutis ex Kunth = Siparuna Aubl. (Siparunac.).

Leonia Ruiz & Pav. Violaceae. 6 trop. S. Am.

Leoniaceae DC. = Violaceae–Leonioïdeae Gilg.

Leonicenia Scop. = Miconia Ruiz & Pav. (Melastomatac.).

Leonicenoa P. & K. = praec.

Leonitis Spach = Leonotis R.Br. (Labiat.).

Leonocassia Britton = Cassia L. (Legumin.).

Leonotis (Pers.) R.Br. in Ait. Labiatae. 40 trop. & S. Afr., 1 pantrop.

Leontia Reichb. = Luutia Neck. = Croton L. (Euphorbiac.).

Leonticaceae (Spach) Airy Shaw. Dicots. 4/14 N. temp. Perennial herbs. Rhiz. tuberous, or occasionally creeping. Ls. mostly rad., simply or 2–3-pinnate, often fleshy, glaucous. Infl. simply or compoundly rac. K 3–9, inner often petaloid; C 6, often nectariform; A 6, dehiscing by 2 valves; G̲ 1, with 2–8 erect basal ovules and short style with small or plicate-dilated stig. Caps. bladdery, indehisc., or gaping above, or the walls evanescent, exposing the large drupe-like seeds. Genera: *Leontice, Gymnospermium, Bongardia, Caulophyllum.* A small fam. related to *Berberidac.* and *Podophyllac.*, but with habit of some *Fumariac.*, esp. *Corydalis* (tuber, fleshy glauc. pinn. ls., etc.).

Leontice L. Leonticaceae. 3–5 SE. Eur. to E. As.

Leontochir Phil. Alstroemeriaceae. 1 Chile. Plac. parietal.

Leontodon Adans. = Taraxacum Weber ex Wigg. (Compos.).

Leontodon L. Compositae. 50 temp. Euras., Medit. to Persia. Very like *Taraxacum.* In *L. hirtus* L. the outer frs. have no pappus.

Leontoglossum Hance = Tetracera L. (Dilleniac.).

Leontondon Robin (sphalm.) = Leontodon L. (Compos.).

Leontonyx Cass. Compositae. 8 S. Afr.

Leontopetaloïdes Boehm. = Tacca J. R. & G. Forst. (Taccac.).

Leontopetalon Mill. = Leontice L. (Leonticac.).

Leontophthalmum Willd. = Colea L. (Compos.).

Leontopodium R.Br. ex Cass. Compositae. 30 mts. of Eur., As. & S. Am. *L. alpinum* Cass. (Edelweiss) is a xero. growing in dense tufts, and covered with woolly hairs. The central florets are ♂, the style remaining, however, to act as pollen-presenter, though it has no stigmas. The outer florets are ♀.

Leonura Usteri ex Steud. = Salvia L. (Labiat.).

Leonuros St-Lag. = seq.

Leonurus L. Labiatae. 14 temp. Euras.

Leonurus Mill. = Leonotis (Pers.) R.Br. in Ait. (Labiat.).

Leopardanthus Blume = Dipodium R.Br. (Orchidac.).

Leopoldia Herb. = Hippeastrum Herb. (Amaryllidac.).

*Leopoldia Parl. = Muscari Mill. (Liliac.).

Leopoldinia Mart. Palmae. 4 Brazil. *L. piassaba* Wallace yields the best piassaba fibre.

Lepachis Rafin. = Echinacea Moench (Compos.).

Lepachys Rafin. = Ratibida Rafin. (Compos.).

Lepactis P. & K. (1) = Lepiactis Rafin. (1837) = Solidago L. (Compos.).

Lepactis P. & K. (2) = Lepiactis Rafin. (1838) = Utricularia L. (Lentibulariac.).

Lepadanthus Ridl. = Ornithoboea Parish ex C. B. Cl. (Gesneriac.).

Lepadena Rafin. = Dichrophyllum Klotzsch & Garcke = Euphorbia L. (Euphorbiac.).

Lepaglaea P. & K. = Lepiaglaia Pierre = Aglaia Lour. (Meliac.).

Lepantes Sw. = Lepanthes Sw. (Orchidac.).

Lepanthes P. & K. = Lepianthes Rafin. = Piper L. (Piperac.).

Lepanthes Sw. Orchidaceae. 100 C. & trop. S. Am., W.I.

22-2

Lepanthopsis Ames. Orchidaceae. 15 trop. Am., W.I.
Lepargochloa Launert. Gramineae. 1 S. trop. Afr.
Lepargyraea Steud. = seq.
Lepargyrea Rafin. = Shepherdia Nutt. (Elaeagnac.).
Lepechinia Willd. Labiatae. 40 Calif. to Argent., Hawaii.
Lepechinella Popov. Boraginaceae. 9 C. As.
Lepedera Rafin. = Lespedeza Michx (Legumin.).
Lepeocercis Trin. = Andropogon L. (Gramin.).
Lepeostegeres Blume. Loranthaceae. 13 W. Malaysia, Celebes.
Leperiza Herb. = Urceolina Reichb. (Amaryllidac.).
Lepervenchea Cordemoy = Angraecum Bory (Orchidac.).
Lepia Desv. = Lepidium L. (Crucif.).
Lepia Hill = Zinnia L. (Compos.).
Lepiactis Rafin. (1837) = Solidago L. (Compos.).
Lepiactis Rafin. (1838) = Utricularia L. (Lentibulariac.).
Lepiagalia Pierre = seq.
Lepiaglaia Pierre = Aglaia Lour. (Meliac.).
Lepiaglia Pierre = praec.
Lepianthes Rafin. = Piper L. (Piperac.).
Lepicaulon Rafin. = Anthericum L. (Liliac.).
Lepicaune Lapeyr. = Crepis L. (Compos.).
Lepicephalus Lag. = Cephalaria Schrad. (Dipsacac.).
Lepichlaena P. & K. = Lepilaena J. Drumm. ex Harv. (Zanichelliac.).
Lepicline Less. = Lepiscline Cass. = Helichrysum Mill. (Compos.).
Lepicochlea Rojas = Coronopus Mill. (Crucif.).
Lepicystis J. Sm. = Polypodium L. (Polypodiac.).
Lepidacanthus C. Presl. Acanthaceae. 2 Brazil.
Lepidadenia Arn. ex Nees = Litsea Lam. (Laurac.).
Lepidagathis Willd. Acanthaceae. 100 trop. & subtrop.
Lepidaglaia Dyer = Lepiaglaia Pierre = Aglaia Lour. (Meliac.).
Lepidalenia P. & K. (sphalm.) = Lepidadenia Arn. ex Nees = Litsea Lam. (Laurac.).
Lepidamphora Zoll. ex Miq. = Dioclea Kunth (Legumin.).
Lepidanche Engelm. = Cuscuta L. (Cuscutac.).
Lepidanthemum Klotzsch = Dissotis Benth. (Melastomatac.).
Lepidanthus Nees = Hypodiscus Nees (Restionac.).
Lepidanthus Nutt. (1837) = Andrachne L. (Euphorbiac.).
Lepidanthus Nutt. (1841) = Matricaria L. (Compos.).
Lepidaploa Cass. = Vernonia Schreb. (Compos.).
Lepidaria Van Tiegh. Loranthaceae. 12 W. Malaysia.
Lepidariaceae Van Tiegh. = Loranthaceae–Elytranthinae Engl.
Lepideilema Trin. = Streptochaeta Schrad. ex Nees (Streptochaetac.).
Lepidella Van Tiegh. = Lepidaria Van Tiegh. (Loranthac.).
Lepiderema Radlk. Sapindaceae. 5 New Guinea, NE. Austr.
Lepidesmia Klatt = Ayapana Spach (Compos.).
Lepidiberis Fourr. = Lepidium L. (Crucif.).
Lepidilema P. & K. = Lepideilema Trin. = Streptochaeta Schrad. ex Nees (Streptochaetac.).

Lepidinella Spach = seq.

Lepidion St-Lag. = seq.

Lepidium L. Cruciferae. 150 cosmop. *L. sativum* L. (Orient) is the garden cress.

Lepidobolus Nees. Restionaceae. 4 S. Austr.

*Lepidobotryaceae Léonard. Dicots. 1/1 trop. Afr. [+2/13 W. Malaysia?] Shrubs, [trees, or climbers,] with alt. 1[–3]-foliol. stip. ls.; lfts. ent., stipellate, pet. and petiolules artic. Fls. reg., ♂ ♀, dioec., [or ♀,] in short strobiloid racemes [or in panic. cymes or elong. rac.]. K 5, imbr., shortly connate below; C 5, imbr.; A 10, on fleshy disk [or disk o], shortly connate [or broadly tubular] below, with dorsifixed anth.; G̲ (3) [or (5)] with 3 [or 5] free or shortly connate styles, and 2 axile collat. [or 1–6 superposed] ov. per loc.; caps. largish, coriac., ellips., septicid., 2(–3)-valved, 1-seeded [or small, fleshy, 5-lobed indehisc., or dehiscing loculicidally with widely spreading valves, few-seeded]; seeds arillate [or (*Sarcotheca*) exarillate], with endosp. Genera: *Lepidobotrys*[, ? *Sarcotheca*, ?*Dapania*]. The divergent characters of *Sarcotheca* and *Dapania*, referred to this fam. by Hutchinson (1959), are indicated by square brackets [ ]. These genera are better referred to the *Averrhoaceae*.

Lepidobotrys Engl. Lepidobotryaceae. 1 trop. Afr.

Lepidocarpa Korth. = Parinari Aubl. (Chrysobalanac.).

Lepidocarpaceae ('-icae') Schultz-Schultzenst. = Proteaceae Juss.

Lepidocarpus Adans. = Protea L. (1771). (Proteac.).

Lepidocarpus P. & K. = Lepidocarpa Korth. = Parinari Aubl. (Chrysobalanac.).

Lepidocarya Korth. ex Miq. = praec.

Lepidocaryaceae O. F. Cook = Palmae–Lepidocaryoïdeae Mart.

Lepidocaryon Spreng. = seq.

Lepidocaryum Mart. Palmae. 8 trop. S. Am.

Lepidocaulon Copel. = Histiopteris (Ag.) J. Sm. (Dennstaedtiac.) (Holttum, *Kew Bull.* 20: 459, 1968).

Lepidoceras Hook. f. Viscaceae (~ Loranthaceae). 1 Peru to Chile.

Lepidocerataceae Van Tiegh. = Viscaceae Miq.

Lepidococca Turcz. = Caperonia A. St-Hil. (Euphorbiac.).

Lepidococcus H. Wendl. & Drude = Mauritiella Burret (Palm.).

Lepidocoma Jungh. = Flemingia Roxb. = Maughania J. St-Hil. (Legumin.).

Lepidocordia Ducke. Ehretiaceae. 1 Brazil.

Lepidocoryphantha Backeb. = Coryphantha (Engelm.) Lem. (Cactac.).

Lepidocroton Klotzsch = Hieronima Fr. Allem. (Euphorbiac.).

Lepidocroton Presl = Chrozophora Neck. ex Juss. (Euphorbiac.).

Lepidogrammitis Ching = Lemmaphyllum Presl (Polypodiac.).

Lepidogyne Blume. Orchidaceae. 3 Malaysia.

Lepidolopha C. Winkler. Compositae. 6 C. As.

Lepidolopsis Poljakov (emend. Tsvelev). Compositae. 1 C. As., Persia, Afghan.

Lepidonema Fisch. & Mey. = Microseris D. Don (Compos.).

Lepidoneuron Fée = Nephrolepis Schott (Oleandrac.).

Lepidonia Blake. Compositae. 1 C. Am. (Guatem.).

Lepidopappus Moç. & Sessé ex DC. = Florestina Cass. (Compos.).

Lepidopelma Klotzsch = Sarcococca Lindl. (Buxac.).

LEPIDOPETALUM

Lepidopetalum Blume. Sapindaceae. 6 Andaman & Nicobar Is., Sumatra, Philipp. Is., Tenimbar & Kei Is., New Guinea, Bismarck Archip.

Lepidopharynx Rusby = Hippeastrum Herb. (Amaryllidac.).

Lepidophorum Neck. ex DC. Compositae. 1 Portugal.

Lepidophyllum Cass. Compositae. 1 Patag.

Lepidophyton Benth. & Hook. f. = seq.

Lepidophytum Hook. f. = Lophophytum Schott & Endl. (Balanophorac.).

Lepidopironia A. Rich. = Tetrapogon Desf. (Gramin.).

Lepidoploa Sch. Bip. (sphalm.) = Lepidaploa Cass. = Vernonia Schreb. (Compos.).

Lepidopogon Tausch = Cylindrocline Cass. (Compos.).

Lepidopteris L. S. Gibbs (sphalm.) = Leptopteris Blume = Gelsemium Juss. (Loganiac.).

Lepidopyronia Benth. = Lepidopironia A. Rich. (Gramin.).

Lepidorrhachis (H. Wendl. & Drude) O. F. Cook. Palmae. 1 Lord Howe I.

Lepidoseris (Reichb.) Fourr. = Crepis L. (Compos.).

Lepidospartum A. Gray. Compositae. 3 SW. U.S.

Lepidosperma Labill. Cyperaceae. 50 S. China, Malay Penins., NW. New Guinea (Waigeo I.), Austr., N.Z. *L. gladiatum* Labill. is the sword-sedge, used to bind sand-dunes in Austr., and as a material for paper-making.

Lepidosperma Schrad. = Schoenus L. (Cyperac.).

Lepidospora F. Muell. = Schoenus L. (Cyperac.).

Lepidostachys Wall. = Aporusa Blume (Euphorbiac.).

Lepidostemon Hassk. = Lepistemon Blume (Convolvulac.).

*Lepidostemon Hook. f. & Thoms. Cruciferae. 1 E. Himalaya.

Lepidostemon Lem. = Keckiella Straw (Scrophulariac.).

Lepidostephanus Bartl. = Achyrachaena Schau. (Compos.).

Lepidostephium Oliv. Compositae. 1 S. Afr.

Lepidostoma Bremek. Rubiaceae. 1 Sumatra.

Lepidothamnus Phil. = Dacrydium Soland. ex Forst. f. (Podocarpac.).

Lepidotheca Nutt. = Matricaria L. (Compos.).

Lepidotis P. Beauv. = Lycopodium L. (Lycopodiac.).

Lepidotosperma Roem. & Schult. = Lepidosperma Labill. (Cyperac.).

Lepidotrichilia (Harms) Leroy. Meliaceae. 4 Madag.

Lepidotrichum Velen. & Bornm. = Aurinia Desf. (Crucif.).

Lepidoturus Baill. = Alchornea Sw. (Euphorbiac.).

Lepidoturus Bojer = ? Acalypha L. (Euphorbiac.).

Lepidozamia Regel. Zamiaceae. 2 E. Austr.

Lepidurus Janchen = Parapholis C. E. Hubbard (Gramin.).

Lepigonum Wahlberg = Spergularia J. & C. Presl (Caryophyllac.).

Lepilaena J. Drumm. ex Harv. Zannichelliaceae. 3 Austr.

Lepimenes Rafin. = Cuscuta L. (Cuscutac.).

Lepinema Rafin. = Enicostema Blume (Gentianac.).

Lepinia Decne. Apocynaceae. 3 Caroline & Solomon Is., Tahiti.

Lepiniopsis Valeton. Apocynaceae. 2 Philipp. & Palau Is., Moluccas.

Lepionopsis Valeton (sphalm.) = praec.

Lepionurus Blume. Opiliaceae. 5–6 E. Himal. to Indoch., Java, New Guinea.

Lepiostegeres Benth. & Hook. f. = Lepeostegeres Bl. (Loranthac.).

Lepiphaia Rafin. = Nevrolis Rafin. = Celosia L. (Amaranthac.).
Lepiphyllum Korth. ex Penzig = Salomonia Lour. (Polygalac.).
Lepipogon Bertol. f. = ? Tricalysia A. Rich. (Rubiac.).
Lepirhiza P. & K. = Leperiza Herb. = Urceolina Reichb. (Amaryllidac.).
Lepirodia Juss. = Lepyrodia R.Br. (Restionac.).
Lepironia Rich. Cyperaceae. 1 Madag., trop. As., Austr., Polynes., *L. mucro-
nata* Rich., cult. in China. The stems are beaten flat and woven into mats, sails
Lepisanthes Blume. Sapindaceae. 40 trop. As.   [(for junks), etc.
Lepiscline Cass. = Helichrysum Mill. (Compos.).
Lepiselinum Presl = Lepisma E. Mey. (Umbellif.).
Lepisia Presl = Tetraria Beauv. (Cyperac.).
Lepisiphon Turcz. = Osteospermum L. (Compos.).
Lepisma E. Mey. = Annesorhiza Cham. & Schlechtd. + Polemannia Eckl. &
Zeyh. + Rhyticarpus Sond. (Umbellif.).
Lepismium Pfeiff. = Rhipsalis Gaertn. (Cactac.).
Lepisorus (J. Sm.) Ching = Pleopeltis Humb. & Bonpl. (Polypodiac.).
Lepisperma Rafin. = Spermolepis Rafin. (Umbellif.).
Lepistachya Zipp. ex Miq. = Mapania Aubl. (Cyperac.).
Lepistemon Blume. Convolvulaceae. 10 trop. Afr. to trop. Austr.
Lepistemonopsis Dammer. Convolvulaceae. 1 trop. E. Afr.
Lepistichaceae Dulac = Cyperaceae Juss.
Lepistoma Blume = Cryptolepis R.Br. (Asclepiadac.).
Lepitoma Torr. ex Steud. = Pleuropogon R.Br. (Gramin.).
Lepiurus Dum. = Lepturus R.Br. (Gramin.).
Leplaea Vermoesen = Guarea L. (Meliac.).
Leporella George. Orchidaceae. 1 W. Austr.
Leposma Blume = Cryptolepis R.Br. (Asclepiadac.).
Lepostegeres Van Tiegh. (sphalm.) = Lepeostegeres Bl. (Loranthac.).
Lepsia Klotzsch = Begonia L. (Begoniac.).
Lepta Lour. = Euodia J. R. & G. Forst. (Rutac.).
Leptacanthus Nees. Acanthaceae. 5 Penins. Ind., Ceylon.
Leptactina Hook. f. Rubiaceae. 25 trop. & S. Afr.
Leptactinia Hook. f. = praec.
Leptadenia R.Br. Asclepiadaceae. 4 trop. Afr., As.
Leptagrostis C. E. Hubbard. Gramineae. 1 trop. Afr. (Abyss.).
Leptalea D. Don ex Hook. & Arn. = Facelis Cass. (Compos.).
Leptaleum DC. Cruciferae. 2 E. Medit. to C. As. & Baluch.
Leptalium Sweet = praec.
Leptalix Rafin. = Fraxinus L. (Oleac.).
Leptaloë Stapf = Aloë L. (Liliac.).
Leptaminium Steud. = seq.
Leptamnium Rafin. = Epifagus Nutt. (Orobanchac.).
Leptandra Nutt. = Veronicastrum Fabr. (Scrophulariac.).
Leptanthe Klotzsch = Macrotomia DC. (Boraginac.).
Leptanthes Wight ex Wall. = Hydrilla Rich. (Hydrocharitac.).
Leptanthis Haw. = Leptasea Haw. = Saxifraga L. (Saxifragac.).
Leptanthus Michx = Heteranthera Ruiz & Pav. (Pontederiac.).
Leptargyreia Schlechtd. = Shepherdia Nutt. (Elaeagnac.).

LEPTARRHENA

Leptarrhena R.Br. Saxifragaceae. 1 Kamtschatka to Rocky Mts.

Leptasea Haw. = Saxifraga L. (Saxifragac.).

Leptaspis R.Br. Gramineae. 5 trop. W. Afr., Masc., Ceylon, Fiji.

Leptatherum Nees = Microstegium Nees ex Lindl. (Gramin.).

Leptaulaceae Van Tiegh. = Icacinaceae Miers.

Leptaulus Benth. Icacinaceae. 5 W. & C. trop. Afr.

× **Leptauminia** Rowley. Liliaceae. Gen. hybr. (Guillauminia × Leptaloë).

Leptaxis Rafin. = Tolmiea Torr. & Gray (Saxifragac.).

Lepteiris Rafin. = Penstemon Mitch. (Scrophulariac.).

Leptemon Rafin. = Crotonopsis Michx (Euphorbiac.).

Lepteranthus Neck. ex Cass. = Centaurea L. (Compos.).

Lepterica N. E. Br. = Scyphogyne Brongn. (Ericac.).

Leptica E. Mey. ex DC. = Gerbera L. ex Cass. (Compos.).

Leptidium Presl = Leptis E. Mey. = Lotononis Eckl. & Zeyh. (Legumin.).

Leptilix Rafin. = Tofieldia Huds. (Liliac.).

Leptilon Rafin. = Erigeron L. (Compos.).

Leptinella Cass. = Cotula L. (Compos.).

Leptis E. Mey. ex Eckl. & Zeyh. = Lotononis Eckl. & Zeyh. (Legumin.).

Leptobaea Benth. Gesneriaceae. 2 E. Himal., 1 Borneo.

Leptobasis Dulac = Sisymbrium L. (Crucif.).

Leptobeaua P. & K. = seq.

Leptoboea C. B. Clarke = Leptobaea Benth. (Gesneriac.).

Leptobotrys Baill. = Tragia L. (Euphorbiac.).

Leptocallis G. Don = Ipomoea L. (Convolvulac.).

Leptocallisia (Benth. & Hook. f.) Pichon. Commelinaceae. 2 Mex. to trop. S. Am., W.I.

Leptocarpaea DC. = Sisymbrium L. (Crucif.).

Leptocarpha DC. Compositae. 1 Andes.

Leptocarpha Endl. = Leptophora Rafin. = Helenium L. (Compos.).

*****Leptocarpus** R. Br. (excl. *Calopsis* Beauv.). Restionaceae. 12 SE. As., Austr., Tasm., N.Z., Chile.

Leptocarpus Willd. ex Link = Tamonea Aubl. (Verbenac.).

Leptocarydion Hochst. ex Benth. & Hook. f. Gramineae. 3 S. Afr.

Leptocarydium auctt. = praec.

Leptocaulis Nutt. ex DC. = Apium L. (Umbellif.).

Leptocentrum Schlechter = Plectrelminthus Rafin. (Orchidac.).

Leptoceras (R.Br.) Lindl. = Caladenia R.Br. (Orchidac.).

Leptoceras R. Fitzg. = Leporella George (Orchidac.).

Leptocercus Rafin. = Lepturus R.Br. (Gramin.).

Leptocereus (A. Berger) Britton & Rose. Cactaceae. 11 W.I.

Leptocereus Rafin. (sphalm.) = Leptocercus Rafin. = Lepturus R.Br. (Gramin.).

Leptochilus Kaulf. Polypodiaceae. 1 India, Malaysia. (In C. Chr., *Ind. Fil.*, 1905, *Bolbitis* and *Lomagramma* were included here.)

Leptochiton Sealy. Amaryllidaceae. 1 Andes.

Leptochlaena Spreng. = Leptolaena Thou. (Sarcolaenac.).

Leptochloa Beauv. Gramineae. 27 trop. & subtrop.

Leptochloopsis Yates. Gramineae. 2 E. N. Am., W.I., N. Andes.

Leptochloris Munro ex Kuntze = Chloris Sw. (Gramin.).
Leptocionium Presl. Hymenophyllaceae. 1 Chile.
Leptocladia F. Buxb. (1951; non J. G. Agardh 1892—Algae) = seq.
Leptocladodia F. Buxb. = Mammillaria Haw. (Cactac.).
Leptocladus Oliver = Mostuea Didr. (Loganiac.).
Leptoclinium Benth. Compositae. 1 Brazil.
Leptoclinium (Nutt.) A. Gray = Garberia A. Gray (Compos.).
Leptocnemia Nutt. ex Torr. & Gray = Cymopterus Rafin. (Umbellif.).
Leptocnide Blume = Pouzolzia Gaudich. (Urticac.).
Leptocodon Lem. Campanulaceae. 1 Himalaya. Pedicel of lat. fl. concrescent
Leptocodon Sond. = Treichelia Vatke (Campanulac.).   [with axis of infl.
Leptocoma Less. = Rhynchospermum Reinw. (Compos.).
Leptocoryphium Nees. Gramineae. 1 warm Am., W.I.
Leptocyamus Benth. = Glycine auctt. (Legumin.).
Leptocytisus Meissn. = Latrobea Meissn. (Legumin.).
Leptodactylon Hook. & Arn. Polemoniaceae. 12 W. N. Am.
Leptodaphne Nees = Ocotea Aubl. (Laurac.).
Leptodermis Wall. Rubiaceae. 30 Himalaya to Japan.
Leptoderris Dunn. Leguminosae. 20 trop. Afr.
Leptodesmia Benth. Leguminosae. 5 Madag., India.
Leptofeddia Diels. Solanaceae. 1 Peru.
Leptoglossis Benth. = Salpiglossis Ruiz & Pav. (Solanac.).
Leptoglottis DC. (∼ Schrankia Willd.). Leguminosae. 28 S. U.S., Mex.
Leptogonum Benth. Polygonaceae. 3 W.I. (S. Domingo).
Leptogramma J. Sm. = Thelypteris Schmid. (Thelypteridac.).
Leptogyna Rafin. = Pluchea Cass. (Compos.).
Leptogyne Less. = Cotula L. (Compos.).
× **Leptolaelia** hort. Orchidaceae. Gen. hybr. (iii) (Laelia × Leptotes).
Leptolaena Thou. Sarcolaenaceae. 12 Madag.
Leptolepia Mett. Dennstaedtiaceae. 2 New Guinea, Austr., N.Z.
Leptolepis Boeck. = Blysmus Panz. ex Schult. + Carex L. (Cyperac.).
Leptolobaceae Dulac = Celastraceae R.Br.
Leptolobium Benth. = Leptocyamus Benth. = Glycine auctt. (Legumin.).
Leptolobium Vog. = Sweetia Spreng. (Legumin.).
Leptoloma Chase. Gramineae. 4 Austr.; 2 S. & E. U.S.
Leptomeria R.Br. Santalaceae. 15 Austr., Tasm.
Leptomischus Drake del Castillo. Rubiaceae. 1 Indoch.
Leptomon Steud. = Leptemon Rafin. = Crotonopsis Michx (Euphorbiac.).
Leptonema Hook. = Dolichostylis Turcz. = Draba L. (Crucif.).
Leptonema A. Juss. Euphorbiaceae. 2 Madag.
Leptonium Griff. = Lepionurus Blume (Opiliac.).
Leptonychia Turcz. Sterculiaceae. 30 trop. Afr., S. India, Burma, W.
  Malaysia (exc. Java), New Guinea.
Leptonychiopsis Ridley. Sterculiaceae. 1 S. Malay Penins.
Leptopaetia Harv. = Tacazzea Decne (Periplocac.).
Leptopeda Rafin. = Leptopoda Nutt. = Helenium L. (Compos.).
Leptopetalum Hook. & Arn. = Hedyotis L. (Rubiac.).
Leptopetion Schott = Biarum Schott (Arac.).

LEPUROPETALON

apex; seeds with scanty endosp. (1 cell thick). Only genus: *Lepuropetalon*.
Long included in *Saxifragac.*, but differs from the more typical genera, and
agrees with *Parnassiac.*, in a number of characters.

Lepuropetalon Ell. Lepuropetalaceae. 1 S. U.S., Mex., Chile.
Lepusa P. & K. = Lipusa Alef. = Phaseolus L. (Legumin.).
Lepyrodia R.Br. Restionaceae. 17 Austr., Tasm., N.Z., Chatham I.
Lepyrodiclis Fenzl. Caryophyllaceae. 3 W. As.
Lepyronia Lestib. = Lepironia Rich. (Cyperac.).
Lepyroxis Beauv. ex Fourn. = Muhlenbergia Schreb. (Gramin.).
Lequeetia Bub. = Limodorum Rich. (Orchidac.).
Lerchea Hall. ex Rüling = Lerchia Zinn = Suaeda Forsk. ex Scop. (Chenopodiac.)
*Lerchea L. Rubiaceae. 6 Sumatra, Java, Lesser Sunda Is.
Lerchenfeldia Schur = Deschampsia Beauv. (Gramin.).
Lerchia Endl. = Lerchea L. (Rubiac.).
Lerchia Reichb. = Leachia Cass. = Coreopsis L. (Compos.).
Lerchia Zinn = Suaeda Forsk. ex Scop. (Chenopodiac.).
Lereschia Boiss. = Pimpinella L. (Umbellif.).
Leretia Vell. = Mappia Jacq. (Icacinac.).
Leria Adans. = Sideritis L. (Labiat.).
Leria DC. = Chaptalia Vent. (Compos.).
Lerisca Schlechtd. = Cryptangium Schrad. (Cyperac.).
Lerouxia Mérat = Lysimachia L. (Primulac.).
Leroyia Cavaco. Rubiaceae. 1 Madag.
Lerrouxia Caballero = Caballeroa Font Quer (Plumbaginac.).
Lescaillea Griseb. Compositae. 2 Cuba.
Leschenaultia R.Br. corr. Benth. Goodeniaceae. 20 Austr. 'In *L. formosa*
R.Br., the insect's proboscis comes in contact with the lower lip of the pollen-
cup (see fam.), opening it and dusting itself with pollen; in the next fl. it places
this pollen on the stigmatic surface which lies outside the pollen-cup.' (Müller.)
Lesemia Rafin. = Salvia L. (Labiat.).
Lesourdia Fourn. = Scleropogon Phil. (Gramin.).
Lespedeza Michx. Leguminosae. 100 Himal. to China & Japan, Austr., temp.
N. Am. Fls. sometimes apetalous, and cleistogamic. *L. striata* Hook. & Arn.
(Japanese clover; As.) is being spread over N. Am. by animal agency. It is
a useful fodder-plant.
Lespedezia Spreng. = praec.
Lesquerella S. Wats. (= Discovium Rafin.). Cruciferae. 40 N. Am.
Lesquereuxia Boiss. & Reut. = Siphonostegia Benth. (Scrophulariac.).
*Lessertia DC. Leguminosae. 60 trop. & S. Afr.
Lessingia Cham. Compositae. 12 Calif., Ariz.
Lessonia Bert. ex Hook. & Arn. = Eryngium L. (Umbellif.).
Lestadia Spach = Laestadia Kunth (Compos.).
Lestibodea Neck. = Lestibudaea Juss. = Dimorphotheca Moench (Compos.).
Lestibondesia Reichb. (sphalm.) = seq.
Lestiboudesia Reichb. = Lestibudesia Thou. = Celosia L. (Amaranthac.).
Lestibudaea Juss. = Dimorphotheca Moench (Compos.).
Lestibudesia Thou. = Celosia L. (Amaranthac.).
Letestua Lecomte. Sapotaceae. 2 W. Equat. Afr.

Letestudoxa Pellegr. Annonaceae. 2 W. Equat. Afr.

Letestuella G. Taylor. Podostemaceae. 2 W. Equat. Afr.

Lethea Nor. = Disporum Salisb. (Liliac.).

Lethedon Spreng. Thymelaeaceae. 10 New Caled., (? 1) New Hebrides, 1 Queensl.

Lethia Forbes & Hemsl. (sphalm.) = Sethia Kunth = Erythroxylum P.Br. (Erythroxylac.).

Leto Phil. = Helogyne Benth. (Compos.).

Letsoma Rafin. = Lettsomia Roxb. = Argyreia Lour. (Convolvulac.).

Letsomia Reichb. = Lettsomia Ruiz & Pav. = Freziera Sw. (Theac.).

Lettowianthus Diels. Annonaceae. 1 trop. E. Afr.

Lettsomia Roxb. = Argyreia Lour. (Convolvulac.).

Lettsomia Ruiz & Pav. = Freziera Sw. (Theac.).

Leucacantha [Dalech.] S. F. Gray corr. Wittst. = Centaurea L. (Compos.).

Leucacantha Nieuwl. & Lunell = praec.

Leucactinia Rydb. Compositae. 1 Mex.

*Leucadendron R.Br. Proteaceae. 80 S. Afr. *L. argenteum* R.Br. (silver-tree) has ls. covered with fine silky hairs, and may be used for painting upon. It has been nearly extirpated. Fl. like *Protea*. The P, when the fr. is ripe, splits into 4 segments, united round the stigma, and acts as a wing.

Leucadendron L. = Protea L. (1771) (Proteac.).

Leucadendrum Salisb. = Leucospermum R.Br. (Proteac.).

Leucadenia Klotzsch ex Baill. = Croton L. (Euphorbiac.).

× **Leucadenia** Schlechter (ii) = × Pseudadenia P. F. Hunt (Orchidac.).

Leucadenium Benth. & Hook. f. = Leucadenia Klotzsch ex Baill. = Croton L. (Euphorbiac.).

Leucaena Benth. Leguminosae. 50 trop. Am., 1 pantrop., 1 Polynesia.

Leucaeria DC. = Leucheria Lag. (Compos.).

Leucalepis Ducke (sphalm.) = Loricalepis Brade (Melastomatac.).

Leucampyx A. Gray ex Benth. & Hook. f. Compositae. 1 SW. U.S. (Colo., New Mex.).

Leucandra Klotzsch = Tragia L. (Euphorbiac.).

Leucandron Steud. = seq.

Leucandrum Neck. = Leucadendron R.Br. (Proteac.).

Leucanotis D. Dietr. = Leuconotis Jack (Apocynac.).

Leucantha S. F. Gray (sphalm.) = Leucacantha S. F. Gray corr. Wittst. = Centaurea L. (Compos.).

Leucantha Zipp. ex Mackl. (nomen). Quid? New Guinea.

Leucanthea Scheele = Bouchetia DC. (Scrophulariac.).

Leucanthemella Tsvelev (~ Chrysanthemum L.). Compositae. 2 SE. Eur., E. As.

Leucanthemum Kuntze = Osmitopsis Cass. (Compos.).

Leucanthemum Mill. (~ Chrysanthemum L.). Compositae. 20 Eur. (esp. mts. C. Eur.) & N. As. *L. vulgare* Lam. (*Chrysanthemum leucanthemum* L.) marguerite or moon-daisy; *L. maximum* (Ram.) DC., etc.

Leucas R.Br. Labiatae. 100 trop. Am., W.I., trop. & S. Afr., Arabia, S. China, Indomal.

Leucasia Rafin. = praec.

Leucaster Choisy. Nyctaginaceae. 1 SE. Brazil.
Leuce Opiz = Populus L. (Salicac.).
Leucea Presl = Leuzea DC. (Compos.).
Leucelene Greene. Compositae. 1 SW. U.S., Mex.
Leuceres Calest. = Endressia J. Gay (Umbellif.).
Leuceria DC. = Leucheria Lag. (Compos.).
× **Leucerminium** Mansf. (ii) = × Pseudinium P. F. Hunt (Orchidac.).
Leucesteria Meissn. = Leycesteria Wall. (Caprifoliac.).
Leuchaeria Less. = seq.
Leucheria Lag. Compositae. 60 S. Andes, Patag. Xerophytes.
Leuchtenbergia Hook. Cactaceae. 1 Mex.
Leuchtenbergiaceae Salm-Dyck ex Pfeiff. = Cactaceae Juss.
Leucipus Rafin. = Dichapetalum Thou. (Dichapetalac.).
Leuciva Rydb. = Iva L. (Compos.).
Leucobarleria Lindau. Acanthaceae. 3 NE. Afr., Arabia.
Leucoblepharis Arn. = Blepharispermum Wight (Compos.).
Leucobotrys Van Tiegh. = Helixanthera Lour. (Loranthac.).
Leucocalantha Barb. Rodr. Bignoniaceae. 1 Amaz. Brazil.
Leucocarpon Endl. = seq.
Leucocarpum A. Rich. = Denhamia Meissn. (Celastrac.).
Leucocarpus D. Don. Scrophulariaceae. 1 trop. Am.
Leucocasia Schott = Colocasia Schott (Arac.).
Leucocephala Roxb. = Eriocaulon L. (Eriocaulac.).
Leucocera Turcz. = Calycera Cav. (Calycerac.).
Leucochlaena P. & K. (1) = Leucolaena R.Br. ex Endl. = Xanthosia Rudge (Umbellif.).
Leucochlaena P. & K. (2) = Leucolena Ridl. (Orchidac.).
Leucochlamys Poepp. ex Engl. = Spathiphyllum Schott (Arac.).
Leucochyle B. D. Jacks. (sphalm.) = Leucohyle Klotzsch (Orchidac.).
Leucocnide Miq. = Debregeasia Gaudich. (Urticac.).
Leucococcus Liebm. = Pouzolzia Gaudich. (Urticac.).
Leucocodon Gardn. Rubiaceae. 1 Ceylon.
Leucocoma Ehrh. (uninom.) = *Eriophorum alpinum* L. = *Trichophorum alpinum* (L.) Pers. (Cyperac.).
Leucocoma Nieuwl. = Thalictrum L. (Ranunculac.).
Leucocoma Rydb. = Trichophorum Pers. (Cyperac.).
Leucocorema Ridley = Trichadenia Thw. (Flacourtiac.).
Leucocoryne Lindl. Alliaceae. 5 Chile.
Leucocraspedum Rydb. = Frasera Walt. (Gentianac.).
Leucocrinum Nutt. ex A. Gray. Liliaceae. 1 SW. U.S.
Leucocroton Griseb. Euphorbiaceae. 20 Cuba, Haiti.
Leucocyclus Boiss. = Anacyclus L. (Compos.).
Leucodendrum P. & K. = Leucadendron L. = Protea L. (Proteac.).
Leucodermis Planch. = Ilex L. (Aquifoliac.).
Leucodesmis Rafin. = Haemanthus L. (Amaryllidac.).
Leucodictyon Dalz. = Galactia P.Br. (Legumin.).
Leucodonium (Reichb.) Opiz = Cerastium L. (Caryophyllac.).
Leucogenes Beauverd. Compositae. 2 N.Z.

Leucoglochin Ehrh. (uninom.) = *Carex leucoglochin* L. (Cyperac.).
Leucoglochin Heuff. = Carex L. (Cyperac.).
Leucohyle Klotzsch. Orchidaceae. 7 C. & trop. S. Am., W.I.
Leucoïum auctt. = Leucojum L. (Amaryllidac.).
Leucoïum Mill. = Matthiola R.Br. (Crucif.).
Leucojaceae Batsch = Amaryllidaceae–Galanthinae Pax.
Leucojum L. Amaryllidaceae. 12 S. Eur., Morocco.
Leucojum Mill. = Matthiola R.Br. (Crucif.).
Leucolaena (DC.) Benth. = Xanthosia Rudge (Umbellif.).
Leucolena Ridley = Didymoplexiella Garay (Orchidac.).
Leucolinum Fourr. = Linum L. (Linac.).
Leucolophus Bremek. Rubiaceae. 3 Malay Penins., Sumatra.
Leucoma B. D. Jacks. (sphalm.) = Leucocoma Ehrh. (uninom.) = *Trichophorum alpinum* (L.) Pers. (Cyperac.).
Leucomalla Phil. = Evolvulus L. (Convolvulac.).
Leucomanes Presl = Pleuromanes Presl (Hymenophyllac.).
Leucomeris Blume ex DC. = Vernonia Schreb. (Compos.).
Leucomeris D. Don = Gochnatia Kunth (Compos.).
Leucomphalos Benth. Leguminosae. 1 trop. W. Afr.
Leucomphalus Benth. = praec.
Leuconocarpus Spruce ex Planch. & Triana = Moronobea Aubl. (Guttif.).
Leuconoë Fourr. = Ranunculus L. (Ranunculac.).
Leuconotis Jack. Apocynaceae. 10 W. Malaysia.
Leuconymphaea Kuntze = Nymphaea Linn. (Nymphaeac.).
Leucophaë Webb & Berth. = Sideritis L. (Labiat.).
Leucophoba Ehrh. (uninom.) = *Juncus niveus* Leers (non L.) = *Luzula nemorosa* (Poll.) E. Mey. (Juncac.).
Leucopholis Gardn. Compositae. 4 S. Brazil.
Leucophora B. D. Jacks. = Leucophoba Ehrh. (uninom.) = *Luzula nemorosa* (Poll.) E. Mey. (Juncac.).
Leucophrys Rendle. Gramineae. 2–3 trop. & S. Afr.
Leucophyllum Humb. & Bonpl. Scrophulariaceae. 14 S. U.S., Mexico.
Leucophylon v. Buch = Leucoxylum Soland. ex Lowe = Heberdenia Banks (Myrsinac.).
Leucophysalis Rydb. = Physalis L. (Solanac.).
Leucophyta R.Br. = Calocephalus R.Br. (Compos.).
Leucopitys Nieuwl. = Pinus L. (Pinac.).
Leucoplocus Endl. = seq.
Leucoploeus Nees = Hypodiscus Nees (Restionac.).
Leucopoa Griseb. Gramineae. 2 C. As. to Himal.
Leucopodon Benth. & Hook. f. = seq.
Leucopodum Gardn. = Chevreulia Cass. (Compos.).
*Leucopogon R.Br. Epacridaceae. 150 Malaysia, Austr., New Caled.
Leucopremna Standley = Pileus Ramírez (Caricac.).
Leucopsidium DC. = Aphanostephus DC. (Compos.).
Leucopsis Baker. Compositae. 12 trop. S. Am.
Leucopsora Rafin. = Cephalaria Schrad. (Dipsacac.).   [thac.).
Leucoraphis T. Anders. = Leucorhaphis Nees = Brillantaisia Beauv. (Acan-

Leucorchis Blume = Didymoplexis Griff. (Orchidac.).
Leucorchis E. Mey. (= Pseudorchis Seguier). Orchidaceae. 3 E. N. Am., Greenland, Iceland, Eur.
Leucorhaphis Nees = Brillantaisia Beauv. (Acanthac.).
× **Leucororchis** Cif. & Giac. (ii) = × Pseudorhiza P. F. Hunt (Orchidac.).
Leucosalpa Scott Elliot. Scrophulariaceae. 3 Madag.
Leucosceptrum Sm. Labiatae. 3 Himalaya, China.
Leucosedum Fourr. = Sedum L. (Crassulac.).
Leucoseris Fourr. = Senecio L. (Compos.).
Leucoseris Nutt. = Malacothrix DC. (Compos.).
Leucosia Thou. = Dichapetalum Thou. (Dichapetalac.).
Leucosidea Eckl. & Zeyh. Rosaceae. 1 S. Afr.
Leucosinapis Spach = Brassica L. (Crucif.).
Leucosmis Benth. = Phaleria Jack (Thymelaeac.).
*****Leucospermum** R.Br. Proteaceae. 40 S. Afr.
Leucosphaera Gilg. Amaranthaceae. 1 SW. Afr.
Leucospora Nutt. Scrophulariaceae. 1 E. N. Am.
Leucostachys Hoffmgg. = Goodyera R.Br. (Orchidac.).
Leucostegane Prain. Leguminosae. 2 Malay Penins., Borneo.
Leucostegia Presl. Davalliaceae. 2 Indomal. to New Hebrides. Terrestrial; vasc. anat. simpler than *Davallia* (Holttum, *Rev. Fl. Malaya*, **2**: 351, 1954).
Leucostele Backeb. = Trichocereus (A. Berger) Riccob. (Cactac.).
Leucostemma Benth. = Stellaria L. (Caryophyllac.).
Leucostemma D. Don = Helichrysum L. (Compos.).
Leucostomon G. Don = Lecostemon Moç. & Sessé = Sloanea L. (Elaeocarpac.).
Leucosyke Zoll. & Mor. Urticaceae. 35 Malaysia, Polynesia.
Leucosyris Greene = Aster L. (Compos.).
× **Leucotella** Schlechter (ii) = × Pseuditella P. F. Hunt (Orchidac.).
Leucothamnus Lindl. = Thomasia J. Gay (Sterculiac.).
Leucothea Moç. & Sessé ex DC. = Saurauia Willd. (Actinidiac.).
Leucothoë D. Don. Ericaceae. 4 E. As., 40 Am.
Leucotodon auct. = Leontodon L. (Compos.).
Leucoxyla Rojas = Quillaja Mol. (Rosac.).
Leucoxylon G. Don = Leucoxylum Blume = Diospyros L. (Ebenac.).
Leucoxylon Rafin. = Tabebuia Gomes ex DC. (Bignoniac.).
Leucoxylum Blume = Diospyros L. (Ebenac.).
Leucoxylum E. Mey. = Ilex L. (Aquifoliac.).
Leucoxylum Soland. ex Lowe = Heberdenia Banks (Myrsinac.)
Leucrinis Rafin. = Leucocrinum Nutt. (Liliac.).
Leucymmaea Benth. = Lencymmoea C. Presl (?Myrtac.).
Leukeria Endl. = Leucheria Lag. (Compos.).
Leukosyke Endl. = Leucosyke Zoll. & Mor. (Urticac.).
Leunisia Phil. Compositae. 1 Chile.
Leuradia Poir. = Lavradia Vell. (Violac.).
Leuranthus Knobl. = Olea L. (Oleac.).
Leurocline S. Moore. Boraginaceae. 3 N. trop. Afr.
Leuwenhoekia Bartl. = Levenhookia R.Br. (Stylidiac.).

Leuzea DC. (~ Centaurea L.). Compositae. 4 Medit.
Leuzia St-Lag. = praec.
Levana Rafin. = Vestia Willd. (Solanac.).
Leveillea Vaniot (1903; non Decne 1839—Algae, nec Fries 1849—Fungi) = Bi-Leveillea Vaniot = Blumea DC. (Compos.).
Levenhoekia Steud. = seq.
Levenhookia R.Br. Stylidiaceae. 8 S. Austr. The labellum is shoe-shaped and at first embraces the column, but if touched it springs downwards.
Levieria Becc. Monimiaceae. 10 Moluccas, New Guinea, Queensl.
Levina Adans. = Prasium L. (Labiat.).
Levisanus Schreb. = Staavia Dahl (Bruniac.).
Levisia Steud. = Lewisia Pursh (Portulacac.).
Levisticum Hill (1) = Ligusticum L. (Umbellif.).
Levisticum Hill (2). Umbelliferae. 3 SW. As.; cult. & nat. N. Am., Eur.
Levretonia Reichb. = Lebretonia Schrank = Pavonia Cav. (Malvac.).
Levya Bur. ex Baill. Bignoniaceae. 1 C. Am.
× **Lewisara** hort. Orchidaceae. Gen. hybr. (iii) (Aërides × Arachnis × Ascocentrum × Vanda).
Lewisia Pursh. Portulacaceae. 20 W. N. Am. *L. rediviva* Pursh (bitter-root), with thick rhiz., fleshy roots and ls., is very xero. Two years' drying will hardly kill it. K 4–8; C 8–16; A ∞.
Lewisi[ac]eae Hook. & Arn. = Portulacaceae–Lewisiinae Franz.
Lexarza La Llave = Myrodia Sw. (Bombacac.).
Lexipyretum Dulac = Gentiana L. (Gentianac.).
Leycephyllum Piper. Leguminosae. 1 C. Am. (Costa Rica).
Leycestera Reichb. = seq.
Leycesteria Wall. Caprifoliaceae. 6 W. Himalaya to SW. China. G (5–8).
× **Leymotrigia** Tsvelev. Gramineae. Gen. hybr. (Leymus × Elytrigia).
Leymus Hochst. = Elymus L. (Gramin.).
Leysera L. = Asteropterus Adans. (Compos.).
Leyseria Neck. = praec.
Leyssera Batsch = praec.
Lhodra Endl. = Lodhra (G. Don) Guillem. = Symplocos Jacq. (Symplocac.).
Lhotzkya Schau. Myrtaceae. 10 Austr.
Liabellum Cabrera = Microliabum Cabr. (Compos.).
Liabellum Rydb. = seq.
Liabum Adans. Compositae. 90 Am., W.I.
× **Liaopsis** hort. Orchidaceae. Gen. hybr. (iii) (Laelia × Laeliopsis).
*****Liatris** Gaertn. ex Schreb. Compositae. 40 N. Am.
Libadion Bub. = Centaurium Hill (Gentianac.).
Libanothamnus Ernst = Espeletia Humb. & Bonpl. (Compos.).
Libanotis Hill (~ Seseli L.). Umbelliferae. 15 temp. Euras.
Libanotis Rafin. = Cistus L. (Cistac.).
Libanotus Stackh. = seq.
Libanus Colebr. = Boswellia Roxb. (Burserac.).
Liberatia Rizzini. Acanthaceae. 1 Brazil.
Liberbaileya Furtado. Palmae. 1 Malay Penins. (Lankawi Is.).
Libertia Dum. = Hosta Tratt. (Liliac.).

LILIACEAE

mode of growth in thickness]. Many are xero., some, e.g. *Aloë* and *Gasteria*, are succulent; [others, e.g. *Phormium*, have hard isobil. ls.; others, e.g. *Dasylirion*, have tuberous stems and narrow ls.;] *Bowiea* only produces leafy shoots in the wet season. [*Smilax*,] *Gloriosa*, etc., are climbing pls.[, the former with peculiar tendrils]. [*Ruscus* exhibits phylloclades.]

Infl. most commonly racemose; fls. with no bracteoles; when the latter occur, the further branching from their axils usually takes a cymose form, especially that of a bostryx, as in *Hemerocallis*. [The apparent umbels of heads of *Allium*, *Agapanthus*, etc., are really cymose.] Sol. term. fls. occur in *Tulipa*, etc. Fls. usu. ♀, reg., pentacyclic, 3-merous (rarely 2, 4, or 5), hypog. P 3+3, free or united, petaloid or sometimes sepaloid; A 3+3 or fewer, rarely more, usually with introrse anthers; G̲ (3), [rarely inf. or semi-inf.,] 3-loc. with axile, or rarely 1-loc. with parietal plac.; ovules usu. ∞, in two rows in each loc., anatr. Fr. usu. capsular, loculic. or septic., sometimes a berry. Seed with straight or curved embryo, in abundant fleshy or cartilaginous, never floury, endosp.

Fls. usu. insect-pollinated. Honey in *Scilla*, [*Allium*,] etc., is secreted by glands in the ovary-wall between the cpls.; in other cases by glands on the bases of the perianth-ls. [*Yucca* (*q.v.*) has a unique pollination-method.]

Economically the *L.* are of no great value. The chief food plants are [*Allium* and] *Asparagus*; [*Phormium*, *Yucca*, and *Sanseverina* yield useful fibre;] [*Smilax*,] *Urginea*, *Aloë*, *Colchicum*, *Veratrum*, etc., are medicinal. [*Xanthorrhoea* and *Dracaena* yield resins;] *Chlorogalum* is used as soap.

(I have here retained Willis's account (above) almost unaltered, but have placed in square brackets the passages referring to genera which have been transferred by Hutchinson (1934) and others to other families. For further information on these groups the relevant entries should be consulted.)

Classification and chief genera (after Engler & Krause):

   I. *Melanthioïdeae* (rhiz., or bulb covered with scale-ls. and with term. infl.; anthers extr. or intr.; caps. loculic. or septic.; fr. never a berry): *Tofieldia*, *Narthecium*, *Veratrum*, *Gloriosa*, *Colchicum*.

  II. *Herrerioïdeae* (tuber, with climbing stem; ls. in tufts; small-flowered racemes at base of these or in panicles at ends of twigs; septic. caps): *Herreria* (only genus).

 III. *Asphodeloïdeae* (rhiz. with radical ls., rarely stem with crown of ls. or leafy branched stem or bulb; infl. usu. term., a simple or cpd. raceme or spike; P or (P); anthers intr.; caps., rarely berry): *Asphodelus*, *Chlorogalum*, *Bowiea*, *Funkia*, *Hemerocallis*, [*Phormium*,] *Kniphofia*, *Aloë*, *Gasteria*, *Haworthia*, [*Aphyllanthes*,|*Lomandra*, *Xanthorrhoea*, *Kingia*].

 [IV. *Allioïdeae* (bulb or short rhiz.; cymose umbel ± enclosed by two broad or rarely narrow ls., sometimes joined; infl. rarely of 1 fl.): *Agapanthus*, *Gagea*, *Allium*, *Brodiaea*.]—See *Alliaceae*.

   V. *Lilioïdeae* (bulb; stem bearing 1 or more ls.; rhaphides present; cover-cells absent, infl. term., racemose; P or (P); anthers intr.; caps. loculic., except in *Calochortus*): *Lilium*, *Fritillaria*, *Tulipa*.

  VI. *Scilloïdeae* (as V, but stem leafless [scapose]; rhaphides o; cover-cells present): *Scilla*, *Ornithogalum*, *Hyacinthus*, *Muscari*.

 [VII. *Dracaenoïdeae* (stem erect with leafy crown, except in *Astelia*; ls.

sometimes leathery, never fleshy; P free or united at base; anthers intr.; berry or caps.): *Yucca, Dasylirion, Dracaena.*]—See *Agavaceae.*
VIII. *Asparagoïdeae* (rhiz. subterranean, sympodial; berry): *Asparagus,* [*Ruscus,*] *Polygonatum, Convallaria*[, *Trillium*].
IX. *Ophiopogonoïdeae* (short rhiz., sometimes with suckers, with narrow or lanceolate radical ls.; P or (P); anthers intr. or semi-intr.; ovary sup. or semi-inf.; fr. with thin pericarp and 1–3 seeds with fleshy coats): [*Sansevieria,*] *Ophiopogon.*
X. *Aletridoïdeae* (short rhiz. with narrow or lanceolate radical ls.; (P); anthers semi-intr.; caps. loculic.; seeds ∞, with thin testa): *Aletris* (only genus).
[XI. *Luzuriagoïdeae* (shrubs or undershrubs with erect or climbing twigs; infl.-twigs usu. many-flowered, cymose, rarely 1-flowered, with scaly bract at base; both whorls of P alike or not; berry with spherical seeds): *Luzuriaga, Lapageria.*]—See *Philesiaceae.*
[XII. *Smilacoïdeae* (climbing shrubs with net-veined l.; fls. small in axillary umbels or racemes or term. panicles; loc. with 1 or 2 orthotr. or semi-anatr. ovules): *Smilax.*]—See *Smilacaceae.*
The *Liliaceae* (*s.str.*) are intimately linked with the *Amaryllidaceae* by means of the *Alliaceae.*

Liliacum Renault = Syringa L. (Oleac.).
Liliago Heist. = Amaryllis L. (Amaryllidac.).
Liliago Presl = Anthericum L. (Liliac.).
Liliastrum Fabr. = Paradisea Mazzuc. (Liliac.).
Lilicella Rich. ex Baill. = Sciaphila Bl. (Triuridac.).
Liliiflorae C. A. Agardh. An order of Monocots. comprising (in Engler) the *Juncac., Stemonac., Liliac., Haemodorac., Amaryllidac., Velloziac., Taccac., Dioscoreac.,* and *Iridac.*
Lilioasphodelus Fabr. = Hemerocallis L. (Liliac.).
Lilio-gladiolus Trew = Gladiolus L. (Iridac.).
Lilio-Hyacinthus Ort. = Scilla L. (Liliac.).
Lilio-narcissus Trew = Amaryllis L. (Amaryllidac.).
Liliorhiza Kellogg = Fritillaria L. (Liliac.).
Lilithia Rafin. = Rhus L. (Anacardiac.).
Lilium L. Liliaceae. 80 N. temp. Herbs with scaly bulbs, leafy stems and usu. large fls. in term. rac. Honey secreted in long grooves at the bases of the P-leaves. The fls. of many spp. are visited by Lepidoptera. *L. martagon* L. gives off its scent at night (cf. *Oenothera*). *L. bulbiferum* L. is reproduced veg. by bulbils in the leaf-axils. In most spp. with hanging fls. the caps. when ripe stands upwards, so that the seeds can only escape when it is shaken.
Lilium-Convallium Moench = Convallaria L. (Liliac.).
Lilloa Speg. = Synandrospadix Engl. (Arac.).
Limacia F. G. Dietr. = Xylosma Forst. f. (Flacourtiac.).
Limacia Lour. Menispermaceae. 3 Indomal.
Limaciopsis Engl. Menispermaceae. 1 trop. Afr.
Limanisa P. & K. = Leimanisa Rafin. = Gentianella Moench (Gentianac.).
Limanthemum P. & K. = Leimanthemum Ritgen = seq.
Limanthium P. & K. = Leimanthium Willd. = Melanthium L. (Liliac.).

×**Limara** hort. Orchidaceae. Gen. hybr. (iii) (Arachnis × Renanthera × Van-
Limatodes Lindl. = Calanthe R.Br. (Orchidac.). [dopsis).
Limatodis Bl. = praec.
×**Limatopreptanthe** hort. (iv) = Calanthe R.Br. (Orchidac.).
Limbaceae Dulac = Campanulaceae–Campanuloïdeae.
Limbarda Adans. = Inula L. (Compos.).
Limborchia Scop. = Coutoubea Aubl. (Gentianac.).
Limeum L. Aïzoaceae. 28 trop. & S. Afr., Arabia, India.
Limia Vand. = Vitex L. (Verbenac.).
Limivasculum Börner = Carex L. (Cyperac.).
Limlia Masam. & Tomiya = Quercus L. (Fagac.).
Limnalsine Rydb. Portulacaceae. 1 W. N. Am.
*****Limnanthaceae** R.Br. Dicots. 2/11 N. Am. Delicate ann. herbs of damp
places, with alt. pinnatif. exstip. ls. and sol. axill. long-pedic. ♂ reg. 3–5-
merous fls. K valv.; C contorted; A 6–10; G̲ 3–5, almost free, with common
gynobasic style, ovules 1 in each loc., ascending, the micropyle facing outwards
and downwards. Fruit separating into 3–5 ± tuberc. nutlets; endosp. o.
Genera: *Limnanthes, Floerkea*. Prob. related to *Geraniac.*; considerable ex-
ternal similarity to *Polemoniac.* and *Hydrophyllac.*
Limnanthemum S. G. Gmel. = Nymphoïdes Seguier (Menyanthac.).
*****Limnanthes** R.Br. Limnanthaceae. 10 Pacific N. Am.
Limnanthes Stokes = Limnanthemum S. G. Gmel. = Nymphoïdes Hill (Men-
Limnanthus Neck. = praec. [yanthac.).
Limnanthus Reichb. = Limnanthes R.Br. (Limnanthac.).
Limnas Ehrh. (uninom.) = *Ophrys paludosa* L. = *Malaxis paludosa* (L.) Sw.
(Orchidac.).
Limnas Trin. Gramineae. 2 C. As. to NE. Sib.
Limnaspidium Fourr. = Veronica L. (Scrophulariac.).
Limnetis Rich. = Spartina Schreb. (Gramin.).
Limnia Haw. = Montia L. (Portulacac.).
Limniboza R. E. Fries. Labiatae. 1 trop. E. Afr.
Limnirion (Reichb.) Opiz = Iris L. (Iridac.).
Limniris Fuss = praec.
Limnobium Rich. Hydrocharitaceae. 3 Am. *L. stoloniferum* (G. F. W. Mey.)
Griseb. (*Trianea bogotensis* Karst.), a small floating plant often cult. It repro-
duces veg. by 'runners' (cf. *Hydrocharis*). Its root-hairs are used to show
circulation of protoplasm. Only the ♀ pl. is known in Eur.
Limnobotrya Rydb. = Ribes L. (Grossulariac.).
Limnocharis Humb. & Bonpl. Limnocharitaceae. 1 trop. S. Am., W.I.
Limnocharis Kunth = Eleocharis R.Br. (Cyperac.).
Limnocharitaceae Takhtadj. Monocots. 4/7 trop. Aquatic herbs, ± intermediate between *Butomac.* and *Hydrocharitac.*, differing from the former in
the ls. differentiated into pet. and lamina, in the presence of latex-tubes, in
the green calycine outer P whorl and delicate non-persistent corolline inner
whorl, in the pollen grains with 4 or more pores, and in the curved or folded
embryo; from the *Hydrocharitac.* principally in the superior ovary, with ovules
scattered over the walls of the carpels. Genera: *Limnocharis, Hydrocleys,
Tenagocharis, Ostenia*.

Limnochloa Beauv. ex Lestib. = Eleocharis R.Br. (Cyperac.).
Limnocitrus Swingle (~ Pleiospermium [Engl.] Swingle). Rutaceae. 1 Indoch., Java.
Limnocrepis Fourr. = Crepis L. (Compos.).
Limnodea L. H. Dewey ex Coult. Gramineae. 1 S. U.S.
Limnogenneton Sch. Bip. = Sigesbeckia L. (Compos.).
Limnogeton Edgew. = Aponogeton Thunb. (Aponogetonac.).
Limnonesis Klotzsch = Pistia L. (Arac.).
Limnopeuce Seguier = Ceratophyllum L. (Ceratophyllac.).
Limnopeuce Zinn = Hippuris L. (Hippuridac.).
***Limnophila** R.Br. Scrophulariaceae. 35 trop. Afr., As., Austr., Pacif.
Limnophyton Miq. Alismataceae. 1–3 trop. Afr., Madag., India to Indoch., Java, Timor.
Limnopoa C. E. Hubbard. Gramineae. 1 S. India.
Limnorchis Rydb. = Platanthera Lindl. (Orchidac.).
Limnosciadium Mathias & Constance. Umbelliferae. 2 S. Cent. U.S.
Limnoseris auctt. = Limonoseris Peterm. = Crepis L. (Compos.).
Limnosipanea Hook. f. Rubiaceae. 7 C. & trop. S. Am.
Limnostachys F. Muell. = Monochoria Presl (Pontederiac.).
Limnoxeranthemum Salzm. ex Steud. = Paepalanthus Mart. (Eriocaulac.).
Limodoraceae Horan. = Orchidaceae Juss.
Limodoron St-Lag.: *s.l.* = Limodorum L.; *s.str.*, as to sole sp. cited = Vanilla Mill. (Orchidac.).
Limodorum Kuntze = Epipactis Zinn + Cephalanthera Rich. + Limodorum Rich. (Orchidac.).
Limodorum L. = Bletia Ruiz & Pav. (Orchidac.).
***Limodorum** Boehm. Orchidaceae. 1 S. Eur. A leafless saprophyte with no chlorophyll (cf. *Epipogium*). The 4 lat. sta. are sometimes fertile.
Limon Mill. = Citrus L. (Rutac.).
Limonaetes Ehrh. (uninom.) = *Carex pallescens* L. (Cyperac.).
Limonanthus Kunth = Leimanthium Willd. = Melanthium L. (Liliac.).
Limonia auctt. = Citropsis (Engl.) Swingle & Kellerm., Swinglea Merr., Feroniella Swingle, etc. (Rutac.).
Limonia Gaertn. = Scolopia Schreb. (Flacourtiac.).
Limonia L. Rutaceae. 1 India to Java, *L. acidissima* L., elephant-apple or wood-apple; wood useful, and yields a gum; fr. ed., used in Japan as a substitute for soap.
Limoniaceae Lincz. (~ Plumbaginaceae Juss. s.l.). Dicots. 14/750, almost cosmop., in maritime conditions or saline deserts. Differs from the *Plumbaginaceae s. str.* in the spreading scarious limb of the calyx, in the stamens being ± adnate to the corolla, and in the styles usually distinct or almost so (connate in *Limoniastrum*). Chief genera: *Acantholimon, Armeria, Goniolimon, Limonium, Limoniastrum.*
Limonias Ehrh. (uninom.) = *Serapias longifolia* L. = *Cephalanthera longifolia* (L.) Fritsch (Orchidac.).
Limoniastrum Moench. Limoniaceae. 10 Medit.
Limoniodes Kuntze = praec.
Limoniopsis Lincz. Plumbaginaceae. 1 E. As. Min., Cauc.

LIMONIUM

*Limonium Mill. Plumbaginaceae. 300 cosmop., esp. Medit. to C. As.,
chiefly in steppes and salt marshes. Infl. cpd., mixed, the total infl. a spike,
the partial a drepanium. Fls. many, e.g. *L. vulgare* Mill., heterostyled like
Primula.

Limonoseris Peterm. = Crepis L. (Compos.).

Limosella L. Scrophulariaceae. 15 cosmop. *L. aquatica* L. multiplies by
runners.

Limosell[ac]eae J. G. Agardh = Scrophulariaceae–Gratioleae Benth.

*Linaceae S. F. Gray. Dicots. 12/290 cosmop. Most are herbs or shrubs
with alt. entire often stip. ls. Infl. cymose, a dichasium or cincinnus, the latter
usu. straightening out very much and looking like a raceme. Fl. ⚥, reg., usu.
5-merous. K 5, quincuncial; C 5, imbr. or conv.; A 5, 10 or more, often with
stds., united at base into a ring; G̲ (2–3–5), multiloc., often with extra partitions projecting from the midribs of the cpls., but not united to the axile plac.;
ovules 1 or 2 per loc., pend., anatr., with the micropyle facing outwards and
upwards. Septic. caps., or drupe. Embryo usu. straight, in fleshy endosp.
Linum (flax, linseed) is economically important. Chief genera: *Hugonia,
Roucheria, Anisadenia, Reinwardtia, Linum.* Through the genus *Anisadenia* the
Linac. are connected with *Plumbaginac.*

Linagrostis Guett. = Eriophorum L. (Cyperac.).

Linanthastrum Ewan. Polemoniaceae. 3 W. U.S.

Linanthus Benth. Polemoniaceae. 50 W. N. Am., Chile.

Linaria Mill. Scrophulariaceae. 150 extra-trop. N. hemisph., esp. Medit. reg.
L. vulgaris Mill. is a perennial, each year's growth arising from an adv. bud
upon the summit of the root. The fl. is closed at the mouth; honey is secreted
by the nectary at the base of the ovary and collects in the spur. The only
visitors are the larger bees, which are able to open the fl., and whose tongues
are long enough to reach the honey. Peloria of the fl. is frequent; a term. fl.
appears upon the raceme and is symmetrical, with 5 spurs upon the C and a
tubular mouth. Sometimes fls. of this type occur all down the raceme.

Linariantha B. L. Burtt & R. M. Smith. Acanthaceae. 1 Borneo.

Linariopsis Welw. Pedaliaceae. 2 W. Equat. & SW. trop. Afr.

Lincania G. Don = Licania Aubl. (Chrysobalanac.).

Linconia L. Bruniaceae. 3 S. Afr.

Lindackera Sieber ex Endl. = Capparis L. (Capparidac.).

Lindackeria Presl. Flacourtiaceae. 18 trop. Am. & Afr.

Lindauea Rendle. Acanthaceae. 1 Somaliland.

Lindbergella Bor. Gramineae. 1 Cyprus.

Lindbergia Bor (1968; non Kindb. 1896—Musci) = praec.

Lindblomia Fries = Platanthera Rich. (Orchidac.).

Lindelofia Lehm. (~ Cynoglossum L.). Boraginaceae. 10 C. As., Afgh.,
Lindenbergia Lehm. Scrophulariaceae. 15 trop. Afr. to E. As.   [Himal.

Lindenia Benth. Rubiaceae. 5 New Caled., Fiji, Mex., C. Am.

Lindenia Mart. & Gal. = Senkenbergia Schau. = Cyphomeris Standl. (Nycta-
Lindeniopiper Trelease. Piperaceae. 2 Mex.   [ginac.].

Lindera Adans. = Ammi L. (Umbellif.).

*Lindera Thunb. Lauraceae. 100 Himal., E. As., W. Malaysia. *L. benzoin*
(L.) Meissn. has aromatic bark (antifebrile).

Lindernia All. Scrophulariaceae. 100 warm, esp. Afr. & As.
Lindheimera A. Gray & Engelm. Compositae. 1 S. U.S.
Lindleya Kunth (1821) = Casearia Jacq. (Flacourtiac.).
***Lindleya** Kunth (1824). Rosaceae. 2 Mex.
Lindleya Nees = Laplacea Kunth (Theac.).
Lindleyaceae J. G. Agardh = Rosaceae–Quillajeae Meissn.
× **Lindleyara** hort. (v) = Hawaiiara hort. (Orchidac.).
Lindleyella Rydb. = Neolindleyella Fedde = Lindleya Kunth (1824) (Rosac.).
Lindleyella Schlechter = Rudolfiella Hoehne (Orchidac.).
Lindmania Mez. Bromeliaceae. 16 S. Am.
Lindnera Reichb. = Tilia L. (Tiliac.).
Lindneria Th. Dur. & Lubbers = Pseudogaltonia Kuntze (Liliac.).
Lindsaea Dryand. ex Sm. (incl. *Schizoloma* Gaud. non Copel.). Lindsaeaceae.
200 trop., S. Afr., Tasmania, N.Z. Small ferns, mostly of forest; branching
of ls. often like *Adiantum*.
Lindsaeaceae Presl. Dicksoniales. 6 gen. trop. & S. temp. Rhiz. creeping,
sometimes protostelic, scales narrow; ls. mostly small (thicket-forming in
Odontosoria); sori submarginal with indus., often fusing laterally, edge of
lamina not reflexed. (Kramer, *Acta Bot. Neerl.* **6**: 98–290, 1957.)
Lindsaya Kaulf. = Lindsaea Dryand. ex Sm. (Lindsaeac.).
Lindsayella Ames & C. Schweinf. Orchidaceae. 1 Panamá.
Lindsaynium Fée = Lindsaea Dryand. ex Sm. (Lindsaeac.).
Lindsayopsis Kuhn = Odontosoria Fée (Lindsaeac.).
Lingelsheimia Pax. Euphorbiaceae. 2 trop. Afr.
Lingnania McClure. Gramineae. 10 S. China, Indoch.
Lingoum Adans. = Pterocarpus L. = Derris Lour. (Legumin.).
Linharea Arruda ex Koster corr. Kosterm. = Ocotea Aubl. (Laurac.).
Linharia Arruda ex Koster = praec.
Linkia Cav. = Persoonia Sm. (Proteac.).
Linkia Pers. = Desfontainia Ruiz & Pav. (Potaliac.).
Linnaea Gronov. ex L. Caprifoliaceae. 1–3 N. circumpolar. Prostrate sub-
shrubs. Sta. 4, didynamous. Ovary covered with glandular hairs. Two loculi
are ∞-ovulate and sterile, the other 1-ovulate and fertile.
Linnaeobreynia Hutch. Capparidaceae. 10 Mex. to trop. S. Am.
Linnaeopsis Engl. Gesneriaceae. 3 trop. E. Afr.
Linnea Neck. = Linnaea Gronov. ex L. (Caprifoliac.).
× **Linneara** hort. (vii) = × Iwanagara hort. (Orchidac.).
Linneusia Rafin. = Linnea Neck. = Linnaea Gronov. ex L. (Caprifoliac.).
Linocalix Lindau. Acanthaceae. 1 W. Equat. Afr.
Linochilus Benth. = Diplostephium Kunth (Compos.).
Linociera Steud. = Linociria Neck. = Haloragis J. R. & G. Forst. (Halora-
gidac.).
***Linociera** Sw. ex Schreb. Oleaceae. 80–100 trop., subtrop.
Linociria Neck. = Haloragis J. R. & G. Forst. (Haloragidac.).
Linodendron Griseb. Thymelaeaceae. 5 Cuba.
Linodes Kuntze = Radiola Roth (Linac.).
Linoma O. F. Cook = Dictyosperma H. Wendl. & Drude (Palm.).
Linophyllum Seguier = Thesium L. (Santalac.).

Linopsis Reichb. = Linum L. (Linac.).

Linospadix Becc. = Paralinospadix Burret (Palm.).

Linospadix H. Wendl. Palmae. 12 New Guinea, Austr.

Linosparton Adans. = Lygeum Loefl. ex L. (Gramin.).

Linospartum Steud. = praec.

Linostachys Klotzsch ex Schlechtd. = Acalypha L. (Euphorbiac.).

Linostigma Klotzsch = Viviania Cav. (Vivianiac.).

Linostoma Wall. ex Endl. Thymaeleaceae. 6 Assam, SE. As., W. Malaysia.

Linostrophum Schrank = Camelina Crantz (Crucif.).

Linostylis Fenzl ex Sond. = Dyschoriste Nees (Acanthac.).

Linosyris Cass. = Crinitaria Cass. (Compos.).

Linosyris Ludw. = Thesium L. (Santalac.).

Linosyris Torr. & Gray = Bigelovia DC. (Compos.).

Linschotenia De Vriese = Dampiera R.Br. (Goodeniac.).

Linschottia Comm. ex Juss. = Homalium Jacq. (Flacourtiac.).

Linscotia Adans. = Limeum L. (Aïzoac.).

Linsecomia Buckl. = Helianthus L. (Compos.).

Lintibularia Gilib. (sphalm.) = Lentibularia Adans. = Utricularia L. (Lentibulariac.).

Lintonia Stapf. Gramineae. 2 trop. E. Afr.

Linum L. Linaceae. 230 temp. & subtrop., esp. Medit. Fls. in sympodial cincinni. Several are heterostyled (dimorphic), e.g. the red-flowered *L. grandiflorum* Desf. Illegitimate pollination in this sp. produces absolutely no seed at all. The seed has a mucilaginous testa which swells on wetting. Flax is the fibre of *Linum usitatissimum* L., obtained by retting off the softer tissues in water; linen is made from it. The shorter fibres form tow, and scraped linen lint. The seeds (linseed) yield an oil by pressure, and the remaining 'cake' (cf. *Gossypium*) is used for cattle-feeding, etc.

Linzia Sch. Bip. = Vernonia Schreb. (Compos.).

Liodendron H. Keng = Drypetes Vahl (Euphorbiac.).

× **Lioponia** hort. Orchidaceae. Gen. hybr. (iii) (Broughtonia × Laeliopsis).

Lioydia Neck. ex Reichb. = Printzia Cass. (Compos.).

Lipandra Moq. = Chenopodium L. (Chenopodiac.).

Lipara Lour. ex Gomes = Canarium L. (Burserac.).

Liparena Poit. ex Leman = Drypetes Vahl (Euphorbiac.).

Liparene Baill. = praec.

Liparia L. Leguminosae. 4 S. Afr.

*****Liparis** Rich. Orchidaceae. 250 cosmop. (except N.Z.).

Liparophyllum Hook. f. Menyanthaceae. 1 Tasm., N.Z.

Lipeocercis Nees = Lepeocercis Trin. = Andropogon L. (Gramin.).

Liphaemus P. & K. = Leiphaimos Cham. & Schlechtd. (Gentianac.).

Liphonoglossa Torr. (sphalm.) = Siphonoglossa Oerst. (Acanthac.).

*****Lipocarpha** R.Br. Cyperaceae. 12 trop. Am., Afr., As.

Lipochaeta DC. Compositae. 35 Hawaii.

Liponeuron Schott, Nym. & Kotschy = Viscaria Roehling (Caryophyllac.).

Lipophragma Schott & Kotschy ex Boiss. = Aëthionema R.Br. (Crucif.).

Lipophyllum Miers = Clusia L. (Guttif.).

Lipostoma D. Don. Rubiaceae. 1 Brazil.

Lipotactes (Blume) Reichb. = Phthirusa Mart. (Loranthac.).
Lipotriche R.Br. = Melanthera Rohr (Compos.).
Lipotriche Less. = Lipochaeta DC. (Compos.).
Lipozygis E. Mey. = Lotononis Eckl. & Zeyh. (Legumin.).
Lippaya Endl. = Dentella J. R. & G. Forst. (Rubiac.).
Lippay[ac]eae Meissn. = Rubiaceae–Hedyotideae Kunth.
Lippia L. Verbenaceae. 220 trop. Am., Afr. The ls. of *L. citriodora* Kunth yield an aromatic oil used in perfumery under the name verbena oil. Some have axillary thorns.
Lippomuellera Kuntze = Agastachys R.Br. (Proteac.).
Lipskya Nevski = Schrenkia Fisch. & Mey. (Umbellif.).
Lipskyella Juzepczuk. Compositae. 1 C. As.
Lipusa Alef. = Phaseolus L. (Legumin.).
Liquidambar L. Altingiaceae. 6 Atlant. N. Am., SW. As. Min., SE. China, Indoch., Formosa. Fls. monoec., apet., the ♂ in upright spikes, the ♀ in heads on pend. stalks. The seeds are easily shaken out in strong winds. Storax (a fragrant balsam) is obtained from all spp., but chiefly from *L. orientalis* Mill. (As. Min.). *L. styraciflua* L., sweet gum, N. Am.; wood useful (satin walnut).
Liquidambaraceae Pfeiff. = Altingiaceae Hayne.
Liquiritia Medik. = Glycyrrhiza L. (Legumin.).
Lirayea Pierre = Afromendoncia Gilg = Mendoncia Vell. ex Vand. (Mendon-
Liriactis Rafin. = Tulipa L. (Liliac.).   [ciac.).
Liriamus Rafin. = Crinum L. (Amaryllidac.).
Lirianthe Spach = Magnolia L. (Magnoliac.).
Liriodendron L. Magnoliaceae. 1 E. N. Am., *L. tulipifera* L., the tulip-tree; 1 China, N. Indoch. The ls. though polymorphic have a very char. shape with truncate or bilobed apex. Fr. a cone of samaroid cpls. *L. tulipifera* yields useful timber (canary whitewood).
Lirio-narcissus Heist. = Amaryllis L. (Amaryllidac.).
Liriope Herb. = Liriopsis Reichb. (Amaryllidac.).
Liriope Lour. Liliaceae. 6 E. As.
Liriope Salisb. = Reineckia Kunth (Liliac.).
Liriopogon Rafin. = Tulipa L. (Liliac.).
Liriopsis Reichb. = Urceolina Reichb. (Amaryllidac.).
Liriopsis Spach = Michelia L. (Magnoliac.).
Liriosma Poepp. & Endl. Olacaceae. 15 trop. S. Am.
Liriothamnus Schlechter. Liliaceae. 2 S. Afr.
Lirium Scop. = Lilium L. (Liliac.).
Lisaea Boiss. Umbelliferae. 5 SW. As.
Lisianthius P.Br. (textu) = seq.
Lisianthus P.Br. (tab.). Gentianaceae. 50 Mex. to trop. S. Am., W.I.
Lisianthus Vell. = Metternichia Mikan (Solanac.).
Lisimachia Neck. = Lysimachia L. (Primulac.).
Lisionotus Reichb. = Lysionotus D. Don (Gesneriac.).
Lissanthe R.Br. Epacridaceae. 4 Austr.
Lissera Fourr. (sphalm.) = Listera Adans. = Genista L. (Legumin.).
Lissocarpa Benth. Ebenaceae. 2 trop. S. Am. G̅! Style simple.
***Lissocarpaceae** Gilg = Ebenaceae Juss.

677

LISSOCARPUS

Lissocarpus P. & K. = Lissocarpa Benth. (Ebenac.).
Lissochilos Bartl. = seq.
Lissochilus R.Br. = Eulophia R.Br. ex Lindl. (Orchidac.).
Lissospermum Bremek. Acanthaceae. 1 Sumatra, Java.
Listera Adans. = Genista L. (Legumin.).
***Listera** R.Br. Orchidaceae. 30 N. temp. The rostellum on being touched ruptures violently and ejects a viscid fluid which cements the pollinia to the insect as in *Epipactis*.
Listeria Neck. ex Rafin. = Oldenlandia L. (Rubiac.).
Listeria Spreng. = Listera R.Br. (Orchidac.).
Listia E. Mey. Leguminosae. 1 S. Afr.
Listrobanthes Bremek. Acanthaceae. 1 E. Himal., Assam.
Listrostachys Reichb. Orchidaceae. 2 trop. Afr., ?Réunion.
Lisyanthus Aubl. = Lisianthus P.Br. (Gentianac.).
Lita Schreb. = Voyria Aubl. (Gentianac.).
Litanthes Lindl. = seq.
Litanthus Harv. Liliaceae. 1 S. Afr.
Litanum Nieuwl. = Talinum Adans. (Portulacac.).
Litchi Sonner. Sapindaceae. 10–12 India, S. China, SE. As., W. Malaysia. *L. chinensis* Sonner. (*litchi* or *leechee*), cult. for ed. fr., a one-seeded nut with fleshy aril.
Lithachne Beauv. Gramineae. 3 C. Am., W.I., Brazil.
Lithagrostis Gaertn. = Coix L. (Gramin.).
Lithanthus Pfeiff. = Litanthus Harv. (Liliac.).
Lithobium Bong. Melastomataceae. 1 Brazil.
Lithocardium Kuntze = Cordia L. (Ehretiac.).
Lithocarpos Targ.-Tozz. ex Steud. = Attalea Kunth (Palm.).
Lithocarpus Blume. Fagaceae. 300 E. & S.E. As., Indomal.
Lithocarpus Bl. ex Pfeiff. = Styrax L. (Styracac.).
Lithocarpus Steud. = Lithocarpos Targ.-Tozz. = Attalea Kunth (Palm.).
Lithocaulon Bally (1959; non Menegh. 1857—gen. foss.) = Pseudolithos Bally
Lithocnide Rafin. = Rousselia Gaudich. (Urticac.).   [(Asclepiadac.).
Lithocnides B. D. Jacks. = praec.
Lithococca Small ex Rydb. Boraginaceae. 1 S. U.S.
Lithodia Blume = Strobilocarpus Klotzsch (Grubbiac.).
Lithodora Griseb. Boraginaceae. 7 NW. France & W. Medit. to As. Min.
Lithodraba Boelcke. Cruciferae. 1 Argent.
Lithofragma Nutt. = Lithophragma (Nutt.) Torr. & Gray (Saxifragac.).
Lithoön Nevski = Astragalus L. (Legumin.).
Lithophila Sw. Amaranthaceae. 15 W.I., Galápagos.
***Lithophragma** (Nutt.) Torr. & Gray. Saxifragaceae. 9 W. N. Am.
Lithophytum T. S. Brandegee. Goetzeaceae. 1 Mex.
Lithoplis Rafin. = Rhamnus L. (Rhamnac.).
Lithops N. E. Brown. Aïzoaceae. 50 S. Afr.
Lithosanthes auctt. = Litosanthes Blume (Rubiac.).
Lithosciadium Turcz. = Cnidium Cuss. (Umbellif.).
Lithospermum L. Boraginaceae. 60 temp.
Lithostegia Ching. Aspidiaceae. 1 E. Himal. to SW. China.

Lithothamnus Zipp. ex Span. = Ehretia L. (Ehretiac.).
Lithoxylon Endl. = Actephila Blume (Euphorbiac.).
Lithraea Miers. Anacardiaceae. 3 S. Am.
Lithrum Huds. = Lythrum L. (Lythrac.).
Litobrochia Presl = Pteris L. (Pteridac.).
Litocarpus L. Bolus = Aptenia N. E. Br. (Aïzoac.).
Litogyne Harv. = Epaltes Cass. (Compos.).
Litonia Pritz. = Littonia Hook. (Liliac.).
Litophila Sw. = Lithophila Sw. = Alternanthera Forsk. (Amaranthac.).
Litorella Aschers. = Littorella Bergius (Plantaginac.).
Litosanthes Blume. Rubiaceae. 1 Formosa, 1 Java, 3 New Guinea.
Litosiphon Pierre ex A. Chev. = Lovoa Harms (Meliac.).
Litrea Phil. = seq.
Litria G. Don = Lithraea Miers (Anacardiac.).
Litrisa Small = Carphephorus Cass. (Compos.).
Litsaea Juss. = seq.
***Litsea** Lam. Lauraceae. 400 warm As. (N. to Korea & Jap.), Austr., Am. Ls. and bark medicinal.
Littaea Brign. ex Tagliabue = Agave L. (Agavac.).
Littanella Roth (sphalm.) = Littorella Bergius (Plantaginac.).
Littlea Dum. (sphalm.) = Littaea Brign. ex Tagliabue = Agave L. (Agavac.).
Littledalea Hemsl. Gramineae. 3 C. As. to W. China.
Littonia Hook. Liliaceae. 8 trop. & S. Afr. Like *Gloriosa*.
Littorella Bergius. Plantaginaceae. 3 spp., 1 N. Am., 1 S. Am., and *L. lacustris* L. (shore-weed) in W., C. & N. Eur. This pl. exhibits two forms, one in water, another on land. The land form has a rosette of narrow ls. about 3 cm. long, which spread out upon the ground and show distinct dorsiventral structure. Fls. in groups of 3, one ♂ on a long stalk between two sessile ♀, which are ripe before the sta. emerge from the former. Both sta. and style are very long and the fls. are wind-pollinated. Fr. a nut. The water form has much larger ls. which grow erect and are cylindrical (centric) in form and internal structure; no fls. are produced, but the plant multiplies largely by runners. It is often mistaken for *Isoëtes*.
Littorella Ehrh. (uninom.) = *Littorella lacustris* L. (Plantaginac.).
Littorellaceae S. F. Gray = Plantaginaceae Juss.
Litwinowia Woronow (~ Euclidium R.Br.). Cruciferae. 2 C. As.
Livistona R.Br. Palmae. 30 Indomal., Austr. Tall trees with fan leaves and panicles of ☿ fls. Fr. a berry.
Lizeron Rafin. = Convolvulus L. (Convolvulac.).
Llagunoa Ruiz & Pav. Sapindaceae. 3 W. trop. S. Am.
Llanosia Blanco = Ternstroemia Mutis ex L. f. (Theac.).
Llavea Lagasca. Cryptogrammataceae. 1 trop. Am.
Llavea Liebm. = Neopringlea S. Wats. (? Flacourtiac.).
Llavea Planch. ex Pfeiff. = Meliosma Blume (Meliosmac.).
Llerasia Triana. Compositae. 11 Colombia, Bolivia.
Llewelynia Pittier. Melastomataceae. 1 Venez.
Lloydia Benth. & Hook. f. = Lioydia Neck. = Printzia Cass. (Compos.).
***Lloydia** Salisb. ex Reichb. Liliaceae. 20 N. temp.

LLOYIDIA

Lloyidia Steud. = praec.

Loasa Adans. Loasaceae. 100 Mex. to temp. S. Am., chiefly mts. of Chile and Peru. They possess stinging hairs. The fls. are generally yellow and face downwards. The nectaries, formed of combined stds. (see fam.), are large and conspicuous. The petals are boat-shapeu and conceal the groups of sta.

***Loasaceae** Spreng. Dicots. 15/250 mostly trop. & subtrop. Am., esp. Andes; 1 gen. & sp. (*Kissenia*) SW. Afr. & Arabia. Mostly herbs, frequently twining, with opp. or alt., rarely stip., ls. The epidermis bears hairs of various kinds; esp. common are grapple-hairs and stinging-hairs. Fls. usu. in cymes, often sympodial, yellow (rarely white or red), ♀, usu. 5-merous. Receptacle deeply hollowed out, so that the fl. is epig. K 5, imbr.; C 5, free or united; A 5–∞. In the genera with ∞ sta. there is much difference as to the arrangement. In *Mentzelia* they are evenly distributed round the style, the outermost in some spp. being sterile. In other gen. it is the antesepalous sta. that are sterile, and in some, e.g. *Loasa*, *Blumenbachia*, 3 or more of the stds. are united to form a large coloured nectary, whose mouth is towards the centre of the fl. and partly obstructed by the other stds. G̅ 1 or more commonly (3–5), with parietal plac.; ovules 1, several, or ∞, anatr., with one integument; style simple. Fr. various, often a caps., sometimes spirally twisted. Endosp. or not. Chief genera: *Mentzelia*, *Eucnide*, *Loasa*, *Caiophora*.

Loasella Baill. = Sympetaleia A. Gray (Loasac.).

Lobadium Rafin. = Rhus L. (Anacardiac.).

Lobake Rafin. = Jacquemontia Choisy (Convolvulac.).

Lobaria Haw. = Saxifraga L. (Saxifragac.).

Lobbia Planch. = Thottea Rottb. (Aristolochiac.).

Lobeira Alexander = Nopalxochia Britt. & Rose (Cactac.).

Lobelia Adans. = Scaevola L. (Goodeniac.).

Lobelia L. Campanulaceae. 200–300 cosmop., mostly trop. & subtrop., esp. Am. The fl. (see fam.) is twisted upon its axis through 180°, and is zygo. The anthers are syngenesious as in *Compositae*, and the style pushes through the tube thus formed, driving the pollen out at the top. Finally it emerges, the stigmas separate, and the ♀ stage begins. (See fam., and cf. *Campanula*, *Phyteuma*, *Jasione* and *Compositae*.)

***Lobeliaceae** R.Br. = Campanulaceae-Lobelioïdeae Engl.

Lobia O. F. Cook. Palmae. 1 C. Am.

Lobirebutia Frič = Lobivia Britt. & Rose (Cactac.).

Lobirota Dulac = Ramonda Rich. (Gesneriac.).

Lobivia Britton & Rose. Cactaceae. 75 S. Andes.

Lobiviopsis Frič = Echinopsis Zucc. (Cactac.).

Lobocarpus Wight & Arn. = Glochidion J. R. & G. Forst. (Euphorbiac.).

Lobogyna P. & K. = seq.

Lobogyne Schlechter = Appendicula Bl. (Orchidac.).

Lobomon Rafin. = Amphicarpaea Ell. (cleistogamous form) (Legumin.).

Lobophyllum F. Muell. = Coldenia L. (Boraginac.).

Lobopogon Schlechtd. = Brachyloma Sond. (Epacridac.).

Loboptera Colla = Columnea L. (Gesneriac.).

Lobostema Spreng. = seq.

Lobostemon Lehm. Boraginaceae. 25–30 S. Afr.

Lobostephanus N. E. Br. Asclepiadaceae. 1 S. Afr.
*Lobularia Desv. Cruciferae. 5 C. Verde & Canary Is., Medit., Arabia.
Locandi Adans. = Samadera Gaertn. = Quassia L. (Simaroubac.).
Locandia Kuntze = praec.
Locardi Steud. = praec.
Locella Van Tiegh. = Taxillus Van Tiegh. (Loranthac.).
Locellaria Welw. = Bauhinia L. (Legumin.).
Lochemia Arn. = Melochia L. (Sterculiac.).
Locheria Neck. = Verbesina L. (Compos.).
Locheria Regel = Achimenes P.Br. (Gesneriac.).
Lochia Balf. f. Caryophyllaceae. 2 Socotra, Abd al Kuri.
Lochmocydia Mart. ex DC. = Cuspidaria DC., Piriadacus Pich., Saldanhaea Bur., *p.p.* (Bignoniac.).
Lochnera Reichb. = Catharanthus G. Don (Apocynac.).
Lochneria Fabr. (uninom.) = *Sisymbrella* Spach sp. (Crucif.).
Lochneria Scop. = Elaeocarpus L. (Elaeocarpac.).
Lockhartia Hook. Orchidaceae. 30 C. Am., trop. S. Am., W.I.
Locusta Medik. = Valerianella Moench (Valerianac.).
Loddigesia Sims = Hypocalyptus Thunb. (Legumin.).
Lodhra (G. Don) Guillem. = Symplocos Jacq. (Symplocac.).
Lodicularia Beauv. = Hemarthria R.Br. (Gramin.).
Lodoicea Comm. ex J. St-Hil. Palmae. 1 Seychelles, *L. maldivica* (Gmel.) Pers., the double coconut or *coco de mer*. Dioec. The fr. is one of the largest known and takes 10 years to ripen. The nut is bilobed. The fr. used to be found floating before the discovery of the tree.
Loeffinglia Augier (sphalm.) = seq.
Loefflingia Neck. = Loeflingia L. (Caryophyllac.).
Loefgrenianthus Hoehne. Orchidaceae. 1 Brazil.
Loeflinga Hedw. f. = seq.
Loeflingia L. Caryophyllaceae. 7 N. Am., Medit.
Loerzingia Airy Shaw. Euphorbiaceae. 1 Sumatra.
Loeselia L. Polemoniaceae. 17 Calif. to Venezuela. Fl. ± zygo.
Loesenera Harms. Leguminosae. 4 trop. W. Afr.
Loeseneriella A. C. Sm. Celastraceae. 16 S. China, SE. As., Indomal., Austr.
Loethainia Heynh. = Wiborgia Thunb. (Legumin.).
Loevigia Karst. & Triana = Monochaetum Naud. (Melastomatac.).
Loewia Urb. Turneraceae. 3 E. trop. Afr.
Loezelia Adans. = Loeselia L. (Polemoniac.).
*Logania R.Br. Loganiaceae. 25 Austr., New Caled., N.Z. Sepal anterior.
Logania J. F. Gmel. = Loghania Scop. = Souroubea Aubl. (Marcgraviac.).
*Loganiaceae Mart. (*s.str.*). Dicots. 7/130 trop., a few warm temp. Trees, shrubs, and herbs, some climbers, with opp. ent. simple stip. ls.; stips. often much reduced. Infl. usu. cymose, various; fls. with br. and bracteoles, usu. reg., ☿, 4–5-merous. Disk small or absent. K (4–5), imbr.; C (4–5), imbr. (rarely contorted); A 4–5, rarely 1, epipet.; G (2), anteropost., rarely semi-inf., 2-loc. (rarely imperfectly so), 1-loc. or more-loc. with simple style and ov. usu. ∞, amphi- or ana-tr. Caps., usu. septicid.; seeds winged or not; endosp.

LOGFIA

Chief genera: *Logania, Gelsemium, Geniostoma, Polypremum, Mostuea.* Nearly allied to *Apocynac., Gentianac., Solanac., Scrophulariac., Rubiac.* (For other groups formerly included in a very heterogeneous 'Loganiac.', see *Antoniac., Buddlejac., Potaliac., Spigeliac., Strychnac.*)

Logfia Cass. = Filago L. (Compos.).

Loghania Scop. = Souroubea Aubl. (Marcgraviac.).

Logia Mutis = Calceolaria L. (Scrophulariac.).

Loheria Merr. Myrsinaceae. 6 Philipp. Is., New Guinea.

Loiseleria Reichb. = Rhododendron L. (Ericac.).

***Loiseleuria** Desv. ex Loisel. Ericaceae. 1 N. circumpolar, *L. procumbens* (L.) Desv. Ls. very wiry, rolled back at margins, reducing transpiration. Fl. reg., protog., opening soon after melting of snow. A 5; G 2-3-loc.

Loiseleuria Reichb. = Loiseleria Reichb. = Rhododendron L. (Ericac.).

Lojaconoa Bobrov (~ Trifolium L.). Leguminosae. 60 W. N. Am., few Andes.

Lojaconoa Gandog. = Festuca L. (Gramin.).

Lolanara Rafin. = Ochrocarpos Thou. = Mammea L. (Guttif.).

Loliolum Krecz. & Bobr. = Nardurus Reichb. (Gramin.).

Lolium L. Gramineae. 12 temp. Euras. Spikelets in 2-ranked spike, and placed edgewise (this distinguishes it from *Triticum, Agropyron,* and *Hordeum,* in the *Hordeëae*). Valuable pasture and fodder.

Lomagramma J. Sm. Lomariopsidaceae. 15 Indomal. to Tahiti. Habit of *Lomariopsis* but veins anast. (Holttum, *Gard. Bull. Str. Settlem.* **9**: 190, 1937).

Lomake Rafin. = Stachytarpheta Vahl (Verbenac.).

Lomandra Labill. Xanthorrhoeaceae. 40 New Guinea, Austr., New Caled. Dioec. P sepaloid, or inner petaloid.

Lomandraceae Lotsy = Xanthorrhoeaceae Dum.

Lomanodia Rafin. (Astronidium A. Gray). Melastomataceae. 35 New Guinea, Pacif.

Lomanthera Rafin. = Tetrazygia Rich. (Melastomatac.).

Lomanthes Rafin. = Phyllanthus L. (Euphorbiac.).

Lomaphlebia J. Sm. = Grammitis Sw. (Grammitidac.).

Lomaresis Rafin. = Ornithogalum L. (Liliac.).

Lomaria Willd. = Blechnum L. (Blechnac.).

Lomaridium Presl = praec.

Lomariobotrys Fée = Stenochlaena J. Sm. (Blechnac.).

Lomariopsidaceae Alston. Aspidiales. Creeping on rocks by streams, or climbers, or epiphytes, with dorsiventral rhizome (except *Elaphoglossum* spp.) and acrostichoid fertile ls. Chief genera: *Egenolfia, Bolbitis, Lomariopsis, Teratophyllum, Lomagramma, Elaphoglossum.* (Holttum, *J. Linn. Soc., Bot.* **53**: 146-9, 1947.)

Lomariopsis Fée. Lomariopsidaceae. 40 pantrop. Rhizome thick, climbing trees in forest to 15 m. or more; ls. simply pinnate, pinnae jointed except terminal one. (Holttum, *Gard. Bull. Str. Settlem.* **5**: 264-77, 1932; *Bull. Misc. Inf., Kew,* **1939**: 613-28, 1940.)

Lomaspora (DC.) Steud. = Arabis L. (Crucif.).

Lomastelma Rafin. = Acmena DC. = Syzygium Gaertn. (Myrtac.).

× **Lomataloë** Guillaumin. Liliaceae. Gen. hybr. (Aloë × Lomatophyllum).

× **Lomateria** Guillaumin. Liliaceae. Gen. hybr. (Gasteria × Lomatophyllum).

*Lomatia R.Br. Proteaceae. 12 E. Austr., Tasm., Chile. Gynophore.

Lomatium Rafin. Umbelliferae. 80 W. N. Am.

Lomatocarum Fisch. & Mey. = Carum L. (Umbellif.).

Lomatogonium A. Br. Gentianaceae. 18 temp. Euras.

Lomatolepis Cass. = Launaea Cass. (Compos.).

Lomatophyllum Willd. (~ Aloë L.). Liliaceae. 14 Madag., Masc. Fr. fleshy, dehisc.

Lomatopodium Fisch. & Mey. = Seseli L. (Umbellif.).

Lomatozona Baker. Compositae. 1 Brazil. Ls. opp. Pappus connate at base.

Lomaxeta Rafin. = Polypteris Nutt. (Compos.).

Lomelosia Rafin. = Scabiosa L. (Dipsacac.).

Lomenia Pourr. = Watsonia Mill. (Iridac.).

Lomentaceae R.Br. = Leguminosae–Caesalpinioïdeae Kunth.

Lomeria Rafin. = Cestrum L. (Solanac.).

Lomilis Rafin. = ?Hamamelis Gronov. ex L. (Hamamelidac.).

Lommelia Willis (sphalm.) = Louvelia Jumelle & Perrier (Palm.).

Lomoplis Rafin. = Mimosa L. (Legumin.).

Lonas Adans. Compositae. 1 SW. Medit.

Lonchanthera Less. ex Baker = Stenachaenium Benth. (Legumin.).

Lonchestigma Dun. = Jaborosa Juss. (Solanac.).

Lonchitis [non L.] Alston (*Ferns W. Trop. Afr.* 33, 1959) = Blotiella R. Tryon (Dennstaedtiac.).

Lonchitis Bub. = Serapias L. (Orchidac.).

Lonchitis L. Dennstaedtiaceae. 2 trop. Am., Afr., Madag. (R. Tryon, *Contr. Gray Herb.* **191**: 94–6, 1962).

*Lonchocarpus Kunth. Leguminosae. 150 trop. Am., W.I., Afr., Austr.

Lonchomera Hook. f. & Thoms. = Mezzettia Becc. (Annonac.).

Lonchophaca Rydb. = Astragalus L. (Legumin.).

Lonchophora Dur. Cruciferae. 1 N. Afr.

Lonchophyllum Ehrh. (uninom.) = *Serapias lonchophyllum* L. = *Cephalanthera longifolia* (L.) Fritsch (Orchidac.).

Lonchostephus Tul. Podostemaceae. 1 Brazil (Amazon).

Lonchostigma P. & K. = Lonchestigma Dunal = Jaborosa Juss. (Solanac.).

*Lonchostoma Wikstr. Bruniaceae. 4 S. Afr.

Lonchostylis Torr. = Rhynchospora Vahl (Cyperac.).

Loncodilis Rafin. = Eriospermum Jacq. (Liliac.).

Loncomelos Rafin. = Ornithogalum L. (Liliac.).

Loncoperis Rafin. = Carex L. (Cyperac.).

Loncostemon Rafin. = Allium L. (Alliac.).

Loncoxis Rafin. = Ornithogalum L. (Liliac.).

Londesia Fisch. & Mey. Chenopodiaceae. 1 C. As.

Londesia Kar. & Kir. ex Moq. = Kirilowia Del. (Chenopodiac.).

Longchampia Willd. = Leysera L. = Asteropterus Adans. (Compos.).

Longetia Baill. = Austrobuxus Miq. (Euphorbiac.).

Longiphylis Thou. (uninom.) = *Bulbophyllum longiflorum* Thou. = *Cirrhopetalum umbellatum* (Forst. f.) Hook. & Arn. (Orchidac.).

Longiviola Gandog. = Viola L. (Violac.).

LONICERA

Lonicera Boehm. = Psittacanthus Mart. (Loranthac.).

Lonicera Gaertn. = Dendrophthoë Mart. (Loranthac.).

Lonicera L. Caprifoliaceae. 200 N. Am. (S. to Mex.), Euras. (S. to N. Afr., Himal., Philipp. Is., & SW. Malaysia). Mostly erect shrubs, a few twining, with opp. frequently connate ls. In the axils of many (e.g. *L. tatarica* L.) are serial buds, of which the lowest gives rise to the fls. usu. in pairs, the central fl. of the small dichasium not being developed. The fl. is frequently zygo., and gives rise to a berry. In some the pair of fls. produces two independent berries, in others the berries fuse into one as they form. Some spp. exhibit the 'fusion' even earlier; and one finds two corollas seated upon what at first glance appears to be a single inf. ovary. Dissection shows that in most cases the two ovaries are side by side, free from one another, in a common hollow axis; in a few cases, however, the union is more complete. The fl. of *L. caprifolium* L. is visited chiefly by hawk-moths (at night). The fl. opens in the evening, the anthers having dehisced shortly before this. The style projects beyond the anthers. The fl. moves into a horiz. position at the same time. At first the style is bent downwards and the sta. form the alighting place for insects. Later on the style moves up to a horiz. position, the sta. shrivel and bend down, and this is complete by the second evening when the next crop of buds is opening. At the same time the fl. has changed from white to yellow. The length of the tube keeps out all but very long-tongued insects.

Lonicer[ac]eae Endl. = Caprifoliaceae Juss.

Loniceroïdes Bullock. Asclepiadaceae. 1 Brazil.

Lontanus Gaertn. = seq.

Lontarus Steck = Borassus L. (Palm.).

Loosa Jacq. = Loasa Adans. (Loasac.).

Lopadocalyx Klotzsch = Olax L. (Olacac.).

Lopanthus Vitm. = Lophanthus J. R. & G. Forst. = Waltheria L. (Sterculiac.).

Lopesia Juss. = seq.

Lopezia Cav. Onagraceae. 17 Mex., C. Am. Fl. zygo. The two upper petals are bent upwards a little way from the base, and at the bend there seems to be a drop of honey. In reality this is a dry glossy piece of hard tissue; like the similar bodies in *Parnassia* it deceives flies. There are real nectaries at the base of the fl. There are two sta., of which the post. only is fertile; it is enclosed at first in the ant. one, which is a spoon-shaped petaloid std. In the early stage of the fl., the style is undeveloped and insects alight on the sta.; later the style grows out into the place first occupied by the sta., which now bends upwards out of the way. In *L. coronata* Andr., etc., there is an upward tension in the sta., a downward in the std., and an explosion occurs when an insect alights.

Lophacma P. & K. = seq.

Lophacme Stapf. Gramineae. 1 S. Afr.

Lophactis Rafin. = Coreopsis L. (Compos.).

Lophalix Rafin. = Alloplectus Mart. (Gesneriac.).

Lophandra D. Don = Erica L. (Ericac.).

***Lophanthera** A. Juss. Malpighiaceae. 4 Brazil. Carpophore.

Lophanthera Rafin. = Sopubia Buch.-Ham. (Scrophulariac.).

Lophanthus Adans. = Nepeta L. (Labiat.).

Lophanthus Benth. = Agastache Clayt. in Gronov. (Labiat.).
Lophanthus J. R. & G. Forst. = Waltheria L. (Sterculiac.).
Lopharina Neck. = Erica L. (Ericac.).
Lophatherum Brongn. Gramineae. 2 E. As., Indomal., trop. Austr.
Lopherina Juss. = Lopharina Neck. = Erica L. (Ericac.).
Lophia Desv. = Alloplectus Mart. (Gesneriac.).
Lophiaris Rafin. = Oncidium Sw. (Orchidac.).
Lophidium Rich. = Schizaea Sm. (Schizaeac.).
Lophiocarpus Miq. = Sagittaria Lam. (Alismatac.).
Lophiocarpus Turcz. (~ Microtea Sw.). Phytolaccaceae(?). 4 S. Afr.
Lophiola Ker-Gawl. Haemodoraceae. 2 Atl. N. Am.
Lophiolaceae Nak. = Haemodoraceae R.Br.
Lophiolepis Cass. = Cirsium Mill. (Compos.).
Lophion Spach = Viola L. (Violac.).
Lophira Banks ex Gaertn. f. Ochnaceae. 2 trop. W. Afr. The fr. of *L. lanceolata* Van Tiegh. ex Keay (*L. alata* auctt., non Banks ex Gaertn. f.) (African oak) has one sep. much, a second less, elongated. The seeds yield an oil on pressure; timber good.
Lophiraceae Endl. = Ochnaceae–Lophireae Gilg.
Lophium Steud. = Lophion Spach = Viola L. (Violac.).
Lophobios Rafin. = Euphorbia L. (Euphorbiac.).
Lophocachrys Koch ex DC. = Hippomarathrum Link (Umbellif.).
Lophocarpinia Burkart. Leguminosae. 1 Paraguay.
Lophocarpus Boeck. = Neolophocarpus E. G. Camus (Cyperac.).
Lophocarpus Link = Froelichia Moench (Amaranthac.).
Lophocarya Nutt. ex Moq. = Obione Gaertn. (Chenopodiac.).
Lophocereus (A. Berger) Britton & Rose. Cactaceae. 4 SW. U.S., Mex.
Lophochlaena Nees = Pleuropogon R.Br. (Gramin.).
Lophochlaena P. & K. = Lopholaena DC. (Compos.).
Lophochloa Reichb. Gramineae. 6 temp. Euras.
Lophoclinium Endl. = Podotheca Cass. (Compos.).
Lophodium Newman = Dryopteris Adans. (Aspidiac.).
Lophoglotis Rafin. = Sophronitis Lindl. (Orchidac.).
Lophogyne Tul. Podostemaceae. 3 E. Brazil, Guiana.
Lopholaena DC. Compositae. 18 trop. & S. Afr.
Lopholepis Decne. Gramineae. 1 S. India, Ceylon.
Lopholepis J. Sm. = Microgramma Presl (Polypodiac.).
Lopholoma Cass. = Centaurea L. (Compos.).
Lophomyrtus Burret. Myrtaceae. 3 N.Z.
Lophopappus Rusby. Compositae. 5 Andes.
Lophopetalum Wight ex Arn. Celastraceae. 4 Indoch., W. Malaysia.
Lophophora Coult. Cactaceae. 2 Texas, New Mex.
Lophophyllum Griff. = Cyclea Arn. (Menispermac.).
Lophophytaceae Horan. = Balanophoraceae–Lophophytoïdeae Harms.
Lophophytum Schott & Endl. Balanophoraceae. 4 trop. S. Am.
Lophopogon Hack. Gramineae. 2 India. Glume 3-toothed.
Lophopteris Griseb. = seq.

LOPHOPTERYS

Lophopterys A. Juss. Malpighiaceae. 3 Guiana.

Lophopteryx Dalla Torre & Harms = praec.

Lophoptilon Gagnep. (nom. subnud.). Compositae. Indochina. Quid?

Lophopyxidaceae (Engl.) H. H. Pfeiff. Dicots. 1/2 Malaysia. Scandent shrubs or small trees, with simple serrul. or crenul. alt. stip. ls., and watch-spring tendrils (modif. infl.). Fls. small, reg., ♂ ♀, monoec., in glomerules on the branches of loose axill. panicles. K 5, valv., v. shortly united, persist.; C 5, much smaller, not contig.; A 5, oppositisep., with filif. fil. and intr. anth., alt. with 5 spreading oppositipet. stds. or glands which in the ♂ fl. are ± adnate to the subtending pet. and in the ♀ fl. are ± concrescent into a 5-lobed disk; G (5–4), ovoid-obl., shallowly ribbed, pubesc., with 5–4 sessile subul. stigs., and 2 pend. apical axile anatr. ov. per loc., each surmounted by an obturator-like appendage. Fr. indehisc., fusiform, 1-loc., 1-seeded, with 5 broad stramineous wings (outline of whole fr. obov. or ellips.); seed oblong, with endosp. Only genus: *Lophopyxis*. Systematic position much disputed, but evidently closely related to *Rhamnaceae–Gouanieae*; cf. habit, foliage, tendrils, tendency to 1-sex. fls. (*Gouania*), valv. K, winged fr. *Lophopyxis* differs principally in having stam. opp. to the sep., and stds. (disk-glands) opp. to the pet., and in the superior (not inf.), 5- (not 2–3-) loc. G, with 2 apical (not 1 basal) ov. per loc.

Lophopyxis Hook. f. Lophopyxidaceae. 2 Malay Penins., Borneo, E. Malaysia, Palau & Solomon Is.

Lophoschoenus Stapf. Cyperaceae. 11 Seychelles, Borneo, New Caled.

Lophosciadium DC. Umbelliferae. 5 S. Eur., W. As.

Lophosoria Presl. Lophosoriaceae. 1 trop. Am. Small tree-fern, hairy as *Dicksonia*.

Lophosoriaceae Pichi-Sermolli. Cyatheales. 2 gen.: *Lophosoria, Metaxya*. Stem short, erect or creeping, not dorsiventral; stem-apex hairy; sori superficial, not indusiate.

Lophospatha Burret. Palmae. 1 Borneo.

Lophospermum D. Don (~ Maurandia Ortega). Scrophulariaceae. 4 Mex. to Venez., W.I.

Lophostachys Pohl. Acanthaceae. 15 C. & trop. S. Am.

Lophostemon Schott = Tristania R.Br. (Myrtac.).

Lophostephus Harv. = Anisotoma Fenzl (Asclepiadac.).

Lophostigma Radlk. Sapindaceae. 1 Bolivia.

Lophostoma Meissn. Thymelaeaceae. 4 N. trop. S. Am.

Lophostylis Hochst. = Securidaca L. (Polygalac.).

Lophotaenia Griseb. = Malabaila Hoffm. (Umbellif.).

Lophothecium Rizzini. Acanthaceae. 1 Brazil.

Lophothele O. F. Cook. Palmae. 1 Mex.

Lophotocarpus Durand (~ Sagittaria L.). Alismataceae. 8 cosmop. (exc. Australasia & Pacif.), mostly Am.

Lophoxera Rafin. = Celosia L. (Amaranthac.).

Lophozonia Turcz. = Nothofagus Bl. (Fagac.).

Lopimia Mart. (~ Pavonia Cav.). Malvaceae. 2 trop. Am.

Lopriorea Schinz. Amaranthaceae. 1 E. Afr.

Lorantea Steud. = Lorentea Ortega = Sanvitalia Gualt. (Compos.).

LORANTHACEAE

*Loranthaceae Juss. (*s.l.*). Dicots. 36/1300 trop. & temp. (Many more genera according to Van Tieghem, Danser, Balle, etc.) An interesting fam. of parasites with green ls. Mostly small semi-parasitic shrubs attached to their hosts by suckers or haustoria—usu. regarded as modified adv. roots. A few root in the earth, e.g. the W. Austr. *Nuytsia*, which grows into a small tree 10 m. high. Most are fairly omnivorous in their choice of hosts, but a few are restricted to one or two. Where the parasitic root joins the host, there is not uncommonly an outgrowth, often of considerable size and complicated in shape. The parasitic root often branches within the tissue of the host, as in mistletoe (*Viscum*). The stem is sympodial, often dichasial, e.g. in *Viscum*, and the ls. usu. evergr. and leathery.

Infl. cymose, the fls. usu. in little groups of 3 (or 2, by abortion of the central fl.). When the fls. are stalked, the bracts of the lateral fls. are always united to their stalks, up to the point of origin of the fl. (see *Viscum* and *Loranthus*). Infl. sometimes in spikes, with the fls. on the internodes as well as on the nodes.

The recept. is hollowed out, and the P springs from its margin. In the *Loranthoïdeae* there is below the P an outgrowth from the axis in the form of a small rim or fringe—the *calyculus*. Some look upon it as a K, many as an outgrowth of the axis itself; and this is perhaps the safest view. P either sepaloid or petaloid. Fls. ♀ or unisexual. Sta. as many as, and (as in *Proteaceae*) opp. and adnate to, the P-leaves. The pollen is often developed in a great number of locelli, separate from one another, though often becoming continuous when mature. Ovary 1-loc., sunk in, and united with, the receptacle, the ovules not differentiated from the placenta, with or without endosp. Embryo-sacs > one, curiously lengthened (cf. *Casuarina*). Fr. a pseudo-berry or -drupe, the fleshy part really the receptacle. Round the seed is a layer of viscin, a very sticky substance.

Classification and chief genera (after Engler & Krause):

I. **Loranthoïdeae.** 'Calyculus' or 2 adnate bracteoles external to perianth; pollen usu. trilobate; fr. with viscous layer outside tepaline vascular strands. [*Loranthaceae s.str.*: 25/850.]

    1. Nuytsiëae (stem with secretory canals and intraxylary phloëm; no calyculus, but 2 adnate bracteoles; fr. dry, 3-winged): *Nuytsia*.

    2. Lorantheae (stem without secretory canals; extraxylary phloëm; calyculus present; fr. drupaceous or baccate): *Amylotheca, Lepeostegeres, Macrosolen, Elytranthe, Loranthus, Phrygilanthus, Struthanthus, Phthirusa, Psittacanthus.*

II. **Viscoïdeae.** No 'calyculus'; pollen spherical; fr. with viscous layer between tepaline and ovarian vascular strands. [See also *Viscaceae*.]

    1. Eremolepideae (fls. in simple spikes or rac.; plac. basal; anth. 4-loc.): *Antidaphne, Eremolepis, Lepidoceras.*

    2. Phoradendreae and 3. Arceuthobiëae (fls. sol. or in groups in axil of persist. bracts, or extra-axill. on the internodes; plac. central; anth. 2- or 1-loc.): *Korthalsella, Dendrophthora, Phoradendron, Ginalloa, Arceuthobium.*

    4. Visceae (as 2 & 3, but plac. basal; anth. with 5–∞ loc.): *Notothixos, Viscum.*

LOXANIA

finally dissolving by autodigestion; A 5, unilat. disposed, with obl. anth.; G (3), with elongate style and 3 lacin. stigs., and ∞ axile anatr. ov. Caps. oblong, with papery walls; seeds globose, with 3-lobed aril. Only genus: *Orchidantha*. Related to *Musaceae* and *Strelitziaceae*.

Loxania Van Tiegh. = Loranthus L. (Loranthac.).
Loxanisa Rafin. = Carex L. (Cyperac.).
Loxanthera Blume. Loranthaceae. 1 W. Malaysia.
Loxanthes Rafin. = Aphyllon Mitch. = Orobanche L. (Orobanchac.).
Loxanthes Salisb. = Nerine Herb. (Amaryllidac.).
Loxanthocereus Backeb. = Borzicactus Riccob. (Cactac.).
Loxanthus Nees = Phlogacanthus Nees (Acanthac.).
Loxidium Vent. = Swainsona Salisb. (Legumin.).
Loxocalyx Hemsl. Labiatae. 2 China.
Loxocarpus R.Br. Gesneriaceae. 15 Malay Penins., Java.
Loxocarya R.Br. Restionaceae. 8 SW. Austr.
Loxococcus H. Wendl. & Drude. Palmae. 1 Ceylon.
Loxodera Launert. Gramineae. 3 S. trop. Afr.
Loxodiscus Hook. f. Sapindaceae. 1 New Caled.
Loxodon Cass. = Chaptalia Vent. (Compos.).
Loxodora Launert (sphalm.) = Loxodera Launert (Gramin.).
Loxogrammaceae Ching = ? Polypodiaceae S. F. Gray
Loxogramme (Blume) Presl. Polypodiaceae(?). 40 Old World trop., esp. Malaysia. (Wilson, *Contr. Gray Herb.* **185**: 114, 1959.)
Loxonia Jack. Gesneriaceae. 1 Sumatra, Japan.
Loxophyllum Blume = praec.
Loxoptera O. E. Schulz. Cruciferae. 1 Peru.
Loxopterygium Hook. f. Anacardiaceae. 5 trop. S. Am.
Loxoscaphe Moore. Aspleniaceae. 4 pantrop.
Loxospermum Hochst. = Trifolium L. (Legumin.).
Loxostachys Peter = Cyrtococcum Stapf (Gramin.).
Loxostemon Hook. f. & Thoms. Cruciferae. 5 E. Himal. to SW. China.
Loxostigma C. B. Clarke. Gesneriaceae. 5 E. Himal. to SW. China.
Loxostylis Spreng. ex Reichb. Anacardiaceae. 1 S. Afr.
Loxothysanus Robinson. Compositae. 3 Mex.
Loxotis R.Br. = Rhynchoglossum Blume (Gesneriac.).
Loxotrema Rafin. = Carex L. (Cyperac.).
Loxsoma R.Br. Loxsomaceae. 1 N.Z.
Loxsomaceae Presl. Loxsomales. Rhiz. hairy; ls. *Davallia*-like; sori as *Trichomanes*. Genera: *Loxsoma, Loxsomopsis*.
Loxsomales. Filicidae. 1 fam.: *Loxsomaceae*.
Loxsomopsis Christ. Loxsomaceae. 3 trop. Am.
Loydia Del. = Beckeropsis Fig. & De Not. (Gramin.).
Lozanella Greenman. Ulmaceae. 2 Mex. to Peru & Bolivia.
Lozania S. Mutis ex Caldas. Lacistemataceae. 7 C. Am. to Peru.
Lubaria Pittier. Rutaceae. 1 Venez.
Lubinia Comm. ex Vent. = Lysimachia L. (Primulac.).
Lucaea Kunth = Arthraxon Beauv. (Gramin.).
Lucaya Britton & Rose = Acacia Mill. (Legumin.).

690

Luchea auctt. (sphalm.) = Luehea Willd. (Tiliac.).
Luchia Steud. = Elodea Michx (Hydrocharitac.).
Lucianea Endl. = Lucinaea DC. (Rubiac.).
Lucilia Cass. Compositae. 20 S. Am.
Luciliodes (Less.) Kuntze = Amphiglossa DC. (Compos.).
Luciliopsis Wedd. Compositae. 4 Andes.
Lucinaea DC. Rubiaceae. 25 Malaysia, New Caled.
Lucinaea Leandro ex Pfeiff. = Anchietea A. St-Hil. (Violac.).
Luciola Sm. = Luzula DC. (Juncac.).
Luckhoffia White & Sloane. Asclepiadaceae. 1 S. Afr.
Luculia Sweet. Rubiaceae. 5 Himal. to SW. China.
Lucuma Molina (~ Pouteria Aubl.). Sapotaceae. 100 Malaysia, Austr., Pacif., trop. Am. The fr. of *L. bifera* Molina is ed.
***Lucya** DC. Rubiaceae. 1 W.I.
Luddemania Reichb. f. = Lueddemannia Reichb. f. (Orchidac.).
Ludia Comm. ex Juss. Flacourtiaceae. 7 E. Afr., Madag., Masc.
Ludisia A. Rich. Orchidaceae. 1 Indochina, Malay Penins.
× **Ludochilus** hort. (v) = × Anoectomaria hort. (Orchidac.).
Ludolfia Adans. = Tetragonia L. (Tetragoniac.).
Ludolfia Willd. = Arundinaria Michx (Gramin.).
***Ludovia** Brongn. Cyclanthaceae. 2 Panamá, trop. S. Am. ♂ fls. as in *Carludovica*, ♀ sunk to stigmas with rudimentary P. Climbers.
Ludovia Pers. = Carludovica Ruiz & Pav. (Cyclanthac.).
Ludovica Vieill. ex Guillaumin = Bikkiopsis Brongn. & Gris (Rubiac.).
Ludovicea Buc'hoz = ? Heisteria Jacq. (Olacac.).
Ludovicia Coss. = Hammatolobium Fenzl (Legumin.).
Ludwigia L. Onagraceae. 75 cosmop., esp. trop. Am.; water and marsh plants. Aerating tissue is well developed (cf. *Sonneratia, Sesbania*). In *L. repens* (L.) Sw., when growing in water, two forms of root develop—ordinary anchorage roots, and erect spongy roots which grow upwards, often till they reach the surface of the water. The bulk of the tissue consists of aerenchyma. In *L. suffruticosa* (L.) Walt. there is an erect stem, whose lower part is covered with aerenchyma if growing in water (cf. *Lycopus*). If the plants be grown on land none of these phenomena appear.
Ludwigiantha (Torr. & Gray) Small = praec.
Lueddemannia Reichb. f. Orchidaceae. 5 W. trop. S. Am.
Luederitzia K. Schum. = Pavonia Cav. (Malvac.).
Luehea F. W. Schmidt = Stilbe Bergius (Stilbac.).
***Luehea** Willd. Tiliaceae. 20 trop. Am., W.I.
Lueheopsis Burret. Tiliaceae. 9 trop. S. Am.
Luerssenia Kuhn. Aspidiaceae. 1 Sumatra.
Luerssenia Kuntze = Cuminum L. (Umbellif.).
Luerssenidendron Domin. Rutaceae. 1 E. Austr.
Luetkea Bong. Rosaceae. 1 W. N. Am.
Luetzelburgia Harms. Leguminosae. 5 Brazil.
Luffa Mill. Cucurbitaceae. 6 trop. *L. cylindrica* (L.) M. Roem. (*L. aegyptiaca* Mill.) furnishes the *loofah* or bath sponge (the vascular bundle net of the
Lugaion Rafin. = Cytisus L., Genista L., etc. (Legumin.). [pericarp).

Lugoa DC. = Gonospermum Less. (Compos.).
Lugonia Wedd. Asclepiadaceae. 3 Peru, Bolivia, Argent.
Luhea DC. = Luehea Willd. (Tiliac.).
Luhea A. DC. = Luehea F. W. Schmidt = Stilbe Bergius (Stilbac.).
Luina Benth. Compositae. 4 NW. N. Am.
× **Luisanda** hort. Orchidaceae. Gen. hybr. (iii) (Luisia × Vanda).
Luisia Gaudich. Orchidaceae. 30 trop. As. to Japan and Polynesia.
Luitkea auct. ex Steud. = Luetkea Bong. (Rosac.).
Luma A. Gray. Myrtaceae. 110 S. Am.
Lumanaja Blanco = Homonoia Lour. (Euphorbiac.).
Lumbricidia Vell. = Andira Lam. (Legumin.).
Lumnitzera Jacq. ex Spreng. = Moschosma Reichb. (Labiat.).
Lumnitzera Willd. Combretaceae. 2 E. Afr. to Malaysia, N. Austr. & Pacific, in mangrove swamps. Fr. floated by ocean currents.
Lunana Endl. (sphalm.) = Lunasia Blanco (Rutac.).
Lunanaea Endl. = seq.
Lunanea DC. = Cola Schott (Sterculiac.).
*****Lunania** Hook. Flacourtiaceae. 18 Mex. to Peru, W.I. (K) in bud.
Lunania Rafin. = Heteranthera Ruiz & Pav. (Pontederiac.).
Lunaria Hill = Botrychium Sw. (Ophioglossac.).
Lunaria L. Cruciferae. 3 C. & SE. Eur.
Lunasia Blanco. Rutaceae. 10 Philipp. Is. & Borneo to E. Malaysia.
Lunathyrium Koidzumi = Diplazium Sw.(?) (Athyriac.).
Lundellia Leonard. Acanthaceae. 1 Mex.
*****Lundia** DC. Bignoniaceae. 15 C. & trop. S. Am., Trinidad.
Lundia Puerari ex DC. = Buchanania Roxb. (Anacardiac.).
Lundia Schumacher & Thonn. = Oncoba Forsk. (Flacourtiac.).
Lunella Nieuwl. = Besseya Rydb. (Scrophulariac.).
Lungia Steud. = Langia Endl. = Hermbstaedtia Reichb. (Amaranthac.).
Luntia Neck. = Luutia Neck. = Croton L. (Euphorbiac.).
Luorea Neck. ex Jaume St-Hil. (sphalm.) = Lourea Jaume St-Hil. corr. Pfeiff. = Maughania Jaume St-Hil. (Legumin.)
Lupatorium Rafin. = Eupatorium L. (Compos.).
Lupinaster Fabr. = Trifolium L. (Legumin.).
Lupinophyllum Gillett ex Hutch. Leguminosae. 1 trop. & S. Afr.
Lupinus L. Leguminosae. 200 Am., Medit. Floral mechanism like *Lotus*. The fr. explodes, its valves twisting spirally. Several spp. are used as fodder.
Lupsia Neck. = Galactites Moench (Compos.).
Lupulaceae Link = Cannabidaceae Endl.
Lupularia (Ser.) Opiz = seq.
Lupulina Noulet = Medicago L. (Legumin.).
Lupulus Mill. = Humulus L. (Cannabidac.).
Luronium Rafin. Alismataceae. 1 Eur., *L. natans* (L.) Rafin.
Luscadium Endl. = Lascadium Rafin. = Croton L. (Euphorbiac.).
Lusekia Opiz = Salix L. (Salicac.).
Lussa Rumph. = Brucea J. S. Mill. (Simaroubac.).
Lussacia Spreng. = Gaylussacia Kunth (Ericac.).

Lussaria Rafin. = Brucea J. S. Mill. (Simaroubac.).
Lustrinia Rafin. = Justicia L. (Acanthac.).
Lusuriaga Pers. = Luzuriaga Ruiz & Pav. (Philesiac.).
Luteola Mill. = Reseda L. (Resedac.).
Luthera Sch. Bip. = Krigia Schreb. (Compos.).
Lutkea Steud. = Luetkea Bong. (Rosac.).
Lutrostylis G. Don = Ehretia P.Br. (Ehretiac.).
Lutzia Gandoger = Alyssoïdes Mill. (Crucif.).
Luutia Neck. (corr. Neck.) = Croton L. (Euphorbiac.).
Luvunga Buch.-Ham. ex Wight & Arn. Rutaceae. 12 Indomal.
Luxembergia auctt. (sphalm.) = seq.
Luxemburgia A. St-Hil. Ochnaceae. 20 Venez., Brazil.
Luxemburgiaceae Van Tiegh. = Ochnaceae–Luxemburgieae Reichb.
Luziola Juss. Gramineae. 10 S. U.S. to trop. S. Am.
Luzonia Elmer. Leguminosae. 1 Philipp. Is.
*Luzula DC. Juncaceae. 80 cosmop., esp. temp. Euras. Rhiz. as in *Juncus*;
ls. usu. flat.
Luzuriaga R.Br. = Geitonoplesium A. Cunn. (Philesiac.).
*Luzuriaga Ruiz & Pav. Philesiaceae. 3 N.Z., Peru to Tierra del Fuego.
Luzuriagaceae Dostál = Philesiaceae Dum.
Lyallia Hook. f. Hectorellaceae. 1 Kerguelen Is.
Lyauteya Maire. Leguminosae. 1 NW. Afr.
Lycapsus Phil. Compositae. 1 Chile (Desventuradas Is.).
Lycaste Lindl. Orchidaceae. 45 C. & trop. S. Am., W.I.
× **Lycastenaria** hort. = seq.
× **Lycasteria** hort. Orchidaceae. Gen. hybr. (iii) (Bifrenaria × Lycaste).
Lychnanthos S. G. Gmel. = Cucubalus L. (Caryophyllac.).
Lychnanthus C. C. Gmel. = praec.
Lychnidea Burm. f. = Manulea L. (Scrophulariac.).
Lychnidea Hill = Phlox L. (Polemoniac.).
Lychnidia Pomel = Lychnis L. (Caryophyllac.).
Lychniothyrsus Lindau. Acanthaceae. 5 Brazil.
Lychnis L. Caryophyllaceae. 12 temp. Euras. Fls. protandrous, suited to
bees and Lepidoptera. The fls. often show the sta. filled with a black or brown
powder, instead of pollen; this is the spores of the fungus *Ustilago violacea*
(Pers.) Fuckel, which are thus distributed from plant to plant, like pollen, by
the visiting insects.
Lychniscabiosa Fabr. (uninom.) = Knautia L. sp. (Dipsacac.).
× **Lychnisilene** Cif. & Giac. Caryophyllaceae. Gen. hybr. (Lychnis × Silene).
Lychnitis Fourr. = Verbascum L. (Scrophulariac.).
Lychnocephaliopsis Sch. Bip. ex Baker = seq.
Lychnocephalus Mart. ex DC. = Lychnophora Mart. (Compos.).
Lychnodiscus Radlk. Sapindaceae. 8 trop. Afr.
Lychnoïdes Fabr. = Silene L. (Caryophyllac.).
Lychnophora Mart. Compositae. 23 S. trop. Brazil.
Lychnophoriopsis Sch. Bip. Compositae. 1 Brazil.
Lycianthes Hassl. (= Parascopolia Baill.). Solanaceae. 200 trop. & temp.
Lycimnia Hance = Melodinus J. R. & G. Forst. (Apocynac.).

LYCIODES

Lyciodes Kuntze = Bumelia Sw. (Sapotac.).
Lycioplesium Miers = Acnistus Schott (Solanac.).
Lyciopsis Schweinf. = Euphorbia L. (Euphorbiac.).
Lyciopsis Spach = Fuchsia L. (Onagrac.).
Lycioserissa Roem. & Schult. = Plectronia L. (Rubiac.).
Lycium L. Solanaceae. 80–90 temp. & subtrop. (45 Am., esp. Argent.). Many
  have thorny twigs; *L. afrum* L. (kaffir thorn) is used for hedges in S. Afr.
  L. barbarum L. often cult. under the name tea-plant.
Lycocarpus O. E. Schulz. Cruciferae. 1 S. Spain.
Lycochloa G. Samuelsson. Gramineae. 1 Syria.
Lycoctonum Fourr. = Aconitum L. (Ranunculac.).
Lycomela Fabr. = Lycopersicon Mill. (Solanac.).
Lycomormium Reichb. f. Orchidaceae. 2 trop. S. Am.
Lycopersicon Mill. (∼ Solanum L.). Solanaceae. 7 Pacif. S. Am., Galápagos.
  L. lycopersicum (L.) Karst. (*L. esculentum* Mill.) is the tomato or love-apple.
Lycopersicum Hill = praec.
Lycopodiaceae Reichb. Lycopodiales. Ls. small with 1 vasc. strand; no
  ligule; sporangia singly at bases of ± modified ls. which are usually in terminal
  strobili (no strobili in *L. selago* L. and allied spp.); spores all alike; prothalli
  monoecious, mostly saprophytic. Genera: *Lycopodium, Phylloglossum*.
Lycopodiales. Lycopsida. 1 fam.: *Lycopodiaceae*.
Lycopodiella Holub = Lycopodium L. (Lycopodiac.).
Lycopodina Bub. & Penz. = seq.
Lycopodiodes Kuntze = seq.
Lycopodioïdes Boehm. = Selaginella Beauv. (Selaginellac.).
Lycopodium L. Lycopodiaceae. 450 trop. & temp. For chromosome num-
  bers, and proposed division into 3 genera, see Löve & Löve, *The Nucleus*, **1**:
  1–10 (1958). For classification as 2 genera, see Herter, *Index Lycopodiorum*
  (Montevideo, 1949).
Lycopsida. Microphyllous pteridophytes, mainly extinct; living members
  included in 3 orders: *Lycopodiales, Selaginellales, Isoëtales*.
Lycopsis L. Boraginaceae. 3 Eur., As.
Lycopus L. Labiatae. 14 N. temp.
Lycoris Herb. Amaryllidaceae. 10 E. Himal. to Japan.
Lycoseris Cass. Compositae. 15 C. Am. to Peru.
Lycotis Hoffmgg. = Arctotis L. (Compos.).
Lycurus Kunth. Gramineae. 3 S. U.S. to N. trop. S. Am.
Lydaea Molina = Kageneckia Ruiz & Pav. (Rosac.).
Lydea Molina = praec.
Lydenburgia N. Robson. Celastraceae. 1 S. Afr.
× **Lyfrenaria** hort. (vii) = × Lycasteria hort. (Orchidac.).
Lygaion Rafin. = Lugaion Rafin. = Genista L. (Legumin.).
Lygeum Loefl. ex L. Gramineae. 1 Medit., *L. spartum* Loefl. ex L., one of the
  esparto-furnishing grasses (cf. *Stipa* and *Ampelodesma*).
Lygeum P. & K. = Lugaion Rafin. = Cytisus L., Genista L., etc. (Legumin.).
Lygia Fasano = Thymelaea L. (Thymelaeac.).
***Lyginia** R.Br. Restionaceae. 1 SW. Austr. Fil. ± connate.
Lygisma Hook. f. Asclepiadaceae. 3 SE. As.

Lygistum P.Br. = Manettia Mutis ex L. (Rubiac.).
Lygodesmia D. Don. Compositae. 12 N. Am.
Lygodiaceae Presl = Schizaeaceae Mart.
Lygodictyon J. Sm. = Lygodium Sw. (Schizaeac.).
Lygodisodea Ruiz & Pav. Rubiaceae. 4 trop. Am.
*Lygodium Sw. Schizaeaceae. 40 trop. & subtrop. Stem a horiz. underground rhizome; ls. with unlimited apical growth and twining axis, which bears short lateral branches each with dormant apex and paired secondary leafy branches; sporangia in a double row on narrow lobes of fertile leaflets.
Lygodysodea Roem. & Schult. = Lygodisodea Ruiz & Pav. (Rubiac.).
Lygodysodeaceae Bartl. = Rubiaceae–Paederieae DC.
Lygoplis Rafin. = seq.
Lygos Adans. = Genista L. (Legumin.).
Lygurus D. Dietr. (sphalm.) = Lycurus Kunth (Gramin.).
Lygustrum Gilib. = Ligustrum L. (Oleac.).
× **Lymanara** hort. Orchidaceae. Gen. hybr. (iii) (Aërides × Arachnis × Renanthera).
Lymnophila Blume = Limnophila R.Br. (Scrophulariac.).
Lyncea Cham. & Schlechtd. = Melasma Bergius (Scrophulariac.).
Lyndenia Miq. = Lijndenia Zoll. & Mor. = Memecylon L. (Memecylac.).
× **Lyonara** hort. (1948) (vii) = × Trichovanda hort. (Orchidac.).
× **Lyonara** hort. (1959). Orchidaceae. Gen. hybr. (iii) (Cattleya × Laelia × Schomburgkia).
Lyonella Rafin. = Polygonella Michx (Polygonac.).
Lyonetia Willk. = Lyonnetia Cass. = Anthemis L. (Compos.).
Lyonettia Endl. = praec.
Lyonia Elliott = Metastelma R.Br. (1809) (Asclepiadac.).
*Lyonia Nutt. Ericaceae. 30 Himal., E. As., N. Am., W.I.
Lyonia Rafin. = Polygonella Michx (Polygonac.).
Lyonia Reichb. = Cassandra D. Don (Ericac.).
Lyonnetia Cass. = Anthemis L. (Compos.).
Lyonothamnus A. Gray. Rosaceae. 1 (variable), islands off southern California.
Lyonsia R.Br. Apocynaceae. 24 Malaysia (Sum., Born., Cel., New Guinea), Austr., New Caled.
Lyonsia Rafin. = Lyonia Ell. = Metastelma R.Br. (Asclepiadac.).
Lyperanthus R.Br. Orchidaceae. 8 Austr., New Caled., N.Z.
Lyperia Benth. = Sutera Roth (Scrophulariac.).
Lyperia Salisb. = Fritillaria L. (Liliac.).
Lyperodendron Willd. ex Meissn. = Coccoloba L. (Polygonac.).
Lyprolepis Steud. = Kyllinga Rottb. (Cyperac.).
Lyraea Lindl. = Bulbophyllum Thou. (Orchidac.).
Lyriloma Schlechtd. (sphalm.) = Lysiloma Benth. (Legumin.).
Lyriodendron DC. = Liriodendron L. (Magnoliac.).
Lyrionotus K. Schum. (sphalm.) = Lysionotus D. Don (Gesneriac.).
Lyrocarpa Hook. & Harv. Cruciferae. 4 California.
Lyrocarpus P. & K. = praec.
Lyroglossa Schlechter. Orchidaceae. 2 Brazil, Trinidad.

LYROLEPIS

Lyrolepis Rech. f. Compositae. 2 Crete, Aegean.
Lysanthe Salisb. = Grevillea R.Br. (Proteac.).
Lysiana Van Tiegh. Loranthaceae. 6 Austr.
Lysianthius Adans. = Lisianthus P.Br. (Gentianac.).
Lysias Salisb. ex Rydb. = Platanthera Rich. (Orchidac.).
Lysicarpus F. Muell. Myrtaceae. 1 Queensland.
Lysichiton Schott. Araceae. 1 E. Sib., Kamch., Sakh., Japan; 1 Pacif. N. Am.
Lysichitum Schott = praec.
Lysichlamys Compton. Compositae. 2 S. Afr.
Lysiclesia A. C. Smith. Ericaceae. 2 Colombia.
Lysidice Hance. Leguminosae. 1 S. China.
Lysiella Rydb. = Platanthera Rich. (Orchidac.).
Lysiloma Benth. Leguminosae. 35 SW. U.S. to trop. S. Am., W.I.
Lysima Medik. = seq.
Lysimachia L. Primulaceae. 200 cosmop., esp. E. As. & N. Am. *L. vulgaris*
L., yellow loosestrife, is said by Müller to occur in two forms, one in sunny
places with large fls. suited to crossing, and one in shady spots with small
self-fert. fls.
Lysimachi[ace]ae Juss. = Primulaceae Vent.
Lysimachiopsis A. A. Heller = seq.
Lysimachusa Pohl = seq.
Lysimandra Reichb. = Lysimachia L. (Primulac.).
Lysimnia Rafin. = Brassavola R.Br. (Orchidac.).
Lysinema R.Br. Epacridaceae. 6 W. Austr.
Lysinotus Low = seq.
Lysionothus D. Dietr. = seq.
Lysionotis G. Don = seq.
Lysionotus D. Don. Gesneriaceae. 20 E. Himal., E. & SE. As.
Lysiopetalum Willis (sphalm.) = seq.
Lysiosepalum F. Muell. Sterculiaceae. 3 W. Austr.
Lysiostyles Benth. Convolvulaceae. 4 C. to N. trop. S. Am.
Lysiphyllum (Benth.) de Wit (~ Bauhinia L.). Leguminosae. 7 India to
Austr.
Lysipoma Spreng. = seq.
Lysipomia Kunth. Campanulaceae. 21 Andes.
Lysis (Baudo) Kuntze = Lysimachia L. (Primulac.).
Lysisepalum P. & K. = Lysiosepalum F. Muell. (Sterculiac.).
Lysistemma Steetz = Vernonia Schreb. (Compos.).
Lysistigma Schott = Taccarum Brongn. (Arac.).
Lysistylis P. & K. = Lysiostyles Benth. (Convolvulac.).
Lyssanthe D. Dietr. = Lissanthe R.Br. (Epacridac.).
Lyssanthe Endl. = Lysanthe Salisb. = Grevillea R.Br. (Proteac.).
× **Lytantholobularia** Svent. Globulariaceae. Gen. hybr. (Globularia ×
Lytanthus).
Lytanthus Wettst. = Globularia L. (Globulariac.).
Lythastrum Hill = Lythrum L. (Lythrac.).
Lythospermum Lucé = Lithospermum L. (Boraginac.).
*Lythraceae Jaume St-Hil. Dicots. 25/550, all zones but frigid. Herbs, shrubs,

or trees; ls. usu. opp., entire, simple, with very small stipules or none. Fls. in racemes, panicles, or dichasial cymes, ♀, reg. or zygo., usu. 4- or 6-merous. The axis ('calyx-tube') is hollow, generally tubular. The sepals are valvate, and frequently possess an epicalyx, formed, as in *Potentilla*, of combined stips. Petals crumpled in bud, sometimes absent. Sta. inserted (often very low down) on calyx-tube, typically twice as many as sepals, but sometimes fewer or ∞. G with simple style and usu. capitate stigma; 2–6-loc., at the base at least, rarely 1-loc. with parietal placentae. Ovules usu. ∞, anatr., ascending. The fls. of *Lythrum* (*q.v.*) and others are heterostyled. Dry fr., usu. capsular. No endosp. A few yield dyes (*Lawsonia*, etc.), or are medicinal.

Chief genera: *Rotala, Lythrum, Cuphea, Diplusodon, Nesaea, Lagerstroemia*.

Lythron St-Lag. = seq.

Lythropsis Welw. ex Koehne = seq.

Lythrum L. Lythraceae. 35 cosmop. The 6-merous fls. are sol. or in small axillary dichasia like *Labiatae*. Each has 12 sta. in two whorls of different length, and the style again is of different length from any of the sta. Three forms of fl. occur (trimorphism), each on a separate pl.; they are distinguished as long-, mid-, and short-styled fls. The diagram illustrates the arrangement of parts (S = stigma, A = anthers, B = base of fl.), as seen in side view. It is evident that an insect visiting the fls. will tend on the whole to transfer pollen from A₂ to S₂, A₁ to S₁, rather than from sta. of one length to style of another, for it will enter these fls. in the same way and to the same depth. The sta. and style project so far that an insect can alight directly upon them. Darwin (*Forms of Flowers*) showed by a long

S₃	A₃	A₃
A₂	S₂	A₂
A₁	A₁	S₁
B	B	B
long-	*mid-*	*short-*
styled	*styled*	*styled*

series of experiments that the best results are obtained by pollinating S₃ from A₃, or S₁ from A₁, etc., i.e. by crossing two plants. The number of seeds thus obtained is much greater and their fertility higher than if S₂ or S₁ be fertilised from A₃, or any other such union be made. Fertilisation of a stigma by sta. of corresponding length Darwin terms 'legitimate', by sta. of a different length 'illegitimate'. The offspring of illegitimate fert. are few, and have the sterility and other sexual characters of hybrids. As in nearly all other heterostyled pls., the longer the sta. the larger the pollen grains, and the longer the style the larger the papillae of the stigma.

Lytocaryum Toledo (~ Syagrus Mart.). Palmae. 3 Brazil.

Lytogomphus Jungh. ex Göpp. = Rhopalocnemis Jungh. (Balanophorac.).

Lytrostylis Wittst. = Lutrostylis G. Don = Ehretia P.Br. (Ehretiac.).

Lytrum Vill. = Lythrum L. (Lythrac.).

M

Maackia Rupr. Leguminosae. 10 E. As.

Maasa Roem. & Schult. (sphalm.) = Maesa Forsk. (Myrsinac.).

Maba J. R. & G. Forst. = Diospyros L. (Ebenac.).

Mabea Aubl. Euphorbiaceae. 50 C. to trop. S. Am., Trinidad.

Mabola Rafin. = Diospyros L. (Ebenac.).

MABURNIA

Maburnia Thou. = Burmannia L. (Burmanniac.).
Macadamia F. Muell. Proteaceae. 1 Madag., 1 Celebes, 5 E. Austr. (nuttree), 3 New Caled. Seeds ed.
Macaglia Rich. ex Vahl = Aspidosperma Mart. & Zucc. (Apocynac.).
Macahanea Aubl. = ? Salacia L. (Celastrac.).
Macairea DC. Melastomataceae. 30 trop. S. Am.
Macananga Reichb. (sphalm.) = Macaranga Thou. (Euphorbiac.).
Macanea Juss. = Macahanea Aubl. = ? Salacia L. (Celastrac.).
Macaranga Thou. Euphorbiaceae. 280 trop. Afr., Madag., Indomal., Austr., Pacif. Many spp. have hollow stems inhabited by ants.
Macarenia v. Royen. Podostemaceae. 1 Colombia.
Macarisia Thou. Rhizophoraceae. 7 Madag.
Macarisiaceae J. G. Agardh corr. Bullock = Macharisi[ac]eae J. G. Agardh = Rhizophoraceae–Macarisieae Baill.
Macarthuria Huegel ex Endl. Aïzoaceae. 4 SW. & SE. Austr.
Macbridea Ell. ex Nutt. Labiatae. 2 SE. U.S.
Macbridea Rafin. = Cynanchum L. (Asclepiadac.).
Macbrideina Standley. Rubiaceae. 1 Peru.
Macclellandia Wight = Pemphis J. R. & G. Forst. (Lythrac.).
Maccoya F. Muell. = Rochelia Reichb. (Boraginac.).
Macdonaldia Gunn ex Lindl. = Thelymitra J. R. & G. Forst. (Orchidac.).
Macdougalia A. A. Heller = Hymenoxys Cass. (Compos.).
Macella C. Koch = Jaegeria Kunth (Compos.).
Maceria DC. ex Meissn. = Ghinia Schreb. (Verbenac.).
Macfadyena A. DC. Bignoniaceae. 4 S. Am., W.I. (Trin. & Tob.).
Macgregoria F. Muell. Stackhousiaceae. 1 E. Austr.
Macgregorianthus Merr. = Enkleia Griff. (Thymelaeac.).
Machadoa Welw. ex Benth. & Hook. f. = Adenia Forssk. (Passiflorac.).
Machaeranthera Nees. Compositae. 25–30 W. N. Am.
Machaerina P. & K. = Macherina Nees = Lepidosperma Labill. (Cyperac.).
Machaerina Vahl. Cyperaceae. 25 trop.
Machaerium Pers. Leguminosae. 150 Mex. to trop. S. Am., W.I. Like *Dalbergia*. Many are lianes, climbing by sensitive lateral shoots, and provided with recurved stipular thorns. Some of the *jacarandá* timbers (rosewoods) are furnished by this gen. (cf. *Dalbergia*).
Machaerocarpus Small (~ Damasonium Mill.). Alismataceae. 1 SW. U.S.
Machaerocereus Britt. & Rose. Cactaceae. 2 Lower Calif.
Machaerophorus Schlechtd. = Mathewsia Hook. & Arn. (Crucif.).
Machairophyllum Schwantes. Aïzoaceae. 9 S. Afr.
Machanaea Steud. = Macahanea Aubl. = ? Salacia L. (Celastrac.).
Machaonia Humb. & Bonpl. Rubiaceae. 30 Mex. to trop. S. Am., W.I.
Macharina Steud. = Machaerina Vahl = Cladium P.Br. (Cyperac.).
Macharisia Planch. ex Hook. f. = Ixonanthes Jack (Ixonanthac.).
Macharisia Spreng. = Macarisia Thou. (Rhizophorac.).
Macharisi[ac]eae J. G. Agardh = Rhizophoraceae–Macarisieae Baill.
Macherina Nees = Lepidosperma Labill. (Cyperac.).
Machilus Nees = Persea Mill. (Laurac.).
Machilus Rumph. = Neolitsea (Benth.) Merr. (Laurac.).

Machlis DC. = Cotula L. (Compos.).
Machura Steud. (sphalm.) = Maclura Nutt. (Morac.).
Macielia Vand. = Cordia L. (Ehretiac.).
Macintyria F. Muell. = Xanthophyllum Roxb. (Xanthophyllac.).
Mackaya Don = Erythropalum Blume (Erythropalac.).
*Mackaya Harv. Acanthaceae. 1 S. Afr.
Mackenia Harv. = Schizoglossum E. Mey. (Asclepiadac.).
Mackenziea Nees. Acanthaceae. 9 India, Ceylon.
Mackinlaya F. Muell. Araliaceae. 12 E. Malaysia, Queensland.
Mackleya Walp. = Macleaya R.Br. (Papaverac.).
Macklotia Pfeiff. = seq.
Macklottia Korth. = Leptospermum J. R. & G. Forst. (Myrtac.).
Maclaya Bernh. = Macleaya R.Br. (Papaverac.).
Macleania Hook. Ericaceae. 45 C. to W. trop. S. Am.
Macleaya R.Br. Papaveraceae. 2 E. As.
Mac-Leaya Benth. & Hook. f. = seq.
Mac-Leayia Montrouz. = Cassia L. (Legumin.).
Macledium Cass. = Dicoma Cass. (Compos.).
Maclelandia Wight = Macclellandia Wight = Pemphis J. R. & G. Forst. (Lythrac.).
Maclenia Dum. = Cattleya Lindl. (Orchidac.).
Macleya Reichb. = Macleaya R.Br. (Papaverac.).
× **Macludrania** André. Moraceae. Gen. hybr. (Cudrania × Maclura).
*Maclura Nutt. Moraceae. 12 warm Am., Afr., As. *M. pomifera* (Rafin.) C. B. Rob. (*M. aurantiaca* Nutt.) (bow-wood or Osage orange), in S. U.S. The tree bears thorns (branches). Fls. dioec., the ♂ in pseudo-racemes, the ♀ in pseudo-heads; individual fls. like *Morus*. After fert. each ♀ fl. produces an achene enclosed in the fleshy P, and at the same time the common recept. swells up into a fleshy mass, so that a large yellow multiple fr. is formed. The wood is used for bows, carriage-poles, etc. The ls. are used for feeding silk-worms.
Maclurea Rafin. = praec.
Macluria Rafin. = praec.
Macnabia Benth. Ericaceae. 1 S. Afr.
Macnemaraea Willem. = ? Hydrangea L. (Hydrangeac.).
Macodes (Bl.) Lindl. Orchidaceae. 10 Malaysia, Solomon Is.
× **Macodisia** hort. (v) = × Macomaria hort. (Orchidac.).
× **Macomaria** hort. Orchidaceae. Gen. hybr. (iii) (Ludisia [Haemaria] × Macodes).
Macoubea Aubl. Apocynaceae. 6 trop. S. Am.
Macoucoua Aubl. = Ilex L. (Aquifoliac.).
Macounastrum Small = Koenigia L. (Polygonac.).
Macowania Oliv. Compositae. 5 S. Afr.
Macphersonia Blume. Sapindaceae. 8 trop. E. Afr., Madag., Comoro Is.
Macqueria Comm. ex Kunth = Fagara L. (Rutac.).
Macquinia Steud. (sphalm.) = Moquinia Spreng. = Loranthus L. (Loranthac.).
Macrachaenium Hook. f. Compositae. 2 Patagonia, Fuegia.
Macradenia R.Br. Orchidaceae. 8 Florida, C. Am., trop. S. Am., Jamaica.

MACROPSIDIUM

Macropsidium Blume = Myrtus L. (Myrtac.).

Macropsychanthus Harms. Leguminosae. 6 Philipp. Is. & E. Malaysia.

Macropteranthes F. Muell. Combretaceae. 4 N. Austr., Queensl.

Macroptilium (Benth.) Urb. (~ Phaseolus L.). Leguminosae. 8 trop. Am., W.I.

Macrorhamnus Baill. Rhamnaceae. 5 Madag.

Macrorhynchus Less. = Troximon Nutt. (Compos.).

Macrorungia C. B. Clarke. Acanthaceae. 2–3 trop. & S. Afr.

Macrosamanea Britton & Rose. Leguminosae. 8 trop. Am.

Macroscapa Kellogg ex Curran = Stropholirion Torr. (Liliac.).

Macroscepis Kunth. Asclepiadaceae. 8 C. & S. trop. Am.

Macroselinum Schur = Peucedanum L. (Umbellif.).

Macrosema Stev. = Astragalus L. (Legumin.).

Macrosepalum Regel & Schmalh. = Sedum L. (Crassulac.).

Macrosiphon Hochst. = Rhamphicarpa Benth. (Scrophulariac.).

Macrosiphon Miq. = Hindsia Benth. (Rubiac.).

Macrosiphonia Muell. Arg. Apocynaceae. 10 SW. U.S. to trop. S. Am. Xero.

Macrosiphonia P. & K. = Macrosyphonia Duby = Dionysia Fenzl (Primulac.).

Macrosolen (Blume) Reichb. Loranthaceae. 40 SE. As., Malaysia.

Macrospermum Steud. = Macrosporum DC. = Sobolewskia M. Bieb. (Crucif.).

Macrosphyra Hook. f. Rubiaceae. 5 trop. Afr.

Macrosporum DC. = Sobolewskia M. Bieb. (Crucif.).

Macrostachya Hochst. ex A. Rich. = Chloris Sw. (Gramin.).

Macrostegia Nees. Acanthaceae. 1 Peru.

Macrostegia Turcz. = Pimelea Banks (Thymelaeac.).

Macrostelia Hochr. Malvaceae. 3 Madag.

Macrostema Pers. = Quamoclit Mill. = Ipomoea L. (Convolvulac.).

Macrostemma Sweet ex Steud. = Fuchsia L. (Onagrac.).

Macrostepis Thou. (uninom.) = *Epidendrum macrostachys* Thou. = *Beclardia macrostachys* (Thou.) A. Rich. (Orchidac.).

Macrostigma Hook. = Stylobasium Desf. (Stylobasiac.).

Macrostigma Kunth = Tupistra Ker-Gawl. (Liliac.).

Macrostoma Griff. = Christensenia Maxon (Kaulfussiac.).

Macrostoma Hedw. (sphalm.) = Macrostema Pers. = Quamoclit Mill. = Ipomoea L. (Convolvulac.).

Macrostomium Blume = Dendrobium Sw. (Orchidac.).

Macrostomum Benth. & Hook. f. = praec.

Macrostomum Holttum (sphalm., sub *Angiopteridaceae*) = Macroglossum Copel. (Angiopteridac.).

Macrostylis Bartl. & Wendl. Rutaceae. 12 S. Afr.

Macrostylis Breda = Corymborkis Thou. (Orchidac.).

Macrosyphonia Duby = Dionysia Fenzl (Primulac.).

Macrosyringion Rothm. Scrophulariaceae. 2 Medit.

Macrothelypteris Ching. Thelypteridaceae. 10 Mascarene Is. to Hawaii. Most spp. have large tripinnatifid fronds; indusia small.

Macrothyrsus Spach = Aesculus L. (Hippocastanac.).

Macrotis Rafin. = Macrotrys Rafin. = Cimicifuga L. (Ranunculac.).

Macrotomia DC. (~ Arnebia Forsk.). Boraginaceae. 6 Medit. to Himalaya.
Macrotonica Steud. (sphalm.) = praec.
Macrotorus Perkins. Monimiaceae. 1 SE. Brazil.
Macrotropis DC. = Ormosia G. Jacks. (Legumin.).
Macrotrullion Klotzsch = Clitoria L. (Legumin.).
Macrotrys Rafin. = Cimicifuga L. (Ranunculac.).
Macrotybus Dulac = Douglasia L. (Primulac.).
Macrotyloma (Wight & Arn.) Verdc. Leguminosae. 25 Afr., As.
Macrotys DC. = Macrotrys Rafin. = Cimicifuga L. (Ranunculac.).
Macroule Pierce. Leguminosae. 1 trop. S. Am.
Macrozamia Miq. Zamiaceae. 14 extra-trop. Austr.
Macrozanonia (Cogn.) Cogn. = Alsomitra (Bl.) M. Roem. (Cucurbitac.).
Macubea Jaume St-Hil. = Macoubea Aubl. (Apocynac.).
Macucua J. F. Gmel. = Macoucoua Aubl. = Ilex L. (Aquifoliac.).
Macuillamia Rafin. = Bacopa Aubl. (Scrophulariac.).
Macuna Scop. = Mucuna Adans. (Legumin.).
Macvaughiella R. M. King & H. Rob. (~ Eupatorium L.). Compositae. 2 Mex., C. Am.
Madacarpus Wight = Senecio L. (Compos.).
Madaractis DC. = Senecio L. (Compos.).
Madaria DC. = Madia Molina (Compos.).
Madariopsis Nutt. = praec.
Madaroglossa DC. = Layia Hook. & Arn. ex DC. (Compos.).
Madarosperma Benth. Asclepiadaceae. 3–4 Brazil.
Maddenia Hook. f. & Thoms. Rosaceae. 4 Himalaya, China.
Madea Soland. ex DC. = Boltonia L'Hérit. (Compos.).
Madhuca J. F. Gmel. Sapotaceae. 85 Indoch., Indomal. (esp. W. Malaysia), Austr. The fls. of *M. longifolia* (L.) Macbride (*mahua, mahwa,* or *mowa,* India) are ed., and the wood is useful.
Madia Molina. Compositae. 20 Pacif. Am. *M. sativa* Mol. (*madi,* Chile; tarweed, U.S.), cult. for the oil from the seed.
Madiola A. St-Hil. (sphalm.) = Modiola Moench (Malvac.).
Madocarpus P. & K. = Madacarpus Wight = Senecio L. (Compos.).
Madorella Nutt. = Madia Molina (Compos.).
Madorius [Rumph.] Kuntze = Calotropis R.Br. (Asclepiadac.).
Madroneila Greene = Monardella Benth. (Labiat.).
Madvigia Liebm. = Cryptanthus Otto & Dietr. (Bromeliac.).
Maecharanthera Pritz. (sphalm.) = Machaeranthera Nees = Aster L. (Compos.).
Maelenia Dumort. = Cattleya Lindl. (Orchidac.).
Maeranthus Benth. & Hook. f. (sphalm.) = Marcanthus Lour. = Mucuna Adans. (Legumin.).
Maerlensia Vell. = Corchorus L. (Liliac.).
Maerua Forsk. Capparidaceae. 100 trop. & S. Afr. to India. The fr. is a berry, constricted between the seeds like a lomentum.
Maesa Forsk. Myrsinaceae. 200 Old World trop.
Maesia B. D. Jacks. (sphalm.) = Meesia Gaertn. = ? Brackenridgea A. Gray (Ochnac.).

MAESOBOTRYA

Maesobotrya Benth. Euphorbiaceae. 20 trop. Afr.

Maesopsis Engl. Rhamnaceae. 1 trop. E. Afr.

Mafekingia Baill. = Raphionacme Harvey (Periplocac.).

Mafureira Bertol. = Trichilia L. (Meliac.).

Maga Urb. Malvaceae. 2 W.I.

Magalhaensia P. & K. (1) = Magallana Cav. (Tropaeolac.).

Magalhaensia P. & K. (2) = Magellania Comm. ex Lam. = Drimys J. R. & G. Forst. (Winterac.).

Magallana Cav. Tropaeolaceae. 2 temp. S. Am.

Magallana Comm. ex DC. = Magellania Comm. ex Lam. = Drimys J. R. & G. Forst. (Winterac.).

Magdalenaea Brade. Scrophulariaceae. 1 SE. Brazil.

Magdaris Rafin. = Trochiscanthes Koch (Umbellif.).

Magellana Poir. = Magallana Cav. (Tropaeolac.).

Magellania Comm. ex Lam. = Drimys J. R. & G. Forst. (Winterac.).

Maghania Steud. = Maughania Jaume St-Hil. (Legumin.).

Magnistipula Engl. Chrysobalanaceae. 16 trop. Afr.

Magnolia L. Magnoliaceae. 80 Himal. to Japan, Borneo & Java; E. N. Am. to Venez., W.I. Trees and shrubs with sheathing stips. covering the bud, and term. fls. P petaloid, except sometimes the outermost ls., and in whorls, usu. large and showy. Sta. and cpls. ∞, on a lengthened torus. Fr. an aggregation of follicles; each dehisces by its dorsal suture, and the seed dangles out of it on a long thread formed by the unravelling of the spiral vessels of the funicle. The outer integ. of the ovule becomes fleshy as it ripens, forming a coloured arilloid seed-coat.

*Magnoliaceae Juss. Dicots. 12/230, temp. & trop. E. As. & Am. Trees and shrubs with alt. simple stipulate ls. Oil passages in parenchyma. Stips. large, enclosing young growth, decid., leaving annular scar round node. Fls. term. or axill., usu. sol., ♀ or ♂ ♀, with decid. spathaceous bracts. P cyclic; A, G spiral. P usu. petaloid; A ∞, hypog.; G usu. ∞, on long torus. Fr. apocarpous (cpls. follicular or samaroid), or a woody or fleshy syncarp; seed album., endosp. not ruminate. Timber often good. Some beetle pollination.

Classification and genera (after Dandy):

1. MAGNOLIËAE (anthers introrse or latrorse; fr. cpls. not samaroid; testa free from endocarp, externally arilloid): *Manglietia, Magnolia, Talauma, Alcimandra, Aromadendron, Pachylarnax, Kmeria, Elmerrillia, Michelia, Paramichelia, Tsoongiodendron.*

2. LIRIODENDREAE (anthers extrorse; fr. cpls. samaroid; testa adherent to endocarp, not arilloid): *Liriodendron.*

Magnusia Klotzsch = Begonia L. (Begoniac.).

Magodendron Vink. Sapotaceae. 1 New Guinea.

Magonaea G. Don = seq.

Magonia A. St-Hil. Sapindaceae. 2 Brazil, Bolivia. Disk of 2 post. lamellae.

Magonia Vell. = Ruprechtia C. A. Mey. (Polygonac.).

Magostan Adans. = Garcinia L. (Guttif.).

Maguirea A. D. Hawkes = Dieffenbachia Schott (Arac.).

Maguireanthus Wurdack. Melastomataceae. 1 Guiana.

Maguireothamnus Steyermark. Rubiaceae. 2 Venez.

Magydaris Koch ex DC. Umbelliferae. 2 Medit.
Mahafalia Jumelle & Perrier. Asclepiadaceae. 1 Madag.
Mahagoni Adans. = Swietenia Jacq. (Meliac.).
Maharanga DC. (~Onosma L.). Boraginaceae. 9 E. Himal. to SW. China.
Mahawoa Schlechter. Asclepiadaceae. 1 Celebes.
Mahea Pierre = Muriea Hartog (Sapotac.).
Mahernia L. = Hermannia L. (Sterculiac.).
×Mahoberberis C. K. Schneider. Berberidaceae. Gen. hybr. (Berberis ×
Mahonia).
Mahoë Hillebr. = Alectryon Gaertn. (Sapindac.).
Mahometa DC. = Monarrhenus Cass. (Compos.).
*Mahonia Nutt. Berberidaceae. 70 Himal. to Japan & Sumatra; N. & C. Am.
Mahurea Aubl. Guttiferae. 8 trop. S. Am. Ls. alt., stip.
Mahya Cordem. Labiatae. 1 Réunion.
Maia Salisb. = seq.
*Maianthemum Weber. Liliaceae. 3 N. temp. Fl. 2-merous, protog., with
2 ls. in middle of infl. axis.
Maidenia Domin = Uldinia J. M. Black (Hydrocotylac.).
Maidenia Rendle. Hydrocharitaceae. 1 NW. Austr.
Maierocactus Rost. = Astrophytum Lem. (Cactac.).
Maieta Aubl. Melastomataceae. 10 C. & trop. S. Am. Heterophyllous. Some
have bladdery outgrowths of the ls., inhabited by ants (cf. Duroia). Fr. ed.
Maihuenia Phil. Cactaceae. 5 Chile, Argent.
Maihueniopsis Speg. (~Opuntia Mill.). Cactaceae. 1 Argent.
Mailelou Adans. = Vitex L. (Verbenac.).
Maillardia Frapp. & Duchartre. Moraceae. 5 Madag., Réunion.
Maillea Parl. Gramineae. 1 Medit. Is. A 2.
Mainea Vell. = Trigonia Aubl. (Trigoniac.).
Maingaya Oliv. Hamamelidaceae. 1 Malay Penins. (Penang, Perak).
Mairania Bub. = Mairrania Neck. ex Desv. = Arctostaphylos Adans. (Ericac.).
Maireana Moq. = Kochia Roth (Chenopodiac.).
Mairella Léveillé = Phelipaea Desf. (Orobanchac.).
Maireria Scop. = Maripa Aubl. (Convolvulac.).
Mairetis I. M. Johnston. Boraginaceae. 1 Canaries, Morocco.
Mairia Nees. Compositae. 15 S. Afr.
Mairrania Neck. ex Desv. = Arctostaphylos Adans. (Ericac.).
Maïs Adans. = Zea L. (Gramin.).
Maizea auctt. = Mayzea Rafin. = praec.
Maizilla Schlechtd. = Paspalum L. (Gramin.).
Maja Klotzsch = Cuphea P.Br. (Lythrac.).
Maja P. & K. = Maia Salisb. = Maianthemum Weber (Liliac.).
Maja Wedd. = Pterygopappus Hook. f. (Compos.).
Majaca P. & K. = Mayaca Aubl. (Mayacac.).
Majana Kuntze = Coleus Lour. (Labiat.).
Majanthemum Kuntze = Convallaria L. (Liliac.).
Majepea Kuntze = Mayepea Aubl. = Linociera Sw. (Oleac.).
Majera Karst. ex Peter = Evolvulus L. (Convolvulac.).
Majeta P. & K. = Maieta Aubl. (Melastomatac.).

Majidea J. Kirk ex Oliv. Sapindaceae. 5 trop. Afr., Madag.

Majodendrum P. & K. = Mayodendron S. Kurz (Bignoniac.).

***Majorana** Mill. = Origanum L. (Labiat.).

× **Majoranamaracus** K. H. Rechinger. Labiatae. Gen. hybr. (Amaracus × Majorana).

Makokoa Baill. = Octolepis Oliver (Thymelaeac.).

Malabaila Hoffm. Umbelliferae. 10 E. Medit. to C. As. & Persia.

Malabaila Tausch = Pleurospermum Hoffm. (Umbellif.).

Malabathris Rafin. = Otanthera Bl. (Melastomatac.).

Malacantha Pierre. Sapotaceae. 1 trop. W. Afr.

Malacarya Rafin. = Spirostylis Rafin. = Thalia L. (Marantac.).

***Malaceae** Small = Mespil[ac]eae Schultz–Schultzenst. = Rosaceae–Pomoïdeae Juss.

Malacha Hassk. = Malachra L. (Malvac.).

Malachadenia Lindl. = Bulbophyllum Thou. (Orchidac.).

Malache B. Vogel = Pavonia Cav. (Malvac.).

Malachium Fries = Myosoton Moench (Caryophyllac.).

Malachochaete Benth. & Hook. f. = Malacochaete Nees = Scirpus L. (Cyperac.).

Malachodendr[ac]eae J. G. Agardh = Theaceae–Schiminae Melch. emend.

Malachodendron Mitch. = Stewartia L. (Theac.). [Sealy.

Malachra L. Malvaceae. 6 warm Am., W.I. No epicalyx.

Malacion St-Lag. = Malachium Fries = Myosoton Moench (Caryophyllac.).

Malacmaea Griseb. = Bunchosia Rich. & Juss. (Malpighiac.).

Malacocarpus Fisch. & Mey. (~ Peganum L.). Zygophyllaceae. 1 C. As.

Malacocarpus Salm-Dyck = Wigginsia D. M. Porter (Cactac.).

Malacocephalus Tausch = Centaurea L. (Compos.).

Malacocera R. H. Anderson. Chenopodiaceae. 2 SE. C. Austr.

Malacochaete Nees = Scirpus L. (Cyperac.).

Malacoïdes Fabr. = Malope L. (Malvac.).

Malacolepis A. A. Heller = Malacothrix DC. (Compos.).

Malacomeles (Decne) Engl. Rosaceae. 2 Mex.

Malacomeris Nutt. = Malacothrix DC. (Compos.).

Malacothamnus Greene. Malvaceae. 20 SW. U.S., Mex., 1 Chile.

Malacothrix DC. Compositae. 20 W. N. Am.

Malacoxylon Jacq. = Cissus L. (Vitidac.).

Malacurus Nevski (~ Elymus L.). Gramineae. 1 C. As.

Malaisia Blanco. Moraceae. 1 E. As., Indoch., Malaysia, Austr., Pacif.

Malanea Aubl. Rubiaceae. 20 trop. S. Am., W.I. Climbing shrubs.

Malanthos Stapf = Hederella Stapf = Catanthera F. Muell. (Melastomatac.).

Malaparius Miq. = ?Pterocarpus L. (Legumin.).

Malapoenna Adans. = Litsea Lam. (Laurac.).

Malasma Scop. (sphalm.) = Melasma Bergius (Scrophulariac.).

Malaspinaea Presl = Aegiceras Gaertn. (Myrsinac.).

Malaxis Soland. ex Sw. Orchidaceae. 300 cosmop., exc. N.Z.

Malbrancia Neck. = Rourea Aubl. (Connarac.).

Malchomia Sang. = seq.

***Malcolmia** R.Br. corr. Spreng. Cruciferae. 35 Medit. to C. As. & Afghan.

Malcomia R.Br. = praec.

Malea Lundell. Ericaceae. 1 Mex.

Malephora N. E. Brown. Aïzoaceae. 9 S. Afr.

Malesherbia Ruiz & Pav. Malesherbiaceae. 27 S. Peru, N. Chile, W. Argent.

*Malesherbiaceae D. Don. Dicots. 1/27 W. S. Am. Herbs or undershrubs with alt. often deeply lobed exstip. ls., often very hairy. Racemes or cymes of ♀ reg. fls. K 5; C 5, imbr.; axis tubular, with central androphore bearing 5 sta. and G̲ (3) with parietal plac. and ∞ anatr. ov.; styles 3–4 below apex of ov. Caps.; no aril. Only genus: *Malesherbia*. Differs from *Passiflorac.* in having the styles more deeply inserted and widely separated; from *Turnerac.* in aestivation of C, and persistent recept.; from both in absence of aril.

Malicope Vitm. (sphalm.) = Melicope J. R. & G. Forst. (Rutac.).

Malidra Rafin. = Syzygium Gaertn. (Myrtac.).

Maliga B. D. Jacks. = seq.

Maligia Rafin. = Allium L. (Alliac.).

Malinvaudia Fourn. Asclepiadaceae. 1 S. Brazil.

Maliortea W. Wats. = Malortiea H. Wendl. (Palm.).

Mallea A. Juss. = Cipadessa Blume (Meliac.).

Malleastrum (Baill.) Leroy. Meliaceae. 11 Madag.

Malleola J. J. Sm. & Schlechter. Orchidaceae. 20 Malaysia.

Mallingtonia Willd. (sphalm.) = Millingtonia L. f. (Bignoniac.).

Mallinoa Coult. Compositae. 1 C. Am.

Mallococca J. R. & G. Forst. = Grewia L. (Tiliac.).

Mallogonum Reichb. = Psammotropha Eckl. (Aïzoac.).

Mallophora Endl. Dicrastylidaceae. 2 W. Austr. Heads. Fls. 4-merous.

Mallophyton Wurdack. Melastomataceae. 1 Venez.

Mallostoma Karst. = Arcytophyllum Willd. ex Schult. (Rubiac.).

Mallota (A. DC.) Willis = Tournefortia L. (Boraginac.).

Mallotonia Britton (~ Tournefortia L.). Boraginaceae. 1 Florida, Mex., W.I.

Mallotopus Franch. & Sav. = Arnica L. (Compos.).

Mallotus Lour. Euphorbiaceae. 2 trop. Afr. & Madag.; 140 E. & SE. As., Indomal. to New Caled. & Fiji, N. & E. Austr. The caps. of *M. philippensis* (Lam.) Muell. Arg. yields *kamala* dye.

Malmea Fries. Annonaceae. 13 S. Mex. to trop. S. Am.

Malnaregam Adans. = Atalantia Corrêa (Rutac.).

Malnerega Rafin. = praec.

Malnereya auctt. = praec.

Malocchia Savi = Canavalia Adans. (Legumin.).

Malocopsis Walp. = Malveopsis Presl = Sphaeralcea A. St-Hil. (Malvac.).

Malope L. Malvaceae. 4 Medit. The 3 ls. of the epicalyx are very large. Cpls. ∞, in vertical rows (see fam.). Cocci indehisc.

Malortiea H. Wendl. (~ Reinhardtia Liebm.). Palmae. 8 C. Am.

Malortieaceae O. F. Cook = Palmae–Areceae Benth. & Hook. f.

Malosma (Nutt.) Rafin. (~ Rhus L.). Anacardiaceae. 1 Calif. & L. Calif.

Malouetia A. DC. Apocynaceae. 25 C. & trop. S. Am., W.I., trop. Afr.

Malouetiella Pichon. Apocynaceae. 2 trop. Afr.

Malparia R. M. King (sphalm.) = seq. (anagram of name of collector, Dr E. Palmer).

MALPERIA

Malperia S. Wats. Compositae. 1 Calif., L. Calif.

Malpighia L. Malpighiaceae. 35 trop. Am., W.I. Erect pls.; some with stinging hairs, some with cleist. fls.

*Malpighiaceae Juss. Dicots. 60/800 trop., esp. S. Am. Shrubs or small trees, usu. climbing, forming a marked feature among the trop. lianes. Stem anatomy peculiar. Ls. usu. opp., entire, stip., frequently gland-dotted; pl. usu. covered with peculiar branched unicellular hairs. Infl. racemose. Fl. ♂, obliquely zygo. K (5), imbr., often with large glands at the base of (outside) the sepals; C 5, petals usually clawed, imbr.; A 5 + 5, obdiplost., often fewer, joined in a ring at the base; anthers opening intr. by longitudinal slits; G (3), obliquely placed in the fl., 3-loc. with axile plac.; one ovule in each loc., pend., semi-anatr., with ventral raphe. Fr. typically a schizocarp breaking into 3 mericarps, but frequently one or more of the loc. abort. The mericarps are often winged, in some cases, e.g. *Heteropterys*, like those of *Acer*. Seed exalbum.

Classification and chief genera (after Niedenzu):

A. PYRAMIDOTORAE (torus pyramidal; mericarps usually winged): *Tetrapteris, Heteropterys, Acridocarpus*.

B. PLANITORAE (torus flat or concave; mericarps not winged): *Malpighia, Bunchosia, Byrsonima*.

Malpighiantha Rojas. Malpighiaceae. 2 Argent.

Malpighiodes Niedenzu = Tetrapteris Cav. + Diplopterys Juss. spp. (Malpighiac.).

× **Maltea** B. Boiv. Gramineae. Gen. hybr. (Phippsia × Puccinellia).

Maltebrunia Kunth. Gramineae. 5 trop. & S. Afr., Madag.

Malteburnia Steud. = praec.

Malthewsia Steud. & Hochst. ex Steud. = Mathewsia Hook. & Arn. (Crucif.).

Malulucban Blanco = Champereia Griff. (Opiliac.).

Malus Mill. Rosaceae. 35 N. temp. The recept. is hollowed out and united to the syncarpous ovary. The fls. are protogynous, and are visited by bees and many other insects. After fert. the fr. becomes a large fleshy pseudocarp (pome), the flesh consisting of the enlarged recept., while the gynaecium forms the core. Several are cultivated for their ed. fruit, e.g. *M. pumila* Mill. (common apple), *M. baccata* (L.) Borkh. (Siberian crab).

Malva L. Malvaceae. 40 N. temp. Fl. of the ordinary type of the fam., with ∞ cpls. *M. sylvestris* L. and *M. neglecta* Wallr. (large and small mallow) afford a contrast in floral mech., etc. Honey is secreted in little pockets in the recept., covered with hairs which exclude rain and very short-tongued insects. The large mallow is very protandr.; the sta. stand up at first in the middle of the fl., and afterwards bend outwards and downwards whilst the styles lengthen and occupy the original positions of the sta. The small mallow has much smaller fls., much less visited by insects; they go through stages similar to those described above, but at the end of the ♀ stage the styles bend downwards, twist in among the anthers and pollinate themselves.

The ls. in autumn may usu. be seen covered with brown spots caused by the fungus *Puccinia malvacearum* Mont. (cf. *Berberis*).

Malvaceae Juss. Dicots. 75/1000, trop. & temp. Herbs, shrubs, or trees, with alt. often palmately lobed stip. ls. Fls. sol. or in cpd. cymose infls. made up of cincinni, ♂, reg., usu. 5-merous. Epicalyx often present; probably an aggre-

gation of bracteoles, but perhaps stipular like that of some *Rosaceae* (*q.v.*).
K 5 or (5), valvate; C 5, conv., the petals usu. asymmetrical; A usu. ∞, owing
to branching of the inner whorl of sta. (the outer is usu. absent), all united
below into a tube which is joined to the petals and at first sight makes the C
appear gamopetalous; the anthers are monothecous (i.e. each = half an anther),
the pollen grains spiny. G (1–∞), frequently (5), multiloc., with axile placentae.
In § 1 a division of the cpls. by horiz. transv. walls occurs, producing vertical
rows of one-ovuled portions. Ovules 1–∞ in each cpl., anatr., usually ascending,
sometimes pend. *Malvaviscus* has a berry, the rest of the order dry fr., either
caps. or schizocarps. Embryo usually curved; endosp. scanty or o. The fls.
are generally protandr. (see *Malva* and *Goethea*). *Gossypium* (cotton), *Hibiscus*,
and others are of economic value. Fls. rarely functionally ♂ ♀, the ♀ with reduced
petals, 1 carpel and 1 ovule (*Plagianthus*).

Classification and chief genera (after Schumann):
A. Cpls. in vert. rows.
   1. MALOPEAE: *Malope, Kitaibelia.*
B. Cpls. in one plane.
   2. MALVEAE (schizocarp; styles as many as cpls.): *Abutilon, Lavatera, Althaea, Malva, Anoda.*
   3. URENEAE (schizocarp; styles twice as many as cpls.): *Urena, Goethea, Pavonia.*
   4. HIBISCEAE (capsule): *Hibiscus, Gossypium.*

× **Malvalthaea** Iljin. Malvaceae. Gen. hybr. (Althaea × Malva).
*Malvastrum A. Gray (emend. Kearney). Malvaceae. 12 trop. & subtrop. Am.
Malvaviscus Adans. Malvaceae. 3 C. & S. Am. Berry.
Malvaviscus Fabr. = Hibiscus + Ketmia + Malvaviscus Adans. (Malvac.).
Malvella Jaub. & Spach (~ Sida L.). Malvaceae. 1 Medit. to Cauc.
Malveola Fabr. = Sida L. (Malvac.).
Malveopsis Presl = Anisodontea Presl (Malvac.).
Malvinda Boehm. = Sida L. (Malvac.).
Malya Opiz = Ventenata Koel. (Gramin.).
Mamboga Blanco = Mitragyna Korth. (Rubiac.).
Mamei Mill. = Mammea L. (Guttif.).
Mamillaria auctt. = Mammillaria Haw. (Cactac.).
Mamillopsis Morr. ex Britt. & Rose. Cactaceae. 2 Mex.
Mammariella J. Shafer (sphalm.) = Mammillaria Haw. (Cactac.).
Mammea L. Guttiferae. 1 trop. Am. & W.I.; 1 trop. Afr.; 20 Madag.;
27 Indomal. & Pacif. *M. americana* L., cult. for ed. fr., the mammee or
S. Domingo apricot. The fls. are used in preparing a liqueur (*eau de Créole*).
Mammilaria Torr. & A. Gray = seq.
*Mammillaria Haw. Cactaceae. 2–300 SW. U.S. to Colombia & Venez.,
W.I. Mostly small plants of very condensed form, often almost spherical in
outline, with well-marked tubercles (mammillae) (see fam.).
Mammilloydia F. Buxb. Cactaceae. 2 Mex.
Mamorea de la Sota. Burmanniaceae. 1 Bolivia.
Mampata Adans. ex Steud. = Parinari Aubl. (Chrysobalanac.).
Manabea Aubl. = Aegiphila Jacq. (Verbenac.).
Manaëlia auctt. = Manoëlia Bowdich = ?Lysimachia L. (Primulac.).

Managa Aubl. = Salacia L. (Celastrac.).

Mananthes Bremek. Acanthaceae. 5 S. China to Java.

Manaosella J. C. Gomes. Bignoniaceae. 1 Brazil.

Mancanilla Mill. = Hippomane L. (Euphorbiac.).

Mançanilla Adans. = praec.

Mancinella Tussac = praec.

Mancoa Rafin. = Hilleria Vell. (Phytolaccac.).

***Mancoa** Wedd. Cruciferae. 7 Mex., Andes.

Mandarus Rafin. = Bauhinia L. (Legumin.).

Mandelorna Steud. = Vetiveria Thou. ex Virey (Gramin.).

Mandevilla Lindl. Apocynaceae. 114 C. & trop. S. Am.

Mandioca Link = Manihot Mill. (Euphorbiac.).

Mandirola Decne = Achimenes P.Br. (Gesneriac.).

Mandonia Hassk. = Neomandonia Hutch. (Commelinac.).

Mandonia Sch. Bip. = Hieracium L. (Compos.).

Mandonia Wedd. = Tridax L. (Compos.).

Mandragora L. Solanaceae. 6 Medit. to Himal. (mandrake).

Manduyta Comm. ex Steud. = Mauduita Comm. ex DC. = Samadera Gaertn. = Quassia L. (Simaroubac.).

Manekia Trelease. Peperomiaceae. 1 W.I. (Hispaniola).

Manettia Adans. = Aridaria N. E. Br. (Aïzoac.).

Manettia Boehm. = Selago L. (Scrophulariac.).

***Manettia** Mutis ex L. Rubiaceae. 130 C. & trop. S. Am., W.I.

Manfreda Salisb. = Agave L. (Agavac.).

Manga Nor. = Mangas Adans. = Mangifera L. (Anacardiac.).

Manganaroa Speg. = Acacia Mill. (Legumin.).

Mangas Adans. = Mangifera L. (Anacardiac.).

Mangenotia Pichon. Periplocaceae. 1 trop. W. Afr.

Mangifera L. Anacardiaceae. 40˜SE. As., Indomal. *M. indica* L. is the mango, everywhere cult. in the trop. for its fr., a large drupe derived from the 1 cpl. of the fl.

Mangium Scop. = seq.

Mangle Adans. = Rhizophora L. (Rhizophorac.).

Mangles DC. = praec.

Manglesia Endl. = Grevillea R.Br. (Proteac.).

Manglesia Lindl. = Beaufortia R.Br. (Myrtac.).

Manglietia Blume. Magnoliaceae. 30 SE. As. to Sumatra. Like *Magnolia*, but cpls. with 4 or more (not 2) ovules.

Manglilla Juss. = Rapanea Aubl. (Myrsinac.).

Mangonia Schott. Araceae. 1 Brazil.

Mangostana [Rumph.] Gaertn. = Garcinia L. (Guttif.).

Manicaria Gaertn. Palmae. 4 trop. Am., W.I.

Manicariaceae O. F. Cook = Palmae–Areceae Benth. & Hook. f.

Manihot Boehm. = Jatropha L. (Euphorbiac.).

Manihot Mill. Euphorbiaceae. 170 SW. U.S. to trop. S. Am. Shrubs and herbs with monoec. fls. *M. glaziovii* Muell. Arg. and other spp. show bud protection well. The petiole of the young leaf curls upwards and inwards, so that the leaf is brought above the bud. *M. esculentus* Crantz (*M. utilissimus*

Pohl), the cassava, manioc or *mandioca*, is a somewhat variable species, possessing many cultivars, some, with high HCN content, being known as the bitter, others, with lower content, as the sweet cassava. Both types are extensively cult. in the trop. for their large tuberous roots, which contain much starch, etc., and form a valuable foodstuff. The bitter cassava is the one more commonly cult.: its poisonous juice is squeezed out, and finally dissipated in the drying. The ground roots form manioc or cassava meal, sometimes called Brazilian arrowroot. By a special mode of preparation, tapioca is prepared from the root. The poisonous juice, evaporated to a syrup and thus rendered harmless, forms an antiseptic, known as *cassareep*, used in preserving meat, etc. *M. glaziovii* is the Ceará rubber; rubber is obtained by tapping the stem of the tree in the usual way. Several other spp. also yield rubber.

*Manilkara Adans. Sapotaceae. 70 trop. *M. zapota* (L.) van Royen, sapodilla plum, cult. for ed. fr.

Maniltoa Scheff. Leguminosae. 20 Malaysia to Fiji (esp. New Guinea).

Manisuris L. Gramineae. 5 India.

Manisuris L. f. = Hackelochloa Kuntze = Rytilix Rafin. (Gramin.).

Manitia Giseke = Globba L. (Zingiberac.).

Manlilia Salisb. = Polyxena Kunth (Liliac.).

Manna D. Don = Alhagi Adans. (Legumin.).

Mannagettaea H. Smith. Orobanchaceae. 3 E. Sib. to W. China.

Mannaphorus Rafin. = Fraxinus L. (Oleac.).

Mannia Hook. f. (1862; nec Opiz 1829—Hepaticae; nec Trevis. 1857—Lichenes) = Pierreodendron Engl. = Quassia L. (Simaroubac.).

Manniella Reichb. f. Orchidaceae. 1 trop. Afr.

Manniophyton Muell. Arg. Euphorbiaceae. 1 trop. Afr. Sympetalous.

Mannoglottis Dur. = Nanoglottis Maxim. (Compos.).

Mannopappus B. D. Jacks. (sphalm.) = Manopappus Sch. Bip. = Helichrysum Mill. (Compos.).

Manochlaenia Börner = Carex L. (Cyperac.).

Manochlamys Aellen. Chenopodiaceae. 1 S. Afr.

Manoëlia Bowdich = ?Lysimachia L. (Primulac.).

Manongarivea Choux. Sapindaceae. 1 Madag.

Manopappus Sch. Bip. = Helichrysum Mill. (Compos.).

Manostachya Bremek. Rubiaceae. 2 trop. Afr.

Manotes Soland. ex Planch. Connaraceae. 11 trop. Afr. (1 Indoch.?).

Manothrix Miers. Apocynaceae. 2 Brazil.

Mansana J. F. Gmel. = Ziziphus Mill. (Rhamnac.).

Mansoa DC. Bignoniaceae. 6 trop. S. Am.

Mansonia J. R. Drumm. Sterculiaceae. 5 W., C. & E. Afr., Assam, Burma. Very disjunct distrib.

Manteia Rafin. = Rubus L. (Rosac.).

Mantisalca Cass. Compositae. 5 N. Afr.

Mantisia Sims. Zingiberaceae. 2–3 Indomal. *M. saltatoria* Sims ('dancing girls') often cult. for its curious fls., borne on separate shoots from the rhiz. At the base is the K, then 3 broad pets., a curiously shaped labellum and 2 filamentous stds., and beyond all the fertile sta. and style.

Mantodda Adans. = Smithia Ait. (Legumin.).

*Manulea L. Scrophulariaceae. 50 S. Afr.

Manuleopsis Thellung. Scrophulariaceae. 2 SW. Afr.

Manungala Blanco = Samadera Gaertn. = Quassia L. (Simaroubac.).

Maoutia Montrouz. = Oxera Labill. (Verbenac.).

Maoutia Wedd. Urticaceae. 15 Indomal., Polynes. No P in the ♀ fl. *M. puya* Wedd. yields good fibre.

Mapa Vell. = Petiveria L. (Phytolaccac.).

Mapania Aubl. Cyperaceae. 80 trop.

Mapaniopsis C. B. Clarke. Cyperaceae. 1 N. Brazil.

Mapira Adans. = Olyra L. (Gramin.).

Mapouria Aubl. (= Psychotria L.?). Rubiaceae. 170 trop.

Mappa A. Juss. = Macaranga Thou. (Euphorbiac.).

Mappia Fabr. (uninom.) = *Sideritis* L. sp. (Labiat.).

Mappia Habl. ex Ledeb. = Crucianella L. (Rubiac.).

*Mappia Jacq. Icacinaceae. 7 C. & trop. S. Am.

Mappia Schreb. = Doliocarpus Roland. (Dilleniac.).

Mappianthus Hand.-Mazz. Icacinaceae. 2 S. China, Borneo.

Maprounea Aubl. Euphorbiaceae. 4 trop. Am., W. Afr.

Mapuria J. F. Gmel. = Mapouria Aubl. (Rubiac.).

Maquineae Mart. [ex Aristotelia macqui L'Hérit.] = Aristoteliaceae Dum. = Elaeocarpaceae DC.

Maquira Aubl. = ? Olmedia Ruiz & Pav. ? Perebea Aubl. (Morac.).

Marah Kellogg. Cucurbitaceae. 7 Pacif. coast, U.S.

Marainophyllum Pohl (nomen). 1 Brazil. Quid?

Maralia Thou. = ? Polyscias J. R. & G. Forst. (Araliac.).

Marama Rafin. = Graptophyllum Nees (Acanthac.).

Maranta L. Marantaceae. 23 trop. Am. The stds. *β, γ* (see fam.) are present in many. The rhiz. of *M. arundinacea* L. furnishes West Indian arrowroot, prepared by grinding and washing to free the starch.

*Marantaceae Petersen. Monocots. 30/400 trop., chiefly Am. Herbaceous perennials of various habit, at once distinguished from the *Zingiberaceae* by the presence of a swollen pulvinus or joint at the junction of petiole and leaf-blade. Ls. 2-ranked, sheathing; one side of the l. is larger than the other and is covered by it when the l. is rolled up on the bud. Fls. usu. upon the leafy shoots, in pairs in the axils of the bracts, either one pair or many (cymose, drepania). The fl. is asymmetric, but in each pair the one is complementary to the other (i.e. like its reflection in a mirror). Fl. ♀, pentacyclic, 3-merous. P 3 + 3, clearly distinguished in most cases into calyx and corolla. As in the allied fams., the A is united to the C. There is one fertile sta., often petaloid, and round it various petaloid structures (cf. carefully *Canna* and *Zingiberaceae*). The staminode α (Eichler system) is repres. by a more or less leathery or callous l. (*Schwielenblatt*); *β* and *γ* are not always present, but are petaloid when they do occur. The same views as to the morphology of these structures have been proposed as in the case of *Canna* (*q.v.*). Ḡ (3), typically 3-loc. 3-ovuled, but commonly 2 of the loc. are abortive and the third contains one ovule; ovule ana-campylo-tropous; style curved and at first enclosed in the *Kapuzen-blatt* or hood. The fl. often has an explosive mechanism. The pollen is shed upon the style, which remains held in the hood. Insects enter upon the staminode α, and in sucking honey (secreted by glands in the septa of the ovary) set free the style, which descends with a sudden shock, touching the insect's back and at the same time showering the pollen upon it (cf. *Genista*). Fr. usu. a loculic. caps., rarely indehiscent. Seed often arillate. Embryo curved, in perisperm. *Maranta* and others furnish arrowroot, etc. Chief genera: *Sarcophrynium, Phrynium, Marantochloa, Calathea, Maranta, Ischnosiphon, Thalia*.

Maranthes Blume. Chrysobalanaceae. 10 trop. Afr.; 1 Malaysia, trop. Austr. The Asiatic sp. occurs also in Panama.

Maranthus Reichb. = praec.

Marantochloa Brongn. ex Gris. Marantaceae. 15 trop. Afr., Masc. (?Comoro Is.).

Marantodes (A. DC.) Kuntze = Labisia Lindl. (Myrsinac.).

Marantopsis Koern. = Stromanthe Sond. (Marantac.).

Marara Karst. = Aïphanes Willd. (Palm.).

Mararungia Scop. = Marurang Adans. = Clerodendrum L. (Verbenac.).

Marasmodes DC. Compositae. 5 S. Afr.

Marathraceae ('-ineae') Dum. = Podostemaceae Rich.

Marathroïdeum Gandog. = Seseli L. (Umbellif.).

Marathrum Humb. & Bonpl. Podostemaceae. 25 C. Am. & NW. trop. S. Am.

Marathrum Link = Seseli L. (Umbellif.).

Marathrum Rafin. = Musineon Rafin. (Umbellif.).

Marattia Sw. Marattiaceae. 60 trop., N.Z. The synangia are paired, along each side of a vein, dehiscing inwards.

Marattiaceae Kaulf. For chars., see *Marattiales*. Genera: *Marattia* Sw., *Protomarattia* Hayata.

Marattiales. 4 fam., trop. & subtrop. Large ferns; stem stout, usu. erect, rarely > 60 cm. long, and seldom branched; strongly dorsiventral in *Danaea*, *Archangiopteris* and *Christensenia*. L., often very large, simply or compoundly pinnate. L.-base with stipular enlargements connected by transverse commissure.

Sori intramarginal on lower side of l.; all have a simple series of sporangia radiately disposed round a central receptacle, linear or point-like as sorus is elongated or circular. The sporangia may be free from one another, or combined to form synangia. Spores all of one kind, giving rise to monoec. prothalli like those of ordinary ferns, but large and capable of somewhat long life.

Classification and families:

Leaf pinnate or bipinnate; veins free:
Sporangia separate · · · · · · · · · · · · · · · · · · · *Angiopteridaceae*
Sporangia united in synangia:
Synangia paired; stem erect, radial · · · · · · · · *Marattiaceae*
Synangia single; stem dorsiventral · · · · · · · · *Danaeaceae*
Leaf palmate; veins reticulate · · · · · · · · · · · *Kaulfussiaceae*

Marattiidae (subclass). 1 order: *Marattiales*.

Marcania Imlay. Acanthaceae. 1 Siam.

Marcanilla Steud. = Mancanilla Mill. = Hippomane L. (Euphorbiac.).

Marcanthus Lour. = Macranthus Lour. corr. Poir. = Mucuna Adans. (Legumin.).

Marcelia Cass. = Anthemis L. (Compos.).

Marcellia Baill. = Marcelliopsis Schinz (Amaranthac.).

Marcellia Mart. ex Choisy = Exogonium Choisy (Convolvulac.).

Marcelliopsis Schinz. Amaranthaceae. 4 trop. Afr.

Marcetella Sventenius = Sanguisorba L. (Rosac.).

Marcetia DC. Melastomataceae. 30 trop. A. Am.

Marcgravia L. Marcgraviaceae. 55 C. & trop. S. Am., W.I. Climbing epiphytic shrubs, with two kinds of shoots—veg., with two-ranked sessile ls. and clasping roots, and flg., with stalked ls., spirally arranged, and ending in a cymose umbel of fls. The central fls. are abortive and their bracts are transformed into pocket-like coloured nectaries with stalks. The fertile fls. stand upside down, the infl. being pendulous. Humming-birds visit the infls. and have been observed to sip nectar from the nectaries, but there appears to be no direct evidence of pollination by their means. Some spp. are certainly self-fertilised and almost cleistogamous.

***Marcgraviaceae** Choisy. Dicots. 5/100 trop. Am. Climbing shrubs, often epiph., with simple alt. exstip. ls., usu. with pend. infls., bracts transformed into brightly coloured nectaries. Fls. ⚥. K 4–5, much imbr.; C (4–5) or free, dropping as a cap; A 3–∞, or (3–∞) epipet.; G originally 1-loc. with parietal plac., later 2–∞-loc. by ingrowth of plac.; ovules ∞, anatr.; style short, with inconspic. or shortly 5-lobed stig. Capsule fleshy-leathery, tardily dehisc., ∞-seeded. Endosp. thin or o. Genera: *Marcgravia, Norantea, Souroubea, Ruyschia, Caracasia*. Related to *Tetrameristac.* and *Theac.*

Marchalanthus Nutt. ex Pfeiff. = Andrachne L. (Euphorbiac.).

Marchandora Pierre = Mangifera L. (Anacardiac.).

Marcielia Steud. = Macielia Vand. = Cordia L. (Ehretiac.).

Marckea A. Rich. = Markea L. C. Rich. (Solanac.).

Marconia Mattei = Pavonia Cav. (Malvac.).

Marcorella Neck. = Colubrina Rich. (Rhamnac.).

Marcuccia Becc. = Enicosanthum Becc. (Annonac.).

Marenga Endl. = Marogna Salisb. = Aframomum K. Schum. (Zingiberac.).

Marenopuntia Backeb. = Opuntia Mill. (Cactac.).

Marenteria Thou. = Uvaria L. (Annonac.).

Maresia Pomel. Cruciferae. 6 Medit. to Caspian & S. Persia.

Mareya Baill. Euphorbiaceae. 3 trop. W. Afr.

Mareyopsis Pax & K. Hoffm. Euphorbiaceae. 1 W. Equat. Afr.

Margacola Buckl. = Trichocoronis A. Gray (Compos.).

Margaranthus Schlechtd. Solanaceae. 3 SW. U.S., Mex.

Margaretta Oliv. Asclepiadaceae. 1 trop. Afr.

Margaripes DC. ex Steud. = Anaphalis DC. (Compos.).

Margaris DC. = Symphoricarpos Boehm. (Caprifoliac.).

Margaris Griseb. = Margaritopsis Wright (Rubiac.).

Margarita Gaudin = Aster L. (Compos.).

Margaritaria L. f. Euphorbiaceae. 10–12 trop.

Margaritaria Opiz = Anaphalis DC. (Compos.).

Margaritolobium Harms. Leguminosae. 1 Venez.

Margaritopsis Wright. Rubiaceae. 4 W.I.

Margarocarpus Wedd. = Pouzolzia Gaudich. (Urticac.).

MARMAROXYLON

Margarospermum (Reichb.) Opiz = Aegonychon S. F. Gray = Lithospermum L. (Boraginac.).

Marggravia Willd. = Marcgravia L. (Marcgraviac.).

Marginaria Bory = Polypodium L. (Polypodiac.).

Marginariopsis C. Chr. Polypodiaceae. 1 trop. Am.

Marginatocereus (Backeb.) Backeb. = Pachycereus (A. Berger) Britton & Rose (Cactac.).

Margotia Boiss. = Elaeoselinum Koch ex DC. (Umbellif.).

× **Margyracaena** Bitter. Rosaceae. Gen. hybr. (Acaena × Margyricarpus).

Margyricarpus Ruiz & Pav. Rosaceae. 10 Andes.

Maria-Antonia Parl. = Crotalaria L. (Legumin.).

Mariacantha Bub. = Silybum Adans. (Compos.).

Marialva Vand. = Tovomita Aubl. (Guttif.).

Marialvaea Mart. = praec.

Mariana Hill = Silybum Adans. (Compos.).

Marianthemum Schrank = Campanula L. (Campanulac.).

Marianthus Hueg. Pittosporaceae. 16 Austr.

Mariarisqueta Guinea = Cheirostylis Bl. (Orchidac.).

Marica Ker-Gawl. = Neomarica Sprague (Iridac.).

Marica Schreb. = Cipura Aubl. (Iridac.).

Mariera Walp. = Moriera Boiss. (Crucif.).

Marignia Comm. ex Kunth = Protium Burm. f. (Burserac.).

Marila Sw. Guttiferae. 15 C. & S. trop. Am., W.I.

Marilaunidium Kuntze = Nama L. (1759) (Hydrophyllac.).

Marina Liebm. Leguminosae. 1 Mex.

Marinellia Bub. = Melampyrum L. (Scrophulariac.).

Maripa Aubl. Convolvulaceae. 25 trop. Am.

Mariposa (Wood) Hoover (~ Calochortus Pursh). Liliaceae. 11 W. N. Am.

Mariscopsis Cherm. Cyperaceae. 1 E. Afr. & Madag. to Indomal.

Mariscus Ehrh. (uninom.) = Schoenus mariscus L. = Cladium mariscus (L.) Pohl (Cyperac.).

Mariscus Gaertn. = Rhynchospora Vahl, Mariscus Vahl, etc. (Cyperac.).

Mariscus Scop. = Cladium P.Br. (Cyperac.).

*****Mariscus** Vahl (~ Cyperus L.). Cyperaceae. 200 trop. & subtrop.

Maritimocereus Akers & Buining corr. Byles = Borzicactus Riccob. (Cactac.).

Maritinocereus Akers & Buining (sphalm.) = praec.

Marizia Gandog. = Daveaua Willk. (Compos.).

Marjorana G. Don = Majorana Mill. = Origanum L. (Labiat.).

Markea L. C. Rich. Solanaceae. 18 trop. Am.

Markhamia Seem. ex Baill. Bignoniaceae. 12 trop. Afr., As.

Markleya Bondar. Palmae. 1 NE. Brazil.

Marlea Roxb. = Alangium Lam. (Alangiac.).

Marlierea Cambess. Myrtaceae. 50 trop. S. Am. Fr. ed.

Marlieriopsis Kiaersk. Myrtaceae. 1 W.I. (S. Domingo).

Marlothia Engl. Rhamnaceae. 1 S. Afr.

Marlothiella H. Wolff. Umbelliferae. 1 S. Afr. (Namaqualand).

Marlothistella Schwantes. Aïzoaceae. 1 S. Afr.

Marmaroxylon Killip. Leguminosae. 1 Brazil.

24 715 A S D

Marmorites Benth. = Nepeta L. (Labiat.).
Marmoritis P. & K. = praec.
Marniera Backeb. = Epiphyllum Haw. (Cactac.).
Marogna Salisb. = Amomum L. (Zingiberac.).
Marojejya Humbert. Palmae. 1 Madag.
Maropsis Pomel = Marrubium L. (Labiat.).
Marottia Rafin. = Hydnocarpus Gaertn. (Flacourtiac.).
Marquartia Hassk. = Pandanus L. f. (Pandanac.).
Marquartia Vog. = Millettia Wight & Arn. (Legumin.).
Marquesia Gilg. Dipterocarpaceae. 4 trop. Afr.
Marquisia A. Rich. = Coprosma J. R. & G. Forst. (Rubiac.).
Marrhubium Delarbre = Marrubium L. (Labiat.).
Marrubiastrum Moench = Sideritis L. (Labiat.).
Marrubiatrum Seguier = Leonurus L. (Labiat.).
Marrubium L. Labiatae. 40 temp. Euras., Medit. *M. vulgare* L., formerly officinal.
Marsana Sonner. = Murraya Koen. ex L. (Rutac.).
Marschallia Bartl. = Marshallia Schreb. (Compos.).
Marsdenia R.Br. Asclepiadaceae. 5–10 trop.
Marsea Adans. = Baccharis L. (Compos.).
Marsesina Rafin. = Capparis L. (Capparidac.).
Marshallfieldia Macbride. Melastomataceae. 1 Peru.
Marshallia J. F. Gmel. = Homalium Jacq. (Flacourtiac.).
*Marshallia Schreb. Compositae. 10 S. U.S.
Marshallocereus Backeb. = Lemaireocereus Britton & Rose (Cactac.).
Marsiglia Rafin. = seq.
Marsilea L. Marsileaceae. 60 trop. & temp. Ls. petiolate with four lobes, resembling those of '4-leaved clover'. In deep water the ls. are floating, in shallow water they stand erect. Some vegetate during the wet season, and pass the dry in the form of sporocarps.
Marsileaceae R.Br. Marsileales. 3/70 trop. & temp. Mature pl. aquatic or amphibious with thin creeping stem. Ls. circinate in vernation like those of ordinary ferns. Sporangia in sporocarps borne at bases of leaves. Each sporocarp is the equivalent of a leaf-segment and encloses several sori, the latter composed of both micro- and mega-sporangia.

Classification:

Leaf filiform	*Pilularia*
Leaf with 2 leaflets	*Regnellidium*
Leaf with 4 leaflets	*Marsilea*

Marsileales. Marsileidae. 1 fam.: *Marsileaceae*.
Marsileidae. Subclass of *Filicopsida*.
Marsilla Rafin. = Marsilea L. (Marsileac.).
Marsippospermum Desv. Juncaceae. 3 N.Z. & Is., temp. S. Am., Falkland Is.
Marssonia Karst. Gesneriaceae. 1 Venez., Trinidad.
Marsupianthes Reichb. = Marsypianthes Mart. ex Benth. (Labiat.).
Marsupiaria Hoehne = Maxillaria Poepp. & Endl. (Orchidac.).
Marsypianthes Mart. ex Benth. Labiatae. 5 Mex. to Parag.
Marsypianthus Bartl. = praec.

Marsypocarpus Neck. = Capsella Medik. (Crucif.).
Marsypopetalum Scheff. Annonaceae. 1 W. Malaysia.
Marsyrocarpus Steud. = Marsypocarpus Neck. = Capsella L. (Crucif.).
Martagon v. Wolf = Lilium L. (Liliac.).
Martensia Giseke = Alpinia L. (Zingiberac.).
Martha Fritz Muell. = Posoqueria Aubl. (Rubiac.).
Marthella Urb. Burmanniaceae. 1 Trinidad.
Martia Benth. = Martiusia Benth. = Martiodendron Gleason (Legumin.).
Martia Lacerda ex J. A. Schmidt = Brunfelsia L. (Solanac.).
Martia Leandro = Clitoria L. (Legumin.).
Martia Spreng. = Hypericum L. (Guttif.).
Martia Valeton = Valetonia Th. Dur. = Pleurisanthes Baill. (Icacinac.).
Martiella Van Tiegh. = Loranthus L. (Loranthac.).
Martinella Baill. Bignoniaceae. 5 C. & trop. S. Am., W.I.
Martinella Léveillé = Neomartinella Pilger (Crucif.).
Martineria Pfeiff. = Martinieria Vell. = Kielmeyera Mart. (Guttif.).
Martinezia Ruiz & Pav. (~ Aïphanes Willd.). Palmae. 10 W.I., trop. S. Am.
Martinia Vaniot = Asteromoea Bl. (Compos.).
Martiniera Guillem. = Wendtia Meyen (Ledocarpac.).
Martinieria Vell. = Kielmeyera Mart. (Guttif.).
Martinieria Walp. = Martiniera Guillem. = Wendtia Meyen (Ledocarpac.).
Martinsia Godr. = Boreava Jaub. & Spach (Crucif.).
Martiodendron Gleason. Leguminosae. 4 trop. S. Am.
Martiusa Benth. & Hook. f. = Martiusia Benth. = praec.
Martiusella Pierre = Chrysophyllum L. (Sapotac.).
Martiusia Benth. = Martiodendron Gleason (Legumin.).
Martiusia Schult. = Clitoria L. (Legumin.).
Martrasia Lag. = Jungia L. f. (Compos.).
Martretia Beille. Euphorbiaceae. 1 trop. Afr.
Martynia L. Martyniaceae. 1 Mex. Fls. with sensitive stigmas like *Mimulus*. The fr. has 2 long curved horns, suited for animal distr.
Martynia Moon = Strobilanthes Bl. (Acanthac.).
*Martyniaceae Stapf. Dicots. 3/13 trop. & subtrop. Am., in dry or coast regions. Herbs, often with tuberous roots, with opp. or alt. ls. and term. racemes of ⚥, 5-merous, zygo. fls. K (5); C (5); A 4 with a std., epipet., didynamous; G (2), 1-loc. with parietal plac., and ∞ or few anatr. ovules. Caps. loculic., the outer pericarp soft and falling off, the inner woody; it is rendered more or less 4-loc. by the union of the T-shaped placentae together and to the endocarp. The tissue at the top of the midrib of each cpl. also becomes woody and forms a projecting spur, usually hooked at the end or curved, and serving for animal distr. Seeds with little endosp. Genera: *Martynia, Craniolaria, Proboscidea*. Close to *Pedaliac.* and *Bignoniac.*
Marubium Roth = Marrubium L. (Labiat.).
Marulea Schrad. ex Moldenke = Chascanum E. Mey. (Verbenac.).
Marum Mill. = Origanum L. (Labiat.).
Marumia Blume = Macrolenes Naud. (Melastomatac.).
Marumia Reinw. ex Blume = Saurauia Willd. (Actinidiac.).

Marungala Blanco = Quassia L. (Simaroubac.).

Marupa Miers = mixture of fls. of Lannea (Anacardiac.) + fruit of ?Quassia (Simaroubac.).

Marurang Adans. = Clerodendrum L. (Verbenac.).

Maruta Cass. = Anthemis L. (Compos.).

Marzaria Rafin. = Macleaya R.Br. (Papaverac.).

Masakia (Nakai) Nakai = Euonymus L. (Celastrac.).

Mascagnia Bert. Malpighiaceae. 60 trop. Am.

Mascalanthus Rafin. = Maschalanthus Nutt. = Andrachne L. (Euphorbiac.).

Mascarena L. H. Bailey. Palmae. 3 Masc.

Mascarenhasia A. DC. Apocynaceae. 10 trop. E. Afr., Madag. *M. elastica* K. Schum. yields rubber.

Maschalanthe Blume = Urophyllum Wall. (Rubiac.

Maschalanthus Nutt. = Andrachne L. (Euphorbiac.).

Maschalocephalus Gilg & K. Schum. Rapateaceae. 1 trop. W. Afr.

Maschalocorymbus Bremek. Rubiaceae. 4 W. Malaysia.

Maschalodesme K. Schum. & Lauterb. Rubiaceae. 2 New Guinea.

Masdevallia Ruiz & Pav. Orchidaceae. 275 Mex. to trop. S. Am.

Masema Dulac = Valerianella Mill. (Valerianac.).

Masoala Jumelle. Palmae. 1 Madag.

Maspeton Rafin. = Opopanax Koch (Umbellif.).

Massangea Éd. Morren. Bromeliaceae. 1 Colombia.

Massartina Maire. Boraginaceae. 1 Sahara.

Massia Balansa. Gramineae. 1 Indoch., Indomal.

Massoia Becc. = Cryptocarya R.Br. (Laurac.).

Massonia Thunb. ex L. f. Liliaceae. 45 S. Afr.

Massounia Thunb. = praec.

Massovia Benth. & Hook. f. = seq.

Massowia C. Koch = Spathiphyllum Schott (Arac.).

Massula Dulac = Typha L. (Typhac.).

Massularia (K. Schum.) Hoyle. Rubiaceae. 1 trop. E. Afr.

Mastacanthus Endl. = Caryopteris Bunge (Verbenac.).

Mastersia Benth. Leguminosae. 3 Himalaya to Celebes.

Mastersiella Gilg-Benedict (~ Hypolaena R.Br.). Restionaceae. 6 S. Afr.

Mastichina Mill. = Thymus L. (Labiat.).

Mastichodendron (Engl.) H. J. Lam. Sapotaceae. 6 Mex., C. Am., W.I.; 1 Colombia; 1 S. Afr.; 1 S. China & Indoch.

Mastichodendron Jacq. ex R. Hedw. = praec.

Mastigloscleria B. D. Jacks. (sphalm.) = Mastigoscleria Nees = Scleria Bergius

Mastigophorus Cass. = Nassauvia Comm. ex Juss. (Compos.).   [(Cyperac.).

Mastigosciadium K. H. Rech. & Kuber. Umbelliferae. 1 Afghan.

Mastigoscleria Nees = Scleria Bergius (Cyperac.).

Mastigostyla I. M. Johnston. Iridaceae. 10 Peru, Argent.

Mastixia Blume. Mastixiaceae. 25 Indomal.

Mastixiaceae Van Tiegh. Dicots. 1/25 Indomal. Trees with alt. or opp. ent. exstip. ls. Fls. reg., ⚥, small, in term. dichot. thyrses, pedic. artic., 2-bracteolate. K (4–5), open; C 4–5, valv., tips inflexed; A 4–5, alternipet., short, anth. intr.; disk fleshy, intrastam.; $\overline{G}$ 1, 1-loc., with 1 pend. epitr. ov. and short conical style with usu. punctif. stig. Fr. an ovoid drupe, endoc. grooved; seed with small embr. in copious fleshy endosp. Only genus: *Mastixia*. Cortical buds present; also resin-canals; no collenchyma in primary bark.

Mastixiodendron Melch. Rubiaceae. 5 New Guinea to Fiji.

Mastostigma Stocks = Glossonema Decne (Apocynac.).

Mastosuke Rafin. = Urostigma Gasp. = Ficus L. (Morac.).

Mastosyce P. & K. = praec.

Mastrucium Cass. = Serratula L. (Compos.).

Mastrutium Endl. = praec.

Masturcium Kitagawa (sphalm.) = praec.

Mastyxia Spach = Mastixia Blume (Cornac.).

Masus G. Don = Mazus Lour. (Scrophulariac.).

Mataiba R. Hedw. = Matayba Aubl. (Sapindac.).

Matalea A. Gray (sphalm.) = Matelea Aubl. (Asclepiadac.).

Matamoria La Llave = Elephantopus L. (Compos.).

Mataxa Spreng. = Lasiospermum Lag. (Compos.).

Matayba Aubl. Sapindaceae. 50 warm Am.

Mateatia Vell. = Sterculia L. (Sterculiac.).

Matelea Aubl. Asclepiadaceae. 130 trop. S. Am.

Matella Bartl. = praec.

Materana Pax & Hoffm. (sphalm.) = Meterana Rafin. = Caperonia A. St-Hil. (Euphorbiac.).

Mathaea Vell. = Schwenckia L. (Solanac.).

Mathea Vell. = praec.

Mathewsia Hook. & Arn. Cruciferae. 6 Peru, Chile.

Mathiasella Constance & C. L. Hitchcock. Umbelliferae. 1 Mex.

Mathieua Klotzsch = Eucharis Planch. (Amaryllidac.).

Mathiola R.Br. = Matthiola R.Br. corr. Spreng. (Crucif.).

Mathiola DC. = Guettarda L. (Rubiac.).

Mathiolaria F. F. Chevallier = Matthiola R.Br. (Crucif.).

Mathurina Balf. f. Turneraceae. 1 Rodrigues I.

Matisia Humb. & Bonpl. Bombacaceae. 35 trop. S. Am.

Matonia R.Br. Matoniaceae. 2 Malay Penins., Sumatra, Borneo, on exposed high ridges and mt. summits. Ls. very rigid, pedately branched. (Bower, *The Ferns*, 2: ch. 25.)

Matonia Rosc. ex Sm. = Elettaria Maton (Zingiberac.).

Matoniaceae Presl. Matoniales. 2/4 Malaysia. Rhizome creeping, containing 2 concentric solenosteles; ls. variously branched; veins free or slightly anastomosing; sporangia large, ± as *Gleichenia*, few in each sorus; sorus covered with umbrella-like indusium. Genera: *Matonia, Phanerosorus*.

Matoniales. Filicidae. 1 fam.: *Matoniaceae*.

Matourea Aubl. = Stemodia L. (Scrophulariac.).

Matpania Gagnep. = Bouea Meissn. (Anacardiac.).

Matrella Pers. = Zoysia Willd. (Gramin.).

Matricaria L. Compositae. 40 Eur., Medit., W. As.; 10 S. Afr.; 2 W. N. Am. Pappus o. Dried fls. of *M. chamomilla* L. officinal.

Matricarioïdes Spach = Tanacetum L. (Compos.).

Matsumurella Makino = Lamium L. (Labiat.).

Matsumuria Hemsl. = Titanotrichum Solereder (Gesneriac.).
***Matteuccia** Todaro. Aspidiaceae. 3 N. temp.
Mattfeldia Urb. Compositae. 1 W.I. (Haiti).
Matthaea Blume. Monimiaceae. 13 Malaysia.
Matthaea P. & K. = Mathaea Vell. = Schwenckia L. (Solanac.).
Matthewsia Reichb. = Mathewsia Hook. & Arn. (Crucif.).
***Matthiola** R.Br. corr. Spreng. Cruciferae. 55 Atl. Is., W. Eur., Medit. to C. As., S. Afr.
Matthiola L. = Guettarda L. (Rubiac.).
Matthisonia Lindl. = seq.
Matthissonia Raddi = Schwenckia L. (Solanac.).
Mattia Schult. = Rindera Pall. (Boraginac.).
Mattiastrum Brand. Boraginaceae. 35 Medit. to Afghan.
Mattiola Sang. = Matthiola R.Br. (Crucif.).
Mattonia Endl. = Matonia Rosc. ex Sm. = Elettaria Maton (Zingiberac.).
Mattuschkaea Schreb. = Perama Aubl. (Rubiac.).
Mattuschkea Batsch = praec.
Mattuschkia J. F. Gmel. = Saururus L. (Saururac.).
Mattuskea Rafin. = Mattuschkaea Schreb. = Perama Aubl. (Rubiac.).
Matucana Britt. & Rose = Borzicactus Riccob. (Cactac.).
Matudacalamus Maekawa. Gramineae. 1 Mex.
Matudaea Lundell. Hamamelidaceae. 1 Mex.
Maturea P. & K. = Matourea Aubl. = Stemodia L. (Scrophulariac.).
Maturna Rafin. = Gomesa R.Br. (Orchidac.).
Mauchartia Neck. = Apium L. (Umbellif.).
Mauchia Kuntze = Bradburia Torr. & Gray (Compos.).
Mauduita Comm. ex DC. = Quassia L. (Simaroubac.).
Mauduyta Endl. = praec.
Maughania N. E. Br. = Maughaniella L. Bolus (Aïzoac.).
Maughania Jaume St-Hil. = Flemingia Roxb. in Ait. (Legumin.).
Maughaniella L. Bolus. Aïzoaceae. 10 S. Afr.
Mauhlia Dahl = Agapanthus L'Hérit. (Alliac.).
Maukschia Heuff. = Carex L. (Cyperac.).
Mauloutchia Warb. = Brochoneura Warb. (Myristicac.).
Maundia F. Muell. Juncaginaceae. 1 Austr. Ovule apical, pend., orthotr.
Maundiaceae Nak. = Juncaginaceae Rich.
Mauneia Thou. = ? Ludia Comm. ex Juss. (Flacourtiac.).
Maurandella (A. Gray) Rothm. Scrophulariaceae. 2–3 SW. U.S., Mex., W.I.
Maurandia auctt. = seq.
Maurandya Ortega. Scrophulariaceae. 10 SW. U.S. to trop. S. Am., W.I. Leaf-climbers with sensitive petioles.
Mauranthe O. F. Cook. Palmae. 1 Brazil.
Mauria Kunth. Anacardiaceae. 20 Andes.
Mauritia L. f. Palmae. 6 trop. Am., W.I. (*moriche*; see Kingsley's *At Last*). They furnish wood, wine, fruit, fibre, etc.
Mauritiella Burret. Palmae. 15 trop. S. Am.
Maurocena Adans. = Maurocenia Mill. (Celastrac.).

Maurocenia Kuntze = Turpinia Vent. (Staphyleac.).

Maurocenia Mill. Celastraceae. 1 S. Afr., *M. capensis* Sond. (Hottentot cherry).

Mausolea Bunge ex Poljakov (~ Artemisia L.). Compositae. 1 C. As., Persia.

Mavaelia Trimen = Podostemum Michx (Podostemac.).

Mavia Bertol. f. = Erythrophleum Afzel. (Legumin.).

Maxburretia Furtado. Palmae. 1 Malay Penins.

Maxia Ö. Nilsson (~ Montia L.). Portulacaceae. 1 W. N. Am.

× **Maxillacaste** hort. Orchidaceae. Gen. hybr. (vi) (Lycaste × Maxillaria).

***Maxillaria** Poepp. & Endl. Orchidaceae. 300 Florida to Argent., W.I.

Maxillaria Ruiz & Pav. = Lycaste Lindl. (Orchidac.).

***Maximiliana** Mart. Palmae. 10 trop. S. Am., W.I.

Maximilianea Mart. & Schrank = Cochlospermum Kunth (Cochlospermac.).

Maximilianea Reichb. = Maximiliana Mart. (Palm.).

Maximiliania Endl. = Maximilianea Mart. & Schrank = Cochlospermum Kunth (Cochlospermac.).

Maximovitzia Benth. & Hook. f. = Maximowiczia Rupr. = Schisandra Michx (Schisandrac.).

Maximowasia Kuntze = Cryptospora Kar. & Kir. (Crucif.).

Maximowiczia Cogn. = Ibervillea Greene (Cucurbitac.).

Maximowiczia Rupr. = Schisandra Michx (Schisandrac.).

Maxonia C. Chr. Aspidiaceae. 1 W.I., S. Am.

Maxwellia Baill. Bombacaceae. 1 New Caled.

Mayaca Aubl. Mayacaceae. 10 SE. U.S. to Parag., W.I.; 1 S. trop. Afr.

***Mayacaceae** Kunth. Monocots. (Farinosae). 1/10 trop. Am. & Afr. Aquatic herbs with alt., simple, ent., lin., non-sheathing ls., shortly bifid at apex. Fls. reg., ⚥, axill., sol., or several subapical, subtended by membr. bracts. K 3, green, subvalv.; C 3, corolline, imbr., shortly clawed; A 3, oppositisep., anth. basifixed, 4-loc., dehisc. by apical pore; G̲ (3), 1-loc., with ∞ biseriate orthotr. ov. on pariet. plac., and simple or shortly 3-fid filif. style. Fr. a 3-valved caps., each valve bearing a median plac.; seeds with strongly retic. testa, with 'embryostega' (cf. *Commelinac.*), and mealy endosp. Only genus: *Mayaca*. Closely related to *Commelinaceae*.

Mayanthemum DC. = Maianthemum Weber (Liliac.).

Mayanthus Rafin. = Smilacina Desf. (Liliac.).

Maycockia A. DC. = Condylocarpon Desf. (Apocynac.).

Maydaceae (Mathieu) Herter = Gramineae–Maydeae Mathieu.

Mayepea Aubl. = Linociera Sw. (Oleac.).

Mayeta Juss. = Maieta Aubl. (Melastomatac.).

Mayna Aubl. Flacourtiaceae. 15 trop. S. Am.

Mayna Schlechtd. = Meyna Roxb. ex Link (Rubiac.).

Mayodendron Kurz. Bignoniaceae. 1 Burma.

Mays Mill. = Zea L. (Gramin.).

Maytenus Molina. Celastraceae. 225 trop.

Mayzea Rafin. = Zea L. (Gramin.).

Mazaea Krug & Urb. (1897; non Bornet & Grunow 1881—Algae) = Neomazaea Urb. (Rubiac.).

Mazeutoxeron Labill. = Correa Andr. (Rutac.).

Mazinna Spach = Mozinna Ort. = Jatropha L. (Euphorbiac.).
Mazus Lour. Scrophulariaceae. 20 E. & SE. As., Indomal., Austr.
Mazzettia Iljin. Compositae. 1 SW. China.
Meadia Mill. = Dodecatheon L. (Primulac.).
Mearnsia Merr. Myrtaceae. 7 Philipp. Is., New Guinea, Solomon Isl., New Caled.
Mebora Steud. = seq.
Meborea Aubl. = Phyllanthus L. (Euphorbiac.).
Mecardonia Ruiz & Pav. Scrophulariaceae. 15 warm Am.
Mechowia Schinz. Amaranthaceae. 1 SW. trop. Afr.
Meckelia (Mart. ex A. Juss.) Griseb. = Spachea A. Juss. (Malpighiac.).
Meclatis Spach = Clematis L. (Ranunculac.).
Mecodium Presl. Hymenophyllaceae. 100 pantrop. & S. temp.
Mecomischus Coss. ex Benth. & Hook. f. Compositae. 1 Algeria.
Meconella Nutt. Papaveraceae. 3–4 Pacif. N. Am.
Meconia Hook. f. & Thoms. = ?praec.
Meconopsis Vig. Papaveraceae. 1 W. Eur., *M. cambrica* Vig.; 42 Himal. to W. China.
Meconostigma Schott = Philodendron Schott (Arac.).
Mecopus Bennett. Leguminosae. 1 SE. As., Java.
Mecosa Blume = Platanthera Rich. (Orchidac.).
Mecoschistum Dulac = Lobelia L. (Campanulac.).
Mecosorus Kl. = Grammitis Sw. (Grammitidac.).
Mecostylis Kurz ex Teijsm. & Binnend. = Macaranga Thou. (Euphorbiac.).
Mecranium Hook. f. Melastomataceae. 20 W.I.
Medea Klotzsch = Croton L. (Euphorbiac.).
Medemia Württemb. ex H. Wendl. Palmae. 2 N. trop. Afr., 1 Madag.
Medeola L. Trilliaceae. 1 N. Am.
Medica Cothenius = Tourrettia Fougeroux (Bignoniac.).
Medica Mill. = seq.
Medicago L. Leguminosae. 100 temp. Euras., Medit., S. Afr. The fl. has an explosive mech. like *Genista* (*q.v.*). The fr. is usu. twisted, often spirally coiled up into a ball or disc, and frequently provided with hooks enabling animal distr. *M. sativa* L. (lucerne or alfalfa), *M. lupulina* L. (nonsuch) and others are useful fodders.
Medicasia Willk. (sphalm.) = Medicusia Moench = Picris L. (Compos.).
Medicia Gardn. & Champ. = Gelsemium Juss. (Loganiac.).
Medicosma Hook. f. Rutaceae. 1 E. Austr.
Medicula Medik. = Medicago L. (Legumin.).
Medicusia Moench = Picris L. (Compos.).
Medinilla Gaudich. Melastomataceae. 400 trop. Afr., Madag., Indomal., Pacif.
Medinillopsis Cogn. = Plethiandra Hook. f. (Melastomatac.).
Mediocactus Britt. & Rose = Hylocereus (A. Berger) Britt. & Rose (Cactac.).
Mediocalcar J. J. Smith. Orchidaceae. 30 New Guinea, Polynesia.
Mediocalcas Willis (sphalm.) = praec.

Mediocereus Frič & Kreuz. = Mediocactus Britt. & Rose = Hylocereus (A. Berger) Britt. & Rose (Cactac.).

Mediolobivia Backeb. = Rebutia K. Schum. (Cactac.).

Mediorebutia Frič = Rebutia K. Schum. (Cactac.).

Medium Opiz = Campanula L. (Campanulac.).

Mediusella (Cavaco) Capuron. Sarcolaenaceae. 1 Madag.

Medora Kunth = Smilacina Desf. (Liliac.).

Medusa Lour. = Rinorea Aubl. (Violac.).

Medusaea Reichb. = Medusea Haw. = Euphorbia L. (Euphorbiac.).

*****Medusagynaceae** Engl. & Gilg. Dicots. 1/1 Seychelles. Shrubs with simple shallowly crenate opp. exstip. ls. Fls. reg., ♀, in small lax opp.-fld. term. panicles. K 5, imbr., decid.; C 5, imbr., decid.; A ∞, with filif. subpersist. fil. and small basifixed anth.; G (17–25), with as many grooves and subapical styles with capit. stigs., and 2 superposed axile ov. per loc., the upper ascending and the lower descending. Fr. an umbrella-shaped multiloc. septic. caps., with winged seeds. Only genus: *Medusagyne*. An ancient relict (now almost extinct), with affinities prob. near the stock of *Theac.*, *Guttif.*, *Myrtac.*, etc.; the char. gynoecium shows an interesting analogy with those of *Eucryphiac.* and *Paracryphiac.*

Medusagyne Baker. Medusagynaceae. 1 Seychelles.

Medusandra Brenan. Medusandraceae. 1 W. Equat. Afr. (Cameroons).

*****Medusandraceae** Brenan. Dicots. 1/1 trop. Afr. Trees, with secretory canals throughout. Ls. alt., simple, crenate, filif.-stip., long-pet., pet. pulvin.-genic. at apex. Fl. small, reg., ♀, in ± dense pendent axill. racemes. K 5, open in bud, persist.; C 5, imbr.; A 5, oppositipet., short, with ± large 4-loc. latrorse anth., stds. 5, oppositisep., elongate, serpentine, long-exserted, densely papillose-pubesc., with abortive term. anth.; G (3), 1-loc., with slender central column, and 6 pend. anatr. ov. attached to roof of ovary close to column, styles 3, remote, short, conical, divergent. Fr. a 3-valved coriaceous caps., silky-fibrous within, subtended by reflexed accresc. K; seed 1, large, pulviniform, often with 6 radiating ribs above, endosp. copious, slightly ruminate. Only genus: *Medusandra*. A very interesting plant, with characters suggesting possible affinities in several directions. The unusual placentation, together with certain features of the anatomy and pollen, has suggested some connection with the *Olacac.* and *Icacinac.* It seems possible, however, that the true affinity of *Medusandrac.* may after all be somewhere near the *Flacourtiac.*, as indeed already indicated by the former association with *Medusandra* of the Flacourtiaceous genus *Soyauxia*. The habit and facies of *Med.* (esp. foliage and infl.) are those of, e.g., *Baccaurea*, *Thecacoris*, *Maesobotrya*, etc., genera currently referred to the *Euphorbiaceae–Phyllanthoïdeae*. These genera are, however, probably not closely related to *Phyllanthus*, etc., but are in some respects intermediate between the *Euphorbiac.* and *Flacourtiac.* The *Medusandrac.* appear to lie somewhere in or near this circle of affinity.

Medusanthera Seem. Icacinaceae. 4–5 Nicobar Is., Philipp. Is. to E. Malaysia & Pacif.

Medusather (Griseb.) Candargy [Kantartzis] = Hordelymus (Jessen) Jessen ex Harz (Gramin.).

Medusea Haw. = Euphorbia L. (Euphorbiac.).

MELACHNA

Melachna Nees = seq.

Melachne Schrad. ex Schult. f. = Gahnia J. R. & G. Forst. (Cyperac.).

Melachrus P. & K. = Melichrus R.Br. (Epacridac.).

Meladendron Molina = Heliotropium L. (Boraginac.).

Meladendron St-Lag. = Melaleuca L. (Myrtac.).   [Austr.

Meladenia Turcz. (~Psoralea L.). Leguminosae. 3 Philipp. Is., New Guinea,

Meladerma Kerr. Periplocaceae. 3 Siam.

Melaenacranis Roem. & Schult. = Melancranis Vahl = Ficinia Schrad. (Cy-

Melalema Hook. f. = Senecio L. (Compos.).   [perac.).

Melaleuca Blanco = Bombax L. (Bombacac.).

*Melaleuca L. Myrtaceae. 1 Indomal., 100 Austr., Pacif. The ls. of *M. leucadendron* L. (Austr., Indomal.) yield cajeput oil. Timber useful.

Melaleuc[ac]eae Reichb. = Lecythidaceae + Myrtaceae p.p.

Melampirum Neck. = Melampyrum L. (Scrophulariac.).

Melampodium L. Compositae. 12 warm Am., esp. Mex.

Melampyraceae Lindl. = Scrophulariaceae–Rhinantheae Juss.

Melampyrum L. Scrophulariaceae. 35 N. temp. Semi-parasites (see fam.).
The fl. has a loose-pollen mechanism; the 4 anthers lie close together and
form a pollen-box; the filaments of the sta. are covered with sharp teeth.

Melanacranis Reichb. = Melancranis Vahl = Ficinia Schrad. (Cyperac.).

Melananthera Michx = Melanthera Rohr (Compos.).

Melananthos Pohl (nomen). 5 Brazil. Quid? (= seq.?).

Melananthus Walp. Solanaceae. 5 C. Am., W.I., Brazil.

Melanaton Rafin. = Thapsia L. (Umbellif.).

Melanchrysum Cass. = Gazania Gaertn. (Compos.).

Melancium Naud. Cucurbitaceae. 1 E. & S. Brazil.

Melancranis Vahl = Ficinia Schrad. (Cyperac.).

Melandrium Roehl. Caryophyllaceae. 100 N. hemisph., mts. trop. S. Am.,
trop. Afr., S. Afr. *M. dioicum* (L.) Coss. & Germ. is dioec. and the ♀ pl. is
stouter and coarser in growth than the ♂.

Melandryum Reichb. = praec.

Melanea Pers. = Malanea Aubl. (Rubiac.).

Melanenthera Link = Melanthera Rohr (Compos.).

Melanidion E. L. Greene = Christolea Cambess. (Crucif.).

Melanium P.Br. = Cuphea P.Br. (Lythrac.).

Melanium Rich. ex DC. = Arthrostema Ruiz & Pav. (Melastomatac.).

Melanix Rafin. = Salix L. (Salicac.).

Melanobatus Greene = Rubus L. (Rosac.).

Melanocarpum Hook. f. = Pleuropetalum Hook. f. (Amaranthac.).

Melanocarya Turcz. = Euonymus L. (Celastrac.).

Melanocenchris Nees. Gramineae. 3 NE. trop. Afr. to India & Ceylon.

Melanochyla Hook. f. Anacardiaceae. 12 W. Malaysia.

Melanococca Blume = Rhus L. (Anacardiac.).

Melanocommia Ridley. Anacardiaceae. 1 Borneo.

Melanodendron DC. Compositae. 1 St Helena. Tree.

Melanodiscus Radlk. Sapindaceae. 3 trop. Afr.

Melanolepis Reichb. f. & Zoll. Euphorbiaceae. 1–2 Formosa, Indoch.,

Melanoleuce St-Lag. = Melaleuca L. (Myrtac.).   [Malaysia, Pacif.

Melanoloma Cass. = Centaurea L. (Compos.).

Melanophylla Baker. Melanophyllaceae. 8 Madag.

Melanophyllaceae Takht. Dicots. 1/8 Madag. Shrubs or small trees, glabr. or almost so, blackening on drying, with alt. ent. or dent. long-pet. exstip. ls., petiole ± sheathing at base. Fls. ⚥, rac. or panic., mostly bibracteolate. K(5), minute; C 5, imbr., ovate or obl., reflexed; A 5, fil. short but slender, anth. elong. oblong, basi- or dorsi-fixed; disk obscure or obsolete; G̅ (2–3), 2–3-loc., with 1 pend. anatr. ovule per loc.; styles 2–3, free, subulate, erect or recurved, stig. linear or punctif. Fruit an ovoid or oblong (2–)3-loc. drupe, with 1 fertile dorsal and 2 sterile ventral loc., the septa marked by external grooves. Only genus: *Melanophylla*.

Melanopsidium Cels. ex Colla = Billiottia DC. (Rubiac.).

Melanopsidium Poit. ex DC. = Alibertia Rich. (Rubiac.).

Melanorrhoea Wall. Anacardiaceae. 20 SE. As., Malay Penins., Borneo. *M. usitata* Wall. (*theetsee*) yields a valuable black varnish, obtained by tapping the stem; the sap turns black on exposure to air.

Melanoschoenos Seguier = Schoenus L. (Cyperac.).

Melanosciadium auctt. = seq.

Melanosciadum de Boissieu. Umbelliferae. 1 W. China.

Melanoselinon Rafin. = seq.

Melanoselinum Hoffm. = Thapsia L. (Umbellif.).

Melanoseris Decne = Lactuca L. (Compos.).

Melanosinapis Schimp. & Spenn. = Brassica L. (Crucif.).

Melanosticta DC. = Hoffmanseggia Cav. (Legumin.).

Melanotis Neck. Inc. sed. (Sympet.-Tubifl.). Schott useful.

Melanoxylon Schott. Leguminosae. 3 trop. S. Am. Timber of *M. brauna*

Melanthaceae R.Br. = Melanthiaceae Batsch = Liliaceae–Melanthiëae Kunth.

Melanthera Rohr (= Synedrella Gaertn.). Compositae. 50 warm Am., trop. Afr., Madag., India.

Melanthes Bl. corr. Bl. = Breynia J. R. & G. Forst. (Euphorbiac.).

Melanthesa Blume (sphalm.) = praec.

Melanthesiopsis Benth. & Hook. f. = praec.

Melanthesopsis Muell. Arg. = praec.

*****Melanthia[ceae]** Batsch = Liliaceae–Melanthiëae Kunth.

Melanthium Kunth = Dipidax Lawson ex Salisb. (Liliac.).

Melanthium L. = Zigadenus Michx (Liliac.).

Melanthium Medik. = Nigella L. (Ranunculac.).

Melanthos P. & K. = Malanthos Stapf = Catanthera F. Muell. (Melastomatac.).

Melanthus Weigel = Melianthus L. (Melianthac.).

Melargyra Rafin. = Spergularia (Pers.) J. & C. Presl (Caryophyllac.).

Melarhiza Kellogg = Wyethia Nutt. (Compos.).

Melasanthus Pohl = Stachytarpheta Vahl (Verbenac.).

Melascus Rafin. = Calonyction Choisy = Ipomoea L. (Convolvulac.).

Melasma Bergius. Scrophulariaceae. 30 trop., exc. Austr. C nearly reg.

Melasphaerula Ker-Gawl. Iridaceae. 1 S. Afr.

Melastoma L. Melastomataceae. 70 S. China, Indomal., Pacif. Sta. very unequal; connective of larger much produced. Fr. blackens mouth.

MELASTOMASTRUM

Melastomastrum Naud. Melastomataceae. 4 trop. Afr.

*Melastomataceae Juss. Dicots. 240/3000 trop. & subtrop. A very natural fam., usu. easy to recognise in veg. condition by the peculiar leaf-veining, etc. They exist under various conditions, and vary much in habit; herbs, shrubs, trees, with usu. erect stem, some climbing, usu. by r., some epiphytic, water, or marsh pls. Stem often 4-angled, ls. usu. decussate, one much larger than other, the latter often withering as it grows older. Ls. usu. simple, the veins diverging from base and converging at apex, and ± the same size, with no true midrib. Phloem in pith. Many myrmecophilous (*Tococa, Maieta,* etc.). A few yield colouring matters.

Infl. cymose in great variety. Fl. usu. ♀, reg. or slightly zygo., very char., easily recognised by the anther appendages. Recept. or K-tube tubular or bell-shaped, usu. ± united with G, sometimes by longitudinal ribs only, often brightly coloured. K 4–5, C 4–5, usu. conv., perig. or epig., both usu. reg.; A often irreg., usu. twice as many as pets., standing when mature in one whorl, bent down in bud with anthers between ov. and recept.; anther-loc. opening by common term. pore, connective developed in various ways and usu. provided with curious appendages, frequently of sickle-like form; G sup. or inf., usu. 4–5-loc., with simple style and stigma, and ∞ anatr. ov. on axile plac. Berry or loculic. caps.; seed exalb., with very small embr., one cot. larger than other.

Classification and chief genera (after Krasser):

[A. Fr. many-seeded. Embryo very small.]

    I. *Melastomatoïdeae* (ovules on slightly projecting plac. in inner angle of loc.): *Tibouchina, Centradenia, Melastoma, Osbeckia, Rhexia, Monochaetum, Microlicia, Medinilla, Leandra, Miconia, Tococa, Maieta.*

    II. *Astronioïdeae* (ov. on basal or parietal plac.): *Astronia,* etc. (caps.), *Kibessia,* etc. (berry).

[B. Fr. a berry, 1–5-seeded. Embryo large.

    III. *Memecyloïdeae*: see *Memecylaceae.*]

Melathallus Pierre = Ilex L. (Aquifoliac.).

Melaxis Smith ex Steud. = Liparis Rich. (Orchidac.).

Melchiora Kobuski (1956; non *Melchioria* Penzig & Sacc. 1897—Fungi) = Balthasaria Verdc. (Theac.).

Meleagrinex Arruda ex Koster = Sapindus L. (Sapindac.).

Melenomphale Rafin. = Melomphis Rafin. = Ornithogalum L. (Liliac.).

Melfona Rafin. = Cuphea P.Br. (Lythrac.).

Melhania Forsk. Sterculiaceae (Malvaceae?). 60 warm Afr., Madag., India (Austr.?). Epicalyx 3, large. Pets. large, flat. A (5), stds. 5.

Melia L. Meliaceae. 2–15 palaeotrop. & subtrop. G to (8). Specific limits very uncertain. Some useful for timber.

*Meliaceae Juss. Dicots. 50/1400 warm. Trees or shrubs, rarely subshrubs, with usu. alt. exstip. pinnate rarely simple ls. without transparent dots, and cymose panicles of ♀ reg. fls. K (3–5) or 3–5; C 3–5(–14), rarely united, imbr., valv. or contorted; A 3–10, usu. united below into a tube, often completely united with anthers sessile on the tube; disk or not; G̲ 2–5-loc., rarely 1-loc. or > 5-loc., with style or not, and 1, 2, or more ov. in each loc., usu. pend. and anatr. with ventral raphe. Caps., berry, or drupe; seeds often winged, usu.

with endosp. Many, e.g. *Swietenia, Khaya*, etc. (mahogany), *Cedrela*, etc. ('cedars,'), yield valuable timber; the seeds of several are used as sources of oil, and others have ed. fr. Ls. of *Chisocheton* usu. show continued apical growth, and a few have epiphyllous infl.

Classification and chief genera (after Engler):

I. **Cedreloïdeae** (sta. free; caps., seeds winged): *Cedrela, Toona.*

II. **Swietenioïdeae** (sta. in a tube; caps., seeds winged, loc. with > 2 ov.): *Swietenia, Khaya, Entandrophragma.*

III. **Melioïdeae** (sta. in a tube; caps., berry or drupe, seeds not winged): *Carapa, Xylocarpus, Melia, Azadirachta, Trichilia, Guarea, Turraea, Chisocheton, Dysoxylum, Aglaia.*

Meliadelpha Radlk. = Dysoxylum Bl. (Meliac.).

Meliandra Ducke = Votomita Aubl. (Memecylac.).

***Melianthaceae** Link. Dicots. 2/15 trop. & S. Afr. Trees and shrubs with alt. pinn. stip. ls., and racemes of ♀ or ♂ ♀ zygo. fls. whose stalks twist through 180° at time of flg. K 5 or (5), sometimes 4 by union of two seps., unequal, imbr.; C 4–5, unequal, clawed, imbr.; disk extra-staminal, annular or variously interrupted; A 5–4, free or united at base, often declinate; G̲ (4–5), 4–5-loc. with 1–4 basal or axile ov. in each; ov. erect or pend., anatr., with ventral or dorsal raphe according as they are erect or pend. Caps.; seed sometimes arillate; endosp. fleshy or horny. Genera: *Melianthus, Bersama.* Related to *Sapindac.*

Melianthus L. Melianthaceae. 6 S. Afr. Post. sep. spurred or saccate. Fls. very rich in honey.

Melica Franch. & Sav. (sphalm.) = Melissa L. (Labiat.).

Melica L. Gramineae. 70 N. & S. temp., exc. Austr. Inf. palea 7–9-nerved. Lodicule 1.

Melichlis auctt. = Meliclis Rafin. = Coryanthes Hook. (Orchidac.).

Melicho Salisb. = Haemanthus L. (Amaryllidac.).

Melichrus R.Br. Epacridaceae. 2 Austr.

Meliclis Rafin. = Coryanthes Hook. (Orchidac.).

Melicocca L. = seq.

Melicoccus P.Br. Sapindaceae. 2 trop. Am., W.I. Anthers extrorse. Fr. ed. *M. bijugatus* Jacq. (W.I.) for timber and fr.

Melicope J. R. & G. Forst. Rutaceae. 70 Indomal., warm Austr., N.Z., Pacif. 4-merous.

Melicopsidium Baill. = Cossinia Comm. ex Lam. (Sapindac.).

Melicytus J. R. & G. Forst. Violaceae. 5 N.Z., Norfolk I., Fiji. Fl. sub-reg. with no claws. Ovules ∞ per plac. Berry.

Melidiscus Rafin. = Cleome L. (Cleomac.).

Melidora Nor. ex Salisb. = Enkianthus Lour. (Ericac.).

Melientha Pierre. Opiliaceae. 2 Indoch., Philipp. Is.

Melilobus Mitch. = Gleditsia L. (Legumin.).

Melilota Medik. = Melilotus Mill. (Legumin.).

Melilothus Hornem. = praec.

Melilotoïdes Fabr. = Trigonella L. (Legumin.).

Melilotus Mill. Leguminosae. 25 temp. & subtrop. Euras., Medit.

Melinia Decne. Asclepiadaceae. 10 S. Am.

MELINIS

Melinis Beauv. Gramineae. 1 trop. S. Am., W.I.; 17 trop. & S. Afr., Madag. Fodder.

Melinonia Brongn. = Pitcairnia L'Hérit. (Bromeliac.).

Melinospermum Walp. = Dichilus DC. (Legumin.).

Melinum Link = Zizania L. (Gramin.).

Melinum Medik. = Salvia L. (Labiat.).

Melioblastis C. Muell. (sphalm.) = Myrioblastus Wall. ex Griff. = Cryptocoryne Fisch. (Arac.).

Meliocarpus Boiss. = Heptaptera Margot & Reuter (Umbellif.).

Meliopsis Reichb. = Fraxinus L. (Oleac.).

Melio-Schinzia K. Schum. = Chisocheton Bl. (Meliac.).

Meliosma Blume. Meliosmaceae. 100 warm As., Am.

Meliosmaceae Endl. Dicots. 2/105 warm As., Am. Evergr. trees and shrubs, with alt. imparipinn. or simple exstip. ls., lfts. opp., ent. or serr. Fls. small, ♀ or polyg.-dioec., in cpd. panicles. K 4–5, imbr. or open, persist.; C 4–5, opp. the seps., imbr., v. unequal, rounded, the 2 inner small and sometimes bifid, or lin.-acute and subvalv.; disk cupular, 3–8-dentate, teeth sometimes bifid; A 5, opp. the pets., v. unequal, free or adnate at base to pets., 2 larger perfect, 3 smaller (opp. larger pets.) without anth., fil. flattened, anth. large, glob., didym., dehisc. by large openings, connective much expanded; G (2–3), 2–3-loc., with 2 axile superposed horiz. or pend. ov. per loc., style straight with simple or 2–3-fid stig., or 2-lobed at apex with lobes stigmatic above. Fr. an obliquely subglob. 1-seeded drupe, with bony endoc. (rarely 2-loc.); seed without endosp., embryo sometimes ± spirally contorted. Genera: *Meliosma*, *Ophiocaryon* (*Phoxanthus*). The usu. assumed relationship with *Sabiac.* appears to require confirmation.

Meliotus Steud. = Melilotus L. (Legumin.).

Meliphlea Zucc. = Phymosia Desv. ex Ham. (Malvac.).

Melisitus Medik. = Melissitus Medik. em. Medik. (Legumin.).

Melissa L. Labiatae. 3 Eur. to C. As. & Persia. *M. officinalis* L. (balm) cult.

Melissitus Medik. emend. Medik. (~ Trigonella L.). Leguminosae. 60 mainly C. As.

Melissophyllon Adans. = seq.

Melissophyllum Hill = Melittis L. (Labiat.).

Melissopsis Sch. Bip. ex Baker = Ageratum L. (Compos.).

Melistaurum J. R. & G. Forst. = Casearia Jacq. (Flacourtiac.).

Melitella Sommier = Crepis L. (Compos.).

Melittacanthus S. Moore. Acanthaceae. 1 Madag.

Melittis L. Labiatae. 1 Eur., *M. melissophyllum* L.

Mella Vand. = Bacopa Aubl. (Scrophulariac.).

Mellera S. Moore. Acanthaceae. 4–5 warm Afr.

Mellichampia A. Gray ex S. Watson. Asclepiadaceae. 2–3 Mex.

Melligo Rafin. = Salvia L. (Labiat.).

Melliniella Harms. Leguminosae. 1 trop. W. Afr.

Melliodendron Hand.-Mazz. Styracaceae. 3 S. & SW. China.

Mellissia Hook. f. Solanaceae. 1 St Helena.

Mellitis Scop. = Melittis L. (Labiat.).

Melloa Bur. Bignoniaceae. 2 trop. Am.

Mellobium A. Juss. = Melolobium Eckl. & Zeyh. (Legumin.).
Mellobium Reichb. = praec.
Melloca Lindl. = Ullucus Lozano (Basellac.).
Melo Mill. = Cucumis L. (Cucurbitac.).
Melocactus Boehm. (1757) = Cereus L. *p.p.* (Cactac.).
Melocactus Boehm. (1760) = Mammillaria Haw. (Cactac.).
Melocactus Link & Otto (= Cactus L. emend. Britton & Rose). Cactaceae.
30 W.I., trop. Am. Ribbed pl. like *Cereus*. Cephalium. Fls. produced
at top.
Melocalamus Benth. = Dinochloa Buese (Gramin.).
Melocanna Trin. Gramineae. 2 Indomal. Berry; seed exalb., ed.
Melocarpus P. & K. = Meliocarpus Boiss. = Prangos Lindl. (Umbellif.).
Melochia L. Sterculiaceae. 54 trop., esp. Am.
Melochia Rottb. Apocynac. vel Asclepiadac. 1 Guiana. Quid?
Melochi[ac]eae J. G. Agardh = Sterculiaceae–Hermannieae Spreng.
Melodinus J. R. & G. Forst. Apocynaceae. 50 Indomal., Austr., Pacif. Scales
in throat.
Melodorum Hook. f. & Thoms. = Fissistigma Griff. (Annonac.).
Melodorum Lour. = Polyalthia sp. + Mitrephora sp. (Annonac.).
Melolobium Eckl. & Zeyh. Leguminosae. 30 S. Afr. K 2-lipped.
Melomphis Rafin. = Ornithogalum L. (Liliac.).
Meloneura Rafin. = Utricularia L. (Lentibulariac.).
Melongena Mill. = Solanum L. (Solanac.).
Melopepo Mill. = Cucurbita L. (Cucurbitac.).
Melorima Rafin. = Fritillaria L. (Liliac.).
Melosmon Rafin. = Teucrium L. (Labiat.).
Melosperma Benth. Scrophulariaceae. 1 Chile. Seed large.
Melospermum Scortech. ex King = Walsura Roxb. (Meliac.).
Melothria L. Cucurbitaceae. 10 New World. Anther-loc. straight.
Melothrianthus Mart. Crovetto. Cucurbitaceae. 1 Brazil.
Melothrix M. Laws. (sphalm.) = Melothria L. (Cucurbitac.).
Melotria P.Br. = praec.
Meltrema Rafin. = Carex L. (Cyperac.).
Meluchia Medik. = Melochia L. (Sterculiac.).
Melvilla A. Anders. = Cuphea P.Br. (Lythrac.).
Memaecylum Mitch. = Epigaea L. (Ericac.).
Memecyclanthus Guillaum. (sphalm.) = Memecylanthus Gilg & Schltr.
(Alseuosmiac.).
Memecylaceae DC. Dicots. 4/360 trop. Glabrous trees and shrubs, with
terete branches and opp. ent. exstip. ls., in habit often closely resembling
Eugenia or *Syzygium* (*Myrtac.*). Infl. usu. axill., with short axes. Floral chars.
± as in *Melastomatac.* (*q.v.*), but $\overline{G}$ (∞–1), when multiloc. with 2–3 axile
ascending ov. per loc., when 1-loc. with few ov. vertic. on free central plac.
Fr. a 1–5-seeded berry; embryo large, with plano-convex or ± leafy cotyledons. Genera: *Memecylon, Mouriri, Votomita, Axinandra*. The fam. is ± intermediate between *Myrtac.* and *Melastomatac.*
Memecylanthus Gilg & Schlechter. Alseuosmiaceae. 1 New Caled.
Memecylon L. Memecylaceae. 300 trop. Afr., As., Austr., Pacif. G 1-loc.

MEMORA

Memora Miers. Bignoniaceae. 25–30 trop. S. Am.
Memorialis Buch.-Ham. ex Wedd. = Hyrtanandra Miq. (Urticac.).
Memycylon Griff. = Memecylon L. (Memecylac.).
Menabea Baill. Periplocaceae. 1 Madag.
Menadena Rafin. = Maxillaria Poepp. & Endl. (Orchidac.).
Menadenium Rafin. ex Cogn. = Zygosepalum Reichb. f. (Orchidac.).
Menais Loefl. Inc. sed. 1 S. Am.
Menalia Nor. = Kibara Endl. (Monimiac.).
Menandra Gronov. = Crocanthemum Spach (Cistac.).
Menanthos St-Lag. = Menyanthes L. (Menyanthac.).
Menaphronocalyx Pohl (nomen). 1 Brazil. Quid?
Menarda Comm. ex A. Juss. = Phyllanthus L. (Euphorbiac.).
Mendevilla Poit. = Mandevilla Lindl. (Apocynac.).
Mendezia DC. = Spilanthes L. (Compos.).
Mendocina Walp. = seq.
Mendoncaea P. & K. = Mendoncia Vell. ex Vand. (Mendonciac.).
Mendoncella A. D. Hawkes. Orchidaceae. 3 C. & trop. S. Am.
Mendoncia Vell. ex Vand. Mendonciaceae. 60 C. & trop. S. Am., trop. Afr.,
Madag.
Mendonciaceae (Lindau) Bremek. Dicots. 2/60 trop. Am., Afr. Twining
shrubs, stems articulated when young. Ls. opp., simple, ent., exstip. Fls. ⚥,
zygo., axill., sometimes in term. rac., with 2 large spathaceous bracteoles.
K reduced, annular, truncate or shortly lobed; C (5), ± hypocrateriform, not
inflated above, contorted; A 4, didynam., std. 1 or 0; disk large, cupular;
G (2), 2-1-loc., with 2 collat. axile ov. per loc., and simple style with small
bilobed stig. Fr. a 1(–2)-loc., 1–2-seeded drupe, with thick bony endoc.; seeds
without endosp.; embryo with twice-folded cots. Genera: *Mendoncia, Gil-
letiella*. Intermediate between *Bignoniac., Pedaliac., Thunbergiac.* and *Acanthac.*
Mendoni Adans. = Gloriosa L. (Liliac.).
Mendoravia Capuron. Leguminosae. 1 Madag.
Mendozia Ruiz & Pav. = Mendoncia Vell. (Mendonciac.).
Meneghinia Endl. = Arnebia Forsk. (Boraginac.).
Meneghinia Vis. = Niphaea Lindl. (Gesneriac.).
Menendezia Britton (~ Tetrazygia Rich.). Melastomataceae. 3 Puerto Rico.
Menepetalum Loes. Celastraceae. 6 New Caled.
Menephora Rafin. = Paphiopedilum Pfitz. (Orchidac.).
Menestoria DC. = Mycetia, Mussaenda, Randia spp. (Rubiac.).
Menestrata Vell. = Persea Mill. (Laurac.).
Menetho Rafin. = Frankenia L. (Frankeniac.).
Mengea Schau. = Amaranthus L. (Amaranthac.).
Menianthes Neck. = Menyanthes L. (Menyanthac.).
Menianthus Gouan = praec.
Menichea Sonner. = Barringtonia J. R. & G. Forst. (Barringtoniac.).
Menicosta D. Dietr. = Meniscosta Blume = Sabia Colebr. (Sabiac.).
Meninia Fua ex Hook. f. = Phlogacanthus Nees (Acanthac.).
Meniocus Desv. = Alyssum L. (Crucif.).
Meniscium Schreb. Thelypteridaceae. 12 trop. Am.
Meniscogyne Gagnep. Urticaceae. 2 Indoch.

Meniscosta Blume = Sabia Colebr. (Sabiac.).

Menisorus Alston = Cyclosorus Link (Thelypteridac.) (*Bol. Soc. Brot.* **30**: 20, 1956).

*****Menispermaceae** Juss. Dicots. 65/350 warm. Mostly twining shrubs, herbs, or trees, a few erect, with alt., usu. simple, exstip. ls., usu. with serial axillary buds. Infl. racemose, with ultimate branching cymose. Fls. unisexual, dioec., often ± caulifloral, rarely brightly coloured, usu. reg. Formula usu. K 3+3, C 3+3, A 3+3, G̲ 3, but many exceptions. K and A often > 6; A often ∞, or united, or in bundles; sometimes only 1 cpl., or more (to 32). Stds. in ♀ fl. various, or 0. Ovules 2 per cpl., soon reduced to 1, ventral, pend., semi-anatr. Fr. drupaceous; endoc. usu. sculptured; endosp. or 0, sometimes ruminate. The fr. usu. curves in development, so that style is no longer term. Classification of the gen. is largely based on structure of the seed. A few medicinal, on account of bitter principle in r. Chief genera: *Chondrodendron, Tiliacora, Triclisia, Anamirta, Coscinium, Tinospora, Jateorhiza, Abuta, Hyperbaena, Hypserpa, Cocculus, Menispermum, Stephania, Cissampelos, Cyclea.* Closely related to *Lardizabalac.*

Menispermum L. Menispermaceae. 3 temp. E. As., Atl. N. Am., Mex. (moonseed). K 4–10, spiral, C 6–9, A 12–24, G 2–4.

Menkea Lehm. Cruciferae. 6 Austr.

Menkenia Bub. = Lathyrus L. (Legumin.). [ringtoniac.).

Mennichea Steud. = Menichea Sonner. = Barringtonia J. R. & G. Forst. (Bar-

Menoceras Lindl. = Vellesia Sm. (Goodeniac.).

Menodora Humb. & Bonpl. Oleaceae. 9 SW. U.S. & Mex., 6 Bolivia to Chile & S. Argent., 2 S. Afr.

Menodoropsis (A. Gray) Small = praec.

Menomphalus Pomel = Centaurea L. (Compos.).

Menonvillea R.Br. ex DC. Cruciferae. 20 Peru, Chile, Argent. Gynophore.

Menophora P. & K. = Menephora Rafin. = Paphiopedilum Pfitz. (Orchidac.).

Menophyla Rafin. = Rumex L. (Polygonac.).

Menopteris Rafin. = Botrychium Swartz (Ophioglossac.).

Menospermum P. & K. = Menispermum L. (Menispermac.).

Menotriche Steetz = Wedelia Jacq. (Compos.).

Mentha L. Labiatae. 25 N. temp., S. Afr., Austr. C sub-reg., 4-merous. A 4 equal. An oil used in medicine is distilled from *M. piperita* L. (peppermint). *M. spicata* (L.) Huds. (garden mint) cult. for flavouring. Many hybrids.

Menthaceae Burnett = Labiatae Juss.

Menthella Pérard = Mentha L. (Labiat.).

Mentocalyx N. E. Brown. Aïzoaceae. 1 S. Afr.

Mentodendron Lundell. Myrtaceae. 1 C. Am. (Guatem.).

Mentzelia L. Loasaceae. 70 trop. & subtrop. Am., W.I. No stinging hairs. In some, outer sta. sterile. G usu. (3).

*****Menyanthaceae** Dum. Dicots. 5/33 N. & S. temp.; trop. SE. As. Aquatic or marsh herbs, with alt. simple (sometimes pelt.) or 3-foliol. exstip. ls. with sheathing pet. Fls. ⚥, reg.; infl. various. K 5 or (5); C (5), valv. or indupl.-valv., margins or interior often fimbr. or barb.; A 5, alternipet. and epipet., with versat. sagitt. anth.; nectaries usu. pres.; G̲ (2), 1-loc., with ∞ ov. on 2 pariet. plac., and simple shortly bifid style. Fr. a 2–4-valved caps., or fleshy indehisc.; seeds few–∞, sometimes winged, with copious endosp. Genera: *Nephrophyllidium, Villarsia, Menyanthes, Nymphoïdes, Liparophyllum.* Near *Gentianac.,* differing *inter alia* by the alt. ls. and valvate aestiv.

Menyanthes L. Menyanthaceae. 1 N. temp., *M. trifoliata* L., a bog pl. with creeping rhiz. and trifoliolate ls. Fls. dimorphic, heterost. (cf. *Primula*). Rhiz. yields bitter tonic.

Menzelia Schreb. = Mentzelia L. (Loasac.).

Menziesia Sm. Ericaceae. 7 N. temp. As., Am.

Menziesiaceae Klotzsch = Ericaceae–Rhododendreae Spreng.

Meon Rafin. = Meum Mill. (Umbellif.).

Meonitis Rafin. = praec.

Meopsis (Calest.) K.-Pol. (∼ Daucus L.). Umbelliferae. 2 W. Medit.

Meoschium Beauv. = Ischaemum L. (Gramin.).

Mephitidia Reinw. ex Blume = Lasianthus Jack (Rubiac.).

Meranthera Van Tiegh. = Loranthus L. (Loranthac.).

Merathrepta Rafin. = Danthonia DC. (Gramin.).

Meratia Cass. = Elvira Cass. (Compos.).

Meratia A. DC. = Moritzia DC. (Boraginac.).

Meratia Loisel. = Chimonanthus Lindl. (Calycanthac.).

Mercadoa Naves = Doryxylon Zoll. (Euphorbiac.).

Merciera A. DC. Campanulaceae. 5 S. Afr. G 1-loc., ov. basal.

Merckia Fisch. ex Cham. & Schlechtd. (1826; non Merkia Borkh. 1792— Hepaticae) = Wilhelmsia Reichb. (Caryophyllac.).

Mercklinia Regel = Hakea Schrad. (Proteac.).

Mercurialis L. Euphorbiaceae. 8 Medit., temp. Euras. to N. Siam. Ls. opp. Dioec. G (2). Veg. repr. by rhiz.

Mercuriastrum Fabr. = Acalypha L. (Euphorbiac.).

Merendera Ram. (∼ Colchicum L.). Liliaceae. 10 Medit. to Afghanistan, Abyssinia. Styles 3.

Meresaldia Bullock. Asclepiadaceae. 1 Venezuela.

Meretricia Néraud (nomen). Rubiaceae. Mauritius. Quid?

Meriana Trew = Watsonia Mill. (Iridac.).

Meriana Vell. = Evolvulus L. (Convolvulac.).

Meriana Vent. = Meriania Sw. (Melastomatac.).

Meriandra Benth. Labiatae. 2 Abyssinia, Himal.

*****Meriania** Sw. Melastomataceae. 50 W.I., trop. S. Am. Connective spurred behind.

Merianthera Kuhlm. Melastomataceae. 1 Brazil.

Mericarpaea Boiss. Rubiaceae. 1 W. As.

Mericocalyx Bamps. Rubiaceae. 2 trop. Afr.

Merida Neck. = Meridiana L. f. = Portulaca L. (Portulacac.).

Meridiana Hill = Gazania Gaertn. (Compos.).

Meridiana L. f. = Portulaca L. (Portulacac.).

Merimea Cambess. = Bergia L. (Elatinac.).

Meringium Presl. Hymenophyllaceae. 60 pantrop. & Austr. Receptacle elongate, indusium as *Hymenophyllum.*

Meringogyne H. Wolff. Umbelliferae. 1 SW. trop. Afr.

Meringurus Murbeck. Gramineae. 1 N. Afr. (Tunis).

Merinthopodium Donn. Smith. Solanaceae. 2 C. Am.

Merinthosorus Copeland. Polypodiaceae. 2 Malaysia (aff. *Aglaomorpha* Schott).

Meriolix Rafin. = Calylophis Spach = Oenothera L. (Onagrac.).

Merione Salisb. = Dioscorea L. (Dioscoreac.).

Merisachne Steud. = Triplasis Beauv. (Gramin.).

Merismia Van Tiegh. = Loranthus L. (Loranthac.).

Merismostigma S. Moore. Rubiaceae. 1 New Caled.

Merista Banks & Soland. ex A. Cunn. = Myrsine L. (Myrsinac.).

Meristostigma A. Dietr. = Lapeyrousia Pourr. (Iridac.).

Meristostylis Klotzsch = ? Kalanchoë Adans. (Crassulac.).

Meristotropis Fisch. & Mey. Leguminosae. 5 C. & SW. As.

Merizadenia Miers = Tabernaemontana L. (Apocynac.).

Merkia Reichb. = Merckia Fisch. ex Cham. & Schlechtd. = Wilhelmsia Reichb. (Caryophyllac.).

Merkusia De Vriese = Scaevola L. (Goodeniac.).

Merleta Rafin. = ? Croton L. (Euphorbiac.).

Merope M. Roem. (~ Atalantia Corrêa). Rutaceae. 1 trop. As.

Merope Wedd. = Gnaphalium L. (Compos.).

Merophragma Dulac = Telephium L. (Aïzoac.).

Merostachys Spreng. Gramineae. 20 C. & trop. S. Am. Climbing bamboos.

Merostela Pierre = Aglaia Lour. (Meliac.).

Merremia Dennst. ex Endl. Convolvulaceae. 80 warm.

Merrettia Soland. ex Engl. = Corynocarpus J. R. & G. Forst. (Corynocarpac.).

Merrillanthus Chun & Tsiang. Asclepiadaceae. 1 S. China (Hainan).

Merrillia Swingle. Rutaceae. 1 Burma, Siam, Malay Penins.

Merrilliodendron Kanehira. Icacinaceae. 1 Philipp., Marianne, Caroline, Bismarck Is.

Merrilliopanax Li. Araliaceae. 2 E. Himal., SW. China.

Merrittia Merr. Compositae. 1 Philipp. Is.

Mertensia Kunth = Celtis L. (Ulmac.).

*Mertensia Roth. Boraginaceae. 50 N. temp., S. to Mex. & Afghan.

Mertensia Willd. = Sticherus Presl (Gleicheniac.).

Mertensiaceae Corda = Gleicheniaceae Gaudich.

Merwia B. Fedtsch. Umbelliferae. 1 C. As. (Transcaspia).

Merxmuellera Conert. Gramineae. 14 SW., S. & S. trop. Afr., Madag.

Meryta J. R. & G. Forst. Araliaceae. 25 New Guinea, Solom. Is., New Caled., N.Z., Polynesia.

Mesadenella Pabst & Garay. Orchidaceae. 2 C. Am., Brazil.

Mesadenia Rafin. (1828) = Frasera Walt. (Gentianac.).

Mesadenia Rafin. (1836) = Senecio L. (Compos.).

Mesadenus Schlechter. Orchidaceae. 5 Florida, W.I., trop. Am.

Mesanchum Dulac = Tillaea L. = Crassula L. (Crassulac.).

Mesandrinia Rafin. = Jatropha L. (Euphorbiac.).

Mesanthemum Koern. Eriocaulaceae. 10 trop. Afr., Madag.

Mesanthus Nees = Cannomoïs Beauv. (Restionac.).

Mesaulosperma van Slooten = Itoa Hemsl. (Flacourtiac.).

Mesechinopsis Y. Ito = Echinopsis Zucc. (Cactac.).

MESECHITES

Mesechites Muell. Arg. Apocynaceae. 10 C. & trop. S. Am., W.I.

Mesembrianthemum Spreng. = Mesembryanthemum L. (Aïzoac.).

Mesembrianthus Rafin. = Mesembryanthus Neck. = praec.

Mesembryaceae ('-bryneae') Dum. = Aïzoaceae J. G. Agardh.

*Mesembryanthem[ac]eae Fenzl = praec.

Mesembryanthemum L., *sensu lato*. Aïzoaceae. 1000 S. Afr., few tropical Afr., Canaries, Mediterranean, Arabia, Austr., 1 Chile, 1 St Helena. Xero. of the most pronounced kind with very succulent ls., usu. closely packed together; the young ls. stand face to face at the growing apex till well grown, and thus protect the young bud. In *M.* (*Conophytum*) *obconellum* Haw. the pairs of ls. are congenitally united into a fleshy body with a little slit in the centre. Several have thorns, sometimes fl.-stalks hardened after the fall of the fl., sometimes branches, as in *M.* (*Eberlanzia*) *spinosum* L. (the leafy branches appear below these in the next year, in the same axils). Fls. usu. term. on the stems, sol. or in dichasia or cincinni. Outer sta. (due to branching) repres. by numerous petaloid stds., having the appearance of a C. The mature ovary is 5-loc. with parietal plac.; this peculiar feature is due to an excessive growth of the peripheral tissue during development, which gradually turns the loculi completely over (cf. *Punica*). Fr. a caps. which opens only in moist air, contrary to the usual wont of capsules. Some, e.g. *M.* (*Carpobrotus*) *edule* L. (Hottentot fig), contain an ed. pulp. *M.* (*Cryophytum*) *crystallinum* L. is the ice-plant, so called because its ls. are covered with small glistening bladder-like hairs.

Mesembryanthemum L., *sensu stricto*, emend. N. E. Brown (*Ruschia* Schwantes). Aïzoaceae. 350 S. Afr.

Mesembryanthes Stokes = praec. (*s.l.*).

Mesembryanthus Neck. = praec.

Mesembryum Adans. = praec.

Mesicera Rafin. = Habenaria Willd. (Orchidac.).

Mesocentron Cass. = Centaurea L. (Compos.).

Mesoceras P. & K. = Mesicera Rafin. = Habenaria Willd. (Orchidac.).

Mesochlaena R.Br. = Sphaerostephanos J. Sm. (Thelypteridac.).

Mesochloa Rafin. = Zephyranthes Herb. (Amaryllidac.).

Mesoclastes Lindl. = Luisia Gaudich. (Orchidac.).

Mesodactylis Wall. = Apostasia Blume (Orchidac.).

Mesodactylus P. & K. = praec.

Mesodetra Rafin. = Helenium L. (Compos.).

Mesodiscus Rafin. = Cryptotaenia DC. (Umbellif.).

Mesogramma DC. = Senecio L. (Compos.).

Mesogyne Engl. (= ?Antiaris Lesch.). Moraceae. 2 trop. Afr.

Mesoligus Rafin. = Aster L. (Compos.).

Mesomelaena Nees = Gymnoschoenus Nees (Cyperac.).

Mesomora Nieuwl. & Lunell = Chamaepericlymenum Hill (Cornac.).

Mesona Blume. Labiatae. 12 Himal. to Formosa & Philipp. Is.; Java.

Mesonephelium Pierre = Nephelium L. (Sapindac.).

Mesoneuris A. Gray = Senecio L. (Compos.).

Mesoneuron Ching (1963; non *Mesonevron* Unger 1856—gen. foss.; nec *Mesonevron* Desf. 1818—Legumin.) = Mesophlebion Holtt. (Thelypteridac.).

Mesopanax R. Viguier = Schefflera, Oreopanax, etc., *p.p.* (Araliac.).

Mesophlebion Holtt. Thelypteridaceae. 45 Malaysia & Pacific.

*Mesoptera Hook. f. Rubiaceae. 1 Malay Penins.

Mesoptera Rafin. = Liparis Rich. (Orchidac.).

Mesoreanthus Greene = Pleiocardia Greene (Crucif.).

Mesosetum Steud. Gramineae. 35 C. & trop. S. Am., W.I.

Mesosorus Hassk. = Sticherus Presl (Gleicheniac.).

Mesosphaerum P.Br. = Hyptis Jacq. (Labiat.).

Mesospinidium Reichb. f. Orchidaceae. 4 C. & trop. S. Am.

Mesostemma Vved. Caryophyllaceae. 1 C. As.

Mesothema Presl = Blechnum L. (Blechnac.).

Mesotricha Shchegl. = Astroloma R.Br. (Epacridac.).

Mespilaceae Schultz-Schultzenst. = Rosaceae–Pomoïdeae Juss.

Mespilodaphne Nees = Ocotea Aubl. (Laurac.).

Mespilophora Neck. = Mespilus L. (*s.l.*, incl. *Aronia* Medik., *Amelanchier* Medik., *Cotoneaster* Medik., etc.) (Rosac.).

Mespilus L. (emend. Medik.). Rosaceae. 1 SE. Eur. to C. As., *M. germanica* L., the medlar (ed. fr.).

Messanthemum Pritz. = Mesanthemum Koern. (Eriocaulac.).

Messermidia Rafin. = seq.

Messerschmidia L. ex Hebenstr. (1763) (= Argusia Boehm.). Boraginaceae. 1 coasts SE. U.S., C. Am. & W.I.; 1 temp. Euras.; 1 islands of Indian & Pacif. oceans.

Messerschmidtia G. Don = praec.

Messersmidia L. (1767) = praec.

Mesterna Adans. = Laëtia Loefl. ex L. (Flacourtiac.).

Mestoklema N. E. Br. Aïzoaceae. 6 S. Afr.

Mestotes Soland. ex DC. = Dichapetalum Thou. (Dichapetalac.).

Mesua L. Guttiferae. 40 Indomal. *M. ferrea* L. (*na* or iron-wood) yields a valuable timber; its fls. are used in perfumery.

Mesynium Rafin. = Linum L. (Linac.).

Meta-aletris Masamune = Aletris L. (Liliac.).

Metabasis DC. = Hypochoeris L. (Compos.).

Metabletaceae Dulac = Portulacaceae Juss.

Metabolos Blume (~ Hedyotis L.). Rubiaceae. 10 trop. As.

Metachilum Lindl. = Appendicula Blume (Orchidac.).

Metachilus P. & K. = praec.

Metachlamydeae Engl. = Sympetalae A. Br. = Gamopetalae Brongn.

Metadina Bakh. f. Naucleaceae (~ Rubiaceae). 1 Java (etc.?).

Metagnanthus Endl. = Athanasia L. (Compos.).

Metagnathus Benth. & Hook. f. (sphalm.) = praec.

Metagonia Nutt. = Vaccinium L. (Ericac.).

Metalasia R.Br. Compositae. 40 S. Afr. Ls. sometimes spirally twisted.

Metalepis Griseb. Asclepiadaceae. 2 W.I.

Metanarthecium Maxim. Liliaceae. 5 Japan, Formosa. Filaments hollow.

Metaplexis R.Br. Asclepiadaceae. 6 E. As.

Metaporana N. E. Br. (~ Bonamia Thou.). Convolvulaceae. 2 trop. Afr., Madag.

Metasequoia Miki. Taxodiaceae. 1 C. China. Repres. of the genus were

METASEQUOIACEAE

discovered in the fossil state (Cretac. to Plioc.) shortly before the discovery of living examples in China. They differ from other *Taxodiac.* in their decussate ls.

Metasequoiaceae Hu & Cheng = Taxodiaceae Neger.

Metasocratea Dugand. Palmae. 1 Colombia.

Metastachydium Airy Shaw. Labiatae. 1 C. As.

Metastachys Knorr. = praec.

Metastachys Van Tiegh. = Loranthus L. (Loranthac.).

Metastelma R.Br. Asclepiadaceae. 100 trop. & subtrop. Am.

Metathelypteris Ching. Thelypteridaceae. 12 trop. As. & Malaysia.

Metathlaspi Krause = Thlaspi L., Aëtheonema R.Br., Iberis L. (Crucif.).

Metatrophis Forest Brown. Moraceae. 1 Polynesia.

Metaxanthus Walp. = Metazanthus Meyen = Senecio L. (Compos.).

Metaxya Presl. Metaxyaceae. 1 S. Am.

Metaxyaceae Pichi-Serm. Cyatheales. 1/1 trop. Am. Rhiz. creeping; fronds simply pinnate.

Metazanthus Meyen = Senecio L. (Compos.).

Metcalfia Conert. Gramineae. 1 Mex.

Meteorina Cass. = Dimorphotheca L. (Compos.).

Meteoromyrtus Gamble. Myrtaceae. 1 S. India.

Meteorus Lour. = Barringtonia J. R. & G. Forst. (Barringtoniac.).

Meterana Rafin. = Caperonia A. St-Hil. (Euphorbiac.).

Meterostachys Nakai (~ Orostachys Fisch.). Crassulaceae. 1 Korea, Japan.

Metharme Phil. ex Engl. Zygophyllaceae. 1 N. Chile.

Methonica Gagneb. = Gloriosa L. (Liliac.).

Methonicaceae Trautv. = Liliaceae–Uvulariëae.

Methorium Schott & Endl. = Helicteres L. (Sterculiac.).

Methyscophyllum Eckl. & Zeyh. = Catha Forsk. (Celastrac.).

Methysticodendron R. E. Schult. = Brugmansia Pers. (Solanac.).

Methysticum Rafin. = Piper L. (Piperac.).

Metopium P.Br. Anacardiaceae. 3 Florida, Mex., W.I. Yields a purging resin (doctor gum) from the stem.

Metrocynia Thou. = Cynometra L. (Legumin.).

Metrodorea A. St-Hil. Rutaceae. 5–6 Brazil.

*Metrosideros Banks ex Gaertn. Myrtaceae. 60 S. Afr., E. Malaysia, Austr., N.Z., Polynes. Some furnish useful timber.

Metroxilon Welw. = seq.

Metroxylon Rottb. Palmae. 15 Siam to Solomon Is., Fiji, New Hebr. Small trees whose stems die after producing their large term. monoec. infls. (cf. *Corypha*, etc.), but form rhiz. branches below. The fr. takes 3 years to ripen. *M. rumphii* Mart. and *M. laeve* Mart. are the sago palms, cult. in Malaya. The trees are cut down when the infl. appears, and the sago is obtained from the pith by crushing and washing.

Metroxylon Spreng. = Raphia Beauv. (Palm.).

Mettenia Griseb. Euphorbiaceae. 6 W.I.

Metteniusa Karst. Alangiaceae. 3 NW. trop. S. Am.

Metteniusaceae Karst. = Alangiaceae DC.

Metternichia Mikan. Solanaceae. 4 C. & trop. S. Am.

Metula Van Tiegh. = Tapinanthus Bl. (Loranthac.).

Metzleria Sond. = Mezleria Presl (Campanulac.).
Meum Mill. Umbelliferae. 1 Eur., *M. athamanticum* Jacq.
Mexianthus B. L. Robinson. Compositae. 1 Mex.
Meyenia Backeb. = Weberbauerocereus Backeb. (Cactac.).
Meyenia Nees (~ Thunbergia L.). Thunbergiaceae (?Pedaliac.). 1 India,
Meyenia P. & K. = Meyna Roxb. ex Link (Rubiac.). [Ceylon.
Meyenia Schlechtd. = Cestrum L. (Solanac.).
Meyera Adans. = Holosteum L. (Caryophyllac.).
Meyera Schreb. = Enydra Lour. (Compos.).
Meyerafra Kuntze = Astephania Oliv. (Compos.).
Meyeria DC. = Calea L. (Compos.).
Meyerophytum Schwantes. Aïzoaceae. 1 S. Afr.
Meyna Roxb. ex Link. Rubiaceae. 11 trop. Afr. & Comoro Is. to SE. As.
Mezereum C. A. Mey. = Daphne L. (Thymelaeac.).
Mezia Kuntze = Neosilvia Pax = Mezilaurus Kuntze ex Taubert (Laurac.).
Mezia Schwacke ex Engl. & Prantl. Malpighiaceae. 1 Brazil.
Meziella Schindler. Haloragidaceae. 1 Austr.
Meziera Baker = seq.
Mezierea Gaudich. = Begonia L. (Begoniac.).
Mezilaurus Kuntze ex Taubert. Lauraceae. 9 trop. S. Am.
Meziothamnus Harms = Abromeitiella Mez (Bromeliac.).
Mezleria Presl (~ Lobelia L.). Campanulaceae. 7 S. Afr., 1 N.Z.
Mezobromelia L. B. Smith. Bromeliaceae. 2 NW. trop. S. Am.
Mezochloa Butzin = Alloteropsis C. Presl (Gramin.).
Mezoneurum auctt. = seq.
Mezonevron Desf. Leguminosae. 30 trop. Afr. & Madag. to Austr. & Pacif.
Mezzettia Becc. Annonaceae. 7 Malay Penins., Borneo.
Mezzettiopsis Ridl. Annonaceae. 1 Borneo.
Miagia Rafin. = Miegia Pers. = Arundinaria Michx (Gramin.).
Miagrum Crantz = Myagrum L. (Crucif.).
Mialisa P. & K. = Meialisa Rafin. = Adriana Gaud. (Euphorbiac.).
Miangis Thou. (uninom.) = *Angraecum parviflorum* Thou. = *Angraecopsis parviflora* (Thou.) Schltr. (Orchidac.).
Miapinon P. & K. = Meiapinon Rafin. = ?Mollugo L. (Aïzoac.).
Mibora Adans. Gramineae. 1 W. Eur., 1 N. Afr.
Micadania R.Br. = Butyrospermum Kotschy (Sapotac.).
Micagrostis d'Anthoine ex Juss. = Mibora Adans. (Gramin.).
Micalia Rafin. = Escobedia Ruiz & Pav. (Scrophulariac.).
Micambe Adans. = Cleome L. (Cleomac.).
Micania D. Dietr. = Mikania Willd. (Compos.).
Michauxia L'Hérit. ex Ait. Campanulaceae. 7 E. Medit., Persia. Fl. 7–10-merous throughout.
Michauxia Neck. = Relhania L'Hérit. (Compos.).
Michauxia P. & K. = Michoxia Vell. = Ternstroemia Mutis ex L. f. (Theac.).
Michauxia Raeuschel (nomen). Hab.? Quid? (Cucurbitac.?).
Michauxia Salisb. = Gordonia Ellis (Theac.).
Michelaria Dum. = Bromus L. (Gramin.).
Michelia Adans. = Pontederia L. (Pontederiac.).

MICHELIA

Michelia Th. Dur. = Lophotocarpus Th. Dur. (Alismatac.).
Michelia Kuntze = Barringtonia J. R. & G. Forst. (Barringtoniac.).
Michelia L. Magnoliaceae. 50 trop. As., China. Differs from *Magnolia* in its axill., not term., fls., and in possessing a gynophore between sta. and cpls. Fr. a lax spike of follicles. Several yield useful timber.
Micheliella Briquet = Hypogon Rafin. (Labiat.).
Micheliopsis H. Kerfg (non Magnolia § Micheliopsis Baill. = Michelia L.) = Magnolia L. (Magnoliac.).
Michelsonia Hauman. Leguminosae. 2 trop. Afr.
Michiea F. Muell. = Coleanthera Shchegl. (Epacridac.).
Micholitzia N. E. Brown. Asclepiadaceae. 1 India.
Michoxia Vell. = Ternstroemia Mutis ex L. f. (Theac.).
***Miconia** Ruiz & Pav. Melastomataceae. 700 trop. Am., W.I.; 1 W. Afr.
Miconiastrum Naud. = Tetrazygia Rich. (Melastomatac.).
Micractis DC. = Sigesbeckia L. (Compos.).
Micradenia Miers = Dipladenia A. DC. (Apocynac.).
Micraea Miers = Ruellia L. (Acanthac.).
Micraeschynanthus Ridley. Gesneriaceae. 1 Malay Penins.
Micragrostis P. & K. = Micagrostis d'Anth. ex Juss. = Mibora Adans. (Gramin.).
Micraira F. Muell. Gramineae. 1 Queensland.
Micrampelis Rafin. = Echinocystis Torr. & Gray (Cucurbitac.).
***Micrandra** Benth. Euphorbiaceae. 14 trop. S. Am.
Micrandra R.Br. = Hevea Aubl. (Euphorbiac.).
Micrangelia Fourr. = Selinum L. (Umbellif.).
Micrantha Dvořák. Cruciferae. 1 Persia.
Micranthea A. Juss. = Micrantheum Desf. (Euphorbiac.).
Micranthea Panch. ex Baill. = Phyllanthus L. (Euphorbiac.).
Micranthe[ace]ae J. G. Agardh = Euphorbiaceae–Porantheroïdeae Pax.
Micranthella Naud. = Tibouchina Aubl. (Melastomatac.).
Micranthemum Endl. = Micrantheum Presl = Trifolium L. (Legumin.).
***Micranthemum** Michx. Scrophulariaceae. 1 E. U.S. to E. S. Am.; 1 Cuba.
Micranthera Choisy = Tovomita Aubl. (Guttif.).
Micranthes Bertol. = Hoslundia Vahl (Labiat.).
Micranthes Haw. = Saxifraga L. (Saxifragac.).
Micrantheum Desf. Euphorbiaceae. 3 Austr.
Micrantheum Presl = Trifolium L. (Legumin.).
Micranthocereus Backeb. Cactaceae. 1 NE. Brazil.
Micranthos St-Lag. = Micranthus (Pers.) Eckl. (Iridac.).
Micranthus Loud. = Marcanthus Lour. = Mucuna Adans. (Legumin.).
***Micranthus** (Pers.) Eckl. Iridaceae. 3 S. Afr.
Micranthus Rafin. = Micranthemum Michx (Scrophulariac.).
Micranthus Roth = Rotala L. (Lythrac.).
Micranthus Wendl. = Phaylopsis Willd. (Acanthac.).
Micrargeria Benth. Scrophulariaceae. 4–5 trop. Afr., India.
Micrargeriella R. E. Fries. Scrophulariaceae. 1 trop. E. Afr.
Micrasepalum Urb. Rubiaceae. 2 W.I.
Micraster Harv. = Brachystelma R.Br. (Asclepiadac.).

Micrauchenia Froel. = Dubyaea DC. (Compos.).
Micrechites Miq. Apocynaceae. 20 S. China, Indomal.
Micrelium Forsk. = Eclipta L. (Compos.).
Micrelus P. & K. = Microëlus Wight & Arn. = Bischofia Blume (Bischofiac.).
Micrembryae Benth. & Hook. f. A 'Series' of Dicots. comprising the fams.
    Piperac., Chloranthac., Myristicac. and Monimiac.
Microbahia Cockerell = Syntrichopappus A. Gray (Compos.).
Microbambus K. Schum. = Guaduella Franch. (Gramin.).
Microberlinia A. Chev. Leguminosae. 2 W. Equat. Afr.
Microbignonia Kraenzlin = Doxantha Miers (Bignoniac.).
Microbiota Komarov (~ Thuja L.). Cupressaceae. 1 E. Siberia.
Microbiotaceae Nak. = Cupressaceae Neger.
Microblepharis M. Roem. = Modecca Lam. (Passiflorac.).
Microbrochis Presl = Tectaria Cav. (Aspidac.).
Microcachrys Hook. f. Podocarpaceae. 1 Tasm.
Microcaelia Hochst. ex A. Rich. = Microcoelia Lindl. (Orchidac.).
Microcala Hoffmgg. & Link. Gentianaceae. 1 Am.; 1 W. Eur., Medit.
Microcalamus Franch. Gramineae. 4 trop. W. Afr.
Microcalamus Gamble = Bambusa Schreb. (Gramin.).
Microcalia A. Rich. = Lagenifera Cass. (Compos.).
Microcardamum O. E. Schulz. Cruciferae. 1 temp. S. Am.
Microcarpaea R.Br. Scrophulariaceae. 1 E. As., Indomal., Austr.
Microcaryum I. M. Johnston. Boraginaceae. 4 C. As., Himal., W. China.
Microcasia Becc. Araceae. 2 Borneo.
Microcephala Pobedim. Compositae. 3 C. As., Persia.
Microcephalum Sch. Bip. ex Klatt = Gymnolomia Kunth (Compos.).
Microcerasus M. Roem. = Prunus L. (Rosac.).
Microchaeta Nutt. = Lipochaeta DC. (Compos.).
Microchaeta Reichb. = Rhynchospora Vahl (Cyperac.).
Microchaete Benth. = Senecio L. (Compos.).
Microcharis Benth. = Indigofera L. (Legumin.).
Microchilus Presl = Erythrodes Blume (Orchidac.).
Microchites Rolfe = Micrechites Miq. (Apocynac.).
Microchlaena Ching. Aspidiaceae. 1 Ceylon, NE. India, W. China. (Bull.
    Fan Mem. Inst. Biol., Bot. Ser. 8: 322, 1938.)
Microchlaena P. & K. = Microlaena R.Br. (Gramin.).
Microchlaena Wight & Arn. = Eriolaena DC. (Sterculiac.).
Microchloa R.Br. Gramineae. 3 Afr., one pantrop.
Microchonea Pierre = Trachelospermum Lem. (Apocynac.).
Microcitrus Swingle. Rutaceae. 5 E. Austr.
Microclisia Benth. = Pleogyne Miers (Menispermac.).
Microcnemum Ung.-Sternb. Chenopodiaceae. 1 Spain.
Micrococca Benth. Euphorbiaceae. 14 trop. Afr., Madag., India, Malaya.
Micrococcus Beckm. = Microcos L. (Tiliac.).
Micrococos Phil. = Jubaea Kunth (Palm.).
Microcodon A. DC. Campanulaceae. 4 S. Afr.
Microcoecia Hook. f. = Elvira Cass. (Compos.).
Microcoelia Hochst. ex Rich. = Angraecum Bory (Orchidac.).

Microcoelia Lindl. Orchidaceae. 27 trop. & S. Afr., Madag.
Microcoelum Burret & Potztal (~ Syagrus Mart.). Palmae. 2 Brazil.
Microcorys R.Br. Labiatae. 16 SW. Austr.
Microcos L. Tiliaceae. 53 Indomal., Fiji.
Microculcas Peter. Araceae. 2 trop. E. Afr.
Microcybe Turcz. Rutaceae. 3 Austr.
Microcycas (Miq.) A. DC. Zamiaceae. 1 Cuba.
Microdactylon T. S. Brandegee. Asclepiadaceae. 1 Mex.
Microderis DC. = Picris L. (Compos.).
Microdesmis Hook. f. Pandaceae. 10 trop. Afr., SE. As., W. Malaysia.
Microdon Choisy. Scrophulariaceae. 6 S. Afr.
Microdonta Nutt. = Heterospermum Cav. (Compos.).
Microdontocharis Baill. = Eucharis Planch. & Linden (Amaryllidac.).
Microdracoïdes Hua. Cyperaceae. 1 trop. Afr. Shrubby Vellozioid pl.; ls. crowded at tips of branches.
Microëlus Wight & Arn. = Bischofia Blume (Bischofiac.).
Microgenetes A. DC. = Phacelia Juss. (Hydrophyllac.).
Microglossa DC. = Conyza L. + Psiadia Jacq. (Compos.).
Microgonium Presl. Hymenophyllaceae. 12 Afr. to Tahiti.
Microgramma Presl. Polypodiaceae. 20 trop. Am., Afr.
Microgyne Cass. = Eriocephalus L. (Compos.).
Microgyne Less. = Vittadinia Rich. (Compos.).
Microgynoecium Hook. f. Chenopodiaceae. 1 Tibet.
Microjambosa Blume = Syzygium Gaertn. (Myrtac.).
Microkentia H. Wendl. ex Benth. & Hook. f. = Basselinia Vieill. (Palm.).
Microkoma Lanessan (sphalm.) = Microloma R.Br. (Asclepiadac.).
Microlaena R.Br. Gramineae. 10 Philipp. Is. & Java to Austr. & N.Z.
Microlaena Wall. = Schillera Reichb. = Eriolaena DC. (Sterculiac.).
Microlecane Sch. Bip. Compositae. 3 W. Afr. & E. trop. Afr.
Microlepia Presl. Dennstaedtiaceae. 45 Old World trop., Japan, N.Z. Distinction from *Dennstaedtia* needs clarification.
Microlepidium F. Muell. = Capsella Medik. (Crucif.).
*****Microlepis** (DC.) Miq. Melastomataceae. 4 S. Brazil.
Microlepis Eichw. = Anabasis L. (Chenopodiac.).
Microlepis Schrad. ex Nees = Cryptangium Schrad. + Lagenocarpus Nees (Cyperac.).
Microlespedeza Makino = Kummerovia Schindl. (Legumin.).
Microliabum Cabrera. Compositae. 1 Argent.
Microlicia D. Don. Melastomataceae. 100 trop. S. Am.
Microlobium Liebm. = Apoplanesia Presl (Legumin.).
Microlobius Presl. Leguminosae. 1 Mex.
Microloma R.Br. Asclepiadaceae. 15 S. Afr.
Microlonchoïdes Candargy [Kantartzis] (~ Centaurea L.). Compositae. 1 Greece.
Microlonchus Cass. (~ Centaurea L.). Compositae. 10 Medit.
Microlophium Fourr. = Polygala L. (Polygalac.).
Microlophopsis Czerep. (~ Centaurea L.). Compositae. 1 Afghanistan.
Microlophus Cass. = Centaurea L. (Compos.).

Microluma Baill. = Lucuma Molina (Sapotac.).
Micromeles Decne = Sorbus L. (Rosac.).
*Micromelum Blume. Rutaceae. 10 Indomal., Pacif.
*Micromeria Benth. (~ Satureia L.). Labiatae. 100 + cosmop. *M. douglasii* Benth. (Calif., etc.) is the *yerba buena* (medicinal).
Micromonolepis Ulbr. Chenopodiaceae. 1 W. U.S.
Micromyrtus Benth. Myrtaceae. 16 Austr.
Micromystria O. E. Schulz. Cruciferae. 2 Austr.
Micronema Schott = Micromeria Benth. (Labiat.).
Micronoma H. Wendl. Palmae. 1 E. Peru.
Micronychia Oliv. Anacardiaceae. 5 Madag.
Micropaegma Pichon = Mussatia Bur. ex Baill. (Bignoniac.).
Micropapyrus Suesseng. Cyperaceae. 1 Brazil.
Micropeplis Bunge (~ Halogeton C. A. Mey.). Chenopodiaceae. 1 C. As.
Micropera Lindl. = Camarotis Lindl. (Orchidac.).
Micropetalon Pers. = Stellaria L. (Caryophyllac.).
Micropetalum Poit. ex Baill. = Amanoa Aubl. (Euphorbiac.).
Micropetalum Spreng. = Micropetalon Pers. = Stellaria L. (Caryophyllac.).
Micropeuce Gordon = Tsuga Carr. (Pinac.).
Microphacos Rydb. = Astragalus L. (Legumin.).
× Microphoenix Naud. ex Carr. Palmae. Gen. hybr. (Chamaerops × Phoenix).
Micropholis (Griseb.) Pierre (~ Sideroxylon L.). Sapotaceae. 40 trop. Am., W.I.
Microphyes Phil. Caryophyllaceae. 2 Chile.
Microphysa Naud. = Microphysca Naud. (Melastomatac.).
Microphysa Schrenk. Rubiaceae. 1 C. As.
Microphysca Naud. (~ Tococa Aubl.). Melastomataceae. 2 trop. S. Am.
Microphyton Fourr. = Trifolium L. (Legumin.).
Micropiper Miq. = Peperomia Ruiz & Pav. (Peperomiac.).
Micropleura Lag. Umbelliferae. 2 Colombia, Chile.
Microplumeria Baill. Apocynaceae. 1 Amaz. Brazil.
Micropodium Mett. = Asplenium L. (Aspleniac.).
Micropodium Reichb. = Brassica L. (Crucif.).
Micropogon Spreng. ex Pfeiff. = ? Microchloa R.Br. (Gramin.).
Micropolypodium Hayata = Xiphopteris Kaulf. (Grammitidac.).
Micropora Dalz. = Micropera Lindl. = Camarotis Lindl. (Orchidac.).
Micropora Hook. f. = Hexapora Hook. f. (Laurac.).
Micropsis DC. Compositae. 5 Chile, Argent., Uruguay.
Microptelea Spach = Ulmus L. (Ulmac.).
Micropteris Desv. = Xiphopteris Kaulf. (Grammitidac.).
Micropterum Schwantes = Cleretum N. E. Br. (Aïzoac.).
Micropteryx Walp. = Erythrina L. (Legumin.).
Micropuntia Daston = Opuntia Mill. (Cactac.).
Micropus ·L. Compositae. 1 W. Medit. to Cauc. & Persia.
Micropyrum (Gaudin) Link. Gramineae. 3 C. Eur., Medit.
Micropyxis Duby = Anagallis L. (Primulac.).
Microrhamnus A. Gray = Condalia Cav. (Rhamnac.).

MICRORHINUM

Microrhinum Fourr. = Chaenorrhinum Lange (Scrophulariac.).
Microrhynchus Less. = Launaea Cass. (Compos.).
Microrphium C. B. Clarke. Gentianaceae. 1 Malay Penins., 1 Philipp. Is.?
Microrynchus Sch. Bip. = Microrhynchus Less. = Launaea Cass. (Compos.).
Microsaccus Blume. Orchidaceae. 6 Indomal.
Microschizaea Reed = Schizaea Sm. (Schizaeac.).
Microschoenus C. B. Clarke. Cyperaceae. 1 W. Himalaya.
Microschwenkia Benth. = Melananthus Walp. (Solanac.).
Microsciadium Boiss. Umbelliferae. 1 Asia Minor.
Microsciadium Hook. f. = Azorella Lam. (Hydrocotylac.).
Microsechium Naud. Cucurbitaceae. 2 Mex.
Microselinum Andrz. (nomen). Umbelliferae. 1 Russia (Ukraine). Quid?
Microsemia Greene. Cruciferae. 1 Calif.
Microsemma Labill. = Lethedon Spreng. (Thymelaeac.).
Microsepala Miq. = Baccaurea Lour. (Euphorbiac.).
Microseris D. Don. Compositae. 14 W. N. Am., 1 Chile; 1 Austr., N.Z.
Microsisymbrium O. E. Schulz. Cruciferae. 6 C. As., Afgh., W. Himal.;
   1 NW. Am.
Microsorium Link. Polypodiaceae. 60 Old World trop., incl. several v.
   common epiphytes. A division into spp. with large sori (*Phymatodes* Presl)
   and small sori seems impracticable.
Microsperma Hook. = Mentzelia L. (Loasac.).
Microspermae Benth. & Hook. f. A 'Series' of Monocots. comprising the
   fams. *Hydrocharitac., Burmanniac.* and *Orchidac.*; used also by Engler, excl.
   Hydrocharitac.
Microspermia Frič = Parodia Speg. (Cactac.).
Microspermum Lag. Compositae. 5 Mex.
Microsplenium Hook. f. = Machaonia Kunth (Rubiac.).
Microstachys A. Juss. = Sebastiania Spreng. (Euphorbiac.).
Microstaphyla Presl. Lomariopsidaceae. 3 trop. Am., St Helena.
Microstegia Presl = Diplazium Sw. (Athyriac.).
Microstegium Nees. Gramineae. 30 trop. & subtrop. Afr. & As.
Microstegnus Presl = Cnemidaria Presl (Cyatheac.).
Microsteira Baker. Malpighiaceae. 25 Madag. Polygamo-dioec.
Microstelma Baill. Asclepiadaceae (inc. sed.). 1–2 Mex. (= ?Gonolobus
Microstemma R.Br. Asclepiadaceae. 2 Austr.   [Michx).
Microstemma Reichb. = Microsemma Labill. = Lethedon Spreng. (Thyme-
   laeac.).
Microstemon Engl. = Pentaspadon Hook. f. (Anacardiac.).
Microstephanus N. E. Br. = Pleurostelma Baillon (Asclepiadac.).
Microstephium Less. = Cryptostemma R.Br. (Compos.).
Microsteris Greene (~ Phlox L.). Polemoniaceae. 1 (polymorph.) Pacif. N.
   & S. Am.
Microstigma Trautv. (~ Matthiola R.Br.). Cruciferae. 1 C. As.
Microstira P. & K. = Microsteira Bak. (Malpighiac.).
Microstrobilus Bremek. Acanthaceae. 4 Sumatra, Java.
Microstrobos Garden & Johnson. Podocarpaceae. 2 E. Austr. (N.S.W.),
   Tasm.

*Microstylis (Nutt.) Eaton = Malaxis Sol. ex Sw. (Orchidac.).
Microsyphus Presl = Alectra Thunb. (Scrophulariac.).
Microtaena Hemsl. = Microtoena Prain (Labiat.).
Microtatorchis Schlechter. Orchidaceae. 30 Malaysia, Polynesia.
Microtea Sw. Chenopodiaceae(?). 10 trop. Am., W.I.
Microterus Presl = Crypsinus Presl (Polypodiac.).
Microthea Juss. = Microtea Sw. (Chenopodiac.?).
Microtheca Schlechter = Cynorkis Thou. (Orchidac.).
Microthouareia Steud. = seq.
Microthuareia Thou. = Thouarea Pers. (Gramin.).
Microtinus Oerst. = Viburnum L. (Caprifoliac.).
Microtis R.Br. Orchidaceae. 10 E. As., Malaysia, Austr., N.Z., Polynesia.
Microtoena Prain. Labiatae. 20 Himal., W. China, Java.
Microtrema Klotzsch = Erica L. (Ericac.).
Microtrichia DC. Compositae. 2 trop. Afr.
Microtrichomanes (Prantl) Copel. Hymenophyllaceae. 10 Madag. to Polynesia & N.Z.
Microtropia Reichb. = Microtropis Wall. (Celastrac.).
Microtropis E. Mey. = Euchlora Eckl. & Zeyh. (Legumin.).
*Microtropis Wall. ex Meissn. Celastraceae. 70 China, Formosa, SE. As., Indomal.; Mex. & C. Am.
Microula Benth. Boraginaceae. 15 Himal., Tibet, W. China.
Microuratea Van Tiegh. = Ouratea Aubl. (Ochnac.).
Mictanthes Rafin. = Myctanthes Rafin. = ?Aster L. (Compos.).
Mida R. Cunn. ex A. Cunn. Santalaceae. 1 N.Z., 1 Juan Fernandez.
Middelbergia Schinz ex Pax = Clutia L. (Euphorbiac.).
Middendorfia Trautv. = Lythrum L. (Lythrac.).
Miediega Bub. = Dorycnium L. (Legumin.).
Miegia Neck. = Hieracium L. (Compos.).
Miegia Pers. = Arundinaria Michx (Gramin.).
Miegia Schreb. = Remirea Aubl. (Gramin.).
Miemianthera P. & K. = Meiemianthera Rafin. = Cytisus L. (Legumin.).
Miena P. & K. = Meiena Rafin. = Dendrophthoë Mart. (Loranthac.).
Mieria La Llave = Schkuhria Roth (Compos.).
Miersia Lindl. Alliaceae. 5 Bolivia, Chile.
Miersiella Urb. Burmanniaceae. 4 trop. S. Am.
Miersiophyton Engl. Menispermaceae. 1 trop. Afr.
Migandra O. F. Cook. Palmae. 1 C. Am.
Mikania Neck. = Perebea Aubl. (Urticac.).
Mikania F. W. Schmidt = Lactuca L. (Compos.).
*Mikania Willd. Compositae. 250 trop. Am., W.I.; 2 S. Afr. Twining herbs or shrubs, with opp. ls.
Mikaniopsis Milne-Redhead. Compositae. 5 trop. Afr.
Mikrobiota Kom. = Microbiota Kom. (Cupressac.).
Mila Britton & Rose. Cactaceae. 12 Peru.
Mildbraedia Pax. Euphorbiaceae. 3 trop. Afr.
Mildbraediochloa Butzin. Gramineae. 1 W. Equat. Afr. (Annobon Is.).
Mildbraediodendron Harms. Leguminosae. 1 trop. Afr.

MILDEA

Mildea Griseb. = Verhuellia Miq. (Piperac.).
Mildea Miq. = Paranephelium Miq. (Sapindac.).
Mildella Trevis. Sinopteridaceae. 1 C. Am.
Milhania Neck. = Ipomoea L. (Convolvulac.).
Milhania Rafin. = Calystegia R.Br. (Convolvulac.).
Miliaceae Burnett = Gramineae–Miliëae + Paniceae.
Miliarium Moench = Milium L. (Gramin.).
Miliastrum Fabr. = Setaria Beauv. (Gramin.).
Milicia Sim = Maclura Nutt. (Morac.).
Milium Adans. = Panicum L. (Gramin.).
Milium L. Gramineae. 3–4 N. temp.
Miliusa Leschen. ex A. DC. Annonaceae. 40 Indomal., Austr.
Milla Cav. Alliaceae. 6 S. U.S. to C. Am.
Milla Vand. = Herpestis Gaertn. (Scrophulariac.).
Millania Zipp. ex Blume = Pemphis J. R. & G. Forst. (Lythrac.).
Millea Standley = Eriotheca Schott & Endl. (Bombacac.).
Millea Willd. = Milla Cav. (Alliac.).
Millefolium Mill. = Achillea L. (Compos.).
Millegrana Adans. = Radiola Hill (Linac.).
Millegrana Juss. ex Turp. = Cypselea Turp. (Aïzoac.).
Millera St-Lag. = seq.
Milleria L. Compositae. 1 Mex., C. Am.
Milletia Meissn. = seq.
Millettia Wight & Arn. Leguminosae. 180 trop. & subtrop. (few Am.).
Milligania Hook. f. (1840) = Gunnera L. (Gunnerac.).
***Milligania** Hook. f. (1853). Liliaceae. 4 Tasmania.
Millina Cass. = Leontodon L. (Compos.).
Millingtonia L. f. Bignoniaceae. 1 SE. As.
Millingtonia Roxb. = Wellingtonia Meissn. = Meliosma Blume (Meliosmac.).
Millingtonia Roxb. ex D. Don = Maughania Jaume St-Hil. (Legumin.).
Millingtoniaceae Wight & Arn. = Meliosmaceae Endl.
Millotia Cass. Compositae. 4 temp. Austr.
Millspaughia Robinson = Gymnopodium Rolfe (Polygonac.).
Milnea Rafin. = ?Ardisia Sw. (Myrsinac.).
Milnea Roxb. = Aglaia Lour. (Meliac.).
× **Milpasia** hort. Orchidaceae. Gen. hybr. (iii) (Aspasia × Miltonia).
× **Milpilia** hort. Orchidaceae. Gen. hybr. (iii) (Miltonia × Trichopilia).
× **Miltassia** hort. Orchidaceae. Gen. hybr. (iii) (Brassia × Miltonia).
Miltianthus Bunge. Zygophyllaceae. 1 C. As.
Miltitzia A. DC. Hydrophyllaceae. 8 Pacif. N. Am.
× **Miltoglossum** hort. (vii) = × Odontonia hort. (Orchidac.).
× **Miltoncidium** hort. (vii) = × Miltonidium hort. (Orchidac.).
× **Miltonguezia** hort. (vii) = × Rodritonia hort. (Orchidac.).
***Miltonia** Lindl. Orchidaceae. 25 C. & trop. S. Am.
× **Miltonidium** hort. Orchidaceae. Gen. hybr. (iii) (Miltonia × Oncidium).
× **Miltonioda** hort. Orchidaceae. Gen. hybr. (iii) (Cochlioda × Miltonia).
× **Miltoniopsis** hort. (iv) = Miltonia Lindl. (Orchidac.).
× **Miltonpasia** hort. (vii) = × Milpasia hort. (Orchidac.).

× **Miltonpilia** hort. (vii) = × Milpilia hort. (Orchidac.).
Miltus Lour. = Gisekia L. (Aïzoac.).
Milula Prain. Alliaceae. 1 E. Himalaya. Infl. rac.!
Mimaecylon St-Lag. = Memecylon L. (Memecylac.).
Mimela Phil. = Leucheria Lag. (Compos.).
Mimetanthe Greene = Mimulus L. (Scrophulariac.).
Mimetes Salisb. Proteaceae. 16 S. Afr.
Mimetophytum L. Bolus. Aïzoaceae. 2 S. Afr.
Mimophytum Greenman. Boraginaceae. 1 Mex.
Mimosa L. Leguminosae. 450–500 trop. & subtrop. Am., a few in Afr. & As.
Mainly herbs and undershrubs, frequently with stipular thorns. *M. pudica* L.
(sensitive plant), now a common trop. weed, has a bipinnate l. with four
secondary petioles. It is exceedingly sensitive, and a touch or shake will make
it move rapidly into the position which it assumes at night. The leaflets move
upwards in pairs, closing against one another, the secondary petioles close up
against one another, and the main petiole drops through about 60°. After a
short time the movements are slowly reversed. They are effected by the aid
of a pulvinus or swollen joint at each point of movement. Each pulvinus can
be made to work independently of the rest by gentle stimulation, and the
propagation of the stimulus from pulvinus to pulvinus may also be seen. The
ribs of the fr. are frequently thorny and are usu. dropped on dehiscence.
***Mimos[ac]eae** R.Br. = Leguminosae–Mimosoïdeae Kunth. (Fls. reg.; K
valv., rarely imbr.; C valv. in bud; A 4–10 or ∞, pollen freq. cpd.; seeds usu.
with lat. areoles; ls. often bipinn.)
Mimosopsis Britton & Rose (~ Mimosa L.). Leguminosae. 25 S. U.S. to
Mimozyganthus Burkart. Leguminosae. 1 Argent.   [C. Am.
Mimulodes (Benth.) Kuntze = Mimetanthe Greene = Mimulus L. (Scrophu-
lariac.).
Mimulopsis Schweinf. Acanthaceae. 30 trop. Afr., Madag.
Mimulus Adans. = Rhinanthus L. (Scrophulariac.).
Mimulus L. Scrophulariaceae. 100 cosmop. (esp. Am.). *M. guttatus* DC.
(yellow monkey-flower) nat. in Eur. *M. moschatus* Dougl. ex Lindl., the musk-
plant, formerly cult. for its scent, is now scentless in Eur. Insects entering the
fl. touch first the stigma, which is sensitive to contact and closes up (cf.
Martynia).
Mimusops L. Sapotaceae. 57 trop. Afr., 1 Malaysia to Pacific. [*M. balata*
Crueg. (*M. globosa* Gaertn.; Guiana) yields a gutta-percha (*balata*). *M. elata*
Allem. is the Brazilian milk tree or *masseranduba*. The timber is hard and
durable, the fr. edible, 'but strangest of all is the vegetable milk, which exudes
in abundance when the bark is cut; it has about the consistence of thick cream'
(Wallace, *Amazon*, ch. 2). These spp. are now referred to *Manilkara* Adans.]
Mina Cerv. (~ Ipomoea L.). Convolvulaceae. 1 Mex. to trop. S. Am.
Minaea La Llave & Lex. = praec.
Minaea Lojacono = Pastorea Tod. ex Bertol. (Crucif.).
Minderera Ram. ex Schrad. = ? Merendera Ram. (Liliac.).
Mindium Adans. = Canarina L. (descr.) + Michauxia L'Hérit. (Campanulac.).
Mindium Rafin. = Canarina L. (Campanulac.).
Minguartia Miers (sphalm.) = Minquartia Aubl. (Olacac.).

MINKELERSIA

Minkelersia Mart. & Gal. Leguminosae. 4 Mex.

Minquartia Aubl. Olacaceae. 3 N. trop. S. Am.

Mintha St-Lag. = seq.

Minthe St-Lag. = Mentha L. (Labiat.).

Minthostachys (Benth.)Griseb.(~ Bystropogon L'Hérit.). Labiatae. 12 Andes.

Mintostachys Spach = praec.

Minuartia Loefl. ex L. Caryophyllaceae. 120 Arctic to Mex., Abyss., Himal.; 1 Chile.

Minuphylis Thou. (uninom.) = *Bulbophyllum minutum* Thou. (Orchidac.).

Minuria DC. Compositae. 8 C. & SE. Austr.

Minuriella Tate = praec.

Minurothamnus DC. Compositae. 1 S. Afr.

Minutalia Fenzl = Antidesma L. (Stilaginac.).

Minutia Vell. = Mayepea Aubl. (Oleac.).

Minyranthes Turcz. = Sigesbeckia L. (Compos.).

Minyria P. & K. = Minuria DC. (Compos.).

Minyrothamnus P. & K. = Minurothamnus DC. (Compos.).

Minythodes Phil. ex Benth. & Hook. f. = Chaetanthera Ruiz & Pav. (Compos.).

Miocarpidium P. & K. = Meiocarpidium Engl. & Diels (Annonac.).

Miocarpus Naud. = Acisanthera P.Br. (Melastomatac.).

Miogyna P. & K. = Meiogyne Miq. (Annonac.).

Mioluma P. & K. = Meioluma Baill. = Micropholis (Griseb.) Pierre (Sapotac.).

Mionandra Griseb. Malpighiaceae. 1 Boliv., Parag., Argent.

Mionectes P. & K. = Meionectes R.Br. = Haloragis J. R. & G. Forst. (Haloragidac.).

Mionula P. & K. = Meionula Rafin. = Utricularia L. (Lentibulariac.).

Mioperis P. & K. = Meioperis Rafin. = Passiflora L. (Passiflorac.).

Mioptrila Rafin. = Cedrela P.Br. (Meliac.) (fruits) + ? Zanthoxylum L. (Rutac.) (leaves & flowers).

Miosperma P. & K. = Meiosperma Rafin. = Justicia L. (Acanthac.).

Miquelia Arn. & Nees = Garnotia Brongn. (Gramin.).

Miquelia Blume = Stauranthera Benth. (Gesneriac.).

Miquelia Meissn. (= Jenkinsia Griff.). Icacinaceae. 5–6 Indoch., Indomal.

Miquelina Van Tiegh. = Macrosolen Bl. (Loranthac.).

Miqueliopuntia Frič = Opuntia Mill. (Cactac.).

Mira Colenso = Mida A. Cunn. ex Endl. (Santalac.).

Mirabellia Bert. ex Baill. = Dysopsis Baill. (Euphorbiac.).

Mirabilidaceae W. R. B. Oliv. = Nyctaginaceae Juss.

Mirabilis L. (incl. *Oxybaphus* L'Hérit.). Nyctaginaceae. 60 Am. At the base of the fl. is an involucre of 5 ls. resembling a K; it is really the bracts of a 3-fld. dich. cyme, of which in most only the central fl. is developed. In some, however, e.g. *M. coccinea* Benth. & Hook. f., the invol. encloses > 1 fl. The fl. opens in the evening and (in *M. jalapa* L. and other spp.) is protog., with ultimate autogamy on withering. The invol. often forms a parachute on the fr. The tuberous roots of *M. jalapa* L. (false jalap, four-o'clock, marvel of Peru) were formerly used as jalap.

Miracyllium P. & K. = Meiracyllium Reichb. f. (Orchidac.).

Miradoria Sch. Bip. ex Benth. & Hook. f. = Microspermum Lag. (Compos.).

Mirandaceltis A. J. Sharp = Gironniera Gaudich. (Ulmac.).
Mirandea Rzedowski. Acanthaceae. 1 Mex.
Mirasolia Sch. Bip. ex Benth. & Hook. f. = Tithonia Desf. ex Juss. + Gymno-
lomia Kunth (Compos.).
Mirbelia Sm. Leguminosae. 25 Austr.
Mirica Nocca = Myrica L. (Myricac.).
Miricacalia Kitamura. Compositae. 7 E. As.
Mirkooa Wight = Ammannia L. (Lythrac.).
Mirmau Adans. = Selaginella Beauv. (Selaginellac.).
Mirmecodia Gaudich. = Myrmecodia Jack (Rubiac.).
Mirobalanus Rumph. = Phyllanthus L. (Euphorbiac.).
Mirobalanus Steud. = Myrobalanus Gaertn. = Terminalia L. (Combretac.).
Miroxilum Blanco = seq.
Miroxylon Scop. = Myroxylon J. R. & G. Forst. = Xylosma G. Forst. [(Fla-
Miroxylum Blanco = praec.   courtiac.).
Mirtana Pierre = Arcangelisia Becc. (Menispermac.).
Misandra Comm. ex Juss. = Gunnera L. (Gunnerac.).
Misandra F. G. Dietr. = Tillandsia L. (Bromeliac.).
Misandropsis Oerst. = seq.
Misanora Urv. (sphalm.) = Misandra Comm. ex Juss. = Gunnera L. (Gun-
Misanteca Cham. & Schlechtd. = Licaria Aubl. (Laurac.).   [nerac.).
Misantheca auctt. = seq.
Misarrhena P. & K. = Meissarrhena R.Br. = Anticharis Endl. (Scrophulariac.).
Miscanthidium Stapf. Gramineae. 7 trop. & S. Afr.
Miscanthus Anderss. Gramineae. 20 trop. & S. Afr. to Japan & Philipp. Is.
Mischanthus Cass. = Mecomischus Coss. ex Benth. & Hook. f. (Compos.).
Mischanthus B. D. Jacks. (in I.K. sub Eulalia Trin.) = Miscanthus Anderss.
(Gramin.).
Mischobulbum Schlechter. Orchidaceae. 7 Himal. to China, Formosa,
Malaysia.
*Mischocarpus Blume. Sapindaceae. 25 China, SE. AS. Indomal., NE. Austr.
Mischocodon Radlk. Sapindaceae. 1 New Guinea.
Mischodon Thw. Euphorbiaceae(?). 1 S. India, Ceylon. Ls. vertic.
Mischogyne Exell. Annonaceae. 2 trop. Afr.
Mischolitzia Wendl. ex Burret (nomen). Palmae. Quid?
Mischolobium P. & K. = Miscolobium Vog. = Dalbergia L. (Legumin.).
Mischopetalum P. & K. = Miscopetalum Haw. = Saxifraga L. (Saxifragac.).
Mischophloeus Scheff. (~ Areca L.). Palmae. 1 Moluccas.
Mischopleura Wernh. ex Ridley = Sericolea Schltr. (Elaeocarpac.).
Mischospora Boeck. = Fimbristylis Vahl (Cyperac.).
Miscodendrum Steud. = Myzodendron Soland. ex DC. (Myzodendrac.).
Miscolobium Vog. = Dalbergia L. (Legumin.).
Miscopetalum Haw. = Saxifraga L. (Saxifragac.).
Misipus Rafin. = Elaeocarpus L. (Elaeocarpac.).
*Misodendraceae auctt. = Myzodendraceae J. G. Agardh.
Misodendron G. Don = seq.
Misodendrum DC. = Myzodendron Soland. ex DC. corr. R.Br. (Myzo-
dendrac.).

MISOPATES

Misopates Rafin. (~Antirrhinum L.). Scrophulariaceae. 2 Medit. to C. Verde Is., Abyss., NW. India.
Missiessia Benth. & Hook. f. = seq.
Missiessya Gaudich. = Leucosyke Zoll. (Urticac.).
Mistralia Fourr. = Daphne L. (Thymelaeac.).
Mistyllus Presl = Trifolium L. (Legumin.).
Mitchella L. Rubiaceae. 2 NE. As. & N. Am. Dimorphic heterostyled. The fls. are in pairs with united ovaries. Occasionally K and C also fuse and give a double ovary surmounted by a 10-lobed K and C (cf. *Lonicera*).
Mitcherlichia Klotzsch = Mitscherlichia Klotzsch = Begonia L. (Begoniac.).
Mitella L. Saxifragaceae. 15 E. Sib., Japan, N. Am. The inconspic. greenish fls. stand in unilateral racemes.
Mitellastra Howell = praec.
Mitellopsis Meissn. = praec.
Mitesia Rafin. = Polygonum L. (Polygonac.).
Mithracarpus Reichb. = Mitracarpum Zucc. (Rubiac.).
Mithridatea Comm. ex Schreb. = Tambourissa Sonner. (Monimiac.).
Mithridatium Adans. = Erythronium L. (Liliac.).
Mitina Adans. = Carlina L. (Compos.).
Mitodendron Walp. = Myzodendron Soland. ex DC. (Myzodendrac.).
Mitolepis Balf. f. Periplocaceae. 2 Socotra.
Mitopetalum Blume = Tainia Blume (Orchidac.).
Mitophyllum Greene. Cruciferae. 1 Calif.
Mitophyllum O. E. Schulz = Rhammatophyllum O. E. Schulz (Crucif.).
Mitostax Rafin. = Prosopis L. (Legumin.).
Mitostemma Mast. Passifloraceae. 3 trop. S. Am.
Mitostigma Blume = Amitostigma Schltr. (Orchidac.).
Mitostigma Decne. Asclepiadaceae. 20 trop. & temp. S. Am.
Mitostylis Rafin. = Cleome L. (Cleomac.).
Mitozus Miers = Echites R.Br. (Apocynac.).
Mitracarpum auctt. = seq.
Mitracarpus Zucc. Rubiaceae. 40 trop. S. Am., W.I., trop. & S. Afr.
Mitracme Schult. = Mitrasacme Labill. (Spigeliac.).
*****Mitragyna** Korth. Naucleaceae (~Rubiac.). 12 trop. Afr., As.
Mitragyne R.Br. = Mitrasacme Labill. (Spigeliac.).
Mitragyne Korth. = Mitragyna Korth. (Naucleac.).
Mitranthes Berg. Myrtaceae. 11 trop. Am., W.I.
Mitranthus Hochst. = Lindernia All. (Scrophulariac.).
*****Mitraria** Cav. Gesneriaceae. 1 Chile.
Mitraria J. F. Gmel. = Barringtonia J. R. & G. Forst. (Barringtoniac.).
Mitrasacme Labill. Spigeliaceae. 35 Formosa, S. China, SE. As., Indomal., Austr., N.Z.
Mitrasacmopsis Jovet. Rubiaceae. 1 Madag.
Mitrastemma Makino = seq.
Mitrastemon Makino. Rafflesiaceae. 1 Japan, 2 Formosa, 1 Indoch., 1 Borneo, 1 Sumatra; 2 Mex., C. Am.
*****Mitrastemonaceae** Mak. = Rafflesiaceae–Mitrastemoneae (Mak.) Hay.
Mitrastigma Harv. = Plectronia L. (Rubiac.).

750

Mitrastylus Alm & T. C. E. Fries. Ericaceae. 2 Madag.
Mitratheca K. Schum. Rubiaceae. 1 trop. Afr.
Mitrella Miq. Annonaceae. 5 Malaysia.
Mitreola Boehm. = Ophiorrhiza L. (Rubiac.).
Mitreola L. ex Schaeffer. Spigeliaceae. 6 Madag., Indomal., Austr., Am.
Mitrephora (Bl.) Hook. f. & Thoms. Annonaceae. 25 trop. SE. As., W. Malaysia.
Mitriostigma Hochst. Rubiaceae. 3 trop. & S. Afr.
Mitrocarpa Torr. ex Steud. = Eleocharis Lestib. (Cyperac.).
Mitrocarpum Hook. = Mitracarpum Zucc. (Rubiac.).
Mitrocarpus P. & K. (1) = Mitracarpum Zucc. (Rubiac.).
Mitrocarpus P. & K. (2) = Mitrocarpa Torr. ex Steud. = Eleocharis Lestib. (Cyperac.).
Mitrocereus (Backeb.) Backeb. emend. Bravo. Cactaceae. 1 Mex.
Mitrogyna P. & K. (1) = Mitragyne R.Br. = Mitrasacme Labill. (Spigeliac.).
Mitrogyna P. & K. (2) = Mitragyna Korth. (Rubiac.).
Mitrophora Neck. ex Rafin. = Fedia Gaertn. (Valerianac.).
Mitrophyllum Schwantes. Aïzoaceae. 23 S. Afr.
Mitropsidium Burret. Myrtaceae. 9 trop. S. Am.
Mitrosacma P. & K. = Mitrasacme Labill. (Spigeliac.).
Mitrosicyos Maxim. = Actinostemma Griff. (Cucurbitac.).
Mitrospora Nees = Rhynchospora Vahl corr. Willd. (Cyperac.).
Mitrostigma P. & K. = Mitrastigma Harv. = Plectronia L. (Rubiac.).
Mitrotheca P. & K. = Mitratheca K. Schum. (Rubiac.).
Mitsa Chapel. ex Benth. = Coleus Lour. (Labiat.).
Mitscherlichia Klotzsch = Begonia L. (Begoniac.).
Mitscherlichia Kunth = Neea Ruiz & Pav. (Nyctaginac.).
Mitwabachloa Phipps = Zonotriche (C. E. Hubbard) Phipps (Gramin.).
Mixandra Pierre ex L. Planch. = Diploknema Pierre (Sapotac.).
Miyakea Miyabe & Tatewaki. Ranunculaceae. 1 Sakhalin.
Miyoshia Makino = Petrosavia Becc. (Petrosaviac.).
Miyoshiaceae Mak. = Petrosaviaceae Hutch.
Mizonia A. Chev. Amaryllidaceae. 1 W. Equat. Afr.
Mizotropis P. & K. = Meizotropis Voigt = Butea Koen. ex Roxb. (Legumin.).
× **Mizutara** hort. Orchidaceae. Gen. hybr. (iii) (Cattleya × Caularthron [Diacrium] × Schomburgkia).
Mkilua Verdc. Annonaceae. 1 trop. E. Afr.
Mnasium Schreb. = Rapatea Aubl. (Rapateac.).
Mnasium Stackhouse = Ensete Bruce (Musac.).
Mnassea Vell. Inc. sed. (? Sapindac.). 1 Brazil.
Mnemion Spach = Viola L. (Violac.).
Mnemosilla Forsk. = Hypecoum L. (Hypecoac.).
Mnesiteon Rafin. = Eclipta L. (Compos.).
Mnesithea Kunth. Gramineae. 5 Indomal.
Mnesitheon Spreng. = Mnesiteon Rafin. = Eclipta L. (Compos.).
Mnianthus Walp. = Terniola Tul. (Podostemac.).
Mniarum J. R. & G. Forst. = Scleranthus L. (Caryophyllac.).
Mniochloa Chase. Gramineae. 2 Cuba.

MNIODES

Mniodes A. Gray. Compositae. 5 Peru.
Mniopsis Mart. & Zucc. Podostemaceae. 5 Brazil.
Mniothamnea (Oliv.) Niedenzu. Bruniaceae. 2 S. Afr.
Mniothamnus Willis = praec.
Mniothamus Dur. & Jacks. = praec.
Moacroton Croizat. Euphorbiaceae. 6 Cuba.
Moacurra Roxb. = Dichapetalum Thou. (Dichapetalac.).
Mobilabium Rupp. Orchidaceae. 1 Queensland.
Mocanera Blanco = Dipterocarpus Gaertn. f. + Anisoptera Korth. (Dipterocarpac.).
Mocanera Juss. = Visnea L. f. (Theac.).
Mocinia DC. = Stifftia Mikan (Compos.).
Mocinna Benth. = Mozinna Ortega = Jatropha L. (Euphorbiac.).
Mocinna Lag. = Calea L. (Compos.).
Mocinna La Llave = Jarilla Rusby (Caricac.).
Mocquerysia Hua. Flacourtiaceae. 1 trop. Afr. Infl. epiphyllous.
Mocquinia Steud. = Moquinia Spreng. = Loranthus L. (Loranthac.).
Modanthos Alef. = Modiola Moench (Malvac.).
Modeca Rafin. = Modesta Rafin. = Ipomoea L. (Convolvulac.).
Modecca Lam. = Adenia Forsk. (Passiflorac.).
Modeccaceae J. G. Agardh = Passifloraceae Juss.
Modeccopsis Griff. = Erythropalum Blume (Erythropalac.).
Modecopsis Griff. = praec.
Modesta Rafin. = Ipomoea L. (Convolvulac.).
Modestia Charadze & Tamamsch. (~ Cirsium Mill.). Compositae. 2–3 C. As.
Modiola Moench. Malvaceae. 1 Am.
Modiolastrum K. Schum. Malvaceae. 5 S. Am.
Modira Rafin. = ? Annona L. (Annonac.).
Moehnia Neck. = Gazania Gaertn. (Compos.).
Moehringella (Franch.) Neumayer = Arenaria L. (Caryophyllac.).
Moehringia L. (~ Arenaria L.). Caryophyllaceae. 20 N. temp.
Moelleria Scop. = Casearia Jacq. (Flacourtiac.).
***Moenchia** Ehrh. Caryophyllaceae. 6 W. & C. Eur., Medit.
Moenchia Medik. = Allium L. (Alliac.).
Moenchia Neck. = Cucubalus L. (Caryophyllac.).
Moenchia Roth = Alyssum L. (Crucif.).
Moenchia Wender. ex Steud. = Paspalum L. (Gramin.).
Moerenhoutia Blume. Orchidaceae. 10 New Guinea, Polynesia.
Moerhingia L. = Moehringia L. (Caryophyllac.).
Moerkensteinia Opiz = Senecio L. (Compos.).
Moeroris Rafin. = Phyllanthus L. (Euphorbiac.).
Moesa Blanco = Maesa Forsk. (Myrsinac.).
Moesslera Reichb. = Tittmannia Brongn. (Bruniac.).
Moghamia Steud. = seq.
Moghania Jaume St-Hil. = Maughania Jaume St-Hil. (Legumin.).
Mogiphanes Mart. = Alternanthera Forsk. (Amaranthac.).
Mogoltavia Korovin. Umbelliferae. 1 C. As.
Mogori Adans. = Jasminum L. (Oleac.).

Mogorium Juss. = praec.
Mohadenium Dur. & Jacks. (sphalm.) = Monadenium Pax (Euphorbiac.).
Mohavea A. Gray. Scrophulariaceae. 2 SW. U.S.
Moheringia Zumaglini (sphalm.) = Moehringia L. (Caryophyllac.).
Mohlana Mart. = Hilleria Vell. (Phytolaccac.).
Mohria Britton = Mohrodendron Britt. = Halesia L. (Styracac.).
Mohria Sw. Schizaeaceae. 3 trop. & S. Afr. Sporangia solitary on under side of ordinary ls., margins turned back over them (cf. *Negripteris*).
Mohriaceae Presl = Schizaeaceae Mart.
Mohrodendron Britton = Halesia L. (Styracac.).
× **Moirara** hort. Orchidaceae. Gen. hybr. (iii) (Phalaenopsis × Renanthera × Vanda).
× **Mokara** hort. Orchidaceae. Gen. hybr. (iii) (Arachnis × Ascocentrum × Vanda).
Mokof Adans. = Ternstroemia Mutis ex L. f. (Theac.).
Mokofua Kuntze = praec.
Moldavica Fabr. = Dracocephalum L. (Labiat.).
Moldenhauera Spreng. = Pyrenacantha Wight (Icacinac.).
Moldenhawera Schrad. Leguminosae. 3 Brazil, Venezuela.
Moldenhaweria Steud. = praec.
Moldenkea Traub. Amaryllidaceae. 1 trop. W. Afr.
Molina Cav. = Hiptage Gaertn. (Malpighiac.).
Molina C. Gay = Dysopsis Baill. (Euphorbiac.).
Molina Giseke = Molinaea Comm. ex Juss. (Sapindac.).
Molina Ruiz & Pav. = Baccharis L. (Compos.).
Molinadendron P. K. Endress. Hamamelidaceae. 3 Mex., C. Am.
Molinaea Bertero = Jubaea Kunth (Palm.).
Molinaea Comm. ex Brongn. = Retanilla Brongn. (Rhamnac.).
Molinaea Comm. ex Juss. Sapindaceae. 10 Madag., Masc.
Molinaea St-Lag. = Molinia Schrank (Gramin.).
Molineria Colla (~ Curculigo Gaertn.). Hypoxidaceae. 7 Indomal.
Molineria Parl. = seq.
Molineriella Rouy = Periballia Trin. (Gramin.).
Molinia Schrank. Gramineae. 2–3 temp. Euras. *M. caerulea* (L.) Moench, char. of wet grass moors.
Moliniera Ball (sphalm.) = Molineria Parl. = Periballia Trin. (Gramin.).
Moliniopsis Gandog. = Cleistogenes Keng (Gramin.).
Moliniopsis Hayata. Gramineae. 3 E. As.
Molium Fourr. = Moly Mill. = Allium L. (Alliac.).
Molkenboeria De Vriese = Scaevola L. (Goodeniac.).
Molle Mill. = Schinus L. (Anacardiac.).
Mollera O. Hoffm. Compositae. 2 trop. Afr.
Mollia J. F. Gmel. = Baeckea L. (Myrtac.).
*****Mollia** Mart. Tiliaceae. 15 trop. S. Am.
Mollia Willd. = Polycarpaea Lam. (Caryophyllac.).
Mollinedia Ruiz & Pav. Monimiaceae. 90 C. & trop. S. Am.
Molloya Meissn. = Grevillea R.Br. (Proteac.).
*****Molluginaceae** Wight = Aïzoaceae–Mollugineae K. Müll.

MOLLUGO

Mollugo Fabr. = Galium L. (Rubiac.).

Mollugo L. Aïzoaceae. 20 trop. & subtrop.

Mollugophytum M. E. Jones = Drymaria Willd. ex Roem. & Schult. (Caryophyllac.).

Molongum Pichon. Apocynaceae. 6 trop. S. Am.

Molopanthera Turcz. Rubiaceae. 1 E. Brazil.

Molopospermum Koch. Umbelliferae. 1 W. Medit.

Molospermum Steud. = praec.

Molpadia Cass. = Buphthalmum L. (Compos.).

Moltkia Lehm. Boraginaceae. 3 N. Italy to N. Greece; 3 As. Min. to NW. Persia.

Moltkiopsis I. M. Johnston. Boraginaceae. 1 N. Afr. to Persia.

Molubda Rafin. = Plumbago L. (Plumbaginac.).

Molucca Mill. = seq.

Moluccella L. Labiatae. 4 Medit. to NW. India.

Molucella Juss. = praec.

Moluchia Medik. = Melochia L. (Sterculiac.).

Moly Mill. = Allium L. (Alliac.).

Molyza Salisb. = praec.

Momisia F. G. Dietr. = Celtis L. (Ulmac.).

Mommsenia Urb. & Ekman. Melastomataceae. 1 Haiti.

Momordica L. Cucurbitaceae. 45 palaeotrop.

Mona Ö. Nilsson. Portulacaceae. 1 Colombia, Venez.

Monacanthus G. Don = Monachanthus Lindl. = Catasetum Rich. (Orchidac.).

Monacather Benth. = Monachather Steud. = Danthonia DC. (Gramin.).

Monachanthus Lindl. = Catasetum Rich. (Orchidac.).

Monachather Steud. = Danthonia DC. (Gramin.).

Monachne Beauv. = Eriochloa Kunth + Panicum L. (Gramin.).

Monachochlamys Baker = Mendoncia Vell. (Mendonciac.).

Monachosoraceae Ching. Aspidiales(?). Stem dictyostelic, radially organized; petiole bases hairy, unicellular hairs on lamina, sori small, lacking indusia, spores trilete. 1 gen., *Monachosorum*. Thought by Christensen, Ching and Hayata to be allied to *Thelypteris*, by Holttum & Copeland to be allied to *Dennstaedtia*, not close to either (*J. Linn. Soc., Bot.* **53**: 158, 1947).

Monachosorella Hayata = seq.

Monachosorum Kunze. Monachosoraceae. 5 E. warm As.

Monachyron Parl. Gramineae. 1 Cape Verde Is.

Monactineirma Bory = Passiflora L. (Passiflorac.).

Monactinocephalus Klatt = Inula L. (Compos.).

Monactis Kunth. Compositae. 4 trop. S. Am.

Monadelphanthus Karst. = Capirona Spruce (Rubiac.).

Monadenia Lindl. Orchidaceae. 20 trop. & S. Afr.

Monadenium Pax. Euphorbiaceae. 47 E. trop. Afr., with outliers in C. & SW. trop. Afr. (Bally, *The Genus Monadenium*, 1961.)

Monadenus Salisb. = Zigadenus Michx (Liliac.).

Monandraira E. Desv. = Deschampsia Beauv. (Gramin.).

Monandriella Engl. Podostemaceae. 1 W. Equat. Afr.

Monandrodendraceae Barkley = Lacistemataceae Mart.

Monandrodendron Mansfeld = Lozania S. Mutis ex Caldas (Lacistematac.).
Monanthella A. Berger = Sedum L. (Crassulac.).
Monanthemum Griseb. = Piptocarpha R.Br. (Compos.).
Monanthemum Scheele = Morisia Gay (Crucif.).
Monanthes Haw. Crassulaceae. 12 Canaries, Morocco.
Monanthium Ehrh. (uninom.) = *Pyrola uniflora* L. = *Moneses uniflora* (L.) Salisb. (Pyrolac.).
Monanthium House = Moneses Salisb. (Pyrolac.).
Monanthochloë Engelm. Gramineae. 1 S. U.S., W.I.
Monanthocitrus Tanaka. Rutaceae. 2 New Guinea.
Monanthotaxis Baill. Annonaceae. 56 trop. Afr.
Monarda L. Labiatae. 12 N. Am., Mex. Sta. 2. Fl. protandrous, visited by bees (and humming-birds in the red spp.). The ls. of some are used medicinally in the form of tea (Oswego tea).
Monardella Benth. Labiatae. 20 W. N. Am.
Monaria Korth. ex Valeton = Erythropalum Bl. (Erythropalac.).
Monarrhenus Cass. Compositae. 3 Madag., Masc.
Monarthrocarpus Merr. Leguminosae. 1 Philipp. Is.
Monastes Rafin. = Centranthus DC. (Valerianac.).
Monastinocephalus Klatt (sphalm.) = Monactinocephalus Klatt = Inula L. (Compos.).
Monathera Rafin. = Ctenium Panz. (Gramin.).
Monavia Adans. = Mimulus L. (Scrophulariac.).
Monbin Mill. = Spondias L. (Anacardiac.).
Mondia Skeels. Periplocaceae. 2 trop. Afr.
Mondo Adans. = Ophiopogon Ker-Gawl. (Liliac.).
Monechma Hochst. Acanthaceae. 60 trop. Afr., 1 India.
Monelasmum Van Tiegh. = Ouratea Aubl. (Ochnac.).
Monelasum Willis (sphalm.) = praec.
Monella Herb. = Cyrtanthus Ait. (Amaryllidac.).
Monelytrum Hackel ex Schinz. Gramineae. 2 SW. Afr.
Monencyanthes A. Gray = Helipterum DC. (Compos.).
Monenteles Labill. = Pterocaulon Ell. (Compos.).
Monerma Beauv. Gramineae. 1 Medit. to Iraq.
Moneses Salisb. Pyrolaceae. 1 boreal & arctic.
Monestes P. & K. = Monoëstes Salisb. = Lachenalia Jacq. (Liliac.).
Monetaria Bronn = Dalbergia L. f. (Legumin.).
Monetia L'Hérit. = Azima L. (Salvadorac.).
Monfetta Neck. = Mouffetta Neck. = Patrinia Juss. (Valerianac.).
Mongesia Miers = seq.
Mongezia Vell. = Symplocos L. (Symplocac.).
Mongorium Desf. = Mogorium Juss. = Jasminum L. (Jasminac.).
Monguia Chapel. ex Baill. = Croton L. (Euphorbiac.).
Moniera P.Br. = Bacopa Aubl. (Scrophulariac.).
Moniera Loefl. (= Ertela Adans.). Rutaceae. 2 trop. S. Am.
Monieria Loefl. = praec.
Monilaria Schwantes. Aïzoaceae. 12 S. Afr.
Monilia S. F. Gray = Molinia Schrank (Gramin.).

MONILIFERA

Monilifera Adans. = Osteospermum L. (Compos.).

Monilistus Rafin. = Populus L. (Salicac.).

Monimia Thou. Monimiaceae. 4 Madag., Masc.

*Monimiaceae Juss. Dicots. 20/150, chiefly S. trop., and esp. in the 'oceanic' floral regions (Madag., Austr., Polynes.). Shrubs and trees, with opp. exstip. leathery evergr. ls., often resiniferous with aromatic scent. Fls. sol. or in cymes, perigynous, commonly unisexual, reg.; often the two sexes differ in the hollowing of the axis. Frequently the bud opens by throwing off the outer ends of the P-leaves as a sort of lid. P 4–∞, simple, or o; A ∞ or few, the anthers intr. or extr., opening by slits; G usu. ∞, sometimes few or 1, each with 1 usu. basal erect (rarely pend.) anatr. ovule. Fr. of achenes, often ± enclosed in or borne on a fleshy recept. Embryo straight, in copious endosp. The fam. forms a connecting link between *Lauraceae* and the '*Ranales*', being allied on one side to *Atherospermataceae* and *L.*, on the other to *Calycanthaceae*. Chief genera: *Hedycarya, Peumus, Tambourissa, Mollinedia, Xymalos*.

Monimiopsis Vieill. ex Perk. = Hedycarya J. R. & G. Forst. (Monimiac.).

Monimopetalum Rehder. Celastraceae. 1 China.

Monina Pers. = Monnina Ruiz & Pav. (Polygalac.).

Monipsis Rafin. = Teucrium L. (Labiat.).

Monium Stapf. Gramineae. 7 W. Afr.

Monixus Finet = Angraecum Bory (Orchidac.):

Monizia Lowe = Melanoselinum Hoffm. (Umbellif.).

Monnella Salisb. = Monella Herb. = Cyrtanthus Ait. (Amaryllidac.).

Monniera Juss. = Moniera Loefl. (Rutac.).

Monniera P. & K. = Moniera P.Br. = Bacopa Aubl. (Scrophulariac.).

Monnieria L. = Moniera Loefl. (Rutac.).

Monnina Ruiz & Pav. Polygalaceae. 150 Mex. to Chile. One of the two cpls. is usu. rudimentary. Fr. indehiscent.

Monnuria Nees & Mart. (sphalm.) = Monnieria L. = Moniera Loefl. (Rutac.).

Monobothrium Hochst. = Swertia L. (Gentianac.).

Monocallis Salisb. = Scilla L. (Liliac.).

Monocardia Pennell = Herpestis Gaertn. f. = Bacopa Aubl. (Scrophulariac.).

Monocarpia Miq. Annonaceae. 1 Siam, W. Malaysia.

Monocaryum R.Br. = Colchicum L. (Liliac.).

Monocelastrus Wang & Tang (~ Celastrus L.). Celastraceae. 2 Himal., W. China.

Monocephalium S. Moore = Pyrenacantha Wight (Icacinac.).

Monocera Ell. = Ctenium Panz. (Gramin.).

Monocera Jack = Elaeocarpus L. (Elaeocarpac.).

Monoceras auct. ex Steud. = praec.

Monoceras Steud. = Velleia Sm. (Goodeniac.).

Monochaete Doell = Gymnopogon Beauv. (Gramin.).

*Monochaetum (DC.) Naud. Melastomataceae. 50 W. trop. Am. Sta. dimorphic. The style, at first bent down, moves slowly up till horiz.

Monochasma Maxim. ex Franch. & Sav. Scrophulariaceae. 4 E. As.

Monochila Spach = Goodenia Sm. (Goodeniac.).

Monochilon Dulac = Teucrium L. (Labiat.).

Monochilus Fisch. & Mey. Verbenaceae. 1 Brazil.

Monochilus Wall. ex Lindl. = Zeuxine Lindl. (Orchidac.).
Monochlaena Cass. = Eriocephalus L. (Compos.).
Monochlaena Gaudich. = Didymochlaena Desv. (Aspidiac.).
Monochlamydeae DC. = Apetalae Juss.
Monochoria C. Presl. Pontederiaceae. 3 NE. Afr. to Manchuria & Austr.
Monochosma Dur. & Jacks. (sphalm.) = Monochasma Maxim. ex Franch. & Sav. (Scrophulariac.).
Monococcus F. Muell. Phytolaccaceae. 1 Austr., New Caled.
Monocodon Salisb. = Fritillaria L. (Liliac.).
Monocosmia Fenzl. Portulacaceae. 1 Chile, Patag.
Monocostus K. Schum. Costaceae. 1 Peru.
Monocotyledones Juss. One of the two great divisions (classes) of *Angiospermae*. Embryo with 1 cotyledon; ls. mostly parallel-nerved; fls. predominantly trimerous. The classification of the Monocots. is less difficult than that of the Dicotyledons, and a comparison should be made of the ways in which it is done in the various systems. There is an evident approach between the Monocots. and Dicots. *via* the *Alismatac.* (Monoc.) and *Ranunculac.* (Dicot.). The similarities can scarcely be fortuitous, and probably indicate real relationship.
Monocyclanthus Keay. Annonaceae. 1 trop. W. Afr.
Monocyclis Wall. ex Voigt = Walsura Roxb. (Meliac.).
Monocymbium Stapf. Gramineae. 4 trop. & S. Afr.
Monocystis Lindl. = Alpinia Roxb. (Zingiberac.).
Monodiella Maire. Gentianaceae. 1 Sahara.
Monodora Dun. Annonaceae. 20 trop. Afr., Madag. Berry with woody epicarp. Seeds of *M. myristica* (Gaertn.) Dun. sometimes used as nutmegs.
Monodoraceae J. G. Agardh = Annonaceae–Monodoroïdeae Diels.
Monodyas (K. Schum.) Kuntze = Halopegia K. Schum. (Zingiberac.).
Monodynamis J. F. Gmel. = Usteria Willd. (Antoniac.).
Monodynamus Pohl = Anacardium L. (Anacardiac.).
Monoëstes Salisb. = Lachenalia Jacq. (Liliac.).
Monogereion Barroso & King. Compositae. 1 NE. Brazil.
Monogonia Presl = Goniopteris Presl (Thelypteridac.).
Monogramma Schkuhr. Vittariaceae. 2 Masc. Is., Malaya. Smallest of all ferns (see also *Vaginularia* Fée).
Monographidium Presl = Cliffortia L. (Rosac.).
Monographis Thou. (uninom.) = *Limodorum concolor* Thou. = *Graphorkis concolor* (Thou.) Kuntze (Orchidac.).
Monogynella Des Moul. = Cuscuta L. (Cuscutac.).
Monolena Triana. Melastomataceae. 6 C. & trop. S. Am.
Monolepis Schrad. Chenopodiaceae. 6 C. & NE. arct. As., N. Am., Patag.
Monolix Rafin. = Callirhoë Nutt. (Malvac.).
Monolophus Wall. = Kaempferia L. + Caulokaempferia Larsen (Zingiberac.).
Monolopia DC. Compositae. 4 Calif.
Monomelangium Hayata = Diplazium Sw. (Athyriac.).
Monomeria Lindl. Orchidaceae. 4 E. Himal., SE. As.
Monomesia Rafin. = Coldenia L. (Ehretiac.).

MONONEURIA

Mononeuria Reichb. = Minuartia L. (Caryophyllac.).
Monoön Miq. = Polyalthia Blume (Annonac.).
Monopanax Regel = Oreopanax Decne & Planch. (Araliac.).
Monopetalanthus Harms. Leguminosae. 10 trop. Afr.
Monophalacrus Cass. = Tessaria Ruiz & Pav. (Compos.).
Monopholis S. F. Blake. Compositae. 4 Ecuador, Peru.
Monophrynium K. Schum. Marantaceae. 3 Philippines, Moluccas.
Monophyllaea R.Br. Gesneriaceae. 15 Malaysia.
Monophyllanthe K. Schum. Marantaceae. 1 Guiana.
Monophyllon Delarbre = Maianthemum Weber (Liliac.).
Monophyllorchis Schlechter. Orchidaceae. 1 Colombia.
Monoplectra Rafin. = Sesbania Scop. (Legumin.).
Monoplegma Piper. Leguminosae. 1 Costa Rica.
Monoploca Bunge = Lepidium L. (Crucif.).
Monopogon J. & C. Presl = Tristachya Nees (Gramin.).
Monoporandra Thw. = Stemonoporus Thw. (Dipterocarpac.).
Monoporidium Van Tiegh. = Ochna L. (Ochnac.).
Monoporina J. Presl = Marila Sw. (Guttif.).
Monoporus A. DC. Myrsinaceae. 8 Madag.
Monopsis Salisb. Campanulaceae. 18 trop. & S. Afr.
Monoptera Sch. Bip. = Chrysanthemum L. (Compos.).
Monopteris Klotzsch ex Radlk. = Matayba Aubl. (Sapindac.).
Monopteryx Spruce. Leguminosae. 3 trop. S. Am.
Monoptilon Torr. & Gray. Compositae. 2 deserts SW. U.S. & N. Mex.
Monopyle Moritz ex Benth. & Hook. f. Gesneriaceae. 18 C. Am. to Peru.
Monopyrena Speg. (= ? Thryothamnus Phil.). Verbenaceae. 1 Patag.
Monorchis Seguier (Herminium Guett.). Orchidaceae. 30 temp. Euras., Siam, Philipp., Java.
Monorchis Ehrh. (uninom.) = *Ophrys monorchis* L. = *Herminium monorchis* (L.) R.Br. (Orchidac.).
Monosalpinx N. Hallé. Rubiaceae. 1 trop. W. Afr.
Monoschisma Brenan. Leguminosae. 2 S. Am.
Monosemeion Rafin. = Amorpha L. (Legumin.).
Monosepalum Schlechter. Orchidaceae. 3 New Guinea.
Monoseris Gaertn. ex Reichb. (nomen). Compositae–Ligulatae. Quid?
Monosis DC. = Vernonia Schreb. (Compos.).
Monosoma Griff. = Carapa Aubl. (Meliac.).
Monospora Hochst. = Trimeria Harv. (Flacourtiac.).
Monostachya Merr. Gramineae. 1 Philipp. Is., New Guinea.
Monostemma Turcz. = Cynanchum L. (Asclepiadac.).
Monostemon Hackel ex Henrard = Briza L. (Gramin.).
Monosteria Rafin. = Hoppea Willd. (Gentianac.).
Monostiche Koern. = Calathea G. F. W. Mey. (Marantac.).
Monostylis Tul. = Apinagia Tul. (Podostemac.).
Monotagma K. Schum. Marantaceae. 20 trop. S. Am.
Monotassa Salisb. = Urginea Steinh. (Liliac.).
Monotaxis Brongn. Euphorbiaceae. 10 Austr.
Monoteles Rafin. = Bauhinia L. (Legumin.).

Monotes A. DC. Dipterocarpaceae. 48 trop. Afr.

Monothactum B. D. Jacks. (sphalm.) = Monochaetum Naud. (Melastomatac.).

Monotheca A. DC. = Reptonia A. DC. (Sapotac.).

Monothecium Hochst. Acanthaceae. 3 trop. Afr., Madag., S. India, Ceylon.

Monothrix Torr. = Perityle Benth. (Compos.).

Monothylacium G. Don = Hoodia Sweet ex Decne (Asclepiadac.).

Monotoca R.Br. Epacridaceae. 8 Austr.

Monotrema Koernicke. Rapateaceae. 3–5 Colombia, Venez., Brazil.

Monotris Lindl. = Holothrix Rich. (Orchidac.).

Monotropa L. Monotropaceae. 5 N. temp. *M. hypopitys* L. (yellow bird's-nest), in pine, birch and beech woods, a yellowish saprophyte with scaly ls. and a short term. raceme of fls. Below the soil is found a very much branched root system, the roots being covered with a superficial mycorhiza by whose aid absorption takes place. Buds are formed adv. upon the roots and lengthen into the flowering shoots.

*****Monotropaceae** Nutt. (~ Pyrolaceae Lindl.). Dicots. 12/21 N. temp., trop. mts. Parasitic ± fleshy herbs devoid of chloroph.; ls. reduced to scales, alt., numerous, uppermost often forming an involucre. Fls. ⚥, reg., rac., capit., or sol. K 2–6 (or indisting. from bracts); C 3–6 or (3–6), imbr. or ± contorted, rarely o; A 6–12, fil. sometimes connate at base, anth. opening by longit. chinks or slits; disk sometimes present; G̲ (4–6), 1–6-loc., with ∞ minute ov. on ax. or pariet. plac.; style various, with capit. stig. Fr. a loculic. caps., with ∞ minute seeds; endosp. copious. Chief genera: *Monotropa, Cheilotheca, Pleuricospora, Monotropastrum.*

Monotropanthum H. Andres. Monotropaceae. 1 Himal. to Japan.

Monotropastrum H. Andres. Monotropaccae. 4 Himal., China, Formosa.

Monotropsis Schweinitz ex Elliott. Monotropaceae. 1 N. Am.

Monoxalis Small = Oxalis L. (Oxalidac.).

Monoxora Wight = Rhodamnia Jack (Myrtac.).

Monroa Torr. = Munroa Torr. corr. Benth. & Hook. f. (Gramin.).

Monrosia Grondona. Polygalaceae. 1 Argent.

Monsonia L. Geraniaceae. 40 Afr., SW. As., NW. India. A 15 in 5 bundles.

Monssonia L. = praec.

Monstera Adans. = Dracontium L. (Arac.).

*****Monstera** Schott. Araceae. 50 trop. Am., W.I. Lianes with ent. or variously perforated, more rarely pinnatifid ls. When very young the l. is entire; then the tissue between the veins ceases to grow rapidly, becomes dry and tears away, thus leaving holes between the ribs; at the edge the marginal part usually breaks, and thus the outermost hole gives rise to a notch in the l., which becomes pinnated. Beginning as a climber the pl. usu. ends as an epiph. with aerial roots to the soil. Fls. ⚥. The fr. of *M. deliciosa* Liebm. is ed.

Montabea Roem. & Schult. (sphalm.) = Moutabea Aubl. (Polygalac.).

Montagnaea DC. = Montañoa Cerv. (Compos.).

Montagueia Baker f. Anacardiaceae. 1 New Caled.

Montalbania Neck. = Clerodendrum L. (Verbenac.).

Montañoa Cerv. Compositae. 50 Mex. to Colombia.

Montbretia DC. = Tritonia Ker-Gawl. (Iridac.).

Montbretiopsis L. Bolus (~ Gladiolus L). Iridaceae. 1 S. Afr.

MONTEIROA

Monteiroa Krapov. Malvaceae. 5 Brazil, Urug., Argent.
Montejacquia Roberty = Jacquemontia Choisy (Convolvulac.).
Montelia A. Gray = Amaranthus L. (Amaranthac.).
Monteverdia A. Rich. = Maytenus Mol. (Celastrac.).
Montezuma (Moç. & Sessé ex) DC. Bombacaceae. 1 Mex.
Montia L. Portulacaceae. 15 N. & S. Am., temp. Euras., mts. trop. Afr.,
Austr. *M. fontana* L., an annual herb, usu. in wet places, with small cymes of
fls. In bad weather or when submerged they become cleistogamic. The stalk
moves like that of *Claytonia* (*q.v.*), and the fr. explodes in the same way. Eaten
as salad.
Montia Mill. = Heliocarpus L. (Tiliac.).
Montiaceae Dum. = Portulacaceae–Montioïdeae Franz.
Montiastrum (A. Gray) Rydb. Portulacaceae. 4 NE. As. to NW. N. Am.
Montinia Thunb. Montiniaceae. 1 S. Afr.
Montiniaceae (Engl.) Nak. Dicots. 2/4 SW. & trop. E. Afr., Madag. Shrubs
or small trees, with opp., subopp. or alt. simple ent. exstip. ls. Fls. ♂ ♀, dioec.,
♂ in few-fld. corymbose term. or axill. cymes, ♀ sol. term. ♂: K (3–5), cupular
or flattened, ent. or shortly 3–5-lobed; C 3–5, imbr., slightly fleshy, decid.;
A 3–5, alternipet., fil. short, thick, anth. rather large, ellipsoid, dorsifixed,
extr.; disk flat, discoid; pistillode o or minute. ♀: K (4–5), adnate to G, limb
shortly tubular, ent. or minutely 4–5-toothed; C 4–5, imbr., fleshy, decid.;
stds. 4–5; disk epig., fleshy, 4–5-angled; Ḡ (2), with short thick persist. style
and 2 large stigs., and 2–6 axile anatr. erect or pend. 1–2-seriate ov. per loc.
Fr. a loculic. caps., or indehisc.; seeds compressed and winged, or subglob.,
endosp. copious or o. Genera: *Montinia, Grevea*. An interesting relict group,
with possible connections with *Celastrac., Cucurbitac., Onagrac.* and *Escal-
loniac.*
Montiopsis Kuntze. Portulacaceae. 1 Bolivia.
Montira Aubl. = Spigelia L. (Spigeliac.).
Montjolya Friesen = Cordia L. (Ehretiac.).
Montolivaea Reichb. f. = Habenaria Willd. (Orchidac.).
Montravelia Montrouz. ex Beauvis. = Deplanchea Vieillard (Bignoniac.).
Montrichardia Crueg. Araceae. 2 trop. Am., W.I.
Montrouzeria Benth. & Hook. f. = seq.
Montrouziera Planch. ex Planch. & Triana. Guttiferae. 5 New Caled.
Monttea C. Gay. Scrophulariaceae. 3 Chile.
Monustes Rafin. = Spiranthes Rich. (Orchidac.).
Monvillea Britton & Rose (~ Cereus Mill.). Cactaceae. 15 trop. S. Am.
Monypus Rafin. = Clematis L. (Ranunculac.).
Moonia Arn. Compositae. 7 Indomal.
Moorcroftia Choisy = Lettsomia Roxb. (Convolvulac.).
Moorea Lemaire = Cortaderia Stapf (Gramin.).
Moorea Rolfe = Neomoorea Rolfe (Orchidac.).
Mooria Montr. = Cloëzia Brongn. & Gris (Myrtac.).
Moparia Britton & Rose = Hoffmanseggia Cav. (Legumin.).
Mopex Lour. ex Gomes = ? Triumfetta L. (Tiliac.).
Mophiganes Steud. (sphalm.) = Mogiphanes Mart. (Amaranthac.).
Moquerysia auctt. = Mocquerysia Hua (Flacourtiac.).

Moquilea Aubl. = Licania Aubl. (Chrysobalanac.).
*Moquinia DC. Compositae. 1 Brazil. Dioecious shrub.
Moquinia Spreng. f. = seq.
Moquiniella Balle. Loranthaceae. 1 S. Afr.
Mora Schomb. ex Benth. Leguminosae. 10 trop. S. Am., W.I.
*Moraceae Link. Dicots. 53/1400, trop. & subtrop., a few temp. Most are trees or shrubs with stip. ls., and with latex. (See *Ficus, Cecropia, Maclura*.) Infl. cymose, usu. in the form of (pseudo-) racemes, spikes, umbels or heads (cf. *Urticaceae*, and paper there cited). Fls. unisexual. P usu. 4 or (4), persistent; A in ♂ = and opp. to P, bent inwards or straight in the bud, anth. not exploding like those of *Urticaceae*; G in ♀ of (2) cpls. of which one is usu. aborted all but the style; ovary 1-loc., sup. to inf.; ovule 1, pend., with micropyle facing upwards, or rarely basal and erect. Fr. an achene or drupe-like; but commonly a multiple fr. arises by union of the frs. of different fls., often complicated by addition of the fleshy common recept. (see *Morus, Ficus, Artocarpus*). Seed with or without endosp.; embryo usu. curved. Many yield useful fruits, e.g. *Morus, Artocarpus, Ficus, Brosimum*, etc.; other important economic plants are *Broussonetia* (paper), *Castilloa* (rubber), *Brosimum* (milk) *Ficus* (caoutchouc, lac, timber, etc.), and others.

Classification and chief genera (after Corner, *Gard. Bull. Str. Settlem.* **19**: 187–252, 1962):

1. FICEAE (fls. in urc. or subglob. recept.; styles not exserted; fil. straight in bud, anth. intr.; ♀ fl. stipit. or sess.; ster. ♀ fls. with gall-wasps): *Ficus.*
2. DORSTENIËAE (fls. not encl., or if encl. styles exserted; anth. extr.; ♀ fl. mostly sess.; infl. bisex., often discoid; ♀ fls. many; fil. inflexed; ov. immersed; seeds small; herbs or suffrut.): *Dorstenia.*
3. BROSIMEAE (as *Dorst.*, but ♀ fl. sol.; fil. straight (rarely infl.); ov. sometimes free; seeds large; woody pls.): *Brosimum, Craterogyne,* etc.
4. MOREAE (infl. unisex., or if bisex. not discoid; ♀ infl. rac. or slenderly spic. or 1-fl.; ov. usu. free; ♂ fl. usu. with pistillode; mostly woody pl.): *Morus, Fatoua, Sorocea, Trophis, Streblus,* etc.
5. OLMEDIËAE (as *Mor.*, but ♀ infl. capit. or stoutly spic.; often immersed in sockets; ♂ fl. usu. without pistillode; ♂ and ♀ infl. similar, discoid, obcon., or urc., with invol., sometimes 1-fld., sometimes syncarp.; fil. 1–8, straight; not spinous): *Olmedia, Pseudolmedia, Olmediopsis, Perebea, Antiaris,* etc.
6. ARTOCARPEAE (as *Olmed.*, but ♂ and ♀ infl. dissimilar, invol. or not; ♀ infl. stoutly spic. to capit.-glob., never 1-fl., syncarp.; ♂ infl. panic., rac., spic. or capit.; stam. 1–4, fil. straight or inflexed; spinous or not): *Maclura, Broussonetia, Malaisia, Artocarpus,* etc.

*Moraea Mill. ex L. Iridaceae. 100 trop. & S. Afr., Masc. The outer integument of the ovule becomes fleshy as it ripens.
Moranda Scop. = Pentapetes L. (Sterculiac.).
Morawetzia Backeb. = Borzicactus Riccob. (Cactac.).
More Gaertn. ex Radlk. = Dimocarpus Lour. (Sapindac.).
Morea Mill. = Moraea Mill. ex L. (Iridac.).
Morelia A. Rich. Rubiaceae. 1 trop. Afr.

MORELLA

Morella Lour. = Myrica L. (Myricac.).

Morelodendron Cavaco & Normand = Pinacopodium Exell & Mendonça (Erythroxylac.).

Morelosia Lex. = Bourreria P.Br. (Boraginac.).

Morelotia Gaudich. Cyperaceae. 2 Hawaii, N.Z.

Morenia Ruiz & Pav. Palmae. 12 Andes.

Morenoa La Llave = Ipomoea L. (Convolvulac.).

Morettia DC. Cruciferae. 4 Morocco to Somal. & Arabia.

Morgagnia Bub. = Simethis Kunth (Liliac.).

Morgania R.Br. Scrophulariaceae. 4 Austr.

Moricanda St-Lag. = seq.

Moricandia DC. Cruciferae. 8 Medit. to Baluchistan.

Moriera Boiss. Cruciferae. 5 C. As., Persia, Afghan.

Morierina Vieill. Rubiaceae. 2 New Caled.

Morilandia Neck. = Cliffortia L. (Rosac.).

Morina L. Morinaceae. 17 SE. Eur. to Himal. & SW. China. Infl. like *Labiatae*.

Morinaceae J. G. Agardh. Dicots. 1/17 temp. Euras. Perenn. herbs with simple stems; ls. opp. or vertic., pinnatif. or spinose-dent., rarely ent. Fls. zygo., in axill. whorls (cf. *Labiatae*), each fl. encl. in a tubular-campan. spinose-margined invol. K bilabiate, lips ent. or 2-lobed, persist.; C (5), bilab., tube usu. slender and exserted; A 2 or 4, didynam.; G̅ 1, with 1 pend. anatr. ov., and slender style with capit. stig. Fr. an achene with thickened ± rugose pericarp. Only genus: *Morina*. Related to *Dipsacac.*

Morinda L. Rubiaceae. 80 trop. Fls. in heads; the ovaries united. Several yield dye-stuffs.

Morindopsis Hook. f. Rubiaceae. 2 SE. As.

Moringa Adans. Moringaceae. 12 NE. & SW. Afr., Madag., Arabia, India. *M. oleifera* Lam. cult. for the oil (ben oil) obtained from the seeds.

*Moringaceae Dum. Dicots. 1/12 Afr. to India. Trees, sometimes with thick stem, containing myrosin-cells; ls. alt., pinn. to tripinn., decid.; stip. 0, or minute, glandular. Fls. ♀, zygo., in many-fld. axill. pan. K 5, imbr.; C 5, imbr., unequal, the lowermost (anticous) sep. the largest, all ± reflexed; A 5+5, the outer (antesep.) staminodial, anth. finally 1-loc.; K, C, and A shortly connate at base into cupular disk; G̲ (3), on short gynoph., 1-loc., with ∞ biseriate pend. anatr. ov. on 3 pariet. plac., and (2–)3(–4) slender term. styles. Fr. an elong. 3–12-angled 3-valved 1-loc. pod-like caps.; seeds large, 3-winged or unwinged, separated by spongy tissue, without endosp. Only genus: *Moringa*. Forms a connecting link between *Capparidaceae* and *Leguminosae–Caesalpinioïdeae*.

Morisea DC. = seq.

Morisia J. Gay. Cruciferae. 1 Sardinia, Corsica.

Morisia Nees = Sphaeroschoenus Arn. = Rhynchospora Vahl (Cyperac.).

Morisina DC. = Morisia J. Gay (Crucif.).

Morisonia L. Capparidaceae. 4 W.I., S. Am.

Moritzia DC. ex Meissn. Boraginaceae. 5 trop. S. Am.

Moritzia Sch. Bip. ex Benth. & Hook. f. = Podocoma Cass. (Compos.).

Morkillia Rose & Painter. Zygophyllaceae. 2 Mex.

Morleya Woodson = Mortoniella Woodson (Apocynac.).

Mormodes Lindl. Orchidaceae. 30 C. & trop. S. Am. The column is twisted to one side, the labellum to the other. The pollinia, with their viscid disk, are violently shot out if an insect touches the articulation of anther to column.

Mormolyca Fenzl. Orchidaceae. 5 C. & trop. S. Am.

Mormolyce auctt. = praec.

Mormolyze auctt. = praec.

Mormoraphis Jack ex Wall. = Arthrophyllum Blume (Araliac.).

Morna Lindl. = Waitzia Wendl. (Compos.).

Morocarpus Boehm. = Blitum L. = Chenopodium L. (Chenopodiac.).

Morocarpus Sieb. & Zucc. = Debregeasia Gaudich. (Urticac.).

Moroea Franch. & Sav. = Moraea Mill. ex L. (Iridac.).

Morolobium Kosterm. (~ Pithecellobium Mart.). Leguminosae. 1 Moluccas.

Morongia Britton (~ Schrankia Willd.). Leguminosae. 10 Am., W.I.

Moronoa Hort. ex Kunze = Morrenia Hort. ex Kunze = Mikania Willd. (Compos.).

Moronobea Aubl. Guttiferae. 7 trop. S. Am.

Moronobeaceae Miers = Guttiferae–Moronobeoïdeae Engl.

Morophorum Neck. = Morus L. (Morac.).

Morphaea Nor. = Fagraea Thunb. (Potaliac.).

Morphixia Ker-Gawl. = Ixia L. (1762) (Iridac.).

Morrenia Hort. ex Kunze = Mikania Willd. (Compos.).

Morrenia Lindl. Asclepiadaceae. 10 trop. & temp. S. Am.

Morrisiella Aellen. Chenopodiaceae. 1 E. Austr. (New S. Wales).

Morsacanthus Rizzini. Acanthaceae. 1 Brazil.

Morstdorffia Steud. = Chirita Buch.-Ham. [+ Dichrotrichum Reinw.] (Gesneriac.).

Mortonia A. Gray. Celastraceae. 8 S. U.S., Mex.

Mortoniella R. E. Woodson. Apocynaceae. 1 C. Am.

Mortoniodendron Standley & Steyerm. Tiliaceae. 5 C. Am.

Morus L. Moraceae. Probably fewer than 10, E. temp. N. Am., SW. U.S. to Andes, trop. Afr., SW. As. to Japan & Java. Fls. monoec. or dioec., the ♂ in catkins, the ♀ in pseudo-spikes, wind-pollinated. Each ovary gives an achene enclosed in the P whose ls. become completely united and fleshy. The whole mass of frs. thus produced on the one spike is closely packed together, giving a multiple fr. like a blackberry (*Rubus*), but of very different morphological nature. The fr. (mulberry) is edible. The leaves of *M. alba* L. (white mulberry), *M. nigra* L. (black mulberry), and others are used for feeding silkworms.

Morysia Cass. = Athanasia L. (Compos.).

Moscaria Pers. = Moscharia Ruiz & Pav. (Compos.).

Moscatella Adans. = Adoxa L. (Adoxac.).

Moscharea I. K. = Moscharia Salisb. = Muscarimia Kostel. ex Los. (Liliac.).

Moscharia Fabr. (uninom.) = *Amberboa* (Pers.) Less. sp. (Compos.).

Moscharia Forsk. = Ajuga L. (Labiat.).

*Moscharia Ruiz. = Pav. Compositae. 1 Chile.

Moscharia Salisb. = Muscarimia Kostel. ex Los. (Liliac.).

Moschatella Scop. = seq.

Moschatellina Mill. = Adoxa L. (Adoxac.).

MOSCHIFERA

Moschifera Molina = Moscharia Ruiz & Pav. (Compos.).

Moschkowitzia Klotzsch = Begonia L. (Begoniac.).

Moschopsis Phil. Calyceraceae. 8 Chile, Patag.

Moschosma Reichb. = Basilicum Moench (Labiat.).

Moschoxylon Meissn. = seq.

Moschoxylum A. Juss. = Trichilia P.Br. (Meliac.).

× **Moscosoara** hort. Orchidaceae. Gen. hybr. (iii) (Broughtonia × Epidendrum × Laeliopsis).

Mosdenia Stent. Gramineae. 1 S. Afr. (Transvaal).

Moseleya Hemsl. = Ellisiophyllum Maxim. (Ellisiophyllac.).

Mosenia Lindm. = Canistrum E. Morr. (Bromeliac.).

Mosenodendron R. E. Fries = Hornschuchia Nees (Annonac.).

Mosenthinia Kuntze = Glaucium L. (Papaverac.).

Mosheovia Eig. Scrophulariaceae. 1 Palestine.

Mosiera Small = Eugenia L. (Myrtac.).

Mosigia Spreng. = Moscharia Ruiz & Pav. (Compos.).

Mosina Adans. = Ortegia L. (Caryophyllac.).

Moskerion Rafin. = Narcissus L. (Amaryllidac.).

Mosla (Benth.) Buch.-Ham. ex Maxim. = Orthodon Benth. & Oliv. (Labiat.).

Mosquitoxylum Krug & Urb. Anacardiaceae. 1 Jamaica (mosquito wood).

Mossia N. E. Br. Aïzoaceae. 2 S. Afr.

Mostacillastrum O. E. Schulz. Cruciferae. 3 Argent.

Mostuea Didr. Loganiaceae. 1 Brazil, 7 trop. Afr., Madag.

Motandra A. DC. Apocynaceae. 10 trop. W. Afr.

Motherwellia F. Muell. Araliaceae. 1 NE. Austr.

Mouffetta Neck. = Patrinia Juss. (Valerianac.).

Mougeotia Kunth = Melochia L. (Sterculiac.).

Moulinsia Cambess. = Erioglossum Blume (Sapindac.).

Moulinsia Rafin. = Aristida L. (Gramin.).

Moullava Adans. ? Leguminosae (= ?Wagatea Dalz.). 1 S. India.

Moultonia Balf. f. & W. W. Smith. Gesneriaceae. 1 Borneo.

Moultonianthus Merr. Euphorbiaceae. 1 Sumatra, Borneo.

Mountnorrisia Szyszył. = Anneslea Wall. (Theac.).

Mourera Aubl. Podostemaceae. 6 N. trop. S. Am.

Mouretia Pitard. Rubiaceae. 1 Indoch.

Mouricou Adans. = Erythrina L. (Legumin.).

Mouriri Aubl. Memecylaceae. 50 C. & trop. S. Am., W.I.

Mouriria Juss. = praec.

Mouririaceae Gardner = Memecylaceae DC.

Mouroucoa Aubl. = Maripa Aubl. (Convolvulac.).

Moussonia Regel = Isoloma Decne (Gesneriac.).

Moutabea Aubl. Polygalaceae. 8 trop. S. Am.

Moutabe[ace]ae Endl. = Polygalaceae–Moutabeëae Chod.

Moutan Reichb. = Paeonia L. (Paeoniac.).

Moutouchi Aubl. = Pterocarpus L. (Legumin.).

Moutouchia Benth. = praec.

Moya Griseb. Celastraceae. 4 Bolivia, Argent.

Mozambe Rafin. = Cadaba Forsk. (Capparidac.).

Mozartia Urb. Myrtaceae. 7 Cuba.

Mozinna Ortega = Jatropha L. (Euphorbiac.).

Mozula Rafin. = Lythrum L. (Lythrac.).

Msuata O. Hoffm. Compositae. 1 trop. Afr.

Muantijamvella Phipps. Gramineae. 1 trop. Afr.

Muantum Pichon. Apocynaceae. 1 Lower Burma & Penins. Siam.

Mucizonia (DC.) Berger. Crassulaceae. 1 Canaries, W. Medit.

Muckia Hassk. = Mukia Arn. (Cucurbitac.).

Mucronea Benth. = Chorizanthe R.Br. (Polygonac.).

***Mucuna** Adans. Leguminosae. 120 trop. & subtrop. Some have irritant hairs on the pods. *M. pruriens* (L.) DC. is the cowage or cowitch, a var. of which is the Florida velvet bean, a useful fodder.

Muehlbergella Feer = Edraianthus DC. (Campanulac.).

***Muehlenbeckia** Meissn. Polygonaceae. 15 New Guinea, Austr., N.Z., W. S. Am. Fls. polyg. or dioec. (For *M. platyclados* Meissn. see *Homocladium* (F. Muell.) L.H. Bailey.)

Muehlenbergia Schreb. (1810) = Muhlenbergia Schreb. (1789) (Gramin.).

***Muellera** L. f. Leguminosae. 3 Mex. to trop. S. Am.

Muelleramra Kuntze = Pterocladon Hook. f. (Melastomatac.).

Muelleranthus Hutch. Leguminosae. 1 C. Austr.

Muellerargia Cogn. Cucurbitaceae. 1 Madag.; 1 Lesser Sunda Is.

Muellerina Van Tiegh. (~ Phrygilanthus Eichl.). Loranthaceae. 4 E. Austr.,

Muellerothamnus Engler = Piptocalyx Oliv. ex Benth. (Trimeniac.). [?N.Z.

Muenchausia Scop. = seq.

Muenchhausia L. ex Murr. = Munchausia L. = Lagerstroemia L. (Lythrac.).

Muenchhusia Fabr. (uninom.) = *Hibiscus* L. sp. (Malvac.).

Muenteria Seem. = Markhamia Seem. ex Baill. (Bignoniac.).

Muenteria Walp. = Picraena Lindl. (Simaroubac.).

Muhlenbergia Schreb. Gramineae. 100 Himal. to Japan, N. Am. to Andes. Some are useful fodder-grasses.

Muilla S. Wats. Alliaceae. 5 SW. U.S., Mex. Rootstock with fibrous covering.

Muiria N. E. Brown. Aïzoaceae. 2 S. Afr.

Muiria C. A. Gardner = seq.

Muiriantha C. A. Gardner. Rutaceae. 1 W. Austr.

× **Muirio-Gibbaeum** Jacobsen. Aïzoaceae. Gen. hybr. (Gibbaeum × Muiria).

Muitis Rafin. = Caucalis L. (Umbellif.).

Mukdenia Koidz. Saxifragaceae. 2 N. China, Manch., Korea.

Mukia Arn. Cucurbitaceae. 4 palaeotrop.

Muldera Miq. = Piper L. (Piperac.).

Mulfordia Rusby = Dimerocostus Kuntze (Costac.).

Mulgedium Cass. = Cicerbita Wallr. (Compos.).

Mulinum Pers. Hydrocotylaceae (~ Umbellif.). 20, char. pl. of southern Andes.

Mullaghera Bub. = Lotus L. (Legumin.).

Mullera Juss. = Muellera L. f. (Legumin.).

Multiovulatae Aquaticae Benth. & Hook. f. A 'Series' of Dicots. comprising the fam. *Podostemaceae* only.

Multiovulatae Terrestres Benth. & Hook. f. A 'Series' of Dicot. fams. comprising the *Nepenthac.*, *Cytinac.* and *Aristolochiac.*

Mumeazalea Makino = Azaleastrum Rydb. = Rhododendron L. (Ericac.).
Munbya Boiss. = Macrotomia DC. (Boraginac.).
Munbya Pomel (~ Psoralea L.). Leguminosae. 2 Algeria.
Munchausia L. = seq.
Munchhausia L. (apud Murr.) = Lagerstroemia L. (Lythrac.).
Münchhusia Fabr. = Hibiscus L. (Malvac.).
Munchusia Rafin. = praec.
Mundia Kunth = Nylandtia Dum. (Polygalac.).
Mundtia auctt. = praec.
Mundubi Adans. = Arachis L. (Legumin.).
Mundulea Benth. Leguminosae. 30 Madag., 1 trop. Afr., S. India, Ceylon.
Mungos Adans. = Ophiorrhiza L. (Rubiac.).
Munnickia Reichb. = Bragantia Lour. (Aristolochiac.).
Munnicksia Dennst. = Hydnocarpus Gaertn. (Flacourtiac.).
Munnozia Ruiz & Pav. (~ Liabum Adans.). Compositae. 20 Andes.
Munroa Torr. corr. Benth. & Hook. f. Gramineae. 1 W. U.S., 4 C. Andes.
Munroidendron Sherff. Araliaceae. 1 Hawaii.
Munronia Wight. Meliaceae. 15 Ceylon to China & W. Malaysia.
Munrozia Steud. = Munnozia Ruiz & Pav. (Compos.).
Muntafara Pichon. Apocynaceae. 1 Madag.
Muntingia L. Elaeocarpaceae. 3 trop. S. Am., W.I.
Munychia Cass. = Felicia Cass. (Compos.).
Munzothamnus P. H. Raven. Compositae. 1 Calif. (Channel Is.: San Clemente).
Muralta Adans. = Clematis L. (Ranunculac.).
Muralta Juss. = seq.
Muraltia [Neck.] Juss. Polygalaceae. 115 S. Afr.
Muratina Maire = Salsola L. (Chenopodiac.).
Murbeckia Urb. & Ekman = Forchhammeria Liebm. (Capparidac.).
Murbeckiella Rothmaler. Cruciferae. 5 SW. Eur. & Alg., 1 Cauc.
Murchisonia N. H. Brittan. Liliaceae. 1 W. Austr.
*Murdannia Royle. Commelinaceae. 50 trop.
Murera Jaume St-Hil. = Mourera Aubl. (Podostemac.).
Muretia Boiss. Umbelliferae. 8 As. Min. to C. As. & Persia.
Murex Kuntze = Pedalium L. (Pedaliac.).
Murexia Mutis apud Alba. Colombia. Quid?
Murianthe (Baill.) Aubrév. Sapotaceae. 1 Cuba.
Muricaria Desv. Cruciferae. 1 N. Afr.
Muricauda Small = Arisaema Mart. (Arac.).
Muricia Lour. = Momordica L. (Cucurbitac.).
Muricococcum Chun & How = Cephalomappa Baill. (Euphorbiac.).
Muriea Hartog. Sapotaceae. 4–5 W.I., trop. E. & SE. Afr.
Muriri J. F. Gmel. = Mouriri Aubl. (Memecylac.).
Muriria Rafin. = praec.
Murocoa Jaume St-Hil. = Mouroucoa Aubl. = Maripa Aubl. (Convolvulac.).
Murraea Koen. ex L. = seq.
*Murraya Koen. ex L., mut. Murr. Rutaceae. 12 E. As., Indomal., Pacif. The timber is useful, and the ls. are used in curries.

Murrinea Rafin. = Baeckea L. (Myrtac.).
Murrithia Zoll. & Mor. = Pimpinella L. (Umbellif.).
Murrya Griff. = Murraya Koen. ex L. (Rutac.).
Murtekias Rafin. = Euphorbia L. (Euphorbiac.).
Murtonia Craib. Leguminosae. 1 Siam.
Murtughas Kuntze = Lagerstroemia L. (Lythrac.).
Murucoa J. F. Gmel. = Mouroucoa Aubl. = Maripa Aubl. + Lettsomia Roxb.
Murucuia Mill. = Passiflora L. (Passiflorac.).   [(Convolvulac.).
Murucuja Guett. = praec.
Murueva Rafin. = Maireria Scop. = Maripa Aubl. (Convolvulac.).
Musa L. Musaceae. 35 palaeotrop. Large to gigantic herbs (to 15 m.) with
rhiz. and 'false' aerial stems (see fam.). The infl. springs from rhiz. and emerges
at the top of the aerial 'stem'. Fls. ∞, in the axils of leathery, often reddish-
coloured bracts, the fruit-forming ♀ fls. at the base of the infl. The sepals and
two ant. petals are joined into a tube, the post. petal is free; there are 5 fertile
sta.; the ovary is 3-loc., with ∞ anatr. ovules. Fr. a longish berry. Seeds with
mealy perisperm. Bracts and fls. usu. decid. by abscission; pollen grains with
granular surface; seeds smaller than those of *Ensete*. *M. paradisiaca* L., the
plantain, with its subsp. *M. sapientum* L., the banana, is one of the most
important food-plants, and is everywhere cult. in the trop. and subtrop.,
yielding much more food per acre than even the potato. The cult. forms are
propagated entirely from the rhiz. and produce no seeds (cf. *Citrus*). About
200 different forms are in cult., and some other spp. are occasionally employed.
There is a vast trade, esp. to the U.S., from C. Am., Jamaica, Canaries, etc.
In Venezuela, etc., alcohol is prepared. The dried frs. are ground to form
plantain-meal. The leaf-stalks of *M. textilis* Née (Philippines, etc.) furnish a
useful fibre, known as Manila hemp or *abaca*. The sp. cult. in the Canaries is
M. nana Lour. (*M. cavendishii* Lambert ex Paxt.) (China).
*Musaceae Juss. Monocots. 2/42 trop. Afr., As., Austr. Gigantic herbs with
usu. freely branching rhiz. from which the many-ranked ls. spring; the sheaths
of the ls. are rolled round one another below, and form what looks like an
aerial stem, attaining in the banana some metres in height. The l. is large and
oval, with a stout midrib, and parallel veins running from it to the edge; it is
rolled up in bud. The edge is easily torn between the bundles, as they do not
join in the same way as in a Dicot.; and so the wind and rain soon reduce the
l. to a very ragged condition. Fls. arranged collaterally in racemes with large
brightly coloured bracts or spathes; ♂ ♀ or ⚥ and zygo., but nearer to the usual
type of Monocotyledonous fl. than those of other *Scitamineae*. P 3 + 3 (5 united
and 1 free), both whorls petaloid; A 3 + 3, the post. sta. often repres. by a std.;
G̅ (3), 3-loc., with 1–∞ ov. in each loc. Fr. a berry. Seed with straight embryo
and mealy perisperm. Fls. rich in honey, and visited by bees and birds.
Genera: *Musa, Ensete*.
Musanga C. Sm. ex R.Br. Urticaceae. 1 trop. Afr. Somewhat intermediate
between *Urticac.* and *Morac.*
Muscadinia Small (~ Vitis L.). Vitidaceae. 2 N. Am., W.I.
Muscari Mill. Liliaceae. 60 Eur., Medit., W. As. Collateral buds in axils.
Upper fls. of the raceme neuter, giving extra conspicuousness to the infl.
(cf. *Centaurea cyanus*).

Muscaria Haw. = Saxifraga L. (Saxifragac.).
Muscarimia Kostel. (~ Muscari Mill.). Liliaceae. 1 As. Min., Cauc.
Muscarius Kuntze = Muscari Mill. (Liliac.).
Muschleria S. Moore. Compositae. 2 trop. Afr.
Muscipula Fourr. = Silene L. (Caryophyllac.).
Museniopsis Coulter & Rose. Umbelliferae. 20 W. U.S., Mex.
Musenium Nutt. = Musineon Rafin. (Umbellif.).
Musgravea F. Muell. Proteaceae. 1 Queensland.
Musidendron Nakai = Phenakospermum Endl. = Ravenala Adans. (Strelitziac.).
Musilia Velen. (~ Odontospermum Moench). Compositae. 1 Arabia.
Musineon Rafin. Umbelliferae. 6 N. Am., Mex.
Mussaenda L. Rubiaceae. 200 palaeotrop. One sepal is large, leafy, and brightly coloured, and helps to make the fl. conspicuous (cf. *Euphorbia, Salvia*).
Mussaendopsis Baill. Rubiaceae. 2 W. Malaysia.
Mussatia Bur. ex Baill. Bignoniaceae. 3 C. & trop. S. Am.
Musschia Dum. Campanulaceae. 2 Madeira. The capsule opens by many transv. slits between the ribs.
Mussinia Willd. = Gazania Gaertn. (Compos.).
Mustelia Cav. ex Steud. = Chusquea Kunth (Gramin.).
Mustelia Spreng. = Stevia Cav. (Compos.).
Musteron Rafin. = Erigeron L. (Compos.).
Mutabea J. F. Gmel. = Moutabea Aubl. (Polygalac.).
Mutafinia Rafin. = Limosella L. (Scrophulariac.).
Mutarda Bernh. = Brassica L. (Crucif.).
Mutelia Gren. ex Mutel = Melissa L. (Labiat.).
Mutellina v. Wolf = Ligusticum L. (Umbellif.).
Mutisia L. f. Compositae. 60 S. Am. Many climbers (a rare habit in *C.*) with ends of leaf-midribs prolonged into tendrils. All are shrubby with large heads of fls.
Mutisiaceae Burnett = Compositae–Mutisiëae Cass.
Mutuchi J. F. Gmel. = Moutouchi Aubl. = Pterocarpus L. (Legumin.).
Mutumocarpon Pohl (nomen). 1 Brazil. Quid?
Muxiria Welw. = Eriosema DC. (Legumin.).
Muza Stokes = Musa L. (Musac.).
Myagropsis Hort. ex O. E. Schulz = Sobolewskia M. Bieb. (Crucif.).
Myagrum L. Cruciferae. 1 Medit. & mid Eur. to Persia.
Myanthe Salisb. = Ornithogalum L. (Liliac.).
Myanthus Lindl. = Catasetum Rich. (Orchidac.).
Myaris Presl = Clausena Burm. f. (Rutac.).
Mycaranthes Blume = Eria Lindl. (Orchidac.).
Mycelis Cass. Compositae. 30 temp. Euras.
Mycerinus A. C. Smith. Ericaceae. 1 Venez.
Mycetanthe Reichb. = Rhizanthes Dum. (Rafflesiac.).
Mycetia Reinw. Rubiaceae. 25 S. China, Indoch., W. Malaysia.
Myconella Sprague = Kremeria Dur. (Compos.).
Myconia Lapeyr. = Ramonda Rich. (Gesneriac.).
Myconia Neck. ex Sch. Bip. = Chrysanthemum L. (Compos.).
Mycostylis Rafin. = Sonneratia L. f. (Sonneratiac.).

Mycropus Gouan = Micropus L. (Compos.).
Mycroseris Hook. & Arn. = Microseris D. Don (Compos.).
Myctanthes Rafin. = Aster L. (Compos.).
Myctirophora Nevski = Astragalus L. (Legumin.).
Mygalurus Link = Festuca L. (Gramin.).
Myginda Jacq. Celastraceae. 15 Mex. to trop. S. Am., W.I.
Mylachne Steud. = seq.
Mylanche Wallr. = Epifagus Nutt. (Orobanchac.).
Mylinum Gaudin = Selinum L. (Umbellif.).
Myllanthus R. S. Cowan. Rutaceae. 3 Venez.
Mylocaryum Willd. = Cliftonia Banks ex Gaertn. f. (Cyrillac.).
Mylosphora Neck. = Singana Aubl. = ? Swartzia Schreb. (Legumin.).
Myobroma Stev. = Astragalus L. (Legumin.).
Myoda Lindl. = Ludisia A. Rich. (Orchidac.).
Myodium Salisb. = Ophrys L. (Orchidac.).
Myodocarpus Brongn. & Gris. Araliaceae. 12 New Caled.
Myogalum Link = Ornithogalum L. (Liliac.).
Myonima Comm. ex Juss. Rubiaceae. 4 Mauritius, Réunion.
***Myoporaceae** R.Br. Dicots. 4/90, chiefly Austr. & S. Pacific Is. (a few in S. Afr., Mauritius, E. As., Hawaii, W.I.). Most are trees or shrubs, with alt. or opp. entire or dent. exstip. ls., often gland-dotted or covered with woolly or glandular or lepidote hairs, frequently very reduced in size. Fls. sol., or in cymose groups, axillary, ♀, usu. zygo. K (5), imbr. or open; C (5), imbr.; A 4, didynamous; anther loculi confluent; G̲ (2), 2-loc. or by segmentation 3–10-loc., in the former case with 1–8, in the latter with 1, pend. anatr. ovule in each loc. Fruit a drupe. Endosperm. Genera: *Myoporum, Oftia, Bontia, Pholidia.*
Myopordon Boiss. Compositae. 2 Persia.
Myoporum Banks & Soland. ex Forst. f. Myoporaceae. 32 Mauritius, E. As., New Guinea, Austr., N.Z., Pacif. *M. laetum* Forst. f. (N.Z.) yields useful timber.
Myopsia Presl = Heterotoma Zucc. (Campanulac.).
Myopsis Benth. & Hook. f. = praec.
Myopteron Spreng. = Alyssum L. (Crucif.).
Myosanthus Desv. = Myosoton Moench (Caryophyllac.).
Myoschilos Ruiz & Pav. Santalaceae. 1 Chile.
Myoseris Link = Crepis L. (Compos.).
Myosotidium Hook. Boraginaceae. 1 Chatham Is.
Myosotis L. Boraginaceae. 50 temp. Euras., mts. trop. Afr., S. Afr., Austr., N.Z. The corolla mouth is nearly closed by scales, and in some there is a coloured ring at the entrance forming a honey guide. The colour of the C changes from pink to blue during anthesis.
Myosotis Mill. = Cerastium L. (Caryophyllac.).
Myosotodon Manning = seq.
Myosoton Moench. Caryophyllaceae. 1 temp. Euras.
Myospyrum Lindl. (sphalm.) = Myxopyrum Bl. (Oleac.).
Myostemma Salisb. = Hippeastrum Herb. (Amaryllidac.).
Myostoma Miers = Thismia Griff. (Burmanniac.).

MYOSURANDRA

Myosurandra Baill. = Myrothamnus Welw. (Myrothamnac.).

Myosuros Adans. = seq.

Myosurus L. Ranunculaceae. 15 N. & S. temp. Recept. much elongated.

Myotoca Griseb. ex Brand = Phlox L. vel Gilia Ruiz & Pav. (Polemoniac.).

Myoxanthus Poepp. & Endl. = Pleurothallis R.Br. (Orchidac.).

Myracrodruon Allem. = Astronium Jacq. (Anacardiac.).

Myrceugenella Kausel. Myrtaceae. 5 Chile.

Myrceugenia Berg. Myrtaceae. 55 temp. S. Am. Ov. ∞ per loc.

Myrcia DC. ex Guillem. Myrtaceae. 500 trop. S. Am., W.I. Ov. 2 per loc. K free or slightly united.

Myrcia Soland. ex Lindl. = Pimenta Lindl. (Myrtac.).

Myrcialeucas Willis (sphalm.) = seq.

Myrcialeucus Roj. Myrtaceae. 1 Argent.

Myrcianthes Berg. Myrtaceae. 6 Argent.

Myrciaria Berg. Myrtaceae. 65 trop. S. Am., W.I. Like *Eugenia*, but ov. 2 per loc. Fr. ed. (*jaboticabá*, etc.).

Myrciariopsis Kausel. Myrtaceae. 1 subtrop. S. Am.

Myria Nor. ex Tul. = Terminalia L. (Combretac.).

Myriachaeta Zoll. & Mor. = Thysanolaena Nees (Gramin.).

Myriactis Less. Compositae. 2–3 Cauc. to Japan & New Guinea. No pappus.

Myriadenus Cass. = Inula L. (Compos.).

Myriadenus Desv. = Zornia J. F. Gmel. (Legumin.).

Myrialepis Becc. ex Hook. f. Palmae. 3 Indoch., Malay Penins., Borneo.

Myriandra Spach = Hypericum L. (Guttif.).

Myriangis Thou. (uninom.) = *Angraecum multiflorum* Thou. (Orchidac.).

Myrianthea Tul. = seq.

Myriantheia Thou. = Homalium Jacq. (Flacourtiac.).

Myrianthemum Gilg. Melastomataceae. 1 trop. W. Afr.

Myrianthus Beauv. Urticaceae. 12 trop. Afr.

Myriaspora DC. Melastomataceae. 2 trop. S. Am. Caulifloral. Calyptra.

Myrica Bub. = Myricaria Desv. (Tamaricac.).

Myrica L. Myricaceae. 35 almost cosmop. (exc. N. Afr., C. & SE. Eur., SW. As., Austr.). *M. cerifera* L. (N. Am., wax-myrtle, bayberry), and others, are sources of wax; the frs. are boiled.

*****Myricaceae** Bl. & Dum. Dicots. 4/40 cosmop. Aromatic trees or shrubs, with alt. or subopp. simple ent. or dent. or pinnatifid stip. or exstip. ls. Fls. ♂ ♀, or ♂ ♀ (*Canacomyrica*), monoec. or dioec., achlam., in short axill. catkin-like simple or cpd. spikes. ♂ usu. with 2 bracteoles; disk annular, sinuate; A (2–)4 (–16), fil. sometimes conn., pistillode rarely present (e.g. *Canac.*); ♀ with 2–4 bracteoles, disk 0, G (2), 1-loc., with 1 erect orthotr. ov., and short style with 2 divaric. branches; ♀ of *Canacomyrica* with 6 perigynous sta., 6-lobed accrescent disk (? perianth), stigmata 2, widened, deeply dichot.-lacin. Fr. a small often waxy-warted drupe with hard endoc., in *Canac.* completely encl. in the accresc. disk; seed without (or in *Canac.* with sparse) endosp. Genera: *Myrica, Gale, Comptonia, Canacomyrica*. Without close relatives; perhaps a distant connection with *Betulac.* The position of *Canac.*, which differs from the other genera anatomically as well as in floral chars., requires further study.

Myricaria Desv. Tamaricaceae. 10 temp. Euras. (A) obdipl., G (3).

Myrinia Lilja = Fuchsia L. (Onagrac.).

Myrioblastus Wall. ex Griff. = Cryptocoryne Fisch. & Mey. (Arac.).

Myriocarpa Benth. Urticaceae. 15 C. to trop. S. Am. Fls. ∞, in catkins.

Myriocephalus Benth. Compositae. 10 temp. Austr. Head 1-fld., cpd.

Myriochaeta P. & K. = Myriachaeta Zoll. & Mor. = Thysanolaena Nees (Gramin.).

Myriocladus Swallen. Gramineae. 2 Venez.

Myriodon Copel. Hymenophyllaceae. 1 New Guinea.

Myriogomphus Didr. = Croton L. (Euphorbiac.).

Myriogyne Less. = Centipeda Lour. (Compos.).

Myriolepis P. & K. = Myrialepis Becc. ex Hook. f. (Palm.).

Myrioneuron R.Br. Rubiaceae. 15 E. Himal., SE. As., W. Malaysia. Heads nodding.

Myriopeltis Welw. ex Hook. f. = Treculia Decne ex Trécul (Morac.).

Myriophillum Neck. = Myriophyllum L. (Haloragidac.).

Myriophyll[ac]eae Schultz-Schultzenst. = Haloragidaceae–Myriophylleae Schindl.

Myriophyllum L. Haloragidaceae. 45 cosmop. Submerged water pls. with usu. whorled much divided ls., borne on shoots that spring from the rhiz.-like creeping stems. Land forms occasionally produced in some. Infl. above water; fls. wind-fert. Hibernation by winter buds as in *Utricularia*.

Myriopteris Fée = Cheilanthes Sw. (Sinopteridac.).

Myriopteron Griff. Periplocaceae. 1 Assam to Malay Penins.

Myriopus Small = Tournefortia L. (Boraginac.).

Myriospora P. & K. = Myriaspora DC. (Melastomatac.).

Myriostachya (Benth.) Hook. f. Gramineae. 1 Ceylon, Bengal, Lower Burma, Malay Penins.

Myriotheca Comm. ex Juss. = Marattia Sw. (Marattiac.).

Myriotriche Turcz. = Abatia Ruiz & Pav. (Flacourtiac.).

Myripnois Bunge. Compositae. 3 N. China. Dioec. Fr. glandular with pappus.

***Myristica** Gronov. Myristicaceae. 120 palaeotrop. Trees with evergr. ls. Berry splits by both sutures, disclosing a large seed—nutmeg—with a branched red aril—mace—around it. The ordinary nutmeg is the seed of *M. fragrans* Houtt. (*M. moschata* Thunb.), of the Moluccas.

***Myristicaceae** R.Br. Dicots. 18/300 trop., esp. As. Trees with simple ent. evergr. exstip. ls., with oil cells, and rac., pan., cymes or heads of small dioec. fls., usu. 3-merous. P (2–)(3)(–5), simple, valv.; A (3–30), fil. united into a column, anth. extr., vertical (adnate to col.) or horizontal (radially spreading); G 1, with 1 basal anatr. ov. and subsess. stig. Fleshy dehisc. fr. Aril. Ruminate endosp. Oils and spices are obtained from *Myristica*, etc. Chief genera: *Myristica, Virola, Horsfieldia, Knema*.

Myrmechis Blume. Orchidaceae. 6 E. As., W. Malaysia.

Myrmecia Schreb. = Tachia Aubl. (Gentianac.).

× **Myrmecocattleya** hort. (iv) = × Schombocattleya hort. (Orchidac.).

Myrmecodendron Britton & Rose = Acacia Mill. (Legumin.).

Myrmecodia Jack. Rubiaceae. 45 Malaysia, trop. Austr., Fiji. Epiph. with leafy stems. The base forms a large tuber, fastened by adv. r., composed of a mass of tissue, chiefly cork, penetrated by numerous communicating galleries

MYRMECOÏDES

and chambers, inhabited by ants. In germin. the hypocotyl swells into a small tuber, in which a phellogen appears, forming cork on the inner side; other phellogens appear later, increasing the number of passages, for the cork falls out. The pl. may be called myrmecophilous, but it is doubtful if the ants perform any service to it. Cf. *Acacia.*

Myrmecoïdes Elmer (sphalm.) = praec.

× **Myrmecolaelia** hort. (iv) = × Schombolaelia hort. (Orchidac.).

Myrmeconauclea Merr. Naucleaceae (~ Rubiac.). 1 Philipp. Is., Borneo.

Myrmecophila (Christ) Nakai = Microsorium Link, *p.p.* (Polypodiac.). (Included in *Lecanopteris* Reinw. by Copeland, *Gen. Fil.*, but differs in scales and sori.)

Myrmecophila Rolfe = Schomburgkia Lindl. (Orchidac.).

Myrmecosicyos C. Jeffrey. Cucurbitaceae. 1 trop. E. Afr. (Kenya).

Myrmecostylum Presl = Meringium Presl (Hymenophyllac.).

Myrmecylon Hook. & Arn. (sphalm.) = Memecylon L. (Memecylac.).

Myrmedoma Becc. Rubiaceae. 2 New Guinea. Epiph.

Myrmedone Dur. & Jacks. (sphalm.) = Myrmidone Mart. ex Meissn. (Melastomatac.).

Myrmephytum Becc. Rubiaceae. 2 Philipp. Is., Celebes. Epiph.

Myrmidone Mart. ex Meissn. Melastomataceae. 4 trop. S. Am. Heterophyllous.

Myrobalan[ac]eae Juss. = Combretaceae–Terminalieae DC.

Myrobalanifera Houttuyn = Terminalia L. (Combretac.).

Myrobalanus Gaertn. = praec.

Myrobroma Salisb. = Vanilla Sw. (Orchidac.).

Myrocarpus Allem. Leguminosae. 4 S. Braz., Paraguay. Ls. translucent-dotted. Yield a balsam like balsam of Peru (see *Myroxylon*).

Myrodendron Schreb. = Houmiri Aubl. (Houmiriac.).

Myrodia Sw. = Quararibea Aubl. (Bombacac.).

Myrosma L. f. Marantaceae. 15 C. & trop. S. Am.

Myrosmodes Reichb. f. = Aa Reichb. f. (Orchidac.).

Myrospermum Jacq. Leguminosae. 1 trop. Am., W.I. Ls. dotted. Fr. winged, indehisc.

*****Myrothamnaceae** Niedenzu. Dicots. 1/2 S. trop. Afr., Madag. Xeroph. resinous shrubs, with opp. flabellate plicate-nerved stip. ls., pet. sheathing. Fls. ♂ ♀, dioec., in erect term. catkin-like spikes. K o; C o; A 4–8, fil. free or shortly conn., anth. large, basifixed, latr., connective produced; G̲ (3), with short free divar. styles and recurved spath. stigs., and ∞ axile (ventral) biser. horiz. anatr. ov. Fr. separating septicid. into 3 ventrally dehiscent follicles, with small seeds, endosp. copious, fleshy. Only genus: *Myrothamnus.* An isolated relict group, prob. related to *Hamamelidac., Cunoniac.,* etc.

Myrothamnus Welw. Myrothamnaceae. 2 S. trop. Afr., Madag.

Myroxylon J. R. & G. Forst. = Xylosma G. Forst. (Flacourtiac.).

*****Myroxylon** L. f. Leguminosae. 2 trop. S. Am. Ls. translucent-dotted. Fr. winged, indehisc. *M. balsamum* (L.) Harms var. *balsamum* yields the medicinal balsam of Tolu, *M.b.* var. *pereirae* (Royle) Baill. the balsam of Peru.

Myroxylon Mutis ex L. f. = Toluifera L. (Legumin.).

Myroxylum P. & K. = praec.

Myroxylum Schreb. = Myroxylon J. R. & G. Forst. = Xylosma G. Forst. (Flacourtiac.)

Myrrha Mitch. = ? Cryptotaenia DC. (Umbellif.).

Myrrhidendron Coulter & Rose. Umbelliferae. 5 C. Am., Colombia.

Myrrhidium (DC.) Eckl. & Zeyh. = Pelargonium L'Hérit. (Geraniac.).

Myrrhina Rupr. = Erodium L'Hérit. (Geraniac.).

Myrrhini[ac]eae Arn. = Myrtaceae–Myrteae DC.

Myrrhinium Schott. Myrtaceae. 5 trop. S. Am. A 4-8. Ed. fr.

Myrrhis Mill. Umbelliferae. 2 Eur., W. As. *M. odorata* Scop. is a pot-herb. Ribs 3-angled.

Myrrhodendrum P. & K. = Myrrhidendron Coult. & Rose (Umbellif.).

Myrrhodes Kuntze = seq.

Myrrhoïdes Fabr. (uninom.) = *Anthriscus* Pers. emend. Hoffm. sp. (Umbellif.).

*Myrsinaceae R.Br. Dicots. 35/1000, trop. & subtrop., S. Afr., N.Z. Trees or shrubs with alt. exstip. ls., often rosetted; usu. leathery, entire, with resin-passages or glands usu. visible to naked eye or lens. Fls. in racemose infls., with 2 bracteoles, ♀ or ♂ ♀, reg., 5-4-merous. K (5-4) or 5-4; C (5-4) or 5-4; A 5-4, epipet., opp. to pets., intr., the stds. in ♀ fl. often almost as large as sta., stds. rarely present in ♂; G, rarely semi-inf., 1-loc., plac. basal or free-central with ∞ or rarely few ov., ½-anatr. or ½-campylotr., sunk in plac. tissue, style and stigma simple. Most ov. usu. abort as fr. ripens. Drupe or berry; embryo straight or slightly curved, endosp. fleshy or horny. Often placed near *Primulac.* on account of the oppositepet. stam. and placentation, but relationship prob. not close. The *Myrsinac.* are very close to the *Theophrastac.,* and prob. more distantly allied to *Sapotac.* and *Theac.*

Classification and chief genera (after Mez in Engler, *Pflanzenr.,* 1901):

I. *Maesoïdeae* (G ½-inf.; fr. many-seeded): *Maesa* (only gen.).

II. *Myrsinoïdeae* (G; fr. 1-seeded):

   1. ARDISIËAE (ov. usu. many in many rows, rarely few): *Aegiceras* (anther-loc. transv. septate, seed exalbum.), *Ardisia* (anth.-loc. v. rarely sept., seed albumin.); *Tapeinosperma* (bony endoc.).

   2. MYRSINEAE (ov. usu. few, in one row): (infl. elongated, loose) *Oncostemon, Embelia* (polypet.); (infl. dense, short-stalked, ± umbellate) *Rapanea,* etc.

Myrsine L. Myrsinaceae. 7 Azores, Afr. to China.

Myrsiphyllum Willd. = Asparagus L. (Liliac.).

Myrssiphylla Rafin. = seq.

Myrstiphylla Rafin. = seq.

Myrstiphyllum P.Br. = Psychotria L. (Rubiac.).

*Myrtaceae Juss. Dicots. 100/3000 warm; chief centres Austr. (§ II) and trop. Am. (§ I). Trees and shrubs, varying from small creepers to the giant *Eucalyptus;* oil glands in ls., bicollateral bundles, phloem in pith. Ls. usu. opp., but alt. in many of § II, exstip., evergr., usu. entire. Fls. usu. in cymes, ♀, reg., perig. or epig., usu. with 2 bracteoles at base; recept. ± hollow and united to G; in § II. 1 the union is not very complete, but in the rest it is so, and the fl. is epig. K 4-5 or (4-5), sometimes thrown off unopened as a lid, usu. quincuncial with second sep. post.; C 4-5, imbr., pets. often nearly circular; A ∞, free or in bundles, rarely definite, usu. bent inwards in bud, connective often with glands above; $\overline{G}$ (or ½-sup.) ∞-1-loc. with usu. 2–∞ anatr. or campylotr. ov. in each, style (usu. long) and stigma simple, plac. usu. axile, rarely parietal. Berry, drupe, caps., or nut; no endosp. Wood usu. hard; several yield useful timber. *Eucalyptus* gives timber, kino, oil; *Eugenia* cloves, etc. Many have ed. fr.

Classification and chief genera:

I. **Myrtoïdeae** (berry, rarely drupe; ls. always opp.):

   1. MYRTEAE: *Myrtus, Psidium, Pimenta, Myrcia, Calyptranthes, Eugenia, Myrciaria, Jambosa, Syzygium.*

II. **Leptospermoïdeae** (dry fr.; ls. opp. or alt.):

   1. LEPTOSPERMEAE (G multiloc., at least when young; loculic. caps.) [embryo straight, anth. versatile]: *Metrosideros* (pets. with narrow base, A free, plac. central, many seeds), *Tristania* (do.; A in bundles opp. pets., G semi-sup.), *Eucalyptus* (pets. with broad base, united, falling as a cap; K-teeth inconspic. or o), *Leptospermum, Callistemon* (sta. much exceeding C, free, K deciduous), *Melaleuca* (do., but sta. in bundles opp. pets., ov. ∞ per loc.), *Baeckea.*

   2. CHAMAELAUCIËAE (1-loc.; 1-seeded nut): *Calycothrix, Chamaelaucium, Darwinia, Verticordia.*

Myrtastrum Burret. Myrtaceae. 1 New Caled.

Myrtekmania Urb. Myrtaceae. 3 Cuba.

Myrtella F. Muell. = Fenzlia Endl. (Myrtac.).

***Myrteola** Berg. Myrtaceae. 12 S. Am. Ed. fr.

× **Myrtgerocactus** Moran. Cactaceae. Gen. hybr. (Myrtillocactus × Bergerocactus).

Myrthoïdes v. Wolf = Syzygium Gaertn. (Myrtac.).

Myrthus Scop. = Myrtus L. (Myrtac.).

Myrtiflorae Endl. A large and somewhat heterogeneous order of Dicots., comprising (in Engler's system) besides the *Myrtac., Lythrac.,* etc. the *Thymelaeac.* et aff., *Nyssac., Alangiac., Hippuridac., Thelygonac.* and *Cynomoriac.*

Myrtilaria Hutch. = Mytilaria Lecomte (Hamamelidac.).

Myrtillocactus Console. Cactaceae. 4 Mex., Guatem.

Myrtillocereus Frič & Kreuz. = praec.

Myrtilloïdes Banks & Soland. ex Hook. = Nothofagus Bl. (Fagac.).

Myrtillus Gilib. = Vaccinium L. (Ericac.).

Myrtiluma Baill. = Micropholis Pierre (Sapotac.).

Myrtinia Nees (sphalm.) = Martynia L. (Martyniac.).

Myrtobium Miq. = Lepidoceras Hook. f. (Loranthac.).

Myrtoleucodendron Kuntze = Melaleuca L. (Myrtac.).

Myrtolobium Chalon = Myrtobium Miq. = Lepidoceras Hook. f. (Loranthac.).

Myrtomera B. C. Stone = Arillastrum Panch. ex Baill. (Myrtac.).

Myrtophyllum Turcz. = Azara Ruiz & Pav. (Flacourtiac.).

***Myrtopsis** Engl. Rutaceae. 8 New Caled.

Myrtopsis O. Hoffm. = Eugenia L. (Myrtac.).

Myrtus L. Myrtaceae. 100 trop. & subtrop., esp. Am. *M. communis* L. (myrtle, W. As.) long nat. in Eur.

Myscolus Cass. = Scolymus L. (Compos.).

Mysicarpus Webb (sphalm.) = Alysicarpus Neck. ex Desv. (Legumin.).

Mysotis Hill (sphalm.) = Myosotis L. (Boraginac.).
Mystacidium Lindl. Orchidaceae. 12 E. trop. & S. Afr.
Mystacinus Rafin. = Helinus E. Mey. (Rhamnac.).
Mystropetalon Harv. Balanophoraceae. 1–3 S. Afr. The pollen suggests the possibility that the genus should be given family rank.
Mystroxylon Eckl. & Zeyh. Celastraceae. 3–8 trop. & S. Afr., Madag.
Mystyllus auct. = Mistyllus Presl = Trifolium L. (Legumin.).
Mytilaria Lecomte. Hamamelidaceae. 1 S. China, Indoch. Fls. ♀︎, densely spicate; K 5–6, imbr.; C 5, fleshy, ligulif.; A 10–13, fil. short, thick, anth. obl., intr., connective broad, trunc.; G (2), semi-inf., styles o, stigs. punctif., ov. 2 per loc. K, C, A very early caduc.; G partly immersed in fleshy axis.
Mytilicoccus Zoll. = Lunasia Blanco (Rutac.).
Myuropteris C. Chr. = Dendroglossa Presl (Polypodiac.).
Myxa (Endl.) Lindl. = Cordia L. (Ehretiac.).
Myxapyrus Hassk. = seq.
Myxopyrum Blume. Oleaceae. 12 Indoch., Indomal. Twiners.
Myxospermum M. Roem. = Glycosmis Corrêa (Rutac.).
Myzodendraceae J. G. Agardh. Dicots. 1/11 S. Am. Small semi-parasitic green shrubs (cf. *Loranthac.*), with alt. green or scale-like ent. or cren. exstip. ls. Fls. v. small, ♂ ♀, dioec., naked, in cpd. spikes. ♂ with 2–3 stam. and centr. disk, anth. term., monothec., dehisc. by term. tangential slit. ♀ with 3 stds. sit. in longit. furrows of the ovary, lengthening into long plumose bristles in fr.; G̲ (3), with 1 pend. ov. per loc., and 3 subsess. stigs., surr. by apical annular 'disk'. Fr. a 3-angled or winged achene, with 3 persist. plumose stds.; seed with endosp., without testa. Only genus: *Myzodendron*. Prob. related to *Santalac.* (esp. *Arjona*).
Myzodendron Soland. ex DC. corr. R.Br. Myzodendraceae. 11 temp. S. Am.; parasitic on *Nothofagus*.
Myzodendrum Soland. ex Forst. f. = praec.
Myzorrhiza Phil. (~ Aphyllon Mitch.). Orobanchaceae. 10 Am.

N

Nabadium Rafin. = Ligusticum L. (Umbellif.).
Nabaluia Ames. Orchidaceae. 1 Borneo.
Nabalus Cass. = Prenanthes L. (Compos.).
Nabea Lehm. = Macnabia Benth. (Ericac.).
Nabelekia Roshev. = Leucopoa Griseb. (Gramin.).
Nabia P. & K. = Nabea Lehm. = Macnabia Benth. (Ericac.).
Nabiasodendron Pitard = Gordonia Ellis (Theac.).
Nablonium Cass. Compositae. 1 Tasmania.
Nachtigalia Schinz ex Engl. = Phaeoptilon Radlk. (Nyctaginac.).
Nacibaea Poir. = seq.
Nacibea Aubl. = Manettia Mutis ex L. (Rubiac.).
Nacrea A. Nelson. Compositae. 1 U.S. (Wyoming).
Naegelia Engl. = Nagelia Lindl. = Malacomeles (Decne) Engl. (Rosac.).
Naegelia Regel = Smithiantha Kuntze (Gesneriac.).

Naegelia Zoll. & Mor. = Gouania L. (Rhamnac.).

Naematospermum Steud. = Nematospermum Rich. = Lacistema Sw. (Lacistematac.).

Nagassari Adans. = Mesua L. (Guttif.).

Nagassarium Rumph. = praec.

Nagatampo Adans. = praec.

Nageia Gaertn. = Myrica L. (Myricac.) + Podocarpus L'Hérit. (Podocarpac.).

Nageia Roxb. = Putranjiva Wall. = Drypetes Vahl (Euphorbiac.).

Nageja auctt. = praecc.

Nagelia Lindl. = Malacomeles (Decne) Engl. (Rosac.).

Nageliella L. O. Williams. Orchidaceae. 3 Mex., C. Am.

Nagelocarpus Bullock. Ericaceae. 2 S. Afr.

Naghas Mirb. ex Steud. = Nagassari Adans. = Mesua L. (Guttif.).

Nahusia Schneev. = Fuchsia L. (Onagrac.).

Naiadothrix Pennell = Benjaminia Mart. ex Benj. = Bacopa Aubl. (Scrophulariac.).

Naias Juss. = Najas L. (Najadac.).

Naiocrene (Torr. & Gray) Rydb. Portulacaceae. 2 Alaska to Mex.

*****Najadaceae** Juss. Monocots. 1/50 cosmop. Freshwater annuals, submerged, with slender stems and subopp. or vertic. usu. toothed linear ls. with sheathing base. Fls. unisexual; ♂ a single anther, term. on the axis and 1- or 4-loc., enclosed in a spathe and 2-lipped P; ♀ fl. G 1, naked or surrounded by a membr. perianth-like organ; ovule 1, basal, anatr.; stigs. 2–4. Fr. an indehisc. nut. Embryo straight; no endosp. Pollination occurs under water as in *Zostera*, but the pollen is spherical. Only genus: *Najas*. Affinities obscure.

Najas L. Najadaceae. 50 cosmop.

× **Nakamotoara** hort. Orchidaceae. Gen. hybr. (iii) (Ascocentrum × Neofinetia × Vanda).

Nalagu Adans. = Leea Royen ex L. (Leeac.).

Naletonia Bremek. Rubiaceae. 1 Guiana.

Nallogia Baill. = Champereia Griff. (Opiliac.).

Nama L (1753) = Hydrolea L. (Hydrophyllac.).

*****Nama** L. (1759). Hydrophyllaceae. 40–50 SW. U.S. to S. Am., W.I.; 1 Hawaii.

Namaquanthus L. Bolus. Aïzoaceae. 1 S. Afr.

Namation Brand. Scrophulariaceae. 1 Mex.

Namibia (Schwantes) Dinter & Schwantes. Aïzoaceae. 3 SW. Afr.

Nananthea DC. Compositae. 1 Corsica, Sardinia; 1 Algeria.

Nananthera Willis (sphalm.) = praec.

Nananthus N. E. Brown. Aïzoaceae. 30 S. Afr.

Nanarepenta Matuda. Dioscoreaceae. 1 Mex.

Nanari Adans. = Canarium L. (Burserac.).

Nanatus Phillips (sphalm.) = Nananthus N. E. Br. (Aïzoac.).

Nandhirob[ac]eae A. St-Hil. = Nhandirob[ac]eae A. St-Hil. corr. Endl. = Cucurbitaceae–Zanonioïdeae C. Jeffr.

Nandina Thunb. Nandinaceae. 1, *N. domestica* Thunb., China, Japan.

Nandinaceae J. G. Agardh. Dicots. 1/1 E. As. Shrubs with alt. bi- or tripinnate exstip. ls.; main joints of larger pinnae and pinnules bulbously swollen at base; lfts. ent., acumin., with arcuate-anast. nerves. Fls. reg., ♀, in ∞-fld.

term. pan. K ∞, spirally arranged, white, strobiloidally imbr. in bud; C 6, larger; nectaries o; A 6, oppositipet., anth. subsess., elong.-ellips., intr. by slits, with broad connective; G̲ 1, with short persist. style and trunc. stig., and 1 lat. pend. ov. Fr. a globose red berry; seed with copious endosp. Only genus: *Nandina*. An isolated type, prob. related to *Berberidac.* and *Podophyllac.*

Nandinaceae Horan. (nom. illegit., non J. G. Ag.) = Berberidaceae Juss.

Nandiroba Adans. = Nhandiroba Adans. = Fevillea L. (Cucurbitac.).

Nangha Zipp. ex Mackl. = ? Artocarpus J. R. & G. Forst. (Morac.).

Nani Adans. = Metrosideros Banks ex Gaertn. (Myrtac.).

Nania Miq. = praec.

Nannoglottis Maxim. Compositae. 8 W. China.

Nannorrhops H. Wendl. Palmae. 4 Arabia to NW. India.

Nannoseris Hedb. Compositae. 1 trop. Afr.

Nannothelypteris Holtt. Thelypteridaceae. 4 Philippines.

Nanochilus K. Schum. Zingiberaceae. 1 Sumatra, 1 New Guinea.

Nanocnide Blume. Urticaceae. 4 E. As.

Nanodea Banks. Santalaceae. 1 S. temp. S. Am.

Nanodes Lindl. = Epidendrum L. (Orchidac.).

Nanoglottis P. & K. = Nannoglottis Maxim. (Compos.).

Nanolirion Benth. Liliaceae. 1 S. Afr.

Nanopetalum Hassk. = Cleistanthus Hook. f. ex Planch. (Euphorbiac.).

Nanophyton Less. Chenopodiaceae. 1 SW. to C. As.

Nanophytum Endl. = praec.

Nanorops P. & K. = Nannorhops H. Wendl. (Palm.).

Nanothamnus T. Thoms. Compositae. 1 Bombay.

Nansiatum Miq. (sphalm.) = Natsiatum Buch.-Ham. (Icacinac.).

Napaea L. Malvaceae. 1 N. Am. Dioec. Fibre from bark.

Napea Crantz = praec.

Napeanthus Gardn. Gesneriaceae. 12 Mex. to trop. S. Am.

Napellus v. Wolf = Aconitum L. (Ranunculac.).

Napeodendron Ridley = Walsura Roxb. (Meliac.).

Napimoga Aubl. = Homalium Jacq. (Flacourtiac.).

Napina Frič = Thelocactus (K. Schum.) Britt. & Rose (Cactac.).

Napoea Hill = Napaea L. (Malvac.).

Napoleona Beauv. = Napoleonaea P. Beauv. ex Fisch. (Napoleonaeac.).

Napoleonaceae P. Beauv. = Napoleonaeaceae P. Beauv. emend.

Napoleonaea P. Beauv. ex Fisch. Napoleonaeaceae. 15 trop. W. Afr. The fl. superficially resembles that of *Passiflora*, owing to the corona of stds. Fr. somewhat like a pomegranate.

Napoleonaeaceae P. Beauv. Dicots. 2/18 W. Afr. Trees or shrubs with ent., alt., exstip. ls. Fls. usu. sol., axill. K 3 imbr. or (5) valv.; C o; A ∞, the outermost staminodial, forming a coloured, many-nerved and -toothed, plicate corona or pseudo-corolla; G (3) or (5), with 2–∞, 2–4-seriate ovules per loc., and short expanded or long filiform style. Fruit a large berry. Genera: *Napoleonaea, Crateranthus*.

Napoleone Robin ex Rafin. = Nelumbo Adans. (Nelumbonac.).

Napus Mill. = Brassica L. (Crucif.).

Naravel Adans. = Atragene L. (Ranunculac.).

NARAVELIA

*Naravelia Adans. mut. DC. = praec.
Narbalia Rafin. = Prenanthes L. (Compos.).
Narcaceae Dulac = Solanaceae Juss.
Narcetis P. & K. = Narketis Rafin. = Lomatogonium R.Br. (Gentianac.).
× **Narcibularia** Wehrh. Amaryllidaceae. Gen. hybr. (Corbularia × Narcissus).
Narcissi Juss. = Amaryllidaceae Jaume St-Hil.
Narcisso-Leucojum Ort. = Leucojum L. (Amaryllidac.).
Narcissulus Fabr. (uninom.) = *Leucojum* L. sp. (Amaryllidac.).
Narcissus L. Amaryllidaceae. 60 Eur., Medit., W. As. Corona well developed, free from the A (see fam.).
Narda Vell. = Strychnos L. (Strychnac.).
Nardina Murr. (sphalm.) = Nandina Thunb. (Nandinac.).
Nardophyllum Hook. & Arn. Compositae. 15 S. Andes.
Nardosmia Cass. (~ Petasites L.). Compositae. 15 temp. Euras., N. Am.
Nardostachys DC. Valerianaceae. 2 Himal., W. China. *N. jatamansi* (D.Don) DC., the spikenard, has very fragrant rhizomes.
Narduretia Hug. del Vill. = Nardurus Reichb. + Vulpia C. C. Gmel. (Gramin.).
Narduroïdes Rouy. Gramineae. 1 Medit.
Nardurus Reichb. Gramineae. 6 W. Eur. to India.
Nardus L. Gramineae. 1 Eur., W. As., *N. stricta* L., common on the drier grass moors. Infl. markedly unilateral.
Narega Rafin. = Randia L. (Rubiac.).
*Naregamia Wight & Arn. Meliaceae. 1 SW. trop. Afr.; 1 Penins. India.
Narenga Burkill. Gramineae. 2 NE. India to Borneo.
Nargedia Bedd. Rubiaceae. 1 Ceylon.
Narica Rafin. = Sarcoglottis Presl (Orchidac.).
Naringi Adans. = Hesperethusa M. Roem.
Narketis Rafin. = Lomatogonium A.Br. (Gentianac.).
Naron Medik. = Moraea Mill. ex L. (Iridac.).
Nartheciaceae Small = Liliaceae–Tofieldieae Kunth.
Narthecium Ehrh. (uninom.) = *Anthericum ossifragum* L. = *Narthecium ossifragum* (L.) Huds. (Liliac.).
Narthecium Gérard = Tofieldia Huds. (Liliac.).
*Narthecium Huds. Liliaceae. 1 Japan; 2–4 N. Am.; 1 NW. Eur., Corsica. *N. ossifragum* (L.) Huds. (Eur.) has a sympodial rhiz. and isobil. ls. The fl. is conspicuous, but contains no honey.
Narthex Falc. = Ferula L. (Umbellif.).
Narukila Adans. = Pontederia L. (Pontederiac.).
Narum Adans. = Uvaria L. (Annonac.).
Naruma Rafin. = praec.
Narvalina Cass. Compositae. 4 W.I., S. Am.   [(Ranunculac.).
Narvelia Link (sphalm.) = Naravelia DC. = Naravel Adans. = Atragene L.
Nashia Millspaugh. Verbenaceae. 7 W.I. Ls. used as tea.
Nasmythia Huds. = Eriocaulon L. (Eriocaulac.).
Nasonia Lindl. = Centropetalum Lindl. (Orchidac.).
Nassauvia Comm. ex Juss. Compositae. 70 S. Andes.
Nassavia Vell. = Allophylus L. (Sapindac.).
Nassawia Lag. = Nassauvia Comm. ex Juss. (Compos.).

Nassella E. Desv. Gramineae. 10 Andes.
Nassovia Batsch = Nassauvia Comm. ex Juss. (Compos.).
Nastanthus Miers = Acarpha Griseb. (Calycerac.).
Nasturtiastrum Gillet & Magne = Lepidium L. (Crucif.).
Nasturtiicarpa Gilli = Calymmatium O. E. Schulz (Crucif.).
Nasturtioïdes Medik. = Lepidium L. (Crucif.).
Nasturtiolum S. F. Gray = Hornungia Reichb. (Crucif.).
Nasturtiolum Medik. = Coronopus Zinn (Crucif.).
Nasturtiopsis Boiss. Cruciferae. 1 N. Afr. to Arabia.
*****Nasturtium** R.Br. (~ Rorippa Scop.). Cruciferae. 6 Eur. to C. As. & Afghan.,
N. Afr., mts. trop. E. Afr., N. Am. *N. officinale* R.Br., *N. microphyllum* Boenn.,
& hybrid (watercress), cult. & almost cosmop.
Nasturtium Mill. = Lepidium L. + Coronopus Zinn (Crucif.).
Nasturtium Zinn = Thlaspi L. (Crucif.).
Nastus Juss. Gramineae. 13 Mascarene Is.
Nastus Lunell = Cenchrus L. (Gramin.).
Natalanthe Sond. = Tricalysia A. Rich. (Rubiac.).
Natalia Hochst. = Bersama Fresen. (Melianthac.).
Nathaliella B. Fedtsch. Scrophulariaceae. 1 C. As.
Nathusia Hochst. = Schrebera Roxb. (Oleac.).
Natrix Moench = Ononis L. (Legumin.).
Natschia Bub. = Nardus L. (Gramin.).
Natsiatopsis Kurz. Icacinaceae. 1 Burma.
Natsiatum Buch.-Ham. Icacinaceae. 1 E. Himalaya to Lower Burma &
Indoch.
Nattamame Banks = Canavalia Adans. (Legumin.).
Nauchea Descourt. = Clitoria L. (Legumin.).
Nauclea Korth. = Neonauclea Merr. (Rubiac.).
Nauclea L. (*s.str.*; *Sarcocephalus* Afzel.). Naucleaceae (~ Rubiac.). 35 trop.
Afr., As., Polynes.
Naucleaceae (DC.) Wernh. (~ Rubiaceae Juss.). Dicots. 10/200 trop. Trees
or shrubs, sometimes scandent, with opp. or vertic. simple ent. stip. ls. (stips.
usu. interpet., caduc.; rarely intrapet., ± persist.). Fls. reg., ⚥, sess. or pedic.,
bracteolate or ebracteolate, in dense globose term. or axill. capit. infl. K (5–4),
valv. or imbr., caduc. or persist., sometimes truncate; C (5–4) valv. or imbr.,
tube long, lobes short; A 5–4, inserted in throat of C; G̅ (2), with slender style
and ent. or bilobed stig., and 1–∞ imbr. ov. per loc., on plac. either pend. from
apex or attached to middle of septum. Fr. a variously dehisc. or more rarely
indehisc. caps., or concrescent into a fleshy syncarp; seeds minute, usu. flattened, with endosp. Chief genera: *Nauclea, Neonauclea, Uncaria, Mitragyna.*
This group of 'Rubiaceae' makes an evident approach, esp. in habit and
vegetative chars., to Combretac. (esp. *Combretum*). As pointed out by
Wernham, it is so distinct from the remainder of *Rubiac.* that it prob. merits
recognition as a distinct fam.
Naucleopsis Miq. Moraceae. 10 C. & trop. S. Am.
Naucorephes Rafin. = Coccoloba L. (Polygonac.).
Naudinia Decne ex Triana = Naudiniella Krasser = Lomanodia Rafin.
(Melastomatac.).

NAUDINIA

*Naudinia Planch. & Lind. Rutaceae. 1 Colombia. Sympetalous.

Naudinia A. Rich. = Tetrazygia Rich. (Melastomatac.).

Naudiniella Krasser = Lomanodia Rafin. (Melastomatac.).

Nauenburgia Willd. = Flaveria Juss. (Compos.).

Nauenia Klotzsch = Lacaena Lindl. (Orchidac.).

Naufraga Constance & Cannon. Hydrocotylaceae. (~ Umbelliferae). 1 Balearic Is.

Naumannia Warb. = Riedelia Oliv. (Zingiberac.).

Naumburgia Moench = Lysimachia L. (Primulac.).

Nauplius Cass. = Odontospermum Neck. ex Sch. Bip. (Compos.).

Nautea Nor. = Tectona L. f. (Verbenac.).

Nauticalyx Hort. ex Loud. = seq.

Nautilocalyx Linden. Gesneriaceae. 14 trop. S. Am.

Nautochilus Bremek. Labiatae. 4 S. Afr. (Transvaal).

Nautonia Decne. Asclepiadaceae. 1 S. Brazil.

Nautophylla Guillaumin (~ Logania R.Br.). Loganiaceae. 1 New Caled.

Navaea Webb & Berth. = Lavatera L. (Malvac.).

Navajoa Croizat = Pediocactus Britt. & Rose (Cactac.).

Navarretia Ruiz & Pav. Polemoniaceae. 30 W. N. Am., 1 Chile & Argent. *N. cotulifolia* Hook. & Arn. is 4-merous; *N. filicaulis* Greene is both loculicid. and septicid.; some dehisce irreg.

Navarria Mutis apud Alba. Colombia. Quid?

Navenia Benth. & Hook. f. = Nauenia Klotzsch = Lacaena Lindl. (Orchidac.).

Navia Schult. f. Bromeliaceae. 60 N. trop. S. Am. (Guayana Highland).

Navicularia Fabr. = Sideritis L. (Labiat.).

Navicularia Raddi = Ichnanthus Beauv. (Gramin.).

Navidura Alef. = Lathyrus L. (Legumin.).

Navipomoea Roberty = Ipomoea L. (Convolvulac.).

Naxiandra (Baill.) Krasser = Axinandra Thw. (Memecylac.).

Nayas Neck. = Najas L. (Najadac.).

Nazia Adans. = Tragus Haller (Gramin.).

Neactelis Rafin. = Helianthus L. (Compos.).

Neaea Juss. = Neea Ruiz & Pav. (Nyctaginac.).

Neaera Salisb. = Stenomesson Herb. (Amaryllidac.).

Nealchornea Huber. Euphorbiaceae. 1 Upper Amazon (Colombia).

Neamyza Van Tiegh. Loranthaceae. 1 N.Z.

Neanotis W. H. Lewis. Rubiaceae. 28 trop. As. & Austr.

Neanthe P.Br. ?Leguminosae (inc. sed.). 1 Jamaica.

Neanthe O. F. Cook = Collinia (Mart.) Liebm. (Palm.).

Neara Soland. ex Seem. = Meryta J. R. & G. Forst. (Araliac.).

Neatostema I. M. Johnston. Boraginaceae. 1 Canary Is., Medit. to Iraq.

Nebasiodendron Pitard (sphalm.) = Nabiasodendron Pitard = Gordonia Ellis (Theac.).

Nebelia Neck. ex Sweet. Bruniaceae. 6 S. Afr.

Neblinaea Maguire & Wurdack. Compositae. 1 Venez.

Neblinanthera Wurdack. Melastomataceae. 1 Venez.

Neblinaria Maguire. Bonnetiaceae. Spp.? Venez.

Neblinathamnus Steyermark. Rubiaceae. 2 Venez.

Nebra Nor. ex Choisy = Neea Ruiz & Pav. (Nyctaginac.).
Nebropsis Rafin. = Aesculus L. (Hippocastanac.).
Nebrownia Kuntze = Philonotion Schott (Arac.).
Necalistis Rafin. = Ficus L. (Morac.).
Necepsia Prain. Euphorbiaceae. 1 trop. Afr.
Nechamandra Planch. = Lagarosiphon Harv. (Hydrocharitac.).
Neckeria J. F. Gmel. = Pollichia Soland. (Caryophyllac.).
Neckeria Scop. = Corydalis Vent. (Fumariac.).
Neckia Korth. Ochnaceae. 9 (or perhaps 1 variable) W. Malaysia.
Necramium Britton. Melastomataceae. 1 Trinidad.
Necranthus Gilli. Orobanchaceae. 1 Turkey.
Nectalisma Fourr. = Luronium Rafin. (Alismatac.).
Nectandra Bergius = Gnidia L. (Thymelaeac.).
*****Nectandra** Roland. ex Rottb. (~ Ocotea Aubl.). Lauraceae. 100 C. to subtrop. S. Am. *N. rodiaei* Hook. (greenheart) and others yield good timber.
Nectandra Roxb. = Linostoma Wall. (Thymelaeac.).
Nectariaceae Dulac = Liliaceae Juss. (*s.str.*).
Nectaripetalum Pohl (nomen). 1 Brazil. Quid?
Nectarobothrium Ledeb. = Lloydia Salisb. (Liliac.).
Nectaropetalaceae Exell & Mendonça = Erythroxylaceae Kunth.
Nectaropetalum Engl. Erythroxylaceae. 6 trop. & S. Afr.
Nectaropetalum P. & K. = Nectaripetalum Pohl (inc. sed.).
Nectaroscilla Parl. = Scilla L. (Liliac.). [to Persia.
Nectaroscordum Lindl. (~ Allium L.). Alliaceae. 2–3 S. France & Sardinia
Nectolis Rafin. = Salix L. (§ Amerina, Viminales, etc.) (Salicac.).
Nectopix Rafin. = Salix L. (§ Vetrix) (Salicac.).
Nectouxia DC. = Morettia DC. (Crucif.).
Nectouxia Kunth. Solanaceae. 1 Mex.
Nectris Schreb. = Cabomba Aubl. (Cabombac.).
Nectusion Rafin. = Salix L. (§ Chamaetia) (Salicac.).
Neea Ruiz & Pav. Nyctaginaceae. 80 Mex. to trop. S. Am., W.I. The ls. of *N. theifera* Oerst. (*caparrosa*) are used as tea, and yield a black dye.
Neeania Rafin. = praec.
Needhamia R.Br. = Needhamiella L. Watson (Epacridac.).
Needhamia Cass. = Narvalina Cass. (Compos.).
Needhamia Scop. = Tephrosia Pers. (Legumin.).
Needhamiella L. Watson. Epacridaceae. 1 SW. Austr.
Neëragrostis Bush (~ Eragrostis Host). Gramineae. 1 S. U.S., Mex., Guiana.
Neerija Roxb. = Elaeodendron Jacq. (Celastrac.).
Neesenbeckia Levyns. Cyperaceae. 1 S. Afr.
*****Neesia** Blume. Bombacaceae. 8 W. Malaysia.
Neesia Mart. ex Meissn. = Funifera Leandro ex C. A. Mey. (Thymelaeac.).
Neesia Spreng. = Otanthus Hoffmgg. & Link (Compos.).
Neesiella Sreemadhavan (1967; non Schiffn. 1893—Bryoph.) = Indoneesiella Sreemadh. (Acanthac.).
Neesiochloa Pilger. Gramineae. 1 Brazil.
Nefflea Spach = Celsia L. (Scrophulariac.).
Nefrakis Rafin. = Brya P.Br. (Legumin.).

NEGRETIA

Negretia Ruiz & Pav. = Mucuna Adans. (Legumin.).

Negria Chiov. = Joannegria Chiov. (Gramin.).

Negria F. Muell. Gesneriaceae. 1 Lord Howe I.

Negripteridaceae Pichi-Serm. Related to *Sinopteris* C. Chr. & Ching, but cells of annulus of sporangium with thickened outer walls, stomium lacking (*Nuovo Giorn. Bot. Ital.* **53**: 129–69, 1946). Only genus: *Negripteris* Pichi-Serm.

Negripteris Pichi-Serm. Negripteridaceae. 1 Ethiopia.

Negundium Rafin. = seq.

Negundo Boehm. (~Acer L.). Aceraceae. 8 N. temp.

Neillia D. Don. Rosaceae. 13 E. Himal. to Korea, Indoch., Sumatra, Java.

Neilliaceae Miq. = Rosaceae–Spiraeëae Juss.

Neilreichia Fenzl = Schistocarpha Less. (Compos.).

Neilreichia B. D. Jacks. (nomen fictum) = Carex L. (Cyperac.).

Neiosperma Rafin. = Ochrosia Juss. (Apocynac.).

Neippergia C. Morr. = Acineta Lindl. (Orchidac.).

Neisandra Rafin. = Hopea Roxb. (Dipterocarpac.).

Neja D. Don = Hysterionica Willd. (Compos.).

Nekemias Rafin. = Ampelopsis Michx (Vitidac.).

Nelanaregam Adans. = Naregamia Wight & Arn. (Melastomatac.).

Nelanaregum Kuntze = praec.

Neleixa Rafin. = Faramea Aubl. (Rubiac.).

Nelensia Poir. = Enslenia Rafin. = Ruellia L. (Acanthac.).

Nelia Schwantes. Aïzoaceae. 4 S. Afr.

Nelipus Rafin. = Utricularia L. (Lentibulariac.).

Nelis Rafin. = Goodyera R.Br. (Orchidac.).

Nelitris Gaertn. = Timonius DC. (Rubiac.).

Nelitris Spreng. = Decaspermum J. R. & G. Forst. (Myrtac.).

Nellica Rafin. = Phyllanthus L. (Euphorbiac.).

Nelmesia Van der Veken. Cyperaceae. 1 trop. Afr.

Nelsia Schinz. Amaranthaceae. 1 S. trop. & S. Afr.

Nelsonia R.Br. Scrophulariaceae or Acanthaceae. 1 trop. Afr., India to Indoch., Malay Penins., Austr.

Neltuma Rafin. = Prosopis L. (Legumin.).

Nelumbicum Rafin. = seq.

Nelumbium Juss. = seq.

Nelumbo Adans. Nelumbonaceae. 2, *N. pentapetala* (Walt.) Fernald (*N. lutea* Pers.), E. U.S. to Colombia, and *N. nucifera* Gaertn. (*N. speciosa* Willd.), As. and NE. Austr. The latter, sometimes supposed to be the sacred Lotus, no longer found in the Nile, is sacred in India, Tibet, China, and was introd. to Egypt about 500 B.C. (cf. *Nymphaea*). The fls., which are v. large and handsome, and the big pelt. slightly hairy ls., stand above the water and do not float upon it. The seeds of *N. nucifera* are used as food in Kashmir, etc.

Nelumbonaceae Dum. Dicots. 1/2 warm As., Austr., Am. Large aquatic perennial herbs. The rhiz. bears 'triads' of leaves; after a long internode comes a scale-l. on the lower side, then one on the upper side, immediately followed by a foliage-l. with ochreate stipule, then a long internode again, and so on. This peculiar leaf-arrangement is unique. From the axil of the second scale-l. springs the fl., from that of the foliage-l. a branch. The fl. has no bracteoles.

The first P-leaf is ant., the second post., then follow 2 lat.; these 4 are sometimes regarded as a K. They are followed by numerous petals and sta., acyclically arranged. In the centre of the fl. stands the obconical G, a large number of cpls. embedded separately in the top of the swollen spongy recept. Each carp. contains 1 pend. ovule. The recept. becomes dry and very light, and the achenes separate from it, as the fruit ripens. It breaks off bodily from the stalk and floats about until decay releases the fruits, which sink to the bottom of the pond. There is no endosp. or perisperm. Only genus: *Nelumbo*. Not closely related to *Nymphaeaceae*, *Euryalaceae*, etc., but perh. some connection with *Podophyllaceae* (cf. *Jeffersonia*, etc.). See Takhtadjian, *Syst. & Phylog. Fl. Pl.*: 89, 1966; also Gupta & Ahuja, *Naturwissenschaften*, **54**: 498, 1967.

Nemacaulis Nutt. Polygonaceae. 1 California.

Nemacladaceae Nutt. = Campanulaceae–Cyphioïdeae Reichb.

Nemacladus Nutt. Campanulaceae. 10 SW. U.S., Mex.

Nemaconia Knowles & Westc. = Ponera Lindl. (Orchidac.).

Nemallosis Rafin. = Mollugo L. (Aïzoac.).

Nemaluma Baill. Sapotaceae. 1 Guiana.

Nemampsis Rafin. = Dracaena L. (Agavac.).

Nemanthera Rafin. = Ipomoea L. vel Merremia Dennst. (Convolvulac.).

Nemastachys Steud. = Pollinia Trin. (Gramin.).

Nemastylis Nutt. Iridaceae. 25 Am.

Nematanthera Miq. = Piper L. (Piperac.).

Nematanthus Nees = Willdenowia Thunb. (Restionac.).

Nematanthus Schrad. Gesneriaceae. 6 Brazil.

Nematoceras Hook. f. = Corybas Salisb. (Orchidac.).

Nematolepis Turcz. Rutaceae. 2 W. Austr.

Nematopera Kunze = Peranema D. Don (Aspidiac.).

Nematophyllum F. Muell. = Templetonia R.Br. (Legumin.).

Nematopoa C. E. Hubbard. Gramineae. 1 S. trop. Afr.

Nematopogon Bureau & K. Schum. = Digomphia Benth. (Bignoniac.).

Nematopteris v. Ald. v. Ros. Grammitidaceae. 1 Borneo.

Nematopus A. Gray = Gnephosis Cass. (Compos.).

Nematopyxis Miq. = Ludwigia L. (Onagrac.).

Nematosciadium H. Wolff. Umbelliferae. 1 Mex.

Nematospermum Rich. = Lacistema Sw. (Lacistematac.).

Nematostemma Choux. Asclepiadaceae. 1 Madag.

Nematostigma A. Dietr. = Libertia Spreng. (Iridac.).

Nematostigma Benth. & Hook. f. = Nemostigma Planch. = Gironniera Gaudich. (Ulmac.).

Nematostylis Hook. f. = Ernestimeyera Kuntze (Rubiac.).

Nemauchenes Cass. = Crepis L. (Compos.).

Nemaulax Rafin. = Albuca L. (Liliac.).

Nemcia Domin (~ Oxylobium Andr.). Leguminosae. 12 W. Austr.

Nemedra A. Juss. = Aglaia Lour. (Meliac.).

Nemelaia Rafin. Inc. sed. (Myrsinaceae?). Hab.?

Nemelataceae Dulac = Urticaceae Juss. (*s.str.*).

Nemepiodon Rafin. = Hymenocallis Salisb. (Amaryllidac.).

Nemepis Rafin. = Cuscuta L. (Cuscutac.).

Nemesia Vent. Scrophulariaceae. 50 S. Afr.
Nemexia Rafin. = Smilax L. (Smilacac.).
Nemexis Rafin. = praec.
Nemia Bergius = Manulea L. (Scrophulariac.).
Nemitis Rafin. = Apteria Nutt. (Burmanniac.).
Nemocharis Beurl. = Scirpus L. (Cyperac.).
Nemochloa Nees = Pleurostachys Brongn. (Cyperac.).
Nemoctis Rafin. = Lachnaea L. (Thymelaeac.).
Nemodaphne Meissn. = Ocotea Aubl. (Laurac.).
Nemodon Griff. = Lepistemon Blume (Convolvulac.).
Nemolapathum Ehrh. (uninom.) = *Rumex nemolapathum* Ehrh. = *R. conglomeratus* Murr. + *R. sanguineus* L. (Polygonac.).
Nemolepis Vilm. = Heliopsis Pers. (Compos.).
Nemopanthes Rafin. = seq.
***Nemopanthus** Rafin. Aquifoliaceae. 1 NE. N. Am.
***Nemophila** Nutt. Hydrophyllaceae. 11 W., 2 SE. N. Am.
Nemopogon Rafin. = Bulbine L. (Liliac.).
Nemorella Ehrh. (uninom.) = *Lysimachia nemorum* L. (Primulac.).
Nemorinia Fourr. = Luzula DC. (Juncac.).
Nemorosa Nieuwl. = Anemone L. (Ranunculac.).
Nemoseris Greene = Rafinesquia Nutt. (Compos.).
Nemostigma Planch. = Gironniera Gaudich. (Ulmac.).
Nemostima Rafin. = Convolvulus L. (Convolvulac.).
Nemostylis Herb. = Nemastylis Nutt. (Iridac.).
Nemostylis Stev. = Phuopsis Benth. & Hook. f. (Rubiac.).
Nemuaron Baill. Atherospermataceae. 2 New Caled.
Nemum Desv. = Scirpus L. (Cyperac.).
Nemuranthes Rafin. = Habenaria Willd. (Orchidac.).
Nenax Gaertn. Rubiaceae. 5 S. Afr.
Nenga H. Wendl. & Drude. Palmae. 4 SE. As., W. Malaysia (exc. Philipp.).
Nengella Becc. Palmae. 7 New Guinea.
Nenningia Opiz = Campanula L. (Campanulac.).
Nenuphar Link = Nuphar Sm. (Nymphaeac.).
Neoabbottia Britton & Rose. Cactaceae. 1 Hispaniola.
Neoalsomitra Hutch. Cucurbitaceae. 12 Indomal., Austr., Polynesia.
Neobaclea Hochr. Malvaceae. 2 temp. S. Am.
Neobaileya Gandog. = Geranium L. (Geraniac.).
Neobakeria Schlechter. Liliaceae. 9 S. Afr.
Neobambus Keng ex P. C. Keng = Arundinaria Michx (Gramin.).
Neobaronia Baker = Phylloxylon Baill. (Legumin.).
Neobartlettia R. M. King & H. Rob. (1971; non Schltr. 1920) = Bartlettina R. M. King & H. Rob. (Compos.).
Neobartlettia Schlechter. Orchidaceae. 6 trop. S. Am.
Neobathiea Schlechter. Orchidaceae. 6 Madag.
Neobaumannia Hutch. & Dalz. Rubiaceae. 1 trop. E. Afr.
Neobeckia Greene = Rorippa Scop. (Crucif.).
Neobeguea Leroy. Meliaceae. 3 Madag.
Neobenthamia Rolfe. Orchidaceae. 1 trop. E. Afr.

Neobertiera Wernham. Rubiaceae. 1 Guiana.
Neobesseya Britton & Rose. Cactaceae. 6 SW. U.S., Mex.
Neobinghamia Backeb. = Haageocereus Backeb. (Cactac.).
Neobiondia Pampan. = Saururus L. (Saururac.).
Neoblakea Standley. Rubiaceae. 1 Venez.
Neoboivinella Aubrév. & Pellegr. = Bequaertiodendron De Wild. (Sapotac.).
Neobolusia Schlechter. Orchidaceae. 4 trop. E. & S. Afr.
Neobotrydium Moldenke = Chenopodium L. (Chenopodiac.).
Neoboutonia Muell. Arg. Euphorbiaceae. 8 trop. Afr.
Neobouteloua Gould. Gramineae. 1 W. C. Argent.
Neoboykinia Hara = Boykinia Nutt. (Saxifragac.).
Neobracea Britton. Apocynaceae. 4 Bahamas, Cuba.
Neobrittonia Hochr. Malvaceae. 1 Mex.
Neobuchia Urban. Bombacaceae. 1 W.I.
Neoburkillia Whitmore. Rutaceae. 1 Malay Penins., Sumatra.
Neobuxbaumia Backeb. (~ Cephalocereus Pfeiff.). Cactaceae. 4 Mex.
Neocabreria R. M. King & H. Rob. (~ Eupatorium L.). Compositae. 3 Braz. to Argent.
Neocaldasia Cuatrec. Compositae. 1 Colombia.
Neocallitropsis Florin. Cupressaceae. 1 New Caled.
Neocalyptrocalyx Hutch. Capparidaceae. 2 trop. S. Am.
Neocardenasia Backeb. = Neoraimondia Britt. & Rose (Cactac.).
Neocarya (DC.) Prance. Chrysobalanaceae. 1 trop. W. Afr.
Neocastela Small = Castela Turp. (Simaroubac.).
Neoceis Cass. = Erechtites Rafin. (Compos.).
Neocentema Schinz. Amaranthaceae. 2 trop. E. Afr.
Neochamaelea (Engl.) Erdtm. Cneoraceae. 1 Canary Is.
Neocheiropteris Christ. Polypodiaceae. 3 Indomal. to Japan.
Neochevaliera A. Chev. & Beille = Chaetocarpus Thw. (Euphorbiac.).
Neochevalierodendron J. Léonard. Leguminosae. 1 W. Equat. Afr.
Neochilenia Backeb. = Nichelia Bullock = Pyrrhocactus (A. Berger) Backeb. & F. M. Knuth (Cactac.).
Neocinnamomum H. Liou (~ Cinnamomum L.). Lauraceae. 8 S. China, Indoch.
Neoclemensia C. E. Carr. Orchidaceae. 1 Borneo.
Neocleome Small = Cleome L. (Cleomac.).
Neoclia Nor. = ? Cyclea Arn. ex Wight (Menispermac.).
Neocogniauxia Schlechter. Orchidaceae. 2 W.I.
Neocollettia Hemsl. Leguminosae. 1 Burma, C. Java, in teak forests. Geocarpic. Two strong roots developing on the gynophore prob. feed the young fr. and perh. help to pull it into the earth.
Neocouma Pierre. Apocynaceae. 2 Brazil.
Neocracca Kuntze. Leguminosae. 1 Bolivia.
Neocuatrecasia R. M. King & H. Rob. (~ Eupatorium L.). Compositae. 4 C. Andes.
Neocussonia Hutch. Araliaceae. 5 trop. & S. Afr., Madag.
Neodawsonia Backeb. = Cephalocereus Pfeiff. (Cactac.).
Neodeutzia (Engl.) Small = Deutzia Thunb. (Hydrangeac.).

Neodielsia Harms. Leguminosae. 1 China.
Neodissochaeta Bakh. f. Melastomataceae. 10+, Penins. Siam, Malaysia.
Neodistemon Babu & Henry. Urticaceae. 1 Indomal.
Neodonnellia Rose (~ Tripogandra Rafin.). Commelinaceae. 1 C. Am.
Neodregia C. H. Wright. Liliaceae. 1 S. Afr.
Neodryas Reichb. f. Orchidaceae. 5 trop. S. Am.
Neodunnia R. Viguier. Leguminosae. 4 Madag.
Neodypsis Baill. Palmae. 15 Madag.
Neoëvansia W. T. Marshall = Peniocereus (A. Berger) Britt. & Rose (Cactac.).
Neofinetia Hu. Orchidaceae. 1 China, Japan.
Neofranciella Guillaumin. Rubiaceae. 1 New Caled.
Neogaerrhinum Rothm. Scrophulariaceae. 2 SW. U.S.
Neogardneria Schlechter. Orchidaceae. 3 Brazil, Guiana.
Neogaya Meissn. = Pachypleurum Ledeb. (Umbellif.).
Neoglaziovia Mez. Bromeliaceae. 2 Brazil.
Neogoetzea Pax = Bridelia Willd. (Euphorbiac.).
Neogoezea Hemsl. Umbelliferae. 3 Mex.
Neogomesia Castañeda = Ariocarpus Scheidw. (Cactac.).   [Afghan.
Neogontscharovia Lincz. Limoniaceae. (~ Plumbaginaceae). 3 C. As.,
Neogoodenia C. A. Gardn. & George. Goodeniaceae. 1 SW. Austr.
Neogossypium Roberty = Gossypium L. (Malvac.).
Neoguillauminia Croizat. Euphorbiaceae. 1 New Caled.
Neogunnia Pax & K. Hoffm. Aïzoaceae. 1 W., S. & E. Austr.
Neogyna Reichb. f. Orchidaceae. 1 N. India, SW. China, Indoch.
Neogyne Pfitzer = praec.
Neohallia Hemsl. Acanthaceae. 1 S. Mex.
Neoharmsia R. Viguier. Leguminosae. 1 Madag.
Neohenricia L. Bolus. Aïzoaceae. 1 S. Afr.
Neohenrya Hemsl. Apocynaceae. 2 China.
Neohickenia Frič = Parodia Speg. (Cactac.).
Neohintonia R. M. King & H. Rob. (~ Eupatorium L.). Compositae. 1 Mex.
Neohopea G. H. S. Wood ex Ashton = Shorea Roxb. ex Gaertn. (Dipterocarpac.).
Neohouzeaua (Camus) Gamble. Gramineae. 4 SE. As.
Neohuberia Ledoux. Lecythidaceae. Spp.? Loc.? Quid?
Neohumbertiella Hochr. Malvaceae. 3 Madag.
Neohusnotia A. Camus = Acroceras Stapf (Gramin.).
Neohyptis J. K. Morton. Labiatae. 1 Angola.
Neojatropha Pax. Euphorbiaceae. 2 E. trop. Afr.
Neojobertia Baill. Bignoniaceae. 1 NE. Brazil.
Neojunghuhnia Koorders = Vaccinium L. (Ericac.).
Neokeithia v. Steenis = Chilocarpus Bl. (Apocynac.).
Neokoehleria Schlechter. Orchidaceae. 4 Peru.
Neokoeleria Ind. Kew. (sphalm.) = praec.
Neolacis Wedd. = Apinagia Tul. (Podostemac.).
Neolauchea Kraenzl. Orchidaceae. 1 Brazil.
Neolehmannia Kraenzl. = Epidendrum L. (Orchidac.).
Neolemaireocereus Backeb. = Stenocereus (A. Berger) Riccob. = Lemaireocereus Britton & Rose (Cactac.).

Neolemonniera Heine. Sapotaceae. 3 trop. Afr.

Neolepisorus Ching. Polypodiaceae. 5–6 Madag., SW. China to W. Malaysia. (*Bull. Fan Mem. Inst.* 10: 11, 1940.)

Neoleptopyrum Hutch. Ranunculaceae. 1 temp. As. (W. Sib. to China & Japan).

Neoleretia Baehni = Nothapodytes Bl. (Icacinac.).

Neoleroya Cavaco. Rubiaceae. 1 Madag.

Neolexis Salisb. = Smilacina Desf. (Liliac.).

Neolindenia Baill. Acanthaceae. 1 Mex.

Neolindleya Kraenzl. = Platanthera Lindl. (Orchidac.).

Neolindleyella Fedde = Lindleya Kunth (Rosac.).

*Neolitsea (Benth.) Merrill (= Bryantea Rafin.). Lauraceae. 80 E. & SE. As., Indomal.

Neolloydia Britton & Rose (~ Thelocactus (K. Schum.) Britt. & Rose). Cactaceae. 8 S. U.S., Mex., Cuba.

Neolobivia Y. Ito = Lobivia Britt. & Rose (Cactac.).

Neolophocarpus E. G. Camus. Cyperaceae. 1 Indoch.

Neolourya L. Rodrig. Liliaceae. 2 Indoch.

Neoluederitzia Schinz. Zygophyllaceae. 1 SW. Afr.

Neoluffa Chakrav. Cucurbitaceae. 1 E. Himal.

Neomacfadya Baill. Bignoniaceae. 1 C. Am., W.I.

Neomacfadyena K. Schum. = praec.

Neomammillaria Britton & Rose = Mammillaria Haw. (Cactac.).

Neomandonia Hutch. Commelinaceae. 1 Mex., C. Am.

Neomanniophyton Pax & K. Hoffm. = Crotonogyne Muell. Arg. (Euphor-

Neomarica Sprague. Iridaceae. 15 trop. Am. [biac.).

Neomartinella Pilger. Cruciferae. 1 China.

Neomazaea Urb. Rubiaceae. 4 W.I.

Neomezia Votsch. Theophrastaceae. 1 W.I. (Cuba).

Neomillspaughia S. F. Blake. Polygonaceae. 2 C. Am.

Neomimosa Britton & Rose = Mimosa L. (Legumin.).

Neomirandea R. M. King & H. Rob. (~ Eupatorium L.) Compositae. 13 Mex. to Equador.

Neomolinia Honda. Gramineae. 3 E. As.

Neomoorea Rolfe. Orchidaceae. 1 Colombia.

Neomphalea Pax & K. Hoffm. = Omphalea L. (Euphorbiac.).

Neomuellera Briquet. Labiatae. 2 SW. Afr.

Neomyrtus Burret. Myrtaceae. 1 N.Z.

Neonauclea Merr. (Nauclea Korth., non L.). Naucleaceae (~ Rubiac.).

Neonelsonia Coulter & Rose. Umbelliferae. 1 Mex. [70 Indomal.

Neonicholsonia Dammer. Palmae. 1 C. Am.

Neoniphopsis Nakai = Pyrrosia Mirbel (Polypodiac.).

Neopalissya Pax. Euphorbiaceae. 1 Madag.

Neopallasia P. Poljakov. Compositae. 1 C. As., Mongol.

Neopanax Allan = Pseudopanax C. Koch (Araliac.).

Neoparrya Mathias. Umbelliferae. 1 SW. U.S.

Neopatersonia Schönland. Liliaceae. 3 S. & SW. Afr.

Neopaxia Ö. Nilss. Portulacaceae. 1 Victoria, Tasm., N.Z.

NEPENTHANDRA

the leaf-midribs. The end of the tendril develops as a rule into a pitcher, with a lid projecting over the mouth, but not closing it except in the young state. The pitcher develops by an invagination of the upper surface of the tip of the l.; the tip takes no part in the development, and the lid grows out below it. The edge of the pitcher is curved inwards; at the entrance are numerous honey-glands, and for some distance below it are other glands, sunk in little pits on the inner surface. Insects attracted by the honey (or by the bright colour) gradually work their way downwards among the glands, and presently get upon the slippery lower part and ultimately into the water at the bottom of the pitcher, where they are drowned. The plant absorbs the products of their decay.

Many are epiphytic. In *N. ampullaria* Jack there are two kinds of ls. (cf. *Cephalotus* and *Triphyophyllum*), some with tendrils and no pitchers; others, as stalked pitchers arranged in a radical rosette.

Fls. dioec., reg., in racemes or with the secondary branching cincinnal; no bracts. P 2+2; in the ♂ fl. A (4–16) in a column; in the ♀ fl. G (4), 4-loc.; ovules ∞, anatr., in many rows. Capsule leathery, loculic. Seeds light, usu. with long hair-like processes at the ends; embryo straight, in fleshy endosp. Genera: *Nepenthes, Anurosperma*.

Nepenthandra S. Moore = Trigonostemon Bl. (Euphorbiac.).
Nepenthes L. Nepenthaceae. 67 Madag., Ceylon, Assam, S. China, Indoch., Malaysia, N. Queensl., New Caled.
Nepeta L. Labiatae. 250 temp. Euras., N. Afr., mts. trop. Afr. Fls. gyno-dioecious.
Nepetaceae Horan. = Labiatae Juss.
Nephelaphyllum Blume. Orchidaceae. 12 India to China and Malaysia.
Nephelium L. Sapindaceae. 35 Burma to Indoch., W. Malaysia. *N. lappaceum* L. (*rambutan*) cult. ed. fr. [For *N. longana* (Lam.) Cambess., see *Euphoria*; for *N. litchi* Cambess., see *Litchi*.]
Nephelochloa Boiss. Gramineae. 1 W. As.
Nephopteris Maxon ex Lellinger. Hemionitidaceae. 1 Colombia.
Nephracis P. & K. = Nefrakis Rafin. = Brya P.Br. (Legumin.).
Nephradenia Decne. Asclepiadaceae. 10 Mex. to Brazil.
Nephraea Hassk. = Nephrea Nor. = Pterocarpus Jacq. (Legumin.).
Nephraeles B. D. Jacks. (sphalm.) = seq.
Nephralles Rafin. = Commelina L. (Commelinac.).
Nephrallus Rafin. = praec.
Nephrandra Willd. = Vitex L. (Verbenac.).
Nephrangis (Schlechter) Summerhayes. Orchidaceae. 1 trop. Afr.
Nephrantera Hassk. = seq.
Nephranthera Hassk. = Renanthera Lour. (Orchidac.).
Nephrea Nor. = Pterocarpus Jacq. (Legumin.).
Nephrocarpus Dammer (~ Basselinia Vieill.). Palmae. 1 New Caled.
Nephrocarya Candargy = Nonea Medik. (Boraginac.).
Nephrocodium Benth. & Hook. f. = seq.
Nephrocodum C. Muell. = seq.
Nephrocoelium Turcz. = Burmannia L. (Burманniac.).
Nephrodesmus Schindler. Leguminosae. 7 New Caled.

NESAEA

Nephrodium Rich. A name variously used for *Dryopteris* Adans. (Aspidiac.), *Thelypteris* Schmid., *Cyclosorus* Link (Thelypteridac.), etc. Typification doubtful.
Nephrogeton Rose ex Pittier (sphalm.)=Niphogeton Schlechtd. (Umbellif.).
Nephroia Lour.=Cocculus DC. (Menispermac.).
Nephroica Miers=praec.
Nephrolepis Schott. Oleandraceae. 30 trop., and Japan, N.Z.; some spp. v. abundant. L.-bearing part of stem short, erect, bearing branching runners which produce new tufts of ls. distally; petioles not jointed at base. *N. acutifolia* (Desv.) Chr. has marginal fusion-sorus.
Nephromedia Kostel.=Trigonella L. (Legumin.).
Nephromeria (Benth.) Schindler. Leguminosae. 11 trop. S. Am.
Nephromischus Klotzsch=Begonia L. (Begoniac.).
Nephropetalum Robinson & Greenman. Sterculiaceae. 1 N. Am.
Nephrophyllidium Gilg. Menyanthaceae. 1 Japan, NW. Am.
Nephrophyllum A. Rich. Convolvulaceae. 1 Abyssinia.
Nephrosis Rich. ex DC.=Drepanocarpus G. F. W. Mey. (Legumin.).
Nephrosperma Balf. f. Palmae. 1 Seychelles.
Nephrostigma Griff.=?Cyathocalyx Champ. (Annonac.).
Nephrostylus Gagnep.=Koilodepas Hassk. (Euphorbiac.).
Nephthytis Schott. Araceae. 5 trop. Afr.
Nepogeton Rose ex Pittier (sphalm.)=Niphogeton Schlechtd. (Umbellif.).
Nepsera Naud. Melastomataceae. 1 trop. S. Am., W.I.
Neptunia Lour. Leguminosae. 11 trop. & subtrop. *N. oleracea* Lour. has a floating stem, rooting at the nodes, and covered by aerenchyma. The ls. are sensitive like those of *Mimosa*. Fls. in heads, the lower ♂, or neuter with [petaloid stds.
Neraudia Gaudich. Urticaceae. 4 Hawaii.
Neretia Moq.=Oreobliton Moq. & Dur. (Chenopodiac.).
Neriacanthus Benth. Acanthaceae. 3 NW. S. Am., W.I.
Neriandra A. DC.=Skytanthus Meyen (Apocynac.).
Nerija Endl.=Neerija Roxb.=Elaeodendron Jacq. (Celastrac.).
*****Nerine** Herb. Amaryllidaceae. 30 S. trop. & S. Afr.
Nerion St-Lag.=Nerium L. (Apocynac.).
Nerissa Rafin.=Ponthieva R.Br. (Orchidac.).
Nerissa Salisb.=Haemanthus L. (Amaryllidac.).
Nerisyrenia Greene. Cruciferae. 5 W. N. Am., Mex.
Nerium L. Apocynaceae. 3 Medit. to Japan. *N. oleander* L. (oleander) has pits on the lower surface of the evergr. ls., in which the stomata are sunk (several in each) and covered with hairs, reducing transpiration. Fls. suited to long-tongued moths.
Nernstia Urb. Rubiaceae. 1 Mex.
Nerophila Naud. Melastomataceae. 1 trop. W. Afr.
*****Nertera** Banks & Soland. ex Gaertn. Rubiaceae. 12 S. China, Formosa, Philipp. Is., Java, Austr., N.Z., Hawaii, temp. S. Am., Tristan da Cunha.
Nerteria Sm.=praec.
*****Nervilia** Comm. ex Gaudich. Orchidaceae. 80 trop. & subtrop. Old World.
Nesaea Comm. ex Juss. Lythraceae. 50 trop. & S. Afr., Madag., Ceylon, Austr., S. Am.

Nescidia A. Rich. = Myonima Comm. ex Juss. (Rubiac.).
Nesiota Hook. f. Rhamnaceae. 1 St Helena.
Neskiza Rafin. = Carex L. (Cyperac.).
Neslea Aschers. = seq.
***Neslia** Desv. Cruciferae. 2 SE. Eur., Medit., SW. As.
Nesobium R. Phil. ex Fuentes = Parietaria L. (Urticac.).
Nesocaryum I. M. Johnston. Boraginaceae. 1 Chile (Desventuradas Is.).
Nesodaphne Hook. f. = Beilschmiedia Nees (Laurac.).
Nesodoxa Calest. = Eremopanax Baill. (Araliac.).
Nesodraba Greene = Draba L. (Crucif.).
Nesoea Wight = Ammannia L. (Lythrac.).   [Seych., Polynesia.
Nesogenes A. DC. Dicrastylidaceae (?). 6 trop. E. Afr., Madag., Masc.,
Nesogordonia Baill. Sterculiaceae. 17 trop. Afr., Madag.
Nesohedyotis (Hook. f.) Bremek. Rubiaceae. 1 St Helena.
Nesoluma Baill. Sapotaceae. 3 Polynesia.
Nesopanax Seem. = Plerandra A. Gray (Araliac.).
Nesopteris Copel. Hymenophyllaceae. 4 Malaysia to Samoa.
Nesoris Rafin. = Pityrogramma Link (Gymnogrammac.).
Nesothamnus Rydb. (~ Perityle Benth.). Compositae. 1 Lower Calif.
Nesphostylis Verdc. Leguminosae. 1 trop. Afr.
Nessea Steud. = Nesaea Comm. ex Juss. (Lythrac.).
Nestegis Rafin. Oleaceae. 4 N.Z.; 1 Hawaii?
Nestlera E. Mey. ex Walp. = Leucosidea Eckl. & Zeyh. (Rosac.).
Nestlera Spreng. Compositae. 18 S. Afr.
Nestlera Willd. ex Steud. = Botelua Lag. (Gramin.).
Nestoria Urb. Bignoniaceae. 1 E. Brazil.
Nestronia Rafin. = Buckleya Torr. (Santalac.).
Nestylix Rafin. = Salix L. (§ Amerina) (Salicac.).
Nesynstylis Rafin. = Strumaria Jacq. (Amaryllidac.).
Netouxia G. Don = Nectouxia Kunth (Solanac.).
Nettlera Rafin. = Cephaëlis Sw. (Rubiac.).
Nettoa Baill. Tiliaceae. 1 Austr.
Neubeckia Alef. = Iris L. (Iridac.).
Neuberia Eckl. = Watsonia Mill. (Iridac.).
Neuburghia Walp. = seq.
Neuburgia Blume. Strychnaceae. 10–12 Philipp., Celebes, New Guinea, Pacif.
Neudorfia Adans. = Nolana L. ex L. f. (Nolanac.).
Neuhofia Stokes = Baeckea L. (Myrtac.).
Neumannia Brongn. = Pitcairnia L'Hérit. (Bromeliac.).
Neumannia A. Rich. = Aphloia (DC.) Benn. (Flacourtiac.) (? Neumanniac.).
Neumanniaceae Van Tiegh. = Flacourtiaceae – Flacourtiëae – Flacourtiinae
Engl. (?). Shrubs with habit ± of *Eurya* (*Theac.*), with alt. simple usu. serr.
exstip. ls., sometimes turning livid blue on drying. Fls. reg., ☿, in axill. fasc.
K 4–5, much imbr.; C o; A ∞, fil. filif., persist., anth. small; G 1, 1-loc., with
1 lat. plac. bearing few biser. ± campylotr. ov., and sess. pelt. stig. Fr. an
indehisc. few-seeded berry; seed ovoid, with hippocrepif. embr. in sparse
endosp. Only genus: *Aphloia* (DC.) Benn. Perhaps some justification for
maintaining the fam. distinct.

Neumayera Reichb. (1841) = Arenaria L. (Caryophyllac.).
Neumayera Reichb. f. (1872) = Niemayera F. Muell. = Apostasia Blume (Orchidac.).
Neuontobotrys O. E. Schulz. Cruciferae. 2 Chile, Argent.
Neuracanthus Nees. Acanthaceae. 20 trop. Afr., Madag., Socotra, Arabia, India.
Neurachne R.Br. Gramineae. 5 Austr.
Neuractis Cass. = Chrysanthellum Rich. (Compos.).
Neurada L. Neuradaceae. 1 E. Medit. to Indian desert.
Neuradaceae J. G. Agardh. Dicots. 3/10 Medit. to India, S. Afr. Prostrate annual tomentose herbs, ± woody below; ls. alt., variously lobed or pinnatif., stips. minute or o. Fls. reg., ⚥, sol., axill., sometimes showy, sometimes with epicalyx of 5 bracteoles. K (5), tube broad and flat, lobes ± valv.; C 5, convol., inserted in throat of K; A 5 + 5, fil. elong., subul., ± persist., anth. small, ovoid; G 3–10, horizont. verticillate, ± connate and adnate to K tube, with 1 pend. ov. per loc., and 3–10 short persist. styles with small capit. stigs. Fr. orbic., depr.-conic, laterally membr.-winged or spinose-muricate, carp. dehisc. ventrally, styles sometimes spinescent, seeds horiz., without endosp. Genera: *Neurada, Neuradopsis, Grielum*. Hitherto associated with *Rosaceae*, but without close relatives there; the pollen also is aberrant. Perh. nearest to *Malvaceae*: cf. leaf-cutting, vertic. carpels, etc. The yellow C of *Grielum* and *Neuradopsis* changes colour in drying to bluish-black, as in e.g. *Althaea ficifolia* Cav. In germination the seeds are retained in the hard discoid fr., which is perforated by the plumule upwards and by the radicle downwards, the fr. remaining as a persistent collar around the hypocotyl.
Neuradopsis Bremek. & Oberm. Neuradaceae. 3 SW. Afr.
Neuras Adans. = Neurada L. (Neuradac.).
Neurelmis Rafin. = ? Pinillosia Ossa vel ? Koehneola Urb. (Compos.).
Neurilis P. & K. = Nevrilis Rafin. = Millingtonia L. f. (Bignoniac.).
Neurocallis Fée. Pteridaceae. 1 trop. Am. Venation as *Acrostichum* L.
Neurocalyx Hook. Rubiaceae. 4–5 S. India, Ceylon.
Neurocarpaea R.Br. = Pentas Benth. (Rubiac.).
Neurocarpaea K. Schum. (sphalm.) = Nodocarpaea A. Gray (Rubiac.).
Neurocarpon Desv. (1820) = seq.
Neurocarpum Desv. (1813) = Clitoria L. (Legumin.).
Neurocarpus P. & K. = Neurocarpaea R.Br. = Pentas Benth. (Rubiac.).
Neurochlaena Less. = Neurolaena R.Br. (Compos.).
Neuroctola Steud. = Nevroctola Rafin. = Uniola L. (Gramin.).
Neurodium Fée = Paltonium Presl (Polypodiac.).
Neurogramma Link = Gymnopteris Bernh. (Gymnogrammac.).
Neurola Rafin. = Sabbatia Adans. (Gentianac.).
Neurolaena R.Br. Compositae. 5 Mex. to trop. S. Am., W.I.
Neurolakis Mattf. Compositae. 1 W. Equat. Afr.
Neurolepis Meissn. Gramineae. 12 trop. S. Am.
Neurolobium Baill. Apocynaceae. 1 Brazil.
Neuroloma Andrz. ex DC. = Parrya R.Br. (Crucif.).
Neuroloma Endl. = Nevroloma Rafin. = Briza L. (Gramin.).
Neuromanes Trevis. = Trichomanes L. (Hymenophyllac.).

NEURONIA

Neuronia D. Don = Oleandra Cav. (Oleandrac.).

Neuropeltis Wall. Convolvulaceae. 12 trop. Afr., W. Penins. India, Indoch., W. Malaysia.

Neuropeltopsis van Ooststr. Convolvulaceae. 1 Borneo.

Neurophyllodes (A. Gray) Degener (~ Geranium L.). Geraniaceae. 1 Hawaii.

Neurophyllum Presl = Trichomanes L. (Hymenophyllac.).

Neurophyllum Torr. & Gray = Peucedanum L. (Umbellif.).

Neuroplatyceros (Endl.) Fée = Platycerium Desv. (Polypodiac.).

Neuropora Comm. ex Endl. = ? Antirhea Comm. ex Juss. (Rubiac.).

Neuropteris Gaud. = Saccoloma Kaulf. (Dennstaedtiac.).

Neuropteris Jack ex Burkill = Neuropeltis Wall. (Convolvulac.).

Neuroscapha Tul. = Lonchocarpus Kunth (Legumin.).

Neurosoria Mett. Sinopteridaceae. 1 trop. Austr.

Neurosorus Trevis. = Coniogramme Fée (Gymnogrammac.).

Neurosperma Rafin. = Nevrosperma Rafin. = Momordica L. (Cucurbitac.).

Neurospermum Bartl. = praec.

Neurotecoma K. Schum. = Spirotecoma Baill. ex Dalla Torre & Harms (Bignoniac.).

Neurotheca Salisb. ex Benth. & Hook. f. Gentianaceae. 10 trop. S. Am., trop. Afr.

Neustanthus Benth. = Pueraria DC. (Legumin.).

Neustruevia Juzepczuk = Pseudomarrubium M. Pop. (Labiat.).

Neuvrada Augier = Neurada L. (Neuradac.).

Neuwiedia Blume. Orchidaceae (~ Apostasiac.). 10 Malaysia.

Neves-Armondia K. Schum. Bignoniaceae. 1 Brazil.

Neviusa Benth. & Hook. f. = seq.

Neviusia A. Gray. Rosaceae. 1 SE. U.S.

Nevosmila Rafin. = Crateva L. (Capparidac.).

Nevreda Dietr. = Neurada L. (Neuradac.).

Nevrilis Rafin. = Millingtonia L. f. (Bignoniac.).

Nevrocarpon Spreng. = Neurocarpum Desv. = Clitoria L. (Legumin.).

Nevroctola Rafin. = Uniola L. (Gramin.).

Nevrola Rafin. = seq.

Nevrolis Rafin. = Celosia L. (Amaranthac.).

Nevroloma Rafin. = Glyceria R.Br. (Gramin.).

Nevroloma Spreng. = Neuroloma Andrz. = Parrya R.Br. (Crucif.).

Nevropora Comm. ex Baill. = Neuropora Comm. ex Endl. = ? Antirhea Comm. ex Juss. (Rubiac.).

Nevrosperma Rafin. = Momordica L. (Cucurbitac.).

Nevskiella (Krecz. & Vved.) Krecz. & Vved. = Bromus L. (Gramin.).

Newberrya Torr. = Hemitomes A. Gray (Monotropac.).

Newbouldia Seem. ex Bur. Bignoniaceae. 1 trop. W. Afr.

Newcastelia F. Muell. Dicrastylidaceae. 8 trop. Austr.

Newcastlia F. Muell. = praec.

Newtonia Baill. Leguminosae. 15 trop. S. Am., trop. Afr.

Newtonia O. Hoffm. = Antunesia O. Hoffm. (Compos.).

Nexilis Rafin. = Rotala L. (Lythrac.).

Neyraudia Hook. f. Gramineae. 2 trop. Afr., Madag., China, Indomal.

Nezera Rafin. = Linum L. (Linac.).

Nhandiroba Adans. = Fevillea L. (Cucurbitac.).

Nhandirobaceae A. St-Hil. ex Endl. = Cucurbitaceae–Zanonioïdeae C. Jeffrey.

Nialel Adans. = Aglaia Lour. (Meliac.).

Niara Dennst. = Ardisia Sw. (Myrsinac.).

Nibbisia Walp. = Nirbisia G. Don = Aconitum L. (Ranunculac.).

Nibo Steud. = Vibo Medik. = Emex Neck. ex Campd. (Polygonac.).

Nibora Rafin. = Gratiola L. (Scrophulariac).

*Nicandra Adans. Solanaceae. 1 Peru, *N. physalodes* ·Scop. Ov. divided in an irreg. way by plac. Berry nearly juiceless, with ∞ seeds, and enclosed in the enlarged K.

Nicandra Schreb. = Potalia Aubl. (Potaliac.).

Nicarago Britt. & Rose = Caesalpinia L. (Legumin.).

Nichelia Bullock = Pyrrhocactus (A. Berger) Britt. & Rose (Cactac.).

Nicholsonia Span. = Nicolsonia DC. (Legumin.).

Nicipe Rafin. = Ornithogalum L. (Liliac.).

Niclouxia Battandier = Lifago Schweinf. & Muschl. (Compos.).

Nicodemia Tenore (~ Buddleja L.). Buddlejaceae. 6 Madag., Mascarene Is.

Nicolaia Horan. Zingiberaceae. 25 Indomal.

Nicolasia S. Moore. Compositae. 8 SW. trop. Afr.

Nicolettia Benth. & Hook. f. = seq.

Nicolletia A. Gray. Compositae. 3 SW. U.S.

Nicolsonia DC. (~ Desmodium Desv.). Leguminosae. 10 trop.

Nicoteba Lindau = Justicia L. (Acanthac.).

Nicotia Opiz = seq.

Nicotiana L. Solanaceae. 21 Austr., Polynes.; 45 extra-trop. N. & S. Am. *N. tabacum* L., cult. in warm countries, esp. U.S., Cuba, Brazil, Egypt, Sumatra, etc., is the tobacco, grown as an annual crop; the ls. are gathered, hung up and slowly dried, then packed in heaps and fermented slightly. Different varieties are grown, and usu. in different places, for cigar, cigarette, and pipe tobacco. *N. rustica* L. and others are also used.

Nicotidendron Griseb. = praec.

Nictanthes All. = Nyctanthes L. (Verbenac.).

Nictitella Rafin. = Chamaecrista Moench = Cassia L. (Legumin.).

Nidema Britton & Millsp. = Epidendrum L. (Orchidac.).

Nidorella Cass. Compositae. 11 trop. & S. Afr.

Nidularium Lem. Bromeliaceae. 22 Brazil.

× Nidumea L. B. Smith. Bromeliaceae. Gen. hybr. (Aechmea × Nidularia).

Nidus Rivinus = Neottia Guett. (Orchidac.).

Nidus-avis Ort. = praec.

Niebuhria DC. = Maerua Forsk. (Capparidac.).

Niebuhria Neck. = Wedelia Jacq. (Compos.).

Niebuhria Scop. = Baltimora L. (Compos.).

Niedenzua Pax = Adenochlaena Baill. (Euphorbiac.).

Niederleinia Hieron. Frankeniaceae. 3 temp. S. Am.

Niedzwedzkia B. Fedtsch. (~ Incarvillea Juss.). Bignoniaceae. 1 C. As.

Niemeyera F. Muell. (1867) = Apostasia Blume (Orchidac.).

*Niemeyera F. Muell. (1870). Sapotaceae. 3 New Guinea, trop. E. Austr.

NIENOKUEA

Nienokuea A. Chev. = Polystachya Hook. (Orchidac.).

Nierembergia Ruiz & Pav. Solanaceae. 35 Mex. to subtrop. S. Am.

Nietneria Klotzsch & Rich. Schomb. ex Benth. & Hook. f. Liliaceae. 1 Venez.,

Nietoa Schaffn. = Hanburia Seem. (Cucurbitac.). [Guiana.

Nigella L. Ranunculaceae. 20 Eur., Medit. to C. As. Annuals. Alt. with the K is an invol. of 5 ls. Within the coloured K are 5–8 nectaries, pocket-like structures with lids which prevent small insects from reaching the honey. The cpls. are more or less completely united but have separate styles; they give a caps. fr. Fl. protandrous.

Nigellaceae J. G. Agardh = Ranunculaceae–Helleboreae DC.

Nigellastrum Fabr. (uninom.) = *Nigella* L. sp. (Ranunculac.).

Nigera Bub. = Caucalis L. (Umbellif.).

× **Nigribicchia** E. G. Camus, Bergon & A. Camus (iv) = × Pseuditella P. F. Hunt (Orchidac.).

Nigrina L. = Melasma Bergius (Scrophulariac.).

Nigrina Thunb. = Chloranthus Sw. (Chloranthac.).

Nigritella Rich. Orchidaceae. 2 Europe.

Nigrolea Nor. = Kibara Endl. (Monimiac.).

× **Nigrorchis** Godfery. Orchidaceae. Gen. hybr. (i) (Nigritella × Orchis).

Nikitinia Iljin. Compositae. 1 C. As.

Nil Medik. = Ipomoea L. (Convolvulac.).

Nilbedousi Augier = Ardisia Sw. (Myrsinac.).

Nilgirianthus Bremek. Acanthaceae. 20 Penins. India.

Nima Buch.-Ham. ex A. Juss. = Picrasma Blume (Simaroubac.).

Nimbo Dennst. = Murraya Koen. ex L. (Rutac.).

Nimiria Prain ex Craib. Leguminosae. 2 Siam.

Nimmoia Wight (1837) = Ammannia L. (Lythrac.).

Nimmoia Wight (1847) = Amoora Roxb. (Meliac.).

Nimmonia Wight (1840) = Nimmoia Wight (1837) = Ammannia L. (Lythrac.).

Nimphaea Neck. = Nymphaea L. (Nymphaeac.).

Nimphea Nocca = praec.

Ninanga Rafin. = Gomphrena L. + Froelichia Moench (Amaranthac.).

Nintooa Sweet = Lonicera L. (Caprifoliac.).

Niobaea Spach = Niobea Willd. ex Schult. f. = Hypoxis L. (Hypoxidac.).

Niobe Salisb. = Hosta Tratt. (Liliac.).

Niobea Willd. ex Schult. f. = Hypoxis L. (Hypoxidac.).

Niopa (Benth.) Britton & Rose = Piptadenia Benth. (Legumin.).

Niota Adans. = Ceropegia L. (Asclepiadac.).

Niota Lam. = Quassia L. (Simaroubac.).

Niotoutt Adans. = Commiphora Jacq. (Burserac.).

Nipa Benth. & Hook. f. (sphalm.) = Nepa Webb (Legumin.).

Nipa Thunb. = Nypa Steck (Palm.).

Nipaceae [Brongn. ex] Chadef. & Emberg. = Palmae–Nypoïdeae Engl.

Niphaea Lindl. Gesneriaceae. 5 Mex., C. Am., W.I.

Niphidium J. Sm. Polypodiaceae. 1 Ecuador.

Niphobolus Kaulf. = Pyrrosia Mirbel (Polypodiac.).

Niphogeton Schlechtd. Umbelliferae. 15 N. Andes.

Niphopsis J. Sm. = Pyrrosia Mirbel (Polypodiac.).

Niphus Rafin. = Aristolochia L. (Aristolochiac.).
Nipponobambusa Muroi. Gramineae. 6 Japan.
Nipponocalamus Nakai (~Pleioblastus Nak., Arundinaria Michx). Gramineae. 70 E. As.
Nipponorchis Masamune = Neofinetia Hu (Orchidac.).
Nirarathamnos Balf. f. Umbelliferae. 1 Socotra.
Nirbisia G. Don = Aconitum L. (Ranunculac.).
Niruri Adans. = Phyllanthus L. (Euphorbiac.).
Niruris Rafin. = praec.
Nirwamia Rafin. = Pellionia Gaudich. (Urticac.).
Nisa Nor. = Homalium Jacq. (Flacourtiac.).
Nisomenes Rafin. = Euphorbia L. (Euphorbiac.).
Nisoralis Rafin. = Helicteres L. (Sterculiac.).
Nispero Aubrév. (~Manilkara Adans.). Sapotaceae. 1 Mex. to trop. S. Am., W. I.
***Nissolia** Jacq. Leguminosae. 12 Mex. to W. trop. S. Am.
Nissolia Mill. = Lathyrus L. (Legumin.).
Nissolius Medik. = Machaerium Pers. (Legumin.).
Nissoloïdes M. E. Jones. Leguminosae. 1 Mex.
Nitelium Cass. = Dicoma Cass. (Compos.).
Nitrapia Pall. (sphalm.) = seq.
Nitraria L. Zygophyllaceae. 7 Sahara & S. Russia to Afghanist. & E. Sib.; 1 SE. Austr.; in salt deserts.
Nitrariaceae Lindl. = Zygophyllaceae–Nitrarioïdeae Engl.
Nitrophila S. Wats. Chenopodiaceae (~Amaranthaceae). 8 W. U.S., Mex., temp. S. Am.
Nivaria Fabr. (uninom.) = Leucojum L. sp. (Amaryllidac.).
Nivenia R.Br. = Paranomus Salisb. (Proteac.).
Nivenia Vent. (~Aristea Soland.). Iridaceae. 8 S. Afr.
Niveophyllum Matuda. Liliaceae. 1 Mex.
Noaea Moq. Chenopodiaceae. 9 W. As.
Noallia Buc'hoz = Eschweilera Mart. (Lecythidac.).
Nobeliodendron O. C. Schmidt = Licaria Aubl. (Laurac.).
× **Nobleara** hort. Orchidaceae. Gen. hybr. (iii) (Aërides × Renanthera × Vanda).
Nobula Adans. = Phyllis L. (Rubiac.).
Nocca Cav. (~Lagascea Cav.). Compositae. 10 Mex.
Noccaea Moench = Thlaspi L. (Crucif.).
Noccaea Willd. = Nocca Cav. (Compos.).
Nochotta S. G. Gmel. = Cicer L. (Legumin.).
Nodocarpaea A. Gray. Rubiaceae. 1 Cuba.
Noëa Boiss. & Bal. = Noaea Moq. (Chenopodiac.).
Nogo Baehni (~Lecomtedoxa [Pierre ex Engl.] Dub.). Sapotaceae. 1(?) trop. Afr.
Nogra Merr. Leguminosae. 3 India, SE. As.
Noisettia Kunth. Violaceae. 1 Peru, Brazil, Guiana.
Noittetia Barb. Rodr. (sphalm.) = praec.
Nolana L. ex L. f. Nolanaceae. 80 Peru to Patag. Many shore plants with fleshy ls.

NOLANACEAE

*Nolanaceae Dum. Dicots. 2/85 W. coast of S. Am. Herbs or low shrubs with simple ls., often covered with glandular hairs. The ls. in the veg. region are alt., but in the infl. portion they become paired in the same way as in *Solanaceae* (*q.v.*). Fls. sol. in the leaf-axils, ⚥, reg. K (5), ± imbr.; C (5), plicate; A 5, alt. with petals; G̲ typically 5, united only in *Alona*, usu. free and divided by irregular transverse constrictions into 5 or 10 portions standing in a row, or by longitudinal and transv. constrictions into 10–30 portions in 2 or 3 rows. The fr. consists of a corresponding number of 1–7-seeded nutlets. Style 1. Seed album. Genera: *Nolana, Alona*. Related to *Solanac.* and *Convolvulac.*

Noldeanthus Knobl. = Jasminum L. (Oleac.).

Nolina Michx. Agavaceae. 30 SW. U.S., Mex. Xero.

Nolinaceae Nak. = Agavaceae–Nolineae Reichb.

Nolinaea Baker = seq.

Nolinea Pers. = Nolina Michx (Agavac.).

Nolinia K. Schum. (sphalm.) = Molinia Schrank (Gramin.).

Nolitangere Rafin. = Impatiens L. (Balsaminac.).

Nolletia Cass. Compositae. 1 Morocco, 10 S. Afr.

Noltea Reichb. Rhamnaceae. 2 S. Afr.

Noltia Eckl. ex Steud. = Selago L. (Scrophulariac.).

Noltia Schumacher & Thonn. = Diospyros L. (Ebenac.).

Nomaphila Blume. Acanthaceae. 12 palaeotrop.

Nomismia Wight & Arn. = Rhynchosia Lour. (Legumin.).

Nomocharis Franch. Liliaceae. 16 Himal. to W. China.

Nomochloa Beauv. = Blysmus Panz. ex Schult. (Cyperac.).

Nomochloa Nees = Pleurostachys Brongn. (Cyperac.).

Nomophila P. & K. = Nomaphila Bl. (Acanthac.).

Nomosa I. M. Johnston. Boraginaceae. 1 Mex.

Nonatelia Aubl. Rubiaceae. 1 trop. Am.

Nonatelia Kuntze = Lasianthus Jack (Rubiac.).

Nonea Medik. Boraginaceae. 35 Medit.

Nonnea Reichb. = praec.

Nonnia St-Lag. = praec.

Nopal Thierry ex Först. & Rümpl. = seq.

Nopalea Salm-Dyck = Opuntia Mill. (Cactac.).

Nopal[e]aceae Burnett = Cactaceae Juss.

Nopalxochia Britton & Rose. Cactaceae. 3 Mex. to Peru.

Norantea Aubl. Marcgraviaceae. 35 trop. Am., W.I. All fls. are fertile, and have saccate nectariferous bracts. Resembles *Philodendron* in habit.

Noratilea Walp. = Nonatelia Aubl. (Rubiac.).

Nordmannia Fisch. & Mey. ex C. A. Mey. = Daphnopsis Mart. & Zucc. (Thymelaeac.).

Nordmannia Ledeb. ex Nordm. = Trachystemon D. Don (Boraginac.).

Norisca Dyer = Norysca Spach = Hypericum L. (Guttif.).

Normanbokea Kladiwa & F. Buxb. Cactaceae. 2 Mex.

Normanbya F. Muell. (~ Ptychosperma Labill.). Palmae. 1 Queensland.

Normandia Hook. f. Rubiaceae. 1 New Caled.

Normania Lowe = Solanum L. (Solanac.).

Norna Wahlenb. = Calypso Salisb. (Orchidac.).
Noronha Thou. ex Kunth = Dypsis Nor. ex Mart. (Palm.).
Noronhaea P. & K. = seq.
Noronhia Stadm. ex Thou. Oleaceae. 40 Madag., Maurit., Comoro Is.
Norrisia Gardn. Antoniaceae. 2 W. Malaysia.
Norta Adans. = Sisymbrium L. (Crucif.).
Nortenia Thou. = Torenia L. (Scrophulariac.).
Northea Hook. f. Sapotaceae. 5 Seychelles.
Northiopsis Kaneh. (~ Manilkara Adans.). Sapotaceae. 1 Caroline Is., Samoa.
Norysca Spach = Hypericum L. (Guttif.).
Nosema Prain. Labiatae. 5 SE. As.
Nostelis Rafin. = Micromeria Benth. (Labiat.).
Nostolachma Th. Dur. Rubiaceae. 10 Indomal.
Notanthera G. Don = Loranthus L. (Loranthac.).
Notaphoebe Pax = Nothaphoebe Bl. (Laurac.).
Notarisia Pestal. ex Cesati = Ricotia L. (Crucif.).
Notelaea Vent. Oleaceae. 7 E. Austr. Hard timber.
Notelea Rafin. = praec.
Noterophila Mart. = Acisanthera P.Br. (Melastomatac.).
Nothaphoebe Bl. Lauraceae. 30 S. China, SE. As., Indomal.
Nothapodytes Blume. Icacinaceae. 4 Ceylon to Formosa, Ryukyu Is. & W. Malaysia.
Nothites Cass. = Stevia Cav. (Compos.).
Nothocalais Greene. Compositae. 4 W. & C. N. Am.
Nothocarpus P. & K. = Nodocarpaea A. Gray (Rubiac.).
Nothocelastrus Blume ex Kuntze = Perrottetia Kunth (Celastrac.).
Nothocestrum A. Gray. Solanaceae. 5 Hawaii.
Nothochelone (A. Gray) Straw. Scrophulariaceae. 1 moist temp. W. N. Am.
Nothochilus Radlk. Scrophulariaceae. 1 Brazil.
Nothochlaena Kaulf. = Notholaena R.Br. = Cheilanthes Sw. (Sinopteridac.).
Nothocnestis Miq. = Kurrimia Wall. ex Thw. = Bhesa Buch.-Ham. ex Arn. (Celastrac.).
Nothocnide Blume. Urticaceae. 5 Malaysia, Bismarcks & Solomon Is.
Nothoderris Blume ex Miq. = Derris Lour. (Legumin.).
Nothofagaceae Kuprianova = Fagaceae–Fagoïdeae Oerst.
*****Nothofagus** Blume. Fagaceae. 35 New Guinea, New Caled., temp. Austr., N.Z., temp. S. Am. *N. cunninghami* (Hook.) Oerst. (myrtle tree), good timber.
Nothoholcus Nash = Notholcus Nash ex Hitchc. = Holcus L. emend. Sw.
Notholaena R.Br. = Cheilanthes Sw. (Sinopteridac.). [(Gramin.).
Notholcus Nash ex Hitchcock = Holcus L. emend. Sw. (Gramin.).
Notholirion Wall. ex Boiss. Liliaceae. 6 Persia to W. China.
Nothomyrcia Kausel. Myrtaceae. 1 Juan Fernandez.
Nothonia Endl. = Notonia DC. (Compos.).
Nothopanax Miq. emend. Harms = Polyscias J. R. & G. Forst. (Araliac.).
*****Nothopegia** Blume. Anacardiaceae. 7 India, Ceylon, Borneo.
Nothopegiopsis Lauterb. Anacardiaceae. 1 New Guinea.
Nothoperanema (Tagawa) Ching. Aspidiaceae. 5 trop. Afr. & As.
Nothophlebia Standley. Rubiaceae. 1 C. Am.

Nurmonia Harms. Meliaceae. 1 S. Afr.

Nutalla Rafin. = seq.

Nuttalia Rafin. = seq.

Nuttalla Rafin. = Nuttallia Rafin. = Mentzelia L. (Loasac.).

Nuttallia Barton = Callirhoë Nutt. (Malvac.).

Nuttallia DC. = Nemopanthus Rafin. (Aquifoliac.).

Nuttallia Rafin. = Mentzelia L. (Loasac.).

Nuttallia Spreng. = Trigonia Aubl. (Trigoniac.).

Nuttallia Torr. & Gray = Osmaronia Greene = Oemleria Reichb. (Rosac.).

Nux Duham. = Juglans L. (Juglandac.).

Nuxia Comm. ex Lam. Buddlejaceae. 40 trop. Afr., Madag., Masc.

Nuxiopsis N. E. Br. ex Engl. = Dobera Juss. (Salvadorac.).

Nuytsia R.Br. Loranthaceae. 1 W. Austr. A small tree, up to 12 m. high, doubtfully parasitic on roots, with alt. lin. ls. Fr. dry, with 3 broad coriaceous wings. Cotyledons 2–4, unequal.

Nuytsiaceae Van Tiegh. = Loranthaceae–Nuytsieae Engl.

Nyachia Small = Paronychia L. (Caryophyllac.).

Nyalel Augier = Nialel Adans. = Aglaia Lour. (Meliac.).

Nyalelia Dennst. = praec.

Nychosma Schlechtd. = Nyctosma Rafin. = Epidendrum L. (Orchidac.).

*__Nyctaginaceae__ Juss. Dicots. (Centrospermae). 30/290, mostly trop. & esp. Am. Trees, shrubs or herbs with opp. (often unequal) or alt. ls. and no stips. Fls. in cymes, ♂ or unisexual, and with much variety. At the base of the fls. are usu. several bracts, often large and coloured. In *Bougainvillea* 3 large conspicuous bracts enclose a group of 3 fls. In *Abronia* the number of bracts and fls. is larger, while in *Mirabilis* there is only one fl. and the involucre resembles a calyx. P usu. (5), petaloid, persistent upon the ripe fr.; usu. the upper part drops away and the fr. remains in the lower part, which is termed the *anthocarp*, and may become glandular, or form an umbrella-like wing, or otherwise serve for seed dispersal. A typically 5, alt. with the P, but often 3, 8, 10 or other number, or raised to 20 or 30 by branching; filaments often of unequal length; G 1, with long style and 1 basal erect ana-campylotr. ov. Achene enclosed in the P. Seed with perisperm. The *N.* are of slight economic value; see *Mirabilis, Neea*, etc. Chief genera: *Mirabilis, Bougainvillea, Pisonia, Neea, Reichenbachia*.

Nyctaginia Choisy. Nyctaginaceae. 1 S. U.S., Mex.

Nyctago Juss. = Mirabilis L. (Nyctaginac.).

Nyctandra Prior = Nectandra Roland. ex Rottb. (Laurac.).

Nyctanth[ac]eae J. G. Agardh = Verbenaceae–Nyctanthoïdeae Airy Shaw.

Nyctanthes L. Verbenaceae. 2 India, Siam, Sumatra, Java.

Nyctanthos St-Lag. = praec.

Nyctelea Scop. = Ellisia L. (Hydrophyllac.).

Nycteranthus Neck. ex Rothm. = Aridaria N. E. Br. (Aïzoac.).

Nycterianthemum auct. ex Haw. = praec.

Nycterinia D. Don = Zaluzianskya F. W. Schmidt (Scrophulariac.).

Nycterisition Ruiz & Pav. = Chrysophyllum L. (Sapotac.).

Nycterium Vent. = Solanum L. (Solanac.).

Nycticalanthus Ducke. Rutaceae. 1 Amaz. Brazil.

Nycticalos auctt. = seq.
Nyctocalos Teijsm. & Binnend. Bignoniaceae. 5 Assam & SW. China to W. Malaysia.
Nyctocereus (A. Berger) Britton & Rose. Cactaceae. 7 Mex., C. Am.
Nyctophyla auctt. = seq.
Nyctophylax Zipp. = Riedelia Oliv. (Zingiberac.).
Nyctosma Rafin. = Epidendrum L. (Orchidac.).
Nylandtia Dum. Polygalaceae. 1 S. Afr.
Nymania Gandog. = Seseli L. (Umbellif.).
Nymania Kuntze (sphalm.) = Nymanina Kuntze = Freesia Klatt (Iridac.).
Nymania S. O. Lindb. Aitoniaceae (see ADDENDA). 1 S. Afr.
Nymania K. Schum. = Phyllanthus L. (Euphorbiac.).
Nymanima Dur. & Jacks. (sphalm.) = seq.
Nymanina Kuntze = Freesia Klatt (Iridac.).
Nymphaea Kuntze = Nuphar Sm. (Nymphaeac.).
*Nymphaea L. Nymphaeaceae. 50 trop. & temp. They grow in shallow water. There is a stout creeping rhiz.; at the tip it is bent up, and bears stip. ls. and fls. on long stalks. The peduncle occupies the position of one of the ls. of the spiral, and there is no bract at its base. The l. is large and floats on the surface; it is nearly circular, entire, and leathery, with stomata, cuticle and palisade tissue on the upper side.

Fl. ⚥, reg., acyclic; floats on the surface. The 4 outermost floral ls. exhibit a peculiar aestivation, the ant. being entirely outside, the post. inside the lat. ls. Most authors regard them as K, but Caspary (Eichler, *Blütendiagr.* **2**: 184) regards the anterior l. as bract adnate to peduncle (cf. *Solanaceae*), the lats. as bracteoles, the post. as a true sep. C ∞, 4 outer alt. with K, 4 inner with these; each of the 8 begins a spiral of pets., usu. 4 in each, alt. approximately with one another and the outer 8, and showing gradual transition to the 50–100 sta. which continue the spirals. K hypog., C and A inserted up sides of G (10–20)-loc. with radiating stigmas on upper surface, and ∞ ovules scattered over the whole carpellary surface (cf. *Butomus*). Fr. a large berry; ∞ seeds, each with spongy aril entangling air bubbles; the seeds float up on dehisc. of fr., and float about till the aril decays. Perisperm round endosp. proper. *N. lotus* L. is considered to have been the original sacred Lotus of Egypt; *Nelumbo* (*q.v.*) was introduced about 500 B.C. (Conard, *The Waterlilies*, Washington, 1905).

*Nymphaeaceae Salisb. (*s.str.*). Dicots. 3/75 cosmop. Perenn. water plants. Rhiz. short, stout, erect and long-lived, or creeping, branching and short-lived. Ls. with basal sinus. K 4 or 5; C ∞, sometimes scale-like and nectarif., passing into A ∞; G (5–35), sup. or semi-inf., ovules ∞ pariet. Fr. a spongy berry, dehiscing by swelling of mucilage surr. the seeds, which have endosp. and perisp., and are sometimes arillate. Genera: *Nymphaea, Nuphar, Ondinea*.
Nymphaeanthe Reichb. = Limnanthemum S. G. Gmel. (Menyanthac.).
Nymphaeum Batsch = ?praec.
Nymphanthus Lour. = Phyllanthus L. (Euphorbiac.).
Nymphea Rafin. = Nymphaea L. (Nymphaeac.).
Nympheanthe Endl. = Nymphaeanthe Reichb. = Limnanthemum S. G. Gmel. (Menyanthac.).

NYMPHODES

Nymphodes Kuntze = seq.

Nymphoïdes Seguier. Menyanthaceae. 20 trop. & temp. *N. nymphaeoïdes* (L.) Britton is a water plant with habit of *Nymphaea*. The infl. appears to spring from the top of the leaf-stalk, but really the floating l. springs from the infl. axis. This is an advance upon the *Nymphaea* construction, as the materials going from l. to seeds have not to travel to the bottom of the pond and up again.

Nymphona Bubani = Nuphar Sm. (Nymphaeac.).

Nymphosanthus Steud. = seq.

Nymphozanthus Rich. = Nuphar Sm. (Nymphaeac.).

Nypa Steck. Nypaceae (~ Palmae). 1 Ceylon, Ganges delta, Malaysia, Marianne & Solomon Is., trop. Austr. *N. fruticans* van Wurmb is a char. pl. of estuarine or swampy mud in brackish or salt water.

Nypaceae (Engl. & Gilg) Tralau (~ Palmae). Monocots. 1/1 coasts trop. As. to Austr. Stem woody, prostrate, branched, creeping. Ls. pinnatisect, palmlike, lfts. plicate. Fls. ♂ ♀, monoec. ♂ infl. encl. in ∞ sheathing spathes; spadix branched, term., erect; ♂ fls. in catkin-like branches of spadix, minute. P 3 + 3, lin., broadened and inflexed at apex, ± imbr.; A 3, fil. short, connate into a column, anth. lin., basifixed; pollen spinose, bisulculate. ♀ fls. in glob. term. head; P 3 + 3, rudimentary; stds. o; G̲ (3), 1-loc., with 1–3 erect basal ov. Fruits in a large cpd. head, each fr. obov., compressed, 1-seeded, with fleshy, fibrous sarcocarp and spongy endoc.; seed with deep unilat. furrow. Only genus: *Nypa*.

Nypha Buch.-Ham. = Nypa Steck (Nypac.).

Nyphar Walp. = Nuphar Sm. (Nymphaeac.).

Nyrophylla Neck. = ?Persea Mill. (Laurac.).

Nyrophyllum Kosterm. (sphalm.) = praec.

Nyssa Gronov. ex L. Nyssaceae. 10 Himal., E. As., W. Malaysia, E. N. Am. *N. sylvatica* Marsh., etc. (N. Am., tupelo, pepperidge, sour or cotton gum-tree, ogeechee lime), yield timber and ed. fr.

*****Nyssaceae** Dum. Dicots. 2/10 E. As., E. N. Am. Trees or shrubs with alt. ent. or dentic. exstip. ls. Fls. reg., polygamo-dioec., the ♂♂ in heads, racemes or umbels, the ♀ and ♂ sol. or in 2–12-fld. heads. K limb minute, 5 (or more)-dentate, or ent., or o; C (4–)5(–8), imbr.; A (8–)10(–16), with elong. subul. fil. and small intr. or latr. anth.; disk large, pulvinif.; G̅ (1–2), 1-loc., with 1 apical pend. anatr. ov., and simple or bifid, erect or spirally coiled style. Fr. drupaceous or shortly subsamaroid, 1-seeded, with endosp. Genera: *Nyssa, Camptotheca*. Related to *Davidiac., Cornac.*, etc.

Nyssanthes R.Br. Amaranthaceae. 2 E. Austr. P 4, two inner smaller.

Nyssopsis Kuntze = Camptotheca Decne (Nyssac.).

Nzidora A. Chev. = Tridesmostemon Engl. (Sapotac.).

O

Oakes-Amesia C. Schweinf. & P. H. Allen. Orchidaceae. 1 Panamá.
Oakesia Tuckerm. = Corema D. Don (Empetrac.).
Oakesia S. Wats. = seq.
Oakesiella Small (~ Uvularia L.). Liliaceae. 3 N. Am.
Oaxacana Rose (sphalm.) = Coaxana Coult. & Rose (Umbellif.).
Oaxacania Robinson & Greenman. Compositae. 1 Mex.
Obaejaca Cass. = Senecio L. (Compos.).
Obbea Hook. f. = Bobea Gaudich. (Rubiac.).
Obeckia Griff. (sphalm.) = Osbeckia L. (Melastomatac.).
Obelanthera Turcz. = Saurauia Willd. (Actinidiac.).
Obeliscaria Cass. = seq.
Obeliscotheca Adans. = Rudbeckia L. (Compos.).
Obelisteca Rafin. = Lepachys Rafin. = Ratibida Rafin. (Compos.).
Obentonia Vell. = Angostura Roem. & Schult. (Rutac.).
Oberna Adans. = Silene L. [subgen. Behen (Moench) Bunge] (Caryophyl-
*****Oberonia** Lindl. Orchidaceae. 330 palaeotrop.   [lac.).
Obesia Haw. = Piaranthus R.Br. (Asclepaidac.).
Obetia Gaudich. Urticaceae. 5 trop. Afr., Madag., Réunion.
Obione Gaertn. (~ Atriplex L.). Chenopodiaceae. 100+, esp. deserts &
steppes N. & S. Am., Medit. to C. As.; few W. & C. Eur., Afr., Austr.
Obistila Rafin. = Anthericum L. (Liliac.).
Obletia Lemonn. ex Rozier = Verbena L. (Verbenac.).
Oblixilis Rafin. = Laportea Gaudich. (Urticac.).
Oboejaca Steud. = Obaejaca Cass. = Senecio L. (Compos.).
Obolaria Kuntze = Linnaea Gronov. ex L. (Caprifoliac.).
Obolaria L. Gentianaceae. 1 E. N. Am. Saprophyte (cf. *Bartonia*) of a
purplish green colour with scale ls.
Obolaria Walt. = Hydrotrida Small = Bacopa Aubl. (Scrophulariac.).
Oboskon Rafin. = Salvia L. (Labiat.).
Obregonia Frič. Cactaceae. 1 Mex.
Obsitila Rafin. = Anthericum L. (Liliac.).
Obularia L. = Obolaria L. (Gentianac.).
Ocalia Klotzsch = Croton L. (Euphorbiac.).
Ocampoa A. Rich. & Gal. = Cranichis Sw. (Orchidac.).
Oceanopapaver Guillaumin. ? Tiliaceae. 1 New Caled. Pet. crumpled in
bud. Fr. elongate, torulose.
Oceanopapaveraceae auct. = ? Tiliaceae Juss.
Oceanorus Small = Zigadenus Michx (Liliac.).
Ocellosia Rafin. = ? Liriodendron L. (Magnoliac.).
Ochagavia Phil. Bromeliaceae. 1 Juan Fernandez.
Ochanostachys Mast. Olacaceae. 1 W. Malaysia.
Ochetocarpus Meyen = Scyphanthus D. Don (Loasac.).
Ochetophila Poepp. ex Reissek = Discaria Hook. (Rhamnac.).
Ochlandra Thw. Gramineae. 12 Madag., India, Ceylon.
Ochlogramma Presl = Diplazium Sm. (Athyriac.).

OCHNA

Ochna L. Ochnaceae. 85 trop. & S. Afr., trop. As. K coloured. Cpls. 3–15, free below, but with a common style. After fert. the style falls and each cpl. gives a drupe, while the recept. becomes fleshy under them.

***Ochnaceae** DC. Dicots. 40/600 trop. Most are trees or shrubs with alt. simple (v. rarely pinnate—*Godoya* sp.) stip. ls., and panicles, racemes or cymes (*Sauvagesia*, etc.) or false umbels of ♀, usu. reg. fls. K 5, free or united at base, imbr.; C 5, rarely 10–12, contorted; A 5, 10, or ∞, hypog. or on an elongated axis; G̲ (2–5), rarely (10–15), often free below with common style (cf. *Apocynaceae*). Ovules 1–2–∞ in each cpl., erect or rarely pend., always with ventral raphe. The axis in some gen. swells and becomes fleshy under the fr., which is usually a cluster of drupes, but sometimes a berry or capsule. Endosp. or not. Chief genera: *Ochna, Ouratea, Sauvagesia, Luxemburgia*.

Ochnella Van Tiegh. = Ochna L. (Ochnac.).

Ochocoa Pierre = Scyphocephalium Warb. (Myristicac.).

Ochoterenaea F. A. Barkley. Anacardiaceae. 1 Colombia.

Ochotonophila Gilli. Caryophyllaceae. 1 Afghanistan.

Ochradenus Delile. Resedaceae. 5 NE. Afr. & Socotra to NW. Penins. India.

Ochrante Walp. = seq.

Ochrantha Beddome = Ochranthe Lindl. = Turpinia Vent. (Staphyleac.).

Ochranthaceae (Lindl.) Endl. = Staphyleaceae–Staphyleoïdeae Pax.

Ochranthe Lindl. = Turpinia Vent. (Staphyleac.).

Ochratellus Pierre ex L. Planch. (sphalm.) = Ochrothallus Pierre (Sapotac.).

Ochreata (Lojac.) Bobrov (∼ Trifolium L.). Leguminosae. 13 E. Afr.

Ochrocarpos Thou. = Mammea L. (Guttif.).

Ochrocarpus auctt. = praec.

Ochrocodon Rydb. = Fritillaria L. (Liliac.).

Ochrolasia Turcz. = Hibbertia Andr. (Dilleniac.).

Ochroma Sw. Bombacaceae. 1 S. Mex. to Bolivia, W.I., *O. pyramidale* (Cav. ex Lam.) Urb. (*O. lagopus* Sw.) (balsa, corkwood). Wood very light. Seeds embedded in hairs.

Ochronelis Rafin. = Helianthus L. (Compos.).

Ochronerium Baill. = Tabernaemontana L. (Apocynac.).

Ochropteris J. Sm. Pteridaceae. 1 Madag., Mascarene Is.

Ochrosia Juss. Apocynaceae. 30 Madag. to Austr., Hawaii & Polynesia.

Ochrosion St-Lag. = praec.

Ochrothallus Pierre (∼ Chrysophyllum L.). Sapotaceae. 3 New Caled.

Ochroxylum Schreb. = Zanthoxylum L. (Rutac.).

Ochrus Mill. = Lathyrus L. (Legumin.).

Ochthocharis Blume. Melastomataceae. 5 Malaysia.

Ochthocloa Edgew. = ?Eleusine Gaertn. (Gramin.).

Ochthocosmus Benth. Ixonanthaceae. 6 trop. Am.

Ochthodium DC. Cruciferae. 1 Greece, As. Min. to Jordan.

Ochtocharis Walp. = Ochthocharis Blume (Melastomatac.).

Ocimum L. Labiatae. 150 trop. & warm temp., esp. Afr. *O. basilicum* L. is the basil, sacred in the Hindu religion (*tulsi*).

Ockea F. G. Dietr. = Adenandra Willd. (Rutac.).

Ockenia Steud. = praec.

Ockia Bartl. & Wendl. = praec.

Oclemena Greene = Aster L. (Compos.).
Oclorosis Rafin. (Iodanthus Torr. & Gray). Cruciferae. 1 Atl. N. Am., 3 Mex.
Ocneron Rafin. = Rolandra Rottb. (Compos.).
Ocotea Aubl. Lauraceae. 3–400 trop. & subtrop. Am., few trop. & S. Afr.,
Mascarenes. *O. bullata* E. Mey. (S. Afr.) yields a useful timber (stinkwood).
Ocreaceae Dulac = Polygonaceae Juss.
Ocrearia Small = Saxifraga L. (Saxifragac.).
Octadenia R.Br. ex Fisch. & Mey. = Alyssum L. (Crucif.).
Octadesmia Benth. = Dilomilis Rafin. (Orchidac.).
Octamyrtus Diels. Myrtaceae. 6 New Guinea.
Octanema Rafin. = Capparis L. (Capparidac.).
Octarillum Lour. = Elaeagnus L. (Elaeagnac.).
Octarrhena Thw. Orchidaceae. 35 Ceylon, Malaysia, Polynesia.
Octas Jack = Ilex L. (cf. *I. cissoïdea* Loes.) (Aquifoliac.).
Octavia DC. = Lasianthus Jack (Rubiac.).
Octelisia Rafin. = Cassia L. (Legumin.).
Octella Rafin. = Melastoma L., etc. (Melastomatac.).
Octhocharis G. Don = Ochthocharis Blume (Melastomatac.).
Octima Rafin. = Populus L. (Salicac.).
Octoceras Bunge. Cruciferae. 1 C. As. to Pers. & Afghan.
Octoclinis F. Muell. = Callitris Vent. (Cupressac.).
Octocnema Van Tiegh. = Octoknema Pierre (Olacac.).
Octodon Thonn. = Borreria G. F. W. Mey. (Rubiac.).
Octogonia Klotzsch = Simocheilus Klotzsch (Ericac.).
Octoknema Pierre. Olacaceae. 6 trop. Afr.
*Octoknemaceae Van Tiegh. = Olacaceae Mirb. (Stellate or fascic. hairs [cf.
Coula]. Fls. ♂ ♀, dioec. K 5, usu. obsolete but sometimes clearly developed;
C 5, valv., persist.; A 5, oppositipet.; stds. present in ♀; disk lobed; Ḡ (3),
almost 3-loc., with 1 apical pend. anatr. ov. per loc., and short thick style with
3 broad reflexed irregularly lobulate lobes; fr. a 1-seeded drupe, seed with
endosp.)
Octolepis Oliv. Thymelaeaceae. 6 trop. Afr.
Octolobus Welw. Sterculiaceae. 5 trop. Afr.
Octomeles Miq. Tetramelaceae. 1–2 Malaysia (exc. Malay Penins., Java &
Lesser Sunda Is.). Ls. & infl. lepidote.
Octomelis auctt. = praec.
Octomeria R.Br. Orchidaceae. 130 C. & trop. S. Am., W.I.
Octomeria D. Don = Eria Lindl. (Orchidac.).
Octomeria Pfeiff. (sphalm.) = Otomeria Benth. (Rubiac.).
Octomeris Naud. = Heterotrichum DC. + Miconia Ruiz & Pav. (Melasto-
Octomeron Robyns. Labiatae. 1 trop. Afr.   [matac.].
Octonum Rafin. = Heterotrichum DC. (Melastomatac.).
Octopera D. Don = Erica L. (Ericac.).
Octopleura Griseb. = Ossaea DC. (Melastomatac.).
Octopleura Spruce ex Prog. = Neurotheca Salisb. (Gentianac.).
Octoplis Rafin. = Gnidia L. (Thymelaeac.).
Octopoma N. E. Brown. Aïzoaceae. 3 S. Afr.
Octosomatium Gagnep. = Trichodesma R.Br. (Boraginac.).

OCTOSPERMUM

Octospermum Airy Shaw. Euphorbiaceae. 1 New Guinea.
Octotheca R. Viguier. Araliaceae. 2 New Caled.
Octotropis Bedd. Rubiaceae. 2 S. India, Burma.
Ocymastrum Kuntze = Centranthus DC. (Valerianac.).
Ocymum Wernischek = Ocimum L. (Labiat.).
Ocyroë Phil. Compositae. 1 Chile.
Odacmis Rafin. = Centella L. (Hydrocotylac.).
Oddoniodendron De Wild. Leguminosae. 2 W. Equat. Afr.
Odicardis Rafin. = Veronica L. (Scrophulariac.).
Odina Netto = Marupa Miers (Simaroubac.).
Odina Roxb. = Lannea A. Rich. (Anacardiac.).
Odisca Rafin. = Colea Boj. ex Meissn. (Bignoniac.).
Ododeca Rafin. = Lythrum L. (Lythrac.).
Odoglossa Rafin. = Coreopsis L. (Compos.).
Odollam Adans. = Cerbera L. (Apocynac.).
Odollamia Rafin. = praec.
Odonectis Rafin. = Isotria Rafin. (Orchidac.).
Odonia Bertol. = Galactia P.Br. (Legumin.).
Odontadenia Benth. Apocynaceae. 30 C. & trop. S. Am., W.I.
Odontandra Willd. ex Roem. & Schult. = Trichilia P.Br. (Meliac.).
Odontandria G. Don = praec.
Odontanthera Wight ex Lindley (nomen). Asclepiadaceae. Quid?
Odontarrhena C. A. Mey. = Alyssum L. (Crucif.).
Odontea Fourr. = Bupleurum L. (Umbellif.).
Odonteilema Turcz. = Acalypha L. (Euphorbiac.).
Odontella Van Tiegh. = Tapinanthus Bl. (Loranthac.).
Odontelytrum Hack. Gramineae. 1 trop. Afr.
Odontilema P. & K. = Odonteilema Turcz. = Acalypha L. (Euphorbiac.).
× **Odontioda** hort. Orchidaceae. Gen. hybr. (iii) (Cochlioda × Odontoglossum).
× **Odontiodonia** hort. (vii) = × Vuylstekeara hort. (Orchidac.).
Odontitella Rothm. Scrophulariaceae. 2 Iberian Penins.
Odontites Ludw. Scrophulariaceae. 30 Medit., W. & S. Eur., W. As. Semi-parasites (see fam.).
Odontites Spreng. = Bupleurum L. (Umbellif.).
× **Odontobrassia** hort. Orchidaceae. Gen. hybr. (iii) (Brassia × Odonto-glossum).
Odontocarpa Neck. ex Rafin. = Valerianella Mill. (Valerianac.).
Odontocarpha DC. = Gutierrezia Lag. (Compos.).
Odontocarya Miers. Menispermaceae. 12 trop. S. Am., W.I.
Odontochilus Blume. Orchidaceae. 21 China, Indomal., Fiji.
× **Odontocidium** hort. Orchidaceae. Gen. hybr. (iii) (Odontoglossum × On-cidium).
Odontocyclus Turcz. = Draba L. (Crucif.).
Odontoglossum Kunth. Orchidaceae. 200 Mex., C. & W. trop. S. Am., Guiana, Jamaica.
Odontoloma Kunth = Pollalesta Kunth (Compos.).
Odontoloma J. Sm. = Lindsaea Dryand. (Lindsaeac.).
Odontolophus Cass. = Centaurea L. (Compos.).

Odontomanes Presl = Trichomanes L. (Hymenophyllac.).
Odontonema Nees. Acanthaceae. 40 Mex. to trop. S. Am., W.I.
Odontonemella Lindau = Pseuderanthemum Radlk. (Acanthac.).
× **Odontonia** hort. Orchidaceae. Gen. hybr. (iii) (Miltonia × Odontoglossum).
Odontonychia Small = Siphonychia Torr. & Gray (Caryophyllac.).
Odontophorus N. E. Brown. Aïzoaceae. 6 S. Afr.
Odontoptera Cass. = Arctotis L. (Compos.).
Odontopteris Bernh. = Lygodium Sw. (Schizaeac.).
Odontorrhynchus M. N. Correa. Orchidaceae. 1 Argentina.
Odontosiphon M. Roem. = Trichilia L. (Meliac.).
Odontosoria Fée. Lindsaeaceae. 11 trop. Am. Thicket-forming spiny ls. of indefinite growth.
Odontospermum Neck. ex Sch. Bip. Compositae. 15 C. Verde Is., Canaries, Medit. *O. pygmaeum* O. Hoffm. is a xero. whose fr.-heads close in dry weather (cf. *Anastatica, Mesembryanthemum*); the seeds only escape in damp weather suitable for germination.
Odontostele Schltr. (nomen). Orchidaceae. Quid?
Odontostelma Rendle. Asclepiadaceae. 1 S. trop. Afr.
Odontostemma Benth. = Arenaria L. (Caryophyllac.).
Odontostemum Baker = Odontostomum Torr. (Tecophilaeac.).
Odontostephana Alexander = Gonolobus Michx (Asclepiadac.).
Odontostigma A. Rich. = Stemmadenia Benth. (Apocynac.).
Odontostigma Zoll. & Mor. = Gymnostachyum Nees (Acanthac.).
Odontostomum Torr. Tecophilaeaceae. 1 California.
Odontostyles auctt. = seq.
Odontostylis Breda = Bulbophyllum Thou. (Orchidac.).
Odontotecoma Bur. & K. Schum. = Adenocalymma sp. + Tabebuia sp., mixtae (Bignoniac.).
Odontotrichum Zucc. (~ Cacalia L.). Compositae. 35 SW. U.S., Mex.
Odontychium K. Schum. = Alpinia Roxb. (§ Dieramalpinia K. Schum.) (Zingiberac.).
× **Odopetalum** hort. Orchidaceae. Gen. hybr. (vi) (Odontoglossum × Zygopetalum). (Doubtful.)
Odoptera Rafin. = Corydalis Vent. (Fumariac.).
Odostelma Rafin. = Passiflora L. (Passiflorac.).
Odostemon Rafin. = Mahonia Nutt. (Berberidac.).
Odostemum Steud. = praec.
Odostima Rafin. = Moneses Salisb. (Pyrolac.).
Odotalon Rafin. = Argithamnia Sw. (Euphorbiac.).
Odotheca Rafin. = Homalium Jacq. (Flacourtiac.).
Odyendea (Pierre) Engl. = Quassia L. (Simaroubac.).
Odyendyea auctt. = praec.
Odyssea Stapf. Gramineae. 3 coasts Red Sea, trop. & SW. Afr.
Oeceoclades Lindl. = Angraecum Bory, etc. (Orchidac.).
Oechmea Jaume St-Hil. = Aechmea Ruiz & Pav. (Bromeliac.).
Oecoeclades Franch. & Sav. (sphalm.) = Oeceoclades Lindl. = Angraecum Bory, etc. (Orchidac.).
Oecopetalum Greenman & C. H. Thompson. Icacinaceae. 3 Mex., C. Am.

OEDEMATOPUS

Oedematopus Planch. & Triana. Guttiferae. 10 trop. Am.

Oedera Crantz = Dracaena Vand. ex L. (Agavac.).

***Oedera** L. Compositae. 6 S. Afr.

Oedibasis K.-Pol. Umbelliferae. 4 C. As.

Oedicephalus Nevski = Astragalus L. (Legumin.).

Oedina Van Tiegh. = Dendrophthoë Mart. (Loranthac.).

Oedipachne Link = Eriochloa Kunth (Gramin.).

Oedmannia Thunb. = Rafnia Thunb. (Legumin.).

Oeginetia Wight = Aeginetia L. (Orobanchac.).

Oegroe B. D. Jacks. (sphalm.) = Ocyroë Phil. (Compos.).

Oehmea F. Buxb. = Mammillaria Harv. (Cactac.).

Oemleria Reichb. Rosaceae. 1 Pacif. N. Am. Like *Prunus*, but with 5 free cpls.

Oenanthe L. Umbelliferae. 40 temp. Euras., mts. trop. Afr.

Oenocarpus Mart. Palmae. 16 trop. S. Am.

Oenone Tul. = Apinagia Tul. (emend. v. Royen) (Podostemac.).

Oenonea Bub. = Melittis L. (Labiat.).

Oenoplea [(Pers.)] 'Michx.' ex Hedw. f. = Berchemia Neck. ex DC. (Rhamnac.).

Oenoplia (Pers.) Roem. & Schult. = praec.

Oenosciadium Pomel (~ Oenanthe L.). Umbelliferae. 1 N. Afr.

Oenostachys Bullock. Iridaceae. 2 Abyss., Uganda.

Oenothera L. Onagraceae. 80 Am. (esp. temp.), W.I. The fls. of *O. biennis* L. emit scent at evening and are visited by nocturnal moths, to which they are suited by the long tubes.

Oenother[ac]eae Endl. = Onagraceae Juss.

Oenotheridium Reiche (~ Godetia Spach). Onagraceae. 1 Chile.

Oenotrichia Copel. Dennstaedtiaceae. 4 New Guinea, Queensl., New Caled.; mixture?

Oeollanthus G. Don (sphalm.) = Aeolanthus Mart. (Labiat.).

***Oeonia** Lindl. Orchidaceae. 7 Mascarene Is.

Oeoniella Schlechter. Orchidaceae. 3 Mascarene Is.

Oerstedella Reichb. f. = Epidendrum L. (Orchidac.).

Oeschinomene Poir. = Aeschynomene L. (Legumin.).

Oeschynomene Rafin. = praec.

Oesculus Neck. = Aesculus L. (Hippocastanac.).

Oethionema Knowles & Westc. = Aëthionema R.Br. (Crucif.).

Oetosis [Neck.] Greene (1900) = Vittaria Sm. (Vittariac.).

Oetosis [Neck.] Kuntze (1891) = mixture: Vittaria (Vittariac.), Taenitis (Gymnogrammac.), Drymoglossum (Polypodiac.), Paltonium (Polypodiac.).

Ofaiston Rafin. Chenopodiaceae. 1 S. Russia to C. As.

Oftia Adans. Myoporaceae. 2 S. Afr.

Ogcerostylus Cass. = Angianthus Wendl. (Compos.).

Ogcodeia Bur. Moraceae. 17 trop. S. Am.

Ogiera Cass. = Eleutheranthera Poit. (Compos.).

Oginetia Wight (sphalm.) = Aeginetia L. (Orobanchac.).

Oglifa Cass. = Filago L. (Compos.).

O-Higgensia Steud. = seq.

Ohigginsia Ruiz & Pav. = Hoffmannia Sw. (Rubiac.).

Ohlendorffia Lehm. = Aptosimum Burch. (Scrophulariac.).
Oïanthus Benth. Asclepiadaceae. 4 India.
Oilapetalum Pohl (nomen). 16 Brazil. Quid?
Oïleus Haw. = Narcissus L. (Amaryllidac.).
Oïonychion Nieuwl. = Viola L. (Violac.).
Oïospermum Less. Compositae. 1 NE. Brazil.
Oisodix Rafin. = Salix L. (Salicac.).
Oïstanthera Markgraf. Apocynaceae. 1 Maurit.
Oïstonema Schlechter. Asclepiadaceae. 1 Borneo.
Okea Steud. = seq.
Okenia F. G. Dietr. = Adenandra Willd. (Rutac.).
Okenia Schlechtd. & Cham. Nyctaginaceae. 1 Florida Keys, Mex.
Okoubaka Pellegr. & Normand. Santalaceae. 2 trop. Afr.
*****Olacaceae** Juss. Dicots. 25/250 trop. Most are shrubs or trees with alt.
entire ls. and small ⚥ (rarely ♂♀) reg. fls. Stellate hairs, resin-canals and/or
latex present in some. K usu. much reduced, often resembling the calyculus
of *Loranthaceae*, often accresc.; C 3-6, valv.; disk sometimes present; A as
many or 2 or 3 times as many as C, anth. rarely dehisc. by valves; G partly sunk
in the disk, or free, rarely G̅, 2-5-loc. at base, 1-loc. above, with free plac. and
1 ovule hanging down into each loc. (occasionally 1-loc. 1-ovuled). Drupe or
nut, one-seeded. Seed with testa and endosp.

    Classification and chief genera (after Engler):

    I. *Anacolosoïdeae* (*'Dysolacoïdeae'*) (ovules with 1 or 2 integs., anatr.
        with dorsal raphe, micropyle facing upward; K often accresc. with
        ripening of fr.; G sup.): *Heisteria, Anacolosa, Strombosia, Ximenia*, etc.
    II. *Olacoïdeae* (ovules without integ., anatr., with micropyle facing
        upwards and dorsal rhaphe; fr. K ± accresc.): *Olax, Liriosma,
        Aptandra*, etc.
    III. *Schoepfioïdeae* (ovules without integ., atropous, pend., micropyle
        facing downwards; G inf.): *Schoepfia*.

    The genus *Octoknema*, now referred to this fam., agrees with the *Anacolo-
soïdeae* in its anatropous ovule with 1 integument, but differs in its inferior
Olamblis Rafin. = Carex L. (Cyperac.).   [ovary.
Olax L. Olacaceae. 55 trop. Afr., Madag., Indomal., Austr.
Olbia Medik. = Lavatera L. (Malvac.).
Oldenburgia Less. Compositae. 3 S. Afr.
Oldenlandia P.Br. = Jussiaea L. (Onagrac.).
Oldenlandia L. Rubiaceae. 300 warm. Some are heterostyled (dimorphic).
Oldfieldia Benth. & Hook. f. Euphorbiaceae (?). 4 trop. Afr. *O. africana*
Benth. & Hook. f., the African oak, gives good timber.
Olea L. Oleaceae. 20 Medit., Afr., Masc., E. As., Indomal., E. Austr., N.Z.,
Polynes. *O. europaea* L. (olive), cult. in Medit. region from early ages. The
wild form has thorny twigs and a small fr., the cult. form (var. *sativa* DC.)
is smooth and has a large drupe with oily flesh. The oil is obtained by bruising
and pressing the fruit. Several yield good timber, e.g. *O. europaea, O. laurifolia*
Lem. (S. Afr.; black ironwood), etc.
*****Oleaceae** Hoffmgg. & Link. Dicots. 29/600 cosmop., esp. temp. & trop. As.
Shrubs and trees usu. with opp. ls., which are exstip., simple or pinnate, often

entire. Serial accessory buds occur in the leaf-axils of many spp. (e.g. *Syringa*) in both flg. and veg. parts. The infl. is racemose or cymose, often bracteolate. Fls. ⚥, rarely unisexual, reg., 2–6-merous, sometimes poly- or a-petalous (*Fraxinus*, etc.). K typically (4), valvate; C (4), valvate or imbr., rarely conv.; A 2, epipet., usu. transv. placed, and alt. with cpls.; no disk; G̲ (2), 2-loc. with 2 anatr. ov. in each loc.; stigma 2-lobed on simple style. Berry, drupe, caps., or schizocarp, with 1–4 seeds. Endosp. or none, embryo straight. *Olea, Fraxinus*, etc., are of economic value. Chief genera: *Jasminum, Schrebera, Fraxinus, Syringa, Osmanthus, Linociera, Olea, Ligustrum, Noronhia.* (For revised classif., see Johnson, *Contrib. N.S.W. Nat. Herb.* **2** (6): 395–418, 1957.)

Oleander Medik. = Nerium L. (Apocynac.).

Oleandra Cav. Oleandraceae. 40 trop. Stem slender, creeping or bushy; ls. simple, jointed to phyllopodia on stem; sori near midrib, dorsal on veins, indus. reniform.

Oleandraceae Ching ex Pichi-Serm. Davalliales. Ls. simple or 1-pinnate, jointed to rhizome exc. *Nephrolepis*; pinnae jointed to rachis; sori terminal or dorsal on veins, rarely confluent, most indusiate. Genera: *Oleandra, Arthropteris, Psammiosorus, Nephrolepis.* [morphous.

Oleandropsis Copeland. Polypodiaceae. 1 New Guinea. Ls. simple, di-

*Olearia Moench. Compositae. 100 New Guinea, Austr., N.Z. Replaces *Aster*, and closely resembles it, but all trees or shrubs.

Oleaster Fabr. (uninom.) = Elaeagnus L. sp. (Elaeagnac.).

Oleicarpon Dwyer. Leguminosae. 1 Costa Rica to Colombia.

Oleicarpus Dwyer = praec.

Oleobachia Hort. ex Mast. = Sterculia L. (Sterculiac.).

Oleoxylon Roxb. = Dipterocarpus Gaertn. f. (Dipterocarpac.).

Olfa Adans. = Isopyrum L. (Ranunculac.).

Olfersia Raddi = Polybotrya Humb. & Bonpl. (Aspidiac.).

Olgaea Iljin. Compositae. 18 C. As. to N. China.

Olgasis Rafin. = Oncidium Sw. (Orchidac.).

Oligacis Rafin. = Rubus L. (Rosac.).

Oligacoce Willd. ex DC. = Valeriana L. (Valerianac.).

Oligactis Cass. = Liabum Adans. (Compos.).

Oligactis Rafin. = Sericocarpus Nees (Compos.).

Oligaerion Cass. = Ursinia Gaertn. (Compos.).

Oligandra Less. (1832). Compositae. 3 trop. S. Am.

Oligandra Less. (1834) = Chenopodium L. (Chenopodiac.).

Oliganthemum F. Muell. Compositae. 8 Austr.

Oliganthera Endl. = Chenopodium L. (Chenopodiac.).

Oliganthes auctt. amer. = Pollalesta Kunth (Compos.).

Oliganthes Cass. Compositae. 9 Madag.

Oligarrhena R.Br. Epacridaceae. 1 W. Austr.

Oligloron Rafin. = Capparis L. (Capparidac.).

Oligobotrya Baker. Liliaceae. 2–3 W. & C. China.

Oligocampia Trevis. = Anisocampium Presl (Athyriac.).

Oligocarpha Cass. = Brachylaena R.Br. (Compos.).

Oligocarpus Less. Compositae. 3 S. Afr.

Oligoceras Gagnep. Euphorbiaceae. 1 Indochina.

Oligochaeta K. Koch = Centaurea L. (Compos.).
Oligocladus Chodat & Wilczek. Umbelliferae. 2 Argentina.
Oligocodon Keay. Rubiaceae. 1 trop. W. Afr.
Oligodora DC. = Athanasia L. (Compos.).
Oligodorella Turcz. = Marasmodes DC. (Compos.).
Oligoglossa DC. = Phymaspermum Less. (Compos.).
Oligogyne DC. = Blainvillea Cass. (Compos.).
Oligogynium Engl. = Nephthytis Schott (Arac.).
Oligolepis Cass. ex DC. = Sphaeranthus L. (Compos.).
Oligolobos Gagnep. = Ottelia Pers. (Hydrocharitac.).
*****Oligomeris** Cambess. Resedaceae. 1 SW. U.S., Mex.; 8 Canaries, N. Afr., S. Afr. to NW India.
Oligonema S. Wats. (1891; non Rostafinsky 1875—Myxomyc.) = Golionema S. Wats. (Compos.).
Oligoneuron Small = Solidago L. (Compos.).
Oligopholis Wight = Christisonia Gardn. (Orobanchac.).
Oligoron Rafin. = Asclepias L. (Asclepiadac.).
Oligoscias Seem. = Polyscias J. R. & G. Forst. (Araliac.).
Oligosma Salisb. = Nothoscordum Kunth (Alliac.).
Oligosmilax Seem. = Heterosmilax Kunth (Smilacac.).
Oligosporus Cass. = Artemisia L. (Compos.).
Oligostemon Benth. = Duparquetia Baill. (Legumin.).
Oligostemon Turcz. = Meliosma Blume (Meliosmac.).
Oligothrix DC. Compositae. 3 trop. & S. Afr.
*****Olinia** Thunb. Oliniaceae. 10 E. trop. & S. Afr.
*****Oliniaceae** Arn. ex Sond. Dicots. 1/10 Afr. Trees or shrubs with simple ent. opp. stip. (!) ls. Fls. ⚥, reg., in term. or axill. cymes. K (5), tubular, limb of 5 minute blunt teeth, or obsolete; C 5, imbr., spath., inserted on margin of K-tube, alt. with 5 small, valvate, ± cucullate, pubesc., coloured scales; A 5, alternipet., inserted immediately below scales and at first hidden and enclosed by them, fil. v. short, connective thickened, alt. with 5 small, thick, subglob., pubesc. stds.; Ḡ (5–4), with 2(–3) superposed pend. axile ov. per loc., and 1 simple style with clavate stig. Fr. a 3–5-loc. false drupe, with 1 seed per loc.; seeds with spiral or convol. endosp. embryo; endosp. o. Only genus: Olinia. Affinities much disputed, but prob. nearly related to *Thymelaeaceae*. The interpretation of the floral whorls has also varied, but the presence in some spp. of 5 quite evident small teeth at the apex of the tubular 'receptacle', *outside* the 5 large spathulate organs (though these teeth are obsolete in other spp.), seems to indicate that the teeth are calycine, and that the apparent petals are in fact such. The coloured scales must then be regarded as ligular structures.
Olisbaea Benth. & Hook. f. = seq.
Olisbea DC. = Mouriri Aubl. (Memecylac.).
Olisca Rafin. = Juncus L. (Juncac.).
Olisia Spach = Stachys L. (Labiat.).
Olivaea Sch. Bip. Compositae. 1 Mex.
Oliveranthus Rose (~ Echeveria DC.). Crassulaceae. 1 Mex.
Oliverella Rose = praec.
Oliverella Van Tiegh. = Tapinanthus Bl. (Loranthac.).

OLIVERIA

Oliveria Vent. Umbelliferae. 1 SW. As.

Oliveriana Reichb. f. Orchidaceae. 1 Colombia.

Oliverodoxa Kuntze = Riedelia Oliver (Zingiberac.).

Oliviera P. & K. = Oliveria Vent. (Umbellif.).

Olmedia Ruiz & Pav. Moraceae. 2 trop. Am.

Olmediella Baill. Flacourtiaceae. 1–2 (of Mex. origin?), only known in cult.

Olmedioperebea Ducke. Moraceae. 2 Amaz. Brazil.

Olmediophaena Karst. Moraceae. 3 trop. S. Am.

Olmediopsis Karst. = Pseudolmedia Tréc. (Morac.).

Olmedoa P. & K. = Olmedia Ruiz & Pav. (Morac.).

Olmedoëlla P. & K. = Olmediella Baill. (Flacourtiac.).

Olmedophaena P. & K. = Olmediophaena Karst. (Morac.).

Olneya A. Gray. Leguminosae. 1 Calif., Mex.

Olofuton Rafin. = Capparis L. (Capparidac.).

Olopetalum Klotzsch = Monsonia L. (Geraniac.).

Olostyla DC. = Holostyla DC. corr. Endl. (Rubiac.).

Olotrema Rafin. = Carex L. (Cyperac.).

Olsynium Rafin. = Sisyrinchium L. (Iridac.).

Oluntos Rafin. = Urostigma Gasp. = Ficus L. (Morac.).

Olus-atrum v. Wolf = Smyrnium L. + Taenidia Drude (Umbellif.).

Olympia Spach = Hypericum L. (Guttif.).

Olymposciadium H. Wolff. Umbelliferae. 1 As. Min. (Bithyn. Olympus).

Olympusa Klotzsch (nomen). Asclepiadaceae. 1 Guiana. Quid?

Olynia Steud. = Olinia Thunb. (Oliniac.).

Olynthia Lindl. = Eugenia L. (Myrtac.).

Olyra L. Gramineae. 20 trop. Am., Afr.

Olythia Steud. = Olynthia Lindl. = Eugenia L. (Myrtac.).

Omalanthus A. Juss. = Homalanthus A. Juss. corr. Reichb. (Euphorbiac.).

Omalanthus Less. = Tanacetum L. (Compos.).

Omalium auct. ex Steud. = Homalium Jacq. (Flacourtiac.).

Omalocarpus Choux. Sapindaceae. 1 Madag.

Omaloclados Hook. f. = Homaloclados Hook. f. = Faramea Aubl. (Rubiac.).

Omalocline Cass. = Crepis L. (Compos.).

Omalotes DC. = Tanacetum L. (Compos.).

Omalotheca Cass. = Gnaphalium L. (Compos.).

Omania S. Moore. Scrophulariaceae. 1 Arabia.

Omanthe O. F. Cook = Chamaedorea Willd. (Palm.).

Ombrocharis Hand.-Mazz. Labiatae. 1 China.

Ombrophytum Poepp. Balanophoraceae. 2–3 E. Peru, W. Brazil, N. Argent.

Omentaria Salisb. = Tulbaghia L. (Alliac.).

Omiltemia Standley. Rubiaceae. 1 Mex.

Ommatodium Lindl. = Pterygodium Sw. (Orchidac.).

Omoea Blume. Orchidaceae. 2 Philippines, Java.

Omolocarpus Neck. = Nyctanthes L. (Verbenac.).

Omonoia Rafin. = Eschscholzia Cham. (Papaverac.).

Omoscleria Nees = Scleria Bergius (Cyperac.).

Omphacarpus Korth. = Microcos L. (Tiliac.).

Omphacomeria (Endl.) A. DC. Santalaceae. 1 SE. Austr.

Omphalandria P.Br. = seq.

*Omphalea L. Euphorbiaceae. 20 trop. Am., Afr., Madag., Indoch., W. Malaysia, Celebes, New Guinea, Queensl., Solomon Is.

Omphalissa Salisb. = Hippeastrum Herb. (Amaryllidac.).

Omphalium Roth = Omphalodes Mill. (Boraginac.).

Omphalobium Gaertn. = Connarus L. (Connarac.).

Omphalobium Jacq. ex DC. = Schotia Jacq. (Legumin.).

Omphalocarpum Beauv. Sapotaceae. 5 trop. W. Afr.

Omphalocarpum Presl ex Dur. = Santalodes Kuntze (Connarac.).

Omphalocarpus P. & K. = praec.

Omphalocaryon Klotzsch = Scyphogyne Brongn. (Ericac.).

Omphalococca Willd. ex Schult. = Aegiphila Jacq. (Verbenac.).

Omphalodes Boerl. (sphalm.) = Omphalopus Naud. (Melastomatac.).

Omphalodes Mill. Boraginaceae. 28 temp. Euras., Mex. The borders of the achenes are inrolled.

Omphalogonus Baill. = Parquetina Baillon (Periplocac.).

Omphalogramma Franch. Primulaceae. 15 Himal., W. China.

Omphalolappula A. Brand. Boraginaceae. 1 temp. Austr.

Omphalopappus O. Hoffm. Compositae. 1 Angola.

Omphalophthalmum Karst. Asclepiadaceae. 1 Colombia.

Omphalopus Naud. Melastomataceae. 1 Sumatra, Java, New Guinea.

Omphalospora Bartl. = Veronica L. (Scrophulariac.).

Omphalostigma Reichb. = Leianthus Griseb. (Gentianac.).

Omphalothalma Pritz. (sphalm.) = Omphalophthalmum Karst. (Asclepiadac.).

Omphalotheca Hassk. = Commelina L. (Commelinac.).

Omphalothrix Maxim. corr. Kom. Scrophulariaceae. 1 NE. As.

Omphalotrix Maxim. = praec.

Onagra Mill. = Oenothera L. (Onagrac.).

*Onagraceae Juss. Dicots. 21/640 temp. & trop. Most are perenn. herbs, a few shrubs or trees, with alt., opp. or whorled ls., usu. simple, rarely stip. Fls. sol. in axils, or in spikes, racemes, or panicles, ⚥, reg. or zygo., usu. 4-merous (2–5). Axis usu. prolonged beyond ovary into a calyx-tube. K 4, valvate; C 4, rarely 0, usu. conv.; A 4+4, or 4−2−1, pollen grains with 3 apertures, often with long viscin threads; G (4), 4-loc., or ½-inf. 2-loc., with axile plac. and ∞–1 anatr. ov.; septa commonly imperfect below; style simple, stigmas 1 or more. Fls. often very protandr., suited to bees or Lepidoptera; cf. *Lopezia*. Fr. usu. a loculic. caps., sometimes nut or berry; endosp. little or 0. Chief genera: *Fuchsia, Lopezia, Circaea, Hauya, Oenothera, Clarkia, Gaura, Ludwigia (Jussiaea), Epilobium.*

Onagrariaceae Dulac = praec. (excl. Circaeëae).

Oncella Van Tiegh. Loranthaceae. 4 trop. Afr.

Oncerostylus P. & K. = Ogcerostylus Cass. = Styloncerus Spreng. = Angianthus

Oncerum Dulac = Silene L. (Caryophyllac.).   [Wendl. (Compos.).

× **Oncidarettia** hort. (vii) = × Oncidettia hort. (Orchidac.).

× **Oncidasia** hort. (vii) = × Aspasium hort. (Orchidac.).

× **Oncidenia** hort. Orchidaceae. Gen. hybr. (iii) (Macradenia × Oncidium).

× **Oncidesa** hort. Orchidaceae. Gen. hybr. (iii) (Gomesa × Oncidium).

× **Oncidettia** hort. Orchidaceae. Gen. hybr. (iii) (Comparettia × Oncidium).

× **Oncidguesia** hort. (vii) = × Rodricidium hort. (Orchidac.).

Oncidiochilus Falc. = Cordylestylis Falc. = Goodyera R.Br. (Orchidac.).

× **Oncidioda** hort. Orchidaceae. Gen. hybr. (iii) (Cochlioda × Oncidium).

Oncidium Sw. Orchidaceae. 350 Florida to temp. S. Am., W.I.

× **Oncidophora** hort. (vii) = × Ornithocidium hort. (Orchidac.).

× **Oncidpilia** hort. Orchidaceae. Gen. hybr. (iii) (Oncidium × Trichopilia).

Oncinema Arnott. Asclepiadaceae. 1 S. Afr.

Oncinocalyx F. Muell. Verbenaceae. 1 Austr.

Oncinotis Benth. Apocynaceae. 25 trop. & S. Afr., Madag.

Oncinus Lour. = Melodinus J. R. & G. Forst. (Apocynac.).

Oncoba Forsk. Flacourtiaceae. 5 trop. Afr.

Oncocalamus (Mann & Wendl.) Mann & Wendl. ex Kerchove. Palmae. 6 trop. Afr. Climbers.

Oncocalyx Van Tiegh. = Odontella Van Tiegh. (Loranthac.).

Oncocarpus A. Gray. Anacardiaceae. 6 Philipp., New Guinea(?), Fiji.

Oncocyclus Siemssen = Iris L. (Iridac.).

Oncodeia Benth. & Hook. f. = Ogcodeia Bur. (Morac.).

Oncodia Lindl. = Brachtia Reichb. f. (Orchidac.).

Oncodostigma Diels. Annonaceae. 3 W. Malaysia, New Guinea, New Hebr.

Oncolon Rafin. = Valerianella Mill. (Valerianac.).

Oncoma Spreng. = Oxera Labill. (Verbenac.).

Oncorhiza Pers. = Oncus Lour. = Dioscorea L. (Dioscoreac.).

Oncorhynchus Lehm. = Orthocarpus Nutt. (Scrophulariac.).

Oncosina Rafin. = Valerianella Mill. (Valerianac.).

Oncosperma Blume. Palmae. 5 Ceylon, Indoch., W. Malaysia. Very thorny.

Oncosporum Putterl. = Marianthus Huegel (Pittosporac.).

Oncostema Rafin. = Scilla L. (Liliac.).

Oncostemma K. Schum. Asclepiadaceae. 1 S. Tomé (W. Equat. Afr.).

Oncostemon Spach = seq.

Oncostemum A. Juss. Myrsinaceae. 100 Madag., Masc. Largest genus endemic to islands only.

Oncostylis Nees = Psilocarya Torr. (Cyperac.).

Oncostylus (Schlechtd.) F. Bolle. Rosaceae. 9 Tasm., N.Z., Auckl. I., temp. S. Am.

Oncotheca Baill. Oncothecaceae. 1 New Caled.

Oncothecaceae Kobuski ex Airy Shaw. Dicots. 1/1 New Caled. Shrub or small tree, glabrous, with alt. simple coriac. oblanc. exstip. ls., ent. or minutely glandular-dentic. towards apex, crowded towards tips of branches. Fls. ⚥, reg., small, in ± narrow axill. thyrses with angular rhachis; bract 1, bracteoles 2, immediately below fl. K 5, orbic., much imbr., persist.; C (5), shortly campan., lobes imbr., rounded; A 5, alternipet., epipet., fil. v. short, anth. basifixed, connective thick, produced into a subulate sharply inflexed appendage; G (5), 5-loc., 5-grooved, with 5 short free recurved styles, and 1 (rarely 2 collat.) pend. ov. per loc. Fr. a small glob. 5-loc. ± woody drupe; ripe seed unknown. Only genus: *Oncotheca*. Affinity prob. with *Theac.*; ls., bract & bracteoles, etc., v. similar. Supposed connection with *Ebenac.* prob. erroneous.

Oncufis Rafin. = Cleome L. (Cleomac.).

Oncus Lour. = Dioscorea L. (Dioscoreac.) [? + Ipomoea L. (Convolvulac.)].
Oncyphis P. & K. = Oncufis Rafin. = Cleome L. (Cleomac.).
Ondetia Benth. Compositae. 1 SW. Afr.
Ondinea den Hartog. Nymphaeaceae. 1 W. Austr.
Onea P. & K. = Onoea Franch. & Sav. = Diarrhena Rafin. (Gramin.).
Onefera Rafin. = Chironia L. (Gentianac.).
Ongokea Pierre. Olacaceaè. 2 W. Equat. Afr.
Oniscophora Wendl. ex Burret = ? Macrophloga Becc. (Palm.).
Onistis Rafin. Quid? Convolvulaceae? Asclepiadaceae? 1 Florida.
Onites Rafin. = Origanum L. (Labiat.).
Onix Medik. = Astragalus L. (Legumin.).
Onixotis Rafin. = Dipidax Laws. (Liliac.).
Onkeripus Rafin. = Xylobium Lindl. (Orchidac.).
Onkerma Rafin. = Carex L. (Cyperac.).
× **Onoara** hort. Orchidaceae. Gen. hybr. (iii) (Ascocentrum × Renanthera × Vanda × Vandopsis).
Onobroma Gaertn. = Carduncellus Adans. (Compos.).
Onobruchus Medik. = seq.
Onobrychis Mill. Leguminosae. 120 Eur., Medit. to C. As. Fl. much like *Trifolium.* Petiole persistent. Ovules collat. *O. viciifolia* Scop. (sainfoin) good
Onochiles Bub. & Penz. = Alkanna Tausch (Boraginac.).   [fodder.
Onochilis Mart. = Nonea Medik. (Boraginac.).
Onoclea L. Aspidiaceae. 1 N. As., N. Am. Fossil in Eocene of Mull (Scot-
Onocleaceae Ching = Aspidiaceae S. F. Gray.   [land).
Onocleopsis F. Ballard. Aspidiaceae. 1 Mex., C. Am.
Onoctonia Naud. = Poteranthera Bong. (Melastomatac.).
Onodontea G. Don = Anodontea Sweet = Alyssum L. (Crucif.).
Onoea Franch. & Sav. = Diarrhena Rafin. (Gramin.).
Onograri[ac]eae Dulac (sphalm.) = Onagrariaceae Dulac = Onagraceae Juss. (excl. Circaeëae).
Onohualcoa Lundell = Mansoa DC. (Bignoniac.).
Ononis L. Leguminosae. 75 Canaries, Medit., Eur. to C. As. Lat. branches sometimes thorny. Rarely > 3 leaflets. Fl. mech. between *Lotus* and *Tri-, folium.* At first upper edges of keel cohere, and pollen is squeezed out at tip; then anthers emerge as in *Trif.*
Onopix Rafin. = Cirsium Mill. (Compos.).
Onopordon Hill = seq.   [brs. thorny.
Onopordum L. Compositae. 40 Eur., N. Afr., W. As. Ls. decurrent. Invol.
Onopyxos Spreng. = Onopix Rafin. = Cirsium Mill. (Compos.).
Onopyxus Bub. = Carduus L. (Compos.).
Onoseris Willd. = Onoseris Willd. emend. DC. + Lycoseris Cass. (Compos.).
Onoseris Willd. emend. DC. Compositae. 25 Mex. to Andes.
Onosma L. Boraginaceae. 150 Medit. to Himal. & China. Root of *O. echioïdes* L. yields a dye (*orsanette*).
Onosm[at]aceae Horan. = Boraginaceae Juss.
Onosmidium Walp. = seq.
Onosmodium Michx. Boraginaceae. 15 N. Am., Mex.
Onosuris Rafin. = Oenothera L. (Onagrac.).

Onotrophe Cass. = Cirsium Mill. (Compos.).
Onuris Phil. Cruciferae. 6 Chile, Patagonia.
Onus Gilli. Acanthaceae. 2 trop. E. Afr.
Onychacanthus Nees = Bravaisia DC. (Acanthac.).
Onychiaceae Dulac = Caryophyllaceae–Caryophylloïdeae.
Onychium Blume = Dendrobium Sw. (Orchidac.).
Onychium Kaulf. Cryptogrammataceae. 6 trop. & subtrop.
Onychium Reinw. = Lecanopteris Reinw. (Polypodiac.).
Onychopetalum R. E. Fries. Annonaceae. 4 Brazil.
Onychosepalum Steud. Restionaceae. 1 SW. Austr.
Onyx Medik. = Astragalus L. (Legumin.).
Oöçarpon Micheli = Ludwigia L. (Onagrac.).
Oöcarpus P. & K. = praec.
Oöchlamys Fée = Thelypteris Schmid. (Thelypteridac.).
Oöclinium DC. = Praxelis Cass. (Compos.).
Oönopsis Greene (~ Haplopappus Cass.). Compositae. 6 N. Am.
Oöphytum N. E. Brown. Aïzoaceae. 2 S. Afr.
Oosterdyckia Boehm. (Cunonia L.). Cunoniaceae. 1, *O. capensis* (L.) Kuntze, S. Afr.; 16 New Caled.
Oosterdykia Kuntze = praec.
Oöthrinax (Bedd.) O. F. Cook = Zombia L. H. Bailey (Palm.).
Opa Lour. = Syzygium Gaertn. (Myrtac.) + Raphiolepis Lindl. (Rosac.).
Opalatoa Aubl. = Apalatoa Aubl. = Crudia Schreb. (Legumin.).
Opanea Rafin. = Rhodamnia Jack (Myrtac.).
Oparanthus Sherff. Compositae. 4 Polynesia.
Opelia Pers. = Opilia Roxb. (Opiliac.).
Opercularia Gaertn. Rubiaceae. 15 Austr.
Operculariaceae Dum. = Rubiaceae–Anthospermeae Cham. & Schlechtd.
Operculicarya H. Perrier. Anacardiaceae. 3 Madag.
Operculina S. Manso. Convolvulaceae. 25 trop. *O. turpethum* (L.) S. Manso yields a drug.
Opetiola Gaertn. = Mariscus Gaertn. (Cyperac.).
Ophelia D. Don ex G. Don = Swertia L. (Gentianac.).
Ophellantha Standley. Euphorbiaceae. 2 C. Am.
Ophelus Lour. = Adansonia L. (Bombacac.).
Ophiala Desv. = Helminthostachys Kaulf. (Ophioglossac.).
Ophianthe Hanst. = Pentarhaphia Lindl. (Gesneriac.).
Ophianthes Rafin. = Chelone L. = Penstemon Schmid. (Scrophulariac.).
Ophiobostryx Skeels = Bowiea Harv. (Liliac.).
Ophiobotrys Gilg (~ Osmelia Thw.). Flacourtiaceae. 1 W. Equat. Afr.
Ophiocaryon Schomb. Meliosmaceae. 2 trop. S. Am. (snakeseed, snakenut).
Ophiocaulon Hook. f. = Adenia Forsk. (Passiflorac.).
Ophiocaulon Rafin. = Chamaecrista Moench = Cassia L. (Legumin.).
Ophiocephalus Wiggins. Scrophulariaceae. 1 Lower Calif.
Ophiocolea H. Perrier. Bignoniaceae. 5 Madag., Comoro Is.
Ophioderma (Bl.) Endl. = Ophioglossum L. (Ophioglossac.).
Ophioglossaceae Presl. Ophioglossales. 4/70 trop., temp. Small herbs, some trop. spp. epiphytic. The l. splits into a dorsal and a ventral part, the former

being the 'sterile' green blade, the latter the 'fertile' sporangiiferous spike, often much branched and containing the sporangia sunk in its tissues. The spores are all of one kind and give rise to subterranean colourless prothalli, living saprophytically. Genera: *Ophioglossum* (sporangia sessile, in two rows, forming a narrow close spike); *Botrychium* (sporangia in small crested clusters forming a long loose spike); *Helminthostachys* (sporangia borne in small groups close to the two sides of the fertile spike); *Rhizoglossum* (fronds dimorphous, spike as *Ophioglossum*).

Ophioglossales. 1 fam.: *Ophioglossaceae.*

Ophioglossidae (subclass). 1 order: *Ophioglossales.*

Ophioglossum L. Ophioglossaceae. 30–50 trop. & temp. Adv. buds are formed on the roots and thus the pl. multiplies veg. The sporangiiferous spike is usu. unbranched, except in *O. palmatum* L. (this sp. and *O. pendulum* L. are epiphytic). High polyploidy occurs.

Ophiolyza Salisb. = Gladiolus L. (Iridac.).

Ophiomeris Miers (~ Thismia Griff.). Burmanniaceae. 10 trop. S. Am.

Ophione Schott = Dracontium L. (Arac.).

***Ophiopogon** Ker-Gawl. Liliaceae. 20 Himal. to Japan & Philipp. Is. The mucilaginous tubers of *O. japonicus* (L. f.) Ker-Gawl. are ed.

Ophiopogon Kunth = Liriope Lour. (Liliac.).

Ophiopogon[ac]eae Endl. = Liliaceae–Ophiopogoneae Endl.

Ophioprason Salisb. = Asphodelus L. (Liliac.).

Ophiopteris Reinw. = Oleandra Cav. (Oleandrac.).

Ophiorrhiza L. Rubiaceae. 150 Indomal.

Ophiorrhiziphyllon Kurz. Acanthaceae. 5 SE. As.

Ophiorrhizophyllum auctt. = praec.

Ophioscorodon Wallr. = Allium L. (Alliac.).

Ophioseris Rafin. = Hieracium L. (Compos.).

Ophiospermae ('-spermes') Vent. = Myrsinaceae R.Br.

Ophiospermataceae Kuntze = praec.

Ophiospermum Reichb. = Ophispermum Lour. = Aquilaria Lam. (Thymelaeac.).

Ophiostachys Delile = Chamaelirium Willd. (Liliac.).

Ophioxylon L. = Rauvolfia L. (Apocynac.).

Ophira Burm. ex L. = Grubbia Bergius (Grubbiac.).

Ophira Lam. = Strobilocarpus Klotzsch (Grubbiac.).

Ophiraceae Reichb. = Grubbiaceae Endl.

Ophiria Becc. = Pinanga Blume (Palm.).

Ophiria Lindl. = Ophira Lam. = Strobilocarpus Klotzsch (Grubbiac.).

Ophiriaceae Arn. = Ophiraceae Reichb. = Grubbiaceae Endl.

Ophismenus Poir. (sphalm.) = Oplismenus Beauv. (Gramin.).

Ophispermum Lour. = Aquilaria Lam. (Thymelaeac.).

Ophiurinella Desv. = Psilurus Trin. (Gramin.).

Ophiuros Gaertn. f. Gramineae. 1 NE. trop. Afr., 6 Indomal., Austr.

Ophiurus R.Br. = praec.

Ophrestia H. M. L. Forbes. Leguminosae. 12 trop. Afr., As.

Ophris Mill. = Listera R.Br. + Epipactis Zinn (Orchidac.).

Ophrydium Schrad. ex Nees = Ophryoscleria Nees = Scleria Bergius (Cyperac.).

ORCHICOELOGLOSSUM

× **Orchicoeloglossum** Aschers. & Graebn. Orchidaceae. Gen. hybr. (vi) (Coeloglossum × Orchis).

*Orchidaceae Juss. Monocots. 735/17,000 cosmop., abundant in trop., rare in arctic regions. They agree in some general features of habit, etc., e.g. they are all perennial herbs, but differ widely in detail, owing to the diversity of conditions in which they exist—terrestrials, epiphytes, saprophytes, etc. Within the trop. they form an important feature of the veg., living chiefly as epiphytes. Most temp. zone forms are terrestrial.

The plant as a whole may be built up in one of three ways: (1) a monopodium, the main axis growing steadily on, year after year, and bearing the fls. on lat. branches; (2) an *acranthous* sympodium, the main axis being composed of annual portions of successive axes, each of which begins with scale ls. and *ends* in an infl.; (3) a *pleuranthous* sympodium, where the infls. are borne on *lateral* axes, the shoot which for the current year continues the main axis stopping short at the end of its growing period, and not ending in an infl.

The saprophytes are few; they have no green ls.; below the soil, in the humus, is a fleshy rhiz., with (*Neottia*) or without roots. It is much branched, and does part or all of the work of absorption. Mycorhiza occurs in most or all. The terrestrial forms are all sympodial, and have usu. a rhiz.; each annual shoot bends up into the leafy shoot of the current year. Many being xero., and all perenn., it becomes a necessity that there should be a storage reservoir to last over the non-veg. period of the year. In a great many this takes the form of a thickened internode of the stem; in many, again, the bud for the next year's growth, i.e. the next part of the sympodium, is laid down at the base of the stem, and from it is developed a thick and fleshy adv. root, forming a large tuber, which lasts over the winter.

Coming lastly to the epiphytes, abundant in the trop., we find great variety. [See Schimper, *Die epiphytische Vegetation Amerikas*.] They are mostly sympodial, but the monopodial *O*. also belong to this group. The exceedingly light seeds and the xero. habit of many *O*. fit them to become epiph. The roots of the epiph. forms are of interest. In the first place, to fasten the pl. to its support there are 'clinging' roots, insensitive to gravity, but negatively heliotropic. The niche between the pl. and its support and the network formed by the roots act as reservoirs for humus, into which project 'absorbing' roots, branches from the others; these are usu. negatively geotropic. Finally the true aerial roots hang down in long festoons. The outer layers of cells (the epidermis and *velamen*) are dead, and act as an insulating water-jacket. Their internal tissue is green (as may be seen on wetting a root) and assimilates. During the dry season a great proportion of the *O*. drop their ls. (though they may flower), and 'hibernate' in the condition of fleshy *pseudobulbs*. One pseudobulb, consisting of one or more thickened stem-internodes, is usu. formed each year. In this, water and other reserves are stored. Those epiphytes which do not form these tubers have fleshy ls. which serve the same purpose; the fleshy-leaved orchids, e.g. *Vanilla*, have usu. a very feebly developed velamen. Some monopodial forms have no green ls. at all, assimilating either by the surface of the stem or by the long aerial roots (*Polyrrhiza*, etc.).

The infls. are racemose, very often spikes, which look like racemes, the long inf. ovary resembling a stalk. The fl. is usu. zygo. (reg. in subfam. *Aposta-*

sioïdeae). There are two kinds of *O.*, with different fls., with 1 and 2 sta. respectively; the great majority are monandrous. P in 2 whorls, epig., petaloid. The post. petal is usu. larger than the rest, and is termed the *labellum*; by the twisting (*resupination*) of the ovary through 180° it comes round to the ant. side of the fl. and forms a landing-place for insects. In many *O.* its structure is exceedingly complex. The essential organs of the fl. are all comprised in a central structure by which the *O.* can be recognised at a glance, viz. the *column*, which consists in the simpler cases of the combined style and sta. (to use the old-fashioned expression; in reality it is very probably an outgrowth of the axis, bearing the anthers and stigmas at the top). In the monandrous forms the column exhibits one anther and two fertile stigmas (often ± confluent), together with a special organ, the *rostellum*, which repres. the third stigma. The single anther is the ant. one of the outer whorl (if we imagine the fl. of *O.* derived from a typical 3-merous fl.); the other two of this whorl are entirely absent, and also all those of the inner whorl. The two fertile stigmas are the post. pair, and the third (ant.) is repres. by the rostellum (in using the terms ant. and post., the resupination is supposed not to have occurred).

The various organs face the labellum and, in the fl. of a simple *O.*, e.g. *Orchis*, can easily be made out. A little above the base are the two stigmas, then above these a projecting point, the rostellum, and above this again, and behind it, forming the apex of the column, is the anther, which shows two lobes. Each is occupied by a *pollinium*, or mass of pollen. Under the microscope the grains of pollen are seen to be tied together in packets by elastic threads; these unite at the base of the pollinium and form a cord, the *caudicle*, which runs down into, and is attached to, part of the rostellum.

The simple construction found in *Orchis*, etc., as thus described, is replaced by much more complex arrangements in many. The labellum itself may be rendered very complex, by the addition of spurs and other outgrowths; often outgrowths of the summit of the receptacle take place, displacing some of the organs. Thus, for example, in *Drymoda* and others, the labellum and the sepals on either side of it are carried forward on an axial protuberance in such a way that the sepals appear to spring from the labellum, the axial growth (*chin*) appearing like the basal part of this organ. Some of these constructions are very elaborate.

Similarly the column shows great variety in structure. One point may be mentioned specially as of importance in some systems of classification. In the simple case of *Orchis*, etc., described above, the *base* of the anther loculi is against the rostellum; such cases are called *basitonic*; in others it is the apex that is next the rostellum (*Oncidium*, etc.), and these are termed *acrotonic*.

So far only monandrous forms have been considered. In *Cypripedium* and its allies the column has 2 anthers, no rostellum, and a simple stigma, composed of the 3 carpellary stigmas. The two sta. belong to the inner whorl, and the sta. which in the monandrous forms is fertile is here repres. by a large std. The stigma is not sticky, but the pollen is, and it is not combined into pollinia.

The ovary is inf. in all *O.*, uniloc., with 3 parietal plac. (exc. *Apostasia*), and ∞ ovules, which do not develop until fert. of the fl. occurs.

The adaptations of orchid flowers to *fertilisation* by insects are endless, and many very complicated. Reference must be made to text-books for the details. No student should omit to read Darwin's *Fertilisation of Orchids*, at least the first two and the last chapters. In it will be found accounts of the mech. of the common gen. A few general points only can be mentioned here. Very few secrete free honey; in most cases the insect has to bite into or drill the tissue for the juice therein contained; this tissue is usu. part of the labellum—often a spur at the base—or the basal part of the column. The pollinia are removed as a rule when the insect is going out of the fl. In most cases the insect in entering displaces the rostellum or some portion of it, and thereby exposes and comes into contact with a sticky mass (due to disorganization of cells formerly living). This becomes cemented to the insect while it is drilling for honey, and as the insect goes out again it takes with it the viscid lump, together with the pollinia, either merely glued to it, or attached by caudicles. In many cases the pollinia are in such a position that when the insect enters the next fl. they will touch the stigmas. In others this is not so, e.g. *Orchis*, where the anthers and stigma are far apart on the column, and in such cases the pollinia, on getting out of the anther, execute a hygroscopic movement which brings them into the proper position on the insect's body to strike the stigmas. Such is the general principle of the orchid mechanism, but the variety in detail is endless.

The *fruit* is a caps., containing usu. a gigantic number of exceedingly small and light seeds, which are well suited to wind distr. (hence, among other causes, the epiph. habit of so many).

The *O.* are favourites in horticulture, and very many gen. are cult. There are many bi- and pluri-generic hybrids; the names of most of those as yet produced, e.g. × *Orchicoeloglossum*, × *Zygocolax*, × *Holttumara*, are included in this book. *Vanilla* is the only orchid of direct economic importance at the present day.

Subdivision of family. The *Orchidaceae* being a very large and natural family with few absolute distinctions between its major subdivisions, many schemes of classification have been suggested. Having considered recent work on this topic we suggest the following practical scheme:

    I. *Apostasioïdeae.* ± regular flowers, shallow labellum, 2–3 anthers, pollen in separate grains.
        1. APOSTASIËAE—as above.
    II. *Cypripedioïdeae.* Irregular flowers, deeply saccate labellum, 2 anthers, usually shield-like staminode, pollen as tetrads in a sticky fluid.
        1. CYPRIPEDIËAE—as above.
    III. *Orchidoïdeae.* Irregular flowers, 1 anther, no staminode, pollen in masses (pollinia).
        1. ORCHIDEAE: granular pollinia, stalk of pollinia part of pollinia (caudicle), viscidium present, base of anther firmly attached to column.
        2. NEOTTIËAE: mealy pollinia, viscidium present, anther deciduous, apex lightly attached to column.
        3. EPIDENDREAE: waxy pollinia, no or poorly developed viscidium, anther deciduous, attached by apex.

4. VANDEAE: horny or waxy pollinia, viscidium present, stalk of pollinia part of rostellum (*stipes*), anther deciduous, attached by apex.

× **Orchidactyla** P. F. Hunt & Summerh. Orchidaceae. Gen. hybr. (i) (Dactylorhiza × Orchis).

× **Orchidactylorhiza** Soó (vii) = praec.

× **Orchidanacamptis** Labrie = × Anacampt-orchis G. Camus (Orchidac.).

Orchidantha N. E. Br. Lowiaceae. 8 S. China, Hainan, Malay Penins., Borneo.

Orchidanthaceae Dostál = Lowiaceae Ridl.

Orchidion Mitch. = Arethusa L. (Orchidac.).

Orchidium Sw. = Calypso Salisb. (Orchidac.).

Orchidocarpum Michx = Asimina Adans. (Annonac.).

Orchidofunckia A. Rich. & Gal. = Cryptarrhena R.Br. (Orchidac.).

Orchidotypus Kränzlin = Pachyphyllum Kunth (Orchidac.).

× **Orchigymnadenia** E. G. Camus. Orchidaceae. Gen. hybr. (i) (Gymnadenia × Orchis).

× **Orchimantoglossum** Aschers. & Graebn. Orchidaceae. Gen. hybr. (i) (Himantoglossum × Orchis).

Orchiodes Kuntze = Goodyera R.Br. (Orchidac.).

Orchiops Salisb. = Lachenalia Jacq. (Liliac.).

Orchipeda Blume = Voacanga Thou. (Apocynac.).

Orchipedum Breda. Orchidaceae. 1 Malay Penins., Java.

× **Orchiplatanthera** E. G. Camus. Orchidaceae. Gen. hybr. (vi) (Orchis × Platanthera).

Orchis L. Orchidaceae. 35 Madeira, temp. Eurasia to India and SW. China.

× **Orchiserapias** E. G. Camus. Orchidaceae. Gen. hybr. (i) (Orchis × Serapias).

Orchites Schur = Traunsteinera Reichb. (Orchidac.).

Orchyllium Barnh. = Utricularia L. (Lentibulariac.).

Orcuttia Vasey. Gramineae. 4 California.

Orcya Vell. = Acanthospermum Schrank (Compos.).

Oreacanthus Benth. Acanthaceae. 3 trop. Afr.

Oreamunoa Oerst. = Oreomunnea Oerst. (Juglandac.).

Oreanthes Benth. Ericaceae. 2 Peru, Ecuador.

Oreanthus Rafin. = Oreotrys Rafin. = Heuchera L. (Saxifragac.).

Oreas Cham. & Schlechtd. = Aphragmus Andrz. ex DC. (Crucif.).

Oreastrum Greene (1896; non Oriastrum Poepp. & Endl. 1842, *q.v.*) = Oreostemma Greene = Aster L. (Compos.).

Orectanthe Maguire & Wurdack. Abolbodaceae. 1 Venez.

Orectospermum Schott (nomen). Leguminosae. Quid?

Oregandra Standley. Rubiaceae. 1 Panamá.

Oreinotinus Oerst. = Viburnum L. (Caprifoliac.).

Oreiostachys Gamble. Gramineae. 5 Malaysia.

Oreithales Schlechtd. Ranunculaceae. 1 Ecuador to Peru.

Orelia Aubl. = Allemanda L. (Apocynac.).

Orellana Kuntze = Bixa L. (Bixac.).

Oreobambos K. Schum. Gramineae. 1 trop. E. Afr.

Oreobatus Rydb. = Rubus L. (Rosac.).

Oreobia Phil. = Jaborosa Juss. (Solanac.).

OREOBLASTUS

Oreoblastus Suslova. Cruciferae. 8 C. As., Tibet, Himal., W. China.

Oreobliton Dur. & Moq. Chenopodiaceae. 2 Algeria.

Oreobolus R.Br. Cyperaceae. 10 mts. Borneo, New Guinea, Austr., N.Z., Polynesia, Andes.

Oreobroma Howell = Calandrinia Kunth + Lewisia Pursh + Montia L. spp. (Portulac.).

Oreocalamus Keng. Gramineae. 2 W. China.

Oreocallis R.Br. Proteaceae. 5 New Guinea, Queensl., New S. Wales, Ecuador, Peru.

Oreocallis Small = Leucothoë D. Don (Ericac.).

Oreocarya Greene. Boraginaceae. 75 Pacif. N. Am.

Oreocaryon Kunze ex K. Schum. = Cruckshanksia Hook. & Arn. (Rubiac.).

Oreocaryum P. & K. = praec.

Oreocereus (A. Berger) Riccob. = Borzicactus Riccob. (Cactac.).

*****Oreocharis** Benth. Gesneriaceae. 20 China, Japan.

Oreocharis Lindl. = Mertensia Roth (Boraginac.).

Oreochloa Link (~ Sesleria Scop.). Gramineae. 4 S. Eur.

Oreochorte K.-Pol. (~ Anthriscus Bernh.). Umbelliferae. 1 China.

Oreochrysum Rydb. = Haplopappus Cass. (Compos.).

Oreocnida B. D. Jacks. = seq.

Oreocnide Miq. Urticaceae. 20 China, Indomal.

Oreocome Edgew. = Selinum L. (Umbellif.).

Oreocosmus Naud. = Tibouchina Aubl. (Melastomatac.).

Oreodaphne Nees & Mart. = Ocotea Aubl. (Laurac.).

Oreodendron C. T. White. Thymelaeaceae. 1 Queensland.

Oreodoxa Kunth & auctt. = Roystonea O. F. Cook (Palm.).

Oreodoxa Willd. = Euterpe Gaertn. [+ Catoblastus H. Wendl.] (Palm.).

Oreogenia I. M. Johnston = Lasiocaryum I. M. Johnst. (Boraginac.).

Oreogrammitis Copeland. Grammitidaceae. 1 Borneo. (Hardly different from *Grammitis* Sw.)

Oreograstis K. Schum. Cyperaceae. 1 trop. E. Afr.

Oreoherzogia W. Vent. Rhamnaceae. 13 C. Eur. & Medit. to Persia.

Oreolirion E. P. Bickn. = Sisyrinchium L. (Iridac.).

Oreomitra Diels. Annonaceae. 1 New Guinea.

Oreomunnea Oerst. Juglandaceae. 3 Mex., C. Am.

Oreomyrrhis Endl. Umbelliferae. 23 Formosa, N. Borneo, New Guinea, SE. Austr., Tasm., N.Z., Mex., C. Am., Andes, Fuegia, Falkl. Is.

Oreonana Jepson. Umbelliferae. 2 Calif.

Oreonesion Raynal. Gentianaceae. 1 W. Equat. Afr.

Oreopanax Decne & Planch. Araliaceae. 120 trop. Am.

Oreophea auctt. ex Steud. = Orophea Blume (Annonac.).

Oreophila D. Don = Hypochoeris L. (Compos.).

Oreophila Nutt. ex Torr. & Gray = Pachystima Rafin. (Celastrac.).

Oreophylax (Endl.) Kusnez. = Gentiana L. (Gentianac.).

Oreophysa (Bunge ex Boiss.) Bornm. Leguminosae. 1 NW. Persia.

Oreophyton O. E. Schulz. Cruciferae. 1 Abyss., Kilimanjaro.

Oreopoa Gandog. = Poa L. (Gramin.).

Oreopogon P. & K. = Oropogon Neck. = Andropogon L. (Gramin.).

Oreopolus Schlechtd. (~ Cruckshanksia Hook. & Arn.). Rubiaceae. 2 Andes, Patag.
Oreoporanthera (Grüning) Hutch. Euphorbiaceae. 1 N.Z.
Oreopteris Holub. Thelypteridaceae. 1 Eur., 2 E. As.
Oreorchis Lindl. Orchidaceae. 14 Himal. to Manchuria & Japan.
Oreorhamnus Ridley = Rhamnus L. (Rhamnac.).
Oreosciadium Wedd. = Apium L. (Umbellif.).
Oreoselinon Rafin. = seq.
Oreoselinum Mill. = Peucedanum L. (Umbellif.).
Oreoselis Rafin. = praec.
Oreoseris DC. = Gerbera L. ex Cass. (Compos.).
Oreosolen Hook. f. Scrophulariaceae. 3 Tibet, Himalaya.
Oreosparte Schlechter. Asclepiadaceae. 1 Celebes.
Oreosphacus Phil. Labiatae. 1 Chile.
Oreosplenium Zahlbr. ex Endl. = Zahlbrucknera Reichb. (Saxifragac.).
Oreostemma Greene = Aster L. (Compos.).
Oreostylidium Berggr. Stylidiaceae. 1 N.Z.
Oreosyce Hook. f. Cucurbitaceae. 1 trop. Afr., Madag.
Oreotelia Rafin. = Seseli L. (Umbellif.).
Oreothamnus P. & K. = Orothamnus Pappe ex Hook. (Proteac.).
Oreothyrsus Lindau. Acanthaceae. 3 Philipp. Is., New Guinea.
Oreotrys Rafin. = Heuchera L. (Saxifragac.).
Oreoxis Rafin. Umbelliferae. 4 SW. U.S.
Oreoxylum P. & K. = Oroxylum Vent. (Bignoniac.).
Orescia Reinw. = Lysimachia L. (Primulac.).
Oresigonia Schlechtd. ex Less. = Culcitium Kunth (Compos.).
Oresigonia Willd. ex Less. = Werneria Kunth (Compos.).
Oresitrophe Bunge. Saxifragaceae. 1 NE. China.
Orestia Ridl. (sphalm.) = seq.
Orestias Ridl. Orchidaceae. 3 trop. Afr.
Orestion Kunze ex Berg = Myrtus L. (Myrtac.).
Orestion Rafin. = Olearia Moench (Compos.).
Orexis Salisb. = Lycoris Herb. (Amaryllidac.).
Orfilea Baill. = Alchornea Sw. (Euphorbiac.).
Orias Dode. Lythraceae. 1 W. China.
Oriastrum Poepp. & Endl. = Chaetanthera Ruiz & Pav. (Compos.).
Oriba Adans. = Anemone L. (Ranunculac.).
Oribasia Moç. & Sessé ex DC. = Werneria Kunth (Compos.).
Oribasia Schreb. = Palicourea Aubl. (Rubiac.).
Oricia Pierre. Rutaceae. 8 trop. & S. Afr.
Oriciopsis Engl. Rutaceae. 1 W. Equat. Afr.
× **Origanomajorana** Domin. Labiatae. Gen. hybr. (Majorana × Origanum).
Origanon St-Lag. = seq.
Origanum L. Labiatae. 15–20 Eur., Medit. to C. As. *O. vulgare* L. used as a flavouring herb. *O. majorana* L. yields oil of marjoram by distillation.
Orimaria Rafin. = Bupleurum L. (Umbellif.).
Orinocoa Rafin. = Deprea Rafin. = Athenaea Sendtn. (Solanac.).
Orinus Hitchcock. Gramineae. 1 Tibet, Kashmir.

ORIOPHORUM

Oriophorum Gunnerus (sphalm.) = Eriophorum L. (Cyperac.).

Orites Banks & Soland. ex Hook. f. = Donatia J. R. & G. Forst. (Donatiac.).

Orites R.Br. Proteaceae. 9 temp. E. Austr., Andes.

Orithalia Blume = Agalmyla Blume (Gesneriac.).

Orithia Bl. ex Dur. = praec.

Orithyia D. Don = Tulipa L. (Liliac.).

Oritina R.Br. = Orites R.Br. (Proteac.).

Oritrephes Ridl. Melastomataceae. 6 Burma, Malay Penins.

Oritrophium (Kunth) Cuatrec. Compositae. 11 Andes.

Orium Desv. = Clypeola L. (Crucif.).

Orixa Thunb. Rutaceae. 1 Japan.

Oriza Franch. & Sav. = Oryza L. (Gramin.).

Orizopsis Rafin. = Oryzopsis Michx (Gramin.).

Orlaya Hoffm. (~Daucus L.). Umbelliferae. 5 SE. Eur., Medit. to C. As.

Orleanesia Barb. Rodr. Orchidaceae. 3 Venez., Brazil.

Orleania Boehm. = Bixa L. (Bixac.).

Orleanisia auctt. = Orleanesia B. Rodr. (Orchidac.).

Orlowia Gueldenst. ex Georgi = Phlomis L. (Labiat.).

Ormenis Cass. Compositae. 10 Medit.

Ormiastis Rafin. = Salvia L. (Labiat.).

Ormilis Rafin. = praec.

Ormiscus Eckl. & Zeyh. = Heliophila Burm. f. (Crucif.).

Ormocarpopsis R. Viguier. Leguminosae. 5 Madag.

*Ormocarpum Beauv. Leguminosae. 30 trop. & subtrop. Afr., Indomal.

Ormoloma Maxon. Lindsaeaceae. 2 trop. Am.

Ormopteris J. Smith. Sinopteridaceae. 1 Brazil.

Ormopterum Schischk. Umbelliferae. 1 C. As.

Ormosciadium Boiss. Umbelliferae. 2 E. As. Min.

*Ormosia G. Jacks. Leguminosae. 50 trop. The seeds of *O. dasycarpa* G. Jacks. (bead or necklace tree) show the same red and black surface as *Abrus precatorius* L.

Ormosiopsis Ducke. Leguminosae. 3 trop. S. Am.

Ormosolenia Tausch = Peucedanum L. (Umbellif.).

Ormostema Rafin. = Dendrobium Sw. (Orchidac.).

Ormycarpus Neck. = Raphanus L. (Crucif.).

Ornanthes Rafin. = Fraxinus L. (Oleac.).

Ornitharium Lindl. & Paxt. = Pteroceras v. Hasselt ex Hassk. (Orchidac.).

Ornithidium R.Br. Orchidaceae. 60 Mex. to trop. S. Am., W.I.

Ornithobaea auctt. = seq.

Ornithoboea Parish ex C. B. Clarke. Gesneriaceae. 8 SE. As.

Ornithocarpa Rose. Cruciferae. 1 Mex.

Ornithocephalochloa Kurz = Thuarea Pers. (Gramin.).

Ornithocephalus Hook. Orchidaceae. 50 Mex. to trop. S. Am.

Ornithochilus (Lindl.) Benth. Orchidaceae. 1 India, Burma, Siam, China.

× **Ornithocidium** Leinig. Orchidaceae. Gen. hybr. (i, iii) (Oncidium × Ornithophora).

Ornithogal[ac]eae Salisb. = Liliaceae–Scilleae Reichb.

Ornithogalon Rafin. = seq.

Ornithogalum L. Liliaceae. 150 temp. Old World.
Ornithogloson Rafin. = seq.
Ornithoglossum Salisb. Liliaceae. 3 S. Afr.
Ornithophora Barb. Rodr. Orchidaceae. 1 Brazil.
Ornithopodioïdes Fabr. (uninom.) = *Coronilla* L. sp. (Legumin.).
Ornithopodium Mill. = Ornithopus L. (Legumin.)
Ornithopteris Bernh. = Anemia Sw. (Schizaeac.).
Ornithopus L. Leguminosae. 10 subtrop. S. Am., trop. Afr., Medit., W. As.
O. sativus Brot. (*seradella, serratella*) affords good fodder.
Ornithorhynchium Röhl. = Euclidium R.Br. (Crucif.).
Ornithosperma Rafin. = Ipomoea L. (Convolvulac.).
Ornithostaphylos Small = Arctostaphylos Adans. (Ericac.).
Ornithoxanthum Link = Gagea Salisb. (Liliac.).
Ornithrophus Boj. ex Engl. = Weinmannia L. (Cunoniac.).
Ornitopus Krock. = Ornithopus L. (Legumin.).
Ornitrope Pers. = seq.
Ornitrophe Comm. ex Juss. = Allophylus L. (Sapindac.).
Ornus Boehm. = Fraxinus L. (Oleac.).
Oroba Medik. = Orobus L. = Lathyrus L. (Legumin.).
***Orobanchaceae** Vent. Dicots. 13/180, chiefly N. temp. Euras.; a few Am.
and trop. All are parasitic herbs with little or no chlorophyll, attached by
suckers formed upon their roots to the roots of other plants (the seeds of
Orobanche only germinate when in contact with a root of a host). For details
see genera. Infl. term., a raceme or spike (exc. *Phelipaea*, which has a sol.
term. fl.). Fl. ♀, zygo. K (2–5), hypog.; C (5), imbr., 2-lipped; A 4, didynamous,
epipet., anthers opening longitudinally; G usu. (2), rarely (3), 1-loc., placentae
parietal, often T-shaped in section or branched, ovules ∞, anatr., style 1.
Loculic. caps.; seeds small, with minute undifferentiated embryo in oily
endosp. Chief genera: *Orobanche, Christisonia, Aeginetia, Cistanche.*
Orobanche L. Orobanchaceae. 140 temp. & subtrop. Parasitic by their roots
upon the roots of other pls.; no green tissue. *O. ramosa* L. is common on hemp,
O. elatior Sutton on *Centaurea*, etc., *O. minor* Sm. on clover. Some are con-
fined to one host, e.g. *O. hederae* Duby to ivy; others are more general in
their attacks.
Orobanche Vell. = Gesneria, Besleria, etc., spp. (Gesneriac.).
Orobanchia Vand. = Alloplectus Mart. (Gesneriac.).
Orobella Presl = Vicia L. (Legumin.).
Orobium Reichb. = Aphragmus Andrz. ex DC. (Crucif.).
Orobium Schrad. ex Nees = Lagenocarpus Nees (Cyperac.).
Orobos St-Lag. = seq.
Orobus L. = Lathyrus L. (Legumin.).
Orochaenactis Coville. Compositae. 1 Calif.
Oroga Rafin. = Origanum L. (Labiat.).
Orogenia S. Wats. Umbelliferae. 2 W. N. Am.
Orollanthus E. Mey. (sphalm.) = Aeolanthus Mart. (Labiat.).
Oronicum S. F. Gray (sphalm.) = Orontium Pers. = Misopates Rafin. (Scrophu-
Orontiaceae Bartl. = Araceae–Orontieae Schott.   [lariac.).
Orontium L. Araceae. 1 Atl. N. Am. Aquatic.

829

Orontium Pers. = Misopates Rafin. (Scrophulariac.).
Oropetium Trin. Gramineae. 5 N. & trop. Afr. to SE. As.
Orophaca Britton = Astragalus L. (Legumin.).
Orophea Blume. Annonaceae. 60 E. As., Indomal.
Orophochilus Lindau. Acanthaceae. 1 Peru.
Orophoma Spruce ex Drude = Mauritia L. f. (Palm.).
Oropogon Neck. = Andropogon L. (Gramin.).
Orospodias Rafin. = Prunus L. (Rosac.).
Orostachys Fisch. (~ Sedum L.). Crassulaceae. 20 temp. As.
Orostachys Steud. (sphalm.) [= Orthostachys Ehrh.] = Elymus L. (Gramin.).
Orothamnus Pappe ex Hook. Proteaceae. 1 S. Afr.
Oroxylon Steud. = seq.
Oroxylum Vent. Bignoniaceae. 2 S. China, SE. As., Indomal.
Oroya Britton & Rose. Cactaceae. 1 Peru (variable).
Orphanidesia Boiss. & Bal. (~ Epigaea L.). Ericaceae. 1 E. As. Min., Cauc.
***Orphium** E. Mey. Gentianaceae. 1 S. Afr.
Orsidice Reichb. f. = Thrixspermum Lour. (Orchidac.).
Orsina Bertol. = Inula L. (Compos.).
Orsinia Bertol. ex DC. = Clibadium L. (Compos.).
Orsopea Rafin. = Opsopea Rafin. = Sterculia L. (Sterculiac.).
Ortachne Nees ex Steud. Gramineae. 2 Chile.
Ortega L. = seq.
Ortegaea Kuntze = seq.
Ortegia L. Caryophyllaceae. 2 Spain, Italy.
Ortegioïdes Soland. ex DC. = Ammannia L. (Lythrac.).
Ortegocactus Alexander. Cactaceae. 1 Mex.
Ortgiesia Regel = Aechmea Ruiz & Pav. (Bromeliac.).
Orthaca Klotzsch = Orthaea Klotzsch corr. Benth. & Hook. f. (Ericac.).
Orthachne D. K. Hughes = seq.
Orthacna P. & K. = Ortachne Nees ex Steud. (Gramin.).
Orthaea Klotzsch corr. Benth. & Hook. Ericaceae. 15 trop. S. Am.
Orthandra Burret. Tiliaceae. 1 C. Am. (Panamá).
Orthandra (Pichon) Pichon = Orthopichonia H. Huber (Apocynac.).
Orthantha (Benth.) Kerner. Scrophulariaceae. 3 Eur., W. As.
Orthanthe Lem. = Gloxinia L'Hérit. (Gesneriac.).
Orthanthera Wight. Asclepiadaceae. 5 Afr., India.
Orthechites Urb. Apocynaceae. 1 Jamaica.
Orthilia Rafin. (~ Pyrola L.). Pyrolaceae. 1–2 circumboreal.
Orthion Standley & Steyerm. Violaceae. 3 C. Am.
Orthiopteris Copel. Dennstaedtiaceae. 9 Malaysia to Fiji, trop. Am. Stock erect, scaly; ls. as *Dennstaedtia*. (Tryon, *Contr. Gray Herb.* **191**: 100–6, 1962, unites with *Saccoloma* Kaulf.)
Orthocarpus Nutt. Scrophulariaceae. 30 W. Am. *O. densiflorus* Benth. in N. Am. appears to exhibit co-ordinated seed dispersal with propagules of introd. host-plant *Hypochoeris glabra* L. (Atsatt, *Science* **149**: 1389–90, 1965).
Orthocentron Cass. = Cirsium Mill. (Compos.).
Orthoceras R.Br. Orchidaceae. 2 E. Austr., N.Z.
Orthochilus Hochst. ex A. Rich. = Eulophia R.Br. ex Lindl. (Orchidac.).

Orthoclada Beauv. Gramineae. 1 S. Mex. to trop. S. Am.; 1 SE. trop. Afr.
Orthodanum E. Mey. = Rhynchosia Lour. (Legumin.). [Ls. petiolate.
Orthodon Benth. & Oliv. Labiatae. 10–20 Himal. to Japan.
Orthogoneuron Gilg. Melastomataceae. 1 trop. Afr.
Orthogramma Presl = Blechnum L. (Blechnac.).
Orthogynium Baill. Menispermaceae. 1 Madag.
Ortholobium Gagnep. = Cylindrokelupha Kosterm. (Legumin.).
Ortholoma Hanst. = Columnea L. (Gesneriac.).
Ortholotus Fourr. = Dorycnium L. (Legumin.).
Orthomene Barneby & Krukoff. Menispermaceae. 4 trop. S. Am.
Orthopappus Gleason. Compositae. 1 trop. Am.
Orthopenthea Rolfe = Disa Bergius (Orchidac.).
Orthopetalum Beer = Pitcairnia L'Hérit. (Bromeliac.).
Orthophytum Beer. Bromeliaceae. 10 Brazil.
Orthopichonia H. Huber. Apocynaceae. 9 trop. W. Afr.
Orthopogon R.Br. = Oplismenus Beauv. (Gramin.).
Orthopterum L. Bolus. Aïzoaceae. 2 S. Afr.
Orthopterygium Hemsl. Julianiaceae. 1 Peru.
Orthoraphium Nees (~ Stipa L.). Gramineae. 2 Himal., Assam, NE. As.
Orthorrhiza Stapf = Diptychocarpus Trautv. (Crucif.).
Orthosanthus Steud. = Orthrosanthus Sweet (Iridac.).
Orthoselis (DC.) Spach = Heliophila Burm. f. ex L. (Crucif.).
Orthosia Decne. Asclepiadaceae. 20 trop. S. Am., W.I.
Orthosiphon Benth. Labiatae. 30 trop. Afr., Madag.; 20 E. As., Indomal.
Orthospermum Opiz = Orthosporum (R.Br.) Kostel. = Chenopodium L. (Chenopodiac.).
Orthosphenia Standley. Celastraceae. 1 Mex.
Orthosporum (R.Br.) Kostel. = Chenopodium L. (Chenopodiac.).
Orthostachys Ehrh. (uninom.) = Elymus europaeus L. = Hordelymus europaeus (L.) Harz (Gramin.)
Orthostachys P. & K. = Ortostachys Fourr. = Stachys L. (Labiat.).
Orthostachys Spach = Orostachys Fisch. (Crassulac.).
Orthostemma Wall. ex Voigt = Pentas Benth. (Rubiac.).
Orthostemon Berg = Feijoa Berg (Myrtac.).
Orthostemon R.Br. = Canscora Lam. (Gentianac.).
Orthotactus Nees. Acanthaceae. 3 trop. S. Am.
Orthotheca Pichon. Bignoniaceae. 1 Brazil.
Orthothecium Schott & Endl. (1832; non *Bruch, Schimp. & Guemb. 1851— Musci) = Helicteres L. (Sterculiac.).
Orthothylax (Hook. f.) Skottsb. Philydraceae. 1 E. Austr.
Orthotropis Benth. = Chorizema Labill. (Legumin.).
Orthrosanthes Rafin. = seq.
Orthrosanthus Sweet. Iridaceae. 5 SW. Austr.; 3 Mex. to trop. S. Am.
Orthurus Juzepczuk. Rosaceae. 2 Medit. to C. As. & Persia.
Ortiga Neck. = Loasa Adans. (Loasac.).
Ortmannia Opiz = Geodorum G. Jacks. (Orchidac.).
Ortostachys Fourr. = Stachys L. (Labiat.).
Orucaria Juss. ex DC. = Drepanocarpus G. F. W. Mey. (Legumin.).

ORUMBELLA

Orumbella Coulter & Rose. Umbelliferae. 1 Alaska.

Orvala L. = Lamium L. (Labiat.).

Orxera Rafin. = Aërides Lour. (Orchidac.).

Orychmophragmus Spach = seq.

Orychophragmos Reichb. = seq.

Orychophragmus Bunge. Cruciferae. 2 C. As., China.

Oryctanthus Eichl. (= Glutago Comm. ex Poir.). Loranthaceae. 20 trop.

Oryctes S. Wats. Solanaceae. 1 SW. U.S.   [Am.

Oryctina Van Tiegh. = Oryctanthus Eichl. (Loranthac.).

Orygia Forsk. Aïzoaceae. 2 N. to SW. Afr., Arabia to Penins. India.

Orymaria Meissn. = Orimaria Rafin. = Bupleurum L. (Umbellif.).

Orypetalum K. Schum. (sphalm.) = Oxypetalum R.Br. (Asclepiadac.).

Orysa Desv. = Oryza L. (Gramin.).

Orythia Endl. = Agalmyla Blume (Gesneriac.).

Oryza L. Gramineae. 25 trop., incl. *O. sativa* L. (rice), one of the chief food plants of the world, and the main subsistence cereal throughout much of Asia. It is cult. in shallow water till nearly ripe, when the water is drained off, there being innumerable varieties suited to different depths and duration of flooding; also occasionally cult. on rain-fed upland soils. The grain in the husk is known as paddy.

Oryzaceae (Kunth) Herter = Gramineae–Oryzeae Kunth.

Oryzidium C. E. Hubbard & Schweickerdt. Gramineae. 1 SW. Afr.

Oryzopsis Michx. Gramineae. 50 N. temp. & subtrop.

Osbeckia L. Melastomataceae. 100 trop. Afr. to Austr.

Osbeckiastrum Naud. = Dissotis Benth. (Melastomatac.).

Osbertia Greene = Erigeron L. + Haplopappus Cass. (Compos.).

Osbornia F. Muell. Myrtaceae. 1 Philipp. Is., Borneo, Bali, New Guinea, NE. Austr.

Oscaria Lilja = Primula L. (Primulac.).

Oschatzia Walp. (~ Azorella Lam.). Hydrocotylaceae (~ Umbellif.). 2 Austr.

Oscularia Schwantes. Aïzoaceae. 5 S. Afr.

Osculisa Rafin. = Carex L. (Cyperac.).

Oserya Tul. & Wedd. Podostemaceae. 6 Mex. to N. trop. S. Am.

Oshimella Masam. & Suzuki = Whytockia W. W. Sm. (Gesneriac.).

Osiodendron Pohl (nomen). 10 Brazil. Quid?

Oskampia Moench = Nonea Medik. (Boraginac.).

Oskampia Rafin. = Tournefortia L. (Boraginac.).

Osmadenia Nutt. = Hemizonia DC. (Compos.).

Osmanthes Rafin. = Dracaena L. (Agavac.).

Osmanthus Lour. Oleaceae. 15 E. & SE. As. *O. fragrans* Lour. has ed. fr., and its ls. are used to perfume tea.

× **Osmarea** Hort. Oleaceae. Gen. hybr. (Osmanthus × Phillyrea).

Osmaronia Greene = Oemleria Reichb. (Rosac.).

Osmaton Rafin. = Carum L. (Umbellif.).

Osmelia Thw. Flacourtiaceae. 1 Ceylon, 3 Malaysia (exc. Java & Lesser Sunda Is.).

× **Osmentara** hort. Orchidaceae. Gen. hybr. (iii) (Broughtonia × Cattleya × Laeliopsis).

Osmhydrophora Barb. Rodr. = Tanaecium Sw. (Bignoniac.).
Osmia Sch. Bip. = Chromolaena DC. (Compos.).
Osmites L. Compositae. 8 S. Afr.
Osmitiphyllum Sch. Bip. = Peyrousea DC. (Compos.).
Osmitopsis Cass. (emend. Bremer). Compositae. 8 S. Afr.
Osmodium Rafin. = Onosmodium Michx (Boraginac.).
Osmoglossum Schlechter. Orchidaceae. 5 C. & trop. S. Am.
Osmohydrophora auctt. = Osmhydrophora Barb. Rodr. = Tanaecium Sw. (Bignoniac.).
Osmorhiza Rafin. Umbelliferae. 15 Cauc. to Himal. & Japan, N. Am., Andes.
Osmoscleria Lindl. = Omoscleria Nees = Scleria Bergius (Cyperac.).
Osmoshiza Rafin. (sphalm.) = Osmorhiza Rafin. (Umbellif.).
Osmothamnus DC. = Rhododendron L. (Ericac.).
Osmoxylon Miq. Araliaceae. 2 Malaysia.
Osmunda L. Osmundaceae. 10 temp. & trop. *O. regalis* L. (royal fern) has a root-stock sometimes a foot high, like the stem of a tree-fern, bearing scale ls. protecting each successive group of green ls. The fronds are large (up to 3 m.); the lower pinnae are veg., the upper are reprod. only and form a sort of panicle. Other spp. have the fertile pinnae on the lower part of the l., others again have separate veg. and reprod. ls.
Osmundaceae R.Br. Osmundales. 3/19, trop. & temp. Short-stemmed ferns, with naked sori. The sporangia are shortly stalked and have an annulus, consisting of a roundish group of cells at one side of the apex; they open by a longitudinal fissure. Genera: *Osmunda* (sori on special pinnae); *Todea* (sori on backs of ordinary pinnae); *Leptopteris* (like *Todea*, but leaf-blade 2 cells thick).
Osmundales. 1 fam.: *Osmundaceae*.
Osmundastrum Presl = Osmunda L. (Osmundac.).
Osmundidae (subclass). 1 order: *Osmundales*.
Osmundopteris (Milde) Small = Botrychium Sw. (Ophioglossac.).
Osmyne Salisb. = Ornithogalum L. (Liliac.).
Ossaea DC. Melastomataceae. 100 trop. Am., W.I.
Ossea Nieuwl. & Lunell = Swida Opiz (Cornac.).
Ossifraga Rumph. = Euphorbia L. (Euphorbiac.).
Ostachyrium Steud. = Otachyrium Nees = Panicum L. (Gramin.).
Osteiza Steud. = Oteiza La Llave = Calea L. (Compos.).
Ostenia Buchenau. Limnocharitaceae. 1 Uruguay.
Osteocarpum F. Muell. = Threlkeldia R.Br. (Chenopodiac.).
Osteocarpus Phil. = Alona Lindl. (Nolanac.).
Osteomeles Lindl. Rosaceae. 15 E. As., Polynesia, C. Am., Andes.
Osteophloeum Warb. Myristicaceae. 1 Amaz. Brazil.
Osteospermum L. Compositae. 70 S. Afr., St Helena.
Osterdamia Neck. = Zoysia Willd. (Gramin.).
Osterdikia Adans. = Oosterdyckia Boehm. (Cunoniac.).
Osterdyckia Reichb. = praec.
Ostericum Hoffm. (~ Angelica L.). Umbelliferae. 1 N. & C. Eur., temp. As.
Ostinia Clairv. = Mespilus L. (*sensu latiss.*, incl. Crataegus & Cotoneaster) (Rosac.). [Borneo.
Ostodes Blume. Euphorbiaceae. 4 E. Himal., SE. As., Sumatra, Java, N.

OSTREARIA

Ostrearia Baill. ex Niedenzu. Hamamelidaceae. 1 Queensland.

Ostreocarpus Rich. ex Endl. = Aspidosperma Mart. & Zucc. (Apocynac.).

Ostrowskia Regel. Campanulaceae. 1 C. As. Ls. whorled.

Ostruthium Link = Peucedanum L. (Umbellif.).

Ostrya Hill = Carpinus L. (Carpinac.).

Ostrya Scop. Carpinaceae. 10 N. temp. (S. to C. Am.). *O. virginiana* (Mill.) K. Koch (lever-wood) furnishes a hard wood.

Ostryocarpus Hook. f. Leguminosae. 8 trop. Afr.

Ostryoderris Dunn = Aganope Miq. (Legumin.).

Ostryodium Desv. = Flemingia Roxb. = Maughania Jaume St-Hil. (Legumin.).

Ostryopsis Decne. Carpinaceae. 2 E. Mongolia, SW. China. Shrubs. A 4–6.

Oswalda Cass. = Clibadium L. (Compos.).

Oswaldia Less. = praec.

Osyricera Blume = Bulbophyllum Thou. (Orchidac.).

Osyriceras P. & K. = praec.

Osyridaceae ['-rinae'] Link = Santalaceae R.Br.

Osyridicarpos A. DC. Santalaceae. 6 trop. & S. Afr.

Osyridicarpus auctt. = praec.

Osyris L. Santalaceae. 6–7 Medit. & Afr. to India.

Otacanthus Lindl. Scrophulariaceae. 4 Brazil.

Otachyrium Nees. Gramineae. 4 trop. S. Am.

Otamplis Rafin. = Cocculus DC. (Menispermac.).

Otandra Salisb. = Geodorum G. Jacks. (Orchidac.).

Otanema Rafin. = Asclepias L. (Asclepiadac.).

Otanthera Blume. Melastomataceae. 15 Nicobars, Malaysia, trop. Austr.

Otanthus Hoffmgg. & Link. Compositae. 1, *O. maritimus* (L.) Hoffmgg. & Link, coasts of W. Eur. & Medit. to Cauc.

Otaria Kunth = Asclepias L. (Asclepiadac.).

Oteiza La Llave = Calea L. (Compos.).

Othake Rafin. (~Palafoxia Lag.). Compositae. 7 S. U.S., Mex.

Othanthera G. Don (sphalm.) = Orthanthera Wight (Asclepiadac.).

Othera Thunb. = Ilex L. (Aquifoliac.).

Otherodendron Makino (~Microtropis Wall. ex Meissn.). Celastraceae. 8 Japan, Ryukyu Is., Formosa.

Othlis Schott = Doliocarpus Roland. (Dilleniac.).

Othocallis Salisb. = Scilla L. (Liliac.).

Othonna L. Compositae. 140–150 trop. & S. Afr. Xero. with swollen roots and often fleshy ls.

Othonnopsis Jaub. & Spach = Hertia Less. (Compos.).

Othostemma Pritz. = Otostemma Blume = Hoya R.Br. (Asclepiadac.).

Othrys Nor. ex Thou. = Crateva L. (Capparidac.).

Otidia Lindl. ex Sweet = Pelargonium L'Hérit. (Geraniac.).

Otilix Rafin. = Solanum L. (Solanac.).

Otillis Gaertn. = Leea L. (Leeac.).

Otiophora Zucc. Rubiaceae. 15 trop. Afr., Madag.

Otites Adans. = Silene L. (Caryophyllac.).

Otoba (DC.) Karst. Myristicaceae. 9 C. & trop. S. Am.

Otocalyx T. S. Brandegee. Rubiaceae. 1 Mex.
Otocarpum Willk. (sphalm.) = Otospermum Willk. = Matricaria L. (Compos.).
Otocarpus Durieu. Cruciferae. 1 Algeria.
Otocephalus Chiov. Rubiaceae. 1 Angola.
Otochilus Lindl. Orchidaceae. 4 E. Himal. to SE. As.
Otochlamys DC. = Cotula L. (Compos.).
Otoglyphis Pomel = praec.
Otolepis Turcz. = Otophora Bl. (Sapindac.).
Otomeria Benth. Rubiaceae. 7 trop. Afr., Madag.
Otonephelium Radlk. Sapindaceae. 1 Penins. India.
× Otonisia hort. Orchidaceae. Gen. hybr. (iii) (Aganisia × Otostylis).
Otonychium Blume = Harpullia Roxb. (Sapindac.).
Otopappus Benth. Compositae. 20 Mex. to Venez.
Otopetalum F. C. Lehm. & Kraenzl. = Kraenzlinella Kuntze (Orchidac.).
Otopetalum Miq. = Micrechites Miq. (Apocynac.).
Otophora Blume. Sapindaceae. 30 Indoch., W. Malaysia.
Otophora P. & K. = Otiophora Zucc. (Rubiac.).
Otophylla Benth. = Tomanthera Rafin. (Scrophulariac.).
Otoptera DC. (~ Vigna Savi). Leguminosae. 2 S. Afr., Madag.
Otosema Benth. = Millettia Wight & Arn. (Legumin.).
Otosma Rafin. = Zantedeschia Spreng. (Arac.).
Otospermum Willk. = Matricaria L. (Compos.).
Otostegia Benth. Labiatae. 20 NE. trop. Afr. to C. As. & Baluchist.
Otostemma Blume = Hoya R.Br. (Asclepiadac.).
Otostylis Schlechter. Orchidaceae. 4 trop. S. Am., Trinidad.
Ototropis Nees = Desmodium Desv. (Legumin.).
Ototropis P. & K. = Oustropis G. Don = Indigofera L. (Legumin.).
Otoxalis Small = Oxalis L. (Oxalidac.).
Ottelia Pers. Hydrocharitaceae. 40 trop. & subtrop.
Ottilis Endl. (sphalm.) = Otillis Gaertn. = Leea L. (Leeac.).
Ottoa Kunth. Umbelliferae. 1 Mex.
Ottochloa Dandy. Gramineae. 6 Indomal., Queensland.
Ottonia Spreng. (~ Piper L.). Piperaceae. 70 trop. S. Am.
Ottoschmidtia Urb. Rubiaceae. 3 W.I.
Ottoschulzia Urb. Icacinaceae. 3 W.I.
Ottosonderia L. Bolus. Aïzoaceae. 2 SW. Afr.
Oubanguia Baill. Scytopetalaceae. 5 trop. Afr.
Ouchemoly Fabr. (nomen). Quid?
Oudemansia Miq. = Helicteres L. (Sterculiac.).
Oudneya R.Br. = Henophyton Coss. & Dur. (Crucif.).
Ougeinia Benth. Leguminosae. 1 India.
*Ouratea Aubl. Ochnaceae. 300 trop.
Ouratella Van Tiegh. = praec.
Ouret Adans. = Aerva Forsk. (Amaranthac.).
Ourisia Comm. ex Juss. Scrophulariaceae. 20 N.Z., Pacif. S. Am.
Ourisianthus Bonati. Scrophulariaceae. 1 Indochina.
Ourouparia Aubl. = Uncaria Schreb. (Naucleac.).
Oustropis G. Don = Indigofera L. (Legumin.).

OUTARDA

Outarda Dum. (sphalm.) = Coutarea Aubl. (Rubiac.).
Outea Aubl. = Macrolobium Schreb. (Legumin.).
Outreya Jaub. & Spach = Jurinea Cass. (Compos.).
Ouvirandra Thou. = Aponogeton L.f. (Aponogetonac.).
Ovaria Fabr. (uninom.) = *Solanum* L. sp. (Ericac.).
Overstratia Deschamps ex R.Br. = Saurauia Willd. (Actinidiac.).
*Ovidia Meissn. Thymelaeaceae. 4 Chile, Patag.?
Ovidia Rafin. = Commelina L. (Commelinac.).
Ovieda L. = Clerodendrum L. (Verbenac.).
Ovieda Spreng. = Lapeyrousia Pourr. (Iridac.).
Ovilla Adans. = Jasione L. (Campanulac.).
Ovostima Rafin. = Aureolaria Rafin. (Scrophulariac.).
Owataria Matsumura = Suregada Roxb. ex Rottl. (Euphorbiac.).
Owenia Hilsenb. ex Meissn. = Oxygonum Burch. (Polygonac.).
Owenia F. Muell. Meliaceae. 6 Queensl., New S. Wales.
*Oxalidaceae R.Br. Dicots. 3/875, mostly trop. & subtrop. Most are perennial herbs with alt. often cpd. (pinn. or digit.) exstip. ls. and rel. large fls., usu. in cymes, $\female$, reg. K 5, imbr., persistent; C 5, imbr. or contorted, free or slightly united; A 10, obdiplost. (i.e. the outer whorl opp. to the petals, the inner to the sepals, and thus the cpls. opp. to the petals, instead of to the sepals, as in diplostemonous fls. with two whorls of sta. in proper alternation), united below, with intr. anthers; G̲ (5) or 5, with free styles, 5-loc., with axile plac.; ovules in 1 or 2 rows in each loc., or few, anatr., with micropyle facing upwards and outwards. Capsule; embryo straight, in fleshy endosp. Genera: *Oxalis, Biophytum, Eichleria*. Allied to *Averrhoac., Lepidobotryac., Hypseocharitac., Linac., Geraniac.*, etc.
Oxalis L. Oxalidaceae. 800 cosmop., chiefly C. & S. Am., S. Afr. *O. acetosella* L. (temp. Euras.) is a small herb with monopodial rhiz. and ternate ls., which sleep at night and in cold weather, the leaflets bending downwards. The fl. is protandr.; the stalk bends downwards and the fl. closes in dull or cold weather. Cleistogamic fls. (cf. *Viola*) occur. Loculic. caps. The seed has a fleshy aril springing from the base. When ripe the cells of the inner layers are extremely turgid, and a small disturbance causes the aril to turn inside out, as one might turn a glove-finger, from U to Ո. This is done instantaneously and the seed is shot off.

Many have bulbous or tuberous stems. Some, e.g. *O. bupleurifolia* A. St-Hil., have phyllodes in place of the ordinary ls. (cf. *Acacia*). Fls. sol. or in cymose infls. Many exhibit trimorphic heterostyled fls.; there are three stocks of pl., one bearing fls. with long styles, and mid- and short-length sta., the others with mid or short styles and correspondingly long and short or long and mid sta. (cf. *Lythrum*). Some produce axillary bulbils; others repr. veg. by underground offshoots. The tubers of *O. deppei* Lodd. (S. Am., Mex.), and others, are used as food. *O. corymbosa* DC., *O. latifolia* Kunth and *O. pes-caprae* L. have become very troublesome weeds in many parts of the world.
Oxalistylis Baill. = Phyllanthus L. (Euphorbiac.).
Oxallis Nor. = Oxalis L. (Oxalidac.).
Oxandra A. Rich. Annonaceae. 22 C. & trop. S. Am., W.I. Wood useful.
Oxanthera Montr. (~ Citrus L.). Rutaceae. 4 New Caled.

Oxera Labill. Verbenaceae. 25 New Caled.
Oxcrostylus Steud. = Ogcerostylus Cass. = Angianthus Wendl. (Compos.).
Oxia Reichb. (sphalm.) = Axia Lour. = Boerhavia L. (Nyctaginac.).
Oxicedrus Garsault = Oxycedrus (Dum.) Hort. ex Carr. = Juniperus L. (Cupressac.).
Oxiceros Lour. = Oxyceros Lour. = Randia L. (Rubiac.).
Oxicoccus Neck. = Oxycoccus Hill = Vaccinium L. (Ericac.).
Oxiphoeria Hort. ex Dum.-Cours. = Humea Sm. (Compos.).
Oxipolis Rafin. = Oxypolis Rafin. (Umbellif.).
Oxis Medik. = Oxys Mill. = Oxalis L. (Oxalidac.).
Oxisma Rafin. = Loreya DC. + Henriettella Naud. (Melastomatac.).
Oxleya A. Cunn. = Flindersia R.Br. (Flindersiac.).
Oxodium Rafin. = Piper L. (Piperac.).
Oxodon Steud. = Oxydon Less. = Chaptalia Vent. (Compos.).
Oxyacantha Medik. = Crataegus L. + Pyracantha M. Roem. (Rosac.).
Oxyacantha Rumph. = ? Carissa L. (Apocynac.).
Oxyacanthus Chevall. = Grossularia Mill. (Grossulariac.).
Oxyadenia Spreng. = Oxydenia Nutt. = Leptochloa Beauv. (Gramin.).
Oxyandra Reichb. = Sloanea L. (Elaeocarpac.).
Oxyanthe Steud. = Phragmites Trin. (Gramin.).
Oxyanthera Brongn. Orchidaceae. 6 Burma to Malaysia.
Oxyanthus DC. Rubiaceae. 50 Afr.
Oxybaphus L'Hérit. ex Willd. (~ Mirabilis L.). Nyctaginaceae. 25 W.
Oxybasis Kar. & Kir. = Chenopodium L. (Chenopodiac.).   [Am.
Oxycarpha Blake. Compositae. 1 Venezuela.
Oxycarpus Lour. = Garcinia L. (Guttif.).
Oxycaryum Nees = Scirpus L. (Cyperac.).
Oxycedrus (Dum.) Hort. ex Carr. = Juniperus L. (Cupressac.).
Oxyceros Lour. = Benkara Adans. = Randia L. (Rubiac.).
Oxychlaena P. & K. = Oxylaena Benth. = Anaglypha DC. (Compos.).
Oxychlamys Schlechter. Gesneriaceae. 1 New Guinea.
Oxychloë Phil. Juncaceae. 7 C. & S. Andes.
Oxycladium F. Muell. = Mirbelia Sm. (Legumin.).
Oxycladus Miers = Monttea C. Gay (Scrophulariac.).
Oxycoca Rafin. = Oxycoccus Hill (Ericac.).
Oxycoccoïdes (Benth. & Hook. f.) Nakai = Vaccinium L. (Ericac.).
Oxycoccos Hedw. f. = seq.
Oxycoccus Hill (~ Vaccinium L.). Ericaceae. 4 N. temp. & arct.
Oxydectes Kuntze = Croton L. (Euphorbiac.).
Oxydendron D. Dietr. = seq.
Oxydendrum DC. Ericaceae. 1 E. U.S., *O. arboreum* (L.) DC. (sorrel tree, sourwood).
Oxydenia Nutt. = Leptochloa Beauv. (Gramin.).
Oxydiastrum Dur. (sphalm.) = Psidiastrum Bello = Eugenia L. (Myrtac.).
Oxydium Benn. = Desmodium Desv. (Legumin.).
Oxydon Less. = Chaptalia Vent. (Compos.).
Oxyglottis (Bunge) Nevski = Astragalus L. (Legumin.).
Oxyglycus Bowd. Quid? W. Afr.

OXYGONIUM

Oxygonium Presl = Diplazium Sw. (Athyriac.).
Oxygonum Burch. Polygonaceae. 30 trop. & S. Afr.
Oxygraphis Bunge. Ranunculaceae. 5 temp. As.
Oxygyne Schlechter. Burmanniaceae. 1 W. Equat. Afr.
Oxylaena Benth. = Anaglypha DC. (Compos.).
Oxylapathon St-Lag. = Rumex L. (Polygonac.).
Oxylepis Benth. = Helenium L. (Compos.).
*****Oxylobium** Andr. Leguminosae. 40 Austr.
Oxylobus Moç. ex DC. Compositae. 5 Mex.
Oxymeris DC. = Leandra Raddi (Melastomatac.).
Oxymitra (Bl.) Hook. f. & Thoms. (1855; non Bischoff ex Lindeb. 1829— Hepaticae) = Friesodielsia v. Steenis (Annonac.).
Oxymitus Presl = Argylia D. Don (Bignoniac.).
Oxymyrrhine Schau. = Baeckea L. (Myrtac.).
Oxymyrsine Bub. = Ruscus L. (Ruscac.).
Oxynemum Rafin. = Lycopodium L. (Lycopodiac.).
Oxynepeta Bunge = Nepeta L. (Labiat.).
Oxynia Nor. = Averrhoa L. (Averrhoac.).
Oxynix B. D. Jacks. (sphalm.) = praec.
Oxyodon DC. = Oxydon Less. = Chaptalia Vent. (Compos.).
Oxyosmyles Spegazz. Boraginaceae. 1 Argentina.
Oxyotis Welw. ex Baker = Aeolanthus Mart. (Labiat.).
Oxypappus Benth. Compositae. 2 Mex.
*****Oxypetalum** R.Br. Asclepiadaceae. 150 Mex., W.I., Brazil.
Oxyphaeria Steud. = seq.
Oxypheria DC. = Oxiphoeria Hort. = Humea Sm. (Compos.).
Oxyphyllum Phil. Compositae. 1 Chile.
Oxypogon Rafin. = Lathyrus L. (Legumin.).
Oxypolis Rafin. Umbelliferae. 7 N. Am.
Oxypteryx Greene = Asclepias L. (Asclepiadac.).
Oxyramphis Wall. ex Meissn. = Lespedeza Michx (Legumin.).
Oxyrhachis Pilger. Gramineae. 1 trop. E. Afr., Madag.
Oxyrhamphis Reichb. = Oxyramphis Wall. ex Meissn. = Lespedeza Michx (Legumin.).
Oxyrhynchus T. S. Brandegee. Leguminosae. 3 Mex., W.I. (India?).
Oxyria Hill. Polygonaceae. 1 N. arctic & subarctic; mts. of temp. Euras. & Calif., *O. digyna* (L.) Hill. Like *Rumex*, but dimerous, and with branching of the outer sta.
Oxys Mill. = Oxalis L. (Oxalidac.).
Oxysepala Wight = Bulbophyllum Thou. (Orchidac.).
Oxysma P. & K. = Oxisma Rafin. = Loreya DC. + Henriettella Naud. (Melastomatac.).
Oxyspermum Eckl. & Zeyh. = Galopina Thunb. (Rubiac.).
Oxyspora DC. Melastomataceae. 20 S. China, Indomal.
Oxystelma R.Br. Asclepiadaceae. 4 Old World tropics.
Oxystemon Planch. & Triana = Clusia L. (Guttif.).
Oxystigma Harms. Leguminosae. 8 trop. Afr.
Oxystophyllum Blume = Dendrobium Sw. (Orchidac.).

Oxystylidaceae Hutch. = Cleomaceae (Pax) Airy Shaw.
Oxystylis Torr. & Frém. Cleomaceae. 1 SW. U.S.
Oxytandrum Neck. = Apeiba Aubl. (Tiliac.).
Oxytenanthera Munro. Gramineae. 20 trop. Afr., E. As., Indomal.
Oxytenia Nutt. Compositae. 1 SW. U.S.
Oxytheca Nutt. Polygonaceae. 9 W. U.S., temp. S. Am.
Oxythece Miq. = Neoxythece Aubrév. & Pellegr. (Sapotac.).
Oxytria Rafin. = Schoenolirion Dur. (Liliac.).
*Oxytropis DC. Leguminosae. 300 N. temp.
Oxyura DC. = Layia Hook. & Arn. (Compos.).
Oxyurus Rafin. (nomen). Quid?
Oyedaea DC. Compositae. 30 C. & trop. S. Am.
Ozandra Rafin. = Melaleuca L. (Myrtac.).
Ozanonia Gandog. = Rosa L. (Rosac.).
Ozanthes Rafin. = Lindera Thunb. (Laurac.).
Ozarthris Rafin. = Viscum L. (Loranthac.).
Oziroë Rafin. = Ornithogalum L. (Liliac.).
Ozodia Wight & Arn. = Foeniculum L. (Umbellif.).
Ozodycus Rafin. = Cucurbita L. (Cucurbitac.).
Ozomelis Rafin. = Mitella L. (Saxifragac.).
Ozophyllum Schreb. = Ticorea Aubl. (Rutac.).
Ozoroa Delile (~ Heeria Meissn.). Anacardiaceae. 40 trop. Afr.
Ozothamnus R.Br. = Helichrysum L. (Compos.).
Ozotis Rafin. = Aesculus L. (Hippocastanac.).
Ozotrix Rafin. = Torilis Adans. (Umbellif.).
Ozoxeta Rafin. = Helicteres L. (Sterculiac.).

P

Pachea Pourr. ex Steud. = Crypsis Ait. (Gramin.).
Pachecoa Standley & Steyerm. Leguminosae. 1 Mex., C. Am.
× Pachgerocereus Moran. Cactaceae. Gen. hybr. (Pachycereus × Bergero-
cactus).
Pachidendron Haw. = Aloë L. (Liliac.).
Pachila Rafin. = Erucaria Gaertn. (Crucif.).
Pachiloma Rafin. = Polytaenia DC. (Umbellif.).
Pachiphillum La Llave & Lex. = Pachyphyllum Kunth (Orchidac.).
Pachira Aubl. (~ Bombax L.). Bombacaceae. 2 trop. Am.
Pachistima Rafin. = Paxistima Rafin. (Celastrac.).
Pachites Lindl. Orchidaceae. 2 S. Afr.
Pachyacris Schlechter ex Bullock. Asclepiadaceae. 10 trop. & S. Afr.
Pachyandra P. & K. = Pachysandra Michx (Buxac.).
Pachyanthus P. & K. = Pachysanthus Presl = Rudgea Salisb. (Rubiac.).
Pachyanthus Rich. Melastomataceae. 25 Cuba, S. Domingo, Colombia.
Pachycalyx Klotzsch = Simocheilus Klotzsch (Ericac.).
Pachycarpus E. Mey. Asclepiadaceae. 30 trop. & S. Afr.
Pachycentria Blume. Melastomataceae. 8 Burma, Malaysia.

PACHYCENTRON

Pachycentron Pomel = Centaurea L. (Compos.).
Pachycereus (A. Berger) Britton & Rose. Cactaceae. 5 Mex.
Pachychaeta Sch. Bip. ex Baker = Ophryosporus Meyen (Compos.).
Pachychilus Blume = Pachystoma Blume (Orchidac.).
Pachychlaena P. & K. = Pachylaena D. Don ex Hook. & Arn. (Compos.).
Pachychlamys Dyer ex Ridl. = Shorea Roxb. (Dipterocarpac.).
Pachycladon Hook. f. Cruciferae. 1 mts. of N.Z.
Pachycormus Coville ex Standl. Anacardiaceae. 1 Lower Calif.
Pachycornia Hook. f. Chenopodiaceae. 2 Austr. Like *Salicornia*.
Pachycornus Willis (sphalm.) = Pachycormus Cov. ex Standl. (Anacardiac.).
Pachyctenium Maire & Pampan. Umbelliferae. 1 Cyrenaica.
Pachydendron Dum. = Pachidendron Haw. = Aloë L. (Liliac.).
Pachyderis Cass. = Pteronia L. (Compos.).
Pachyderma Blume = Olea L. (Oleac.).
Pachydesmia Gleason. Melastomataceae. 1 Colombia.
Pachydiscus Gilg & Schlechter = Periomphale Baill. (Alseuosmiac.).
Pachyelasma Harms. Leguminosae. 1 W. Afr.
Pachyglossum Decne = Oxypetalum R.Br. (Asclepiadac.).
Pachygone Miers. Menispermaceae. 12 China, SE. As., Indomal., Austr., Pacif. Exalb.
Pachygraphea P. & K. = Pachyraphea Presl = Aspalathus L. (Legumin.).
Pachylaena D. Don ex Hook. & Arn. Compositae. 2 Chilean Andes.
Pachylarnax Dandy. Magnoliaceae. 2 Assam, Indoch., Malay Penins. Fr. of few united cpls. forming a woody loculic. caps.
Pachylepis Brongn. = Parolinia Endl. = Widdringtonia Endl. (Cupressac.).
Pachylepis Less. = Crepis L. (Compos.).
Pachylobium (Benth.) Willis = Dioclea Kunth (Legumin.).
Pachylobus G. Don = Dacryodes Vahl (Burserac.).
Pachyloma v. d. Bosch = Craspedophyllum (Presl) Copel. (Hymenophyllac.).
Pachyloma DC. Melastomataceae. 3 Brazil. Connective with auricles in front, appendages behind.
Pachyloma P. & K. = Pachiloma Rafin. = Polytaenia DC. (Umbellif.).
Pachyloma Spach = Pfundia Opiz ex Nevski = Hericinia Fourr. (Ranunculac.).
Pachylophis auctt. = seq.
Pachylophus Spach = Oenothera L. (Onagrac.).
Pachymeria Benth. = Meriania Sw. (Melastomatac.).
Pachymitra Nees = Rhynchospora Vahl (Cyperac.).
Pachymitus O. E. Schulz. Cruciferae. 2 Austr.
Pachyne Salisb. = Phaius Lour. (Orchidac.).
Pachynema R.Br. ex DC. Dilleniaceae. 4 N. Austr. Phylloclades. Sta. thick below.
Pachyneurum Bunge. Cruciferae. 1 C. As.
Pachynocarpus Hook. f. = Vatica L. (Dipterocarpac.).
Pachypharynx Aellen. Chenopodiaceae. 2 W. & S. Austr.
Pachyphragma (DC.) Reichb. Cruciferae. 1 Armenia, Cauc.
Pachyphyllum Kunth. Orchidaceae. 12 W. trop. S. Am.
Pachyphytum Link, Klotzsch & Otto. Crassulaceae. 12 Mex.
Pachyplectron Schlechter. Orchidaceae. 2 New Caled.

Pachypleuria Presl = Humata Cav. (Davalliac.).
Pachypleurum Ledeb. Umbelliferae. 1 mts. C. Eur., 3 arct. Euras. to mts. C. As.
Pachypodanthium Engl. & Diels. Annonaceae. 4 trop. W. Afr.
Pachypodium Lindl. Apocynaceae. 20 Afr., Madag.
Pachypodium Nutt. ex Torr. & Gray = Thelypodium Endl. + Pleurophragma Rydb. (Crucif.).
Pachypodium Webb & Berth. = Sisymbrium L. (Crucif.).
Pachyptera DC. Bignoniaceae. 1 C. & N. trop. S. Am., Trinidad.
Pachypteris Kar. & Kir. (1842; non Brongn. 1829—gen. foss.) = seq.
Pachypterygium Bunge. Cruciferae. 9 C. As. to Persia & Afghan. Fr. margin thick.
Pachyra A. St-Hil. & Naud. = Pachira Aubl. (Bombacac.).
× **Pachyrantia** Walth. Crassulaceae. Gen. hybr. (Courantia × Pachyphytum).
Pachyraphea Presl = Aspalathus L. (Legumin.).
Pachyrhiza B. D. Jacks. (sphalm.) = Pachyrhizus Rich. ex DC. (Legumin.).
Pachyrhizanthe (Schlechter) Nakai = Cymbidium Sw. (Orchidac.).
*****Pachyrhizus** Rich. ex DC. Leguminosae. 6 trop. Tall twining herbs, cult. for the ed. tuberous root (yam-bean).
Pachyrhynchus DC. Compositae. 1 S. Afr. Fr. glandular with pappus.
Pachyrrhizos Spreng. = Pachyrhizus Rich. ex DC. (Legumin.).
Pachysa D. Don = Erica L. (Ericac.).
Pachysandra Michx. Buxaceae. 3 E. As., 1 E. U.S.
Pachysandraceae J. G. Agardh = Buxaceae Loisel.
Pachysanthus Presl = Rudgea Salisb. (Rubiac.).
Pachysolen Phil. = Nolana L. (Nolanac.).
Pachystachys Nees. Acanthaceae. 5 trop. Am., W.I.
Pachystegia Cheeseman. Compositae. 1 N.Z.
Pachystela Pierre ex Radlk. Sapotaceae. 4 trop. Afr.
Pachystele Schlechter. Orchidaceae. 6 Mex., C. Am.
Pachystemon Blume = Macaranga Thou. (Euphorbiac.).
Pachystigma Hochst. Rubiaceae. 11 warm Afr.
Pachystigma Hook. = Peltostigma Walp. (Rutac.).
Pachystigma Meissn. = seq.
Pachystima Rafin. = Paxistima Rafin. (Celastrac.).
Pachystoma Blume. Orchidaceae. 11 China, Indomal., N. Austr., New Caled.
Pachystrobilus Bremek. Acanthaceae. 2 Java.
Pachystroma (Klotzsch) Muell. Arg. Euphorbiaceae. 1 S. Brazil.
Pachystylidium Pax & K. Hoffm. Euphorbiaceae. 1 Penins. India, Siam, Indoch., Philipp. Is., Java.
Pachystylum Eckl. & Zeyh. = Heliophila Burm. f. (Crucif.).
Pachystylus K. Schum. Rubiaceae. 2 New Guinea.
Pachysurus Steetz = Calocephalus R.Br. (Compos.).
Pachythamnus (R. M. King & H. Rob.) R. M. King & H. Rob. (~ Eupatorium L.). Compositae. 1 Mex., C. Am.
Pachythelia Steetz = Epaltes Cass. (Compos.).
Pachythrix Hook. f. = Pleurophyllum Hook. f. (Compos.).

PACHYTROPHE

Pachytrophe Bur. Moraceae. 2 Madag. G oblique, style excentric. L. form variable.

Pachyurus P. & K. = Pachysurus Steetz = Calocephalus R.Br. (Compos.).

× **Pachyveria** Hort. ex Haage & Schmidt. Crassulaceae. Gen. hybr. (Echeveria × Pachyphytum).

Pacoseroca Adans. = Amomum L. (Zingiberac.).

Pacouria Aubl. Apocynaceae. 20 trop. S. Am., trop. & S. Afr., Madag.

Pacourina Aubl. Compositae. 1 trop. S. Am. Aquatic. Head sessile. Ed. ls.

Pacourinopsis Cass. = praec.

Pacurina Rafin. = Messerschmidia L. ex Hebenstr. (Boraginac.).

Padbruggea Miq. Leguminosae. 5 Siam, Malay Penins., Java.

Padia Zoll. & Mor. = Oryza L. (Gramin.).

Padostemon Griff. (sphalm.) = Podostemum Michx (Podostemac.).

Padota Adans. = Marrubium L. (Labiat.).

Padus Mill. (~ Prunus L.). Rosaceae. 110 Euras., N. & S. Am.

***Paederia** L. Rubiaceae. 50 trop. Climbing shrubs. Ls. sometimes whorled.

Paederota L. = Veronica L. (Scrophulariac.).

Paederotella (E. Wulff) Kemul.-Nath. (~ Veronica L.). Scrophulariaceae. 2 As. Min., Cauc.

Paedicalyx Pierre ex Pitard. Rubiaceae. 1 Hainan, Indoch.

Paennaea Meerb. = Penaea L. (Penaeac.).

Paënoë P. & K. = Panoë Adans. = Vateria L. (Dipterocarpac.).

Paeonia L. Paeoniaceae. 33 temp. Euras., W. N. Am. *P. officinalis* L. has tuberous r., large fls. with much honey, slight cohesion of cpls., follicle with red seeds, protog. fls. closing at night.

***Paeoniaceae** Rudolphi. Dicots. 1/33 N. temp. Perenn. rhizomatous herbs, occasionally shrubby, roots sometimes tuberous. Ls. alt., biternate, exstip. Fls. large, reg., ♀. K 5, rounded or subfoliac., much imbr., persist.; C 5(-10), large, imbr.; A ∞, centrifugal, with oblong extr. anth.; disk fleshy, of separate glands or forming a large subglob. envelope surrounding the G; G 2-5, large, free, ± fleshy, atten. into thick falcate stig., bearing ∞ biser. ov. Fr. of 2-5 large leathery ventr. dehisc. follicles; seeds large, at first red, later black and shining, arillate, with copious endosp. Only genus: *Paeonia*. Related to *Ranunculac.*, but differing in the possession of a disk, in the centrifugal dehiscence of the stamens, in the large arillate seeds, and in the distinctive anatomy. The habit should be compared with that of *Romneya* (*Papaverac.*). The fls. of some spp. are beetle-pollinated (cantharophilous).

***Paepalanthus** Kunth. Eriocaulaceae. 485 warm S. Am., W.I.

Paësia A. St-Hil. Dennstaedtiaceae. 12 Malaysia to Tahiti, N.Z., trop. Am. Spreading sun-ferns; like *Hypolepis*, but true indusium present.

Pagaea Griseb. Gentianaceae. 6 trop. S. Am.

Pagamea Aubl. Rubiaceae. 20 trop. S. Am.

Pagameopsis Steyerm. Rubiaceae. 2 Venez.

Pagapate Sonnerat = Sonneratia L. f. (Sonneratiac.).

Pagella Schönland. Crassulaceae. 1 S. Afr. Remarkable small annual; stem at first turbinate, finally discoid, with fls. on upper surface.

Pagerea Pierre ex Laness. (sphalm.) = Sageraea Dalz. (Annonac.).

Pageria Juss. = Lapageria Ruiz & Pav. (Philesiac.).

Pageria Rafin. = seq.
Pagesia Rafin. Scrophulariaceae. 15–20 S. U.S. to trop. S. Am.
Pagetia F. Muell. Rutaceae. 2 Queensland.
Pagiantha Markgraf. Apocynaceae. 20 Indomal., Pacif.
Pahudia Miq. = Afzelia Sm. (Legumin.).
Paillotia Gandog. = Erodium L..(Geraniac.).
Painteria Britton & Rose. Leguminosae. 4 Mex., 1 Ceylon.
Paiva Vell. = Sabicea Aubl. (Rubiac.).
Paivaea Berg. Myrtaceae. 1 SE. Brazil.
Paivaea P. & K. = Paiva Vell. = Sabicea Aubl. (Rubiac.).
Paivaeusa Welw. = Oldfieldia Hook. (Euphorbiac.?).
Pajanelia DC. Bignoniaceae. 1 Indomal.
Pala Juss. = Alstonia R.Br. (Apocynac.).
Paladelpha Pichon. Apocynaceae. 1 W. Malaysia.
Palaeconringia E. H. L. Krause = Erysimum L. (Crucif.).
Palaeno Rafin. = Campanula L. (Campanulac.).
Palafoxia Lag. Compositae. 2 S. U.S., N. Mex.
Palala Kuntze = Myristica L. s.l. (Myristicac.).
Palala Rumph. = Horsfieldia Willd. (Myristicac.).   [L. (Euphorbiac.).
Palamostigma Benth. & Hook. f. = Palanostigma Mart. ex Klotzsch = Croton
Palandra O. F. Cook (~ Phytelephas Ruiz & Pav.). Palmae. 1 Ecuador.
Palanostigma Mart. ex Klotzsch = Croton L. (Euphorbiac.).
Palaoea Kanehira = Tristiropsis Radlk. (Sapindac.).
Palaquium Blanco. Sapotaceae. 115 +, Formosa, SE. As., Indomal., Solomon
Is. P. gutta (Hook.) Burck was formerly the chief source of gutta-percha, but
it is now extinct exc. in cult., and gutta is obtained from other spp. and from
Payena leerii (Teijsm. & Binnend.) Kurz, etc. The trees are cut down or ringed
and the milky latex coagulates, forming gutta-percha.
Palaua Cav. Malvaceae. 15 Andes.
Palaua Ruiz & Pav. = Saurauia Willd. (Actinidiac.).
Palava Juss. = Palaua Cav. (Malvac.).
Palava Pers. = Palaua Ruiz & Pav. = Saurauia Willd. (Actinidiac.).
Palavia Ruiz & Pav. ex Ort. = Calyxhymenia Ort. = Mirabilis L. (Nyctaginac.).
Palavia Schreb. = Palaua Cav. (Malvac.).
Paleaepappus Cabr. Compositae. 1 Patag.
Paleista Rafin. = Eclipta L. (Compos.).
Palenga Thw. = Putranjiva Wall. = Drypetes Vahl (Euphorbiac.).
Palenia Phil. (~ Heterothalamus Less.). Compositae. 1 Chile.
Paleolaria Cass. = Palafoxia Lag. (Compos.).
Paletuviera Thou. ex DC. = Bruguiera Lam. (Rhizophorac.).
Paletuvieraceae Lam. ex Kuntze = Rhizophoraceae R.Br.
Paleya Cass. = Crepis L. (Compos.).
Palhinhaea Franco & Vasconc. = Lycopodium L. p.p. (Lycopodiac.).
Paliavana Vell. ex Vand. Gesneriaceae. 5 Brazil.
Palicourea Aubl. Rubiaceae. 200 trop. Am., W.I.
Palicuria Rafin. = praec.
Palilia Allam. ex L. = Heliconia L. (Heliconiac.).
Palimbia Bess. Umbelliferae. 1 S. & E. Russia to C. As.

PALINDAN

Palindan Blanco ex P. & K. = Orania Zipp. (Palm.).

Palinetes Salisb. = Ammocharis Herb. (Amaryllidac.).

Paliris Dum. = Liparis Rich. (Orchidac.).

***Palisota** Reichb. Commelinaceae. 25 trop. W. Afr.

Palissya Baill. (1858; non Endl. 1847—gen. foss. Conif.) = Neopalissya Pax (Euphorbiac.).

Paliuros St-Lag. = seq.

Paliurus Mill. Rhamnaceae. 8 S. Eur. to Japan. In some spp. the stip. thorns are both straight; in others (e.g. *P. spina-christi* Mill., Christ's thorn) one is straight and the other recurved. The fr. has a horizontal wing, developed at the base of the style after fert.

Palladia Lam. = ? Escallonia Mutis ex L. f. (Escalloniac.).

Palladia Moench = Lysimachia L. (Primulac.).

Pallasia Houtt. = Calodendrum Thunb. (Rutac.).

Pallasia Klotzsch. Rubiaceae. 1 Guiana.

Pallasia L'Hérit. = Encelia Adans. (Compos.).

Pallasia L. f. = Calligonum L. (Polygonac.).

Pallasia Scop. = Crypsis Ait. (Gramin.).

Pallastema Salisb. = Albuca L. (Liliac.).

Pallavia Vell. = Pisonia L. (Nyctaginac.).

Pallavicinia De Not. = Cyphomandra Sendtn. (Solanac.).

***Pallenis** (Cass.) Cass. (~ Asteriscus Moench). Compositae. 4 Medit. to C. As.

Palma Mill. = Palmae variae (incl. Phoenix), Dracaena (Agavac.), Cycas (Cycadac.), etc.

***Palmae** Juss. (nom. altern. **Arecaceae** Schultz-Schultzenst.). 217/2500 trop. & subtrop.; most gen. well localized, chief exceptions being *Cocos nucifera* L. and *Elaeis guineënsis* Jacq. The palms form a char. feature of trop. veg. The veg. hab. is familiar—a crown of ls. at the end of an unbranched stem (*Hyphaene* is sometimes branched). The stem exhibits various forms; some palms, e.g., *Nypa*, *Phytelephas*, have a short rhiz. or stock bearing 'radical' leaves and often branching below ground; some, e.g. *Geonoma, Calamus, Desmoncus*, have a thin reed-like stem with long internodes (the two latter are climbers); others again have a tall stem with a crown of ls. at the top. The stem is often covered with the remains of old leaf-sheaths, or is thorny. Its height may reach 45–60 m. in some, and it grows slowly in thickness. At the base the stem is usu. conically thickened or provided with buttress roots; this gives the necessary mechanical rigidity. The stems of *Cocos* and other palms are curved instead of straight; this appears to be due to reaction to light.

The l. is very characteristic; the only closely similar l. is that of *Carludovica*, though those of Cycads and some tree-ferns have a superficial likeness. Some have palmate (fan) ls., some pinnate (feather) ls., but this structure arises by a development unlike that which gives rise to these forms in Dicots. and more like that in *Araceae*. The apparently cpd. condition is attained through the formation of splits in the originally continuous tissue. The l. is usu. very large, and at the base of the petiole is a sheath, which makes a firmer attachment to the stem than a mere articulation. The sheath contains many bundles of fibres, which remain after the decay of the softer tissues. The pinnae are folded where they meet the main stalk of the l., sometimes upwards (induplicate, V in section), sometimes downwards (reduplicate, ∧ in section); these chars. are important in classification. The l. emerges from the bud in an almost vertical line and thus escapes excessive radiation and transpiration. The palms are pronounced sun plants, and show xero. chars. in their ls. The leaf surface is glossy with a thick cuticle, and is rarely arranged ⊥ to the incident rays. Often the l. is corrugated, or placed at an angle by the twisting or upward slope of the stalk; sometimes the leaflets slope upwards, and so on.

Infl. usu. very large and much branched. In *Corypha* and others it is term., its production being a mark of the end of the life of the plant (monocarpic; cf. *Agave*), but usu. it is axillary; sometimes in the axils of the current ls., sometimes lower on the stem. The branching is racemose and the fls. are often embedded in the axis; the whole is often termed a *spadix*. It is enclosed in a spathe of several ls. and emerges when the fls. are ready to open. Some are dioec., some monoec., in the latter case often with the fls. in groups (small dichasia) of 3, one ♀ between two ♂.

The fl. has usu. the formula P 3 + 3, A 3 + 3, G̱ 3 or (3). P homochlam., varying in texture. G (3), 1-loc. or 3-loc., with 3, or sometimes 1, anatr. ovules (rarely semi-anatr., or orthotr.). Some are wind-pollinated, others are entomoph.

Fr. a berry or drupe; in the latter case the endocarp usu. united to the seed. Fr. in § III covered with dry woody scales. Endosperm large; in date, vegetable ivory, etc., it is very hard, the non-nitrogenous storage-material taking the form of cellulose, deposited upon the cell walls. In germ. the cot. lengthens and pushes out the radicle, and then the plumule grows out of the sheathing cotyledon.

Economically, the *P*. are very important, furnishing many of the necessaries of life in the tropics, etc. Some (e.g. *Elaeis*, *Cocos*) are valuable sources of vegetable fats, used for cooking-fats, margarine, soap, etc. Many have ed. fr. or seed, e.g. date (*Phoenix*) and coconut (*Cocos*); the stems contain much starch as reserve food, esp. in those spp. which save up for a great terminal infl., e.g. *Metroxylon* (sago), *Caryota*, etc.; the rush of sap to the infl., esp. in the cases just mentioned, is great; and by tapping the stem great quantities of sugar-containing fluid may be obtained and utilised, either directly as a source of sugar or indirectly to make intoxicating drinks by fermentation. The bud of ls. at the top of the stem is sometimes used as cabbage, but of course its removal kills the tree. The stems are used in building, but do not yield plank-timber; the ls. in thatching and basket-making, and for hats, mats, etc.; the fibres of the leaf-sheaths or sometimes of the pericarp (e.g. *Cocos*) are used for ropes, etc.; other *P*. furnish wax (*Copernicia*), vegetable ivory (*Phytelephas*, etc.), betel-nuts (*Areca*), etc.

Classification and chief genera (after Drude):

A. Perianth 6-partite, enclosing the fruit after fertilisation.

    I. **Coryphoïdeae** (spadix loosely branched, often a prolix panicle; fls. single or in long rows flowering from above; cpls. 3, or loosely united, separating after fert.; berry; fan or feather ls., induplicate):

        1. PHOENICEAE (feather ls.): *Phoenix*.

        2. SABALEAE (fan ls.): *Chamaerops, Rhapis, Corypha, Livistona, Sabal, Copernicia*.

II. *Borassoïdeae* (spadix simple or little-branched with thick cylindrical twigs; fls. markedly diclinous-dimorphic, invested with bracts, the ♂ in 1–∞ cincinni in grooves of the twigs; cpls. (3), fully united, producing a one-seeded drupe; fan ls. induplicate):

    1. BORASSEAE: *Hyphaene, Borassus, Lodoicea.*

III. *Lepidocaryoïdeae* (spadix branched once or more in a 2-ranked arrangement; fls. in cincinni or 2-ranked spikes with bracts and bracteoles round them; cpls. (3), fast united, covered with scales; fr. 1-seeded, covered with hard scales; feather or fan ls., reduplicate):

    1. MAURITIËAE (fan ls.): *Mauritia.*

    2. METROXYLEAE (feather ls.): *Raphia, Metroxylon, Calamus.*

IV. *Arecoïdeae (Ceroxyloïdeae)* (spadix simple or once or several times branched; fls. diclinous, usu. dimorphic; when dioec., sol. with rudimentary bracts; when monoec., usu. in cymes of 3 fls., 2 being ♂ and 1 ♀, or rarely ∞ ♂ and 1 at the end of the row being ♀; cpls. (3), 3- 2- 1-loc.; fruit smooth, not scaly; feather ls.):

    1. ARECEAE (berry fr.): *Caryota, Arenga, Leopoldinia, Iriartea, Ceroxylon, Chamaedorea, Roystonea, Euterpe, Kentia, Areca.*

    2. COCOËAE (drupe fr.): *Elaeis, Attalea, Cocos, Bactris, Desmoncus.*

B. Perianth rudimentary in ♂ or ♀. Fruit in dense heads.

V. *Phytelephantoïdeae* (♂ fl. with ∞ free sta.; ♀ with P; endosp. ivory-like): *Phytelephas* (only gen.).

VI. *Nypoïdeae* (♂ with (3) sta.; ♀ naked; woody endocarp): *Nypa* (only gen.).

The above classification has been somewhat modified by Burret (1953) and Beccari & Pichi-Sermolli (1918–56). The tribe *Cocoëae* is raised to the rank of subfamily (*Cocoïdeae*), whilst the *Phytelephantoïdeae* are (by Burret) merged into the *Arecoïdeae*. Anatomical evidence (Tomlinson, 1961) indicates that there are certainly no grounds for assuming a close relationship between *Phytel.* and *Nypoïdeae.*

Palma-filix Adans. = Zamia L. (Zamiac.).

Palmangis Thou. (uninom.) = *Angraecum palmiforme* Thou. (Orchidac.).

Palmerella A. Gray. Campanulaceae. 2 Mex., California.

Palmeria F. Muell. Monimiaceae. 20 E. Malaysia, Austr.

Palmerocassia Britton = Cassia L. (Legumin.).

Palmervandenbroeckia L. S. Gibbs. Araliaceae. 1 W. New Guinea.

Palmia Endl. = Hewittia Wight & Arn. (Convolvulac.).

Palmifolia Kuntze = Zamia L. (Zamiac.).

Palmijuncus Kuntze = Calamus L. (Palm.).

Palmofilix P. & K. = Palma-filix Adans. = Zamia L. (Zamiac.).

Palmoglossum Klotzsch ex Reichb. f. = Pleurothallis R.Br. (Orchidac.).

Palmolmedia Ducke. Moraceae. 1 Brazil.

Palmonaria Boiss. (sphalm.) = Pulmonaria L. (Boraginac.).

Palmorchis Barb. Rodr. Orchidaceae. 2 trop. S. Am., W.I.

Palmstruckia Retz. f. = Chaenostoma Benth. (Scrophulariac.).

Palmstruckia Sond. = Thlaspeocarpa C. A. Smith (Crucif.).

Paloue Aubl. Leguminosae. 4 trop. S. Am.

Palovea Juss. = praec.
Palovea Rafin. = Sabicea Aubl. (Rubiac.).
Paloveopsis R. S. Cowan. Leguminosae. 1 Guiana.
Paltonium Presl. Polypodiaceae. 1 trop. Am.
Paltoria Ruiz & Pav. = Ilex L. (Aquifoliac.).
Paludana Giseke = Amomum L. (Zingiberac.).
Paludana Salisb. = seq.
Paludaria Salisb. = Colchicum L. (Liliac.).
Paluëa P. & K. = Paloue Aubl. (Legumin.).
Palumbina Reichb. f. Orchidaceae. 1 Guatem.
Palura (G. Don) Buch.-Ham. ex Miers = Symplocos L. (Symplocac.).
Pamburus Swingle (~ Atalantia Corrêa). Rutaceae. 1 S. India, Ceylon.
Pamea Aubl. = Terminalia L. (Combretac.).
Pamianthe Stapf. Amaryllidaceae. 3 Andes.
Pamphalea Lag. corr. DC. Compositae. 8 subtrop. & temp. S. Am.
Pamphilia Mart. Styracaceae. 3 Brazil.
Pamplethantha Bremek. Rubiaceae. 3 trop. Afr.
Panacea Mitch. = Panax L. (Araliac.).
Panaetia Cass. = Podolepis Labill. (Compos.).
Panargyrum D. Don = seq.
Panargyrus Lag. = Nassauvia Juss. (Compos.).
Panax Hill = Opopanax Koch (Umbellif.).
Panax L. Araliaceae. 8 trop. & E. As., N. Am. Herbs. Ginseng from *P. schinseng* Nees (*P. ginseng* Mey.).
Panaxus St-Lag. = praec.
Pancalum Ehrh. (uninom.) = *Hypericum pulchrum* L. (Guttif.).
Panchera P. & K. = Pancheria Montr. = Ixora L. (Rubiac.).
Pancheria Brongn. & Gris. Cunoniaceae. 25 New Caled.
Pancheria Montr. = Ixora L. (Rubiac.).
Panchezia B. D. Jacks. (sphalm.) = praec.
Panciatica Picciv. = Cadia Forsk. (Legumin.).
Pancicia Vis. = Pimpinella L. (Umbellif.).
Pancovia Fabr. (uninom.) = *Potentilla palustris* (L.) Scop. (Rosac.).
***Pancovia** Willd. Sapindaceae. 10–12 trop. Afr.
Pancratiaceae Horan. = Amaryllidaceae Jaume St-Hil.
Pancratio-Crinum Herb. ex Steud. = Crinum L. (Amaryllidac.).
Pancratium L. Amaryllidaceae. 15 Medit. to trop. As. & trop. Afr.
Panctenis Rafin. = Aureolaria Rafin. (Scrophulariac.).
Panda Pierre. Pandaceae. 1 trop. W. Afr. Large drupe with massive stony endocarp, containing 3 one-seeded chambers, finally dehiscing by valves at germination.
Pandaca Nor. ex Thou. = Tabernaemontana L. (Apocynac.).
Pandacastrum Pichon. Apocynaceae. 1 Madag.
***Pandaceae** Pierre. Dicots. 4/28 trop. Afr. & As. Dioecious trees. Ls. alt., distich., simple, often serrate, stip., on shoots which bear buds in their 'axils'. Fls. reg., ♂ ♀, in axill. fascicles (*Microdesmis*) or cymes (*Centroplacus*) or in terminal (*Galearia*) or cauliflorous (*Galearia* and *Panda*) racemiform thyrses. K (5), imbr. or open; C 5, imbr. or valv.; A 5–10–15, sometimes unequal,

anth. intr.; small stds. sometimes present (*Centroplacus*); disk small or 0, rarely larger (*Centroplacus* ♀); G (2–5), with 1 or rarely 2 (*Centroplacus*) apical pend. orthotr. or anatr. ov. per loc., without obturator; and short style with 2–5(–10)-lobed stig. Fr. a drupe, sometimes flattened, more rarely a caps. (*Centroplacus*); endoc. ± bony or stony, thin to very thick and massive, usu. variously tuberculate or muricate or pitted or ridged, sometimes dehisc. by valves; seeds usu. flattened-concave, more rarely ovoid (*Centroplacus*), with endosp. Genera: *Panda, Galearia, Microdesmis, Centroplacus*.

Pandamus Rafin. = Pandanus L. f. (Pandanac.).

*Pandanaceae R.Br. Monocots. 3/700, char. pl. of the Old World tropics, but a few warm temp. Mostly sea-coast or marsh pls. with tall stems supported upon aerial roots, frequently branched; buds are found in all axils, and the branching appears dichotomous; some are climbers. The aerial roots have marked root-caps of membranous texture. Ls. in 3-ranked phyllotaxy, but stem usu. twisted so that they appear to run in well-marked spirals, whence the name of screw-pines. Ls. parallel-veined, long, and narrow, with open sheath and usu. thorny margin; generally sharply bent downwards at the middle, and corrugated like a palm l.

Infl. term., with a few bract-like ls. at the base going gradually over into the foliage ls., usu. a racemose spadix with neither bracts nor bracteoles to the individual fls., which are somewhat difficult to make out. The ♂ fls. in spp. of *Freycinetia* have a rudimentary G, but in the rest of the fam. they have not. The floral axis of the ♂ fl. bears a number of sta., arranged in a racemose or umbel-like manner upon it. The G in the ♀ fl. of ∞ cpls. in a ring, 1-loc. or ∞-loc., the union being ± complete, or it may be reduced, even to 1 cpl., or to a row of cpls. arranged transv. Stigmas sessile. Ovules anatr. Berry or multi-loc. drupe, often containing hollow spaces which aid it in floating. Seed with oily endosp. The plants yield thatch, etc. Genera: *Freycinetia* (infl. capitulate or spicate; fls. sessile; berry); *Pandanus* (infl. as last; fls. sessile; drupe); *Sararanga* (infl. paniculate; fls. pedicelled; drupe).

Pandanophyllum Hassk. = Mapania Aubl. (Cyperac.).

*Pandanus L. f. Pandanaceae. 600 palaeotrop. (screw-pines). Trees with flying-buttress roots. Fls. in large heads, enclosed in spathes. ♂ of ∞ sta., arranged in various ways upon the axis, ♀ of 1–∞ cpls., free or united. Each gives a drupe containing as many seeds as cpls. Seeds album. The pericarp is rich in fibres. The frs. of some are cooked and eaten, e.g. *P. leram* Jones, the Nicobar bread-fruit. The ls. of many are used for weaving, e.g. *P. spurius* Miq., cultivar 'Putat', which is cult. in Java (cf. H. St John, *Pacif. Sci.* **19**: 232–5, 1965). Several have sweetly scented fls. or ls. which are used for ornament and otherwise in the East.

Panderia Fisch. & Mey. Chenopodiaceae. 2 SW. & C. As.

Pandiaka (Moq.) Hook. f. Amaranthaceae. 20 trop. & S. Afr.

Pandora Nor. ex Thou. = Rhodolaena Thou. (Sarcolaenac.).

Pandorea Spach. Bignoniaceae. 8 E. Malaysia, C. Austr.

Paneguia Rafin. = Sisyrinchium L. (Iridac.).

Paneion Lunell = Poa L. (Gramin.).

Panel Adans. = Terminalia L. (Combretac.).

Panemata Rafin. = Gymnostachyum Nees (Acanthac.).

Panetos Rafin. = Hedyotis L. + Arcytophyllum Willd. ex Schult. & Schult. f. (Rubiac.).

Pangi[ac]eae Bl. ex Endl. = Flacourtiaceae–Pangieae Clos.

Pangium Reinw. Flacourtiaceae. 1 Malaysia, Bismarck Archip., Palau Is. The seeds of *P. edule* Reinw. are eaten after long soaking to dissipate the hydrocyanic acid which they, like all parts, contain.

Panhopia Nor. ex Muell. Arg. = Panopia Nor. ex Thou. = Macaranga Thou. (Euphorbiac.).

Panicaceae (R.Br.) Herter = Gramineae–Paniceae R.Br.

Panicastrella Moench = Echinaria Desf. (Gramin.).

Panicularia Cotta = Thyrsopteris Kunze (Dicksoniac.).

Panicularia Fabr. (uninom.) = *Glyceria* R.Br. sp. (Gramin.).

Paniculum Ard. = Oplismenus Beauv. (Gramin.).

Panicum L. Gramineae. 500 trop. & warm temp. The spikelets are 2-flowered. Some *P.*, known as millets, are important cereals, extensively cult. in India, S. Eur., etc., e.g. *P. miliaceum* L., the 'proso' or common millet, *P. sumatrense* Roth ex Roem. & Schult. (*P. miliare* Lam.), the little millet, and other minor spp. Many are important fodder plants, *P. maximum* Jacq. (trop.; Guinea grass), *P. molle* Sw. (trop. Am.; Mauritius grass), etc. Many are distributed by animals, for the joints of the stem will grow after passing the alimentary canal.

Paniopsis Rafin. = Inula L. (Compos.).

Panios Adans. = Erigeron L. (Compos.).

***Panisea** (Lindl.) Lindl. Orchidaceae. 4 NE. India, Indoch.

Panisia Rafin. = Cassia L. (Legumin.).

Panke Mol. = Gunnera L. (Gunnerac.).

Panke Willd. = Francoa Cav. (Francoac.).

Pankea Oerst. = Panke Mol. = Gunnera L. (Gunnerac.).

Panninia Th. Dur. (sphalm.) = Fanninia Harv. (Asclepiadac.).

Panoë Adans. = Vateria L. (Dipterocarpac.).

Panope Rafin. = Lippia L. (Verbenac.).

Panopia Nor. ex Thou. = Macaranga Thou. (Euphorbiac.).

Panopsis Salisb. Proteaceae. 20 trop. Am.

Panoxis Rafin. = Hebe Comm. ex Juss. (Scrophulariac.).

Panphalea Lag. = Pamphalea Lag. corr. DC. (Compos.).

Panslowia Wight ex Pfeiff. = Kadsura Juss. (Schisandrac.).

Panstenum Rafin. = Allium L. (Alliac.).

Panstrepis Rafin. = Coryanthes Hook. (Orchidac.).

Pantacantha Spegazz. Solanaceae. 1 Patag.

Pantadenia Gagnep. Euphorbiaceae. 1 Indoch.

Pantasachme Endl. (sphalm.) = Pentasachme Wall. ex Wight, *q.v.* (Asclepiadac.).

Pantathera Phil. Gramineae. 1 Juan Fernandez.

Panterpa Miers = Petastoma Miers (Bignoniac.).

Panthocarpa Rafin. = Acacia Willd. (Legumin.).

Pantlingia Prain = Stigmatodactylus Maxim. (Orchidac.).

Pantocsekia Griseb. = Convolvulus L. (Convolvulac.) (infected with the smut *Thecaphora* Fingerh.).

849

PANTORRHYNCHUS

Pantorrhynchus Murb. = Trachystoma O. E. Schulz (Crucif.).
Panulia Baill. (~ Ligusticum L.). Umbelliferae. 3 Chile.
Panurea Spruce ex Benth. & Hook. f. Leguminosae. 1 N. Brazil.
Panza Salisb. = Narcissus L. (Amaryllidac.).
Panzera Willd. = Eperua Aubl. (Legumin.).
Panzeria J. F. Gmel. = Lycium L. (Solanac.).
Panzeria Moench (~ Leonurus L.). Labiatae. 5 W. Siberia to Mongolia.
Paolia Chiov. Rubiaceae. 1 Somaliland.
Paoluccia Gandog. = Lathyrus L. (Legumin.).
Papas Opiz = Battata Hill = Solanum L. (Solanac.).
Papaver L. Papaveraceae. 100 Eur., As., S. Afr., Austr., Am. The fls. nod
in bud, not by their own weight, but by more rapid growth of one side of
the stalk. Ovary crowned by a sessile rayed stigma, each lobe of which stands
over a placenta instead of as usual over a midrib. This is commonly explained
by supposing each actual ray of the stigma to be formed of one half of each of
two adjacent stigmas. The fl. of most contains no honey, and is homogamous;
both cross- and self-pollination usually occur with insect visits. Fr. a round
or cylindr. caps., opening by pores under the eaves of the roof formed by the
dry stigmas, so that the seeds are protected from rain and can only escape
when the capsule is shaken by strong winds or other agencies (censer mechanism,
cf. *Aconitum*). *P. somniferum* L. is the opium poppy; the drug is obtained by
cutting notches in the half-ripened capsules, from which the latex exudes and
hardens. The seeds of this and other spp. yield an oil on pressure.
*Papaveraceae Juss. (*s.str.*). Dicots. 26/200, chiefly N. temp. Most are herbs
with alt. ls., and contain latex. *Dendromecon* and *Bocconia* spp. are shrubs or
small trees. Fls. sol. or in racemes, or in dichasia with cincinnal tendency, reg.,
♀, hypog. (exc. *Eschscholzia*). K 2 (united in *Eschscholzia*), caducous; C 2 + 2,
rolled or crumpled in bud; A ∞; G (2–∞), 1-loc., with parietal plac., which
in *Papaver*, etc., project into the loc.; in *Platystemon* G 6–20, almost distinct,
torulose, sometimes separating in fruit. Ovules generally ∞, anatr. or slightly
campylotr. Fr. a septic. caps., or one opening by pores, or a nut, rarely of
∞ free carpels; seeds with oily endosp., and small embryo. The fls. are mostly
large and conspicuous, but many contain no honey and are visited by pollen-
seeking insects; they are often protandr. The order is of little economic value;
see *Papaver*.

Chief genera: *Platystemon, Romneya, Eschscholzia, Sanguinaria, Chelidonium,
Bocconia, Glaucium, Meconopsis, Argemone, Papaver*.
Papaya Mill. = Carica L. (Caricac.).
Papayaceae Blume = Caricaceae Dum.
Papeda Hassk. = Citrus L. (Rutac.).
Paphia Seem. (~ Agapetes D. Don ex G. Don). Ericaceae. 15 New Guinea,
Queensl., Fiji.
Paphinia Lindl. Orchidaceae. 5 trop. S. Am.
*Paphiopedilum Pfitz. Orchidaceae. 50 trop. As. to Solomon Is.
*Papilionaceae Giseke = Leguminosae–Papilionoïdeae DC. (Fls. usu. zygo.;
K imbr.; C imbr.-descending [lower petal innermost] in bud; A usu. 10, pollen
simple; seeds without areoles; ls. pinn., digit., trifoliol. or simple.)
× **Papilionanda** hort. (v) = seq.

850

Parapottsia Miq. = Pottsia Hook. & Arn. (Apocynac.).
Paraprotium Cuatrec. Burseraceae. 1 Colombia.
Parapyrenaria H. T. Chang (~ Pyrenaria Bl.). Theaceae. 1 S. China (Hainan).
Parapyrola Miq. = Epigaea L. (Ericac.).
Paraqueiba Scop. = Poraqueiba Aubl. (Icacinac.).
Paraquilegia J. R. Drumm. & Hutch. Ranunculaceae. 8 C. As. to Afghan.
× **Pararachnis** hort. (iv) = × Arachnopsis hort. (Orchidac.).
× **Pararenanthera** hort. (iv) = × Renanthopsis hort. (Orchidac.).   [W. Afr.
Pararistolochia Hutch. & Dalz. (~ Aristolochia L.). Aristolochiaceae. 16 trop.
Parartabotrys Miq. = Xylopia L. (Annonac.).
Parartocarpus Baill. Moraceae. 2 Penins. Siam, Malaysia (exc. Lesser Sunda
Is.) to Solomon Is.
Pararuellia Bremek. & Nannenga-Bremek. Acanthaceae. 6 Indoch., W.
Parasamanea Kosterm. Leguminosae. 1 Borneo.   [Malaysia.
Parasarcochilus Dockr. Orchidaceae. 3 E. Austr.
Parascheelea Dugand. Palmae. 2 trop. S. Am.
Parascopolia Baill. (Lycianthes Baill.). Solanaceae. 200 trop. & temp.
Paraselinum Wolff. Umbelliferae. 1 Peru.
Parasenecio W. W. Smith & Small. Compositae. 1 W. China.
Parashorea Kurz. Dipterocarpaceae. 11 SE. As., W. Malaysia. *P. malaänonan*
(Blanco) Merr. yields the timber known as 'Borneo white seraya'.
Parasia P. & K. = Parrasia Rafin. = Belmontia E. Mey. (Gentianac.).
Parasitipomoea Hayata = Ipomoea L. (Convolvulac.).
Parasorus v. Ald. v. Ros. Davalliaceae. 1 Moluccas (Ternate). Ls. simple;
sori confluent along edges, otherwise as *Humata*.
Paraspalathus Presl = Aspalathus L. (Legumin.).
Parasponia Miq. Ulmaceae. 6 Malaysia, Polynesia.
Parastemon A. DC. Chrysobalanaceae. 2 Malaysia.
Parastranthus G. Don = Lobelia L. (Campanulac.).
Parastrephia Nutt. Compositae. 5 Andes.
Parastriga Mildbr. Scrophulariaceae. 1 trop. Afr.
Parastrobilanthes Bremek. Acanthaceae. 4 Sumatra, Java.
Parastyrax W. W. Smith. Styracaceae. 1 Burma.
Parasympagis Bremek. Acanthaceae. 3 Burma, Siam.
Parasyringa W. W. Smith (~ Ligustrum L.). Oleaceae. 1 SW. China.
Parasystasia Baill. = Asystasia Bl. (Acanthac.).
Paratecoma Kuhlm. Bignoniaceae. 1 Brazil.
Paratephrosia Domin. Leguminosae. 1 C. Austr.
Parathelypteris Ching emend. Holtt. Thelypteridaceae. 10 SE. As.; N. Am.?
Paratheria Griseb. Gramineae. 2 W.I., trop. S. Am., trop. Afr., Madag.
Parathesis Hook. f. Myrsinaceae. 40 Mex. to trop. S. Am., Cuba.
Parathyrium Holttum = Lunathyrium Koidz. (Athyriac.).
Paratriaina Bremek. Rubiaceae. 1 Madag.
Paratrophis Blume = Streblus Lour. (Morac.).   [(Araliac.).
Paratropia DC. = Heptapleurum Gaertn. = Schefflera J. R. & G. Forst.
Paravallaris Pierre (~ Kibatalia G. Don). Apocynaceae. 4 Siam, Indoch.,
Malay Penins.
× **Paravanda** hort. (iv) = × Vandaenopsis hort. (Orchidac.).

PARAVANDANTHERA

× **Paravandanthera** hort. (iv) = × Moirara hort. (Orchidac.).

Paravinia Hassk. = Praravinia Korth. (Rubiac.).

Paravitex Fletcher. Verbenaceae. 1 Siam.

× **Pardancanda** Lenz. Iridaceae. Gen. hybr. (Belamcanda × Pardanthopsis).

Pardanthopsis (Hance) Lenz (~ Iris L.). Iridaceae. 1 temp. As.

Pardanthus Ker-Gawl. = Belamcanda Adans. (Iridac.).

Pardinia Herb. = Hydrotaenia Lindl. + Tigridia Juss. (Iridac.).

Pardisium Burm. f. = Gerbera L. ex Cass. (Compos.).

Parduyna Salisb. Liliaceae. 2 E. Austr.

Parechites Miq. = Trachelospermum Lem. (Apocynac.).

Parectenium P. Beauv. corr. Stapf. Gramineae. 1 Austr.

Pareira Lour. ex Gomes = Vitis L. (Vitidac.).

Parenterolobium Kosterm. Leguminosae. 1 Borneo.

Parentucellia Viv. Scrophulariaceae. 4 Atl. Eur., Medit. to C. As. & Persia.

Parestia Presl = Davallia Sm. (Davalliac.).

Pareugenia Turrill = Syzygium Gaertn. (Myrtac.).

Parexuris Nakai & Maekawa = Sciaphila Bl. (Triuridac.).

Parfonsia Scop. (sphalm.) = Parsonsia P.Br. = Cuphea P.Br. (Lythrac.).

Parhabenaria Gagnep. Orchidaceae. 1 Indochina.

Pariana Aubl. Gramineae. 34 trop. S. Am.

Parianaceae (Hack.) Nak. = Gramineae–Parianeae (Hack.) C. E. Hubb.

Pariatica P. & K. = seq.

Pariaticu Adans. = Nyctanthes L. (Verbenac.).

Parid[ac]eae Dum. = Trilliaceae Lindl.

Parietales Lindl. A large order of Dicots., including amongst others (in Engler's system) the *Dilleniac.*, *Theac.*, *Guttiferae*, *Tamaricac.*, *Cistac.*, *Violac.*, *Flacourtiac.*, *Passiflorac.*, *Loasac.*, *Begoniac.* and *Ancistrocladac.*

Parietaria L. Urticaceae. 30 temp. & trop. (in Malaysia rare, and mts. only). Fls. mostly ⚥ (unlike most of the fam.), in little cymes in the l. axils. According to Eichler the first fl. is ♀, the bulk of the cyme ⚥, and the last fls. ♂. The ⚥ fls. are exceedingly protog., the style protruding from the bud; the sta. develop later, exploding when ripe like those of the nettle (*Urtica*), but by this time the stigma is incapable of fert., and usu. the style has dropped off, so that at first glance the fl. looks as if ♂.

Parilax Rafin. = Parillax Rafin. = Smilax L. (Smilacac.).

Parilia Dennst. = Elaeodendron Jacq. (Celastrac.).

Parilium Gaertn. = Nyctanthes L. (Verbenac.).

Parillax Rafin. = Smilax L. (Smilacac.).

Pariltaria Burm. f. (sphalm.) = Parietaria L. (Urticac.).

Parinari Aubl. Chrysobalanaceae. 60 trop. Some have ed. seed.

Parinarium Juss. = praec.

Paripon Voigt. Palmae (inc. sed.). 1, habitat?

Paris L. Trilliaceae. 20 N. palaeotemp. *P. quadrifolia* L. has monopodial rhiz. and aerial stem with whorl of 4 or more net-veined ls.; the aerial stems are formed, not annually, but at irreg. periods. P 4(or more)-merous, as well as the other whorls; in *P. quadrifolia* the sepals alt. with the foliage-ls. The fls. of this sp. are very protog., and colour and scent attract flies.

Parisetta Augier = praec.

Parishella A. Gray. Campanulaceae. 1 California.
Parishia Hook. f. Anacardiaceae. 12 Burma, W. Malaysia.
Parita Scop. = seq.
Pariti Adans. (~ Hibiscus L.). Malvaceae. 25 trop.
Paritium A. Juss. = praec.
Parivoa Aubl. = Eperua Aubl. (Legumin.).
Parkeria Hook. = Ceratopteris Brongn. (Parkeriac.).
Parkeriaceae Hook. Pteridales. Only genus: *Ceratopteris*.
Parkia R.Br. Leguminosae. 40 trop. Fls. in heads of which either the upper
or lower fls. are male or neuter; pollinated by bats. The seeds of *P. africana*
R.Br. (trop. Afr.) are eaten.
Parkinsonia L. Leguminosae. 2 trop. Am., S. Afr.
Parlatorea Barb. Rodr. = Sanderella Kuntze (Orchidac.).
Parlatoria Boiss. Cruciferae. 2 SW. As.
Parmena Greene = Rubus L. (Rosac.).
Parmentiera DC. Bignoniaceae. 8 Mex. to Colombia. *P. cerifera* Seem.,
used as fodder, has caulifloral frs. which look like candles.
Parmentiera Rafin. = Solanum L. (Solanac.).
Parnassia L. Parnassiaceae. 50 N. temp., chiefly in upland bogs.
*****Parnassiaceae** S. F. Gray. Dicots. 1/50 N. temp. Perenn. herbs with thick
rootstock. Ls. radic. and alt., simple, ent., petiolate, often cordate, exstip.,
with mostly basal nerves, and tannin-containing cells in epidermis. Fls. reg.,
⚥, sol., on long stems. Floral axis hollowed out and united to the base of the
ovary. K 5, imbr.; C 5, imbr., sometimes fimbr.; A 5, and 5 alternating
stds.; G (3–4) or half-inf., 1-loc., with large projecting parietal plac. and
∞ ov. The fl. is protandr., the anthers in turn dehiscing just above the pistil
and then moving outwards. Stds. opp. to the petals. Each has a solid nectar-
secreting base, and ends above in a candelabra-like structure, each branch of
which is terminated by a yellow knob, glistening in the sun and looking like
a drop of honey. Flies are deceived by this appearance, and have been seen
licking the knobs. Only genus: *Parnassia*.
Parochetus Buch.-Ham. ex D. Don. Leguminosae. 1 mts. of trop. Afr. & As.
It has cleistogamic and chasmog. fls.
Parodia Speg. Cactaceae. 35 trop. & subtrop. S. Am.
Parodianthus Troncoso. Verbenaceae. 1 Argentina.
Parodiella J. & C. Reeder (1968; non Speg. 1880—Fungi) = Lorenzochloa
J. & C. Reeder (Gramin.).
Parodiodendron Hunziker. Euphorbiaceae. 1 Argent.
Parodiodoxa O. E. Schulz. Cruciferae. 1 high mts. N. Argentina.
Parolinia Engl. = Widdringtonia Endl. (Cupressac.).
Parolinia Webb. Cruciferae. 1 Canaries.
Paronychia Hill = Erophila DC. (Crucif.).
Paronychia Mill. Caryophyllaceae. 50 cosmop. The small axillary fls. are
concealed by the stipules.
Paronychi[ac]eae A. St-Hil. = Caryophyllaceae–Paronychioïdeae Vierh.
Parophiorrhiza C. B. Clarke = Mitreola L. (Loganiac.).
Paropsia Nor. ex Thou. Flacourtiaceae. 10 trop. Afr., Madag.; 1 Sumatra,
Malay Penins.

PAROPSIACEAE

Paropsiaceae Dum. = Flacourtiaceae–Paropsieae DC.
Paropsiopsis Engl. Flacourtiaceae. 7 trop. W. Afr.
Paropyrum Ulbr. Ranunculaceae. 1 west C. As.
Parosela Cav. = Dalea Juss. (Legumin.).
Paroxygraphis W. W. Smith. Ranunculaceae. 1 E. Himal.
Parquetina Baill. Periplocaceae. 1 W. Equat. Afr.
Parqui Adans. = Cestrum L. (Solanac.).
Parrasia Greene = Nerisyrenia Greene (Crucif.).
Parrasia Rafin. = Belmontia E. Mey. = Sebaea Sol. ex R.Br. (Gentianac.).
Parria Steud. = Parrya R.Br. (Crucif.).
Parrotia C. A. Mey. Hamamelidaceae. 1 Persia. Fl. ♂, apet.
Parrotia Walp. (sphalm.) = Barrotia Gaudich. = Pandanus L. f. (Pandanac.).
Parrotiaceae Horan. = Hamamelidaceae R.Br.
Parrotiopsis (Niedenzu) Schneider. Hamamelidaceae. 1 W. Himal.
Parrya R.Br. Cruciferae. 25 C. & N. temp. As., N. Am.
Parryella Torr. & Gray. Leguminosae. 1 Mex.
Parryodes Jafri. Cruciferae. 1 S. Tibet, E. Himal.
Parryopsis Botschantsev. Cruciferae. 1 Tibet.
Parsana Parsa & Maleki. Urticaceae. 1 Persia.
Parsonsia P.Br. = Cuphea P.Br. (Lythrac.).
***Parsonsia** R.Br. Apocynaceae. 100 S. China, SE. As., Indomal., Austr., N.Z., Polynes.
Partheniaceae Link = Compositae–Heliantheae Cass.
Partheniastrum Fabr. = Parthenium L. (Compos.).
Parthenice A. Gr. Compositae. 1 SW. U.S., Mex.
Parthenium L. Compositae. 15 Am., W.I. Br. attached to fr.
***Parthenocissus** Planch. Vitidaceae. 15 temp. As., Am. *P. tricuspidata* (Sieb. & Zucc.) Planch. and *P. quinquefolia* (L.) Planch. are the Virginia creepers.
Parthenopsis Kellogg = Venegasia DC. (Compos.).
Parthenostachys Fourr. = Ornithogalum L. (Liliac.).
Parthenoxylon Blume = Cinnamomum Schaeff. (Laurac.).
Parvatia Decne (~ Stauntonia Wall.). Lardizabalaceae. 3 Assam, China.
Parviopuntia Soulaire = Opuntia Mill. (Cactac.).
Parvisedum R. T. Clausen (~ Sedum L.). Crassulaceae. 3–4 Calif.
Parvotrisetum Chrtek. Gramineae. 1 N. Italy to Greece.
Paryphantha Schau. = Thryptomene Endl. (Myrtac.).
Paryphanthe Benth. = praec.
Paryphosphaera Karst. = Parkia R.Br. (Legumin.).
Pasaccardoa Kuntze. Compositae. 4 trop. Afr.
Pasania (Miq.) Oerst. = Lithocarpus Bl. (Fagac.).
Pasaniopsis Kudo = Castanopsis Spach (Fagac.).
Pascalia Ortega (~ Wedelia Jacq.). Compositae. 1 Chile.
Paschanthus Burch. = Adenia Forsk. (Passiflorac.).
Paschira G. Kuntze = Pachira Aubl. (Bombacac.).
Pasina Adans. = Horminum L. (Labiat.).
Pasithea D. Don. Liliaceae. 1 Peru, Chile.
Paspalanthium Desv. = Paspalum L. (Gramin.).

Paspalidium Stapf. Gramineae. 20 warm, esp. Old World.

Paspalum L. Gramineae. 250 warm lands. In temp. Am. they form a large proportion of the pasture of the campos, pampas, etc. Good fodder. *P. scrobiculatum* L. (Kodo millet) cult. in India.

Paspalus Fluegge = praec.

Passaea Adans. (~ Ononis L.). Leguminosae. 1 Medit.

Passaea Baill. = Bernardia Mill. (Euphorbiac.).

Passalia Soland. ex R.Br. = Rinorea Aubl. (Violac.).

Passaveria Mart. & Eichl. = Ecclinusa Mart. (Sapotac.).

Passerina L. Thymelaeaceae. 15 S. Afr. G 1-loc. C o.

Passiflora L. Passifloraceae. 500 chiefly Am.; a few in As. & Austr., 1 in Madag. Climbing pl. with axillary tendrils. Some have curious bilobed ls. (crescentic or swallow-tailed in shape), the centre lobe not developing. At the base of the leaf-stalk there are usu. extra-floral nectaries. The fls. spring from the same axils as the tendrils, sol. or in small cymes; the bract is usu. 'adnate' to the peduncle. The recept. is hollowed into a cup, bearing on its margin 5 sepals, 5 petals, and a number of effigurations of the axis—thread-like petaloid bodies, forming a dense mass (the corona) round the central androphore, at whose apex is borne the ovary. Five sta. spring from the androphore at the base of the ovary, and are bent downwards at first; afterwards the styles bend down also. Honey is secreted at the base of the androphore. Fr. a berry; seed enveloped in a fleshy aril. Several have ed. fr., e.g. *P. quadrangularis* L., the granadilla (trop. Am.), *P. maliformis* L., the sweet calabash (W.I.), *P. laurifolia* L., the water-lemon, *P. edulis* Sims (passion fruit), etc.

***Passifloraceae** Juss. Dicots. 12/600 trop. & warm temp. Shrubs and herbs, mostly climbers with axillary tendrils, and with alt. stip. ls. Fls. ♀ or unisexual, reg. Recept. of various shapes, often hollowed and frequently with a central andro- or gynophore; usu. term. by outgrowths, often of petaloid or staminodial appearance, forming the *corona*. K 3–5, imbr.; C 3–5, imbr. or open, or o; A 3–5(–10), anth. versat.; G̲ (3), 1-loc. with parietal plac. and several or ∞ anatr. ov.; style 1, simple or branched, or 3–5 separate styles. Caps. or berry. Seed with fleshy aril and endosp. Chief genera: *Passiflora, Adenia, Tryphostemma, Deïdamia.*

Passoura Aubl. = Rinorea Aubl. (Violac.).

Passovia Karst. ex Klotzsch = Loranthus Linn. (Loranthac.).

Passowia Karst. = praec.

Pastinaca L. Umbelliferae. 15 temp. Euras. *P. sativa* L. is the parsnip, a biennial, often cult. for ed. root.

Pastinacopsis Golosk. Umbelliferae. 1 C. As.

Pastoraea Tod. = seq.

Pastorea Tod. ex Bertol. (~ Bivonaea DC.). Cruciferae. 4 W. & C. Medit.

Patabea Aubl. = Ixora L. (Rubiac.).

Patagnana Steud. = Petagnana J. F. Gmel. = Smithia Ait. (Legumin.).

Patagonia Dur. & Jacks. (sphalm.) = Patagonium Schrank = Adesmia DC. (Legumin.).

Patagonica Boehm. = Patagonula L. (Boraginac.).

Patagonium E. Mey. = Aeschynomene L. (Legumin.).

PATAGONIUM

Patagonium Schrank = Adesmia DC. (Legumin.).
Patagonula L. Ehretiaceae. 2 Brazil, Argentina. Good timber.
Patagua Poepp. ex Baill. = Orites R.Br. (Proteac.).
Patagua Poepp. ex Reiche = Villaresia Ruiz & Pav. (Icacinac.).
Patamogeton Honck. (sphalm.) = Potamogeton L. (Potamogetonac.).
Patania Presl = Dennstaedtia Bernh. (Dennstaedtiac.).
Patascoya Urb. Theaceae. 1 Colombia.
***Patersonia** R.Br. Iridaceae. 3 mts. Philipp. Is., Borneo, New Guinea, 17 Austr., Tasmania.
Patersonia Poir. = Pattersonia J. F. Gmel. = Ruellia L. (Acanthac.).
Pathersonia Poir. = praec.
Patientia Rafin. = Rumex L. (Polygonac.).
Patima Aubl. Rubiaceae. 2 Guiana.
Patinoa Cuatrec. Bombacaceae. 2 trop. S. Am.
Patis Ohwi (~ Stipa L.). Gramineae. 1 Korea.
Patisna Jack ex Burkill = Urophyllum Wall. (Rubiac.).
Patmaceae Schultz-Schultzenst. = Rafflesiaceae Dum. [Rafflesia patma Bl.].
Patonia Wight = Xylopia L. (Annonac.).
Patosia Buchen. Juncaceae. 2 Chile, Argent.
***Patrinia** Juss. Valerianaceae. 20 C. As. & Himal. to E. As.
Patrinia Rafin. = Vexibia Rafin. = Sophora L. (Legumin.).
Patrisia Rich. = Ryania Vahl (Flacourtiac.).
Patrisia Rohr ex Steud. = Dichapetalum Thou. (Dichapetalac.).
Patrisiaceae Mart. = Flacourtiaceae–Casearieae Reichb.
Patrocles Salisb. = Narcissus L. (Amaryllidac.).
Patsjotti Adans. = Strumpfia Jacq. (Rubiac.).
Pattalias W. Wats. Asclepiadaceae. 2 SW. U.S., Mex.
Pattara Adans. = Embelia Burm. f. (Myrsinac.).
Pattersonia J. F. Gmel. = Ruellia L. (Acanthac.).
Pattonia Wight = Grammatophyllum Blume (Orchidac.).
× **Pattoniheadia** hort. Orchidaceae. Gen. hybr. (vi; nomen nugax) (Bromheadia × Pattonia).
Patulix Rafin. = Clerodendrum L. (Verbenac.).
Patya Neck. = Verbena L. (Verbenac.).
Paua Caballero. Compositae. 1 Morocco.
Paua Gandog. = Torilis Adans. (Umbellif.).
Pauella Ramamurthy & Sebastine = Theriophonum Bl. (Arac.).
Pauia Deb & Dutta. Solanaceae. 1 India.
Pauladolphia Börner = Acetosella (Meissn.) Fourr. (Polygonac.).
Pauldopia v. Steenis. Bignoniaceae. 1 NE. India, SW. China, SE. As.
Pauletia Cav. = Bauhinia L. (Legumin.).
Paulinia Th. Dur. (sphalm.) = Roulinia Decne = Rouliniella Vail (Asclepiadac.).
Paullinia L. Sapindaceae. 180 warm Am. Lianes with watch-spring tendrils. Caps. often winged. *P. cupana* Kunth (*guarana*) cult. in Brazil; seeds used like cacao.
Paulomagnusia Kuntze = Micranthus Eckl. (Iridac.).
Paulo-Wilhelmia Hochst. (Jan. 1844) = Mollugo L. (Aïzoac.).
Paulo-Wilhelmia Hochst. (late 1844) = Eremomastax Lindau (Acanthac.).

Paulownia Sieb. & Zucc. Scrophulariaceae. 17 E. As. Trees (rare in *Scroph.*). Extremely close to *Catalpa* Scop. (*Bignoniac.*), only differing in chars. of G (plac., style, stig., seeds). Anat. chars. identical. The two genera evidently stand close to the common stock of the two fams.

Paulowniaceae Nak. = Scrophulariaceae–Cheloneae Benth.

Paulseniella Briquet. Labiatae. 1 C. As. (Pamirs).

Pauridia Harv. Hypoxidaceae. 1 S. Afr.

Pauridiantha Hook. f. Rubiaceae. 25 trop. Afr.

Paurolepis S. Moore. Compositae. 1 S. trop. Afr.

Paurotis O. F. Cook = Acoelorrhaphe H. Wendl. (Palm.).

Pausandra Radlk. Euphorbiaceae. 12 C. & trop. S. Am.

Pausia Rafin. (1836) = Thymelaea Endl. (Thymelaeac.).

Pausia Rafin. (1838) = Cartrema Rafin. = Osmanthus Lour. (Oleac.).

Pausinystalia Pierre ex Beille. Rubiaceae. 13 trop. W. Afr.

Pautsauvia Juss. = Alangium Lam. (Alangiac.).

Pavate Adans. = seq.

Pavetta L. Rubiaceae. 400 palaeotrop. The ls. of many have little warts inhabited by bacterial colonies.

Pavia Mill. = Aesculus L. (Hippocastanac.).

Paviaceae Horan. = Hippocastanaceae DC.

Paviana Rafin. = Pavia Mill. = Aesculus L. (Hippocastanac.).

Pavieasia Pierre. Sapindaceae. 1 Indoch.

Pavinda Thunb. ex Bartl. = Audouinia Brongn. (Bruniac.).

*Pavonia Cav. Malvaceae. 200 trop. & subtrop. There are 5 cpls. and 10 styles, 5 of these corresponding to cpls. which abort in development. The cpls. are hooked in fr.

Pavonia Domb. ex Lam. = Cordia L. (Ehretiac.).

Pavonia Ruiz & Pav. = Laurelia Juss. (Monimiac.).

Pawia Kuntze = Pavia Mill. = Aesculus L. (Hippocastanac.).

Paxia Gilg. Connaraceae. 9 trop. W. Afr.

Paxia Herter = Paxiuscula Herter (Euphorbiac.).

Paxia Ö. Nilss. (non Gilg 1892, nec Herter 1931!) = Neopaxia Ö. Nilss. (Portulacac.).

Paxiactes Rafin. = Heracleum L. (Umbellif.).

Paxiodendron Engl. = Xymalos Baill. (Trimeniac.).

Paxistima Rafin. Celastraceae. 5 N. Am. G semi-inf.

Paxiuscula Herter. Euphorbiaceae. 4 S. Brazil, Urug.

Paxtonia Lindl. = Spathoglottis Bl. (Orchidac.).

Payanelia C. B. Clarke = Pajanelia DC. (Bignoniac.).

Payena A. DC. Sapotaceae. 16 SE. As., W. Malaysia. *P. leerii* (Teysm. & Binnend.) Kurz yields a good gutta-percha (see *Palaquium*), known as *gutta sundek*.

*Payera Baill. Rubiaceae. 1 Madag.

Payeria Baill. = Quivisia Comm. ex Juss. = Turraea L. (Meliac.).

Paypayrola Aubl. Violaceae. 7 trop. S. Am. Tree. Stam. tube.

Payrola Juss. = praec.

Pearcea Regel. Gesneriaceae. 2 Ecuador.

Pearsonia Dümmer. Leguminosae. 12 S. Afr.

PEAUTIA

Peautia Comm. ex Pfeiff. = Hydrangea L. (Hydrangeac.).
Peccana Rafin. = Euphorbia L. (Euphorbiac.).
Pechea Pourr. ex Kunth = Crypsis Ait. (Gramin.).
Pechea Steud. (sphalm.) = Peckia Vell. = Cybianthus Mart. (Myrsinac.).
Pecheya Scop. = Coussarea Aubl. (Rubiac.).
Pechuelia Kuntze = Selago L. (Scrophulariac.).
Pechuel-Loeschea O. Hoffm. Compositae. 2 SW. Afr.
Peckelia Hutch. = Peekelia Harms (Legumin.).
Peckeya Rafin. = Pecheya Scop. = Coussarea Aubl. (Rubiac.).
Peckia Vell. = Cybianthus Mart. (Myrsinac.).
Peckoltia Fourn. Asclepiadaceae. 1 Brazil.
× **Pectabenaria** hort. Orchidaceae. Gen. hybr. (iii) (Habenaria × Pecteilis).
Pectangis Thou. (uninom.) = Angraecum pectinatum Thou. (Orchidac.).
Pectanisia Rafin. = Reseda L. (Resedac.).
Pectantia Rafin. = Pectiantia Rafin. = Mitella L. (Saxifragac.).
Pecteilis Rafin. Orchidaceae. 4 E. As., Indomal.
Pecten Lam. = Scandix L. (Umbellif.).
Pectianthia Rydb. = seq.
Pectiantia Rafin. = Mitella L. (Saxifragac.).
Pectidium Less. = Pectis L. (Compos.).
Pectidopsis DC. = praec.
Pectinaria Bernh. = Scandix L. (Umbellif.).
Pectinaria Cordem. = Angraecum Bory (Orchidac.).
Pectinaria Hack. = Eremochloa Buese (Gramin.).
*****Pectinaria** Haw. Asclepiadaceae. 7 S. Afr.
Pectinastrum Cass. = Centaurea L. (Compos.).
Pectinea Gaertn. = Erythrospermum Lam. (Flacourtiac.).
Pectinella J. M. Black = Amphibolis C. Agardh (Cymodoceac.).
Pectis L. Compositae. 70 S. U.S. to Brazil, W.I., Galáp.
Pectocarya DC. ex Meissn. Boraginaceae. 10 Pacif. Am.
Pectophyllum Reichb. = seq.
Pectophytum Kunth = Azorella Lam. (Hydrocotylac.).
*****Pedaliaceae** R.Br. Dicots. 12/50 trop. & S. Afr., Madag., Indomal. Mostly shore and desert plants. Herbs or rarely shrubs with opp. ls. and glandular hairs. Fls. sol. or in cymes (usu. 3-flowered), with glands (metamorphosed fls.) at the base of the stalks, ⚥, zygo. K (5); C (5); A 4, didynamous, with a post. std.; G (2), with long style and 2 stigmas, 2–4-loc. or apparently 1-loc., often with false septa; ovules 1–∞ per loc., on axile plac. Caps. or nut, often with hooks. Embryo straight; endosp. thin. *Sesamum* is economically important. Chief genera: *Pedalium, Sesamum, Pterodiscus, Uncarina, Josephinia*. Near *Martyniaceae*. The chief distinctions lie in the placentation, fruit, calyx, and glandular hairs.
Pedaliodiscus Ihlenf. Pedaliaceae. 1 trop. E. Afr.
Pedaliophyton Engl. = Pterodiscus Hook. (Pedaliac.).
Pedalium Adans. = Atraphaxis L. (Polygonac.).
Pedalium Royen ex L. Pedaliaceae. 1 trop. Afr., Madag., trop. As.
Pedastis Rafin. = Cayratia Juss. (Vitidac.).
Peddiea Harv. Thymelaeaceae. 20 trop. & S. Afr.

862

Pederia Nor. = Paederia L. (Rubiac.).
Pederlea Rafin. = Acnistus Schott (Solanac.).
Pederota Scop. = Paederota L. = Veronica L. (Scrophulariac.).
Pedicellaria Schrank = Gynandropsis DC. (Cleomac.).
Pedicellia Lour. = Mischocarpus Bl. (Sapindac.).
Pedicularidaceae ('-lares') Juss. = Scrophulariaceae–Rhinantheae Spreng.
Pedicularis L. Scrophulariaceae. 500 N. hemisph., esp. on mts. of C. &
  E. As. Semi-parasites with loose-pollen fls., fert. by humble-bees, etc.
*****Pedilanthus** Neck. ex Poit. Euphorbiaceae. 14 Florida & Mex. to trop.
  S. Am., W.I.
Pedilea Lindl. = Malaxis Soland. ex Sw. (Orchidac.).
Pedilochilus Schlechter. Orchidaceae. 18 New Guinea, Solomon Is.
Pedilonia Presl = Wachendorfia L. (Hypoxidac.).
Pedilonum Blume = Dendrobium Sw. (Orchidac.).
Pedina Stev. = Astragalus L. (Legumin.).
Pedinogyne A. Brand. Boraginaceae. 1 E. Himal.
Pedinopetalum Urb. & H. Wolff. Umbelliferae. 1 S. Domingo.
Pediocactus Britton & Rose. Cactaceae. 6–7 W. U.S.
Pediomelum Rydb. (~ Psoralea L.). Leguminosae. 25 S. U.S., Mex.
Pedochelus Wight (sphalm.) = Podochilus Blume (Orchidac.).
Pedrosia Lowe = Lotus L. (Legumin.).
Peekelia Harms. Leguminosae. 1 New Guinea.
Peekeliodendron Sleum. = Merrilliodendron Kanehira (Icacinac.).
Peekeliopanax Harms. Araliaceae. 1 New Guinea, Bismarck Archip., Solomon
  Is.
Peersia L. Bolus. Aïzoaceae. 2 S. Afr.
Pegaeophyton Hayek & Hand.-Mazz. Cruciferae. 2 Himal. to W. China.
Pegamea Vitm. (sphalm.) = Pagamea Aubl. (Rubiac.).
Peganaceae Van Tiegh. = Zygophyllaceae–Peganoïdeae Engl.
Peganon St-Lag. = seq.
Peganum L. Zygophyllaceae. 5–6 Medit. to Mongolia, S. U.S., Mex. The
  seeds of *P. harmala* L. yield 'Turkey red'.
Pegesia Steud. = Pagesia Rafin. (Scrophulariac.).
Pegia Colebr. Anacardiaceae. 4 E. Himalaya to S. China & Philipp. Is.
Peglera Bolus = Nectaropetalum Engl. (Erythroxylac.).
Pegolettia Cass. Compositae. 12 N., trop., & S. Afr., Madag.
Pegolletia Less. = praec.
Pehria Sprague. Lythraceae. 1 Colombia, Venez.
Peiranisia Rafin. = Cassia L. (Legumin.).
Peirescia Zucc. = Pereskia Mill. (Cactac.).
Peireskia P. & K. = Pereskia Vell. = Hippocratea L. (Celastrac.).
Peireskia Steud. = Pereskia Mill. (Cactac.).
Peireskiopsis Vaupel = Pereskiopsis Britton & Rose (Cactac.).
Peixotoa A. Juss. Malpighiaceae. 11 Brazil.
Pekea Aubl. = Caryocar L. (Caryocarac.).
Pekia Steud. = praec.
Pelae Adans. = Xanthophyllum Roxb. (Xanthophyllac.).
Pelagatia O. E. Schulz. Cruciferae. 1 Peru.

PELAGONDENDRON

Pelagodendron Seem. = Randia L. (Rubiac.).

Pelagodoxa Becc. Palmae. 2 Marquesas.

Pelaphia Banks & Soland. ex A. Cunn. = Coprosma J. R. & G. Forst. (Rubiac.).

Pelaphoïdes Banks & Soland. ex Cheesem. = praec.

Pelargonion St-Lag. = seq.

Pelargonium L'Hérit. Geraniaceae. 250 trop. & esp. S. Afr.; 1 each Canaries, St Helena, Tristan, E. Medit., S. Arabia, S. India, Austr., N.Z. The so-called 'geranium' of greenhouses, etc. In many the base of the stem is tuberous. An oil, used as a substitute for otto of roses, is distilled in Algeria from *P. odoratissimum* Ait. The partial infl. (unlike that of *Geranium* and *Erodium*, *qq.v.*) is a dichasial umbel.

Pelatantheria Ridl. Orchidaceae. 8 India to Formosa, Malay Penins.

Pelea A. Gray. Rutaceae. 75 Pacif., mostly Hawaii.

Pelecinus Mill. = Biserrula L. (Legumin.).

Pelecostemon Leonard. Acanthaceae. 1 Colombia.

Pelecynthis E. Mey. = Rafnia Thunb. (Legumin.).

Pelecyphora Ehrenb. Cactaceae. 2 Mex.

*****Pelexia** Poit. ex Lindl. Orchidaceae. 40 Mex. to temp. S. Am., W.I.

Pelianthus E. Mey. ex Moq. = Hermbstaedtia Reichb. (Amaranthac.).

Pelidnia Barnh. = Utricularia L. (Lentibulariac.).

Peliosanth[ac]eae Salisb. = Liliaceae–Peliosantheae Reichb.

Peliosanthes Andr. Liliaceae. 15 Himal. to Formosa & Java.

Peliostomum E. Mey. Scrophulariaceae. 7 trop. & S. Afr.

Peliotes Harv. & Sond. = seq.

Peliotis E. Mey. = Lonchostoma Wikstr. (Thymelaeac.).

Peliotus E. Mey. = praec.

Pella Gaertn. = Ficus L. (Morac.).

Pellacalyx Korth. Rhizophoraceae. 7–8 SE. As., W. Malaysia, N. Celebes.

*****Pellaea** Link. Sinopteridaceae. 80 trop. & subtrop.

Pellaeopsis J. Sm. = praec.

Pellea André = Pellionia Gaudich. (Urticac.).

Pellegrinia Sleum. (emend. A. C. Sm.). Ericaceae. 5 Andes.

Pellegriniodendron J. Léonard. Leguminosae. 1 trop. Afr.

Pelleteria Poir. = seq.

Pelletiera A. St-Hil. Primulaceae. 1 extratrop. S. Am.

Pelliceria Planch. & Triana ex Benth. & Hook. f. corr. Pl. & Tr. Pelliceriaceae. 1 Costa Rica to Ecuador.

Pelliceriaceae (Planch. & Triana) Beauvis. Dicots. 1/1 trop. Am. Mangrove trees, with buttresses but not stilt-roots. Ls. alt., simple, ent., decurrent, slightly asymm. (as *Bonnetiac.* and *Foetidiac.*), fringed with decid. gland. teeth. Fls. reg., ☿, sol., axill., showy, with 2 long sheathing bracteoles. K 5, short, much imbr.; C 5, elong.-obl., imbr., spreading widely at anthesis; A 5, alternipet., with subul. fil. adpr. to grooves of G, and elong.-lin. basifixed sagitt. latero-extr. anth.; G (2), conic-cylindric, 10-grooved, atten. into smooth style with small bifid stig., and 1 axile ov. per loc., the loc. sometimes confluent above. Fr. large (to 13 cm. diam.), turbinate or onion-shaped, many-ridged, pustulate, 1-seeded, indehisc. Ripe seed consisting of a large naked embr. without endosp. or testa; cotyledons ± hemisph., filling cavity of fr., and en-

closing the conspic. red plumule, the radicle penetrating the beak of fr. (base
of style), through which germination takes place. Only genus: *Pelliceria*. An
interesting plant, clearly related to *Bonnetiac.*, *Foetidiac.*, *Marcgraviac.*, etc.,
but with the mangrove habit of *Rhizophorac.*
Pelliciera Planch. & Triana ex Benth. & Hook. f. (sphalm.) = Pelliceria Planch.
& Triana (Pelliceriac.).
Pellinia Molina = Eucryphia Cav. (Eucryphiac.).
*Pellionia Gaudich. Urticaceae. 50 trop. & E. As., Polynes. *P. umbellata*
Wedd. (Japan) has the brs. of the ♂ fls. united to form an invol.
Pellocalyx P. & K. = Pellacalyx Korth. (Rhizophorac.).
Pelma Finet = Bulbophyllum Thou. (Orchidac.).
Pelonastes Hook. f. = Myriophyllum L. (Haloragidac.).
Peloria Adans. = Linaria Mill. (Scrophulariac.).
Pelotris Rafin. = Muscari Mill. (Liliac.).
Pelozia Rose. Onagraceae. 2 Mex.
Pelta Dulac = Zannichellia L. (Zannichelliac.).
Peltactila Rafin. = Daucus L. (Umbellif.).
Peltaea (Gürke) Standley (~ Pavonia Cav.). Malvaceae. 4 trop. Am., W.I.
*Peltandra Rafin. Araceae. 4 S. U.S. & Atl. N. Am
Peltandra Wight = Phyllanthus L. (Euphorbiac.).
*Peltanthera Benth. Buddlejaceae. 2 C Am., Peru.
Peltanthera Roth = Vallaris Burm. f. (Apocynac.).
Peltanatheria P. & K. = Pelatantheria Ridl. (Orchidac.).
Peltapteris Link. Lomariopsidaceae. 3 trop. Am. (Morton, *Amer. Fern Journ.*
45: 11–14, 1955).
Peltaria DC. = Wiborgia Thunb. (Legumin.).
Peltaria Jacq. Cruciferae. 7 E. Medit. to C. As. & Persia.
Peltariopsis (Boiss.) N. Busch. Cruciferae. 3 Cauc., N. Persia.
Peltastes R. E. Woodson. Apocynaceae. 7 C. & trop. S. Am.
Pelticalyx Griff. = ? Desmos Lour. (Annonac.).
Peltidium Zollikofer = Chondrilla L. (Compos.).
Peltimela Rafin. = Glossostigma Arn. (Scrophulariac.).
Peltiphyllum Engl. Saxifragaceae. 1 Pacif. U.S.
Peltispermum Moq. = Anthochlamys Fenzl ex Endl. (Chenopodiac.).
Peltoboykinia (Engl.) Hara (~ Boykinia Nutt.). Saxifragaceae. 1 Japan.
Peltobractea Rusby. Malvaceae. 1 Bolivia.
Peltobryon Klotzsch = Piper L. (Piperac.).
Peltocalyx P. & K. = Pelticalyx Griff. = ? Desmos Lour. (Annonac.).
Peltocarpus Zipp. (nomen). Quid? New Guinea.
Peltochlaena Fée = Stigmatopteris C. Chr., § Peltochlaena (Fée) (Aspidiac.).
Peltodon Pohl. Labiatae. 6 Brazil, Paraguay.
Peltogyne Vogel. Leguminosae. 25 trop. S. Am. Timber, dye.
Peltomesa Rafin. = Struthanthus Mart. (Loranthac.).
Peltophora Benth. & Hook. f. (sphalm.) = Peltophorus Desv. = Manisuris L.
(Gramin.).
Peltophoropsis Chiov. Leguminosae. 1 NE. trop. Afr.
*Peltophorum (Vogel) Benth. Leguminosae. 12 trop.
Peltophorus Desv. = Manisuris L. (Gramin.).

Peltophyllum Gardn. = Triuris Miers (Triuridac.).
Peltopsis Rafin. = Potamogeton L. (Potamogetonac.).
Peltospermum Benth. = Sacosperma G. Tayl. (Rubiac.).
Peltospermum DC. = Aspidosperma Mart. & Zucc. (Apocynac.).
Peltospermum P. & K. = Peltispermum Moq. = Anthochlamys Fenzl ex Endl. (Chenopodiac.).
Peltostegia Turcz. = Kosteletzkya C. Presl (Malvac.).
Peltostigma Walp. Rutaceae. 3 Mex., C. Am., W.I.
Pelucha S. Wats. Compositae. 1 Lower Calif.
Pemphis J. R. & G. Forst. Lythraceae. 2 trop. Afr. & Madag. to Pacific. *P. acidula* J. R. & G. Forst., on palaeotrop. coasts, esp. on beaches that are washing away. *P. madagascariensis* (Baker) Koehne, mts. SW. Madag.
Penaea L. Penaeaceae. 6 S. Afr.
***Penaeaceae** Guillemin. Dicots. 5/25 SW. S. Afr. Shrubby xero. of ericoid habit, with opp. ent. evergr. ls., stip. minute or o. Fls. reg., ♀, axill., sol. or paired, the br. often coloured. Recept. hollow, tubular. K (4), valv., persist.; C o; A 4, alternisep., fil. short, inserted in throat of K, anth. intr.; disk o; G̲ (4), 4-loc., with 2–4 anatr. ov. per loc. (ascend. or descend. or both); style simple, with 4-lobed stig. Fr. a loculic. caps., often 4-seeded; endosp. o. Genera: *Penaea, Saltera, Brachysiphon, Endonema, Glischrocolla.* Prob. related to *Lythraceae.*
Penanthes Vell. (sphalm.) = Prenanthes L. (Compos.).
Penar-valli Adans. = Zanonia L. (Cucurbitac.).
Penarvallia P. & K. = praec.
Pendiphylis Thou. (uninom.) = Bulbophyllum pendulum Thou. (Orchidac.).
Pendulina Willk. = Diplotaxis DC. (Crucif.).
Penelopeia Urb. Cucurbitaceae. 1 Hispaniola.
Penianthus Miers. Menispermaceae. 6 trop. W. Afr.
Penicillanthemum Vieill. = Hugonia L. (Linac.).
Penicillaria Willd. = Pennisetum Rich. (Gramin.).
Peniculifera Ridley = Trigonopleura Hook. f. (Euphorbiac.).
Peniculus Swallen = Mesosetum Steud. (Gramin.).
Peniocereus (A. Berger) Britton & Rose. Cactaceae. 7 SW. U.S., Mex.
Peniophyllum Pennell (~ Oenothera L.). Onagraceae. 1 S. U.S.
Pennantia J. R. & G. Forst. Icacinaceae. 4 Austr., N.Z., Norfolk I.
Pennanti[ac]eae J. G. Agardh = Icacinaceae Miers.
Pennellia Nieuwl. Cruciferae. 5 C. & S. Am.
Pennellianthus Crosswh. & Kawano (~ Penstemon Schmid.). Scrophulariaceae. 1 Japan.
Pennilabium J. J. Smith. Orchidaceae. 5 Siam, Malay Penins., Java.
Pennisetum Rich. Gramineae. 130 warm. Involucre as in *Cenchrus. P. typhoïdeum* Rich., the pearl, spiked, or bulrush millet, is extensively cult. in India.
Penslemon Rafin. = seq.
Penstemon Schmid. Scrophulariaceae. 1 NE. As., 250 N. Am. (esp. W. U.S.), 1 C. Am. Post. sta. repres. by a large std. which is bent down to the lower side of the C (cf. *Scrophularia*).
Penstemum Rafin. = praec.
Pentabothra Hook. f. Asclepiadaceae. 1 Assam.

PENTADIPLANDRACEAE

Pentabrachion Muell. Arg. = Microdesmis Planch. (Pandac.).
Pentabrachium Muell. Arg. = praec.
Pentacaelium Franch. & Sav. = Pentacoelium Sieb. & Zucc. = Myoporum Banks & Soland. ex Forst. f. (Myoporac.).
Pentacaena Bartl. = Cardionema DC. (Caryophyllac.).
Pentacalia Cass. = Senecio L. (Compos.).
Pentacarpaea Hiern. Rubiaceae. 1 trop. Afr.
Pentacarpus P. & K. = praec.
Pentacarya DC. ex Meissn. = Heliotropium L. (Boraginac.).
Pentace Hassk. Tiliaceae. 25 SE. As., W. Malaysia.
***Pentaceras** Hook. f. Rutaceae. 1 E. Austr.
Pentaceras Roem. & Schult. = seq.
Pentaceros G. F. W. Mey. = Byttneria Loefl. (Sterculiac.).
Pentachaeta Nutt. = Chaetopappa DC. (Compos.).
Pentachlaena Perrier. Sarcolaenaceae. 1 Madag.
Pentachondra R.Br. Epacridaceae. 4 Victoria, Tasm., N.Z.
Pentaclathra Endl. = Polyclathra Bertol. (Cucurbitac.).
Pentaclethra Benth. Leguminosae. 3 trop. Am., Afr.
Pentacme A. DC. Dipterocarpaceae. 3 SE. As., Philipp. Is.
Pentacnida P. & K. = Pentocnide Rafin. = Pouzolzia Gaudich. (Urticac.).
Pentacocca Turcz. = Ochthocosmus Benth. (Ixonanthac.).
Pentacoelium Sieb. & Zucc. = Myoporum Banks & Soland. ex Forst. f. (Myoporac.).
Pentacraspedon Steud. = Amphipogon R.Br. (Gramin.).
Pentacrophys A. Gray = Acleisanthes A. Gray (Nyctaginac.).
Pentacrostigma K. Afzel. Convolvulaceae. 1 Madag.
Pentacrypta Lehm. = Arracacia Bancr. (Umbellif.).
Pentactina Nakai. Rosaceae. 1 Korea.
Pentacyphus Schlechter. Asclepiadaceae. 1 Peru.
Pentadactylon Gaertn. f. = Persoonia Sm. (Proteac.).
Pentadenia Hanst. = Columnea L. (Gesneriac.).
Pentadesma Sabine. Guttiferae. 4 trop. Afr., Seychelles. Incl. *P. butyracea* Sabine, the tallow or butter tree. The fr. yields a greasy juice used as butter.
Pentadesmos Spruce ex Planch. & Triana = Moronobea Aubl. (Guttif.).
Pentadiplandra Baill. (1886). Pentadiplandraceae. 1–2 trop. Afr.
Pentadiplandra (Baill. 1879 pro sect.) Kuntze = Dizygotheca N. E. Br. (Araliac.).
Pentadiplandraceae Hutch. & Dalz. Dicots. 1/2 Afr. Large shrubs or climbers; ls. alt., simple, exstip. Fls. in abbrev. axill. rac., on rel. long pedic., polygam., opening at an early stage and enlarging after opening. K 5, valv.; C 5, imbr., loosely connivent or coherent at thickened scale-like base or claw, limb free, lanc., thin, acuminate; A 9–13, inserted within a thick fleshy cupular disk, fil. filif., anth. small, basifixed; G̲ (3–5), shortly stipit., ovules several, axile, 2–3-seriate in each loc., style 1, elongate, 5-lobed at apex (10 filiform staminodes in ♀ fls., pistillode with abort. ov. and no style in ♂ fls.). Fr. a globose brownish-scaly berry, with numerous small seeds. Only genus: *Pentadiplandra*. Probably related to *Capparidac*.

Pentadynamis R.Br. Leguminosae. 1 S. Austr.

Pentaglossum Forsk. = Lythrum L. (Lythrac.).

Pentaglottis Tausch (~ Anchusa L.). Boraginaceae. 1 SW. Eur.

Pentaglottis Wall. = Melhania Forsk. (Sterculiac.).

Pentagonanthus Bullock. Periplocaceae. 2 trop. Afr.

Pentagonaster Klotzsch = Kunzea Reichb. (Myrtac.).

*Pentagonia Benth. Rubiaceae. 20 trop. Am.

*Pentagonia Fabr. (uninom.) = Nicandra Adans. sp. (Solanac.).

Pentagonia Kuntze = Specularia Fabr. (Campanulac.).

Pentagonium Schau. = Philibertia Kunth (Asclepiadac.).

Pentagonocarpus Parl. = Kosteletzkya Presl (Malvac.).

Pentake Rafin. = Cuscuta L. (Cuscutac.).

Pentalepis F. Muell. = Chrysogonum L. (Compos.).

Pentalinon Voigt = Prestonia R.Br. (Apocynac.).

Pentaloba Lour. = Rinorea Aubl. (Violac.).

Pentaloncha Hook. f. Rubiaceae. 3 trop. W. Afr.

Pentalophus A. DC. = Lithospermum L. (Boraginac.).

Pentamera Willis (sphalm.) = Pentanura Bl. (Periplocac.).

Pentamerea Baill. = seq.

Pentameria Klotzsch ex Baill. = Bridelia Willd. (Euphorbiac.).

Pentameris Beauv. Gramineae. 5–8 S. Afr.

Pentameris E. Mey. = Pavonia Cav. (Malvac.).

Pentamerista Maguire. Bonnetiaceae. Spp.? Venez.

Pentamorpha Scheidw. = Erythrochiton Nees & Mart. (Rutac.).

Pentanema Cass. (~ Vicoa Cass.). Compositae. 2 Afr.; 12 Medit. to C. As. & Himal.; 1 Java.

Pentanisia Harv. Rubiaceae. 16 trop. Afr.

Pentanome Moç. & Sessé ex DC. = Zanthoxylum L. (Rutac.).

Pentanopsis Rendle. Rubiaceae. 1 Somaliland.

Pentanthus Hook. & Arn. = Paracalia Cuatrec. (Compos.).

Pentanthus Less. = Nassauvia Comm. ex Juss. (Compos.).

Pentanthus Rafin. = Jacquemontia Choisy (Convolvulac.).

Pentanura Blume. Periplocaceae. 2 Burma, Sumatra.

Pentapanax Seem. Araliaceae. 15 Himal. to Formosa, Java, Queensl., S. Am.

Pentapeltis Bunge = Xanthosia Rudge (Umbellif.).

Pentapera Klotzsch (~ Erica L.). Ericaceae. 1 Sicily.

Pentapetes L. Sterculiaceae (Malvaceae?). 1 Indomal.

Pentapetiopsis auctt. = Pentopetiopsis Costant. & Gall. (Periplocac.).

Pentaphalangium Warb. Guttiferae. 7 Caroline Is., Moluccas, New Guinea, Solomon Is.

Pentaphiltrum Reichb. = Physalis L. (Solanac.).

Pentaphorus D. Don = Gochnatia Kunth (Compos.).

Pentaphragma Wall. ex G. Don. Pentaphragmataceae. 30 S. China, Indoch., Malaysia.

Pentaphragma Zucc. ex Reichb. = Araujia Brot. (Asclepiadac.).

*Pentaphragmataceae J. G. Agardh. Dicots. 1/30 SE. As., Malaysia. Perenn. ± succul. herbs, often with multicell. hairs. Ls. alt., simple, usu. ± asymm., sinuate-dentic. or subent., exstip. Fls. reg., ♀ or rarely ♂ ♀, in

axill., often dense and scorpioid, acropetal cymes, bracts ± conspic., membr.
K 5, imbr., unequal, persist.; C (5) or 5 (rarely 4), valv., usu. fleshy or cartilag.
(rarely delicate), persist.; A 5, alternipet., fil. persist.(!), anth. basifixed, intr.;
G̲ (2) (actually only adnate to K tube by means of longit. septa formed by
continuation of fil., leaving 5 intervening nectarif. pits), with short thick style
and massive stig., and ∞ pend. anatr. ov. on bifid axile plac. Fr. baccate,
indehisc.; seeds minute, with copious endosp. Only genus: *Pentaphragma*.
Anat. and asymm. ls. suggest affinity with *Begoniac.*; supposed connection
with *Campanulac.* v. doubtful.
*__Pentaphylacaceae__ Engl. Dicots. 1/2 SE. As. Shrubs or trees with alt. ent.
exstip. ls., and small reg. ♂ fls. arranged pseudo-racemosely on twigs which
often continue as leafy shoots. K 5, imbr.; C 5, imbr., thickish, coherent at
base with sta.; A 5, alternipet., fil. thickish, flattened, anth.-cells 2, free, with
apiculate term. pore; disk o; G̲ (5), with 2 collat. pend. ov. per loc., and simple
persist. style with minutely 5-lobed stig. Caps. loculic.; seeds with slight
endosp. and arcuate embr. Only genus: *Pentaphylax*. Related to *Theac.*
Pentaphylax Gardn. & Champ. Pentaphylacaceae. 1–2 S. China to Malay
Penins. & Sumatra.
Pentaphylloïdes Duham. = Potentilla L. (Rosac.).
Pentaphyllon Pers. = Trifolium L. (Legumin.).
Pentaphyllum Hill = Potentilla L. (Rosac.).
Pentaplaris L. O. Williams & Standley. Tiliaceae. 1 C. Am.
Pentaple Reichb. = Cerastium L. (Caryophyllac.).
Pentapleura Hand.-Mazz. Labiatae. 1 Kurdistan.
Pentapogon R.Br. Gramineae. 1 Victoria, Tasmania.
Pentaptelion Turcz. = Leucopogon R.Br. (Epacridac.).
Pentaptera Roxb. = Terminalia L. (Combretac.).
Pentapteris Haller = seq.
Pentapterophyllon Hill = Myriophyllum L. (Haloragidac.).
Pentapterophyllum Fabr. = praec.
Pentapterygium Klotzsch = Agapetes D. Don ex G. Don (Ericac.).
Pentaptilon Pritzel. Goodeniaceae. 1 W. Austr.
Pentaptychaceae Dulac = Plumbaginaceae Juss.
Pentapyxis Hook. f. = Leycesteria Wall. (Caprifoliac.).
Pentarhaphia Lindl. Gesneriaceae. 20 trop. Am., W.I.
Pentarhizidium Hayata = Matteuccia Todaro (Aspidiac.).
Pentaria M. Roem. = Passiflora L. (Passiflorac.).
Pentarrhaphis Kunth. Gramineae. 3 Mex. to Colombia.
Pentarrhinum E. Mey. Asclepiadaceae. 1–2 trop. & S. Afr.
Pentas Benth. Rubiaceae. 50 Afr., Madag.
Pentasachme Wall. ex Wight (sphalm.) = seq.   [SE. As.
Pentasacme Wall. ex Wight, corr. D. Don. Asclepiadaceae. 8 Himal., E. &
Pentaschistis (Nees) Spach. Gramineae. 60 Afr., Madag.
Pentascyphus Radlk. Sapindaceae. 1 Guiana.
Pentaspadon Hook. f. Anacardiaceae. 5 Malaysia, Solomon Is.
Pentaspatella Gleason. Ochnaceae. 1 Venez.
Pentastachya Hochst. ex Steud. = Pennisetum Rich. (Gramin.).
Pentastemon Batsch = Penstemon Schmid. (Scrophulariac.).

PENTASTEMONODISCUS

Pentastemonodiscus K. H. Rechinger. Caryophyllaceae. 1 Afghan.
Pentastemum Steud. = Pentastemon Batsch = Penstemon Schmid. (Scrophulariac.).
Pentasticha Turcz. (~ Fuirena Rottb.). Cyperaceae. 1 trop. Afr., Madag.
Pentastira Ridley = Dichapetalum Thou. (Dichapetalac.).
Pentataenium Tamamsch. Umbelliferae. 2 As. Min., Cauc., Iran.
Pentataphrus Schlechtd. = Astroloma R.Br. (Epacridac.).
Pentataxis D. Don = Helichrysum L. (Compos.).
Pentatherum Nábělek (~ Agrostis L.). Gramineae. 7 Cauc. to C. As.
Pentathymelaea Lecomte. Thymelaeaceae. 1 E. Tibet.
Pentatrichia Klatt. Compositae. 5 SW. Afr.
Pentatropis R.Br. Asclepiadaceae. 2 Afr., Masc., Orient, India & Ceylon; 4 Austr.
Penteca Rafin. = Croton L. (Euphorbiac.).
Pentelesia Rafin. = Arrabidaea DC. (Bignoniac.).
Pentena Rafin. = Scabiosa L. (Dipsacac.).
Penthea Lindl. = Disa Bergius (Orchidac.).
Penthea Spach = Barnadesia Mutis (Compos.).
Pentheriella O. Hoffm. & Muschler = Heteromma Benth. (Compos.).
***Penthoraceae** Van Tiegh. Dicots. 1/3 E. As., E. N. Am. Perenn. rhiz. herbs, with alt. simple serr. exstip. ls. Fls. small, reg., ⚥, in term. secund cymes. K 5(–8), valv., persist.; C 5(–8), inconspic., or more often 0; A 5 + 5, fil. filif., anth. obl., basifixed; G̲ 5–8, united half-way and slightly sunk in recept., recurved, each with short style and capit. stig., and ∞ ov. on a thick ventr. plac. Fr. depressed, of 5(–8) follicles circumscissile above their union, containing ∞ scobiform papillose seeds. Only genus: *Penthorum*. Somewhat intermediate between *Saxifragac. (s.str.)* and *Crassulac.*
Penthorum Gronov. ex L. Penthoraceae. 1–3 E. As., Indoch., Atlant. N. Am.
Penthysa Rafin. = Lobostemon Lehm. (Boraginac.).
Pentilium Rafin. (nomen). Liliaceae. Quid?
Pentiphragma Hook. (sphalm.) = Pentaphragma Wall. ex G. Don (Pentaphragmatac.).
Pentisea Lindl. = Caladenia R.Br. (Orchidac.).
Pentitdis Zipp. ex Blume = Opilia Roxb. (Opiliac.).
Pentlandia Herb. = Urceolina Reichb. (Amaryllidac.).
Pentochna Van Tiegh. = Ochna L. (Ochnac.).
Pentocnide Rafin. = Pouzolzia Gaudich. (Urticac.).
Pentodon Ehrenb. ex Boiss. = Kochia Roth (Chenopodiac.).
Pentodon Hochst. Rubiaceae. 2 SE. U.S., C. Am., W.I., trop. Afr., Arabia, Seychelles.
Pentopetia Decne. Periplocaceae. 10 Madag.
Pentopetiopsis Costantin & Gallaud. Periplocaceae. 1 Madag.
Pentostemon Rafin. = Penstemon Schmid. (Scrophulariac.).
Pentrias Benth. & Hook. f. = seq.
Pentrius Rafin. = Amaranthus L. (Amaranthac.).
Pentropis Rafin. = Campanula L. (Campanulac.).
Pentsteira Griff. = Torenia L. (Scrophulariac.).
Pentstemon Ait. = Penstemon Schmid. (Scrophulariac.).

Pentstemonacanthus Nees. Acanthaceae. 1 Brazil.

Pentstemonopsis Rydb. = Chionophila Benth. (Scrophulariac.).

Pentstemum Steud. = Penstemon Schmid. (Scrophulariac.).

Pentsteria Griff. (sphalm.) = Pentsteira Griff. = Torenia L. (Scrophulariac.).

Pentstira P. & K. = praec.

Pentulops Rafin. = Xylobium Lindl. (Orchidac.).

Pentzia Thunb. Compositae. 35 S. Afr., few trop. & N. Afr.

Peperidia Kostel. = Piper L. (Piperac.).

Peperidia Reichb. = Chloranthus Sw. (Chloranthac.).

Peperidium Lindl. = Renealmia L. (Zingiberac.).

Peperomia Ruiz & Pav. Peperomiaceae. 1000 +, trop. & subtrop., esp. Am. Many are epiph. with creeping stems, adv. roots and fleshy ls. (water tissue under the upper epidermis).

Peperomiaceae (Miq.) Wettst. Dicots. 4/1000 trop. Succulent herbs or subshrubs, with alt., opp., or vertic., exstip. ls. Fls. ♂, in axill. or term., sol. or aggreg. spikes, with succul. bracts. P 0; A 2, lateral, anth. finally confluent; G 1, with 1 basal orthotr. ov.; stig. simple (v. rarely divided), pilose. Seed with mealy perisp. Genera: *Peperomia, Verhuellia, Manekia, Piperanthera.* Near *Piperac.*, but differing in anatomy, pollen, absence of stips., etc.

Pepinia Brongn. ex E. André = Pitcairnia L'Hérit. (Bromeliac.).

Peplidium Delile. Scrophulariaceae. 2 N. & trop. Afr. to Austr.

Peplis L. Lythraceae. 3 N. temp., in wet places. *P. portula* L. is a little annual herb, very like *Montia fontana*, with minute hexamerous fls. Self-fert. by the bending inwards of the sta. over the stigma. Fr. biloc. (the partition does not come up to the very apex) with many seeds, but indehisc. When submerged the pl. has a more etiolated structure and becomes perennial.

Peplonia Decne. Asclepiadaceae. 3 Brazil.

Pepo Mill. = Cucurbita L. (Cucurbitac.).

Peponia Naud. = Peponium Engl. (Cucurbitac.).

Peponidium (Baill.) Arènes. Rubiaceae. 20 Madag., Comoro Is.

Peponiella Kuntze = seq.

Peponium Engl. Cucurbitaceae. 20 Afr., Madag.

Peponopsis Naud. Cucurbitaceae. 1 Mex.

Pera Mutis. Peraceae (~ Euphorbiaceae). 40 Mex. to trop. S. Am. W.I.

Peracarpa Hook. f. & Thoms. Campanulaceae. 1 Himal. to Japan & Formosa, Philipp. Is.

Peraceae (Baill.) Benth. ex Klotzsch. Dicots. 1/40 trop. Am., W.I. Trees or shrubs, with stellate or lepidote or occasionally simple indumentum, more rarely glabrous. Ls. alt. (v. rarely opp.), simple, ent., shortly pet., with minute or obsolete stips. Fls. ♂♀, monoec. or dioec., in uni- or bisex. 3–4-flowered involucrate capitula; invol. of 1–2 small free outer and 2 larger variously connate spathaceous inner bracts. ♂ fl.: K 4–6, united into a lobed, dentate, or ent. cupule, or 0; C 0; disk 0; A 2–5, ± united below into a column, anth. intr. or extr.; pistillode sometimes present. ♀ fl.: K 0; C 0; disk 0; G (3), styles forming a subsess. pelt. or 3-lobed disc; ov. 1 per loc., apical, pend. Fruit a tardily dehisc. 3-loc. drupiform caps., with woody endoc. and fleshy or spongy mesocarp, epicarp smooth and much wrinkled when dry, valves splitting along midline but not springing back elastically, usu. remaining attached at their base to

pedic., central column slender, splitting longit. into 3; seeds with caruncle, testa smooth, shining, cots. broad, flat, endosp. fleshy. Only genus: *Pera*. Long included in the *Euphorbiaceae*, but probably distinct.

Perakanthus Robyns. Rubiaceae. 2 Malay Penins.

Peraltea Kunth = Brongniartia Kunth (Legumin.).

Perama Aubl. Rubiaceae. 6 trop. S. Am., W.I.

Peramibus Rafin. = Rudbeckia L. (Compos.).

Peramium Salisb. ex Britton & Brown = Goodyera R.Br. (Orchidac.).

Peranema D. Don. Aspidiaceae. 2 NE. India, Malaysia.

Peranem[at]aceae Presl = Aspidiaceae S. F. Gray.

Perantha Craib = Oreocharis Benth. (Gesneriac.).

Peraphora Miers = Cyclea Arn. (Menispermac.).

Peraphyllum Nutt. ex Torr. & Gray. Rosaceae. 1 W. U.S.

Peratanthe Urb. Rubiaceae. 2 Cuba, Haiti.

Peraxilla Van Tiegh. Loranthaceae. 2 N.Z.

Percepier Moench = Aphanes L. (Rosac.).

Perdicesca Provanch. = Mitchella L. (Rubiac.).

Perdicesea Delam., Ren. & Card. = praec.

Perdici[ac]eae Link = Compositae–Mutisieae Cass.

Perdicium L. = Gerbera L. ex Cass. + Trixis P.Br. (Compos.).

Perebea Aubl. Moraceae. 20 trop. Am.

Pereilema J. & C. Presl. Gramineae. 4 Mex. to trop. S. Am.

Pereiria Lindl. = Coscinium Colebr. (Menispermac.).

Perella Van Tiegh. (~ Peraxilla Van Tiegh.). Loranthaceae. 4 Austr.

Peremis Rafin. = Passiflora L. (Passiflorac.).

Perenema Ching = Peranema D. Don (Aspidiac.).

Perenem[at]aceae Ching = Peranem[at]aceae Presl = Aspidiaceae S. F. Gray.

Perenideboles Goyena. Acanthaceae. 1 Nicaragua.

Perepusa Steud. = Prepusa Mart. (Gentianac.).

Perescia Lem. = seq.

Pereskia Mill. Cactaceae. 20 Mex. to trop. S. Am., W.I. Leafy plants (see fam.). *P. aculeata* Mill. climbs with recurved spines.

Pereskia Vell. = Hippocratea L. (Celastrac.).

Pereskiopsis Britton & Rose. Cactaceae. 17 Mex., C. Am.

Pereuphora Hoffmgg. = Serratula L. (Compos.).

Perezia Lag. Compositae. 90 S. U.S. to Patag.

Pereziopsis Coulter. Compositae. 1 C. Am.

Perfoliata Fourr. = Bupleurum L. (Umbellif.).

Perfoliata Kuntze = Hermas L. (Hydrocotylac.).

Perfolisa Rafin. = Bupleurum L. (Umbellif.).

Perfonon Rafin. = Rhamnus L. (Rhamnac.).

Pergamena Finet = Dactylostalix Reichb. f. (Orchidac.).

Pergularia L. Asclepiadaceae. 3–5 Afr. & Madag. to India.

Periandra Cambess. = Thylacospermum Fenzl (Caryophyllac.).

Periandra Mart. ex Benth. Leguminosae. 8 C. Am., W.I., Brazil.

Perianthomega Bur. ex Baill. Bignoniaceae. 1 C. Brazil.

Perianthopodus S. Manso = Cayaponia S. Manso (Cucurbitac.).

Periarrabidaea A. Sampaio. Bignoniaceae. 2 Amaz. Brazil.

Peribaea Lindl. = Periboea Kunth = Hyacinthus L. (Liliac.).
Periballanthus Franch. & Sav. = Polygonatum Adans. (Liliac.).
Periballia Trin. = Deschampsia Beauv. (Gramin.).
Periblema DC. = Boutonia DC. (Acanthac.).
Periblepharis Van Tiegh. = Luxemburgia A. St-Hil. (Ochnac.).
Periboea Kunth = Hyacinthus L. (Liliac.).
Pericalia Cass. (~ Senecio L.). Compositae. 6 Mex. to trop. S. Am.
Pericallis Webb & Berth. = Senecio L. (Compos.).
Pericalymma Endl. = Leptospermum J. R. & G. Forst. (Myrtac.).
Pericalymna Meissn. = praec.
Pericalypta Benoist. Acanthaceae. 1 Madag.
Pericampylus Miers. Menispermaceae. 7 E. Himal. & S. China to W. Malaysia & Moluccas.
Pericaulon Rafin. = Baptisia Vent. (Legumin.).
Perichasma Miers = Stephania Lour. (Menispermac.).
Perichlaena Baill. Bignoniaceae. 1 Madag.
Pericla Rafin. = Cleome L. (Cleomac.).
Periclesia A. C. Smith = Ceratostema Juss. (Ericac.).
Periclistia Benth. = Paypayrola Aubl. (Violac.).
Periclyma Rafin. = seq.
Periclymenum Mill. = Lonicera L. (Caprifoliac.).
Pericodia Rafin. = Passiflora L. (Passiflorac.).
Pericome A. Gray. Compositae. 4 SW. U.S., Mex.
Pericopsis Thw. Leguminosae. 5 trop. Afr., yielding Afrormosia wood; 1 Ceylon, yielding a pretty cabinet wood (*nedun*); 1 Palau & Caroline Is.
Perictenia Miers. Apocynaceae. 1 Peru.
Pericycla Blume = Licuala van Wurmb (Palm.).
Perideraea Webb = Ormenis Cass. (Compos.).
Perideridia Reichb. Umbelliferae. 10 N. Am.
*__*Peridiscaceae__ Kuhlm. Dicots. 2/2 trop. S. Am. A small group, clearly falling within the *Tiliaceae–Flacourtiaceae–Euphorbiaceae* circle of affinity, differing from the great majority in having monothecous anthers, a group of apical pendulous ovules in a unilocular ovary, and seeds without endosperm and with a curved embryo. It is interesting that *Medusandraceae*, another 'peripheral' group of the same affinity, also possesses the character of a group of apical pendulous ovules; there is, however, probably no close affinity with *Peridiscac.* There is nothing distinctive in the facies and general habit of the present family. Genera: *Peridiscus, Whittonia.*
Peridiscus Benth. Peridiscaceae. 1 Amaz. Brazil, Venezuela.
Peridium Schott = Pera Mutis (Perac.).
Perieilema Benth. & Hook. f. = Pereilema J. & C. Presl (Gramin.).
Periestes Baill. Acanthaceae. 2 Madag., Comoros.
Perieteris Rafin. = Nicotiana L. (Solanac.).
Perigaria Span. = Planchonia Bl. (Barringtoniac.).
Periglossum Decne. Asclepiadaceae. 4 S. Afr.
Perihema Rafin. = Haemanthus L. (Liliac.).
Perihemia B. D. Jacks. = praec.
Perijea (Tul.) Tul. = Fagara L. (Rutac.).

PERILEPTA

Perilepta Bremek. Acanthaceae. 8 India to China & Indoch.
Perilimnastes Ridley. Melastomataceae. 1 Malay Penins.
Perilla L. Labiatae. 4–6 India to Japan.
Perillula Maxim. Labiatae. 1 Japan.
Periloba Rafin. = Nolana L. ex L. f. (Nolanac.).
Perilomia Kunth = Scutellaria L. (Labiat.).
Perima Rafin. = Entada Adans. (Legumin.).
Perimenium Steud. = Perymenium Schrad. (Compos.).
Perinerion Baill. = Baissea A. DC. (Apocynac.).
Perinka Rafin. = seq.
Perinkara Adans. = Elaeocarpus L. (Elaeocarpac.).
Periomphale Baill. Alseuosmiaceae. 2 New Caled.
Peripea Steud. = Piripea Aubl. = Buchnera L. (Scrophulariac.).
Peripentadenia L. S. Smith. Elaeocarpaceae. 1 Queensland.
Peripeplus Pierre. Rubiaceae. 1 W. Equat. Afr.
Peripetasma Ridley = Dioscorea L. (Dioscoreac.).
Periphanes Salisb. Amaryllidaceae. 15 S. Afr.
Periphas Rafin. = Convolvulus L. (Convolvulac.).
Periphragmos Ruiz & Pav. = Cantua Juss. (Polemoniac.).
Periphyllium Gandog. = Paronychia L. (Caryophyllac.).
Periplexis Wall. = Drypetes Vahl (Euphorbiac.).
***Periploca** L. Periplocaceae. 10 N. & trop. Afr., Orient, E. As.
Periplocaceae Schltr. Dicots. 45–50/200 trop. & warm temp. Old World,
esp. trop. Afr. Perenn. laticif. herbs or shrubs with wiry or softly woody
stems, erect, scrambling or twining; rootstock a fleshy or woody tuber. Ls.
opp., ent., linear to very broad, penninerved; stip. o, but sometimes a nodal
stipular annulus, which may become enlarged, indurated and variously dissected. Infl. a term. or lat. 1–2–∞-fld. cyme, never umbelliform or racemiform;
bracts and bracteoles minute. Fls. ⚥, but sometimes functionally unisexual
and then dioecious; K 5, tube short or obsolescent, segs. valv. or imbr., but
opening very early; C (5), tube short or sometimes as long as or longer than
lobes; lobes usu. contorted, rarely valv. or imbr.; corona of 5 free lobes of
various form arising from base of stamen filaments, rarely absent or reduced
to minute tubercles; A 5, fil. free from each other, inserted at or below throat
of corolla; anthers coherent and appressed to the expanded style-head, introrse,
discharging granular pollen (in tetrads) on to spathulate pollen-carriers arising
from the style-head and alternating with the anthers. G 2, free from each other
but united through the expanded style-head; stigmatic surfaces concealed by
the pollen-carriers; ov. ∞, multiseriate on a single adaxial plac. Fruit of 2(–1)
divaric. or reflexed sessile ± fusif. or linear follicles, membr. or woody, finally
dehiscent along adaxial suture; seeds flat, often winged, crowned with coma
of silky hairs; endosp. present, embryo straight, almost as long as seed, cots.
flat.

Distinguished from *Asclepiadac.* by the free stamens, spathulate pollen-
carriers and granular pollen. The pollination mechanism is unique: pollen is
discharged on to the spoon-shaped carriers and removed by visiting insects,
to whose heads the glandular base of the carrier adheres. Cross-pollination is
ensured by the concealment of the stigmatic surfaces by the carriers.

Periptera DC. Malvaceae. 4 Mex.
Peripteris Rafin. = Pteris L. (Pteridac.).
Peripterygia Loes. Celastraceae. 1 New Caled.
Peripterygiaceae F. N. Williams = Cardiopterygaceae Bl. corr. Van Tiegh.
Peripterygium Hassk. Cardiopterygaceae. 3 Assam, SE. As., Malaysia, Queensland.
Perispermum Degener (1932; non Heydrich 1901—Algae) (~ Bonamia Thou.). Convolvulaceae. 2 Hawaii.
Perissandra Gagnep. = Vatica L. (Dipterocarpac.).
Perissocoeleum Mathias & Constance. Umbelliferae. 4 Colombia.
Perissolobus N. E. Brown. Aïzoaceae. 1 S. Afr.
Perissus Miers = Asthotheca Miers ex Planch. & Triana = Clusia L. (Guttif.).
Peristeira Hook. f. = Pentsteira Griff. = Torenia L. (Scrophulariac.).
Peristera Eckl. & Zeyh. = Pelargonium L'Hérit. (Geraniac.).
Peristera Endl. = Peristeria Hook. (Orchidac.).
Peristeranthus T. E. Hunt. Orchidaceae. 1 Austr.
Peristeria Hook. Orchidaceae. 9 trop. S. Am.
Peristethium Van Tiegh. = Loranthus L. (Loranthac.).
Peristima Rafin. = Heliotropium L. (Boraginac.).
Peristrophe Nees. Acanthaceae. 30 warm Afr. to E. Malaysia.
Peristylus Blume. Orchidaceae. 60 China, Formosa, India to Austr., Polynesia.
Peritassa Miers. Celastraceae. 14 trop. S. Am., Tobago.
Perithrix Pierre = Batesanthus N. E. Br. (Asclepiadac.).
Peritoma DC. (~ Cleome L.). Cleomaceae. 6 SW. U.S., Mex.
Peritomia G. Don = Perilomia Kunth (Labiat.).
Peritris Rafin. = Arnica L. (Compos.).
Peritropa Presl ex Reichb. (nomen). Leguminosae. Quid?
Perittium Vog. = Melanoxylon Schott (Legumin.).
Perittostema I. M. Johnston. Boraginaceae. 1 Mex.
Perityle Benth. Compositae. 25 SW. U.S., Mex.
Perizoma (Miers) Lindl. = Salpichroa Miers (Solanac.).
Perizomanthus Pursh = Dicentra Borckh. corr. Bernh. (Fumariac.).
Perlaria Fabr. (uninom.) = *Aegilops* L. sp. (Gramin.).
Perlarius Kuntze = Pipturus Wedd. (Urticac.).
Perlebia DC. = Heptaptera Margot & Reuter (Umbellif.).
Perlebia Mart. = Bauhinia L. (Legumin.).
***Pernettya** Gaudich. Ericaceae. 20 Tasm., N.Z., Mex. to temp. S. Am., Galáp., Falkl. Is. Probably to be united with *Gaultheria* L.
Pernettyopsis King & Gamble. Ericaceae. 1 Malay Penins.
Pernetya Scop. = Canarina L. (Campanulac.).
Peroa Pers. = Perojoa Cav. = Leucopogon R.Br. (Epacridac.).
Perobachne J. & C. Presl = Anthistiria L. f. (Gramin.).
Perocarpa Feer = Peracarpa Hook. f. & Thoms. (Campanulac.).
Perojoa Cav. = Leucopogon R.Br. (Epacridac.).
Peronema Jack. Verbenaceae. 1 Lower Burma, Siam, Malay Penins., Sumatra, Borneo.
Peronia R.Br. = Sarcosperma Hook. f. (Sarcospermatac.).
Peronia De la Roche ex DC. = Thalia L. (Marantac.).

PERONIACEAE

Peroniaceae Dostál = Sarcospermataceae H. J. Lam.
Perostema Raeusch. = Nectandra Roland. (Laurac.).
Perostis P. Beauv. = seq. [awned.
Perotis Ait. Gramineae. 10 trop. Afr., S. India, Ceylon, E. As., Austr. Glumes
Perotriche Cass. Compositae. 1 S. Afr.
Perottetia P. & K. = Perrottetia DC. = Desmodium Desv. (Legumin.).
Perovskia Karelin. Labiatae. 7 NE. Persia & C. As. to Baluchist. & NW. Himal.
Perpensum Burm. f. = Gunnera L. (Gunnerac.).
Perplexia Iljin. Compositae. 2 C. As., Persia.
Perralderia Coss. Compositae. 4 NW. Afr.
× **Perreiraara** hort. Orchidaceae. Gen. hybr. (iii) (Aërides × Rhynochostylis × Vanda).
Perreymondia Barnéoud = Schizopetalon Sims (Crucif.).
Perriera Courchet. Simaroubaceae. 1 Madag.
Perrieranthus Hochr. Malvaceae. 1 Madag.
Perrierastrum Guillaumin. Labiatae. 1 Madag.
Perrierbambus A. Camus. Gramineae. 2 Madag.
Perrieriella Schlechter. Orchidaceae. 1 Madag.
Perrierodendron Cavaco. Sarcolaenaceae. 1 Madag.
Perrierophytum Hochr. Malvaceae. 9 Madag.
Perrottetia DC. = Desmodium Desv. (Legumin.).
Perrottetia Kunth. Celastraceae. 20 E. As., Malaysia, Austr., Hawaii, Mex. to Colombia.
***Persea** Mill. Lauraceae. 150 trop. The fr. of *P. americana* Mill. (*aguacate*, *avocado*, alligator pear, *palta*) is ed.
Perseaceae Horan. = Lauraceae Juss.
Persica Mill. = Amygdalus L. (Rosac.).
Persicana Scop. = seq.
Persicaria Mill. (~ Polygonum L.). Polygonaceae. 150 N. temp., W.I., trop.
Persicaria Neck. = Atraphaxis L. (Polygonac.). [S. Am.
Persimon Rafin. = Diospyros L. (Ebenac.).
Personaceae Dulac = Scrophulariaceae Juss.
Personaria Lam. = Gorteria L. (Compos.).
Personatae Vent. = Scrophulariaceae Juss.
Personia Rafin. = Persoonia Michx = Marshallia Schreb. (Compos.).
Personula Rafin. = Utricularia L. (Lentibulariac.).
Persoonia Michx = Marshallia Schreb. (Compos.).
***Persoonia** Sm. Proteaceae. 60 Austr., N.Z.
Persoonia Willd. = Carapa Aubl. (Meliac.).
Perspicillum Fabr. (uninom.) = *Biscutella* L. sp. (Crucif.).
Pertya Sch. Bip. Compositae. 10 Afghanistan to Japan.
Perula Rafin. = Urostigma Gasp. = Ficus L. (Morac.).
Perula Schreb. = Pera Mutis (Perac.).
Perularia Lindl. = Tulotis Rafin. (Orchidac.).
Perulifera A. Camus. Gramineae. 1 Madag.
Peruvocereus Akers = Haageocereus Backeb. (Cactac.).
Pervillaea Decne = Toxocarpus Wight & Arn. (Asclepiadac.).
Pervinca Mill. = Vinca L. (Apocynac.).

Perxo Rafin. = Moschosma Reichb. (Labiat.).
Perymeniopsis Sch. Bip. ex Klatt = Gymnolomia Kunth (Compos.).
Perymenium Schrad. Compositae. 50 Mex. to Peru.
Perytis Rafin. = Cyclobalanopsis (Endl.) Oerst. = Quercus L. (Fagac.).
× Pescarhyncha hort. Orchidaceae. Gen. hybr. (iii) (Chondrorhyncha ×
Pescatoria).
× Pescatobollea hort. Orchidaceae. Gen. hybr. (iii) (Bollea × Pescatoria).
Pescatorea Reichb. f. (1869) = seq.
Pescatoria Reichb. f. (1852). Orchidaceae. 17 Costa Rica, W. trop. S. Am.
Peschiera A. DC. (~ Tabernaemontana L.). Apocynaceae. 25 trop. Am., W.I.
× Pescoranthes hort. Orchidaceae. Gen. hybr. (iii) (Cochleanthes × Pes-
catoria).
Pesomeria Lindl. = Phaius Lour. (Orchidac.).
Pessopteris Underwood & Maxon. Polypodiaceae. 1 trop. Am.
Pessularia Salisb. = Anthericum L. (Liliac.).
Pestallozzia Willis (sphalm.) = seq.
Pestalozzia Moritzi = Gynostemma Blume (Cucurbitac.).
Petagna Endl. = Petagnia Rafin. = Solanum L. (Solanac.).
Petagnaea Caruel = Petagnia Guss. (Umbellif.).
Petagnana J. F. Gmel. = Smithia Ait. (Legumin.).
Petagnia Guss. Umbelliferae. 1 Sicily.
Petagnia Rafin. = Solanum L. (Solanac.).
Petagniana Rafin. = Petagnana J. F. Gmel. = Smithia Ait. (Legumin.).
Petagomoa Bremek. Rubiaceae. 7 trop. S. Am.
Petalacte D. Don. Compositae. 2 S. Afr.
Petalactella N. E. Br. = Ifloga Cass. (Compos.).
Petaladenium Ducke. Leguminosae. 1 Amaz. Brazil.
Petalandra Hassk. = Hopea Roxb. (Dipterocarpac.).
Petalandra F. Muell. ex Boiss. = Euphorbia L. (Euphorbiac.).
Petalanisia Rafin. = Hypericum L. (Guttif.).
Petalanthera Nees = Ocotea Aubl. (Laurac.).
Petalanthera Nutt. = Cevallia Lag. (Loasac.).
Petalanthera Rafin. = Justicia L. (Acanthac.).
Petalepis Rafin. = Lepuropetalon Ell. (Lepuropetalac.).
Petalidium Nees. Acanthaceae. 35 trop. & S. Afr.; 1 W. Himal. & W.
Penins. India.
Petalinia Becc. = Ochanostachys Mast. (Olacac.).
Petalocaryum Pierre ex A. Chev. Olacaceae. 1 W. Equat. Afr.
Petalocentrum Schlechter = Sigmatostalix Reichb. f. (Orchidac.).
Petalochilus R. S. Rogers. Orchidaceae. 2 N.Z.
Petalodactylis Arènes = Cassipourea Aubl. (Rhizophorac.).
Petalodiscus Baill. = Savia Willd. (Euphorbiac.).
Petalogyne F. Muell. = Petalostylis R.Br. (Legumin.).
Petalolepis Cass. = Helichrysum Mill. (Compos.).
Petalolepis Less. = Petalacte D. Don (Compos.).
Petalolophus K. Schum. Annonaceae. 1 NE. New Guinea.
Petaloma Rafin. ex Boiss. = Euphorbia L. (Euphorbiac.).
Petaloma Roxb. = Lumnitzera Willd. (Combretac.).

PETALOMA

Petaloma Sw. = Mouriri Aubl. (Memecylac.).
Petalonema Gilg (1897; non Correns 1889—Cyanophyceae) = Neopetalonema Brenan (Melastomatac.).
Petalonema Peter (1928) = Impatiens L. (Balsaminac.).
Petalonema Schlechter (1915) = Quisumbingia Merr. (Asclepiadac.).
Petalonyx A. Gray. Loasaceae. 5 SW. U.S., Mex.
Petalopogon Reiss. = Phylica L. (Rhamnac.).
Petalosteira Rafin. = Tiarella L. (Saxifragac.).
Petalostelma Fourn. Asclepiadaceae. 2 Brazil.
Petalostemma R.Br. = Glossonema Decne (Asclepiadac.).
*****Petalostemon** Michx mut. Pers. Leguminosae. 50 N. Am.
Petalostemum Michx = praec.
Petalostigma F. Muell. Euphorbiaceae (?). 7 New Guinea, Austr.
Petalostima Rafin. = Wahlenbergia Schrad. ex Roth (Campanulac.).
Petalostyles Benth. = seq.
Petalostylis R.Br. Leguminosae. 3 Austr.
Petalostylis Lindl. = Petasostylis Griseb. = Leianthus Griseb. (Gentianac.).
Petalotoma DC. = Carallia Roxb. (Rhizophorac.).
Petaloxis Rafin. = Dichorisandra Mikan (Commelinac.).
Petalvitemon Rafin. = Petalostemon Michx (Legumin.).
Petamenes Salisb. Iridaceae. 24 trop. & S. Afr.
Petasioïdes Vitm. = Petesioïdes Jacq. = Wallenia Sw. (Myrsinac.).
Petasites Mill. Compositae. 5 W. Eur. to C. As. *P. hybridus* (L.) Gaertn., Mey. & Scherb. spreads largely by rhiz. It is dioecious (cf. *Tussilago*, its close ally). The ♂ head has about 30 fls. with the usual mech. of *Compositae*, the style acting as pollen-presenter, though the ovary is not fertile. Occasionally a few ♀ fls. are found. The ♀ head consists of about 150 ♀ fls. surrounding 1–3 ♂ fls. Only the male fls. secrete honey. In *P. fragrans* (Vill.) C. Presl the fls., which appear in February, are fragrant.
Petasitis Mill. (sphalm.) = praec.
Petasostylis Griseb. = Leianthus Griseb. (Gentianac.).
Petastoma Miers. Bignoniaceae. 12 trop. Am.
Petasula Nor. = Schefflera J. R. & G. Forst. + Trevesia Vis. (Araliac.).
Petchia Livera. Apocynaceae. 1 Ceylon.
Petelotia Gagnep. (1928; non Patouill. 1924—Fungi) = seq.
Petelotiella Gagnep. Urticaceae. 1 Indoch.
Petenaea Lundell. Elaeocarpaceae. 1 C. Am. (Guatem.).
Peteravenia R. M. King & H. Rob. (~ Eupatorium). Compositae. 5 Mex., C. Am.
Peteria A. Gray. Leguminosae. 3 SW. U.S.
Petermannia Klotzsch = Begonia L. (Begoniac.).
*****Petermannia** F. Muell. Philesiaceae. 1 New S. Wales.
Petermannia Reichb. = Cycloloma Moq. (Chenopodiac.).
*****Petermanniaceae** Hutch. = Philesiaceae Dum. (Woody climbers with ± prickly stem and alt. ent. exstip. ls., the venules retic. between the main arcuate-parallel nerves. Fls. reg., ☿, in small lax leaf-opp. cymes, these sometimes modif. into branched tendrils. P 6, spreading or reflexed; A 6, with erect fil. and oblong extr. anth.; G̅ (3), 1-loc., with slender style and capit.

stig., and ∞ ov. on 3 pariet. plac. Fr. a ∞-seeded berry. Only genus: *Peter-mannia* F. Muell.)

Peterodendron Sleum. Flacourtiaceae. 1 trop. E. Afr.

Petersia Klotzsch = Capparis L. (Capparidac.).

Petersia Welw. ex Benth. & Hook. f. = Petersianthus Merr. = Combretodendron A. Chev. (Barringtoniac.).

Petersianthus Merr. = praec.

Petesia P. Br. = Rondeletia L. (Rubiac.).

Petesiodes Kuntze = seq.

Petesioïdes Jacq. = Wallenia Sw. (Myrsinac.).

Petilium Ludw. = Fritillaria L. (Liliac.).

Petitia J. Gay = Xatardia Meissn. (Umbellif.).

Petitia Jacq. Verbenaceae. 2 W.I.

Petitia Neck. = Hibiscus L. (Malvac.).

Petitmenginia Bonati. Scrophulariaceae. 2 S. China, Indoch.

Petiveria L. Phytolaccaceae. 1 warm Am., W.I. Polymorphic.

Petiver[iac]eae C. A. Agardh = Phytolaccaceae–Rivineae Reichb.

Petkovia Stefanoff (~ Campanula L.). Campanulaceae. 1 Balkan Penins.

Petlomelia Nieuwl. = Fraxinus L. (Oleac.).

Petopentia Bullock. Asclepiadaceae. 1 S. Afr.

Petracanthus Nees = Gymnostachyum Nees (Acanthac.).

Petradoria Greene (~ Solidago L.). Compositae. 1 SW. U.S.

Petradosia Dur. & Jacks. (sphalm.) = praec.

Petraea Juss. = Petrea L. (Verbenac.).

Petraeovitex Oliv. Verbenaceae. 7 Malaysia (exc. Java & Lesser Sunda Is.), N.Z., Pacif.

Petramnia Rafin. = Pentarhaphia Lindl. (Gesneriac.).

Petrantha DC. (sphalm.) = Tetrantha Poit. = Riencourtia Cass. (Compos.).

Petranthe Salisb. = Scilla L. (Liliac.).

Petrea L. Verbenaceae. 30 trop. Am., W.I. Climbers.

Petreaceae J. G. Agardh = Verbenaceae–Petreëae Briq.

Petriella Zotov = Ehrharta Thunb. (Gramin.).

Petrina Phipps. Gramineae. 3 S. trop. Afr.

Petrobium Bong. = Lithobium Bong. (Melastomatac.).

Petrobium R.Br. Compositae. 1 St Helena.

Petrocallis R.Br. Cruciferae. 2 mts. S. Eur., N. Persia.

Petrocarvi Tausch = Athamantha L. (Umbellif.).

Petrocarya Schreb. = Parinari Aubl. (Chrysobalanac.).

Petrocodon Hance. Gesneriaceae. 2 China.

Petrocoma Rupr. = Silene L. (Caryophyllac.).

Petrocoptis A.Br. ex Endl. Caryophyllaceae. 6 Pyrenees.

Petrocosmea Oliv. Gesneriaceae. 15 China.

Petrodora Fourr. = Veronica L. (Scrophulariac.).

Petrodoxa Anthony = Beccarinda Kuntze (Gesneriac.).

Petroëdmondia Tamamsh. Umbelliferae. Spp.? Hab.?

Petrogenia I. M. Johnston. Convolvulaceae. 1 Mex.

Petrogeton Eckl. & Zeyh. = Crassula L. (Crassulac.).

Petrollinia Chiov. Compositae. 1 NE. trop. Afr.

Petromarula Vent. ex Hedw. f. Campanulaceae. 1 Crete.

Petromecon Greene = Eschscholzia Cham. (Papaverac.).

Petronia Barb. Rodr. = Batemannia Lindl. (Orchidac.

Petronia Jungh. (sphalm.) = Pteronia L. (Compos.).

Petronymphe H. E. Moore. Amaryllidaceae. 1 Mex.

Petrophila R.Br. Proteaceae. 35 Austr.

Petrophile Knight = praec.

Petrophiloïdes Bowerbank ex Reid & Chandler = Platycarya Sieb. & Zucc. (Juglandac.).

Petrophyes Webb & Berth. = Monanthes Haw. (Crassulac.).

Petrophyton Rydb. = seq.

Petrophytum (Nutt. ex Torr. & Gray) Rydb. Rosaceae. 4 N. Am.

Petrorhagia (Ser.) Link. Caryophyllaceae. 20 Medit.

Petrosavia Becc. Petrosaviaceae. 3 E. As., W. Malaysia.

*****Petrosaviaceae** Hutch. Monocots. 1/3 E. As., W. Malaysia. Small colourless herbaceous forest-floor saprophytes, with simple erect stem and ls. reduced to scales. Fls. reg., ⚥, in a term. sometimes corymbif. rac. P 3 + 3, persist., the outer narrow, the inner broad-ovate, sometimes with a nectary; A 3 + 3, epitep., fil. persist., anth. dorsifixed, intr.; G̲ 3, very shortly united below and with the P, with short styles and small subcapit. stig., and ∞ biser. ov. on ventr. (axile) plac. Fr. of 3 spreading ventr. dehisc. follicles; seeds with endosp. Only genus: *Petrosavia* (*Protolirion*). Prob. related to *Liliaceae–Melanthioïdeae–Tofieldieae*.

Petrosciadium Edgew. = Pimpinella L. (Umbellif.).

Petroselinum Hill. Umbelliferae. 5 Eur., Medit. *P. crispum* (Mill.) Nym. ex A. W. Hill is the parsley, cult. as condiment.

Petrosilene Fourr. = Silene L. (Caryophyllac.).

Petrosimonia Bunge. Chenopodiaceae. 12 Greece to C. As.

Petrostylis Pritz. (sphalm.) = Pterostylis R.Br. (Orchidac.).

Petrotheca Steud. = Tetratheca Sm. (Tremandrac.).

Petrusia Baill. = Zygophyllum L. (Zygophyllac.).

Pettera Reichb. = Arenaria L. (Caryophyllac.).

*****Petteria** C. Presl. Leguminosae. 1 Balkan Penins.

Pettospermum Roxb. = Pittosporum Banks ex Soland. (Pittosporac.).

Petunga DC. Rubiaceae. 10 India to W. Malaysia.

Petunia Juss. Solanaceae. 40 S. & warm N. Am.

Peuce Rich. = Picea + Abies + Cedrus + Larix (Pinac.).

Peucedanon Rafin. = seq.

Peucedanum L. Umbelliferae. 120 temp. Euras., trop. & S. Afr. *P. officinale* L. is the sulphur-root used in veterinary practice; *P. ostruthium* (L.) Koch is also used.

Peuceluma Baill. = Lucuma Molina (Sapotac.).

Peucephyllum A. Gray. Compositae. 1 SW. U.S.

Peudanum Dingl. (sphalm.) = Peucedanum L. (Umbellif.).

Peumus Molina = Boldu Adans. (Monimiac.) + Cryptocarya R.Br. (Laurac.).

*****Peumus** Molina emend. Pers. Monimiaceae. 1 Chile, *P. boldus* Molina, the *boldo*. Wood hard; bark yields dye; fr. ed.

Peurousea Steud. = Peyrousea DC. (Compos.).

Peutalis Rafin. = Polygonum L. (Polygonac.).
Peuteron Rafin. = Pleuteron Rafin. = Capparis L. (Capparidac.).
Pevraea Comm. ex Juss. = Poivrea Comm. ex Thou. = Combretum L. (Combretac.).
Peyritschia Fourn. ex Benth. & Hook. f. = Deschampsia Beauv. (Gramin.).
*Peyrousea DC. Compositae. 2 S. Afr.
Peyrousia Poir. = Lapeyrousia Pourr. (Iridac.).
Peyrusa Rich. ex Dun. = Hornemannia Vahl (Ericac.).
Peyssonelia Boiv. ex Webb & Berth. = Cytisus L. (Legumin.).
Pezisicarpus Vernet. Apocynaceae. 1 Indoch.
Pfaffia Mart. Amaranthaceae. 50 warm S. Am.
Pfeiffera Salm-Dyck (~ Rhipsalis Gaertn.). Cactaceae. 1 Bolivia, Argent.
Pfeifferago Kuntze = Codia J. R. & G. Forst. (Cunoniac.).
Pfeifferia Buching. = Cuscuta L. (Cuscutac.).
Pfundia Opiz ex Nevski = Hericinia Fourr. (Ranunculac.).
× **Phabletia** hort. Orchidaceae. Gen. hybr. (vi) (Bletia × Phaius).
Phaca L. = Astragalus L. (Legumin.).
Phacelia Juss. Hydrophyllaceae. 200 N. Am., Andes. The fl. is a bee-flower with honey secreted below the ovary and guarded by stipule-like flaps at the base of the sta. The large-flowered spp. are highly protandrous. The anther as it dehisces turns inside out.
Phacellanthus Klotzsch ex Kunth = Picramnia Sw. (Simaroubac.).
Phacellanthus Sieb. & Zucc. Orobanchaceae. 1 E. Sib., China, Japan.
Phacellanthus Steud. ex Zoll. & Mor. = Gahnia J. R. & G. Forst. (Cyperac.).
Phacellaria Benth. Santalaceae. 7 SE. As.
Phacellaria Willd. ex Steud. = Chloris Sw. (Gramin.).
Phacellothrix F. Muell. Compositae. 1 E. trop. Austr.
Phacelophrynium K. Schum. Marantaceae. 9 Nicobar Is., Malaysia.
Phacelura Benth. = seq.
Phacelurus Griseb. Gramineae. 4 warm E. Afr., As.
Phacocapnos Bernh. Fumariaceae. 3 S. Afr.
Phacolobus P. & K. = Fakeloba Rafin. = Anthyllis L. (Legumin.).
Phacomene Rydb. = Astragalus L. (Legumin.).
Phacopsis Rydb. = praec.
Phacosperma Haw. = Calandrinia Kunth (Portulacac.).
Phadrosanthus Neck. = Oncidium Sw. (Orchidac.).
Phaeanthus Hook. f. & Thoms. Annonaceae. 20 S. India, SE. As., Malaysia.
Phaeanthus P. & K. = Phaianthes Rafin. = Moraea Mill. ex L. (Iridac.).
Phaecasium Cass. = Crepis L. (Compos.).
Phaedra Klotzsch ex Endl. = Bernardia Mill. (Euphorbiac.).
Phaedranassa Herb. Amaryllidaceae. 6 Andes.
*Phaedranthus Miers. Bignoniaceae. 1 Mex.
Phaedrosanthus P. & K. [Phadrosanthus Neck.] = Epidendrum L., Cattleya Lindl., etc. (Orchidac.).
Phaelypaea P.Br. = Stemodia L. (Scrophulariac.).
Phaenanthoecium C. E. Hubbard. Gramineae. 1 mts. NE. trop. Afr.
Phaeneilema Brückn. = Murdannia Royle (Commelinac.).
Phaenicanthus Thw. = Premna L. (Verbenac.).

PHAENICAULIS

Phaenicaulis Greene = Phoenicaulis Nutt. (Crucif.).

Phaenix Hill = Phoenix L. (Palm.).

Phaenixopus Cass. = Lactuca L. (Compos.).

Phaenocodon Salisb. = Lapageria Ruiz & Pav. (Philesiac.).

Phaenocoma D. Don. Compositae. 1 S. Afr.

Phaenohoffmannia Kuntze. Leguminosae. 10 S. Afr.

Phaenomeria Steud. = Phaeomeria Lindl. ex K. Schum. = Nicolaia Horan. (Zingiberac.).

Phaenopoda Cass. = Podotheca Cass. (Compos.).

Phaenopus DC. = Phaenixopus Cass. = Lactuca L. (Compos.).

Phaenopyrum M. Roem. = Crataegus L. (Rosac.).

Phaenopyrum Schrad. ex Nees = Lagenocarpus Nees (Cyperac.).

Phaenosperma Munro ex Benth. Gramineae. 1 China.

Phaenostoma Steud. (sphalm.) = Chaenostoma Benth. = Sutera Roth (Scrophulariac.).

Phaeocarpus Mart. = Magonia A. St-Hil. (Sapindac.).

Phaeocephalum Ehrh. (uninom.) = Schoenus fuscus L. = Rhynchospora fusca (L.) Vahl (Cyperac.).

Phaeocephalum House = Rhynchospora Vahl (Cyperac.).

Phaeocephalus S. Moore. Compositae. 1 S. Afr.

Phaeocles Salisb. = Caruelia Parl. = Ornithogalum L. (Liliac.).

Phaeocordylis Griff. = Rhopalocnemis Jungh. (Balanophorac.).

Phaeolorum Ehrh. (uninom.) = Carex flacca Schreb. (Cyperac.).

Phaeomeria Lindl. ex K. Schum. = Nicolaia Horan. (Zingiberac.).

Phaeoneuron Gilg. Melastomataceae. 6 trop. Afr.

Phaeonychium O. E. Schulz. Cruciferae. 2 C. As., W. Tibet, Himal.

Phaeopappus Boiss. (~ Centaurea L.). Compositae. 50 SW. As.

Phaeophleps P. & K. = Phaiophleps Rafin. (Iridac.).

Phaeopsis Nutt. ex Benth. = Stenogyne Benth. (Labiat.).

Phaeoptilon Engl. = seq.

Phaeoptilum Radlk. Nyctaginaceae. 1 SW. Afr. Habit of Lycium (Solanac.).

Phaeosperma P. & K. = Phaiosperma Rafin. = Polytaenia DC. (Umbellif.).

Phaeosphaerion Hassk. Commelinaceae. 6 trop. S. Am.

Phaeosphaeriona B. D. Jacks. (sphalm.) = praec.

Phaeospherion auctt. = praec.

Phaeospheriona Willis (sphalm.) = praec.

Phaeostemma Fourn. Asclepiadaceae. 5 trop. S. Am.

Phaeostoma Spach = Clarkia Pursh (Onagrac.).

Phaëthusa Gaertn. = Verbesina L. (Compos.).

Phaëthusia Rafin. = praec.

Phaëtusa Schreb. = praec.

Phaeus P. & K. = Phaius Lour. (Orchidac.).

Phagnalon Cass. Compositae. 40 Canaries & Medit. to C. As. & Himalaya.

Phaianthes Rafin. = Moraea Mill. ex L. (Iridac.).

Phainantha Gleason. Melastomataceae. 1 trop. S. Am.

× **Phaiocalanthe** hort. Orchidaceae. Gen. hybr. (iii) (Calanthe × Phaius).

× **Phaiocalanthodes** hort. (iv) = praec.   [Phaius).

× **Phaiocymbidium** hort. Orchidaceae. Gen. hybr. (iii) (Cymbidium ×

× **Phaiolimatopreptanthe** hort. (iv) = × Phaiocalanthe hort. (Orchidac.).

Phaiophleps Rafin. Iridaceae. 8 Andes.

× **Phaiopreptanthe** hort. (iv) = × Phaiocalanthe hort. (Orchidac.).

Phaiosperma Rafin. = Polytaenia DC. (Umbellif.).

Phaius Lour. Orchidaceae. 50 trop. Afr., Masc., trop. As., Austr., Polynesia (introd. and now widespread in Panamá and W.I.).

Phajus Lindl. = praec.

Phakellanthus Steud. = Phacellanthus Sieb. & Zucc. = Gahnia J. R. & G. Forst. (Cyperac.). [C. As.

Phalacrachena Iljin (~ Centaurea L.). Compositae. 2 SE. Russia & Cauc., **Phalacraea** DC. = Piqueria Cav. (Compos.).

Phalacrocarpum Willk. (~ Chrysanthemum L.). Compositae. 1 Spain, Portugal.

Phalacrocarpus (Boiss.) Van Tiegh. = Cephalaria Schrad. (Dipsacac.).

Phalacroderis DC. = Rodigia Spreng. = Crepis L. (Compos.).

Phalacrodiscus Less. = Chrysanthemum L. (Compos.).

Phalacroglossum Sch. Bip. = Chrysanthemum L. (Compos.).

Phalacroloma Cass. = Erigeron L. (Compos.).

Phalacromesus Cass. = Tessaria Ruiz & Pav. (Compos.).

Phalacros Wenzig = Crataegus L. (Rosac.).

Phalacroseris A. Gray. Compositae. 1 California.

× **Phalaenetia** hort. Orchidaceae. Gen. hybr. (iii) (Neofinetia × Phalaenopsis).

Phalaenopsis Blume. Orchidaceae. 35 China, Indomal., Queensland.

× **Phalaërianda** hort. Orchidaceae. Gen. hybr. (iii) (Aërides × Phalaenopsis × Vanda).

× **Phalandopsis** hort. Orchidaceae. Gen. hybr. (iii) (Phalaenopsis × Vandopsis).

Phalanganthus Schrank = Anthericum L. (Liliac.).

Phalangion St-Lag. = Phalangium Mill. = Anthericum L. (Liliac.).

Phalangites Bub. = praec.

Phalangium Adans. = Urginea Steinheil (Liliac.).

Phalangium Burm. f. = Melasphaerula Ker-Gawl. (Iridac.).

Phalangium Kuntze = Bulbine L. (Liliac.).

Phalangium Mill. = Anthericum L. (Liliac.).

× **Phalanthe** hort. = × Phaiocalanthe hort. (Orchidac.).

Phalaridaceae Burnett = Gramineae–Pharideae Link.

Phalaridantha St-Lag. = Phalaris L. (Gramin.).

Phalaridium Nees = Dissanthelium Trin. (Gramin.).

Phalaris L. Gramineae. 20 N. & S. temp. *P. canariensis* L. (canary grass) seeds are used for cage-birds.

Phalaroïdes v. Wolf = Phalaris L. + Phleum L. (Gramin.).

Phaleria Jack. Thymelaeaceae. 20 Ceylon, SE. As., Indomal., Austr., Pacif.

Phaleri[ac]eae Meissn. = Thymelaeaceae–Phalarieae Domke.

Phalerocarpus G. Don = Gaultheria L. (Ericac.).

Phallaria Schumach. & Thonn. = Plectronia L. (Rubiac.).

Phallerocarpus G. Don (sphalm.) = Phalerocarpus G. Don = Gaultheria L. (Ericac.).

Phalocallis Herb. = Cypella Herb. (Iridac.).

Phalodallis Dur. & Jacks. (sphalm.) = praec.
Phaloë Dum. = Sagina L. (Caryophyllac.).
Phalolepis Cass. = Centaurea L. (Compos.).
Phalona Dum. = Falona Adans. = Cynosurus L. (Gramin.).
Phanera Lour. (~ Bauhinia L.). Leguminosae. 60 trop. As. to Austr.
Phanerandra Shchegl. = Leucopogon R.Br. (Epacridac.).
Phaneranthera DC. ex Meissn. = Nonea Medik. (Boraginac.).
Phanerocalyx S. Moore = Heisteria Jacq. (Olacac.).
Phanerodiscus Cavaco. Olacaceae. 1 Madag.
Phanerogonocarpus Cavaco. Monimiaceae. 2 Madag.
Phanerophlebia Presl (incl. Cyrtomium Presl). Aspidiaceae. 20 trop. As.,
Hawaii, trop. Am., S. Afr. *P. falcata* (L. f.) Copel. cult. as house plant.
Phanerophlebiopsis Ching. Aspidiaceae. 4 C. China.
Phanerosorus Copel. Matoniaceae. 2 Borneo, New Guinea. On limestone;
ls. pendent, their short lateral branches often ending in dormant apex, ± as
Phanerotaenia St John = Polytaenia DC. (Umbellif.). [*Lygodium*.
Phania DC. Compositae. 4 W.I.
Phaniasia Blume ex Miq. = Gymnadenia R.Br. (Orchidac.).
Phanopyrum (Rafin.) Nash = Panicum L. (Gramin.).
Phanrangia Tardieu = Mangifera L. (Anacardiac.).
Phantis Adans. = Atalantia Corrêa (Rutac.).
Pharaceae (Stapf) Herter = Gramineae–Phareae Stapf.
*Pharbitis Choisy = Diatremis Rafin. = Ipomoea L. (Convolvulac.).
Pharetranthus F. W. Klatt = Petrobium R.Br. (Compos.).
Pharetrella Salisb. = Cyanella L. (Tecophilaeac.).
Pharia Steud. = Phania DC. (Compos.).
Pharium Herb. = Bessera Schult. f. (Liliac.).
Pharmacaceae Dulac = Ranunculaceae Juss.
Pharmacosycea Miq. = Ficus L. (Morac.).
Pharmacum Kuntze = Astronia Nor. ex Bl. (Melastomatac.).
Pharnaceum L. Aïzoaceae. 20 S. Afr.
Pharseophora Miers = Memora Miers (Bignoniac.).
Pharus P.Br. Gramineae. 7–8 trop. Am., W.I.
Phasellus Medik. = Phaseolus L. (Legumin.).
Phaseolaceae Ponce de León & Alvares = Fabaceae Reichb.
Phaseolodes Kuntze = Millettia Wight & Arn. (Legumin.).
Phaseoloïdes Duham. = Wisteria Nutt. (Legumin.).
Phaseolus L. Leguminosae. 200–240 trop. & subtrop., chiefly Am. Fl. mech.
like *Vicia*, but complicated by the spiral coiling of the keel with the enclosed
style. *P. coccineus* L. (Mexico) is the scarlet runner; *P. vulgaris* L. the French
or kidney bean; *P. lunatus* L. the Lima or duffin bean, similarly used in the
trop.; *P. acutifolius* A. Gray the tepary of the SW. U.S.
Phaulanthus Ridl. = Anerincleistus Korth. (Melastomatac.).
*Phaulopsis Willd. corr. Spreng. Acanthaceae. 20 trop. Afr., Masc., India.
Phaulothamnus A. Gray. Achatocarpaceae. 1 N. Mex.
Phaylopsis Willd. = Phaulopsis Willd. corr. Spreng. (Acanthac.).
Phebalium Vent. Rutaceae. 40 Austr., N.Z.
Pheboantha Reichb. (sphalm.) = Phleboantha Tausch = Ajuga L. (Labiat.).

Phebolitis DC. (sphalm.) = Phlebolithis Gaertn. = ? Mimusops L. (Sapotac.).
Phedimus Rafin. = Sedum L. (Crassulac.).
Phegopteris (Presl) Fée, emend. Ching. Thelypteridaceae. 3 N. Temp.
(Formerly used for an unnatural mixture of all ferns of *Aspidiales* lacking in-
Phegopyrum Peterm. = Fagopyrum Mill. (Polygonac.). [dusia.)
Phegos St-Lag. = Fagus L. (Fagac.).
Pheidochloa S. T. Blake. Gramineae. 1 New Guinea, 1 Queensland.
Phelandrium Neck. = Phellandrium L. = Oenanthe L. (Umbellif.).
Pheliandra Werderm. = Solanum L. (Solanac.).
Phelima Nor. = Horsfieldia Willd. (Myristicac.).
Phelipaca Fourr. (sphalm.) = seq.
Phelipaea Desf. Orobanchaceae. 4 SW. As.
Phelipaea P. & K. (1) = Phaelypaea P.Br. = Stemodia L. (Scrophulariac.).
Phelipaea P. & K. (2) = Phelypaea Boehm. = Aeginetia L. (Orobanchac.).
Phelipaea P. & K. (3) = Phelypea Thunb. = Cytinus L. (Rafflesiac.).
Phelipanche Pomel = Orobanche L. (Orobanchac.).
Phelipea Pers. = Phelipaea Desf. (Orobanchac.).
Phellandrium L. = Oenanthe L. (Umbellif.).
Phellandryum Gilib. = praec.
Phellinaceae ['-neaceae'] (Loes.) Takht. Dicots. 1/12 New Caled. Trees or
shrubs; ls. alt., simple, evergr., exstip., ± crowded towards tips of branches.
Fls. dioec., in spicate or panic. infl. K 4–6, small, open; C 4–6, free, fleshy, valv.,
with a small inflexed apiculus. A 4–6, free, alternipet. (stds. present in ♀ fl.);
G̲ (2–5), with sessile lobed stig., and 1 pend. hemitr. or slightly campylotr.
apotr. ovule per loc. (pistillode present in ♂ fl.). Fruit a drupe with 2–5 stones;
seeds with copious endosp. Only genus: *Phelline*.
Phelline Labill. Phellinaceae. 12 New Caled.
Phellocarpus Benth. = Pterocarpus L. (Legumin.).
Phellodendron Rupr. Rutaceae. 10 E. As.
Phelloderma Miers = Priva Adans. (Verbenac.).
Phellolophium Baker. Umbelliferae. 1 Madag.
Phellopterus Benth. = Glehnia F. Schmidt (Umbellif.).
Phellopterus (Torr. & Gray) Coult. & Rose (~ Cymopterus Rafin.). Um-
belliferae. 5 SW. U.S.
Phellosperma Britton & Rose = Mammillaria Haw. (Cactac.).
Phelpsiella Maguire. Rapateaceae. 1 Venez.
Phelypaea Boehm. = Aeginetia L. (Orobanchac.).
Phelypaea D. Don = Phelipaea Desf. (Orobanchac.).
Phelypaeaceae Horan. = Orobanchaceae Vent.
Phelypea Adans. = Phelypaea Boehm. = Aeginetia L. (Orobanchac.).
Phelypea Thunb. = Cytinus L. (Rafflesiac.).
Phemeranthus Rafin. = Talinum Adans. (Portulacac.).
Phemoranthus Rafin. = praec.
Phenakosperum Endl. (sphalm.) = seq.
Phenakospermum Endl. corr. Miq. (~ Ravenala Adans.). Strelitziaceae.
2 E. trop. S. Am.
Phenax Wedd. Urticaceae. 25 Mex. to trop. S. Am., W.I.; 2 Madag.
Phenianthus Rafin. = Lonicera L. (Caprifoliac.).

PHENOPUS

Phenopus Hook. f. = Phaenopus DC. = Lactuca L. (Compos.).

Phenotrichis Steud. = Pherotrichis Decne (Asclepiadac.).

Pherelobus Phillips (sphalm.) = seq.

Pherolobus N. E. Brown. Aïzoaceae. 1 S. Afr.

Pherosphaera Archer = Diselma Hook. f. (Cupressac.) *p.p.* + Microcachrys Hook. f. (Podocarpac.) *p.p.*

Pherosphaera Hook. f. = Microstrobos Garden & Johnson (Podocarpac.).

Pherotrichis Decne. Asclepiadaceae. 2 Mex.

Phialacanthus Benth. Acanthaceae. 5 E. Himal. to Malay Penins.

Phialanthus Griseb. Rubiaceae. 10 W.I.

Phialis Spreng. = Eriophyllum Lag. (Compos.).

Phialocarpus Deflers = Kedrostis Medik. (Cucurbitac.).

Phialodiscus Radlk. Sapindaceae. 8 trop. Afr.

Phidiasia Urb. Acanthaceae. 1 Cuba.

Philacanthus B. D. Jacks. (sphalm.) = Phialacanthus Benth. (Acanthac.).

Philacra Dwyer. Ochnaceae. 3 trop. S. Am.

Philactis Schrad. Compositae. 3 S. Mex., Guatem.

Philadelphaceae D. Don. Dicots. 7/135 S. Eur. to E. As. & N. Am., S. to Philipp. Is. & C. Am. Shrubs or small trees, usu. with ± stellate pubesc. Ls. simple, serr. or ent., opp. or vertic., more rarely alt., exstip., usu. decid. Fls. reg., ♀, in term. cymes or rac., rarely sol. K (5–4), valv. or imbr.; C 5–7, valv., imbr. or contorted; A 4–∞, fil. sometimes broad with elong. lat. teeth at apex, sometimes conn. at base; G̲ to G̅ (3–5–7), rarely 1-loc., with 3–5–⌐ usu. free styles, and ∞ or rarely sol. ov. on axile or rarely pariet. plac. F a loculic. caps., seeds with fleshy endosp. Genera: *Jamesia, Fendlera, Fendlerella, Whipplea, Carpenteria, Philadelphus, Deutzia.* Closely related to *Hydrangeac.*

Philadelphus L. Philadelphaceae. 75 N. temp., esp. E. As. Shrubs with opp. ls.; the buds arise closely protected by the l.-bases through which in many they have to break. Fls. conspicuous, strongly scented, protogynous. Sta. 20–40; ovary inf., usu. 4-loc.

× **Philageria** Mast. Philesiaceae. Gen. hybr. (Lapageria × Philesia).

Philaginopsis Walp. = Filaginopsis Torr. & Gray = Diaperia Nutt. (Compos.).

Philagonia Blume = Evodia J. R. & G. Forst. (Rutac.).

Philammos Stev. = Astragalus L. (Legumin.).

Philastrea Pierre = Munronia Wight (Meliac.).

Philbornea Hallier. Linaceae. 2 Borneo, Philippines.

Philenoptera Fenzl = Lonchocarpus Kunth (Legumin.).

Phileozera Buckl. = Actinella Nutt. (Compos.).

Philesia Comm. ex Juss. Philesiaceae. 1 S. Chile, a much-branched shrub with petioled, 1-nerved, rolled-back ls., not easily recognised as a Monocot.

***Philesiaceae** Dum. Monocots. 8/10 S. hemisph. Shrubs or climbers, sometimes epiph., with alt. simple exstip. ls. Fls. reg., ♀, large, pend., term. or axill., sol. or in various infl. P 3 + 3 or (3 + 3), sim. or dissim., caduc.; A 3 + 3, variously connate or adnate; anth. large, obl., dorsifixed, intr.; G̲ (3), 3-loc. or 1-loc., with ∞ to few ov. on pariet. plac., and a simple style with 3-lobed or capit. stig. Fr. a berry. Genera: *Philesia, Luzuriaga, Lapageria, Eustrephus, Elachanthera, Geitonoplesium, Behnia, Petermannia.* Near *Alstroemeriaceae.*

PHILYDRACEAE

Philetaeria Liebm. = Fouquieria Kunth (Fouquieriac.).
Philexia Rafin. = Lythrum L. (Lythrac.).
Philgamia Baill. Malpighiaceae. 4 Madag.
Philibertella Vail = Funastrum Fourn. (Asclepiadac.).
Philibertia Kunth. Asclepiadaceae. 40 Am., W.I.
Philippia Klotzsch (~Blaeria L.). Ericaceae. 80 trop. & S. Afr., Madag., Mascarenes.
Philippiamra Kuntze. Portulacaceae. 4 N. Chile.
Philippicereus Backeb. = Eulychnia Phil. (Cactac.).
Philippiella Spegazz. Caryophyllaceae. 1 Patag.
Philippimalva Kuntze = Tetraptera Phil. = Gaya Kunth (Malvac.).
Philippinaea Schlechter & Ames. Orchidaceae. 1 Philippines.
Philippodendraceae Endl. = Malvaceae–Malveae–Sidinae K. Schum.
Philippodendron Endl. = seq.
Philippodendrum Poit. = Plagianthus J. R. & G. Forst. (Malvac.).
Phillipsia Rolfe = Satanocrater Schweinf. (Acanthac.).
Phillyrea L. Oleaceae. 4 Madeira, Medit., to N. Persia.
Phillyrophyllum O. Hoffm. = Philyrophyllum O. Hoffm. corr. Schinz (Compos.).
Philocrena Bong. = Tristicha Thou. (Podostemac.).
Philocrenaceae Bong. = Podostemaceae Rich.
***Philodendron** Schott. Araceae. 275 warm Am., W.I. Usu. shrubs, usu. climbing, often epiph., with both clasping roots and aerial roots reaching the soil (see fam.). The latter sometimes twine as they descend. The pinnation of the l. is due to a delayed development of the portions between the ribs, and not to a process such as occurs in *Monstera* (*q.v.*). Monoecious.
Philodendrum Schott = praec.
Philodice Mart. Eriocaulaceae. 2 Brazil.
Philoglossa DC. Compositae. 1 Peru, Ecuador.
Philogyne Salisb. = Narcissus L. (Amaryllidac.).
Philomeda Nor. ex Thou. = Ouratea L. (Ochnac.).
Philonomia DC. ex Meissn. = Macromeria D. Don (Boraginac.).
Philonotion Schott = Schismatoglottis Zoll. & Mor. (Arac.).
Philopodium Hort. = Muehlenbeckia Meissn. (Polygonac.).
Philostemon Rafin. = Rhus L. (Anacardiac.).
Philostemum Steud. = praec.
Philostizus Cass. = Centaurea L. (Compos.).
Philotheca Rudge. Rutaceae. 6 Austr.
Philotria Rafin. = Elodea Michx (Hydrocharitac.).
Philoxerus R.Br. Amaranthaceae. 10 coasts trop. Austr., warm Am., W.I., W. Afr.
Philyca Boehm. = Phylica L. (Rhamnac.).
***Philydraceae** Link. Monocots. 4/5 Indomal., Austr. Erect rhiz. herbs, with 2-ranked sheathing narrow ls., and fls. in simple or cpd. spikes, ⚥, zygo. P homochlam., outer 1+(2) post., inner 2 anter. only developed; A 1, anter., with flattened fil., cells of anth. sometimes twisted; G (3), with 1 style, axile or parietal plac., and ∞ anatr. ov. Caps.; endosp. Genera: *Philydrum, Helmholtzia, Pritzelia, Orthothylax.*

887

PHILYDRELLA

Philydrella Caruel. Philydraceae. 1 W. Austr.

Philydrum Banks ex Gaertn. Philydraceae. 1 E. As., Malaysia, Austr.

Philyra Klotzsch. Euphorbiaceae. 1 S. Brazil, Paraguay.

Philyrophyllum O. Hoffm. corr. Schinz. Compositae. 1 Transvaal, Bechuanaland.

Phinaea Benth. Gesneriaceae. 6 Mex. to Colombia. G semi-inf.

Phippsia (Trin.) R.Br. Gramineae. 1 arctic circumpolar, 2 mts. temp. S. Am.

Phisalis Nocca = Physalis L. (Solanac.).

Phitopis Hook. f. Rubiaceae. 2 Peru. Densely hairy trees.

Phlebaceae Dulac = Ruscaceae + Trilliaceae + Liliaceae–Polygonateae & –Convallarieae.

Phlebanthe P. & K. = Phleboanthe Tausch = Ajuga L. (Labiat.).

Phlebanthe Reichb. = seq.

Phlebanthia Reichb. = Minuartia L. (Caryophyllac.).

Phlebidia Lindl. = Disa Bergius (Orchidac.).

Phlebiogonium Fée = Tectaria Cav. (Aspidiac.).

Phlebiophragmus O. E. Schulz. Cruciferae. 1 Peru.

Phlebiophyllum v. d. Bosch = Polyphlebium Copel. (Hymenophyllac.).

Phleboanthe Tausch = Ajuga L. (Labiat.).

Phlebocalymna Griff. ex Miers = Sphenostemon Baill. (Sphenostemonac.).

Phlebocarya R.Br. Haemodoraceae. 3 W. Austr. G 1-loc.

Phlebochiton Wall. = Pegia Colebr. (Anacardiac.).

Phlebodium (R.Br.) J. Smith. Polypodiaceae. 10 trop. Am. *P. aureum* (L.) J. Sm. much cult. in several vars.

Phlebolithis Gaertn. = ?Mimusops L. (Sapotac.).

Phlebolobium O. E. Schulz. Cruciferae. 1 Falkland Is.

Phlebophyllum Nees. Acanthaceae. 8 Penins. India.

Phlebosporium Hassk. = seq.

Phlebosporum Jungh. = Lespedeza Michx (Legumin.).

Phlebotaenia Griseb. (~ Polygala L.). Polygalaceae. 3 Cuba, Puerto Rico.

Phledinium Spach = Delphinium L. (Ranunculac.).

Phlegmariurus [Herter] Holub = Lycopodium L. (Lycopodiac.).

Phlegmatospermum O. E. Schulz. Cruciferae. 5 Austr.

Phleobanthe Ledeb. (sphalm.) = Phleboanthe Tausch = Ajuga L. (Labiat.).

Phleoïdes Ehrh. (uninom.) = *Phalaris phleoïdes* L. = *Phleum phleoïdes* (L.) Simonk. (Gramin.).

Phleum L. Gramineae. 15 temp. Euras., N. Am. to Mex., temp. S. Am. Inf. palea shorter than awned glume. *P. pratense* L. (timothy grass) valuable fodder.

Phloeodicarpus Bess. = Phlojodicarpus Turcz. ex Ledeb. (Umbellif.).

Phloeophila Hoehne & Schlechter. Orchidaceae. 2 Brazil.

Phloga Nor. ex Hook. f. Palmae. 2 Madag. Ls. with whorled pinnae.

Phlogacanthus Nees. Acanthaceae. 30 Indomal. A 2.

Phlogella Baill. (~ Chrysalidocarpus H. Wendl.). Palmae. 1 Comoro Is.

Phloiodicarpus Reichb. = seq.

Phlojodicarpus Turcz. ex Ledeb. Umbelliferae. 2–4 Siberia.

Phlomidopsis Link = Phlomis L. (Labiat.). [& W. Himal.

Phlomidoschema (Benth.) Vved. (~ Stachys L.). Labiatae. 1 Persia to C. As.

Phlomis L. Labiatae. 100+, N. palaeotemp. The helmet-like upper lip of fl. is raised by a visiting insect. Style branches differ.

Phlomitis Reichb. ex T. Nees = praec.

Phlomoïdes Moench = praec.

Phlomostachys C. Koch = Pitcairnia L'Hérit. (Bromeliac.).

Phlox L. Polemoniaceae. 1 NE. As., 66 N. Am., Mex. Sta. inserted at unequal

Phlyarodoxa S. Moore = Ligustrum L. (Oleac.). [heights.

Phlyctidocarpa Cannon & Theobald. Umbelliferae. 1 SW. Afr.

Phoberos Lour. = Scolopia Schreb. (Flacourtiac.).

Phocea Seem. = Macaranga Thou. (Euphorbiac.).

Phoebanthus Blake (~ Helianthella Torr. & Gray). Compositae. 2 N. Am.

Phoebe Nees. Lauraceae. 70 Indomal., trop. Am., W.I. (The American spp. hitherto referred to *Ph.* should be removed to *Persea* and *Cinnamomum*, fide Kostermans.)

Phoenicaceae Schultz-Schultzenst. = Palmae–Phoeniceae Spreng.

Phoenicanthemum (Blume) Reichb. = Helixanthera Lour. (Loranthac.).

Phoenicanthus Alston. Annonaceae. 1 Ceylon.

Phoenicanthus P. & K. = Phaenicanthus Thw. = Premna L. (Verbenac.).

Phoenicaulis Nutt. Cruciferae. 2 Pacif. N. Am.

Phoenicimon Ridley = Glycosmis Corrêa (Rutac.).

Phoenicocissus Mart. ex Meissn. = Lundia DC. (Bignoniac.).

Phoenicophorium H. Wendl. Palmae. 1 Seychelles.

Phoenicopus Spach = Phaenixopus Cass. = Lactuca L. (Compos.).

Phoenicoseris (Skottsb.) Skottsb. Compositae. 3 Juan Fernandez.

Phoenicosperma Miq. = Sloanea L. (Elaeocarpac.).

Phoenicospermum B. D. Jacks. (sphalm.) = praec.

Phoenix Haller = Chrysopogon Trin. (Gramin.).

Phoenix L. Palmae. 17 warm Afr., As., incl. *P. dactylifera* L. (date, N. Afr., SW. As.). Columnar stem covered with old l.-bases; ls. pinnate. Dioec.; ♀ spadix fert. by hanging a ♂ over it. Berry; usu. 1 cpl. only ripens; seeds with hard cellulose endosp. Yields fr., wine, sugar (cf. *Cocos*), etc. Hats, mats, thatch, etc., from leaves.

Phoenixopus Reichb. = Phaenixopus Cass. = Lactuca L. (Compos.).

Phoenocoma G. Don (sphalm.) = Phaenocoma D. Don (Compos.).

Phoenopus Nym. (sphalm.) = Phaenopus DC. = Lactuca L. (Compos.).

Pholacilla Griseb. = Trichilia L. (Meliac.).

Pholidandra Neck. = Raputia Aubl. (Rutac.).

Pholidia R.Br. Myoporaceae. 45 Austr.

Pholidiopsis F. Muell. = praec.

Pholidocarpus Blume. Palmae. 7 Indoch., W. Malaysia, Moluccas. Fr. papillose, hairy.

Pholidophyllum Vis. = Cryptanthus Otto & Dietr. (Bromeliac.).

Pholidostachys H. Wendl. ex Benth. & Hook. f. Palmae. 2 C. Am., Colombia.

Pholidota Lindl. Orchidaceae. 55 China & India to Australia & Polynesia.

Pholisma Nutt. ex Hook. Lennoaceae. 2 S. Calif. & Lower Calif. In sandy ground, on roots of *Hymenoclea* (Compos.) or *Croton* (Euphorbiac.).

Pholistoma Lilja (~ Nemophila Nutt.). Hydrophyllaceae. 3 SW. U.S., Lower Calif.

PHOLIURUS

Pholiurus Trin. Gramineae. 1 SE. Eur.

Pholomphis Rafin. = Miconia Ruiz & Pav. (Melastomatac.).

Phoniphora Neck. = Phoenix L. (Palm.).

Phonus Hill = Carthamus L. (Compos.).

Phoradendron Nutt. Viscaceae (~ Loranthaceae). 190 Am., W.I.

Phormangis Schlechter = Ancistrorhynchus Finet (Orchidac.).

Phormiaceae J. G. Agardh = Agavaceae J. G. Agardh.

Phormium J. R. & G. Forst. Agavaceae. 2 N.Z., Norfolk I. Ls. isobilat. The ls. of *P. tenax* J. R. & G. Forst. furnish N.Z. flax.

Phornothamnus Baker. Melastomataceae. 1 Madag.

Phorodendrum P. & K. = Phoradendron Nutt. (Loranthac.).

Phorolobus Desv. = Cryptogramma R.Br. (Cryptogrammac.).

Phosanthus Rafin. = Isertia Schreb. (Rubiac.).

Photinia Lindl. Rosaceae. 60 Himal. to Japan & Sumatra, N. Am.

Photinopteris J. Sm. Polypodiaceae. 1 Malaysia.

Phoxanthus Benth. = Ophiocaryon Schomb. (Meliosmac.).

Phragmanthera Van Tiegh. = Tapinanthus Bl. (Loranthac.).

× **Phragmipaphiopedilum** hort. (vii) = × Phragmipaphium hort. (Orchidac.).

× **Phragmipaphium** hort. Orchidaceae. Gen. hybr. (iii) (Paphiopedilum × Phragmipedium).

Phragmipedilum Rolfe = seq.

Phragmipedium Rolfe. Orchidaceae. 11 trop. S. Am.

Phragmites Adans. Gramineae. 3 cosmop. *P. communis* Trin. is the common reed. It forms floating fens at the Danube mouth. It has a creeping rhiz. and tall upright stem with a dense panicle of spikelets. The lowest fl. of the spikelet is ♂, the rest ♀. A few cm. above the leaf-sheath are three transverse dents in the l. (' *Teufelsbiss*'); these are due to pressure at the time when the rolled-up blade is still in the sheaths of older ls.

Phragmites Allam. ? Cyperaceae. Quid?

Phragmites Trin. = Gynerium Kunth (Gramin.).

Phragmocarpidium Krapov. Malvaceae. 1 Brazil.

Phragmocassia Britton & Rose = Cassia L. (Legumin.).

Phragmopedilum Pfitzer = Phragmipedium Rolfe (Orchidac.).

Phragmorchis L. O. Williams. Orchidaceae. 1 Philippines.

Phragmotheca Cuatrec. Bombacaceae. 1 Colombia.

Phreatia Lindl. Orchidaceae. 190 NE. India, China & Formosa to Polynesia, Austr., N.Z.

Phrenanthes Wigg. = Prenanthes L. (Compos.).

Phrissocarpus Miers. Apocynaceae. 1 S. Am.

Phrodus Miers. Solanaceae. 4 Chile.

Phryganocydia Mart. ex Baill. Bignoniaceae. 3 trop. S. Am.

Phryganthus Baker (sphalm.) = Phyganthus Poepp. & Endl. = Tecophilaea Bert. (Tecophilaeac.).

Phrygia S. F. Gray = Centaurea L. (Compos.).

Phrygilanthus Eichl. Loranthaceae. 8 Philipp. Is., E. New Guinea, E. Austr.; 40 C. & trop. S. Am.

Phrygiobureaua Kuntze = Phryganocydia Mart. ex Baill. (Bignoniac.).

Phryma Forsk. = Priva Adans. (Verbenac.).

Phryma L. Phrymataceae. 1–2 India to Japan, E. N. Am. [Name perh. in error for, or altered from, Gr. *phyrma*, mixture, defilement, etc.]
*Phrymataceae Schauer. Dicots. 1/2 E. As., E. N. Am. Herbs with opp. ls. and term. spikes of small zygo. fls. K (5), bilabiate, with hooked teeth; C (5), bilabiate; A 4, didynamous; G̲ 1 with 1 erect orthotr. ov. Only genus: *Phryma*. Doubtfully worth separating as a fam. from *Verbenaceae*. The chief distinction is in the erect orthotr. ovule; no known transitions between this and the anatr. ov. of the *Verb*.
Phryna (Boiss.) Pax & K. Hoffm. = Phrynella Pax & K. Hoffm. (Caryophyllac.).
Phryne Bubani. Cruciferae. 5 mts. of S. Eur., Caucasus.
Phrynella Pax & K. Hoffm. Caryophyllaceae. 1 As. Min.
Phrynium Loefl. ex Kuntze = Heteranthera Ruiz & Pav. (Pontederiac.).
*Phrynium Willd. Marantaceae. 30 trop. Afr., Indomal. Aril.
Phtheirospermum Bunge ex Fisch. & Mey. Scrophulariaceae. 7 E. As.
Phtheirotheca Maxim. ex Regel = Caulophyllum Michx (Leonticac.).
Phthirusa Mart. (= Hemitria Rafin.). Loranthaceae. 60 trop. Am.
Phtirium Rafin. = Delphinium L. (Ranunculac.).
Phu Ludw. = Valeriana L. (Valerianac.).
Phucagrostis Cavol. = Zostera L. (Zosterac.) + Cymodocea Koenig (Cymodoceac.).
Phuodendron Graebn. Valerianaceae. 1 Brazil.
Phuopsis (Griseb.) Benth. & Hook. f. Rubiaceae. 1 Caucasus, E. As. Min., NW. Persia.
Phusicarpos Poir. = Hovea R.Br. (Legumin.).
Phycagrostis P. & K. = Phucagrostis Cavol. = Zostera L. (Zosterac.) + Cymodocea Koenig (Cymodoceac.).
Phycella Lindl. = Hippeastrum Herb. (Amaryllidac.).
Phycoschoenus (Aschers.) Nakai = Cymodocea Koenig (Cymodoceac.).
Phyganthus Poepp. & Endl. = Tecophilaea Bert. (Tecophilaeac.).
Phygelius E. Mey. Scrophulariaceae. 2 S. Afr.
Phyla Lour. (~ Lippia L.). Verbenaceae. 10 N. & C. Am.
Phylacanthus Benth. = Angelonia Kunth (Scrophulariac.).
Phylacium Bennett. Leguminosae. 2 SE. As., Malay Penins., Java.
Phylanthera Nor. = Hypobathrum Bl. (Rubiac.).
Phylanthus Murr. (sphalm.) = Phyllanthus L. (Euphorbiac.).
Phylax Nor. = Polygala L. (Polygalac.).
Phylepidum Rafin. = Phyllepidum Rafin. = Polygonella Michx (Polygonac.).
Phylesiaceae Dum. = Philesiaceae Dum. (corr.).
Phylica L. Rhamnaceae. 150 S. Afr., Madag., Tristan da Cunha. Mostly xero. shrubs, often of heath-like habit with revolute ls.
Phylicaceae J. G. Agardh = Rhamnaceae Juss.
Phylidr[ac]eae Lindl. = Philydraceae Link.
Phylidrum Willd. = Philydrum Banks ex Gaertn. (Philydrac.).
Phylirastrum (Pierre) Pierre = Caloncoba Gilg (Flacourtiac.).
Phyllacantha Hook. f. Rubiaceae. 1 Cuba.
Phyllachne J. R. & G. Forst. Stylidiaceae. 4 Tasm., N.Z., temp. S. Am.
Phyllactinia Benth. (1873; non Lévl. 1851—Fungi) = Pasaccardoa Kuntze (Compos.).

PHYLLACTIS

Phyllactis Pers. Valerianaceae. 25 Mex., W. S. Am. to Patag.

Phyllactis Steud. = Philactis Schrad. (Compos.).

Phyllagathis Blume. Melastomataceae. 20 S. China to W. Malaysia.

Phyllamphora Lour. = Nepenthes L. (Nepenthac.).

Phyllanoa Croizat. Euphorbiaceae. 1 Colombia.

Phyllanth[ac]eae J. G. Agardh = Euphorbiaceae–Phyllantheae Bl.

Phyllanthera Blume. Periplocaceae. 2 Malay Penins., Java.

Phyllantherum Rafin. = Trillium L. (Trilliac.).

Phyllanthidea Didr. = Andrachne L. (Euphorbiac.).

Phyllanthodendron Hemsl. (~ Phyllanthus L.). Euphorbiaceae. 12 S. China, Indoch., Siam, Malay Penins.

Phyllanthos St-Lag. = seq.

Phyllanthus L. Euphorbiaceae. 600 trop. & subtrop., exc. Eur. & N. As. The trop. Am. § *Xylophylla* has flat green phylloclades bearing fls. on the margins. The ultimate shoots in § *Phyllanthus (Euphyllanthus)* look like pinnate ls. In *P. cyclanthera* Baill. (Cuba) the ♂ fl. has its 3 sta. united into a synandrium with ring-like anther at top. *P. fluitans* Muell. Arg. (S. Am.) is a remarkable free-floating aquatic with the habit of *Salvinia*, unique in *Euphorbiac.*

Phyllapophysis Mansfeld. Melastomataceae. 1 New Guinea.

Phyllarthron DC. Bignoniaceae. 13 Madag., Comoro Is. The l. is reduced to a jointed winged petiole.

Phyllarthus Neck. = ?Opuntia Mill. (Cactac.).

Phyllaurea Lour. = Codiaeum Juss. (Euphorbiac.).

Phyllepidum Rafin. = Polygonella Michx (Polygonac.).

Phyllimena Blume ex DC. = Enydra Lour. (Compos.).

Phyllirea Duham. = Phillyrea L. (Oleac.).

Phyllis L. Rubiaceae. 2 Canaries, Madeira.

Phyllitis Hill (syn. *Scolopendrium* Adanson). Aspleniaceae. 8 trop. & subtrop. Copeland includes in *Asplenium* L.

Phyllitis Rafin. = Pteris L. (Pteridac.).

Phylloboea Benth. Gesneriaceae. 3 Burma, China, Malay Penins.

Phyllobolus N. E. Br. Aïzoaceae. 4 S. Afr.

Phyllobotrium Willis (sphalm.) = seq.

Phyllobotryon Muell. Arg. Flacourtiaceae. 2 trop. Afr. Infl. epiphyllous.

Phyllobotrys Fourr. = Genista L. (Legumin.).

Phyllobotryum Muell. Arg. = Phyllobotryon Muell. Arg. (Flacourtiac.).

Phyllobryon Miq. = Peperomia Ruiz & Pav. (Peperomiac.).

Phyllocactus Link = Epiphyllum Haw., etc. (Cactac.).

Phyllocalymma Benth. = Angianthus Wendl. (Compos.).

Phyllocalyx Berg = Eugenia L. (Myrtac.).

Phyllocalyx A. Rich. = Crotalaria L. (Legumin.).

Phyllocara Guşul. Boraginaceae. 1 Cauc., Armenia.

Phyllocarpa Nutt. ex Moq. = Atriplex L. (Chenopodiac.).

Phyllocarpus Riedel ex Endl. Leguminosae. 2 C. & trop. S. Am.

Phyllocasia Reichb. = Xanthosoma Schott (Arac.).

Phyllocephalium Miq. = seq.

Phyllocephalum Blume = Centratherum Cass. (Compos.).

× **Phyllocereus** Knebel = × Heliphyllum Rowley (Cactac.).

Phyllocereus Miq. = Epiphyllum Haw., etc. (Cactac.).

× **Phyllocereus** Worsley (nom. illegit.). Cactaceae. Gen. hybr. ([Epiphyllum × Nopalxochia × Selenicereus] × Heliocereus).

Phyllocharis Diels (1917; non Fée 1824—Lichenes) = Ruthiella v. Steenis (Campanulac.).

Phyllochilium Cabrera = Chiliophyllum R. Phil. (Compos.).

Phyllochlamys Bur. = Streblus Lour. (Morac.).

Phyllocladaceae (Pilger) Core (~ Podocarpac.). Gymnosp. 1/7 Malaysia, Tasm., N.Z. Close to *Podocarpaceae* (*q.v.*), but differing conspicuously in the production of phyllocladoid short shoots, which may be flabellate, lobed or dentate, in the axils of minute scale-like ls.; in the absence of the epimatium or ephimatium (a liguliform outgrowth of the carpel, variously associated with the ovule, present in all *Podocarpac.* except *Microstrobos*); and in the arillate seed. Only genus: *Phyllocladus*.

*****Phyllocladus** Rich. ex Mirb. Phyllocladaceae. 7 Philippine Islands, Borneo, Moluccas, New Guinea, Tasm., N.Z. (celery pine). The 'short shoots' are represented by flat green leaf-like structures—phylloclades—whose stem-nature is easily recognised by their position in the axils of the scale ls. on the 'long shoots'. The edges of the phylloclades also bear scales. The fls. (mon- or dioec.) occupy the position of phylloclades. Each cpl. has one axillary erect ovule. The seed has a small basal aril. The timber is useful; the bark of *P. trichomanoïdes* D. Don ex A. Cunn. is used for tanning.

Phylloclinium Baill. Flacourtiaceae. 2 trop. Afr. Infl. epiphyllous.

Phyllocomos Mast. Restionaceae. 1 S. Afr.

Phyllocoryne Hook. f. = Scybalium Schott & Endl. (Balanophorac.).

Phyllocosmus Klotzsch (~ Ochthocosmus Benth.). Ixonanthaceae. 8 trop. Afr.

Phyllocrater Wernham (~ Oldenlandia L.). Rubiaceae. 1 Borneo.

Phylloctenium Baill. Bignoniaceae. 2 Madag.

Phyllocyclus Kurz = Canscora Lam. (Gentianac.).

Phyllocytisus Fourn. = Cytisus L. (Legumin.).

Phyllodes Lour. = Phrynium Willd. (Marantac.).

Phyllodesmis Van Tiegh. = Taxillus Van Tiegh. (Loranthac.).

Phyllodium Desv. Leguminosae. 9 SE. As.

Phyllodoce Link = Acacia Mill. (Legumin.).

Phyllodoce Salisb. Ericaceae. 7 N. circumpolar & temp.

Phyllodolon Salisb. = Allium L. (Alliac.).

Phyllogeiton (Weberb.) Herzog (~ Berchemia Neck. ex DC.). Rhamnaceae. 2 trop. & S. Afr.

Phylloglossaceae Kunze = Lycopodiaceae Reichb.

Phylloglossum Kunze. Lycopodiaceae. 1 Austr., N.Z., *P. drummondii* Kunze. Small plant producing only a single strobilus.

Phylloglottis Salisb. = Eriospermum Jacq. (Liliac.).

Phyllogonum Coville (1893; non *Phyllogonium* Brid. 1827—Musci) = Gilmania Coville (Polygonac.).

Phyllolobium Fisch. = Astragalus L. (Legumin.).

Phylloma Ker-Gawl. = Lomatophyllum Willd. (Liliac.).

Phyllomatia Benth. = Rhynchosia Lour. (Legumin.).

Phyllomelia Griseb. Rubiaceae. 1 Cuba.

PHYLLOMERIA

Phyllomeria Griseb. (sphalm.) = praec.

Phyllomphax Schlechter = Brachycorythis Lindl. (Orchidac.).

Phyllonoma Willd. ex Schult. Dulongiaceae. 8 Mex. to Peru.

Phyllonomaceae Rusby = Dulongiaceae J. G. Agardh.

Phyllopappus Walp. = Microseris D. Don (Compos.).

Phyllophiorhiza Kuntze = Ophiorhiziphyllon S. Kurz (Acanthac.).

Phyllophyton Kudo = Nepeta L. (Labiat.).

Phyllopodium Benth. Scrophulariaceae. 15 S. Afr.

Phyllopus DC. = Henriettea DC. (Melastomatac.).

Phyllorachis Trimen = Phyllorhachis Trimen corr. Benth. (Gramin.).

Phyllorchis Thou. = Phyllorkis Thou. = Bulbophyllum Thou. (Orchidac.).

Phyllorhachis Trimen corr. Benth. Gramineae. 1 S. trop. Afr.

Phyllorkis Thou. = Bulbophyllum Thou. (Orchidac.).

Phylloschoenus Fourr. = Juncus L. (Juncac.).

Phylloscirpus C. B. Clarke. Cyperaceae. 1 Argentina.

× **Phylloselenicereus** Hort. = × Seleniphyllum Rowley (Cactac.).

Phyllosma Bolus. Rutaceae. 1 S. Afr.

Phyllospadix Hook. Zosteraceae. 5 Japan, Pacif. coast N. Am.

*****Phyllostachys** Sieb. et Zucc. Gramineae. 40 Himal. to Japan. The stripped stems are whangee canes.

Phyllostachys Torr. ex Steud. = Carex L. (Cyperac.).

Phyllostegia Benth. Labiatae. 25 Hawaii, Tahiti.

Phyllostelidium Beauverd. Compositae. 2 Andes.

Phyllostema Neck. = Simaba Aubl. = Quassia L. (Simaroubac.).

Phyllostemonodaphne Kosterm. Lauraceae. 1 Brazil.

Phyllostephanus Van Tiegh. = Loranthus L. (Loranthac.).

Phyllostylon Capanema ex Benth. & Hook. f. Ulmaceae. 3 W.I. to Paraguay. Good timber.

Phyllota Benth. Leguminosae. 10 Austr.

Phyllotaenium André = Xanthosoma Schott (Arac.).

Phyllotephrum Gandog. = Clinopodium L. (Labiat.).

× **Phyllothamnus** C. K. Schneider. Ericaceae. Gen. hybr. (Phyllodoce × Rhodothamnus).

Phyllotheca Nutt. ex Moq. = Atriplex L. (Chenopodiac.).

Phyllotrichum Thorel ex Lecomte. Sapindaceae. 1 Indoch.

Phylloxylon Baill. Leguminosae. 4 Madag., Mauritius. Timber hard.

Phylloxys auct. ex Moq. = Cornulaca Delile(?) (Chenopodiac.).

Phyllymena Blume ex Miq. = Phyllimena Blume ex DC. = Enydra Lour. (Compos.).

Phyllyrea G. Don = Phillyrea L. (Oleac.).

Phylocarpos Rafin. = Physocarpus (Cambess.) Maxim. (Rosac.).

Phyloma Gmel. = Cymbaria L. (Scrophulariac.).

Phymaspermum Less. Compositae. 9 S. Afr.

Phymatanthus Sweet = Pelargonium L'Hérit. (Geraniac.).

Phymatarum M. Hotta. Araceae. 2 Borneo.

Phymatidium Lindl. Orchidaceae. 5 Brazil.

Phymatis E. Mey. = Carum L. (Umbellif.).

Phymatocarpus F. Muell. Myrtaceae. 2 W. Austr.

Phymatodes Presl = Microsorium Link (Polypodiac.).
Phymatopsis J. Sm. = Crypsinus Presl (Polypodiac.).
Phymosia Desv. ex Ham. Malvaceae. 8 Mex., Guatem., W.I.
Phyodina Rafin. (~ Tradescantia L.). Commelinaceae. 5 Florida, C. & trop. S. Am.
Phyrrheima Hassk. (sphalm.) = Pyrrheima Hassk. = Siderasis Rafin. (Commelinac.).
Physa Thou. = Glinus L. (Aïzoac.).
Physacanthus Benth. Acanthaceae. 5 trop. Afr.
Physaliastrum Makino. Solanaceae. 4 E. As.
Physalidium Fenzl. Cruciferae. 2 Persia.
Physalis L. Solanaceae. 100 cosmop., esp. Am. The berry of *P. alkekengi* L. (winter cherry) is edible, also that of *P. peruviana* L. (strawberry or gooseberry tomato, Cape gooseberry). It is enclosed in the bladdery persistent calyx, which becomes orange red.
Physalobium Steud. = Physolobium Benth. = Kennedya Vent. (Legumin.).
Physalodes Boehm. = Nicandra Adans. (Solanac.).
Physaloïdes Moench = Withania Pauq. (Solanac.).
Physandra Botsch. Chenopodiaceae. 1 C. As.
Physanthemum Klotzsch = Courbonia Brongn. = Maerua Forsk. (Capparidac.).
Physanthera Bert. ex Steud. = Rodriguezia Ruiz & Pav. (Orchidac.).
Physanthyllis Boiss. = Anthyllis L. (Legumin.).
Physaria (Nutt.) A. Gray. Cruciferae. 15 Pacif. N. Am.
Physarus Steud. = Physurus Rich. (Orchidac.).
Physcium P. & K. = Physkium Lour. = Vallisneria L. (Hydrocharitac.).
Physedra Hook. f. = Coccinia Wight & Arn. (Cucurbitac.).
Physematium Kaulf. = Woodsia R.Br. (Aspidiac.).
Physena Nor. ex Thou. Capparidaceae. 2 Madag.
Physenaceae Takhtadjian (nomen) = Capparidaceae–Stixeae p.p. (?).
Physetobasis Hassk. Apocynaceae. 1 Java.
Physianthus Mart. = Araujia Brot. (Asclepiadac.).
Physicarpos DC. = Phusicarpos Poir. = Hovea R.Br. (Legumin.).
Physichilus Nees = Hygrophila R.Br. (Acanthac.).
Physidium Schrad. = Angelonia Humb. & Bonpl. (Scrophulariac.).
Physiglochis Neck. = Carex L. (Cyperac.).
Physinga Lindl. = Epidendrum L. (Orchidac.).
Physiphora Soland. ex DC. = Rinorea Aubl. (Violac.).
Physkium Lour. = Vallisneria L. (Hydrocharitac.).
Physocalycium Vest = Bryophyllum Salisb. (Crassulac.).
Physocalymma Pohl. Lythraceae. 1 trop. S. Am. Timber valuable.
Physocalymna DC. = praec.
Physocalyx Pohl. Scrophulariaceae. 2 Brazil.
Physocardamum Hedge. Leguminosae. 1 E. Turkey.
Physocarpa Rafin. = Physocarpus (Cambess.) Maxim. (Rosac.).
Physocarpon Neck. = Lychnis L. (Caryophyllac.).
Physocarpum Bercht. & Presl = Sumnera Nieuwl. = Thalictrum L. (Ranunculac.).

PHYSOCARPUS

*Physocarpus (Cambess.) Rafin. corr. Maxim. Rosaceae. 10 NE. As., N. Am. G usu. 5. Fr. 2-valved.

Physocarpus P. & K. = Phusicarpos Poir. = Hovea R.Br. (Legumin.).

Physocaulis (DC.) Tausch. Umbelliferae. 1 Medit.

Physocaulos Fiori & Paol. = praec.

Physocaulus Koch = praec.

Physoceras Schlechter. Orchidaceae. 7 Madag.

Physocheilus Nutt. ex Benth. = Orthocarpus Nutt. (Labiat.).

Physochilus P. & K. = Physichilus Nees = Hygrophila R.Br. (Acanthac.).

Physochlaena C. Koch = seq.

Physochlaina G. Don. Solanaceae. 10 C. As. to Himal. & China.

Physoclaina Boiss. = praec.

Physocodon Turcz. = Melochia L. (Sterculiac.).

Physodeira Hanst. = Episcia Mart. (Gesneriac.).

Physodia Salisb. = Urginea Steinh. (Liliac.).

Physodium Presl. Sterculiaceae. 2 Mex.

Physogeton Jaub. & Spach = Halanthium C. Koch (Chenopodiac.).

Physoglochin P. & K. = Physiglochis Neck. = Carex L. (Cyperac.).

Physokentia Becc. Palmae. 3–4 Fiji, New Hebrides.

Physolepidion Schrenk = Lepidium L. (Crucif.).

Physolepidium Endl. = praec.

Physoleucas Jaub. & Spach. Labiatae. 1 Arabia.

Physolobium Benth. = Kennedya Vent. (Legumin.).

Physolophium Turcz. = Angelica L. (Umbellif.).

Physolychnis Rupr. = Lychnis L. (Caryophyllac.).

Physondra Rafin. = Phaca L. (Legumin.).

Physophora Link = Physospermum Cusson (Umbellif.).

Physophora P. & K. = Physiphora Soland. ex R.Br. = Rinorea Aubl. (Violac.)

Physoplexis (Endl.) Schur = Synotoma (G. Don) R. Schulz (Campanulac.).

Physopodium Desv. = ? Combretum L. (Combretac.).

Physopsis Turcz. Dicrastylidaceae. 2 W. & S. Austr.

Physoptychis Boiss. Cruciferae. 2 E. As. Min., NW. Persia.

Physopyrum Popov. Polygonaceae. 1 C. As.

Physorhyncus Hook. f. & Anders. = seq.

Physorrhynchus Hook. Cruciferae. 2 S. Persia to NW. India.

Physosiphon Lindl. Orchidaceae. 20 Mex. to trop. S. Am.

Physospermopsis H. Wolff. Umbelliferae. 9 Himal. to W. China.

Physospermum Cusson. Umbelliferae. 6 temp. Euras.

Physospermum Lag. = Pleurospermum Hoffm. (Umbellif.).

Physostegia Benth. Labiatae. 15 N. Am.

Physostelma Wight. Asclepiadaceae. 6 SE. As.

Physostemon Mart. (~ Cleome L.). Cleomaceae. 7 Mex., trop. S. Am. Post. fil. with swelling at apex.

Physostigma Balf. Leguminosae. 5 trop. Afr. The keel is spurred. *P. venenosum* Balf. is the ordeal bean of Calabar.

Physothallis Garay. Orchidaceae. 1 Ecuador.

Physotheca Rafin. = Physocarpus (Cambess.) Maxim. (Rosac.).

Physotrichia Hiern. Umbelliferae. 10 trop. Afr.

Physurus Rich. = Erythrodes Bl. (Orchidac.).

Phytarrhiza Vis. = Tillandsia L. (Bromeliac.).

Phytelephant[ac]eae Mart. = Palmae–Phytelephantoïdeae Drude.

Phytelephas Ruiz & Pav. Palmae. 15 trop. Am. Like *Nypa*, widely different from other palms; with possible affinities to *Pandanaceae* and *Cyclanthaceae*. Short-stemmed with large pinnate rad. ls., and dioec. infls. ♂ infl. a sausage-shaped spadix; the fl. has an irreg. P. and ∞ sta. with long filaments. ♀ spadix simple with spathe of several ls., and about 6 fls.; the fl. has an irreg. P (an outer whorl of 3 and inner of 5–10 longer ls.), numerous stds. and usu. a 5-loc. ov. with long style and stigmas. Each fl. gives a berry, and the actual fr. consists of 6 or more of these united together. The outer coat is hard, with woody protuberances. Each partial fr. contains several seeds; the endosp. (cellulose) is very hard (vegetable ivory) and is used for turning into billiard balls, etc. [Compare this fr. with those of *Pandanus* and *Carludovica*.]

Phytelephasi[ac]eae [Brongn. ex] Chadef. & Emberg. = Palmae–Phytelephantoïdeae Drude.

Phyteuma L. Campanulaceae. 40 Medit., Eur., As. Fl. mech. interesting (see fam.). The fls. are small, and massed together in heads. A tube is formed by the coherence of the tips of the long thin petals, within which the anthers are held. The style pushes up through this and drives the pollen gradually out at the end, where it is exposed to insects. Finally the style emerges, the stigmas open and the petals separate and fall back. [Compare with *Campanula*, *Jasione*

Phyteuma Lour. = Sambucus L. (Sambucac.). [and *Compositae*.]

Phyteumoïdes Smeathm. ex DC. = Virecta Sm. (Rubiac.).

Phyteumopsis Juss. ex Poir. = Marshallia Schreb. (Compos.).

Phytholacca Brot. = Phytolacca L. (Phytolaccac.).

Phytocren[ac]eae R.Br. = Icacinaceae–Phytocreneae Engl.

Phytocrene Wall. Icacinaceae. 11 SE. As., Malaysia. Twining shrubs with very large vessels in the stem. If the stem be cut a quantity of water escapes, which according to Wallich was drunk by the natives of Martaban (Lower Burma). Fls. dioec.

Phytolacca L. Phytolaccaceae. 35 trop. & subtrop. Herbs with fleshy roots, or shrubs or trees. Fls. reg.; P 5, A 10–30, G 7–16 or (7–16); in the latter case fr. a berry, in the former an aggregate of achenes or drupes.

***Phytolaccaceae** R.Br. Dicots. (Centrospermae). 12/100, chiefly trop. Am. & S. Afr. Herbs, shrubs, or trees, with alt. ent. rarely stip. ls., and racemose or cymose infls. of regular inconspic. ♂ fls. P 4–5; A 4–5 or more (to ∞); G, rarely G̅, 1–∞ or (1–∞), ovules 1 in each cpl., amphi- or campylotropous. Drupe or nut, rarely capsule. Seed with perisperm, often arillate. The fls. exhibit great variety in structure, owing to branching of sta. and different numbers and arrangements of cpls. Chief genera: *Seguieria*, *Rivina*, *Schindleria*, *Hilleria*, *Ercilla*, *Phytolacca*.

Phytosalpinx Lunell = Lycopus L. (Labiat.).

Phytoxis Molina = Sphacele Benth. (Labiat.).

Phytoxys Spreng. = praec.

Piaggiaea Chiov. = Wrightia R.Br. (Apocynac.).

Piaradena Rafin. = Salvia L. (Labiat.).

Piaranthus R.Br. Asclepiadaceae. 16 S. Afr.

Piarimula Rafin. = Lippia L. (Verbenac.).

Piarophyla Rafin. = Bergenia Moench (Saxifragac.).

Piaropus Rafin. = Eichhornia Kunth (Pontederiac.).

Picardaea Urb. Rubiaceae. 2 Cuba, Haiti.

Picardenia Steud. = Picradenia Hook. = Actinella Nutt. (Compos.).

Piccia Neck. = Moronobea Aubl. (Guttif.).

Picconia DC. Oleaceae. 2 Canaries, Madeira, Azores.

Picea A. Dietr. Pinaceae. 50 N. temp., esp. E. As. Long shoots only with needle ls. Fls. single. Cones ripening in one year. *P. abies* (L.) Karst., the Norway spruce or spruce-fir, found in Eur. from the Pyrenees to 68° N., furnishes valuable wood, resin, and turpentine. *P. glauca* (Moench) Voss (white spruce), *P. smithiana* (Wall.) Boiss. (Himalayan spruce), and others, are also valuable.

Pichleria Stapf & Wettst. Umbelliferae. 1 W. Persia.

Pichonia Pierre = Lucuma Mol. (Sapotac.).

Pickeringia Nutt. (1834) = Ardisia Sw. (Myrsinac.).

***Pickeringia** Nutt. ex Torr. & Gray (1840). Leguminosae. 1 Calif.

Picnocomon Wallr. ex DC. = Cephalaria Schrad. (Dipsacac.).

Picnomon Adans. = Cirsium Mill. (Compos.).

Picotia Roem. & Schult. = Omphalodes Moench (Boraginac.).

Picradenia Hook. = Hymenoxys Cass. (Compos.).

Picradeniopsis Rydb. (~ Bahia Lag.). Compositae. 3 N. Am.

Picraena Lindl. = Aeschrion Vell. (Simaroubac.).

Picraena Stev. = Astragalus L. (Legumin.).

Picralima Pierre. Apocynaceae. 1 W. Equat. Afr.

***Picramnia** Sw. Simaroubaceae. 55 Mex. to trop. S. Am., W.I.

Picranenia Endl. = Picraena Lindl. = Aeschrion Vell. (Simaroubac.).

Picrasma Blume. Simaroubaceae. 6 W. Himal. to Japan, Malaysia, Fiji. The bitter wood and bark are used as a substitute for quassia. Ls. stip.! (rare in the fam.).

Picrella Baill. = ? Euodia J. R. & G. Forst. (Rutac.).

Picreus Juss. = Pycreus Beauv. (Cyperac.).

Picria Benth. & Hook. f. = Picrium Schreb. = Coutoubea Aubl. (Gentianac.).

Picria Lour. = Curanga Juss. (Scrophulariac.).

Picricarya Dennst. = Olea L. (Oleac.).

Picridium Desf. = Reichardia Roth (Compos.).

Picrina Reichb. ex Steud. = seq.

Picris L. Compositae. 40–50 temp. Euras., Medit., Abyss.

Picrita Schumacher = Aeschrion Vell. (Simaroubac.).

Picrium Schreb. = Coutoubea Aubl. (Gentianac.).

Picrocardia Radlk. = Soulamea Lam. (Simaroubac.).

Picrococcus Nutt. = Vaccinium L. (Ericac.).

Picrodendraceae Small. Dicots. 1/3 W.I. Decid. trees with alt. long-pet. digitately 3-foliol. stip. ls.; lfts. ent., petiolules artic.; stips. minute, setif., caduc. or persist. Fls. ♂ ♀, dioec. ♂ infl. of strict catkin-like pseudo-rac. or narrow thyrses, precoc. in axils of fallen ls. ♂ fl. very shortly pedic., subtended by 3–7 imbr. bracts and/or bracteoles at apex of pedic.; K 0; C 0; A ∞, in-

serted on hemisph. recept.; fil. short, anth. ellips., latr. to extr., slightly pubesc.
towards apex; pistillode o. ♀ fl. sol., axill., on long strict pedunc. with shortly
cupular apical expansion; K (or invol. ?) 4–5, unequal, lanc., subvalv. or
scarcely imbr., sometimes obscurely toothed or ciliate; C o; stds. o; G̲ (not
G̲!) (2), with long style and 2 thick subul. spreading stigs., and 2 pend. apical
axile anatr. ov. per loc., with obturators. Fr. a ± glob. drupe, with thin
fleshy orange pericarp containing ∞ vesicles filled with bitter juice and 1–2-
seeded obscurely 4-angled indehisc. endocarp; seed glob., cots. massive, much
corrugated, endosp. o. The family should prob. be restricted to the genus
Picrodendron; the other 5 genera mentioned in ed. 7 of this Dictionary, p. 875,
should perhaps constitute a separate family, intermediate between *Picrodendrac.*
and *Euphorbiac.*

Picrodendron Planch. Picrodendraceae. 3 W.I.
Picroderma Thorel & Gagnep. = Trichilia P.Br. (Meliac.).
Picrolemma Hook. f. Simaroubaceae. 3 E. Peru, Amaz. Brazil.
Picrophloeus Blume = Fagraea Thunb. (Potaliac.).
Picrophyta F. Muell. = Goodenia Sm. (Goodeniac.).
Picrorhiza Royle ex Benth. Scrophulariaceae. 2 W. Himal.
Picrorrhiza Wittst. = praec.
Picrosia D. Don. Compositae. 2 warm S. Am.
Picrothamnus Nutt. = Artemisia L. (Compos.).
Picroxylon Warb. = Eurycoma Jack (Simaroubac.).
Pictetia DC. Leguminosae. 6 Mex., W.I.
Piddingtonia A. DC. = Pratia Gaudich. (Campanulac.).
Pierardia P. & K. = Pirarda Adans. = Ethulia L. (Compos.).
Pierardia Rafin. = Dendrobium Sw. (Orchidac.).
Pierardia Roxb. = Baccaurea Lour. (Euphorbiac.).
Piercea Mill. = Rivina L. (Phytolaccac.).
Pieridia Reichb. = seq.
Pieris D. Don. Ericaceae. 10 E. As., N. Am.
Pierotia Blume = Ixonanthes Jack (Ixonanthac.).
Pierranthus Bonati. Scrophulariaceae. 1 Indoch.
Pierrea Hance = Homalium Jacq. (Flacourtiac.).
*Pierrea Heim = Hopea Roxb. (Dipterocarpac.).
Pierreanthus Willis = Pierranthus Bonati (Scrophulariac.).
Pierredmondia Tamamsh. (? sphalm.) = Petroëdmondia Tamamsh. (Umbellif.).
Pierreocarpus Ridley ex Symington = Hopea Roxb. (Dipterocarpac.).
Pierreodendron A. Chev. = Letestua Lecomte (Sapotac.).
Pierreodendron Engl. = Quassia L. (Simaroubac.).
Pierrina Engl. Scytopetalaceae. 2 W. Equat. Afr.
Pietrosia Nyárády (~ Hieracium L.). Compositae. 1 Roumania.
Pigafetta Adans. = Eranthemum L. (Acanthac.). [(Palm.).
*Pigafetta (Bl.)ʃMart. ex Becc. mut. Hook. f. = Pigafettia (Bl.) Mart. ex Becc.
Pigafettaea P. & K. (1) = Pigafetta Adans. = Eranthemum L. (Acanthac.).
Pigafettaea P. & K. (2) = seq.
Pigafettia (Bl.) Mart. ex Becc. Palmae. 3 E. Malaysia.
Pigea DC. = Hybanthus Jacq. (Violac.).

PIGEUM

Pigeum Laness. = Pygeum Gaertn. = Lauro-Cerasus Duham. (Rosac.).

Pikria G. Don = Picria Lour. = Curanga Juss. (Scrophulariac.).

Pilanthus Poit. ex Endl. = Centrosema DC. (Legumin.).

Pilasia Rafin. = Urginea Steinh. (Liliac.).

Pilderia Klotzsch = Begonia L. (Begoniac.).

*Pilea Lindl. Urticaceae. 400 trop. *P. microphylla* (L.) Liebm., 'artillery plant', so called from the puffs of pollen ejected by the exploding sta. (cf. *Urtica*).

Pileanthus Labill. Myrtaceae. 3 W. Austr.

Pileocalyx Gasparr. = Cucurbita L. (Cucurbitac.).

Pileocalyx P. & K. = Piliocalyx Brongn. & Gris (Myrtac.).

Pileostegia Hook. f. & Thoms. Hydrangeaceae. 3 Himal. to Formosa.

Pileostegia Turcz. = Ilex L. (Aquifoliac.).

Pileostigma B. D. Jacks. = Piliostigma Hochst. (Legumin.).

Piletocarpus Hassk. = Aneilema R.Br. (Commelinac.).

Pileus Ramírez (~ Jacaratia Endl.). Caricaceae. 1–2 Mex.

Pilgerochloa Eig. Gramineae. 1 Asia Minor.

Pilgerodendron Florin (~ Libocedrus Endl.). Cupressaceae. 1 S. Chile.

Pilidiostigma Burret. Myrtaceae. 4 Austr.

Pilinophytum Klotzsch = Croton L. (Euphorbiac.).

*Piliocalyx Brongn. & Gris. Myrtaceae. 8 New Caled., Fiji.

Piliosanthes Hassk. = Peliosanthes Andr. (Liliac.).

*Piliostigma Hochst. Leguminosae. 3, drier (monsoon) regions trop. Afr., Indomal. & Queensland.

Pilitis Lindl. = Richea R.Br. (Epacridac.).

Pillansia L. Bolus. Iridaceae. 1 S. Afr.

Pillera Endl. = Mucuna Adans. (Legumin.).

Piloblephis Rafin. = Satureia L. (Labiat.).

Pilocanthus B. W. Benson & Backeb. = Pediocactus Britt. & Rose (Cactac.).

Pilocarp[ac]eae J. G. Agardh = Rutaceae–Cusparieae–Pilocarpinae Engl.

Pilocarpus Vahl. Rutaceae. 22 trop. Am., W.I. The leaves of *P. pennatifolius* Lem. are the officinal 'folia Jaborandi'.

Pilocereus Lem. = Cephalocereus Pfeiff. (Cactac.).

Pilocopiapoa Ritter. Cactaceae. 1 Chile.

Pilogyne Gagnep. = Myrsine L. (Myrsinac.).

Pilogyne Schrad. = Zehneria Endl. (Cucurbitac.).

Piloisa B. D. Jacks. = seq.

Piloisia Rafin. = Cordia L. (Ehretiac.).

Pilophora Jacq. = Manicaria Gaertn. (Palm.).

Pilophyllum Schlechter. Orchidaceae. 1 W. Malaysia.

Pilopleura Schischk. Umbelliferae. 1 C. As.

Pilopsis Ito = Arthrocereus A. Berger (Cactac.).

Pilopus Rafin. = Phyla Lour. (Verbenac.).

Pilorea Rafin. = Wahlenbergia Schrad. ex Roth (Campanulac.).

Pilosanthus Steud. = Psilosanthus Neck. = Liatris Schreb. (Compos.).

Pilosella Hill (~ Hieracium L.). Compositae. 200 Eur., SW. As. to W. Sib.

Pilosella Kostel. = Arabidopsis Heynh. (Crucif.).

Pilosella F. W. & C. H. Sch. Bip. = Hieracium L. (Compos.).

Piloselloïdes (Less.) C. Jeffrey. Compositae. 1 trop. Afr., Madag., trop. As.; 1 S. Afr.

Pilosocereus Byles & Rowl. = Cephalocereus Pfeiff., etc. (Cactac.).

Pilosperma Planch. & Triana. Guttiferae. 2 Colombia.

Pilostachys B. D. Jacks. = seq.

Pilostaxis Rafin. = Polygala L. (Polygalac.).

Pilostemon Iljin. Compositae. 2 C. As. & Afghanist. (? Iran) to W. China.

Pilostigma Costantin = Costantina Bullock (Asclepiadac.).

Pilostigma Van Tiegh. (~ Amyema Van Tiegh.). Loranthaceae. 3 Austr.

Pilostyles Guill. Rafflesiaceae. 18 S. U.S. to trop. S. Am.; 2 trop. Afr.; 1 Persia; 1 W. Austr.

Pilotheca T. L. Mitch. (sphalm.) = Philotheca Rudge (Rutac.).

Pilothecium (Kiaersk.) Kausel. Myrtaceae. 5 Brazil.

Pilotrichum Hook. f. & T. Anders. (sphalm.) = Ptilotrichum C. A. Mey. (Crucif.).

Pilouratea Van Tiegh. = Ouratea L. (Ochnac.).

Pilularia Ehrh. (uninom.) = Pilularia globulifera L. (Marsileac.).

Pilularia L. Marsileaceae. 6 N. & S. temp. *P. globulifera* L., the pillwort, on the margins of lakes. The pea-shaped sporocarp, borne on the ventral side of a l.-stalk, has a hard outer coat, and consists of four sori, each containing micro- and megasporangia. Life-history like *Marsilea*.

Pilulariaceae Dumort. = Marsileaceae R.Br.

Pilumna Lindl. = Trichopilia Lindl. (Orchidac.).

Pimecaria Rafin. = Ximenia L. (Olacac.).

Pimela Lour. = Canarium L. (Burserac.).

Pimelaea Kuntze = Pimelea Banks & Soland. (Thymelaeac.).

Pimelandra A. DC. = Ardisia Sw. (Myrsinac.).

*Pimelea Banks & Soland. Thymelaeaceae. 80 N. Philipp. Is., Lesser Sunda Is., SE. New Guinea, Austr., Tasm., N.Z. Fls. in heads, protandrous.

Pimeledendrum Hassk. = Pimelodendron Hassk. (Euphorbiac.).

Pimeleodendron Muell. Arg. = seq.

Pimelodendron Hassk. Euphorbiaceae. 6–8 Malaysia.

Pimenta Lindl. Myrtaceae. 18 trop. Am., W.I. The unripe frs. of *P. officinalis* Lindl., rapidly dried, form allspice.

Pimentelea Willis = seq.

Pimentelia Wedd. Rubiaceae. 1 Peru.

Pimentella Walp. = praec.

Pimentus Rafin. = Pimenta Lindl., Eugenia L., Melaleuca L., etc. (Myrtac.).

Pimia Seem. Sterculiaceae. 1 Fiji.

Pimpinele St-Lag. = Pimpinella L. (Umbellif.).

Pimpinella L. Umbelliferae. 150 Euras., Afr.; 1 Pacif. N. Am., few S. Am. *P. anisum* L. (Medit., anise), fr. (aniseed) used in flavouring.

Pimpinella Seguier = Poterium L. (Rosac.).

Pinacantha Gilli. Umbelliferae. 1 Afghanistan.

*Pinaceae Lindl. Gymnosp. (Conif.). 10/250 N. hemisph., S. to Sumatra, Java, C. Am. & W.I. Trees, rarely prostrate or creeping shrubs, with acic. spirally arranged ls.; either long shoots only, or long and short shoots developed; long shoots with scale-ls. or needle-ls. Fls. naked, usu. monoec. ♂ fls. usu.

PINACOPODIUM

with env. of scales at base; A ∞, pollen-sacs 2, completely adnate to the anther-scale below; pollen mostly with bladders. ♀ fl.-cones with ∞ spirally arrd. bract-scales, bearing on their upper surface the ± free flat ovuliferous scale, which bears on its upper (inner) surface 2 anatr. ov. with 1 integ. Fr. cones woody, closed until ripe, mostly composed of the much enlarged ovulif. scales, the bract-scales also enlarged but narrower and thinner than the ovulif. scales, or reduced and hidden; seeds mostly unilaterally winged; embr. with several cots. Genera: *Abies, Keteleeria, Pseudotsuga, Tsuga, Picea, Pseudolarix, Larix, Cedrus, Pinus, Cathaya.*

Pinacopodium Exell & Mendonça. Erythroxylaceae. 2 trop. Afr.
Pinalia Lindl. = Eria Lindl. (Orchidac.).
Pinanga Blume. Palmae. 115 SE. As., Indomal.
Pinarda Vell. = Micranthemum Michx (Scrophulariac.).
Pinardia Cass. = Chrysanthemum L. (Compos.).
Pinardia Neck. = Aster L. (Compos.).
Pinaria (DC.) Reichb. = Matthiola R.Br. (Crucif.).
Pinaropappus Less. Compositae. 7 S. U.S., Mex.
Pinarophyllon T. S. Brandegee. Rubiaceae. 2 Mex., C. Am.
Pinasgelon Rafin. = Cnidium Cusson (Umbellif.).
Pincecnitia Hort. ex Lem. = seq.
Pincenectia Hort. ex Lem. = seq.
Pincenectitia Hort. ex Lem. = Nolina Michx (Agavac.).
Pincenictitia Baker = praec.
Pincinectia Hort. ex Lem. = praec.
Pinckneya Michx. Rubiaceae. 1 S. U.S. Cinchonin in bark.
Pindarea Barb. Rodr. = Attalea Kunth (Palm.).
Pinea v. Wolf = Pinus L. (Pinac.).
Pineda Ruiz & Pav. Flacourtiaceae. 1 N. Andes.
Pinelea Willis = seq.
Pinelia Lindl. Orchidaceae. 2 Brazil.
Pinellia Tenore. Araceae. 7 China, Japan.
Pingraea Cass. = Baccharis L. (Compos.).
Pinguicula L. Lentibulariaceae. 46 N. Temp. (exc. W. & C. U.S.A. & C. As.), S.E. U.S.A., C. Am., W.I., N. Andes, Chile, T. del Fuego, Himal. In *P. vulgaris* L. (N. temp. & circumpolar, exc. NE. As.) the rad. ls. are covered with glands, some sessile, some on stalks, secreting a sticky fluid to which small insects adhere. Rain washes them against the edge of the l., which is slightly upturned: when stimulated by the presence of proteid bodies it rolls over upon itself and encloses them, and then the sessile glands secrete a ferment, digest the prey, and absorb the products, after which the l. unrolls again.
Pinguiculaceae Dum. = Lentibulariaceae Rich.
Pinguin Adans. = Bromelia L. (Bromeliac.).
Pinillosia de la Ossa. Compositae. 2 Cuba.
Pinknea Pers. = Pinckneya Michx (Rubiac.).
Pinkneya Rafin. = praec.
Pinonia Gaudich. = Cibotium Kaulf. (Dicksoniac.).
Pinosia Urb. Caryophyllaceae. 1 Cuba.
Pinsona auctt. = Pinzona Mart. & Zucc. (Dilleniac.).

Pintoa C. Gay. Zygophyllaceae. 1 Chile.

Pinus L. Pinaceae. 70–100 N. temp. and on mts. in the N. trop. Evergr., resinous trees with both long and short shoots (see *Coniferae*). If a tree be examined in winter the main axes will be found each with a group of buds at the end, one term., the rest lat., covered with resinous scale ls. Each gives rise in spring to a 'long shoot' or shoot of unlimited growth; the term. bud continues the main axis of all, forming a year's growth before branching in a similar way again. The large branches thus form rough whorls marking each year's growth. On the stem of a long shoot no green ls. are directly borne, but only scales, first the bud scales above mentioned and then others in whose axils arise the 'short shoots', or shoots of limited growth. Each of these has a few scale ls. at the base of a very short stem and ends with 2 or more green ls. of needle shape. When there are two, their upper flat sides face one another. These needle ls. exhibit xero. characters; they are thick in proportion to surface exposed, they have a very stout epidermis with a hypoderm of thick-walled tissue under it, and the stomata are placed at the bottom of deep pits; the intercellular spaces too are very small.

The infls. take the form of the familiar cones, the ♂ grouped together in spikes. Each infl., whether ♂ or ♀, occupies the position of a short shoot and is of limited growth—an axis with a few scale ls. below bearing a number of fertile scales (sporophylls). In the ♂ there are many fertile scales, each with two pollen-sacs on the under side; the pollen is loose and powdery, and each grain has two bladdery expansions of the cuticle helping it to float in the air. In the ♀, the bract scales are very small, but the ovuliferous scales, which show at the outside of the cone, are very large, and each bears two ovules at its base, with the micropyles facing the axis. The ♀ cones take 2–3 years before the seeds are ripe. In May of the first year, the first stage may be seen—young cones, about 1 cm. long, in the position of short shoots near the tip of the lengthening axis. The ovules are not ripe for fert. In June (the time varies from year to year according to season) pollination takes place. The ♂ cones shed their pollen in great quantities, so that in a pine forest the air is often full of it (if it rains, the phenomenon of 'showers of sulphur' may occur), and the wind carries it about. At the same time the ovuliferous scales spread apart. If a grain falls between two of them it slips down to the micropyle of an ovule, where it becomes held by the sticky fluid then exuding. After a short time the ♀ cones close up again. The pollen grain is brought into contact with the nucellus by the drying up of the mucilage; it forms a short pollen-tube, and then a resting period comes on. Next year in May or June the ♀ cone has become a fat green body about 3 cm. long, with the ovules ready for fert.; the pollen-tubes now recommence growth and reach the ova. Then in the third year the cone is mature—a hard woody cone containing the seeds between the scales. Each seed contains an embryo with a whorl of cots., embedded in rich endosp., and has a hard testa. To the end of this is attached a thin membranous wing, derived from the ovuliferous scale. In dry weather the cone opens and the seeds are blown away. In germ. the seed is lifted up above the earth by the growing plant and the cots. remain inside the testa till the reserves are exhausted. They are green whilst in the seed, though in darkness—an exception to the rule that chlorophyll requires light for its formation. During the first year no short shoots are formed, and the seedling has green ls. borne directly on the main stem.

The pines are amongst the most valuable of all plants and are cultivated on an enormous scale, chiefly for their timber, which is easily worked, and resinous products. The resin renders the timber very resistant to decay, etc. Some of the more important spp. are mentioned in the following conspectus:

I. Sect. STROBUS Spach. Base of scale-leaves non-decurrent; vascular bundle of needle-leaves single. 'Soft Pines.' (Timber relatively soft with little resin.)

    a. Scales of female cone with terminal boss (umbo) (about 12 spp.). *P. strobus* L., the Weymouth pine (E. N. Am.), formerly an important timber tree; *P. lambertiana* Douglas, the giant sugar pine of the western U.S.; *P. wallichiana* A. B. Jackson, the Bhutan pine (Afghanistan to Nepal, Bhutan); etc.

    b. Scales of female cone with dorsal umbo (about 10 spp.). *P. cembroïdes* Zucc., Mexican nut pine, with edible seeds; etc.

II. Sect. PINUS (Sect. PINASTER Endl.). Base of scale-leaves decurrent; vascular bundle of needle-leaves double. 'Hard Pines.' (Timber relatively hard, often resinous.)

    a. Seeds with an effective wing.

      (i) Cones symmetrical, opening once only; pits of wood-ray cells large. (About 15–20 spp., nearly all of the Old World.) *P. sylvestris* L., the Scots pine, the only native British sp., occurs in Europe to 68° N., in Asia to 66° N. and as far south as Spain and Italy (alpine). The wood (yellow deal) is largely used in the arts; turpentine is obtained by tapping the tree. The resin exudes and is distilled; the distillate is oil of turpentine, the remainder rosin. Tar and pitch are correspondingly the products of destructive distillation in closed chambers. *P. mugo* Turra is a shrubby decumbent sp., Pyrenees to Caucasus. *P. nigra* Arn. var. *maritima* (Aiton) Melville (S. Europe) is the Corsican pine.

      (ii) Cones symmetrical, opening once only; pits of wood-ray cells small. (About 12–15 spp., exclusively of the New World.) *P. taeda* L. (loblolly pine, southern U.S.) yields turpentine. *P. palustris* Mill. (longleaf or pitch pine, SE. U.S.) yields timber and turpentine. Other spp., e.g. *P. ponderosa* Dougl. ex Lawson and *P. rigida* Mill., are also known by the name of pitch pine.

      (iii) Cones ± asymmetrical, persistent in various degrees, opening periodically (serotinous). (About 12 spp., nearly all of the New World.) An Old World representative is *P. pinaster* Aiton, the maritime or cluster pine of the Mediterranean region. It is a valuable tree, growing well near the sea, and large areas of the Landes of S. France are planted with it. It furnishes much of the turpentine, etc., in use. *P. radiata* D. Don, California, is much planted for its valuable timber.

    b. Seed-wing ineffective.

      P. pinea L. (Medit.), the umbrella pine, has large edible seeds (*pignons*) and the seed wing is scarcely developed.

Pinzona Mart. & Zucc. (~ Doliocarpus Rol. or Curatella L.). Dilleniaceae. 1–2 trop. S. Am., W.I.
Pioctonon Rafin. = Heliotropium L. (Boraginac.).
Piofontia Cuatrec. Compositae. 1 Colombia.
Pionandra Miers = Cyphomandra Mart. (Solanac.).
Pionocarpus Blake. Compositae. 1 Mex.
Piora Koster. Compositae. 1 New Guinea.
Pioriza Rafin. = Picriza Rafin. = Gentiana L. (Gentianac.).
Piotes Soland. ex Britt. = Augea Thunb. (Zygophyllac.).
Pipaceae Dulac = Aristolochiaceae Juss.
Pipalia Stokes = Litsea Lam. (Laurac.).
Piparea Aubl. = Casearia Jacq. (Flacourtiac.).
Piper L. Piperaceae. 2000 trop. Mostly climbing shrubs (peppers). Fls. in sympodial spikes, the bracts closely appressed to the axis. Fr. a berry. That of *P. nigrum* L., gathered before ripe and dried, forms a black peppercorn; or if the outside be removed by maceration, a white one. Pepper is chiefly cult. in Malaya. *P. cubeba* L. f. is the cubebs, *P. betle* L. the betel pepper (see *Areca*).
***Piperaceae** C. A. Agardh. Dicots. 4/2000 trop. Shrubs or climbers or small trees, with sympodial growth and enlarged nodes. Ls. alt., simple, ent., stip.; stips. adnate, caduc. or rarely obsol.; taste and odour pungent. Fls. naked, ♀ or ♂ ♀, in dense lf.-opp. (more rarely axill.) sol. (rarely umbell.) spikes. P 0; A 1–10, with opp. discrete loc.; G̲ (1–4), 1-loc., with 1 basal orthotr. ov., and 1–5 short stigs. Fr. a small berry; seed with much mealy perisp. Genera: *Piper, Trianaeopiper, Ottonia, Pothomorphe.* [See also *Peperomiac.*]
Piperanthera C. DC. (~ Peperomia Ruiz & Pav.). Peperomiaceae. 1 W.I.
Piperella Presl = Micromeria Benth. (Labiat.). [(S. Domingo).
Piperi St-Lag. = Piper L. (Piperac.).
Piperia Rydb. = Platanthera Lindl. (Orchidac.).
Piperiphorum Neck. = Piper L. (Piperac.).
Piperodendron Fabr. (uninom.) = Schinus L. sp. (Anacardiac.).
Piperomia Pritz. (sphalm.) = Peperomia Ruiz & Pav. (Peperomiac.).
Piperonia Pritz. (sphalm.) = praec.
Pippenalia McVaugh. Compositae. 1 Mex.
Pipseva Rafin. = Chimaphila Pursh (Pyrolac.).
Piptadenia Benth. (1840). Leguminosae. 11 Mex. to trop. S. Am.
Piptadenia Benth. (1841) = Parapiptadenia Brenan (Legumin.).
Piptadeniastrum Brenan. Leguminosae. 1 trop. Afr.
Piptadeniopsis Burkart. Leguminosae. 1 S. trop. S. Am.
Piptandra Turcz. = Scholtzia Schau. (Myrtac.).
Piptanthocereus (A. Berger) Riccob. = Cereus Mill. (Cactac.).
Piptanthus Sweet. Leguminosae. 8 C. As., Himalaya, China.
Piptatherum Beauv. = Oryzopsis Michx (Gramin.).
Piptocalyx Oliv. ex Benth. Trimeniaceae. 2 New Guinea, E. Austr. (New S. Wales).
Piptocalyx Torr. = Greeneocharis Guerke & Harms (Boraginac.).
Piptocarpha R.Br. Compositae. 50 C. to trop. S. Am., W.I.
Piptocarpha Hook. & Arn. = Chuquiraga Juss. (Compos.).
Piptocelus C. Presl = ? Schinus L. (Anacardiac.).

PIPTOCEPHALUM

Piptocephalum Sch. Bip. = Catananche L. (Compos.).
Piptoceras Cass. = Centaurea L. (Compos.).
Piptochaetium J. Presl. Gramineae. 20 temp. S. Am.
Piptochlaena P. & K. = Piptolaena Harv. = Voacanga Thou. (Apocynac.).
Piptochlamys C. A. Mey. = Thymelaea Endl. (Thymelaeac.).
Piptoclaina G. Don = Heliotropium L. (Boraginac.).
Piptocoma Cass. Compositae. 2 S. Domingo.
Piptocoma Less. = Lychnophora Mart. (Compos.).
Piptolaena Harv. = Voacanga Thou. (Apocynac.).
Piptolepis Benth. = Forestiera Poir. (Oleac.).
*****Piptolepis** Sch. Bip. Compositae. 8 Brazil.
Piptomeris Turcz. = Jacksonia R.Br. (Legumin.).
Piptophyllum C. E. Hubbard. Gramineae. 1 trop. Afr.
Piptopogon Cass. = Hypochoeris L. (Compos.).
Piptoptera Bunge. Chenopodiaceae. 1 C. As.
Piptosaccos Turcz. = Dysoxylum Bl. (Meliac.).
Piptospatha N. E. Br. Araceae. 10 W. Malaysia.
Piptostachya (C. E. Hubbard) Phipps. Gramineae. 1 trop. Afr.
Piptostegia Hoffmgg. = Operculina S. Manso (Convolvulac.).
Piptostemma Spach = Nassauvia Comm. ex Juss. (Compos.).
Piptostemma Turcz. = Angianthus Wendl. (Compos.).
Piptostemum Steud. = Piptostemma Spach = Nassauvia Comm. ex Juss. (Compos.).
Piptostigma Oliv. Annonaceae. 15 trop. Afr.
Piptostylis Dalz. = Clausena Burm. f. (Rutac.).
Piptothrix A. Gray. Compositae. 5 Mex.
Pipturus Wedd. Urticaceae. 40 Mascarenes to Austr. & Polynesia.
Piqueria Cav. Compositae. 20 Mex. to Andes.
Piqueriopsis R. M. King. Compositae. 1 Mex.
Piquetia N. E. Br. = Kensitia Fedde (Aïzoac.).
Piquetia (Pierre) H. Hallier = Camellia L. (Theac.).
Piranhea Baill. Euphorbiaceae(?). 1–2 Venez., Brazil, Guiana.
Pirarda Adans. = Ethulia L. f. (Compos.).
Piratinera Aubl. Moraceae. 12 trop. S. Am.
Pirazzia Chiov. = Matthiola R.Br. (Crucif.).
Pircunia Bert. = Phytolacca L. (Phytolaccac.).
Pirea Durand = Nasturtium R.Br. = Rorippa Scop. (Crucif.).
Pirenia C. Koch = Pyrenia Clairv. = Pyrus L. (*sens. lat.*) (Rosac.).
Piresia Swallen. Gramineae. 2 trop. S. Am., Trinidad.
Piresodendron Aubrév. Sapotaceae. 1 Brazil.
Piriadacus Pichon. Bignoniaceae. 1 E. Brazil.
Pirigara Aubl. = Gustavia L. (Lecythidac.).
Pirigarda C. B. Clarke (sphalm.) = praec.
Piringa Juss. = Gardenia Ellis (Rubiac.).
Piripea Aubl. = Buchnera L. (Scrophulariac.).
Piriqueta Aubl. Turneraceae. 20 warm Am., Afr., Madag.
Piritanera R. H. Schomb. = Piratinera Aubl. (Morac.).
+**Pirocydonia** H. Winkler ex Daniel = +Pyronia Hort. (Rosac.).

Pirola Neck. = Pyrola L. (Pyrolac.).
Pirolaceae auctt. = Pyrolaceae Lindl.
Pironneaua Benth. & Hook. f. = seq.
Pironneauella Kuntze = seq.
Pironneava Gaudich. Bromeliaceae. 1 SE. Brazil.
Pirophorum Neck. = Pyrus L. (*sens. lat.*) (Rosac.).
Pirottantha Speg. = Plathymenia Benth. (Legumin.).
Pirroneana Benth. & Hook. f. (sphalm.) = Pironneava Gaudich. (Bromeliac.).
Pirus auctt. = Pyrus L. (Rosac.).
Pisaura Bonato = Lopezia Cav. (Onagrac.).
Piscaria Piper. Euphorbiaceae. 2 NW. U.S.
*****Piscidia** L. Leguminosae. 10 Florida, Mex., W.I.
Piscipula Loefl. = praec.
Pisonia L. Nyctaginaceae. 50 trop. & subtrop. Fls. usu. unisexual. The anthocarp is glandular and is one of the few frs. which are able to cling to feathers. On some Pacific islands birds and even reptiles are ensnared and disabled and eventually killed by the masses of viscid anthocarps.
Pisonia Rottb. = Diospyros L. (Ebenac.).
Pisoniaceae J. G. Agardh = Nyctaginaceae–Pisonieae Benth. & Hook. f.
Pisoniella Standley. Nyctaginaceae. 1 warm Am.
Pisophaca Rydb. = Astragalus L. (Legumin.).
Pisosperma Sond. = Kedrostis Medik. (Cucurbitac.).
Pistacia L. Pistaciaceae. 10 Medit. to Afghan.; SE. & E. As. to Malaysia; S. U.S., Mex., Guatem. *P. terebinthus* L. yields Chian turpentine; *P. lentiscus* L., mastic. Fr. of *P. vera* L. (pistachio nuts) ed.
Pistaciaceae (Marchand) Caruel. Dicots. 1/10 Medit. to Afghan., E. As. to Malaysia, warm N. Am. Resinous trees or shrubs, with alt., pinn. (rarely trifoliol. or simple), exstip. ls. Fls. small, reg., ♂ ♀, dioec., in ∞-fld. panicles. ♂ fls. subtended by a bract and 2 bracteoles; K(?) 2–1–0; C 0; A 3–5, fil. v. short, adnate to disk, anth. large, ovoid, basifixed, latrorse; pistillode small or 0. ♀ fls. with bract and 2 bracteoles; K(?) 2–5, small, scarious; stds. 0; disk minute or 0; G̲ (3), 1-loc., with 1 ov. suspended from a basal funicle, and short simple style with 3 elong.-obl. spreading stigs. Fr. an oblique ± compressed drupe, with thin exocarp and bony endoc.; seed without endosp. Only genus: *Pistacia*. Differs from *Anacardiac.*, with which it is usu. associated, in the dioec., prob. naked fls., and in the structure of the pollen. The supposed K is prob. bracteal in nature. Cf. *Picrodendrac.*
Pistaciopsis Engl. = Haplocoelum Radlk. (Sapindac.).
Pistaciovitex Kuntze = Aglaia Lour. (Meliac.).
Pistia L. Araceae. 1 trop. & subtrop., *P. stratiotes* L., a floating water plant, rarely anchored by its roots, and often blown about by wind. It is of sympodial structure, but the internodes remain short and bear a rosette of large ls.; these sleep at night, moving upwards from the nearly horiz. day position. The continuation shoots of the sympodium are axillary, but beside each l. arises a stolon which grows out along the water and gives rise to a new pl. The infl. is small and bisexual; above is a whorl of ♂ fls., each with a synandrium of 2 sta.; below is a ♀ fl. of 1 cpl. Both are naked. *P.* is a link between *Araceae* and *Lemnaceae* (*q.v.*). Cf. Arber, *Water Plants*.

PISTIACEAE

Pistiaceae C. A. Agardh = Cytinus L. (Rafflesiac.) + Nepenthes L. (Nepenthac.) + Pistia L. (Arac.).

Pistiaceae Dum. = Araceae–Pistioïdeae Engl.

Pistolochia Bernh. = Corydalis Vent. (Fumariac.).

Pistolochia Rafin. = Aristolochia L. (Aristolochiac.).

Pistolochiaceae ('-inae') Link = Aristolochiaceae Juss.

Pistorinia DC. = Cotyledon L. (Crassulac.).

Pisum L. Leguminosae. 6 Medit., W. As., incl. *P. sativum* L. (pea). The fl. mech. resembles that of *Lathyrus*.

Pitardia Battand. Labiatae. 2 NW. Afr.

Pitavia Molina. Rutaceae. 1 Chile.

Pitavia Nutt. ex Torr. & Gray = Cneoridium Hook. f. (Simaroubac.).

Pitcairinia Regel (sphalm.) = Pitcairnia L'Hérit. (Bromeliac.).

Pitcairnia J. R. & G. Forst. = Pennantia J. R. & G. Forst. (Icacinac.).

*Pitcairnia L'Hérit. Bromeliaceae. 250 trop. Am., W.I.; 1 W. Afr. (Senegal). Most are terrestrial; many form stolons at the base.

Pitcarnia J. F. Gmel. = praec.

Pitcheria Nutt. = Rhynchosia Lour. (Legumin.).

Pitchia auct. (sphalm.) = Fitchia Hook. f. (Compos.).

*Pithecellobium Mart. Leguminosae. 200 trop. Stipules often thorny. Fr. often coiled like *Medicago* (whence name = 'monkey's ear-ring'). *P. saman* Benth. (trop. S. Am.) is the 'rain tree', so called because of a legend that it was always raining under the branches. The ejections of juice by the cicadas are responsible for this (cf. *Acer, Andira*). It shows sleep movement of ls. well.

Pithecoctenium Mart. ex Meissn. Bignoniaceae. 7 Mex. to trop. S. Am., W.I.

Pithecodendron Speg. = Acacia Mill. (Legumin.).

Pithecolobium auctt. (sphalm.) = Pithecellobium Mart. (Legumin.).

Pithecoseris Mart. Compositae. 1 N. Brazil.

Pithecoxanium Corrêa de Mello apud Stellfeld = Clytostoma Miers ex Bur. (Bignoniac.).

Pithecurus Willd. ex Kunth = Andropogon L. (Gramin.).

Pithocarpa Lindl. Compositae. 4 W. Austr.

Pithodes O. F. Cook. Palmae. 1 W.I. (Hispaniola).

Pithosillum Cass. = Senecio L. (Compos.).

Pithuranthos DC. = Pituranthos Viv. (Umbellif.).

Pitraea Turcz. Verbenaceae. 1 S. Am.

Pittiera Cogn. = Polyclathra Bertol. (Cucurbitac.).

Pittierella Schlechter = Cryptocentrum Benth. (Orchidac.).

Pittierothamnus Steyerm. Rubiaceae. 1 Venez.

Pittonia Mill. = Tournefortia L. (Boraginac.).

Pittoniotis Griseb. = Antirhea Comm. ex Juss. (Rubiac.).

*Pittosporaceae R.Br. Dicots. 9/200, trop. Afr. to Pacif.; 8 gen. endem. in Australia. Trees or shrubs, often climbing, with alt., leathery, evergr., usu. entire, exstip. ls. Resin is present in large quantity in passages at the outer side of the bast. Fls. ⚥, reg.; K, C and A 5-merous, C often with erect claws; sta. hypog.; $\underline{G}$ (2)(–3–5), forming a 1-loc. or multiloc. ovary with parietal or axile plac., and 2-ranked ∞ anatr. ov.; style simple. Loculic. caps. or berry, with album. seeds, which are often black and shining and embedded in viscid pulp. Chief genera: *Pittosporum, Marianthus, Billardiera, Sollya.* Relationships somewhat obscure, but prob. some connection with *Escalloniaceae.*

Pittosporoïdes Soland. ex Gaertn. = Pittosporum Banks ex Soland. (Pittosporac.).

Pittosporopsis Craib. Icacinaceae. 1 SE. As.

*Pittosporum Banks ex Soland. apud Gaertn. Pittosporaceae. 150 trop. & subtrop. Afr., As., Austr., N.Z., Pacif. The seeds of some are sticky. Some yield useful timber.

Pittunia Miers (sphalm.) = Petunia Juss. (Solanac.).

Pitumba Aubl. = Casearia Jacq. (Flacourtiac.).

Pituranthos Viv. Umbelliferae. 10 Morocco to Syria, 2 S. Afr.

Pitygentias Gilg. Gentianaceae. 2 Peru.

Pityopsis Nutt. = Chrysopsis Ell. (Compos.).

Pityopus Small. Monotropaceae. 1 Pacif. U.S.

Pityothamnus Small = Asimina Adans. (Annonac.).

Pityphyllum Schlechter. Orchidaceae. 1 Colombia.

Pityranthe Thw. = Diplodiscus Turcz. (Tiliac.).

Pityranthes Willis = Pituranthos Viv. (Umbellif.).

Pityranthus Mart. = Alternanthera Forsk. (Amaranthac.).

Pityranthus H. Wolff = Pituranthos Viv. (Umbellif.).

Pityrocarpa (Benth.) Britton & Rose = Piptadenia Benth. (1840) (Legumin.).

Pityrodia R.Br. Dicrastylidaceae. 25 Austr.

Pityrogramma Link. Hemionitidaceae. 40 trop. Am., Afr. Ls. covered beneath with white or yellow waxy powder ('golden fern'). *P. calomelanos* (L.) Link, orig. in trop. Am., is a weed in trop. As.

Pityrophyllum Beer = Tillandsia L. (Bromeliac.).

Pityrosperma Sieb. & Zucc. = Cimicifuga L. (Ranunculac.).

Piuttia Mattei = Thalictrum L. (Ranunculac.).

Pivonneava Hook. f. (sphalm.) = Pironneava Gaudich. (Bromeliac.).

Pixidaria Schott = ? Lecythis Loefl. (Lecythidac.).

Pixydaria Schott = praec.

Placea Miers. Amaryllidaceae. 5 Chile.

Placocarpa Hook. f. Rubiaceae. 1 Mex.

Placodiscus Radlk. Sapindaceae. 14 trop. Afr.

Placodium Benth. & Hook. f. (sphalm.) = Plocama Ait. (Rubiac.).

Placolobium Miq. = Ormosia G. Jacks. (Legumin.).

Placoma J. F. Gmel. = Plocama Ait. (Rubiac.).

Placopoda Balf. f. Rubiaceae. 1 Socotra.

Placospermum C. T. White & Francis. Proteaceae. 1 Queensland.

Placostigma Blume = Podochilus Blume (Orchidac.).

Placseptalia Espinosa = Ochagavia Phil. (Bromeliac.).

Placus Lour. = Blumea DC. (Compos.).

Pladaroxylon Hook. f. = Senecio L. (Compos.).

Pladera Soland. ex Roxb. = Canscora Lam. (Gentianac.).

Plaea Pers. = Pleea Michx (Liliac.).

Plaesiantha Hook. f. = Pellacalyx Korth. (Rhizophorac.).

Plaesianthera Livera. Acanthaceae. 1 Ceylon.

Plagiacanthus Nees = Dianthera L. (Acanthac.).

PLAGIANTH[AC]EAE

Plagianth[ac]eae J. G. Agardh = Malvaceae–Malveae–Sidinae K. Schum.
Plagianthera Reichb. f. & Zoll. = Mallotus Lour. (Euphorbiac.).
Plagianthus J. R. & G. Forst. Malvaceae. 15 Austr., N.Z. *P. betulinus* A. Cunn. (N.Z.; lace-bark) good timber.
Plagidia Rafin. = Paronychia Mill. (Caryophyllac.).
Plagielytrum P. & K. = Plagiolytrum Nees = Tripogon Roth (Gramin.).
Plagiobasis Schrenk. Compositae. 3 C. As.
Plagiobothrys Fisch. & Mey. Boraginaceae. 100 Pacif. Am.
Plagiocarpus Benth. Leguminosae. 1 trop. Austr.
Plagiocaryum Willd. ex Steud. (nomen). 1 Brazil. Quid?
Plagioceltis Mildbr. ex Baehni. Ulmaceae. 1 Peru.
Plagiocheilus Arn. Compositae. 6–7 S. Am.
Plagiochilus Lindl. = praec.
Plagiochloa Adamson & Sprague. Gramineae. 7 S. Afr.
Plagiogyria (Kunze) Mett. Plagiogyriaceae. 36 E. As., Am., in forest on mt. ridges.
Plagiogyriaceae Bower. Plagiogyriales. Young ls. covered with mucilage through which pneumathodes project; ls. simply pinnate; fertile pinnae narrow, acrostichoid; annulus complete, oblique. 1 gen.: *Plagiogyria*.
Plagiogyriales. Filicidae. 1 fam.: *Plagiogyriaceae*.
Plagiolirion Baker = Urceolina Reichb. (Amaryllidac.).
Plagioloba Reichb. = Hesperis L. (Crucif.).
Plagiolobium Sweet = Hovea R.Br. (Legumin.).
Plagiolophus Greenman. Compositae. 1 Mex. (Yucatan).
Plagiolytrum Nees = Tripogon Roth (Gramin.).
Plagion St-Lag. = Plagius L'Hérit. ex DC. = Chrysanthemum L. (Compos.).
Plagiopetalum Rehder. Melastomataceae. 5 China.
Plagiophyllum Schlechtd. = Centradenia G. Don (Melastomatac.).
Plagiopoda Spach = Grevillea R.Br. (Proteac.).
Plagiopteraceae Airy Shaw. Dicots. 1/1 SE. As. Climbing shrubs of Combretaceous or Malpighiaceous habit, shortly ferrugineous-stellate-pubesc. Ls. opp., simple, ent., membr., exstip., condupl. in vernation. Fls. small, reg., ⚥, in ∞-fld. subumbellif. groups in thyrsif. infls., v. fragrant. K (3–)4(–5), small, shortly united below, open in aestiv.; C (3–)4(–5), valv., externally pubesc., revolute; A ∞, ± biser., on small disk, fil. elong., filif., subclav., anth. small, 4-loc., ± horizont., transv. dehisc.; G (3), with elong. subul. shortly 3-lobed style, and 2 collat. erect basal anatr. ov. per loc. Fr. a turbinate finally septicid. caps., with 3 apical spreading spatulate wings. Only genus: *Plagiopteron*. An interesting plant, 'presenting a curious mixture of characters' (Griffith). General appearance of a *Combretum*; fruit somewhat Malpighiaceous; perhaps some relationship with *Verbenac.–Caryopteridoïdeae* and *Symphorematac.* Stamens, anther-dehiscence and ovule-arrangement very distinctive. Latex-canals in ls. and branches.
Plagiopteron Griff. Plagiopteraceae. 1 Lower Burma.
Plagiorhegma Maxim. = Jeffersonia Barton (Podophyllac.).
Plagiorrhiza (Pierre) H. Hallier = Mesua L. (Guttif.).
Plagioscyphus Radlk. Sapindaceae. 2 Madag.
Plagiosetum Benth. Gramineae. 1 Austr.

Plagiosiphon Harms. Leguminosae. 5 trop. Afr.

Plagiospermum Oliv. (1886; non Cleve 1868—Algae) (~ Prinsepia Royle). Rosaceae. 2 N. China.

Plagiospermum Pierre = Styrax L. (Styracac.).

Plagiostachys Ridl. Zingiberaceae. 15 W. Malaysia.

Plagiostemon Klotzsch = Simocheilus Klotzsch (Ericac.).

Plagiostigma Presl = Aspalathus L. (Legumin.).

Plagiostigma Zucc. = Ficus L. (Morac.).

Plagiostyles Pierre. Euphorbiaceae. 1 trop. Afr.

Plagiotaxis Wall. ex Kuntze = Chukrasia A. Juss. (Meliac.).

Plagiotheca Chiov. Acanthaceae. 1 trop. E. Afr.

Plagistra Rafin. = Aristolochia L. (Aristolochiac.).

Plagius L'Hérit. ex DC. = Chrysanthemum L. (Compos.).

Planaltoa Taubert. Compositae. 1 C. Brazil.

Plananthes P. Beauv. = Lycopodium L. (Lycopodiac.).

Planarium Desv. = Chaetocalyx DC. (Legumin.).

Planchonella Pierre (~ Xantolis Rafin.). Sapotaceae. 100 Seychelles, Indoch., Malaysia, trop. Austr., N.Z., Pacif.; 2 S. Am.

Planchonella Van Tiegh. = Godoya Ruiz & Pav. (Ochnac.).

Planchonia Blume. Barringtoniaceae. 8 Andamans to N. & NE. Austr.

Planchonia Dun. = Salpichroa Miers (Solanac.).

Planchonia J. Gay ex Benth. & Hook. f. = Polycarpaea Lam. (Caryophyllac.).

Plancia Neck. = Leontodon L. (Compos.).

Planera Giseke = Costus L. (Costac.).

Planera J. F. Gmel. Ulmaceae. 1 S. U.S., *P. aquatica* J. F. Gmel., a useful timber tree.

Planetanthemum (Endl.) Kuntze = Pseuderanthemum Radlk. (Acanthac.).

Planodes Greene. Cruciferae. 1 S. U.S.

Planotia Munro. Gramineae. 8 trop. S. Am.

Plantaginaceae Juss. Dicots. 3/270 cosmop. Annual or perennial herbs; ls. without distinction into stalk and blade, exstip. Fls. usu. in heads or spikes, inconspic., usu. ⚥, reg., without bracteoles, wind- or partly insect-fert. K (4), imbr., diagonal; C (4), imbr., membranous; A 4, with very long filaments and versatile anthers containing much powdery pollen; G̲ usu. (2), 2-loc., with 1–∞ semi-anatr. ov. on axile plac. Fr. a membranous caps., opening with a lid cut off by a peripheral dehiscence, or sometimes a nut surrounded by the persistent calyx. Embryo straight, in fleshy endosp. Genera: *Plantago, Littorella, Bougueria.* The fl. is usu. regarded as derived from a 5-merous type in the same way as that of *Veronica*, and there are good grounds (incl. entomological, viz. the food-plants of certain Lepidoptera and Coleoptera) for believing that the *P.* are in fact allied to *Scrophulariaceae.*

Plantaginastrum Fabr. (uninom.) = *Alisma plantago* L. (Alismatac.).

Plantaginella Fourr. = Plantago L. (Plantaginac.).

Plantaginella Hill = Limosella L. (Scrophulariac.).

Plantago L. Plantaginaceae. 265 cosmop. The 5 following Eur. spp. are representative of the genus. *P. major* L. (greater plantain) is a perennial with a thick root and a rosette of large erect ls., in whose axils arise the infls. (spikes). Fls. protog., the stigmas protruding from the bud; the sta. appear later. Wind-pollination is the rule, but insects sometimes visit them for pollen. The fruit-spikes are often given as food to cage-birds. *P. media* L. (hoary plantain) shows similar general features, but the ls. lie flat on the ground (hence it is troublesome as a weed); they exhibit the 3/8 phyllotaxy very clearly. The fl. is more conspicuous than *P. major* and has a pleasant scent, and, though primarily wind-pollinated, is largely visited for pollen. It is sometimes gynodioec. (cf. *Labiatae*). *P. lanceolata* L. (ribwort plantain) has narrow erect ls., and fls. also gynodioec.; the C-segments have a brown midrib. *P. coronopus* L. (buck's-horn plantain) is xero. with hairy ls., growing in sandy places. Many S. Am. spp. also show marked xero. characters—dense tufting, small hairy ls., often grooved on the lower surface (cf. *Ericaceae*), etc. *P. maritima* L. (the seaside plantain) has linear fleshy ls.; it is also found at high levels by streams in mts., though rarely in the intermediate regions.

The seeds of many swell up when wetted and become mucilaginous (cf. *Linum*). Those of *P. psyllium* L. (Medit. to India) are used in silk and cotton manufacture; they, and those of *P. ovata* Forsk. (Orient), have also been used in medicine.

Plantia Herb. = Hexaglottis Vent. (Iridac.).

Plantinia Bub. = Phleum L. (Gramin.).

Plappertia Reichb. = Dichapetalum Thou. (Dichapetalac.).

Plarodrigoa Looser. Malvaceae. 5 temp. S. Am.

Plaso Adans. = Butea Roxb. ex Willd. (Legumin.).

Plastolaena Pierre ex A. Chev. = Schumanniophyton Harms (Rubiac.).

*Platanaceae Dum. Dicots. 1/10 N. temp. Large decid. trees, with alt. simple palmatifid and palmatinerved (rarely elliptic-oblong and penninerved) dentate stip. ls. The bark of the trunk scales off in large flakes, leaving a smooth surface. The axillary bud is developed within the base of the petiole, which fits over it like an extinguisher. Ls. below and young parts often clothed with an evanescent or ± persistent felt of stellate or simple hairs. The stips. are usu. large and conspic., sometimes membr., and are united round the stem. Infl. consisting of pendulous strings of up to 12 dense globose sess. or pedunc. heads of fls., each infl. wholly ♂ or wholly ♀, monoecious, anemoph. The structure of the individual fls. has received conflicting interpretations at the hands of equally eminent authorities. Some claim to find a basic tetramerous pattern: K 4, C 4, A 4, G̲ 4, with various modifications and reductions; others that the fls. are essentially naked, but subtended by various small scales, presumably of bracteal origin. The genus needs the systematic investigation of all the spp. from this standpoint. It seems at least probable that the ♂ fl. consists of A 3–4, subsessile, the anth. ± cuneiform or obpyramidal, biloc., with expanded peltate connective. ♀ fl.: G̲ 6–9, free, subsess., subulate, with 1(–2) pend. orthotr. ov., and elongate linear style with unilateral stig., usu. ± hooked at apex. (3–4 pistillodes or stds. sometimes present in ♂ and ♀ fls. respectively.) Fr. a globose head of ± turbinate achenes or caryopses, with persist. styles, each fr. surrounded by a tuft of long bristly pappose hairs, and containing 1 linear seed with sparse endosp. The 'pappus' originates from the accresc. P or scales. The ls. are sometimes opposite on vigorous young shoots. Only genus: *Platanus*. Affinities disputed, but perhaps remotely with *Altingiac.*

Platanaria S. F. Gray = Sparganium L. (Sparganiac.).

Platanocarpum Korth. = Sarcocephalus Afzel. (Naucleac.).

Platanocarpus Korth. = praec.

Platanocephalus Crantz (Anthocephalus A. Rich.). Naucleaceae (~ Rubiaceae). 3 Indomal.

Platanos St-Lag. = Platanus L. (Platanac.).

*****Platanthera** Rich. Orchidaceae. 200 temp. & trop. Euras., N. Afr., N. & C. Am.

Platanus L. Platanaceae. 1 (*P. orientalis* L.) SE. Eur. to N. Persia; 1 (*P. kerrii* Gagnep.) Indoch., remarkable for elliptic-oblong unlobed pinnate-veined ls. and 10–12-headed ♀ infl.); 1 (*P. occidentalis* L.) E. N. Am., and 7 SW. U.S. and Mex. The hybrid × *P. hispanica* Muenchh. (*P. orientalis* × *P. occidentalis*) widely planted.

Platcalaria W. T. Stearn (sphalm.) = Platolaria Rafin. = Anemopaegma Mart. ex Meissn. (Bignoniac.).

Platea Blume. Icacinaceae. 5 Assam, SE. As., Malaysia.

Plateana Salisb. = Narcissus L. (Amaryllidac.).

Plateilema Cockerell (~ Actinella Pers.). Compositae. 1 Mex.

Platenia Karst. = Syagrus Mart. (Palm.).

Plathymenia Benth. Leguminosae. 3 Brazil, Argent. Good timber.

Platolaria Rafin. = Anemopaegma Mart. ex Meissn. (Bignoniac.).

Platoloma Rafin. Cruciferae (quid?). 1 W. U.S.

Platonia Kunth = Planotia Munro (Gramin.).

*****Platonia** Mart. Guttiferae. 1–2 Brazil. Fr. ed.

Platonia Rafin. (1808) = Phyla Lour. (Verbenac.).

Platonia Rafin. (1810) = Helianthemum Mill. (Cistac.).

Platorheedia Roj. = ? Rheedia L. (Guttif.).

Platostoma Beauv. Labiatae. 4 trop. Afr., W. Penins. India.

Platunum A. Juss. = Holmskioldia Retz. (Verbenac.).

Platyadenia B. L. Burtt. Gesneriaceae. 1 Borneo (Sarawak).

Platycalyx N. E. Br. = Eremia D. Don (Ericac.).

Platycapnos Bernh. Fumariaceae. 3–4 Canaries, W. Medit.

Platycarpha Less. Compositae. 5 trop. & S. Afr.

Platycarpidium F. Muell. = Platysace Bunge (Umbellif.).

Platycarpum Humb. & Bonpl. Henriqueziaceae. 10 N. trop. S. Am.

Platycarya Sieb. & Zucc. Juglandaceae. 2 E. As.

Platycaryaceae Nak. = Juglandaceae Kunth.

Platycelyphium Harms. Leguminosae. 1 trop. Afr.

Platycentrum Klotzsch = Begonia L. (Begoniac.).

Platycentrum Naud. Melastomataceae. 1 Guiana.

Platyceriaceae Ching = Polypodiaceae S. F. Gray.

Platycerium Desv. Polypodiaceae. 17 Afr., Malaysia, Austr., S. Am. (staghorn ferns), epiph., or on steep rock surfaces. The rhiz. is short and bears alt. ls. of two kinds. The young ls. are protected by stellate hairs. Of the two kinds of ls., the one stands ± erect (the 'mantle' l.), the other is pend., usu. much branched, and bears the sporangia in particular areas (according to sp.) on its lower surface. Two types of mantle ls. occur, repres. in *P. grande* J. Sm. and *P. bifurcatum* (Cav.) C. Chr. In the former the base of the l. clings closely to the supporting trunk, whilst the upper part spreads out and makes a niche in

PLATYCHAETA

which humus collects; in this the roots ramify and absorb food. In the latter
the whole of the mantle l. clings to the support, and the only humus-supply
is that furnished by the decay of old mantle ls. and perhaps of the tree bark;
it grows in great colonies, owing to adv. budding from the r.

Platychaeta Boiss. (~Pulicaria Gaertn.). Compositae. 8 S. Arabia to NW.
India (Sind).

Platychaete Bornm. = praec.

Platycheilus Cass. = Trixis P.Br. (Compos.).

Platychilum Delaun. = Hovea R.Br. (Legumin.).

Platychilus P. & K. = Platycheilus Cass. = Trixis P.Br. (Compos.).

Platycladus Spach = Thuja L. (Cupressac.).

Platyclinis Benth. = Dendrochilum Blume (Orchidac.).

Platyclinium T. Moore = Begonia L. (Begoniac.).

Platycodon A. DC. Campanulaceae. 1 NE. As.

Platycodon Reichb. = Daucus L. (Umbellif.).

Platycoryne Reichb. f. Orchidaceae. 17 trop. Afr., Madag.

Platycraspedum O. E. Schulz. Cruciferae. 1 E. Tibet.

Platycrater Sieb. & Zucc. Hydrangeaceae. 1 Japan.

Platycyamus Benth. Leguminosae. 2 Brazil.

Platydaucon Reichb. = Daucus L. (Umbellif.).

Platydesma H. Mann. Rutaceae. 6 Hawaii.

Platyelasma Kitagawa = Elsholtzia Willd. (Labiat.).

Platyestes Salisb. = Lachenalia Jacq. (Liliac.).

Platyglottis L. O. Williams. Orchidaceae. 1 Panamá.

Platygonia Naud. = Trichosanthes L. (Cucurbitac.).

Platygyna Mercier. Euphorbiaceae. 5 Cuba.

Platygyne Howard = praec.

Platyhymenia Walp. = Plathymenia Benth. (Legumin.).

Platykeleba N. E. Br. Asclepiadaceae. 1 Madag.

Platylepis Kunth = Ascolepis Nees (Cyperac.).

***Platylepis** A. Rich. Orchidaceae. 10 trop. & S. Afr., Madag., Masc.,
Seychelles.

Platylobeae Muell. Arg. An inclusive term denoting all *Euphorbiaceae* with
cotyledons much broader than the radicle, as opposed to the *Stenolobeae* (*q.v.*).
Probably an unnatural distinction for major groups in this fam.

Platylobium Sm. Leguminosae. 3 Austr.

Platyloma Merr. (sphalm.) = Platoloma Rafin. (Crucif.).

Platyloma J. Sm. = Pellaea Link (Sinopteridac.).

Platylophus Cass. = Centaurea L. (Compos.).

***Platylophus** D. Don. Cunoniaceae. 1 S. Afr.

Platyluma Baill. = Micropholis Pierre (Sapotac.).

Platymerium Bartl. ex DC. = Hypobathrum Bl. (Rubiac.).

Platymetra Nor. ex Salisb. = Tupistra Ker-Gawl. (Liliac.).

Platymetr[ac]eae Salisb. = Aspidistr[ac]eae J. G. Agardh = Liliaceae–Aspidistrinae Engl.

Platymiscium Vog. Leguminosae. 30 Mex. to trop. S. Am. Good timber.

Platymitium Warb. = Dobera Juss. (Salvadorac.).

Platymitra Boerl. Annonaceae. 2 Siam, Malay Penins., Java.

914

Platymitrium Willis (sphalm.) = Platymitium Warb. = Dobera Juss. (Salva-
Platynema Schrad. = Mertensia Roth (Boraginac.). [dorac.).
Platynema Wight & Arn. = Tristellateia Thou. (Malpighiac.).
Platyosprion Maxim. = Cladrastis Rafin. (Legumin.).
Platypetalum R.Br. = Braya Sternb. & Hoppe (Crucif.).
Platypholis Maxim. Orobanchaceae. 1 Bonin Is. (Japan).
Platypodium Vog. Leguminosae. 3 Panamá, Brazil.
Platyptelea J. Drumm. ex Harv. = Aphanopetalum Endl. (Cunoniac.).
Platypteris Kunth = Verbesina L. (Compos.).
Platypterocarpus Dunkley & Brenan. Celastraceae. 1 trop. E. Afr.
Platypus Small & Nash = Eulophia R.Br. ex Lindl. (Orchidac.).
Platyraphium Cass. = Lamyra Cass. (Compos.).
Platyrhaphe Miq. = Pimpinella L. (Umbellif.).
Platyrhiza Barb. Rodr. Orchidaceae. 1 Brazil.
Platyrhodon (Decne) Hurst (~ Rosa L.). Rosaceae. 1 E. As.
Platysace Bunge. Hydrocotylaceae (~ Umbellif.). 20 Austr.
Platyschkuhria (A. Gray) Rydb. (~ Bahia Lag.). Compositae. 1–3 W. U.S.
Platysema Benth. = Centrosema DC. (Legumin.).
Platysepalum Welw. ex Baker. Leguminosae. 12 trop. Afr.
Platysma Blume = Podochilus Blume (Orchidac.).
Platysperma Reichb. = Platyspermum Hoffm. = Daucus L. (Umbellif.).
Platyspermation Guillaumin. Rutaceae(!). 1 New Caled. G or semi-inf.!
Platyspermum Hoffm. = Daucus L. (Umbellif.).
Platyspermum Hook. = Idahoa A. Nels. & Macbr. (Crucif.).
Platystachys C. Koch = Tillandsia L. (Bromeliac.).
Platystele Schlechter. Orchidaceae. 7 Mex. to trop. S. Am.
Platystemma Wall. Gesneriaceae. 1 Himalaya.
Platystemon Benth. Papaveraceae. 60 (or 1 variable?) W. N. Am. Ls. in
apparent whorls (see fam.).
Platystephium Gardn. = Egletes Cass. (Compos.).
Platystigma Benth. Papaveraceae. 9 Pacif. N. Am. Ls. as in *Platystemon.*
Platystigma R.Br. ex Benth. & Hook. f. = Platea Bl. (Icacinac.).
Platystoma Benth. & Hook. f. = Platostoma Beauv. (Labiat.).
Platystylis Lindl. = Liparis Rich. (Orchidac.).
Platystylis Sweet = Lathyrus L. (Legumin.).
Platytaenia Kuhn = Taenitis Willd. (Gymnogrammac.).
Platytaenia Nevski & Vved. Umbelliferae. 7–8 C. As.
Platythea O. F. Cook. Palmae. 1 Mex.
Platytheca Steetz. Tremandraceae. 1 W. Austr.
Platythyra N. E. Br. Aïzoaceae. 1 S. Afr.
Platytinospora Diels. Menispermaceae. 1 W. Equat. Afr.
Platyzamia Zucc. = Dioön Lindl. (Zamiac.).
Platyzoma R.Br. Hemionitidaceae (see A. F. Tryon, *Rhodora*, **63**: 91, 1961).
1 NE. Austr.
Platyzomataceae Nakai = Hemionitidaceae Pichi-Serm.
Platzchaeta Sch. Bip. (sphalm.) = Platychaeta Boiss. (Compos.).
Plazaea P. & K. = Plazia Ruiz & Pav. (Compos.).
Plazeria Steud. = seq.

PLAZERIUM

Plazerium Willd. ex Kunth = Saccharum L. (Gramin.).

Plazia Ruiz & Pav. Compositae. 8 S. Andes, Argentina.

Pleconax Rafin. = Silene L. (Caryophyllac.).

Plecosorus Fée = Polystichum Roth (Aspidiac.).

Plecospermum Tréc. Moraceae. 2 Burma, Andamans.

Plecostigma Turcz. = Gagea Salisb. (Liliac.).

Plectaneia Thou. Apocynaceae. 14 Madag.

Plectanthera Mart. = Luxemburgia A. St-Hil. (Ochnac.).

Plectis O. F. Cook. Palmae. 1 C. Am.

Plectocephalus D. Don = Centaurea L. (Compos.).

Plectocomia Mart. & Blume. Palmae. 1 Assam to S. China & W. Malaysia. Climbers like *Calamus* with hooked ls.

Plectocomiopsis Becc. Palmae. 8 SE As., Malay Penins.

Plectogyne Link = Aspidistra Ker-Gawl. (Liliac.).

Plectoma Rafin. = Utricularia L. (Lentibulariac.).

Plectomirtha W. R. B. Oliv. = Pennantia J. R. & G. Forst. (Icacinac.).

Plectopoma Hanst. = Gloxinia L'Hérit. (Gesneriac.).

Plectopteris Fée = Calymmodon Presl (Grammitidac.).

Plectorrhiza Dockr. Orchidaceae. 2 E. Austr., 1 Lord Howe I.

Plectrachne Henrard. Gramineae. 11 Austr.

Plectranthastrum T. C. E. Fries = Alvesia Welw. (Labiat.).

Plectranthera Benth. & Hook. f. (sphalm.) = Plectanthera Mart. = Luxemburgia A. St-Hil. (Ochnac.).

Plectranthrastrum Willis (sphalm.) = Plectranthastrum T. C. E. Fries (Labiat.) = Alvesia Welw. (Labiat.).

*Plectranthus L'Hérit. Labiatae. 250 trop. Afr. to Japan, Malaysia, Austr., Pacif.

Plectreca Rafin. = Vernonia Schreb. (Compos.).

Plectrelminthes Merr. (sphalm.) = seq.

Plectrelminthus Rafin. Orchidaceae. 2 trop. W. Afr., Comoro Is.

Plectritis DC. Valerianaceae. 15 W. U.S., Chile.

Plectrocarpa Gillies ex Hook. Zygophyllaceae. 3 temp. S. Am.

Plectronema Rafin. = Zephyranthes Herb. (Amaryllidac.).

Plectronia auctt. = Canthium Lam. (Rubiac.).

Plectronia Buching. ex Krauss = Olinia Thunb. (Oliniac.).

Plectronia L. = Canthium Lam. (Rubiac.) + Olinia Thunb. (Oliniac.).

Plectronia Lour. = Acanthopanax Decne & Planch. (Araliac.).

Plectroniella Robyns. Rubiaceae. 2 trop. Afr.

Plectrophora H. C. Focke. Orchidaceae. 5 Brazil, Guiana, Trinidad.

Plectrornis Rafin. ex Lunell = Delphinium L. (Ranunculac.).

Plectrotropis Schumach. & Thonn. = Vigna Savi (Legumin.).

Plectrurus Rafin. = Tipularia Nutt. (Orchidac.).

Pleea Michx. Liliaceae. 1 SE. U.S.

Plegerina B. D. Jacks. (sphalm.) = Pleragina Arruda ex Koster = Licania Aubl. + Couepia Aubl. (Chrysobalanac.).

Plegmatolemma Bremek. Acanthaceae. 2 Siam.

Plegorhiza Molina = Limonium Mill. (Plumbaginac.).

Pleiacanthus Rydb. (~ Lygodesmia D. Don). Compositae. 1 W. U.S.

Pleiadelphia Stapf = Elymandra Stapf (Gramin.).
Pleianthemum K. Schum. ex A. Chev. = Desplatsia Bocq. (Tiliac.).
Pleiarina Rafin. = Salix L. (Salicac.).
Pleienta Rafin. = Sabatia Adans. (Gentianac.).
Pleimeris Rafin. = Gardenia Ellis (Rubiac.).
Pleioblastus Nakai (~ Arundinaria Rich.). Gramineae. 100 E. As.
Pleiocardia Greene. Cruciferae. 6 Calif.
Pleiocarpa Benth. Apocynaceae. 3 trop. Afr.
Pleiocarpidia K. Schum. Rubiaceae. 27 W. Malaysia.
Pleioceras Baill. Apocynaceae. 3 trop. W. Afr.
Pleiochasia Barnh. = Utricularia L. (Lentibulariac.).
Pleiochiton Naud. ex A. Gray. Melastomataceae. 9 S. Brazil.
Pleiococca F. Muell. Rutaceae. 1 E. Austr.
Pleiocraterium Bremek. Rubiaceae. 4 S. India, Ceylon, Sumatra.
Pleiodon Reichb. = Polyodon Kunth = Botelua Lag. (Gramin.).
Pleiogyne C. Koch = Cotula L. (Compos.).
Pleiogynium Engl. Anacardiaceae. 2 Philippines, Lesser Sunda Is., New Guinea, Queensland.
Pleiokirkia R. Capuron. Simaroubaceae. 1 Madag.
Pleiomeris A. DC. Myrsinaceae. 1 Canaries, Madeira.
Pleïone D. Don. Orchidaceae. 10 India to Formosa & Siam.
Pleioneura K. H. Rechinger. Caryophyllaceae. 2 C. As., Afghan., W. Himal.
Pleiophaca F. Muell. ex Baill. = Archidendron F. Muell. (Legumin.).
Pleiosepalum Hand.-Mazz. Rosaceae. 1 SW. China.
Pleiosepalum C. E. Moss = Krauseola Pax & Hoffm. (Caryophyllac.).
Pleiosmilax Seem. = Smilax L. (Smilacac.).
Pleiospermium (Engl.) Swingle. Rutaceae. 7 SE. As., W. Malaysia.
Pleiospermum auctt. = praec.
Pleiospilos N. E. Brown. Aïzoaceae. 38 S. Afr. (Cape Prov.).
Pleiospora Harv. (1859; nec *Pleospora* Rabenhorst 1851—Fungi) = Phaeno-hoffmannia Kuntze (Legumin.).
Pleiostachya K. Schum. Marantaceae. 2 C. Am., Ecuador.
Pleiostachyopiper Trelease. Piperaceae. 1 Amaz. Brazil.
Pleiostemon Sond. (~ Securinega Juss.). Euphorbiaceae. 1 S. Afr.
Pleiotaenia Coulter & Rose. Umbelliferae. 1 W. N. Am.
Pleiotaxis Steetz. Compositae. 25 trop. Afr.
Pleisolirion Rafin. = Paradisea Mazzuc. (Liliac.).
Plemasium Presl = Osmunda L. (Osmundac.).
Plenckia Moç. & Sessé ex DC. = Choisya Kunth (Rutac.).
Plenckia Rafin. = Glinus L. (Aïzoac.).
*****Plenckia** Reissek. Celastraceae. 4 S. Am.
Pleocarphus D. Don = Jungia L. f. (Compos.).
Pleocarpus Walp. = praec.
Pleocaulus Bremek. Acanthaceae. 3 Penins. India.
Pleocnemia Presl. Aspidiaceae. 15 Malaysia. (Included in *Tectaria*, Copeland 1947.) Large ferns, distinct in anat. and sinus-teeth (Holttum, *Reinwardtia*, 1: 171–189, 1951).
Pleodendron Van Tiegh. Canellaceae. 2 W.I.

PLEODIPOROCHNA

Pleodiporochna Van Tiegh. = Ochna L. (Ochnac.).

Pleogyne Miers. Menispermaceae. 1 trop. E. Austr.

Pleomele Salisb. = Dracaena Vand. ex L. (Agavac.).

Pleonanthus Ehrh. (uninom.) = *Dianthus prolifer* L. = *Kohlrauschia prolifera* (L.) Kunth (Caryophyllac.).

Pleonotoma Miers. Bignoniaceae. 12 trop. Am., Trinidad.

Pleopadium Rafin. = Croton L. (Euphorbiac.).

Pleopeltis Humb. & Bonpl. (incl. *Lepisorus* Ching). Polypodiaceae. 40 pantrop., but few in Malaysia–Polynesia. Ls. bearing peltate scales. (Beddome & v. Ald. v. Ros. included here almost all spp. of *Polypodiaceae* with reticulate venation and round sori.)

Pleopetalum Van Tiegh. = Ochna L. (Ochnac.).

Pleopogon Nutt. = Lycurus Kunth (Gramin.).

Pleorothyrium Endl. (sphalm.) = Pleurothyrium Nees = Ocotea Aubl. (Laurac.).

Pleotheca Wall. = Spiradiclis Blume (Rubiac.).

Pleouratea Van Tiegh. = Ouratea Aubl. (Ochnac.).

Pleragina Arruda ex Koster = Licania Aubl. + Couepia Aubl. (Chrysobalanac.).

Plerandra A. Gray. Araliaceae. 14 New Guinea to Fiji.

Plerandropsis R. Viguier = ? Trevesia Vis. (Araliac.).

Pleroma D. Don = Tibouchina Aubl. (Melastomatac.).

Plesiagopus Rafin. = Ipomoea L. (Convolvulac.).

Plesiatropha Pierre ex Hutch. = Mildbraedia Pax (Euphorbiac.).

Plesilia Rafin. = Stylisma Rafin. = Breweria R.Br. (Convolvulac.).

Plesiopsora Rafin. = Scabiosa L. (Dipsacac.).

Plesisa Rafin. = Utricularia L. (Lentibulariac.).

Plesmonium Schott. Araceae. 1 N. India.

Plethadenia Urb. Rutaceae. 2 W.I.

Plethiandra Hook. f. Melastomataceae. 9 Malay Penins., Borneo.

Plethiosphace (Benth.) Opiz = Salvia L. (Labiat.).

Plethostephia Miers = Cordia L. (Ehretiac.).

Plethyrsis Rafin. = ? Richardia L. (Rubiac.).

Plettkea Mattf. Caryophyllaceae. 4 Andes.

Pleudia Rafin. = Salvia L. (Labiat.).

Pleurachne Schrad. = Ficinia Schrad. (Cyperac.).

Pleuradena Rafin. (1833) = Euphorbia L. (Euphorbiac.)

Pleuradenia B. D. Jacks. (sphalm.) = praec.

Pleuradenia Rafin. (1825) = Collinsonia L. (Labiat.).

Pleurandra Labill. = Hibbertia Andr. (Dilleniac.).

Pleurandra Rafin. = Pleurostemon Rafin. = Gaura L. (Onagrac.).

Pleurandropsis Baill. Rutaceae. 1 W. Austr.

Pleurandros St-Lag. = Pleurandra Labill. = Hibbertia Andr. (Dilleniac.).

Pleuranthe Salisb. = Protea L. (Proteac.).

Pleuranthemum (Pichon) Pichon. Apocynaceae. 1 trop. Afr.

Pleuranthium Benth. = Epidendrum L. (Orchidac.).

Pleuranthodendron L. O. Williams. ? Flacourtiaceae. 1 Mex. to trop. S. Am.

Pleuranthodes Weberb. Rhamnaceae. 2 Hawaii.

Pleuranthus Rich. ex Pers. = Dulichium Pers. (Cyperac.).
Pleuraphis Torr. (~ Hilaria Kunth). Gramineae. 3 SW. U.S., Mex.
Pleurastis Rafin. = Lycoris Herb. (Amaryllidac.).
Pleureia Rafin. = Psychotria L. (Rubiac.).
Pleuremidis Rafin. = Thunbergia Retz. (Thunbergiac.).
Pleurendotria Rafin. = Lithophragma (Nutt.) Torr. & Gray (Saxifragac.).
Pleurenodon Rafin. = Hypericum L. (Guttif.).
Pleuriarum Nakai = Arisaema Mart. (Arac.).
Pleuricospora A. Gray. Monotropaceae. 2 Pacif. N. Am.
Pleuridium (Presl) Fée (1852; non Bridel 1819—Musci) = Pessopteris Underw. & Moxon (Polypodiac.).
Pleurima Rafin. = Campanula L. (Campanulac.).
Pleurimaria B. D. Jacks. = Plurimaria Rafin. = Blackstonia Huds. (Gentianac.).
Pleuripetalum Th. Dur. (sphalm.) = Eburopetalum Becc. = Anaxagorea A. St-Hil. (Annonac.).
Pleurisanthaceae Van Tiegh. = Icacinaceae Miers.
Pleurisanthes Baill. Icacinaceae. 5 trop. S. Am.
Pleuroblepharis Baill. Acanthaceae. 1 Madag.
Pleuroblepharon Kunze ex Reichb. (nomen). Orchidaceae. Quid?
Pleurobotryum Barb. Rodr. Orchidaceae. 2 Brazil.
Pleurocalyptus Brongn. & Gris. Myrtaceae. 1 New Caled.
Pleurocarpaea Benth. Compositae. 1 trop. Austr.
Pleurocarpus Klotzsch = Rhyssocarpus Endl. (Rubiac.).
Pleurochaenia Griseb. = Miconia Ruiz & Pav. (Melastomatac.).
Pleurocitrus Tanaka. Rutaceae. 1 Queensland.
Pleurocoffea Baill. Rubiaceae. 1 Madag.
Pleurocoronis R. M. King & H. Rob. (~ Eupatorium L.). Compositae. 3 SW. U.S., NW. Mex.
Pleuroderris Maxon. Aspidiaceae. 1 C. Am. Probably inter-generic hybrid, Dictyoxiphium × Tectaria.
Pleurodesmia Arn. = Schumacheria Vahl (Dilleniac.).
Pleurodiscus Pierre ex A. Chev. = Laccodiscus Radlk. (Sapindac.).
Pleurofossa Nakai ex H. Ito = seq.
Pleurogramme (Bl.) Presl = Monogramma Schkuhr (Vittariac.).
Pleurogyna Eschsch. ex Cham. & Schlechtd. Gentianaceae. 15 temp. Euras., 1 N. Am.
Pleurogyne Griseb. = praec.
Pleurogynella Ikonn. Gentianaceae. 4 mts. C. As.
Pleurolobus Jaume St-Hil. = Desmodium Desv. (Legumin.).
Pleuromanes Presl. Hymenophyllaceae. 3 Ceylon to Tahiti.
Pleuromenes Rafin. = Prosopis L. (Legumin.).
Pleuropappus F. Muell. = Angianthus Wendl. (Compos.).
Pleuropetalon Blume = Chariessa Miq. (Icacinac.).
Pleuropetalum Benth. & Hook. f. = praec.
Pleuropetalum Hook. f. Amaranthaceae. 5 C. & trop. S. Am., Galápagos.
Pleurophora D. Don. Lythraceae. 7 S. Am.
Pleurophragma Rydb. Cruciferae. 4 Pacif. U.S.

Pleurophyllum Hook. f. Compositae. 3, islands to S. of N.Z.
Pleurophyllum Mart. ex K. Schum. = Warszewiczia Klotzsch (Rubiac.).
Pleuroplitis Trin. = Arthraxon Beauv. (Gramin.).
Pleuropogon R.Br. Gramineae. 5 W. U.S., 1 circumpolar.
Pleuropsa Merr. (sphalm.) = Pleurospa Rafin. (Arac.).
Pleuropterantha Franch. Amaranthaceae. 2 Somaliland.
Pleuropteropyrum H. Gross (= Aconogonum [Meissn.] Reichb.). Polygonaceae. 15 N. As., Japan, N. Am.
Pleuropterus Turcz. = Polygonum L. (Polygonac.).
Pleuroraphis P. & K. = Pleuraphis Torr. (Gramin.).
Pleuroridgea Van Tiegh. = Brackenridgea A. Gray (Ochnac.).
Pleurosoriopsis Fomin. Hemionitidaceae. 1 China, Japan.
Pleurosorus Fée. Aspleniaceae. 3 Spain, N. Afr., Austr., Chile.
Pleurospa Rafin. (Montrichardia Crueg.). Araceae. 2 trop. Am., W.I.
Pleurospermopsis C. Norman. Umbelliferae. 1 E. Himal.
Pleurospermum Hoffm. Umbelliferae. *S.str.* 3, *s.l.* 80, temp. Euras.
Pleurostachys Brongn. Cyperaceae. 50 S. Am.
Pleurostelma Baillon. Asclepiadaceae. 1 E. Afr., Madag., Aldabra I.
Pleurostelma Schlechter = Schlechterella K. Schum. (Periplocac.).
Pleurostemon Rafin. = Gaura L. (Onagrac.).
Pleurostena Rafin. = Polygonum L. (Polygonac.).
Pleurostigma Hochst. = Bouchea Cham. (Verbenac.).
Pleurostima Rafin. = Barbacenia Vand. (Velloziac.).
Pleurostylia Wight & Arn. Celastraceae. 3–4 trop. & S. Afr., Madag., Masc., India, SE. As.
Pleurostylis Walp. = praec.
Pleurotaenia Hohenack. ex Benth. & Hook. f. = Peucedanum L. (Umbellif.).
Pleurothallis R.Br. Orchidaceae. 1000 Mex. to trop. S. Am., W.I.
Pleurothallopsis C. Porto & Brade. Orchidaceae. 1 Brazil.
Pleurothyrium (~ Ocotea Aubl.). Lauraceae. 1 Peru.
Pleuteron Rafin. = Capparis L. (Capparidac.).
Plexaure Endl. = Phreatia Lindl. (Orchidac.).
Plexinium Rafin. = Androcymbium Willd. (Liliac.).
Plexipus Rafin. = Bouchea Cham. (Verbenac.).
Plexistena Rafin. = Allium L. (Alliac.).
Pliarina P. & K. = Pleiarina Rafin. = Salix L. (Salicac.).
Plicangis Thou. (uninom.) = *Angraecum implicatum* Thou. (Orchidac.).
Plicosepalus Van Tiegh. Loranthaceae. 1 S. trop. Afr.
Plicouratea Van Tiegh. = Ouratea Aubl. (Ochnac.).
Plicula Rafin. = Acnistus Schott (Solanac.).
Plienta P. & K. = Pleienta Rafin. = Sabatia Adans. (Gentianac.).
Plinia Blanco = Mesua L. (Guttif.).
Plinia L. Myrtaceae. 30 W.I., trop. S. Am.
Plinthanthesis Steud. = Danthonia DC. (Gramin.).
Plinthine Reichb. = Arenaria L. (Caryophyllac.).
Plinthocroma Dulac = Azalea auctt. = Rhododendron L. (Ericac.).
Plinthus Fenzl. Aïzoaceae. 5–6 S. Afr.
Pliocarpidia P. & K. = Pleiocarpidia K. Schum. (Rubiac.).

Pliodon P. & K. = Pleiodon Reichb. = Botelua Lag. (Gramin.).
Pliogyna P. & K. = Pleogyne Miers (Menispermac.).
Pliogynopsis Kuntze = Pleiogynium Engl. (Anacardiac.).
Pliophaca P. & K. = Pleiophaca F. Muell. ex Baill. = Archidendron F. Muell. (Legumin.).
Ploca Lour. ex Gomes = Lourea Neck. ex Desv. (Legumin.).
Plocaglottis Steud. = Plocoglottis Blume (Orchidac.).
Plocama Ait. Rubiaceae. 1 Canaries.
Plocandra E. Mey. = Chironia L. (Gentianac.).
Plocaniophyllon T. S. Brandegee. Rubiaceae. 1 Mex.
Plocoglottis Blume. Orchidaceae. 30 Malaysia (?China, ?Indochina), Solomon Is.
Plocosperma Benth. Plocospermataceae. 3 Mex., Guatem.
Plocospermataceae Hutch. Dicots. 1/3 Mex., C. Am. Trees or shrubs with opp. or subvertic. simple ent. sometimes emarg. exstip. ls. Fls. reg., ⚥, axill., 1–4 together. K (5–6), imbr. or open, small; C (5–6), tube broadly infundibulif., lobes rounded, imbr.; A 5–6, epipet., unequal, anth. dorsifixed, ± latrorse; disk small; G̲ (2), 1-loc., stipitate, with 2 parietal plac., the plac. bearing either 1 basal erect ov. each, 2 basal ov. each, or 2 basal. ov. on one and 2 sub-apical pend. ov. on the other; style shortly twice bifid, with 4 small clavate stigs. Fr. an elongate-fusiform ribbed 2-valved caps.; seed usu. sol., linear, terete, with an apical tuft of long hairs, endosp. sparse, fleshy. Only genus: *Plocosperma*. An interesting group, combining features of *Apocynac.*, *Loganiac.*, *Convolvulac.* and *Ehretiac.*
Plocostemma Blume = Hoya R.Br. (Asclepiadac.).
Plocostigma Benth. (sphalm.) = Placostigma Bl. (Orchidac.).
Plocostigma P. & K. = Plokiostigma Schuch. = Stackhousia Sm. (Stackhousiac.).
Ploearium P. & K. = Ploiarium Korth. (Bonnetiac.).
Ploesslia Endl. = Boswellia Roxb. (Burserac.).
Ploiarium Korthals. Bonnetiaceae. 3 SE. As., W. Malaysia, Moluccas, New Guinea.
Ploionixus Van Tiegh. ex Lecomte = Viscum L. (Viscac.).
Plokiostigma Schuchardt = Stackhousia Sm. (Stackhousiac.).
Plostaxis Rafin. = Polygala L. (Polygalac.).
Plotea J. F. Gmel. = Plotia Adans. = Salvadora L. (Salvadorac.).
Plothirium Rafin. = Delphinium L. (Ranunculac.).
Plotia Adans. = ? Salvadora L. (Salvadorac.).
Plotia Neck. = Embelia Burm. f. (Myrsinac.).
Plotia Schreb. ex Steud. = Poa L. (Gramin.).
Plottzia Arn. = Paronychia Juss. (Caryophyllac.).
Pluchea Cass. Compositae. 50 trop. & subtrop.
Pluchia Vell. = Diclidanthera Mart. (Polygalac.).
Pluechea Zoll. & Mor. = Pluchea Cass. (Compos.).
Plukenetia L. Euphorbiaceae. 10 warm Am.; 1 Madag.(?).
Pluknetia Boehm. = praec.
Plumaria Fabr. (uninom.) = Eriophorum L. sp. (Cyperac.).
Plumaria Opiz = Dianthus L. (Caryophyllac.).

Plumbagella Spach (~Plumbago L.). Plumbaginaceae. 1 C. As.

Plumbagidium Spach = Plumbago L. (Plumbaginac.).

*Plumbaginaceae Juss.(*s.l.*). Dicots. 19/775 cosmop., but esp. on salt steppes and sea coast. Perennial herbs or shrubs (sometimes scandent) with alt. simple exstip. ls., on whose surface water glands occur, or sometimes chalk glands (cf. *Saxifraga*). Infl. of various types, racemose and cymose (see *Plumbago*, *Ceratostigma, Limonium, Armeria*), bracteolate. Fls. reg., ♀, 5-merous, the odd sepal post. K persistent; C often nearly polypetalous, conv.; A 5, epipet. and oppositipet.; G̲ (5), 1-loc., with basal placenta, and one anatr. ov., whose funicle curves up to the top of the loc. and causes the micropyle to be directed upwards; styles or stigmas 5. Nut; embryo straight, in floury endosp. Chief genera: *Plumbago, Ceratostigma, Acantholimon, Armeria, Limonium, Limoniastrum*. Prob. marginally related to the *Centrospermae* (though the pollen is not Centrospermous); connected with the *Linaceae* through the Linaceous genus *Anisadenia*. Distinguished from *Primulaceae* by the sol. ovule and free styles.

Plumbaginaceae Juss. (*s.str.*). Dicots. 4/24 warm dry regions of the world. Differ from *Limoniaceae* in the erect lobes or teeth of the calyx, in the stamens being free (adnate to C in *Ceratostigma*), and in the long connation of the styles. Genera: *Plumbago, Plumbagella, Ceratostigma, Dyerophytum*.

Plumbaginella Ledeb. = Plumbagella Spach (Plumbaginac.).

Plumbago L. Plumbaginaceae. 12 warm regions. Racemose infl. K with glandular hairs, aiding seed dispersal.

Plumea Lunan = Guarea L. (Meliac.).

Plumeria L. Apocynaceae. 7 warm Am. Hybrid swarms are known in Haiti. *P. rubra* L. ('frangipani') widely cult. in Old World tropics; the flowers are offered in Buddhist temples.

Plumeriaceae Horan. = Apocynaceae Juss.

Plumeriopsis Rusby & R. E. Woodson = Ahouai Mill. (Apocynac.).

Plummera A. Gray. Compositae. 2 SW. U.S. (Arizona).

Plumosipappus Czerep. Compositae. 1 As. Min.

Pluridens Neck. = Bidens L. (Compos.).

Plurimaria Rafin. = Blackstonia Huds. (Gentianac.).

Plutarchia A. C. Smith. Ericaceae. 12 N. Andes.

Plutea Nor. = ?Lansium Corrêa (Meliac.).

Plutonia Nor. = Phaleria Jack (Thymelaeac.).

Pneumaria Hill = Mertensia Roth (Boraginac.).

Pneumatopteris Nakai, emend. Holtt. Thelypteridaceae. 50 Afr., SE. As. to Hawaii. Related to *Sphaerostephanos* J. Sm., but fronds and scales sparsely hairy, never with spherical glands.

Pneumonanthe Gleditsch = Gentiana L. (Gentianac.).

Pneumonanthopsis Miq. = Voyria Aubl. (Gentianac.).

Poa L. Gramineae. 300 cosmop. Many are useful pasture-grasses.

*Poaceae Barnhart: *see* **Gramineae** Juss. (*nom. altern.*).

Poacynum Baill. Apocynaceae. 3 C. As. Ls. alt.

Poaephyllum Ridley. Orchidaceae. 4 Malaysia.

Poagris Rafin. = Poa L. (Gramin.).

Poagrostis Stapf. Gramineae. 1 S. Afr.

Pogogyne Benth. Labiatae. 5 California, S. Oregon.
Pogoina B. Grant (sphalm.) = Pogonia Griff. (Orchidac.).
Pogomesia Rafin. = Tinantia Scheidw. (Commelinac.).
Pogonachne Bor. Gramineae. 1 India (Bombay).
Pogonanthera Blume. Melastomataceae. 1–4 Malaysia.
Pogonanthera Spach = Scaevola L. (Goodeniac.).
Pogonanthus Montr. = Morinda L. (Rubiac.).
Pogonarthria Stapf. Gramineae. 5 trop. & S. Afr.
Pogonatherum Beauv. Gramineae. 2 India to Japan. Sta. 2.
Pogonatum Steud. (sphalm.) = praec.
Pogonella Salisb. = Simethis Kunth (Liliac.).
Pogonema Rafin. = Zephyranthes Herb. (Amaryllidac.)
Pogonetes Lindl. = Scaevola L. (Goodeniac.).
Pogonia Andr. = Myoporum Banks & Soland. (Myoporac.).
Pogonia Juss. Orchidaceae. 50 India to E. As., Canada to trop. S. Am., W.I.
Pogoniopsis Reichb. f. Orchidaceae. 2 Brazil.
Pogonitis Reichb. = Anthyllis L. (Legumin.).
Pogonochloa C. E. Hubbard. Gramineae. 1 S. trop. Afr.
Pogonolepis Steetz = Angianthus Wendl. (Compos.).
Pogonolobus F. Muell. = Coelospermum Blume (Rubiac.).
Pogononeura Napper. Gramineae. 1 trop. E. Afr.
Pogonophora Miers. Euphorbiaceae (?). 1–2 trop. S. Am.; 1 Equat. Afr. An anomalous genus; some features suggest *Icacinaceae*.
Pogonophyllum Didrichs. = Micrandra Benth. (Euphorbiac.).
Pogonopsis J. &' C. Presl = Pogonatherum Beauv. (Gramin.).
Pogonopus Klotzsch. Rubiaceae. 3 C. & S. Am.
Pogonorhynchus Crueg. = Miconia Ruiz & Pav. (Melastomatac.).
Pogonospermum Hochst. = Monechma Hochst. (Acanthac.).
Pogonostemon Hassk. = Pogostemon Desf. (Labiat.).
Pogonostigma Boiss. = Tephrosia Pers. (Legumin.).
Pogonostylis Bertol. = Fimbristylis Vahl (Cyperac.).
Pogonotrophe Miq. = Ficus L. (Morac.).
Pogonura DC. ex Lindl. = Perezia Lag. (Compos.).
Pogopetalum Benth. = Emmotum Desv. (Icacinac.).
Pogospermum Brongn. = Catopsis Griseb. (Bromeliac.).
Pogostemon Desf. Labiatae. 40 China, Indomal. *P. patchouly* Pellet. yields the well-known perfume by distillation.
Pogostoma Schrad. = Capraria L. (Scrophulariac.).
Pohlana Leandro = Zanthoxylum L. (Rutac.).
Pohliella Engl. Podostemaceae. 2 W. Equat. Afr.
Poicilanthe Schltr. Orchidaceae. Nomen. Quid?
Poicilla Griseb. Asclepiadaceae. 1 Cuba.
Poicillopsis Schlechter ex Rendle. Asclepiadaceae. 6 S. Domingo, Cuba.
Poïdium Nees = Poa L. (Gramin.).
Poikadenia Ell. = Psoralea L. (Legumin.).
Poikilacanthus Lindau. Acanthaceae. 9 C. & S. Am.
Poikilogyne E. G. Baker. Melastomataceae. 12 Borneo, New Guinea.
Poikilospermum Zipp. ex Miq. Urticaceae. 20 E. Himal. to S. China &

Malaysia. Somewhat intermediate between *Urticac.* and *Morac.* The ls. of *P. suaveolens* (Bl.) Merr. possess water-secreting glands.

Poilanedora Gagnep. Capparidaceae [?]. 1 Indoch.

Poilania Gagnep. = Epaltes Cass. (Compos.).

Poilaniella Gagnep. Euphorbiaceae. 1 Indoch.

Poincettia Klotzsch & Garcke = Poinsettia R. Grah. = Euphorbia L. (Euphor-

Poincia Neck. = seq.   [biac.).

Poinciana auctt. = Delonix Rafin. (Legumin.).

Poinciana L. = Caesalpinia L. (Legumin.).

Poincianella Britton & Rose = Caesalpinia L. (Legumin.).

Poinsettia R. Grah. (~ Euphorbia L.). Euphorbiaceae. 11-12 E. U.S. to N. Argent. (esp. Mex.).

Poiretia Cav. = Sprengelia Sm. (Epacridac.).

Poiretia J. F. Gmel. = Houstonia Gronov. ex L. (Rubiac.).

Poiretia Sm. = Hovea R.Br. (Legumin.).

*Poiretia Vent. Leguminosae. 7 trop. Am.

Poissonella Pierre = Lucuma Molina (Sapotac.).

Poissonia Baill. Leguminosae. 3 Andes.

Poitaea DC. = seq.

Poitea Vent. Leguminosae. 6 W.I.

Poivraea auctt. = seq.

Poivrea Comm. ex Thou. = Combretum L. (Combretac.).

Pokornya Montr. = Lumnitzera Willd. (Combretac.).

Polakia Stapf. Labiatae. 1 Persia.

Polakiastrum Nakai = Salvia L. (Labiat.).

Polakowskia Pittier. Cucurbitaceae. 1 Costa Rica.

Polameia Reichb. (sphalm.) = Potameia Thou. (Laurac.).

Polanina Rafin. = seq.

Polanisia Rafin. Cleomaceae. 7 temp. N. Am.

Polanysia Rafin. = praec.

Polaskia Backeb. Cactaceae. 1 Mex.

Polathera Rafin. = seq.

Polatherus Rafin. = Gaillardia Fouger. (Compos.).

Polemannia Bergius ex Schlechtd. = Dipcadi Medik. (Liliac.).

*Polemannia Eckl. & Zeyh. Umbelliferae. 3-4 S. Afr.

Polembrium Steud. = seq.

Polembryon Benth. & Hook. f. = seq.

Polembryum A. Juss. = Esenbeckia Kunth (Rutac.).

*Polemoniaceae Juss. Dicots. 15/300, chiefly N. Am.; a few in Chile, Peru, Eur., N. As. Herbs (rarely shrubby below), glabrous or shortly hairy, with usu. opp. exstip. ls. Fls. in cymes (sometimes condensed into involucrate heads), ⚥, reg. or slightly zygo., with or without bracteoles. K (5), valvate or imbr., persistent; C (5), bell-, funnel-, or plate-shaped, usu. conv.; A 5, epipet., alt. with petals; G̲ (3) or rarely (2-5), on a disk, multiloc., with simple style ± lobed at tip. Ovules 1-∞ in each loc., axile, anatr., sessile. Fr. usu. a loculic. caps. Embryo straight, in endosp. Chief genera: *Cantua, Phlox, Collomia, Gilia, Polemonium*.

Polemoniella A. A. Heller = seq.

POLYANTHINA

Polemonium L. Polemoniaceae. 50 temp. Euras., N. Am., Mex., 2 Chile.
Polevansia de Winter. Gramineae. 1 SW. Afr.
Polgidon Rafin. = Chaerophyllum L. (Umbellif.).
Polia Lour. = Polycarpaea Lam. (Caryophyllac.).
Polia Tenore = Cypella Herb. (Iridac.).
Polianthes L. Agavaceae. 13 Mex., Trinidad.
Policarpaea Lam. = Polycarpaea Lam. (Caryophyllac.).
Poligala Neck. = Polygala L. (Polygalac.).
Poligonum Neck. = Polygonum L. (Polygonac.).
Poliodendron Webb & Berth. = Teucrium L. (Labiat.).
Poliomintha A. Gray. Labiatae. 4 SW. U.S., Mex.
Poliophyton O. E. Schulz. Cruciferae. 1 S. U.S., NE. Mex.
Poliothyrsis Oliv. Flacourtiaceae. 1 China.
Polium Mill. = Teucrium L. (Labiat.).
Polium Stokes = Polia Lour. = Polycarpaea Lam. (Caryophyllac.).
Pollalesta Kunth. Compositae. 24 Costa Rica to N. Braz. & Peru.
Pollia Thunb. Commelinaceae. 16 palaeotrop.
Pollichia Medik. = Trichodesma R.Br. (Boraginac.).
Pollichia Schrank = Galeobdolon Adans. (Labiat.).
*****Pollichia** (Soland. in) Ait. Caryophyllaceae. 1 trop. & S. Afr., Arabia.
Pollinia Spreng. = Chrysopogon Trin., etc. (Gramin.).
Pollinia Trin. = Microstegium Nees (Gramin.).
Pollinidium Stapf ex Haines = Eulaliopsis Honda (Gramin.).
Polliniopsis Hayata. Gramineae. 1 Formosa.
Pollinirhiza Dulac = Listera R.Br. (Orchidac.).
Poloa DC. = Pulicaria Gaertn. (Compos.).
Polpoda C. Presl. Aïzoaceae. 1 S. Afr.
Polpodaceae Nak. = Aïzoaceae–Limeëae–Adenogramminae K. Müll.
Poltolobium Presl = Andira Lam. (Legumin.).
Polyacantha S. F. Gray = Centaurea L. (Compos.).
Polyacantha Hill = Carduus L. + Cirsium Mill. (Compos.).
Polyacanthus Presl = Gymnosporia Benth. & Hook. f. (Celastrac.).
Polyachyrus Lag. Compositae. 20 Peru, Chile.
Polyactidium DC. = seq.
Polyactis Less. = Erigeron L. (Compos.).
Polyactium Eckl. & Zeyh. = Pelargonium L'Hérit. (Geraniac.).
Polyadelphaceae Dulac = Hypericaceae Juss.
Polyadenia Nees = Lindera Thunb. (Laurac.).
Polyadoa Stapf. Apocynaceae. 4 trop. Afr.
Polyalthia Blume. Annonaceae. 120 palaeotrop., esp. Indomal.
Polyandra Carlos Leal. Euphorbiaceae. 1 Brazil.
Polyandrococos Barb. Rodr. Palmae. 1 Brazil.
Polyanthemum Bub. = Leucojum L. (Amaryllidac.).
Polyanthemum Medik. = Armeria Willd. (Plumbaginac.).
Polyantherix Nees = Elymus L. (Gramin.).
Polyanthes Hill = Polianthes L. (Agavac.).
Polyanthes Jacq. = Polyxena Kunth (Liliac.). [Am. to Peru.
Polyanthina R. M. King & H. Rob. (~Eupatorium L.). Compositae. 1 C.

929

POLYANTHUS

Polyanthus auctt. = Polianthes L. (Agavac.).
Polyarrhena Cass. = Felicia Cass. (Compos.).
Polyaster Hook. f. Rutaceae. 1–2 Mex.
Polyaulax Backer. Annonaceae. 1 Java, S. New Guinea.
Polybaea Klotzsch ex Benth. & Hook. f. = Polyboea Klotzsch (1851) = Cavendishia Lindl. (Ericac.).
Polyboea Klotzsch ex Endl. (1850) = Bernardia Mill. (Euphorbiac.).
Polyboea Klotzsch (1851) = Cavendishia Lindl. (Ericac.).
Polybotrya Humb. & Bonpl. Aspidiaceae. 25 trop. Am. Large scandent ferns; ls. dimorphous.
Polycalymma F. Muell. & Sond. = Myriocephalus Benth. (Compos.).
Polycampium Presl = Pyrrosia Mirbel (Polypodiac.).
Polycandia Steud. = Polycardia Juss. (Celastrac.).
Polycantha Hill = Polyacantha Hill = Carduus L. + Cirsium Mill. (Compos.).
Polycardia Juss. Celastraceae. 9 Madag.
Polycarena Benth. Scrophulariaceae. 40 S. Afr.
Polycarpa Linden ex Carr. = Idesia Maxim. (Flacourtiac.).
Polycarpa Loefl. = Polycarpon L. (Caryophyllac.).
*****Polycarpaea** Lam. Caryophyllaceae. 50 trop. & subtrop. (few in Am.).
Polycarpea Pomel = praec.
Polycarpia Webb & Berth. = praec.
Polycarpoea Lam. = praec.
Polycarpon Loefl. ex L. Caryophyllaceae. 16 cosmop.
Polycenia Choisy = Hebenstretia L. (Scrophulariac.).
Polycephalium Engl. Icacinaceae. 2 trop. Afr.
Polycephalos Forsk. = Sphaeranthus L. (Compos.).
Polyceratocarpus Engl. & Diels. Annonaceae. 5 trop. Afr.
Polychaetia Less. = Nestlera Spreng. (Compos.).
Polychaetia Tausch ex Less. = Tolpis Adans. (Compos.).
Polychilos Breda = Phalaenopsis Bl. (Orchidac.).
Polychisma C. Muell. (sphalm.) = Polyschisma Turcz. = Pelargonium L'Hérit.
Polychlaena G. Don = Melochia L. (Sterculiac.). [(Geraniac.).
Polychlaena Garcke = Hibiscus L. (Malvac.).
Polychnemum Zumaglini (sphalm.) = Polycnemum L. (Chenopodiac.).
Polychroa Lour. = Pellionia Gaudich. (Urticac.).
Polychrysum (Tsvelev) Kovalevsk. Compositae. 1 C. As.
Polyclados Phil. = Lepidophyllum Cass. (Compos.).
Polyclathra Bertol. Cucurbitaceae. 4 (7?) Mex., Guatem.
Polycline Oliv. = Athroisma DC. (Compos.).
Polyclita A. C. Smith. Ericaceae. 1 Bolivia.
Polyclonos Rafin. = Orobanche L. (Orobanchac.).
Polycnemon F. Muell. = seq.
Polycnemum L. Chenopodiaceae (~ Amaranthaceae). 6–7 C. & S. Eur., Medit. to C. As. Structure of the fr. curious, a ridge developing at its apex after fert.
Polycocca Hill = Selaginella Beauv. (Selaginellac.).
Polycodium Rafin. (~ Vaccinium L.). Ericaceae. 20 SE. U.S.
Polycoelium A. DC. = Myoporum Banks & Soland. (Myoporac.).

Polycoryne Keay. Rubiaceae. 1 trop. W. Afr.

Polyctenium Greene (~ Smelowskia C. A. Mey.). Cruciferae. 1 NW. U.S.

Polycycliska Ridley. Rubiaceae. 1 Sumatra (Mentawi Is.).

Polycycnis Reichb. f. Orchidaceae. 11 Costa Rica, trop. S. Am.

Polycyema Voigt = Clausena Burm. f. (Rutac.).

Polycyrtus Schlechtd. = Ferula L. (Umbellif.).

Polydendris Thou. (uninom.) = Dendrobium polystachion Thou. = Polystachya mauritiana Spreng. (Orchidac.).

Polydiclis Miers = Nicotiana L. (Solanac.).

Polydictyum Presl = Tectaria Cav. (Aspidiac.).

Polydontia Blume = Pygeum Gaertn. = Lauro-Cerasus Duham. (Rosac.).

Polydora Fenzl = Vernonia Schreb. (Compos.).

Polydragma Hook. f. = Spathiostemon Bl. (Euphorbiac.).

Polyechma Hochst. = Hygrophila R.Br. (Acanthac.).

Polyembrium Schott ex Steud. = seq.

Polyembryum Schott ex Steud. = Polembryum A. Juss. = Esenbeckia Kunth (Rutac.).

Polygala L. Polygalaceae. 5–600 cosmop., exc. N.Z., Polynes., and arctic zone. A few have stipular thorns. The essential organs in most spp. are enclosed by the keel, and emerge from it, as in Leguminosae, when it is depressed by a visiting insect. In *P. vulgaris* L. the fls., which owe their conspicuousness to the two coloured sepals, occur in three colours, red, white, and blue, usu. on different plants but sometimes on the same. *P. senega* L. (Senega snake-root) in N. Am. is medicinal.

*****Polygalaceae** Juss. Dicots. 12/800 cosmop. exc. N.Z., Polynes., and arctic zone. Herbs, shrubs, or small trees, with simple entire alt. opp. or whorled usu. exstip. ls.; the stipules when present are usu. thorny or scaly. Infl. a raceme, spike, or panicle, with bracts and bracteoles. Fl. diplochlam., medially zygo. K usu. 5, rarely (5), the 2 inner sepals (alae) often large and petaloid; C 5, rarely all present, usu. only 3—the lowest and two upper—± joined to sta.-tube, the median ant. petal keel-like and often with a term. brush; A in two 5-merous whorls, usu. only 8, or 7, 5, 4 or 3, usu. united below into an open tube; G (5–2), usu. (2), antero-post., 2-loc. with 1 anatr. pend. ov. in each loc. (rarely 1-loc. with ∞ ov.). Caps., nut, or drupe. Endosp. or not. The fl. mech., like the structure, resembles that found in many Leguminosae, to which the fam. is certainly related. (Cf. also Krameriac.) Chief genera: Polygala, Securidaca, Bredemeyera, Monnina, Muraltia. (See also Xanthophyllaceae.)

Polygaloïdes Haller (~ Polygala L.). Polygalaceae. 25 cosmop. (exc. Australas.).

Polyglochin Ehrh. (uninom.) = Carex dioica L. (Cyperac.).

*****Polygonaceae** Juss. Dicots. 40/800, chiefly N. temp.; a few trop., arctic, and southern. Most are herbs, but some shrubs, and a few trees (Triplaris). The ls. (exc. Eriogoneae) possess a peculiar sheathing stipule or ocrea clasping the stem above the leaf-base. This forms a char. feature of the fam. The infl. is primarily racemose, but the partial infls. usu. cymose. (See Eriogonum.) Fls. ⚥, reg., cyclic or acyclic. The former have usu. the formula

POLYGONANTHACEAE

P 3+3, homochlamydeous; A 3+3; G (3); but many vary from this type. *Oxyria* is 2-merous; others, e.g. *Eriogonum, Rheum,* have branching of the outer sta. The acyclic fls. have P 5, arranged according to 2/5 phyllotaxy (e.g. *Polygonum*), A 5–8, G (3). Ovary 1-loc. with 1 erect orthotr. ov. and 3 styles. Fls. pollinated by wind or by insects. Fr. almost always a triangular nut, with smooth exterior. The seed contains an excentric curved or straight embryo surrounded by mealy endosp., sometimes ruminate. The fruits are usually wind-distributed; the persistent P usu. forms a membranous wing. Others are provided with hooks.

Classification and chief genera (after Dammer):

A. Flower cyclic; endosp. not ruminate.
  I. **Rumicoïdeae:**
    1. ERIOGONEAE (no ocrea): *Chorizanthe, Eriogonum.*
    2. RUMICEAE (ocreate): *Rumex, Rheum, Oxyria.*
B. Acyclic (except a few *Coccoloboïdeae*).
  II. **Polygonoïdeae** (endosp. not ruminate):
    1. ATRAPHAXIDEAE (shrubs): *Calligonum.*
    2. POLYGONEAE (herbs): *Polygonum, Fagopyrum.*
  III. **Coccoloboïdeae** (ruminate):
    1. COCCOLOBEAE (usu. ♀): *Muehlenbeckia, Coccoloba.*
    2. TRIPLARIDEAE (usu. dioec.): *Triplaris.*

There is undoubtedly some affinity between this fam. and *Plumbaginac.*; both fams. can be reckoned among 'peripheral Centrospermae'.

Polygonanthaceae (Croiz.) Croiz. = Anisophylleaceae Ridl.

Polygonanthus Ducke. Anisophylleaceae. 2 Amaz. Brazil.

Polygonastrum Moench = Smilacina Desf. (Liliac.).

Polygonat[ac]eae Salisb. = Liliaceae–Polygonateae Benth.

Polygonatum Mill. Liliaceae. 50 N. temp. There is a sympodial fleshy rhizome, upon which the annual shoots leave curious seal-like marks when they die away. Infl. unilat.; fl. homogamous, bee-pollinated.

Polygonatum Zinn = Convallaria L. (*s.l.,* incl. Polygonatum Mill.) (Liliac.).

Polygonella Michx (~ Polygonum L.). Polygonaceae. 10 N. Am.

Polygonifolia Fabr. (uninom.) = Corrigiola L. sp. (Caryophyllac.).

Polygonoïdes Ort. = Calligonum L. (Polygonac.).

Polygonon St-Lag. = Polygonum L. (Polygonac.).

× **Polygonorumex** Weill. Polygonaceae. Gen. hybr. (Polygonum × Rumex).

Polygonum L. Polygonaceae. *S.l.,* 300 cosmop., but esp. temp.; *s.str.* (i.e. sects. *Avicularia, Duravia, Pseudomollia* and *Tephis*), 50 cosmop. Herbaceous. Some are xero., some water plants (e.g. *P. amphibium* L., which may however be found almost as often on land, where its ls. have not the stalks of the water form). The fls. are in spikes and panicles (the partial infl. is cymose). Fls. ♀, acyclic, usu. with a coloured 5-leaved P and about 8 sta. Honey is secreted at the base of the sta., and the fls. are visited by insects, but in varying degree. Cleistog. fls. are found under the ochrea in *P. aviculare* L., etc. In *P. viviparum* L. many of the fls. are replaced by bulbils in the lower part of the infl. (cf. [*Lilium, Allium*].

Polygyne Phil. = Eclipta L. (Compos.).

Polylepis Ruiz & Pav. Rosaceae. 35 trop. S. Am.

Polylobium Eckl. & Zeyh. = Lotononis DC. (Legumin.).

932

Polylophium Boiss. Umbelliferae. 5 W. As.

Polylychnis Bremek. Acanthaceae. 2 Guiana.

Polymeria R.Br. Convolvulaceae. 8–10 Timor, Queensl., New Caled.

Polymita N. E. Brown. Aïzoaceae. 1 S. Afr.

Polymnia L. Compositae. 20 warm Am. Disk florets pedicellate.

Polymniastrum Lam. (*P. variabilis* Lam.) = praec.

Polymniastrum Small (*P. uvedalia* [L.] Small) = Smallanthus Mackenzie (Compos.).

Polymorpha Fabr. (uninom.) = *Salvia* L. sp. (Labiat.).

Polyneura Peter = Panicum L. (Gramin.).

Polynome Salisb. = Dioscorea L. (Dioscoreac.).

Polyochnella Van Tiegh. = Ochna L. (Ochnac.).

Polyodon Kunth = Botelua Lag. (Gramin.).

Polyodontia Meissn. = Polydontia Blume = Pygeum Gaertn. = Lauro-Cerasus Duham. (Rosac.).

Polyosma Blume. Escalloniaceae. 60 E. Himal. to trop. Austr.

Polyosmaceae Blume = Escalloniaceae–Polyosmeae [sphalm. '-mateae'] Engl.

Polyosus Lour. = Polyozus Lour. = Canthium Lam. + Psychotria L. (Rubiac.).

Polyothyris Koord. (sphalm.) = seq.

Polyothyrsis Koord. = Poliothyrsis Oliv. (Flacourtiac.).

Polyotidium Garay. Orchidaceae. 1 Colombia.

Polyotus Nutt. = Asclepias L. (Asclepiadac.).

Polyouratea Van Tiegh. = Ouratea Aubl. (Ochnac.).

Polyozus Blume = Psychotria L. (Rubiac.).

Polyozus Lour. = Canthium Lam. + Psychotria L. (Rubiac.).

Polypappus Less. = Baccharis L. (Compos.).

Polypappus Nutt. = Tessaria Ruiz & Pav. (Compos.).

Polypara Lour. = Houttuynia Thunb. (Saururac.).

Polypetalae Juss. An assemblage of Dicot. fams. in which polypetaly (choripetaly) predominates. The *P.* together with the *Apetalae* constitute the subclass *Archichlamydeae* of Engler.

Polypetalia Hort. = Prunus L. (Rosac.).

Polyphema Lour. = Artocarpus J. R. & G. Forst. (Morac.).

Polyphlebium Copel. Hymenophyllaceae. 1 Austr., N.Z.

Polyphragmon Desf. = Timonius DC. (Rubiac.).

Polyplethia (Griff.) Van Tiegh. = Balanophora J. R. & G. Forst. (Balanophorac.).

Polypleurella Engl. Podostemaceae. 1 Siam.

Polypleurum (Taylor ex Tul., corr.) Warming. Podostemaceae. 1 trop. W. Afr., 6–7 S. India, Ceylon.

Polypodiaceae S. F. Gray (*s.str.*). Polypodiales. ± 50 genera (many formerly included in *Polypodium* L.), cosmop., esp. in wet tropics, almost all epiphytes. Rhiz. dorsiventral, covered with peltate scales, pet. jointed to it; venation usually reticulate with free veins in areoles; sori mostly at vein-junctions, round, exindusiate, or spreading along veins or acrostichoid; spores·monolete. (Copeland, *Gen. Fil.*, includes *Grammitidaceae*, here separated.)

Polypodiales. Filicidae. Families: *Dipteridaceae*, *Cheiropleuriaceae*, *Polypodiaceae*, *Grammitidaceae*.

Polypodiopsis Carr. = ?Bauprea Brongn. & Gris (Proteac.).
Polypodiopsis Copel. = seq.
Polypodiopteris Reed. Polypodiaceae. 3 Borneo.
Polypodium L. Polypodiaceae. 75 cosmop. (Copel., *Gen. Fil.*); limits of genus need study. Hooker adopted an artificial definition, and included all ferns with naked ± round superficial sori, many quite unrelated to *Polypodium vulgare* L.; Christensen (*Ind. Fil.*) removed the unrelated spp. but still maintained a very comprehensive genus, later broken up by Copeland and Ching.
Polypogon Desf. Gramineae. 15 trop. & warm temp.
× **Polypogonagrostis** (Aschers. & Graebn.) Maire & Weiller = × Agropogon P. Fourn. (Gramin.).
*****Polypompholyx** Lehm. Lentibulariaceae. 2 Austr.
Polyporandra Becc. Icacinaceae. 1 Moluccas, New Guinea.
Polypremum Adans. = Valerianella Moench (Valerianac.).
Polypremum L. Loganiaceae. 1 N. Am., W.I., Colombia.
Polypsecadium O. E. Schulz. Cruciferae. 1 C. Andes.
Polypteris Nutt. (~Palafoxia Lag.). Compositae. 1 SE. U.S. (Georgia, Florida).
Polyradicion Garay. Orchidaceae. 2 trop. N. Am. (Florida), W.I.
Polyrhaphis Lindl. = Pappophorum Schreb. (Gramin.).
Polyrrhiza Pfitz. = Dendrophylax Reichb. f. + Polyradicion Garay (Orchidac.).
Polyscalis Wall. = Cyathula Lour. (Amaranthac.).
Polyscelis Hook. f. = praec.
Polyschemone Schott, Nym. & Kotschy = Silene L. (Caryophyllac.).
Polyschisma Turcz. = Pelargonium L'Hérit. (Geraniac.).
Polyschistis J. & C. Presl = Pentarrhaphis Kunth (Gramin.).
Polyscias J. R. & G. Forst. Araliaceae. 80 palaeotrop.
Polysolenia Hook. f. Rubiaceae. 1 Assam.
Polyspatha Benth. Commelinaceae. 3 trop. W. Afr.
Polysphaeria Hook. f. Rubiaceae. 12 trop. E. Afr.
Polyspora Sweet = Gordonia Ellis (Theac.).
*****Polystachya** Hook. Orchidaceae. 210 Florida to trop. S. Am., W.I.; trop. & S. Afr., Masc.; India, Ceylon, SW. China, Philipp.
Polystemma Decne. Asclepiadaceae. 3 Mex., C. Am.
Polystemon D. Don = Belangera Cambess. = Lamanonia Vell. (Cunoniac.).
Polystemonanthus Harms. Leguminosae. 1 trop. W. Afr.
Polystepis Thou. (uninom.) = *Epidendrum polystachys* Thou. = *Oeoniella polystachys* (Thou.) Schltr. (Orchidac.).
Polystichopsis (J. Sm.) C. Chr. Aspidiaceae. 4 trop. Am. (Morton, *Am. Fern Journ.* **50**: 155, 1960).
Polystichopsis Holttum = Arachniodes Bl. (Aspidiac.).
*****Polystichum** Roth. Aspidiaceae. 135 cosmop.
Polystigma Meissn. = Byronia Endl. = Ilex L. (Aquifoliac.).
Polystorthia Blume = Pygeum Gaertn. = Lauro-Cerasus Duham. (Rosac.).
Polystylus Hasselt ex Hassk. = Phalaenopsis Blume (Orchidac.).
Polytaenia DC. Umbelliferae. 2 N. Am.
Polytaenium Desvaux. Vittariaceae. 10 trop. Am.

Polytaxis Bunge = Jurinea Cass. (Compos.).
Polytemia Rafin. = seq.
Polytenia Rafin. = Polytaenia DC. (Umbellif.).
Polytepalum Suesseng. & Beyerle. Caryophyllaceae. 1 Angola.
Polythecandra Planch. & Triana = Clusia L. (Guttif.).
Polythecanthum Van Tiegh. = Ochna L. (Ochnac.).
Polythecium Van Tiegh. = Ochna L. (Ochnac.).
Polythrix Nees = Crossandra Salisb. (Acanthac.).
Polythysania Hanst. = Alloplectus Mart. (Gesneriac.).
Polytoca R.Br. Gramineae. 6 Indomal.
Polytoma Lour. ex Gomes = Bletilla Reichb. f. + Aërides Lour. (Orchidac.).
Polytrema C. B. Clarke. Acanthaceae. 10 Indoch., W. Malaysia.
Polytrias Hack. Gramineae. 1 Java.
Polytropia Presl = Rhynchosia Lour. (Legumin.).
Polytropis B. D. Jacks. (sphalm.) = praec.
Polyura Hook. f. Rubiaceae. 1 Assam.
Polyxena Kunth. Liliaceae. 10 S. Afr.
Polyzone Endl. = Darwinia Rudge (Myrtac.).
Polyzygus Dalzell. Umbelliferae. 1 S. India.
Pomaceae S. F. Gray = Rosaceae–Pomoïdeae Juss.
Pomaderris Labill. Rhamnaceae. 45 Austr., N.Z.
Pomangium Reinw. = Argostemma Wall. (Rubiac.).
Pomaria Cav. = Caesalpinia L. (Legumin.).
Pomasterion Miq. = Actinostemma Griff. (Cucurbitac.).
Pomataphytum Jones = Cheilanthes Sw. (Sinopteridac.).
Pomatiderris Roem. & Schult. = Pomatoderris Roem. & Schult. = Pomaderris Labill. (Rhamnac.).
Pomatium Gaertn. f. = Bertiera Aubl. (Rubiac.).
Pomatium Nees & Mart. ex Lindl. = Ocotea Aubl. (Laurac.).
Pomatocalpa Breda, Kuhl & v. Hasselt. Orchidaceae. 60 China, Formosa, Indomal., Austr., Solomons, Samoa.
Pomatoderris Roem. & Schult. = Pomaderris Labill. (Rhamnac.).
Pomatosace Maxim. Primulaceae. 1 NW. China.
Pomatostoma Stapf. Melastomataceae. 1 Sumatra, 4 Borneo.
Pomatotheca F. Muell. = Trianthema L. (Aïzoac.).
Pomax Soland. ex DC. Rubiaceae. 1 E. Austr.
Pomazota Ridl. Rubiaceae. 13 W. Malaysia.
Pombalia Vand. = Hybanthus Jacq. (Violac.).
Pombea Caldas (nomen). 1 Ecuador. Quid?
Pomelia Durando ex Pomel = Daucus L. (Umbellif.).
Pomereula Dombey ex DC. = Miconia Ruiz & Pav. (Melastomatac.).
Pometia J. R. & G. Forst. Sapindaceae. 10 Indomal.
Pometia Vell. = Neopometia Aubrév. (Sapotac.).
Pometia Willd. = Allophylus L. (Sapindac.).
Pommereschea Wittmack. Zingiberaceae. 2 Burma.
Pommereschia Dur. & Jacks. = praec.
Pommereulla L. f. Gramineae. 1 S. India, Ceylon.   [(Melastomatac.).
Pommereullia P. & K. = Pomereula Dombey ex DC. = Miconia Ruiz & Pav.

935

POMPADOURA

Pompadoura Buc'hoz ex DC. = Calycanthus L. (Calycanthac.).
Pomphidea Miers = Ravenia Sw. (Rutac.).
Pompila Nor. = Sterculia L. (Sterculiac).
Ponaea Bub. = Carpesium L. (Compos.).
Ponaea Schreb. = Toulicia Aubl. (Sapindac.).
Ponapea Becc. Palmae. 4 Caroline Is.
Ponaria Rafin. = Veronica L. (Scrophulariac.).
Ponceletia R.Br. = Sprengelia Sm. (Epacridac.).
Ponceletia Thou. = Psammophila Schult. = Spartina Schreb. (Gramin.).
Poncirus Rafin. (~ Citrus L.). Rutaceae. 1 N. China.
Ponera Lindl. Orchidaceae. 8 Mex. to trop. S. Am.
Ponerorchis Reichb. f. = Gymnadenia R.Br. (Orchidac.).
Pongam Adans. = seq.
*Pongamia Adans. mut. Vent. Leguminosae. 1 Indomal.
Pongamiopsis R. Viguier. Leguminosae. 2 Madag.
Pongati Adans. = Sphenoclea Gaertn. (Sphenocleac.).
Pongati[ac]eae Endl. = Sphenocleaceae Lindl.
Pongatium Adans. mut. Juss. = Pongati Adans. = Sphenoclea Gaertn. (Spheno-
Pongelia Rafin. = Dolichandrone Fenzl (Bignoniac.).   [cleac.).
Pongelion Adans. = Ailanthus Desf. (Simaroubac.).
Pongelium Scop. = praec.
Pongonia Grant (sphalm.) = Pogonia Griff. (Orchidac.).
Ponista Rafin. = Saxifraga L. (Saxifragac.).
Ponna Boehm. = Calophyllum L. (Guttif.).
Pontaletsje Adans. (sphalm.) = Poutaletsje Adans. = Hedyotis L. (Rubiac.).
Pontania Lem. = Brachysema R.Br. (Legumin.).
Pontederas Hoffmgg. = seq.   [Lythrum).
Pontederia L. Pontederiaceae. 4 Am. Fls. trimorphic, heterostyled (cf.
*Pontederiaceae Kunth. Monocots. (Farinosae). 7/30 trop. Water plants,
floating or rooted, of sympodial structure, the successive axes ending in infls.
(sympodial cymose pseudo-racemes). Often, e.g. in Eichhornia, the axillary
shoot is adnate to the main shoot from which it springs. Sometimes extra
branches are formed, and the axis of the infl. is often pushed to one side so
that it appears to spring from a leaf-sheath. Fls. reg., more rarely zygo.; P (3 + 3),
persistent; A 6-3-1, epitepalous, anth. occasionally dehisc. by pores; G (3),
3-loc. with ∞ anatr. ovules, or 1-loc. with 1 ovule; style long, stigma entire or
slightly lobed. Capsule or nut. Embryo central in the seed, scarcely, or not,
shorter than the rich mealy endosp. Genera: Eichhornia, Pontederia, Mono-
choria, Heteranthera, Reussia, Hydrothrix, Scholleropsis.
Pontesia Vell. = ? Riencourtia Cass. (Compos.).
Ponthieva R.Br. Orchidaceae. 60 SE. U.S. to Argent., W.I.
Pontia Bub. = Chrysanthemum L. (Compos.).
Pontinia Fries = Silene L. (Caryophyllac.).
Pontopidana Scop. = Couroupita Aubl. (Lecythidac.).
Pontoppidana Steud. = praec.
Pontya A. Chev. = Bosqueia Thou. (Morac.).
Poortmannia Drake = Trianaea Planch. & Lind. (Solanac.).
Pootia Dennst. = Canscora Lam. (Gentianac.).

Pootia Miq. = Voacanga Thou. (Apocynac.).
Poponax Rafin. = Acacia Mill. (Legumin.).
Popoviocodonia Fedorov. Campanulaceae. 2 Far E. Siberia, Sakhalin.
Popoviolimon Lincz. Limoniaceae. (~ Plumbaginaceae). 1 C. As.
Popowia auctt. afr. = Monanthotaxis Baill. (Annonac.).
Popowia Endl. Annonaceae. 50 trop. As. to trop. Austr.
Poppia Carr. ex Vilm. = Poppya Neck. ex M. Roem. = Luffa Mill. (Cucurbitac.).
Poppigia Hook. & Arn. = Poeppigia Bert. = Rhaphithamnus Miers (Verbenac.).
Poppya Neck. ex M. Roem. = Luffa L. (Cucurbitac.).
Populago Mill. = Caltha L. (Ranunculac.).
Populina Baill. Acanthaceae. 2 Madag.
Populus L. Salicaceae. 35 N. temp. Buds covered by several imbr. scales.
Ls. usu. long-pet., often broad, deltoid or cordate. Fls. anemoph., usu. appearing in advance of ls.; scale (or bract) toothed or lacin.; P reduced to a
cupular disk; nectary o; A 4–∞; G resembling that of *Salix*; caps. 2(–4)-valved.
Most spp. are quick-growing and are cult. for paper-pulp, for timber for
packing-cases, matches, etc. The resinous-aromatic *P. trichocarpa* Torr. &
Gray ex Hook. and allies are freq. grown as ornamentals.
Porana Burm. f. Convolvulaceae. 3 Afr., 20 SE. As. & Indomal., 1 Austr.
(1 Am.?)
Poranaceae J. G. Agardh = Convolvulaceae–Convolvulinae Peter.
Poranopsis Roberty = Porana Burm. f. (Convolvulac.).
Poranthera Rafin. = Sorghastrum Nash (Gramin.).
Poranthera Rudge. Euphorbiaceae. 10 Austr.
Porantheraceae (Pax) Hurusawa = Euphorbiaceae–Porantheroïdeae Pax.
Poraqueiba Aubl. Icacinaceae. 3 trop. S. Am.
Poraresia Gleason = Pogonophora Miers (? Euphorbiac.).
Porcelia Ruiz & Pav. Annonaceae. 5 C. & trop. S. Am.
Porcellites Cass. = Hypochoeris L. (Compos.).
Porfiria Bödeker = Mammillaria Haw. (Cactac.).
Porfuris Rafin. = Nemesia Vent. (Scrophulariac.).
Porliera Pers. = seq.
Porlieria Ruiz & Pav. Zygophyllaceae. 6 Mex., Andes. The leaflets of *P.
hygrometrica* R. & P. spread out horiz. in the day-time, but at night fold up in
pairs. Timber useful.
Porocarpus Gaertn. = Timonius DC. (Rubiac.).
Porochna Van Tiegh. = Ochna L. (Ochnac.).
Porocillaea Miers (nomen). Chile. Quid?
Porocystis Radlk. Sapindaceae. 2 trop. S. Am.
Porodittia G. Don. Scrophulariaceae. 1 Peru.
Porolabium Tang & Wang. Orchidaceae. 1 Mongolia.
Porophyllum Guett. Compositae. 50 warm Am.
Porosectaceae Dulac = Loranthaceae Juss.
Porospermum F. Muell. Araliaceae. 1 NE. Austr.
Porostema Schreb. = Ocotea Aubl. (Laurac.).
Porotheca K. Schum. = Chlaenandra Miq. (Menispermac.).
Porothrinax H. Wendl. ex Griseb. = Thrinax L. (Palm.).
Porotiddia Walp. = Porodittia G. Don (Scrophulariac.).

Porpa Blume = Triumfetta L. (Tiliac.).

Porpax Lindl. Orchidaceae. 10 India to Siam.

Porpax Salisb. = Aspidistra Ker-Gawl. (Liliac.).

Porphyra Lour. = Callicarpa L. (Verbenac.).

Porphyranthus Engl. = Panda Pierre (Pandac.).

Porphyrocodon Hook. f. = Cardamine L. (Crucif.).

Porphyrocoma Scheidw. Acanthaceae. 2 trop. S. Am.

Porphyrodesme Schlechter. Orchidaceae. 1 New Guinea.

Porphyroglottis Ridl. Orchidaceae. 1 Malay Penins., Borneo.

Porphyroscias Miq. Umbelliferae. 2 W. China, Japan.

Porphyrospatha Engl. Araceae. 3 C. Am., Colombia.

Porphyrostachys Reichb. f. Orchidaceae. 1 Ecuador, Peru.

Porphyrostemma Benth. ex Oliv. Compositae. 3 trop. Afr.

Porroglossum Schlechter. Orchidaceae. 5 trop. S. Am.

Porroteranthe Steud. = Glyceria R.Br. (Gramin.).

Porrum Mill. = Allium L. (Alliac.).

Portaea Tenore = Juanulloa Ruiz & Pav. (Solanac.).

Portalesia Meyen = Nassauvia Juss. (Compos.).

Portea Brongn. & C. Koch. Bromeliaceae. 6 Brazil.

Portenschlagia Tratt. = Elaeodendron Jacq. f. (Celastrac.).

Portenschlagia Vis. = seq.

Portenschlagiella Tutin. Umbelliferae. 1 Dalmatia.

Porterandia Ridley. Rubiaceae. 5 trop. Afr., 9 W. Malaysia.

Porteranthus Britton = Gillenia Moench (Rosac.).

Porterella Torr. (~ Laurentia Adans.). Campanulaceae. 1 W. U.S.

Porteresia Tateoka (~ Oryza L.). Gramineae. 1–2 Indomal.

Porteria Hook. = Valeriana L. (Valerianac.).

Portesia Cav. = Trichilia P.Br. (Meliac.).

Portlandia P.Br. Rubiaceae. 25 Mex., C. Am., W.I.

Portlandia Ellis = Gardenia L. (Rubiac.).

Portula Hill = Peplis L. (Lythrac.).

Portulaca L. Portulacaceae. 200 trop. & subtrop. The fl. has a semi-inf. ovary and 4–∞ sta. It remains closed in bad weather. The sta. of *P. oleracea* L. are sensitive to contact and move toward the side touched.

*****Portulacaceae** Juss. Dicots. (Centrospermae). 19/580, cosmop., but esp. Am. Most are annual herbs, often with fleshy opp. or alt. ls., and with stipules (sometimes repres. by axillary bundles of hairs). Fls. usu. in cymes (often dich. with tendency to cincinni), reg., ♀. K 2, the lower sepal (usu. ant.) overlapping the upper (the two are often regarded as bracteoles); C 5, often with satiny surface; A 5 + 5, or 5 opp. the petals, or 4–6–∞; G̱ (2–8), usu. (3), (G̱ in *Portulaca*), 1-loc., with several stigmas and 2–∞ campylotr. ov. on a central basal plac. The fls. secrete honey and are mostly insect-pollinated. Caps. with album. seeds; that of *Claytonia* and *Montia* is explosive; embryo more or less curved round the perisperm. Chief genera: *Portulaca, Lewisia, Calandrinia, Talinum, Anacampseros, Claytonia, Montia.* Closely related to *Cactaceae* and *Aïzoaceae.*

Portulacaria Jacq. Portulacaceae. 2 S. Afr.

Portulacastrum Juss. ex Medik. = Trianthema L. (Aïzoac.).

Portuna Nutt. = Pieris D. Don (Ericac.).

Posadaea Cogn. Cucurbitaceae. 1 C. & S. Am., W.I.

*Posidonia Koenig. Posidoniaceae. 1 coasts Medit., 2 Austr. Stems used for packing glass.

*Posidoniaceae (Kunth) Lotsy. Monocots. 1/3 Medit., Austr. Submerged marine perenn. rhiz. herbs, with lin. ent. or serr. alt. exstip. ls., sheathing and ligulate at base, which is fibr. and persist. Fls. ♀ ♂, in 3–12-fld. spicate infls. borne in the axils of foliaceous bracts at the summit of an erect scapose stem; floral bracts o. P 3, squamiform, caduc., or o; A 3–4, anth. large, extr., sessile, thecae caducous, widely separated on broad thick apically produced persist. connective; pollen thread-like (cf. *Zosterac.*); G 1, with sess. stig., and 1 ventr. campylotr. ov. Fr. a detached and floating drupe, indehisc., or the fleshy exoc. splitting from the base into 2–3 valves, whilst the thin endoc. splits shortly at the apex, permitting the emergence of the plumule of the germinating seed; seed without testa(?) or endosp. Only genus: *Posidonia*.

Posidonion St-Lag. = Posidonia Koenig (Posidoniac.).

Poskea Vatke. Globulariaceae. 2 Somaliland, Socotra.

Posoqueria Aubl. Rubiaceae. 15 C. & trop. S. Am., W.I.

Posoria Rafin. = praec.

Possira Aubl. = Swartzia Schreb. (Legumin.).

Possiria Steud. = praec.

Possura Steud. = praec.

Postia Boiss. & Blanche. Compositae. 4 Syria, Persia.

Postuera Rafin. = Notelaea Vent. (Oleac.).

Potalia Aubl. Potaliaceae. 1 trop. S. Am.

Potaliaceae Mart. Dicots. 4/70 trop. Trees or shrubs with opp. simple ent. or spinose-dent. often thickly coriac. ls., connected at base by transv. line or stipule-like sheath. Fls. reg., ♀, sometimes large and showy, in few–many-fld. bracteate cymes, or sol. and term. K 4–5, connate below, imbr.; C (5–16), infundib. or hypocrat., lobes contorted; A 5–16, epipet., often exserted; disk fleshy; G (2), rarely (3–5), with simple style and capit. stig., and ∞ ov. on pariet. or centrally coalescent plac. Fr. a berry; seeds with endosp., unwinged. Genera: *Fagraea, Anthocleista, Potalia, Desfontainia.* Differs from *Loganiac.* (*s.str.*) chiefly in the contorted C and baccate fr. with unwinged seeds; from the *Gesneriaceae–Cyrtandreae* (some of which it approaches rather closely) in the reg., non-zygo. fls. with completely developed androec.

Potameae Juss. = Potamogetonaceae Juss.

Potameia Thou. Lauraceae. 19 Madag., 1 E. Himal., 1 S. China (Hainan).

Potamica Poiret = praec.

Potamisia M'Murtrie (sphalm.) = Polanisia Rafin. (Cleomac.).

Potamobryon Liebm. = Tristicha Thou. (Podostemac.).

Potamocharis Rottb. = Mammea L. (Guttif.).

Potamochloa Griff. = Hygroryza Nees (Gramin.).

Potamoganos Sandwith. Bignoniaceae. 1 Guiana.

Potamogeton L. Potamogetonaceae. 100 cosmop. Water pls. with creeping sympodial rhiz. and erect leafy branches; ls. all submerged or some floating. A series of types occurs, beginning with the floating spp. and ending with the narrow-leaved submerged ones. There can be no doubt of the origin of the *P.*

POTAMOGETON

from land pls., and Schenck looks upon *P. natans* L. as the sp. least modified to suit a water existence, i.e. the nearest to the ancestral type. The upper ls. are ovate, leathery, and floating; the lower submerged, sometimes linear. Then come such as *P. gramineus* L., where the submerged ls. are all narrow. Next *P. lucens* L., *P. crispus* L., etc., with all the ls. lanceolate and submerged. Then in *P. obtusifolius* Mert. & Koch, *P. pusillus* L., etc., the leaves are narrow and of a long ribbon shape. *P. trichoïdes* Cham. & Schlechtd. represents the most highly modified type (easily studied in a herbarium).

Hibernation in various ways: some green all winter; *P. natans*, etc., die down, leaving rhiz.; *P. pectinatus* L. forms tubers on special branches; *P. crispus* and others form winter buds with broad ls. (not closely packed); *P. obtusifolius* ordinary winter buds. Fl. protog., wind-fert. The outer layer of pericarp contains air, so that the achene floats and may be distr.

Potamogeton Walt. = Myriophyllum L. (Haloragidac.).

*Potamogetonaceae Dum. Monocots. 2/100, cosmop. Submerged or floating freshwater perenn. herbs. Usu. creeping stem or rhiz., mono- or sympodial, attached to soil by adv. r., with erect branches upwards into water, usu. with ribbon ls., usu. alt. in ½ phyllotaxy. Base sheathing; within sheath are the small scales (*squamulae intravaginales*) which occur in most *Helobiëae*. Infl. of pedunc. axill. spikes. Fl. ⚥, reg.; P 4, free, rounded, shortly clawed, valv.; A 4, anth. extr., sess. on claws of P; G 4, free, with 1 campylotr. basal-ventr. ov. in each. Fr. a group of 1–4 small drupes or ach.; embryo usu. with well-developed hypocotyl; no endosp. Genera: *Potamogeton, Groenlandia.*

Potamogetum Clairv. = Potamogeton L. (Potamogetonac.).

Potamogiton Rafin. = praec.

Potamophila R.Br. Gramineae. 1 Austr.

Potamophila Schrank = Microtea Sw. (Chenopodiac.?).

Potamopithys Seub. = seq.

Potamopitys Adans. = Elatine L. (Elatinac.).

Potamotheca Ph. & K. = Pomatotheca F. Muell. = Trianthema L. (Aïzoac.).

Potamoxylon Rafin. = Tabebuia Gomes ex DC. (Bignoniac.).

Potaninia Maxim. Rosaceae. 1 Mongolia. Epicalyx. K, C, A, 3; G 1.

Potarophytum Sandwith. Rapateaceae. 1 Guiana.

Potentilla L. Rosaceae. 500 nearly cosmop., chiefly N. temp. & arctic. Herbs, usu. with creeping stems rooting at nodes (veg. repr.). Fl. with epicalyx of small green ls., outside and alt. with the seps. (the stips. of seps. united in pairs; often one or more may be seen with two lobes, or fully divided). Fls. homogamous, visited by flies; honey secreted by ring-shaped nectary within sta.

Potentill[ac]eae Trautv. = Rosaceae–Potentilleae Juss.

Potentillopsis Opiz = Potentilla L. (Rosac.).

Poteranthera Bong. Melastomataceae. 1 trop. S. Am. Fl. 5-merous.

Poteridium Spach = Sanguisorba L. (Rosac.).

Poterion St-Lag. = seq.

Poterium L. (~ Sanguisorba L.). Rosaceae. 25 temp. Euras. Fl. ♂ ♀ monoec. or polyg., anemoph.; K green; A ∞. (For *P. spinosum* L., sometimes erroneously treated as the type-species of the genus, see *Sarcopoterium* Spach.)

Pothoïdium Schott. Araceae. 1 Philipp. Is., Celebes, Moluccas.

Pothomorpha Willis (sphalm.) = seq.
Pothomorphe Miq. (~ Piper L.). Piperaceae. 10 trop.
Pothos Adans. = Polianthes L. (Agavac.).
Pothos L. Araceae. 75 Madag., Indomal. Monopodial (see fam.). Stem climbing, with adv. roots. The buds break through the axils, so that the branching seems infra-axillary. Fl. ♂. P 3+3.
Pothuava Gaudich. = Aechmea Ruiz & Pav. (Bromeliac.).
Potima R. Hedw. = Faramea Aubl. (Rubiac.).
× **Potinara** hort. Orchidaceae. Gen. hybr. (iii) (Brassavola × Cattleya × Laelia × Sophronitis).
Pottingeria Prain. Celastraceae (?). 1 Assam, NE. Burma, NW. Siam. Ls. stip.! Venation like *Cinnamomum.*
Pottsia Hook. & Arn. Apocynaceae. 4 India, SE. As., Java.
Pouchetia A. Rich. Rubiaceae. 6 trop. Afr.
Poulsenia Eggers. Moraceae. 1 Ecuador.
Pounguia Benoist. Thunbergiaceae. 1 W. Equat. Afr.
Poupartia Comm. ex Juss. Anacardiaceae. 12 trop., esp. Madag., Masc.
Pourouma Aubl. Urticaceae. 50 C. & trop. S. Am. Ed. fr.
Pourretia Ruiz & Pav. = Puya Mol. (Bromeliac.).
Pourretia Willd. = Cavanillesia Ruiz & Pav. (Malvac.).
Pourthiaea Decne = Photinia Lindl. (Rosac.).
Pouslowia Wight. Inc. sed. S. India, Ceylon.
Poutaletsje Adans. = Hedyotis L. (Rubiac.).
Pouteria Aubl. Sapotaceae. 50 trop. Am.
Pouzolzia Gaudich. Urticaceae. 50 trop. Am., trop. & S. Afr., trop. As. The root of *P. tuberosa* Wight is eaten in India.
Pozoa Hook. f. = Schizeilema Domin (Umbellif.).
Pozoa Lag. Hydrocotylaceae (~ Umbellif.). 2 Andes of Chile & Argent.
Pozopsis Hook. = Huanaca Cav. (Hydrocotylac.).
Pradosia Liais. Sapotaceae. 12 trop. S. Am. Hard wood.
Praealstonia Miers = Symplocos L. (Symplocac.).
Praecereus F. Buxb. Cactaceae. 6 trop. S. Am.
Praecitrullus Pangalo. Cucurbitaceae. 1 India.
Praenanthes Hook. = Prenanthes L. (Compos.).
Praesepium Spreng. ex Reichb. (nomen). Quid? (Rosac.? Umbellif.?)
Praetoria Baill. = Pipturus Wedd. (Urticac.).
Prageluria N. E. Br. = Telosma Coville (Asclepiadac.).
Pragmatropa Pierre = Euonymus L. (Celastrac.).
Pragmotessara Pierre = Euonymus L. (Celastrac.).
× **Prago-Aureolobivia** hort., × **Prago-Chamaecereus** hort., etc. Cactaceae. Gen. hybr., involving Chamaecereus, Echinopsis, Lobivia, etc.
Prainea King ex Hook. f. Moraceae. 7 Malaysia.
Prangos Lindl. Umbelliferae. 30 Medit. to C. As.
Prantleia Mez = Orthophytum Beer (Bromeliac.).
Praravinia Korth. Rubiaceae. 50 Borneo, Philipp. Is., Celebes.
Prasanthea Decne = Paliavana Vand. (Gesneriac.).
Prascoenum P. & K. = Praskoinon Rafin. = Allium L. (Alliac.).
Prasiteles Salisb. = Narcissus L. (Amaryllidac.).

PRASIUM

Prasium L. Labiatae. 1 Medit.
Praskoinon Rafin. = Allium L. (Alliac.).
Prasopepon Naud. = Cucurbitella Walp. corr. Hook. f. (Cucurbitac.).
Prasophyllum R.Br. Orchidaceae. 70 Austr., N.Z.
Prasoxylon M. Roem. = Dysoxylum Blume (Meliac.).
Pratia Gaudich. (~ Lobelia L.). Campanulaceae. 35 S. Am., trop. Afr. (1), trop. As., Austr., N.Z., Pacif.
Praticola Ehrh. (uninom.) = *Thalictrum simplex* L. (Ranunculac.).
Pravinaria Bremek. Rubiaceae. 2 Borneo.
Praxeliopsis G. M. Barroso. Compositae. 1 Brazil.
Praxelis Cass. (~ Eupatorium L.). Compositae. 13 trop. S. Am.
Preauxia Sch. Bip. = Chrysanthemum L. (Compos.).
Preissia Opiz = Avena L. (Gramin.).
Premna L. Verbenaceae. 200 trop. & subtrop. Afr., As.
Prenanthella Rydb. (~ Prenanthes L.). Compositae. 1 NW. Am.
Prenanthenia Svent. Compositae. 1 Canary Is.
Prenanthes L. Compositae. 40 N. Am., Canaries, trop. Afr., temp. & trop. As.
Prenia N. E. Brown. Aïzoaceae. 4 S. Afr.
Preonanthus Ehrh. (uninom.) = *Anemone alpina* L. (Ranunculac.).
Preonanthus (DC.) Schur = Anemone L. (Ranunculac.).
Prepodesma N. E. Brown. Aïzoaceae. 2 S. Afr.
Preptanthe Reichb. f. = Calanthe R.Br. (Orchidac.).
Prepusa Mart. Gentianaceae. 5 Brazil.
Prescottia Lindl. Orchidaceae. 22 Florida, C. & trop. S. Am., W.I.
Preslaea Mart. = Heliotropium L. (Boraginac.).
Preslea G. Don = Preslia Opiz (Labiat.).
Preslea Spreng. = Preslaea Mart. = Heliotropium L. (Boraginac.).
Preslia Opiz (1819) = Woodsia R.Br. (Aspidiac.).
Preslia Opiz (1824). Labiatae. 1 W. Medit.
Prestelia Sch. Bip. ex Benth. & Hook. f. = Eremanthus L. (Compos.).
Prestinaria Sch. Bip. ex Hochst. = Coreopsis L. (Compos.).
*****Prestoea** Hook. f. (~ Euterpe Gaertn.). Palmae. 12 Panamá, Colombia, W.I.
*****Prestonia** R.Br. Apocynaceae. 65 C. & trop. S. Am., W.I.
Prestonia Scop. = Abutilon L. (Malvac.).
Prestoniopsis Muell. Arg. = Dipladenia DC. (Apocynac.).
Pretrea J. Gay = Dicerocaryum Boj. (Pedaliac.).
Pretreothamnus Engl. = Josephinia Vent. (Pedaliac.).
Preussiella Gilg. Melastomataceae. 2 trop. W. Afr.
Preussiodora Keay. Rubiaceae. 1 trop. W. Afr.
Prevoita Steud. = Prevotia Adans. = Cerastium L. (Caryophyllac.).
Prevostea Choisy = Calycobolus Willd. ex Roem. & Schult. (Convolvulac.).
Prevotia Adans. = Cerastium L. (Caryophyllac.).
Priamosia Urb. Flacourtiaceae. 1 S. Domingo.
Prianthes Pritz. = Prionanthes Schrank = Trixis P.Br. (Compos.).
Prickothamnus Nutt. ex Baill. = Pickeringia Nutt. ex Torr. & Gray (Legumin.)
Pridania Gagnep. (= ? Pycnarrhena Miers). Menispermaceae. 2 Indoch.
Priestleya DC. Leguminosae. 20 S. Afr.
Priestleya Moç. & Sessé ex DC. = Montañoa Cerv. (Compos.).

942

PRIMULACEAE

Prieurea DC. = Ludwigia L. (Onagrac.).
Prieurella Pierre (~ Ecclinusa Mart.). Sapotaceae. 3 trop. S. Am.
Prieuria Benth. & Hook. f. = Prieurea DC. = Ludwigia L. (Onagrac.).
Primula Kuntze = Androsace L. (Primulac.).
Primula L. Primulaceae. 500 N. hemisph., chiefly in hilly districts. A few elsewhere, e.g. *P. farinosa* L. var. *magellanica* Hook. at the Str. of Magellan. The rhizome is a sympodium, each joint terminating in an infl. In some spp. this consists of successive whorls of fls. arranged up a long stalk, e.g. *P. japonica* A. Gray. A few of the better known spp. are: *P. sinensis* Sabine, the Chinese primrose; *P. japonica*; *P. elatior* (L.) Hill, the oxlip; *P. vulgaris* Huds. (*P. acaulis* (L.) Hill), the primrose; *P. veris* L. (*P. officinalis* Jacq.), the cowslip; *P. farinosa* L.; *P. auricula* L., the auricula, with its many forms. A great many hybrids occur. In the double-crowned cowslip the K has become petaloid, so that the fl. looks as if it had two Cs, one within the other.

The fls. are dimorphic, heterostyled. On one pl. are long-styled fls., with sta. halfway up the tube and the stigma at its mouth; on another plant are short-styled fls., with stigma halfway up and anthers at the mouth. The depth and narrowness of the tube suit the fl. to bees or butterflies, and these tend to carry pollen from long sta. to long style or from short to short. These 'legitimate' pollinations (see *Lythrum*), which are at the same time crossings, are the only ones which produce a full complement of fertile seed.

The fl. stalks in umbellate forms, e.g. cowslip, stand close and erect till the fls. open, then spread out, and close up again as the fr. ripens; thus the caps. is held erect and the seeds must be shaken out.

*Primulaceae Vent. Dicots. 20/1000 cosmop., but esp. N. temp. Herbaceous pls., commonly perenn., with rhiz. or tubers; ls. opp. or alt., exstip. Fls. often borne on scapes, which when > 1-flowered are term.; they are usually actinom., ⚥, often heterostyled, and 5-merous, without bracteoles, the odd (4th) sepal post. K (5), persistent; C (5), or 5, or 0 (*Glaux*); A 5, epipet. and opp. the pets.; occasionally 5 stds. alt. with the pets.; anthers intr. The presence of the stds. here (as in *Myrsinac.*) explains antepet. position of sta. as due to abortion of originally outer whorl. Formerly much discussion, esp. after Pfeffer's discovery of development of C from backs of sta. G or semi-inf., syncp. with free-central plac., typically of 5 cpls., but this is not easily proved, as there are no partitions (cf. *Caryophyllac.*), and style and stigma are simple; ov. usu. ∞, spirally or in whorls on plac., semi-anatr. Caps. fr., 5-valved, usu. dehisc. by teeth at tip, one opp. each sep. Seeds few or many; embryo small, in fleshy or horny endosp. Fl. often heterostyled (cf. *Primula*, also *Hottonia, Glaux, Androsace*). No econ. value.

Classification and chief genera:

1. PRIMULEAE (ANDROSACEAE) (G; C imbr.): *Primula* (C not reflexed, lobes entire or bifid, sta. on tube, ov. ∞), *Dionysia* (do.; ov. few), *Androsace* (do.; tube very short), *Soldanella* (do.; lobes fimbriate, caps. operculate), *Hottonia* (water pl.), *Dodecatheon* (C reflexed).
2. CYCLAMINEAE (G; C conv.; pl. with tubers): *Cyclamen* only.
3. LYSIMACHIËAE (G; C conv.; no tubers): *Lysimachia* (caps. with valves; 5-merous), *Trientalis* (do.; 7-merous), *Glaux* (do.; apet.), *Anagallis* (caps. with lid).

PRIMULARIA

4. SAMOLEAE (G semi-inf.; C imbr.): *Samolus* only.
Probably some relationship with *Centrospermae*, esp. *Caryophyllac.*; also with *Diapensiac.* (cf. *Soldanella* with *Schizocodon*); and possibly even with *Ochnac.* (cf. *Lysimachia* spp. with *Sauvagesia*, etc.). Supposed affinity with *Myrsinac.* doubtful.

Primularia Brenan. Melastomataceae. 1 trop. E. Afr.

Primulidium Spach = Primula L. (Primulac.).

Primulina Hance. Gesneriaceae. 1 China.

Princea Dubard & Dop. Rubiaceae. 1 Madag.

Principes Endl. An order of Monocots. consisting (in Engler's system) of the fam. *Palmae* only.

Principina Uitt. Cyperaceae. 1 W. Equat. Afr. (Principe I.).

Pringlea Anders. ex Hook. f. Cruciferae. 1, *P. antiscorbutica* R.Br. (Kerguelen cabbage), on Kerguelen and Crozet Is. Habit of cabbage (*Brassica*), but fls. on lat. axes. 'Winged insects cannot exist, because at every flight they run the risk of being drowned...the pl. has become modified for fert. by wind... exserted anthers...long filiform stigmatic papillae...still retains traces of... entomophilous ancestors...usu. devoid of pets., it occurs abundantly in shaded places with pets." (Müller.) No septum in ovary.

Pringleochloa Scribner. Gramineae. 1 Mex.

Pringleophytum A. Gray = Berginia Harv. (Acanthac.).

Prinodia Griseb. = seq.

Prinoïdes (DC.) Willis = seq.

Prinos Gronov. ex L. = Ilex L. (Aquifoliac.).

Prinsepia Royle. Rosaceae. 3–4 Himal. to N. China & Formosa. Style lat. Oil from seeds.

*Printzia Cass. Compositae. 8 S. Afr.

Prinus P. & K. = Prinos L. = Ilex L. (Aquifoliac.).

Prionachne Nees = Prionanthium Desv. (Gramin.).

Prionanthes Schrank = Trixis P.Br. (Compos.).

Prionanthium Desv. Gramineae. 3 S. Afr.

Prionantium Desv. = praec.

Prionites Pritz. (sphalm.) = Prionotes R.Br. (Epacridac.).

Prionitis Adans. = Falcaria Bernh. (Umbellif.).

Prionitis Oerst. = Barleria L. (Acanthac.).

Prionium E. Mey. Juncaceae. 1 S. Afr., *P. serratum* (Thunb.) Drège, the *palmiet*, a shrubby aloe-like plant with a stem 1–2 metres high, covered with the fibrous remains of old ls. It grows on the edges of streams, sometimes almost blocking them up. Veg. propagation takes place by formation of runners. Adv. roots form between the ls. [Buchenau in *Bibl. Bot.* 5, Heft 27, 1893.]

Prionolepis Poepp. & Endl. = Liabum Adans. (Compos.).

Prionophyllum C. Koch. Bromeliaceae. 2 S. Brazil, Urug.

Prionoplectus Oerst. = Alloplectus Mart. (Gesneriac.).

Prionopsis Nutt. = Haplopappus Cass. (Compos.).

Prionoschoenus (Reichb.) Kuntze = Prionium E. Mey. (Juncac.).

Prionosciadium S. Wats. Umbelliferae. 25 Mex. to Ecuad.

Prionosepalum Steud. = Chaetanthus R.Br. (Restionac.).

Prionostachys Hassk. = Aneilema R.Br. (Commelinac.).

Prionostemma Miers. Celastraceae. 1 Panamá, trop. S. Am., Trinidad.
Prionotaceae ('-tidaceae') Hutch. = Epacridaceae–Prionoteae.
Prionotes R.Br. Epacridaceae. 1 Tasmania.
Prionotis Benth. & Hook. f. = praec.
Prionotrichon Botsch. & Vved. Cruciferae. 4 C. As., Afghan.
Priopetalon Rafin. = Alstroemeria L. (Alstroemeriac.).
Prioria Griseb. Leguminosae. 1 Panamá, Jamaica.
Priotropis Wight & Arn. = Crotalaria L. (Legumin.).
Prisciana Rafin. = Heliophila Burm. f. ex L. (Crucif.).
Prismatanthus Hook. & Arn. = Siphonostegia Benth. (Scrophulariac.).
Prismatocarpus L'Hérit. Campanulaceae. 30 S. (mostly SW.) Afr., 1 S. trop. Afr.
Prismatomeris Thw. Rubiaceae. 25 Ceylon, E. Himal. to S. China & W. Malaysia.
Prismocarpa Rafin. = Prismatocarpus L'Hérit. (Campanulac.).
Prismophylis Thou. (uninom.) = Bulbophyllum prismaticum Thou. (Orchidac.).
Pristidia Thw. = Gaertnera Lam. (Rubiac.).
Pristiglottis Cretzoiu & J. J. Smith. Orchidaceae. 13 Sumatra, Borneo, Java, New Guinea, Solomon Is., New Hebrid., New Caled., Samoa.
Pristimera Miers. Celastraceae. 14 E. As., Indomal., C. & trop. S. Am., W.I.
Pristocarpha E. Mey. ex DC. = Athanasia L. (Compos.).
***Pritchardia** Seem. & H. Wendl. (1861; non Unger ex Endl. 1842, gen. foss.). Palmae. 35 Hawaii, Fiji, etc.
Pritchardiopsis Becc. Palmae. 1 New Caled.
Pritzelago Kuntze = Hutchinsia R.Br. (Crucif.).
Pritzelia Klotzsch = Begonia L. (Begoniac.).
Pritzelia F. Muell. = Philydrella Caruel (Philydrac.).
Pritzelia Schau. = Scholtzia Schau. (Myrtac.).
Pritzelia Walp. = Trachymene Rudge (Umbellif.).
Priva Adans. Verbenaceae. 20 trop. & subtrop. The ls. of *P. echinata* Juss. are used as tea; tubers of *P. laevis* Juss. ed.
Probatea Rafin. = Asarina Moench (Scrophulariac.).
Problastes Reinw. = Lumnitzera Willd. (Combretac.).
Probletostemon K. Schum. = Tricalysia A. Rich. (Rubiac.).
Proboscella Van Tiegh. = Ochna L. (Ochnac.).
Proboscidea Schmidel. Martyniaceae. 9 warm Am.
Proboscidia Rich. ex DC. = Rhynchanthera DC. (Melastomatac.).
Probosciphora Neck. = Rhynchocorys Griseb. (Scrophulariac.).
Procephalium P. & K. = Proscephaleium Korth. (Rubiac.).
Prochnyanthes S. Wats. Agavaceae. 3 Mex.
Prochynanthes Bak. = praec.
Prockia P.Br. ex L. Flacourtiaceae. 3 trop. Am., W.I.
Prockiaceae D. Don = Flacourtiaceae–Scolopieae–Prockiinae Gilg.
Prockiopsis Baill. Flacourtiaceae. 1 Madag.
Proclesia Klotzsch = Cavendishia Lindl. (Ericac.).
Procopiania Guşul. (~ Symphytum L.). Boraginaceae. 1–2 Greece (Euboea), Crete. [Symphytum].
× **Procopiphytum** Pawłowski. Boraginaceae. Gen. hybr. (Procopiania ×

PROCRASSULA

Procrassula Griseb. = Sedum L. (Crassulac.).

Procris Comm. ex Juss. Urticaceae. 20 palaeotrop.

Proferea Presl = (?) Cyclosorus Link (Thelypteridac.).

Proïneia Ehrh. (uninom.) = *Aira praecox* L. (Gramin.).

Proïphys Herb. = Eurycles Salisb. (Amaryllidac.).

Prolongoa Boiss. = Chrysanthemum L. (Compos.).

Promenaea Lindl. Orchidaceae. 6 Brazil.

Pronacron Cass. = Melampodium L. (Compos.).

Pronaya Huegel. Pittosporaceae. 1 W. Austr.

Pronephrium Presl emend. Holtt. Thelypteridaceae. 60–70 trop. As. & Malaysia to Fiji. Fronds simply pinnate with entire or slightly lobed pinnae, basal pinnae not reduced, veins anast. (Syn. *Abacopteris* Fée.)

Prosanerpis S. F. Blake. Melastomataceae. 2 C. Am.

Prosaptia Presl. Grammitidaceae. 20 trop. As. & Polynesia. Distinction from *Ctenopteris* not clear.

Prosartema Gagnep. = Trigonostemon Bl. (Euphorbiac.).

Prosartes D. Don = Disporum Salisb. (Liliac.).

Proscephaleium Korth. Rubiaceae. 1 Java.

Proscephalium Benth. & Hook. f. = praec.

Proselia D. Don = Chaetanthera Ruiz & Pav. (Compos.).

Proselias Stev. = Astragalus L. (Legumin.).

Proserpinaca L. Haloragidaceae. 4 N. Am. 3-merous.

Prosopanche de Bary. Hydnoraceae. 6 Paraguay, Argentina.

Prosopia Reichb. = Pedicularis L. (Scrophulariac.).

Prosopidastrum Burkart. Leguminosae. 1 Mex.

Prosopis L. Leguminosae. 40 warm Am.; 1 trop. Afr.; 2 Cauc. to W. Penins. India. Some xero., without ls., many thorny, the thorns being epidermal, or metamorphosed branches or stipules. *P. juliflora* (Sw.) DC. (trop. Am.) is the *mezquit* tree (fodder, etc.). *P. alba* Griseb. has sweet succulent pods (*algaroba blanca*), used as food.

Prosopostelma Baill. Asclepiadaceae (inc. sed.). 1 trop. W. Afr.

Prosorus Dalz. = Margaritaria L. f. (Euphorbiac.).

Prospero Salisb. = Scilla L. (Liliac.).

Prosphysis Dulac = Festuca L. (Gramin.).

Prosphytochloa Schweickerdt. Gramineae. 1 S. Afr.

Prosporus Thw. (sphalm.) = Prosorus Dalz. = Margaritaria L. f. (Euphorbiac.).

Prostanthera Labill. Labiatae. 50 Austr.

Prostea Cambess. = Pometia J. R. & G. Forst. (Sapindac.).

Prosthechea Knowles & Westc. = Epidendrum L. (Orchidac.).

Prosthecidiscus Donn. Smith. Asclepiadaceae. 1 C. Am.

Prosthesia Blume = Rinorea Aubl. (Violac.).

Protamomum Ridl. = Orchidantha N. E. Br. (Lowiac.).

Protangiopteris Hayata = Archangiopteris Chr. (Angiopteridac.).

Protanthera Rafin. (nomen). Liliaceae. 2 N. Am. Quid?

Protarum Engl. Araceae. 1 Seychelles.

Protea L. (1753) = Leucadendron R.Br. (Proteac.).

***Protea** L. (1771). Proteaceae. 130 trop. & S. Afr. Fls. in showy heads, often with coloured bracts.

*Proteaceae Juss. Dicots. 62/1050 trop. As., Malaysia, Austr., New Caled., N.Z.; trop. S. Am., Chile; mts. trop. Afr., S. Afr., Madag. The great majority live in regions where there is annually a long dry season. Correlated with this is the fact that they are mostly xero. The primitive members of the fam., however (*Helicia*, etc.), are largely rain-forest trees; the xerophily is therefore prob. secondary. Nearly all are shrubs or trees with ent. or much-divided exstip. ls., which have commonly a thick cuticle and often a covering of hairs, further checking transpiration. The fls. are borne in rac., spikes, heads, etc., and are often very showy; many have their pollen freely exposed, though they are not wind-fert.—a peculiarity perh. connected with their life in a dry climate (cf. the acacias of Austr.). The pollen is often shed on to non-receptive parts of the style (often specially modif.), whence it is transferred to the pollinators. The fls. are usu. ⚥, often zygo. P (4), corolline, valv., the tep. commonly bent or rolled back when open, and often alternating with 4 (3–2) 'glands' or 'scales' (?petals); A 4, opp. to and often inserted on the tepals, sometimes with only the anth. free; G 1, with ∞–1 ov., which may be sub-basal or sub-apical, anatr. or hemitr. or orthotr., the micropyle always directed downwards. Style term., long, often bent inwards, sometimes wiry. Fr. a follicle, drupe or nut; seed exalbum. (exc. in *Bellendena*). The ovary is sometimes borne on a gynophore, and at its base are commonly nectarial outgrowths. The fls. are protandrous and adapted to pollination by birds or insects. (For classif., etc., cf. Johnson & Briggs, *Austr. J. Bot.* **11**: 21–61, 1963.)

Classification and chief genera:

    I. *Proteoïdeae (Persoonioïdeae)* (fls. single in axils of bracts; ovules 1–2-few; drupe or nut, 1-seeded): *Persoonia, Protea, Leucadendron.*

    II. *Grevilleoïdeae* (fls. in pairs; ovules 2–several; fr. usu. dehisc., ∞–1-seeded): *Grevillea, Hakea, Banksia, Helicia, Roupala.*

Proteinia (Ser.) Reichb. = Saponaria L. (Caryophyllac.).
Proteinophallus Hook. f. = Amorphophallus Blume (Arac.).
Proteocarpus Börner = Carex L. (Cyperac.).
Proteopsis Mart. & Zucc. ex DC. Compositae. 5 campos of S. Brazil.
Proterpia Rafin. = Tabebuia Gomes ex DC. (Bignoniac.).
Protionopsis Blume = Commiphora Jacq. (Burserac.).
*Protium Burm. f. Burseraceae. 90 Madag. to Malaysia & trop. Am.
Protium Wight & Arn. = Commiphora Jacq. (Burserac.).
Protocamusia Gandog. = Buphthalmum L. (Compos.).
Protocyatheaceae Bower = Lophosoriaceae Pichi-Serm.
Protocyrtandra Hosokawa. Gesneriaceae. 1 Formosa.
Protohopea Miers = Symplocos L. (Symplocac.).
Protolepis Steud. (sphalm.) = Proteopsis Mart. & Zucc. ex DC. (Compos.).
Protolindsaya Copel. = Tapeinidium (Presl) C. Chr. (Lindsaeac.).
Protoliriaceae Mak. = Petrosaviaceae Hutch.
Protolirion Ridl. = Petrosavia Becc. (Petrosaviac.).
Protomarattia Hayata. Marattiaceae. 1 N. Vietnam. (Pichi-Sermolli, *Webbia*, **12**: 350–2, 1957.)
Protomegabaria Hutch. Euphorbiaceae. 2 trop. W. Afr.
Protonopsis Pfeiff. = Protionopsis Bl. = Commiphora Jacq. (Burserac.).
Protorhus Engl. Anacardiaceae. 1 SW. Afr., 20 Madag.

PROTOSCHWENKIA

Protoschwenkia Solereder. Solanaceae. 1 Bolivia.
Protowoodsia Ching = Woodsia R.Br. (Aspidiac.).
Proustia Lag. (1811). Compositae. 15 W.I., Andes, temp. S. Am.
Proustia Lag. (1807) ex DC. (1830) = Actinotus Labill. (Umbellif.).
Provancheria B. Boiv. Caryophyllaceae. 2 temp. N. Am.
Provenzalia Adans. = Calla L. (Arac.).
Prozetia Neck. = Pouteria Aubl. (Sapotac.).
Prozopsis C. Muell. (sphalm.) = Pozopsis Hook. = Huanaca Cav. (Umbellif.).
Prumnopitys Phil. = Podocarpus L'Hérit. ex Pers. (Podocarpac.).
Prunaceae Burnett = Amygdalaceae Bartl. = Rosaceae–Amygdaleae Juss.
Prunella L. Labiatae. 7 temp. Euras., NW. Afr., N. Am. In *P. vulgaris* L.
the fr. K is closed and points up in dry air, but opens and stands horiz. in damp.
Prunellopsis Kudo = praec.
Prunophora Neck. = Prunus L. (Rosac.).
Prunopsis André = Louiseania Carr. (Rosac.).
Prunus L. Rosaceae. In the widest sense (incl. *Amygdalus, Cerasus, Padus, Lauro-Cerasus, Pygeum,* etc.), 430 cosmop.; in the strict sense (including only the plums, apricots, etc., subgen. *Prunophora* Focke), 36 N. temp. The fl. buds are laid down in August or September of the preceding year. There is 1 cpl., which gives rise to a drupe, while the hollow recept. usu. falls away. Many spp. are cult. for their fr., e.g. *P. (Prunophora) armeniaca* L. (apricot), *P. (Prunophora) domestica* L. (plum, prune), *P. (Amygdalus) amygdalus* Batsch (almond), *P. (Amygdalus) persica* (L.) Batsch (peach, with its smooth-fruited variety the nectarine), *P. (Cerasus) cerasus* L. (cherry), etc. *P. (Laurocerasus) laurocerasus* L. is the cherry laurel; it has extra-floral nectaries on the backs of the ls., showing as brownish patches against the midribs. The spines of some spp. are axillary, as in *Crataegus*.
Pryona Miq. = Crudia Schreb. (Legumin.).
Przewalskia Maxim. Solanaceae. 2 C. As.
Psacadocalymma Bremek. = Stethoma Rafin. (Acanthac.).
Psacadopaepale Bremek. Acanthaceae. 2 Sumatra.
Psacalium Cass. (~ Senecio L.). Compositae. 20 warm Am.
Psalina Rafin. = Gentiana L. (Gentianac.).
Psamma Beauv. = Ammophila Host (Gramin.).
Psammagrostis C. A. Gardn. & C. E. Hubb. Gramineae. 1 W. Austr.
Psammanthe Hance = Sesuvium L. (Aïzoac.).
Psammanthe Reichb. = Rhodalsine J. Gay = Minuartia L. (Caryophyllac.).
Psammetes Hepper. Scrophulariaceae. 1 trop. W. Afr.
Psammiosorus C. Chr. Oleandraceae. 1 Madag.
Psammisia Klotzsch. Ericaceae. 50 C. Am., N. & W. trop. S. Am.
Psammochloa Hitchcock. Gramineae. 1 Mongolia.
Psammocorchorus Reichb. = Corchorus L. (Tiliac.).
Psammogeton Edgew. Umbelliferae. 5 C. Am. to Persia & W. Himal.
Psammogonum Nieuwl. = Polygonella Michx (Polygonac.).
Psammomoya Diels & Loesener. Celastraceae. 2 Austr.
Psammophila Fourr. = Gypsophila L. (Caryophyllac.).
Psammophila Schult. = Spartina Schreb. (Gramin.).
Psammophora Dinter & Schwantes. Aïzoaceae. 5 S. Afr.

Psammoseris Boiss. & Reut. = Crepis L. (Compos.).
Psammostachys Presl = Striga Lour. (Scrophulariac.).
Psammotropha Eckl. & Zeyh. Aïzoaceae. 8 S. Afr.
Psammotrophe Benth. & Hook. f. = praec.
Psanacetum Neck. = Tanacetum L. (Compos.).
Psanchum Neck. = Cynanchum L. (Asclepiadac.).
Psatherips Rafin. = Salix L. (Salicac.).
Psathura Comm. ex Juss. Rubiaceae. 8 Madag., Mascarenes.
Psathurochaeta DC. = Melanthera Rohr (Compos.).
Psathyra Spreng. = Psathura Comm. ex Juss. (Rubiac.).
Psathyranthus Ule. Loranthaceae. 1 W. Amaz. Brazil.
Psathyrochaeta P. & K. = Psathurochaeta DC. = Melanthera Rohr (Compos.).
Psathyrodes Willis (sphalm.) = Psathyrotes A. Gray (Compos.).
Psathyrostachys Nevski = Hordeum L. (Gramin.).
Psathyrotes A. Gray. Compositae. 4 W. U.S., N. Mex.
Psathyrotopsis Rydb. Compositae. 1 Mex.
Psatura Poir. = Psathura Comm. ex Juss. (Rubiac.).
Psedera Necker ex Greene = Ampelopsis Michx, Parthenocissus Planch., etc. (Vitidac.).
Psednotrichia Hiern. Compositae. 2 trop. & S. Afr.
Psedomelia Neck. = Bromelia L. (Bromeliac.).
Pselium Lour. = Pericampylus Miers + Stephania Lour. (Menispermac.).
Psephellus Cass. = Centaurea L. (Compos.).
Pseudabutilon R. E. Fries. Malvaceae. 17 warm Am.
Pseudacacia Moench = Pseudo-Acacia Duham. = Robinia L. (Legumin.).
Pseudacanthopale Benoist. Acanthaceae. 1 SE. trop. Afr.
Pseudacoridium Ames. Orchidaceae. 1 Philipp. Is.
Pseudactis S. Moore. Compositae. 1 trop. Afr.
× **Pseudadenia** P. F. Hunt. Orchidaceae. Gen. hybr. (i) (Gymnadenia × Pseudorchis).
Pseudaechmanthera Bremek. Acanthaceae. 1 Himalaya.
Pseudaegiphila Rusby. Verbenaceae. 1 W. Amaz. Brazil.
Pseudaegle Miq. = Poncirus Rafin. (Rutac.).
Pseudagrostistachys Pax & K. Hoffm. Euphorbiaceae. 3 trop. Afr.
Pseudais Decne = Phaleria Jack (Thymelaeac.).
Pseudalangium F. Muell. = Alangium Lam. (Alangiac.).
Pseudalbizzia Britton & Rose = Albizia Durazz. (Legumin.).
Pseudaleia Thou. = Olax L. (Olacac.).
Pseudaleioïdes Thou. = praec.
Pseudaleiopsis Reichb. = praec.
Pseudalepyrum Dandy = Centrolepis Labill. (Centrolepidac.).
Pseudalomia Zoll. & Mor. = ? Ethulia L. f. (Compos.).
Pseudalthenia (Graebn.) Nakai = Zannichellia L. (Zannichelliac.).
Pseudammi Wolff. Umbelliferae. 1 C. As.
Pseudanamomis Kausel. Myrtaceae. 1 Puerto Rico.
Pseudananas (Hassler) Harms. Bromeliaceae. 1 Paraguay, N. Argent.
Pseudanastatica (Boiss.) Lemée = Pseudoanastatica (Boiss.) Grossh. = Clypeola L. (Crucif.).

PSEUDANCHUSA

Pseudanchusa (A. DC.) Kuntze = Lindelofia Lehm. (Boraginac.).
Pseudannona (Baill.) Safford. Annonaceae. 2 Mauritius (& Madag.?).
Pseudanth[ac]eae Endl. = Micrantheaceae J. G. Agardh = Euphorbiaceae–Porantheroïdeae Pax.
Pseudanthistiria Hook. f. Gramineae. 4 India, Siam, S. China.
Pseudanthus Sieb. ex Spreng. Euphorbiaceae. 7 Austr.
Pseudanthus Wight = Nothosaerva Wight (Amaranthac.).
Pseudarabidella O. E. Schulz. Cruciferae. 1 S. Austr.
Pseudarrhenatherum Rouy (~ Arrhenatherum Beauv.). Gramineae. 1–2 W. France, NW. Spain, N. Portugal.
Pseudartabotrys Pellegr. Annonaceae. 1 W. Equat. Afr.
Pseudarthria Wight & Arn. Leguminosae. 6 trop. & S. Afr., 1 trop. As.
Pseudatalaya Baill. = Atalaya Blume (Sapindac.).
Pseudathyrium Newman = Athyrium Roth (Athyriac.).
Pseudechinolaena Stapf. Gramineae. 1 trop.
Pseudechinopepon (Cogn.) Kuntze = Vaseyanthus Cogn. (Cucurbitac.).
Pseudehretia Turcz. = Ilex L. (Aquifoliac.).
Pseudelephantopus Rohr. Compositae. 1–3 C. & trop. S. Am.
Pseudellipanthus Schellenb. Connaraceae. 2 Borneo.
Pseudeminia Verdc. Leguminosae. 4 trop. Afr.
Pseudephedranthus Aristeg. Annonaceae. 1 Brazil, Venez.
Pseudepidendrum Reichb. f. = Epidendrum L. (Orchidac.).
Pseuderanthemum Radlk. Acanthaceae. 120 trop.
Pseuderemostachys Popov. Labiatae. 1 C. As.
Pseuderia Schlechter. Orchidaceae. 20 Malaysia, Polynesia.
Pseuderiopsis Reichb. f. = Eriopsis Lindl. (Orchidac.).
Pseudernestia P. & K. = Pseudoërnestia (Cogn.) Krasser (Melastomatac.).
Pseuderucaria (Boiss.) O. E. Schulz. Cruciferae. 3 Morocco to Palestine.
Pseudetalon Rafin. = Pseudopetalon Rafin. (Rutac.).
Pseudeugenia Legrand & Mattos. Myrtaceae. 1 Brazil.
Pseudeugenia P. & K. = Pseudoeugenia Scortechini = Syzygium Gaertn. (Myrtac.).
Pseudevax DC. ex Pomel = Evax Gaertn. (Compos.).
Pseudibatia Malme. Asclepiadaceae. 15 S. Am.
Pseudima Radlk. Sapindaceae. 3 C. & N. trop. S. Am.
× **Pseudinium** P. F. Hunt. Orchidaceae. Gen. hybr. (i) (Herminium × Pseudorchis).
Pseudiosma A. Juss. Quid? 1 Indoch.
Pseudipomoea Roberty = Ipomoea L. (Convolvulac.).
Pseudiris P. & K. = Pseudo-Iris Medik. = Iris L. (Iridac.).
Pseuditea Hassk. = Pittosporum Banks ex Soland. (Pittosporac.).
× **Pseuditella** P. F. Hunt. Orchidaceae. Gen. hybr. (i) (Nigritella × Pseudorchis).
Pseudixora Miq. = Randia L., etc. (Rubiac.).
Pseudixus Hayata = Korthalsella Van Tiegh. (Viscac.).
Pseudo-Acacia Duham. = Robinia L. (Legumin.).
Pseudoampelopsis Planch. = Ampelopsis Planch. (Vitidac.).
Pseudoanastatica (Boiss.) Grossheim = Clypeola L. (Crucif.).

Pseudoarrenatherum Holub = Pseudarrenatherum Rouy (Gramin.).
Pseudobaccharis Cabrera = Psila Phil. (Compos.).
Pseudobaeckea Niedenzu. Bruniaceae. 4 S. Afr.
Pseudobahia (A. Gray) Rydb. Compositae. 3 Calif.
Pseudobarleria T. Anders. = Petalidium Nees (Acanthac.).
Pseudobarleria Oerst. = Barleria L. (Acanthac.).
Pseudobartlettia Rydb. Compositae. 1 SW. U.S., Mex.
Pseudobasilicum Steud. = Pseudo-brasilium Adans. = Picramnia Sw. (Simaroubac.).
Pseudobastardia Hassler = Herissantia Medik. (Malvac.).
Pseudoberlinia Duvign. Leguminosae. 4 trop. Afr.
Pseudobersama Verdc. Meliaceae. 1 trop. E. Afr.
Pseudobesleria Oerst. = Besleria L. (Acanthac.).
Pseudobetckea (Hoeck) Lincz. Valerianaceae. 1 Cauc.
Pseudoblepharis Baill. = Sclerochiton Harv. (Acanthac.).
Pseudoboivinella Aubrév. & Pellegr. Sapotaceae. 3 trop. Afr.
Pseudobombax Dugand. Bombacaceae. 20 Mex. to trop. S. Am.
Pseudobotrys Moeser. Icacinaceae. 2 New Guinea.
Pseudobrachiaria Launert. Gramineae. 1 trop. & S. Afr.
Pseudo-brasilium Adans. = Picramnia Sw. (Simaroubac.).
Pseudobravoa Rose. Agavaceae. 1 Mex.
Pseudobraya Korshinsky = Draba L. (Crucif.).
Pseudobrazzeia Engl. = Brazzeia Baill. (Scytopetalac.).
Pseudobromus K. Schum. Gramineae. 8 E. trop. & S. Afr., Madag.
Pseudobulbostylis Nutt. = ? Bulbostylis DC. = Brickellia Ell. (Compos.).
Pseudocadia Harms = Xanthocercis Baill. (Legumin.).
Pseudocalymma A. Sampaio & Kuhlm. Bignoniaceae. 4 trop. Am., W.I.
Pseudocalyx Radlk. Thunbergiaceae. 4 trop. Afr., Madag.
Pseudocamelina (Boiss.) N. Busch. Cruciferae. 6 Persia.
Pseudocapsicum Medik. = Solanum L. (Solanac.).
Pseudocarapa Hemsl. Meliaceae. 5–6 Ceylon, New Guinea, E. Austr., New Caled.
Pseudocarex Miq. = Carex L. (Cyperac.).
Pseudocarpidium Millsp. Verbenaceae. 8 W.I.
Pseudocarum C. Norman. Umbelliferae. 2 trop. E. Afr.
Pseudocaryophyllus Berg. Myrtaceae. 17 trop. S. Am.
Pseudocassia Britton & Rose = Cassia L. (Legumin.).
Pseudocassine Bredell = Crocoxylon Eckl. & Zeyh. (Celastrac.).
Pseudocedrela Harms. Meliaceae. 6 trop. Afr.
Pseudocentema Chiov. = Centema Hook. f. (Amaranthac.).
Pseudocentrum Lindl. Orchidaceae. 8 C. & trop. S. Am., W.I.
Pseudo-chaenomeles Carr. = Chaenomeles Lindl. (Rosac.).
Pseudochaetochloa Hitchcock. Gramineae. 1 W. Austr.
Pseudochimarrhis Ducke. Rubiaceae. 3 trop. S. Am.
Pseudochrosia Blume = Ochrosia Juss. (?) (Apocynac.).
Pseudocimum Bremek. = Endostemon N. E. Br. (Labiat.).
Pseudocinchona A. Chev. ex E. Perrot (~ Corynanthe Welw.). Rubiaceae. 4 trop. Afr.

PSEUDOCIONE

Pseudocione Mart. ex Engl. = Thyrsodium Salzm. ex Benth. (Anacardiac.).

Pseudocladia Pierre = Lucuma Molina (Sapotac.).

Pseudoclappia Rydb. Compositae. 1 SW. U.S.

Pseudoclausia Popov. Cruciferae. 1 C. As.

Pseudoclinium Kuntze = Leptoclinium Gardn. ex Benth. (Compos.).

Pseudocoix A. Camus. Gramineae. 1 Madag.

Pseudoconnarus Radlk. Connaraceae. 6 trop. S. Am.

Pseudoconyza Cuatrec. Compositae. 1 Florida & Mex. to Bolivia.

Pseudocopaiva Britton & Wilson = Guibourtia Benn. (Legumin.).

Pseudocorchorus Capuron. Tiliaceae. 6 Madag.

Pseudocroton Muell. Arg. Euphorbiaceae. 1 C. Am.

Pseudocryptocarya Teschn. = Cryptocarya R.Br. (Laurac.).

Pseudoctomeria Kraenzl. Orchidaceae. 1 Costa Rica.

Pseudocunila Brade. Labiatae. 1 Brazil.

Pseudocyclanthera Mart. Crovetto. Cucurbitaceae. 1 S. Am.

Pseudocylosorus Ching. Thelypteridaceae. 10 SE. As. Free-veined, but perhaps = Pneumatopteris Nak.

Pseudocydonia C. K. Schneider = Chaenomeles Lindl. (Rosac.).

Pseudocymopterus Coulter & Rose. Umbelliferae. 7 SW. U.S.

Pseudocynometra (Wight & Arn.) Kuntze = Maniltoa Scheff. (Legumin.).

Pseudocyperus Steud. = Fimbristylis Vahl (Cyperac.).

Pseudocystopteris Ching = Athyrium Roth (Athyriac.).

Pseudocytisus Kuntze = Vella DC. (Crucif.).

Pseudodanthonia Bor & C. E. Hubbard. Gramineae. 1 Himal.

Pseudodatura v. Zijp = Brugmansia Bl. (Solanac.).

Pseudodesmos Spruce ex Engl. = Moronobea Aubl. (Guttif.).

Pseudodichanthium Bor. Gramineae. 1 W. Penins. India.

Pseudodicliptera Benoist. Acanthaceae. 2 Madag.

Pseudodictamnus Fabr. (uninom.) = Ballota L. sp. (Labiat.).

Pseudodigera Chiov. Amaranthaceae. 1 Somalia.

Pseudodiphryllum Nevski. Orchidaceae. 1 Siberia, Japan.

Pseudodissochaetą Nayar. Melastomataceae. 4 N. India to Hainan.

Pseudodracontium N. E. Br. Araceae. 7 Siam, Indoch.

Pseudodrynaria C. Chr. Polypodiaceae. 1 NE. India to Formosa. Habit of Drynariopsis.

Pseudoëchinocereus Buining = Borzicactus Riccob. (Cactac.).

Pseudo-elephantopus Steud. = Pseudelephantopus Rohr (Compos.).

Pseudoëntada Britton & Rose = Entada Adans. (Legumin.).

Pseudo-eranthemum Radlk. = Pseuderanthemum Radlk. (Acanthac.).

Pseudoëriosema Hauman. Leguminosae. 6 trop. Afr.

Pseudoërnestia (Cogn.) Krasser. Melastomataceae. 1 Venezuela.

Pseudoëspostoa Backeb. = Espostoa Britt. & Rose (Cactac.).

Pseudoëugenia Scortech. = Aphanomyrtus Miq. = Syzygium Gaertn. (Myrtac.).

Pseudoëurya Yamamoto. Theaceae. 1 Formosa. (= ?Eurya Thunb.)

Pseudoëurystyles Hoehne. Orchidaceae. 3 Brazil.

Pseudoëverardia Gilly. Cyperaceae. 1 Venez.

Pseudofortuynia Hedge. Cruciferae. 1 Persia.

Pseudofumaria Medik. = Corydalis Vent. (Fumariac.).

Pseudogaltonia Kuntze. Liliaceae. 1 S. Afr.

Pseudogardenia Keay. Rubiaceae. 1 trop. Afr.

Pseudogardneria Raciborski = Gardneria Wall. (Strychnac.).

Pseudoglochidion Gamble = Glochidion J. R. & G. Forst. (Euphorbiac.).

Pseudoglycine F. J. Hermann = Ophrestia H. M. L. Forbes (Legumin.).

Pseudognaphalium Kirpich. Compositae. 1 C. Am.

Pseudognidia Phillips. Thymelaeaceae. 1 S. Afr.

Pseudogomphrena R. E. Fries. Amaranthaceae. 1 Brazil.

Pseudogoodyera Schlechter. Orchidaceae. 1 Cuba.

Pseudo-gunnera Oerst. = Gunnera L. (Gunnerac.).

Pseudogynoxys (Greenm.) Cabrera. Compositae. 16 trop. S. Am.

Pseudohamelia Wernham. Rubiaceae. 1 Andes.

Pseudohandelia Tsvelev (~ Tanacetum L.). Compositae. 1 C. As., Persia, Afghan.

Pseudohomalomena A. D. Hawkes = Zantedeschia Spreng. (Arac.).

Pseudohydrosme Engl. Araceae. 2 trop. W. Afr.

Pseudo-Iris Medik. = Iris L. (Iridac.).

Pseudolabatia Aubrév. & Pellegr. Sapotaceae. 5 Brazil.

Pseudolachnostylis Pax. Euphorbiaceae(?). 5 trop. & S. Afr.

Pseudolaelia C. Porto & Brade = Schomburgkia Lindl. (Orchidac.).

Pseudolarix Gordon. Pinaceae. 2 China. *P. amabilis* (Nelson) Rehd., the golden larch. Like *Larix*, but distinguished chiefly by the deciduous cone-scales.

Pseudolasiacis (A. Camus) A. Camus. Gramineae. 4 Madag.

Pseudolinosyris Novopokr. Compositae. 3 C. As.

Pseudoliparis Finet = Malaxis Soland. ex Sw. (Orchidac.).

Pseudolitchi Danguy & Choux. Sapindaceae. 1 Madag.

Pseudolithos Bally. Asclepiadaceae. 2 NE. trop. Afr.

Pseudolitsea Yang = Litsea Lam. (Laurac.).

Pseudolmedia Karst. = Olmediophaena Karst. (Morac.).

Pseudolmedia Tréc. Moraceae. 25 C. & trop. S. Am., W.I.

Pseudolobelia A. Chev. = Torenia L. (Scrophulariac.).

Pseudolobivia (Backeb.) Backeb. = Echinopsis Zucc. (Cactac.).

Pseudolopezia Rose. Onagraceae. 1 Mex.

Pseudolophanthus Levin (~ Glechoma L.). Labiatae. 3 Tibet, Himal., Yunnan.

Pseudolotus K. H. Rechinger = Lotus L. (Legumin.).

Pseudoludovia Harling. Cyclanthaceae. 1 Colombia.

Pseudolysimachion (Koch) Opiz (~ Veronica L.). Scrophulariaceae. 30 N. temp.

Pseudolysimachion Opiz = Veronica L. (Scrophulariac.).

Pseudomachaerium Hassler. Leguminosae. 1 Paraguay.

Pseudomacodes Rolfe = Macodes Bl. (Orchidac.).

Pseudomacrolobium Hauman. Leguminosae. 1 trop. Afr.

Pseudomammillaria F. Buxb. = Mammillaria Haw. (Cactác.).

Pseudomarrubium M. Popov (= ? Neustruevia Juzepczuk). Labiatae. 1 C. As.

Pseudomarsdenia Baill. Asclepiadaceae. 5 trop. Am.   [Matricaria).

× **Pseudomatricaria** Domin. Compositae. Gen. hybr. (Chrysanthemum ×

PSEUDOMAXILLARIA

Pseudomaxillaria Hoehne = Ornithidium R.Br. (Orchidac.).

Pseudomertensia H. Riedl. Boraginaceae. 8 Persia to Tibet & Himal.

Pseudomitrocereus Bravo & F. Buxb. = Pachycereus (A. Berger) Britton & Rose (Cactac.).

Pseudomorus Bur. = Streblus Lour. (Morac.).

Pseudomuscari Garbari & Greuter. Liliaceae. 7 (?) Caucasus.

Pseudomussaenda Wernham. Rubiaceae. 7 trop. Afr.

Pseudomyrcianthes Kausel. Myrtaceae. 10 SW. Brazil to N. Argent.

Pseudonemacladus McVaugh. Campanulaceae. 1 Mex.

Pseudonephelium Radlk. = Dimocarpus Lour. (Sapindac.).

Pseudonesohedyotis Tennant. Rubiaceae. 1 trop. E. Afr.

Pseudonopalxochia Backeb. = Nopalxochia Britt. & Rose (Cactac.).

Pseudopachystela Aubrév. & Pellegr. Sapotaceae. 2 trop. Afr.

Pseudopaegma Urb. Bignoniaceae. 6 trop. S. Am.

Pseudopanax C. Koch. Araliaceae. 6 N.Z., temp. S. Am.

Pseudopancovia Pellegr. Sapindaceae. 1 W. Equat. Afr.

Pseudoparis H. Perrier. Commelinaceae. 2 Madag.

Pseudopavonia Hassler. Malvaceae. 1 Paraguay.

Pseudopectinaria Lavr. Asclepiadaceae. 1 Somalia.

Pseudopentatropis Costantin. Asclepiadaceae. 1 Indochina.

Pseudopeponidium Homolle ex Arènes. Rubiaceae. 6 Madag.

Pseudopetalon Rafin. = Zanthoxylum L. (Rutac.).

Pseudophacelurus A. Camus = Phacelurus Griseb. (Gramin.).

Pseudophegopteris Ching. Thelypteridaceae. 20 Afr., SE. As. to Hawaii. Most spp. bipinnate, all exindusiate.

Pseudophoenicaceae O. F. Cook = Palmae–Areceae Lindl.

Pseudophoenix H. Wendl. & Drude. Palmae. 2–4 Florida, Caribbean shores, Hispaniola.

Pseudopholidia A. DC. = Pholidia R.Br. (Myoporac.).

Pseudopilocereus F. Buxb. Cactaceae. 22 trop. S. Am.

Pseudopinanga Burret (~Pinanga Bl.). Palmae. 15 Formosa, Philippines, Borneo, Celebes (? 1 Sumatra).

Pseudopipturus Skottsb. = Nothocnide Bl. (Urticac.).

Pseudoplantago Suesseng. Amaranthaceae. 1 Argentina.

Pseudoplatanthera Rouy = × Gymplatanthera Camus (1906) (Orchidac.).

Pseudopogonatherum A. Camus. Gramineae. 2 trop. As.

Pseudoprimula (Pax) O. Schwarz. Primulaceae. 15 C. As. & Afghan. to E. As. & Philipp. Is.

Pseudoprosopis Harms. Leguminosae. 4 trop. Afr.

Pseudoprotorhus H. Perrier. Anacardiaceae. 1 Madag.

Pseudopteris Baill. Sapindaceae. 1 Madag.

Pseudopteryxia Rydb. = Pseudocymopterus Coult. & Rose (Umbellif.).

Pseudopyxis Miq. Rubiaceae. 2 Japan.

Pseudorachicallis P. & K. = Pseudrachicallis Karst. = Mallostoma Karst.

Pseudoraphis Griff. Gramineae. 7 India & Japan to Austr.   [(Rubiac.).

Pseudorchis S. F. Gray = Liparis Rich. (Orchidac.).

Pseudorchis Seguier (Leucorchis E. Mey.). Orchidaceae. 3 E. N. Am., Greenland, Iceland, Eur.

PSEUDOSTENOMESSON

Pseudoreoxis Rydb. = Pseudocymopterus Coult. & Rose. (Umbellif.).
Pseudorhachicallis Benth. & Hook. f. = Pseudrachicallis Karst. = Mallostoma Karst. (Rubiac.).
Pseudorhipsalis Britton & Rose = Disocactus Lindl. (Cactac.).
× **Pseudorhiza** P. F. Hunt. Orchidaceae. Gen. hybr. (i) (Dactylorhiza × Pseudorchis).
Pseudorlaya Murb. Umbelliferae. 3 W. Eur., Medit.
Pseudorobanche Rouy = Alectra Thunb. (Scrophulariac.).
Pseudorontium (A. Gray) Rothm. Scrophulariaceae. 1 SW. U.S.
Pseudoruellia Benoist. Acanthaceae. 1 Madag.
Pseudoryza Griff. = ? Leersia Sw. (Gramin.).
Pseudosabicea Hallé. Rubiaceae. 9 trop. Afr.
Pseudosalacia Codd. Celastraceae. 1 S. Afr. (Natal).
Pseudosamanea Harms. Leguminosae. 1 Panamá to Venez. & Ecuador.
Pseudosantalum Kuntze = Osmoxylon Miq. (Araliac.).
Pseudosantalum [Sloane ex] Mill. = Caesalpinia L. (Legumin.).
Pseudosarcolobus Costantin. Asclepiadaceae. 1 Indochina.
Pseudosasa Makino = Sasa Mak. & Shibata (Gramin.).
Pseudosassafras Lecomte (~ Sassafras Boehm.). Lauraceae. 1 China.
Pseudosbeckia A. & R. Fernandes. Melastomataceae. 1 trop. E. Afr.
Pseudosciadium Baill. Araliaceae. 1 New Caled.
Pseudoscolopia Gilg. Flacourtiaceae. 1 S. Afr.
Pseudoscolopia Phillips = praec.
Pseudoscordum Herb. = Nothoscordum Kunth (Alliac.).
Pseudosecale (sphalm. 'Psendo-') (Godr.) Degen = Dasypyrum (Coss. & Durieu) Candargy [Kantartzis] = Haynaldia Schur (Gramin.).
Pseudosedum (Boiss.) Berger. Crassulaceae. 10 C. As.
Pseudoselinum C. Norman. Umbelliferae. 1 Angola.
Pseudosempervivum (Boiss.) Grossheim = Cochlearia L. (Crucif.).
Pseudosericocoma Cavaco. Amaranthaceae. 1 SW. & S. Afr.
Pseudoseris Baill. = Gerbera L. ex Cass. (Compos.).
Pseudosicydium Harms. Cucurbitaceae. 1 Peru, Bolivia.
Pseudosindora Symington (~ Copaïfera L.). Leguminosae. 1 Borneo.
Pseudosmelia Sleum. Flacourtiaceae. 1 Moluccas.
Pseudosmilax Hayata. Smilacaceae. 2 Formosa.
Pseudosmodingium Engl. Anacardiaceae. 7 Mex.
Pseudosophora Sweet = Sophora L. (Legumin.).
Pseudosopubia Engl. Scrophulariaceae. 7 trop. Afr.
Pseudosorghum A. Camus. Gramineae. 1 trop. As.
Pseudosorocea Baill. = Sorocea A. St-Hil. + Acanthinophyllum Fr. Allem. (Morac.).
Pseudospermum S. F. Gray = Physospermum Cuss. (Umbellif.).
Pseudospigelia Klett (~ Spigelia L.). Loganiaceae. 1 C. & trop. S. Am.
Pseudospondias Engl. Anacardiaceae. 2 W. & C. trop. Afr.
Pseudostachyum Munro. Gramineae. 1 E. Himal. to Burma.
Pseudostelis Schlechter. Orchidaceae. 3 Brazil.
Pseudostellaria Pax. Caryophyllaceae. 15 C. As. & Afghanistan to Japan.
Pseudostenomesson Velarde. Amaryllidaceae. 2 Andes.

955

PSEUDOSTENOSIPHONIUM

Pseudostenosiphonium Lindau. Acanthaceae. 9 Ceylon.

Pseudostonium Kuntze = praec.

Pseudostreblus Bur. = Streblus Lour. (Morac.).

Pseudostreptogyne A. Camus. Gramineae. 1 Réunion.

Pseudostriga Bonati. Scrophulariaceae. 1 Indoch.

Pseudostrophis Dur. & Jacks. (sphalm.) = Pseudotrophis Warb. = Streblus Lour. (Morac.).

Pseudotaenidia Mackenzie. Umbelliferae. 1 N. Am.

Pseudotaxus Cheng. Taxaceae. 1 China.

Pseudotectaria Tard. Aspidiaceae. 2 Madag. (*Not. Syst.* **15**: 87, 1955).

Pseudotephrocactus Frič & Schelle = Opuntia Mill. (Cactac.).

Pseudotragia Pax = Pterococcus Hassk. (Euphorbiac.).

Pseudotreculia (Baill.) B. D. Jacks. = Treculia Decne ex Tréc. (Morac.).

Pseudotrewia Miq. = Wetria Baill. (Euphorbiac.).

Pseudotrimezia R. C. Foster. Iridaceae. 1 Brazil.

Pseudotrophis Warb. = Streblus Lour. (Morac.).

Pseudotsuga Carr. Pinaceae. 7 E. As., W. N. Am., incl. *P. menziesii* (Mirb.) Franco (*P. taxifolia* [Poir.] Britton), the Douglas fir of W. N. Am., useful for masts, etc.

Pseudourceolina Vargas. Amaryllidaceae. 1 Peru.

Pseudovesicaria (Boiss.) Rupr. Cruciferae. 1 Caucasus.

Pseudovigna (Harms) Verdc. Leguminosae. 1 trop. Afr.

Pseudovossia A. Camus. Gramineae. 1 Indochina.

Pseudovouapa Britton & Killip. Leguminosae. 1 Colombia.

Pseudoweinmannia Engl. Cunoniaceae. 2 Queensland.

Pseudowillughbeia Markgraf. Apocynaceae. 1 NE. New Guinea.

Pseudowintera Dandy. Winteraceae. 3 N.Z.

Pseudowolffia den Hartog & van der Plas. Lemnaceae. 3 N. & C. Afr.

Pseudoxalis Rose = Oxalis L. (Oxalidac.).

Pseudoxandra R. E. Fries. Annonaceae. 6 trop. S. Am.

Pseudozoysia Chiov. Gramineae. 1 Somalia.

Pseudozygocactus Backeb. = Hatiora Britt. & Rose (Cactac.).

Pseudrachicallis Karst. = Mallostoma Karst. (Rubiac.).

Pseuduvaria Miq. Annonaceae. 17 SE. As., W. Malaysia, 1 New Guinea.

Pseusmagennetus Ruschenb. = Marsdenia R. Br. (Asclepiadac.).

Pseva Rafin. = Chimaphila Pursh (Pyrolac.).

Psiadia Jacq. Compositae. 60 trop. Afr., St Helena, Madag., Masc., Socotra.

Psiadiella Humbert. Compositae. 1 Madag.

Psidiastrum Bello = Eugenia L. (Myrtac.).

Psidiomyrtus Guillaumin = Rhodomyrtus Reichb. (Myrtac.).

Psidiopsis Berg. Myrtaceae. 1 Venezuela. Ed. fr.

Psidium L. Myrtaceae. 140 trop. Am., W.I. Many yield ed. fr., e.g. *P. guajava* L., the guava.

Psiguria Neck. ex Arn. Cucurbitaceae. 15 trop. Am.

Psila Phil. Compositae. 12 Andes.

Psilachaenia P. & K. = seq.

Psilachenia Benth. = Psilochenia Nutt. = Crepis L. (Compos.).

Psilactis A. Gray. Compositae. 5 SW. U.S., N. Mex.

Psilaea Miq. = Linostoma Wall. ex Endl. (Thymelaeac.).
Psilantha (C. Koch) Tsvelev. Gramineae. 1 E. Medit. & S. Russia to C. As.
Psilanthele auctt. = Oplonia Rafin. (Acanthac.).
Psilanthele Lindau. Acanthaceae. 1 Ecuador.
Psilanthopsis A. Chev. Rubiaceae. 1 Angola.
***Psilanthus** Hook. f. Rubiaceae. 1 trop. W. Afr.
Psilanthus [Juss.] (DC.) M. Roem. = Passiflora L. (Passiflorac.).
Psilarabis Fourr. = Arabis L. (Crucif.).
Psilathera Link (~ Sesleria Scop.). Gramineae. 1 S. Bavaria, N. Italy, Austria.
Psilobium Jack = Acranthera Arn. ex Meissn. (Rubiac.).
Psilocarphus Nutt. Compositae. 5 W. U.S., 1 W. temp. S. Am.
Psilocarpus Pritz. = Psophocarpus Neck. ex DC. (Legumin.).
Psilocarya Torr. = Rhynchospora Vahl (Cyperac.).
Psilocaulon N. E. Brown. Aïzoaceae. 75 S. Afr.
Psilochenia Nutt. = Crepis L. (Compos.).
Psilochilus Barb. Rodr. = Pogonia Juss. (Orchidac.).
Psilochlaena Walp. = Psilochenia Nutt. = Crepis L. (Compos.).
Psilochloa Launert. Gramineae. 1 SW. Afr.
Psilodigera Suesseng. = Saltia R.Br. ex Moq. (Amaranthac.).
Psilodochea Presl = Angiopteris Hoffm. (Angiopteridac.).
Psiloësthes Benoist. Acanthaceae. 1 Indoch.
Psilogramme Kuhn emend. Underw. = Eriosorus Fée (Gymnogrammac.).
Psilogyne DC. = Vitex L. (Verbenac.).
Psilolaemus I. M. Johnston. Boraginaceae. 1 Mex.
Psilolepus Presl = Aspalathus L. (Legumin.).
Psilonema C. A. Mey. = Alyssum L. (Crucif.).
Psilopeganum Hemsl. Rutaceae. 1 C. China.
Psilopogon Hochst. = Arthraxon Beauv. (Gramin.).
Psilopogon Phil. = Picrosia D. Don (Compos.).
Psilopsis Neck. = Lamium L. (Labiat.).
Psilorhegma (Vogel) Britton & Rose = Cassia L. (Legumin.).
Psilosanthus Neck. = Liatris Schreb. (Compos.).
Psilosiphon Welw. ex Baker = Lapeyrousia Pourr. (Iridac.).
Psilosolena Presl = Peddiea Harv. (Thymelaeac.).
Psilostachys Hochst. = Psilotrichum Bl. (Amaranthac.).
Psilostachys Steud. = Dimeria R.Br. (Gramin.).
Psilostachys Turcz. = Cleidion Blume (Euphorbiac.).
Psilostemon DC. = Trachystemon D. Don (Boraginac.).
Psilostoma Klotzsch = Plectronia L. (Rubiac.).
Psilostrophe DC. Compositae. 6 S. U.S., N. Mex.
Psilotaceae Eichler. Psilotales. Epiphytes or in rock crevices; rhizome rootless, bearing small bulbils; green stems branched, bearing minute ls. and bilobed sporophylls each with a 3-locular sporangium. Only genus: *Psilotum*.
Psilotales. 2 fam.: *Psilotaceae, Tmesipteridaceae*.
Psilothamnus DC. = Gamolepis Less. (Compos.).
Psilothonna E. Mey. ex DC. Compositae. 5 S. Afr.
Psilotopsida. One order: *Psilotales*.
Psilotrichium Hassk. = seq.

PSILOTRICHUM

Psilotrichum Blume. Amaranthaceae. 27 NE., trop. E., & SE. Afr., Madag., India, SE. As., Malaysia.

Psilotum Sw. Psilotaceae. 3 trop. & subtrop.

Psiloxylaceae Croizat. Dicots. 1/1 Masc. Trees with white bark and alt. simple ent. gland-dotted exstip. ls. with intramarg. nerve. Fls. reg., ♂ ♀, dioec., or polygam., in small axill. rac. or pan. K 5(-6), imbr., persist.; C 5(-6), imbr., caduc., ± coriac., punct., shortly clawed; A 10(-12), inserted on perig. disk, fil. subul., anth. versat., ovoid, intr.; G (3-4), with 3-4 subsess., thick, fleshy, flattened, obtuse, spreading and reflexed style-lobes, persist. on the fr., and ∞ anatr. ov. on axile plac. Fr. a small punctate many-seeded berry; embryo fleshy, endosp. o. Only genus: *Psiloxylon* Thou. ex Tul. Closely related to *Myrtac.*, differing in the wholly superior ovary, the thick reflexed persist. style-lobes, and the large anthers.

Psiloxylon Thou. ex Tul. Psiloxylaceae. 1 Mauritius, Réunion.

Psilurus Trin. Gramineae. 1 S. Eur. to Afghanistan.

Psistina Rafin. = Helianthemum Mill. (Cistac.).

Psistus Neck. = praec.

Psithyrisma Herb. = Symphyostemon Miers (Iridac.).

Psittacanthaceae Nak. = Loranthaceae–Psittacanthinae Engl.

Psittacanthus Mart. Loranthaceae. 50 trop. Am.

Psittacaria Fabr. (uninom.) = *Amaranthus* L. sp. (Amaranthac.).

Psittacoglossum La Llave & Lex. = Maxillaria Poepp. & Endl. (Orchidac.).

Psittacoschoenus Nees = Gahnia J. R. & G. Forst. (Cyperac.).

Psittaglossum P. & K. = Psittacoglossum La Llave & Lex. = Maxillaria Poepp. & Endl. (Orchidac.).

Psolanum Neck. = Solanum L. (Solanac.).

Psomiocarpa Presl. Aspidiaceae. 1 Philipp. Is.

Psophiza Rafin. = Aristolochia L. (Aristolochiac.).

*Psophocarpus Neck. ex DC. Leguminosae. 10 trop. Afr., Masc. *P. tetragonolobus* (L.) DC. and others have ed. pods.

Psora Hill = Centaurea L. (Compos.).

Psoralea L. Leguminosae. 130 trop. & subtrop. *P. esculenta* Pursh (N. Am. prairie turnip) has ed. tuberous root.

Psoralidium Rydb. (~ Psoralea L.). Leguminosae. 15 W. N. Am.

Psorobates Willis (sphalm.) = seq.

Psorobatus Rydb. = Dalea L. (Legumin.).

Psorodendron Rydb. = Dalea L. (Legumin.).

Psorophytum Spach = Hypericum L. (Guttif.).

Psorospermum Spach. Guttiferae. 40–45 trop. Afr., Madag.

Psorothamnus Rydb. = Dalea L. (Legumin.).

Psychanthus Rafin. = Polygala L. (Polygalac.).

Psychanthus Ridley = Alpinia Roxb. (Zingiberac.).

Psychechilus Breda = Zeuxine Lindl. (Orchidac.).

Psychilis Rafin. = Epidendrum L. (Orchidac.).

Psychine Desf. Cruciferae. 1 N. Afr.

Psychochilus P. & K. = Psychechilus Breda = Zeuxine Lindl. (Orchidac.).

Psychodendron Walp. ('Wall.') ex Voigt = Bischofia Bl. (Bischofiac.).

Psycholobium Blume ex Burck = Mucuna Adans. (Legumin.).

Psychopsis Nutt. ex Greene = Hosackia Dougl. (Legumin.).
Psychopsis Rafin. = Oncidium Sw. (Orchidac.).
Psychosperma Dum. = Ptychosperma Labill. (Palm.).
Psychothria L. = seq.
*Psychotria L. Rubiaceae. 700 warm. Some heterostyled. Many have infl.-axis brightly coloured.
Psychotrophum P.Br. = praec.
Psychridium Stev. = Astragalus L. (Legumin.).
Psychrobatia Greene = Ametron Rafin. = Rubus L. (Rosac.).
Psychrogeton Boiss. (~ Aster L.). Compositae. 29 SW. & C. As.
Psychrophila Bercht. & Presl = Caltha L. (Ranunculac.).
Psychrophyton Beauverd˙ (~ Raoulia Hook. f.). Compositae. 9 N.Z.
Psycothria L. (sphalm.) = Psychotria L. (Rubiac.).
Psycrophila Rafin. = Psychrophila Bercht. & Presl = Caltha L. (Ranunculac.).
Psydaranta Neck. ex Rafin. = Calathea G. F. W. Mey. (Marantac.).
Psydarantha Steud. = praec.
Psydax Steud. = seq.
Psydrax Gaertn. = Canthium Lam. (Rubiac.).
Psygmium Presl = Aglaomorpha Schott (Polypodiac.).
Psylliaceae Horan. = Plantaginaceae Juss.
Psylliostachys (Jaub. & Spach) Nevski. Plumbaginaceae. 6 C. As.
Psyllium Mill. = Plantago L. (Plantaginac.).
Psyllocarpus Mart. Rubiaceae. 5 Brazil.
Psyllocarpus Pohl ex DC. = Declieuxia Kunth (Rubiac.).
Psyllophora Ehrh. (uninom.) = Carex pulicaris L. (Cyperac.).
Psyllophora Heuffel = Carex L. (Cyperac.).
Psyllothamnus Oliv. = Sphaerocoma T. Anders. (Caryophyllac.).
Psylostachys Oerst. = Chamaedorea Willd. (Palm.).
Psyloxylon Thou. ex Gaudich. = Psiloxylon Thou. ex Tul. (Psiloxylac.).
Psythirhisma Lindl. = Psithyrisma Herb. = Symphyostemon Miers (Iridac.).
Ptacoseia Ehrh. (uninom.) = Carex leporina L. (Cyperac.).
Ptaeroxylaceae Sonder. Dicots. 2/3 (? + 1/8) trop. & S. Afr., Madag. Aromatic trees or shrubs, with alt. or opp. pinn. exstip. ls.; lfts. ent., oblique, opp. or alt. Fls. reg., polyg.-dioec., small, in small axill. cymes or false rac. K 4–5, shortly conn., ± imbr.; C 4–5, imbr. or valv.; A 4–5, alternipet.; disk present or o; G̲ (2–5), with 2–5 short connate styles with capit. stigs., and 1–3 campylotr. ov. per loc. Fr. a coriac. samaroid caps., of 2–5 follicles, dehisc. by inner suture; seeds 2–5, winged above, exarillate; embr. curved or folded, endosp. o, or thin and fleshy. Genera: Ptaeroxylon, Cedrelopsis, ?Kirkia. Somewhat intermediate between Meliac. and Rutac. (Cf. Flindersiac.).
Ptaeroxylon Eckl. & Zeyh. Ptaeroxylaceae. 1 S. Afr., P. utile E. & Z. (sneezewood, Cape mahogany); timber.
Ptarmica Mill. = Achillea L. (Compos.).
Ptelandra Triana = Macrolenes Naud. (Melastomatac.).
Ptelea L. Rutaceae. 3 U.S., Mex. Fls. monoec. Fr. winged (cf. Ulmus).
Pteleaceae Kunth = Rutaceae–Toddalioïdeae–Pteleïnae Engl.
Pteleocarpa Oliv. Ehretiaceae. 2 W. Malaysia.
Pteleodendron K. Schum. (sphalm.) = Pleodendron Van Tiegh. (Canellac.).

PTELEOPSIS

Pteleopsis Engl. Combretaceae. 8 trop. Afr.

Ptelidium Thou. Celastraceae. 2 Madag.

Pteracanthus (Nees) Bremek. Acanthaceae. 20 Himalaya, Assam.

Pterachaenia (Benth.) Lipschitz. Compositae. 2 Afghan., Baluch.

Pterachne Schrad. ex Nees = Ascolepis Nees (Cyperac.).

Pteralyxia K. Schum. Apocynaceae. 3 Hawaii.

Pterandra A. Juss. Malpighiaceae. 5 trop. Am. ͼ

Pteranthera Blume = Vatica L. (Dipterocarpac.).

Pteranthus Forsk. Caryophyllaceae. 2 Morocco to Persia, Malta, Cyprus.

Pteraton Rafin. = Bupleurum L. (Umbellif.).

Pteretis Rafin. = Matteuccia Todaro (Aspidiac.).

Pterichis Lindl. Orchidaceae. 13 trop. S. Am.

Pteridaceae (S. F. Gray) Gaudich. Pteridales. Here confined to *Pteris* L. & immediate allies. Chief genera: *Pteris, Acrostichum*.

Pteridales. Filicidae. 8 fams.; the more highly evolved ferns which are thought to be nearest to *Schizaeaceae*. Fams.: *Pteridaceae, Negripteridaceae, Sinopteridaceae, Cryptogrammataceae, Parkeriaceae, Hemionitidaceae* (*Gymnogrammaceae*), *Adiantaceae, Vittariaceae, Actiniopteridaceae*.

Pteridanetium Copel. = Anetium (Kunze) Splitg. (Vittariac.).

Pteridella Mett. = Pellaea Link (Sinopteridac.).

Pteridium Rafin. = Pteris L. (Pteridac.).

***Pteridium** Scop. Dennstaedtiaceae. 1 cosmop., *P. aquilinum* (L.) Kuhn, the bracken: subsp. *aquilinum*, N. hemisph. & Africa; subsp. *caudatum* (L.) Bonap., S. hemisph.; overlap in W. Indies & Malaysia (R. M. Tryon, *Rhodora*, **43**: 1–67, 1941).

Pteridoblechnum Hennipman. Blechnaceae. 1 N. Queensland.

Pteridocalyx Wernham. Rubiaceae. 2 Guiana.

Pteridophyllaceae (Murb.) Sugiura ex Nak. Dicots. 1/1 Japan. Perenn. rhizomatous stemless herbs. Ls. all radic., oblong-oblanc., but deeply and reg. pinnatif., somewhat fern-like (e.g. *Blechnum*), exstip., surr. at the base by several large orbic. membr. cataphylls. Scape erect, simple or v. slightly branched, bearing ± distant 2-fld. fascicles of fls. on slender pedicels. K 2; C 4, ent., subequal, white; A 4, in pairs opp. to the 2 outer C-members, with v. short fil. and obl. anth.; G̲ (2), 1-loc., flattened, with 2–4 basal anatr. ov. and elong. style with 2 recurved commissural stigs. Fr. siliculiform, dehiscing by 2 valves. Only genus: *Pteridophyllum*. An interesting endemic, sometimes associated taxonomically with *Hypecoum*, but only remotely related. Cf. the foliage of *Acrotrema thwaitesii* Hook. f. & Thoms. (*Dilleniac.*).

Pteridophyllum Sieb. & Zucc. Pteridophyllaceae. 1 Japan.

Pteridophyllum Thw. = Filicium Thw. (Sapindac.).

Pteridophyta. Here divided into 4 Classes: *Lycopsida, Sphenopsida, Psilotopsida, Filicopsida*.

Pteridrys C. Chr. & Ching. Aspidiaceae. 8 trop. As. & Malaysia.

Pterigeron (DC.) Benth. = Oliganthemum F. Muell. (Compos.).

Pterigium Corrêa = Dipterocarpus Gaertn. (Dipterocarpac.).

Pteriglyphis Fée = Diplazium Sw. (Athyriac.).

Pterigostachyum Nees ex Steud. = Pterygostachyum Nees ex Steud. = Dimeria R.Br. (Gramin.).

Pterilema Reinw. = Engelhardtia Leschen. (Juglandac.).
Pterilis Raf. (1819) = Matteuccia Todaro (Aspidiac.).
Pterilis Raf. (1830) = Pteris L. (Pteridac.).
Pterinodes Kuntze = Matteuccia Todaro (Aspidiac.).
Pteriphis Rafin. = Aristolochia L. (Aristolochiac.).
Pteris L. Pteridaceae. 250 cosmop. (For obs. on cytol. & hybr., see Walker, *Evolution*, **12**: 81–92, 1958; **16**: 27–43, 1962.)
Pteris Scop. = Dryopteris Adans. (Aspidiac.).
Pterisanth[ac]eae J. G. Agardh = Vitidaceae Juss.
Pterisanthes Blume. Vitidaceae. 20 Burma, W. Malaysia.
Pterium Desv. = Lamarkia Moench (Gramin.).
Pternandra Jack. Melastomataceae. 1 (variable) SE. As., Malaysia. Berry. Calyptra o.
Pternix Hill = Carduus L. (Compos.).
Pternix Rafin. = Silybum Adans. (Compos.).
Pternopetalum Franch. Umbelliferae. 27 Tibet, E. As.
Pterobesleria Morton. Gesneriaceae. 2 Venez.
Pterocactus K. Schum. Cactaceae. 6–7 Argentina.
Pterocalymma Turcz. = Lagerstroemia L. (Lythrac.).
Pterocalymna Benth. & Hook. f. (sphalm.) = praec.
Pterocalyx Schrenk = Alexandra Bunge (Chenopodiac.).
Pterocarpos St-Lag. = Pterocarpus Jacq. (Legumin.).
Pterocarpus Bergius = Ecastaphyllum P.Br. (Legumin.).
Pterocarpus Burm. = Brya P.Br. (Legumin.).
Pterocarpus Jacq. Leguminosae. 100 trop. Fruit winged. Several, esp. *P. marsupium* Roxb., furnish *kino*, an astringent resin. *P. santalinus* L. f. yields red sandalwood.
Pterocarpus L. = Derris Lour. (Legumin.).
Pterocarya Kunth. Juglandaceae. 10 Caucasus to Japan.
Pterocarya Nutt. ex Moq. = Atriplex L. (Chenopodiac.).
Pterocaryaceae Nak. = Juglandaceae Kunth.
Pterocassia Britton & Rose = Cassia L. (Legumin.).
Pterocaulon Ell. Compositae. 25 Madag., warm As., Austr., Am.
Pterocelastrus Meissn. Celastraceae. 6 S. Afr.
Pteroceltis Maxim. Ulmaceae. 1 N. China, Mongolia.
Pterocephalus Adans. Dipsacaceae. 25 Medit. to C. As., Himal. & W. China; trop. Afr.
Pteroceras van Hasselt ex Hassk. Orchidaceae. 30 NE. India, SE. As., Malaysia.
Pterocereus MacDougall & Miranda. Cactaceae. 1 Mex.
Pterochaeta Steetz = Waitzia Wendl. (Compos.).
Pterochaete Arn. ex Boeck. = Rhynchospora Vahl (Cyperac.).
Pterochaete Boiss. = Pulicaria Gaertn. (Compos.).
Pterochilus Hook. & Arn. = Malaxis Soland. ex Sw. (Orchidac.).
Pterochiton Torr. = Atriplex L. (Chenopodiac.).
Pterochlaena Chiov. = Alloteropsis Presl (Gramin.).
Pterochlamys Fisch. ex Endl. = Panderia Fisch. & Mey. (Chenopodiac.).
Pterochlamys Roberty = Hildebrandtia Vatke (Convolvulac.).

PTEROCHLORIS

Pterochloris A. Camus = Chloris Sw. (Gramin.).
Pterochrosia Baill. = Cerberiopsis Vieill. (Apocynac.).
Pterocissus Urb. & Ekm. Vitidaceae. 1 W.I. (Haiti).
Pterocladis Lamb. ex G. Don = Baccharis L. (Compos.).
Pterocladon Hook. f. Melastomataceae. 1 Peru.
Pterocladum Triana = praec.
*****Pterococcus** Hassk. (~ Plukenetia L.). Euphorbiaceae. 1 E. Himal., SE. As.,
W. Malaysia, Moluccas.
Pterococcus Pall. = Calligonum L. (Polygonac.).
Pterocoelion Turcz. = Berrya Roxb. (Tiliac.).
× **Pterocottia** hort. Orchidaceae. Gen. hybr. (vi; nomen nugax) (Prescottia ×
Pterostylis).
Pterocyclus Klotzsch. Umbelliferae. 3 Himal., W. China.
Pterocymbium R.Br. Sterculiaceae. 15 SE. As., Malaysia, Fiji.
Pterocyperus (Peterm.) Opiz = Cyperus L. (Cyperac.).
Pterodiscus Hook. Pedaliaceae. 15 trop. & S. Afr.
Pterodon Vog. Leguminosae. 4 Brazil, Bolivia.
Pterogastra Naud. Melastomataceae. 4 N. trop. S. Am.
Pteroglossa Schlechter. Orchidaceae. 2 Brazil, Argent.
Pteroglossaspis Reichb. f. Orchidaceae. 5 Florida, Cuba, Argent., trop. Afr.
Pteroglossis Miers = Reyesia Clos = Salpiglossis Ruiz & Pav. (Solanac.).
Pterogyne Schrad. ex Nees = Ascolepis Nees (Cyperac.).
Pterogyne Tul. Leguminosae. 1 Brazil.
Pterolepis (DC.) Miq. Melastomataceae. 35 trop. Am.
Pterolepis Endl. = Osbeckia L. (Melastomatac.).
Pterolepis Schrad. = Scirpus L. (Cyperac.).
Pterolobium Andrz. ex DC. = Pachyphragma (DC.) Reichb. (Crucif.).
*****Pterolobium** R.Br. ex Wight & Arn. Leguminosae. 1 trop. Afr.; 10 trop.
& subtrop. As., Austr.
Pteroloma Desv. ex Benth. (~ Desmodium Desv.). Leguminosae. 5 SE. As.
Pteroloma Hochst. & Steud. = Dipterygium Decne (Crucif.).
Pterolophus Cass. = Centaurea L. (Compos.).
Pteromarathrum Koch ex DC. = Prangos Lindl. (Umbellif.).
Pteromimosa Britton = Mimosa L. (Legumin.).
Pteromischus Pichon = Crescentia L. (Bignoniac.).
Pteronema Pierre = Spondias L. (Anacardiac.).
Pteroneuron Fée = Humata Cav. (Davalliac.).
Pteroneuron Meissn. = seq.
Pteroneurum DC. = Cardamine L. (Crucif.).
*****Pteronia** L. Compositae. 75 trop. & S. Afr., Madag., W. Austr.
Pteropappus Pritz. = Pterygopappus Hook. f. (Compos.).
Pteropavonia Mattei = Pavonia Cav. (Malvac.).
Pteropepon Cogn. Cucurbitaceae. 3 Brazil, Argentina.
Pteropetalum Pax = Euadenia Oliv. (Capparidac.).
Pterophacos Rydb. = Astragalus L. (Legumin.).
Pterophora Harv. = Dregea E. Mey. (Asclepiadac.).
Pterophora L. = Pteronia L. (Compos.).
Pterophorus Boehm. = praec.

Pterophylla D. Don = Weinmannia L. (Cunoniac.).
Pterophyllus 'Senilis' [Nelson] = Ginkgo L. (Ginkgoac.).
Pterophyton Cass. = Actinomeris Nutt. (Compos.).
Pteropodium DC. = Jacaranda Juss. (Bignoniac.).
Pteropodium Willd. ex Steud. = Deyeuxia Beauv. (Gramin.).
Pteropogon DC. = Helipterum DC. (Compos.).
Pteropogon Fenzl = Facelis Cass. (Compos.).
Pteropogon Neck. = Scabiosa L. (Dipsacac.).
Pteropsis Desv. = Drymoglossum Presl (Polypodiac.).
Pteroptychia Bremek. Acanthaceae. 5 Indoch., W. Malaysia.
Pteropyrum Jaub. & Spach. Polygonaceae. 5 SW. As.
Pterorhachis Harms. Meliaceae. 1 W. Equat. Afr.
Pteroscleria Nees. Cyperaceae. 2 N. trop. S. Am., Trinidad.
Pteroselinum Reichb. = Peucedanum L. (Umbellif.).
Pterosenecio Sch. Bip. ex Baker = Senecio L. (Compos.).
Pterosicyos T. S. Brandegee. Cucurbitaceae. 1 Mex.
Pterosiphon Turcz. = Cedrela L. (Meliac.).
Pterospartum (Spach) K. Koch = Chamaespartium Adans. (Legumin.).
Pterospermadendron Kuntze = Pterospermum Schreb. (Sterculiac.).
Pterospermopsis Arènes = Macarisia Thou. (Rhizophorac.).
*Pterospermum Schreb. Sterculiaceae. 40 E. Himal., SE. As., W. Malaysia.
Pterospora Nutt. Monotropaceae. 1 N. Am.
Pterosporopsis Kellogg = Sarcodes Torr. (Monotropac.).
Pterostegia Fisch. & Mey. Polygonaceae. 1 Pacif. U.S., NW. Mex.
Pterostelma Wight = Hoya R.Br. (Asclepiadac.).
Pterostemma Lehm. & Kraenzl. Orchidaceae. 1 Colombia.
Pterostemon Schau. Pterostemonaceae. 2 Mex.
*Pterostemonaceae (Engl.) Small. Dicots. 1/2 Mex. Much-branched shrubs with alt. simple dentate stip. ls., shining glutinous-resinous above, softly pubesc. below; stips. small, decid. Fls. in few-fld. corymbose cymes, reg., ⚥. K 5, valv.; C 5, imbr., conspic., pubesc., ± persist.; A 5 + 5, the outer fert., with broad fil. dentic. at apex and ovoid intr. anth., the inner stds. with narrow fil. and no anth.; disk o; G̲ (5), with simple shortly 5-lobed style, and 4-6 ascending axile ov. per loc. Fr. a septic. woody caps.; seeds few, with cartilag. testa, atten. at either end. Only genus: *Pterostemon*. Related to *Philadelphaceae*.
Pterostephanus Kellogg = Anisocoma Torr. (Compos.).
Pterostigma Benth. = Adenosma R.Br. (Scrophulariac.).
*Pterostylis R.Br. Orchidaceae. 95 New Guinea, Austr., N.Z., New Caled.
Pterostyrax Sieb. & Zucc. Styracaceae. 7 Burma to Japan.
Pterota P.Br. = Fagara L. (Rutac.).
Pterotaberna Stapf. Apocynaceae. 1 trop. W. Afr.
Pterotheca Cass. = Crepis L. (Compos.).
Pterotheca Presl = Rhynchospora Vahl (Cyperac.).
Pterothrix DC. Compositae. 5 S. Afr.
Pterotrichis Reichb. = Pherotrichis Decne = Lachnostoma Kunth (Asclepiadac.).
Pterotropia Hillebrand = Dipanax Seem. (Araliac.).
Pterotropis Fourr. = Thlaspi L. (Crucif.).

PTEROTUM

Pterotum Lour. = ?Rourea Aubl. (Connarac.).
Pteroxygonum Dammer & Diels. Polygonaceae. 1 China.
Pteroxylon Hook. f. = Ptaeroxylon Eckl. & Zeyh. (Ptaeroxylac.).
Pterozonium Fée. Hemionitidaceae. 4 trop. S. Am.
Pterygiella Oliv. Scrophulariaceae. 3 China.
Pterygiosperma O. E. Schulz. Cruciferae. 1 Patag.
Pterygium Endl. = Pterigium Corrêa = Dipterocarpus Gaertn. (Dipterocarpac.).
Pterygocalyx Maxim. = Crawfurdia Wall. = Gentiana L. (Gentianac.).
Pterygocarpus Hochst. = Dregea E. Mey. (Asclepiadac.).
Pterygodium Sw. Orchidaceae. 18 E. trop. & S. Afr.
Pterygolepis Reichb. = Pterolepis Schrad. = Scirpus L. (Cyperac.).
Pterygoloma Hanst. = Alloplectus Mart. (Gesneriac.).
Pterygopappus Hook. f. Compositae. 1 Tasmania.
Pterygopleurum Kitagawa. Umbelliferae. 1 Korea, Japan.
Pterygopodium Harms. Leguminosae. 2 trop. Afr.
Pterygostachyum Nees ex Steud. = Dimeria R.Br. (Gramin.).
Pterygota Schott & Endl. Sterculiaceae. 20 trop., chiefly Old World.
Pterypodium Reichb. f. = Pterygodium Sw. (Orchidac.).
Pteryxia Nutt. Umbelliferae. 5 W. N. Am.
Ptichochilus Benth. = Ptychochilus Schau. = Tropidia Lindl. (Orchidac.).
Ptilagrostis Griseb. (~ Stipa L.). Gramineae. 7 C. & NE. As.
Ptilanthelium Steud. Cyperaceae. 2 Austr.
Ptilanthus Gleason. Melastomataceae. 1 Colombia.
Ptilepida Rafin. = Actinella Nutt. (Compos.).
Ptileris Rafin. = Erechtites Rafin. (Compos.).
Ptilimnium Rafin. Umbelliferae. 10 N. Am.
Ptilina Nutt. ex Torr. & Gray = Didiplis Rafin. (Lythrac.).
Ptilium Pers. (sphalm.) = Petilium L. = Fritillaria L. (Liliac.).
Ptilocalais A. Gray ex Greene = Microseris D. Don (Compos.).
Ptilocalyx Torr. & Gray = Coldenia L. (Ehretiac.).
Ptilochaeta Nees = Rhynchospora Vahl (Cyperac.).
*****Ptilochaeta** Turcz. Malpighiaceae. 5 subtrop. S. Am.
Ptilocnema D. Don = Pholidota Lindl. (Orchidac.).
Ptilomeris Nutt. = Baeria Fisch. & Mey. (Compos.).
Ptiloneilema Steud. = Melanocenchris Nees (Gramin.).
Ptilonella Nutt. = Blepharipappus Hook. (Compos.).
Ptilonema Hook. f. (sphalm.) = Ptiloneilema Steud. = Melanocenchris Nees (Gramin.).
Ptilonilema P. & K. = praec.
Ptilophora A. Gray = Microseris D. Don (Compos.).
Ptilophyllum v. d. Bosch = Trichomanes L. (Hymenophyllac.).
Ptilophyllum Rafin. = Myriophyllum L. (Haloragidac.).
Ptilopteris Hance = Monachosorum Kunze (Monachosorac.).
Ptiloria Rafin. = Stephanomeria Nutt. (Compos.).
Ptilosciadium Steud. = Rhynchospora Vahl (Cyperac.).
Ptilosia Tausch = Picris L. (Compos.).
Ptilostemon Cass. (~ Cirsium Mill.). Compositae. 15 Medit. to C. As.

Ptilostemum Steud. = praec.
Ptilostephium Kunth = Tridax L. (Compos.).
Ptilotrichum C. A. Mey. (~Alyssum L.). Cruciferae. 15 Medit. to C. As.
Ptilotum Dulac = Dryas L. (Rosac.).
Ptilotus R.Br. Amaranthaceae. 100 Austr., Tasm.
Ptilurus D. Don = Leuceria Lag. (Compos.).
Ptosimopappus Boiss. = Centaurea L. (Compos.).
Ptyas Salisb. = Aloë (Liliac.).
Ptycanthera Decne. Asclepiadaceae. 1 S. Domingo.
Ptychandra Scheff. Palmae. 7 Moluccas, New Guinea.
Ptychanthera P. & K. = Ptycanthera Decne (Asclepiadac.).
Ptychocarpa (R.Br.) Spach = Grevillea R.Br. (Proteac.).
Ptychocarpus Hilsenberg ex Sieber = Melochia L. (Sterculiac.).
Ptychocarpus Kuhlm. = Neoptychocarpus Buchheim (Flacourtiac.).
Ptychocarya R.Br. ex Wall. = Scirpodendron Zipp. ex Kurz (Cyperac.).
Ptychocaryum Kuntze ex H. Pfeiff. = praec.
Ptychocentrum Benth. = Rhynchosia Lour. (Legumin.).
Ptychochilus Schau. = Tropidia Lindl. (Orchidac.).
Ptychococcus Becc. Palmae. 8 E. Malaysia.
Ptychodea Willd. ex Cham. & Schlechtd. = Sipanea Aubl. (Rubiac.).
Ptychodon Klotzsch ex Reichb. = Lafoënsia Vand. (Lythrac.).
Ptychogyne Pfitz. Orchidaceae. 3 Malay Penins., Java, Celebes.
Ptycholepis Griseb. ex Lechler = Blepharodon Decne (Asclepiadac.).
Ptycholobium Harms. Leguminosae. 2 trop. Afr.
Ptychomeria Benth. (~Gymnosiphon Blume). Burmanniaceae. 20 trop. Am.
Ptychopetalum Benth. Olacaceae. 2 trop. S. Am., 5 trop. Afr.
Ptychophyllum Presl = Meringium Presl (Hymenophyllac.).
Ptychopyxis Miq. Euphorbiaceae. 13 Siam, Indoch., W. Malaysia, E. New Guinea.
Ptychoraphis Becc. = Rhopaloblaste Scheff. (Palm.).
Ptychosema Benth. Leguminosae. 2 W. & C. Austr.
Ptychosperma Labill. Palmae. 38 Kei Is., New Guinea, Bismarck & Solomon Is., N. Austr. Fls. in threes, 2 ♂♂ and 1 ♀.
Ptychostigma Hochst. = Galiniera Delile (Rubiac.).
Ptychostoma P. & K. = Ptyxostoma Vahl = Lonchostoma Wikstr. (Bruniac.).
Ptychostylus Van Tiegh. = Loranthus L. (Loranthac.).
Ptychotis Koch. Umbelliferae. 1 C. & S. Eur.
Ptyssiglottis T. Anders. Acanthaceae. 30 Ceylon, Indoch., W. Malaysia.
Ptyxostoma Vahl = Lonchostoma Wikstr. (Bruniac.).
Pubeta L. = Duroia L. f. (Rubiac.).
Pubilaria Rafin. = Simethis Kunth (Liliac.).
Pubistylus Thothathri. Rubiaceae. 1 Andaman Is.
Publicaria Deflers (sphalm.) = Pulicaria Gaertn. (Compos.).
*****Puccinellia** Parl. Gramineae. 100 N. temp., S. Afr.
Puccionia Chiov. Cruciferae (~Capparidac.). 1 Somalia.
× **Pucciphippsia** Tsvelev. Gramineae. Gen. hybr. (Phippsia × Puccinellia).
Pucedanum Hill (sphalm.) = Peucedanum L. (Umbellif.).
Puelia Franch. Gramineae. 5 trop. Afr.

PUERARIA

Pueraria DC. Leguminosae. 35 Himal. to Japan, SE. As., Malaysia, Pacif. Fibre. Medicinal roots.

Pugetia Gandog. = Rosa L. (Rosac.).

Pugionella Salisb. = Strumaria Jacq. (Amaryllidac.).

Pugionium Gaertn. Cruciferae. 5 Mongolia.

Pugiopappus A. Gray ex Torr. = Coreopsis L. (Compos.).

Puja Molina = Puya Molina (Bromeliac.).

Pukanthus Rafin. = Grabowskia Schlechtd. (Solanac.).

Pukateria Raoul = Griselinia J. R. & G. Forst. (Cornac.).

Pulassarium Kuntze = Alyxia Banks (Apocynac.).

Pulcheria Comm. ex Moewes = Polycardia Juss. (Celastrac.).

Pulcheria Nor. = Kadsura Juss. (Schisandrac.).

Pulchia Steud. = Pluchia Vell. = Diclidanthera Mart. (Polygalac.).

Pulegium Mill. = Mentha L. (Labiat.).

Pulicaria Gaertn. Compositae. 50–60 temp. & warm Euras., trop. & S. Afr.

Puliculum Stapf ex Haines = Pseudopogonatherum A. Camus (Gramin.).

Pullea Schlechter. Cunoniaceae. 8 New Guinea, Fiji.

Pullipes Rafin. = Caucalis L. (Umbellif.).

Pullipuntu Ruiz = Phytelephas Ruiz & Pav. + Yarina O. F. Cook (Palm.).

Pulmonaria L. Boraginaceae. 10 Eur. *P. officinalis* L. (lungwort, formerly officinal) and *P. angustifolia* L. have dimorphic heterostyled fls. which change from red to blue as they grow older (see fam.).

Pulpaceae Dulac = Grossulariaceae Lam. & DC.

Pulsatilla Mill. (~ Anemone L.). Ranunculaceae. 30 temp. Euras.

Pultenaea Sm. Leguminosae. 90 Austr.

Pultenea A. St-Hil. = praec.

Pulteneja Hoffmgg. = praec.

Pulteneya P. & K. = praec.

Pultnaea R. Grah. (sphalm.) = praec.

Pultoria Rafin. = Paltoria Ruiz & Pav. = Ilex L. (Aquifoliac.).

Pulvinaria Fourn. Asclepiadaceae. 1 Brazil.

Pumilea P.Br. = Turnera L. (Turnerac.).

Pumilo Schlechtd. = Rutidosis DC. (Compos.).

Puncticularia N. E. Br. ex Lemée = seq.

Punctilaria Lemée (sphalm.) = seq.

Punctillaria N. E. Brown = Pleiospilos N. E. Br. (Aïzoac.).

Punduana Steetz = Vernonia Schreb. (Compos.).

Puneeria Stocks = Withania Pauquier (Solanac.).

Pungamia Lam. = Pongamia Adans. mut. Vent. (Legumin.).

Punica L. Punicaceae. 2, one endem. in Socotra, the other, *P. granatum* L., the pomegranate, Balkans to Himalaya, and cult.

*Punicaceae Horan. Dicots. 2 SE. Eur. to Himal., and Socotra. Shrubs, sometimes spiny, with terete branches; ls. opp. or subopp., simple, ent., exstip. The young twigs have four wings, composed simply of epidermis and cortical parenchyma; these are early thrown off. Fls. sol. or fascic., axill., ☿, reg., perig. K 5–8, valvate; C 5–8, imbr.; A ∞; G (∞), adnate to receptacle. The mature ovary has a peculiar structure, due to a development like that in *Mesembryanthemum*. Two whorls of cpls. with basal plac. are laid down, and then a peripheral growth tilts them up from ‖ · ‖ to = · = so that two layers of loculi are formed and the placentation appears to be parietal. Ovules ∞, anatr. The arrangement is also seen in the fr., commonly termed a berry, but not strictly so. The pericarp (axial in part) is leathery, and the fleshy inner part round the seeds is really the outer layers of the seed coats. No endosp. Only genus: *Punica*. Near *Lythrac.* and *Sonneratiac.*

Punicella Turcz. = Balaustion Hook. (Myrtac.).

Punjuba Britton & Rose = Pithecellobium Mart. (Legumin.).

Pupal Adans. = seq.

***Pupalia** Adans. mut. Juss. Amaranthaceae. 10 trop. Afr., Madag., India.

Pupartia P. & K. = Poupartia Comm. ex Juss. (Anacardiac.).

Pupilla Rizzini. Acanthaceae. 2 trop. S. Am.

Puralia Ham. ex Reichb. (nomen). Quid?

Puraria Wall. = Pueraria DC. (Legumin.).

Purchia Dum. = Purshia Spreng. = Onosmodium Michx (Boraginac.).

Purdiaea Planch. Cyrillaceae. 10 Cuba, 1 C. Am., 1 NW. S. Am.

Purdieanthus Gilg. Gentianaceae. 1 Colombia.

Purga Schiede ex Zucc. = Exogonium Choisy (Convolvulac.).

Purgosea Haw. = Crassula L. (Crassulac.).

Purgosia G. Don = praec.

Purkayasthaea Purkay. = Beilschmiedia Nees (Laurac.).

Purkinjia Presl = Ardisia Sw. (Myrsinac.).

Purpurabenis Thou. (uninom.) = *Habenaria purpurea* Thou. = *Cynorkis purpurea* (Thou.) Kraenzl. (Orchidac.).

Purpurella Naud. = Tibouchina Aubl. (Melastomatac.).

Purpureostemon Gugerli. Myrtaceae. 1 New Caled.

Purpurocynis Thou. (uninom.) = *Cynosorchis purpurascens* Thou. = *Cynorkis purpurascens* Thou. (Orchidac.).

Purpusia T. S. Brandegee. Rosaceae. 1 SW. U.S.

Purshia DC. Rosaceae. 2 Pacif. N. Am.

Purshia Dennst. = Centranthera R.Br. (Scrophulariac.).

Purshia Rafin. = Burshia Rafin. = Myriophyllum L. (Haloragidac.).

Purshia Spreng. = Onosmodium Michx (Boraginac.).

Puruma Jaume St-Hil. = Pourouma Aubl. (Morac.).

Pusaetha Kuntze = Entada Adans. (Legumin.).

Puschkinia Adams. Liliaceae. 2 W. As.

× **Puschkiscilla** Cif. & Giac. Liliaceae. Gen. hybr. (Puschkinia × Scilla).

Pusiphylis Thou. (uninom.) = *Bulbophyllum pusillum* Thou. (Orchidac.).

Putoria Pers. Rubiaceae. 3 Medit.

Putranjiva Wall. = Drypetes Vahl (Euphorbiac.).

Putranjiv[ac]eae Endl. = Euphorbiaceae–Drypetinae Pax.

Putterlickia Endl. Celastraceae. 2 S. Afr.

Putzeysia Klotzsch = Begonia L. (Begoniac.).

Putzeysia Planch. & Linden = Aesculus L. (Hippocastanac.).

Puya Molina. Bromeliaceae. 120 Andes. Some 3 m. high, with thick stem.

Pycanthus P. & K. = Pukanthus Rafin. = Grabowskia Schlechtd. (Solanac.).

Pychnanthemum G. Don (sphalm.) = Pycnanthemum Michx (Labiat.).

Pychnostachys G. Don (sphalm.) = Pycnostachys Hook. (Labiat.).

PYCNANDRA

Pycnandra Benth. Sapotaceae. 5 New Caled.

***Pycnanthemum** Michx. Labiatae. 17 N. Am.

Pycnanthes Rafin. = praec.

Pycnanthus Warb. Myristicaceae. 8 trop. Afr.

Pycnarrhena Miers ex Hook. f. & Thoms. Menispermaceae. 25 SE. As., Indomal., Austr.

Pycnobolus Willd. ex O. E. Schulz = Eudema Humb. & Bonpl. (Crucif.).

Pycnobotrya Benth. Apocynaceae. 2 trop. Afr.

Pycnobregma Baill. Asclepiadaceae. 1 Colombia.

Pycnocephalum DC. = Eremanthus Less. (Compos.).

Pycnocoma Benth. Euphorbiaceae. 14 trop. Afr., Madag., Masc.

Pycnocomon Hoffmgg. & Link (~ Scabiosa L.). Dipsacaceae. 2 Medit.

Pycnocomon St-Lag. = Picnomon Adans. = Cirsium Mill. (Compos.).

Pycnocomon Wallr. = Cephalaria Schrad. (Dipsacac.).

Pycnocomum Link = Pycnocomon Hoffmgg. & Link (Dipsacac.).

Pycnocomus Hill = Calcitrapoïdes Fabr. = Centaurea L. (Compos.).

Pycnocycla Lindl. Umbelliferae. 10 trop. W. Afr. to NW. India.

Pycnodoria Presl = Pteris L. (Pteridac.).

Pycnolachne Turcz. = Lachnostachys Hook. (Verbenac.).

Pycnoloma C. Chr. Polypodiaceae. 3 Malay Penins., Borneo.

Pycnoneurum Decne. Asclepiadaceae. 2 Madag.

Pycnophyllopsis Skottsb. Caryophyllaceae. 2 Andes (Boliv., Patag.).

Pycnophyllum Remy. Caryophyllaceae. 17 Andes.

Pycnoplinthus O. E. Schulz. Cruciferae. 1 Himal.

Pycnopteris Moore = Dryopteris Adans. (Aspidiac.).

Pycnorhachis Benth. Asclepiadaceae. 1 Malay Penins.

Pycnosandra Blume = Drypetes Vahl (Euphorbiac.).

Pycnosorus Benth. = Craspedia J. R. & G. Forst. (Compos.).

Pycnospatha Thorel ex Gagnep. Araceae. 2 Siam, Indoch.

Pycnosphace Rydb. = Salvia L. (Labiat.).

Pycnosphaera Gilg. Gentianaceae. 5 trop. Afr.

Pycnospora R.Br. ex Wight & Arn. Leguminosae. 1 trop. Afr., India, SE. As., S. China, Formosa, Philipp., Java, E. Malaysia, NE. Austr.

Pycnostachys Hook. Labiatae. 37 trop. & S. Afr., Madag.

Pycnostelma Decne. Asclepiadaceae. 2–3 China.

Pycnostylis Pierre. Menispermaceae. 2 trop. E. Afr., Madag.

Pycnothryx M. E. Jones = Drudeophytum Coulter & Rose (Umbellif.).

Pycnothymus Small = Satureia L. (Labiat.).

Pycreus Beauv. Cyperaceae. 100 trop. & warm temp.

Pygeum Gaertn. = Lauro-Cerasus Duham. (Rosac.).

Pygmaea Hook. f.(~ Veronica L.). Scrophulariaceae. 1 Tasm., 6 N.Z.

Pygmaeocereus Johns. & Backeb. (~ Haageocereus Backeb.). Cactaceae. 1 Peru.

Pygmaeopremna Merr. Verbenaceae. 6 India, SE. As., Philipp. Is., E. Malaysia, Austr. Probably pyrogenous forms of *Premna*.

Pygmaeorchis Brade. Orchidaceae. 1 Brazil.

Pygmaeothamnus Robyns. Rubiaceae. 4 trop. & S. Afr.

Pygmea J. Buchanan = Pygmaea Hook. f. (Scrophulariac.).
Pylostachya Rafin. = Polygala L. (Polygalac.).
Pynaertia De Willd. = Anopyxis (Pierre) Engl. (Rhizophorac.).
Pynaertiodendron De Wild. Leguminosae. 2 trop. Afr.
Pynanthemum Rafin. = Pycnanthemum Michx (Labiat.).
Pyracantha M. Roem. Rosaceae. 10 SE. Eur. to C. China & Indoch.
Pyraceae Burnett = Pomaceae S. F. Gray = Rosaceae–Pomoïdeae Juss.
× **Pyracomeles** Rehder. Rosaceae. Gen. hybr. (Osteomeles × Pyracantha).
Pyragma Nor. = Stelechocarpus Hook. f. & Thoms. (Annonac.).
Pyragra Bremek. Rubiaceae. 2 Madag.
Pyramia Cham. Melastomataceae. 3 S. Brazil.
Pyramidanthe Miq. Annonaceae. 1 Siam, W. Malaysia.
Pyramidium Boiss. (1853; non Bridel 1826—Musci) = Veselskya Opiz (Crucif.).
Pyramidocarpus Oliv. = Dasylepis Oliv. (Flacourtiac.).
Pyramidoptera Boiss. Umbelliferaᵔ. 1 Afghanistan.
Pyramidostylium Mart. = Salacia L. (Celastrac.).
Pyrarda Cass. = Grangea Adans. (Compos.).
× **Pyraria** A. Chev. = × Sorbopyrus C. K. Schneid. (Rosac.).
Pyrecnia Nor. ex Hassk. = seq.
Pyreenia Nor. = Laportea Gaudich. (Urticac.).
*****Pyrenacantha** Hook. ex Wight. Icacinaceae. 20 trop. & S. Afr., Madag.; 1 SE. As., 1 Philipp. K o!
Pyrenaceae Vent. = Verbenaceae Juss.
Pyrenaria Blume. Theaceae. 20 SE. As., W. Malaysia.
Pyrenia Clairv. = Pyrus L. (*sensu latiss.*) (Rosac.)
Pyrenoglyphis Karst. (~ Bactris Jacq.). Palmae. 30 trop. Am.
Pyrethraria Pers. ex Steud. = Cotula L. (Compos.).
Pyrethrum Medik. = Spilanthes Jacq. (Compos.).
Pyrethrum Zinn (~ Chrysanthemum L.). Compositae. 100 temp. Euras.
Pyretrum Burm. Inc sed. (?Compos.). 1 W.I.
Pyrgosea Eckl. & Zeyh. = Purgosea Haw. = Crassula L. (Crassulac.).
Pyrgus Lour. = Ardisia Sw. (Myrsinac.).
Pyriluma (Baill.) Aubrév. Sapotaceé. 2 New Caled.
+ **Pyro-cydonia** Guillaumin = + Pirocydonia H. Winkler ex Daniel = + Pyronia Hort. (Rosac.).
Pyrogennema Lunell = Chamaenerion Boehm. emend. S. F. Gray (Onagrac.).
Pyrola L. Pyrolaceae. 20 N. temp. Evergreens with creeping stocks. *P.* (*Moneses*) *uniflora* L. has adv. buds on the roots, and a solitary term. fl. The fls. of *P. minor* L. are in racemes, pend., without disks. There is no honey; the stigma projects beyond the anthers, but pollen may at last fall upon it from them. *P. rotundifolia* L. is similar. The seeds of *P.* are very light and are distr. by wind.
*****Pyrolaceae** Dum. Dicots. 3/30, cold N. temp. & arctic. Evergreen perenn. slightly woody herbs with sympodial growth from rhiz.; ls. alt. or vertic., ent. or dent., exstip. The infl. is term.; it may be a true raceme (*Pyrola*), or a cyme, or sol., leafless or with scaly bracts. Fl. ☿, reg.; K 5–4, imbr.; C 5–4, rarely shortly united; A 10–8, obdiplost.; G (5–4). The petals and sta. are often at

the edge of a nectariferous disk. Anthers intr., opening by apical pores; pollen in tetrads. Cpls. opp. petals; ovary imperfectly 5-4-loc. Style simple, sometimes declinate; ovules minute, ∞, anatr., on thick fleshy plac. Caps. loculic. Seeds ∞, small, in loose testa. Embryo of few cells, without differentiation of cotyledons. Genera: *Chimaphila*, *Pyrola*, *Moneses*. Somewhat intermediate between *Ericac.* and *Monotropac.*

Pyrolirion Herb. Amaryllidaceae. 6 Andes.

+Pyronia Hort. (?Veitch). Rosaceae. Gen. hybr. asex. (Cydonia+Pyrus). (See *Journ. Hered.* 7: 416, 1916.)

Pyrophorum DC.=Pirophorum Neck.=Pyrus L. (*sens. lat.*) (Rosac.).

Pyropsis Hort. ex Fisch., Mey. & Avé-Lallem.=Madia Molina (Compos.).

Pyrospermum Miq.=Kurrimia Wall. ex Thw.=Bhesa Buch.-Ham. (Celastrac.).

Pyrostegia C. Presl. Bignoniaceae. 5 trop. S. Am.

Pyrostoma G. F. W. Mey.=Vitex L. (Verbenac.).

Pyrostria Comm. ex Juss. Rubiaceae. 10 Madag., Mauritius, Rodrigues.

Pyrostria Roxb.=Timonius DC. (Rubiac.).

Pyrotheca Steud. (sphalm.)=Gyrotheca Salisb.=Lachnanthes Ell. (Haemodorac.).

Pyrrhanthera Zotov. Gramineae. 1 N.Z.

Pyrrhanthus Jack=Lumnitzera Willd. (Combretac.).

Pyrrheima Hassk.=Siderasis Rafin. (Commelinac.).

Pyrrhocactus (A. Berger) Backeb. & F. M. Knuth (~Neoporteria Britt. & Rose). Cactaceae. 6 Argent., Chile(?).

Pyrrhocoma Walp.=Pyrrocoma Hook.=Haplopappus Cass. (Compos.).

*****Pyrrhopappus** DC. Compositae. 8 N. Am.

Pyrrhopappus A. Rich.=Lactuca L. (Compos.).

Pyrrhosa Endl.=Horsfieldia Willd. (Myristicac.).

Pyrrhotrichia Wight & Arn.=Eriosema DC. (Legumin.).

Pyrrocoma Hook.=Haplopappus Cass. (Compos.).

Pyrrorhiza Maguire & Wurdack. Haemodoraceae. 1 Venez.

Pyrrosia Mirbel. Polypodiaceae. 100 Old World trop., temp. NE. As. Ls. simple, fleshy, covered with stellate hairs. (*Cyclophorus* Desv. of C. Chr., *Ind. Fil.*)

Pyrrothrix Bremek. Acanthaceae. 10 Assam to Sumatra.

Pyrsonota Ridley=Sericolea Schlechter (Elaeocarpac.).

Pyrularia Michx. Santalaceae. 4 Himalaya, China, SE. U.S.

Pyrus L. Rosaceae. 30 temp. Euras. *P. communis* L. (pear), cult. for fruit, which differs from that of *Malus* (apple) in possession of stone-cells in the flesh. Several vars. are self-sterile.

Pythagorea Lour.=Homalium Jacq. (Flacourtiac.).

Pythagorea Rafin.=Lythrum L. (Lythrac.).

Pythion Mart.=Amorphophallus Blume (Arac.).

Pythius B. D. Jacks. (sphalm.) [Tithymalis subgen. Pythiusa Rafin.] = Euphorbia L. (Euphorbiac.).

Pythonium Schott=Thomsonia Wall. (Arac.).

Pyxa Nor.=Costus L. (Costac.).

Pyxidanthera Michx. Diapensiaceae. 1-2 E. U.S.

Pyxidanthera Muehlenb. = Lepuropetalon Ell. (Lepuropetalac.).
Pyxidanthus Naud. = Blakea P.Br. (Melastomatac.).
Pyxidaria Gled. = Trichomanes L., *sens. lat.* (no binomial) (Hymenophyllac.).
Pyxidaria Kuntze = Lindernia All. (Scrophulariac.).
Pyxidaria Schott = ? Lecythis Loefl. (Lecythidac.).
Pyxidiaceae Dulac = Plantaginaceae Juss.
Pyxidium Moq. = Amaranthus L. (Amaranthac.).
Pyxipoma Fenzl = Sesuvium L. (Aïzoac.).

Q

Qaeria Rafin. = Queria Loefl. = Minuartia L. (Caryophyllac.).
Quadrania Nor. = ? Kopsia Bl. (Apocynac.).
Quadrasia Elmer = Claoxylon Juss. (Euphorbiac.).
Quadrella J. S. Presl = Capparis L. (Capparidac.).
Quadria Mutis = Vismia Vand. (Guttif.).
Quadria Ruiz & Pav. = Gevuina Molina (Proteac.).
Quadriala Sieb. & Zucc. = Buckleya Torr. (Santalac.).
Quadricasaea R. E. Woodson. Apocynaceae. 2 Colombia.
Quadricosta Dulac = Ludwigia L. (Onagrac.).
Quadrifaria Manetti ex Gord. = Araucaria Juss. (Araucariac.).
Quadripterygium Tardieu. Celastraceae. 1 Indoch.
Quaiacum Scop. (sphalm.) = Guaiacum L. (Zygophyllac.).
Qualea Aubl. Vochysiaceae. 60 trop. S. Am. G̲; plac. not thickened.
Quamasia Rafin. = Camassia Lindl. (Liliac.).
Quamassia B. D. Jacks. = praec.
Quamoclidion Choisy. Nyctaginaceae. 1 Mex.
Quamoclit Mill. = Ipomoea L. (Convolvulac.).
Quamoclita Rafin. = praec.
Quamoclitia Lowe = praec.
Quamoclitium P. & K. = Quamoclidion Choisy (Nyctaginac.).
Quamoctita Rafin. = Quamoclit Mill. = Ipomoea L. (Convolvulac.).
Quapira auct. ex Pfeiff. = Guapira Aubl. = Pisonia L. (Nyctaginac.).
Quapoia auctt. = seq.
Quapoja Batsch = seq.
Quapoya Aubl. Guttiferae. 3 Guiana, Peru.
Quaqua N. E. Br. = Caralluma R.Br. (Asclepiadac.).
Quararibea Aubl. Bombacaceae. 50 C. & trop. S. Am.
Quarena Rafin. = Cordia L. (Ehretiac.).
Quartinia Endl. = Rotala L. (Lythrac.).
Quartinia A. Rich. = Pterolobium R.Br. (Legumin.).
Quassia L. Simaroubaceae. 40 trop., esp. Am. *Q. amara* L. (Am.) is the source of quassia wood.
Quaternella Ehrh. (uninom.) = *Moenchia quaternella* Ehrh. = *M. erecta* (L.) Gaertn., Mey. & Scherb. (Caryophyllac.).
Quebitea Aubl. = Piper L. (Piperac.). [diac.).
Quebrachia Griseb. (1874) (nomen) = Loxopterygium Hook. f. (Anacar-

QUEBRACHIA

Quebrachia Griseb. (1879) = Schinopsis Engl. (Anacardiac.).

Queenslandiella Domin. Cyperaceae. 1 trop. E. Afr., Madag., Maurit., India, Ceylon, Borneo, Java, Timor, Queensland.

Quekettia Lindl. Orchidaceae. 6 Brazil, Guiana, Trinidad.

Quelchia N. E. Br. Compositae. 4 Venez., Guiana(?).

Queltia Salisb. = Narcissus L. (Amaryllidac.).

Quélusia Vand. = Fuchsia L. (Onagrac.).

Quercaceae ('-ineae') Juss. = Fagaceae–Quercoïdeae Oerst.

Quercifilix Copel. Aspidiaceae. 1 Ceylon, Malaysia. Hybridizes with *Tectaria decurrens* (Presl) Copel. in Ceylon.

Quercus L. (*s.str.*). Fagaceae. 450 N. Am. to W. trop. S. Am., temp. & subtrop. Euras., N. Afr. The oaks are evergreen or deciduous trees, in the latter case esp. with well-developed winter buds. The cupule contains 1 ♀ fl. only (see fam.), and forms the acorn-cup at the base of the nut in fr. The ♂ fls. are usu. sol. in pend. catkins. Anemoph. Many are important economic plants. Among the most noteworthy are: *Q. aegilops* L. (E. Eur., W. As.), whose cupules and unripe acorns, known as *valonia*, are used in tanning; *Q. alba* L. (N. Am.), the white or Quebec oak (timber); *Q. cerris* L. (Eur., W. As.), the Turkey oak (timber); *Q. ilex* L. (Medit.), the holm, holly or evergreen oak (timber, bark for tanning); *Q. robur* L. and *Q. petraea* (Mattuschka) Liebl. (Eur., W. As.), yielding timber and tan bark; *Q. suber* L. (Medit.), the cork oak, whose bark, stripped off in thick layers and flattened, forms ordinary cork; *Q. tinctoria* Bartr. (N. Am.), whose bark (quercitron bark) forms a yellow

Queria L. = Minuartia L. (Caryophyllac.).   [dye; and many others.

Quesnelia Gaudich. Bromeliaceae. 12 trop. S. Am.

Queteletia Blume = Orchipedum Breda (Orchidac.).

Quetia Gandog. = Chaerophyllum L. (Umbellif.).

Quetzalia Lundell. Celastraceae. 9 C. & trop. S. Am.

Quezelia Scholz. Cruciferae. 1 Sahara.

Quiabentia Britton & Rose. Cactaceae. 5 trop. S. Am.

Quiducia Gagnep. Rubiaceae. 1 Indoch.

Quiina Aubl. Quiinaceae. 35 trop. S. Am.

*****Quiinaceae** Engl. Dicots. 4/50 trop. S. Am. Trees or shrubs, sometimes scandent, with opp. or vertic. simple or pinnatif. ent. or cren. many-nerved ('satiny') stip. ls.; stips. interpet., 1–4 pairs. Fls. small, reg., ♀ or ♂ ♀, in rac. or pan. K 4–5, imbr.; C 4–5–8, imbr. or contorted; A 15–30–∞ (160–170), free or adnate to C below; G̲ 3 or (2–3) or (7–11), with as many filif. finally reflexed styles with obliquely pelt. stigs., and 2 axile erect or ascending ov. per loc. Fr. either baccate, finally dehiscing by valves, usu. 1-loc., 1(–4)-seeded, or of 3 free 1-seeded carpels; seed often with densely velutinous covering, endosp. o. Genera: *Quiina, Touroulia, Lacunaria, Froësia.* Somewhat intermediate between *Guttiferae* and *Ochnaceae.*

Quilamum Blanco = Crypteronia Blume (Crypteroniac.).

Quilesia Blanco = Dichapetalum Thou. (Dichapetalac.).

Quiliusa Hook. f. = Quelusia Vand. = Fuchsia L. (Onagrac.).

Quillaia Mol. = seq.

Quillaja Molina. Rosaceae. 3 temp. S. Am. *Q. saponaria* Molina is the soaptree of Chile; the powdered bark lathers with water.

Quillaj[ac]eae D. Don = Rosaceae–Quillajeae Meissn.
Quinaria Lour. = Clausena Burm. f. (Rutac.).
Quinaria Rafin. = Parthenocissus Planch. (Vitidac.).
Quinasis Rafin. = Polylepis Ruiz & Pav. (Rosac.).
Quinata Medik. = Machaerium Pers. (Legumin.).
Quinchamala Willd. = seq.
*Quinchamalium Juss. Santalaceae. 25 Andes.
Quinchamalium Mol. Quid? 1 Chile.
Quincula Rafin. = Physalis L. (Solanac.).
Quinetia Cass. Compositae. 1 W. Austr.
Quinio Schlechtd. = Cocculus DC. (Menispermac.).
Quinquedula Nor. = Litsea Lam. (Laurac.) + ?
Quinquefolium Seguier = Potentilla L (Rosac.).·
Quinquelobus Benj. = Benjaminia Mart. ex Benj. = Bacopa Aubl. (Scrophulariac.).
Quinquelocularia C. Koch = Campanula L. (Campanulac.).
Quinquina Boehm. = Cinchona L. (Rubiac.).
Quinsonia Montr. = Pittosporum Banks ex Soland. (Pittosporac.).
Quintilia Endl. = Stauranthera Benth. (Gesneriac.).
Quintinia A. DC. Saxifragaceae. 20 Philipp. Is., New Guinea, Austr., New Caled., N.Z.
Quirina Rafin. = Cuphea P.Br. (Lythrac.).
Quirivelia Poir. = Ichnocarpus R.Br. (Apocynac.).
Quirosia Blanco = Crotalaria L. (Legumin.).
Quisqualis L. Combretaceae. 17 trop. & S. Afr., Indomal. *Q. indica* L. is erect below, ± twining above, with alt. ls. Fl. shoots with opp. ls.
Quisumbingia Merrill. Asclepiadaceae. 1 Philipp. Is.
Quivisia Comm. ex Juss. = Turraea L. (Meliac.).
Quivisiantha Willis (sphalm.) = seq.
Quivisianthe Baill. = ? Trichilia P.Br. (Meliac.).
Quoya Gaudich. = Pityrodia R.Br. (Verbenac.).
Quoya Néraud (nomen). Quid? (Rhamnac.?). 1 Mauritius.
Qveria L. = Queria L. = Minuartia L. (Caryophyllac.).

R

Raaltema Mus. Lugd. ex C. B. Clarke = Boea Comm. ex Lam. (Gesneriac.).
Rabarbarum P. & K. = Rhabarbarum Adans. = Rheum L. (Polygonac.).
Rabdadenia P. & K. = Rhabdadenia Muell. Arg. (Apocynac.).
Rabdia P. & K. = Rhabdia Mart. = Rotula Lour. (Ehretiac.).
Rabdochloa Beauv. = Leptochloa Beauv. (Gramin.).
Rabdosia Hassk. = Plectranthus L'Hérit. (Labiat.).
Rabelaisia Planch. = Lunasia Blanco (Rutac.).
Rabenhorstia Reichb. = Berzelia Brongn. (Bruniac.).
Rabiea N. E. Br. Aïzoaceae. 9 S. Afr.
Racapa M. Roem. = Carapa Aubl. (Meliac.).
Racaria Aubl. = Talisia Aubl. (Sapindac.).

RACEMARIA

Racemaria Rafin. = Smilacina Desf. (Liliac.).
Racemobambos Holttum. Gramineae. 5 Malay Penins., Borneo.
Rachia Klotzsch = Begonia L. (Begoniac.).
Rachicallis DC. = Rhachicallis DC. corr. Spach (Rubiac.).
Racka J. F. Gmel. = Avicennia L. (Avicenniac.).
Raclathris Rafin. = Rochelia Reichb. (Boraginac.).
Racletia Adans. = ? Reaumuria L. (Tamaricac.).
Racoma Willd. ex Steud. = Rocama Forsk. = Trianthema L. (Aïzoac.).
Racoubea Aubl. = Homalium Jacq. (Flacourtiac.).
Racua J. F. Gmel. = Racka J. F. Gmel. = Avicennia L. (Avicenniac.).
Radackia Cham. & Endl. (nomen). Leguminosae. 1 Marshall Is. Quid?
Radamaea Benth. Scrophulariaceae. 5 Madag.
Raddia Bertol. Gramineae. 10 Mexico.
Raddia DC. ex Miers = Raddisia Leandro = Salacia L. (Celastrac.).
Raddia Pieri = Crypsis Ait. (Gramin.).
Raddia P. & K. = Radia A. Rich. ex Kunth = Barbacenia Vand. (Velloziac.).
Raddiella Swallen. Gramineae. 3 trop. S. Am., Trinidad.
Raddisia Leandro = Salacia L. (Celastrac.).
Rademachia Steud. = Radermachia Thunb. = Artocarpus Forsk. (Morac.).
Radermachera Zoll. & Mor. Bignoniaceae. 40 India to China, Philipp. Is., Celebes, Java.
Radermachia B. D. Jacks. (sphalm.) = praec.
Radermachia Thunb. = Artocarpus Forsk. (Morac.).
Radia Nor. = Mimusops L. (Sapotac.).
Radia A. Rich. ex Kunth = Barbacenia Vand. (Velloziac.).
Radiana Rafin. = Cypselea Turp. (Aïzoac.).
Radiata Medik. = Medicago L. (Legumin.).
Radiaxaceae Dulac = Cornaceae Dum.
Radicula Hill = Nasturtium R.Br. = Rorippa Scop. (Crucif.).
Radinocion Ridl. = Aërangis Reichb. f. (Orchidac.).
Radinosiphon N. E. Br. Iridaceae. 5 trop. & S. Afr.
Radiola Hill. Linaceae. 1 Eur., N. Afr., temp. As., *R. linoïdes* Roth. Infl. a dichasial cyme.
Radiusia Reichb. = Sophora L. (Legumin.).
Radlkofera Gilg. Sapindaceae. 1 trop. Afr.
Radlkoferella Pierre = Lucuma Mol. (Sapotac.).
Radlkoferotoma Kuntze. Compositae. 3 S. Brazil, Urug. Hairy shrubs with
Radojitskya Turcz. = Lachnaea L. (Thymelaeac.). [opp. ls.
Radyera Bullock. Malvaceae. 1 S. Afr., 1 southern Austr.
Raffenaldia Godr. Cruciferae. 2 Morocco, Algeria.
Rafflesia R.Br. Rafflesiaceae. 12 W. Malaysia; parasitic on roots of *Vitidaceae*.
R. arnoldi R.Br. has a colossal fl. 45 cm. across and weighing 7 kg. It smells like putrid meat, and is visited by carrion flies.
***Rafflesiaceae** Dum. Dicots. 8/50 trop. Usu. fleshy herbaceous root- or stem-parasites, whose veg. organs are reduced to what is practically a *mycelium* like that of a true fungus, viz. a network of fine cellular threads ramifying in the tissues of the host. The fls. appear above ground, developing as adv. shoots upon the mycelium. They are usu. unisexual, dioec. or monoec., rarely ⚥,

sometimes of enormous size, reg., haplochlam. P (4–5); A ∞ on a column;
G̅ (rarely G̲) (4–10–20), with parietal plac., or ∞ twisted loc. Berry. Endosp.
Genera: *Rafflesia, Sapria, Rhizanthes; Apodanthes, Pilostyles; Mitrastemon;*
Cytinus, Bdallophyton.
Rafia Bory = Raphia Beauv. (Palm.).
*Rafinesquia Nutt. Compositae. 2 SW. U.S.
Rafinesquia Rafin. (1837/1) = Hosackia Dougl. (Legumin.).
Rafinesquia Rafin. (1837/2) = Clinopodium L. (Labiat.).
Rafinesquia Rafin. (1838) = Jacaranda Juss. (Bignoniac.).
Rafnia Thunb. Leguminosae. 32 S. Afr.
Ragadiolus P. & K. = Rhagadiolus Zinn (Compos.).
Ragala Pierre (~ Chrysophyllum L.). Sapotaceae. 3 trop. S. Am.
Ragatelus Presl = Trichomanes L. (Hymenophyllac.).
Ragenium Gandog. = Geranium L. (Geraniac.).
Ragiopteris Presl = Onoclea L. (Aspidiac.).
Raiania Scop. = Rajania L. (Dioscoreac.).
Raillarda Endl. = Raillardia Spreng. = Railliardia Gaudich. (Compos.).
Raillardella (A. Gray) Benth. Compositae. 5 W. U.S.
Raillardia Spreng. = Railliardia Gaudich. (Compos.).
Raillardiopsis Rydb. = Raillardella (A. Gray) Benth. (Compos.).
Railliarda DC. = seq.
Railliardia Gaudich. Compositae. 20 Hawaii.
× **Railliautia** Sherff. Compositae. Gen. hybr. (Dubautia × Railliardia).
Raimannia Rose = Oenothera L. (Onagrac.).
Raimondia Safford. Annonaceae. 2 Colombia to Ecuador.
Raimondianthus Harms. Leguminosae. 1 Peru.
Rainiera Greene = Luina Benth. (Compos.).
Raja Burm. = seq.
Rajania L. Dioscoreaceae. 25 W.I. Caps. with 1 wing.
Rajania Walt. = Brunnichia Banks (Polygonac.).
Raleighia Gardn. = Abatia Ruiz & Pav. (Flacourtiac.).
Ramangis Thou. (uninom.) = *Angraecum ramosum* Thou. (Orchidac.).
Ramatuela Kunth. Combretaceae. 3 trop. Am.
Ramelia Baill. Euphorbiaceae. 1 New Caled.
Rameya Baill. Menispermaceae. 2 Madag., Comoro Is.
Ramirezella Rose. Leguminosae. 9 Mex., C. Am.
Ramirezia A. Rich. = Poeppigia Presl (Legumin.).
Ramischia Opiz ex Garcke = Orthilia Rafin. (Pyrolac.).
Ramisia Glaziou ex Baill. Nyctaginaceae. 1 SE. Brazil.
Ramium Kuntze = Boehmeria Jacq. (Urticac.).
Ramona Greene = Audibertia Benth. = Salvia L. (Labiat.).
*Ramonda Rich. Gesneriaceae. 4 Pyrenees, Balkan Penins. Fl. almost reg. A 5.
Ramondia Jaume St-Hil. = praec.
Ramondia Mirbel = Lygodium Sw. (Schizaeac.).
Ramondiaceae Godr. & Gren. ex Godr. = Gesneriaceae Nees.
Ramonia P. & K. = Ramona Greene = Salvia L. (Labiat.).
Ramonia Schlechter = Hexadesmia Brongn. (Orchidac.).
Ramorinoa Spreng. Leguminosae. 1 Argentina.

RAMOSIA

Ramosia Merr. = Centotheca Desv. (Gramin.).
Ramostigmaceae Dulac = Empetraceae S. F. Gray.
Ramotha Rafin. = Xyris Gronov. ex L. (Xyridac.).
Ramphicarpa Reichb. = Rhamphicarpa Benth. (Scrophulariac.).
Ramphidia Miq. = Rhamphidia Lindl. = Hetaeria Blume (Orchidac.).
Ramphocarpus Neck. = Geranium L. (Geraniac.).
Rampholepis Stapf (sphalm.) = Rhampholepis Stapf (Gramin.).
Ramphospermum Andrz. ex Reichb. = Rhamphospermum Reichb. = Sinapis L. (Crucif.).
Rampinia C. B. Clarke = Herpetospermum Wall. ex Hook. f. (Cucurbitac.).
Ramsaia W. Anders. ex R.Br. = Bauera Banks ex Andr. (Bauerac.).
Ramsdenia Britton = Phyllanthus L. (Euphorbiac.).
Ramspekia Scop. = Posoqueria Aubl. (Rubiac.).
Ramtilla DC. = Guizotia Cass. (Compos.).
Ramusia E. Mey. = Asystasia Blume (Acanthac.).
Ramusia Nees = Peristrophe Nees (Acanthac.).
Ranales Lindl. An order of Dicots. centring around the *Ranunculac.*, *Magnoliac.*, and presumed allies. [1 Indoch., Malay Penins.
Ranalisma Stapf (~Echinodorus Rich.). Alismataceae. 1 trop. W. Afr.;
Ranapalus Kellogg = Herpestis Gaertn. f. = Bacopa Aubl. (Scrophulariac.).
Ranaria Cham. = Bacopa Aubl. (Scrophulariac.).
Rancagua Poepp. & Endl. = Lasthenia Cass. (Compos.).
Randalia Beauv. ex Desv. = Eriocaulon L. (Eriocaulac.).
Randia L. Rubiaceae. 200–300 trop. The two ls. at a node are often unequal and one frequently aborts early. Thorns often occur. In *R. dumetorum* Lam. the thorn arises in the axil above the bud, and is carried up by intercalary growth.
Randonia Coss. Resedaceae. 3 N. Afr., Somal., Arabia.
Ranevea L. H. Bailey. Palmae. 1 Comoro Is.
Rangaëris (Schltr.) Summerh. Orchidaceae. 6 trop. & S. Afr.
Rangia Griseb. (sphalm.) = Randia L. (Rubiac.).
Rangium Juss. = Forsythia Vahl (Oleac.).
Ranisia Salisb. = Gladiolus L. (Iridac.).
Ranugia (Schlechtd.) P. & K. = Dieudonnaea Cogn. (Cucurbitac.).
Ranula Fourr. = Ranunculus L. (Ranunculac.).
*Ranunculaceae Juss. Dicots. 50/1900 chiefly N. temp. Most are herbaceous perennials with rhiz., usually of condensed (rootstock) form, and always sympodial. Each year's shoot ends in an infl. and a bud is formed in the axil of one of the ls. at the base, which forms the next year's growth. In most the primary root soon dies away, and adventit. roots are formed from the stem; often (e.g. *Aconitum, Ranunculus* spp.) these swell up into tubers holding reserve materials. Ls. usu. exstip., more rarely stip. (*Trollius, Caltha, Thalictrum, Ranunculus*, etc.), usu. alt., with sheathing bases and often very much divided. The chief exceptions to the above general statements, and special cases of interest, are described under the genera, e.g. *Helleborus, Eranthis, Clematis, Ranunculus*.

The infl. is typically determinate; in *Anemone* spp., *Eranthis*, etc., a single term. fl. is produced. More often a cymose branching occurs, the buds in the axils of the ls. below the term. fl. developing in descending order. In *Nigella* spp. and others, after the term. fl. is formed, the buds below develop in ascending order, so that a raceme with an end fl. is formed; in *Aconitum*, etc., the same thing occurs, but the term. fl. rarely develops. In *Nigella, Anemone*, etc., there is an invol. of green leaves below the fl., usually alt. with the K. Fl. itself typically spiral upon a ± elongated recept., but frequently the ls. of the P in whorls; usu. reg. and ♀. The P usu. petaloid; rarely (e.g. *Ranunculus*) a true K and C. Frequently there occur nectaries of various patterns between the P proper and the sta.; these are usu. considered as modified petals, but it is as probable that they are derived directly from sta. An interesting series of transitions may be seen by comparing the following fls.: *Caltha* (honey secreted by cpls., 'calyx' present, nothing between it and sta.); *Helleborus* or *Eranthis* (honey secreted in little tubular 'petals'); *Nigella* (ditto, but 'petals' with a small leafy end); *Ranunculus auricomus* ('petals' distinct and coloured, with pocket-like nectary at base); *R. acris*, etc. (petals large, nectary at base). In *Aconitum* and *Delphinium* the fl. is zygo. The sta. are usu. ∞ and spiral, the anthers extr.; the cpls. ∞, apocarpous, spiral, with either one basal or several ventral anatr. ovules. In *Nigella* the cpls. are united; in *Actaea* there is only 1, as in *Podophyllaceae, Berberidaceae*, etc.

As a rule the fls. are protandrous, and the sta., as their anthers open, bend outwards from the centre. A series of fls. showing various grades of adaptation to insects may be found, e.g. *Clematis* (pollen fl.), *Ranunculus* (actinomorphic, honey scarcely concealed at all), *Nigella* (honey in little closed cavities), *Aquilegia* (honey in long spurs), *Delphinium* (ditto, but zygo. also), etc.

Fr. a group of achenes or follicles (caps. in *Nigella*, berry in *Actaea*); seeds with minute embryo and oily endosp. The *R.* are mostly poisonous; a few, e.g. *Aconitum*, are or have been medicinal.

Classification and chief genera (adapted from Hutchinson, 1923):

I. **Helleboroïdeae** (carpels with more than 1 ovule; fruits follicular or baccate):
   1. HELLEBOREAE (fls. actinomorphic): *Trollius, Helleborus, Eranthis, Caltha, Coptis, Isopyrum, Aquilegia, Nigella, Actaea.*
   2. DELPHINIËAE (fls. zygomorphic): *Delphinium, Aconitum.*

II. **Ranunculoïdeae** (carpels with 1 ovule; fruit a bunch of dry achenes, very rarely baccate):
   3. RANUNCULEAE (leaves alt., sep. imbr.; fls. not subtended by invol. of ls.; K mostly caducous; C mostly present): *Myosurus, Callianthemum, Ranunculus, Adonis, Knowltonia, Thalictrum.*
   4. ANEMONEAE (ls. alt., sep. imbr.; fls. subtended by an invol. or ls. usu. remote from K, the latter mostly coloured and persist. at anthesis; C o): *Anemone, Hepatica, Barneoudia.*
   5. CLEMATIDEAE (ls. opp., sep. indupl.-valv. or rarely partly imbr., C o or repr. by outer stds.): *Clematopsis, Clematis, Naravelia.*

Ranunculastrum Fabr. (uninom.) = *Trollius* L. sp. (Ranunculac.).
Ranunculastrum Fourr. = seq.
Ranunculus L. Ranunculaceae. 400 cosmop., temp. & cold, trop. mts. *R. ficaria* L. (pilewort or celandine) has tuberous roots, one formed at the base of each axillary bud; these may give rise by separation to new plants. *R. aquatilis* L.

(water crowfoot) has a floating stem bearing ls. which are of two kinds (heterophylly), the submerged ls. being much divided into linear segments, whilst the floating ls. are merely lobed. *R. repens* L. (creeping buttercup or crowfoot) has creeping runners which root at the nodes and give rise to new pls. *R. acris* L. and *R. bulbosus* L. are other common Eur. spp.; the latter has the base of the stem thickened for storage. Fls. in cymes, reg., with well-marked K and C, protandrous and visited by a miscellaneous lot of insects. Honey is secreted in little pockets at the base of the petals.

Ranzania T. Ito. Podophyllaceae. 1 Japan.

Raoulia Hook. f. Compositae. 25 New Guinea, Austr., N.Z. Woolly herbs forming dense tufted whitish masses ('vegetable sheep').

Raouliopsis Blake. Compositae. 2 Andes.

Rapa Mill. = Brassica L. (Crucif.).

Rapanea Aubl. (~ Myrsine L.). Myrsinaceae. 200 trop., subtrop.

Rapatea Aubl. Rapateaceae. 20 trop. S. Am.

*Rapateaceae Dum. Monocots. (Farinosae). 16/80 trop. S. Am., W. Afr. Perenn. herbs, from ± fleshy rootstocks; ls. lin. or gladiate, usu. twisted through 90°, from condupl. sheaths. Fls. reg., ⚥, in term. heads of spikelets encl. in 2 large spathes, each spikelet of ∞ bracts and 1 term. fl. K 3, imbr., rigid; C 3, usu. connate at base, convol.; A 3 + 3, fil. ± conn. or adnate to C, anth. basifixed, 4- or 2-celled, dehisc. by 1, 2 or 4 apical pores; G (3), ± 3-loc., with 1 to several axile to basal anatr. ov. per loc. Fr. a loculic. caps.; seed with copious mealy endosp.

Classification and chief genera (after Maguire):

  I. **Saxofridericioïdeae** (carpels with several ovules, on axile or septal plac.; seeds ± prismatic, pyramidal, or lunate): *Saxofridericia, Stegolepis, Schoenocephalium.*

  II. **Rapateoïdeae** (carpels 1-ovulate; plac. basal or sub-basal axile; seeds oval or oblong): *Rapatea, Cephalostemon, Monotrema.*

Raphanaceae Horan. = Cruciferae Juss.

Raphanis Moench = Armoracia Gaertn., Mey. & Scherb. (Crucif.).

Raphanistrocarpus (Baill.) Pax = Momordica L. (Cucurbitac.).

Raphanistrum Mill. = Raphanus L. (Crucif.).

Raphanocarpus Hook. f. = Momordica L. (Cucurbitac.).

Raphanopsis Welw. = Oxygonum Burch. (Polygonac.).

Raphanus L. Cruciferae. 8 W. & C. Eur., Medit. to C. As. Pods jointed between seeds (lomentose). *R. sativus* L. (radish) with root-storage.

Raphelingia Dum. = Ornithogalum L. (Liliac.).

Raphia Beauv. Palmae. 30 trop. & S. Afr., Madag. *R. hookeri* Mann & Wendl. is the wine palm, *R. vinifera* Beauv. the bamboo palm, of W. Afr. *R. taedigera* Mart. (= ? *R. vinifera*) in S. Am. (? introd.). Spadix monoec.; the bracts have a curious sheathing form. In *R. farinifera* (Gaertn.) Hyl. (*R. ruffia* [Jacq.] Mart.) (W. Afr. to Madag.) roots develop between the dead leaf-bases; they curve upwards and are said to act as respiratory organs.

Raphiacme K. Schum. = Raphionacme Harvey (Periplocac.).

Raphidiocystis Hook. f. Cucurbitaceae. 7 trop. Afr., Madag.

Raphidophora Hassk. = Rhaphidophora Schott (Arac.).

Raphidophyllum Hochst. = Sopubia Buch.-Ham. (Scrophulariac.).

Raphidospora Reichb. = Rhaphidospora Nees = Justicia L. (Acanthac.)
Raphiocarpus Chun. Gesneriaceae. 1 S. China.
Raphiodon Benth. = Rhaphiodon Schau. = Hyptis Jacq. (Labiat.).
Raphiolepis Lindl. = Rhaphiolepis Lindl. corr. Poir. (Rosac.).
Raphionacme Harv. Periplocaceae. 20 trop. & S. Afr.
Raphione Salisb. = Allium L. (Alliac.).
Raphiostyles Benth. & Hook. f. = Rhaphiostylis Planch. (Icacinac.).
Raphis Beauv. = Rhaphis Lour. = Chrysopogon Trin. (Gramin.).
Raphisanthe Lilja = Blumenbachia Schreb. (Loasac.).
Raphistemma Wall. Asclepiadaceae. 2 Indomal.
Raphithamnus Dalla Torre & Harms = Rhaphithamnus Miers (Verbenac.).
Rapicactus F. Buxb. & Oehme (~ Neolloydia Britt. & Rose). Cactaceae. 2 Mex.
Rapinia Lour. = Sphenoclea Gaertn. (Sphenocleac.).
Rapinia Montrouz. = Neorapinia Moldenke (Verbenac.).
× **Rapistrella** Pomel. Cruciferae. Gen. hybr. (Cordylocarpus × Rapistrum).
× **Rapistrosymbrium** P. Fourn. ex Madiot. Cruciferae. Gen. hybr. (Rapistrum × Sisymbrium).
Rapistrum Bergeret = Calepina Adans. (Crucif.).
Rapistrum R.Br. = Ochthodium DC. (Crucif.).
***Rapistrum** Crantz. Cruciferae. 3 C. Eur., Medit., W. As.
Rapistrum Fabr. = Neslia Desv. (Crucif.).
Rapistrum Medik. = Crambe L. (Crucif.).
Rapistrum Mill. (nomen) = Sinapis L. (Crucif.).
Rapistrum Scop. = Rapistrum Crantz (Crucif.).
Rapolocarpus Boj. (sphalm.) = Ropalocarpus Boj. (Sphaerosepalac.).
Rapona Baill. Convolvulaceae. 1 Madag.
Rapourea Reichb. = Ropourea Aubl. = Diospyros L. (Ebenac.).
Raptostylus P. & K. = Rhaptostylum Humb. & Bonpl. = Heisteria Jacq.
Rapum Hill = Rapa Mill. = Brassica L. (Crucif.). [(Olacac.).
Rapunculus Fourr. = Campanula L. (Campanulac.).
Rapunculus Mill. = Phyteuma L. (Campanulac.).
Rapuntia Chevallier = Campanula L. (Campanulac.).
Rapuntium Mill. = Lobelia L. (Campanulac.).
Rapuntium P. & K. = Rapuntia Chevallier = Campanula L. (Campanulac.).
Raputia Aubl. Rutaceae. 10 W.I., trop. S. Am.
Raram Adans. = Cenchrus L. (Gramin.).
Raritebe Wernh. Rubiaceae. 2 Colombia.
Raspailia Endl. = Raspalia Brongn. (Bruniac.).
Raspailia J. & C. Presl = Polypogon Desf. (Gramin.).
Raspalia Brongn. Bruniaceae. 16 S. Afr.
Rassia Neck. = Gentiana L. (Gentianac.).
Ratabida Loud. = Ratibida Rafin. (Compos.).
Ratanhia Rafin. = Krameria Loefl. (Krameriac.).
Rathbunia Britton & Rose. Cactaceae. 4 Mex.
Rathea Karst. (1858) = Synechanthus H. Wendl. (Palm.).
Rathea Karst. (1860) = Passiflora L. (Passiflorac.).
Rathkea Schumach. & Thonn. = Ormocarpum Beauv. (Legumin.).

Ratibida Rafin. Compositae. 6 N. Am., Mex.

Ratonia DC. = Matayba Aubl. (Sapindac.).

Ratopitys Carr. (sphalm.) = Raxopitys 'Senilis' = Cunninghamia R.Br. ex Rich. (Taxodiac.).

Rattraya Phipps. Gramineae. 1 S. trop. Afr.

Ratzeburgia Kunth. Gramineae. 1 Burma.

Rauhia Traub. Amaryllidaceae. 1 Peru.

Rauhocereus Backeb. (~ Browningia Britt. & Rose). Cactaceae. 1 N. Peru.

Rauia Nees & Mart. Rutaceae. 3 trop. S. Am.

Raukana Seem. = Pseudopanax C. Koch (Araliac.).

Raulinoa R. S. Cowan. Rutaceae. 1 Brazil.

Raulinoreitzia R. M. King & H. Rob. (~ Eupatorium L.). Compositae. 2 Boliv. & Braz. to Argent.

Raussinia Neck. = Pachira Aubl. (Bombacac.).

Rautanenia Buchenau. Alismataceae. 1 SW. Afr.

Rauvolfia L. Apocynaceae. 100 trop. Ls. often in whorls of 3 or 4.

Rauwenhoffia Scheff. Annonaceae. 5 Siam, Indoch., Malay Penins., New Guinea, E. Austr.

Rauwolfia L. = Rauvolfia L. (Apocynac.).

Rauwolfia Ruiz & Pav. = Citharexylum Mill. (Verbenac.).

Ravenala Adans. Strelitziaceae. 1 Madag., *R. madagascariensis* J. F. Gmel., the traveller's tree, so-called because the water that accumulates in the leaf-bases has been used for drinking in cases of necessity. It may be found by piercing the base with a knife. The pl. has a true sub-aerial stem, which bears large 2-ranked ls. giving it a peculiar fan-like appearance.

Ravenea H. Wendl. Palmae. 9 Madag., Comoro Is.

Ravenia Bouché = praec.

Ravenia Vell. Rutaceae. 18 C. & trop. S. Am., W.I.

Raveniopsis Gleason. Rutaceae. 2 Venez.

Ravensara Sonnerat. Lauraceae. 18 Madag. *R. aromatica* J. F. Gmel. is the Madagascar clove (fr. a spice).

Ravia Schult. = Rauia Nees & Mart. (Rutac.).

Ravinia P. & K. = Ravenia Vell. (Rutac.).

Ravnia Oerst. Rubiaceae. 4 C. Am.

Rawsonia Harv. & Sond. Flacourtiaceae. 7 trop. & S. Afr.

Raxamaris Rafin. = Soulamea Lam. (Simaroubac.).

Raxopitys 'Senilis' [Nelson] = Cunninghamia R.Br. ex Rich. (Taxodiac.).

Rayania Meissn. = Rajania Walt. = Brunnichia Banks (Polygonac.).

Rayania Rafin. = Rajania L. (Dioscoreac.).

Rayeria Gaudich. = Alona Lindl. (Nolanac.).

Raynaudetia Bub. = Telephium L. (Caryophyllac.).

Raynia Rafin. = Rajania L. (Dioscoreac.).

Razisea Oerst. Acanthaceae. 1 C. Am.

Razoumowskia Hoffm. = Arceuthobium Bieb. (Viscac.).

Razoumowskiaceae Van Tiegh. = Loranthaceae–Viscoïdeae–Arceuthobieae

Razulia Rafin. = Angelica L. (Umbellif.). [Engl.]

Razumovia Spreng. (1805) = Humea Sm. (Compos.).

Razumovia Spreng. (1807) = Centranthera R.Br. (Scrophulariac.).

Rea Bertol. ex Decne. Compositae. 3 Juan Fernandez.
Readea Gillespie. Rubiaceae. 3 Fiji.
Reana Brign. = Euchlaena Schrad. (Gramin.).
Reaumurea Steud. = seq.
Reaumuria Hasselq. ex L. Tamaricaceae. 20 E. Medit. to C. As. &
Baluchistan. Halophytes.
Reaumuri[ac]eae Ehrenb. = Tamaricaceae–Reamurieae Reichb.
Rebentischia Opiz = Trisetum Pers. (Gramin.).
Rebis Spach = Ribes L. (Grossulariac.).
Reboudia Coss. & Dur. ex Coss. Cruciferae. 2 N. Afr., SE. Medit.
Reboulea Kunth (1830; non Rebouillia Raddi 1818 = Reboullia Raddi 1820
= Reboulia Nees 1846—Hepaticae) = Sphenopholis Scribn. (Gramin.).
Rebsamenia Conzatti = Robinsonella Rose & Baker (Malvac.).
Rebulobivia Frič = Rebutia K. Schum. (Cactac.).   [Rose (Cactac.).
Rebulobivia Frič & Schelle ex Backeb. & F. M. Knuth = Lobivia Britton &
Rebutia K. Schum. Cactaceae. 4 Bolivia, Argent.
× Recchara hort. Orchidaceae. Gen. hybr. (iii) (Brassavola × Cattleya ×
Laelia × Schomburgkia).
Recchia Moç. & Sessé ex DC. Simaroubaceae. 2 Mex. Ls. stip.! (rare in fam.).
× Recchiara hort. (vii) = × Recchara hort. (Orchidac.).
Receveura Vell. = Hypericum L. (Guttif.).
Rechsteinera Kuntze = seq.
*Rechsteineria Regel. Gesneriaceae. 75 Mex. to N. Argent.
Recordia Moldenke. Verbenaceae. 1 Bolivia.
Recordoxylon Ducke. Leguminosae. 2 Brazil.
Rectangis Thou. (uninom.) = Angraecum rectum Thou. = Jumellea recta (Thou.)
Schltr. (Orchidac.).
Rectanthera Degener (~ Callisia L.). Commelinaceae. 2 Mex., C. Am.
Rectomitra Blume = Kibessia DC. (Melastomatac.).
Rectophylis Thou. (uninom.) = Bulbophyllum erectum Thou. (Orchidac.).
Rectophyllum P. & K. = Rhektophyllum N. E. Br. (Arac.).
Redfieldia Vasey. Gramineae. 1 SW. U.S.
Redia Casar. = Cleidion Blume (Euphorbiac.).
Redoutea Vent. = Cienfuegosia Cav. (Malvac.).
Redowskia Cham. & Schlechtd. Cruciferae. 1 NE. Siberia (Yakutia).
Redutea Pers. = Redoutea Vent. = Cienfuegosia Cav. (Malvac.).
Reederochloa Soderstrom & Decker. Gramineae. 1 Mex. Dioecious.
Reedia F. Muell. Cyperaceae. 1 SW. Austr.
Reediella Pichi-Sermolli. Hymenophyllaceae. 5 Malaysia to Tahiti & N.Z.
Reedrollinsia J. W. Walker. Annonaceae. 1 Mex.
Reesia Ewart = Polycarpaea Lam. (Caryophyllac.).
Reevesia Lindl. Sterculiaceae. 15 Himalaya to Formosa.
Regelia Hort. ex H. Wendl. = Verschaffeltia H. Wendl. (Palm.).
Regelia Lem. = Aregelia Kuntze (Bromeliac.).
Regelia Schau. Myrtaceae. 4 W. Austr.
Reggeria Rafin. = Gagea Salisb. (Liliac.).
Regia Loud. ex C. DC. = Juglans L. (Juglandac.).
Regina Buc'hoz = Bontia L. (Myoporac.).

REGMUS

Regmus Dulac = Circaea L. (Onagrac.).
Regnaldia Baill. = Chaetocarpus Thw. (Euphorbiac.).
Regnellia Barb. Rodr. = Bletia Ruiz & Pav. (Orchidac.).
Regnellidium Lindman. Marsileaceae. 1 S. Brazil.
Rehdera Moldenke. Verbenaceae. 3 C. Am.
Rehderodendron Hu. Styracaceae. 9 China, Indoch.
Rehderophoenix Burret. Palmae. 1 Solomon Is.
*Rehmannia Libosch. ex Fisch. & Mey. Gesneriaceae? Scrophulariaceae?
    10 E. As.
Reichardia Dennst. = Tabernaemontana L. (Apocynac.).
Reichardia Roth (1787). Compositae. 10 Medit.
Reichardia Roth (1800) = Maurandia Ortega (Scrophulariac.).
Reichardia Roth (1821) = Pterolobium R.Br. (Legumin.).
Reicheëlla Pax. Caryophyllaceae. 1 Chile.
Reicheia Kausel. Myrtaceae. 1 temp. S. Am.
Reichelea A. W. Benn. = seq.
Reichelia Schreb. = Hydrolea L. (Hydrophyllac.).
Reichembachanthus B. D. Jacks. (sphalm.) = seq.
Reichenbachanthus Barb. Rodr. Orchidaceae. 3 Costa Rica, W.I., trop.
    S. Am.
× **Reichenbachara** hort. (v) = × Opsisanda hort. (Orchidac.).
Reichenbachia Spreng. Nyctaginaceae. 2 trop. S. Am.
Reichenheimia Klotzsch = Begonia L. (Begoniac.).
Reicheocactus Backeb. = Pyrrhocactus (A. Berger) Backeb. & F. M. Knuth
    (Cactac.).
Reichertia Karst. = Schultesia Mart. (Gentianac.).
Reidia Wight = Eriococcus Hassk. = Phyllanthus L. (Euphorbiac.).
Reifferscheidia Presl = Dillenia L. (Dilleniac.).
Reigera Opiz = Bolboschoenus Palla (Cyperac.).
Reilia Steud. Inc. sed. (?Juncac. ?Eriocaulac.). 1 Argentina.
Reimaria Fluegge. Gramineae. 5 warm Am.
Reimarochloa Hitchcock = praec.
Reimbolea Debeaux = Echinaria Desf. (Gramin.).
*Reineckea Kunth. Liliaceae. 2 E. As.
Reineckia auctt. = praec.
Reineckia Karst. = Synechanthus H. Wendl. (Palm.).
Reinera Dennst. = Leptadenia R.Br. (Asclepiadac.).
Reineria Moench = Tephrosia Pers. (Legumin.).
Reinhardtia Liebm. Palmae. 8 Mex., C. Am.
Reinia Franch. & Sav. = Itea Gronov. ex L. (Iteac.).
× **Reinikkaara** hort. (v) = × Christieara hort. (Orchidac.).
Reinwardtia Blume ex Nees = Saurauia Willd. (Actinidiac.).
Reinwardtia Dum. Linaceae. 2 N. India, China.
Reinwardtia Korth. = Ternstroemia Mutis ex L. f. (Theac.).
Reinwardtia Spreng. = Breweria R.Br. (Convolvulac.).
Reinwardtiodendron Koorders. Meliaceae. 1 E. Borneo, Philipp., Celebes,
    Moluccas, W. New Guinea.
Reissantia Hallé. Celastraceae. 7 trop. Afr., As.

*Restio Rottb. Restionaceae. 120 S. Afr., Madag., Austr. Assimilation is performed by the green stems, the ls. being reduced to sheaths.

*Restionaceae R.Br. Monocots. (Farinosae). 28/320, mostly in S. Afr. and Austr., a few in N.Z., Chile and Indochina. Xero., usu. of tufted growth, with the general habit of *Juncus*; below ground is a rhiz. with scale-ls., giving off erect cylindrical shoots bearing sheathing ls. (very rarely with ligules), which have a short blade, or sometimes none; assim. is also performed by the stem. Fls. dioec. (rarely monoec. or ♀), reg., in spikelets. P in two whorls, but single members often absent; A 3 or 2, opp. to the inner perianth-ls.; G̲ (3–1), 1–3-loc., with 1 pend. orthotr. ov. per loc. Caps. or nut. Embryo lens-shaped, in mealy endosp. Chief genera: *Elegia, Chondropetalum, Restio, Lepyrodia, Thamnochortus, Leptocarpus, Willdenowia.*

Restrepia Kunth. Orchidaceae. 60 Mex. to trop. S. Am.

Restrepiella Garay & Dunsterville. Orchidaceae. 8 C. & N. trop. S. Am.

Resupinaria Rafin. = Sesbania Scop. (Legumin.).

Retalaceae Dulac = Pyrolaceae Lindl.

Retama [Rafin.] Boiss. = Lygos Adans. = Genista L. (Legumin.).

Retamilia Miers = seq.

Retanilla Brongn. Rhamnaceae. 2 Peru, Chile.

Retinaria Gaertn. = Gouania L. (Rhamnac.).

Retiniphyllum Humb. & Bonpl. Rubiaceae. 20 trop. S. Am.

Retinispora Sieb. & Zucc. = Chamaecyparis Spach (Cupressac.).

Retinodendron Korth. = Vatica L. (Dipterocarpac.).

Retinodendropsis Heim = praec.   [(Legumin.).

Retinophleum Benth. & Hook. f. = Rhetinophloeum Karst. = Cercidium Tul.

Retinospora Carr. Cupressaceae. A 'form genus'. Seedlings of many spp. of the genera *Chamaecyparis, Cupressus, Thuja*, etc., exhibit, instead of the decussate appressed ls. of the mature plant, spreading needle-ls. (often in whorls of 4) like those of *Abies*, etc. (cf. *Pinus, Acacia*, etc.). If now these young seedlings be used as offsets, the new pls. thus formed retain throughout life this form of foliage; and pls. are thus obtained of totally different habit from that usual in these genera. To these juvenile forms the name *R.* was at

Retrosepalaceae Dulac = Violaceae Lam. & DC.   [one time given.

Rettbergia Raddi = Chusquea Kunth (Gramin.).

Retzia Thunb. Retziaceae. 1 S. Afr.

Retziaceae Bartl. Dicots. 1/1 S. Afr. Small shrub with virgate stems, bare below, densely leafy above. Ls. lin., ent., exstip., with revolute margins, arranged in whorls of 3 on young shoots, of 4 on older stems. Fls. sol., axill., forming a leafy spike. K (5), unequal; C (5)(–7), long-tubular, orange-red, villous without, lobes short, purplish-black, indupl.-valv.; A 5(–7), alternipet., inserted at mouth of C, anth. shortly sagitt.; disk small, hypog.; G̲ (2), ± constricted ½-way, with long exserted filif. style and shortly bifid stig., and 2–3 erect axile anatr. ov. per loc. in lower part. Fr. a 2-valved septic. caps., valves bifid, each part with 1 pend. seed; endosp. copious. Only genus: *Retzia*. An isolated genus, showing possible connections with *Solanac., Scrophulariac.*, and *Loganiales*; it differs from all fams. in the last-named group (except the very different *Buddlejac.*) in the absence of intraxylary phloem.

Reussia Dennst. = Paederia L. (Rubiac.).

REUSSIA

*Reussia Endl. Pontederiaceae. 3 S. Am.
Reutealis Airy Shaw. Euphorbiaceae. 1 Philipp. Is.
Reutera Boiss. Umbelliferae. 10 As. Min. to C. As. & Afghan.
Revatophyllum Roehl. (sphalm.) = Ceratophyllum L. (Ceratophyllac.).
Reveesia Walp. = Reevesia Lindl. (Sterculiac.).
Reverchonia Gandog. = Armeria Willd. (Plumbaginac.).
Reverchonia A. Gray. Euphorbiaceae. 1 S. U.S., Mex.
Reya Kuntze = Burchardia R.Br. (Liliac.).
Reyesia Clos = Salpiglossis Ruiz & Pav. (Solanac.).
Reymondia H. Karst. = Pleurothallis R.Br. (Orchidac.).
Reynandia B. D. Jacks. (sphalm.) = seq.
Reynaudia Kunth. Gramineae. 1 Cuba, S. Domingo.
Reynoldsia A. Gray. Araliaceae. 14 Polynesia.
Reynosia Griseb. Rhamnaceae. 1 Florida, 15 W.I. Endosp. ruminate.
Reynoutria Houtt. (~ Polygonum L.). Polygonaceae. 15 temp. As.
Rhabarbarum Fabr. = Rheum L. (Polygonac.).
Rhabdadenia Muell. Arg. Apocynaceae. 4 Florida, C. & trop. S. Am., W.I.
Rhabdia Mart. = Rotula Lour. (Ehretiac.).
Rhabdocalyx Lindl. = Cordia L. (Ehretiac.).
Rhabdocaulon (Benth.) Epling. Labiatae. 7 trop. S. Am.
Rhabdochloa Kunth = Rabdochloa Beauv. = Leptochloa Beauv. (Gramin.).
Rhabdocrinum Reichb. = Lloydia Salisb. (Liliac.).
Rhabdodendraceae (Engl.) Prance. Dicots. (Centrospermae). 1/4 N. trop. S. Am. Tall, fastigiate or subpyramidal shrubs or small trees. Ls. alt., simple, ent., elongate, variable, gland-dotted; stips. small or obsol. Fls. in supra-axill. racemose pan., with thick rhachis, fls. ± distant. K turbinate, (5), or with ent. annular rim; C 5, oblong, slightly imbr. below, ± valv. at apex, caducous; A ∞, with v. short flattened finally persist. and recurved fil. and elongate lin. 4-locell. anthers, emarg. at apex; G 1, with 1 basal anatr. ov. and 1 thickened gynobasic style with unilat. (dorsal) decurr. stig. Fr. ± glob., 1-seeded, borne in concave recept. at apex of expanded woody pedic., with thin ± fleshy exocarp and thin ± woody endocarp; endosperm o. Only genus: *Rhabdodendron*. Affinities obscure; formerly included in *Rutac.* (with which it has lysigenous oil-glands in common), but differing in the wood-anat., the ∞ sta. with short persist. fil. and elongate linear basifixed non-versatile anth., the absence of a disk, and the 1-loc. 1-ovulate ovary with gynobasic style. The wood-anat. shows significant similarities to that of *Phytolaccac.* (*s.l.*), and this affinity is supported (*inter alia*) by the elongate 'Phytolaccaceous' anthers. The *Rhabd.* are perhaps to be regarded as a primitive survival of the stock from which the otherwise relatively specialised Centrospermous groups were differentiated.
Rhabdodendron Gilg & Pilger. Rhabdodendraceae. 4 N. Brazil, Guianas.
Rhabdophyllum Van Tiegh. (~ Ouratea Aubl.). Ochnaceae. 25 trop. Afr.
Rhabdosciadium Boiss. Umbelliferae. 5 Persia, Kurdist.
Rhabdostigma Hook. f. = Kraussia Harv. (Rubiac.).
Rhabdothamnopsis Hemsl. Gesneriaceae. 1 China.
Rhabdothamnus A. Cunn. Gesneriaceae. 1 N.Z.
Rhabdotheca Cass. = Launaea Cass. (Compos.).

Rhachicallis DC. corr. Spach. Rubiaceae. 1 W.I.
Rhachidosorus Ching = Athyrium Roth (Athyriac.).
Rhachidospermum Vasey = Jouvea Fourn. (Gramin.).
Rhacodiscus Lindau. Acanthaceae. 4 trop. S. Am.
Rhacoma Adans. = Leuzea DC. (Compos.).
Rhacoma P.Br. ex L. = Crossopetalum P.Br. (Celastrac.).
Rhadamanthus Salisb. Liliaceae. 2–3 S. Afr.
Rhadinocarpus Vog. = Chaetocalyx DC. (Legumin.).
Rhadinopus S. Moore. Rubiaceae. 1 New Guinea.
Rhadinothamnus P. G. Wilson. Rutaceae. 1 W. Austr.
Rhadiola Savi = Radiola L. (Linac.).
Rhaeo C. B. Clarke (sphalm.) = Rhoeo Hance (Commelinac.).
Rhaesteria Summerh. Orchidaceae. 1 trop. E. Afr.
Rhagadiolus Scop. Compositae. 1 Medit. Fr. linear, without pappus, completely enwrapped in an involucral bract.
Rhaganus E. Mey. = Bersama Fresen. (Melianthac.).
Rhagodia R.Br. Chenopodiaceae. 12 Austr.
Rhamindium Sarg. (sphalm.) = Rhamnidium Reissek (Rhamnac.).
Rhammatophyllum O. E. Schulz. Cruciferae. 2 C. As.
Rham-Moluma Baill. (sphalm.) = Rhamnoluma Baill. = Lucuma Mol. (Sapotac.).
***Rhamnaceae** Juss. Dicots. 58/900 cosmop. Mostly trees or shrubs, often climbing (by aid of hooks in *Ventilago*, tendrils in *Gouania*, etc., twining stems in *Berchemia*); thorns occur in some, and especially in *Colletia*, etc. (*q.v.*). In these pls. too, serial buds occur in the l.-axils. Ls. simple, usu. with stips., never lobed or divided. Infl. cymose, usu. a corymb.

Fls. inconspic., ♀ or rarely unisexual, reg., sometimes apet. Recept. hollow, free from or united to the ovary. K 5–4, valvate; C 5–4, usu. small, often strongly concave, frequently clawed at base; A 5–4, alt. with sepals, usu. enclosed by the petals, at any rate at first; disk usu. well developed, intrastaminal; G free or ± united to recept., 3–2-(rarely by abortion 1-)loc. (sometimes 4- or typically 1-loc.); in each loc. 1 (rarely 2) basal ovule with downwardly directed micropyle; style simple or divided. Fr. dry, splitting into dehisc. or indehisc. mericarps, or a drupe with 1 or several stones, or a nut. Endosp. little or none. Many of the dry frs. show special adaptations for windcarriage, e.g. *Paliurus*, *Ventilago*. Related to *Vitidaceae*, from which they are chiefly distinguished by the small petals, the recept., the endocarp and simple ls.; they also approach *Celastraceae*, the chief distinction being the antepetalous sta. Few are of economic value; see *Ziziphus*, *Rhamnus*, *Hovenia*. Chief genera: *Ventilago*; *Paliurus*, *Ziziphus*; *Rhamnus*, *Hovenia*, *Ceanothus*, *Phylica*; *Colletia*; *Gouania*.

Rhamnella Miq. Rhamnaceae. 10 W. Himal. to Japan, New Guinea, Queensl., Fiji.
Rhamnicastrum Kuntze = Scolopia Schreb. (Flacourtiac.).
Rhamnidium Reissek. Rhamnaceae. 12 trop. S. Am., W.I.
Rhamnobrina H. Perrier = Colubrina Rich. ex Brongn. (Rhamnac.).
Rhamnoïdes Mill. = Hippophaë L. (Elaeagnac.).
Rhamnoluma Baill. = Lucuma Molina (Sapotac.).

Rhamnoneuron Gilg. Thymelaeaceae. 1 Indoch.

Rhamnopsis Reichb. = Flacourtia Comm. ex L'Hérit. (Flacourtiac.).

Rhamnos St-Lag. = seq.

Rhamnus L. Rhamnaceae. Excl. *Frangula* Mill., 110; incl. *Frangula*, 160; cosmop. Shrubs with alt. or opp. ls. and small cymose clusters of fls. Two subgenera. To subgen. *Rhamnus* (*Eurhamnus*) (fls. usu. 4-merous, polyg. or dioec.) belong *R. alaternus* L. (Medit.) and *R. cathartica* (Eur., As., Medit.), whose berries are purgative; the juice of the fr. is mixed with alum and evaporated, thus forming the paint known as sap-green; also *R. infectoria* L. (mts. of S. Eur.) whose berries (*graines d'Avignon* or 'yellow berries') yield useful green and yellow dye-stuffs, and *R. chlorophora* Decne from whose bark the Chinese prepare the dye known as 'Chinese green indigo' used in dyeing silk (*R. utilis* Decne is also employed). To subgen. *Frangula* (fls. usu. 5-merous, ♀), belong *R. frangula* L. (*Frangula alnus* Mill.) (Eur., As., N. Afr.) whose bark is officinal (cathartic) and whose wood forms one of the best charcoals; *R. purshiana* DC., in N. Am., whose bark (*cascara sagrada*) is largely used as a cathartic; etc.

Rhamphicarpa Benth. Scrophulariaceae. 30 Afr., Madag., India, New Guinea, Austr.

Rhamphidia Lindl. = Hetaeria Blume (Orchidac.).

Rhamphocarya Kuang = Annamocarya A. Chev. (Juglandac.).

Rhamphogyne S. Moore. Compositae. 1 Rodrigues.

Rhampholepis Stapf = Sacciolepis Nash (Gramin.).

Rhamphospermum Andrz. ex Bess. = Sinapis L. (Crucif.).

Rhanterium Desf. Compositae. 3 NW. Afr. to Baluchistan.

Rhaphanistrocarpus auctt. = Raphanistrocarpus Pax (Cucurbitac.).

Rhaphanocarpus auctt. = Raphanocarpus Hook. f. (Cucurbitac.).

Rhaphanos St-Lag. = seq.

Rhaphanus auctt. = Raphanus L. (Crucif.).

Rhaphedospera Wight (sphalm.) = Rhaphidospora Nees = Justicia L. (Acanthac.).

Rhaphia auct. ex Steud. = Raphia Beauv. (Palm.).

Rhaphiacme K. Schum. = Raphionacme Harv. (Asclepiadac.).

Rhaphidanthe Hiern ex Gürke = Diospyros L. (Ebenac.).

Rhaphidiocystis Hook. f. = Raphidiocystis Hook. f. (Cucurbitac.).

Rhaphidophora Hassk. Araceae. 100 Indomal., New Caled. Sympodial climbing stems with clasping roots and pend. aerial roots. The pinnation of the ls. originates like that in *Monstera*, i.e. by long holes arising between the ribs, and the margin finally breaking. Fls. ♀.

Rhaphidophyllum Benth. = Raphidophyllum Hochst. = Sopubia Buch.-Ham. (Scrophulariac.).

Rhaphidophyton Iljin. Chenopodiaceae. 1 C. As.

Rhaphidorhynchus Finet = Microcoelia Lindl., Aërangis Reichb. f., Tridactyle Schltr., Calyptrochilum Kraenzl., Beclardia A. Rich., etc. (Orchidac.).

Rhaphidosperma G. Don = seq.

Rhaphidospora Nees. Acanthaceae. 12 trop. Afr., W. Malaysia.

Rhaphidura Bremek. Rubiaceae. 1 Borneo.

Rhaphiodon Schau. = Hyptis Jacq. (Labiat.).

*Rhaphiolepis Lindl. corr. Poir. Rosaceae. 15 subtrop. E. As.
Rhaphionacme C. Muell. = Raphionacme Harv. (Asclepiadac.).
Rhaphiophallus Schott = Amorphophallus Bl. (Arac.).
Rhaphiostylis Planch. ex Benth. Icacinaceae. 10 trop. W. Afr.
Rhaphis Lour. = Chrysopogon Trin. (Gramin.).
Rhaphis Walp. (sphalm.) = Rhapis L. f. (Palm.).
Rhaphispermum Benth. Scrophulariaceae. 1 Madag.
Rhaphistemma Meissn. = Raphistemma Walt. (Asclepiadac.).
Rhaphistemum Walp. = praec.
Rhaphitamnus B. D. Jacks. (sphalm.) = seq.
Rhaphithamnus Miers. Verbenaceae. 2 Chile, Argent., Juan Fernandez.
Rhapidophyllum H. Wendl. & Drude. Palmae. 1 SE. U.S.
Rhapidospora Reichb. = Rhaphidospora Nees (Acanthac.).
Rhapis L. f. Palmae. 15 S. China to Java.
Rhapontica Hill = seq.
Rhaponticum Ludw. (~ Centaurea L.). Compositae. 20 W. & C. As.
Rhaptocalymma Börner = Carex L. (Cyperac.).
Rhaptocarpus Miers = Echites P.Br. (Apocynac.).
Rhaptomeris Miers = Cyclea Arn. (Menispermac.).
Rhaptonema Miers. Menispermaceae. 6 Madag.
Rhaptopetalaceae Pierre ex Van Tiegh. = Scytopetalaceae Engl.
Rhaptopetalum Oliv. Scytopetalaceae. 6 trop. Afr.
Rhaptostylum Humb. & Bonpl. = Heisteria Jacq. (Olacac.).
Rhazya Decne. Apocynaceae. 1 Grecian Thrace, NW. As. Min.; 1 Arabia to
Rhea Endl. = Rea Bertol. ex Decne (Compos.). [NW. India.
Rheedia L. Guttiferae. 45 C. & trop. S. Am., W.I., Madag.
Rheithrophyllum Hassk. (1844) = seq.
Rheitrophyllum Hassk. (1842) = Aeschynanthus Jack (Gesneriac.).
Rhektophyllum N. E. Br. Araceae. 1 trop. W. Afr.
Rheopteris Alston = Aspleniopsis Mett. (Hemionitidac.).
Rhesa Walp. = Bhesa Buch.-Ham. ex Arn. (Celastrac.).
Rhetinodendron Meissn. Compositae. 1 Juan Fernandez.
Rhetinolepis Coss. = Anthemis L. (Compos.).
Rhetinophloeum Karst. = Cercidium Tul. (Legumin.).
Rhetinosperma Radlk. Meliaceae. 1 Queensland.
Rheum L. Polygonaceae. 50 temp. & subtrop. As. Fls. like *Rumex,* but coloured and entomophilous, though they exhibit traces of anemophily in very large stigmas (cf. *Poterium,* etc.). *R. officinale* Baill. furnishes medicinal rhubarb; *R. rhaponticum* L. is the rhubarb used as a vegetable.
Rhexia L. Melastomataceae. 10 E. U.S.
Rhigicarya Miers = seq.
Rhigiocarya Miers corr. Miers. Menispermaceae. 3 trop. W. Afr.
Rhigiophyllum Hochst. Campanulaceae. 1 S. Afr.
Rhigiothamnus Spach = Dicoma Cass. (Compos.).
Rhigospira Miers = Tabernaemontana L. (Apocynac.).
Rhigozum Burch. Bignoniaceae. 6 trop. & S. Afr., Madag.
Rhinacanthus Nees. Acanthaceae. 15 trop. Afr., Madag., Socotra, E. As., Indomal.

989

RHINACTINA

Rhinactina Less. = Rhinactinidia Novopokr. (Compos.).
Rhinactina Willd. = Jungia L. f. (Compos.).
Rhinactinidia Novopokr. (~ Aster L.). Compositae. 3 C. As.
Rhinanthaceae ('-oïdeae') Vent. = Scrophulariaceae–Rhinanthoïdeae Link.
Rhinanthera Blume = Scolopia Schreb. (Flacourtiac.).
Rhinanthus L. Scrophulariaceae. 50 temp. Euras., N. Am.; common in damp pastures. Semi-parasites with loose pollen fls. (see fam.).
Rhinchoglossum Blume = Rhynchoglossum Blume corr. G. Don (Gesneriac.).
Rhinchosia Zoll. & Mor. = Rhynchosia Lour. (Legumin.).
Rhincospora Gaudich. = Rhynchospora Vahl (Cyperac.).
Rhinephyllum N. E. Brown. Aïzoaceae. 14 S. Afr.
Rhinerrhiza Rupp. Orchidaceae. 1 Queensl., New S. Wales.
Rhiniachne Hochst. ex Steud. = Thelepogon Roth (Gramin.).
Rhinium Schreb. = Tetracera L. (Dilleniac.).
Rhinocarpus Bert. ex Kunth = Anacardium L. (Anacardiac.).
× **Rhinochilus** hort. Orchidaceae. Gen. hybr. (iii) (Rhinorrhiza × Sarcochilus).
Rhinoglossum Pritz. = Rhynchoglossum Blume (Gesneriac.).
Rhinolobium Arn. = Lagarinthus E. Meyer (Asclepiadac.).
Rhinopetalum Fisch. ex Alex. = Fritillaria L. (Liliac.).
Rhinopterys Niedenzu emend. Ndz. Malpighiaceae. 3 trop. W. Afr.
Rhinopteryx Niedenzu = praec.
Rhinostegia Turcz. = Thesium L. (Santalac.).
Rhinostigma Miq. = Garcinia L. (Guttif.).
Rhipidantha Bremek. Rubiaceae. 1 trop. Afr.
Rhipidia Markgraf. Apocynaceae. 1 Amaz. Brazil.
Rhipidodendrum Willd. = Aloë L. (Liliac.).
Rhipidoglossum Schlechter = Diaphananthe Schltr. (Orchidac.).
Rhipidopteris Schott = Peltapteris Link (Lomariopsidac.).
Rhipidostigma Hassk. = Diospyros L. (Ebenac.).
Rhipogonum J. R. & G. Forst. corr. Spreng. Liliaceae. 7 E. New Guinea, E. Austr., N.Z.
Rhipsalidopsis Britton & Rose. Cactaceae. 2 S. Brazil.
***Rhipsalis** Gaertn. Cactaceae. 60 W.I., Brazil, Argentina. *R. baccifera* (J. Mill.) W. T. Stearn is found in Africa, the Mascarene Is. & Ceylon but is quite possibly introduced.
× **Rhipsaphyllopsis** Werderm. = [Epiphyllopsis (= Rhipsalidopsis) ×] Rhipsalidopsis Britt. & Rose (Cactac.).
Rhizaëris Rafin. = Laguncularia Gaertn. f. (Combretac.).
Rhizakenia Rafin. = Limnobium Rich. (Hydrocharitac.).
Rhizanota Lour. ex Gomes = Corchorus L. (Tiliac.).
Rhizanthella R. S. Rogers. Orchidaceae. 1 W. Austr.
Rhizanthemum Van Tiegh. = Amyema Van Tiegh. (Loranthac.).
× **Rhizanthera** P. F. Hunt & Summerh. Orchidaceae. Gen. hybr. (i) (Dactylorhiza × Platanthera).
Rhizanthes Dumort. Rafflesiaceae. 1–2 W. Malaysia.
Rhizemys Rafin. = Testudinaria Salisb. = Dioscorea L. (Dioscoreac).
Rhizirideum (G. Don) Fourr. = Allium L. (Alliac.).

Rhizium Dulac = Elatine L. (Elatinac.).

Rhizobol[ac]eae DC. = Caryocaraceae Szyszyl.

Rhizobolus Gaertn. ex Schreb. = Caryocar L. (Caryocarac.).

Rhizobotrya Tausch. Cruciferae. 1 S. Tyrol.

Rhizocarpaceae Dulac = Salviniaceae + Marsileaceae.

Rhizocephalum Wedd. Campanulaceae. 5 Andes.

Rhizocephalus Boiss. Gramineae. 2 SW. & C. As.

Rhizocorallon Gagneb. = Corallorhiza Chatel. (Orchidac.).

Rhizoëtes D. Meyer ex Rauh = Stylites Amstutz (Isoëtac.).

Rhizoglossum Presl. Ophioglossaceae. 1 S. Afr.

Rhizogum Reichb. = Rhigozum Burch. (Bignoniac.).

Rhizomonanthes Danser. Loranthaceae. 3 New Guinea.

Rhizophora L. Rhizophoraceae. 7 trop. coasts. These mangroves are moderate-sized trees with a great development of roots from the stem and branches. On the sub-aerial parts of the roots are large lenticels, probably serving in the same way as the aerenchyma of *Bruguiera*, etc. The seed germinates upon the tree; the hypocotyl projects at the micropyle and grows rapidly. The bark is used for tanning, yielding a substance known as cutch (cf. *Acacia*).

*Rhizophoraceae R.Br. Dicots. 16/120 trop., mostly Old World. Trees, often of 'mangrove' habit, with simple ent. opp. stip. (more rarely alt. exstip.) occasionally dimorphic or tripli–quintupli-nerved ls.; stips. conspic., inter-pet., caduc. Fls. ♀ (more rarely ♂♀), reg., hypo- to epi-gynous in axill. cymes or (sometimes serial) rac., rarely sol. K 3–16, valv., persist.; C 3–16, usu. clawed and apically lacin.; A 8–∞, inserted on outer edge of perig. or epig. disk, anth. intr., usu. 4-loc., rarely multilocell. or dehisc. by a valve (stds. in ♀ fls. sometimes adn. to pet.); G or Ḡ (2–12), with usu. 2 anatr. pend. ov. per loc. Fr. a slightly soft berry or drupe, or dry and indehisc., rarely a dehisc. caps. or winged; seeds sometimes arillate, viviparous in mangrove spp., endosp. fleshy or o. Chief genera: *Rhizophora, Bruguiera, Cassipourea, Carallia, Anisophyllea*. Relationships with *Combretac., Rubiac.* and *Elaeocarpac.*

Rhizosperma Meyen = Azolla Lam. (Azollac.).

Rhizotaechia auctt. = Rhysotoechia Radlk. (Sapindac.).

Rhoanthus Rafin. = Mentzelia L. (Loasac.).

Rhodactinea Gardn. = Barnadesia Mutis (Compos.).

Rhodactinia Hook. f. = praec.

Rhodalix Rafin. = Spiraea L. (Rosac.).

Rhodalsine J. Gay = Minuartia L. (Caryophyllac.).

Rhodamnia Jack. Myrtaceae. 20 SE. As., Malaysia, E. Austr., New Caled.

Rhodanthe Lindl. = Helipterum DC. (Compos.).

Rhodax Spach = Helianthemum Mill. (Cistac.).

× **Rhodazalea** Hort. Ericaceae. Gen. hybr. (Azalea × Rhododendron).

Rhodea Endl. = Rohdea Roth (Liliac.).

Rhodia Adans. = seq.

Rhodiola L. (~ Sedum L.). Crassulaceae. 50 N. temp.

Rhodiola Lour. = Cardiospermum L. (Sapindac.).

× **Rhodobranthus** Traub. Amaryllidaceae. Gen. hybr. (Habranthus × Rhodophiala).

Rhodocactus (A. Berger) F. M. Knuth = Pereskia Mill. (Cactac.).

RHODOCALYX

Rhodocalyx Muell. Arg. Apocynaceae. 1 campos of Brazil. K coloured.

Rhodochiton Zucc. Scrophulariaceae. 1 Mex., *R. atrosanguineus* (Zucc.) Rothm. (*R. volubile* Zucc.), a twiner with sensitive petioles (cf. *Clematis*).

Rhodochlaena Spreng. = Rhodolaena Thou. (Sarcolaenac.).

Rhodochlamys Schau. = Salvia L. (Labiat.).

× **Rhodocinium** Avrorin. Ericaceae. Gen. hybr. (Rhodococcum × Vac-
Rhodocistus Spach = Cistus L. (Cistac.).   [cinium).

Rhodoclada Baker = Asteropeia Thou. (Asteropeiac.).

Rhodococcum (Rupr.) Avrorin = Vaccinium L. (Ericac.).

Rhodocodon Baker. Liliaceae. 8 Madag.

Rhodocolea Baill. Bignoniaceae. 6 Madag.

Rhodocoma Nees = Restio L. (Restionac.).

Rhododendraceae ('-dra') Juss. = Ericaceae–Rhododendroïdeae Endl.

Rhododendron L. Ericaceae. 5–600 N. temp., etc. 'One sp. (*R. lochae* F. Muell.) is found in trop. Austr.; the greatest richness of spp. is in E. As., from S. China to the Himal. and Japan; a second and lesser abundance is found in temp. N. Am., and a few spp. in the arctic regions. 4 spp. in Mid. & S. Eur., 5 in Caucasus.' (Drude.) Over 250 in Malaysia. Shrubs and small trees with usu. leathery ls.; the ls. of § *Azalea* last one year, those of the other subgenera usu. more. Large winter buds are formed covered with scale ls.; the larger and stouter ones contain infls., the slender ones merely ls. The branch bearing an infl. is continued by the formation of a bud in one of the upper axils. Some of the Asiatic spp. are epiph. The C is slightly zygo., and the sta. and styles bend upwards to touch the under surface of a visiting insect. *R. ferrugineum* L. (alpine rose) is protandr. and visited by humble-bees.

Rhododendros Adans. = Andromeda + Kalmia + Chamaedaphne + Rhododendron, etc. (Ericac.).

Rhododendrum L. = Rhododendron L. (Ericac.).

Rhododon Epling. Labiatae. 1 S. U.S. (Texas).

Rhodogeron Griseb. Compositae. 1 Cuba.

Rhodognaphalon (Ulbr.) Roberty, emend. A. Robyns. Bombacaceae. 7 trop.

Rhodognaphalopsis A. Robyns. Bombacaceae. 9 trop. Am.   [Afr.

Rhodohypoxis Nel. Hypoxidaceae. 2 SE. Afr.

Rhodolaena Thou. Sarcolaenaceae. 5 Madag.

Rhodolaenaceae Bullock = Sarcolaenaceae Caruel.

Rhodoleia Champ. ex Hook. Rhodoleiaceae. 1–7 Upper Burma & S. China; Malay Penins. & Sumatra.

Rhodoleiaceae (Harms) Nak. Dicots. 1/1(–7) SE. As., W. Malaysia. Evergreen trees or shrubs of *Rhododendron*-like habit; bud-scales large, membr., glauc.; ls. alt., ent., coriac., exstip., glaucous below, long-pet., crowded toward ends of branchlets. Fls. ♀, in 5–10-fld. pend. axill. capitula; capit. surr. by ∞ broadly imbr. rounded toment. bracts. K (5), minute; C 0–4, imbr., only the outer ones in the capit. developed, spath., clawed, exserted, red; A (7–)10(–11), fil. elong., surr. by small nectarif. glands, anth. obl., basifixed, latrorse, with apic. connective; G (2), semi-inf., sometimes glaucous, 2-1-loc., with ∞ biseriate axile ± horiz. ov., and 2 long slender erect caduc. styles with minute stigs. Fr. a head of 5–10 woody biloc. 4-valved caps., connate below, with 1 fert. and 20–40 ster. angular or flattened seeds, with copious endosp. Only

RHOÏPTELEACEAE

genus: *Rhodoleia*. The flowers are bird-pollinated (*Zosteropidae, Nectariniidae,* etc.). The exstipulate, coriaceous ls., remarkable bracteate infl. (simulating a *Camellia* flower), and ornithophily seem to justify family rank for this genus. The ls. and venation are unlike anything in *Hamamelidac.* (in which it has been usu. included), but are suggestive of *Magnoliac., Theac.* (e.g. *Ternstroemia), Daphniphyllac.,* or *Pygeum (Rosac.).*

Rhodolirion Phil. = Hippeastrum Herb. (Amaryllidac.).
Rhodomyrtus Reichb. Myrtaceae. 20 S. India, Ceylon, Siam, Philipp. Is., New Guinea, New Caled., Austr.
Rhodophiala Presl = Hippeastrum Herb. (Amaryllidac.).
Rhodophora Neck. = Rosa L. (Rosac.).
*Rhodopis Urb. Leguminosae. 1 W.I.
Rhodopsis (Endl.) Reichb. = Rosa L. (Rosac.).
Rhodopsis (Ledeb.) Dippel = Hulthemia Dum. (Rosac.).
Rhodopsis Lilja = Calandrinia Kunth (Portulacac.).
Rhodoptera Rafin. = Rumex L. (Polygonac.).
Rhodora L. = Rhododendron L. (Ericac.).
Rhodoraceae Vent. = Ericaceae–Rhododendroïdeae Endl.
Rhodorhiza Webb = Convolvulus L. (Convolvulac.).
Rhodormis Rafin. = Salvia L. (Labiat.).
Rhodorrhiza Webb & Berth. = Rhodorhiza Webb = Convolvulus L. (Convolvulac.).
Rhodosciadium S. Wats. Umbelliferae. 12 Mex.
Rhodosepala Baker. Melastomataceae. 3 Madag.
Rhodoseris Turcz. = Onoseris DC. (Compos.).
Rhodospatha Poepp. & Endl. Araceae. 25 C. & trop. S. Am.
Rhodosphaera Engl. Anacardiaceae. 1 E. Austr.
Rhodostachys Phil. (~ Ochagavia Phil.). Bromeliaceae. 3 Chile.
Rhodostoma Scheidw. = Palicourea Aubl. (Rubiac.).
Rhodothamnus Lindl. & Paxt. = Rhododendron L. (Ericac.).
*Rhodothamnus Reichb. Ericaceae. 1 E. Alps, 1 NE. As. Min.
Rhodotyp[ac]eae J. G. Agardh = Rosaceae–Kerrieae Engl.
Rhodotypos Sieb. & Zucc. Rosaceae. 1 Japan, *R. scandens* (Thunb.) Mak. (*R. kerrioïdes* S. & Z.). It has opp. ls., found in no other member of the fam., except in seedlings of *Prunus.* There is an epicalyx (see *Potentilla).*
Rhodoxylon Rafin. = Convolvulus L. (Convolvulac.).
Rhodusia Vasilch. (~ Medicago L.). Leguminosae. 1 Canary Is., Medit.
Rhoea St-Lag. = Punica L. (Punicac.).
Rhoeadales Bartl. corr. Engl. An order of Dicots. comprising (in Engler's system) the fams. *Papaverac., Capparidac., Crucif., Tovariac., Resedac., Moringac.* and *Bretschneiderac.*
Rhoeidium Greene = Rhus L. (Anacardiac.).
Rhoeo Hance. Commelinaceae. 1 Mex., C. Am., W.I.
Rhoiacarpos A. DC. Santalaceae. 1 S. Afr.
Rhoicarpos B. D. Jacks. = praec.
Rhoïcissus Planch. Vitidaceae. 12 trop. & S. Afr.
Rhoïptelea Diels & Hand.-Mazz. Rhoïptelaceae. 1 S. China, Indoch.
*Rhoïpteleaceae Hand.-Mazz. Dicots. 1/1 S. China, Indoch. Tall trees;

993

branchlets densely lentic. and glandular-aromatic; ls. alt., imparipinn., stip.; lfts. alt., dentic.; stips. small, caduc. Fls. ♂ ♀, reg., small, in groups of 3 (♀ ♂ ♀) in loose axill. and term. tassel-like panicles. ♂ fl. short-pedic., ♀ fls. sess., ster. K 4, imbr., persist., membr.; C o; disk o; A 6, fil. v. short, persist., anth. basifixed, short; G̲ (2), 2-loc., with 1 empty and 1 1-ov. loc., vesic.-pilose; ov. attached to septum, ascending, semi-anatr.; stigs. 2, free. Fr. a woody tuberc. flattened orbic. 2-winged nut, wings chartac., emarg. at apex between the 2 persist. stigs., seed without endosp. Only genus: *Rhoïptelea*. Perh. related to *Juglandac.*, but wood structure similar to *Acerac.* (cf. *Dipteronia?*).

Rhombifolium Rich. ex DC. = Clitoria L. (Legumin.).

Rhombochlamys Lindau. Acanthaceae. 2 Colombia.

Rhomboda Lindl. = Zeuxine Lindl. (Orchidac.).

Rhombolobium Rich. ex Kunth = Clitoria L. (Legumin.).

Rhombolythrum Link. Gramineae. 1 Chile.

Rhombophyllum Schwantes. Aïzoaceae. 3 S. Afr.

Rhombospora Korth. = Greenia Wight & Arn. (Rubiac.).

Rhoögeton Leeuwenb. Gesneriaceae. 3 Venez., Guiana.

Rhopala Schreb. = Roupala Aubl. (Proteac.).

Rhopalandria Stapf = Dioscoreophyllum Engl. (Menispermac.).

Rhopalephora Hassk. = Aneilema R.Br. (Commelinac.).

Rhopaloblaste Scheff. Palmae. 7 Nicobars, Malaya, Moluccas, New Guinea, Solomons.

Rhopalobrachium Schlechter & Krause. Rubiaceae. 2 New Caled. (1 Caroline Is.?).

Rhopalocarpaceae Hemsl. = Sphaerosepalaceae Van Tiegh.

Rhopalocarpus Boj. Sphaerosepalaceae. 13 Madag.

Rhopalocarpus Teijsm. & Binnend. ex Miq. = Anaxagorea A. St-Hil. (Annonac.).

Rhopalocnemis Jungh. Balanophoraceae. 1 Madag.; 1 C. & E. Himal., Assam, Sumatra to Moluccas.

Rhopalocyclus Schwantes (~ Leipoldtia L. Bolus). Aïzoaceae. 2 S. Afr.

Rhopalopilia Pierre. Opiliaceae. 10 trop. Afr., Madag.

Rhopalopodium Ulbr. Ranunculaceae. 7 N. Andes.

Rhopalosciadium K. H. Rechinger. Umbelliferae. 1 Persia.

Rhopalostigma Phil. = Phrodus Miers (Solanac.).

Rhopalostigma Schott = Asterostigma Fisch. & Mey. (Arac.).

Rhopalostylis Klotzsch ex Baill. = Dalechampia L. (Euphorbiac.).

Rhopalostylis H. Wendl. & Drude. Palmae. 3 N.Z., Norfolk I., Kermadec I.

Rhopalota N. E. Brown. Crassulaceae. 1 S. Afr.

Rhophostemon Wittstein = Roptrostemon Bl. = Nervilia Comm. ex Gaudich. (Orchidac.).

Rhopium Schreb. = Phyllanthus L. (Euphorbiac.).

Rhoradendron Griseb. (sphalm.) = Phoradendron Nutt. (Loranthac.).

Rhouancou Augier = Rouhamon Aubl. = Strychnos L. (Strychnac.).

Rhuacophila Blume = Dianella Lam. (Liliac.).

Rhus L. Anacardiaceae. 250 subtrop. & warm temp. *R. coriaria* L. is the sumac (S. Eur.); its ls., ground fine, are used for tanning and dyeing. *R. toxicodendron*

L. (N. Am., poison-ivy) climbs like ivy. Its juice produces ulcerations or erysipelas. *R. vernicifera* DC. is the lacquer-tree. Japan lacquer is obtained from notches in the stem. *R. succedanea* L. is the wax-tree of Japan; its crushed berries yield wax.

Rhuyschiana Adans. = Dracocephalum L. (Labiat.).

Rhyacophila Hochst. = Quartinia Endl. (Lythrac.).

Rhyditospermum Walp. = Rhytidospermum Sch. Bip. = Matricaria L. (Com-

Rhynchadenia A. Rich. = Macradenia R.Br. (Orchidac.). [pos.).

Rhynchandra Reichb. = seq.

Rhynchanthera Blume = Corymborkis Thou. (Orchidac.).

***Rhynchanthera** DC. Melastomataceae. 40 C. & trop. S. Am.

× **Rhynchanthera** hort. (vii) = × Renanstylis hort. (Orchidac.).

Rhynchanthus Hook. f. Zingiberaceae. 5–6 Burma, SW. China, W. Malaysia.

Rhynchelythrum Nees = seq.

Rhynchelytrum Nees corr. Endl. Gramineae. 37 trop. Afr., Madag., Arabia to Indoch. *Rh. repens* (Willd.) C. E. Hubb. (*Tricholaena rosea* Nees) cult. for

Rhynchium Dulac = Vicia L. (Legumin.). [dry bouquets.

Rhynchocalyx Oliv. Lythraceae. 1 Natal.

Rhynchocarpa Backer ex K. Heyne = Dansera van Steenis = Dialium L. (Legumin.).

Rhynchocarpa Becc. = Burretiokentia Pichi-Serm. (Palm.).

Rhynchocarpa Schrad. ex Endl. = Kedrostis Medik. (Cucurbitac.).

Rhynchocarpus Less. = Relhania L'Hérit. (Compos.).

Rhynchocarpus Reinw. ex Blume = Cyrtandra J. R. & G. Forst. (Gesneriac.).

× **Rhynchocentrum** hort. Orchidaceae. Gen. hybr. (iii) (Ascocentrum × Rhynchostylis).

***Rhynchocorys** Griseb. Scrophulariaceae. 3 S. Eur. to Persia.

Rhynchodia Benth. Apocynaceae. 8 India & S. China to Java.

Rhynchodium C. Presl = Psoralea L. (Legumin.).

Rhynchoglossum Blume. Gesneriaceae. 6 Formosa, Indomal. Markedly anisophyllous, with heterodromous fls.

Rhyncholacis Tul. Podostemaceae. 25 N. trop. S. Am.

Rhyncholaelia Schlechter. Orchidaceae. 2 Mex., C. Am., Panamá.

Rhyncholepis Miq. = Piper L. (Piperac.).

Rhyncholepsis C. DC. (sphalm.) = praec.

× **Rhynchonopsis** hort. Orchidaceae. Gen. hybr. (iii) (Phalaenopsis × Rhynchostylis).

Rhynchopappus Dulac = Crepis L. (Compos.).

Rhynchopera Börner = Carex L. (Cyperac.).

Rhynchopera Klotzsch = Pleurothallis R.Br. (Orchidac.).

Rhynchopetalum Fresen. = Lobelia L. (Campanulac.).

Rhynchophora Arènes. Malpighiaceae. 1 Madag.

Rhynchophorum (Miq.) Small = Peperomia Ruiz & Pav. (Peperomiac.).

Rhynchophreatia Schlechter. Orchidaceae. 5 Micronesia, New Guinea.

Rhynchopsidium DC. = Relhania L'Hérit. (Compos.).

Rhynchopyle Engl. = Piptospatha N.E.Br. (Arac.).

× **Rhynchorades** hort. (sphalm.) = seq.

× **Rhynchorides** hort. Orchidaceae. Gen. hybr. (iii) (Aërides × Rhynchostylis).

RHYNCHORYZA

Rhynchoryza Baill. = Oryza L. (Gramin.).

*Rhynchosia Lour. Leguminosae. 300 trop. & subtrop., esp. Am. & Afr.

Rhynchosinapis Hayek. Cruciferae. 8 W. Eur. to Alps, NW. Afr., Canary Is.

Rhynchospermum Lindl. = Rhyncospermum A. DC. = Rhynchodia Benth. (Apocynac.).

Rhynchospermum Reinw. Compositae. 2 Himalaya to Japan.

*Rhynchospora Vahl corr. Willd. Cyperaceae. 200 cosmop., esp. trop.

Rhynchostele Reichb. f. = Leochilus Knowles & Westc. (Orchidac.).

Rhynchostemon Steetz = Thomasia J. Gay (Sterculiac.).

Rhynchostigma Benth. Asclepiadaceae. 1 trop. W. Afr.

Rhynchostylis Blume corr. Hassk. Orchidaceae. 15 China, Indomal.

Rhynchostylis Tausch = Chaerophyllum L. (Umbellif.).

Rhynchotechum Blume. Gesneriaceae. 12 Ceylon, S. China, Indomal.

Rhynchotheca Ruiz & Pav. Ledocarpaceae (?). 1 Andes. C o. Palynological research suggests that this genus should be transferred from the *Ledocarpaceae* to the *Vivianiaceae*. (Bortenschlager, *Grana Palynol.* 7: 420–1, 1967.) It differs from *Viviania* in its spinose habit, free imbricate sepals and lack of corolla.

Rhynchothec[ac]eae J. G. Agardh = Ledocarpaceae Meyen (?).

Rhynchotheha Pers. (sphalm.) = Rhynchotheca Ruiz & Pav. (Ledocarpac.).

Rhynchotropis Harms. Leguminosae. 2 trop. Afr.

× **Rhynchovanda** hort. Orchidaceae. Gen. hybr. (iii) (Rhynchostylis × Vanda).

× **Rhynchovandanthe** hort. (vii) = praec.

× **Rhynchovola** hort. Orchidaceae. Gen. hybr. (iii) (Brassavola × Rhyncholaelia).

Rhyncosia Webb = Rhynchosia Lour. (Legumin.).

Rhyncospermum A. DC. = Rhynchodia Benth. (Apocynac.).

Rhyncospora auctt. = Rhynchospora Vahl corr. Willd. (Cyperac.).

Rhyncostylis Steud. = Rhynchostylis Blume (Orchidac.).

Rhyncothecum A. DC. (sphalm.) = Rhynchotechum Blume (Gesneriac.).

Rhyncothelia auctt. = Rhynchotheca Ruiz & Pav. (Ledocarpac.).

× **Rhyndoropsis** hort. Orchidaceae. Gen. hybr. (iii) (Doritis × Phalaenopsis × Rhynchostylis).

Rhynea DC. = Cassinia R.Br. (Compos.).

Rhynea Scop. = Mesua L. (Guttif.).

Rhynospermum Walp. = Rhyncospermum A. DC. = Rhynchodia Benth. (Apocynac.).

Rhyparia Hassk. = Ryparosa Blume (Flacourtiac.).

Rhysolepis Blake. Compositae. 2 Mex.

Rhysopterus Coulter & Rose. Umbelliferae. 1 N. Am.

Rhysospermum Gaertn. f. = Notelaea Vent. (Oleac.).

Rhysotoechia Radlk. Sapindaceae. 13–14 Philipp. Is. & Borneo to Queensland.

Rhyssocarpus Endl. = Billiotia DC. (Rubiac.).

Rhyssolobium E. Mey. Asclepiadaceae. 1 S. Afr.

*Rhyssopteris Rickett & Stafleu (sphalm.?) = seq.

Rhyssopterys Blume ex A. Juss. corr. Wittst. Malpighiaceae. 6 Philipp., Java, E. Malaysia, Queensl., New Caled.

Rhyssopteryx Dalla Torre & Harms = praec.
Rhyssostelma Decne. Asclepiadaceae. 1 Argentina.
Rhytachne Desv. Gramineae. 10 trop. Afr., Madag.
Rhyticalymma Bremek. Acanthaceae. 5 W. Malaysia.
Rhyticarpus Sond. (= Anginon Rafin.). Umbelliferae. 3 S. Afr.
Rhyticarum Boerl. (sphalm.) = seq.
Rhyticaryum Becc. corr. Becc. Icacinaceae. 12 New Guinea.
Rhyticocos Becc. Palmae. 1 W.I.
Rhytidachne K. Schum. = Rhytachne Desv. (Gramin.).
Rhytidandra A. Gray = Alangium Lam. (Alangiac.).
Rhytidanthe Benth. = Leptorhynchos Less. (Compos.).
Rhytidanthera [Planch.] Van Tiegh. Ochnaceae. 6 Colombia.
Rhytidocaryum K. Schum. & Lauterb. = Rhyticaryum Becc. (Icacinac.).
Rhytidocaulon Bally. Asclepiadaceae. 2 NE. trop. Afr.
Rhytidolobus auctt. = Rytidolobus Dulac = Hyacinthus L. (Liliac.).
Rhytidomene Rydb. = Psoralea L. (Legumin.).
Rhytidophyllum Mart. Gesneriaceae.' 20 W.I.
Rhytidosolen Van Tiegh. = Arthrosolen C. A. Mey. (Thymelaeac.).
Rhytidospermum Sch. Bip. = Tripleurospermum Sch. Bip. (Compos.).
Rhytidosporum F. Muell. ex Hook. f. = Marianthus Hueg. (Pittosporac.).
Rhytidostylis Reichb. = Rytidostylis Hook. & Arn. (Cucurbitac.).
Rhytidotus Hook. f. = Rytidotus Hook. f. (Rubiac.).
Rhytiglossa Nees = Dianthera Gronov. ex L. = Justicia L. (Acanthac.).
Rhytileucoma F. Muell. = Chilocarpus Blume (Apocynac.).
Rhytis Lour. = Antidesma L. (Stilaginac.).
Rhytispermum Link = Aegonychon S. F. Gray = Lithospermum L. (Boraginac.).
Rhyttiglossa T. Anders. = Rhytiglossa Nees = Justicia L. (Acanthac.).
Riana Aubl. = Rinorea Aubl. (Violac.).
Ribeirea Fr. Allem. = Schoepfia Schreb. (Olacac.).
Ribeirea Arruda ex Koster = Hancornia Gomes (Apocynac.).
Ribeiria Willis (sphalm.) = praec.
Ribes L. Grossulariaceae. 150 N. temp. and Andine. Shrubs, often with spines (emergences), and with racemes of fls. on 'short shoots'. G̅ with two parietal plac. Fls. usu. homogamous, with self-pollination in default of insect visits. *R. alpinum* L. is dioec. In *R. sanguineum* Pursh (N. Am.) ('flowering currant') the petals change from white to pink as fls. grow older, and in *R. aureum* Pursh from yellow to carmine (cf. *Fumaria, Boraginaceae*). *R. rubrum* L. is the red, *R. nigrum* L. the black currant, *R. grossularia* L. the gooseberry, all cult. for fr.
Ribesi[ac]eae A. Rich. = Grossulariaceae DC.
Ribesiodes Kuntze = Embelia Burm. f. (Myrsinac.).
Ribesioïdes auctt. = praec.
Ribesium Medik. = Ribes L. (Grossulariac.).
Ricardia Adans. = Richardia L. (Rubiac.).
Ricaurtea Triana = Doliocarpus Roland. (Dilleniac.).
Richaeia Thou. = Cassipourea Aubl. (Rhizophorac.).
Richardella Pierre = Lucuma Molina (Sapotac.).

RICHARDIA

Richardia Kunth = Zantedeschia Spreng. (Arac.).
Richardia Lindl. (sphalm.) = Reichardia Roth = Picridium Desf. (Compos.).
Richardia L. Rubiaceae. 10 trop. S. Am.
Richardsiella Elffers & Kennedy-O'Byrne. Gramineae. 1 S. trop. Afr. (N. Rhod.).
Richardsonia Kunth = Richardia L. (Rubiac.).
__*Richea__ R.Br. Epacridaceae. 10 Victoria, Tasmania.
Richea Kuntze = Richaeia Thou. = Cassipourea Aubl. (Rhizophorac.).
Richea Labill. = Craspedia J. R. & G. Forst. (Compos.).
Richella A. Gray. Annonaceae. 1 Borneo, 1 New Caled., 1 Fiji.
Richeopsis Arènes = Scolopia Schreb. (Rhizophorac.).
Richeria Vahl. Euphorbiaceae. 7 trop. S. Am.
Richeriella Pax & K. Hoffm. Euphorbiaceae. 2 Hainan, Philipp. Is., Borneo, Malay Penins.
Richetia Heim = Balanocarpus Bedd. (Dipterocarpac.).
Richiaea Benth. & Hook. f. = Richaeia Thou. = Cassipourea Aubl. (Rhizo-
Richiea G. Don = Ritchiea R.Br. (Capparidac.).   [phorac.).
Richtera Reichb. = Anneslea Wall. (Theac.).
Richterago Kuntze = Gochnatia Kunth (Compos.).
Richteria Kar. & Kir. = Chrysanthemum L. (Compos.).
Richthofenia Hosseus = Sapria Griff. (Rafflesiac.).
Ricinaceae Barkley = Euphorbiaceae Juss.
Ricinella Muell. Arg. = Adelia L. (Euphorbiac.).
Ricinocarpaceae (Pax) Hurusawa = Euphorbiaceae–Ricinocarpoïdeae Pax.
Ricinocarpodendron Amm. ex Boehm. = ? Dysoxylum Bl. (Meliac.).
Ricinocarpos Desf. Euphorbiaceae. 15 Austr., Tasm., 1 New Caled.
Ricinocarpus A. Juss. = praec.
Ricinocarpus Kuntze = Acalypha L. (Euphorbiac.).
Ricinodendron Muell. Arg. Euphorbiaceae. 2 trop. & SW. Afr.
Ricinoïdes Mill. = Jatropha L. (Euphorbiac.).
Ricinoïdes Moench = Chrozophora Neck. ex Juss. (Euphorbiac.).
Ricinophyllum Pall. ex Ledeb. = Fatsia Decne & Planch. (Araliac.).
Ricinus L. Euphorbiaceae. 1 trop. Afr., As., *R. communis* L. (castor-oil), a shrub in trop., a herb when adventive in temp. regions. Monoec. The ♂ fl. has much-branched sta. The fr. explodes into the separate cpls., which at the same time open and drop the seeds. The seed is rich in oil, used medicinally and as a lubricant.
Ricoila Rénéalm ex Rafin. = Gentiana L. (Gentianac.).
Ricophora Mill. = Dioscorea L. (Dioscoreac.).
__*Ricotia__ L. Cruciferae. 9 E. Medit.
Ridan Adans. = Actinomeris Nutt. (Compos.)
Ridania Kuntze = praec.
Riddelia Rafin. = Riddellia Rafin. = Melochia L. (Sterculiac.).
Riddellia Nutt. Compositae. 2 SW. U.S., Mex.
Riddellia Rafin. = Melochia L. (Sterculiac.).
Ridelia Spach = Riedelia Cham. = Lantana L. (Verbenac.).
Ridleia Endl. = Riedlea Vent. = Melochia L. (Sterculiac.).
Ridleya (Hook. f.) Pfitz. = Thrixspermum Lour. (Orchidac.).

Ridleya K. Schum. = Risleya King & Pantl. (Orchidac.).
× Ridleyara hort. Orchidaceae. Gen. hybr. (iii) (Arachnis × Trichoglottis × Vanda).
Ridleyella Schlechter. Orchidaceae. 1 New Guinea.
Ridleyinda Kuntze = Isoptera Scheff. ex Burck (Dipterocarpac.).
Ridolfia Moris = Carum L. (Umbellif.).
Riedelia Cham. = Lantana L. (Verbenac.).
Riedelia Meissn. = Satyria Klotzsch (Ericac.).
*Riedelia Oliv. Zingiberaceae. 50 Malaysia. Lat. stds. unequal, or 1.
Riedelia Trin. ex Kunth = Arundinella Raddi (Gramin.).
Riedeliella Harms. Leguminosae. 2 Brazil, Paraguay.
Riedlea Mirbel = Onoclea L. (Aspidiac.)
Riedlea Vent. = Melochia L. (Sterculiac.).
Riedleia DC. = praec.
Riedleja Hassk. = praec.
Riedlia Dum. = praec.
Riencourtia Cass. corr. Cass. Compositae. 8 N. trop. S. Am.
Riencurtia Cass. = praec.
Riesenbachia C. Presl. Onagraceae. 1 Mex.
Riessia Klotzsch = Begonia L. (Begoniac.).
Rigidella Lindl. Iridaceae. 4 Mex., Guatemala, Peru.
Rigiocarya P. & K. = Rhigiocarya Miers (Menispermac.).
Rigiolepis Hook. f. (~Vaccinium L.). Ericaceae. 16 W. Malaysia.
Rigiopappus A. Gray. Compositae. 1 SW. U.S.
Rigiophyllum (Less.) Spach = Relhania L'Hérit. (Compos.).
Rigiophyllum P. & K. = Rhigiophyllum Hochst. (Campanulac.).
Rigiostachys Planch. = Recchia Moç. & Sessé ex DC. (Simaroubac.).
Rigocarpus Neck. = Rytidiostylis Hook. & Arn. (Cucurbitac.).
Rigospira P. & K. = Rhigospira Miers = Tabernaemontana L. (Apocynac.).
Rigozum P. & K. = Rhigozum Burch. (Bignoniac.).
Rima Sonner. = Artocarpus J. R. & G. Forst. (Morac.).
Rimacola Rupp. Orchidaceae. 1 New S. Wales.
Rimaria L. Bolus = Vanheerdia L. Bolus (Aïzoac.).
Rimaria N. E. Br. Aïzoaceae. 10 S. Afr.
Rinanthus Gilib. = Rhinanthus L. (Scrophulariac.).
Rindera Pall. Boraginaceae. 25 Medit. to C. As.
Ringentiarum Nakai = Arisaema Mart. (Arac.).
Rinopodium Salisb. = Scilla L. (Liliac.).
Rinorea Aubl. Violaceae. 340 warm. C sub-reg.
Rinoreocarpus Ducke. Violaceae. 1 Amaz. Brazil.
Rinxostylis Rafin. = Cissus L. (Vitidac.).
Rinzia Schau. = Baeckea L. (Myrtac.).
Riocreuxia Decne. Asclepiadaceae. 5 trop. & S. Afr.; 1 Nepal.
Riparia Rafin. = Ripasia Rafin. = Baptisia Vent. (Legumin.).
Ripartia Gandog. = Rosa L. (Rosac.).
Ripasia Rafin. = Baptisia Vent. (Legumin.).
Ripidium Bernh. = Schizaea Sm. (Schizaeac.).
Ripidium Trin. = Erianthus Michx (Gramin.).

RIPIDODENDRUM

Ripidodendrum P. & K. =Rhipidodendrum Willd. =Aloë L. (Liliac.).
Ripidostigma P. & K. =Rhipidostigma Hassk. =Diospyros L. (Ebenac.).
Ripogonum J. R. & G. Forst. =Rhipogonum J. R. & G. Forst. corr. Spreng. (Liliac.).
Ripsalis P. & K. =Rhipsalis Gaertn. (Cactac.).
Ripselaxis Rafin. =Salix L. (Salicac.).
Ripsoctis Rafin. =Salix L. (Salicac.).
Riqueria Pers. =seq.
Riqueuria Ruiz & Pav. Rubiaceae (inc. sed.). 1 Peru.
Riseleya Hemsl. =Drypetes Vahl (Euphorbiac.).
Risleya King & Pantling. Orchidaceae. 1 Himalaya, W. China.
Rissoa Arn. =Atalantia Corrêa (Rutac.).
Ritaia King & Pantling =Ceratostylis Bl. (Orchidac.).
Ritchiea R.Br. ex G. Don. Capparidaceae. 50 trop. Afr. A ∞.
Ritchieophyton Pax =Givotia Griff. (Euphorbiac.).
Rithrophyllum P. & K. =Rheitrophyllum Hassk. =Aeschynanthus Jack (Gesneriac.).
Ritinophora Neck. =Amyris P.Br. (Burserac.).
Ritonia Benoist. Acanthaceae. 3 Madag.
Ritron[ac]eae Hoffmgg. & Link =Compositae–Cynareae–Echinopinae O. Hoffm. [*Ritro* Lobel].
Rittenasia Rafin. =Menispermum L. (Menispermac.).
Rittera Rafin. =Centranthus DC. (Valerianac.).
Rittera Schreb. =Swartzia Schreb. (Legumin.)
Ritterocereus Backeb. =Lemaireocereus Britt. & Rose (Cactac.).
Rivea Choisy. Convolvulaceae. 5 India, SE. As.
Riveria Kunth =Swartzia Schreb. (Legumin.).
Rivina L. Phytolaccaceae. 3 trop. Am. P 4, A 4 or 8, G 1. Berry.
Rivin[ac]eae J. G. Agardh =Phytolaccaceae R.Br.
Rivinia L. =Rivina L. (Phytolaccac.).
Rivinia Mill. =Trichostigma A. Rich. (Phytolaccac.).
Rivinoïdes Afzel. ex Prain =Erythrococca Benth. (Euphorbiac.).
Rixea C. Morr. =Tropaeolum L. (Tropaeolac.).
Rixia Lindl. =praec.
Rizoa Cav. =Gardoquia Ruiz & Pav. (Labiat.).
Robbairea Boiss. (~Polycarpaea Lam.). Caryophyllaceae. 2 N. Afr. to Arabia.
Robbia A. DC. =Malouetia A. DC. (Apocynac.).
Robergia Roxb. =Pegia Colebr. (Anacardiac.).
Robergia Schreb. =Rourea Aubl. (Connarac.).
Roberta St-Lag. =seq.
Robertia DC. =Hypochoeris L. (Compos.).
Robertia Mérat =Eranthis Salisb. (Ranunculac.).
Robertia Rich. ex Carr. =Phyllocladus L. C. & A. Rich. (Phyllocladac.).
Robertia Scop. =Bumelia Sw. (Sapotac.).
Robertiella Hanks =seq.
Robertium Picard =Geranium L. (Geraniac.).
Robertsia Endl. =Sideroxylon L. (Sapotac.).

Robertsonia Haw. = Saxifraga L. (Saxifragac.).
Robeschia Hochst. ex Fourn. Cruciferae. 1 Sinai Penins. to Afghan.
Robina Aubl. = seq.
Robinia L. Leguminosae. 20 E. N. Am., Mex. *R. pseud-acacia* L. (false acacia, locust). Stipules thorny. The leaflets move upwards in hot or dry air. The horiz. shoots branch in one plane, while the upright show radial symmetry. The base of the petiole forms a cap protecting a series of axillary buds.
Robiniaceae ('-aceae') Welw. = Leguminosae–Papilionoïdeae DC.
Robinsonella Rose & Baker f. Malvaceae. 7 Mex., C. Am.
*Robinsonia DC. Compositae. 6 Juan Fernandez.
Robinsonia Scop. = Quiina Aubl. (Quiinac.).
Robinsoniodendron Merr. Urticaceae. 1 Moluccas.
Robiquetia Gaudich. Orchidaceae. 20 India, SE. As., Malaysia, Solomons, Fiji.
Roborowskia Batalin. Fumariaceae. 1 C. As.
Robsonia Reichb. = Ribes L. (Grossulariac.).
Robynsia Drapiez = Bravoa Lex. = Polianthes L. (Amaryllidac.).
*Robynsia Hutch. (= ?Rytigynia Bl.). Rubiaceae. 1 trop. W. Afr.
Robynsia Mart. & Gal. = Pachyrhizus Rich. (Legumin.).
Robynsiella Suesseng. Amaranthaceae. 1 trop. Afr.
Robynsiochloa Jacques-Félix. Gramineae. 1 trop. Afr.
Robynsiophyton R. Wilczek. Leguminosae. 1 trop. Afr.
Rocama Forsk. = Trianthema L. (Aïzoac.).
Roccardia Neck. = Helipterum DC. (Compos.).
Roccardia Rafin. = Staehelina L. (Compos.).
*Rochea DC. Crassulaceae. 4 S. Afr.
Rochea Salisb. = Geissorhiza Ker-Gawl. (Iridac.).
Rochea Scop. = Aeschynomene L. (Legumin.).
Rochefortia Sw. Ehretiaceae. 12 C. Am., Colombia, W.I.
*Rochelia Reichb. Boraginaceae. 20 Medit. to Austr.
Rochelia Roem. & Schult. = Echinospermum Sw. (Boraginac.).
Rochetia Delile = Trichilia P.Br. (Meliac.).
Rochonia DC. Compositae. 4 Madag.
Rockia Heimerl. Nyctaginaceae. 1 Hawaii.
Rockinghamia Airy Shaw. Euphorbiaceae. 1 NE. Queensland.
Rodatia Rafin. = Beloperone Nees (Acanthac.).
Rodentiophila Ritter (nomen). Cactaceae. 1 Chile. Quid?
Rodetia Moq. = Bosea L. (Amaranthac.).
Rodgersia A. Gray. Saxifragaceae. 6 E. As.
Rodigia Spreng. = Crepis L. (Compos.).
Rodora Adans. = Rhodora L. = Rhododendron L. (Ericac.).
× Rodrassia hort. Orchidaceae. Gen. hybr. (iii) (Brassia × Rodriguezia).
× Rodrenia hort. (vii) = × Rodridenia hort. (Orchidac.).
× Rodrettia hort. Orchidaceae. Gen. hybr. (iii) (Comparettia × Rodriguezia).
× Rodricidium hort. Orchidaceae. Gen. hybr. (iii) (Oncidium × Rodriguezia).
× Rodridenia hort. Orchidaceae. Gen. hybr. (iii) (Macradenia × Rodriguezia).
Rodriguezia Ruiz & Pav. Orchidaceae. 30 trop. S. Am., W.I.
Rodrigueziella Kuntze. Orchidaceae. 2 Brazil.

RODRIGUEZIOPSIS

Rodrigueziopsis Schlechter. Orchidaceae. 2 Brazil.

× **Rodriopsis** hort. Orchidaceae. Gen. hybr. (iii) (Ionopsis × Rodriguezia).

× **Rodritonia** hort. Orchidaceae. Gen. hybr. (iii) (Miltonia × Rodriguezia).

Rodschiedia Dennst. = ? Callicarpa L. (Verbenac.).

Rodschiedia Gaertn., Mey. & Scherb. = Capsella Medik. (Crucif.).

Rodschiedia Miq. = Securidaca L. (Polygalac.).

Rodwaya F. Muell. = Thismia Griff. (Burmanniac.).

Roëa Hueg. ex Benth. = Sphaerolobium Sm. (Legumin.).

Roebelia Engel = Calyptrogyne H. Wendl. vel Welfia H. Wendl. (Palm.).

Roegneria C. Koch (~ Agropyron J. Gaertn.). Gramineae. 60 temp. Euras.

Roehlingia Dennst. = Tetracera L. (Dilleniac.).

Roehlingia Roepert = Eranthis Salisb. (Ranunculac.).

Roëla Scop. = Roëlla L. (Campanulac.).

Roëlana Comm. ex DC. = Erythroxylum P.Br. (Erythroxylac.).

Roëlla L. Campanulaceae. 25 S. Afr.

Roëllana Comm. ex Lam. = Erythroxylum P.Br. (Erythroxylac.).

Roëlloïdes Banks ex A. DC. = Prismatocarpus L'Hérit. (Campanulac.).

Roelpinia Scop. (sphalm.) = Koelpinia Scop. = Acronychia J. R. & G. Forst. (Rutac.).

Roemera Tratt. = Steriphoma Spreng. (Capparidac.).

Roemeria DC. = praec.

Roemeria Dennst. = Scaevola L. (Goodeniac.).

Roemeria Medik. Papaveraceae. 7 Medit. to C. As. & Afghanistan.

Roemeria Moench = Amaranthus L. (Amaranthac.).

Roemeria Thunb. = Heeria Meissn. (Anacardiac.) + Myrsine L. (Myrsinac.) + Sideroxylon L. (Sapotac.).

Roemeria Zea ex Roem. & Schult. = Diarrhena Rafin. (Gramin.).

Roentgenia Urb. Bignoniaceae. 2 N. trop. S. Am.

Roepera A. Juss. = Zygophyllum L. (Zygophyllac.).

Roeperia F. Muell. = Justago Kuntze (Capparidac.).

Roeperia Spreng. = Ricinocarpus Desf. (Euphorbiac.).

Roeperocharis Reichb. f. Orchidaceae. 5 trop. Afr.

Roeslinia Moench = Chironia L. (Gentianac.).

Roettlera P. & K. (1) = Rottlera Willd. (1804) = Mallotus Lour. (Euphorbiac.).

Roettlera P. & K. (2) = Rottlera Willd. (1797) = Trewia L. (Euphorbiac.).

Roettlera Vahl = Didymocarpus Wall. (Gesneriac.).

Roezlia Hort. = Furcraea Vent. (Amaryllidac.).

Roezlia Regel = Monochaetum Naud. (Melastomatac.).

Roezliella Schlechter. Orchidaceae. 5 Colombia.

Rogeonella A. Chev. = Afrosersalisia A. Chev. (Sapotac.).

Rogeria J. Gay. Pedaliaceae. 6 Brazil, trop. & S. Afr.

Rogiera Planch. = Rondeletia L. (Rubiac.).

Rohdea Roth. Liliaceae. 3 E. As. Said to be fert. by snails crawling over the fls.

Rohria Schreb. = Tapura Aubl. (Dichapetalac.).

Rohria Vahl = Berkheya Ehrh. (Compos.).

Roia Scop. = Swietenia Jacq. (Meliac.).

Roigia Britton = Phyllanthus L. (Euphorbiac.).

Rojasia Malme. Asclepiadaceae. 1 Brazil.
Rojasianthe Standley & Steyerm. Compositae. 1 C. Am.
Rojasiophyton Hassler = Xylophragma Sprague (Bignoniac.).
Rojoc Adans. = Morinda Linn. (Rubiac.).
Rokejeka Forsk. = Gypsophila L. (Caryophyllac.).
Rolandra Rottb. Compositae. 1 trop. S. Am.
Roldana La Llave = Senecio L. (Compos.).
Rolfea Zahlbr. = Palmorchis Rodr. (Orchidac.).
× **Rolfeara** hort. Orchidaceae. Gen. hybr. (iii) (Cattleya × Rhyncholaelia [Brassavola hort.] × Sophronitis).
Rolfeëlla Schlechter. Orchidaceae. 1 Madag.
Rolfinkia Zenk. = Centratherum Cass. (Compos.).
Rollandia Gaudich. Campanulaceae. 15 Hawaii.
Rollinia A. St-Hil. Annonaceae. 65 C. Am. to Argent., W.I. Some have ed. fr.
Rolliniopsis Safford. Annonaceae. 5 Brazil.
Rolofa Adans. = Glinus L. (Aïzoac.).
Romana Vell. = Buddleja L. (Buddlejac.).
Romanesia Gandog. = Antirrhinum L. (Scrophulariac.).
Romanoa Trevisan. Euphorbiaceae. 1 E. Brazil.
Romanowia Hort. Sand. = Ptychosperma Labill. (Palm.).
Romanschulzia O. E. Schulz. Cruciferae. 8 Mex., C. Am.
Romanzoffia Cham. Hydrophyllaceae. 4 Alaska to Calif. (?NE. Siberia).
Romboda P. & K. = Rhomboda Lindl. = Hetaeria Bl. (Orchidac.).
Rombolobium P. & K. = Rhombolobium Rich. ex Kunth = Clitoria L. (Legumin.).
Rombut Adans. = Cassytha L. (Laurac.).
Romeroa Dugand. Bignoniaceae. 1 Colombia.
Romneya Harv. Papaveraceae. 2 California, NW. Mex.
Romovia Muell. Arg. = Ronnowia Buc'hoz = Omphalea L. (Euphorbiac.).
Rompelia K.-Pol. = Angelica L. (Umbellif.).
Romualdea Triana & Planch. = Cuervea Triana ex Miers (Celastrac.).
*****Romulea** Maratti. Iridaceae. 90 Eur., Medit., S. Afr.
Ronabea Aubl. = Psychotria L. (Rubiac.).
Ronabia St-Lag. = praec.
Roncelia Willk. = Roucela Dum. = Campanula L. + Wahlenbergia Schrad. ex Roth (Campanulac.).
Ronconia Rafin. = Ammannia L. (Lythrac.).
Rondachine Bosc = Brasenia Schreb. (Cabombac.).
Rondeletia L. Rubiaceae. 120 warm Am., W.I.
Rondonanthus Herzog. Eriocaulaceae. 2 Venez., Guiana.
Ronnbergia E. Morr. & André. Bromeliaceae. 5 trop. S. Am.
Ronnowia Buc'hoz = Omphalea L. (Euphorbiac.).
Ronoria Augier = Rinorea Aubl. (Violac.).
Roodia N. E. Brown. Aïzoaceae. 3 S. Afr.
Rooksbya (Backeb.) Backeb. = Neobuxbaumia Backeb. (Cactac.).
Rooseveltia O. F. Cook. Palmae. 1 Costa Rica (I. del Coco).
Ropala J. F. Gmel. = Roupala Aubl. (Proteac.).

Ropalocarpus Boj. = Rhopalocarpus Boj. (Sphaerosepalac.).

Ropalon Rafin. = Nuphar Sm. (Nymphaeac.).

Ropalopetalum Griff. = Artabotrys R.Br. (Annonac.).

Ropalophora P. & K. = Rhopalephora Hassk. = Aneilema R.Br. (Commelinac.).

Rophostemon Endl. = Roptrostemon Blume = Nervilia Comm. ex Gaudich. (Orchidac.).

Rophostemum Reichb. = praec.

Ropourea Aubl. = Diospyros L. (Ebenac.).

Roptrostemon Blume = Nervilia Comm. ex Gaudich. (Orchidac.).

Roraimanthus Gleason = Sauvagesia L. (Ochnac.).

Roram Endl. = Raram Adans. = Cenchrus L. (Gramin.).

Rorella Hill = Drosera L. (Droserac.).

Rorella Rafin. = Drosophyllum Link (Droserac.).

Rorida J. F. Gmel. = Cleome L. (Cleomac.).

Roridula Burm. f. ex L. Roridulaceae. 2 S. Afr.

Roridula Forsk. = Rorida J. F. Gmel. = Cleome L. (Cleomac.).

*Roridulaceae Engl. & Gilg (~ Byblidaceae Domin). Dicots. 1/2 S. Afr. Insectivorous shrublets, bearing stalked capit. viscous glands of various lengths on stems, ls. and K. Ls. alt., elong. lin.-lanc., ent. or pinnatif., circinate in vernation, exstip. Fls. reg., ♀, sol., axill., bibracteolate. K 5, connate at base, imbr., persist.; C 5, imbr., broad-ellipt., acute; A 5, alternipet., anth. with basal thickening, basifixed, inverted in bud, later erect, dehisc. by apic. pore or short slit; G (3), with short simple style and small capit. stig., and 1–several pend. axile ov. per loc. Fr. a 3-valved loculic. caps.; seeds rather large, with crustaceous areolate testa and copious fleshy endosp. Only genus: *Roridula*. Perhaps some affinity with *Byblidac.* and *Pittosporac.*; great superficial similarity to *Droserac.* Pollination by small *Heteroptera*; the anthers are sensitive, and upon being touched by a suitable visitor spring suddenly upright, scattering their pollen over it.

Roripa Adans. = seq.   [mts.

Rorippa Scop. Cruciferae. 70 almost cosmop., esp. N. & S. temp., and trop.

Rosa L. Rosaceae. 250 N. temp., and trop. mts. The thorns are epidermal appendages. The fr. (hip) consists of a number of achenes enclosed in the fleshy recept. which closes over them after fert. The fl. of *R. canina* L. is a pollen fl. *R. centifolia* L. is the form from which the cabbage rose is derived; numerous forms of this and other spp. are cult. Otto (attar) of roses is distilled mainly from *R. damascena* Mill., cult. in the Balkans. (E. Willmot, *The genus Rosa*, London, 1914; Baker, Revised Classification of Roses, in *J. Linn. Soc., Bot.* **35**: 70, 1905; Boulenger, *Bull. Jard. Bot. Brux.* **9–14**, 1924–37, *passim*.)

*Rosaceae Juss. Dicots. 100/2000, cosmop. Trees, shrubs and herbs, usu. perenn.; ls. alt. (exc. *Rhodotypos*), simple or cpd., usu. stip., the stipules often adnate to the petiole. Veg. repr. in various ways, but esp. by creeping stems— runners as in *Fragaria*, or suckers as in *Rubus idaeus* L. Fls. term., in racemose or cymose infls. of various types; great variety of forms. Receptacle generally ± hollowed, so that various degrees of perigyny occur. Frequently there is a central protuberance bearing the cpls., even in the forms with very much hollowed recept. In a few cases (subfam. II) the cpls. are united to the recept. and fully inf. The recept. often forms a part of the fr. Fl. usu. ♀ and actinom.

K 5, often with an epicalyx of outer and smaller ls. (see *Potentilla*), usu. imbr.; C 5, usu. imbr.; A 2, 3 or 4 times as many as petals, or ∞, bent inwards in bud; G usu. apocarpous and sup., rarely syncarpous or inf.; cpls. as many or 2 or 3 times as many as petals, or ∞ or 1–4. Ovules anatr., usu. 2 in each cpl. Style often lat. or basal. Fr. various, dry or fleshy; often an aggregate of achenes (*Potentilla*) or drupes (*Rubus*), or a single drupe (*Prunus*), or pome (*Malus*), and so on (cf. genera, esp. those mentioned, and *Fragaria, Geum, Rosa, Poterium*). Seed usu. exalbum.

Fls. in general of simple type, with slightly concealed honey and sta., usu. protandrous. *Poterium* spp. are anemoph.

Ls. or lfts. usu. ± toothed, but entire in *Cotoneaster, Osteomeles*, some *Cliffortia, Spiraea*, etc.

Classification and chief genera (after Focke). Closely related to *Saxifragaceae*, some genera being almost arbitrarily placed in one or the other; perhaps also allied to *Calycanthaceae, Myrtaceae* (floral diagram of *M.* practically the same as that of § II), and *Chrysobalanaceae*. See also *Biebersteiniaceae*.

I. **Spiraeoïdeae** (cpls. 12–1, usu. 5–2, whorled, neither on special carpophore nor sunk in recept., with 2 or more ovules in each; fr. usu. dehisc.; sta. on broad base, tapering upwards; stipules often absent):
1. SPIRAEEAE (follicle, seeds not winged): *Spiraea*.
2. QUILLAJEAE (follicle, seeds winged): *Quillaja*.
3. HOLODISCEAE (achene): *Holodiscus* (only genus).

II. **Pyroïdeae (Pomoïdeae)** (cpls. 5–2, united to inner wall of recept., usually syncarpous; axis fleshy in fruit; stipules):
1. PYREAE ('Pomariae'): *Pyrus, Malus, Sorbus*.

III. **Rosoïdeae** (cpls. ∞ or rarely 1 on carpophore, sometimes enclosed in axis in fr.; fr. 1-seeded, indehisc.):
1. KERRIEAE (stips. distinct; axis not forming part of fr.; sta. tapering upwards from broad base; cpls. few, whorled; sta. ∞): *Rhodotypos, Kerria*.
2. POTENTILLEAE (as last, but cpls. usu. ∞, in a head, or rarely few and then sta. also few):
   a. Rubinae (drupes, no epicalyx): *Rubus*.
   b. Potentillinae (achenes, seed pend.; usu. epicalyx): *Fragaria Potentilla*.
   c. Dryadinae (as *b*, seed erect): *Geum, Dryas*.
3. CERCOCARPEAE (stipules slightly developed; torus tubular; cpl. 1; achene): *Adenostoma, Purshia*.
4. ULMARIEAE (torus flat or nearly so; sta. with narrow base): *Ulmaria*.
5. SANGUISORBEAE (torus cup-like enclosing cpls., hardening in fr.; cpls. 2 or more): *Alchemilla, Agrimonia, Poterium*.
6. ROSEAE (torus cup-like or tubular, enclosing ∞ cpls., and fleshy in [fr.): *Rosa*.

[IV. **Neuradoïdeae**: see *Neuradaceae*.]
V. **Prunoïdeae** (cpl. 1, rarely 2–5, free of torus; drupe; trees with simple ls.; style almost term.; ovules pend.; fls. reg.).
1. PRUNEAE: *Nuttallia, Prunus*.
[VI. **Chrysobalanoïdeae**: see *Chrysobalanaceae*.]

ROSALESIA

Rosalesia La Llave = Brickellia Ell. (Compos.).

Rosanovia Benth. & Hook. f. = seq.

Rosanowia Regel = Sinningia Nees (Gesneriac.).

Rosanthus Small = Gaudichaudia Kunth (Malpighiac.).

Rosaura Nor. = Ardisia Sw. (Myrsinac.).

Roscheria H. Wendl. Palmae. 1 Seychelles.

Roscia D. Dietr. (sphalm.) = Boscia Thunb. = Toddalia Juss. (Rutac.).

Roscoea Roxb. = Sphenodesma Jack (Symphorematac.).

Roscoea Sm. Zingiberaceae. 15 Himal. to China. *R. purpurea* Sm. has a zygo. fl. with two lips. Insects landing on the lower and probing are obstructed by two projecting spikes from the lower end of the anther; pressure on these brings the anther (with the stigma, which projects beyond it) down upon the insect's back. Protandrous. (Cf. *Sálvia.*)

Roscyna Spach = Hypericum L. (Guttif.).

Rosea Fabr. (uninom.) = *Rhodiola* L. sp. (Crassulac.).

Rosea Klotzsch = Neorosea N. Hallé (Rubiac.).

Rosea Mart. = Iresine P.Br. (Amaranthac.).

Roseanthus Cogn. = Polyclathra Bertol. (Cucurbitac.).

Roseia Frič = Ancistrocactus Britt. & Rose (Cactac.).

Rosenbachia Regel = Ajuga L. (Labiat.).

Rosenbergia Oerst. = Cobaea Cav. (Cobaeac.).

Rosenbergiodendron Fagerl. Rubiaceae. 3 trop. S. Am.

Rosenia Thunb. Compositae. 1 S. Afr. Shrub. Ls. opp.

Rosenstockia Copel. Hymenophyllaceae. 1 New Caled.

Roseocactus A. Berger = Ariocarpus Scheidw. (Cactac.).

Roseocereus (Backeb.) Backeb. = Trichocereus Britt. & Rose (Cactac.).

Roseodendron Miranda. Bignoniaceae. 3 C. & trop. S. Am.

Roshevitzia Tsvelev = Diandrochloa de Winter (Gramin.).

Rosilla Less. (~ Dyssodia Cav.). Compositae. 1 Mex.

Roslinia G. Don = Roeslinia Moench = Chironia L. (Gentianac.).

Roslinia Neck. = Justicia L. (Acanthac.).

Rosmarinus L. Labiatae. 3 Medit. *R. officinalis* L. (rosemary), a xero. shrub with ls. rolled back and stomata in hairy grooves on lower side (cf. *Ericaceae, Empetrum*). Oil of rosemary is employed in perfumery, etc.

Rospidios A. DC. = Diospyros L. (Ebenac.).

Rossatis Thou. (uninom.) = *Satyrium rosellatum* Thou. = *Habenaria rosellata* (Thou.) Schltr. (Orchidac.).

Rossenia Vell. = Angostura Roem. & Schult. (Rutac.).

Rossina Steud. (sphalm.) = Possira Aubl. = Swartzia Schreb. (Legumin.).

Rossittia Ewart = Hibbertia Andr. (Dilleniac.).

Rossmaesslera Reichb. = Gilia Ruiz & Pav. (Polemoniac.).

Rossmannia Klotzsch = Begonia L. (Begoniac.).

Rossolis Adans. = Drosera L. (Droserac.).

Rostellaria Gaertn. f. = ? Bumelia Sw. (Sapotac.).

Rostellaria Nees = seq. [1 Queensland.

Rostellularia Reichb. Acanthaceae. 1 Abyss., 20 trop. & subtrop. As.,

Rostkovia Desv. Juncaceae. 1 N.Z., Campbell I., T. del Fuego, Falkland Is., S. Georgia; 1 Tristan da Cunha.

Rostraceae Dulac = Geraniaceae Juss.
Rostraria Trin. = Trisetum Pers. (Gramin.).
Rostrinucula Kudo = Elsholtzia Willd. (Labiat.).
Rosularia (DC.) Stapf. Crassulaceae. 25 E. Medit. to C. As.
Rotala L. Lythraceae. 50 trop. & subtrop., in wet places.
Rotang Adans. = seq.
Rotanga Boehm. = Calamus L. (Palm.).
Rotantha Baker = Lawsonia L. (Lythrac.).
Rotantha Small = Campanula L. (Campanulac.).
Rotbolla Zumagl. = Rottboellia L. f. (Gramin.).
Roterbe Klatt = Botherbe Steud. = Calydorea Herb. (Iridac.).
Rotheca Rafin. = Clerodendrum L. (Verbenac.).
Rotheria Meyen = Cruckshanksia Hook. & Arn. (Rubiac.)
Rothia Borckh. = Mibora Adans. (Gramin.).
Rothia Lam. = Hymenopappus L'Hérit. (Compos.).
*Rothia Pers. Leguminosae. 2 trop. Afr., As., Austr.
Rothia Schreb. = Andryala L. (Compos.).
Rothmaleria Font Quer. Compositae. 1 S. Spain.
Rothmannia Thunb. Rubiaceae. 20 trop. & S. Afr.
Rothrockia A. Gray. Asclepiadaceae. 3 SW. U.S.
Rotmannia Neck. = Eperua Aubl. (Legumin.).
Rottboelia Scop. = Ximenia L. (Olacac.).
Rottboella Murr. = Rottboellia L. f. (Gramin.).
Rottboellia Host = Parapholis C. E. Hubb. + Pholiurus Trin. (Gramin.).
*Rottboellia L. f. Gramineae. 4 trop. & subtrop. Afr., As.
Rottlera Roem. & Schult. = Roettlera Vahl = Didymocarpus Wall. (Gesneriac.)
Rottlera Willd. (1797) = Trewia L. (Euphorbiac.).
Rottlera Willd. (1804) = Mallotus Lour. (Euphorbiac.).
Rotula Lour. Ehretiaceae. 3 E. Brazil, trop. Afr., Indomal. Char. shrub of rocky places by streams, often submerged for weeks; branches v. tough.
Roubieva Moq. = Chenopodium L. (Chenopodiac.).
Roucela Dum. = Campanula L. + Wahlenbergia Schrad. ex Roth (Campanulac.).
Rouchera H. Hallier = seq.
Roucheria Planch. Linaceae. 8 trop. S. Am.
Rouhamon Aubl. = Strychnos L. (Strychnac.).
Roulinia Brongn. = Nolina Michx (Liliac.).
Rouliniella Vail. Asclepiadaceae. 9 trop. Am.
Roumea DC. = Rumea Poit. = Xylosma G. Forst. (Flacourtiac.).
Roumea Wall. ex Meissn. = Daphne L. (Thymelaeac.).
Roupala Aubl. Proteaceae. 50 C. & S. trop. Am.
Roupalia T. Moores & Ayres = seq.
Roupallia Hassk. = Roupellia Wall. & Hook. ex Benth. = Strophanthus DC. (Apocynac.).
Roupelina Pichon = Roupellina (Baill.) Pichon (Apocynac.).
Roupellia Wall. & Hook. ex Benth. = Strophanthus DC. (Apocynac.).
Roupellina (Baill.) Pichon. Apocynaceae. 1 Madag.

*Rourea Aubl. Connaraceae. 80–90 trop. Am., Afr., Madag., SE. As., Malaysia, trop. Austr., Pacif.

Roureopsis Planch. Connaraceae. 2 trop. Afr.; 8 SE. As., W. Malaysia.

Rouseauvia Boj. = Roussea Smith (Brexiac.).

Roussaea DC. = Roussea Sm. (Brexiac.).

Roussea L. = Russelia Jacq. (Scrophulariac.).

Roussea L. ex B. D. Jacks. = Russelia L. f. = Bistella Adans. (Vahliac.).

Roussea Smith. Brexiaceae. 1 Mauritius.

Rousseaceae DC. = Brexiaceae Lindl.

Rousseauxia DC. Melastomataceae. 1 Madag.

Rousselia Gaudich. Urticaceae. 2 C. Am., Colombia, W.I.

Roussinia Gaudich. = Pandanus L. f. (Pandanac.).

Roussoa Roem. & Schult. = Roussea Sm. (Escalloniac.).

Rouxia Husn. = × Elytrordeum Hyl. (Gramin.).

Rouya Coincy. Umbelliferae. 1 Corsica, N. Afr.

Rovillia Bub. = Polycnemum L. (Chenopodiac.).

Roxburghia Banks = Stemona Lour. (Stemonac.).

Roxburghia Koen. ex Roxb. = Olax Schreb. (Olacac.).

Roxburghiaceae Wall. & Lindl. = Stemonaceae Franch. & Sav.

Roycea C. A. Gardner. Chenopodiaceae. 2 W. Austr.

Roydsia Roxb. = Stixis Lour. (Capparidac.).

Royena L. = Diospyros L. (Ebenac.).

Roylea Nees ex Steud. = Melanocenchrus Nees (Gramin.).

Roylea Wall. Labiatae. 1 Himalaya.

Roystonea O. F. Cook. Palmae. 17 Florida, C. & trop. S. Am., W.I. Monoec.; fls. in groups of 3; a ♀ between two ♂♂. *R. oleracea* (Jacq.) O. F. Cook is the cabbage palm; the young head of ls. is cut out and eaten. The fr. yields an oil, and a form of sago is obtained from the stem (see *Metroxylon*). The ls. are used for thatch, etc. *R. regia* (Kunth) O. F. Cook is the royal palm. Both are extensively used for avenues.

Rrynchoryza B. D. Jacks. (sphalm.) = Rhynchoryza Baill. = Oryza L. (Gramin.)

Ruagea Karst. Meliaceae. 20 NW. trop. S. Am.

Rubacer Rydb. = Rubus L. (Rosac.).

Rubachia Berg = Marlierea Cambess. (Myrtac.).

Rubentia Boj. ex Steud. = Toddalia Juss. (Rutac.).

Rubentia Comm. ex Juss. = Elaeodendron Jacq. f. (Celastrac.).

Rubeola Hill = Sherardia L. (Rubiac.).

Rubeola Mill. = Crucianella L. (Rubiac.).

Rubia L. Rubiaceae. 60 W. & C. Eur., Medit., E. trop. & S. Afr., temp. As., Himal., Mex. to trop. S. Am. *R. tinctorum* L. is the madder, formerly cult. for its dye (alizarin), which is now prepared artificially.

*Rubiaceae Juss. Dicots. 500/6000, one of the largest fams. of pls. Most are trop., but a number (esp. *Rubieae*) are temp., and *Galium* has even a few arctic spp. Trees, shrubs and herbs with decussate entire or rarely toothed stip. ls. The stips. exhibit great variety of form; they stand either between the petioles (*interpetiolar*) or between the petiole and the axis (*intrapetiolar*), and are frequently united to one another and to the petioles, so that a sheath is formed round the stem. The two stips.—one from each l.—that stand side by side are usu. united, and in the *Rubieae* are leaf-like, and often as large as the ordinary ls.; a char. appearance is thus produced, the plants seeming to have whorls of ls.; and it is only by noting the axillary buds that a clue is obtained to the real state of affairs. The number of organs—ls. and stips.—in a whorl varies from 4 upwards, according to the amount of 'fusion' or 'branching' of the stips. The simplest case is a whorl of 6, each leaf having 2 separate stips.; if the stips. be united in pairs, a whorl of 4 results; if each stip. be branched into two, we get a whorl of 10, and, if the centre pair of half-stips. on either side be united, a whorl of 8.

Several are myrmecophilous, e.g. *Myrmecodia, Cuviera, Duroia, Hydnophytum.*

Infl. typically cymose. Sol. term. fls. rare; small dichasia more frequent; most common case a much branched cymose panicle.

Fl. usu. ⚥, reg., epig., 4- or 5-merous. K 4–5, epig., often almost absent, usu. open in aestivation, sometimes with one sepal larger than the rest and brightly coloured (*Mussaenda,* etc.); C (4–5), valvate, conv., or imbr.; A 4–5, alt. with petals, epipet.; Ḡ (v. rarely G̲—*Gaertnera*) (2), rarely (1–∞), 2-loc. with 1–∞ anatr. ov. in each loc.; ov. erect, pend., or horiz.; style simple; stigma capitate or lobed. Caps. (septi- or loculi-cidal), berry, or schizocarp. Embryo small, in rich endosp.

Most have conspic. insect-pollinated fls. Many trop. spp. have bee- and Lepidoptera-fls. with long tubes. The *Rubieae* have small fls. with freely exposed or slightly concealed honey, chiefly visited by flies. Honey usu. secreted by an epig. nectary round base of style. Heterostylism is common, and dioecism sometimes occurs.

Several are of economic importance, e.g. *Cinchona,* which yields quinine, while many of its allies have also valuable alkakoids, *Coffea* (coffee), *Uragoga* (ipecacuanha), *Rubia,* etc.

Classification and chief genera (after Schumann):
  I. **Cinchonoïdeae** (ovules ∞ in each loculus).
    A. **Cinchoninae** (fruit dry):
      α. Fls. solitary or in decussate panicles.
        a. Fl. regular; seed not winged; C valvate.
      1. CONDAMINEËAE: *Condaminea.*
      2. OLDENLANDIËAE: *Oldenlandia, Houstonia, Pentas.*
        b. As a, but C imbr. or conv.
      3. RONDELETIËAE: *Rondeletia.*
        [c. As a, but C 2-lipped.
      4. HENRIQUEZIËAE: see *Henriqueziaceae.*]
        d. As b, but seed winged.
      5. CINCHONEAE: *Cinchona, Bouvardia, Cosmibuena.*
        β. Fls. in heads.
      6. NAUCLEËAE: *Uncaria, Nauclea* [see also *Naucleaceae*].
    B. **Gardeniinae** (fruit fleshy):
      7. MUSSAENDEAE (C valvate): *Mussaenda.*
      8. GARDENIËAE (C imbr. or conv.): *Randia, Gardenia, Posoqueria, Duroia.*
  II. **Rubioïdeae (Coffeoïdeae)** (ovules 1 in each loculus).

RUBINA

A. **Guettardinae** (ovule pendulous; micropyle facing upwards):
  1. VANGUERIËAE: *Plectronia, Cuviera.*
  2. GUETTARDEAE: *Guettarda.*
  3. CHIOCOCCEAE: *Chiococca.*
B. **Rubiinae (Psychotriinae)** (ovule ascending; micropyle facing downwards):
  α. C convolute.
  4. IXOREAE: *Coffea, Ixora, Pavetta.*
  β. C valvate.
    a. Ovules inserted at base of loculus.
  5. PSYCHOTRIËAE: *Psychotria, Rudgea, Uragoga, Lasianthus, Myrmecodia.*
  6. PAEDERIËAE: *Paederia.*
  7. ANTHOSPERMEAE: *Nertera, Coprosma, Mitchella.*
  8. COUSSAREËAE: *Faramea.*
    b. Ovules on septum.
  9. MORINDEAE (stips. undivided, not leafy; trees and shrubs): *Morinda.*
  10. SPERMACOCEAE (stips. divided; shrubs and undershrubs): *Borreria.*
  11. RUBIËAE (GALIËAE) (stips. leafy; herbs): *Sherardia, Crucianella, Asperula, Galium, Rubia.*

The following outline revised classification has been proposed by Verdcourt (*Bull. Jard. Bot. Brux.* **28**: 209–90, 1958):

I. *Rubioïdeae.* Rhaphides present; seeds albuminous.—Trees, shrubs, or (mostly) herbs; aestiv. usu. valv.; indum. of stem and ls. often septate. Heterostyly frequent. Ovary 1–∞-loc. Ovules sol. or ∞, erect or pend. Fr. dry or succ., dehisc. or not. Stips. freq. fimbr. Style usu. with 2 lin. lobes, stigmatic within. Cells of testa rarely pitted.—Tribes PSYCHOTRIËAE, COUSSAREËAE, MORINDEAE, SCHRADEREAE, CRATERISPERMEAE, KNOXIËAE (all containing aluminium accumulators); PAEDERIËAE, COCCO-CYPSELEAE, ARGOSTEMMATEAE, OPHIORRHIZEAE, HAMELIËAE, CRUCKSHANKSIËAE, HEDYOTIDEAE, ANTHOSPERMEAE, SPERMACOCEAE, RUBIËAE.

II. *Cinchonoïdeae.* Rhaphides 0; seeds albuminous. Hairs never truly septate.—Trees and shrubs, rarely herbs. Aestiv. various. Indum. usu. of thick-walled non-sept. or incompletely sept. hairs. No complete heterostyly. Ovary 1–∞-loc., with 1–∞ axile or pend. ov. per loc. Fr. dry or succ., dehisc. or indehisc. Stips. usu. ent. Style fusif., capit. or with lin. lobes. Cells of testa of first 4 tribes usu. pitted.—Tribes NAUCLEËAE, CINCHONEAE, RONDELETIËAE (incl. CONDAMINEËAE), MUSSAENDEAE, CATESBAEËAE, GARDENIËAE, IXOREAE, RETINIPHYLLEAE, ALBERTEAE, VANGUERIËAE, CHIOCOCCEAE.

III. *Guettardoïdeae.* Rhaphides 0; seeds exalbuminous (or with traces only).—Trees or shrubs. Aestiv. imbr. or valv. Indum. not truly septate. Heterostyly unknown, but style sometimes grows as fl. expands. Ovary 2–∞-loc., with 1 sol. pend. ov. per loc. Fr. drupac. or with woody putamen, with 2 pyrenes, rarely dicoccous. Seeds with little or no endosp. Testa not sculptured or irreg. retic.—Tribe GUETTARDEAE.

Rubina Nor. = Antidesma L. (Stilaginac.).
Rubioïdes Soland. ex Gaertn. = Opercularia Gaertn. (Rubiac.).

Rubiteucris Kudo = Teucrium L. (Labiat.).

Rubus L. Rosaceae. 250 cosmop., esp. N. temp.; ± 3000 critical segregates and apomictic forms of *R. fruticosus* L. have also been recognised. Fls. conspic.; honey secreted by a ring-shaped nectary upon the hollowed axis just within the insertion of the sta. Fls. homogamous, visited by many insects, including bees. Fr. an aggregate of drupes. *R. chamaemorus* L., the cloudberry (arctic, Scotland), has creeping underground stems by means of which a large veg. repr. is carried on. Fls. sol., term. and unisexual, occasionally ♀. *R. idaeus* L. (raspberry) multiplies largely by suckers—stems which grow out horiz. beneath the soil to some distance, then turn up and give rise to new pls. which flower in their second year. *R. fruticosus* L. (a general specific name for the ∞ ± apomictic variants of the common bramble or blackberry) is a hook-climber (the hooks being emergences) sprawling over the surrounding vegetation. Branches which reach the soil often take root there and grow up into new plants. *R. caesius* L. (dewberry) has fr. covered with bloom (wax) like grapes. *R. occidentalis* L. is the black-cap raspberry or thimbleberry of N. Am. *R. australis* Forst. f. has the blades of the leaves reduced to the minimum. Many spp. and vars. of blackberry, raspberry, etc., are cult. for ed. fr. The loganberry, a form which appeared in 1881 in the grounds of Judge Logan at Santa Cruz, Calif., is usu. supposed to be a hybrid, but this is disputed.

Ruckeria DC. Compositae. 3 S. Afr.

Ruckia Regel = Rhodostachys Phil. (Bromeliac.).

Rudbeckia Adans. = Conocarpus L. (Combretac.).

Rudbeckia L. Compositae. 25 N. Am.

Ruddia Yakovl. (~ Ormosia G. Jacks.). Leguminosae. 1 India, SE. As., S. China.

Rudelia B. D. Jacks. (sphalm.) = Riedelia Oliv. (Zingiberac.).

Rudella Loes. (sphalm.) = praec.

Rudgea Salisb. Rubiaceae. 150 C. & trop. S. Am., W.I. Some are heterostyled.

Rudicularia Moç. & Sessé ex Ramírez = Semeiandra Hook. & Arn. (Onagrac.).

Rudolfiella Hoehne. Orchidaceae. 9 W.I., trop. S. Am.

Rudolphia Medik. = Malpighia L. (Malpighiac.).

Rudolphia Willd. = Neorudolphia Britton (Legumin.).

Rudolpho-Roemeria Steud. ex Hochst. = Kniphofia Moench (Liliac.).

Rudua Maekawa (~ Phaseolus L., Vigna Savi). Leguminosae. 2 trop.

Ruehssia Karst. Asclepiadaceae. 5 trop. S. Am.   [As.

Ruelingia Ehrh. = Anacampseros L. (Portulacac.).

Ruelingia F. Muell. = Rulingia R.Br. (Sterculiac.).

Ruellia L. (emend. Bremek.). Acanthaceae. 5 trop. & subtrop. Am. The capsule explodes. The seeds possess surface hairs which, when wetted, swell and adhere to the soil.

Ruellia Nees = Hemigraphis Nees emend. T. Anders. (Acanthac.).

Ruelliola Baill. Acanthaceae. 1 Madag.

Ruelliopsis C. B. Clarke. Acanthaceae. 2–3 trop. & S. Afr.

Rueppelia A. Rich. = Aeschynomene L. (Legumin.).

Rufacer Small = Acer L. (Acerac.).

Rugelia Shuttlew. ex Chapm. = Senecio L. (Compos.).

RUGENDASIA

Rugendasia Schiede ex Schlechtd. = Weldenia Schult. f. (Commelinac.).

Rugenia Neck. = Eugenia L. (Myrtac.).

Ruhamon P. & K. = Rouhamon Aubl. = Strychnos L. (Strychnac.).

Ruizia Cav. Sterculiaceae (Malvaceae?). 3 Réunion.

Ruizia Mutis apud Alba. Colombia. Quid?

Ruizia Ruiz & Pav. = Peumus Molina (*q.v.*) (Laurac. + Monimiac.).

Ruizodendron R. E. Fries. Annonaceae. 1 Peru, Bolivia.

Ruizterania Marcano-Berti (~ Qualea Aubl.). Vochysiaceae. 18 trop. S. Am.

Rulac Adans. = Acer L. (Acerac.).

***Rulingia** R.Br. Sterculiaceae. 20 Madag., Austr.

Rumea Poit. = Xylosma G. Forst. (Flacourtiac.).

Rumex L. Polygonaceae. *S.l.*, 200; *s.str.* (i.e. subgen. *Lapathum*), 170; cosmop., esp. N. temp. Fls. of the type usual in the fam., wind.-fert., with large stigmas (cf. *Rheum*). Some have adv. shoots upon the roots, e.g. *R. acetosella* L. *R. hydrolapathum* Huds. is said to produce aerating roots like a mangrove. The roots of *R. hymenosepalus* Torr. (NW. N. Am.), the *canaigré*, are used for tanning.

Rumfordia DC. Compositae. 12 Mex., C. Am.

Rumia Hoffm. Umbelliferae. 1 Crimea.

Rumicaceae ('-cineae') Dum. = Polygonaceae–Rumiceae Reichb.

Rumicastrum Ulbr. Chenopodiaceae. 1 SW. Austr.

Rumicicarpus Chiov. Tiliaceae. 1 Somalia.

Ruminia Parl. = Leucojum L. (Amaryllidac.).

Rumohra Ching *p.p. max.*, excl. *R. adiantiformis* (Forst.) Ching = Arachniodes Bl. (Aspidiac.).

Rumohra Raddi. Davalliaceae. 1 warm S. hemisph. (Holttum, *Rev. Fl. Malaya*, **2**: 484, 1954).

Rumpfia L. = seq.

Rumphia L. = ? Cordia L. (Ehretiac.). vel ? Croton L. (Euphorbiac.).

Rumputris Rafin. = Cassytha L. (Laurac.).

× **Rumrillara** hort. Orchidaceae. Gen. hybr. (iii) (Ascocentrum × Neofinetia × Rhynchostylis).

Runcina Allemand = Cenchrus L. (Gramin.).

Rungia Nees. Acanthaceae. 50 trop. Afr., India, SE. As., W. Malaysia, Celebes, New Guinea.

Runyonia Rose. Agavaceae. 1 S. U.S. (Texas), Mexico.

Rupala Vahl = Roupala Aubl. (Proteac.).

Rupalleya Morière = Stropholirion Torr. (Liliac.).

Rupicapnos Pomel. Fumariaceae. 32 Spain, N. Afr.

Rupicola Maiden & Betche. Epacridaceae. 2 Austr.

Rupifraga (Sternb.) Rafin. = Saxifraga L. (Saxifragac.).

Ruppalleya Krause (sphalm.) = Rupalleya Morière = Stropholirion Torr. (Liliac.).

Ruppelia Baker = Rueppelia A. Rich. = Aeschynomene L. (Legumin.).

Ruppia L. Ruppiaceae. 2 temp. & subtrop., in salt or brackish water. Slender swimming pl., with the habit of a *Potamogeton*. The fls. are borne just at the surface of the water, where fert. occurs by floating pollen. Each infl. of 2 fls. not enclosed in the spathe at the flowering time.

Ruppiaceae Horan. =Potamogetonaceae Dum.

*Ruppiaceae (Kunth) Hutch. Monocots. 1/2 temp. & subtrop. Submerged slender aquatic herbs of brackish water, with opp. or alt. lin. sheathing ls. Fls. reg., ♀, small, in short term. subumbell. rac., ebract. P o; A 2, fil. short and broad, anth. with renif. loc., extr.; G 4, free, with sess. stig. and 1 pend. campylotr. ov. in each. Fr. of 4 long-stipit. subdrupac. ach. borne on elong. often spirally coiled pedunc. Only genus: *Ruppia*. Prob. related to *Potamogetonac.*

Ruprechtia C. A. Mey. Polygonaceae. 17 Mex. to N. Argent. & Urug., Trinidad.

Ruprechtia Opiz = Thalictrum L. (Ranunculac.).

Ruprechtia Reichb. = Plinthus Fenzl (Aïzoac.).

Rurea P. & K. = Rourea Aubl. (Connarac.).

Rureopsis P. & K. = Roureopsis Planch. (Connarac.).

Rusbya Britton. Ericaceae. 1 Bolivia.

Rusbyanthus Gilg. Gentianaceae. 1 Bolivia.

Rusbyella Rolfe. Orchidaceae. 1 Bolivia.

Ruscaceae Spreng. Monocots. 3/9 W. & C. Eur., Medit. reg. Erect or scandent shrubs, with ls. reduced to small membr. scales, bearing in their axils flattened leaf-like branches (cladodes); these may have a pungent apex, and bear infls. on margins or upper or lower surface, or more rarely the infls. are in short term. rac. free from the cladodes. Fls. reg., ♀ or ♂ ♀ and dioec. P 3 + 3, free or ± connate, unequal, sometimes with fleshy corona; A 6 or 3, fil. connate into a short tube or column, anth. sess., extr. (tube without anth. present in ♀ fls.); G (3), 3- or 1-loc., with 2 collat. orthotr. or ± anatr. ov. per loc. (pistillode sometimes pres. in ♂ fls.). Fr. a berry, with 1 glob. or 2 hemisph. seeds, with endosp. Genera: *Danaë, Semele, Ruscus*. Related to *Liliaceae–Asparageae*, but habit and stam. tube distinctive.

Ruschia Schwantes (=Mesembryanthemum L. emend. N. E. Br.). Aïzoaceae. 350 S. Afr.

Ruschianthemum Friedrich. Aïzoaceae. 1 SW. Afr.

Ruschianthus L. Bolus. Aïzoaceae. 1 SW. Afr.

Ruscus L. Ruscaceae. 7 Madeira, W. & C. Eur., Medit. to Persia. *R. aculeatus* L., butcher's broom, a small shrub. In the axils of scale-ls. stand leaf-like phylloclades; half-way up each is another scale-l., in whose axil stands the fl.

Ruspolia Lindau. Acanthaceae. 4 trop. Afr.

Russea J. F. Gmel. = Roussea Sm. (Escalloniac.).

Russeggera Endl. = Lepidagathis Willd. (Acanthac.).

Russelia Jacq. Scrophulariaceae. 40 Mex. to trop. S. Am. *R. juncea* Zucc. is xero. with much reduced ls. and pendulous green stems. Shoots sometimes appear under cult. with broad ls. (perhaps a reversion to an ancestral type).

Russelia Koen. ex Roxb. = Ormocarpum Beauv. (Legumin.).

Russelia L. f. = Bistella Adans. (Vahliac.).

Russellodendron Britton & Rose = Caesalpinia L. (Legumin.).

Russeria Buek (sphalm.) = Bursera Jacq. (Burserac.).

Russowia C. Winkler. Compositae. 1 C. As.

Rustia Klotzsch. Rubiaceae. 12 Mex. to trop. S. Am., W.I.

RUTA

Ruta L. (excl. Haplophyllum A. Juss.). Rutaceae. 7 Canaries, etc., Medit. reg. to SW. As. *R. graveolens* L., rue, is a strongly smelling shrub, owing to the presence in the ls., etc., of an ethereal oil. The terminal fl. of the infl. is 5-merous, the lat. fls. 4-merous. The sta. lie in pairs in the boat-like petals; one by one they bend upwards over the stigma, dehisce and fall back; when all have done this, the stigma ripens, and finally the sta. again move up and effect self-fert. Chiefly visited by small flies. Rue is a narcotic and stimulant.

*Rutaceae Juss. Dicots. 150/900 trop. & temp., esp. S. Afr. and Austr. Most are shrubs and trees, often xero., frequently of heath-like habit (e.g. *Diosma*). Ls. alt. or opp., exstip., usu. cpd., with glandular dots, often aromatic. In many *Aurantioïdeae* there are short shoots whose ls. are reduced to thorns (cf. *Cactaceae*). Infl. of various forms, usu. cymose, v. rarely epiphyllous (*Erythrochiton* sp.). Fl. ♀, rarely ♂ ♀, reg. or zygo., 5–4-merous (see *Ruta*), with a large disk below G. K or (K) 5 or 4, odd sepal post.; C 5 or 4, imbr.; A 10 or 8, obdiplost., or 5, 3, 2, or ∞, with intr. anthers; G̲, rarely G̅ or semi-inf. (*Platyspermation*), (5 or 4), rarely (3–1) or (∞), often free at base and united above by the style (cf. *Apocynac.*), multiloc.; ov. 2–∞ or 1 in each loc., anatr. with ventral rhaphe and micropyle facing upwards. Fr. various; schizocarps, drupes, berries, etc. Seeds without endosp.

Several *R.* are or have been used in medicine, chiefly on account of the oils they contain, e.g. *Ruta*, *Galipea*, *Toddalia*, etc. *Citrus* yields important fruits (orange, etc.).

Classification and chief genera (after Engler). The groups of *R.* differ considerably among themselves, and several of them have been regarded as independent fams. The relationships to allied fams. are thus given by Engler:

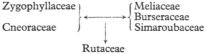

Zygophyllaceae ⎤　　　⎧ Meliaceae
　　　　　　　 ⎬ ←————→ ⎨ Burseraceae
Cneoraceae 　⎦　　　　⎩ Simaroubaceae
　　　　　　　 ↓
　　　　　Rutaceae

    I. **Rutoïdeae.** Cpls. usu. 4–5, rarely 3–1, or > 5, often only united by the style, and ± divided when ripe; loculic. dehiscence usu. with separation of the endocarp; rarely 4–1 fleshy drupes.

       1. ZANTHOXYLEAE (woody plants, usu. with small, greenish, reg., often unisexual fls.; cpl. rarely with > 2 ovules; embryo with flat cots. in endosp.): *Zanthoxylum, Fagara, Choisya.*

       2. RUTEAE (herbs or undershrubs, rarely shrubs, with moderate-sized ♀ fls., sometimes slightly zygo.; cpls. usu. with > 2 ov.; endosp.): *Ruta, Dictamnus.*

       3. BORONIËAE (undershrubs and shrubs, with reg. usu. ♀ fls.; endosp. fleshy, otherwise as 2): *Boronia, Eriostemon, Correa.*

       4. DIOSMEAE (undershrubs and shrubs, rarely trees, with simple ls.; exalbum., embryo usu. straight with fleshy cots.): *Calodendron, Adenandra, Diosma.*

       5. CUSPARIËAE (shrubs and trees with reg. or zygo. fls.; endosp. little or 0; embryo curved, with radicle between cots.): *Almeidea, Galipea, Cusparia.*

    II. **Dictyolomatoïdeae.** Cpls. with several ov., united only at base; fr. with separating endoc., 3–4-seeded.

I. DICTYOLOMATEAE (fls. reg., haplostemonous; sta. with bifid pubescent scales at base; trees with bipinnate ls.): *Dictyoloma*.

III. *Spathelioïdeae*. Cpls. (3), each with 2 pend. ov.; drupe winged; secretory cells and lysigenous oil-glands at margins of ls.

   I. SPATHELIËAE: *Spathelia*.

IV. *Toddalioïdeae*. Cpls. (5–2) or 1, each with 2–1 ov.; drupe or dry winged fr.; endosp. or 0; ls. and bark with lysigenous oil-glands.

   I. TODDALIËAE: *Ptelea, Toddalia, Skimmia*.

V. *Aurantioïdeae*. Berry, often with periderm, and with pulp derived from sappy emergences of cpl. wall. Seeds exalbum., often with 2 or more embryos. Lysigenous oil-glands.

   I. AURANTIËAE: *Glycosmis, Limonia, Atalantia, Feronia, Aegle, Citrus*.

[See also *Flindersiaceae, Rhabdodendraceae*.]

Rutaea M. Roem. = Turraea L. (Meliac.).
Rutamuraria Ort. = Asplenium L. (Aspleniac.).
Rutaria Webb ex Benth. & Hook. f. = Ruta L. (Rutac.).
Ruteria Medik. = Psoralea L. (Legumin.).
Ruthalicia C. Jeffrey. Cucurbitaceae. 2 trop. W. Afr.
Ruthea Bolle (~ Lichtensteinia Cham. & Schlechtd.). Umbelliferae. 3 Canaries, St Helena, S. Afr.
Ruthiella v. Steenis. Campanulaceae. 4 New Guinea. Infl. epiphyllous.
Ruthrum Hill = Echinops L. (Compos.).
Rutica Neck. = Urtica L. (spp. dioec.) (Urticac.).
Rutidea DC. Rubiaceae. 30 trop. Afr., Madag.
Rutidochlamys Sond. = Podolepis Labill. (Compos.).
Rutidosis DC. Compositae. 10 Austr.
Rutilia Vell. [Ls. alt. 3-fol.; K (5), C 5, A 8, G?] Sapindaceae? 1 Brazil.
Rutosma A. Gray = Thamnosma Torr. & Frém. (Rutac.).
Ruttya Harv. Acanthaceae. 6 trop. & S. Afr., Madag.
× **Ruttyruspolia** Meeuse & de Wet. Acanthaceae. Gen. hybr. (Ruspolia × Ruttya).
Ruyschia Fabr. = Ruyschiana Mill. = Dracocephalum L. (Labiat.).
Ruyschia Jacq. Marcgraviaceae. 10 C. & trop. S. Am.
Ruyschiana Mill. = Dracocephalum L. (Labiat.).
Ryacophila P. & K. = Rhuacophila Bl. = Dianella Lam. (Liliac.).
Ryanaea DC. = seq.
*****Ryania** Vahl. Flacourtiaceae. 8 N. trop. S. Am., Trinidad.
Ryckia Balf. f. = Rykia De Vriese = Pandanus L. f. (Pandanac.).
Rydbergia Greene = Actinella Nutt. (Compos.).
Rydbergiella Fedde & Sydow ex Rydb. = Astragalus L. (Legumin.).
Ryditophyllum Walp. = Rytidophyllum Mart. (Gesneriac.).
Ryditostylis Walp. = Rytidostylis Hook. & Arn. (Cucurbitac.).
Rykia De Vriese = Pandanus L. f. (Pandanac.).
Rylstonea R. T. Baker. Myrtaceae. 1 E. Austr.
Rymandra Salisb. = Knightia R.Br. (Proteac.).
Rymia Endl. = Euclea Murr. (Ebenac.).
Ryncholeucaena Britton & Rose. Leguminosae. 1 S. U.S. (Texas), Mex.
Rynchosia Macfad. = Rhynchosia Lour. (Legumin.).

RYNCHOSPERMUM

Rynchospermum P. & K. = Rhyncospermum A. DC. = Rhynchodia Benth. (Apocynac.).

Rynchospora Vahl = Rhynchospora Vahl corr. Willd. (Cyperac.).

Rynchostylis Blume = Rhynchostylis Blume corr. Hassk. (Orchidac.).

Rynchostylis P. & K. = Rinxostylis Rafin. = Cissus L. (Vitidac.).

Ryncospora auctt. = Rhynchospora Vahl corr. Willd. (Cyperac.).

Ryparia Blume = seq.

Ryparosa Blume. Flacourtiaceae. 18 Andaman & Nicobar Is., W. Malaysia, N. New Guinea.

Rysodium Stev. = Astragalus L. (Legumin.).

Ryssopteris Hassk. = seq.

Ryssopterys Blume ex A. Juss. = Rhyssopterys Bl. ex A. Juss. corr. Wittst. (Malpighiac.).

Ryssosciadium Kuntze = Rhysopterus Coulter & Rose (Umbellif.).

Ryssotoechia Kuntze = Rhysotoechia Radlk. (Sapindac.).

Rytachne Endl. = Rhytachne Desv. (Gramin.).

Ryticaryum Becc. = Rhyticaryum Becc. corr. Becc. (Icacinac.).

Rytidea Spreng. = Rutidea DC. (Compos.).

Rytidocarpus Coss. Cruciferae. 1 Morocco.

Rytidochlamys P. & K. = Rutidochlamys Sond. = Podolepis Labill. (Compos.).

Rytidolobus Dulac = Hyacinthus L. (Liliac.). [pos.).

Rytidoloma Turcz. = Dictyanthus Decne (Asclepiadac.).

Rytidophyllum Mart. = Rhytidophyllum Mart. (Gesneriac.).

Rytidosperma Steud. = Deschampsia Beauv. (Gramin.).

Rytidostylis Hook. & Arn. Cucurbitaceae. 5 trop. Am., W.I.

Rytidotus Hook. f. = Bobea Gaudich. (Rubiac.).

Rytiglossa Steud. = Rhytiglossa Nees = Dianthera L. (Acanthac.).

Rytigynia Blume. Rubiaceae. 70 trop. Afr.

Rytilix Rafin. Gramineae. 1 tropics, to S. China & S. U.S.

S

Saba (Pichon) Pichon. Apocynaceae. 3 trop. Afr., Madag., Comoro Is.

Sabadilla Brandt & Ratzeb. = Schoenocaulon A. Gray (Liliac.).

Sabadilla Rafin. = praec.

Sabal Adans. Palmae. 25 warmer Am., W.I. *S. palmetto* Lodd. (palmetto or thatch palm), etc., ls. are used for thatching; the wood is also useful.

Sabalaceae ('-ineae') Schultz–Schultzenst. = Palmae–Sabaleae Mart.

Sabatia Adans. Gentianaceae. 20 S. & E. U.S., Mex., W.I. (rose pinks).

Sabaudia Buscal. & Muschl. Labiatae. 2 NE. & SE. trop. Afr.

Sabaudiella Chiov. Convolvulaceae. 1 NE. trop. Afr.

Sabazia Cass. Compositae. 10 Mex., C. Am.

Sabbata Vell. Compositae (inc. sed.). 2 Brazil.

Sabbatia Moench = Micromeria Benth. (Labiat.).

Sabbatia P. & K. = Sabbata Vell. (Compos.).

Sabbatia Salisb. = Sabatia Adans. (Gentianac.).

Sabdariffa Kostel. = Hibiscus L. (Malvac.).

Sabia Colebr. Sabiaceae. 55 E. & SE. As., Indomal., Solomon Is.
*Sabiaceae Bl. Dicots. 1/55 India & E. As. to Solomons. Decid. or evergr.
scandent shrubs, more rarely erect, with alt. simple ent. membr. to coriac.
exstip. ls., margin usu. cartilag., lat. nerves usu. conspic. arc.-anastom. Fls.
reg., ⚥, small or very small, usu. in axill. few–many-fld. pedunc. cymes, more
rarely in a small pan. or simple rac. K 5, imbr., small; C (4–)5(–6), imbr.,
larger, sometimes thickish, opposite to the sep.; A (4–)5(–6), opposite the pet.
and attached to their base, with ovoid or obl. extr. or intr. anth.; disk annular
or cupular (sometimes tumid), or dentate or lobed, the lobes sometimes bearing
indurated ± discoid glands; G (2), ovoid to conical, with 2 term. erect ±
coherent styles and simple stigs., and 2 collat. or superposed horiz. semi-anatr.
ov. per loc. Fr. of 2 flattened dorsally gibbous drupac. carp., styles becoming
adaxially sub-basal and persist. as a beak on each carp.; exocarp fleshy, endocarp
crustac. and conspic. sculptured or pitted; seeds with little or no endosp.
Only genus: *Sabia*. An interesting group, showing possible or probable connections with several others, e.g. *Menispermac.*, *Icacinac.*, *Meliosmac.*, etc. The
opposition of the members of the K, C and A is a most unusual feature, but
can probably be derived from the Menispermaceous type of flower. Minute
obscure reddish gland-dots occurring in the ls. and various parts of the fl.
recall those found in *Myrsinac.*
Sabicea Aubl. Rubiaceae. 130 trop. Am., Afr., Madag.
Sabina Mill. = Juniperus L. (Cupressac.).
Sabinea DC. Leguminosae. 3 W.I.
Sabinella Nakai = Juniperus L. (Cupressac.).
Sabouraea Léandri = Talinella Baill. (?Portulacac.).
Saboureauea auct. = praec.
Sabsab Adans. = Paspalum L. (Gramin.).
Sabularia Small (sphalm.) = seq.
Sabulina Reichb. = Minuartia L. (Caryophyllac.).
Sacaglottis G. Don = Sacoglottis Mart. (Houmiriac.).
Saccaceae Dulac = Nymphaeaceae Salisb.
× **Saccanthera** hort. Orchidaceae. Gen. hybr. (vi) (Renanthera × Saccola-
bium).
Saccanthus Herzog = Basistemon Turcz. (Scrophulariac.).
Saccardophytum Speg. Solanaceae. 2 Patag.
Saccarum Sanguin. = Saccharum L. (Gramin.).
Saccellium Humb. & Bonpl. Ehretiaceae. 3 trop. S. Am.
Saccharaceae Burnett = Gramineae–Andropogoneae.
Saccharifera Stokes = Saccharum L. (Gramin.).
Saccharodendron Nieuwl. = Acer L. (Acerac.).
Saccharophorum Neck. = Saccharum L. (Gramin.).
Saccharum L. Gramineae. 5 trop., subtrop., incl. *S. officinarum* L. (sugar
cane), a native(?) of trop. E. As., now cult. in most warm regions, esp. Cuba,
Java, Hawaii, etc. From the rhiz. there spring each year shoots which may
reach 3.5–4.5 m. and a thickness of 5 cm.; the outer tissues have much silica in
their cell-walls. The infl. is a dense woolly spike, the spikelets being obscured
by long hairs from the callus. The cult. form has always been veg. propagated
(pieces of the haulm, each bearing a bud, are planted), but all plantations now

use named (or numbered) cultivars which have been developed by planned hybridization involving seedling production. The sugar is contained in the soft central tissues of the stem; the canes are cut before flowering and crushed between rollers to extract the juice; afterwards it is boiled down under reduced pressure and laid out to crystallize.

Sacchrosphendamus Nieuwl. = Sachrosphendamnus Nieuwl. = Acer L.

Saccia Naudin. Convolvulaceae. 1 Bolivia. Quid? [(Acerac.).

Saccidium Lindl. = Holothrix Rich. (Orchidac.).

Saccilabium Rottb. = Nepeta L. (Labiat.).

Sacciolepis Nash. Gramineae. 30 trop. & subtrop.

*****Saccocalyx** Coss. & Dur. (~ Satureia L.). Labiatae. 1 NW. Afr.

Saccocalyx Stev. = Astragalus L. (Legumin.).

Saccochilus Blume = Saccolabium Blume (Orchidac.).

Saccoglossum Schlechter. Orchidaceae. 2 New Guinea.

Saccoglottis Endl. = Sacoglottis Mart. (Houmiriac.).

Saccolabiopsis J. J. Smith. Orchidaceae. 1 Java.

*****Saccolabium** Blume. Orchidaceae. 6 Malaysia.

Saccolabium P. & K. = Saccilabium Rottb. = Nepeta L. (Labiat.).

Saccolena Gleason. Melastomataceae. 1 Colombia.

Saccolepis Nash = Sacciolepis Nash (Gramin.).

Saccoloma Kaulf. Dennstaedtiaceae. 1 trop. Am. Rhiz. scaly, not dorsiventral; ls. simply pinnate. (See *Orthiopteris* Copel.)

Sacconia Endl. = Chione DC. (Rubiac.).

Saccolaria Kuhlmann = Biovularia Kamienski = Utricularia L. (Lentibulariac.).

Saccopetalum Bennett = Miliusa Leschen. (Annonac.).

Saccoplectus Oerst. = Alloplectus Mart. (Gesneriac.).

Saccostoma Wall. ex Voigt = ? Anisochilus Wall. or Coleus Lour. or Plectranthus L'Hérit. (Labiat.).

× **Saccovanda** hort. (vii) = × Sanda hort. (Orchidac.).

Saccularia Kellogg. Scrophulariaceae. 3 NW. Mex. (Lower Calif.).

Sacculina Bosser = Utricularia L. (Lentibulariac.).

Saccus Kuntze = Artocarpus J. R. & G. Forst. (Morac.).

Sacellium Spreng. = Saccellium Humb. & Bonpl. (Ehretiac.).

Sacharum Scop. = Saccharum L. (Gramin.).

Sachrosphendamnus Nieuwl. = Acer L. (Acerac.).

Sachsia Griseb. Compositae. 4 Florida, Cuba, Bahamas.

Sacleuxia Baill. Asclepiadaceae. 2 trop. E. Afr.

Sacodon Rafin. = Cypripedium L. (Orchidac.).

Sacoglottis Mart. Houmiriaceae. 8 C. to trop. S. Am., 1 W. Afr.

Sacoila Rafin. = Stenorhynchos Rich. (Orchidac.).

Sacosperma G. Taylor. Rubiaceae. 2 trop. Afr.

Sacranthus Endl. = Sairanthus G. Don = Nicotiana L. (Solanac.).

Sacrosphendamus Willis (sphalm.) = Sachrosphendamnus Nieuwl. = Acer L. (Acerac.).

Sacropteryx Radlk. (sphalm.) = Sarcopteryx Radlk. (Sapindac.).

Sadiria Mez. Myrsinaceae. 5 E. Himal., Assam.

Sadleria Kaulf. Blechnaceae. 6 Hawaii.

Sadrum Soland. ex Baill. = Pyrenacantha Wight (Icacinac.).

Sadymia Griseb. = Samyda L. (Flacourtiac.).
Saelanthus Forsk. = Cissus L. (Vitidac.).
Saeranthus P. & K. = Sairanthus G. Don = Nicotiana L. (Solanac.).
Saerocarpus P. & K. = Sairocarpus Nutt. ex DC. = Antirrhinum L. (Scrophulariac.).
Saffordia Maxon. Hemionitidaceae (?). 1 Peru.
Saffordiella Merr. Myrtaceae. 1 Philipp. Is.
Safran Medik. = Crocus L. (Iridac.).
Sagapenon Rafin. = Physospermum Cusson (Umbellif.).
Sagenia Presl = Tectaria Cav. (Aspidiac.).
Sageraea Dalzell. Annonaceae. 9 India & Ceylon to W. Malaysia.
Sageretia Brongn. Rhamnaceae. 35 As. Min. & Somalia to Formosa, S. U.S. to trop. S. Am.
Sagina L. Caryophyllaceae. 20–30 N. temp., S. to E. Afr. mts., Himal., New Guinea, Andes. Small herbs with inconspic., sometimes apet. fls.
Sagitta Guett. = seq.
Sagittaria L. (incl. *Lophotocarpus* Dur.). Alismataceae. 20 cosmop., esp. Am., incl. *S. sagittifolia* L. (arrow-head) in Eur., a water plant with a short rhiz. bearing ls. of various types, the number of each kind depending on the depth of the water, etc. The fully submerged ls. are ribbon-shaped, the floating ones have an ovate blade, whilst those (usually the majority) that project above water are arrow-shaped (sagittate). In the axils are formed the 'renewal' shoots which last over the winter, short branches which burrow into the mud and swell up at the ends, each into a large bud whose central axis is swollen with reserve-materials; in spring this develops into a new plant. The diclinous racemose infl. projects above water; the ♀ fls. are lower down than the ♂. The ♂ contains ∞ sta., the ♀ ∞ cpls.
Sagittipetalum Merr. = Carallia Roxb. (Rhizophorac.).
Saglorithys Rizzini. Acanthaceae. 7 Brazil.
Sagmen Hill = Centaurea L. (Compos.).
Sagoaceae ('-goïneae') Schultz-Schultzenst. = Palmae–Metroxyleae Drude.
Sagonea Aubl. = Hydrolea L. (Hydrophyllac.).
Sagotanthus Van Tiegh. = Chaunochiton Benth. (Olacac.).
*Sagotia Baill. Euphorbiaceae. 2 N. Brazil, Guiana.
Sagotia Duchass. & Walp. = Desmodium Desv. (Legumin.).
Sagraea DC. = Clidemia D. Don (Melastomatac.).
Saguaster Kuntze = Drymophloeus Zipp. (Palm.).
Saguerus Steck = Arenga Labill. (Palm.).
Sagus Gaertn. = Raphia P. Beauv. (Palm.).
Sagus Steck = Metroxylon Rottb. (Palm.).
Sahagunia Liebm. Moraceae. 3 trop. Am. *S. strepitans* Engl. (Brazil) yields good timber.
Saheria Fenzl ex Durand = Maerua Forsk. (Capparidac.).
Sahlbergia Reichb. = Salhbergia Neck. = Gardenia Ellis (Rubiac.).
Saintlegeria Cordem. = Chloranthus Sw. (Chloranthac.).
Saintmorysia Endl. = Athanasia L. (Compos.).
Saintpaulia H. Wendl. Gesneriaceae. 12 trop. E. Afr., incl. *S. ionantha* H. Wendl. (Afr. violet). The fl. is like that of *Exacum*, with similar dimorphic

symmetry. In some the style projects to the left over the C, in others to the right (cf. *Exacum*, *Cassia*).

Saintpauliopsis Staner = Staurogyne Wall. (Acanthac. or Scrophulariac.).

Saintpierrea Germain de Saint Pierre = Rosa L. (Rosac.).

Saiothra Rafin. = Sarothra L. = Hypericum L. (Guttif.).

Saipania Hosokawa = Croton L. (Euphorbiac.).

Sairanthus G. Don = Nicotiana L. (Solanac.).

Sairocarpus Nutt. ex A. DC. = Antirrhinum L. (Scrophulariac.).

Saivala Jones = Blyxa Nor. (Hydrocharitac.).

Sajorium Endl. = Plukenetia L. (Euphorbiac.).

Sakakia Nakai = Eurya Thunb. (Theac.).

Sakersia Hook. f. = Dichaetanthera Endl. (Melastomatac.).

Sakoanala R. Viguier. Leguminosae. 2 Madag.

Salabertia Neck. = Tapiria Juss. (Anacardiac.).

Salacca Reinw. Palmae. 10 Indomal. Fr. ed.

Salacia L. Celastraceae. 200 trop., often lianes with dimorphic branches, one form suited for climbing.

Salacicratea Loes. = praec.

Salacighia Loes. Celastraceae. 1 W. Equat. Afr.

Salaciopsis Baker f. Celastraceae. 5 New Caled.

Salacistis Reichb. f. = Hetaeria Bl. or Goodyera R.Br. (Orchidac.).

Salakka Reinw. ex Blume = Salacca Reinw. (Palm.).

Salaxid[ac]eae J. G. Agardh = Ericaceae–Salaxideae Benth.

Salaxis Salisb. Ericaceae. 11 S. Afr.

Salazaria Torr. Labiatae. 1 SW. U.S., Mex.

Salazia Dur. & Jacks. (sphalm.) = Sabazia Cass. (Compos.).

Salceda Blanco = Camellia L. (Theac.).

Saldanha Vell. = Hillia Jacq. (Rubiac.).

Saldanhaea Bur. Bignoniaceae. 5 trop. S. Am.

Saldanhaea P. & K. = Saldanha Vell. = Hillia Jacq. (Rubiac.).

Saldania Sim = Ormocarpum Beauv. (Legumin.).

Saldinia A. Rich. Rubiaceae. 2 Madag.

Salgada Blanco = Cryptocarya R.Br. (Laurac.).

Salhbergia Neck. = Gardenia Ellis (Rubiac.).

Salica Hill = Lythastrum Hill = Lythrum L. (Lythrac.).

*****Salicaceae** Mirbel. Dicots. 3/530, chiefly N. temp. Trees, shrubs and sub-shrubs. Ls. simple, usu. alt., stip. Fls. ♂ ♀, dioec., in catkins, often precocious, each fl. subtended by a scale (or bract). P o or repr. by a cupular disk or small nectary; A 2–30, free or connate, anth. 2-loc.; G (2), 1-loc., ± flask-shaped, with ∞ anatr. ov. on 2–4 pariet. or basal plac. Fr. a 2–4-valved caps.; seeds exalbum., covered with silky hairs arising from the funicle. Genera: *Populus*, *Salix*, *Chosenia*. Many hybrids. Incidence of rust fungi (*Uredinales*) suggests affinity with *Flacourtiaceae* (Holm, *Nytt Mag. Bot.* **16**: 147–50, 1969).

Salicaria Mill. = Lythrum L. (Lythrac.).

Salicaria Moench = Nesaea Comm. ex Juss. (Lythrac.).

Salicari[ace]ae Juss. = Lythraceae Jaume St-Hil.

Salicornia L. Chenopodiaceae. 35 temp. & trop., on sea-coasts and inland salt-pans. *S. herbacea* L. (*s.l.*) (saltwort) cosmop. Succulent herbs, with the habit

SALPIGLOSSIDACEAE

of a cactus, leafless and with jointed nodes. Fls. in groups of 3 or more, one
group sunk in the tissue on either side of each internode. P fleshy; sta. 1 or 2.
Salicorni[ac]eae J. G. Agardh = Chenopodiaceae–Salicornioïdeae Ulbr.
Salimori Adans. = Cordia L. (Ehretiac.).
Saliola Schwag. ex Demid. = ? Salsola L. (Chenopodiac.).
Salisburia Sm. = Ginkgo L. (Ginkgoac.).
Salisburiaceae Link = Ginkgoaceae Engl.
Salisburiana Wood (sphalm.) = Salisburia Sm. = Ginkgo L. (Ginkgoac.).
Salisburya Hoffmgg. = praec.
Salisburyaceae Kuntze = Salisburiaceae Link = Ginkgoaceae Engl.
Salisia Lindl. = Kunzea Reichb. (Myrtac.).
Salisia Panch. ex Brongn. & Gris = Xanthostemon F. Muell. (Myrtac.).
Salisia Regel = Gloxinia L'Hérit. (Gesneriac.).
Saliunca Rafin. = Valerianella Mill. (Valerianac.).
Salix L. Salicaceae. 500, chiefly N. temp.; often difficult to identify on account
of frequent hybridization. Trees, shrubs or subshrubs. Bud covered with a
single calyptrate scale. Ls. simple, alt. or occasionally subopp., usu. short-
stalked, often conspic. stipulate. Fls. entomophilous, with an entire scale (or
bract); nectaries 1–4; stamens usually 2, sometimes 1–12. Style 1; stigmas 2,
often divided. Capsule with 2 valves. Few willows are of economic importance,
though the osiers (chiefly *S. purpurea* L. and *S. viminalis* L.) are used in
basket-making. Cricket-bats are made from the wood of *S. alba* L. var.
caerulea (Sm.) Sm. *S. babylonica* L., *S. daphnoïdes* Vill. and many others are
valued as ornamentals.
Salizaria A. Gray (sphalm.) = Salazaria Torr. (Labiat.).
Salkea Steud. = seq.
Salken Adans. = Derris Lour. (Legumin.).
Salloa Walp. (sphalm.) = Galloa Hassk. = Cocculus DC. (Menispermac.).
Salmalia Schott & Endl. = Bombax L. (Bombacac.).
Salmasia Bub. = Aira L. (Gramin.).
Salmasia Reichb. (sphalm.) = Salmalia Schott & Endl. (Bombacac.).
Salmasia Schreb. = Tachibota Aubl. = Hirtella L. (Chrysobalanac.).
*****Salmea** DC. Compositae. 12 Mex., C. Am., W.I.
Salmeopsis Benth. Compositae. 1 S. Brazil, Paraguay.
Salmia Cav. = Sanseverinia Petagna (Agavac.).
Salmia Willd. = Carludovica Ruiz & Pav. (Cyclanthac.).
Salmiopuntia Frič = Opuntia Mill. (Cactac.).
Salmonea Vahl = Salomonia Lour. (Polygalac.).
Salmonia Scop. = Vochysia Juss. (Vochysiac.).
Saloa Stuntz = Blumenbachia Schrad. (Loasac.).
Salomonia Fabr. (uninom.) = *Polygonatum* Mill. sp. (Liliac.).
*****Salomonia** Lour. Polygalaceae. 8 Indomal., Austr. Some are parasitic.
Salpianthus Humb. & Bonpl. Nyctaginaceae. 1–4 Mex., C. Am.
Salpichlaena J. Sm. Blechnaceae. 1 trop. Am. Rachis of ls. scandent, twining.
Salpichroa Miers. Solanaceae. 25 warm Am.
Salpichroma Miers = praec.
Salpiglaena Klotzsch = Salpichlaena J. Sm. (Blechnac.).
Salpiglossidaceae Hutch. = Solanaceae–Salpiglossideae.

SALPIGLOSSIS

Salpiglossis Ruiz & Pav. Solanaceae. 18 S. Am.

Salpiglottis Hort. ex C. Koch = praec.

Salpinctes R. E. Woodson. Apocynaceae. 2 Venez.

Salpinga Mart. ex DC. Melastomataceae. 6 trop. S. Am.

Salpingacanthus S. Moore = Ruellia L. (Acanthac.).

Salpingantha Hook. corr. Hort. ex Lem. Acanthaceae. 2 Jamaica.

Salpingia (Torr. & Gray) Raimann = Oenothera L. (Onagrac.).

Salpingoglottis C. Koch = Salpiglossis Ruiz & Pav. (Solanac.).

Salpingolobivia Y. Ito = Echinopsis Zucc., etc. (Cactac.).

Salpingostylis Small = Ixia L. (Iridac.).

Salpinxantha Hook. corr. Urb. = Salpingantha Hook. corr. Hort. ex Lem. (Acanthac.).

Salpixantha Hook. = praec.

Salpixanthus Lindl. = praec.

Salsa Feuillée ex Ruiz & Pav. = Herreria Ruiz & Pav. (Liliac.).

Salsola L. Chenopodiaceae. 150 cosmop., maritime or on salt steppes. *S. kali* L. (glasswort), a very fleshy plant with ls. ending in spines. The var. *tenuifolia* Tausch (*S. pestifera* A. Nels.) ('Russian thistle') has since about 1900 become a pest of agriculture in N. Am.

Salsolaceae Moq.-Tand. = Chenopodiaceae Vent.

Saltera Bullock. Penaeaceae. 1 S. Afr.

Saltia R.Br. (1814) = Cometes L. (Caryophyllac.).

Saltia R.Br. ex Moq. (1849). Amaranthaceae. 1 S. Arabia.

Saltzwedelia Gaertn., Mey. & Scherb. = Genistella Moench (Legumin.).

Salutiaea Colla = Achimenes P.Br. (Gesneriac.).

Salutiea Griseb. = praec.

Saluzzia Colla ex auctt. = praec.

Salvadora L. Salvadoraceae. 4–5 warm Afr., As. *S. persica* L. has been thought by some (prob. erroneously) to be the mustard of the Bible. Its ls. taste like mustard.

*Salvadoraceae Lindl. Dicots. 3/12 dry hot regions (often coastal or saline) of Afr., Madag., As. Trees, shrubs or scramblers, sometimes spinose (*Azima*), with opp. ent. stip. ls.; whole plant usu. olive grey in colour. Fls. ♂ or polyg.-dioec., reg. K (2–4), imbr. or ± valv.; C 4–5 or (4–5), imbr. or contorted, usu. with teeth or glands on inner side; A 4–5, alternipet., free or epipet., fil. sometimes conn. into a tube; G (2), 1–2-loc., with 1–2 erect anatr. ov. per loc., and with short style and bifid or subent. stig. Fr. a 1-seeded berry or drupe; seed exalbum. Genera: *Azima, Dobera, Salvadora*. Relationships doubtful; perhaps some connection with *Avicenniaceae*.

Salvadoropsis H. Perrier. Celastraceae. 1 Madag.

Salvertia A. St-Hil. Vochysiaceae. 1 campos of S. Brazil. Stigma lat.

Salvetia Pohl (nomen). Brazil. Quid?

Salvia L. Labiatae. 700 trop. & temp. The sta. are reduced to 2 (the ant.), each of which has a sort of T-shape, the connective of the versatile anther being greatly elongated. The stalks of the sta. stand up together across the mouth of the fl., and a bee, in pushing down towards the honey, comes into contact with the inner end of the anther and, raising it, causes the outer to descend upon its back and dust it with pollen. In some forms of *S*. both ends of the lever bear fertile anthers; but in most the useless half-anther at the inner end is aborted, and the outer half of the connective is much longer than the inner (compare *S. officinalis* L. with *S. pratensis* L.). The fl. is protandrous, and in the later stage the style bends down and places the stigma in position to be touched first by an entering insect. Some have coloured bracts at the top of the infl., adding to its conspicuousness. *S. officinalis* L. (Medit.) is the garden sage.

Salviacanthus Lindau = Justicia L. (Acanthac.).

Salviastrum Fabr. (uninom.) = *Tarchonanthus* L. sp. (Compos.).

Salviastrum Scheele (~ Salvia L.). Labiatae. 4 S. U.S.

Salvinia Seguier. Salviniaceae. 10 trop. & warm temp., incl. *S. natans* (L.) All.

Salviniaceae Dum. Salviniales. Free-floating ferns on water; at each node is a whorl of three ls., two floating ls. and a submerged l. There are no roots, their function being performed by the finely divided submerged ls. The sporocarps are borne several together as outgrowths from the base of a submerged l. The microspores germinate inside the sporangium, the prothalli emerging through its wall as fine tubes, at the end of which the antheridia form. Megaspores produce prothalli bearing archegonia. 1 gen.: *Salvinia*.

Salviniales. Salviniidae. 2 fams.: *Salviniaceae, Azollaceae.*

Salviniidae. Subclass of Filicopsida.

Salweenia E. G. Baker. Leguminosae. 1 SE. Tibet.

Salzmannia DC. Rubiaceae. 1 E. Brazil.

Salzwedelia O. F. Lang = Saltzwedelia Gaertn., Mey. & Scherb. = Genista L. (Legumin.).

*****Samadera** Gaertn. = Quassia L. (Simaroubac.).

Samaipaticereus Cárdenas. Cactaceae. 2 Bolivia.

Samama Kuntze = Anthocephalus A. Rich. (Naucleac.).

Samandura Baill. = Samadera Gaertn. = Quassia L. (Simaroubac.).

Samanea (Benth.) Merr. Leguminosae. 20 Mex. to trop. S. Am., trop. Afr.

Samara L. = Embelia Burm. f. (Myrsinac.).

Samara Sw. = Myrsine L. (Myrsinac.).

Samaraceae Dulac = Ulmaceae Mirb.

Samaroceltis Poiss. = Phyllostylon Capan. (Urticac.).

Samaropyxis Miq. = Hymenocardia Wall. (Hymenocardiac.).

Samarpses Rafin. = Fraxinus L. (Oleac.).

Samba Roberty = Triplochiton K. Schum. (Sterculiac.).

Sambirania Tard. Lindsaeaceae. 2 Madag. (*Mém. Inst. Sci. Madag.* B. 7: 36, fig. 1, 1–9, 1956.)

Sambucaceae Link. Dicots. 1/40 cosmop. Intermediate between *Staphyleac.* and *Caprifoliac.*, differing from the former in the connate, sometimes valv. pet., inf. ovary and baccate fr.; from the latter in the pinnate, stip. ls. (the stips. sometimes clustered or transformed into glands), with stom. on upper as well as lower surface, in the extrorse anthers and sometimes valv. pet., and in many anat. features (cf. Schwerin in *Mitt. Deutsch. Dendrol. Ges.* 1920). K, C, A 3–5, $\overline{G}$ (3–5); fr. a berry-like drupe with 3–5 1-seeded pyrenes. Only genus: *Sambucus*.

Sambucus L. Sambucaceae. 40 cosmop. (exc. Amazonia, Afr. [1 on E. Afr. mts.], Arabia, Penins. Ind., W. Austr., Pacif.). Lenticels show clearly in the

SAOUARI

Saouari Aubl. = Caryocar L. (Caryocarac.).

Saphesia N. E. Brown. Aïzoaceae. 1 S. Afr.

*Sapindaceae Juss. Dicots. 150/2000 trop. & subtrop. 5 gen. (*Serjania*, *Paullinia*, etc.) with 300 spp. are lianes, the rest erect trees or shrubs. The lianes climb by tendrils, which are metam. infl.-axes and are usu. branched or sometimes watch-spring-like; their stems often show peculiar internal anatomy. Ls. alt., stip. in the climbing spp., usu. cpd., pinnate; in the climbing spp. there is usu. a true term. leaflet, but not in the erect; in these one of the last pair of leaflets often becomes term., so that the l. is asymmetric. The tissues of the plants usually contain resinous or latex-like secretions in special cells. The infl. is cymose, usu. a cincinnus, with bracts and bracteoles.

Fl. unisexual (the sta. are apparently well developed in the ♀, so that it is easily mistaken for ♂, but the pollen is useless, and the anthers do not open), generally monoec., reg. or often obliquely zygo., 5- or 4-merous. K usu. 5, rarely (5), imbr. or rarely valvate or open, sometimes apparently 4-merous by union of 2 sepals; C usu. 5, imbr., with well-marked disk between it and the sta.; A usu. 5+5 in one whorl, often with 2 absent, more rarely 5, 4, or ∞, inserted within or rarely upon the disk round the rudimentary ovary; G in ♀ fl. usu. (3), 3-loc., with term. style; ovules usu. 1 in each loc., ascending, with ventral raphe. Fr. a caps., nut, berry, drupe, schizocarp, or samara, usu. large, often red; seed often arillate, with no endosp.; embryo usu. curved.

Many *S.* are of economic value; several yield valuable timber; *Nephelium*, *Litchi* and others furnish ed. fr.

Classification and chief genera (after Radlkofer):

I. **Sapindoïdeae** ('*Eusapindaceae*') (ov. sol. in loc., erect or ascending, micropyle down):
   i. 'Nomophyllae' (ls. usu. imparipinn.; disk oblique): *Serjania*, *Paullinia*.
   ii. 'Anomophyllae' (ls. usu. paripinn.; disk annular): *Sapindus*, *Talisia*, *Schleichera*, *Litchi*, *Nephelium*, *Pappea*, *Cupania*, *Blighia*.

II. **Dodonaeoïdeae** ('*Dyssapindaceae*') (ov. usu. 2 or several in each loc., in the first case erect or pend., in the second horiz., rarely 1 pend. with micropyle up): *Koelreuteria*, *Dodonaea*, *Harpullia*.

Sapindopsis How & Ho. Sapindaceae. 1 S. China (Hainan).

Sapindus L. Sapindaceae. 13 trop. & subtrop. As., Pacif. (not Austr.), Am. The berries of *S. saponaria* L. (Am.) contain saponin, form a lather with water, and may be used as soap.

Sapiopsis Muell. Arg. = Sapium P.Br. (Euphorbiac.).

Sapium P.Br. Euphorbiaceae. 120 trop. & subtrop. (in Am. S. to Patag.). Seeds of *S. sebiferum* Roxb. (tallow-tree, China) are coated with fat; they also yield an oil by pressure. Some yield rubber (Bolivian, Colombian).

Saponaceae Vent. = Sapindaceae Juss.

Saponaria L. Caryophyllaceae. 30 temp. Euras., chiefly Medit. The leaves of *S. officinalis* L. (soapwort) lather if rubbed with water. Fls. protandrous,

Saposhnikovia Schischk. Umbelliferae. 1 NE. As.   [butterfly-visited.

Sapota Mill. = Achras L. (Sapotac.).

*Sapotaceae Juss. Dicots. 35–75 ill-defined gen., 800 spp., trop. Mostly trees with entire leathery ls., sometimes stip. They are commonly hairy with

SARCHOCHILUS

2-shanked hairs, and contain secretory passages in pith, cortex and ls. Fls. sol.
or in cymose bunches in the l. axils or on old stems, bracteolate, ♂, reg. or not.
K 2+2, 3+3, 4+4, or 5; C usu. equal in number to sepals, and alt. with the
K as a whole, rarely in 2 whorls. In *Mimusopeae* the petals have dorsal
appendages like themselves, giving the appearance of more than one whorl.
Sta. in 2 or 3 whorls, but frequently the outer staminodial or absent; anthers
commonly extr. G, syncarpous, multiloc.; cpls. = or twice the number of sta.
in a whorl, or more; ovules at base of axile placenta, one in each loc., anatr.
with micropyle facing down; style simple. Berry, the flesh sometimes
sclerenchymatous near the surface. Seeds few or one, usually album.; endosp.
oily; testa hard and rich in tannin. Many *S.* furnish useful products, esp.
gutta-percha and balata; see all gen. below. A fam. of conspic. economic value.

Classification and chief genera (after Engler):
1. SAPOTEAE (petals without appendages): *Pouteria, Madhuca, Payena, Palaquium, Achras, Butyrospermum, Sideroxylon, Chrysophyllum.*
2. MIMUSOPEAE (petals with appendages—see above): *Mimusops, Manilkara.*

× **Sappanara** hort. Orchidaceae. Gen. hybr. (iii) (Arachnis × Phalaenopsis × Renanthera).

Sapphoa Urb. Acanthaceae. 1 Cuba.
Sapranthus Seem. Annonaceae. 9 Mex., C. Am.
Sapria Griff. Rafflesiaceae. 1-2 Assam to Indoch.
Saprosma Blume. Rubiaceae. 30 SE. As., Indomal.
Sapucaya Knuth. Lecythidaceae. 1 Brazil.
Saraca L. Leguminosae. 20 trop. As. Young shoots pend. (cf. *Amherstia, Brownea*). Fls. (scented at night) used as temple offerings.
Saracenia Spreng. = Sarracenia L. (Sarraceniac.).
Saracha Ruiz & Pav. Solanaceae. 20 Mex. to Peru.
Saragodra Hort. ex Steud. = Suregada Roxb. ex Rottl. (Euphorbiac.).
Sarana Fisch. ex Baker = Fritillaria L. (Liliac.).
Saranthe (Regel & Koern.) Eichl. Marantaceae. 10 Brazil.
Sararanga Hemsl. Pandanaceae. 2 Philipp. & Solomon Is.
Sarathrochilus auctt. (sphalm.) = Sarothrochilus Schltr. = Trichoglottis Bl. (Orchidac.).
Sarawakodendron Ding Hou. Celastraceae. 1 Borneo. Ls. alt.; seeds with filamentous aril.
Sarazina Rafin. = Sarracenia L. (Sarraceniac.).
Sarcandra Gardn. Chloranthaceae. 3 E. As., Indomal. Stem without
Sarcanthemum Cass. = Psiadia Jacq. (Compos.). [vessels.
Sarcanthera Rafin. = Gymnostachyum Nees (Acanthac.).
Sarcanthidion Baill. (~ Citronella D. Don). Icacinaceae. 1 New Caled.
Sarcanthidium Endl. & Prantl = praec.
Sarcanthus Anderss. = Heliotropium L. (Boraginac.).
Sarcanthus Lindl. Orchidaceae. 100 trop. As., Austr.
Sarcathria Rafin. = Salicornia L. (Chenopodiac.).
Sarcaulis B. D. Jacks. (sphalm.) = seq.
Sarcaulus Radlk. Sapotaceae. 2 Venez., N. Brazil, Guiana.
Sarcheta Rafin. (nomen). Compositae. Quid?
Sarchochilus Vidal = Sarcochilus R.Br. (Orchidac.).

1027

SARCINANTHUS

Sarcinanthus Oerst. = Carludovica Ruiz & Pav. (Cyclanthac.).

Sarcobatus Nees. Chenopodiaceae. 1–2 N. Am.

Sarcobatus K. Schum. (sphalm.) = Sarcolobus R.Br. (Asclepiadac.).

Sarcobodium Beer = Bulbophyllum Thou. (Orchidac.).

Sarcobotrya R. Viguier = Kotschya Endl. (Legumin.).

Sarcoca Rafin. = Phytolacca L. (Phytolaccac.).

Sarcocalyx Walp. = Aspalathus L. (Legumin.).

Sarcocalyx Zipp. = Exocarpus Labill. (Santalac.).

Sarcocampsa Miers = Peritassa Miers (Celastrac.).

Sarcocapnos DC. Fumariaceae. 4 W. Medit.

Sarcocarpon Blume = Kadsura Juss. (Schisandrac.).

Sarcocaulon (DC.) Sweet. Geraniaceae. 12 SW. Afr. Xero.; fleshy stems. When the l. falls the base of the petiole hardens to a thorn.

Sarcoccaceae [sic] Dulac = Coriariaceae DC.

× **Sarcocentrum** hort. Orchidaceae. Gen. hybr. (iii) (Ascocentrum × Sarcochilus).

Sarcocephalus Afzel. ex R.Br. = Nauclea L., s.str. (Naucleac.).

Sarcochilus R.Br. Orchidaceae. 12 New Guinea, Solomons, Austr., Polynesia.

Sarcochlaena Spreng. = Sarcolaena Thou. (Sarcolaenac.).

Sarcochlamys Gaudich. Urticaceae. 1 Indomal.

Sarcoclinium Wight = Agrostistachys Dalz. (Euphorbiac.).

Sarcococca Lindl. Buxaceae. 16–20 Himal. to C. China, Indoch., Hainan, Formosa, N. Philipp. Is.; S. India, Ceylon; Sumatra, Java.

Sarcocodon N. E. Br. = Caralluma R.Br. (Asclepiadac.).

Sarcocolla Boehm. = Penaea L. (Penaeac.).

Sarcocolla Kunth = Saltera Bullock (Penaeac.).

Sarcocordylis Wall. = Balanophora J. R. & G. Forst. (Balanophorac.).

Sarcocyphula Harv. = Cynanchum L. (Asclepiadac.).

Sarcodactilis Gaertn. f. = Citrus L. (Rugac.).

Sarcodes Torr. Monotropaceae. 1 W. U.S.

Sarcodiscaceae Dulac = Rutaceae Juss.

Sarcodiscus Griff. = Kibara Endl. (Monimiac.).

Sarcodiscus Mart. ex Miq. = Sorocea A. St-Hil. (Morac.).

Sarcodium Pers. = Sarcodum Lour. = Clianthus Banks & Soland. (Legumin.).

Sarcodraba Gilg & Muschler. Cruciferae. 4 Andes.

Sarcodum Lour. = Clianthus Banks & Soland. (Legumin.).

Sarcoglossum Beer = Cirrhaea Lindl. (Orchidac.).

Sarcoglottis Presl. Orchidaceae. 17 Mex. to trop. S. Am., W.I.

Sarcogonum G. Don = Muehlenbeckia Meïssn. (Polygonac.).

Sarcolaena Thou. Sarcolaenaceae. 10 Madag.

***Sarcolaenaceae** Caruel. Dicots. 8/40 Madag. Trees or shrubs, with simple ent. alt. stip. ls., stips. often large, similar to *Ficus*, extra- or intra-pet. Fls. reg., ♀, singly or 2 together in an invol. of various form, in cymose infl. K 3–5, imbr., equal or unequal; C 5–6, large, contorted; disk (or stds.) sometimes present; A ∞, more rarely 5–10, sometimes fasic., anth. intr. or extr., basifixed or dorsif.; G̲ (1–5), 1–5-loc., with few to several, basal, apical or axile, ascend. or descend. anatr. ov. per loc., style usu. thick, ± elong., usu. with lobed stig. Fr. a several-seeded loculic. caps., or 1-seeded and indehisc., sometimes encl.

in a woody sac, often surr. by lignified bracts or a cupule; seeds with fleshy or horny endosp. Genera: *Sarcolaena, Leptolaena, Xyloölaena, Perrierodendron, Eremolaena, Schizolaena, Rhodolaena, Pentachlaena.*

Sarcolemma Griseb. ex Lorentz = Sarcostemma R.Br. (Asclepiadac.).

Sarcolipes Eckl. & Zeyh. = Crassula L. (Crassulac.).

Sarcolobus R.Br. Asclepiadaceae. 15 SE. As., Malaysia.

Sarcolophium Troupin. Menispermaceae. 1 trop. Afr.

Sarcomelicope Engl. Rutaceae. 2 New Caled.

Sarcomeris Naud. = Pachyanthus A. Rich. (Melastomatac.).

Sarcomorphis Boj. ex Moq. = Salsola L. (Chenopodiac.).

Sarcomphalium Dulac = Hyacinthus L. (Liliac.).

Sarcomphalodes (DC.) Kuntze = Noltea Reichb. (Rhamnac.).

Sarcomphalus P.Br. = Ziziphus Mill. (Rhamnac.).

× **Sarconopsis** hort. Orchidaceae. Gen. hybr. (iii) (Phalaenopsis × Sarcochilus).

Sarcoperis Rafin. = Campelia Rich. (Commelinac.).

Sarcopetalum F. Muell. Menispermaceae. 1 E. Austr.

Sarcophagophilus Dinter. Asclepiadaceae. 2 SW. Afr.

Sarcophrynium K. Schum. Marantaceae. 5 trop. Afr.

Sarcophyllum E. Mey. = Lebeckia Thunb. (Legumin.).

Sarcophyllum Willd. = seq.

Sarcophyllus Thunb. = Aspalathus L. (Legumin.).

Sarcophysa Miers = Juanulloa Ruiz & Pav. (Solanac.).

Sarcophyta auctt. = Sarcophyte Sparrm. (Balanophorac.).

Sarcophytaceae (Engl.) Van Tiegh. = Balanophoraceae–Sarcophytoïdeae Engl.

Sarcophyte Sparrm. Balanophoraceae. 2 trop. E. Afr., N. Transvaal.

Sarcopilea Urb. Urticaceae. 1 S. Domingo.

Sarcopodaceae Gagnep. = Santalaceae–Anthoboleae (Dum.) Endl.

Sarcopodium Lindl. (1853; non Ehrenb. ex Brongn. 1824—Fungi) = Epigeneium Gagnep. (Orchidac.).

Sarcopoterium Spach. Rosaceae. 1 Italy, E. Medit.

Sarcopteryx Radlk. Sapindaceae. 8 E. Malaysia, Austr.

Sarcopus Gagnep. = Exocarpus Labill. (Santalac.).

Sarcopygme Setchell & Christophersen. Rubiaceae. 5 Samoa.

Sarcopyramis Wall. Melastomataceae. 6 Himal. & S. China to W. Malaysia.

Sarcorhachis Trelease. Piperaceae. 4 C. & trop. S. Am.

× **Sarcorhiza** hort. (vii) = × Rhinochilus hort. (Orchidac.).

Sarcorhyna Presl = Rostellaria Gaertn. f. = ?Bumelia Sw. (Sapotac.).

Sarcorhynchus Schlechter. Orchidaceae. 3 trop. Afr.

Sarcorrhiza Bullock. Periplocaceae. 1 trop. Afr.

Sarcoryna P. & K. = Sarcorhyna Presl = Rostellaria Gaertn. f. = ?Bumelia Sw. (Sapotac.).

Sarcosiphon Blume = Thismia Griff. (Burmanniac.).

Sarcosperma Hook. f. Sarcospermataceae. 6–10 E. Himal. to S. & E. China, scattered in W. Malaysia & Moluccas.

*Sarcospermataceae H. J. Lam. Dicots. 1/6 SE. As., Malaysia. Trees or shrubs, sometimes laticiferous, with simple ent. opp. or subopp. stip. ls., stips. small, caduc. Fls. small, reg., ☿, singly or in groups on axill. pan. or rac. K 5, imbr.; C (5), tube short, lobes imbr., spreading, rounded; A 5 + 5, epipet., the outer staminodial, fil. short, anth. basifixed, latero-extr.; G (1-2), with short stout style and ± trunc. stig., and 1 basal-axile apotr. ascend. ov. per loc. Fr. drupac., 1-2-seeded; endosp. o. Only genus: *Sarcosperma.* Closely related to *Sapotac.*

Sarcospermum Reinw. ex De Vriese = Gunnera L. (Gunnerac.).

Sarcostachys Juss. = Stachytarpheta Vahl (Verbenac.).

Sarcostemma R.Br. Asclepiadaceae. 10 trop. & subtrop. Old World. Leafless xero. with slightly fleshy stems.

Sarcostigma Wight & Arn. Icacinaceae. 6 India to W. Malaysia.

Sarcostigmataceae Van Tiegh. = Icacinaceae–Sarcostigmateae Miers.

Sarcostoma Blume. Orchidaceae. 3 Malay Penins., Java, Celebes.

Sarcostyles Presl ex DC. = Hydrangea L. (Hydrangeac.).

Sarcothalamicae Schultz-Schultzenst. = fam. heterogen., incl. Monimiac., Morac., etc.

Sarcotheca Blume. Averrhoaceae (~ Oxalidac.). 11 W. Malaysia, Celebes.

Sarcotheca Kuntze = Sarotheca Nees = Justicia L. (Acanthac.).

Sarcotheca Turcz. = Schinus L. (Anacardiac.).

× **Sarcothera** hort. Orchidaceae. Gen. hybr. (iii) (Pteroceras [Sarcochilus] × Renanthera).

Sarcotoechia Radlk. Sapindaceae. 2 Austr.

× **Sarcovanda** hort. Orchidaceae. Gen. hybr. (iii) (Sarcochilus × Vanda).

Sarcoyucca (Engelm.) Lindinger. Agavaceae. 13 S. U.S., Mex., W.I.

Sarcozona J. M. Black. Aïzoaceae. 1 S. & E. Austr.

Sarcozygium Bunge = Zygophyllum L. (Zygophyllac.).

Sardinia Vell. = Guettarda L. (Rubiac.).

Sardonula Rafin. = Ranunculus L. (Ranunculac.).

Sarga Ewart = Chrysopogon Trin. (Gramin.).

*****Sargentia** S. Wats. Rutaceae 1 Mex. Fr. ed.

Sargentia Wendl. & Drude ex Salomon = Pseudophoenix Wendl. (Palm.).

Sargentodoxa Rehder & Wilson. Sargentodoxaceae. 1 C. China.

Sargentodoxaceae Stapf. Dicots. 1/1 China. Scandent or scrambling decid. shrubs, with alt. 3-foliol. (sometimes 3-lobed or ent.) exstip. ls. Fls. reg., ♂ ♀, dioec., rac., small, greenish-yellow, long-pedic. K 3 + 3, imbr.; C 3 + 3, squami-form, minute; A 3 + 3, fil. short, anth. obl., extr., thecae separated by produced connective (in ♀ fl. 6 stds.); G ∞, free, imbr., on ± cylindr. torus, with subulate styles, each carp. with 1 ventr. pend. anatr. ov. Fr. formed of enlarged torus bearing ∞ stipit. berry-like carpels, black and pruinose when ripe; seed with fleshy endosp. Only genus: *Sargentodoxa.* An interesting type, combining features of *Lardizabalac.* and *Schisandrac.*

Sariava Reinw. = Symplocos L. (Symplocac.).

Sariawa auctt. = praec.

Saribus Blume = Livistona R.Br. (Palm.).

Sarinia O. F. Cook. Palmae. 1 trop. S. Am.

Sarissus Gaertn. = Hydrophylax L. f. (Rubiac.).

Saritaea Dugand. Bignoniaceae. 1 Colombia.

Sarlina Guillaumin = ? Linociera Sw. (Oleac.).

Sarmasikia Bub. = Cynanchum L. (Asclepiadac.).

Sarmentaceae Schultz-Schultzenst. = Dioscoreaceae R.Br. + Liliaceae Juss.
Sarmentaceae Sonnenb. = Smilacaceae + Liliaceae–Convallariëae, etc. [*p.p.*
Sarmentaceae Vent. = Vitidaceae Juss.
Sarmentaria Naud. = Adelobotrys DC. (Melastomatac.).
***Sarmienta** Ruiz & Pav. Gesneriaceae. 1 Chile.
Sarmienta Sieb. ex Baill. = Cnestis Juss. (Connarac.).
Sarna Karst. = Pilostyles Guillemin (Rafflesiac.).
Sarojusticia Bremek. Acanthaceae. 1 N. & C. Austr.
Sarosanthera Korth. = Adinandra Jack (Theac.).
Sarotes Lindl. = Guichenotia J. Gay (Sterculiac.).
***Sarothamnus** Wimm. (~ Cytisus L.). Leguminosae. 10 Atl. Is., Eur., W. Sib., *S. scoparius* (L.) Wimm. ex Koch, the broom. The ls. are reduced to scales and assim. is chiefly performed by the stems. The fl. has an explosive mech., in general like *Genista* (*q.v.*), but different in detail. The style is very long and there are two lengths of sta., so that pollen is shed near the tip of the keel (where also is the stigma) and also about half-way along its upper side. When an insect alights on the fl. (there is no honey), the keel begins to split from the base towards the tip, and presently the pollen of the short sta. is shot out upon the lower surface of the visitor; immediately afterwards, the split having reached the tip, the other pollen and the style spring violently out and strike the insect on the back. As the stigma touches first there is thus a chance of a cross, if the insect bear any pollen. Afterwards the style bends right round and the stigma occupies a position just above the short sta., so that another chance of cross-fert. is afforded if other insects visit the fl. (in most exploding fls. there is only the one chance). The fr. explodes by a twisting of the valves.
Sarotheca Nees = Justicia L. (Acanthac.).
Sarothra L. = Hypericum L. (Guttif.).
Sarothrochilus Schlechter = Trichoglottis Blume (Orchidac.).
Sarothrostachys Klotzsch = Sebastiania Spreng. (Euphorbiac.).
Sarpedonia Rafin. = Ranunculus L. (Ranunculac.).
Sarracena L. = seq.
Sarracenia L. Sarraceniaceae. 10 Atl. N. Am. (side-saddle fl.), in sunny marshy places. Low herbs with rosettes of rad. ls.; each l. is repres. by a long narrow pitcher with a flat green wing of tissue on the ventral side, serving chiefly for assim. The general structure of the pitcher is similar to that found in *Nepenthes* (*Nepenthac.*); it has a fixed lid projecting over the mouth, and the lip is usu. turned down inwards. The mouth bears numerous honey-glands; below these comes the 'slide-zone', then the zone of hairs (cf. *Nepenthes*), and at the bottom is water in which the insects are drowned. The pitchers are often brightly coloured. In *S.* the entire l. is a pitcher, while in *Nepenthes* it is only part of the l., and in *Cephalotus* (*Cephalotac.*) only certain ls. Many hybrids.
***Sarraceniaceae** Dum. Dicots. 3/17 Atl. & Pacif. N. Am., N. trop. S. Am. Insectivorous pitcher-plants (see gen.), with rosettes of rad. ls. and rather large reg. ⚥ fls. subtended by 1–3 bracteoles. K (4–)5(–6), persist.; C=K, or 0; A ∞ or 12–15, with ± short fil. and obl. intr. anth.; G̲ (5–6) or (3), with ∞ anatr. ov. on inrolled cpl.-walls. Fr. a loculic. ∞-seeded caps.; endosp. fleshy. Genera: *Sarracenia* (fl. sol.; G̲ (5), style with large umbrella-shaped apical expansion; top of pitcher erect, with ovate 'lid'); *Darlingtonia* (fl. sol.;

G (5), style 5-partite at apex; top of pitcher recurved, with fish-tail-shaped appendage); *Heliamphora* (fls. rac.; C o; G (3), style ± trunc., pitchers with minute appendiculiform 'lid'). Related to *Nepenthac.*

Sarracha Reichb. = Saracha Ruiz & Pav. (Solanac.).

Sarratia Moq. = Amaranthus L. (Amaranthac.).

Sarrazinia Hoffmgg. = Sarracenia L. (Sarraceniac.).

Sarsaparilla Kuntze = Smilax L. (Smilacac.).

Sartidia de Winter. Gramineae. 3 trop. & S. Afr.

Sartoria Boiss. = Onobrychis L. (Legumin.).

Sartwellia A. Gray. Compositae. 3 S. U.S., Mex.

Saruma Oliv. Aristolochiaceae. 1 NW. to SW. China.

Sarumaceae (O. C. Schmidt) Nak. = Aristolochiaceae–Asaroïdeae–Sarumeae

Sasa Makino & Shibata. Gramineae. 200 E. As.   [O. C. Schmidt.

Sasaëlla Makino (~ Arundinaria Michx). Gramineae. 50 E. As.

Sasali Adans. = Microcos L. (Tiliac.).

Sasamorpha Nakai. Gramineae. 15 E. As.

Sasanqua Nees = Camellia L. (Theac.).

Sassa Bruce ex J. F. Gmel. = Acacia Willd. (Legumin.).

Sassafras Bercht. & Presl = Lindera Thunb. (Laurac.).

Sassafras Trew. Lauraceae. 2 China & Formosa; 1 Canada to Florida, *S. albidum* (Nutt.) Nees (*S. officinale* Nees & Eberm.). The wood and bark of *S. albidum* yield oil of sassafras, used in medicine.

Sassafridium Meissn. = Ocotea Aubl. (Laurac.).

Sassea Klotzsch = Begonia L. (Begoniac.).

Sassia Molina = Oxalis L. (Oxalidac.).

Satakentia H. E. Moore. Palmae. 1 Ryukyu Is.

Satania Nor. = Flacourtia Comm. ex L'Hérit. (Flacourtiac.).

Satanocrater Schweinf. Acanthaceae. 4 trop. Afr.

Sataria Rafin. = Oxypolis Rafin. (Umbellif.).

Saterna Nor. = Rauvolfia L. + Urceola Roxb. (Apocynac.).

Satirium Neck. = Satyrium L. (Orchidac.).

Satorchis Thou. = seq.

Satorkis Thou. = Habenaria Willd. + Cynorkis Thou. + Benthamia A. Rich.

Sattadia Fourn. Asclepiadaceae. 2 Brazil.   [(Orchidac.).

Saturegia Leers = seq.

Satureia auctt. = seq.

Satureja L. Labiatae. *S.l.* 200; *s.str.* 30, temp. & warm regions. Fls. gynodioec. *S. hortensis* L. and *S. montana* L. (summer and winter savories), cult. condiments.

Saturiastrum Fourr. = praec.

Saturna B. D. Jacks. = Saterna Nor. = Rauvolfia L. + Urceola Roxb. (Apocynac.).

Saturnia Maratti = Allium L. (Alliac.).

Satyria Klotzsch. Ericaceae. 25 C. & trop. S. Am.

Satyridium Lindl. Orchidaceae. 1 S. Afr.

Satyrium L. = Coeloglossum Hartm. (Orchidac.).

***Satyrium** Sw. Orchidaceae. 115 trop. & S. Afr., Masc., India, Tibet, China.

Saubinetia Remy = Verbesina L. (Compos.).
Saueria Klotzsch = Begonia L. (Begoniac.).
Saugetia Hitchcock & Chase. Gramineae. 2 W.I.
Saul Roxb. ex Wight & Arn. = Shorea L. (Dipterocarpac.).
Saulcya Michon = Odontospermum Neck. ex Sch. Bip. (Compos.).
Saundersia Reichb. f. Orchidaceae. 1 Brazil.
Saurauia Willd. Actinidiaceae. 300 trop. As., Am.
*__Saurauiaceae__ J. G. Agardh corr. Hutch. = Actinidiaceae–Saurauiëae DC.
Saurauja auctt. = Saurauia Willd. (Actinidiac.).
Saurauj[ac]eae J. G. Agardh = Saurauiaceae (q.v.).
Sauravia Spreng. = Saurauia Willd. (Actinidiac.).
Saurobroma Rafin. = Celtis L. (Ulmac.).
Sauroglossum Lindl. Orchidaceae. 4 trop. & temp. S. Am.
Sauromatum Schott. Araceae. 6 trop. Afr. to W. Malaysia. Ls. pedate (cymosely branched).
Sauropus Blume. Euphorbiaceae. 40 SE. As., Indomal.
*__Saururaceae__ A. Rich. Dicots. 5/7 E. As., N. Am. Perenn. herbs, with alt. ent. simple stip. ls., stips. adnate to pet.; upper ls. often coloured or forming an invol. of bracts. Fls. small, reg., ♂, densely or laxly spicate or rac. K o; C o; A 3, 6, or 8, free or ± adn. to G, with extr. or intr. anth.; G̲ or G̅, 3–4 or (3–4), when free with 2–4 orthotr. ov. per carp., when syncarp. with 6–8 or 1 ov. on pariet. plac. Fr. of sep. dehisc. cocci, or (when syncarp.) indehisc. or dehisc. apically, sometimes glochidiate; seeds with scanty endosp. and copious mealy perisp. Genera: *Saururus, Gymnotheca, Houttuynia, Anemopsis, Circaeocarpus.* Related to *Piperac.* and *Chloranthac.*; often much similarity to *Polygonac.*
Saururopsis Turcz. = seq.
Saururus L. Saururaceae. 1 E. As., Philipp. Is., 1 E. U.S. Bog pls. with spikes of fls., bract usu. adnate to axis of its fl.
Saururus Mill. = Piper L. (Piperac.).
*__Saussurea__ DC. Compositae. 1 Eur., 400 temp. As., 1 Austr., 1 W. N. Am.; esp. on mts. In the European *S. alpina* (L.) DC. the fls. are sweet-scented, an unusual char. in *Compos.* Many spp. have 3 cpls. Extensive hybridism has been reported.
Saussurea Salisb. = Hosta Tratt. (Liliac.).
Saussuria Moench = Nepeta L. (Labiat.).
Saussuria St-Lag. = Saussurea DC. (Compos.).
Sautiera Decne. Acanthaceae. 1 Timor.
Sauvagea Adans. = seq.
Sauvagesia L. Ochnaceae. 30 trop., esp. trop. S. Am. 5 fertile sta., surrounded by ∞ stds. Cpls. 3. This gen. and its relatives (*Luxemburgiëae*) form a transition to *Violac.*
Sauvagesiaceae Dum. = Ochnaceae–Luxemburgiëae Reichb.
Sauvagia St-Lag. = Sauvagesia L. (Ochnac.).
Sauvalella Willis (sphalm.) = Sauvallella Rydb. (Legumin.).
Sauvallea Wright (~ Commelina L.). Commelinaceae. 1 Cuba.
Sauvallella Rydb. Leguminosae. 1 Cuba.
Sava Adans. = Onosma L. (Boraginac.).

SAVASTANA

Savastana Rafin. = Savastania Scop. = Tibouchina Aubl. (Melastomatac.).
Savastana Schrank = Hierochloë S. G. Gmel. (Gramin.).
Savastania Scop. = Tibouchina Aubl. (Melastomatac.).
Savastonia Neck. ex Steud. = praec.
Savia Rafin. = Amphicarpaea Ell. (Legumin.).
Savia Willd. Euphorbiaceae. 2 S. U.S., 12 W.I., 2 S. Brazil; 1 S. Afr., 14 Madag.
Savignya DC. Cruciferae. 2 Morocco to Afghan.
Saviniona Webb & Berth. = Lavatera L. (Malvac.).
Saviona Pritz. (sphalm.) = praec.
Saxegothaea Eichl. = seq.
Saxe-Gothaea Lindl. Podocarpaceae. 1 Andes of Patag. Forms a connecting link between *Podocarpaceae* and *Araucariaceae* (Erdtman, *Pollen Morph. & Pl. Taxon.* 3: 73, 1965).
Saxegothea Benth. = praec.
Saxe-Gothea Gay = praec.
Saxicolella Engl. Podostemaceae. 2 trop. W. Afr.
Saxifraga L. Saxifragaceae. 370 N. temp., Arctic, Andes, chiefly alpine. Most show xero. chars., such as tufted growth, close packing of ls. (esp. well shown in *S. oppositifolia* L.), succulence, hairiness, etc. Many are veg. repr. by offsets, or (e.g. *S. granulata* L.) by bulbils in the lower leaf-axils. Many exhibit chalk-glands at the tips or edges of the ls. (e.g. *S. oppositifolia* at the tip); these are water-pores with nectary-like tissue beneath, secreting water containing chalk in solution. As the water evaporates, the chalk forms an incrustation. Fls. usu. in dich. cymes with cincinnus tendency. Every stage occurs from hypogyny to epigyny. Honey only partially concealed; fls. visited by miscellaneous insects. Most are protandrous. A few, e.g. *S. stolonifera* Curtis, have zygo. fls. See Engler, *Pflanzenreich* IV, **117** (1916–19).
*****Saxifragaceae** Juss. Dicots. 30/580, chiefly N. temp.; a few S. temp., and trop. mts. Perenn. (rarely ann.) herbs, with usu. alt., exstip. ls. Many alpine and arctic forms of xero. habit; many hygrophilous. Infl. of various kinds, both racemose and cymose. Fl. usu. ⚥, reg., cyclic, 5-merous (exc. cpls.). Recept. flat or hollowed to various depths, so that sta. and P may be peri- or epi-gynous. K usu. 5, imbr. or valv.; C 5, imbr. or valv., sometimes (5) or 0; A usu. 5 + 5, obdiplost.; cpls. rarely free and as many as petals, usually fewer and joined below, often 2; plac. axile, with several rows of anatr. ov.; styles as many as cpls. Fls. mostly protandrous. Capsule. Seed with rich endosp. round a small embryo. Chief genera: *Astilbe, Bergenia, Saxifraga, Heuchera, Mitella, Chrysosplenium*. Much in common with *Ranunculac.*
Saxifragella Engl. Saxifragaceae. 2 Antarctic S. Am.
Saxifragites Gagnep. = Distylium Sieb. & Zucc. (Hamamelidac.).
Saxifragodes D. M. Moore. Saxifragaceae. 1 Tierra del Fuego.
Saxifragopsis Small. Saxifragaceae. 1 Pacif. U.S.
Saxiglossum Ching. Polypodiaceae. 1 China.
Saxofridericia R. Schomb. Rapateaceae. 9 trop. S. Am.
Saxogothaea Dalla Torre & Harms = Saxe-Gothaea Lindl. (Podocarpac.).
Sayera P. & K. = seq.
Sayeria Kraenzl. = Dendrobium Sw. (Orchidac.).

Scabiosa L. (*s.str.*). Dipsacaceae. 100 temp. Euras., Medit.; mts. E. Afr., S. Afr.

Scabiosella Van Tiegh. (~ Scabiosa L.). Dipsacaceae. 2 SE. Eur.

Scabrita L. = Nyctanthes L. (Verbenac.).

Scadianus Rafin. = Crinum L. (Amaryllidac.).

Scadiasis Rafin. = Angelica L. (Umbellif.).

Scadoxus Rafin. = Haemanthus L. (Amaryllidac.).

Scaduakintos Rafin. = Brodiaea Sm. (Alliac.).

***Scaevola** L. Goodeniaceae. 80–100 trop. & subtrop., esp. Austr., Polynes. *S. taccada* (Gaertn.) Roxb. and *S. plumieri* Vahl, widely distr. in trop. beach jungle, furnish a kind of rice paper; the pith is squeezed flat.

Scaevol[ac]eae Lindl. = Goodeniaceae R.Br.

Scalesia Arn. Compositae. 20 Galápagos.

Scalia Sims = Podolepis Labill. (Compos.).

Scaligera Adans. = Aspalathus L. (Legumin.).

***Scaligeria** DC. Umbelliferae. 22 E. Medit. to C. As.

Scaliopsis Walp. = Podolepis Labill. (Compos.).

Scambopus O. E. Schulz. Cruciferae. 2 E. & S. Austr.

Scammonea Rafin. = Convolvulus L. (Convolvulac.).

Scandalida Adans. = Tetragonolobus Scop. = Lotus L. (Legumin.).

Scandederis Thou. (uninom.) = Neottia scandens Thou. = Bulbophyllum (Hederorkis) sp. (non B. scandens Kraenzl.) (Orchidac.).

Scandia J .W. Dawson. Umbelliferae. 2 N.Z.

Scandicium Thell. Umbelliferae. 1 Medit.

Scandivepres Loes. Celastraceae. 1 Mex.

Scandix L. Umbelliferae. 15–20 Eur., Medit. The ripe mericarps separate with a jerk.

Scandix Mol. = Erodium L'Hérit. (Geraniac.).

Scapha Nor. = Saurauia Willd. (Actinidiac.) + ?

Scaphespermum Edgew. = Eriocycla Lindl. (Umbellif.).

Scaphiophora Schlechter. Burmanniaceae. 2 Philipp. Is., New Guinea.

Scaphispatha Brongn. Araceae. 1 Bolivia.

Scaphium Endl. Sterculiaceae. 6 W. Malaysia.

Scaphium P. & K. = Skaphium Miq. = Xanthophyllum Roxb. (Xanthophyllac.).

Scaphocalyx Ridley. Flacourtiaceae. 2 Malay Penins.

Scaphochlamys Baker. Zingiberaceae. ?30 Indomal.

Scaphoglottis Dur. & Jacks. = Scaphyglottis Poepp. & Endl. (Orchidac.).

Scaphopetalum Mast. Sterculiaceae. 15 trop. Afr.

Scaphosepalum Pfitz. Orchidaceae. 25 C. Am. to W. trop. S. Am.

Scaphospatha P. & K. = Scaphispatha Brongn. ex Schott (Arac.).

Scaphospermum Korovin. Umbelliferae. 1 C. As.

Scaphospermum P. & K. = Scaphespermum Edgew. = Eriocycla Lindl. (Umbellif.).

Scaphula R. N. Parker = Anisoptera Korth. (Dipterocarpac.).

Scaphyglottis Poepp. & Endl. Orchidaceae. 35 Mex. to trop. S. Am., W.I.

Scaredederis Thou. (sphalm.) = Scandederis Thou. (Orchidac.), *q.v.*

Scariola F. W. Schmidt = Lactuca L. (Compos.).

Scassellatia Chiov. Anacardiaceae. 1 Somalia.

SCATOHYACINTHUS

Scatohyacinthus P. & K. = Scaduakintos Rafin. = Brodiaea Sm. (Alliac.).

Scatoxis P. & K. = Scadoxus Rafin. = Haemanthus L. (Amaryllidac.).

Sceletium N. E. Brown. Aïzoaceae. 20 S. Afr.

Scelochilus Klotzsch. Orchidaceae. 15 C. Am., trop. S. Am.

Scepa Lindl. = Aporusa Blume (Euphorbiac.).

Scepaceae Lindl. = Euphorbiaceae–'Antidesminae' Pax, *p.p.*, excl. *Antidesma*! (Somewhat intermediate between *Euphorbiac.* and *Flacourtiac.*; perhaps some grounds for recognition as distinct fam.)

Scepanium Ehrh. (uninom.) = *Pedicularis sceptrum-carolinum* L. (Scrophulariac.).

Scepasma Blume = Phyllanthus L. (Euphorbiac.).

Scepinia Neck. = Pteronia L. (Compos.).

Scepocarpus Wedd. = Urera Gaudich. (Urticac.).

Scepseothamnus Cham. = Alibertia A. Rich. (Rubiac.).

Scepsothamnus Steud. = praec.

Sceptranthes R. Grah. = Cooperia Herb. (Amaryllidac.).

Sceptranthus Benth. & Hook. f. = praec.

Sceptridium Lyon = Botrychium Sw. (Ophioglossac.).

Sceptrocnide Maxim. = Laportea Gaudich. (Urticac.).

Sceura Forsk. = Avicennia L. (Avicenniac.).

Scevola Rafin. = Scaevola L. (Goodeniac.).

Schachtia Karst. = Duroia L. f. (Rubiac.).

Schaeffera Schreb. = seq.

Schaefferia Jacq. Celastraceae. 16 S. U.S. to trop. S. Am., W.I.

Schaeffnera Benth. & Hook. f. (sphalm.) = Schaffnera Sch. Bip. = Dicoma Cass. (Compos.).

Schaenfeldia Edgew. (sphalm.) = Schoenefeldia Kunth (Gramin.).

Schaenolaena Lindl. = Schoenolaena Bunge = Xanthosia Rudge (Umbellif.).

Schaenoprasum Franch. & Sav. = Schoenoprasum Kunth = Allium L. (Alliac.).

Schaenus Gouan = Schoenus L. (Cyperac.).

Schaetzelia Sch. Bip. = Schaetzellia Sch. Bip. (1850) (Compos.).

Schaetzellia Klotzsch = Onoseris DC. (Compos.).

Schaetzellia Sch. Bip. (1849) = Hinterhubera Sch. Bip. (Compos.).

Schaetzellia Sch. Bip. (1850). Compositae. 2 Mex., C. Am.

Schaffnera Benth. = Schaffnerella Nash (Gramin.).

Schaffnera Sch. Bip. = Dicoma Cass. (Compos.).

Schaffnerella Nash. Gramineae. 1 Mex.

Schaffneria Fée. Aspleniaceae. 1 C. Am.

Schanginia C. A. Mey. = Suaeda Forsk. (Chenopodiac.).

Schanginia Sievers ex Pall. = Hololachne Ehrenb. (Tamaricac.).

Schaphespermum Edgew. (sphalm.) = Scaphespermum Edgew. = Seseli L. (Umbellif.).

Schauera Nees = Endlicheria Nees (Laurac.).

Schaueria Hassk. = Hyptis Jacq. (Labiat.).

Schaueria Meissn. = Endlicheria Nees (Laurac.).

*Schaueria Nees. Acanthaceae.' 8 trop. Am.

Scheadendron Bertol. f. = Sheadendron Bertol. f. = Combretum L. (Combretac.).

Schedonnardus Steud. Gramineae. 1 S. U.S.
Schedonorus Beauv. = Bromus L. + Festuca L. (Gramin.).
Scheelea Karst. Palmae. 40 trop. Am.
Scheeria Seem. = Achimenes P.Br. (Gesneriac.).
Schefferella Pierre = Payena A. DC. (Sapotac.).
Schefferomitra Diels. Annonaceae. 1 New Guinea.
Scheffieldia Scop. = Sheffieldia J. R. & G. Forst. = Samolus L. (Primulac.).
*Schefflera J. R. & G. Forst. Araliaceae. 200 trop. & subtrop.
Schefflerodendron Harms. Leguminosae. 4 trop. Afr.
Scheffleropsis Ridley. Araliaceae. 2 Siam, Malay Penins. $\overline{G}$ (8).
Scheidweileria Klotzsch = Begonia L. (Begoniac.).
Schelhameria Fabr. (uninom.) = Matthiola R.Br. sp. (Crucif.).
*Schelhammera R.Br. Liliaceae. 3 E. Austr.
Schelhammeria Moench = Carex L. (Cyperac.).
Schellanderia Francisci = Synotoma (G. Don) R. Schulz (Campanulac.).
Schellenbergia C. E. Parkinson. Connaraceae. 1 Lower Burma.
Schellingia Steud. = Aegopogon Beauv. = Amphipogon R.Br. (Gramin.).
Schellolepis J. Sm. = Goniophlebium (Bl.) Presl (Polypodiac.).
Schelveria Nees & Mart. = Angelonia Humb. & Bonpl. (Scrophulariac.).
Schenckia K. Schum. Rubiaceae. 1 S. Brazil.
Schenkia Griseb. = Centaurium Hill (Gentianac.).
Schenodorus Beauv. (sphalm.) = Schedonorus Beauv. = Festuca L. + Bromus L. (Gramin.).
Scheperia Rafin. = seq.
Schepperia Neck. = Cadaba Forsk. (Capparidac.).
Scherardia Neck. = Sherardia L. (Rubiac.).
Schetti Adans. = Ixora L. (Rubiac.).
Scheuchleria Heynh. = ? Vernonia L. (Compos.).
Scheuchzera St-Lag. = seq.
Scheuchzeria L. Scheuchzeriaceae. 1–2 N. temp. & arctic.
*Scheuchzeriaceae Rudolphi. Monocots. 1/2 N. temp. Slender perenn. herbs of Sphagnum bogs; ls. alt., lin., sheathing, with ligules or squamulae intravaginales in axils (cf. Potamogetonac.). Fls. reg., ⚥, in term. bract. racemes, greenish, anemoph., protog. P 3 + 3, homochlam.; A 3 + 3, with basifixed extr. anth.; G 6 or 3, shortly united below, with sess. stigs., and 2–few basal erect anatr. ov. per carp. Fr. a schizocarp, with free divaricate 1–2-seeded ventrally dehisc. follicles; seeds exalbum. Only genus: Scheuchzeria.
Scheutzia Gandog. = Rosa L. (Rosac.).
Schickendantzia Pax. Alstroemeriaceae. 2 Argentina.
Schickendantzia Speg. = seq.
Schickendantziella Speg. Liliaceae. 1 Argentina.
Schidiomyrtus Schau. = Baeckea L. (Myrtac.).
Schidospermum Griseb. ex Lechl. = Fosterella L. B. Smith (Liliac.).
Schieckia Benth. & Hook. f. = Schiekia Meissn. (Haemodorac.).
Schieckia Karst. = Celastrus L. (Celastrac.).
Schiedea Bartl. = Richardsonia Kunth (Rubiac.).
Schiedea Cham. & Schlechtd. Caryophyllaceae. 24 Hawaii.
Schiedea A. Rich. = Tertrea DC. (Rubiac.).

SCHIEDEËLLA

Schiedeëlla Schlechter. Orchidaceae. 7 S. U.S., Mex.
Schiedeophytum H. Wolff = Donnellsmithia Coult. & Rose (Umbellif.).
Schiedia Willis (sphalm.) = Schiedea Cham. & Schlechtd. (Caryophyllac.).
Schiekea Walp. = Schieckia Karst. = Celastrus L. (Celastrac.).
Schiekia Meissn. Haemodoraceae. 1 trop. S. Am.
Schievereckia Nym. = Schivereckia Andrz. ex DC. (Crucif.).
Schillera Reichb. = Eriolaena DC. (Sterculiac.).
Schilleria Kunth = Piper L. (Piperac.).
Schillingia Verdc. Convolvulaceae. 1 Nepal.
Schima Reinw. ex Blume. Theaceae. 15 E. Himal. to Formosa, Bonin & Ryukyu Is., W. Malaysia. *S. wallichii* (DC.) Choisy, good timber.
Schima auct. ex Steud. (sphalm.) = Sehima Forsk. = Ischaemum L. (Gramin.)
Schimmelia Holmes = Amyris P.Br. (Rutac.).
Schimpera Steud. & Hochst. ex Endl. Cruciferae. 2 E. Medit. to S. Persia.
Schimperella H. Wolff. Umbelliferae. 2 mts. E. Afr.
Schimperina Van Tiegh. = Tapinanthus Bl. (Loranthac.).
Schindleria H. Walter. Phytolaccaceae. 6 Peru, Bolivia.
Schinnongia Schrank = ? Hypoxis L. (Hypoxidac.).
Schinocarpus K. Schum. = Selinocarpus A. Gray (Nyctaginac.).
Schinopsis Engl. Anacardiaceae. 7 S. Am. The wood (*quebracho*) is hard and rich in tannin; used for tanning.
Schinus L. Anacardiaceae. 30 Mex. to Argent. *S. molle* L. yields American mastic (resin); cult. for shade, etc. (pepper-tree).
Schinzafra Kuntze = Thamnea Soland. ex Brongn. (Bruniac.).
Schinzia Dennst. = ? Rinorea Aubl. (Violac.).
Schinziella Gilg. Gentianaceae. 2 trop. Afr.
Schippia Burret. Palmae. 1 C. Am.
Schisachyrium Munro = Schizachyrium Nees = Andropogon L. (Gramin.).
*****Schisandra** Michx. Schisandraceae. 25 trop. & warm temp. As., E. N. Am.
*****Schisandraceae** Bl. Dicots. 2/47 E. As., W. Malaysia, SE. U.S. Climbing shrubs with alt. simple exstip. sometimes gland-dotted ls. Fls. reg., ♂ ♀, monoec. or dioec., sol., axill. P 9–15, spiral, imbr.; A 4–∞, short, fil. ± united below, sometimes forming a fleshy glob. mass, anth. small, extr. to intr.; G 12–∞, on a ± elong. column or torus, with 2–5–(11) collat. or superposed anatr. ov. per loc. Fr. an aggr. of ∞ sess. drupe-like carp. on the modified torus; seeds 1–5, flattened, with copious oily endosp. Genera: *Schisandra, Kadsura*. Related to *Magnoliac.*, etc.; fruit analogous to that of *Sargentodoxac.*
Schisanthes Haw. = Narcissus L. (Amaryllidac.).
Schischkinia Iljin. Compositae. 1 Persia to Baluchistan.
Schischkiniella v. Steenis (~ Silene L.). Caryophyllaceae. 1 Turkey.
Schismaceras P. & K. = Schismoceras Presl = Dendrobium Sw. (Orchidac.).
Schismatoclada Baker. Rubiaceae. 20 Madag.
Schismatoclaea Willis (sphalm.) = praec.
Schismatoglottis Zoll. & Mor. Araceae. 100 Malaysia; 1 N. trop. S. Am. At top of spadix, above the ♂ fls., are sterile fls. consisting of stds.
Schismatopera Klotzsch = Pera Mutis (Perac.).
Schismaxon Steud. = Xyris Gronov. ex L. (Xyridac.).
Schismocarpus Blake. Loasaceae. 2 Mex.

Schismoceras Presl = Dendrobium Sw. (Orchidac.).
Schismus Beauv. Gramineae. 5 Afr., Medit. to NW. India.
Schistachne Figari & De Not. = Aristida L. (Gramin.).
Schistanthe Kunze = Alonsoa Ruiz & Pav. (Scrophulariac.).
Schistocarpaea F. Muell. Rhamnaceae. 1 Queensland.
Schistocarpha Less. Compositae. 15 Mex. to Peru.
Schistocarpia Pritz. (sphalm.) = praec.
Schistocaryum Franch. Boraginaceae. 1 SW. China.
Schistocodon Schau. = Toxocarpus Wight & Arn. (Asclepiadac.).
Schistogyne Hook. & Arn. Asclepiadaceae. 12 S. Am.
Schistonema Schlechter. Asclepiadaceae. 1 Peru.
Schistophragma Benth. Scrophulariaceae. 5 Am.
Schistostemon (Urb.) Cuatrec. Houmiriaceae. 7 trop. S. Am.
Schistostephium Less. Compositae. 12 trop. & S. Afr.
Schistostigma Lauterb. = Cleistanthus Hook. f. ex Planch. (Euphorbiac.).
Schistotylus Dockr. Orchidaceae. 1 New S. Wales.
Schivereckia Andrz. ex DC. Cruciferae. 5 N. Russia to Balkan Penins. & As. Min.
Schiverekia Reichb. = praec.
Schiwereckia Andrz. ex DC. = praec.
Schizachne Hack. Gramineae. 1 arct. Eur., NE. Siberia, Japan, temp. N. Am., mts. SW. U.S.
Schizachyrium Nees. Gramineae. 50 trop.
***Schizaea** Sm. Schizaeaceae. 30 trop., also temp. (N. Am., all S. continents). Sporangia in a double row on lower surface of each of the reduced fertile leaflets; branching of l. dichotomous.
Schizaeaceae Mart. Schizaeales. 4/150, trop. & a few subtrop. or temp. *Lygodium* is a leaf-climber. The sporangia are borne (exc. in *Mohria*) on special pinnae or parts of pinnae. The sporangia are large, sessile, usu. without indusium; at the apex is a cap-like annulus, and the sporangium dehisces longitudinally. Genera: *Schizaea, Anemia, Lygodium, Mohria.*
Schizaeales. Filicidae. 1 fam.: *Schizaeaceae* (also 4 fossil fams.).
Schizandra DC. = Schisandra Michx (Schisandrac.).
Schizandraceae auctt. = Schisandraceae Bl.
Schizangium Bartl. ex DC. = Mitracarpum Zucc. (Rubiac.).
Schizanthera Turcz. = Miconia Ruiz & Pav. (Melastomatac.).
Schizanthes Endl. = Schisanthes Haw. = Narcissus L. (Amaryllidac.).
Schizanthoseddera (Roberty) Roberty = Seddera Hochst. (Convolvulac.).
Schizanthus Ruiz & Pav. Solanaceae. 15 Chile. Fl. zygo.; stalk curved, and the two really upper petals form the lower lip which is 3–4-lobed, while the lat. petals are 4-lobed and the lowest petal forms the simple or slightly 2-lobed upper lip. Sta. 4, 2 fertile and 2 staminodial. Fl. like the papilionate *Leguminosae* (cf. *Collinsia*), and fert. in a similar way, usu. by an explosive movement (cf. *Genista*).
Schizeilema Domin. Hydrocotylaceae (~ Umbellif.). 18 Austr., N.Z.
Schizenterospermum Homolle ex Arènes. Rubiaceae. 4 Madag.
Schizobasis Baker. Liliaceae. 9 trop. & S. Afr.
Schizobasopsis Macbride = Bowiea Hook. f. & Haw. (Liliac.).

SCHIZOCAENA

Schizocaena J. Sm. (incl. *Gymnosphaera* sect. 3, Copel., *Gen. Fil.* 99, 1947). (=*Cyathea* subg. *Sphaeropteris* sect. *Schizocaena*, Holttum, *Fl. Males.* II, I, 1963.) Cyatheaceae. 25 Malaysia; small tree-ferns of forest (excl. *C. sinuata* Hook. & Grev., *C. hookeri* Thw. of Ceylon).

Schizocalomyrtus Kausel. Myrtaceae. 1 trop. S. Am.

Schizocalyx Berg = praec.

Schizocalyx Hochst. = Dobera Juss. (Salvadorac.).

Schizocalyx Scheele = Origanum L. (Labiat.).

***Schizocalyx** Wedd. Rubiaceae. 2 Colombia.

Schizocapsa Hance. Taccaceae. 1 SE. China. Scarcely more than a section of *Tacca*.

Schizocardia A. C. Smith & Standl. = Purdiaea Planch. (Cyrillac.).

Schizocarphus van der Merwe. Liliaceae. 4 S. Afr.

Schizocarpum Schrad. Cucurbitaceae. 4 (6?) Mex.

Schizocarya Spach = Gaura L. (Onagrac.).

Schizocasia Schott = Xenophya Schott (Arac.).

Schizocentron Meissn. = Heterocentron Hook. & Arn. (Melastomatac.).

Schizochilus Sond. Orchidaceae. 26 trop. & S. Afr.

Schizochiton Spreng. = Chisocheton Blume (Meliac.).

Schizochlaena Spreng. = Schizolaena Thou. (Sarcolaenac.).

Schizochlaenaceae Wettst. = Sarcolaenaceae Caruel.

Schizococcus Eastw. (~ Arctostaphylos Adans.). Ericaceae. 4 W. U.S.

Schizocodon Sieb. & Zucc. Diapensiaceae. 2 Japan.

Schizocolea Bremek. Rubiaceae. 1 trop. W. Afr.

Schizodium Lindl. Orchidaceae. 9 S. Afr.

Schizoglossum E. Mey. Asclepiadaceae. 50 trop. & S. Afr.

Schizogramma Link (nom. dub.) = ? Hemionitis L. (Gymnogrammac.).

Schizogyna Willis = seq.

Schizogyne Cass. = Inula L. (Compos.).

Schizogyne Ehrenb. ex Pax = Acalypha L. (Euphorbiac.).

Schizoica Alef. = Napaea L. (Malvac.).

Schizojacquemontia (Roberty) Roberty = Jacquemontia Choisy (Convolvulac.).

Schizolaena Thou. Sarcolaenaceae. 7 Madag.

Schizolegna Alston = Lindsaea Dryand. (Lindsaeac.). (See Kramer, *Acta Bot. Neerl.* 6: 102, 1957.)

Schizolepis Schrad. ex Nees = Scleria Bergius (Cyperac.).

Schizolepton Fée = Taenitis Willd. (Gymnogrammac.).

Schizolobium Vog. Leguminosae. 5 C. Am. to Brazil.

Schizoloma Gaud. = Lindsaea Dryand. (Lindsaeac.).

Schizoloma *sensu* Copel. (*Gen. Fil.* 55, 1947) = Taenitis Willd. (Gymnogrammac.), *q.v.*

Schizomeria D. Don. Cunoniaceae. 18 Moluccas, New Guinea, Queensland. Drupe.

Schizomeryta R. Viguier. Araliaceae. 1 New Caled.

Schizomussaenda Li. Rubiaceae. 1 SE. As.

Schizonepeta Briq. Labiatae. 3 temp. As.

Schizonephos Griff. = Piper L. (Piperac.).

Schizonotus A. Gray = Solanoa Greene (Asclepiadac.).
Schizonotus Lindl. ex Wail. = Sorbaria (Ser. ex DC.) A.Br. (Rosac.).
Schizonotus Rafin. = Holodiscus Maxim. (Rosac.).
Schizopepon Maxim. Cucurbitaceae. 3 N. India & China. Fls. sometimes ♀.
Schizopetalon Sims. Cruciferae. 8 Chile.
Schizopetalum DC. = praec.
Schizophragma Sieb. & Zucc. Hydrangeaceae. 8 Himal., E. As.
Schizophyllum Nutt. = Lipochaeta DC. (Compos.).
Schizopleura Endl. = Beaufortia R.Br. (Myrtac.).
Schizopogon Reichb. = Andropogon L. (Gramin.).
Schizopremna Baill. = Faradaya F. Muell. (Verbenac.).
Schizopsera Turcz. = Schizoptera Turcz. corr. C. Muell. (Compos.).
Schizopsis Bur. = Tynnanthus Miers (Bignoniac.).
Schizoptera Turcz. corr. C. Muell. Compositae. 3 Mex., Ecuador.
Schizoscyphus Taub. = Maniltoa Scheff. (Legumin.).
Schizoseddera Roberty = Seddera Hochst. (Convolvulac.).
Schizosepala G. M. Barroso. Scrophulariaceae. 1 Brazil.
Schizosiphon K. Schum. Leguminosae. 1 New Guinea.
Schizospatha Furtado. Palmae. 1 New Guinea.
Schizospermum Boiv. ex Baill. = Cremaspora Benth. (Rubiac.).
Schizostachyum Nees. Gramineae. 35 E. As.
Schizostege Hillebr. = Pteris L. (Pteridac.) (Wagner, *Bull. Torr. Bot. Cl.* **76**: 444–61, 1949).
Schizostegopsis (sphalm. '-stegeopsis') Copel. Pteridaceae. 2 Philippines (Mindanao) (Copel., *Fern Fl. Philipp.* 147, 1958).
Schizostemma Decne = Oxypetalum R.Br. (Asclepiadac.).
Schizostephanus Hochst. ex K. Schum. = Cynanchum L. (Asclepiadac.).
Schizostigma Arn. (1839). Rubiaceae. 1 Ceylon. Ovary 5–7-loc.
Schizostigma Arn. (1841) = Cucurbitella Walp. corr. Hook. f. (Cucurbitac.).
Schizostylis Backh. & Harv. Iridaceae. 2 S. Afr.
Schizotechium Reichb. = Stellaria L. (Caryophyllac.).
Schizotheca Ehrenb. ex Solms = Thalassia Soland. ex Koen. (Hydrocharitac.).
Schizotheca Lindl. = Atriplex L. (Chenopodiac.).
Schizothrinax Wendl. = ? Thrinax L. f. (Palm.).
Schizotrichia Benth. Compositae. 1 Peru.
Schizozygia Baill. Apocynaceae. 1 trop. E. Afr.
Schkuhria Moench = Sigesbeckia L. (Compos.).
*Schkuhria Roth. Compositae. 10 warm Am.
Schlagintweitia Griseb. = Hieracium L. (Compos.).
Schlagintweitiella Ulbr. Ranunculaceae. 2 Tibet, W. China.
Schlechtendahlia Benth. & Hook. f. (sphalm.) = Schlechtendalia Willd. = Adenophyllum Pers. (Compos.).
*Schlechtendalia Less. Compositae. 1 Brazil, Urug., Argent. (variable). Habit somewhat *Eryngium*-like, with opp. or whorled lin. ls. Pollen indicates affinity with *Chuquiraga* Juss.
Schlechtendalia Spreng. = Mollia Mart. (Tiliac.).
Schlechtendalia Willd. = Adenophyllum Pers. (Compos.).

SCHLECHTERA

Schlechtera P. & K. = Schlechteria Bolus (Crucif.).
Schlechteranthus Schwantes. Aïzoaceae. 1 SW. Afr.
× **Schlechterara** hort. (v) = × Ascocenda hort. (Orchidac.).
Schlechterella Hoehne = Rudolfiella Hoehne (Orchidac.).
Schlechterella K. Schum. Periplocaceae. 1 E. Afr.
Schlechteria Bolus. Cruciferae. 1 S. Afr.
Schlechteria Mast. = Phyllocomos Mast. (Restionac.).
Schlechterina Harms. Passifloraceae. 1 trop. E. Afr.
Schlechterosciadium H. Wolff. Umbelliferae. 1 S. Afr.
Schlegelia Miq. Bignoniaceae. 18 trop. S. Am., W.I.
*****Schleichera** Willd. Sapindaceae. 1 Indomal., *S. oleosa* (Lour.) Oken
(Ceylon oak). Useful timber; aril of seed ed.; oil expressed from seed itself.
Furnishes the best lac (Mirzapore lac).
Schleidenia Endl. = Heliotropium L. (Boraginac.).
Schleinitzia Warb. (~ Prosopis L.). Leguminosae. 1 Moluccas to Tahiti.
Schleranthus L. = Scleranthus L. (Caryophyllac.).
Schlerochloa Parl. = Sclerochloa Beauv. (Gramin.).
Schleropelta Buckley = Hilaria Kunth (Gramin.).
Schliebenia Mildbr. Acanthaceae. 2 trop. E. Afr.
Schlimia Regel = Lisianthus Aubl. (Gentianac.).
Schlimmia Planch. & Linden. Orchidaceae. 4 N. Andes.
Schlosseria Ellis = Styrax L. (Styracac.).
Schlosseria Garden (nomen). Palmae. Quid?
Schlosseria Mill. ex Steud. = Coccoloba L. (Polygonac.).
Schlosseria Vukot. = Peucedanum L. (Umbellif.).
Schlumbergera Lemaire. Cactaceae. 2–5 Brazil.
Schlumbergera E. Morr. (sphalm.) = seq.
Schlumbergeria E. Morr. (~ Guzmania Ruiz & Pav.). Bromeliaceae. 50 S.
Schmalhausenia C. Winkler. Compositae. 2 C. As.   [Am., W.I.
Schmaltzia Steud. = seq.
Schmalzia Desv. (~ Rhus L.). Anacardiaceae. 50 N. Am.
Schmardaea Karst. Meliaceae. 2 trop. S. Am.
Schmidelia Boehm. = Ehretia P.Br. (Ehretiac.).
Schmidelia L. = Allophylus L. (Sapindac.).
Schmidia Wight = Thunbergia L. (Thunbergiac.).
Schmidtia Moench = Tolpis Adans. (Compos.).
*****Schmidtia** Steud. Gramineae. 4–5 trop. & S. Afr., C. Verde Is.
Schmidtia Tratt. = Coleanthus Seidl (Gramin.).
Schmidtottia Urb. Rubiaceae. 10 Cuba.
Schmiedelia Murr. = Schmidelia L. (Sapindac.).
Schmiedtia Rafin. = Schmidtia Tratt. = Coleanthus Seidl (Gramin.).
Schnabelia Hand.-Mazz. Verbenaceae. 1 SW. China. Both chasmog. and
cleistog. fls. produced.
Schnella Raddi (~ Bauhinia L.). Leguminosae. 30 Mex. to trop. S. Am.
Schnittspahnia Reichb. = Polyalthia Bl. (Annonac.).
Schnittspahnia Sch. Bip. = Landtia Less. (Compos.).
Schnitzleinia Walp. = Schnizleinia Steud. ex Hochst. = Vellozia Vand.
(Velloziac.).

Schnizleinia Mart. ex Engl. = Emmotum Desv. (Icacinac.).
Schnizleinia Steud. (1840) = Boissiera Hochst. (Gramin.).
Schnizleinia Steud. (1841) = Trochiscanthes Koch (Umbellif.).
Schnizleinia Steud. ex Hochst. (1844) = Vellozia Vand. (Velloziac.).
Schobera Scop. = Heliotropium L. (Boraginac.).
Schoberia C. A. Mey. = Suaeda Forsk. (Chenopodiac.).
Schoebera Neck. = Schobera Scop. = Heliotropium L. (Boraginac.).
Schoedonardus Scribn. (sphalm.) = Schedonnardus Steud. (Gramin.).
Schoenanthus Adans. = Ischaemum L. (Gramin.).
Schoenanthus auct. ex Just (sphalm.) = Ichnanthus Beauv. (Gramin.).
Schoenefeldia Kunth. Gramineae. 2 trop. Afr., As.
Schoenia Steetz. Compositae. 1 temp. Austr.
Schoenidium Nees = Ficinia Schrad. (Cyperac.).
Schoenissa Salisb. = Allium L. (Alliac.).
Schoenlandia Cornu = Cyanastrum Oliv. (Tecophilaeac.).
Schoenleinia Klotzsch (1843) = Bathysa Presl (Rubiac.).
Schoenleinia Klotzsch ex Lindl. (1847) = Ponthieva R.Br. (Orchidac.).
Schoenobiblos Endl. = seq.
Schoenobiblus Mart. Thymelaeaceae. 7 trop. S. Am., W.I.
Schoenocaulon A. Gray. Liliaceae. 10 Florida to Peru. Veratrin from seeds.
Schoenocephalium Seub. Rapateaceae. 5 Colombia, Venez.
Schoenochlaena P. & K. = Schoenolaena Bunge (Hydrocotylac.).
Schoenocrambe Greene = Sisymbrium L. (Crucif.).
Schoenodendron Engl. = Microdracoïdes Hua (Cyperac.).
Schoenodorus Roem. & Schult. = Schedonorus Beauv. (Gramin.).
Schoenodum Labill. = Lyginia R.Br. + Leptocarpus R.Br. (Restionac.).
Schoenolaena Bunge. Hydrocotylaceae (~ Umbellif.). 2 W. Austr.
*Schoenolirion Torr. ex Durand. Liliaceae. 1 Calif., 3 SE. U.S.
Schoenomorphus Thorel ex Gagnep. Orchidaceae (~ Apostasiac.). 1 Indoch. (Laos).
*Schoenoplectus (Reichb.) Palla (~ Scirpus L.). Cyperaceae. 25 N. temp.
Schoenoprasum Kunth = Allium L. (Alliac.).
Schoenopsis Beauv. = Tetraria Beauv. (Cyperac.).
Schoenorchis Blume. Orchidaceae. 20 China, Indomal., Solomons, Fiji.
Schoenoxiphium Nees (~ Kobresia Willd.). Cyperaceae. 15 S. Afr., Madag., C. As., Sumatra.
Schoenus L. Cyperaceae. 100, esp. S. Afr., Austr., N.Z.; a few Eur. to C. As., Malaysia, Am.
Schoepfia Schreb. Olacaceae. 35 trop.
Schoepfiaceae Blume = Olacaceae–Schoepfioïdeae Engl.
Schoepfianthus Engl. ex De Wild. (nomen). Olacaceae. 1 trop. Afr. Quid?
Schoepfiopsis Miers = Schoepfia Schreb. (Olacac.).
Scholera Hook. f. (sphalm.) = Schollia Jacq. f. = Hoya R.Br. (Asclepiadac.).
Schollera Rohr = Microtea Sw. (Chenopodiac.?).
Schollera Roth = Oxycoccus L. (Ericac.).
Schollera Schreb. = Heteranthera Ruiz & Pav. (Pontederiac.).
Scholleropsis H. Perrier. Pontederiaceae. 1 Madag.

SCHOLLIA

Schollia Jacq. f. = Hoya R.Br. (Asclepiadac.).

Scholtzia Schau. Myrtaceae. 12 Austr.

× **Schombavola** hort. Orchidaceae. Gen. hybr. (iii) (Brassavola × Schomburgkia).

× **Schombletia** hort. Orchidaceae. Gen. hybr. (vi) (Bletia × Schomburgkia).

× **Schombobrassavola** hort. (vii) = × Schombavola hort. (Orchidac.).

× **Schombocattleya** hort. Orchidaceae. Gen. hybr. (iii) (Cattleya × Schom-

× **Schombodiacrium** hort. (vii) = × Diaschomburgkia hort. (Orchidac.).

× **Schomboëpidendrum** hort. Orchidaceae. Gen. hybr. (iii) (Epidendrum × Schomburgkia).

× **Schombolaelia** hort. Orchidaceae. Gen. hybr. (iii) (Laelia × Schomburgkia)

× **Schombolaeliocattleya** hort. (vii) = × Lyonara hort. (1959) (Orchidac.).

× **Schombonia** hort. Orchidaceae. Gen. hybr. (iii) (Broughtonia × Schomburgkia).

× **Schombonitis** hort. Orchidaceae. Gen. hybr. (iii) (Schomburgkia × Sophronitis).

× **Schombotonia** hort. Orchidaceae. Gen. hybr. (iii) (Broughtonia × Schomburgkia).

Schomburghia DC. = Geissopappus Benth. (Compos.).

Schomburgkia Benth. & Hook. f. = praec.

Schomburgkia Lindl. Orchidaceae. 17 Mex., C. Am., W.I., trop. S. Am.

× **Schomburgkiocattleya** hort. (vii) = × Schombocattleya hort. (Orchidac.).

× **Schomcattleya** hort. (vii) = praec.

× **Schomocattleya** hort. (vii) = × Schombocattleya hort. (Orchidac.).

Schonlandia L. Bolus = Corpuscularia Schwant. (Aïzoac.).

Schorigeram Adans. = Tragia L. (Euphorbiac.).

Schortia Hort. ex Vilm. = Actinolepis DC. (Compos.).

*****Schotia** Jacq. Leguminosae. 18 trop. & S. Afr.

Schotiaria (DC.) Kuntze = Griffonia Baill. (Legumin.).

Schousbea Rafin. = Schousboea Willd. = Cacoucia Aubl. (Combretac.).

Schousboea Nicotra = Stipa L. (Gramin.).

Schousboea Schumach. & Thonn. = Alchornea Sw. (Euphorbiac.).

Schousboea Willd. = Cacoucia Aubl. (Combretac.).

Schoutenia Korth. Tiliaceae. 8 Siam, Indoch., ? Borneo, Java, Lesser Sunda

Schoutensia Endl. = Pittosporum Banks ex Soland. (Pittosporac.).   [Is.

*****Schouwia** DC. Cruciferae. 2 Sahara to Arabia.

Schouwia Schrad. = Goethea Nees & Mart. (Malvac.).

Schowia Sweet (sphalm.) = praec.

*****Schradera** Vahl. Rubiaceae. 15 trop. S. Am., W.I.

Schradera Willd. = Croton L. (Euphorbiac.).

Schraderia Fabr. ex Medik. = Arischrada Pobed. (Labiat.).

Schrameckia Danguy. Monimiaceae. 1 Madag.

Schrammia Britton & Rose = Hoffmanseggia Cav. (Legumin.).

Schranckia Scop. ex J. F. Gmel. = Goupia Aubl. (Goupiac.).

Schranckiastrum Hassl. Leguminosae. 1 Paraguay.

Schrankia Medik. = Rapistrum Crantz (Crucif.).

*****Schrankia** Willd. Leguminosae. 30 warm Am.

Schrankiastrum Willis (sphalm.) = Schranckiastrum Hassl. (Legumin.).

Schrebera L. = Cuscuta L. (Cuscutac.).
Schrebera L. ex Schreb. = Myrica L. (Myricac.). + Cuscuta L. (Cuscutac.).
Schrebera Retz. = Elaeodendron Jacq. f. (Celastrac.).
*Schrebera Roxb. Oleaceae. 1 S. Am., 25 trop. Afr., India, 1 SE. As., 1 Borneo.
Schrebera Thunb. = Hartogia L. f. (Celastrac.).
Schreiberia Steud. (sphalm.) = seq.
Schreibersia Pohl = Augusta Pohl (Rubiac.).
Schrenckia Benth. & Hook. f. (sphalm.) = seq.
Schrenkia Fisch. & Mey. Umbelliferae. 7 C. As.
Schroeterella Briquet (1926; non Herzog 1916—Bryoph.) = Neoschroetera Briq. (Zygophyllac.).
Schrophularia Medik. (sphalm.) = Scrophularia L. (Scrophulariac.).
Schstschurowskia Willis (sphalm.) = seq.
Schtschurowskia Regel & Schmalh. Umbelliferae. 2 C. As.
Schubea Pax = Cola Schott & Endl. (Sterculiac.) (leaf) + Trichoscypha Hook. f. (Anacardiac.) (infl.).
Schuberta St-Lag. = seq.
Schubertia Blume ex DC. = Harmsiopanax Warb. (Araliac.).
*Schubertia Mart. Asclepiadaceae. 6 S. Am.
Schubertia Mirb. = Taxodium Rich. (Taxodiac.).
Schudia Molina ex C. Gay = Osmorhiza Rafin. (Umbellif.).
Schuebleria Mart. = Curtia Cham. & Schlechtd. (Gentianac.).
Schuechia Endl. = Qualea Aubl. (Vochysiac.).
Schuenkia Rafin. = Schwenckia L. (Solanac.).
Schuermannia F. Muell. = Darwinia Rudge (Myrtac.).
Schufia Spach = Fuchsia L. (Onagrac.).
*Schultesia Mart. Gentianaceae. 20 trop. Am., Afr.
Schultesia Roth = Wahlenbergia Schrad. ex Roth (Campanulac.).
Schultesia Schrad. = Gomphrena L. (Amaranthac.).
Schultesia Spreng. = Chloris Sw. (Gramin.).
Schultesiophytum Harling. Cyclanthaceae. 1 NW. trop. S. Am.
Schultzia Nees = Herpetacanthus Nees (Acanthac.).
Schultzia Rafin. = Obolaria L. (Gentianac.).
*Schultzia Spreng. Umbelliferae. 2 C. As., ?2 NW. India.
Schultzia Wall. = Cortia DC. (Umbellif.).
Schumacheria Spreng. = Wormskioldia Thonn. (Turnerac.).
Schumacheria Vahl. Dilleniaceae. 3 Ceylon.
Schumannia Kuntze. Umbelliferae. 1 Persia, C. As.
Schumannianthus Gagnep. (= Clinogyne Salisb. ex Benth., as to type). Marantaceae. 2 Ceylon, Indomal.
Schumanniophyton Harms. Rubiaceae. 5 trop. Afr.
Schumeria Iljin. Compositae. 6 E. Medit. to C. As.
Schunda-Pana Adans. = Caryota L. (Palm.).
Schuurmansia Blume. Ochnaceae. 3 Borneo & Philippines to Solomon Is.
Schuurmansiella H. Hallier. Ochnaceae. 1 Borneo.
Schwabea Endl. Acanthaceae. 5 trop. Afr.
Schwackaea Cogn. Melastomataceae. 1 Mex., C. Am.

SCHWAEGRICHENIA

Schwaegrichenia Reichb. = Tetragastris Gaertn. (Burserac.).

Schwaegrichenia Spreng. = Anigozanthus Labill. (Haemodorac.).

Schwalbea L. Scrophulariaceae. 2 E. N. Am.

Schwannia Endl. Malpighiaceae. 6 warm S. Am. Usu. 6 fertile sta.

Schwantesia L. Bolus = Monilaria Schw. + Conophytum Schw. (Aïzoac.).

Schwantesia Dinter. Aïzoaceae. 10 S. Afr.

Schwartzia Vell. = Norantea Aubl. (Marcgraviac.).

Schwartzkopffia Kraenzl. Orchidaceae. 2 trop. Afr.

Schwarzia Vell. = Schwartzia Vell. = Norantea Aubl. (Marcgraviac.).

Schweiggera Mart. = Renggeria Meissn. (Guttif.).

Schweiggera E. Mey. ex Baker = Gladiolus L. (Iridac.).

Schweiggeria Spreng. Violaceae. 2 Mex., Brazil.

Schweinfurthafra Kuntze = Glyphaea Hook. f. (Tiliac.).

Schweinfurthia A. Br. Scrophulariaceae. 3 NE. Afr. to NW. India.

Schweinitzia Elliott = Monotropsis Schwein. (Monotropac.).

Schwenckea P. & K. = seq.

Schwenckia Royen ex L. Solanaceae. 25 C. & trop. S. Am.; 1 trop. W. Afr.

Schwendenera K. Schum. Rubiaceae. 1 SE. Brazil.

Schwenkfelda Schreb. = Sabicea Aubl. (Rubiac.).

Schwenkfeldia Willd. = praec.

Schwenkia L. (sphalm.) = Schwenckia Royen ex L. (Solanac.).

Schwenkiopsis Dammer. Solanaceae. 1 Andes.

Schwerinia Karst. = Meriania Sw. (Melastomatac.).

Schweyckerta C. C. Gmel. = Nymphoïdes Hill (Menyanthac.).

Schweykerta Griseb. = praec.

Schychowskya Endl. = Laportea Gaudich. (Urticac.).

Sciacassia Britton = Cassia L. (Legumin.).

Sciadiara Rafin. = Convolvulus L. (Convolvulac.).

Sciadicarpus Hassk. = Kibara Endl. (Monimiac.).

Sciadiodaphne Reichb. = Umbellularia Nees (Laurac.).

Sciadioseris Kunze = Senecio L. (Compos.).

Sciadiphyllum Hassk. = Sciadophyllum P.Br. (Araliac.).

Sciadocalyx Regel = Isoloma Decne (Gesneriac.).

Sciadocarpus P. & K. = Sciadicarpus Hassk. = Kibara Endl. (Monimiac.).

Sciadocephala Mattf. Compositae. 1 Ecuador.

Sciadodendron Griseb. Araliaceae. 1 C. Am. to Colombia, Haiti.

Sciadonardus Steud. = Gymnopogon Beauv. (Gramin.).

Sciadopanax Seem. Araliaceae. 1 Madag.

Sciadophila Phil. = Rhamnus L. (Rhamnac.).

Sciadophyllum P.Br. corr. Reichb. (~ Schefflera J. R. & G. Forst.). Araliaceae. 30 trop. S. Am., W.I.

Sciadopityaceae (Pilger) J. Doyle = Taxodiaceae Neger.

Sciadopitys Sieb. & Zucc. Taxodiaceae. 1 Japan, *S. verticillata* Sieb. & Zucc., the parasol-pine or umbrella-pine, planted round temples. Long shoots with small scale-ls., bearing in their axils (acc. to one interpr.) 'needle'-like short shoots, or cladodes, crowded into whorls; acc. to other authors these 'cladodes' represent pairs of true needle-ls. fused laterally by their margins. Cone-scales with strongly developed scale-pulvinus(?). Certain features of the ♂ cells in the pollen tube and of the archegonia of the ♀ gametophyte are similar to those of some *Podocarpac.* or *Pinac.*, whilst the chromosome number ($n = 10$) differs from that of other *Taxodiac.* The cones take 2 years to ripen. Wood useful.

Sciadoseris C. Muell. = Sciadioseris Kunze = Senecio L. (Compos.).

Sciadotaenia Benth. = seq.

Sciadotenia Miers. Menispermaceae. 15 trop. S. Am.

Sciaphila Blume. Triuridaceae. 50 trop. (Giesen in Engl. *Pflanzenr.* IV. 18 (Triuridac.): 30–71, 1938).

Sciaphyllum Bremek. Acanthaceae. 1 cult. (origin unknown).

Scilla L. Liliaceae. 80 temp. Euras., S. Afr., few trop. Afr.

Scillaceae Lotsy = Liliaceae–Scilloïdeae K. Krause.

Scillopsis Lem. = Lachenalia Jacq. (Liliac.).

Scindapsus Schott. Araceae. 40 S. China, SE. As., Indomal.

Sciobia Reichb. = Sciophila Gaudich. = Procris Comm. ex Juss. (Urticac.).

Sciodaphyllum P.Br. = Sciadophyllum P.Br. corr. Reichb. (Araliac.).

Sciophila Gaudich. = Procris Comm. ex Juss. (Urticac.).

Sciophila P. & K. = Skiophila Hanst. = Columnea L. (Gesneriac.).

Sciophila Wibel = Maianthemum Weber (Liliac.).

Sciophylla Heller (sphalm.) = praec.

Sciothamnus Endl. = Peucedanum L. (Umbellif.).

Scirpaceae Burnett = Cyperaceae–Scirpoïdeae (excl. Cyperus).

Scirpidium Nees = Eleocharis R.Br. (Cyperac.).

Scirpobambus (A. Rich.) P. & K. = Oxytenanthera Munro (Gramin.).

Scirpo-cyperus Seguier = Scirpus L. (*s.str.*) (Cyperac.).

Scirpodendron Engl. (sphalm.) = Schoenodendron Engl. = Microdracoïdes Hua (Cyperac.).

Scirpodendron Zippel ex Kurz. Cyperaceae. 1 Indomal., Austr., Polynesia. Perhaps the most 'primitive' member of the family.

Scirpoïdes Seguier = seq.

Scirpus L. Cyperaceae. 300 cosmop., char. of wet moors, bogs and marshes. Stem usu. erect and angular, bearing 3 ranks of ls. reduced to sheaths; stem performs assim. The stem often arises from a creeping rhiz., which sometimes emits shoots ending in tubers like potatoes. The racemose many-flowered spikelets are aggregated into a terminal tuft. Fl. ⚥, with 6 P-scales in two whorls; in many spp. protog.; in all wind-pollinated. *S. lacustris* L. (the true 'bulrush') is used for matting, chair-seats, etc.

Scirrhophorus Turcz. = Skirrhophorus DC. = Angianthus Wendl. (Compos.).

Scitamineae R.Br. = Zingiberales Hutch., incl. *Musaceae, Strelitziac., Heliconiac., Lowiac., Marantac., Cannac.* and *Zingiberac.*

Sciuris Nees & Mart. = Galipea Aubl. (Rutac.).

Sciuris Schreb. = Raputia Aubl. (Rutac.).

Sciurus D. Dietr. = praec.

Scizanthus Pers. (sphalm.) = Schizanthus Ruiz & Pav. (Solanac.).

Sclaeranthus Thunb. (sphalm.) = Scleranthus L. (Caryophyllac.).

Sclaraea Steud. = seq.

Sclarea Mill. = Salvia L. (Labiat.).

Sclareastrum Fabr. (uninom.) = *Salvia* L. sp. (Labiat.).

Sclepsion Rafin. ex Wedd. [cf. Selepsion Rafin.] = Laportea Gaudich. (Urticac.).

SCLERACHNE

Sclerachne R.Br. Gramineae. 1 Java, Timor.

Sclerachne Torr. ex Trin. = Thurberia Benth. (Gramin.).

Sclerandrium Stapf & C. E. Hubbard. Gramineae. 3 New Guinea, NE. Austr.

Scleranthaceae Bartl. = Caryophyllaceae–Paronychioïdeae Fenzl.

Scleranthera Pichon. Apocynaceae. 2 SE. As., Malay Penins.

Scleranthopsis Rech. f. Caryophyllaceae. 1 Afghan.

Scleranthus L. Caryophyllaceae. 10 Eur., As., Afr., Austr. Fls. apet., self-fert.

Scleria Bergius. Cyperaceae. 200 trop. & subtrop.

Sclerobasis Cass. = Senecio L. (Compos.).

Sclerobassia Ulbr. (~ Bassia All.). Chenopodiaceae. 1 W. Austr.

Scleroblitum Ulbr. Chenopodiaceae. 1 SE. Austr.

Sclerocactus Britton & Rose. Cactaceae. 6 SW. U.S.

Sclerocalyx Nees = Gymnacanthus Nees (Acanthac.).

Sclerocarpa Sond. (sphalm.) = Sclerocarya Hochst. (Anacardiac.).

Sclerocarpus Jacq. (~ Madia Mol.). Compositae. 15 S. U.S. to Colombia, trop. Afr.

Sclerocarya Hochst. Anacardiaceae. 5 trop. & S. Afr. Ed. fr.

Sclerocaryopsis A. Brand. Boraginaceae. 1 N. Afr. to C. As.

Sclerocephalus Boiss. Caryophyllaceae. 1 C. Verde & Canary Is. to Persia, a char. plant.

Sclerochaetium Nees = Tetraria Beauv. (Cyperac.).

Sclerochiton Harv. Acanthaceae. 15 trop. & S. Afr.

Sclerochlaena auctt. = Scleroölaena Baill. = Xyloölaena Baill. (Sarcolaenac.).

Sclerochlaena P. & K. = Sclerolaena R.Br. (Chenopodiac.).

Sclerochlamys F. Muell. = Kochia Roth (Chenopodiac.).

Sclerochloa Beauv. Gramineae. 1 S. Eur., W. As.

Sclerochloa Reichb. = Festuca L. (Gramin.).

Sclerochorton Boiss. Umbelliferae. 2 SE. Eur. to Persia.

Sclerocladus Rafin. = Bumelia Sw. (Sapotac.).

Sclerococcus Bartl. = Hedyotis L. (Rubiac.).

Sclerocroton Hochst. = Excoecaria L. (Euphorbiac.).

Sclerocyathium Prokh. = Euphorbia L. (Euphorbiac.).

Sclerodactylon Stapf. Gramineae. 1 Madag.

Sclerodeyeuxia (Stapf) Pilger (~ Deyeuxia Clar.). Gramineae. 1 New Guinea.

Sclerodictyon Pierre = Dictyophleba Pierre (Apocynac.).

Scleroglossum v. Ald. v. Ros. Grammitidaceae. 6 Ceylon, Malaysia, Micronesia, Queensld. (To be united with *Cochlidium* Kaulf.?)

Sclerolaena Boivin ex A. Camus = Boivinella A. Camus (Gramin.).

Sclerolaena R.Br. (~ Bassia All.). Chenopodiaceae. 20 Austr.

Scleroleima Hook. f. = Abrotanella Cass. (Compos.).

Sclerolepis Cass. Compositae. 1 W. U.S.

Sclerolepis Monn. = Rodigia Spreng. = Crepis L. (Compos.).

Sclerolinon C. M. Rogers. Linaceae. 1 W. U.S.

Sclerolobium Vog. Leguminosae. 30 trop. S. Am.

Scleromelum K. Schum. & Lauterb. = Scleropyrum Arn. (Santalac.).

Scleromitrion Wight & Arn. = Hedyotis L. (Rubiac.).

Scleromphalos Griff. = Withania Pauq. (Solanac.).
Scleronema Benth. Bombacaceae. 5 trop. S. Am.
Scleronema Brongn. & Gris = Xeronema Brongn. (Liliac.).
Scleroölaena Baill. = Xyloölaena Baill. (Sarcolaenac.).
Scleroön Benth. = Petitia Jacq. (Verbenac.).
Scleropelta auctt. = Schleropelta Buckley = Hilaria Kunth (Gramin.).
Sclerophylacaceae Miers = Solanaceae Juss.
Sclerophylax Miers. Solanaceae. 12 Argent., Paraguay, Uruguay. Succulent halophytes; ls. opp.
Sclerophyllum Gaudin = Phaecasium Cass. = Crepis L. (Compos.).
Sclerophyllum Griff. = Indoryza A. N. Henry & Roy = Porteresia Tateoka (Gramin.).
Sclerophyrum Hieron. (sphalm.) = Scleropyrum Arn. (Santalac.).
Scleropoa Griseb. = Catapodium Link (Gramin.).
Scleropogon Phil. Gramineae. 1 SW. U.S., Mex., Argent.
Scleropteris Scheidw. = Cirrhaea Lindl. (Orchidac.).
Scleropus Schrad. = Amaranthus L. (Amaranthac.).
Scleropyron Endl. = seq.
***Scleropyrum** Arn. Santalaceae. 6 Indomal.
Sclerorhachis (K. H. Rech.) K. H. Rech. Compositae. 2 Persia.
Sclerosciadium Koch = Capnophyllum Gaertn. (Umbellif.).
Sclerosia Klotzsch (nomen). Ochnaceae. 1 Guiana. Quid?
Sclerosiphon Nevski = Iris L. (Iridac.).
Sclerosperma G. Mann & H. Wendl. Palmae. 2-3 trop. W. Afr.
Sclerostachya (Hack.) A. Camus. Gramineae. 3 Assam, Indoch., Malay Penins.
Sclerostachyum Stapf ex Ridley (sphalm.) = praec.
Sclerostemma Schott ex Roem. & Schult. = Scabiosa L. (Dipsacac.).
Sclerostephane Chiov. Compositae. 2 Somalia.
Sclerostylis Blume = Atalantia Corrêa (Rutac.).
Sclerothamnus R.Br. = Eutaxia R.Br. (Legumin.).
Sclerothamnus Fedde (sphalm.) = Selerothamnus Harms = Hesperothamnus Brandegee (Legumin.).
Sclerotheca A. DC. Campanulaceae. 4 Cook Is., Society Is.
Sclerothrix C. Presl. Loasaceae. 1 Mex. to trop. S. Am. 4-merous. Stds. 6-10.
Sclerotiaria Korovin. Umbelliferae. 1 C. As.
Sclerotriaria Czerep. (sphalm.) = praec.
Sclerotris Rafin. (nomen). Florida. Quid?
Scleroxylon Bertol. = ? Chrysophyllum L. (Sapotac.).
Scleroxylon Steud. = seq.
Scleroxylum Willd. = Myrsine L. (Myrsinac.).
Sclerozus Rafin. = Bumelia Sw. (Sapotac.).
Scobedia Labill. ex Steud. (nomen). Labiatae. Quid?
Scobia Nor. = Lagerstroemia L. (Lythrac.).
Scobinaria Seibert = Arrabidaea DC. (Bignoniac.).
Scoliaxon Payson. Cruciferae. 1 NE. Mex.
Scoliochilus Reichb. f. = Appendicula Bl. (Orchidac.).

Scoliopus Torr. Trilliaceae. 2 W. N. Am.
Scoliosorus Moore. Vittariaceae. 1 C. Am.
Scoliotheca Baill. Gesneriaceae. 1 Colombia.
Scolleropsis Hutch. = Scholleropsis H. Perr. (Pontederiac.).
Scolobus Rafin. = Thermopsis R.Br. (Legumin.).
*Scolochloa Link. Gramineae. 2 N. temp.
Scolochloa Mert. & Koch = Arundo L. (Gramin.).
Scolodia Rafin. = Cassia L. (Legumin.).
Scolodrys Rafin. = Quercus L. (Fagac.).
Scolopacium Eckl. & Zeyh. = Erodium L'Hérit. (Geraniac.).
Scolopendrium Adans. = Phyllitis Hill (Aspleniac.).
*Scolopia Schreb. Flacourtiaceae. 45 trop. & S. Afr., As., Austr. S. Afr. spp. yield timber.
Scolopospermum Hemsl. = Scolospermum Less. = Baltimora L. (Compos.).
Scolosanthes Willis (sphalm.) = seq.
Scolosanthus Vahl. Rubiaceae. 15 W.I.
Scolosperma Rafin. = Cleome L. (Cleomac.).
Scolospermum Less. = Baltimora L. (Compos.).
Scolymanthus Willd. ex DC. = Perezia Lag. (Compos.).
Scolymocephalus Kuntze = Protea L. (Proteac.).
Scolymus L. Compositae. 3 Medit.
Scoparebutia Frič & Kreuz. = Lobivia Britt. & Rose (Cactac.).
Scoparia L. Scrophulariaceae. 20 trop. Am.
Scopelogena L. Bolus. Aïzoaceae. 2 S. Afr.
Scopola Jacq. = Scopolia Jacq. corr. Link (Solanac.).
Scopolia Adans. = Ricotia L. (Crucif.).
Scopolia J. R. & G. Forst. = Griselinia Forst. f. (Cornac.).
*Scopolia Jacq. corr. Link. Solanaceae. 6 C. & S. Eur. to India & Japan.
Scopolia Lam. (sphalm.) = Scolopia Schreb. (Flacourtiac.).
Scopolia L. f. = Daphne L. (Thymelaeac.).
Scopolia Sm. = Toddalia Juss. (Rutac.).
Scopolina Schult. = Scopolia Jacq. (Solanac.).
Scopularia Lindl. = Holothrix Rich. (Orchidac.).
Scopulophila M. E. Jones. Caryophyllaceae. 1 SW. U.S.
Scorbion Rafin. [sphalm. pro Scordion] = seq.
Scordium Mill. = Teucrium L. (Labiat.).
Scoria Rafin. = Hicoria Rafin. = Carya Nutt. (Juglandac.).
Scorias Rafin. = praec.   [(Sapindac.).
Scorodendron Pierre (sphalm.) = Scorododendron Bl. = Lepisanthes Bl.
Scorodocarpaceae Van Tiegh. = Olacaceae–Anacoloseae Engl.
Scorodocarpus Becc. Olacaceae. 1 W. Malaysia.
Scorododendron Blume = Lepisanthes Bl. (Sapindac.).
Scorodon Fourr. = Allium L. (Alliac.).
Scorodonia Hill = Teucrium L. (Labiat.).
Scorodophloeus Harms. Leguminosae. 2 trop. Afr.
Scorodosma Bunge = Ferula L. (Umbellif.).
Scorodoxylum Nees = Ruellia L. (Acanthac.).
Scorpaena Nor. (nomen). Orchidaceae (terrestres). 2 Java. Quid?

Scorpia Ewart & Petrie = Corchorus L. (Tiliac.).

Scorpiaceae Dulac = Boraginaceae Juss.

Scorpianthes Rafin. = Heliotropium L. (Boraginac.).

Scorpioïdes Gilib. = Myosotis L. (Boraginac.).

Scorpioïdes Hill = Scorpiurus L. (Legumin.).

Scorpiothyrsus Li. Melastomataceae. 1 Indoch., 5 S. China (Hainan).

Scorpiurus Fabr. (uninom.) = Heliotropium L. sp. (Boraginac.).

Scorpiurus Haller = Myosotis L. (Boraginac.).

Scorpiurus L. Leguminosae. 8 Medit to Cauc. Pod twisted, indehisc.

Scorpius Loisel. = praec.

Scorpius Medik. = Coronilla L. (Legumin.).

Scorpius Moench = Genista L. (Legumin.).

Scortechinia Hook. f. (1887; non Saccardo 1885—Fungi) = Neoscortechinia Pax (Euphorbiac.).

Scorzonella Nutt. = Microseris D. Don (Compos.).

Scorzonera L. Compositae. 150 Medit., C. Eur. to C. As. Roots of *S. hispanica* L., etc. are eaten as a vegetable.

Scorzoneroïdes Moench = Leontodon L. (Compos.).

Scotanthus Naud. = Gymnopetalum Arn. (Cucurbitac.).

Scotanum Adans. = Ficaria Guett. = Ranunculus L. (Ranunculac.)

Scotia Thunb. = Schotia Jacq. (Legumin.).

Scottea DC. = Scottia R.Br. = Bossiaea Vent. (Legumin.).

Scottellia Oliv. Flacourtiaceae. 10 trop. Afr.

Scottia R.Br. = Bossiaea Vent. (Legumin.).

Scovitzia Walp. = Szowitsia Fisch. & Mey. (Umbellif.).

Scribaea Borkh. = Cucubalus L. (Caryophyllac.).

Scribneria Hackel. Gramineae. 1 Pacif. U.S.

Scrobicaria Cass. = Gynoxys Cass. (Compos.).

Scrobicularia Mansfeld. Melastomataceae. 1 New Guinea.

Scrofella Maxim. Scrophulariaceae. 1 NW. China.

Scrofularia Spreng. = seq.

Scrophularia L. Scrophulariaceae. 300 temp. Euras.; 12 N. & trop. Am. Perenn. herbs with opp. ls., which on the lat. twigs are commonly anisophyllous. Fls. in tall infls. whose primary branching is racemose; the lat. shoots are dichasial. Sta. and style arranged along the lower lip of the C (upper usual in such fls.). The posterior sta., usu. absent in the fam., is repres. by a std. Fl. markedly protog., largely visited by wasps.

*Scrophulariaceae Juss. 220/3000 cosmop. Most are herbs and undershrubs, a few shrubs or trees (e.g. *Paulownia*), with alt., opp., or whorled exstip. ls. Many exhibit interesting features in the veg. organs. Several are climbers (e.g. *Maurandia*, *Rhodochiton*, etc.). The 'Veronicas' (*Hebe*) of N.Z. are xero., with resemblance in habit to certain *Coniferae*. A number of spp. in III. 2 and 3 (below), e.g. *Euphrasia*, *Bartsia*, *Pedicularis*, grow in swampy grassland and are parasitic by their roots upon the roots of the grasses. Suckers are formed at the points of contact, in spring; they absorb food till the summer, and later absorb organic compounds from the dead parts of the host, and function for storage of reserve-materials. The plants possess green ls. of their own, and so are able to assimilate.

SCROPHULARIACEAE

Infl. racemose or cymose, in the former case usu. a spike or raceme, axillary or term. (every variety in spp. of *Veronica*). Sol. axillary fls. in many, e.g. *Linaria*. Cymose infls. usu. dichasia, often united into complex corymbs, etc. Bracts and bracteoles usu. present. In *Castilleja* the upper ls. and bracts brightly coloured.

Fl. ♂, zygo., sometimes nearly reg. (*Verbascum*, etc.); considerable variety in structure, as may be seen from floral diagrams. The bulk of the family show the *Linaria* type. K (5), of various aestivations; C (5), median zygo., often 2-lipped; A 4 (sometimes 2), didynamous, epipet., the post. sta. sometimes repres. by a std. (e.g. in *Scrophularia* and *Penstemon*). *Verbascum* and its allies have an actinom. C and 5 sta.; *Veronica* (*q.v.*) shows 4 sepals (the post. one of the typical 5 absent), 4 petals (the post. pair of the 5 united), and 2 sta., the C rotate. Other variations occur in the *Selagineae*, etc. Below the ovary is a honey-secreting disk. G (2), medially placed (not obliquely as in *Solanaceae*), 2-loc., with axile plac.; ov. usu. ∞, less commonly few (e.g. *Veronica*, etc.), anatr.; style simple or bilobed. Fr. surrounded below by the persistent K, usu. a capsule (dehisc. in various ways) or a berry. Seeds usu. numerous, small, with endosp. Embryo straight or slightly curved.

Most have fls. ± adapted to insect-visits. Müller divides them into 4 types: (1) the *Verbascum* or *Veronica* type (see gen.), with open fl. and short tube (bees and flies); (2) the *Scrophularia* type (wasps); (3) the *Digitalis* and *Linaria* type, with long wide tubes and the essential organs so placed as to touch the back of the insect (bees); and (4) the *Euphrasia* type, or 'loose-pollen' fl., where the pollen is loose and powdery, and the anthers (protected by upper lip) have spines, etc., so that they may be shaken upon the entrance of the insect which thus receives a shower of pollen. The fls. are seldom markedly dichogamous, but the stigma usu. projects beyond the sta. so as to be first touched. Most are capable of self-fert. in default of visits. For further details see gen.

In *Linaria*, *Digitalis*, etc. (*q.v.*) there sometimes appears a terminal fl. to the raceme, and this exhibits *peloria*, having a symmetrical C with (in *Linaria*) spurs to all the petals (cf. *Ruta*, or compare *Aquilegia* with *Delphinium*).

A number are or have been officinal, e.g. *Digitalis*; most are poisonous.

Classification and chief genera (after von Wettstein):

A. Two post. C-teeth (or upper lip) cover lat. teeth in bud.
  I. **Verbascoïdeae ('Pseudosolaneae')** (all ls. usu. alt.; 5 sta. often present):
    1. VERBASCEAE (C with very short tube or none, rotate or shortly campanulate): *Verbascum*, *Celsia*.
    2. APTOSIMEAE (C with long tube): *Aptosimum*.
  II. **Scrophularioïdeae (Antirrhinoïdeae)** (lower ls. at least opp.; the 5th sta. wanting or staminodial):
    α. C 2-lipped; lower lip concave, bladder-like.
    2. CALCEOLARIËAE: *Calceolaria*.
    β. C almost actinom., or 2-lipped with flat or convex lips.
    2. HEMIMERIDEAE (dehisc. caps.; C spurred or saccate at base, with no tube): *Alonsoa*.
    3. ANTIRRHINEAE (as 2, but with tube): *Linaria*, *Antirrhinum*, *Maurandia*, *Rhodochiton*.

4. SCROPHULARIËAE (CHELONEAE) (dehisc. caps. or many-seeded berry; C not spurred or saccate; infl. cymose, cpd.): *Russelia, Wightia, Collinsia, Scrophularia, Chelone, Penstemon, Paulownia*.

5. MANULEAE (dehisc. caps.; C as in 4; infl. not cymose; usu. simple; anthers finally 1-loc.): *Zaluzianskia, Sutera*.

6. GRATIOLEAE (as 5, but anthers finally 2-loc.): *Mimulus, Gratiola, Torenia*.

7. SELAGINEAE (drupe or indehisc. few-seeded caps.): *Hebenstretia, Selago*.

B. Two post. teeth (or upper lip) of C covered in bud by one or both of the lat. teeth.

III. *Rhinanthoïdeae*:

α. C-teeth all flat and divergent, or the 2 upper erect.

1. DIGITALIDEAE (anther-loc. finally united at tip; 2 upper C-lobes often erect; not paras.): *Veronica, Digitalis*.

2. GERARDIËAE (anther-loc. always separate, one often reduced; C-lobes all flat, divergent; often paras.): *Gerardia*.

β. 2 upper C-teeth form a helmet-like upper lip. Often paras.

3. RHINANTHEAE: *Castilleja, Melampyrum, Tozzia, Euphrasia, Bartsia, Pedicularis, Rhinanthus*.

Scrophularioïdes Forst. f. = Premna L. (Verbenac.).

× **Scrophulari-verbascum** P. Fourn. Scrophulariaceae. Gen. hybr. (Scrophularia × Verbascum).

Scrotalaria Ser. ex Pfeiff. = Teucrium L. (Labiat.).

Scubalia Nor. = Lasianthus Jack (Rubiac.).

Scubulon Rafin. = Lycopersicon Mill. (Solanac.).

Sculeria Rafin. = Vancouveria Decne (Berberidac.).

Sculertia K. Schum. (sphalm.) = Seubertia Kunth (Liliac.).

Scuria Rafin. = Carex L. (Cyperac.).

Scurrula L. Loranthaceae. 50 SE. As., W. Malaysia.

Scutachne Hitchc. & Chase. Gramineae. 2 Cuba.

Scutellaria L. Labiatae. 300 cosmop., exc. S. Afr.

Scutellariaceae Caruel = Labiatae–Scutellarioïdeae Briq.

***Scutia** (DC.) Comm. ex Brongn. Rhamnaceae. 9 trop. Am., trop. & S. Afr., India, Indoch.

Scuticaria Lindl. Orchidaceae. 3 trop. S. Am.

Scutinanthe Thw. Burseraceae. 2 Ceylon, S. Burma, W. Malaysia, Celebes.

Scutis Endl. = Scutia (DC.) Comm. ex Brongn. (Rhamnac.).

Scutula Lour. = Memecylon L. (Memecylac.).

Scybalium Schott & Endl. Balanophoraceae. 4 trop. S. Am., W.I.

Scynopsole Reichb. (sphalm.) = Cynopsole Endl. = Balanophora J. R. & G. Forst. (Balanophorac.).

Scyphaea Presl = Marila Sw. (Guttif.).

Scyphanthus D. Don. Loasaceae. 2 Chile.

Scypharia Miers = Scutia Comm. ex Brongn. (Rhamnac.).

Scyphellandra Thw. (~Rinorea Aubl.). Violaceae. 1 Ceylon, 3 Siam, Indoch.

Scyphiphora Gaertn. f. Rubiaceae. 1 sea-coasts Indomal., Austr.

SCYPHOCEPHALIUM

Scyphocephalium Warb. Myristicaceae. 3 trop. W. Afr.
Scyphochlamys Balf. f. Rubiaceae. 1 Rodrigues I.
Scyphocoronis A. Gray. Compositae. 1 W. Austr.
Scyphofilix Thou. (gen. dub.) = ? Microlepia Presl (Dennstaedtiac.).
Scyphoglottis Pritz. = Scaphyglottis Poepp. & Endl. (Orchidac.).
Scyphogyne Brongn. Ericaceae. 47 S. Afr.
Scypholepia J. Sm. = Microlepia Presl (Dennstaedtiac.).
Scyphonychium Radlk. Sapindaceae. 1 NE. Brazil.
Scyphopetalum Hiern = Paranephelium Miq. (Sapindac.).
Scyphophora P. & K. = Scyphiphora Gaertn. f. (Rubiac.).
Scyphopteris Rafin. = Scyphofilix Thou. = ? Microlepia Presl (Dennstaedtiac.).
Scyphostachys Thw. Rubiaceae. 2 Ceylon.
Scyphostegia Stapf. Scyphostegiaceae. 1 NW. Borneo.
*****Scyphostegiaceae** Hutch. Dicots. 1/1 Borneo. Small trees with soft wood and slender lenticellate branchlets, usu. ± anfractuose and strongly 3–4-angled when young. Ls. simple, alt., distich., closely serrul., shortly pet., with v. small caduc. stip., and close transverse ('rhamnaceous') tertiary venation. Fls. reg., ♂ ♀, dioec., in term. elongate racemiform pan. usu. leafy below; pan.-branches bare below, bearing in upper part 2 (in ♀ infl.) to 12 (in ♂ infl.) concentric infundib. or tubular bracts, the upper bracts each shortly exserted from the lower, with truncate margins, each bract subtending a single fl.; pedic. of ♂ fl. flattened, 2-nerved. P (3 + 3), lobes imbr., rounded; A (3), opposite the inner tepals, fil. united into a column with clavate common apical connective, anth. 4-loc., oblong, extrorse, col. surr. at base by 3 short fleshy glands opposite to the sta.; disk o. G (± 10), large, globose, fleshy, with a thick apical discoid stig. with a narrow central ostiole; ovules ∞, on a slightly convex basal receptacle, erect, anatr., cylindr.–fusiform, stipitate, puberulous, surrounded at base by a fleshy collar-like aril(loid) (membranous when dry). Fruit a fleshy capsule, ultimately dehiscing into 8–12 reflexed segments, enclosing ∞ dry seeds with large embryo; endosp. scanty or o. Only genus: *Scyphostegia*. The morphology of the parts of the flower has given rise to much discussion, but it now seems clear that the above interpretation is the correct one, and that the affinity of the plant is with the *Parietales*, and especially with the *Flacourtiaceae*. See van Heel, *Blumea*, **15**: 107–25, 1967.
Scyphostelma Baill. Asclepiadaceae. 1 Colombia.
Scyphostigma M. Roem. = Turraea L. (Meliac.).
Scyphostrychnos S. Moore. Strychnaceae. 2 trop. Afr.
Scyphosyce Baill. Moraceae. 3 trop. Afr.
Scyphularia Fée. Davalliaceae. 8 Malaysia. Small epiphytes; ls. simply pinnate.
Scyrtocarpa Miers = seq.
Scyrtocarpus Miers = Barberina Vell. = Symplocos L. (Symplocac.).
Scytala E. Mey. ex DC. = Oldenburgia Less. (Compos.).
Scytalanthus Schau. = Skytanthus Meyen (Apocynac.).
Scytalia Gaertn. = Nephelium L. (Sapindac.).
Scytalis E. Mey. = Vigna Savi (Legumin.).
Scytanthus T. Anders. ex Benth. = Thomandersia Baill. (Acanthac.).
Scytanthus Hook. = Hoodia Sweet (Asclepiadac.).

Scytanthus Liebm. = Bdallophyton Eichl. (Rafflesiac.).
Scytanthus P. & K. = Skytanthus Meyen (Apocynac.).
Scytodephyllum Pohl (nomen). 1 Brazil. Quid?
*****Scytopetalaceae** Engl. Dicots. 5/20 trop. Afr. Trees or shrubs (sometimes scandent?), with simple alt. often distich. ent. or dent. exstip. ls., often asymm. at base. Fls. reg., ♂, often long-pedic., in axill. or term. pan. or rac., or in fascicles on old wood. K (3–4), or more often patellif. or cupulif. with ent. margin, persist.; C (3–16), valv., sometimes thick, reflexed at anthesis, or sometimes not separating and then falling as an entire cap; disk annular, inconspic.; A ∞, 3–6-ser., sometimes connate below, anth. sometimes dehisc. by apical pores; G̲ (3–8), loc. sometimes incomplete above, with simple style and small stig., and 2–6(–∞) biseriate axile pend. anatr. ov. per loc. Fr. usu. a tardily dehisc. caps., more rarely ± drupaceous, 1-loc., 1–8-seeded; seeds sometimes with a covering of agglutinated mucilag. hairs; endosp. copious, sometimes ruminate. Genera: *Oubanguia, Scytopetalum, Rhaptopetalum, Brazzeia, Pierrina.* Affinities prob. with *Olacaceae*, rather than with *Malvales* as often suggested.
Scytopetalum Pierre ex Engl. Scytopetalaceae. 3 trop. W. Afr.
Scytophyllum Eckl. & Zeyh. = Elaeodendron Jacq. f. (Celastrac.).
Scytopteris Presl = Pyrrosia Mirbel (Polypodiac.).
Sczegleëwia Turcz. (1858) = Pterospermum Schreb. (Sterculiac.).
Sczegleëwia Turcz. (1863) = Symphorema Roxb. (Symphorematac.).
Sczukinia Turcz. = Swertia L. (Gentianac.).
Seaforthia R.Br. = Ptychosperma Labill. (Palm.).
Seala Adans. = Pectis L. (Compos.).
Searsia F. A. Barkley = Terminthia Bernh. (Anacardiac.).
Sebaea Soland. ex R.Br. Gentianaceae. 100 warm Afr., Madag., India, Austr., N.Z.
Sebastiana Benth. & Hook. f. = Sebastiania Bertol. = Chrysanthellum Rich. (Compos.).
Sebastiana Spreng. = Sebastiania Spreng. (Euphorbiac.).
Sebastiania Bertol. = Chrysanthellum Rich. (Compos.).
Sebastiania Spreng. Euphorbiaceae. 95 trop. Am., Atl. U.S.; 1 trop. W. Afr., India to S. China & Austr.; 3 W. Malaysia.
Sebastiano-Schaueria Nees. Acanthaceae. 1 Brazil.
Sebeekia Steud. = seq.
Sebeokia Neck. = Gentiana L. (Gentianac.).
Sebertia Pierre. Sapotaceae. 2 New Caled.
Sebesten Adans. = seq.
Sebestena Boehm. = Cordia L. (Ehretiac.).
Sebesten[ace]ae Vent. = Ehretiaceae Mart.
Sebicea Pierre ex Diels = Tiliacora Colebr. (Menispermac.).
Sebifera Lour. = Litsea Lam. (Laurac.).
Sebipira Mart. = Bowdichia Kunth (Legumin.).
Sebizia Mart. = Mappia Jacq. (Icacinac.).
Sebophora Neck. = Myristica L. (Myristicac.).
Seborium Rafin. = Sapium P.Br. (Euphorbiac.).
Sebschauera Kuntze = Sebastiano-Schaueria Nees (Acanthac.).
Secale L. Gramineae. 4 Medit., E. Eur. to C. As.; 1 S. Afr. *S. cereale* L.,

SECALIDIUM

the rye, is largely cult. in N. Eur. as a cereal, forming a staple food. There are no well-marked races. The hardy winter ryes are the best. Also used as fodder. Cf. Schiemann, *Weizen, Roggen, Gerste*, 60–70 (1948).
Secalidium Schur = Agropyron Gaertn. (Gramin.).
Secamone R.Br. Asclepiadaceae. 100 palaeotrop.
Secamonopsis Jumelle. Asclepiadaceae. 1 Madag.
Sechiopsis Naud. Cucurbitaceae. 1 Mex.
*Sechium P.Br. Cucurbitaceae. 1 trop. Am., *S. edule* (Jacq.) Sw., cult. for its ed. fr. (*chocho*), containing one enormous seed.
Secondatia A. DC. Apocynaceae. 7 trop. S. Am., Jamaica.
Secretania Muell. Arg. = Minquartia Aubl. (Olacac.).
Secula Small = Aeschynomene L. (Legumin.).
Securidaca L. (1753) = Dalbergia L. f. (Legumin.).
*Securidaca L. (1759). Polygalaceae. 80 trop., exc. Austr. Climbers.
Securidaca Mill. = Securigera DC. = Coronilla L. (Legumin.).
Securidaea Turcz. (sphalm.) = praec.
*Securigera DC. = Coronilla L. (Legumin.).
Securilla Gaertn. ex Steud. = praec.
Securina Medik. = praec.
*Securinega Comm. ex Juss. Euphorbiaceae. 25 temp. & subtrop.
Sedaceae Barkley = Crassulaceae Juss.
Sedastrum Rose = Sedum L. (Crassulac.).
Seddera Hochst. Convolvulaceae. 20 trop. & S. Afr., Madag., Arabia.
Seddera Hochst. & Steud. ex Moq. = Saltia R.Br. ex Moq. (Amaranthac.).
Sedderopsis Roberty = Seddera Hochst. (Convolvulac.).
Sedella Britton & Rose = Parvisedum R. T. Clausen (Crassulac.).
Sedella Fourr. = Sedum L. (Crassulac.).
× **Sedeveria** Walth. Crassulaceae. Gen. hybr. (Echeveria × Sedum).
Sedgwickia Griff. = Altingia Noronha (Altingiac.).
Sedopsis (Engl.) Exell & Mendonçá. Portulacaceae. 5 trop. Afr.
Sedum Adans. = Sempervivum L. (Crassulac.).
Sedum L. Crassulaceae. 600 N. temp., 1 Peru. Fleshy-leaved xero.
Seemannantha Alef. = Tephrosia Pers. (Legumin.).
Seemannaralia R. Viguier. Araliaceae. 1 S. Afr.
Seemannia Hook. = Pentagonia Benth. (Rubiac.).
*Seemannia Regel. Gesneriaceae. 10 Peru, Bolivia.
Seetzenia R.Br. Zygophyllaceae. 2 N. & S. Afr., Arabia, NW. India, in deserts.
Segeretia G. Don (sphalm.) = Sageretia Brongn. (Rhamnac.).
Segetella Desv. (~ Spergularia J. & C. Presl). Caryophyllaceae. 3 N. temp.
Segregatia A. Wood (sphalm.) = Sageretia Brongn. (Rhamnac.).
Seguiera Adans. = Seguieria Loefl. (Phytolaccac.).
Seguiera Kuntze = Blackstonia Huds. (Gentianac.).
Seguiera Reichb. ex Oliver = Combretum Loefl. (Combretac.).
Seguieria Loefl. Phytolaccaceae. 30 trop. S. Am. Ls. leathery; stipules thorny. Powerful odour of garlic. Cpl. 1. Fr. a samara.
Seguieriaceae Nak. = Phytolaccaceae–Rivineae Reichb.
Seguinum Rafin. = Dieffenbachia Schott (Arac.).

Segurola Larrañaga = ? Aeschynomene L., etc. (Legumin.).
Sehima Forsk. Gramineae. 7 warm Afr., India, Austr.
Seidelia Baill. Euphorbiaceae. 2 S. Afr.
Seidlia Kostel. = Vatica L. (Dipterocarpac.).
Seidlia Opiz = Scirpus L. (Cyperac.).
Seidlitzia Bunge ex Boiss. Chenopodiaceae. 8 Canary Is. & Medit. to C. As. & Persia.
Sekika Medik. = Saxifraga L. (Saxifragac.).
*__Selaginaceae__ Choisy = Scrophulariaceae–Selagineae Reichb.
Selaginastrum Schinz & Thellung = Antherothamnus N. E. Br. (Scrophulariac.).
*__Selaginella__ Beauv. Selaginellaceae. 700 chiefly trop.; a few temp., e.g. *S. selaginoïdes* (L.) Link on boggy hillsides in Eur.; most in damp places, esp. in forests, but a few xero.
Selaginellaceae Mett. Selaginellales. Ls. small, 1-veined, usually of 2 kinds, in 4 ranks; a minute ligule at base of each l., sporangia containing either microspores or megaspores; prothalli very small, unisexual. Only gen.: *Selaginella*.
Selaginellales. Lycopsida. 1 fam.: *Selaginellaceae.*
Selaginoïdes Boehm. = Selaginella P. Beauv. (Selaginellac.).
Selago Adans. = Camphorosma L. + Polycnemum L. (Chenopodiac.).
Selago Boehm. = Lycopodium L. (Lycopodiac.).
Selago P.Br. = Selaginella Beauv. (Selaginellac.).
Selago L. Scrophulariaceae. 150 trop. & S. Afr.
Selas Spreng. = Gela Lour. = Acronychia J. R. & G. Forst. (Rutac.).
Selatium D. Don ex G. Don = Gentiana L. (Gentianac.).
Selbya M. Roem. = Aglaia Lour. (Meliac.).
Selenia Nutt. Cruciferae. 6 S. U.S., Mex.
Selenicereus (A. Berger) Britton & Rose. Cactaceae. 25 S. U.S. to C. Am., N. coasts of S. Am.; 1 Urug. & Argent. *S. grandiflorus* (Mill.) Britt. & Rose is one of the night-flowering cacti, whose sweetly scented flowers open in the evening and wither before morning.
Selenidium (Kunze) Fee = Dennstaedtia Bernh. (Dennstaedtiac.).
× **Selenipanthes** hort. Orchidaceae. Gen. hybr. (vi; nomen nugax) (Lepanthes × Selenipedium).
Selenipedilum Pfitz. = seq.
Selenipedium Reichb. f. Orchidaceae. 3 trop. S. Am., W.I.
× **Seleniphyllum** hort. Cactaceae. Gen. hybr. (Epiphyllum × Selenicereus).
Selenocarpaea Eckl. & Zeyh. = Heliophila Burm. f. (Crucif.).
Selenocera Zipp. ex Span. = Mitreola L. ex Schaeff. (Loganiac.).
× **Selenocypripedium** hort. (vi) = × Cysepedium hort. (Orchidac.).
Selenodesmium (Prantl) Copel. Hymenophyllaceae. 10 pantrop. & S. hemisph.
Selenogyne DC. = Solenogyne Cass. = Lagenophora Cass. (Compos.).
Selenothamnus Melville. Malvaceae. 2 Austr.
Selepsion Rafin. = Urtica L. (Urticac.).
Selera Ulbr. Malvaceae. 1 Mex.
Seleranthus Hill (sphalm.) = Scleranthus L. (Caryophyllac.).
Seleria Boeck. (sphalm.) = Scleria Bergius (Cyperac.).
Selerothamnus Harms = Hesperothamnus T. S. Brandegee (Legumin.).

SELINOCARPUS

Selinocarpus A. Gray. Nyctaginaceae. 8 SW. U.S., Mex.
Selinon Adans.=Apium L. (Umbellif.).
Selinon Rafin.=Selinum L. (Umbellif.).
Selinopsis Coss. & Dur. ex Munby=Carum L. (Umbellif.).
Selinum L. (1753)=Peucedanum L. (Umbellif.).
*Selinum L. (1762). Umbelliferae. 4 Scand. & C. Eur. to C. As.
Selkirkia Hemsl. Boraginaceae. 1 Juan Fernandez.
Selleola Urb. Caryophyllaceae. 1 Haiti.
Selleophytum Urb. Compositae. 1 Haiti.
Selliera Cav. Goodeniaceae. 2 Austr., N.Z., temp. S. Am. Fr. indehisc.
Selliguea Bory. Polypodiaceae. 5 Malaysia, Polynesia. *S. feei* Bory terrestrial near craters of volcanoes in Java. (See *Crypsinus* Presl.)
*Selloa Kunth. Compositae. 1 Mex.
Selloa Spreng.=Gymnosperma Less. (Compos.).
Sellocharis Taub. Leguminosae. 1 SE. Brazil.
Sellowia Roth ex Roem. & Schult.=Ammannia L. (Lythrac.).
Sellunia Alef.=Vicia L. (Legumin.).
Selmation Th. Dur. (sphalm.)=Stelmation Fourn. (Asclepiadac.).
Selnorition Rafin.=Rubus L. (Rosac.).
Selonia Regel=Eremurus Bieb. (Liliac.).
Selwynia F. Muell.=Cocculus DC. (Menispermac.).
Selysia Cogn. Cucurbitaceae. 2 Brazil, Colombia.
Semaquilegia P. & K.=Semiaquilegia Makino (Ranunculac.).
Semarilla Rafin.=Gymnosporia Benth. & Hook. f. (Celastrac.).
Semarillaria Ruiz & Pav.=Paullinia L. (Sapindac.).
Semecarpos St-Lag.=seq.
Semecarpus L. f. Anacardiaceae. 50 Indomal., Micron., Solomon Is. The young fr. yields a black resin used as marking ink, etc.
Semeiandra Hook. & Arn. Onagraceae. 1 Mex.
Semeiocardium Hassk.=Polygala L. (Polygalac.).
Semeiocardium Zoll. Balsaminaceae. 1 Indonesia (Madura, Kangean Is.), on calcareous rocks.
Semeionotis Schott=Dalbergia L. (Legumin.).
Semeiostachys Drobov=Agropyron Gaertn. (Gramin.).
Semele Kunth. Ruscaceae. 5 Canaries, Madeira. *S. androgyna* (L.) Kunth is a climbing shrub with leaf-like phylloclades in the axils of scale-ls. Fls. in little cymes (cf. *Asparagus*) on edges of phylloclades. The new shoots rise from the soil, and reach some length before the lat. branches, bearing the phylloclades, begin to unfold.
Semenovia Regel & Herder=Heracleum L. (Umbellif.).
Semetor Rafin.=Derris Lour. (Legumin.).
Semetum Rafin.=Lepidium L. (Crucif.).
Semiaquilegia Makino. Ranunculaceae. 7 E. As.
Semiarundinaria Makino ex Nak. Gramineae. 20 E. As.
Semibegoniella C. DC. Begoniaceae. 2 Ecuador.
Semicipium Pierre=Mimusops L. (Sapotac.).
Semicirculaceae Dulac=Monotropaceae Nutt.
Semidopsis Zumaglini=Duschekia Opiz (Betulac.).

Semilta Rafin. = Croton L. (Euphorbiac.).
Semiphajus Gagnep. = Eulophia R.Br. ex Lindl. (Orchidac.).
Semiramisia Klotzsch. Ericaceae. 7 N. Andes.
Semnanthe N. E. Brown. Aïzoaceae. 1 S. Afr.
Semnos Rafin. = Chilianthus Burch. (Buddlejac.).
Semnostachya Bremek. Acanthaceae. 9 W. Malaysia.
Semnothyrsus Bremek. Acanthaceae. 1 Sumatra.
Semonvillea J. Gay (~ Limeum L.). Aïzoaceae. 3 trop. & S. Afr.
Semperviv[ace]ae Juss. = Crassulaceae DC.
Sempervivella Stapf. Crassulaceae. 4 W. Himalaya.
Sempervivum L. Crassulaceae. 25 mts. S. Eur. to Cauc. Xero. with fleshy
ls. and veg. repr. by offsets. *S. tectorum* L. (houseleek), planted on cottages
to keep slates in position.
Senacia Comm. ex Lam. = Pittosporum (Pittosporac.) + Maytenus (Celastrac.)
+ Celastrus, etc.
Senacia Comm. ex Lam. emend. Thou. = Pittosporum Banks ex Soland.
Senaea Taub. Gentianaceae. 2 Brazil. [(Pittosporac.).
Senapea Aubl. = ? Passiflora L. (Passiflorac.).
Senckenbergia Gaertn., Mey. & Scherb. = Lepidium L. (Crucif.).
Senckenbergia P. & K. (2) = Senkebergia Neck. = Mendoncia Vell. (Mendonciac.).
Senckenbergia P. & K. (3) = Senkenbergia Schau. = Cyphomeris Standl.
(Nyctaginac.).
Senebiera DC. = Coronopus Zinn (Crucif.).
Senebiera P. & K. = Senneberia Neck. = Ocotea Aubl. (Laurac.).
× **Senecillicacalia** Kitamura. Compositae. Gen. hybr. (Cacalia × Senecillis).
Senecillis Gaertn. = Senecio L. (Compos.).
Senecio L. Compositae. 2–3000 cosmop. The gen. includes pls. of most various
habit. Some are climbers, e.g. *S. macroglossus* DC. (S. Afr.), which is remarkably like ivy. Many are xero., some with fleshy ls., others with fleshy stems,
others with hairy or inrolled ls. The fls. of *S. vulgaris* L. (groundsel) are regularly self-fert. and are very inconspic.; there are normally no ray-florets. In
S. jacobaea L. (ragwort) there are ray-florets, and the conspic. fls. are largely
visited by insects. The fleshy stems of *S. (Kleinia) articulatus* Sch. Bip. (S. Afr.)
separate at the joints and grow into new pls. Over a dozen remarkable tree-
species (*S. johnstonii* Oliv., etc.) at high altitudes on E. African volcanoes.
Senecionaceae ('-idaceae') Bessey = Compositae–Senecioneae Cass.
Seneciunculus Opiz = Senecio L. (Compos.).
Senefeldera Mart. Euphorbiaceae. 10 trop. S. Am.
Senefelderopsis Steyerm. Euphorbiaceae. 2 NW. trop. S. Am.
Senega Spach = Polygala L. (Polygalac.).
Senegalia Rafin. = Acacia Mill. (Legumin.).
Senegaria Rafin. = Polygala L. (Polygalac.).
Seneico Hill (sphalm.) = Senecio L. (Compos.).
Senftenbergia Klotzsch & Karst. ex Klotzsch = Langsdorffia Mart. (Balanophorac.).
Senisetum Honda = Agrostis L. (Gramin.).
Senites Adans. = Zeugites P.Br. (Gramin.).

SENKEBERGIA

Senkebergia Neck. = Mendoncia Vell. (Mendonciac.).

Senkenbergia Reichb. = Senckenbergia Gaertn., Mey. & Scherb. = Lepidium L. (Crucif.).

Senkenbergia Schau. = Cyphomeris Standl. (Nyctaginac.).

Senna Mill. = Cassia L. (Legumin.).

Senneberia Neck. = Ocotea Aubl. (Laurac.).

Sennebiera Willd. = Senebiera DC. = Coronopus Zinn (Crucif.).

Sennefeldera Endl. = Senefeldera Mart. (Euphorbiac.).

Sennenia Pau ex Sennen = Trisetum Pers. (Gramin.).

Sennia Chiov. Sapindaceae. 1 Somalia.

Senniella Aellen. Chenopodiaceae. 1 N., W. & S. Austr.

Senra Cav. Malvaceae. 3 E. Afr., Arabia.

Senraea Willd. = praec.

Sensitiva Rafin. = Mimosa L. (Legumin.).

Sentis Comm. ex Brongn. = Scutia (DC.) Comm. ex Brongn. (Rhamnac.).

Sentis F. Muell. = Pholidia R.Br. (Myoporac.).

Sepalosaccus Schlechter. Orchidaceae. 1 Costa Rica, Ecuador.

Sepalosiphon Schlechter. Orchidaceae. 1 New Guinea.

Separotheca Waterf. Commelinaceae. 1 Mex.

Sepikea Schlechter. Gesneriaceae. 1 New Guinea.

Septacanthus Wight (sphalm.) = Leptacanthus Nees = Strobilanthes Blume (Acanthac.).

Septas L. = Crassula L. (Crassulac.).

Septas Lour. = Brami Adans. = Bacopa Aubl. (Scrophulariac.).

Septilia Rafin. = praec.

Septimetula Van Tiegh. = Tapinanthus Bl. (Loranthac.).

Septina Nor. = ? Litsea Lam. (Laurac.).

Septogarcinia Kosterm. Guttiferae. 1 E. Malaysia (Sumbawa).

Septotheca Ulbr. Bombacaceae. 1 Peru.

Septulina Van Tiegh. = Taxillus Van Tiegh. (Loranthac.).

*Sequoia Endl. Taxodiaceae. *S. sempervirens* Endl., the redwood, even taller than the mammoth tree (*Sequoiadendron, q.v.*), though not so thick (up to 102 m. high and 8·5 m. thick), and valued for its timber.

Sequoiadendron Buchholz. Taxodiaceae. 1 Calif., *S. giganteum* (Lindl.) Buchholz, the mammoth tree, discovered in the Sierra Nevada in 1850. The tallest is 96 m., the thickest 10·5 m. (Sargent; cf. *Eucalyptus*); the age of the largest is about 1500 years. In some museums are sections of a tree cut down in 1882 and showing 1335 annual rings.

Serangium W. Wood ex Salisb. = Monstera Adans. (Arac.).

Seraphyta Fisch. & Mey. Orchidaceae. 1 Mex., C. Am., W.I.

*Serapias L. Orchidaceae. 10 Azores, Medit.

Serapiastrum Kuntze = praec.

× **Serapicamptis** Godfery. Orchidaceae. Gen. hybr. (i) (Anacamptis × Serapias).

× **Serapirhiza** Potůček. Orchidaceae. Gen. hybr. (i) (Dactylorhiza × Serapias).

Serena Rafin. = Haemanthus L. (Amaryllidac.).

Serenaea Hook. f. (sphalm.) = seq.

Serenoa Hook. f. Palmae. 1 SE. U.S.

Sererea Rafin. = Phaedranthus Miers (Bignoniac.).
Seretoberlinia Duvign. Leguminosae. 1 trop. Afr.
Sergia Fedorov. Campanulaceae. 2 C. As.
Sergilus Gaertn. = Baccharis L. (Compos.).
Serialbizzia Kosterm. Leguminosae. 2 Indoch., W. Malaysia, Celebes.
Seriana Willd. = seq.
Seriania Schumacher = Serjania Mill. (Sapindac.).
Serianthes Benth. (1) = Albizia Durazz. (Legumin.).
Serianthes Benth. (2). Leguminosae. 11 Malaysia, Polynesia.
Sericandra Rafin. = Albizia Durazz. (Legumin.).
Sericeocassia Britton = Cassia L. (Legumin.).
× **Sericobonia** André. Acanthaceae. Gen. hybr. (Libonia × Sericographis).
Sericocactus Y. Ito = Notocactus (A. Berger) Backeb. & F. M. Knuth (Cactac.).
Sericocalyx Bremek. Acanthaceae. 15 S. China, SE. As., Indomal.
Sericocarpus Nees. Compositae. 5 U.S.
Sericocoma Fenzl. Amaranthaceae. *S.str.* 6, *s.l.* 25, trop. & S. Afr.
Sericocomopsis Schinz. Amaranthaceae. 4 E. trop. Afr.
Sericodes A. Gray. Zygophyllaceae. 1 N. Mex.
Sericographis Nees (~ Jacobinia Moric.). Acanthaceae. 30 Mex. to trop. S. Am.
Sericola Rafin. = Miconia Ruiz & Pav. (Melastomatac.).
Sericolea Schlechter. Elaeocarpaceae. 20 mts. of New Guinea.
Sericoma Hochst. (sphalm.) = Sericocoma Fenzl (Amaranthac.).
Sericorema (Hook. f.) Lopriore. Amaranthaceae. 3 trop. Afr., Madag.
Sericospora Nees. Acanthaceae (inc. sed.). 1 Antilles.
Sericostachys Gilg & Lopriore. Amaranthaceae. 1–2 trop. Afr.
Sericostoma Stocks. Boraginaceae. 8 trop. E. & NE. Afr. to NW. India.
Sericotheca Rafin. = Holodiscus Maxim. (Rosac.).
Sericrostis Rafin. = Muhlenbergia Schreb. (Gramin.).
Sericura Hassk. = Pennisetum Rich. (Gramin.).
Seridia Juss. = Centaurea L. (Compos.).
Serigrostis Steud. = Sericrostis Rafin. = Muhlenbergia Schreb. (Gramin.).
Seringea F. Muell. = seq.
***Seringia** J. Gay. Sterculiaceae. 1 New Guinea, E. Austr.
Seringia Spreng. = Ptelidium Thou. (Celastrac.).
Serinia Rafin. = Krigia Schreb. (Compos.).
Seriola L. = Hypochoeris L. (Compos.).
Seriphidium (Bess.) P. Poljakov (~ Artemisia L.). Compositae. 60 temp. As.
Seriphium L. = Stoebe L. (Compos.).
Seris Less. = Richterago Kuntze = Gochnatia Kunth (Compos.).
Seris Willd. = Onoseris DC. Compositae.
Serissa Comm. ex Juss. Rubiaceae. 1–3 E. As., cult. medicinal.
Serjania Mill. Sapindaceae. 215 S. U.S. to trop. S. Am. Lianes with watch-spring tendrils and stip. ls. Fr. a 3-winged schizocarp.
Serjania Vell. Inc. sed. (Diandria Monogynia). ?Verbenac. ?Acanthac. 1 Brazil.
Serophyton Benth. = Argithamnia Sw. (Euphorbiac.).
Serpaea Gardn. = Dimerostemma Cass. + Oyedaea DC. (Compos.).

Serpicula L. = Laurembergia Bergius (Haloragidac.).
Serpicula L. f. = Hydrilla Rich. (Hydrocharitac.).
Serpicula Pursh = Elodea Michx (Hydrocharitac.).
Serpillaria Fabr. (uninom.) = *Illecebrum verticillatum* L. (Caryophyllac.).
Serpyllopsis v. d. Bosch. Hymenophyllaceae. 1 temp. S. Am.
Serpyllum Mill. = Thymus L. (Labiat.).
Serra Cav. (sphalm.) = Senra Cav. (Malvac.).
Serraea Spreng. = praec.
Serrafalcus Parl. = Bromus L. (Gramin.).
Serraria Adans. = Serruria Adans. corr. Salisb. (Proteac.).
Serrastylis Rolfe = Macradenia R.Br. (Orchidac.).
Serratula L. Compositae. 70 Eur. to Japan. *S. tinctoria* L. is dioec.
Serresia Montr. Inc. sed. 1 New Caled.
Serronia Gaudich. = Piper L. (Piperac.).
Serrulata DC. (sphalm.) = Serratula L. (Compos.).
Serruria Adans. corr. Salisb. Proteaceae. 50 S. Afr.
Sersalisia R.Br. = Lucuma Mol. + Planchonella Pierre (Sapotac.).
Sertifera Lindl. Orchidaceae. 2 trop. S. Am.
Sertuernera Mart. = Pfaffia Mart. (Amaranthac.).
Sertula L. = Melilotus L. (Legumin.).
Seruneum Kuntze = Wedelia Jacq. (Compos.).
Serveria Neck. = Doliocarpus Roland. (Dilleniac.).
Sesamaceae Horan. = Pedaliaceae R.Br.
Sesamella Reichb. = seq.
Sesamodes Kuntze = seq.
Sesamoïdes Ortega. Resedaceae. 1–2 W. Medit.
Sesamopteris DC. ex Meissn. = Sesamum L. (Pedaliac.).
Sesamothamnus Welw. Pedaliaceae. 6 trop. Afr.
Sesamum Adans. = Martynia L. (Martyniac.).
Sesamum L. Pedaliaceae. 30 trop. & S. Afr., As. *S. indicum* L. largely cult.
  in India, etc., for the oil from seeds (*gingili*, sesame, etc.).
Sesban Adans. = seq.
Sesbana R.Br. = seq.
*Sesbania Adans. mut. Scop. Leguminosae. 50 trop. & subtrop. *S. cannabina*
  (Retz.) Pers. (*S. aculeata* [Willd.] Poir.) is a marsh plant, giving off floating roots
  from the base of the stem, covered with spongy aërenchyma (cf. *Neptunia*).
Seseli L. Umbelliferae. 80 Eur. to C. As.
Seselinia G. Beck = praec.
Seselopsis Schischk. Umbelliferae. 1 C. As.
Seshagiria Ansari & Hemadri. Asclepiadaceae. 1 W. Penins. India.
Seslera St-Lag. = Sesleria Scop. (Gramin.).
Sesleria Nutt. = Buchloë Engelm. (Gramin.).
Sesleria Scop. Gramineae. 33 Eur., W. As.
Sesleriella Deyl (~ Sesleria Scop.). Gramineae. 2 N. Italy, Austria.
Sesquicella Alef. = Callirhoë Nutt. (Malvac.).
Sessea Ruiz & Pav. Solanaceae. 30 Andes.
Sesseopsis Hassl. Solanaceae. 2 subtrop. S. Am.
Sessilanthera Molseed & Cruden. Iridaceae. 2 Mex., Guatem.

Sessleria Spreng. = Sesleria Scop. (Gramin.).
Sestinia Boiss. = Hymenocrater Fisch. & Mey. (Labiat.).
Sestinia Boiss. & Hohen. = Wendlandia Bartl. (Rubiac.).
Sestinia Rafin. = Agrimonia L. (Rosac.).
Sestochilos Breda = Bulbophyllum Thou. (Orchidac.).
Sestochilus P. & K. = praec.
Sesuviaceae Horan. = Aïzoaceae J. G. Agardh.
Sesuvium L. Aïzoaceae. 8 trop. & subtrop. coasts. Halophytes.
Setachna Dulac = Centaurea L. (Compos.).
*Setaria Beauv. Gramineae. 140 trop. & warm temp. *S. italica* (L.) Beauv.
(Italian millet) is cult. as a cereal from S. Eur. to Japan.
Setariopsis Scribner ex Millsp. Gramineae. 2 Mex.
Setchellanthus T. S. Brandegee. Capparidaceae (inc. sed.). 1 Mex. Petals
blue!
Setcreasea K. Schum. & Sydow. Commelinaceae. 9 S. U.S., Mex.
Sethia Kunth = Erythroxylum P.Br. (Erythroxylac.).
Seticereus Backeb. = Borzicactus Riccob. (Cactac.).
Seticleistocactus Backeb. (~ Cleistocactus Lem.). Cactaceae. 1 Bolivia.
× Setidenmoza Hort. Cactaceae. Gen. hybr. (Denmoza × Seticereus).
Setiechinopsis (Backeb.) de Haas = Arthrocereus A. Berger (Cactac.).
Setilobus Baill. Bignoniaceae. 3 Brazil.
Setirebutia Frič & Kreuzinger = Rebutia K. Schum. (Cactac.).
Setiscapella Barnhart = Utricularia L. (Lentibulariac.).
Setosa Ewart = Chamaeraphis R.Br. (Gramin.).
Setouratea Van Tiegh. = Ouratea Aubl. (Ochnac.).
Seubertia Kunth = Brodiaea Sm. (Alliac.).
Seubertia H. C. Wats. = Bellis L. (Compos.).
Seutera Reichb. (~ Vincetoxicum Moench). Asclepiadaceae. 2 coasts NE. As.,
Pacif. N. Am.
Sevada Moq. (~ Suaeda Forsk.). Chenopodiaceae. 1 Abyss., Somal., Arabia.
Severinia Tenore (~ Atalantia Corrêa). Rutaceae. 7 E. As., Indomal.
Sewerzowia Regel & Schmalh. = Astragalus L. (Legumin.).
Sexglumaceae Dulac = Juncaceae Juss.
Sexilia Rafin. = Polygala L. (Polygalac.).
Seychellaria Hemsl. Triuridaceae. 3 Madag., Seychelles.
*Seymeria Pursh. Scrophulariaceae. 25 S. U.S., Mex.
Seymouria Sweet = Pelargonium L'Hérit. (Geraniac.).
Seyrigia Keraudren. Cucurbitaceae. 4 Madag.
Shae[ace]ae Bertol. f. = Combretaceae R.Br.
Shafera Greenman. Compositae. 1 Cuba.
Shaferocharis Urb. Rubiaceae. 1 Cuba.
Shaferodendron Gilly. Sapotaceae. 2 Cuba.
Shakua Boj. = Spondias L. (Anacardiac.).
Shallonium Rafin. = Gaultheria L. (Ericac.).
Shantzia Lewton = Azanza Alef. (Malvac.).
Shawia J. R. & G. Forst. = Olearia Moench (Compos.).
Sheadendron Bertol. f. = Combretum L. (Combretac.).
Sheareria S. Moore. Compositae. 1 E. China.

SHEFFIELDIA

Sheffieldia J. R. & G. Forst. = Samolus L. (Primulac.).

Sheperdia Rafin. = seq.

***Shepherdia** Nutt. Elaeagnaceae. 3 N. Am. Recept. fleshy in fr. Fr. of *S. argentea* Nutt. (buffalo-berry) ed.

Sherarda St-Lag. = Sherardia L. (Rubiac.).

Sherardia Boehm. = Glinus L. (Aïzoac.).

Sherardia L. Rubiaceae. 1 Eur., W. As., N. Afr., *S. arvensis* L.

Sherardia Mill. = Stachytarpheta Vahl (Verbenac.).

Sherbournea G. Don. Rubiaceae. 10 trop. Afr. K conv.

Sherwoodia House = Shortia Torr. & Gray (Diapensiac.).

Shibataea Makino. Gramineae. 6 E. As.

Shibateranthis Nakai = Eranthis Salisb. (Ranunculac.).

Shicola M. Roem. = Eriobotrya Lindl. (Rosac.).

× **Shigeuraara** hort. Orchidaceae. Gen. hybr. (iii) (Ascocentrum × Ascoglossum × Renanthera × Vanda).

Shiia Makino = Castanopsis (D. Don) Spach (Fagac.).

Shinnersia R. M. King & H. Rob. Compositae. 1 S. U.S. (Texas), Mex.

× **Shipmanara** hort. Orchidaceae. Gen. hybr. (iii) (Broughtonia × Caularthron [Diacrium] × Schomburgkia).

Shirakia Hurusawa = Sapium P.Br. (Euphorbiac.).

Shirleyopanax Domin = Kissodendron Seem. (Araliac.).

Shishindenia Makino ex Koidz. = Retinospora Carr. (Cupressac.).

Shiuyinghua Paclt. Scrophulariaceae. 1 C. China. Near *Paulownia* Sieb. & Zucc.

Shorea Roxb. ex Gaertn. Dipterocarpaceae. 180 Ceylon to S. China, W. Malaysia, Moluccas, Lesser Sunda Is. Many yield useful timber, such as *S. robusta* Roxb. (*sal*) of India, and *meranti, luan* or *seraya*, obtained from several spp. in the Malay Arch. The fruit of *S. macrophylla* (De Vriese) Ashton and related Bornean spp., called *illipe* or *engkabang* nuts, yield ed. fat used as a substitute for cocoa butter in chocolate manufacture. Some yield resin (*dammar*) used in varnishes.

Shoreaceae Barkley = Dipterocarpaceae Bl.

Shortia Rafin. = Arabis L. (Crucif.).

***Shortia** Torr. & Gray. Diapensiaceae. 7–8 SW. China, Formosa, Japan; 1 SE. U.S.

Shortiopsis Hayata = praec.

Shultzia Rafin. = Schultzia Rafin. = Obolaria L. (Gentianac.).

Shuria Hort. ex Hérincq = Achimenes P.Br. (Gesneriac.).

Shutereia Choisy = Hewittia Wight (Convolvulac.).

***Shuteria** Wight & Arn. Leguminosae. 10 trop. Afr. to E. As. & Java.

Shuttelworthia Steud. = seq.

Shuttleworthia Meissn. = Uwarowia Bunge = Verbena L. (Verbenac.).

Siagonanthus Poepp. & Endl. = Ornithidium R.Br. (Orchidac.).

Siagonanthus Pohl ex Engler = Emmotum Desv. ex Ham. (Icacinac.).

Siagonarrhen Mart. ex J. A. Schmidt = Hyptis Jacq. (Labiat.).

Sialita Rafin. = Syalita Adans. = Dillenia L. (Dilleniac.).

Sialodes Eckl. & Zeyh. = Galenia L. (Aïzoac.).

Sibaldia L. = Sibbaldia L. = Potentilla L. (Rosac.).

Sibangea Oliv. = Drypetes Vahl (Euphorbiac.).
Sibara Greene. Cruciferae. 11 N. Mex. & Calif. to E. U.S.
Sibbaldia L. (~ Potentilla L.). Rosaceae. 20 temp. Euras. to Himal.
Sibbaldianthe Juzepczuk. Rosaceae. 2 temp. As.
Sibbaldiopsis Rydb. = Potentilla L. (Rosac.).
Sibertia Steud. (sphalm.) = Libertia Lej. = Bromus L. (Gramin.).
Sibiraea Maxim. (= Eleiosina Rafin.). Rosaceae. 5 Balkans, C. As., W. China.
Sibthorpia L. Scrophulariaceae. 5 C. & S. Am., Azores, Madeira, W. & S.
Eur., mts. trop. Afr.
Sibthorpiaceae D. Don = Scrophulariaceae–Digitalideae Benth. *p.p.*
Sibtorpia Scop. = Sibthorpia L. (Scrophulariac.).
Siburatia Thou. = Maesa Forsk. (Myrsinac.).
Sicana Naud. Cucurbitaceae. 2 trop. Am., W.I. Fr. ed.
Sicelium P.Br. = Coccocypselum P.Br. (Rubiac.).
Sickingia Willd. = Simira Aubl. (Rubiac.).
Sicklera M. Roem. = Poechia Opiz = Murraya Koen. ex L. (Rutac.).
Sicklera Sendtn. = Brachistus Miers (Solanac.).
Sickmannia Nees = Ficinia Schrad. (Cyperac.).
Sicrea H. Hallier. Tiliaceae. 1 Siam, Indoch.
Sicydium A. Gray = Ibervillea Greene (Cucurbitac.).
Sicydium Schlechtd. Cucurbitaceae. 1 trop. Am.
Sicyocarpus Boj. = Dregea E. Mey. (Asclepiadac.).
Sicyocaulis Wiggins. Cucurbitaceae. 1 Galápagos.
Sicyocodon Feer = Campanula L. (Campanulac.).
Sicyodes Ludw. = seq.
Sicyoïdes Mill. = Sicyos L. (Cucurbitac.).
Sicyomorpha Miers = Peritassa Miers (Celastrac.).
Sicyos L. Cucurbitaceae. 15 Hawaii, Polynes., Austr., trop. Am.
Sicyosperma A. Gray. Cucurbitaceae. 1 S. U.S.
Sicyus Clements = Sicyos L. (Cucurbitac.).
Sida L. Malvaceae. 200 all warm regions, esp. Am. Ov. 1 per loc., pena.
Schizocarp.
Sidalcea A. Gray. Malvaceae. 25 W. N. Am.
Sidanoda Wooton & Standley = Anoda Cav. (Malvac.).
Sidastrum E. G. Baker = Sida L. (Malvac.).
Side St-Lag. = Sida L. (Malvac.).
Sideranthus 'Fraser' [Nutt.] ex Nees = Haplopappus Cass. (Compos.).
Siderasis Rafin. Commelinaceae. 1 trop. S. Am.
Sideria Ewart & Petrie. Malvaceae. 1 N. Austr. Quid?
Sideritis L. Labiatae. 100 N. temp. Euras.
Siderobombyx Bremek. Rubiaceae. 1 N. Borneo.
Siderocarpos Small = Acacia Mill. (Legumin.).
Siderocarpus Pierre ex L. Planch. = Planchonella Pierre (Sapotac.).
Siderocarpus Willis (sphalm.) = Siderocarpos Small = Acacia L. (Legumin.).
Siderodendron Roem. & Schult. = seq.
Siderodendrum Schreb. = Sideroxyloïdes Jacq. = Ixora L. (Rubiac.).
Sideropogon Pichon. Bignoniaceae. 1 Brazil.
Sideroxyloïdes Jacq. = Ixora L. (Rubiac.).

SIDEROXYLON

Sideroxylon L. Sapotaceae. 100 trop. Some ironwoods.

Sideroxylum Salisb. = praec.

Sidopsis Rydb. = Sida L. (Malvac.).

***Siebera** J. Gay. Compositae. 1–2 W. As.

Siebera Hoppe = Minuartia L. (Caryophyllac.).

Siebera P. & K. = Sieberia Spreng. (Orchidac.), *q.v.*

Siebera Presl = Anredera Juss. (Basellac.).

Siebera Reichb. = Platysace Bunge (Umbellif.).

Sieberia Spreng. = Platanthera Lindl., Nigritella Rich., Coeloglossum Hartm., Leucorchis E. Mey., etc. (Orchidac.).

Sieboldia Heynh. = Simethis Kunth (Liliac.).

Sieboldia Hoffmgg. = Clematis L. (Ranunculac.).

Siegesbeckia auctt. = Sigesbeckia L. (Compos.).

Siegfriedia C. A. Gardner. Rhamnaceae. 1 W. Austr.

Sieglingia Bernh. (~ Danthonia DC.). Gramineae. 1 Madeira, Algeria, Eur., As. Min.

Siemensia Urb. Rubiaceae. 1 Cuba.

Siemssenia Steetz = Podolepis Labill. (Compos.).

Sieruela Rafin. = Cleome L. (Cleomac.).

Sievekingia Reichb. f. Orchidaceae. 5 trop. Am.

Sieversia Willd. Rosaceae. 2 NE. As., Aleutian Is.

Siflora Rafin. = Sison L. (Umbellif.).

Sigesbeckia L. Compositae. 6 trop. & warm temp. Heads small, with invol. of 5 bracts, covered with very sticky glandular hairs, aiding in distr., the whole head breaking off.

Sigilaria Rafin. = Sigillaria Rafin. = Smilacina Desf. (Liliac.).

Sigillabenis Thou. (uninom.) = *Habenaria sigillum* Thou. (Orchidac.).

Sigillaria Rafin. = Smilacina Desf. (Liliac.).

Sigillum Montandon = Polygonatum Adans. (Liliac.).

Sigmatanthus Huber ex Ducke = Raputia Aubl. (Rutac.).

Sigmatochilus Rolfe = Pholidota Lindl. (Orchidac.).

Sigmatogyne Pfitzer. Orchidaceae. 2 E. Himal., Indoch.

Sigmatophyllum D. Dietr. = Stigmaphyllum A. Juss. (Malpighiac.).

Sigmatosiphon Engl. = Sesamothamnus Welw. (Pedaliac.).

Sigmatostalix Reichb. f. Orchidaceae. 16 C. & trop. S. Am.

Sigmodostyles Meissn. = Rhynchosia Lour. (Legumin.).

Sikira Rafin. = Chaerophyllum L. (Umbellif.).

Silamnus Rafin. = ? Cephalanthus L. (Naucleac.).

Silaum Mill. Umbelliferae. 10 temp. Euras.

Silaus Bernh. = praec.

Silen[ac]eae Bartl. = Caryophyllaceae–Silenoïdeae A. Br.

Silenanthe Griseb. & Schenk = Silene L. (Caryophyllac.).

Silene L. Caryophyllaceae. 500 N. temp., esp. Medit. Fls. of many adapted to butterflies, e.g. *S. acaulis* L. (moss campion), a tufted alpine; others to moths, e.g. *S. vulgaris* (Moench) Garcke (bladder campion), which emits scent at night.

Silenopsis Willk. = Petrocoptis A.Br. (Caryophyllac.).

Siler Mill. Umbelliferae. 1 Eur., Siberia.

Sileriana Urb. & Loes. (sphalm.) = Jacquinia L. (Theophrastac.).

Silerium Rafin. = Trochiscanthes Koch (Umbellif.).

Silicularia Compton. Cruciferae. 1 S. Afr.

Siliqua Duham. = Ceratonia L. (Legumin.).

Siliquamomum Baill. Zingiberaceae. 1 Indoch.

Siliquaria Forsk. = Cleome L. (Cleomac.).

Siliquastrum Duham. = Cercis L. (Legumin.).

Sillybum Hassk. = Silybum Adans. (Compos.).

Siloxerus Labill. = Angianthus Wendl. (Compos.).

Silphion St-Lag. = Silphium L. (Compos.).

Silphiosperma Steetz = Brachycome Cass. (Compos.).

Silphium L. Compositae. 15 E. U.S. *S. laciniatum* L. is the 'compass plant' of the prairies. In an exposed position its ls. turn their edges to N. and S. and avoid the mid-day radiation. (Cf. *Lactuca*.)

Silvaea Hook. & Arn. = Trigonostemon Blume (Euphorbiac.).

Silvaea Meissn. = Silvia Allem. = Mezilaurus Taub. (Laurac.).

Silvaea Phil. = Philippiamra Kuntze (Portulacac.)

Silvalismis (sphalm. '-alisimis') Thou. (uninom.) = *Centrosis? plantaginea* Thou. = *Calanthe sylvatica* (Thou.) Lindl. (Orchidac.).

Silvia Allem. = Mezia Kuntze = Neosilvia Pax = Mezilaurus Taub. (Laurac.).

Silvia Benth. = Silviella Pennell (Scrophulariac.).

Silvia Vell. = Escobedia Ruiz & Pav. (Scrophulariac.).

Silvianthus Hook. f. Carlemanniaceae. 2 Assam, S. China, Indoch.

Silviella Pennell. Scrophulariaceae. 2 Mex.

Silvinula Pennell = Bacopa Aubl. (Scrophulariac.).

Silvorchis J. J. Smith. Orchidaceae. 1 Java.

Silybon St-Lag. = seq.

*__Silybum__ Adans. Compositae. 2 Medit. *S. marianum* (L.) Gaertn. (milk-thistle) is now widely distributed over the pampas of S. Am., where it was introduced.

Silymbrium Neck. (sphalm.) = Sisymbrium L. (Crucit.).

Simaba Aubl. = Quassia L. (Simaroubac.).

Simabaceae Horan. = fam. heterogen., incl. Ochnac., Simaroubac., Staphyleac., and Caryocarac.

*__Simarouba__ Aubl. = Quassia L. (Simaroubac.).

*__Simaroubaceae__ DC. Dicots. 20/120 trop. & subtrop. Shrubs and trees with alt. pinnate or simple usu. exstip. ls., never gland-dotted. Fls. small, reg., ⚥, often ∞, in axillary cpd. pan. or cymose spikes. K and C 3–7-merous. K free or more often united; C imbr. or rarely valv.; disk between sta. and ovary ring- or cup-like, sometimes enlarged into a gynophore; A twice as many as petals, obdiplost., often with scales at the base; G̲ (4–5) or fewer, often free below and united by the style or stigma; ovules usu. 1 in each loc. as in *Rutaceae*. Schizocarp or caps.; endosp. thin or none; embryo with thick cots. A few yield useful timber. Many have bitter bark, sometimes officinal. Chief genera: *Quassia* (*Simarouba*), *Harrisonia*, *Brucea*, *Ailanthus*, *Soulamea*. An ill-defined assemblage, hardly differing significantly from the *Rutaceae* except in the absence of pellucid glands in the ls. Ls. stipulate in *Picrasma, Recchia, Cadellia, Guilfoylia*.

Simaruba auctt. = Simarouba Aubl. = Quassia L. (Simaroubac.).

SIMARUBA

Simaruba Boehm. = Bursera Jacq. ex L. (Burserac.).

Simarubopsis Engl. = Pierreodendron Engl. = Quassia L. (Simaroubac.).

Simblocline DC. = Diplostephium Kunth (Compos.).

Simbuleta Forsk. = Anarrhinum Desf. (Scrophulariac.).

Simenia Szabó (~ Cephalaria Schrad., ~ Dipsacus L.). Dipsacaceae. 1 N. trop. Afr. (Ethiopia).

*Simethis Kunth. Liliaceae. 1 W. & S. Eur., N. Afr.

Simira Aubl. Rubiaceae. 35 Mex. to trop. S. Am. Some medicinal.

Simira Rafin. = Scilla L. (Liliac.).

Simirestis N. Hallé. Celastraceae. 21 trop. Afr.

Simlera Bub. = Leontopodium R.Br. (Compos.).

Simmondsia Nutt. Simmondsiaceae. 1 California, L. Calif.

Simmondsiaceae (Pax) Van Tiegh. Dicots. 1/1 N. Am. Rigid divaric. branched shrub, shortly pubesc., with opp. simple ent. exstip. ls. Fls. reg., ♂ ♀, dioec., on short axill. pedunc., the ♂ small, in cernuous capit. clusters, the ♀ large, pend., sol. (more rarely in pend. 2–7-fld. rac.). ♂: K (4–)5(–6), imbr., fimbr.; C o; A 10–12, imbr., inserted ± distantly on flat recept., fil. almost o, anth. large, obl. ♀: K (4–)5(–6), ± ventric. below, much imbr., acumin.; C o; G (3), with 3 free reflexed subul. papill. styles, and 1 apical pend. anatr. ov. per loc. Fr. an ovoid loculic. caps., with shiny brown coriac. pericarp and 1–3 seeds; endosp. scanty or o. Only genus: *Simmondsia*. Usu. included in *Buxaceae*, but differing in the pentamery, the numerous sta., the solitary ovule in each loculus, the pollen, and the anomalous wood-structure, consisting of several concentric rings of vasc. strands; besides other anat. features. The anat. and pollen have much in common with *Centrospermae*, but it is difficult to envisage an actual relationship on grounds of general morphology. The most probable affinity would seem to be with the *Monimiaceae* (*s.str.*), from which *Simmondsia* differs principally in its syncarpous gynoecium and fruit, and in the scanty endosp. An unusual feature of *S.* is the storage and mobilisation of waxes in the seed, instead of the more usual carbohydrates, proteins or fats.

Simocheilus Klotzsch (~ Eremia D. Don). Ericaceae. 40 S. Afr.

Simonenium Willis (sphalm.) = Sinomenium Diels (Menispermac.).

Simonisia Nees = Beloperone Nees (Acanthac.).

Simonsia Kuntze (sphalm.) = Simonisia Nees = Beloperone Nees (Acanthac.).

Simphitum Neck. = Symphytum L. (Boraginac.).

Simplicia T. Kirk. Gramineae. 1 N.Z.

Simplocarpus F. Schmidt = Symplocarpus Salisb. (Arac.).

Simplocos Lex. = Symplocos L. (Symplocac.).

Simpsonia O. F. Cook. Palmae. 1 S. Florida, W.I.

Simsia R.Br. = Stirlingia Endl. (Proteac.).

Simsia Pers. Compositae. 40 trop. Am., Jamaica.

Simsimum Bernh. = Sesamum L. (Pedaliac.).

Sinapi Dulac = Rhynchosinapis Hayek (Crucif.).

Sinapi Mill. = Sinapis L. (Crucif.).

Sinapidendron Lowe. Cruciferae. 5–6 Madeira, Canary Is.

Sinapis L. Cruciferae. 10 Medit., Eur. *S. arvensis* L. (charlock) is an abundant weed of cult.; cornfields are often yellow with it in summer. *S. alba* L. is the white mustard.

SINOPTERIS

Sinapistrum Chevall. = praec.
Sinapistrum Medik. = Gynandropsis DC. (Cleomac.).
Sinapistrum Mill. = Cleome L. (Cleomac.).
Sinapodendron Ball = Sinapidendron Lowe (Crucif.).
Sinarundinaria Nak. = Sinoarundinaria Ohwi (Gramin.).
Sincarpia Tenore = Syncarpia Tenore (Myrtac.).
Sinclairea Sch. Bip. = seq.
Sinclairia Hook. & Arn. = Liabum Adans. (Compos.).
Sinclairiopsis Rydb. = praec.
Sincoraea Ule. Bromeliaceae. 1 NE. Brazil.
Sindechites Oliv. Apocynaceae. 3 C. & SW. China, Siam.
Sindora Miq. Leguminosae. 1 trop. Afr.; 20 SE. As., Hainan, W. Malaysia, Celebes, Moluccas.
Sindoropsis J. Léonard. Leguminosae. 1 trop. Afr.
Sindroa Jumelle. Palmae. 1 Madag.
Singana Aubl. ?Leguminosae (inc. sed.). 1 Guiana.
Singlingia Benth. (sphalm.) = Sieglingia Bern. (Gramin.).
Sinia Diels. Ochnaceae. 1 SE. China.
Sinistrophorum Schrank ex Endl. = Camelina Crantz (Crucif.).
Sinningia Nees. Gesneriaceae. 20 Brazil. Tuberous plants. S. speciosa Hiern and others are the 'Gloxinias' of cultivation. When ls. are planted on the soil, a new pl. arises from the base of the petiole by budding (cf. Streptocarpus, Begonia).
Sinoarundinaria Ohwi (~ Arundinaria Michx). Gramineae. 23 E. As.
Sinobambusa Makino. Gramineae. 8 E. As.
Sinoboea Chun = Ornithoboea Parish ex C. B. Cl. (Gesneriac.).
Sinocalamus McClure. Gramineae. 8 E. As., India.
Sinocalycanthus (Cheng & S. Y. Chang) Cheng & S. Y. Chang = Calycanthus
Sinocarum H. Wolff. Umbelliferae. 8 China.   [L. (Calycanthac.).
Sinochasea Keng. Gramineae. 1 China.
Sinocrassula Berger. Crassulaceae. 5 Himalaya, W. China.
Sinodielsia H. Wolff. Umbelliferae. 1 SW. China.
Sinodolichos Verdc. Leguminosae. 2 Burma, S. China.
Sinofranchetia (Diels) Hemsl. Lardizabalaceae. 1 W. China.
Sinoga S. T. Blake. Myrtaceae. 1 Queensland.
Sinojackia Hu. Styracaceae. 3 S. China.
Sinojohnstonia Hu. Boraginaceae. 1 W. China.
Sinolimprichtia H. Wolff. Umbelliferae. 1 E. Tibet.
Sinomalus Koidz. = Malus L. (Rosac.).
Sinomenium Diels. Menispermaceae. 1 E. As.
Sinomerrillia Hu. Celastraceae. 1 SW. China.
Sinopanax Li. Araliaceae. 1 Formosa.
Sinopteridaceae Koidz. Pteridales. Ferns with sori at ends of veins, single or confluent, protected by reflexed margin. Chief genera: Cheilanthes, Aleuritopteris, Pellaea, Doryopteris.
Sinopteris C. Chr. & Ching. Sinopteridaceae. 2 China. L.-form as Aleuritopteris Fée; sporangia large, sessile, usually single at ends of veins; annulus of many cells.

Sinopyrenaria Hu = Pyrenaria Bl. (Theac.).

Sinosideroxylon (Engl.) Aubrév. Sapotaceae. 3 Indoch., S. China.

Sinowilsonia Hemsl. Hamamelidaceae. 1 C. China.

Sinthroblastes Bremek. Acanthaceae. 1 Timor.

Sioja Buch.-Ham. ex Lindl. = Peripterygium Hassk. (Cardiopterygac.).

Siolmatra Baill. Cucurbitaceae. 3 S. Am.

Sion Adans. = Sium L. (Umbellif.).

Siona Salisb. = Dichopogon Kunth (Liliac.).

Sipanea Aubl. Rubiaceae. 17 trop. S. Am.

Sipaneopsis Steyerm. Rubiaceae. 6 NW. trop. S. Am.

Sipania Seem. = Limnosipanea Hook. f. (Rubiac.).

Sipapoa Maguire. Malpighiaceae. 9 Venez.

Siparuna Aubl. Siparunaceae. 150 Mex. to trop. S. Am., W.I. Fr. fig-like.

Siparunaceae (A. DC.) Schodde. Dicots. 3/160 trop. Am., W.I., W. Afr. As *Monimiaceae* (*s. str.*), but vessel perforation plates simple, wood fibres usu. non-septate, wood-rays narrow, wood parenchyma apotracheal, node trace single, arc-shaped; floral velum present; anthers valvate, pollen-grains granular, intectate; ovule erect; fruiting hypanthium fleshy, perigynous. Genera: *Siparuna, Bracteanthus, Glossocalyx.* (Schodde, *Taxon*, **19**: 325, 1970.)

Siphanthemum Van Tiegh. = Psittacanthus Mart. (Loranthac.).

Siphanthera Pohl. Melastomataceae. 20 Brazil, Guiana.

Siphantheropsis Brade. Melastomataceae. 1 Brazil.

Sipharissa P. & K. = Sypharissa Salisb. = Urginea Steinheil (Liliac.).

Siphaulax Rafin. = Nicotiana L. (Solanac.).

Siphidia Rafin. = seq.

Siphisia Rafin. = Aristolochia L. (Aristolochiac.).

Siphoboea Baill. = Clerodendrum L. (Verbenac.).

Siphocampylus Pohl. Campanulaceae. 215 trop. Am., W.I.

Siphocodon Turcz. Campanulaceae. 2–3 S. Afr.

Siphocolea Baill. = Stereospermum Cham. (Bignoniac.).

Siphocranion Kudo = Hancea Hemsl. = Hanceola Kudo (Labiat.).

Siphokentia Burret. Palmae. 2 Moluccas.

Siphomeris Boj. = Lecontea A. Rich. (Rubiac.).

Siphonacanthus Nees = Ruellia L. (Acanthac.).

Siphonandra Klotzsch. Ericaceae. 3 Andes.

Siphonandra Turcz. = Chiococca L. (Rubiac.).

Siphonandraceae Klotzsch = Ericaceae–Vaccinieae Reichb.

Siphonandrium K. Schum. Rubiaceae. 1 New Guinea.

Siphonanthera Pohl (nomen). 3 Brazil. Quid?

Siphonanthus L. = Clerodendrum L. (Verbenac.).

Siphonanthus Schreb. ex Baill. = Hevea Aubl. (Euphorbiac.).

Siphonella (A. Gray) A. A. Heller = Leptodactylon Hook. & Arn. (Polemoniac.).

Siphonella Small. Valerianaceae. 2 N. Am.

Siphonema Rafin. = Nierembergia Ruiz & Pav. (Solanac.).

Siphoneranthemum Kuntze = Pseuderanthemum Radlk. (Acanthac.).

Siphoneugenia Berg (~Eugenia L.). Myrtaceae. 1 trop. S. Am. Polymorphic.

Siphonia Benth. = Lindenia Benth. (Rubiac.).
Siphonia Rich. = Hevea Aubl. (Euphorbiac.).
Siphonidium Armstr. = Euphrasia L. (Scrophulariac.).
Siphoniopsis Karst. = Cola Schott (Sterculiac.).
Siphonochilus Wood & Franks. Zingiberaceae. 1 Natal.
Siphonodendron Metc. & Chalk (sphalm.) = Siphonodon Griff. (Siphonodontac.).
Siphonodon auct. (sphalm.) = Siphocodon Turcz. (Campanulac.).
Siphonodon Griff. Siphonodontaceae (~ Celastraceae). 5-6 SE. As., Malaysia NE. Austr.
*Siphonodontaceae Gagnep. & Tardieu. Dicots. 1/5 SE. As. to Austr. Trees or climbing shrubs, as *Celastrac.*, but ls. alt., K 5 C 5 A 5; G (± 10), each loc. divided horizontally into 2 superposed 1-ovulate locelli (the upper ov. ascending, the lower pend.), the whole embedded in a large hemisph. disk, with a central channel or pit lined with 5 stigmatic lines ending in 5 apical stigmatic tufts and almost filled by a narrow column resembling a style, which just reaches the orifice; the sta. bend over and are closely adpressed to the disk, the anth. forming a 5-rayed star at the apex. Fr. large, pyriform to glob., ± crustaceous, indehisc., with ± 20 radiating bony pyrenes; embryo with large cots., in bony endosp. Only genus: *Siphonodon*. Structure of gynoecium very curious, but possibly only an extreme modif. of *Celastrac.* Some anat. features similar to *Flacourtiac.*
Siphonoglossa Oerst. Acanthaceae. 15 Mex. to trop. S. Am., trop. & S. Afr.
Siphonogyne Cass. = Eriocephalus L. (Compos.).
Siphonosmanthus Stapf. Oleaceae. 3 E. As.
Siphonostegia Benth. Scrophulariaceae. 1 As. Min., 2 E. As.
Siphonostelma Schlechter. Asclepiadaceae. 1 SW. Afr.
Siphonostema Griseb. = Ceratostema Juss. (Ericac.).
Siphonostoma Benth. & Hook. f. (sphalm.) = praec.
Siphonostylis W. Schulze (~Iris L.). Iridaceae. 3 Medit. reg.
*Siphonychia Torr. & A. Gray. Caryophyllaceae. 6 Atl. N. Am.
Siphostigma B. D. Jacks. = seq.
Siphostima Rafin. = Cyanotis D. Don (Commelinac.).
Siphotoma Rafin. = Hymenocallis Salisb. (Amaryllidac.).
Siphotoxis Boj. ex Benth. = Achyrospermum Blume (Labiat.).
Siphotria Rafin. = Renealmia L. f. (Zingiberac.).
Siphyalis Rafin. = Polygonatum Mill. (Liliac.).
Sipolisia Glaziou = Proteopsis Mart. & Zucc. ex DC. (Compos.).
Siponima A. DC. = Ciponima Aubl. = Symplocos Jacq. (Symplocac.).
Siraitia Merr. Cucurbitaceae. 1 Sumatra.
Siraitos Rafin. = Chionographis Maxim. (Liliac.).
Sirhookera Kuntze. Orchidaceae. 2 S. India, Ceylon.
Sirium L. = Santalum L. (Santalac.).
Sirmuellera Kuntze = Banksia L. f. (Proteac.).
Siryrinchium Rafin. (sphalm.) = Sisyrinchium L. (Iridac.).
Sisarum Mill. = Sium L. (Umbellif.).
Sisimbryum Clairv. = Sisymbrium L. (Crucif.).
Sismondaea Delponte = Dioscorea L. (Dioscoreac.).

SISON

Sison L. Umbelliferae. 2 Eur., Medit.
Sison Wahlenb. = Apium L. (Umbellif.).
Sisymbrella Spach. Cruciferae. 5 C. France, W. & C. Medit.
Sisymbrianthus Chevall. = Rorippa Scop. (Crucif.).
Sisymbrion St-Lag. = Sisymbrium L. (Crucif.).
Sisymbriopsis Botschantsev & Tsvelev. Cruciferae. 1 C. As.
Sisymbrium L. Cruciferae. 90 temp. Euras., Medit., S. Afr., N. Am., Andes.
Sisyndite E. Mey. Zygophyllaceae. 1 S. Afr.
Sisyranthus E. Mey. Asclepiadaceae. 12 S. trop. & S. Afr.
Sisyrinchium Eckl. = Aristea Soland. (Iridac.).
Sisyrinchium L. Iridaceae. 100 Am., W.I.
Sisyrinchium Mill. = Gynandriris Parl. (Iridac.).
Sisyrocarpum Klotzsch = seq.
Sisyrocarpus Klotzsch = Capanea Decne (Gesneriac.).
Sisyrolepis Radlk. = Delpya Pierre (Sapindac.).
Sitanion Rafin. Gramineae. 4 W. N. Am. Hybrids frequent.
Sitella L. H. Bailey. Sterculiaceae. 1 N. trop. S. Am.
Sitilias Rafin. = Pyrrhopappus DC. (Compos.).
Sitobolium Desv. (sphalm.) = Sitolobium Desv. corr. J. Sm. = Dennstaedtia
  Bernh. (Dennstaedtiac.).
Sitocodium Salisb. = Camassia Lindl. (Liliac.).
Sitodium Parkinson = Artocarpus J. R. & G. Forst. (Morac.).
Sitolobium Desv. corr. J. Sm. = Dennstaedtia Bernh. (Dennstaedtiac.).
× **Sitordeum** Bowden. Gramineae. Gen. hybr. (Hordeum × Sitanion).
Sitospelos Adans. = Elymus L. (Gramin.).
Sium L. Umbelliferae. 10–15 cosmop., exc. S. Am., Austr. *S. sisarum* L.
Siumis Rafin. = praec.   [(skirret) cult. for tuberous roots.
Sivetes Rafin. = Isoëtes L. (Isoëtac.).
Sixalix Rafin. = Scabiosa L. (Dipsacac.).
Sizygium Duch. = Syzygium Gaertn. (Myrtac.).
Skalnika Pohl (nomen). 1 Brazil. Quid?
Skaphium Miq. = Xanthophyllum Roxb. (Xanthophyllac.).
Skiatophytum L. Bolus. Aïzoaceae. 1 S. Afr.
Skidanthera Rafin. = Elaeocarpus L. (Elaeocarpac.).
Skilla Rafin. = Scilla L. (Liliac.).
Skimmi Adans. = Illicium L. (Illiciac.).
*****Skimmia** Thunb. Rutaceae. 7–8 Himal., E. As., Philipp. Is. Dioecious.
Skinnera J. R. & G. Forst. = Fuchsia L. (Onagrac.).
Skinneria Choisy = Merremia Dennst. (Convolvulac.).
Skiophila Hanst. = Episcia Mart. (Gesneriac.).
Skirrhophorus DC. ex Lindl. = Angianthus Wendl. (Compos.).
Skirrophorus C. Muell. = praec.
Skizima Rafin. = Astelia Banks & Soland. (Liliac.).
Skofitzia Hassk. & Kanitz = Tradescantia L. (Commelinac.).
Skoinolon Rafin. = Schoenocaulon A. Gray (Liliac.).
Skolemora Arruda = Andira Lam. (Legumin.).
Skoliopteris Cuatrec. Malpighiaceae. 1 Colombia.
Skoliostigma Lauterb. = Spondias L. (Anacardiac.).

SMILACACEAE

Skottsbergiella Epling = Cuminia Colla (Labiat.).
Skutchia Pax & K. Hoffm. = Trophis P.Br. (Morac.).
Skytalanthus Endl. = seq.
Skytanthus Meyen. Apocynaceae. 3 Brazil, Chile.
Slackia Griff. (1848) = Decaisnea Hook. f. & Thoms. (Lardizabalac.).
Slackia Griff. (1854a) = Beccarinda Kuntze (Gesneriac.).
Slackia Griff. (1854b) = Iguanura Blume (Palm.).
Sladenia Kurz. Sladeniaceae. 1 Burma, Yunnan, Siam.
Sladeniaceae (Gilg & Werderm.) Airy Shaw. Dicots. 1/1 SE. As. Trees with simple alt. cren.-serr. exstip. ls. Fls. ⚥, in short axill. dichas. cymes. K 5, much imbr., persist. C 5, imbr., as long as sep., v. slightly conn. at base. A 10(–13), 1-ser., v. slightly adnate to C at base; fil. broad, ± ovate, thick, abruptly atten. above; anth. ± obl., basifixed, loc. divaric. (± sagitt.) at base, bifid at apex, setulose, introrse by means of poriform slits. G̲ (3), ovoid, 10-ribbed below, 20-ribbed above, narrowed into a thick shortly trifid style, ribs thickened and glandular below, simulating a disk; ovules 2 per loc., collat., pend., anatr., with a long beaked appendage near funicle. Fruit (not yet known in mature state) dry, seated on a broad base, with crustac. endoc. and papery exoc., perhaps eventually splitting septicid. into 3 cocci. Only genus: *Sladenia*. Near *Theac.*, but differing from the great majority in its dichasial cymes, and in its dry ribbed fruit recalling *Lophira* (*Ochnac.*) or *Pelliceria* (*Pelliceriac.*).
Slateria Desv. = Ophiopogon Ker-Gawl. (Liliac.).
Sleumerodendron Virot. Proteaceae. 1 New Caled.
Slevogtia Reichb. = Enicostema Blume (Gentianac.).
Sloana Adans. = seq.
Sloanea L. Elaeocarpaceae. 120 trop. As. & Am.
Sloanea Loefl. = Apeiba Aubl. (Tiliac.).
Sloania St-Lag. = Sloanea L. (Elaeocarpac.).
Sloetia Teijsm. & Binnend. ex Kurz = Streblus Lour. (Morac.).
Sloetiopsis Engl. Moraceae. 1 trop. E. Afr.
Sloteria auct. ex Steud. = Slateria Desv. = Ophiopogon Ker-Gawl. (Liliac.).
Smallanthus Mackenzie (~ Polymnia L.). Compositae. 1 E. U.S.
Smallia Nieuwl. = Pteroglossaspis Reichb. f. (Orchidac.).
Smeathmannia Soland. ex R.Br. Flacourtiaceae. 2 trop. W. Afr.
Smegmadermos Ruiz & Pav. = Quillaja Molina (Rosac.).
Smegmaria Willd. = praec.
Smegmathamnium Fenzl ex Reichb. = Saponaria L. (Caryophyllac.).
Smelophyllum Radlk. Sapindaceae. 1 S. Afr.
Smelowskia C. A. Mey. Cruciferae. 10 temp. As., Afghan., Pacif. N. Am.
Smicrostigma N. E. Brown. Aïzoaceae. 1 S. Afr.
Smidetia Rafin. = Schmidtia Tratt. = Coleanthus Seidl (Gramin.).
*****Smilacaceae** Vent. Monocots. 4/375 trop. & temp. Mostly climbing rhiz. shrubs, stems often prickly, with alt. or opp. 3-nerved ± coriac. ls. and stipular or leaf-sheath tendrils. Fls. reg., mostly ♂ ♀ dioec., rarely ⚥, in axill. rac., spikes or umb. P 3+3, rarely (3+3); A usu. 3+3 or (3+3), anth. intr. (♀ fl. with stds.); G̲ (3), with recurved stigs., and 1–2 pend. orthotr. or semi-anatr. ov. per loc. (♂ fl. without pistillode). Fr. a 1–3-seeded berry; endosp. hard. Genera: *Smilax, Rhipogonum, Heterosmilax, Pseudosmilax*.

1073

SMILACINA

*Smilacina Desf. Liliaceae. 25 Himal. & E. As. to N. & C. Am.
Smilax L. Smilacaceae. 350 trop. & subtrop. Most are climbing shrubs with net-veined ls. At base of ls. spring two tendrils, one on either side, usu. regarded as modified stips., though these organs scarcely occur in Monocots. Stems often furnished with recurved hooks which aid in climbing. Fls. dioec., in umbels. The dried roots of several S. Am. spp. form sarsaparilla.
Smirnowia Bunge. Leguminosae. 1 C. As.
Smirus Rafin. = ? Simira Aubl. (Rubiac.).
× **Smithara** hort. (v) = × Nakamotoara hort. (Orchidac.).
× **Smithennis** Hammond (nom. illegit.) = × Gloxinantha R. E. Lee (Gesneriac.).
*Smithia Ait. Leguminosae. 70 trop. Afr., As., incl. 3 Malaysia & Queensland.
Smithia J. F. Gmel. = Humbertia Lam. (Humbertiac.).
Smithia Scop. = Clusia L. (Guttif.).
Smithiantha Kuntze. Gesneriaceae. 8 Mex. They form subterranean runners,
Smithiella Dunn. Urticaceae. 1 E. Himalaya.   [covered with scale-ls.
Smithiodendron Hu = Broussonetia L'Hérit. (Morac.).
Smithorchis Tang & Wang. Orchidaceae. 1 China.
Smitinandia Holttum. Orchidaceae. 1 Siam, Indoch.
Smodingium E. Mey. Anacardiaceae. 1 S. Afr.
Smyrniaceae Burnett = Umbelliferae-Smyrnieae, etc.
Smyrniopsis Boiss. Umbelliferae. 4 E. Medit. to Persia.
Smyrnium L. Umbelliferae. 8 Eur., Medit. S. olus-atrum L. (alexanders) formerly used like celery.
Smythea Seem. Rhamnaceae. 7 Burma, Malaysia, Polynesia. Exalb.
Snowdenia C. E. Hubbard. Gramineae. 4 trop. E. Afr.
Soala Blanco = Cyathocalyx Champ. ex Hook. f. & Thoms. (Annonac.).
Soaresia Allem. = Clarisia Ruiz & Pav. (Urticac.).
*Soaresia Sch. Bip. Compositae. 1 campos of S. Brazil.
Sobennikoffia Schlechter. Orchidaceae. 4 Mascarene Is.
Soberbaea D. Dietr. (sphalm.) = Sowerbaea Sm. (Liliac.).
Sobisco Merr. (sphalm.) = seq.
Sobiso Rafin. = Salvia L. (Labiat.).
Sobolewskia M. Bieb. Cruciferae. 4 Crimea, As. Min., Cauc.
× **Sobraleya** hort. Orchidaceae. Gen. hybr. (vi) (Cattleya × Sobralia).
Sobralia Ruiz & Pav. Orchidaceae. 90 Mex. to trop. S. Am.
Sobreyra Ruiz & Pav. = Enydra Lour. (Compos.).
Sobrya Pers. = praec.
Socotora Balf. f. Periplocaceae. 1 NE. trop. Afr., Socotra.
Socotranthus Kuntze. Periplocaceae. 1 Socotra.
Socratea Karst. Palmae. 12 N. trop. S. Am.
Socratesia Klotzsch = Cavendishia Lindl. (Ericac.).
Socratina S. Balle. Loranthaceae. 2 Madag.
Soda Fourr. = Salsola L. (Chenopodiac.).
Sodada Forsk. = Capparis L. (Capparidac.).
Soderstromia Morton. Gramineae. 1 Mex.
Sodiroa André. Bromeliacaae. 8 Colombia, Ecuador.
Sodiroëlla Schlechter. Orchidaceae. 1 Ecuador.

Soehrensia (Backeb.) Backeb. Cactaceae. 7 NW. Argent.
Soelanthus Rafin. = Saelanthus Forsk. = Cissus L. (Vitidac.).
Soemmeringia Mart. Leguminosae. 1 NE. Brazil.
Sogalgina Cass. = Tridax L. (Compos.).
Sogaligna Steud. = praec.
Sogerianthe Danser. Loranthaceae. 5 New Guinea, Solomon Is.
Sohnreyia Krause = Spathelia L. (Rutac.).
Sohnsia Airy Shaw. Gramineae. 1 Mex. Dioecious.
Sohrea Steud. = Shorea Roxb. (Dipterocarpac.).
Soja Moench = Glycine L. (Legumin.).
Sokolofia Rafin. = Salix L. (Salicac.).
Solaenacanthus Oerst. = Ruellia L. (Acanthac.).
*Solanaceae Juss. Dicots. 90/2000 +, trop. & temp.; chief centre C. & S. Am.
where there are 40 local gen.; in Eur. and As. only §§ 2 & 3 are repres. Herbs
shrubs or small trees; ls. in the non-flowering part usu. alt., but in the infl.
portion alt. or in pairs; the arrangement in pairs is due to the mode of branching
and adnation. In *Datura* the branching is dichasial, and the bracts are adnate
to their axillary shoots up to the point at which the next branches arise, so that
α looks like the bracteole of 2, rather than its bract. In *Atropa* the branching
is cincinnal, one of the two branches at a node remaining undeveloped, and
the bract is again adnate to its axillary branch. Of the pair of ls. thus found
at any node, one is usu. smaller than the other. In *Solanum*, etc., further
complications occur.

Fls. sol. or in cymes, ☿, sometimes zygo. K (5), persistent; C (5), of various
forms, rarely 2-lipped, usually folded and conv.; A 5, or fewer in zygo. fls., alt.
with petals, epipet., often opening by pores; G (2), obliquely placed in the fl.
(the post. cpl. to the right, the ant. to the left, when shown in a floral diagram),
2-loc., sometimes with secondary divisions (e.g. *Datura*), upon a hypog. disk;
ov. 1–∞ in each loc., anatr. or slightly amphitr., on axile plac. (most often the
plac. are swollen and the ov. numerous); style simple, with 2-lobed stigma.
Berry or caps. Embryo curved or straight, in endosp. Fls. conspic., insect-
visited; some, e.g. *Nicotiana*, suited to Lepidoptera. A few are economically
important, e.g. *Solanum* (potato), *Nicotiana* (tobacco), *Lycopersicon*, *Capsicum*,
etc.; *Datura*, *Atropa*, etc. are medicinal; several are favourites in horticulture.

The S. are nearly related to *Scrophulariaceae*, the most general distinction
being the oblique ovary; this, however, is by no means easily made out, and
the zygomorphism of the fl. is most often used as a distinction. Certain genera
of S. are nearly related to various *Boraginaceae*, *Gesneriaceae*, *Nolanaceae*, etc.,
and it is possible that the S. are not really a simple monophyletic family; they
occupy a middle place in the *Tubiflorae* between those with actinom. and
those with zygom. fls.

Classification and chief genera (after von Wettstein):

A. Embryo clearly curved, through more than a semicircle. All 5 sta. fertile,
equal or only slightly different in length.

   1. NICANDREAE (ovary 3–5-loc., the walls of the loc. dividing the
placentae irregularly): *Nicandra* (only genus).

   2. SOLANEAE (ovary 2-loc.): *Lycium*, *Atropa*, *Hyoscyamus*, *Physalis*,
Capsicum, *Solanum*, *Lycopersicon*, *Mandragora*.

3. DATUREAE (ovary 4-loc., the walls dividing the placentae equally): *Datura*, *Solandra* (only genera).

B. Embryo straight or slightly curved (less than a semicircle).

4. CESTREAE (all 5 sta. fertile): *Cestrum*, *Nicotiana*, *Petunia*.

5. SALPIGLOSSIDEAE (2 or 4 sta. fertile, of different lengths): *Salpiglossis*, *Schizanthus*.

Solanandra Pers. (sphalm.) = Solenandria Beauv. ex Vent. = Galax Rafin. (Diapensiac.).

Solanastrum Fabr. (uninom.) = *Solanum* L. sp. (Solanac.).

Solandera Kuntze = Solandra Sw. (Solanac.).

Solandra L. = Centella L. (Hydrocotylac.).

Solandra Murr. = Hibiscus L. (Malvac.).

*****Solandra** Sw. Solanaceae. 10 Mex. to trop. S. Am.

Solanecio (Sch. Bip.) Walp. = Senecio L. (Compos.).

Solanoa Greene (~ Asclepias L.). Asclepiadaceae. 1 Calif.

Solanoana Kuntze = praec.

Solanocharis Bitter = Solanum L. (Solanac.).

Solanoïdes Mill. = Rivina L. (Phytolaccac.).

Solanopsis Börner = Solanum tuberosum L. + Lycopersicon Mill. (Solanac.).

Solanopteris Copel. corr. Copel. Polypodiaceae. 1 W. trop. S. Am. Fronds dimorphous; tubers present. (*Am. Fern J.* **41**: 75, 128, 1951.)

Solanum L. (incl. *Lycopersicon* Mill.). Solanaceae. 1700 trop. & temp. In *S. dulcamara* L. (bittersweet, nightshade) and *S. nigrum* L. the fls. are small, with a cone of anthers opening at the tip as in *Borago*. *S. tuberosum* L. (S. Am.) is the potato. From the axils of the lowest ls. there spring branches which grow horiz. underground and swell up at the ends into tubers (potatoes). That these are stem structures is shown by their origin and by their possession of buds— the 'eyes'. Each eye is a small bud in the axil of an aborted l. (repres. by a semicircular rim). When the parent plant dies down in autumn the tubers become detached, and in the next season they form new plants by the development of the eyes, at the expense of the starch and other reserves stored in the tuber. By heaping earth against the stem, so as to cover more of the leaf-axils, more of the axillary shoots are made to become tuber-bearing; hence the value of ridging potatoes. *S. lycopersicum* L. (Am.) is the tomato, cult. for ed. fr. *S. melongena* L., the egg-plant, is cult. in warm countries for ed. fr.

Solaria Phil. Alliaceae. 2 Chile.

Soldanella L. Primulaceae. 11 mts. of S. & C. Eur. The fls. expand at very low temperatures, often coming up through the snow; they have a mechanism like that of *Erica*.

Soldevilla Lag. = Hispidella Barnad. (Compos.).

Solea Spreng. = Hybanthus Jacq. (Violac.).

Soleirolia Gaudich. Urticaceae. 1 Corsica, Sardinia.

Solena Lour. Cucurbitaceae. 1 trop. As. & Malaysia.

Solena Willd. = Posoqueria Aubl. (Rubiac.).

Solenacanthus C. Muell. = Solaenacanthus Oerst. = Ruellia L. (Acanthac.).

Solenachne Steud. (~ Spartina Schreb.). Gramineae. 1 temp. S. Am.

Solenandra Benth. & Hook. f. = Solenandria Beauv. ex Vent. = Galax Rafin. (Diapensiac.).

SOLIVA

Solenandra Hook. f. = Exostema (Pers.) Rich. (Rubiac.).
Solenandra (Reissek) Kuntze = Stenanthemum Reiss. (Rhamnac.).
Solenandria Beauv. ex Vent. = Galax Rafin. (Diapensiac.).
Solenangis Schlechter. Orchidaceae. 5 trop. Afr.
Solenantha G. Don = Hymenanthera R.Br. (Violac.).
Solenanthus Ledeb. (~ Cynoglossum L.). Boraginaceae. 15 Medit. to C. As. & Afghan.
Solenanthus Steud. ex Klatt = Acidanthera Hochst. (Iridac.).
Solenarium Dulac = Gagea Salisb. (Liliac.).
Solenidium Lindl. Orchidaceae. 4 Costa Rica, trop. S. Am.
Solenipedium Beer (sphalm.) = Selenipedium Reichb. f. (Orchidac.).
Soleniscia DC. = Styphelia Sm. (Epacridac.).
Solenisia Steud. = praec.
Solenixora Baill. Rubiaceae. 1 Madag.
Solenocalyx Van Tiegh. = Psittacanthus Mart. (Loranthac.).
Solenocarpus Wight & Arn. = Spondias L. (Anacardiac.).
Solenocentrum Schlechter. Orchidaceae. 1 Costa Rica.
Solenochasma Fenzl = Dicliptera Juss. (Acanthac.).
Solenogyne Cass. (1827) = Eriocephalus L. (Compos.).
Solenogyne Cass. (1828) = Lagenifera Cass. (Compos.).
Solenolantana (Nakai) Nakai = Viburnum L. (Caprifoliac.).
Solenomeles Dur. & Jacks. (sphalm.) = seq.
Solenomelus Miers. Iridaceae. 2 Chile.
Solenophora Benth. Gesneriaceae. 10 Mex., C. Am. Ḡ.
Solenophyllum Nutt. ex Baill. = Monanthochloë Engelm. (Gramin.).
Solenopsis Presl = Laurentia Michx (Campanulac.).
Solenopteris Copel. = Solanopteris Copel. corr. Copel. (Polypodiac.).
Solenoruellia Baill. Acanthaceae. 1 Mex.
Solenospermum Zoll. Celastraceae. 20 SE. As., Malaysia.
Solenostemma Hayne. Asclepiadaceae. 1 Egypt, Arabia.
Solenostemon Thonn. Labiatae. 10 trop. Afr.
Solenosterigma Klotzsch ex K. Krause = Philodendron Schott (Arac.).
Solenostigma Endl. = Celtis L. (Ulmac.).
Solenostigma Klotzsch ex Walp. = Retzia Thunb. (Retziac.).
Solenostyles Hort. ex Pasq. (nomen). Acanthaceae. Quid?
Solenotheca Nutt. = Tagetes L. (Compos.).
Solenotinus Spach = Viburnum L. (Caprifoliac.).
Solenotus Stev. = Astragalus L. (Legumin.).
Solfia Rechinger = Drymophloeus Zipp. (Palm.).
Solia Nor. = Premna L. (Verbenac.).
Solidago L. Compositae. 100 Am.; 1 Eur., S. virgaurea L., the golden-rod.
Solidago Mill. = Senecio L. (Compos.).
× Solidaster Wehrhahn. Compositae. Gen. hybr. (Aster × Solidago).
Soliera Clos (1849; non Solieria Agardh 1842—Algae) = Kurtzamra Kuntze (Labiat.).
Solisia Britton & Rose = Mammillaria Haw. (Cactac.).
Soliva Ruiz & Pav. Compositae. 8 S. Am.

1077

Sollya Lindl. Pittosporaceae. 2 W. Austr. Twiners; fls. bright blue.
Solmsia Baill. Thymelaeaceae. 2 New Caled.
Solmsiella Borbás = Capsella Medik. (Crucif.).
Solms-Laubachia Muschler. Cruciferae. 9 Himal., Tibet, SW. China.
Solonia Urb. Myrsinaceae. 1 Cuba.
Solori Adans. = Derris Lour. (Legumin.).
Solstitiaria Hill = Centaurea L. (Compos.).
Soltmannia [Klotzsch] ex Naud. = Miconia Ruiz & Pav. (Melastomatac.).
Solulus Kuntze = Ormocarpum Beauv. (Legumin.).
Somalia Oliv. = Barleria L. (Acanthac.).
Somera Salisb. = Scilla L. (Liliac.).
Somerauera Hoppe = Minuartia L. (Caryophyllac.).
Sommea Bory = Acicarpha Juss. (Calycerac.).
Sommera Schlechtd. Rubiaceae. 12 Mex. to trop. S. Am.
Sommerauera Endl. = Somerauera Hoppe = Minuartia L. (Caryophyllac.).
Sommerfeldtia Schumacher = Drepanocarpus G. F. W. Mey. (Legumin.).
*Sommerfeltia Less. Compositae. 1 S. Braz., Urug., Argent. Char. pl.
Sommeringia Lindl. = Soemmeringia Mart. (Legumin.).
Sommiera Benth. & Hook. f. = seq.
Sommieria Becc. Palmae. 3 New Guinea.
Somphocarya Torr. ex Steud. = Scirpus L. (Cyperac.).
Somphoxylon Eichl. Menispermaceae. 6 trop. S. Am.
Sonchidium Pomel = Sonchoseris Fourr. = Sonchus L. (Compos.).
Sonchos St-Lag. = Sonchus L. (Compos.).
Sonchoseris Fourr. = Sonchus L. (Compos.).
Sonchus L. Compositae. 50 Euras., Medit., Atl. Is., trop. Afr.
× **Sonchustenia** Svent. Compositae. Gen. hybr. (Sonchus × Sventenia).
Sondaria Dennst. ?Rhamnaceae (inc. sed.). 1 S. India.
Sondera Lehm. = Drosera L. (Droserac.).
Sonderina H. Wolff. Umbelliferae. 6 S. Afr.
Sonderothamnus R. Dahlgren. Penaeaceae. 2 S. Afr.
*Sonerila Roxb. Melastomataceae. 175 warm As.
Soninnia Kostel. = Sonninia Reichb. = Diplolepis R.Br. (Asclepiadac.).
Sonnea Greene. Compositae. 6 Pacif. N. Am.
Sonneratia Comm. ex Endl. = Celastrus L. (Celastrac.).
*Sonneratia L. f. Sonneratiaceae. 5 coasts trop. E. Afr. & Madag. to Hainan
& Ryukyu Is., Micronesia, Malaysia, New Hebrides & Solomons, N. Austr.,
New Caled. Mangroves with the general habit of *Rhizophoraceae*. Aërial roots
spring vertically out of the mud, arising as lat., negatively geotropic branches
upon the ordinary roots; they are provided with aërenchyma, and give out into
the uppermost layers of the constantly renewed silt fine horizontal 'nutrition
roots', which also perform a respiratory function.
*Sonneratiaceae Engl. & Gilg. Dicots. 2/7 trop. E. Afr. & As. to Austr. & W.
Pacif. Trees or shrubs with simple ent. opp. coriac. exstip. ls. Fls. reg., conspic.,
♀ or ♂ ♀, in 1–3-fld. term. cymes or corymbs. K (4–8), valv., acute, coriac.,
often coloured inside, persist.; C 4–8, sometimes broad and wrinkled, or o;
A 12–∞, fil. filif., inflexed in bud, anth. medifixed; G̲ (4–15), seated on broad
base, with long robust simple style and capit. stig., and multiloc. ov. with

∞ axile ov. Caps. or berry; ∞ seeds, without endosp. Genera: *Sonneratia*, *Duabanga*. Near *Lythrac.*, *Crypteroniac.* and *Punicac.*

Sonninia Reichb. = Diplolepis R.Br. (Asclepiadac.).

Sonraya Engl. (sphalm.) = seq.

Sonzaya Marchand = Canarium L. (Burserac.).

Sonzeya Engl. (sphalm.) = praec.

Sooja Siebold = Cassia L. (Legumin.).

Sophandra Meissn. (sphalm.) = Lophandra D. Don = Erica L. (Ericac.).

Sophia Adans. = Descurainia Webb & Berth. (Crucif.).

Sophia L. = Pachira Aubl. (Bombacac.).

Sophiopsis O. E. Schulz. Cruciferae. 5 mts. of C. As.

Sophisteques Comm. ex Endl. = Ochna L. (Ochnac.).

Sophoclesia Klotzsch = Sphyrospermum Poepp. & Endl. (Ericac.).

Sophonodon Miq. (sphalm.) = Siphonodon Griff. (Siphonodontac.).

Sophora L. Leguminosae. 50 trop. & warm temp. Winter-buds naked. The wood is very hard.

Sophorocapnos Turcz. = Corydalis Vent. (Fumariac.).

× **Sophrobroughtonia** hort. Orchidaceae. Gen. hybr. (iii) (Broughtonia × Sophronitis).

× **Sophrocatlaelia** hort. (vii) = × Sophrolaeliocattleya hort. (Orchidac.).

× **Sophrocattleya** hort. Orchidaceae. Gen. hybr. (iii) (Cattleya × Sophronitis).

× **Sophrolaelia** hort. Orchidaceae. Gen. hybr. (iii) (Laelia × Sophronitis).

× **Sophrolaeliocattleya** hort. Orchidaceae. Gen. hybr. (iii) (Cattleya × Laelia × Sophronitis).

× **Sophroleya** hort. (vii) = × Sophrocattleya hort. (Orchidac.).

Sophronanthe Benth. Scrophulariaceae. 1 SE. U.S.

Sophronia Lichtenst. ex Roem. & Schult. = Lapeirousia Pourr. (Iridac.).

Sophronia Lindl. = Sophronitis Lindl. (Orchidac.).

Sophronitella Schlechter. Orchidaceae. 1 Brazil.

Sophronitis Lindl. Orchidaceae. 6 Brazil.

× **Sophrovola** hort. (vii) = × Brassophronitis hort. (Orchidac.).

Sopropis Britton & Rose = Prosopis L. (Legumin.).

Sopubia Buch.-Ham. Scrophulariaceae. 40 trop. & S. Afr., Madag., Himal. to Indoch. & Formosa; 1 Queensl.

Soramia Aubl. = Doliocarpus Roland. (Dilleniac.).

Soranthe Salisb. = Sorocephalus R.Br. (Proteac.).

Soranthus Ledeb. Umbelliferae. 1 C. As.

***Sorbaria** (Ser. ex DC.) A. Br. Rosaceae. 10 C. & E. As., N. Am.

× **Sorbaronia** C. K. Schneider. Rosaceae. Gen. hybr. (Aronia × Sorbus).

× **Sorbocotoneaster** Pojark. Rosaceae. Gen. hybr. (Cotoneaster × Sorbus).

× **Sorbopyrus** C. K. Schneider. Rosaceae. Gen. hybr. (Pyrus × Sorbus).

Sorbus L. Rosaceae. 100 N. temp. Many apomicts.

Sorema Lindl. = Periloba Rafin. = Nolana L. ex L. f. (Nolanac.).

Sorghastrum Nash. Gramineae. 12 trop. & warm temp. Am., trop. Afr.

***Sorghum** Moench. Gramineae. 60 trop. & subtrop. *S. vulgare* Pers. (millet or guinea corn), largely cult. in Medit., etc., as a cereal. From the haulm of the var. *saccharatum* Koern. sugar is sometimes prepared.

Sorgum Adans. = Holcus L. (Gramin.).
Sorgum Kuntze = Andropogon L. (Gramin.).
Soria Adans. = Euclidium R.Br. (Crucif.).
Soridium Miers. Triuridaceae. 2 N. trop. S. Am.
Sorindeia Thou. Anacardiaceae. 50 trop. Afr., Madag.
Sorindeiopsis Engl. = praec.
Sorocea A. St-Hil. Moraceae. 22 C. Am. to Argent.
***Sorocephalus** R.Br. Proteaceae. 13 S. Afr.
Sorolepidium Christ = Polystichum Roth (Aspidiac.).
Soromanes Fée = Polybotrya Humb. & Bonpl. (Aspidiac.).
Soroseris Stebbins. Compositae. 8 Himal., Tibet, W. China.
Sorostachys Steud. = Cyperus Michx (Cyperac.).
Sosnovskya Takht. Compositae. 2 Cauc.
Sotor Fenzl = Kigelia DC. (Bignoniac.).
Sotrophola Buch.-Ham. = Chukrasia A. Juss. (Meliac.).
Sotularia Rafin. = Lagerstroemia L. (Lythrac.).
Souari Endl. = Saouari Aubl. = Caryocar L. (Caryocarac.).
Soubeyrania Neck. = Barleria L. (Acanthac.).
Soulamea Lam. Simaroubaceae. 10 Borneo, Moluccas, New Guinea to Fiji.
Soulame[ace]ae Endl. = Simaroubaceae–Soulameëae Engl.
Soulangia Brongn. = Phylica L. (Rhamnac.).
Souleyetia Gaudich. = Pandanus L. f. (Pandanac.).
Souliea Franch. Ranunculaceae. 1 China.
Souroubea Aubl. Marcgraviaceae. 25 C. & trop. S. Am., W.I. G 5-loc.
Southwellia Salisb. = Sterculia L. (Sterculiac.).
Souza Vell. = Sisyrinchium L. (Iridac.).
Sovara Rafin. = Tovara Adans. = Polygonum L. (Polygonac.).
Sowerbaea Smith. Liliaceae. 4 Austr.
Soya Benth. = Soja Moench = Glycine L. (Legumin.).
Soyauxia Oliv. Flacourtiaceae (~ Medusandrac.). 7 trop. W. Afr.
Soyauxiaceae Barkley = Flacourtiaceae DC.
Soyera St-Lag. = seq.
Soyeria Monn. = Crepis L. (Compos.).
Soymida A. Juss. Meliaceae. 1 Indomal. Astringent bark. Wood.
Spachea A. Juss. Malpighiaceae. 10 W.I., trop. S. Am.
Spachelodes Kimura = Hypericum L. (Guttif.).
Spachia Lilja = Fuchsia L. (Onagrac.).
Spadactis Cass. = Atractylis L. (Compos.).
Spadicaceae Dulac = Araceae Juss.
Spadonia Less. (1832; non Fries 1817 – Fungi) = Moquinia DC. (Compos.).
Spadostyles Benth. = Pultenaea Sm. (Legumin.).
Spaendoncea Desf. = Cadia Forsk. (Legumin.).
Spaetalumeae Wyeth & Nuttall = Lewisi[ac]eae Hook. & Arn. = Portulacaceae–
Lewisiinae Franz. ['Spoet'lum of the Sailish or Flat-Head Indians', Wyeth &
Nutt. in *J. Nat. Acad. Sci. Philad.* 7: 24, 1834.]
Spalanthus Walp. = Sphalanthus Jack = Quisqualis L. (Combretac.).
Spallanzania DC. = Mussaenda DC. (Rubiac.).
Spallanzania Neck. = Gustavia L. (Lecythidac.).

Spallanzania Pollini = Aremonia Neck. (Rosac.).
Spananthe Jacq. Hydrocotylaceae (~ Umbellif.). 1 trop. Am., W.I. Annual,
Spaniopappus B. L. Robinson. Compositae. 5 Cuba. [to 2 m. high.
Spanioptilon Less. = Cirsium Mill. (Compos.).
Spanizium Griseb. = Saponaria L. (Caryophyllac.).
Spanoghea Blume = Alectryon Gaertn. (Sapindac.).
Spanotrichum E. Mey. ex DC. = Osmites L. (Compos.).
× **Sparanthera** Cif. & Giac. Iridaceae. Gen. hybr. (Sparaxis × Streptanthera).
Sparattanthelium Mart. Gyrocarpaceae. 15 trop. S. Am.
Sparattosperma Mart. ex Meissn. Bignoniaceae. 2 trop. S. Am.
Sparattosyce Bur. Moraceae. 2 New Caled. [dac.).
Sparattothamnella v. Steenis (sphalm.) = Spartothamnella Briq. (Dicrastyli-
Sparaxis Ker-Gawl. Iridaceae. 4–5 S. Afr.
*****Sparganiaceae** Schultz-Schultzenst. 1/20 N. temp., Austr., N.Z. Perenn.
aquatic herbs from creeping rhiz.; stem usu. projecting above water; ls. elong.
± lin., erect or floating, sheathing below. Fls. ♂ ♀, monoec., in spherical heads,
♂ usu. higher up and more crowded than ♀. P 3–6, scaly, sepaloid; (♂) A 3–6
alt. with P when equal in number; anth. obl., basifixed; (♀) <u>G</u> 1 or (2), with one
ov. pend. near base of G, with micropyle up. Fr. drupaceous with narrow
obconic base, in muricate glob. heads; seed with mealy endosp. Only genus:
Sparganium. Rather closely related to *Typhac.*
Sparganion Adans. = seq. [*Acorus.*)
Sparganium L. Sparganiaceae. 20 N. temp., Austr., N.Z. (Cf. note under
Sparganophoros Adans. = seq.
Sparganophorus (sphalm. 'Spharg-') Boehm. = Struchium P.Br. (Compos.).
Spargonaphoros Adans. (sphalm.) = praec.
Sparmannia Buc'hoz = Rehmannia Libosch. (Scrophulariac.).
*****Sparmannia** L. f. Tiliaceae. 7 trop. & S. Afr., Madag. Fls in cymose umbels
(recognized by centrifugal order of opening). K 4, 2 sometimes petaloid.
Sta. ∞ with ∞ stds., sensitive, moving outward when touched (cf. *Helian-
themum*); no androphore.
Sparmanniaceae J. G. Agardh = Tiliaceae–Tilieae Reichb.
Sparrmania L. ex B. D. Jacks. = Melanthium L. (Liliac.).
Spartianthus Link = Spartium L. (Legumin.).
Spartidium Pomel = Genista L. (Legumin.).
Spartina Schreb. Gramineae. 16 mostly temp. Am., few coasts Eur. & Afr.,
Tristan da Cunha. Halophytes. *S.* × *townsendi* H. & J. Groves (*S. alterniflora*
Lois. × *S. maritima* (Curt.) Fernald), and the amphidiploid, are still spreading
on the S. coasts of England.
Spartinaceae Burnett = Gramineae–Spartineae + Eragrostideae p.p. + Chlori-
deae p.p.
Spartium Duham. = Genista L. (Legumin.).
Spartium L. Leguminosae. 1 Medit., *S. junceum* L. (Spanish broom), like
Sarothamnus in habit. Exstip. ls. with one leaflet. Fls. explosive like *Genista*;
they yield yellow dye, the pl. fibre.
Spartochloa C. E. Hubbard. Gramineae. 1 W. Austr.
Spartocysus Willk. & Lange (sphalm.) = seq.
Spartocytisus Webb & Berth. = Cytisus L. (Legumin.).

SPARTOTAMNUS

Spartotamnus Webb & Berth. ex Presl = praec.

Spartothamnella Briq. Dicrastylidaceae. 3 Austr. Switch-plants. Fr. a succ.

Spartothamnus A. Cunn. ex Walp. = praec. [drupe.

Spartothamnus Walp. = Spartotamnus Webb & Berth. = Spartocytisus Webb & Berth. = Cytisus L. (Legumin.).

Spatalanthus Sweet = Romulea Maratt. (Iridac.).

Spatalla Salisb. Proteaceae. 21 S. Afr. Invol. 1–4-fld. Fl. slightly zygo.

Spatallopsis Phillips. Proteaceae. 5 S. Afr.

Spatanthus Juss. = Spathanthus Desv. (Rapateac.).

Spatela Adans. = Spathelia L. (Rutac.).

Spatellaria Reichb. = Spathularia A. St-Hil. = Amphirrhox Spreng. (Violac.).

Spatha P. & K. = Spathe P.Br. = Spathelia L. (Rutac.).

Spathacanthus Baill. Acanthaceae. 5 Mex., C. Am. (K) in pairs.

Spathaceae Dulac = Iridaceae Juss.

Spathandra Guill. & Perr. = Memecylon L. (Memecylac.).

Spathandus Steud. = Spathanthus Desv. (Rapateac.).

Spathantheum Schott. Araceae. 2 Bolivia. G 6–8-loc.

Spathanthus Desv. Rapateaceae. 2 N. trop. S. Am. G 1-loc., 1-ovuled. Ls. 1·5 m. long. Spike one-sided.

Spathe P.Br. = seq.

*****Spathelia** L. Rutaceae. 20 W.I., N. S. Am.

Spathepteris Presl = Anemia Sw. (Schizaeac.).

Spathestigma Hook. & Arn. = Adenosma R.Br. (Scrophulariac.).

Spathia Ewart. Gramineae. 1 N. Austr.

Spathicalyx J. C. Gomes. Bignoniaceae. Brazil.

Spathicarpa Hook. Araceae. 7 trop. S. Am. Spadix adnate to spathe, monoec. Down centre 1–3 rows of ♂ fls. (stalked synandria); at sides ♀ fls., each of a bottle-shaped G with stds.

Spathichlamys R. N. Parker. Rubiaceae. 1 Burma.

Spathidolepis Schlechter. Asclepiadaceae. 1 New Guinea.

Spathiflorae Engl. An order of Monocots. comprising the *Arac.* and *Lemnac.*

Spathiger Small = Epidendrum L. (Orchidac.).

Spathionema Taub. Leguminosae. 1 trop. Afr.

Spathiostemon Blume. Euphorbiaceae. 2 L. Siam, W. Malaysia, New Guinea.

Spathipappus Tsvelev (∼ Chrysanthemum L.). Compositae. 2 C. As., Afghan., N. India.

Spathiphyllopsis Teijsm. & Binnend. = seq.

Spathiphyllum Schott. Araceae. 3 Philipp., Palau Is., Moluccas, New Guinea, Bismarcks & Solomons; 35 C. & trop. S. Am. Spathe partly adnate to spadix.

Spathirachis Klotzsch ex Klatt = Sisyrinchium L. (Iridac.).

Spathium Edgew. = Aponogeton Thunb. (Aponogetonac.).

Spathium Lour. = Saururus L. (Saururac.).

Spathocarpus P. & K. = Spathicarpa Hook. (Arac.).

Spathodea Beauv. Bignoniaceae. 2 trop. Afr. Large water-pores on backs of leaflets near midrib. In *S. campanulata* Beauv. the K is inflated and water secreted between it and the C.

Spathodeopsis Dop. Bignoniaceae. 2 Indoch.

Spathodithyros Hassk. = Commelina L. (Commelinac.).
Spathoglottis Blume. Orchidaceae. 46 China, Indomal., Austr., Solomon Is., Fiji.
Spatholirion Ridl. Commelinaceae. 2 Siam, Indoch., S. China; 1 sp. climbing, 1 erect.
Spatholobus Hassk. Leguminosae. 40 Himal. to China, W. Malaysia.
× **Spathophaius** hort. Orchidaceae. Gen. hybr. (vi) (Phaius × Spathoglottis).
Spathophyllopsis P. & K. = Spathiphyllopsis Teijsm. & Binnend. = seq.
Spathophyllum P. & K. = Spathiphyllum Schott (Arac.).
Spathorachis P. & K. = Spathirachis Klotzsch ex Klatt = Sisyrinchium L. (Iridac.).
Spathoscaphe Oerst. = Chamaedorea Willd. (Palm.).
Spathostigma P. & K. = Spathestigma Hook. & Arn. = Adenosma R.Br. (Scrophulariac.).
Spathotecoma Bur. = Newbouldia Seem. ex Bur. (Bignoniac.).
Spathula Fourr. = Iris L. (Iridac.).
Spathularia DC. = Spatularia Haw. = Saxifraga L. (Saxifragac.).
Spathularia A. St-Hil. = Amphirrhox Spreng. (Violac.).
Spathulopetalum Chiov. = Caralluma R.Br. (Asclepiadac.).
Spathyema Rafin. = Symplocarpus Salisb. (Arac.).
Spatularia Haw. = Saxifraga L. (Saxifragac.).
Spatulima Rafin. = Lathyrus L. (Legumin.).
Specklinia Lindl. = Pleurothallis R.Br. (Orchidac.).
Specularia A. DC. = Legousia Durande (Campanulac.).
Speea Loes. Alliaceae. 1–2 Chile.
Spegazzinia Backeb. = Weingartia Werderm. = Gymnocalycium Pfeiff. (Cactac.).
Speirantha Baker. Liliaceae. 1 E. China.
Speiranthes Hassk. = Spiranthes Rich. (Orchidac.).
Speirema Hook. f. & Thoms. = Pratia Gaudich. (Campanulac.).
Speirodela S. Wats. = Spirodela Schleid. (Lemnac.).
Speirostyla Baker = Christiana DC. (Tiliac.).
Spelta v. Wolf = Triticum L. (Gramin.).
Spencera Stapf = seq.
Spenceria Trimen. Rosaceae. 2 W. China.
Spennera Mart. ex DC. = Aciotis D. Don (Melastomatac.).
Spenocarpus B. D. Jacks. (sphalm.) = Sphenocarpus Wall. = Magnolia L. (Magnoliac.).
Spenotoma G. Don = Sphenotoma R.Br. = Dracophyllum Labill. (Epacridac.)
Speranskia Baill. Euphorbiaceae. 3 China. Rhizome-herbs.
Spergella Reichb. = Sagina L. (Caryophyllac.).
Spergula L. Caryophyllaceae. 5 temp. *S. arvensis* L., corn-spurrey, a general weed. The axillary shoots do not lengthen their internodes, so that the ls. seem to be tufted. Fls. in cymes, gynomonoec. or gynodioec. Sometimes used as fodder.
***Spergularia** (Pers.) J. & C. Presl. Caryophyllaceae. 40 cosmop., mostly halophytes. *S. rubra* (L.) J. & C. Presl, sand-spurrey.
Spergulastrum Rich. in Michx = Stellaria L. (Caryophyllac.).
Spergulus Brot. ex Steud. = Drosophyllum Link (Droserac.).

Sperlingia Vahl = Hoya R.Br. (Asclepiadac.).
Spermabolus Teijsm. & Binnend. = Anaxagorea A. St-Hil. (Annonac.).
Spermachiton Llanos = Sporobolus R.Br. (Gramin.).
Spermacoce L. Rubiaceae. 100 warm Am.
Spermacoceodes Kuntze = praec.
Spermacon Rafin. = praec.
Spermadictyon Roxb. = Hamiltonia Roxb. (Rubiac.).
Spermadon P. & K. = Spermodon Beauv. ex Lestib. = Rhynchospora Vahl (Cyperac.).
Spermaphyllum P. & K. = Spermophylla Neck. = Ursinia Gaertn. (Compos.).
Spermatochiton Pilger = Spermachiton Llanos = Sporobolus R.Br. (Gramin.).
Spermatococe Clem. = Spermacoce L. (Rubiac.).
Spermatolepis Clem. = Spermolepis Brongn. & Gris = Arillastrum Panch. ex Baill. = Myrtomera B. C. Stone = Stereocaryum Burret (Myrtac.).
Spermatura Reichb. = Osmorhiza Rafin. (Umbellif.).
Spermaulaxen Rafin. = Polygonum L. (Polygonac.).
Spermaxyron Steud. = seq.
Spermaxyrum Labill. = Olax L. (Olacac.).
Spermodon Beauv. ex Lestib. = Rhynchospora Vahl (Cyperac.).
Spermolepis Brongn. & Gris = Stereocaryum Burret (Myrtac.).
Spermolepis Rafin. Umbelliferae. 1 Hawaii, 4 N. Am., 1 Argent.
Spermophylla Neck. = Ursinia Gaertn. (Compos.).
Sphacanthus Benoist. Acanthaceae. 2 Madag.
***Sphacele** Benth. Labiatae. 25 Mex., Venez., Andes, 1 SE. Brazil.
Sphacophyllum Benth. = Epallage DC. (Compos.).
Sphacopsis Briq. = Salvia L. (Labiat.).
Sphaenodesma Schau. = Sphenodesme Jack (Symphorematac.).
Sphaeradenia Harling. Cyclanthaceae. 38 C. Am. & NW. trop. S. Am.
Sphaeralcea A. St-Hil. Malvaceae. 60 warm Am., S. Afr.
Sphaeranthoïdes A. Cunn. ex DC. = Pterocaulon Ell. (Compos.).
Sphaeranthus L. Compositae. 40 Afr. (not NW.), Madag., Iraq to Persia, India, SE. As., W. Malaysia, Celebes, NE. Austr.
Sphaerella Bub. = Airopsis Desv. (Gramin.).
Sphaereupatorium Kuntze. Compositae. 1 Bolivia.
Sphaeridiophora Benth. & Hook. f. (sphalm.) = seq.
Sphaeridiophorum Desv. = Indigofera L. (Legumin.).
Sphaerine Herb. = Bomarea Mirb. (Alstroemeriac.).
Sphaeritis Eckl. & Zeyh. = Crassula L. (Crassulac.).
Sphaerium Kuntze = Coix L. (Gramin.).
Sphaerocardamum Schau. Cruciferae. 1 Mex.
Sphaerocarpos J. F. Gmel. (1791; non Boehm. 1760—Hepat.) = Globba L. (Zingiberac.).
Sphaerocarpum Steud. (sphalm.) = Sphaerocaryum Nees ex Hook. f. (Gramin.).
Sphaerocarpus Batsch. Compositae. Quid?
Sphaerocarpus Fabr. (uninom.) = *Neslia* Desv. sp. (Crucif.).
Sphaerocarpus Steud. (sphalm.) = Sphenocarpus Rich. = Laguncularia Gaertn. (Combretac.).

Sphaerocarya Dalz. ex DC. = Strombosia Bl. (Olacac.).
Sphaerocarya Wall. = Pyrularia Michx (Santalac.).
Sphaerocaryum Nees ex Hook. f. Gramineae. 1 India to S. China & Formosa; Malay Penins., Banka.
Sphaerocephala Hill = Centaurea L. (Compos.).
Sphaerocephalus Kuntze = Echinops L. (Compos.).
Sphaerocephalus Lag. ex DC. = Nassauvia Juss. (Compos.).
Sphaerochloa Beauv. ex Desv. = Eriocaulon L. (Eriocaulac.).
Sphaerocionium Presl. Hymenophyllaceae. 50 pantrop. & S. hemisph.
Sphaeroclinium Sch. Bip. = Matricaria L. (Compos.).
Sphaerocodon Benth. Asclepiadaceae. 2 Afr.
Sphaerocoma T. Anders. Caryophyllaceae. 2 NE. Sudan, S. Arabia, Persia.
Sphaerocoryne Scheff. ex Ridley = Melodorum Lour. (Annonac.).
Sphaerodendron Seem. = Cussonia Thunb. (Araliac.).
Sphaerodia Dulac = Marsilea L. (Marsileac.).
Sphaerodiscus Nakai = Euonymus L. (Celastrac.).
Sphaerogyne Naud. = Tococa Aubl. (Melastomatac.).
Sphaerolobium Sm. Leguminosae. 12 Austr.
Sphaeroma (DC.) Schlechtd. = Phymosia Desv. ex Ham. (Malvac.).
Sphaeromariscus E. G. Camus. Cyperaceae. 1 Indoch.
Sphaeromeria Nutt. = Tanacetum L. (Compos.).
Sphaeromorphaea DC. = Epaltes Cass. (Compos.).
Sphaerophora Blume = Morinda L. (Rubiac.).
Sphaerophora Sch. Bip. = Eremanthus Less. (Compos.).
Sphaerophyllum auct. = Sphacophyllum Benth. = Epallage DC. (Compos.).
Sphaerophysa DC. Leguminosae. 2 Cauc. to N. China. Halophytes.
Sphaeropteris Bernh. = Cyathea subg. Sphaeropteris Holttum (Cyatheac.). Mostly large tree-ferns of trop. Asia & Pacific; type is *C. medullaris* (Forst.) Sw., of N.Z.
Sphaeropteris Wall. = Peranema D. Don (Aspidiac.).
Sphaeropus Boeck. = Scleria Bergius (Cyperac.).
Sphaerorhizon Hook. f. = Scybalium Schott (Balanophorac.).
Sphaerosacme Wall. ex Roem. = Lansium Corrêa (Meliac.).
Sphaeroschoenus Arn. = Rhynchospora Vahl (Cyperac.).
Sphaerosepalaceae Van Tiegh. Dicots. 2/14 Madag. Trees and shrubs with alternate simple entire sometimes trinerved stipulate ls.; stipules intrapet., caduc., leaving prominent scar. Fls. reg., ⚥, in term. and axill. pan. of umbellif. cymes. K 4 (rarely 3 + 3), unequal, much imbr.; C 4 (rarely 3), unequal, imbr., slightly clawed, densely streaked with short resinous lines; A ∞, fil. filif., ± connate at base, resin-dotted, anth. small, 2-loc., loc. widely sep.; disk (interior to A) large, cupular, wrinkled, dentic.; G (2), rarely (3), 2-loc., and deeply 2-lobed, with simple genic. style (arising between the lobes) and entire stig., and ± 3 erect basal anatr. ov. per loc. Fr. globose or didymous, densely muricate, 1-(rarely 2-)seeded; seed large, renif., with horny ± ruminate endosp. and minute embr. with foliac. cots. Genera: *Rhopalocarpus* (*Sphaerosepalum*), *Dialyceras*. An interesting group, apparently related to the Malaysian *Gonystylus* (*Thymelaeac.*): the twigs and bark, l. texture and venation (exc. in the 3-nerved *Rhop.* spp.), infl., disk, fil. and anth., ovary, and

slender 'kinked' style are remarkably similar. Chief differences lie in the exstip. ls., pentamerous fls., valv., not imbr., K, absence of true C, much divided *extra*-staminal disk (formerly regarded as C), and usu. dehiscent, never didymous or muricate, fr. of *Gonystylus*.

Sphaerosepalum Baker = Rhopalocarpus Boj. (Sphaerosepalac.).

Sphaerosicyos Hook. f. = Lagenaria Ser. (Cucurbitac.).

Sphaerospora Klatt = Acidanthera Hochst. (Iridac.).

Sphaerospora Sweet = Gladiolus L. (Iridac.).

Sphaerostachys Miq. = Piper L. (Piperac.).

Sphaerostema Blume = Schisandra Michx (Schisandrac.).

Sphaerostemma Reichb. (sphalm.) = praec.

Sphaerostephanaceae Ching = Thelypteridaceae Ching.

Sphaerostephanos J. Smith emend. Holtt. Thelypteridaceae. 120 Madag. to Tahiti. Fronds bipinnatifid with ± numerous reduced lower pinnae; veins usu. anast.; spherical glands often present on lamina, indusia and sporangia.

Sphaerostichum Presl = Pyrrosia Mirbel (Polypodiac.).

Sphaerostigma Fisch. & Mey. = Camissonia Link (Onagrac.).

Sphaerostylis Baill. Euphorbiaceae. 8 trop. E. Afr., Madag., W. Malaysia.

Sphaerotele Link = Urceolina Reichb. (Amaryllidac.).

Sphaerotele Presl = Stenomesson Herb. (Amaryllidac.).

Sphaerothalamus Hook. f. = Polyalthia Bl. (Annonac.).

Sphaerotheca Cham. & Schlechtd. = Conobea Aubl. (Scrophulariac.).

Sphaerothele Benth. & Hook. f. = Sphaerotele Presl = Stenomesson Herb. (Amaryllidac.).

Sphaerothylax Bischoff ex Krauss. Podostemaceae. 3 S. trop. Afr.

Sphaerula W. Anders. ex Hook. f. = Acaena L. (Rosac.).

Sphagneticola O. Hoffm. Compositae. 1 SE. Brazil.

Sphalanthus Jack = Quisqualis L. (Combretac.).

Sphallerocarpus Bess. Umbelliferae. 1 temp. S. As.

Sphalmanthus N. E. Brown. Aïzoaceae. 17 S. Afr.

Sphanellolepis Cogn. (sphalm.) = seq.

Sphanellopsis Steud. ex Naud. = Adelobotrys DC. (Melastomatac.).

Spharganophorus Boehm. (sphalm.) = Sparganophorus Boehm. = Struchium P.Br. (Compos.).

Sphedamnocarpus Planch. ex Benth. & Hook. f. Malpighiaceae. 20 trop. & **Sphenandra** Benth. Scrophulariaceae. 2 S. Afr.   [S. Afr., Madag.

Sphenantha Schrad. = ? Cucurbita L. (Cucurbitac.).

Sphenanthera Hassk. = Begonia L. (Begoniac.).

Sphendamnocarpus Baker (sphalm.) = Sphedamnocarpus Planch. ex Benth. & Hook. f. (Malpighiac.).

Spheneria Kuhlm. Gramineae. 1 trop. S. Am.

Sphenista Rafin. = Hirtella L. (Chrysobalanac.).

Sphenocarpus Korovin. Umbelliferae. 1 C. As.

Sphenocarpus Rich. = Laguncularia Gaertn. (Combretac.).

Sphenocarpus Wall. = Magnolia L. (Magnoliac.).

Sphenocentrum Pierre. Menispermaceae. 1 trop. W. Afr.

★Sphenoclea Gaertn. Sphenocleaceae. 1 pantrop. (orig. in Africa?), 1 W. Afr. *S. zeylanica* Gaertn. a common weed in Asiat. rice-fields.

Spondylococcos Mitch. = Callicarpa L. (Verbenac.).
Spondylococcus Reichb. = Sphondylococca Willd. = Bergia L. (Elatinac.).
Spongiosyndesmus Gilli. Umbelliferae. 1 Afghanistan.
Spongopyrena Van Tiegh. = Ouratea Aubl. (Ochnac.).
Spongostemma (Reichb.) Reichb. = Scabiosa L. (Dipsacac.).
Spongotrichum Nees = Olearia Moench (Compos.).
Sponia Comm. ex Lam. = Trema Lour. (Ulmac.).
Sporabolus Hassk. = Sporobolus R.Br. (Gramin.).
Sporadanthus F. Muell. ex J. Buch. Restionaceae. 1 N.Z. & Chatham I., in salt-marshes & peat-swamps.
Sporichloë Pilger (sphalm.) = Spirochloë Lunell = Schedonnardus Steud. (Gramin.).
Sporledera Bernh. = Ceratotheca Endl. (Pedaliac.).
Sporobolaceae (Stapf) Herter = Gramineae–Sporoboleae Stapf.
Sporobolus R.Br. Gramineae. 150 trop. & warm temp.
Sporoxeia W. W. Smith. Melastomataceae. 1 Burma.
Sportella Hance = Pyracantha M. Roem. (Rosac.).
Spraguea Torr. Portulacaceae. 10 W. N. Am.
Spragueanella Balle. Loranthaceae. 1 trop. E. Afr.
Spreckelia Fabr. = seq.
× **Sprekanthus** Traub. Amaryllidaceae. Gen. hybr. (Habranthus × Sprekelia).
Sprekelia Heist. Amaryllidaceae. 1 Mex., S. formosissima Herb.
Sprengalia Steud. = Sprengelia Sm. (Epacridac.).
Sprengelia Schult. = Melhania Forsk. (Sterculiac.).
Sprengelia Sm. Epacridaceae. 3 SE. Austr., Tasmania.
Sprengeria Greene = Lepidium L. (Crucif.).
Springalia DC. (sphalm.) = Sprengelia Sm. (Epacridac.).
Springia Heurck & Muell. Arg. = Ichnocarpus R.Br. (Apocynac.).
Springula Nor. = Hygrophila R.Br. (Acanthac.).
Sprucea Benth. = Simira Aubl. (Rubiac.).
Spruceanthus Sleum. (1936; non Verdoorn 1934—Hepaticae) = Neosprucea Sleum. (Flacourtiac.).
Sprucella Pierre = Micropholis (Griseb.) Pierre (Sapotac.).
Sprucina Niedenzu = Diplopterys Juss. (Malpighiac.).
Sprunera Sch. Bip. ex Hochst. = Sphaeranthus L. (Compos.).
Sprunnera Sch. Bip. = Codonocephalum Fenzl (Compos.).
Spryginia Popov = Orychophragmus Bunge (Crucif.).
Spuriodaucus C. Norman. Umbelliferae. 3 trop. Afr.
Spurionucaceae Dulac = Ambrosiaceae Dum.
Spuriopimpinella (Boissieu) Kitagawa = Pimpinella L. (Umbellif.).
Spyridanthus Wittst. = Spiridanthus Fenzl ex Endl. = Monolopia DC.
Spyridium Fenzl. Rhamnaceae. 30 temp. Austr. [(Compos.).
Squamaria Ludw. = Lathraea L. (Orobanchac.).
Squamataxus 'Senilis' [Nelson] = Saxe-Gothaea Lindl. (Podocarpac.).
Squamellaria Becc. Rubiaceae. 2 Fiji.
Squibbia Rafin. = Sesuvium L. (Aïzoac.).
Squilla Steinh. = Urginea Steinh. (Liliac.).
Sredinskya (Stein) Fedorov (~ Primula L.). Primulaceae. 1 Cauc.

Srutanthus Pritz. (sphalm.) = seq.
Sruthanthus DC. (sphalm.) = Struthanthus Mart. (Loranthac.).
Staavia Dahl. Bruniaceae. 10 S. Afr.
Staberoha Kunth. Restionaceae. 8 S. Afr.
Stachiopsis Ikonnikov-Galitzky = Stachyopsis Popov & Vved. (Labiat.).
Stachis Neck. = Stachys L. (Labiat.).
Stachyacanthus Nees (= ?Eranthemum L.). Acanthaceae. 1 Brazil.
Stachyanthemum Klotzsch = Cyrilla Garden ex L. (Cyrillac.).
Stachyanthus Blume = Phyllorkis Thou. = Bulbophyllum Thou. (Orchidac.).
Stachyanthus DC. = Eremanthus Less. (Compos.).
*****Stachyanthus** Engl. Icacinaceae. 4 trop. Afr.
Stachyarpagophora Gomez de la Maza = Achyranthes L. (Amaranthac.).
Stachyarrhena Hook. f. Rubiaceae. 8 trop. Am.
Stachycarpus Van Tiegh. = Podocarpus L'Hérit. ex Pers. (Podocarpac.).
Stachycephalum Sch. Bip. ex Benth. Compositae. 1 Mex., 1 Argent.
Stachychrysum Boj. = Piptadenia Benth. (Legumin.).
Stachycnida P. & K. = Stachyocnide Bl. = Pouzolzia Gaudich. (Urticac.).
Stachycrater Turcz. = Osmelia Thw. (Flacourtiac.).
Stachydeoma Small (~ Hedeoma Pers.). Labiatae. 4 S. U.S.
Stachydesma Willis (sphalm.) = praec.
Stachygynandrum P. Beauv. = Selaginella P. Beauv. (Selaginellac.).
Stachyobium Reichb. f. = Dendrobium Sw. (Orchidac.).
Stachyocnide Blume = Pouzolzia Gaudich. (Urticac.).
Stachyococcus Standley. Rubiaceae. 1 Peru.
Stachyophorbe Liebm. (~ Chamaedorea Willd.). Palmae. 6 Mex., trop.
Stachyopogon Klotzsch = Aletris L. (Liliac.).   [Am.
Stachyopsis Popov & Vved. Labiatae. 3 C. As.
Stachyothyrsus Harms. Leguminosae. 2 trop. Afr.
Stachyphrynium K. Schum. Marantaceae. 12 India, SE. As., W. Malaysia.
Stachyphyllum Van Tiegh. = Antidaphne Poepp. & Endl. (Loranthac.).
Stachypogon P. & K. = Stachyopogon Klotzsch = Aletris L. (Liliac.).
Stachys L. Labiatae. 300 N. & S. trop. & subtrop. (exc. Austr., N.Z.); trop.
mts. Tubers of *S. sieboldi* Miq. ed. (Japanese artichokes, *crosnes du Japon*).
Stachystemon Planch. Euphorbiaceae. 3 W. Austr.
Stachytarpha Link = seq.
Stachytarpheia auct. = seq.
*****Stachytarpheta** Vahl. Verbenaceae. 100 Am., some spp. widely dispersed
as weeds in tropics. Ls. of *S. dichotoma* Vahl (*S. jamaïcensis* Gardn.) sometimes
used as tea.
Stachythyrsus P. & K. = Stachyothyrsus Harms (Legumin.).
*****Stachyuraceae** J. G. Agardh. Dicots. 1/10 E. As. Small trees or shrubs,
sometimes climbing, with alt. simple serrul. stip. ls., stips. lin.-lanc., caduc.
Fls. reg., ♀ or polygamous, in axill. pendent spikes or rac., usu. precocious,
each fl. subtended by 2 connate bracteoles. K 4, much imbr.; C 4, imbr.,
connivent; A 4 + 4, with subul. fil. and small intr. anth.; G̲ (4), with short
simple style and capit. stig., and ∞ axile ov. Fr. a 4-loc. ∞-seeded berry;
seeds small, arillate, with fleshy endosp. Only genus: *Stachyurus*. Relationships with *Actinidiac.* and *Flacourtiac.*

Stachyurus Sieb. & Zucc. Stachyuraceae. 10 Himal. to Formosa & Japan.
Stachyus St-Lag. = Stachys L. (Labiat.).
Stackhousia Sm. Stackhousiaceae. 25 Austr., N.Z., 1 extending NW. to Sumatra, Philippines and Micronesia.
*Stackhousiaceae R.Br. Dicots. 3/27 Malaysia, Austr., N.Z. Ann. or perenn. ± xeroph. rhiz. herbs with alt. simple ent. exstip. ls. Fls. reg., ☿, in rac. or cymose (rarely umbellate) infls. K (5), imbr.; C 5, perig., long-clawed, claws mostly connate in middle portion, lobes imbr.; A 5, usu. 3 long and 2 short; G (2–5), 2-5-loc., with 2–5 free or connate styles, and 1 erect anatr. ov. in each loc. Fr. a schizocarp of 2–5 indehisc. 1-seeded cocci. Perhaps related to *Scrophulariac.–Selagineae.*
   Classification and genera (after Mattfeld):
   I. *Macgregorioïdeae* (pet. free, spatulate; sta. equal, fil. v. short, anth. with apic. append., pollen smooth; carp. 5, style with discoid collar beneath stig. lobes; fls. ebract.): *Macgregoria.*
   II. *Stackhousioïdeae* (claws of pet. free below, connate into a tube above; sta. unequal, 3 long, 2 short, fil. elong., anth. obtuse or shortly mucr. at apex, pollen lamellate-areolate; carp. 3, rarely 2 or 5; style without collar; fls. with 2 transv. bracteoles): *Stackhousia, Tripterococcus.*
Stadmannia Lam. Sapindaceae. 1 trop. E. Afr., Madag., Masc.
Stadtmannia Walp. = praec.
Staebe Hill = Centaurea L. (Compos.).
Staebe Juss. = Stoebe L. (Compos.).
Staeblorhiza Dur. (sphalm.) = Streblorrhiza Endl. (Legumin.).
Staehelina L. Compositae. 6 Medit.
Staehelina Rafin. = Helipterum DC. (Compos.).
Staehelinia Haller = Bartsia L. (Scrophulariac.).
Staehelinoïdes Loefl. = Ludwigia L. (Onagrac.).
Staëlia Cham. & Schlechtd. Rubiaceae. 12 trop. S. Am.
Staflinus Rafin. = Daucus L. (Umbellif.).
Stagmaria Jack = Gluta L. (Anacardiac.).
Stahelia Jonker. Gentianaceae. 1 N. Brazil, Guiana.
Stahlia Bello. Leguminosae. 1 W.I. (Puerto Rico). Good timber.
Stahlianthus Kuntze. Zingiberaceae. 6–7 E. Himal., SE. As., Hainan.
Stalagmites Miq. = Cratoxylum Bl. (Guttif.).
Stalagmites Spreng. = seq.
Stalagmitis Murr. = Garcinia L. (Guttif.).
Stammarium Willd. ex DC. = Pectis L. (Compos.).
Standleya Brade. Rubiaceae. 4 Brazil.
Standleyacanthus Leonard. Acanthaceae. 1 C. Am. (Costa Rica).
Standleyanthus R. M. King & H. Rob. (~ Eupatorium L.). Compositae. 1 C. Am. (Costa Rica).
× **Stanfieldara** hort. Orchidaceae. Gen. hybr. (iii) (Epidendrum × Laelia × Sophronitis).
Stanfieldia Small = Haplopappus Cass. (Compos.).
Stanfieldiella Brenan. Commelinaceae. 4 trop. Afr.
Stanfordia S. Wats. Cruciferae. 1 California.
Stangea Graebn. Valerianaceae. 7 Peru, Argent.

STANGERIA

Stangeria T. Moore. Stangeriaceae. 1 SE. Afr., *S. eriopus* (Kunze) Nash
(*S. paradoxa* T. Moore).
Stangeriaceae (Pilger) L. A. S. Johnson. Gymnosp. (Cycadales). 1/1 S. Afr.
Fern-like cycads with general chars. of *Zamiaceae* (*q.v.*), but pinnae of ls.
penninerved, with definite midrib and ∞ transv. parallel dichot.-branched lat. ´
nerves, convol. in vernation. Sporophylls imbr. but in almost vert. rows.
Caudex subterranean, naked (frond-bases decid.). Only genus: *Stangeria*.
Stanggeria Stevens (sphalm.) = Stangeria T. Moore (Stangeriac.).
Stanhopea Frost ex Hook. Orchidaceae. 45 C. & trop. S. Am., W.I.
Stanhopeastrum Reichb. f. = praec.
Stanleya Nutt. Cruciferae. 8 W. U.S.
Stanle[yace]ae Nutt. = Cruciferae–Stanleyeae O. E. Schulz. (Approaching
Cleomaceae in the spreading K, long-exserted A, and long slender gynophore.)
Stanleyella Rydb. Cruciferae. 2 SW. U.S., Mex.
Stannia Karst. = Posoqueria Aubl. (Rubiac.).
Stapelia L. Asclepiadaceae. 75 trop. & S. Afr., carrion-flowers. Like the
Cacti and the fleshy Euphorbias they inhabit semi-arid regions, and exhibit
similar swollen stems, the ls. reduced to deciduous scales, standing in 4 ranks
corresponding to the usual l. arrangement in the fam. The green tissue occupies
the periphery of the stem, and the centre is full of water-storage cells. Fls.
sometimes large with dull red or dark maroon colour and carrion smell,
attracting flies. Corona ± cupular, with an inner ring of 5 entire or divided
horns of varied form.
Stapeliaceae Horan. = Asclepiadaceae R.Br.
Stapelianthus Choux ex A. White & Sloane. Asclepiadaceae. 2 Madag.
Stapeliopsis Choux (1931) = praec.
Stapeliopsis Phillips (1932) = Stultitia Phillips (Asclepiadac.).
Stapeliopsis Pillans (1928). Asclepiadaceae. 1 S. Afr.
Stapfia Davy (1898; non Chodat 1897—Chlorophyceae) = Davyella Hack.
= Neostapfia Davy (Gramin.).
Stapfiella Gilg. Turneraceae. 5 trop. Afr.
Stapfiola Kuntze (Desmostachya Stapf). Gramineae. 1 NE. trop. Afr. to
Stapfiophyton Li. Melastomataceae. 4 S. China.   [India.
Staphidiastrum Naud. = Sagraea DC. (Melastomatac.).
Staphidium Naud. = Clidemia D. Don (Melastomatac.).
Staphilea Medik. = Staphylea L. (Staphyleac.).
Staphisagria Hill = Staphysagria (DC.) Spach = Delphinium L. (Ranunculac.).
Staphylea L. Staphyleaceae. 10 N. temp.
*****Staphyleaceae** (DC.) Lindl. Dicots. 5/60, trop. As. & Am., N. temp.
Trees and shrubs with opp. or alt. imparipinn. (rarely unifoliol.) stip. (rarely
exstip.) ls.; lfts. serr. Fls. reg., ♀ or polygam., rarely dioec., sometimes v. small
in small (but sometimes ∞-fld.) pan. K 5 or (5), imbr.; C 5, imbr.; A 5,
alternipet., fil. sometimes much flattened, anth. dorsif., intr.; disk usu. conspic.
and ± lobed, sometimes o; G (3), rarely (2) or (4), or 3–4 free (*Euscaphis*),
with 1–few axile anatr. ov. per loc.; styles variously free to completely united.
Fr. either caps., of ± free follicles, or indehisc., ± drupac. or baccate; seeds
1–few, with fleshy or horny endosp. Possible relationships with *Sambucac.*,
Cunoniac. and *Bischofiac.*

Classification and genera:

I. **Staphyleoïdeae** (ls. opp., often 3-foliol., rarely 1-foliol.; stips. and stipels present at least when young; K ± free, never connate into a tube; disk usu. v. conspic.; carp. free, or only partly, never wholly, connate; ov. usu. > 1, often several, per carp.; fr. dehisc. or indehisc.; wood prosench. with bordered pits): *Euscaphis, Staphylea, Turpinia.*

II. **Tapiscioïdeae** (ls. alt., usu. 3–10-jugate, never 1-foliol.; stips. and stipels not always present even on young ls.; K ± connate into a tube; disk small or o; carp. completely united into an unlobed ovary with stylar column; ov. only 1–2 in entire ovary; fr. always indehisc., drupac. or baccate; wood prosench. with simple pits): *Tapiscia, Huertea.* (The *Tapiscioïdeae* should perhaps be treated as a distinct fam.)

Staphylis St-Lag. = Staphylea L. (Staphyleac.).
Staphylium Dum. (nomen) = ? Daucus L. (Umbellif.).
Staphyllaea Scop. = Staphylea L. (Staphyleac.).
Staphyllodendron Scop. = seq.
Staphylodendron Mill. = Staphylea L. (Staphyleac.).
Staphylodendrum Moench = praec.
Staphylorhodos Turcz. = ? Azara Ruiz & Pav. (Flacourtiac.).
Staphylosyce Hook. f. = Coccinia Wight & Arn. (Cucurbitac.).
Staphysagria (DC.) Spach = Delphinium L. (Ranunculac.).
Staphysora Pierre = Maesobotrya Benth. (Euphorbiac.).
Starbia Thou. = Alectra Thunb. (Scrophulariac.).
Starkea Willd. = Liabum Adans. (Compos.).
Starkia Steud. = praec.
Stathmostelma K. Schum. Asclepiadaceae. 12 trop. Afr.
Staticaceae ('-inae') Hoffmgg & Link ex S. F. Gray = Plumbaginaceae–Staticeae Reichb.
Statice L. = Armeria Willd. + Limonium Mill. (Plumbaginac.).
Staudtia Warb. Myristicaceae. 2–3 W. Afr.
Stauntonia DC. Lardizabalaceae. 15 Burma to Formosa & Japan.
Stauracanthus Link (~ Ulex L.). Leguminosae. 6 Iberian Penins.
× **Staurachnis** hort. (vii) = × Trichachnis hort. (Orchidac.).
× **Stauranda** hort. (vii) = × Trichovanda hort. (Orchidac.).
Stauranthera Benth. Gesneriaceae. 10 SE. As., W. Malaysia.
Stauranthus Liebm. Rutaceae. 1 S. Mex.
Stauregton Fourr. (sphalm.) = Staurogeton Reichb. = Lemna L. (Lemnac.).
Stauritis Reichb. f. = Phalaenopsis Blume (Orchidac.).
Staurochilus Ridl. Orchidaceae. 2 Siam, W. Malaysia.
Staurochlamys Baker. Compositae. 1 N. Brazil.
Staurogeton Reichb. = Lemna L. (Lemnac.).
Stauroglottis Schau. = Phalaenopsis Blume (Orchidac.).
Staurogyne Wall. Acanthaceae or Scrophulariaceae. 80 trop., esp. W. Malaysia.
The genus is almost exactly intermediate between the two families.
Staurogynopsis Mangenot & Aké Assi = praec.
Stauromatum Endl. (sphalm.) = Sauromatum Schott (Arac.).
Staurophragma Fisch. & Mey. Scrophulariaceae. 1 As. Min.

STAUROPSIS

Stauropsis Reichb. f. = Trichoglottis Bl. (Orchidac.).
Staurospermum Thonn. = Mitracarpum Zucc. (Rubiac.).
Staurostigma Scheidw. = Asterostigma Fisch. & Mey. (Arac.).
Staurothylax Griff. = Cicca L. (Euphorbiac.).
Staurothyrax Griff. = praec.
Stavia Thunb. = Staavia Dahl (Bruniac.).
Stawellia F. Muell. Liliaceae. 2 SW. Austr.
Stayneria L. Bolus. Aïzoaceae. 1 S. Afr.
Stearodendron Engl. = Allanblackia Oliv. (Guttif.).
Stebbinsia Lipschitz. Compositae. 1 C. As.
Stechmannia DC. = Jurinea Cass. (Compos.).
Stechys Boiss. (sphalm.) = Stachys L. (Labiat.).
Steegia Steud. = Stegia DC. (Malvac.).
Steenhamera Kostel. (sphalm.) = seq.
Steenhammera Reichb. = Mertensia Roth (Boraginac.).
Steenisia Bakh. f. (~ Neurocalyx Hook.). Rubiaceae. 5 Borneo.
Steenisia Kuprianova = Nothofagus (Bl.) Oerst. (Fagac.).
Steerbeckia J. F. Gmel. = Sterbeckia Schreb. = Singana Aubl. (inc. sed.).
Steetzia Sond. = Olearia Moench (Compos.).
Stefaninia Chiov. = ? Reseda L. (Resedac.).
Stefanoffia H. Wolff. Umbelliferae. 1 C. Medit.
Steffensia Kunth = Piper L. (Piperac.).
Stegania R.Br. = Blechnum L. (Blechnac.).
Steganotaenia Hochst. Umbelliferae. 2 trop. Afr.
Steganotropis Lehm. = Centrosema Benth. (Legumin.).
Steganotus Cass. = Arctotis L. (Compos.).
Steganthera Perk. Monimiaceae. 28 E. Malaysia.
Steganthus Knobl. = Olea L. (Oleac.).
Stegastrum Van Tiegh. = Lepeostegeres Bl. (Loranthac.).
Stegia DC. (~ Lavatera L.). Malvaceae. 1 Medit.
Stegitrio P. & K. = seq.
Stegitris Rafin. = Halimium (Dunal) Spach (Cistac.).
Stegnocarpus Torr. = Coldenia L. (Ehretiac.).
Stegnogramma Blume. Thelypteridaceae. 4 trop. As.
Stegnogramma Fourn. = Bommeria Fourn. (Gymnogrammac.).
Stegnosperma Benth. Stegnospermataceae. 3 L. Calif. to C. Am., W.I.
Stegnospermataceae (H. Walter) Nak. Dicots. (Centrosperm.). 1/3 Mex.,
C. Am., W.I. Shrubs of dry or coastal regions, sometimes scandent, with alt.
simple ent. fleshy grey-green membr.-margined exstip. ls. Fls. reg., ⚥, in
term. bract. racemif. thyrses or axill. cymes; pedic. thickened upwards,
bibracteol. at base. K 5, imbr., persist.; C 5, imbr., suborbic., slightly clawed;
A 10, fil. subul., connate at base, persist., anth. dorsif., intr.; G (3–5), 1-loc.,
with 3–5 short recurved stigs., and 3–5 basal erect amphitr. ov. around a central
column. Fr. a coriac. 3–5-valved caps.; seeds 1–5, almost covered by large
fleshy aril; embryo curved; endosp. mealy. Only genus: *Stegnosperma*. Closely
related to *Phytolaccac.*, of which it could be regarded as a ± primitive member.
Stegolepis Klotzsch ex Koern. Rapateaceae. 23 N. trop. S. Am.
Stegonotus P. & K. = Steganotus Cass. = Arctotis Gaertn. (Compos.).

Stegosia Lour. = Rottboellia L. f. (Gramin.).
Steigeria Muell. = Baloghia Endl. (Euphorbiac.).
Steinbachiella Harms. Leguminosae. 1 Bolivia.
Steinchisma Rafin. = Panicum L. (Gramin.).
Steineria Klotzsch = Begonia L. (Begoniac.).
Steinhauera [Presl pro gen. foss.] P. & K. = Sequoia Endl. + Sequoiadendron Buchholz (Taxodiac.).
Steinheilia Decne. Asclepiadaceae. 1 Arabia.
Steinitzia Gandog. = Anthemis L. (Compos.).
Steinmannia Opiz = Rumex L. (Polygonac.).
Steinmannia Phil. f. = Garaventia Looser (Alliac.).
Steinreitera Opiz = Thesium L. (Santalac.).
Steinschisma Steud. = Steinchisma Rafin. = Panicum L. (Gramin.).
Steirachne Ekman. Gramineae. 1 NE. Brazil.
Steiractinia Blake. Compositae. 18 Colombia, Ecuador.
Steiractis DC. = Olearia Moench (Compos.).
Steiractis Rafin. = Layia Hook. & Arn. ex DC. (Compos.).
Steiranisia Rafin. = Heterisia Rafin. = Micranthes Haw. = Saxifraga L. (Saxi-
Steirema Benth. & Hook. f. (sphalm.) = seq.   [fragac.).
Steiremis Rafin. = Telanthera R.Br. (Amaranthac.).
Steirexa Rafin. = Trichopus Gaertn. (Trichopodac.).
Steireya B. D. Jacks. (sphalm.) = praec.
Steirocoma Reichb. = Dicoma Cass. (Compos.).
Steiroctis Rafin. = Lachnaea L. + Cryptadenia Meissn. (Thymelaeac.).
Steirodiscus Less. = Psilothonna E. Mey. ex DC. (Compos.).
Steiroglossa DC. = Brachycome Cass. (Compos.).
Steironema Rafin. = Lysimachia L. (Primulac.).
Steiropteris C. Chr. = Thelypteris Schmid. (Thelypteridac.).
Steirosanchezia Lindau. Acanthaceae. 1 Peru.
Steirostemon Phil. = Samolus L. (Primulac.).
Steirotis Rafin. (Struthanthus Mart.). Loranthaceae. 75 trop. Am.
Stekhovia De Vriese = Goodenia Sm. (Goodeniac.).
Stelanthes Stokes = Marlea Roxb. = Alangium Lam. (Alangiac.).
Stelechanteria Thou. ex Baill. = Drypetes Vahl (Euphorbiac.).
Stelechantha Bremek. Rubiaceae. 1 Angola.
Stelechocarpus Hook. f. & Thoms. Annonaceae. 5 Siam, Malaysia. Fr. ed.
Stelechospermum Blume = Mischocarpus Bl. (Sapindac.).
Steleostemma Schlechter. Asclepiadaceae. 1 Bolivia.
Stelephuros Adans. = Phleum L. (Gramin.).
Stelestylis Drude. Cyclanthaceae. 4 N. trop. S. Am.
Stelin Bub. = Viscum L. (Loranthac.).
Stelis Loefl. = Oryctanthus (Griseb.) Eichl. + Struthanthus Mart. (Loranthac.).
*Stelis Sw. Orchidaceae. 270 trop. Am., W.I.
Stelitaceae Dulac = Primulaceae Vent.
Stella Medik. = Astragalus L. (Legumin.).
Stellandria Brickell = Schisandra Michx (Schisandrac.).
Stellara Fisch. ex Reut. = Boschniakia C. A. Mey. (Orobanchac.).
Stellaria Hill = Corispermum L. (Chenopodiac.).

STELLARIA

Stellaria L. Caryophyllaceae. 120 cosmop. *S. media* (L.) Vill. has small homogamous fls. that fert. themselves in absence of insects; it flowers all the year, and in winter (?on account of weak light, cold, etc.) is often cleistogamic. The number of sta. is most often 3, but varies a good deal. The fls. of *S. graminea* L. are larger and protandr., but with autogamy, whilst in *S. holostea* L. the fls. are still larger and very protandr. with little self-fert.

Stellaria Seguier = Callitriche L. (Callitrichac.).

Stellariaceae Dum. = Caryophyllaceae Juss.

Stellariaceae MacMill. = Callitrichaceae Link.

Stellarioïdes Medik. = ? Anthericum L. (Liliac.).

Stellariopsis Rydb. = Potentilla L. (Rosac.).

Stellaris Fabr. = Scilla L. (Liliac.).

Stellaris Moench = Scilla L. + Ornithogalum L. + Gagea Salisb. (Liliac.).

Stellaster Fabr. (uninom.) = *Scilla* L. sp. (Liliac.).

Stellatae Batsch = Rubiaceae–Rubioïdeae–Rubieae.

Stellera L. = Wikstroemia Endl. + Dendrostellera Van Tiegh. (Thymelaeac.).

Stellera Turcz. = Swertia L. (Gentianac.).

Stelleropsis Pobedim. Thymelaeaceae. 18 W. & C. As.

Stellia Nor. = Tarenna Gaertn. (Rubiac.).

Stellilabium Schlechter. Orchidaceae. 2 Colombia, Peru.

Stellimia Rafin. = Pectis L. (Compos.).

Stellina Bubani = Callitriche L. (Callitrichac.).

Stellix Nor. = Psychotria L. (Rubiac.).

Stellorkis Thou. = Nervilia Comm. ex Gaudich. (Orchidac.).

Stellularia Benth. Scrophulariaceae. 2 trop. Afr.

Stellularia Hill = Stellaria L. (Caryophyllac.).

Stelmagonum Baill. Asclepiadaceae. 2 trop. Am.

Stelmanis Rafin. (1836) = Heterotheca Cass. (Compos.).

Stelmanis Rafin. (1840) = Anistelma Rafin. = Hedyotis L. (Rubiac.).

Stelmation Fourn. (~ Metastelma R.Br.). Asclepiadaceae. 1 Brazil.

Stelmatocodon Schlechter. Asclepiadaceae (inc. sed.). 1 Bolivia.

Stelmatogonum K. Schum. = Stelmagonum Baill. (Asclepiadac.).

Stelmesus Rafin. = Allium L. (Alliac.).

Stelmotis Rafin. = Stelmanis Rafin. = Anistelma Rafin. = Hedyotis L. (Rubiac.).

Stelophurus P. & K. = Stelephuros Adans. = Phleum L. (Gramin.).

Stelostylis P. & K. = Stelestylis Drude (Cyclanthac.).

Stemeiena Rafin. = Krameria Loefl. (Krameriac.).

Stemmacantha Cass. = Cirsium Mill. (Compos.).

Stemmadenia Benth. Apocynaceae. 20 Mex. to Ecuad.

Stemmatella Wedd. ex Benth. & Hook. f. = Galinsoga Ruiz & Pav. (Compos.).

Stemmatium Phil. = Tristagma Poepp. (Liliac.).

Stemmatodaphne Gamble. Lauraceae. 1 Malay Penins., Borneo.

Stemmatophyllum Van Tiegh. = Amyema Van Tiegh. (Loranthac.).

Stemmatophysum Steud. (sphalm.) = seq.

Stemmatosiphon Meissn. = seq.

Stemmatosiphum Pohl = Symplocos L. (Symplocac.).

Stemmatospermum Beauv. = Nastus Juss. (Gramin.).

Stemmodontia Cass. = Wedelia Jacq. (Compos.).

*Stemodia L. Scrophulariaceae. 30 trop.

Stemodiacra P.Br. = praec.

Stemodiopsis Engl. Scrophulariaceae. 10 trop. Afr., Madag.

Stemodoxis Rafin. = Allium L. (Alliac.).

Stemona Lour. Stemonaceae. 25 E. As., Indomal., N. Austr.

Stemonacanthus Nees = Ruellia L. (Acanthac.).

*Stemonaceae Engler. Monocots. 1/25 E. As., Indomal., N. Austr. Perenn. erect or climbing rhiz. herbs, often with fascic. tubers. Ls. alt., opp. or vertic., pet., often *Smilax*-like, with parallel main nerves and ∞ closely parallel cross-veins; sometimes reduced to scales. Fls. reg., ☿, rarely ♂♀, sometimes precoc., in axill. few-fld. cymes, sometimes unpleasantly scented. P 2+2, sepaloid or petaloid; A 2+2, with short fil. conn. below, and linear anth. with long-produced linear-lanc. sometimes lamellate connective, free from anther above (fil. sometimes red with green connective); G (2), 1-loc., compressed, with small subsess. stig., and few to several basal orthotr. ov.; funicles sometimes long and ± pilose. Fr. an ovoid 2-valved caps.; seeds with endosp., sometimes arillate(?), often dependent on long funicles. Only genus: *Stemona*. Habit recalling *Smilax*, *Polygonatum* or even *Asclepiadaceae*. Prob. closely related to *Liliaceae–Asparagoïdeae–Polygonateae*. (See also *Croomiaceae*.)

Stemone Franch. & Sav. = Stemona Lour. (Stemonac.).

Stemonix Rafin. = Eurycles Salisb. (Amaryllidac.).

Stemonocoleus Harms. Leguminosae. 1 W. Equat. Afr.

Stemonoporus Thw. Dipterocarpaceae. 14 Ceylon.

Stemonurus Blume (1825, emend. Kuntze, Engl., Howard, etc.). Icacinaceae. 12 Indomal.

Stemonurus Blume (1849, sensu Becc., Valet., Sleum., etc.) = Urandra Thw. (Icacinac.).

Stemoptera Miers = Apteria Nutt. (Burmanniac.).

Stemotis Rafin. = Rhododendron L. (Ericac.).

Stemotria Wettst. & Harms. Scrophulariaceae. 1 Peru.

Stenachaenium Benth. Compositae. 4 S. Brazil, Argentina.

Stenactis Cass. = Erigeron L. (Compos.).

Stenadenium Pax. Euphorbiaceae. 1 trop. E. Afr.

Stenandriopsis S. Moore. Acanthaceae. 10 Madag.

*Stenandrium Nees. Acanthaceae. 30 warm Am.

Stenanona Standley. Annonaceae. 2 C. Am. (Costa Rica, Panamá).

Stenanthella Rydb. Liliaceae. 2 Sakhalin, Pacif. N. Am.

Stenanthemum Reissek = Cryptandra Sm. (Rhamnac.).

Stenanthera R.Br. = Astroloma R.Br. (Epacridac.).

Stenanthera Engl. & Diels = Neostenanthera Exell (Annonac.).

*Stenanthium Kunth. Liliaceae. 1 E. U.S.

Stenanthus Oerst. ex Hanst. = Columnea L. (Gesneriac.).

Stenaphia A. Rich. (sphalm.) = Stephania Lour. (Menispermac.).

Stenaria Rafin. = Houstonia L. (Rubiac.).

Stenarrhena D. Don = Salvia L. (Labiat.).

Stengelia Neck. = Mourera Aubl. (Podostemac.).

Stengelia Sch. Bip. = Vernonia Schreb. (Compos.).

Stenhammaria Nym. = Steenhammeria Reichb. = Mertensia Roth (Boraginac.).

STENIA

Stenia Lindl. Orchidaceae. 2 Trinidad, N. trop. S. Am.
Stenocactus A. Berger = Echinofossulocactus Lawr. (Cactac.).
Stenocaelium Benth. & Hook. f. = Stenocoelium Ledeb. (Umbellif.).
Stenocalyx Berg = Eugenia L. (Myrtac.).
Stenocalyx Turcz. = Diplopterys Juss. (Malpighiac.).
Stenocarpha Blake. Compositae. 1 Mex.
*Stenocarpus R.Br. Proteaceae. 25 E. Austr., New Caled.
Stenocephalum Sch. Bip. = Vernonia Schreb. (Compos.).
Stenocereus (A. Berger) Riccob. = Lemaireocereus Britt. & Rose (Cactac.).
Stenochasma Griff. = Hornstedtia Retz. (Zingiberac.).
Stenochasma Miq. Urticaceae. 1 Sumatra.
Stenochilum Willd. ex Cham. & Schlechtd. = Lamourouxia Kunth (Scrophulariac.).
Stenochilus R.Br. Myoporaceae. 15 Austr.
Stenochilus P. & K. = Stenochilum Willd. ex Cham. & Schlechtd. = Lamourouxia Kunth (Scrophulariac.).
Stenochlaena J. Sm. Blechnaceae(?). 5 Afr. to Pacific. Superficially like Lomariopsis and Teratophyllum, but distinct in anatomy and spores (Holttum, Gard. Bull. Str. Settlem. 5: 251–63, 1932).
Stenochloa Nutt. = Dissanthelium Trin. (Gramin.).
Stenocline DC. Compositae. 13 Brazil, S. Afr., Madag.
Stenocoelium Ledeb. Umbelliferae. 2 C. As.
Stenocoryne Lindl. Orchidaceae. 10 Brazil, Guiana.
Stenodiptera Kozo-Poljansky = Caropodium Stapf & Wettst. (Umbellif.).
Stenodiscus Reissek = Spyridium Fenzl (Rhamnac.).
Stenodon Naud. Melastomataceae. 2 S. Brazil.
Stenodraba O. E. Schulz. Cruciferae. 6 S. Andes.
Stenodrepanum Harms. Leguminosae. 1 Argentina.
Stenofestuca (Honda) Nakai = Bromus L. (Gramin.).
Stenogastra Hanst. = Sinningia Nees (Gesneriac.).
Stenoglossum Kunth = Epidendrum L. (Orchidac.).
Stenoglottis Lindl. Orchidaceae. 4 trop. & S. Afr.
Stenogonum Nutt. = Eriogonum Michx (Polygonac.).
*Stenogyne Benth. Labiatae. 25 Hawaii.
Stenogyne Cass. = Eriocephalus L. (Compos.).
Stenolepia v. Ald. v. Ros. Aspidiaceae. 1 Malaysia.
Stenolirion Baker (~ Crinum L.). Amaryllidaceae. 1 trop. E. Afr.
Stenolobeae Muell. Arg. In the Euphorbiaceae, an inclusive term denoting those (Australian) genera in which the cotyledons are narrow, no wider than the radicle. These genera should probably be distributed among the other tribes of Euph. (vide Pax & Hoffmann in Engl. Pflanzenfam. ed. 2, 19c: 29–30, 1931).
Stenolobium Benth. = Calopogonium Desv. (Legumin.).
Stenolobium D. Don = Tecoma Juss. (Bignoniac.).
Stenolobus Presl = Davallia Sm. (Davalliac.).
Stenoloma Fée = Sphenomeris Maxon (Lindsaeac.).
Stenolophus Cass. = Centaurea L. (Compos.).
Stenomeria Turcz. Asclepiadaceae. 3 Colombia.

*Stenomeridaceae J. G. Agardh = Dioscoreaceae R.Br.

Stenomeris Planch. Dioscoreaceae. 2 W. Malaysia.

Stenomesson Herb. Amaryllidaceae. 20 trop. Am.

Stenonema Hook. ex Benth. & Hook. f. = Dolichostylis Turcz. = Draba L. (Crucif.).

Stenonia Baill. = Stenoniella Kuntze = Cleistanthus Hook. f. ex Planch. (Euphorbiac.).

Stenonia Didr. = Ditaxis Vahl ex Juss. (Euphorbiac.).

Stenoniella Kuntze = Cleistanthus Hook. f. ex Planch. (Euphorbiac.).

Stenopadus Blake. Compositae. 13 NW. trop. S. Am.

Stenopetalum R.Br. ex DC. Cruciferae. 8 W. & S. Austr.

Stenophragma Čelak. = Arabidopsis Heynh. (Crucif.).

Stenophyllum Sch. Bip. ex Benth. & Hook. f. = Calea L. (Compos.).

Stenophyllus Rafin. = Bulbostylis Kunth (Cyperac.).

Stenopolen Rafin. = Stenia Lindl. (Orchidac.).

Stenoptera Presl. Orchidaceae. 5 trop. S. Am.

Stenorhynchus Lindl. = seq.

Stenorhyncus Lindl. = seq.

Stenorrhynchium Reichb. = seq.

Stenorrhynchos Rich. ex Spreng. Orchidaceae. 45 SE. U.S. to temp.

Stenorrhynchus Reichb. = praec. [S. Am.

Stenorynchus Rich. (nomen) = praec.

Stenoschista Bremek. Acanthaceae. 1 trop. W. Afr.

Stenoselenium Popov (sphalm.) = Stenosolenium Turcz. (Boraginac.).

Stenosemia Presl. Aspidiaceae. 2 Malaysia, Solomon Is. Acrostichoid deriv. of *Tectaria* Cav.

Stenosemis E. Mey. ex Harv. & Sond. = Annesorhiza Cham. & Schlechtd. (Umbellif.).

Stenosiphanthus A. Sampaio = Arrabidaea DC. (Bignoniac.).

Stenosiphon Spach. Onagraceae. 1 SW. U.S.

Stenosiphonium Nees. Acanthaceae. 6 Penins. Ind., Ceylon.

Stenosolen (Muell. Arg.) Markgraf. Apocynaceae. 5 trop. S. Am.

Stenosolenium Turcz. Boraginaceae. 1 C. As.

Stenospermation Schott. Araceae. 25 trop. Am., sub-andine.

Stenospermatium Schott = praec.

Stenospermum Sweet = Metrosideros Banks (Myrtac.).

Stenostachys Turcz. = ? Hystrix Moench (Gramin.).

Stenostelma Schlechter. Asclepiadaceae. 5 S. trop. & S. Afr.

Stenostephanus Nees. Acanthaceae. 6 trop. S. Am.

Stenostomum Gaertn. f. = Antirhea Comm. ex Juss. (Rubiac.).

Stenotaenia Boiss. (~ Heracleum L.). Umbelliferae. 5–6 As. Min., Persia.

Stenotaphrum Trin. Gramineae. 7 trop. & subtrop. *S. secundatum* (Walt.) Kuntze is useful for binding drift-sand (cf. *Ammophila*).

Stenotheca Monn. = Hieracium L. (Compos.).

Stenothyrsus C. B. Clarke. Acanthaceae. 1 Malay Penins.

Stenotium Presl ex Steud. = Lobelia L. (Campanulac.).

Stenotopsis Rydb. = Haplopappus Cass. (Compos.).

Stenotropis Hassk. = Erythrina L. (Legumin.).

Stenotus Nutt. = Haplopappus Cass. (Compos.).
Stenouratea Van Tiegh. = Ouratea Aubl. (Ochnac.).
Stenurus Salisb. = Biarum Schott (Arac.).
Stephalea Rafin. = Campanula L. (Campanulac.).
Stephanachne Keng. Gramineae. 1 China.
Stephanandra Sieb. & Zucc. Rosaceae. 4 E. As.
Stephanangaceae Dulac = Valerianaceae Lam. & DC.
Stephananthus Lehm. = Baccharis L. (Compos.).
Stephania Kuntze = Astephania Oliver (Compos.).
Stephania Lour. Menispermaceae. 40 trop. Afr., As., Austr.
Stephania Willd. = Steriphoma Spreng. (Capparidac.).
Stephaniscus Van Tiegh. = Tapinanthus Bl. (Loranthac.).
Stephanium Schreb. = Palicourea Aubl. (Rubiac.).
Stephanocarpus Spach = Cistus L. (Cistac.).
Stephanocaryum Popov. Boraginaceae. 1 C. As.
Stephanocereus A. Berger. Cactaceae. 1 Brazil.
Stephanochilus Coss. & Dur. ex Benth. & Hook. f. = Volutarella Cass.
Stephanococcus Bremek. Rubiaceae. 1 trop. Afr.   [(Compos.).
Stephanocoma Less. = Berkheya Ehrh. (Compos.).
Stephanodaphne Baill. Thymelaeaceae. 8–9 Madag., Comoro Is.
Stephanodoria Greene (~ Xanthocephalum Willd.). Compositae. 1 Mex.
Stephanogastra Karst. & Triana = Centronia D. Don (Melastomatac.).
Stephanogyna P. & K. = Stephegyne Korth. = Mitragyna Korth. (Rubiac.).
Stephanolepis S. Moore. Compositae. 1 trop. Afr.
Stephanolirion Baker = Tristagma Poepp. (Liliac.).
Stephanoluma Baill. = Sideroxylon L. (Sapotac.).
***Stephanomeria** Nutt. Compositae. 15 W. N. Am.
Stephanopappus Less. = Nestlera Spreng. (Compos.).
Stephanopholis Blake. Compositae. 1 Mex.
Stephanophorum Dulac = Narcissus L. (Amaryllidac.).
Stephanophyllum Guill. = Paepalanthus Mart. (Eriocaulac.).
Stephanophysum Pohl. Acanthaceae. 5 trop. Am.
Stephanopodium Poepp. & Endl. Dichapetalaceae. 7 trop. S. Am.
Stephanorossia Chiov. Umbelliferae. 2 E. trop. Afr.
Stephanosiphon Boiv. ex C. DC. = Turraea L. (Meliac.).
Stephanostachys Klotzsch ex Oerst. = Chamaedorea Willd. (Palm.).
Stephanostegia Baill. Apocynaceae. 3 Madag.
Stephanostema K. Schum. Apocynaceae. 1 Zanzibar.
Stephanotella Fourn. Asclepiadaceae. 2 trop. S. Am.
Stephanotis Thou. Asclepiadaceae. 5 Madag.
Stephanotrichum Naud. = Clidemia D. Don (Melastomatac.).
Stephegyne Korth. = Mitragyna Korth. (Naucleac.).
Stephensonia Hort. = Stevensonia J. Dunc. ex Balf. f. = Phoenicophorium H. Wendl. (Palm.).
Steptium Boiss. = Streptium Roxb. = Priva Adans. (Verbenac.).
Steptorhamphus Bunge. Compositae. 7 SW. to C. As.
Stera Ewart = Cratystylis S. Moore (Compos.).
Sterbeckia Schreb. = Singana Aubl. (inc. sed.).

Sterculia L. Sterculiaceae. 300 trop. Fls. ♂ ♀; C o.

*Sterculiaceae Vent. Dicots. 60/700, chiefly trop. Trees, shrubs, or herbs, sometimes lianes, with alt. simple or digit. stip. ls.; petiole often pulvinate and geniculate at apex. Fls. in complex cymes, ♀ or ♂ ♀, usu. reg., 5-merous. K (3–5), valvate, with no epicalyx; C 5, often absent or small, contorted; A in two whorls, the outer staminodial or o, the inner often branched, free or ± united into a tube, anthers 2–loc.; G usu. (5), sometimes (1–4) or (10–12), with 2–∞ axile anatr. ovules in each, with the micropyle outwards; style simple, lobed, rarely free to base. Fruit various, often a schizocarp. Endosperm fleshy or thin or o. *Cola* and *Theobroma* (cacao) are economically important. Intimately related to *Tiliac.*, *Malvac.*, *Bombacac.*, etc., on the one hand, and to *Euphorbiac.* on the other. The fam. prob. requires considerable re-casting. Chief genera: *Dombeya, Hermannia, Melochia, Byttneria, Theobroma, Helicteres, Sterculia, Cola.*

Stereimis Rafin. = Steiremis Rafin. = Alternanthera Forsk. (Amaranthac.).

Stereocarpus Hallier = Camellia L. (Theac.).

Stereocaryum Burret. Myrtaceae. 3 New Caled.

Stereochilus Lindl. = Sarcanthus Lindl. (Orchidac.).

Stereochlaena Hackel. Gramineae. 1 trop. E. Afr.

Stereoderma Blume = Olea L. (Oleac.).

Stereosandra Blume. Orchidaceae. 1 Indoch., W. Malaysia.

Stereosanthus Franch. = Nannoglottis Maxim. (Compos.).

Stereospermum Cham. Bignoniaceae. 24 trop. Afr., As.

Stereoxylon Ruiz & Pav. = Escallonia Mutis (Escalloniac.).

Sterigma DC. = Sterigmostemum M. Bieb. (Crucif.).

Sterigmanthe Klotzsch & Garcke = Euphorbia L. (Euphorbiac.).

Sterigmapetalum Kuhlm. Rhizophoraceae. 3 N. trop. S. Am.

Sterigmostemon Poir. = seq.

Sterigmostemum M. Bieb. Cruciferae. 8–9 SW. to C. As.

Steripha Banks ex Gaertn. = Dichondra J. R. & G. Forst. (Convolvulac.).

Steriphe Phil. = Haplopappus Cass. (Compos.).

*Steriphoma Spreng. Capparidaceae. 8 C. Am. to Peru, Trinidad.

Steris Adans. = Lychnis L. (Caryophyllac.).

Steris L. = Hydrolea L. (Hydrophyllac.).

Sterisia Rafin. = praec.

Sternbeckia Pers. (sphalm.) = Sterbeckia Schreb. = Singana Aubl. (inc. sed.).

Sternbergia Waldst. & Kit. Amaryllidaceae. 8 E. Medit. to Cauc.

Sterrhymenia Griseb. = Sclerophylax Miers (Solanac.).

Sterropetalum N. E. Brown. Aïzoaceae. 1 SW. Afr.

Stethoma Rafin. Acanthaceae. 5 trop. S. Am.

Stetsonia Britton & Rose. Cactaceae. 1 Argentina.

Steudelago Kuntze = Exostema (Pers.) Rich. (Rubiac.).

Steudelella Honda = Sphaerocarya Nees ex Hook. f. (Gramin.).

Steudelia Mart. = Leonia Ruiz & Pav. (Violac.).

Steudelia Presl = Adenogramma Reichb. (Aïzoac.).

Steudelia Spreng. = Erythroxylum P.Br. (Erythroxylac.).

Steudnera C. Koch. Araceae. 8 Himal., SE. As., Malay Penins.

Stevena Andrz. ex DC. = Alyssum L. (Crucif.).

STEVENIA

Stevenia Adams & Fisch. Cruciferae. 3 Siberia, Mongolia, Korea.
Steveniella Schlechter. Orchidaceae. 1 S. Russia (Crimea), W. As.
Stevenorchis Wankow & Kraenzl. = Neottiella Schltr (Orchidac.).
Stevensia Poit. Rubiaceae. 6 W.I.
Stevensonia J. Dunc. ex Balf. f. = Phoenicophorium H. Wendl. (Palm.).
Stevia Cav. Compositae. 150 trop. & subtrop. Am.
Steviopsis R. M. King & H. Rob. (~ Eupatorium L.). Compositae. 1 Mex.
Stevogtia Neck. = Convolvulus L. (Convolvulac.).
Stevogtia Rafin. = Phacelia Juss. (Hydrophyllac.).
Stewartia L. Theaceae. 10 E. As., E. U.S.
Stewartiella Nasir. Umbelliferae. 1 W. Pakistan.
Steyermarkia Standley. Rubiaceae. 1 C. Am.
Steyermarkina R. M. King & H. Rob. (~ Eupatorium L.). Compositae. 4 Venez., Braz.
Sthaelina Lag. (sphalm.) = Staehelina L. (Compos.).
Stibadotheca Klotzsch = Begonia L. (Begoniac.).
Stibas Comm. ex DC. = Levenhookia R.Br. (Campanulac.).
Stibasia Presl = Marattia Sw. (Marattiac.).
Stiburus Stapf. Gramineae. 2 S. Afr.
Sticherus Presl. Gleicheniaceae. 100 pantrop. & S. hemisph. Branches of ls. always pseudo-dichot., rhiz. scaly.
Stichianthus Valeton. Rubiaceae. 3 Borneo.
Stichoneuron Hook. f. Croomiaceae. 2 Assam, Siam, Malay Penins.
Stichophyllum Phil. = Pycnophyllum Remy (Caryophyllac.).
Stichorchis Thou. = Liparis Rich. (Orchidac.).
Stickmannia Neck. = Dichorisandra Mikan (Commelinac.).
Stictocardia Hallier f. Convolvulaceae. 6–7 trop., esp. Malaysia.
Stictophyllum Edgew. = Tricholepis DC. (Compos.).
Stiefia Medik. = Salvia L. (Labiat.).
*Stifftia Mikan. Compositae. 7 N. trop. S. Am. Shrubs. Very large fls. (for
Stiftia Cass. = praec.   [*Compos.*).
Stiftia Pohl ex Nees = Ebermaiera Nees (Acanthac.).
Stigmanthus Lour. = Morinda L. (Rubiac.).
Stigmaphyllon A. Juss. Malpighiaceae. 60–70 trop. Am., W.I.
Stigmarosa Hook. f. & Thoms. (sphalm.) = seq.
Stigmarota Lour. = Flacourtia Comm. ex L'Hérit. (Flacourtiac.).
Stigmatanthus Roem. & Schult. = Stigmanthus Lour. = Morinda L. (Rubiac.).
Stigmatella Eig. Cruciferae. 1 Palestine.
Stigmatocarpum L. Bolus = Dorotheanthus Schwant. (Aïzoac.).
Stigmatococca Willd. = Ardisia Sw. (Myrsinac.).
Stigmatodactylus Maxim. ex Makino. Orchidaceae. 4 India, Japan, Java, Celebes.
Stigmatophyllon Meissn. = Stigmaphyllon A. Juss. (Malpighiac.).
Stigmatophyllum Spach = praec.
Stigmatopteris C. Chr. Aspidiaceae. 26 trop. Am.
Stigmatorhynchus Schlechter. Asclepiadaceae. 2 trop. E. & SW. Afr.
Stigmatotheca Sch. Bip. = Chrysanthemum L. (Compos.).
Stilaginella Tul. = Hieronyma Allem. (Euphorbiac.).

Stilaginaceae C. A. Agardh. Dicots. 1/170 trop. & subtrop. Afr. & As. Shrubs or small trees, with alt. ent. simple shortly pet. distich. conspicuously stip. ls., with conspic. looping venation. Fls. ♂ ♀, dioec., v. small, in axill. or term. spikes, rac., or true panicles. K (3–5[-8]), imbr. or open; C o; disk rel. large and conspic., variously formed, lobes free or united, extra-staminal or -ovarial, or surrounding each sta.; A [2]3–5[6], oppositisep., fil. elong., filif., inflexed in bud, anth. bilobed, borne on short thick connect., thecae ± divergent, almost free. (In ♀ fl. stds. o.) G 1, with 2 apical pend. anatr. ov., and short term. or subterm. style with 2–4 short spreading stigs. (In ♂ fl. a small cylindr. pistillode, or o.) Fr. an ovoid or flattened, often oblique drupe, with often lateral persist. style, and conspic. foveolate-retic. endocarp; seeds 1(-2), with fleshy endosp. Only genus: *Antidesma* (*Stilago*). Long included in the *Euphorbiac.*, this distinctive genus appears to have no close relatives in that fam. The small drupaceous fruits, with their characteristic foveolate-reticulate often flattened endocarps, differ widely from anything found in the *Euph.*, but show much similarity to those of several genera of *Icacinaceae* (*Rhyticaryum, Natsiatum, Iodes,* etc.); the leaf venation is also v. similar to that of *Rhyticaryum,* etc. There are, however, also obvious and profound differences from *Icacinac.* The genus *Antidesma* may be regarded as to some extent intermediate between *Icacinac.* and *Euphorbiac.*, and fully merits the family rank given it by Agardh, Lindley and others.

Stilago L. = Antidesma L. (Stilaginac.).

*Stilbaceae Kunth. Dicots. 5/12 S. Afr. Shrubs with densely leafy branches; indumentum not stellate; ls. vertic., lin. or acic., ent., exstip. Fls. zygo., ⚥, in dense term. short or elong. spikes, with leafy bracts. K (5), sometimes bilab., valv. or subimbr.; C (4–5), reg. or zygo., lobes imbr.; A 4, alternipet. and epipet., fil. filif., anth. loc. sometimes diverg. downwards; disk o; G̲ (2), with filif. style and punctif. stig., and 1 erect anatr. ov. per loc. Fr. a dehisc. caps., or indehisc.; seeds with endosp. Genera: *Stilbe, Campylostachys, Xeroplana, Eurylobium, Euthystachys.* Closely related to *Dicrastylidac.* and *Verbenac.*

Stilbanthus Hook. f. Amaranthaceae. 1 Himalaya.

Stilbe Bergius. Stilbaceae. 8 S. Afr.

Stilbeaceae Bullock = Stilbaceae Kunth.

Stilbocarpa A. Gray. Araliaceae. 2 N.Z. islands. Herbs.

Stilingia Rafin. = Stillingia L. (Euphorbiac.).

Stilla W. Young = ? Scilla L. = ? Camassia Lindl. (Liliac.).

Stillengia Torr. (sphalm.) = Stillingia L. (Euphorbiac.).

Stillingfleetia Boj. = Sapium P.Br. (Euphorbiac.).

Stillingia Garden ex L. Euphorbiaceae. 1–2 Masc., 1 E. Malaysia, Fiji; 30 warm Am.

Stilopus Hook. = Stylypus Rafin. = Geum L. (Rosac.).

Stilpnogyne DC. Compositae. 1 S. Afr.

Stilpnolepis I. M. Krascheninn. Compositae. 1 Mongolia.

Stilpnopappus Mart. ex DC. Compositae. 20 trop. S. Am.

Stilpnophleum Nevski = Deyeuxia Clarion (Gramin.).

Stilpnophyllum Hook. f. Rubiaceae. 1 Peru.

Stilpnophytum Less. Compositae. 4 S. Afr. (karroo).

Stimegas Rafin. = Paphiopedilum Pfitz. (Orchidac.).

Stimenes Rafin. = Nierembergia Ruiz & Pav. (Solanac.).
Stimomphis Rafin. = Salpiglossis Ruiz & Pav. (Solanac.).
Stimoryne Rafin. = Petunia Juss. (Solanac.).
Stimpsonia Wright. Primulaceae. 1 E. As.
Stingana B. D. Jacks. (sphalm.) = Singana Aubl. (inc. sed.).
Stipa L. Gramineae. 300 trop. & temp., usu. xero. *S. pennata* L. (feather grass, steppes) and others have ls. which roll inwards when the air is dry, covering the stomata and green tissue (which are on the upper side only) and exposing only the impermeable lower surface. The awn of the fr. is long, ending in a long feather, and hygroscopic, curling up when dry and uncurling when damp. The fr. is thin and sharply pointed, with backward-pointing hairs on the tip. As in *Erodium*, the awn when damped uncurls and, if the point of the fr. be on the soil and the feather be entangled with other objects, drives the fr. into the soil. When the air dries the feather is drawn down, not the fr. up. *S. tenacissima* L. (N. Afr.) is the esparto grass, from which paper is extensively made.
Stipaceae Burnett = Gramineae–Stipeae (Kunth) Nees.
Stipagrostis Nees = Aristida L. (Gramin.).
Stipecoma Muell. Arg. Apocynaceae. 1 Brazil.
Stipellaria Benth. = Alchornea Sw. (Euphorbiac.).
Stipellaria Klotzsch (nomen). Leguminosae. Quid?
Stiphonia Hemsl. = Styphonia Nutt. = Rhus L. (Anacardiac.).
Stipocoma P. & K. = Stipecoma Muell. Arg. (Apocynac.).
× **Stiporyzopsis** B. L. Johnson & Rogler. Gramineae. Gen. hybr. (Oryzopsis × Stipa).
Stiptanthus (Benth.) Briquet. Labiatae. 1 E. Himal., Assam.
Stipularia Beauv. = Sabicea Aubl. (Rubiac.).
Stipularia Delpino = Piuttia Mattei = Thalictrum L. (Ranunculac.).
Stipularia Haw. = Spergularia (Pers.) J. & C. Presl (Caryophyllac.).
Stipulicida Rich. in Michx. Caryophyllaceae. 1 SE. U.S.
Stiractis P. & K. = Steiractis Rafin. = Layia Hook. & Arn. ex DC. (Compos.).
Stiradotheca Klotzsch (sphalm.) = Stibadotheca Klotzsch = Begonia L. (Begoniac.).
Stiranisia P. & K. = Steiranisia Rafin. = Saxifraga L. (Saxifragac.).
Stireja P. & K. = Steireya B. D. Jacks. = Steirexa Rafin. = Trichopus Gaertn. (Trichopodac.).
Stiremis P. & K. = Steiremis Rafin. = Telanthera R.Br. (Amaranthac.).
Stirlingia Endl. Proteaceae. 6 Austr.
Stiroctis P. & K. = Steiroctis Rafin. = Lachnaea L. + Cryptadenia Meissn. (Thymelaeac.).
Stirodiscus P. & K. = Steirodiscus Less. = Psilothonna E. Mey. ex DC. (Compos.).
Stiroglossa P. & K. = Steiroglossa DC. = Brachycome Cass. (Compos.).
Stironema P. & K. = Steironema Rafin. = Lysimachia L. (Primulac.).
Stironeurum Radlk. ex De Wild. & Dur. Sapotaceae. 1 trop. Afr.
Stirostemon P. & K. = Steirostemon Phil. = Samolus L. (Primulac.).
Stirotis P. & K. = Steirotis Rafin. (Loranthac.).
Stissera Giseke = Curcuma L. (Zingiberac.).

Stissera Kuntze = Stapelia L. (Asclepiadac.).
Stisseria Fabr. (uninom.) = *Stapelia* L. sp. (Asclepiadac.).
Stisseria Scop. = Mimusops L. (Sapotac.).
Stixis Lour. Capparidaceae. 7 E. Himal. to Indoch. & Hainan, W. Malaysia, Lesser Sunda Is.
Stiza E. Mey. = Lebeckia Thunb. (Legumin.).
Stizolobium P.Br. = Mucuna Adans. (Legumin.).
Stizolophus Cass. = Centaurea L. (Compos.).
Stizophyllum Miers. Bignoniaceae. 3–4 trop. Am.
Stobaea Thunb. = Berkheya Ehrh. (Compos.).
Stoberia Dinter & Schwantes. Aïzoaceae. 3 SW. Afr.
Stocksia Benth. Sapindaceae. 1 E. Persia, Afghan., Baluch.
Stoebe L. Compositae. 40 trop. & E. Afr., Madag., Masc.
Stoeberia Dinter & Schwantes. Aïzoaceae. 3 S. Afr.
Stoechadomentha Kuntze = Adenosma R.Br. (Scrophulariac.).
Stoechas Gueldenst. = Helichrysum L. (Compos.).
Stoechas Mill. = Lavandula L. (Labiat.).
Stoechas Rumph. = Adenosma R.Br. (Scrophulariac.).
Stoehelina Benth. = Staehelina Crantz = Bartsia L. (Scrophulariac.).
Stoerkea Baker = seq.
Stoerkia Crantz = Dracaena Vand. ex L. (Agavac.).
Stoibrax Rafin. = Carum L. (Umbellif.).
Stokesia L'Hérit. Compositae. 1 SE. U.S.
Stolidia Baill. = Badula Juss. (Myrsinac.).
Stollaea Schlechter = Caldcluvia D. Don (Cunoniac.).
Stolzia Schlechter. Orchidaceae. 10 trop. Afr.
Stomadena Rafin. = Ipomoea L. (Convolvulac.).
Stomandra Standley. Rubiaceae. 1 C. Am.
Stomarrhena DC. = Astroloma R.Br. (Epacridac.).
Stomatanthes R. M. King & H. Rob. (~ Eupatorium L.). Compositae. 1 trop. Afr.
Stomatechium B. D. Jacks. (sphalm.) = Stomotechium Lehm. = Anchusa L. (Boraginac.).
Stomatium Schwantes. Aïzoaceae. 40 S. Afr.
Stomatocalyx Muell. Arg. = Pimelodendron Hassk. (Euphorbiac.).
Stomatochaeta (Blake) Maguire & Wurdack. Compositae. 4 Venez., Guiana
Stomatostemma N. E. Br. Periplocaceae. 1 SE. trop. Afr., S. Afr.
Stomatotechium Spach = Stomotechium Lehm. = Anchusa L. (Boraginac.).
Stomoisia Rafin. = Utricularia L. (Lentibulariac.).
Stomotechium Lehm. = Anchusa L. (Boraginac.).
Stonckenya Rafin. = Honkenya Ehrh. (Caryophyllac.).
Stonesia G. Taylor. Podostemaceae. 3 trop. W. Afr.
Stongylocaryum Burret (sphalm.) = Strongylocaryum Burret (Palm.).
Stooria Neck. = Lobelia L. (Campanulac.).
Stopinaca Rafin. = Polygonella Michx (Polygonac.).
Storckiella Seem. Leguminosae. 3 New Caled., Fiji.
Stormesia Kickx f. = Asplenium L. (Aspleniac.).
Stormia S. Moore = Cardiopetalum Schlechtd. (Annonac.).

STORTHOCALYX

Storthocalyx Radlk. Sapindaceae. 4 New Caled.
Strabonia DC. = Pulicaria Gaertn. (Compos.).
Stracheya Benth. Leguminosae. 1 Tibet.
Strailia Th. Dur. Lecythidaceae. 1 Amaz. Brazil. Known from a single fallen flower only!
Strakaea Presl = Apama Lam. (Aristolochiac.).
Stramonium Mill. = Datura L. (Solanac.).
Strangalis Dulac = Hirschfeldia Moench (Crucif.).
Strangea Meissn. Proteaceae. 3 W. Austr., 1 E. Austr.
Strangeveia Baker = Strangweja Bertol. = Hyacinthus L. (Liliac.).
Strangula Nor. = Ardisia Sw. (Myrsinac.).
Strangwaysia P. & K. (1) = Strangweja Bertol. = Hyacinthus L. (Liliac.).
Strangwaysia P. & K. (2) = Stranvaesia Lindl. (Rosac.).
Strangweia Baker = seq.
Strangweja Bertol. = Hyacinthus L. (Liliac.).
Strangweya Benth. & Hook. f. = praec.
Strania Nor. = Canarium L. (Burserac.).
Stranvaesia Lindl. Rosaceae. 10 Himal. to China, Formosa, Philipp. Is.
Strasburgeria Baill. Strasburgeriaceae. 1 New Caled.
*Strasburgeriaceae Van Tiegh. Dicots. 1/1 New Caled. Trees with large obov.-spath. alt. remotely dent. stip. ls., stips. paired, connate, intrapet. Fls. reg., ⚥, sol., axill., short-pedic. K 8–10, much imbr., spirally arranged, outermost smallest, persist.; C 5, imbr., thick, venose; A 10, 1-ser., with stout subul. fil. and large obl. dorsif. versat. intr. anth.; disk thick, annular, sinuous, 10-lobed; G̲ (5), 10-ribbed, narrowed into subul. style with small capit. stig., and with 2 superposed descending axile ov. per loc. Fr. large, glob., indehisc., corky-woody, with 1–2 seeds per loc., or fr. 1-seeded; seeds trigonous, with fleshy endosp. Only genus: *Strasburgeria*. An interesting isolated survival from the primitive reservoir of forms that may have given rise to the *Brexiac.*, *Theac.*, etc.
Strateuma Rafin. = Zeuxine Lindl. (Orchidac.).
Strateuma Salisb. = Orchis L. (Orchidac.).
Stratioites Gilib. (sphalm.) = Stratiotes L. (Hydrocharitac.).
Stratiot[ac]eae Link = Hydrocharitaceae–Stratioteae Reichb.
Stratiotes L. Hydrocharitaceae. 1 Eur., *S. aloïdes* L. (water-soldier). Short stem bearing roots and a number of aloe-like ls. with toothed edges. In the summer it floats up to the surface and bears the (dioec.) fls. It sinks in autumn. It gives off numerous axillary shoots with big buds at the ends, and these grow into young plants, which become free and sink to the bottom, where they remain over winter.
Straussia A. Gray = Psychotria L. (Rubiac.).
Straussiella Hausskn. Cruciferae. 1 Persia.
Stravadia Pers. = seq.
Stravadium Juss. = Barringtonia J. R. & G. Forst. (Barringtoniac.).
Strebanthus Rafin. = Streblanthus Rafin. = Eryngium L. (Umbellif.).
Streblacanthus Kuntze. Acanthaceae. 6 C. Am., 1 Bolivia.
Streblanthera Steud. = Trichodesma R.Br. (Boraginac.).
Streblanthus Rafin. = Eryngium L. (Umbellif.).

Streblidia Link = Schoenus L. (Cyperac.).
Streblina Rafin. = Nyssa L. (Nyssac.).
Streblocarpus Arn. = Maerua Forsk. (Capparidac.).
Streblochaeta Benth. & Hook. f. = seq.
Streblochaete Hochst. ex A. Rich. Gramineae. 1 trop. Afr., Java, Lombok, Philippine Is., always on mountains.
Streblorrhiza Endl. (~ Clianthus Soland.). Leguminosae. 1 Norfolk I. (Phillip I.). Extinct?.
Streblosa Korth. Rubiaceae. 25 W. Malaysia.
Streblosiopsis Valet. Rubiaceae. 1 Borneo.
Streblus Lour. Moraceae. 22 Madag. (?Afr.), SE. As., Indomal. Used for paper in Siam.
Streckera Sch. Bip. = Leontodon L. (Compos.).
Streleskia Hook. f. = Wahlenbergia Schrad. ex Roth (Campanulac.).
Strelitsia Thunb. = seq.
Strelitzia Dryand. Strelitziaceae. 5 S. Afr. Pollen matted together into a mass by means of a network of filaments.
***Strelitziaceae** (K. Schum.) Hutch. Monocots. 3/7 trop. S. Am., S. Afr., Madag. Perenn. rhiz. herbs or banana-like trees, with distich. long-pet. ls. Fls. large, ♂, in term. or lat. long-pedunc. cincinni, encl. in large cymbif. bract. P 3+3, the outer ± equal, of the inner the 2 lateral developed unilaterally, closely apposed, forming a conspic. sagittate struct., the median short, cymbiform. A 5(–6), with long rigid fil. and lin. anth. Ḡ (3), with ∞ axile anatr. ov., and slender rigid style with trifid stig. Fr. a woody loculic. ∞-seeded caps.; seeds arillate. Genera: *Ravenala, Phenakospermum, Strelitzia*.
Strempelia A. Rich. Rubiaceae. 5 trop. S. Am.
Strempeliopsis Benth. Apocynaceae. 2 Cuba, Jamaica.
Strepalon Rafin. = Streptalon Rafin. = Hypericum L. (Guttif.).
Strephium Schrad. ex Nees = Olyra L. (Gramin.).
Strephonema Hook. f. Combretaceae. 6 trop. W. Afr.
Strephonemataceae (Benth. & Hook. f.) Venkat. & Prak. Rao = Combretaceae–Strephonematoïdeae Benth. & Hook. f.
Strepsanthera Rafin. = Anthurium Schott (Arac.).
Strepsia Steud. = Tillandsia L. (Bromeliac.).
Strepsiloba Rafin. = seq.
Strepsilobus Rafin. = Entada Adans. (Legumin.).
Strepsimela Rafin. = Helixanthera Lour. (Loranthac.).
Strepsiphigla Krause (sphalm.) = Strepsiphyla Rafin. = Drimia Jacq. (Liliac.)
Strepsiphus Rafin. = Peristrophe Nees (Acanthac.).
Strepsiphyla Rafin. = Drimia Jacq. (Liliac.).
Streptachne R.Br. Gramineae. 1 Queensland.
Streptachne Kunth = Aristida L. (Gramin.).
Streptalon Rafin. = Hypericum L. (Guttif.).
Streptanthella Rydb. Cruciferae. 1 W. U.S.
Streptanthera Sweet. Iridaceae. 2 S. Afr.
Streptanthus Nutt. Cruciferae. 14 W. & S. U.S., Mex.
Streptia Rich. ex Hook. f. = Streptogyne Beauv. (Gramin.).
Streptilon Rafin. = Geum L. (Rosac.).

STREPTIMA

Streptima Rafin. = Frankenia L. (Frankeniac.).
Streptium Roxb. = Priva Adans. (Verbenac.).
Streptocalyx Beer. Bromeliaceae. 12 trop. Am.
Streptocarpus Lindl. Gesneriaceae. 132 trop. & S. Afr., Madag. In *S. polyanthus* Hook., etc., the embryo in the exalbum. seed has 2 cots. and a hypocotyl, but no plumule or radicle; the hypocotyl enters the soil, swells up at the end and develops absorbent hairs; presently, however, roots (adv.) form above the swelling, which dies off. One of the cots. continues to grow, while the other dies. Thus the young pl. consists of a large green cot. with few adv. roots. The cot. continues to grow, and reaches considerable size. Finally the infl. arises as a bud from the base of the petiole, and leafy shoots may also arise (Cf. the artificial repr. of *Sinningia*.)
Streptocaulon Wight & Arn. Periplocaceae. 5 Indomal.
Streptochaeta Schrad. Streptochaetaceae (~ Gramineae). 2 Brazil, Ecuador.
Streptochaetaceae (C. E. Hubb.) Nak. Monocots. 2 trop. S. Am. Perenn. forest 'grasses' with broad ovate-ellipt. l.-blades contracted into a short petiole-like base, with ∞ nerves and cross-veins. Fls. ♀, in 1-fld. spikelets arranged spirally and ± distantly in sol. term. rac., the spikelets falling entire; glumes 4–5, small, truncate, dentic., irreg. arranged; lemma much longer than gl., tapering into v. long helically twisted awn; palea shorter than lemma, deeply bifid; lodicules 3, large, coriaceous, exceeding palea. A 6, sometimes connate at base; G with sol. elong. style and 3 short stigs. Caryopsis elongate, free. Only genus: *Streptochaeta*.
Streptodesmia A. Gray = Adesmia DC. (Legumin.).
Streptoglossa Steetz ex F. Muell. = Oliganthemum F. Muell. (Compos.).
Streptogloxinia hort. = Sinningia Nees (Gesneriac.).
× **Streptogloxinia** Rodigas. Gesneriaceae. Gen. hybr. (Gloxinia × Streptocarpus).
Streptogyna Beauv. Gramineae. 1 Mex. to trop. S. Am.; 1 trop. Afr., S. India, Ceylon.
Streptogyne Poir. = praec.
Streptolirion Edgew. Commelinaceae. 1 E. Himal. to Indoch. & Korea. Climber.
Streptoloma Bunge. Cruciferae. 1 C. As. to Afghan.
Streptolophus Hughes. Gramineae. 1 Angola.
Streptomanes K. Schum. Periplocaceae. 1 New Guinea.
Streptopetalum Hochst. Turneraceae. 3–4 trop. & S. Afr.
Streptopus Rich. in Michx. Liliaceae. 10 N. temp. Euras. (exc. W. Eur.) & Am., S. to Himal. & S. U.S.
Streptorhamphus Regel (sphalm.) = Steptorhamphus Bunge (Compos.).
Streptosema Presl = Aspalathus L. (Legumin.).
Streptosiphon Mildbr. Acanthaceae. 1 trop. E. Afr.
Streptosolen Miers. Solanaceae. 1 trop. S. Am.
Streptostachys Desv. Gramineae. 1 N. trop. S. Am., Trinidad.
Streptostigma Regel = Cacabus Bernh. (Solanac.).
Streptostigma Thw. = Harpullia Roxb. (Sapindac.).
Streptothamnus F. Muell. Flacourtiaceae. 2 New S. Wales.
Streptotrachelus Greenm. Apocynaceae. 1 Mex.

Streptylis Rafin. = Murdannia Royle (Commelinac.).
Striangis Thou. (uninom.) = *Angraecum striatum* Thou. (Orchidac.).
Stricklandia Baker. Amaryllidaceae. 1 Ecuador.
Striga Lour. Scrophulariaceae. 40 trop. & S. Afr., As., Austr. Semiparasites
Strigilia Cav. = Styrax L. (Styracac.). [like *Rhinanthus.*
Strigina Engl. Scrophulariaceae. 1 trop. Afr.
Strigosella Boiss. Cruciferae. 23 W. Medit. to C. China.
Striolaria Ducke. Rubiaceae. 1 Amaz. Brazil.
Strobidia Miq. = Alpinia Roxb. (Zingiberac.).
Strobila G. Don = Arnebia Forsk. (Boraginac.).
Strobila Nor. = Nicolaia Horan. (Zingiberac.).
Strobilacanthus Griseb. Acanthaceae. 1 Panamá.
Strobilaceae Dulac = Cannabidaceae Endl.
Strobilaceae Reichb. = Coniferae Juss.
Strobilanthes Blume. Acanthaceae. 250 Madag., trop. As. Many occur gregariously in vast numbers, forming almost the sole undergrowth in forests. They fl. simultaneously and die down. Some, e.g. *S. anisophyllus* T. Anders., show marked anisophylly. The stigma is sensitive to contact (cf. *Mimulus*); when touched it moves downwards, and becomes pressed against the lower lip of the fl.
Strobilanthopsis Léveillé (sphalm.) = Strobilanthes Blume (Acanthac.).
Strobilanthopsis S. Moore. Acanthaceae. 5 trop. Afr.
Strobilanthos St-Lag. = seq.
Strobilanthus Reichb. = Strobilanthes Bl. (Acanthac.).
Strobilocarpos Benth. & Hook. f. = seq.
Strobilocarpus Klotzsch (~ Grubbia Bergius). Grubbiaceae. 1–2 S. Afr.
Strobilopanax R. Viguier. Araliaceae. 3 New Caled.
Strobilorhachis Klotzsch = Aphelandra R.Br. (Acanthac.).
Strobocalyx Sch. Bip. = Vernonia Schreb. (Compos.).
Strobon Rafin. = Cistus L. + Halimium (Dunal) Spach (Cistac.).
Strobopetalum N. E. Br. Asclepiadaceae. 1 NE. trop. Afr., Arabia.
Strobus (Endl.) Opiz = Pinus L. (Pinac.).
Stroemeria Roxb. (sphalm.) = seq.
Stroemia Vahl = Cadaba Forsk. (Capparidac.).
Stroganowia Kar. & Kir. Cruciferae. 13 C. As.
Strogylodon Dur. & Jacks. (sphalm.) = Strongylodon Vog. (Legumin.).
Stromadendrum Pav. ex Bur. = Broussonetia L'Hérit. (Morac.).
Stromanthe Sond. Marantaceae. 13 trop. Am. Fls. antidromous.
Stromatocactus Karw. ex Rümpler = Ariocarpus Scheidw. (Cactac.).
Stromatocarpus Rümpler (sphalm.) = praec.
Stromatopteridaceae (Nak.) Bierhorst. Gleicheniales. 1/1 New Caled., *Stromatopteris.* Shows some similarities to *Psilotaceae.* L.-bases branch below ground. (Bierhorst, *Phytomorph.* **18**: 232–68, 1968.)
Stromatopteris Mett. Stromatopteridaceae. 1 New Caled.
Strombocactus Britton & Rose. Cactaceae. 1 Mex.
Strombocarpa A. Gray = Prosopis L. (Legumin.).
Strombocarpus Benth. & Hook. f. = praec.
Strombodurus Willd. ex Steud. = Pentarrhaphis Kunth (Gramin.).

STROMBOSIA

Strombosia Blume. Olacaceae. 17 trop. Afr., India, Ceylon, Burma, W. Malaysia.

Strombosiaceae Van Tiegh. = Olacaceae–Anacoloseae Engl.

Strombosiopsis Engl. Olacaceae. 2 trop. Afr.

Strongylocalyx Blume = Syzygium Gaertn. (Myrtac.).

Strongylocaryum Burret. Palmae. 3 Solomon Is.

Strongylodon Vog. Leguminosae. 20 Madag., Masc., Philipp. Is., New Guinea, Pacif.

Strongyloma DC. = Nassauvia Comm. ex Juss. (Compos.).

Strongylomopsis Spegazz. = praec.

Strongylosperma Less. = Cotula L. (Compos.).

Stropha Nor. = Chloranthus Sw. (Chloranthac.).

Strophacanthus Lindau. Acanthaceae. 6 E. Himal., Burma, Java, E. Malaysia.

Strophades Boiss. = Erysimum L. (Crucif.).

Strophanthus DC. Apocynaceae. 60 trop. Afr., Madag., Indomal. Free parts of petals long, strap-shaped, often twisted; follicles divergent when ripe. The seeds of *S. hispidus* DC. (S. Afr.) furnish the drug strophanthin.

Strophioblachia Boerlage. Euphorbiaceae. 2 S. China (Hainan), Indoch., Philipp. Is., Celebes.

Strophiodiscus Choux. Sapindaceae. 1 Madag.

Strophiostoma Turcz. = Myosotis L. (Boraginac.).

Strophis Salisb. = Dioscorea L. (Dioscoreac.).

Strophium Dulac = Moehringia L. (Caryophyllac.).

Strophocactus Britton & Rose. Cactaceae. 1 Brazil.

Strophocaulos Small = Convolvulus L. (Convolvulac.).

Strophocereus Frič & Kreuz. = Strophocactus Britt. & Rose (Cactac.).

Stropholirion Torr. Alliaceae. 1 Calif.

Strophopappus DC. = Stilpnopappus Mart. (Compos.).

Strophostyles Ell. = Phaseolus L. (Legumin.).

Strophostyles E. Mey. = Vigna Savi (Legumin.).

Strotheria B.L. Turner. Compositae. 1 Mex.

Struchium P.Br. Compositae. 1 trop. Am., W.I., trop. Afr.

Struckeria Steud. = seq.

Strukeria Vell. = Vochysia Juss. (Vochysiac.).

Strumaria Jacq. Amaryllidaceae. 8 S. Afr.

Strumar[iac]eae Salisb. = Amaryllidaceae–Haemantheae Kunth.

Strumarium Rafin. = Xanthium L. (Compos.).

Strumpfia Jacq. Rubiaceae. 1 W.I.

Strusiola Rafin. = Struthiola L. (Thymelaeac.).

Struthanthus Mart. (= Steirotis Rafin.). Loranthaceae. 75 trop. Am.

Struthia Boehm. = Gnidia L. (Thymelaeac.).

*****Struthiola** L. Thymelaeaceae. 40 trop. & S. Afr.

Struthiolopsis Phillips. Thymelaeaceae. 1 S. Afr.

Struthiopteris Scop. (also Willd.) = Matteuccia Todaro (Aspidiac.).

Struthiopteris Weis = Blechnum L. (Blechnac.).

Struthopteris Bernh. = Osmunda L. (Osmundac.).

Struvea Reichb. = Torreya Arn. (Taxac.).

Strychnaceae Link. Dicots. 4/250 trop. & subtrop. Trees or shrubs, some-times spinous, often climbing by axill. tendrils. Ls. opp., simple, ent., most often parallel-3–5-nerved, exstip. Fls. reg., ♀, usu. in cymes, rarely sol. K (4–5) or 4–5, imbr.; C (4–5), valv.; A 4–5, intr.; G̲ (2), occasionally 1-loc., with short style and capit. or 2-lobed stig., and ∞–1 ov. per loc. Fr. a drupe or berry, sometimes large; seeds with endosp. Genera: *Strychnos, Scypho-strychnos, Gardneria, Neuburgia*. Near *Loganiac., Antoniac., Potaliac.*, etc.

Strychnodaphne Nees = Ocotea Aubl. (Laurac.).

Strychnopsis Baill. Menispermaceae. 1 Madag.

Strychnos L. Strychnaceae. 200 trop. Some, e.g. *S. nux-vomica* L. (India, Ceylon), are erect trees, others are climbing shrubs, with curious hook-tendrils. The hook is a modified axillary shoot; the l. in whose axil it arises usu. becomes a scale-l. If the hook catch upon a support it twines close round it and thickens and lignifies (cf. *Clematis*). Other spp. have axillary thorns. A few have a 1-loc. ovary with free-central placenta. Fr. a berry; the flesh is harmless, but the seeds are exceedingly poisonous, owing to the presence of strychnine in the seed-coats. From these seeds the alkaloid is chiefly obtained. *S. toxifera* Schomb. (S. Am.) yields the famous *wourali* or *curare* poison, with which the S. Am. Indians poison their arrows; it is obtained from the bark by scraping and maceration in water. The seeds of *S. potatorum* L. f. (clearing nut) are used to purify water for drinking. They are rubbed on the inside of the vessel, and cause precipitation.

Strychnus P. & K. = Strychnos L. (Strychnac.).

Stryphnodendron Mart. Leguminosae. 15 trop. Am.

Strzeleckya F. Muell. = Flindersia R.Br. (Flindersiac.).

Stuartia L'Hérit. = Stewartia L. (Theac.).

Stuartina Sond. Compositae. 2 S. & E. Austr.

Stubendorffia Schrenk. Cruciferae. 5 C. As., Afghan.

Stuckenia Börner = Potamogeton L. (Potamogetonac.).

Stuckertia Kuntze. Asclepiadaceae. 1 S. Am.

Stuckertiella Beauverd. Compositae. 2 Argentina.

Stuebelia Pax = Belencita Karst. (Capparidac.).

Stuhlmannia Taub. Leguminosae. 1 trop. E. Afr.

Stultitia Phillips. Asclepiadaceae. 2 S. Afr.

Stupa Aschers. = Stipa L. (Gramin.).

Sturmia Gaertn. f. = Antirhea Comm. ex Juss. (Rubiac.).

Sturmia Hoppe = Mibora Adans. (Gramin.).

Sturmia Reichb. = Liparis Rich. (Orchidac.).

Sturtia R.Br. = Gossypium L. (Malvac.).

Stutzeria F. Muell. = Pullea Schlechter (Cunoniac.).

Styasasia S. Moore = Asystasia Bl. (Acanthac.).

Stychophyllum Phil. (sphalm.) = Pycnophyllum Remy (Caryophyllac.).

Stygiaria Ehrh. (uninom.) = *Juncus stygius* L. (Juncac.).

Stygiopsis Gandog. = Juncus L. (Juncac.).

Stygnanthe Hanst. = Columnea L. (Gesneriac.).

Stylago Salisb. = Strumaria Jacq. (Liliac.).

Stylagrostis Mez = Deyeuxia Clar. (Gramin.).

Stylandra Nutt. = Podostigma Ell. (Asclepiadac.).

STYLANTHUS

Stylanthus Reichb. f. & Zoll. = Mallotus Lour. (Euphorbiac.).

Stylaptera Benth. & Hook. f. (sphalm.) = seq.

Stylapterus A. Juss. = Penaea L. (Penaeac.).

Stylarthropus Baill. = Whitfieldia Hook. (Acanthac.).

Stylbocarpa Decne & Planch. = Stilbocarpa Decne (Araliac.).

Styledium Andr. (sphalm.) = Stylidium Sw. (Stylidiac.).

Stylesia Nutt. = Bahia Lag. (Compos.).

Styleurodon Rafin. = Stylodon Rafin. (Verbenac.).

Stylexia Rafin. = Caylusea A. St-Hil. (Resedac.).

*Stylidiaceae R.Br. Dicots. 5/150 trop. As., Austr., N.Z., temp. S. Am. Small herbs or undershrubs, ± xero., without latex. Ls. simple, exstip., almost grass-like, often in rad. rosettes with fls. on a scape; successive rosettes may be separated by a slightly leafy piece of stem. Rosettes sometimes almost bulbous, with aerial roots. Fls. in racemes or cymes, ♀ or ♂ ♀, usu. zygo, K 5 or (5), often covered with stalked viscid glands, odd sep. post.; C (5), the ant. pet. (*labellum*) often different from the rest; A 2 (post.-lat.), rarely 3, united with style to form a gynostemium (cf. *Orchidac.*, *Asclepiadac.*), anthers extr.; G̅ (2), usu. 2-loc., but sometimes the post. loc. aborted; ov. ∞, anatr., variously attached to septum or central plac. Caps.; seeds minute, ∞–few; fleshy endosp. Genera: *Levenhookia*, *Phyllachne*, *Stylidium*, *Oreostylidium*, *Forstera*. Cf. Erickson, *Triggerplants* (Perth, W. Austr., 1958).

Stylidium Lour. = Pautsauvia Juss. = Alangium Lam. (Alangiac.).

*Stylidium Sw. ex Willd. Stylidiaceae. 136 SE. As., Austr., N.Z. (trigger-plants). Some have an irritable gynostemium. It bends over to one side, and may be released by a touch, when it springs over to the other. These periodic movements go on for some time.

Stylimnus Rafin. = Pluchea Cass. (Compos.).

Stylipus Rafin. = Stylypus Rafin. = Geum L. (Rosac.).

Stylis Poir. = Alangium Lam. (Alangiac.).

Stylisma Rafin. Convolvulaceae. 6 S. & E. U.S.

Stylista Rafin. = Cleome L. (Cleomac.).

Stylites E. Amstutz emend. Rauh. Isoëtaceae. 2 Peruvian Andes. Differs from *Isoëtes* L. in elongate stem and unbranched roots. (Rauh & Falk, *Sitzungsb. Heidelb. Akad. Wiss.* 1959: Abh. 1, 2.)

Stylobasiaceae J. G. Agardh. Dicots. 1/2 SW. Austr. Small shrubs, with alt. simple ent. lin. or obl. coriac. ls., stips. minute or o. Fls. reg., ♀ or polyg., sol., axill., forming leafy rac. K (5), turb.-campan., persist., lobes imbr.; C o; disk o; A 10, fil. filif., persist., anth. large, obl., basif., extr., exserted (♀ fl. with long filif. stds.); G 1, subglob., with filif. ± sigmoid gynobasic style with large term. pelt. granular-glandular stig., and 2 basal erect ov. (♂ fl. with small pistillode). Fr. a bony nut or dry drupe, ± surr. by the enlarged K; seed 1, with v. sparse endosp. Only genus: *Stylobasium*. Perhaps related to *Surianac.*, as already suggested by Agardh (1858); prob. no connection with *Chrysobalanac.* (Cf. Prance, *Bull. Jard. Bot. Brux.* **35**: 435–48, 1965.)

Stylobasium Desf. Stylobasiaceae. 2 SW. Austr.

Styloceras Juss. Stylocerataceae. 3 trop. Andes (Colombia to Bolivia). Fruit ed., with pleasant taste and odour.

Stylocerataceae Baill. 1/3 W. trop. S. Am. Glabrous trees, with simple alt.

ent. coriac. shining exstip. ls., with few usu. steeply ascending lat. nerves, incised above, prominent below, subtriplinerved at base. Fls. ♂ ♀, monoec. or dioec., in short spicate unisex. or when monoec. sometimes bisex. axill. infl., or the ♀ sol. ♂ fl. subtended by a small triang. bract; K o; C o; A 6–30, sess. or subsess. on lower part of bract, crowded, anth. thick, unequal, biloc., intr. ♀ fl. pedicellate, pedic. bearing several bracts; K (apical bracts?) 3–5, much imbr.; G̲ (2–3), loc. bilocell., with 1 pend. anatr. ov. in each locellus; styles 2–3, elong., free (rarely united below into a column), divaricate, with long decurrent ventr. grooved stigs., recurved at tip. Fr. drupaceous, ± fleshy, indehisc., crowned with persist. styles (usu. widely separated at base); seeds with copious fleshy endosp. Only genus: *Styloceras.* Related to *Buxac.*, but differing in the naked ♂ fls. with ∞ ± sess. anth. borne on a sol. bract, and in the loculi of the ovary completely divided by secondary longit. septa. The ♂ 'fls.' appear to be simple androphylls, as in *Didymeles*, to which *Styloceras* is prob. also related.

Stylochiton Lepr. Araceae. 21 warm Afr. The monoec. infl. remains below the ground, only the tip protruding and opening.

Stylocline Nutt. Compositae. 6 SW. U.S., N. Mex.; 1 Afghanistan.

Styloconus Baill. Haemodoraceae. 1 SW. Austr.

Stylocoryna Cav. = Tarenna Gaertn. (Rubiac.).

Stylocoryne Wight & Arn. = praec.

Stylodiscus Benn. = Bischofia Blume (Bischofiac.).

Stylodon Rafin. (∼ Verbena L.). Verbenaceae. 1 SE. U.S.

Styloglossum Breda = Calanthe R.Br. (Orchidac.).

Stylogyne A. DC. Myrsinaceae. 50 trop. Am., W.I.

Stylolepis Lehm. Compositae. 1 Austr.

Styloma O. F. Cook = Eupritchardia Kuntze = Pritchardia Seem. & H. Wendl. (Palm.).

Stylomecon Benth. (sphalm.) = Hylomecon Maxim. = Stylophorum Nutt. (Papaverac.).

Stylomecon G. Taylor (∼ Meconopsis Vig.). Papaveraceae. 1 Calif.

Styloncerus Labill. corr. Spreng. = Siloxerus Labill. = Angianthus Wendl. (Compos.).

Stylonema (DC.) Kuntze = Syrenia Andrz. ex DC. (Crucif.).

Stylopappus Nutt. = Troximon Nutt. (Compos.).

Stylophorum Nutt. Papaveraceae. 2 E. As., 1 Atl. N. Am.

Stylophyllum Britton & Rose (∼ Dudleya Britt. & Rose). Crassulaceae. 7 Calif., Lower Calif.

Stylopus Hook. = Stylypus Rafin. = Geum L. (Rosac.).

Stylosanthes Sw. Leguminosae. 50 trop. & subtrop. Am., Afr., trop. As., incl. 2 Java & Lesser Sunda Is.

Stylosiphonia T. S. Brandegee. Rubiaceae. 2 Mex., C. Am.

Stylotrichium Mattf. Compositae. 2 NE. Brazil.

Stylurus Rafin. = Ranunculus L. (Ranunculac.).

Stylurus Salisb. (∼ Grevillea R.Br.). Proteaceae. 4 Austr.

Stylvianthes Rafin. = Stylosanthes Sw. (Legumin.).

Stylypus Rafin. = Geum L. (Rosac.).

Stypa Garcke = Stipa L. (Gramin.).

SUMNERA

Sumnera Nieuwl. = Thalictrum L. (Ranunculac.).
Sunania Rafin. = Antenoron Rafin. (Polygonac.).
Sunaptea Griff. = Vatica L. (Dipterocarpac.).
Sunapteopsis Heim = Stemonoporus Thw. = Vateria L. (Dipterocarpac.).
Sunipia Buch.-Ham. ex Lindl. Orchidaceae. 15 India, SE. As., Formosa.
Superbangis Thou. (uninom.) = *Angraecum superbum* Thou. (Orchidac.).
Suprago Gaertn. = Vernonia Schreb. (Compos.).
Supushpa Suryanarayana. Acanthaceae 1 W. Penins. India.
Suregada Roxb. ex Rottl. (*Gelonium* Roxb. ex Willd.). Euphorbiaceae. 40
(many ill-defined) trop. & S. Afr., Madag., SE. As. to Formosa, Indomal. Ls.
pellucid-punctate, flexible when dry; infl. lf.-opp., gummy when young.
Surenus Kuntze = Cedrela P.Br. (Meliac.).
Suriana Domb. & Cav. ex D. Don = Ercilla A. Juss. (Phytolaccac.).
Suriana L. Surianaceae. 1 coasts Atl. trop. Am., trop. E. Afr., Madag., Masc.,
Is. of Indian Ocean, Ceylon to Malay Penins., Formosa, Philipp., E. Malaysia,
NE. Austr., Pacif. The distribution should be compared with that of *Pisonia
grandis* R.Br. (Nyctaginac.).
*****Surianaceae** Arn. Dicots. 1/1 trop. coasts. Shrubs, rarely ± arboresc., with
densely grey-puberulous branchlets, and crowded simple ent. alt. lin.-spath.
exstip. ls. Fls. reg., ♀, in few-fld. axill. cymes with large bracts and bracteoles.
K 5, connate at base, imbr., acum., persist.; C 5, shortly clawed, imbr.;
A 5 + 5, the inner (oppositipet.) shorter or ster. or abortive, fil. subul., pilose
below, anth. small, medifixed, intr.; disk inconspic.; G 5, free, pilose, each
carp. with filif. gynobasic style and scarcely capit. stig., and 2 collat. subbasal
anatr. ov. Fr. of 5–3 free 1-seeded drupac. carp., embryo hippocrepiform,
without endosp. Only genus: *Suriana*. Some relationship with *Simaroubac.*
and *Stylobasiac.* (Cf. Gutzwiller, Engl., *Bot. Jahrb.* 8: 1–49, 1961.)
Suringaria Pierre = Symplocos Jacq. (Symplocac.).
Surubea Jaume St-Hil. = Souroubea Aubl. = Ruyschia Jacq. (Marcgraviac.).
Surwala M. Roem. = Walsura Roxb. (Meliac.).
Susanna Phillips. Compositae. 3 S. Afr.
Susarium Phil. = Symphyostemon Miers (Iridac.).
Sussea Gaudich. = Pandanus L. f. (Pandanac.).
Sussodia Buch.-Ham. ex D. Don = Colebrookia Sm. (Labiat.).
Susum Blume = Hanguana Bl. (Hanguanac.).
Sutera Hort. ex Steud. = Lessertia DC. (Legumin.).
Sutera Roth (1807). Scrophulariaceae. 130 trop. & S. Afr., Canaries.
Sutera Roth (1821) = Jamesbrittenia Kuntze (Scrophulariac.).
Suteria DC. = Psychotria L. (Rubiac.).
*****Sutherlandia** R.Br. ex Ait. Leguminosae. 6 S. Afr.
Sutherlandia J. F. Gmel. = Heritiera Dryand. (Sterculiac.).
Sutrina Lindl. Orchidaceae. 1 Peru.
Suttonia Mez = seq.
Suttonia A. Rich. = Rapanea Aubl. (Myrsinac.).
Suzukia Kudo. Labiatae. 2 Ryukyu Is., Formosa.
Suzygium P.Br. = Calyptranthes Sw. (Myrtac.).
Svenhedinia Urb. = Talauma Juss. (Magnoliac.).
Svensonia Moldenke. Verbenaceae. 4 trop. E. Afr., Arabia, India, Ceylon.

Sventenia Font Quer. Compositae. 1 Canary Is.
Svida Small = Swida Opiz (Cornac.).
Svitramia Cham. Melastomataceae. 1 S. Brazil.
Svjda Opiz = Swida Opiz (Cornac.).
Swainsona Salisb. Leguminosae. 60 + Austr., N.Z.
Swainsonia Spreng. = praec.
Swallenia Soderstr. & Decker. Gramineae. 1 Calif.
Swammerdamia DC. = Helichrysum L. (Compos.).
Swanalloia Hort. ex Walp. = Juanulloa Ruiz & Pav. (Solanac.).
Swantia Alef. = Vicia L. (Legumin.).
Swartzia J. F. Gmel. = Solandra Sw. (Solanac.).
*****Swartzia** Schreb. Leguminosae. 100 trop. Am., Afr.
Swartzi[ac]eae Bartl. = Leguminosae–Caesalpinioïdeae–Swartziëae DC.
Swarzia Retz. = Hellenia Retz. = Costus L. Costac.).
Sweertia Koch = Swertia L. (Gentianac.)
Sweertia P. & K. = Swertia Boehm. = Tolpis L. (Compos.).
× **Sweetara** hort. (iv) = × Yapara hort. (Orchidac.).
Sweetia DC. = Galactia P.Br. (Legumin.).
*****Sweetia** Spreng. Leguminosae. 12 S. Am.
Sweetiopsis Chodat & Hassl. = Riedeliella Harms (Legumin.).
Swertia Boehm. = Tolpis Adans. (Compos.).
Swertia L. Gentianaceae. 100 N. Am., Euras., Afr., Madag. Ls. sometimes
alt. The corolla segments each bear 2 nectaries on the upper side, consisting of
little pits covered with hairs.
Swertopsis Makino = praec.
Swertya Steud. = Swertia Boehm. = Tolpis L. (Compos.).
Swida Opiz. Cornaceae. 36 N. temp., 3 Mex., 1 N. Andes.
Swietenia Jacq. Meliaceae. 7–8 trop. Am., W.I., incl. *S. mahagoni* Jacq., the
mahogany, a valuable timber tree.
Swinburnia Ewart = Neotysonia Dalla Torre & Harms (Compos.).
Swingera Dunal = Zwingera Hofer = Nolana L. (Nolanac.).
Swinglea Merr. Rutaceae. 1 Philipp. Is.
Swintonia Griff. Anacardiaceae. 15 SE. As., W. Malaysia. Pets. accrescent
and persistent, forming wings to fr.
Swjda Opiz = Swida Opiz (Cornac.).
Swynnertonia S. Moore. Asclepiadaceae. 1 Rhodesia.
Syagrus Mart. Palmae. 50 trop. Am.
Syalita Adans. = Dillenia L. (Dilleniac.).
Syama Jones = Pupalia Juss. (Amaranthac.).
Sycamorus Oliv. = Sycomorus Gasp. = Ficus L. (Morac.).
Sychinium Desv. = Dorstenia L. (Urticac.).
Sychnosepalum Eichl. = Sciadotaenia Miers (Menispermac.).
Sycios Medik. = Sicyos L. (Cucurbitac.).
Sycocarpus Britton = Guarea L. (Meliac.).
Sycodendron Rojas = Ficus L. (Morac.).
Sycodium Pomel = Anvillea DC. (Compos.).
Sycoïdeae Link = Moraceae–Ficeae + Artocarpeae + Dorsteniëae.
Sycomorphe Miq. = Ficus L. (Morac.).

SYCOMORUS

Sycomorus Gasp. = Ficus L. (Morac.).

× **Sycoparrotia** P. Endr. & Anliker. Hamamelidaceae. Gen. hybr. (Parrotia × Sycopsis).

Sycophila Welw. ex Van Tiegh. = Helixanthera Lour. (Loranthac.).

Sycopsis Oliv. Hamamelidaceae. 7 Assam to C. & S. China, Philippines, Celebes, New Guinea.

Syderitis All. = Sideritis L. (Labiat.).

× **Sydneya** Traub. Amaryllidaceae. Gen. hybr. (Habranthus × Zephyranthes).

Syena Schreb. = Mayaca Aubl. (Mayacac.).

Sykesia Arn. = Gaertnera Lam. (Rubiac.).

Sykoraea Opiz = Campanula L. (Campanulac.).

Sylitra E. Mey. = Ptycholobium Harms (Legumin.).

Syllepis Fourn. ex Benth. & Hook. f. = Imperata Cyr. (Gramin.).

Syllisium Endl. = seq.

Syllysium Meyen & Schau. = Syzygium Gaertn. (Myrtac.).

Sylvalismis Dalla Torre & Harms = seq.

Sylvalismus P. & K. = Silvalismis Thou. (uninom.) = Calanthe sylvatica (Thou.) Lindl. (Orchidac.).

Sylvia Lindl. = Silvia Benth. (Scrophulariac.).

Sylvorchis Schlechter = Silvorchis J. J. Sm. (Orchidac.).

Symbasiandra Willd. ex Steud. = Hilaria Kunth (Gramin.).

Symbegonia Warb. Begoniaceae. 12 New Guinea.

Symblomeria Nutt. = Albertinia Spreng. (Compos.).

Symbolanthus G. Don. Gentianaceae. 20–25 C. Am. to Bolivia.

Symbryon Griseb. = Lunania Hook. (Flacourtiac.).

Symea Baker = Solaria Phil. (Liliac.).

Symethus Rafin. = Convolvulus L. (Convolvulac.).

Symingtonia van Steenis = Exbucklandia R. W. Brown (Hamamelidac.).

Symmeria Benth. Polygonaceae. 1 N. trop. S. Am., trop. W. Afr.

Symmeria Hook. f. = Synmeria Nimmo = Habenaria Willd. (Orchidac.).

Symmetria Blume = Carallia Roxb. (Rhizophorac.).

Sympachne Steud. = Symphachne Beauv. = Eriocaulon L. (Eriocaulac.).

Sympagis (Nees) Bremek. Acanthaceae. 5 E. Himal., Assam.

Sympegma Bunge. Chenopodiaceae. 1 C. As.

Sympetalae A.Br. (Metachlamydeae Engl.) = Gamopetalae Brongn.

Sympetalandra Stapf. Leguminosae. 1 Borneo.

Sympetaleia A. Gray. Loasaceae. 2 SW. U.S., NW. Mex.

Symphachne Beauv. ex Desv. = Eriocaulon L. (Eriocaulac.).

Symphiandra Steud. = Symphyandra DC. (Campanulac.).

Symphionema R.Br. = Symphyonema R.Br. corr. Spreng. (Proteac.).

Symphipappus Klatt = Cadiscus E. Mey. ex DC. (Compos.).

Symphitum Neck. = Symphytum L. (Boraginac.).

Symphocoronis Dur. (sphalm.) = Scyphocoronis A. Gray (Compos.).

× **Symphodontioda** hort. (v) = × Odontioda hort. (Orchidac.).

× **Symphodontoglossum** hort. (v) = × Odontioda hort. (Orchidac.).

× **Symphodontonia** hort. (v) = × Vuylstekeara hort. (Orchidac.).

Symphonia L. f. Guttiferae. 2 Colombia, 1 trop. Am. & Afr., 18 Madag.

Symphoniaceae Presl = Guttiferae Juss.

Symphoranthus Mitch. = Polypremum L. (Loganiac.).

Symphoranthera Dur. & Jacks. = Synphoranthera Boj. = Dialypetalum Benth.

Symphorema Roxb. Symphoremataceae. 4 Indomal. [(Campanulac.).

Symphoremataceae Van Tiegh. Dicots. 3/34 trop. Am., Afr., As. Large scandent shrubs, with opp. simple ent. or toothed exstip. ls., sometimes stellate-tomentose. Fls. ♀, zygo., sess., in 3–7(–9)-fld. capit. cymes, often aggr. into large term. pan., each cyme with an invol. of 6 equal coloured 'bracts' (2 bracts and 4 bracteoles), which are ± membr., accresc. and persist. K (4–5–8), ± accresc. or inflated in fr., teeth open or scarcely valv.; C (5–16), lobes imbr., subequal or bilab.; A 4–16, epipet., incl. or exserted; G (2), imperf. 4-loc., with 4 orthotr. ov. pend. from the free centr. plac., style filif., shortly bifid. Fr. a small dry 1–4-seeded drupe, ± incl. in the K; endosp. o. Genera: *Symphorema, Sphenodesme, Congea.* Related to *Verbenac.,* but differing *inter alia* in the free-central placentation.

Symphoria Pers. = seq.

Symphoricarpa Neck. = seq.

Symphoricarpos Duham. Caprifoliaceae. 1 C. China, 17 N. Am. The pend. fls. are fert. chiefly by bees and wasps.

Symphoricarpus Kunth = praec.

Symphostemon Hiern. Labiatae. 2 Angola.

Symphyachna P. & K. = Symphachne Beauv. ex Desv. = Eriocaulon L. (Eriocaulac.).

Symphyandra A. DC. Campanulaceae. 10 E. Medit. to Cauc. & C. As. The pend. caps. opens at the base (cf. *Campanula*).

Symphydolon Salisb. = Gladiolus L. (Iridac.).

Symphyecarpon Pohl (nomen). 8 Brazil. Quid?

× **Symphyglossonia** hort. (v) = × Miltonioda hort. (Orchidac.).

*****Symphyglossum** Schlechter. Orchidaceae. 2 trop. S. Am.

Symphyllanthus Vahl = Dichapetalum Thou. (Dichapetalac.).

Symphyllarion Gagnep. Rubiaceae. 1 Indoch.

Symphyllia Baill. = Epiprinus Griff. (Euphorbiac.).

Symphyllium P. & K. = Synphyllium Griff. = Curanga Juss. (Scrophulariac.).

Symphyllocarpus Maxim. Compositae. 1 E. Siberia, Manchuria.

Symphyllochlamys Willis (sphalm.) = Symphyochlamys Gürke (Malvac.).

Symphyllophyton Gilg. Gentianaceae. 2 Brazil.

Symphyloma Steud. = Symphyoloma C. A. Mey. (Umbellif.).

Symphyobasis Krause. Goodeniaceae. 2 W. Austr.

Symphyochaeta (DC.) Skottsb. Compositae. 1 Chile, Juan Fernandez.

Symphyochlamys Gürke. Malvaceae. 1 NE. trop. Afr.

Symphyodolon Baker = Symphydolon Salisb. = Gladiolus L. (Iridac.).

Symphyogyne Burret (1940; non *Symphyogyna* Nees & Montagne 1836— Hepaticae) = Liberbaileya Furtado + Maxburretia Furtado (Palm.).

Symphyoloma C. A. Mey. Umbelliferae. 1 Caucasus, at high altitudes.

Symphyomera Hook. f. = Cotula L. (Compos.).

Symphyomyrtus Schau. = Eucalyptus L'Hérit. (Myrtac.).

Symphyonema R.Br. corr. Spreng. Proteaceae. 2 New S. Wales.

Symphyopappus P. & K. = Symphipappus Klatt = Cadiscus E. Mey. ex DC. (Compos.).

SYMPHYOPAPPUS

Symphyopappus Turcz. Compositae. 12 campos of S. Brazil.
Symphyopetalon J. Drumm. ex Harv. = Nematolepis Turcz. (Rutac.).
Symphyosepalum Hand.-Mazz. Orchidaceae. 1 China.
Symphyostemon Klotzsch = Cleome L. (Cleomac.).
Symphyostemon Miers ex Klatt = Phaiophleps Rafin. (Iridac.).
Symphyotrichum Nees = Aster L. (Compos.).
Symphysia Presl. Ericaceae. 1 W.I.
Symphysicarpus Hassk. = Heterostemma Wight & Arn. (Asclepiadac.).
Symphysodaphne A. Rich. = Licaria Aubl. (Laurac.).
Symphytonema Schlechter. Periplocaceae. 3 Madag.
Symphytosiphon Harms = Trichilia P.Br. (Meliac.).
Symphytum L. Boraginaceae. 25 Eur., Medit. to Cauc. (comfrey). *S. tuberosum* L. has tubers like those of potato. The pend. fls. are bee-visited; the entrance to the honey is narrowed by the C scales. Mech. of fl. as in *Borago*. Some cult. for fodder, e.g. *S. asperrimum* Donn.
Sympieza Lichtenst. ex Roem. & Schult. Ericaceae. 8 S. Afr.
Symplectochilus Lindau. Acanthaceae. 2 trop. Afr., Madag.
Sympleura Miers = Barberina Vell. = Symplocos Jacq. (Symplocac.).
***Symplocaceae** Desf. 2/500 trop. & subtrop. (exc. Afr.). Shrubs and trees with alt. simple exstip. often leathery ls., and racemed bracteolate ⚥ reg. fls. K (5), imbr.; C (5) or (5+5), imbr.; A 5 or 5+5 or 5+5+5 or more, epipet. or free of C; anthers round or ovoid; G (2-5), inf. or semi-inf., with 2-4 anatr. pend. ov. on an axile plac. in each loc. Style simple, stigma capitate or lobed. Fr. drupaceous, one seed in each loc. of the stone. Embryo straight or curved, in endosp. Genera: *Symplocos, Cordyloblaste*. Closely related to *Theaceae*, and scarcely differing except in the rac. infl. and inferior ov. (the latter occurs also in the Theaceous [Ternstroemiaceous *s.str.*] genera *Anneslea* and *Symplococarpon*). There is no close relationship with *Styracaceae*.
***Symplocarpus** Salisb. Araceae. 1, *S. foetidus* (L.) Salisb., the skunk-cabbage, NE. As., Japan, Atl. N. Am.
Symplococarpon Airy Shaw. Theaceae. 9 Mex. to Colombia. G̅!
Symplocos Jacq. Symplocaceae. 350 trop. & subtrop. As., Austr., Polynesia, Am. Many aluminium-accumulators.
Sympodium C. Koch = Carum L. (Umbellif.).
Symptera P. & K. = Synptera Llanos = Trichoglottis Bl. (Orchidac.).
Synactila Rafin. = Passiflora L. (Passiflorac.).
Synadena Rafin. = Phalaenopsis Blume (Orchidac.).
Synadenium Boiss. Euphorbiaceae. 15 warm Afr., Madag., Masc.
Synaecia Pritz. (sphalm.) = Synoecia Miq. = Ficus L. (Morac.).
Synaedrys Lindl. = Lithocarpus Bl. (Fagac.).
Synallodia Rafin. = Swertia L. (Gentianac.).
Synammia Presl. Polypodiaceae. 1 Chile.
Synandra Nutt. Labiatae. 1 E. U.S.
Synandra Schrad. = Stenandrium Nees (Acanthac.).
Synandrina Standley & L. O. Williams. Flacourtiaceae. 1 C. Am.
***Synandrodaphne** Gilg. Thymelaeaceae. 1 W. Equat. Afr.
Synandrodaphne Meissn. = Ocotea Aubl. (Laurac.).
Synandrogyne Buchet. Araceae. 1 Madag.

Synandropus A. C. Smith. Menispermaceae. 1 NE. Brazil.
Synandrospadix Engl. Araceae. 1 N. Argentina.
Synanthae Engl. An order of Monocots. consisting of the fam. *Cyclanthaceae*
Synantheraceae Cass. = Compositae Giseke. [only.
Synantherias Schott = Amorphophallus Bl. (Arac.).
Synaphe Dulac = Catapodium Link (Gramin.).
Synaphea R.Br. Proteaceae. 8 W. Austr.
Synaphlebium J. Sm. = Lindsaea Dryand. (Lindsaeac.).
Synapisma Steud. = Synaspisma Endl. = Codiaeum Juss. (Euphorbiac.).
Synapsis Griseb. Bignoniaceae. 1 Cuba.
Synaptantha Hook. f. Rubiaceae. 1 subtrop. Austr.
Synaptanthe Willis (sphalm.) = praec.
Synaptanthera K. Schum. = praec.
Synaptea Griff. corr. Kurz = Sunaptea Griff. = Vatica L. (Dipterocarpac.).
Synapteopsis P. & K. = Sunapteopsis Heim = Stemonoporus Thw. = Vateria L. (Dipterocarpac.).
Synaptera Willis (sphalm.) = Synaptea Griff. = Vatica L. (Dipterocarpac.).
Synaptolepis Oliv. Thymelaeaceae. 8 trop. Afr., Madag.
Synaptophyllum N. E. Brown. Aïzoaceae. 2 SW. Afr.
Synardisia (Mez) Lundell. Myrsinaceae. 1 C. Am.
Synarrhena Fisch. & Mey. = Mimusops L. (Sapotac.).
Synarrhena F. Muell. = Saurauia Willd. (Actinidiac.).
Synarthron Benth. & Hook. f. = seq.
Synarthrum Cass. = Senecio L. (Compos.).
Synaspisma Endl. = Codiaeum Juss. (Euphorbiac.).
Synassa Lindl. Orchidaceae. 1 Peru.
Syncalathium Lipschitz. Compositae. 1 Tibet.
Syncarpha DC. = Helipterum DC. (Compos.).
Syncarpia Tenore. Myrtaceae. 5 Queensland.
Syncephalantha Bartl. Compositae. 3 Mex., C. Am.
Syncephalanthus Benth. & Hook. f. = praec.
Syncephalum DC. Compositae. 7 Madag.
Synchaeta Kirp. (~ Gnaphalium L.). Compositae. 1 N. temp.
Synchodendron Boj. ex DC. (?sphalm. pro Sychnodendron) = Brachylaena
Synchoriste Baill. Acanthaceae. 1 Madag. [R.Br. (Compos.).
Synclisia Benth. Menispermaceae. 3 trop. Afr.
Syncodium Rafin. = Myogalum Link = Ornithogalum L. (Liliac.).
Syncodon Fourr. = Campanula L. (Campanulac.).
Syncolostemon E. Mey. Labiatae. 8 S. Afr.
Syncretocarpus Blake. Compositae. 2 Peru. Xerophytes.
Syndechites Dur. & Jacks. = Sindechites Oliv. (Apocynac.).
Syndesmanthus Klotzsch = Scyphogyne Brongn. (Ericac.).
Syndesmis Wall. = Gluta L. (Anacardiac.).
Syndesmon Hoffmgg. = Anemonella Spach (Ranunculac.).
Syndiaspermaceae Dulac = Orobanchaceae Vent.
Syndiclis Hook. f. = Potameia Thou. (Laurac.).
Syndyophyllum Lauterb. & K. Schum. Euphorbiaceae. 1 Sumatra, Borneo, New Guinea.

SYNECHANTHACEAE

Synechanthaceae O. F. Cook = Palmae–Chamaedoreëae Benth. & Hook. f.
Synechanthus H. Wendl. Palmae. 6 Mex. to trop. S. Am.
Synedrella Gaertn. (Melanthera Rohr). Compositae. 50 warm Am., trop.
Afr., Madag., India.
Synedrellopsis Hieron. & Kuntze. Compositae. 1 Argentina.
Syneilesis Maxim. (~ Cacalia L.). Compositae. 5 E. As.
Synelcosciadium Boiss. Umbelliferae. 1 Syria.
Synema Dulac = Mercurialis L. (Euphorbiac.).
Synepilaena Baill. Gesneriaceae. 1 Colombia.
Synexemia Rafin. = Andrachne L. (Euphorbiac.).
Syngeneticae Horan. = Compositae Giseke. [Madag.
Syngonanthus Ruhl. Eriocaulaceae. 195 C. & trop. S. Am., W.I., trop. Afr.,
Syngonium Schott. Araceae. 20 C. & trop. S. Am., W.I. Climbers with
cymes of monoec. spadices. Synandrous.
Syngramma J. Sm. Hemionitidaceae. 20 Malaysia to Fiji.
Syngrammatopsis Alston. Hemionitidaceae. 3 S. Am. (*Mutisia* 7: 7, 1952).
Synima Radlk. Sapindaceae. 1 Austr.
Synisoön Baill. Rubiaceae. 1 Guiana.
Synmeria Nimmo = Habenaria Willd. (Orchidac.).
Synnema Benth. Acanthaceae. 20 palaeotrop.
Synnetia Sweet (sphalm.) = seq.
Synnotia Sweet. Iridaceae. 5 S. Afr.
Synnottia Baker (sphalm.) = praec.
Synochlamys Fée = Pellaea Link (Sinopteridac.).
Synodon Rafin. = Conostegia D. Don (Melastomatac.).
Synoecia Miq. = Ficus L. (Morac.).
Synoliga Rafin. = Xyris L. (Xyridac.).
Synoplectris Rafin. = Sarcoglottis Presl (Orchidac.).
Synoptera Rafin. = Miconia Ruiz & Pav. (Melastomatac.).
Synosma Rafin. = Cacalia L. (Compos.).
Synostemon F. Muell. (~ Phyllanthus L.). Euphorbiaceae. 1 Madag. to SE.
As. & Indomal., 12 Austr.
Synotoma (G. Don) R. Schulz. Campanulaceae. 1 SE. Alps.
Synoum A. Juss. Meliaceae. 2 Austr.
Synphoranthera Boj. = Dialypetalum Benth. (Campanulac.).
Synphyllium Griff. = Curanga Juss. (Scrophulariac.).
Synptera Llanos = Trichoglottis Blume (Orchidac.).
Synsepalum (A. DC.) Daniell. Sapotaceae. 10 trop. Afr.
Synsiphon Regel = Colchicum L. (Liliac.).
Synstemon Botsch. Cruciferae. 1 NW. China.
Synstemon Taub. = Synostemon F. Muell. (Euphorbiac.).
Synstima Rafin. = Ilex L. (Aquifoliac.).
Syntherisma Walt. = Digitaria Fabr. (Gramin.).
Synthlipsis A. Gray. Cruciferae. 3 S. U.S., Mex.
Synthyris Benth. Scrophulariaceae. 15 mts. W. N. Am.
Syntriandrum Engl. Menispermaceae. 4 W. trop. Afr.
Syntrichopappus A. Gray. Compositae. 2 SW. U.S.
Syntrinema H. Pfeiff. Cyperaceae. 1 Brazil.

Syntrophe G. Ehrenb. ex Muell. Arg. = Caylusea A. St-Hil. (Resedac.).
Synurus Iljin. Compositae. 6 temp. E. As.
Synzistachium Rafin. = Heliotropium L. (Boraginac.).
Synzyganthera Ruiz & Pav. = Lacistema Sw. (Lacistematac.).
Syoctonum Bernh. = Chenopodium L. (§ Orthosporum C. A. Mey.) (Chenopodiac.).
Syorhynchium Hoffmgg. (sphalm.) = Sisyrinchium L. (Iridac.).
Sypharissa Salisb. = Urginea Steinh. (Liliac.).
Syphocampylos Hort. Belg. ex Hook. = Siphocampylus Pohl (Campanulac.).
Syphomeris Steud. = Siphomeris Boj. = Grewia Juss. (Tiliac.).
Syreitschikovia Pavlov. Compositae. 2 C. As
Syrenia Andrz. ex Besser. Cruciferae. 10 E. Eur. to W. Siberia.
Syreniopsis H. P. Fuchs = Acachmena H. P. Fuchs (Crucif.).
Syrenopsis Jaub. & Spach. Cruciferae. 2 As. Min., Kurdist.
Syringa L. Oleaceae. 30 SE. Eur. to E. As. *S. vulgaris* L. is the common lilac. Serial accessory buds in axils. Well-marked false dichotomy; the term. bud usu. fails to develop each spring and the two nearest lat. buds continue the growth. Winter buds scaly; the scales secrete a gummy substance as the bud elongates. Fls. in thyrses, each branch with a term fl. Seeds flat, slightly
Syringa Mill. = Philadelphus L. (Philadelphac.).   [winged.
Syringaceae Horan. = Oleaceae Hoffmgg. & Link.
Syringantha Standley. Rubiaceae. 1 Mex.
Syringidium Lindau. Acanthaceae. 1 Colombia.
Syringodea D. Don = Erica L. (Ericac.).
*Syringodea Hook. f. Iridaceae. 8 S. Afr.
Syringodium Kütz. Cymodoceaceae. 1 coasts of Caribbean; 1 Indian & W. Pacif. oceans.
Syringosma Mart. ex Reichb. = Forsteronia G. F. W. Mey. (Apocynac.).
Syrium Steud. = Sirium Schreb. = Santalum L. (Santalac.).
Syrmatium Vog. = Hosackia Dougl. (Legumin.).
Syrrheonema Miers. Menispermaceae. 3 trop. W. Afr.
Syrrhonema Miers = praec.
Sysepalum P. & K. = Synsepalum (A. DC.) Baill. (Sapotac.).
Sysimbrium Pall. = Sisymbrium L. (Crucif.).
Sysiphon P. & K. = Synsiphon Regel = Colchicum L. (Liliac.).
Sysirinchium Rafin. = Sisyrinchium L. (Iridac.).
Syspone Griseb. = Genistella Ort. (Legumin.).
Systeloglossum Schlechter. Orchidaceae. 1 Costa Rica.
Systemon P. & K. = Synostemon F. Muell. (Euphorbiac.).
Systemon Regel = Galipea Aubl. (Rutac.).
Systemonodaphne Mez. Lauraceae. 2 trop. S. Am.
Systigma P. & K. = Synstima Rafin. = Ilex L. (Aquifoliac.).
Systrepha Burch. = Ceropegia L. (Asclepiadac.).
Systrephia Benth. & Hook. f. = praec.
Syziganthus Steud. = Gahnia J. R. & G. Forst. (Cyperac.).
Syzistachyum P. & K. = Synzistachium Rafin. = Heliotropium L. (Boraginac.).
Syzyganthera P. & K. = Synzyganthera Ruiz & Pav. = Lacistema Sw. (Lacistematac.).

SYZYGIOPSIS

Syzygiopsis Ducke. Sapotaceae. 2 trop. S. Am.

*Syzygium Gaertn. Myrtaceae. 500 palaeotrop. For distinctions from *Eugenia* L., see R. Schmid, A resolution of the *Eugenia–Syzygium* controversy, *Amer. J. Bot.* **59**: 423–36, 1972. The dried fl. buds of *S. aromaticum* (L.) Merr. form the spice 'cloves'.

Szechenyia Kanitz = Gagea Salisb. (Liliac.).

Szeglewia C. Muell. (sphalm.) = Sczegleëwia Turcz. = Symphorema Roxb. (Symphorematac.).

Szovitsia Fisch. & Mey. Umbelliferae. 1 Transcauc., Armenia, NW. Persia.

T

Tabacina Reichb. = seq.

Tabacum [Gilib.] Opiz = Nicotiana L. (Solanac.).

Tabacus Moench = praec.

Tabascina Baill. Acanthaceae. 1 Mex.

Tabebuia Gomes ex DC. Bignoniaceae. 100 Mex. to N. Argent., W.I.

Taberna Miers (~ Tabernaemontana L.). Apocynaceae. 1 trop. Am.

Tabernaemontana L. Apocynaceae. 100 trop.

Tabernanthe Baill. Apocynaceae. 7 trop. Afr.

Tabernaria Rafin. = Tabernaemontana L. (Apocynac.).

Tabraca Nor. = ?Polyalthia Bl. (Annonac.).

Tacamahaca Mill. = Populus L. (Salicac.).

Tacazzea Decne. Periplocaceae. 4 trop. & S. Afr.

*Tacca J. R. & G. Forst. Taccaceae. 30 trop., esp. SE. As. East Indian arrowroot is made from the rhiz. of *T. leontopodioïdes* (L.) Kuntze and other spp.

*Taccaceae Dum. Monocots. 2/31 trop. Perenn. herbs from creeping or tub. rhiz. Ls. rad., broad elliptic or much lobed (resembling *Arac.*), often longstalked. Fls. ☿, reg., in involucrate cymose umbels on scapes, bracts broad or elong.-lin. P (3 + 3), tube shortly and broadly cupular, segs. broad, spreading, imbr., persist.; A 3 + 3, epitep., fil. short, ± petaloid, anth. intr.; Ḡ (3), 1-loc., with short style and 3 reflexed ± petaloid stigs., and ∞ anatr. ov. on pariet. plac. Fr. a ∞-seeded berry, rarely caps. (*Schizocapsa*); endosp. copious, ± cartilag. Genera: *Tacca, Schizocapsa*.

Taccarum Brongn. Araceae. 6 trop. S. Am.

Tachia Aubl. Gentianaceae. 4 Guiana, Brazil.

Tachia Pers. = Tachigalia Aubl. (Legumin.).

Tachiadenus Griseb. Gentianaceae. 7 Madag.

Tachibota Aubl. = Hirtella L. (Chrysobalanac.).

Tachigalea Griseb. (sphalm.) = Taligalea Aubl. = Amasonia L. f. (Verbenac.).

Tachigalia Aubl. Leguminosae. 25 C. & trop. S. Am.

Tachites Soland. ex Gaertn. = Melicytus J. R. & G. Forst. (Violac.).

Tachytes Steud. = praec.

Tacinga Britton & Rose. Cactaceae. 2 NE. Brazil.

Tacoanthus Baill. Acanthaceae. 1 Bolivia.

Tacsonia Juss. = Passiflora L. (Passiflorac.).

Taeckholmia Boulos. Compositae. 7 Canary Is.

Taenais Salisb. = Crinum L. (Amaryllidac.).
Taenia P. & K. = Tainia Bl. (Orchidac.).
Taeniandra Bremek. Acanthaceae. 1 Penins. India.
Taenianthera Burret. Palmae. 10 trop. S. Am.
Taeniatherum Nevski. Gramineae. 2 Medit. to NW. India.
Taenidia (Torr. & Gray) Drude. Umbelliferae. 1 E. U.S.
Taenidium Targ.-Tozz. = Posidonia Koen. (Posidoniac.).
Taeniocarpum Desv. = Pachyrhizus Rich. (Legumin.).
Taeniochlaena Hook. f. = Roureopsis Planch. (Connarac.).
Taeniola Salisb. = Ornithogalum L. (Liliac.).
Taenionema P. & K. = Tainionema Schltr. (Asclepiadac.).
Taeniopetalum Vis. = Peucedanum L. (Umbellif.).
Taeniophyllum Blume. Orchidaceae. 120 trop. Afr. to Japan and Tahiti,
Taenioplehrum Dur. & Jacks. (sphalm.) = seq.   [Australasia.
Taeniopleurum Coulter & Rose. Umbelliferae. 1 NW. U.S.
Taeniopsis J. Sm. = Vittaria Sm. (Vittariac.).
Taeniopteris Hook. = Vittaria Sm. (Vittariac.).
Taeniorrhiza Summerhayes. Orchidaceae. 1 W. Equat. Afr. (Gabon).
Taeniosapium Muell. Arg. = Excoecaria L. (Euphorbiac.).
Taeniostema Spach = Crocanthemum Spach (Cistac.).
Taenitis Willd. ex Spreng. Hemionitidaceae. 15 Ceylon, S. India, SE. As.,
    Malaysia, Queensl., Pacif. (Holttum, *Blumea*, **16**: 87–95, 1968).
Taenosapium Benth. & Hook. f. = Taeniosapium Muell. Arg. = Excoecaria L.
    (Euphorbiac.).
Taetsia Medik. = Cordyline Comm. ex Juss. (Liliac.).
Tafalla D. Don = Loricaria Wedd. (Compos.).
Tafalla Ruiz & Pav. = Hedyosmum Sw. (Chloranthac.).
Tafallaea Kuntze = praec.
Tagera Rafin. = Chamaecrista Moench = Cassia L. (Legumin.).
Tagetes L. Compositae. 50 warm Am.
Taguaria Rafin. = Gaiadendron G. Don (Loranthac.).
Tahitia Burret (= ?Berrya Roxb.).Tiliaceae. 1 Society Is. (Tahiti).
Tainia Blume. Orchidaceae. 25 China, Formosa, Indomal.
Tainionema Schlechter. Asclepiadaceae. 1 W.I. (S. Domingo).
Tainiopsis Hayata = Tainia Blume (Orchidac.).
Tainiopsis Schlechter. Orchidaceae. 1 NE. India to China and Indoch.
Taitonia Yamamoto. Labiatae. 1 Formosa.
Taiwania Hayata. Taxodiaceae. 1 Manchuria, 2 SW. China, Formosa.
Taiwaniaceae Hay. = Taxodiaceae Neger.
Taiwanites Hayata = Taiwania Hayata (Taxodiac.).
Takasagoya Kimura = Hypericum L. (Guttif.).
Takeikadzuchia Kitag. & Kitam. = Olgaea Iljin (Compos.).
Tala Blanco = Limnophila R.Br. (Scrophulariac.).
Talanelis Rafin. = Campanula L. (Campanulac.).
Talangninia Chapel. ex DC. = Randia L. (Rubiac.).
Talasium Spreng. = Thalasium Spreng. = Panicum L. (Gramin.).
Talassia Korov. Umbelliferae. 2 C. As.
Talauma Juss. Magnoliaceae. 50 E. Himal., SE. As., Malaysia, Mex. to trop.

TALBOTIA

S. Am., W.I. Like *Magnolia*, but fr. indehisc. or the upper portions of the cpls. breaking off from the persistent lower portions, from which the seeds dangle on silky threads as in *Magnolia*.

Talbotia Balf. = Vellozia Vand. (Velloziac.).

Talbotia S. Moore = Afrofittonia Lindau (Acanthac.).

Talbotiella E. G. Baker. Leguminosae. 3 trop. W. Afr.

Talechium Hill (sphalm.) = Trachelium Hill = Campanula L. (Campanulac.).

Talguenea Miers. Rhamnaceae. 1 Chile.

Tali Adans. = Connarus L. (Connarac.).

Taliera Mart. = Corypha L. (Palm.).

Taligalea Aubl. = Amasonia L. f. (Verbenac.).

Talinaria T. S. Brandegee = Talinum Adans. (Portulacac.).

Talinella Baill. Portulacaceae(?). 2 Madag.

Talinium Rafin. = Talinum Adans. (Portulacac.).

Talinopsis A. Gray. Portulacaceae. 1 S. U.S., Mex.

Talinum Adans. Portulacaceae. 50 warm Am. (esp. Mex.), Afr., As.

Talipulia Rafin. = Aneilema R.Br. (Commelinac.).

Talisia Aubl. Sapindaceae. 50 C. & trop. S. Am., Trinidad.

Talisiopsis Radlk. = Zanha Hiern (Sapindac.).

Talmella Dur. (sphalm.) = Talinella Baill. (? Portulacac.).

Talpa Rafin. = Catalpa Scop. (Bignoniac.).

Talpinaria Karst. = Pleurothallis R.Br. (Orchidac.).

Tamaceae S. F. Gray = Dioscoreaceae R.Br.

Tamala Rafin. = Persea Mill. (Laurac.).

Tamara Roxb. ex Steud. = Nelumbo Adans. (Nelumbonac.).

*Tamaricaceae Link. Dicots. 4/120 temp. & subtrop. Desert, steppe, and shore pls. Shrubs or herbs with alt. simple ent. exstip. ls., often heath-like. Fls. sol. or in racemose infls., ebracteolate, ⚥, reg., hypog. K (4–5); C 4–5, imbr.; A 4–5, 8–10 or ∞, on a disk; anth. extr. or intr.; G̲ (4–5) or (2), 1-loc. Styles usu. free. Ovules ∞ or few, on basal-parietal plac., ascending, anatr. Caps. Seeds hairy. Embryo straight; endosp. or not. Genera: *Reaumuria*, *Hololachna*, *Tamarix*, *Myricaria*. A 'peripheral Centrosperm' fam., related to *Frankeniac.*, *Fouquieriac.*, etc.

Tamarindus L. Leguminosae. 1 trop. Afr.(?), *T. indica* L., the tamarind, largely cultivated in the trop. for its ed. fruit (the part eaten is the pulp round the seeds; it is also officinal). The 2 ant. pets. are reduced to bristles, and the sta. (3 fert. + 4 minute ster.) united below into an anticous band (v. unusual in *Caesalpinioïdeae*). The wood is useful.

Tamariscaceae ('-ineae') A. St-Hil. = Tamaricaceae S. F. Gray.

Tamariscus Mill. = seq.

Tamarix L. Tamaricaceae. 54 W. Eur., Medit., to India & N. China. Halophytes. *T. mannifera* Ehrenb. (Egypt to Afghanistan) produces, owing to the punctures of the scale-insects *Trabutina mannipara* Ehrenb. & *Najacoccus serpentinus* Green (Bodenheimer, *Bibl. Archaeol.* **10**: 944, 1947), the manna of the Bedouins, a white substance which falls from the twigs.

Tamatavia Hook. f. = Chapelieria A. Rich. (Rubiac.).

Tamaulipa R. M. King & H. Rob. (~ Eupatorium L.). Compositae. 1 Texas, NE. Mex.

Tamayoa Badillo. Compositae. 1 Venez.
Tambourissa Sonner. Monimiaceae. 25 Madag., Mascarenes.
Tammsia Karst. Rubiaceae. 1 Colombia.
Tamnus Mill. = Tamus L. (Dioscoreac.).
Tamonea Aubl. (1) = Miconia Ruiz & Pav. (Melastomatac.).
Tamonea Aubl. (2) = Ghinia Schreb. (Verbenac.).
Tamonopsis Griseb. = Lantana L. (Verbenac.).
Tampoa Aubl. = Salacia L. (Celastrac.).
Tamus L. Dioscoreaceae. 5 Canaries, Madeira, Eur., Medit. Climbing plants, hibernating by tubers formed by a lat. outgrowth of the first two internodes of the stem.
Tanacetopsis (Tsvel.) Kovalevsk. (~ Cancrinia Kar. & Kir.). Compositae. 9 C. As.
Tanacetum L. (~ Chrysanthemum L.). Compositae. 50–60 N. temp. T. vulgare L. (tansy) cult. as a popular remedy.
Tanaecium Sw. Bignoniaceae. 7 C. & trop. S. Am., W.I.
Tanaesium Rafin. = praec.
× Tanakara hort. (1947) (vii) = × Opsisanda hort. (Orchidac.).
× Tanakara (sphalm. '-aria') hort. (1952) (vii) = × Phalaërianda hort.
× Tanakaria hort. (sphalm.) = praec.   [(Orchidac.).
Tanakea Franch. & Sav. Saxifragaceae. 2 China, Japan.
Tanaosolen N. E. Brown. Iridaceae. 2 S. Afr.
Tanarius Kuntze = Macaranga Thou. (Euphorbiac.).
Tanaxion Rafin. = Pluchea Cass. (Compos.).
Tandonia Baill. = Tannodia Baill. (Euphorbiac.).
Tandonia Moq. = Anredera Juss. (Basellac.).
Tangaraca Adans. = Hamelia Jacq. (Rubiac.).
Tanghekolli Adans. = Crinum L. (Liliac.).
Tanghinia Thou. (~ Cerbera L.). Apocynaceae. 1 Madag.
Tangtsinia S. C. Chen. Orchidaceae. 1 SW. China.
Tanibouca Aubl. = Terminalia L. (Combretac.).
Tankervillia Link = Phaius Lour. (Orchidac.).
Tannodia Baill. Euphorbiaceae. 3 trop. Afr., Comoro Is.
Tanroujou Juss. = Hymenaea L. (Legumin.).
Tantalus Nor. ex Thou. = Sarcolaena Thou. (Sarcolaenac.).
Tanulepis Balf. f. Periplocaceae. 5 Madag., Rodrigues I.
Taonabo Aubl. = Ternstroemia Mutis ex L. f. (Theac.).
Tapagomea Kuntze = Tapogomea Aubl. = Cephaëlis Sw. (Rubiac.).
Tapanava Adans. = Pothos L. (Arac.).
Tapanawa Hassk. = praec.
Tapanhuacanga Vell. ex Vand. Quid? (Rubiac.? Thymelaeac.?) 1 Brazil.
Tapeinaegle Herb. = Tapeinanthus Herb. (Amaryllidac.).
Tapeinanthus Boiss. ex Benth. = Thuspeinanta Dur. (Labiat.).
Tapeinanthus Herb. Amaryllidaceae. 1 W. Medit.
Tapeinia Comm. ex Juss. Iridaceae. 2 S. Chile, Patag.
Tapeinidium (Presl) C. Chr. Lindsaeaceae. 14 trop. As., Polynes.
*Tapeinocheilos Miq. Costaceae. 20 Moluccas, New Guinea, Bismarck
Tapeinochilos auctt. = praec.   [Arch., Queensland.

TAPEINOCHILUS

Tapeinochilus Benth. & Hook. f. = praec.

Tapeinoglossum Schlechter. Orchidaceae. 2 New Guinea.

Tapeinophallus Baill. = Amorphophallus Blume (Arac.).

Tapeinosperma Hook. f. Myrsinaceae. 40 New Guinea, Queensland, New Hebrid., New Caled., Fiji.

Tapeinostelma Schlechter = Brachystelma R.Br. (Asclepiadac.).

Tapeinostemon Benth. Gentianaceae. 1 N. Brazil, Guiana.

Tapeinotes DC. = Sinningia Nees (Gesneriac.).

Taphrogiton Montand. = Scirpus L. (Cyperac.).

Taphrospermum C. A. Mey. Cruciferae. 2 C. As.

Tapia Mill. = Crateva L. (Capparidac.).

Tapina Mart. = Sinningia Nees (Gesneriac.).

Tapinaegle P. & K. = Tapeinaegle Herb. = Tapeinanthus Herb. (Amaryllidac.).

Tapinanthus (Blume) Reichb. Loranthaceae. 250 trop. Afr.

Tapinanthus P. & K. (1) = Tapeinanthus Herb. (Amaryllidac.).

Tapinanthus P. & K. (2) = Tapeinanthus Boiss. ex Benth. = Thuspeinanta Dur. (Labiat.).

Tapinia P. & K. = Tapeinia Comm. ex Juss. (Iridac.).

Tapinia Steud. (sphalm.) = Tapirira Aubl. (Anacardiac.).

Tapinocarpus Dalz. = Theriophonum Blume (Arac.).

Tapinochilus P. & K. = Tapeinocheilos Miq. (Costac.).

Tapinopentas Bremek. Rubiaceae. 3 trop. Afr.

Tapinophallus P. & K. = Tapeinophallus Baill. = Amorphophallus Bl. (Arac.).

Tapinosperma P. & K. = Tapeinosperma Hook. f. (Myrsinac.).

Tapinostemma Van Tiegh. Loranthaceae. 4 trop. Afr.

Tapiphyllum Robyns. Rubiaceae. 18 trop. Afr.

Tapiria Juss. = seq.

Tapirira Aubl. Anacardiaceae. 15 Mex. to S. Am.

Tapirocarpus Sagot. Burseraceae. 1 Guiana.

Tapiscia Oliv. Staphyleaceae(?). 1 China.

Tapogamea Rafin. = seq.

Tapogomea Aubl. = Cephaëlis Sw. (Rubiac.).

Tapoïdes Airy Shaw. Euphorbiaceae. 1 Borneo.

Tapomana Adans. = Connarus L. (Connarac.).

Tapura Aubl. Dichapetalaceae. 20 trop. S. Am., W.I.; 4 trop. Afr.

Tara Molina = Caesalpinia L. (Legumin.).

Tarachia Presl = Asplenium L. (Aspleniac.).

Taraktogenos Hassk. = Hydnocarpus Gaertn. (Flacourtiac.).

Taralea Aubl. Leguminosae. 8 trop. S. Am.

Taramea Rafin. = Faramea Aubl. (Rubiac.).

Tarasa Phil. Malvaceae. 25 Andes.

Taravalia Greene (~Ptelea L.). Rutaceae. 1 NW. Mex. (L. Calif.).

Taraxaconastrum Guett. = Hyoseris L. (Compos.).

Taraxaconoïdes Guett. = Leontodon L. (Compos.).

*Taraxacum Weber. Compositae. 60 mostly N. temp.; 2 temp. S. Am. *T. officinale* Weber (dandelion), almost cosmop. The thick primary root is perenn. and crowned by a very short sympodial stem; each year a new bud is formed on the leafy axis, to come into active growth in the following year.

The roots as they grow to maturity contract and thus drag the stem downwards so that it never rises much above the soil. If the root be cut through, a callus forms over the wound, and from this adv. shoots develop. The fl. mech., etc., are of the usu. type of the fam., and show the final autogamy very clearly. Hundreds of apomictic forms have been described.

Taraxacum Zinn = Leontodon L. (Compos.).

Taraxia Nutt. ex Torr. & Gray = Camissonia Link (Onagrac.).

Tarchonanthus L. Compositae. 2 Mex.; 4 S. Afr. The wood of *T. camphoratus* L. (S. Afr.) is used for musical instruments.

Tardavel Adans. = Borreria G. F. W. Mey. (Rubiac.).

Tardiella Gagnep. = Casearia Jacq. (Flacourtiac.).

Tarenaya Rafin. = Cleome L. (Cleomac.).

Tarenna Gaertn. (= Chomelia L.). Rubiaceae. 370 trop. Afr., Madag., Sey-

Tarigidia Stent. Gramineae. 1 S. Afr. [chelles, trop. As., Austr.

Tariri Aubl. = Picramnia Sw. (Simaroubac.).

Tarpheta Rafin. = Stachytarpheta Vahl (Verbenac.).

Tarphochlamys Bremek. Acanthaceae. 1 Assam.

Tarrietia Blume = Heritiera Dryand. (Sterculiac.).

Tarsina Nor. = Lepeostegeres Bl. (Loranthac.).

Tartagalia Capurro = Eriotheca Schott & Endl. (Bombacac.).

Tartonia Rafin. = Thymelaea Mill. (Thymelaeac.).

Taschneria Presl = Crepidomanes Presl (Hymenophyllac.).

Tashiroa Willis (sphalm.) = seq.

Tashiroea Matsumura. Melastomataceae. 1 E. China, 2 Ryukyu Is.

Tasmannia R.Br. = Drimys J. R. & G. Forst. (Magnoliac.).

Tasoba Rafin. = Polygonum L. (Polygonac.).

Tassadia Decne. Asclepiadaceae. 25 S. Am.

Tassia Rich. = Tachigalia Juss. (Legumin.).

Tatea F. Muell. = Pygmaeopremna Merr. = Premna L. (Verbenac.).

Tatea Seem. = Bikkia Reinw. (Rubiac.).

Tateanthus Gleason. Melastomataceae. 1 Venez.

Tatina Rafin. = Ilex L. (Aquifoliac.).

Tattia Scop. = Homalium Jacq. (Flacourtiac.).

Taubertia K. Schum. ex Taub. Menispermaceae. 1 Brazil.

Taumastos Rafin. = Libertia Spreng. (Iridac.).

Tauroceras Britton & Rose = Acacia Mill. (Legumin.).

Taurophthalmum Duchassaing ex Griseb. = Dioclea Kunth (Legumin.).

Taurostalix Reichb. f. = Bulbophyllum Thou. (Orchidac.).

Taurrettia Raeusch. = Tourrettia Juss. (Bignoniac.).

Tauscheria Fisch. ex DC. Cruciferae. 2 C. As.

Tauschia Preissler = Symphysia C. B. Presl (Ericac.).

*Tauschia Schlechtd. Umbelliferae. 20 W. U.S. to C. Am.

Tavalla Pers. = Tafalla Ruiz & Pav. = Hedyosmum Sw. (Chloranthac.).

Tavaresia Welw. = Decabelone Decne (Asclepiadac.).

Tavernaria Reichb. = seq.

Taverniera DC. Leguminosae. 7 NE. Afr. to NW. India.

Taveunia Burret. Palmae. 1 Fiji.

Tavomyta Vitm. = Tovomita Aubl. (Guttif.).

TAXACEAE

Taxaceae S. F. Gray. Gymnosp. (Taxales). 5/20 N. hemisph., S. to Celebes and Mex.; 1 in New Caled. Much branched trees or shrubs with acic., lin., or lin.-lanc., often asymm. ls. ♂ fls. sol. or in small spikes (*Austrotaxus*) in the l. axils; sta. with 2–8 sporangia. ♀ fls. on small axillary shoots, surr. at the base by pairs of scale-ls.; ovule 1, term., with 1 integ.; seed ± covered by an aril. Genera: *Taxus, Pseudotaxus, Torreya, Austrotaxus, Amentotaxus.*

Taxales Sahni. Gymnospermae. An order containing the single fam. *Taxaceae.*

Taxanthema Neck. = Statice L. (Plumbaginac.).

Taxillus Van Tiegh. Loranthaceae. 60 S. Afr., Madag., Masc., S. China, Indoch., W. Malaysia.

Taxodiaceae Warming. Gymnosp. (Conif.). 10/16 E. As., Tasm., N. Am. Medium-sized to giant trees, with squamif., acic., or cultrate ls.; sympodial structure in *Taxodium*; long and short shoots usu. not distinguished, but in *Sciadopitys* (*q.v.*) whorls of broadly aciculif. cladodes are borne in axils of scale-ls. on long shoots. ♂ fls. sol., term. or axill., or capit., or in panic. infl. (*Taxodium*); fertile scales with 2–9 pollen-sacs; pollen without bladders. ♀ cones sol., term., bract-scales & ovulif. scales ∞, spirally arranged, partly or completely fused; ov. 2–9, erect or anatr. Fr. cones ± woody, glob., dehisc., the scales dentic. or cren.; seeds often narrowly winged. Genera: *Sciadopitys, Sequoia, Sequoiadendron, Metasequoia, Taxodium, Glyptostrobus, Cryptomeria, Athrotaxis, Taiwania, Cunninghamia.*

Taxodium Rich. Taxodiaceae. 3 SE. U.S., Mex., *T. distichum* (L.) Rich., *T. ascendens* Brongn., *T. mucronatum* Ten. (swamp-cypresses). In the first, esp. in swampy ground, curious 'knees' are formed, hollow cylindrical or spherical outgrowths projecting upwards from the roots, believed to be aerating organs (cf. *Sonneratia*).

Taxotrophis Blume = Streblus Lour. (Morac.).

Taxus L. Taxaceae. 10 N. temp., S. to Himal., Philipp., Celebes, and Mex.; incl. *T. baccata* L., the common yew. No short shoots, but the ls. of the spreading branches arrange themselves ± closely in two rows with their upper surfaces nearly in one plane, a dorsiventral structure to the shoot. Fls. dioec., sol. in the axils of the ls. of the preceding year. The ♂ has a few scale-ls. below and about 8 or 10 sta., each of which is shield-shaped with a number of pollen-sacs on the axial side of the shield arranged round its stalk like the sporangia in *Equisetum*. The ♀ has a rather complex structure. The primary axis bears scale-ls. only. In the axil of one of the uppermost of these arises a shoot, continuing the line of the first axis and bearing 3 pairs of scales and a term. ovule. This is orthotr. with one integument, and develops into a seed surrounded by a cup-shaped, red, fleshy aril.

The wood of the yew is valuable; in the middle ages it was the chief material used in making bows. The ls. are very poisonous, but the aril is harmless. Birds swallow it, and thus distrib. the seeds.

Tayloriophyton Nayar. Melastomataceae. 2 Malay Penins., Borneo.

Tayotum Blanco = Geniostoma J. R. & G. Forst. (Loganiac.).

Tchihatchewia Boiss. Cruciferae. 1 Armenia.

Teclea Delile. Rutaceae. 30 trop. Afr., Comoro Is., Madag.

Tecleopsis Hoyle & Leakey. Rutaceae. 1 trop. E. Afr.

Tecmarsis DC. = Vernonia Schreb. (Compos.).

Tecoma Juss. Bignoniaceae. 16 Florida, W.I., Mex. to Argent.

Tecomanthe Baill. (~ Campsis Lour.). Bignoniaceae. 17 Moluccas to Queensl. & N.Z.

Tecomaria Spach. Bignoniaceae. 2 trop. E. & S. Afr.

Tecomella Seem. Bignoniaceae. 1 SW. As., Arabia.

Tecophilaea Bertol. ex Colla. Tecophilaeaceae. 2 Chile.

*Tecophilaeaceae Leybold. Monocots. 6/22 Pacif. N. & S. Am., C. & S. Afr. Perenn. herbs with corms or tubers. Ls. alt., mostly rad., of various forms. Fls. reg. or slightly zygo. (esp. androec.), ♀, in bract. rac. or pan. P 3+3, sometimes shortly connate, imbr.; A 3+3, all perfect or 2–3 staminodial, epitep., anth. obl., intr., opening by term. pore, rarely by longit. slit, connective often produced at both ends, base sometimes enlarged or calcarate; G (3), ± semi-inf., with filif. style and small ± 3-lobed stig., and 2–∞ 2-ser. axile anatr. ov. per loc. Fr. a loculic. caps., seeds 1–∞, with fleshy endosp. Genera: *Conanthera, Odontostomum, Cyanastrum, Cyanella, Zephyra, Tecophilaea*. Somewhat intermediate between *Liliac.* and *Iridac.*

Tecophilea Herb. = Tecophilaea Bertol. ex Colla (Tecophilaeac.).

Tectaria Cav. Aspidiaceae. 200 pantrop. (*Aspidium* of C. Chr., *Ind. Fil.*, 1905.) Segregated genera with acrostichoid fertile ls. are *Stenosemia, Tectaridium, Quercifilix, Hemigramma*.

Tectaridium Copel. Aspidiaceae. 2 Philipp. Is.

Tecticornia Hook. f. Chenopodiaceae. 1 New Guinea, NW. Austr. to Queensl.

*Tectona L. f. Verbenaceae. 3 Indomal. *T. grandis* L. f. is the teak, cult. in India, Java, etc., for its timber, which is very hard and durable; enormous quantities are used for shipbuilding, etc. There are two areas of teak, in Peninsular India and in Burma; it grows in deciduous forest, but not gregariously. The wood sinks in water unless thoroughly dried; this is effected in India by the process of 'girdling', which consists in removing a ring of bark and sap-wood from the tree near the base. It soon dies, and is left standing for two years.

Tectonia Spreng. = praec.

Tecunumania Standley & Steyerm. Cucurbitaceae. 1 C. Am. (Guatem.).

Teedea P. & K. = seq.

Teedia Rudolphi. Scrophulariaceae. 4 S. Afr.

Teesdalea Aschers. = seq.

Teesdalia R.Br. Cruciferae. 2 Eur., Medit. Raceme at first corymbose, elongating as flowering proceeds.

Teesdaliopsis (Willk.) Rothm. (~ Iberis L.). Cruciferae. 1 Spain.

Teganium Schmid. = Nolana L. ex L. f. (Nolanac.).

Teganocharis Hochst. (sphalm.) = Tenagocharis Hochst. (Limnocharitac.).

Tegneria Lilja = Calandrinia Kunth (Portulacac.).

Tegularia Reinw. = Didymochlaena Desv. (Aspidiac.).

Teichmeyeria Scop. = Gustavia L. (Lecythidac.).

Teichostemma R.Br. = Vernonia Schreb. (Compos.).

Teijsmannia auctt. = Teyssmannia Reichb. f. & Zoll. = Johannesteijsmannia H. E. Moore (Palm.).

Teijsmannia P. & K. = Teysmannia Miq. = Pottsia Hook. & Arn. (Apocynac.).

TEIJSMANNIODENDRON

Teijsmanniodendron Koorders. Verbenaceae. 14 S. Indoch., Malaysia (exc. Java & Lesser Sunda Is.).
Teinosolen Hook. f. Rubiaceae. 4 Andes.
Teinostachyum Munro. Gramineae. 6 India, Ceylon, Burma.
Tekel Adans. = Libertia Spreng. (Iridac.).
Tekelia Kuntze = praec.
Tekelia Scop. = Argania Roem. & Schult. (Sapotac.).
Tektona L. f. = Tectona (? sphalm.) L. f. (Verbenac.).
Telanthera R.Br. = Alternanthera Forsk. (Amaranthac.).
Telectadium Baill. Periplocaceae. 3 Indoch.
Teleiandra Nees & Meyen = Ocotea Aubl. (Laurac.).
Teleianthera Endl. = Telanthera R.Br. = Alternanthera Forsk. (Amaranthac.).
Telekia Baumg. Compositae. 2 C. Eur. to Cauc.
Telelophus Dulac = Seseli L. (Umbellif.).
Telemachia Urb. Celastraceae. 1 Trinidad.
Teleozoma R.Br. = Ceratopteris Brongn. (Parkeriac.).
Telephiaceae Link = Aïzoaceae–Telephiëae Bartl.
Telephiastrum Fabr. = Anacampseros L. (Portulacac.).
Telephioïdes Ort. = Andrachne L. (Euphorbiac.).
Telephium Hill = Sedum L. (Crassulac.).
Telephium L. Caryophyllaceae. 5 Medit., 1 Madag.
Telesia Rafin. = Zexmenia La Llave & Lex. (Compos.).
Telesilla Klotzsch (nomen). Asclepiadaceae. 1 Guiana. Quid?
Telesmia Rafin. = Salix L. (Sect. Amerina Dum.) (Salicac.).
Telesonix Rafin. = Boykinia Nutt. (Saxifragac.).
Telestria Rafin. = Bauhinia L. (Legumin.).
Telfairia Hook. Cucurbitaceae. 2 trop. Afr. *T. pedata* (Sm. ex Sims) Hook. is cult. for its seeds, which are ed. and also yield oil.
Telfairia Newm. ex Hook. = Buettneria Loefl. (Buettneriac.).
Telina Dur. & Jacks. (sphalm.) = Teline Medik. = Chamaespartium Adans. (Legumin.).
Telina E. Mey. = Lotononis Eckl. & Zeyh. (Legumin.).
Telinaria Presl = seq.
Teline Medik. = Chamaespartium Adans. (Legumin.).
Teliostachya Nees. Acanthaceae. 10 trop. S. Am.
Telipodus Rafin. = Philodendron Schott (Arac.).
Telipogon Mutis ex Kunth. Orchidaceae. 80 Costa Rica, W. trop. S. Am.
Telis Kuntze = Trigonella L. (Legumin.).
Telitoxicum Moldenke. Menispermaceae. 6 trop. S. Am.
Tellima R.Br. Saxifragaceae. 1 Alaska to Calif.
Telmatophace Schleid. = Lemna L. (Lemnac.).
Telmatophila Ehrh. (uninom.) = Scheuchzeria palustris L. (Scheuchzeriac.).
Telmatophila Mart. ex Baker. Compositae. 1 Brazil.
Telmatosphace Ball (sphalm.) = Telmatophace Schleid. = Lemna L. (Lemnac.).
Telminostelma Fourn. Asclepiadaceae. 1 Brazil.
Telmissa Fenzl = Sedum L. (Crassulac.).
Telogyne Baill. = Trigonostemon Blume (Euphorbiac.).
Telopaea [Soland. ex] Parkinson = Aleurites J. R. & G. Forst. (Euphorbiac.).

*Telopea R.Br. Proteaceae. 4 E. Austr., Tasmania.

Telopea Soland. ex Baill. = Aleurites J. R. & G. Forst. (Euphorbiac.).

Telophyllum Van Tiegh. = Myzodendron Soland. ex DC. (Myzodendrac.).

Telopogon Mutis ex Spreng. = Telipogon Mutis ex Kunth (Orchidac.).

Telosma Coville. Asclepiadaceae. 10 Old World tropics.

Telotia Pierre = Pycnarrhena Miers (Menispermac.).

Teloxis Reichb. = seq.

Teloxys Moq. Chenopodiaceae. 1 C. As. to Manchuria.

Telukrama Rafin. = Thelycrania (Dum.) Fourr. = Swida Opiz (Cornac.).

Tema Adans. = Setaria Beauv. (Gramin.).

Temenia O. F. Cook = Maximiliana Mart. (Palm.).

Temminckia De Vriese = Scaevola L. (Goodeniac.).

Temnadenia Miers. Apocynaceae. 4 trop. S. Am.

Temnemis Rafin. = Carex L. (Cyperac.).

Temnocalyx Robyns. Rubiaceae. 4 trop. E. Afr.

Temnocydia Mart. ex DC. = Bignonia L. sens. latiss. (incl. Memora, Cydista, etc.) (Bignoniac.).

Temnolepis Baker = Epallage DC. (Compos.).

Temnopteryx Hook. f. Rubiaceae. 1 trop. W. Afr.

Templetonia R.Br. Leguminosae. 10 Austr.

Temu Berg. Myrtaceae. 2 Chile.

Temus Molina = ? Drimys J. R. & G. Forst. (Winterac.).

Tenageia Ehrh. (uninom.) = Juncus tenageia L. (Juncac.).

Tenageia (Reichb.) Reichb. = Juncus L. (Juncac.).

Tenagocharis Hochst. Limnocharitaceae. 1 trop. Afr., India, SE. As., Java, Madura, N. & NE. Austr.

Tenaris E. Mey. Asclepiadaceae. 2-3 trop. & S. Afr.

Tendana Reichb. f. = Micromeria Benth. (Labiat.).

Tengia Chun. Gesneriaceae. 1 SW. China.

Tenicroa Rafin. = Urginea Steinh. (Liliac.).

Tenorea Colla = Trixis P.Br. (Compos.).

Tenorea Gasp. = Ficus L. (Morac.).

Tenorea C. Koch = Tenoria Spreng. = Bupleurum L. (Umbellif.).

Tenorea Rafin. = Zanthoxylum L. (Rutac.).

Tenoria Dehnh. & Giord. = Hygrophila R.Br. (Acanthac.).

Tenoria Spreng. = Bupleurum L. (Umbellif.).

× Tenranara hort. (vii) = × Fujiwaraära hort. (Orchidac.).

× Teohara hort. Orchidaceae. Gen. hybr. (iii) (Arachnis × Renanthera × Vanda × Vandopsis).

Teonongia Stapf = Streblus Lour. (Morac.).

Tepesia Gaertn. f. = Hamelia Jacq. (Rubiac.).

Tephea Delile = Olinia Thunb. (Oliniac.).

Tephis Adans. = Polygonum L. (Polygonac.).

Tephranthus Neck. = Phyllanthus L. (Euphorbiac.).

Tephras E. Mey. ex Harv. & Sond. = Galenia L. (Aïzoac.).

Tephrocactus Lem. = Opuntia Mill. (Cactac.).

Tephroseris Reichb. = Senecio L. (Compos.).

*Tephrosia Pers. Leguminosae. 300 trop. & subtrop.

TEPHROTHAMNUS

Tephrothamnus Sch. Bip. = Vernonia Schreb. (Compos.).
Tephrothamnus Sweet = Argyrolobium Eckl. & Zeyh. (Legumin.).
Tepion Adans. = Verbesina L. (Compos.).
Tepso Rafin. = Bupleurum L. (Umbellif.).
Tepualia Griseb. Myrtaceae. 1 Chile. Hard wood.
Tepuia Camp. Ericaceae. 5 Venez.
Teramnus P.Br. Leguminosae. 15 trop.
Terana La Llave. Compositae (inc. sed.). 1 Mex.
Terania Berland. = Leucophyllum Humb. & Bonpl. (Scrophulariac.).
Teratophyllum Mett. Lomariopsidaceae. 9 Malaysia. Young plants on bases
of tree-trunks produce distinctive bathyphylls; higher-climbing part of stem
bears larger simply pinnate acrophylls.
Terauchia Nakai. Liliaceae. 1 Korea.
Terebint[h]aceae Juss. = Anacardiac. + Pistaciac. + Connarac. + Burserac. +
Rutac. + Simaroubac., etc.
Terebinthina Kuntze = Limnophila R.Br. (Scrophulariac.).
Terebinthus Mill. = Pistacia L. (Pistaciac.) + Bursera L. (Burserac.).
Terebraria Sessé ex Kuntze. Rubiaceae. 2 W.I.
Tereianthes Rafin. = Reseda L. (Resedac.).
Tereianthus Fourr. = praec.
Tereietra Rafin. = Calonyction Choisy = Ipomoea L. (Convolvulac.).
Tereiphas Rafin. = Scabiosa L. (Dipsacac.).
Teremis Rafin. = Lycium L. (Solanac.).
Terepis Rafin. = Salvia L. (Labiat.).
Terera Domb. ex Naud. = Miconia Ruiz & Pav. (Melastomatac.).
*****Terminalia** L. Combretaceae. 250 trop. The fr. of many are winged (see
fam.). Those of *T. chebula* Retz. and others (myrobalans) are used in dyeing
and tanning, and also in medicine. The seed of *T. catappa* L. is ed. (country
almond). *T. glabra* Wight & Arn. has aerating roots. The bark is burnt for
lime. Good timber; tan from bark.
Terminaliaceae Jaume St-Hil. = Combretaceae R.Br.
Terminaliopsis Danguy. Combretaceae. 2 Madag.'
Terminalis Kuntze = Cordyline Adans. (Liliac.).
Terminalis Medik. = Dracaena Vand. ex L. (Liliac.).
Terminthia Bernh. (~ Rhus L.). Anacardiaceae. 70 S. Afr. to India.
Terminthodia Ridley. Rutaceae. 1 Malay Penins., 7 New Guinea.
Terminthos St-Lag. = Pistacia L. (Pistaciac.).
Termontis Rafin. (1815) = Linaria Mill. (Scrophulariac.).
Termontis Rafin. (1840) = Antirrhinum L. (Scrophulariac.).
Ternatea Mill. = Clitoria L. (Legumin.).
Terniola Tul. = Lawia Tul. (Podostemac.).
*****Ternstroemia** Mutis ex L. f. Theaceae. 100 trop. (2 Afr., 1 Queensl.).
Ternstroemiaceae Mirb. (~ Theaceae Mirb., *q.v.*). Dicots. 10/350 trop. (v.
few Afr.). Closely rel. to *Camelliaceae*, but anthers mostly elongate and non-
versatile; fruit a berry, or dry and indehiscent; embryo curved or rarely almost
straight. Chief genera: *Ternstroemia, Anneslea, Adinandra, Eurya.*
Ternstroemiopsis Urb. Theaceae. 1 Hawaii.
Terobera Steud. = Cladium P.Br. (Cyperac.).

Terogia Rafin. = Ortegia Loefl. ex L. (Caryophyllac.).
Terpnanthus Nees & Mart. = Spiranthera A. St-Hil. (Rutac.).
Terpnophyllum Thw. = Garcinia L. (Guttif.).
Terranea Colla = Erigeron L. (Compos.).
Terrelia auctt. (sphalm.) = seq.
Terrella Nevski = seq.
Terrellia Lunell = Elymus L. (Gramin.).
Terrentia Vell. = Ichthyothere Mart. (Compos.).
Tersonia Moq. Gyrostemonaceae. 2 W. Austr.
Tertrea DC. = Machaonia Kunth (Rubiac.).
Tertria Schrank = Polygala L. (Polygalac.).
Terua Standley & F. J. Hermann. Leguminosae. 1 C. As.
Teruncius Lunell = Thlaspi L. (Crucif.).
Terustroemia Jack (sphalm.) = Ternstroemia L. (Theac.).
Tesmannia Willis (sphalm.) = Tessmannia Harms (Legumin.).
Tesota C. Muell. = Olneya A. Gray (Legumin.).
Tessarandra Miers. Oleaceae. 1 Brazil.
Tessaranthium Kellogg = Frasera Walt. (Gentianac.).
Tessaria Ruiz & Pav. Compositae. 4-5 SW. U.S. to Argentina.
Tessenia Bub. = Erigeron L. (Compos.).
Tesserantherum Curran = seq.
Tesseranthium Pritz. (sphalm.) = Tessaranthium Kellogg = Frasera Walt. (Gentianac.).
Tessiera DC. = Spermacoce L. (Rubiac.).
Tessmannia Harms. Leguminosae. 12 trop. W. Afr.
Tessmanniacanthus Mildbraed. Acanthaceae. 1 E. Peru.
Tessmannianthus Markgraf. Melastomataceae. 1 E. Peru.
Tessmanniodoxa Burret. Palmae. 2 trop. S. Am.
Tessmanniophoenix Burret = Chelyocarpus Dammer (Palm.).
Testudinaria Salisb. = Dioscorea L. (Dioscoreac.).
Testudipes Markgraf. Apocynaceae. 1 Assam, Burma.
Testulea Pellegr. Ochnaceae. 1 trop. Afr.
Teta Roxb. = Peliosanthes Andr. (Liliac.).
Tetanosia Rich. ex M. Roem. = Opilia Roxb. (Opiliac.).
Tetaris Lindl. (in Chesn.) = Arnebia Forsk. (Boraginac.).
Tetilla DC. Francoaceae. 1 Chile.
Tetraberlinia (Harms) Hauman. Leguminosae. 1 W. Equat. Afr.
Tetracanthus A. Rich. Compositae. 1 Cuba.
Tetracanthus Wright ex Griseb. = Pinillosia Ossa (Compos.).
Tetracarpaea Benth. (sphalm.) = Tetracrypta Gardn. = Anisophyllea R.Br. ex Sabine (Anisophylleac.).
Tetracarpaea Hook. f. Tetracarpaeaceae. 1 Tasmania.
Tetracarpaeaceae (Engl.) Nak. Dicots. 1/1 Tasm. Low shrubs with alt. simple doubly serr. evergr. strongly incised-nervose exstip. ls., lobulate-serr. when young. Fls. reg., ⚥, in erect persist. bract. rac. K 4-5, imbr., v. shortly connate, small; C 4, imbr., orbic., erose, clawed, caduc.; A 4+4, fil. filif., anth. ellipt. basif. ± latr.; G̲ 4, free, fusif., substipit., with small subsess. stig. and ∞ ov. on a ventr. plac. Fr. of 4 fusif. ventr. dehisc. follicles;

TETRACARPIDIUM

seeds ∞, with fleshy endosp. Only genus: *Tetracarpaea*. Closely related to *Escalloniac.*, differing chiefly in the tetram. fls., two whorls of sta., and free carpels.

Tetracarpidium Pax. Euphorbiaceae. 1 W. Equat. Afr.

Tetracarpum Moench = Schkuhria Roth (Compos.).

Tetracarpus P. & K. = Tetracarpaea Hook. f. (Tetracarpaeac.).

Tetracarya Dur. = Tretocarya Maxim. = Microula Benth. (Boraginac.).

Tetracellion Turcz. ex Fisch. & Mey. = Tetrapoma Turcz. = Rorippa Scop. (Crucif.).

*****Tetracentraceae** Van Tiegh. Dicots. 1/1 SE. As. Decid. trees, with slender branches bearing short shoots, these marked with crowded concentr. scars of fallen ls. and perulae, and bearing 1 subterm. l. and 1 infl. Ls. alt., cordate-ov., dent., palmately curvinerved from base, stips. adnate to pet. Fls. reg., ⚥, v. small, in groups of 4 on slender catkin-like spikes. K 4, imbr.; C o; A 4, oppositisep., fil. filif., anth. small, basif., latr.; G̲ (4), free above, with several ov. per loc. on 2 axile plac., and 4 free erect subul. styles with minute stigs. Fr. a 4-loc. loculic. caps., styles now externally subbasal by great overgrowth of ventr. region of each carp.; seeds few, with oily endosp. Only genus: *Tetracentron*. Perhaps related to *Trochodendraceae*.

Tetracentron Oliv. Tetracentraceae. 1 NE. India, Burma, SW. China.

Tetracera L. Dilleniaceae. 40 trop. Aril branched.

Tetraceras P. & K. = praec.

Tetraceras Webb = seq.

Tetraceratium (DC.) Kuntze = Tetracme Bunge (Crucif.).

Tetrachaete Chiovenda. Gramineae. 1 Eritrea, Arabia.

Tetracheilos Lehm. = Acacia Mill. (Legumin.).

Tetrachne Nees. Gramineae. 1 S. Afr.

Tetrachondra Petrie ex Oliv. Tetrachondraceae. 2 N.Z., Patag.

Tetrachondraceae Skottsb. Dicots. 1/2 N.Z. Patag. Small branched creeping herbs with habit of *Crassula* § *Tillaea*, with prostr. rooting stems emitting short erect leafy branches. Ls. opp., with flat connate petioles, simple, exstip., ciliate towards base, minutely dentic. distally, ± fleshy, obscurely gland-dotted. Fl. reg., ⚥, term. or axill., sol. K (4), imbr.(?); C (4), imbr., subrotate; A 4, alternipet. and epipet., anth. dorsif., intr.; G̲ (4), 4-lobed to base, with long or short slender gynobasic style and inconspic. stig., and 1 erect anatr. ov. per loc. Fr. of 4 basally attached 1-seeded setulose nutlets; seeds with copious endosp. Only genus: *Tetrachondra*. Probably near *Labiatae* (cf. *Mentha*).

Tetrachyron Schlechtd. = Calea L. (Compos.).

Tetraclea A. Gray. Labiatae. 2 S. U.S., Mex.

Tetraclinidaceae Hay. = Cupressaceae Neger.

Tetraclinis Mast. Cupressaceae. 1 S. Spain to Tunis, Malta, *T. articulata* (Vahl) Mast., the source of Arar wood and sandarach resin or pounce.

Tetraclis Hiern = Diospyros L. (Ebenac.).

Tetracma P. & K. = seq.

Tetracme Bunge. Cruciferae. 8 E. Medit. to C. As. & Baluchistan.

Tetracmidion Korshinsky = praec.

Tetracoccus Engelm. ex Parry. Euphorbiaceae. 4–5 Calif., Ariz., Lower Calif.

Tetracocyne Turcz. = Patrisia Rich. (Flacourtiac.).

Tetracronia Pierre = Glycosmis Corrêa (Rutac.).

Tetracrypta Gardn. & Champ. = Anisophyllea R.Br. ex Sabine (Anisophylleac.).

Tetractinostigma Hassk. = Aporusa Blume (Euphorbiac.).

Tetractis DC. = Tetractys Spreng. (inc. sed.).

Tetractomia Hook. f. Rutaceae. 15 Malaysia.

Tetractys Spreng. Inc. sed. (? Crassulac. ? Rutac.). 1 S. Afr.

Tetracustelma Baill. Asclepiadaceae. 2 Mex.

Tetradapa Osbeck = Erythrina L. (Legumin.).

Tetradema Schlechter = Agalmyla Blume (Gesneriac.).

Tetradenia Benth. Labiatae. 3 Madag.

Tetradenia Nees = Neolitsea (Benth.) Merr. (Laurac.).

Tetradia R.Br. = Pterygota Schott & Endl. (Sterculiac.).

Tetradia Thou. ex Tul. = Tetrataxis Hook. f. (Lythrac.).

Tetradiclis Stev. ex Bieb. Zygophyllaceae. 1 SE. Russia & E. Medit. to C. As. & Persia.

Tetradium Dulac = Sedum L. (Crassulac.).

Tetradium Lour. = Euodia J. R. & G. Forst. (Rutac.).

Tetradoa Pichon. Apocynaceae. 2 trop. W. Afr.

Tetradyas Danser. Loranthaceae. 1 New Guinea.

Tetradymia DC. Compositae. 8 W. N. Am.

Tetradynamae Reichb. = Cruciferae Juss.

Tetraëdrocarpus Schwartz. Boraginaceae. 1 Arabia.

Tetraena Maxim. Zygophyllaceae (?). 1 Mongolia.

Tetraeugenia Merr. = Syzygium Gaertn. (Myrtac.).

Tetragamestus Reichb. f. Orchidaceae. 3 C. & trop. S. Am.

Tetragastris Gaertn. Burseraceae. 8 C. Am. to Guiana, W.I. *T. balsamifera* (Sw.) Kuntze (Antilles) is known as pig's balsam, on account of a legend that wounded pigs rub against the trees to heal wounds with the resin.

Tetraglochidion K. Schum. = Glochidion J. R. & G. Forst. (Euphorbiac.).

Tetraglochidium Bremek. Acanthaceae. 8 W. Malaysia.

Tetraglochin Kunze ex Poepp. (~ Margyricarpus Ruiz & Pav.). Rosaceae. 8 Andes, temp. S. Am.

Tetraglossa Bedd. = Cleidion Blume (Euphorbiac.).

Tetragocyanis Thou. (uninom.) = *Epidendrum tetragonum* Thou. = *Phaius tetragonus* (Thou.) Reichb. f. (Orchidac.).

Tetragoga Bremek. Acanthaceae. 2 Assam, Sumatra.

Tetragompha Bremek. Acanthaceae. 2 Sumatra.

Tetragonanthus S. G. Gmel. = Halenia Borkh. (Gentianac.).

Tetragonella Miq. = seq.

Tetragonia L. Tetragoniaceae. 50–60 Afr., E. As., Austr., N.Z., temp. S. Am. Sometimes 2 fls. stand one above the other in the same axil. From the fr. thorny projections grow out which may bear fls. (an argument for the axial nature of the inf. ovary). *T. tetragonioïdes* (Pall.) Kuntze (*T. expansa* Murr.) is often used as a vegetable (New Zealand spinach).

***Tetragoniaceae** Link. Dicots. (Centrospermae). 2/60 mostly S. hemisph. Fleshy herbs or subshrubs, prostr. or scrambling, with simple ent. alt. exstip. ls. Fls. axill., sol. or few, or in small racemiform cymes, ⚥ or ♂ ♀, reg. P (3–5);

TETRAGONOBOLUS

A 1–∞, inserted on P-tube, sometimes fascic.; G̅ (3–8), rarely (1–2), with 1 pend. ov. per loc., and as many subul. styles as loculi. Fr. indehisc., ± drupaceous, sometimes winged, cornute or spiny. Genera: *Tetragonia, Tribulocarpus.* Closely related to *Aïzoaceae.*

Tetragonobolus Scop. (sphalm.) = Tetragonolobus Scop. = Lotus L. (Legumin.).

Tetragonocalamus Nakai. Gramineae. 2 E. As.

Tetragonocarpos Mill. = Tetragonia L. (Aïzoac.).

Tetragonocarpus Hassk. = Marsdenia R.Br. (Asclepiadac.).

*Tetragonolobus Scop. = Lotus L. (Legumin.).

Tetragonosperma Scheele = seq.

Tetragonotheca L. Compositae. 3–4 S. U.S., Mex.

Tetragyne Miq. = Microdesmis Hook. f. (Pandac.).

Tetrahit Adans. = Sideritis L. (Labiat.).

Tetrahit Gérard = Stachys L. (Labiat.).

Tetrahitum Hoffmgg. & Link = praec.

Tetraith Bub. = praec.

× **Tetralaelia** (sphalm. ' -laenia ') hort. Orchidaceae.̈ Gen. hybr. (vi) (Laelia × Tetramicra).

Tetralepis Steud. = Cyathochaeta Nees (Cyperac.).

× **Tetraliopsis** hort. Orchidaceae. Gen. hybr. (iii) (Laeliopsis × Tetramicra).

*Tetralix Griseb. Tiliaceae. 2 Cuba.

Tetralix Hill = Cirsium Mill. (Compos.).

Tetralix Zinn = Erica L. (Ericac.).

Tetralobus A. DC. = Polypompholyx Lehm. (Lentibulariac.).

Tetralocularia O'Donell. Convolvulaceae. 1 Colombia.

Tetralopha Hook. f. Rubiaceae. 5 Borneo, Philipp. Is. Style 0.

Tetralyx Hill = Tetralix Hill = Cirsium Mill. (Compos.).

Tetramelaceae (Warb.) Airy Shaw. Dicots. 2/2 trop. As. Large or very large trees, often buttressed. Ls. alt., simple, cordate-ovate (sometimes obliquely), ent. or dent., pubesc. or lepidote, long-pet., exstip., palmately 3–5-nerved. Fls. in term. fasc. of precocious-flg. panicles, or in long sol. axill. pend. spikes. Fls. ♂ ♀, dioec. K (4) or (6–8), valv., sometimes unequal. C 6–8, valv., inserted on K, or 0. A 4 or 6–8, fil. elong., anth. short-ovate, or large, oblong, recurved (stds. in ♀ fl. 0). G̅ (4) or (6–8), 1-loc., with 4–8 pariet. plac. bearing ∞ anatr. ov.; styles 4–8, thick, short or long, free, with oblique or capit. stig. Fr. a dehisc. caps., the inner crustaceous-horny ovary-wall sometimes breaking through the adherent K-tube and spreading stellately in 6–8 valves. Seeds minute. Genera: *Tetrameles, Octomeles.* Some relationship with *Lythrac., Sonneratiac.,* etc.; connection with *Datiscac.* uncertain.

Tetrameles R.Br. Tetramelaceae. 1 Indoch., Indomal. (exc. Borneo, Philipp. & Moluccas), in drier forests. Ls. decid. Fls. precocious.

Tetrameranthus R. E. Fries. Annonaceae. 2 trop. S. Am.

Tetrameris Naud. = Comolia DC. (Melastomatac.).

Tetramerista Miq. Tetrameristaceae. 2–3 Sumatra, Malay Penins., Borneo.

Tetrameristaceae (H. Hallier) Hutch. Dicots. 1/3 W. Malaysia. Large trees or shrubs of acid soils, peat swamps, etc., with alt. simple ent. sess. asymm. oblanc. coriac. exstip. ls., sometimes auric. or sagitt. at base, lower surface usu.

with 2 irreg. rows of black 'glandulae hypophyllae' (cf. Marcgraviac.) of uncertain function. Fls. reg., ☿, in axill. long-pedunc. but abbrev. conspic. bract. rac., sometimes almost umbellif., bracts strongly reflexed, pedic. bibracteol. at apex, persist. K 2+2, imbr., persist.; C 4, imbr., persist.; A 4, alternipet., fil. subul., anth. obl. with sep. loc., sagitt. and glandular at base, dorsif. but ultimately reversed and extr.; G (4), with 1 basal erect anatr. ov. per loc., style subul. with 4-lobul. stig. Fr. a dry glob. ± coriac. 4-seeded berry. Only genus: *Tetramerista*. Related to *Marcgraviac.*, *Bonnetiac.*, *Foetidiac.*, and *Pelliceriac.*; habit of *Ploiarium* (*Bonnetiac.*).

Tetramerium Gaertn. f. = Faramea Aubl. (Rubiac.).

*Tetramerium Nees. Acanthaceae. 23 Mex., C. Am.

Tetramicra Lindl. Orchidaceae. 10 W.I.

Tetramolopium Nees. Compositae. 25 New Guinea, Hawaii.

Tetramorphaea DC. = Centaurea L. (Compos.).

Tetramorphandra Baill. = Hibbertia Andr. (Dilleniac.).

Tetramyxis Gagnep. = Spondias L. (Anacardiac.).

Tetrandra Miq. = Tournefortia L. (Boraginac.).

*Tetranema Benth. Scrophulariaceae. 3 Mex.

Tetranema Sweet = Desmodium Desv. (Legumin.).

Tetraneuris Greene = Hymenoxys Cass. (Compos.).

Tetrantha Poit. ex DC. = Riencourtia Cass. (Compos.).

Tetranthera Jacq. = Litsea Lam. (Laurac.).

Tetranthus Sw. Compositae. 4 W.I.

Tetraotis Reinw. = Enydra Lour. (Compos.).

Tetrapanax Harms = Hoplopanax (Torr. & Gray) Miq. (Araliac.).

Tetrapanax K. Koch. Araliaceae. 1 S. China, Formosa, *T. papyriferum* (Hook.) K. Koch, the rice-paper tree. The pith is split into thin sheets and **Tetrapasma** G. Don = Discaria Hook. (Rhamnac.). [pressed.

Tetrapathaea Reichb. Passifloraceae. 1 N.Z.

Tetrapeltis Wall. ex Lindl. = Otochilus Lindl. (Orchidac.).

Tetraperone Urb. Compositae. 1 Cuba.

Tetrapetalum Miq. Annonaceae. 2 Borneo. [(Gramin.).

Tetraphis auct. (sphalm., 1899; non Hedw. 1782—Musci) = Triraphis R.Br.

Tetraphyla Reichb. = Tetraphyle Eckl. & Zeyh. = Sedum L. (Crassulac.).

Tetraphylax De Vriese = Goodenia Sm. (Goodeniac.).

Tetraphyle Eckl. & Zeyh. = Crassula L. (Crassulac.).

Tetraphyllaster Gilg. Melastomataceae. 1 trop. W. Afr.

Tetraphyllum Griff. ex C. B. Cl. Gesneriaceae. 2 NE. India, Siam.

Tetraphysa Schlechter. Asclepiadaceae. 1 Colombia.

Tetrapilus Lour. Oleaceae. 10–15 SE. As.

Tetraplacus Radlk. = Otacanthus Lindl. (Scrophulariac.).

Tetraplandra Baill. Euphorbiaceae. 5 Brazil.

Tetraplasandra A. Gray. Araliaceae. 1 Philipp. Is., 1 Celebes, 2 New Guinea & Solomon Is., 20 Hawaii.

Tetraplasia Rehder. Rubiaceae. 4 Formosa, Ryukyu Is.

Tetraplasium Kunze = Tetilla DC. (Francoac.).

Tetrapleura Benth. Leguminosae. 2 trop. Afr.

Tetrapleura Parl. ex Webb = Tornabenia Parl. (Umbellif.).

TETRAPODENIA

Tetrapodenia Gleason. Malpighiaceae. 1 Guiana.

Tetrapogon Desf. Gramineae. 5–6 Medit. to India, trop. & S. Afr.

Tetrapoma Turcz. ex Fisch. & Mey. = Rorippa Scop. (Crucif.).

Tetrapora Schau. = Baeckea L. (Myrtac.).

Tetraptera Miers ex Lindl. = Burmannia L. (Burmanniac.).

Tetraptera Phil. = Gaya Kunth (Malvac.).

Tetrapteris Cav. = Tetrapterys Cav. emend. A. Juss. (Malpighiac.).

Tetrapterocarpon Humbert. Leguminosae. 1 Madag.

Tetrapterygium Fisch. & Mey. = Sameraria Desv. (Crucif.).

Tetrapterys Cav. emend. A. Juss. Malpighiaceae. 80 Mex. to trop. S. Am., W.I.

Tetrapteryx Dalla Torre & Harms = praec.

Tetraracus Klotzsch ex Engler = Tapirira Aubl. (Anacardiac.).

Tetrardisia Mez. Myrsinaceae. 3 Java, Queensland.

Tetrarhaphis Miers = Oxytheca Nutt. (Polygonac.).

Tetraria Beauv. Cyperaceae. 38 S. Afr., 1 trop. E. Afr.; 1 Borneo; 4 Austr.

Tetrariopsis C. B. Clarke. Cyperaceae. 1 Austr.

Tetrarrhena R.Br. Gramineae. 4 Austr.

Tetraselago Junell. Scrophulariaceae. 1 S. Afr.

Tetrasida Ulbr. Malvaceae. 1 Peru.

Tetrasiphon Urb. Celastraceae. 1 W.I. (Jamaica).

Tetrasperma Steud. = Tetrapasma G. Don = Discaria Hook. (Rhamnac.).

Tetraspidium Baker. Scrophulariaceae. 1 Madag.

Tetraspis Chiov. = Kirkia Oliv. (? Ptaeroxylac.).

Tetraspora Miq. = Baeckea L. (Myrtac.).

Tetrastemma Diels ex H. Winkler = Uvariopsis Engl. (Annonac.).

Tetrastemon Hook. & Arn. = Myrrhinium Schott (Myrtac.).

Tetrastichella Pichon = Arrabidaea DC. (Bignoniac.).

Tetrastigma Planch. Vitidaceae. 90 SE. As., Indomal., Austr.

Tetrastigma K. Schum. = Schumanniophytum Harms (Rubiac.).

Tetrastylidiaceae Van Tiegh. = Olacaceae–Anacoloseae Engl.

Tetrastylidium Engl. Olacaceae. 3 S. Brazil.

Tetrastylis Barb. Rodr. = Passiflora L. (Passiflorac.).

Tetrasynandra Perkins. Monimiaceae. 3 Austr.

Tetrataenium (DC.) Manden. (∼ Heracleum L.). Umbelliferae. Spp.? C. As. to Ceylon.

Tetrataxis Hook. f. Lythraceae. 1 Mauritius.

Tetrateleia Arwidsson = seq.

Tetratelia Sond. Cleomaceae. 2 SE. trop. Afr.

Tetrathalamus Lauterb. Winteraceae. 1 New Guinea.

Tetratheca Sm. Tremandraceae. 20 extratrop. Austr.

Tetrathecaceae R.Br. = Tremandraceae DC.

Tetrathylacium Poepp. & Endl. Flacourtiaceae. 4–5 C. & W. trop. S. Am.

Tetrathyrium Benth. Hamamelidaceae. 1 Hongkong.

Tetratome Poepp. & Endl. = Mollinedia Ruiz & Pav. (Monimiac.).

× **Tetratonia** hort. Orchidaceae. Gen. hybr. (iii) (Broughtonia × Tetramicra).

Tetraulacium Turcz. Scrophulariaceae. 1 Brazil.

Tetrazygia Rich. Melastomataceae. 30 W.I.

Tetrazygos Rich. ex DC. = Charianthus D. Don (Melastomatac.).

Tetreilema Turcz. = Frankenia L. (Frankeniac.).

Tetrixus Van Tiegh. ex Lecomte = Viscum L. (Viscac.).

Tetrodea Rafin. = Amphicarpaea Ell. mut. DC. (Legumin.).

Tetrodes Rafin. = praec.

Tetrodus Cass. = Helenium L. (Compos.).

Tetroncium Willd. Juncaginaceae. 1 temp. S. Am. (Strait of Magellan). Fl. 2-merous; fr. sharply defl. (cf. *Carex pulicaris*).

Tetrorchidium Poepp. & Endl. Euphorbiaceae. 16 C. & trop. S. Am., W.I., trop. W. Afr.

Tetrorhiza Rafin. ex Jacks. = Tretorhiza Adans. = Gentiana L. (Gentianac.).

Tetrorum Rose = Sedum L. (Crassulac.).

Tetrouratea Van Tiegh. = Ouratea Aubl. (Ochnac.).

Teucridium Hook. f. Verbenaceae. 1 N.Z.

Teucrion St-Lag. = seq.

Teucrium L. Labiatae. 300 cosmop., esp. Medit. Fl. with small upper lip, protandr. with movement of style and sta.

Teuscheria Garay. Orchidaceae. 3 trop. S. Am.

Texiera Jaub. & Spach = Glastaria Boiss. (Crucif.).

Textoria Miq. = Dendropanax Decne (Araliac.).

Teyleria (Backer) Backer. Leguminosae. 1 Hainan, Java.

Teysmannia Miq. (1857) = Pottsia Hook. & Arn. (Apocynac.).

Teysmannia Miq. (1859) = Teyssmannia Reichb. f. & Zoll. = Johannesteijsmannia H. E. Moore (Palm.).

Teysmanniodendron Koord. = Teijsmanniodendron Koord. (Verbenac.).

Teyssmannia Reichb. f. & Zoll. = Johannesteijsmannia H. E. Moore (Palm.).

Thacla Spach = Caltha L. (Ranunculac.).

Thacombauia Seem. = Flacourtia Comm. ex L'Hérit. (Flacourtiac.).

Thalamia Spreng. = Phyllocladus Rich. (Phyllocladac.).

Thalamiflorae DC. A 'Series' of Dicots. comprising (in Bentham & Hooker) the orders *Ranales, Parietales, Polygalinae, Caryophyllinae, Guttiferales* and *Malvales*.

Thalasium Spreng. = Panicum L. (Gramin.).

Thalassia Soland. ex Koenig. Hydrocharitaceae. 1 coasts of Caribbean; 1 Ind. & Pacif. oceans. Submerged marine aquatics.

Thalassiaceae (Aschers. & Gürke) Nak. = Hydrocharitaceae–Thalassioïdeae Aschers. & Gürke.

Thalassiophila Denizot = Thalassia Soland. ex Koen. (Hydrocharitac.).

Thalassodendron den Hartog. Cymodoceaceae. 1 coasts Red Sea, W. Indian Ocean, E. Malaysia, Queensl.; 1 extratrop. W. Austr.

Thalesia Mart. ex Pfeiff. = Sweetia Spreng. (Legumin.).

Thalesia Rafin. = Aphyllon Mitch. = Orobanche L. (Orobanchac.).

Thalestris Rizzini. Acanthaceae. 1 Brazil.

Thalia L. Marantaceae. 11 trop. Am., Afr. The std. β (see fam.) is present.

Thalianthus Klotzsch = Myrosma L. f. (Marantac.).

Thalictrella A. Rich. = Isopyrum L. (Ranunculac.).

Thalictrodes Kuntze = Cimicifuga L. (Ranunculac.).

Thalictrum L. Ranunculaceae. 150 N. temp., trop. S. Am., trop. & S. Afr. Fls. small; P sepaloid or slightly coloured and soon falling. Some are visited

THELIGONUM

Theligonum L. Theligonaceae. 3 Canaries, Medit., SW. China, Japan. Myrmecochorous. The suggestion is made by Wunderlich in *Österr. Bot. Zeit.* **119**: 329–94, 1971, that *Theligonum* should be referred to the Rubiaceae.
Thelira Thou. = Parinari Aubl. (Chrysobalanac.).
Thellungia Stapf. Gramineae. 1 E. Austr.
Thellungiella O. E. Schulz. Cruciferae. 2 C. & NE. As., W. N. Am.
Thelmatophace Godr. = Telmatophace Schleid. = Lemna L. (Lemnac.).
Thelocactus (K. Schum.) Britton & Rose. Cactaceae. 17 S. U.S., Mex.
Thelocarpus P. & K. = Thelecarpus Van Tiegh. = Tapinanthus Bl. (Loranthac.).
Thelocephala Y. Ito = Pyrrhocactus (A. Berger) Backeb. & F. M. Knuth (Cactac.).
Thelomastus Frič = Thelocactus (K. Schum.) Britt. & Rose + Echinomastus Britt. & Rose (Cactac.).
Thelophytum P. & K. = Theleophyton (Hook. f.) Moq. (Chenopodiac.).
Thelopogon P. & K. = Thelepogon Roth ex Roem. & Schult. (Gramin.).
Thelosperma P. & K. = Thelesperma Less. (Compos.).
Thelychiton Endl. = Dendrobium Sw. (Orchidac.).
Thelycrania (Dumort.) Fourr. = Swida Opiz (Cornac.).
Thelygonaceae auctt. = Theligonaceae Dum.
Thelygonum Schreb. = Theligonum L. (Cynocrambac.).
Thelymitra J. R. & G. Forst. Orchidaceae. 45 Malaysia, Austr., N.Z.
Thelypetalum Gagnep. = Leptopus Decne (Euphorbiac.).
Thelypodiopsis Rydb. (~ Thelypodium Endl.). Cruciferae. 14 W. U.S.
Thelypodium Endl. Cruciferae. *S.str.*, 1 W. U.S.; *s.lat.*, 45 W. U.S., Mex.
Thelypogon Mutis ex Spreng. = Telipogon Kunth (Orchidac.).
Thelypteridaceae Ching. Aspidiales. Terrestrial ferns, distinct in scales, hairs and anatomy from *Aspidiaceae* (Holttum, *J. Linn. Soc., Bot.* **53**: 130–3, 1947). Chief genera: *Thelypteris, Cyclosorus, Goniopteris.* In this edition of the *Dictionary* the Old World members of this family are arranged in 23 genera, according to Holttum, *Blumea*, **17**: 5–32, 1969 and **19**: 17–52, 1971.
Thelypteris Schmid. Thelypteridaceae. Now restricted to c. 4 spp., cosmop. Type species: *Acrostichum thelypteris* L. = *Thelypteris palustris* Schott. Copeland's distinction between *Thelypteris* and *Cyclosorus* (veins free or anastomosing) gives an unnatural division, and part of *Thelypteris* as so distinguished should be united to *Cyclosorus*; this leaves a polymorphic *Thelypteris* which needs further study. See Morton, *Amer. Fern Journ.* **53**: 149–54, 1963, and Ching, *Acta Phytotax. Sinica*, **8**: 289–335, 1963, for two quite different solutions.
Thelyra DC. = Hirtella L. (Chrysobalanac.).
Thelysia Salisb. = Iris L. (Iridac.).
Thelythamnos Spreng. f. = Ursinia Gaertn. (Compos.).
Themeda Forsk. Gramineae. 10 warm Afr., As. *Th. triandra* Forsk. covers very large areas in trop. and S. Afr.
Themid[ac]eae Salisb. = Alliaceae J. G. Agardh.
Themis Salisb. = Brodiaea Sm. (Alliac.).
Themistoclesia Klotzsch. Ericaceae. 30 N. Andes.
Thenardia Kunth. Apocynaceae. 6 Mex., C. Am.
Thenardia Moç. & Sessé ex DC. = Rhynchanthera DC. (Melastomatac.).

Theobroma L. Sterculiaceae. 30 trop. Am., incl. *Th. cacao* L., *Th. pentagona* Bernoulli, and others, producing cacao, cocoa, or chocolate. Young ls. red, pend. Fls. on old wood. Pets. cap-like at base. Stds. almost petal-like. Fr. large, tough, indehisc., berry-like, with nearly exalb. seeds, which yield cocoa after roasting, etc.; cocoa butter by pressing seeds.
Theobrom[atac]eae J. G. Agardh = Sterculiaceae–Buettneriëae Kunth.
Theobromodes Kuntze = Theobroma L. (Sterculiac.).
Theodora Medik. = Schotia Jacq. (Legumin.).
Theodorea Cass. = Saussurea DC. (Compos.).
Theodorea Barb. Rodr. = Rodrigueziella Kuntze (Orchidac.).
Theodoria Neck. = Sterculia L. (Sterculiac.).
Theodoricea Buc'hoz. Quid?
Theophrasta L. Theophrastaceae. 2 W.I. Thorny scales on upper part of stem. Serial buds in axils, fls. in axils of scales on these.
Theophrastaceae Link. Dicots. 5/110 trop. Am., W.I. Trees or shrubs, with alt. usu. pseudo-vertic. ent. or spinose-serr. exstip. ls., often crowded at apex of stems, sometimes pungent. Fls. reg., ♀ or ♂ ♀, rac. or fascic., rarely sol., usu. term. K 4–5, sometimes shortly connate, imbr., persist.; C (4–5), imbr. (K and C gland-dotted or streaked); A 4–5, oppositipet., epipet. near base of C-tube, fil. sometimes connate into a tube, anth. extr., with produced connective; stds. 4–5, alternipet., sometimes petaloid; G 1-loc., with usu. thick style and discoid or conical stig., and stipit. free-central plac. bearing ∞ anatr. ov. immersed in mucilage. Fr. baccate or drupac., sometimes almost dry, indehisc. Seeds ∞–few, rarely 1; endosp. copious, horny. Genera: *Theophrasta, Neomezia, Deherainia, Clavija, Jacquinia.* Related to *Myrsinac.*
Theophroseris Andrae = Tephroseris Reichb. = Senecio L. (Compos.).
Theopsis Nakai = Camellia L. (Theac.).
Theopyxis Griseb. = Lysimachia L. (Primulac.).
Therebina Nor. = Pilea Lindl. (Urticac.).
Therefon Rafin. = Therofon Rafin. = Boykinia Nutt. (Saxifragac.).
Thereianthus G. J. Lewis. Iridaceae. 8 S. Afr.
Theresa Clos = Perilomia Kunth (Labiat.).
Theresia C. Koch = Fritillaria L. (Liliac.).
Theriophonum Blume. Araceae. 6 Penins. Ind., Ceylon.
Thermia Nutt. = Thermopsis R.Br. (Legumin.).
Therminthos St-Lag. (sphalm.) = Terminthos St-Lag. = Pistacia L. (Pistaciac.).
Thermophila Miers = Salacia L. (Celastrac.).
Thermopsis R.Br. Leguminosae. 30 C. As. & Himal. to E. U.S. Rhiz. herbs.
Therofon Rafin. = Boykinia Nutt. (Saxifragac.).
Therogeron DC. = Minuria DC. (Compos.).
Therolepta Rafin. = Marshallia Schreb. (Compos.).
Therophon Rydb. = Therofon Rafin. = Boykinia Nutt. (Saxifragac.).
Therophonum P. & K. = praec.
Theropogon Maxim. Liliaceae. 1 Himalaya.
Therorhodion Small (~ Rhododendron L.). Ericaceae. 3 NE. As., NW. Am.
Thesidium Sonder. Santalaceae. 7–8 S. Afr.
Thesion St-Lag. = seq.
Thesium L. Santalaceae. 325 Eur., Afr., As. (to Philipp. Is. & Lesser Sunda

THESPESIA

Is.), Austr. Herbaceous root-parasites with yellow-green ls. Fls. ♀, in racemes. Bract adnate to peduncle, and with the 2 bracteoles forming a sort of involucre.

*Thespesia Soland. ex Corrêa. Malvaceae. 15 trop. *T. populnea* (L.) Soland. Corr. a common strand plant.

Thespesiopsis Exell & Hillc. Malvaceae. 1 SE. trop. Afr.

Thespesocarpus Pierre = Diospyros L. (Ebenac.).

Thespidium F. Muell. Compositae. 1 trop. Austr.

Thespis DC. Compositae. 1 Nepal to Burma; 2 Indoch.(?).

Thevenotia DC. Compositae. 2 SW. As.

Thevetia Adans. = Cerbera L. (Apocynac.).

*Thevetia [L.] Juss. ex Endl. Apocynaceae. 9 trop. Am., W.I. Ls. alt.

Thevetia Vell. = ? Aeschrion Vell. (Simaroubac.).

Thevetiana Kuntze = praec.

Theyga Molina = Thiga Molina = Laurelia Juss. (Monimiac.).

Theyodis A. Rich. = Oldenlandia L. (Rubiac.).

Thezera Rafin. = ? Rhus L. (Anacardiac.).

Thibaudia Ruiz & Pav. Ericaceae. 60 trop. Am.

Thicuania Rafin. = Dendrobium Sw. (Orchidac.).

Thiebautia Colla = Bletia Ruiz & Pav. (Orchidac.).

Thieleodoxa Cham. Rubiaceae. 5 Brazil, Guiana. Fr. ed.

Thiersia Baill. = Faramea Aubl. (Rubiac.).

Thiga Molina = Laurelia Juss. (Monimiac.).

Thilachium Lour. Capparidaceae. 15 trop. E. Afr., Madag., Masc.

Thilakium Lour. = praec.

Thilcum Molina = Tilco Adans. = Fuchsia L. (Onagrac.).

Thillaea Sang. = Tillaea L. = Crassula L. (Crassulac.).

Thiloa Eichl. Combretaceae. 3 trop. S. Am.

Thimus Neck. = Thymus L. (Labiat.).

Thinobia Phil. = Nardophyllum Hook. & Arn. (Compos.).

Thinogeton Benth. = Cacabus Bernh. (Solanac.).

Thinouia Planch. & Triana. Sapindaceae. 12 warm S. Am.

Thiodia Benn. = Laëtia Loefl. ex L. (Flacourtiac.).

Thiodia Griseb. = Zuelania A. Rich. (Flacourtiac.).

Thiollierea Montr. = Randia L. (Rubiac.).

Thisantha Eckl. & Zeyh. = Tillaea L. = Crassula L. (Crassulac.).

Thisbe Falc. = Herminium L. (Orchidac.).

Thiseltonia Hemsl. Compositae. 1 W. Austr.

Thismia Griff. Burmanniaceae. 25 trop. Saprophytes.

*Thismiaceae J. G. Agardh = Burmanniaceae–Thismiëae Miers.

Thium Steud. = Tium Medik. = Astragalus L. (Legumin.).

Thladiantha Bunge. Cucurbitaceae. 15 E. As. to Malaysia. Climbing herbs with root-tubers.

Thladianthopsis Cogn. = praec.

Thlasidia Rafin. = Scabiosa L. (Dipsacac.).

Thlaspeocarpa C. A. Smith. Cruciferae. 2 S. Afr.

Thlaspi L. Cruciferae. 60 mostly N. temp. (& mts.) Euras.; few N. & S. Am.

Thlaspidea Opiz = praec.

Thlaspidium Bub. = praec.

Thlaspidium Mill. = Biscutella L. (Crucif.).
Thlaspidium Spach = Lepidium L. (Crucif.).
Thlaspius St-Lag. = Thlaspi L. (Crucif.).
Thlipsocarpus Kunze = Hyoseris L. (Compos.).
Thoa Aubl. = Gnetum L. (Gnetac.).
Thoaceae Agardh = Gnetaceae Lindl.
Thodaya Compton. Compositae. 2 S. Afr.
Thollonia Baill. = Icacina A. Juss. (Icacinac.).
Thomandersia Baill. Acanthaceae (?Pedaliac.). 6 trop. Afr.
Thomasia J. Gay. Sterculiaceae. 25 Austr.
Thomassetia Hemsl. = Brexia Nor. ex Thou. (Brexiac.).
Thommasinia Steud. = Tommasinia Bertol. = Peucedanum L. (Umbellif.).
Thompsonella Britton & Rose (~ Echeveria DC.). Crassulaceae. 2 Mex.
Thompsonia R.Br. = Deïdamia Nor. ex Thou. (Passiflorac.).
Thompsonia Steud. = seq.
Thomsonia Wall. Araceae. 1 Himalaya, Assam.
Thonnera De Wild. = Uvariopsis Engl. (Annonac.).
Thonningia Vahl. Balanophoraceae. 5 (or 1?) trop. Afr., 1 Madag.
Thora Hill = Ranunculus L. (Ranunculac.).
Thoracocarpus Harling. Cyclanthaceae. 1 trop. S. Am.
Thoracosperma Klotzsch = Eremia D. Don (Ericac.).
Thoracostachyum Kurz. Cyperaceae. 2 Seychelles; 3 Malaysia to Polynesia.
Thoraea Gandog. = Pimpinella L. (Umbellif.).
Thorea Briq. (1902; non Bory 1808—Algae) = Thorella Briq. (Umbellif.).
Thorea Rouy (1913) = Pseudarrhenatherum Rouy = Arrhenatherum Beauv. (Gramin.).
Thoreldora Pierre. Rutaceae. 1 Indoch.
*Thorelia Gagnep. = Thoreliella C. Y. Wu = Camchaya Gagnep. (Compos.).
Thorelia Hance = Tristania R.Br. (Myrtac.).
Thoreliella C. Y. Wu = Camchaya Gagnep. (Compos.).
Thorella Briquet. Umbelliferae. 1 Eur.
Thoreochloa Holub = Pseudarrenatherum Rouy (Gramin.).
Thornbera Rydb. = Dalea L. (Legumin.).
Thorncroftia N. E. Br. Labiatae. 1 S. Afr. (Transvaal).
Thorntonia Reichb. (Kosteletzkya C. Presl). Malvaceae. 30 N. Am., Mex., trop. & S. Afr., Madag.
Thorvaldsenia Liebm. = Chysis Lindl. (Orchidac.).
Thorwaldsenia auct. = praec.
Thottea Rottb. Aristolochiaceae. 9 W. Malaysia (exc. Java).
Thouarea Kunth = Thuarea Pers. (Gramin.).
Thouarsia P. & K. = praec.
Thouarsia Vent. ex DC. = Psiadia Jacq. (Compos.).
Thouarsiora Homolle ex Arènes. Rubiaceae. 1 Madag.
Thouinia Comm. ex Planch. = Vitis L. (Vitidac.).
Thouinia Domb. ex DC. = Lardizabala Ruiz & Pav. (Lardizabalac.).
Thouinia L. f. = Linociera Sw. (Oleac.).
*Thouinia Poit. Sapindaceae. 28 Mex., C. Am., W.I. Lianes.
Thouinia Sm. = Humbertia Lam. (Humbertiac.).

THOUINIDIUM

Thouinidium Radlk. Sapindaceae. 7 Mex., C. Am., W.I.

Thouvenotia Danguy = Beilschmiedia Nees (Laurac.).

Thozetia F. Muell. ex Benth. Asclepiadaceae. 1 Austr.

Thrasya Kunth. Gramineae. 15 trop. S. Am., Trinidad.

Thrasyopsis L. Parodi. Gramineae. 2 Brazil.

Thraulococcus Radlk. Sapindaceae. 2 India, Ceylon.

Threlkeldia R.Br. Chenopodiaceae. 5 Austr.

Thrica S. F. Gray = Thrincia Roth = Leontodon L. (Compos.).

Thrinax L. f. ex Sw. Palmae. 12 W.I. (thatch-palm). The ls. are used for roofing, and the plants also yield useful fibre.

Thrincia Roth = Leontodon L. (Compos.).

Thrincoma O. F. Cook = Coccothrinax Sargent (Palm.).

Thringis O. F. Cook = Coccothrinax Sargent (Palm.).

Thrixanthocereus Backeb. = Espostoa Britt. & Rose (Cactac.).

Thrixgyne Keng = Duthiea Hack. (Gramin.).

Thrixia Dulac = Leontodon L. (Compos.).

Thrixspermum Lour. Orchidaceae. 100 SE. As. to Formosa, Indomal., Austr., Polynesia.

Thryallis L. Malpighiaceae. 12 warm Am.

*Thryallis Mart. Malpighiaceae. 3 warm S. Am. K umbrella-like after fert.

Thrycocephalum Steud. = seq.

Thryocephalon J. R. & G. Forst. = Kyllinga Rottb. (Cyperac.).

Thryothamnus Phil. Verbenaceae. 1 Chile. (Cf. *Junellia* Mold.)

*Thryptomene Endl. Myrtaceae. 25 Austr., esp. W.

Thuarea Pers. Gramineae. 1 Madag.; 1 Indomal., on the coast.

Thuessinkia Korth. ex Miq. = Caryota L. (Palm.).

Thuia Scop. = Thuja L. (Cupressac.).

Thuiacarpus Benth. & Hook. f. = seq.

Thuiaecarpus Trautv. = Juniperus L. (Cupressac.).

Thuinia Rafin. = Thouinia L. f. = Linociera Sw. (Oleac.).

Thuiopsis Endl. = Thujopsis Sieb. & Zucc. = Thuja L. (Cupressac.).

Thuja L. Cupressaceae. 5 China, Japan, N. Am. *Th. occidentalis* L. is the American, *Th. plicata* D. Don the Western, *Th. orientalis* L. the Chinese Arbor-vitae. The ls. are small and closely appressed to the stems, which show dorsiventral symmetry. Cones of 3 or 4 pairs of scales, the uppermost sterile and often united to form the *columella*, the lowest also often sterile. The timber of the American species is valuable, esp. that of *Th. plicata*, under the name Western Red Cedar.

Thujaecarpus Trautv. = Thuiaecarpus Trautv. = Juniperus L. (Cupressac.).

Thujocarpus P. & K. = praec.

Thujopsis P. & K. = Thyopsis Wedd. = Tafalla D. Don (Compos.).

Thujopsis Sieb. & Zucc. Cupressaceae. 1 Japan, *Th. dolabrata* Sieb. & Zucc., 'Hiba Arbor-vitae'.

Thumbergia Poit. (sphalm.) = Thunbergia Retz. (Acanthac.).

Thumung Koen. = Zingiber Boehm. (Zingiberac.).

Thunbergia Montin = Gardenia Ellis (Rubiac.).

*Thunbergia Retz. Thunbergiaceae. 200 palaeotrop. The bracteoles enclose the K and tube of the C and are often united post.

Thunbergiaceae Van Tiegh. Dicots. 4/205 trop. Herbs or shrubs, with artic. often twining stems. Ls. opp., simple, ent. or hast., exstip. Fls. ♀, reg. or slightly zygo., axill. or in term. rac., with 2 large spathaceous bracteoles. K reduced, truncate or shortly lobed or 5–16-dentate; C (5), ± hypocraterif., often inflated above, contorted; A 4, didynam., std. minute or o; disk large, annular; G̲ (2), with 2 collat. ov. per loc., and simple style with large infundibulif. or bilobed stig. Fr. a globose biloc. loculic. caps. with massive ensiform beak; seeds hemisph., with excavated ventral face; endosp. o; embryo with slightly curved cots. Genera: *Thunbergia, Pseudocalyx, Meyenia, Pounguia.* Intermediate between *Bignoniac., Pedaliac., Mendonciac.* and *Acanthac.*

Thunbergianthus Engl. Scrophulariaceae. 1 S. Tomé, 1 trop. E. Afr.

Thunbergiella H. Wolff. Umbelliferae. 1 S. Afr.

Thunbergiopsis Engl. (sphalm.) = Thunbergianthus Engl. (Scrophulariac.).

Thunia Reichb. f. Orchidaceae. 6 India, SE. As.

Thuranthos C. H. Wright. Liliaceae. 1–3 S. Afr.

Thuraria Molina. Inc. sed. 1 Chile.

Thuraria Nutt. = Grindelia Willd. (Compos.).

Thurberia Benth. = Limnodea L. H. Dewey (Gramin.).

Thurberia A. Gray = Gossypium L. (Malvac.).

Thurnhausera Pohl ex G. Don = Curtia Cham. & Schlechtd. (Gentianac.).

Thurnheyssera Mart. ex Meissn. = Symmeria Benth. (Polygonac.).

Thurnia Hook. f. Thurniaceae. 2–3 Venez., Guiana.

*Thurniaceae Engl. Monocots. 1/3 trop. S. Am. Perenn. herbs with triquetrous stems and narrow sometimes spinose-serr. sheathing ls. Fls. reg., ♀, very small, in dense glob. heads subtended by several ± leafy bracts. P 6, narrow, persist.; A 6, much exserted, anth. basifixed; G̲ (3), 3-loc., fusif., with 3 filif. stigs., and 1–∞ axile anatr. ov. per loc. Fr. a 3-seeded loculic. caps., seeds with endosp. Only genus: *Thurnia.* Anatomical evidence does not support suggested affinity with *Rapateac.* or *Juncac.* (Cutler, *Kew Bull.* **19**: 431–41, 1965).

Thurovia Rose. Compositae. 1 S. U.S.

Thurya Boiss. & Bal. Caryophyllaceae. 1 Asia Minor.

Thuspeinanta Durand. Labiatae. 2 C. As., Persia, Afghan.

Thuspeinantha auctt. = praec.

Thuya L. = Thuja L. (Cupressac.).

Thuyopsis Parl. = Thujopsis Sieb. & Zucc. (Cupressac.).

Thya Adans. = Thuja L. (Cupressac.).

Thyana Ham. = Thouinia Poit. (Sapindac.).

Thyarea Benth. = Thuarea Pers. (Gramin.).

Thyella Rafin. = Jacquemontia Choisy (Convolvulac.).

Thyia Aschers. = Thuja L. (Cupressac.).

Thylacantha Nees & Mart. = Angelonia Humb. & Bonpl. (Scrophulariac.).

Thylacanthus Tul. Leguminosae. 1 Amaz. Brazil.

Thylachium DC. = Thilachium Lour. (Capparidac.).

Thylacis Gagnep. = Thrixspermum Lour. (Orchidac.).

Thylacitis Adans. = Gentiana L. (Gentianac.).

Thylacitis Rafin. = Centaurium Hill (Gentianac.).

Thylacium Spreng. = Thilachium Lour. (Capparidac.).

THYLACODRABA

Thylacodraba (Nábělek) O. E. Schulz (~Draba L.). Cruciferae. 1 As. Min.

Thylacophora Ridley (~Riedelia Oliv.). Zingiberaceae. 1 W. New Guinea.

Thylacopteris Kunze ex Mett. Polypodiaceae. 2 Malaysia.

Thylacospermum Fenzl. Caryophyllaceae. 1 C. As., Himal., W. China.

Thylactitis Steud. = Thylacitis Adans. = Gentiana L. (Gentianac.).

Thylax Rafin. = Zanthoxylum L. (Rutac.).

Thylaxus Rafin. = praec.

Thyloceras Steud. = Styloceras A. Juss. (Styloceratac.).

Thyloglossa Nees = Tyloglossa Hochst. (Acanthac.).

Thylostemon Kunkel (sphalm.) = Tylostemon Engl. = Beilschmiedia Nees (Laurac.).

Thymalis P. & K. = Tumalis Rafin. = Euphorbia L. (Euphorbiac.).

Thymbra L. Labiatae. 2 SE. Eur., SW. As.

Thymbra Mill. = Satureja L. (Labiat.).

Thymelaea Endl. Thymelaeaceae. 20 Medit., temp. As.

*Thymelaea Mill. (et Adans., Scop., All.) = Daphne L. (Thymelaeac.).

*Thymelaeaceae Juss. Dicots. 50/500 temp. & trop., esp. in Afr. Most are shrubs, some trees or lianes, few herbs, with tough fibrous bast, and entire alt. or opp. exstip. ls. Infl. basically racemose, often much condensed, capit. or fascic. Fl. usu. ♀, reg., 4–5-merous. Recept. much hollowed, usu. forming a deep tube of leafy consistence ('calyx-tube'); outgrowths of the axis are sometimes found at the base of the tube round the ovary. K petaloid, like the tube, imbr. or subvalv.; C conspic., imbr., or small or o; disk membr. or ± fleshy, ann., cupluar, or variously lobed or lacin.; A 4–5 or 8–10, sometimes ∞, rarely 4, inserted on edge of tube, more rarely at base of recept.; G̲ (1–2), rarely 3–8), with as many loc., each loc. with 1 axile or pariet. pend. anatr. ov. with ventral raphe; style simple, sometimes with small 'parastyles' at base. Achene, berry, or drupe, often enclosed in the persistent recept.; a few have caps. Seed carunculate or arillate. Embryo straight; endosp. little or none. Chief genera: *Gonystylus, Aquilaria, Gnidia, Lasiosiphon, Thymelaea, Daphne, Pimelea, Wikstroemia, Peddiea.* The family is a very natural one, with no very close affinities, but prob. connected through the subfam. *Gonystyloïdeae* with the *Sphaerosepalaceae* of Madagascar.

Thymelina Hoffmgg. = Gnidia L. (Thymelaeac.).

Thymium P. & K. = Tumion Rafin. = Torreya Arn. (Taxac.).

Thymophylla Lag. Compositae. 25 S. U.S., Mex., C. Am.

Thymophyllum Benth. & Hook. f. = praec.

*Thymopsis Benth. Compositae. 2 W.I.

Thymopsis Jaub. & Spach = Hypericum L. (Guttif.).

Thymos St-Lag. = seq.

Thymus L. Labiatae. 3–400 temp. Euras. Fls. gynodioec. with marked protandry. *T. vulgaris* L. (garden thyme) used in flavouring.

Thyopsis Wedd. = Tafalla D. Don (Compos.).

Thypha Costa = Typha L. (Typhac.).

Thyrasperma N. E. Brown = Apatesia N. E. Br. (Aïzoac.).

Thyridachne C. E. Hubbard. Gramineae. 1 trop. Afr.

Thyridiaceae Dulac = Orchidaceae Juss.

Thyridocalyx Bremek. Rubiaceae. 1 Madag.

Thyridostachyum Nees = Mnesithea Kunth (Gramin.).

Thyrocarpus Hance. Boraginaceae. 3 China.

Thyroma Miers = Aspidosperma Mart. & Zucc. (Apocynac.).

Thyrophora Neck. = Gentiana L. (Gentianac.).

Thyrsacanthus Nees = Odontonema Nees (Acanthac.).

Thyrsanthella (Baill.) Pichon. Apocynaceae. 1 SE. U.S.

Thyrsanthema Neck. = Chaptalia Vent. (Compos.).

Thyrsanthemum Pichon. Commelinaceae. 1 Mex.

Thyrsanthera Pierre ex Gagnep. Euphorbiaceae. 1 Indoch.

Thyrsanthus Benth. = Forsteronia G. F. W. Mey. (Apocynac.).

Thyrsanthus Ell. = Wisteria Nutt. (Legumin.).

Thyrsanthus Schrank = Lysimachia L. (Primulac.).

Thyrsia Stapf. Gramineae. 3–4 trop. Afr., NE. India, SE. As.

Thyrsine Gled. = Cytinus L. (Rafflesiac.).

Thyrsodium Salzm. ex Benth. Anacardiaceae. 7–8 trop. S. Am., W. Afr.

Thyrsopteridaceae Presl. Regarded by some authors as monotypic; if united to *Dicksoniaceae*, the name *Thyrsopteridaceae* has priority. Fronds dimorphic.

Thyrsopteridales Kunkel. Filicidae. 1 fam., *Thyrsopteridaceae*.

Thyrsopteris Kunze. Thyrsopteridaceae. 1 Juan Fernandez.

Thyrsosalacia Loes. Celastraceae. 1–2 W. Equat. Afr.

Thyrsosma Rafin. = Viburnum L. (Caprifoliac.).

Thyrsostachys Gamble. Gramineae. 2 Assam, Burma, Siam.

Thysamus Reichb. (sphalm.) = Thysanus Lour. = Cnestis Juss. (Connarac.).

Thysanachne Presl = Arundinella Raddi (Gramin.).

Thysanella A. Gray ex Engelm. & Gray (~ Polygonella Michx, ~ Polygonum L.). Polygonaceae. 2 SE. U.S.

Thysanella Salisb. = Thysanotus R.Br. (Liliac.).

Thysanobotrya v. Ald. v. Ros. = Cyathea Sm. (Cyatheac.).

Thysanocarex Börner = Carex L. (Cyperac.).

Thysanocarpus Hook. Cruciferae. 4–5 W. U.S.

Thysanochilus Falc. = ? Cyrtopera Lindl. = Eulophia R.Br. (Orchidac.).

Thysanoglossa C. Porto & Brade. Orchidaceae. 2 Brazil.

Thysanolaena Nees. Gramineae. 1 trop. As. (tiger-grass).

Thysanosoria Gepp. Lomariopsidaceae. 1 New Guinea. Like *Lomariopsis* but not acrostichoid.

Thysanospermum Champ. ex Benth. = Coptosapelta Korth. (Rubiac.).

Thysanostemon Maguire. Guttiferae. 2 Guiana.

Thysanostigma Imlay. Acanthaceae. 1 Siam.

***Thysanotus** R.Br. Liliaceae. 1 Siam, Indoch.; 1 S. China, Philipp., New Guinea, Queensl.; 30 Austr.

Thysantha Hook. = Thisantha Eckl. & Zeyh. = Tillaea L. = Crassula L. (Crassulac.).

Thysanurus O. Hoffm. Compositae. 1 Angola.

Thysanus Lour. = Cnestis Juss. (Connarac.).

Thyselium Rafin. = Thysselinum Hoffm. = Peucedanum L. (Umbellif.).

Thysselinum Adans. = Selinum L. (Umbellif.).

Thysselinum Hoffm. = Peucedanum L. (Umbellif.).

THYSSELINUM

Thysselinum Moench = Pleurospermum Hoffm. (Umbellif.).

Tianschaniella B. Fedtsch. & Popov. Boraginaceae. 1 C. As.

Tiaranthus Herb. = Pancratium L. (Amaryllidac.).

Tiarella L. Saxifragaceae. 5 Himal. & E. As., Pacif. & Atl. N. Am.

Tiaridium Lehm. = Heliotropium L. (Boraginac.).

Tiarrhena (Maxim.) Nakai = Miscanthus Anders. (Gramin.).

Tibestina Maire. Compositae. 1 Sahara.

Tibouchina Aubl. Melastomataceae. 200 + trop. Am.

Tibouchinopsis Markgraf. Melastomataceae. 1 NE. Brazil.

Tibuchina Rafin. = Tibouchina Aubl. (Melastomatac.).

Ticanto Adans. = Caesalpinia L. (Legumin.).

Ticorea Aubl. Rutaceae. 3 Guiana, Brazil.

Ticorea A. St-Hil. = Galipea Aubl. (Rutac.).

Tidestromia Standley. Amaranthaceae. 7 SW. U.S., Mex.

Tiedemannia DC. = Oxypolis Rafin. (Umbellif.).

Tiedmannia Torr. & Gray (sphalm.) = praec.

Tieghemella Pierre. Sapotaceae. 2 trop. W. Afr.

Tieghemia Balle. Loranthaceae. 2 trop. & S. Afr.

Tieghemopanax Viguier (~ Polyscias J. R. & G. Forst.). Araliaceae. 35 New Hebrid., New Caled., Austr.

Tienmuia Hu. Orobanchaceae. 1 E. China.

Tigarea Aubl. = Doliocarpus Roland. + Tetracera L. (Dilleniac.).

Tigarea Pursh = Purshia DC. (Rosac.).

Tiglium Klotzsch = Croton L. (Euphorbiac.).

Tigridia Juss. Iridaceae. 12 Mex. to Chile. In *T. pavonia* (L.) Ker-Gawl. (tiger flower) the fls. only last 8–12 hours.

Tikalia Lundell. Sapindaceae. 1 C. Am. (Guatem.).

Tilco Adans. = Fuchsia L. (Onagrac.).

Tikusta Rafin. = Tupistra Ker-Gawl. (Liliac.).

Tildenia Miq. = Peperomia Ruiz & Pav. (Peperomiac.).

Tilecarpus K. Schum. (sphalm.) = Tylecarpus Engl. = Medusanthera Seem.

Tilesia G. F. W. Mey. = Wulffia Neck. ex Cass. (Compos.). [(Icacinac.).

Tilesia Thunb. ex Steud. = Gladiolus L. (Iridac.).

Tilia L. Tiliaceae. 50 N. temp., S. to Indoch. & Mex. Note leaf-mosaic (see fam.). The upper surfaces of the ls. are usu. covered with honey-dew (see *Acer*) due to abundant aphids on under-surfaces of ls. above. Fls. in little cymes, arising from axils of ls. of current year; the axillary growing-point elongates transversely, giving rise to two buds, one of which forms the infl., the other the buds of next year's growth. The further development of the infl. is complex, but throughout there occurs 'adnation' of bracts to the axes arising in their axils, particularly noticeable in the first l. of the infl.-axis, which forms a wing, covering the fls. Honey is secreted at the base of the sepals. Fls. protandrous, dependent upon insects for fert.; largely visited by bees, etc., and a valuable source of honey. Fr. a nut. Endosp. very oily. The wood of *T.* × *europaea* L. (lime) and of *T. americana* L. (bass-wood) is useful. The inner fibre of the bark (bass) is very useful for tying.

★Tiliaceae Juss. Dicots. 50/450, trop. & temp., chiefly SE. As. and Brazil. Trees or shrubs, rarely herbs, with alt. stip. ls., often showing well-marked

2-ranked arrangement. In the trees, the shoots spread out horiz. and the insertions of the ls. are upon the upper half, so that the divergence is not $\frac{1}{2}$. The end bud of the branch does not develop in the next year. Frequently the l. is asymmetrical, with the larger side towards the branch. In the herbs, the ls. are in two ranks diverging at a right angle; torsion of the l. occurs later on and produces a dorsiventrality. The infl. is always, at least after the first branching, cymose, and often very complex, e.g. in *Tilia* and *Triumfetta* (*q.v.*). Fl. usu. ⚥, reg., 5–4-merous, sometimes with epicalyx. K 5 or (5), valvate; C 5, rarely 0, often glandular at base; A usu. ∞, free or united in groups, inserted at base of petals or on androphore, with dithecous anthers; G 2–∞-loc., with 1–∞ ov. in each; ov. usu. ascending, ± anatr.; style simple, with capitate or lobed stigma. Fr. usu. caps. or schizocarp; endosp. The *T.* yield useful timber, jute (*Corchorus*) and other fibre. The most constant distinction from *Malvaceae* is in the dithecous anthers. Stellate pubescence frequent. Chief genera: *Corchorus*, *Sparmannia*, *Tilia*, *Grewia*, *Microcos*, *Triumfetta*.

Tiliacora Colebr. Menispermaceae. 25 trop. Afr., Indomal.

Tilingia Regel & Tiling = Ligusticum L. (Umbellif.).

Tilioïdes Medik. = Tilia L. (Tiliac.).

Tillaea L. (~ Crassula L.). Crassulaceae. 40 almost cosmop.

Tillaeastrum Britton = praec.

Tillandsia L. Bromeliaceae. 500 warm Am., 1 W. Afr. Some resemble the rest of the fam.—epiphytes with rosette habit—while others, especially *T. usneoïdes* L. (long moss, Spanish moss, old man's beard, vegetable horsehair), show a different habit, hanging in long grey festoons from the branches of trees, looking rather like a lichen (esp. *Usnea*). At the base, each of the pendent stems is wound round its support, and as the apex grows on downwards the older parts die away, leaving the axile strand of sclerenchyma (the 'horsehair'). The whole pl. is thickly covered with the usual scaly hairs for absorbing the water trickling over it. It has no storage reservoir for water at all. The fls. appear but rarely. The pl. is largely distributed from tree to tree by the wind. Birds also use it for nesting and thus carry it about. It is used like horsehair.

Tillandsi[ac]eae Juss. = Bromeliaceae–Tillandsioïdeae Harms.

Tillea L. = Tillaea L. = Crassula L. (Crassulac.).

Tillia St-Lag. = praec.

Tilliandsia Michx = Tillandsia L. (Bromeliac.).

Tilmia O. F. Cook = Aïphanes Willd. (Palm.).

Tilocarpus auctt. = Tylecarpus Engl. = Medusanthera Seem. (Icacinac.).

Timaeosia Klotzsch = Gypsophila L. (Caryophyllac.).

Timandra Klotzsch = Croton L. (Euphorbiac.).

Timanthea Salisb. = Baltimora L. (Compos.).

Timbalia Clos = Pyracantha M. Roem. (Rosac.).

Timbuleta Steud. = Simbuleta Forsk. = Anarrhinum Desf. (Scrophulariac.).

Timeroya Benth. & Hook. f. = seq.

Timeroyea Montr. = Calpidia Thou. (Nyctaginac.).

Timmia J. F. Gmel. (1791; non Hedw. 1789—Musci) = Cyrtanthus Ait. (Amaryllidac.).

*Timonius DC. Rubiaceae. 2 Seychelles, Mauritius; 1 Ceylon; 1 Andaman Is.; 150 Malaysia (esp. New Guinea), Austr., Pacif.

TIMORON

Timoron Rafin. = Capnophyllum Gaertn. (Umbellif.).
Timotocia Moldenke = Casselia Nees & Mart. (Verbenac.).
Timouria Roshev. (~Achnatherum Beauv.). Gramineae. 1 C. As.
Tina Blume = Harpullia Roxb. (Sapindac.).
Tina Roem. & Schult. Sapindaceae. 16 Madag.
Tinaea Boiss. = Tinea Bivona = Neotinea Reichb. f. (Orchidac.).
Tinaea Garzia = Lamarkia Moench (Gramin.).
Tinantia Dumort. = Cyphella Herb. (Iridac.).
Tinantia Mart. & Gal. = Cyphomeris Standl. (Nyctaginac.).
*Tinantia Scheidw. Commelinaceae. 8 Mex. to trop. S. Am., W.I.
Tinda Reichb. = Streblus Lour. (Morac.).
Tinea Bivona = Neotinea Reichb. f. (Orchidac.).
Tinea Spreng. = Prockia P.Br. ex L. (Tiliac.).
Tineoa P. & K. (1) = Tinea Biv. = Neotinea Reichb. f. (Orchidac.).
Tineoa P. & K. (2) = Tinaea Garzia = Lamarkia Moench (Gramin.).
Tineoa P. & K. (3) = Tinea Spreng. = Prockia P.Br. ex L. (Flacourtiac.).
Tinguarra Parl. (~Athamanta L.). Umbelliferae. 1 Canaries, 1 Sicily.
Tingulong Rumph. = Protium Burm. f. (Burserac.).
Tingulonga Kuntze = praec.
Tiniaria (Meissn.) Reichb. = Fallopia Adans. (Polygonac.).
Tinnea Kotschy & Peyr. Labiatae. 30 trop. Afr.
Tinnea Vatke = Cyclocheilon Oliv. (Dicrastylidac.?).
Tinnia Nor. = Leea L. (Leeac.).
Tinomiscium Miers. Menispermaceae. 8 SE. As., Indomal.
Tinopsis Radlk. = Tina Roem. & Schult. (Sapindac.).
Tinosolen P. & K. = Teinosolen Hook. f. (Rubiac.).
Tinospora Miers. Menispermaceae. 40 trop. Afr., SE. As., Indomal., Austr.
Tinostachyum P. & K. = Teinostachyum Munro (Gramin.).
Tintinabulum Rydb. = Gilia Ruiz & Pav. (Polemoniac.).
Tintinnabularia R. E. Woodson. Apocynaceae. 1 C. Am.
Tinus Kuntze = Ardisia Sw. (Myrsinac.).
Tinus L. (1754) (nomen) = Premna L. (Verbenac.).
Tinus L. (1759) = Clethra Gronov. ex L. (Clethrac.).
Tinus Mill. = Viburnum L. (Caprifoliac.).
Tipalia Dennst. = Zanthoxylum L. (Rutac.).
Tipha Neck. = Typha L. (Typhac.).
Tiphogeton Ehrh. (uninom.) = Isnardia palustris L. = Ludwigia palustris (L.) Ell. (Onagrac.).
Tipuana Benth. Leguminosae. 1 trop. S. Am.
Tipularia Nutt. Orchidaceae. 4 Himal. to Japan, E. U.S.
Tiquilia Pers. = Coldenia L. (Ehretiac.).
Tiquiliopsis A. A. Heller = praec.
Tirania Pierre. Capparidaceae. 1 Indoch.
Tirasekia G. Don (sphalm.) = Jirasekia F. W. Schmidt = Anagallis L. (Primulac.).
Tiricta Rafin. = Daucus L. (Umbellif.).
Tirpitzia H. Hallier. Linaceae. 2 SW. China, Indoch.
Tirtalia Rafin. = Ipomoea L. (Convolvulac.).

Tirucalia Rafin. = Euphorbia L. (Euphorbiac.).

Tirucalla Rafin. = praec.

Tischleria Schwantes. Aïzoaceae. 1 S. Afr.

Tisonia Baill. Flacourtiaceae (?). 14 Madag. Pollen aberrant.

Tissa Adans. = Spergularia (Pers.) J. & C. Presl (Caryophyllac.).

Tisserantia Humbert. Compositae. 1 trop. Afr.

Tisserantiella Mimeur = Thyridachne C. E. Hubbard (Gramin.).

Tisserantiodoxa Aubrév. & Pellegr. (~ Bequaertiodendron De Wild.). Sapotaceae. 1 W. Equat. Afr.

Tisserantodendron Sillans = Fernandoa Welw. ex Seem. (Bignoniac.).

Tita Scop. = Cassipourea Aubl. (Rhizophorac.).

Titania Endl. = Oberonia Lindl. (Orchidac.).

Titanopsis Schwantes. Aïzoaceae. 8 S. Afr.

Titanotrichum Solereder. Gesneriaceae. 1 Formosa.

Titelbachia Klotzsch = Tittelbachia Klotzsch = Begonia L. (Begoniac.).

Tithonia Desf. ex Juss. Compositae. 10 Mex., C. Am., W.I. *T. diversifolia* A. Gray (Mexican sunflower) now a common weed in trop. As.

Tithonia Kuntze = Rivina L. (Phytolaccac.).

Tithonia Raeusch. = Rudbeckia L. (Compos.).

Tithymalaceae ('-oïdeae') Vent. = Euphorbiaceae Juss.

Tithymalis Rafin. = Tithymalus Seguier = Euphorbia L. (Euphorbiac.).

Tithymalodes Kuntze = seq.

Tithymaloïdes Ortega = Pedilanthus Neck. ex Poit. (Euphorbiac.).

Tithymalopsis Klotzsch & Garcke = Euphorbia L. (Euphorbiac.).

*Tithymalus Gaertn. (~ Euphorbia L.). Euphorbiaceae. 200 N. temp., Andes.

Tithymalus Mill. = Pedilanthus Neck. ex Poit. (Euphorbiac.).

Tithymalus Seguier = Euphorbia L. (Euphorbiac.).

Titragyne Salisb. = Rohdea Roth (Liliac.).

Tittelbachia Klotzsch = Begonia L. (Begoniac.).

*Tittmannia Brongn. Bruniaceae. 3–4 S. Afr.

Tittmannia Reichb. = Lindernia All. (Scrophulariac.).

Tityrus Salisb. = Narcissus L. (Amaryllidac.).

Tium Medik. = Astragalus L. (Legumin.).

Tjongina Adans. = Baeckea L. (Myrtac.).

Tjutsjau Rumph. = Salvia L. (Labiat.).

Tmesipteridaceae Reimers. Psilotales. Epiphytes on the trunks of tree ferns, or terrestrial; rhizome rootless; ls. simple, sporophylls deeply bilobed; sporangia 2-locular; prothallus saprophytic. (*Victorian Nat.* 71: 97–9, 1954.) Only gen.: *Tmesipteris.*

Tmesipteris Bernh. Tmesipteridaceae. 7 Austr., N.Z., Polynes.

Toanabo DC. (sphalm.) = Taonabo Aubl. = Ternstroemia L. f. (Theac.).

Tobagoa Urb. Rubiaceae. 1 W.I. (Tobago).

Tobaphes Phil. (sphalm.) = Jobaphes Phil. = Plazia Ruiz & Pav. (Compos.).

Tobinia Desv. = Fagara L. (Rutac.).

Tobira Adans. = Pittosporum Banks ex Soland. (Pittosporac.).

Tobium Rafin. = Poterium L. (Rosac.).

Tococa Aubl. Melastomataceae. 50 trop. S. Am.

Tocoyena Aubl. Rubiaceae. 20 Mex., trop. S. Am., W.I.

TODAROA

Todaroa Parl. Umbelliferae. 1 Canary Is.

Todaroa A. Rich. & Gal. = Campylocentron Benth. (Orchidac.).

Toddalia auctt. = Teclea Del. + Vepris Comm. ex A. Juss. (Rutac.).

*****Toddalia** Juss. Rutaceae. 1 trop. Afr., Madag., trop. As.

Toddaliopsis Engl. Rutaceae. 2–4 trop. Afr.

Todda-pana Adans. = Cycas L. (Cycadac.).

Toddavaddi Kuntze = Biophytum DC. (Oxalidac.).

Toddavaddia Kuntze = praec.

Todea Willd. Osmundaceae. 1 S. Afr., Austr., N.Z., *T. barbara* (L.) Moore.

Toechima Radlk. Sapindaceae. 8 New Guinea, Austr.

Toffieldia Schrank = seq.

Tofielda Pers. = seq.

Tofieldia Huds. Liliaceae. 20 N. temp., Venez., Guiana, Andes. *T. calyculata* (L.) Wahlenb., *T. pusilla* (Michx) Pers. and other spp. have a 3-lobed invol. (*calyculus*) beneath the K.

× **Toisochosenia** Kimura. Salicaceae. Gen. hybr. (Chosenia × Toisusu).

Toisusu Kimura = Salix L. (Salicac.).

Tokoyena Rich. ex Steud. = Tocoyena Aubl. (Rubiac.).

Tola Wedd. ex Benth. & Hook. f. = Lepidophyllum Cass. (Compos.).

Tolbonia Kuntze = Calotis R.Br. (Compos.).

Tollatia Endl. = Layia Hook. & Arn. (Compos.).

Tolmiaea Buek = seq.

Tolmiea Hook. = Cladothamnus Bong. (Ericac.).

*****Tolmiea** Torr. & Gray. Saxifragaceae. 1 Pacif. N. Am., *T. menziesii* (Pursh) Torr. & Gray. Adv. buds on upper part of petiole. Axial cup split down ant. side. Pets. thread-like; only 3 post. sta. occur.

Tolpis Adans. Compositae. 20 Azores, Canaries, Medit., S. to Abyss. & Somal.

Toluifera L. = Myroxylon L. f. (Legumin.).

Toluifera Lour. = Loureira Meissn. = Glycosmis Corrêa (Rutac.).

Tolumnia Rafin. = Oncidium Sw. (Orchidac.).

Tolypanthus (Blume) Reichb. Loranthaceae. 4 India, Ceylon, SE. China.

Tolypeuma E. Mey. = Nesaea Comm. ex Juss. (Lythrac.).

Tomantea Steud. = seq.

Tomanthea DC. (~ Centaurea L.). Compositae. 10 As. Min. to C. As.

Tomanthera Rafin. (~ Gerardia L.). Scrophulariaceae. 2 temp. & arct. N. Am.

Tomaris Rafin. = Rumex L. (Polygonac.).

Tombea Brongn. & Gris = Chiratia Montrouz. = Sonneratia L. f. (Sonneratiac.).

Tomex Forsk. = Dobera Juss. (Salvadorac.).

Tomex L. = Callicarpa L. (Verbenac.).

Tomex Thunb. = Litsea Lam. (Laurac.).

Tomiephyllum Fourr. = Scrophularia L. (Scrophulariac.).

Tomilix Rafin. = Macranthera Torr. (Scrophulariac.).

Tomista Rafin. = Florestina Cass. (Compos.).

Tommasinia Bertol. (~ Angelica L., ~ Peucedanum L.). Umbelliferae. 1 Alps to Balkans.

Tomodon Rafin. = Hymenocallis Salisb. (Amaryllidac.).

Tomostima Rafin. = Draba L. (Crucif.).

Tomostina Willis (sphalm.) = praec.

Tomostoma Merr. (sphalm.) = praec.

Tomostylis Montr. = Crossostyles J. R. & G. Forst. (Rhizophorac.).

Tomotris Rafin. = Corymborkis Thou. + Tropidia Lindb. (Orchidac.).

Tomoxis Rafin. = Ornithogalum L. (Liliac.).

Tonabea Juss. = Taonabo Aubl. = Ternstroemia L. f. (Theac.).

Tonalanthus T. S. Brandegee = Calea L. (Compos.).

Tonca Rich. = Bertholletia Humb. & Bonpl. (Lecythidac.).

Tondin Vitm. = Paullinia L. (Sapindac.).

Tonduzia Boeck. ex Tonduz = Durandia Boeck. (Cyperac.).

Tonduzia Pittier. Apocynaceae. 7 C. Am. to Brazil.

Tonella Nutt. ex A. Gray. Scrophulariaceae. 3 W. U.S.

Tonestus A. Nelson = Haplopappus Cass. (Compos.).

Tongoloa H. Wolff. Umbelliferae. 14 W. Himal., Tibet, W. China.

Tonguea Endl. = Sisymbrium L. (Crucif.).

Tonina Aubl. Eriocaulaceae. 1 C. & trop. S. Am., W.I.

Tonningia Neck. = Cyanotis D. Don (Commelinac.).

Tonsella Schreb. = Tontelea Aubl. (Celastrac.).

Tonshia Buch.-Ham. ex D. Don = Saurauia Willd. (Actinidiac.).

Tontanea Aubl. = Coccocypselum P.Br. (Rubiac.).

Tontelea Aubl. Celastraceae. 30 C. & trop. S. Am.

Tooldia Lehm. ex Wittst. (nomen). Quid?

Toona M. Roem. (~ Cedrela P.Br.). Meliaceae. 15 trop. As., Austr.

Topeinostemon C. Muell. (sphalm.) = Tapeinostemon Benth. (Gentianac.).

Topiaris Rafin. = Cordia L. (Ehretiac.).

Topobea Aubl. Melastomataceae. 40–50 warm Am. Ed. fr.

Toppingia O. & I. Degener = Pseudophegopteris Ching (Thelypteridac.).

Toquera Rafin. = Cordia L. (Ehretiac.).

Torcula Nor. = Pithecellobium Mart. (Legumin.).

Tordylioïdes Wall. ex DC. = Heracleum L. (Umbellif.).

Tordyliopsis DC. = praec.

Tordylium L. Umbelliferae. 6 Eur., N. Afr., SW. As.

Toreala B. D. Jacks. (sphalm.) = Torcula Nor. = Pithecellobium Mart. (Legumin.).

Torenia L. Scrophulariaceae. 50 trop.

Toresia Pers. = Torresia Ruiz & Pav. = Hierochloë R.Br. (Gramin.).

Torfasadis Rafin. = Euphorbia L. (Euphorbiac.).

Torfosidis B. D. Jacks. (sphalm.) = praec.

Torgesia Bornm. = Crypsis Ait. (Gramin.).

Toricellia DC. Toricelliaceae. 3 E. Himalaya, W. China.

Toricelliaceae (Wang.) Hu (corr.). Dicots. 1/3 Himal., China. Small trees with thick branches and broad pith. Ls. alt., ± palmatilobed, ent. or coarsely toothed, palmately nerved, long-pet., pet. broadly sheathing at base, exstip. Fls. reg., ♂ ♀, small, in lax ∞-fld. pend. thyrses. ♂: K (5), lobes open, ± unequal; C 5, indupl.-valv., apex elong., inflexed; A 5, alternipet., fil. short, anth. basifixed, latero-intr.; disk ± flat; pistillode of 1–3 subul. processes. ♀: K (3–5), minutely toothed; C o; A (stds.) o; disk inconspic.; Ḡ (3–4), with 3 thick subul. erect or divaric. persist. stigs., and 1 apical pend. anatr. ov. per loc. (often partly ster.), funicle thickened to form an obturator. Fr. an

TORILIS

obliquely ovoid 3–4-loc. 1-seeded drupe; seed lin., curved, with fleshy endosp. Only genus: *Toricellia*. Intermediate between *Cornac*. and *Araliac*.; anatomically closer to latter.

Torilis Adans. Umbelliferae. 15 Canaries, Medit. to E. As.
Tormentilla L. = Potentilla L. (Rosac.).
Torminalis Medik. = Sorbus L. (Rosac.).
Torminaria Opiz = praec.
Tornabenea Parl. ex Webb. Umbelliferae. 3 Cape Verde Is.
Tornabenia Benth. & Hook. f. = praec.
Tornelia Gutierrez ex Schlechtd. = Monstera Adans. (Arac.).
Torpesia M. Roem. = Trichilia L. (Meliac.).
Torralbasia Krug & Urb. Celastraceae. 2 W.I.
Torrcya Rafin. (sphalm.) = Torreya Rafin. (1818) = Synandra Nutt. (Labiat.).
Torrentia Vell. = Ichthyothere Mart. (Compos.).
Torrenticola Domin. Podostemaceae. 1 New Guinea, Queensland.
Torresea Allem. Leguminosae. 2 Brazil.
Torresia Ruiz & Pav. = Hierochloë R.Br. (Gramin.).
Torresia Willis (sphalm.) = Torresea Allem. (Legumin.).
*Torreya Arn. Taxaceae. 6 E. As., Calif., Florida. The timber is useful.
Torreya Croom ex Meissn. = Croomia Torr. ex Torr. & Gray (Croomiac.).
Torreya Eaton = Mentzelia L. (Loasac.).
Torreya Rafin. (1818) = Synandra Nutt. (Labiat.).
Torreya Rafin. (1819) = Pycreus Beauv. = Cyperus L. (Cyperac.).
Torreya Spreng. = Clerodendrum L. (Verbenac.).
Torreyaceae Nak. = Taxaceae Spreng.
Torreyochloa Church = Glyceria L. (Gramin.).
Torricellia Harms ex Diels = Toricellia DC. (Toricelliac.).
Torricelliaceae Hu = Toricelliaceae Hu corr. Airy Shaw.
Torrubia Vell. (1825; non Léveillé ex Tul. 1865—Fungi) = Pisonia L. (Nyctaginac.).
Tortipes Small = Uvularia L. (Liliac.).
Tortuella Urb. Rubiaceae. 1 S. Domingo (Haiti).
Tortula Roxb. ex Willd. (1800; non Hedw. 1782—Musci) = Priva Adans. (Verbenac.).
Torularia O. E. Schulz. Cruciferae. 16 Medit. to C. As. & Afghan.
Torulinium Desv. = Cyperus L. (Cyperac.).
Torymenes Salisb. = Amomum L. (Zingiberac.).
Tosagris Beauv. = Muhlenbergia Schreb. (Gramin.).
Toubaouate Kunkel (sphalm.) = seq.
Toubasuate Aubrév. & Pellegr. Leguminosae. 1 trop. Afr.
Touchardia Gaudich. Urticaceae. 1 Hawaii.
Touchiroa Aubl. = Crudia Schreb. (Legumin.).
Toulichiba Adans. = Ormosia G. Jacks. (Legumin.).
Toulicia Aubl. Sapindaceae. 14 trop. S. Am.
Touloucouna M. Roem. = Carapa Aubl. (Meliac.).
Toumboa Naud. = Tumboa Welw. = Welwitschia Hook. f. (Welwitschiac.).
Toumeya Britton & Rose = Pediocactus Britt. & Rose (Cactac.).
Tounatea Aubl. = Swartzia Schreb. (Legumin.).

Tournefortia L. Boraginaceae. 150 trop. & subtrop. Trees and shrubs.
Tournefortiopsis Rusby. Rubiaceae. 3 Andes.
Tournesol Adans. = Chrozophora Neck. ex Juss. (Euphorbiac.).
Tournesolia Scop. = praec.
Tourneuxia Coss. Compositae. 1 Algeria.
Tournonia Moq. Basellaceae. 1 Colombia.
Tourolia Stokes = Touroulia Aubl. (Quiinac.).
Touroubea Steud. = Souroubea Aubl. (Marcgraviac.).
Touroulia Aubl. Quiinaceae. 4 trop. S. Am.
Tourretia Fougeroux = seq.
***Tourrettia** Fougeroux corr. DC. Bignoniaceae. 1 C. Am. to Peru.
Toussaintia Boutique. Annonaceae. 1 trop. Afr.
Touterea Eaton & Wright = Mentzelia L. (Loasac.).
Touteria Willis (sphalm.) = praec.
Tovara Adans. = Antenoron Rafin. (Polygonac.).
Tovaria Neck. = Smilacina Desf. (Liliac.).
***Tovaria** Ruiz & Pav. Tovariaceae. 1–2 Mex., Jamaica, trop. Andes.
***Tovariaceae** Pax. Dicots. 1/2 warm Am. Shrubs or ann. herbs, with alt.
trifoliol. stip. ls., lfts. ent.; plant smelling of *Apium* or *Cestrum* when fresh,
and of coumarin when dried. Fls. reg., ⚥, in loose elong. ∞-fld. term. rac.
K 8, imbr.; C 8, shortly clawed, imbr.; A 8, oppositisep., with lanceol. long-
papill. fil. and basifixed intr. anth.; G̲ (6–8), with short style and pelt. lobul.
stig., and ∞ ov. on thickened spongy plac. Fr. a glob. green slender-pedic.
∞-seeded berry, mucilaginous when young; pericarp membr.; seeds with
curved embr. and sparse endosp. Only genus: *Tovaria*. Related to *Cleomac*.
Tovomia Pers. = seq.
Tovomita Aubl. Guttiferae. 60 C. & trop. S. Am.
Tovomitidium Ducke. Guttiferae. 2 Brazil.
Tovomitopsis Planch. & Triana (~ Chrysochlamys Poepp.). Guttiferae. 10
Townsendia Hook. Compositae. 21 W. N. Am., Mex.   [trop. Am.
Townsonia Cheeseman. Orchidaceae. 2 Tasmania, N.Z.
Toxanthera Endl. ex Grüning = Monotaxis Brongn. (Euphorbiac.).
Toxanthera Hook. f. = Kedrostis Medik. (Cucurbitac.).
Toxanthes Turcz. Compositae. 3 W. & S. Austr.
Toxanthus Benth. = praec.
Toxicaria Aepnel ex Steud. = Antiaris Leschen. (Morac.).
Toxicaria Schreb. = Strychnos L. (Strychnac.).
Toxicodendron auctt. = Toxicodendrum Thunb. = Hyaenanche Lamb. & Vahl
(Euphorbiac.).
Toxicodendron Mill. (~ Rhus L.). Anacardiaceae. 15 E. As., N. & S. Am.
Toxicodendrum Gaertn. = Allophylus L. (Sapindac.).
Toxicodendrum Thunb. = Hyaenanche Lamb. & Vahl (Euphorbiac.).
Toxicophlaea Harv. = Acokanthera G. Don (Apocynac.).
Toxicophloea Lindl. = praec.
Toxicoscordion Rydb. (~ Zigadenus Michx). Liliaceae. 10 N. Am.
Toxina Nor. = Allophylus L. (Sapindac.).
Toxocarpus Wight & Arn. Asclepiadaceae. 40 trop. Afr., Madag., Masc.,
SE. As., Indomal.

TOXOPHOENIX

Toxophoenix Schott = Astrocaryum G. F. W. Mey. (Palm.).
Toxopteris Trevis. = Syngramma J. Sm. (Gymnogrammac.).
Toxopus Rafin. = Macranthera Torr. (Scrophulariac.).
Toxosiphon Baill. = Erythrochiton Nees & Mart. (Rutac.).
Toxostigma A. Rich. = Arnebia Forsk. (Boraginac.).
Toxotrophis Planch. (sphalm.) = Taxotrophis Blume (Morac.).
Toxotropis Turcz. = Corynella DC. (Legumin.).
Toxylon Rafin. = Maclura Nutt. (Morac.).
Toxylus Rafin. = praec.
Tozzettia Parl. = Fritillaria L. (Liliac.).
Tozzettia Savi = Alopecurus L. (Gramin.).
Tozzia L. Scrophulariaceae. 1–2 Alps, Carpathians. Semi-parasites, with loose-pollen fls. (see fam.).
Tracanthelium Kit. ex Schur = Phyteuma L. (Campanulac.).
Tracaulon Rafin. = Polygonum L. (Polygonac.).
Trachelanthus Klotzsch = Begonia L. (Begoniac.).
Trachelanthus Kunze. Boraginaceae. 4 Persia to C. As.
Trachelioïdes Opiz = Campanula L. (Campanulac.).
Tracheliopsis Buser. Campanulaceae. 5 Medit.
Tracheliopsis Opiz = Campanula L. (Campanulac.).
Trachelium Hill = Campanula L. (Campanulac.).
Trachelium L. Campanulaceae. 7 Medit.
Trachelocarpus C. Muell. = Begonia L. (Begoniac.).
Trachelosiphon Schlechter = Eurystyles Wawra (Orchidac.).
Trachelospermum Lem. Apocynaceae. 30 India to Japan, SE. U.S.
Trachinema Rafin. = Anthericum L. (Liliac.).
Trachodes D. Don = Sonchus L. (Compos.).
Trachomitum Woodson. Apocynaceae. 6 S. Russia to China.
Trachopyron Rafin. = Fagopyrum Mill. (Polygonac.).
Trachyandra Kunth = Anthericum L. (Liliac.).
Trachycalymma (K. Schum.) Bullock. Asclepiadaceae. 4 trop. Afr.
Trachycarpus H. Wendl. Palmae. 8 Himal., E. As.
Trachycaryon Klotzsch = Adriana Gaudich. (Euphorbiac.).
× **Trachycnemum** Maire & G. Samuelsson. Cruciferae. Gen. hybr. (Ceratocnemum × Trachystoma).
Trachydium Lindl. Umbelliferae. *S.str.* 10, *s.l.* 40, C. As. & Persia to W. China.
Trachylobium Hayne. Leguminosae. 1 trop. E. Afr., Madag., Maurit. Yields copal resin, which is dug up from the soil near the roots or in a half-fossilised condition from places where trees once existed.
Trachyloma Pfeiff. = seq.
Trachylomia Nees = Scleria Bergius (Cyperac.).
Trachymarathrum Tausch = Hippomarathrum Link (Umbellif.).
Trachymene DC. = Platysace Bunge (Umbellif.).
Trachymene Rudge. Hydrocotylaceae (~ Umbelliferae). 40 Philipp., Borneo, E. Malaysia, Austr., New Caled., Fiji.
Trachynia Link (~ Brachypodium Beauv.). Gramineae. 1 Medit. to W.
Trachynotia Michx = Spartina Schreb. (Gramin.). [Pakistan.

Trachyozus Reichb. = Trachys Pers. (Gramin.).
Trachyphrynium Benth. Marantaceae. 1 trop. Afr.
Trachyphytum Nutt. ex Torr. & Gray = Mentzelia L. (Loasac.).
Trachypleurum Reichb. = Bupleurum L. (Umbellif.).
Trachypoa Bub. = Dactylis L. (Gramin.).
Trachypogon Nees. Gramineae. 10 trop. Am., trop. & S. Afr., Madag.
Trachypteris André ex Christ. Hemionitidaceae. 2 S. Am., Galápagos, Madag.
Trachypyrum P. & K. = Trachopyron Rafin. = Fagopyrum Mill. (Polygonac.).
Trachyrhynchium Nees = Cladium P.Br. (Cyperac.).
Trachyrhyngium Kunth = praec.
Trachys Pers. Gramineae. 1 S. India & Burma, esp. coasts.
Trachysciadium Eckl. & Zeyh. = Chamarea Eckl. & Zeyh. (Umbellif.).
Trachysperma Rafin. = Nymphoïdes Hill (Menyanthac.).
*****Trachyspermum** Link. Umbelliferae. 20 trop. & NE. Afr. to C. As., India & W. China.
Trachystachys A. Dietr. = Trachys Pers. (Gramin.).
Trachystella Steud. = Trachytella DC. = Tetracera L. (Dilleniac.).
Trachystemma auctt. = seq.
Trachystemon D. Don. Boraginaceae. 2 Medit.
Trachystigma C. B. Clarke. Gesneriaceae. 1 trop. Afr.
Trachystoma O. E. Schulz. Cruciferae. 3 Morocco.
Trachystylis S. T. Blake. Cyperaceae. 2 Queensland.
Trachytella DC. = Tetracera L. (Dilleniac.).
Trachytheca Nutt. ex Benth. = Eriogonum Michx (Polygonac.).
Trachythece Pierre. Sapotaceae. 1 W. Equat. Afr. Quid?
Tractema Rafin. = Scilla L. (Liliac.).
Tractocopevodia Raiz. & Narayanasw. Rutaceae. 1 Burma.
Tracyanthus Small = Zigadenus Michx (Liliac.).
Tracyina Blake. Compositae. 1 Calif.
Tradescantella Small = Phyodina Rafin. (Commelinac.).
Tradescantia L. Commelinaceae. 60 N. Am. to trop. S. Am. 6 perfect sta. covered with hairs. Protandr. Infl. a cincinnus.
Traevia Neck. = Trewia L. (Euphorbiac.).
Tragacantha Mill. = Astragalus L. (Legumin.).
Traganopsis Maire & Wilczek. Chenopodiaceae. 1 Morocco.
Tragantha Endl. (sphalm.) = seq.
Traganthes Wallr. = Eupatorium L. (Compos.).
Traganthus Klotzsch = Bernardia Mill. (Euphorbiac.).
Traganum Delile. Chenopodiaceae. 3 Canaries, Morocco & Sahara to [Arabia.
Tragia L. Euphorbiaceae. 100 trop., subtrop.
Tragiella Pax & K. Hoffm. = Sphaerostylis Baill. (Euphorbiac.).
Tragiola Small & Pennell. Scrophulariaceae. 1 E. & S. U.S.
Tragiopsis Karst. = Sebastiania Spreng. (Euphorbiac.).
Tragiopsis Pomel. Umbelliferae. 4 Medit., S. Afr.
Tragium Spreng. = Pimpinella L. (Umbellif.).
Tragoceras Spreng. = seq.
Tragoceros Kunth. Compositae. 5 Mex., C. Am. The C of ♀ fl. becomes rigid after fert., and forms a double hook upon the fr.

TRAGOLINUM

Tragolinum Rafin. = Pimpinella L. (Umbellif.).

Tragopogon L. Compositae. 50 temp. Euras.; 1 S. Afr. The fl.-heads of the Eur. spp. close at midday, whence the English name of 'John-go-to-bed-at-noon'. *T. porrifolius* L. (salsify) sometimes grown as a vegetable.

Tragopogonodes Kuntze = Urospermum Scop. (Compos.).

Tragopyrum Bieb. = Atraphaxis L. (Polygonac.).

Tragoriganum Gronov. = Satureja L. (Labiat.).

Tragoselinum Mill. = Pimpinella L. (Umbellif.).

Tragosma C. A. Mey. ex Ledeb. = Cymbocarpum DC. (Umbellif.).

Tragularia Koen. ex Roxb. = Pisonia L. (Nyctaginac.).

*Tragus Haller. Gramineae. 6 warm Afr., 1 pantrop.

Tragus Panz. = Festuca L. (Gramin.).

Traillia Lindl. ex Endl. = Schimpera Hochst. (Crucif.).

Trailliaedoxa W. W. Smith & Forrest. Rubiaceae. 1 SW. China.

Trallesia Zumaglini = Matricaria L. (Compos.).

Tralliana Lour. = Colubrina Rich. (Rhamnac.).

Trambis Rafin. = Phlomis L. (Labiat.).

Tramoia Schwacke & Taub. ex Glaziou (nomen). Urticaceae. Brazil. Quid?

Trankenia Thunb. (sphalm.) = Frankenia L. (Frankeniac.).

Trapa L. Trapaceae. 30 (or 1 polymorphic) C. & SE. Eur., temp. & trop. As. & Afr.

*Trapaceae Dum. 1/30 warm Euras. & Afr. Floating aquatic herbs, with opp. or ternate pinnatisect root-like submerged ls., and rosulate long-pet. rhomboid proximally ent. distally dent. floating ls., pet. with an ellipsoid air-filled 'float' midway, exstip. Fls. reg., ☿, sol., axill., on short pubesc. pedic. K (4), valv., sometimes spinose; C 4, imbr.; A 4, alternipet., fil. short, anth. intr.; disk cupular, angular, lobulate; G (2), semi-inf., with ± elong. simple style and capit. stig., and 1 apical axile pend. anatr. ov. per loc. Fr. a ± large woody or bony ± turbinate variously sculptured or cornute or spinose 'nut' or false drupe, 1-loc., 1-seeded; seed large, without endosp., edible; cots. v. unequal; germination through apical pore left after fall of style. Only genus: *Trapa*. Closely related to *Onagraceae*.

Trapaulos Reichb. = Traupalos Rafin. = Hydrangea L. (Hydrangeac.).

Trapella Oliv. Trapellaceae. 1–2 E. As.

Trapellaceae (F. W. Oliver) Honda & Sakisake. Dicots. 1/2 E. As. Aquatic perenn. herbs with creeping rhiz., stems ± floating. Ls. opp., simple, dent., exstip., the lower oblong, remotely serr., the upper (floating) deltoid-rotundate, crenate. Fls. zygo., ☿, sol., axill., chasmogam. from floating ls., cleistogam. from submerged ls., usu. only 1 fl. devel. at each node. K 5, imbr., persist.; C (5), slightly bilab. (2 upper, 3 lower lobes), upper lip exterior, lobes rounded, imbr.; A 2, + 2 stds., fert. sta. latero-post., stds. antic., epipet., fil. filif., anth. biloc., seated on large pelt. connective, included; G̅ (2), postic. loc. with 2 apic. axile pend. anatr. ov., antic. loc. abortive and empty, style slender with uneq. bilab. stig. Fr. narrow-elong., 1-seeded, indehisc., crowned with 5 spreading rigid appendages below the K: 3 elongate, slender, uncinate, 2 short, subul., spinose; seed with thin endosp. Only genus: *Trapella*. Prob. closely related to *Scrophulariaceae–Gratioleae*, with which it agrees in the aquatic habitat, androecium, and pollen; differing in the inf. 2-ovulate ovary and

TREMANDRACEAE

appendaged indehisc. fr. In the last char. it shows much similarity to the *Pedaliac.*, but the geogr. distr., habitat and pollen argue against a close relationship.

Trasera Rafin. = Frasera Walt. (Gentianac.).
Trasi Lestib. = seq.
Trasis Beauv. = Cladium P.Br. (Cyperac.).
Trasus S. F. Gray = Carex L. (Cyperac.).
Trattenikia Pers. = Marshallia Schreb. (Compos.).
Trattinickia auctt. = seq.
Trattinickya A. Juss. = seq.
Trattinnickia Willd. Burseraceae. 11 trop. S. Am.
Traubia Mold. Amaryllidaceae. 1 Chile.
Traunsteinera Reichb. Orchidaceae. 2 Eur., Medit.
Traupalos Rafin. = Hydrangea L. (Hydrangeac.).
Trautvetteria Fisch. & Mey. Ranunculaceae. 1 E. As., Pacif. N. Am., Mex.
Traversia Hook. f. = Senecio L. (Compos.).
Traxara Rafin. = Lobostemon Lehm. (Boraginac.).
Traxilisa Rafin. = Tetracera L. (Dilleniac.).
Traxilum Rafin. = Ehretia L. (Ehretiac.).
Trecacoris Pritz. (sphalm.) = Thecacoris A. Juss. (Euphorbiac.).
Trechonaetes Miers. Solanaceae. 3 Chile.
Treculia Decne ex Tréc. Moraceae. 12 trop. Afr., Madag. The seeds of *T. africana* Decne (*okwa*) are ground into meal.
Treichelia Vatke. Campanulaceae. 1 S. Afr.
Treisia Haw. = Euphorbia L. (Euphorbiac.).
Treisteria Griff. = Curanga Juss. (Scrophulariac.).
Treleasea Rose (1899; non *Treleasia* Spegazzini 1896—Fungi) = Setcreasea K. Schum. & Sydow (Commelinac.).
Trema Lour. Ulmaceae. 30 trop. & subtrop.
Tremacanthus S. Moore. Acanthaceae. 1 Brazil.
Tremacron Craib. Gesneriaceae. 3 SW. China.
Tremandra R.Br. ex DC. Tremandraceae. 2 W. Austr.
Tremandr[ac]eae R.Br., nom. illegit. (*Tremandra* tunc nondum descripta) = seq.
***Tremandraceae** DC. Dicots. 3/25 Austr. Small slender subshrubs, or subherbaceous, stems occas. winged and/or leafless. Ls. alt., opp. or vertic., simple, ent. or dent., often narrow, exstip., often glandular, sometimes stell.-toment. Fls. reg., ⚥, sol., axill., on slender pedic. K (3-)4-5, valv.; C (3-)4-5, indupl.-valv.; disk rarely present (*Tremandra*), annular, lobulate; A (6-)8-10, fil. v. short, anth. elong., ± beaked, basifixed, opening by 1 apic. pore; G (2), with slender style and inconspic. stig., and 1-2 apical pend. anatr. ov. per loc. Fr. a compressed 2-loc. loculicid. (sometimes also septicid.) caps.; seeds with copious endosp., sometimes pilose, often with large chalazal appendage (?aril). Genera: *Tremandra, Tetratheca, Platytheca*. Relationships much disputed; often associated with *Pittosporac.* or *Polygalac.* Habit of *Tetratheca* spp. suggestive of some *Rutac.* (*Boronia*, etc.), or *Ericac.* (*Erica, Vaccinium*, etc.; cf. also anthers!).

1165

38-2

TREMANTHERA

Tremanthera P. & K. = Trematanthera F. Muell. = Saurauia Willd. (Actinidiac.).

Tremanthus Pers. = Styrax L. (Styracac.).

Tremasperma Rafin. = Calonyction Choisy (Convolvulac.).

Tremastelma Rafin. Dipsacaceae. 1 Balkan Penins., E. Medit.

Trematanthera F. Muell. = Saurauia Willd. (Actinidiac.).

Trematocarpus Zahlbr. (~ Lobelia L.). Campanulaceae. 4 Hawaii.

Trematolobelia Zahlbr. ex Rock = praec.

Trematosperma Urb. = Pyrenacantha Hook. ex Wight (Icacinac.).

Trembleya DC. Melastomataceae. 14 S. Brazil.

Tremolsia Gandog. = Atractylis L. (Compos.).

Tremotis Rafin. (Synoecia Miq.) = Ficus L. (Morac.).

Tremula Dum. = Populus L. (Salicac.).

Tremularia Fabr. (uninom.) = *Briza* L. sp. (Gramin.).

Trendelenburgia Klotzsch = Begonia L. (Begoniac.).

Trentepohlia Boeck. (1858) = Cyperus L. (Cyperac.).

Trentepohlia Roth (1800; non *Mart. 1817—Algae) = Heliophila Burm. f. (Crucif.).

Trepnanthus Steud. = Terpnanthus Nees & Mart. = Spiranthera A. St-Hil. (Rutac.).

Trepodandra Durand (sphalm.) = Tripodandra Baill. (Menispermac.).

Trepocarpus Nutt. ex DC. Umbelliferae. 1 S. U.S.

Tresanthera Karst. Rubiaceae. 2 Venezuela, W.I. (Tobago).

Tresteira Hook. f. (sphalm.) = Treisteria Griff. = Curanga Juss. (Scrophulariac.).

Tretocarya Maxim. = Microula Benth. (Boraginac.).

Tretorhiza Adans. = Gentiana L. (Gentianac.).

Treubania Van Tiegh. = Decaisnina Van Tiegh. (Loranthac.).

Treubaniaceae Van Tiegh. = Loranthaceae–Elytranthinae Engl.

Treubella Pierre = Palaquium Blanco (Sapotac.).

Treubella Van Tiegh. = Treubania Van Tiegh. = Decaisnina Van Tiegh. (Loranthac.).

Treubellaceae Van Tiegh. = Treubaniaceae Van Tiegh. = Loranthaceae–Elytranthinae Engl.

Treubia Pierre ex Boerl. = Lophopyxis Hook. f. (Lophopyxidac.).

Treuia Stokes = Trewia L. (Euphorbiac.).

Treutlera Hook. f. Asclepiadaceae. 1 E. Himal.

Trevauxia Steud. = Trevouxia Scop. = Luffa L. (Cucurbitac.).

Trevesia Vis. Araliaceae. 10 Indomal., Pacif.

Trevia L. = Trewia L. (Euphorbiac.).

Treviaceae Bullock = Trewiaceae Lindl.

Trevirana Willd. = Achimenes P.Br. (Gesneriac.).

Trevirania Heynh. = Psychotria L. (Rubiac.).

Trevirania Roth = Lindernia All. (Scrophulariac.).

Trevirania Spreng. = Trevirana Willd. = Achimenes P.Br. (Gesneriac.).

Trevoa Miers ex Hook. Rhamnaceae. 6 Andes.

× **Trevorara** hort. Orchidaceae. Gen. hybr. (iii) (Arachnis × Phalaenopsis × Vanda).

Trevoria F. C. Lehmann. Orchidaceae. 2 Colombia, Ecuador.
Trevouxia Scop. = Luffa L. (Cucurbitac.).
Trewia L. Euphorbiaceae. 2 W. Himal. & Ceylon to SE. As. & Hainan; scarce in W. Malaysia. Fr. a drupe.
Trewiaceae Lindl. = Euphorbiaceae–Mercurialinae–Trewiiformes Pax & Hoffm., *p.p.*
Triachne Cass. = Nassauvia Comm. ex Juss. (Compos.).
Triachyrium Benth. = seq.
Triachyrum Hochst. = Sporobolus R.Br. (Gramin.).
Triacis Griseb. = Turnera L. (Turnerac.).
Triacma van Hass. ex Miq. = Hoya R.Br. (Asclepiadac.).
Triactina Hook. f. & Thoms. = Sedum L. (Crassulac.).
Triadenia Miq. = Trachelospermum Lem. (Apocynac.).
Triadenia Spach = Hypericum L. (Guttif.).
Triadenum Rafin. Guttiferae. 6–10 Assam, temp. E. As., E. N. Am.
Triadica Lour. = Sapium P.Br. (Euphorbiac.).
Triaena Kunth = Botelua Lag. (Gramin.).
Triaenacanthus Nees corr. Bremek. Acanthaceae. 1 Assam.
Triaenanthus Nees (sphalm.) = praec.
Triaenophora Solereder. Scrophulariaceae. 2 China.
Triaina Kunth = Triaena Kunth = Botelua Lag. (Gramin.).
Triainolepis Hook. f. Rubiaceae. 2 trop. E. Afr., Madag.
Triallosia Rafin. = Lachenalia Jacq. (Liliac.).
Trianaea Planch. & Linden. Solanaceae. 3 N. Andes.
Trianaeopiper Trelease. Piperaceae. 18 N. Andes.
Triandrophora Schwarz = Cleome L. (Cleomac.).
Trianea Karst. = Limnobium Rich. (Hydrocharitac.).
Triangis Thou. (uninom.) = *Angraecum triquetrum* Thou. (Orchidac.).
Trianoptiles Fenzl. Cyperaceae. 3 S. Afr.
Trianosperma Mart. = Cayaponia S. Manso (Cucurbitac.).
Triantha Baker = Tofieldia Huds. (Liliac.).
Trianthaea Spach = Vernonia Schreb. (Compos.).
Trianthella House = Triantha Baker = Tofieldia Huds. (Liliac.).
Trianthema L. Aïzoaceae. 20 trop. & subtrop. Afr., As., & esp. Austr.; 1 trop. Am.
Trianthema Spreng. ex Turcz. = Adenocline Turcz. (Euphorbiac.).
Trianthera Wettst. (1891; non Conwentz 1886—gen. foss.) = Stemotria Wettst. & Harms (Scrophulariac.).
Trianthium Desv. = Chrysopogon Trin. (Gramin.).
Trianthus Hook. f. = Nassauvia Comm. ex Juss. (Compos.).
Triarthron Baill. = Phthirusa Mart. (Loranthac.).
Trias Lindl. Orchidaceae. 6 SE. As.
Triascidium Benth. & Hook. f. (sphalm.) = Trisciadium Phil. = Huanaca Cav. (Umbellif.).
Triasekia G. Don (sphalm.) = Jirasekia F. W. Schmidt = Anagallis L. (Primulac.).
Triaspis Burchell. Malpighiaceae. 18 trop. & S. Afr.
Triathera Desv. = Botelua Lag. (Gramin.).

TRIATHERA

Triathera Roth ex Roem. & Schult. = Tripogon Roem. & Schult. (Gramin.).

Triatherus Rafin. = Ctenium Panz. (Gramin.).

Triavenopsis Candargy [Kantartzis] = Duthiea Hack. (Gramin.).

Tribelaceae (Engl.) Airy Shaw. Dicots. 1/1 temp. S. Am. Prostr. glabr. shrublet, with robust terete flexuous stems, emitting short ascending densely leafy branches. Ls. alt., simple, sess., semiamplex., ent., but apically minutely tridentate, thickish, smooth, ± glauc. above, obscurely nerved, exstip. Fls. reg., ☿, sol. at apex of branchlets, subsess., ebract. K 5, small, shortly conn. below, imbr., persist.; C 5, rel. large, ellipt., thickish, slightly clawed, contorted; A 5, alternipet., fil. subul., anth. small, ovoid, extrorse; disk o; G̲ (3), with short simple style and capit. 3-lobed stig., and ∞ axile anatr. ov. Fr. a small ∞-seeded loculic. caps., borne on short erect pedic., epicarp coriac., endoc. crustac., valves finally separating from axis, to which the shining black seeds remain attached long after dehiscence of caps.; seeds with fleshy and oily endosp. Only genus: *Tribeles*. Somewhat intermediate between *Pittosporac.* and *Escalloniac.*, differing from both in the obscurely nerved glaucescent ls., contorted petals, extrorse anthers, and persistent seed-bearing column of the fr. It differs further from *Pittosporac.* in the absence of resin canals, and from *Escalloniac.* in the absence of a disk. J. D. Hooker reports a 'bitter-sweet' taste to the stems.

Tribeles Phil. Tribelaceae. 1 temp. S. Am.

Triblemma R.Br. ex DC. = Bertolonia Raddi (Melastomatac.).

Tribolacis Griseb. = Turnera L. (Turnerac.).

Tribolium Desv. = Lasiochloa Kunth (Gramin.).

Tribonanthes Endl. Haemodoraceae. 5 SW. Austr.

Tribrachia Lindl. = Bulbophyllum Thou. (Orchidac.).

Tribrachium Benth. & Hook. f. = praec.

Tribrachya Korth. = Rennellia Korth. (Rubiac.).

Tribrachys Champ. ex Thw. = Thismia Griff. (Burmanniac.).

Tribroma O. F. Cook (~ Theobroma L.). Sterculiaceae. 1 trop. S. Am.

Tribula Hill = ? Caucalis L. + Athamanta L. (Umbellif.).

Tribulaceae Trautv. = Zygophyllaceae R.Br.

Tribulastrum B. Juss. ex Pfeiff. = Neurada L. (Neuradac.).

Tribulocarpus S. Moore. Tetragoniaceae. 1 SW. Afr.

Tribuloïdes Seguier = Trapa L. (Trapac.).

Tribulopis R.Br. = Tribulus L. (Zygophyllac.).

Tribulopsis F. Muell. = praec.

Tribulus L. Zygophyllaceae. 20 trop. & subtrop. (caltrops). The mericarps have sharp rigid spines which may stick into the foot of an animal. Each contains 3–5 seeds, and is divided by cross walls which develop after fert.

Tricalistra Ridley (= ? Tupistra Ker-Gawl.). Liliaceae. 1 Malay Penins.

Tricalysia A. Rich. Rubiaceae. 100 trop. Afr., Madag., few Indomal.

Tricardia Torr. Hydrophyllaceae. 1 SW. U.S.

Tricarium Lour. = Phyllanthus L. (Euphorbiac.).

Tricarpelema J. K. Morton. Commelinaceae. 1 E. Himal.

Tricaryum Spreng. = Tricarium Lour. = Phyllanthus L. (Euphorbiac.).

Tricatus Pritz. (sphalm.) = Tricratus L'Hérit. = Abronia Juss. (Nyctaginac.).

Tricentrum DC. = Comolia DC. (Melastomatac.).

Tricera Schreb. = Buxus L. (Buxac.).
Triceraia Willd. ex Roem. & Schult. = Turpinia Vent. (Staphyleac.).
Triceras Andrz. = Matthiola R.Br. (Crucif.).
Triceras P. & K. = Triceros Griff. = Gomphogyne Griff. (Cucurbitac.).
Triceras Wittst. = Triceros Lour. = Turpinia Vent. (Staphyleac.).
Tricerastes Presl = Datisca L. (Datiscac.).
Triceratella Brenan. Commelinaceae. 1 S. trop. Afr.
Triceratia A. Rich. = Sicydium Schlechtd. (Cucurbitac.).
Triceratorhynchus Summerhayes. Orchidaceae. 1 trop. E. Afr.
Tricercandra A. Gray = Chloranthus Sw. (Chloranthac.).
Tricerma Liebm. = Maytenus Molina (Celastrac.).
Triceros Griff. = Gomphogyne Griff. (Cucurbitac.).
Triceros Lour. = Turpinia Vent. (Staphyleac.).
Trichacanthus Zoll. & Mor. Acanthaceae. 1 Java.
Trichachne Nees. Gramineae. 15 Austr., warm Am.
× Trichachnis hort. (vii) = × Arachnoglottis hort. (Orchidac.).
Trichadenia Thw. Flacourtiaceae. 1 Ceylon, 1 Philipp. Is., New Guinea.
Trichaeta Beauv. = Trisetum Pers. (Gramin.).
Trichaetolepis Rydb. = Adenophyllum Pers. (Compos.).
Trichandrum Neck. = Helichrysum Mill. (Compos.).
Trichantha Hook. Gesneriaceae. 12 Colombia, Ecuador.
Trichantha Karst. & Triana = Breweria R.Br. (Convolvulac.).
Trichanthemis Regel & Schmalh. Compositae. 7 C. As.
Trichanthera Ehrenb. = Hermannia L. (Sterculiac.).
Trichanthera Kunth. Acanthaceae. 2 NW. trop. S. Am.
Trichanthodium Sond. & F. Muell. = Gnephosis Cass. (Compos.).
Trichanthus Phil. = Jaborosa Juss. (Solanac.).
Tricharis Salisb. = Dipcadi Medik. (Liliac.).
Trichasma Walp. = Argyrolobium Eckl. & Zeyh. (Legumin.).
Trichasterophyllum Willd. ex Link = Crocanthemum Spach (Cistac.).
Trichaurus Arn. = Tamarix L. (Tamaricac.).
Trichelostylis Lestib. = Fimbristylis Vahl (Cyperac.).
Trichera Schrad. = Knautia L. (Dipsacac.).
Tricherostigma Boiss. (sphalm.) = Trichosterigma Klotzsch & Garcke = Euphorbia L. (Euphorbiac.).
*Trichilia P.Br. Meliaceae. 300 Mex. to trop. S. Am., W.I., trop. Afr.
Trichinium R.Br. = Ptilotus R.Br. (Amaranthac.).
Trichiocarpa (Hook.) J. Sm. = Cionidium Moore (Aspidiac.).
Trichiogramme Kuhn = Syngramma J. Sm. (Gymnogrammac.).
Trichlisperma Rafin. = Polygala L. (Polygalac.).
Trichlora Baker. Alliaceae. 1 Peru.
Trichloris Fourn. = Chloris Sw. (Gramin.).
Trichoa Pers. = Abuta Aubl. (Menispermac.).
Trichoballia Presl = Tetraria Beauv. (Cyperac.).
Trichobasis Turcz. = Conothamnus Lindl. (Myrtac.).
*Trichocalyx Balf. f. Acanthaceae. 2 Socotra.
Trichocalyx Schau. = Calythrix Labill. (Myrtac.).
Trichocarpus Neck. = Prunus L. (Rosac.).

TRICHOCARPUS

Trichocarpus Schreb. = Sloanea L. (Elaeocarpac.).
Trichocarya Miq. = Angelesia Korth. (= Licania Aubl.) + Diemenia Korth. (= Parastemon A. DC.) (Chrysobalanac.).
Trichocaulon N. E. Br. Asclepiadaceae. 25 Afr.
Trichocentrum Poepp. & Endl. Orchidaceae. 30 C. & trop. S. Am.
Trichocephalum Schur = Virga Hill (Dipsacac.).
Trichocephalus Brongn. = Phylica L. (Rhamnac.).
Trichoceras Spreng. = Trichoceros Kunth (Orchidac.).
Trichocereus (Berger) Riccob. Cactaceae. 40 subtrop. & temp. S. Am.
Trichoceros Kunth. Orchidaceae. 4 trop. S. Am.
Trichochaeta Steud. = Rhynchospora Vahl (Cyperac.).
Trichochilus Ames = Dipodium R.Br. (Orchidac.).
Trichochiton Komarov (~ Cryptospora Kar. & Kir.). Cruciferae. 2 C. As.
Trichochlaena P. & K. = Tricholaena Schrad. (Gramin.).
Trichochloa Beauv. = Muhlenbergia Schreb. (Gramin.).
× **Trichocidium** hort. Orchidaceae. Gen. hybr. (iii) (Oncidium × Trichocentrum).
Trichocladus Pers. Hamamelidaceae. 6–8 trop. & S. Afr.
Trichocline Cass. Compositae. 30 S. Am.
Trichocoronis A. Gray. Compositae. 3 SW. U.S., Mex.
Trichocoryne S. F. Blake. Compositae. 1 Mex.
Trichocrepis Vis. = Pterotheca Cass. = Crepis L. (Compos.).
Trichocyamos Yakovl. (~ Ormosia G. Jacks.). Leguminosae. 4 S. China, Hainan, Hongk.
Trichocyclus N. E. Br. = Brownanthus Schwant. (Aïzoac.).
Trichocyclus Dulac = Woodsia R.Br. (Aspidiac.).
***Trichodesma** R.Br. Boraginaceae. 35 trop. & subtrop. Afr., As., Austr.
Trichodia Griff. = Paropsia Nor. ex Thou. (Passiflorac.).
Trichodiadema Schwantes. Aïzoaceae. 26 S. Afr.
Trichodiclida Cerv. = Blepharidachne Hack. (Gramin.).
Trichodium Michx = Agrostis L. (Gramin.).
Trichodon Benth. (sphalm.) = Trichoön Roth = Phragmites Trin. (Gramin.).
Trichodoum Beauv. ex Taub. = Dioclea Kunth (Legumin.).
Trichodrymonia Oerst. = Episcia Mart. (Gesneriac.).
Trichodypsis Baill. (~ Dypsis Nor.). Palmae. 3 Madag.
× **Trichoëchinopsis** Hort. Cactaceae. Gen. hybr. (Echinopsis × Trichocereus).
Trichogalium Fourr. = Galium L. (Rubiac.).   [cereus).
Trichogamia Boehm. = seq.
Trichogamila P.Br. Inc. sed. 1 Jamaica.
Trichoglottis Blume. Orchidaceae. 60 Formosa, Indomal., Polynesia.
Trichogonia Gardn. Compositae. 20 trop. S. Am.
Trichogyne Less. = Ifloga Cass. (Compos.).
Tricholaena Schrad. Gramineae. 8 Canaries, Medit., Afr., Madag. (*For T. rosea* Nees see *Rhynchelytrum* Nees.)
Tricholaser Gilli. Umbelliferae. 1 Afghanistan.
Tricholepis DC. Compositae. 15 C. As., Himal., Burma.
Tricholobos Turcz. = Sisymbrium L. (Crucif.).
Tricholobus Blume = Connarus L. (Connarac.).

Tricholoma Benth. (1846; non *[Fr.] Kummer 1871—Fungi) = Glossostigma Arn. (Scrophulariac.).

Tricholophus Spach = Polygala L. (Polygalac.).

Tricholeptus Gandog. = Daucus L. (Umbellif.).

Trichomaneaceae [sic] Kunkel = Hymenophyllaceae Gaudich. *p.p.*

Trichomanes L. Hymenophyllaceae. 25 trop. & subtrop. Am. [*T. radicans* Sw. = *Vandenboschia radicans* (Sw.) Copel.]

Trichomanes Scop. = Asplenium L. (Aspleniac.).

Trichomaria Steud. = Tricomaria Gill. (Malpighiac.).

Trichomema S. F. Gray (sphalm.) = seq.

Trichonema Ker-Gawl. = Romulea Maratti (Iridac.).

Trichoneura Anderss. Gramineae. 1 S. U.S., 1 Peru, 1 Galápagos; 6 trop. Afr.

Trichoneuron Ching. Probably Aspidiaceae. 1 SW. China.

Trichoön Roth = Phragmites Trin. (Gramin.).

× Trichopasia hort. Orchidaceae. Gen. hybr. (vi) (Aspasia × Trichopilia).

Trichopetalon Rafin. = seq.

Trichopetalum Lindl. = Bottionea Colla (Liliac.).

Trichophorum Pers. = Scirpus L. (Cyperac.).

Trichophyllum Ehrh. (uninom.) = *Scirpus acicularis* L. = *Eleocharis acicularis* (L.) Roem. & Schult. (Cyperac.).

Trichophyllum House = Eleocharis R.Br. (Cyperac.).

Trichophyllum Nutt. = Bahia Lag. (Compos.).

Trichopilia Lindl. Orchidaceae. 30 Mex. to trop. S. Am.

*Trichopodaceae Hutch. Monocots. 1/1 S. India, Ceylon, Malay Penins. Small erect rigid glabrous tufted perenn. herb, with wiry roots. Stem angular, bearing a single apparently terminal l. Ls. very variable, petiolate, rigid, lin.-lanc. to triang.-ovate, base cuneate to deeply cordate, 5–9-nerved, with persist. ovate-lanc. stip. Fls. small, reg., ⚥, on elong. filif. pedic., fascic. at base of l. P 3 + 3, tube campan., green, segs. ovate-lanc., spreading, dark brown-purple, persist.; A 3 + 3, adnate to base of P-segs., fil. v. short, anth. short, broad, thecae divaric., connective with lanc. apical append. and short bifid basal process. G (3), with 2 superposed ov. per loc., and v. short style with 3 short bifid reflexed stigs. Fruit a trigonous obovoid pendulous berry with thick pericarp and 3 thick wings; seeds oblong, rugose, dorsally grooved, testa thin, embryo minute in cartilaginous endosp. Only genus: *Trichopus*. The anatomy supports the segregation of this genus (but not of *Avetra*) from the *Dioscoreac.*

Trichopodium Lindl. = Trichopus Gaertn. (Trichopodac.).

Trichopodium Presl = Dalea L. (Legumin.).

Trichopteria Lindl. (sphalm.) = Trichopteryx Nees (Gramin.).

Trichopteris Neck. = Scabiosa L. (Dipsacac.).

Trichopteris Presl = Cyathea Sm. (Cyatheac.).

Trichopterya Lindl. (sphalm.) = seq.

Trichopteryx Nees. Gramineae. 7 trop. & S. Afr., Madag.

Trichoptilium A. Gray. Compositae. 1 SW. U.S.

Trichopus Gaertn. Trichopodaceae. 1 S. India, Ceylon, Malay Penins.

Trichorhiza Lindl. ex Steud. = Luisia Gaudich. (Orchidac.).

Trichoryne F. Muell. (sphalm.) = Tricoryne R.Br. (Liliac.).

Trichosacme Zucc. Asclepiadaceae. 1 Mex.

TRICHOSANCHEZIA

Trichosanchezia Mildbraed. Acanthaceae. 1 E. Peru.
Trichosandra Decne. Asclepiadaceae. 1 Mauritius.
Trichosantha Steud. (sphalm.) [Trichosathera Ehrh.] = Stipa L. (Gramin.).
Trichosanthes L. Cucurbitaceae. 15 Indomal., Austr.
Trichosathera Ehrh. (uninom.) = Stipa capillata L. (Gramin.).
Trichoschoenus J. Raynal. Cyperaceae. 1 Madag.
Trichoscypha Hook. f. Anacardiaceae. 50 trop. Afr.
Trichoseris Poepp. & Endl. (sphalm.) = Trochoseris Poepp. & Endl. = Troximon Nutt. (Compos.).
Trichoseris Vis. = Pterotheca Cass. = Crepis L. (Compos.).
Trichosia Blume = Trichotosia Blume (Orchidac.).
Trichosiphon Schott & Endl. = Sterculia L. (Sterculiac.).
Trichosma Lindl. = Eria Lindl. (Orchidac.).
Trichosorus Liebm. = Lophosoria Presl (Lophosoriac.).
Trichospermum Beauv. ex Cass. = Parthenium L. (Compos.).
Trichospermum Blume. Tiliaceae. 20 Nicobar Is., Malaysia, Pacif.
Trichospira Kunth. Compositae. 1 trop. Am.
Trichosporum D. Don = Aeschynanthus Jack (Gesneriac.).
***Trichostachys** Hook. f. Rubiaceae. 10 trop. Afr.
Trichostachys Welw. = Faurea Harv. (Proteac.).
Trichostegia Turcz. = Athrixia Ker-Gawl. (Compos.).
Trichostelma Baill. Asclepiadaceae. 3 Mex., Guatem.
Trichostema Gronov. ex L. Labiatae. 16 N. Am.
Trichostemma R.Br. = Vernonia Schreb. (Compos.).
Trichostemma Cass. = Wedelia Jacq. (Compos.).
Trichostemma L. = Trichostema Gronov. ex L. (Labiat.).
Trichostemum Rafin. = praec.
Trichostephania Tardieu. Sterculiaceae. 1 Indoch.
Trichostephanus Gilg. Flacourtiaceae. 1 W. Equat. Afr.
Trichostephium Cass. = seq.
Trichostephus Cass. = Wedelia Jacq. (Compos.).
Trichosterigma Klotzsch & Garcke = Euphorbia L. (Euphorbiac.).
Trichostigma A. Rich. Phytolaccaceae. 4 trop. Am.
Trichostomanthemum Domin. Apocynaceae. 1 Austr.
Trichotaenia Yamazaki. Scrophulariaceae. 2 Indoch.
Trichothalamus Spreng. = Potentilla L. (Rosac.).
Trichotheca (Niedenzu) Willis = Byrsonima Rich. (Malpighiac.).
Trichotolinum O. E. Schulz. Cruciferae. 1 Patag.
Trichotosia Blume. Orchidaceae. 80 NE. India, SE. As., Malaysia, Solomons.
Trichouratea Van Tiegh. = Ouratea Aubl. (Ochnaceae).
× **Trichovanda** hort. Orchidaceae. Gen. hybr. (iii) (Trichoglottis × Vanda).
Trichovaselia Van Tiegh. (= ?Elvasia DC.). Ochnaceae. 1 Venezuela.
Trichroa Rafin. = Rhamnus L. (Rhamnac.).
Trichrysus Rafin. = Helleborus L. (Ranunculac.).
Trichymenia Rydb. = Hymenothrix A. Gray (Compos.).
Triclanthera Rafin. = Crateva L. (Capparidac.).
Tricliceras Thonn. ex DC. = Wormskioldia Thonn. (Turnerac.).
Triclinium Rafin. = Sanicula L. (Umbellif.).

Triclis Hall. = Mollugo L. (Aïzoac.).
Triclisia Benth. Menispermaceae. 25 trop. Afr., Madag.
Triclisperma Rafin. = Polygala L. (Polygalac.).
Triclissa Salisb. = Kniphofia Moench (Liliac.).
Triclocladus Hutch. (sphalm.) = Trichocladus Pers. (Hamamelidac.).
Tricoccae Batsch = Euphorbiaceae Juss.
Tricoilendus Rafin. = Indigofera L. (Legumin.).
Tricomaria Gill. Malpighiaceae. 1 Argentina.
Tricomariopsis Dubard = Sphedamnocarpus Planch. ex Benth. & Hook. f.
Tricondylus Salisb. = Lomatia R.Br. (Proteac.). [(Malpighiac.).
Tricoryne R.Br. Liliaceae. 6 Austr.
Tricoscypha Engl. (sphalm.) = Trichoscypha Hook. f. (Anacardiac.).
Tricostularia Nees. Cyperaceae. 1 Ceylon, Siam, Malaysia, 3 Austr.
Tricratus L'Hérit. ex Willd. = Abronia Juss. (Nyctaginac.).
Tricuspidaria Ruiz & Pav. = Crinodendron Mol. (Elaeocarpac.).
Tricuspis Beauv. = Tridens Roem. & Schult. (Gramin.).
Tricuspis Pers. = Tricuspidaria Ruiz & Pav. = Crinodendron Mol. (Elaeocarpac.).
Tricycla Cav. (~ Bougainvillea Comm. ex Juss.). Nyctaginaceae. 1 Andes.
Tricyclandra Keraudren. Cucurbitaceae. 1 Madag.
***Tricyrtis** Wall. Liliaceae. 10 Himal., E. As.
Tridachne Liebm. ex Lindl. & Paxt. = Notylia Lindl. (Orchidac.).
Tridactyle Schlechter. Orchidaceae. 35 trop. & S. Afr.
Tridactylina Sch. Bip. (~ Chrysanthemum L.). Compositae. 1 E. Siberia, N. Mong.(?).
Tridactylites Haw. = Saxifraga L. (Saxifragac.).
Tridalia Nor. = Abroma Jacq. (Sterculiac.).
Tridaps Comm. ex Endl. = Artocarpus J. R. & G. Forst. (Morac.).
Tridax L. Compositae. 26 Mex. to trop. S. Am.
Triddenum Rafin. = Triadenum Rafin. (Guttif.).
Tridens Roem. & Schult. Gramineae. 16 N. Am.
Tridentea Haw. = Stapelia L. (Asclepiadac.).
Tridermia Rafin. = Grewia L. (Tiliac.).
Tridesmis Lour. = Croton L. (Euphorbiac.).
Tridesmis Spach = Cratoxylon Blume (Guttif.).
Trideşmostemon Engl. Sapotaceae. 1 W. Equat. Afr.
Tridesmus Steud. = Tridesmis Lour. = Croton L. (Euphorbiac.).
Tridia Korth. = Hypericum L. (Guttif.).
Tridianisia Baill. Icacinaceae. 1 Madag.
Tridimeris Baill. Annonaceae. 1 Mex.
Tridophyllum Neck. = Potentilla L. (Rosac.).
Tridynamia Gagnep. = Porana Burm. f. (Convolvulac.).
Tridynia Rafin. = Lysimachia L. (Primulac.).
Tridyra Steud. = praec.
Triendilix Rafin. = Glycine L. (Legumin.).
Trientalis L. Primulaceae. 3–4 N. temp. Rhiz. with erect stem bearing about 4–7 ls. in a tuft and a few heptamerous fls.
Triexastima Rafin. = Heteranthera Ruiz & Pav. (Pontederiac.).

TRIFAX

Trifax Nor. = Reissantia Hallé (Celastrac.).

Trifidacanthus Merr. Leguminosae. 1 Philipp. Is.

Trifillium Medik. = Triphyllum Medik. = Medicago L. (Legumin.).

Trifoliada Rojas. Oxalidaceae. 2 Argent.

Trifoliastrum Moench = Trigonella L. (Legumin.).

Trifolium L. Leguminosae. 300 temp. & subtrop. (not SE. As. or Austr.) (clover, trefoil, shamrock). The fl. has the simplest mechanism in the fam., the sta. and style emerging as the keel is depressed by an insect resting on the wings, and returning when it is released. The fls. of *T. repens* L. (white clover) are an important source of honey; those of *T. pratense* L. (red clover) are too long-tubed for hive-bees and are visited by humble-bees. *T. subterraneum* L. has two kinds of infl., one normal, the other becoming subterranean. Only 3 or 4 of its fls. develop, the rest forming grapnels (each sepal forming a reflexed hook): the stalk of the infl. bends downwards and gradually forces the fls. under the earth, where the frs. ripen (cf. *Arachis*). *T. badium* Schreb. has a wing upon the fr. formed by the persistent C, *T. fragiferum* L. a bladdery 'wing' formed by the K. The clovers are important pasture and hay plants; among the chief are *T. repens* and *T. pratense*, *T. hybridum* L. (alsike), etc.

Trifurcaria Endl. = seq.

Trifurcia Herb. = Herbertia Sweet (Iridac.).

Trigastrotheca F. Muell. = Mollugo L. (Aïzoac.).

Trigella Salisb. = Cyanella L. (Tecophilaeac.).

Triglochin L. Juncaginaceae. 15 cosmop., esp. Austr. & temp. S. Am., in freshwater- or salt-marshes. Tufted herbs with leafless flg. scapes ending in ebracteate spikes or racemes. Ls. linear, fleshy in the maritime spp. P 3 + 3; A 3 + 3; G (3 + 3), or sometimes 3 with abortive cpls. between the fertile. Fl. protog., wind-pollinated. The pollen collects in the hollowed bases of the P-leaves. The ripe cpls. surround a central beak (cf. *Geranium*), and are prolonged outwards at the base into long sharp spines, by whose means, breaking away from the beak, they may be animal-distr.

Triglochin[ac]eae Dum. = Juncaginaceae Rich.

Triglossum Fisch. = Arundinaria Michx (Gramin.).

Trigoglottis auct. = Trichoglottis Blume (Orchidac.)

Trigonachras Radlk. Sapindaceae. 1 Malay Penins., 8 Philippines.

Trigonanthera E. André = Peperomia Ruiz & Pav. (Peperomiac.).

Trigonanthus Korth. ex Hook. f. = Ceratostylis Bl. (Orchidac.).

Trigonea Parl. = Nectaroscordum Lindl. (Alliac.).

Trigonella L. Leguminosae. *S.l.*, 135 Medit., Eur., As., S. Afr., Austr.; *s.str.* (excl. *Melissitus* Medik.), 75. *T. foenum-graecum* L. (fenugreek) is sometimes cult. in E. Medit. for flavouring (ground seeds), and for veterinary medicine. The fls. of *T. aschersoniana* Urban bury themselves like those of *Arachis*.

Trigonia Aubl. Trigoniaceae. 30 C. & trop. S. Am.

*Trigoniaceae Endl. Dicots. 4/35 Madag., W. Malaysia, trop. Am. Shrubs, often climbing, more rarely trees, with simple ent. alt. or opp. stip. or exstip. ls. (stips. of opp. ls. sometimes conn.). Fls. ⚥, obliquely zygo., in term. or axill. rac. or thyrses, rarely in 3-fld. cymes. K (5), imbr.; C 5 or 3, unequal, imbr. or contorted, rarely valv. or subimbr.; A 5–12 (usu. incl. 3–6 stds.), fil.

conn. into longer or shorter stam. tube, split posticously, anth. intr.; diskglands 1–3, usu. adjoining split; G (3), 3- or 1-loc., with simple style and capit. stig., and ∞–1 biser. axile pend. or erect ov. per loc. Fr. usu. a septic. caps., sometimes winged, rarely a samara; seeds usu. long-pilose, rarely glabr., with or rarely without endosp. Genera: *Trigonia*, *Lightia*, *Trigoniastrum*, *Humbertiodendron*. Closely related to *Polygalac.*; possibly connected also with *Chrysobalanac.* and *Sapindac.*

*Trigoniastrum Miq. Trigoniaceae. 1 W. Malaysia (exc. Philipp.).

Trigonidium Lindl. Orchidaceae. 14 C. & trop. S. Am.

Trigonis Jacq. = Cupania L. (Sapindac.).

Trigonobalanus Forman. Fagaceae. 2 N. Siam, Malay Penins., Borneo, Celebes. Ls. in 1 sp. in whorls of 3. Nuts resembling those of *Fagus*.

Trigonocapnos Schlechter. Fumariaceae. 1 S. Afr.

Trigonocarpaea Steud. = Trigonocarpus Wall. = Kokoona Thw. (Celastrac.).

Trigonocarpus Bert. ex Steud. = Chorizanthe R.Br. (Polygonac.).

Trigonocarpus Vell. = Cupania L. (Sapindac.).

Trigonocarpus Wall. = Kokoona Thw. (Celastrac.).

Trigonocaryum Trautv. Boraginaceae. 1 Caucasus.

Trigonochlamys Hook. f. = Santiria Bl. (Burserac.).

Trigonopleura Hook. f. Euphorbiaceae. 1 W. Malaysia (exc. Java). Close resemblance in habit, and certainly related, to *Casearia* (Flacourtiac.).

Trigonopterum Steetz ex Anderss. = Lipochaete DC. (Compos.).

Trigonopyren Bremek. Rubiaceae. 9 Madag., Comoro Is.

Trigonosciadium Boiss. Umbelliferae. 3 Turkey, Iraq, W. Iran.

Trigonospermum Less. Compositae. 5 S. Mex.

Trigonospora Holtt. Thelypteridaceae. 8 SE. As.

*Trigonostemon Blume (corr. Bl.). Euphorbiaceae. 40–50 Ceylon, E. Himal., SE. As., W. Malaysia, ?Fiji. Perhaps some relationship with *Dichapetalac.*; cf. the bifid petals of *T. diplopetalus* Thw. (Ceylon), their purple colour, and the foliage of some spp.

Trigonotheca Hochst. = Catha Forsk. (Celastrac.).

Trigonotheca Sch. Bip. = Melanthera Rohr (Compos.).

Trigonotis Stev. Boraginaceae. 40 C. As. & Himal. to Japan, Borneo, New Guinea. Two New Guinea species have 8–10 nutlets (*Zoelleria* Warb.).

Trigostemon Blume = Trigonostemon Blume corr. Bl. (Euphorbiac.).

Triguera Cav. (1785) = Hibiscus L. (Malvac.).

*Triguera Cav. (1786). Solanaceae. 1 S. Spain, Algeria.

Trigula Nor. = Clematis L. (Ranunculac.).

Trigynaea Schlechtd. Annonaceae. 5 trop. S. Am.

Trigynaea auctt. = Unonopsis R. E. Fries (Annonac.).

Trigyneia Reichb. = Trigynaea Schlechtd. (Annonac.).

Trigynia Jacques-Félix corr. A. W. Hill. Melastomataceae. 1 W. Equat. Afr.

Trihesperus Herb. = Anthericum L. (Liliac.).

Trihexastigma P. & K. = Triexastima Rafin. = Heteranthera Ruiz & Pav. (Pontederiac.).

Trikalis Rafin. = Suaeda Forsk. (Chenopodiac.).

Trikeraia Bor. Gramineae. 1 W. Himal., Tibet.

Trilepidea Van Tiegh. Loranthaceae. 1 N.Z. (North I.).

TRILEPIS

Trilepis Nees. Cyperaceae. 5 Guiana, E. Brazil. Erect woody-stemmed pl., with tristichous persist. ligulate ls.

Trilepisium Thou. Inc. sed. (?Thymelaeac.). 1 Madag.

Triliena Rafin. = Acnistus Schott (Solanac.).

Trilisa Cass. = Carphephorus Cass. (Compos.).

Trilix L. = Prockia P.Br. ex L. (Flacourtiac.).

Trillesianthus Pierre ex A. Chev. = Marquesia Gilg (Dipterocarpac.).

*Trilliaceae Lindl. Monocots. 4/53 temp. Euras., N. Am. Perenn. rhiz. herbs, with simple erect stems, and opp. or vertic. simple ent. ± retic.-veined ls. Fls. reg., ♀, term., sol. or several umbellate, rather large. K (2–)3–5, imbr. or ± contorted; C (2–)3–5, ± imbr., sometimes linear; K and C decid. or persist.; A (2–)3–5, with ± flattened fil. and elong. anth., sometimes with produced connective; G̲ (3–10), with 3–10 loc. and axile plac., or sometimes 1-loc. with pariet. plac., ovules ∞, styles 3–5, free or ± conn. Fr. a berry or fleshy loculic. caps.; seeds with endosp. Genera: *Trillium, Paris, Medeola, Scoliopus.* Closely related to *Liliac.*

Trillidium Kunth = seq.

Trillium L. Trilliaceae. 30 W. Himal. to Japan & Kamtsch., N. Am.

Trilobachne Schenck ex Henrard. Gramineae. 1 W. Penins. India.

Trilobulina Rafin. = Utricularia L. (Lentibulariac.).

Trilocularia Schlechter = Balanops Baill. (Balanopac.).

Trilomisa Rafin. = Begonia L. (Begoniac.).

Trilophus Fisch. = Menispermum L. (Menispermac.).

Trilophus Lestib. = Kaempferia L. (Zingiberac.).

Trilopus Adans. = Hamamelis L. (Hamamelidac.).

Trima Nor. = Mycetia Reinw. (Rubiac.).

Trimeiandra Rafin. = Lonchostoma Wikstr. (Bruniac.).

Trimelopter Rafin. = Ornithogalum L. (Liliac.).

Trimenia Seem. Trimeniaceae. 3 Celebes, Moluccas, New Guinea, New Caled., Fiji, Samoa, Marquesas.

*Trimeniaceae (Perk. & Gilg) Gibbs. Dicots. 2/5 E. Malaysia, E. Austr., Pacif. Trees or shrubs, sometimes scrambling, with simple opp. pet. dent. gland-dotted exstip. ls.; young parts often densely rufous-pubesc. Fls. small, ♀ or ♂ ♀ or polyg.-dioec., bracteate, in axill. or term. rac. or pan. Fl. axis not differentiated into pedic. and recept., bearing an ellipsoid strobilus of ∞ small decuss. or spiral brown ± chaffy gland-dotted bracteoles (?tepals), much imbr., early caduc.; C o; A 10–20, 2-3-ser., with slender (long or short) fil. and elong. latr. or subintr. anth. with shortly produced linguiform connective; G̲ 1(-2), ± ridged, with sess. strongly papill. ± lobed pulvinate stig., and 1 pend. anatr. ov. per loc. Fr. berry-like, usu. oblique, red or black when ripe; seed with hard, thick, smooth or ± ridged testa, and copious endosp. Genera: *Trimenia, Piptocalyx.* Bark of *Trim.* said to smell of peppermint. For arr. of bracteoles or tepals, cf. *Nandina (Nandinac.).*

Trimeranthes Cass. = Sigesbeckia L. (Compos.).

Trimeranthus Karst. = Chaetolepis Miq. (Melastomatac.).

Trimeria Harv. Flacourtiaceae. 5 trop. & S. Afr.

Trimeris Presl = Lobelia L. (Campanulac.).

Trimerisma Presl = Platylophus D. Don (Cunoniac.).

Trimeriza Lindl. = Apama Lam. (Aristolochiac.).
Trimerocalyx (Murb.) Murb. Scrophulariaceae. 1 N. Afr.
Trimetra Moç. ex DC. = Borrichia Adans. (Compos.).
Trimeza Salisb. = seq.
Trimezia Salisb. ex Herb. Iridaceae. 5 Mex. to trop. S. Am., W.I.
Trimista Rafin. = Mirabilis L. (Nyctaginac.).
Trimorpha Cass. = Erigeron L. (Compos.).
Trimorphaea Cass. = praec.
Trimorphandra Brongn. & Gris = Hibbertia Andr. (Dilleniac.).
Trimorphoea Benth. & Hook. f. = Trimorphaea Cass. = Erigeron L. (Compos.).
Trimorphopetalum Baker (~ Impatiens L.). Balsaminaceae. 1 Madag.
Trinacte Gaertn. (Jungia L. f.). Compositae. 30 Mex., C. Am., Andes.
Trinax D. Dietr. = Thrinax L. f. (Palm.).
Trinchinettia Endl. = Geissopappus Benth. (Compos.).
Trinciatella Adans. = Hyoseris L. (Compos.).
Trineuria Presl = Aspalathus L. (Legumin.).
Trineuron Hook. f. = Abrotanella Cass. (Compos.).
Tringa Roxb. (sphalm.) = Tunga Roxb. = Hypolytrum Rich. (Cyperac.).
***Trinia** Hoffm. Umbelliferae. 12 Eur., Medit. to C. As.
Triniella Calest. = praec.
Triniochloa Hitchcock. Gramineae. 4 Mex. to Ecuador.
Triniusa Steud. = Bromus L. (Gramin.).
Trinogeton Walp. = Thinogeton Benth. = Cacabus Bernh. (Solanac.).
Triodanis Rafin. Campanulaceae. 7 N. Am., 1 Medit.
Triodia R.Br. Gramineae. 35 Austr.
Triodica Steud. = Triadica Lour. = Sapium P.Br. (Euphorbiac.).
Triodoglossum Bullock. Periplocaceae. 1 trop. Afr.
Triodon DC. = Diodia L. (Rubiac.).
Triodon Rich. = Rhynchospora Vahl (Cyperac.).
Triodris Thou. (uninom.) = Dryopeia tripetaloïdes Thou. = Disperis tripetaloïdes (Thou.) Lindl. (Orchidac.).
Triodus Rafin. = Carex L. (Cyperac.).
Triolaena Dur. & Jacks. = seq.
Triolena Naud. Melastomataceae. 20 Mex. to W. trop. Am.
Triomma Hook. f. Burseraceae. 1 W. Malaysia (exc. Philipp.).
Trionaea Medik. = Hibiscus L. (Malvac.).
Trionfettaria P. & K. = Triumfettaria Reichb. = Triumfetta L. (Tiliac.).
Trionfettia P. & K. = Triumfetta L. (Tiliac.).
Trionum L. ex Schaeff. = Hibiscus L. (Malvac.).
Triopteris L. = seq.
Triopterys L. emend. Juss. Malpighiaceae. 3 trop. Am., W.I.
Triopteryx Dalla Torre & Harms = praec.
Trioptolemea Benth. = Triptolemea Mart. = Dalbergia L. (Legumin.).
Triorchis Agosti = Spiranthes Rich. (Orchidac.).
Triorchos Small & Nash = Pteroglossaspis Reichb. f. (Orchidac.).
Triostemon Benth. & Hook. f. (sphalm.) = Triosteum L. (Caprifoliac.).
Triosteon Adans. = seq.
Triosteospermum Mill. = seq.

TRIOSTEUM

Triosteum L. Caprifoliaceae. 6 Himal., E. As., 4 N. Am.

Tripagandra Rafin. = Tripogandra Rafin. (Commelinac.).

Tripentas Casp. = Hypericum L. (Guttif.).

Tripetalanthus A. Chev. = Plagiosiphon Harms (Legumin.).

Tripetaleia Sieb. & Zucc. Ericaceae. 2 Japan.

Tripetalum P. & K. = Tripetelus Lindl. = Sambucus L. (Sambucac.).

Tripetalum K. Schum. Guttiferae. 1 New Guinea.

Tripetelus Lindl. = Sambucus L. (Sambucac.).

Tripha Nor. = Mischocarpus Bl. vel Guioa Cav. (Sapindac.).

Triphaca Lour. = Sterculia L. (Sterculiac.).

Triphalia Banks & Sol. ex Hook. f. = Aristotelia L'Hérit. (Elaeocarpac.).

Triphasia Lour. Rutaceae. 2 trop. As., Philipp. Is.

Triphelia R.Br. ex Endl. = Actinodium Schau. (Myrtac.).

Triphlebia Baker = Diplora Baker (Aspleniac.).

Triphlebia Stapf = Stiburus Stapf (Gramin.).

Triphora Nutt. Orchidaceae. 13 N., C. & trop. S. Am., W.I.

Triphylleion Suesseng. Umbelliferae. 1 C. Am.

Triphyllocynis Thou. (uninom.) = Cynorkis triphylla Thou. (Orchidac.).

Triphylloïdes Moench = Trifolium L. (Legumin.).

Triphyllum Medik. = Medicago L. (Legumin.).

Triphyophyllaceae Emberger = Dioncophyllaceae Airy Shaw.

Triphyophyllum Airy Shaw. Dioncophyllaceae. 1 Sierra Leone, Liberia, Ivory Coast. Ls. of 3 kinds: (1) on the long shoots, ± small, biuncinate at apex; (2) on the short axillary shoots, larger, without hooks; (3) on vigorous sterile branches, elongate and ± reduced to the midrib, circinate in vernation and covered with ∞ stalked and sessile glands (cf. *Drosophyllum*).

Triphysaria Fisch. & Mey. = Orthocarpus Nutt. (Scrophulariac.).

Tripinna Lour. = Vitex L. (Verbenac.).

Tripinnaria Pers. = praec.

Triplachne Link. Gramineae. 1 Sicily.

Tripladenia D. Don = Kreysigia Reichb. (Liliac.).

Triplandra Rafin. = Croton L. (Euphorbiac.).

Triplandron Benth. = Clusia L. (Guttif.).

Triplarina Rafin. = Baeckea L. (Myrtac.).

Triplaris Loefl. ex L. Polygonaceae. 25 trop. S. Am. Trees. All are said to harbour ants in their hollow stems (cf. *Cecropia*). Fls. cyclic (see fam.), dioecious. The 3 outer P-leaves grow into long wings which project beyond the fr. and prob. aid in distribution.

Triplasandra Seem. = Tetraplasandra A. Gray (Araliac.).

Triplasis Beauv. Gramineae. 2 SE. U.S.

Triplateia Bartl. = Hymenella Moç. & Sessé = Minuartia L. (Caryophyllac.).

Triplathera Endl. = Botelua Lag. (Gramin.).

Triplectrum D. Don ex Wight & Arn. = Medinilla Gaudich. (Melastomatac.).

Tripleura Lindl. = Zeuxine Lindl. (Orchidac.).

Tripleurospermum Sch. Bip. (~ Matricaria L.). Compositae. 30 N. temp.

Triplima Rafin. = Carex L. (Cyperac.).

Triplisomeris Aubrév. & Pellegr. Leguminosae. 3 trop. Afr.

Triplobus Rafin. = Sterculia L. (Sterculiac.).

Triplocentron Cass. = Centaurea L. (Compos.).

Triplocephalum O. Hoffm. Compositae. 1 trop. E. Afr.

Triplochiton Alef. = Hibiscus L. (Malvac.).

Triplochiton K. Schum. Sterculiaceae. 2–3 trop. Afr.

Triplochitonaceae K. Schum. = Sterculiaceae–Mansoniëae Prain.

Triplochlamys Ulbr. Malvaceae. 5 trop. S. Am.

Triplolepis Turcz. = Streptocaulon Wight & Arn. (Asclepiadac.).

Triplomeia Rafin. = Licaria Aubl. (Laurac.).

Triplopetalum Nyárády = Alyssum L. (Crucif.).

Triplopogon Bor. Gramineae. 1 W. Penins. India.

Triplorhiza Ehrh. (uninom.) = Satyrium albidum L. = Leucorchis albida (L.) E. Mey. ex Schur (Orchidac.).

Triplosperma G. Don = Ceropegia L. (Asclepiadac.).

Triplostegia Wall. ex DC. Triplostegiaceae. 2 E. Himal., S. China, Formosa, Celebes, New Guinea.

Triplostegiaceae (Höck) Bobrov ex Airy Shaw. Dicots. 1/2 SE. As., E. Malaysia. Perenn. rhiz. herbs with habit of *Verbena*, rhiz. sometimes with fusif. tubers. Ls. opp., dent. to pinnatifid, exstip., mostly basal. Infl. a ± few-fld. term. thyrse, branches glandular. Fls. small, ♀, almost reg., each subtended by 2 opp. narrow bracts, an outer epicalyx of 4 further conspic. capit.-gland. persist. bracts, connate at base and uncinate-cuspid. at apex, and an 8-ribbed urceolate persist. inner epicalyx. K (5), minute. C (5), infundib., caduc., lobes imbr., subequal. A 4, epipet. and alternipet. G̅ (3), 1 cell fertile, with 1 pend. ov., the other 2 cells abortive and evanescent; style simple, slender, with small capit. stig. Fr. 1-seeded, surr. by persist. indurated epicalyces; seed with endosp. Only genus: *Triplostegia*. Intermediate between *Valerianac.* and *Dipsacac.*

Triplotaxis Hutch. Compositae. 3 trop. Afr.

Tripodandra Baill. Menispermaceae. 1 Madag.

Tripodanthera M. Roem. = Gymnopetalum Arn. (Cucurbitac.).

Tripodanthus Van Tiegh. = Loranthus L. (Loranthac.).

Tripodion Medik. = Anthyllis L. (Legumin.).

Tripogandra Rafin. Commelinaceae. 20 trop. Am.

Tripogon Roem. & Schult. Gramineae. 20 trop. Afr., As.

Tripolion Rafin. = seq.

Tripolium Nees (~ Aster L.). Compositae. 1 temp. Euras., N. Afr., N. Am.

Tripsacum L. Gramineae. 7 warm Am. *T. dactyloïdes* L., is a fodder; it is like *Euchlaena*, but with ♂ and ♀ fls. in same infl.

Tripsilina Rafin. = Passiflora L. (Passiflorac.).

Tripterachaenium Kuntze = Tripteris Less. (Compos.).

Tripteranthus Wall. ex Miers = seq.

Tripterella Michx = Burmannia L. (Burmanniac.).

Tripterell[ac]eae Dum. = Burmanniaceae Bl.

*****Tripteris** Less. Compositae. 40 S. Afr. to Arabia. Fr. 3-winged.

Tripteris Thunb. (sphalm.) = Triopterys L. emend. Juss. (Malpighiac.).

Tripterium Bercht. & Presl = Thalictrum L. (Ranunculac.).

Tripterocalyx (Torr.) Hook. = Abronia Juss. (Nyctaginac.).

Tripterocarpus Meissn. = Bridgesia Bert. (Sapindac.).

TRIPTEROCOCCUS

Tripterococcus Endl. Stackhousiaceae. 1 NW. Austr.
Tripterodendron Radlk. Sapindaceae. 1 Brazil.
Tripterospermum Blume = Crawfurdia Wall. = Gentiana L. (Gentianac.).
Tripterygium Hook. f. Celastraceae. 4–5 E. As.
Triptilion Ruiz & Pav. Compositae. 16 Chile.
Triptilium DC. = praec.
Triptilodiscus Turcz. = Helipterum DC. (Compos.).
Triptolemaea Walp. = seq.
Triptolemea Mart. = Dalbergia L. (Legumin.).
Triptorella Ritgen = Tripterella Michx = Burmannia L. (Burmanniac.).
Triquetra Medik. = Astragalus L. (Legumin.).
Triquiliopsis Rydb. = Tiquiliopsis A. A. Heller = Coldenia L. (Ehretiac.).
Triraphis R.Br. Gramineae. 10 trop. & S. Afr., Austr.
Triraphis Nees = Pentaschistis Stapf (Gramin.).
Trisacarpis Rafin. = Hippeastrum Herb. (Amaryllidac.).
Trisanthus Lour. = Centella L. (Hydrocotylac.).
Triscaphis Gagnep. = Picrasma Blume (Simaroubac.).
Triscenia Griseb. Gramineae. 1 Cuba.
Trischidium Tul. = Swartzia Schreb. (Legumin.).
Trisciadia Hook. f. = Coelospermum Bl. (Rubiac.).
Trisciadium Phil. = Huanaca Cav. (Umbellif.).
Triscyphus Taub. ex Warm. Burmanniaceae. 1 SE. Brazil.
Trisecus Willd. ex Roem. & Schult. 1 Venez. Quid?
Trisema Hook. f. = Hibbertia Andr. (Dilleniac.).
Trisepalum C. B. Clarke. Gesneriaceae. 3 Burma.
Trisetaria Forsk. Gramineae. 2 E. Medit.
Trisetarium Poir. = Trisetum Pers. (Gramin.).
Trisetobromus Nevski. Gramineae. 1 Chile.
× **Trisetokoehleria** Tsvelev. Gramineae. Gen. hybr. (Koeleria × Trisetum).
Trisetum Pers. Gramineae. 75 N. & S. temp. *T. flavescens* (L.) Beauv. is a good forage grass.
Trisiola Rafin. = Distichlis Rafin. (Gramin.).
Trismeria Fée. Hemionitidaceae. 1 trop. Am.
Trismeriaceae Kunkel = Gymnogrammaceae Ching.
Trispermium Hill = Selaginella Beauv. (Selaginellac.).
Tristachya Nees. Gramineae. 25 trop. Am., Afr., Madag.
Tristagma Poepp. & Endl. Alliaceae. 5 Chile, Patag.
Tristania R.Br. Myrtaceae. 50 Malaysia, Queensl., New Caled., Fiji. Ls. alt. G semi-sup.
Tristania Poir. = Spartina Schreb. (Gramin.).
Tristaniopsis Brongn. & Gris = Tristania R.Br. (Myrtac.).
Tristegis Nees = Melinis Beauv. (Gramin.).
Tristellateia Thou. Malpighiaceae. 1 trop. E. Afr., 20 Madag.; 1 SE. As. to Formosa, Malaysia, Queensl., New Caled.
Tristemma Juss. Melastomataceae. 20 trop. Afr., Madag., Masc.
Tristemon Klotzsch = Scyphogyne Brongn. (Ericac.).
Tristemon Rafin. (1819) = Triglochin L. (Juncaginac.).
Tristemon Rafin. (1838) = Juncus L. (Juncac.).

Tristemon Scheele = Cucurbita L. (Cucurbitac.). [W. Afr.

Tristemonanthus Loes. (~ Campylostemon Welw.). Celastraceae. 2 trop.

Tristeria Hook. f. (sphalm.) = Treisteria Griff. = Curanga Juss. (Scrophulariac.).

Tristerix Mart. = Macrosolen Bl. (Loranthac.).

Tristicha Thou. Podostemaceae. 2 trop. Am., Afr., Madag., Masc., India, Ceylon.

Tristichaceae Willis = Podostemaceae–Tristichoïdeae Engl. + Weddellinoïdeae Engl. [5/10 trop. Herbs of rapid water in hill streams, with creeping threadlike roots giving off (exc. *Lawia*, where the primary axis is flattened into a thallus and gives off) large numbers of secondary shoots with minute delicate simple exstip. ls. P 3–5 or (3–5), reg., sepaloid; A 3, 5, ∞, or 1; G (2–3), 2–3-loc. with ∞ anatr. ov. Caps. Genera: *Tristicha, Dalzellia, Terniola, Lawia, Weddellina*.]

Tristichocalyx Miers = Legnephora Miers (Menispermac.).

Tristichopsis A. Chev. = Tristicha Thou. (Podostemac.).

Tristira Radlk. Sapindaceae. 4 Philippines, Celebes, Moluccas.

Tristiropsis Radlk. Sapindaceae. 2–3 Philipp., Borneo, Mariannes, New Guinea, Solomons, Christmas I.

Tristylea Jord. & Fourr. = Saxifraga L. (Saxifragac.).

Tristylium Turcz. = Cleyera DC. (Theac.).

Tristylopsis Kaneh. & Hatus. (sphalm.) = Tristiropsis Radlk. (Sapindac.).

Trisyngyne Baill. = Nothofagus Bl. (Fagac.).

Tritaenicum Turcz. = Asteriscium Cham. & Schlechtd. (Umbellif.).

Tritaxis Baill. = Trigonostemon Bl. (Euphorbiac.).

Tritelandra Rafin. = Epidendrum L. (Orchidac.).

Triteleia Dougl. ex Lindl. (~ Brodiaea Sm.). Alliaceae. 16 W. Am.

Triteleiopsis Hoover (~ Brodiaea Sm.). Alliaceae. 1 W. U.S.

Triteleya Phil. = Triteleia Dougl. ex Lindl. (Alliac.).

Tritheca Miq. = Ammannia L. (Lythrac.).

Trithecanthera Van Tiegh. Loranthaceae. 4 Malay Penins., Borneo.

Trithrinax Mart. Palmae. 5 S. Am. Fls. ♀.

Trithuria Hook. f. Centrolepidaceae. 4 W. Austr., Tasm., N.Z.

Trithyrocarpus Hassk. = Commelina L. (Commelinac.).

× **Triticosecale** Wittmack. Gramineae. Gen. hybr. (Secale × Triticum).

Triticum L. Gramineae. About 20 Eur., Medit., W. As. The cultivated species fall into 3 groups according to chromosome number. The diploid species (*T. monococcum* L.) is apparently derived from the wild *T. boeoticum* Boiss. The tetraploid species (*T. dicoccon* Schrank, *T. durum* Desf., *T. polonicum* L., *T. turgidum* L.) are probably selected from several wild species, of which the commonest is *T. dicoccoïdes* Koern.; the latter seems to be an allotetraploid derived from *T. monococcum* and *Aegilops speltoïdes* Tausch. The hexaploid species (*T. spelta* L., *T. aestivum* L.) are not represented in the wild, but have almost certainly arisen from the crossing of an allotetraploid wheat with *Aegilops squarrosa* L. Many of the so-called species would be better ranked as cultivars. Cf. Schiemann, *Weizen, Roggen, Gerste*, 1–59 (1948).

Tritillaria Rafin. = Fritillaria L. (Liliac.).

Tritoma Ker-Gawl. = Kniphofia Moench (Liliac.).

Tritomanthe Link = praec.

TRITOMIUM

Tritomium Link = praec.

Tritomodon Turcz. = Enkianthus Lour. (Ericac.).

Tritomopterys Niedenzu = Gaudichaudia Kunth (Malpighiac.).

Tritonia Ker-Gawl. Iridaceae. 55 trop. & S. Afr.

Tritoniopsis L. Bolus. Iridaceae. 14 S. Afr.

Tritonixia Klatt = Tritonia Ker-Gawl. (Iridac.).

Tritophus Lestib. (sphalm.) = Trilophus Lestib. = Kaempferia L. (Zingiberac.).

× **Tritordeum** Aschers. & Graebn. Gramineae. Gen. hybr. (Hordeum × Triticum).

Tritriela Rafin. = Ornithogalum L. (Liliac.).

Triumfetta L. Tiliaceae. 150 trop. Herbs or shrubs, often with extra-floral nectaries at base of ls. On each internode of infl. are usu. at least three 3-flowered dichasial cymes. The first and oldest is opp. to the l.; the rest stand alt. right and left between the first and the l. Fruit with hooked spines (animal distr.).

Triumfettaria Reichb. = praec.

Triunila Rafin. = Uniola L. (Gramin.).

Triuranthera Backer. Melastomataceae. 2 Borneo, Sumatra, Java.

*****Triuridaceae** Gardn. Monocots. 7/80 trop. Am., Afr., As. Small reddish, purplish or ± colourless saprophytes, with scale ls. Fls. small, reg., ♀ or ♂ ♀ (monoec. or dioec.), ± long-pedic., in bracteate racemes. P 3–10, corolline, valv., equal or unequal, often with an apical knob or tuft of hairs. A 2–6, on flat or convex recept., fil. short, anth. 2–3–4-loc., sometimes transv. dehisc. (stds. in ♀ fl. o); G ∞, immersed in recept., each with term. or lat. or basal style, stig. subul. or clav. or penicill., and 1 erect anatr. ov. with much endosp. Genera: *Triuris, Hexuris, Seychellaria, Andruris, Hyalisma, Sciaphila, Soridium.* A very isolated fam., with possible affinity to the *Alismatac.*

Triuris Miers. Triuridaceae. 1 Guatemala, Brazil.

Triurocodon Schlechter. Burmanniaceae. 2 Brazil.

Trivalvaria Miq. Annonaceae. 5 Assam, Burma, Siam, W. Malaysia (exc. Philipp.).

Trivolvulus Moç. & Sessé ex Choisy = Ipomoea L. (Convolvulac.).

Trixago Haller = Stachys L. (Labiat.).

Trixago Rafin. = Teucrium L. (Labiat.).

Trixago Stev. = Bellardia All. (Scrophulariac.).

Trixanthera Rafin. = Trichanthera Kunth (Acanthac.).

Trixapias Rafin. = Utricularia L. (Lentibulariac.).

Trixella Fourr. = Stachys L. (Labiat.).

Trixis Adans. = Proserpinaca L. (Haloragidac.).

Trixis P.Br. Compositae. 50–60 SW. U.S. to Chile.

Trixis Sw. = Clibadium L. (Compos.).

Trixostis Rafin. = Aristida L. (Gramin.).

Trizeuxis Lindl. Orchidaceae. 1 Costa Rica, trop. S. Am.

Trocdaris Rafin. = Carum L. (Umbellif.).

Trochera Rich. = Ehrharta Thunb. (Gramin.).

Trochetia DC. Sterculiaceae (Malvac.?). 8 St Helena, Madag., Mauritius, Réunion.

Trochilocactus Lindinger = Disocactus Lindl. (Cactac.).

Trochisandra Bedd. = Kurrimia Wall. (Celastrac.).

Trochiscanthes Koch. Umbelliferae. 1 S. Eur.

Trochiscanthos St-Lag. = praec.

Trochiscus O. E. Schulz. Cruciferae. 1 NE. India.

Trochocarpa R.Br. Epacridaceae. 14 N. Borneo, Celebes, New Guinea, E. Austr., Tasm.

Trochocephalus (Mert. & Koch) Opiz = Scabiosa L. (Dipsacac.).

Trochocodon Candargy. Campanulaceae. 1 Greece.

*****Trochodendraceae** [Seem. ex] Prantl. Dicots. 1/1 Japan to Formosa. Trees or shrubs, with simple serr. long-pet. pseudo-vertic. coriac. exstip. ls., branchlets term. by a conspic. perulate bud. Fls. reg. or slightly asymm., ♀, in term. racemiform pleiochasial cymes, pedic. with bract and sev. bracteoles, expanded above into subconical torus. P o(?); A ∞, 3–4-ser., fil. filif., anth. basifixed, mucron., latr.; G ∞, 1-ser., laterally coalesc., with short recurved ventr. stigmatic style, and ∞ biser. anatr. ov. near ventr. suture. Fr. a ring of coalesc. ventr. dehisc. ∞-seeded follicles; seeds with oily endosp. Only genus: *Trochodendron*, but perhaps also *Paracryphia* (*Paracryphiac.*, *q.v.*). Perhaps some relationship with *Eupteleac.* Facies 'Araliaceous'.

Trochodendron Sieb. & Zucc. Trochodendraceae. 1 Japan, Ryukyu Is., Formosa, *T. aralioïdes* Sieb. & Zucc.

Trochomeria Hook. f. Cucurbitaceae. 10 Afr.

Trochomeriopsis Cogn. Cucurbitaceae. 1 Madag.

Trochopteris Gardn. = Anemia Sw. (Schizaeac.).

Trochoseris Poepp. & Endl. = Troximon Nutt. (Compos.).

Trochostigma Sieb. & Zucc. = Actinidia Lindl. (Actinidiac.).

Trogostolon Copel. Davalliaceae. 2 W. China, Philippines.

Trollius L. Ranunculaceae. 25 N. temp. & arctic. The 'sepals' completely cover in the fl. Fl. homogamous, regularly self-fert.

Trommsdorffia Mart. = Iresine P.Br. (Amaranthac.).

Trommsdorfia Bernh. = Hypochoeris L. (Compos.).

Tromotriche Haw. = Stapelia L. (Asclepiadac.).

Tromsdorffia Benth. & Hook. f. (sphalm.) = Trommsdorffia Mart. = Iresine P.Br. (Amaranthac.).

Tromsdorffia Blume = Liebigia Endl. = Morstdorffia Steud. = Chirita Buch.-Ham. [+ Dichrotrichum Reinw.] (Gesneriac.).

Tromsdorffia R.Br. = Dichrotrichum Reinw. (Gesneriac.).

Tronicena Steud. = Aeginetia L. (Orobanchac.).

Troniceus Miq. (sphalm.) = praec.

Troostwyckia Benth. & Hook. f. = seq.

Troostwykia Miq. = Castanola Llanos = Agelaea Soland. ex Planch. (Connarac.).

*****Tropaeolaceae** DC. Dicots. 2/92 Mex. to temp. S. Am. Somewhat succulent sometimes tuberous herbs with watery juice, climbing by sensitive petioles. Ls. alt. (rarely opp.), long-pet., often pelt., ent. or variously lobed or palmate, stip. (stips. sometimes obsol.). Fls. ± zygo., ♀, sol., axill., showy. K (5), produced into a short or long spur below, lobes imbr. or valv.; C 5, the 2 upper often different from 3 lower (which may sometimes be wanting), clawed, inserted on K, variously lobed or lacin., imbr.; A 8, fil. filif., anth. small,

TROPAEOLUM

basif., latr.; <u>G</u> (3), 3-loc., with simple style and 3-lobed stig., and 1 axile pend. anatr. ov. per loc. Fr. of 3 indehisc. cocci, rarely baccate; seeds without endosp. Genera: *Tropaeolum*, *Magallana*. All parts of the plants contain a mustard-oil, as in *Crucif.*, etc.; hence the pop. name 'nasturtium'. Affinities possibly with *Geraniac.*, *Limnanthac.* and *Sapindac.*

Tropaeolum L. Tropaeolaceae. 90 Mex. to temp. S. Am.
Tropalanthe S. Moore = Pycnandra Benth. (Sapotac.).
Tropentis Rafin. = Seseli L. (Umbellif.).
Tropexa Rafin. = Aristolochia L. (Aristolochiac.).
Trophaeastrum Sparre. Tropaeolaceae. 1 Patag.
Trophaeum Kuntze = Tropaeolum L. (Tropaeolac.).
Trophianthus Scheidw. = Aspasia Lindl. (Orchidac.).
*Trophis P.Br. Moraceae. *S.lat.*, 11 Madag., Malaysia, trop. Am., W.I.; *s.str.*, 4 Mex. & C. Am.
Trophisomia Rojas = ? Sorocea A. St-Hil. (Morac.).
Trophospermum Walp. = Taphrospermum C. A. Mey. (Crucif.).
Tropidia Lindl. Orchidaceae. 22 India to Formosa, Malaysia and Polynesia; Florida, C. Am., W.I.
Tropidocarpum Hook. Cruciferae. 2 Calif.
Tropidolepis Tausch = Chiliotrichum Cass. (Compos.).
Tropidopetalum Turcz. = Bouea Meissn. (Anacardiac.).
Tropilis Rafin. = Dendrobium Sw. (Orchidac.).
Tropitia Pichon = Tropitria Rafin. = Tradescantia L. (Commelinac.).
Tropitoma Rafin. = Desmodium Desv. (Legumin.).
Tropitria Rafin. = Tradescantia L. (Commelinac.).
Tropocarpa D. Don ex Meissn. = Orites R.Br. (Proteac.).
Tros Haw. = Narcissus L. (Amaryllidac.).
Troschelia Klotzsch & Schomb. = Schiekia Meissn. (Haemodorac.).
Trotula Comm. ex DC. = Nesaea Comm. ex Juss. (Lythrac.).
Trouettea Pierre ex Aubrév. = seq.
Trouettia Pierre ex Baill. (~ Chrysophyllum L.). Sapotaceae. 4 New Caled.
Troxilanthes Rafin. = Polygonatum Mill. (Liliac.).
Troximon Gaertn. = Krigia Schreb. + Scorzonera L. (Compos.).
Troximon Nutt. = Agoseris Rafin. (Compos.).
Troxirum Rafin. = Peperomia Ruiz & Pav. (Peperomiac.).
Troxistemon Rafin. = Hymenocallis Salisb. (Amaryllidac.).
Trozelia Rafin. = Acnistus Schott (Solanac.).
Truellum Houttuyn = Polygonum L. (Polygonac.).
Trujanoa La Llave = ? Pseudosmodingium Engl. ? Rhus L. (Anacardiac.).
Trukia Kanehira = Randia L. (Rubiac.).
Truncaria DC. = Miconia Ruiz & Pav. (Melastomatac.).
Tryallis C. Muell. = Thryallis Mart. (Malpighiac.).
Trybliocalyx Lindau. Acanthaceae. 1 C. Am.
Trychinolepis B. L. Robinson. Compositae. 1 Peru.
Tryginia Jacques-Félix = Trigynia Jacques-Félix corr. A. W. Hill (Melastomatac.).
Trygonanthus Endl. ex Steud. = Loranthus L. (Loranthac.).
Trymalium Fenzl. Rhamnaceae. 11 Austr.

centre are about 40 ♂ fls., surrounded by about 300 ♀. The ♂ retain the style, as usual, to act as pollen-presenter, but it has no stigmas. Honey is secreted in the ♂ fls., but not in the ♀. The ♀ fls., being the outer ones, are ripe before the ♂, and self-fert. is almost impossible.

Tutcheria Dunn = Pyrenaria Bl. (Theac.).
Tutuca Molina. Ericaceae(?). 1 Chile. Quid?
Tuyamaea Yamazaki. Scrophulariaceae. Spp.? E. As.
Tweedia Hook. & Arn. Asclepiadaceae. 1 temp. S. Am.
Tydaea Decne = Kohleria Regel (Gesneriac.).
Tydea C. Muell. = praec.
Tylacantha Endl. = Thylacantha Nees & Mart. = Angelonia Humb. & Bonpl. (Scrophulariac.).
Tylachenia P. & K. = Tulakenia Rafin. = Jurinea Cass. (Compos.).
Tylachium Gilg = Thilachium Lour. (Capparidac.).
Tylanthus Reissek = Phylica L. (Rhamnac.).
Tylecarpus Engl. = Medusanthera Seem. (Icacinac.).
Tyleria Gleason. Ochnaceae. 4 Venez.
Tyleropappus Greenm. Compositae. 1 Venez.
Tylexis P. & K. = Tulexis Rafin. = Brassavola R.Br. (Orchidac.).
Tylisma P. & K. = Tulisma Rafin. = Corytholoma (Decne) Benth. (Gesneriac.).
Tylista P. & K. = Tulista Rafin. = Haworthia Duval (Liliac.).
Tylloma D. Don = Chaetanthera Ruiz & Pav. (Compos.).
Tylocarpus P. & K. (1) = Tylecarpus Engl. = Medusanthera Seem. (Icacinac.).
Tylocarpus P. & K. (2) = Tulocarpus Hook. & Arn. = Guardiola Cerv. ex Humb. & Bonpl. (Compos.).
Tylocarya Nelmes (~ Fimbristylis Vahl). Cyperaceae. 1 Siam.
Tylochilus Nees = Cyrtopodium R.Br. (Orchidac.).
Tyloclinia P. & K. = Tuloclinia Rafin. = Metalasia R.Br. (Compos.).
Tyloderma Miers = Hylenaea Miers (Celastrac.).
Tyloglossa Hochst. = Justicia L. (Acanthac.).
Tylomium Presl = Lobelia L. (Campanulac.).
Tylopetalum Barneby & Krukoff. Menispermaceae. 1 Colombia.
Tylophora R.Br. Asclepiadaceae. 50 palaeotrop. & S. Afr.
Tylophoropsis N. E. Br. Asclepiadaceae. 2 E. Afr.
Tylophus P. & K. = Tulophos Rafin. = Triteleia Dougl. ex Lindl. (Alliac.).
Tylopsacas Leeuwenb. Gesneriaceae. 1 trop. Am.
Tylorima P. & K. = Tulorima Rafin. = Saxifraga L. (Saxifragac.).
Tylosema (Schweinf.) Torre & Hillc. Leguminosae. 1 trop. Afr.
Tylosepalum Kurz ex Teijsm. & Binnend. = Trigonostemon Blume (Euphor-
Tylostemon Engl. = Beilschmiedia Nees (Laurac.). [biac.].
Tylosperma Botsch. Rosaceae. 1 C. As., N. Persia.
Tylosperma Leeuwenb. = Tylopsacas Leeuwenb. (Gesneriac.).
Tylostigma Schlechter. Orchidaceae. 3 Madag.
Tylostylis Blume = Eria Lindl. (Orchidac.).
Tylothrasya Doell = Panicum L. (Gramin.).
Tylotis P. & K. = Tulotis Rafin. (Orchidac.).
Tympananthe Hassk. = Dictyanthus Decne (Asclepiadac.).
Tynanthus Miers = seq.

TYNNANTHUS

Tynnanthus Miers corr. K. Schum. Bignoniaceae. 12 C. & trop. S. Am., W.I.

Tynus J. S. Presl = Tinus Mill. = Viburnum L. (Caprifoliac.).

Typha L. Typhaceae. 10–20 temp. & trop., in ponds & still water.

***Typhaceae** Juss. Monocots. 1/10 temp. & trop. The lower part of the stem is a thick rhiz.; the upper projects high out of the water (ls. 2-ranked) and bears the infl., a dense cylindr. spike, divided into two parts, the upper ♂ (usu. yellow), the lower ♀ (brown). Fls. surr. by elong. threads or spath. ± forked scales (=P?): ♂ fl. of 2–5 sta., fil. free or conn., anth. lin.-obl., basifixed, the connective projecting beyond the anthers; pollen in tetrads; ♀ of 1 cpl. with 1 pend. ov., micropyle towards the base or ventral side of the ovary. Fl. anemoph. Achenes covered by the long threads or scales mentioned, which aid in distr. Seed album.; embryo straight. Only genus: *Typha*. Rather closely related to *Sparganiac.*

Typhalea Neck. = Pavonia Cav. (Malvac.).

Typhodes P. & K. = seq.

Typhoïdes Moench = Phalaris L. (Gramin.).

Typhonium Schott. Araceae. 30 SE. As., Indomal.

Typhonodorum Schott. Araceae. 1 trop. E. Afr., Madag., Masc.

Tyria Klotzsch ex Endl. (1850) = Bernardia Mill. (Euphorbiac.).

Tyria Klotzsch (1851) = Macleania Hook. (Ericac.).

Tyrimnus Cass. Compositae. 1 S. Eur., W. As.

Tysonia Bolus. Boraginaceae. 1 SE. Afr.

Tysonia F. Muell. = Swinburnia Ewart = Neotysonia Dalla Torre & Harms (Compos.).

Tyssacia Steud. = Tussacia Reichb. = Chrysothemis Decne (Gesneriac.).

Tytonia G. Don = Hydrocera Blume (Balsaminac.).

Tytthostemma Nevski. Caryophyllaceae. 1 C. As., E. Persia.

Tzellemtinia Chiov. Rhamnaceae. 1 NE. trop. Afr.

U

Uapaca Baill. Uapacaceae. 50 trop. Afr., Madag.

Uapacaceae (Muell. Arg.) Airy Shaw. Dicots. 1/50 trop. Afr., Madag. Trees or shrubs, ± pachycaulous, with alt. simple ent. ± cuneate-obov. stip. ls. crowded towards ends of branches. Fls. ♂ ♀, dioec. ♂ fls. crowded in dense glob. capit. pedunc. infl., surr. by calycine invol. of 5–10 large much imbr. bracts: K (5–6), ± campan. or turbin., lobed or dent.; C 0; disk 0; A 4–6, oppositisep., anth. shortly ov., intr.; sometimes 4–6 pilose stds. alt. with stamens; pistillode large, obconic, infundib., or pileif., sometimes lobed. ♀ fls. sol., pedunc., surr. by calycine invol. as ♂: K much reduced, sinuate or truncate; C 0; disk 0; stds. 0; G̲ (2–)3(–4), with thick much lacin. recurved styles, and 2 apical pend. ov. per loc. Fruit drupaceous, with tough ± fleshy exocarp, and 3 dorsally 2-sulcate pyrenes; seeds with fleshy endosp. and flat cots. Only genus: *Uapaca*. Hitherto included in *Euphorbiac.*, but foliage, infl. and anat. aberrant. Perh. some connection with *Anacardiac., Picrodendrac., Pistaciac.*, etc.

Ubiaea J. Gay = Landtia Less. (Compos.).

ULLUCACEAE

Ubidium Rafin. = seq.
Ubium J. F. Gmel. = Dioscorea L. (Dioscoreac.).
Ubochea Baill. Verbenaceae. 1 Cape Verde Is.
Ucacea Cass. = Blainvillea Cass. (Compos.).
Ucacou Adans. = Synedrella Gaertn. (Compos.).
Uchi P. & K. = Uxi Almeida Pinto = Sacoglottis Mart. (Houmiriac.).
Ucnopsolen A. W. Hill (sphalm.) = seq.
Ucnopsolon Rafin. = Lindernia All. (Scrophulariac.).
Ucria Pfeiff. (*nomen casu fictum!*) = Ambrosinia Bassi (Arac.).
Ucriana Spreng. = Augusta Pohl + Tocoyena Aubl. (Rubiac.).
Ucriana Willd. = Tocoyena Aubl. (Rubiac.).
Udani Adans. = Quisqualis L. (Combretac.).
Udora Nutt. = Anacharis Rich. (Hydrocharitac.).
Udoza Rafin. = praec.
Udrastina Rafin. = Laportea Gaudich. (Urticac.).
Uebelinia Hochst. Caryophyllaceae. 10 trop. Afr.
Uebelmannia Buining. Cactaceae. 2 Brazil.
Uechtritzia Freyn. Compositae. 3 Armenia, C.As., W. Himal.
Uffenbachia Fabr. (uninom.) = *Uvularia* L. sp. (Liliac.).
Ugamia Pavlov. Compositae. 1 C. As.
Ugena Cav. = Lygodium Sw. (Schizaeac.).
Ugni Turcz. Myrtaceae. 15 Mex., Andes. Ed. fr.
Ugona Adans. = Hugonia L. (Linac.).
Uhdea Kunth = Montañoa La Llave (Compos.).
Uitenia Nor. = Erioglossum Blume (Sapindac.).
Uittienia v. Steenis = Dialium L. (Legumin.).
Uladendron Marcano-Berti. Malvaceae. 1 Venez.
Ulantha Hook. = Chloraea Lindl. (Orchidac.).
Ulanthia Rafin. = praec.
Ulbrichia Urb. Malvaceae. 1 W.I. (S. Domingo).
Uldinia J. M. Black. Hydrocotylaceae (~ Umbellif.). 1 C. Austr.
Ulea C. B. Clarke ex H. Pfeiff. = Exochogyne C. B. Clarke (Cyperac.).
Ulea-flos A. W. Hill (sphalm.) = praec.
Uleanthus Harms. Leguminosae. 1 Amaz. Brazil.
Ulearum Engl. Araceae. 1 Upper Amazon valley.
Uleiorchis Hoehne. Orchidaceae. 1 Brazil.
Uleophytum Hieron. Compositae. 1 Peru. Climber.
Ulex L. Leguminosae. 20 W. Eur., N. Afr. *U. europaeus* L., *U. minor* Roth (*U. nanus* T. F. Forst.) and *U. gallii* Planch. (gorse, furze, or whin) cover large areas, esp. on heaths. The ls. are reduced in size, and many branches reduced to green spines (xerophytism). The fls. explode like *Genista*, and the fr. explodes by the twisting up of its valves in dry air. The seedlings show interesting transition stages from the usual compound ls. seen in the family to the needle-ls. of the mature pl. (cf. *Acacia*).
Ulina Opiz = Inula L. (Compos.).
Ulleria Bremek. Acanthaceae. 4 C. & S. Am.
Ulloa Pers. = Juanulloa Ruiz & Pav. (Solanac.).
Ullucaceae Nak. = Basellaceae Moq.

ULLUCUS

Ullucus Caldas. Basellaceae. 1 Andes, *U. tuberosus* Caldas. Differs from other *Basellac.* in production of potato-like tubers from rhiz., prostrate, scarcely climbing habit, bibracteol. fls., long-caud. petals, anth. dehisc. by apic. pores, columnar style with capit. stig., and bacc. fr. The tubers are used as food.

*Ulmaceae Mirb. Dicots. 15/200 trop. & temp. Trees or shrubs with sympodial stems, bearing 2-ranked, simple, ent., dent. or lobul., often asymm. ls. with stips. Fls. usu. in cymose clusters, ♀ or ♂ ♀. P 4–8, free or united, sepaloid, according to Engler theoretically belonging to two whorls; A 4–8, opp. the perianth-ls., in two whorls; G rudimentary in ♂ fl., in the ♀ of (2) cpls., sometimes 2-loc. but usu. 1-loc., the second loc. aborting; ov. 1 per loc., anatr. or amphitr., pend., style linear or bifid. Nut, samara or drupe. Seed usu. with no endosp. The wood of many is useful. Chief genera: *Ulmus, Celtis, Trema, Gironniera, Aphananthe, Ampelocera.*

Ulmaria Mill. = Filipendula Mill. (Rosac.).

Ulmari[ace]ae S. F. Gray = Rosaceae–Ulmariëae Meissn.

Ulmarronia Friesen = Cordia L. (Ehretiac.).

Ulmus L. Ulmaceae. 45 N. temp., S. to Himal., Indoch., & Mex. Growth sympodial, the term. bud being suppressed. Ls. asymmetrical, one side larger than the other (cf. *Begonia*). The fls. are ♀ and usu. come out before the ls. as little reddish tufts, each a short axis with a number of ls., beginning 2-ranked at the base and going over to 5-ranked above. (Some autumn-flowering spp., e.g. *U. parvifolia* Jacq., have fls. on leafy branches.) There are no fls. in the axils of the lowest 10 or 12 ls.; in the axils of the upper ls. are fls. arranged in small dich. cymes (cf. *Betulaceae*), which are reduced, in *U. procera* Salisb. and others, to the one central fl. Each fl. has P 4–8 and as many sta. with 1-loc. ovary. Fr. a samara. Elms produce valuable timber.

Uloma Rafin. = Rhodocolea Baill. (Bignoniac.).

Uloptera Fenzl = Ferula L. (Umbellif.).

Ulospermum Link = Capnophyllum Gaertn. (Umbellif.).

Ulostoma D. Don = Gentiana L. (Gentianac.).

Ulricia Jacq. ex Steud. = Lepechinia Willd. (Labiat.).

Ulticona Rafin. = Hebecladus Miers (Solanac.).

Ultragossypium Roberty. Malvaceae. 1 N. trop. Afr. (cult.).

Ulugbekia Zakirov. Boraginaceae. 1 C. As.

Uluxia Juss. = Columellia Ruiz & Pav. (Columelliac.).

Ulva Adans. (1763; non L. 1753—Algae) = Carex L. (Cyperac.).

Umari Adans. = Geoffroea Jacq. (Legumin.).

Umbellales Benth. & Hook. f. = Umbelliflorae Bartl.

Umbellifera Honigberger = ? Ligusticum L. (Umbellif.).

*Umbelliferae Juss. (nom. altern. **Apiaceae** Lindl.). Dicots. 275/2850, cosmop., chiefly N. temp. Mostly herbs with stout stems, hollow internodes, and alt. exstip. sheathing ls. with their blades much divided pinnately. A few, e.g. *Hydrocotyle* and *Bupleurum*, have entire ls. Infl. usu. a cpd. umbel. At the top of the stalk of each partial umbel (umbellule), an invol. of bracts is often found (the bracts of the outer fls.), and a similar larger invol. often occurs at the top of the main stalk bearing the cpd. umbel; the latter is sometimes termed the involucre in contradistinction to the *involucels* of the partial umbels. A term. fl. often occurs, e.g. in *Daucus*, or rarely a tuft of penicillate hairs

(*Artedia*). In a number of genera (e.g. *Astrantia, Hydrocotyle*) simple umbels occur, cymose in type (as the non-centripetal order of opening of fls. shows) and often arranged in cymose groupings, e.g. in *Sanicula. Eryngium* has a cymose head. Some spp. of *Xanthosia* and *Azorella* have such cymose infls. reduced to single fls., and these infls. have commonly invols. of bracts. In spite of superfic. similarity the species are generally well characterized, but the delimitation of genera presents considerable difficulties.

Fl. usu. ♀ and reg. (see below), epig. K 5, usu. very small, the odd sepal post.; C 5 (rarely o), usu. white or yellow; A 5, intr.; $\overline{G}$ (2), antero-post., 2-loc.; in each loc. one pend. ovule, anatr., with ventral raphe. On top of the ovary is an epig. disk, prolonged into two short styles set on variously and characteristically developed swollen bases (*stylopodia*).

The massing of the fls. into dense infls. makes them conspic. (cf. *Compositae*), and this is aided by the zygomorphism of the C often seen; the outer petals of the outer fls. are drawn out (cf. *Cruciferae*) so as to form a sort of ray. Honey is secreted by the disk; it is accessible to all insects. The chief visitors are flies; beetles and Hymenoptera are also frequent. Fls. very protandrous, the ♂ stage being most commonly over before the ♀ begins.

The ovary ripens into a very char. fruit, a dry schizocarp, which generally splits down a septum (commissure) between the cpls. into 2 *mericarps*, each containing one seed. The two are generally held together at first by a thin stalk (*carpophore*) running up between them. The structure of the pericarp is of great importance in determining the gen. It is nearly always necessary to havè ripe fr. in order to identify one of the U. The shape is often important; the outer surface of each mericarp has generally 5 projecting *primary ridges*, two of which (the *lateral ridges*) are at the edges where the splitting takes place. Between these are sometimes *secondary ridges*, 4 to each mericarp. In the furrows are often found oil-cavities (seen as small openings in cross-section) known as *vittae*. The presence or absence of crystals in the pericarp is of taxonomic importance. The seed is often united to the pericarp; it is album. with small embryo in oily endosp., which is usu. cartilaginous in texture. The shape of the endosp. as seen in cross-section is of importance; it may be crescentic, or ventrally grooved, or concave on ventral side, and with the margins plane or variously involute. The fr. often shows adaptations for distr.; in many (e.g. *Heracleum* and allies) the mericarp is thin and flat, suited to wind-carriage; in others (e.g. *Daucus*) it has hooks. See also *Scandix*.

Many U. are economically useful, but as a rule they are poisonous. See *Daucus* (carrot), *Pastinaca* (parsnip), *Apium* (celery), *Crithmum* (samphire), *Foeniculum* (fennel), *Archangelica, Carum, Ferula, Pimpinella, Coriandrum, Petroselinum*, etc.

Classification and chief genera (after Engler):

I. **Hydrocotyloïdeae** (stipules present; fr. with no free carpophore, and woody endocarp; vittae none or in main ribs). [See also *Hydrocotylaceae*.]

    1. HYDROCOTYLEAE (fr. with narrow commissure, lat. flattened): *Hydrocotyle, Azorella*.

    2. MULINEAE (fr. with flattened or rounded back; S. hemisph.): *Bowlesia*.

UMBELLIFLORAE

II. *Saniculoïdeae* (stipules absent; endocarp soft, exocarp rarely smooth; style long with capitate stigmas, surrounded by ring-like disk; vittae various).

1. SANICULEAE (ov. 2-loc.; fr. 2-seeded, with broad commissure; vittae distinct): *Eryngium, Astrantia, Sanicula.*
2. LAGOECIËAE (ov. 1-loc.; fr. 1-seeded; vittae indistinct): *Lagoecia.*

III. *Apioïdeae* (stipules absent; endocarp soft, sometimes hardened by subepidermal fibre layers; style on apex of disk; vittae various).

A. Primary ridges projecting, the lat. sometimes wing-like; no secondary ridges.

a. Secondary umbels each with 1 or few ♀ fls. surrounded by ♂♂.
   1. ECHINOPHOREAE (fr. enclosed by hardened stalks of ♂ fls.): *Echinophora.*
b. Fls. all ♀, or irreg. polygamous.
   α. Seed at commissure deeply forked or hollow.
   2. SCANDICEAE (parenchyma around carpophore with crystal layer): *Chaerophyllum, Anthriscus, Torilis.*
   3. CORIANDREAE (without crystal layer; fr. ovate-spherical, nut-like, rarely long, with woody subepidermal layer): *Coriandrum.*
   4. SMYRNIËAE (narrow commissure, mericarps rounded outwards): *Smyrnium, Conium.*
   β. Seeds flattened at commissure.
   5. AMMIËAE (primary ridges all alike; seed semicircular in section): *Bupleurum, Apium, Petroselinum, Carum, Pimpinella, Seseli, Foeniculum, Oenanthe, Ligusticum.*
   6. PEUCEDANEAE (lat. ribs much broader, forming wings; seed narrow in section): *Angelica, Ferula, Peucedanum, Pastinaca.*
B. Lat. ridges equal or larger than primary; vittae in furrows or under secondary ridges.
   7. LASERPITIËAE (secondary ridges very marked, often extended into broad undivided or wavy wings): *Laserpitium, Thapsia.*
   8. DAUCEAE (ribs with spines): *Daucus.*

Umbelliflorae Bartl. An order of Dicots. comprising (in Engler's system) the *Araliac., Umbellif.* and *Cornac.*

Umbellulanthus S. Moore = Triaspis Burch. (Malpighiac.).

Umbellularia Nutt. Lauraceae. 1 Calif., *U. californica* Nutt., the California olive, with useful timber.

Umbilicaria Fabr. (uninom.) = *Omphalodes* Mill. sp. (Boraginac.).

Umbilicaria Pers. (1805; non [Ach.] Hoffm. 1789—Lichenes) = Cotyledon L. (Crassulac.).

Umbilicus DC. Crassulaceae. 18 Medit.

Umbraculum Kuntze = Aegiceras Gaertn. (Myrsinac.).

Umsema Rafin. = Unisema Rafin. = Pontederia L. (Pontederiac.).

Umtiza Sim. Leguminosae. 1 S. Afr.

Unamia Greene = Aster L. (Compos.).

Unannea Steud. = seq.

Unanuea Ruiz & Pav. ex Pennell. Scrophulariaceae. 2 Colombia, Ecuador.

Uncaria Burch. = Harpagophytum DC. (Pedaliac.).

*Uncaria Schreb. Naucleaceae (~Rubiac.). 60 trop. They climb by hooks, which are metam. infl.-axes, and sensitive to continued contact; after clasping they enlarge and become woody. *U. gambir* (Hunter) Roxb. (gambier; Malaysia) is a valuable source of tan.

Uncarina (Baill.) Stapf. Pedaliaceae. 5 Madag.

Uncariopsis Karst. = Schradera Vahl (Rubiac.).

Uncasia Greene = Eupatorium L. (Compos.).

Uncifera Lindl. Orchidaceae. 5 India, Siam, Malay Penins.

Uncina C. A. Mey. = Uncinia Pers. (Cyperac.).

Uncinaria Reichb. = Uncaria Schreb. (Rubiac.).

Uncinia Pers. Cyperaceae. 35 mts. Borneo, New Guinea, Austr., N.Z.; Mex. & Venez. to temp. S. Am.; antarct. Is. The axis of origin of the fl. projects beyond the utricle in the form of a long hook, serving as a means of dispersal for the fr.

Uncinus Raeusch. = Oncinus Lour. = Melodinus J. R. & G. Forst. (Apocynac.).

Unedo Hoffmgg. & Link = Arbutus L. (Ericac.).

Ungeria Nees ex C. B. Clarke = Cyperus L. (Cyperac.).

Ungeria Schott & Endl. Sterculiaceae. 1 E. Austr., Norfolk I.

Ungernia Bunge. Amaryllidaceae. 6 Persia, C. As.

Ungnadia Endl. Sapindaceae. 1 S. U.S., Mex. (Mexican buckeye).

Unguacha Hochst. = Strychnos L. (Strychnac.).

Ungula Barlow. Loranthaceae. 1 New Guinea.

Ungulipetalum Moldenke. Menispermaceae. 1 Brazil.

Unifolium Ludw. = Maianthemum Weber (Liliac.).

Unigenes E. Wimm. Campanulaceae. 1 S. Afr.

Uniola L. Gramineae. 2 N. Am., W.I. Useful pasture.

Unisema Rafin. = Pontederia L. (Pontederiac.).

Unisexuales Benth. & Hook. f. A 'Series' of Dicots. comprising the fams. *Euphorbiac.*, *Balanopac.*, *Urticac.*, *Platanac.*, *Leitneriac.*, *Juglandac.*, *Myricac.*, *Casuarinac.* and *Cupuliferae.*

Unjala Reinw. ex Blume = Schefflera J. R. & G. Forst. (Araliac.).

Unona Hook. f. & Thoms. = Desmos Lour. + Dasymaschalon (Hook. f. & Thoms.) Dalla Torre & Harms (Annonac.).

Unona L. f. = Xylopia L. (Annonac.).

Unonopsis R. E. Fries. Annonaceae. 27 Mex. to trop. S. Am., W.I.

Unxia Bert. ex Colla = Blennosperma Less. (Compos.).

Unxia Kunth = Villanova Lag. (Compos.).

Unxia L. f. Compositae. 2 Panama, N. trop. S. Am.

Upata Adans. = Avicennia L. (Avicenniac.).

Upoda Adans. = Hypoxis L. (Hypoxidac.).

Upopion Rafin. = Thaspium Nutt. (Umbellif.).

Upoxis Adans. = Gagea L. (Hypoxidac.).

Upudalia Rafin. (Daedalacanthus T. Anders.). Acanthaceae. 15 Penins. India, Ceylon, E. Himal. to Lower Burma.

Upuna Symington. Dipterocarpaceae. 1 Borneo. Aril. Useful timber.

Upuntia Rafin. = Opuntia Mill. (Cactac.).

Urachne Trin. = Oryzopsis Michx. (Gramin.).

Uragoga Baill. = Cephaëlis Sw. (Rubiac.).

URALEPIS

Uralepis Rafin. = seq.

Uralepsis Nutt. = Triplasis Beauv. (Gramin.).

Urananthus Benth. = Eustoma Salisb. (Gentianac.).

Urandra Thw. Icacinaceae. 17 Indomal.

Urania DC. (sphalm.) = Uraria Desv. (Legumin.).

Urania. Schreb. = Ravenala Adans. (Strelitziac.).

Uranodactylus Gilli. Cruciferae. 2 C. As., Afghan., Himal.

Uranostachys Fourr. = Veronica L. (Scrophulariac.).

Uranthera Naud. = Acisanthera P.Br. (Melastomatac.).

Uranthera Pax & K. Hoffm. = Phyllanthodendron Hemsl. (Euphorbiac.).

Uranthera Rafin. = Justicia L. (Acanthac.).

Uranthoecium Stapf. Gramineae. 1 New S. Wales.

Uraria Desv. Leguminosae. 20 trop. Afr., SE. As., Formosa, Indomal., N. Austr., Pacif.

Urariopsis Schindl. Leguminosae. 1 SE. As.

Uraspermum Nutt. = Osmorhiza Rafin. (Umbellif.).

Uratea J. F. Gme. = Ouratea Aubl. (Ochnac.).

Uratella P. & K. = Ouratella Van Tiegh. = Ouratea Aubl. (Ochnac.).

Urbananthus R. M. King & H. Rob. (~ Eupatorium L.). Compositae. 2 W.I.

Urbanella Pierre = Lucuma Molina (Sapotac.).

***Urbania** Phil. Verbenaceae. 2 Chile.

Urbania Vatke = Lyperia Benth. (Scrophulariac.).

Urbaniella Dusén ex Melch. = Urbanolophium Melch. (Bignoniac.).

Urbanisol Kuntze = Tithonia Desf. ex Juss. (Compos.).

Urbanodendron Mez. Lauraceae. 1 E. Brazil.

Urbanodoxa Muschler = Cremolobus DC. (Crucif.).

Urbanoguarea Harms. Meliaceae. 1 W.I. (Santo Domingo).

Urbanolophium Melch. Bignoniaceae. 2 Brazil.

Urbanosciadium H. Wolff. Umbelliferae. 1 Peru.

Urbinella Greenman. Compositae. 1 Mex.

Urbinia Rose = Echeveria DC. (Crassulac.).

× **Urbiphytum** Gossot. Crassulaceae. Gen. hybr. (Pachyphytum × Urbinia).

× **Urceocharis** Mast. Amaryllidaceae. Gen. hybr. (Eucharis × Urceolina).

***Urceola** Roxb. Apocynaceae. 15 Burma, W. Malaysia.

Urceola Vand. Inc. sed. (? Rubiac. ? Loganiac.). 1 Brazil.

Urceolaria F. G. Dietr. (sphalm.) = Utricularia L. (Lentibulariac.).

Urceolaria Herb. = Urceolina Reichb. (Amaryllidac.).

Urceolaria Molina = Sarmienta Ruiz & Pav. (Gesneriac.).

Urceolaria Willd. ex Cothenius = Schradera Vahl (Rubiac.).

***Urceolina** Reichb. Amaryllidaceae. 5 Andes.

Urechites Muell. Arg. Apocynaceae. 2 Florida, C. Am., W.I.

Urelytrum Hack. Gramineae. 18 trop. & S. Afr.

Urena L. Malvaceae. 6 trop. & subtrop. Schizocarp, the individual cpls. provided with hooks. Useful fibre.

Urera Gaudich. Urticaceae. 35 Hawaii, warm Am., trop. & S. Afr., Madag. Stinging hairs powerful. Achene enclosed in persistent fleshy P (pseudo-berry).

Ureskinnera P. & K. = Uroskinnera Lindl. (Scrophulariac.).

Uretia P. & K. = Ouret Adans. = Aerva Forsk. (Amaranthac.).
Uretia Rafin. = Aerva Forsk. + Digera Forsk. (Amaranthac.).
Urginea Steinh. Liliaceae. 100 Medit., Afr., India. *U. maritima* (L.) Baker
(*U. scilla* Steinh., squill), large bulbs used in medicine.
Urgineopsis Compton. Liliaceae. 1–2 S. Afr.
Uribea Dugand & Romero. Leguminosae. 1 Costa Rica, Colombia.
Uricola Boerl. (sphalm.) = Urceola Roxb. (Apocynac.).
Urinaria Medik. = Phyllanthus L. (Euphorbiac.).
Urmenetia Phil. Compositae. 1 N. Chile.
Urnectis Rafin. = Salix L. (Salicac.).
Urnularia Stapf. Apocynaceae. 6 W. Malaysia.
Urobotrya Stapf (~ Opilia Roxb.). Opiliaceae. 2 trop. Afr., 2 SE. As.
Urocarpidium Ulbr. Malvaceae. 11 Mex., C. Andes, Galápagos.
Urocarpus J. Drumm. ex Harv. = Asterolasia F. Muell. (Rutac.).
Urochlaena Nees. Gramineae. 2 S. Afr.
Urochloa Beauv. Gramineae. 25 trop. Afr., As.
Urochondra C. E. Hubbard. Gramineae. 1 NE. trop. Afr.
Urodesmium Naud. = Pachyloma DC. (Melastomatac.).
Urodon Turcz. = Pultenaea Sm. (Legumin.).
Urogentias Gilg & Gilg-Benedict. Gentianaceae. 1 trop. E. Afr.
Urolepis (A. DC.) R. M. King & H. Rob. (~ Eupatorium L.). Compositae.
1 Bolivia & Braz. to Argent.
Uromorus Bur. = Streblus Lour. (Morac.).
Uromyrtus Burret. Myrtaceae. 10 New Caled.
Uropappus Nutt. = Microseris D. Don (Compos.).
Uropedilum Pfitz. = seq.
Uropedium Lindl. = Phragmipedium Rolfe (Orchidac.).
Uropetalon Ker-Gawl. = Dipcadi Medik. (Liliac.).
Uropetalum Burch. = praec.
Urophyllon Salisb. = Ornithogalum L. (Liliac.).
Urophyllum C. Koch (sphalm.) = Urospatha Schott (Arac.).
Urophyllum Wall. Rubiaceae. 150 trop. Afr., trop. As. to Japan & New
Urophysa Ulbr. Ranunculaceae. 2 China.   [Guinea.
Uroskinnera Lindl. Scrophulariaceae. 4 Mex., C. Am.
Urospatha Schott. Araceae. 20 C. & trop. S. Am. Rhiz. spongy.
Urospermum auct. ex Steud. = Uraspermum Nutt. = Osmorhiza Rafin.
(Umbellif.).
Urospermum Scop. Compositae. 2 Medit.
Urostachyaceae Rothm. = Lycopodiaceae Reichb.
Urostachys Herter = Lycopodium L. (Lycopodiac.).
Urostelma Bunge = Metaplexis R.Br. (Asclepiadac.).
Urostephanus Robinson & Greenman. Asclepiadaceae. 1 Mex.
Urostigma Gasp. = Ficus L. (Morac.).
Urostylis Meissn. = Layia Hook. & Arn. ex DC. (Compos.).
Urotheca Gilg. Melastomataceae. 1 trop. E. Afr.
Ursinea Willis (sphalm.) = seq.
*****Ursinia** Gaertn. Compositae. 80 S. Afr., 1 Abyssinia.
Ursiniopsis Phillips. Compositae. 3 S. Afr.

URTICA

Urtica L. Urticaceae. 50 mostly N. temp., few trop. & S. temp. (stinging nettle). Herbs with opp. ls. and stips. (sometimes united in pairs between the petioles, as in *Rubiaceae*), usu. covered with stinging hairs. The various types of infl. are well shown in 3 Eur. spp. In general the infl. is a dich. cyme with tendency to cincinnus by preference of the β-bracteole. In *U. pilulifera* L. (Roman nettle) the ♂ and ♀ infls. spring side by side from each node, the ♂ catkin-like, the ♀ a pseudo-head. In *U. urens* L. (small nettle) a panicle is formed containing both ♂ and ♀ fls. In *U. dioica* L. (large or common nettle) there is a panicle, but each sex is confined to its own plant. P 4; A 4, opp. to P leaves. The sta. are bent down inwards in the bud, and when ripe spring violently upwards and bend out of the fl., the anther turning inside out, so that the loose powdery pollen is ejected as a little cloud, and may be borne by wind to the stigma. The ♀ fl. has a 1-loc., 1-ovuled ovary with a large brush-like stigma. Achene enclosed in the persistent P. Young tops can be eaten like spinach. Useful fibre from stems.

*Urticaceae Juss. Dicots. 45/550 trop. & temp. Most are herbs or undershrubs, with no latex, and with alt. or opp. stip. ls. Infl. cymose, often 'condensed' into pseudo-heads, etc. Fls. usu. unisexual and reg. P 4–5, free or united, sepaloid; sta. as many, straight (*Conocephaleae*) or bent down inwards in bud and exploding when ripe; G 1-loc. with 1 erect basal orthotr. ov. (sometimes apical in *Conocephaleae*) and 1 style. Achene. Seed usu. with rich oily endosp.; embryo small, straight. *Boehmeria, Urtica, Maoutia* and others are used as sources of fibre.

Classification and chief genera (after Engler):
A. With stinging hairs. P (4–5) in ♀.
    1. URTICEAE (UREREAE): *Urtica, Urera, Laportea.*
B. No stinging hairs.
    2. PROCRIDEAE (P of ♀ 3-merous, stigma paintbrush-like): *Pilea, Pellionia, Elatostema.*
    3. BOEHMERIEAE (♂ usu. with 4–5 sta.; no invol.): *Boehmeria, Maoutia.*
    4. PARIETARIEAE (P present; bracts often united in invol.): *Parietaria.*
    5. FORSKOHLEEAE (♂ fl. reduced to 1 sta.): *Forskohlea.*
    6. CONOCEPHALEAE (fil. in ♂ fl. straight): *Poikilospermum, Cecropia.*
Ulmaceae are distinguished by infl., aestivation of sta., and ovule; *Moraceae* by presence of latex, and also usu. by pend. ovule, curved embryo, etc.

Urticastrum Fabr. = Laportea Gaudich. (Urticac.).

Urucu Adans. = Bixa L. (Bixac.).

Uruparia Rafin. = Ourouparia Aubl. = Uncaria Schreb. (Naucleac.).

Urvillaea DC. = seq.

Urvillea Kunth. Sapindaceae. 13 warm Am. Lianes like *Serjania*

Usionis Rafin. = Salix L. (Salicac.).

Usoricum Lunell = Oenothera L. (Onagrac.).

Usteria Cav. = Maurandya Ortega (Scrophulariac.).

Usteria Dennst. = Acalypha L. (Euphorbiac.).

Usteria Medik. = Endymion Dum. (Liliac.).

Usteria Willd. Antoniaceae. 1 trop. W. Afr.

Usubis Burm. f. = Allophylus L. (Sapindac.).

Utahia Britton & Rose = Pediocactus Britt. & Rose (Cactac.).

Utania G. Don = Fagraea Thunb. (Potaliac.).

Utea J. St-Hil. = Outea Aubl. = Macrolobium Schreb. (Legumin.).

Uterveria Bertol. = Capparis L. (Capparidac.).

Utleria Bedd. ex Benth. Periplocaceae. 1 S. India.

Utricularia L. Lentibulariaceae. 120 trop. & temp., the latter all aquatic. The morphology is interesting, for the usual distinctions drawn between root, stem and ls. cannot be applied here. *U. vulgaris* L. is a submerged water pl. with finely divided ls.; it never has roots, even in the embryo. The fls. project above water on short shoots, and there are also short shoots with small ls., which arise from the main axis and grow upwards to the surface. Upon the ordinary submerged ls. are borne the bladders, remarkable hollow structures with trap-door entrances. Small Crustacea and other animals coming in contact with a sensitive trigger mechanism on the door cause it to open inwards, the bladder at the same time suddenly expanding. The animal is thus sucked into the bladder, the door of which then closes. The plant takes up the products of the decay of the organism thus captured; it is very doubtful whether any special ferment is secreted. Other spp. are land pls. with peculiar runners, which develop in the moss or other substratum on which they grow, and there bear the bladders. Others again, e.g. *U. alpina* Jacq., are epiph. with water storage in tuberous branches. The ls. of all these forms are simple. Goebel (*Pflanzenbiol. Sch.*) has investigated the development of *U.* and finds that all these parts —ls., bladders, runners, water shoots, erect shoots, etc.—are practically equivalent to one another, and that the same rudiment at the growing-point may give rise to any one of them, or they may themselves change from one to another type. Similarly on germ. a number of spirally arranged primary ls. are produced, and then one or two water shoots appear lat. on the growing-point, bearing no direct relation to the ls. in position, but apparently homologous with them. 'Like *Genlisea*, *U.* possessed originally a leaf-rosette, ending with an infl., and consisting partly of bladders. Then were added the swimming water shoots or (in land forms) runners, which though externally unlike leaves (since they develop indefinitely and produce leaves and infls.) yet are originally homologous with them.' For further details see Goebel, *loc. cit.* Hibernation in temp. spp. by winter buds full of reserves, which drop off and sink.

Utriculariaceae ('-ulinae') Hoffmgg. & Link = Lentibulariaceae Rich.

Utsetela Pellegr. Moraceae. 1 W. Equat. Afr.

Uva Kuntze = seq.

Uvaria L. Annonaceae. 150 trop. Afr., Madag., Indomal., Austr. Mostly lianes with recurved hooks (infl.-axes). The connective of the anther is usu. leafy.

Uvariastrum Engl. Annonaceae. 7 trop. Afr.

Uvariella Ridley = Uvaria L. (Annonac.).

Uvariodendron (Engl. & Diels) R. E. Fries. Annonaceae. 12 trop. Afr. (mostly W.).

Uvariopsis Engl. ex Engl. & Diels. Annonaceae. 12 trop. W. Afr.

Uva-Ursi Duham. = Arctostaphylos Adans. (Ericac.).

Uvedalia R.Br. = Mimulus L. (Scrophulariac.).

Uvifera Kuntze = Coccoloba L. (Polygonac.).

Uvirandra Jaume St-Hil. = Ouvirandra Thou. = Aponogeton Thunb. (Aponogetonac.).

UVULANA

Uvulana Rafin. = seq.
Uvularia L. Liliaceae. 4 E. N. Am.
Uvulari[ac]eae Kunth = Liliaceae–Melanthioïdeae–Uvulariëae Endl.
Uwarowia Bunge = Verbena L. (Verbenac.).
Uxi Almeida Pinto = Sacoglottis Mart. (Houmiriac.).

V

Vaccaria v. Wolf. Caryophyllaceae. 4 C. & E. Eur., Medít., temp. As.
***Vacciniaceae** S. F. Gray = Ericaceae–Gaultheriëae Drude.
Vacciniopsis Rusby. Ericaceae. 2 Bolivia.
Vaccinium L. (incl. *Oxycoccus* Hill). Ericaceae. 3–400 N. temp., trop. mts.
(exc. Afr.), Andes, S. Afr., Madag. The Eur. *V. myrtillus* L., the whortle-,
bil- or blae-berry, is common in hilly districts. *V. uliginosum* L. at high levels.
Both have deciduous ls. and blue berries. *V. vitis-idaea* L., the cow- or whin-
berry (often called cranberry by error), also a mountain sp., evergr., with red
berries. *V. oxycoccos* L., the cranberry, in mountain bogs, a slender trailing
evergr., with l. edges rolled back. The fls. resemble *Erica*, both in structure
and mech., but ov. inf.; largely visited by humble-bees. The red fleshy fr. is ed.
(used for jams, etc.) and is much distr. by birds.
Vachellia Wight & Arn. = Acacia Mill. (Legumin.).
Vachendorfia Adans. = Wachendorfia Burm. (Hypoxidac.).
Vada-Kodi Adans. = Justicia L. (Acanthac.).
Vadia O. F. Cook = Chamaedorea Willd. (Palm.).
Vagaria Herb. Amaryllidaceae. 3 Morocco, 1 Syria.
Vaginaria Kunth = praec.
Vaginaria Pers. = Fuirena Rottb. (Cyperac.).
Vaginularia Fée. Vittariaceae. 6 Ceylon to Fiji (often incl. in *Monogramma*
Schkuhr).
Vagnera Adans. = Smilacina Desf. (Liliac.).
Vahadenia Stapf. Apocynaceae. 2 trop. W. Afr.
Vahea Lam. = Landolphia Beauv. (Apocynac.).
Vahlbergella Blytt = Wahlbergella Fries = Melandrium Roehl. (Caryophyllac.).
Vahlenbergella Pax & Hoffm. (sphalm.) = praec.
Vahlia Dahl = Dombeya Cav. (Sterculiac.).
***Vahlia** Thunb. = Bistella Adans. (Vahliac.).
Vahliaceae (Reichb.) Dandy. Dicots. 1/5 trop. & S. Afr. to NW. India. Erect
branched sometimes glandular ann. or bienn. herbs, with opp. simple ent.
exstip. ls. Fls. reg., ♀, in pairs in sympod. cymose infls. K 5, valv.; C 5, imbr.
or open; A 5, alternipet., with subul. fil. and dorsif. intr. anth.; disk epig.,
inconspic.; G (2–3), 1-loc., with ∞ ov. on 2–3 thick pend. flattened orbic.
apic. plac., and 2–3 thick divaric. styles with capit. stigs. Fr. a ± glob. apically
dehisc. caps., with ∞ minute appendaged seeds. Only genus: *Bistella*. Somewhat intermediate between *Saxifragac.* and *Rubiac.–Oldenlandiëae.*
Vahlodea Fries (~Deschampsia Beauv.). Gramineae. 3–4 NE. As.
Vailia Rusby. Asclepiadaceae. 1 Bolivia.
Vaillanta Rafin. = seq.

Vaillantia Neck. ex Hoffm. = Valantia L. (Rubiac.).
Vainilla Salisb. = Vanilla Juss. (Orchidac.).
Valantia L. Rubiaceae. 3–4 Canary Is. to N. Persia.
Valarum Schur = Sisymbrium L. (Crucif.).
Valbomia Rafin. = Wahlbomia Thunb. = Tetracera L. (Dilleniac.).
Valcarcelia Lag. (nomen). Leguminosae. Quid?
Valcarcella Steud. = praec.
Valdesia Ruiz & Pav. = Blakea P.Br. (Melastomatac.).
Valdia Boehm. = Ovieda L. = Clerodendrum L. (Verbenac.).
Valdivia Remy. Escalloniaceae. 1 Chile.
Valentiana Rafin. 'Caprifoliaceae'. 1 'Abyssinia'. Quid?
Valentina Speg. = Valentiniella Speg. (Boraginac.).
Valentinia Fabr. (uninom.) = Maianthemum Weber sp. (Liliac.).
Valentinia Neck. = Tachigalia Aubl. (Legumin.).
Valentinia Raeusch. = Eystathes Lour. = Xanthophyllum Roxb. (Xanthophyllac.).
Valentinia Sw. = Casearia Jacq. (Flacourtiac.).
Valentiniella Speg. Boraginaceae. 1 Patag.
Valenzuela B. D. Jacks. = Valenzuelia S. Mutis ex Caldas = Picramnia Sw. (Simaroubac.).
Valenzuelia Bertero ex Cambess. Sapindaceae. 2 Chile, Argentina.
Valenzuelia S. Mutis ex Caldas = Picramnia Sw. (Simaroubac.).
Valeranda Neck. = Orphium E. Mey. (Gentianac.).
Valerandia Dur. & Jacks. = praec.
Valeria Minod. Scrophulariaceae. 1 E. Brazil.
Valeriana L. Valerianaceae. 200+ Euras., S. Afr., temp. N. Am., Andes. Fls. protandr. The K forms a pappus upon the fr.
*Valerianaceae Batsch. Dicots. 13/400 Eur., As., Afr., Am. Herbs (rarely shrubs) with opp. or rosul. exstip. ls. and dich. branching. Fls. in cymose panicles, etc., ♂ or ♂ ♀, zygo., usu. 5-merous. K sup., little developed at time of flowering, afterwards often forming a pappus as in *Compositae*; C usu. (5), often spurred at base, lobes imbr.; A 1–4, epipet., alt. with petals; anthers intr.; G̅ (3); only 1 loc. is fertile, and contains 1 pend. anatr. ov.; style simple, slender. Achene. Seed exalbum. Chief genera: *Valerianella, Valeriana, Centranthus, Patrinia, Phyllactis, Plectritis*.
Valerianella Mill. Valerianaceae. 80 W. Eur. to C. As & Afghan. Seed-dispersal mech. various. In *V. rimosa* Bast. the sterile loc. of the fr., in *V. vesicaria* Moench the K, is inflated; in *V. discoidea* Loisel. the K forms a parachute, whilst in others it is provided with hooks.
Valerianodes Dur. & Jacks. = seq.
Valerianoïdes Medik. = Stachytarpheta Vahl (Verbenac.).
Valerianopsis C. A. Muell. = Valeriana L. (Valerianac.).
Valerioa Standley & Steyerm. Solanaceae. 1 C. Am.
Valetonia Durand = Pleurisanthes Baill. (Icacinac.).
Validallium Small = Allium L. (Alliac.).
Valikaha Adans. = Memecylon L. (Memecylac.).
Valisneria Scop. = Vallisneria L. (Hydrocharitac.).
Valkera Stokes = Walkera Schreb. = Ouratea Aubl. (Ochnac.).

VALLANTHUS

Vallanthus Cif. & Giac. Amaryllidaceae. Gen. hybr. (Cyrtanthus × Vallota).

Vallantia A. Dietr. = Valantia L. (Rubiac.).

Vallariopsis R. E. Woodson. Apocynaceae. 1 W. Malaysia.

Vallaris Burm. f. Apocynaceae. 10 India to SE. As., Philipp. & Malay Penins.

Vallaris Rafin. = Euphorbia L. (Euphorbiac.).

Vallea Mutis ex L. f. Elaeocarpaceae. 1 Colombia to Bolivia.

Vallesia Ruiz & Pav. Apocynaceae. 10 Florida to Argentina.

Valliera Ruiz & Pav. (apud Lopez). Tiliaceae? Flacourtiaceae? 1 Ecuador.

Vallifilix Thou. = Lygodium Sw. (Schizaeac.).

Vallisneria L. Hydrocharitaceae. 6-10 trop. & subtrop. *V. spiralis* L. (Eur.) a dioec. submerged water pl. with ribbon ls. ♂ fls. in dense spikes enclosed in spathes; when ready to open the fls. break off and float up to the surface, where they open. ♀ fl. sol. on very long stalk, which brings it to the surface; it has a green P, inf. ov. and 3 large stigmas. Pollination occurs on the surface (cf. *Elodea*); and after it the stalk curls up into a close spiral, drawing the young fr. to the bottom to ripen. Veg. repr. by runners, rooting at the ends.

Vallisneriaceae Dum. = Hydrocharitaceae–Vallisneriëae Endl.

*****Vallota** Salisb. ex Herb. Amaryllidaceae. 1 S. Afr.

Vallota Steud. = Valota Adans. = Trichachne Nees (Gramin.).

Vallotia P. & K. = praec.

Valoradia Hochst. = Ceratostigma Bunge (Plumbaginac.).

Valota Adans. = Trichachne Nees (Gramin.).

Valota Dum. = Vallota Salisb. ex Herb. (Amaryllidac.).

Valsonica Scop. = Watsonia Mill. (Iridac.).

Valteta Rafin. = Iochroma Benth. (Solanac.).

Valvanthera C. T. White. ?Hernandiaceae. 1 Queensland.

Valvaria Seringe = Clematis L. (Ranunculac.).

Valvinterlobus Dulac = Wahlenbergia Schrad. ex Roth (Campanulac.).

Vanalphimia Leschen. ex DC. = Saurauia Willd. (Actinidiac.).

Vanalpighmia Steud. = praec.

Vananthes Willis (sphalm.) = Vauanthes Haw. (Crassulac.).

Vanasta Rafin. = Vanessa Rafin. = Manettia Mutis ex L. (Rubiac.).

× **Vancampe** hort. Orchidaceae. Gen. hybr. (iii) (Acampe × Vanda).

Vanclevea Greene (~ Grindelia Willd.). Compositae. 1 U.S.

Vancouveria C. Morr. & Decne. Berberidaceae. 3 Pacif. N. Am.

Vanda R.Br. Orchidaceae. 60 China, Indomal., Marianne Is.

× **Vandachnanthe** hort. (v) = × Aranda hort. (Orchidac.).

× **Vandachnis** hort. Orchidaceae. Gen. hybr. (iii) (Arachnis × Vandopsis).

× **Vandachostylis** (sphalm. 'Vandaco-') hort. (iii) = × Rhynchovanda hort. (Orchidac.).

× **Vandacostylis** hort. (sphalm.) = praec.

× **Vandaecum** hort. Orchidaceae. Gen. hybr. (vi) (Angraecum × Vanda).

× **Vandaenopsis** hort. Orchidaceae. Gen. hybr. (iii) (Phalaenopsis × Vanda).

Vandalea Fourr. = Sisymbrium L. (Crucif.).

× **Vandaeopsis** hort. (vii) = praec.

× **Vandanthe** hort. (v) = Vanda R.Br. (Orchidac.).

× **Vandanopsis** hort. (vii) = × Vandaenopsis hort. (Orchidac.).

× **Vandantherides** hort. (v) = × Aëridovanda hort. (Orchidac.).

1202

VARGASIA

× **Vandarachnis** hort. (vii) = × Aranda hort. (Orchidac.).
Vandasia Domin. Leguminosae. 1 Queensland.
× **Vandathera** hort. (vii) = × Renantanda hort. (Orchidac.).
Vandea Griff. (sphalm.) = Vanda R.Br. (Orchidac.).
Vandellia P.Br. ex L. = Lindernia All. (Scrophulariac.).
Vandenboschia Copel. Hymenophyllaceae. 25 cosmop. (type sp. *Trichomanes radicans* Sw.).
Vandera Rafin. = ? Croton L. (Euphorbiac.).
Vanderystia De Wild. = Ituridendron De Wild. (Sapotac.).
Vandesia Salisb. = Bomarea Mirb. (Alstroemeriac.).
× **Vandofinetia** hort. Orchidaceae. Gen. hybr. (iii) (Neofinetia × Vanda).
× **Vandopsides** hort. Orchidaceae. Gen. hybr. (iii) (Aërides × Vandopsis Pfitz.).
× **Vandopsis** hort. (vii) = × Vandaenopsis hort. (Orchidac.).
Vandopsis Pfitz. Orchidaceae. 21 China, Formosa, NE. India to Polynesia.
× **Vandopsisvanda** hort. (vii) = × Opsisanda hort. (Orchidac.).
× **Vandoritis** hort. Orchidaceae. Gen. hybr. (iii) (Doritis × Vanda).
Vanessa Rafin. = Manettia Mutis ex L. (Rubiac.).
× **Vanglossum** hort. Orchidaceae. Gen. hybr. (iii) (Ascoglossum × Vanda).
Vangueria Comm. ex Juss. Rubiaceae. 27 trop. Afr., Madag.
Vangueriopsis Robyns ex Good. Rubiaceae. 18 trop. Afr.
Vanhallia Schult. f. = Apama Lam. (Aristolochiac.).
Vanheerdia L. Bolus. Aïzoaceae. 4 S. Afr.
Vanhouttea Lem. Gesneriaceae. 4 Brazil.
Vaniera Jaume St-Hil. = seq.
Vanieria Lour. = Cudrania Tréc. (Morac.).
Vanieria Montr. = Trisema Hook. f. = Hibbertia Andr. (Dilleniac.).
Vanilla Mill. Orchidaceae. 90 trop. & subtrop. The spice 'vanilla' is obtained from the pods of spp. of this genus.
Vanillaceae Lindl. = Orchidaceae–Epidendreae–Vanillinae Benth. (This group contains perhaps the most 'primitive' or least specialised members of the
Vanillophorum Neck. = Vanilla Mill. (Orchidac.). 　　　　[*Orchidac.*)
Vanillosma Spach = Piptocarpha R.Br. (Compos.).
Vanillosmopsis Sch. Bip. Compositae. 7 Brazil.
Vaniotia Léveillé = Petrocosmea Oliv. (Gesneriac.).
Vanoverberghia Merr. Zingiberaceae. 1 Philippines.
Van-Royena Aubrév. Sapotaceae. 1 Queensland.
Vantanea Aubl. Houmiriaceae. 14 C. to trop. S. Am.
Vanzijlia L. Bolus. Aïzoaceae. 3 S. Afr.
Varangevillea Willis (sphalm.) = Varengevillea Baill. = Rhodocolea Baill. (Bignoniac.) + Vitex L. (Verbenac.).
Varasia Phil. = Gentiana L. (Gentianac.).
Vareca Gaertn. = Casearia Jacq. (Flacourtiac.).
Vareca Roxb. = Rinorea Aubl. (Violac.).
Varengevillea Baill. = Rhodocolea Baill. (Bignoniac.) + Vitex L. (Verbenac.).
Varennea DC. = Eysenhardtia Kunth (Legumin.).
Vargasia Bert. ex Spreng. = Thouinia Poit. (Sapindac.).
Vargasia DC. = Galinsoga Ruiz & Pav. (Compos.).

Vargasia Ernst = Caracasia Szyszyl. (Marcgraviac.).
Vargasiella C. Schweinf. Orchidaceae. 2 Venezuela, Peru.
Varilla A. Gray. Compositae. 2 S. U.S., Mex.
Varinga Rafin. = Ficus L. (Morac.).
Variphylis Thou. (uninom.) = *Bulbophyllum variegatum* Thou. (Orchidac.).
Varneria L. = Gardenia Ellis (Rubiac.).
Varonthe Juss. ex Reichb. = Physena Nor. ex Thou. (Capparidac.).
Varronia P.Br. = Cordia L. (Ehretiac.).
Varroniopsis Friesen = praec.
Vartheimia Benth. & Hook. f. = seq.
Varthemia DC. Compositae. 4 E. Medit. to C. As. & NW. India.
Vasargia Steud. = Vargasia DC. = Galinsoga Ruiz & Pav. (Compos.).
Vascoa DC. (1824) = Mundia Kunth = Nylandtia Dum. (Polygalac.).
Vascoa DC. (1825) = Rafnia Thunb. (Legumin.).
Vasconcellea A. St-Hil. = Carica L. (Caricac.).
Vasconcellia Mart. = Arrabidaea DC. (Bignoniac.).
Vasconcellosia Caruel = Carica L. (Caricac.).
Vasconella Regel = praec.
× **Vascostylis** hort. Orchidaceae. Gen. hybr. (iii) (Ascocentrum × Rhynchostylis × Vanda).
Vaselia Van Tiegh. = Elvasia DC. (Ochnac.).
Vaseya Thurb. = Muhlenbergia Schreb. (Gramin.).
Vaseyanthus Cogn. Cucurbitaceae. 2 Calif., Mex.
Vaseyochloa Hitchcock. Gramineae. 1 S. U.S. (Texas).
Vasivaea Baill. Tiliaceae. 2 Peru, Brazil.
Vasovulaceae Dulac = Aquifoliaceae Bartl.
Vasquezia Phil. = Villanova Lag. (Compos.).
Vassobia Rusby. Solanaceae. 1–2 Bolivia.
Vatairea Aubl. Leguminosae. 7 trop. S. Am.
Vataireopsis Ducke. Leguminosae. 3 Brazil.
Vateria L. Dipterocarpaceae. 21 Seychelles, S. India, Ceylon. *V. indica* L. yields a gum resin (Indian copal, white dammar).
Vateriopsis Heim. Dipterocarpaceae. 1 Seychelles.
Vatica L. Dipterocarpaceae. 76 S. India, Ceylon, Siam, Indoch., Hainan, Malaysia (exc. Lesser Sunda Is.). A few yield resins or useful timber.
Vatkea Hildebr. & O. Hoffm. = Martynia L. (Martyniac.).
Vatovaea Chiov. Leguminosae. 1 NE. trop. Afr.
Vatricania Backeb. = Espostoa Britt. & Rose (Cactac.).
Vauanthes Haw. Crassulaceae. 1 S. Afr.
Vaughania S. Moore. Leguminosae. 1 Madag.
× **Vaughnara** hort. Orchidaceae. Gen. hybr. (iii) (Brassavola × Cattleya × Epidendrum).
*****Vaupelia** Brand. Boraginaceae. 8 trop. Afr.
Vaupellia Griseb. = Pentarhaphia Lindl. (Gesneriac.).
Vaupesia R. E. Schultes. Euphorbiaceae. 1 Colombia, W. Brazil.
Vauquelinia Corrêa ex Humb. & Bonpl. Rosaceae. 8–10 SW. U.S., Mex.
Vausagesia Baill. Ochnaceae. 2 S. trop. Afr.
Vauthiera A. Rich. = Cladium P.Br. (Cyperac.).

Vavaea Benth. Meliaceae. 4 Philipp., N. Borneo, Caroline Is., Java, New
Vavanga Rohr = Vangueria Juss. (Rubiac.). [Guinea, Pacif.
Vavara Benoist. Acanthaceae. 1 Madag.
Vavilovia Fedorov. Leguminosae. 1 Cauc., Persia.
Vazea Fr. Allem. ex Mart. (nomen). Olacaceae. 1 Brazil. Quid?
Vazquezia Phil. = Vasquezia Phil. = Villanova Lag. (Compos.).
Veatchia A. Gray = Pachycormus Coville ex Standl. (Anacardiac.).
Veatchia Kellogg = Hesperoscordum Lindl. (Alliac.).
Veconcibea Pax & K. Hoffm. Euphorbiaceae. 2 trop. Am.
Vedela Adans. = Ardisia Sw. (Myrsinac.).
Veeresia Monachino & Moldenke. Sterculiaceae. 1 Mex.
Vegaea Urb. Myrsinaceae. 1 W.I. (S. Domingo).
Vegelia Neck. = Weigela Thunb. (Caprifoliac.).
Veitchia Lindl. = Picea A. Dietr. (Pinac.).
*****Veitchia** H. Wendl. Palmae. 1 Philipp., 8 New Caled., New Hebrides, Fiji.
Velaea DC. = Tauschia Schlechtd. (Umbellif.).
Velaga Adans. = Pentapetes L. (Sterculiac.).
Velarum Reichb. = Sisymbrium L. (Crucif.).
Velasquezia Pritz. = Vellasquezia Bertol. = Triplaris Loefl. (Polygonac.).
Velezia L. Caryophyllaceae. 4 Medit. to Afghanistan.
Velheimia Scop. (sphalm.) = Veltheimia Gled. (Liliac.).
Vella DC. Cruciferae. 4 W. Medit. Thorns = stems.
Vella L. = Vella DC. + Carrichtera Adans. (Crucif.).
Vellasquezia Bertol. = Triplaris Loefl. ex L. (Polygonac.).
Vellea D. Dietr. ex Steud. = Velaea DC. = Tauschia Schlechtd. (Umbellif.).
Velleia Sm. Goodeniaceae. 20 Austr. Ovary ± sup.
Velleruca Pomel = Eruca Mill. (Crucif.).
Velleya Walp. = Velleia Sm. (Goodeniac.).
Vellosia Spreng. = seq. [Arabia.
Vellozia Vand. Velloziaceae. 100 trop. Am. (esp. campos), trop. Afr., Madag.,
*****Velloziaceae** Endl. Monocots. 4–5/250 trop. Am., Afr., Madag., Arabia.
Xero., chiefly of rocky places or dry campos. Perenn. with woody, dichot.
branched stems and ls. in rosettes (cf. *Aloë*). Upper parts of stems clothed with
fibrous sheaths of old ls., lower parts with adv. roots. The stem is thin, but its
coating of roots may be inches deep. Water poured over the roots disappears as
if into a sponge, and the pl. is thus able to supply itself from dew, etc., during the
dry season. The ls. also are xero., often pungent. Fls. sol., term., reg. P 3 + 3,
corolline; A 3 + 3, or ∞, in 6 bundles. G̅ (3), 3-loc., with axile plac. in the form
of lamellae, ± peltately widened or thickened at the outer edge. Ovules ∞.
Style slender, with capit. stig. or 3-lobed. Caps. woody, often capitate-
verrucose, septicid. or irreg. dehisc. Endosp. Genera: *Vellozia* (*Talbotia*),
Barbacenia, Xerophyta, Barbaceniopsis.
Velloziella Baill. Scrophulariaceae. 3 Brazil.
Vellozoa Lem. = Vellozia Vand. (Velloziac.).
Velophylla Benj. Clarke ex Durand. Podostemaceae (gen. dub.). 1 Brazil.
Velpeaulia Gaudich. = Dolia Lindl. = Nolana L. (Nolanac.).
Veltheimia Gleditsch. Liliaceae. 6 S. Afr.
Veltis Adans. = ? Centaurea L. (Compos.).

VELVETIA

Velvetia Van Tiegh. = Loranthus L. (Loranthac.).

Velvitsia Hiern. Scrophulariaceae. 1 trop. Afr.

Vemonia Edgew. (sphalm.) = Vernonia Schreb. (Compos.).

Venana Lam. = Brexia Nor. ex Thou. (Brexiac.).

Venatris Rafin. = Aster L. (Compos.).

Vendredia Baill. = Rhetinodendron Meissn. (Compos.).

Venegasia DC. Compositae. 1 California, NW. Mex.

Venegazia Benth. & Hook. f. = praec.

Venelia Comm. ex Endl. = Erythroxylum P.Br. (Erythroxylac.).

Venidium Less. Compositae. 20–30 S. Afr.

Veniera Salisb. = Narcissus L. (Amaryllidac.).

Venilia Fourr. = Scrophularia L. (Scrophulariac.).

Ventana Macbride (sphalm.) = Vantanea Aubl. (Houmiriac.).

*****Ventenata** Koel. Gramineae. 5 S. Eur., Medit. to Caspian.

Ventenatia Beauv. = Oncoba Forsk. (Flacourtiac.).

Ventenatia Cav. = Vintenatia Cav. = Astroloma R.Br. + Melichrus R.Br. (Epa-

Ventenatia Sm. = Stylidium Sw. (Stylidiac.).   [cridac.).

Ventenatia Tratt. = Pedilanthus Neck. ex Poit. (Euphorbiac.).

Ventenatum Leschen. ex Reichb. = Diplolaena R.Br. (Rutac.).

Ventilago Gaertn. Rhamnaceae. 1 trop. Afr., 1 Madag., 35 India & China to New Guinea, Austr. & Pacif. Some climb by hooks. Fr. with wing on upper end, formed from style after fert.

Ventraceae Dulac = Cucurbitaceae Juss.

Veprecella Naud. Melastomataceae. 20 Madag.

Vepris Comm. ex A. Juss. Rutaceae. 10 trop. & S. Afr., 30 Madag., Masc., **Veratr[ac]eae** C. A. Agardh = Liliaceae–Melanthioïdeae Engl.   [1 S. India.

Veratrilla (Baill.) Franch. Gentianaceae. 2 E. Himal., Assam, W. China.

Veratronia Miq. = Hanguana Bl. (Hanguanac.).

Veratrum L. Liliaceae. 25 N. temp. Rhiz. with leafy stem and racemes, lower fls. ♀, but upper usu. ♂ by abortion (andro-monoecism). Sometimes pls. occur with ♂ fls. only. Protandr. Seeds with membranous border. Veratrin is obtained from the rhiz.; that of *V. album* L. is known as white hellebore root.

Verbascaceae Bonnier = Scrophulariaceae–Verbascoïdeae.

Verbascum L. Scrophulariaceae. 360 N. temp. Euras. Large perenn. herbs with stout tap-roots, wrinkled like *Taraxacum*. Infl. primarily racemose, but lat. fls. often replaced by condensed dichasia (cf. *Labiatae*). Fls. visited for pollen by bees and drone-flies. Those of several formerly officinal (*flores Verbasci*).

Verbena L. Verbenaceae. 250 trop. & temp. Am.; 2–3 Old World. *V. officinalis* L., the vervain, was formerly in great repute as a remedy in eye-diseases, its bright-eyed C, like that of *Euphrasia*, being supposed, under the old doctrine of signatures, to indicate its virtues in that direction.

Verbena Rumph. = Aerva Forsk. (Amaranthac.).

*****Verbenaceae** Jaume St-Hil. 75/3000 almost all trop. & subtrop. Herbs, shrubs or trees; many lianes, e.g. spp. of *Lantana, Clerodendrum, Vitex*; xero. also frequent, often armed with thorns. Ls. usu. opp., rarely whorled or alt., entire or divided, exstip. Infl. racemose or cymose, in the former case most often a spike or head, often with an invol. of coloured bracts. The cymes usu. dich.

with a cincinnal tendency (cf. *Caryophyllaceae*); sometimes they also form heads.

Fl. usu. ♀, zygo., usu. 5-merous. K (5) [or (4–8)], hypog.; C (5), usu. with narrow tube, rarely campanulate, often 2-lipped; A 4, didynamous, rarely 5 or 2, or of equal length, alt. with C-lobes, with intr. anthers; G usu. (2), rarely (4) or (5), usu. 4-lobed, originally 2- (or more-)loc., but very early divided into 4 (or more) loc. by the formation of a 'false' septum in each loc. (cf. *Labiatae*); plac. axile, with 2 ov. per cpl. (i.e. 1 in each loc. after septation); ovules ana- to ortho-tr., basal, lat. or pend., but always with the micropyle directed downwards. Style term., rarely ± sunk between lobes of ovary (contrast *Labiatae*); stigma usu. lobed. Fr. generally a drupe, more rarely a caps. or schizocarp. Seed exalbum.

Several are useful as sources of timber, e.g. *Tectona* (teak). See also *Lippia*, *Priva*, *Clerodendrum*, etc., for other economic uses.

Classification and chief genera (after Moldenke):

A. Infl. racemose or spicate.
  I. *Verbenoïdeae*: *Verbena*, *Lantana*, *Lippia*, *Priva*, *Petrea*, *Citharexylum*.
B. Infl. of cymose type. Cymes often united into panicles, corymbs, etc.; if axillary, often reduced to 1 fl.
  II. *Viticoïdeae* (fr. drupac.; fls. sometimes reg.): *Callicarpa*, *Tectona*, *Vitex*, *Clerodendrum*.
  III. *Nyctanthoïdeae* (fr. a 2-loc., 2-valved, 2-seeded caps.; fls. reg. or almost so): *Nyctanthes*, *Dimetra*.
  IV. *Caryopteridoïdeae* (fr. capsule-like, 4-valved, the valves falling so as to take the stones with them or loosen them from the placental axis; or 1-celled and indehiscent): *Caryopteris*, *Teijsmanniodendron*.

Verbenastrum Lippi ex Del. = Capraria L. (Scrophulariac.).
Verbenoxylum Troncoso. Verbenaceae. 1 S. Brazil.
Verbesina L, Compositae. 150 warm Am.
Verbiascum Fenzl (sphalm.) = Verbascum L. (Scrophulariac.)
Verdcourtia Wilczek = Dipogon Liebm. (Legumin.)
Verdickia De Wild. Liliaceae. 1 trop. Afr.
Verea Willd. (sphalm.) = seq.
Vereia Andr. = Kalanchoë Adans. (Crassulac.).
Verena Minod. Scrophulariaceae. 1 Paraguay.
Verhuellia Miq. Peperomiaceae. 3 W.I.
Verinea Merino = Melica L. (Gramin.).
Verinea Pomel = Asphodelus L. (Liliac.).
Verlangia Neck. = Argania Roem. & Schult. (Sapotac.).
Verlotia Fourn. Asclepiadaceae. 5 Brazil.
Vermicularia Moench = Stachytarpheta Vahl (Verbenac.).
Vermifrux J. B. Gillett. Leguminosae. 1 Sudan, Ethiopia, SW. Arabia.
Vermifuga Ruiz & Pav. = Flaveria Juss. (Compos.).
Verminiaria Hort. (sphalm.) = Viminaria Sm. (Legumin.).
Vermoneta Comm. ex Juss. = Homalium Jacq. (Flacourtiac.).
Vermonia Edgew. (sphalm.) = Vernonia Schreb. (Compos.).
Vermontea Steud. = Vermoneta Comm. ex Juss. = Homalium Jacq. (Flacourtiac.).

VERNASOLIS

Vernasolis Rafin. = Coreopsis L. (Compos.).

Vernic[ac]eae Link = Anacardiaceae R.Br. + Burseraceae Kunth + Rutaceae-Toddalioïdeae–Amyridinae Engl.

Vernicia Lour. (~Aleurites J. R. & G. Forst.). Euphorbiaceae. 3 E. As. *V. montana* Lour. and *V. fordii* (Hemsl.) Airy Shaw give tung or China woodoil, a drying oil.

Verniseckia Steud. = Wernisekia Scop. = Houmiri Aubl. (Houmiriac.).

Vernix Adans. = Toxicodendron Mill. = Rhus L. (Anacardiac.).

Vernonella Sond. = Vernonia Schreb. (Compos.).

***Vernonia** Schreb. Compositae. 1000 Am., Afr., As., Austr., very common

Vernoniaceae Bessey = Compositae–Vernonieae Cass.   [in grassy places.

Vernoniopsis Humbert. Compositae. 1 Madag.

Veronica L. Scrophulariaceae. 300 mostly N. temp., many alpine; few S. temp. & trop. mts. The post. sepal of the 5 typical of *Scrophulariac.* is absent, and the two post. petals are united into one large one, so that the P is 4-merous. The 2 sta. and style project horiz. from the rotate C. A small percentage of fls. exhibit a different number of parts (e.g. 5 petals). The fert. of the fl. in *V. chamaedrys* L., a common Eur. sp., is performed chiefly by drone-flies. The style projects over the lower petal, while the two sta. project lat. Honey is secreted at the base of the ovary and concealed by the hairs at the mouth of the short tube. Insects alighting on the lower petal touch the style and grasp the bases of the sta., thus causing the anthers to move inwards and dust them with pollen. The peduncles stand close up against the main stem of the raceme whilst the fls. are in bud, diverge as the fls. open, and again close up as they wither. Caps. with a few flattened seeds suited to wind-distr. In *V. arvensis* L., and some spp. that live in damp places, the capsule merely cracks as it dries and only opens so far as to allow the seeds to escape when thoroughly wetted; the seeds then become slimy (cf. *Linum*).

Veronicaceae Horan. = Scrophulariaceae Juss.

Veronicastrum Moench. Scrophulariaceae. 1 temp. NE. As., 1 temp. NE. N. Am.

Veronicastrum Opiz (non Moench) = Veronica L. (Scrophulariac.).

Veronicella Fabr. (uninom.) = *Veronica* L. sp. (Scrophulariac.).

× **Veronicena** Moldenke. ?Gen. hybr. interfamil. [?Verbena (Verbenac.) × Veronica (Scrophulariac.)]. Requires confirmation.

Verreauxia Benth. Goodeniaceae. 3 W. Austr. Nut.

Verrucaria Medik. (1787; non Scop. 1777, nec Weber 1780, nec Schrad. 1794 —Lichenes) = Tournefortia L. (Boraginac.).

Verrucifera N. E. Brown. Aïzoaceae. 3 S. Afr.

Verrucularia A. Juss. Malpighiaceae. 1 E. Brazil.

Verschaffeltia H. Wendl. Palmae. 1 Seychelles.

Versteegia Valeton. Rubiaceae. 2 New Guinea.

Versteggia Willis (sphalm.) = praec.

Verticillaceae Dulac = Hippuridaceae Link.

Verticillaria Ruiz & Pav. = Rheedia L. (Guttif.).

Verticillatae Engl. An order of Dicots. comprising the fam. *Casuarinaceae*

***Verticordia** DC. Myrtaceae. 40 Austr., esp. W.   [only.

Verulamia DC. ex Poir. = Pavetta L. (Rubiac.).

VICIA

Verutina Cass. = Centaurea L. (Compos.).
Verzinum Rafin. = Thermopsis R.Br. (Legumin.).
Vesalea Mart. & Gal. = Abelia R.Br. (Caprifoliac.).
Veselskya Opiz. Cruciferae. 1 Afghan.
Veseyochloa Phipps. Gramineae. 1 trop. E. Afr.
Vesicarex Steyerm. Cyperaceae. 1 Venez.
Vesicaria Adans. (~ Alyssoïdes Mill.). Cruciferae. 2–3 Alps, Balkans, As. Min.
Vesicarpa Rydb. = Artemisia L. (Compos.).
Vesiculina Rafin. = Utricularia L. (Lentibulariac.).
Veslingia Fabr. (uninom.) = Aïzoön L. sp. (Aïzoac.).
Veslingia Vis. = Guizotia Cass. (Compos.).
Vespuccia Parl. = Hydrocleys Rich. (Alismatac.).
Vesqueatoma Pierre. Sapotaceae. 1 Guiana. Quid?
Vesquella Heim = Stemonoporus Thw. (Dipterocarpac.).
Vesselowskya Pampanini. Cunoniaceae. 1 E. Austr., 1 New Caled.
Vestia Willd. Solanaceae. 1 Chile.
Vetiveria Bory. Gramineae. 10 trop. Afr., As., Austr., incl. *V. zizanioïdes* (L.) Nash (*Andropogon squarrosus* L. f., *A. muricatus* Retz.), the khus-khus, whose roots are woven into fragrant mats, baskets, fans, etc., which give off scent when sprinkled with water.
Vetrix Rafin. = Salix L. (Salicac.).
Vexibia Rafin. = Sophora L. (Legumin.).
Vexillabium Maekawa. Orchidaceae. 2 Korea, Japan.
Vexillaria Hoffmgg. ex Benth. = Centrosema DC. (Legumin.).
Vexillaria Rafin. = Clitoria L. (Legumin.).
Vexillifera Ducke = Dussia Krug & Urb. (Legumin.).
Vialia Vis. = Melhania Forsk. (Sterculiac.).
Vibexia Rafin. = Vexibia Rafin. = Sophora L. (Legumin.)
Vibo Medik. = Emex Neck. ex Campd. (Polygonac.).
Vibones Rafin. = Rumex L. (Polygonac.).
Viborgia Moench = Tubocytisus (DC.) Fourr. = Cytisus L. (Legumin.).
Viborgia Spreng. = Wiborgia Roth = Galinsoga Ruiz & Pav. (Compos.).
Viborquia Ortega = Eysenhardtia Kunth (Legumin.).
Viburnaceae ('-nideae') Dum. = Viburnum L. (Caprifoliac.) + Sambucus L. (Sambucac.).
Viburnum L. Caprifoliaceae. 200 temp. & subtrop., esp. As., N. Am.; 16 Malaysia (exc. New Guinea). Winter buds of some naked, i.e. with no scale-ls. The outer fls. of the cymose corymb are neuter in some, e.g. *V. opulus* L. (guelder-rose), having a large C, but at cost of essential organs. In the cult. guelder-rose all the fls. are neuter. *V. tinus* L. (laurustinus), S. Eur., widely [cult.
Vicarya Stocks (nomen). Malvaceae. 1 NW. India. Quid?
Vicarya Wall. ex Voigt = Myriopteron Griff. (Periplocac.).
Vicatia DC. Umbelliferae. 5 Himalaya, W. China.
Vicentia Allem. = Terminalia L. (Combretac.).
Vicia L. Leguminosae. 150 N. temp., and S. Am. (vetch, tare). Most are climbers with leaf-tendrils. Fl. mech. typical of many L. Pollen early shed by anthers into apex of keel; upon style, below stigma, is a brush of hairs which carries out the pollen when keel is depressed (see fam.). *V. sativa* L.

VICIACEAE

and many other vetches are valuable fodder pls.; *V. faba* L. is the broad bean, with its many vars.

Viciaceae Dostál = Leguminosae–Papilionoïdeae DC.

Vicilla Schur = Vicia L. (Legumin.).

Vicioïdes Moench = Vicia L. (Legumin.).

Vicoa Cass. Compositae. 12 NE. trop. Afr. to C. As. & NW. India.

Vicq-aziria Buc'hoz = Gurania Cogn. (Cucurbitac.).

Victoria Schomb. Euryalaceae. 2–3 trop. S. Am. *V. amazonica* (Poepp.) Sow. is the giant water-lily of the Amazon; it has the habit of *Nymphaea*, but is of enormous size. The floating ls. may be 2 m. across; the edge is turned up to a height of several cm., and on the lower side the ribs project and are armed with spines. Fl. like *Nymphaea* but fully epig. Fr. also similar; the seeds contain both endo- and peri-sperm. They are roasted and eaten in Brazil. The plant is now cult.; it was discovered in 1801, but not brought into general notice till 1837. Pollination is by comparatively large beetles.

Victorinia León (~ Jatropha L.). Euphorbiaceae. 2 Cuba.

Victoriperrea Hombr. & Jacquinot ex Decne = Freycinetia Gaudich. (Pandanac.).

Vidalia F.-Villar = Mesua L. (Guttif.).

Vidoricum Kuntze̅ = Bassia Koen. ex L. = Madhuca Gmel. (Sapotac.).

Vieillardia Brongn. & Gris = Timeroyea Montr. = Calpidia Thou. (Nyctaginac.).

Vieillardia Montr. = Castanospermum A. Cunn. (Legumin.).

Vieillardorchis Kraenzl. = Goodyera R.Br. (Orchidac.).

Viellardia Benth. & Hook. f. (1865) (sphalm.) = Vieillardia Montr. = Castanospermum A. Cunn. (Legumin.).

Viellardia Benth. & Hook. f. (1880) (sphalm.) = Vieillardia Brongn. & Gris = Calpidia Thou. (Nyctaginac.).

Vieraea Sch. Bip. Compositae. 1 Canary Is.

Viereya Steud. (1) = Vireya Blume = Rhododendron L. (Ericac.).

Viereya Steud. (2) = Vireya Rafin. = Alloplectus Mart. (Gesneriac.).

Vierhapperia Hand.-Mazz. = Nannoglottis Maxim. (Compos.).

Vieria Webb & Berth. = Vieraea Sch. Bip. (Compos.).

Vieusseuxia De la Roche = Moraea Mill. ex L. (Iridac.).

Vigethia W. A. Weber. Compositae. 1 Mex.

Vigia Vell. = Fragariopsis A. St-Hil. (Euphorbiac.).

Vigiera Benth. & Hook. f. = seq.

Vigieria Vell. = Escallonia L. (Escalloniac.).

Vigineixia Pomel = Picris L. (Compos.).

Vigna Savi. Leguminosae. 80–100 trop. (esp. Afr. & As.). *V. unguiculata* L. (*V. sinensis* Endl.) is the horse-gram, cherry-bean or cow-pea (trop. As.); pods eaten like French beans, or used as horse- or cattle-fodder. *V. catjang* Endl. (blackeye pea) is also cult. *V. mungo* (L.) Hepper and related spp. (formerly placed in *Phaseolus*), incl. the green and black grams, rice bean, etc., are highly valued as animal and human food in India and E. As.

Vignaldia A. Rich. = Pentas Benth. (Rubiac.).

Vignantha Schur = Carex L. (Cyperac.).

Vignaudia Schweinf. = Vignaldia A. Rich. = Pentas Benth. (Rubiac.).

Vignea Beauv. = Carex L. (Cyperac.).
Vigneopsis De Wild. = Psophocarpus Neck. ex DC. (Legumin.).
Vignidula Börner = Carex L. (Cyperac.).
Vigolina Poir. = Galinsoga Ruiz & Pav. (Compos.).
Viguiera Kunth. Compositae. 150 warm Am., W.I.
Viguierella A. Camus. Gramineae. 1 Madag.
Vilaria Guett. = Berardia Vill. (Compos.).
Vilbouchevitchia A. Chev. = Alafia Thou. (Apocynac.).
Vilfa Adans. = Agrostis L. (Gramin.).
Vilfa Beauv. = Sporobolus R.Br. (Gramin.).
Vilfagrostis A.Br. & Aschers. ex Doell = Eragrostis P. Beauv. (Gramin.).
Villadia Rose. Crassulaceae. 25–30 Mex., Andes.
Villamilla Ruiz & Pav. = Trichostigma A. Rich. (Phytolaccac.).
Villamillia Lopez = praec.
*Villanova Lag. Compositae. 10 Chile to Mex.
Villanova Ortega = Parthenium L. (Compos.). [(Euphorbiac.).
Villanova Pourr. ex Cutanda = Colmeiroa Reut. = Securinega Comm. ex Juss.
Villaresia Ruiz & Pav. (1794). Inc. sed. (? Celastrac.). 1 Peru.
Villaresia Ruiz & Pav. (1802) = Citronella D. Don (Icacinac.).
Villaresiopsis Sleum. Icacinaceae. 1 Peru.
Villarezia Roem. & Schult. = Villaresia Ruiz & Pav. (Icacinac.).
Villaria Bally = Villarsia Vent. (Menyanthac.).
Villaria DC. = Vilaria Guett. = Berardia Vill. (Compos.).
*Villaria Rolfe. Rubiaceae. 5 Philipp. Is.
Villaria Schreb. Quid? (Rutac.?). Hab.?
Villarsia J. F. Gmel. = Nymphoïdes Seguier (Menyanthac.).
Villarsia Neck. = Cabomba Aubl. (Cabombac.).
Villarsia P. & K. = Vilaria Guett. = Berardia Vill. (Compos.).
Villarsia Sm. = Villaria Schreb. (? Rutac.).
*Villarsia Vent. Menyanthaceae. 1 S. Afr., 9 Austr. (The water plant often known under this name is a *Nymphoïdes*.)
Villebrunea Gaudich. Urticaceae. 12 Ceylon to Japan. *V. integrifolia* Gaudich.
Villebrunia Willis = praec. [yields a good fibre.
Villemetia Moq. = Willemetia Maerkl. = Chenolea Thunb. (Chenopodiac.).
Villocuspis (A. DC.) Aubrév. & Pellegr. Sapotaceae. 5 Brazil.
Villosogastris Thou. (uninom.) = Limodorum villosum Thou. = Phaius villosus (Thou.) Bl. (Orchidac.).
Villouratea Van Tiegh. = Ouratea Aubl. (Ochnac.).
Vilmorinia DC. = Poitea Vent. (Legumin.).
Vilobia Strother. Compositae. 1 Bolivia.
Vilshenica Thou. ex Reichb. (nomen). Quid?
Vimen P.Br. = Hyperbaena Miers ex Benth. (Menispermac.).
Vimen Rafin. = Salix L. (Salicac.).
Viminaria Sm. Leguminosae. 1 Austr.
Vinca L. Apocynaceae. 5 Eur., N. Afr., W. As. *V. minor* L. and *V. major* L., the periwinkles, widely naturalised. The anthers stand above the stigmatic disc, but the stigma itself is on the outer surface, so that self-fert. is not caused as the insect's tongue enters the fl.

VINCACEAE

Vinc[ace]ae S. F. Gray = Apocynaceae Juss.

Vincentella Pierre. Sapotaceae. 4 trop. & S. Afr.

Vincentia Boj. = Vinticena Steud. (Tiliac.).

Vincentia Gaudich. (~ Cladium P.Br.). Cyperaceae. 10 Masc. to Polynes.

Vincetoxicopsis Costantin. Asclepiadaceae. 1 Indochina.

Vincetoxicum Walt. = Gonolobus Michx (Asclepiadac.).

Vincetoxicum v. Wolf. Asclepiadaceae. 10–20 temp. Euras.

Vindasia Benoist. Acanthaceae. 1 Madag.

Vindicta Rafin. = Aceranthus Morr. & Decne = Epimedium L. (Berberidac.).

Vinsonia Gaudich. = Pandanus L. f. (Pandanac.).

Vintenatia Cav. = Astroloma R.Br. + Melichrus R.Br. (Epacridac.).

Vintera Humb. & Bonpl. = Wintera Humb. & Bonpl. = Drimys J. R. & G. Forst.
(Winterac.). [Arabia, India.

Vinticena Steud. Tiliaceae. 30–35 trop. & S. Afr., Madag., Masc., Socotra,

Viola L. Violaceae. 500 cosmop., chiefly N. temp., but many Andine. *V. odorata*
L. and *V. canina* L., the sweet and dog violets, *V. tricolor* L., the pansy or
heart's-ease, and others, are well known. Herbs with large stips., on which
glands sometimes occur. Fls. usu. one in each axil; sometimes (e.g. *V. tricolor*)
a veg. shoot arises above the fl. in the same axil. The intr. anthers form a close
ring round the ovary, below the style, which ends in a variously shaped head
on whose ant. surface is the stigma, often a hollow pocket. The lower pet.
forms a landing-place and is often prolonged backwards into a spur, in which
collects honey, secreted by processes projecting into it from the lower sta.
Honey guides show as streaks upon the C leading to nectaries. These fls. are
as a rule incapable of self-fert. In *V. tricolor* the pollen is shed on to the ant.
pet., and the lower edge of the stigma is guarded by a flap which the insect,
when withdrawing, closes; and thus the fl.'s own pollen does not reach the
stigma. The small-flowered species *V. arvensis* Murr. has not this flap and
fertilises itself. In *V. odorata* the stigma is merely the bent-over end of the
style, and is first touched as the insect enters. The size, colour, etc., of the flower
of this sp. and of *V. canina* render them suited to bees.

In many, e.g. *V. canina*, *V. odorata*, *V. sylvestris* Lam., the fls. are rarely
visited, and little seed is set. They usu. flower early in the season; later on
appears a second form of fl. on the same pl. These are the *cleistogamic* fls.,
which never open, but set seed by self-fert. In *V. canina* this fl. looks like a
bud; the seps. remain shut, there are 5 minute pets., 2 ant. sta. with anthers
containing a little pollen (only enough for fert.—there is no waste as in open
fls.) and 3 other abortive sta.; pistil much as usual. The anthers are closely
appressed to the stigma; the pollen-grains germinate within them, and the
tubes burrow through the anther-walls into the stigma. *V. odorata* has similar
fls., but with all 5 sta. fertile. The production of these fls. ensures the setting
of seed.

Fr. a 3-valved capsule; seeds very hard and slippery. One plac. with its
seeds remains attached to each valve; as this dries it bends upwards into a
U-shape, squeezing the seeds against one another and shooting them out (cf.
Claytonia, Buxus).

*Violaceae Batsch. Dicots. 22/900 cosmop. Annual or perennial herbs, or
shrubs. Ls. alt., stip., usu. undivided. Fls. ⚥, usu. zygo., in usu. racemose

infls., bracteolate, 1 or 2 in each axil. K 5, imbr. or open, persistent; C 5, hypog., usu. zygo., the ant. petal often spurred to hold the honey, with descending aestivation; A 5, alt. with petals, hypog., free or conn., often forming a cylinder round the ovary; fil. very short, the 2 anticous often appendaged or spurred, anth. intr., connective usu. with membranous prolongation; G̱ (3), 1-loc., with 1–∞ anatr. ov. on each of the parietal plac. Style simple, often sigmoid, or thickened upwards; stig. very various, usu. ± truncate, sometimes appendic. Fr. a 3-valved loculic. caps., rarely indehisc., baccate or nut-like. Endosp. Chief genera: *Rinorea, Hymenanthera, Paypayrola, Hybanthus, Anchietea, Viola, Leonia.* Closest relationship prob. with *Flacourtiac.*, but some affinity with *Resedac.* also probable.

Violaeoïdes Michx ex DC. = Noisettia Kunth (Violac.).

Violaria P. & K. = Talauma Juss. (Magnoliac.).

Vionaea Neck. = Leucadendron Bergius (Proteac.).

Viorna Reichb. = Clematis L. (Ranunculac.).

Viposia Lundell. Celastraceae. 1 Argent.

Viraea Vahl ex Benth. & Hook. f. = Virea Adans. = Leontodon L. (Compos.).

Viraya Gaudich. = Waitzia Wendl. (Compos.).

Virchowia Schenk = Ilysanthes Rafin. (Scrophulariac.).

Virdika Adans. = Albuca L. (Liliac.).

Virea Adans. = Leontodon L. (Compos.).

Virecta Afzel. ex Sm. Rubiaceae. 3 trop. Afr.

Virecta L. f. = Sipanea Aubl. (Rubiac.).

Virectaria Bremek. Rubiaceae. 7 trop. Afr.

Vireya Blume = Rhododendron L. (Ericac.).

Vireya P. & K. = Viraya Gaudich. = Waitzia Wendl. (Compos.).

Vireya Rafin. = Alloplectus Mart. (Gesneriac.).

Virga Hill (~ Cephalaria Schrad., ~ Dipsacus L.). Dipsacaceae. 2–3 temp. Euras. *V. pilosa* (L.) Hill is the shepherd's rod.

Virgaria Rafin. ex DC. = Aster L. (Compos.).

Virgilia L'Hérit. = Gaillardia Fouger. (Compos.).

*Virgilia Poir. Leguminosae. 1–2 S. Afr. Useful wood.

Virgularia Ruiz & Pav. = Gerardia L. (Scrophulariac.).

Virgulus Rafin. = Aster L. (Compos.).

Viridivia J. H. Hemsl. & Verdc. Passifloraceae. 1 trop. Afr. (N. Rhod.).

Virletia Sch. Bip. ex Benth. & Hook. f. = Bahia Lag. (Compos.).

Virola Aubl. Myristicaceae. 60 C. & trop. S. Am.

Viscaceae Miq. (~ Loranthaceae Juss.). Dicots. 11/450 cosmop., esp. trop. & subtrop. Semiparasitic shrubs on branches of trees; haustorial attachment without runners. Ls. usu. opp. (more rarely alt. or 0), with ± parallel nerves. Fls. minute, in small cymes or rac., monochlam. P (?K) 2–4, valv.; A 2–4, opp. the tepals, free or adnate, anthers 1–∞-loc., opening by pores; G̱1, with short plac. column containing sporogenous cells; ovules 0; embryo-sac single, devel. from one cell of a dyad, confined to plac. column or extending into adjacent ovary tissue. Fr. baccate, the viscous layer within the vasc. bundles; cleavage of zygote usu. transv.; suspensor v. short or 0. Chief genera: *Phoradendron, Dendrophthora, Korthalsella, Arceuthobium, Viscum.* (Cf. Dixit, *Bull. Bot. Surv. India* **4**: 49–55, 1962; Barlow, *Proc. Linn. Soc. N.S.W.* **89**: 268–72, 1965.)

VISCAGO

Viscago Zinn = Cucubalus L. + Silene L. (Caryophyllac.).

Viscainoa Greene. Zygophyllaceae. 1–2 Lower Calif.

Viscaria Commers. apud Danser = Korthalsella Van Tiegh. (Viscac.).

Viscaria Roehl. (~ Lychnis L.). Caryophyllaceae. 5 N. temp.

Viscoïdes Jacq. = Psychotria L. (Rubiac.).

Viscum L. Viscaceae (~ Loranthaceae). 60–70 warm Old World; few temp. Euras., incl. *V. album* L. (mistletoe), a semiparasitic shrubby evergr., growing on apple, hawthorn, oak, etc., and drawing nourishment from its host by suckers. It is repeatedly branched in a dichot. manner, the central stalk usu. ending in an infl. Each branch bears two green leathery ls., and repres. a year's growth. The unisexual dioec. fls. are in groups of three. No calyculus. Sta. completely fused to the P-leaf. Pollen-sacs very numerous. Ovary as usual. The fls. secrete honey and are visited by flies. Pseudo-berry. The layer of viscin prevents the bird that eats the berry from swallowing the seed, which it scrapes off its bill on to a branch, where it adheres and germinates.

Visena Schult. = seq.

Visenia Houtt. = Melochia L. (Sterculiac.).

Visiania DC. = Ligustrum L. (Oleac.).

Visiania Gasp. = Ficus L. (Morac.).

Visinia Turcz. (sphalm.) = seq.   [2 E. trop. Afr.

*****Vismia** Vand. Guttiferae. 30 Mex. to trop. S. Am.; 2–3 W. trop. Afr.,

Vismianthus Mildbr. Connaraceae. 1 trop. E. Afr.

Visnaga Mill. = Ammi L. (Umbellif.).

Visnea L. f. Theaceae. 1 Canaries.

Visnea Steud. ex Endl. = Barbacenia Vand. (Velloziac.).

Vissadali Adans. = Knoxia L. (Rubiac.).

Vistnu Adans. = Evolvulus L. (Convolvulac.).

Vitaceae: see **Vitidaceae**.

Vitaeda Börner = Ampelopsis Michx (Vitidac.).

Vitaliana Sesier (~ Douglasia Lindl.). Primulaceae. 5 mts. S. & E. Spain, Pyrenees, Alps.

Vitellaria Gaertn. f. = Butyrospermum Kotschy (Sapotac.).

Vitellariopsis (Baill.) Dubard. Sapotaceae. 5 E. trop. & S. Afr.

Vitenia Nor. ex Cambess. = Uitenia Nor. = Erioglossum Blume (Sapindac.).

Vitex L. Verbenaceae. 250 trop. & temp.

Viticaceae ('-ices') Juss. = Verbenaceae Juss.

Viticastrum Presl = Sphenodesme Jack (Symphorematac.).

Viticella Mitch. = Galax L. = Nemophila Nutt. (Hydrophyllac.).

Viticella Moench = Clematis L. (Ranunculac.).

Viticena Benth. (sphalm.) = Vinticena Steud. (Tiliac.).

Viticipremna Lam. Verbenaceae. 2 Java, Philipp. Is., New Guinea, Admiralty Is., Bismarck Arch.

*****Vitidaceae** Juss. Dicots. 12/700, mostly trop. & subtrop. Climbing or rarely erect shrubs; stems usu. sympod., bearing tendrils (modif. shoots or infl.) which may end in discoid suckers. Ls. alt., simple or cpd., usu. distich., stip. Infl. cymose, usu. complex, usu. leaf-opposed; axis occasionally flattened and expanded (*Pterisanthes*); bracteoles present. Fl. reg., ♂ or ♀ ♀, dioec. or polyg.-monoec. K (4–5), small and cup-like, very slightly lobed; C 4–5, valvate, often

united at the tips and falling off as a hood upon the opening of the bud; A 4–5, opp. to the petals, at the base of a hypog. disk, with intr. anthers; G usu. (2), rarely 3–6, multiloc., with usu. 2 collat. anatr. ov., erect, with ventral raphe; style long or short, stig. inconspic., rarely 4-lobed (*Tetrastigma*). Berry. Endosp.; embryo straight. *Vitis* is economically important. Chief genera: *Vitis, Cissus, Ampelocissus, Tetrastigma*. Related to *Leeac.* and *Rhamnac.*
Vitiphoenix Becc. (~Veitchia H. Wendl.). Palmae. 9 Fiji.
Vitis L. Vitidaceae. 60–70 N. hemisph. The vines are climbing pls., with tendrils which repres. modified infls.; the stem is usu. regarded as a sympodium, each axis in turn ending in a tendril, but there has been much argument upon the subject. The tendril may attach itself by the ordinary coiling method, or may be negatively phototropic and thus force its way into the crevices of the support; in these crevices the tips of the tendrils form large balls of tissue, the outer parts of which become mucilaginous and cement the tendril to its support. *V. vinifera* L. (Orient, NW. India) is the grape-vine, cult. in most warm countries. Over 25 million metric tons of wine are made every year. When dried the fruits form raisins; the sultana raisin is a seedless var. The currants of commerce are the fruit of the Corinthian variety ('currant' is a corruption of Corinth). *V. aestivalis* Michx (summer-grape) and *V. labrusca* L. (fox-grape) are N. Am. spp. which have been largely introduced into Eur., as they resist the attacks of the dreaded insect, *Phylloxera*, better than the Eur. spp. (For 'Virginia creepers' cf. *Parthenocissus*.)
Vitis-Idaea Seguier = Vaccinium L. (Ericac.).
Vitmania Turra ex Cav. = Mirabilis L. (Nyctaginac.).
Vitmannia Vahl = Quassia L. (Simaroubac.).
Vitmannia Wight & Arn. = Noltea Reichb. (Rhamnac.).
Vittadenia Steud. = seq.
Vittadinia A. Rich. Compositae. 8 New Guinea, Austr., New Caled., N.Z., S. Am. (Australian daisy).
Vittaria Sm. Vittariaceae. 50 trop. & subtrop.
Vittariaceae Presl. Pteridales. Small epiphytes; scales clathrate; ls. simple; sori elongate along veins, exindusiate but often in a groove, protected when young by paraphyses; spores trilete. Chief genera: *Vittaria, Antrophyum.*
Vittmannia Endl. = Vitmania Turra ex Cav. = Mirabilis L. (Nyctaginac.).
Viviana Merr. = Viviania Rafin. = Guettarda L. (Rubiac.).
Viviania Cav. Vivianiaceae. 30 S. Brazil, Chile.
Viviania Colla = Billiottia DC. (Rubiac.).
Viviania Rafin. = Guettarda L. (Rubiac.).
Viviania Willd. ex Less. = Liabum Adans. (Compos.).
Vivianiaceae Klotzsch. Dicots. 2/30 S. Am. Branched woody herbs or shrublets, with opp. or vertic. simple ent. or crenate exstip. ls., often white-toment. below. Fls. reg., ☿, loosely or closely cymose, often long-pedic. K (5–4) or 5–4, parallel-veined, valv. or imbr.; C 5–4, contorted, or o; disk-glands 5, alternipet., ent. or 2-lobed; A 10–8, eq. or uneq.; G (3–2), with 3–2 conn. or almost free stigs., and 2 superposed axile ov. (1 ascend., 1 pend.). Fr. a loculic. 3–2-valved caps., seeds 1 or 2 per loc., with fleshy endosp.; embryo much curved. Genera: *Viviania, Rhynchotheca*. Habit sometimes of *Helianthemum (Cistac.)*. *Rhynchotheca* Ruiz & Pav. should probably be added to this

VLADIMIRIA

family. The family belongs in the *Centrospermae*, near *Caryophyllaceae* and *Amaranthaceae*. (Bortenschlager, *Grana Palyn.* 7: 420–1, 1967).

Vladimiria Iljin. Compositae. 12 SW. China.

Vlamingia De Vriese = Ionidium Vent. (Violac.).

Vlechia Rafin. = seq.

Vleckia Rafin. = Agastache Clayt. in Gronov. (Labiat.).

Voacanga Thou. Apocynaceae. 25 trop. Afr., Madag., Malaysia.

Voandzeia Thou. Leguminosae. 1 trop. Afr., Madag., *V. subterranea* Thou.; it buries its young fr. like *Arachis*. The seed is ed. and the pl. is largely cult. (Bambarra groundnut).

Vochisia Juss. = seq.

Vochy Aubl. = seq.

Vochya Vell. ex Vand. = seq.

*Vochysia Aubl. mut. Poir. Vochysiaceae. 105 C. & trop. S. Am.

*Vochysiaceae A. St-Hil. Dicots. 6/200 trop. Am., W. Afr. Trees or shrubs, rarely herbs, with opp. or vertic. rarely alt. ent. simple stip. or exstip. ls. Fls. ♀, obliquely zygo., usu. arranged in a cpd. rac. (pan.) of cincinni. K 5, connate at base, lobes imbr., postic. lobe often spurred; C usu. 3–1, unequal, rarely 5, subequal, contorted, perig. or epig.; A 1 fertile, intr., + 2–4 stds.; G̲ (3), with 1–∞ epitr. axile ov. per loc., or G̅ 1, with 2 lateral ov.; style simple, stig. small. Fr. a loculic. caps., or samaroid and indehisc.; seeds often winged, sometimes pilose; endosp. o. Genera: *Vochysia, Callisthene, Qualea, Salvertia, Erisma, Erismadelphus*. Rather closely related to *Trigoniac.* and *Polygalac.*

Voelckeria Klotzsch & Karst. = Ternstroemia L. (Theac.).

Vogelia J. F. Gmel. = Burmannia L. (Burmanniac.).

Vogelia Lam. = Dyerophytum Kuntze (Plumbaginac.).

Vogelia Medik. = Neslia Desv. (Crucif.).

Vogelocassia Britton = Cassia L. (Legumin.).

Voglera Gaertn., Mey. & Scherb. = Genista L. (Legumin.).

Voharanga Costantin & Bois. Asclepiadaceae. 1 Madag.

Vohemaria Buchenau. Asclepiadaceae. 2 Madag.

Vohiria Juss. = Voyria Aubl. (Gentianac.).

Voigtia Klotzsch = Bathysa Presl (Rubiac.).

Voigtia Roth = Andryala L. (Compos.).

Voigtia Spreng. = Barnadesia Mutis (Compos.).

Voladeria Benoist. Juncaceae (?). 1 Ecuador.

Volataceae Dulac = Aceraceae Juss.

Volckameria Fabr. (uninom.) = *Cedronella* Moench sp. (Labiat.).

Volkamera P. & K. = Volkameria Burm. f. = Capparis L. (Capparidac.).

Volkameria P.Br. = Clethra L. (Clethrac.).

Volkameria Burm. f. = Capparis L. (Capparidac.).

Volkameria L. = Clerodendrum L. (Verbenac.).

Volkensia O. Hoffm. Compositae. 8 trop. E. Afr.

Volkensiella H. Wolff. Umbelliferae. 1 trop. E. Afr.

Volkensinia Schinz. Amaranthaceae. 2 trop. E. Afr.

Volkensiophyton Lindau = Lepidagathis Lindau (Acanthac.).

Volkensteinia Van Tiegh. = Wolkensteinia Regel = Ouratea Aubl. (Ochnac.).

Volkiella Merxm. & Czech. Cyperaceae. 1 SW. Afr.

Volkmannia Jacq. = Clerodendrum L. (Verbenac.).
Volubilis Catesby = Vanilla L. (Orchidac.).
Volucrepis Thou. (uninom.) = Epidendrum volucre Thou. = Oeonia volucris (Thou.) Spreng. (Orchidac.).
Volutarella Cass. = seq.
Volutaria Cass. = Amberboi Adans. (Compos.).
Volutella Forsk. = Cassytha L. (Laurac.).
Volvulopsis Roberty = Evolvulus L. (Convolvulac.).
Volvulus Medik. = Calystegia R.Br. (Convolvulac.).
Vonitra Becc. Palmae. 5 Madag.
Vonroemeria J. J. Smith = Octarrhena Thw. (Orchidac.).
Vormia Adans. = Selago L. (Scrophulariac.).
Vorstia Adans. = Thryallis L (Malpighiac.).
Vosacan Adans. = Helianthus L. (Compos.).
Vossia Adans. = Glottiphyllum Haw. (Aïzoac.).
*Vossia Wall. & Griff. Gramineae. 1 trop. & SW. Afr.; Bengal, Assam, Burma.
A swimming grass, which with Saccharum spontaneum L. makes the great grass bars of the Nile.
Vossianthus Kuntze = Sparmannia L. (Tiliac.).
Votomita Aubl. Memecylaceae. 5 trop. S. Am.
Vouacapoua Aubl. = Andira Lam. (Legumin.).
Vouapa Aubl. = Macrolobium Schreb. (Legumin.).
Vouarana Aubl. Sapindaceae. 1 Guiana, N. Brazil.
Vouay Aubl. = Geonoma Willd. (Palm.).
Voucapoua Steud. = Vouacapoua Aubl. = Andira Lam. (Legumin.).
Voyara Aubl. = Capparis L. (Capparidac.).
Voyria Aubl. Gentianaceae. 8 Panamá, N. trop. S. Am., trop. W. Afr.
Voyriella Miq. Gentianaceae. 2 Guiana, N. Brazil.
Vrena Nor. = Urena L. (Malvac.).
Vriesea Hassk. = Lindernia All. (Scrophulariac.).
*Vriesea Lindl. mut. Beer. Bromeliaceae. 190 trop. Am.
Vrieseida Rojas. Bromeliaceae. 1 Argent.
Vriesia Lindl. = praec.
Vroedea Bub. = Glaux L. (Primulac.).
Vrolickia Steud. = seq.
Vrolikia Spreng. = Heteranthia Nees (Scrophulariac.).
Vrydagzenia Benth. & Hook. f. = seq.
Vrydagzynea Blume. Orchidaceae. 40 NE. India to Formosa & Malaysia, Polynesia.
Vuacapua Kuntze = Vouacapoua Aubl. = Andira Lam. (Legumin.).
Vuapa Kuntze = Vouapa Aubl. = Macrolobium Schreb. (Legumin.).
Vulneraria Mill. = Anthyllis L. (Legumin.).
Vulpia C. C. Gmel. (~ Festuca L.). Gramineae. 25–30 temp., esp. Medit. & Pacif. N. & S. Am.
Vulpiella (Battand. & Trab.) Andreánszky. Gramineae. 1 Medit.
Vulvaria Bub. = Chenopodium L. (Chenopodiac.).
× Vuylstekeara hort. Orchidaceae. Gen. hybr. (iii) (Cochlioda × Miltonia × Odontoglossum).

VVEDENSKYA

Vvedenskya Korovin. Umbelliferae. 1 C. As.
Vvedenskyella Botschantsev. Cruciferae. 2 C. As., Kashmir, Kashgaria.
Vyenomus Presl = Euonymus L. (Celastrac.).

W

Wacchendorfia Burm. f. (sphalm.) = seq.
Wachendorfia Burm. Haemodoraceae. 25 Afr. Transv. zygomorphism in fl., but not obvious on account of twisting of stalk.
Wachendorfia Loefl. = Callisia Loefl. (Commelinac.).
Wadapus Rafin. = Gomphrena L. (Amaranthac.).
Waddingtonia Phil. = Petunia Juss. (Solanac.).
Wadea Rafin. = Cestrum L. (Solanac.).
Wagatea Dalz. Leguminosae. 1 SW. India.
Wageneria Klotzsch = Begonia L. (Begoniac.).
Wagnera P. & K. = Vagnera Adans. = Smilacina Desf. (Liliac.).
Wagneria Klotzsch = Wageneria Klotzsch = Begonia L. (Begoniac.).
Wagneria Lem. = Diervilla L. (Caprifoliac.).
Wahabia Fenzl = Barleria L. (Acanthac.).
Wahlbergella Fries = Melandrium Roehl. (Caryophyllac.).
Wahlbomia Thunb. = Tetracera L. (Dilleniac.).
Wahlenbergia Blume = Tarenna Gaertn. (Rubiac.).
Wahlenbergia R.Br. ex Wall. = Dichapetalum Thou. (Dichapetalac.).
*****Wahlenbergia** Schrad. ex Roth. Campanulaceae. 150+ chiefly S. temp. Fl. like *Campanula*. Capsule loculic. (the chief difference between these two gen.).
Wahlenbergia Schum. = Enydra Lour. (Compos.).
Wailesia Lindl. = Dipodium R.Br. (Orchidac.).
Waitzia Reichb. = Tritonia Ker-Gawl. (Iridac.).
Waitzia Wendl. Compositae. 6 temp. W. & S. Austr.
Wakilia Gilli. Cruciferae. 1 Afghanistan.
Walafrida E. Mey. Scrophulariaceae. 30–40 trop. & S. Afr., Madag.
Walcottia F. Muell. = Lachnostachys Hook. (Dicrastylidac.).
Walcuffa J. F. Gmel. = Dombeya Cav. (Sterculiac.).
Waldeckia Klotzsch ex Rich. Schomb. = Hirtella L. (Chrysobalanac.).
Waldemaria Klotzsch = Rhododendron L. (Ericac.).
Waldensia Lavy. Fam.? 1 NW. Italy. Quid?
Waldheimia Kar. & Kir. Compositae. 8 C. As., Himal.
Waldschmidia Weber = Nymphoïdes Hill (Menyanthac.).
Waldschmidtia Bluff & Fingerh. = praec.
Waldschmidtia Scop. = Crudia Schreb. (Legumin.).
Waldsteinia Willd. Rosaceae. 6 N. temp.
Walidda (A. DC.) Pichon. Apocynaceae. 1 Ceylon.
Walkera Schreb. = Ouratea Aubl. (Ochnac.).
Walkeria A. Chev. = Nogo Baehni (Sapotac.).
Walkeria Mill. ex Ehret = Nolana L. ex L. f. (Nolanac.).
Walkuffa Bruce ex Steud. = Walcuffa J. F. Gmel. = Dombeya Cav. (Sterculiac.).

Wallacea Spruce. Ochnaceae. 3 Amaz. Brazil.
Wallaceaceae Van Tiegh. = Ochnaceae–? Luxembergiëae Planch.
Wallaceodendron Koorders. Leguminosae. 1 Celebes.
*****Wallenia** Sw. Myrsinaceae. 25 W.I.
Walleniella P. Wils. Myrsinaceae. 1 Cuba.
Walleria J. Kirk. Liliaceae. 1–5 trop. & S. Afr., Madag.
Wallia Alef. = Juglans L. (Juglandac.).
Wallichia DC. = Eriolaena DC. (Sterculiac.).
Wallichia Reinw. ex Blume = Urophyllum Wall. (Rubiac.).
Wallichia Roxb. Palmae. 6 E. Himal. to S. China.
Wallinia Moq. = Lophiocarpus Turcz. (? Phytolaccac.).
Wallisia (Regel 1869) E. Morr. (1870) = Tillandsia L. (Bromeliac.).
Wallisia Regel (1875) = Lisianthus Aubl. (Gentianac.).
Wallrothia Roth = Vitex L. (Verbenac.).
Wallrothia Spreng. = Seseli L. (Umbellif.).
Walnewa Hort. (sphalm.) = Waluewa Regel = Leochilus Knowles & Westc. (Orchidac.).
*****Walpersia** Harv. & Sond. = Phyllota DC. (Legumin.).
Walpersia Meissn. ex Krauss = Rhynchosia Lour. (Legumin.).
Walpersia Reissek = Phylica L. (Rhamnac.).
Walsura Roxb. Meliaceae. 30–40 Himal. & Ceylon to S. China, Andamans, Indoch., W. Malaysia, Celebes.
Walteria Scop. = Waltheria L. (Sterculiac.).
Walteria A. St-Hil. = Vateria L. (Dipterocarpac.).
Walteriana Fraser ex Endl. = Cliftonia Banks (Cyrillac.).
Waltheria L. Sterculiaceae. 50 trop. Am., W.I.; 1 S. Rhodesia, 1 Madag., 1 Malay Penins., 1 Formosa.
Waluewa Regel = Leochilus Knowles & Westc. (Orchidac.).
Wangenheimia F. G. Dietr. = Gilibertia Ruiz & Pav. (Araliac.).
Wangenheimia Moench. Gramineae. 1 Spain, N. Afr.
Wangerinia Franz. Portulacaceae. 1 Chile.
*****Warburgia** Engl. Canellaceae. 3 trop. E. Afr.
Warburgina Eig. Rubiaceae. 1 Syria, Palestine.
Warburtonia F. Muell. = Hibbertia Andr. (Dilleniac.).
Warczewitzia Skinner = Catasetum Rich. (Orchidac.).
Wardaster Small. Compositae. 1 W. China.
Wardenia King. Araliaceae. 1 Malay Penins.
Warea C. B. Clarke = Biswarea Cogn. (Cucurbitac.).
Warea Nutt. Cruciferae. 4 SE. U.S. Near *Stanleya* Nutt. (*q.v.*), and approaching *Cleomaceae.*
Waria Aubl. = Uvaria L. (Annonac.).
Warionia Benth. & Coss. Compositae. 1 NW. Sahara.
Warmingia Engl. (1874, a) = Ticorea Aubl. (Rutac.).
Warmingia Engl. (1874, b) = Spondias L. (Anacardiac.).
*****Warmingia** Reichb. f. Orchidaceae. 2 Brazil.
× **Warneara** hort. Orchidaceae. Gen. hybr. (iii) (Comparettia × Oncidium × Rodriguezia).
Warneckea Gilg = Memecylon L. (Memecylac.).

WARNERA

Warnera Mill. = Warneria Mill. = Hydrastis L. (Hydrastidac.).

Warneria Ellis = Varneria L. = Gardenia Ellis (Rubiac.).

Warneria Mill. = Hydrastis L. (Hydrastidac.).

Warneria Mill. ex L. = Watsonia Mill. (Iridac.).

Warpuria Stapf. Acanthaceae. 2 Madag.

Warrea Lindl. Orchidaceae. 3 trop. S. Am.

Warreëlla Schlechter. Orchidaceae. 1 Colombia.

Warscewiczella Reichb. f. = Cochleanthes Rafin. (Orchidac.).

Warscewiczia P. & K. = Warczewitzia Skinner = Catasetum Rich. (Orchidac.).

Warscewiczia auctt. = Warszewiczia Klotzsch (Rubiac.).

Warszewiczella Benth. & Hook. f. = Warscewiczella Reichb. f. = Cochleanthes Rafin. (Orchidac.).

Warszewiczia Klotzsch. Rubiaceae. 4 trop. Am., W.I.

Warthemia Boiss. = Varthemia DC. = Iphiona Cass. (Compos.).

Wartmannia Muell. Arg. = Homalanthus A. Juss. (Euphorbiac.).

Wasabia Matsumura = Eutrema R.Br. (Crucif.).

Wasatchia M. E. Jones = Hesperochloa Rydb. (Gramin.).

Washingtonia Rafin. = Osmorhiza Rafin. (Umbellif.).

*****Washingtonia** H. Wendl. (~ Pritchardia Seem. & H. Wendl.). Palmae. 2 S. Calif., Arizona, & N. Lower Calif.

Washingtonia Winsl. = Sequoiadendron Buchholz (Taxodiac.).

Watsonamra Kuntze = Pentagonia Benth. (Rubiac.).

Watsonia Boehm. = Byttneria Loefl. (Byttneriac.).

*****Watsonia** Mill. Iridaceae. 60–70 S. Afr., Madag.

Wattakaka Hassk. = Dregea E. Mey. (Asclepiadac.).

Weatherbya Copel. = Lemmaphyllum Presl (Polypodiac.). (Donk, *Reinwardtia*, 2: 409, 1954.)

Webbia DC. = Vernonia Schreb. (Compos.).

Webbia Ruiz & Pav. ex Engl. = Dictyoloma Juss. (Rutac.).

Webbia Sch. Bip. = Conyza L. (Compos.).

Webbia Spach = Huebneria Reichb. = Hypericum L. (Guttif.).

Webera Cramer = Plectronia L. (Rubiac.).

Webera J. F. Gmel. = Bellucia Neck. ex Naud. (Melastomatac.).

Webera Schreb. = Tarenna Gaertn. (Rubiac.).

Weberbauera Gilg & Muschler. Cruciferae. 2 Andes.

Weberbauerella Ulbrich. Leguminosae. 2 Peru.

Weberbaueriella Ferreyra = Chucoa Cabr. (Compos.).

Weberbauerocereus Backeb. Cactaceae. 5 S. Peru, N. Chile(?).

Weberiopuntia Frič = Opuntia Mill. (Cactac.).

Weberocereus Britton & Rose. Cactaceae. 3 C. Am.

Websteria S. H. Wright. Cyperaceae. 1 trop. (rare).

Weddellina Tul. Podostemaceae. 1 N. trop. S. Am., *W. squamulosa* Tul. Roots ± flattened, with haptera; shoots borne at their edges of two kinds: veg. to 75 cm. long and much branched, and short unbranched flowering ones. Between the branches of the long shoots are branches of limited growth, as in *Tristicha.*

Wedela Steud. = Vedela Adans. = Ardisia Sw. (Myrsinac.).

*****Wedelia** Jacq. Compositae. 70 trop. & warm temp.

Wedelia Loefl. = Allionia Loefl. (Nyctaginac.).

Wedelia P. & K. = Vedela Adans. = Ardisia Sw. (Myrsinac.).

Wedeliella Cockerell = Allionia Loefl. (Nyctaginac.).

Wedeliopsis Planch. ex Benth. = Dissotis Benth. (Melastomatac.).

Wehlia F. Muell. Myrtaceae. 5 W. Austr.

Weigela Thunb. (~ Diervilla Adans.). Caprifoliaceae. 12 E. As. Fls. suited to bees; change colour after fert.

Weigelastrum (Nakai) Nakai = praec.

Weigelia Pers. = Weigela Thunb. (Caprifoliac.).

Weigeltia A. DC. Myrsinaceae. 35 W.I., Panamá, trop. S. Am.

Weigeltia Reichb. (nomen). Leguminosae (aff. Copaïfera L.). Quid?

Weihea Eckl. = Geissorhiza Ker-Gawl. (Iridac.).

Weihea Reichb. = Burtonia R.Br. (Legumin.).

*****Weihea** Spreng. = Cassipourea Aubl. (Rhizophorac.).

Weihea Spreng. ex Eichl. = Phthirusa Mart. (Loranthac.).

Weilbachia Klotzsch & Oerst. = Begonia L. (Begoniac.).

Weingaertneria Bernh. = Corynephorus Beauv. (Gramin.).

Weingartia Werderm. = Gymnocalycium Pfeiff. (Cactac.).

Weingartneria Benth. (sphalm.) = Weingaertneria Bernh. = Corynephorus Beauv. (Gramin.).

*****Weinmannia** L. Cunoniaceae. 170 Madag., Masc., Malaysia, Pacif., N.Z., Andes (Mex. to Chile).

Weinreichia Reichb. = Pterocarpus L. (Legumin.).

Weitenwebera Opiz = Campanula L. (Campanulac.).

Weldena Pohl ex K. Schum. = Abutilon L. (Malvac.).

Weldenia Reichb. = ? Hibbertia Andr. (Dilleniac.).

Weldenia Schult. Commelinaceae. 1 Mex.

Welezia Neck. = Velezia L. (Caryophyllac.).

Welfia H. Wendl. Palmae. 2 C. Am.

Wellingtonia Lindl. = Sequoiadendron Buchholz (Taxodiac.).

Wellingtonia Meissn. = Meliosma Blume (Meliosmac.).

Wellingtoniaceae Meissn. = Millingtoniaceae Wight & Arn. = Meliosmaceae Endl.

Wellstedia Balf. f. Wellstediaceae. 2 SW. Afr., Somalia, Socotra.

Wellstediaceae (Pilger) Novák. Dicots. 1/2 SW. & NE. Afr., Socotra. Low woody herbs or shrublets, with alt. simple ent. densely adpr. grey-strigose exstip. ls. Fls. reg., ☿, small, sol., axill., on the lateral branches often forming dense subscorpioid cymes. K (4), open or subimbr.; C (4), ± imbr., brick-red; A 4, alternipet. and epipet., fil. short, anth. intr.; disk 0; G̲ (2), compressed, 2-loc., with 1 pend. anatr. ov. per loc., and a simple shortly bifid style and small stigs. Fr. a compressed broadly obcordate 1–2-seeded loculic. caps.; seeds comose, without endosp. Only genus: *Wellstedia*. Related to *Boraginac.*, differing in the tetram. fls. and loculic. caps.

*****Welwitschia** Hook. f. (*Tumboa* Welw.). Welwitschiaceae. 1, *W. mirabilis* Hook. f. (*Tumboa bainesii* Hook. f., *nom. provis.*; *T. strobilifera* Welw. ex Hook. f., pro synon.; *W. bainesii* (Hook. f.) Carr.), a remarkable plant discovered by Baines in Damaraland in SW. trop. Afr., and shortly afterwards by Welwitsch in Mossâmedes, and described by Hooker in *Trans. Linn. Soc.* **24**: 1–48, 1863

WELWITSCHIA

(*q.v.*). The plant grows for at least a century, and probably much longer. Its native climate is a markedly desert one, with a mere trifle of rainfall, the bulk of the moisture being derived from sea fogs, which cause a heavy deposit of dew. Seeds are produced in large quantities, and being enclosed in the winged P are blown about, and germinate in the occasional wet years.

Welwitschia P. & K. = Velvitsia Hiern (Scrophulariac.).

Welwitschia Reichb. = Gilia Ruiz & Pav. (Polemoniac.).

*****Welwitschiaceae** (Engl.) Markgr. Gymnosp. (Gnetales). 1/1 SW. Afr. Long-lived woody perennials, with v. short stout truncated stem, two-lobed above, almost circular in section, and narrowing downwards into a stout tap-root. At the edges of the two lobes are two grooves, from each of which springs a broad-based oblong l. These ls. are the first pair after the cots. and are the only ls. the plant ever has; they go on growing at the base throughout its life, wearing away at the tips and often becoming torn down to the base into long ribbons. The stem continues to grow in thickness, and exhibits concentric grooves upon the top surface. In the outer (younger) of these grooves the fls. appear, in cpd. dichasia of small (♂) or larger (♀) spikes; they are covered by bracts which become bright red after fert. The fls. are dioec., and are produced annually. Pollination by insects (*Heteroptera*). In the ♂, there is a P of 2 + 2 ls., the outer whorl transv. to the bract; sta. 6, united below, with 3-loc. anthers; gynoecium rudimentary, but with the integument of the ovule looking like a style and stigma. In the ♀, the perianth-ls. are fused into a tube, and are equivalent to the two outer ls. of the ♂; there is no trace of sta. Ovule 1, erect, with the integument drawn out beyond it. Seed with endosp. and perisperm, enclosed in the P which becomes winged. (See *Gymnospermae*, and Pearson in *Phil. Trans. R. Soc. B*, **198**: 291, 1906.) Only genus: *Welwitschia*.

Welwitschiella Engl. = Welwitschiina Engl. = Triclisia Benth. (Menispermac.).

Welwitschiella O. Hoffm. Compositae. 1 Angola.

Welwitschiina Engl. = Triclisia Benth. (Menispermac.).

Wenchengia C. Y. Wu & S. Chow. Labiatae. 1 S. China (Hainan).

Wendelboa van Soest. Compositae. 1 W. Pakistan.

Wenderothia Schlechtd. = Canavalia DC. (Legumin.).

Wendia Hoffm. = Heracleum L. (Umbellif.).

*****Wendlandia** Bartl. ex DC. Rubiaceae. 70 India, SE. As., Formosa, Malaysia, Queensl. Ls. sometimes whorled.

Wendlandia Willd. = Cocculus DC. (Menispermac.).

Wendlandiella Dammer. Palmae. 3 Peru, Brazil.

Wendtia Ledeb. = Wendia Hoffm. = Heracleum L. (Umbellif.).

*****Wendtia** Meyen. Ledocarpaceae. 3 Chile, Argent.

Wensea Wendl. = Pogostemon Desf. (Labiat.).

Wenzelia Merr. Rutaceae. 9 Philipp. Is., New Guinea, Solomon Is., Hawaii.

Wepferia Fabr. (uninom.) = *Aethusa cynapium* L. (Umbellif.).

Wercklea Pittier & Standley. Malvaceae. 2 Costa Rica.

Werckleocereus Britton & Rose = Weberocereus Britt. & Rose (Cactac.).

Werdermannia O. E. Schulz. Cruciferae. 3–4 N. Chile.

Wernera Kuntze = seq.

Werneria Kunth. Compositae. 40 Andes.

Wernhamia S. Moore. Rubiaceae. 1 Bolivia.

Wernisekia Scop. = Houmiri Aubl. (Houmiriac.).
Werrinuwa Heyne = Guizotia Cass. (Compos.).
Westia Vahl = Berlinia Soland. ex Hook. f. + Afzelia Sm. (Legumin.).
Westonia Spreng. = Rothia Pers. (Legumin.).
Westringia Sm. Labiatae. 22 Austr.
Wetria Baill. Euphorbiaceae. 1 Lower Burma, Penins. Siam, W. Malaysia, New Guinea.
Wetriaria (Muell. Arg.) Kuntze = Argomuellera Pax (Euphorbiac.).
Wettinella O. F. Cook & Doyle = seq.
Wettinia Poepp. ex Endl. Palmae. 5 W. trop. S. Am.
Wettiniicarpus Burret. Palmae. 2 Colombia.
Wettsteinia Petrak (1910; non Schiffner 1898—Hepaticae) = Olgaea Iljin (Compos.).
Wettsteiniola Suesseng. Podostemaceae. 2 Brazil.
Wheelera Schreb. = ? Geissospermum Allem. or Forsteronia G. F. W. Mey. (Apocynac.).
Wheelerella G. B. Grant = Greeneocharis Gürke & Harms (Boraginac.).
Whipplea Torr. Philadelphaceae. 1 Pacif. U.S.
Whitefieldia Nees (sphalm.) = Whitfieldia Hook. (Acanthac.).
Whiteheadia Harv. Liliaceae. 1 S. Afr.
Whiteochloa C. E. Hubbard. Gramineae. 1 Queensland.
Whiteodendron van Steenis. Myrtaceae. 1 Borneo.
White-Sloanea Chiov. Asclepiadaceae. 1 Somaliland.
Whitfieldia Hook. Acanthaceae. 15 trop. Afr.
Whitfordia Elmer (1910; non Murrill 1908—Fungi) = seq.
Whitfordiodendron Elmer. Leguminosae. 9 Formosa, China, W. Malaysia.
Whitia Blume = Cyrtandra J. R. & G. Forst. (Gesneriac.).
Whitlavia Harv. = Phacelia Juss. (Hydrophyllac.).
Whitleya Sweet = Scopolia Jacq. (Solanac.).
Whitmorea Sleumer. Icacinaceae. 1 Solomon Is.
Whitneya A. Gray. Compositae. 1 California.
Whittonia Sandwith. Flacourtiaceae (~ Peridiscac.). 1 Guiana.
Whytockia W. W. Smith. Gesneriaceae. 1 W. China, 1 Formosa.
Wiasemskya Klotzsch (nomen). Rubiaceae (aff. Rondeletia L. & Sommera Schlechtd.). 1 Colombia. Quid?
Wibelia Bernh. = Davallia Sm. (Davalliac.).
Wibelia Fée = Tapeinidium Presl (Lindsaeac.).
Wibelia Gaertn., Mey. & Scherb. = Crepis L. (Compos.).
Wibelia Pers. = Paypayrola Aubl. (Violac.).
Wibelia Roehl. = Chondrilla L. (Compos.).
Wiborgia Kuntze = Viborquia Ort. = Eysenhardtia Kunth (Legumin.).
Wiborgia P. & K. = Viborgia Moench = Cytisus L. (Legumin.).
Wiborgia Roth = Galinsoga Ruiz & Pav. (Compos.).
***Wiborgia** Thunb. Leguminosae. 7 S. Afr.
Wichuraea Nees = Cryptandra Sm. (Rhamnac.).
Wichuraea M. Roem. = Bomarea Mirb. (Alstroemeriac.).
Wichurea Benth. & Hook. f. = Wichuraea Nees = Cryptandra Sm. (Rhamnac.).

Wickstroemia Reichb. = Wikstroemia Schrad. = Laplacea Kunth (Theac.).
Widdringtonia Endl. Cupressaceae. 5 trop. & S. Afr.
Widgrenia Malme. Asclepiadaceae. 1 Brazil.
Wiedemannia Fisch. & Mey. Labiatae. 3 As. Min., Cauc.
Wiegmannia Hochst. & Steud. ex Steud. = Maerua Forsk. (Capparidac.).
Wiegmannia Meyen = Kadua Cham. & Schlechtd. (Rubiac.).
Wielandia Baill. Euphorbiaceae. 1 Seychelles.
Wierzbickia Reichb. = Minuartia L. (Caryophyllac.).
Wiesbauria Gandog. = Viola L. (Violac.).
Wiesneria M. Micheli. Alismataceae. 4 trop. Afr., Madag., India.
Wiestia Boiss. = Boissiera Hochst. (Gramin.).
Wiestia Sch. Bip. = Lactuca L. (Compos.).
Wiganda St-Lag. = seq.
***Wigandia** Kunth. Hydrophyllaceae. 6 Mex. to Peru, W.I.
Wigandia Neck. = Disparago Gaertn. (Compos.).
Wiggersia Gaertn., Mey. & Scherb. = Vicia L. (Legumin.).
Wigginsia D. M. Porter (~Notocactus [K. Schum.] Backeb. & F. M. Knuth).
Cactaceae. 13 trop. S. Am. Stem not jointed; fls. apical in mass of hairs.
Wightia Spreng. ex DC. = Centratherum Cass. (Compos.).
Wightia Wall. Scrophulariaceae. 2–3 E. Himal. to SE. As., W. Malaysia (exc.
Philipp.), Sumbawa, Flores. Epiph. shrubs, later becoming independent trees.
Wigmannia Walp. (sphalm.) = Wiegmannia Meyen = Kadua Cham. &
Schlechtd. (Rubiac.).
***Wikstroemia** Endl. Thymelaeaceae. 70 S. China, Indoch., Austr., Pacif.
Some are parthenogenetic.
Wikstroemia Schrad. = Laplacea Kunth (Theac.).
Wikstroemia Spreng. = Eupatorium L. (Compos.).
Wilberforcia Hook. f. ex Planch. = Bonamia Thou. (Convolvulac.).
Wilbrandia Presl = ?Cordia L. (Ehretiac.).
Wilbrandia S. Manso. Cucurbitaceae. 2 trop. S. Am.
Wilckea Scop. = Vitex L. (Verbenac.).
Wilckia Scop. = Malcolmia R.Br. corr. Spreng. (Crucif.).
Wilcoxia Britton & Rose. Cactaceae. 7–8 SW. U.S., Mex.
Wildemaniodoxa Aubrév. & Pellegr. Sapotaceae. 1 trop. Afr.
Wildenowia Thunb. = Willdenowia Thunb. (Restionac.).
Wildpretina Kuntze = Ixanthus Griseb. (Gentianac.).
Wildungenia Wender. = Sinningia Nees (Gesneriac.).
Wilhelminia Hochr. = Hibiscus L. (Malvacac.).
Wilhelmsia C. Koch = Koeleria Pers. (Gramin.).
Wilhelmsia Reichb. (Merckia Fisch. ex Cham. & Schlechtd.). Caryo-
phyllaceae. 1 arct. E. As., NW. Am.
Wilibalda Sternb. = Coleanthus Seidl (Gramin.).
Wilibald-Schmidtia Seidel [sic] ex Princ. Friedr. v. Sachsen = Sieglingia
Bernh. (Gramin.).
Wilkea P. & K. = Wilckea Scop. = Vitex L. (Verbenac.).
Wilkesia A. Gray. Compositae. 2 Hawaii. Small trees.
Wilkia F. Muell. = Wilckia Scop. = Malcolmia R.Br. corr. Spreng. (Crucif.).
Wilkiea F. Muell. Monimiaceae. 6 E. Austr.

Willardia Rose. Leguminosae. 6 Mex. Timber.
Willbleibia Herter = Willkommia Hack. (Gramin.).
Willdenovia J. F. Gmel. = Rondeletia L. (Rubiac.).
Willdenovia Thunb. = Willdenowia Thunb. (Restionac.).
Willdenowa Cav. = Adenophyllum Pers. (Compos.).
Willdenowia Steud. (1) = praec.
Willdenowia Steud. (2) = Willdenovia J.F. Gmel. = Rondeletia L. (Rubiac.).
Willdenowia Thunb. Restionaceae. 15 S. Afr. The stems of some are used in making brooms.
Willemetia Brongn. = Noltea Reichb. (Rhamnac.).
Willemetia Maerkl. = Bassia All. + Kochia Roth (Chenopodiac.).
Willemetia Neck. = Chondrilla L. (Compos.).
Williamia Baill. = Phyllanthus L. (Euphorbiac.).
Williamsia Merr. Rubiaceae. 15 Philipp., N. Borneo.
Willibalda Steud. = Wilibalda Sternb. = Coleanthus Seidl (Gramin.).
Willichia Mutis ex L. = Sibthorpia L. (Scrophulariac.).
Willisellus S. F. Gray = Elatine L. (Elatinac.).
Willisia Warm. Podostemaceae. 1 S. India. There is a small thallus, with closely crowded erect shoots bearing 4 closely packed ranks of scaly s., and ribbon-like ls. at the tips. Each shoot bears one fl. (Cf. Willis in *Ann. Perad.* 1: 369, 1902.)
Willkommia Hackel. Gramineae. 1 S. U.S., 1 temp. S. Am.; 3 trop. & SW. Afr.
Willkommia Sch. Bip. ex Nym. = Senecio L. (Compos.).
Willoughbeia Hook. f. = Willughbeia Roxb. (Apocynac.).
Willoughbya Kuntze = Willugbaeya Neck. = Mikania Willd. (Compos.).
Willrussellia A. Chev. = Pitcairnia L'Hér. (Bromeliac.).
Willugbaeya Neck. = Mikania Willd. (Compos.).
Willughbeia Klotzsch = Landolphia Beauv. (Apocynac.).
*****Willughbeia** Roxb. Apocynaceae. 25 Indomal. Some, e.g. *W. edulis* Roxb. (Assam to Borneo), and *W. firma* Bl. (Java, etc.), contain rubber in their latex, and are used as sources of rubber.
Willughbeia Scop. = Ambelania Aubl. + Pacouria Aubl. (Apocynac.).
Willughbeiaceae ('-bejieae') J. G. Agardh = Apocynaceae–Plumieroïdeae–Arduineae K. Schum.
Wilmattea Britton & Rose = Hylocereus (A. Berger) Britt. & Rose (Cactac.).
Wilmattia Willis = praec.
× **Wilsonara** hort. Orchidaceae. Gen. hybr. (iii) (Cochlioda × Odontoglossum × Oncidium).
Wilsonia R.Br. Convolvulaceae. 4 Austr.
Wilsonia Hook. = Dipyrena Hook. (Verbenac.).
Wilsonia Rafin. = ? Epacris J. R. & G. Forst. (Epacridac.).
Wimmera P. & K. = Wimmeria Schlechtd. (Celastrac.).
Wimmeria Nees ex Meissn. = Beilschmiedia Nees (Laurac.).
Wimmeria Schlechtd. Celastraceae. 14 Mex., C. Am.
Winchia A. DC. Apocynaceae. 2 SE. As.
Windmannia P.Br. = Weinmannia L. (Cunionac.).
Windsoria Nutt. = Triodia R.Br. (Gramin.).

WINDSORINA

Windsorina Gleason. Rapateaceae. 1 Guiana.
Winklera P. & K. = Winkleria Reichb. = Mertensia Roth (Boraginac.).
Winklera Regel = Uranodactylus Gilli (Crucif.).
Winklerella Engl. Podostemaceae. 1 W. Equat. Afr.
Winkleria Reichb. = Mertensia Roth (Boraginac.).
Wintera G. Forst. ex Van Tiegh. = Pseudowintera Dandy (Winterac.).
Wintera Murr. = Drimys J. R. & G. Forst. (Winterac.).
*****Winteraceae** Lindl. Dicots. 7/120 Malaysia to Pacif., E. Austr., N.Z., C. &
S. Am. Trees or shrubs, with alt. or subvertic. simple ent. gland-dotted exstip.
ls. Fls. reg., ♀, rarely polyg., cymose or fascic. K 2–6, free, rarely totally
concresc., valv.; C 4–∞, 1–3-ser., imbr.; A 15–∞, 2–5-ser., anth. intr. or
sometimes extr.; G̲ 1–several, free or partly or wholly conn., ± 1-ser., with
1–∞ ventr. or axile anatr. ov. per loc., styles short or o. Fr. of dehisc. follicles,
or baccate; seeds with copious endosp. Genera: *Drimys, Pseudowintera, Bubbia,
Belliolum, Exospermum, Zygogynum, Tetrathalamus.* Related to *Illiciac., Dege-
neriac., Magnoliac.*
Winterana L. = Canella P.Br. (Canellac.).
Winterana Soland. ex Medik. = Drimys J. R. & G. Forst. (Winterac.).
Winteranaceae Warb. = Canellaceae Mart.
Winterania L. = Winterana L. = Canella P.Br. (Canellac.).
Winterania P. & K. = Winterana Soland. ex Medik. = Drimys J. R. & G. Forst.
(Winterac.).
Winteria F. Ritter = Hildewintera F. Ritter (Cactac.).
Winterlia Dennst. = ? Hesperethusa M. Roem. (Rutac.).
Winterlia Moench = Ilex L. (Aquifoliac.).
Winterlia Spreng. = Ammannia L. (Lythrac.).
Winterocereus Backeb. = Hildewintera F. Ritter (Cactac.).
Wirtgenia H. Andres = Andresia Sleum. (Monotropac.).
Wirtgenia Jungh. ex Hassk. = Spondias L. + Lannea A. Rich. (Anacardiac.).
Wirtgenia Nees ex Doell = Paspalum L. (Gramin.).
Wirtgenia Sch. Bip. = Aspilia Thou. (Compos.).
Wisenia J. F. Gmel. = Visenia Houtt. = Melochia L. (Sterculiac.).
Wislizenia Engelm. Cleomaceae. 1 (very polymorphic) SW. U.S., Mex.
Wisneria M. Micheli (sphalm.) = Wiesneria M. Micheli (Alismatac.).
Wissadula Medik. Malvaceae. 40 trop. (esp. Am.).
Wissmannia Burret. Palmae. 1 Somalia, S. Arabia.
Wistaria Spreng. = seq.
*****Wisteria** Nutt. Leguminosae. 10 E. As., E. N. Am. *W. chinensis* DC. (China)
is a climbing shrub with sweet-scented fls. Floral mech. like *Trifolium.* The
pods explode violently.
*****Withania** Pauquy. Solanaceae. 10 S. Am., S. Afr., Canaries, Medit. to India.
W. coagulans Dun. is used in India in preparing cheese.
Witharia Reichb. (sphalm.) = praec.
Witheringia L'Hérit. = Bassovia Aubl. (Solanac.).
Witheringia Miers = Athenaea Sendtn. (Solanac.).
× **Withnerara** hort. Orchidaceae. Gen. hybr. (iii) (Aspasia × Miltonia ×
Odontoglossum × Oncidium).
Witsenia Thunb. Iridaceae. 1 S. Afr.

Wittea Kunth = Downingia Torr. (Campanulac.).
Wittelsbachia Mart. = Cochlospermum Kunth (Cochlospermac.).
Wittia K. Schum. (~ Disocactus Lindl.). Cactaceae. 2 Panamá, NW. trop.
S. Am.
Wittmackanthus Kuntze = Pallasia Klotzsch (Rubiac.).
Wittmackia Mez. Bromeliaceae. 5 W.I., E. trop. S. Am.
Wittrockia Lindm. Bromeliaceae. 6 Brazil.
Wittsteinia F. Muell. Epacridaceae (?). 1 mts. SE. Austr. G̅. Berry.
Woehleria Griseb. Amarantuaceae. 1 Cuba.
Woikoia Baehni. Sapotaceae. 1(?) New Guinea.
Wokoia Baehni (sphalm.) = praec.
Wolffia Horkel ex Schleid. Lemnaceae. 10 trop. & temp. *W. arrhiza* (L.)
Hork. ex Wimm. is the smallest known flowering plant.
Wolffiaceae (Engl.) Nak. = Lemnaceae–Wolffioïdeae Engl.
Wolffiella (Hegelm.) Hegelm. Lemnaceae. 8 trop. Am., Afr.
Wolffiopsis den Hartog & v. d. Plas. Lemnaceae. 1 trop. Am. & Afr.
Wolfia Dennst. = ?Renealmia L. f. (Zingiberac.).
Wolfia Kunth (sphalm.) = Wolffia Horkel ex Schleid. (Lemnac.).
Wolfia P. & K. = Orchidantha N. E. Br. (Lowiac.).
Wolfia Schreb. = Casearia Jacq. (Flacourtiac.).
Wolfia Spreng. Inc. sed. (?Menispermac.). 1 Brazil.
Wolkensteinia Regel = Ouratea Aubl. (Ochnac.).
Wollastonia DC. ex Decne = Wedelia Jacq. (Compos.).
Woodburnia Prain. Araliaceae. 1 Burma.
Woodfordia Salisb. Lythraceae. 1 Abyss.; 1 (*W. floribunda* Salisb.) Madag.,
India, Ceylon, China, Sumatra to Timor.
Woodia Schlechter. Asclepiadaceae. 3 S. Afr.
Woodiella Merr. Annonaceae. 1 Borneo.
Woodier Roxb. ex Kostel. = Odina Roxb. = Lannea A. Rich. (Anacardiac.).
Woodrowia Stapf (~ Dimeria R.Br.). Gramineae. 1 W. Penins. India. Infl.
sharply deflexed and infl. branches convolute in fruit.
Woodsia R.Br. Aspidiaceae. 40 alpine & arctic, also S. Am. & S. Afr.
Woodsiaceae Ching = Aspidiaceae S. F. Gray.
Woodsonia L. H. Bailey (~ Neonicholsonia Dammer). Palmae. 1 Panamá.
Woodvillea DC. = Erigeron L. (Compos.).
Woodwardia Sm. Blechnaceae. 12 S. Eur. to Japan, W. N. Am.
Wooleya L. Bolus. Aïzoaceae. 1 SW. Afr.
Woollsia F. Muell. = Lysinema R.Br. (Epacridac.).
Wootonella Standley. Compositae. 1 N. Am.
Wootonia Greene. Compositae. 1 SW. U.S.
Worcesterianthus Merr. = Microdesmis Hook. f. (Pandac.).
Wormia P. & K. = Vormia Adans. = Selago L. (Scrophulariac.).
Wormia Rottb. = Dillenia L. (Dilleniac.).
Wormia Vahl = Ancistrocladus Wall. (Ancistrocladac.).
Wormskioldia Schumacher & Thonn. (? 1829; non Spreng. 1827—Algae).
Turneraceae. 10 trop. & S. Afr.
Woronowia Juzepczuk (~ Sieversia Willd.). Rosaceae. 1 Caucasus.
Worrnia J. F. Gmel. (sphalm.) = Wormia Rottb. = Dillenia L. (Dilleniac.).

WORSLEYA

Worsleya (Traub) Traub = Hippeastrum L. (Amaryllidac.).
Worsleya W. Watson = Hippeastrum L. (Amaryllidac.).
Woytkowskia Woodson. Apocynaceae. 1 Peru.
Wredowia Eckl. = Aristea Soland. (Iridac.).
× **Wrefordara** hort. (v) = × Burkillara hort. (Orchidac.).
Wrenciala A. Gray = Plagianthus J. R. & G. Forst. (Malvac.).
Wrightea Roxb. = Wallichia Roxb. (Palm.).
Wrightea Tussac = Meriania Sw. (Melastomatac.).
Wrightia R.Br. Apocynaceae. 23 trop. Afr., As., Austr.
Wrightia Soland. ex Naud. = Meriania Sw. (Melastomatac.).
Wrixonia F. Muell. Labiatae. 1 W. Austr.
Wuerschmittia Sch. Bip. ex Hochst. = Melanthera Rohr (Compos.).
Wuerthia Regel = Ixia L. (Iridac.).
Wulfenia Jacq. Scrophulariaceae. 3 SE. Eur., W. Himal.
Wulffia Neck. ex Cass. Compositae. 4 W.I., S. Am.
Wulfhorstia C. DC. = Entandrophragma C. DC. (Meliac.).
Wullschlaegelia Reichb. f. Orchidaceae. 3 trop. S. Am., W.I.
Wunderlichia Riedel. Compositae. 4 Brazil.
Wunschmannia Urb. Bignoniaceae. 1 Haiti.
Wurdackia Moldenke. Eriocaulaceae. 1 Venez.
Wurfbaeinia Steud. = seq.
Wurfbainia Giseke = Amomum L. (Zingiberac.).
Wurmbaea Steud. = seq.
Wurmbea Thunb. Liliaceae. 8 trop. & S. Afr., W. Austr.
Wurmschnittia(!) Benth. = Wuerschmittia Sch. Bip. ex Hochst. = Melanthera Rohr (Compos.).
Wurtzia Baill. = Phyllanthus L. (Euphorbiac.).
Wycliffea Ewart & Petrie = Glinus L. (Aïzoac.).
Wydlera P. & K. = Wydleria Fisch. & Trautv. = Apium L. (Umbellif.).
Wydleria DC. = Carum L. (Umbellif.).
Wydleria Fisch. & Trautv. = Apium L. (Umbellif.).
Wyethia Nutt. Compositae. 14 W. N. Am.
Wylia Hoffm. = Scandix L. (Umbellif.).
Wyomingia A. Nelson = Erigeron L. (Compos.).

X

Xaiasme Rafin. = Stellera L. (Thymelaeac.).
Xalkitis Rafin. = Bindera Rafin. = ? Aster L. (Compos.).
Xamacrista Rafin. = Chamaecrista Moench = Cassia L. (Legumin.).
Xamesike Rafin. = Chamaesyce S. F. Gray = Euphorbia L. (Euphorbiac.).
Xamesuke Rafin. = praec.
Xamilenis Rafin. = Silene L. (Caryophyllac.).
Xananthes Rafin. = Utricularia L. (Lentibulariac.).
Xanthaea Reichb. = Centaurium Hill (Gentianac.).
Xanthanthos St-Lag. = Anthoxanthum L. (Gramin.).
Xanthe Schreb. = Clusia L. (Guttif.).

Xantheranthemum Lindau. Acanthaceae. 1 Peru.

Xanthidium Delpino = Franseria Cav. (Compos.).

Xanthisma DC. Compositae. 1 S. U.S.

Xanthium L. Compositae. 30 cosmop. They have been so widely distr. by man (unintentionally) that it is hard to discover their native place. Fls. in unisexual heads, single or in axillary cymes, the ♂ at the ends of the branches. The ♀ head has 2 fls., enclosed in a prickly gamophyllous invol., only the styles projecting from it through openings in the two horns of the invol. The frs. are enclosed in the hard woody invol., which is covered with hooks and well suited to animal-distr. One sp. has gradually spread in this way from the East of Europe. 'In 1828 it was brought into Wallachia by the Cossack horses, whose manes and tails were covered with the burrs. It travelled in Hungarian wool, and in cattle from the same region, to Regensburg, and on to Hamburg, appearing here and there on the way.' Strenuous laws for its extirpation have been enforced in South Africa, where at one time it had become so common as seriously to impair the value of the wool.

Xantho Remy = Lasthenia Cass. (Compos.).

Xanthocephalum Willd. Compositae. 20 S. U.S. to C. Mex.

Xanthoceras Bunge. Sapindaceae. 2 N. China. Ed. seed.

Xanthocercis Baill. Leguminosae. 2 trop. E. Afr., Madag.

Xanthochrysum Turcz. = Helichrysum Mill. (Compos.).

Xanthochymus Roxb. = Garcinia L. (Guttif.).

Xanthocoma Kunth = Xanthocephalum Willd. (Compos.).

Xanthocromyon Karst. = Trimeza Salisb. (Iridac.).

Xanthogalum Avé-Lallem. Umbelliferae. 3 As. Min., Cauc., Persia.

Xantholepis Willd. ex Less. = Cacosmia Kunth (Compos.).

Xantholinum Reichb. = Linum L. (Linac.).

Xanthomyrtus Diels. Myrtaceae. 25 New Guinea, New Caled.

Xanthonanthos St-Lag. = Anthoxanthum L. (Gramin.).

Xanthopappus C. Winkler. Compositae. 2 NW. China.

Xanthophthalmum Sch. Bip. = Chrysanthemum L. (Compos.).

Xanthophyllaceae (Chodat) Gagnep. Dicots. 1/60 Indomal. Trees with alt. simple coriac. exstip. ls. Infl. axill. or term., rac. or panic. Fls. ☿, zygo. K 5, imbr., the 2 inner slightly longer; C 5–4, imbr., unequal, sometimes clawed, the lowermost within, folded, forming a 'keel' (not crested); A 8, fil. free or ± adnate to claw of pet., ± inflated and pubesc. below, anth. intr., pubesc. below; disk present; G (2), stipit., 1-loc., with 2 pariet. plac., each bearing 2–∞ anatr. ov., and simple style with small capit. bilobed stig. Fr. globose, fibrous-fleshy or dry, indehisc., 1-seeded; endosp. o. Only genus: *Xanthophyllum*. Intermediate between *Polygalaceae* and *Leguminosae–Caesalpinioïdeae*; nearer to the former in floral structure, but making an extremely close approach to certain genera of the latter in habit, etc. It differs notably from both in its strong aluminium-accumulating tendency (shown by the yellow-green colour of the ls. of many spp. on drying).

Xanthophyllon St-Lag. (sphalm.) = Xanthoxylon Spreng. = Zanthoxylum L. (Rutac.).

***Xanthophyllum** Roxb. Xanthophyllaceae. 60 Indomal. Many aluminium-accumulators.

40-2

XANTHOPHYTOPSIS

Xanthophytopsis Pitard. Rubiaceae. 2 S. China, Indoch.

Xanthophytum Reinw. Rubiaceae. 15 SE. As., Malaysia to Fiji.

Xanthopsis C. Koch = Centaurea L. (Compos.).

Xanthorhiza Marshall. Ranunculaceae. 1 Atl. N. Am.

Xanthorrhoea Sm. Xanthorrhoeaceae. 15 Austr. The best known is *X. hastilis* R.Br., the grass-tree or black-boy, a char. plant of the Austr. veg. It has the habit of an *Aloë* or *Dasylirion*, with a long bulrush-like spike of fls. (really cymose as may be seen from the many bracts on the individual fl.-stalks). P sepaloid. From the bases of the old leaves trickles a resin, used in making varnish, sealing-wax, etc.

*Xanthorrhoeaceae Dum. Monocots. 8/66 Austr., New Caled., N.Z. Stout woody ± rhiz. perennials; stem sometimes tall and little branched; ls. simple, ± lin., sometimes pungent, often sheathing. Fls. reg., ♀ or ♂ ♀, spicate, panic. or capit., usu. dry and glumaceous (cf. *Juncac.*), often persist. P 3 + 3 (sometimes shortly conn.); A 3 + 3 (outer ± free, inner usu. adn. to inner tep.), anth. basif. or versat., intr. or latr.; G̲ (3), 3-loc. with axile plac. or 1-loc. with basal plac., ov. 1–3–few per loc. (erect when basal); styles free or ± completely conn. Fr. a loculic. caps. or indehisc. nut; seeds 1–few, with hard endosp. Genera: *Baxteria, Calectasia, Xanthorrhoea, Chamaexeros, Acanthocarpus, Dasypogon, Lomandra* (by far the largest), *Kingia*. Near *Liliac., Agavac.* and *Juncac.*

Xanthoselinum Schur = Peucedanum L. (Umbellif.).

Xanthosia Rudge. Hydrocotylaceae (~ Umbellif.). 15 Austr. The umbels in some are reduced to single fls.

Xanthosoma Schott. Araceae. 45 Mex. to trop. S. Am., W.I. Large herbs. *X. appendiculatum* Schott has a pocket at the back of the leaf due to a tangential division of the embryonic leaf. Fls. monoec., naked; synandria. Some spp. (*X. jacquinii* Schott, etc.) cult. for ed. rhiz. (*yautia*), like *Colocasia*.

Xanthostachya Bremek. Acanthaceae. 2 Lesser Sunda Is.

Xanthostemon F. Muell. Myrtaceae. 40 Philipp. Is., E. Malaysia, N. & NE. Austr., New Caled.

Xanthoxalis Small = Oxalis L. (Oxalidac.).

Xanthoxylaceae Nees & Mart. = Rutaceae–Zanthoxyleae Benth. & Hook. f.

Xanthoxylon Spreng. = seq.

Xanthoxylum J. F. Gmel. = Zanthoxylum L. (Rutac.).

Xantium Gilib. = Xanthium L. (Compos.).

Xantolis Rafin. Sapotaceae. 14 S. India, SE. As., N. Philipp. Is.

Xantonnea Pierre ex Pitard. Rubiaceae. 3 Siam, Indochina.

Xantonneopsis Pitard. Rubiaceae. 1 Indochina.

Xantophtalmum Sang. = Xanthophthalmum Sch. Bip. = Chrysanthemum L. (Compos.).

Xaritonia Rafin. = Oncidium Sw. (Orchidac.).

Xatardia Meissn. Umbelliferae. 1 Pyrenees.

Xatartia St-Lag. = praec.

Xatatia Bub. = praec.

Xaveria Endl. = Anemonopsis Sieb. & Zucc. (Ranunculac.).

Xeilyathum Rafin. = Oncidium Sw. (Orchidac.).

Xenacanthus Bremek. Acanthaceae. 4 Penins. India.

Xeniatrum Salisb. = Clintonia Rafin. (Liliac.).

Xenismia DC. = Oligocarpus Less. (Compos.).
Xenocarpus Cass. = Cineraria L. emend. Less. (Compos.).
Xenochloa Lichtenstein = ?Danthonia DC. (Gramin.).
Xenodendron K. Schum. & Lauterb. = Acmena DC. = Syzygium Gaertn. (Myrtac.).
Xenophonta Benth. & Hook. f. = seq.
Xenophontia Vell. = Barnadesia Mutis (Compos.).
Xenophya Schott. Araceae. 2 Moluccas to Bismarcks.
Xenopoma Willd. = Micromeria Benth. (Labiat.).
Xeodolon Salisb. = Scilla L. (Liliac.).
Xeracina Rafin. = Adelobotrys DC. (Melastomatac.).
Xeractis Oliv. = Xerotia Oliv. (Caryophyllac.).
Xeraea Kuntze = Gomphrena L. (Amaranthac.).
Xeraenanthus Mart. ex Koehne = Pleurophora D. Don (Lythrac.).
Xeralis Rafin. = Charachera Forsk. = ?Lantana L. (Verbenac.), vel Acanthac.?
Xeralsine Fourr. = Minuartia L. (Caryophyllac.).
Xerandra Rafin. = Iresine P.Br. (Amaranthac.).
Xeranthemum L. Compositae. 6 Medit. to SW. As.
Xeranthium Lepekh. = Xanthium L. (Compos.).
Xeranthus Miers = Grahamia Gill. (Portulacac.).
Xeregathis Rafin. = Baccharis L. (Compos.).
Xeria Presl ex Rohrb. = Pycnophyllum Remy (Caryophyllac.).
Xeris Medik. = Iris L. (Iridac.).
Xeroaloysia Troncoso. Verbenaceae. 1 Argentina.
Xerobius Cass. = Egletes Cass. (Compos.).
Xerobotrys Nutt. = Arctostaphylos Adans. (Ericac.).
***Xerocarpa** Lam. Verbenaceae. 1 New Guinea.
Xerocarpa Spach = Scaevola L. (Goodeniac.).
Xerocarpus Guill. & Perr. = Rothia Pers. (Legumin.).
Xerocassia Britton & Rose = Cassia L. (Legumin.).
Xerochlamys Baker. Sarcolaenaceae. 16 Madag.
Xerochloa R.Br. Gramineae. 4 Siam, Java, Austr.
Xerocladia Harv. Leguminosae. 1 S. Afr.
Xerococcus Oerst. Rubiaceae. 1 C. Am.
Xerodanthia Phipps. Gramineae. 2 Egypt & Sudan to Sind & Baluchistan.
Xerodera Fourr. = Ranunculus L. (Ranunculac.).
Xeroderris Roberty. Leguminosae. 1 savannahs, trop. Afr.
Xerodraba Skottsb. Cruciferae. 6 S. Patag.
Xerogona Rafin. = Passiflora L. (Passiflorac.).
Xerololophus B. D. Jacks. (sphalm.) = seq.
Xerolophus Dulac = Thesium L. (Santalac.).
Xeroloma Cass. = Xeranthemum L. (Compos.).
Xeromalon Rafin. = Crataegus L. (Rosac.).
Xeromphis Rafin. = ?Catunaregam Adans. ex v. Wolf = Randia L. (Rubiac.).
Xeronema Brongn. & Gris. Liliaceae. 2 New Caled., N. N.Z. (Poor Knights Is.).
Xeropappus Wall. = Dicoma Cass. (Compos.).
Xeropetalon Hook. = Viviania Cav. (Vivianiac.).
Xeropetalum Delile = Dombeya Cav. (Sterculiac.).

XEROPETALUM

Xeropetalum Reichb. = Dillwynia Sm. (Legumin.).

Xerophyllum Michx. Liliaceae. 3 N. Am.

Xerophylum Rafin. = praec.

Xerophysa Stev. = Astragalus L. (Legumin.).

Xerophyta Juss. (~Barbacenia Vand.). Velloziaceae. 55 S. Am., trop. Afr., Madag.

Xeroplana Briq. Stilbaceae. 1 S. Afr.

Xerorchis Schlechter. Orchidaceae. 2 trop. S. Am.

Xerosicyos Humbert. Cucurbitaceae. 2 Madag.

Xerosiphon Turcz. = Gomphrena L. (Amaranthac.).

Xerosollya Turcz. = Sollya Lindl. (Pittosporac.).

Xerospermum Blume. Sapindaceae. 20–25 Assam, SE. As., W. Malaysia.

Xerotaceae ('-ideae') Endl. = Xanthorrhoeaceae Dum.

Xerotecoma J. C. Gomes. Bignoniaceae. 1 NE. Brazil.

Xerotes R.Br. = Lomandra Labill. (Xanthorrhoeac.).

Xerothamnella C. T. White. Acanthaceae. 1 Queensland.

Xerothamnus DC. = Osteospermum L. (Compos.).

Xerotia Oliv. Caryophyllaceae. 1 Arabia.

Xerotis Hoffmgg. = Xerotes R.Br. = Lomandra Labill. (Xanthorrhoeac.).

Xerotium Bluff & Fingerh. = Filago L. (Compos.).

Xestaea Griseb. = Schultesia Mart. (Gentianac.).

Xetola Rafin. = Cephalaria Schrad. (Dipsacac.).

Xetoligus Rafin. = Stevia Cav. (Compos.).

Xilophia Augier = seq.

Xilopia Juss. = Xylopia L. (Annonac.).

Ximenesia Cav. = Verbesina L. (Compos.).

Ximenia L. Olacaceae. 10–15 trop. Am., trop. & S. Afr., trop. As., Austr.
X. americana L. yields good wood.

Ximeniaceae Van Tiegh. = Olacaceae–Anacolosoïdeae–Ximeniëae Engl.

Xiphagrostis Coville = Miscanthus Anders. (Gramin.).

Xiphidiaceae Dum. = Hypoxidaceae R.Br.

Xiphidium Loefl. ex Aubl. Haemodoraceae. 1–2 trop. Am., W.I.

Xiphion Mill. = Iris L. (Iridac.).

Xiphium Mill. = praec.

Xiphizusa Reichb. f. = Bulbophyllum Thou. (Orchidac.).

Xiphocarpus Presl = Tephrosia Pers. (Legumin.).

Xiphochaeta Poepp. & Endl. = Stilpnopappus Mart. ex DC. (Compos.).

Xiphocoma Stev. = Ranunculus L. (Ranunculac.).

Xiphodendron Rafin. = ? Yucca L. (Agavac.).

Xipholepis Steetz = Vernonia Schreb. (Compos.).

Xiphophyllum Ehrh. (uninom.) = *Serapias xiphophyllum* L. = *Cephalanthera longifolia* (L.) Fritsch (Orchidac.).

Xiphopteris Kaulf. Grammitidaceae. 50 pantrop. Distinction from *Grammitis* Sw. and *Ctenopteris* Bl. not clear.

Xiphosium Griff. = Eria Lindl. (Orchidac.).

Xiphostylis Gasp. = Trigonella L. (Legumin.).

Xiphotheca Eckl. & Zeyh. = Priestleya DC. (Legumin.).

Xiris Rafin. = Xyris Gronov. ex L. (Xyridac.).

Xolantha Rafin. = Helianthemum Mill. (Cistac.).
Xolanthes Rafin. = praec.
Xolemia Rafin. = Gentiana L. (Gentianac.).
Xolisma Rafin. = Lyonia Nutt. (Ericac.).
Xolocotzia Miranda. Verbenaceae. 1 Mex.
Xoxylon Rafin. = Toxylon Rafin. = Maclura Nutt. (Morac.).
Xuaresia Pers. = seq.
Xuarezia Ruiz & Pav. = Capraria L. (Scrophulariac.).
Xuris Adans. = Iris L. (Iridac.).
Xyladenius Desv. = Banara Aubl. (Flacourtiac.).
Xylanche G. Beck. Orobanchaceae. 2 Himalaya, China, Formosa.
Xylanthema Neck. = Cirsium Mill. (Compos.).
Xylanthemum Tsvelev (~ Chrysanthemum L.). Compositae. 5 C. As., NE. Persia, Afghan.
Xylia Benth. (= Esclerona Rafin.). Leguminosae. 15 trop. Afr., Madag., trop. As. Good timber.
Xylinabaria Pierre. Apocynaceae. 2 Indochina, 2 Java.
Xylinabariopsis Pitard. Apocynaceae. 2 Indoch.
Xylobium Lindl. Orchidaceae. 33 C. & trop. S. Am., W.I.
Xylocalyx Balf. f. Scrophulariaceae. 4 Somal., Socotra.
Xylocarpus Koen. Meliaceae. 3 coasts trop. E. Afr., Ceylon, Malaysia, N. Austr., Pacif.
Xylochlaena Dalla Torre & Harms (sphalm.) = Xyloölaena Baill. = Scleroölaena Baill. (Sarcolaenac.).
Xylochlamys Domin = Amyema Van Tiegh. (Loranthac.).
Xylociste Adans. = Xylocyste P.Br. (fam.?).
Xylococcus R.Br. ex Britt. & S. Moore = Petalostigma F. Muell. (? Euphorbiac.).
Xylococcus Nutt. (~ Arctostaphylos Adans.). Ericaceae. 1 S. Calif., Lower Calif.
Xylocyste P.Br. Jamaica. Quid?
Xylolaena Baill. (sphalm.) = Xyloölaena Baill. = Scleroölaena Baill. (Sarcolaenac.).
Xylolobus Kuntze = Xylia Benth. (Legumin.).
Xylomelum Sm. Proteaceae. 4 Austr. The fruits are known as wooden pears, being of the size of a large pear, and looking ed. at first glance. Inside is a thick wall of woody tissue enveloping the winged seeds. It splits along the post. side.
Xylon Kuntze = Bombax L. (Malvac.).
Xylon 'L.' = Ceiba Mill. (Bombacac.).
Xylon Mill. = Gossypium L. (Malvac.).
Xylonagra Donn. Sm. & Rose. Onagraceae. 1 Lower Calif.
Xylonymus Kalkm. Celastraceae. 1 W. New Guinea.
Xyloölaena Baill. Sarcolaenaceae. 1 Madag.
Xylophacos Rydb. = Astragalus L. (Legumin.).
Xylophragma Sprague. Bignoniaceae. 4 trop. Am., Trinidad.
Xylophylla L. = Phyllanthus L. (spp. with phylloclades) (Euphorbiac.) + Exocarpos Labill. (Santalac.).
Xylophyllos Kuntze = Exocarpos Labill. (Santalac.).

*Xylopia L. Annonaceae. 100–150 trop., esp. Afr. Fr. used as peppers.

Xylopiastrum Roberty = Uvaria L. (Annonac.).

Xylopicron Adans. = seq.

Xylopicrum P.Br. = Xylopia L. (Annonac.).

Xylopleurum Spach = Oenothera L. (Onagrac.).

Xylorhiza Nutt. = Machaeranthera Nees (Compos.).

Xylorhiza Salisb. = Allium L. (Alliac.).

*Xylosma G. Forst. Flacourtiaceae. 100 warm. Usu. dioec. Often axillary thorns. Seps. ± united.

Xylosma Harv. = Xymalos Baill. (Monimiac.).

Xylosteon Mill. = Lonicera L. (Caprifoliac.).

Xylosteum auctt. = praec.

Xylotheca Hochst. Flacourtiaceae. 10 trop. Afr., Madag.

Xylothermia Greene = Pickeringia Nutt. ex Torr. & Gray (Legumin.).

Xymalobium Steud. = Xysmalobium R.Br. (Asclepiadac.).

Xymalos Baill. Monimiaceae. 1–2 trop. & S. Afr.

Xynophylla Montr. [= Xylophylla L.] = Exocarpos Labill. (Santalac.).

Xyochlaena Stapf = Tricholaena Schrad. (Gramin.).

Xyphanthus Rafin. = Erythrina L. (Legumin.).

Xypherus Rafin. = Amphicarpaea Ell. mut. DC. (Legumin.).

Xyphidium Neck. = Xiphidium Loefl. ex Aubl. (Hypoxidac.).

Xyphidium Steud. = Xiphion Mill. = Iris L. (Iridac.).

Xyphion Medik. = praec.

Xyphostylis Rafin. = Canna L. (Cannac.).

*Xyridaceae C. A. Agardh. Monocots. (Farinosae). 2/250 trop. & subtrop., mostly Am. Mostly marsh plants, herbaceous, tufted, with radical distichous sheathing ls. and densely bracteate spikes or heads of ⚥ fls. P heterochlam. K 3, zygo., the lat. sepals small, the ant. large, membr., enclosing the corolla; C 3, yellow; A 3, epipet., the outer whorl absent or repres. by stds.; pollen espinose; G̲ (3), 1-loc., with parietal or free basal plac. and ∞ orthotr. ov.; styles 3, inappendic. Caps. Seeds subglob., vert. striate, usu. biapic. Embryo small, in mealy endosp. Genera: *Xyris, Achlyphila.* (Cf. *Abolbodaceae.*)

Xyridanthe Lindl. = Helipterum DC. (Compos.).

Xyridion Fourr. = Iris L. (Iridac.).

Xyridium Tausch ex Steud. = praec.

Xyridopsis Welw. ex O. Hoffm. = Oligothrix DC. (Compos.).

Xyris Gronov. ex L. Xyridaceae. 250 trop. & subtrop.

Xyroïdes Thou. = praec.

Xyropleurum auct. ex Steud. = Xylopleurum Spach = Oenothera L. (Onagrac.).

Xyropterix Kramer. Lindsaeaceae. 1 Borneo. (*Acta Bot. Neerl.* **6**: 599, 1957.)

Xysmalobium R.Br. Asclepiadaceae. 1 trop. & S. Afr.

Xystidium Trin. = Perotis Ait. (Gramin.).

Xystris Schreb. Inc. sed. [K (5), C (5), A 5, G̲ (10), styles 2.] Quid?

Xystrolobos Gagnep. = Ottelia Pers. (Hydrocharitac.).

Xystrolobus Willis = praec.

Y

Yabea K.-Pol. = Caucalis L. (Umbellif.).

Yadakea Makino (~ Arundinaria Michx). Gramineae. 2 Japan.

Yadakeya Makino = praec.

× **Yamadara** Hort. Orchidaceae. Gen. hybr. (Cattleya × Epidendrum × Laelia × Rhyncholaelia [Brassavola Hort.]).

Yamala Rafin. = Heuchera L. (Saxifragac.).

Yangapa Rafin. = Gardenia Ellis (Rubiac.).

Yangua Spruce = Cybistax Mart. ex Meissn. (Bignoniac.).

Yarima Burret (sphalm.) = seq.

Yarina O. F. Cook (= ? Phytelephas Ruiz & Pav.). Palmae. 1 Peru.

Yatabea Maxim. ex Yatabe = Ranzania T. Ito (Podophyllac.).

Yaundea Schellenb. ex De Wild. = Jaundea Gilg (Connarac.).

Yeatesia Small. Acanthaceae. 1–2 SE. U.S.

Yermoloffia Bélang. = Lagochilus Bunge (Labiat.).

Yervamora Kuntze = Bosea L. (Amaranthac.).

Ygramela Rafin. = Limosella L. (Scrophulariac.).

Ygramelta Rafin. = praec.

Ymnostema Neck. = Lobelia L. (Campanulac.).

Ymnostemma Steud. = praec.

Ynesa O. F. Cook. Palmae. 1 Ecuador.

Yoania Maxim. Orchidaceae. 2 NE. India, Japan, ?N.Z.

Yodes Kurz = Iodes Bl. (Icacinac.).

Yolanda Hoehne. Orchidaceae. 1 Brazil.

Yongsonia Young = Fothergilla L. (Hamamelidac.).

Youngia Cass. Compositae. 35–40 temp. & trop. As.

Ypomaea Robin = Ipomoea L. (Convolvulac.).

Ypsilandra Franch. Liliaceae. 5 Tibet, Burma, W. China.

Ypsilopus Summerhayes. Orchidaceae. 3 trop. E. Afr.

Ystia Compère (~ Schizachyrium Nees). Gramineae. 1 trop. Afr.

Yuca Rafin. = Yucca L. (Agavac.).

Yucaratonia Burkart. Leguminosae. 1 Ecuad., Peru.

Yucca L. Agavaceae. 40 S. U.S., Mex., W.I. Stem short, growing in thickness, and branching occasionally (cf. *Dracaena*); at the end is a rosette of fleshy and pointed ls. Fls. large, white, in panicle. Remarkable mode of pollination (for details and figures see Riley in *3rd Ann. Rep. Missouri Bot. Gard.* 99–158, tt. 34–43, 1892). This is one of the few cases of mutual dependence and adaptation of the single fl. and a single insect—a small Tineid moth belonging to the genus *Pronuba* (*Prodoxidae*). The fl. emits its perfume esp. at night, and is then visited by the moths. The female has a long ovipositor with which she can penetrate the tissue of the ovary of the fl., and possesses peculiar prehensile, spinous, maxillary tentacles confined to the genus. She begins soon after dark, collecting a load of pollen, and shaping it into a pellet about thrice as large as her head. She then flies to another fl. and deposits a few eggs in the ovary, piercing its wall with her ovipositor. Having done this, she climbs to the top and presses the ball of pollen into the stigma. The ovules are thus fertilised, and are so numerous that there are plenty for the larvae to feed upon and also to reproduce the plant.

The setting of abundant, apparently good seed has, however, been reported in *Y. aloïfolia* L. cultivated at Lucknow, India, with no sign of the seeds being eaten or damaged. The pollinator was not observed.

The leaves of *Y. filamentosa* L. and other spp. furnish an excellent fibre (cf. *Agave*).

Yuccaceae J. G. Agardh = Agavaceae J. G. Agardh.

Yucea Rafin. = Yucca L. (Liliac.).

Yulania Spach = Magnolia L. (Magnoliac.).

Yunckeria Lundell. Myrsinaceae. 3 Mex., C. Am.

Yungasocereus Ritter (nomen). Cactaceae. 1 Bolivia(?). Quid?

Yunnanea Hu = ? Camellia L. (Theac.).

Yunquea Skottsb. Compositae. 1 Juan Fernandez.

Yushania P. C. Keng. Gramineae. 1 Formosa, Philippines (Luzon).

Yushunia Kamikoti = Sassafras Boehm. (Laurac.).

Yuyba (Barb.-Rodr.) L. H. Bailey (~ Bactris Jacq.). Palmae. 30–40 trop. S. Am.

Yvesia A. Camus. Gramineae. 1 Madag.

Z

Zaa Baill. = Phyllarthron DC. (Bignoniac.).

Zabelia (Rehder) Makino. Caprifoliaceae. 15 E. As. Anat. and pollen support separation from *Abelia*.

Zacateza Bullock. Periplocaceae. 1 trop. Afr.

Zacintha Mill. = Crepis L. (Compos.).

Zacintha Vell. = Clavija Ruiz & Pav. (Theophrastac.).

Zacyntha Adans. = Zacintha Mill. = Crepis L. (Compos.).

Zaga Rafin. = Adenanthera L. (Legumin.).

Zahlbruckera Steud. = Zahlbrucknera Reichb. (Saxifragac.).

Zahlbrucknera Pohl ex Nees = Hygrophila R.Br., Ebermaiera Nees, etc. (Acanthac.).

Zahlbrucknera Reichb. (~ Saxifraga L.). Saxifragaceae. 1 SE. Eur.

Zala Lour. = Pistia L. (Arac.).

Zalacca Reinw. ex Bl. = Salacca Reinw. (Palm.).

Zalaccella Becc. Palmae. 1 Indoch.

Zaleia Steud. = seq.

Zaleja Burm. f. = seq.

Zaleya Burm. f. Aïzoaceae. 3 trop. Afr., India, Ceylon, Lesser Sunda Is., Austr.

Zalitea Rafin. = Euphorbia L. (Euphorbiac.).

Zallia Roxb. = Zaleya Burm. f. = Trianthema L. (Aïzoac.).

Zalmaria B. D. Jacks. (sphalm.) = Zamaria Rafin. = Rondeletia L. (Rubiac.).

Zalucania Steud. = Zaluzania Pers. (Compos.).

Zaluzania Comm. ex Gaertn. f. = Bertiera Aubl. (Rubiac.).

Zaluzania Pers. Compositae. 12–15 Mex.

ZANTHOXYLUM

Zaluzanskia Neck. = Zaluzianskia Neck. = Marsilea L. (Marsileac.).
Zaluziana Link = Zaluzania Pers. (Compos.).
Zaluzianskia Benth. & Hook. f. = Zaluzianskya F. W. Schmidt (Scrophulariac.).
Zaluzianskia Neck. = Marsilea L. (Marsileac.).
Zaluzianskya F. W. Schmidt. Scrophulariaceae. 35 S. Afr., 1 mts. trop. E. Afr.
Zamaria Rafin. = Rondeletia L. (Rubiac.).
*Zamia L. Zamiaceae. 30–40 trop. Am., W.I.
Zamiaceae Reichb. Gymnosp. (Cycadales). 8/80 trop. & warm temp. Austr., Am., Afr. A family comprising the bulk of the Cycads. Pinnae (or pinnules) straight and imbr. (but not convol.) in vernation (frond circinate as a whole in *Ceratozamia*), with ∞ ± parallel longit. nerves, dichot.-branched near base. ♂ and ♀ sporophylls in determinate cones, the ♀ scale-like, ± pelt. with thickened and laterally expanded end, on the axis-facing margins of which the 2 (sometimes 3 in *Lepidozamia*, rarely 3 or more in other genera) inward-facing ('inverted') ovules are inserted. Genera: *Lepidozamia, Macrozamia, Encephalartos, Dioön, Microcycas, Ceratozamia, Zamia, Bowenia*. (See also *Cycadales*.)
Zamioculcas Schott. Araceae. 1 trop. E. Afr. Ls. pinnate.
Zamzela Rafin. = Hirtella L. (Chrysobalanac.).
Zanha Hiern. Sapindaceae. 1–2 trop. Afr.
Zanichelia Gilib. = seq.
Zanichellia Roth = seq.
Zannichallia Reut. = seq.
Zannichellia L. Zannichelliaceae. 1 cosmop., *Z. palustris* L.; 1 S. Afr. (SW. Cape Prov.). Fls. monoec.; ♀ term.; from the axil of its lower bracteole springs the ♂. From the axil of the upper a new branch may arise, bearing ♀ and ♂ fls. again. The ♂ fl. consists of 1 or 2 sta., the ♀ of usu. 4 cpls., surrounded by a small cup-like P. Pollination under water.
*Zannichelliaceae Dum. Monocots. 3/6 cosmop. Submerged aq. herbs, of fresh or brackish water. Ls. alt. or opp., lin., sheathing (sometimes reduced to sheath), sheath usu. ligulate. Fls. ♂♀, monoec. or dioec., axill., sol. or cymose. P a small cup, or a few small scales, or 0; A 1–3, sometimes connate; anth. 1- or 2-loc., pollen globose; G 1–9, free, sometimes stipit., each with a simple or 2–3-lobed style and conspic. stig., and 1 pend. ov. Fr. carpels achaenioid, indehisc., endosp. o. Genera: *Zannichellia, Althenia, Lepilaena.*
Zanonia Cram. = Campelia Rich. (Commelinac.).
Zanonia L. Cucurbitaceae. 1 Indomal., *Z. indica* L., with winged seeds.
Zanoniaceae Dumort. = Cucurbitaceae–Zanonioïdeae (Dum.) C. Jeffrey.
Zantedeschia C. Koch = Schismatoglottis Zoll. & Mor. (Arac.).
*Zantedeschia Spreng. Araceae. 8–9 temp. & subtrop. S. Afr., few trop. Afr.
Zanthorhiza L'Hérit. = Xanthorhiza Marsh. (Ranunculac.).
Zanthoxilon Franch. & Sav. = Zanthoxylum L. (Rutac.).
Zanthoxylaceae Nees & Mart. corr. Bartl. = Rutaceae–Zanthoxyleae Benth. & Hook. f.
Zanthoxylon Walt. = seq.
Zanthoxylum L. (sens. Engl.). Rutaceae. 20–30 temp. & subtrop. E. As.,

1237

Philipp., E. Malaysia, N. Am. *Z. piperitum* DC., the Japan pepper, yields fr. used as a condiment. The bark of *Z. fraxineum* Willd. (prickly ash or toothache-tree) is used in Am. as a remedy for toothache. Some yield good timber.

Zanthyrsis Rafin. = Sophora L. (Legumin.).

Zantorrhiza Steud. = Xanthorhiza Marsh. (Ranunculac.).

Zapamia Steud. = seq.

Zapania Lam. = Lippia L. (Verbenac.).

Zapateria Pau = Ballota L. (Labiat.).

Zappania Zuccagni = Zapania Lam. = Lippia L. (Verbenac.).

Zarabellia Cass. = Melampodium L. (Compos.).

Zarabellia Neck. = Berkheya Ehrh. (Compos.).

Zarcoa Llanos = Glochidion J. R. & G. Forst. (Euphorbiac.).

Zatarendia Rafin. = Origanum L. (Labiat.).

Zatarhendi Forsk. = Plectranthus L'Hérit. (Labiat.).

Zataria Boiss. Labiatae. 1 Persia, Afghanistan.

Zauscheria Steud. = seq.

Zauschneria Presl. Onagraceae. 4 W. U.S., NW. Mex.

Zazintha Boehm. = Zacintha L. (Compos.).

Zea L. Gramineae. 1 sp., *Z. mays* L., maize or Indian corn, originally from trop. South and Central America, now cult. in most trop. & subtrop. regions. A tall annual grass, with term. ♂ infl. (the 'tassel'), and ♀ infls. in the axils of the foliage ls. ♂ spikelets 2-flowered, in pairs, one sessile and the other pedicelled. The ♀ infl. ('ear') is enveloped by large leaf-sheaths, from the tip of which protrude the long pendent stigmas ('silks'). ♀ spikelets paired, both members of the pair sessile, and borne upon the surface of a woody 'cob', which is the axis of a single raceme with extremely short internodes. Glumes and lower floret poorly developed, and concealed beneath the large grain when mature.

Probably derived from a *Tripsacum*-like ancestor. The earliest known archaeological specimens are some 5600 years old. They resemble the smaller popcorns in size, but with the grains enveloped by long glumes as in present-day popcorns, and with a short terminal portion bearing ♂ spikelets. *Euchlaena,* for long regarded as an ancestor of maize, now appears to have originated from the crossing of *Zea* with *Tripsacum*.

A most important cereal; it is termed 'corn' in the U.S., in the same way as wheat in England and oats in Scotland. The grain is made into flour (Indian meal) or cooked without grinding; green corn (unripe cobs) forms a favourite vegetable; the ls. are useful as fodder, the dry cobs as firing; the spathes are used in paper-making, and so on.

Zebrina Schnizl. Commelinaceae. 4–5 S. U.S. to C. Am.

Zederachia Fabr. (uninom.) = *Melia* L. sp. (Meliac.).

Zederbauera H. P. Fuchs. Cruciferae. 1 SW. As.

Zedoaria Rafin. = Curcuma L. (Zingiberac.).

Zeduba Ham. ex Meissn. = Calanthe R.Br. (Orchidac.).

Zehneria Endl. Cucurbitaceae. 30 palaeotrop.

Zehntnerella Britton & Rose. Cactaceae. 1 NE. Brazil.

Zeia Lunell = Agropyron L. (Gramin.).

Zeiba Rafin. = Ceiba L. (Bombacac.).

Zelea Hort. ex Tenore = Carapa Aubl. (Meliac.).
Zeliauros Rafin. = ? Veronica L. (Scrophulariac.).
*****Zelkova** Spach. Ulmaceae. 6–7 E. Medit., Cauc., E. As. Timber valuable.
Zelmira Rafin. = Calathea G. F. W. Mey. (Marantac.).
Zelonops Rafin. = Phoenix L. (Palm.).
Zemisne Degener & Sherff. Compositae. 1 Hawaii.
Zenia Chun. Leguminosae. 1 S. China.
Zenkerella Taub. Leguminosae. 6 trop. Afr.
Zenkeria Arn. = Apuleia Mart. (Legumin.).
Zenkeria Reichb. = Parmentiera DC. (Bignoniac.).
Zenkeria Trin. Gramineae. 3 India, Ceylon.
Zenkerina Engl. = Staurogyne Wall. (Acanthac. or Scrophulariac.).
Zenkerodendron Gilg ex Jablonszky = Cleistanthus Hook. f. ex Planch. (Euphorbiac.).
Zenkerophytum Engl. ex Diels. Menispermaceae. 1 trop. W. Afr.
Zenobia D. Don. Ericaceae. 1 E. N. Am.
Zenopogon Link = Anthyllis L. (Legumin.).
Zeocriton v. Wolf = Hordeum L. (Gramin.).
Zephiranthes Rafin. = Zephyranthes Herb. (Amaryllidac.).
Zephyra D. Don. Tecophilaeaceae. 1 Chile.
Zephyranthaceae Salisb. = Amaryllidaceae–Zephyranthinae Pax.
Zephyranthella Pax. Amaryllidaceae. 1 Argentina.
*****Zephyranthes** Herb. Amaryllidaceae. 35–40 warm Am., W.I.
Zeravschania Korovin. Umbelliferae. 1 C. As.
Zerdana Boiss. Cruciferae. 1 mts. of Persia.
Zerna Panz. (~ Bromus L.). Gramineae. 30–40 N. temp.
Zerumbet Garsault = Kaempferia L. (Zingiberac.).
Zerumbet Lestib. = Zingiber Boehm. (Zingiberac.).
Zerumbet Wendl. (Catimbium Holtt., non Juss.). Zingiberaceae. 5 Malay
Zerumbeth Retz. = Curcuma L. (Zingiberac.).   [Penins.
Zetagyne Ridl. Orchidaceae. 1 Indoch.
Zetocapnia Link & Otto = Coetocapnia Link & Otto = Polianthes L. (Amarylli-
Zeugandra P. H. Davis. Campanulaceae. 1 Persia.   [dac.).
*****Zeugites** P.Br. Gramineae. 10 Mex. to Venez., W.I.
Zeuktophyllum N. E. Br. Aïzoaceae. 1 S. Afr.
Zeuxanthe Ridl. Rubiaceae. 3 Borneo.
Zeuxina Summerh. (sphalm.) = seq.
*****Zeuxine** Lindl. Orchidaceae. 76 trop. & subtrop. Old World.
Zexmenia La Llave. Compositae. 80 trop. & subtrop. Am.
Zeydora Lour. ex Gomes = Pueraria DC. (Legumin.).
Zeyhera DC. = Zeyheria Mart. (Bignoniac.).
Zeyhera Less. = Zeyheria Spreng. f. = Geigeria Griesselich (Compos.).
Zeyherella (Engl.) Aubrév. & Pellegr. = Bequaertiodendron De Wild. (Sapotac.).
Zeyheria Mart. Bignoniaceae. 2 Brazil.
Zeyheria Spreng. f. = Geigeria Griesselich (Compos.).
Zeylanidium (Tul.) Engl. Podostemaceae. 2–3 S. India, Ceylon, Assam, Burma.

ZEZYPHOÏDES

Zezyphoïdes Parkinson (sphalm.). = Zizyphoïdes Soland. ex Drake = Alphitonia Reissek ex Endl. (Rhamnac).

Zhumeria K. H. Rech. & Wendelbo. Labiatae. 1 Persia.

Zichia Steud. = seq.

Zichya Hueg. = Kennedya Vent. (Legumin.).

Ziegera Rafin. = Miconia Ruiz & Pav. (Melastomatac.).

Zieria Sm. Rutaceae. 15 E. Austr.

Zieridium Baill. Rutaceae. 3 New Caled.

Ziervoglia Neck. = Cynanchum L. (Asclepiadac.).

Zietenia Gled. = Stachys L. (Labiat.).

Zigadenus Michx. Liliaceae. 15 C. & E. Siberia, E. As., N. & C. Am.

Zigara Rafin. = Bupleurum L. (Umbellif.).

Zigmaloba Rafin. = Acacia Mill. (Legumin.).

Zilla Forsk. Cruciferae. 3 N. Afr. to Arabia.

Zimapania Engl. & Pax = Jatropha L. (Euphorbiac.).

Zimmermannia Pax. Euphorbiaceae. 4 trop. E. Afr.

Zingania A. Chev. = Didelotia Baill. (Legumin.).

Zingeria P. Smirn. Gramineae. 2 SE. Russia to As. Min., Syria & Iraq.

*Zingiber Boehm. Zingiberaceae. 80–90 E. As., Indomal., N. Austr. Labellum large; opp. to it are the style and the petaloid fertile sta. The stigma has many rays. *Z. officinale* Rosc. is the ginger; it is always reprod. by veg. methods, and is quite sterile (cf. *Musa*). Largely cult.; the rhiz. are dug up and killed in boiling water. According to whether the rind is or is not scraped off, the product is known as 'scraped' or 'coated' ginger.

*Zingiberaceae Lindl. (*s.str.*). Monocots. 45/700 trop., chiefly Indomal. Perenn. aromatic herbs, usu. with sympodial fleshy rhiz., often with tuberous roots. Aerial stem, if any, short; sometimes an apparent stem is formed as in *Musa* by the rolled-up leaf-sheaths. Ls. 2-ranked, with short stalks and sheathing bases. At the top of the sheath is a char. ligule (cf. *Gramineae*). Fls. in racemes, heads, or cymes. Their morphology has been much discussed. Bracteole often sheathing. K (3), the odd one ant. C 3, usu. different in colour and texture from the outer P-leaves. Of the possible 6 members of the A (two whorls), the post. one of the inner whorl is present as a fertile epipet. sta., and the other two of this whorl are united to form the petaloid labellum (not equivalent to that of Orchids), which may be 2- or 3-lobed; the ant. sta. of the outer whorl is always absent; the other two may be absent (as in *Renealmia*) or may be present as large leafy stds. right and left of the fertile sta. (cf. with *Cannaceae* and *Marantaceae*). Anther-loc. without apical appendage. G̅ (3), 3-1-loc., with ∞ 2-4-seriate anatr. or semi-anatr. ov. Fr. usu. a loculic. caps. Seeds with perisperm. The fam. contains several economic plants; see *Curcuma*, *Alpinia*, *Zingiber*, *Amomum*, *Elettaria*. Chief genera: *Hedychium*, *Kaempferia*, *Curcuma*, *Globba*, *Zingiber*, *Amomum*, *Renealmia*, *Alpinia*. (See also *Costaceae*.)

*Zinnia L. Compositae. 20 S. U.S. to Brazil & Chile. Ls. opp. or whorled. Fr. winged.

Zinowiewia Turcz. Celastraceae. 9 Mex. to Venez.

Zinziber Mill. = Zingiber Boehm. (Zingiberac.).

Zipania Pers. (sphalm.) = Zapania Scop. = Lippia L. (Verbenac.).

Zippelia Blume (~ Piper L.). Piperaceae. 1 Java.

Zippelia Reichb. = Rhizanthes Dum. (Rafflesiac.).
Zizania Gronov. ex L. Gramineae. 1 NE. India, Burma, E. As.; 2 N. Am.
Z. aquatica L., 'wild rice', formerly staple food of N. Am. Indians and still use by Chippewa tribe; also important as game-bird attractant.
Zizaniopsis Doell & Aschers. Gramineae. 2 SE. U.S., 2 trop. S. Am.
Zizia Koch. Umbelliferae. 3 N. Am.
Zizia Pfeiff. = Zizzia Roth = Draba L. + Alyssum L. (Crucif.).
Zizifora Adans. = seq.
Ziziforum Caruel = seq.
Ziziphora L. Labiatae. 25 Medit. to C. As. & Afghan.
Ziziphus Mill. Rhamnaceae. 100 trop. Am., Afr., Medit., Indomal., Austr. Stips. often repres. by thorns; one is sometimes recurved whilst the other is straight (cf. *Paliurus*); occasionally only one is developed. *Z. chloroxylon* (L.) Oliv. (cogwood; Jamaica) hard tough wood. Fr. of many ed.; those of *Z. lotus* (L.) Willd. (Medit.) are said to be [the lotus fruits of antiquity; those of *Z. jujuba* Mill. (*Z. vulgaris* Lam.) (E. Medit.) are known as French jujubes; those of *Z. joazeiro* Mart. are used in Brazil as fodder. *Z. spina-christi* (L.) Willd. is said to have furnished the crown of thorns (cf. *Paliurus*).
Zizyphoïdes Soland. ex Drake = Alphitonia Reissek ex Endl. (Rhamnac.).
Zizyphon St-Lag. = Ziziphus Mill. (Rhamnac.).
Zizyphora Dum. = Ziziphora L. (Labiat.).
Zizyphus Adans. = Ziziphus Mill. (Rhamnac.).
Zizzia Roth = Draba L. + Alyssum L. (Crucif.).
Zoduba Buch.-Ham. ex D. Don = Calanthe R.Br. (Orchidac.).
Zoëgea L. Compositae. 6 SW. & C. As.
Zoelleria Warb. = Trigonotis Stev. (Boraginac.).
Zoisia Aschers. & Graebn. = Zoysia Willd. (Gramin.).
Zollernia Maximil. & Nees. Leguminosae. 10 C. & trop. S. Am.
Zollikoferia DC. = Launaea Cass. (Compos.)
Zollikoferia Nees = Chondrilla L. (Compos.).
***Zollingeria** Kurz. Sapindaceae. 2 SE. As.
Zollingeria Sch. Bip. = Rhynchospermum Reinw. (Compos.).
Zomacarpus P. & K. = Zomicarpa Schott (Arac.).
Zombia L. H. Bailey. Palmae. 1 W.I. (Haiti).
Zombiana Baill. = Rotula Lour. (Ehretiac.).
Zombitsia Keraudren. Cucurbitaceae. 1 Madag.
Zomicarpa Schott. Araceae. 3 S. Brazil.
Zomicarpella N. E. Br. Araceae. 1 Colombia.
Zonablephis Rafin. = Acanthus L. (Acanthac.).
Zonanthemis Greene = Hemizonia DC. (Compos.).
Zonanthus Griseb. Gentianaceae. 1 Cuba.
Zonaria Steud. (sphalm.) = Zornia J. F. Gmel. (Legumin.).
Zonotriche (C. E. Hubbard) Phipps. Gramineae. 1 trop. Afr.
Zoöphora Bernh. = Orchis L. (Orchidac.).
Zoöphthalmum P.Br. = Mucuna Adans. (Legumin.).
Zornia J. F. Gmel. Leguminosae. 75 trop., esp. Am.
Zornia Moench = Lallemantia Fisch. (Legumin.).
Zoroxus Rafin. = Polygala L. (Polygalac.).

ZOSIMA

Zosima Hoffm. Umbelliferae. 10 W. As.

Zosima Phil. = Philibertia Kunth (Asclepiadac.).

Zosimia M. Bieb. = Zosima Hoffm. (Umbellif.).

Zoster St-Lag. = Zostera L. (Zosterac.).

Zostera Cavol. = Posidonia Koenig (Posidoniac.).

Zostera L. Zosteraceae. 12 subtrop., temp., subarct., subantarct., in shallow salt water on gently sloping shores. The pl. is largely used for packing glass, stuffing cushions, etc. esp. in Venice.

*Zosteraceae Dum. 3/18 extratrop. coasts. The lower part of the stem creeps, rooting as it advances, and has monopodial branching; the branches grow upwards and exhibit sympodial branching, complicated by union of axillary shoot to main shoot for some distance above its point of origin. This is most easily seen in the infl. region; the branching is that of a rhipidium, but shoot 2, which springs from the axil of a l. on shoot 1, is adnate to 1 up to the point at which the first l. is borne on 2; this l. occupies the angle between the two shoots where they separate. Shoot 1 (and 2, 3, etc., successively) is pushed aside and bears an infl. Ls. long, linear, sheathing at base.

Infl. a flattened spadix, enclosed at flowering time in a spathe (the sheath of the uppermost l.). This is open down one side, and on the corresponding side of the spadix the fls. are borne, the essential organs forming two vertical rows, each composed of a cpl. and a sta. alt. On the outer side of the spadix next the sta. is often a small l. (*retinaculum* of systematic works). The midrib of the cpl. faces outwards. Each cpl. contains one ovule and has two flat stigmas. The sta. consists of two half anthers, joined by a small connective. It is difficult to decide what is the actual 'flower' in these pls.; the usual view is that each sta. with the cpl. on the same level forms a fl., the retinaculum representing the bract.

Fert. peculiar, *Z.* being among the most completely modified of all water pls. from the ancestral land-pl. type. Fl. submerged like the rest of the pl. The pollen grains are long threads of the same specific gravity as salt water, so that when discharged they float freely at any depth. The stigmas are very large, and thus have a good chance of catching some of the grains. The whole mech. is similar in principle to that of a wind-fert. pl. Fr. an achene. In winter it hibernates without any special modification. Genera: *Phyllospadix, Zostera, Heterozostera.*

Zosterella Small = Heteranthera Ruiz & Pav. (Pontederiac.).

Zosterospermum Beauv. = Rhynchospora Vahl (Cyperac.).

Zosterostylis Blume = Cryptostylis R.Br. (Orchidac.).

Zouchia Rafin. = Pancratium L. (Amaryllidac.).

Zoutpansbergia Hutch. Compositae. 1 S. Afr.

Zoydia Pers. = seq.

*Zoysia Willd. Gramineae. 10 Mascarene Is. to N.Z.

Zozima DC. = Zosima Hoffm. (Umbellif.).

Zozimia DC. = praec.

Zschokkea Muell. Arg. Apocynaceae. 15 trop. S. Am.

Zschokkia Benth. & Hook. f. = praec.

Zubiaea Gandog. = Daucus L. (Umbellif.).

Zucca Comm. ex Juss. = Momordica L. (Cucurbitac.).

*Zuccagnia Cav. Leguminosae. 1 Chile.
Zuccagnia Thunb. = Dipcadi Medik. (Liliac.).
*Zuccarinia Blume. Rubiaceae. 2 Sumatra, Java.
Zuccarinia Maerklin = ? Rehmannia Libosch. ex Fisch. & Mey. (Scrophulariac.).
Zuccarinia Spreng. = Jackia Wall. (Rubiac.).
Zucchellia Decne = Raphionacme Harv. (Asclepiadac.).
Zuckertia Baill. = Tragia L. (Euphorbiac.).
Zuckia Standley. Chenopodiaceae. 1 SW. U.S.
Zuelania A. Rich. Flacourtiaceae. 5 C. Am., W.I.
Zugilus Rafin. = Ostrya Scop. (Carpinac.).
Zulatia Neck. = Miconia Ruiz & Pav. (Melastomatac.).
Zurloa Tenore = Carapa Aubl. (Meliac.).
Zwaardekronia Korth. = Psychotria L. (Rubiac.).
Zwackhia Sendtner (1858; non Körber 1855—Lichenes) = Halacsya Dörfl. (Boraginac.).
Zwingera Hofer = Nolana L. ex L. f. (Nolanac.).
Zwingera Schreb. = Quassia L. (Simaroubac.).
Zwingeria Fabr. (uninom.) = Ziziphora L. sp. (Labiat.).
Zycona Kuntze = Allendea La Llave (Compos.).
Zygadenus Endl. = Zigadenus Michx (Liliac.).
Zygalchemilla Rydb. = Alchemilla L. (Rosac.).
Zyganthera N. E. Br. = Pseudohydrosme Engl. (Arac.).
Zygella S. Moore = Cypella Herb. (Iridac.).
Zygia Benth. & Hook. f. (sphalm.) = Zygis Desv. = Micromeria Benth. (Labiat.).
Zygia P.Br. Leguminosae. 20 Malaysia, Austr., trop. Am.
Zygia Kosterm. = Pithecellobium Mart. (Legumin.).
Zygia Walp. = Albizia Durazz. (Legumin.).
Zygilus P. & K. = Zugilus Rafin. = Ostrya L. (Carpinac.).
Zygis Desv. = Micromeria Benth. (Labiat.).
× Zygobatemannia (sphalm. '-mania') hort. Orchidaceae. Gen. hybr. (iii) (Batemannia × Zygopetalum).
Zygocactus K. Schum. = Schlumbergera Lem. (Cactac.).
× Zygocaste hort. Orchidaceae. Gen. hybr. (iii) (Lycaste × Zygopetalum).
× Zygocella hort. Orchidaceae. Gen. hybr. (iii) (Mendoncella × Zygopetalum)
Zygocereus Frič & Kreuz. = Zygocactus K. Schum. = Schlumbergera Lem. (Cactac.).
Zygochloa S. T. Blake. Gramineae. 1 C. Austr.
× Zygocidium hort. Orchidaceae. Gen. hybr. (vi) (Oncidium × Zygopetalum).
× Zygocolax hort. Orchidaceae. Gen. hybr. (iii) (Colax × Zygopetalum).
× Zygodendrum hort. Orchidaceae. Gen. hybr. (vi) (Epidendrum × Zygopetalum).
Zygodia Benth. Apocynaceae. 5 trop. Afr.
Zygoglossum Reinw. ex Blume = Cirrhopetalum Lindl. (Orchidac.).
Zygogonum Hutch. (sphalm.) = seq.
Zygogynum Baill. Winteraceae. 5 New Caled.
× Zygolax hort. (vii) = × Zygocolax hort. (Orchidac.).

ZYGOLEPIS

Zygolepis Turcz. = Arytera Blume (Sapindac.).

× **Zygomena** hort. Orchidaceae. Gen. hybr. (iii) (Zygopetalum × Zygosepalum [Menadenium]).

Zygomenes Salisb. = Amischophacelus Rao & Kammathy (Commelinac.).

Zygomeris Moç. & Sessé ex DC. = Amicia Kunth (Legumin.).

Zygonerion Baill. Apocynaceae. 1 Angola.

× **Zygonisia** hort. Orchidaceae. Gen. hybr. (vi) (Aganisia × Zygopetalum).

Zygoön Hiern. Rubiaceae. 1 S. trop. Afr.

Zygopeltis Fenzl ex Endl. = Heldreichia Boiss. (Crucif.).

Zygopetalum Hook. Orchidaceae. 20 C. & trop. S. Am., Trinidad.

***Zygophyllaceae** R.Br. Dicots. 25/240 trop. & subtrop. Xero- or halophytes; most are woody perennials. Ls. opp. (rarely alt.), 2–3–∞-foliol., usu. ± fleshy, leathery or hairy, stip. (stips. sometimes spinesc.). Fls. reg., ⚥, in cymose infl. K 5–4, imbr., rarely valv.; C 5–4, rarely 0, imbr. or contorted, rarely valv.; disk usu. pres.; A 5–10–15, obdiplost., fil. often with ligular appendages, anth. dorsif., latrorse; G (5), rarely (2–12), often angled or winged, with 1–∞ axile pend. ov. per loc., and gradually atten. into a simple style with lobed or capit. stig. Fr. usu. a loculic. and/or septic. caps., or of indehisc. cocci, rarely baccate (*Peganum*) or drupac. (*Nitraria*); seeds with or without endosp.

Classification and chief genera (after Engler):

A. Fruit a caps., or of separate cocci, rarely baccate.
    I. **Peganoïdeae** (ls. alt., multifid; fr. a glob. loculic. caps., loc. ∞-seeded, rarely baccate): *Peganum*.
    II. **Morkillioïdeae** (*Chitonioïdeae*) (ls. alt., simple or imparipinn.; fr. a caps., or of cocci): *Morkillia, Sericodes*.
    III. **Tetradiclidoïdeae** (lower ls. opp., upper ls. alt.; stam. 3–4; ov. 3–4-lobed, each loc. incompletely 3-locell., with 4 ovules in the central and 1 ov. in the lateral locelli): *Tetradiclis*.
    IV. **Augeoïdeae** (ls. opp.; ovary of 10 carpels; ls. clavate, exstip.): *Augea*.
    V. **Zygophylloïdeae** (ls. opp., simple or ternate or paripinn., stip.; ovary of 5 or fewer carpels, rarely of 10–12 [*Kallstroemia*]): *Fagonia, Guaiacum, Zygophyllum, Tribulus, Kallstroemia.*
B. Fr. drupaceous.
    VI. **Nitrarioïdeae** (ls. alt., simple; ovary 3-loc., not lobed, with 1 pend. ov. per loc.; drupe 1-seeded): *Nitraria*.
    [VII. **Balanitoïdeae**: see *Balanitaceae*.]

The *Z.* are closely related to *Rutac.* and perhaps other fams. Pollen very diverse, suggesting need for review of classification.

Zygophyllidium (Boiss.) Small = Euphorbia L. (Euphorbiac.).

Zygophyllon St-Lag. = seq.

Zygophyllum L. Zygophyllaceae. 100 Medit. to C. As., S. Afr., Austr., in deserts and steppes. Ls. and twigs fleshy.

× **Zygorhyncha** hort. Orchidaceae. Gen. hybr. (iii) (Chondrorhyncha × Zygopetalum).

Zygoruellia Baill. Acanthaceae. 1 Madag.

Zygosepalum Reichb. f. Orchidaceae. 5 trop. S. Am.

Zygosicyos Humbert. Cucurbitaceae. 2 Madag.

Zygospermum Thw. ex Baill. = Prosorus Dalz. = Margaritaria L. f. (Euphorbiac.).

Zygostates Lindl. Orchidaceae. 4 Brazil.

Zygostelma Benth. Periplocaceae. 1 Siam.

Zygostelma Fourn. = Lagoa Dur. (Asclepiadac.).

Zygostemma Van Tiegh. = Scabiosa L. (Dipsacac.).

Zygostigma Griseb. Gentianaceae. 2 Brazil, Argentina.

× **Zygostylis** hort. Orchidaceae. Gen. hybr. (iii) (Otostylis × Zygopetalum).

Zygotritonia Mildbraed. Iridaceae. 6–7 trop. Afr.

Zymum Nor. ex Thou. = Tristellateia Thou. (Malpighiac.).

Zyrphelis Cass. = Mairia Nees (Compos.).

Zyzophyllum Salisb. = Zygophyllum L. (Zygophyllac.).

Zyzygium Brongn. = Syzygium Gaertn. (Myrtac.).

KEY
TO THE FAMILIES OF
FLOWERING PLANTS

BASED ON ENGLER'S CLASSIFICATION AS GIVEN IN
DIE NATÜRLICHEN PFLANZENFAMILIEN
AND REVISED IN HIS *SYLLABUS*, ED. 7

MONOCOTYLEDONEAE

[Embryo with one cot.; stem with closed bundles; ls. usu. ‖-veined; fls. usu. 3-merous.]

A. Orders with predominant variability in number of floral parts (Orders 1–7):
  a. *Typically achlamydeous fls. appear (Orders 1–4).*
    α. Fls. usu. naked. Great variability in number of sta. and cpls.
  1. PANDANALES. Marsh herbs, or trees, with linear ls., and cpd. heads or spikes of naked, haplo- or homo-chlamydeous ♂ ♀ fls. P bract-like, A ∞–1, G ∞–1. Endosp.
    β. Naked fls. occur, but also all stages from achlam. to heterochlam. fls., and from hypog. to epig. Number of essential organs definite or not (Orders 2, 3).
  2. HELOBIAE. Water or marsh pls. with scales in axils, and cyclic or hemicyclic fls. P in 0, 1, or 2 whorls, homo- or hetero-chlam., hypog. or epig. A ∞–1, G ∞–1, free or united. Endosp. little or none.
    1. *Potamogetonineae:* fls. hypog., achlam., haplo-, or homo-chlam. (fams. 1–4).

  2. *Alismatineae:* fls. hypog., usu. heterochlam.; ov. on ventral suture (fam. 5).

  3. *Butomineae:* fls. hypog. or epig., usu. heterochlam.; ov. on inner surface of cpls. (fams. 6, 7).

  3. TRIURIDALES. Saprophytes with scale ls. and small long-stalked homochlam. ♀ or ♂ ♀ fls. P 3–8, valv., petaloid; ♂ A 3, 4, or 6; ♀, 2 stds., G̲ ∞, each with 1 basal ov.; ∞ styles. Pericarp thick. Endosp.

MONOCOTYLEDONEAE

1

1. **Typhaceae:** rhiz. herbs with linear 2-ranked ls. and cylindrical spikes of naked fls., ♀ below, ♂ above; A 2–5, G̲ 1 on hairy axis with 1 pend. ov.; nutlet, with album. seed.

2. **Pandanaceae:** woody pls., sometimes climbing, with 3-ranked ls. and term. or racemed spikes of ♂ ♀ fls., ♂ of ∞ sta. racemed or umbelled on short or long axis, ♀ of (∞–1) cpls. with sessile stigs. and ∞–1 ov.; heads of berries or drupes; endosp. oily.

3. **Sparganiaceae:** rhiz. herbs with 2-ranked ls. and fls. in ♂ ♀ heads, ♀ heads lower. P 3–6, sepaloid, A 3–6, G̲ (1–2), each with 1 pend. ov.; fr. drupaceous; endosp. floury.

2

1. **Potamogetonaceae:** submerged or floating herbs of fresh or salt water, with usu. 2-ranked ls. and sol. or spiked ♀ or ♀ ♂ reg. fls. P usually 0, A 4–1, G̲ 4–1 each with 1 pend. ov.; fr. 1-seeded.

2. **Najadaceae:** submerged herbs with opp. linear toothed ls. and ♂ ♀ fls., ♂ P 2, A 1 term.; ♀ P 1 or 0, G̲ 1, with 1 basal anatr. ov.

3. **Aponogetonaceae:** tuber-rhiz. water herbs with submerged or floating ls. and spikes (in caducous spathes) of ♀ reg. fls. P 3–1 petaloid, A 3+3 or more, G̲ 3–6; fr. leathery, seeds 2 or ∞, exalb.

4. **Scheuchzeriaceae:** marsh herbs with narrow ls. and racemes or spikes of ♀ or ♂ ♀ reg. fls.; P usu. 3+3, homochlam., bract-like, A 3+3, G̲ 3+3 sometimes united, outer often absent, 1 or 2 anatr. ov. in each. [*Lilaea* ♀ ♂ ♀, A 1, G 1.]

5. **Alismataceae:** water or marsh herbs with rad. ls., latex, and much branched infl. of reg. heterochlam. ♀ or ♂ ♀ fls.; K 3, C 3, A 6–∞ or 3, G̲ 6–∞ with 1–∞ anatr. ov. and 6–∞ styles; no endosp.

6. **Butomaceae:** water and marsh herbs; latex; usu. ± umbel-like cymose infl. of reg. usu. heterochlam. ♀ fls.; K 3, C 3, A 9–∞; G̲ 6–∞, often united below, with ∞ ov. on inner surface; follicles.

7. **Hydrocharitaceae:** salt or fresh water pls. with alt. or whorled ls. and sol. or cymose-paniculate fls. enclosed in 1 or 2 bracts, usu. heterochlam., reg., 3-merous, usu. ♂ ♀; A in 1–5 whorls, inner often stds., G̅ (2–15), 1-loc. with parietal plac. and ∞ ov.; fr. irreg. dehisc. with ∞ seeds. Exalb.

3

1. **Triuridaceae.**

γ. **Fls. usu. naked. Number of sta. rarely indefinite.**
4. GLUMIFLORAE. Usu. herbs, with naked fls. (rarely with trichome-like or true P) covered by bracts (glumes). G̲ 1-loc. with 1 ov.

b. *Fls. rarely naked, and then usu. by reduction, and accompanied by spathes of bracts; A and G commonly definite, but also frequently ∞ sta. and > 3 cpls.*
5. PRINCIPES. Tree-like or woody pls., sometimes climbing, with fan or feather ls., and reg. usu. ♂ ♀ fls. in spikes (usu. compound) or spike-like racemes, usu. in spathe; P 3 + 3, A 3 + 3, or 3, 9, or ∞, G 3 or (3), usu. with 1 ov. each; berry or drupe; endosp. rich.
6. SYNANTHAE. Often palm-like pls., climbers, or large herbs with ♂ ♀ fls. alternating over surface of spike, ♂ naked or with thick short P and 6–∞ sta.; ♀ naked or with 4 fleshy scaly P and long thread-like std. in front of each, G (2) or (4) with 2 or 4 plac. and ∞ ov.; the 1-loc. ovaries sunk in spike and united; multiple fr. with ∞ seeds; endosp.
7. SPATHIFLORAE. Herbs, or woody, sometimes climbing, rarely forming erect stem, usu. sympodial; fls. cyclic, haplo- or homo-chlam. or naked by abortion, 3–2-merous, ⚥ or ♂ ♀ often reduced to 1 sta. or cpl., in simple spikes (spadix), ± enclosed in bract (spathe).
B. Fls. typically 5-cyclic, whorls typically iso-, usu. 3-merous, rarely more or 2-merous (Orders 8–11).
    a. *Fls. homo- to hetero-chlam., rarely naked; P still often bract-like; hypogyny and actinomorphy the rule (Orders 8, 9).*
8. FARINOSAE. Usu. herbs, rarely with stout stem; fls. cyclic, homo- or hetero-chlam., 3–2-merous, usu. P 3 + 3, A 3 + 3, G (3), one whorl of A sometimes wanting, or all reduced to 1; ov. usu. orthotr.; endosp. mealy.
    1. *Flagellariineae:* P homochlam., bracteoid, hypog.; ov. anatr. (fam. 1).
    2. *Enantioblastae:* P various, hypog.; ov. orthotr. (fams. 2–6).

4

1. *Gramineae:* herbs, rarely woody, with jointed stem and alt. 2-ranked ls. with split sheath and ligule, and panicle or spike-like infls. of small ♀ rarely ♂ ♀ naked fls. in spikelets, each beginning with 1 or more glumes, then paleae with axillary fls.; A usu. 3, G̲ with 1 ov., micropyle facing down; stigs. 2, 3, or 1; caryopsis with rich endosp.

2. *Cyperaceae:* herbs with usu. 3-angled stem and 3-ranked ls. with closed sheath; fls. in spikelets or cymes united to large infls., naked, ♀ or ♂ ♀, A usu. 3–1, G̲ (3–2), styles 3–2, 1-loc. with 1 basal anatr. ov.; nut; endosp.

5

1. *Palmae.*

6

1. *Cyclanthaceae.*

7

1. *Araceae:* tuberous herbs, sometimes woody, or lianes, with ♀ or ♂ ♀ fls. in same spike, often with spathe; fl. 2–3-merous or reduced to 1 sta. or cpl.; fr. usu. berry; outer seed-coat fleshy.

2. *Lemnaceae:* free-swimming water pls. usu. with no ls. and naked ♂ ♀ fls., ♂ of 1 sta., ♀ of 1 cpl. with 1–6 basal erect ov.; endosp. thin.

8

1. *Flagellariaceae:* pls. sometimes climbing, with long many-veined ls. and small, ♀ or ♂ ♀, 3-merous, reg. fls. in cpd. term. panicles; P bract-like, G̲ (3) 3-loc. each with 1 ov.; fr. 3-loc. or with 3–1 stones; endosp.

2. *Restionaceae:* rush-like xero. or marsh herbs with creeping rhiz. and 2-ranked bracts or scale ls. on stem; fls. in spikes in axils of bracts, usu. ♂ ♀ reg.; P 3–2 + 3–2 sepaloid, A 3–2, G̲ (3–1) with 3–1 styles, 3–1-loc. with 1 ov. in each; caps. or nut; endosp.

3. *Centrolepidaceae:* usu. marsh pls. with ♀ or ♂ ♀ fls., naked or with 1–3 hair-like brs.; A 1–2, G̲ (1–∞) each with 1 pend. ov.

4. *Mayacaceae:* marsh pls. with alt. linear ls. and sol. or umbelled ♀ reg. heterochlam. 3-merous fls.; K 3, C 3, A 3, G̲ (3), style 1 with 3 stigs., 1-loc. with parietal plac. and few ov.; caps. 3-valved.

5. *Xyridaceae:* perenn. herbs with long narrow ls. and axill. spikes of ♀ heterochlam. 3-merous fls.; K ·|· with 2 smaller lat. ls., C (3) with tube, A 3 epipet., with sometimes 3 outer stds., G̲ (3), 1-loc. with ∞ ov.; caps. 3-valved; endosp.

3. *Bromeliineae:* P usu. heterochlam., hypog. to epig.; ov. anatr. (fams. 7–9).

4. *Commelinineae:* P heterochlam.; part of A often stds. or wanting (fam. 10).

5. *Pontederiineae:* P homochlam., petaloid, united (fams. 11, 12).

6. *Philydrineae:* P petaloid, the outer ls. larger than inner, the 2 post. of outer whorl united, the post. of inner whorl aborted (fam. 13).

9. LILIIFLORAE. As last, but endosp. fleshy or oily; ov. usu. anatr.; fls. usu. 3-merous, rarely 2, 4, or more.

1. *Juncineae:* P homochlam., bracteoid; endosp. mealy with starch (fam. 1).

2. *Liliineae:* P homochlam., rarely bracteoid, usu. petaloid, rarely heterochlam.; endosp. without starch; inner whorl of A present (fams. 2–8).

vi

6. *Eriocaulaceae:* perenn. herbs with long linear ls. and involucrate heads of fls. on long stalks, ♂ ♀, reg. or ·l·, heterochlam., 2–3-merous, sta. usu. in 1 whorl, G̲ (2–3), 2–3-loc. with 1 pend. ov. in each; caps.; endosp.

7. *Thurniaceae:* perenn. herbs with narrow ls. and bracteate heads of ♀ reg. homochlam. 3-merous fls. on scapes; P 3 + 3, A 6, G̲ (3), 3-loc. with 1–∞ ov. in each; caps.; endosp.

8. *Rapateaceae:* perenn. herbs with 2-ranked narrow ls.; infl. term. with 2 large spathes encl. head of spikelets, each of ∞ brs. and term. ♀ reg. 3-merous heterochlam. fl.; K (3), C (3), A 3 + 3, G̲ (3), 3-loc. with ∞–1 ov. in each; caps.; endosp.

9. *Bromeliaceae:* herbs, often epiph., with alt. usu. rad. ls. and spikes or panicles of usu. ♀ reg. heterochlam. 3-merous fls.; K 3, C or (C) 3, A 3 + 3, G (3), sup. to inf., 3-loc. with ∞ ov.; berry or caps.; endosp.

10. *Commelinaceae:* herbs with jointed stems, alt. sheathing ls. and cymes of blue or violet ♀ reg. or ·l· heterochlam. 3-merous fls.; K 3, C 3 rarely united, A 3 + 3, G̲ (3–2), style 1, 3–2-loc. with few ov.; caps.; endosp.

11. *Pontederiaceae:* water pls. often with 2-ranked ls. and spicate ♀ ·l· fls.; P 3 + 3 with long tube, A 3 + 3, 3, or 1, on tube, G̲ (3) with 1 style, 3-loc. with ∞ ov. or 1-loc. with 1; caps. or nut; endosp.

12. *Cyanastraceae:* herbs with tuber or rhiz. and raceme or panicle of ♀ reg. 3-merous fls.; (P) with short tube, A (6), G̅ (3), with 1 style, 3-loc. with 2 ov. in each; caps. 1-seeded; perisp.

13. *Philydraceae:* herbs with 2-ranked narrow ls. and spikes of homochlam. 3-merous ♀ ·l· fls.; sta. 1 ant., G̲ (3) with 1 style, 3- or 1-loc. with ∞ ov.; caps.; endosp.

9

1. *Juncaceae:* perenn. herbs with narrow usu. rad. ls. and many-fld. infl. of homochlam. 3-merous ♀ reg. fls.; P sepaloid, A 6 or 3, G̲ (3), style 1 with 3 stigs., 1–3-loc. each with 1–8 ov.; caps.; endosp.

2. *Stemonaceae:* perenn. herbs with rhiz. and often climbing stem and axillary infls. of homochlam. ♀ reg. 2-merous fls.; P sepaloid, G̲ (2), 1-loc.; caps.; endosp.

3. *Liliaceae:* herbs with rhiz. or bulbs, shrubs, or trees with infl. of usu. racemose type, of usu. homochlam. ♀ reg. usu. 3-merous fls.; P or (P) 3 + 3, petaloid, A 3 + 3, G̲ to G̅ (2–)3(–5)-loc.; fr. various; endosp. fleshy or cartilaginous.

4. *Haemodoraceae:* perenn. herbs with 2-ranked ls. and simple or cpd. infl. of ♀ reg. or ·l· fls.; P 3 + 3, A 3, G̲ to G̅, 3-loc. with few ov.; caps.

5. *Amaryllidaceae:* herbs or shrubs of various habit and cymose infl. on scape, of ♀ reg. or ·l· fls.; P 3 + 3 petaloid, A 3 + 3 usu. intr., often with stipular corona, G̅ (3), rarely ½-inf., 3-loc. with ∞ ov.; caps. or berry.

6. *Velloziaceae:* herbs or shrubs with linear crowded ls. and term. sol. ♀ reg. 3-merous fls. on long stalks; P petaloid, A 6 or 6 bundles, G̅ (3), 3-loc. with ∞ ov. on lamellar plac.; caps.; endosp.

3. *Iridineae:* as last, but inner sta. aborted (fam. 9).

b. *Fls. homo-(petaloid) or hetero-chlam., epig., usu.* ·|· (*Orders* 10, 11).
10. SCITAMINEAE. Trop. herbs, sometimes very large or woody, with cyclic, homo- or hetero-chlam. usu. ·|· 3-merous fls.; A typically 3 + 3, but often with great reduction, even to 1 sta., $\overline{G}$ usu. 3-loc. with large ov.; usu. aril, peri- and endosp.

11. MICROSPERMAE. Fls. cyclic, homo- or hetero-chlam., 3-merous, typically diplostemonous, but commonly with great reduction in A, $\overline{G}$ 3- or 1-loc. with ∞ small ov.; endosp. or o.
   1. *Burmanniineae:* fls. usu. reg.; endosp.
   2. *Gynandrae:* fls. always ·|· ; no endosp.

7. **Taccaceae:** perenn. herbs with tubers and large entire or cymosely branched ls., and cymose umbels of ☿ reg. fls. with long thread-like brs.; P 3 + 3, petaloid, A 3 + 3, Ḡ (3), 1-loc. with parietal plac., 6 petaloid stigs. and ∞ ov.; caps. or berry; endosp.

8. **Dioscoreaceae:** climbing herbs with usu. tuberous rhiz. and alt. or opp. often sagittate ls.; fls. in racemes, homochlam., ☿ or ♂ ♀, reg.; P sepaloid, usu. united, with tube, A 3 + 3, inner sometimes stds., Ḡ (3), 3- or 1-loc.; usu. with 2 ov. to each, styles 3 or 6; caps. or berry; endosp.

9. **Iridaceae:** perenn. herbs or undershrubs with equitant ls. and term. cymose infl. of ☿ reg. or ·|· fls.; P 3 + 3 homo- or hetero-chlam., A 3 extr., Ḡ (3), 3-loc. with 3 styles sometimes divided and leafy, rarely 1-loc., ov. ∞; caps.; endosp.

10

1. **Musaceae:** very large herbs with 'false' stem, or trees, with cpd. infl. with large often petaloid brs. and ☿ or ♂ ♀, ·|· homo- or hetero-chlam. fls.; P 3 + 3, petaloid, often united, A 3 + 2 and std., Ḡ (3), 3-loc. with 1–∞ ov. in each; berry or caps.; endosp. and perisp.

2. **Zingiberaceae:** perenn. herbs with tuberous rhiz. and lanc. petiolate ls., with ligule, and simple or cpd. infls. of usu. ☿ ·|· fls.; K (3), C (3) forming tube below, A 1 (of inner whorl, with labellum opp. to it of 2 inner stds., and sometimes 2 outer stds.), Ḡ (3), usu. 3-loc. with ∞ ov.; caps.; usu. aril; endosp. and perisp.

3. **Cannaceae:** perenn. herbs with large ls. and cpd. infl. of showy heterochlam. ☿ asymmetric fls.; K 3, C (3), A 1–5, only half of 1 inner sta. fertile, the other half, and rest, petaloid stds., Ḡ (3), 3-loc. with ∞ ov.; caps.; endosp. and perisp.

4. **Marantaceae:** perenn. herbs with 2-ranked ls. with pulvinus at end of stalk, and heterochlam. ☿ asymmetric fls.; P 3 + 3, A 4–5, only 1 inner half fertile, as in last, the 2 other inner and 1–2 outer petaloid (1 inner usu. hood-like), Ḡ (3), 3-loc. or 1-loc. by suppression, with 1 ov. in each; caps.; aril; endosp. and perisp.

11

1. **Burmanniaceae:** green or saproph. herbs with sol. or cymose fls.; P (3 + 3) or 3 + 3, A 3 + 3 or 3, Ḡ (3), 3- or 1-loc.; caps., ∞ seeds; endosp.

2. **Orchidaceae:** perenn. herbs of various form, often epiph. with pseudobulbs, and ☿, ·|·, usu. resupinated, homo- or hetero-chlam. fls.; P 3 + 3, A 1 or 2, united with style to form a column; pollen in tetrads usu. united to pollinia; Ḡ (3), 1-loc., stigmas 3, the third usu. rudimentary or forming a rostellum, ovules ∞; caps.; no endosp.

DICOTS. VERTICILLATAE—BALANOP[SID]ALES
A; B; a
DICOTYLEDONEAE

[Embryo with two cots.; stem with open bundles; ls. usu. net-veined; fls. usu. not 3-merous.]

Archichlamydeae (Orders 1–30)

(Fls. achlam., haplochlam., or diplochlam., usu. polypet., rarely sympet. or apet.)

A. Ov. with 20 or more embryo sacs, and chalazogamic fert. (Order 1).

1. VERTICILLATAE. Woody pls. of Equisetum habit; ♂ fls. in catkin-like spikes, ♀ in heads, at end of twigs; ♂ with 2 median bract-like P and a central sta., ♀ naked, G (2) with 2 thread-like stigmas, 2-loc., the post. sterile, the other with 2–4 erect orthotr. ov.; fr. indehisc.; no endosp.

B. Ov. usu. with only 1 embryo sac (Orders 2–30).
   a. Fls. naked or with haplochlam. bract-like P (Orders 2–12).

2. PIPERALES. Ls. simple, stip. or not, and spikes of small achlam. or haplochlam. ♀ or ♂ ♀ fls.; A 1–10, G 1–4, free or united.

3. SALICALES. Woody with simple alt. stip. ls. and spikes of dioec. achlam. fls., disk cup-like or reduced to scales; A 2–∞, G̲ (2), 1-loc. with parietal plac. and ∞ anatr. ov.; caps. with ∞ seeds, seeds small with basal tuft of hairs and no endosp.

4. GARRYALES. Woody pls. with opp. evergr. ls. and fls. in catkin-like panicles, ♂ ♀; ♂ P 4, A 4, ♀ naked, G̲ (2–3), 1-loc. with 2 ov.; endosp.

5. MYRICALES. Woody, usu. with simple ls. and fls. in simple, rarely cpd. spikes, ♂ ♀ achlam., sometimes with bracts at base; A 2–16, usu. 4, G̲ (2), 1-loc. with 1 basal orthotr. ov. and 2 stigs.; porogamous; drupe with waxy exocarp; no endosp.

6. BALANOP[SID]ALES. Woody with simple ls.; ♂ fls. in spikes, haplochlam, ♀ sol. surrounded by ∞ scaly bracts; G̲ (2), imperfectly 2-loc. each with 2 ascending ov.; drupe.

x

DICOTYLEDONEAE

Archichlamydeae

1

1. *Casuarinaceae.*

2

1. *Saururaceae:* herbs with alt. ls. and spikes of achlam. ♂ fls.; A 6 or fewer, G (3–4) or 3–4, plac. parietal, ov. 2–∞; endo- and perisp.
2. *Piperaceae:* herbs and shrubs with alt. ls. of biting taste, and spikes, etc. of ♀ or ♂ ♀ achlam. fls.; A 1–10, G (1–4), 1-loc. with 1 basal ov.; endo- and perisp.
3. *Chloranthaceae:* herbs or woody pls. with opp. stip. ls. and spikes or cymes of ♀ or ♂ ♀ fls., sometimes with sepaloid P; A (1 or 3) united to ovary, G 1 with 1 pend. ov.; peri- and endosp.
4. *Lacistemaceae:* shrubs with 2-ranked lanc. exstip. ls. and spikes of minute ♀ fls., naked or with sepaloid P; A 1, G (2–3) plac. parietal, with 1–2 pend. ov. on each; caps. 1-seeded; endosp.

3

1. *Salicaceae.*

4

1. *Garryaceae.*

5

1. *Myricaceae.*

6

1. *Balanop[sid]aceae.*

7. LEITNERIALES. Woody with alt. entire ls. and spikes of dioec. fls.; ♂ achlam., A 3–12, ♀ haplochlam., P of small scaly united ls., G̲ 1 with long style and 1 amphitr. ov.; drupe; thin endosp.

8. JUGLANDALES. Woody with alt. usu. pinnate exstip. ls. and spikes of achlam. or haplochlam. ♂ ♀ fls.; A 3–40, G̅ (2), 1-loc. with 1 basal orthotr. ov.; chalazogamic; fr. drupe or nut-like; no endosp.

9. BATIDALES. Coast shrubs with opp. fleshy ls. and panicles of spikes; fls. ♂ ♀, ♂ with cup-like P and A 4, ♀ naked, originally 2-loc. with 2 ov. in each, divided by false septum, all ♀ fls. in spike concrescent; aggregate fr.; no endosp.

10. JULIANIALES. Woody with alt. usu. pinnate exstip. ls. and dioec. fls.; ♂ ∞ in ± dense panicle, P, A, 6–8, ♀ in fours at end of downward-directed spike, naked, G 1-loc. with 1 ov. on broad hollowed funicle; no endosp.

11. FAGALES. Woody with alt. stip. ls. and fls. in simple or cymose spikes, cyclic, homochlam., rarely naked, usu. monoec.; A opp. P, G̅ (2–6) each with 1–2 ov.; fr. nut-like, seed 1; no endosp.

12. URTICALES. Herbs, shrubs, trees with alt. or opp. stip. ls. and cymose infls. of cyclic homochlam. rarely haplochlam. or naked usu. reg. ⚥ or ♂ ♀ fls., usu. 2 + 2 rarely 2 + 3-merous; sta. before P, G̲ (2–1) with 1 ov.; drupe or nut.

b. *Usu. with sepaloid or petaloid P, rarely heterochlam.* (*Orders* 13–16).

13. PROTEALES. Woody with alt. exstip. ls. and spikes or racemes of cyclic homo-(apparently haplo-)chlam. 2 + 2-merous ⚥ or ♂ ♀ reg. or ·|· fls.; P petaloid; sta. anteposed and usu. adherent to P, G̲ 1; fr. various, no endosp.

7

1. *Leitneriaceae.*

8

1. *Juglandaceae.*

9

1. *Batidaceae.*

10

1. *Julianiaceae.*

11

1. *Betulaceae:* shrubs and trees with alt. simple ls. with caducous stips. and monoec. anemoph. fls. in catkins, typically 3 fls. per axil; P sepaloid or 0, A 2–10, $\overline{G}$ (2), 2-loc. each with 1 pend. ov.; nut; no endosp.
2. *Fagaceae:* trees, rarely shrubs, with simple ls. and caducous scaly stips., and usu. catkins or small spikes of ♂ ♀ fls.; P sepaloid (4–7), A 4–7 or 8–14, $\overline{G}$ usu. (3), 3-loc., 3-styled, each with 2 pend. ov.; nut; no endosp.

12

1. *Ulmaceae:* trees and shrubs with 2-ranked simple stip. ls. and axill. cymes of homochlam. ♀ or ♂ ♀ fls.; P 4–5, sepaloid, A 4–5 or 8–10, $\underline{G}$ (2), styles 2, usu. 1-loc. with 1 pend. ov.; nut or drupe; usu. no endosp.
2. *Moraceae:* usu. trees and shrubs with stip. ls., latex, and cymes of small ♂ ♀ fls., often head-like; P usu. 4 or (4), persistent, rarely 0, A as many, opp. P, $\underline{G}$ (2), 1-loc. with usu. 1 pend. ov.; nut or drupe; endosp. or not.
3. *Urticaceae:* usu. herbs with opp. or alt. stip. ls., no latex, and cymose infls. of small homochlam. usu. ♂ ♀ fls.; P usu. 4–5, A 4–5 opp. P, bent inwards in bud and exploding, $\underline{G}$ 1-loc. with 1 basal ov. and 1 style; nut or drupe; endosp.

13

1. *Proteaceae.*

14. SANTALALES. Herbs, shrubs, trees, often paras., with cyclic, usu. homochlam. fls.; A anteposed, in 1 or 2 whorls, $\overline{G}$, rarely $\underline{G}$ (2–3), rarely 1, each with 1 pend. ov. (or ov. not differentiated).

   1. *Santalineae:* ov. differentiated from plac., often without integ. (fams. 1–6).

   2. *Loranthineae:* ov. usu. not differentiated (fam. 7).

   3. *Balanophorineae:* plac. central with pend. ov. with no integ.; chlorophyll-less paras. (fam. 8).

15. ARISTOLOCHIALES. Fls. cyclic, homo- or haplo-chlam., reg. or ·⊦· ; P petaloid. G usu. inf. 3–6-loc. with axile plac., or 1-loc. with parietal, and ∞ ov.

16. POLYGONALES. Ls. usu. ocreate, fls. haplo- to hetero-chlam., ♀, reg.; $\underline{G}$ 1-loc. with usu. 1 basal erect ov.; nut; endosp.

   c. *P haplochlam., sepaloid or petaloid, sometimes heterochlam. (Order 17).*
17. CENTROSPERMAE. Usu. herbs with spiral or cyclic homo- or hetero-chlam. fls.; A usu. = and opp. P, but also ∞–1, $\underline{G}$ (∞–1) or free, rarely $\overline{G}$, usu. 1-loc. with ∞–1 campylotr. ov.; perisperm.

14

1. *Myzodendraceae:* semiparas. shrublets with alt. ls. and minute naked ♂ ♀ fls.; A 2–3–1 with monothecous anthers, $\overline{\text{G}}$ (3) with axile plac. and 3 ov.; fr. with 3 feathery bristles in angles.

2. *Santalaceae:* semiparas. herbs, shrubs, trees with opp. or alt. ls. and small ♀ or ♂ ♀ homochlam. fls. with perig. or epig. disk; P usu. 2+2 or 2+3, A as many, inserted on P, $\overline{\text{G}}$ 1-loc. with axile plac. and 1–3 ov.; nut or drupe, 1-seeded; endosp.

3. *Opiliaceae:* fls. ♀ heterochlam. with slight seam-like K; $\overline{\text{G}}$ with 1 ov. with no integument.

4. *Grubbiaceae:* trees or shrubs with opp. leathery ls. and small ♀ reg. fls.; P 4 sepaloid, A 4+4, $\overline{\text{G}}$ (2), 2-loc. below when young, later 1-loc. with 2 pend. ov. on central plac.; drupe; oily endosp.

5. *Olacaceae:* trees and shrubs with usu. alt. entire ls. and small ♀ reg. fls.; K 4–6, very small, C 4–6, A as many or 2–3 times as many, $\underline{\text{G}}$ (2–5), 2–5-loc. at base, 1-loc. above, with 1 ov. pend. into each loc.; 1-seeded drupe or nut; endosp.

6. *Octoknem[at]aceae:* woody with alt. ls. and ♀ fls.; P 2+3, A 2+3, anteposed, $\overline{\text{G}}$ 1-loc. with 3 pend. ov.; drupe 1-seeded.

7. *Loranthaceae:* woody semiparas., usu. on trees, with usu. reg. 2–3-merous, usu. homochlam. ♀ or ♂ ♀ fls.; P in two whorls, A as many, $\overline{\text{G}}$ 1-loc. usu. without differentiation of ov. and plac.; layer of viscin round seed; endosp.

8. *Balanophoraceae:* fleshy root paras. with tuberous rhiz. from which stems rise endog., and small fls. in spikes or heads, homochlam. or naked, usu. ♂ ♀; P in ♂ 3–4 (2–8), united below, A as many or 1–2; P in ♀ usu. o; $\overline{\text{G}}$ (1–2), rarely (3–5); nut or drupe; endosp.

15

1. *Aristolochiaceae:* herbs or climbing shrubs with alt. exstip. ls. and homochlam. ♀ reg. or ·⎸· fls.; P usu. (3), petaloid, A 6–36, free or united with style, $\overline{\text{G}}$, rarely $\underline{\text{G}}$, 4–6-loc. with ∞ ov.; caps.; endosp.

2. *Rafflesiaceae:* thalloid parasites, shoots very short with term. fl. or raceme, usu. ♂ ♀, reg. haplochlam.; P (4–5), A ∞ on column, $\overline{\text{G}}$ (4–6–8) with parietal plac. or ∞ twisted loc.; berry with ∞ seeds; endosp.

3. *Hydnoraceae:* thalloid paras. with ♀ reg. fls.; P (3–4), fleshy, A 3–4, epiphyllous, $\overline{\text{G}}$ (3) with parietal plac. and ∞ ov.; berry; endo- and perisp.

16

1. *Polygonaceae.*

17

1. *Chenopodiaceae:* usu. herbs with alt. often fleshy ls. and cymose infls. of small reg. homochlam. ♀ or ♂ ♀ fls.; P (5) or less, imbr. sepaloid, A as many, anteposed, bent inwards in bud, $\underline{\text{G}}$ (2) 1-loc. with 1 basal ov.; nut; endosp.

DICOTS. CENTROSPERMAE—RANALES
B; c; d; α

1. *Chenopodiineae:* P bracteoid, not > 5, A anteposed; ovule usu. 1 (fams. 1, 2).

2. *Phytolaccineae:* P haplo- to hetero-chlam., tending to cyclic; A sometimes ∞, G sometimes little united (fams. 3–6).

3. *Portulacineae:* P heterochlam.; K 2, C 4–5 (fams. 7, 8).

4. *Caryophyllineae:* P heterochlam., K = C; fls. cyclic, sometimes with no C (fam. 9).

d. *Fls. usu. heterochlam.* (*Orders* 18–30).
   α. Apocarpy and hypogyny the rule; perig. and epig. fls. only in Lauraceae and Hernandiaceae (Order 18).
18. RANALES. Herbs or woody pls. with spiral, spirocyclic, or cyclic, usu. haplo- or hetero-chlam. rarely achlam. reg. or ·|· fls.; A usu. ∞, G ∞–1, rarely united.
   1. *Nymphaeineae:* fls. various, usu. spiral; ov. (exc. in 2) usu. ∞ on inner surface of cpls.; mostly water plants (fams. 1, 2).
   2. *Trochodendrineae:* fls. naked, spirocyclic; ov. on ventral suture; no oil cells (fams. 3, 4).

   3. *Ranunculineae:* fls. with P, spiral to cyclic; ov. on ventral suture; no oil cells (fams. 5–8).

2. *Amaranthaceae:* herbs or shrubs with opp. or alt. exstip. ls. and small haplochlam. usu. ♂ reg. fls. in cymose or cpd. infls.; P 4–5 or (4–5) usu. sepaloid, A 1–5 anteposed and ± united below, G̲ (2–3), 1-loc. with ∞–1 ov.; nut; endosp.

3. *Nyctaginaceae:* herbs or woody, with opp. exstip. ls. and cymose ♂ or ♂ ♀ reg. fls. with bracts, sometimes united or petaloid, at base; P (5) petaloid, lower part persistent on fr.; A typically 5 (1–30), G̲ 1 with 1 basal erect ov.; achene; perisp.

4. *Cynocrambaceae:* herbs with fleshy stip. ls., the lower opp., and ♂ ♀ fls., ♂ P 2–5, A 10–30, ♀ P (3–4), G̲ 1, 1 ov.; drupe; endosp.

5. *Phytolaccaceae:* herbs or woody, with racemes or cymes of reg. usu. ♂ fls.; P usu. 4–5, A 4–5 or ∞, G̲ (rarely G̅) 1–∞, free or united, 1 ov. in each; drupe or nut, rarely caps.; perisp.

6. *Aïzoaceae:* herbs or undershrubs with thread-like or fleshy opp. or alt. exstip. ls. and cymose infls. of ♂ reg. fls.; P 4–5 or (4–5), A 5 (3–∞), the outer petaloid stds., G̲ or G̅ (2–∞) with ∞ ov., usu. 2–∞-loc.; caps.; perisp.

7. *Portulacaceae:* herbs or undershrubs with fleshy ls. and often hair-like stips., and cymes of reg. ♂ fls.; K usu. 2, C 4–5, A 5 or 5 + 5, or fewer or ∞, G̲ or semi-inf. (3–5) 1-loc. with 2–∞ ov. on basal plac.; caps.; endosp.

8. *Basellaceae:* twining herbs with ♂ reg. fls.; K 2, C 5 united below, A 5, anteposed, G̲ (3), 1-loc. with 1 basal ov.; nut; endosp.

9. *Caryophyllaceae:* herbs or undershrubs with entire usu. opp. ls. and cymose panicles of usu. reg. ♂ fls.; K 5 or (5), C 5 or 0, A 5 or 10, G̲ (5–2), 1-loc. usu. with free-central plac., ov. 1–∞; caps. or berry; endosp.

18

1. *Nymphaeaceae:* water or marsh pls. with usu. submerged or swimming ls. and sol. reg. ♂ fls.; axis often hollowed; P 6–∞, A 6–∞, G̲ or G̅ 3–∞ or (3–∞), each with 1–∞ ov.; endosp. or o.

2. *Ceratophyllaceae:* submerged water pls. with whorls of 4 ls. and sol. ♂ ♀ axillary reg. fls.; P 9–12 sepaloid, A 12–16, G̲ 1 with 1 pend. ov.; nut; endosp.

3. *Trochodendraceae:* woody with alt. exstip. ls. and sol. or racemed naked ♂ or ♂ ♀ fls.; A ∞, G̲ 5–∞ with ∞–1 ov.; endosp.

4. *Cercidiphyllaceae:* woody with opp. stip. ls. and sol. dioec. fls.; A ∞ spiral, G̲ 2–5, stalked, with ∞ ov.; follicles; endosp.

5. *Ranunculaceae:* usu. herbs, often with divided ls. and usu. ♂ reg. rarely ·|· or fully cyclic fls.; P often haplochlam., usu. petaloid, rarely K, C, A usu. ∞, G̲ ∞–1 rarely united, with ∞–1 ov.; follicle, caps., or ach., rarely berry; endosp. oily.

6. *Lardizabalaceae:* climbing shrubs with cpd. ls. and sol. or racemed ♂ or ♂ ♀ reg. fls.; P 3 + 3 usu. with two whorls of honey-ls., A 3 + 3, G̲ 3 or more with ∞ ov.; berry; endosp.

7. *Berberidaceae:* herbs or shrubs with simple or cpd. ls. and ♂ reg. homo- or hetero-chlam. 3-2-merous fls.; P in 2–4 whorls, often with 2 whorls of honey-ls., A in two, G̲ 1, rarely more, with ∞–1 ov.; berry; endosp.

4. *Magnoliineae:* fls. with P, spiral to cyclic; ov. on ventral suture; oil cells (fams. 9–18).

β. **Syncarpy and hypogyny the rule (Orders 19, 20).**
19. RHOEADALES. Usu. herbs with racemes of fls., cyclic (exc. sometimes the A), heterochlam., rarely homochlam. or apet., hypog., reg. or ·|· ; G (∞–2), ov. with 2 integ.
  1. *Rhoeadineae:* fls. heterochlam, K usu. 2 (fam. 1).
  2. *Capparidineae:* fls. heterochlam., K usu. 4 or more (fams. 2–4).

3. *Resedineae:* fls. heterochlam., spirocyclic (fam. 5).

8. *Menispermaceae:* climbing shrubs with usu. alt. simple ls. and small usu. reg. ♂ ♀ fls.; K, C, A usu. each 2 whorls, G͟ ∞–3–1 each with 1 ov.; drupe; endosp. or 0.

9. *Magnoliaceae:* woody pls. with alt. simple ls. and usu. sol. reg. heterochlam. ♀ or ♂ ♀ fls.; P usu. petaloid; A ∞; G͟ usu. ∞, rarely united; endosp.

10. *Calycanthaceae:* shrubs with opp. simple ls. and ♀ fls. with hollowed recept.; P ∞, petaloid, A 10–30, G͟ ∞ each with 2 ov.; achenes enclosed in axis; endosp. little.

11. *Lactoridaceae:* shrub with haplochlam. cyclic fls.; P 3, A 3 + 3, G͟ 3.

12. *Annonaceae:* woody pls. with entire exstip. ls. and usu. showy ♀ reg. heterochlam. fls.; P 3 + 3 + 3, A ∞ spiral, G ∞–1; berry; endosp. ruminate.

13. *Eupomatiaceae:* fl. deeply perig., naked; A ∞, G ∞.

14. *Myristicaceae:* woody pls. with evergr. simple ls. and axill. racemes of ♂ ♀ reg. cyclic fls.; P (3), A (3–18) extr., G͟ 1 with 1 basal ov.; fr. fleshy dehisc.; aril; endosp. ruminate.

15. *Gomortegaceae:* tree with opp. evergr. ls. and racemes of ♀ fls.; P 7, A 2–3, G (2–3), with 1 pend. ov. in each; drupe; endosp.

16. *Monimiaceae:* woody pls. with usu. opp. exstip. ls. and sol. or cymose infls. of ♀ or ♂ ♀ reg. or ·|· fls.; P often perig. or epig., 4–∞ or 0, A ∞ or few, G͟ ∞ each with 1 ov.; achene; endosp.

17. *Lauraceae:* woody with leathery alt. exstip. ls., and oil cavities in tissues; infl. various, of 3-merous reg. ♀ or ♂ ♀ fls. with ± concave axis; P homochlam. in 2 whorls, A in 3 or 4, one sometimes stds., anthers opening by valves, G͟ (3), 1-loc. with 1 pend. ov.; berry sometimes encl. in fleshy axis; no endosp.

18. *Hernandiaceae:* woody with alt. exstip. ls., and oil passages; and ♀ or ♂ ♀ reg. homochlam. fls.; P 4–10, A in whorl before outer P, G͞ 1-loc. with 1 pend. anatr. ov.; fr. winged; no endosp.

19

1. *Papaveraceae:* usu. herbs with alt. ls. and latex, and reg. or ·|· ♀ fls.; K 2, C 4, rarely 6 or more, or 0, A ∞–4–2 (branched), G͟ (2–16) with parietal plac. and ∞ ov., or 1 basal; caps.; oily endosp.

2. *Capparidaceae:* herbs and shrubs with alt. ls. and bracteate racemes of ♀ reg. or ·|· fls., axis usu. elongated below A or G; K, C 4, A ∞–6–4, G͟ (2–several), 1-loc. or more with ∞ ov.; caps., berry or drupe; no endosp.

3. *Cruciferae:* herbs with alt. exstip. ls. and simple or branched hairs, and ebracteate racemes of ♀ reg. fls.; K 2 + 2, C 4 diagonal, A 2 (short) + 2 + 2 (long), G͟ (2), 1-loc., with 'spurious' partition; usu. siliqua; no endosp.

4. *Tovariaceae:* herbs with ternate ls. and term. racemes of ♀ reg. fls.; K, C, A 8, G͟ (6–8) with plac. reaching centre, and ∞ ov.; berry; endosp. thin.

5. *Resedaceae:* herbs with alt. stip. ls. and racemes of ♀ ·|· fls., with post. disk; K 4–8, C 0–8, A 3–10, G͟ (2–6) open above, 1-loc. with 1–∞ ov.; caps.; no endosp.

4. *Moringineae:* fls. homochlam., cyclic (fam. 6).

20. SARRACENIALES. Herbs with usu. alt. insectivorous ls. and spirocyclic to cyclic homo- or hetero-chlam. hypog. reg. fls.; G̲ (3–5) with parietal or axile plac. and 3–∞ ov.; endosp.

γ. **Apocarpy and hypogyny occur, but perigyny is commoner; syncarpy and epigyny also common (Order 21).**

21. ROSALES. Fls. cyclic, rarely spirocyclic, heterochlam. rarely apet., hypog. to epig., reg. or ⊹ ; G or (G) sometimes with thick plac. and ∞ ov.

1. *Podostemineae:* submerged trop. water pls. of alga- or lichen-like form (fams. 1–3).

2. *Saxifragineae:* G = or fewer than C; endosp. usu. rich (fams. 4–13).

6. *Moringaceae:* trees with pinnate exstip. ls. and panicles of ☿ ·ǀ· fls.; K, C, A 5, and 5 stds., G̲ (3) on short gynophore, with parietal plac. and ∞ ov.; caps.; no endosp.

20

1. *Sarraceniaceae:* herbs with pitcher ls. and scapes with sol. or racemed ☿ reg. fls.; K 8–5, C 5, A ∞, G̲ (5–3) 5- or 3-loc. with ∞ ov.; caps.; endosp.
2. *Nepenthaceae:* climbers with alt. ls., the lower with pitchers, the upper tendrilled, and racemes or panicles of ♂ ♀ reg. fls., P 2+2 homochlam., A (4–16), G̲ (4), 4-loc. with ∞ ov.; caps.; endosp.
3. *Droseraceae:* herbs usu. with alt. ls., usu. inrolled in bud, and with sticky glands, and cymose ☿ reg. fls.; K, C 5–4, A 5–4–20, G̲ (5–3), 1-loc. with ∞–3 ov.; caps.; endosp.

21

1. *Podostemaceae:* herbs (usu. trop.) of rushing water with reg. or ·ǀ· ☿ achlam. fls.; A ∞–1 free or united, G̲ (2), 2–1-loc. with thick central plac. and ∞ or few anatr. ov.; caps.
2. *Tristichaceae:* as last, with reg. or slightly ·ǀ· homochlam. ☿ fls.; P 3–5 sepaloid, A as many, or 4–5 times as many, or 2–1, G̲ (2–3), 2–3-loc. with ∞ ov. on thick central plac.; caps.
3. *Hydrostachyaceae:* herbs (S. Afr.) of running water with spikes of dioec. naked fls.; ♂ of 1 sta., ♀ of (2) cpls. with ∞ ov.; caps.
4. *Crassulaceae:* succulent exstip. herbs or undershrubs, usu. with cymose infl. of reg. ☿ 3–30-merous fls.; C or (C), A obdipl. or in one whorl, G̲ sometimes slightly united, with ∞ ov.; follicles; endosp.
5. *Cephalotaceae:* perenn. herbs with some pitcher ls. and panicles of ☿ reg. fls.; P 6, A 6, G̲ 6 with 1–2 basal ov.; follicles; endosp.
6. *Saxifragaceae:* herbs, shrubs or trees with usu. alt. ls. and various infl. of usu. ∞ ☿ reg. (rarely ·ǀ·) fls. with convex, flat or concave axis; A usu. obdipl. or = C, G = C or fewer, with usu. free styles, 2–1 (rarely 5)-loc. with swollen plac. and ∞ ov. in several ranks, sup. or inf.; caps. or berry; endosp.
7. *Pittosporaceae:* woody, sometimes climbing, with alt. ls. and resin passages, and ☿ reg. 5-merous fls.; G̲ (2 or more) 1–5-loc. with parietal or axile plac. and 2-ranked ∞ anatr. ov., and simple style; caps. or berry; endosp.
8. *Brunelliaceae:* woody with opp. or whorled ls. and panicles of small ♂ ♀ 4–5–7-merous diplost. fls.; K valv., C 0, G̲ 5–2 each with 2 pend. ov.; follicle-caps.; endosp.
9. *Cunoniaceae:* woody with opp. or whorled stip. ls.; like 6, but ov. in 2 ranks.
10. *Myrothamnaceae:* small shrubs with opp. fan-folded ls. and spikes of ♂ ♀ reg. achlam. fls.; A 4–8, G̲ (4–3); caps. septicidal; endosp.

xxi

3. *Rosineae:* G ∞–1; ov. with 2 integ.; endosp. little or 0 (fams. 14–18).

δ. Fls. usu. with 5 or 4 whorls; apocarpy and isomery appear, but syncarpy and oligomery of G̲ are the rule (Orders 22–26).

22. PANDALES. Fls. cyclic, heterochlam., dioec. G̲ (3), each with 1 pend. orthotr. ov.; drupe.

23. GERANIALES. Fls. cyclic, heterochlam., apet. or naked, usu. 5-merous; A various, G̲ (5–2), rarely more, often separating when ripe, usu. with 2–1 rarely ∞ ov., pend. with ventral raphe and micropyle up, or, when > 1 present, some with dorsal raphe and micropyle down.

1. *Geraniineae:* fls. heterochlam. rarely apet., usu. reg. and obdipl., rarely haplostemonous and in ·|· fls. usu. abortion of some sta.; anthers opening longitud., G iso- or oligo-merous; ov. with 2 integ. (fams. 1–12).

A. No secretory cells or passages (fams. 1–7).

11. **Bruniaceae:** heath-like undershrubs with alt. exstip. ls. and cpd. spikes, racemes and heads of usu. reg. and perig. ♂ fls.; K, C, A 5, G̲ (3–2) each with 3–4 ov. or 1 with 1; caps.; aril; endosp.
12. **Hamamelidaceae:** woody with usu. alt. stip. ls. and spikes or heads of ♀ or ♂ ♀ reg. heterochlam. apet. or naked fls. surrounded by brs.; K, C, A 4–5, G̲ (2) with 1–∞ pend. ov.; caps.; endosp.
13. **Eucommiaceae:** trees with alt. exstip. ls. and latex, and naked ♂ ♀ reg. fls.; A 6–10, G̲ (2), one aborting, with 2 pend. ov.; samara; endosp.
14. **Platanaceae:** woody with alt. 3–5-lobed stip. ls. and pend. spherical heads of ♂ ♀ reg. fls.; K, C, A 3–8, G̲ usu. 1, free, with 1–2 ov.; caryopsis; endosp.
15. **Crossosomataceae:** shrubs with small stiff grey-green ls. and sol. fls.; like Rosaceae–Spiraeoideae, but seeds kidney-shaped; aril; endosp.
16. **Rosaceae:** herbs, shrubs, or trees with alt. usu. stip. ls. and reg. (rarely ⋅|⋅) 5 (3–8 or more)-merous fls.; axis flat or hollowed; K 5, C 5 or 0, A 2–4 or more times as many, bent inwards in bud, G = K or 2–3 times as many, or ∞, rarely 1–4, free or united to hollow axis, usu. 1-loc. with 2 ov. per cpl.; follicle, achene, drupe or pome; endosp. thin or 0.
17. **Connaraceae:** usu. climbing shrubs, rarely trees, with alt. exstip. ls. and panicles of reg. ♀ or ♂ ♀ fls.; K 5 or (5) persist., C 5, A 5+5, G̲ usu. 5, rarely 4 or 1, each with 2 ov.; one follicle with 1 seed; aril; endosp. or none.
18. **Leguminosae:** trees, shrubs, or herbs, usu. with alt. stip. ls. and racemes of reg. or ⋅|⋅ usu. ♀ fls.; K 5, C 5, A 5+5 or more, G usu. 1, rarely 2–5–15, with ∞ ov.; pod or indehisc. fr.; endosp. usu. none.

22

1. **Pandaceae.**

23

1. **Geraniaceae:** herbs with lobed or divided ls., stip. or not, and ♀ usu. reg. 5-merous fls.; A 10–15, sometimes only 5 fertile, G̲ (5–2) usu. with 1–2, rarely 2–∞ ov. per cpl.; schizocarp, rarely caps.; endosp.
2. **Oxalidaceae:** usu. herbs with alt. cpd. stip. or exstip. ls. and ♀ reg. 5-merous fls. with no disk; A 10 obdiplost., united at base, G̲ (5) with ∞–1 ov.; caps. or berry; endosp.
3. **Tropaeolaceae:** usu. climbers with sensitive petioles, stip. or not, and ♀ 5-merous ⋅|⋅ fls., with axis prolonged into post. spur; A 8, G̲ 3-loc. with 1 ov. in each; schizocarp; no endosp.
4. **Linaceae:** herbs or woody with alt. ls., stip. or not, and ♀ reg. 5–4-merous fls. with no disk; A 5–20 united below, G̲ 5–4 (or fewer)-loc. with 1–2 ov. in each and often with extra partitions; caps. or drupe; endosp.

B. Secretory cells or passages (in 10 sometimes only in pith and bark) (fams. 8–12).

2. *Malpighiineae:* as last, but fls. obliquely ·l·, at least in G; ls. often opp. (fams. 13–15).

3. *Polygalineae:* fls. reg. or ·l· with two whorls of sta.; anthers opening by pores, G (2) median (fams. 16, 17).

4. *Dichapetalineae:* fls. reg. or ·l· with 1 whorl of sta.; C or (C), ov. with 1 integ., seed sometimes with caruncle (fam. 18).

5. *Humiriaceae:* woody with alt. stip. ls. and reg. ♂ 5-merous fls. with cup-shaped disk; A 10–∞, G (5) each with 1–2 ov.; drupe; endosp.

6. *Erythroxylaceae:* woody with alt. simple stip. ls. and 5-merous ♀ reg. fls., heterostyled with no disk; C with appendages on inner side, A 10, united in tube at base, G(3–4), 3–4-loc., but only 1 fertile, with 1–2 ov.; drupe; endosp.

7. *Zygophyllaceae:* usu. shrubby with opp. often pinnate stip. ls. and cymes or cpd. infls. of reg. ♀ 5–4-merous fls. with disk or gynophore; A 10–8, rarely 15, often with united basal appendages, G (5–4) or more with 1–∞ ov.; usu. caps. or schizocarp; endosp. or o.

8. *Cneoraceae:* shrubs with alt. narrow leathery exstip. ls., oil cells, and single or cymose reg. ♀ 3–4-merous fls. with disk; A 3–4, G (3–4), lobed, each with 2 ov., style 1; schizocarp.

9. *Rutaceae:* usu. woody with alt. or opp. simple or cpd. exstip. ls. and reg. or ·l· usu. ♀ 5–4-merous fls. with disk; A obdipl. or 5–4–3–2, rarely ∞, G (5–4) rarely (3–1 or ∞) with ∞–2 ov.; fr. various; endosp. or none.

10. *Simaroubaceae:* woody pls. with bitter bark, alt. or opp. usu. pinnate exstip. ls. and reg. usu. ♂ ♀ 5–4-merous fls. with disk; A 10, 5, or ∞, G (5) or fewer; fr. various; endosp. thin or none.

11. *Burseraceae:* woody pls. with alt. usu. cpd. ls., resin-passages and small reg. usu. ♂ ♀ 5–3-merous fls. with disk; A obdipl. or 5, G (5–3) each usu. with 2 ov.; style 1; drupe or caps.; no endosp.

12. *Meliaceae:* woody pls. usu. with pinnate exstip. ls. and usu. ♀ reg. fls. in cymose panicles; axis rounded or with effigurations; K, C sometimes united, A usu. in tube, obdipl. or 5, G (5) or fewer, multiloc. with 1–2 rarely more ov. in each, and 1 style; fr. various; endosp. or o.

13. *Malpighiaceae:* woody usu. climbing pls. with opp. stip. ls. and ♀ obdipl. 5-merous fls. with convex or flat axis, sometimes with gynophore; K (5), often with nectaries, C 5 usu. clawed, A 5+5, often some aborted, G usu. (3), each with 1 ov.; schizocarp, nut or drupe; no endosp.

14. *Trigoniaceae:* woody often climbing pls. with alt. or opp. ls., stip. or not, and ♀ obliquely ·l· 5-merous fls.; K (5), C 5–3 often very unequal, A 5–6–10, ± united in tube at base, G (3) with ∞–2 ov. each; caps.; endosp. or not.

15. *Vochysiaceae:* woody, rarely herbs, with opp. or whorled simple ls., stip. or not, and ♀ obliquely ·l· fls.; K (5), one often spurred, C usu. 3–1, perig. or epig., A 1 and stds., G or G̅ (3) each with ∞–2 ov.; fr. indehisc. or caps.; no endosp.

16. *Tremandraceae:* shrubs with entire or toothed ls. and sol. axillary 4–5 (rarely 3)-merous ♀ reg. fls.; K free, C valv., A in 2 whorls, G (2) with 1–2 ov. each; caps.; endosp.

17. *Polygalaceae:* herbs, shrubs, or trees with simple entire usu. alt. exstip. ls. and racemes, spikes, or panicles of ♀ ·l· fls.; K usu. 5, 2 larger and petaloid, C 3, 1 often keel-like, A (4+4) or fewer, usu. united below, G usu. (2), 2-loc. with 1 ov. in each; caps., nut or drupe; endosp. or o.

18. *Dichapetalaceae:* woody, often lianes, with entire stip. ls. and small ♀ or ♂ ♀ usu. reg. fls. with disk or scales; K 5 or (5), C 5 or (5), often bifid, A 5, sometimes united to C, G (2–3) each with 2 ov.; drupe; no endosp.

5. *Tricoccae:* fls. reg. ♂ ♀, often much reduced; G (3) each with 2–1 ov. with 2 integ.; usu. caruncle (fam. 19).

6. *Callitrichineae:* herbs, often submerged, with crowded ls. and small axillary monoec. naked fls.; ♂ with term. sta., ♀ with 2 transv. cpls. divided into 4, with 1 ov. in each section; fr. of 4 nutlets; endosp. (fam. 20).

24. SAPINDALES. Usu. woody; as last, but ov. in reversed position, pend. with dorsal raphe and micropyle up, or erect with ventral raphe and micropyle down.

   1. *Buxineae:* haplochlam.; ov. with 2 integs. (fam. 1).

   2. *Empetrineae:* heterochlam., cpls. each with 1 erect ov. with 1 integ., united till ripe; shrubs (fam. 2).

   3. *Coriariineae:* heterochlam., cpls. each with 1 pend. ov. with 2 integ., finally free; shrubs (fam. 3).

   4. *Limnanthineae:* heterochlam., cpls. each with 1 erect ov. with 1 integ., finally free; herbs (fam. 4).

   5. *Anacardiineae:* heterochlam., rarely apet., reg.; G usu. oligomerous; woody with resin passages (fam. 5).

   6. *Celastrineae:* fls. heterochlam., reg., with 2 or 1 whorls of A; G most often oligomerous (fams. 6–14).

19. *Euphorbiaceae:* herbs, shrubs, and trees, with alt. or opp. usu. stip. ls., often latex, and cpd. infls. of ♂ ♀ reg. usu. 5-merous fls.; P usu. in 1 whorl, or 0, A 1–∞ free or united or branched, G̲ usu. (3), 3-loc. with 2-lobed styles, and 1–2 pend. anatr. ov. in each, with ventral raphe and micropyle usu. with caruncle; usu. schizocarp-caps.; endosp.

20. *Callitrichaceae.*

24

1. *Buxaceae:* woody pls. with entire evergr. exstip. ls. and reg. ♂ ♀ apet. or naked fls., sol. or in racemose infls.; A 4–∞, G̲ (3) or (2–4) each with 2–1 ov.; caps. or drupe; endosp.

2. *Empetraceae:* ericoid shrubs with linear exstip. grooved ls. and heads of small ♂ ♀ reg. fls.; K, C, A 2–3, G̲ (2–9); drupe; no caruncle.

3. *Coriariaceae:* woody pls. with opp. or whorled exstip. ls. and axillary or racemed ♀ or ♂ ♀ reg. fls.; K, C 5, A 5 + 5, G̲ 5–8; schizocarp, endosp.

4. *Limnanthaceae:* annuals with alt. exstip. ls. and sol. axillary ♀ reg. 5–3-merous fls.; K, C 5–3, A 10–6, G̲ (5–3), with 1 ov. in each, separating when ripe; no endosp.

5. *Anacardiaceae:* woody pls. with alt. exstip. not gland-dotted ls. and ∞ fls. in panicles, typically 5-merous, hypog. to epig.; A 10–5 or other number, G (3–1) rarely (5), each with 1 anatr. ov., often only one fertile; drupe, no endosp.

6. *Cyrillaceae:* woody pls. with evergr. ls., and racemes of small ♀ reg. 5-merous fls.; K, C sometimes united, A in 2 whorls, G̲ (5–2)-loc. each with 1 ov.; endosp.

7. *Pentaphylacaceae:* woody pls. with alt. leathery ls. and small ♀ reg. fls. in racemes below ls.; K, C, A 5, G̲ (5) each with 2 pend. ov.; caps.; endosp.

8. *Corynocarpaceae:* woody pls. with alt. leathery ls. and small ♀ fls. in panicles; inner sta. stds., G̲ (2), 1 fertile with 1 pend. ov.; drupe; no endosp.

9. *Aquifoliaceae:* woody pls. with alt. evergr. simple ls., stips. small or none, and dioec., reg., 4–more-merous ♂ ♀ fls. in cymose umbels; K, C 4, A 4 often epipet., G̲ (4–6) or more, each with 1–2 pend. ov.; drupe with several stones; endosp.

10. *Celastraceae:* woody pls. with simple opp. or alt. ls., sometimes stip., and small ♀ reg. 4–5-merous fls. in cymose umbels; A 4–5 on edge of disk, G̲ (2–5) each with ∞–1 ov.; caps. or berry; often aril; endosp. or not.

11. *Hippocrateaceae:* woody pls., often climbing, with opp. or alt. simple ls., stips. small or none, and small, ♀ reg. fls. in cymose umbels; K, C 5, A 3, rarely 5, G̲ (3) each with ∞–2 ov.; berry or 3-winged fr.; no endosp.

12. *Salvadoraceae:* woody pls. with opp. simple ls. and sometimes bristle-like stips., and panicles of ♀ or ♂ ♀ reg. fls.; K (4–2), C 4–5 or (4–5), A 4–5, G̲ (2), 1–2-loc. with 1–2 basal ov. in each; berry or drupe, usu. 1-seeded; no endosp.

7. *Icacinineae:* fls. heterochlam., reg., with 1 whorl of sta. before K; G usu. 1, integ. 1, fr. 1-seeded (fam. 15).

8. *Sapindineae:* fls. heterochlam., typically with 2 whorls of sta., but with aborted sta. and cpls., reg. or obliquely ·⊢·; ov. with 2 integs. (fams. 16–18).

9. *Sabiineae:* fls. heterochlam., sta. before pets. (fam. 19).

10. *Melianthineae:* fls. heterochlam., ·⊢·, with 1, rarely 2, whorls of sta. with free anthers (fam. 20).

11. *Balsaminineae:* as last, but anthers united (fam. 21).

25. RHAMNALES. Fls. cyclic, diplochlam., sometimes apet., with 1 whorl of sta. before pets., reg.; G (5–2) each with 1–2 ascending ov. with dorsal, lat., or ventral raphe and 2 integs.

26. MALVALES. Fls. cyclic, exc. sometimes the A, heterochlam., rarely apet., usu. ♀ and reg.; K, C usu. 5-merous, K usu. valv., A ∞ or in 2 whorls, the inner branched, G (2–∞) each with 1–∞ anatr. ov. with 2 integs.

1. *Elaeocarpineae:* K ± free, anthers dithecous with pores; no mucilage cells (fam. 1).

2. *Chlaenineae:* K free, imbr., A enclosed by a cup, anthers dithecous with slits; mucilage cells often present (fam. 2).

13. *Stackhousiaceae:* herbs with alt. exstip. ls. and spikes or cymes of fl.; K, C, A 5, G̲ (2–5)-loc. each with 1 erect ov.; schizocarp; endosp.

14. *Staphyleaceae:* woody pls. with opp. or alt. usu. pinn. stip. ls. and panicles or racemes of fls.; K, C 5, A 5, outside disk, G̲ (2–3), free above with ∞–few pend. ov.; caps.; endosp.

15. *Icacinaceae:* woody pls. some climbing, usu. with alt. exstip. ls. and small ♀ or ♂ ♀ reg. fls.; K, C, A 5–4, G̲ (3), usu. 1 only with 2 pend. ov.; drupe; endosp.

16. *Aceraceae:* trees with opp. exstip. ls. and small reg. ♀ ♂ ♀ fls. in spikes, racemes or panicles; axis disk-like or concave; K, C, A 4–10, G̲ (2) each with 2 ov.; fr. with 1-seeded samaras; no endosp.

17. *Hippocastanaceae:* trees with opp. palmate exstip. ls. and cymose racemes of ·l· ♀ ♂ ♀ fls.; K (5), C 4–5, A 5–8, G̲ (3)-loc. each with 2 ov.; caps. 3–1-loc. usu. 1-seeded; no endosp.

18. *Sapindaceae:* woody pls. with alt. ls. and usu. ·l· ♀ ♂ ♀ fls. with extra-staminal disk; K 5, C 5–3 or 0, often with scales, A usu. 8, rarely 10, 5, or ∞, G̲ (2–3) each usu. with 1 ov.; caps., drupe, nut, or schizocarp; no endosp.

19. *Sabiaceae:* woody pls., often climbers, with alt. exstip. ls. and small ♀ or ♀ ♂ ♀ fls. in racemes or cymose racemes; K (2–5), C 4–5, A 5 antepetalous, G̲ (2–3) each with 2 ov.; fr. 1-loc., 1-seeded; no endosp.

20. *Melianthaceae:* woody pls. with alt. usu. pinnate ls., stip. or not, and racemes of ♀ ·l· fls.; K, C 5, A 5–4, rarely 10, unequal or partly united, G̲ (4–5) each with ∞–1 ov.; caps.; aril or not; endosp.

21. *Balsaminaceae:* herbs with watery translucent stems and alt. usu. exstip. ls., and ♀ ·l· fls.; K 5, the 2 ant. often small or aborted, C 5, the lat. ones united in pairs, A (5), G̲ (5)-loc. each with ∞ ov.; caps. usu. explosive; no endosp.

25

1. *Rhamnaceae:* woody pls., rarely herbs, often climbing, with simple stip. ls. and small greenish or yellowish fls. often in axillary cymose infls.; K 5–4, C 5–4 small, or 0, A 5–4, G̲ to G̅ (5–2) with 1 ov. in each.; dry fr. or drupe; endosp. little or none.

2. *Vitidaceae:* climbing shrubs often with tendrils opp. ls.; like preceding, but berry; C valv., often united above and falling as a whole, G̲ (2–8); endosp.

26

1. *Elaeocarpaceae:* woody pls. with simple stip. ls. and ♀ 5–4-merous fls.; A ∞, G̲ (2–8) with ∞ ov. and 1 style, 2–∞-loc., rarely 1-loc.; caps., rarely drupe; sometimes aril; endosp.

2. *Chlaenaceae:* woody pls. with alt. stip. ls. and ♀ reg. fls.; K 5, C 5–6, A 10–∞, G̲ (3) each with 2 ov.; caps.; endosp.

3. *Malvineae:* K rarely imbr., usu. valv; mucilage cells (fams. 3–7).

4. *Scytopetalineae:* seps. united into dish-like K (fam. 8).

ε. **Fls. spirocyclic or in 5–4 whorls; apocarpy only in lower forms, syncarpy the rule, often with a sinking of G in axis (Orders 27, 28).**

27. PARIETALES. Fls. spirocyclic or cyclic, often A and G ∞, heterochlam., rarely apet., hypog. to epig.; G ± united, often with parietal plac. which may touch in centre, very rarely with basal ov.

1. *Theineae:* G free on convex or flat axis; endosp. oily (fams. 1–9).

3. *Gonystylaceae:* shrubs with alt. entire exstip. ls. and cymose panicles of ♀ reg. fls.; K 5–4, C 5–4, usu. divided, A ∞, anthers dithecous, G̲ (5–3), each with 1 pend. ov.; berry; no endosp.

4. *Tiliaceae:* usu. woody pls. with alt. stip. ls. and ♀ reg. fls.; K 5, C 5 or o, A ∞ rarely to 10, free or in bundles, anthers dithecous, G̲ (2–∞), each with 1–∞ ov., 2–∞-loc.; fr. various; endosp.

5. *Malvaceae:* herbs, shrubs, or trees with simple or lobed stip. ls. and ♀ usu. conspic. fls., sol., or in infls.; K 5, often with epicalyx, C 5, conv., A usu. ∞ in 2 whorls, united in a tube below, monothecous, with thorny pollen, G̲ (5–∞) each with 1–∞ ov.; styles as many or twice; caps. or schizocarp; endosp.

6. *Bombacaceae:* woody pls. with entire or palmate stip. ls. and often conspic. fls.; like last, but anthers with 1, 2 or more loc. and smooth pollen; G̲ (2–5) with 2–∞ ov., seeds sometimes enclosed in hairs from pericarp; endosp. thin or o.

7. *Sterculiaceae:* trees, shrubs and herbs with alt. simple or cpd. stip. ls. and complex infls. of ♀ or ♂ ♀ fls.; (K), C conv. or o, A in 2 whorls, the outer stds., the inner often branched, all ± united, anthers 2-loc.; often androgynophore; G̲ usu. (5), antepet., each with 2–∞ ov.; usu. schizocarp; endosp.

8. *Scytopetalaceae:* woody pls. with alt. leathery ls. and bunches or racemes of long-stalked fls.; K dish-like, C 3–7 valv., A ∞, G (4–6), each with 2–6 pend. ov.; fr. woody or drupe, 1-seeded.

27

1. *Dilleniaceae:* woody, sometimes climbing, rarely herbs, with usu. entire alt. evergr. ls., stip. or not, and usu. ♀ reg. yellow or white fls.; K 3–∞, C 5–3, A ∞, rarely 10 or less, G̲ ∞–1, each with 1–∞ ov.; fr. dehisc. or not; aril; endosp.

2. *Eucryphiaceae:* woody with evergr. opp. stip. ls. and sol. axillary ♀ reg. white fls.; K, C 4, A ∞, G̲ (5–18) each with ∞ pend. ov., becoming free on ripening; seed winged; endosp.

3. *Ochnaceae:* woody, or undershrubs, with evergr. stip. ls., usu. with ‖ lat. nerves, and panicles of showy usu. yellow ♀ reg. (rarely ·|·) fls., axis often enlarging after flg.; K 4–10, C 5, rarely 4–10, A 5–10–∞, sometimes stds., G̲ (2–5–10) with one style, often free below, with ∞–1 erect or pend. ov.; endosp. or o.

4. *Caryocaraceae:* woody with ternate evergr. stip. ls. and term. racemes of ♀ reg. fls.; K (5), C (5), A ∞, G̲ (4–8–20) rarely (1–3) each loc. with 1 pend. ov.; schizocarp; endosp. thin or o.

5. *Marcgraviaceae:* woody, often climbing and epiph., with simple exstip. ls. and racemes of ♀ reg. fls., the brs. metam. into hollow nectaries; K 4–5, C (4–5), A 3–6–∞, G̲ (5) or (2–8–∞) with ∞ ov. on originally parietal plac. afterwards meeting in centre; caps.; no endosp.

2. *Tamaricineae:* G free on flat axis; endosp. starchy or none, C free, A in whorls, or if ∞ in bundles (fams. 10–12).

3. *Fouquieriineae:* as last, but endosp. oily, and (C) (fam. 13).

4. *Cistineae:* G free on flat or convex axis; endosp. starchy, C free, A ∞ not in bundles (fams. 14, 15).

5. *Cochlospermineae:* as last, but endosp. of kidney-shaped seed oily (fam. 16).

6. *Flacourtiineae:* G free on convex axis, or in tubular axis rarely united at sides to G; endosp. oily (fams. 17–24).

6. **Quiinaceae:** woody with shining evergr. stip. ls. and racemes or panicles of ♀ ♂ ♀ reg. fls.; K, C 4–5, A 15–30, G̲ (2–3) or (7), each with 2 axile ov.; berry, seeds felted.

7. **Theaceae:** woody with simple usu. alt. exstip. ls. and ♀ reg. fls.; K 5–7, C 5–9, sometimes united below, A ∞–5, sometimes in bundles, G̲ (3–5) or (2–∞) with ∞–1 ov. in each on axile plac.; caps.; endosp. or 0.

8. **Guttiferae:** woody, rarely herbs, with simple usu. opp. rarely stip. ls., resin passages, and ♀ or ♂ ♀ reg. fls.; A ∞–4, often partly stds. and united in groups, G̲ (3–5) or (1–15) with ∞–1 ov.; caps., berry, or drupe; no endosp.

9. **Dipterocarpaceae:** trees with alt. evergr. stip. ls., resin passages, and panicles of ♀ reg. fls.; K 5 (2, 3 or all lengthening to wings on the fr.), C 5 free or united, A ∞ or 15–10–5, G̲ (3–1) each with ∞–2 ov.; fr. usu. 1-seeded indehisc.; no endosp.

10. **Elatinaceae:** undershrubs or herbs, often water pls., with opp. or whorled stip. ls. and small ♀ reg. fls., axillary or in cymes, K, C 2–5, A 2–5 or 4–10, G̲ (2–5) with ∞ axill. ov.; caps.; endosp. thin or 0.

11. **Frankeniaceae:** undershrubs or herbs with small opp. exstip. ls. and term. or cymed ♀ reg. 4–6-merous fls.; (K), C with ligule, A usu. 6, sometimes ∞, free or united below, G̲ (4–2) with ∞ erect ov. on parietal plac.; caps.; endosp.

12. **Tamaricaceae:** shrubs or herbs with small alt. exstip. ls. and ♀ reg. 4–6-merous fls.; A as many or twice as many as C, or ∞ in groups, G̲ (5–2) with ∞ ascending ov. on basal plac.; style divided; caps.; seed hairy; endosperm or none.

13. **Fouquieriaceae:** shrubs with decid. ls. and thorny midrib, and racemes or panicles of showy ♀ reg. fls.; K 5, C (5), A 10–15, G̲ (3), each with 4–6 ov.; seeds hairy or winged; endosp.

14. **Cistaceae:** herbs and shrubs with usu. opp. ls. with glandular hairs and ethereal oil, and ♀ reg. fls.; K 5–3, C 5–3–0, A ∞, G̲ (3–10) with ∞ or 2 ov. on parietal plac.; caps.; endosp.

15. **Bixaceae:** woody pls. with alt. simple ls. and showy ♀ reg. fls. in panicles; K, C 5, A ∞, G̲ (2) each with ∞ ov. on parietal plac.; style 1; caps.; endosp.

16. **Cochlospermaceae:** woody, usu. with lobed or cpd. ls. and showy ♀ reg. or ·|· fls. in racemes or panicles; K, C 4–5, A ∞, G̲ (3–5) each with ∞ ov. on parietal or almost central plac.; caps.; endosp.

17. **Winteranaceae:** woody pls. with alt. exstip. ls. and cymose umbels of ♀ reg. fls.; K 4–5, C 4–5 or 0, A (20 or fewer), G̲ (2–5) with 2–∞ ov. on parietal plac.; berry; endosp.

18. **Violaceae:** herbs, or woody, with alt. stip. ls. and ♀ reg. or ·|· fls.; K, C, A 5, G̲ (3), each with 1–∞ ov. on parietal plac.; caps. or berry; endosp.

19. **Flacourtiaceae:** usu. woody pls. with alt. stip. simple ls., and ♀ or ♂ ♀ reg. fls.; K 2–15, C 15–0, A usu. ∞, G̲ or semi-inf. (2–10) usu. with ∞ ov. on parietal plac.; berry or caps.; often aril; endosp.

20. **Stachyuraceae:** small shrubs with alt. ls. and racemes of small ♀ or polyg. reg. fls.; K, C 4, A 8, G̲ (4) with ∞ ov.; berry; aril; endosp.

7. *Papayineae:* G free in tubular or bell-shaped axis; endosp. oily; latex (fam. 25).

8. *Loasineae:* G sunk in and united to axis; endosp. oily, rarely none (fam. 26).

9. *Datiscineae:* G sunk in and united to axis; endosp. thin, embryo oily; fls. in racemes (fam. 27).

10. *Begoniineae:* as last, but no endosp.; fls. in dichasia or scorpioid cymes (fam. 28).

11. *Ancistrocladineae:* G sunk in and united to axis, 1-loc. with 1 basal ov.; endosp. ruminate, starchy (fam. 29).

28. OPUNTIALES. Succulents, usu. without ls., often thorny, with hemicyclic, heterochlam., ♂ reg. or rarely ·⊢ fls.; K, C, A ∞, on tubular axis, and G̅ (4–∞), 1-loc. with ∞ ov. on parietal plac.; berry-like fr. with ∞ seeds; endosp. little or none.

ζ. **Fls. cyclic; G usu. sunk in hollow axis, and usu. united thereto (Orders 29, 30).**

29. MYRTIFLORAE. Herbs or woody pls., with cyclic heterochlam., rarely apet. or ·⊢ fls. with concave axis; A in 1 or 2 whorls, sometimes branched and in bundles, G (2–∞) usu. united to axis, rarely 1 free.

1. *Thymelaeineae:* woody pls., rarely herbs, with simple ls.; fls. with dish or tubular axis (at least in ♂ and ♀), reg. with (2–4) cpls. free of axis (fams. 1–5).

21. **Turneraceae:** herbs, shrubs or trees with alt. ls., stip. or not, and axillary or racemed or cymed ♀ reg. fls. with tubular axis; K, C, A 5, G̲ (3), each with 3–8 ov. on parietal plac.; style divided; caps.; aril; endosp.

22. **Malesherbiaceae:** herbs or undershrubs with alt. exstip. usu. very hairy ls. and racemes or cymes of ♀ reg. 5-merous fls. with tubular axis and gynophore; A 5, concrescent with gynophore, G̲ (3) with ∞ ov. on parietal plac.; caps.; no aril.

23. **Passifloraceae:** herbs or shrubs, often climbing by tendrils, with simple usu. palmately lobed ls., stip. or not, and fls. sol. or in racemes or cymes, reg., ♀ or ♂ ♀, with axis often ± tubular ending in effigurations; K, C 5, rarely 3–8, A usu. 5 or 4–8, rarely ∞, united to prolongation of axis, G̲ (3–5) usu. with ∞ ov. on parietal plac.; caps. or berry; usu. aril and endosp.

24. **Achariaceae:** herbs or undershrubs with simple or lobed ls. and single fls. or few in an axil, ♂ ♀ reg., 3–5-merous; K, C, A 3–5, G̲ as last; caps.; endosp.

25. **Caricaceae:** woody pls. with simple or cpd. exstip. ls. and axillary infls. of ♂ ♀ reg. fls. with hollow axis; K 5, C (5) in long tube in ♂, short in ♀; A 5+5, G̲ (3–5) with ∞ ov. on parietal plac.; berry; endosp.

26. **Loasaceae:** herbs, rarely shrubs, sometimes twining, with alt. or opp. exstip. ls. and often stinging hairs, and ♀ fls.; K 5 (rarely 4–7), C 5, rarely united, often boat-shaped, A ∞, those before K often transformed into nectaries, G̅ (3–7) each with 1–∞ ov., usu. on parietal plac.; caps. sometimes spirally twisted; endosp.

27. **Datiscaceae:** herbs or shrubs with exstip. ls. and racemes of small usu. ♂ ♀ fls.; ♂ K 3–9, C 0–4–9, A 4–25, ♀ and ♀ P 3–8, G̅ (3–8) with parietal plac. and 8 ov.; caps.; endosp. slight.

28. **Begoniaceae:** herbs or undershrubs with alt. asymmetric stip. ls. and dichasia or cymes of ♂ ♀ fls.; ♂ K 2, rarely 5, C 2–6 or 0, A ∞, ♀ P 5–2 or 3+3 or 8, G̅ (3), rarely (4–5) with ∞ ov. on parietal plac. often united in middle; caps.; no endosp.

29. **Ancistrocladaceae:** lianes with lanc. ls. and racemes or panicles of ♀ reg. fls.; K 5, C 5, slightly united below, A 5–10, G̅ (3), only 1 loc. with 1 basal ov.; nut; endosp.

28

1. **Cactaceae.**

29

1. **Geissolomataceae:** shrubs with opp. evergr. ls. and sol. axillary ♀ fls.; K 4, imbr., C 0, A 4+4, G̲ (4), each with 2 pend. ov.; 1 style; caps.; endosp.

2. **Penaeaceae:** shrubs with small opp. ls. and sol. axillary ♀ reg. fls.; K 4, valv., C 0, A 4, G̲ (4), each with 2–4 erect ov.; 1 style; caps.; no endosp.

2. *Myrtineae:* herbs or woody pls. with alt. or opp. ls. and fls. with tubular axis and (2–∞) cpls. usu. united to axis; ov. with 1 integ. (fams. 6–17).

3. *Oliniaceae:* shrubs with opp. leathery ls. and small ♂ fls. in cymose umbels at ends of twigs; K 4–5, petaloid, C 4–5, smaller, A 4–5, anteposed, Ḡ (3–5) each with 2–3 axile ov.; 1 style; drupe; no endosp.

4. *Thymelaeaceae:* shrubs and trees, rarely herbs, with entire alt. or opp. exstip. ls. and sol. or racemed or spiked ♀ fls. with cup-like or tubular axis; K 5–4, C 5–4–0, A 5–4 or 10–8, Ḡ (5–2) or 1, each with 1 pend. ov.; 1 style; endosp. or o.

5. *Elaeagnaceae:* woody with alt. or opp. entire ls. and fls. as last, ♀ or ♂ ♀ with flat or cup-shaped axis; K 4, C usu. o, A 4 or 8, Ḡ 1 with 1 ascending ov.; nut; endosp. little or none.

6. *Lythraceae:* herbs and shrubs with simple entire usu. opp. stip. ls. and racemes, panicles, or dichasia of ♀, reg. or ·l·, 3–16- usu. 4–6-merous fls. with hollow or tubular axis; K valv., C sometimes o, A twice as many or 1–∞, Ḡ (2–6), 2–6- rarely 1-loc. each with ∞–2 ov.; caps.; no endosp.

7. *Sonneratiaceae:* woody pls. with opp. exstip. ls. and ♀ or ♂ ♀ reg. fls. with bell-shaped axis; K 4–8, C 4–8 or o, A ∞, Ḡ (4–15) united to hollow axis, with 1 style, 4–15-loc. with ∞ ov.; caps. or berry-like fr.; no endosp.

8. *Punicaceae:* woody pls. with entire ls. and showy axillary ♀ reg. fls. with top-shaped axis; K, C 5–7, A ∞, G (∞) in superposed whorls with ∞ ov. united to axis, 1 style; berry-like fr.; no endosp.

9. *Lecythidaceae:* woody pls. with alt. entire exstip. ls. and ♂ fls. with hollow axis; K usu. 4–6, C 4–6, rarely more or o, A ∞, ± united at base, bent inwards in bud, Ḡ (2–6) each with ∞–1 ov.; style 1; fleshy or woody fr.; no endosp.

10. *Rhizophoraceae:* woody pls. usu. with opp. stip. ls. and usu. ♀ reg. fls., sol. or in cymose infls., hypog. to epig.; K 3–16, usu. 4–8, C as many or o, A 8–∞, G usu. (2–5), rarely 6, each with 2–4–∞ pend. axile ov.; fr. usu. with 1 seed per loc., sometimes viviparous; endosp. or o.

11. *Nyssaceae:* shrubs with alt. exstip. ls. and small ♀ or ♂ ♀ fls. usu. with hollow axis, the ♂ in racemes, the ♀ sol.; K 5 or more, C usu. 5, valv., or o, A twice as many, Ḡ usu. 1-loc. rarely 6–10-loc. with 1 ov. in each; drupe; endosp.

12. *Alangiaceae:* shrubs with alt. ls. and umbels of ♀ fls.; K (4–10), C 4–10, narrow, valv., A 4–10 or 2–4 times as many, Ḡ 1–2-loc. with 1 pend. ov. in each; fr. drupaceous with 1 seed; endosp.

13. *Combretaceae:* woody, often climbing, with opp. entire exstip. ls. and racemes of ♀ or ♂ ♀ reg. fls.; K, C, 4–3, rarely 6–8 (C may be o), A 4–5–8–10, rarely ∞, Ḡ 1-loc. with 2–6 pend. ov.; fr. leathery or drupaceous, often winged; no endosp.

14. *Myrtaceae:* woody with opp. or alt. entire exstip. ls. and ♀ reg. fls.; K, C usu. 4–5, A ∞ sometimes in bundles, Ḡ (2–5–∞)-loc., each with ∞–1 ov.; style 1; fr. various; no endosp.

15. *Melastomataceae:* herbs or woody pls. with opp. or whorled exstip. ls. with often 3–9 equal nerves, and showy ♀ reg. 3–∞-merous fls. with hollow axis; K = C, A twice as many, anthers usu. opening by pores, connective usu. with appendages, (G) usu. = K, free or united to axis, 1 style; seeds ∞ in caps. or berry; no endosp.

3. *Hippuridineae:* fls. epig. with 1 sta.; 1 cpl. with 1 ov. and no integ. (fam. 18).

4. *Cynomoriineae:* root paras. with epig. fls. with 1 sta., cpl. with 1 ov., with 1 integ. (fam. 19).

30. UMBELLIFLORAE. Fls. usu. in umbels, cyclic, heterochlam.; usu. with 1 whorl of sta., epig., 4–5-, rarely ∞-merous, ☿ reg.; $\overline{G}$ (5–1) or (∞) each with 1 (rarely 2) pend. ov. with 1 integ., rich endosp.

Sympetalae (Orders 1–10)
(fls. usu. sympetalous)

A. Fls. sometimes polypetalous; 2 or 1 whorls of sta.; usu. hypog., rarely epig. (Orders 1–3).

1. ERICALES. Woody pls. or herbs with simple ls. and ☿ usu. reg. 5–4-merous fls.; C usu. united, A hypog. or epig., rarely united to pets. at base, obdipl. or whorl before C not developed, G 2–∞, usu. before C when equal in number, sup. to inf., ov. with 1 integ.

16. **Onagraceae:** usu. herbs with opp. or alt. exstip. ls. and axillary or racemed ♀ usu. reg. fls. with tubular axis; K 2–4, rarely more, C 2–4 or more or 0, A usu. 4–8, G̅ usu. (4), each with 1–∞ ov.; 1 style; caps., nut, or berry; endosp. little or 0.

17. **Haloragidaceae:** herbs often of marsh or water, with inconspic. reg. 4–1-merous ♀ or ♂ ♀ fls.; C often 0, A twice or less, G̅ (4), rarely 1; fr. nut- or drupe-like; endosp.

18. **Hippuridaceae:** water pls. with whorled ls. and inconspic. apet. fls., G̅ 1 with 1 style and 1 pend. ov.

19. **Cynomoriaceae:** paras. with rhiz. and ♀ or ♂ ♀ fls.; ♂ with 1 epig. sta., ♀ with 1 pend. ov.

30

1. **Araliaceae:** woody pls., rarely herbs, with usu. alt. often much divided ls., commonly stip., and oil passages, and usu. 5 (3–∞)-merous fls. in heads, umbels, or spikes, often in cpd. infls.; K sometimes indistinct, A = C, G̅ (∞–1); fr. berry- or drupe-like with ∞–1 stones; endosp.

2. **Umbelliferae:** herbs with tap root or rhiz., hollow stem, and alt. usu. much divided, sheathing exstip. ls., and usu. ♀ reg. small 5-merous fls. in umbels, simple or cpd.; K (often indistinguishable), C, A 5, G̅ (2) with two styles on swollen style base; schizocarp, the mericarps on a carpophore, each usu. with 5 ribs, often with vittae between; oily endosp.

3. **Cornaceae:** trees or shrubs with opp. or alt. usu. entire exstip. ls. and umbels, panicles or heads of small, sometimes ♂ ♀, reg. 4–5–∞-merous fls.; A = or 2–4 times as many as C, G̅ (4–1) with epig. disk and usu. 1 ov. each; fr. 1–4-loc. with 1–4 seeds.

Sympetalae

I

1. **Clethraceae:** woody with alt. ls. and racemes of ♀ reg. 5-merous obdipl. fls.; C free, A 10 hypog., G̲ (3) each with ∞ ov.; style long with 3 stigs.; caps. 3-valved; endosp.

2. **Pyrolaceae:** evergr. or saprophytic herbs with alt. ls. and ♀ reg. 5-merous obdipl. fls., sol. or in racemes; C free or united, A hypog., G̲ (5–4) with ∞ ov. in each; caps. loculic.; endosp. fleshy.

3. **Lennoaceae:** root paras. with ∞ ♀ reg. 5–∞-merous fls.; A = C, G̲ (6–14) each with 2 ov. and false partition; drupe with 12–28 stones; endosp.

4. **Ericaceae:** usu. undershrubs or shrubs with alt., opp. or whorled usu. evergr. ls. and single or racemed ♀ 5–4-merous obdipl. fls.; C rarely free, inserted with sta. on disk, anther loc. often with projections, pollen in tetrads, (G) sup. or inf. with axile plac. each with 1–∞ ov., style 1 with capitate stig.; berry, drupe, caps.; endosp.

5. **Epacridaceae:** shrubs or undershrubs with stiff entire sess. alt. ls. and usu. racemes of ♀ reg. 5–4-merous fls.; K, (C), A = C, epipet. or at base of hypog. disk, thecae with common slit, G̲ usu. (5) each with 1–∞ ov. on axile plac., style 1 with capitate stig.; caps. or drupe; endosp.

2. PRIMULALES. Fls. ⚥ or ♂ ♀, reg., rarely ·⊦·, 5(rarely 4–∞)-merous, usu. with 1 whorl of antepet. sta., rarely also 5 opp. K; C usu. united, G apparently as many as C, sup. to inf., 1-loc. with ∞–1 ov. on basal or free-central plac.

3. PLUMBAGINALES. Shrubs, undershrubs or herbs with simple ls., often with water- or chalk-secreting glands and cpd. infl. of ⚥ fls.; C or (C), A in 1 whorl, G̲ (5) with 5 stigs., 1-loc. with 1 ov.; endosp. starchy.

B. Fls. sympet. only; sta. sometimes ∞, usu. in 3–2 whorls; fl. usu. hypog. (Order 4).

4. EBENALES. Woody pls. with simple ls.; (C), A in 2–3 whorls, or in 1 by abortion, rarely ∞, G with axile plac. and several loc. with 1 or few ov. in each.

   1. *Sapotineae:* G̲ completely divided into loc., each with 1 ascending ov. with 1 integ. (fam. 1).

   2. *Diospyrineae:* G or ½-inf. not chambered above; ov. with 2 integs. (fams. 2–4).

C. Sympetaly the rule; sta. always in 1 whorl; union of cpls. sometimes small; usu. hypogyny (Order 5).

5. CONTORTAE. Woody pls. or herbs with usu. opp. simple exstip. ls. and usu. 5 (rarely 2–6)-merous fls.; usu. (C), rarely C or none, usu. conv., with as many or fewer sta. usu. epipet. at base of C, and G (2).

   1. *Oleineae:* A 2, ov. with 1 integ. (fam. 1).

   2. *Gentianineae:* A=C, G̲ 1–2-loc. usu. with ∞ ov. on axile or parietal plac. with each 1 integ. (fams. 2–5).

6. *Diapensiaceae:* undershrubs or woody herbs with ♀ reg. fls.; K 5 or (5), C (5), A 10 obdipl., or 5, G̲ (3) each with ∞ ov. on axile plac., style 1; caps.; endosp.

2

1. *Theophrastaceae:* woody with alt. exstip. ls. often crowded at ends of stem or branches, and ♀ or ♂ ♀ reg. rarely ·|· fls.; K 5, C (5), A 5+5 stds., G̲ 1-loc. with ∞ ov. on free-central or basal plac.; drupe with ∞–2 seeds; endosp.
2. *Myrsinaceae:* woody with often evergr. entire alt. exstip. ls., and ♀ or ♂ ♀ reg. fls.; K 5, C (5), A 5 rarely with 5 stds., G̲ to G̅, 1-loc. with ∞ ov. on basal or free-central plac.; style 1; drupe with 1 or few seeds; endosp.
3. *Primulaceae:* herbs with usu. alt. exstip. ls. and ♀ reg. rarely ·|· fls.; K (5), C (5), A 5, epipet., anteposed, and rarely 5 stds., G̲ rarely ½-inf., 1-loc. with ∞ ov. on free-central plac.; caps.; endosp.

3

1. *Plumbaginaceae.*

4

1. *Sapotaceae:* woody with simple alt. ls., secretory passages, and usu. ♀ fls.; K 4–8 in two whorls, (C) as many in 1 whorl, or twice in 2, sometimes with lat. or dorsal appendages, A in 2 or 3 whorls, outer sometimes stds., (G̲) as many (or twice) as 1 whorl of sta., each with 1 basal or axile ov.; style 1; berry; endosp. or 0.
2. *Ebenaceae:* trees with entire alt. rarely opp. exstip. ls., and usu. ♂ ♀ fls., sol. or in few-fld. umbels, 3–more-merous; K persistent, C usu. conv., A as many, or 2–more times as many, free or united in bundles, G̲ (2–16) each with 1–2 pend. ov.; berry with 1 or few seeds; endosp. often ruminate.
3. *Symplocaceae:* woody pls. with alt. exstip. ls. and ♀ 5-merous fls.; C = or twice K, ± united, A epipet. in 1–3 whorls, G̅ sometimes ½-sup.(5–2) each with 2–4 pend. ov.; style 1; drupe; endosp.
4. *Styracaceae:* woody pls. with simple alt. ls. with stellate or scaly hairs, and small or smallish ♀ fls.; K, C (5–4), A 10–8 united at base or rarely into tube, G̲, rarely ½-inf. (5–3) each with 1 or few ov., 3–5-loc. below, 1-loc. above; drupe, indehisc. fr. or caps., with 1 or few seeds; endosp.

5

1. *Oleaceae:* woody, sometimes climbing, rarely herbs, with opp. or whorled simple or pinnate exstip. ls., and cpd. infls. of ♀ or ♂ ♀ reg. 2–6-merous fls.; C 4–5–6 or 0, free or united, imbr. or valv., A 2 epipet. or hypog., G̲ (2) each usu. with 2, rarely 1 or 4–8 axile ov.; caps., berry or drupe; endosp. or 0.
2. *Loganiaceae:* woody, rarely herbs, with opp. or whorled often stip. ls. and cymose umbels of ♀ or ♂ ♀ reg. fls.; K usu. imbr., C (4–5–∞), valv., imbr., or conv., A = C or 1, G̲ (2) rarely more with ∞–1 axile ov. and 1 style; caps.; endosp.

D. Fls. always sympetalous, with 1 whorl of sta., often ·|·, with usu. 2 median cpls. fully united (Orders 6–10).

   a. *K, C hypogynous,* with few exceptions (*Orders* 6, 7).

  6. TUBIFLORAE. Usu. herbs, fls. typically with 4 isomerous whorls or usu. with oligomerous G, and if ·|· also oligomerous A; sta. epipet., ov. with 1 integ.

    1. *Convolvulineae:* ls. usu. alt., fls. usu. reg.; cpls. with few or 2 ov. with micropyle downwards; fr. rarely 4 nutlets (fams. 1, 2).

    2. *Boraginineae:* as last, but micropyle facing upwards; caps. or drupe, or 4 nutlets (fams. 3, 4).

    3. *Verbenineae:* ls. usu. opp. or whorled, fls. usu. ·|· ; cpls. with 2, rarely 1, ov.; fr. drupe or drupe-like, or 4 nutlets (fams. 5, 6).

    4. *Solanineae:* fls. ·|· or reg. usu. 5-merous; A 5–4–2, G̲ rarely (5), usu. (2) with usu. ∞, rarely 2–1 ov.; fr. usu. caps., never splitting to base, rarely berry or drupe (fams. 7–17).

     A. Fr. splitting into 5 or many mericarps (fam. 7).

     B. Fr. 2-, rarely 5–∞-loc., or 1-loc. (fams. 8–17).

     1. Vascular bundles bicollateral (fam. 8).

3. *Gentianaceae:* herbs, rarely shrubs, with opp. entire exstip. ls. and cymose infls. of usu. ♀ reg. 4–5-merous fls.; K or (K), (C) usu. conv., A as many, G (2) usu. with ∞ ov. in 1-loc. ovary; caps.; endosp.

4. *Apocynaceae:* woody or herbs with simple usu. opp. entire ls., and latex, and cymose infls. of ♀ reg. 5–4-merous fls.; (C) usu. conv., A epipet., G (2) often only united by style; fr. various, endosperm thin or 0.

5. *Asclepiadaceae:* herbs or shrubby, often climbing, some succulent, with opp. or whorled, rarely alt. exstip. ls., and ♀ reg. fls. sol. or in cymose umbels; K 5, C (5), usu. conv., sometimes with appendages forming a corona, A 5 usu. united below, usu. with appendages forming a corona, pollen usu. in pollinia with translators, G (2) enclosed in sta. tube, with ∞ rarely few or 1 pend. ov., united by style above; fr. 2 follicles, seeds usu. hairy; endosp.

6

1. *Convolvulaceae:* usu. herbs with alt. ls., often twining, usu. with large ♀ reg. 5–4-merous fls.; A epipet., G (2) rarely (3–5) each with 2 basal erect ov. on axile plac.; caps.; endosp.

2. *Polemoniaceae:* usu. herbs with alt. or opp. exstip. ls. and ♀ usu. reg. 5-merous fls.; C usu. conv., G (3) rarely (2) or (5) each with ∞–1 erect ov.; caps.; endosp.

3. *Hydrophyllaceae:* herbs with alt. rarely opp. ls. and scorpioid cymes of ♀ reg. 5-merous fls.; A 5, G (2) each with ∞–2 sessile or pend. ov.; caps.; endosp.

4. *Boraginaceae:* herbs or woody pls., often roughly hairy, with usu. alt. simple ls., and scorpioid cymes of ♀ reg. 5 (rarely more)-merous fls.; G (2) each with 2 ov., 2-loc., usu. with false septum; fr. drupaceous or of 4 nutlets; endosp. or none.

5. *Verbenaceae:* herbs or woody pls. with usu. opp. or whorled entire or divided ls. and cymes orrac. of ♀ usu. ·l· 5–4 (rarely more)-merous fls.; (K), (C) often 2-lipped, A usu. 4 didynamous, or 2, G (2) rarely more, each with 2 ov., usu. 4-loc. by formation of secondary septa, style 1; drupe or schizocarp; usu. no endosp.

6. *Labiatae:* herbs or shrubs with decussate or whorled exstip. ls. and cymose infls. often condensed in the axils into seeming whorls of ♀ ·l· 5-merous fls.; K (5), C usu. 2-lipped, A 4 didynamous or 2 with or without 2 stds., G (2) each with 2 erect ov., infolded between them; fr. of 4 nutlets; endosp. little or none.

7. *Nolanaceae:* herbs or undershrubs with alt. ls. and sol. or racemed ♀ reg. fls.; K, C (5), A 5, G 5 with ∞ ov., divided by long. or transv. constrictions into 1–7-ovuled sections; endosp.

8. *Solanaceae:* herbs or shrubs with alt. ls. and term. sol. or cymosely umbelled ♀ usu. reg. 5-merous fls.; A 5, G (2) obliquely placed, each with ∞–1 ov. on axile plac., style 1; berry or caps.; endosp.

xliii

2. Vascular bundles collateral (fams. 9–17).

i. G 2-loc. with ∞ to few ov. (fams. 9–11).

ii. G 1-loc. with ± parietal plac. and ∞ ov. (fams. 12–15).

iii. G rarely 2-loc., usu. 1-loc. with basal central plac. and ∞ ov. (fam. 16).

iv. G 2- or 1-loc., in each 1 pend. ov., or 1 pend. ov. only (fam. 17).

5. *Acanthineae:* fls. usu. ⊹, typically 5-merous; A 4 or 2, G̲ (2) with usu. ∞ ov.; caps. loculicidal to very base (fam. 18).

6. *Myoporineae:* woody with alt. or opp. ls. and fls. reg. or ⊹, 5-merous; G̲ (2) later 4-loc., each with 2–4–∞ ov., or (2–∞) each with 1 pend. ov. with micropyle upwards; drupe; endosp. thin or none (fam. 19).

7. *Phrymineae:* herbs, fls. ⊹; G̲ 1 with 1 orthotr. ascending ov. (fam. 20).

7. PLANTAGINALES. Usu. herbs, rarely shrubby, with usu. alt. ls.; fls. ⚥ or ♂ ♀, reg., 4-merous; K (4), C (4), membranous, A 4, epipet., G̲ (2) or 1, 4–1-loc. with few or 1 anatr. ov.; caps. or nut; endosp.

9. *Scrophulariaceae:* herbs or shrubs, rarely trees, with alt. opp. or whorled ls., and variously arranged, never term., ♀, ± ·|·, 5-merous fls.; A usu. 4 or 2, G (2) median with each ∞ or few ov. on axile plac., and 1 style; caps. or berry; endosp.

10. *Bignoniaceae:* woody pls. often climbing, with usu. opp. often cpd. ls. and showy ♀ ·|· 5-merous fls., often in cpd. infls.; A 4 or 2, sometimes with 3–1 stds., G (2) median with ∞ ov., 2- or 1-loc., style 1; caps. or fleshy fr.; no endosp.

11. *Pedaliaceae:* herbs with glandular hairs and opp. ls. (sometimes alt. above) and axillary or racemed ♀ ·|· 5-merous fls.; A 4 or 2, G (2), rarely (3–4) or G̅, each with ∞ ov., 2–4-loc. transv. divided with axile plac.; caps. or nut; thin endosp.

12. *Martyniaceae:* as last, but fls. rac.; anther thecae spurred; G with 2 bilobed parietal plac.; caps.; thin endosp.

13. *Orobanchaceae:* paras. herbs with scale-ls. and term. or racemed ♀ ·|· 5-merous fls.; C 2-lipped, A 4 didynamous, G (2), rarely (3), each with 2 parietal plac. sometimes united in middle, and ∞ ov., 1 style; caps.; endosp.

14. *Gesneriaceae:* herbs or woody pls. with opp. simple ls. and showy sol. or cymosely umbelled ♀ ·|· 5-merous fls.; C 2-lipped, A 4 or 2 with sometimes 1–3 stds., G to G̅ (2), 1-loc. with parietal plac. and ∞ ov.; caps. or berry; endosp. or not.

15. *Columelliaceae:* woody pls. with opp. entire ls. and cymose umbels of ♀ nearly reg. 5–8-merous fls.; A 2, G (2) with ∞ ov. on 2 parietal bilobed plac.; caps. 4-valved; endosp.

16. *Lentibulariaceae:* herbs, usu. of water or damp ground, with ♀ ·|· 5-merous fls.; C 2-lipped, A usu. 2, G (2) usu. 1-loc. with basal free plac. and ∞ ov.; caps. 2–4-valved, ∞- or 1-seeded; no endosp.

17. *Globulariaceae:* herbs with rad. ls. and spherical heads or spikes of ♀ 5-merous ·|· fls.; A 4 or 2, G (2) 1-loc. each with 1 ov., or 1 ov. only, 1 style; 1-seeded nut; endosp.

18. *Acanthaceae:* herbs or shrubs with opp. ls. and spikes, racemes or cymose umbels of ♀ ·|· 5-merous fls.; K free or united, C reg. or ·|·, A 4 or 2, sometimes with 1–3 stds., G (2) median, each with ∞–2 ov.; caps. loculicidal to very base; seeds usu. with no endosp. and with jaculators.

19. *Myoporaceae.*

20. *Phrymaceae.*

7

1. *Plantaginaceae.*

b. *P epigynous (Orders* 8–10).

α. **Sta. free (Order 8).**

8. RUBIALES. Woody pls. or herbs with opp. usu. simple ls. and usu. reg. 5-4-merous fls.; $\overline{G}$ 1 (or more)-loc., each with ∞-1 anatr. ov.

A. Sta. = C segments (fams. 1–3).

B. Sta. fewer than C segments, $\overline{G}$ always with only 1 fertile loc. and 1 pend. ov. (fams. 4, 5).

β. **Sta. close together or partly united (Orders 9, 10).**

9. CUCURBITALES. Fls. typically 5-merous, usu. ♂ ♀ reg., with cup-like axis; A 5 free, at edge of axis, or each 2 united, or all 5 in a central synandrium, $\overline{G}$ usu. (3), 3-loc. usu. with ∞ ov. and usu. forked stigs.; fr. berry-like; no endosp.

10. CAMPANULATAE. Usu. herbs, rarely woody, with typically 5-merous fls. with 1 whorl of sta. and usu. fewer cpls.; anthers with 2-loc. thecae, often united, $\overline{G}$ or $\underline{G}$ with several loc. and ∞-1 ov. in each, or 1-loc. with 1 ov.

8

1. *Rubiaceae:* herbs or woody pls. with decussate entire ls. and interpetiolar stips. sometimes = ls., and usu. ♀ reg. fls. in cymes often condensed to heads, 5–4 (rarely more)-merous; K usu. open, C valv. or conv., G̅ (2) each with 1–∞ ov., style 1; fr. various; endosp. or o.

2. *Caprifoliaceae:* woody with opp. usu. exstip. ls. and ♀ reg. or ·l· 5-merous fls.; (C), G̅ (2–5) each with 1–∞ axile pend. ov.; fr. usu. berry- or drupe-like; endosp.

3. *Adoxaceae:* rhiz. herb; stems with 2 opp. ls. and 5–7-fld. cyme of ♀ homo-chlam. fls. (or with aborted K); term. fl. 4 (5)-, lat. 5 (6)-merous, all with 2 bracteoles; A 4–5–6 split to base, G (3–4–5) ½-inf. each with 1 pend. ov.; drupe; endosp.

4. *Valerianaceae:* herbs, rarely shrubby, with opp. exstip. ls. and cymose umbels or heads of ♀ or ♂ ♀ fls. without plane of symmetry; K indistinct in fl., later enlarging to pappus, C (5) or (3–4), often spurred at base, A 1–4, G̅ (3), 1 developed with 1 pend. ov.; style 1; no endosp.

5. *Dipsacaceae:* herbs or undershrubs with opp. exstip. ls. and cymose heads or umbels of ♀ usu. ·l· fls. with epicalyx; A 4 or fewer, G̅ (2), 1-loc. with 1 pend. ov. and 1 style; endosp.

9

1. *Cucurbitaceae.*

10

1. *Campanulaceae:* herbs or woody pls. usu. with alt. exstip. ls., latex, and often showy ♀ reg. or ·l· 5-merous fls.; C usu. united, A free or united with intr. anthers, G̅ usu. (2–5) with ∞ ov., rarely 1-loc., style 1; fr. caps. or berry-like; endosp.

2. *Goodeniaceae:* herbs or shrubs with simple ls. and ♀ usu. ·l· 5-merous fls.; A free or epipet., G usu. inf., 2- rarely 1-loc. with 1–2 or many ov. in each; style with pollen cup; fr. caps.-like; endosp.

3. *Brunoniaceae:* herb with rad. entire exstip. ls., and blue ♀ reg. 5-merous fls. in heads; C cylindrical, A 5 with united anthers, G̲ 1, 1-loc., style simple with pollen cup; no endosp.

4. *Stylidiaceae:* herb with simple exstip. ls. and ♀ or ♂ ♀ usu. ·l· 5-merous fls.; C usu. united, A 3–2 free or united to style, with extr. anthers, G̅ (2) 2- or 1-loc.; fr. septicidal or indehisc.; endosp.

5. *Calyceraceae:* herbs or undershrubs with alt. exstip. ls. and ♀ or ♂ ♀ reg. or ·l· 4–6-merous fls. in heads surrounded by bracts; A united but anthers free, G̅ 1-loc. with 1 pend. ov.; style 1; little endosp.

6. *Compositae:* herbs, shrubs or rarely trees with usu. alt., rarely opp., ls. and ♀ or ♂ ♀ reg. or ·l· 5-merous fls. in heads or short spikes, with invol.; K usu. repres. by hairs of pappus, C often ·l·, 2-lipped or strap-shaped, A at base epipet., anthers intr. united, G̅ (2) median, 1-loc. with 1 erect ov., and 1 style with 2 stigs.; achene; no endosp.

SYSTEM OF BENTHAM AND HOOKER
1862–93

I. DICOTYLEDONES (as above)

I. **Polypetalae** (fl. usually with two whorls of perianth, the inner polyphyllous: exceptions as in Engler's system):

SERIES I. THALAMIFLORAE. Sepals usu. distinct and separate, free from ovary; petals 1-, 2- to ∞-seriate, hypog.; sta. hypog., rarely inserted on a short or long torus or on a disk; ovary superior.

Order 1. *Ranales* (sta. rarely definite; cpls. free or immersed in torus, very rarely united; micropyle usu. inferior; embryo minute in fleshy albumen):

1. Ranunculaceae. 2. Dilleniaceae. 3. Calycanthaceae. 4. Magnoliaceae. 5. Anonaceae. 6. Menispermaceae. 7. Berberideae. 8. Nymphaeaceae.

Order 2. *Parietales* (sta. definite or ∞; cpls. united into a 1-loc. ovary with parietal placentae, rarely spuriously 2- or more-loc. by prolongation of placentae):

9. Sarraceniaceae. 10. Papaveraceae. 11. Cruciferae. 12. Capparideae. 13. Resedaceae. 14. Cistineae. 15. Violarieae. 16. Canellaceae. 17. Bixineae.

Order 3. *Polygalinae* (K and C 5, rarely 4 or 3; sta. as many or twice as many as petals; ovary 2-, rarely 1- or more-loc.; endosperm fleshy, rarely absent; herbs or shrubs with exstip. ls.):

18. Pittosporeae. 19. Tremandreae. 20. Polygaleae. 21. Vochysiaceae.

Order 4. *Caryophyllinae* (fl. regular; K 2–5, rarely 6; petals usu. as many; sta. as many or twice as many, rarely more or fewer; ovary 1-loc. or imperfectly 2–5-loc.; placenta free-central, rarely parietal; embryo usu. curved in floury albumen):

22. Frankeniaceae. 23. Caryophylleae. 24. Portulaceae. 25. Tamariscineae.

Order 5. *Guttiferales* (fl. regular; K and C usu. 4–5, imbr.; sta. usu. ∞; ovary 3–∞-loc., rarely 2-loc. or of 1 cpl.; placentae on inner angles of loculi):

26. Elatineae. 27. Hypericineae. 28. Guttiferae. 29. Ternstroemiaceae. 30. Dipterocarpeae. 31. Chlaenaceae.

Order 6. *Malvales* (fl. rarely irregular; K 5, rarely 2–4, free or united, valvate or imbr.; petals as many or 0; sta. usu. ∞, monadelphous; ovary 3–∞-loc., rarely of 1 cpl.; ovules in inner angles of loculi):

32. Malvaceae. 33. Sterculiaceae. 34. Tiliaceae.

SERIES II. DISCIFLORAE. Sepals distinct or united, free or adnate to ovary; disk usu. conspicuous as a ring or cushion, or spread over the base of the calyx-tube, or confluent with the base of the ovary, or broken up into glands; sta. usu. definite, inserted upon or at the outer or inner base of the disk; ovary superior.

Order 7. *Geraniales* (fl. often irregular; disk usu. annular, adnate to the sta. or reduced to glands, rarely 0; ovary of several cpls., syncarpous or sub-apocarpous; ovules 1–2, rarely ∞, ascending or pendulous; raphe usu. ventral):

35. Lineae. 36. Humiriaceae. 37. Malpighiaceae. 38. Zygophylleae. 39. Geraniaceae. 40. Rutaceae. 41. Simarubeae. 42. Ochnaceae. 43. Burseraceae. 44. Meliaceae. 45. Chailletiaceae.

Order 8. *Olacales* (fl. regular, ♂ or unisex.; calyx small; disk free, cupular or annular, rarely glandular or 0; ovary entire, 1–∞-loc.; ovules 1–3 in each loc., pend.; raphe dorsal, integ. confluent with the nucellus; endosp. usu. copious, fleshy; embryo small; shrubs or trees; leaves alt., simple, exstip.):

46. Olacineae. 47. Ilicineae. 48. Cyrilleae.

Order 9. *Celastrales* (fl. regular, ♂; corolla hypo- or peri-gynous; disk tumid, adnate to base of calyx-tube or lining it; sta. = petals or fewer, rarely twice as many, perig. or inserted outside the disk or on its edge; ovary usu. entire; ovules 1–2 in each loc., erect with ventral raphe; leaves simple, except in fam. 52):

49. Celastrineae. 50. Stackhousieae. 51. Rhamneae. 52. Ampelideae.

Order 10. *Sapindales* (fl. often irregular and unisex.; disk tumid, adnate to base of calyx or lining its tube; sta. perig. or inserted upon the disk or between it and the ovary, usu. definite; ovary entire, lobed or apocarpous; ovules 1–2 in each loc. usu. ascending with a ventral raphe, or reversed, or pend. from a basal funicle, rarely ∞ horizontal; seed usu. exalb.; embryo often curved or crumpled; shrubs or trees, ls. usu. compound):

53. Sapindaceae. 54. Sabiaceae. 55. Anacardiaceae.

Anomalous fams. or rather genera:

56. Coriarieae. 57. Moringeae.

SERIES III. CALYCIFLORAE. Sepals united, rarely free, often adnate to ovary; petals 1-seriate, peri- or epi-gynous; disk adnate to base of calyx, rarely tumid or raised into a torus or gynophore; sta. perig., usu. inserted on or beneath the outer margin of the disk; ovary often inferior.

Order 11. *Rosales* (fl. usu. ♂, regular or irregular; cpls. 1 or more, usu. quite free in bud, sometimes variously united afterwards with the calyx-tube or enclosed in the swollen top of the peduncle; styles usu. distinct):

58. Connaraceae. 59. Leguminosae. 60. Rosaceae. 61. Saxifrageae. 62. Crassulaceae. 63. Droseraceae. 64. Hamamelideae. 65. Bruniaceae. 66. Halorageae.

Order 12. *Myrtales* (fl. regular or sub-regular, usu. ♂; ovary syncarpous, usu. inferior; style undivided, or very rarely styles free; placentae axile or apical, rarely basal; ls. simple, usu. quite entire, rarely 3-foliolate in fam. 68):

67. Rhizophoraceae. 68. Combretaceae. 69. Myrtaceae. 70. Melastomaceae. 71. Lythrarieae. 72. Onagrarieae.

Order 13. *Passiflorales* (fl. usu. regular, ♂ or unisex.; ovary usu. inferior, syncarpous, 1-loc. with parietal placentae, sometimes 3- or more-loc. by the produced placentae; styles free or connate):

73. Samydaceae. 74. Loaseae. 75. Turneraceae. 76. Passifloreae. 77. Cucurbitaceae. 78. Begoniaceae. 79. Datisceae.

Order 14. *Ficoidales* (fl. regular or sub-regular; ovary syncarpous, inferior to superior, 1-loc. with parietal, or 2–8-loc. with axile or basal placentae; embryo curved, with endosp., or cyclical, or oblique with no endosp.):

80. Cacteae. 81. Ficoideae.

Order 15. *Umbellales* (fl. regular, usu. ♂; sta. usu. definite; ovary inferior,

1–2–∞-loc.; ovules solitary, pend. in each loc. from its apex; styles free or united at base; seeds with endosp.; embryo usu. minute):
82. Umbelliferae. 83. Araliaceae. 84. Cornaceae.

II. **Gamopetalae** (fl. usu. with two whorls of perianth, the inner gamophyllous; exceptions as in Engler's system):
SERIES I. INFERAE. Ovary inferior; sta. usu. as many as corolla-lobes.
Order 1. Rubiales (fl. regular or irregular; sta. epipet.; ovary 2–∞-loc., with 1–∞ ovules in each loc.):
85. Caprifoliaceae. 86. Rubiaceae.
Order 2. Asterales (fl. regular or irregular; sta. epipet.; ovary 1-loc., 1-ovuled, sometimes > 1-loc. but with only 1 ovule):
87. Valerianeae. 88. Dipsaceae. 89. Calycereae. 90. Compositae.
Order 3. Campanales (fl. usu. irregular; sta. usu. epig.; ovary 2–6-loc., with usu. ∞ ovules in each loc.):
91. Stylidieae. 92. Goodenovieae. 93. Campanulaceae.
SERIES II. HETEROMERAE. Ovary usu. superior; sta. epipet. or free from corolla, opp. or alt. to its segments, or twice as many, or ∞; cpls. > 2.
Order 4. Ericales (fl. usu. regular and hypog.; sta. as many or twice as many as petals; ovary 1–∞-loc. with 1–∞ ovules in each loc.; seeds minute):
94. Ericaceae. 95. Vaccinieae. 96. Monotropeae. 97. Epacrideae. 98. Diapensiaceae. 99. Lennoaceae.
Order 5. Primulales (corolla usu. regular and hypog., sta. usu. = and opp. to corolla-lobes; ovary 1-loc. with free-central or basal placenta and 1–∞ ovules):
100. Plumbagineae. 101. Primulaceae. 102. Myrsineae.
Order 6. Ebenales (corolla usu. hypog.; sta. usu. more than corolla-lobes, or, if as many, then opposite to them, except in 103, often ∞; ovary 2–∞-loc.; ovules usu. few; trees or shrubs):
103. Sapotaceae. 104. Ebenaceae. 105. Styraceae.
SERIES III. BICARPELLATAE. Ovary usu. superior; sta. as many as or fewer than corolla-lobes, alt. to them; cpls. 2, rarely 1 or 3.
Order 7. Gentianales (corolla regular, hypog.; sta. epipet.; ls. generally opp.):
106. Oleaceae. 107. Salvadoraceae. 108. Apocynaceae. 109. Asclepiadaceae. 110. Loganiaceae. 111. Gentianaceae.
Order 8. Polemoniales (corolla regular, hypog.; sta. = corolla-lobes, epipet.; ovary 1–5-loc.; ls. generally alt.):
112. Polemoniaceae. 113. Hydrophyllaceae. 114. Boragineae. 115. Convolvulaceae. 116. Solanaceae.
Order 9. Personales (fl. usu. very irregular; corolla hypog., often 2-lipped; sta. generally fewer than corolla-lobes, usu. 4, didynamous, or 2; ovary 1–2- or rarely 4-loc.; ovules usu. ∞):
117. Scrophularineae. 118. Orobanchaceae. 119. Lentibularieae. 120. Columelliaceae. 121. Gesneraceae. 122. Bignoniaceae. 123. Pedalineae. 124. Acanthaceae.
Order 10. Lamiales (corolla usu. 2-lipped, hypog., rarely regular; sta. as in preceding; ovary 2–4-loc.; ovules solitary in loc., or rarely > 1 in fams. 125 and 127; fruit a drupe or nutlets):
125. Myoporineae. 126. Selagineae. 127. Verbenaceae. 128. Labiatae.

1

Anomalous Fam.
129. Plantagineae.

III. **Monochlamydeae or Incompletae** (fl. usu. with one whorl of perianth, commonly sepaloid, or none):

SERIES I. CURVEMBRYAE. Terrestrial plants with usu. ☿ fls.; sta. generally = perianth-segments; ovule usu. solitary; embryo curved in floury endosp.
130. Nyctagineae. 131. Illecebraceae. 132. Amarantaceae. 133. Chenopodiaceae. 134. Phytolaccaceae. 135. Batideae. 136. Polygonaceae.

SERIES II. MULTIOVULATAE AQUATICAE. Aquatic plants with syncarpous ovary and ∞ ovules.
137. Podostemaceae.

SERIES III. MULTIOVULATAE TERRESTRES. Terrestrial plants with syncarpous ovary and ∞ ovules.
138. Nepenthaceae. 139. Cytinaceae. 140. Aristolochieae.

SERIES IV. MICREMBRYAE. Ovary syn- or apo-carpous; ovules usu. solitary; embryo very small, surrounded by endosp.
141. Piperaceae. 142. Chloranthaceae. 143. Myristiceae. 144. Monimiaceae.

SERIES V. DAPHNALES. Ovary usu. of 1 cpl.; ovules solitary or few; perianth perfect, sepaloid, or in 1 or 2 whorls; sta. perig.
145. Laurineae. 146. Proteaceae. 147. Thymelaeaceae. 148. Penaeaceae. 149. Elaeagnaceae.

SERIES VI. ACHLAMYDOSPOREAE. Ovary 1-loc., 1–3-ovuled; ovules not apparent till after fert.; seed with endosp., but no testa, adnate to receptacle or pericarp.
150. Loranthaceae. 151. Santalaceae. 152. Balanophoreae.

SERIES VII. UNISEXUALES. Fls. unisex.; ovary syncarpous or of 1 cpl.; ovules solitary or 2 per cpl.; endosp. or none; perianth sepaloid or much reduced or absent.
153. Euphorbiaceae. 154. Balanopseae. 155. Urticaceae. 156. Platanaceae. 157. Leitnerieae. 158. Juglandeae. 159. Myricaceae. 160. Casuarineae. 161. Cupuliferae.

SERIES VIII. ANOMALOUS FAMILIES. Unisex. fams. of doubtful or unknown affinities.
162. Salicaceae. 163. Lacistemaceae. 164. Empetraceae. 165. Ceratophylleae.

II. MONOCOTYLEDONES (as in Engler)

SERIES I. MICROSPERMAE. Inner perianth petaloid; ovary inferior with 3 parietal or rarely axile placentae; seeds minute, exalb.
169. Hydrocharideae. 170. Burmanniaceae. 171. Orchideae.

SERIES II. EPIGYNAE. Perianth partly petaloid; ovary usu. inferior; endosp. abundant.
172. Scitamineae. 173. Bromeliaceae. 174. Haemodoraceae. 175. Irideae. 176. Amaryllideae. 177. Taccaceae. 178. Dioscoreaceae.

SERIES III. CORONARIEAE. Inner perianth petaloid; ovary usu. free, superior; endosp. abundant.
179. Roxburghiaceae. 180. Liliaceae. 181. Pontederiaceae. 182. Philydraceae. 183. Xyrideae. 184. Mayacaceae. 185. Commelinaceae. 186. Rapateaceae.

SERIES IV. CALYCINAE. Perianth sepaloid, herbaceous or membranous; ovary, etc. as in III.

187. Flagellarieae. 188. Juncaceae. 189. Palmae.

SERIES V. NUDIFLORAE. Perianth none, or represented by hairs or scales; cpl. 1 or several syncarpous; ovary superior; ovules 1–∞; endosp. usu. present.

190. Pandaneae. 191. Cyclanthaceae. 192. Typhaceae. 193. Aroideae. 194. Lemnaceae.

SERIES VI. APOCARPAE. Perianth in 1 or 2 whorls, or none; ovary superior, apocarpous; no endosp.

195. Triurideae. 196. Alismaceae. 197. Naiadaceae.

SERIES VII. GLUMACEAE. Fls. solitary, sessile in the axils of bracts and arranged in heads or spikelets with bracts; perianth of scales, or none; ovary usu. 1-loc., 1-ovuled; endosp.

198. Eriocauleae. 199. Centrolepideae. 200. Restiaceae. 201. Cyperaceae. 202. Gramineae.

FAMILY EQUIVALENTS*
IN THIS EDITION OF THE *DICTIONARY*,
IN ENGLER'S *SYLLABUS* (ED. 12),
AND IN THE *GENERA PLANTARUM*.

Willis ed. 8 (1973)	*Engler & Melchior* (1954–64)	*Bentham & Hooker* (1862–83)
Abolbodaceae	Xyridaceae	Xyrideae
Acanthaceae	Acanthaceae	Acanthaceae
Aceraceae	Aceraceae	Sapindaceae
Achariaceae	Achariaceae	Passifloreae
Achatocarpaceae	Achatocarpaceae	Amarantaceae
Actinidiaceae	Actinidiaceae	Ternstroemiaceae
Adoxaceae	Adoxaceae	Caprifoliaceae
Aegialitidaceae	Plumbaginaceae	Plumbagineae
Aextoxicaceae	Aextoxicaceae	Euphorbiaceae
Agavaceae	Agavaceae	Amaryllideae
Agdestidaceae	Phytolaccaceae	Phytolaccaceae (genus anomalum)
Aitoniaceae	?	Sapindaceae
Aïzoaceae	Aïzoaceae	Ficoïdeae
Akaniaceae	Akaniaceae	Sapindaceae
Alangiaceae	Alangiaceae	Cornaceae
Alismataceae	Alismataceae	Alismaceae
Alliaceae	Liliaceae	Liliaceae
Alseuosmiaceae	'Probably do not belong to the family [Caprifoliaceae]'	Caprifoliaceae
Alstroemeriaceae	Liliaceae	Amaryllideae
Altingiaceae	Hamamelidaceae	Hamamelideae
Amaranthaceae	Amaranthaceae	Amarantaceae
Amaryllidaceae	Amaryllidaceae	Amaryllideae
Amborellaceae	Amborellaceae	Monimiaceae
Anacardiaceae	Anacardiaceae	Anacardiaceae
Anarthriaceae	Restionaceae	Restiaceae
Ancistrocladaceae	Ancistrocladaceae	Dipterocarpeae
Androstachydaceae	[Euphorbiaceae?]	—
Anisophylleaceae	Rhizophoraceae	Rhizophoreae

* These do not, of course, imply exact coextensive equivalence, but only that the 'Willis' family was *included in* the corresponding families of Engler & Melchior or Bentham & Hooker.

Willis ed. 8	Engler & Melchior	Bentham & Hooker
Annonaceae	Annonaceae	Anonaceae
Anomochloaceae	Gramineae	Gramineae
Aphyllanthaceae	Liliaceae	Liliaceae
Apocynaceae	Apocynaceae	Apocynaceae
Aponogetonaceae	Aponogetonaceae	Naiadaceae
Apostasiaceae	Orchidaceae	Orchideae
Aquifoliaceae	Aquifoliaceae	Ilicineae
Araceae	Araceae	Aroïdeae
Araliaceae	Araliaceae	Araliaceae
Araucariaceae	Araucariaceae	Coniferae
Aristolochiaceae	Aristolochiaceae	Aristolochiaceae
Asclepiadaceae	Asclepiadaceae	Asclepiadeae
Asteranthaceae	Lecythidaceae	Myrtaceae
Asteropeiaceae	Theaceae	Samydaceae (genus anomalum)
Atherospermataceae	Monimiaceae	Monimiaceae
Aucubaceae	Cornaceae	Cornaceae
Austrobaileyaceae	Austrobaileyaceae	—
Averrhoaceae	Oxalidaceae	Geraniaceae
Avicenniaceae	Verbenaceae	Verbenaceae
Balanitaceae	Zygophyllaceae	Simarubeae
Balanopaceae	Balanopaceae	Balanopseae
Balanophoraceae	Balanophoraceae	Balanophoreae
Balsaminaceae	Balsaminaceae	Geraniaceae
Bambusaceae	Gramineae	Gramineae
Barbeuiaceae	Phytolaccaceae	Phytolaccaceae
Barbeyaceae	Ulmaceae	—
Barclayaceae	Nymphaeaceae	Nymphaeaceae
Barringtoniaceae	Lecythidaceae	Myrtaceae
Basellaceae	Basellaceae	Chenopodiaceae
Batidaceae	Bataceae	Batideae
Baueraceae	Saxifragaceae	Saxifrageae (genus anomalum)
Begoniaceae	Begoniaceae	Begoniaceae
Berberidaceae	Berberidaceae	Berberideae
Betulaceae	Betulaceae	Cupuliferae
Biebersteiniaceae	Geraniaceae	Geraniaceae
Bignoniaceae	Bignoniaceae	Bignoniaceae
Bischofiaceae	[Euphorbiaceae?]	Euphorbiaceae
Bixaceae	Bixaceae	Bixineae
Blepharocaryaceae	[Anacardiaceae?]	—
Boerlagellaceae	Sapotaceae	
Bombacaceae	Bombacaceae	Malvaceae
Bonnetiaceae	Theaceae	Ternstroemiaceae
Boraginaceae	Boraginaceae	Boragineae
Bretschneideraceae	Bretschneideraceae	—
Brexiaceae	Saxifragaceae	Saxifrageae

Willis ed. 8	Engler & Melchior	Bentham & Hooker
Bromeliaceae	Bromeliaceae	Bromeliaceae
Brunelliaceae	Brunelliaceae	Simarubeae
Bruniaceae	Bruniaceae	Bruniaceae
Brunoniaceae	Brunoniaceae	Goodenoviëae
Buddlejaceae	Buddlejaceae	Loganiaceae
Burmanniaceae	Burmanniaceae	Burmanniaceae
Burseraceae	Burseraceae	Burseraceae
Butomaceae	Butomaceae	Alismaceae
Buxaceae	Buxaceae	Euphorbiaceae
Byblidaceae	Byblidaceae	Droseraceae
Cabombaceae	Nymphaeaceae	Nymphaeaceae
Cactaceae	Cactaceae	Cacteae
Callitrichaceae	Callitrichaceae	Halorageae
Calycanthaceae	Calycanthaceae	Calycanthaceae
Calyceraceae	Calyceraceae	Calycereae
Camelliaceae	Theaceae	Ternstroemiaceae
Campanulaceae	Campanulaceae	Campanulaceae
Canellaceae	Canellaceae	Canellaceae
Cannabidaceae	Moraceae	Urticaceae
Cannaceae	Cannaceae	Scitamineae
Canotiaceae	?	Rosaceae
Capparidaceae	Capparaceae	Capparideae
Caprifoliaceae	Caprifoliaceae	Caprifoliaceae
Cardiopterygaceae	Cardiopteridaceae	Olacineae (genus valde anomalum)
Caricaceae	Caricaceae	Passifloreae
Carlemanniaceae	Caprifoliaceae	Rubiaceae
Carpinaceae	Betulaceae	Cupuliferae
Cartonemataceae	Commelinaceae	Commelinaceae
Caryocaraceae	Caryocaraceae	Ternstroemiaceae
Caryophyllaceae	Caryophyllaceae	Caryophylleae
Casuarinaceae	Casuarinaceae	Casuarineae
Celastraceae	Celastraceae	Celastrineae
Centrolepidaceae	Centrolepidaceae	Centrolepideae
Cephalotaceae	Cephalotaceae	Saxifrageae (genus anomalum)
Cephalotaxaceae	Cephalotaxaceae	Coniferae
Ceratophyllaceae	Ceratophyllaceae	Ceratophylleae
Cercidiphyllaceae	Cercidiphyllaceae	—
Chenopodiaceae	Chenopodiaceae	Chenopodiaceae
Chloranthaceae	Chloranthaceae	Chloranthaceae
Chrysobalanaceae	Chrysobalanaceae	Rosaceae
Circaeasteraceae	Ranunculaceae	Chloranthaceae
Cistaceae	Cistaceae	Cistineae
Cleomaceae	Capparaceae	Capparideae
Clethraceae	Clethraceae	Ericaceae (genus anomalum)

Willis ed. 8	Engler & Melchior	Bentham & Hooker
Cneoraceae	Cneoraceae	Simarubeae
Cobaeaceae	Polemoniaceae	Polemoniaceae
Cochlospermaceae	Cochlospermaceae	Bixineae
Columelliaceae	Columelliaceae	Columelliaceae
Combretaceae	Combretaceae	Combretaceae
Commelinaceae	Commelinaceae	Commelinaceae
Compositae	Compositae	Compositae
Connaraceae	Connaraceae	Connaraceae
Convolvulaceae	Convolvulaceae	Convolvulaceae
Coriariaceae	Coriariaceae	Coriariëae
Coridaceae	Primulaceae	Primulaceae
Cornaceae	Cornaceae	Cornaceae
Corsiaceae	Corsiaceae	Burmanniaceae
Corylaceae	Betulaceae	Cupuliferae
Corynocarpaceae	Corynocarpaceae	Anacardiaceae
Costaceae	Zingiberaceae	Scitamineae
Crassulaceae	Crassulaceae	Crassulaceae
Croomiaceae	Stemonaceae	Roxburghiaceae
Crossosomataceae	Crossosomataceae	Dilleniaceae (?)
Cruciferae	Cruciferae	Cruciferae
Crypteroniaceae	Crypteroniaceae	Lythrariëae
Ctenolophonaceae	Linaceae	—
Cucurbitaceae	Cucurbitaceae	Cucurbitaceae
Cunoniaceae	Cunoniaceae	Saxifrageae
Cupressaceae	Cupressaceae	Coniferae
Curtisiaceae	Cornaceae	Cornaceae
Cuscutaceae	Convolvulaceae	Convolvulaceae
Cycadaceae	Cycadaceae	Cycadaceae
Cyclanthaceae	Cyclanthaceae	Cyclanthaceae
Cymodoceaceae	Zannichelliaceae	Naiadaceae
Cynomoriaceae	Cynomoriaceae	Balanophoreae
Cyperaceae	Cyperaceae	Cyperaceae
Cyrillaceae	Cyrillaceae	Cyrilleae
Daphniphyllaceae	Daphniphyllaceae	Euphorbiaceae
Datiscaceae	Datiscaceae	Datisceae
Davidiaceae	Davidiaceae	—
Davidsoniaceae	Davidsoniaceae	—
Degeneriaceae	Degeneriaceae	—
Dialypetalanthaceae	Dialypetalanthaceae	—
Diapensiaceae	Diapensiaceae	Diapensiaceae
Dichapetalaceae	Dichapetalaceae	Chailletiaceae
Dicrastylidaceae	Verbenaceae	Verbenaceae
Didiereaceae	Didiereaceae	—
Didymelaceae	Didymelaceae	[Leitnerieae ??]
Diegodendraceae	—	—
Dilleniaceae	Dilleniaceae	Dilleniaceae
Dioncophyllaceae	Dioncophyllaceae	—